The Ocean and Cryosphere in a Changing Climate

Special Report of the Intergovernmental Panel on Climate Change

Edited by

Hans-Otto Pörtner

Working Group II Co-Chair

Debra C. Roberts

Working Group II Co-Chair

Valérie Masson-Delmotte

Working Group I Co-Chair

Panmao Zhai

Working Group I Co-Chair

Melinda Tignor

Head of WGII TSU

Elvira Poloczanska

Science Advisor to the WGII
Co-Chairs and TSU

Katja Mintenbeck

Director of Science

Andrés Alegría

Graphics Officer

Maike Nicolai

Communications Officer

Andrew Okem

Science Officer

Jan Petzold

Science Officer

Bardhyl Rama

Director of Operations

Nora M. Weyer

Science Officer

Working Group II Technical Support Unit

CAMBRIDGE
UNIVERSITY PRESS

University Printing House, Cambridge CB2 8BS, United Kingdom
One Liberty Plaza, 20th Floor, New York, NY 10006, USA
477 Williamstown Road, Port Melbourne, VIC 3207, Australia
314–321, 3rd Floor, Plot 3, Splendor Forum, Jasola District Centre, New Delhi – 110025, India
103 Penang Road, #05-06/07, Visioncrest Commercial, Singapore 238467

Cambridge University Press is part of the University of Cambridge.
It furthers the University's mission by disseminating knowledge in the pursuit of education,
learning, and research at the highest international levels of excellence.

www.cambridge.org
Information on this title: www.cambridge.org/9781009157971
DOI: 10.1017/9781009157964

First published 2022

Printed in the United Kingdom by TJ Books Limited, Padstow Cornwall

A catalogue record for this publication is available from the British Library.

ISBN 978-1-009-15797-1 Paperback

Cambridge University Press has no responsibility for the persistence or accuracy of URLs for external or
third-party internet websites referred to in this publication and does not guarantee that any content on
such websites is, or will remain, accurate or appropriate.

This report should be cited as:
IPCC, 2019: *IPCC Special Report on the Ocean and Cryosphere in a Changing Climate* [H.-O. Pörtner, D.C. Roberts,
V. Masson-Delmotte, P. Zhai, M. Tignor, E. Poloczanska, K. Mintenbeck, A. Alegría, M. Nicolai, A. Okem, J. Petzold,
B. Rama, N.M. Weyer (eds.)]. Cambridge University Press, Cambridge, UK and New York, NY, USA, 755 pp.
https://doi.org/10.1017/9781009157964.

The designations employed and the presentation of material on maps do not imply the expression of any opinion
whatsoever on the part of the Intergovernmental Panel on Climate Change concerning the legal status of any
country, territory, city or area or of its authorities, or concerning the delimitation of its frontiers or boundaries.

Front and back cover artwork and layout by Stefanie Langsdorf

Contents

Foreword
and Preface

Foreword

This *IPCC Special Report on the Ocean and Cryosphere in a Changing Climate* (SROCC), is the third Special Report to be produced in the Intergovernmental Panel on Climate Change's (IPCC) Sixth Assessment Report (AR6) cycle. SROCC is unique because – for the first time – the IPCC has produced an in-depth report examining the farthest corners of the Earth – from the highest mountains and remote polar regions to the deepest oceans. The report finds that even and especially in these places, human-caused climate change is evident. These changes show that the world's ocean and cryosphere have been 'taking the heat' for climate change for decades. The consequences for nature and humanity are sweeping and severe. This report highlights the urgency of timely, ambitious, coordinated, and enduring action.

SROCC was jointly prepared by Working Groups I and II, and provides the latest state of knowledge on the ocean and cryosphere in a changing climate. It focuses on changes to mountain cryosphere, polar regions and ecosystems, sea level rise and coastal extremes, ocean and marine life, as well as providing key information to enable action at all scales and to manage risks and build resilience through adaptation for the benefit of ecosystems and human societies. The report highlights the observed and projected changes in the ocean and cryosphere, the associated impacts and risks for human societies and ecosystems, as well as assessing a range of response options and adaption measures. SROCC clearly presents the level of risks and the limits of adaptation for high emission scenarios and thereby the benefits of ambitious and effective adaptation for sustainable development. It highlights the importance of irreversible and committed changes on timescales of decades to centuries. It stresses the urgency of near-term action to reduce risks also by reducing emissions of greenhouse gases, strengthening findings from the SR15 and SRCCL reports.

The IPCC provides policymakers with regular scientific assessments on climate change, its impacts and risks, as well as adaptation and mitigation options. Since it was established jointly in 1988 by the World Meteorological Organisation and the United Nations Environment Programme, the IPCC has produced a series of Assessment Reports, Special Reports, Technical Papers and Methodological Reports and other products that have become the gold standard scientific resource on climate change issues for policymakers.

SROCC was made possible thanks to the commitment and dedication of hundreds of experts worldwide, representing a wide range of disciplines. We express our deep gratitude to all Coordinating Lead Authors, Lead Authors, Contributing Authors, Review Editors, Chapter Scientists and Expert and Government Reviewers who devoted their time and effort to make the Special Report on the Ocean and Cryosphere in a Changing Change possible. We would like to thank the staff of the Working Group Technical Support Units and the IPCC Secretariat for their dedication and professionalism.

We are also grateful to the governments that supported their scientists' participation in developing this report and that contributed to the IPCC Trust Fund to provide for the essential participation of experts from developing countries and countries with economies in transition. We would like to express our appreciation to the Principality of Monaco for hosting the SROCC Scoping Meeting, to the Governments of Fiji, Ecuador, China, and the Russian Federation for hosting Lead Author Meetings, and to the Principality of Monaco for hosting the Second Joint Session of Working Group I and Working Group II. Our thanks also to the Government of France for funding the Technical Support Unit of Working Group I, and to the Government of Germany and the Governments of Norway and New Zealand, for funding the Technical Support Unit of Working Group II. We also acknowledge the Government of Norway's generous contribution in support of the development of the graphics for SROCC Summary for Policymakers, and the support of the Prince Albert II of Monaco Foundation and the Fondation de France for an additional post in the Working Group II Technical Support Unit.

We especially wish to thank the IPCC Chair, Hoesung Lee, the IPCC Vice-Chairs Ko Barrett, Thelma Krug, and Youba Sokona for their guidance, and the Co-Chairs of Working Groups II and I Hans-Otto Pörtner, Debra Roberts, Valérie Masson-Delmotte and Panmao Zhai for their inspired leadership throughout the process.

Petteri Taalas
Secretary-General
World Meteorological Organisation

Inger Andersen
Executive Director
United Nations Environment Programme

Preface

This IPCC Special Report on the Ocean and Cryosphere in a Changing Climate (SROCC), is the third Special Report to be produced in the Intergovernmental Panel on Climate Change's (IPCC) Sixth Assessment Report (AR6) cycle. Its findings reinforce those of the two earlier Special Reports, the IPCC Special Report on Global Warming of 1.5°C and the IPCC Special Report on Climate Change and Land. The report was jointly prepared by Working Groups I and II, with the Working Group II Technical Support Unit leading the operational production. It was prepared following IPCC principles and procedures. This Special Report builds upon the IPCC's Fifth Assessment Report (AR5) in 2013–2014 and on relevant research published in the scientific, technical and socio-economic literature. The report sits alongside other related reports from other UN Bodies, including Intergovernmental Science Policy Platform on Biodiversity and Ecosystem Services (IPBES) Global Assessment Report on Biodiversity and Ecosystem Services.

Scope of the Report

The IPCC SROCC responds to proposals for Special Reports from governments and observer organisations provided at the start of the IPCC AR6 cycle. It assesses the observed and projected changes to the ocean and cryosphere and their associated impacts and risks, with a focus on resilience, risk management response options, and adaptation measures, considering both their potential and limitations. SROCC brings together knowledge on physical and biogeochemical changes, the interplay with ecosystem changes, and the implications for human communities. The report was produced with careful attention to other assessments, with the aim of achieving coherence and complementarity, as well as providing an updated assessment of the current state of knowledge. The Special Report considered literature accepted for publication up to 15 May 2019.

Structure of the Report

This report consists of a short Summary for Policymakers, a Technical Summary, six Chapters, an Integrative Cross-Chapter Box, four Annexes, as well as online Supplementary Material.

Chapter 1: Framing and Context introduces the reader to the structure of the report and the content presented in more detail in subsequent chapters. It highlights the role of the ocean and cryosphere in the Earth system, assessment of climate impacts and future risks for ecosystems and human societies from the high mountains to the deep ocean, the knowledge systems informing responses to climate change. as well as the capacities of governance and institutions to implement such responses, and it highlights key concepts and terms as well as linkages between chapters.

Chapter 2: High Mountain Areas provides a wide-ranging assessment of the observed and projected cryosphere (including snow, glaciers, permafrost, lake and river ice) changes in high mountain areas, as well as associated impacts, risks, and adaptation measures related to natural and human systems.

Chapter 3: Polar Regions presents the state of knowledge concerning changes in the Arctic and Antarctic oceans and marine and land cryosphere, how they are affected by climate change, and projections for the future. It assesses impacts of individual and interacting polar system changes, as well as response options to reduce risk and build resilience in the polar regions.

Chapter 4: Sea Level Rise and Implications for Low-lying Islands, Coasts and Communities assesses past and future contributions of various processes to global, regional and extreme sea level changes, the associated risks, and response options and pathways to resilience and sustainable development.

Chapter 5: Changing Ocean, Marine Ecosystems, and Dependent Communities focuses on observations of climate-related trends, impacts and adaptation, projected changes and associated risks, as well as the response options to enhance resilience.

Chapter 6: Extremes, Abrupt Changes and Managing Risks assesses extreme as well as abrupt or irreversible changes in the ocean and cryosphere including recent anomalous extreme events, compound risk, cascading effects, their impacts on human and natural systems, and sustainable and resilient risk management strategies.

Finally, the *Integrative Cross-Chapter Box on Low-Lying Islands and Coasts* highlights the key assessment findings relating to low lying islands and coasts. It includes summary information on the critical climate-related drivers, their observed and projected impacts on related geographies and major sectors, and responses, including adaptation strategies in practice.

The Process

The IPCC SROCC was prepared in accordance with the principles and procedures established by the IPCC and represents the combined efforts of leading experts in the field of climate change. A scoping meeting for SROCC was held in Monaco in December 2016, and the final outline was agreed by the Panel at its 45th Session in March 2017 in Guadalajara, Mexico. Governments and IPCC observer organisations nominated more than 500 experts for the chapter team. The team of 14 Coordinating Lead Authors, 75 Lead Authors, and 15 Review Editors were selected by Working Groups I and II Bureaux. In addition, 222 Contributing Authors were invited by the chapter teams to provide scientific and technical information in the form of text, graphs or data. The report drafts prepared by the authors were subject to two rounds of formal review and revision followed by a final round of government comments on the Summary for Policymakers. The enthusiastic participation of the scientific community and governments to the review process resulted in over 31,000 written review comments, submitted by 824 expert reviewers and 43 governments. The Review Editors for the chapters monitored the review process to ensure that all substantive review comments received appropriate consideration. The Summary for Policymakers was approved at the Second Joint Session of Working Groups I and II, and the Summary for Policymakers and the underlying chapters were then accepted by the IPCC at its 51st Session in September 2019 in Monaco.

Acknowledgements

We express our deepest appreciation for the expertise and commitment shown by the Coordinating Lead Authors and Lead Authors throughout the process. They were ably helped by the many Contributing Authors who served on SROCC. The Review Editors were critical in assisting the author teams and ensuring the integrity of the review process. We are grateful to the Chapter Scientists who supported the chapter teams in the delivery of the report. We would also like to thank all the expert and government reviewers who submitted comments on the drafts.

The production of the report was guided by members of the IPCC Bureau. We would like to thank our colleagues who supported and advised us in the development of the report: Working Group I Vice-Chairs Edvin Aldrian, Fatima Driouech, Gregory Flato, Jan Fuglestvedt, Muhammad I. Tariq, Carolina Vera, Noureddine Yassaa; Working Group II Vice-Chairs Andreas Fischlin, Mark Howden, Carlos Méndez, Joy Jacqueline Pereira, Roberto A. Sánchez-Rodríguez, Sergey Semenov, Pius Yanda, and Taha M. Zatari; and Working Group III Vice-Chair Amjad Abdulla. Our appreciation also goes to Ko Barrett, Vice Chair of IPCC, who served as champion for the report and ably supported us from scoping through approval.

Our sincere thanks go to the hosts and organizers of the Scoping Meeting, the four Lead Author Meetings, and the Joint Working Group and IPCC Sessions. We gratefully acknowledge the support from the Principality of Monaco and the Prince Albert II of Monaco Foundation, the Government of Fiji and the University of the South Pacific, the Government of Ecuador, the Government of China and the Chinese Academy of Sciences, the Government of the Russian Federation and Kazan Federal University, and the Principality of Monaco and the Prince Albert II of Monaco Foundation. The support provided by many governments as well as through the IPCC Trust Fund for the many experts participating in the process is also noted with appreciation.

The staff of the IPCC Secretariat based in Geneva provided a wide range of support for which we would like to thank Abdalah Mokssit, Secretary of the IPCC, and his colleagues: Kerstin Stendahl, Jonathan Lynn, Sophie Schlingemann, Jesbin Baidya, Laura Biagioni, Annie Courtin, Oksana Ekzarkho, Judith Ewa, Joelle Fernandez, Andrea Papucides Bach, Nina Peeva, Mxolisi Shongwe, and Werani Zabula. Thanks are due to Elhousseine Gouaini and Adriana Oskarsson who served as the conference officers for the 51st Session of the IPCC.

The report production was managed by the Technical Support Unit of IPCC Working Group II, through the generous financial support of the German Federal Ministry for Education and Research and the Alfred Wegener Institute Hemholtz Centre for Polar and Marine Research. The Prince Albert II of Monaco Foundation and the Fondation de France provided financial support for a scientific staff member to the Working Group II Technical Support Unit, while the Norwegian Environment Agency enabled additional graphics support for the Summary for Policymakers. Additional funding from the Governments of Norway and New Zealand support the Working Group II Technical Support Unit office in Durban, South Africa. Without the support of all these bodies this report would not have been possible.

This report could not have been prepared without the dedication, commitment, and professionalism of the members of the Working Group II Technical Support Unit: Melinda Tignor, Elvira Poloczanska, Katja Mintenbeck, Andrés Alegría, Marlies Craig, Anka Freund, Stefanie Langsdorf, Philisiwe Manqele, Maike Nicolai, Andrew Okem, Jan Petzold, Bardhyl Rama, Jussi Savolainen, Stefan Weisfeld, Nora Weyer, and Mallou. Our warmest thanks go to the collegial and collaborative support provided by Sarah Connors, Melissa Gomis, Robin Matthews, Clotilde Péan, Anna Pirani, and Rong Yu from the WGI Technical Support Unit, and Katie Kissick from the WGIII Technical Support Unit. In addition, the following contributions are gratefully acknowledged: Martin Künsting (graphics support for the Summary for Policymakers), David Dokken (approval session support), Naomi Stewart (copyedit), Marilyn Anderson (index), and Soapbox (layout).

And a final, special thank you to the colleagues, family and friends who supported us through the many long hours and days spent at home and away from home while producing this report.

Hans-Otto Pörtner
IPCC Working Group II Co-Chair

Debra C. Roberts
IPCC Working Group II Co-Chair

Valérie Masson-Delmotte
IPCC Working Group I Co-Chair

Panmao Zhai
IPCC Working Group I Co-Chair

Summary for Policymakers

Summary
for Policymakers

Drafting Authors:

Nerilie Abram (Australia), Carolina Adler (Switzerland/Australia), Nathaniel L. Bindoff (Australia), Lijing Cheng (China), So-Min Cheong (Republic of Korea), William W. L. Cheung (Canada), Matthew Collins (UK), Chris Derksen (Canada), Alexey Ekaykin (Russian Federation), Thomas Frölicher (Switzerland), Matthias Garschagen (Germany), Jean-Pierre Gattuso (France), Bruce Glavovic (New Zealand), Stephan Gruber (Canada/Germany), Valeria Guinder (Argentina), Robert Hallberg (USA), Sherilee Harper (Canada), Nathalie Hilmi (Monaco/France), Jochen Hinkel (Germany), Yukiko Hirabayashi (Japan), Regine Hock (USA), Anne Hollowed (USA), Helene Jacot Des Combes (Fiji), James Kairo (Kenya), Alexandre K. Magnan (France), Valérie Masson-Delmotte (France), J.B. Robin Matthews (UK), Kathleen McInnes (Australia), Michael Meredith (UK), Katja Mintenbeck (Germany), Samuel Morin (France), Andrew Okem (South Africa/Nigeria), Michael Oppenheimer (USA), Ben Orlove (USA), Jan Petzold (Germany), Anna Pirani (Italy), Elvira Poloczanska (UK/Australia), Hans-Otto Pörtner (Germany), Anjal Prakash (Nepal/India), Golam Rasul (Nepal), Evelia Rivera-Arriaga (Mexico), Debra C. Roberts (South Africa), Edward A.G. Schuur (USA), Zita Sebesvari (Hungary/Germany), Martin Sommerkorn (Norway/Germany), Michael Sutherland (Trinidad and Tobago), Alessandro Tagliabue (UK), Roderik Van De Wal (Netherlands), Phil Williamson (UK), Rong Yu (China), Panmao Zhai (China)

Draft Contributing Authors:

Andrés Alegría (Honduras), Robert M. DeConto (USA), Andreas Fischlin (Switzerland), Shengping He (Norway/China), Miriam Jackson (Norway), Martin Künsting (Germany), Erwin Lambert (Netherlands), Pierre-Marie Lefeuvre (Norway/France), Alexander Milner (UK), Jess Melbourne-Thomas (Australia), Benoit Meyssignac (France), Maike Nicolai (Germany), Hamish Pritchard (UK), Heidi Steltzer (USA), Nora M. Weyer (Germany)

This Summary for Policymakers should be cited as:

IPCC, 2019: Summary for Policymakers. In: *IPCC Special Report on the Ocean and Cryosphere in a Changing Climate* [H.-O. Pörtner, D.C. Roberts, V. Masson-Delmotte, P. Zhai, M. Tignor, E. Poloczanska, K. Mintenbeck, A. Alegría, M. Nicolai, A. Okem, J. Petzold, B. Rama, N.M. Weyer (eds.)]. Cambridge University Press, Cambridge, UK and New York, NY, USA, pp. 3–35. https://doi.org/10.1017/9781009157964.001.

Introduction

This Special Report on the Ocean and Cryosphere[1] in a Changing Climate (SROCC) was prepared following an IPCC Panel decision in 2016 to prepare three Special Reports during the Sixth Assessment Cycle[2]. By assessing new scientific literature[3], the SROCC[4] responds to government and observer organization proposals. The SROCC follows the other two Special Reports on Global Warming of 1.5°C (SR1.5) and on Climate Change and Land (SRCCL)[5] and the Intergovernmental Science Policy Platform on Biodiversity and Ecosystem Services (IPBES) Global Assessment Report on Biodiversity and Ecosystem Services.

This Summary for Policymakers (SPM) compiles key findings of the report and is structured in three parts: SPM.A: Observed Changes and Impacts, SPM.B: Projected Changes and Risks, and SPM.C: Implementing Responses to Ocean and Cryosphere Change. To assist navigation of the SPM, icons indicate where content can be found. Confidence in key findings is reported using IPCC calibrated language[6] and the underlying scientific basis for each key finding is indicated by references to sections of the underlying report.

Key of icons to indicate content

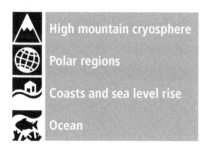

High mountain cryosphere

Polar regions

Coasts and sea level rise

Ocean

---

[1] The cryosphere is defined in this report (Annex I: Glossary) as the components of the Earth System at and below the land and ocean surface that are frozen, including snow cover, glaciers, ice sheets, ice shelves, icebergs, sea ice, lake ice, river ice, permafrost, and seasonally frozen ground.

[2] The decision to prepare a Special Report on Climate Change and Oceans and the Cryosphere was made at the Forty-Third Session of the IPCC in Nairobi, Kenya, 11–13 April 2016.

[3] Cut-off dates: 15 October 2018 for manuscript submission, 15 May 2019 for acceptance for publication.

[4] The SROCC is produced under the scientific leadership of Working Group I and Working Group II. In line with the approved outline, mitigation options (Working Group III) are not assessed with the exception of the mitigation potential of blue carbon (coastal ecosystems).

[5] The full titles of these two Special Reports are: "Global Warming of 1.5°C. An IPCC special report on the impacts of global warming of 1.5°C above pre-industrial levels and related global greenhouse gas emission pathways, in the context of strengthening the global response to the threat of climate change, sustainable development, and efforts to eradicate poverty"; "Climate Change and Land: an IPCC special report on climate change, desertification, land degradation, sustainable land management, food security, and greenhouse gas fluxes in terrestrial ecosystems".

[6] Each finding is grounded in an evaluation of underlying evidence and agreement. A level of confidence is expressed using five qualifiers: very low, low, medium, high and very high, and typeset in italics, e.g., *medium confidence*. The following terms have been used to indicate the assessed likelihood of an outcome or a result: virtually certain 99–100% probability, very likely 90–100%, likely 66–100%, about as likely as not 33–66%, unlikely 0–33%, very unlikely 0–10%, exceptionally unlikely 0–1%. Assessed likelihood is typeset in italics, e.g., *very likely*. This is consistent with AR5 and the other AR6 Special Reports. Additional terms (extremely likely 95–100%, more likely than not >50–100%, more unlikely than likely 0–<50%, extremely unlikely 0–5%) are used when appropriate. This Report also uses the term '*likely* range' or '*very likely* range' to indicate that the assessed likelihood of an outcome lies within the 17–83% or 5–95% probability range. {1.9.2, Figure 1.4}

Startup Box | The Importance of the Ocean and Cryosphere for People

All people on Earth depend directly or indirectly on the ocean and cryosphere. The global ocean covers 71% of the Earth surface and contains about 97% of the Earth's water. The cryosphere refers to frozen components of the Earth system[1]. Around 10% of Earth's land area is covered by glaciers or ice sheets. The ocean and cryosphere support unique habitats, and are interconnected with other components of the climate system through global exchange of water, energy and carbon. The projected responses of the ocean and cryosphere to past and current human-induced greenhouse gas emissions and ongoing global warming include climate feedbacks, changes over decades to millennia that cannot be avoided, thresholds of abrupt change, and irreversibility. {Box 1.1, 1.2}

Human communities in close connection with coastal environments, small islands (including Small Island Developing States, SIDS), polar areas and high mountains[7] are particularly exposed to ocean and cryosphere change, such as sea level rise, extreme sea level and shrinking cryosphere. Other communities further from the coast are also exposed to changes in the ocean, such as through extreme weather events. Today, around 4 million people live permanently in the Arctic region, of whom 10% are Indigenous. The low-lying coastal zone[8] is currently home to around 680 million people (nearly 10% of the 2010 global population), projected to reach more than one billion by 2050. SIDS are home to 65 million people. Around 670 million people (nearly 10% of the 2010 global population), including Indigenous peoples, live in high mountain regions in all continents except Antarctica. In high mountain regions, population is projected to reach between 740 and 840 million by 2050 (about 8.4–8.7% of the projected global population). {1.1, 2.1, 3.1, Cross-Chapter Box 9, Figure 2.1}

In addition to their role within the climate system, such as the uptake and redistribution of natural and anthropogenic carbon dioxide (CO_2) and heat, as well as ecosystem support, services provided to people by the ocean and/or cryosphere include food and water supply, renewable energy, and benefits for health and well-being, cultural values, tourism, trade, and transport. The state of the ocean and cryosphere interacts with each aspect of sustainability reflected in the United Nations Sustainable Development Goals (SDGs). {1.1, 1.2, 1.5}

---

[7]   High mountain areas include all mountain regions where glaciers, snow or permafrost are prominent features of the landscape. For a list of high mountain regions covered in this report, see Chapter 2. Population in high mountain regions is calculated for areas less than 100 kilometres from glaciers or permafrost in high mountain areas assessed in this report. {2.1} Projections for 2050 give the range of population in these regions across all five of the Shared Socioeconomic Pathways. {Cross-Chapter Box 1 in Chapter 1}

[8]   Population in the low elevation coastal zone is calculated for land areas connected to the coast, including small island states, that are less than 10 metres above sea level. {Cross-Chapter Box 9} Projections for 2050 give the range of population in these regions across all five of the Shared Socioeconomic Pathways. {Cross-Chapter Box 1 in Chapter 1}

A. Observed Changes and Impacts

Observed Physical Changes

A.1 **Over the last decades, global warming has led to widespread shrinking of the cryosphere, with mass loss from ice sheets and glaciers (*very high confidence*), reductions in snow cover (*high confidence*) and Arctic sea ice extent and thickness (*very high confidence*), and increased permafrost temperature (*very high confidence*). {2.2, 3.2, 3.3, 3.4, Figures SPM.1, SPM.2}**

A.1.1   Ice sheets and glaciers worldwide have lost mass (*very high confidence*). Between 2006 and 2015, the Greenland Ice Sheet[9] lost ice mass at an average rate of 278 ± 11 Gt yr^{-1} (equivalent to 0.77 ± 0.03 mm yr^{-1} of global sea level rise)[10], mostly due to surface melting (*high confidence*). In 2006–2015, the Antarctic Ice Sheet lost mass at an average rate of 155 ± 19 Gt yr^{-1} (0.43 ± 0.05 mm yr^{-1}), mostly due to rapid thinning and retreat of major outlet glaciers draining the West Antarctic Ice Sheet (*very high confidence*). Glaciers worldwide outside Greenland and Antarctica lost mass at an average rate of 220 ± 30 Gt yr^{-1} (equivalent to 0.61 ± 0.08 mm yr^{-1} sea level rise) in 2006–2015. {3.3.1, 4.2.3, Appendix 2.A, Figure SPM.1}

A.1.2   Arctic June snow cover extent on land declined by $13.4 \pm 5.4\%$ per decade from 1967 to 2018, a total loss of approximately 2.5 million km^2, predominantly due to surface air temperature increase (*high confidence*). In nearly all high mountain areas, the depth, extent and duration of snow cover have declined over recent decades, especially at lower elevation (*high confidence*). {2.2.2, 3.4.1, Figure SPM.1}

A.1.3   Permafrost temperatures have increased to record high levels (1980s–present) (*very high confidence*) including the recent increase by $0.29°C \pm 0.12°C$ from 2007 to 2016 averaged across polar and high mountain regions globally. Arctic and boreal permafrost contain 1460–1600 Gt organic carbon, almost twice the carbon in the atmosphere (*medium confidence*). There is *medium evidence* with *low agreement* whether northern permafrost regions are currently releasing additional net methane and CO_2 due to thaw. Permafrost thaw and glacier retreat have decreased the stability of high mountain slopes (*high confidence*). {2.2.4, 2.3.2, 3.4.1, 3.4.3, Figure SPM.1}

A.1.4   Between 1979 and 2018, Arctic sea ice extent has *very likely* decreased for all months of the year. September sea ice reductions are *very likely* $12.8 \pm 2.3\%$ per decade. These sea ice changes in September are *likely* unprecedented for at least 1000 years. Arctic sea ice has thinned, concurrent with a transition to younger ice: between 1979 and 2018, the areal proportion of multi-year ice at least five years old has declined by approximately 90% (*very high confidence*). Feedbacks from the loss of summer sea ice and spring snow cover on land have contributed to amplified warming in the Arctic (*high confidence*) where surface air temperature *likely* increased by more than double the global average over the last two decades. Changes in Arctic sea ice have the potential to influence mid-latitude weather (*medium confidence*), but there is *low confidence* in the detection of this influence for specific weather types. Antarctic sea ice extent overall has had no statistically significant trend (1979–2018) due to contrasting regional signals and large interannual variability (*high confidence*). {3.2.1, 6.3.1, Box 3.1, Box 3.2, SPM A.1.2, Figures SPM.1, SPM.2}

---

[9]   Including peripheral glaciers.

[10]   360 Gt ice corresponds to 1 mm of global mean sea level.

Past and future changes in the ocean and cryosphere

Historical changes (observed and modelled) and projections under RCP2.6 and RCP8.5 for key indicators

Historical (observed) | Historical (modelled) | Projected (RCP2.6) | Projected (RCP8.5)

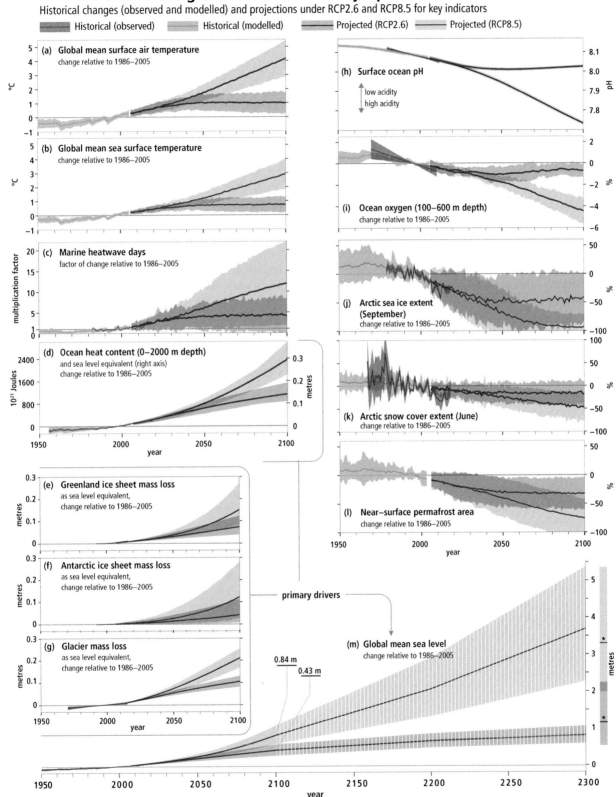

Figure SPM.1 | Observed and modelled historical changes in the ocean and cryosphere since 1950[11], and projected future changes under low (RCP2.6) and high (RCP8.5) greenhouse gas emissions scenarios. {Box SPM.1}

---

11   This does not imply that the changes started in 1950. Changes in some variables have occurred since the pre-industrial period.

SPM

Figure SPM.1 (continued): Changes are shown for: **(a)** Global mean surface air temperature change with *likely* range. {Box SPM.1, Cross-Chapter Box 1 in Chapter 1} **Ocean-related changes** with *very likely* ranges for **(b)** Global mean sea surface temperature change {Box 5.1, 5.2.2}; **(c)** Change factor in surface ocean marine heatwave days {6.4.1}; **(d)** Global ocean heat content change (0–2000 m depth). An approximate steric sea level equivalent is shown with the right axis by multiplying the ocean heat content by the global-mean thermal expansion coefficient ($\varepsilon \approx 0.125$ m per 10^{24} Joules)[12] for observed warming since 1970 {Figure 5.1}. **(h)** Global mean surface pH (on the total scale). Assessed observational trends are compiled from open ocean time series sites longer than 15 years {Box 5.1, Figure 5.6, 5.2.2}; and **(i)** Global mean ocean oxygen change (100–600 m depth). Assessed observational trends span 1970–2010 centered on 1996 {Figure 5.8, 5.2.2}. **Sea level changes** with *likely* ranges for **(m)** Global mean sea level change. Hashed shading reflects *low confidence* in sea level projections beyond 2100 and bars at 2300 reflect expert elicitation on the range of possible sea level change {4.2.3, Figure 4.2}; and components from **(e,f)** Greenland and Antarctic ice sheet mass loss {3.3.1}; and **(g)** Glacier mass loss {Cross-Chapter Box 6 in Chapter 2, Table 4.1}. Further **cryosphere-related changes** with *very likely* ranges for **(j)** Arctic sea ice extent change for September[13] {3.2.1, 3.2.2 Figure 3.3}; **(k)** Arctic snow cover change for June (land areas north of 60°N) {3.4.1, 3.4.2, Figure 3.10}; and **(l)** Change in near-surface (within 3–4 m) permafrost area in the Northern Hemisphere {3.4.1, 3.4.2, Figure 3.10}. Assessments of projected changes under the intermediate RCP4.5 and RCP6.0 scenarios are not available for all variables considered here, but where available can be found in the underlying report. {For RCP4.5 see: 2.2.2, Cross-Chapter Box 6 in Chapter 2, 3.2.2, 3.4.2, 4.2.3, for RCP6.0 see Cross-Chapter Box 1 in Chapter 1}

Box SPM.1 | Use of Climate Change Scenarios in SROCC

Assessments of projected future changes in this report are based largely on CMIP5[14] climate model projections using Representative Concentration Pathways (RCPs). RCPs are scenarios that include time series of emissions and concentrations of the full suite of greenhouse gases (GHGs) and aerosols and chemically active gases, as well as land use / land cover. RCPs provide only one set of many possible scenarios that would lead to different levels of global warming. {Annex I: Glossary}

This report uses mainly RCP2.6 and RCP8.5 in its assessment, reflecting the available literature. RCP2.6 represents a low greenhouse gas emissions, high mitigation future, that in CMIP5 simulations gives a two in three chance of limiting global warming to below 2°C by 2100[15]. By contrast, RCP8.5 is a high greenhouse gas emissions scenario in the absence of policies to combat climate change, leading to continued and sustained growth in atmospheric greenhouse gas concentrations. Compared to the total set of RCPs, RCP8.5 corresponds to the pathway with the highest greenhouse gas emissions. The underlying chapters also reference other scenarios, including RCP4.5 and RCP6.0 that have intermediate levels of greenhouse gas emissions and result in intermediate levels of warming. {Annex I: Glossary, Cross-Chapter Box 1 in Chapter 1}

Table SPM.1 provides estimates of total warming since the pre-industrial period under four different RCPs for key assessment intervals used in SROCC. The warming from the 1850–1900 period until 1986–2005 has been assessed as 0.63°C (0.57°C to 0.69°C *likely* range) using observations of near-surface air temperature over the ocean and over land.[16] Consistent with the approach in AR5, modelled future changes in global mean surface air temperature relative to 1986–2005 are added to this observed warming. {Cross-Chapter Box 1 in Chapter 1}

Table SPM.1 | Projected global mean surface temperature change relative to 1850–1900 for two time periods under four RCPs[15] {Cross-Chapter Box 1 in Chapter 1}

Scenario	Near-term: 2031–2050		End-of-century: 2081–2100	
	Mean (°C)	*Likely* range (°C)	Mean (°C)	*Likely* range (°C)
RCP2.6	1.6	1.1 to 2.0	1.6	0.9 to 2.4
RCP4.5	1.7	1.3 to 2.2	2.5	1.7 to 3.3
RCP6.0	1.6	1.2 to 2.0	2.9	2.0 to 3.8
RCP8.5	2.0	1.5 to 2.4	4.3	3.2 to 5.4

---

[12] This scaling factor (global-mean ocean expansion as sea level rise in metres per unit heat) varies by about 10% between different models, and it will systematically increase by about 10% by 2100 under RCP8.5 forcing due to ocean warming increasing the average thermal expansion coefficient. {4.2.1, 4.2.2, 5.2.2}

[13] Antarctic sea ice is not shown here due to *low confidence* in future projections. {3.2.2}

[14] CMIP5 is Phase 5 of the Coupled Model Intercomparison Project (Annex I: Glossary).

[15] A pathway with lower emissions (RCP1.9), which would correspond to a lower level of projected warming than RCP2.6, was not part of CMIP5.

[16] In some instances this report assesses changes relative to 2006–2015. The warming from the 1850–1900 period until 2006–2015 has been assessed as 0.87°C (0.75 to 0.99°C *likely* range). {Cross-Chapter Box 1 in Chapter 1}

A.2 It is *virtually certain* that the global ocean has warmed unabated since 1970 and has taken up more than 90% of the excess heat in the climate system (*high confidence*). Since 1993, the rate of ocean warming has more than doubled (*likely*). Marine heatwaves have *very likely* doubled in frequency since 1982 and are increasing in intensity (*very high confidence*). By absorbing more CO_2, the ocean has undergone increasing surface acidification (*virtually certain*). A loss of oxygen has occurred from the surface to 1000 m (*medium confidence*). {1.4, 3.2, 5.2, 6.4, 6.7, Figures SPM.1, SPM.2}

A.2.1. The ocean warming trend documented in the IPCC Fifth Assessment Report (AR5) has continued. Since 1993 the rate of ocean warming and thus heat uptake has more than doubled (*likely*) from 3.22 ± 1.61 ZJ yr^{-1} (0–700 m depth) and 0.97 ± 0.64 ZJ yr^{-1} (700–2000 m) between 1969 and 1993, to 6.28 ± 0.48 ZJ yr^{-1} (0–700 m) and 3.86 ± 2.09 ZJ yr^{-1} (700–2000 m) between 1993 and 2017[17], and is attributed to anthropogenic forcing (*very likely*). {1.4.1, 5.2.2, Table 5.1, Figure SPM.1}

A.2.2 The Southern Ocean accounted for 35–43% of the total heat gain in the upper 2000 m global ocean between 1970 and 2017 (*high confidence*). Its share increased to 45–62% between 2005 and 2017 (*high confidence*). The deep ocean below 2000 m has warmed since 1992 (*likely*), especially in the Southern Ocean. {1.4, 3.2.1, 5.2.2, Table 5.1, Figure SPM.2}

A.2.3 Globally, marine heat-related events have increased; marine heatwaves[18], defined when the daily sea surface temperature exceeds the local 99th percentile over the period 1982 to 2016, have doubled in frequency and have become longer-lasting, more intense and more extensive (*very likely*). It is *very likely* that between 84–90% of marine heatwaves that occurred between 2006 and 2015 are attributable to the anthropogenic temperature increase. {Table 6.2, 6.4, Figures SPM.1, SPM.2}

A.2.4 Density stratification[19] has increased in the upper 200 m of the ocean since 1970 (*very likely*). Observed surface ocean warming and high latitude addition of freshwater are making the surface ocean less dense relative to deeper parts of the ocean (*high confidence*) and inhibiting mixing between surface and deeper waters (*high confidence*). The mean stratification of the upper 200 m has increased by 2.3 ± 0.1% (*very likely* range) from the 1971–1990 average to the 1998–2017 average. {5.2.2}

A.2.5 The ocean has taken up between 20–30% (*very likely*) of total anthropogenic CO_2 emissions since the 1980s causing further ocean acidification. Open ocean surface pH has declined by a *very likely* range of 0.017–0.027 pH units per decade since the late 1980s[20], with the decline in surface ocean pH *very likely* to have already emerged from background natural variability for more than 95% of the ocean surface area. {3.2.1, 5.2.2, Box 5.1, Figures SPM.1, SPM.2}

---

[17] ZJ is Zettajoule and is equal to 10^{21} Joules. Warming the entire ocean by 1°C requires about 5500 ZJ; 144 ZJ would warm the top 100 m by about 1°C.

[18] A marine heatwave is a period of extreme warm near-sea surface temperature that persists for days to months and can extend up to thousands of kilometres (Annex I: Glossary).

[19] In this report density stratification is defined as the density contrast between shallower and deeper layers. Increased stratification reduces the vertical exchange of heat, salinity, oxygen, carbon, and nutrients.

[20] Based on in-situ records longer than fifteen years.

A.2.6 Datasets spanning 1970–2010 show that the open ocean has lost oxygen by a *very likely* range of 0.5–3.3% over the upper 1000 m, alongside a *likely* expansion of the volume of oxygen minimum zones by 3–8% (*medium confidence*). Oxygen loss is primarily due to increasing ocean stratification, changing ventilation and biogeochemistry (*high confidence*). {5.2.2, Figures SPM.1, SPM.2}

A.2.7 Observations, both in situ (2004–2017) and based on sea surface temperature reconstructions, indicate that the Atlantic Meridional Overturning Circulation (AMOC)[21] has weakened relative to 1850–1900 (*medium confidence*). There is insufficient data to quantify the magnitude of the weakening, or to properly attribute it to anthropogenic forcing due to the limited length of the observational record. Although attribution is currently not possible, CMIP5 model simulations of the period 1850–2015, on average, exhibit a weakening AMOC when driven by anthropogenic forcing. {6.7}

A.3 Global mean sea level (GMSL) is rising, with acceleration in recent decades due to increasing rates of ice loss from the Greenland and Antarctic ice sheets (*very high confidence*), as well as continued glacier mass loss and ocean thermal expansion. Increases in tropical cyclone winds and rainfall, and increases in extreme waves, combined with relative sea level rise, exacerbate extreme sea level events and coastal hazards (*high confidence*). {3.3, 4.2, 6.2, 6.3, 6.8, Figures SPM.1, SPM.2, SPM.4, SPM.5}

A.3.1 Total GMSL rise for 1902–2015 is 0.16 m (*likely* range 0.12–0.21 m). The rate of GMSL rise for 2006–2015 of 3.6 mm yr⁻¹ (3.1–4.1 mm yr⁻¹, *very likely* range), is unprecedented over the last century (*high confidence*), and about 2.5 times the rate for 1901–1990 of 1.4 mm yr⁻¹ (0.8– 2.0 mm yr⁻¹, *very likely* range). The sum of ice sheet and glacier contributions over the period 2006–2015 is the dominant source of sea level rise (1.8 mm yr⁻¹, *very likely* range 1.7–1.9 mm yr⁻¹), exceeding the effect of thermal expansion of ocean water (1.4 mm yr⁻¹, *very likely* range 1.1–1.7 mm yr⁻¹)[22] (*very high confidence*). The dominant cause of global mean sea level rise since 1970 is anthropogenic forcing (*high confidence*). {4.2.1, 4.2.2, Figure SPM.1}

A.3.2 Sea level rise has accelerated (*extremely likely*) due to the combined increased ice loss from the Greenland and Antarctic ice sheets (*very high confidence*). Mass loss from the Antarctic ice sheet over the period 2007–2016 tripled relative to 1997–2006. For Greenland, mass loss doubled over the same period (*likely, medium confidence*). {3.3.1, Figures SPM.1, SPM.2, SPM A.1.1}

A.3.3 Acceleration of ice flow and retreat in Antarctica, which has the potential to lead to sea level rise of several metres within a few centuries, is observed in the Amundsen Sea Embayment of West Antarctica and in Wilkes Land, East Antarctica (*very high confidence*). These changes may be the onset of an irreversible[23] ice sheet instability. Uncertainty related to the onset of ice sheet instability arises from limited observations, inadequate model representation of ice sheet processes, and limited understanding of the complex interactions between the atmosphere, ocean and the ice sheet. {3.3.1, Cross-Chapter Box 8 in Chapter 3, 4.2.3}

A.3.4 Sea level rise is not globally uniform and varies regionally. Regional differences, within ±30% of the global mean sea level rise, result from land ice loss and variations in ocean warming and circulation. Differences from the global mean can be greater in areas of rapid vertical land movement including from local human activities (e.g. extraction of groundwater). (*high confidence*) {4.2.2, 5.2.2, 6.2.2, 6.3.1, 6.8.2, Figure SPM.2}

---

[21] The Atlantic Meridional Overturning Circulation (AMOC) is the main current system in the South and North Atlantic Oceans (Annex I: Glossary).

[22] The total rate of sea level rise is greater than the sum of cryosphere and ocean contributions due to uncertainties in the estimate of landwater storage change.

[23] The recovery time scale is hundreds to thousands of years (Annex I: Glossary).

A.3.5 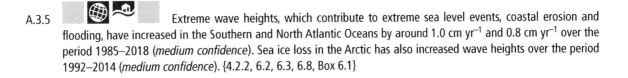 Extreme wave heights, which contribute to extreme sea level events, coastal erosion and flooding, have increased in the Southern and North Atlantic Oceans by around 1.0 cm yr^{-1} and 0.8 cm yr^{-1} over the period 1985–2018 (*medium confidence*). Sea ice loss in the Arctic has also increased wave heights over the period 1992–2014 (*medium confidence*). {4.2.2, 6.2, 6.3, 6.8, Box 6.1}

A.3.6 Anthropogenic climate change has increased observed precipitation (*medium confidence*), winds (*low confidence*), and extreme sea level events (*high confidence*) associated with some tropical cyclones, which has increased intensity of multiple extreme events and associated cascading impacts (*high confidence*). Anthropogenic climate change may have contributed to a poleward migration of maximum tropical cyclone intensity in the western North Pacific in recent decades related to anthropogenically-forced tropical expansion (*low confidence*). There is emerging evidence for an increase in annual global proportion of Category 4 or 5 tropical cyclones in recent decades (*low confidence*). {6.2, Table 6.2, 6.3, 6.8, Box 6.1}

Observed Impacts on Ecosystems

A.4   Cryospheric and associated hydrological changes have impacted terrestrial and freshwater species and ecosystems in high mountain and polar regions through the appearance of land previously covered by ice, changes in snow cover, and thawing permafrost. These changes have contributed to changing the seasonal activities, abundance and distribution of ecologically, culturally, and economically important plant and animal species, ecological disturbances, and ecosystem functioning. (*high confidence*) {2.3.2, 2.3.3, 3.4.1, 3.4.3, Box 3.4, Figure SPM.2}

A.4.1 Over the last century some species of plants and animals have increased in abundance, shifted their range, and established in new areas as glaciers receded and the snow-free season lengthened (*high confidence*). Together with warming, these changes have increased locally the number of species in high mountains, as lower-elevation species migrate upslope (*very high confidence*). Some cold-adapted or snow-dependent species have declined in abundance, increasing their risk of extinction, notably on mountain summits (*high confidence*). In polar and mountain regions, many species have altered seasonal activities especially in late winter and spring (*high confidence*). {2.3.3, Box 3.4}

A.4.2 Increased wildfire and abrupt permafrost thaw, as well as changes in Arctic and mountain hydrology have altered frequency and intensity of ecosystem disturbances (*high confidence*). This has included positive and negative impacts on vegetation and wildlife such as reindeer and salmon (*high confidence*). {2.3.3, 3.4.1, 3.4.3}

A.4.3 Across tundra, satellite observations show an overall greening, often indicative of increased plant productivity (*high confidence*). Some browning areas in tundra and boreal forest are indicative that productivity has decreased (*high confidence*). These changes have negatively affected provisioning, regulating and cultural ecosystem services, with also some transient positive impacts for provisioning services, in both high mountains (*medium confidence*) and polar regions (*high confidence*). {2.3.1, 2.3.3, 3.4.1, 3.4.3, Annex I: Glossary}

A.5 **Since about 1950 many marine species across various groups have undergone shifts in geographical range and seasonal activities in response to ocean warming, sea ice change and biogeochemical changes, such as oxygen loss, to their habitats (*high confidence*). This has resulted in shifts in species composition, abundance and biomass production of ecosystems, from the equator to the poles. Altered interactions between species have caused cascading impacts on ecosystem structure and functioning (*medium confidence*). In some marine ecosystems species are impacted by both the effects of fishing and climate changes (*medium confidence*). {3.2.3, 3.2.4, Box 3.4, 5.2.3, 5.3, 5.4.1, Figure SPM.2}**

A.5.1 Rates of poleward shifts in distributions across different marine species since the 1950s are 52 ± 33 km per decade and 29 ± 16 km per decade (*very likely* ranges) for organisms in the epipelagic (upper 200 m from sea surface) and seafloor ecosystems, respectively. The rate and direction of observed shifts in distributions are shaped by local temperature, oxygen, and ocean currents across depth, latitudinal and longitudinal gradients (*high confidence*). Warming-induced species range expansions have led to altered ecosystem structure and functioning such as in the North Atlantic, Northeast Pacific and Arctic (*medium confidence*). {5.2.3, 5.3.2, 5.3.6, Box 3.4, Figure SPM.2}

A.5.2 In recent decades, Arctic net primary production has increased in ice-free waters (*high confidence*) and spring phytoplankton blooms are occurring earlier in the year in response to sea ice change and nutrient availability with spatially variable positive and negative consequences for marine ecosystems (*medium confidence*). In the Antarctic, such changes are spatially heterogeneous and have been associated with rapid local environmental change, including retreating glaciers and sea ice change (*medium confidence*). Changes in the seasonal activities, production and distribution of some Arctic zooplankton and a southward shift in the distribution of the Antarctic krill population in the South Atlantic are associated with climate-linked environmental changes (*medium confidence*). In polar regions, ice associated marine mammals and seabirds have experienced habitat contraction linked to sea ice changes (*high confidence*) and impacts on foraging success due to climate impacts on prey distributions (*medium confidence*). Cascading effects of multiple climate-related drivers on polar zooplankton have affected food web structure and function, biodiversity as well as fisheries (*high confidence*). {3.2.3, 3.2.4, Box 3.4, 5.2.3, Figure SPM.2}

A.5.3 Eastern Boundary Upwelling Systems (EBUS) are amongst the most productive ocean ecosystems. Increasing ocean acidification and oxygen loss are negatively impacting two of the four major upwelling systems: the California Current and Humboldt Current (*high confidence*). Ocean acidification and decrease in oxygen level in the California Current upwelling system have altered ecosystem structure, with direct negative impacts on biomass production and species composition (*medium confidence*). {Box 5.3, Figure SPM.2}

A.5.4 Ocean warming in the 20th century and beyond has contributed to an overall decrease in maximum catch potential (*medium confidence*), compounding the impacts from overfishing for some fish stocks (*high confidence*). In many regions, declines in the abundance of fish and shellfish stocks due to direct and indirect effects of global warming and biogeochemical changes have already contributed to reduced fisheries catches (*high confidence*). In some areas, changing ocean conditions have contributed to the expansion of suitable habitat and/or increases in the abundance of some species (*high confidence*). These changes have been accompanied by changes in species composition of fisheries catches since the 1970s in many ecosystems (*medium confidence*). {3.2.3, 5.4.1, Figure SPM.2}

A.6 **Coastal ecosystems are affected by ocean warming, including intensified marine heatwaves, acidification, loss of oxygen, salinity intrusion and sea level rise, in combination with adverse effects from human activities on ocean and land (*high confidence*). Impacts are already observed on habitat area and biodiversity, as well as ecosystem functioning and services (*high confidence*). {4.3.2, 4.3.3, 5.3, 5.4.1, 6.4.2, Figure SPM.2}**

A.6.1   Vegetated coastal ecosystems protect the coastline from storms and erosion and help buffer the impacts of sea level rise. Nearly 50% of coastal wetlands have been lost over the last 100 years, as a result of the combined effects of localised human pressures, sea level rise, warming and extreme climate events (*high confidence*). Vegetated coastal ecosystems are important carbon stores; their loss is responsible for the current release of 0.04–1.46 GtC yr^{-1} (*medium confidence*). In response to warming, distribution ranges of seagrass meadows and kelp forests are expanding at high latitudes and contracting at low latitudes since the late 1970s (*high confidence*), and in some areas episodic losses occur following heatwaves (*medium confidence*). Large-scale mangrove mortality that is related to warming since the 1960s has been partially offset by their encroachment into subtropical saltmarshes as a result of increase in temperature, causing the loss of open areas with herbaceous plants that provide food and habitat for dependent fauna (*high confidence*). {4.3.3, 5.3.2, 5.3.6, 5.4.1, 5.5.1, Figure SPM.2}

A.6.2   Increased sea water intrusion in estuaries due to sea level rise has driven upstream redistribution of marine species (*medium confidence*) and caused a reduction of suitable habitats for estuarine communities (*medium confidence*). Increased nutrient and organic matter loads in estuaries since the 1970s from intensive human development and riverine loads have exacerbated the stimulating effects of ocean warming on bacterial respiration, leading to expansion of low oxygen areas (*high confidence*). {5.3.1}

A.6.3   The impacts of sea level rise on coastal ecosystems include habitat contraction, geographical shift of associated species, and loss of biodiversity and ecosystem functionality. Impacts are exacerbated by direct human disturbances, and where anthropogenic barriers prevent landward shift of marshes and mangroves (termed coastal squeeze) (*high confidence*). Depending on local geomorphology and sediment supply, marshes and mangroves can grow vertically at rates equal to or greater than current mean sea level rise (*high confidence*). {4.3.2, 4.3.3, 5.3.2, 5.3.7, 5.4.1}

A.6.4   Warm-water coral reefs and rocky shores dominated by immobile, calcifying (e.g., shell and skeleton producing) organisms such as corals, barnacles and mussels, are currently impacted by extreme temperatures and ocean acidification (*high confidence*). Marine heatwaves have already resulted in large-scale coral bleaching events at increasing frequency (*very high confidence*) causing worldwide reef degradation since 1997, and recovery is slow (more than 15 years) if it occurs (*high confidence*). Prolonged periods of high environmental temperature and dehydration of the organisms pose high risk to rocky shore ecosystems (*high confidence*). {SR1.5; 5.3.4, 5.3.5, 6.4.2, Figure SPM.2}

Observed regional impacts from changes in the ocean and the cryosphere

¹ Eastern Boundary Upwelling Systems (Benguela Current, Canary Current, California Current, and Humboldt Current); {Box 5.3}

² including Hindu Kush, Karakoram, Hengduan Shan, and Tien Shan; ³ tropical Andes, Mexico, eastern Africa, and Indonesia; ⁴ includes Finland, Norway, and Sweden; ⁵ includes adjacent areas in Yukon Territory and British Columbia, Canada; ⁶ Migration refers to an increase or decrease in net migration, not to beneficial/adverse value.

Figure SPM.2 | Synthesis of observed regional hazards and impacts in ocean[24] (top) and high mountain and polar land regions (bottom) assessed in SROCC. For each region, physical changes, impacts on key ecosystems, and impacts on human systems and ecosystem function and services are shown. For physical changes, yellow/green refers to an increase/decrease, respectively, in amount or frequency of the measured variable. For impacts on ecosystems, human systems and ecosystems services blue or red depicts whether an observed impact is positive (beneficial) or negative (adverse), respectively, to the given system or service. Cells assigned 'increase and decrease' indicate that within that region, both increase and decrease of physical changes are found, but are not necessarily equal; the same holds for cells showing 'positive and negative' attributable impacts. For ocean regions, the confidence level refers to the confidence in attributing observed changes to changes in greenhouse gas forcing for physical changes and to climate change for ecosystem, human systems, and ecosystem services. For high mountain and polar land regions, the level of confidence in attributing physical changes and impacts at least partly to a change in the cryosphere is shown. No assessment means: not applicable, not assessed at regional scale, or the evidence is insufficient for assessment. The physical changes in the ocean are defined as: Temperature change in 0–700 m layer of the ocean except for Southern Ocean (0–2000 m) and Arctic Ocean (upper mixed layer and major inflowing branches); Oxygen in the 0–1200 m layer or oxygen minimum layer; Ocean pH as surface pH (decreasing pH corresponds to increasing ocean acidification). Ecosystems in the ocean: Coral refers to warm-water coral reefs and cold-water corals. The 'upper water column' category refers to epipelagic zone for all ocean regions except Polar Regions, where the impacts on some pelagic organisms in open water deeper than the upper 200 m were included. Coastal wetland includes salt marshes, mangroves and seagrasses. Kelp forests are habitats of a specific group of macroalgae. Rocky shores are coastal habitats dominated by immobile calcified organisms such as mussels and barnacles. Deep sea is seafloor ecosystems that are 3000–6000 m deep. Sea-ice associated includes ecosystems in, on and below sea ice. Habitat services refer to supporting structures and services (e.g., habitat, biodiversity, primary production). Coastal Carbon Sequestration refers to the uptake and storage of carbon by coastal blue carbon ecosystems. Ecosystems on Land: Tundra refers to tundra and alpine meadows, and includes terrestrial Antarctic ecosystems.

---

24 Marginal seas are not assessed individually as ocean regions in this report.

Figure SPM.2 (continued): Migration refers to an increase or decrease in net migration, not to beneficial/adverse value. Impacts on tourism refer to the operating conditions for the tourism sector. Cultural services include cultural identity, sense of home, and spiritual, intrinsic and aesthetic values, as well as contributions from glacier archaeology. The underlying information is given for land regions in tables SM2.6, SM2.7, SM2.8, SM3.8, SM3.9, and SM3.10, and for ocean regions in tables SM5.10, SM5.11, SM3.8, SM3.9, and SM3.10. {2.3.1, 2.3.2, 2.3.3, 2.3.4, 2.3.5, 2.3.6, 2.3.7, Figure 2.1, 3.2.1, 3.2.3, 3.2.4, 3.3.3, 3.4.1, 3.4.3, 3.5.2, Box 3.4, 4.2.2, 5.2.2, 5.2.3, 5.3.3, 5.4, 5.6, Figure 5.24, Box 5.3}

Observed Impacts on People and Ecosystem Services

A.7   **Since the mid-20th century, the shrinking cryosphere in the Arctic and high mountain areas has led to predominantly negative impacts on food security, water resources, water quality, livelihoods, health and well-being, infrastructure, transportation, tourism and recreation, as well as culture of human societies, particularly for Indigenous peoples (*high confidence*). Costs and benefits have been unequally distributed across populations and regions. Adaptation efforts have benefited from the inclusion of Indigenous knowledge and local knowledge (*high confidence*). {1.1, 1.5, 1.6.2, 2.3, 2.4, 3.4, 3.5, Figure SPM.2}**

A.7.1   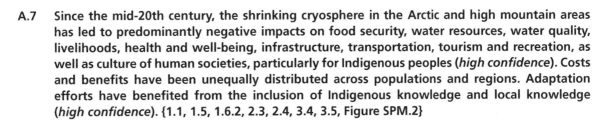 Food and water security have been negatively impacted by changes in snow cover, lake and river ice, and permafrost in many Arctic regions (*high confidence*). These changes have disrupted access to, and food availability within, herding, hunting, fishing, and gathering areas, harming the livelihoods and cultural identity of Arctic residents including Indigenous populations (*high confidence*). Glacier retreat and snow cover changes have contributed to localized declines in agricultural yields in some high mountain regions, including Hindu Kush Himalaya and the tropical Andes (*medium confidence*). {2.3.1, 2.3.7, Box 2.4, 3.4.1, 3.4.2, 3.4.3, 3.5.2, Figure SPM.2}

A.7.2   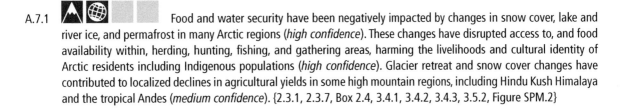 In the Arctic, negative impacts of cryosphere change on human health have included increased risk of food- and waterborne diseases, malnutrition, injury, and mental health challenges especially among Indigenous peoples (*high confidence*). In some high mountain areas, water quality has been affected by contaminants, particularly mercury, released from melting glaciers and thawing permafrost (*medium confidence*). Health-related adaptation efforts in the Arctic range from local to international in scale, and successes have been underpinned by Indigenous knowledge (*high confidence*). {1.8, Cross-Chapter Box 4 in Chapter 1, 2.3.1, 3.4.3}

A.7.3   Arctic residents, especially Indigenous peoples, have adjusted the timing of activities to respond to changes in seasonality and safety of land, ice, and snow travel conditions. Municipalities and industry are beginning to address infrastructure failures associated with flooding and thawing permafrost and some coastal communities have planned for relocation (*high confidence*). Limited funding, skills, capacity, and institutional support to engage meaningfully in planning processes have challenged adaptation (*high confidence*). {3.5.2, 3.5.4, Cross-Chapter Box 9}

A.7.4   Summertime Arctic ship-based transportation (including tourism) increased over the past two decades concurrent with sea ice reductions (*high confidence*). This has implications for global trade and economies linked to traditional shipping corridors, and poses risks to Arctic marine ecosystems and coastal communities (*high confidence*), such as from invasive species and local pollution. {3.2.1, 3.2.4, 3.5.4, 5.4.2, Figure SPM.2}

A.7.5   In past decades, exposure of people and infrastructure to natural hazards has increased due to growing population, tourism and socioeconomic development (*high confidence*). Some disasters have been linked to changes in the cryosphere, for example in the Andes, high mountain Asia, Caucasus and European Alps (*medium confidence*). {2.3.2, Figure SPM.2}

A.7.6   Changes in snow and glaciers have changed the amount and seasonality of runoff and water resources in snow dominated and glacier-fed river basins (*very high confidence*). Hydropower facilities have experienced changes in seasonality and both increases and decreases in water input from high mountain areas, for example, in central Europe, Iceland, Western USA/Canada, and tropical Andes (*medium confidence*). However, there is only *limited evidence* of resulting impacts on operations and energy production. {SPM B.1.4, 2.3.1}

A.7.7   High mountain aesthetic and cultural aspects have been negatively impacted by glacier and snow cover decline (e.g. in the Himalaya, East Africa, the tropical Andes) (*medium confidence*). Tourism and recreation, including ski and glacier tourism, hiking, and mountaineering, have also been negatively impacted in many mountain regions (*medium confidence*). In some places, artificial snowmaking has reduced negative impacts on ski tourism (*medium confidence*). {2.3.5, 2.3.6, Figure SPM.2}

A.8   **Changes in the ocean have impacted marine ecosystems and ecosystem services with regionally diverse outcomes, challenging their governance (*high confidence*). Both positive and negative impacts result for food security through fisheries (*medium confidence*), local cultures and livelihoods (*medium confidence*), and tourism and recreation (*medium confidence*). The impacts on ecosystem services have negative consequences for health and well-being (*medium confidence*), and for Indigenous peoples and local communities dependent on fisheries (*high confidence*). {1.1, 1.5, 3.2.1, 5.4.1, 5.4.2, Figure SPM.2}**

A.8.1   Warming-induced changes in the spatial distribution and abundance of some fish and shellfish stocks have had positive and negative impacts on catches, economic benefits, livelihoods, and local culture (*high confidence*). There are negative consequences for Indigenous peoples and local communities that are dependent on fisheries (*high confidence*). Shifts in species distributions and abundance has challenged international and national ocean and fisheries governance, including in the Arctic, North Atlantic and Pacific, in terms of regulating fishing to secure ecosystem integrity and sharing of resources between fishing entities (*high confidence*). {3.2.4, 3.5.3, 5.4.2, 5.5.2, Figure SPM.2}

A.8.2   Harmful algal blooms display range expansion and increased frequency in coastal areas since the 1980s in response to both climatic and non-climatic drivers such as increased riverine nutrients run-off (*high confidence*). The observed trends in harmful algal blooms are attributed partly to the effects of ocean warming, marine heatwaves, oxygen loss, eutrophication and pollution (*high confidence*). Harmful algal blooms have had negative impacts on food security, tourism, local economy, and human health (*high confidence*). The human communities who are more vulnerable to these biological hazards are those in areas without sustained monitoring programs and dedicated early warning systems for harmful algal blooms (*medium confidence*). {Box 5.4, 5.4.2, 6.4.2}

A.9   **Coastal communities are exposed to multiple climate-related hazards, including tropical cyclones, extreme sea levels and flooding, marine heatwaves, sea ice loss, and permafrost thaw (*high confidence*). A diversity of responses has been implemented worldwide, mostly after extreme events, but also some in anticipation of future sea level rise, e.g., in the case of large infrastructure. {3.2.4, 3.4.3, 4.3.2, 4.3.3, 4.3.4, 4.4.2, 5.4.2, 6.2, 6.4.2, 6.8, Box 6.1, Cross Chapter Box 9, Figure SPM.5}**

A.9.1   Attribution of current coastal impacts on people to sea level rise remains difficult in most locations since impacts were exacerbated by human-induced non-climatic drivers, such as land subsidence (e.g., groundwater extraction), pollution, habitat degradation, reef and sand mining (*high confidence*). {4.3.2, 4.3.3}

A.9.2   Coastal protection through hard measures, such as dikes, seawalls, and surge barriers, is widespread in many coastal cities and deltas. Ecosystem-based and hybrid approaches combining ecosystems and built infrastructure are becoming more popular worldwide. Coastal advance, which refers to the creation of new land by building seawards (e.g., land reclamation), has a long history in most areas where there are dense coastal populations and a shortage of land. Coastal retreat, which refers to the removal of human occupation of coastal areas, is also observed, but is generally restricted to small human communities or occurs to create coastal wetland habitat. The effectiveness of the responses to sea level rise are assessed in Figure SPM.5. {3.5.3, 4.3.3, 4.4.2, 6.3.3, 6.9.1, Cross-Chapter Box 9}

B. Projected Changes and Risks

Projected Physical Changes[25]

B.1   **Global-scale glacier mass loss, permafrost thaw, and decline in snow cover and Arctic sea ice extent are projected to continue in the near-term (2031–2050) due to surface air temperature increases (*high confidence*), with unavoidable consequences for river runoff and local hazards (*high confidence*). The Greenland and Antarctic Ice Sheets are projected to lose mass at an increasing rate throughout the 21st century and beyond (*high confidence*). The rates and magnitudes of these cryospheric changes are projected to increase further in the second half of the 21st century in a high greenhouse gas emissions scenario (*high confidence*). Strong reductions in greenhouse gas emissions in the coming decades are projected to reduce further changes after 2050 (*high confidence*). {2.2, 2.3, Cross-Chapter Box 6 in Chapter 2, 3.3, 3.4, Figure SPM.1, SPM Box SPM.1}**

B.1.1   Projected glacier mass reductions between 2015 and 2100 (excluding the ice sheets) range from 18 ± 7% (*likely* range) for RCP2.6 to 36 ± 11% (*likely* range) for RCP8.5, corresponding to a sea level contribution of 94 ± 25 mm (*likely* range) sea level equivalent for RCP2.6, and 200 ± 44 mm (*likely* range) for RCP8.5 (*medium confidence*). Regions with mostly smaller glaciers (e.g., Central Europe, Caucasus, North Asia, Scandinavia, tropical Andes, Mexico, eastern Africa and Indonesia), are projected to lose more than 80% of their current ice mass by 2100 under RCP8.5 (*medium confidence*), and many glaciers are projected to disappear regardless of future emissions (*very high confidence*). {Cross-Chapter Box 6 in Chapter 2, Figure SPM.1}

B.1.2   In 2100, the Greenland Ice Sheet's projected contribution to GMSL rise is 0.07 m (0.04–0.12 m, *likely* range) under RCP2.6, and 0.15 m (0.08–0.27 m, *likely* range) under RCP8.5. In 2100, the Antarctic Ice Sheet is projected to contribute 0.04 m (0.01–0.11 m, *likely* range) under RCP2.6, and 0.12 m (0.03–0.28 m, *likely* range) under RCP8.5. The Greenland Ice Sheet is currently contributing more to sea level rise than the Antarctic Ice Sheet (*high confidence*), but Antarctica could become a larger contributor by the end of the 21st century as a consequence of rapid retreat (*low confidence*). Beyond 2100, increasing divergence between Greenland and Antarctica's relative contributions to GMSL rise under RCP8.5 has important consequences for the pace of relative sea level rise in the Northern Hemisphere. {3.3.1, 4.2.3, 4.2.5, 4.3.3, Cross-Chapter Box 8 in Chapter 3, Figure SPM.1}

B.1.3   Arctic autumn and spring snow cover are projected to decrease by 5–10%, relative to 1986–2005, in the near-term (2031–2050), followed by no further losses under RCP2.6, but an additional 15–25% loss by the end of century under RCP8.5 (*high confidence*). In high mountain areas, projected decreases in low elevation mean winter snow depth, compared to 1986–2005, are *likely* 10–40% by 2031–2050, regardless of emissions scenario (*high confidence*). For 2081–2100, this projected decrease is *likely* 10–40% for RCP2.6 and 50–90% for RCP8.5. {2.2.2, 3.3.2, 3.4.2, Figure SPM.1}

---

[25]   This report primarily uses RCP2.6 and RCP8.5 for the following reasons: These scenarios largely represent the assessed range for the topics covered in this report; they largely represent what is covered in the assessed literature, based on CMIP5; and they allow a consistent narrative about projected changes. RCP4.5 and RCP6.0 are not available for all topics addressed in the report. {Box SPM.1}

B.1.4 Widespread permafrost thaw is projected for this century (*very high confidence*) and beyond. By 2100, projected near-surface (within 3–4 m) permafrost area shows a decrease of 24 ± 16% (*likely* range) for RCP2.6 and 69 ± 20% (*likely* range) for RCP8.5. The RCP8.5 scenario leads to the cumulative release of tens to hundreds of billions of tons (GtC) of permafrost carbon as CO_2[26] and methane to the atmosphere by 2100 with the potential to exacerbate climate change (*medium confidence*). Lower emissions scenarios dampen the response of carbon emissions from the permafrost region (*high confidence*). Methane contributes a small fraction of the total additional carbon release but is significant because of its higher warming potential. Increased plant growth is projected to replenish soil carbon in part, but will not match carbon releases over the long term (*medium confidence*). {2.2.4, 3.4.2, 3.4.3, Figure SPM.1, Cross-Chapter Box 5 in Chapter 1}

B.1.5 In many high mountain areas, glacier retreat and permafrost thaw are projected to further decrease the stability of slopes, and the number and area of glacier lakes will continue to increase (*high confidence*). Floods due to glacier lake outburst or rain-on-snow, landslides and snow avalanches, are projected to occur also in new locations or different seasons (*high confidence*). {2.3.2}

B.1.6 River runoff in snow-dominated or glacier-fed high mountain basins is projected to change regardless of emissions scenario (*very high confidence*), with increases in average winter runoff (*high confidence*) and earlier spring peaks (*very high confidence*). In all emissions scenarios, average annual and summer runoff from glaciers are projected to peak at or before the end of the 21st century (*high confidence*), e.g., around mid-century in High Mountain Asia, followed by a decline in glacier runoff. In regions with little glacier cover (e.g., tropical Andes, European Alps) most glaciers have already passed this peak (*high confidence*). Projected declines in glacier runoff by 2100 (RCP8.5) can reduce basin runoff by 10% or more in at least one month of the melt season in several large river basins, especially in High Mountain Asia during the dry season (*low confidence*). {2.3.1}

B.1.7 Arctic sea ice loss is projected to continue through mid-century, with differences thereafter depending on the magnitude of global warming: for stabilised global warming of 1.5°C the annual probability of a sea ice-free September by the end of century is approximately 1%, which rises to 10–35% for stabilised global warming of 2°C (*high confidence*). There is *low confidence* in projections for Antarctic sea ice. {3.2.2, Figure SPM.1}

B.2 Over the 21st century, the ocean is projected to transition to unprecedented conditions with increased temperatures (*virtually certain*), greater upper ocean stratification (*very likely*), further acidification (*virtually certain*), oxygen decline (*medium confidence*), and altered net primary production (*low confidence*). Marine heatwaves (*very high confidence*) and extreme El Niño and La Niña events (*medium confidence*) are projected to become more frequent. The Atlantic Meridional Overturning Circulation (AMOC) is projected to weaken (*very likely*). The rates and magnitudes of these changes will be smaller under scenarios with low greenhouse gas emissions (*very likely*). {3.2, 5.2, 6.4, 6.5, 6.7, Box 5.1, Figures SPM.1, SPM.3}

B.2.1 The ocean will continue to warm throughout the 21st century (*virtually certain*). By 2100, the top 2000 m of the ocean are projected to take up 5–7 times more heat under RCP8.5 (or 2–4 times more under RCP2.6) than the observed accumulated ocean heat uptake since 1970 (*very likely*). The annual mean density stratification[19] of the top 200 m, averaged between 60°S and 60°N, is projected to increase by 12–30% for RCP8.5 and 1–9% for RCP2.6, for 2081–2100 relative to 1986–2005 (*very likely*), inhibiting vertical nutrient, carbon and oxygen fluxes. {5.2.2, Figure SPM.1}

---

[26] For context, total annual anthropogenic CO_2 emissions were 10.8 ± 0.8 GtC yr^{-1} (39.6 ± 2.9 GtCO$_2$ yr^{-1}) on average over the period 2008–2017. Total annual anthropogenic methane emissions were 0.35 ± 0.01 GtCH$_4$ yr^{-1}, on average over the period 2003–2012. {5.5.1}

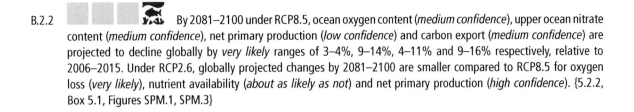

B.2.2 By 2081–2100 under RCP8.5, ocean oxygen content (*medium confidence*), upper ocean nitrate content (*medium confidence*), net primary production (*low confidence*) and carbon export (*medium confidence*) are projected to decline globally by *very likely* ranges of 3–4%, 9–14%, 4–11% and 9–16% respectively, relative to 2006–2015. Under RCP2.6, globally projected changes by 2081–2100 are smaller compared to RCP8.5 for oxygen loss (*very likely*), nutrient availability (*about as likely as not*) and net primary production (*high confidence*). {5.2.2, Box 5.1, Figures SPM.1, SPM.3}

B.2.3 Continued carbon uptake by the ocean by 2100 is *virtually certain* to exacerbate ocean acidification. Open ocean surface pH is projected to decrease by around 0.3 pH units by 2081–2100, relative to 2006–2015, under RCP8.5 (*virtually certain*). For RCP8.5, there are elevated risks for keystone aragonite shell-forming species due to crossing an aragonite stability threshold year-round in the Polar and sub-Polar Oceans by 2081–2100 (*very likely*). For RCP2.6, these conditions will be avoided this century (*very likely*), but some eastern boundary upwelling systems are projected to remain vulnerable (*high confidence*). {3.2.3, 5.2.2, Box 5.1, Box 5.3, Figure SPM.1}

B.2.4 Climate conditions, unprecedented since the preindustrial period, are developing in the ocean, elevating risks for open ocean ecosystems. Surface acidification and warming have already emerged in the historical period (*very likely*). Oxygen loss between 100 and 600 m depth is projected to emerge over 59–80% of the ocean area by 2031–2050 under RCP8.5 (*very likely*). The projected time of emergence for five primary drivers of marine ecosystem change (surface warming and acidification, oxygen loss, nitrate content and net primary production change) are all prior to 2100 for over 60% of the ocean area under RCP8.5 and over 30% under RCP2.6 (*very likely*). {Annex I: Glossary, Box 5.1, Box 5.1 Figure 1}

B.2.5 Marine heatwaves are projected to further increase in frequency, duration, spatial extent and intensity (maximum temperature) (*very high confidence*). Climate models project increases in the frequency of marine heatwaves by 2081–2100, relative to 1850–1900, by approximately 50 times under RCP8.5 and 20 times under RCP2.6 (*medium confidence*). The largest increases in frequency are projected for the Arctic and the tropical oceans (*medium confidence*). The intensity of marine heatwaves is projected to increase about 10-fold under RCP8.5 by 2081–2100, relative to 1850–1900 (*medium confidence*). {6.4, Figure SPM.1}

B.2.6 Extreme El Niño and La Niña events are projected to *likely* increase in frequency in the 21st century and to *likely* intensify existing hazards, with drier or wetter responses in several regions across the globe. Extreme El Niño events are projected to occur about as twice as often under both RCP2.6 and RCP8.5 in the 21st century when compared to the 20th century (*medium confidence*). Projections indicate that extreme Indian Ocean Dipole events also increase in frequency (*low confidence*). {6.5, Figures 6.5, 6.6}

B.2.7 The AMOC is projected to weaken in the 21st century under all RCPs (*very likely*), although a collapse is *very unlikely* (*medium confidence*). Based on CMIP5 projections, by 2300, an AMOC collapse is *about as likely as not* for high emissions scenarios and *very unlikely* for lower ones (*medium confidence*). Any substantial weakening of the AMOC is projected to cause a decrease in marine productivity in the North Atlantic (*medium confidence*), more storms in Northern Europe (*medium confidence*), less Sahelian summer rainfall (*high confidence*) and South Asian summer rainfall (*medium confidence*), a reduced number of tropical cyclones in the Atlantic (*medium confidence*), and an increase in regional sea level along the northeast coast of North America (*medium confidence*). Such changes would be in addition to the global warming signal. {6.7, Figures 6.8–6.10}

B.3 Sea level continues to rise at an increasing rate. Extreme sea level events that are historically rare (once per century in the recent past) are projected to occur frequently (at least once per year) at many locations by 2050 in all RCP scenarios, especially in tropical regions (*high confidence*). The increasing frequency of high water levels can have severe impacts in many locations depending on exposure (*high confidence*). Sea level rise is projected to continue beyond 2100 in all RCP scenarios. For a high emissions scenario (RCP8.5), projections of global sea level rise by 2100 are greater than in AR5 due to a larger contribution from the Antarctic Ice Sheet (*medium confidence*). In coming centuries under RCP8.5, sea-level rise is projected to exceed rates of several centimetres per year resulting in multi-metre rise (*medium confidence*), while for RCP2.6 sea level rise is projected to be limited to around 1 m in 2300 (*low confidence*). Extreme sea levels and coastal hazards will be exacerbated by projected increases in tropical cyclone intensity and precipitation (*high confidence*). Projected changes in waves and tides vary locally in whether they amplify or ameliorate these hazards (*medium confidence*). {Cross-Chapter Box 5 in Chapter 1, Cross-Chapter Box 8 in Chapter 3, 4.1, 4.2, 5.2.2, 6.3.1, Figures SPM.1, SPM.4, SPM.5}

B.3.1 The global mean sea level (GMSL) rise under RCP2.6 is projected to be 0.39 m (0.26–0.53 m, *likely* range) for the period 2081–2100, and 0.43 m (0.29–0.59 m, *likely* range) in 2100 with respect to 1986–2005. For RCP8.5, the corresponding GMSL rise is 0.71 m (0.51–0.92 m, *likely* range) for 2081–2100 and 0.84 m (0.61–1.10 m, *likely* range) in 2100. Mean sea level rise projections are higher by 0.1 m compared to AR5 under RCP8.5 in 2100, and the *likely* range extends beyond 1 m in 2100 due to a larger projected ice loss from the Antarctic Ice Sheet (*medium confidence*). The uncertainty at the end of the century is mainly determined by the ice sheets, especially in Antarctica. {4.2.3, Figures SPM.1, SPM.5}

B.3.2 Sea level projections show regional differences around GMSL. Processes not driven by recent climate change, such as local subsidence caused by natural processes and human activities, are important to relative sea level changes at the coast (*high confidence*). While the relative importance of climate-driven sea level rise is projected to increase over time, local processes need to be considered for projections and impacts of sea level (*high confidence*). {SPM A.3.4, 4.2.1, 4.2.2, Figure SPM.5}

B.3.3 The rate of global mean sea level rise is projected to reach 15 mm yr^{-1} (10–20 mm yr^{-1}, *likely* range) under RCP8.5 in 2100, and to exceed several centimetres per year in the 22nd century. Under RCP2.6, the rate is projected to reach 4 mm yr^{-1} (2–6 mm yr^{-1}, *likely* range) in 2100. Model studies indicate multi-meter rise in sea level by 2300 (2.3–5.4 m for RCP8.5 and 0.6–1.07 m under RCP2.6) (*low confidence*), indicating the importance of reduced emissions for limiting sea level rise. Processes controlling the timing of future ice-shelf loss and the extent of ice sheet instabilities could increase Antarctica's contribution to sea level rise to values substantially higher than the *likely* range on century and longer time-scales (*low confidence*). Considering the consequences of sea level rise that a collapse of parts of the Antarctic Ice Sheet entails, this high impact risk merits attention. {Cross-Chapter Box 5 in Chapter 1, Cross-Chapter Box 8 in Chapter 3, 4.1, 4.2.3}

B.3.4 Global mean sea level rise will cause the frequency of extreme sea level events at most locations to increase. Local sea levels that historically occurred once per century (historical centennial events) are projected to occur at least annually at most locations by 2100 under all RCP scenarios (*high confidence*). Many low-lying megacities and small islands (including SIDS) are projected to experience historical centennial events at least annually by 2050 under RCP2.6, RCP4.5 and RCP8.5. The year when the historical centennial event becomes an annual event in the mid-latitudes occurs soonest in RCP8.5, next in RCP4.5 and latest in RCP2.6. The increasing frequency of high water levels can have severe impacts in many locations depending on the level of exposure (*high confidence*). {4.2.3, 6.3, Figures SPM.4, SPM.5}

B.3.5   Significant wave heights (the average height from trough to crest of the highest one-third of waves) are projected to increase across the Southern Ocean and tropical eastern Pacific (*high confidence*) and Baltic Sea (*medium confidence*) and decrease over the North Atlantic and Mediterranean Sea under RCP8.5 (*high confidence*). Coastal tidal amplitudes and patterns are projected to change due to sea level rise and coastal adaptation measures (*very likely*). Projected changes in waves arising from changes in weather patterns, and changes in tides due to sea level rise, can locally enhance or ameliorate coastal hazards (*medium confidence*). {6.3.1, 5.2.2}

B.3.6   The average intensity of tropical cyclones, the proportion of Category 4 and 5 tropical cyclones and the associated average precipitation rates are projected to increase for a 2°C global temperature rise above any baseline period (*medium confidence*). Rising mean sea levels will contribute to higher extreme sea levels associated with tropical cyclones (*very high confidence*). Coastal hazards will be exacerbated by an increase in the average intensity, magnitude of storm surge and precipitation rates of tropical cyclones. There are greater increases projected under RCP8.5 than under RCP2.6 from around mid-century to 2100 (*medium confidence*). There is *low confidence* in changes in the future frequency of tropical cyclones at the global scale. {6.3.1}

Projected Risks for Ecosystems

B.4   **Future land cryosphere changes will continue to alter terrestrial and freshwater ecosystems in high mountain and polar regions with major shifts in species distributions resulting in changes in ecosystem structure and functioning, and eventual loss of globally unique biodiversity (*medium confidence*). Wildfire is projected to increase significantly for the rest of this century across most tundra and boreal regions, and also in some mountain regions (*medium confidence*). {2.3.3, Box 3.4, 3.4.3}**

B.4.1   In high mountain regions, further upslope migration by lower-elevation species, range contractions, and increased mortality will lead to population declines of many alpine species, especially glacier- or snow-dependent species (*high confidence*), with local and eventual global species loss (*medium confidence*). The persistence of alpine species and sustaining ecosystem services depends on appropriate conservation and adaptation measures (*high confidence*). {2.3.3}

B.4.2   On Arctic land, a loss of globally unique biodiversity is projected as limited refugia exist for some High-Arctic species and hence they are outcompeted by more temperate species (*medium confidence*). Woody shrubs and trees are projected to expand to cover 24–52% of Arctic tundra by 2050 (*medium confidence*). The boreal forest is projected to expand at its northern edge, while diminishing at its southern edge where it is replaced by lower biomass woodland/shrublands *(medium confidence)*. {3.4.3, Box 3.4}

B.4.3   Permafrost thaw and decrease in snow will affect Arctic and mountain hydrology and wildfire, with impacts on vegetation and wildlife *(medium confidence)*. About 20% of Arctic land permafrost is vulnerable to abrupt permafrost thaw and ground subsidence, which is projected to increase small lake area by over 50% by 2100 for RCP8.5 (*medium confidence*). Even as the overall regional water cycle is projected to intensify, including increased precipitation, evapotranspiration, and river discharge to the Arctic Ocean, decreases in snow and permafrost may lead to soil drying with consequences for ecosystem productivity and disturbances (*medium confidence*). Wildfire is projected to increase for the rest of this century across most tundra and boreal regions, and also in some mountain regions, while interactions between climate and shifting vegetation will influence future fire intensity and frequency (*medium confidence*). {2.3.3, 3.4.1, 3.4.2, 3.4.3, SPM B.1}

B.5 **A decrease in global biomass of marine animal communities, their production, and fisheries catch potential, and a shift in species composition are projected over the 21st century in ocean ecosystems from the surface to the deep seafloor under all emission scenarios (*medium confidence*). The rate and magnitude of decline are projected to be highest in the tropics (*high confidence*), whereas impacts remain diverse in polar regions (*medium confidence*) and increase for high emissions scenarios. Ocean acidification (*medium confidence*), oxygen loss (*medium confidence*) and reduced sea ice extent (*medium confidence*) as well as non-climatic human activities (*medium confidence*) have the potential to exacerbate these warming-induced ecosystem impacts. {3.2.3, 3.3.3, 5.2.2, 5.2.3, 5.2.4, 5.4.1, Figure SPM.3}**

B.5.1 Projected ocean warming and changes in net primary production alter biomass, production and community structure of marine ecosystems. The global-scale biomass of marine animals across the foodweb is projected to decrease by $15.0 \pm 5.9\%$ (*very likely* range) and the maximum catch potential of fisheries by 20.5–24.1% by the end of the 21st century relative to 1986–2005 under RCP8.5 (*medium confidence*). These changes are projected to be *very likely* three to four times larger under RCP8.5 than RCP2.6. {3.2.3, 3.3.3, 5.2.2, 5.2.3, 5.4.1, Figure SPM.3}

B.5.2 Under enhanced stratification reduced nutrient supply is projected to cause tropical ocean net primary production to decline by 7–16% (*very likely* range) for RCP8.5 by 2081–2100 (*medium confidence*). In tropical regions, marine animal biomass and production are projected to decrease more than the global average under all emissions scenarios in the 21st century (*high confidence*). Warming and sea ice changes are projected to increase marine net primary production in the Arctic (*medium confidence*) and around Antarctica (*low confidence*), modified by changing nutrient supply due to shifts in upwelling and stratification. Globally, the sinking flux of organic matter from the upper ocean is projected to decrease, linked largely due to changes in net primary production (*high confidence*). As a result, 95% or more of the deep sea (3000–6000 m depth) seafloor area and cold-water coral ecosystems are projected to experience declines in benthic biomass under RCP8.5 (*medium confidence*). {3.2.3, 5.2.2. 5.2.4, Figure SPM.1}

B.5.3 Warming, ocean acidification, reduced seasonal sea ice extent and continued loss of multi-year sea ice are projected to impact polar marine ecosystems through direct and indirect effects on habitats, populations and their viability (*medium confidence*). The geographical range of Arctic marine species, including marine mammals, birds and fish is projected to contract, while the range of some sub-Arctic fish communities is projected to expand, further increasing pressure on high-Arctic species (*medium confidence*). In the Southern Ocean, the habitat of Antarctic krill, a key prey species for penguins, seals and whales, is projected to contract southwards under both RCP2.6 and RCP8.5 (*medium confidence*). {3.2.2, 3.2.3, 5.2.3}

B.5.4 Ocean warming, oxygen loss, acidification and a decrease in flux of organic carbon from the surface to the deep ocean are projected to harm habitat-forming cold-water corals, which support high biodiversity, partly through decreased calcification, increased dissolution of skeletons, and bioerosion (*medium confidence*). Vulnerability and risks are highest where and when temperature and oxygen conditions both reach values outside species' tolerance ranges (*medium confidence*). {Box 5.2, Figure SPM.3}

Projected changes, impacts and risks for ocean ecosystems as a result of climate change

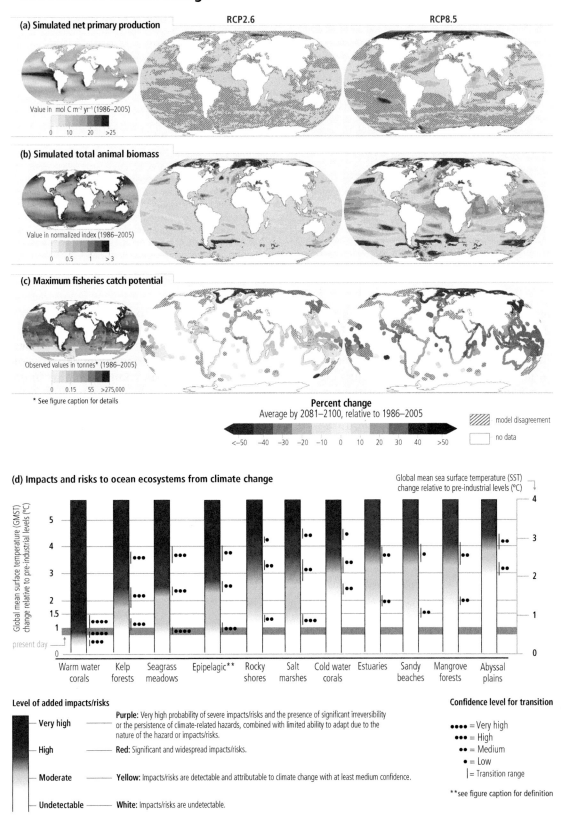

Figure SPM.3 | Projected changes, impacts and risks for ocean regions and ecosystems.

Figure SPM.3 (continued): **(a)** depth integrated net primary production (NPP from CMIP5[27]), **(b)** total animal biomass (depth integrated, including fishes and invertebrates from FISHMIP[28]), **(c)** maximum fisheries catch potential and **(d)** impacts and risks for coastal and open ocean ecosystems. The three left panels represent the simulated (a,b) and observed (c) mean values for the recent past (1986–2005), the middle and right panels represent projected changes (%) by 2081–2100 relative to recent past under low (RCP2.6) and high (RCP8.5) greenhouse gas emissions scenario {Box SPM.1}, respectively. Total animal biomass in the recent past (b, left panel) represents the projected total animal biomass by each spatial pixel relative to the global average. (c) *Average observed fisheries catch in the recent past (based on data from the Sea Around Us global fisheries database); projected changes in maximum fisheries catch potential in shelf seas are based on the average outputs from two fisheries and marine ecosystem models. To indicate areas of model inconsistency, shaded areas represent regions where models disagree in the direction of change for more than: (a) and (b) 3 out of 10 model projections, and (c) one out of two models. Although unshaded, the projected change in the Arctic and Antarctic regions in (b) total animal biomass and (c) fisheries catch potential have *low confidence* due to uncertainties associated with modelling multiple interacting drivers and ecosystem responses. Projections presented in (b) and (c) are driven by changes in ocean physical and biogeochemical conditions e.g., temperature, oxygen level, and net primary production projected from CMIP5 Earth system models. **The epipelagic refers to the uppermost part of the ocean with depth <200 m from the surface where there is enough sunlight to allow photosynthesis. (d) Assessment of risks for coastal and open ocean ecosystems based on observed and projected climate impacts on ecosystem structure, functioning and biodiversity. Impacts and risks are shown in relation to changes in Global Mean Surface Temperature (GMST) relative to pre-industrial level. Since assessments of risks and impacts are based on global mean Sea Surface Temperature (SST), the corresponding SST levels are shown[29]. The assessment of risk transitions is described in Chapter 5 Sections 5.2, 5.3, 5.2.5 and 5.3.7 and Supplementary Materials SM5.3, Table SM5.6, Table SM5.8 and other parts of the underlying report. The figure indicates assessed risks at approximate warming levels and increasing climate-related hazards in the ocean: ocean warming, acidification, deoxygenation, increased density stratification, changes in carbon fluxes, sea level rise, and increased frequency and/or intensity of extreme events. The assessment considers the natural adaptive capacity of the ecosystems, their exposure and vulnerability. Impact and risk levels do not consider risk reduction strategies such as human interventions, or future changes in non-climatic drivers. Risks for ecosystems were assessed by considering biological, biogeochemical, geomorphological and physical aspects. Higher risks associated with compound effects of climate hazards include habitat and biodiversity loss, changes in species composition and distribution ranges, and impacts/risks on ecosystem structure and functioning, including changes in animal/plant biomass and density, productivity, carbon fluxes, and sediment transport. As part of the assessment, literature was compiled and data extracted into a summary table. A multi-round expert elicitation process was undertaken with independent evaluation of threshold judgement, and a final consensus discussion. Further information on methods and underlying literature can be found in Chapter 5, Sections 5.2 and 5.3 and Supplementary Material. {3.2.3, 3.2.4, 5.2, 5.3, 5.2.5, 5.3.7, SM5.6, SM5.8, Figure 5.16, Cross Chapter Box 1 in Chapter 1 Table CCB1}

B.6 **Risks of severe impacts on biodiversity, structure and function of coastal ecosystems are projected to be higher for elevated temperatures under high compared to low emissions scenarios in the 21st century and beyond. Projected ecosystem responses include losses of species habitat and diversity, and degradation of ecosystem functions. The capacity of organisms and ecosystems to adjust and adapt is higher at lower emissions scenarios (*high confidence*). For sensitive ecosystems such as seagrass meadows and kelp forests, high risks are projected if global warming exceeds 2°C above pre-industrial temperature, combined with other climate-related hazards (*high confidence*). Warm-water corals are at high risk already and are projected to transition to very high risk even if global warming is limited to 1.5°C (*very high confidence*). {4.3.3, 5.3, 5.5, Figure SPM.3}**

B.6.1 All coastal ecosystems assessed are projected to face increasing risk level, from moderate to high risk under RCP2.6 to high to very high risk under RCP8.5 by 2100. Intertidal rocky shore ecosystems are projected to be at very high risk by 2100 under RCP8.5 (*medium confidence*) due to exposure to warming, especially during marine heatwaves, as well as to acidification, sea level rise, loss of calcifying species and biodiversity (*high confidence*). Ocean acidification challenges these ecosystems and further limits their habitat suitability (*medium confidence*) by inhibiting recovery through reduced calcification and enhanced bioerosion. The decline of kelp forests is projected to continue in temperate regions due to warming, particularly under the projected intensification of marine heatwaves, with high risk of local extinctions under RCP8.5 (*medium confidence*). {5.3, 5.3.5, 5.3.6, 5.3.7, 6.4.2, Figure SPM.3}

B.6.2 Seagrass meadows and saltmarshes and associated carbon stores are at moderate risk at 1.5°C global warming and increase with further warming (*medium confidence*). Globally, 20–90% of current coastal wetlands are projected to be lost by 2100, depending on projected sea level rise, regional differences and wetland types, especially where vertical growth is already constrained by reduced sediment supply and landward migration is constrained by steep topography or human modification of shorelines (*high confidence*). {4.3.3, 5.3.2, Figure SPM.3, SPM A.6.1}

---

[27] NPP is estimated from the Coupled Models Intercomparison Project 5 (CMIP5).

[28] Total animal biomass is from the Fisheries and Marine Ecosystem Models Intercomparison Project (FISHMIP).

[29] The conversion between GMST and SST is based on a scaling factor of 1.44 derived from changes in an ensemble of RCP8.5 simulations; this scaling factor has an uncertainty of about 4% due to differences between the RCP2.6 and RCP8.5 scenarios. {Table SPM.1}

B.6.3   Ocean warming, sea level rise and tidal changes are projected to expand salinization and hypoxia in estuaries (*high confidence*) with high risks for some biota leading to migration, reduced survival, and local extinction under high emission scenarios (*medium confidence*). These impacts are projected to be more pronounced in more vulnerable eutrophic and shallow estuaries with low tidal range in temperate and high latitude regions (*medium confidence*). {5.2.2, 5.3.1, Figure SPM.3}

B.6.4   Almost all warm-water coral reefs are projected to suffer significant losses of area and local extinctions, even if global warming is limited to 1.5°C (*high confidence*). The species composition and diversity of remaining reef communities is projected to differ from present-day reefs (*very high confidence*). {5.3.4, 5.4.1, Figure SPM.3}

Projected Risks for People and Ecosystem Services

B.7 Future cryosphere changes on land are projected to affect water resources and their uses, such as hydropower (*high confidence*) and irrigated agriculture in and downstream of high mountain areas (*medium confidence*), as well as livelihoods in the Arctic (*medium confidence*). Changes in floods, avalanches, landslides, and ground destabilization are projected to increase risk for infrastructure, cultural, tourism, and recreational assets (*medium confidence*). {2.3, 2.3.1, 3.4.3}

B.7.1   Disaster risks to human settlements and livelihood options in high mountain areas and the Arctic are expected to increase (*medium confidence*), due to future changes in hazards such as floods, fires, landslides, avalanches, unreliable ice and snow conditions, and increased exposure of people and infrastructure (*high confidence*). Current engineered risk reduction approaches are projected to be less effective as hazards change in character (*medium confidence*). Significant risk reduction and adaptation strategies help avoid increased impacts from mountain flood and landslide hazards as exposure and vulnerability are increasing in many mountain regions during this century (*high confidence*). {2.3.2, 3.4.3, 3.5.2}

B.7.2   Permafrost thaw-induced subsidence of the land surface is projected to impact overlying urban and rural communication and transportation infrastructure in the Arctic and in high mountain areas (*medium confidence*). The majority of Arctic infrastructure is located in regions where permafrost thaw is projected to intensify by mid-century. Retrofitting and redesigning infrastructure has the potential to halve the costs arising from permafrost thaw and related climate-change impacts by 2100 (*medium confidence*). {2.3.4, 3.4.1, 3.4.3}

B.7.3   High mountain tourism, recreation and cultural assets are projected to be negatively affected by future cryospheric changes (*high confidence*). Current snowmaking technologies are projected to be less effective in reducing risks to ski tourism in a warmer climate in most parts of Europe, North America, and Japan, in particular at 2°C global warming and beyond (*high confidence*). {2.3.5, 2.3.6}

B.8 **Future shifts in fish distribution and decreases in their abundance and fisheries catch potential due to climate change are projected to affect income, livelihoods, and food security of marine resource-dependent communities** (*medium confidence*). **Long-term loss and degradation of marine ecosystems compromises the ocean's role in cultural, recreational, and intrinsic values important for human identity and well-being** (*medium confidence*). {3.2.4, 3.4.3, 5.4.1, 5.4.2, 6.4}

B.8.1 Projected geographical shifts and decreases of global marine animal biomass and fish catch potential are more pronounced under RCP8.5 relative to RCP2.6 elevating the risk for income and livelihoods of dependent human communities, particularly in areas that are economically vulnerable (*medium confidence*). The projected redistribution of resources and abundance increases the risk of conflicts among fisheries, authorities or communities (*medium confidence*). Challenges to fisheries governance are widespread under RCP8.5 with regional hotspots such as the Arctic and tropical Pacific Ocean (*medium confidence*). {3.5.2, 5.4.1, 5.4.2, 5.5.2, 5.5.3, 6.4.2, Figure SPM.3}

B.8.2 The decline in warm-water coral reefs is projected to greatly compromise the services they provide to society, such as food provision (*high confidence*), coastal protection (*high confidence*) and tourism (*medium confidence*). Increases in the risks for seafood security (*medium confidence*) associated with decreases in seafood availability are projected to elevate the risk to nutritional health in some communities highly dependent on seafood (*medium confidence*), such as those in the Arctic, West Africa, and Small Island Developing States. Such impacts compound any risks from other shifts in diets and food systems caused by social and economic changes and climate change over land (*medium confidence*). {3.4.3, 5.4.2, 6.4.2}

B.8.3 Global warming compromises seafood safety (*medium confidence*) through human exposure to elevated bioaccumulation of persistent organic pollutants and mercury in marine plants and animals (*medium confidence*), increasing prevalence of waterborne *Vibrio* pathogens (*medium confidence*), and heightened likelihood of harmful algal blooms (*medium confidence*). These risks are projected to be particularly large for human communities with high consumption of seafood, including coastal Indigenous communities (*medium confidence*), and for economic sectors such as fisheries, aquaculture, and tourism (*high confidence*). {3.4.3, 5.4.2, Box 5.3}

B.8.4 Climate change impacts on marine ecosystems and their services put key cultural dimensions of lives and livelihoods at risk (*medium confidence*), including through shifts in the distribution or abundance of harvested species and diminished access to fishing or hunting areas. This includes potentially rapid and irreversible loss of culture and local knowledge and Indigenous knowledge, and negative impacts on traditional diets and food security, aesthetic aspects, and marine recreational activities (*medium confidence*). {3.4.3, 3.5.3, 5.4.2}

B.9 **Increased mean and extreme sea level, alongside ocean warming and acidification, are projected to exacerbate risks for human communities in low-lying coastal areas (*high confidence*). In Arctic human communities without rapid land uplift, and in urban atoll islands, risks are projected to be moderate to high even under a low emissions scenario (RCP2.6) (*medium confidence*), including reaching adaptation limits (*high confidence*). Under a high emissions scenario (RCP8.5), delta regions and resource rich coastal cities are projected to experience moderate to high risk levels after 2050 under current adaptation (*medium confidence*). Ambitious adaptation including transformative governance is expected to reduce risk (*high confidence*), but with context-specific benefits. {4.3.3, 4.3.4, SM4.3, 6.9.2, Cross-Chapter Box 9, Figure SPM.5}**

B.9.1   In the absence of more ambitious adaptation efforts compared to today, and under current trends of increasing exposure and vulnerability of coastal communities, risks, such as erosion and land loss, flooding, salinization, and cascading impacts due to mean sea level rise and extreme events are projected to significantly increase throughout this century under all greenhouse gas emissions scenarios (*very high confidence*). Under the same assumptions, annual coastal flood damages are projected to increase by 2–3 orders of magnitude by 2100 compared to today (*high confidence*). {4.3.3, 4.3.4, Box 6.1, 6.8, SM.4.3, Figures SPM.4, SPM.5}

B.9.2   High to very high risks are approached for vulnerable communities in coral reef environments, urban atoll islands and low-lying Arctic locations from sea level rise well before the end of this century in case of high emissions scenarios. This entails adaptation limits being reached, which are the points at which an actor's objectives (or system needs) cannot be secured from intolerable risks through adaptive actions (*high confidence*). Reaching adaptation limits (e.g., biophysical, geographical, financial, technical, social, political, and institutional) depends on the emissions scenario and context-specific risk tolerance, and is projected to expand to more areas beyond 2100, due to the long-term commitment of sea level rise (*medium confidence*). Some island nations are *likely* to become uninhabitable due to climate-related ocean and cryosphere change (*medium confidence*), but habitability thresholds remain extremely difficult to assess. {4.3.4, 4.4.2, 4.4.3, 5.5.2, Cross-Chapter Box 9, SM.4.3, SPM C.1, Glossary, Figure SPM.5}

B.9.3   Globally, a slower rate of climate-related ocean and cryosphere change provides greater adaptation opportunities (*high confidence*). While there is *high confidence* that ambitious adaptation, including governance for transformative change, has the potential to reduce risks in many locations, such benefits can vary between locations. At global scale, coastal protection can reduce flood risk by 2–3 orders of magnitude during the 21st century, but depends on investments on the order of tens to several hundreds of billions of US$ per year (*high confidence*). While such investments are generally cost efficient for densely populated urban areas, rural and poorer areas may be challenged to afford such investments with relative annual costs for some small island states amounting to several percent of GDP (*high confidence*). Even with major adaptation efforts, residual risks and associated losses are projected to occur (*medium confidence*), but context-specific limits to adaptation and residual risks remain difficult to assess. {4.1.3, 4.2.2.4, 4.3.1, 4.3.2, 4.3.4., 4.4.3, 6.9.1, 6.9.2, Cross-Chapter Boxes 1–2 in Chapter 1, SM.4.3, Figure SPM.5}

Extreme sea level events

Due to projected global mean sea level (GMSL) rise, local sea levels that historically occurred once per century (historical centennial events, HCEs) are projected to become at least annual events at most locations during the 21st century. The height of a HCE varies widely, and depending on the level of exposure can already cause severe impacts. Impacts can continue to increase with rising frequency of HCEs.

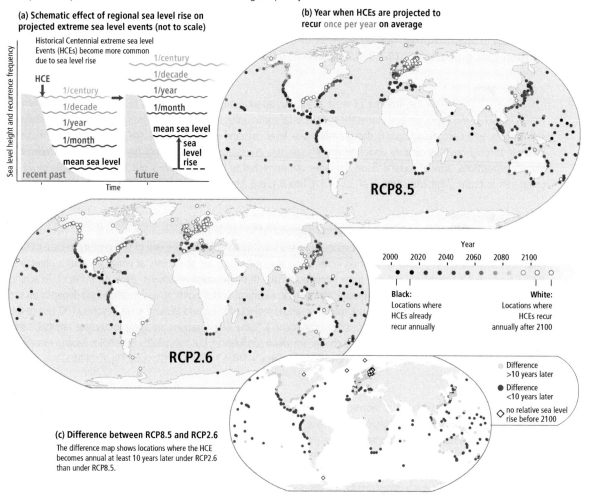

(a) Schematic effect of regional sea level rise on projected extreme sea level events (not to scale)

(b) Year when HCEs are projected to recur once per year on average

RCP8.5

RCP2.6

(c) Difference between RCP8.5 and RCP2.6
The difference map shows locations where the HCE becomes annual at least 10 years later under RCP2.6 than under RCP8.5.

Figure SPM.4 | The effect of regional sea level rise on extreme sea level events at coastal locations. **(a)** Schematic illustration of extreme sea level events and their average recurrence in the recent past (1986–2005) and the future. As a consequence of mean sea level rise, local sea levels that historically occurred once per century (historical centennial events, HCEs) are projected to recur more frequently in the future. **(b)** The year in which HCEs are expected to recur once per year on average under RCP8.5 and RCP2.6, at the 439 individual coastal locations where the observational record is sufficient. The absence of a circle indicates an inability to perform an assessment due to a lack of data but does not indicate absence of exposure and risk. The darker the circle, the earlier this transition is expected. The *likely* range is ±10 years for locations where this transition is expected before 2100. White circles (33% of locations under RCP2.6 and 10% under RCP8.5) indicate that HCEs are not expected to recur once per year before 2100. **(c)** An indication at which locations this transition of HCEs to annual events is projected to occur more than 10 years later under RCP2.6 compared to RCP8.5. As the scenarios lead to small differences by 2050 in many locations results are not shown here for RCP4.5 but they are available in Chapter 4. {4.2.3, Figure 4.10, Figure 4.12}

C. Implementing Responses to Ocean and Cryosphere Change

Challenges

C.1 **Impacts of climate-related changes in the ocean and cryosphere increasingly challenge current governance efforts to develop and implement adaptation responses from local to global scales, and in some cases pushing them to their limits. People with the highest exposure and vulnerability are often those with lowest capacity to respond (*high confidence*). {1.5, 1.7, Cross-Chapter Boxes 2–3 in Chapter 1, 2.3.1, 2.3.2, 2.3.3, 2.4, 3.2.4, 3.4.3, 3.5.2, 3.5.3, 4.1, 4.3.3, 4.4.3, 5.5.2, 5.5.3, 6.9}**

C.1.1 The temporal scales of climate change impacts in ocean and cryosphere and their societal consequences operate on time horizons which are longer than those of governance arrangements (e.g., planning cycles, public and corporate decision making cycles, and financial instruments). Such temporal differences challenge the ability of societies to adequately prepare for and respond to long-term changes including shifts in the frequency and intensity of extreme events (*high confidence*). Examples include changing landslides and floods in high mountain regions and risks to important species and ecosystems in the Arctic, as well as to low-lying nations and islands, small island nations, other coastal regions and to coral reef ecosystems. {2.3.2, 3.5.2, 3.5.4, 4.4.3, 5.2, 5.3, 5.4, 5.5.1, 5.5.2, 5.5.3, 6.9}

C.1.2 Governance arrangements (e.g., marine protected areas, spatial plans and water management systems) are, in many contexts, too fragmented across administrative boundaries and sectors to provide integrated responses to the increasing and cascading risks from climate-related changes in the ocean and/or cryosphere (*high confidence*). The capacity of governance systems in polar and ocean regions to respond to climate change impacts has strengthened recently, but this development is not sufficiently rapid or robust to adequately address the scale of increasing projected risks (*high confidence*). In high mountains, coastal regions and small islands, there are also difficulties in coordinating climate adaptation responses, due to the many interactions of climatic and non-climatic risk drivers (such as inaccessibility, demographic and settlement trends, or land subsidence caused by local activities) across scales, sectors and policy domains (*high confidence*). {2.3.1, 3.5.3, 4.4.3, 5.4.2, 5.5.2, 5.5.3, Box 5.6, 6.9, Cross-Chapter Box 3 in Chapter 1}

C.1.3 There are a broad range of identified barriers and limits for adaptation to climate change in ecosystems (*high confidence*). Limitations include the space that ecosystems require, non-climatic drivers and human impacts that need to be addressed as part of the adaptation response, the lowering of adaptive capacity of ecosystems because of climate change, and the slower ecosystem recovery rates relative to the recurrence of climate impacts, availability of technology, knowledge and financial support, and existing governance arrangements (*medium confidence*). {3.5.4, 5.5.2}

C.1.4 Financial, technological, institutional and other barriers exist for implementing responses to current and projected negative impacts of climate-related changes in the ocean and cryosphere, impeding resilience building and risk reduction measures (*high confidence*). Whether such barriers reduce adaptation effectiveness or correspond to adaptation limits depends on context specific circumstances, the rate and scale of climate changes and on the ability of societies to turn their adaptive capacity into effective adaptation responses. Adaptive capacity continues to differ between as well as within communities and societies (*high confidence*). People with highest exposure and vulnerability to current and future hazards from ocean and cryosphere changes are often also those with lowest adaptive capacity, particularly in low-lying islands and coasts, Arctic and high mountain regions with development challenges (*high confidence*). {2.3.1, 2.3.2, 2.3.7, Box 2.4, 3.5.2, 4.3.4, 4.4.2, 4.4.3, 5.5.2, 6.9, Cross-Chapter Boxes 2 and 3 in Chapter 1, Cross-Chapter Box 9}

Strengthening Response Options

C.2 **The far-reaching services and options provided by ocean and cryosphere-related ecosystems can be supported by protection, restoration, precautionary ecosystem-based management of renewable resource use, and the reduction of pollution and other stressors (*high confidence*). Integrated water management (*medium confidence*) and ecosystem-based adaptation (*high confidence*) approaches lower climate risks locally and provide multiple societal benefits. However, ecological, financial, institutional and governance constraints for such actions exist (*high confidence*), and in many contexts ecosystem-based adaptation will only be effective under the lowest levels of warming (*high confidence*). {2.3.1, 2.3.3, 3.2.4, 3.5.2, 3.5.4, 4.4.2, 5.2.2, 5.4.2, 5.5.1, 5.5.2, Figure SPM.5}**

C.2.1   Networks of protected areas help maintain ecosystem services, including carbon uptake and storage, and enable future ecosystem-based adaptation options by facilitating the poleward and altitudinal movements of species, populations, and ecosystems that occur in response to warming and sea level rise (*medium confidence*). Geographic barriers, ecosystem degradation, habitat fragmentation and barriers to regional cooperation limit the potential for such networks to support future species range shifts in marine, high mountain and polar land regions (*high confidence*). {2.3.3, 3.2.3, 3.3.2, 3.5.4, 5.5.2, Box 3.4}

C.2.2   Terrestrial and marine habitat restoration, and ecosystem management tools such as assisted species relocation and coral gardening, can be locally effective in enhancing ecosystem-based adaptation (*high confidence*). Such actions are most successful when they are community-supported, are science-based whilst also using local knowledge and Indigenous knowledge, have long-term support that includes the reduction or removal of non-climatic stressors, and under the lowest levels of warming (*high confidence*). For example, coral reef restoration options may be ineffective if global warming exceeds 1.5°C, because corals are already at high risk (*very high confidence*) at current levels of warming. {2.3.3, 4.4.2, 5.3.7, 5.5.1, 5.5.2, Box 5.5, Figure SPM.3}

C.2.3   Strengthening precautionary approaches, such as rebuilding overexploited or depleted fisheries, and responsiveness of existing fisheries management strategies reduces negative climate change impacts on fisheries, with benefits for regional economies and livelihoods (*medium confidence*). Fisheries management that regularly assesses and updates measures over time, informed by assessments of future ecosystem trends, reduces risks for fisheries (*medium confidence*) but has limited ability to address ecosystem change. {3.2.4, 3.5.2, 5.4.2, 5.5.2, 5.5.3, Figure SPM.5}

C.2.4   Restoration of vegetated coastal ecosystems, such as mangroves, tidal marshes and seagrass meadows (coastal 'blue carbon' ecosystems), could provide climate change mitigation through increased carbon uptake and storage of around 0.5% of current global emissions annually (*medium confidence*). Improved protection and management can reduce carbon emissions from these ecosystems. Together, these actions also have multiple other benefits, such as providing storm protection, improving water quality, and benefiting biodiversity and fisheries (*high confidence*). Improving the quantification of carbon storage and greenhouse gas fluxes of these coastal ecosystems will reduce current uncertainties around measurement, reporting and verification (*high confidence*). {Box 4.3, 5.4, 5.5.1, 5.5.2, Annex I: Glossary}

C.2.5   Ocean renewable energy can support climate change mitigation, and can comprise energy extraction from offshore winds, tides, waves, thermal and salinity gradient and algal biofuels. The emerging demand for alternative energy sources is expected to generate economic opportunities for the ocean renewable energy sector (*high confidence*), although their potential may also be affected by climate change (*low confidence*). {5.4.2, 5.5.1, Figure 5.23}

C.2.6 Integrated water management approaches across multiple scales can be effective at addressing impacts and leveraging opportunities from cryosphere changes in high mountain areas. These approaches also support water resource management through the development and optimization of multi-purpose storage and release of water from reservoirs (*medium confidence*), with consideration of potentially negative impacts to ecosystems and communities. Diversification of tourism activities throughout the year supports adaptation in high mountain economies (*medium confidence*). {2.3.1, 2.3.5}

C.3 **Coastal communities face challenging choices in crafting context-specific and integrated responses to sea level rise that balance costs, benefits and trade-offs of available options and that can be adjusted over time (*high confidence*). All types of options, including protection, accommodation, ecosystem-based adaptation, coastal advance and retreat, wherever possible, can play important roles in such integrated responses (*high confidence*). {4.4.2, 4.4.3, 4.4.4, 6.9.1, Cross-Chapter Box 9, Figure SPM.5}**

C.3.1 The higher the sea levels rise, the more challenging is coastal protection, mainly due to economic, financial and social barriers rather than due to technical limits (*high confidence*). In the coming decades, reducing local drivers of exposure and vulnerability such as coastal urbanization and human-induced subsidence constitute effective responses (*high confidence*). Where space is limited, and the value of exposed assets is high (e.g., in cities), hard protection (e.g., dikes) is *likely* to be a cost-efficient response option during the 21st century taking into account the specifics of the context (*high confidence*), but resource-limited areas may not be able to afford such investments. Where space is available, ecosystem-based adaptation can reduce coastal risk and provide multiple other benefits such as carbon storage, improved water quality, biodiversity conservation and livelihood support (*medium confidence*). {4.3.2, 4.4.2, Box 4.1, Cross-Chapter Box 9, Figure SPM.5}

C.3.2 Some coastal accommodation measures, such as early warning systems and flood-proofing of buildings, are often both low cost and highly cost-efficient under current sea levels (*high confidence*). Under projected sea level rise and increase in coastal hazards some of these measures become less effective unless combined with other measures (*high confidence*). All types of options, including protection, accommodation, ecosystem-based adaptation, coastal advance and planned relocation, if alternative localities are available, can play important roles in such integrated responses (*high confidence*). Where the community affected is small, or in the aftermath of a disaster, reducing risk by coastal planned relocations is worth considering if safe alternative localities are available. Such planned relocation can be socially, culturally, financially and politically constrained (*very high confidence*). {4.4.2, Box 4.1, Cross-Chapter Box 9, SPM B.3}

C.3.3 Responses to sea level rise and associated risk reduction present society with profound governance challenges, resulting from the uncertainty about the magnitude and rate of future sea level rise, vexing trade-offs between societal goals (e.g., safety, conservation, economic development, intra- and inter-generational equity), limited resources, and conflicting interests and values among diverse stakeholders (*high confidence*). These challenges can be eased using locally appropriate combinations of decision analysis, land-use planning, public participation, diverse knowledge systems and conflict resolution approaches that are adjusted over time as circumstances change (*high confidence*). {Cross-Chapter Box 5 in Chapter 1, 4.4.3, 4.4.4, 6.9}

C.3.4 Despite the large uncertainties about the magnitude and rate of post 2050 sea level rise, many coastal decisions with time horizons of decades to over a century are being made now (e.g., critical infrastructure, coastal protection works, city planning) and can be improved by taking relative sea level rise into account, favouring flexible responses (i.e., those that can be adapted over time) supported by monitoring systems for early warning signals, periodically adjusting decisions (i.e., adaptive decision making), using robust decision-making approaches, expert judgement, scenario-building, and multiple knowledge systems (*high confidence*). The sea level rise range that needs to be considered for planning and implementing coastal responses depends on the risk tolerance of stakeholders. Stakeholders with higher risk tolerance (e.g., those planning for investments that can be very easily adapted to unforeseen conditions) often prefer to use the *likely* range of projections, while stakeholders with a lower risk tolerance (e.g., those deciding on critical infrastructure) also consider global and local mean sea level above the upper end of the *likely* range (globally 1.1 m under RCP8.5 by 2100) and from methods characterised by lower confidence such as from expert elicitation. {1.8.1, 1.9.2, 4.2.3, 4.4.4, Figure 4.2, Cross-Chapter Box 5 in Chapter 1, Figure SPM.5, SPM B.3}

Sea level rise risk and responses

The term response is used here instead of adaptation because some responses, such as retreat, may or may not be considered to be adaptation.

(a) Risk in 2100 under different sea level rise and response scenarios

Risk for illustrative geographies based on mean sea level changes *(medium confidence)*

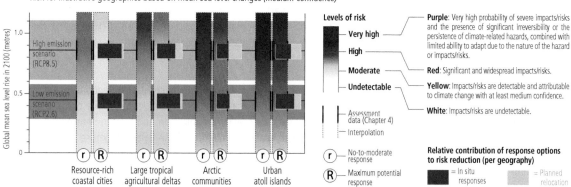

In this assessment, the term response refers to in situ responses to sea level rise (hard engineered coastal defenses, restoration of degraded ecosystems, subsidence limitation) and planned relocation. Planned relocation in this assessment refers to proactive managed retreat or resettlement only at a local scale, and according to the specificities of a particular context (e.g., in urban atoll islands: within the island, in a neighbouring island or in artificially raised islands). Forced displacement and international migration are not considered in this assessment.

The illustrative geographies are based on a limited number of case studies well covered by the peer reviewed literature. The realisation of risk will depend on context specifities.

Sea level rise scenarios: RCP4.5 and RCP6.0 are not considered in this risk assessment because the literature underpinning this assessment is only available for RCP2.6 and RCP8.5.

(b) Benefits of responses to sea level rise and mitigation

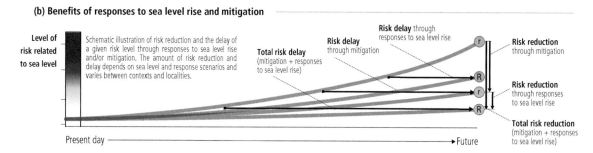

Figure SPM.5 | a, b

SPM

(c) Responses to rising mean and extreme sea levels

The table illustrates responses and their characteristics. It is not exhaustive. Whether a response is applicable depends on geography and context.

Confidence levels (assessed for effectiveness): ●●●● = Very High   ●●●= High   ●● = Medium   ● = Low

Responses		Potential effectiveness in terms of reducing sea level rise (SLR) risks (technical/biophysical limits)	Advantages (beyond risk reduction)	Co–benefits	Drawbacks	Economic efficiency	Governance challenges
Hard protection		Up to multiple metres of SLR {4.4.2.2.4} ●●●	Predictable levels of safety {4.4.2.2.4}	Multifunctional dikes such as for recreation, or other land use {4.4.2.2.5}	Destruction of habitat through coastal squeeze, flooding & erosion downdrift, lock–in, disastrous consequence in case of defence failure {4.3.2.4, 4.4.2.2.5}	High if the value of assets behind protection is high, as found in many urban and densely populated coastal areas {4.4.2.2.7}	Often unaffordable for poorer areas. Conflicts between objectives (e.g., conservation, safety and tourism), conflicts about the distribution of public budgets, lack of finance {4.3.3.2, 4.4.2.2.6}
Sediment–based protection		Effective but depends on sediment availability {4.4.2.2.4} ●●●	High flexibility {4.4.2.2.4}	Preservation of beaches for recreation/tourism {4.4.2.2.5}	Destruction of habitat, where sediment is sourced {4.4.2.2.5}	High if tourism revenues are high {4.4.2.2.7}	Conflicts about the distribution of public budgets {4.4.2.2.6}
Ecosystem based adaptation	Coral conservation / Coral restoration	Effective up to 0.5 cm yr⁻¹ SLR. ●● Strongly limited by ocean warming and acidification. Constrained at 1.5°C warming and lost at 2°C at many places. {4.3.3.5.2, 4.4.2.3.2, 5.3.4} ●●●	Opportunity for community involvement, {4.4.2.3.1}	Habitat gain, biodiversity, carbon sequestration, income from tourism, enhanced fishery productivity, improved water quality. Provision of food, medicine, fuel, wood and cultural benefits {4.4.2.3.5}	Long–term effectiveness depends on ocean warming, acidification and emission scenarios {4.3.3.5.2., 4.4.2.3.2}	Limited evidence on benefit–cost ratios; Depends on population density and the availability of land {4.4.2.3.7}	Permits for implementation are difficult to obtain. Lack of finance. Lack of enforcement of conservation policies. EbA options dismissed due to short–term economic interest, availability of land {4.4.2.3.6}
	Wetland conservation (Marshes, Mangroves)	Effective up to 0.5–1 cm yr⁻¹ SLR, ●● decreased at 2°C {4.3.3.5.1, 4.4.2.3.2, 5.3.7} ●●●			Safety levels less predictable, development benefits not realized {4.4.2.3.5, 4.4.2.3.2}		
	Wetland restoration (Marshes, Mangroves)				Safety levels less predictable, a lot of land required, barriers for landward expan–sion of ecosystems has to be removed {4.4.2.3.5, 4.4.2.3.2}		
Coastal advance		Up to multiple metres of SLR {4.4.2.2.4} ●●●	Predictable levels of safety {4.4.2.2.4}	Generates land and land sale revenues that can be used to finance adaptation {4.4.2.4.5}	Groundwater salinisa–tion, enhanced erosion and loss of coastal ecosystems and habitat {4.4.2.4.5}	Very high if land prices are high as found in many urban coasts {4.4.2.4.7}	Often unaffordable for poorer areas. Social conflicts with regards to access and distribution of new land {4.4.2.4.6}
Coastal accommodation (Flood–proofing buildings, early warning systems for flood events, etc.)		Very effective for small SLR {4.4.2.5.4} ●●●	Mature technology; sediments deposited during floods can raise elevation {4.4.2.5.5}	Maintains landscape connectivity {4.4.2.5.5}	Does not prevent flooding/impacts {4.4.2.5.5}	Very high for early warning systems and building–scale measures {4.4.2.5.7}	Early warning systems require effective insti–tutional arrangements {4.4.2.6.6}
Retreat	Planned relocation	Effective if alternative safe localities are available {4.4.2.6.4} ●●●	Sea level risks at origin can be eliminated {4.4.2.6.4}	Access to improved services (health, education, housing), job opportunities and economic growth {4.4.2.6.5}	Loss of social cohesion, cultural identity and well–being. Depressed services (health, education, housing), job opportunities and economic growth {4.4.2.6.5}	Limited evidence {4.4.2.6.7}	Reconciling the divergent interests arising from relocating people from point of origin and destination {4.4.2.6.6}
	Forced displacement	Addresses only immediate risk at place of origin	Not applicable	Not applicable	Range from loss of life to loss of livelihoods and sovereignty {4.4.2.6.5}	Not applicable	Raises complex humanitarian questions on livelihoods, human rights and equity {4.4.2.6.6}

(d) Choosing and enabling sea level rise responses

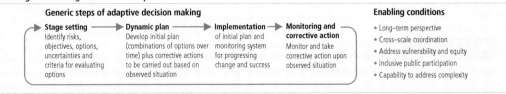

Generic steps of adaptive decision making				Enabling conditions

Stage setting → **Dynamic plan** → **Implementation** → **Monitoring and corrective action**

Stage setting: Identify risks, objectives, options, uncertainties and criteria for evaluating options

Dynamic plan: Develop initial plan (combinations of options over time) plus corrective actions to be carried out based on observed situation

Implementation: of initial plan and monitoring system for progressing change and success

Monitoring and corrective action: Monitor and take corrective action upon observed situation

Enabling conditions
- Long–term perspective
- Cross–scale coordination
- Address vulnerability and equity
- Inclusive public participation
- Capability to address complexity

Figure SPM.5 | c, d

Figure SPM.5 | Sea level rise risks and responses. The term response is used here instead of adaptation because some responses, such as retreat, may or may not be considered to be adaptation. **(a)** shows the combined risk of coastal flooding, erosion and salinization for illustrative geographies in 2100, due to changing mean and extreme sea levels under RCP2.6 and RCP8.5 and under two response scenarios. Risks under RCPs 4.5 and 6.0 were not assessed due to a lack of literature for the assessed geographies. The assessment does not account for changes in extreme sea level beyond those directly induced by mean sea level rise; risk levels could increase if other changes in extreme sea levels were considered (e.g., due to changes in cyclone intensity). Panel a) considers a socioeconomic scenario with relatively stable coastal population density over the century. {SM4.3.2} Risks to illustrative geographies have been assessed based on relative sea level changes projected for a set of specific examples: New York City, Shanghai and Rotterdam for resource-rich coastal cities covering a wide range of response experiences; South Tarawa, Fongafale and Male' for urban atoll islands; Mekong and Ganges-Brahmaputra-Meghna for large tropical agricultural deltas; and Bykovskiy, Shishmaref, Kivalina, Tuktoyaktuk and Shingle Point for Arctic communities located in regions remote from rapid glacio-isostatic adjustment. {4.2, 4.3.4, SM4.2} The assessment distinguishes between two contrasting response scenarios. "No-to-moderate response" describes efforts as of today (i.e., no further significant action or new types of actions). "Maximum potential response" represents a combination of responses implemented to their full extent and thus significant additional efforts compared to today, assuming minimal financial, social and political barriers. The assessment has been conducted for each sea level rise and response scenario, as indicated by the burning embers in the figure; in-between risk levels are interpolated. {4.3.3} The assessment criteria include exposure and vulnerability (density of assets, level of degradation of terrestrial and marine buffer ecosystems), coastal hazards (flooding, shoreline erosion, salinization), in-situ responses (hard engineered coastal defenses, ecosystem restoration or creation of new natural buffers areas, and subsidence management) and planned relocation. Planned relocation refers to managed retreat or resettlement as described in Chapter 4, i.e., proactive and local-scale measures to reduce risk by relocating people, assets and infrastructure. Forced displacement is not considered in this assessment. Panel **(a)** also highlights the relative contributions of in-situ responses and planned relocation to the total risk reduction. **(b)** schematically illustrates the risk reduction (vertical arrows) and risk delay (horizontal arrows) through mitigation and/or responses to sea level rise. **(c)** summarizes and assesses responses to sea level rise in terms of their effectiveness, costs, co-benefits, drawbacks, economic efficiency and associated governance challenges. {4.4.2} **(d)** presents generic steps of an adaptive decision-making approach, as well as key enabling conditions for responses to sea level rise. {4.4.4, 4.4.5}

Enabling Conditions

C.4 **Enabling climate resilience and sustainable development depends critically on urgent and ambitious emissions reductions coupled with coordinated sustained and increasingly ambitious adaptation actions (*very high confidence*). Key enablers for implementing effective responses to climate-related changes in the ocean and cryosphere include intensifying cooperation and coordination among governing authorities across spatial scales and planning horizons. Education and climate literacy, monitoring and forecasting, use of all available knowledge sources, sharing of data, information and knowledge, finance, addressing social vulnerability and equity, and institutional support are also essential. Such investments enable capacity-building, social learning, and participation in context-specific adaptation, as well as the negotiation of trade-offs and realisation of co-benefits in reducing short-term risks and building long-term resilience and sustainability. (*high confidence*). This report reflects the state of science for ocean and cryosphere for low levels of global warming (1.5°C), as also assessed in earlier IPCC and IPBES reports. {1.1, 1.5, 1.8.3, 2.3.1, 2.3.2, 2.4, Figure 2.7, 2.5, 3.5.2, 3.5.4, 4.4, 5.2.2, Box 5.3, 5.4.2, 5.5.2, 6.4.3, 6.5.3, 6.8, 6.9, Cross-Chapter Box 9, Figure SPM.5}**

C.4.1 In light of observed and projected changes in the ocean and cryosphere, many nations will face challenges to adapt, even with ambitious mitigation (*very high confidence*). In a high emissions scenario, many ocean- and cryosphere-dependent communities are projected to face adaptation limits (e.g. biophysical, geographical, financial, technical, social, political and institutional) during the second half of the 21st century. Low emission pathways, for comparison, limit the risks from ocean and cryosphere changes in this century and beyond and enable more effective responses (*high confidence*), whilst also creating co-benefits. Profound economic and institutional transformative change will enable Climate Resilient Development Pathways in the ocean and cryosphere context (*high confidence*). {1.1, 1.4–1.7, Cross-Chapter Boxes 1–3 in Chapter 1, 2.3.1, 2.4, Box 3.2, Figure 3.4, Cross-Chapter Box 7 in Chapter 3, 3.4.3, 4.2.2, 4.2.3, 4.3.4, 4.4.2, 4.4.3, 4.4.6, 5.4.2, 5.5.3, 6.9.2, Cross-Chapter Box 9, Figure SPM.5}

C.4.2 Intensifying cooperation and coordination among governing authorities across scales, jurisdictions, sectors, policy domains and planning horizons can enable effective responses to changes in the ocean, cryosphere and to sea level rise (*high confidence*). Regional cooperation, including treaties and conventions, can support adaptation action; however, the extent to which responding to impacts and losses arising from changes in the ocean and cryosphere is enabled through regional policy frameworks is currently limited *(high confidence)*. Institutional arrangements that provide strong multiscale linkages with local and Indigenous communities benefit adaptation *(high confidence)*. Coordination and complementarity between national and transboundary regional policies can support efforts to address risks to resource security and management, such as water and fisheries *(medium confidence)*. {2.3.1, 2.3.2, 2.4, Box 2.4, 2.5, 3.5.2, 3.5.3, 3.5.4, 4.4.4, 4.4.5, Table 4.9, 5.5.2, 6.9.2}

C.4.3 Experience to date – for example, in responding to sea level rise, water-related risks in some high mountains, and climate change risks in the Arctic – also reveal the enabling influence of taking a long-term perspective when making short-term decisions, explicitly accounting for uncertainty of context-specific risks beyond 2050 *(high confidence)*, and building governance capabilities to tackle complex risks *(medium confidence)*. {2.3.1, 3.5.4, 4.4.4, 4.4.5, Table 4.9, 5.5.2, 6.9, Figure SPM.5}

C.4.4 Investments in education and capacity building at various levels and scales facilitates social learning and long-term capability for context-specific responses to reduce risk and enhance resilience *(high confidence)*. Specific activities include utilization of multiple knowledge systems and regional climate information into decision making, and the engagement of local communities, Indigenous peoples, and relevant stakeholders in adaptive governance arrangements and planning frameworks *(medium confidence)*. Promotion of climate literacy and drawing on local, Indigenous and scientific knowledge systems enables public awareness, understanding and social learning about locality-specific risk and response potential *(high confidence)*. Such investments can develop, and in many cases transform existing institutions and enable informed, interactive and adaptive governance arrangements *(high confidence)*. {1.8.3, 2.3.2, Figure 2.7, Box 2.4, 2.4, 3.5.2, 3.5.4, 4.4.4, 4.4.5, Table 4.9, 5.5.2, 6.9}

C.4.5 Context-specific monitoring and forecasting of changes in the ocean and the cryosphere informs adaptation planning and implementation, and facilitates robust decisions on trade-offs between short- and long-term gains *(medium confidence)*. Sustained long-term monitoring, sharing of data, information and knowledge and improved context-specific forecasts, including early warning systems to predict more extreme El Niño/La Niña events, tropical cyclones, and marine heatwaves, help to manage negative impacts from ocean changes such as losses in fisheries, and adverse impacts on human health, food security, agriculture, coral reefs, aquaculture, wildfire, tourism, conservation, drought and flood *(high confidence)*. {2.4, 2.5, 3.5.2, 4.4.4, 5.5.2, 6.3.1, 6.3.3, 6.4.3, 6.5.3, 6.9}

C.4.6 Prioritising measures to address social vulnerability and equity underpins efforts to promote fair and just climate resilience and sustainable development *(high confidence)*, and can be helped by creating safe community settings for meaningful public participation, deliberation and conflict resolution *(medium confidence)*. {Box 2.4, 4.4.4, 4.4.5, Table 4.9, Figure SPM.5}

C.4.7 This assessment of the ocean and cryosphere in a changing climate reveals the benefits of ambitious mitigation and effective adaptation for sustainable development and, conversely, the escalating costs and risks of delayed action. The potential to chart Climate Resilient Development Pathways varies within and among ocean, high mountain and polar land regions. Realising this potential depends on transformative change. This highlights the urgency of prioritising timely, ambitious, coordinated and enduring action *(very high confidence)*. {1.1, 1.8, Cross-Chapter Box 1 in Chapter 1, 2.3, 2.4, 3.5, 4.2.1, 4.2.2, 4.3.4, 4.4, Table 4.9, 5.5, 6.9, Cross-Chapter Box 9, Figure SPM.5}

Technical
Summary

Technical Summary

Editorial Team:

Hans-Otto Pörtner (Germany), Debra C. Roberts (South Africa), Valerie Masson-Delmotte (France), Panmao Zhai (China), Elvira Poloczanska (United Kingdom/Australia), Katja Mintenbeck (Germany), Melinda Tignor (USA), Andrés Alegría (Honduras), Maike Nicolai (Germany), Andrew Okem (Nigeria), Jan Petzold (Germany), Bard Rama (Kosovo), Nora M. Weyer (Germany)

Authors:

Amro Abd-Elgawad (Egypt), Nerilie Abram (Australia), Carolina Adler (Switzerland/Australia), Andrés Alegría (Honduras), Javier Arístegui (Spain), Nathaniel L. Bindoff (Australia), Laurens Bouwer (Netherlands), Bolívar Cáceres (Ecuador), Rongshuo Cai (China), Sandra Cassotta (Denmark), Lijing Cheng (China), So-Min Cheong (Republic of Korea), William W. L. Cheung (Canada), Maria Paz Chidichimo (Argentina), Miguel Cifuentes-Jara (Costa Rica), Matthew Collins (United Kingdom), Susan Crate (USA), Rob Deconto (USA), Chris Derksen (Canada), Alexey Ekaykin (Russian Federation), Hiroyuki Enomoto (Japan), Thomas Frölicher (Switzerland), Matthias Garschagen (Germany), Jean-Pierre Gattuso (France), Tuhin Ghosh (India), Bruce Glavovic (New Zealand), Nicolas Gruber (Switzerland), Stephan Gruber (Canada/Germany), Valeria A. Guinder (Argentina), Robert Hallberg (USA), Sherilee Harper (Canada), John Hay (Cook Islands), Nathalie Hilmi (France), Jochen Hinkel (Germany), Yukiko Hirabayashi (Japan), Regine Hock (USA), Elisabeth Holland (Fiji), Anne Hollowed (USA), Federico Isla (Argentina), Miriam Jackson (Norway), Hélène Jacot Des Combes (Fiji), Nianzhi Jiao (China), Andreas Kääb (Norway), James G. Kairo (Kenya), Shichang Kang (China), Md Saiful Karim (Australia), Gary Kofinas (USA), Roxy Mathew Koll (India), Raphael Martin Kudela (USA), Stanislav Kutuzov (Russian Federation), Lisa Levin (USA), Iñigo Losada (Spain), Andrew Mackintosh (New Zealand), Alexandre K. Magnan (France), Ben Marzeion (Germany), Valerie Masson-Delmotte (France), Robin Matthews (United Kingdom), Kathleen McInnes (Australia), Jess Melbourne-Thomas (Australia), Michael Meredith (United Kingdom), Benoit Meyssignac (France), Alexander Milner (United Kingdom), Katja Mintenbeck (Germany), Ulf Molau (Sweden), Samuel Morin (France), Mônica M.C. Muelbert (Brazil), Maike Nicolai (Germany), Sean O'Donoghue (South Africa), Andrew Okem (Nigeria), Michael Oppenheimer (USA), Ben Orlove (USA), Geir Ottersen (Norway), Jan Petzold (Germany), Anna Pirani (Italy), Hans-Otto Pörtner (Germany), Elvira Poloczanska (United Kingdom/Australia), Anjal Prakash (India), Hamish Pritchard (United Kingdom), Sara R. Purca Cuicapusa (Peru), Golam Rasul (Nepal), Beate Ratter (Germany),

Jake Rice (Canada), Baruch Rinkevich (Israel), Evelia Rivera-Arriaga (Mexico), Debra C. Roberts (South Africa), Karina von Schuckmann (France), Ted Schuur (USA), Zita Sebesvari (Hungary), Martin Sommerkorn (Norway/Germany), Konrad Steffen (Switzerland), Heidi Steltzer (USA), Toshio Suga (Japan), Raden Dwi Susanto (Indonesia), Michael Sutherland (Trinidad and Tobago), Didier Swingedouw (France), Alessandro Tagliabue (United Kingdom), Lourdes Tibig (Philippines), Roderik van de Wal (Netherlands), Phillip Williamson (United Kingdom), Rong Yu (China), Panmao Zhai (China)

Review Editors:

Amjad Abdulla (Maldives), Ayako Abe-Ouchi (Japan), Oleg Anisimov (Russian Federation), Manuel Barange (South Africa), Gregory Flato (Canada), Kapil Gupta (India), Marcelino Hernández González (Cuba), Georg Kaser (Austria), Aditi Mukherji (Nepal), Joy Pereira (Malaysia), Monika Rhein (Germany), David Schoeman (Australia), Brad Seibel (USA), Carol Turley (United Kingdom), Cunde Xiao (China)

Chapter Scientists:

Maya Buchanan (USA), Axel Durand (Australia), Bethany Ellis (Australia), Shengping He (Norway/China), Jules Kajtar (United Kingdom), Pierre-Marie Lefeuvre (Norway/France), Santosh Nepal (Nepal), Avash Pandey (Nepal), Victoria Peck (United Kingdom)

Additional Graphics Support:

Martin Künstig (Germany), Stefanie Langsdorf (Germany)

This Technical Summary should be cited as:
IPCC, 2019: Technical Summary [H.-O. Pörtner, D.C. Roberts, V. Masson-Delmotte, P. Zhai, E. Poloczanska, K. Mintenbeck, M. Tignor, A. Alegría, M. Nicolai, A. Okem, J. Petzold, B. Rama, N.M. Weyer (eds.)]. In: *IPCC Special Report on the Ocean and Cryosphere in a Changing Climate* [H.- O. Pörtner, D.C. Roberts, V. Masson-Delmotte, P. Zhai, M. Tignor, E. Poloczanska, K. Mintenbeck, A. Alegría, M. Nicolai, A. Okem, J. Petzold, B. Rama, N.M. Weyer (eds.)]. Cambridge University Press, Cambridge, UK and New York, NY, USA, pp. 39–69. https://doi.org/10.1017/9781009157964.002

Table of contents

TS

TS.0   Introduction

This Technical Summary of the *IPCC Special Report on Ocean and Cryosphere in a Changing Climate (SROCC)* consists of the Executive Summaries of all chapters (1–6) of the Special Report, the Executive Summary from the Integrative Cross-Chapter Box on Low-Lying Islands and Coasts, and supporting figures drawn from the chapters and the Summary for Policymakers. The Technical Summary follows the structure of the Report (Table TS.1).

Section TS.1 (Chapter 1) introduces important key concepts, summarizes the characteristics and interconnection of ocean and cryosphere and highlights their importance in the earth system and for human societies in the light of climate change. TS.2 (Chapter 2) assesses changes in high mountain cryosphere and their impacts on local mountain communities and far beyond. TS.3 (Chapter 3) evaluates the state of knowledge concerning changes and impacts in the Arctic and Antarctic ocean and cryosphere systems, including challenges and opportunities for societies. TS.4 (Chapter 4) focusses on regional and global changes in sea level, the associated risk to low-lying islands, coasts and human settlements, and response options. TS.5 (Chapter 5) assesses changes in the ocean and marine ecosystems, including risks to ecosystem services and vulnerability of the dependent communities. TS.6 (Chapter 6) examines extremes and abrupt or irreversible changes in the ocean and cryosphere in a changing climate, and identifies sustainable and resilient risk management strategies. All chapters and their Executive Summaries build on findings since the *IPCC Fifth Assessment Report (AR5)* and, whenever applicable, outcomes of the *IPCC Special Report on Global Warming of 1.5°C (SR15)*.

SROCC uses IPCC calibrated language[1] for the communication of confidence in the assessment process (see Chapter 1 and references therein). This calibrated language is designed to consistently evaluate and communicate uncertainties that arise from incomplete knowledge due to a lack of information, or from disagreement about what is known or even knowable. The IPCC calibrated language uses qualitative expressions of confidence based on the robustness of evidence for a finding, and (where possible) uses quantitative expressions to describe the likelihood of a finding (Figure TS.1).

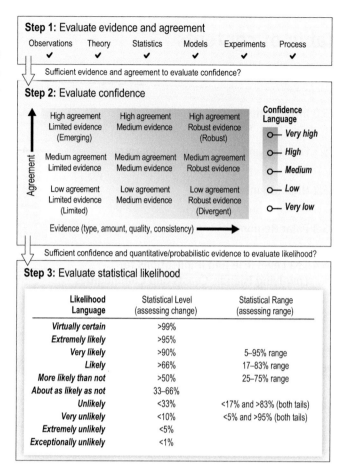

Figure TS.1 | Schematic of the IPCC usage of calibrated language (for more details see Section 1.9.2, Figure 1.4 and Cross-Chapter Box 5 in Chapter 1).

References to chapter sections, boxes, cross-chapter boxes as well as to figures and tables are provided in curly brackets {} at the end of each statement below.

Table TS.1 | Structure of the Technical Summary (TS) and Chapters included in the IPCC Special Report on Ocean and Cryosphere in a Changing Climate (SROCC).

TS.1	Chapter 1: Framing and Context of the Report
TS.2	Chapter 2: High Mountain Areas
TS.3	Chapter 3: Polar Regions
TS.4	Chapter 4: Sea Level Rise and Implications for Low-Lying Islands, Coasts and Communities
TS.5	Chapter 5: Changing Ocean, Marine Ecosystems, and Dependent Communities
TS.6	Chapter 6: Extremes, Abrupt Changes and Managing Risks
TS.7	Integrative Cross-Chapter Box: Low-lying Islands and Coasts

---

[1] Each finding is grounded in an evaluation of underlying evidence and agreement. The summary terms for evidence are: limited, medium or robust. For agreement, they are low, medium or high. In many cases, a synthesis of evidence and agreement supports an assignment of confidence. A level of confidence is expressed using five qualifiers: very low, low, medium, high and very high, and typeset in italics, e.g., *medium confidence*. The following terms have been used to indicate the assessed likelihood of an outcome or a result: virtually certain 99–100% probability, very likely 90–100%, likely 66–100%, about as likely as not 33–66%, unlikely 0–33%, very unlikely 0–10%, exceptionally unlikely 0–1%. Additional terms (extremely likely 95–100%, more likely than not >50–100%, more unlikely than likely 0–<50%, extremely unlikely 0–5%) may also be used when appropriate. Assessed likelihood is typeset in italics, e.g., *very likely*. This Report also uses the term 'likely range' or 'very likely range' to indicate that the assessed likelihood of an outcome lies within the 17–83% or 5–95% probability range. For more details see Chapter 1, Section 1.9.2 and Figure 1.4.

TS.1 Framing and Context of the Report

This special report assesses new knowledge since the IPCC 5th Assessment Report (AR5) and the Special Report on Global Warming of 1.5°C (SR15) on how the ocean and cryosphere have and are expected to change with ongoing global warming, the risks and opportunities these changes bring to ecosystems and people, and mitigation, adaptation and governance options for reducing future risks. Chapter 1 provides context on the importance of the ocean and cryosphere, and the framework for the assessments in subsequent chapters of the report.

All people on Earth depend directly or indirectly on the ocean and cryosphere. The fundamental roles of the ocean and cryosphere in the Earth system include the uptake and redistribution of anthropogenic carbon dioxide and heat by the ocean, as well as their crucial involvement of in the hydrological cycle. The cryosphere also amplifies climate changes through snow, ice and permafrost feedbacks. Services provided to people by the ocean and/or cryosphere include food and freshwater, renewable energy, health and wellbeing, cultural values, trade and transport. {1.1, 1.2, 1.5, Figure TS.2}

Sustainable development is at risk from emerging and intensifying ocean and cryosphere changes. Ocean and cryosphere changes interact with each of the United Nations Sustainable Development Goals (SDGs). Progress on climate action (SDG 13) would reduce risks to aspects of sustainable development

that are fundamentally linked to the ocean and cryosphere and the services they provide (*high confidence*). Progress on achieving the SDGs can contribute to reducing the exposure or vulnerabilities of people and communities to the risks of ocean and cryosphere change (*medium confidence*). {1.1}

Communities living in close connection with polar, mountain, and coastal environments are particularly exposed to the current and future hazards of ocean and cryosphere change. Coasts are home to approximately 28% of the global population, including around 11% living on land less than 10 m above sea level. Almost 10% of the global population lives in the Arctic or high mountain regions. People in these regions face the greatest exposure to ocean and cryosphere change, and poor and marginalised people here are particularly vulnerable to climate-related hazards and risks (*very high confidence*). The adaptive capacity of people, communities and nations is shaped by social, political, cultural, economic, technological, institutional, geographical and demographic factors. {1.1, 1.5, 1.6, Cross-Chapter Box 2 in Chapter 1}

Ocean and cryosphere changes are pervasive and observed from high mountains, to the polar regions, to coasts, and into the deep ocean. AR5 assessed that the ocean is warming (0 to 700 m: *virtually certain*; 700 to 2000 m: *likely*), sea level is rising (*high confidence*), and ocean acidity is increasing (*high confidence*). Most glaciers are shrinking (*high confidence*), the Greenland and Antarctic ice sheets are losing mass (*high confidence*), sea ice extent in the Arctic is decreasing (*very high confidence*), Northern Hemisphere

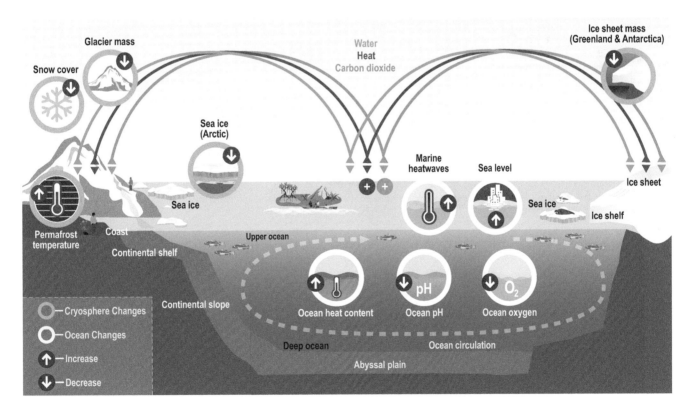

Figure TS.2 | Schematic illustration of key components and changes of the ocean and cryosphere, and their linkages in the Earth system through the global exchange of heat, water, and carbon (Section 1.2). Climate change-related effects (increase/decrease indicated by arrows in pictograms) in the ocean include sea level rise, increasing ocean heat content and marine heat waves, increasing ocean oxygen loss and ocean acidification (Section 1.4.1). Changes in the cryosphere include the decline of Arctic sea ice extent, Antarctic and Greenland ice sheet mass loss, glacier mass loss, permafrost thaw, and decreasing snow cover extent (Section 1.4.2). For illustration purposes, a few examples of where humans directly interact with ocean and cryosphere are shown (for more details see Box 1.1).

Past and future changes in the ocean and cryosphere

Historical changes (observed and modelled) and projections under RCP2.6 and RCP8.5 for key indicators

Figure TS.3

Figure TS.3 | Observed and modelled historical changes in the ocean and cryosphere since 1950[2], and projected future changes under low (RCP2.6) and high (RCP8.5) greenhouse gas emissions scenarios. Changes are shown for: **(a)** Global mean surface air temperature change with likely range. {Box SPM.1, Cross-Chapter Box 1 in Chapter 1} Ocean-related changes with very likely ranges for **(b)** Global mean sea surface temperature change {Box 5.1, 5.2.2}; **(c)** Change factor in surface ocean marine heatwave days {6.4.1}; **(d)** Global ocean heat content change (0–2000 m depth). An approximate steric sea level equivalent is shown with the right axis by multiplying the ocean heat content by the global-mean thermal expansion coefficient ($\varepsilon \approx 0.125$ m per 1024 Joules[3]) for observed warming since 1970 {Figure 5.1}; **(h)** Global mean surface pH (on the total scale). Assessed observational trends are compiled from open ocean time series sites longer than 15 years {Box 5.1, Figure 5.6, 5.2.2}; and **(i)** Global mean ocean oxygen change (100–600 m depth). Assessed observational trends span 1970–2010 centered on 1996 {Figure 5.8, 5.2.2}. Sea level changes with likely ranges for **(m)** Global mean sea level change. Hashed shading reflects low confidence in sea level projections beyond 2100 and bars at 2300 reflect expert elicitation on the range of possible sea level change {4.2.3, Figure 4.2}; and components from **(e,f)** Greenland and Antarctic ice sheet mass loss {3.3.1}; and **(g)** Glacier mass loss {Cross-Chapter Box 6 in Chapter 2, Table 4.1}. Further cryosphere-related changes with very likely ranges for **(j)** Arctic sea ice extent change for September[4] {3.2.1, 3.2.2 Figure 3.3}; **(k)** Arctic snow cover change for June (land areas north of 60°N) {3.4.1, 3.4.2, Figure 3.10}; and **(l)** Change in near-surface (within 3–4 m) permafrost area in the Northern Hemisphere {3.4.1, 3.4.2, Figure 3.10}. Assessments of projected changes under the intermediate RCP4.5 and RCP6.0 scenarios are not available for all variables considered here, but where available can be found in the underlying report. {For RCP4.5 see: 2.2.2, Cross-Chapter Box 6 in Chapter 2, 3.2.2, 3.4.2, 4.2.3, for RCP6.0 see Cross-Chapter Box 1 in Chapter 1}

snow cover is decreasing (*very high confidence*), and permafrost temperatures are increasing (*high confidence*). Improvements since AR5 in observation systems, techniques, reconstructions and model developments, have advanced scientific characterisation and understanding of ocean and cryosphere change, including in previously identified areas of concern such as ice sheets and Atlantic Meridional Overturning Circulation (AMOC). {1.1, 1.4, 1.8.1, Figure TS.3}

Evidence and understanding of the human causes of climate warming, and of associated ocean and cryosphere changes, has increased over the past 30 years of IPCC assessments (*very high confidence*). Human activities are estimated to have caused approximately 1.0°C of global warming above pre-industrial levels (SR15). Areas of concern in earlier IPCC reports, such as the expected acceleration of sea level rise, are now observed (*high confidence*). Evidence for expected slow-down of AMOC is emerging in sustained observations and from long-term palaeoclimate reconstructions (*medium confidence*), and may be related with anthropogenic forcing according to model simulations, although this remains to be properly attributed. Significant sea level rise contributions from Antarctic ice sheet mass loss (*very high confidence*), which earlier reports did not expect to manifest this century, are already being observed. {1.1, 1.4}

Ocean and cryosphere changes and risks by the end-of-century (2081–2100) will be larger under high greenhouse gas emission scenarios, compared with low emission scenarios (*very high confidence*). Projections and assessments of future climate, ocean and cryosphere changes in the Special Report on the Ocean and Cryosphere in a Changing Climate (SROCC) are commonly based on coordinated climate model experiments from the Coupled Model Intercomparison Project Phase 5 (CMIP5) forced with Representative Concentration Pathways (RCPs) of future radiative forcing. Current emissions continue to grow at a rate consistent with a high emission future without effective climate change mitigation policies (referred to as RCP8.5). The SROCC assessment contrasts this high greenhouse gas emission future with a low greenhouse gas emission, high mitigation future (referred to as RCP2.6) that gives a two in three chance of limiting warming by the end of the century to less than 2°C above pre-industrial. {Cross-Chapter Box 1 in Chapter 1, Table TS.2}

Characteristics of ocean and cryosphere change include thresholds of abrupt change, long-term changes that cannot be avoided, and irreversibility (*high confidence*). Ocean warming, acidification and deoxygenation, ice sheet and glacier mass loss, and permafrost degradation are expected to be irreversible on time scales relevant to human societies and ecosystems. Long response times of decades to millennia mean that the ocean and cryosphere are committed to long-term change even after atmospheric greenhouse gas concentrations and radiative forcing stabilise (*high confidence*). Ice-melt or the thawing of permafrost involve thresholds (state changes) that allow for abrupt, nonlinear responses to ongoing climate warming (*high confidence*). These characteristics of ocean and cryosphere change pose risks and challenges to adaptation. {1.1, Box 1.1, 1.3}

Societies will be exposed, and challenged to adapt, to changes in the ocean and cryosphere even if current and future efforts to reduce greenhouse gas emissions keep global warming well below 2°C (*very high confidence*). Ocean and cryosphere-related mitigation and adaptation measures include options that address the causes of climate change, support biological and ecological adaptation, or enhance societal adaptation. Most ocean-based local mitigation and adaptation measures have limited effectiveness to mitigate climate change and reduce its consequences at the global scale, but are useful to implement because they address local risks, often have co-benefits such as biodiversity conservation, and have few adverse side effects. Effective mitigation at a global scale will reduce the need and cost of adaptation, and reduce the risks of surpassing limits to adaptation. Ocean-based carbon dioxide removal at the global scale has potentially large negative ecosystem consequences. {1.6.1, 1.6.2, Cross-Chapter Box 2 in Chapter 1, Figure TS.4}

The scale and cross-boundary dimensions of changes in the ocean and cryosphere challenge the ability of communities, cultures and nations to respond effectively within existing governance frameworks (*high confidence*). Profound economic and institutional transformations are needed if climate-resilient development is to be achieved (*high confidence*). Changes in the ocean and cryosphere, the ecosystem services that they provide,

---

[2]   This does not imply that the changes started in 1950. Changes in some variables have occurred since the pre-industrial period.

[3]   This scaling factor (global-mean ocean expansion as sea level rise in metres per unit heat) varies by about 10% between different models, and it will systematically increase by about 10% by 2100 under RCP8.5 forcing due to ocean warming increasing the average thermal expansion coefficient. {4.2.1, 4.2.2, 5.2.2}

[4]   Antarctic sea ice is not shown here due to low confidence in future projections. {3.2.2}

Table TS.2 | Projected change in global mean surface air temperature and key ocean variables for the *near-term* (2031–2050) and *end-of-century* (2081–2100) relative to the *recent past* (1986–2005) reference period from CMIP5. Small differences in the projections given here compared with AR5 reflect differences in the number of models available now compared to at the time of the AR5 assessment (for more details see Cross-Chapter Box 1 in Chapter 1).

	Scenario	Near-term: 2031–2050		End-of-century: 2081–2100	
		Mean	5–95% range	Mean	5–95% range
Global Mean Surface Air Temperature (°C)[a]	RCP2.6	0.9	0.5–1.4	1.0	0.3–1.7
	RCP4.5	1.1	0.7–1.5	1.8	1.0–2.6
	RCP6.0	1.0	0.5–1.4	2.3	1.4–3.2
	RCP8.5	1.4	0.9–1.8	3.7	2.6–4.8
Global Mean Sea Surface Temperature (°C)[b] (Section 5.2.5)	RCP2.6	0.64	0.33–0.96	0.73	0.20–1.27
	RCP8.5	0.95	0.60–1.29	2.58	1.64–3.51
Surface pH (units)[b] (Section 5.2.2.3)	RCP2.6	–0.072	–0.072 to –0.072	–0.065	–0.065 to –0.066
	RCP8.5	–0.108	–0.106 to –0.110	–0.315	–0.313 to –0.317
Dissolved Oxygen (100–600 m) (% change) (Section 5.2.2.4)[b]	RCP2.6	–0.9	–0.3 to –1.5	–0.6	0.0 to –1.2
	RCP8.5	–1.4	–1.0 to –1.8	–3.9	–2.9 to –5.0

Notes:

[a] Calculated following the same procedure as the IPCC 5th Assessment Report (AR5). The 5–95% model range of global mean surface air temperature across CMIP5 projections was assessed in AR5 as the *likely* range, after accounting for additional uncertainties or different levels of confidence in models.

[b] The 5–95% model range for global mean sea surface temperature, surface pH and dissolved oxygen (100–600 m) as referred to in the SROCC assessment as the *very likely* range (see also Chapter 1, Section 1.9.2, Figure 1.4).

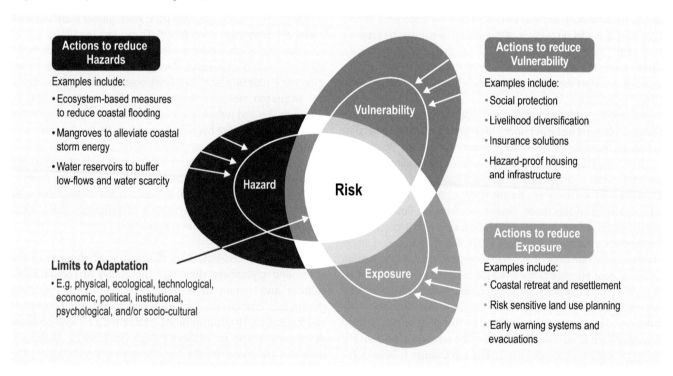

Figure TS.4 | There are options for risk reduction through adaptation. Adaptation can reduce risk by addressing one or more of the three risk factors: vulnerability, exposure, and/or hazard. The reduction of vulnerability, exposure, and/or hazard potential can be achieved through different policy and action choices over time until limits to adaptation might be reached. The figure builds on the conceptual framework of risk used in AR5 (for more details see Cross-Chapter Box 2 in Chapter 1).

the drivers of those changes, and the risks to marine, coastal, polar and mountain ecosystems, occur on spatial and temporal scales that may not align within existing governance structures and practices (*medium confidence*). This report highlights the requirements for transformative governance, international and transboundary cooperation, and greater empowerment of local communities in the governance of the ocean, coasts, and cryosphere in a changing climate. {1.5, 1.7, Cross-Chapter Box 2 in Chapter 1, Cross-Chapter Box 3 in Chapter 1}

Robust assessments of ocean and cryosphere change, and the development of context-specific governance and response options, depend on utilising and strengthening all available knowledge systems (*high confidence*). Scientific knowledge from observations, models and syntheses provides global to local scale understandings of climate change (*very high confidence*). Indigenous knowledge (IK) and local knowledge (LK) provide context-specific and socio-culturally relevant understandings for effective responses and policies (*medium confidence*). Education and climate literacy enable climate action and adaptation (*high confidence*). {1.8, Cross-Chapter Box 4 in Chapter 1}

Long-term sustained observations and continued modelling are critical for detecting, understanding and predicting ocean and cryosphere change, providing the knowledge to inform risk assessments and adaptation planning (*high confidence*). Knowledge gaps exist in scientific knowledge for important regions, parameters and processes of ocean and cryosphere change, including for physically plausible, high impact changes like high end sea level rise scenarios that would be costly if realised without effective adaptation planning and even then may exceed limits to adaptation. Means such as expert judgement, scenario building, and invoking multiple lines of evidence enable comprehensive risk assessments even in cases of uncertain future ocean and cryosphere changes. {1.8.1, 1.9.2, Cross-Chapter Box 5 in Chapter 1}

TS.2 High Mountain Areas

The cryosphere (including, snow, glaciers, permafrost, lake and river ice) is an integral element of high mountain regions, which are home to roughly 10% of the global population. Widespread cryosphere changes affect physical, biological and human systems in the mountains and surrounding lowlands, with impacts evident even in the ocean. Building on the IPCC's 5th Assessment Report (AR5), this chapter assesses new evidence on observed recent and projected changes in the mountain cryosphere as well as associated impacts, risks and adaptation measures related to natural and human systems. Impacts in response to climate changes independently of changes in the cryosphere are not assessed in this chapter. Polar mountains are included in Chapter 3, except those in Alaska and adjacent Yukon, Iceland and Scandinavia, which are included in this chapter.

Observations of cryospheric changes, impacts, and adaptation in high mountain areas

Observations show general decline in low-elevation snow cover (*high confidence*) glaciers (*very high confidence*) and permafrost (*high confidence*) due to climate change in recent decades. Snow cover duration has declined in nearly all regions, especially at lower elevations, on average by 5 days per decade, with a *likely* range from 0–10 days per decade. Low elevation snow depth and extent have declined, although year-to-year variation is high. Mass change of glaciers in all mountain regions (excluding the Canadian and Russian Arctic, Svalbard, Greenland and Antarctica) was *very likely* -490 ± 100 kg m^{-2} yr^{-1} (-123 ± 24 Gt yr^{-1}) in 2006–2015. Regionally averaged mass budgets were *likely* most negative (less than -850 kg m^{-2} yr^{-1}) in the southern Andes, Caucasus and the European Alps/Pyrenees, and least negative in High Mountain Asia (-150 ± 110 kg m^{-2} yr^{-1}) but variations within regions are strong. Between 3.6–5.2 million km^2 are underlain by permafrost in the eleven high mountain regions covered in this chapter corresponding to 27–29% of the global permafrost area (*medium confidence*). Sparse and unevenly distributed measurements show an increase in permafrost temperature (*high confidence*), for example, by 0.19°C $\pm$ 0.05°C on average for about 28 locations in the European Alps, Scandinavia, Canada and Asia during the past decade. Other observations reveal decreasing permafrost thickness and loss of ice in the ground. {2.2.2, 2.2.3, 2.2.4, Figure TS.5}

Glacier, snow and permafrost decline has altered the frequency, magnitude and location of most related natural hazards (*high confidence*). Exposure of people and infrastructure to natural hazards has increased due to growing population, tourism and socioeconomic development (*high confidence*). Glacier retreat and permafrost thaw have decreased the stability of mountain slopes and the integrity of infrastructure (*high confidence*). The number and area of glacier lakes has increased in most regions in recent decades (*high confidence*), but there is only *limited evidence* that the frequency of glacier lake outburst floods (GLOF) has changed. In some regions, snow avalanches involving wet snow have increased (*medium confidence*), and rain-on-snow floods have decreased at low elevations in spring and increased at high elevations in winter (*medium confidence*). The number and extent of wildfires have increased in the Western USA partly due to early snowmelt (*medium confidence*). {2.3.2, 2.3.3}

Changes in snow and glaciers have changed the amount and seasonality of runoff in snow-dominated and glacier-fed river basins (*very high confidence*) with local impacts on water resources and agriculture (*medium confidence*). Winter runoff has increased in recent decades due to more precipitation falling as rain (*high confidence*). In some glacier-fed rivers, summer and annual runoff have increased due to intensified glacier melt, but decreased where glacier melt water has lessened as glacier area shrinks. Decreases were observed especially in regions dominated by small glaciers, such as the European Alps (*medium confidence*). Glacier retreat and snow cover changes have contributed to localized declines in agricultural yields in some high mountain regions, including the Hindu Kush Himalaya and the tropical Andes (*medium confidence*).

There is *limited evidence* of impacts on operation and productivity of hydropower facilities resulting from changes in seasonality and both increases and decreases in water input, for example, in the European Alps, Iceland, Western Canada and USA, and the tropical Andes. {2.3.1}

Species composition and abundance have markedly changed in high mountain ecosystems in recent decades (*very high confidence*), partly due to changes in the cryosphere (*high confidence*). Habitats for establishment by formerly absent species have opened up or been altered by reduced snow cover (*high confidence*), retreating glaciers (*very high confidence*), and thawing of permafrost (*medium confidence*). Reductions in glacier and snow cover have directly altered the structure of many freshwater communities (*high confidence*). Reduced snow cover has negatively impacted the reproductive fitness of some snow-dependent plant and animal species, including foraging and predator-prey relationships of mammals (*high confidence*). Upslope migration of individual species, mostly due to warming and to a lesser extent due to cryosphere-related changes, has often increased local species richness (*very high confidence*). Some cold-adapted species, including endemics, in terrestrial and freshwater communities have declined in abundance (*high confidence*). While the plant productivity has generally increased, the actual impact on provisioning, regulating and cultural ecosystem services varies greatly (*high confidence*). {2.3.3}

Tourism and recreation activities such as skiing, glacier tourism and mountaineering have been negatively impacted by declining snow cover, glaciers and permafrost (*medium confidence*). In several regions, worsening route safety has reduced mountaineering opportunities (*medium confidence*). Variability and decline in natural snow cover have compromised the operation of low-elevation ski resorts (*high confidence*). Glacier and snow decline have impacted aesthetic, spiritual and other cultural aspects of mountain landscapes (*medium confidence*), reducing the well-being of people (e.g., in the Himalaya, eastern Africa, and the tropical Andes). {2.3.5, 2.3.6}

Adaptation in agriculture, tourism and drinking water supply has aimed to reduce the impacts of cryosphere change (*medium confidence*), though there is *limited evidence* on their effectiveness owing to a lack of formal evaluations, or technical, financial and institutional barriers to implementation. In some places, artificial snowmaking has reduced the negative impacts on ski tourism (*medium confidence*). Release and storage of water from reservoirs according to sectoral needs (agriculture, drinking water, ecosystems) has reduced the impact of seasonal variability on runoff (*medium confidence*). {2.3.1, 2.3.5}

Future projections of cryospheric changes, their impacts and risks, and adaptation in high mountain areas

Snow cover, glaciers and permafrost are projected to continue to decline in almost all regions throughout the 21st century (*high confidence*). Compared to 1986–2005, low elevation snow depth will *likely* decrease by 10–40% for 2031–2050, regardless of

Representative Concentration Pathway (RCP) and for 2081–2100, *likely* by 10–40% for RCP2.6 and by 50–90% for RCP8.5. Projected glacier mass reductions between 2015–2100 are *likely* 22–44% for RCP2.6 and 37–57% for RCP8.5. In regions with mostly smaller glaciers and relatively little ice cover (e.g., European Alps, Pyrenees, Caucasus, North Asia, Scandinavia, tropical Andes, Mexico, eastern Africa and Indonesia), glaciers will lose more than 80% of their current mass by 2100 under RCP8.5 (*medium confidence*), and many glaciers will disappear regardless emission scenario (*very high confidence*). Permafrost thaw and degradation will increase during the 21st century (*very high confidence*) but quantitative projections are scarce. {2.2.2, 2.2.3, 2.2.4}

Most types of natural hazards are projected to change in frequency, magnitude and areas affected as the cryosphere continues to decline (*high confidence*). Glacier retreat and permafrost thaw are projected to decrease the stability of mountain slopes and increase the number and area of glacier lakes (*high confidence*). Resulting landslides and floods, and cascading events, will also emerge where there is no record of previous events (*high confidence*). Snow avalanches are projected to decline in number and runout distance at lower elevation, and avalanches involving wet snow even in winter will occur more frequently (*medium confidence*). Rain-on-snow floods will occur earlier in spring and later in autumn, and be more frequent at higher elevations and less frequent at lower elevations (*high confidence*). {2.3.2, 2.3.3}

River runoff in snow dominated and glacier-fed river basins will change further in amount and seasonality in response to projected snow cover and glacier decline (*very high confidence*) with negative impacts on agriculture, hydropower and water quality in some regions (*medium confidence*). The average winter snowmelt runoff is projected to increase (*high confidence*), and spring peaks to occur earlier (*very high confidence*). Projected trends in annual runoff vary substantially among regions, and can even be opposite in direction, but there is *high confidence* that in all regions average annual runoff from glaciers will have reached a peak that will be followed by declining runoff at the latest by the end of the 21st century. Declining runoff is expected to reduce the productivity of irrigated agriculture in some regions (*medium confidence*). Hydropower operations will increasingly be impacted by altered amount and seasonality of water supply from snow and glacier melt (*high confidence*). The release of heavy metals, particularly mercury, and other legacy contaminants currently stored in glaciers and permafrost, is projected to reduce water quality for freshwater biota, household use and irrigation (*medium confidence*). {2.3.1}

Current trends in cryosphere-related changes in high mountain ecosystems are expected to continue and impacts to intensify (*very high confidence*). While high mountains will provide new and greater habitat area, including refugia for lowland species, both range expansion and shrinkage are projected, and at high elevations this will lead to population declines (*high confidence*). The latter increases the risk of local extinctions, in particular for freshwater cold-adapted species (*medium confidence*). Without genetic plasticity and/or behavioural shifts, cryospheric changes will continue to negatively impact endemic and native species, such as some coldwater fish (e.g., trout) and species whose traits directly depend on snow (e.g., snowshoe hares) or many large mammals (*medium confidence*). The survival of such species will depend on appropriate conservation and adaptation measures (*medium confidence*). Many projected ecological changes will alter ecosystem services (*high confidence*), affecting ecological disturbances (e.g., fire, rock fall, slope erosion) with considerable impacts on people (*medium confidence*). {2.3.3}

Cultural assets, such as snow- and ice-covered peaks in many UNESCO World Heritage sites, and tourism and recreation activities, are expected to be negatively affected by future cryospheric change in many regions (*high confidence*). Current snowmaking technologies are projected to be less effective in a warmer climate in reducing risks to ski tourism in most parts of Europe, North America and Japan, in particular at 2°C global warming and beyond (*high confidence*). Diversification through year-round activities supports adaptation of tourism under future climate change (*medium confidence*). {2.3.5, 2.3.6}

Enablers and response options to promote adaptation and sustainable development in high mountain areas

The already committed and unavoidable climate change affecting all cryosphere elements, irrespective of the emission scenario, points to integrated adaptation planning to support and enhance water availability, access, and management (*medium confidence*). Integrated management approaches for water across all scales, in particular for energy, agriculture, ecosystems and drinking water supply, can be effective at dealing with impacts from changes in the cryosphere. These approaches also offer opportunities to support social-ecological systems, through the development and optimisation of storage and the release of water from reservoirs (*medium confidence*), while being cognisant of potential negative implications for some ecosystems. Success in implementing such management options depends on the participation of relevant stakeholders, including affected communities, diverse knowledge and adequate tools for monitoring and projecting future conditions, and financial and institutional resources to support planning and implementation (*medium confidence*). {2.3.1, 2.3.3, 2.4}

Effective governance is a key enabler for reducing disaster risk, considering relevant exposure factors such as planning, zoning, and urbanisation pressures, as well as vulnerability factors such as poverty, which can challenge efforts towards resilience and sustainable development for communities (*medium confidence*). Reducing losses to disasters depend on integrated and coordinated approaches to account for the hazards concerned, the degree of exposure, and existing vulnerabilities. Diverse knowledge that includes community and multi-stakeholder experience with past impacts complements scientific knowledge to anticipate future risks. {Cross-Chapter Box 2 in Chapter 1, 2.3.2, 2.4}

International cooperation, treaties and conventions exist for some mountain regions and transboundary river basins with potential to support adaptation action. However, there is *limited evidence* on the extent to which impacts and losses arising from changes in the cryosphere are specifically monitored and addressed in these frameworks. A wide range of institutional arrangements and practices have emerged over the past three decades that respond to a shared global mountain agenda and specific regional priorities. There is potential to strengthen them to also respond to climate-related cryosphere risks and open opportunities for development through adaptation (*limited evidence, high agreement*). The Sustainable Development Goals (SDGs), Sendai Framework and Paris Agreement have directed some attention in mountain-specific research and practice towards the monitoring and reporting on targets and indicators specified therein. {2.3.1, 2.4}

Figure TS.5 (next pages) | Synthesis of observed regional hazards and impacts in ocean[5] (top) and high mountain and polar land regions (bottom) assessed in SROCC. The same data are shown in different formats in **(a)** and **(b)**. For each region, physical changes, impacts on key ecosystems, and impacts on human systems and ecosystem function and services are shown. For physical changes, yellow/green refers to an increase/decrease, respectively, in amount or frequency of the measured variable. For impacts on ecosystems, human systems and ecosystems services blue or red depicts whether an observed impact is positive (beneficial) or negative (adverse), respectively, to the given system or service. Cells assigned 'increase and decrease' indicate that within that region, both increase and decrease of physical changes are found, but are not necessarily equal; the same holds for cells showing 'positive and negative' attributable impacts. For ocean regions, the confidence level refers to the confidence in attributing observed changes to changes in greenhouse gas forcing for physical changes and to climate change for ecosystem, human systems, and ecosystem services. For high mountain and polar land regions, the level of confidence in attributing physical changes and impacts at least partly to a change in the cryosphere is shown. No assessment means: not applicable, not assessed at regional scale, or the evidence is insufficient for assessment. The physical changes in the ocean are defined as: Temperature change in 0–700 m layer of the ocean except for Southern Ocean (0–2000 m) and Arctic Ocean (upper mixed layer and major inflowing branches); Oxygen in the 0–1200 m layer or oxygen minimum layer; Ocean pH as surface pH (decreasing pH corresponds to increasing ocean acidification). Ecosystems in the ocean: Coral refers to warm-water coral reefs and cold-water corals. The 'upper water column' category refers to epipelagic zone for all ocean regions except Polar Regions, where the impacts on some pelagic organisms in open water deeper than the upper 200 m were included. Coastal wetland includes salt marshes, mangroves and seagrasses. Kelp forests are habitats of a specific group of macroalgae. Rocky shores are coastal habitats dominated by immobile calcified organisms such as mussels and barnacles. Deep sea is seafloor ecosystems that are 3000–6000 m deep. Sea-ice associated includes ecosystems in, on and below sea ice. Habitat services refer to supporting structures and services (e.g., habitat, biodiversity, primary production). Coastal Carbon Sequestration refers to the uptake and storage of carbon by coastal blue carbon ecosystems. Ecosystems on Land: Tundra refers to tundra and alpine meadows, and includes terrestrial Antarctic ecosystems. Migration refers to an increase or decrease in net migration, not to beneficial/adverse value. Impacts on tourism refer to the operating conditions for the tourism sector. Cultural services refer to cultural identity, sense of home, and spiritual, intrinsic and aesthetic values, as well as contributions from glacier archaeology. The underlying information is given for land regions in tables SM2.6, SM2.7, SM2.8, SM3.8, SM3.9, and SM3.10, and for ocean regions in tables SM5.10, SM5.11, SM3.8, SM3.9, and SM3.10. {2.3.1, 2.3.2, 2.3.3, 2.3.4, 2.3.5, 2.3.6, 2.3.7, Figure 2.1, 3.2.1, 3.2.3, 3.2.4, 3.3.3, 3.4.1, 3.4.3, 3.5.2, Box 3.4, 4.2.2, 5.2.2, 5.2.3, 5.3.3, 5.4, 5.6, Figure 5.24, Box 5.3]

---

[5]   Marginal seas are not assessed individually as ocean regions in this report.

(a)

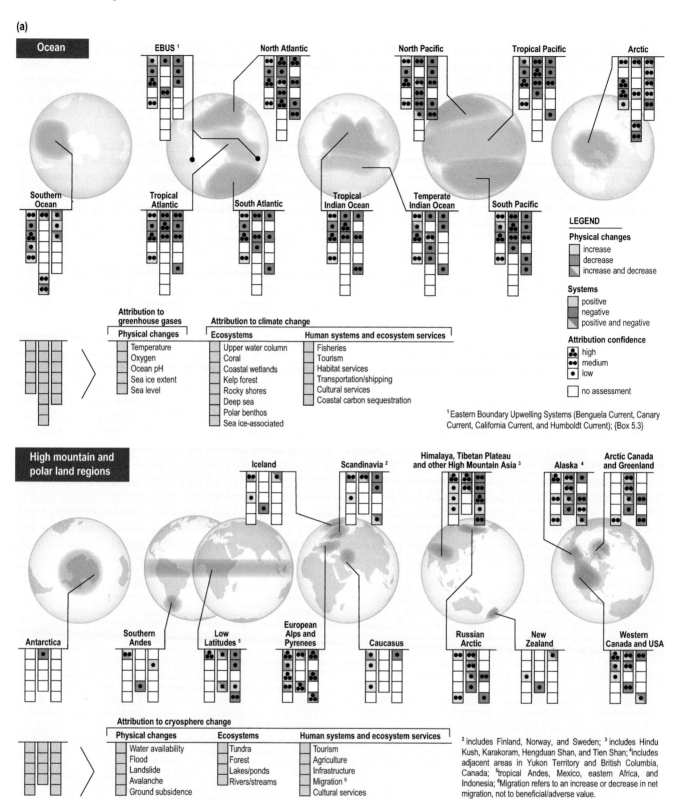

Figure TS.5 | a

(b)

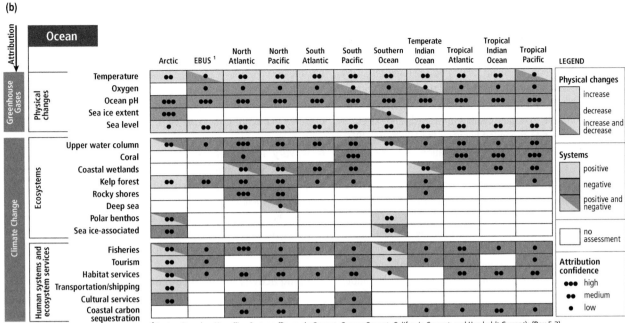

Figure TS.5 | b

TS.3 Polar Regions

This chapter assesses the state of physical, biological and social knowledge concerning the Arctic and Antarctic ocean and cryosphere, how they are affected by climate change, and how they will evolve in future. Concurrently, it assesses the local, regional and global consequences and impacts of individual and interacting polar system changes, and it assesses response options to reduce risk and build resilience in the polar regions. Key findings are:

The polar regions are losing ice, and their oceans are changing rapidly. The consequences of this polar transition extend to the whole planet, and are affecting people in multiple ways.

Arctic surface air temperature has *likely* **increased by more than double the global average over the last two decades, with feedbacks from loss of sea ice and snow cover contributing to the amplified warming.** For each of the five years since the IPCC 5th Asesssment Report (AR5) (2014–2018), Arctic annual surface air temperature exceeded that of any year since 1900. During the winters (January to March) of 2016 and 2018, surface temperatures in the central Arctic were 6°C above the 1981–2010 average, contributing to unprecedented regional sea ice absence. These trends and extremes provide *medium evidence* with *high agreement* of the contemporary coupled atmosphere-cryosphere system moving well outside the 20th century envelope. {Box 3.1, 3.2.1.1}

The Arctic and Southern Oceans are continuing to remove carbon dioxide from the atmosphere and to acidify (*high confidence*). There is *medium confidence* that the amount of CO_2 drawn into the Southern Ocean from the atmosphere has experienced significant decadal variations since the 1980s. Rates of calcification (by which marine organisms form hard skeletons and shells) declined in the Southern Ocean by $3.9 \pm 1.3\%$ between 1998 and 2014. In the Arctic Ocean, the area corrosive to organisms that form shells and skeletons using the mineral aragonite expanded between the 1990s and 2010, with instances of extreme aragonite undersaturation. {3.2.1.2.4}

Both polar oceans have continued to warm in recent years, with the Southern Ocean being disproportionately and increasingly important in global ocean heat increase (*high confidence*). Over large sectors of the seasonally ice-free Arctic, summer upper mixed layer temperatures increased at around $0.5°C$ per decade during 1982–2017, primarily associated with increased absorbed solar radiation accompanying sea ice loss, and the inflow of ocean heat from lower latitude increased since the 2000s (*high confidence*). During 1970–2017, the Southern Ocean south of 30°S accounted for 35–43% of the global ocean heat gain in the upper 2000 m (*high confidence*), despite occupying ~25% of the global ocean area. In recent years (2005–2017), the Southern Ocean was responsible for an increased proportion of the global ocean heat increase (45–62%) (*high confidence*). {3.2.1.2.1, Figure TS.5}

Climate-induced changes in seasonal sea ice extent and thickness and ocean stratification are altering marine primary production (*high confidence*), with impacts on ecosystems (*medium confidence*). Changes in the timing, duration and intensity of primary production have occurred in both polar oceans, with marked regional or local variability (*high confidence*). In the Antarctic, such changes have been associated with locally-rapid environmental change, including retreating glaciers and sea ice change (*medium confidence*). In the Arctic, changes in primary production have affected regional species composition, spatial distribution, and abundance of many marine species, impacting ecosystem structure (*medium confidence*). {3.2.1, 3.2.3, 3.2.4}

In both polar regions, climate-induced changes in ocean and sea ice, together with human introduction of non-native species, have expanded the range of temperate species and contracted the range of polar fish and ice-associated species (*high confidence*). Commercially and ecologically important fish stocks like Atlantic cod, haddock and mackerel have expanded their spatial distributions northwards many hundreds of kilometres, and increased their abundance. In some Arctic areas, such expansions have affected the whole fish community, leading to higher competition and predation on smaller sized fish species, while some commercial fisheries have benefited. There has been a southward shift in the distribution of Antarctic krill in the South Atlantic, the main area for the krill fishery (*medium confidence*). These changes are altering biodiversity in polar marine ecosystems (*medium confidence*). {3.2.3, Box 3.4}

Arctic sea ice extent continues to decline in all months of the year (*very high confidence*); the strongest reductions in September (*very likely* $-12.8 \pm 2.3\%$ per decade; 1979–2018)

are unprecedented in at least 1000 years (*medium confidence*). Arctic sea ice has thinned, concurrent with a shift to younger ice: since 1979, the areal proportion of thick ice at least 5 years old has declined by approximately 90% (*very high confidence*). Approximately half the observed sea ice loss is attributable to increased atmospheric greenhouse gas concentrations (*medium confidence*). Changes in Arctic sea ice have potential to influence mid-latitude weather on timescales of weeks to months (*low to medium confidence*). {3.2.1.1, Box 3.2}

It is *very likely* that Antarctic sea ice cover exhibits no significant trend over the period of satellite observations (1979–2018). While the drivers of historical decadal variability are known with *medium confidence*, there is currently *limited evidence* and *low agreement* concerning causes of the strong recent decrease (2016–2018), and *low confidence* in the ability of current-generation climate models to reproduce and explain the observations. {3.2.1.1}

Shipping activity during the Arctic summer increased over the past two decades in regions for which there is information, concurrent with reductions in sea ice extent (*high confidence*). Transit times across the Northern Sea Route have shortened due to lighter ice conditions, and while long-term, pan-Arctic datasets are incomplete, the distance travelled by ships in Arctic Canada nearly tripled during 1990–2015 (*high confidence*). Greater levels of Arctic ship-based transportation and tourism have socioeconomic and political implications for global trade, northern nations, and economies linked to traditional shipping corridors; they will also exacerbate region specific risks for marine ecosystems and coastal communities if further action to develop and adequately implement regulations does not keep pace with increased shipping (*high confidence*). {3.2.1.1, 3.2.4.2, 3.2.4.3, 3.4.3.3.2, 3.5.2.7}

Permafrost temperatures have increased to record high levels (*very high confidence*), but there is *medium evidence* and *low agreement* that this warming is currently causing northern permafrost regions to release additional methane and carbon dioxide. During 2007–2016, continuous-zone permafrost temperatures in the Arctic and Antarctic increased by $0.39 \pm 0.15°C$ and $0.37 \pm 0.10°C$ respectively. Arctic and boreal permafrost region soils contain 1460–1600 Gt organic carbon (*medium confidence*). Changes in permafrost influence global climate through emissions of carbon dioxide and methane released from the microbial breakdown of organic carbon, or the release of trapped methane. {3.4.1, 3.4.3}

Climate-related changes to Arctic hydrology, wildfire and abrupt thaw are occurring (*high confidence*), with impacts on vegetation and water and food security. Snow and lake ice cover has declined, with June snow extent decreasing $13.4 \pm 5.4\%$ per decade (1967–2018) (*high confidence*). Runoff into the Arctic Ocean increased for Eurasian and North American rivers by $3.3 \pm 1.6\%$ and $2.0 \pm 1.8\%$ respectively (1976–2017; *medium confidence*). Area burned and frequency of fires (including extreme fires) are unprecedented over the last 10,000 years (*high confidence*). There has been an overall greening of the tundra biome, but also browning in some regions of tundra and boreal forest, and changes in the abundance and distribution of animals including reindeer and salmon (*high confidence*). Together, these impact access to (and food availability within) herding, hunting, fishing, forage and gathering areas, affecting the livelihood, health and cultural identity of residents including Indigenous peoples (*high confidence*). {3.4.1, 3.4.3, 3.5.2}

Limited knowledge, financial resources, human capital and organisational capacity are constraining adaptation in many human sectors in the Arctic (*high confidence*). Harvesters of renewable resources are adjusting timing of activities to changes in seasonality and less safe ice travel conditions. Municipalities and industry are addressing infrastructure failures associated with flooding and thawing permafrost, and coastal communities and cooperating agencies are in some cases planning for relocation (*high confidence*). In spite of these adaptations, many groups are making decisions without adequate knowledge to forecast near- and long-term conditions, and without the funding, skills and institutional support to engage fully in planning processes (*high confidence*). {3.5.2, 3.5.4, Cross-Chapter Box 9}

It is *extremely likely* that the rapid ice loss from the Greenland and Antarctic ice sheets during the early 21st century has increased into the near present day, adding to the ice sheet contribution to global sea level rise. From Greenland, the 2012–2016 ice losses (-247 ± 15 Gt yr^{-1}) were similar to those from 2002 to 2011 (-263 ± 21 Gt yr^{-1}) and *extremely likely* greater than from 1992 to 2001 (-8 ± 82 Gt yr^{-1}). Summer melting of the Greenland Ice Sheet (GIS) has increased since the 1990s (*very high confidence*) to a level unprecedented over at least the last 350 years, and two-to-fivefold the pre-industrial level (*medium confidence*). From Antarctica, the 2012–2016 losses (-199 ± 26 Gt yr^{-1}) were *extremely likely* greater than those from 2002 to 2011 (-82 ± 27 Gt yr^{-1}) and *likely* greater than from 1992 to 2001 (-51 ± 73 Gt yr^{-1}). Antarctic ice loss is dominated by acceleration, retreat and rapid thinning of major West Antarctic Ice Sheet (WAIS) outlet glaciers (*very high confidence*), driven by melting of ice shelves by warm ocean waters (*high confidence*). The combined sea level rise contribution from both ice sheets for 2012–2016 was 1.2 ± 0.1 mm yr^{-1}, a 29% increase on the 2002–2011 contribution and a ~700% increase on the 1992–2001 period. {3.3.1}

Mass loss from Arctic glaciers (-212 ± 29 Gt yr^{-1}) during 2006–2015 contributed to sea level rise at a similar rate (0.6 ± 0.1 mm yr^{-1}) to the GIS (*high confidence*). Over the same period in Antarctic and subantarctic regions, glaciers separate from the ice sheets changed mass by -11 ± 108 Gt yr^{-1} (*low confidence*). {2.2.3, 3.3.2}

There is *limited evidence* and *high agreement* that recent Antarctic Ice Sheet (AIS) mass losses could be irreversible over decades to millennia. Rapid mass loss due to glacier flow acceleration in the Amundsen Sea Embayment (ASE) of West Antarctica and in Wilkes Land, East Antarctica, may indicate the beginning of Marine Ice Sheet Instability (MISI), but observational data are not yet sufficient to determine whether these changes mark the beginning of irreversible retreat. {3.3.1, Cross-Chapter Box 8 in Chapter 3, 4.2.3.1.2}

The polar regions will be profoundly different in future compared with today, and the degree and nature of that difference will depend strongly on the rate and magnitude of global climatic change[6]. This will challenge adaptation responses regionally and worldwide.

It is *very likely* that projected Arctic warming will result in continued loss of sea ice and snow on land, and reductions in the mass of glaciers. Important differences in the trajectories of loss emerge from 2050 onwards, depending on mitigation measures taken (*high confidence*). For stabilised global warming of 1.5°C, an approximately 1% chance of a given September being sea ice free at the end of century is projected; for stabilised warming at a 2°C increase, this rises to 10–35% (*high confidence*). The potential for reduced (further 5–10%) but stabilised Arctic autumn and spring snow extent by mid-century for Representative Concentration Pathway (RCP)2.6 contrasts with continued loss under RCP8.5 (a further 15–25% reduction to end of century) (*high confidence*). Projected mass reductions for polar glaciers between 2015 and 2100 range from $16 \pm 7\%$ for RCP2.6 to $33 \pm 11\%$ for RCP8.5 (*medium confidence*). {3.2.2, 3.3.2, 3.4.2, Cross-Chapter Box 6 in Chapter 2}

Both polar oceans will be increasingly affected by CO$_2$ uptake, causing conditions corrosive for calcium carbonate shell-producing organisms (*high confidence*), with associated impacts on marine organisms and ecosystems (*medium confidence*). It is *very likely* that both the Southern Ocean and the Arctic Ocean will experience year-round conditions of surface water undersaturation for mineral forms of calcium carbonate by 2100 under RCP8.5; under RCP2.6 the extent of undersaturated waters are reduced markedly. Imperfect representation of local processes and sea ice interaction in global climate models limit the ability to project the response of specific polar areas and the precise timing of undersaturation at seasonal scales. Differences in sensitivity and the scope for adaptation to projected levels of ocean acidification exist across a broad range of marine species groups. {3.2.1, 3.2.2.3, 3.2.3}

Future climate-induced changes in the polar oceans, sea ice, snow and permafrost will drive habitat and biome shifts, with associated changes in the ranges and abundance of ecologically important species (*medium confidence*). Projected shifts will include further habitat contraction and changes in abundance for polar species, including marine mammals, birds, fish, and Antarctic krill (*medium confidence*). Projected range expansion of subarctic marine species will increase pressure for high-Arctic species (*medium confidence*), with regionally variable impacts. Continued loss of Arctic multi-year sea ice will affect ice-related and pelagic primary production (*high confidence*), with impacts for whole ice-associated, seafloor and open ocean ecosystems. On Arctic land, projections indicate a loss of globally unique biodiversity as some high Arctic species will be outcompeted by more temperate species and very limited refugia exist (*medium confidence*). Woody shrubs and trees are projected to expand, covering 24–52% of the current tundra region by 2050. {3.2.2.1, 3.2.3, 3.2.3.1, Box 3.4, 3.4.2, 3.4.3}

---

[6]   Projections for ice sheets and glaciers in the polar regions are summarized in Chapters 4 and 2, respectively.

The projected effects of climate-induced stressors on polar marine ecosystems present risks for commercial and subsistence fisheries with implications for regional economies, cultures and the global supply of fish, shellfish, and Antarctic krill (*high confidence*). Future impacts for linked human systems depend on the level of mitigation and especially the responsiveness of precautionary management approaches (*medium confidence*). Polar regions support several of the world's largest commercial fisheries. Specific impacts on the stocks and economic value in both regions will depend on future climate change and on the strategies employed to manage the effects on stocks and ecosystems (*medium confidence*). Under high emission scenarios current management strategies of some high-value stocks may not sustain current catch levels in the future (*low confidence*); this exemplifies the limits to the ability of existing natural resource management frameworks to address ecosystem change. Adaptive management that combines annual measures and within-season provisions informed by assessments of future ecosystem trends reduces the risks of negative climate change impacts on polar fisheries (*medium confidence*). {3.2.4, 3.5.2, 3.5.4}

Widespread disappearance of Arctic near-surface permafrost is projected to occur this century as a result of warming (*very high confidence*), with important consequences for global climate. By 2100, near-surface permafrost area will decrease by 2–66% for RCP2.6 and 30–99% for RCP8.5. This is projected to release 10s to 100s of billions of tons (Gt C), up to as much as 240 Gt C, of permafrost carbon as carbon dioxide and methane to the atmosphere with the potential to accelerate climate change. Methane will contribute a small proportion of these additional carbon emissions, on the order of $0.01–0.06$ Gt CH_4 yr^{-1}, but could contribute 40–70% of the total permafrost-affected radiative forcing because of its higher warming potential. There is *medium evidence* but with *low agreement* whether the level and timing of increased plant growth and replenishment of soil will compensate these permafrost carbon losses. {3.4.2, 3.4.3}

Projected permafrost thaw and decrease in snow will affect Arctic hydrology and wildfire, with impacts on vegetation and human infrastructure (*medium confidence*). About 20% of Arctic land permafrost is vulnerable to abrupt permafrost thaw and ground subsidence, which is expected to increase small lake area by over 50% by 2100 for RCP8.5 (*medium confidence*). Even as the overall regional water cycle intensifies, including increased precipitation, evapotranspiration, and river discharge to the Arctic Ocean, decreases in snow and permafrost may lead to soil drying (*medium confidence*). Fire is projected to increase for the rest of this century across most tundra and boreal regions, while interactions between climate and shifting vegetation will influence future fire intensity and frequency (*medium confidence*). By 2050, 70% of Arctic infrastructure is located in regions at risk from permafrost thaw and subsidence; adaptation measures taken in advance could reduce costs arising from thaw and other climate change related impacts such as increased flooding, precipitation, and freeze-thaw events by half (*medium confidence*). {3.4.1, 3.4.2, 3.4.3, 3.5.2}

Response options exist that can ameliorate the impacts of polar change, build resilience and allow time for effective mitigation measures. Institutional barriers presently limit their efficacy.

Responding to climate change in polar regions will be more effective if attention to reducing immediate risks (short-term adaptation) is concurrent with long-term planning that builds resilience to address expected and unexpected impacts (*high confidence*). Emphasis on short-term adaptation to specific problems will ultimately not succeed in reducing the risks and vulnerabilities to society given the scale, complexity and uncertainty of climate change. Moving toward a dual focus of short- and long-term adaptation involves knowledge co-production, linking knowledge with decision making and implementing ecosystem-based stewardship, which involves the transformation of many existing institutions (*high confidence*). {3.5.4}

Innovative tools and practices in polar resource management and planning show strong potential in improving society's capacity to respond to climate change (*high confidence*). Networks of protected areas, participatory scenario analysis, decision support systems, community-based ecological monitoring that draws on local and indigenous knowledge, and self assessments of community resilience contribute to strategic plans for sustaining biodiversity and limit risk to human livelihoods and wellbeing. Such practices are most effective when linked closely to the policy process. Experimenting, assessing, and continually refining practices while strengthening the links with decision making has the potential to ready society for the expected and unexpected impacts of climate change (*high confidence*). {3.5.1, 3.5.2, 3.5.4}

Institutional arrangements that provide for strong multiscale linkages with Arctic local communities can benefit from including indigenous knowledge and local knowledge in the formulation of adaptation strategies (*high confidence*). The tightly coupled relationship of northern local communities and their environment provide an opportunity to better understand climate change and its effects, support adaptation and limit unintended consequences. Enabling conditions for the involvement of local communities in climate adaptation planning include investments in human capital, engagement processes for knowledge co-production and systems of adaptive governance. {3.5.3}

The capacity of governance systems in polar regions to respond to climate change has strengthened recently, but the development of these systems is not sufficiently rapid or robust to address the challenges and risks to societies posed by projected changes (*high confidence*). Human responses to climate change in the polar regions occur in a fragmented governance landscape. Climate change, new polar interests from outside the regions, and an increasingly active role played by informal organisations are compelling stronger coordination and integration between different levels and sectors of governance. The governance landscape is currently not sufficiently equipped to address cascading risks and uncertainty in an integrated and precautionary way within existing legal and policy frameworks (*high confidence*). {3.5.3, 3.5.4}

TS.4   Sea Level Rise and Implications for Low-lying Islands, Coasts and Communities

This chapter assesses past and future contributions to global, regional and extreme sea level changes, associated risk to low-lying islands, coasts, cities, and settlements, and response options and pathways to resilience and sustainable development along the coast.

Observations

Global mean sea level (GMSL) is rising (*virtually certain*) and accelerating (*high confidence*)[7]. The sum of glacier and ice sheet contributions is now the dominant source of GMSL rise (*very high confidence*). GMSL from tide gauges and altimetry observations increased from 1.4 mm yr^{-1} over the period 1901–1990 to 2.1 mm yr^{-1} over the period 1970–2015 to 3.2 mm yr^{-1} over the period 1993–2015 to 3.6 mm yr^{-1} over the period 2006–2015 (*high confidence*). The dominant cause of GMSL rise since 1970 is anthropogenic forcing (*high confidence*). {4.2.2.1.1, 4.2.2.2}

GMSL was considerably higher than today during past climate states that were warmer than pre-industrial, including the Last Interglacial (LIG; 129–116 ka), when global mean surface temperature was 0.5°C–1.0°C warmer, and the mid-Pliocene Warm Period (mPWP; ~3.3 to 3.0 million years ago), 2°C–4°C warmer. Despite the modest global warmth of the Last Interglacial, GMSL was *likely* 6–9 m higher, mainly due to contributions from the Greenland and Antarctic ice sheets (GIS and AIS, respectively), and *unlikely* more than 10m higher (*medium confidence*). Based on new understanding about geological constraints since the IPCC 5th Assessment Report (AR5), 25 m is a plausible upper bound on GMSL during the mPWP (*low confidence*). Ongoing uncertainties in palaeo sea level reconstructions and modelling hamper conclusions regarding the total magnitudes and rates of past sea level rise (SLR). Furthermore, the long (multi-millennial) time scales of these past climate and sea level changes, and regional climate influences from changes in Earth's orbital configuration and climate system feedbacks, lead to *low confidence* in direct comparisons with near-term future changes. {Cross-Chapter Box 5 in Chapter 1, 4.2.2, 4.2.2.1, 4.2.2.5, SM 4.1}

Non-climatic anthropogenic drivers, including recent and historical demographic and settlement trends and anthropogenic subsidence, have played an important role in increasing low-lying coastal communities' exposure and vulnerability to SLR and extreme sea level (ESL) events (*very high confidence*). In coastal deltas, for example, these drivers have altered freshwater and sediment availability (*high confidence*). In low-lying coastal areas more broadly, human-induced changes can be rapid and modify coastlines over short periods of time, outpacing the effects of SLR (*high confidence*). Adaptation can be undertaken in the short- to medium-term by targeting local drivers of exposure and vulnerability, notwithstanding uncertainty about local SLR impacts in coming decades and beyond (*high confidence*). {4.2.2.4, 4.3.1, 4.3.2.2, 4.3.2.3}

Coastal ecosystems are already impacted by the combination of SLR, other climate-related ocean changes, and adverse effects from human activities on ocean and land (*high confidence*). Attributing such impacts to SLR, however, remains challenging due to the influence of other climate-related and non-climatic drivers such as infrastructure development and human-induced habitat degradation (*high confidence*). Coastal ecosystems, including saltmarshes, mangroves, vegetated dunes and sandy beaches, can build vertically and expand laterally in response to SLR, though this capacity varies across sites (*high confidence*). These ecosystems provide important services that include coastal protection and habitat for diverse biota. However, as a consequence of human actions that fragment wetland habitats and restrict landward migration, coastal ecosystems progressively lose their ability to adapt to climate-induced changes and provide ecosystem services, including acting as protective barriers (*high confidence*). {4.3.2.3}

Coastal risk is dynamic and increased by widely observed changes in coastal infrastructure, community livelihoods, agriculture and habitability (*high confidence*). As with coastal ecosystems, attribution of observed changes and associated risk to SLR remains challenging. Drivers and processes inhibiting attribution include demographic, resource and land use changes and anthropogenic subsidence. {4.3.3, 4.3.4}

A diversity of adaptation responses to coastal impacts and risks have been implemented around the world, but mostly as a reaction to current coastal risk or experienced disasters (*high confidence*). Hard coastal protection measures (dikes, embankments, sea walls and surge barriers) are widespread, providing predictable levels of safety in northwest Europe, East Asia, and around many coastal cities and deltas. Ecosystem-based adaptation (EbA) is continuing to gain traction worldwide, providing multiple co-benefits, but there is still *low agreement* on its cost and long-term effectiveness. Advance, which refers to the creation of new land by building into the sea (e.g., land reclamation), has a long history in most areas where there are dense coastal populations. Accommodation measures, such as early warning systems (EWS) for ESL events, are widespread. Retreat is observed but largely restricted to small communities or carried out for the purpose of creating new wetland habitat. {4.4.2.3, 4.4.2.4, 4.4.2.5}

Projections

Future rise in GMSL caused by thermal expansion, melting of glaciers and ice sheets and land water storage changes, is strongly dependent on which Representative Concentration Pathway (RCP) emission scenario is followed. SLR at the end

---

[7]   Statements about uncertainty in Section 4.2 are contingent upon the RCP or other emissions assumptions that accompany them. In Section 4.4, the entirety of information facing a decision maker is taken into consideration, including the unknown path of future emissions, in assessing uncertainty. Depending on which perspective is chosen, uncertainty may or may not be characterised as 'deep'.

of the century is projected to be faster under all scenarios, including those compatible with achieving the long-term temperature goal set out in the Paris Agreement. GMSL will rise between 0.43 m (0.29–0.59 m, *likely* range; RCP2.6) and 0.84 m (0.61–1.10 m, *likely* range; RCP8.5) by 2100 (*medium confidence*) relative to 1986–2005. Beyond 2100, sea level will continue to rise for centuries due to continuing deep ocean heat uptake and mass loss of the GIS and AIS and will remain elevated for thousands of years (*high confidence*). Under RCP8.5, estimates for 2100 are higher and the uncertainty range larger than in AR5. Antarctica could contribute up to 28 cm of SLR (RCP8.5, upper end of *likely* range) by the end of the century (*medium confidence*). Estimates of SLR higher than the *likely* range are also provided here for decision makers with low risk tolerance. {SR1.5, 4.1, 4.2.3.2, 4.2.3.5}

Under RCP8.5, the rate of SLR will be 15 mm yr^{-1} (10–20 mm yr^{-1}, *likely* range) in 2100, and could exceed several cm yr^{-1} in the 22nd century. These high rates challenge the implementation of adaptation measures that involve a long lead time, but this has not yet been studied in detail. {4.2.3.2, 4.4.2.2.3}

Processes controlling the timing of future ice shelf loss and the spatial extent of ice sheet instabilities could increase Antarctica's contribution to SLR to values higher than the *likely* range on century and longer time scales (*low confidence*). Evolution of the AIS beyond the end of the 21st century is characterized by deep uncertainty as ice sheet models lack realistic representations of some of the underlying physical processes. The few model studies available addressing time scales of centuries to millennia indicate multi-metre (2.3–5.4 m) rise in sea level for RCP8.5 (*low confidence*). There is *low confidence* in threshold temperatures for ice sheet instabilities and the rates of GMSL rise they can produce. {Cross-Chapter Box 5 in Chapter 1, Cross-Chapter Box 8 in Chapter 3, and Sections 4.1, 4.2.3.1.1, 4.2.3.1.2, 4.2.3.6}

Sea level rise is not globally uniform and varies regionally. Thermal expansion, ocean dynamics and land ice loss contributions will generate regional departures of about ±30% around the GMSL rise. Differences from the global mean can be greater than ±30% in areas of rapid vertical land movements, including those caused by local anthropogenic factors such as groundwater extraction (*high confidence*). Subsidence caused by human activities is currently the most important cause of relative sea level (RSL) change in many delta regions. While the comparative importance of climate-driven RSL rise will increase over time, these findings on anthropogenic subsidence imply that a consideration of local processes is critical for projections of sea level impacts at local scales (*high confidence*). {4.2.1.6, 4.2.2.4}

Due to projected GMSL rise, ESLs that are historically rare (for example, today's hundred-year event) will become common by 2100 under all RCPs (*high confidence*). Many low-lying cities and small islands at most latitudes will experience such events annually by 2050. Greenhouse gas (GHG) mitigation envisioned in low-emission scenarios (e.g., RCP2.6) is expected to sharply reduce but not eliminate risk to low-lying coasts and islands from SLR and ESL events. Low-emission scenarios lead to slower rates of SLR and

allow for a wider range of adaptation options. For the first half of the 21st century differences in ESL events among the scenarios are small, facilitating adaptation planning. {4.2.2.5, 4.2.3.4, Figure TS.6}

Non-climatic anthropogenic drivers will continue to increase the exposure and vulnerability of coastal communities to future SLR and ESL events in the absence of major adaptation efforts compared to today (*high confidence*). {4.3.4, Cross-Chapter Box 9}

The expected impacts of SLR on coastal ecosystems over the course of the century include habitat contraction, loss of functionality and biodiversity, and lateral and inland migration. Impacts will be exacerbated in cases of land reclamation and where anthropogenic barriers prevent inland migration of marshes and mangroves and limit the availability and relocation of sediment (*high confidence*). Under favourable conditions, marshes and mangroves have been found to keep pace with fast rates of SLR (e.g., >10 mm yr^{-1}), but this capacity varies significantly depending on factors such as wave exposure of the location, tidal range, sediment trapping, overall sediment availability and coastal squeeze (*high confidence*). {4.3.3.5.1}

In the absence of adaptation, more intense and frequent ESL events, together with trends in coastal development will increase expected annual flood damages by 2–3 orders of magnitude by 2100 (*high confidence*). However, well designed coastal protection is very effective in reducing expected damages and cost efficient for urban and densely populated regions, but generally unaffordable for rural and poorer areas (*high confidence*). Effective protection requires investments on the order of tens to several hundreds of billions of USD yr^{-1} globally (*high confidence*). While investments are generally cost efficient for densely populated and urban areas (*high confidence*), rural and poorer areas will be challenged to afford such investments with relative annual costs for some small island states amounting to several percent of GDP (*high confidence*). Even with well-designed hard protection, the risk of possibly disastrous consequences in the event of failure of defences remains. {4.3.4, 4.4.2.2, 4.4.3.2, Cross-Chapter Box 9}

Risk related to SLR (including erosion, flooding and salinisation) is expected to significantly increase by the end of this century along all low-lying coasts in the absence of major additional adaptation efforts (*very high confidence*). While only urban atoll islands and some Arctic communities are expected to experience moderate to high risk relative to today in a low emission pathway, almost high to very high risks are expected in all low-lying coastal settings at the upper end of the *likely* range for high emission pathways (*medium confidence*). However, the transition from moderate to high and from high to very high risk will vary from one coastal setting to another (*high confidence*). While a slower rate of SLR enables greater opportunities for adapting, adaptation benefits are also expected to vary between coastal settings. Although ambitious adaptation will not necessarily eradicate end-century SLR risk (*medium confidence*), it will help to buy time in many locations and therefore help to lay a robust foundation for adaptation beyond 2100. {4.1.3, 4.3.4, Box 4.1, SM4.2}

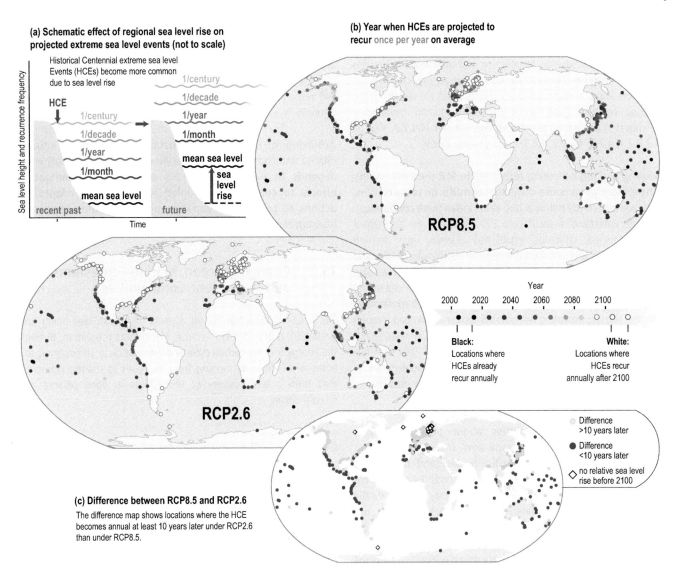

Figure TS.6 | The effect of regional sea level rise on extreme sea level events at coastal locations. Due to projected global mean sea level (GMSL) rise, local sea levels that historically occurred once per century (historical centennial events, HCEs) are projected to become at least annual events at most locations during the 21st century. The height of a HCE varies widely, and depending on the level of exposure can already cause severe impacts. Impacts can continue to increase with rising frequency of HCEs. **(a)** Schematic illustration of extreme sea level events and their average recurrence in the recent past (1986–2005) and the future. As a consequence of mean sea level rise, HCEs are projected to recur more frequently in the future. **(b)** The year in which HCEs are expected to recur once per year on average under RCP8.5 and RCP2.6, at the 439 individual coastal locations where the observational record is sufficient. The absence of a circle indicates an inability to perform an assessment due to a lack of data but does not indicate absence of exposure and risk. The darker the circle, the earlier this transition is expected. The likely range is ±10 years for locations where this transition is expected before 2100. White circles (33% of locations under RCP2.6 and 10% under RCP8.5) indicate that HCEs are not expected to recur once per year before 2100. **(c)** An indication at which locations this transition of HCEs to annual events is projected to occur more than 10 years later under RCP2.6 compared to RCP8.5. As the scenarios lead to small differences by 2050 in many locations results are not shown here for RCP4.5 but they are available in Chapter 4. {4.2.3, Figure 4.10, Figure 4.12}

Choosing and Implementing Responses

All types of responses to SLR, including protection, accommodation, EbA, advance and retreat, have important and synergistic roles to play in an integrated and sequenced response to SLR (*high confidence*). Hard protection and advance (building into the sea) are economically efficient in most urban contexts facing land scarcity (*high confidence*), but can lead to increased exposure in the long term. Where sufficient space is available, EbA can both reduce coastal risks and provide multiple other benefits (*medium confidence*). Accommodation such as flood proofing buildings and EWS for ESL events are often both low-cost and highly cost-efficient in all contexts (*high confidence*). Where coastal

risks are already high, and population size and density are low, or in the aftermath of a coastal disaster, retreat may be especially effective, albeit socially, culturally and politically challenging. {4.4.2.2, 4.4.2.3, 4.4.2.4, 4.4.2.5, 4.4.2.6, 4.4.3}

Technical limits to hard protection are expected to be reached under high emission scenarios (RCP8.5) beyond 2100 (*high confidence*) and biophysical limits to EbA may arise during the 21st century, but economic and social barriers arise well before the end of the century (*medium confidence*). Economic challenges to hard protection increase with higher sea levels and will make adaptation unaffordable before technical limits are reached (*high confidence*). Drivers other than SLR are expected to contribute more to biophysical limits of EbA. For corals, limits may be reached during this century, due to ocean acidification and ocean warming, and for tidal wetlands due to pollution and infrastructure limiting their inland migration. Limits to accommodation are expected to occur well before limits to protection occur. Limits to retreat are uncertain, reflecting research gaps. Social barriers (including governance challenges) to adaptation are already encountered. {4.4.2.2, 4.4.2.3, 4.4.2.3.2, 4.4.2.5, 4.4.2.6, 4.4.3, Cross-Chapter Box 9}

Choosing and implementing responses to SLR presents society with profound governance challenges and difficult social choices, which are inherently political and value laden (*high confidence*). The large uncertainties about post 2050 SLR, and the substantial impact expected, challenge established planning and decision making practises and introduce the need for coordination within and between governance levels and policy domains. SLR responses also raise equity concerns about marginalising those most vulnerable and could potentially spark or compound social conflict (*high confidence*). Choosing and implementing responses is further challenged through a lack of resources, vexing trade-offs between safety, conservation and economic development, multiple ways of framing the 'sea level rise problem', power relations, and various coastal stakeholders having conflicting interests in the future development of heavily used coastal zones (*high confidence*). {4.4.2, 4.4.3}

Despite the large uncertainties about post 2050 SLR, adaptation decisions can be made now, facilitated by using decision analysis methods specifically designed to address uncertainty (*high confidence*). These methods favour flexible responses (i.e., those that can be adapted over time) and periodically adjusted decisions (i.e., adaptive decision making). They use robustness criteria (i.e., effectiveness across a range of circumstances) for evaluating alternative responses instead of standard expected utility criteria (*high confidence*). One example is adaptation pathway analysis, which has emerged as a low-cost tool to assess long-term coastal responses as sequences of adaptive decisions in the face of dynamic coastal risk characterised by deep uncertainty (*medium evidence, high agreement*). The range of SLR to be considered in decisions depends on the risk tolerance of stakeholders, with stakeholders whose risk tolerance is low also considering SLR higher than the *likely* range. {4.1, 4.4.4.3}

Adaptation experience to date demonstrates that using a locally appropriate combination of decision analysis, land use planning, public participation and conflict resolution approaches can help to address the governance challenges faced in responding to SLR (*high confidence*). Effective SLR responses depend, first, on taking a long-term perspective when making short-term decisions, explicitly accounting for uncertainty of locality-specific risks beyond 2050 (*high confidence*), and building governance capabilities to tackle the complexity of SLR risk (*medium evidence, high agreement*). Second, improved coordination of SLR responses across scales, sectors and policy domains can help to address SLR impacts and risk (*high confidence*). Third, prioritising consideration of social vulnerability and equity underpins efforts to promote fair and just climate resilience and sustainable development (*high confidence*) and can be helped by creating safe community arenas for meaningful public deliberation and conflict resolution (*medium evidence, high agreement*). Finally, public awareness and understanding about SLR risks and responses can be improved by drawing on local, indigenous and scientific knowledge systems, together with social learning about locality-specific SLR risk and response potential (*high confidence*). {4.4.4.2, 4.4.5, Table 4.9, Figure TS.7}

Achieving the United Nations Sustainable Development Goals (SDGs) and charting Climate Resilient Development Pathways depends in part on ambitious and sustained mitigation efforts to contain SLR coupled with effective adaptation actions to reduce SLR impacts and risk (*medium evidence, high agreement*).

TS.5 Changing Ocean, Marine Ecosystems, and Dependent Communities

The ocean is essential for all aspects of human well-being and livelihood. It provides key services like climate regulation, through the energy budget, carbon cycle and nutrient cycle. The ocean is the home of biodiversity ranging from microbes to marine mammals that form a wide variety of ecosystems in open pelagic and coastal ocean.

Observations: Climate-related trends, impacts, adaptation

Carbon emissions from human activities are causing ocean warming, acidification and oxygen loss with some evidence of changes in nutrient cycling and primary production. The warming ocean is affecting marine organisms at multiple trophic levels, impacting fisheries with implications for food production and human communities. Concerns regarding the effectiveness of existing ocean and fisheries governance have already been reported, highlighting the need for timely mitigation and adaptation responses.

The ocean has warmed unabated since 2005, continuing the clear multi-decadal ocean warming trends documented in the IPCC Fifth Assessment Report (AR5). The warming trend is further confirmed by the improved ocean temperature measurements over the last decade. The 0–700 m and 700–2000 m layers of the ocean have warmed at rates of 5.31 ± 0.48 and 4.02 ± 0.97 ZJ yr^{-1} from 2005 to 2017. The long-term trend for 0–700 m and 700–2000 m layers have warmed 4.35 ± 0.8 and 2.25 ± 0.64 ZJ yr^{-1} from between the averages of 1971–1990 and 1998–2017 and is attributed to anthropogenic influences. It is *likely* the ocean warming has continued in the abyssal and deep ocean below 2000 m (southern hemisphere and Southern Ocean). {1.8.1, 1.2, 5.2.2}

It is *likely* that the rate of ocean warming has increased since 1993. The 0–700 m and 700–2000 m layers of the ocean have warmed by 3.22 ± 1.61 ZJ and 0.97 ± 0.64 ZJ from 1969 to 1993, and 6.28 ± 0.48 ZJ and 3.86 ± 2.09 ZJ from 1993 to 2017. This represents at least a two-fold increase in heat uptake. {Table 5.1, 5.2.2}

The upper ocean is *very likely* to have been stratifying since 1970. Observed warming and high-latitude freshening are making the surface ocean less dense over time relative to the deeper ocean (*high confidence*) and inhibiting the exchange between surface and deep waters. The upper 200 m stratification increase is in the *very likely* range of between 2.18–2.42% from 1970 to 2017. {5.2.2}

Multiple datasets and models show that the rate of ocean uptake of atmospheric CO_2 has continued to strengthen in the recent two decades in response to the increasing concentration of CO_2 in the atmosphere. The *very likely* range for ocean uptake is between 20–30% of total anthropogenic emissions in the recent two decades. Evidence is growing that the ocean carbon sink is dynamic on decadal timescales, especially in the Southern Ocean, which has affected the total global ocean carbon sink (*medium confidence*). {5.2.2.3}

The ocean is continuing to acidify in response to ongoing ocean carbon uptake. The open ocean surface water pH is observed to be declining (*virtually certain*) by a *very likely* range of 0.017–0.027 pH units per decade since the late 1980s across individual time series observations longer than 15 years. The anthropogenic pH signal is *very likely* to have emerged for three-quarters of the near-surface open ocean prior to 1950 and it is *very likely* that over 95% of the near surface open ocean has already been affected. These changes in pH have reduced the stability of mineral forms of calcium carbonate due to a lowering of carbonate ion concentrations, most notably in the upwelling and high-latitude regions of the ocean. {5.2.2.3, Box 5.1}

There is a growing consensus that the open ocean is losing oxygen overall with a *very likely* loss of 0.5–3.3% between 1970–2010 from the ocean surface to 1000 m (*medium confidence*). Globally, the oxygen loss due to warming is reinforced by

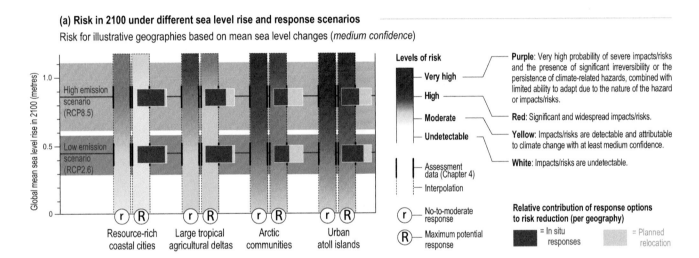

(a) Risk in 2100 under different sea level rise and response scenarios

Risk for illustrative geographies based on mean sea level changes (*medium confidence*)

In this assessment, the term response refers to in situ responses to sea level rise (hard engineered coastal defenses, restoration of degraded ecosystems, subsidence limitation) and planned relocation. Planned relocation in this assessment refers to proactive managed retreat or resettlement only at a local scale, and according to the specificities of a particular context (e.g., in urban atoll islands: within the island, in a neighbouring island or in artificially raised islands). Forced displacement and international migration are not considered in this assessment.

The illustrative geographies are based on a limited number of case studies well covered by the peer reviewed literature. The realisation of risk will depend on context specifities.

Sea level rise scenarios: RCP4.5 and RCP6.0 are not considered in this risk assessment because the literature underpinning this assessment is only available for RCP2.6 and RCP8.5.

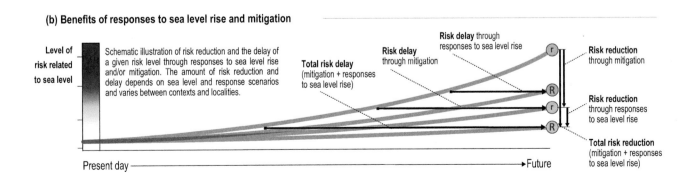

(b) Benefits of responses to sea level rise and mitigation

Figure TS.7 | a, b

(c) Responses to rising mean and extreme sea levels

The table illustrates responses and their characteristics. It is not exhaustive. Whether a response is applicable depends on geography and context.

Confidence levels (assessed for effectiveness): •••• = Very High   ••• = High   •• = Medium   • = Low

Responses		Potential effectiveness in terms of reducing sea level rise (SLR) risks (technical/biophysical limits)	Advantages (beyond risk reduction)	Co–benefits	Drawbacks	Economic efficiency	Governance challenges
Hard protection		Up to multiple metres of SLR {4.4.2.2.4} •••	Predictable levels of safety {4.4.2.2.4}	Multifunctional dikes such as for recreation, or other land use {4.4.2.2.5}	Destruction of habitat through coastal squeeze, flooding & erosion downdrift, lock-in, disastrous consequence in case of defence failure {4.3.2.4, 4.4.2.2.5}	High if the value of assets behind protection is high, as found in many urban and densely populated coastal areas {4.4.2.2.7}	Often unaffordable for poorer areas. Conflicts between objectives (e.g., conservation, safety and tourism), conflicts about the distribution of public budgets, lack of finance {4.3.3.2, 4.4.2.2.6}
Sediment-based protection		Effective but depends on sediment availability {4.4.2.2.4} •••	High flexibility {4.4.2.2.4}	Preservation of beaches for recreation/ tourism {4.4.2.2.5}	Destruction of habitat, where sediment is sourced {4.4.2.2.5}	High if tourism revenues are high {4.4.2.2.7}	Conflicts about the distribution of public budgets {4.4.2.2.6}
Ecosystem based adaptation	Coral conservation	Effective up to 0.5 cm yr⁻¹ SLR. •• Strongly limited by ocean warming and acidification. Constrained at 1.5°C warming and lost at 2°C at many places. {4.3.3.5.2, 4.4.2.3.2, 5.3.4} •••	Opportunity for community involvement, {4.4.2.3.1}	Habitat gain, biodiversity, carbon sequestration, income from tourism, enhanced fishery productivity, improved water quality. Provision of food, medicine, fuel, wood and cultural benefits {4.4.2.3.5}	Long-term effectiveness depends on ocean warming, acidification and emission scenarios {4.3.3.5.2., 4.4.2.3.2}	Limited evidence on benefit-cost ratios; Depends on population density and the availability of land {4.4.2.3.7}	Permits for implementation are difficult to obtain. Lack of finance. Lack of enforcement of conservation policies. EbA options dismissed due to short-term economic interest, availability of land {4.4.2.3.6}
	Coral restoration						
	Wetland conservation (Marshes, Mangroves)	Effective up to 0.5–1 cm yr⁻¹ SLR, •• decreased at 2°C {4.3.3.5.1, 4.4.2.3.2, 5.3.7} •••			Safety levels less predictable, development benefits not realized {4.4.2.3.5, 4.4.2.3.2}		
	Wetland restoration (Marshes, Mangroves)				Safety levels less predictable, a lot of land required, barriers for landward expansion of ecosystems has to be removed {4.4.2.3.5, 4.4.2.3.2}		
Coastal advance		Up to multiple metres of SLR {4.4.2.2.4} •••	Predictable levels of safety {4.4.2.2.4}	Generates land and land sale revenues that can be used to finance adaptation {4.4.2.4.5}	Groundwater salinisation, enhanced erosion and loss of coastal ecosystems and habitat {4.4.2.4.5}	Very high if land prices are high as found in many urban coasts {4.4.2.4.7}	Often unaffordable for poorer areas. Social conflicts with regards to access and distribution of new land {4.4.2.4.6}
Coastal accommodation (Flood-proofing buildings, early warning systems for flood events, etc.)		Very effective for small SLR {4.4.2.5.4} •••	Mature technology; sediments deposited during floods can raise elevation {4.4.2.5.5}	Maintains landscape connectivity {4.4.2.5.5}	Does not prevent flooding/impacts {4.4.2.5.5}	Very high for early warning systems and building-scale measures {4.4.2.5.7}	Early warning systems require effective institutional arrangements {4.4.2.6.6}
Retreat	Planned relocation	Effective if alternative safe localities are available {4.4.2.6.4} •••	Sea level risks at origin can be eliminated {4.4.2.6.4}	Access to improved services (health, education, housing), job opportunities and economic growth {4.4.2.6.5}	Loss of social cohesion, cultural identity and well-being. Depressed services (health, education, housing), job opportunities and economic growth {4.4.2.6.5}	Limited evidence {4.4.2.6.7}	Reconciling the divergent interests arising from relocating people from point of origin and destination {4.4.2.6.6}
	Forced displacement	Addresses only immediate risk at place of origin	Not applicable	Not applicable	Range from loss of life to loss of livelihoods and sovereignty {4.4.2.6.5}	Not applicable	Raises complex humanitarian questions on livelihoods, human rights and equity {4.4.2.6.6}

(d) Choosing and enabling sea level rise responses

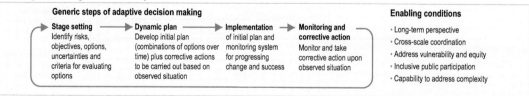

Generic steps of adaptive decision making

➤ **Stage setting**
Identify risks, objectives, options, uncertainties and criteria for evaluating options

➤ **Dynamic plan**
Develop initial plan (combinations of options over time) plus corrective actions to be carried out based on observed situation

➤ **Implementation**
of initial plan and monitoring system for progressing change and success

➤ **Monitoring and corrective action**
Monitor and take corrective action upon observed situation

Enabling conditions
· Long-term perspective
· Cross-scale coordination
· Address vulnerability and equity
· Inclusive public participation
· Capability to address complexity

Figure TS.7 | c, d

Figure TS.7 | Sea level rise risks and responses. The term response is used here instead of adaptation because some responses, such as retreat, may or may not be considered to be adaptation. **(a)** shows the combined risk of coastal flooding, erosion and salinization for illustrative geographies in 2100, due to changing mean and extreme sea levels under RCP2.6 and RCP8.5 and under two response scenarios. Risks under RCPs 4.5 and 6.0 were not assessed due to a lack of literature for the assessed geographies. The assessment does not account for changes in extreme sea level beyond those directly induced by mean sea level rise; risk levels could increase if other changes in extreme sea levels were considered (e.g., due to changes in cyclone intensity). Panel **(a)** considers a socioeconomic scenario with relatively stable coastal population density over the century. {SM4.3.2} Risks to illustrative geographies have been assessed based on relative sea level changes projected for a set of specific examples: New York City, Shanghai and Rotterdam for resource-rich coastal cities covering a wide range of response experiences; South Tarawa, Fongafale and Male' for urban atoll islands; Mekong and Ganges-Brahmaputra-Meghna for large tropical agricultural deltas; and Bykovskiy, Shishmaref, Kivalina, Tuktoyaktuk and Shingle Point for Arctic communities located in regions remote from rapid glacio-isostatic adjustment. {4.2, 4.3.4, SM4.2} The assessment distinguishes between two contrasting response scenarios. "No-to-moderate response" describes efforts as of today (i.e., no further significant action or new types of actions). "Maximum potential response" represents a combination of responses implemented to their full extent and thus significant additional efforts compared to today, assuming minimal financial, social and political barriers. The assessment has been conducted for each sea level rise and response scenario, as indicated by the burning embers in the figure; in-between risk levels are interpolated. {4.3.3} The assessment criteria include exposure and vulnerability (density of assets, level of degradation of terrestrial and marine buffer ecosystems), coastal hazards (flooding, shoreline erosion, salinization), in-situ responses (hard engineered coastal defenses, ecosystem restoration or creation of new natural buffers areas, and subsidence management) and planned relocation. Planned relocation refers to managed retreat or resettlement as described in Chapter 4, i.e., proactive and local-scale measures to reduce risk by relocating people, assets and infrastructure. Forced displacement is not considered in this assessment. Panel **(a)** also highlights the relative contributions of in-situ responses and planned relocation to the total risk reduction. **(b)** schematically illustrates the risk reduction (vertical arrows) and risk delay (horizontal arrows) through mitigation and/or responses to sea level rise. **(c)** summarizes and assesses responses to sea level rise in terms of their effectiveness, costs, co-benefits, drawbacks, economic efficiency and associated governance challenges. {4.4.2} **(d)** presents generic steps of an adaptive decision-making approach, as well as key enabling conditions for responses to sea level rise. {4.4.4, 4.4.5}

other processes associated with ocean physics and biogeochemistry, which cause the majority of the observed oxygen decline (*high confidence*). The oxygen minimum zones (OMZs) are expanding by a *very likely* range of 3–8%, most notably in the tropical oceans, but there is substantial decadal variability that affects the attribution of the overall oxygen declines to human activity in tropical regions (*high confidence*). {5.2.2.4, Figure TS.3, Figure TS.5}

In response to ocean warming and increased stratification, open ocean nutrient cycles are being perturbed and there is *high confidence* that this is having a regionally variable impact on primary producers. There is currently *low confidence* in appraising past open ocean productivity trends, including those determined by satellites, due to newly identified region-specific drivers of microbial growth and the lack of corroborating *in situ* time series datasets. {5.2.2.5, 5.2.2.6}

Ocean warming has contributed to observed changes in biogeography of organisms ranging from phytoplankton to marine mammals (*high confidence*), consequently changing community composition (*high confidence*), and in some cases, altering interactions between organisms (*medium confidence*). Observed rate of range shifts since the 1950s and *its very likely range* are estimated to be 51.5 ± 33.3 km per decade and 29.0 ± 15.5 km per decade for organisms in the epipelagic and seafloor ecosystems, respectively. The direction of the majority of the shifts of epipelagic organisms are consistent with a response to warming (*high confidence*). {5.2.3, 5.3}

Warming-induced range expansion of tropical species to higher latitudes has led to increased grazing on some coral reefs, rocky reefs, seagrass meadows and epipelagic ecosystems, leading to altered ecosystem structure (*medium confidence*). Warming, sea level rise (SLR) and enhanced loads of nutrients and sediments in deltas have contributed to salinisation and deoxygenation in estuaries (*high confidence*), and have caused upstream redistribution of benthic and pelagic species according to their tolerance limits (*medium confidence*). {5.3.4, 5.3.5, 5.3.6, 5.2.3}

Fisheries catches and their composition in many regions are already impacted by the effects of warming and changing primary production on growth, reproduction and survival of fish stocks (*high confidence*). Ocean warming and changes in primary production in the 20th century are related to changes in productivity of many fish stocks (*high confidence*), with an average decrease of approximately 3% per decade in population replenishment and 4.1% (*very likely range* of 9.0% decline to 0.3% increase) in maximum catch potential (*robust evidence, low agreement* between fish stocks, *medium confidence*). Species composition of fisheries catches since the 1970s in many shelf seas ecosystems of the world is increasing dominated by warm water species (*medium confidence*). {5.2.3, 5.4.1}

Warming-induced changes in spatial distribution and abundance of fish stocks have already challenged the management of some important fisheries and their economic benefits (*high confidence*). For existing international and national ocean and fisheries governance, there are concerns about the reduced effectiveness to achieve mandated ecological, economic, and social objectives because of observed climate impacts on fisheries resources (*high confidence*). {5.4.2, 5.5.2}

Coastal ecosystems are observed to be under stress from ocean warming and SLR that are exacerbated by non-climatic pressures from human activities on ocean and land (*high confidence*). Global wetland area has declined by nearly 50% relative to pre-industrial level as a result of warming, SLR, extreme climate events and other human impacts (*medium confidence*). Warming related mangrove encroachment into subtropical salt marshes has been observed in the past 50 years (*high confidence*). Distributions of seagrass meadows and kelp forests are contracting at low-latitudes that is attributable to warming (*high confidence*), and in some areas a loss of 36–43% following heat waves (*medium confidence*). Inundation, coastline erosion and salinisation are causing inland shifts in plant species distributions, which has been accelerating in the last decades (*medium confidence*). Warming has increased the frequency of large-scale coral bleaching events, causing worldwide reef degradation since 1997–1998 with cases of shifts to algal-dominated reefs (*high confidence*). Sessile calcified organisms (e.g., barnacles and mussels) in intertidal rocky shores are highly sensitive to extreme temperature events and acidification (*high confidence*), a reduction in their biodiversity and abundance have been observed in naturally-acidified rocky reef ecosystems (*medium confidence*). Increased nutrient and organic matter loads in estuaries since the 1970s have exacerbated the effects of warming on bacterial respiration and eutrophication, leading to expansion of hypoxic areas (*high confidence*). {5.3.1, 5.3.2, 5.3.4, 5.3.6}

Coastal and near-shore ecosystems including salt marshes, mangroves and vegetated dunes in sandy beaches have a varying capacity to build vertically and expand laterally in response to SLR. These ecosystems provide important services including coastal protection, carbon sequestration and habitat for diverse biota (*high confidence*). The carbon emission associated with the loss of vegetated coastal ecosystems is estimated to be 0.04–1.46 Gt C yr^{-1} (*high confidence*). The natural capacity of ecosystems to adapt to climate impacts may be limited by human activities that fragment wetland habitats and restrict landward migration (*high confidence*). {5.3.2, 5.3.3, 5.4.1, 5.5.1}

Three out of the four major Eastern Boundary Upwelling Systems (EBUS) have shown large-scale wind intensification in the past 60 years (*high confidence*). However, the interaction of coastal warming and local winds may have affected upwelling strength, with the direction of changes varies between and within EBUS (*low confidence*). Increasing trends in ocean acidification in the California Current EBUS and deoxygenation in California Current and Humboldt Current EBUS are observed in the last few decades (*high confidence*), although there is *low confidence* to distinguish anthropogenic forcing from internal climate variability. The expanding California EBUS OMZ has altered ecosystem structure and fisheries catches (*medium confidence*). {Box 5.3}

Since the early 1980s, the occurrence of harmful algal blooms (HABs) and pathogenic organisms (e.g., *Vibrio*) has increased in coastal areas in response to warming, deoxygenation and eutrophication, with negative impacts on food provisioning, tourism, the economy and human health (*high confidence*). These impacts depend on species-specific responses to the interactive effects of climate change and other human drivers (e.g., pollution). Human communities in poorly monitored areas are among the most vulnerable to these biological hazards (*medium confidence*). {Box 5.4, 5.4.2}

Many frameworks for climate resilient coastal adaptation have been developed since AR5, with substantial variations in approach between and within countries, and across development status (*high confidence*). Few studies have assessed the success of implementing these frameworks due to the time-lag between implementation, monitoring, evaluation and reporting (*medium confidence*). {5.5.2}

Projections: scenarios and time horizons

Climate models project significant changes in the ocean state over the coming century. Under the high emissions scenario (Representative Concentration Pathway (RCP)8.5) the impacts by 2090 are substantially larger and more widespread than for the low emissions scenario (RCP2.6) throughout the surface and deep ocean, including: warming (*virtually certain*); ocean acidification (*virtually certain*); decreased stability of mineral forms of calcite (*virtually certain*); oxygen loss (*very likely*); reduced near-surface nutrients (*likely as not*); decreased net primary productivity (*high confidence*); reduced fish production (*likely*) and loss of key ecosystems services (*medium confidence*) that are important for human well-being and sustainable development. {5.2.2, Box 5.1, 5.2.3, 5.2.4, 5.4}

By 2100 the ocean is *very likely* to warm by 2 to 4 times as much for low emissions (RCP2.6) and 5 to 7 times as much for the high emissions scenario (RCP8.5) compared with the observed changes since 1970. The 0–2000 m layer of the ocean is projected to warm by a further 2150 ZJ (*very likely* range 1710–2790 ZJ) between 2017 and 2100 for the RCP8.5 scenario. The 0–2000 m layer is projected to warm by 900 ZJ (*very likely* range 650–1340 ZJ) by 2100 for the RCP2.6 scenario, and the overall warming of the ocean will continue this century even after radiative forcing and mean surface temperatures stabilise (*high confidence*). {5.2.2.2}

The upper ocean will continue to stratify. By the end of the century the annual mean stratification of the top 200 m (averaged between 60°S–60°N relative to the 1986–2005 period) is projected to increase in the *very likely* range of 1–9% and 12–30% for RCP2.6 and RCP8.5 respectively. {5.2.2.2}

It is *very likely* that the majority of coastal regions will experience statistically significant changes in tidal amplitudes over the course of the 21st century. The sign and amplitude of local changes to tides are *very likely* to be impacted by both human coastal adaptation measures and climate drivers. {5.2.2.2.3}

It is *virtually certain* that surface ocean pH will decline, by 0.036–0.042 or 0.287–0.29 pH units by 2081–2100, relative to 2006–2015, for the RCP2.6 or RCP8.5 scenarios, respectively. These pH changes are *very likely* to cause the Arctic and Southern Oceans, as well as the North Pacific and Northwestern Atlantic Oceans to become corrosive for the major mineral forms of calcium carbonate under RCP8.5, but these changes are *virtually certain* to be avoided under the RCP2.6 scenario. There is increasing evidence of an increase in the seasonal exposure to acidified conditions in the future (*high confidence*), with a *very likely* increase in the amplitude of seasonal cycle of hydrogen iron concentrations of 71–90% by 2100, relative to 2000 for the RCP8.5 scenario, especially at high latitudes. {5.2.2.3}

Oxygen is projected to decline further. Globally, the oxygen content of the ocean is *very likely* to decline by 3.2–3.7% by 2081–2100, relative to 2006–2015, for the RCP8.5 scenario or by 1.6–2.0% for the RCP2.6 scenario. The volume of the oceans OMZ is projected to grow by a *very likely* range of 7.0 ± 5.6% by 2100 during the RCP8.5 scenario, relative to 1850–1900. The climate signal of oxygen loss will *very likely* emerge from the historical climate by 2050 with a *very likely* range of 59–80% of ocean area being affected by 2031–2050 and rising with a *very likely* range of 79–91% by 2081–2100 (RCP8.5 emissions scenario). The emergence of oxygen loss is *very likely* smaller in area for the RCP2.6 scenario in the 21st century and by 2090 the emerged area is declining. {5.2.2.4, Box 5.1 Figure 1}

Overall, nitrate concentrations in the upper 100 m are *very likely* to decline by 9–14% across CMIP5 models by 2081–2100, relative to 2006–2015, in response to increased stratification for RCP8.5, with *medium confidence* in these projections due to the *limited evidence* of past changes that can be robustly understood and reproduced by models. There is *low confidence* regarding projected increases in surface ocean iron levels due to systemic uncertainties in these models. {5.2.2.5}

Climate models project that net primary productivity will *very likely* decline by 4–11% for RCP8.5 by 2081–2100, relative to 2006–2015. The decline is due to the combined effects of warming, stratification, light, nutrients and predation and will show regional variations between low and high latitudes (*low confidence*). The tropical ocean NPP will *very likely* decline by 7–16% for RCP8.5, with *medium confidence* as there are improved constraints from historical variability in this region. Globally, the sinking flux of organic matter from the upper ocean into the ocean interior is *very likely* to decrease by 9–16% for RCP8.5 in response to increased stratification and reduced nutrient supply, especially in tropical regions (*medium confidence*), which will reduce organic carbon supply to deep sea ecosystems (*high confidence*). The reduction in food supply to the

deep sea is projected to lead to a 5–6% reduction in biomass of benthic biota over more than 97% of the abyssal seafloor by 2100 (*medium confidence*). {5.2.2.6, 5.2.4.2, Figure TS.8}

New ocean states for a broad suite of climate indices will progressively emerge over a substantial fractions of the ocean in the coming century (relative to past internal ocean variability), with Earth System Models (ESMs) showing an ordered emergence of first pH, followed by sea surface temperature (SST), interior oxygen, upper ocean nutrient levels and finally net primary production (NPP). The anthropogenic pH signal has *very likely* emerged for three quarters of the ocean prior to 1950, with little difference between scenarios. Oxygen changes will *very likely* emerge over 59–80% of the ocean area by 2031–2050 and rises to 79–91% by 2081–2100 (RCP8.5 emissions scenario). The projected time of emergence for five primary drivers of marine ecosystem change (surface warming and acidification, oxygen loss, nitrate content and net primary production change) are all prior to 2100 for over 60% of the ocean area under RCP8.5 and over 30% under RCP2.6 (*very likely*). {Box 5.1, Box 5.1 Figure 1}

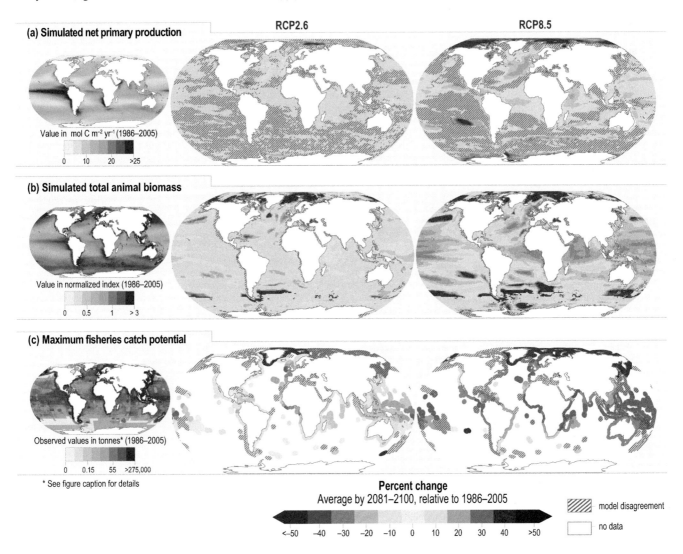

Figure TS.8 | a, b , c

(d) Impacts and risks to ocean ecosystems from climate change

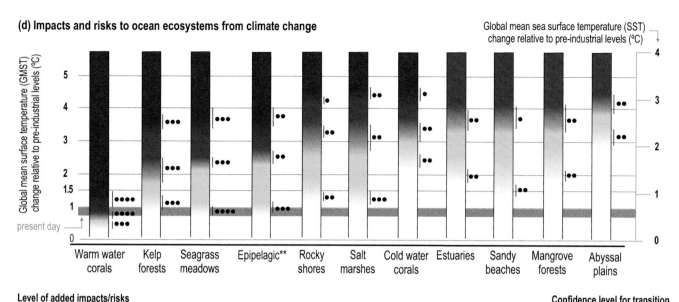

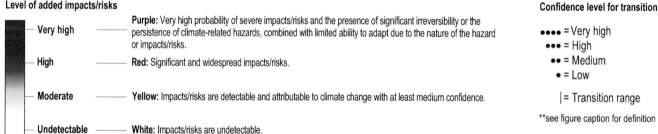

Figure TS.8 | Projected changes, impacts and risks for ocean regions and ecosystems. **(a)** depth integrated net primary production (NPP from CMIP5)[8], **(b)** total animal biomass (depth integrated, including fishes and invertebrates from FISHMIP)[9], **(c)** maximum fisheries catch potential and **(d)** impacts and risks for coastal and open ocean ecosystems. The three left panels represent the simulated **(a,b)** and observed **(c)** mean values for the recent past (1986–2005), the middle and right panels represent projected changes (%) by 2081–2100 relative to recent past under low (RCP2.6) and high (RCP8.5) greenhouse gas emissions scenario (see Table TS.2), respectively. Total animal biomass in the recent past (b, left panel) represents the projected total animal biomass by each spatial pixel relative to the global average. **(c)** *Average observed fisheries catch in the recent past (based on data from the Sea Around Us global fisheries database); projected changes in maximum fisheries catch potential in shelf seas are based on the average outputs from two fisheries and marine ecosystem models. To indicate areas of model inconsistency, shaded areas represent regions where models disagree in the direction of change for more than: **(a)** and **(b)** 3 out of 10 model projections, and **(c)** one out of two models. Although unshaded, the projected change in the Arctic and Antarctic regions in **(b)** total animal biomass and **(c)** fisheries catch potential have low confidence due to uncertainties associated with modelling multiple interacting drivers and ecosystem responses. Projections presented in **(b)** and **(c)** are driven by changes in ocean physical and biogeochemical conditions e.g., temperature, oxygen level, and net primary production projected from CMIP5 Earth system models. ******The epipelagic refers to the uppermost part of the ocean with depth <200 m from the surface where there is enough sunlight to allow photosynthesis. **(d)** Assessment of risks for coastal and open ocean ecosystems based on observed and projected climate impacts on ecosystem structure, functioning and biodiversity. Impacts and risks are shown in relation to changes in Global Mean Surface Temperature (GMST) relative to pre-industrial level. Since assessments of risks and impacts are based on global mean Sea Surface Temperature (SST), the corresponding SST levels are shown[10]. The assessment of risk transitions is described in Chapter 5 Sections 5.2, 5.3, 5.2.5 and 5.3.7 and Supplementary Materials SM5.3, Table SM5.6, Table SM5.8 and other parts of the underlying report. The figure indicates assessed risks at approximate warming levels and increasing climate-related hazards in the ocean: ocean warming, acidification, deoxygenation, increased density stratification, changes in carbon fluxes, sea level rise, and increased frequency and/or intensity of extreme events. The assessment considers the natural adaptive capacity of the ecosystems, their exposure and vulnerability. Impact and risk levels do not consider risk reduction strategies such as human interventions, or future changes in non-climatic drivers. Risks for ecosystems were assessed by considering biological, biogeochemical, geomorphological and physical aspects. Higher risks associated with compound effects of climate hazards include habitat and biodiversity loss, changes in species composition and distribution ranges, and impacts/risks on ecosystem structure and functioning, including changes in animal/plant biomass and density, productivity, carbon fluxes, and sediment transport. As part of the assessment, literature was compiled and data extracted into a summary table. A multi-round expert elicitation process was undertaken with independent evaluation of threshold judgement, and a final consensus discussion. Further information on methods and underlying literature can be found in Chapter 5, Sections 5.2 and 5.3 and Supplementary Material. {3.2.3, 3.2.4, 5.2, 5.3, 5.2.5, 5.3.7, SM5.6, SM5.8, Figure 5.16, Cross Chapter Box 1 in Chapter 1 Table CCB1}

---

[8]   NPP is estimated from the Coupled Models Intercomparison Project 5 (CMIP5).

[9]   Total animal biomass is from the Fisheries and Marine Ecosystem Models Intercomparison Project (FISHMIP).

[10]   The conversion between GMST and SST is based on a scaling factor of 1.44 derived from changes in an ensemble of RCP8.5 simulations; this scaling factor has an uncertainty of about 4% due to differences between the RCP2.6 and RCP8.5 scenarios. {Table SPM.1}

Simulated ocean warming and changes in NPP during the 21st century are projected to alter community structure of marine organisms (*high confidence*), reduce global marine animal biomass (*medium confidence*) and the maximum potential catches of fish stocks (*medium confidence*) with regional differences in the direction and magnitude of changes (*high confidence*). The global biomass of marine animals, including those that contribute to fisheries, is projected to decrease with a *very likely* range under RCP2.6 and RCP8.5 of 4.3 ± 2.0% and 15.0 ± 5.9%, respectively, by 2080–2099 relative to 1986–2005. The maximum catch potential is projected to decrease by 3.4% to 6.4% (RCP2.6) and 20.5% to 24.1% (RCP8.5) in the 21st century. {5.4.1}

Projected decreases in global marine animal biomass and fish catch potential could elevate the risk of impacts on income, livelihood and food security of the dependent human communities (*medium confidence*). Projected climate change impacts on fisheries also increase the risk of potential conflicts among fishery area users and authorities or between two different communities within the same country (*medium confidence*), exacerbated through competing resource exploitation from international actors and mal-adapted policies (*low confidence*). {5.2.3, 5.4, 5.5.3}

Projected decrease in upper ocean export of organic carbon to the deep seafloor is expected to result in a loss of animal biomass on the deep seafloor by 5.2–17.6% by 2090–2100 compared to the present (2006–2015) under RCP8.5 with regional variations (*medium confidence*). Some increases are projected in the polar regions, due to enhanced stratification in the surface ocean, reduced primary production and shifts towards small phytoplankton (*medium confidence*). The projected impacts on biomass in the abyssal seafloor are larger under RCP8.5 than RCP4.5 (*very likely*). The increase in climatic hazards beyond thresholds of tolerance of deep sea organisms will increase the risk of loss of biodiversity and impacts on functioning of deep water column and seafloor that is important to support ecosystem services, such as carbon sequestration (*medium confidence*). {5.2.4}

Structure and functions of all types of coastal ecosystems will continue to be at moderate to high risk under the RCP2.6 scenario (*medium confidence*) and will face high to very high risk under the RCP8.5 scenario (*high confidence*) by 2100. Seagrass meadows (*high confidence*) and kelp forests (*high confidence*) will face moderate to high risk at temperature above 1.5°C global sea surface warming. Coral reefs will face very high risk at temperatures 1.5°C of global sea surface warming (*very high confidence*). Intertidal rocky shores are also expected to be at very high risk (transition above 3°C) under the RCP8.5 scenario (*medium confidence*). These ecosystems have low to moderate adaptive capacity, as they are highly sensitive to ocean temperatures and acidification. The ecosystems with moderate to high risk (transition above 1.8°C) under future emissions scenarios are mangrove forests, sandy beaches, estuaries and salt marshes (*medium confidence*). Estuaries and sandy beaches are subject to highly dynamic hydrological and geomorphological processes, giving them more natural adaptive capacity to climate hazards. In these systems, sediment relocation, soil accretion and landward expansion of vegetation may initially mitigate against flooding and habitat loss, but salt marshes in particular will be at very high risk in the context of SLR and extreme climate-driven erosion under RCP8.5. {5.3, Figure 5.16}

Expected coastal ecosystem responses over the 21st century are habitat contraction, migration and loss of biodiversity and functionality. Pervasive human coastal disturbances will limit natural ecosystem adaptation to climate hazards (*high confidence*). Global coastal wetlands will lose between 20–90% of their area depending on emissions scenario with impacts on their contributions to carbon sequestration and coastal protection (*high confidence*). Kelp forests at low-latitudes and temperate seagrass meadows will continue to retreat as a result of intensified extreme temperatures, and their low dispersal ability will elevate the risk of local extinction under RCP8.5 (*high confidence*). Intertidal rocky shores will continue to be affected by ocean acidification, warming, and extreme heat exposure during low tide emersion, causing reduction of calcareous species and loss of ecosystem biodiversity and complexity shifting towards algae dominated habitats (*high confidence*). Salinisation and expansion of hypoxic conditions will intensify in eutrophic estuaries, especially in mid and high latitudes with microtidal regimes (*high confidence*). Sandy beach ecosystems will increasingly be at risk of eroding, reducing the habitable area for dependent organisms (*high confidence*). {5.3, 5.4.1}

Almost all coral reefs will degrade from their current state, even if global warming remains below 2°C (*very high confidence*), and the remaining shallow coral reef communities will differ in species composition and diversity from present reefs (*very high confidence*). These declines in coral reef health will greatly diminish the services they provide to society, such as food provision (*high confidence*), coastal protection (*high confidence*) and tourism (*medium confidence*). {5.3.4, 5.4.1}

Multiple hazards of warming, deoxygenation, aragonite under-saturation and decrease in flux of organic carbon from the surface ocean will decrease calcification and exacerbate the bioerosion and dissolution of the non-living component of cold water coral. Habitat-forming, cold water corals will be vulnerable where temperature and oxygen exceed the species' thresholds (*medium confidence*). Reduced particulate food supply is projected to be experienced by 95% of cold water coral ecosystems by 2100 under RCP8.5 relative to the present, leading to a *very likely* range of 8.6 ± 2% biomass loss (*medium confidence*). {5.2.4, Box 5.2}

Anthropogenic changes in EBUS will emerge primarily in the second half of the 21st century (*medium confidence*). EBUS will be impacted by climate change in different ways, with strong regional variability with consequences for fisheries, recreation and climate regulation (*medium confidence*). The Pacific EBUS are projected to have calcium carbonate undersaturation in surface waters within a few decades under RCP8.5 (*high confidence*); combined with warming and decreasing oxygen levels, this will increase the impacts on shellfish larvae, benthic invertebrates and demersal fishes (*high confidence*) and related fisheries and aquaculture (*medium confidence*). The inherent natural variability of EBUS, together with uncertainties in present and future trends in the intensity and seasonality of upwelling, coastal warming and stratification, primary production and biogeochemistry of source waters poses large challenges in projecting the response of EBUS to climate change and to the adaptation of governance of biodiversity conservation and living marine resources in EBUS (*high confidence*). {Box 5.3}

Climate change impacts on ecosystems and their goods and services threatens key cultural dimensions of lives and livelihoods. These threats include erosion of Indigenous and non-indigenous culture, their knowledge about the ocean and knowledge transmission, reduced access to traditional food, loss of opportunities for aesthetic and spiritual appreciation of the ecosystems, and marine recreational activities (*medium confidence*). Ultimately, these can lead to the loss of part of people's cultural identity and values beyond the rate at which identify and values can be adjusted or substituted (*medium confidence*). {5.4.2}

Climate change increases the exposure and bioaccumulation of contaminants such as persistent organic pollutants and mercury (*medium confidence*), and their risk of impacts on marine ecosystems and seafood safety (*high agreement, medium evidence, medium confidence*). Such risks are particularly large for top predators and for human communities that have high consumption on these organisms, including coastal Indigenous communities (*medium confidence*). {5.4.2}

Shifting distributions of fish stocks between governance jurisdictions will increase the risk of potential conflicts among fishery area users and authorities or between two different communities within the same country (*medium confidence*). These fishery governance related risks are widespread under high emissions scenarios with regional hotspots (*medium confidence*), and highlight the limits of existing natural resource management frameworks for addressing ecosystem change (*high confidence*). {5.2.5, 5.4.2.1.3, 5.5, 5.5.2}

Response options to enhance resilience

There is clear evidence for observed climate change impacts throughout the ocean with consequences for human communities and require options to reduce risks and impacts. Coastal blue carbon can contribute to mitigation for many nations but its global scope is modest (offset of <2% of current emissions) (*likely*). Some ocean indices are expected to emerge earlier than others (e.g., warming, acidification and effects on fish stocks) and could therefore be used to prioritise planning and building resilience. The survival of some keystone ecosystems (e.g., coral reefs) are at risk, while governance structures are not well-matched to the spatial and temporal scale of climate change impacts on ocean systems. Ecosystem restoration may be able to locally reduce climate risks (*medium confidence*) but at relatively high cost and effectiveness limited to low emissions scenarios and to less sensitive systems (*high confidence*). {5.2, 5.3, 5.4, 5.5}

Coastal blue carbon ecosystems, such as mangroves, salt marshes and seagrasses, can help reduce the risks and impacts of climate change, with multiple co-benefits. Some 151 countries around the world contain at least one of these coastal blue carbon ecosystems and 71 countries contain all three. Below-ground carbon storage in vegetated marine habitats can be up to 1000 tC ha^{-1}, much higher than most terrestrial ecosystems (*high confidence*). Successful implementation of measures to maintain and promote carbon storage in such coastal ecosystems could assist several countries in achieving a balance between emissions and removals of greenhouse gases (*medium confidence*). Conservation of these habitats would also sustain the wide range of ecosystem services they provide and assist with climate adaptation through improving critical habitats for biodiversity, enhancing local fisheries production, and protecting coastal communities from SLR and storm events (*high confidence*). The climate mitigation effectiveness of other natural carbon removal processes in coastal waters, such as seaweed ecosystems and proposed non-biological marine CO_2 removal methods, are smaller or currently have higher associated uncertainties. Seaweed aquaculture warrants further research attention. {5.5.1.1, 5.5.1.1, 5.5.1, 5.5.2, 5.5.1.1.3, 5.5.1.1.4}

The potential climatic benefits of blue carbon ecosystems can only be a very modest addition to, and not a replacement for, the very rapid reduction of greenhouse gas emissions. The maximum global mitigation benefits of cost-effective coastal wetland restoration is *unlikely* to be more than 2% of current total emissions from all sources. Nevertheless, the protection and enhancement of coastal blue carbon can be an important contribution to both mitigation and adaptation at the national scale. The feasibility of climate mitigation by open ocean fertilisation of productivity is limited to negligible, due to the likely decadal-scale return to the atmosphere of nearly all the extra carbon removed, associated difficulties in carbon accounting, risks of unintended side effects and low acceptability. Other human interventions to enhance marine carbon uptake, for example, ocean alkalinisation (enhanced weathering), would also have governance challenges, with the increased risk of undesirable ecological consequences (*high confidence*). {5.5.1.2}

Socioinstitutional adaptation responses are more frequently reported in the literature than ecosystem-based and built infrastructure approaches. Hard engineering responses are more effective when supported by ecosystem-based adaptation (EbA) approaches (*high agreement*), and both approaches are enhanced by combining with socioinstitutional approaches for adaptation (*high confidence*). Stakeholder engagement is necessary (*robust evidence, high agreement*). {5.5.2}

EbA is a cost-effective coastal protection tool that can have many co-benefits, including supporting livelihoods, contributing to carbon sequestration and the provision of a range of other valuable ecosystem services (*high confidence*). Such adaptation does, however, assume that the climate can be stabilised. Under changing climatic conditions there are limits to the effectiveness of ecosystem-based adaptation, and these limits are currently difficult to determine. {5.5.2.1}

Socioinstitutional adaptation responses, including community-based adaptation, capacity-building, participatory processes, institutional support for adaptation planning and support mechanisms for communities are important tools to address climate change impacts (*high confidence*). For fisheries management, improving coordination of integrated coastal management and marine protected areas (MPAs) have emerged in the literature as important adaptation governance responses (*robust evidence, medium agreement*). {5.5.2.2, 5.5.2.6}

Observed widespread decline in warm water corals has led to the consideration of alternative restoration approaches to enhance climate resilience. Approaches, such as 'coral reef gardening' have been tested, and ecological engineering and other approaches such as assisted evolution, colonisation and chimerism are being researched for reef restoration. However, the effectiveness of these approaches to increase resilience to climate stressors and their large-scale implementation for reef restoration will be limited unless warming and ocean acidification are rapidly controlled (*high confidence*). {Box 5.5, 5.5.2}

Existing ocean governance structures are already facing multi-dimensional, scale-related challenges because of climate change. This trend of increasing complexity will continue (*high confidence*). The mechanisms for the governance of marine Areas Beyond National Jurisdiction (ABNJ), such as ocean acidification, would benefit from further development (*high confidence*). There is also scope to increase the overall effectiveness of international and national ocean governance regimes by increasing cooperation, integration and widening participation (*medium confidence*). Diverse adaptations of ocean related governance are being tried, and some are producing promising results. However, rigorous evaluation is needed of the effectiveness of these adaptations in achieving their goals. {5.5.3}

There are a broad range of identified barriers and limits for adaptation to climate change in ecosystems and human systems (*high confidence*). Limitations include the space that ecosystems require, non-climatic drivers and human impacts that need to be addressed as part of the adaptation response, the lowering of adaptive capacity of ecosystems because of climate change, and the slower ecosystem recovery rates relative to the recurrence of climate impacts, availability of technology, knowledge and financial support and existing governance structures (*medium confidence*). {5.5.2}

TS.6 Extremes, Abrupt Changes and Managing Risks

This chapter assesses extremes and abrupt or irreversible changes in the ocean and cryosphere in a changing climate, to identify regional hot spots, cascading effects, their impacts on human and natural systems, and sustainable and resilient risk management strategies. It is not comprehensive in terms of the systems assessed and some information on extremes, abrupt and irreversible changes, in particular for the cryosphere, may be found in other chapters.

Ongoing and Emerging Changes in the Ocean and Cryosphere, and their Impacts on Ecosystems and Human Societies

Anthropogenic climate change has increased observed precipitation (*medium confidence*), winds (*low confidence*), and extreme sea level events (*high confidence*) associated with some tropical cyclones, which has increased intensity of multiple extreme events and associated cascading impacts (*high confidence*). Anthropogenic climate change may have contributed to a poleward migration of maximum tropical cyclone intensity in the western North Pacific in recent decades related to anthropogenically-forced tropical expansion (*low confidence*). There is emerging evidence for an increase in the annual global proportion of Category 4 or 5 tropical cyclones in recent decades (*low confidence*). {6.3, Table 6.2, Figure 6.2, Box 6.1}

Changes in Arctic sea ice have the potential to influence mid-latitude weather (*medium confidence*), but there is *low confidence* in the detection of this influence for specific weather types. {6.3}

Extreme wave heights, which contribute to extreme sea level events, coastal erosion and flooding, have increased in the Southern and North Atlantic Oceans by around 1.0 cm yr^{-1} and 0.8 cm yr^{-1} over the period 1985–2018 (*medium confidence*). Sea ice loss in the Arctic has also increased wave heights over the period 1992–2014 (*medium confidence*). {6.3}

Marine heatwaves (MHWs), periods of extremely high ocean temperatures, have negatively impacted marine organisms and ecosystems in all ocean basins over the last two decades, including critical foundation species such as corals, seagrasses and kelps (*very high confidence*). Globally, marine heat related events have increased; marine heatwaves, defined when the daily sea surface temperature exceeds the local 99th percentile over the period 1982 to 2016, have doubled in frequency and have become longer-lasting, more intense and more extensive (*very likely*). It is *very likely* that between 84–90% of marine heatwaves that occurred between 2006 and 2015 are attributable to the anthropogenic temperature increase. {6.4, Figures 6.3, 6.4}

Both palaeoclimate and modern observations suggest that the strongest El Niño and La Niña events since the pre-industrial period have occurred during the last fifty years (*medium confidence*). There have been three occurrences of extreme El Niño events during the modern observational period (1982–1983, 1997–1998, 2015–2016), all characterised by pronounced rainfall in the normally dry equatorial East Pacific. There have been two occurrences of extreme La Niña (1988–1989, 1998–1999). El Niño and La Niña variability during the last 50 years is unusually high compared with average variability during the last millennium. {6.5, Figure 6.5}

The equatorial Pacific trade wind system experienced an unprecedented intensification during 2001–2014, resulting in enhanced ocean heat transport from the Pacific to the Indian

Ocean, influencing the rate of global temperature change (*medium confidence*). In the last two decades, total water transport from the Pacific to the Indian Ocean by the Indonesian Throughflow (ITF), and the Indian Ocean to Atlantic Ocean has increased (*high confidence*). Increased ITF has been linked to Pacific cooling trends and basin-wide warming trends in the Indian Ocean. Pacific sea surface temperature (SST) cooling trends and strengthened trade winds have been linked to an anomalously warm tropical Atlantic. {6.6, Figure 6.7}

Observations, both in situ (2004–2017) and based on sea surface temperature reconstructions, indicate that the Atlantic Meridional Overturning Circulation (AMOC) has weakened relative to 1850–1900 (*medium confidence*). There is insufficient data to quantify the magnitude of the weakening, or to properly attribute it to anthropogenic forcing due to the limited length of the observational record. Although attribution is currently not possible, CMIP5 model simulations of the period 1850–2015, on average, exhibit a weakening AMOC when driven by anthropogenic forcing. {6.7, Figure 6.8}

Climate change is modifying multiple types of climate-related events or hazards in terms of occurrence, intensity and periodicity. It increases the likelihood of compound hazards that comprise simultaneously or sequentially occurring events to cause extreme impacts in natural and human systems. Compound events in turn trigger cascading impacts (*high confidence*). Three case studies are presented in the chapter, (i) Tasmania's Summer of 2015–2016, (ii) The Coral Triangle and (ii) Hurricanes of 2017. {6.8, Box 6.1}

Projections of Ocean and Cryosphere Change and Hazards to Ecosystems and Human Society Under Low and High Emission Futures

The average intensity of tropical cyclones, the proportion of Category 4 and 5 tropical cyclones and the associated average precipitation rates are projected to increase for a 2°C global temperature rise above any baseline period (*medium confidence*). Rising mean sea levels will contribute to higher extreme sea levels associated with tropical cyclones (*very high confidence*). Coastal hazards will be exacerbated by an increase in the average intensity, magnitude of storm surge and precipitation rates of tropical cyclones. There are greater increases projected under RCP8.5 than under RCP2.6 from around mid-century to 2100 (*medium confidence*). There is *low confidence* in changes in the future frequency of tropical cyclones at the global scale. {6.3.1}

Significant wave heights (the average height from trough to crest of the highest one-third of waves) are projected to increase across the Southern Ocean and tropical eastern Pacific (*high confidence*) and Baltic Sea (*medium confidence*) and decrease over the North Atlantic and Mediterranean Sea under RCP8.5 (*high confidence*). Coastal tidal amplitudes and patterns are projected to change due to sea level rise and coastal adaptation measures (*very likely*). Projected changes in waves arising from changes in weather patterns, and changes in tides due to sea level rise, can locally enhance or ameliorate coastal hazards (*medium confidence*). {6.3.1, 5.2.2}

Marine heatwaves are projected to further increase in frequency, duration, spatial extent and intensity (maximum temperature) (*very high confidence*). Climate models project increases in the frequency of marine heatwaves by 2081–2100, relative to 1850–1900, by approximately 50 times under RCP8.5 and 20 times under RCP2.6 (*medium confidence*). The largest increases in frequency are projected for the Arctic and the tropical oceans (*medium confidence*). The intensity of marine heatwaves is projected to increase about 10-fold under RCP8.5 by 2081–2100, relative to 1850–1900 (*medium confidence*). {6.4}

Extreme El Niño and La Niña events are projected to *likely* increase in frequency in the 21st century and to *likely* intensify existing hazards, with drier or wetter responses in several regions across the globe. Extreme El Niño events are projected to occur about as twice as often under both RCP2.6 and RCP8.5 in the 21st century when compared to the 20th century (*medium confidence*). Projections indicate that extreme Indian Ocean Dipole events also increase in frequency (*low confidence*). {6.5, Figures 6.5, 6.6}

Lack of long-term sustained Indian and Pacific Ocean observations, and inadequacies in the ability of climate models to simulate the magnitude of trade wind decadal variability and the inter-ocean link, mean there is *low confidence* in future projections of the trade wind system. {6.6, Figure 6.7}

The AMOC will *very likely* weaken over the 21st century (*high confidence*), although a collapse is *very unlikely* (*medium confidence*). Nevertheless, a substantial weakening of the AMOC remains a physically plausible scenario. Such a weakening would strongly impact natural and human systems, leading to a decrease in marine productivity in the North Atlantic, more winter storms in Europe, a reduction in Sahelian and South Asian summer rainfall, a decrease in the number of TCs in the Atlantic, and an increase in regional sea level around the Atlantic especially along the northeast coast of North America (*medium confidence*). Such impacts would be superimposed on the global warming signal. {6.7, Figure 6.8}

Impacts from further changes in TCs and ETCs, MHWs, extreme El Niño and La Niña events and other extremes will exceed the limits of resilience and adaptation of ecosystems and people, leading to unavoidable loss and damage (*medium confidence*). {6.9.2}

Strengthening the Global Responses in the Context of Sustainable Development Goals (SDGs) and Charting Climate Resilient Development Pathways for Oceans and Cryosphere

There is *medium confidence* that including extremes and abrupt changes, such as AMOC weakening, ice sheet collapse (West Antarctic Ice Sheet (WAIS) and Greenland Ice Sheet (GIS)), leads to a several-fold increase in the cost of carbon emissions (*medium confidence*). If carbon emissions decline, the risk of extremes and abrupt changes are reduced, creating co-benefits. {6.8.6}

For TCs and ETCs, investment in disaster risk reduction, flood management (ecosystem and engineered) and early warning systems decreases economic loss (*medium confidence*), but such investments may be hindered by limited local capacities, such as increased losses and mortality from extreme winds and storm surges in less developed countries despite adaptation efforts. There is emerging evidence of increasing risks for locations impacted by unprecedented storm trajectories (*low confidence*). Managing the risk from such changing storm trajectories and intensity proves challenging because of the difficulties of early warning and its receptivity by the affected population (*high confidence*). {6.3, 6.9}

Limiting global warming would reduce the risk of impacts of MHWs, but critical thresholds for some ecosystems (e.g., kelp forests, coral reefs) will be reached at relatively low levels of future global warming (*high confidence*). Early warning systems, producing skillful forecasts of MHWs, can further help to reduce the vulnerability in the areas of fisheries, tourism and conservation, but are yet unproven at large scale (*medium confidence*). {6.4}

Sustained long-term monitoring and improved forecasts can be used in managing the risks of extreme El Niño and La Niña events associated with human health, agriculture, fisheries, coral reefs, aquaculture, wildfire, drought and flood management (*high confidence*). {6.5}

Extreme change in the trade wind system and its impacts on global variability, biogeochemistry, ecosystems and society have not been adequately understood and represent significant knowledge gaps. {6.6}

By 2300, an AMOC collapse is *as likely as not* for high emission pathways and *very unlikely* for lower ones, highlighting that an AMOC collapse can be avoided in the long term by CO_2 mitigation (*medium confidence*). Nevertheless, the human impact of these physical changes have not been sufficiently quantified and there are considerable knowledge gaps in adaptation responses to a substantial AMOC weakening. {6.7}

The ratio between risk reduction investment and reduction of damages of extreme events varies. Investing in preparation and prevention against the impacts from extreme events is *very likely* less than the cost of impacts and recovery (*medium confidence*). Coupling insurance mechanisms with risk reduction measures can enhance the cost-effectiveness of adapting to climate change (*medium confidence*). {6.9}

Climate change adaptation and disaster risk reduction require capacity building and an integrated approach to ensure trade-offs between short- and long-term gains in dealing with the uncertainty of increasing extreme events, abrupt changes and cascading impacts at different geographic scales (*high confidence*). {6.9}

Limiting the risk from the impact of extreme events and abrupt changes leads to successful adaptation to climate change with the presence of well-coordinated climate-affected sectors and disaster management relevant agencies (*high confidence*). Transformative governance inclusive of successful integration of disaster risk management (DRM) and climate change adaptation, empowerment of vulnerable groups, and accountability of governmental decisions promotes climate-resilient development pathways (*high confidence*). {6.9}

TS.7 Low-lying Islands and Coasts (Integrative Cross-Chapter Box)

Ocean and cryosphere changes already impact Low-Lying Islands and Coasts (LLIC), including Small Island Developing States (SIDS), with cascading and compounding risks. Disproportionately higher risks are expected in the course of the 21st century. Reinforcing the findings of the IPCC Special Report on Global Warming of 1.5°C, vulnerable human communities, especially those in coral reef environments and polar regions, may exceed adaptation limits well before the end of this century and even in a low greenhouse gas emission pathway (*high confidence*). Depending on the effectiveness of 21st century mitigation and adaptation pathways under all emission scenarios, most of the low-lying regions around the world may face adaptation limits beyond 2100, due to the long-term commitment of sea level rise (*medium confidence*). LLIC host around 11% of the global population, generate about 14% of the global Gross Domestic Product and comprise many world cultural heritage sites. LLIC already experience climate-related ocean and cryosphere changes (*high confidence*), and they share both commonalities in their exposure and vulnerability to climate change (e.g., low elevation, human disturbances to terrestrial and marine ecosystems), and context-specificities (e.g., variable ecosystem climate sensitivities and risk perceptions by populations). Options to adapt to rising seas, e.g., range from hard engineering to ecosystem-based measures, and from securing current settings to relocating people, built assets and activities. Effective combinations of measures vary across geographies (cities and megacities, small islands, deltas and Arctic coasts), and reflect the scale of observed and projected impacts, ecosystems' and societies' adaptive capacity, and the existence of transformational governance (*high confidence*) {Sections 3.5.3, 4.4.2 to 4.4.5, 5.5.2, 6.8, 6.9, Cross-Chapter Box 2 in Chapter 1}.

TS

Chapters

Framing and Context of the Report

Coordinating Lead Authors:

Nerilie Abram (Australia), Jean-Pierre Gattuso (France), Anjal Prakash (Nepal/India)

Lead Authors:

Lijing Cheng (China), Maria Paz Chidichimo (Argentina), Susan Crate (USA), Hiroyuki Enomoto (Japan), Matthias Garschagen (Germany), Nicolas Gruber (Switzerland), Sherilee Harper (Canada), Elisabeth Holland (Fiji), Raphael Martin Kudela (USA), Jake Rice (Canada), Konrad Steffen (Switzerland), Karina von Schuckmann (France)

Contributing Authors:

Nathaniel Bindoff (Australia), Sinead Collins (UK), Rebecca Colvin (Australia), Daniel Farinotti (Switzerland), Nathalie Hilmi (France/Monaco), Jochen Hinkel (Germany), Regine Hock (USA), Alexandre Magnan (France), Michael Meredith (UK), Avash Pandey (Nepal), Mandira Singh Shrestha (Nepal), Anna Sinisalo (Nepal/Finland), Catherine Sutherland (South Africa), Phillip Williamson (UK)

Review Editors:

Monika Rhein (Germany), David Schoeman (Australia)

Chapter Scientists:

Avash Pandey (Nepal), Bethany Ellis (Australia)

This chapter should be cited as:

Abram, N., J.-P. Gattuso, A. Prakash, L. Cheng, M.P. Chidichimo, S. Crate, H. Enomoto, M. Garschagen, N. Gruber, S. Harper, E. Holland, R.M. Kudela, J. Rice, K. Steffen, and K. von Schuckmann, 2019: Framing and Context of the Report. In: *IPCC Special Report on the Ocean and Cryosphere in a Changing Climate* [H.-O. Pörtner, D.C. Roberts, V. Masson-Delmotte, P. Zhai, M. Tignor, E. Poloczanska, K. Mintenbeck, A. Alegría, M. Nicolai, A. Okem, J. Petzold, B. Rama, N.M. Weyer (eds.)]. Cambridge University Press, Cambridge, UK and New York, NY, USA, pp. 73–129. https://doi.org/10.1017/9781009157964.003.

Table of contents

Executive Summary

This special report assesses new knowledge since the IPCC 5th Assessment Report (AR5) and the Special Report on Global Warming of 1.5°C (SR15) on how the ocean and cryosphere have and are expected to change with ongoing global warming, the risks and opportunities these changes bring to ecosystems and people, and mitigation, adaptation and governance options for reducing future risks. Chapter 1 provides context on the importance of the ocean and cryosphere, and the framework for the assessments in subsequent chapters of the report.

All people on Earth depend directly or indirectly on the ocean and cryosphere. The fundamental roles of the ocean and cryosphere in the Earth system include the uptake and redistribution of anthropogenic carbon dioxide and heat by the ocean, as well as their crucial involvement of in the hydrological cycle. The cryosphere also amplifies climate changes through snow, ice and permafrost feedbacks. Services provided to people by the ocean and/or cryosphere include food and freshwater, renewable energy, health and wellbeing, cultural values, trade and transport. {1.1, 1.2, 1.5}

Sustainable development is at risk from emerging and intensifying ocean and cryosphere changes. Ocean and cryosphere changes interact with each of the United Nations Sustainable Development Goals (SDGs). Progress on climate action (SDG 13) would reduce risks to aspects of sustainable development that are fundamentally linked to the ocean and cryosphere and the services they provide (*high confidence*[1]). Progress on achieving the SDGs can contribute to reducing the exposure or vulnerabilities of people and communities to the risks of ocean and cryosphere change (*medium confidence*). {1.1}

Communities living in close connection with polar, mountain, and coastal environments are particularly exposed to the current and future hazards of ocean and cryosphere change. Coasts are home to approximately 28% of the global population, including around 11% living on land less than 10 m above sea level. Almost 10% of the global population lives in the Arctic or high mountain regions. People in these regions face the greatest exposure to ocean and cryosphere change, and poor and marginalised people here are particularly vulnerable to climate-related hazards and risks (*very high confidence*). The adaptive capacity of people, communities and nations is shaped by social, political, cultural, economic, technological, institutional, geographical and demographic factors. {1.1, 1.5, 1.6, Cross-Chapter Box 2 in Chapter 1}

Ocean and cryosphere changes are pervasive and observed from high mountains, to the polar regions, to coasts, and into the deep ocean. AR5 assessed that the ocean is warming (0 to 700 m: *virtually certain*[2]; 700 to 2,000 m: *likely*), sea level is rising (*high confidence*), and ocean acidity is increasing (*high confidence*). Most glaciers are shrinking (*high confidence*), the Greenland and Antarctic ice sheets are losing mass (*high confidence*), sea ice extent in the Arctic is decreasing (*very high confidence*), Northern Hemisphere snow cover is decreasing (*very high confidence*), and permafrost temperatures are increasing (*high confidence*). Improvements since AR5 in observation systems, techniques, reconstructions and model developments, have advanced scientific characterisation and understanding of ocean and cryosphere change, including in previously identified areas of concern such as ice sheets and Atlantic Meridional Overturning Circulation (AMOC). {1.1, 1.4, 1.8.1}

Evidence and understanding of the human causes of climate warming, and of associated ocean and cryosphere changes, has increased over the past 30 years of IPCC assessments (*very high confidence*). Human activities are estimated to have caused approximately 1.0°C of global warming above pre-industrial levels (SR15). Areas of concern in earlier IPCC reports, such as the expected acceleration of sea level rise, are now observed (*high confidence*). Evidence for expected slow-down of AMOC is emerging in sustained observations and from long-term palaeoclimate reconstructions (*medium confidence*), and may be related with anthropogenic forcing according to model simulations, although this remains to be properly attributed. Significant sea level rise contributions from Antarctic ice sheet mass loss (*very high confidence*), which earlier reports did not expect to manifest this century, are already being observed. {1.1, 1.4}

Ocean and cryosphere changes and risks by the end-of-century (2081–2100) will be larger under high greenhouse gas emission scenarios, compared with low emission scenarios (*very high confidence*). Projections and assessments of future climate, ocean and cryosphere changes in the Special Report on the Ocean and Cryosphere in a Changing Climate (SROCC) are commonly based on coordinated climate model experiments from the Coupled Model Intercomparison Project Phase 5 (CMIP5) forced with Representative Concentration Pathways (RCPs) of future radiative forcing. Current emissions continue to grow at a rate consistent with a high emission future without effective climate change mitigation policies (referred to as RCP8.5). The SROCC assessment contrasts this high greenhouse gas emission future with a low greenhouse gas emission, high mitigation future (referred to as RCP2.6) that gives a two in three chance of limiting warming by the end of the century to less than 2°C above pre-industrial. {Cross-Chapter Box 1 in Chapter 1}

---

[1]   In this report, the following summary terms are used to describe the available evidence: limited, medium, or robust; and for the degree of agreement: low, medium or high. A level of confidence is expressed using five qualifiers: very low, low, medium, high and very high, and typeset in italics, for example, *medium confidence*. For a given evidence and agreement statement, different confidence levels can be assigned, but increasing levels of evidence and degrees of agreement are correlated with increasing confidence (see Section 1.9.2 and Figure 1.4 for more details).

[2]   In this report, the following terms have been used to indicate the assessed likelihood of an outcome or a result: Virtually certain 99–100% probability, Very likely 90–100%, Likely 66–100%, About as likely as not 33–66%, Unlikely 0–33%, Very unlikely 0–10%, and Exceptionally unlikely 0–1%. Additional terms (Extremely likely: 95–100%, More likely than not >50–100%, and Extremely unlikely 0–5%) may also be used when appropriate. Assessed likelihood is typeset in italics, for example, *very likely* (see Section 1.9.2 and Figure 1.4 for more details). This Report also uses the term '*likely* range' to indicate that the assessed likelihood of an outcome lies within the 17–83% probability range.

Characteristics of ocean and cryosphere change include thresholds of abrupt change, long-term changes that cannot be avoided, and irreversibility (*high confidence*). Ocean warming, acidification and deoxygenation, ice sheet and glacier mass loss, and permafrost degradation are expected to be irreversible on time scales relevant to human societies and ecosystems. Long response times of decades to millennia mean that the ocean and cryosphere are committed to long-term change even after atmospheric greenhouse gas concentrations and radiative forcing stabilise (*high confidence*). Ice-melt or the thawing of permafrost involve thresholds (state changes) that allow for abrupt, nonlinear responses to ongoing climate warming (*high confidence*). These characteristics of ocean and cryosphere change pose risks and challenges to adaptation. {1.1, Box 1.1, 1.3}

Societies will be exposed, and challenged to adapt, to changes in the ocean and cryosphere even if current and future efforts to reduce greenhouse gas emissions keep global warming well below 2°C (*very high confidence*). Ocean and cryosphere-related mitigation and adaptation measures include options that address the causes of climate change, support biological and ecological adaptation, or enhance societal adaptation. Most ocean-based local mitigation and adaptation measures have limited effectiveness to mitigate climate change and reduce its consequences at the global scale, but are useful to implement because they address local risks, often have co-benefits such as biodiversity conservation, and have few adverse side effects. Effective mitigation at a global scale will reduce the need and cost of adaptation, and reduce the risks of surpassing limits to adaptation. Ocean-based carbon dioxide removal at the global scale has potentially large negative ecosystem consequences. {1.6.1, 1.6.2, Cross-Chapter Box 2 in Chapter 1}

The scale and cross-boundary dimensions of changes in the ocean and cryosphere challenge the ability of communities, cultures and nations to respond effectively within existing governance frameworks (*high confidence*). Profound economic and institutional transformations are needed if climate-resilient development is to be achieved (*high confidence*). Changes in the ocean and cryosphere, the ecosystem services that they provide, the drivers of those changes, and the risks to marine, coastal, polar and mountain ecosystems, occur on spatial and temporal scales that may not align within existing governance structures and practices (*medium confidence*). This report highlights the requirements for transformative governance, international and transboundary cooperation, and greater empowerment of local communities in the governance of the ocean, coasts, and cryosphere in a changing climate. {1.5, 1.7, Cross-Chapter Box 2 in Chapter 1, Cross-Chapter Box 3 in Chapter 1}

Robust assessments of ocean and cryosphere change, and the development of context-specific governance and response options, depend on utilising and strengthening all available knowledge systems (*high confidence*). Scientific knowledge from observations, models and syntheses provides global to local scale understandings of climate change (*very high confidence*). Indigenous knowledge (IK) and local knowledge (LK) provide context-specific and socio-culturally relevant understandings for effective responses and policies (*medium confidence*). Education and climate literacy enable climate action and adaptation (*high confidence*). {1.8, Cross-Chapter Box 4 in Chapter 1}

Long-term sustained observations and continued modelling are critical for detecting, understanding and predicting ocean and cryosphere change, providing the knowledge to inform risk assessments and adaptation planning (*high confidence*). Knowledge gaps exist in scientific knowledge for important regions, parameters and processes of ocean and cryosphere change, including for physically plausible, high impact changes like high end sea level rise scenarios that would be costly if realised without effective adaptation planning and even then may exceed limits to adaptation. Means such as expert judgement, scenario building, and invoking multiple lines of evidence enable comprehensive risk assessments even in cases of uncertain future ocean and cryosphere changes. {1.8.1, 1.9.2; Cross-Chapter Box 5 in Chapter 1}

1.1 Why this Special Report?

All people depend directly or indirectly on the ocean and cryosphere (see FAQ1.1). Coasts are the most densely populated areas on Earth. As of 2010, 28% of the global population (1.9 billion people) were living in areas less than 100 km from the coastline and less than 100 m above sea level, including 17 major cities which are each home to more than 5 million people (Kummu et al., 2016). Small Island Developing States are together home to around 65 million people (UN, 2015a). The low elevation coastal zone (land less than 10 m above sea level), where people and infrastructure are most exposed to coastal hazards, is currently home to around 11% of the global population (around 680 million people), and by 2050 the population in this zone is projected to grow to more than one billion under all Shared Socioeconomic Pathways (SSPs) (Section 4.3.3.2; Merkens et al., 2016; O'Neill et al., 2017). In 2010, approximately 4 million people lived in the Arctic (Section 3.5.1), and an increase of only 4% is projected for 2030 (Heleniak, 2014) compared to 16–23% for the global population increase (O'Neill et al., 2017). Almost 10% of the global population (around 670 million people) lived in high mountain regions in 2010, and by 2050 the population in these regions is expected to grow to between 736–844 million across the SSPs (Section 2.1). For people living in close contact with the ocean and cryosphere, these systems provide essential livelihoods, food security, well-being and cultural identity, but are also a source of hazards (Sections 1.5.1, 1.5.2).

Even people living far from the ocean or cryosphere depend on these systems. Snow and glacier melt from high mountains helps to sustain the rivers that deliver water resources to downstream populations (Kaser et al., 2010; Sharma et al., 2019). In the Indus and Ganges river basins, for example, snow and glacier melt provides enough water to grow food crops to sustain a balanced diet for 38 million people, and supports the livelihoods of 129 million farmers (Biemans et al., 2019). The ocean and cryosphere regulate global climate and weather; the ocean is the primary source of rain and snowfall needed to sustain life on land, and uptake of heat and carbon into the ocean has so far limited the magnitude of anthropogenic warming experienced at the Earth's surface (Section 1.2). The ocean's biosphere is responsible for about half of the primary production on Earth, and around 17% of the non-grain protein in human diets is derived from the ocean (FAO, 2018). Communities far from the coast can also be exposed to changes in the ocean through extreme weather events. Ocean and cryosphere changes can result in differing consequences and benefits on local to global scales; for example, declining sea ice in the Arctic is allowing access to shorter international shipping routes but restricting traditional sea ice based travel for Arctic communities.

Human activities are estimated to have so far caused approximately 1°C of global warming (0.8°C–1.2°C likely range; above pre-industrial levels; IPCC, 2018). The IPCC Fifth Assessment Report (AR5) concluded that, 'Warming of the climate system is unequivocal, and since the 1950s, many of the observed changes are unprecedented over decades to millennia. The atmosphere and ocean have warmed, the amounts of snow and ice have diminished, sea level has risen, and the concentrations of greenhouse gases have increased' (IPCC, 2013). Subsequently, Parties to the Paris Agreement aimed to strengthen the global response to the threats of climate change, including by 'holding the increase in global average temperature to well below 2°C above pre-industrial levels and pursuing efforts to limit the temperature increase to 1.5°C' (UNFCCC, 2015).

Pervasive ocean and cryosphere changes that are already being caused by human-induced climate change are observed from high mountains, to the polar regions, to coasts and into the deep reaches of the ocean. Changes by the end of this century are expected to be larger under high greenhouse gas emission futures compared with low-emission futures (Cross-Chapter Box 1 in Chapter 1), and inaction on reducing emissions will have large economic costs. If human impacts on the ocean continue unabated, declines in ocean health and services are projected to cost the global economy 428 billion USD yr^{-1} by 2050, and 1.979 trillion USD yr^{-1} by 2100. Alternatively, steps to reduce these impacts could save more than a trillion dollars USD yr^{-1} by 2100 (Ackerman, 2013). Similarly, sea level rise scenarios of 25 to 123 cm by 2100 without adaptation are expected to see 0.2–4.6% of the global population impacted by coastal flooding annually, with average annual losses amounting to 0.3–9.3% of global GDP. Investment in adaptation reduces by 2 to 3 orders of magnitude the number of people flooded and the losses caused (Hinkel et al., 2014).

The United Nations 2030 SDGs (UN, 2015b) are all connected to varying extents with the ocean and cryosphere (see FAQ1.2). Climate action (SDG 13) would limit future ocean and cryosphere changes (*high confidence*; Cross-Chapter Box 1 in Chapter 1, Figure 1.5, Chapter 2 to 6), and would reduce risks to SDGs that are fundamentally linked to the ocean and cryosphere, including life below water, and clean water and sanitation. (Sections 2.4, 4.4, 5.4; Szabo et al., 2016; LeBlanc et al., 2017; Singh et al., 2018; Visbeck, 2018; Wymann von Dach et al., 2018; Kulonen, Accepted). Other goals for sustainable development depend on the services the ocean and cryosphere provide or are impacted by ocean and cryosphere change; including, life on land, health and wellbeing, eradicating poverty and hunger, economic growth, clean energy, infrastructure, and sustainable cities and communities. Progress on the other SDGs (education, gender equality, reduced inequalities, responsible consumption, strong institutions, and partnerships for the goals) are important for reducing the vulnerability of people and communities to the risks of ocean and cryosphere changes (Section 1.5; 2.3), and for supporting mitigation and adaptation responses (Sections 1.6, 1.7 and 1.8.3; *medium confidence*).

The characteristics of ocean and cryosphere change (Section 1.3) present particular challenges to climate-resilient development pathways (CRDPs). Ocean acidification and deoxygenation, ice sheet and glacier mass loss, and permafrost degradation are expected to be irreversible on time scales relevant to human societies and ecosystems (Lenton et al., 2008; Solomon et al., 2009; Frölicher and Joos, 2010; Cai et al., 2016; Kopp et al., 2016). Ocean and cryosphere changes also have the potential to worsen anthropogenic climate change, globally and regionally; for example, by additional greenhouse gas emissions released through permafrost thaw that would intensify anthropogenic climate change globally, or by increasing the absorption of solar radiation through snow and ice loss in the Arctic that is causing regional climate to warm at more than twice the global rate (AMAP, 2017; Steffen et al., 2018). Ocean and cryosphere changes place particular pressures on the adaptive capacities of cultures who maintain centuries to millennia-old relationships to the planet's polar, mountain, and coastal environments, as well as on cities, states and nations whose territorial boundaries are being transformed by ongoing sea level rise (Gerrard and Wannier, 2013). The scale and cross-boundary dimensions of changes in the ocean and cryosphere challenge the ability of current local, regional and international governance structures to respond (Section 1.7). Profound economic and institutional transformations are needed if climate-resilient development is to be achieved, including ambitious mitigation efforts to avoid the risks of large-scale and abrupt ocean and cryosphere changes.

The commissioning of this IPCC special report recognises the interconnected ways in which the ocean and cryosphere are expected to change in a warming climate. SROCC assesses new knowledge since AR5 and provides an integrated approach across IPCC working groups I and II, linking physical changes with their ecological and human impacts, and the strategies to respond and adapt to future risks. It is one of three special reports being produced by the IPCC during its Sixth Assessment Cycle (in addition to the three working groups' main assessment reports). The concurrent IPCC Special Report on Climate Change and Land (released August 2019) links to SROCC where terrestrial environments and their habitability interact closely with the ocean or cryosphere, such as in mountain, Arctic, and coastal regions. SR15 concluded that human-induced warming will reach 1.5°C between 2030–2052 if it continues to increase at the current rate (*high confidence*), and that there are widespread benefits to human and natural systems of limiting warming to 1.5°C compared with 2°C or more (*high confidence*; IPCC, 2018).

Box 1.1 | Major Components and Characteristics of the Ocean and Cryosphere

Ocean

The global ocean is the interconnected body of saline water that encompasses polar to equatorial climate zones and covers 71% of the Earth surface. It includes the Arctic, Pacific, Atlantic, Indian and Southern Oceans, as well as their marginal seas. The ocean contains about 97% of the Earth's water, supplies 99% of the Earth's biologically habitable space, and provides roughly half of the primary production on Earth.

Coasts are where ocean and land processes interact, and includes coastal cities, deltas, estuaries, and other coastal ecosystems such as mangrove forests. Low elevation coastal zones (less than 10 m above sea level) are densely populated and particularly exposed to hazards from the ocean (Chapters 4 to 6, Cross-Chapter Box 9). Moving into the ocean, the continental shelf represents the shallow ocean areas (depth <200 m) that surround continents and islands, before the seafloor descends at the continental slope into the deep ocean. The edge of the continental shelf is often used to identify the coastal ocean from the open ocean. Ocean depth and distance from the coast may influence the governance and economic access that applies to ocean areas (Cross-Chapter Box 3 in Chapter 1).

The average depth of the global ocean is about 3,700 m, with a maximum depth of more than 10,000 m. The ocean is vertically stratified with less dense water sitting above more dense layers, determined by the seawater temperature, salinity and pressure. The surface of the ocean is in direct contact with the atmosphere, except for sea ice covered regions. Sunlight penetrates the water column and supports primary production (by phytoplankton) down to 50–200 m depth (epipelagic zone). Atmospheric-driven mixing occurs from the sea surface and into the mesopelagic zone (200–1,000 m). The distinction between the upper ocean and deep ocean depends on the processes being considered.

The ocean is a fundamental climate regulator on seasonal to millennial time scales. Seawater has a heat capacity four times larger than air and holds vast quantities of dissolved carbon. Heat, water, and biogeochemically relevant gases (e.g., O_2 and CO_2) exchange at the air-sea interface, and ocean currents and mixing caused by winds, tides, wave dynamics, density differences and turbulence redistribute these throughout the global ocean (Box 1.1, Figure 1).

Cryosphere

The cryosphere refers to frozen components of the Earth system that are at or below the land and ocean surface. These include snow, glaciers, ice sheets, ice shelves, icebergs, sea ice, lake ice, river ice, permafrost and seasonally frozen ground. Cryosphere is widespread in polar regions (Chapter 3) and high mountains (Chapter 2), and changes in the cryosphere can have far-reaching and even global impacts (Chapters 2 to 6, Cross-Chapter Box 9).

Snow is common in polar and mountain regions. It can ultimately either melt seasonally or transform into ice layers that build glaciers and ice sheets. Snow feeds groundwater and river runoff together with glacier melt causes natural hazards (avalanches, rain-on-snow flood events) and is a critical economic resource for hydropower and tourism. Snow plays a major role in maintaining high mountain and Arctic ecosystems, affects the Earth's energy budget by reflecting solar radiation (albedo effect), and influences the temperature of underlying permafrost.

Box 1.1 (continued)

Ice sheets and glaciers are land-based ice, built up by accumulating snowfall on their surface. Presently, around 10% of Earth's land area is covered by glaciers or ice sheets, which in total hold about 69% of Earth's freshwater (Gleick, 1996). Ice sheets and glaciers flow, and at their margins ice and/or melt water is discharged into lakes, rivers or the ocean. The largest ice bodies on Earth are the Greenland and Antarctic ice sheets. Marine-based sections of ice sheets (e.g., West Antarctic Ice Sheet) sit upon bedrock that largely lies below sea level and are in contact with ocean heat, making them vulnerable to rapid and irreversible ice loss. Ice sheets and glaciers that lose more ice than they accumulate contribute to global sea level rise.

Ice shelves are extensions of ice sheets and glaciers that float in the surrounding ocean. The transition between the grounded part of an ice sheet and a floating ice shelf is called the grounding line. Changes in ice shelf size do not directly contribute to sea level rise, but buttressing of ice shelves restrict the flow of land-based ice past the grounding line into the ocean.

Sea ice forms from freezing of seawater, and sea ice on the ocean surface is further thickened by snow accumulation. Sea ice may be discontinuous pieces moved on the ocean surface by wind and currents (pack ice), or a motionless sheet attached to the coast or to ice shelves (fast ice). Sea ice provides many critical functions: it provides essential habitat for polar species and supports the livelihoods of people in the Arctic (including Indigenous peoples); regulates climate by reflecting solar radiation; inhibits ocean-atmosphere exchange of heat, momentum and gases (including CO_2); supports global deep ocean circulation via dense (cold and salty) water formation; and aids or hinders transportation and travel routes in the polar regions.

Permafrost is ground (soil or rock containing ice and frozen organic material) that remains at or below 0°C for at least two consecutive years. It occurs on land in polar and high mountain areas, and also as submarine permafrost in shallow parts of the Arctic and Southern Oceans. Permafrost thickness ranges from less than 1 m to greater than 1,000 m. It usually occurs beneath an active layer, which thaws and freezes annually. Unlike glaciers and snow, the spatial distribution and temporal changes of permafrost cannot easily be observed. Permafrost thaw can cause hazards, including ground subsidence or landslides, and influence global climate through emissions of greenhouse gases from microbial breakdown of previously frozen organic carbon.

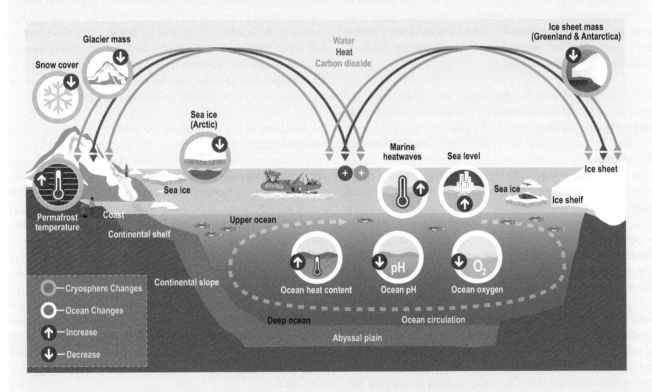

Box 1.1, Figure 1 | Schematic illustration of key components and changes of the ocean and cryosphere, and their linkages in the Earth system through the movement of heat, water, and carbon dioxide (Section 1.2). Climate change-related effects in the ocean include sea level rise, increasing ocean heat content and marine heat waves, ocean deoxygenation, and ocean acidification (Section 1.4.1). Changes in the cryosphere include the decline of Arctic sea ice extent, Antarctic and Greenland ice sheet mass loss, glacier mass loss, permafrost thaw and decreasing snow cover extent (Section 1.4.2). For illustration purposes, a few examples of where humans directly interact with ocean and cryosphere are shown.

1.2 Role of the Ocean and Cryosphere in the Earth System

1.2.1 Ocean and Cryosphere in Earth's Energy, Water and Biogeochemical Cycles

The ocean and cryosphere play a key role in the Earth system. Powered by the Sun's energy, large quantities of energy, water and biogeochemical elements (predominantly carbon, nitrogen, oxygen and hydrogen) are exchanged between all components of the Earth system, including between the ocean and cryosphere (Box 1.1, Figure 1).

During an equilibrium (stable) climate state, the amount of incoming solar energy is balanced by an equal amount of outgoing radiation at the top of Earth's atmosphere (Hansen et al., 2011). At the Earth's surface energy from the Sun is transformed into various forms (heat, potential, latent, kinetic, and chemical), that drive weather systems in the atmosphere and currents in the ocean, fuel photosynthesis on land and in the ocean, and fundamentally determine the climate (Trenberth et al., 2014). The ocean has a large capacity to store and release heat, and the Earth's energy budget can be effectively monitored through the heat content of the ocean on time scales longer than one year (Palmer and McNeall, 2014; von Schuckmann et al., 2016; Cheng et al., 2018). The large heat capacity of the ocean leads to different characteristics of the ocean response to external forcings compared with the atmosphere (Sections 1.3, 1.4). The reflective properties of snow and ice also play an important role in regulating climate via the albedo effect. Increased amounts of solar energy are absorbed when snow or ice are replaced by less reflective land or ocean surfaces, resulting in a climate change feedback responsible for amplified changes.

Water is exchanged between the ocean, atmosphere, land and cryosphere as part of the hydrological cycle driven by solar heating (Box 1.1, Figure 1; Trenberth et al., 2007; Lagerloef et al., 2010; Durack et al., 2016). Evaporation from the surface ocean is the main source of water in the atmosphere, which is moved back to the Earth's surface as precipitation (Gimeno et al., 2012). The hydrological cycle is closed by the eventual return of water to the ocean by rivers, streams, and groundwater flow, and through ice discharge and melting of ice sheets and glaciers (Yu, 2018). Hydrological extremes related to the ocean include floods from extreme rainfall (including tropical cyclones) or ocean circulation-related droughts (Sections 6.3, 6.5), while cryosphere-related flooding can be caused by rapid snow melt and melt water discharge events (Sections 2.3, 3.4).

Ninety-two percent of the carbon on Earth that is not locked up in geological reservoirs (e.g., in sedimentary rocks or coal, oil and gas reservoirs) resides in the ocean (Sarmiento and Gruber, 2002). Most of this is in the form of dissolved inorganic carbon, some of which readily exchanges with CO_2 in the overlying atmosphere. This represents a major control on atmospheric CO_2 and makes the ocean and its carbon cycle one of the most important climate regulators in the Earth system, especially on time scales of a few hundred years and more (Sigman and Boyle, 2000; Berner and Kothavala, 2001). The ocean also contains as much organic carbon (mostly in the form of dissolved organic matter) as the total vegetation on land (Jiao et al., 2010; Hansell, 2013). Primary production in the ocean, which is as large as that on land (Field et al., 1998), fuels complex food-webs that provide essential food for people.

Ocean circulation and mixing redistribute heat and carbon over large distances and depths (Delworth et al., 2017). The ocean moves heat laterally from the tropics towards polar regions (Rhines et al., 2008). Vertical redistribution of heat and carbon occurs where warm, low-density surface ocean waters transform into cool high-density waters that sink to deeper layers of the ocean (Talley, 2013), taking high carbon concentrations with them (Gruber et al., 2019). Driven by winds, ocean circulation also brings cold water up from deep layers (upwelling) in some regions, allowing heat, oxygen and carbon exchange between the deep ocean and the atmosphere (Oschlies et al., 2018; Shi et al., 2018) and fuelling biological production (Sarmiento and Gruber, 2006).

1.2.2 Interactions Between the Ocean and Cryosphere

The ocean and cryosphere are interconnected in a multitude of ways (Box 1.1, Figure 1). Evaporation from the ocean provides snowfall that builds and sustains the ice sheets and glaciers that store large amounts of frozen water on land (Section 4.2.1). The vast ice sheets in Antarctica and Greenland currently hold about 66 m of potential global sea level rise (Fretwell et al., 2013), although the loss of a large fraction of this potential would require millennia of ice sheet retreat. Ocean temperature and sea level affect ice sheet, glacier and ice shelf stability in places where the base of ice bodies are in direct contact with ocean water (Section 3.3.1). The nonlinear response of ice-melt to ocean temperature changes means that even slight increases in ocean temperature have the potential to rapidly melt and destabilise large sections of an ice sheet or ice shelf (Section 3.3.1.5).

The formation of sea ice leads to the production of dense ocean water that contributes to the deep ocean circulation (Section 3.3.3.2). Paleoclimate evidence and modelling indicates that releases of large amounts of glacier and ice sheet melt water into the surface ocean can disrupt deep overturning circulation of the ocean, causing global climate impacts (Knutti et al., 2004; Golledge et al., 2019). Ice sheet melt water in the Antarctic may cause changes in surface ocean salinity, stratification and circulation, that feedback to generate further ocean-driven melting of marine-based ice sheets (Golledge et al., 2019) and promote sea ice formation (Purich et al., 2018). The cryosphere and ocean further link through the movement of biogeochemical nutrients. For example, iron accumulated in sea ice during winter is released to the ocean during the spring and summer melt, helping to fuel ocean productivity in the seasonal sea ice zone (Tagliabue et al., 2017). Nutrient rich sediments delivered by glaciers further connect cryosphere processes to ocean productivity (Arrigo et al., 2017).

1.3 Time Scales, Thresholds and Detection of Ocean and Cryosphere Change

It takes hundreds of years to millennia for the entire deep ocean to turn over (Matsumoto, 2007; Gebbie and Huybers, 2012), while renewal of the large ice sheets requires many thousands of years (Huybrechts and de Wolde, 1999). Long response times mean that the deep ocean and the large ice sheets tend to lag behind in their response to the rapidly changing climate at Earth's surface, and that they will continue to change even after radiative forcing stabilises (e.g., Golledge et al., 2015; Figure 1.1a). Such 'committed' changes mean that some ocean and cryosphere changes are essentially irreversible on time scales relevant to human societies (decades to centuries), even in the presence of immediate action to limit further global warming (e.g., Section 4.2.3.5).

While some aspects of the ocean and cryosphere might respond in a linear (i.e., directly proportional) manner to a perturbation by some external forcing, this may change fundamentally when critical thresholds are reached. A very important example for such a threshold is the transition from frozen water to liquid water at around 0°C that can lead to rapid acceleration of ice-melt or permafrost thaw (e.g., Abram et al., 2013; Trusel et al., 2018). Such thresholds often act as tipping points, as they are associated with rapid and abrupt changes even when the underlying forcing changes gradually (Figure 1.1a, 1.1c). Tipping elements include, for example, the collapse of the ocean's large-scale overturning circulation in the Atlantic (Section 6.7), or the collapse of the West Antarctic Ice Sheet though a process called marine ice sheet instability (Cross-Chapter Box 8 in Chapter 3; Lenton et al., 2008). Potential ocean and cryosphere tipping elements form part of the scientific case for efforts to limit climate warming to well below 2°C (IPCC, 2018).

Anthropogenically forced change occurs against a backdrop of substantial natural variability (Figure 1.1b). The anthropogenic signal is already detectable in global surface air temperature and several other climate variables, including ocean temperature and salinity (IPCC, 2014), but short observational records and large year-to-year variability mean that formal detection is not yet the case for many expected ocean and cryosphere changes (Jones et al., 2016). 'Time of Emergence' refers to the time when anthropogenic change signals emerge from the background noise of natural variability in a pre-defined reference period Hawkins and Sutton, 2012; (Figure 1.1b; Section 5.2, Box 5.1). For some variables, (e.g., for those associated with ocean acidification), the current signals emerge from this natural variability within a few decades, whereas for others, such as primary production and expected Antarctic-wide sea ice decline, the signal may not emerge for many more decades even under high emission scenarios (Collins et al., 2013; Keller et al., 2014; Rodgers et al., 2015; Frölicher et al., 2016; Jones et al., 2016).

'Detection and Attribution' assesses evidence for past changes in the ocean and cryosphere, relative to normal/reference-interval conditions (*detection*), and the extent to which these changes have been caused by anthropogenic climate change or by other factors (*attribution*) (Bindoff et al., 2013; Cramer et al., 2014; Knutson et al., 2017; Figure 1.1d). Reliable detection and attribution is fundamental to our understanding of the scientific basis of climate change (Hegerl et al., 2010). For example, the main attribution conclusion of the IPCC 4th Assessment Report (AR4), in other words, that 'most of the observed increase in global average temperatures since the mid-20th century is *very likely* due to the observed increase in anthropogenic greenhouse gas concentrations', has had a strong impact on climate policy (Petersen, 2011). In AR5 this attribution statement was elevated to '*extremely likely*' (Bindoff et al., 2013). Statistical approaches for attribution often involve using contrasting forcing scenarios in climate model experiments to detect the forcing that best explains an observed change (Figure 1.1d). In addition to passing the statistical test, a successful attribution also requires a firm process understanding. Confident attribution remains challenging though, especially when there are multiple or confounding factors that influence the state of a system (Hegerl et al., 2010). Particular challenges to detection and attribution in the ocean and cryosphere include the often short observational records (Section 1.8.1.1, Figure 1.3), which are particularly confounding given the long adjustment time scales to anthropogenic forcing of many properties of interest.

Extreme climate events (e.g., marine heatwaves or storm surges) push a system to near or beyond the ends of its normally observed range (Seneviratne et al., 2012; Figure 1.1b; Chapter 6;). Extremes can be very costly in terms of loss of life, ecosystem destruction, and economic damage. In a system affected by climate change, the recurrence and intensity of these extreme events can change much faster and have greater impacts than changes of the average system state (Easterling et al., 2000; Parmesan et al., 2000; Hughes et al., 2018). Of particular concern are 'compound events', when the joint probability of two or more properties of a system is extreme at the same time or closely connected in time and space (Cross-Chapter Box 5 in Chapter 1; Sections 4.3.4, 6.8). Such a compound event is given, for example, when marine heatwaves co-occur with very low nutrient levels in the ocean potentially resulting in extreme impacts (Bond et al., 2015). The interconnectedness of the ocean and cryosphere (Section 1.2.2) can also lead to cascading effects where changes in one element trigger secondary changes in completely different but connected elements of the systems, including its socioeconomic aspects. (Figure 1.1e). An example is the large change in ocean productivity triggered by the changes in circulation and iron inputs induced by the large outflow of melt waters from Greenland (Kanna et al., 2018). New methodologies for attributing extreme events and the risks they bring to climate change have emerged since AR5 (Trenberth et al., 2015; Stott et al., 2016; Kirchmeier-Young et al., 2017; Otto, 2017), especially also for the attribution of individual events through an assessment of the fraction of attributable risk (Figure 1.1f).

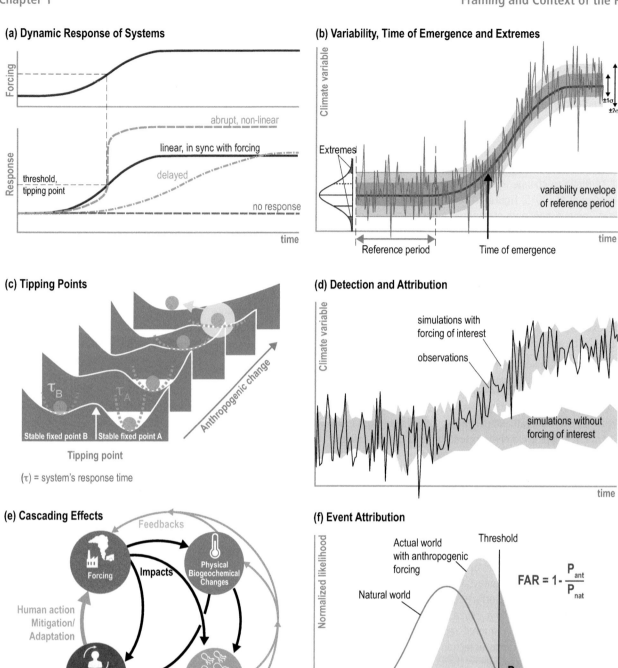

(a) Dynamic Response of Systems

(b) Variability, Time of Emergence and Extremes

(c) Tipping Points

(τ) = system's response time

(d) Detection and Attribution

(e) Cascading Effects

(f) Event Attribution

$$FAR = 1 - \frac{P_{ant}}{P_{nat}}$$

Figure 1.1 | Schematic of key concepts associated with changes in the ocean and cryosphere. **(a)** Differing responses of systems to gradual forcing (e.g., linear, delayed, abrupt, nonlinear). **(b)** Evolution of a dynamical system in time, revealing both natural (unforced) variability and a response to a new (e.g., anthropogenic) forcing. Key concepts include (i) the time of emergence and (ii) extreme events near or beyond the observed range of variability. **(c)** Tipping points and the change of their behaviour through time in response to, for example, anthropogenic change (adapted from Lenton et al., 2008). The two minima represent two stable fixed points, separated by a maximum representing an unstable fixed point, acting as a tipping point. The ball represents the state of the system with the red dash line indicating the stability of the fixed point and the system's response time to small perturbations. **(d)** Detection and attribution, i.e., the statistical framework used to determine whether a change occurs or not (detection), and whether this detected change is caused by a particular set of forcings (e.g., greenhouse gases) (attribution). **(e)** Cascading effects, where changes in one part of a system inevitably affect the state in another, and so forth, ultimately affecting the state of the entire system. These cascading effects can also trigger feedbacks, altering the forcing. **(f)** Event attribution and fraction of attributable risk. The blue (orange) probability density function shows the likelihood of the occurrence of a particular value of a climate variable of interest under natural (present = including anthropogenic forcing) conditions. The corresponding areas above the threshold indicate the probabilities P_{nat} and P_{ant} of exceedance of this threshold. The fraction of attributable risk (given by FAR = 1 − P_{ant}/P_{nat}) indicates the likelihood that a particular event has occurred as a consequence of anthropogenic change (adapted from Stott et al., 2016).

1.4 Changes in the Ocean and Cryosphere

Earth's climate, ocean and cryosphere vary across a wide range of time scales. This includes the seasonal growth and melting of sea ice, interannual variation of ocean temperature caused by the El Niño-Southern Oscillation and ice age cycles across tens to hundreds of thousands of years.

Climate variability can arise from internally generated (i.e., unforced) fluctuations in the climate system. Variability can also occur in response to external forcings, including volcanic eruptions, changes in the Earth's orbit around the Sun, oscillations in solar activity and changing atmospheric greenhouse gas concentrations.

Since the onset of the industrial revolution, human activities have had a strong impact on the climate system, including the ocean and cryosphere. Human activities have altered the external forcings acting on Earth's climate (Myhre et al., 2013) by changes in land use (albedo), and changes in atmospheric aerosols (e.g., soot) from the burning of biomass and fossil fuels. Most significantly, human activities have led to an accumulation of greenhouse gases (including CO_2) in the atmosphere as a result of the burning of fossil fuels, cement production, agriculture and land use change. In 2016, the global average atmospheric CO_2 concentration crossed 400 parts per million, a level Earth's atmosphere did not experience for at least the past 800,000 years and possibly much longer (Lüthi et al., 2008; Fischer et al., 2018). These anthropogenic forcings have not only warmed the ocean and begun to melt the cryosphere, but have also led to widespread biogeochemical changes driven by the oceanic uptake of anthropogenic CO_2 from the atmosphere (IPCC, 2013).

It is now nearly three decades since the first assessment report of the IPCC, and over that time evidence and confidence in observed and projected ocean and cryosphere changes have grown (*very high confidence*; Table SM1.1). Confidence in climate warming and its anthropogenic causes has increased across assessment cycles; robust detection was not yet possible in 1990, but has been characterised as unequivocal since AR4 in 2007. Projections of near-term warming rates in early reports have been realised over the subsequent decades, while projections have tended to err on the side of caution for sea level rise and ocean heat uptake that have developed faster than predicted (Brysse et al., 2013; Section 4.2, 5.2). Areas of concern in early reports which were expected but not observable are now emerging. The expected acceleration of sea level rise is now observed with *high confidence* (Section 4.2). There is emerging evidence in sustained observations and from long-term palaeoclimate reconstructions for the expected slow-down of AMOC (*medium confidence*), although this remains to be properly attributed (Section 6.7). Significant sea level rise contributions from Antarctic ice sheet mass loss (*very high confidence*), which earlier reports did not expect to manifest this century, are already being observed (Section 3.3.1). Other newly emergent characteristics of ocean and cryosphere change (e.g., marine heat waves; Section 6.4) are assessed for the first time in SROCC.

AR5 (IPCC, 2013; IPCC, 2014) provides ample evidence of profound and pervasive changes in the ocean and cryosphere (Sections 1.4.1, 1.4.2), and along with the recent SR15 report (IPCC, 2018), is the point of departure for the updated assessments made in SROCC.

1.4.1 Observed and Projected Changes in the Ocean

Increasing greenhouse gases in the atmosphere cause heat uptake in the Earth system (Section 1.2) and as reported since 1970, there is *high confidence*[3] that the majority (more than 90%) of the extra thermal energy in the Earth's system is stored in the global ocean (IPCC, 2013). Mean ocean surface temperature has increased since the 1970s at a rate of 0.11 (0.09–0.13)°C per decade (*high confidence*), and forms part of a long-term warming of the surface ocean since the mid-19th century. The upper ocean (0–700 m, *virtually certain*) and intermediate ocean (700 to 2,000 m, *likely*) have warmed since the 1970s. Ocean heat uptake has continued unabated since AR5 (Sections 3.2.1.2.1, 5.2), increasing the risk of marine heat waves and other extreme events (Section 6.4). During the 21st century, ocean warming is projected to continue even if anthropogenic greenhouse gas emissions cease (Sections 1.3, 5.2). The global water cycle has been altered, resulting in substantial regional changes in sea surface salinity (*high confidence*; Rhein et al., 2013), which is expected to continue in the future (Sections 5.2.2, 6.3, 6.5).

The rate of sea level rise since the mid-19th century has been larger than the mean rate of the previous two millennia (*high confidence*). Over the period 1901 to 2010, global mean sea level rose by 0.19 (0.17–0.21) m (*high confidence*) (Church et al., 2013; Table SM1.1). Sea level rise continues due to freshwater added to the ocean by melting of glaciers and ice sheets, and as a result of ocean expansion due to continuous ocean warming, with a projected acceleration and century to millennial-scale commitments for ongoing rise (Section 4.2.3). In SROCC, recent developments of ice sheet modelling are assessed (Sections 1.8, 4.3, Cross-Chapter Box 8 in Chapter 3) and the projected sea level rise at the end of 21st century is higher than reported in AR5 but with a larger uncertainty range (Sections 4.2.3.2, 4.2.3.3).

By 2011, the ocean had taken up about 30 ± 7% of the anthropogenic CO_2 that had been released to the atmosphere since the industrial revolution (Ciais et al., 2013; Section 5.2). In response, ocean pH decreased by 0.1 since the beginning of the industrial era (*high confidence*), corresponding to an increase in acidity of 26% (Table SM1.1) and leading to both positive and negative biological and ecological impacts (*high confidence*) (Gattuso et al., 2014). Evidence is increasing that the ocean's oxygen content is declining (Oschlies et al., 2018). AR5 did not come to a final conclusion with regard to potential long-term changes in ocean productivity due to short observational records and divergent scientific evidence (Boyd et al., 2014; Section 5.2.2). Ocean acidification and deoxygenation are projected to continue over the next century with *high confidence* (Sections 3.2.2.3, 5.2.2).

---

[3]   Confidence/likelihood statements in Sections 1.4.1 and 1.4.2 derived from AR5 and SR15, unless otherwise specified.

1.4.2 Observed and Projected Changes in the Cryosphere

Changes in the cryosphere documented in AR5 included the widespread retreat of glaciers (*high confidence*), mass loss from the Greenland and Antarctic ice sheets (*high confidence*) and declining extents of Arctic sea ice (*very high confidence*) and Northern Hemisphere spring snow cover (*very high confidence*; IPCC, 2013; Vaughan et al., 2013).

A particularly rapid change in Earth's cryosphere has been the decrease in Arctic sea ice extent in all seasons (Section 3.2.1.1). AR5 assessed that there was *medium confidence* that a nearly ice-free summer Arctic Ocean is *likely* to occur before mid-century under a high emissions future (IPCC, 2013), and SR15 assessed that ice-free summers are projected to occur at least once per century at 1.5°C of warming, and at least once per decade at 2°C of warming above pre-industrial levels (IPCC, 2018). Sea ice thickness is decreasing further in the Northern Hemisphere and older ice that has survived multiple summers is rapidly disappearing; most sea ice in the Arctic is now 'first year' ice that grows in the autumn and winter but melts during the spring and summer (AMAP, 2017).

AR5 assessed that the annual mean loss from the Greenland ice sheet *very likely* substantially increased from 34 (-6–74) Gt yr⁻¹ (billion tonnes yr⁻¹) over the period 1992–2001, to 215 (157–274) Gt yr⁻¹ over the period 2002–2011 (IPCC, 2013). The average rate of ice loss from the Antarctic ice sheet also *likely* increased from 30 (-37–97) Gt yr⁻¹ over the period 1992–2001, to 147 (72–221) Gt yr⁻¹ over the period 2002–2011 (IPCC, 2013). The average rate of ice loss from glaciers around the world (excluding glaciers on the periphery of the ice sheets), was *very likely* 226 (91–361) Gt yr⁻¹ over the period 1971–2009, and 275 (140–410) Gt yr⁻¹ over the period 1993–2009 (IPCC, 2013). The Greenland and Antarctic ice sheets are continuing to lose mass at an accelerating rate (Section 3.3) and glaciers are continuing to lose mass worldwide (Section 2.2.3, Cross-Chapter Box 6 in Chapter 2). Confidence in the quantification of glacier and ice sheet mass loss has increased across successive IPCC reports (Table SM1.1) due to the development of remote sensing observational methods (Section 1.8.1).

Changes in seasonal snow are best documented for the Northern Hemisphere. AR5 reported that the extent of snow cover has decreased since the mid-20th century (*very high confidence*). Negative trends in both snow depth and duration are also detected with station observations (*medium confidence*), although results depend on elevation and observational period (Section 2.2.2). AR5 assessed that permafrost temperatures have increased in most regions since the early 1980s (*high confidence*), and the rate of increase has varied regionally (IPCC, 2013). Methane and carbon dioxide release from soil organic carbon is projected to continue in high mountain and polar regions (Box 2.2), and SROCC has used multiple lines of evidence to reduce uncertainty in permafrost change assessments (Cross-Chapter Box 5 in Chapter 1, Section 3.4.3.1.1).

Cross-Chapter Box 1 | Scenarios, Pathways and Reference Periods

Authors: Nerilie Abram (Australia), William Cheung (Canada), Lijing Cheng (China), Thomas Frölicher (Switzerland), Mathias Hauser (Switzerland), Shengping He (Norway/China), Anne Hollowed (USA), Ben Marzeion (Germany), Samuel Morin (France), Anna Pirani (Italy), Didier Swingedouw (France)

Introduction

Assessing the future risks and opportunities that climate change will bring for the ocean and cryosphere, and for their dependent ecosystems and human communities, is a main objective of this report. However, the future is inherently uncertain. A well-established methodological approach that the Special Report on the Ocean and Cryosphere in a Changing Climate (SROCC) report uses to assess the future under these uncertainties is through scenario analysis (Kainuma et al., 2018). The ultimate physical driver of the ocean and cryosphere changes that SROCC assesses are greenhouse gas emissions, while the exposure to hazards and the future risks to natural and human systems are also shaped social, economic and governance factors (Cross-Chapter Box 2 in Chapter 1; Section 1.5). This Cross-Chapter Box introduces the main scenarios that are used in the SROCC assessment. Examples of key climate change indicators in the atmosphere and ocean projected under future greenhouse gas emission scenarios are also provided (Table CB1.1).

Scenarios and pathways

Scenarios are a plausible description of how the future may develop based on a coherent and internally consistent set of assumptions about key driving forces and relationships. *Pathways* refer to the temporal evolution of natural and/or human systems towards a future state. In SROCC, assessments of future change frequently use climate model projections forced by pathways of future radiative forcing changes related to different socioeconomic scenarios.

Representative Concentration Pathways (RCPs) are a set of time series of plausible future concentrations of greenhouse gases, aerosols and chemically active gases, as well as land use changes (Moss et al., 2008; Moss et al., 2010; van Vuuren et al., 2011a; Figure SM1.1). The word representative signifies that each RCP provides only one of many possible pathways that would lead to the specific radiative forcing characteristics. The term pathway emphasises the fact that not only the long-term concentration levels, but also the trajectory taken over time to reach that outcome are of interest.

Cross-Chapter Box 1 (continued)

Four RCPs were used for projections of the future climate in the Coupled Model Intercomparison Project Phase 5 (CMIP5) (Taylor et al., 2012). They are identified by their approximate anthropogenic radiative forcing (in W m^{-2}, relative to 1750) by 2100: RCP2.6, RCP4.5, RCP6.0, and RCP8.5 (Figure SM1.1). RCP8.5 is a high greenhouse gas emission scenario without effective climate change mitigation policies, leading to continued and sustained growth in atmospheric greenhouse gas concentrations (Riahi et al., 2011). RCP2.6 represents a low greenhouse gas emission, high mitigation future that gives a two in three chance of limiting global atmospheric surface warming to below 2°C by the end of the century (van Vuuren et al., 2011b; Collins et al., 2013; Allen et al., 2018). Achieving the RCP2.6 pathway would require implementation of negative emissions technologies at a not-yet-proven scale to remove greenhouse gases from the air, in addition to other mitigation strategies such as energy from sustainable sources and existing nature-based strategies (Gasser et al., 2015; Sanderson et al., 2016; Royal Society, 2018; National Academies of Sciences, 2019). An even more stringent RCP1.9 pathway is considered most compatible with limiting global warming to below 1.5°C, called a 1.5°C-consistent pathway in the Special Report on Global Warming of 1.5°C (SR15; O'Neill et al., 2016; IPCC, 2018), and will be assessed in the IPCC 6th Assessment Report (AR6) using projections of Phase 6 of the Coupled Model Intercomparison Project (CMIP6). Global fossil CO_2 emissions rose more than 2% in 2018 and 1.6% in 2017, after a temporary slowdown in emissions from 2014–2016. Current emissions continue to grow in line with the RCP8.5 trajectory (Peters et al., 2012; Le Quéré et al., 2018).

In SROCC, the CMIP5 simulations forced with RCPs are used extensively to assess future ocean and cryosphere changes. In particular, RCP2.6 and RCP8.5 are used to contrast the possible outcomes of low-emission versus high-emission futures, respectively (Table CB1.1). In some cases the SROCC assessments use literature that is based on the earlier Special Report on Emission Scenarios (SRES; IPCC, 2000), and details of these and their approximate RCP equivalents are provided in Tables SM1.3 and SM1.4.

Shared Socioeconomic Pathways (SSPs) complement the RCPs with varying socioeconomic challenges to adaptation and mitigation (e.g., population, economic growth, education, urbanisation and the rate of technological development; O'Neill et al., 2017). The SSPs describe five alternative socioeconomic futures comprising: sustainable development (SSP1), middle-of-the-road development (SSP2), regional rivalry (SSP3), inequality (SSP4), and fossil-fuelled development (SSP5; Figure SM1.1; Kriegler et al., 2016; Riahi et al., 2017). The RCPs set plausible pathways for greenhouse gas concentrations and the climate changes that could occur, and the SSPs set the stage on which reductions in emissions will or will not be achieved within the context of the underlying socioeconomic characteristics and shared policy assumptions of that world. The combination of SSP-based socioeconomic scenarios and RCP-based climate projections provides an integrative frame for climate impact and policy analysis. The SSPs will be included in the CMIP6 simulations to be assessed in AR6 (O'Neill et al., 2016). In SROCC, the SSPs are used only for contextualising estimates from the literature on varying future populations in regions exposed to ocean and cryosphere changes.

Baselines and reference intervals

A baseline provides a reference period from which changes can be evaluated. In the context of anthropogenic climate change, the baseline should ideally approximate the 'pre-industrial' conditions before significant human influences on the climate began. The IPCC 5th Assessment Report (AR5) and SR15 (Allen et al., 2018) use 1850–1900 as the *pre-industrial* baseline for assessing historical and future climate change. Atmospheric greenhouse gas concentrations and global surface temperatures had already begun to rise in this interval from early industrialisation (Abram et al., 2016; Hawkins et al., 2017; Schurer et al., 2017). However, the scarcity of reliable climate observations represents a major challenge for quantifying earlier pre-industrial states (Hawkins et al., 2017). To maintain consistency across IPCC reports, the 1850–1900 *pre-industrial* baseline is used wherever possible in SROCC, recognising that this is a compromise between data coverage and representativeness of typical pre-industrial conditions.

In SROCC, the 1986–2005 reference interval used in AR5 is referred to as the *recent past*, and a 2006–2015 reference is used for *present day*, consistent with SR15 (Allen et al., 2018). The 2006–2015 reference interval incorporates near-global upper ocean data coverage and reasonably comprehensive remote-sensing cryosphere data (Section 1.8.1), and aligns this report with a more current reference than the 1986–2005 reference adopted by AR5. This 10-year *present day* period is short relative to natural variability. However, at this decadal scale the bias in the *present day* interval due to natural variability is generally small compared to differences between *present day* conditions and the *pre-industrial* baseline. There is also no indication of global average surface temperature in either 1986–2005 or 2006–2015 being substantially biased by short-term variability (Allen et al., 2018), consistent with the AR5 finding that each of the last three decades has been successively warmer at the Earth's surface than any preceding decade since 1850 (IPCC, 2013).

SROCC commonly provides future change assessments for two key intervals: A *near term* interval of 2031–2050 is comparable to a single generation time scale from present day, and incorporates the interval when global warming is *likely* to reach 1.5°C if warming continues at the current rate (IPCC, 2018). An *end-of-century* interval of 2081–2100 represents the average climate

Table CB1.1 | Projected change in global mean surface air temperature and key ocean variables for the *near term* (2031–2050) and *end-of-century* (2081–2100) relative to the *recent past* (1986–2005) reference period from CMIP5. See Table SM1.2 for the list of CMIP5 models and ensemble member used for calculating these projections. Small differences in the projections given here compared with AR5 (e.g., Table 12.2 in Collins et al., 2013) reflect differences in the number of models available now compared to at the time of the AR5 assessment (Table SM1.2).

		Near term: 2031–2050		End-of-century: 2081–2100	
	Scenario	Mean	5–95% range	Mean	5–95% range
Global Mean Surface Air temperature (°C) [a]	RCP2.6	0.9	0.5 to 1.4	1.0	0.3 to 1.7
	RCP4.5	1.1	0.7 to 1.5	1.8	1.0 to 2.6
	RCP6.0	1.0	0.5 to 1.4	2.3	1.4 to 3.2
	RCP8.5	1.4	0.9 to 1.8	3.7	2.6 to 4.8
Global Mean Sea Surface Temperature (°C) [b] (Section 5.2.5)	RCP2.6	0.64	0.33 to 0.96	0.73	0.20 to 1.27
	RCP8.5	0.95	0.60 to 1.29	2.58	1.64 to 3.51
Surface pH (units) [b] (Section 5.2.2.3)	RCP2.6	-0.072	-0.072 to -0.072	-0.065	-0.065 to -0.066
	RCP8.5	-0.108	-0.106 to -0.110	-0.315	-0.313 to -0.317
Dissolved Oxygen (100–600 m) (% change) (Section 5.2.2.4) [b]	RCP2.6	-0.9	-0.3 to -1.5	-0.6	0.0 to -1.2
	RCP8.5	-1.4	-1.0 to -1.8	-3.9	-2.9 to -5.0

Notes:

[a] Calculated following the same procedure as the IPCC 5th Assessment Report (AR5) (Table 12.2 in Collins et al., 2013). The 5–95% model range of global mean surface air temperature across CMIP5 projections was assessed in AR5 as the *likely* range, after accounting for additional uncertainties or different levels of confidence in models.

[b] The 5–95% model range for global mean sea surface temperature, surface pH and dissolved oxygen (100–600 m) as referred to in the SROCC assessment as the *very likely* range (Section 1.9.2, Figure 1.4).

conditions reached at the end of the standard CMIP5 future climate simulations and is relevant to long-term infrastructure planning and climate-resilient development pathways (CRDPs) (Cross-Chapter Box 2 in Chapter 1). In some cases where committed changes exist over multi-century time scales, such as the assessment of future sea level rise (Section 4.3.2) or deep ocean oxygen changes (Section 5.2.4.2, Table 5.5), SROCC also considers model evidence for *long-term* changes beyond the end of the current century.

Key indicators of future ocean and cryosphere change
Table CB1.1 compiles information on key indicators of climate change in the atmosphere and ocean. This information is given for different RCPs and for changes in the *near term* and *end-of-century* assessment intervals, relative to the *recent past*, noting that this does not capture changes that have already taken place since the *pre-industrial* baseline. SR15 assessed that global mean surface warming from the *pre-industrial* (1850–1900) to the *recent past* (1986–2005) reference period was 0.63°C (*likely* range of 0.57°C–0.69°C), and during the *present day* interval (2006–2015) was 0.87°C (*likely* range of 0.75°C–0.99°C) higher than the average over the 1850–1900 *pre-industrial* period (*very high confidence*; IPCC, 2018).

These key climate and ocean change indicators allow for some harmonisation of the risk assessments in the chapters of SROCC. Projections of future change across a wider range of ocean and cryosphere components is also provided in Figure 1.5. Ocean and cryosphere changes and risks by the end-of-century (2081–2100) are expected to be larger under high greenhouse gas emission scenarios, compared with low greenhouse gas emission scenarios (*very high confidence*) (Table CB1.1, Figure 1.5).

1.5   Risk and Impacts Related to Ocean and Cryosphere Change

SROCC assesses the risks (i.e., potential for adverse consequences) and impacts (i.e., manifested risk) resulting from climate-related changes in the ocean and cryosphere. Knowledge on risk is essential for conceiving and implementing adequate responses. Cross-Chapter Box 2 in Chapter 1 introduces key concepts of risk, adaptation, resilience and transformation, and explains why and how they matter for this report.

In SROCC, the term 'natural system' describes the biological and physical components of the environment, independent of human involvement but potentially affected by human activities. 'Natural systems' may refer to portions of the total system without necessarily considering all its components (e.g., an ocean upwelling system). Throughout the assessment usage of 'natural system' does not imply a system unaltered by human activities.

'Human systems' include physiological, health, socio-cultural, belief, technological, economic, food, political, and legal systems, among others. Humans have depended upon the Earth's ocean (WOA, 2016; IPBES, 2018b) and cryosphere (AMAP, 2011; Hovelsrud et al., 2011; Watt-Cloutier, 2018) for many millennia (Redman, 1999). Contemporary human populations still depend directly on elements of the ocean and cryosphere, and the ecosystem services they provide, but at a much larger scale and with greater environmental impact than in pre-industrial times (Inniss and Simcock, 2017).

An ecosystem is a functional unit consisting of living organisms, their non-living environment, and the interactions within and between them. Ecosystems can be nested within other ecosystems and their scale can range from very small to the entire biosphere. Today, most ecosystems either contain humans as key organisms, or are influenced by the effects of human activities in their environment. In SROCC, a social-ecological system describes the combined system and all of its sub-components and refers specifically to the interaction of natural and human systems.

The ocean and cryosphere are unique systems that have intrinsic value, including the ecosystems and biodiversity they support. Frameworks of Ecosystem Services and Nature's Contributions to People are both used within SROCC to assess the impacts of changes in the ocean and cryosphere on humans directly, and through changes to the ecosystems that support human life and civilisations (Sections 2.3, 3.4.3.2, 4.3.3.5, 5.4, 6.4, 6.5, 6.8). The Millennium Ecosystem Assessment (MEA, 2005) established a conceptual Ecosystem Services framework between biodiversity, human well-being, and drivers of change. This framework highlights that natural systems provide vital life-support services to humans and the planet, including direct material services (e.g., food, timber), non-material services (e.g., cultural continuity, health), and many services that regulate environmental status (e.g., soil formation, water purification). This framework supports decision-making by quantifying benefits for valuation and trade-off analyses. The Ecosystem Services framework has been challenged as monetising the relationships of people with nature, and undervaluing small-scale livelihoods, cultural values and other considerations that contribute little to global commerce (Díaz et al., 2018). More recent frameworks, such as Nature's Contributions to People (Díaz et al., 2018), used in the Intergovernmental Platform on Biodiversity and Ecosystem Services assessments (IPBES), aim to better encompass the non-commercial ways that nature contributes to human quality of life.

Cross-Chapter Box 2 | Key Concepts of Risk, Adaptation, Resilience and Transformation

Authors: Matthias Garschagen (Germany), Carolina Adler (Switzerland/Australia), Susie Crate (USA), Hélène Jacot Des Combes (Fiji/France), Bruce Glavovic (New Zealand/South Africa), Sherilee Harper (Canada), Elisabeth Holland (Fiji/USA), Gary Kofinas (USA), Sean O'Donoghue (South Africa), Ben Orlove (USA), Zita Sebesvari (Hungary/Germany), Martin Sommerkorn (Norway/Germany)

This box introduces key concepts used in the Special Report on the Ocean and Cryosphere in a Changing Climate (SROCC) in relation to risk, adaptation, resilience, and transformation. Building on an assessment of the current literature, it provides a conceptual framing for the report and for the assessments within its chapters. Full definitions of key terms are provided in the SROCC Annex I: Glossary.

Risk and adaptation

SROCC considers risk from climate change-related effects on the ocean and cryosphere as the result of the interaction between: (1) environmental hazards triggered by climate change, (2) exposure of humans, infrastructure and ecosystems to those hazards, and (3) systems' vulnerabilities. *Risk* refers to the potential for adverse consequences, and *impacts* refer to materialised effects of climate change. Next to assessing risk and impacts specifically resulting from climate change-related effects on the ocean, coast and cryosphere, SROCC is also concerned with the options to reduce climate-related risk.

Beyond mitigation, adaptation is a key avenue to reduce risk (Section 1.6). Adaptation can also include exploiting new opportunities; however, this box focuses on risk, and thus, the latter is not discussed in detail here. Adaptation efforts link into the causal fabric of risk by reducing existing and future vulnerability, exposure, and/or (where possible) hazards (Figure CB2.1). Addressing the different risk components (hazards, exposure and vulnerability) involves assessing and selecting options for policy.

Cross-Chapter Box 2 (continued)

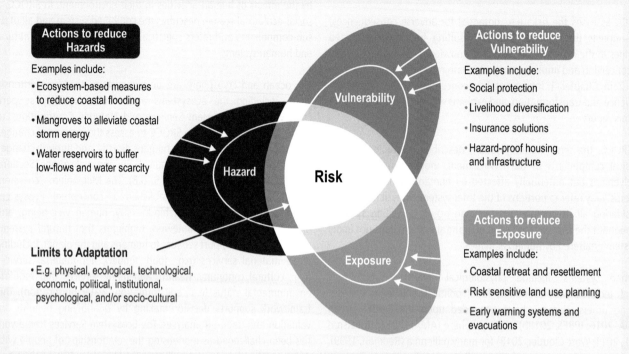

Figure CB2.1 | There are options for risk reduction through adaptation. Adaptation can reduce risk by addressing one or more of the three risk factors: vulnerability, exposure, and/or hazard. The reduction of vulnerability, exposure, and/or hazard potential can be achieved through different policy and action choices over time until limits to adaptation might be reached. The figure builds on the conceptual framework of risk used in the IPCC 5th Assessment Report (AR5) (Oppenheimer et al., 2014).

and action. Such decision-making entails evaluation of the effectiveness, efficiency, efficacy and acceptance of actions. Adaptation responses are more effective when they promote resilience to climate change, consider plausible futures and unexpected events, strengthen essential or desired characteristics as well as values of the responding system and/or make adjustments to avoid unsustainable pathways (*high agreement, medium evidence*; Section 2.3; Box 2.4; 4.4.4; 4.4.5).

Adaptation requires adaptive capacity, which for human systems includes assets (financial, physical, and/or ecological), capital (social and institutional), knowledge and technical know-how (Klein et al., 2014). The extent of adaptive capacity determines adaptation potential, but does not necessarily translate into effective adaptation if awareness of the need to act, the willingness to act and/or the cooperation needed to act is lacking (*high confidence*; Sections 2.3; Box 2.4; 4.3.2.6.3; 5.5.2.4).

There are limits to adaptation, which include, for example, physical, ecological, technological, economic, political, institutional, psychological and/or socio-cultural aspects (*medium evidence, high agreement*) (Dow et al., 2013; Barnett et al., 2014; Klein et al., 2014). For example, the ability to adapt to sea level rise depends, in part, on the elevation of the low-lying islands and coasts in question, but also on the capacity to successfully negotiate protection or relocation measures socially and politically (Cross-Chapter Box 9, also see Section 6.4.3 for a wider overview). Limits to adaptation are sometimes considered as something different from barriers to adaptation. Barriers can in principle be overcome if adaptive capacity is available (e.g., where funding is made available), even though overcoming barriers is often hard in reality, particularly for resource-poor communities and countries (*high confidence*; Section 4.4.3). Limits to adaptation are reached when adaptation no longer allows an actor or ecosystem to secure valued objectives or key functions from intolerable risks (Section 4.4.2; Dow et al., 2013). Defining tolerable risks and key system functions is, therefore, of central importance for the assessment of limits to adaptation.

Residual risks (i.e., the risk that endures following adaptation and risk reduction efforts) remain even where adaptation is possible (*very high confidence*; Chapters 2–6; Section 6.3.2; Table 6.2). Residual risks have bearing on the emerging debate about loss and damage (Huq et al., 2013; Warner and van der Geest, 2013; Boyd et al., 2017; Djalante et al., 2018; Mechler et al., 2018; Roy et al., 2018). This report addresses loss and damage in relation to slow onset processes, including ocean changes (Section 5.4.2.3), and sea level rise (Section 4.3), and glacier retreat (Section 2.3.6), and polar cryosphere changes (Section 3.4.3.3.4), as well as rapid onset hazards such as tropical cyclones (Chapter 6). The assessment encompasses non-economic losses, including the impacts on intrinsic and

Cross-Chapter Box 2 (continued)

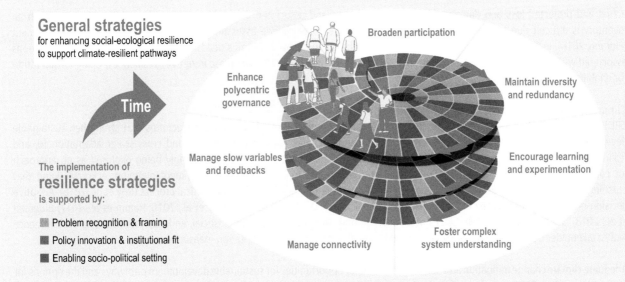

Figure CB2.2 | General strategies for enhancing social-ecological resilience to support climate-resilient pathways have been identified. The seven strategies are adapted from synthesis papers by Biggs et al. (2012) and Quinlan et al. (2016), the illustration of the climate-resilient development pathway (CRDP) builds on Figure SPM9 in the IPCC 5th Assessment Report (AR5) (IPCC, 2014).

spiritual attributes with which high mountain societies value their landscapes (Section 2.3.5); the interconnected relationship with, and reliance upon, the land, water and ice for culture, livelihoods and wellbeing in the Arctic (Section 3.4.3.3); and cultural heritage and displacement addressed in the Cross-Chapter Box on low-lying islands and coasts (Cross-Chapter Box 9; Burkett, 2016; Markham et al., 2016; Tschakert et al., 2017; Huggel et al., 2018).

Building resilience
Addressing climate change-related risk, impacts (including extreme events and shocks) and trade-offs together with shaping the trajectories of social and ecological systems is facilitated by considering resilience (Biggs et al., 2012; Quinlan et al., 2016). In SROCC, resilience is understood as the capacity of interconnected social, economic and ecological systems to cope with disturbances by reorganising in ways that maintain their essential function, structure, and identity (Walker et al., 2004). Resilience may be considered as a positive attribute of a system and an aspirational goal when it contributes to the capacity for adaptation and learning without changing the structure, function, and identity of the system (Walker et al., 2004; Steiner, 2015). Alternately, resilience may be used descriptively as a system property that is neither good nor bad (Walker et al., 2004; Chapin et al., 2009; Weichselgartner and Kelman, 2014). For example, a system can be highly resilient in keeping its unfavoured attributes, such as poverty or institutional rigidity (Carpenter and Brock, 2008). Critics of the resilience concept warn that the application of resilience to social systems is problematic when the responsibility for resilience building is shifted onto the shoulders of vulnerable and resource-poor populations (e.g., Chandler, 2013; Reid, 2013; Rigg and Oven, 2015; Tierney, 2015; Olsson et al., 2017).

Applying the concept of resilience in mitigation and adaptation planning builds the capacity of a social-ecological system to navigate anticipated changes and unexpected events (Biggs et al., 2012; Varma et al., 2014; Sud et al., 2015). Resilience also emphasises social-ecological system dynamics, including the possibility of crossing critical thresholds and experiencing a regime shift (i.e., state change). Seven general strategies for building social-ecological resilience have been identified (Figure CB2.2; Ostrom, 2010; Biggs et al., 2012; Quinlan et al., 2016). The concept of resilience also allows analysts, accessors of risk and decision makers to recognise how climate-change related risks often cannot be fully avoided or alleviated despite adaptation. For SROCC, this is especially relevant along low-lying coasts, high mountain areas and the polar regions (*medium evidence, high agreement*; Sections 2.3; 2.4; 3.5, 6.8, 6.9).

Many efforts are underway to apply resilience thinking in assessments, management practices, policy making and the day-to-day practices of affected sectors and local communities. For example, leaders of the Pacific small island developing states use the Framework for Resilient Development in the Pacific, which integrates climate change and disaster risk management (Pacific Community, 2016; Cross-Chapter Box 9). In the Philippines, a new framework has been developed to conduct full inventories of

Cross-Chapter Box 2 (continued)

actual and projected loss and damage due to climate change and associated disasters such as from cyclones. Creating such an inventory is difficult due to the disconnect between tools for climate change assessment and those for post disaster assessment (Florano, 2018). In Arctic Alaska, evaluative frameworks are being applied to determine needs, responsibilities, and alternative actions associated with coastal village relocations (Bronen, 2015; Cross-Chapter Box 9). In all these initiatives, resilience is a key consideration for enabling CRDPs.

Climate-resilient development pathways (CRDPs)
CRDPs are a relatively new concept to describe climate change mitigation and adaptation trajectories that strengthen sustainable development and efforts to eradicate poverty and reduce inequalities while promoting fair and cross-scalar adaptation to, and resilience in, a changing climate (Kainuma et al., 2018; Roy et al., 2018). CRDPs are increasingly being explored as an approach for combining scientific assessments, stakeholder participation, and forward-looking development planning, acknowledging that pursuing CRDP is not only a technical challenge of risk management but also a social and political process (Roy et al., 2018). Adaptive decision-making over time is key to CRDPs (Haasnoot et al., 2013; Wise et al., 2014; Fazey et al., 2016; Ramm et al., 2017; Bloemen et al., 2018; Lawrence et al., 2018). CRDPs accommodate both the interacting cultural, social, and ecosystem factors that influence multi-stakeholder decision making processes, and the overall sustainability of adaptation measures.

Adequate climate change mitigation and adaptation allows for opportunities for sustainable development pathways and the options for resilience building. CRDPs involve series of mitigation and adaptation choices over time, balancing short-term and long-term goals and accommodating newly available knowledge (Denton et al., 2014). The CRDPs approach has been successfully used, for example, in urban, remote and disadvantaged communities, and can showcase the potential to counter maladaptive choices (e.g., Barnett et al., 2014; Butler et al., 2014; Maru et al., 2014). CRDPs aim to establish narratives of hope and opportunity that can extend beyond risk reduction and coping (Amundsen et al., 2018). Although climate change impacts on the ocean and cryosphere elicit many emotions, including fear, anger, despair and apathy (Cunsolo Willox et al., 2013; Cunsolo and Landman, 2017; Cunsolo and Ellis, 2018), narratives of hope are critical in provoking motivation, creative thinking and behavioural changes in response to climate change (Myers et al., 2012; Smith and Leiserowitz, 2014; Feldman and Hart, 2016; Feldman and Hart, 2018; Prescott and Logan, 2018; Section 1.8.3).

Much of the adaptation and resilience literature published since AR5 highlights the need for transformations that enable effective climate change mitigation (most notably, to decarbonise the economy) (Riahi et al., 2017), and support adaptation (e.g., Pelling et al., 2015; Few et al., 2017). Transformation becomes particularly relevant when existing mitigation and adaptation practices cannot reduce risks and impacts to an acceptable level. Transformative adaptation, therefore, involves fundamental modifications of policies, policy making processes, institutions, human behaviour and cultural values (Pelling et al., 2015; Solecki et al., 2017). Successful transformation requires attention to conditions that allow for such changes, including timing (e.g., windows of opportunity), social readiness (e.g., some level of willingness) and resources to act (e.g., trust, human skill and financial resources; Kofinas et al., 2013; Moore et al., 2014). Examples related to SROCC include shifting from a paradigm of protection reliant on seawalls to living with saltwater as a response to coastal flooding in rural areas (Renaud et al., 2015) or involving fundamental risk management changes in coastal megacities, including retreat (Solecki et al., 2017). Transformation in changing ocean and cryosphere contexts can be fostered by transdisciplinary collaboration between actors in science, government, the private sector, civil society and affected communities (Padmanabhan, 2017; Cross-Chapter Box 3 in Chapter 1; Cross-Chapter Box 4 in Chapter 1).

1.5.1   Hazards and Opportunities for Natural Systems, Ecosystems, and Human Systems

Hazards faced by marine and coastal organisms, and the ecosystem services they provide are generally dependent on future greenhouse gas emission pathways, with moderate likelihood under a low-emission future, but high to very high likelihood under higher emission scenarios (*very high confidence*) (Mora et al., 2013; Gattuso et al., 2015). Hazards to marine ecosystems assessed in AR5 (IPCC, 2014) included degradation of coral reefs (*high confidence*), ocean deoxygenation (*medium confidence*) and ocean acidification (*high confidence*). Shifts in the ranges of plankton and fish were identified with *high confidence* regionally, but with uncertain trends globally.

SROCC provides more evidence for global shifts in the distribution of marine organisms, and in how the phenology of animals is responding to ocean change (Sections 3.2.3, 5.2). The signature of climate change is now detected in almost all marine ecosystems. Similar trends of changing habitat due to climate change are reported for the cryosphere (Sections 2.2, 3.4.3.2). The risk of irreversible loss of many marine and coastal ecosystems increases with global warming, especially at 2°C or more (*high confidence*; IPCC, 2018). Risk also increases for habitat displacements, both poleward (Section 3.2.4) and to greater ocean depths (Section 5.2.4), or habitat reductions, such as that caused by glacier retreat (Section 2.2.3).

Changes in the ocean and cryosphere bring hazards that affect the health, wellbeing, safety and security of populations in coastal, mountain and polar environments (Section 2.3.5, 3.4.3, 4.3.2). Some impacts are direct, such as sea level rise or coastal erosion that can displace coastal residents (4.3.2.3, 4.4.2.6, Box 4.1). Other effects are indirect; for example, rising ocean temperatures have led to increases in maximum wind speed and rainfall rates in tropical cyclones (Section 6.3), creating hazards with severe consequences for natural and human systems (Sections 4.3, 6.2, 6.3, 6.8). The multiple category 4 and 5 Atlantic hurricanes in 2017 caused the loss of over 3300 lives and more than 350 billion USD in economic damages (Cross-Chapter Box 9; Andrade et al., 2018; Murakami et al., 2018; NOAA, 2018). In mountain regions, glacial lake outburst floods have caused severe impacts on lives, livelihoods and infrastructure that often extend beyond the directly affected areas (Section 2.3.2 and 6.2.2). Some hazards related to ocean and cryosphere change involve abrupt and irreversible changes (Section 1.3), which generate sometimes unpredictable risks, and multiple hazards can coincide to greatly elevate the total risk (Section 6.8.2). For example, combinations of thawing permafrost, sea level rise, loss of sea ice, ocean surface waves and extreme weather events (Thomson and Rogers, 2014; Ford et al., 2017) have damaged Arctic infrastructure (e.g., buildings, roads) (AMAP, 2015; AMAP, 2017), impacted reindeer husbandry livelihoods for Sami and other Arctic Indigenous peoples and impeded access to hunting grounds, other communities and travel routes fundamental to the livelihoods, food security and wellbeing of Inuit and other Northern cultures (Section 3.4.3). In some Arctic regions, tipping points may have already been reached such that adaptive practices can no longer work (Section 3.5).

Climate change impacts on the ocean and cryosphere can also present opportunities, in at least the near- and medium-term. For example, in Nepal warming of high mountain environments and accelerated melting of snow and ice have extended the growing season and crop yields in some regions (Section 2.3; Gaire et al., 2015; Merrey et al., 2018), while tourism and shipping has increased in the Arctic with loss of sea ice (Section 3.2.4). Moreover, rising ocean temperatures redistribute the global fish population, allowing new fishing opportunities while reducing some established fisheries (Bell et al., 2011; Fenichel et al., 2016; Section 5.4). To gain from new opportunities, while also avoiding or mitigating new or increasing hazards, it is necessary to be aware of trade-offs between risks and benefits to understand who is and is not benefiting. For example, opportunities can involve trade-offs with mitigation and/or SDGs (Section 3.5.2), and the balance of economic costs and benefits may differ substantially between the near-term and long-term future (Section 5.4.2.2).

1.5.2 Exposure of Natural Systems, Ecosystems, and Human Systems

Exposure to hazards in cryosphere systems occur in the immediate vicinity of cryosphere components, and at regional to global scales where cryosphere changes link to other natural systems. For example, decreasing Arctic sea ice increases exposure for organisms that depend upon habitats provided by sea ice, but also has far-reaching impacts

through the resulting direct albedo feedback and amplification of Arctic climate warming (e.g., Pistone et al., 2014) that then locally increases surface melting of the Greenland ice sheet (Liu et al., 2016; Stroeve et al., 2017). Additionally, ice loss from ice sheets contribute to the global-scale exposure of sea level rise, and more local-scale modifications and losses of coastal habitats and ecosystems (Sections 3.2.3 and 4.3.3.5). Interactions within and between natural systems also influence the spatial reach of risks associated with cryosphere change. Permafrost degradation, for example, interacts with ecosystems and climate on various spatial and temporal scales, and feedbacks from these interactions range from local impacts on topography, hydrology and biology, to global-scale impacts via biogeochemical cycling (e.g., methane release) on climate (Sections 2.2, 2.3, 3.4; Kokelj et al., 2015; Grosse et al., 2016).

Exposure to climate change risk exists for virtually all coastal organisms, habitats and ecosystems (Section 5.2), through processes such as inundation and salinisation (Section 4.3), ocean acidification and deoxygenation (Sections 3.2.3, 5.2.3), increasing marine heatwaves (Section 6.4.1.2), and increases in harmful algal blooms and invasive species (Glibert et al., 2014; Gobler et al., 2017; Townhill et al., 2017; Box 5.3). Aggregate impacts of multiple drivers are dramatically altering ecosystem structure and function in the coastal and open ocean (Boyd et al., 2015; Deutsch et al., 2015; Przeslawski et al., 2015), such as coral reefs under increasing pressure from both rising ocean temperature and acidification (Section 5.3.4). Increasing exposure to climate change hazards in open ocean natural systems includes ocean acidification (O'Neill et al., 2017; Section 5.2.3), changes in ocean ventilation, deoxygenation (Shepherd et al., 2017; Breitburg et al., 2018; Section 5.2.2.4), increased cyclone and flood risk (Section 6.3.3) and an increase in extreme El Niño and La Niña events (Section. 6.5.1). Heat content is rapidly increasing within the ocean (Section 5.2.2) and marine heat waves are becoming more frequent across the world ocean (Section 6.4.1).

People who live close to the ocean and/or cryosphere, or depend directly on their resources for livelihoods, are particularly exposed to climate change impacts and hazards (*very high confidence*) (Barange et al., 2014; Romero-Lankao et al., 2014; AMAP, 2015). These exposures can result in infrastructure damage and failure (Sections 2.3.1.3, 3.4.3, 3.5., 4.3.2), loss of habitability (Sections 2.3.7, 3.4.3, 3.5, 4.3.3), changes in air quality (Section 6.5.2), proliferation of disease vectors (Sections 3.4.3.2.2, 5.4.2.1.1), increased morbidity and mortality due to injury, infectious disease, heat stress, and mental health and wellness challenges (Section 3.4.3.3), compromised food and water security (Sections 2.3.1, 3.4.3.3, 4.3.3.6, 5.4.2.1, 6.8.4), degradation of ecosystem services (Sections 2.3.1.2, 2.3.3.4, 4.3.3, 5.4.1, 6.4.2.3), economic and non-economic impacts due to reduced production and social network system disruption (Section 2.3.7), conflict (Sections 2.3.1.14, 3.5) and widespread human migration (Sections 2.3.7, 4.4.3.5; Oppenheimer et al., 2014; van Ruijven et al., 2014; AMAP, 2015; Cunsolo and Ellis, 2018).

This report documents how people residing in coastal and cryosphere regions are already exposed to climate change hazards, and which of these hazards are projected to increase in the future. For example, mountain communities have been exposed to increased rockfall, rock avalanches and landslides due to permafrost degradation and glacier shrinkage, and to changes in snow avalanche type and seasonal timing (Section 2.3.1). Cryosphere changes that can impact water availability in mountain regions and for downstream populations (Sections 2.3.1, 2.3.4, 2.3.5) have implications for drinking water, irrigation, livestock grazing, hydropower production and tourism (Section 2.3). Some declining mountain glaciers hold sacred and symbolic meanings for local communities who will experience spiritual losses (Section 2.3.4, 2.3.5, and 2.3.6). Exposures to extreme warming, and continued sea ice and permafrost loss in the Arctic, challenge Indigenous communities with close interdependent relationships of economy, lifestyles, cultural identity, self-sufficiency, Indigenous knowledge, health and wellbeing with the Arctic cryosphere (Section 3.4.3, 3.5).

The population living in low elevation coastal zones (land less than 10 m above sea level) is projected to increase to more than one billion by 2050 (Section 4.3.2.2). These people and communities are particularly exposed to future sea level rise, rising ocean temperature (including marine heat waves; Section 6.4), enhanced coastal erosion, increasing wind, wave height, storm intensity and ocean acidification (Section 4.3.4). These exposures bring associated risks for livelihoods linked to fisheries, tourism and trade, as well as loss of life, damaged assets, and disruption of basic services including safe water supplies, sanitation, energy and transportation networks (Chapters 4, 5, and 6; Cross-Chapter Box 9).

1.5.3   Vulnerabilities in Natural Systems, Ecosystems, and Human Systems

Direct and indirect risks to natural systems are influenced by vulnerability to climate change as well as deterioration of ecosystem services. For example, about half of species assessed on the northeast United States continental shelf exhibited high to very high climate vulnerability due to temperature preferences and changes in habitat space (Hare et al., 2016), with corresponding northward range shifts for many species (Kleisner et al., 2017) and increased vulnerability for organisms or ecosystems unable to migrate or evolve at the rate required to adapt to ocean and cryosphere changes (Miller et al., 2018). Non-climatic pressures also magnify the vulnerability of ocean and cryosphere ecosystems to climate-related changes, such as overfishing, coastal development, and pollution, including plastic pollution (Halpern et al., 2008; Halpern et al., 2015; IPBES, 2018a; IPBES, 2018b; IPBES, 2018c; IPBES, 2018d). Conventional (fossil fuel-based) plastics produced in 2015 accounted for 3.8% of global CO_2 emissions and could reach up to 15% by 2050 (Zheng and Suh, 2019).

The vulnerability of mountain, Arctic and coastal communities is affected by social, political, historical, cultural, economic, institutional, environmental, geographical and/or demographic factors such as gender, age, race, class, caste, Indigeneity and disability (Thomas et al., 2019; Sections 2.3.6 and 3.5; Cross-Chapter Box 9). Disparities and inequities in such factors may result in social exclusion, inequalities and non-climatic challenges to health and wellbeing, economic development and basic human rights (Adger et al., 2014; Olsson et al., 2014; Smith et al., 2014). Those less advantaged often also have reduced access to and control over the social, financial, technological and environmental resources that are required for adaptation and transformation (Oppenheimer et al., 2014; AMAP, 2015), thus limiting options for coping and adapting to change (Hijioka et al., 2014). However, even populations with greater wealth and privilege can be vulnerable to some climate change risks (Cardona et al., 2012; Smith et al., 2014), especially if sources of wealth and wellbeing depend upon established infrastructure that is poorly suited to ocean or cryosphere change.

Institutions and governance can shape vulnerability and adaptive capacity, and it can be challenging for weak governance structures to respond effectively to extreme or persistent climate change hazards (Sections 6.4 and 6.9; Cross-Chapter Box 3 in Chapter 1; Berrang-Ford et al., 2014; Hijioka et al., 2014). Furthermore, populations can be negatively impacted by inappropriate climate change mitigation and/or adaptation policies, particularly ones that further marginalise their knowledge, culture, values and livelihoods (Field et al., 2014; Cross-Chapter Box 4 in Chapter 1).

Vulnerability is not static in place and time, nor homogeneously experienced. The vulnerabilities of individuals, groups, and populations to climate change is dynamic and diverse, and reflects changing societal and environmental conditions (Thomas et al., 2019). SROCC examines vulnerability following the conceptual definition presented in Cross-Chapter Box 2 in Chapter 1, and vulnerability in human systems is treated in relative rather than absolute terms.

1.6   Addressing the Causes and Consequences of Climate Change for the Ocean and Cryosphere

Effective and ambitious mitigation of climate change would be required to meet the temperature goal of the Paris Agreement (UNFCCC, 2015; IPCC, 2018). Similarly, effective and ambitious adaptation to climate change impacts on the ocean and cryosphere is necessary to enable CRDPs that minimise residual risk, and loss and damage (*very high confidence*; Cross-Chapter Box 2 in Chapter 1; IPCC, 2018). *Mitigation* refers to human actions to limit climate change by reducing the emissions and enhancing the sinks of greenhouse gases. *Adaptation* refers to processes of adjustment by natural or human systems to actual or expected climate and its effects, intended to moderate harm or exploit beneficial opportunities. The presidency of the 23rd Conference Of the Parties of United Nations Framework Convention on Climate Change (UNFCCC) introduced the oceans pathway into the climate solution space, acknowledging both the importance of the ocean in the climate system and that ocean commitments for adaptation and mitigation are available through Nationally Determined Contributions under the UNFCCC (Gallo et al., 2017).

1.6.1   Mitigation and Adaptation Options in the Ocean and Cryosphere

Mitigation and adaptation pathways to avoid dangerous anthropogenic interference with the climate system (United Nations,

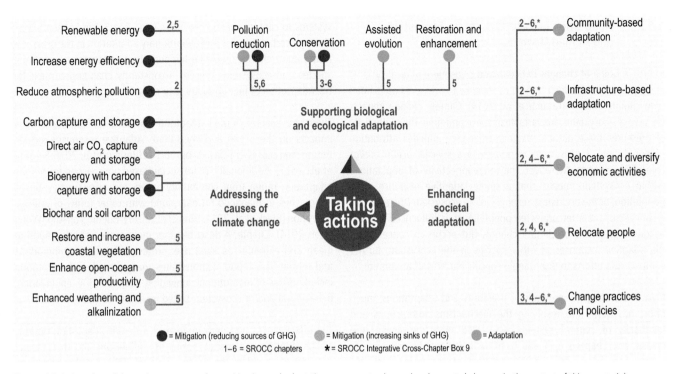

Figure 1.2 | Overview of the main ocean-cryosphere mitigation and adaptation measures to observed and expected changes in the context of this report. A longer description of these measures are given in SM1.3. Solar radiation management techniques are omitted because they are covered in other IPCC 6th Assessment Report (AR6) products. Governance and enabling conditions are implicitly embedded in all mitigation and adaptation measures. Some governance-based measures (e.g., institutional arrangements) are not included in this figure but are covered in Cross-Chapter Box 3 in Chapter 1 and in Chapters 2 to 6. GHG: greenhouse gases. Modified from Gattuso et al. (2018).

1992) are considered in SR15 (IPCC, 2018). SROCC assesses several ocean and cryosphere-specific measures for mitigation and adaptation including options for to address the causes of climate change, support biological and ecological adaptation, and enhance societal adaptation (Figure 1.2). Other measures have been proposed, including solar radiation management and several other forms of carbon dioxide removal, but these are not addressed in SROCC as they are covered in other products of the IPCC Sixth Assessment Cycle (SR15 and AR6 Working Group III) and are outside the scope of SROCC. SROCC does assess indirect mitigation measures that involve the ocean and the cryosphere (Figure 1.2) by supporting biological and ecological adaptation, such as through reducing nutrient and organic carbon pollution (which moderates ocean acidification in eutrophied areas) and conservation (which preserves biodiversity and habitats) in coastal regions (Billé et al., 2013).

A literature-based expert assessment shows that ocean-related mitigation measures have trade-offs, with the greatest benefits derived by combining global and local measures (*high confidence*; Gattuso et al., 2018). Local measures, such as pollution reduction and conservation, provide significant co-benefits and few adverse side effects (*high confidence*; Sections 5.5.1, 5.5.2). They can be relatively rapidly implemented, but are generally less effective in addressing the global problem (*high confidence*; Sections 5.5.1, 5.5.2). Likewise, local efforts to decrease air pollution near mountain glaciers and other cryosphere components, for example reducing black carbon emissions, can bring regional-scale benefits for health and in reducing snow and ice-melt (Shindell et al., 2012; Box 2.2).

Well-chosen human interventions can enhance the adaptive capacity of natural systems to climate change. Such interventions through manipulating an ecosystem's structural or functional properties (e.g., restoration of mangroves) may minimise climate change pressures, enhance natural resilience and/or re-direct ecosystem responses to reduce cascading risks on societies. In human systems, adaptation can involve both infrastructure (e.g., enhanced sea defences) and community-based action (e.g., changes in policies and practices). Adaptation options to ongoing climate change are most effective when considered together with mitigation strategies because there are limits to effective adaptation, mitigation actions can make adaptation more difficult, and some adaptation measures may increase greenhouse gas emissions.

Adaptation and mitigation decisions are connected with economic concerns. In SROCC, two main economic approaches are used. The first comprises the Total Economic Value method and the valuation of ecosystem services. SROCC considers the paradigm of sustainable development, and the linkages between climate impacts on ecosystem services (Section 5.4.1) and the consequences on SDGs including food security or poverty eradication (Section 5.4.2). The second economic approach used are formal decision analysis methods, which help to identify options (also called alternatives) that perform best or well with regards to given objectives. These methods include cost-benefit analysis, multi-criteria analysis and robust decision-making and are specifically relevant for appraising long-term investment decisions in the context of coastal adaptation (Section 4.4.4.6).

1.6.2   Adaptation in Natural Systems, Ecosystems, and Human Systems

In AR5, a range of changes in ocean and cryosphere natural systems were linked with *medium* to *high confidence* to pressures associated with climate change (Cramer et al., 2014). Climate change impacts on natural ecosystems are variable in space and time. The multiplicity of pressures these natural systems experience impedes attribution of population or ecosystem responses to a specific ocean and/or cryosphere change. Moreover, the interconnectivity of populations within ecosystems means that a single 'adaptive response' of a population, or the aggregate response of an ecosystem (the adaptive responses of the interconnected populations), is influenced not just by direct pressures of climate change, but occurs in concert with the adaptive responses of other species in the ecosystem, further complicating efforts to disentangle specific patterns of adaptation.

Notwithstanding the network of pressures and adaptations, much effort has gone into resolving the mechanisms, interactions and feedbacks of natural systems associated with the ocean and cryosphere. Chapters 4, 5 and 6, as well as Cross-Chapter Box 9, assess new knowledge on the adaptive responses of wetlands, coral reefs, other coastal habitats, and the populations of marine organisms encountering ocean-based risks, including. Likewise, Chapters 2 and 3 describe emerging knowledge on how ecosystems in high-mountain and polar areas are adapting to cryosphere decline.

AR5 and SR15 have highlighted the importance of evolutionary adaptation as a component of how populations adapt to climate change pressures (e.g., Pörtner et al., 2014; Hoegh-Guldberg et al., 2018). Acclimatisation (variation in morphology, physiology or behaviour) can result from changes in gene expression but does not involve change in the underlying DNA sequence. Responses related to acclimatisation can occur both within single generations and over several generations. In contrast, evolution requires changes in the genetic composition of a population over multiple generations; for example, by differential survival or fecundity of different genotypes (Sunday et al., 2014). Adaptive evolution is the subset of evolution attributable to natural selection, and natural selection may lead to populations becoming more fit (Sunday et al., 2014) or extend the range of environments where populations persist (van Oppen et al., 2015). The efficacy of natural selection is affected by population size (Charlesworth, 2009), standing genetic variation, the ability of a population to generate novel genetic variation, migration rates and the frequency of genetic recombination (Rice, 2002). Many studies have shown evolution of traits within and across life stages of populations (Pespeni et al., 2013; Hinners et al., 2017), but there are fewer studies on how evolutionary change can impact ecosystem or community function, and whether trait evolution is stable (Schaum and Collins, 2014). Although acclimatisation and evolutionary adaptation are separate processes, they influence each other, and both adaptive and maladaptive variation of traits can facilitate evolution (Schaum and Collins, 2014; Ghalambor et al., 2015). Natural evolutionary adaptation may be challenged by the speed and magnitude of current ocean and cryosphere changes, but emerging studies investigate how human actions may assist evolutionary adaptation and thereby possibly enhance the resilience of natural systems to climate change pressures (e.g., Box 5.4 in Section 5.5.2). Through acclimatisation and evolutionary adaptation to the pressures from climate change (and all other persistent pressures), populations, species and ecosystems present a constantly changing context for the adaptation of human systems to climate change.

There are several human adaptation options for climate change impacts on the ocean and cryosphere. Adaptive responses include nature- and ecosystem-based approaches (Renaud et al., 2016; Serpetti et al., 2017). Additionally, more social-based approaches for human adaptation range from community-based and infrastructure-based approaches to managed retreat, along with other forms of internal migration (Black et al., 2011; Hino et al., 2017). Building on AR5 (Wong et al., 2014), Chapter 4 describes four main modes of adaptation to mean and extreme sea level rise: protect, advance, accommodate, and retreat. This report demonstrates that all modes of adaptation include mixes of institutional, individual, socio-cultural, engineering, behavioural and/or ecosystem-based measures (e.g., Section 4.4.2).

The effectiveness and performance of different adaptation options across spatial and social scales is influenced by their social acceptance, political feasibility, cost-efficiency, co-benefits and trade-offs (Jones et al., 2012; Adger et al., 2013; Eriksen et al., 2015). Scientific evaluation of past successes and future options, including understanding barriers, limits, risks and opportunities, are complex and inadequately researched (Magnan and Ribera, 2016). In the end, adaptation priorities will depend on multiple parameters including the extent and rate of climate change, the risk attitudes and social preferences of individuals and institutions (and the returns they may gain) (Adger et al., 2009; Brügger et al., 2015; Evans et al., 2016; Neef et al., 2018) and access to finances, technology, capacity and other resources (Berrang-Ford et al., 2014; Eisenack et al., 2014).

Since AR5, transformational adaptation (i.e., the need for fundamental changes in private and public institutions and flexible decision-making processes to face climate change consequences) has been increasingly studied (Cross-Chapter Box 2 in Chapter 1). The recent literature documents how societies, institutions, and/or individuals increasingly assume a readiness to engage in transformative change, via their acceptance and promotion of fundamental alterations in natural or human systems (Klinsky et al., 2016). People living in and near coastal, mountain and polar environments often pioneer these types of transformations, since they are at the forefront of ocean and cryosphere change (e.g., Solecki et al., 2017). Community led and indigenous led adaptation research continues to burgeon (Ayers and Forsyth, 2009; David-Chavez and Gavin, 2018), especially in many mountain (Section 2.3.2.3), Arctic (Section 3.5), and coastal (Section 4.4.4.4, 4.4.5.4, Cross-Chapter Box 9) areas, and demonstrate potential for enabling transformational adaptation (Dodman and Mitlin, 2013; Chung Tiam Fook, 2017). Similarly, the concepts of scenario planning and 'adaptation pathway' design have expanded since AR5, especially in the context of development planning for coastal and delta regions (Section 4.4, Cross-Chapter Box 9; Wise et al., 2014; Maier et al., 2016; Bloemen et al., 2018; Flynn et al., 2018; Frame et al., 2018; Lawrence et al., 2018).

1.7   Governance and Institutions

SROCC conceptualises governance as deciding, managing, implementing and monitoring policies in the context of ocean and cryosphere change. Institutions are defined as formal and informal social rules that shape human behaviour (Roggero et al., 2017). Governance guides how different actors negotiate, mediate their interests and share their rights and responsibilities (Forino et al., 2015; See SROCC Annex I: Glossary and Cross-Chapter Box 3 in Chapter 1 for definition). Governance and institutions interface with climate and social-ecological change process across local, regional and global scales (Fischer et al., 2015; Pahl-Wostl, 2019).

SROCC explores how the interlinked social-ecological systems affect challenge current governance systems in the context of ocean and cryosphere change. These challenges include three aspects. First, the scale of changes to ocean and cryosphere properties driven by global warming, and in the ecosystems they support and services they provide, are poorly matched to existing scales of governance (Sections 2.2.2.1; 2.3.1.3; 3.2.1; 3.5.3). Second, the nature of changes in ecosystem services resulting from changes in ocean and cryosphere properties, including services provided to humans living far from the mountains and coasts, are poorly matched to existing institutions and processes of governance (Section 4.4.4). Third, many possible governance responses to these challenges could be of limited or diminished effectiveness unless they are coordinated on scales beyond that of currently available governance options (Section 6.9.2; Box 5.5).

Hydrological processes in the high mountain cryosphere connect through upstream and downstream areas of river basins (Molden et al., 2016; Chen et al., 2018), including floodplains and deltaic regions (Kilroy, 2015; Cross-Chapter Box 3 in Chapter 1). These cross boundary linkages challenge local-scale governance and institutions that determine how the river-based ecosystem services that sustain food, water and energy are used and distributed (Rasul, 2014; Warner, 2016; Lele et al., 2018; Pahl-Wostl et al., 2018). Small Island States face rising seas that threaten habitability of their homeland and the possibility of losing their nation-state, cultural identity and voices in international governance (Gerrard and Wannier, 2013; Philip, 2018; Section 1.4, Cross-Chapter Box 9), highlighting the need for transboundary components to governance.

These governance challenges cannot be met without working across multiple organisations and institutions, bringing varying capacities, frameworks and spatial extents (Cross-Chapter Box 3 in Chapter 1). Progress in governance for ocean and cryosphere change will require filling gaps in legal frameworks (Amsler, 2016), aligning spatial mismatches (Eriksen et al., 2015; Young, 2016; Cosens et al., 2018), improving the ability for nations to cooperate effectively (Downie and Williams, 2018; Hall and Persson, 2018) and integrating across divided policy domains, most notably of climate change adaptation and disaster risk reduction (e.g., where slow sea level change also alters the implications for civil defense planning and the management of extreme events; Mysiak et al., 2018).

Harmonising local, regional and global governance structures would provide an overarching policy framework for action and allocation of necessary resources for adaptation. Coordinating the top-down and bottom-up governance processes (Bisaro and Hinkel, 2016; Sabel and Victor, 2017; Homsy et al., 2019) to increase effectiveness of responses, mobilise and equitably distribute adequate resources and access private and public sector capabilities requires a polycentric approach to governance (Ostrom, 2010; Jordan et al., 2015). Polycentric governance connotes a complex form of governance with multiple centres of decision making working with some degree of autonomy (Carlisle and Gruby, 2017; Baldwin et al., 2018; Mewhirter et al., 2018; Hamilton and Lubell, 2019).

Cross-Chapter Box 3 | Governance of the Ocean, Coasts and the Cryosphere under Climate Change

Authors: Anjal Prakash (Nepal/India), Sandra Cassotta (Denmark), Bruce Glavovic (New Zealand/South Africa), Jochen Hinkel (Germany), Elisabeth Holland (Fiji/USA), Md Saiful Karim (Australia/Bangladesh), Ben Orlove (USA), Beate Ratter (Germany), Jake Rice (Canada), Evelia Rivera-Arriaga (Mexico), Catherine Sutherland (South Africa)

This Cross-Chapter Box outlines governance and associated institutional challenges and emerging solutions relevant to the ocean, coasts and cryosphere in a changing climate. It illustrates these through three cases: (Case 1) multi-level interactions in Ocean and Arctic governance; (Case 2) mountain governance; and (Case 3) coastal risk governance. Governance refers to how political, social, economic and environmental systems and their interactions are governed or 'steered' by establishing and modifying institutional and organisational arrangements which regulate social processes, mitigate conflicts and realise mutual gains (North, 1990; Pierre and Peters, 2000; Paavola, 2007). Institutions are formal and informal rules and norms, constructed and held in common by social actors that guide, constrain and shape human interactions (North, 1990; Ostrom, 2005). Formal institutions include constitutions, laws, policies and contracts, while informal institutions include customs, social norms and taboos. Both administrative or state government structures and indigenous or traditional governance structures govern the ocean, coasts and cryosphere.

Understanding governance in a changing climate
The Special Report on the Ocean and Cryosphere in a Changing Climate (SROCC), together with the Special Report on Global Warming of 1.5°C (SR15) (IPCC, 2018), highlights the critical role of governance in implementing effective climate adaptation. Chapter 2 explores local community institutions offering autonomous adaptation in the Alps, Andes, Himalayas and other mountain

Cross-Chapter Box 3 (continued)

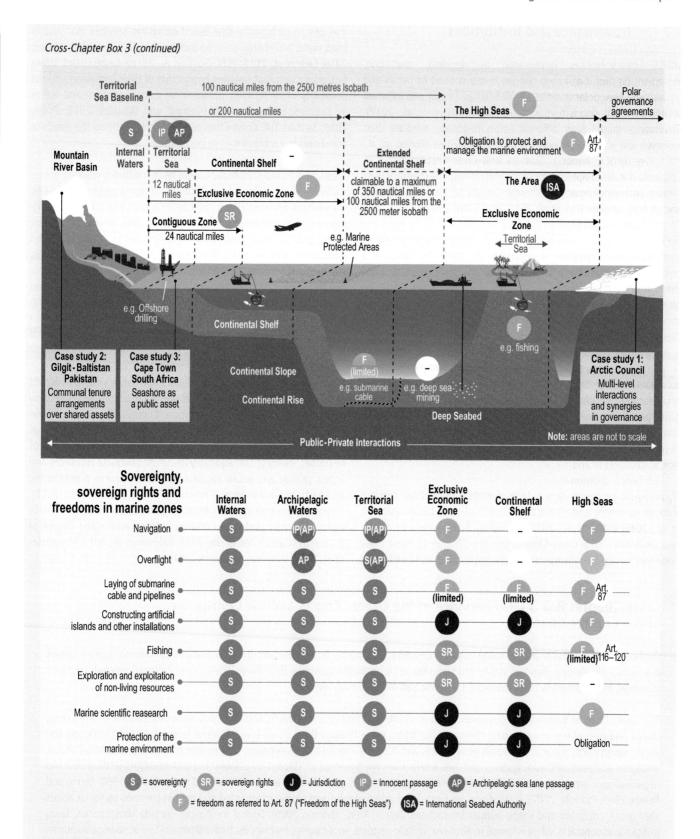

Figure CB3.1 | Spatial distribution of multi-faceted governance arrangements for the ocean, coasts and cryosphere (Panel A) sovereignty, sovereign rights, jurisdictions and freedoms defined for different ocean zones and sea by the United Nations Convention on the Law of the Sea (UNCLOS) (Panel B). *Figure CB3.1 is designed to be illustrative and is not comprehensive of all governance arrangements for the ocean, coasts and cryosphere.*

Cross-Chapter Box 3 (continued)

regions (Section 2.4), focusing on the need for transboundary cooperation to support water governance and mitigate conflict. Chapter 3 explores how polar governance system facilitate building resilient pathways, knowledge co-production, social learning, adaptation and power-sharing with Indigenous Peoples at the regional level. This would help in increasing international cooperation in multi-level governance arenas to strengthen responses supporting adaptation in socio-ecological systems (Section 3.5.4). Chapter 4 illustrates how sea level rise governance attempts to address conflicting interests in coastal development, risk management and adaptation with a diversity of governance contexts and degrees of community participation, with a focus on equity concerns and inevitable trade-offs (Section 4.4). Chapter 5 includes a review of existing international legal regimes for addressing ocean warming, acidification and deoxygenation impacts on social-ecological systems and considers ways to facilitate appropriate responses to ocean change (Sections 5.4, 5.5). Chapter 6 explores the issues of credibility, trust, and reliability in government that arise from promoting 'paying the costs of preparedness and prevention' as an alternative to 'bearing the costs of loss and damage' (Section 6.9).

Climate change challenges existing governance arrangements in a variety of ways. First, there are complex interconnections between climate change and other processes that influence the ocean, coasts and cryosphere, making it difficult to untangle climate governance from other governance efforts. Second, the time frames for societal decision-making and government terms are mismatched with the long-term commitment of climate change. Third, governance choices have to be made in the face of uncertainty about the rate and scale of change that will occur in the medium to long-term (Cross-Chapter Box 5 in Chapter 1). Lastly, climate change progressively alters the environment and hence requires continual innovation and adjustment of governance arrangements (Bisaro and Hinkel, 2016; Roggero et al., 2018). Novel transboundary interactions and conflicts are emerging as well as new multi-level governance structures for international and regional cooperation, strengthening shared decision-making among States and other actors (Case 1). The prospects of 'disappearing states', glacier retreat and increasing water scarcity are resulting in States redefining complex water-sharing agreements (Case 2). Coastal risk is escalating, which may require participatory governance responses and the co-production of knowledge at the local scale (Case 3; see also Cross-Chapter Box 9).

Governance, exercised through legal, administrative and other social processes, is essential to prevent, mitigate and adapt to the challenges and risks posed by a changing climate. These governance processes determine roles in the exercising of power and hence decision-making (Graham et al., 2003). Governance may be an act of governments (e.g., passing laws or providing

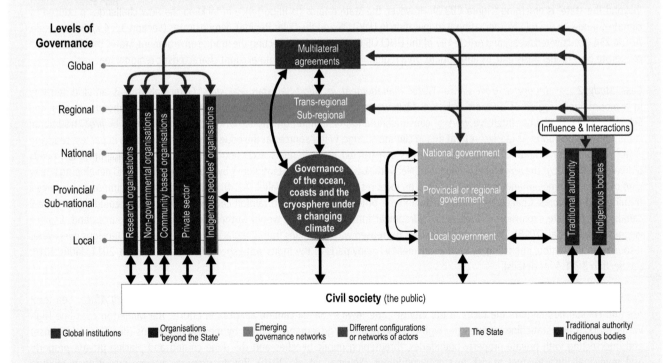

Figure CB3.2 | Interactions and emergence of network governance arrangements for the ocean, coasts and cryosphere across different scales. Adapted from Sommerkorn and Nilsson (2015).

Cross-Chapter Box 3 (continued)

incentives or information such that citizens can respond more effectively to climate change); private sector actions (e.g., insurance); a co-operative effort among local actors governing themselves through customary law (e.g., by establishing entitlements or norms regulating the common use of scarce resources); a collaborative multi-level effort involving multiple actors (state, private and civil society; e.g., UNFCCC); or a multi-national effort (e.g., Antarctic Treaty; see Figure CB3.2). The complexities of governance arrangements in the ocean, coasts and cryosphere (Figure CB3.1), and the interactions and emergence of relationships between different governance actors in multiple configurations across various spatial scales (Figure CB3.2) are illustrated below.

Case Study 1 – Multi-level Interactions and Synergies in Governance. The UN Convention on the Law of the Sea (UNCLOS) and the changing Arctic: Climate-change induced sea level rise (Section 4.2), could shift the boundaries and territory of some coastal states, changing the areas where their coastal rights are applied under UNCLOS. In extreme cases, inundation from sea level rise might lead to loss of territory and sovereignty, the disappearance of islands and the loss of international maritime jurisdiction subject to maritime claim. These challenges have limited opportunities for recourse in international law and it remains unclear what adequate responses from an international law perspective would be (Vidas et al., 2015; Andreone, 2017; Mayer and Crépeau, 2017; Chircop et al., 2018). While specific legal arrangements and instruments of environmental protection are in place at a regional, sub-regional and national level, they are insufficient to address the new challenges sea level rise brings. Institutional responses to the geopolitical transformation caused by climate change, such as through the Arctic Council (AC) and the 'Law of the Sea' are still evolving. Similar to many international agreements, UNCLOS 'Law of the Sea' provisions for enforcement, compliance, monitoring and dispute settlement mechanisms are not comprehensive, and commonly depend on further, detailed law-making by state parties, acting through competent international organizations (Vidas, 2000; Karim, 2015; De Lucia, 2017; Grip, 2017). Shifts from traditional state-based practices of international law to multi-level and informal governance structures that involve state and non-state actors (including Indigenous Peoples) may address these challenges (*medium confidence*; Cassotta, 2012; Shadian, 2014; Young, 2016; Andreone, 2017). The AC is a regionally focused governance structure blending new forms of formal and informal multi-level regional cooperation (Young, 2016). The soft law mechanisms employed draw upon best available practice and standards from multiple knowledge systems (Cassotta and Mazza, 2015; Pincus and Ali, 2015) in an attempt to respond to the ocean's global, trans-regional and national climate challenges (Section 3.5.4.2). Reconfiguration and restructuring of the AC has been proposed in order to address emerging trans-regional and global problems (*high confidence*; Baker and Yeager, 2015; Pincus and Ali, 2015; Young, 2016). Within the existing scope, the AC has amplified the voice of Arctic people affected by the impacts of climate change and mobilized action (Koivurova, 2016). The influence of actors 'beyond the state' is emerging (Figure CB3.2). However, the state retains its importance in tackling the new challenges produced by climate change, as the role of international cooperation in UNCLOS and the Polar Regions demonstrates (Section 3.5.4.2). For example, Article 234 (Ice-covered areas) and Article 197 of the UNCLOS Convention in protecting the marine environment, states that 'States shall cooperate on a global basis and, as appropriate, on a regional basis [...] taking into account characteristic regional features'.

Case Study 2 – Mountain Governance: Water management in Gilgit-Baltistan, Pakistan. Gilgit-Baltistan is an arid territory in a mountainous region of northern Pakistan. Melt water fed streams supply irrigation water for rural livelihoods (Nüsser and Schmidt, 2017). The labour intensive work of constructing and maintaining gravity-fed irrigation canals is done by *jirga*, traditional community associations. As glaciers retreat due to climate change, water sources at the edge of glaciers have been impacted, reducing water available for irrigation. In response, villagers constructed new channels accessing more distant water for irrigation needs (Parveen et al., 2015). The Aga Khan Development Network supported this substantial task by providing funding and developing a new kind of cross-scale governance network, drawing on local residents for staff (Walter, 2014), and strengthening community resources, training and networks. Challenges remain, including the potential for increased rainfall causing landslides that could damage new canals, and possible expansion of Pakistan's hydropower infrastructure that would further diminish water resources and displace villages (Shaikh et al., 2015). On a geopolitical scale, decreased water supplies from the glaciers could exacerbate tensions over water resources in the region, impacting water management in many parts of the Indus watershed (Uprety and Salman, 2011; Jamir, 2016; see Section 2.3.1.4 for details).

Case Study 3 – Coastal Governance: Risk management for sea level changes in the City of Cape Town, South Africa. Sea level rise and coastal flooding are the focus of the City of Cape Town's coastal climate adaptation efforts. The Milnerton coastline High Water Mark, a non-static line marking the high tide, is creating a governance conflict by moving landwards (due to sea level rise) and intersecting with private property boundaries, threatening public beaches and the dune cordon and placing private property and municipal infrastructure at risk in storm conditions (Sowman et al., 2016). Private property owners are using a mixture of formal, ad hoc, and in some cases illegal, coastal barrier measures to protect their assets from sea level and storm risks, but these are creating additional erosion impacts on the coastline. Legally, the City of Cape Town is not responsible for remediating private land impacted by coastal erosion (Smith et al., 2016). However, city officials feel compelled to take action for the common good using a progressive, multi-stakeholder participatory approach. This involves opening up opportunities for dialogue and co-producing knowledge, instead of a purely legalistic and state-centric compliance approach (Colenbrander et al., 2015). The city's actions are both mindful of international frameworks on climate change and responsive to national and provincial legislation and policy. A major challenge that remains is how to navigate the power struggles that will be triggered by this consultative process, as different actors define and negotiate their interests, roles and responsibilities (see Section 4.4.3; Table 4.9).

Conclusions

These cases illustrate four important points. First, new governance challenges are emerging due to climate change, including: disruptions to long-established cultures, livelihoods and even territorial sovereignty (Case 1); changes in the accessibility and availability of vital resources (Case 2); and the blurring of public and private boundaries of risk and responsibility through accelerated coastal erosion (Case 3; Figure CB3.1). Second, new governance arrangements are emerging to address these challenges, including participatory and networked structures linking formal and informal networks, and involving state, private sector, indigenous and civil society actors in different configurations (Figure CB3.2). Third, climate governance is a complex, contested and unfolding process, with governance actors and networks having to learn from experience, to innovate and develop context-relevant arrangements that can be adjusted in the face of ongoing change. Lastly, there is no single climate governance panacea for the ocean, coasts and cryosphere. Empirical evidence on which governance arrangements work well in which context is still limited, but 'good governance' norms indicate the importance of inclusivity, fairness, deliberation, reflexivity, responsiveness, social learning, the co-production of knowledge and respect for ethnic and cultural diversity.

1.8 Knowledge Systems for Understanding and Responding to Change

Assessments of how climate change interacts with the planet and people are largely based on scientific knowledge from observations, theories, modelling and synthesis to understand physical and ecological systems (Section 1.8.1), societies (e.g., Cross-Chapter Box 2 in Chapter 1, Section 1.5) and institutions (e.g., Cross-Chapter Box 3 in Chapter 1). However, humans integrate information from multiple sources to observe and interact with their environment, respond to changes, and solve problems. Accordingly, SROCC also recognises the importance of Indigenous knowledge and local knowledge in understanding and responding to changes in the ocean and cryosphere (Sections 1.8.2, 1.8.3; Cross-Chapter Box 4 in Chapter 1).

1.8.1 Scientific Knowledge

1.8.1.1 Ocean and Cryosphere Observations

Long-term sustained observations are critical for detecting and understanding the processes of ocean and cryosphere change (Rhein et al., 2013; Vaughan et al., 2013). Scientific knowledge of the ocean and cryosphere has increased through time and geographical space (Figure 1.3). *In situ* ocean subsurface temperature and salinity observations have increased in spatial and temporal coverage since the middle of the 19th century (Abraham et al., 2013), and near global coverage (60°S–60°N) of the upper 2,000 m has been achieved since 2007 due to the international Argo network (Riser et al., 2016; Figure 1.3). Improved data quality and data analysis techniques have reduced uncertainties in global ocean heat uptake estimates (Sections 1.4.1, 5.2.2). In addition to providing deep ocean

measurements, repeated hydrographic physical and biogeochemical observations since AR5 have led to improved estimates of ocean carbon uptake and ocean deoxygenation (Sections 1.4.1, 5.2.2.3, 5.2.2.4). Targeted observational programmes have improved scientific knowledge for specific regions and physical processes of particular concern in a warming climate, including the Greenland and West Antarctic ice sheets (Section 3.3), and the AMOC (Section 6.7). Ocean and cryosphere mass changes and sea level studies have benefited from sustained or newly implemented satellite-based remote sensing technologies, complemented by *in situ* data such as tide gauges measurements (Sections 3.3, 4.2; Dowell et al., 2013; Raup et al., 2015; PSMSL, 2016). Glacier length measurements in some locations go back many centuries (Figure 1.3), but it is the systematic high resolution satellite monitoring of a large number of the world's glaciers since the late 1970s that has improved global assessments of glacier mass loss (Sections 2.2.3, 3.3.2).

Limitations in knowledge of ocean and cryosphere change remain, creating knowledge gaps for the SROCC assessment. Ocean and cryosphere datasets are frequently short, and do not always span the key IPCC assessment time intervals (Cross-Chapter Box 1 in Chapter 1), so for many parameters the full magnitude of changes since the pre-industrial period is not observed (Figure 1.3). The brevity of ocean and cryosphere measurements also means that some expected changes cannot yet be detected with confidence in direct observations (e.g., Antarctic sea ice loss in Section 3.2.1, AMOC weakening in Section 6.7.1), or other observed changes cannot yet be robustly attributed to anthropogenic factors (e.g., ice sheet mass loss in Section 3.3.1). Observations for many key ocean variables (Bojinski et al., 2014), such as ocean currents, surface heat fluxes, oxygen, inorganic carbon, subsurface salinity, phytoplankton biomass and diversity, etc., do not yet have global coverage or have not reached the required density or accuracy for detection of change. Some ocean and cryosphere areas remain difficult to observe systematically, for example, the ocean under sea ice, subsurface permafrost, high mountain areas, marginal seas, coastal areas (Section 4.2.2.3) and ocean boundary currents (Hu and Sprintall, 2016), basin interconnections (Section 6.6) and the Southern Ocean (Sections 3.2, 5.2.2). Measurements that reflect ecosystem change are often location or species specific, and assessments of long-term ocean ecosystem changes are currently only feasible for a limited subset of variables, for example coral reef health (e.g., coral reef health) (Section 5.3; Miloslavich et al., 2018). The deep ocean below 2,000 m is still rarely observed (Talley et al., 2016), limiting (for example) the accurate estimate of deep ocean heat uptake and, consequently the full magnitude of Earth's energy imbalance (e.g., von Schuckmann et al., 2016; Johnson et al., 2018; Sections 1.2, 1.4, 5.2.2).

1.8.1.2   Reanalysis Products

Advances have been made over the past decade in developing more reliable and more highly resolved ocean and atmosphere reanalysis products. Reanalysis products combine observational data with numerical models through data assimilation to produce physically consistent, and spatially complete ocean and climate products (Balmaseda et al., 2015; Lellouche et al., 2018; Storto et al., 2018; Zuo et al., 2018). Ocean reanalyses are widely used to understand changes in physical properties (Section 3.2.1, 5.2), extremes (Sections 6.3 to 6.6), circulation (Section 6.6, 6.7) and to provide climate diagnostics (Wunsch et al., 2009; Balmaseda et al., 2013; Hu and Sprintall, 2016; Carton et al., 2018). Reanalysis products are used in SROCC for assessing climate change process that cause changes in the ocean and cryosphere (e.g., Sections 2.2.1, 3.2.1, 3.3.1, 3.4.1, 5.2.2, 6.3.1, 6.6.1, 6.7.1). Improvements in reanalysis products provide more realistic forcing for regional models, which are used for assessing regional ocean and cryosphere changes that cannot be resolved in global-scale models (e.g., Section 2.2.1; Mazloff et al., 2010; Fenty et al., 2017). The weather forecasts, and seasonal to decadal predictions building on reanalysis products have important applications in the early warning systems that reduce risk and aid human adaptation to extreme events (Sections 6.3.4, 6.4.3, 6.5.3, 6.7.3, 6.8.5).

1.8.1.3   Model Simulation Data

Models are numerical approximations of the Earth system that allow hypotheses about the mechanisms of ocean and cryosphere change to be tested, support attribution of observed changes to specific forcings (Section 1.3), and are the best available information for assessing future change (Figure 1.3). General Circulation Models (GCMs) typically simulate the atmosphere, ocean, sea ice, and land surface, and sometimes also incorporate terrestrial and marine ecosystems. Earth System Models (ESM) are climate models that explicitly include the carbon cycle and may include additional components (e.g., atmospheric chemistry, ice sheets, dynamic vegetation, nitrogen cycle, but also urban or crop models). The systematic set of global-scale model experiments (Taylor et al., 2012) used in SROCC were produced by CMIP5 (Cross-Chapter Box 1 in Chapter 1), including both GCMs and ESMs.

Models may differ in their spatial resolution, and in the extent to which processes are explicitly represented or approximated (parameterised). Model output can be biased due to uncertainties in their physical equations or parameterisations, specification of initial conditions, knowledge of external forcing factors, and unaccounted processes and feedbacks (Hawkins and Sutton, 2009; Deser et al., 2012; Gupta et al., 2013; Lin et al., 2016). Since AR5 there have been advances in modelling the dynamical processes of the Greenland and Antarctica ice sheets, leading to better representation of the range of potential future sea level rise scenarios (Sections 4.2.3). Downscaling, including the use of regional models, makes it possible to improve the spatial resolution of model output in order to better resolve past and future climate change in specific areas, such as high mountains and coastal seas (e.g., Sections 2.2.2, 3.2.3, 3.5.4, 4.2.2, 6.3.1). For biological processes, such as nutrient levels and organic matter production, model uncertainty at regional scales is the main issue limiting confidence in future projections (Sections 5.3, 5.7). While model projections of range shifts for fishes agree with theory and observations, at a regional scale there are known deficiencies in the ways models represent the impacts of ocean variables such as temperature and productivity (Sections 5.2.3, 5.7).

1.8.1.4   Palaeoclimate Data

Palaeoclimate data provide a way to establish the nature of ocean and cryosphere changes prior to direct measurements (Figure 1.3), including natural variability and early anthropogenic climate change (Masson-Delmotte et al., 2013; Abram et al., 2016). Palaeoclimate records utilise the accumulation of physical, chemical or biological properties within natural archives that are related to climate at the time the archive formed. Commonly used palaeoclimate evidence for ocean and cryosphere change comes from marine and lake sediments, ice layers and bubbles, tree growth rings, past shorelines and shallow reef deposits. In many mountain areas, centuries to millennia of palaeoclimate information is now being lost through widespread melting of glacier ice (Cross-Chapter Box 6 in Chapter 2). Palaeoclimate data are spatially limited (Figure 1.3), but often represent regional to global-scale climate patterns, either individually or as syntheses of networks of data (PAGES2K Consortium, 2017).

Palaeoclimate data provide evidence for multi-metre global sea level rises and shifts in climate zones and ocean ecosystems during past warm climate states where temperatures were similar to those expected later this century (Hansen et al., 2016; Fischer et al., 2018; Section 4.2.2). Palaeoclimate reconstructions give context to recent ocean and cryosphere changes that are unusual in the context of variability over past centuries to millennia, including acceleration in Greenland and Antarctic Peninsula ice-melt (Section 3.3.1), declining Arctic sea ice (Section 3.2.1), and emerging evidence for a slowdown of AMOC (Section 6.7.1). Assessments of climate model performance across a wider-range of climate states than is possible using direct observations alone also draws on palaeoclimate data (Flato et al., 2013), and since AR5 important progress has been made to calibrate modelled ice sheet processes and future sea level rise based on palaeoclimate evidence (Cross-Chapter Box 8 in Chapter 3).

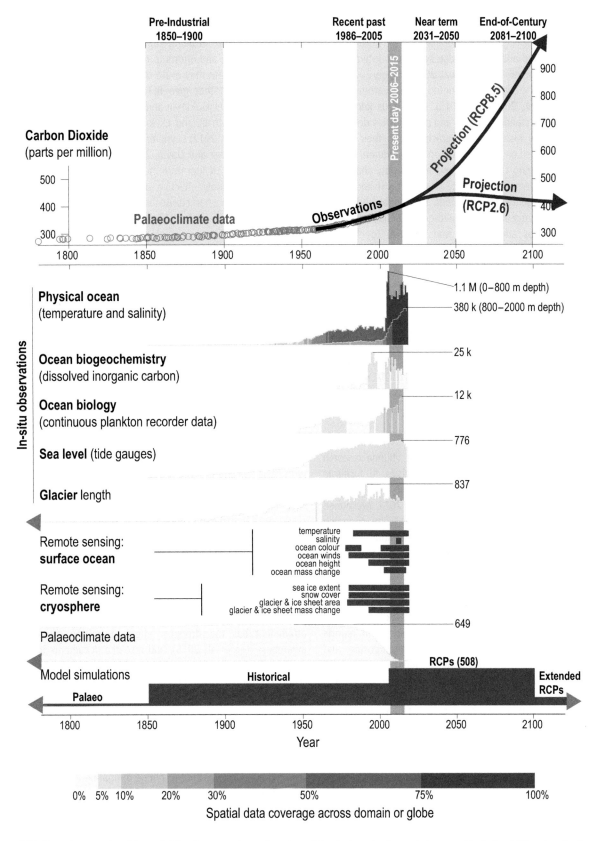

Figure 1.3 | Illustrative examples of the availability of ocean and cryosphere data relative to the major time periods assessed in the Special Report on the Oceans and Cryosphere in a Changing Climate (SROCC). Upper panel; observed (Keeling et al., 1976) and reconstructed (Bereiter et al., 2015) atmospheric CO₂ concentrations, as well as the Representative Concentration Pathways (RCPs) of CO₂ for low (RCP2.6) and high (RCP8.5) future emission scenarios (van Vuuren et al., 2011a; Cross-Chapter Box 1 in Chapter 1). Lower panel; illustrative examples of data availability for the ocean and cryosphere (Section 1.8.1; Taylor et al., 2012; Boyer et al., 2013; Dowell et al., 2013; McQuatters-Gollop et al., 2015; Raup et al., 2015; Olsen et al., 2016; PSMSL, 2016; PAGES2K Consortium, 2017; WGMS, 2017). The amount of data available through time is shown by the heights of the time series for observational data, palaeoclimate data and model simulations, expressed relative to the maximum annual data availability (maximum values given on plot; M = million, k = thousand). Spatial coverage of data across the globe or the relevant domain is shown by colour scale. See SM1.4 for further details.

1.8.2   Indigenous Knowledge and Local Knowledge

Humans create, use, and adapt knowledge systems to interact with their environment (Agrawal, 1995; Escobar, 2001; Sillitoe, 2007), and to observe and respond to change (Huntington, 2000; Gearheard et al., 2013; Maldonado et al., 2016; Yeh, 2016). Indigenous knowledge (IK) refers to the understandings, skills, and philosophies developed by societies with long histories of interaction with their natural surroundings. It is passed on from generation to generation, flexible, and adaptive in changing conditions, and increasingly challenged in the context of contemporary climate change. Local knowledge (LK) is what non-Indigenous communities, both rural and urban, use on a daily and lifelong basis. It is multi-generational, embedded in community practices and cultures and adaptive to changing conditions (FAO, 2018). Each chapter of SROCC cites examples of IK and LK related to ocean and cryosphere change.

IK and LK stand on their own, and also enrich and complement each other and scientific knowledge. For example, Australian Aboriginal groups' Indigenous oral history provides empirical corroboration of the sea level rise 7,000 years ago (Nunn and Reid, 2016), and their seasonal calendars direct hunting, fishing, planting, conservation and detection of unusual changes today (Green et al., 2010). LK works in tandem with scientific knowledge, for example, as coastal Australian communities consider the impacts and trade-offs of sea level rise (O'Neill and Graham, 2016).

Both IK and LK are increasingly used in climate change research and policy efforts to engage affected communities to facilitate site-specific understandings of, and responses to, the local effects of climate change (Hiwasaki et al., 2014; Hou et al., 2017; Mekonnen et al., 2017). IK and LK enrich CRDPs particularly by engaging multiple stakeholders and the diversity of socioeconomic, cultural and linguistic contexts of populations affected by changes in the ocean and cryosphere (Cross-Chapter Box 4 in Chapter 1).

Global environmental assessments increasingly recognise the importance of IK and LK (Thaman et al., 2013; Beck et al., 2014; Díaz et al., 2015). References to IK in IPCC assessment reports increased 60% from AR4 to AR5, and highlighted the exposures and vulnerabilities of Indigenous populations to climate change risks related to socioeconomic status, resource-based dependence and geographic location (Ford et al., 2016a). All four IPBES assessments in 2018 (IPBES, 2018a; IPBES, 2018b; IPBES, 2018c; IPBES, 2018d) engaged IK and LK (Díaz et al., 2015; Roué and Molnar, 2017; Díaz et al., 2018). Peer-reviewed research on IK and LK is burgeoning (Savo et al., 2016), providing information that can guide responses and inform policy (Huntington, 2011; Nakashima et al., 2012; Lavrillier and Gabyshev, 2018). However, most global assessments still fail to incorporate 'the plurality and heterogeneity of worldviews' (Obermeister, 2017), resulting 'in a partial understanding of core issues that limits the potential for locally and culturally appropriate adaptation responses' (Ford et al., 2016b).

IK and LK provide case specific information that may not be easily extrapolated to the scales of disturbance that humans exert on natural systems (Wohling, 2009). Some forms of IK and LK are also not amenable to being captured in peer-reviewed articles or published reports, and efforts to translate IK and LK into qualitative or quantitative data may mute the multidimensional, dynamic and nuanced features that give IK and LK meaning (DeWalt, 1994; Roncoli et al., 2009; Goldman and Lovell, 2017). Nonetheless, efforts to collaborate with IK and LK knowledge holders (Baptiste et al., 2017; Karki et al., 2017; Lavrillier and Gabyshev, 2017; Roué et al., 2017; David-Chavez and Gavin, 2018) and to systematically assess published IK and LK literature in parallel with scientific knowledge result in increasingly effective usage of the multiple knowledge systems to better characterise and address ocean and cryosphere change (Huntington et al., 2017; Nalau et al., 2018; Ford et al., 2019).

1.8.3   The Role of Knowledge in People's Responses to Climate, Ocean and Cryosphere Change

To hold global average temperature to well below 2°C above pre-industrial levels, substantial changes in the day-to-day activities of individuals, families, communities, the private sector, and governance bodies will be required (Ostrom, 2010; Creutzig et al., 2018). Enabling these changes at a meaningful societal scale requires sensitivity to communities and their use of multiple knowledge systems to best motivate effective responses to the risks and opportunities posed by climate change (*medium confidence*) (1.8.2, Cross-Chapter Box 4 in Chapter 1). Meaningful engagement of people and communities with climate change information depends on that information cohering with their perception of how the world works (Crate and Fedorov, 2013). The values and identities people hold affect how acceptable they find the behavioural changes, technological solutions and governance that climate change action requires (Moser, 2016).

Education and climate literacy contribute to climate change action and adaptation (*high confidence*). Although public understanding of humanity's role in both causing and abating climate change has increased in the last decade (Milfont et al., 2017), levels of climate concern vary greatly globally (Lee et al., 2015). Educational attainment has the strongest effect on raising climate change awareness (Lee et al., 2015), and research documents the value of evidence-based climate change education, particularly during formal schooling (Motta, 2018). People further understand climate change as a serious threat when they experience it in their lives and have knowledge of its human causes (Lee et al., 2015; Shi et al., 2016). Education and tailored climate communication strategies that are respectful of people's values and identity can aid acceptance and implementation of the local to global-scale approaches and policies required for effective climate change mitigation and adaptation (Shi et al., 2016; Anisimov and Orttung, 2018; Sections 3.5.4, 4.4), while also supporting CRDPs (see also Cross-Chapter Box 2 in Chapter 1, and FAQ1.2).

Human psychology complicates engagement with climate change, due to complex social factors, including values (Corner et al., 2014), identity (Unsworth and Fielding, 2014), ideology (Smith and Mayer, 2019) and the framing of climate messaging. Additionally, psychology effects adaptation actions, motivated by perceptions that others are already adapting, avoidance of an unpleasant state of mind, feelings of self-efficacy and belief in the efficacy of the adaptation action (van Valkengoed and Steg, 2019). Better understandings of the psychological implications across diverse communities and social and political contexts will facilitate a just transition of both emissions reduction and adaptation (Schlosberg et al., 2017). Impacts of climate change on natural and human environments (e.g., extreme weather) or human-caused modifications to the environment (e.g., adaptation) will raise further psychological challenges. This includes psychological impacts to the emotional wellbeing of people adversely affected by climate change (Ogunbode et al., 2018), resulting in solastalgia (Albrecht et al., 2007), a distress akin to homesickness while in their home environment (McNamara and Westoby, 2011).

Cross-Chapter Box 4 | Indigenous Knowledge and Local Knowledge in Ocean and Cryosphere Change

Authors: Susan Crate (USA), William Cheung (Canada), Bruce Glavovic (New Zealand), Sherilee Harper (Canada), Hélène Jacot Des Combes (Fiji/France), Monica Ell Kanayuk (Canada), Ben Orlove (USA), Joanna Petrasek MacDonald (Canada), Anjal Prakash (Nepal/India), Jake Rice (Canada), Pasang Yangjee Sherpa (Nepal), Martin Sommerkorn (Norway/Germany)

Introduction

This Cross-Chapter Box describes how Indigenous knowledge (IK) and local knowledge (LK) are different and unique sources of knowledge, which are critical to observing, responding to, and governing the ocean and cryosphere in a changing climate (See SROCC Annex I: Glossary for definitions). International organisations recognise the importance of IK and LK in global assessments, including UN Environment, UNDP, UNESCO, IPBES, and the World Bank. IK and LK are referenced throughout SROCC, understanding that many climate change impacts affect, and will require responses from, local communities (both Indigenous and non-Indigenous) who maintain a close connection with the ocean and/or cryosphere.

Attention to IK and LK in understanding global change is relatively recent, but important (*high confidence*). For instance, in 1980, Alaskan Inuit formed the Alaska Eskimo Whaling Commission in response to the International Whaling Commission's science that underestimated the Bowhead whale population and, in 1977, banned whaling as a result (Huntington, 1992). The Commission facilitated an improved population count using a study design based on IK, which indicated a harvestable population (Huntington, 2000). There are various approaches for utilising multiple knowledge systems. For example, the Mi'kmaw Elders' concept of Two Eyed Seeing: which is 'learning to see from one eye with the strengths of Indigenous knowledges, and from the other eye with the strengths of Western [scientific] knowledges, and to use both together, for the benefit of all' (Bartlett et al., 2012), to preserve the distinctiveness of each, while allowing for fuller understandings and actions (Bartlett et al., 2012: 334).

Knowledge Co-production

Scientific knowledge, IK and LK can complement one another by engaging both quantitative data and qualitative information, including people's observations, responses and values (Huntington, 2000; Crate and Fedorov, 2013; Burnham et al., 2016; Figure CB4.1). However, this process of knowledge co-production is complex (Jasanoff, 2004) and IK and LK possess uncertainties of a different nature from those of scientific knowledge (Kahneman and Egan, 2011), often resulting in the dominance of scientific knowledge over IK and LK in policy, governance and management (Mistry and Berardi, 2016). Working across disciplines (interdisciplinarity; Strang, 2009), and/or engaging multiple stakeholders (transdisciplinarity; Klenk and Meehan, 2015; Crate et al., 2017), are approaches used to bridge knowledge systems. The use of all knowledge relevant to a specific challenge can involve approaches such as: scenario building across stakeholder groups to capture the multiple ways people perceive their environment and act within it (Klenk and Meehan, 2015); knowledge co-production to achieve collaborative management efforts (Armitage et al., 2011); and working with communities to identify shared values and perceptions that enable context-specific adaptation strategies (Grunblatt and Alessa, 2017). Broad stakeholder engagement, including affected communities, Indigenous Peoples, local and regional representatives, policy makers, managers, interest groups and organisations, has the potential to effectively use all relevant knowledge (Obermeister, 2017) and produce results that reduce the disproportionate influence that formally educated and economically advantaged groups often exert in scientific assessments (Castree et al., 2014).

Figure CB4.1 | Knowledge co-production using scientific knowledge, Indigenous knowledge (IK) and/or local knowledge (LK) to create new understandings for decision making. Panels A, B, and C represent the use of one, two, and three knowledge systems, respectively, illustrating co-production moments in time (collars). Panel A represents a context which uses one knowledge system, for example, of IK used by Indigenous peoples; or of LK used by farmers, fishers and rural or urban inhabitants; or of scientific knowledge used in contexts where substantial human presence is lacking. Panel B depicts the use of two knowledge systems, as described in this Cross-Chapter Box in the case of Bowhead whale population counts and in Himalayan flood management. Panel C illustrates the use of all three knowledge systems, as in the Pacific case in this Cross-Chapter Box. Each collar represents how making use of knowledge from different systems is a matter of both identifying available knowledge across systems and of knowledge holder deliberations. In these processes, learning takes place on how to relate knowledge from different systems for the purpose of improved decisions and solutions. Knowledge from different systems can enrich the body of relevant knowledge while continuing independently or can be combined to co-produce new knowledge.

Contributions to SROCC

Observations, responses, and governance are three important contributions that IK and LK make in ocean and cryosphere change:

Observations: IK and LK observations document glacier and sea ice dynamics, permafrost dynamics, coastal processes, etc. (Sections 2.3.2.2.2, 2.5, 3.2.2, 3.4.1.1, 3.4.1.1, 3.4.1.2, 4.3.2.4.2, 5.2.3 and Box 2.4), and how they interact with social-cultural factors (West and Hovelsrud, 2010). Researchers have begun documenting IK and LK observations only recently (Sections 2.3.1.1, 3.2, 3.4, 3.5, Box 4.4, 5.4.2.2.1).

Responses: Either IK or LK alone (Yager, 2015), or used with scientific knowledge (Nüsser and Schmidt, 2017) inform responses (Sections 2.3.1.3.2, 2.3.2.2.2, 3.5.2, 3.5.4, 4.4.2, Box 4.4, 5.5.2, 6.8.4, 6.9.2). Utilising multiple knowledge systems requires continued development, accumulation, and transmission of IK, LK and scientific knowledge towards understanding the ecological and cultural context of diverse peoples (Crate and Fedorov, 2013; Jones et al., 2016), resulting in the incorporation of relevant priorities and contexts into adaptation responses (Sections 3.5.2, 3.5.4, 4.4.4, 5.5.2, 6.8.4, 6.9.2, Box 2.3).

Governance: Using IK and LK in climate decision and policy making includes customary Indigenous and local institutions (Karlsson and Hovelsrud, 2015), as in the case when Indigenous communities are engaged in an integrated approach for disaster risk reduction in response to cryosphere hazards (Carey et al., 2015). The effective engagement of communities and stakeholders in decisions requires using the multiple knowledge systems available (Chilisa, 2011; Sections 2.3.1.3.2, 2.3.2.3, 3.5.4, 4.4.4, Table 4.4, 5.5.2, 6.8.4, 6.9.2; Sections 2.3.1.3.2, 2.3.2.3, 3.5.4, 4.4.4, Table 4.9, 5.5.2, 6.8.4, 6.9.2).

Cross-Chapter Box 4 (continued)

Examples from regions covered in this report

IK and LK in the Pacific: Historically, Pacific communities, who depend on marine resources for essential protein (Pratchett et al., 2011), use LK for management systems to determine access to, and closure of, fishing grounds, the latter to respect community deaths, sacred sites, and customary feasts. Today a hybrid system, Locally Managed Marine Protected Areas (LMMAs), is common and integrates local governance with NGO or government agency interventions (Jupiter et al., 2014). The expected benefits of these management systems support climate change adaptation through sustainable resource management (Roberts et al., 2017) and mitigation through improved carbon storage (Vierros, 2017). The challenges to wider use include both how to upscale LMMAs (Roberts et al., 2017; Vierros, 2017), and how to assess them as climate change adaptation and mitigation solutions (Rohe et al., 2017; Section 5.4).

IK and Pikialasorsuaq: Pikialasorsuaq (North Water Polynya), in Baffin Bay, is the Arctic's largest polynya, or area of open water surrounded by ice, and is also one of the most biologically productive regions in the Arctic (Barber et al., 2001). Adjacent Inuit communities depend on Pikialasorsuaq for their food security and subsistence economy (Hastrup et al., 2018). They use Qaujimajatuqangit, an IK system, in daily and seasonal activities (ICC, 2017). The sea ice bridge north of the Pikialasorsuaq is no longer forming as reliably as in the past, resulting in a polynya that is geographically and seasonally less defined (Ryan and Münchow, 2017). In response, the Inuit Circumpolar Council initiated the Pikialasorsuaq Commission who formed an Inuit-led management authority to (1) oversee monitoring and research to conserve the polynya's living resources; (2) identify an Indigenous Protected Area, to include the polynya and dependent communities; and (3) establish a free travel zone for Inuit across the Pikialasorsuaq region (ICC, 2017; Box 3.2).

LK in the Alps: Mountain guides and other local residents engaged in supporting mountain tourism draw on LK for livelihood management. A study at Mont Blanc lists specific cryosphere changes which they have observed, including glacial shrinkage and reduction in ice and snow cover. As a result, the categorisation of the difficulty of a number of routes has changed, and the timing of the climbing season has shifted earlier (Mourey and Ravanel, 2017; Section 2.3.5).

LK to manage flooding: Climate change is increasing glacial melt water and rain-induced disasters in the Himalayan region and affected communities in China, Nepal, and India use LK to adapt (Nadeem et al., 2012). For instance, rains upstream in Gandaki (Nepal) flood downstream areas of Bihar, India. Local communities' knowledge of forecasting floods has evolved over time through the complexities of caste, class, gender and ecological flux, and is critical to flood forecasting and disaster risk reduction. Local communities manage risk by using a diverse set of knowledge, including phenomenological (e.g., river sound), ecological (e.g., red ant movement) and riverine (e.g., river colour) indicators, alongside meteorological and official information (Acharya and Prakash, 2018; Section 2.3.2.3).

Knowledge Holders' Recommendations for Utilising IK and LK in Assessment Reports

Perspectives from the Himalayas: IK and LK holders in the Himalayas have conducted long-term systematic observations in these remote areas for centuries. Contemporary IK details change in phenology, weather patterns, and flora and fauna species, which enriches scientific knowledge of glacial retreat and potential glacial lake outbursts (Sherpa, 2014). The scientific community can close many knowledge gaps by engaging IK and LK holders as counterparts. Suggestions towards this objective are to work with affected communities to elicit their knowledge of change, especially IK and LK holders with more specialised knowledge (farmers, herders, mountain guides, etc.), and use location- and culture-specific approaches to share scientific knowledge and use it with IK and LK.

Perspectives from the Inuit Circumpolar Council (ICC), Canada: Engaging Inuit as partners across all climate research disciplines ensures that Inuit knowledge and priorities guide research, monitoring, and the reporting of results in Inuit homeland. Doing so enhances the effectiveness, impact, and usefulness of global assessments, and ensures that Inuit knowledge is appropriately reported in assessments. Inuit seek to achieve self-determination in all aspects of research carried out in Inuit homeland (e.g., Nickels et al., 2005). Inuit actively produce and use climate research (e.g., ITK, 2005; ICC, 2015) and lead approaches to address climate challenges spurred by great incentive to develop innovative solutions. Engaging Inuit representative organisations and governments as partners in research recognises that the best available knowledge includes IK, enabling more robust climate research that in turn informs climate policy. When interpreted and applied properly, IK comes directly from research by Inuit and from an Inuit perspective (ICC, 2018). This can be achieved by working with Inuit on scoping and methodology for assessments and supporting inclusion of Inuit experts in research, analysis, and results dissemination.

1.9   Approaches Taken in this Special Report

1.9.1   Methodologies Relevant to this Report

SROCC assesses literature on ocean and cryosphere change and associated impacts and responses, focusing on advances in knowledge since AR5. The literature used is primarily published, peer-reviewed scientific, social science and humanities research. In some cases, grey literature sources (for example, published reports from governments, industry, research institutes and non-government organisations) are used where there are important gaps in available peer-reviewed literature. It is recognised that published knowledge from many parts of the world most vulnerable to ocean and cryosphere change is still limited (Czerniewicz et al., 2017).

Where possible, SROCC draws upon established methodologies and/or frameworks. Cross-Chapter Boxes in Chapter 1 address methodologies used for projections of future change (Cross-Chapter Box 1 in Chapter 1), for assessing and reducing risk (Cross-Chapter Box 2 in Chapter 1), for governance options relevant to a problem or region (Cross-Chapter Box 3 in Chapter 1), and for using IK and LK (Cross-Chapter Box 4 in Chapter 1). It is recognised in the assessment process that multiple and non-static factors determine human vulnerabilities to climate change impacts, and that ecosystems provide essential services that have both commercial and non-commercial value (Section 1.5). Economic methods are also important in SROCC, for estimating the economic value of natural systems, and for aiding decision-making around mitigation and adaptation strategies (Section 1.6).

1.9.2   Communication of Confidence in Assessment Findings

SROCC uses calibrated language for the communication of confidence in the assessment process (Mastrandrea et al., 2010; Mach et al., 2017). Calibrated language is designed to consistently evaluate and communicate uncertainties that arise from incomplete knowledge due to a lack of information, or from disagreement about what is known or even knowable. The IPCC calibrated language uses qualitative expressions of confidence based on the robustness of evidence for a finding, and (where possible) uses quantitative expressions to describe the likelihood of a finding (Figure 1.4).

Qualitative expressions (*confidence scale*) describe the validity of a finding based on the type, amount, quality and consistency of evidence, and the degree of agreement between different lines of evidence (Figure 1.4, step 2). Evidence includes all knowledge sources, including IK and LK where available. *Very high* and *high* confidence findings are those that are supported by multiple lines of robust evidence with high agreement. *Low* or *very low* confidence describe findings for which there is limited evidence and/or low agreement among different lines of evidence, and are only presented in SROCC if they address a major topic of concern.

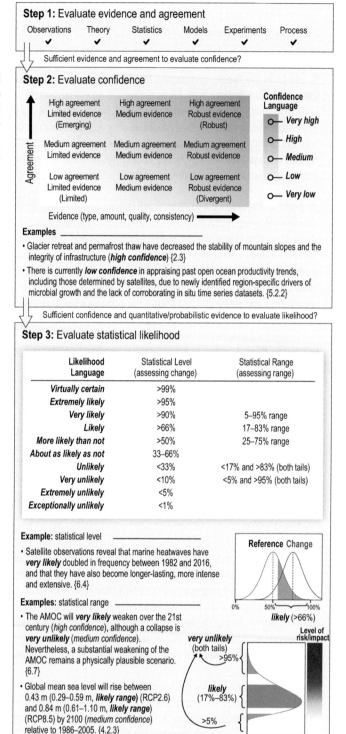

Figure 1.4 | Schematic of the IPCC usage of calibrated language, with examples of confidence and likelihood statements from this report. Figure developed after Mastrandrea et al. (2010), Mach et al. (2017) and Sutton (2018).

Quantitative expressions (*likelihood scale*) are used when sufficient data and confidence exists for findings to be assigned a quantitative or probabilistic estimate (Figure 1.4, step 3). In the scientific literature, a finding is often said to be significant if it has a likelihood exceeding 95% confidence. Using calibrated IPCC language, this level of statistical confidence would be termed *extremely likely*. Lower levels of likelihood than those derived numerically can be assigned by expert judgement to take into account structural or measurement uncertainties within the products or data used to determine the probabilistic estimates (e.g., Table CB1.1). Likelihood statements may be used to describe how climate changes relate to the ends of distribution functions, such as in detection and attribution studies that assess the likelihood that an observed climate change or event is different to a reference climate state (Section 1.3). In other situations, likelihood statements refer to the central region across a distribution of possibilities. Examples are the estimates of future changes based on large ensembles of climate model simulations, where the central 66% of estimates across the ensemble (i.e., the 17–83% range) would be termed a *likely* range (Figure 1.4, step 3).

It is increasingly recognised that effective risk management requires assessments not just of 'what is most likely' but also of 'how bad things could get' (Mach et al., 2017; Weaver et al., 2017; Xu and Ramanathan, 2017; Spratt and Dunlop, 2018; Sutton, 2018). In response to the need to reframe policy relevant assessments according to risk (Section 1.5; Mach et al., 2016; Weaver et al., 2017; Sutton, 2018), an effort is made in SROCC to report on potential changes for which there is low scientific confidence or a low likelihood of occurrence, but that would have large impacts if realised (Mach et al., 2017). In some cases where evidence is limited or emerging, phenomena may instead be discussed according to physically plausible scenarios of impact (e.g., Table 6.1).

In some cases, *deep uncertainty* (Cross-Chapter Box 5 in Chapter 1) may exist in current scientific assessments of the processes, rate, timing, magnitude, and consequences of future ocean and cryosphere changes. This includes physically plausible high-impact changes, such as high-end sea level rise scenarios that would be costly if realised without effective adaptation planning and even then may exceed limits to adaptation. Means such as expert judgement, scenario building, and invoking multiple lines of evidence enable comprehensive risk assessments even in cases of uncertain future ocean and cryosphere changes.

Cross-Chapter Box 5 | Confidence and Deep Uncertainty

Authors: Carolina Adler (Switzerland/Australia), Michael Oppenheimer (USA), Nerilie Abram (Australia), Kathleen McInnes (Australia) and Ted Schuur (USA)

Definition and Context

Characterising, assessing and managing risks to climate change involves dealing with inherent uncertainties. Uncertainties can lead to complex decision-making situations for managers and policymakers tasked with risk management, particularly where decisions relate to possibilities assessed as having low or unknown confidence/likelihood, yet would have high impacts if realised. While uncertainty can be quantitatively or qualitatively assessed (Section 1.9.2; Figure 1.4), a situation of *deep uncertainty* exists when experts or stakeholders do not know or cannot agree on: (1) appropriate conceptual models that describe relationships among key driving forces in a system; (2) the probability distributions used to represent uncertainty about key variables and parameters; and/or, (3) how to weigh and value desirable alternative outcomes (adapted from Lempert et al., 2003; Marchau et al., 2019b).

The concept of deep uncertainty has been debated and addressed in the literature for some time, with diverse terminology used. Terms such as great uncertainty (Hansson and Hirsch Hadorn, 2017), contested uncertain knowledge (Douglas and Wildavsky, 1983), ambiguity (Ellsberg, 1961) and Knightian uncertainty (Knight, 1921) among others, are also present in the literature to refer to the multiple components of uncertainty that need to be accounted for in decision making. The purpose of this Cross-Chapter Box is to constructively engage with the concept of deep uncertainty, by first providing some context for how the IPCC has dealt with deep uncertainty in the past. This is followed by examples of cases from the ocean and cryosphere assessments in the Special Report on the Oceans and Cryosphere in a Changing Climate (SROCC), where deep uncertainty has been addressed to advance assessment of risks and their management.

How has the IPCC and other literature dealt with deep uncertainty?

The IPCC assessment process provides instances of how deep uncertainty can manifest. In assessing the scientific evidence for anthropogenic climate change, and its influence on the Earth system in the past and future, IPCC assessments can identify areas where a large range of possibilities exist in the scientific literature or where knowledge of the underlying processes and responses is lacking. Existing guidelines to ensure consistent treatment of uncertainties by IPCC author teams (Mastrandrea et al., 2010; Section 1.9.2) may not be sufficient to ensure the desired consistency or guide robust findings when conditions of deep uncertainty are present (Adler and Hirsch Hadorn, 2014).

The IPCC, and earlier assessments, encountered deep uncertainty when evaluating numerous aspects of the climate change problem. Examining these cases sheds light on approaches to quantifying and reducing deep uncertainty. An assessment by the US National Academy of Sciences (Charney et al., 1979; commonly referred to as the Charney Report) provides a classic example. Evaluating climate sensitivity to a doubling of carbon dioxide concentration, and developing a probability distribution for it, was challenging because only two 3-D climate models and a handful of model variants and realisations were available. The panel invoked three strategies to eliminate some of these simulations: (1) Using multiple lines of evidence to complement the limited model results; (2) estimating the consequences of poor or absent model representations of certain physical processes (particularly cumulus convection, high-altitude cloud formation, and non-cloud entrainment); and, (3) evaluating mismatches between model results and observations. This triage yielded 'probable bounds' of 2°C–3.5°C on climate sensitivity. The panel then invoked expert judgment (Box 12.2 in Collins et al., 2013) to broaden the range to 3 ± 1.5°C, with 3°C referred to as the 'most probable value'. The panel did not report its confidence in these judgments.

The literature has expanded greatly since, allowing successive IPCC assessments to refine the approach taken in the Charney report. By the IPCC 5th Assessment Report (AR5), four lines of evidence (from instrumental records, palaeoclimate data, model intercomparison of sensitivity, and model-climatology comparisons) were assessed to determine that 'Equilibrium climate sensitivity is *likely* in the range 1.5°C–4.5°C (*high confidence*), *extremely unlikely* less than 1°C (*high confidence*), and *very unlikely* greater than 6°C (*medium confidence*)' (Box 12.2 in Collins et al., 2013). The Charney report began the process of convergence of opinion around a single probability range (essentially, category (2) in the definition of deep uncertainty, above), at least for sensitivity arising from fast feedbacks captured by global climate models (Hansen et al., 2007). Subsequent assessments increased confidence, eliminating deep uncertainty about this part of the sensitivity problem over a wide range of probability.

Cases of Deep Uncertainty from SROCC

Case A: Permafrost carbon and greenhouse gas emissions. AR5 reported the estimated size of the organic carbon pool stored frozen in permafrost zone soils, but uncertainty estimates were not available (Tarnocai et al., 2009; Ciais et al., 2013). AR5 further reported that future greenhouse gas emissions (CO_2 only) from permafrost were the most uncertain biogeochemical feedback on climate of the ten factors quantified (Figure 6.20 in Ciais et al., 2013). However, the *low confidence* assigned to permafrost was not due to few studies, but rather to divergence on the conceptual framework relating changes in permafrost carbon and future greenhouse gas emissions, as well as the probability distribution of key variables. Most large-scale carbon climate models still lack key landscape-level mechanisms that are known to abruptly thaw permafrost and expose organic carbon to decomposition, and many do not include mechanisms needed to differentiate the release of methane versus carbon dioxide with their very different global warming potentials. Studies since AR5 on potential methane release from laboratory soil incubations (Schädel et al., 2016; Knoblauch et al., 2018), actual methane release from the Siberian shallow Arctic ocean shelves (Shakhova et al., 2013; Thornton et al., 2016), changes in permafrost carbon stocks from the Last Glacial Maximum until present (Ciais et al., 2011; Lindgren et al., 2018) and potential carbon uptake by future plant growth (Qian et al., 2010; McGuire et al., 2018) have widened rather than narrowed the uncertainty range (Section 3.4.3.1.1). Accounting for greenhouse gas release from polar and high mountain (Box 2.2) permafrost, introduces an element of deep uncertainty when determining emissions pathways consistent with Article 2 of the Paris Agreement (Comyn-Platt et al., 2018). With stakeholder needs in mind, scientists have been actively engaged in narrowing this uncertainty by using multiple lines of evidence, expert judgment, and joint evaluation of observations and models. As a result, SROCC has reduced uncertainty and introduced confidence assessments across some but not all components of this problem (Section 3.4.3.1.1.).

Case B: Antarctic ice sheet and sea level rise. Dynamical ice loss from Antarctica (Cross-Chapter Box 8 in Chapter 3) provides an example of lack of knowledge about processes, and disagreement about appropriate models and probability distributions for representing uncertainty (categories (1) and (2) in the definition of deep uncertainty). AR5 used a statistical model and expert judgment to reduce uncertainty compared to AR4 (Church et al., 2013). Based on modelling of marine ice sheet processes after AR5, SROCC has further reduced uncertainty in the Antarctic contribution to sea level rise. The *likely* range including the potential contribution of marine ice sheet instability is quantified as 0.02–0.23 m for 2081–2100 (and 0.03–0.28 m for 2100) compared to 1986–2005 under RCP8.5 (*medium confidence*). However, the magnitude of additional rise beyond 2100, and the probability of greater sea level rise than that included in the *likely* range before 2100, are characterised by deep uncertainty (Section 4.2.3).

Cross-Chapter Box 5 (continued)

Policy makers at various levels of governance are considering adaptation investments (e.g., hard infrastructure, retreat, and nature-based defences) for multi-decadal time horizons that consider projection uncertainty (Sections 4.4.2, 4.4.3). For example, extreme sea levels (e.g., the local 'hundred-year flood') now occurring during storms that are historically rare are projected to become annual events by 2100 or sooner at many low-lying coastal locations (Section 4.4.3). Sea level rise exceeding the *likely* range, or an alternate pathway to the assumed climate change scenario (e.g., which RCP is used in risk estimation), could alter these projections and both factors are characterised by deep uncertainty. Among the strategies used to reduce deep uncertainty in these cases are formal and informal elicitation of expert judgment to project ice sheet behaviour (Horton et al., 2014; Bamber et al., 2019), and development of plausible sea level rise scenarios, including extreme cases (Sections 4.2.3, 4.4.5.3). Frameworks for risk management under deep uncertainty in the context of time lags between commitment to ice sheet losses and emissions mitigation, and between coastal adaptation planning and implementation, are currently emerging in the literature (Section 4.4.5.3.4).

Case C: Compound risks and cascading impacts. Compound risks and cascading impacts (Section 6.1, 6.8, Figure 1.1, Figure 6.1) arise from multiple coincident or sequential hazards (Zscheischler et al., 2018). Compound risks are an example of deep uncertainty because their rarity means that there is often a lack of data or modelling to characterise the risks statistically under present conditions or future changes (Gallina et al., 2016), and there is the potential that climate elements could cross tipping points (e.g., Cai et al., 2016). Nevertheless, effective risk reduction strategies can be developed without knowing the statistical likelihoods of such events by acknowledging the possibility that an event can occur (Dessai et al., 2009). Such strategies are typically well hedged against a variety of different futures and adjustable through time in response to emerging information (Lempert et al., 2010). Case studies are useful for raising awareness of the possibility of compound events and provide valuable learnings for decision makers in the form of analogues (McLeman and Hunter, 2010). They can provide a basis for devising scenarios to stress test systems in other regions for the purposes of understanding and reducing risk. The case study describing the ocean, climate and weather events in the Australian state of Tasmania in 2015/2016 (Box 6.1) provides such an example. It led to compound risks that could not have been estimated due to deep uncertainty. The total cost of the co-occurring fires, floods and marine heat wave to the state government was estimated at about 300 million USD, and impacts on the food, energy and manufacturing sectors reduced Tasmania's anticipated economic growth by approximately half (Eslake, 2016). In the aftermath of this event, the government increased funding to relevant agencies responsible for flood and bushfire management and independent reviews have recommended major policy reforms that are now under consideration (Blake et al., 2017; Tasmanian Climate Change Office, 2017).

What can we learn from SROCC cases in addressing deep uncertainty?
Using the adapted definition as a framing concept for deep uncertainty (see also Annex 1: Glossary), we find that each of the three cases described in this Cross-Chapter Box involve at least one of the three ways that deep uncertainty can manifest. In Case A, incomplete knowledge on relationships and key drivers and feedbacks (category 1), coupled with broadened probability distributions in post-AR5 literature (category 2), are key reasons for deep uncertainty. In Case B, the inability to characterise the probability of marine ice sheet instability due to a lack of adequate models resulting in divergent views on the probability of ice loss lead to deep uncertainty (categories 1 and 2). In Case C, the Australian example provides insights on the inadequacy of models or previous experience for estimating risk of multiple simultaneous extreme events, contributing to the exhaustion of resources which were then insufficient to meet the need for emergency response. This case also points to the complex task of addressing multiple simultaneous extreme events, and the multiple ways of valuing preferred outcomes in reducing future losses (category 3).

The three cases validate the continued iterative process required to meaningfully engage with deep uncertainty in situations of risk, through means such as elicitation, deliberation and application of expert judgement, scenario building and invoking multiple lines of evidence. These approaches demonstrate feasible ways to address or even reduce deep uncertainty in complex decision situations (see also Marchau et al., 2019a), considering that possible obstacles and time investment needed to address deep uncertainty, should not be underestimated.

1.10   Integrated Storyline of this Special Report

The chapters that follow in this special report are framed around geographies or climatic processes where the ocean and/or cryosphere are particularly important for ecosystems and people. The chapter order follows the movement of water from Earth's shrinking mountain and polar cryosphere into our rising and warming ocean.

Chapter 2 assesses *High Mountain* areas outside of the polar regions, where glaciers, snow and/or permafrost are common. Chapter 3 moves to the *Polar Regions* of the northern and southern high latitudes, which are characterised by vast stores of frozen water in ice sheets, glaciers, ice shelves, sea ice and permafrost, and by the interaction of these cryosphere elements and the polar oceans. Chapter 4 examines *Sea Level Rise* and the hazards this brings to *Low-Lying Regions, Coasts and Communities*. Chapter 5 focuses on the *Changing Ocean*, with a particular focus on how climate change impacts on the ocean are altering *Marine Ecosystems* and affecting *Dependent Communities*. Chapter 6 is dedicated to assessing *Extremes* and *Abrupt Events*, and reflects the potential for rapid and possibly irreversible changes in Earth's ocean and cryosphere, and the challenges this brings to *Managing Risk*. The multitude ways in which *Low-Lying Islands and Coasts* are exposed and vulnerable to the impacts of ocean and cryosphere change, along with resilience and adaptation strategies, opportunities and governance options specific to these settings, is highlighted in integrative Cross-Chapter Box 9.

This report does not attempt to assess all aspects of the ocean and cryosphere in a changing climate. Examples of research themes that will be covered elsewhere in the IPCC Sixth Assessment Cycle and not SROCC include: assessments of ocean and cryosphere changes in the CMIP6 experiments (AR6), cryosphere changes outside of polar and high mountain regions (e.g., snow cover in temperate and low altitude settings; AR6), and a thorough assessment of mitigation options for reducing climate change impacts (SR15, AR6 WGIII).

Each chapter of SROCC presents an integrated storyline on the ocean and/or cryosphere in a changing climate. The chapter assessments each present evidence of the pervasive changes that are already underway in the ocean and cryosphere (Figure 1.5). The impacts that physical changes in the ocean and cryosphere have had on ecosystems and people are assessed, along with lessons learned from adaptation measures that have already been employed to avoid adverse impacts. The assessments of future change in the ocean and cryosphere demonstrate the growing and accelerating changes projected for the future and identify the reduced impacts and risks that choices for a low greenhouse gas emission future would have compared with a high emission future (Figure 1.5). Potential adaptation strategies to reduce future risks to ecosystems and people are assessed, including identifying where limits to adaptation may be exceeded. The local- to global-scale responses for charting CRDPs are also assessed.

Figure 1.5 | (right) Observed and modelled historical changes in the ocean and cryosphere since 1950, and projected future changes under low (RCP2.6) and high (RCP8.5) greenhouse gas emissions scenarios (Cross-Chapter Box 1 in Chapter 1). -Changes are shown for: **(a)** Global mean surface air temperature change with *likely* range (Cross-Chapter Box 1 in Chapter 1). **Ocean-related changes** with *very likely* ranges for **(b)** Global mean sea surface temperature change (Box 5.1, Section 5.2.2); **(c)** Change factor in surface ocean marine heatwave days (6.4.1); **(d)** Global ocean heat content change (0–2000 m depth). An approximate steric sea level equivalent is shown with the right axis by multiplying the ocean heat content by the global-mean thermal expansion coefficient ($\varepsilon \approx 0.125$ m per 10^{24} Joules) for observed warming since 1970 (Figure 5.1); **(h)** Global mean surface pH (on the total scale). Assessed observational trends are compiled from open ocean time series sites longer than 15 years (Box 5.1, Figure 5.6, Section 5.2.2); and **(i)** Global mean ocean oxygen change (100–600 m depth). Assessed observational trends span 1970–2010 centered on 1996 (Figure 5.8, Section 5.2.2). **Sea-level changes** with *likely* ranges for **(m)** Global mean sea level change. Hashed shading reflects *low confidence* in sea level projections beyond 2100 and bars at 2300 reflect expert elicitation on the range of possible sea level change (Section 4.2.3, Figure 4.2); and components from **(e,f)** Greenland and Antarctic ice sheet mass loss (Section 3.3.1); and **(g)** Glacier mass loss (Cross-Chapter Box 6 in Chapter 2, Table 4.1). Further **cryosphere-related changes** with *very likely* ranges for **(j)** Arctic sea ice extent change for September (Sections 3.2.1, 3.2.2 Figure 3.3); **(k)** Arctic snow cover change for June (land areas north of 60°N) (Sections 3.4.1, 3.4.2, Figure 3.10); and **(l)** Change in near-surface (within 3–4 m) permafrost area in the Northern Hemisphere (Sections 3.4.1, 3.4.2, Figure 3.10).

Past and future changes in the ocean and cryosphere

Historical changes (observed and modelled) and projections under RCP2.6 and RCP8.5 for key indicators

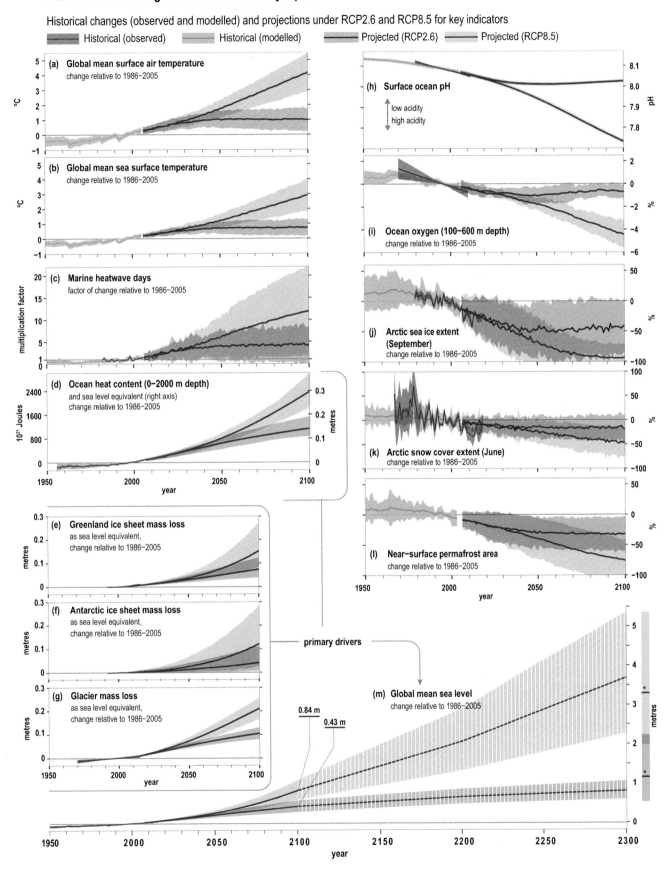

1

FAQ 1.1 | How do changes in the ocean and cryosphere affect our life on planet Earth?

The ocean and cryosphere regulate the climate and weather on Earth, provide food and water, support economies, trade and transportation, shape cultures and influence our well-being. Many of the recent changes in Earth's ocean and cryosphere are the result of human activities and have consequences on everyone's life. Deep cuts in greenhouse gas emissions will reduce negative impacts on billions of people and help them adapt to changes in their environment. Improving education and combining scientific knowledge with Indigenous knowledge and local knowledge helps communities to further address the challenges ahead.

The ocean and cryosphere – a collective name for the frozen parts of the Earth – are essential to the climate and life giving processes on our planet.

Changes in the ocean and cryosphere occur naturally, but the speed, magnitude, and pervasiveness of the global changes happening right now have not been observed for millennia or longer. Evidence shows that the majority of ocean and cryosphere changes observed in the past few decades are the result of human influences on Earth's climate.

Every one of us benefits from the role of the ocean and cryosphere in regulating climate and weather. The ocean has absorbed about a third of the carbon dioxide humans have emitted from the burning of fossil fuels since the Industrial Revolution, and the majority (more than 90%) of the extra heat within the Earth system. In this way, the ocean has slowed the warming humans and ecosystems have experienced on land. The reflective surface of snow and ice reduce the amount of the sun's energy that is absorbed on Earth. This effect diminishes as snow and ice melts, contributing to amplified temperature rise across the Arctic. The ocean and cryosphere also sustain life giving water resources, by rain and snow that come from the ocean, and by melt water from snow and glaciers in mountain and polar regions.

Nearly two billion people live near the coast, and around 800 million on land less than 10 m above sea level. The ocean directly supports the food, economies, cultures and well-being of coastal populations (see FAQ 1.2). The livelihoods of many more are tied closely to the ocean through food, trade, and transportation. Fish and shellfish contribute about 17% of the non-grain protein in human diets and shipping transports at least 80% of international imports and exports. But the ocean also brings hazards to coastal populations and infrastructure, and particularly to low-lying coasts. These populations are increasingly exposed to tropical cyclones, marine heat waves, sea level rise, coastal flooding and saltwater incursion into groundwater resources.

In high mountains and the Arctic, around 700 million people live in close contact with the cryosphere. These people, including many Indigenous Peoples, depend on snow, glaciers and sea ice for their livelihoods, food and water security, travel and transport, and cultures (see FAQ 1.2). They are also exposed to hazards as the cryosphere changes, including flood outbursts, landslides and coastal erosion. Changes in the polar and high mountain regions also have far-reaching consequences for people in other parts of the world (see FAQ 3.1).

Warming of the climate system leads to sea level rise. Melt from glaciers and ice sheets is adding to the amount of water in the ocean, and the heat being absorbed by the ocean is causing it to expand and take up more space. Today's sea level is already about 20 cm higher than in 1900. Sea level will continue to rise for centuries to millennia because the ocean system reacts slowly. Even if global warming were to be halted, it would take centuries or more to halt ice sheet melt and ocean warming.

Enhanced warming in the Arctic and in high mountains is causing rapid surface melt of glaciers and the Greenland ice sheet. Thawing of permafrost is destabilising soils, human infrastructure, and Arctic coasts, and has the potential to release vast quantities of methane and carbon dioxide into the atmosphere that will further exacerbate climate change. Widespread loss of sea ice in the Arctic is opening up new routes for shipping, but at the same time is reducing habitats for key species and affecting the livelihoods of Indigenous cultures. In Antarctica, glacier and ice sheet loss is occurring particularly quickly in places where ice is in direct contact with warm ocean water, further contributing to sea level rise.

Ocean ecosystems are threatened globally by three major climate change-induced stressors: warming, loss of oxygen and acidification. Marine heat waves are occurring everywhere across the surface ocean, and are becoming more frequent and more intense as the ocean warms. These are causing disease and mass-mortality that put, for example, coral reefs and fish populations at risk. Marine heat waves last much longer than the heat waves experienced on land, and are particularly harmful for organisms that cannot move away from areas of warm water.

Warming of the ocean reduces not only the amount of oxygen it can hold, but also tend to stratify it. As a result, less oxygen is transported to depth, where it is needed to support ocean life. Dissolved carbon dioxide that has been taken up by the ocean reacts with water molecules to increase the acidity of seawater. This makes the water more corrosive for marine organisms that build their shells and structures out of mineral carbonates, such as corals, shellfish and plankton. These climate-change stressors occur alongside other human-driven impacts, such as overfishing, excessive nutrient loads (eutrophication), and plastic pollution. If human impacts on the ocean continue unabated, declines in ocean health and services are projected to cost the global economy 428 billion USD yr^{-1} by 2050, and 1.979 trillion USD yr^{-1} by 2100.

The speed and intensity of the future risks and impacts from ocean and cryosphere change depend critically on future greenhouse gas emissions. The more these emissions can be curbed, the more the changes in the ocean and cryosphere can be slowed and limited, reducing future risks and impacts. But humankind is also exposed to the effects of changes triggered by past emissions, including sea level rise that will continue for centuries to come. Improving education and using scientific knowledge alongside local knowledge and Indigenous knowledge can support the development of context-specific options that help communities to adapt to inevitable changes and respond to challenges ahead.

FAQ 1.2 | How will changes in the ocean and cryosphere affect meeting the Sustainable Development Goals?

Ocean and cryosphere change affect our ability to meet the United Nations Sustainable Development Goals (SDGs). Progress on the SDGs support climate action that will reduce future ocean and cryosphere change, and as well as the adaptation responses to unavoidable changes. There are also trade-offs between SDGs and measures that help communities to adjust to their changing environment, but limiting greenhouse gas emissions opens more options for effective adaptation and sustainable development.

The SDGs were adopted by the United Nations in 2015 to support action for people, planet and prosperity (FAQ 1.2, Figure 1). The 17 goals and their 169 targets strive to end poverty and hunger, protect the planet and reduce gender, social and economic inequities by 2030.

SDG 13 (Climate Action) explicitly recognises that changing climatic conditions are a global concern. Climate change is already causing pervasive changes in Earth's ocean and cryosphere (FAQ 1.1). These changes are impacting food, water and health securities, with consequences for achieving SDG 2 (Zero Hunger), SDG 3 (Good Health and Well-Being), SDG 6 (Clean Water and Sanitation), and SDG 1 (No Poverty). Climate change impacts on Earth's ocean and cryosphere also affect the environmental goals for SDG 14 (Life below Water) and SDG 15 (Life on Land), with additional implications for many of the other SDGs.

SDG 6 (Clean Water and Sanitation) will be affected by ocean and cryosphere changes. Melting mountain glaciers bring an initial increase in water, but as glaciers continue to shrink so too will the essential water they provide to millions of mountain dwellers, downstream communities, and cities. These populations also depend on water flow from the high mountains for drinking, sanitation, and irrigation, and for SDG 7 (Affordable and Clean Energy). Water security is also threatened by changes in the magnitude and seasonality of rainfall, driven by rising ocean temperatures, which increases the risk of severe storms and flooding in some regions, or the risk of more severe or more frequent droughts in other regions. Among other effects, ongoing sea level rise is allowing salt water to intrude further inland, contaminating drinking water and irrigation sources for some coastal populations. Actions to address these threats will likely require new infrastructure to manage rain, melt water, and river flow, in order to make water supplies more reliable. These actions would also benefit SDG 3 (Good Health and Well-Being) by reducing the risk of flooding and negative health outcomes posed by extreme rainfall and outbursts of glacial melt.

Climate change impacts on the ocean and cryosphere also have many implications for progress on food security that is addressed in SDG 2 (Zero Hunger). Changes in rainfall patterns caused by ocean warming will increase aridity in some areas and bring more (or more intense) rainfall to others. In mountain regions, these changes bring varying challenges for maintaining reliable crops and livestock production. Some adaptation opportunities might be found in developing strains of crops and livestock better adapted to the future climate conditions, but this response option is also challenged by the rapid rate of climate change. In the Arctic, very rapidly warming temperatures, diminishing sea ice, reduced snow cover and degradation of permafrost are restricting the habitats and migration patterns of important food sources (SDG 2 Zero Hunger), including reindeer and several marine mammals (SDG 15 Life on Land; SDG 14 Life below Water), resulting in reduced hunting opportunities for staple foods that many northern Indigenous communities depend upon.

Rising temperatures, and changes in ocean nutrients, acidity and salinity are altering SDG 14 (Life Below Water). The productivity and distributions of some fish species are changing in ways that alter availability of fish to long-established fisheries, whereas the range of fish populations may move to become available in some new coastal and open ocean areas.

Ocean changes are of concern for small island developing states and coastal cities and communities. Beyond possible reductions in marine food supply and related risks for SDG 2 (Zero Hunger), their lives, livelihoods and well-being are also threatened in ways that are linked to several SDGs, including SDG 3 (Good Health and Wellbeing), SDG 8 (Decent Work and Economic Growth), SDG 9 (Industry, Innovation, and Infrastructure), and SDG 11 (Sustainable Cities and Communities). For example, sea level rise and warming oceans can cause inundation of coastal homes and infrastructure, more powerful tropical storms, declines in established economies such as tourism and losses of cultural heritage and identity. Improved community and coastal infrastructure can help to adapt to these changes, and more effective and faster disaster responses from health sectors and other emergency services can assist the populations who experience these impacts. In some situations, the most appropriate responses may involve relocation of critical services and, in some cases, communities; and for some populations, migration away from their homeland may become the only viable response.

FAQ 1.2, Figure 1 | The United Nations 2030 Sustainable Development Goals (SDGs).

Without transformative adaptation and mitigation, climate change could undermine progress towards achieving the 2030 SDGs, and make it more difficult to implement CRDPs in the longer term. Reducing global warming (mitigation) provides the best possibility to limit the speed and extent of ocean and cryosphere change and give more options for effective adaptation and sustainable development. Progress on SDG 4 (Quality Education), SDG 5 (Gender Equality) and SDG 10 (Reduced Inequalities) can moderate the vulnerabilities that shape people's risk to ocean and cryosphere change, while SDG 12 (Responsible Consumption and Production), SDG 16 (Peace, Justice and Institutions) and SDG 17 (Partnerships for the Goals) will help to facilitate the scales of adaptation and mitigation responses required to achieve sustainable development. Investment in social and physical infrastructure that supports adaptation to inevitable ocean and cryosphere changes will enable people to participate in initiatives to achieve the SDGs. Current and past IPCC efforts have focused on identifying CRDPs. Such adaptation and mitigation strategies, supported by adequate investments, and understanding the potential for SDG initiatives to increase the exposure or vulnerability of the activities to climate change hazards, could also constitute pathways for progress on the SDGs.

Acknowledgements

Thanks are due to Mohd Abdul Fahad, Mohamed Khamla, Jelto von Schuckmann and Debabrat Sukla for their assistance with early versions of figures and graphics.

References

Abraham, J.P. et al., 2013: A review of global ocean temperature observations: Implications for ocean heat content estimates and climate change. *Rev. Geophysics*, **51**(3), 450–483, doi:10.1002/rog.20022.

Abram, N.J. et al., 2016: Early onset of industrial-era warming across the oceans and continents. *Nature*, **536**(7617), 411–418, doi:10.1038/nature19082.

Abram, N.J. et al., 2013: Acceleration of snow melt in an Antarctic Peninsula ice core during the twentieth century. *Nat. Geosci.*, **6**(5), 404–411, doi:10.1038/ngeo1787.

Acharya, A. and A. Prakash, 2018: When the river talks to its people: local knowledge-based flood forecasting in Gandak River basin, India. *Environmental Development*, **31**, 55–67, doi:10.1016/j.envdev.2018.12.003.

Ackerman, F., 2013: Valuing the ocean environment. In: *Managing ocean environments in a changing climate: sustainability and economic perspectives* [Noone, K.J., U.R. Sumaila and R.J. Diaz (eds.)]. Elsevier, Amsterdam, pp. 243–275. ISBN 9780124076686.

Adger, W.N. et al., 2013: Cultural dimensions of climate change impacts and adaptation. *Nat. Clim. Change*, **3**(2), 112–117, doi:10.1038/Nclimate1666.

Adger, W.N. et al., 2009: Are there social limits to adaptation to climate change? *Clim. Change*, **93**(3–4), 335–354, doi:10.1007/s10584-008-9520-z.

Adger, W.N. et al., 2014: Human security. In: Climate Change 2014: Impacts, Adaptation, and Vulnerability. Part A: Global and Sectoral Aspects. Contribution of Working Group II to the Fifth Assessment Report of the Intergovernmental Panel on Climate Change [Field, C.B., V.R. Barros, D.J. Dokken, K.J. Mach, M.D. Mastrandrea, T.E. Biller, M. Chatterjee, K.L. Ebi, Y.O. Estrada, R.C. Genova, B. Girma, E.S. Kissel, A.N. Levy, S. MacCracken, P.R. Mastrandrea and L.L. White (eds.)]. Cambridge University Press, Cambridge, United Kingdom and New York, NY, USA, 793–832.

Adler, C.E. and G. Hirsch Hadorn, 2014: The IPCC and treatment of uncertainties: Topics and sources of dissensus. *WIREs. Clim. Change*, **5**(5), 663–676, doi:10.1002/wcc.297.

Agrawal, A., 1995: Dismantling the divide between indigenous and scientific knowledge. *Dev. Change*, **26**(3), 413–439, doi:10.1111/j.1467-7660.1995.tb00560.x.

Albrecht, G. et al., 2007: Solastalgia: the distress caused by environmental change. *Australas. Psych.*, **15**, S95-S98, doi:10.1080/10398560701701288.

Allen, M.R. et al., 2018: Framing and Context. In: Global Warming of 1.5°C. An IPCC Special Report on the impacts of global warming of 1.5°C above pre-industrial levels and related global greenhouse gas emission pathways, in the context of strengthening the global response to the threat of climate change, sustainable development, and efforts to eradicate poverty [Masson-Delmotte, V., P. Zhai, H.-O. Portner, D. Roberts, J. Skea, P.R. Shukla, A. Pirani, W. Moufouma-Okia, C. Péan, R. Pidcock, S. Connors, J.B.R. Matthews, Y. Chen, X. Zhou, M.I. Gomis, E. Lonnoy, T. Maycock, M. Tignor and T. Waterfield (eds.)]. World Meteorological Organization, Geneva, Switzerland, 49–91.

AMAP, 2011: *Snow, Water, Ice and Permafrost in the Arctic (SWIPA): Climate Change and the Cryosphere*. Arctic Monitoring and Assessment Programme (AMAP), Oslo, Norway, 538 pp.

AMAP, 2015: *AMAP Assessment 2015: Human Health in the Arctic*. Arctic Monitoring and Assessment Programme (AMAP), Oslo, Norway, 165 pp.

AMAP, 2017: *Snow, Water, Ice and Permafrost in the Arctic (SWIPA) 2017*. Arctic Monitoring and Assessment Programme (AMAP), Oslo, Norway, 269 pp.

Amsler, L.B., 2016: Collaborative governance: Integrating management, politics, and law. *Public Adm. Rev.*, **76**(5), 700–711, doi:10.1111/puar.12605.

Amundsen, H. et al., 2018: Local governments as drivers for societal transformation: Towards the 1.5°C ambition. *Curr. Opin. Environ. Sustain.*, **31**, 23–29, doi:10.1016/j.cosust.2017.12.004.

Andrade, E. et al., 2018: *Project Report: Ascertainment of the estimated excess mortality from Hurricane Maria in Puerto Rico*. Milken Institute School of Public Health of The George Washington University, Washington D.C., 69 pp.

Andreone, G., 2017: *The Future of the Law of the Sea: Bridging Gaps Between National, Individual and Common Interests*. Springer International Publishing, Cham, Switzerland. ISBN 9783319512747.

Anisimov, O.A. and R. Orttung, 2018: Climate change in Northern Russia through the prism of public perception. *Ambio*, **48**(6), 661–671, doi:10.1007/s13280-018-1096-x.

Armitage, D. et al., 2011: Co-management and the co-production of knowledge: Learning to adapt in Canada's Arctic. *Global Environ. Change*, **21**(3), 995–1004, doi:10.1016/j.gloenvcha.2011.04.006.

Arrigo, K.R. et al., 2017: Melting glaciers stimulate large summer phytoplankton blooms in southwest Greenland waters. *Geophys. Res. Letters*, **44**(12), 6278–6285, doi:10.1002/2017GL073583.

Ayers, J. and T. Forsyth, 2009: Community-Based adaptation to climate change. *Environment: Science and Policy for Sustainable Development*, **51**(4), 22–31, doi:10.1093/acrefore/9780190228620.013.602.

Baker, B. and B. Yeager, 2015: Coordinated ocean stewardship in the Arctic: Needs, challenges and possible models for an Arctic Ocean coordinating agreement. *Transnational Environmental Law*, **4**(2), 359–394, doi:10.1017/S2047102515000151.

Baldwin, E., P. McCord, J. Dell'Angelo and T. Evans, 2018: Collective action in a polycentric water governance system. *Environmental Policy and Governance*, **28**(4), 212–222, doi:10.1002/eet.1810.

Balmaseda, M.A. et al., 2015: The Ocean Reanalyses Intercomparison Project (ORA-IP). *J. Oper. Oceanogr.*, **8**(sup1), s80-s97, doi:10.1080/1755876X.2015.1022329.

Balmaseda, M.A., K.E. Trenberth and E. Källén, 2013: Distinctive climate signals in reanalysis of global ocean heat content. *Geophys. Res. Letters*, **40**(9), 1754–1759, doi:10.1002/grl.50382.

Bamber, J.W. et al., 2019: Ice sheet contributions to future sea level rise from structured expert judgement. *Proc. Natl. Acad. Sci. U.S.A.* **116**(23), 11195–11200, doi:10.1073/pnas.1817205116.

Baptiste, B., D. Pacheco, M. Carneiro da Cunha and S. Diaz, 2017: *Knowing our lands and resources: Indigenous and local knowledge of biodiversity and ecosystem services in the Americas*. Knowledges of Nature 11, UNESCO, Paris, 176 pp.

Barange, M. et al., 2014: Impacts of climate change on marine ecosystem production in societies dependent on fisheries. *Nat. Clim. Change*, **4**(3), 211–216, doi:10.1038/nclimate2119.

Barber, D. et al., 2001: Physical processes within the North Water (NOW) polynya. *Atmos. Ocean*, **39**(3), 163–166, doi:10.1080/07055900.2001.96 49673.

Barnett, J. et al., 2014: A local coastal adaptation pathway. *Nat. Clim. Change*, **4**(12), 1103–1108, doi:10.1038/nclimate2383.

Bartlett, C., M. Marshall and A. Marshall, 2012: Two-eyed seeing and other lessons learned within a co-learning journey of bringing together indigenous and mainstream knowledges and ways of knowing. *J. Environ. Stud. Sci.*, **2**(4), 331–340, doi:10.1007/s13412-012-0086-8.

Beck, S. et al., 2014: Towards a reflexive turn in the governance of global environmental expertise the cases of the IPCC and the IPBES. *GAIA.*, **23**(2), 80–87, doi:10.14512/gaia.23.2.4.

Bell, J.D., J.E. Johnson and A.J. Hobday, 2011: *Vulnerability of tropical Pacific fisheries and aquaculture to climate change*. Secretariat of the Pacific Community, Noumea, New Caledonia, 925 pp.

Bereiter, B. et al., 2015: Revision of the EPICA Dome C CO_2 record from 800 to 600 kyr before present. *Geophys. Res. Letters*, **42**(2), 542–549, doi:10.1002/2014GL061957.

Berner, R.A. and Z. Kothavala, 2001: GEOCARB III: A revised model of atmospheric CO_2 over Phanerozoic time. *Am. J. Sci.*, **301**(2), 182–204, doi:10.2475/ajs.301.2.182.

Berrang-Ford, L. et al., 2014: What drives national adaptation? A global assessment. *Clim. Change*, **124**(1–2), 441–450, doi:10.1007/s10584-014-1078-3.

Biemans, H. et al., 2019: Importance of snow and glacier meltwater for agriculture on the Indo-Gangetic Plain. *Nature Sustainability*, **2**, 594–601, doi: 10.1038/s41893-019-0305-3.

Biggs, R. et al., 2012: Toward principles for enhancing the resilience of ecosystem services. *Annu. Rev. Environ. Resour.*, **37**(1), 421–448, doi:10.1146/annurev-environ-051211-123836.

Billé, R. et al., 2013: Taking action against ocean acidification: a review of management and policy options. *Environ. Manage.*, **52**(4), 761–779, doi:10.1007/s00267-013-0132-7.

Bindoff, N.L. and P.A. Stott, 2013: Detection and Attribution of Climate Change: from Global to Regional. In: Climate Change 2013: The Physical Science Basis. Contribution of Working Group I to the Fifth Assessment Report of the Intergovernmental Panel on Climate Change [Stocker, T.F., D. Qin, G.-K. Plattner, M. Tignor, S.K. Allen, J. Boschung, A. Nauels, Y. Xia, V. Bex and P.M. Midgley (eds.)]. Cambridge University Press, Cambridge, United Kingdom and New York, NY, USA, 867–952.

Bisaro, A. and J. Hinkel, 2016: Governance of social dilemmas in climate change adaptation. *Nat. Clim. Change*, **6**(4), 354–359, doi:10.1038/nclimate2936.

Black, R., S.R. Bennett, S.M. Thomas and J.R. Beddington, 2011: Climate change: Migration as adaptation. *Nature*, **478**(7370), 447–449, doi:10.1038/478477a.

Blake, M., P. Considine and B. Edmons, 2017: *Report of the independent review into the Tasmanian floods of June and July 2016: Shared responsibility, resilience and adaptation*. Department of Premier and Cabinet (Tasmania), Hobart, Tasmania, 138 [Available at: www.dpac.tas.gov.au/__data/assets/pdf_file/0015/332610/floodreview.pdf].

Bloemen, P., M. Van Der Steen and Z. Van Der Wal, 2018: Designing a century ahead: climate change adaptation in the Dutch Delta. *Policy and Society*, **38**(1), 58–76, doi:10.1080/14494035.2018.1513731.

Bojinski, S. et al., 2014: The concept of essential climate variables in support of climate research, applications, and policy. *Bull. Am. Meteorol. Soc.*, **95**(9), 1431–1443, doi:10.1175/BAMS-D-13-00047.1.

Bond, N.A., M.F. Cronin, H. Freeland and N. Mantua, 2015: Causes and impacts of the 2014 warm anomaly in the NE Pacific. *Geophys. Res. Letters*, **42**, 3414–3420, doi:10.1002/2015GL063306.

Boyd, E. et al., 2017: A typology of loss and damage perspectives. *Nat. Clim. Change*, **7**(10), 723–729, doi:10.1038/nclimate3389.

Boyd, P.W., S.T. Lennartz, D.M. Glover and S.C. Doney, 2015: Biological ramifications of climate-change-mediated oceanic multi-stressors. *Nat. Clim. Change*, **5**(1), 71–79, doi:10.1038/nclimate2441.

Boyd, P.W., S. Sundby and H.-O. Pörtner, 2014: Cross-chapter box on net primary production in the ocean. In: Climate Change 2014: Impacts, Adaptation, and Vulnerability. Part A: Global and Sectoral Aspects. Contribution of Working Group II to the Fifth Assessment Report of the Intergovernmental Panel on Climate Change [Field, C.B., V.R. Barros, D.J. Dokken, K.J. Mach, M.D. Mastrandrea, T.E. Bilir, M. Chatterjee, K.L. Ebi, Y.O. Estrada, R.C. Genova, B. Girma, E.S. Kissel, A.N. Levy, S. MacCracken, P.R. Mastrandrea and L.L. White (eds.)]. Cambridge University Press, Cambridge, United Kingdom and New York, NY, USA, 133–136.

Boyer, T.P. et al., 2013: World Ocean Database 2013.[Levitus, S. and A. Mishonov (eds.)]. Silver Spring, MD, NOAA Atlas, 209 pp.

Breitburg, D. et al., 2018: Declining oxygen in the global ocean and coastal waters. *Science*, **359**(6371), eaam7240, doi:10.1126/science.aam7240.

Bronen, R., 2015: Climate-induced community relocations: Using integrated social-ecological assessments to foster adaptation and resilience. *Ecol. Soc.*, **20**(3), 36, doi:10.5751/ES-07801-200336.

Brügger, A. et al., 2015: Psychological responses to the proximity of climate change. *Nat. Clim. Change*, **5**(12), 1031–1037, doi:10.1038/nclimate2760.

Brysse, K., N. Oreskes, J. O'Reilly and M. Oppenheimer, 2013: Climate change prediction: Erring on the side of least drama? *Global Environ. Change*, **23**(1), 327–337, doi:10.1016/j.gloenvcha.2012.10.008.

Burkett, M., 2016: Reading between the red lines: Loss and damage and the Paris outcome. *Climate Law*, **6**(1–2), 118–129, doi:10.1163/18786561-00601008.

Burnham, M., Z. Ma and B. Zhang, 2016: Making sense of climate change: Hybrid epistemologies, socio-natural assemblages and smallholder knowledge. *Area*, **48**(1), 18–26, doi:10.1111/area.12150.

Butler, J. et al., 2014: Framing the application of adaptation pathways for rural livelihoods and global change in eastern Indonesian islands. *Global Environ. Change*, **28**, 368–382, doi:10.1016/j.gloenvcha.2013.12.004.

Cai, Y., T.M. Lenton and T.S. Lontzek, 2016: Risk of multiple interacting tipping points should encourage rapid CO_2 emission reduction. *Nat. Clim. Change*, **6**(5), 520–525, doi:10.1038/nclimate2964.

Cardona, O.D. et al., 2012: Determinants of risk: Exposure and vulnerability. In: Managing the risks of extreme events and disasters to advance climate change adaptation – A special report of Working Groups I and II of the Intergovernmental Panel on Climate Change (IPCC) [Field, C.B., V. Barros, T.F. Stocker, D. Qin, D.J. Dokken, K.L. Ebi, M.D. Mastrandrea, K.J. Mach, G.-K. Plattner, S.K. Allen, M. Tignor and P.M. Midgley (eds.)]. Cambridge University Press, Cambridge, United Kingdom, 65–108.

Carey, M. et al., 2015: Integrated approaches to adaptation and disaster risk reduction in dynamic socio-cryospheric systems. In: *Snow and ice-related hazards, risks and disasters* [Shroder, J.F., C. Whiteman and W. Haeberli (eds.)]. Elsevier, pp. 219–261. ISBN 9780123948496.

Carlisle, K. and R.L. Gruby, 2017: Polycentric systems of governance: a theoretical model for the commons. *Policy Stud. J.*, **26**, doi:10.1111/psj.12212.

Carpenter, S.R. and W.A. Brock, 2008: Adaptive capacity and traps. *Ecol. Soc.*, **13**(2), 40.

Carton, J.A., G.A. Chepurin and L. Chen, 2018: SODA3: A new ocean climate reanalysis. *J. Clim.*, **31**, 6967–6983, doi:10.1175/JCLI-D-18-0149.1.

Cassotta, S., 2012: *Environmental Damage Liability Problems in a Multilevel Context: The Case of the Environmental Liability Directive*. Energy and environmental law & policy series: Supanational and comparative aspects, Kluwer Law International, 280 pp.

Cassotta, S. and M. Mazza, 2015: Balancing *de jure* and *de facto* Arctic environmental law applied to the oil and gas industry: Linking indigenous rights, social impact assessment and business in Greenland. *The Yearbook of Polar Law Online*, **6**(1), 63–119, doi:10.1163/1876-8814_004.

Castree, N. et al., 2014: Changing the intellectual climate. *Nat. Clim. Change*, **4**(9), 763–768, doi:10.1038/nclimate2339.

Chandler, D., 2013: Resilience and the autotelic subject: Toward a critique of the societalization of security. *Int. Political Sociol.*, **7**(2), 210–226, doi:10.1111/ips.12018.

Chapin, F.S., G.P. Kofinas and C. Folke, 2009: *Principles of Ecosystem Stewardship: Resilience-Based Natural Resource Management in a Changing World*. Springer-Verlag, New York, 401 pp.

Charlesworth, B., 2009: Effective population size and patterns of molecular evolution and variation. *Nat. Rev. Genet.*, **10**(3), 195–205, doi:10.1038/nrg2526.

Charney, J.G. et al., 1979: *Carbon Dioxide and Climate: A Scientific Assessment. Report of an Ad Hoc Study Group on Carbon Dioxide and Climate*. National Academy of Sciences, Washington, DC., 22 pp.

Chen, Y., Z. Li, G. Fang and W. Li, 2018: Large hydrological processes changes in the transboundary rivers of Central Asia. *J. Geophys. Res. Atmos.*, **123**(10), 5059–5069, doi:10.1029/2017jd028184.

Cheng, L. et al., 2018: Taking the pulse of the planet. *Eos*, **98**, 14–16, doi:10.1029/2017EO081839.

Chilisa, B., 2011: *Indigenous Research Methodologies*. SAGE Publications, Los Angeles, 368 pp.

Chircop, A., S. Coffen-Smout and M.L. McConnell, 2018: *Ocean Yearbook 32*. Ocean Yearbook, Brill Nijhoff, 381 pp. ISBN: 9789004367005.

Chung Tiam Fook, T., 2017: Transformational processes for community-focused adaptation and social change: A synthesis. *Clim. Dev.*, **9**(1), 5–21, doi:10.1080/17565529.2015.1086294.

Church, J.A. et al., 2013: Sea Level Change. In: Climate Change 2013: The Physical Science Basis. Contribution of Working Group I to the Fifth Assessment Report of the Intergovernmental Panel on Climate Change [Stocker, T.F., D. Qin, G.K. Plattner, M. Tignor, S.K. Allen, J. Boschung, A. Nauels, Y. Xia, V. Bex and P.M. Midgley (eds.)]. Cambridge University Press, Cambridge, United Kingdom and New York, NY, USA, 1137–1216.

Ciais, P. et al., 2013: Carbon and Other Biogeochemical Cycles. In: Climate Change 2013: The Physical Science Basis. Contribution of Working Group I to the Fifth Assessment Report of the Intergovernmental Panel on Climate Change [Stocker, T.F., D. Qin, G.K. Plattner, M. Tignor, S.K. Allen, J. Boschung, A. Nauels, Y. Xia, V. Bex and P.M. Midgley (eds.)]. Cambridge University Press, Cambridge, United Kingdom and New York, NY, USA.

Ciais, P. et al., 2011: Large inert carbon pool in the terrestrial biosphere during the Last Glacial Maximum. *Nat. Geosci.*, **5**, 74–79, doi:10.1038/ngeo1324.

Colenbrander, D., A. Cartwright and A. Taylor, 2015: Drawing a line in the sand: Managing coastal risks in the City of Cape Town. *S. Afr. Geogr. J.*, **97**(1), 1–17, doi:10.1080/03736245.2014.924865.

Collins, M. et al., 2013: Long-term Climate Change: Projections, Commitments and Irreversibility. In: Climate Change 2013: The Physical Science Basis. Contribution of Working Group I to the Fifth Assessment Report of the Intergovernmental Panel on Climate Change [Stocker, T.F., D. Qin, G.-K. Plattner, M. Tignor, S.K. Allen, J. Boschung, A. Nauels, Y. Xia, V. Bex and P.M. Midgley (eds.)]. Cambridge University Press, Cambridge, United Kingdom and New York, NY, USA, 1029–1136.

Comyn-Platt, E. et al., 2018: Carbon budgets for 1.5 and 2°C targets lowered by natural wetland and permafrost feedbacks. *Nat. Geosci.*, **11**(8), 568–573, doi:10.1038/s41561-018-0174-9.

Corner, A., E. Markowitz and N. Pidgeon, 2014: Public engagement with climate change: The role of human values. *WIREs Clim. Change*, **5**(3), 411–422, doi:10.1002/wcc.269.

Cosens, B.A., L. Gunderson and B.C. Chaffin, 2018: Introduction to the Special Feature Practicing Panarchy: Assessing legal flexibility, ecological resilience, and adaptive governance in regional water systems experiencing rapid environmental change. *Ecol. Soc.*, **23**(1), 4, doi:10.5751/ES-09524-230104.

Cramer, W. et al., 2014: Detection and attribution of observed impacts. In: Climate Change 2014: Impacts, Adaptation, and Vulnerability. Part A: Global and Sectoral Aspects. Contribution of Working Group II to the Fifth Assessment Report of the Intergovernmental Panel on Climate Change [Field, C.B., V.R. Barros, D.J. Dokken, K.J. Mach, M.D. Mastrandrea, T.E. Bilir, M. Chatterjee, K.L. Ebi, Y.O. Estrada, R.C. Genova, B. Girma, E.S. Kissel, A.N. Levy, S. MacCracken, P.R. Mastrandrea and L.L. White (eds.)]. Cambridge University Press, Cambridge, United Kingdom and New York, NY, USA, 979–1037.

Crate, S. et al., 2017: Permafrost livelihoods: A transdisciplinary review and analysis of thermokarst-based systems of Indigenous land use. *Anthropocene*, **18**, 89–104, doi:10.1016/j.ancene.2017.06.001.

Crate, S.A. and A.N. Fedorov, 2013: A methodological model for exchanging local and scientific climate change knowledge in northeastern Siberia. *Arctic*, **66**(3), 338–350, doi:10.14430/arctic4312.

Creutzig, F. et al., 2018: Towards demand-side solutions for mitigating climate change. *Nat. Clim. Change*, **8**(4), 260–263, doi:10.1038/s41558-018-0121-1.

Cunsolo, A. and N.R. Ellis, 2018: Ecological grief as a mental health response to climate change-related loss. *Nat. Clim. Change*, **8**(4), 275–281, doi:10.1038/s41558-018-0092-2.

Cunsolo, A. and K. Landman, 2017: *Mourning Nature: Hope at the Heart of Ecological Loss and Grief*. McGill-Queen's University Press, Montreal-Kingston, 360 pp.

Cunsolo Willox, A. et al., 2013: The land enriches the soul: On climatic and environmental change, affect, and emotional health and well-being in Rigolet, Nunatsiavut, Canada. *Emote. Space Soc.*, **6**, 14–24, doi:10.1016/j.emospa.2011.08.005.

Czerniewicz, L., S. Goodier and R. Morrell, 2017: Southern knowledge online? Climate change research discoverability and communication practices. *Inf. Commun. Soc.*, **20**(3), 386–405, doi:10.1080/1369118X.2016.1168473.

David-Chavez, D.M. and M.C. Gavin, 2018: A global assessment of Indigenous community engagement in climate research. *Environ. Res. Letters*, **13**(12), 123005, doi:10.1088/1748-9326/aaf300.

De Lucia, V., 2017: The Arctic environment and the BBNJ negotiations. Special rules for special circumstances? *Mar. Policy*, **86**, 234–240, doi:10.1016/j.marpol.2017.09.011.

Delworth, T.L. et al., 2017: The central role of ocean dynamics in connecting the North Atlantic Oscillation to the extratropical component of the Atlantic Multidecadal Oscillation. *J. Clim.*, **30**(10), 3789–3805, doi:10.1175/JCLI-D-16-0358.1.

Denton, F. et al., 2014: Climate-resilient pathways: adaptation, mitigation, and sustainable development. In: Climate Change 2014: Impacts, Adaptation, and Vulnerability. Part A: Global and Sectoral Aspects. Contribution of Working Group II to the Fifth Assessment Report of the Intergovernmental Panel on Climate Change [Field, C.B., V. Barros, D.J. Dokken, K.J. Mach, M.D. Mastrandrea, T.E. Bilir, M. Chatterjee, K.L. Ebi, Y.O. Estrada, R.C. Genova, B. Girma, E.S. Kissel, A.N. Levy, S. MacCracken, P.R. Mastrandrea and L.L. White (eds.)]. Cambridge University Press, Cambridge, United Kingdom and New York, NY, USA, 1101–1131.

Deser, C., A. Phillips, V. Bourdette and H. Teng, 2012: Uncertainty in climate change projections: The role of internal variability. *Clim. Dyn.*, **38**(3–4), 527–546, doi:10.1007/s00382-010-0977-x.

Dessai, S., M. Hulme, R. Lempert and R.A. Pielke Jr, 2009: Climate prediction: A limit to adaptation? In: *Adapting to Climate Change: Thresholds, Values, Governance* [Adger, W.N., I. Lorenzoni and K. O'Brien (eds.)]. Cambridge University Press, pp. 64–78. ISBN: 9780521764858.

Deutsch, C. et al., 2015: Climate change tightens a metabolic constraint on marine habitats. *Science*, **348**(6239), 1132–1135, doi:10.1126/science.aaa1605.

DeWalt, B., 1994: Using indigenous knowledge to improve agriculture and natural resource management. *Hum. Organ.*, **53**(2), 123–131, doi:10.17730/humo.53.2.ku60563817m03n73.

Díaz, S. et al., 2015: A Rosetta Stone for nature's benefits to people. *PLoS Biol.*, **13**(1), e1002040, doi:10.1371/journal.pbio.1002040.

Díaz, S. et al., 2018: Assessing nature's contributions to people. *Science*, **359**(6373), 270–272, doi:10.1126/science.aap8826.

Djalante, R. et al., 2018: Cross Chapter Box 12: Residual risks, limits to adaptation and loss and damage. In: Global Warming of 1.5°C. An IPCC Special Report on the impacts of global warming of 1.5°C above pre-industrial levels and related global greenhouse gas emission pathways, in the context of strengthening the global response to the threat of climate change, sustainable development, and efforts to eradicate poverty [Masson-Delmotte, V., P. Zhai, H.O. Pörtner, D. Roberts, J. Skea, P.R. Shukla, A. Pirani, W. Moufouma-Okia, C. Péan, R. Pidcock, S. Connors, J.B.R. Matthews, Y. Chen, X. Zhou, M.I. Gomis, E. Lonnoy, T. Maycock, M. Tignor and T. Waterfield (eds.)]. World Meteorological Organization, Geneva, Switzerland.

Dodman, D. and M. Mitlin, 2013: Challenges for community-based adaptation: Discovering the potential for transformation. *J. Int. Dev.*, **25**(5), 640–659, doi:10.1002/jid.1772.

Douglas, M. and A. Wildavsky, 1983: Risk and Culture. *J. Policy Anal. Manage.*, **2**(2), 221, doi:10.2307/3323308.

Dow, K. et al., 2013: Limits to adaptation. *Nat. Clim. Change*, **3**(4), 305–307, doi:10.1038/nclimate1847.

Dowell, M. et al., 2013: *Strategy towards an architecture for climate monitoring from space*. Committee on Earth Observation Satellites, pp 39. [Available at: www.wmo.int/pages/prog/sat/.../ARCH_strategy-climate-architecture-space.pdf].

Downie, C. and M. Williams, 2018: After the Paris Agreement: What Role for the BRICS in Global Climate Governance? *Glob. Policy*, **9**(3), 398–407, doi:10.1111/1758-5899.12550.

Durack, P.J., T. Lee, N.T. Vinogradova and D. Stammer, 2016: Keeping the lights on for global ocean salinity observation. *Nat. Clim. Change*, **6**, 228–231, doi:10.1038/nclimate2946.

Easterling, D.R. et al., 2000: Climate extremes: Observations, modeling, and impacts. *Science*, **289**(5487), 2068–2074, doi:10.1126/science.289.5487.2068.

Eisenack, K. et al., 2014: Explaining and overcoming barriers to climate change adaptation. *Nat. Clim. Change*, **4**(10), 867–872, doi:10.1038/nclimate2350.

Ellsberg, D., 1961: Risk, ambiguity, and the Savage axioms. *Q.J. Econ.*, **75**(4), 643–669, doi:10.2307/1884324.

Eriksen, S.H., A.J. Nightingale and H. Eakin, 2015: Reframing adaptation: The political nature of climate change adaptation. *Global Environ. Change*, **35**, 523–533, doi:10.1016/j.gloenvcha.2015.09.014.

Escobar, A., 2001: Culture sits in places: Reflections on globalism and subaltern strategies of localization. *Political Geogr*, **20**, 139–174, doi:10.1016/S0962-6298(00)00064-0.

Eslake, S., 2016: *Tasmania Report 2016*. Tasmania Chamber of Commerce and Industry, Hobart, Tasmania, 92 [Available at: www.tcci.com.au/getattachment/Events/Tasmania-Report-2016/Tasmania-Report-2016-FINAL.pdf.aspx].

Evans, L.S. et al., 2016: Structural and psycho-social limits to climate change adaptation in the Great Barrier Reef Region. *PloS ONE*, **11**(3), e0150575, doi:10.1371/journal.pone.0150575.

FAO, 2018: *The State of World Fisheries and Aquaculture 2018 – Meeting the sustainable development goals*. The State of World Fisheries and Aquaculture, FAO, Rome, Italy, 210 pp.

Fazey, I. et al., 2016: Past and future adaptation pathways. *Clim. Dev.*, **8**(1), 26–44, doi:10.1080/17565529.2014.989192.

Feldman, L. and P. Hart, 2018: Is there any hope? How climate change news imagery and text influence audience emotions and support for climate mitigation policies. *Risk Anal.*, **38**(3), 585–602, doi:10.1111/risa.12868.

Feldman, L. and P.S. Hart, 2016: Using political efficacy messages to increase climate activism: The mediating role of emotions. *Sci. Commun.*, **38**(1), 99–127, doi:10.1177/1075547015617941.

Fenichel, E.P. et al., 2016: Wealth reallocation and sustainability under climate change. *Nat. Clim. Change*, **6**(3), 237–244, doi:10.1038/nclimate2871.

Fenty, I., D. Menemenlis and H. Zhang, 2017: Global coupled sea ice-ocean state estimation. *Clim. Dyn.*, **49**(3), 931–956, doi:10.1007/s00382-015-2796-6.

Few, R. et al., 2017: Transformation, adaptation and development: relating concepts to practice. *Palgrave Commun.*, **3**, doi:10.1057/palcomms.2017.92.

Field, C.B. et al., 2014: Technical summary. In: Climate Change 2014: Impacts, Adaptation, and Vulnerability. Part A: Global and Sectoral Aspects. Contribution of Working Group II to the Fifth Assessment Report of the Intergovernmental Panel on Climate Change [Field, C.B., V.R. Barros, D.J. Dokken, K.J. Mach, M.D. Mastrandrea, T.E. Biller, M. Chatterjee, K.L. Ebi, Y.O. Estrada, R.C. Genova, B. Girma, E.S. Kissel, A.N. Levy, S. MacCracken, P.R. Mastrandrea and L.L. White (eds.)]. Cambridge University Press, Cambridge, United Kingdom and New York, NY, USA, 35–94.

Field, C.B., M.J. Behrenfeld, J.T. Randerson and P. Falkowski, 1998: Primary production of the biosphere: Integrating terrestrial and oceanic components. *Science*, **281**(5374), 237–240, doi:10.1126/science.281.5374.237.

Fischer, H. et al., 2018: Palaeoclimate constraints on the impact of 2°C anthropogenic warming and beyond. *Nat. Geosci.*, **11**, 474–485, doi:10.1038/s41561-018-0146-0.

Fischer, J. et al., 2015: Advancing sustainability through mainstreaming a social–ecological systems perspective. *Curr. Opin. Environ. Sustain.*, **14**, 144–149, doi:10.1016/j.cosust.2015.06.002.

Flato, G. et al., 2013: Evaluation of climate models In: Climate Change 2013: The Physical Science Basis. Contribution of Working Group I to the Fifth Assessment Report of the Intergovernmental Panel on Climate Change [Stocker, T.F., D. Qin, G.-K. Plattner, M. Tignor, S.K. Allen, J. Boschung, A. Nauels, Y. Xia, V. Bex and P.M. Midgley (eds.)]. Cambridge University Press, Cambridge, United Kingdom and New York, NY, USA, 741–866.

Florano, E.R., 2018: Integrated Loss and Damage–Climate Change Adaptation–Disaster Risk Reduction Framework: The Case of the Philippines. In: *Resilience* [Zommers, Z. (ed.)]. Elsevier, Amsterdam, The Netherlands, pp. 317–324.

Flynn, M. et al., 2018: Participatory scenario planning and climate change impacts, adaptation and vulnerability research in the Arctic. *Environmental Science & Policy*, **79**, 45–53, doi:10.1016/j.envsci.2017.10.012.

Ford, J.D. et al., 2016a: Including indigenous knowledge and experience in IPCC assessment reports. *Nat. Clim. Change*, **6** (4), 349–353, doi:10.1038/nclimate2954.

Ford, J.D. et al., 2019: Changing access to ice, land and water in Arctic communities. *Nat. Clim. Change*, **9**, 335–339, doi:10.1038/s41558-019-0435-7.

Ford, J.D., N. Couture, T. Bell and D.G. Clark, 2017: Climate change and Canada's north coast: Research trends, progress, and future directions. *Environmental Reviews*, **26**(1), 82–92, doi:10.1139/er-2017-0027.

Ford, J.D. et al., 2016b: Adaptation and indigenous peoples in the United Nations framework convention on climate change. *Clim. Change*, **139**(3–4), 429–443.

Forino, G., J. von Meding and G.J. Brewer, 2015: A conceptual governance framework for climate change adaptation and disaster risk reduction integration. *Int. J. Disaster Risk Sc.*, **6**(4), 372–384, doi:10.1007/s13753-015-0076-z.

Frame, B. et al., 2018: Adapting global shared socio-economic pathways for national and local scenarios. *Clim. Risk Manage.*, **21**, 39–51, doi:10.1016/j.crm.2018.05.001.

Fretwell, P. et al., 2013: Bedmap2: Improved ice bed, surface and thickness datasets for Antarctica. *The Cryosphere*, **7**, 375–393, doi:10.5194/tc-7-375-2013.

Frölicher, T.L. and F. Joos, 2010: Reversible and irreversible impacts of greenhouse gas emissions in multi-century projections with the NCAR global coupled carbon cycle-climate model. *Clim. Dyn.*, **35**(7–8), 1439–1459, doi:10.1007/s00382-009-0727-0.

Frölicher, T.L., K.B. Rodgers, C.A. Stock and W. Cheung, 2016: Sources of uncertainties in 21st century projections of potential ocean ecosystem stressors. *Global Biogeochem. Cycles*, **30**(8), 1224–1243, doi:10.1002/2015GB005338.

Gaire, K., R. Beilin and F. Miller, 2015: Withdrawing, resisting, maintaining and adapting: Food security and vulnerability in Jumla, Nepal. *Reg. Environ. Change*, **15**(8), 1667–1678, doi:10.1007/s10113-014-0724-7.

Gallina, V. et al., 2016: A review of multi-risk methodologies for natural hazards: Consequences and challenges for a climate change impact assessment. *J. Environ. Manage.*, **168**, 123–132, doi:10.1016/j.jenvman.2015.11.011.

Gallo, N.D., D.G. Victor and L.A. Levin, 2017: Ocean commitments under the Paris Agreement. *Nat. Clim. Change*, **7**, 833–838, doi:10.1038/nclimate3422.

Gasser, T. et al., 2015: Negative emissions physically needed to keep global warming below 2°C. *Nat. Commun.*, **6**, 7958, doi:10.1038/ncomms8958.

Gattuso, J.-P. et al., 2014: Ocean Acidification. In: Climate Change 2014: Impacts, Adaptation, and Vulnerability. Part A: Global and Sectoral Aspects. Contribution of Working Group II to the Fifth Assessment Report of the Intergovernmental Panel on Climate Change [Field, C.B., V.R. Barros, D.J. Dokken, K.J. Mach, M.D. Mastrandrea, T.E. Biller, M. Chatterjee, K.L. Ebi, Y.O. Estrada, R.C. Genova, B. Girma, E.S. Kissel, A.N. Levy, S. MacCracken, P.R. Mastrandrea and L.L. White (eds.)]. Cambridge University Press, Cambridge, United Kingdom and New York, NY, USA, 129–131.

Gattuso, J.-P. et al., 2015: Contrasting futures for ocean and society from different anthropogenic CO_2 emissions scenarios. *Science*, **349**(6243), 45–55, doi:10.1126/science.aac4722.

Gattuso, J.-P. et al., 2018: Ocean solutions to address climate change and its effects on marine ecosystems. *Front. Mar. Sci.*, **5**(337), doi:10.3389/fmars.2018.00337.

Gearheard, S.F. et al., 2013: *The Meaning of Ice: People and Sea Ice in Three Arctic Communities*. International Polar Institute Press Hanover, New Hampshire, 366 pp. ISBN: 9780996193856.

Gebbie, G. and P. Huybers, 2012: The mean age of ocean waters inferred from radiocarbon observations: Sensitivity to surface sources and accounting for mixing histories. *J. Phys. Oceanogr.*, **42**(2), 291–305, doi:10.1175/JPO-D-11-043.1.

Gerrard, M.B. and G.E. Wannier, 2013: *Threatened Island Nations: Legal implications of rising seas and a changing climate*. Cambridge University Press, Cambridge, 627 pp. ISBN: 978113919877.

Ghalambor, C.K. et al., 2015: Non-adaptive plasticity potentiates rapid adaptive evolution of gene expression in nature. *Nature*, **525**(7569), 372–375, doi:10.1038/nature15256.

Gimeno, L. et al., 2012: Oceanic and terrestrial sources of continental precipitation. *Rev. Geophys.*, **50**(4), RG4003, doi:10.1029/2012RG000389.

Gleick, P.H., 1996: Water resources. In: *Encyclopedia of Climate and Weather* [Schneider, S.H. (ed.)]. Oxford University Press, New York, pp. 817–823. ISBN 978-0816063505.

Glibert, P.M. et al., 2014: Vulnerability of coastal ecosystems to changes in harmful algal bloom distribution in response to climate change: Projections based on model analysis. *Glob. Chang. Biol.*, **20**(12), 3845–3858, doi:10.1111/gcb.12662.

Gobler, C.J. et al., 2017: Ocean warming since 1982 has expanded the niche of toxic algal blooms in the North Atlantic and North Pacific oceans. Proc. Natl. Acad. Sci. U.S.A., **114**(19), 4975–4980, doi:10.1073/pnas.1619575114.

Goldman, M.J. and E. Lovell, 2017: Indigenous Technical Knowledge. In: *International Encyclopedia of Geography: People, the Earth, Environment and Technology* [Richardson, D., N. Castree, M.F. Goodchild, A. Kobayashi, W. Liu and R.A. Marston (eds.)]. Wiley, pp. 1–4. ISBN 9781118786352.

Golledge, N.R. et al., 2019: Global environmental consequences of twenty-first-century ice-sheet melt. *Nature*, **566**, 65–72, doi:10.1038/s41586-019-0889-9.

Golledge, N.R. et al., 2015: The multi-millennial Antarctic commitment to future sea-level rise. *Nature*, **526**(7573), 421–425, doi:10.1038/nature15706.

Graham, J., T.W. Plumptre and B. Amos, 2003: *Principles for Good Governance in the 21st Century*. IOG Policy Briefs, Institute on Governance Ottawa, 9 pp.

Green, D., J. Billy and A. Tapim, 2010: Indigenous Australians' knowledge of weather and climate. *Clim. Change*, **100**(2), 337–354, doi:10.1007/s10584-010-9803-z.

Grip, K., 2017: International marine environmental governance: A review. *Ambio*, **46**(4), 413–427, doi:10.1007/s13280-016-0847-9.

Grosse, G. et al., 2016: Changing permafrost in a warming world and feedbacks to the Earth system. *Environ. Res. Letters*, **11**(4), 04021, doi:10.1088/1748-9326/11/4/040201.

Gruber, N. et al., 2019: The oceanic sink for anthropogenic CO_2 from 1994 to 2007. *Science*, **363**(6432), 1193–1199, doi:10.1126/science.aau5153.

Grunblatt, J. and L. Alessa, 2017: Role of perception in determining adaptive capacity: communities adapting to environmental change. *Sustain. Sci.*, **12**(1), 3–13, doi:10.1007/s11625-016-0394-0.

Gupta, A.S., N.C. Jourdain, J.N. Brown and D. Monselesan, 2013: Climate drift in the CMIP5 models. *J. Clim.*, **26**(21), 8597–8615, doi:10.1175/JCLI_D_12_00521.1.

Haasnoot, M., J.H. Kwakkel, W.E. Walker and J. ter Maat, 2013: Dynamic adaptive policy pathways: A method for crafting robust decisions for a deeply uncertain world. *Global Environ. Change*, **23**(2), 485–498, doi:10.1016/j.gloenvcha.2012.12.006.

Hall, N. and Å. Persson, 2018: Global climate adaptation governance: Why is it not legally binding? *Eur. J. Int. Relat.*, **24**(3), 540–566, doi:10.1177/1354066117725157.

Halpern, B.S. et al., 2015: Spatial and temporal changes in cumulative human impacts on the world's ocean. *Nat. Commun.*, **6**, 7615, doi:10.1038/ncomms8615.

Halpern, B.S. et al., 2008: A global map of human impact on marine ecosystems. *Science*, **319**(5865), 948–952, doi:10.1126/science.1149345.

Hamilton, M.L. and M. Lubell, 2019: Climate change adaptation, social capital, and the performance of polycentric governance institutions. *Clim. Change*, **152**(3–4), 307–326, doi:10.1007/s10584-019-02380-2.

Hansell, D.A., 2013: Recalcitrant dissolved organic carbon fractions. *Annu. Rev. Mar. Sci.*, **5**(1), 421–445, doi:10.1146/annurev-marine-120710-100757.

Hansen, J. et al., 2016: Ice melt, sea level rise and superstorms: Evidence from paleoclimate data, climate modeling, and modern observations that 2°C global warming could be dangerous. *Atmos. Chem. Phys.*, **16**(6), 3761–3812, doi:10.5194/acp-16-3761-2016.

Hansen, J. et al., 2007: Climate change and trace gases. *Philosophical Transactions of the Royal Society A: Mathematical, Physical and Engineering Sciences*, **365**(1856), 1925–1954, doi:10.1098/rsta.2007.2052.

Hansen, J., M. Sato, P. Kharecha and K. von Schuckmann, 2011: Earth's energy imbalance and implications. *Atmos. Chem. Phys.*, **11**(24), 13421–13449, doi:10.5194/acp-11-13421-2011.

Hansson, S.O. and G. Hirsch Hadorn, 2017: Argument-based decision support for risk analysis. *J. Risk Res.*, **21**(12), 1449–1464, doi:10.1080/13669877.2017.1313767.

Hare, J.A. et al., 2016: A vulnerability assessment of fish and invertebrates to climate change on the Northeast US Continental Shelf. *PloS ONE*, **11**(2), e0146756, doi:10.1371/journal.pone.0146756.

Hastrup, K., A. Mosbech and B. Grønnow, 2018: Introducing the North Water: Histories of exploration, ice dynamics, living resources, and human settlement in the Thule Region. *Ambio*, **47**(2), 162–174, doi:10.1007/s13280-018-1030-2.

Hawkins, E. et al., 2017: Estimating changes in global temperature since the preindustrial period. *Bull. Am. Meteorol. Soc.*, **98**(9), 1841–1856, doi:10.1175/BAMS-D-16-0007.1.

Hawkins, E. and R. Sutton, 2009: The potential to narrow uncertainty in regional climate predictions. *Bull. Am. Meteorol. Soc.*, **90**(8), 1095–1108, doi:10.1175/2009BAMS2607.1.

Hawkins, E. and R. Sutton, 2012: Time of emergence of climate signals. *Geophys. Res. Letters*, **39**(1), L01702, doi:10.1029/2011GL050087.

Hegerl, G.C. et al., 2010: Good practice guidance paper on detection and attribution related to anthropogenic climate change. In: Meeting Report of the Intergovernmental Panel on Climate Change Expert Meeting on Detection and Attribution of Anthropogenic Climate Change [Stocker, T.F., C.B. Field, D. Qin, V. Barros, G.-K. Plattner, M. Tignor, P.M. Midgley and K.L. Ebi (eds.)]. IPCC Working Group I Technical Support Unit, University of Bern, Bern, Switzerland.

Heleniak, T., 2014: *Arctic Populations and Migration*. Arctic Human Development Report: Regional Processes and Global Linkages, Nordic Council of Ministers, Copenhagen. 53–104 pp.

Hijioka, Y. et al., 2014: Asia. In: Climate Change 2014: Impacts, Adaptation, and Vulnerability. Part B: Regional Aspects: Working Group II Contribution to the Fifth Assessment Report of the Intergovernmental Panel on Climate Change [Barros, V.R., C.B. Field, D.J. Dokken, M.D. Mastrandrea, K.J. Mach, T.E. Bilir, M. Chatterjee, K.L. Ebi, Y.O. Estrada, R.C. Genova, B. Girma, E.S. Kissel, A.N. Levy, S. MacCracken, P.R. Mastrandrea and L.L. White (eds.)]. Cambridge University Press, Cambridge, United Kingdom and New York, NY, USA, 1327–1370.

Hinkel, J. et al., 2014: Coastal flood damage and adaptation costs under 21st century sea-level rise. *Proc. Natl. Acad. Sci. U.S.A.*, **111**(9), 3292–3297, doi:10.1073/pnas.1222469111.

Hinners, J., A. Kremp and I. Hense, 2017: Evolution in temperature-dependent phytoplankton traits revealed from a sediment archive: Do reaction norms tell the whole story? *Proc. R. Soc. Lond. B.*, **284**(1864), 20171888, doi:10.1098/rspb.2017.1888.

Hino, M., C.B. Field and K.J. Mach, 2017: Managed retreat as a response to natural hazard risk. *Nat. Clim. Change*, **7**(5), 364–370, doi:10.1038/nclimate3252.

Hiwasaki, L., E. Luna and S.R. Syamsidik, 2014: *Local and indigenous knowledge for community resilience: Hydro-meteorological disaster risk reduction and climate change adaptation in coastal and small island communities*. UNESCO, Jakarta, 60 pp. ISBN: 9786029416114.

Hoegh-Guldberg, O. et al., 2018: Impacts of 1.5°C global warming on natural and human systems. In: Global Warming of 1.5°C. An IPCC Special Report on the impacts of global warming of 1.5°C above pre-industrial levels and related global greenhouse gas emission pathways, in the context of strengthening the global response to the threat of climate change, sustainable development, and efforts to eradicate poverty [Masson-Delmotte, V., P. Zhai, H.O. Pörtner, D. Roberts, J. Skea, P.R. Shukla, A. Pirani, W. Moufouma-Okia, C. Péan, R. Pidcock, S. Connors, J.B.R. Matthews, Y. Chen, X. Zhou, M.I. Gomis, E. Lonnoy, T. Maycock, M. Tignor and T. Waterfield (eds.)]. World Meteorological Organization, Geneva, Switzerland.

Homsy, G.C., Z. Liu and M.E. Warner, 2019: Multilevel governance: Framing the integration of top-down and bottom-up policymaking. *Int. J. Public Adm.*, **42**(7), 572–582.

Horton, B.P., S. Rahmstorf, S.E. Engelhart and A.C. Kemp, 2014: Expert assessment of sea-level rise by AD 2100 and AD 2300. *Quaternary Sci. Rev.*, **84**, 1–6, doi:10.1016/j.quascirev.2013.11.002.

Hou, L., J. Huang and J. Wang, 2017: Early warning information, farmers' perceptions of, and adaptations to drought in China. *Clim. Change*, **141**(2), 197–212, doi:10.1007/s10584-017-1900-9.

Hovelsrud, G.K., B. Poppel, B. Van Oort and J.D. Reist, 2011: Arctic societies, cultures, and peoples in a changing cryosphere. *Ambio*, **40**, 100–110, doi:10.1007/s13280-011-0219-4.

Hu, S. and J. Sprintall, 2016: Interannual variability of the Indonesian Throughflow: The salinity effect. *J. Geophys. Res. Oceans*, **121**(4), 2596–2615, doi:10.1002/2015JC011495.

Huggel, C. et al., 2019: Loss and damage in the mountain cryosphere. *Reg. Environ. Change*, 19(5), 1387–1399, doi:10.1007/s10113-018-1385-8.

Hughes, T.P. et al., 2018: Spatial and temporal patterns of mass bleaching of corals in the Anthropocene. *Science*, **359**(6371), 80–83, doi:10.1126/science.aan8048.

Huntington, H.P., 1992: The Alaska Eskimo Whaling Commission and other cooperative marine mammal management organizations in northern Alaska. *Polar Rec.*, **28**(165), 119–126, doi:10.1017/S0032247400013413.

Huntington, H.P., 2000: Using traditional ecological knowledge in science: methods and applications. *Ecological Applications*, **10**(5), 1270–1274, doi:10.1890/1051-0761(2000)010[1270:UTEKIS]2.0.CO;2.

Huntington, H.P., 2011: Arctic science: The local perspective. *Nature*, **478**(7368), 182–183, doi:10.1038/478182a.

Huntington, H.P. et al., 2017: How small communities respond to environmental change: Patterns from tropical to polar ecosystems. *Ecol. Soc.*, **22**(3), 9.

Huq, S., E. Roberts and A. Fenton, 2013: Loss and damage. *Nat. Clim. Change*, **3**, 947–949, doi:10.1038/nclimate2026.

Huybrechts, P. and J. de Wolde, 1999: The dynamic response of the Greenland and Antarctic ice sheets to multiple-century climatic warming. *J. Clim.*, **12**(8), 2169–2188, doi:10.1175/1520-0442(1999)012<2169:TDROTG>2.0.CO;2.

ICC, 2015: *Inuit Circumpolar Council, Alaskan Inuit Food Security Conceptual Framework: How to assess the Arctic from an Inuit perspective*. Alaska [Available at: https://iccalaska.org/wp-icc/wp-content/uploads/2016/05/Food-Security-Full-Technical-Report.pdf].

ICC. 2018: People of the Ice Bridge: The Future of the Pikialasorsuaq. [Available at: http://pikialasorsuaq.org/en/, accessed 23 March].

ICC, 2018: *Utqiaġvik Declaration*. Inuit Circumpolar Council, Utqiaġvik, Alaska [Available at: www.inuitcircumpolar.com/uploads/3/0/5/4/30542564/2018_utqiagvik_declaration_-_final.pdf].

Inniss, L. and A. Simcock, 2017: *The First Global Integrated Marine Assessment: World Ocean Assessment I*. World Ocean Assessment, Cambridge University Press, Cambridge, United Kingdom, 973 pp. ISBN: 9781108186148.

IPBES, 2018a: The IPBES regional assessment report on biodiversity and ecosystem services for Africa.[Archer, E., L. Dziba, K.J. Mulongoy, M.A. Maoela and M. Walters (eds.)]. Secretariat of the Intergovernmental Science-Policy Platform on Biodiversity and Ecosystem services, Bonn, Germany, 492 pp. ISBN: 9783947851003.

IPBES, 2018b: The IPBES regional assessment report on biodiversity and ecosystem services for Asia and the Pacific.[Senaratna Sellamuttu, S., M. Karki, N. Moriwake and S. Okayasu (eds.)] [Karki, M., S. Senaratna Sellamuttu, S. Okayasu and W. Suzuki (eds.)]. Secretariat of the Intergovernmental Science-Policy Platform on Biodiversity and Ecosystem services, Bonn, Germany, 612 pp. ISBN: 9783947851027.

IPBES, 2018c: The IPBES regional assessment report on biodiversity and ecosystem services for Europe and Central Asia.[Rounsevell, M., M. Fischer, A. Torre-Marin Rando and A. Mader (eds.)]. Secretariat of the Intergovernmental Science-Policy Platform on Biodiversity and Ecosystem services, Bonn, Germany, 892 pp. ISBN: 9783947851034.

IPBES, 2018d: The IPBES regional assessment report on biodiversity and ecosystem services for the Americas. [Rice, J., C.S. Seixas, M.E. Zaccagnini, M. Bedoya-Gaitán and N. Valderrama (eds.)]. Secretariat of the Intergovernmental Science-Policy Platform on Biodiversity and Ecosystem Services, Bonn, Germany, 656 pp. ISBN: 9783947851010.

IPCC, 2000: *Emissions Scenarios* [Nakicenovic, N. and R. Swart (eds.)]. Cambridge University Press, Cambridge, United Kingdom, 570 pp.

IPCC, 2013: Summary for Policymakers. In: Climate Change 2013: The Physical Science Basis. Contribution of Working Group I to the Fifth Assessment Report of the Intergovernmental Panel on Climate Change [Stocker, T.F., D. Qin, G.-K. Platter, M. Tignor, S.K. Allen, J. Boschung, A. Nauels, Y. Xia, V. Bex and P.M. Midgley (eds.)]. Cambridge University Press, Cambridge, United Kingdom and New York, NY, USA.

IPCC, 2014: Summary for Policymakers. In: Climate Change 2014: Impacts, Adaptation, and Vulnerability. Part A: Global and Sectoral Aspects. Contribution of Working Group II to the Fifth Assessment Report of the Intergovernmental Panel on Climate Change [Field, C.B., V. Barros, D.J. Dokken, K.J. Mach, M.D. Mastrandrea, T.E. Bilir, M. Chatterjee, K.L. Ebi, Y.O. Estrada, R.C. Genova, B. Girma, E.S. Kissel, A.N. Levy, S. MacCracken,

M.P.R. and L.L. White (eds.)]. Cambridge University Press, Cambridge, United Kingdom and New York, NY, USA, 1–32.

IPCC, 2018: Summary for Policymakers. In: Global Warming of 1.5°C. An IPCC Special Report on the impacts of global warming of 1.5°C above pre-industrial levels and related global greenhouse gas emission pathways, in the context of strengthening the global response to the threat of climate change, sustainable development, and efforts to eradicate poverty [Masson-Delmotte, V., P. Zhai, H.O. Pörtner, D. Roberts, J. Skea, P.R. Shukla, A. Pirani, W. Moufouma-Okia, C. Péan, R. Pidcock, S. Connors, J.B.R. Matthews, Y. Chen, X. Zhou, M.I. Gomis, E. Lonnoy, T. Maycock, M. Tignor and T. Waterfield (eds.)]. World Meteorological Organization, Geneva, Switzerland, 32 pp.

ITK, 2005: *Putting the human face on climate change: Perspectives from Inuit in Canada*. Inuit Tapiriit Kanatami, Nasivvik Centre for Inuit Health and Changing Environments at Université Laval and the Ajunnginiq Centre at the National Aboriginal Health Organization, Unikkaaqatigiit [Available at: www.itk.ca/wp-content/uploads/2016/07/unikkaaqatigiit01–1.pdf.].

Jamir, O., 2016: Understanding India-Pakistan water politics since the signing of the Indus Water Treaty. *Water Policy*, **18**(5), 1070–1087, doi:https://doi.org/10.2166/wp.2016.185.

Jasanoff, S., 2004: *States of knowledge: The co-production of science and the social order*. Routledge, London and New York, 332 pp. ISBN: 9780415403290.

Jiao, N. et al., 2010: Microbial production of recalcitrant dissolved organic matter: Long-term carbon storage in the global ocean. *Nat. Rev. Microbiol.*, **8**(8), 593–599, doi:10.1038/nrmicro2386.

Johnson, G.C. et al., 2018: Ocean heat content. In: *State of the Climate 2017* [Blunden, J., D.S. Arndt and G. Hartifled (eds.)]. *Bull. Am. Meteorol. Soc*, **99**(8), S72-S77.

Jones, H.P., D.G. Hole and E.S. Zavaleta, 2012: Harnessing nature to help people adapt to climate change. *Nat. Clim. Change*, **2**(7), 504–509, doi:10.1038/nclimate1463.

Jones, J.M. et al., 2016: Assessing recent trends in high-latitude Southern Hemisphere surface climate. *Nat. Clim. Change*, **6**(10), 917–926, doi:10.1038/nclimate3103.

Jordan, A.J. et al., 2015: Emergence of polycentric climate governance and its future prospects. *Nat. Clim. Change*, **5**(11), 977–982, doi:10.1038/nclimate2725.

Jupiter, S.D. et al., 2014: Locally-managed marine areas: Multiple objectives and diverse strategies. *Pac. Conserv. Biol.*, **20**(2), 165–179, doi:10.1071/PC140165.

Kahneman, D. and P. Egan, 2011: *Thinking, fast and slow*. Farrar, Straus and Giroux, New York, 499 pp. ISBN: 9780141033570.

Kainuma, M. et al., 2018: Cross-Chapter Box 1: Scenarios and Pathways. In: Global Warming of 1.5°C. An IPCC Special Report on the impacts of global warming of 1.5°C above pre-industrial levels and related global greenhouse gas emission pathways, in the context of strengthening the global response to the threat of climate change, sustainable development, and efforts to eradicate poverty. [Masson-Delmotte, V., P. Zhai, H.O. Pörtner, D. Roberts, J. Skea, P.R. Shukla, A. Pirani, W. Moufouma-Okia, C. Péan, R. Pidcock, S. Connors, J.B.R. Matthews, Y. Chen, X. Zhou, M.I. Gomis, E. Lonnoy, T. Maycock, M. Tignor and T. Waterfield (eds.)]. World Meteorological Organization, Geneva, Switzerland, 62–64.

Kanna, N. et al., 2018: Upwelling of macronutrients and dissolved inorganic carbon by a subglacial freshwater driven plume in Bowdoin Fjord, northwestern Greenland. *J. Geophys. Res. Biogeosci.*, **123**(5), 1666–1682, doi:10.1029/2017JG004248.

Karim, M.S., 2015: *Prevention of pollution of the marine environment from vessels: The potential and limits of the International Maritime Organisation*. Springer. ISBN: 9783319106083.

Karki, M. et al., 2017: *Knowing our lands and resources: Indigenous and local knowledge and practices related to biodiversity and ecosystem*

services in Asia. Knowledges of Nature 10, UNESCO, Paris, 212 pp. ISBN: 9789231002106.

Karlsson, M. and G.K. Hovelsrud, 2015: Local collective action: Adaptation to coastal erosion in the Monkey River Village, Belize. *Global Environ. Change*, **32**, 96–107, doi:10.1016/j.gloenvcha.2015.03.002.

Kaser, G., M. Großhauser and B. Marzeion, 2010: Contribution potential of glaciers to water availability in different climate regimes. *Proc. Natl. Acad. Sci. U.S.A.*, **107**, 20223–20227, doi:10.1073/pnas.1008162107.

Keeling, C.D. et al., 1976: Atmospheric carbon dioxide variations at Mauna Loa observatory, Hawaii. *Tellus*, **28**(6), 538–551, doi:10.1111/j.2153-3490.1976.tb00701.x.

Keller, K.M., F. Joos and C. Raible, 2014: Time of emergence of trends in ocean biogeochemistry. *Biogeosciences*, **11**(13), 3647–3659, doi:10.5194/bg-11-3647-2014.

Kilroy, G., 2015: A review of the biophysical impacts of climate change in three hotspot regions in Africa and Asia. *Reg. Environ. Change*, **15** (5), 771–782, doi:10.1007/s10113-014-0709-6.

Kirchmeier-Young, M.C., F.W. Zwiers and N.P. Gillett, 2017: Attribution of extreme events in Arctic sea ice extent. *J. Clim.*, **30** (2), 553–571, doi:10.1175/JCLI-D-16-0412.1.

Klein, R.J.T. et al., 2014: Adaptation opportunities, constraints, and limits. In: Climate Change 2014: Impacts, Adaptation, and Vulnerability. Part A: Global and Sectoral Aspects. Contribution of Working Group II to the Fifth Assessment Report of the Intergovernmental Panel on Climate Change [Field, C.B., V.R. Barros, D.J. Dokken, K.J. Mach, M.D. Mastrandrea, T.E. Bilir, M. Chatterjee, K.L. Ebi, Y.O. Estrada, R.C. Genova, B. Girma, E.S. Kissel, A.N. Levy, S. MacCracken, P.R. Mastrandrea and L.L. White (eds.)]. Cambridge University Press, Cambridge, United Kingdom and New York, NY, USA, 899–943.

Kleisner, K.M. et al., 2017: Marine species distribution shifts on the US Northeast Continental Shelf under continued ocean warming. *Prog. Oceanogr.*, **153**, 24–36, doi:10.1016/j.pocean.2017.04.001.

Klenk, N. and K. Meehan, 2015: Climate change and transdisciplinary science: Problematizing the integration imperative. *Environmental Science & Policy*, **54**, 160–167, doi:10.1016/j.envsci.2015.05.017.

Klinsky, S. et al., 2016: Why equity is fundamental in climate change policy research. *Global Environ. Change*, **44**, 170–173, doi:10.1016/j.gloenvcha.2016.08.002.

Knight, F.H., 1921: *Risk, uncertainty and profit*. University of Chicago Press, Chicago, 381 pp. ISBN: 9781614276395.

Knoblauch, C. et al., 2018: Methane production as key to the greenhouse gas budget of thawing permafrost. *Nat. Clim. Change*, **8**(4), 309, doi:10.1038/s41558-018-0095-z.

Knutson, T. et al., 2017: Detection and attribution of climate change. In: *Climate Science Special Report: Fourth National Climate Assessment*, Volume I.U.S. Global Change Research Program, Washington DC, USA, pp.114–132.

Knutti, R.F., T.F. Stocker and A. Timmermann, 2004: Strong hemispheric coupling of glacial climate through freshwater discharge and ocean circulation. *Nature*, **430**(7002), 851–856, doi:10.1038/nature02786.

Kofinas, G. et al., 2013: Adaptive and transformative capacity. In: *Arctic Resilience Interim Report to the Arctic Council*. Stockholm Environment Institute and Stockholm Resilience Centre, Stockholm, pp. 71–91. ISBN: 9789186125424.

Koivurova, T., 2016: Arctic resources: Exploitation of natural resources in the arctic from the perspective of international law. In: *Research Handbook on International Law and Natural Resources* [Morgera, E. and K. Kulovesi (eds.)]. Edward Elgar Publishing, pp. 349–367. ISBN: 9781783478323.

Kokelj, S.V. et al., 2015: Increased precipitation drives mega slump development and destabilization of ice-rich permafrost terrain, northwestern Canada. *Glob. Planet. Change*, **129**, 56–68, doi:10.1016/j.gloplacha.2015.02.008.

Kopp, R.E., R.L. Shwom, G. Wagner and J. Yuan, 2016: Tipping elements and climate–economic shocks: Pathways toward integrated assessment. *Earth's Future*, **4**(8), 346–372, doi:10.1002/2016EF000362.

Kriegler, E. et al., 2016: Fossil-fueled development (SSP5): an energy and resource intensive scenario for the 21st century. *Global Environ. Change*, **42**, 297–315, doi:10.1016/j.gloenvcha.2016.05.015.

Kulonen, K., Adler, C., Bracher, C., and S. Wymann von Dach, 2019: Spatial context matters for monitoring and reporting on SDGs: Reflections based on research in mountain regions. *GAIA*, **28**(2), 90–94(5), doi.org/10.14512/gaia.28.2.5.

Kummu, M. et al., 2016: Over the hills and further away from coast: global geospatial patterns of human and environment over the 20th–21st centuries. *Environ. Res. Letters*, **11**(3), 034010, doi:10.1088/1748-9326/11/3/034010.

Lagerloef, G., R. Schmitt, J. Schanze and H.-Y. Kao, 2010: The ocean and the global water cycle. *Oceanography*, **23**(4), 82–93, doi:10.5670/oceanog.2010.07.

Lavrillier, A. and S. Gabyshev, 2017: *An Arctic indigenous knowledge system of landscape, climate, and human interactions*. Evenki Reindeer Herders and Hunters, Studies in Social and Cultural Anthropology, SEC Publications, Kulturstiftung Sibirien, Fürstenberg/Havel, Germany, 467 pp. ISBN: 9783942883313.

Lavrillier, A. and S. Gabyshev, 2018: An emic science of climate: A reindeer evenki environmental knowledge and the notion of an extreme process of change. In: *Human-environment relationships in Siberia and Northeast China: Skills, rituals, mobility and politics among the Tungus peoples* [Lavrillier, A., A. Dumont and D. Brandisauskas (eds.)]. EMSCAT, 49.

Lawrence, J. et al., 2018: National guidance for adapting to coastal hazards and sea-level rise: Anticipating change, when and how to change pathway. *Environmental Science & Policy*, **82**, 100–107, doi:10.1016/j.envsci.2018.01.012.

Le Quéré, C. et al., 2018: Global Carbon Budget 2018. *Earth Syst. Sci. Data*, **10**(4), doi:10.5194/essd-10-2141-2018.

LeBlanc, D., C. Freire and M. Vierros, 2017: *Mapping the linkages between oceans and other Sustainable Development Goals*. DESA Working Paper No. 149, ST/ESA/2017/DWP/149 [Available at: www.un.org/en/development/desa/pa].

Lee, T.M. et al., 2015: Predictors of public climate change awareness and risk perception around the world. *Nat. Clim. Change*, **5**(11), 1014–1020, doi:10.1038/nclimate2728.

Lele, S., V. Srinivasan, B.K. Thomas and P. Jamwal, 2018: Adapting to climate change in rapidly urbanizing river basins: Insights from a multiple-concerns, multiple-stressors, and multi-level approach. *Water Int.*, **43**(2), 281–304, doi:10.1080/02508060.2017.1416442.

Lellouche, J. et al., 2018: Recent updates on the Copernicus Marine Service global ocean monitoring and forecasting real-time 1/12 high resolution system. *Ocean Sci.*, **14**(5), 1093–1126, doi:10.5194/os-2018-15.

Lempert, R.J., S.W. Popper and S.C. Bankes, 2003: *Shaping the next one hundred years: New methods for quantitative, long-term policy analysis*. RAND Corporation, Santa Monica, CA, 186 pp. ISBN: 9780833032751.

Lempert, R.J., S.W. Popper and S.C. Bankes, 2010: Robust Decision Making: Coping with Uncertainty. *The Futurist*, **44**(1), 47–48.

Lenton, T.M. et al., 2008: Tipping elements in the Earth's climate system. *Proc. Natl. Acad. Sci. U.S.A.*, **105**(6), 1786–1793, doi:10.1073/pnas.0705414105.

Lin, P.-F. et al., 2016: Long-term surface air temperature trend and the possible impact on historical warming in CMIP5 models. *Atmos. Ocean Sci. Letters*, **9**(3), 153–161, doi:10.1080/16742834.2016.1159911.

Lindgren, A., G. Hugelius and P. Kuhry, 2018: Extensive loss of past permafrost carbon but a net accumulation into present-day soils. *Nature*, **560** (7717), 219–222, doi:10.1038/s41586-018-0371-0.

Liu, J. et al., 2016: Has Arctic sea ice loss contributed to increased surface melting of the Greenland ice sheet? *J. Clim.*, **29**(9), 3373–3386, doi:10.1175/JCLI-D-15-0391.1.

Lüthi, D. et al., 2008: High-resolution carbon dioxide concentration record 650,000–800,000 years before present. *Nature*, **453**, 379–382, doi:10.1038/nature06949.

Mach, K.J., M.D. Mastrandrea, T.E. Bilir and C.B. Field, 2016: Understanding and responding to danger from climate change: The role of key risks in the IPCC AR5. *Clim. Change*, **136**(3–4), 427–444, doi:10.1007/s10584-016-1645-x.

Mach, K.J., M.D. Mastrandrea, P.T. Freeman and C.B. Field, 2017: Unleashing expert judgment in assessment. *Global Environ. Change*, **44**, 1–14, doi:10.1016/j.gloenvcha.2017.02.005.

Magnan, A.K. and T. Ribera, 2016: Global adaptation after Paris. *Science*, **352**(6291), 1280–1282, doi:10.1126/science.aaf5002.

Maier, H.R. et al., 2016: An uncertain future, deep uncertainty, scenarios, robustness and adaptation: How do they fit together? *Environ. Model. Soft.*, **81**, 154–164, doi:10.1016/j.envsoft.2016.03.014.

Maldonado, J. et al., 2016: Engagement with indigenous peoples and honoring traditional knowledge systems. *Clim. Change*, **135**(1), 111–126, doi:10.1007/s10584-015-1535-7.

Marchau, V.A.W.J., W.E. Walker, P.J.T.M. Bloemen and S.W. Popper, Eds., 2019a: Decision making under deep uncertainty: from theory to practice. Springer International Publishing, Cham, Switzerland, 400 pp. ISBN: 9783030052522.

Marchau, V.A.W.J., W.E. Walker, P.J.T.M. Bloemen and S.W. Popper, 2019b: Introduction. In: *Decision making under deep uncertainty: From theory to practice* [Marchau, V.A.W.J., W.E. Walker, P.J.T.M. Bloemen and S.W. Popper (eds.)]. Springer International Publishing, Cham, Switzerland, pp. 1–20. ISBN: 9783030052522.

Markham, A., E. Osipova, K. Lafrenz Samuels and A. Caldas, 2016: *World heritage and tourism in a changing climate*. United Nations Environment Programme, Nairobi, Kenya and United Nations Educational, Scientific and Cultural Organization, Paris, France, 104 pp. ISBN: 9789231001529.

Maru, Y.T. et al., 2014: A linked vulnerability and resilience framework for adaptation pathways in remote disadvantaged communities. *Global Environ. Change*, **28**, 337–350, doi:10.1016/j.gloenvcha.2013.12.007.

Masson-Delmotte, V. et al., 2013: Information from paleoclimate archives. In: Climate Change 2013: The Physical Science Basis. Contribution of Working Group I to the Fifth Assessment Report of the Intergovernmental Panel on Climate Change [Stocker, T.F., D. Qin, G.-K. Plattner, M. Tignor, S.K. Allen, J. Boschung, A. Nauels, Y. Xia, V. Bex and P.M. Midgley (eds.)]. Cambridge University Press, Cambridge, United Kingdom and New York, NY, USA, 383–464.

Mastrandrea, M.C. et al., 2010: *Guidance Note for Lead Authors of the IPCC Fifth Assessment Report on Consistent Treatment of Uncertainties*. Intergovernmental Panel on Climate Change (IPCC) [Available at: www.ipcc.ch].

Matsumoto, K., 2007: Radiocarbon-based circulation age of the world oceans. *J. Geophys. Res. Oceans*, **112**, C09004, doi:10.1029/2007JC004095.

Mayer, B. and F. Crépeau, 2017: *Research Handbook on Climate Change, Migration and the Law*. Edward Elgar Publishing, Cheltenham, UK, Northampton, MA, USA. ISBN: 9781785366581.

Mazloff, M.R., P. Heimbach and C. Wunsch, 2010: An eddy-permitting Southern Ocean state estimate. *J. Phys. Oceanogr.*, **40**(5), 880–899, doi:10.1175/2009JPO4236.1.

McGuire, A.D. et al., 2018: Dependence of the evolution of carbon dynamics in the northern permafrost region on the trajectory of climate change. *Proc. Natl. Acad. Sci. U.S.A.S*, **115**(15, 3882–3887, doi:10.1073/pnas.1719903115.

McLeman, R.A. and L.M. Hunter, 2010: Migration in the context of vulnerability and adaptation to climate change: Insights from analogues. *WIRes. Clim. Change*, **1**(3), 450–461, doi:10.1002/wcc.51.

McNamara, K.E. and R. Westoby, 2011: Solastalgia and the gendered nature of climate change: An example from Erub Island, Torres Strait. *EcoHealth*, **8**(2), 233–236, doi:10.1007/s10393-011-0698-6.

McQuatters-Gollop, A. et al., 2015: The Continuous Plankton Recorder survey: How can long-term phytoplankton datasets contribute to the assessment of Good Environmental Status? *Estuarine, Coastal and Shelf Science*, **162**, 88–97, doi:10.1016/j.ecss.2015.05.010.

MEA, 2005: *Ecosystems and human well-being: Synthesis*. World Resources Institute, New Island Press, Washington, DC, 155 pp. ISBN: 9781597260404.

Mechler, R. et al., 2018: *Loss and damage from climate change*. Springer Berlin Heidelberg, New York NY. 557 pp. ISBN: 9783319720258.

Mekonnen, Z., H. Kassa, T. Woldeamanuel and Z. Asfaw, 2019: Analysis of observed and perceived climate change and variability in Arsi Negele District, Ethiopia. *Environ. Dev. Sustain.*, **20**(3), 1191–1212, doi:10.1007/s10668-017-9934-8.

Merkens, J.L., L. Reimann, J. Hinkel and A.T. Vafeidis, 2016: Gridded population projections for the coastal zone under the Shared Socioeconomic Pathways. *Glob. Planet. Change*, **145**, 57–66, doi:10.1016/j.gloplacha.2016.08.009.

Merrey, D.J. et al., 2018: Evolving high altitude livelihoods and climate change: A study from Rasuwa District, Nepal. *Food Security*, **10**(4), 1055–1071, doi:10.1007/s12571-018-0827-y.

Mewhirter, J., M. Lubell and R. Berardo, 2018: Institutional externalities and actor performance in polycentric governance systems. *Environmental Policy and Governance*, **28**(4), 295–307, doi:10.1002/eet.1816.

Milfont, T.L., M.S. Wilson and C.G. Sibley, 2017: The public's belief in climate change and its human cause are increasing over time. *PLoS ONE*, **12** (3), e0174246, doi:10.1371/journal.pone.0174246.

Miller, D.D. et al., 2018: Adaptation strategies to climate change in marine systems. *Glob. Chang. Biol.*, **24**(1), e1-e14, doi:10.1111/gcb.13829.

Miloslavich, P. et al., 2018: Essential ocean variables for global sustained observations of biodiversity and ecosystem changes. *Glob. Chang. Biol.*, **24**(6), 2416–2433, doi:10.1111/gcb.14108.

Mistry, J. and A. Berardi, 2016: Bridging indigenous and scientific knowledge. *Science*, **352**(6291), 1274–1275, doi:10.1126/science.aaf1160.

Molden, D.J., A.B. Shrestha, S. Nepal and W.W. Immerzeel, 2016: Downstream implications of climate change in the Himalayas. In: *Water security, climate change and sustainable development*. Springer, Singapore, pp. 65–82. ISBN: 9789812879769.

Moore, M.-L. et al., 2014: Studying the complexity of change: Toward an analytical framework for understanding deliberate social-ecological transformations. *Ecol. Soc.*, **19**(4), 54, doi:10.5751/ES-06966-190454.

Mora, C. et al., 2013: The projected timing of climate departure from recent variability. *Nature*, **502**(7470), 183–187, doi:10.1038/nature12540.

Moser, S.C., 2016: Reflections on climate change communication research and practice in the second decade of the 21st century: What more is there to say? *WIRes Clim. Change*, **7**(3), 345–369, doi:10.1002/wcc.403.

Moss, R. et al., 2008: *Towards new scenarios for analysis of emissions, climate change, impacts and response strategies*. IPCC Expert Meeting Report, 25 pp.

Moss, R.H. et al., 2010: The next generation of scenarios for climate change research and assessment. *Nature*, **463**, 747–756, doi:10.1038/nature08823 www.nature.com/articles/nature08823#supplementary-information.

Motta, M., 2018: The enduring effect of scientific interest on trust in climate scientists in the United States. *Nat. Clim. Change*, **8**(6), 485–488, doi:10.1038/s41558-018-0126-9.

Mourey, J. and L. Ravanel, 2017: Evolution of Access Routes to High Mountain Refuges of the Mer de Glace Basin (Mont Blanc Massif, France): An Example of Adapting to Climate Change Effects in the Alpine High Mountains. *Journal of Alpine Research | Revue de géographie alpine*, **105**(4), doi:10.4000/rga.3790.

Murakami, H. et al., 2018: Dominant effect of relative tropical Atlantic warming on major hurricane occurrence. *Science*, **362**(6416), 794–799, doi:10.1126/science.aat6711.

Myers, T.A., M.C. Nisbet, E.W. Maibach and A.A. Leiserowitz, 2012: A public health frame arouses hopeful emotions about climate change. *Clim. Change*, **113**(3–4), 1105–1112, doi:10.1007/s10584-012-0513-6.

Myhre, G. et al., 2013: Anthropogenic and natural radiative forcing. In: Climate Change 2013: The Physical Science Basis. Contribution of Working Group I to the Fifth Assessment Report of the Intergovernmental Panel on Climate Change [Stocker, T.F., D. Qin, G.-K. Plattner, M. Tignor, S.K. Allen, J. Boschung, A. Nauels, Y. Xia, V. Bex and P.M. Midgley (eds.)]. Cambridge University Press, Cambridge, United Kingdom and New York, NY, USA, 658–740.

Mysiak, J. et al., 2018: Brief communication: Strengthening coherence between climate change adaptation and disaster risk reduction. *Nat. Hazar Earth Sys. Sci.*, **18**(11), 3137–3143, doi:10.5194/nhess-18-3137-2018.

Nadeem, S., I. Elahi, A. Hadi and I. Uddin, 2012: *Traditional knowledge and local institutions support adaptation to water-induced hazards in Chitral, Pakistan*. ICIMOD, Kathmandu, 51 pp.

Nakashima, D. et al., 2012: *Weathering uncertainty: Traditional knowledge for climate change assessment and adaptation*. United Nations Educational, Scientific and Cultural Organization and United Nations University, Paris/Darwin. 120 pp. ISBN: 9789230010683.

Nalau, J. et al., 2018: The role of indigenous and traditional knowledge in ecosystem-based adaptation: A review of the literature and case studies from the Pacific Islands. *Weather Clim. Soc.*, **10**(4), 851–865, doi:10.1175/WCAS-D-18-0032.1.

National Academies of Sciences, Engineering, and Medicine, 2019: *Negative Emissions Technologies and Reliable Sequestration: A Research Agenda*. The National Academies Press, Washington, DC. ISBN: 9780309484527.

Neef, A. et al., 2018: Climate adaptation strategies in Fiji: The role of social norms and cultural values. *World Dev.*, **107**, 125–137, doi:10.1016/j.worlddev.2018.02.029.

Nickels, S., C. Furgal, M. Buell and H. Moquin, 2005: *Putting the human face on climate change: Perspectives from Inuit in Canada*. Joint publication of Inuit Tapiriit Kanatami, Nasivvik Centre for Inuit Health and Changing Environments at Université Laval and the Ajunnginiq Centre at the National Aboriginal Health Organization, Ottawa, 123 pp. ISBN: 9780969977414.

NOAA. 2018: National Centers for Environmental Information, 2018: State of the Climate: Hurricanes and Tropical Storms for Annual 2017. [Available at: www.ncdc.noaa.gov/sotc/tropical-cyclones/201713, accessed September 13].

North, D.C., 1990: A transaction cost theory of politics. *J. Theor. Polit.*, **2**(4), 355–367, doi:10.1177/0951692890002004001.

Nunn, P.D. and N.J. Reid, 2016: Aboriginal memories of inundation of the Australian coast dating from more than 7000 years ago. *Aust. Geogr.*, **47**(1), 11–47, doi:10.1080/00049182.2015.1077539.

Nüsser, M. and S. Schmidt, 2017: Nanga Parbat revisited: Evolution and dynamics of sociohydrological interactions in the Northwestern Himalaya. *Ann. Am. Assoc. Geogr.*, **107**(2), 403–415, doi:10.1080/24694452.2016.1235495.

O'Neill, B.C. et al., 2017: IPCC reasons for concern regarding climate change risks. *Nat. Clim. Change*, **7**(1), 28–37, doi:10.1038/nclimate3179.

O'Neill, B.C. et al., 2016: The scenario model intercomparison project (ScenarioMIP) for CMIP6. *Geosci. Model Dev.*, **9**(9), 3461–3482, doi:10.5194/gmd-9-3461-2016.

O'Neill, S.J. and S. Graham, 2016: (En)visioning place-based adaptation to sea-level rise. *Geography and Environment*, **3**(2), 1–16, doi:10.1002/geo2.28.

O'Neill, B.C. et al., 2017: The roads ahead: Narratives for shared socioeconomic pathways describing world futures in the 21st century. *Global Environ. Change*, **42**, 169–180, doi:10.1016/j.gloenvcha.2015.01.004.

Obermeister, N., 2017: From dichotomy to duality: Addressing interdisciplinary epistemological barriers to inclusive knowledge governance in global environmental assessments. *Environmental Science & Policy*, **68**, 80–86, doi:10.1016/j.envsci.2016.11.010.

Ogunbode, C.A. et al., 2018: The resilience paradox: Flooding experience, coping and climate change mitigation intentions. *Clim. Policy*, **19**(6), 703–715, doi:10.1080/14693062.2018.1560242.

Olsen, A. et al., 2016: The Global Ocean Data Analysis Project version 2 (GLODAPv2) – an internally consistent data product for the world ocean. *Earth Syst. Sci. Data*, **8**(2), 297–323, doi:10.5194/essd-8-297-2016.

Olsson, L. et al., 2017: Why resilience is unappealing to social science: Theoretical and empirical investigations of the scientific use of resilience. In: *The Routledge Handbook of International Resilience* [Chandler, D. and J. Coaffe (eds.)]. Routledge. 402 pp. ISBN: 9781138784321.

Olsson, L. et al., 2014: Livelihoods and poverty. In: Climate Change 2014: Impacts, Adaptation, and Vulnerability. Part A: Global and Sectoral Aspects. Contribution of Working Group II to the Fifth Assessment Report of the Intergovernmental Panel on Climate Change [Field, C.B., V.R. Barros, D.J. Dokken, K.J. Mach, M.D. Mastrandrea, T.E. Biller, M. Chatterjee, K.L. Ebi, Y.O. Estrada, R.C. Genova, B. Girma, E.S. Kissel, A.N. Levy, S. MacCracken, P.R. Mastrandrea and L.L. White (eds.)]. Cambridge University Press, Cambridge, United Kingdom and New York, NY, USA, 793–832.

Oppenheimer, M. et al., 2014: Emergent Risks and Key Vulnerabilities. In: Climate Change 2014: Impacts, Adaptation, and Vulnerability. Part A: Global and Sectoral Aspects. Contribution of Working Group II to the Fifth Assessment Report of the Intergovernmental Panel on Climate Change [Field, C.B., V.R. Barros, D.J. Dokken, K.J. Mach, M.D. Mastrandrea, T.E. Biller, M. Chatterjee, K.L. Ebi, Y.O. Estrada, R.C. Genova, B. Girma, E.S. Kissel, A.N. Levy, S. MacCracken, P.R. Mastrandrea and L.L. White (eds.)]. Cambridge University Press, Cambridge, United Kingdom and New York, NY, USA, 1039–1099.

Oschlies, A., P. Brandt, L. Stramma and S. Schmidtko, 2018: Drivers and mechanisms of ocean deoxygenation. *Nat. Geosci.*, **11**, 467–473, doi:10.1038/s41561-018-0152-2.

Ostrom, E., 2005: *Understanding institutional diversity*. Princeton University Press, Princeton, NJ, 355 pp. ISBN: 9780691122380.

Ostrom, E., 2010: Polycentric systems for coping with collective action and global environmental change. *Global Environ. Change*, **20**(4), 550–557, doi:10.1016/j.gloenvcha.2010.07.004.

Otto, F.E., 2017: Attribution of weather and climate events. *Annu. Rev. Environ. Resour.*, **42**, 627–646, doi:10.1146/annurev-environ-102016-060847.

Paavola, J., 2007: Institutions and environmental governance: A reconceptualization. *Ecol. Econ.*, **63**(1), 93–103, doi:10.1016/j.ecolecon.2006.09.026.

Pacific Community, 2016: *Framework for resilient development in the Pacific: An integrated approach to address climate change and disaster risk management (FRDP)*. Pacific Community, Suva, Fiji, 32 pp. [Available at: http://gsd.spc.int/frdp/assets/FRDP_2016_Resilient_Dev_pacific.pdf].

Padmanabhan, M., 2017: *Transdisciplinary Research and Sustainability: Collaboration, Innovation and Transformation*. Routledge. 443 pp. ISBN: 9781138216402.

PAGES2K Consortium, 2017: A global multiproxy database for temperature reconstructions of the Common Era. *Scientific Data*, **4**, 170088, doi:10.1038/sdata.2017.88.

Pahl-Wostl, C., 2019: The role of governance modes and meta-governance in the transformation towards sustainable water governance. *Environmental Science & Policy*, **91**, 6–16, doi:10.1016/j.envsci.2018.10.008.

Pahl-Wostl, C., A. Bhaduri and A. Bruns, 2018: Editorial special issue: The Nexus of water, energy and food – An environmental governance perspective. *Environmental Science & Policy*, **90**, 161–163, doi:10.1016/j.envsci.2018.06.021.

Palmer, M.D. and D.J. McNeall, 2014: Internal variability of Earth's energy budget simulated by CMIP5 climate models. *Environ. Res. Letters*, **9**(3), 034016, doi:10.1088/1748-9326/9/3/034016.

Parmesan, C., T.L. Root and M.R. Willig, 2000: Impacts of extreme weather and climate on terrestrial biota. *Bull. Am. Meteorol. Soc.*, **81**(3), 443–450, doi:10.1175/1520-0477(2000)081<0443:IOEWAC>2.3.CO;2.

Parveen, S., M. Winiger, S. Schmidt and M. Nüsser, 2015: Irrigation in Upper Hunza: Evolution of socio-hydrological interactions in the Karakoram, northern Pakistan. *Erdkunde*, **69**(1), 69–85, doi:10.3112/erdkunde.2015.01.05.

Pelling, M., K. O'Brien and D. Matyas, 2015: Adaptation and transformation. *Clim. Change*, **133**, 113–127, doi:10.1007/s10584-014-1303-0.

Pespeni, M.H. et al., 2013: Evolutionary change during experimental ocean acidification. *Proc. Natl. Acad. Sci. U.S.A.*, **110**(17), 6937–6942, doi:10.1073/pnas.1220673110.

Peters, G.P. et al., 2012: The challenge to keep global warming below 2°C. *Nat. Clim. Change*, **3**(1), 4–6, doi:10.1038/nclimate1783.

Petersen, A.C., 2011: Climate simulation, uncertainty, and policy advice–the case of the IPCC. In: *Climate Change and Policy: The Calculability of Climate Change and the Challenge of Uncertainty* [Gramelsberger, G. and J. Feichter (eds.)]. Springer, Berlin, Heidelberg, pp. 91–111. ISBN: 9783642177002.

Philip, T., 2018: Climate change displacement and migration: An analysis of the current international legal regime's deficiency, proposed solutions and a way forward for Australia. *Melbourne Journal of International Law*, **19**, 27 pp.

Pierre, J. and G. Peters, 2000: *Governance, Politics, and the State*. St. Martin's Press. 240 pp. ISBN: 9780333718483.

Pincus, R. and S.H. Ali, 2015: *Diplomacy on Ice: Energy and the Environment in the Arctic and Antarctic*. Yale University Press, 304 pp. ISBN: 9780300205169.

Pistone, K., I. Eisenman and V. Ramanathan, 2014: Observational determination of albedo decrease caused by vanishing Arctic sea ice. *Proc. Natl. Acad. Sci. U.S.A.*, **111**(9), 3322–3326, doi:10.1073/pnas.1318201111.

Pörtner, H.-O. et al., 2014: Ocean Systems. In: Climate Change 2014: Impacts, Adaptation, and Vulnerability. Part A: Global and Sectoral Aspects. Contribution of Working Group II to the Fifth Assessment Report of the Intergovernmental Panel on Climate Change [Field, C.B., V.R. Barros, D.J. Dokken, K.J. Mach, M.D. Mastrandrea, T.E. Biller, M. Chatterjee, K.L. Ebi, Y.O. Estrada, R.C. Genova, B. Girma, E.S. Kissel, A.N. Levy, S. MacCracken, P.R. Mastrandrea and L.L. White (eds.)]. Cambridge University Press, Cambridge, United Kingdom and New York, NY, USA, 411–484.

Pratchett, M.S. et al., 2011: Vulnerability of coastal fisheries in the tropical Pacific to climate change. In: *Vulnerability of Tropical Pacific Fisheries and Aquaculture to Climate Change* [Bell, J.D., J.E. Johnson and A.J. Hobday (eds.)]. Secretariat of the Pacific Community, Noumea, pp. 493–576. ISBN: 9789820005082.

Prescott, S.L. and A.C. Logan, 2018: Larger than life: Injecting hope into the planetary health paradigm. *Challenges*, **9**(13), 1–27, doi:10.3390/challe9010013.

Przeslawski, R., M. Byrne and C. Mellin, 2015: A review and meta-analysis of the effects of multiple abiotic stressors on marine embryos and larvae. *Glob. Chang. Biol.*, **21** (6), 2122–2140, doi:10.1111/gcb.12833.

PSMSL. Permanent Service for Mean Sea Level. [Available at: www.psmsl.org].

Purich, A. et al., 2018: Impacts of broad-scale surface freshening of the Southern Ocean in a coupled climate model. *J. Clim.*, **31**(7), 2613–2632, doi:10.1175/JCLI-D-17-0092.1.

Qian, H., R. Joseh and N. Zeng, 2010: Enhanced terrestrial carbon uptake in the Northern High Latitudes in the 21st century from the Coupled Carbon Cycle Climate Model Intercomparison Project model projections. *Glob. Chang. Biol.*, **16**(2), 641–656, doi:10.1111/j.1365-2486.2009.01989.x.

Quinlan, A.E., M. Berbés-Blázquez, L.J. Haider and G.D. Peterson, 2016: Measuring and assessing resilience: Broadening understanding through multiple disciplinary perspectives. *J. Appl. Ecol.*, **53**(3), 677–687, doi:10.1111/1365-2664.12550.

Ramm, T.D., C.J. White, A.H.C. Chan and C.S. Watson, 2017: A review of methodologies applied in Australian practice to evaulate long-term coastal adaptation options. *Clim. Risk Manage.*, **17**, 35–51, doi:10.1016/j.crm.2017.06.005.

Rasul, G., 2014: Food, water, and energy security in South Asia: A nexus perspective from the Hindu Kush Himalayan region. *Environmental Science & Policy*, **39**, 35–48, doi:10.1016/j.envsci.2014.01.010.

Raup, B.H., L.M. Andreassen, T. Bolch and S. Bevan, 2015: Remote Sensing of Glaciers. In: *Remote Sensing of the Cryosphere* [Tedesco, M. (ed.)]. Wiley Blackwell, pp. 123–156. ISBN: 9781118368855.

Redman, C.L., 1999: *Human Impact on Ancient Environments*. University of Arizona Press, Tuscon, 256 pp. ISBN: 9780816519637.

Reid, J., 2013: Interrogating the neoliberal biopolitics of the sustainable development-resilience nexus. *Int. Political Sociol.*, **7**(4), 353–367, doi:10.1111/ips.12028.

Renaud, F.G. et al., 2015: Resilience and shifts in agro-ecosystems facing increasing sea-level rise and salinity intrusion in Ben Tre Province, Mekong Delta. *Clim. Change*, **133**(1), 69–84, doi:10.1007/s10584-014-1113-4.

Renaud, F.G., U. Nehren, K. Sudmeier-Rieux and M. Estrella, 2016: Developments and Opportunities for Ecosystem-Based Disaster Risk Reduction and Climate Change Adaptation. In: *Ecosystem-Based Disaster Risk Reduction and Adaptation in Practice* [Renaud, F.G., K. Sudmeier-Rieux, M. Estrella and U. Nehren (eds.)]. Springer International Publishing, Switzerland, pp.1–20. ISBN: 9783319436319.

Rhein, M. et al., 2013: Observations: Ocean. In: Climate Change 2013: The Physical Science Basis. Contribution of Working Group I to the Fifth Assessment Report of the Intergovernmental Panel on Climate Change [Stocker, T.F., D. Qin, G.-K. Plattner, M. Tignor, S.K. Allen, J. Boschung, A. Nauels, Y. Xia, V.B. And and P.M. Midgley (eds.)]. Cambridge University Press, Cambridge, United Kingdom and New York, NY, USA, pp. 255–315.

Rhines, P., S. Häkkinen and S.A. Josey, 2008: Is oceanic heat transport significant in the climate system? In: *Arctic-Subarctic Ocean Fluxes* [Dickson, R.R., J. Meincke and P. Rhines (eds.)]. Springer, Dordrecht, pp. 87–109. ISBN: 9781402067730.

Riahi, K. et al., 2011: RCP 8.5 – a scenario of comparatively high greenhouse gas emissions. *Clim. Change*, **109**, doi:10.1007/s10584-011-0149-y.

Riahi, K. et al., 2017: The Shared Socioeconomic Pathways and their energy, land use, and greenhouse gas emissions implications: An overview. *Global Environ. Change*, **42**, 153–168, doi:10.1016/j.gloenvcha.2016.05.009.

Rice, W.R., 2002: Evolution of sex: Experimental tests of the adaptive significance of sexual recombination. *Nat. Rev. Genet.*, **3**(4), 241–251, doi:10.1038/nrg760.

Rigg, J. and K. Oven, 2015: Building liberal resilience? A critical review from developing rural Asia. *Global Environ. Change*, **32**, 175–186, doi:10.1016/j.gloenvcha.2015.03.007.

Riser, S.C. et al., 2016: Fifteen years of ocean observations with the global Argo array. *Nat. Clim. Change*, **6**(2), 145–153, doi:10.1038/nclimate2872.

Roberts, C.M. et al., 2017: Marine reserves can mitigate and promote adaptation to climate change. *Proc. Natl. Acad. Sci. U.S.A.*, **114**, 6167–6175, doi:10.1073/pnas.1701262114.

Rodgers, K., J. Lin and T. Frölicher, 2015: Emergence of multiple ocean ecosystem drivers in a large ensemble suite with an Earth system model. *Biogeosciences*, **12**(11), 3301–3320, doi:10.5194/bg-12-3301-2015.

Roggero, M., A. Bisaro and S. Villamayor-Tomas, 2017: Institutions in the climate adaptation literature: A systematic literature review through the lens of the Institutional Analysis and Development framework. *Journal of Institutional Economics*, **14**(3), 423–448, doi:10.1017/S1744137417000376.

Roggero, M. et al., 2018: Introduction to the special issue on adapting institutions to climate change. *Journal of Institutional Economics*, **14**(3), 409–422, doi:10.1017/S1744137417000649.

Rohe, J.R., S. Aswani, A. Schlüter and S.C. Ferse, 2017: Multiple drivers of local (non-) compliance in community-ased marine resource management: case studies from the South Pacific. *Front. Mar. Sci.*, **4**, 172, doi:10.3389/fmars.2017.00172.

Romero-Lankao, P. et al., 2014: North America. In: Climate Change 2014: Impacts, Adaptation, and Vulnerability. Part B: Regional Aspects: Working Group II Contribution to the Fifth Assessment Report of the Intergovernmental Panel on Climate Change [Field, C.B., V.R. Barros, D.J. Dokken, K.J. Mach, M.D. Mastrandrea, T.E. Biller, M. Chatterjee, K.L. Ebi, Y.O. Estrada, R.C. Genova, B. Girma, E.S. Kissel, A.N. Levy, S. MacCracken, P.R. Mastrandrea and L.L. White (eds.)]. Cambridge University Press, Cambridge, United Kingdom and New York, NY, USA, 1439–1498.

Roncoli, C., T. Crane and B. Orlove, 2009: Fielding Climate Change in Cultural Anthropology. In: *Anthropology and Climate Change: From Encounters to Actions* [Crate, S. and M. Nuttall (eds.)]. Left Coast Press, Walnut Creek, CA, pp. 87–115. ISBN: 9781598743340.

Roué, M., N. Césard, N.C. Adou Yao and A. Oteng-Yeboah, 2017: *Knowing our lands and resources: Indigenous and local knowledge of biodiversity and ecosystem services in Africa*. Knowledges of Nature 8, UNESCO, Paris. 156 pp. ISBN: 9789231002083.

Roué, M. and Z. Molnar, 2017: *Knowing our lands and resources: indigenous and local knowledge of biodiversity and ecosystem services in Europe and Central Asia*. Knowledges of Nature 9, UNESCO, Paris. 148 pp. ISBN: 9789231002106.

Roy, J., P. Tschakert, H. Waisman and S.A. Halim, 2018: Sustainable development, poverty eradication and reducing inequalities. In: Global Warming of 1.5°C. An IPCC Special Report on the impacts of global warming of 1.5°C above pre-industrial levels and related global greenhouse gas emission pathways, in the context of strengthening the global response to the threat of climate change, sustainable development, and efforts to eradicate poverty [Masson-Delmotte, V., P. Zhai, H.O. Pörtner, D. Roberts, J. Skea, P.R. Shukla, A. Pirani, W. Moufouma-Okia, C. Péan, R. Pidcock, S. Connors, J.B.R. Matthews, Y. Chen, X. Zhou, M.I. Gomis, E. Lonnoy, T. Maycock, M. Tignor and T. Waterfield (eds.)]. World Meteorological Organization, Geneva, Switzerland.

Royal Society, 2018: *Greenhouse Gas Removal*. Royal Academy of Engineering, London, UK [Available at: https://royalsociety.org/~/media/policy/projects/greenhouse-gas-removal/royal-society-greenhouse-gas-removal-report-2018.pdf].

Ryan, P.A. and A. Münchow, 2017: Sea ice draft observations in Nares Strait from 2003 to 2012. *J. Geophys. Res. Oceans*, **122**(4), 3057–3080, doi:10.1002/2016JC011966.

Sabel, C.F. and D.G. Victor, 2017: Governing global problems under uncertainty: Making bottom-up climate policy work. *Clim. Change*, **144**(1), 15–27.

Sanderson, B.M., B.C. O'Neill and C. Tebaldi, 2016: What would it take to achieve the Paris temperature targets? *Geophys. Res. Letters*, **43**(13), 7133–7142, doi:10.1002/2016GL069563.

Sarmiento, J.L. and N. Gruber, 2002: Sinks for anthropogenic carbon. *Physics Today*, **55**(8), 30–36, doi:10.1063/1.1510279.

Sarmiento, J.L. and N. Gruber, 2006: *Ocean Biogeochemical Dynamics*. Princeton University Press, Princeton, New Jersey, 528 pp. ISBN: 9780691017075.

Savo, V. et al., 2016: Observations of climate change among subsistence-oriented communities around the world. *Nat. Clim. Change*, **6**(5), 462–473, doi:10.1038/nclimate2958.

Schädel, C. et al., 2016: Potential carbon emissions dominated by carbon dioxide from thawed permafrost soils. *Nat. Clim. Change*, **6**, 950–953, doi:10.1038/nclimate3054.

Schaum, C.E. and S. Collins, 2014: Plasticity predicts evolution in a marine alga. *Proc. R. Soc. Lond. B.*, **281**(1793), 20141486, doi:10.1098/rspb.2014.1486.

Schlosberg, D., L.B. Collins and S. Niemeyer, 2017: Adaptation policy and community discourse: Risk, vulnerability, and just transformation. *Environ. Politics*, **26**(3), 413–437, doi:10.1080/09644016.2017.1287628.

Schurer, A.P. et al., 2017: Importance of the pre-industrial baseline for likelihood of exceeding Paris goals. *Nat. Clim. Change*, **7**(8), 563–567, doi:10.1038/nclimate3345.

Seneviratne, S.I. et al., 2012: Summary for Policymakers. In: Managing the Risks of Extreme Events and Disasters to Advance Climate Change Adaptation. A Special Report of Working Groups I and II of the Intergovernmental Panel on Climate Change (IPCC). [Field, C.B., V. Barros, T.F. Stocker, D. Qin, D.J. Dokken, K.L. Ebi, M.D. Mastrandrea, K.J. Mach, G.-K. Plattner, S.K. Allen, M. Tignor and P.M. Midgley (eds.)]. Cambridge University Press, Cambridge, United Kingdom and New York, NY, USA, 109–230.

Serpetti, N. et al., 2017: Impact of ocean warming on sustainable fisheries management informs the ecosystem approach to fisheries. *Sci. Rep.*, **7**(13438), 15, doi:10.1038/s41598-017-13220-7.

Shadian, J.M., 2014: *The politics of Arctic sovereignty: oil, ice, and Inuit governance*. Routledge.

Shaikh, F., Q. Ji and Y. Fan, 2015: The diagnosis of an electricity crisis and alternative energy development in Pakistan. *Renew. Sust. Energ. Rev.*, **52**, 1172–1185, doi:10.1016/j.rser.2015.08.009.

Shakhova, N. et al., 2013: Ebullition and storm-induced methane release from the East Siberian Arctic Shelf. *Nat. Geosci.*, **7**, 64–70, doi:10.1038/ngeo2007.

Sharma, E. et al., 2019: Introduction. In: *The Hindu Kush Himalaya Assessment – Mountains, Climate Change, Sustainability and People* [Wester, P., A. Mishra, A. Mukherji and A.B. Shrestha (eds.)]. SpringerNature, Dordrecht. 627 pp. ISBN: 9783319950518.

Shepherd, J.G., P.G. Brewer, A. Oschlies and A.J.Watson, 2017: Ocean ventilation and deoxygenation in a warming world: Introduction and overview. *Phil. Trans. Roy. Soc.*, **375**, 20170240, doi:10.1098/rsta.2017.0240.

Sherpa, P., 2014: Climate change, perceptions, and social heterogeneity in Pharak, Mount Everest region of Nepal. *Hum. Organ.*, **73**(2), 153–161, doi:10.17730/humo.73.2.94q43152111733t6.

Shi, J., V.H.M. Visschers, M. Siegrist and J. Arvai, 2016: Knowledge as a driver of public perceptions about climate change reassessed. *Nat. Clim. Change*, **6**(8), 759–762, doi:10.1038/nclimate2997.

Shi, J.-R., S.-P. Xie and L.D. Talley, 2018: Evolving relative importance of the southern ocean and North Atlantic in anthropogenic ocean heat uptake. *J. Clim.*, **31**(18), 7459–7479, doi:10.1175/JCLI-D-18-0170.1.

Shindell, D. et al., 2012: Simultaneously mitigating near-term climate change and improving human health and food security. *Science*, **335**(6065), 183–189, doi:10.1126/science.1210026.

Sigman, D. and E. Boyle, 2000: Glacial/interglacial variations in carbon dioxide. *Nature*, **407**, 859–869, doi:10.1038/35038000.

Sillitoe, P., 2007: *Local science vs. global science: Approaches to indigenous knowledge in international development*. Studies in Environmental Anthropology and Ethnobiology, Berghahn Books, New York, Oxford. 300 pp. ISBN: 9781845456481.

Singh, G.G. et al., 2018: A rapid assessment of co-benefits and trade-offs among Sustainable Development Goals. *Mar. Policy*, **93**, 223–231, doi:10.1016/j.marpol.2017.05.030.

Smith, E.K. and A. Mayer, 2019: Anomalous Anglophones? Contours of free market ideology, political polarization, and climate change attitudes in English-speaking countries, Western European and post-Communist states. *Clim. Change*, **152**(1), 17–34, doi:10.1007/s10584-018-2332-x.

Smith, K.R. et al., 2014: Human health: impacts, adaptation, and co-benefits. In: Climate Change 2014: Impacts, Adaptation, and Vulnerability. Part A: Global and Sectoral Aspects. Contribution of Working Group II to the Fifth Assessment Report of the Intergovernmental Panel on Climate Change [Field, C.B., V.R. Barros, D.J. Dokken, K.J. Mach, M.D. Mastrandrea, T.E. Biller, M. Chatterjee, K.L. Ebi, Y.O. Estrada, R.C. Genova, B. Girma, E.S. Kissel, A.N. Levy, S. MacCracken, P.R. Mastrandrea and L.L. White (eds.)]. Cambridge University Press, Cambridge, United Kingdom and New York, NY, USA, 709–754.

Smith, N. and A. Leiserowitz, 2014: The role of emotion in global warming policy support and opposition. *Risk Anal.*, **34**(5), 937–948, doi:10.1111/risa.12140.

Smith, N.D., S.I.F. Ndlovu and R.W. Summers, 2016: *Milnerton Coast Legal Review: Legal Issues Relevant to the City of Cape Town's Coastal Erosion Management Strategy (for the portion of the Milnerton Coast that forms the study area), Prepared for ERMD, City of Cape Town*. ERMD, Cape Town.

Solecki, W., M. Pelling and M. Garschagen, 2017: Transitions between risk management regimes in cities. *Ecol. Soc.*, **22**(2), 38, doi:10.5751/ES-09102-220238.

Solomon, S., G.-K. Plattner, R. Knutti and P. Friedlingstein, 2009: Irreversible climate change due to carbon dioxide emissions. Proc. Natl. Acad. Sci. U.S.A., **106**(6), 1704–1709, doi:10.1073/pnas.0812721106.

Sommerkorn, M. and A.E. Nilsson, 2015: Governance of Arctic ecosystem services. In: *The Economics of Biodiversity and Ecosystem Services TEEB Scoping Study for the Arctic*, ed. CAFF (Akureyri, Iceland: Conservation of Arctic Flora and Fauna, 2015), ISBN: 9789935431462.

Sowman, M., D. Scott and C. Sutherland, 2016: *Governance and Social Justice Position Paper: Milnerton Beach, Prepared for ERMD, City of Cape Town*. ERMD, Cape Town.

Spratt, D. and I. Dunlop, 2018: *What lies beneath: The understatement of existential climate risk*. Breakthrough-National Centre for Climate Restoration, Melbourne, Australia, 40 pp.

Steffen, W. et al., 2018: Trajectories of the Earth System in the Anthropocene. *Proc. Natl. Acad. Sci. U.S.A.*, **115**(33), 8252–8259, doi:10.1073/pnas.1810141115.

Steiner, C.E., 2015: A sea of warriors: Performing an identity of resilience and empowerment in the face of climate change in the Pacific. *The Contemporary Pacific*, **27**(1), 147–180, doi:10.1353/cp.2015.0002.

Storto, A. et al., 2018: Extending an oceanographic variational scheme to allow for affordable hybrid and four-dimensional data assimilation. *Ocean Model.*, **128**, 67–86, doi:10.1016/j.ocemod.2018.06.005.

Stott, P.A. et al., 2016: Attribution of extreme weather and climate-related events. *WIRes Clim. Change*, **7**(1), 23–41, doi:10.1002/wcc.380.

Strang, V., 2009: Integrating the social and natural sciences in environmental research: A discussion paper. *Environ. Dev. Sustain.*, **11**(1), 1–18, doi:10.1007/s10668-007-9095-2.

Stroeve, J.C. et al., 2017: Investigating the local-scale influence of sea ice on Greenland surface melt. *The Cryosphere*, **11**(5), 2363–2381, doi:10.5194/tc-2017-65.

Sud, R., A. Mishra, N. Varma and S. Bhadwal, 2015: Adaptation policy and practice in densely populated glacier-fed river basins of South Asia: A systematic review. *Reg. Environ. Change*, **15** (5), 825–836, doi:10.1007/s10113-014-0711-z.

Sunday, J.M. et al., 2014: Evolution in an acidifying ocean. *Trends Ecol. Evol.*, **29**(2), 117–125, doi:10.1016/j.tree.2013.11.001.

Sutton, R.T., 2018: Ideas: A simple proposal to improve the contribution of IPCC WG1 to the assessment and communication of climate change risks. *Earth Syst. Dynam.*, **9**, 1155–1158, doi:10.5194/esd-2018-36.

Szabo, S. et al., 2016: Making SDGs work for climate change hotspots. *Environment: Science Policy for Sustainable Development*, **58** (6), 24–33, doi:10.1080/00139157.2016.1209016.

Tagliabue, A. et al., 2017: The integral role of iron in ocean biogeochemistry. *Nature*, **543**(7643), 51–59, doi:10.1038/nature21058.

Talley, L.D., 2013: Closure of the global overturning circulation through the Indian, Pacific, and Southern Oceans: Schematics and transports. *Oceanography*, **26**(1), 80–97, doi:10.5670/oceanog.2013.07.

Talley, L.D. et al., 2016: Changes in ocean heat, carbon content and ventilation: A review of the first decade of GO-SHIP global repeat hydrography. *Annu. Rev. Mar. Sci.*, **8**(1), 185–215, doi:10.1146/annurev-marine-052915-100829.

Tarnocai, C. et al., 2009: Soil organic carbon pools in the northern circumpolar permafrost region. *Global Biogeochem. Cycles*, **23**(2), 1–11, doi:10.1029/2008GB003327.

Tasmanian Climate Change Office, 2017: *Tasmanian Wilderness and World Heritage Area Bushfire and Climate Change Research Project: Tasmanian Government's Response*. Tasmanian Climate Change Office Department of Premier and Cabinet, 49 pp. ISBN: 9780724656456.

Taylor, K.E., R.J. Stouffer and G.A. Meehl, 2012: An overview of CMIP5 and the experiment design. *Bull. Am. Meteorol. Soc.*, **93**(4), 485–498, doi:10.1175/BAMS-D-11-00094.1.

Thaman, R. et al., 2013: *The contribution of indigenous and local knowledge systems to IPBES: Building synergies with science*. IPBES Expert Meeting Report, UNESCO/UNU, Paris, UNESCO, 49 pp.

Thomas, K. et al., 2019: Explaining differential vulnerability to climate change: A social science review. *WIRes Clim. Change*, **10**(2), 1–18, doi:10.1002/wcc.565.

Thomson, J. and W.E. Rogers, 2014: Swell and sea in the emerging Arctic Ocean. *Geophys. Res. Letters*, **41**(9), 3136–3140, doi:10.1002/2014GL059983.

Thornton, B.F. et al., 2016: Methane fluxes from the sea to the atmosphere across the Siberian shelf seas. *Geophys. Res. Letters*, **43**(11), 5869–5877, doi:10.1002/2016GL068977.

Tierney, K., 2015: Resilience and the neoliberal project: Discourses, critiques, practices – and Katrina. *Am. Behav. Sci.*, **59**(10), 1327–1342, doi:10.1177/0002764215591187.

Townhill, B. et al., 2017: Non-native marine species in north-west Europe: Developing an approach to assess future spread using regional downscaled climate projections. *Aquat. Conserv.*, **27**(5), 1035–1050, doi:10.1002/aqc.2764.

Trenberth, K.E., J.T. Fasullo and M.A. Balmaseda, 2014: Earth's Energy Imbalance. *J. Clim.*, **27** (9), 3129–3144, doi:10.1175/JCLI-D-13-00294.1.

Trenberth, K.E., J.T. Fasullo and T.G. Shepherd, 2015: Attribution of climate extreme events. *Nat. Clim. Change*, **5**, 725–730, doi:10.1038/nclimate2657.

Trenberth, K.E. et al., 2007: Estimates of the global water budget and its annual cycle using observational and model data. *J. Hydrometeorol.*, **8**, 758–769, doi:10.1175/JHM600.1.

Trusel, L.D. et al., 2018: Nonlinear rise in Greenland runoff in response to post-industrial Arctic warming. *Nature*, **564** (7734), 104–108, doi:10.1038/s41586-018-0752-4.

Tschakert, P. et al., 2017: Climate change and loss, as if people mattered: Values, places, and experiences. *WIRes Clim. Change*, **8**(5), e476, doi:10.1002/wcc.476.

UN, 1992: *United Nations Framework Convention on Climate Change*. FCCC/INFORMAL/84, United Nations, New York, 24.

UN, 2015a: *Small Island Developing States in Numbers, Climate Change*. Edition 2015. United Nations, New York, [Available at: http://unohrlls.org/small-island-developing-states-in-numbers-2015/].

UN, 2015b: *United Nations General Assembly. Transforming our world: The 2030 Agenda for Sustainable Development*. **A/RES/70/1** [Available at: www.un.org/sustainabledevelopment/development-agenda/].

UNFCCC, 2015: *The Paris Agreement*. (FCCC/CP/2015/10/Add.1).

Unsworth, K.L. and K.S. Fielding, 2014: It's political: How the salience of one's political identity changes climate change beliefs and policy support. *Global Environ. Change*, **27**, 131–137, doi:10.1016/j.gloenvcha.2014.05.002.

Uprety, K. and S.M. Salman, 2011: Legal aspects of sharing and management of transboundary waters in South Asia: Preventing conflicts and promoting cooperation. *Hydrolog. Sci. J.*, **56**(4), 641–661, doi:10.1080/02626667.2011.576252.

van Oppen, M.J.H., J.K. Oliver, H.M. Putnam and R.D. Gates, 2015: Building coral reef resilience through assisted evolution. Proc. Natl. Acad. Sci. U.S.A., **112**, 2307–2313, doi:10.1073/pnas.1422301112.

van Ruijven, B.J. et al., 2014: Enhancing the relevance of Shared Socioeconomic Pathways for climate change impacts, adaptation and vulnerability research. *Clim. Change*, **122**(3), 481–494, doi:10.1007/s10584-013-0931-0.

van Valkengoed, A.M. and L. Steg, 2019: Meta-analyses of factors motivating climate change adaptation behaviour. *Nat. Clim. Change*, **9** (2), 158–163, doi:10.1038/s41558-018-0371-y.

van Vuuren, D.P. et al., 2011a: The representative concentration pathways: An overview. *Clim. Change*, **109** (1–2), 5–31, doi:10.1007/s10584-011-0148-z.

van Vuuren, D.P. et al., 2011b: RCP2.6: exploring the possibility to keep global mean temperature increase below 2°C. *Clim. Change*, **109**, 95–116, doi:10.1007/s10584-011-0152-3.

Varma, N. et al., 2014: Climate change, disasters and development: Testing the waters for adaptive governance in India. *Vision: The Journal of Business Perspective*, **18**(4), 327–338, doi:10.1177/0972262914551664.

Vaughan, D.G. et al., 2013: Observations: Cryosphere. In: Climate Change 2013: The Physical Science Basis. Contribution of Working Group I to the Fifth Assessment Report of the Intergovernmental Panel on Climate Change [Stocker, T.F., D. Qin, G.-K. Plattner, M. Tignor, S.K. Allen, J. Boschung,

A. Nauels, Y. Xia, V. Bex and P.M. Midgley (eds.)]. Cambridge University Press, Cambridge, United Kingdom and New York, NY, USA, 317–382.

Vidas, D., 2000: *Protecting the Polar marine environment: Law and policy for pollution prevention*. Cambridge University Press, Cambridge, United Kingdom, 276 pp. ISBN: 0521663113.

Vidas, D., D. Freestone and J. McAdam, 2015: International law and sea level rise: The new ILA committee. *ILSA Journal of International & Comparative Law*, **21**(2), 397–408.

Vierros, M., 2017: Communities and blue carbon: The role of traditional management systems in providing benefits for carbon storage, biodiversity conservation and livelihoods. *Clim. Change*, **140**(1), 89–100, doi:10.1007/s10584-013-0920-3.

Visbeck, M., 2018: Ocean science research is key for a sustainable future. *Nature Commun.*, **9**(1), 690, doi:10.1038/s41467-018-03158-3.

von Schuckmann, K. et al., 2016: An imperative to monitor Earth's energy imbalance. *Nat. Clim. Change*, **6**, 138–144, doi:10.1038/nclimate2876.

Walker, B., C.S. Holling, S. Carpenter and A. Kinzig, 2004: Resilience, adaptability and transformability in social–ecological systems. *Ecol. Soc.*, **9**(2), 5.

Walter, A.M., 2014: Changing Gilgit-Baltistan: Perceptions of the recent history and the role of community activism. *Ethnoscripts*, **16**(1), 31–49.

Warner, J.F., 2016: Of river linkage and issue linkage: Transboundary conflict and cooperation on the River Meuse. *Globalizations*, **13**(6), 741–766.

Warner, K. and K. van der Geest, 2013: Loss and damage from climate change: Local-level evidence from nine vulnerable countries. *Int. J. Global Warm.*, **5**(4), 367–386, doi:10.1504/IJGW.2013.057289.

Watt-Cloutier, S., 2018: *The right to be cold: One woman's fight to protect the Arctic and save the planet from climate change*. University of Minnesota Press, Minneapolis; London, 368 pp. ISBN: 9780670067107.

Weaver, C. et al., 2017: Reframing climate change assessments around risk: Recommendations for the US National Climate Assessment. *Environ. Res. Letters*, **12** (8), 080201, doi:10.1088/1748-9326/aa7494.

Weichselgartner, J. and I. Kelman, 2014: Geographies of resilience: Challenges and opportunities of a descriptive concept. *Prog. Hum. Geogr.*, **39**(3), 249–267, doi:10.1177/0309132513518834.

West, J.J. and G.K. Hovelsrud, 2010: Cross-scale adaptation challenges in the coastal fisheries: Findings from Lebesby, Northern Norway. *Arctic*, **63**(3), 338–354, doi:10.2307/20799601.

WGMS, 2017: *Fluctuations of Glaciers Database*. World Glacier Monitoring Service, Zurich, Switzerland [Available at: http://dx.doi.org/10.5904/wgms-fog-2017–10].

Wise, R.M. et al., 2014: Reconceptualising adaptation to climate change as part of pathways of change and response. *Global Environ. Change*, **28**, 325–336, doi:10.1016/j.gloenvcha.2013.12.002.

WOA, 2016: *The First Global Integrated Marine Assessment: World Ocean Assessment I*. Cambridge University Press, Cambridge, United Kingdom, 973 pp. ISBN: 9781108186148.

Wohling, M., 2009: The problem of scale in indigenous knowledge: A perspective from northern Australia. *Ecol. Soc.*, **14**(1), 1.

Wong, P.P. et al., 2014: Coastal systems and low-lying areas. In: Climate Change 2014: Impacts, Adaptation, and Vulnerability. Part A: Global and Sectoral Aspects. Contribution of Working Group II to the Fifth Assessment Report of the Intergovernmental Panel on Climate Change [Field, C.B., V.R. Barros, D.J. Dokken, K.J. Mach, M.D. Mastrandrea, T.E. Bilir, M. Chatterji, K.L. Ebi, Y.O. Estrada, R.C. Genova, B. Girma, E.S. Kissel, A.N. Levy, S. MacCracken, P.R. Mastrandrea and L.L. White (eds.)]. Cambridge University Press, Cambridge, United Kingdom and New York, NY, USA, 361–409.

Wunsch, C., P. Heimbach, R. Ponte and I. Fukumori, 2009: The global general circulation of the ocean estimated by the ECCO-Consortium. *Oceanography*, **22**(2), 88–103, doi:10.5670/oceanog.2009.41.

Wymann von Dach, S. et al., 2018: *Leaving no one in mountains behind: localizing the SDGs for resilience of mountain people and ecosystems*. Issue Brief 2018: Sustainable Mountain Development Centre for Development and Environment and Mountain Research Initiative, with Bern Open Publishing (BOP), Bern, Switzerland. 12 pp.

Xu, Y. and V. Ramanathan, 2017: Well below 2°C: Mitigation strategies for avoiding dangerous to catastrophic climate changes. *PNAS*, **114**(39), 10315–10323, doi:10.1073/pnas.1618481114.

Yager, K., 2015: Satellite imagery and community perceptions of climate change impacts and landscape change. In: *Climate Cultures: Anthropological Perspectives on Climate Change* [Jessica, B. and M.R. Dove (eds.)]. Yale University Press, New Haven, pp. 146–168. ISBN: 9780300198812.

Yeh, E.T., 2016: How can experience of local residents be "knowledge"? Challenges in interdisciplinary climate change research. *Area*, **48**(1), 34–40, doi:10.1111/area.12189.

Young, O.R., 2016: The shifting landscape of Arctic politics: Implications for international cooperation. *The Polar Journal*, **6**(2), 209–223, doi:10.1080/2154896X.2016.1253823.

Yu, L., 2018: Global air–sea fluxes of heat, fresh water, and momentum: Energy budget closure and unanswered questions. *Annu. Rev. Mar. Sci.*, **11**, 227–248, doi:10.1146/annurev-marine-010816-060704.

Zheng, J. and S. Suh, 2019: Strategies to reduce the global carbon footprint of plastics. *Nat. Clim. Change*, **9**, 374–378, doi:10.1038/s41558-019-0459-z.

Zscheischler, J. et al., 2018: Future climate risk from compound events. *Nat. Clim. Change*, **8**(6), 469–477, doi:10.1038/s41558-018-0156-3.

Zuo, H., M.A. Balmaseda, K. Mogensen and S. Tietsche, 2018: *OCEAN5: The ECMWF Ocean Reanalysis System and its Real-Time analysis component.* ECMWF Tech Memo 823.

2

High Mountain Areas

Coordinating Lead Authors:
Regine Hock (USA), Golam Rasul (Nepal)

Lead Authors:
Carolina Adler (Switzerland/Australia), Bolívar Cáceres (Ecuador), Stephan Gruber (Canada/Germany), Yukiko Hirabayashi (Japan), Miriam Jackson (Norway), Andreas Kääb (Norway), Shichang Kang (China), Stanislav Kutuzov (Russian Federation), Alexander Milner (UK), Ulf Molau (Sweden), Samuel Morin (France), Ben Orlove (USA), Heidi Steltzer (USA)

Contributing Authors:
Simon Allen (Switzerland), Lukas Arenson (Canada), Soumyadeep Banerjee (India), Iestyn Barr (UK), Roxana Bórquez (Chile), Lee Brown (UK), Bin Cao (China), Mark Carey (USA), Graham Cogley (Canada), Andreas Fischlin (Switzerland), Alex de Sherbinin (USA), Nicolas Eckert (France), Marten Geertsema (Canada), Marca Hagenstad (USA), Martin Honsberg (Germany), Eran Hood (USA), Matthias Huss (Switzerland), Elizabeth Jimenez Zamora (Bolivia), Sven Kotlarski (Switzerland), Pierre-Marie Lefeuvre (Norway/France), Juan Ignacio López Moreno (Spain), Jessica Lundquist (USA), Graham McDowell (Canada), Scott Mills (USA), Cuicui Mou (China), Santosh Nepal (Nepal), Jeannette Noetzli (Switzerland), Elisa Palazzi (Italy), Nick Pepin (UK), Christian Rixen (Switzerland), Maria Shahgedanova (UK), S. McKenzie Skiles (USA), Christian Vincent (France), Daniel Viviroli (Switzerland), Gesa Weyhenmeyer (Sweden), Pasang Yangjee Sherpa (Nepal/USA), Nora M. Weyer (Germany), Bert Wouters (Netherlands), Teppei J. Yasunari (Japan), Qinglong You (China), Yangjiang Zhang (China)

Review Editors:
Georg Kaser (Austria), Aditi Mukherji (Nepal/India)

Chapter Scientist:
Pierre-Marie Lefeuvre (Norway/France), Santosh Nepal (Nepal)

This chapter should be cited as:
Hock, R., G. Rasul, C. Adler, B. Cáceres, S. Gruber, Y. Hirabayashi, M. Jackson, A. Kääb, S. Kang, S. Kutuzov, Al. Milner, U. Molau, S. Morin, B. Orlove, and H. Steltzer, 2019: High Mountain Areas. In: *IPCC Special Report on the Ocean and Cryosphere in a Changing Climate* [H.-O. Pörtner, D.C. Roberts, V. Masson-Delmotte, P. Zhai, M. Tignor, E. Poloczanska, K. Mintenbeck, A. Alegría, M. Nicolai, A. Okem, J. Petzold, B. Rama, N.M. Weyer (eds.)]. Cambridge University Press, Cambridge, UK and New York, NY, USA, pp. 131–202. https://doi.org/10.1017/9781009157964.004.

Table of contents

2

Executive Summary

The cryosphere (including, snow, glaciers, permafrost, lake and river ice) is an integral element of high mountain regions, which are home to roughly 10% of the global population. Widespread cryosphere changes affect physical, biological and human systems in the mountains and surrounding lowlands, with impacts evident even in the ocean. Building on the IPCC's 5th Assessment Report (AR5), this chapter assesses new evidence on observed recent and projected changes in the mountain cryosphere as well as associated impacts, risks and adaptation measures related to natural and human systems. Impacts in response to climate changes independently of changes in the cryosphere are not assessed in this chapter. Polar mountains are included in Chapter 3, except those in Alaska and adjacent Yukon, Iceland and Scandinavia, which are included in this chapter.

Observations of cryospheric changes, impacts, and adaptation in high mountain areas

Observations show general decline in low-elevation snow cover (*high confidence*[1]), glaciers (*very high confidence*) and permafrost (*high confidence*) due to climate change in recent decades. Snow cover duration has declined in nearly all regions, especially at lower elevations, on average by 5 days per decade, with a *likely*[2] range from 0–10 days per decade. Low elevation snow depth and extent have declined, although year-to-year variation is high. Mass change of glaciers in all mountain regions (excluding the Canadian and Russian Arctic, Svalbard, Greenland and Antarctica) was *very likely* -490 ± 100 kg m^{-2} yr^{-1} (-123 ± 24 Gt yr^{-1}) in 2006–2015. Regionally averaged mass budgets were *likely* most negative (less than -850 kg m^{-2} yr^{-1}) in the southern Andes, Caucasus and the European Alps/Pyrenees, and least negative in High Mountain Asia (-150 ± 110 kg m^{-2} yr^{-1}) but variations within regions are strong. Between 3.6–5.2 million km^2 are underlain by permafrost in the eleven high mountain regions covered in this chapter corresponding to 27–29% of the global permafrost area (*medium confidence*). Sparse and unevenly distributed measurements show an increase in permafrost temperature (*high confidence*), for example, by 0.19°C ± 0.05°C on average for about 28 locations in the European Alps, Scandinavia, Canada and Asia during the past decade. Other observations reveal decreasing permafrost thickness and loss of ice in the ground. {2.2.2, 2.2.3, 2.2.4}

Glacier, snow and permafrost decline has altered the frequency, magnitude and location of most related natural hazards (*high confidence*). Exposure of people and infrastructure to natural hazards has increased due to growing population, tourism and socioeconomic development (*high confidence*). Glacier retreat and permafrost thaw have decreased the stability of mountain slopes and the integrity of infrastructure (*high confidence*). The number and area of glacier lakes has increased in most regions in recent decades (*high confidence*), but there is only *limited evidence* that the frequency of glacier lake outburst floods (GLOF) has changed. In some regions, snow avalanches involving wet snow have increased (*medium confidence*), and rain-on-snow floods have decreased at low elevations in spring and increased at high elevations in winter (*medium confidence*). The number and extent of wildfires have increased in the Western USA partly due to early snowmelt (*medium confidence*). {2.3.2, 2.3.3}

Changes in snow and glaciers have changed the amount and seasonality of runoff in snow-dominated and glacier-fed river basins (*very high confidence*) with local impacts on water resources and agriculture (*medium confidence*). Winter runoff has increased in recent decades due to more precipitation falling as rain (*high confidence*). In some glacier-fed rivers, summer and annual runoff have increased due to intensified glacier melt, but decreased where glacier melt water has lessened as glacier area shrinks. Decreases were observed especially in regions dominated by small glaciers, such as the European Alps (*medium confidence*). Glacier retreat and snow cover changes have contributed to localized declines in agricultural yields in some high mountain regions, including the Hindu Kush Himalaya and the tropical Andes (*medium confidence*). There is *limited evidence* of impacts on operation and productivity of hydropower facilities resulting from changes in seasonality and both increases and decreases in water input, for example, in the European Alps, Iceland, Western Canada and USA, and the tropical Andes. {2.3.1}

Species composition and abundance have markedly changed in high mountain ecosystems in recent decades (*very high confidence*), partly due to changes in the cryosphere (*high confidence*). Habitats for establishment by formerly absent species have opened up or been altered by reduced snow cover (*high confidence*), retreating glaciers (*very high confidence*), and thawing of permafrost (*medium confidence*). Reductions in glacier and snow cover have directly altered the structure of many freshwater communities (*high confidence*). Reduced snow cover has negatively impacted the reproductive fitness of some snow-dependent plant and animal species, including foraging and predator-prey relationships of mammals (*high confidence*). Upslope migration of individual species, mostly due to warming and to a lesser extent due to cryosphere-related changes, has often increased local species richness (*very high confidence*). Some cold-adapted species, including endemics, in terrestrial and freshwater communities have declined

---

[1] In this report, the following summary terms are used to describe the available evidence: limited, medium, or robust; and for the degree of agreement: low, medium or high. A level of confidence is expressed using five qualifiers: very low, low, medium, high and very high, and typeset in italics, for example, *medium confidence*. For a given evidence and agreement statement, different confidence levels can be assigned, but increasing levels of evidence and degrees of agreement are correlated with increasing confidence (see Section 1.9.2 and Figure 1.4 for more details).

[2] In this report, the following terms have been used to indicate the assessed likelihood of an outcome or a result: Virtually certain 99–100% probability, Very likely 90–100%, Likely 66–100%, About as likely as not 33–66%, Unlikely 0–33%, Very unlikely 0–10% and Exceptionally unlikely 0–1%. Additional terms (Extremely likely: 95–100%, More likely than not >50–100% and Extremely unlikely 0–5%) may also be used when appropriate. Assessed likelihood is typeset in italics, for example, *very likely* (see Section 1.9.2 and Figure 1.4 for more details). This Report also uses the term '*likely* range' to indicate that the assessed likelihood of an outcome lies within the 17–83% probability range.

in abundance (*high confidence*). While the plant productivity has generally increased, the actual impact on provisioning, regulating and cultural ecosystem services varies greatly (*high confidence*). {2.3.3}

Tourism and recreation activities such as skiing, glacier tourism and mountaineering have been negatively impacted by declining snow cover, glaciers and permafrost (*medium confidence*). In several regions, worsening route safety has reduced mountaineering opportunities (*medium confidence*). Variability and decline in natural snow cover have compromised the operation of low-elevation ski resorts (*high confidence*). Glacier and snow decline have impacted aesthetic, spiritual and other cultural aspects of mountain landscapes (*medium confidence*), reducing the well-being of people (e.g., in the Himalaya, eastern Africa, and the tropical Andes). {2.3.5, 2.3.6}

Adaptation in agriculture, tourism and drinking water supply has aimed to reduce the impacts of cryosphere change (*medium confidence*), though there is *limited evidence* on their effectiveness owing to a lack of formal evaluations, or technical, financial and institutional barriers to implementation. In some places, artificial snowmaking has reduced the negative impacts on ski tourism (*medium confidence*). Release and storage of water from reservoirs according to sectoral needs (agriculture, drinking water, ecosystems) has reduced the impact of seasonal variability on runoff (*medium confidence*). {2.3.1, 2.3.5}

Future projections of cryospheric changes, their impacts and risks, and adaptation in high mountain areas

Snow cover, glaciers and permafrost are projected to continue to decline in almost all regions throughout the 21st century (*high confidence*). Compared to 1986–2005, low elevation snow depth will *likely* decrease by 10–40% for 2031–2050, regardless of Representative Concentration Pathway (RCP) and for 2081–2100, *likely* by 10–40 % for RCP2.6 and by 50–90% for RCP8.5. Projected glacier mass reductions between 2015–2100 are *likely* 22–44% for RCP2.6 and 37–57% for RCP8.5. In regions with mostly smaller glaciers and relatively little ice cover (e.g., European Alps, Pyrenees, Caucasus, North Asia, Scandinavia, tropical Andes, Mexico, eastern Africa and Indonesia), glaciers will lose more than 80% of their current mass by 2100 under RCP8.5 (*medium confidence*), and many glaciers will disappear regardless emission scenario (*very high confidence*). Permafrost thaw and degradation will increase during the 21st century (*very high confidence*) but quantitative projections are scarce. {2.2.2, 2.2.3, 2.2.4}

Most types of natural hazards are projected to change in frequency, magnitude and areas affected as the cryosphere continues to decline (*high confidence*). Glacier retreat and permafrost thaw are projected to decrease the stability of mountain slopes and increase the number and area of glacier lakes (*high confidence*). Resulting landslides and floods, and cascading events, will also emerge where there is no record of previous events (*high confidence*). Snow avalanches are projected to decline in number and runout distance at lower elevation, and avalanches involving wet

snow even in winter will occur more frequently (*medium confidence*). Rain-on-snow floods will occur earlier in spring and later in autumn, and be more frequent at higher elevations and less frequent at lower elevations (*high confidence*). {2.3.2, 2.3.3}

River runoff in snow dominated and glacier-fed river basins will change further in amount and seasonality in response to projected snow cover and glacier decline (*very high confidence*) with negative impacts on agriculture, hydropower and water quality in some regions (*medium confidence*). The average winter snowmelt runoff is projected to increase (*high confidence*), and spring peaks to occur earlier (*very high confidence*). Projected trends in annual runoff vary substantially among regions, and can even be opposite in direction, but there is *high confidence* that in all regions average annual runoff from glaciers will have reached a peak that will be followed by declining runoff at the latest by the end of the 21st century. Declining runoff is expected to reduce the productivity of irrigated agriculture in some regions (*medium confidence*). Hydropower operations will increasingly be impacted by altered amount and seasonality of water supply from snow and glacier melt (*high confidence*). The release of heavy metals, particularly mercury, and other legacy contaminants currently stored in glaciers and permafrost, is projected to reduce water quality for freshwater biota, household use and irrigation (*medium confidence*). {2.3.1}

Current trends in cryosphere-related changes in high mountain ecosystems are expected to continue and impacts to intensify (*very high confidence*). While high mountains will provide new and greater habitat area, including refugia for lowland species, both range expansion and shrinkage are projected, and at high elevations this will lead to population declines (*high confidence*). The latter increases the risk of local extinctions, in particular for freshwater cold-adapted species (*medium confidence*). Without genetic plasticity and/or behavioural shifts, cryospheric changes will continue to negatively impact endemic and native species, such as some coldwater fish (e.g., trout) and species whose traits directly depend on snow (e.g., snowshoe hares) or many large mammals (*medium confidence*). The survival of such species will depend on appropriate conservation and adaptation measures (*medium confidence*). Many projected ecological changes will alter ecosystem services (*high confidence*), affecting ecological disturbances (e.g., fire, rock fall, slope erosion) with considerable impacts on people (*medium confidence*). {2.3.3}

Cultural assets, such as snow- and ice-covered peaks in many UNESCO World Heritage sites, and tourism and recreation activities, are expected to be negatively affected by future cryospheric change in many regions (*high confidence*). Current snowmaking technologies are projected to be less effective in a warmer climate in reducing risks to ski tourism in most parts of Europe, North America and Japan, in particular at 2°C global warming and beyond (*high confidence*). Diversification through year-round activities supports adaptation of tourism under future climate change (*medium confidence*). {2.3.5, 2.3.6}

Enablers and response options to promote adaptation and sustainable development in high mountain areas

The already committed and unavoidable climate change affecting all cryosphere elements, irrespective of the emission scenario, points to integrated adaptation planning to support and enhance water availability, access, and management (*medium confidence*). Integrated management approaches for water across all scales, in particular for energy, agriculture, ecosystems and drinking water supply, can be effective at dealing with impacts from changes in the cryosphere. These approaches also offer opportunities to support social-ecological systems, through the development and optimisation of storage and the release of water from reservoirs (*medium confidence*), while being cognisant of potential negative implications for some ecosystems. Success in implementing such management options depends on the participation of relevant stakeholders, including affected communities, diverse knowledge and adequate tools for monitoring and projecting future conditions, and financial and institutional resources to support planning and implementation (*medium confidence*). {2.3.1, 2.3.3, 2.4}

Effective governance is a key enabler for reducing disaster risk, considering relevant exposure factors such as planning, zoning, and urbanisation pressures, as well as vulnerability factors such as poverty, which can challenge efforts towards resilience and sustainable development for communities (*medium confidence*). Reducing losses to disasters depend on integrated and coordinated approaches to account for the hazards concerned, the degree of exposure, and existing vulnerabilities. Diverse knowledge that includes community and multi-stakeholder experience with past impacts complements scientific knowledge to anticipate future risks. {CCB-1, 2.3.2, 2.4}

International cooperation, treaties and conventions exist for some mountain regions and transboundary river basins with potential to support adaptation action. However, there is *limited evidence* on the extent to which impacts and losses arising from changes in the cryosphere are specifically monitored and addressed in these frameworks. A wide range of institutional arrangements and practices have emerged over the past three decades that respond to a shared global mountain agenda and specific regional priorities. There is potential to strengthen them to also respond to climate-related cryosphere risks and open opportunities for development through adaptation (*limited evidence, high agreement*). The Sustainable Development Goals (SDGs), Sendai Framework and Paris Agreement have directed some attention in mountain-specific research and practice towards the monitoring and reporting on targets and indicators specified therein. {2.3.1, 2.4}

2.1   Introduction

High mountain regions share common features, including rugged terrain, a low-temperature climate regime, steep slopes and institutional and spatial remoteness. These features are often linked to physical and social-ecological processes that, although not unique to mountain regions, typify many of the special aspects of these regions. Due to their higher elevation compared with the surrounding landscape, mountains often feature cryosphere components, such as glaciers, snow cover and permafrost, with a significant influence on surrounding lowland areas even far from the mountains (Huggel et al., 2015a). Hence the mountain cryosphere plays a major role in large parts of the world. Considering the close relationship between mountains and the cryosphere, high mountain areas are addressed in a dedicated chapter within this special report. Almost 10%

(671 million people) of the global population lived in high mountain regions in 2010, based on gridded population data (Jones and O'Neill, 2016) and a distance of less than 100 km from glaciers or permafrost located in mountains areas as defined in Figure 2.1. This population is expected to grow to 736–844 million across the shared socioeconomic pathways by 2050 (Gao, 2019). Many people living outside of mountain areas and not included in these numbers are also affected by changes in the mountain cryosphere.

This chapter assesses recent and projected changes in glaciers, snow cover, permafrost and lake and river ice in high mountain areas, their drivers, as well as their impact on the different services provided by the cryosphere and related adaptation, with a focus on literature published after AR5. The assessment of cryospheric change is focused on recent decades rather than a perspective over a longer period,

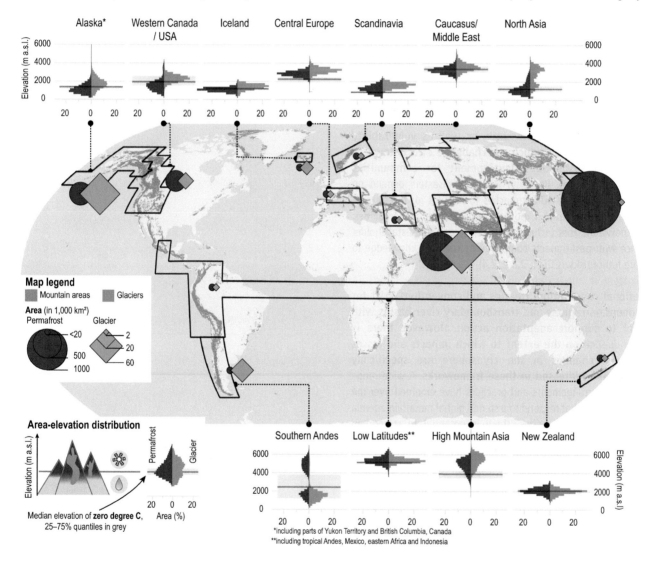

Figure 2.1 | Distribution of mountain areas (orange shading) and glaciers (blue) as well as regional summary statistics for glaciers and permafrost in mountains. Mountains are distinguished based on a ruggedness index (>3.5), a logarithmically scaled measure of relative relief (Gruber, 2012). Eleven distinct regions with glaciers, generally corresponding to the primary regions in the Randolph Glacier Inventory, RGI v6.0 (RGI Consortium, 2017) are outlined, although some cryosphere related impacts presented in this chapter may go beyond these regions. Region names correspond to those in the RGI. Diamonds represent regional glacier area (RGI 6.0) and circles the permafrost area in all mountains within each region boundary (Obu et al., 2019). Histograms for each region show glacier and permafrost area in 200 m elevation bins as a percentage of total regional glacier and permafrost area, respectively. Also shown is the median elevation of the annual mean 0°C free-atmosphere isotherm calculated from the ERA-5 re-analysis of the European Centre for Medium Range Weather Forecasts over each region's mountain area for the period 2006–2015, with 25–75% quantiles in grey. The annual 0°C isotherm elevation roughly separates the areas where precipitation predominantly falls as snow and rain. Areas above and below this elevation are loosely referred to as high and low elevations, respectively, in this chapter.

and future changes spanning the 21st century. A palaeo-perspective is covered in IPCC Sixth Assessment Report (AR6) Working Group I contribution on 'The Physical Science Basis'. High mountain areas, as discussed here, include all mountain regions where glaciers, snow or permafrost are prominent features of the landscape, without a strict and quantitative demarcation, but with a focus on distinct regions (Figure 2.1). Mountain regions located in the polar regions are considered in Chapter 3 except those in Iceland, Scandinavia and Alaska and parts of adjacent Yukon Territory and British Columbia, which are included in this chapter. Many changes in the mountain environment are not solely or directly related to climate change induced changes in the cryosphere, but to other direct or indirect effects of climate change, or to other consequences of socioeconomic development. Consistent with the scope of this report with a focus on the ocean and the cryosphere, this section deals primarily with the impacts that can at least partially be attributed to cryosphere changes. Even though other drivers may be the dominant driver of change in many cases, they are not considered explicitly in this chapter, although unambiguous attribution to cryosphere changes is often difficult.

2.2 Changes in the Mountain Cryosphere

2.2.1 Atmospheric Drivers of Changes in the Mountain Cryosphere

Past changes of surface air temperature and precipitation in high mountain areas have been documented by *in situ* observations and regional reanalyses (Table SM2.2 and Table SM2.4). However, mountain observation networks do not always follow standard measurement procedures (Oyler et al., 2015; Nitu et al., 2018) and are often insufficiently dense to capture fine-scale changes (Lawrimore et al., 2011) and the underlying larger scale patterns. Future changes are projected using General Circulation Models (GCMs) or Regional Climate Models (RCMs) or simplified versions thereof (e.g., Gutmann et al., 2016), used to represent processes at play in a dynamically consistent manner, and to relate mountain changes to larger-scale atmospheric forcing based on physical principles. Existing mountain-specific model studies typically cover individual mountain ranges, and there is currently no initiative found, such as model intercomparisons or coordinated model experiments, which specifically and comprehensively addresses high mountain meteorology and climate globally. This makes it difficult to provide a globally uniform assessment.

2.2.1.1 Surface Air Temperature

Mountain surface air temperature observations in Western North America, European Alps and High Mountain Asia show warming over recent decades at an average rate of 0.3°C per decade, with a *likely* range of ± 0.2°C, thereby outpacing the global warming rate 0.2 ± 0.1°C per decade (IPCC, 2018). Underlying data from global and regional studies are compiled in Table SM2.2, and Figure 2.2 provides a synthesis on mountain warming trends, mostly based on studies using *in situ* observations. Local warming rates depend on the season (*high confidence*). For example, in the European Alps,

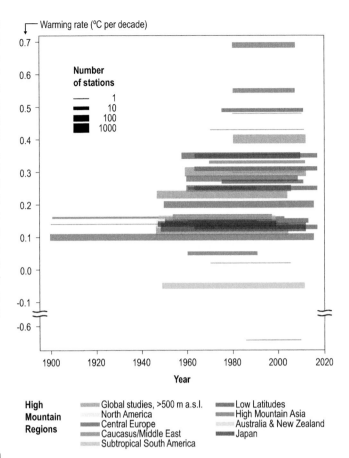

Figure 2.2 | Synthesis of trends in mean annual surface air temperature in mountain regions, based on 4672 observation stations (partly overlapping) aggregated in 38 datasets reported in 19 studies. Each line refers to a warming rate from one dataset, calculated over the time period indicated by the extent of the line. Colours indicate mountain region (Figure 2.1), and line thickness the number of observation stations used. Detailed references are found in Table SM2.2, which also provides additional information on trends for individual seasons and other temperature indicators (daily minimum or maximum temperature).

warming has been found to be more pronounced in summer and spring (Auer et al., 2007; Ceppi et al., 2012), while on the Tibetan Plateau warming is stronger in winter (Liu et al., 2009; You et al., 2010). Studies comparing observations at lower and higher elevation at the global scale indicate that warming is generally enhanced above 500 m above sea level (a.s.l.) (e.g., Wang et al., 2016a; Qixiang et al., 2018, Table SM2.2). At the local and regional scale, evidence for elevation dependent warming, i.e., that the warming rate is different across elevation bands, is scattered and sometimes contradictory (Box 2.1). On the Tibetan Plateau, evidence based on combining *in situ* observations (often scarce at high elevation) with remote sensing and modelling approaches, indicates that warming is amplified around 4,000 m a.s.l., but not above 5,000 m a.s.l. (Qin et al., 2009; Gao et al., 2018). Studies in the Italian Alps (Tudoroiu et al., 2016) and Southern Himalaya (Nepal, 2016) have shown higher warming at lower elevation. Evidence from Western North and South America is conflicting (Table SM2.2). In other regions, evidence to assess whether warming varies with elevation is insufficient. In summary, there is *medium evidence* (*medium agreement*) that surface warming is different across elevation bands. Observed changes also depend on the type of temperature indicator: changes in daily mean, minimum

Box 2.1 | Does Atmospheric Warming in the Mountains Depend on Elevation?

In mountain regions, surface air temperature generally tends to decrease with increasing elevation thus directly impacting how much of the precipitation falls as snow as opposed to rain. Therefore, changes in air temperature have different consequences for snow cover, permafrost and glaciers at different elevations. A number of studies have reported that trends in air temperature vary with elevation, a phenomenon referred to as elevation dependent warming (EDW; Pepin et al., 2015, and references therein), with potential consequences beyond those of uniform warming. EDW does not imply that warming is larger at higher elevation, and smaller at lower elevation, but it means that the warming rate (e.g., in °C per decade) is not the same across all elevation bands. Although this concept has received wide attention in recent years, the manifestation of EDW varies by region, season and temperature indicator (e.g., daily mean, minimum or maximum temperature), meaning that a uniform pattern does not exist. The identification of the underlying driving mechanisms for EDW and how they combine is complex.

Several physical processes contribute to EDW, and quantifying their relative contributions has remained largely elusive (Minder et al., 2018; Palazzi et al., 2019). Some of the processes identified are similar to those explaining the amplified warming in the polar regions (Chapter 3). For example, the sensitivity of temperature to radiative forcing is increased at low temperatures common in both polar and mountain environments (Ohmura, 2012). Because the relationship between specific humidity and downwelling radiation is nonlinear, in a dry and cold atmosphere found at high elevation, any increase in atmospheric humidity due to temperature increase drives disproportionately large warming (Rangwala et al., 2013; Chen et al., 2014). Snow-albedo feedback plays an important role where the snow cover is in decline (Pepin and Lundquist, 2008; Scherrer et al., 2012), increasing the absorption of solar radiation which in turn leads to increased surface air temperature and further snowmelt. Other processes are specific to the mountain environment. Especially in the tropics, warming can be enhanced at higher elevation by a reduction of the vertical temperature gradient, due to increased latent heat release above the condensation level, favored in a warmer and moister atmosphere (Held and Soden, 2006). The cooling effect of aerosols, which also cause solar dimming, is more pronounced at low elevation and reduced at high elevation (Zeng et al., 2015). While many mechanisms suggest that warming should be enhanced at high elevation, observed and simulated EDW patterns are usually more complex (Pepin et al., 2015, and references therein). Numerical simulations by global and regional climate models, which show EDW, need to be considered carefully because of intrinsic limitations due to potentially incomplete understanding and implementation of relevant physical processes, in addition to coarse grid spacing with respect to mountainous topography (Ménégoz et al., 2014; Winter et al., 2017).

and maximum temperature can display contrasting patterns depending on region, season and elevation (Table SM2.2).

Attribution studies for changes in surface air temperature specifically in mountain regions are rare. Bonfils et al. (2008) and Dileepkumar et al. (2018) demonstrated that anthropogenic greenhouse gas emissions are the dominant factor in the recent temperature increases, partially compensated by other anthropogenic factors (land use change and aerosol emissions for Western USA and Western Himalaya, respectively). These findings are consistent with conclusions of AR5 regarding anthropogenic effects (Bindoff et al., 2013). It is thus *likely* that anthropogenic influence is the main contributor to surface temperature increases in high mountain regions since the mid-20th century, amplified by regional feedbacks.

Until the mid-21st century, regardless of the climate scenario (Cross-Chapter Box 1 in Chapter 1), surface air temperature is projected to continue increasing (*very high confidence*) at an average rate of 0.3°C per decade, with a *likely* range of ±0.2°C per decade, locally even more in some regions, generally outpacing global warming rates (0.2 ± 0.1°C per decade; IPCC, 2018) (*high confidence*). Beyond mid-21st century, atmospheric warming in mountains will be stronger under a high greenhouse gas emission scenario (RCP8.5) and will stabilise at mid-21st levels under a low greenhouse gas emission scenario (RCP2.6), similar to global change patterns (*very high confidence*). The warming rate will result from the

combination of regional (*high confidence*) and elevation-dependent (*medium confidence*) enhancement factors. Underlying evidence of future projections from global and regional studies is provided in Table SM2.3. Figure 2.3 provides examples of regional climate projections of surface air temperature, as a function of elevation and season (winter and summer) in North America (Rocky Mountains), South America (Subtropical Central Andes), Europe (European Alps) and High Mountain Asia (Hindu Kush, Karakoram, Himalaya), based on global and regional climate projections.

2.2.1.2   Rainfall and Snowfall

Past precipitation changes are less well quantified than temperature changes and are often more heterogeneous, even within mountain regions (Hartmann and Andresky, 2013). Regional patterns are characterised by decadal variability (Mankin and Diffenbaugh, 2015) and influenced by shifts in large-scale atmospheric circulation (e.g., in Alaska; Winski et al., 2017). While mountain regions do not exhibit clear direction of trends in annual precipitation over the past decades (*medium confidence* that there is no trend), snowfall has decreased, at least in part due to higher temperatures, especially at lower elevation (Table SM2.4, *high confidence*).

Future projections of annual precipitation indicate increases of the order of 5 to 20% over the 21st century in many mountain regions, including the Hindu Kush and Himalaya, East Asia, eastern Africa, the European Alps and the Carpathian region, and decreases in the Mediterranean and the Southern Andes (*medium confidence*, Table SM2.5). Changes in the frequency and intensity of extreme precipitation events vary according to season and region. For example, across the Himalayan-Tibetan Plateau mountains, the frequency and intensity of extreme rainfall events are projected to increase throughout the 21st century, particularly during the summer monsoon (Panday et al., 2015; Sanjay et al., 2017). This suggests a transition toward more episodic and intense monsoonal precipitation, especially in the easternmost part of the Himalayan chain (Palazzi et al., 2013). Increases in winter precipitation extremes are projected in the European Alps (Rajczak and Schär, 2017). At lower elevation, near term (2031–2050) and end of century (2081–2100) projections of snowfall all indicate a decrease, for all greenhouse gas emission scenarios (*very high confidence*). At higher elevation, where temperature increase is insufficient to affect rain/snow partitioning, total winter precipitation increases can lead to increased snowfall (e.g., Kapnick and Delworth, 2013; O'Gorman, 2014) (*medium confidence*).

2.2.1.3   Other Meteorological Variables

Atmospheric humidity, incoming shortwave and longwave radiation, and near-surface wind speed and direction also influence the high mountain cryosphere. Detecting their changes and associated effects on the cryosphere is even more challenging than for surface air temperature and precipitation, both from an observation and modelling standpoint. Therefore, most simulation studies of cryosphere changes are mainly driven by temperature and precipitation (see, e.g., Beniston et al., 2018, and references therein).

Atmospheric moisture content, which is generally increasing in a warming atmosphere (Stocker et al., 2013), affects latent and longwave heat fluxes (Armstrong and Brun, 2008) with implications for the timing and rate of snow and ice ablation, and in some areas changes in atmospheric moisture content could be a significant driver of cryosphere change (Harpold and Brooks, 2018). Short-lived climate forcers, such as sulphur and black carbon aerosols (You et al., 2013), reduce the amount of solar radiation reaching the surface, with potential impacts on snow and ice ablation. Solar brightening caused by declining anthropogenic aerosols in Europe since the 1980s was

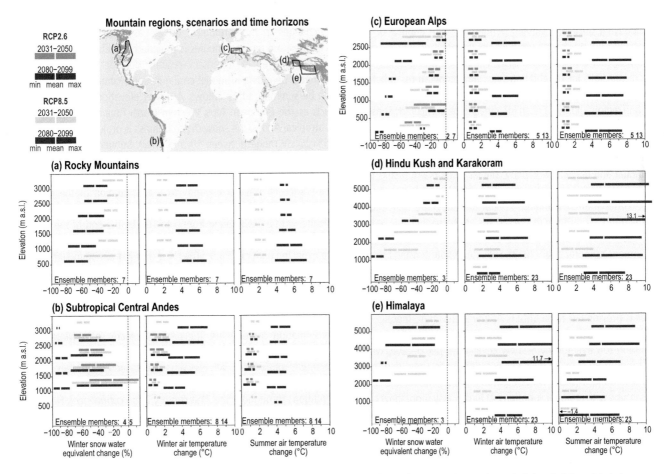

Figure 2.3 | Projected change (1986–2005 to 2031–2050 and 2080–2099) of mean winter (December to May; June to August in Subtropical Central Andes) snow water equivalent, winter air temperature and summer air temperature (June to August; December to February in Subtropical Central Andes) in five high mountain regions for RCP8.5 (all regions) and RCP2.6 (European Alps and Subtropical Central Andes). Changes are averaged over 500 m (a,b,c) and 1,000 m (d,e) elevation bands. The numbers in the lower right of each panel reflect the number of simulations (note that not all models provide snow water equivalent). For the Rocky Mountains, data from NA-CORDEX RCMs (25 km grid spacing) driven by Coupled Model Intercomparison Project Phase 5 (CMIP5) General Circulation Models (GCMs) were used (Mearns et al., 2017). For the European Alps, data from EURO-CORDEX RCMs (12 km grid spacing) driven by CMIP5 GCMs were used (Jacob et al., 2014). For the other regions, CMIP5 GCMs were used: Zazulie (2016) and Zazulie et al. (2018) for the Subtropical Central Andes, and Terzago et al. (2014) and Palazzi et al. (2017) for the Hindu Kush and Karakoram and Himalaya. The list of models used is provided in Table SM2.8.

shown to have only a minor effect on atmospheric warming at high elevation (Philipona, 2013), and effects on the cryosphere were not specifically discussed.

Wind controls preferential deposition of precipitation, post-depositional snow drift and affects ablation of snow and glaciers through turbulent fluxes. Near-surface wind speed has decreased on the Tibetan Plateau between the 1970s and early 2000s, and stabilised or increased slightly thereafter (Yang et al., 2014a; Kuang and Jiao, 2016). This is consistent with existing evidence for a decrease in near-surface wind speed on mid-latitude continental areas since the mid-20th century (Hartmann et al., 2013). In general, the literature on past and future changes of near-surface wind patterns in mountain areas is very limited.

2.2.2   Snow Cover

Snow on the ground is an essential and widespread component of the mountain cryosphere. It affects mountain ecosystems and plays a major role for mass movement and floods in the mountains. It plays a key role in nourishing glaciers and provides an insulating and reflective cover at their surface. It influences the thermal regime of the underlying ground, including permafrost, with implications for ecosystems. Climate change modifies key variables driving the onset and development of the snow cover (e.g., solid precipitation), and those responsible for its ablation (e.g., air temperature, radiation). The snow cover, especially in low-lying and mid-elevation areas of mountain regions, has long been identified to be particularly sensitive to climate change.

The mountain snow cover is characterised by a very strong interannual and decadal variability, similar to its main driving force solid precipitation (Lafaysse et al., 2014; Mankin and Diffenbaugh, 2015). Observations spanning several decades are required to quantify trends. Long-term *in situ* records are scarce in some regions of the world, particularly in High Mountain Asia, Northern Asia and South America (Rohrer et al., 2013). Satellite remote sensing provides new capabilities for monitoring mountain snow cover on regional scales. The satellite record length is often insufficient to assess trends (Bormann et al., 2018). Evidence of past changes from regional studies is provided in Table SM2.6. At lower elevation, there is *high confidence* that the mountain snow cover has generally declined in duration (on average by 5 snow cover days per decade, with a *likely* range from 0 to 10 days per decade), mean snow depth and accumulated mass (snow water equivalent) since the middle of the 20th century, with regional variations. At higher elevation, snow cover trends are generally insignificant (*medium confidence*) or unknown.

Most of the snow cover changes can be attributed, at lower elevation, to more precipitation falling as liquid precipitation (rain) and to increases in melt at all elevations, mostly due to changes in atmospheric forcings, especially increased air temperature (Kapnick and Hall, 2012; Marty et al., 2017) which in turn are attributed to anthropogenic forcings at a larger scale (Section 2.2.1). Formal anthropogenic attribution studies provide similar conclusions in Western North America (Pierce et al., 2008; Najafi et al., 2017).

Assessing the impact of the deposition of short-lived climate forcers on snow cover changes is an emerging issue (Skiles et al., 2018 and references therein). This concerns light absorbing particles, in particular, which include deposited aerosols such as black carbon, organic carbon and mineral dust, or microbial growth (Qian et al., 2015), although the role of the latter has not been specifically quantified. Due to their seasonally variable deposition flux and impact, and mostly episodic nature in case of dust deposition (Kaspari et al., 2014; Di Mauro et al., 2015), light absorbing particles contribute to interannual fluctuations of seasonal snowmelt rate (Painter et al., 2018) (*medium evidence, high agreement*). There is *limited evidence* (*medium agreement*) that increases in black carbon deposition from anthropogenic and biomass burning sources have contributed to snow cover decline in High Mountain Asia (Li et al., 2016; Zhang et al., 2018) and South America (Molina et al., 2015).

Projected changes of mountain snow cover are studied based on climate model experiments, either directly from GCM or RCM output, or following downscaling and the use of snowpack models. These projections generally do not specifically account for future changes in the deposition rate of light absorbing particles on snow (or, if so, simple approaches have been used hitherto; e.g., Deems et al., 2013), so that future changes in snow conditions are mostly driven by changes in meteorological drivers assessed in Section 2.2.1. Evidence from regional studies is provided in Table SM2.7. Although existing studies in mountain regions do not use homogenous reference periods and model configurations, common future trends can be summarised as follows. At lower elevation in many regions such as the European Alps, Western North America, Himalaya and subtropical Andes, the snow depth or mass is projected to decline by 25% (*likely* range between 10 and 40%), between the recent past period (1986–2005) and the near future (2031–2050), regardless of the greenhouse gas emission scenario (Cross-Chapter Box 1 in Chapter 1). This corresponds to a continuation of the ongoing decrease in annual snow cover duration (on average 5 days per decade, with a *likely* range from 0 to 10). By the end of the century (2081–2100), reductions of up to 80% (*likely* range from 50 to 90%) are expected under RCP8.5, 50% (*likely* range from 30 to 70%) under RCP4.5 and 30% (*likely* range from 10 to 40%) under RCP2.6. At higher elevations, projected reductions are smaller (*high confidence*), as temperature increases at higher elevations affect the ablation component of snow mass evolution, rather than both the onset and accumulation components. The projected increase in winter snow accumulation may result in a net increase in winter snow mass (*medium confidence*). All elevation levels and mountain regions are projected to exhibit sustained interannual variability of snow conditions throughout the 21st century (*high confidence*). Figure 2.3 provides projections of temperature and snow cover in mountain areas in Europe, High Mountain Asia (Hindu Kush, Karakoram and Himalaya), North America (Rocky Mountains) and South America (sub-tropical Central Andes), illustrating how changes vary with elevation, season, region, future time period and climate scenario.

2.2.3   Glaciers

The high mountain areas considered in this chapter (Figure 2.1), including all glacier regions in the world except those in Antarctica, Greenland, the Canadian and Russian Arctic, and Svalbard (which are covered in Chapter 3) include ~170,000 glaciers covering an area of ~250,000 km² (RGI Consortium, 2017) with a total ice volume of 87 ± 15 mm sea level equivalent (Farinotti et al., 2019). These glaciers span an elevation range from sea level, for example in south-east Alaska, to >8,000 m a.s.l. in the Himalaya and Karakoram, and occupy diverse climatic regions. Their mass budget is determined largely by the balance between snow accumulation and melt at the glacier surface, driven primarily by atmospheric conditions. Rapid changes in mountain glaciers have multiple impacts for social-ecological systems, affecting not only biophysical properties such as runoff volume and sediment fluxes in glacier-fed rivers, glacier related hazards, and global sea level (Chapter 4) but also ecosystems and human livelihoods, socioeconomic activities and sectors such as agriculture and tourism, as well as other intrinsic assets such as cultural values. While glaciers worldwide have experienced considerable fluctuations throughout the Holocene driven by multidecadal variations of solar and volcanic activity, and changes in atmospheric circulation

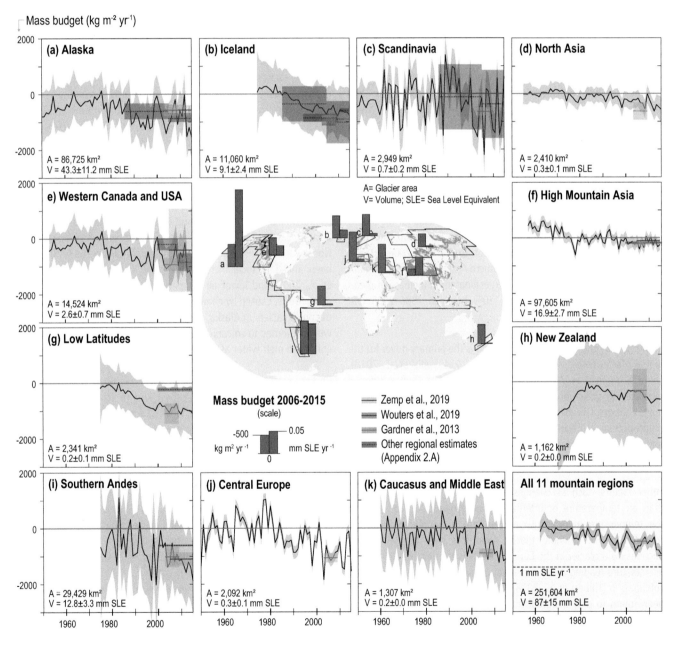

Figure 2.4 | Glacier mass budgets for the eleven mountain regions assessed in this Chapter (Figure 2.1) and these regions combined. Mass budgets for the remaining polar regions are shown in Chapter 3, Figure 3.8. Regional time series of annual mass change are based on glaciological and geodetic balances (Zemp et al., 2019). Superimposed are multi-year averages by Wouters et al. (2019) based on the Gravity Recovery and Climate Experiment (GRACE), only shown for the regions with glacier area >3,000 km². Estimates by Gardner et al. (2013) were used in the IPCC 5th Assessment Report (AR5). Additional regional estimates available in some regions and shown here are listed in Table 2.A.1. Annual and time-averaged mass-budget estimates include the errors reported in each study. Glacier areas (A) and volumes (V) are based on RGI Consortium (2017) and Farinotti et al. (2019), respectively. Red and blue bars on map refer to regional budgets averaged over the period 2006–2015 in units of kg m⁻² yr⁻¹ and mm sea level equivalent (SLE) yr⁻¹, respectively, and are derived from each region's available mass-balance estimates (Appendix 2.A).

(Solomina et al., 2016), this section focuses on observed glacier changes during recent decades and changes projected for the 21st century (Cross-Chapter Box 6 in Chapter 2).

Satellite and *in situ* observations of changes in glacier area, length and mass show a globally largely coherent picture of mountain glacier recession in the last decades (Zemp et al., 2015), although annual variability and regional differences are large (Figure 2.4; *very high confidence*). The global trend is statistically significant despite considerable interannual and regional variations (Medwedeff and Roe, 2017). Since AR5's global 2003–2009 estimate based on Gardner et al. (2013), several new estimates of global-scale glacier mass budgets have emerged using largely improved data coverage and methods (Bamber et al., 2018; Wouters et al., 2019; Zemp et al., 2019).

These estimates combined with available regional estimates (Table 2.A.1) indicate that the glacier mass budget of all mountain regions (excluding Antarctica, Greenland, the Canadian and Russian Arctic, and Svalbard) was *very likely* -490 ± 100 kg m^{-2} yr^{-1} (-123 ± 24 Gt yr^{-1}) during the period 2006–2015 with most negative averages (less than -850 kg m^{-2} yr^{-1}) in the Southern Andes, Caucasus/Middle East, European Alps and Pyrenees. High Mountain Asia shows the least negative mass budget (-150 ± 110 kg m^{-2} yr^{-1}, Figure 2.4), but variations within the region are large with most negative regional balance estimates in Nyainqentanglha, Tibet (-620 ± 230 kg m^{-2} yr^{-1}) and slightly positive balances in the Kunlun Mountains for the period 2000–2016 (Brun et al., 2017). Due to large ice extent, the total mass loss and corresponding contribution to sea level 2006–2015 is largest in Alaska, followed by the Southern Andes and High Mountain Asia (Table 2.A.1). Zemp et al. (2019) estimated an increase in mean global-scale glacier mass loss by ~30% between 1986–2005 and 2006–2015.

It is *very likely* that atmospheric warming is the primary driver for the global glacier recession (Marzeion et al., 2014; Vuille et al., 2018). There is *limited evidence* (*high agreement*) that human-induced increases in greenhouse gases have contributed to the observed mass changes (Hirabayashi et al., 2016). It was estimated that the anthropogenic fraction of mass loss of all glaciers outside Greenland and Antarctica increased from 25 ± 35% during 1851–2010 to 69 ± 24% during 1991–2010 (Marzeion et al., 2014).

Other factors, such as changes in meteorological variables other than air temperature or internal glacier dynamics, have modified the temperature-induced glacier response in some regions (*high confidence*). For example, glacier mass loss over the last seven decades on a glacier in the European Alps was intensified by higher air moisture leading to increased longwave irradiance and reduced sublimation (Thibert et al., 2018). Changes in air moisture have also been found to play a significant role in past glacier mass changes in eastern Africa (Prinz et al., 2016), while an increase in shortwave radiation due to reduced cloud cover contributed to an acceleration in glacier recession in the Caucasus (Toropov et al., 2019). In the Tien Shan mountains changes in atmospheric circulation in the North Atlantic and North Pacific in the 1970s resulted in an abrupt reduction in precipitation and thus snow accumulation, amplifying temperature-induced glacier mass loss (Duethmann et al., 2015).

Deposition of light absorbing particles, growth of algae and bacteria and local amplification phenomena such as the enhancement of particles concentration due to surface snow and ice melt, and cryoconite holes, have been shown to enhance ice melt (e.g., Ginot et al., 2014; Zhang et al., 2017; Williamson et al., 2019) but there is *limited evidence* and *low agreement* that long-term changes in glacier mass are linked to light absorbing particles (Painter et al., 2013; Sigl et al., 2018). Debris cover can modulate glacier melt but there is *limited evidence* on its role in recent glacier changes (Gardelle et al., 2012; Pelliccciotti et al., 2015). Rapid retreat of calving outlet glaciers in Patagonia was attributed to changes in glacier dynamics (Sakakibara and Sugiyama, 2014).

Departing from this global trend of glacier recession, a small fraction of glaciers have gained mass or advanced in some regions mostly due to internal glacier dynamics or, in some cases, locally restricted climatic causes. For example, in Alaska 36 marine-terminating glaciers exhibited a complex pattern of periods of significant retreat and advance during 1948–2012, highly variable in time and lacking coherent regional behaviour (McNabb and Hock, 2014). These fluctuations can be explained by internal retreat-advance cycles typical of tidewater glaciers that are largely independent of climate (Brinkerhoff et al., 2017). Irregular and spatially inconsistent glacier advances, for example, in Alaska, Iceland and Karakoram, have been associated with surge-type flow instabilities largely independent of changes in climate (Sevestre and Benn, 2015; Bhambri et al., 2017; Section 2.3.2). Regional scale glacier mass gain and advances in Norway in the 1990s and in New Zealand between 1983–2008 have been linked to local increases in snow precipitation (Andreassen et al., 2005) and lower air temperatures (Mackintosh et al., 2017), respectively, caused by changes in atmospheric circulation. Advances of some glaciers in Alaska, the Andes, Kamchatka and the Caucasus were attributed to volcanic activity causing flow acceleration through enhanced melt water at the ice-bed interface (Barr et al., 2018).

Region averaged glacier mass budgets have been nearly balanced in the Karakoram since at least the 1970s (Bolch et al., 2017; Zhou et al., 2017; Azam et al., 2018), while slightly positive balances since 2000 have been reported in the western Kunlun Shan, eastern Pamir, and the central and northern Karakoram mountains (Gardelle et al., 2013; Brun et al., 2017; Lin et al., 2017; Berthier and Brun, 2019). This anomalous behavior has been related to specific mechanisms countering the effects of atmospheric warming, for example, an increase in cloudiness (Bashir et al., 2017) and snowfall (Kapnick et al., 2014) spatially heterogeneous glacier mass balance sensitivity (Sakai and Fujita, 2017), feedbacks due to intensified lowland irrigation (de Kok et al., 2018), and changes in summer atmospheric circulation (Forsythe et al., 2017).

There is *medium evidence* (*high agreement*) that recent glacier mass changes have modified glacier flow. A study covering all glaciers in High Mountain Asia showed glacier slowdown for regions with negative mass budgets since the 1970s and slightly accelerated glacier flow for Karakoram and West Kunlun regions where mass budgets were close to balance (Dehecq et al., 2019). Waechter et al. (2015) report reduced flow velocities in the St. Elias Mountains in North America, especially in areas of rapid ice thinning near glacier termini. In contrast Mouginot and Rignot (2015) found complex ice flow patterns with simultaneous acceleration and deceleration for glaciers of the Patagonian Icefield as well as large interannual variability during the last three decades concurrent with general thinning of the ice field.

Cross-Chapter Box 6 | Glacier Projections in Polar and High Mountain Regions

Authors: Regine Hock (USA), Andrew Mackintosh (Australia/New Zealand), Ben Marzeion (Germany)

Century-scale projections for all glaciers on Earth including those around the periphery of Greenland and Antarctica are presented here. Projections of the Greenland and Antarctic ice sheets are presented in Chapter 4. Future changes in glacier mass have global implications through their contribution to sea level change (Chapter 4) and local implication, for example, by affecting freshwater resources (Section 2.3.1). Glacier decline can also lead to loss of palaeoclimate information contained in glacier ice (Thompson et al., 2017).

AR5 included projections of 21st century glacier evolution from four process-based global-scale glacier models (Slangen and Van De Wal, 2011; Marzeion et al., 2012; Giesen and Oerlemans, 2013; Radić et al., 2014). Results have since been updated (Bliss et al., 2014; Slangen et al., 2017; Hock et al., 2019) using new glacier inventory data and/or climate projections, and projections from two additional models have been presented (Hirabayashi et al., 2013; Huss and Hock, 2015). These six models were driven by climate projections from 8–21 General Circulation Models (GCMs) from the CMIP5 (Taylor et al., 2012) forced by various RCPs, and results are systematically compared in Hock et al. (2019).

Based on these studies there is *high confidence* that glaciers in polar and high mountain regions will lose substantial mass by the end of the century. Results indicate global glacier mass losses by 2100 relative to 2015 of 18% (*likely* range 11–25%) (mean of all projections with range referring to ± one standard deviation) for scenario RCP2.6 and 36% (*likely* range 26–47%) for RCP8.5, but relative mass reductions vary greatly between regions (Figure CB6.1). Projected end-of-century mean mass losses relative to 2015 tend to be largest in mountain regions dominated by smaller glaciers and relatively little ice cover, exceeding on average 80%, for example, the European Alps, Pyrenees, Caucasus/Middle East, Low Latitudes and North Asia for RCP8.5 (see Figure 2.1 for region definitions). While these glaciers' contribution to sea level is negligible their large relative mass losses have implications for streamflow (Section 2.3.1, FAQ 2.1).

The magnitude and timing of these projected mass losses is assigned *medium confidence* because the projections have been carried out using relatively simple models calibrated with limited observations in some regions and diverging initial glacier volumes. For example, mass loss by iceberg calving and subaqueous melt processes that can be particularly important components of glacier mass budgets in polar regions (McNabb et al., 2015) have only been included in one global-scale study (Huss and Hock, 2015). In addition instability mechanisms that can cause rapid glacier retreat and mass loss are not considered (Dunse et al., 2015; Sevestre et al., 2018; Willis et al., 2018).

The projected global-scale relative mass losses 2015–2100 correspond to a sea level contribution of 94 (*likely* range 69–119) mm sea level equivalent (SLE) corresponding to an average rate of 1.1 (*likely* range 0.8–1.4) mm SLE yr^{-1} for RCP2.6, and 200 (*likely* range 156 to 240) mm SLE, a rate of 2.4 (*likely* range 1.8–2.8) mm SLE yr^{-1} for RCP8.5, in addition to the sea level contribution from the Greenland and Antarctic ice sheets (Chapter 4). Averages refer to the mean and ranges to ± one standard deviation of all simulations. For RCP2.6, rates increase only slightly until approximately year 2040 with a steady decline thereafter, as glaciers retreat to higher elevations and reach new equilibrium. In contrast, for RCP8.5, the sea level contribution from glaciers increases steadily for most of the century, reaching an average maximum rate exceeding 3 mm SLE yr^{-1} (Hock et al., 2019). For both RCPs the polar regions are the largest contributors with projected mass reductions by 2100 relative to 2015 combined for the Antarctic periphery, Arctic Canada, the Greenland periphery, Iceland, Russian Arctic, Scandinavia and Svalbard ranging from 16% (*likely* range 9 to 23%) for RCP2.6 to 33% (*likely* range 22 to 44%) for RCP8.5. Due to extensive ice cover, these regions make up roughly 80% of the global sea level contribution from glaciers by 2100. The global projections are similar to those reported in AR5 for the period 2081–2100 relative to 1986–2005, if differences in period length and domain are accounted for (AR5's glacier estimates excluded the Antarctic periphery). The eleven mountain regions covered in Chapter 2 are *likely* to lose 22–44% of their glacier mass by 2100 relative to 2015 for RCP2.6 and 37–57% for RCP8.5. Worldwide many glaciers are expected to disappear by 2100 regardless emission scenario, especially in regions with smaller glaciers (*very high confidence*) (Cullen et al. 2013; Rabatel et al., 2013; Huss and Fischer, 2016; Rabatel et al., 2017).

Cross-Chapter Box 6 (continued)

The global-scale projections (Figure CB6.1) are consistent with results from regional-scale studies using more sophisticated models. Kraaijenbrink et al. (2017) projected mass losses for all glaciers in High Mountain Asia of 64 ± 5% (RCP8.5) by the end of the century (2071–2100) compared to 1996–2015. A high-resolution regional glaciation model including ice dynamics indicated that by 2100 glacier volume in western Canada will shrink by ~70% (RCP2.6) to ~90% (RCP8.5) relative to 2005 (Clarke et al., 2015). Zekollari et al. (2019) projected that the glaciers in the European Alps will largely disappear by 2100 (94 ± 4% mass loss relative to 2017) for RCP8.5, while projected mass losses are 63 ± 11% for RCP2.6.

AR5 concluded with *high confidence* that due to a pronounced imbalance between current glacier mass and climate, glaciers are expected to further recede even in the absence of further climate change. Studies since AR5 agree and provide further evidence (Mernild et al., 2013; Marzeion et al., 2018).

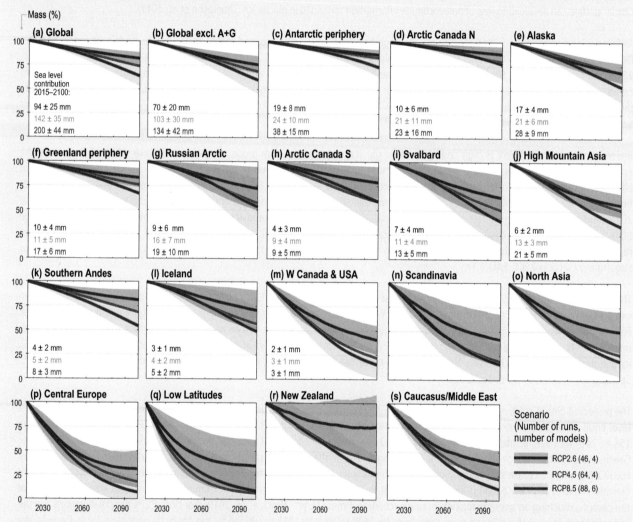

Figure CB6.1 | Projected glacier mass evolution between 2015 and 2100 relative to each region's glacier mass in 2015 (100%) based on three Representative Concentration Pathways (RCP) emission scenarios (Cross-Chapter Box 1 in Chapter 1). Thick lines show the averages of 46 to 88 model projections based on four to six glacier models for the same RCP, and the shading marks ± 1 standard deviation (not shown for RCP4.5 for better readability). Global projections are shown excluding and including the Antarctic (A) and Greenland (G) periphery. Regional sea level contributions are given for three RCPs for all regions with >0.5 mm sea level equivalent (SLE) between 2015–2100. The Low Latitudes region includes the glaciers in the tropical Andes, Mexico, eastern Africa and Indonesia. Region Alaska includes adjacent glaciers in the Yukon and British Columbia, Canada. Regions are sorted by glacier volume according to Farinotti et al. (2019). Data based on Marzeion et al. (2012); Giesen and Oerlemans (2013); Hirabayashi et al. (2013); Bliss et al. (2014); Huss and Hock (2015); Slangen et al. (2017). Modified from Hock et al. (2019).

2.2.4 Permafrost

This section assesses permafrost, but not seasonally frozen ground, in high mountain areas. As mountains also exist in polar areas, some overlap exists between this section and Chapter 3. Observations of permafrost are scarce (Tables 2.1 and 2.2, PERMOS, 2016; Bolch et al., 2018) and unevenly distributed among and within mountain regions. Unlike glaciers and snow, permafrost is a subsurface phenomenon that cannot easily be observed remotely. As a consequence, its distribution and change are less understood than for glaciers or snow, and in many mountain regions it can only be inferred (Gruber et al., 2017). Permafrost thaw and degradation impact people via runoff and water quality (Section 2.3.1), hazards and infrastructure (Section 2.3.2) and greenhouse gas emissions (Box 2.2).

AR5 and IPCC's Special Report on 'Managing the Risks of Extreme Events and Disasters to Advance Climate Change Adaptation' (SREX) assessed permafrost change globally, but not separately for mountains. AR5 concluded that permafrost temperatures had increased in most regions since the early 1980s (*high confidence*), although warming rates varied regionally, and attributed this warming to increased air temperature and changes in snow cover (*high confidence*). The temperature increase for colder permafrost was generally greater than for warmer permafrost (*high confidence*). SREX found a *likely* warming of permafrost in recent decades and expressed *high confidence* that its temperatures will continue to increase. AR5 found decreases of northern high-latitude near surface permafrost for 2016–2035 to be *very likely* and a general retreat of permafrost extent for the end of the 21st century and beyond to be *virtually certain*. While some permafrost phenomena, methods of observation and scale issues in scenario simulations are specific to mountainous terrain, the basic mechanisms connecting climate and permafrost are the same in mountains and polar regions.

Between 3.6–5.2 million km^2 are underlain by permafrost in the eleven high mountain regions outlined in Figure 2.1 (*medium confidence*) based on data from two modelling studies (Gruber, 2012; Obu et al., 2019). For comparison, this is 14–21 times the area of glaciers (Section 2.2.3) in these regions (Figure 2.1) or 27–29% of the global permafrost area. The distribution of permafrost in mountains is spatially highly heterogeneous, as shown in detailed regional modelling studies (Boeckli et al., 2012; Bonnaventure et al., 2012; Westermann et al., 2015; Azócar et al., 2017; Zou et al., 2017).

Permafrost in the European Alps, Scandinavia, Canada, Mongolia, the Tien Shan and the Tibetan Plateau has warmed during recent decades and some observations reveal ground-ice loss and permafrost degradation (*high confidence*). The heterogeneity of mountain environments and scarcity of long-term observations challenge the quantification of representative regional or global warming rates. A recent analysis finds that permafrost at 28 mountain locations in the European Alps, Scandinavia, Canada as well as High Mountain Asia and North Asia, warmed on average by 0.19 ± 0.05°C per decade between 2007–2016 (Biskaborn et al., 2019). Over longer periods, observations in the European Alps, Scandinavia, Mongolia, the Tien Shan and the Tibetan Plateau (see also Cao et al., 2018) show general warming (Table 2.1, Figure 2.5) and degradation of permafrost at

individual sites (e.g., Phillips et al., 2009). Permafrost close to 0°C warms at a lower rate than colder permafrost because ground-ice melt slows warming. Similarly, bedrock warms faster than debris or soil because of low ice content. For example, several European bedrock sites (Table 2.1) have warmed rapidly, by up to 1°C per decade, during the past two decades. By contrast, total warming of 0.5°C–0.8°C has been inferred for the second half of the 20th century based on thermal gradients at depth in an ensemble of European bedrock sites (Isaksen et al., 2001; Harris et al., 2003). Warming has been shown to accelerate at sites in Scandinavia (Isaksen et al., 2007) and in mountains globally within the past decade (Biskaborn et al., 2019). During recent decades, rates of permafrost warming in the European Alps and Scandinavia exceeded values of the late 20th century (*limited evidence, high agreement*).

The observed thickness of the active layer (see Annex I: Glossary), the layer of ground above permafrost subject to annual thawing and freezing, increased in the European Alps, Scandinavia (Christiansen et al., 2010), and on the Tibetan Plateau during the past few decades (Table 2.2), indicating permafrost degradation. Geophysical monitoring in the European Alps during approximately the past 15 years revealed increasing subsurface liquid water content (Hilbich et al., 2008; Bodin et al., 2009; PERMOS, 2016), indicating gradual ground-ice loss.

During recent decades, the velocity of rock glaciers in the European Alps exceeded values of the late 20th century (*limited evidence, high agreement*). Some rock glaciers, that is, masses of ice-rich debris that show evidence of past or present movement, show increasing velocity as a transient response to warming and water input, although continued permafrost degradation would eventually inactivate them (Ikeda and Matsuoka, 2002). Rock glacier velocities observed in the European Alps in the 1990s were on the order of a few decimetres per year and during approximately the past 15 years they often were about 2–10 times higher (Bodin et al., 2009; Lugon and Stoffel, 2010; PERMOS, 2016). Destabilisation, including collapse and rapid acceleration, has been documented (Delaloye et al., 2010; Buchli et al., 2013; Bodin et al., 2016). One particularly long time series shows velocities around 1960 just slightly lower than during recent years (Hartl et al., 2016). In contrast to nearby glaciers, no clear change in rock glacier velocity or elevation was detected at a site in the Andes between 1955–1996 (Bodin et al., 2010). The majority of similar landforms investigated in the Alaska Brooks Range increased their velocity since the 1950s, while few others slowed down (Darrow et al., 2016).

Decadal-scale permafrost warming and degradation are driven by air temperature increase and additionally affected by changes in snow cover, vegetation and soil moisture. Bedrock locations, especially when steep and free of snow, produce the most direct signal of climate change on the ground thermal regime (Smith and Riseborough, 1996), increasing the confidence in attribution. Periods of cooling, one or few years long, have been observed and attributed to extraordinary low-snow conditions (PERMOS, 2016). Extreme increases of active-layer thickness often correspond with summer heat waves (PERMOS, 2016) and permafrost degradation can be accelerated by water percolation (Luethi et al., 2017). Similarity and synchronicity of interannual to decadal velocity changes of rock glaciers within the European Alps (Bodin et al., 2009; Delaloye et al., 2010) and the Tien Shan (Sorg et al., 2015), suggest common regional forcing such as summer air temperature or snow cover.

Because air temperature is the major driver of permafrost change, permafrost in high mountain regions is expected to undergo increasing thaw and degradation during the 21st century, with stronger consequences expected for higher greenhouse gas emission scenarios (*very high confidence*). Scenario simulations for the Tibetan Plateau until 2100 estimate permafrost area to be strongly reduced, for example by 22–64% for RCP2.6 and RCP8.5 and a spatial resolution of 0.5° (Lu et al., 2017). Such coarse-scale studies (Guo et al., 2012; Slater and Lawrence, 2013; Guo and Wang, 2016), however, are of limited use in quantifying changes and informing impact studies in steep terrain due to inadequate representation of topography (Fiddes and Gruber, 2012). Fine-scale simulations, on the other hand, are local or regional, limited in areal extent and differ widely in their representation of climate change and permafrost. They reveal regional and elevational differences of warming and degradation (Bonnaventure and Lewkowicz, 2011; Hipp et al., 2012; Farbrot et al., 2013) as well as warming rates that differ between locations (Marmy et al., 2016) and seasons (Marmy et al., 2013). While structural differences in simulations preclude a quantitative summary, these studies agree on increasing warming and thaw of permafrost for the 21st century and reveal increased loss of permafrost under stronger atmospheric warming (Chadburn et al., 2017). Permafrost thaw at depth is slow but can be accelerated by mountain peaks warming from multiple sides (Noetzli and Gruber, 2009) and deep percolation of water (Hasler et al., 2011). Near Mont Blanc in the European Alps, narrow peaks below 3,850 m a.s.l. may lose permafrost entirely under RCP8.5 by the end of the 21st century (Magnin et al., 2017). As ground-ice from permafrost usually melts slower than glacier ice, some mountain regions will transition from having abundant glaciers to having few and small glaciers but large areas of permafrost that is thawing (Haeberli et al., 2017).

Table 2.1 | Observed changes in permafrost mean annual ground temperature (MAGT) in mountain regions. Values are based on individual boreholes or ensembles of several boreholes. The MAGT refers to the last year in a period and is taken from a depth of 10–20 m unless the borehole is shallower. Region names refer to Figure 2.1. Numbers in brackets indicate how many sites are summarised for a particular surface type and area; the underscored value is an average. Elevation is metres above sea level (m a.s.l.).

Region	Elevation [m a.s.l.]	Surface Type	Period	MAGT [°C]	MAGT trend [°C per decade]	Reference
Global	>1,000	various (28)	2006–2017	not specified	0.2 ± 0.05	Biskaborn et al. (2019)
European Alps	2,500–3,000	debris or coarse blocks (>10)	1987–2005	>-3	0.0–0.2	PERMOS (2016)
			2006–2017	>-3	0.0–0.6	Noetzli et al. (2018)
	3,500–4,000	bedrock (4)	2008–2017	>-5.5	0.0–1.0	Pogliotti et al. (2015) Magnin et al. (2015) Noetzli et al. (2018)
Scandinavia	1,402–1,505	moraine (3)	1999–2009	0 to -0.5	0.0–0.2	Isaksen et al. (2011)
	1,500–1,894	bedrock (2)	1999–2009	-2.7	0.5	Christiansen et al. (2010)
High Mountain Asia (Tien Shan)	~3,330	bare soil (2)	1974–2009	-0.5 to -0.1	0.3–0.6	Zhao et al. (2010)
	3,500	meadow (1)	1992–2011	-1.1	0.4	Liu et al. (2017)
High Mountain Asia (Tibetan Plateau)	~4,650	meadow (6)	2002–2012	-1.52 to -0.41	0.08–0.24	Wu et al. (2015)
	~4,650	steppe (3)	2002–2012	-0.79 to -0.17		Wu et al. (2015)
	~4,650	bare soil (1)	2003–2012	-0.22	0.15	Wu et al. (2015)
	4,500–5,000	unknown (6)	2002–2011	-1.5 to -0.16	0.08–0.24	Peng et al. (2015)
North Asia (Mongolia)	1,350–2,050	steppe (6)	2000–2009	-0.06 to -1.54	0.2–0.3	Zhao et al. (2010)

Table 2.2 | Observed changes of active-layer thickness (ALT) in mountain regions. Numbers in brackets indicate how many sites are summarised for a particular surface type and area. Region names refer to Figure 2.1. Elevation is metres above sea level (m a.s.l.).

Region	Elevation [m a.s.l.]	Surface Type	Period	ALT in last year [m]	ALT trend [cm per decade]	Reference
Scandinavia	353–507	peatland (9)	1978–2006 1997–2006	~0.65–0.85	7–13 13–20	Åkerman and Johansson (2008)
European Alps	2,500–2,910	bedrock (4)	2000–2014	4.2–5.2	10–100	PERMOS (2016)
High Mountain Asia (Tien Shan)	3,500	meadow (1)	1992–2011	1.70	19	Liu et al. (2017)
High Mountain Asia (Tibetan Plateau)	4,629–4,665	meadow (6)	2002–2012	2.11–2.3	34.8–45.7	Wu et al. (2015
	4,638–4,645	steppe (3)	2002–2012	2.54–3.03	39.6–67.2	Wu et al. (2015
	4,635	bare soil (1)	2002–2012	3.38	18.9	Wu et al. (2015
	4,848	meadow	2006–2014	1.92–2.72	15.2–54	Lin et al. (2016)

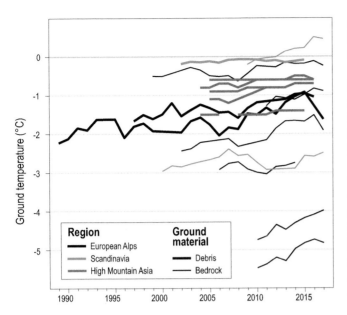

Figure 2.5 | Mean annual ground temperature from boreholes in debris and bedrock in the European Alps, Scandinavia and High Mountain Asia. Temperatures differ between locations and warming trends can be interspersed by short periods of cooling. One location shows degrading of permafrost. Overall, the number of observed boreholes is small and most records are short. The depth of measurements is approximately 10 m, and years without sufficient data are omitted (Noetzli et al., 2018).

2.2.5 Lake and River Ice

Based on *limited evidence*, AR5 reported shorter seasonal ice cover duration during the past decades (*low confidence*), however, did not specifically address changes in mountain lakes and rivers. Observations of extent, timing, duration and thickness of lake and river ice rely mostly on *in situ* measurements (e.g., Sharma et al., 2019) and, increasingly on remote sensing (Duguay et al., 2014). Lake and river ice studies focusing specifically on mountain regions are rare but observations from lakes in the European Alps, Scandinavia and the Tibetan Plateau show highly variable trends in ice cover duration during the past decades.

For example, Cai et al. (2019) reported shorter ice cover duration for 40 lakes and longer duration for 18 lakes on the Tibetan Plateau during the period 2000–2017. Similarly, using microwave remote sensing, Du et al. (2017) found shorter ice cover duration for 43 out of 71 lakes >50 km² including lakes on the Tibetan Plateau during 2002–2015, but only five of these had statistically significant trends ($p < 0.05$), due to large interannual variability. The variable trends in the duration of lake ice cover on the Tibetan Plateau between 2002–2015 corresponded to variable trends in surface water temperatures. Of 52 study lakes in this region, 31 lakes showed a mean warming rate of 0.055 ± 0.033°C yr⁻¹, and 21 lakes showed a mean cooling rate of -0.053 ± 0.038°C yr⁻¹ during 2001–2012 (Zhang et al., 2014). Kainz et al. (2017) reported a significant ($p < 0.05$) increase in the interannual variability in ice cover duration for a subalpine lake in Austria during 1921–2015 in addition to a significant trend in later freeze on, earlier ice break up and shorter ice cover duration.

A significant ($p < 0.05$) trend towards shorter ice cover duration was found for another Austrian alpine lake during 1972–2015 (Niedrist et al., 2018).

Highly variable trends were also found in the timing and magnitude of river ice jams during 1903–2015, as reported by Rokaya et al. (2018) for Canadian rivers, including rivers in the mountains. Most of the variability in river ice trends could be explained by variable water flow, in particular due to flow regulation.

There is *high confidence* that air temperature and solar radiation are the most important drivers to explain observed changes of lake ice dynamics (Sharma et al., 2019). In mountainous regions where the interannual variability in ice cover duration is high, additional drivers become important, for example, morphometry, wind exposure, salinity, and hydrology, in particular hydrological processes driven by glaciers (Kropáček et al., 2013; Song et al., 2014; Yao et al., 2016; Gou et al., 2017). Despite high spatial and temporal variability in lake and river ice cover dynamics in mountain regions there is *limited evidence* (*high agreement*) that further air temperature increases will result in a general trend towards later freezing, earlier break-up, and shorter ice cover duration in the future (Gebre et al., 2014; Du et al., 2017).

Overall, there is only *limited evidence* on changes in lake and river ice specifically in the mountains, indicating a trend, but not universally, towards shorter lake ice cover duration consistent with increased water temperature.

Box 2.2 | Local, Regional and Global Climate Feedbacks Involving the Mountain Cryosphere

The cryosphere interacts with the environment and contributes to several climate feedbacks, most notably ones involving the snow cover, referred to as the snow albedo feedback. The presence or absence of snow on the ground drives profound changes in the energy budget of land surfaces, hence influencing the physical state of the overlying atmosphere (Armstrong and Brun, 2008). The reduction of snow on the ground, potentially amplified by aerosol deposition and modulated by interactions with the vegetation, increases the absorption of incoming solar radiation and leads to atmospheric warming. In mountain regions, this positive feedback loop mostly operates at the local scale and is seasonally variable, with most visible effects at the beginning and end of the snow season (Scherrer et al., 2012). Examples of other mechanisms contributing to local feedbacks are introduced in Box 2.1. At the regional scale, feedbacks associated with deposition of light absorbing particles and enhanced snow albedo feedback were shown to induce surface air warming (locally up to 2°C) (Ménégoz et al., 2014) with accelerated snow cover reduction (Ji, 2016; Xu et al., 2016), and may also influence the Asian monsoon system (Yasunari et al., 2015). However, many of these studies have considered so-called rapid adjustments, without changes in large-scale atmospheric circulation patterns, because they used regional or global models constrained by large-scale synoptic fields. In summary, regional climate feedbacks involving the high mountain cryosphere, particularly the snow albedo feedback, have only been detected in large mountain regions such as the Himalaya, using global and regional climate models (*medium confidence*).

Global-scale climate feedbacks from the cryosphere remain largely unexplored with respect to the proportion originating from high mountains. Although mountain topography affects global climate (e.g., Naiman et al., 2017), there is little evidence for mountain-cryosphere specific feedbacks, largely because of the limited spatial extent of the mountain cryosphere. The most relevant feedback probably relates to permafrost in mountains, which contain about 28% of the global permafrost area (Section 2.2.4). Organic carbon stored in permafrost can be decayed following thaw and transferred to the atmosphere as carbon dioxide or methane (Schuur et al., 2015). This self-reinforcing effect accelerates the pace of climate change and operates in polar (Section 3.4.1.2.3) and mountain areas alike (Mu et al., 2017; Sun et al., 2018a). In contrast to polar areas, however, there is *limited evidence* and *low agreement* on the total amount of permafrost carbon in mountains because of differences in upscaling and difficulties to distinguish permafrost and seasonally frozen soils due to the lack of data. For example, on the Tibetan Plateau, the top 3 m of permafrost are estimated to contain about 15 petagrams (Ding et al., 2016) and mountain soils with permafrost globally are estimated to contain approximately 66 petagrams of organic carbon (Bockheim and Munroe, 2014). At the same time, there is *limited evidence* and *high agreement* that the average density (kg C m^{-2}) of permafrost carbon in mountains is lower than in other areas. For example, densities of soil organic carbon are low in the sub-arctic Ural (Dymov et al., 2015) and 1–2 orders of magnitude lower in subarctic Sweden (Fuchs et al., 2015) in comparison to lowland permafrost, and 50% lower in mountains than in steppe-tundra in Siberia and Alaska (Zimov et al., 2006). Some mechanisms of soil carbon decay and transfer to the atmosphere in mountains are similar to those in lowland areas, for example collapse following thaw in peatlands (Mu et al., 2016; Mamet et al., 2017), and some are specific to areas with steep slopes, for example drainage of water from thawing permafrost leading to soil aeration (Dymov et al., 2015). There is no global-scale analysis of the climate feedback from permafrost in mountains. Given that projections indicate increasing thaw and degradation of permafrost in mountains during the 21st century (*very high confidence*) (Section 2.2.4), a corresponding increase in greenhouse gas emissions can be anticipated but is not quantified.

2.3   Mountain Social-Ecological Systems: Impacts, Risks and Human Responses

2.3.1   Water Resources

The mountain cryosphere is an important source of freshwater in the mountains themselves and in downstream regions. The runoff per unit area generated in mountains is on average approximately twice as high as in lowlands (Viviroli et al., 2011) making mountains a significant source of fresh water in sustaining ecosystem and supporting livelihoods in and far beyond the mountain ranges themselves. The presence of snow, glaciers, and permafrost generally exert a strong control on the amount, timing and biogeochemical properties of runoff (FAQ 2.1). Changes to the cryosphere due to climate change can alter freshwater availability with direct consequences for human populations and ecosystems.

2.3.1.1   Changes in River Runoff

AR5 reported increased winter flows and a shift in timing towards earlier spring snowmelt runoff peaks during previous decades (*robust evidence, high agreement*). In glacier-fed river basins, it was projected that melt water yields from glaciers will increase for decades in many regions but then decline (*very high confidence*). These findings have been further supported and refined by a wealth of new studies since AR5.

Recent studies indicate considerable changes in the seasonality of runoff in snow and glacier dominated river basins (*very high confidence*; Table SM2.9). Several studies have reported an increase in average winter runoff over the past decades, for example in Western Canada (Moyer et al., 2016), the European Alps (Bocchiola, 2014; Bard et al., 2015) and Norway (Fleming and Dahlke, 2014), due to more precipitation falling as rain under warmer conditions. Summer runoff has been observed to decrease in basins, for example in Western Canada (Brahney et al., 2017) and the European Alps (Bocchiola, 2014), but to increase in several basins in High Mountain Asia (Mukhopadhyay and Khan, 2014; Duethmann et al., 2015; Reggiani and Rientjes, 2015; Engelhardt et al., 2017). Both increases, for example, in Alaska (Beamer et al., 2016) and the Tien Shan (Wang et al., 2015; Chen et al., 2016), and decreases, for example, in Western Canada (Brahney et al., 2017) have also been found for average annual runoff. In Western Austria, Kormann et al. (2015) detected an increase in annual flow at high elevations and a decrease at low elevations between 1980–2010.

These contrasting trends for summer and annual runoff often result from spatially variable changes in the contribution of glacier and snow melt. As glaciers shrink, annual glacier runoff typically first increases, until a turning point, often called 'peak water' is reached, upon which runoff declines (FAQ 2.1). There is *robust evidence* and *high agreement* that peak water in glacier-fed rivers has already passed with annual runoff declining especially in mountain regions with predominantly smaller glaciers, for example, in the tropical Andes (Frans et al., 2015; Polk et al., 2017), Western Canada (Fleming and Dahlke, 2014; Brahney et al., 2017) and the Swiss Alps (Huss and Fischer, 2016). A global modelling study (Huss and Hock, 2018) suggests that peak water has been reached before 2019 for 82–95% of the glacier area in the tropical Andes, 40–49% in Western Canada and USA, and 55–67% in Central Europe (including European Alps and Pyrenees) and the Caucasus (Figure 2.6).

Projections indicate a continued increase in winter runoff in many snow and/or glacier-fed rivers over the 21st century (*high confidence*) regardless of the climate scenario, for example, in North America (Schnorbus et al., 2014; Sultana and Choi, 2018), the European Alps (Addor et al., 2014; Bosshard et al., 2014), Scotland (Capell et al., 2014) and High Mountain Asia (Kriegel et al., 2013) due to increased winter snowmelt and more precipitation falling as rain in addition to increases in precipitation in some basins (Table SM2.9). There is *robust evidence* (*high agreement*) that summer runoff will decline over the 21st century in many basins for all emission scenarios, for example, in Western Canada and USA (Shrestha et al., 2017), the European Alps (Jenicek et al., 2018), High Mountain Asia (Prasch et al., 2013; Engelhardt et al., 2017) and the tropical Andes (Baraer et al., 2012), due to less snowfall and decreases in glacier melt after peak water. A global-scale projection suggests that decline in glacier runoff by 2100 (RCP8.5) may reduce basin runoff by 10% or more in at least one month of the melt season in several large river basins, especially in High Mountain Asia during dry seasons, despite glacier cover of less than a few percent (Huss and Hock, 2018).

Projected changes in annual runoff in glacier dominated basins are complex including increases and decreases over the 21st century for all scenarios depending on the time period and the timing of peak water (*high confidence*) (Figure 2.6). Local and regional-scale projections in High Mountain Asia, the European Alps, and Western Canada and USA suggest that peak water will generally be reached before or around the middle of the century. These finding are consistent with results from global-scale modelling of glacier runoff

(Bliss et al., 2014; Huss and Hock, 2018) indicating generally earlier peak water in regions with little ice cover and smaller glaciers (e.g., Low Latitudes, European Alps and Pyrenees, and the Caucasus) and later peak water in regions with extensive ice cover and large glaciers (e.g., Alaska, Southern Andes). In some regions (e.g., Iceland) peak water from most glacier area is projected to occur earlier for RCP2.6 than RCP8.5, caused by decreasing glacier runoff as glaciers find a new equilibrium. In contrast melt-driven glacier runoff continues to rise for the higher emission scenario. There is *very high confidence* that spring peak runoff in many snow-dominated basins around the world will occur earlier in the year, up to several weeks, by the end of the century caused by earlier snowmelt (e.g., Coppola et al., 2014; Bard et al., 2015; Yucel et al., 2015; Islam et al., 2017; Sultana and Choi, 2018).

In addition to changes in ice and snow melt, changes in other variables such as precipitation and evapotranspiration due to atmospheric warming or vegetation change affect runoff amounts and timing (e.g., Bocchiola, 2014; Lutz et al., 2016). Changes in melt water from ice and snow often dominates the runoff response to climate change at higher elevations, while changes in precipitation and evapotranspiration become increasingly important at lower elevations (Kormann et al., 2015). Permafrost thaw may affect runoff by releasing water from ground ice (Jones et al., 2018) and indirectly by changing hydrological pathways or ground water recharge as permafrost degrades (Lamontagne-Hallé et al., 2018). The relative importance of runoff from thawing permafrost compared to runoff from melting glaciers is expected to be greatest in arid areas where permafrost tends to be more abundant (Gruber et al., 2017). Because glaciers react more rapidly to climate change than permafrost, runoff in some mountain landscapes may become increasingly affected by permafrost thaw in the future (Jones et al., 2018).

In summary, there is *very high confidence* that glacier and snow cover decline have affected and will continue to change the amounts and seasonality of river runoff in many snow-dominated and/or glacier-fed river basins. The average winter runoff is expected to increase (*high confidence*), and spring peak maxima will occur earlier (*very high confidence*). Although observed and projected trends in annual runoff vary substantially among regions and can even be opposite in sign, there is *high confidence* that average annual runoff from glaciers will have reached a peak, with declining runoff thereafter, at the latest by the end of the 21st century in all regions regardless emission scenario. The projected changes in runoff are expected to affect downstream water management, related hazards and ecosystems (Section 2.3.2, 2.3.3).

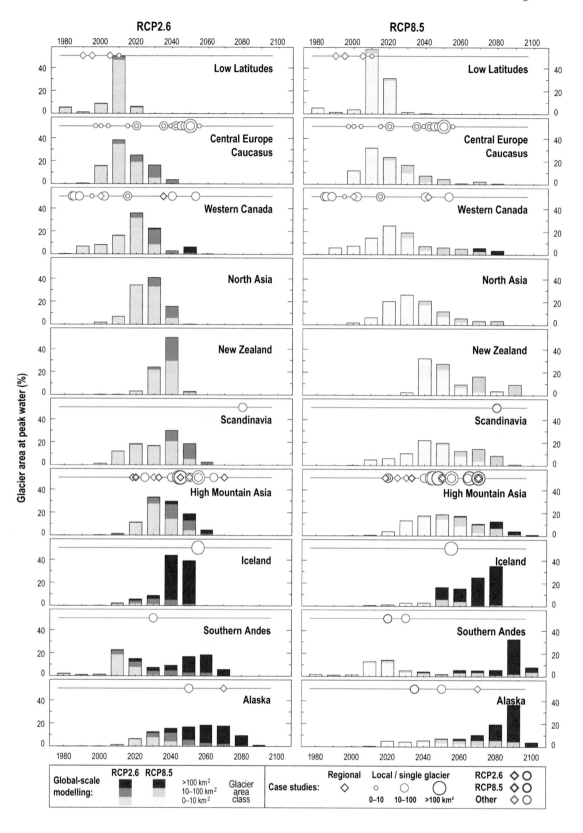

Figure 2.6 | Timing of peak water from glaciers in different regions (Figure 2.1) under two emission scenarios for Representative Concentration Pathways RCP2.6 and RCP8.5. Peak water refers to the year when annual runoff from the initially glacier-covered area will start to decrease due to glacier shrinkage after a period of melt induced increase. The bars are based on Huss and Hock (2018) who used a global glacier model to compute the runoff of all individual glaciers in a region until year 2100 based on 14 General Circulation Models (GCMs). Depicted is the area of all glaciers that fall into the same 10-year peak water interval expressed as a percentage of each region's total glacier area, i.e., all bars for the same RCP sum up to 100% glacier area. Shadings of the bars distinguish different glacier sizes indicating a tendency for peak water to occur later for larger glaciers. Circles/diamonds mark timing of peak water from individual case studies based on observations or modelling (Table SM2.10). Circles refer to results from individual glaciers regardless of size or a collection of glaciers covering <150 km² in total, while diamonds refer to regional-scale results from a collection of glaciers with >150 km² glacier coverage. Case studies based on observations or scenarios other than RCP2.6 and RCP8.5 are shown in both the left and right set of panels.

Frequently Asked Questions

FAQ 2.1 | How does glacier shrinkage affect river runoff further downhill?

Glaciers supply water that supports human communities both close to the glacier and far away from the glacier, for example for agriculture or drinking water. Rising temperatures cause mountain glaciers to melt and change the water availability. At first, as the glacier melts, more water runs downhill away from the glacier. However, as the glacier shrinks, the water supply will diminish and farms, villages and cities might lose a valuable water source.

Melting glaciers can affect river runoff, and thus freshwater resources available to human communities, not only close to the glacier but also far from mountain areas. As glaciers shrink in response to a warmer climate, water is released from long-term glacial storage. At first, glacier runoff increases because the glacier melts faster and more water flows downhill from the glacier. However, there will be a turning point after several years or decades, often called 'peak water', after which glacier runoff and hence its contribution to river flow downstream will decline (FAQ 2.1; Figure 1). Peak water runoff from glaciers can exceed the amount of initial yearly runoff by 50% or more. This excess water can be used in different ways, such as for hydropower or irrigation. After the turning point, this additional water decreases steadily as the glacier continues to shrink, and eventually stops when the glacier has disappeared, or retreated to higher elevations where it is still cold enough for the glacier to survive. As a result, communities downstream lose this valuable additional source of water. Total amounts of river runoff will then depend mainly on rainfall, snowmelt, ground water and evaporation.

Furthermore, glacier decline can change the timing in the year and day when the most water is available in rivers that collect water from glaciers. In mid- or high latitudes, glacier runoff is greatest in the summer, when the glacier ice continues to melt after the winter snow has disappeared, and greatest during the day when air temperature and solar radiation are at their highest (FAQ 2.1, Figure 1). As peak water occurs, more intense glacier melt rates also increase these daily runoff maxima significantly. In tropical areas, such as parts of the Andes, seasonal air temperature variations are small, and alternating wet and dry seasons are the main control on the amount and timing of glacier runoff throughout the year.

The effects of glaciers on river runoff further downhill depend on the distance from the glacier. Close to the glaciers (e.g., within several kilometres), initial increases in yearly glacier runoff until peak water followed by decreases can affect water supply considerably, and larger peaks in daily runoff from the glaciers can cause floods. Further away from the glaciers the impact of glacier shrinkage on total river runoff tends to become small or negligible. However, the melt water from glaciers in the mountains can be an important source of water in hot and dry years or seasons when river runoff would otherwise be low, and thereby also reducing variability in total river runoff from year to year, even hundreds of kilometres away from the glaciers. Other components of the water cycle such as rainfall, evaporation, groundwater and snowmelt can compensate or strengthen the effects of changes in glacier runoff as the climate changes.

2

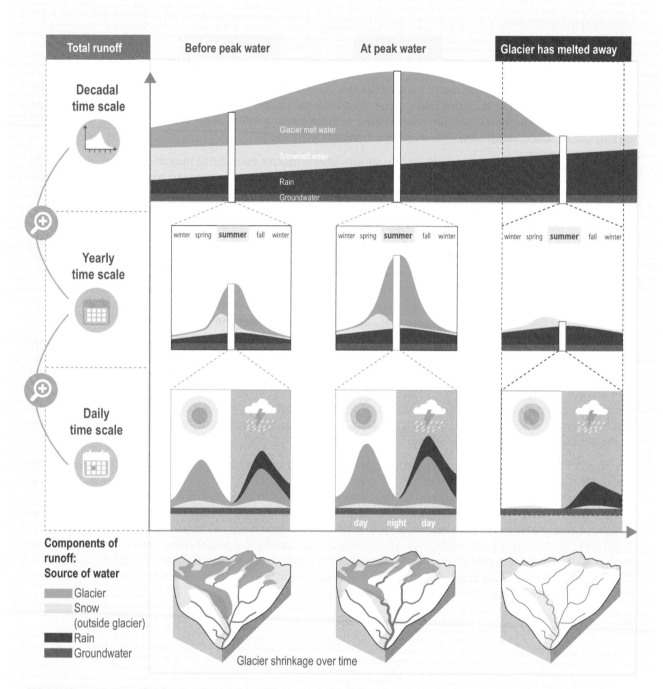

FAQ 2.1, Figure 1 | A simplified overview of changes in runoff from a river basin with large (e.g., >50%) glacier cover as the glaciers shrink, showing the relative amounts of water from different sources – glaciers, snow (outside the glacier), rain and groundwater. Three different time scales are shown: annual runoff from the entire basin (upper panel); runoff variations over one year (middle panel) and variations during a sunny then a rainy summer day (lower panel). Note that seasonal and daily runoff variations are different before, during and after peak flow. The glacier's initial negative annual mass budget becomes more negative over time until eventually the glacier has melted away. This is a simplified figure so permafrost is not addressed specifically and the exact partitioning between the different sources of water will vary between river basins.

2.3.1.2   Water Quality

Glacier decline can influence water quality by accelerating the release of stored anthropogenic legacy pollutants, with impacts to downstream ecosystem services. These legacy pollutants notably include persistent organic pollutants (POPs), particularly polychlorinated biphenyls (PCBs) and dichlorodiphenyl-trichloroethane (DDT), polycyclic aromatic hydrocarbons, and heavy metals (Hodson, 2014) and are associated with the deposition and release of black carbon. There is *limited evidence* that some of these pollutants found in surface waters in the Gangetic Plain during the dry season originate from Himalayan glaciers (Sharma et al., 2015), and glaciers in the European Alps store the largest known quantity of POPs in the Northern Hemisphere (Milner et al., 2017). Although their use has declined or ceased worldwide, PCBs have been detected in runoff from glacier melt due to the lag time of release from glaciers (Li et al., 2017). Glaciers also represent the most unstable stores of DDT in European and other mountain areas flanking large urban centres and glacier derived DDT is still accumulating in lake sediments downstream from glaciers (Bogdal et al., 2010). However, bioflocculation (the aggregation of dispersed organic particles by the action of organisms) can increase the residence time of these contaminants stored in glaciers thereby reducing their overall toxicity to freshwater ecosystems (Langford et al., 2010). Overall the effect on freshwater ecosystems of these contaminants is estimated to be low (*medium confidence*) (Milner et al., 2017).

Of the heavy metals, mercury is of particular concern and an estimated 2.5 tonnes has been released by glaciers to downstream ecosystems across the Tibetan Plateau over the last 40 years (Zhang et al., 2012). Mercury in glacial silt, originating from grinding of rocks as the glacier flows over them, can be as large or larger than the mercury flux from melting ice due to anthropogenic sources deposited on the glacier (Zdanowicz et al., 2013). Both glacier erosion and atmospheric deposition contributed to the high rates of total mercury export found in a glacierised watershed in coastal Alaska (Vermilyea et al., 2017) and mercury output is predicted to increase in glacierised mountain catchments (Sun et al., 2017; Sun et al., 2018b) (*medium confidence*). However, a key issue is how much of this glacier-derived mercury, largely in the particulate form, is converted to toxic methyl mercury downstream. Methyl mercury can be incorporated into aquatic food webs in glacier streams (Nagorski et al., 2014) and bio-magnify up the food chain (Lavoie et al., 2013). Water originating from rock glaciers can also contribute other heavy metals that exceed guideline values for drinking water quality (Thies et al., 2013). In addition, permafrost degradation can enhance the release of other trace elements (e.g., aluminium, manganese and nickel) (Colombo et al., 2018). Indeed, projections indicate that all scenarios of future climate change will enhance the mobilisation of metals in metamorphic mountain catchments (Zaharescu et al., 2016). The release of toxic contaminants, particularly where glacial melt waters are used for irrigation and drinking water in the Himalayas and the Andes, is potentially harmful to human health both now and in the future (Hodson, 2014) (*medium confidence*).

Soluble reactive phosphorus concentrations in rivers downstream of glaciers are predicted to decrease with declining glacier coverage (Hood et al., 2009) as a large percentage is associated with glacier-derived suspended sediment (Hawkings et al., 2016). In contrast, dissolved organic carbon (DOC), dissolved inorganic nitrogen and dissolved organic nitrogen concentrations in pro-glacial rivers is projected to increase this century due to glacier shrinkage (Hood et al., 2015; Milner et al., 2017) (*robust evidence, medium agreement*). Globally, mountain glaciers are estimated to release about 0.8 Tera g C yr^{-1} (Li et al., 2018) of highly bioavailable DOC that may be incorporated into downstream food webs (Fellman et al., 2015; Hood et al., 2015). Loss rates of DOC from glaciers in the high mountains of the Tibetan Plateau were estimated to be ~0.19 Tera g C yr^{-1}, (Li et al., 2018) higher than other regions suggesting that DOC is released more efficiently from Asian mountain glaciers (Liu et al., 2016). Glacier DOC losses are expected to accelerate as they shrink, leading to

a cumulative annual loss of roughly 15 Tera g C yr^{-1} of glacial DOC by 2050 from melting glaciers and ice sheets (Hood et al., 2015). Permafrost degradation is also a major and increasing source of bioavailable DOC (Abbott et al., 2014; Aiken et al., 2014). Major ions calcium, magnesium, sulphate and nitrate (Colombo et al., 2018) are also released by permafrost degradation as well as acid drainage leaching into alpine lakes (Ilyashuk et al., 2018).

Increasing water temperature has been reported in some high mountain streams (e.g., Groll et al., 2015; Isaak et al., 2016) due to decreases in glacial runoff, producing changes in water quality and species richness (Section 2.3.3). In contrast, water temperature in regions with extensive glacier cover are expected to show a transient decline, due to an enhanced cooling effect from increased glacial melt water (Fellman et al., 2014).

In summary, changes in the mountain cryosphere will cause significant shifts in downstream nutrients (DOC, nitrogen, phosphorus) and influence water quality through increases in heavy metals, particularly mercury, and other legacy contaminants (*medium evidence, high agreement*) posing a potential threat to human health. These threats are more focused where glaciers are subject to substantial pollutant loads such as High Mountain Asia and Europe, rather than areas like Alaska and Canada.

2.3.1.3   Key Impacts and Vulnerability

2.3.1.3.1 *Hydropower*

Hydropower comprises about 16% of electricity generation globally but close to 100%, in many mountainous countries (Hamududu and Killingtveit, 2012; IHA, 2018). It represents a significant source of revenue for mountainous regions (Gaudard et al., 2016). Due to the dependence on water resources as key input, hydropower operations are expected to be affected by changes in runoff from glaciers and snow cover (Section 2.3.1.1, FAQ 2.1). Both increases and decreases in annual and/or seasonal water input to hydropower facilities have been recorded in several high mountain regions, for example, in Switzerland (Hänggi and Weingartner, 2012; Schaefli et al., 2019), Canada (Jost et al., 2012; Jost and Weber, 2013), Iceland (Einarsson and Jónsson, 2010) and High Mountain Asia (Ali et al., 2018). However, there is only *limited evidence* (*medium agreement*) that changes in runoff have led to changes in hydropower plant operation. For example, in Iceland, the National Power Company observed in 2005 that flows into their energy system were greater than historical flows. By incorporating the most recent runoff data into strategies for reservoir management it was possible to increase production capacity (Braun and Fournier, 2016).

There is *robust evidence* (*medium agreement*) that water input to hydropower facilities will change in the future due to cryosphere-related impacts on runoff (Section 2.3.1.1). For example, in the Skagit river basin in British Columbia and Northern Washington (Lee et al., 2016) and in California (Madani and Lund, 2010) projections (SRES A1B) show more runoff in winter and less in summer. In India, snow and glacier runoff to hydropower plants is projected to decline in several basins (Ali et al., 2018). In some cases, catchments that are close together are projected to evolve in contrasting directions in terms of runoff, for example in the European Alps (Gaudard et al., 2013; Gaudard et al., 2014). Increased runoff due to changes in the cryosphere will increase the risk of overflows (non-productive discharge), particularly during winter and spring melt, with the greatest impacts on run-of-river power plants (e.g., in Canada; Minville et al., 2010; Warren and Lemmen, 2014) (*medium confidence*).

There is *medium evidence* (*high agreement*) that changes in glacier- and moraine-dammed lakes, and changes in sediment supply will affect hydropower generation (Colonia et al., 2017; Hauer et al., 2018). Many glacier lakes have increased in volume, and can damage hydropower infrastructure when they empty suddenly (Engeset et al., 2005; Jackson and Ragulina, 2014; Carrivick and Tweed, 2016) (Section 2.3.2; Figure 2.7). If large enough, hydropower reservoirs can reduce the downstream negative impacts of changes in the cryosphere by storing and providing freshwater during hot, dry periods or by alleviating the effects of glacier floods (Jackson and Ragulina, 2014; Colonia et al., 2017). In mountain rivers, sediment volume and type depend on connectivity between hillslopes and the valley floor (Carrivick et al., 2013), glacier activity (Lane et al., 2017) and on water runoff regime feedbacks with river channel dynamics (Schmidt and Morche, 2006). An increase in suspended sediment loading under current reservoir operating policies is projected for some hydropower facilities, for example, in British Columbia and Northern Washington (Lee et al., 2016).

Only a few studies have addressed the economic effects on hydropower due directly to changes in the cryosphere. For example in Peru, Vergara et al. (2007) studied the effect of both reduced glacier runoff and runoff with no glacier input once the glaciers have completely melted for the Cañón del Pato hydropower plant in Peru, and found an economic cost of between 5–20 million USD yr^{-1}, with the lower figure for the cost of energy paid to the producer and the higher figure the society cost. Costs calculated for all of Peru, where ~80% of electricity comes from hydropower range from 60–212 million USD yr^{-1}. If the cost of rationing energy is considered, the national cost is estimated as 1,500 million USD yr^{-1}.

Other factors than changes in the cryosphere, such as market policies and regulation, may have greater significance for socioeconomic development of hydropower in the future (Section 2.3.1.4, Gaudard et al., 2016). Hence, despite the efforts of hydropower agencies and regulatory bodies to quantify changes or to develop possible adaptation strategies (IHA, 2018), only a few organisations are incorporating current knowledge of climate change into their investment planning. The World Bank uses a decision tree approach to identify potential vulnerabilities in a hydropower project incurred from key uncertain factors and their combinations (Bonzanigo et al., 2015).

2.3.1.3.2 Agriculture

High mountains have supported agricultural livelihoods for centuries. Rural communities are dependent on adequate levels of soil moisture at planting time, derived in part in many cases from irrigation water which includes glacier and snowmelt water; as a result, they

are exposed to risk which stems from cryosphere changes (*high confidence*) (Figure 2.8). The relative poverty of many mountain communities contributes to their vulnerability to the impacts of these cryosphere changes (McDowell et al., 2014; Carey et al., 2017; Rasul and Molden, 2019) (*medium evidence, high agreement*). Glacier and snowmelt water contribute irrigation water to adjacent lowlands as well. Pastoralism, an important livelihood strategy in mountain regions, is also impacted by cryosphere changes, but described in Section 2.3.7.

There is *medium evidence* (*medium agreement*) that reduction in streamflow due to glacier retreat or reduced snow cover has led to reduced water availability for irrigation of crops and declining agricultural yields in several mountain areas (Table SM2.11), for example in the tropical Andes (e.g., Bury et al., 2011) and High Mountain Asia (e.g., Nüsser and Schmidt, 2017). In the Southern Andes, increased streamflow in the Elqui River in Chile, due to glacier retreat or changing snow cover, has led to increased water availability for irrigation and increased agricultural yields (Young et al., 2010).

In addition to the effects on agriculture of changing availability of irrigation water, reductions in snow cover can also impact agriculture through its direct effects on soil moisture, as reported for Nepal, where lesser snow cover has led to the drying of soils and lower yields of potatoes and fodder (Smadja et al., 2015). Agriculture in high mountain areas is sensitive to other climatic drivers as well. Rising air temperatures increase crop evapotranspiration, thus increasing water demand for crop production to maintain optimal yield (Beniston and Stoffel, 2014). They are also associated with upslope movement of cropping zones, which favours some farmers in high mountain areas, who are increasingly able to cultivate new crops, such as onions, garlic and apples in Nepal (Huntington et al., 2017; Hussain et al., 2018), and maize in Ecuador (Skarbø and VanderMolen, 2014). Dry spells and unseasonal frosts have also impacted agriculture in Peru (Bury et al., 2011).

Adaptation activities in mountain agriculture related at least partially to cryospheric changes are detailed in Table SM2.12 and their geographic spread shown in Figure 2.9. Agriculture in these areas is sensitive to non-climate drivers as well, such as market forces and political pressures (Montana et al., 2016; Sietz and Feola, 2016; Figueroa-Armijos and Valdivia, 2017) and shifts in water governance (Rasmussen, 2016). The majority of the adaptation activities are autonomous, though some are planned or carried out with support from national governments, non-governmental organisations (NGOs), or international aid organisations. Though many studies report on benefits from these activities which accrue to community members as increased harvests and income, systematic evaluations of these adaptation strategies are generally lacking. A range of factors, discussed below, place barriers which limit the scale and scope of these activities in the mountain agricultural sector, including a lack of finance and technical knowledge, low adaptive capacity within communities, ill-equipped state organisations, ambiguous property rights and inadequate institutional and market support (*medium evidence, high agreement*). Section 2.3.7 examines two other responses to decreasing irrigation water: wage labour migration, which often serves as an adaptation strategy, and displacement of entire communities, an indication of the limits to adaptation – this displacement is also due in some cases to natural hazards.

To cope with the reduced water supplies, planted areas have been reduced in a number of different places in Nepal (Gentle and Maraseni, 2012; Sujakhu et al., 2016). Adaptation responses within irrigation systems include the adoption of new irrigation technologies or upgrading existing technologies, adopting water conservation measures, water rationing, constructing water storage infrastructure, and change in cropping patterns (Rasul et al., 2019; Figure 2.9). Water delivery technologies which reduce loss are adopted in Chile (Young et al., 2010) and Peru (Orlove et al., 2019). Similarly, greenhouses have been adopted in Nepal (Konchar et al., 2015) to reduce evapotranspiration and frost damage, though limited access to finance is a barrier to these activities. Box 2.3 describes innovative irrigation practices in India. Local pastoral communities have responded to these challenges with techniques broadly similar to those in agricultural settings by expanding irrigation facilities, for example, in Switzerland (Fuhrer et al., 2014). In addition to adopting new technologies, some water users make investments to tap more distant sources of irrigation water. Cross-Chapter Box 3 in Chapter 1 discusses such efforts in Northern Pakistan, where landslides, associated with cryosphere change, have also damaged irrigation systems.

The adoption of new crops and varieties is an adaptation response found in several regions. Farmers in northwest India have increased production of lentils and vegetables, which provide important nutrients to the local diet, with support from government watershed improvement programs which help address decreased availability of irrigation water, though stringent requirements for participation in the programs have limited access by poor households to this assistance (Dame and Nüsser, 2011). Farmers who rely on irrigation in the Naryn River basin in Kyrgyzstan have shifted from the water intensive fruits and vegetables to fodder crops such as barley and alfalfa, which are more profitable. Upstream communities, with greater access to water and more active local institutions, are more willing to experiment with new crops than those further downstream (Hill et al., 2017). In other areas, crop choices also reflect responses to rising temperatures along with new market opportunities such as the demand for fresh vegetables by tourists in Nepal (Konchar et al., 2015; Dangi et al., 2018) and the demand for roses in urban areas in Peru (SENASA, 2017). Indigenous knowledge and local knowledge (Cross-Chapter Box 4 in Chapter 1), access to local and regional seed supply networks, proximity to agricultural extension and support services also facilitate the adoption of new crops (Skarbø and VanderMolen, 2014).

Local institutions and embedded social relations play a vital role in enabling mountain communities to respond to the impacts of climate driven cryosphere change. Indigenous pastoral communities who have tapped into new water sources to irrigate new areas have also strengthened the control of access to existing irrigated pastures in Peru (Postigo, 2014) and Bolivia (Yager, 2015). In an example of indigenous populations in the USA, two tribes who share a large reservation in the Northern Rockies rely on rivers which receive glacier melt water to irrigate pasture, and maintain fisheries, domestic water supplies, and traditional ceremonial practices. Tribal water managers have sought to install infrastructure to promote more efficient water use and protect fisheries, but these efforts have been impeded by land and water governance institutions in the region and by a history of social marginalisation (McNeeley, 2017).

High mountain communities have sought new financial resources from wage labour (Section 2.3.7), tourism (Mukherji et al., 2019) and government sources to support adaptation activities. Local water user associations in Kyrgyzstan and Tajikistan have adopted less water intensive crops and reorganised the use and maintenance of irrigation systems, investing government relief payments after floods (Stucker et al., 2012). Similar measures are reported from India and Pakistan (Dame and Mankelow, 2010; Clouse, 2016; Nüsser and Schmidt, 2017), Nepal (McDowell et al., 2013) and Peru (Postigo, 2014). In contrast, fewer adaptation measures have been adopted in Uzbekistan, due to low levels of capital availability and to agricultural policies, including centralised water management, crop production quotas and weak agricultural extension, which limit the response capacity of farmers (Aleksandrova et al., 2014).

Lowland agricultural areas which receive irrigation water from rivers fed by glacier melt and snowmelt are projected to face negative impacts in some regions (*limited evidence, high agreement*). In the Rhone basin in Switzerland, many irrigated pasture areas are projected to face water deficits by 2050, under the A1B scenario (Fuhrer et al., 2014; Cross Chapter Box 1 in Chapter 1). For California and the southwestern USA, a shift to peak snowmelt earlier in the year would create more frequent floods, and a reduced ability of existing reservoirs to store water by 2050 under RCP8.5 (Pagán et al., 2016) and by 2100 under RCP2.6, RCP4.5 and RCP8.5 (Pathak et al., 2018). The economic values of these losses have been estimated at 10.8–48.6 billion USD by around 2050 (Sturm et al., 2017). A similar transition to runoff peaks earlier in the year by 2100 under RCP2.6, RCP4.5 and RCP8.5, creating challenges for management of irrigation water, has been reported for the countries in central Asia which are dependent on snow cover and glaciers of the Tien Shan (Xenarios et al., 2018). In India and Pakistan, where over 100 million farmers receive irrigation from the Indus and Ganges Rivers, which also have significant inputs from glaciers and snowmelt, also face risks of decreasing water supplies from cryosphere change by 2100 (Biemans et al., 2019; Rasul and Molden, 2019).

Box 2.3 | Local Responses to Water Shortage in northwest India

Agriculture in Ladakh, a cold arid mountain region (~100,000 km^2) in the western Himalaya of India with median elevation of 3,350 m a.s.l. and mean annual precipitation of less than 100 mm, is highly dependent on streams for irrigation in the agricultural season in the spring and summer (Nüsser et al., 2012; Barrett and Bosak, 2018). Glaciers in Ladakh, largely located at 5,000–6,000 m a.s.l. and small in size have retreated at least since the late 1960s although less pronounced than in many other Himalayan regions (Chudley et al., 2017; Schmidt and Nüsser, 2017). However, the effect of glaciers on streamflow in Ladakh is poorly constrained, and measurements on changes in runoff and snow cover are lacking (Nüsser et al., 2018).

To cope with seasonal water scarcity at critical times for irrigation, villagers in the region have developed four types of artificial ice reservoirs: basins, cascades, diversions and a form known locally as ice stupas. All these types of ice reservoirs capture water in the autumn and winter, allowing it to freeze, and hold it until spring, when it melts and flows down to fields (Clouse et al., 2017; Nüsser et al., 2018). In this way, they retain a previously unused portion of the annual flow and facilitate its use to supplement the decreased flow in the following spring (Vince, 2009; Shaheen, 2016). Frozen basins are formed from water which is conveyed across a slope through channels and check dams to shaded surface depressions near the villages. Cascades and diversions direct water to pass over stone walls, slowing its movement and allowing it to freeze. Ice stupas direct water through pipes into fountains, where it freezes into conical shapes (Box 2.3 Figure 1). These techniques use local materials and draw on local knowledge (Nüsser and Baghel, 2016).

A study examined 14 ice reservoirs, including ice stupas, and concluded that they serve as 'site-specific water conservation strategies; and that they can be regarded as appropriate local technologies to reduce seasonal water scarcity at critical times (Nüsser et al., 2018). It listed the benefits of ice reservoirs as improved water availability in spring, reduction of seasonal water scarcity and resulting crop failure risks, and the possibility of growing cash crops. However, the study questioned their usefulness as a long-term adaptation strategy, because their operation depends on winter runoff and freeze-thaw cycles, both of which are sensitive to interannual variability, and often deviate from the optimum range required for effective functioning of the reservoirs. It also raised questions about the financial costs and labour requirements, which vary across the four types of ice reservoirs.

Box 2.3, Figure 1 | Ice stupas in Ladakh, India (Photo: Padma Rigzin)

2.3.1.3.3 Drinking water supply

Only a few studies provide detailed empirical assessments of the effects of cryosphere change on the amounts of drinking water supply. Decreases in drinking water supplies due to reduced glacier and snowmelt water have been reported for rural areas in the Nepal Himalaya (McDowell et al., 2013; Dangi et al., 2018), but the tropical Andes have received the most attention, including both urban conglomerates and some rural areas, where water resources are especially vulnerable to climate change due to water scarcity and increased demands (Chevallier et al., 2011; Somers et al., 2018), amidst rapidly retreating glaciers (Burns and Nolin, 2014).

The contribution of glacier water to the water supply of La Paz, Bolivia, between 1963–2006 was assessed at 15% annually and 27% during the dry season (Soruco et al., 2015), though rising as high as 86% during extreme drought months (Buytaert and De Bièvre, 2012). Despite a 50% area loss, the glacier retreat has not contributed to reduced water supplies for the city, because increased melt rates have compensated for reductions in glacier volume. However, for a complete disappearance of the glaciers, assuming no change in precipitation, a reduction in annual runoff by 12% and 24% in the dry season was projected (Soruco et al., 2015) similar to reductions projected by 2050 under a RCP8.5 scenario for a basin in southern Peru (Drenkhan et al., 2019). Huaraz and Huancayo in Peru are other cities with high average contribution of melt water to surface water resources (up to ~20%; Buytaert et al., 2017) and rapid glacier retreat in their headwaters (Rabatel et al., 2013).

Overall, risks to water security and related vulnerabilities are highly heterogeneous varying even at small spatial scales with populations closer to the glaciers being more vulnerable, especially during dry months and droughts (Buytaert et al., 2017; Mark et al., 2017). A regional-scale modelling study including all of Bolivia, Ecuador and Peru (Buytaert et al., 2017) estimated that roughly 390,000 domestic water users, mostly in Peru, rely on a high (>25%) long-term average contribution from glacier melt, with this number rising to almost 4 million in the driest month of a drought year. Despite *high confidence* in declining longer-term melt water contributions from glaciers in the tropical Andes (Figure CB6.1), major uncertainties remain how these will affect future human water use. Regional-scale water balance simulations forced by multi-model climate projections (Buytaert and De Bièvre, 2012), suggest a relatively limited effect of glacier retreat on water supply in four major cities (Bogota, La Paz, Lima, Quito) due to the dominance of human factors influencing water supply (Carey et al., 2014; Mark et al., 2017; Vuille et al., 2018), though uncertainties are large. Population growth and limited funding for infrastructure maintenance exacerbate water scarcity, though water managers have established programs in Quito and in Huancayo and the Santa and Vilcanota basins (Peru) to improve water management through innovations in grey infrastructure and ecosystem-based adaptations (Buytaert and De Bièvre, 2012; Buytaert et al., 2017; Somers et al., 2018).

In summary, there is *limited evidence* (*medium agreement*) that glacier decline places increased risks to drinking water supply. In the Andes, future increases in water demand due to population growth and other socioeconomic stressors are expected to outpace the impact of climate change induced changes on water availability regardless the emission scenario.

2.3.1.4 Water Governance and Response Measures

Cryospheric changes induced by climate change, and their effects on hydrological regime and water availability, bear relevance for the management and governance of water as a resource for communities and ecosystems (Hill, 2013; Beniston and Stoffel, 2014; Carey et al., 2017), particularly in areas where snow and ice contribute significantly to river runoff (*medium confidence*) (Section 2.3.1.1). However, water availability is one aspect relevant for water management and governance, given that multiple and diverse decision making contexts and governance approaches and strategies can influence how the water resource is accessed and distributed (*medium confidence*) (De Stefano et al., 2010; Beniston and Stoffel, 2014).

A key risk factor that influences how water is managed and governed, rests on existing and unresolved conflicts that may or may not necessarily arise exclusively from demands over shared water resources, raising tensions within and across borders in river basins influenced by snow and glacier melt (Valdés-Pineda et al., 2014; Bocchiola et al., 2017). For example, in Central Asia, competing demand for water for hydropower and irrigation between upstream and downstream countries has raised tensions (Bernauer and Siegfried, 2012; Bocchiola et al., 2017). Similarly, competing demand for water is also reported in Chile (Valdés-Pineda et al., 2014) and in Peru (Vuille, 2013; Drenkhan et al., 2015). Since AR5, some studies have examined the impacts and risks related to projections of cryosphere-related changes in streamflow in transboundary basins in the 21st century, and suggest that these changes create barriers in effectively managing water in some settings (*medium confidence*). For instance, within the transnational Indus River basin, climate change impacts may reduce streamflow by the end of this century, thus putting pressure on established water sharing arrangements between nations (Jamir, 2016) and subnational administrative units (Yang et al., 2014b). In this basin, management efforts may be hampered by current legal and regulatory frameworks for evaluating new dams, which do not take into account changes in streamflow that may result from climate change (Raman, 2018). Within the transnational Syr Darya and Amu Darya basins in Central Asia, competition for water between multiple uses, exacerbated by reductions in flow later in this century, may hamper future coordination (Reyer et al., 2017; Yu et al., 2019). However, other evidence from Central Asia suggests that relative water scarcity may not be the only factor to exacerbate conflict in this region (Hummel, 2017). Overall, there is *medium confidence* in the ability to meet future water demands in some mountain regions, given the combined uncertainties associated with accurate projections of water supply in terms of availability and the diverse sociocultural and political contexts in which decisions on water access and distribution are taken.

Since AR5, several studies highlight that integrated water management approaches, focused on the multipurpose use of water that includes water released from the cryosphere, are important as adaptation measures, particularly for sectors reliant on this water source to sustain energy production, agriculture, ecosystems and drinking water supply (Figure 2.9). These measures, backed by effective governance arrangements to support them, demonstrate an ability to address increasing challenges to water availability arising from climate change in the mountain cryosphere, providing co-benefits through the optimisation of storage and the release of water from high mountain reservoirs (*medium confidence*). Studies in Switzerland (e.g., Haeberli et al., 2016; Brunner et al., 2019), Peru (e.g., Barriga Delgado et al., 2018; Drenkhan et al., 2019), Central Asia (Jalilov et al., 2018) and Himalaya (Molden et al., 2014; Biemans et al., 2019) highlight the potential of water reservoirs in high mountains, including new reservoirs located in former glacier beds, alleviating seasonal water scarcity for multiple water usages. However, concerns are also raised in the environmental literature about their actual and potential negative impacts on local ecosystems and biodiversity hotspots, such as wetlands and peat bogs, which have been reported for small high mountain reservoirs, for example, in the European Alps (Evette et al., 2011) and for large dam construction projects in High Mountain Asia (e.g., Dharmadhikary, 2008).

Transboundary cooperation at regional scales are reported to further support efforts that address the potential risks to water resources in terms of its availability and its access and distribution governance (Dinar et al., 2016). Furthermore, the UN 2030 Agenda and its Sustainable Development Goals (SDGs) (UN, 2015) may offer additional prospects to strengthen water governance under a changing cryosphere, given that monitoring and reporting on key water-related targets and indicators, and their interaction across other SDGs, direct attention to the provision of water as a key condition for development (Section 2.4). However, there is *limited evidence* to date to assess their effectiveness on an evidentiary basis.

2.3.2   Landslide, Avalanche and Flood Hazards

High mountains are particularly prone to hazards related to snow, ice and permafrost as these elements exert key controls on mountain slope stability (Haeberli and Whiteman, 2015). This section assesses knowledge gained since previous IPCC reports, in particular SREX (e.g., Seneviratne et al., 2012), and AR5 Working Group II (Cramer et al., 2014). In this section, observed and projected changes in hazards are covered first, followed by exposure, vulnerability and resulting impacts and risks, and finally disaster risk reduction and adaptation. Cryospheric hazards that constitute tipping points are also listed in Table 6.1 in Chapter 6.

Hazards assessed in this section range from localised effects on mountain slopes and adjacent valley floors (distance reach of up to several kilometres) to events reaching far into major valleys and even surrounding lowlands (reach of tens to hundreds of kilometres), and include cascading events. Changes in the cryosphere due to climate change influence the frequency and magnitude of hazards, the processes involved, and the locations exposed to the hazards (Figure 2.7). Natural hazards and associated disasters are sporadic by nature, and vulnerability and exposure exhibit strong geographic variations. Assessments of change are based not only

on direct evidence, but also on laboratory experiments, theoretical considerations and calculations, and numerical modelling.

2.3.2.1   Observed and Projected Changes

2.3.2.1.1 Unstable slopes, landslides and glacier instabilities

Permafrost degradation and thaw as well as increased water flow into frozen slopes can increase the rate of movement of frozen debris bodies and lower their surface due to loss of ground ice (subsidence). Such processes affected engineered structures such as buildings, hazard protection structures, roads, or rail lines in all high mountains during recent decades (Section 2.3.4). Movement of frozen slopes and ground subsidence/heave are strongly related to ground temperature, ice content, and water input (Wirz et al., 2016; Kenner et al., 2017). Where massive ground ice gets exposed, retrogressive thaw erosion develops (Niu et al., 2012). The creep of rock glaciers (frozen debris tongues that slowly deform under gravity) is in principle expected to accelerate in response to rising ground temperatures, until substantial volumetric ice contents have melted out (Kääb et al., 2007; Arenson et al., 2015a). As documented for instance for sites in the European Alps and Scandinavia for recent years to decades, rock glaciers replenished debris flow starting zones at their fronts, so that the intensified material supply associated with accelerated movement (Section 2.2.4) contributed to increased debris flow activity (higher frequency, larger magnitudes) or slope destabilisation (Stoffel and Graf, 2015; Wirz et al., 2016; Kummert et al., 2017; Eriksen et al., 2018).

There is *high confidence* that the frequency of rocks detaching and falling from steep slopes (rock fall) has increased within zones of degrading permafrost over the past half-century, for instance in high mountains in North America, New Zealand, and Europe (Allen et al., 2011; Ravanel and Deline, 2011; Fischer et al., 2012; Coe et al., 2017). Compared to the SREX and AR5 reports, the confidence in this finding increased. Available field evidence agrees with theoretical considerations and calculations that permafrost thaw increases the likelihood of rock fall (and also rock avalanches, which have larger volumes compared to rock falls) (Gruber and Haeberli, 2007; Krautblatter et al., 2013). These conclusions are also supported by observed ice in the detachment zone of previous events in North America, Iceland and Europe (Geertsema et al., 2006; Phillips et al., 2017; Sæmundsson et al., 2018). Summer heat waves have in recent years triggered rock instability with delays of only a few days or weeks in the European Alps (Allen and Huggel, 2013; Ravanel et al., 2017). This is in line with theoretical considerations about fast thaw of ice filled frozen fractures in bedrock (Hasler et al., 2011) and other climate impacts on rock stability, such as from large temperature variations (Luethi et al., 2015). Similarly, permafrost thaw increased the frequency and volumes of landslides from frozen sediments in many mountain regions in recent decades (Wei et al., 2006; Ravanel et al., 2010; Lacelle et al., 2015). At lower elevations in the French Alps, though, climate driven changes such as a reduction in number of freezing days are projected to lead to a reduction in debris flows (Jomelli et al., 2009).

A range of slope instability types was found to be connected to glacier retreat (Allen et al., 2011; Evans and Delaney, 2015). Debris left behind by retreating glaciers (moraines) slid or collapsed, or formed fast flowing water-debris mixtures (debris flows) in recent decades, for instance in the European and New Zealand Alps (Zimmermann and Haeberli, 1992; Blair, 1994; Curry et al., 2006; Eichel et al., 2018). Over decades to millennia, or even longer, rock slopes adjacent to or formerly covered by glaciers, became unstable and in some cases, eventually collapsed. Related landslide activity increased in recently deglacierised zones in most high mountains (Korup et al., 2012; McColl, 2012; Deline et al., 2015; Kos et al., 2016; Serrano et al., 2018). For example, according to Cloutier et al. (2017) more than two-thirds of the large landslides that occurred in Northern British Columbia between 1973–2003, occurred on cirque walls that have been exposed after glacier retreat from the mid-19th century on. Ice-rich permafrost environments following glacial retreat enhanced slope mass movements (Oliva and Ruiz-Fernández, 2015). At lower elevations, re-vegetation and rise of tree limit are able to stabilise shallow slope instabilities (Curry et al., 2006). Overall, there is *high confidence* that glacier retreat in general has in most high mountains destabilised adjacent debris and rock slopes over time scales from years to millennia, but robust statistics about current trends in this development are lacking. This finding reconfirms, and for some processes increases confidence in related findings from the SREX and AR5 reports.

Ice break-off and subsequent ice avalanches are natural processes at steep glacier fronts. How climate driven changes in geometry and thermal regime of such glaciers influenced ice avalanche hazards over years to decades depended strongly on local conditions, as shown for the European Alps (Fischer et al., 2013; Faillettaz et al., 2015). The few available observations are insufficient to detect trends. Where steep glaciers are frozen to bedrock, there is, however, *medium evidence* and *high agreement* from observations in the European Alps and from numerical simulations that failures of large parts of these glaciers were and will be facilitated in the future due to an increase in basal ice temperature (Fischer et al., 2013; Faillettaz et al., 2015; Gilbert et al., 2015).

In some regions, glacier surges constitute a recurring hazard, due to widespread, quasi-periodic and substantial increases in glacier speed over a period of a few months to years, often accompanied by glacier advance (Harrison et al., 2015; Sevestre and Benn, 2015). In a number of cases, mostly in North America and High Mountain Asia (Bevington and Copland, 2014; Round et al., 2017; Steiner et al., 2018), surge-related glacier advances dammed rivers, causing major floods. In rare cases, glacier surges directly inundated agricultural land and damaged infrastructure (Shangguan et al., 2016). Sevestre and Benn (2015) suggest that surging operates within a climatic envelope of temperature and precipitation conditions, and that shifts in these conditions can modify surge frequencies and magnitudes. Some glaciers have reduced or stopped surge activity, or are projected to do so within decades, as a consequence of negative glacier mass balances (Eisen et al., 2001; Kienholz et al., 2017). For such cases, related hazards can also be expected to decrease. In contrast, intensive or increased surge activity (Hewitt, 2007; Gardelle et al., 2012; Yasuda and Furuya, 2015) occurred in a region on and

around the Western Tibetan plateau which exhibited balanced or even positive glacier mass budgets in recent decades (Brun et al., 2017). Enhanced melt water production was suggested to be able to trigger or enhance surge-type instability, in particular for glaciers that contain ice both at the melting point and considerably below (Dunse et al., 2015; Yasuda and Furuya, 2015; Nuth et al., 2019).

A rare type of glacier instability with large volumes (in the order of 10–100 million m^3) and high mobility (up to 200–300 km/h) results from the complete collapse of large sections of low-angle valley glaciers and subsequent combined ice/rock/debris avalanches. The largest of such glacier collapses have been reported in the Caucasus Mountains in 2002 (Kolka Glacier, ~130 fatalities) (Huggel et al., 2005; Evans et al., 2009), and in the Aru Range in Tibet in 2016 (twin glacier collapses with 9 fatalities) (Kääb et al., 2018). Although there is no evidence that climate change has played a direct role in the 2002 event, changes in glacier mass balance, water input into the glaciers, and the frozen regime of the glacier beds were involved in the 2016 collapses and at least partly linked with climate change (Gilbert et al., 2018). Besides the 2016 Tibet cases, it is unknown if such massive and rare collapse-like glacier instabilities can be attributed to climate change.

2.3.2.1.2 Snow avalanches

Snow avalanches can occur either spontaneously due to meteorological factors such as loading by snowfall or liquid water infiltration following, for example, surface melt or rain-on-snow, or can be triggered by the passage of people in avalanche terrain, the impact of falling ice or rocks, or by explosives used for avalanche control (Schweizer et al., 2003). There is no published evidence found that addresses the links between climate change and accidental avalanches triggered by recreationists or workers. Changes in snow cover characteristics are expected to induce changes in spontaneous avalanche activity including changes in friction and flow regime (Naaim et al., 2013; Steinkogler et al., 2014).

Ballesteros-Cánovas et al. (2018) reported increased avalanche activity in some slopes of the Western Indian Himalaya over the past decades related to increased frequency of wet-snow conditions. In the European Alps, avalanche numbers and runout distance have decreased where snow depth decreased and air temperature increased (Teich et al., 2012; Eckert et al., 2013). In the European Alps and Tatras mountains, over past decades, there has been a decrease in avalanche mass and run-out distance and a decrease in avalanches with a powder part; avalanche numbers decreased below 2,000 m a.s.l., and increased above (Eckert et al., 2013; Lavigne et al., 2015; Gadek et al., 2017). A positive trend in the proportion of avalanches involving wet snow in December through February was shown for the last decades (Pielmeier et al., 2013; Naaim et al., 2016). Land use and land cover changes also contributed to changes in avalanches (García-Hernández et al., 2017; Giacona et al., 2018). Correlations between avalanche activity and the El Niño-Southern Oscillation (ENSO) were identified from 1950–2011 in North and South America but there was no significant temporal trend reported for avalanche activity (McClung, 2013). Mostly inconclusive results were reported by Sinickas et al. (2015) and Bellaire et al. (2016) regarding the

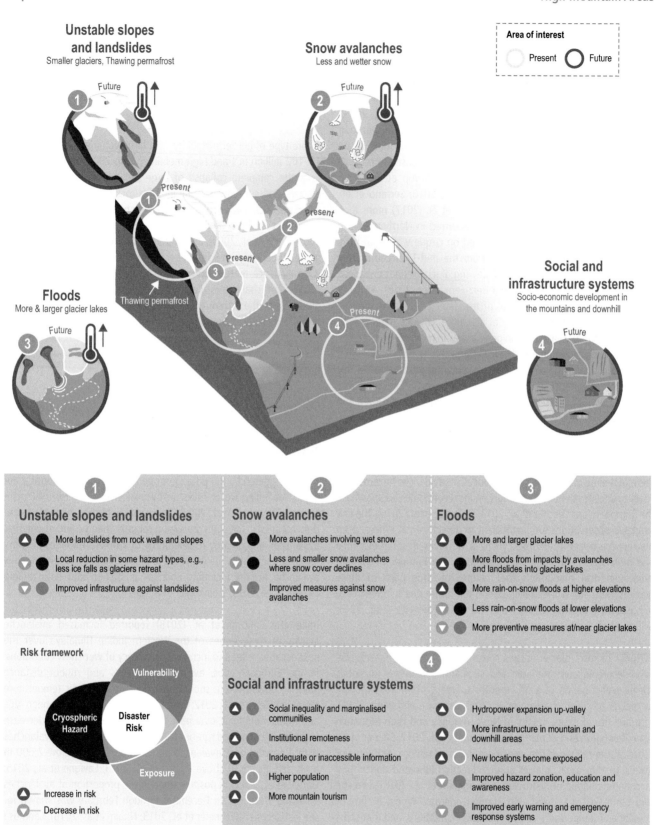

Figure 2.7 | Anticipated changes in high mountain hazards under climate change, driven by changes in snow cover, glaciers and permafrost, overlay changes in the exposure and vulnerability of individuals, communities, and mountain infrastructure.

relationship between avalanche activity, climate change and disaster risk reduction activities in North America. In summary, in particular in Europe, there is *medium confidence* in an increase in avalanche activity involving wet snow, and a decrease in the size and run-out distance of snow avalanches over the past decades.

Future projections mostly indicate an overall decrease in snow depth and snow cover duration at lower elevation (Section 2.2.2), but the probability of occurrence of occasionally large snow precipitation events is projected to remain possible throughout most of the 21st century (Section 2.2.1). Castebrunet et al. (2014) estimated an overall 20 and 30% decrease of natural avalanche activity in the French Alps for the mid and end of the 21st century, respectively, under A1B scenario, compared to the reference period 1960–1990. Katsuyama et al. (2017) reached similar conclusions for Northern Japan, and Lazar and Williams (2008) for North America. Avalanches involving wet snow are projected to occur more frequently during the winter at all elevations due to surface melt or rain-on-snow (e.g., Castebrunet et al., 2014, for the French Alps), and the overall number and runout distance of snow avalanches is projected to decrease in regions and elevations experiencing significant reduction in snow cover (Mock et al., 2017). In summary, there is *medium evidence* and *high agreement* that observed changes in avalanches in mountain regions will be exacerbated in the future, with generally a decrease in hazard at lower elevation, and mixed changes at higher elevation (increase in avalanches involving wet snow, no clear direction of trend for overall avalanche activity).

2.3.2.1.3 Floods

Glacier-related floods, including floods from lake outbursts (GLOFs), are documented for most glacierised mountain ranges and are among the most far-reaching glacier hazards. Past events affected areas tens to hundreds of kilometres downstream (Carrivick and Tweed, 2016). Retreating glaciers produced lakes at their fronts in many high mountain regions in recent decades (Frey et al., 2010; Gardelle et al., 2011; Loriaux and Casassa, 2013). Lake systems in High Mountain Asia also often developed on the surface of downwasting, low-slope glaciers where they coalesced from temporally variable supraglacial lakes (Benn et al., 2012; Narama et al., 2017). Corroborating SREX and AR5 findings, there is *high confidence* that current global glacier shrinkage caused new lakes to form and existing lakes to grow in most regions, for instance in South America, High mountain Asia and Europe (Loriaux and Casassa, 2013; Paul and Mölg, 2014; Zhang et al., 2015; Buckel et al., 2018). Exceptions occurred and are expected to occur in the future for few lakes where evaporation, runoff and reduced melt water influx in total led to a negative water balance (Sun et al., 2018a). Also, advancing glaciers temporarily dammed rivers, lake sections, or fjords (Stearns et al., 2015), for instance through surging (Round et al., 2017), causing particularly large floods once the ice dams breached. Outbursts from water bodies in and under glaciers are able to cause floods similar to those from surface lakes but little is known about the processes involved and any trends under climate change. In some cases, the glacier thermal regime played a role so that climate driven changes in thermal regime are expected to alter the hazard potential, depending on local conditions (Gilbert et al., 2012). Another source of large water bodies under

glaciers and subsequent floods has been subglacial volcanic activity (Section 2.3.2.1.4). There is also *high confidence* that the number and area of glacier lakes will continue to increase in most regions in the coming decades, and new lakes will develop closer to steep and potentially unstable mountain walls where lake outbursts can be more easily triggered by the impact of landslides (Frey et al., 2010; ICIMOD, 2011; Allen et al., 2016a; Linsbauer et al., 2016; Colonia et al., 2017; Haeberli et al., 2017).

In contrast to the number and size of glacier lakes, trends in the number of glacier-related floods are not well known for recent decades (Carrivick and Tweed, 2016; Harrison et al., 2018), although a number of periods of increased and decreased flood activity have been documented for individual glaciers in North America and Greenland, spanning decades (Geertsema and Clague, 2005; Russell et al., 2011). A decrease in moraine-dammed glacier lake outburst floods in recent decades suggests a response of lake outburst activity being delayed by some decades with respect to glacier retreat (Harrison et al., 2018) but inventories might significantly underestimate the number of events (Veh et al., 2018). For the Himalaya, Veh et al. (2019)[161] found no increase in the number of glacier lake outburst floods since the late 1980s. The degradation of permafrost and the melting of ice buried in lake dams have been shown to lower dam stability and contribute to outburst floods in many high mountain regions (Fujita et al., 2013; Erokhin et al., 2017; Narama et al., 2017).

Floods originating from the combination of rapidly melting snow and intense rainfall, referred to as rain-on-snow events, are some of the most damaging floods in mountain areas (Pomeroy et al., 2016; Il Jeong and Sushama, 2018). The hydrological response of a catchment to a rain-on-snow event depends on the characteristics of the precipitation event, but also on turbulent fluxes driven by wind and humidity, which typically provide most of the melting energy during such events (Pomeroy et al., 2016), and the state of the snowpack, in particular the liquid water content (Würzer et al., 2016). An increase in the occurrence of rain-on-snow events in high-elevation zones, and a decrease at the lowest elevations have been reported (Western USA, 1949–2003, McCabe et al. (2007); Oregon, 1986–2010, Surfleet and Tullos (2013); Switzerland, 1972–2016, Moran-Tejéda et al. (2016), central Europe, 1950–2010, Freudiger et al. (2014)). These trends are consistent with studies carried out at the scale of the Northern Hemisphere (Putkonen and Roe, 2003; Ye et al., 2008; Cohen et al., 2015). There are no studies found on this topic in Africa and South America. In summary, evidence since AR5 suggests that rain-on-snow events have increased over the last decades at high elevations, particularly during transition periods from autumn to winter and winter to spring (*medium confidence*). The occurrence of rain-on-snow events has decreased over the last decade in low-elevation or low-latitude areas due to a decreasing duration of the snowpack, except for the coldest months of the year (*medium confidence*).

Il Jeong and Sushama (2018) projected an increase in rain-on-snow events in winter and a decrease in spring, for the period 2041–2070 (RCP4.5 and RCP8.5) in North America, corroborated by Musselman et al. (2018). Their frequency in the Swiss Alps is projected to increase at elevations higher than 2,000 m a.s.l. (SRES A1B, 2025, 2055, and 2085) (Beniston and Stoffel, 2016). This study showed that the number of rain-on-snow events may increase by 50% with a regional temperature increase of 2°C to 4°C, and decrease with a temperature rise exceeding 4°C. In Alaska, an overall increase of rain-on-snow events is projected, however with a projected decline in the southwestern/southern region (Bieniek et al., 2018). In summary, evidence since AR5 suggests that the frequency of rain-on-snow events is projected to increase and occur earlier in spring and later in autumn at higher elevation and to decrease at lower elevation (*high confidence*).

2.3.2.1.4 Combined hazards and cascading events

The largest mountain disasters in terms of reach, damage and lives lost that involve ice, snow and permafrost occurred through a combination or chain of processes. New evidence since SREX and AR5 has strengthened these findings (Anacona et al., 2015a; Evans and Delaney, 2015). Some process chains occur frequently, while others are rare, specific to local circumstances and difficult to anticipate. Glacier lake outbursts were in many mountain regions and over recent decades documented to have been triggered by impact waves from snow-, ice- or rock-avalanches, landslides, iceberg calving events, or by temporary blockage of surface or subsurface drainage channels (Benn et al., 2012; Narama et al., 2017). Rock-slope instability and catastrophic failure along fjords caused tsunamis (Hermanns et al., 2014; Roberts et al., 2014). For instance, a landslide generated wave in 2015 at Taan Fjord, Alaska, ran up 193 m on the opposite slope and then travelled more than 20 km down the fjord (Higman et al., 2018). Earthquakes have been a starting point for different types of cascading events, for instance by causing snow-, ice- or rock-avalanches, and landslides (van der Woerd et al., 2004; Podolskiy et al., 2010; Cook and Butz, 2013; Sæmundsson et al., 2018). Glaciers and their moraines, including morainic lake dams, seem however, not particularly prone to earthquake triggered failure (Kargel et al., 2016).

Landslides and rock avalanches in glacier environments were often documented to entrain snow and ice that fluidise, and incorporate additional loose glacial sediments or water bodies, thereby multiplying their mobility, volume and reach (Schneider et al., 2011; Evans and Delaney, 2015). Rock avalanches onto glaciers triggered glacier advances in recent decades, for instance in North America, New Zealand and Europe, mainly through reducing surface melt (Deline, 2009; Reznichenko et al., 2011; Menounos et al., 2013). In glacier covered frozen rock walls, particularly complex thermal, mechanical, hydraulic and hydrologic interactions between steep glaciers, frozen rock and its ice content, and unfrozen rock sections lead to combined rock/ice instabilities that are difficult to observe and anticipate (Harris et al., 2009; Fischer et al., 2013; Ravanel et al., 2017). There is *limited evidence* of observed direct event chains to project future trends. However, from the observed and projected degradation of permafrost, shrinkage of glaciers and increase in glacier lakes it is reasonable to assume that event chains involving these could increase in frequency or magnitude, and that accordingly hazard zones could expand.

Volcanoes covered by snow and ice often produce substantial melt water during eruptions. This typically results in floods and/or lahars (mixtures of melt water and volcanic debris) which can be exceptionally violent and cause large-scale loss of life and destruction to infrastructure (Barr et al., 2018). The most devastating example from recent history occurred in 1985, when the medium-sized eruption of Nevado del Ruiz volcano, Colombia, produced lahars that killed more than 23,000 people some 70 km downstream (Pierson et al., 1990). Hazards associated with ice and snow-clad volcanoes have been reported mostly from the Cordilleras of the Americas, but also from the Aleutian arc (USA), Mexico, Kamchatka (Russia), Japan, New Zealand and Iceland (Seynova et al., 2017). In particular, under Icelandic glaciers, volcanic activity and eruptions melted large amounts of ice and caused especially large floods if water accumulated underneath the glacier (Björnsson, 2003; Seneviratne et al., 2012). There is *medium confidence* that the overall hazard related to floods and lahars from ice- and snow-clad volcanoes will gradually diminish over years-to-decades as glaciers and seasonal snow cover continue to decrease under climate change (Aguilera et al., 2004; Barr et al., 2018). On the other hand, shrinkage of glaciers may uncover steep slopes of unconsolidated volcanic sediments, thus decreasing in the future the resistance of these volcano flanks to heavy rain fall and increasing the hazard from related debris flows (Vallance, 2005). In summary, future changes in snow and ice are expected to modify the impacts of volcanic activity of snow- and ice-clad volcanoes (*high confidence*) although in complex and locally variable ways and at a variety of time scales (Barr et al., 2018; Swindles et al., 2018).

2.3.2.2 Exposure, Vulnerability and Impacts

2.3.2.2.1 Changes in exposure

Confirming findings from SREX, there is *high confidence* that the exposure of people and infrastructure to cryosphere hazards in high mountain regions has increased over recent decades, and this trend is expected to continue in the future (Figure 2.7). In some regions, tourism development has increased exposure, where often weakly regulated expansion of infrastructure such as roads, trails, and overnight lodging brought more visitors into remote valleys and exposed sites (Gardner et al., 2002; Uniyal, 2013). As an example for the consequences of increased exposure, many of the more than 350 fatalities resulting from the 2015 earthquake triggered snow-ice avalanche in Langtang, Nepal, were foreign trekkers and their local guides (Kargel et al., 2016). Further, several thousand religious pilgrims were killed during the 2013 Kedarnath glacier flood disaster (State of Uttarakhand, Northern India) (Kala, 2014). The expansion of hydropower (Section 2.3.1) is another key factor, and in the Himalaya alone, up to two-thirds of the current and planned hydropower projects are located in the path of potential glacier floods (Schwanghart et al., 2016). Changes in exposure of local communities, for instance, through emigration driven by climate change related threats (Grau and Aide, 2007; Gosai and Sulewski, 2014), or increased connectivity and quality of life in urban centres (Tiwari and Joshi, 2015), are complex and vary regionally. The effects of changes in exposure on labour migration and relocation of entire communities are discussed in Section 2.3.7.

2.3.2.2.2 Changes in vulnerability

Considering the wide ranging social, economic, and institutional factors that enable communities to adequately prepare for, respond to and recover from climate change impacts (Cutter and Morath, 2013), there is *limited evidence* and *high agreement* that mountain communities, particularly within developing countries, are highly vulnerable to the adverse effects of enhanced cryosphere hazards. There are few studies that have systematically investigated the vulnerability of mountain communities to natural hazards (Carey et al., 2017). Coping capacities to withstand impacts from natural hazards in mountain communities are constrained due to a number of reasons. Fundamental weather and climate information is lacking to support both short-term early warning for imminent disasters, and long-term adaptation planning (Rohrer et al., 2013; Xenarios et al., 2018). Communities may be politically and socially marginalised (Marston, 2008). Incomes are typically lower and opportunities for livelihood diversification restricted (McDowell et al., 2013). Emergency responders can have difficulties accessing remote mountain valleys after disasters strike (Sati and Gahalaut, 2013). Cultural or social ties to the land can limit freedom of movement (Oliver-Smith, 1996). Conversely, there is evidence that some mountain communities exhibit enhanced levels of resilience, drawing on long-standing experience, and Indigenous knowledge and local knowledge (Cross-Chapter Box 4 in Chapter 1) gained over many centuries of living with extremes of climate and related disasters (Gardner and Dekens, 2006). In the absence of sufficient data, few studies have considered temporal trends in vulnerability (Huggel et al., 2015a).

2.3.2.2.3 Impacts on livelihoods

Empirical evidence from past events shows that cryosphere related landslides and floods can have severe impacts on lives and livelihoods, often extending far beyond the directly affected region, and persisting for several years. Glacier lake outburst floods alone have over the past two centuries directly caused at least 400 deaths in Europe, 5,745 deaths in South America, and 6,300 deaths in Asia (Carrivick and Tweed, 2016), although these numbers are heavily skewed by individual large events occurring in Huaraz and Yungay, Peru (Carey, 2005) and Kedarnath, India (Allen et al., 2016b).

Economic losses associated with these events are incurred through two pathways. The first consists of direct losses due to the disasters, and the second includes indirect costs from the additional risk and loss of potential opportunities, or from additional investment that would be necessary to manage or adapt to the challenges brought about by the cryosphere changes. Nationwide economic impacts from glacier floods have been greatest in Nepal and Bhutan (Carrivick and Tweed, 2016). The disruption of vital transportation corridors that can impact trading of goods and services (Gupta and Sah, 2008; Khanal et al., 2015), and the loss of earnings from tourism can represent significant far-reaching and long-lasting impacts (Nothiger and Elsasser, 2004; IHCAP, 2017). The Dig Tsho flood in the Khumbu Himal of Nepal in 1985 damaged a hydropower plant and other properties, with estimated economic losses of 500 million USD (Shrestha et al., 2010). Less tangible, but equally important impacts concern the cultural and social disruption resulting from temporary or permanent evacuation (Oliver-Smith, 1979). According to the International Disaster – Emergency Events Database (EM-DAT), over the period 1985–2014, absolute economic losses in mountain regions from all flood and mass movements (including non-cryosphere origins) were highest in the Hindu Kush Himalaya region (45 billion USD), followed by the European Alps (7 billion USD), and the Andes (3 billion USD) (Stäubli et al., 2018). For example, a project to dig a channel in Tsho Rolpa glacier in Nepal that lowered a glacial lake cost 3 million USD in 2000 (Bajracharya, 2010), and similar measures have been taken at Imja Tsho Lake in Nepal in 2016 (Cuellar and McKinney, 2017). Other impacts are related to drinking and irrigation water and livelihoods (Section 2.3.1). In summary, there is *high confidence* that in the context of mountain flood and landslide hazards, exposure, and vulnerability growing in the coming century, significant risk reduction and adaptation strategies will be required to avoid increased impacts.

2.3.2.3 Disaster Risk Reduction and Adaptation

There is *medium confidence* that applying an integrative socioecological risk perspective to flood, avalanche and landslide hazards in high mountain regions paves the way for adaptation strategies that can best address the underlying components of hazard, exposure and vulnerability (Carey et al., 2014; McDowell and Koppes, 2017; Allen et al., 2018; Vaidya et al., 2019). Some degree of adaptation action has been identified in a number of countries with glacier covered mountain ranges, mostly in the form of reactive responses (rather than formal anticipatory plans) to high mountain hazards (Xenarios et al., 2018; McDowell et al., 2019) (Figure 2.9). However, scientific literature reflecting on lessons learned from adaptation efforts generally remains scarce. Specifically for flood and landslide hazards, adaptation strategies that were applied include: hard engineering solutions such as lowering of glacier lake levels, channel engineering, or slope stabilisation that reduce the hazard potential; nature-based solutions such as revegetation efforts to stabilise hazard prone slopes or channels; hazard and risk mapping as a basis for land zoning and early warning systems that reduce potential exposure; various community level interventions to develop disaster response programmes, build local capacities and reduce vulnerability. For example, there is a long tradition of engineered responses to reduce glacier flood risk, most notably beginning in the mid-20th century in Peru (Box 2.4), Italian and Swiss Alps (Haeberli et al., 2001), and more recently in the Himalaya (Ives et al., 2010). There is no published evidence that avalanche risk management, through defence structures design and norms, control measures and warning systems, has been modified as an adaptation to climate change, over the past decades. Projected changes in avalanche character bear potential reductions of the effectiveness of current approaches for infrastructure design and avalanche risk management (Ancey and Bain, 2015).

Early warning systems necessitate strong local engagement and capacity building to ensure communities know how to prepare for and respond to emergencies, and to ensure the long-term sustainability of any such project. In Pakistan and Chile, for instance, glacier flood warnings, evacuation and post-disaster relief have largely been community led (Ashraf et al., 2012; Anacona et al., 2015b).

Cutter et al. (2012) highlight the post-recovery and reconstruction period as an opportunity to build new resilience and adaptive capacities. Ziegler et al. (2014) exemplify consequences when such process is rushed or poorly supported by appropriate long-term planning, as illustrated following the 2013 Kedarnath glacier flood disaster, where guest houses and even schools were being rebuilt in the same exposed locations, driven by short-term perspectives. As changes in the mountain cryosphere, together with socioeconomic, cultural and political developments are producing conditions beyond historical precedent, related responses are suggested to include forward-thinking planning and anticipation of emerging risks and opportunities (Haeberli et al., 2016).

Researchers, policymakers, international donors and local communities do not always agree on the timing of disaster risk reduction projects and programs, impeding full coordination (Huggel et al., 2015b; Allen et al., 2018). Several authors highlight the value of improved evidential basis to underpin adaptation planning. Thereby, transdisciplinary and cross-regional collaboration that places human societies at the centre of studies provides a basis for more effective and sustainable adaptation strategies (McDowell et al., 2014; Carey et al., 2017; McDowell et al., 2019; Vaidya et al., 2019).

In summary, the evidence from regions affected by cryospheric floods, avalanches and landslides generally confirms the findings from the SREX report (Chapter 3), including the requirement for multi-pronged approaches customised to local circumstances, integration of Indigenous knowledge and local knowledge (Cross-Chapter Box 4 in Chapter 1) together with improved scientific understanding and technical capacities, strong local participation and early engagement in the process, and high-level communication and exchange between all actors. Particularly for mountain regions, there is *high confidence* that integration of knowledge and practices across natural and social sciences, and the humanities, is most efficient in addressing complex hazards and risks related to glaciers, snow, and permafrost.

Box 2.4 | Challenges to Farmers and Local Population Related to Shrinkages in the Cryosphere: Cordillera Blanca, Peru

The Cordillera Blanca of Peru contains most of the glaciers in the tropics, and its glacier coverage declined significantly in the recent past (Burns and Nolin, 2014; Mark et al., 2017). Since the 1940s, glacier hazards have killed thousands (Carey, 2005) and remain threatening. Glacier wastage has also reduced river runoff in most of its basins in recent decades, particularly in the dry season (Baraer et al., 2012; Vuille et al., 2018). Residents living adjacent to the Cordillera Blanca have long recognised this glacier shrinkage, including rural populations living near glaciers and urban residents worried about glacier lake floods and glacier landslides (Jurt et al., 2015; Walter, 2017). Glacier hazards and the glacier runoff variability increase exposure and uncertainty while diminishing adaptive capacity (Rasmussen, 2016).

Cordillera Blanca residents' risk of glacier-related disasters is amplified by intersecting physical and societal factors. Cryosphere hazards include expanding or newly forming glacial lakes, slope instability, and other consequences of rising temperatures, and precipitation changes (Emmer et al., 2016; Colonia et al., 2017; Haeberli et al., 2017). Human vulnerability to these hazards is conditioned by factors such as poverty, limited political influence and resources, minimal access to education and healthcare, and weak government institutions (Hegglin and Huggel, 2008; Carey et al., 2012; Lynch, 2012; Carey et al., 2014; Heikkinen, 2017). Early warning systems have been, or are being, installed at glacial lakes Laguna 513 and Palcacocha to protect populations (Muñoz et al., 2016). Laguna 513 was lowered by 20 m for outburst prevention in the early 1990s but nonetheless caused a destructive flood in 2010, though much smaller and less destructive than a flood that would have been expected without previous lake mitigation works (Carey et al., 2012; Schneider et al., 2014). An early warning system was subsequently installed, but some local residents destroyed it in 2017 due to political, social and cultural conflicts (Fraser, 2017). The nearby Lake Palcacocha also threatens populations (Wegner, 2014; Somos-Valenzuela et al., 2016). The usefulness for ground-level education and communication regarding advanced early warning systems has been demonstrated in Peru (Muñoz et al., 2016).

Vulnerability to hydrologic variability and declining glacier runoff is also shaped by intertwining human and biophysical drivers playing out in dynamic hydro-social systems (Bury et al., 2013; Rasmussen 2016; Drenkhan et al., 2015; Carey et al., 2017). Water security is influenced by both water availability (supply from glaciers) as well as by water distribution, which is affected by factors such as water laws and policies, global demand for agricultural products grown in the lower Santa River basin, energy demands and hydroelectricity production, potable water usage, and livelihood transformations over time (Carey et al., 2014; Vuille et al., 2018). In some cases, the formation of new glacial lakes can create opportunities as well as hazards, such as new tourist attractions and reservoirs of water, thereby showing how socioeconomic and geophysical forces intersect in complex ways (Colonia et al., 2017).

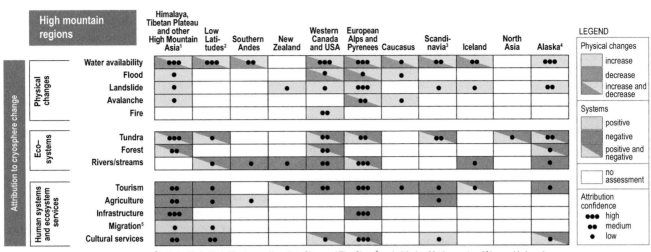

¹ includes Hindu Kush, Karakoram, Hengduan Shan, and Tien Shan; ² tropical Andes, Mexico, eastern Africa, and Indonesia;
³ includes Finland, Norway, and Sweden; ⁴ includes adjacent areas in Yukon Territory and British Columbia, Canada;
⁵ Migration refers to an increase or decrease in net migration, not to beneficial/adverse value.

Figure 2.8 | Synthesis of observed physical changes and impacts on ecosystems and human systems and ecosystems services in eleven high mountain regions over past decades that can at least partly be attributed to changes in the cryosphere. Only observations documented in the scientific literature are shown, but impacts may also be experienced elsewhere. For physical changes yellow/green refers to an increase/decrease, respectively, in amount or frequency of the measured variable. For impacts on ecosystems and human systems and ecosystems services blue or red depicts whether an observed impact is positive (beneficial) or negative (adverse). Cells assigned 'increase and decrease' indicate that within that region both increase and decrease of physical changes are found, but are not necessarily equal; the same holds for cells showing 'positive and negative' impacts. Confidence levels refer to confidence in attribution to cryospheric changes. No assessment means: not applicable, not assessed at regional scale, or the evidence is insufficient for assessment. Tundra refers to tundra and alpine meadows. Migration refers to an increase and decrease in net migration, not beneficial/adverse value. Impacts on tourism refer to the operating conditions for the tourism sector. Cultural services include cultural identity, sense of home, intrinsic and aesthetic values, as well as contributions from glacier archaeology. Figure is based on observed impacts listed in Table SM2.11.

2.3.3   Ecosystems

Widespread climate driven ecological changes have occurred in high mountain ecosystems over the past century. Those impacts were assessed in a dedicated manner only in earlier IPCC assessments (Beniston and Fox, 1996; Gitay et al., 2001; Fischlin et al., 2007) but not in AR5 (Settele et al., 2014). Two of the most evident changes include range shifts of plants and animals in Central Europe and the Himalaya but also for other mountain regions (e.g., Morueta-Holme et al., 2015; Evangelista et al., 2016; Freeman et al., 2018; Liang et al., 2018; You et al., 2018; He et al., 2019), and increases in species richness on mountain summits (Khamis et al., 2016; Fell et al., 2017; Steinbauer et al., 2018) of which some have accelerated during recent decades (e.g., Steinbauer et al., 2018), though slowing over the past ten years in Austria (e.g., Lamprecht et al., 2018). While many changes in freshwater communities have been directly attributed to changes in the cryosphere (Jacobsen et al., 2012; Milner et al., 2017), separating the direct influence of atmospheric warming from the influence of concomitant cryospheric change and independent biotic processes has been often challenging for terrestrial ecosystems (Grytnes et al., 2014; Lesica and Crone, 2016; Frei et al., 2018; Lamprecht et al., 2018). Changing climate in high mountains places further stress on biota, which are already impacted by land use and its change, direct exploitation, and pollutants (Díaz et al., 2019; Wester et al., 2019). Species are required to shift their behaviours, including seasonal aspects, and distributional ranges to track suitable climate conditions (Settele et al., 2014). In the Special Report on Global Warming of 1.5°C (SR15), climate change scenarios exceeding mean global warming of 1.5° C relative to preindustrial levels have been estimated to lead to major impacts on species

abundances, community structure, and ecosystem functioning in high mountain areas (Hoegh-Guldberg et al., 2018). The size and isolation of mountain habitats (Steinbauer et al., 2016; Cotto et al., 2017), which may vary strongly with the topography of mountain ridges (Elsen and Tingley, 2015; Graae et al., 2018), affects critically the survival of species as they migrate across mountain ranges, increasing in general the risks for many species from climate change (Settele et al., 2014; Dobrowski and Parks, 2016).

2.3.3.1   Terrestrial Biota

The cryosphere can play a critical role in moderating and driving how species respond to climate change in high mountains (*high confidence*). Many mountain plant and animal species have changed abundances and migrated upslope while expanding or contracting their ranges over the past decades to century, whereas others show no change (Morueta-Holme et al., 2015; Suding et al., 2015; Lesica and Crone, 2016; Fadrique et al., 2018; Freeman et al., 2018; Rumpf et al., 2018; Johnston et al., 2019; Rumpf et al., 2019) (*medium agreement, robust evidence*). These responses are often linked directly to warming, yet a changing cryosphere, for example, in the form of decreasing snow thickness or altered seasonality of snow (e.g., Matteodo et al., 2016; Kirkpatrick et al., 2017; Amagai et al., 2018; Wu et al., 2018) or indirectly leading to changes in soil moisture (Harpold and Molotch, 2015), can play a significant role for growth, fitness and survival of many species (e.g., Grytnes et al., 2014; Winkler et al., 2016) (*medium evidence, high agreement*).

Cryospheric changes were found to be beneficial for some plant species and for ecosystems in some regions, improving a number of ecosystem services, such as by provisioning new habitat for endemic plant species and increasing plant productivity (*high confidence*). Decreasing snow cover duration, glacier retreat and permafrost thaw have already and will over coming decades allow plant species, including some endemic species, to increase their abundance and extend their range in many mountain ranges (Yang et al., 2010a; Grytnes et al., 2014; Elsen and Tingley, 2015; Dolezal et al., 2016; Wang et al., 2016b; D'Amico et al., 2017; Liang et al., 2018; Yang et al., 2018; You et al., 2018; He et al., 2019). Over recent decades, plant colonisation after glacier retreat has been swift, for example, at many sites with favourable soils in the European Alps (Matthews and Vater, 2015; Fickert and Grüninger, 2018) or has even accelerated compared to 100 years ago (Fickert et al., 2016). At other sites of the European Alps (D'Amico et al., 2017) and in other mountain ranges (e.g., Andes and Alaska; Darcy et al., 2018; Zimmer et al., 2018) the rate of colonisation remains slow due to soil type, soil formation and phosphorous limitation (Darcy et al., 2018). In Bhutan, snowlines have ascended and new plant species have established themselves in these areas, yet despite range expansion and increased productivity, yak herders describe impacts on the ecosystem services as mostly negative (Wangchuk and Wangdi, 2018). Earlier snowmelt often leads to earlier plant growth and, provided there is sufficient water, including from underlying permafrost, plant productivity has increased in many alpine regions (e.g., Williams et al., 2015; Yang et al., 2018). Decreased snow cover duration has led to colonisation of snowbed communities by wide-ranging species in several regions, for example, in the Australian Alps (Pickering et al., 2014), though this can lead to declines in the abundance of resident species, for example, in the Swiss Alps (Matteodo et al., 2016).

Cryospheric change in high mountains directly harms some plant species and ecosystems in some regions, degrading a number of ecosystem services, such as maintaining regional and global biodiversity, and some provisioning services, for example, fodder or wood production, in terms of timing and magnitude (*high confidence*). In mountains, microrefugia (a local environment different from surrounding areas) and isolation have contributed to high plant endemism that increases with elevation (Steinbauer et al., 2016; Zhang and Zhang, 2017; Muellner-Riehl, 2019). Microrefugia may enable alpine species to persist if global warming remains below 2°C relative pre-industrial levels (Scherrer and Körner, 2011; Hannah et al., 2014; Graae et al., 2018) (*medium evidence, medium agreement*). Yet, where glaciers have been retreating over recent decades, cool microrefugia have shifted location or decreased in extent (Gentili et al., 2015). In regions with insufficient summer precipitation, earlier snowmelt and absence of permafrost lead to insufficient water supply during the growing season, and consequently an earlier end of peak season, altered species composition, and a decline in greenness or productivity (Trujillo et al., 2012; Sloat et al., 2015; Williams et al., 2015; Yang et al., 2018) (*medium evidence, high agreement*). Across elevations, alpine-restricted species show greater sensitivity to the timing of snowmelt than wide ranging species (Lesica, 2014; Winkler et al., 2018), and though the cause is often not known, some alpine-restricted species have declined in abundance or disappeared in regions with distinctive flora (Evangelista et al., 2016; Giménez-Benavides et al., 2018; Lamprecht et al., 2018; Panetta et al., 2018) (*medium evidence, high agreement*).

The shrinking cryosphere represents a loss of critical habitat for wildlife that depend on snow and ice cover, affecting well-known and unique high-elevation species (*high confidence*). Areas with seasonal snow and glaciers are essential habitat for birds and mammals within mountain ecosystems for foraging, relief from climate stress, food caching and nesting grounds (Hall et al., 2016; Rosvold, 2016) (*robust evidence*). Above 5,000 m a.s.l. in Peru, there was recently a first observation of bird nesting for which its nesting may be glacier obligate (Hardy et al., 2018). The insulated and thermally stable region under the snow at the soil-snow interface, termed the subnivean, has been affected by changing snowpack, limiting winter activity and decreasing population growth for some mountain animals, including frogs, rodents and small carnivores (Penczykowski et al., 2017; Zuckerberg and Pauli, 2018; Kissel et al., 2019) (*medium evidence*). Many mountain animals have been observed to change their behaviour in a subtle manner, for example., in foraging or hunting behaviour, due to cryospheric changes (e.g., Rosvold, 2016; Büntgen et al., 2017; Mahoney et al., 2018) (*medium evidence, high agreement*). In the Canadian Rocky Mountains, grizzly bears have moved to new snow free habitat after emerging in spring from hibernation to dig for forage, which may increase the risk of human-bear encounters (Berman et al., 2019). In the US Central Rocky Mountains, migratory herbivores, such as elk, moose and bison, track newly emergent vegetation that greens soon after snowmelt (Merkle et al., 2016). For elk, this was found to increase fat gain (Middleton et al., 2018). Due to loss of snow patches that increase surface water and thus insect abundance, some mammal species, for example, reindeer and ibex, have changed their foraging behaviour to evade the biting insects with negative impacts on reproductive fitness (Vors and Boyce, 2009; Büntgen et al., 2017).

Many endemic plant and animal species including mammals and invertebrates in high mountain regions are vulnerable to further decreasing snow cover duration, such as later onset of snow accumulation and/or earlier snowmelt (*high confidence*) (Williams et al., 2015; Slatyer et al., 2017). Winter-white animals for which coat or plumage colour is cued by day length will confront more days with brown snowless ground (Zimova et al., 2018), which has already contributed to range contractions for several species, including hares and ptarmigan (Imperio et al., 2013; Sultaire et al., 2016; Pedersen et al., 2017) (*robust evidence*). Under all climate scenarios, the duration of this camouflage mismatch will increase, enhancing predation rates thereby decreasing populations of coat-colour changing species (e.g., 24% decrease by late century under RCP8.5 for snowshoe hares; Zimova et al., 2016; see also Atmeh et al., 2018) (*medium evidence, high agreement*). For roe deer (Plard et al., 2014) and mountain goats (White et al., 2017), climate driven changes in snowmelt duration and summer temperatures will reduce survival considerably under RCP4.5 and RCP8.5 scenarios (*medium evidence, high agreement*).

2

2.3.3.2   Freshwater Biota

Biota in mountain freshwater ecosystems is affected by cryospheric change through alterations in both the quantity and timing of runoff from glaciers and snowmelt. Where melt water from glaciers decreases, river flows have become more variable, with water temperature and overall channel stability increasing and habitats becoming less complex (Giersch et al., 2017; Milner et al., 2017) (*medium evidence, medium agreement*).

Analysis of three invertebrate datasets from tropical (Ecuador), temperate (Italian Alps) and sub-Arctic (Iceland) alpine regions indicates that a number of cold-adapted species have decreased in abundance below a threshold of watershed glacier cover varying from 19–32%. With complete loss of the glaciers, 11–38% of the regional species will be lost (Jacobsen et al., 2012; Milner et al., 2017) (*medium confidence*). As evidenced in Europe (Pyrenees, Italian Alps) and North America (Rocky Mountains) (Brown et al., 2007; Giersch et al., 2015; Giersch et al., 2017; Lencioni, 2018) the loss of these invertebrates – many of them endemic – as glacier runoff decreases and transitions to a regime more dominated by snowmelt leading to a reduction in turnover between and within stream reaches (beta diversity) and regional (gamma) diversity (*very high confidence*). Regional genetic diversity within individual riverine invertebrate species in mountain headwater areas has decreased with the loss of environmental heterogeneity (Giersch et al., 2017), as decreasing glacier runoff reduces the isolation of individuals permitting a greater degree of genetic intermixing (Finn et al., 2013; Finn et al., 2016; Jordan et al., 2016; Hotaling et al., 2018) (*medium evidence, high agreement*). However, local (alpha) diversity, dominated by generalist species of invertebrates and algae, has increased (Khamis et al., 2016; Fell et al., 2017; Brown et al., 2018) (*very high confidence*) in certain regions as species move upstream, although not in the Andes, where downstream migration has been observed (Jacobsen et al., 2014; Cauvy-Fraunié et al., 2016).

Many climate variables influence fisheries, through both direct and indirect pathways. The key variables linked to cryospheric change include: changes in air and water temperature, precipitation, nutrient levels and ice cover (Stenseth et al., 2003). A shrinking cryosphere has significantly affected cold mountain resident salmonids (e.g., brook trout, *Salvelinus fontinalis*), causing further migration upstream in summer thereby shrinking their range (Hari et al., 2006; Eby et al., 2014; Young et al., 2018). Within the Yanamarey watershed of the Cordillera Blanca in Peru, fish stocks have either declined markedly or have become extinct in many streams, possibly due to seasonal reductions of fish habitat in the upper watershed resulting from glacier recession (Bury et al., 2011; Vuille et al., 2018). In contrast, glacier recession in the mountains of coastal Alaska and to a lesser extent the Pacific northwest have created a large number of new stream systems that have been, and could continue to be with further glacier retreat, colonised from the sea by salmon species that contribute to both commercial and sport fisheries (Milner et al., 2017; Schoen et al., 2017) (*medium confidence*). Changes in water temperature will vary seasonally, and a potential decreased frequency of rain-on-snow events in winter compared to rain-on-ground would increase water temperature, benefiting overwintering survival (Leach

and Moore, 2014). Increased water temperature remaining below thermal tolerance limits for fish and occurring earlier in the year can benefit overall fish growth and increase fitness (Comola et al., 2015) (*medium evidence, medium agreement*).

In the future, increased primary production dominated by diatoms and golden algae will occur in streams as glacier runoff decreases, although some cold-tolerant diatom species will be lost, resulting in a decrease in regional diversity (Fell et al., 2017; Fell et al., 2018). Reduced glacier runoff is projected to improve water clarity in many mountain lakes, increasing biotic diversity and the abundance of bacterial and algal communities and thus primary production (Peter and Sommaruga, 2016) (*limited evidence*). Extinction of range-restricted prey species may increase as more favourable conditions facilitate the upstream movement of large bodied invertebrate predators (Khamis et al., 2015) (*medium confidence*). Modelling studies indicate a reduction in the range of native species, notably trout, in mountain streams, (Papadaki et al., 2016; Vigano et al., 2016; Young et al., 2018) (*medium evidence, high agreement*), which will potentially impact sport fisheries. In northwest North America, where salmon are important in native subsistence as well as commercial and sport fisheries, all species will potentially be affected by reductions in glacial runoff from mountain glaciers over time (Milner et al., 2017; Schoen et al., 2017), particularly in larger systems where migratory corridors to spawning grounds are reduced (*medium confidence*).

In summary, cryospheric change will alter freshwater communities with increases in local biodiversity but range shrinkage and extinctions for some species causes regional biodiversity to decrease (*robust evidence, medium agreement, i.e., high confidence*).

2.3.3.3   Ecosystem Services and Adaptation

The trend to a higher productivity in high mountain ecosystems due to a warmer environment and cryospheric changes, affects provisioning and regulating services (*high confidence*). Due to earlier snowmelt, the growing season has begun earlier, for example, on the Tibetan Plateau, and in the Swiss Alps (Wang et al., 2017; Xie et al., 2018), and in some regions earlier growth has been linked to greater plant production or greater net ecosystem production, possibly affecting carbon uptake (Scholz et al., 2018; Wang et al., 2018; Wu et al., 2018). In other areas productivity has decreased, despite a longer growing season, for example, in the US Rocky Mountains, US Sierra Nevada Mountains, Swiss Alps, and Tibetan Plateau (Arnold et al., 2014; Sloat et al., 2015; Wang et al., 2017; De Boeck et al., 2018; Knowles et al., 2018) (*robust evidence, medium agreement*). Changed productivity of the vegetation in turn can affect the timing, quantity and quality of water supply, a critical regulating service ecosystems play in high mountain areas (Goulden and Bales, 2014; Hubbard et al., 2018) (*medium confidence*). Permafrost degradation has dramatically changed some alpine ecosystems through altered soil temperature and permeability, decreasing the climate regulating service of a vast region and leading to lowered ground water and new and shrinking lakes on the Tibetan Plateau (Jin et al., 2009; Yang et al., 2010b; Shen et al., 2018) (*medium evidence, high agreement*).

Ecosystems and their services are vulnerable to changes in the intensity and/or the frequency of an ecological disturbance that exceed the previous range of variation (Johnstone et al., 2016; Camac et al., 2017; Fairman et al., 2017; cf. 3.4.3.2 Ecosystems and their Services) (*high confidence*). For example, in the Western USA, mountain ecosystems are experiencing an increase in the number and extent of wildfires, which have been attributed to many factors, including climate factors such as earlier snowmelt and vapour-pressure deficit (Settele et al., 2014; Westerling, 2016; Kitzberger et al., 2017; Littell, 2018; Littell et al., 2018). Similarly, landslides and floods in many areas have been attributed to cryospheric changes (Section 2.3.2). Disturbances can feedback and diminish many of the ecosystem services such as provisioning, regulating and cultural services (Millar and Stephenson, 2015; McDowell and Koppes, 2017; Mcdowell et al., 2018; Murphy et al., 2018; Maxwell et al., 2019). Consistent with AR5 findings (Settele et al., 2014) the capacity of many freshwater and terrestrial mountain species to adapt naturally to climate change is projected to be exceeded for high warming levels, leading to species migration across mountain ranges or loss with consequences for many ecosystem services (Elsen and Tingley, 2015; Dobrowski and Parks, 2016; Pecl et al., 2017; Rumpf et al., 2019) (*robust evidence, medium agreement*, i.e., *high confidence*). Although the adaptive potential of aquatic biota to projected changes in glacial runoff is not fully understood (Lencioni et al., 2015), dispersion and phenotypic plasticity together with additional microrefugia formation due to cryospheric changes, is expected to help threatened species to better adapt, perhaps even in the long term (Shama and Robinson, 2009). Likewise, traits shaped by climate and with high genetically-based standing variation may be used to spatially identify, map and manage global 'hotspots' for evolutionary rescue from climate change (Jones et al., 2018; Mills et al., 2018). Nature conservation increases the potential for mitigating adverse effects on many of these ecosystem services, including those that are essential for the support of the livelihoods and the culture of mountain peoples, including economical aspects such as recreation and tourism (e.g., Palomo, 2017; Elsen et al., 2018; Wester et al., 2019) (*medium confidence*).

2.3.4   Infrastructure and Mining

There is *high confidence* that permafrost thaw has had negative impacts on the integrity of infrastructure in high mountain areas. Like in polar regions (Section 3.4.3.3.4), the local effects of infrastructure together with climate change degraded permafrost beneath and around structures (Dall'Amico et al., 2011; Doré et al., 2016) Infrastructure on permafrost in the European Alps, mostly found near mountain summits but not in major valleys, has been destabilised by permafrost thaw, including mountain stations in France and Austria (Ravanel et al., 2013; Keuschnig et al., 2015; Duvillard et al., 2019) as well as avalanche defence structures (Phillips and Margreth, 2008) and a ski lift (Phillips et al., 2007) in Switzerland. On the Tibetan Plateau, deformation or damage has been found on roads (Yu et al., 2013; Chai et al., 2018), power transmission infrastructure (Guo et al., 2016) and around an oil pipeline (Yu et al., 2016). For infrastructure on permafrost, engineering practices suitable for polar and high mountain environments (Doré et al., 2016) as well as

specific for steep terrain (Bommer et al., 2010) have been developed to support adaptation.

In some mountain regions, glacier retreat and related processes of change in the cryosphere have afforded greater accessibility for extractive industries and related activities to mine minerals and metals (*medium confidence*). Accelerated glacier shrinkage and retreat have been reported to facilitate mining activities in Chile, Argentina and Peru (Brenning, 2008; Brenning and Azócar, 2010; Anacona et al., 2018), and Kyrgyzstan (Kronenberg, 2013; Petrakov et al., 2016), which also interact with and have consequences for other social, cultural, economic, political and legal measures, where climate change impacts also play a role (Brenning and Azócar, 2010; Evans et al., 2016; Khadim, 2016; Anacona et al., 2018). However, negative impacts due to cryosphere changes may also occur. One study projects that reductions in glacier melt water and snowmelt in the watershed in the Chilean Andes will lead to a reduction of water supply to a copper mine by 2075–2100 of 28% under scenario A2 and of 6% under B2; construction of infrastructure to draw water from other sources will cost between 16–137 million USD (Correa-Ibanez et al., 2018).

Conversely, there is also evidence suggesting that some of these mining activities affect glaciers locally, and the mountain environment around them, further altering glacier dynamics, glacier structure and permafrost degradation. This is due mainly to excavation, extraction, and use of explosives (Brenning, 2008; Brenning and Azócar, 2010; Kronenberg, 2013), and deposition of dust and other mine waste material close to or top of glaciers during extraction and transportation (Brenning, 2008; Torgoev and Omorov, 2014; Arenson et al., 2015b; Jamieson et al., 2015). These activities have reportedly generated slope instabilities (Brenning, 2008; Brenning and Azócar, 2010; Torgoev and Omorov, 2014), glacier mass loss due to enhanced surface melt from dust and debris deposition (Torgoev and Omorov, 2014; Arenson et al., 2015b; Petrakov et al., 2016), and even glacier advance by several kilometres (Jamieson et al., 2015), although their impact is considered less than that reported for changes in glaciers due to climatic change (*limited evidence, medium agreement*). Glacier Protection Laws and similar measures have been introduced in countries such as Chile and Argentina to address these impacts (Khadim, 2016; Anacona et al., 2018; Navarro et al., 2018). In addition, the United Nations Human Rights Council passed a declaration in 2018 to "protect and restore water-related ecosystems" in mountain areas as elsewhere from contamination by mining (UNHRC, 2018); however, evidence on the effectiveness of these measures remains inconclusive.

2.3.5   Tourism and Recreation

The mountain cryosphere provides important aesthetic, cultural, and recreational services to society (Xiao et al., 2015). These services support tourism, providing economic contributions and livelihood options to mountain communities and beyond. The relevant changes in the cryosphere affecting mountain tourism and recreation include shorter seasons of snow cover, more winter precipitation falling as rain instead of snow, and declining glaciers and permafrost

(Sections 2.2.1, 2.2.2, 2.2.3 and 2.2.4). Downhill skiing, the most popular form of snow recreation, occurs in 67 countries (Vanat, 2018). The Alps in Europe support the largest ski industry (Vanat, 2018). In Europe, the growth of alpine skiing and winter tourism after 1930 brought major economic growth to alpine regions and transformed winter sports into a multi-billion USD industry (Denning, 2014). Sixteen percent of skier visits occur in the USA, where expenditures from all recreational snow sports generated more than 695,000 jobs and 72.7 billion USD in trip-related spending in 2016 (Outdoor Industry Association, 2017). While the number of ski resorts in the USA has been decreasing since the 1980s, China added 57 new ski resorts in 2017 (Vanat, 2018). Although the bulk of economic activity is held within mountain communities, supply chains for production of ski equipment and apparel span the globe. Steiger et al. (2017) point out that Asia, Africa and South America are underrepresented in the ski tourism literature, and Africa and the Middle East are not significant markets from a ski tourism perspective.

Skiing's reliance on favourable atmospheric and snow conditions make it particularly vulnerable to climate change (Arent et al., 2014; Hoegh-Guldberg et al., 2018). Snow reliability, although not universally defined, quantifies whether the snow cover is sufficient for ski resorts operations. Depending on the context, it focuses on specific periods of the winter season, and may account for interannual variability and/or for snow management (Steiger et al., 2017). The effects of less snow, due to strong correlation between snow cover and skier visits, cost the USA economy 1 billion USD and 17,400 jobs per year between 2001–2016 in years of less seasonal snow (Hagenstad et al., 2018). Efforts to reduce climate change impacts and risks to economic losses focus on increased snowmaking, such as artificial production of snow (Steiger et al., 2017), summertime slope preparation (Pintaldi et al., 2017), grooming (Steiger et al., 2017), and snow farming, that is, storage of snow (Grünewald et al., 2018). The effectiveness of snow management methods as adaptation to long-term climate change depends on sufficiently low air temperature conditions needed for snowmaking, water and energy availability, compliance with environmental regulations (de Jong, 2015), and ability to pay for investment and operating costs. When these requirements are met, evidence over the past decades shows that snow management methods have generally proven efficient in reducing the impact of reduced natural snow cover duration for many resorts (Dawson and Scott, 2013; Hopkins and Maclean, 2014; Steiger et al., 2017; Spandre et al., 2019a). The number of skier visits was found to be 39% less sensitive to natural snow variations in Swiss ski resorts with 30% areal snowmaking coverage (representing the national average), compared to resorts without snowmaking (Gonseth, 2013). In some regions, many resorts (mostly smaller, low-elevation resorts) have closed due to unfavourable snow conditions brought on by climate change and/or the associated need for large capital investments for snowmaking capacities (e.g., in northeast USA; Beaudin and Huang, 2014). To offset loss in ski tourism revenue, a key adaptation strategy is diversification, offering other non-snow recreation options such as mountain biking, mountain coasters and alpine slides, indoor climbing walls and water parks, festivals and other special events (Figure 2.9; Hagenstad et al., 2018; Da Silva et al., 2019).

In the near term (2031–2050) and regardless of the greenhouse gas emission scenario, risks to snow reliability exist for many resorts, especially at lower elevation, although snow reliability is projected to be maintained at many resorts in North America (Wobus et al., 2017) and in the European Alps, Pyrenees and Scandinavia (Marke et al., 2015; Steiger et al., 2017; Scott et al., 2019; Spandre et al., 2019a; Spandre et al., 2019b). At the end of the century (2081–2100), under RCP8.5, snow reliability is projected to be unviable for most ski resorts under current operating practices in North America, the European Alps and Pyrenees, Scandinavia and Japan, with some exceptions at high elevation or high-latitudes (Steiger et al., 2017; Wobus et al., 2017; Suzuki-Parker et al., 2018; Scott et al., 2019; Spandre et al., 2019a; Spandre et al., 2019b). Only few studies have used RCP2.6 in the context of ski tourism, and results indicate that the risks at the end of the century (2081–2100) are expected to be similar to the near term impacts (2031–2050) for RCP8.5 (Scott et al., 2019; Spandre et al., 2019a).

The projected economic losses reported in the literature include an annual loss in hotel revenues of EUR 560 million (2012 value) in Europe, compared to the period 1971–2000 under a 2°C global warming scenario (Damm et al., 2017). This estimate includes population projections but does not account for snow management. In the USA, Wobus et al. (2017) estimate annual revenue losses from tickets (skiing) and day fees (cross country skiing and snowmobiling) due to reduced snow season length, will range from 340–780 million USD in 2050 for RCP4.5 and RCP8.5, respectively, and from 130 million to 2 billion USD in 2090 for RCP4.5 and RCP8.5 respectively, taking into account snow management and population projections. Total economic losses from these studies would be much higher if all costs were included (costs for tickets, transport, lodging, food and equipment). Regardless of the climate scenario, as risk of financial unviability increases, there are reported expectations that companies would need to forecast when their assets may become stranded assets and require devaluation or conversion to liabilities, and report this on their balance sheets (Caldecott et al., 2016). Economic impacts are projected to occur in other snow-based winter activities including events (e.g., ski races) and other recreation activities such as cross-country skiing, snowshoeing, backcountry skiing, ice climbing, sledding, snowmobiling and snow tubing. By 2050, 13 (out of 21) prior Olympic Winter Games locations are projected to exhibit adequate snow reliability under RCP2.6, and 10 under RCP8.5. By 2080, the number decreases to 12 and 8, respectively (Scott et al., 2018). Even for cities remaining cold enough to host ski competitions, costs are projected to rise for making and stockpiling snow, as was the case in Sochi, Russia in 2014 and Vancouver, Canada in 2010 (Scott et al., 2018), and preserving race courses through salting (Hagenstad et al., 2018).

In summer, cryosphere changes are impacting glacier-related activities (hiking, sightseeing, skiing, climbing and mountaineering) (Figure 2.8). In recent years, several ski resorts operating on glaciers have ceased summer operations due to unfavourable snow conditions and excessive operating costs (e.g., Falk, 2016). Snow management and snowmaking are increasingly used on glaciers (Fischer et al., 2016). Glacier retreat has led to increased moraine instability which can compromise hiker and climber safety along established trails and common access routes, for example, in Iceland (Welling et al., 2019), though it has made some areas in the Peruvian Andes more accessible to trekkers (Vuille et al., 2018). In response, some hiking routes have been adjusted and ladders and fixed anchors installed (Duvillard et al., 2015; Mourey and Ravanel, 2017). As permafrost thaws, rock falls on and off glaciers are increasingly observed, threatening the safety of hikers and mountaineers, for example, in Switzerland (Temme, 2015) and New Zealand (Purdie et al., 2015). Glacier retreat and permafrost thaw have induced major changes to iconic mountaineering routes in the Mont Blanc area, European Alps with impacts on mountaineering practices, such as shifts in suitable climbing seasons, and reduced route safety (Mourey and Ravanel, 2017; Mourey et al., 2019). Cryosphere decline has also reduced opportunities for ice climbing and reduced attractions for summer trekking in the Cascade Mountains, USA (Orlove et al., 2019). In response to these impacts, tour companies have shifted to new sites, diversified to offer other activities or simply reduced their activities (Furunes and Mykletun, 2012) (Figure 2.9). Steps to improve consultation and participatory approaches to understand risk perception and design joint action

between affected communities, authorities and operators, are evident, for example, in Iceland (Welling et al., 2019). In some cases, new opportunities are presented such as marketing 'climate change tourism' where visitors are attracted by 'last chance' opportunities to view a glacier; for example, in New Zealand (Stewart et al., 2016), in China (Wang et al., 2010) or through changing landscapes such as new lakes, for instance in Iceland (Þórhallsdóttir and Ólafsson, 2017), or to view the loss of a glacier, for example, in the Bolivian Andes (Kaenzig et al., 2016). The opening of a trekking route promoting this opportunity created tensions between a National Park and a local indigenous community in the Peruvian Andes over the management and allocation of revenue from the route (Rasmussen, 2019). The consequences of ongoing and future glacier retreat are projected to negatively impact trekking and mountaineering in the Himalaya (Watson and King, 2018). Reduced snow cover has also negatively impacted trekking in the Himalaya, since tourists find the mountains

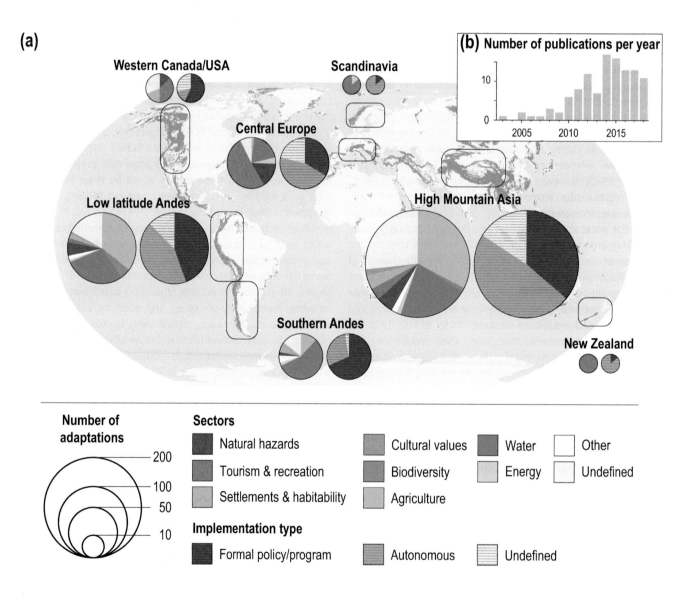

Figure 2.9 | **(a)** Documented number of individual adaptation actions distributed across seven of the high mountain regions addressed in this Chapter, with pie charts indicating the number of adaptation measures for sectors addressed in this chapter (left pie chart), and the relative proportion of these classified as either 'formal', 'autonomous' or 'undefined' (right pie chart). Note that for regions with less than five reported adaptation measures were excluded from the figure (i.e., Caucasus, Iceland and Alaska), however these are detailed in Table SM2.9. **(b)** Number of publications reported in the assessed literature over time. In some cases, multiple adaptation measures are discussed in a single publication (Table SM2.9).

less attractive as a destination, and the reduced water availability affects the ability of hotels and campsites to serve visitors (Becken et al., 2013).

In summary, financial risks to mountain communities that depend on tourism for income, are high and include losses to revenues generated from recreation primarily in the winter season. Adaptation to cryosphere change for ski tourism focuses on snowmaking and is expected to be moderately effective for many locations in the near term (2031–2050), but it is unlikely to substantially reduce the risks in most locations in the longer term (end of century) (*high confidence*). Determining the extent to which glacier retreat and permafrost thaw impact upon overall visitor numbers in summer tourism, and how any losses or increased costs are offset by opportunities, is inconclusive. Furthermore, tourism is also impacted by cryospheric change that impacts on water resources availability, increasing competition for its use (Section 2.3.1.3).

2.3.6   Cultural Values and Human Well-being

Cryosphere changes also impact cultural values, which are held by populations in high mountains and other regions around the world; these impacts often harm human well-being (Tschakert et al., 2019) (*medium evidence, high agreement*). Cultural values were covered extensively in AR5, with particular emphasis on small island states and the Arctic; the research on cultural values in high mountain regions is relatively new. Out of a total of 247 UNESCO World Heritage natural sites recognised for their outstanding universal value, 46 sites include glaciers within their boundaries, where the presence of glaciers is stated among the principal reason (5 sites), or secondary reason (28 sites), for World Heritage inscription; complete glacier extinction is projected by 2100 in 8 to 21 of these sites, under RCP2.6 and RCP8.5 scenarios, respectively, compromising the outstanding universal value placed on these sites, which have been inscribed at least partly for their exceptional glaciers (Bosson et al., 2019). UNESCO defines "outstanding universal value" as "cultural and/or natural significance which is so exceptional as to transcend national boundaries and to be of common importance for present and future generations of all humanity" (UNESCO, 2012). Furthermore, in recognising the importance of the cultural and intangible value placed by communities on aspects of their surrounding environment, such as those afforded by cryosphere elements in the high mountains, cultural values are mentioned under the workplan of the Warsaw International Mechanism as a specific work area under 'Non-economic loss and damage' (UNFCCC Secretariat, 2014; Serdeczny, 2019).

Cultural values include spiritual, intrinsic and existence values, as well as aesthetic dimensions, which are also an element of tourism and recreation (Section 2.3.5), though they focus more directly on ties to sacred beings or to inherent rights of entities to exist. However, these values overlap, since the visual appeal of natural landscapes links with a sense of the immensity of mountain landscapes, glaciers and fresh snow (Paden et al., 2013; Gagné et al., 2014). Moreover, different stakeholders, such as local communities, tourists and policymakers, may place different emphasis on specific cultural values (Schirpke et al., 2016). For the indigenous Manangi community of the

Annapurna Conservation Area of Nepal, the loss of glaciers which they have observed threatens their ethnic identity (Konchar et al., 2015). Villagers in the Italian Alps also report that glacier retreat weakens their identity (Jurt et al., 2015).

Spiritual and intrinsic values in high mountain regions often, but not exclusively, rest on deeply held religious beliefs and other local customs (*medium evidence, high agreement*). Some communities understand mountains through a religious framework (Bernbaum, 2006). In settings as diverse as the Peruvian Andes, the Nepal Himalaya, the European Alps, the North Cascades (USA), Mount Kilimanjaro and the Hengduan Mountains of southwest China, local populations view glacier retreat as the product of their failure to show respect to sacred beings or to follow proper conduct. Experiencing deep concern that they have disturbed cosmic order, they seek to behave in closer accord with established traditions; they anticipate that the retreat will continue, leading to further environmental degradation and to the decline of natural and social orders – a prospect which causes them distress (Becken et al., 2013; Gagné et al., 2014; Allison, 2015). In the USA, the snow covered peaks of the Cascades have also evoked a deep sense of awe and majesty, and an obligation to protect them (Carroll, 2012; Duntley, 2015). Similar views are found in the Italian Alps, where villagers speak of treating glacier peaks with "respect," and state that glacier retreat is due, at least in part, to humans "disturbing" the glaciers (Brugger et al., 2013), resulting in an emotion which Albrecht et al. (2007) termed solastalgia, a kind of deep environmental distress or ecological grief (Cunsolo and Ellis, 2018).

Glacier retreat threatens the Indigenous knowledge and local knowledge of populations in mountain regions; this knowledge constitutes a cultural service to wider society by contributing to scientific understanding of glaciers (Cross-Chapter Box 4 in Chapter 1). Though this knowledge is dynamic, and records previous states of glaciers, it has been undermined by the complete disappearance of glaciers in a local area (Rhoades et al., 2008). This knowledge of glaciers is often tied to religious beliefs and practices. It is based on direct observation, stories passed down from one generation to another within community, placenames, locations of structures and other sources (Gagné et al., 2014). Residents of mountain areas can provide dates for previous locations of glacier fronts, sometimes documenting these locations through the presence of structures (Brugger et al., 2013). Much like other cases of data from citizen science (Theobald et al., 2015), their observations often overlap with the record of instrumental observations (Deng et al., 2012), and can significantly extend this record (Mark et al., 2010).

An additional cultural value is the contribution of glaciers to the understanding of human history. Glacier retreat has supported the increase of knowledge of past societies by providing access to archaeological materials and other cultural resources that had previously been covered by ice. The discovery of Oetzi, a mummified Bronze Age man whose remains were discovered in 1991 in the Alps near the Italian-Austrian border, marked the beginning of scientific research with such materials (Putzer and Festi, 2014). Subsequent papers described objects that were uncovered in retreating glaciers and shrinking ice patches in the Wrangell-Saint Elias Range (Dixon et al., 2005), the Rocky Mountains (Lee, 2012) and Norway (Bjørgo et al., 2016). This field provides new insight into human cultural history and contributes to global awareness of climate change (Dixon et al., 2014). Though climate change permits the discovery of new artefacts and sites, it also threatens these objects and places, since they become newly exposed to harsh weather (Callanan, 2016).

2.3.7   Migration, Habitability and Livelihoods

High mountain communities have historically included mobility in their sets of livelihood strategies, as a means to gain access to production zones at different elevations within mountain zones and in lowland areas, and as a response to the strong seasonality of agricultural and pastoral livelihoods. Cryosphere changes in high mountain areas have influenced human mobility and migration during this century by altering water availability and increasing exposure to mass movements and floods and other cryospheric induced disasters (Figure 2.7) (Barnett et al., 2005; Carey et al., 2017; Rasul and Molden, 2019). These changes affect three forms of human mobility: transhumant pastoralism, temporary or permanent wage labour migration and displacement, in which entire communities resettle in new areas.

Transhumant pastoralism, involving movements between summer and winter pastures, is a centuries old practice in high mountain areas (Lozny, 2013). In High Mountain Asia and other regions, it is declining due to both climatic factors, including changes in snow distribution and glaciers, and to non-climatic factors, and is projected to continue declining, at least in the short term (*medium evidence, high agreement*). The changes in snow and glaciers adversely affect herders at their summer residences and winter camps in the Himalaya (Namgay et al., 2014) and in Scandinavian mountains (Mallory and Boyce, 2018). Reduced winter snowfall has led to poorer pasture quality in Nepal (Gentle and Maraseni, 2012) and India (Ingty, 2017). Other climate change impacts, including erratic snowfall patterns and a decrease in rainfall, are perceived by herders in Afghanistan, Nepal and Pakistan to have resulted in vegetation of lower quality and quantity (Shaoliang et al., 2012; Joshi et al., 2013; Gentle and Thwaites, 2016). Heavy snowfall incidents in winter caused deaths of a large number of livestock in northern Pakistan in 2009 (Shaoliang et al., 2012). Herders in Nepal reported of water scarcity in traditional water sources along migration routes (Gentle and Thwaites, 2016). Increased glacier melt water has caused lakes on the Tibetan Plateau to increase in size, covering pasture areas and leading pastoralists to alter their patterns of seasonal movement (Nyima and Hopping, 2019). However, rising temperatures, with associated effects on snow cover, have some positive impacts. Seasonal migration from winter to summer pastures start earlier in Northern Pakistan, and residence in summer pasture lasts longer (Joshi et al., 2013), as it does in Afghanistan (Shaoliang et al., 2012).

Wage labour migration is also a centuries old practice in the Himalaya, the Andes and the European Alps (Macfarlane, 1976; Cole, 1985; Viazzo, 1989). Studies show that migration is a second-order effect of cryosphere changes, since the first-order effects, a decrease in agricultural production (Section 2.3.1.3.2), have led to increased

wage labour migration to provide supplementary income in a number of regions (*medium evidence, high agreement*). Wage labour migration linked to cryosphere changes occurs on several time scales, including short-term, long-term and permanent migration, and on different spatial scales. Though migration usually takes place within the country of origin, and sometimes within the region, cases of international migration have also been recorded (Merrey et al., 2018). The studies since AR5 on migration driven by cryosphere changes are concentrated in High Mountain Asia and the Andes, supporting the finding, reported in AR5 Working Group II (Section 12.7), that stress on livelihoods is an important driver of climate change induced migration. The research on such migration also supports the finding in SR15 (Section 4.3.5.6) that migration can have mixed outcomes on reducing socioeconomic vulnerability, since cases of increase and of reduction of vulnerability are both found in migration from high mountain regions that is driven by cryosphere changes.

Changing water availability, mass movements and floods are cryosphere processes which drive wage labour migration (*medium evidence, high agreement*). A debris flow in central Nepal in 2014, in a region where landslides have increased in recent decades, led more than half the households to migrate for months (van der Geest and Schindler, 2016). In the Santa River drainage, Peru, rural populations have declined 10% between 1970–2000, and the area of several major subsistence crops also declined (Bury et al., 2013). Research in this region suggests that seasonal wage labour migration from small basins within the main Santa basin is largest in the small drainages in which glacier retreat has reduced melt water flow most significantly; where this process is not as acute, and streamflow is less reduced, migration rates are lower (Wrathall et al., 2014). A study from a region in the central Peruvian Andes shows that the residents of the villages that have the highest dependence on glacier melt water travel further and stay away longer than the residents of the villages where glacier melt water forms a smaller portion of stream flow (Milan and Ho, 2014). However, the inverse relation between reliance on cryosphere-related water sources and migration was noted in a case in the Naryn River drainage in Kyrgyzstan, where the villages that are more dependent on glacier melt water had lower, rather than higher, rates of wage labour migration than the villages which were less dependent on it; the villages with lower rates of such migration also had more efficient water management institutions than the others (Hill et al., 2017). Several studies, which project cryosphere-related emigration to continue in the short term, emphasise decreased water availability, due to glacier retreat as a driver in Kyrgyzstan (Chandonnet et al., 2016) and Peru (Oliver-Smith, 2014), and to reduced snow cover in Nepal (Prasain, 2018). In most cases, climate is only one of several drivers (employment opportunities and better educational and health services in lowland areas are others).

Several studies show that wage labour migration is more frequent among young adults than among other age groups, supporting the observation in AR5 that climate change migrants worldwide are concentrated in this age (*limited evidence, high agreement*). This age-specific pattern is found in a valley in Northern Pakistan in which agriculture relies on glacier melt water for irrigation; as river flow decreases, the returns to agricultural labour have declined, and emigration has increased, particularly among the youth, who are assigned, by local cultural practices, to carry out the heaviest work (Parveen et al., 2015). Emigration has increased in recent decades from two valleys in highland Bolivia which rely on glacier melt water, as water supplies have declined, though other factors also contribute to emigration, including land fragmentation, increasing household needs for income, the lack of local wage-labour opportunities and an interest among the young in educational opportunities located in cities (Brandt et al., 2016). In Nepal, young members of high-elevation pastoral households impacted by cryosphere change have been increasingly engaged in tourism and labour migration since 2000 (Shaoliang et al., 2012); similar responses are reported for Sikkim in the Indian Himalaya (Ingty, 2017). A recent study documents the inter-generational dynamics of emigration from a livestock raising community in the Peruvian Andes, where glacier retreat has led to reduced streamflow that supports crucial dry season pasture (Alata et al., 2018). Though people 50 years old or older in this community are accustomed to living in the high pasture zones, younger people use livestock raising as a means of accumulating capital. They sell off their animals and move to towns at lower elevations. This loss of young adults has reduced the capacity of households to undertake the most demanding tasks, particularly in periods of inclement weather, accelerating the decline of herding. As a result, the human and animal populations of the communities are shrinking.

Recent research on cryosphere driven migration shows some cases of complex livelihood interactions or feedback loops, in which migration is not merely a result of changes in agricultural livelihoods, but also has impacts, either positive or negative, on these livelihoods (*medium confidence*). In some instances, the different livelihood strategies complement each other to support income and well-being. A review of migration in the Himalaya and Hindu Kush found that households that participated in labour migration and received remittances had improved adaptive capacity, and lowered exposure to natural hazards (Banerjee et al., 2018). In other cases, the households and communities, which undertake wage labour migration, encounter conflicts or incompatibilities between migration and agricultural livelihoods. Sustainable management of land, water and other resources is highly labour intensive, and hence labour mobility constrains and limits the adoption of sustainable practices (Gilles et al., 2013). Moreover, the labour available to a household is differentiated by age. In Northern Pakistan, where cryosphere changes are reducing streamflow the emigration of young people has led to a decline not only in the labour in fields and orchards, but also a decline in the maintenance of irrigation infrastructure, leading to an overall reduction of the agricultural livelihoods in the community (Parveen et al., 2015).

In addition to affecting pastoral transhumance and increasing wage labour migration, cryosphere changes impact human mobility by creating cases of displacement. These cases differ from wage labour migration because they involve entire communities. As a result, they are irreversible, unlike cases in which individuals undertake long-term or permanent migration from their communities but retain the possibility of returning, because, for example, some relatives or former neighbours have remained in place. In this way, these cases of displacement represent cryosphere driven challenges to habitability. Though natural hazards have historically led some communities

to relocate, cryosphere changes have contributed to instances of displacement. Unreliable water availability and increased risks of natural hazards are responsible for resettlement of villages in certain high mountain areas (McDonald, 1989; Parveen et al., 2015). A village in Western Nepal moved to lower elevation after decreasing snowfall reduced the flow of water in the river on which their pastoralism and agriculture depended (Barnett et al., 2005). Three villages in Nepal faced severe declines in agricultural and pastoral livelihoods because decreased snow cover led to reduced soil moisture and to the drying up of springs, which were the historical source of irrigation water; in conjunction with an international non-governmental organisation (INGO), the residents planned a move to a lower area (Prasain, 2018).

The issue of habitability arises in the cases, mentioned above, of communities that relocate after floods or debris flows destroy houses and irrigation infrastructure, or damage fields and pastures. It occurs as well in the cases of households with extensive long-term migration, where agricultural and pastoral livelihoods are undermined by reduced water supply caused by cryospheric change (Barnett et al., 2005). In addition, the loss of cultural values, including spiritual and intrinsic values (Section 2.3.6), can contribute to decisions to migrate (Kaenzig, 2015). Combined with the patterns of permanent emigration, this issue of habitability raises the issue of limits to adaptation in mountain areas (Huggel et al., 2019). Projections of decreased streamflow by 2100 in watersheds with strong glacier melt water components in Asia, Europe, and North and South America (Section 2.3.1.1) indicate that threats to habitability may continue through this period and affect the endeavours of achieving the SDGs in developing countries (Rasul et al., 2019).

2.4 International Policy Frameworks and Pathways to Sustainable Development

The governance of key resources that are affected by climate-related changes in the cryosphere, such as water, is a relevant aspect for climate resilient sustainable development in mountains at the catchment level (Section 2.3.1.4). In this section, we address broader policy frameworks that are expected to shape a solution space through global action. An important development since AR5, at the global level, is the adoption of key frameworks that include the Paris Agreement (UNFCCC, 2015), UN 2030 Agenda and its SDGs (UN, 2015), and the Sendai Framework for Disaster Risk Reduction (UNISDR, 2015), which call for integrated and coordinated climate adaptation action that is also relevant for and applicable in mountain regions.

In international climate policy, the importance of averting, minimising and addressing loss and damage associated with adverse impacts of climate change is articulated in the Paris Agreement under Article 8, more specifically (UNFCCC, 2015). However, despite evident impacts of climate change on the mountain cryosphere (Section 2.3.2), there is *limited evidence* or reference in the literature to loss and damage for mountains, globally (Huggel et al., 2019). With already committed and unavoidable climate change, its effects on the high mountain cryosphere (Section 2.2) and related impacts and risks (Section 2.3), substantial adverse effects are expected in the coming decades (Huggel et al., 2019), especially at high emission scenarios, which renders this issue a relevant aspect for planning climate resilient development in mountains. At least in one region, a concrete example for responding to and translating the Paris Agreement in a transboundary mountain setting, is reported. In 2015, through policy measures afforded by the Alpine Convention for the European Alps, the ministers for the environment of the Alpine countries established the Alpine Climate Board, who at the XV Alpine Conference in April 2019, presented a climate target system that includes strategic targets for 'climate-resilient Alps' (Hojesky et al., 2019). The implementation and monitoring of these initiatives, however, remains to be assessed on an evidentiary basis. Furthermore, mechanisms afforded through the workplan of the Warsaw International Mechanism, specifically its work area under 'Non-economic loss and damage', are prospects relevant to address impacts to cultural and intrinsic values associated with losses in the high mountain cryosphere (UNFCCC Secretariat, 2014; Serdeczny, 2019).

Monitoring and reporting on progress towards sustainable development through the implementation of the SDGs (UN, 2015) is receiving some research attention in the context of mountain regions (Rasul and Tripura, 2016; Gratzer and Keeton, 2017; Bracher et al., 2018; Wymann von Dach et al., 2018; Kulonen et al., 2019; Mishra et al., 2019), noting key mountain specific considerations to improve the conditions under which the SDGs may serve a purpose in the mountain context. For example, previous research has identified a need for disaggregated data for SDG indicators and targets at subnational scales, with relevant area units that are both within country boundaries and/or across borders in transboundary settings (Rasul and Tripura, 2016; Bracher et al., 2018; Wymann von Dach et al., 2018). Furthermore, the use of non-standardised proxy data can further limit the potential for comparisons between countries and within regions (Bracher et al., 2018; Kulonen et al., 2019). On substance, assessments of the economic performance of livelihood options, combined with robust socioeconomic data for mountain systems, are still lacking in many parts of the world, compromising the ability for meaningful comparison and aggregation of data and knowledge for monitoring and reporting on progress of SDGs at regional or global scales (Gratzer and Keeton, 2017).

Disasters associated with natural hazards in high mountains are placing many communities and their potential for sustainable development at risk (Wymann von Dach et al., 2017; Keiler and Fuchs, 2018; Vaidya et al., 2019). The Sendai Framework for Disaster Risk Reduction 2015–2030 (UNISDR, 2015) offers a global policy framework under which risks, including climate change, can be accounted for and addressed at national scales. However, there is *limited evidence* in monitoring and reporting on progress on targets therein (Wymann von Dach et al., 2017), particularly in systematically reporting on root causes of disasters in high mountains and associated compounded risks and cascading impacts, and even more so when accounting for impacts related to climate change. Technical guidelines available for the high mountain context provide complementary means to monitor and report on the effectiveness of measures to reduce associated risks with changes in the cryosphere (e.g., GAPHAZ, 2017).

Other relevant frameworks include the Convention Concerning the Protection of the World Cultural and Natural Heritage, enacted to protect the planet's most significant and irreplaceable places from loss or damage (UNESCO, 1972). In it, conservation strategies are listed that aim at preserving natural and cultural heritage across regions, including sites that contain glaciers (Section 2.3.6), and are suggested as means to further support efforts towards the promotion of knowledge, collective cultural memory and climate policy (Bosson et al., 2019).

Overall, there are promising prospects through international policy frameworks to support governance and adaptation to climate-related changes in the mountain cryosphere whilst addressing sustainable development, with evidence suggesting that treaties and conventions are relevant enablers to support cooperation and implementation at the mountain region scale (Dinar et al., 2016). However, there is *limited evidence* to systematically assess for effectiveness in addressing specific challenges posed by changes in the mountain cryosphere, globally.

2.5   Key Gaps in Knowledge and Prospects

Impacts associated with climate-related changes in the high mountain cryosphere are evident in the observations reported in this chapter (Section 2.3). However, uncertainties remain with detection and attribution of key atmospheric drivers that influence much of these climate-related changes (Section 2.2.1), due to limited spatial density and/or temporal extent of observation records at high elevations. For example, trends in total or solid precipitation at high elevation remain highly uncertain, due to intrinsic uncertainties with *in situ* observation methods, and large natural variability. There are clear knowledge gaps in the distribution and characteristics of cryospheric variables, in particular the extent and ice content of permafrost in mountains, but also current glacier ice volumes, trends in lake and river ice, and the spatial and temporal variation of snow cover. These knowledge gaps persist despite a wealth of new data since AR5 especially from Earth observation satellites, which overcome much of the remoteness and inaccessibility of high mountains yet still face challenges for observations in mountains such as dealing with cloud cover and rugged terrain. Along with improved capacities to generate and integrate diverse observation data, initiatives such as citizen science (e.g., Dickerson-Lange et al., 2016; Wikstrom Jones et al., 2018) or Indigenous knowledge and local knowledge (Section 1.8.2, Cross Chapter Box 4 in Chapter 1) can also complement some observations that are based on conventional instruments and models. Radiative forcing effects of light absorbing particles, and understanding their spatiotemporal dynamics, is a key knowledge gap for the attribution of changes in high mountain snow and glaciers, and the understanding of regional feedbacks (Section 2.2.2, Box 2.2).

These observational knowledge gaps currently impede efforts to quantify trends, and to calibrate and evaluate models that simulate the past and future evolution of the cryosphere and its impacts. Specific uncertainties are associated with projections of future climate change trends at high elevations due mostly to current limits in regional climate models and downscaling methods to capture the subtle interplays between large-scale climate change and local phenomena influenced by complex topography and high relief (Section 2.2.1). Coarse-scale simulations of future permafrost conditions in high mountains are fraught with difficulties in capturing fine-scale variation of topography, surface cover and near-surface materials (Section 2.2.4). Improved cross-disciplinary studies bringing together current observation and modelling approaches in each specific field hold potential to contribute to addressing these gaps in the future.

Experiences with changes in water availability, and with changes in frequency and/or magnitude of natural hazards, demonstrate the relevance of integrated approaches to understand past impacts and prepare for future risks, where exposure and the underpinning existing vulnerabilities of mountain socioecological systems influence the extent of these impacts (Section 2.3.2.3). However, there is insufficient understanding of the effects of cryospheric change on some natural hazards such as glacier outburst floods and on infrastructure, for example for transportation. Increased wildfire risk with a shrinking cryosphere is an uncertainty both spatially and temporally and with consequent effects on mountain ecosystems, particularly with respect to soil carbon and potential biome shifts. Overall, few studies have taken a comprehensive risk approach to systematically characterise and compare magnitude and extent of past impacts and future risks across high mountain regions, including compound risks and cascading impacts where instances of deep uncertainty in responses and outcomes may arise (Cross Chapter Box 5 in Chapter 1). Furthermore, a key knowledge gap is the

capacity to economically quantify cryosphere-specific impacts and potential risks.

With ecosystems, particularly the terrestrial component, uncertainty exists at which community changes can be directly linked to cryospheric change as distinct from those due to atmospheric warming. In some cases, the changes can be linked, for example, where a receding glacier creates new habitat, but rising air temperature allow some species to establish that would not otherwise be able to. A major research gap is in our understanding of the fate of legacy pollutants such as mercury downstream of their release from glaciers and permafrost in terms of quantity and regional differences, freshwater sinks, and potential effects to ecosystems and human health. Similarly, the effect of permafrost thaw on water quality and ecosystems due to the increasing release of natural heavy metals and nutrients represents a gap in knowledge.

While adaptation measures are reported for high mountain cryosphere changes (Figure 2.9 a), it stands as a relatively new and developing area of research since AR5 (Figure 2.9 b), with particular gaps in terms of systematically evaluating their cost-benefits and long-term effectiveness as 'fit-for-purpose' solutions in the mountain context. Improved inter-comparability of successful adaptation cases, including the transferability of evidence for how adaptation can address both climate change and sustainable development objectives in different mountain regions, are prospects to support an evidentiary basis for future assessments of adaptation to cryosphere changes in the high mountains (Adler et al., 2019; McDowell et al., 2019).

Acknowledgements

We acknowledge the kind contributions of Matvey Debolskiy (Unversity of Alaska Fairbanks, USA), Florian Hanzer (University of Innsbruck, Austria), Andreas F. Prein (National Center for Atmospheric Research, Boulder, CO, USA), Silvia Terzago (Institute of Atmospheric Sciences and Climate, National Research Council, Torino, Italy) and Natalia Zazulie (CONICET/University of Buenos Aires, Argentina) who contributed to drafting figures.

References

Aas, K.S. et al., 2016: The climatic mass balance of Svalbard glaciers: a 10-year simulation with a coupled atmosphere-glacier mass balance model. *The Cryosphere*, **10**(3), 1089–1104, doi:10.5194/tc-10-1089-2016.

Abbott, B.W. et al., 2014: Elevated dissolved organic carbon biodegradability from thawing and collapsing permafrost. *J. Geophys. Res-Biogeo.*, **119**(10), 2049–2063, doi:10.1002/2014JG002678.

Addor, N. et al., 2014: Robust changes and sources of uncertainty in the projected hydrological regimes of Swiss catchments. *Water Resour. Res.*, **50**(10), 7541–7562, doi:10.1002/2014wr015549.

Adler, C., C. Huggel, B. Orlove and A. Nolin, 2019: Climate change in the mountain cryosphere: impacts and responses. *Reg. Environ. Change*, **19**(5), 1225–1228, doi:10.1007/s10113-019-01507-6.

Aguilera, E., M.T. Pareschi, M. Rosi and G. Zanchetta, 2004: Risk from lahars in the northern valleys of Cotopaxi volcano (Ecuador). *Nat. Hazards*, **33**(2), 161–189, doi:10.1023/B:NHAZ.0000037037.03155.23.

Aiken, G.R. et al., 2014: Influences of glacier melt and permafrost thaw on the age of dissolved organic carbon in the Yukon River basin. *Global Biogeochem. Cy.*, **28**(5), 525–537, doi:10.1002/2013GB004764.

Åkerman, H.J. and M. Johansson, 2008: Thawing permafrost and thicker active layers in sub-arctic Sweden. *Permafrost Periglac.*, **19**(3), 279–292, doi:10.1002/ppp.626.

Alata, E., B. Fuentealba, J. Recharte, B. Fuentealba and J. Recharte, 2018: El despoblamiento de la Puna: efectos del cambio climático y otros factores. Revista Kawsaypacha, 2, 49–69, doi:10.18800/kawsaypacha.201802.003.

Albrecht, G. et al., 2007: Solastalgia: the distress caused by environmental change. *Australas. Psychiatry*, **15**(1), S95-S98, doi:10.1080/10398560701701288.

Aleksandrova, M., J.P.A. Lamers, C. Martius and B. Tischbein, 2014: Rural vulnerability to environmental change in the irrigated lowlands of Central Asia and options for policy-makers: A review. *Environ. Sci. Policy*, **41**, 77–88, doi:10.1016/j.envsci.2014.03.001.

Ali, S.A., S. Aadhar, H.L. Shah and V. Mishra, 2018: Projected increase in hydropower production in India under climate change. *Sci. Rep.*, **8**(1), 12450, doi:10.1038/s41598-018-30489-4.

Allen, S. and C. Huggel, 2013: Extremely warm temperatures as a potential cause of recent high mountain rockfall. *Glob. Planet. Change*, **107**, 59–69, doi:10.1016/j.gloplacha.2013.04.007.

Allen, S.K. et al., 2018: Translating the concept of climate risk into an assessment framework to inform adaptation planning: Insights from a pilot study of flood risk in Himachal Pradesh, Northern India. *Environ. Sci. Policy*, **87**, 1–10, doi:10.1016/j.envsci.2018.05.013.

Allen, S.K., S.C. Cox and I.F. Owens, 2011: Rock avalanches and other landslides in the central Southern Alps of New Zealand: a regional study considering possible climate change impacts. *Landslides*, **8**(1), 33–48, doi:10.1007/s10346-010-0222-z.

Allen, S.K. et al., 2016a: Glacial lake outburst flood risk in Himachal Pradesh, India: an integrative and anticipatory approach considering current and future threats. *Nat. Hazards*, **84**(3), 1741–1763. doi:10.1007/s11069-016-2511-x.

Allen, S.K. et al., 2016b: Lake outburst and debris flow disaster at Kedarnath, June 2013: hydrometeorological triggering and topographic predisposition. *Landslides*, **13**(6), 1479–1491, doi:10.1007/s10346-015-0584-3.

Allison, E.A., 2015: The spiritual significance of glaciers in an age of climate change. *WiRes. Clim. Change*, **6**(5), 493–508, doi:10.1002/wcc.354.

Amagai, Y., G. Kudo and K. Sato, 2018: Changes in alpine plant communities under climate change: Dynamics of snow-meadow vegetation in northern Japan over the last 40 years. *Applied Vegetation Science*, **21**, 561–571. doi:10.1111/avsc.12387.

Anacona, P.I. et al., 2018: Glacier protection laws: Potential conflicts in managing glacial hazards and adapting to climate change. *Ambio*, **47**(8), 835–845, doi:10.1007/s13280-018-1043-x.

Anacona, P.I., A. Mackintosh, and K.P. Norton, K.P., 2015a: Hazardous processes and events from glacier and permafrost areas: lessons from the Chilean and Argentinean Andes. *Earth Surf. Process. Landf,.*, **40**(1), 2–21, doi:10.1002/esp.3524.

Anacona, P.I., A. Mackintosh and K. Norton, 2015b: Reconstruction of a glacial lake outburst flood (GLOF) in the Engaño valley, chilean patagonia: Lessons for GLOF risk management. *Sci. Total Environ.*, **527–528**, 1–11, doi:10.1016/j.scitotenv.2015.04.096.

Ancey, C. and V. Bain, 2015: Dynamics of glide avalanches and snow gliding. *Rev. Geophys.*, **53**(3), 745–784, doi:10.1002/2015RG000491.

Andreassen, L.M. et al., 2005: Glacier mass-balance and length variation in Norway. *Ann. Glaciol.*, **42**, 317–325, doi:10.3189/172756405781812826.

Arenson, L.U., W. Colgan and H.P. Marshall, 2015a: Chapter 2 – Physical, Thermal, and Mechanical Properties of Snow, Ice, and Permafrost. In: *Snow and Ice-Related Hazards, Risks, and Disasters* [Shroder, J.F., W. Haeberli and C. Whiteman (eds.)]. Academic Press, Boston, pp. 35–75. ISBN 9780123948496.

Arenson, L.U., M. Jakob and P. Wainstein, 2015b: Effects of dust deposition on glacier ablation and runoff at the Pascua-Lama Mining Project, Chile and Argentina. In: *Engineering Geology for Society and Territory – Volume 1: Climate Change and Engineering Geology* [Lollino, G., A. Manconi, J. Clague, W. Shan and M. Chiarle (eds.)], Springer International Publishing, Cham, pp. 27–32. ISBN 9783319093000.

Arent, D.J. et al., 2014: Key economic sectors and services. In: *Climate Change 2014 Impacts, Adaptation and Vulnerability: Part A: Global and Sectoral Aspects* [Field, C.B., V.R. Barros, D.J. Dokken, K.J. Mach and M.D. Mastrandrea (eds.)]. Cambridge University Press, Cambridge, pp. 659–708. ISBN 9781107415379.

Armstrong, R.L. and E. Brun, 2008: *Snow and climate: physical processes, surface energy exchange and modelling.* Cambridge University Press, Cambridge, 256 pp. ISBN 9780521854542.

Arnold, C., T.A. Ghezzehei and A.A. Berhe, 2014: Early spring, severe frost events, and drought induce rapid carbon loss in high elevation meadows. *PLOS ONE*, **9**(9), e106058, doi:10.1371/journal.pone.0106058.

Ashraf, A., R. Naz and R. Roohi, 2012: Glacial lake outburst flood hazards in Hindukush, Karakoram and Himalayan ranges of Pakistan: Implications and risk analysis. *Geomat. Nat. Haz. Risk*, **3**(2), 113–132, doi:10.1080/19475705.2011.615344.

Atmeh, K., A. Andruszkiewicz and K. Zub, 2018: Climate change is affecting mortality of weasels due to camouflage mismatch. *Sci Rep*, **8** (7648), doi:10.1038/s41598-018-26057-5.

Auer, I. et al., 2007: HISTALP – historical instrumental climatological surface time series of the Greater Alpine Region. *Int. J. Climatol.*, **27** (1), 17–46, doi:10.1002/joc.1377.

Azam, M.F. et al., 2018: Review of the status and mass changes of Himalayan-Karakoram glaciers. *J. Glaciol.*, **64**(243), 61–74, doi:10.1017/jog.2017.86.

Azócar, G.F., A. Brenning and X. Bodin, 2017: Permafrost distribution modelling in the semi-arid Chilean Andes. *The Cryosphere*, **11** (2), 877–890, doi:10.5194/tc-11-877-2017.

Bajracharya, S.R., 2010: Glacial Lake Outburst Flood Disaster Risk Reduction Activities in Nepal. *International Journal of Erosion Control Engineering*, **3**(1), 92–101, doi:10.13101/ijece.3.92.

Ballesteros-Cánovas, J.A. et al., 2018: Climate warming enhances snow avalanche risk in the Western Himalayas. *PNAS*, **115** (13), 3410–3415, doi:10.1073/pnas.1716913115.

Bamber, J.L., R.M. Westaway, B. Marzeion and B. Wouters, 2018: The land ice contribution to sea level during the satellite era. *Environ. Res. Lett.*, **13**(6), 063008, doi:10.1088/1748-9326/aac2f0.

Banerjee, S., R. Black, A. Mishra and D. Kniveton, 2018: Assessing vulnerability of remittance-recipient and non-recipient households in rural communities affected by extreme weather events: Case studies from south-west China and northeast India. *Popul. Space Place*, **25**(2), e2157, doi:10.1002/psp.2157.

Baraer, M. et al., 2012: Glacier recession and water resources in Peru's Cordillera Blanca. *J. Glaciol.*, **58**(207), 134–150, doi:10.3189/2012JoG11J186.

Bard, A. et al., 2015: Trends in the hydrologic regime of Alpine rivers. *J. Hydrol.*, **529**, 1823–1837, doi:10.1016/j.jhydrol.2015.07.052.

Barnett, T.P., J.C. Adam and D.P. Lettenmaier, 2005: Potential impacts of a warming climate on water availability in snow-dominated regions. *Nature*, **438**(7066), 303–309, doi:10.1038/nature04141.

Barr, I.D. et al., 2018: Volcanic impacts on modern glaciers: A global synthesis. *Earth-Sci. Rev.*, **182**, 186–203, doi:10.1016/j.earscirev.2018.04.008.

Barrett, K. and K. Bosak, 2018: The role of place in adapting to climate change: A case study from Ladakh, Western Himalayas. *Sustainability*, **10**(4), 898, doi:10.3390/su10040898.

Barriga Delgado, L.M., F. Drenkhan and C. Huggel, 2018: Proyectos multipropósito para la gestión de recursos hídricos en los Andes tropicalesplanteamientos generales basados en proceso participativo [Multi-purpose Projects for Water Resources Management in the Tropical Andes: Participatory-based approaches]. *Espacio y Desarrollo*, **32**, 7–28, doi:10.18800/espacioydesarrollo.201802.001.

Bashir, F., X. Zeng, H. Gupta and P. Hazenberg, 2017: A Hydrometeorological Perspective on the Karakoram Anomaly Using Unique Valley-Based Synoptic Weather Observations. *Geophys. Res. Lett.*, **44**(20), 10,410–10,478, doi:10.1002/2017GL075284.

Beamer, J.P., D.F. Hill, A.A. Arendt and G.E. Liston, 2016: High-resolution modeling of coastal freshwater discharge and glacier mass balance in the Gulf of Alaska watershed. *Water Resour. Res.*, **52**(5), 3888–3909, doi:10.1002/2015WR018457.

Beaudin, L. and J.C. Huang, 2014: Weather conditions and outdoor recreation: A study of New England ski areas. *Ecol. Econ.*, **106**, 56–68, doi:10.1016/j.ecolecon.2014.07.011.

Becken, S., A.K. Lama and S. Espiner, 2013: The cultural context of climate change impacts: Perceptions among community members in the Annapurna Conservation Area, Nepal. *Environ. Dev.*, **8**, 22–37, doi:10.1016/J.ENVDEV.2013.05.007.

Bellaire, S. et al., 2016: Analysis of long-term weather, snow and avalanche data at Glacier National Park, B.C., Canada. *Cold Reg. Sci. Technol.*, **121**, 118–125, doi:10.1016/j.coldregions.2015.10.010.

Beniston, M. et al., 2018: The European mountain cryosphere: a review of its current state, trends, and future challenges. *The Cryosphere*, **12**(2), 759–794, doi:10.5194/tc-12-759-2018.

Beniston, M. and D.G. Fox, 1996: Impacts of climate change on mountain regions. In: Climate change 1995 – Impacts, adaptations and mitigation of climate change: scientific-technical analysis. Contribution of Working Group II to the Second Assessment Report of the Intergovernmental Panel of Climate Change. [Watson, R., M.C. Zinyowera and R.H. Moss (eds.)]. Cambridge University Press, Cambridge, United Kingdom and New York, NY, WGII, 191–213.

Beniston, M. and M. Stoffel, 2014: Assessing the impacts of climatic change on mountain water resources. *Sci. Total Environ.*, **493**, 1129–1137, doi:10.1016/j.scitotenv.2013.11.122.

Beniston, M. and M. Stoffel, 2016: Rain-on-snow events, floods and climate change in the Alps: Events may increase with warming up to 4°C and decrease thereafter. *Sci. Total Environ.*, **571**, 228–236, doi:10.1016/j.scitotenv.2016.07.146.

Benn, D.I. et al., 2012: Response of debris-covered glaciers in the Mount Everest region to recent warming, and implications for outburst flood hazards. *Earth-Sci. Rev.*, **114**(1–2), 156–174, doi:10.1016/j.earscirev.2012.03.008.

Berman, E E., N.C. Coops, S.P. Kearney and G.B. Stenhouse, 2019: Grizzly bear response to fine spatial and temporal scale spring snow cover in Western Alberta. *PLOS ONE*, **14**(4), e0215243. doi:10.1371/journal.pone.0215243.

Bernauer, T. and T. Siegfried, 2012: Climate change and international water conflict in Central Asia. *J. Peace Res.*, **49**(1), 227–239, doi:10.1177/0022343311425843.

Bernbaum, E., 2006: Sacred mountains: Themes and teachings. *Mt. Res. Dev.*, **26**(4), 304–309, doi:10.1659/0276-4741(2006)26[304:smtat]2.0.co;2.

Berthier, E. and F. Brun, 2019: Karakoram geodetic glacier mass balances between 2008 and 2016: persistence of the anomaly and influence of a large rock avalanche on Siachen Glacier. *J. Glaciol.*, **65**(251), 494–507, doi:10.1017/jog.2019.32.

Bevington, A. and L. Copland, 2014: Characteristics of the last five surges of Lowell Glacier, Yukon, Canada, since 1948. *J. Glaciol.*, **60**(219), 113–123, doi:10.3189/2014JoG13J134.

Bhambri, R., K. Hewitt, P. Kawishwar and B. Pratap, 2017: Surge-type and surge-modified glaciers in the Karakoram. *Sci. Rep.*, **7**, 15391, doi:10.1038/s41598-017-15473-8.

Biemans, H. et al., 2019: Importance of snow and glacier melt water for agriculture on the Indo-Gangetic Plain. *Nat. Sustain.* **2**(7).

Bieniek, P.A. et al., 2018: Assessment of Alaska rain-on-snow events using dynamical downscaling. *J. Appl. Meteorol. Climatol.*, **57**(8), 1847–1863, doi:10.1175/JAMC-D-17-0276.1.

Bindoff, N.L. et al., 2013: Chapter 10 – Detection and attribution of climate change: From global to regional. In: Climate Change 2013: The Physical Science Basis. IPCC Working Group I Contribution to AR5 [Stocker, T.F., G.K. Plattner, M. Tignor, S.K. Allen, J. Boschung, A. Nauels, Y. Xia, V. Bex and P.M. Midgley (eds.)]. Cambridge University Press, Cambridge, United Kingdom and New York, NY, USA, Cambridge. 867–952.

Biskaborn, B.K. et al., 2019: Permafrost is warming at a global scale. *Nat. Commun.*, **10**(1), 264, doi:10.1038/s41467-018-08240-4.

Bjørgo, T. et al., 2016: Fragments of a Late Iron Age Sledge Melted Out of the Vossaskavlen Snowdrift Glacier in Western Norway. *Journal of Glacial Archaeology*, **2**(0), 73–81, doi:10.1558/jga.v2i1.27719.

Björnsson, H., 2003: Subglacial lakes and jökulhlaups in Iceland. *Glob. Planet. Change*, **35**(3–4), 255–271, doi:10.1016/S0921-8181(02)00130-3.

Björnsson, H. and F. Pálsson, 2008: Icelandic glaciers. *Jökull*, **58**, 365–386.

Björnsson, H. et al., 2013: Contribution of Icelandic ice caps to sea level rise: Trends and variability since the Little Ice Age. *Geophys. Res. Lett.*, **40**(8), 1546–1550, doi:10.1002/grl.50278.

Blair, R.W., 1994: Moraine and valley wall collapse due to rapid deglaciation in Mount Cook National Park, New Zealand. *Mt. Res. Dev.*, **14**(4), 347–358, doi:10.2307/3673731.

Bliss, A., R. Hock and V. Radić, 2014: Global response of glacier runoff to twenty-first century climate change. *J. Geophys. Res-Earth.*, **119**(4), 717–730, doi:10.1002/2013JF002931.

Bocchiola, D., 2014: Long term (1921–2011) hydrological regime of Alpine catchments in Northern Italy. *Adv. Water Resour.*, **70**, 51–64, doi:10.1016/j.advwatres.2014.04.017.

Bocchiola, D., M.G. Pelosi and A. Soncini, 2017: Effects of hydrological changes on cooperation in transnational catchments: the case of the Syr Darya. *Water Int.*, **42**(7), 852–873, doi:10.1080/02508060.2017.1376568.

Bockheim, J.G. and J.S. Munroe, 2014: Organic Carbon Pools and Genesis of Alpine Soils with Permafrost: A Review. *Arct. Antarct. Alp. Res.*, **46**, 987–1006, doi:10.1657/1938-4246-46.4.987.

Bodin, X. et al., 2016: The 2006 Collapse of the Bérard Rock Glacier (Southern French Alps). *Permafrost Periglac.*, **28**(1), 209–223, doi:10.1002/ppp.1887.

Bodin, X., F. Rojas and A. Brenning, 2010: Status and evolution of the cryosphere in the Andes of Santiago (Chile, 33.5°S.). *Geomorphology*, **118**, 453–464, doi:10.1016/j.geomorph.2010.02.016.

Bodin, X. et al., 2009: Two Decades of Responses (1986–2006) to Climate by the Laurichard Rock Glacier, French Alps. *Permafrost Periglac.*, **20**(4), 331–344, doi:10.1002/ppp.665.

Boeckli, L., A. Brenning, S. Gruber and J. Noetzli, 2012: Permafrost distribution in the European Alps: calculation and evaluation of an index map and summary statistics. *The Cryosphere*, **6**(4), 807–820, doi:10.5194/tc-6-807-2012.

Bogdal, C. et al., 2010: Release of legacy pollutants from melting glaciers: Model evidence and conceptual understanding. *Environ. Sci. Technol.*, **44**(11), 4063–4069, doi:10.1021/es903007h.

Bolch, T., T. Pieczonka, K. Mukherjee and J. Shea, 2017: Brief communication: Glaciers in the Hunza catchment (Karakoram) have been nearly in balance since the 1970s. *The Cryosphere*, **11** (1), 531–539, doi:10.5194/tc-11-531-2017.

Bolch, T. et al., 2018: *Status and Change of the Cryosphere in the Extended Hindu Kush Himalaya Region*. The Hindu Kush Himalaya Assessment Mountains, Climate Change, Sustainability and People, Springer Nature, Switzerland. ISBN 9783319922874.

Bommer, C., M. Phillips and L.U. Arenson, 2010: Practical recommendations for planning, constructing and maintaining infrastructure in mountain permafrost. *Permafrost Periglac.*, **21**(1), 97–104, doi:10.1002/ppp.679.

Bonfils, C. et al., 2008: Detection and attribution of temperature changes in the mountainous Western United States. *J. Clim.*, **21**(23), 6404–6424, doi:10.1175/2008JCLI2397.1.

Bonnaventure, P.P. and A.G. Lewkowicz, 2011: Modelling climate change effects on the spatial distribution of mountain permafrost at three sites in northwest Canada. *Clim. Change*, **105**(1–2), 293–312, doi:10.1007/s10584-010-9818-5.

Bonnaventure, P.P., A.G. Lewkowicz, M. Kremer and M.C. Sawada, 2012: A Permafrost Probability Model for the Southern Yukon and Northern British Columbia, Canada. *Permafrost Periglac.*, **23**(1), 52–68, doi:10.1002/ppp.1733.

Bonzanigo, L. et al., 2015: *South Asia investment decision making in hydropower: decision tree case study of the Upper Arun Hydropower Project and Koshi Basin Hydropower Development in Nepal*. Report No. AUS11077. World Bank, Washington, D.C., 127 pp.

Bormann, K.J., R.D. Brown, C. Derksen and T.H. Painter, 2018: Estimating snow cover trends from space. *Nat. Clim. Change*, **8**(11), 924, doi:10.1038/s41558-018-0318-3.

Bosshard, T., S. Kotlarski, M. Zappa and C. Schär, 2014: Hydrological climate-impact projections for the Rhine River: GCM–RCM uncertainty and separate temperature and precipitation effects. *J. Hydrometeorol.*, **15**(2), 697–713, doi:10.1175/JHM-D-12-098.1.

Bosson, J.-B., M. Huss and E. Osipova, 2019: Disappearing World Heritage Glaciers as a Keystone of Nature Conservation in a Changing Climate. *Earth's Future*, **7** (4), 469–479, doi:10.1029/2018ef001139.

Box, J.E. et al., 2018: Global sea-level contribution from Arctic land ice: 1971–2017. *Environ. Res. Lett.*, **13**(12), 125012, doi:10.1088/1748-9326/aaf2ed.

Bracher, C.P., S.W. von Dach and C. Adler, 2018: *Challenges and Opportunities in Assessing Sustainable Mountain Development Using the UN Sustainable Development Goals*. Universitat Bern, Bern, 42 pp.

Brahney, J. et al., 2017: Evidence for a climate-driven hydrologic regime shift in the Canadian Columbia Basin. *Can. Water. Resour. J.* **42**(2), 179–192, doi:10.1080/07011784.2016.1268933.

Brandt, R., R. Kaenzig and S. Lachmuth, 2016: Migration as a risk management strategy in the context of climate change: Evidence from the Bolivian Andes. In: Global Migration Issues, IOMS(6) [Milan, A., B. Schraven, K. Warner and N. Cascone (eds.)]. Springer International Publishing Ag, Cham, 43–61.

Braun, M. and E. Fournier, 2016: *Adaptation Case Studies in the Energy Sector – Overcoming Barriers to Adaptation, Report presented to Climate Change Impacts and Adaptation Division*. Natural Resources Canada, pp. 114. ISBN 9782923292229.

Braun, M.H. et al., 2019: Constraining glacier elevation and mass changes in South America. *Nat. Clim. Change*, **9**(2), 130–136, doi:10.1038/s41558-018-0375-7.

Brenning, A., 2008: The impact of mining on rock glaciers and glaciers. In: Darkening peaks: glacier retreat, science, and society [Orlove, B.S., E. Weigandt and B. Luckman (eds.)]. University of California Press, Berkely, 196–205.

Brenning, A. and G.F. Azócar, 2010: Minería y glaciares rocosos: impactos ambientales, antecedentes políticos y legales, y perspectivas futuras. *Revista de geografía Norte Grande*, **47**, 143–158, doi:10.4067/S0718-34022010000300008.

Brinkerhoff, D., M. Truffer and A. Aschwanden, 2017: Sediment transport drives tidewater glacier periodicity. *Nat. Commun.*, **8**(1), 90, doi:10.1038/s41467-017-00095-5.

Brown, L.E., D.M. Hannah and A.M. Milner, 2007: Vulnerability of alpine stream biodiversity to shrinking glaciers and snowpacks. *Glob. Change Biol*, **13**(5), 958–966, doi:10.1111/j.1365-2486.2007.01341.x.

Brown, L.E. et al., 2018: Functional diversity and community assembly of river invertebrates show globally consistent responses to decreasing glacier cover. *Nat. Ecol. Evol.*, **2**(2), 325–333, doi:10.1038/s41559-017-0426-x.

Brugger, J., K.W. Dunbar, C. Jurt and B. Orlove, 2013: Climates of anxiety: Comparing experience of glacier retreat across three mountain regions. *Emote. Space Soc.*, **6**, 4–13, doi:10.1016/j.emospa.2012.05.001.

Brun, F. et al., 2017: A spatially resolved estimate of High Mountain Asia glacier mass balances, 2000–2016. *Nat. Geosci.*, **10**(9), 668–673, doi:10.1038/NGEO2999.

Brunner, M.I. et al., 2019: Present and future water scarcity in Switzerland: Potential for alleviation through reservoirs and lakes. *Sci. Total Environ.*, **666**, 1033–1047, doi:10.1016/j.scitotenv.2019.02.169.

Buchli, T. et al., 2013: Characterization and monitoring of the furggwanghorn rock glacier, Turtmann Valley, Switzerland: Results from 2010 to 2012. *Vadose Zone J.*, **12**, doi:10.2136/vzj2012.0067.

Buckel, J., J.C. Otto, G. Prasicek and M. Keuschnig, 2018: Glacial lakes in Austria – Distribution and formation since the Little Ice Age. *Glob. Planet. Change*, **164**, 39–51, doi:10.1016/j.gloplacha.2018.03.003.

Büntgen, U. et al., 2017: Elevational range shifts in four mountain ungulate species from the Swiss Alps. *Ecosphere*, **8**(4), e01761, doi:10.1002/ecs2.1761.

Burns, P. and A. Nolin, 2014: Using atmospherically-corrected Landsat imagery to measure glacier area change in the Cordillera Blanca, Peru from 1987 to 2010. *Remote Sens. Environ.*, **140**, 165–178, doi:10.1016/j.rse.2013.08.026.

Bury, J. et al., 2013: New geographies of water and climate change in Peru: Coupled natural and social transformations in the Santa River Watershed. *Ann. Am. Assoc. Geogr.*, **103** (2), 363–374, doi:10.1080/00045608.2013.754665.

Bury, J.T. et al., 2011: Glacier recession and human vulnerability in the Yanamarey watershed of the Cordillera Blanca, Peru. *Clim. Change*, **105** (1–2), 179–206, doi:10.1007/s10584-010-9870-1.

Buytaert, W. and B. De Bièvre, 2012: Water for cities: The impact of climate change and demographic growth in the tropical Andes. *Water Resour. Res.*, **48** (8), 897, doi:10.1029/2011WR011755.

Buytaert, W. et al., 2017: Glacial melt content of water use in the tropical Andes. *Environ. Res. Lett.*, **12**, 1–8, doi:10.1088/1748-9326/aa926c.

Cai, Y. et al., 2019: Variations of lake ice phenology on the Tibetan Plateau From 2001 to 2017 based on MODIS Data. *J. Geophys. Res. Atmos.*, **124**(2), 825–843. doi:10.1029/2018jd028993.

Caldecott, B. et al., 2016: *Stranded Assets: A Climate Risk Challenge* [Rios, A.R. (ed.)]. Inter-American Development Bank. [Available at: https://publications.iadb.org/handle/11319/7946]. Accessed 05/08/2019.

Callanan, M., 2016: Managing frozen heritage: Some challenges and responses. *Quaternary Int.*, **402**, 72–79, doi:10.1016/j.quaint.2015.10.067.

Camac, J.S. et al., 2017: Climatic warming strengthens a positive feedback between alpine shrubs and fire. *Global Change Biol.*, **23**(8), 3249-s3258, doi:10.1111/gcb.13614.

2

Cao, B. et al., 2018: Thermal characteristics and recent changes of permafrost in the upper reaches of the Heihe River Basin, Western China. *J. Geophys. Res-Atmos.*, **123**(15), 7935–7949, doi:10.1029/2018JD028442.

Capell, R., D. Tetzlaff, R. Essery and C. Soulsby, 2014: Projecting climate change impacts on stream flow regimes with tracer-aided runoff models-preliminary assessment of heterogeneity at the mesoscale. *Hydrol. Process.*, **28**(3), 545–558, doi:10.1002/hyp.9612.

Carey, M., 2005: Living and dying with glaciers: people's historical vulnerability to avalanches and outburst floods in Peru. *Glob. Planet. Change*, **47**(2–4), 122–134, doi:10.1016/j.gloplacha.2004.10.007.

Carey, M. et al., 2014: Toward hydro-social modeling: Merging human variables and the social sciences with climate-glacier runoff models (Santa River, Peru). *J. Hydrol.*, **518**, 60–70, doi:10.1016/j.jhydrol.2013.11.006.

Carey, M., A. French and E. O'Brien, 2012: Unintended effects of technology on climate change adaptation: An historical analysis of water conflicts below Andean Glaciers. *J. Hist. Geogr.*, **38** (2), 181–191, doi:10.1016/j.jhg.2011.12.002.

Carey, M. et al., 2017: Impacts of glacier recession and declining melt water on mountain societies. *Ann Am. Assoc. Geogr.*, **107**(2), 350–359, doi:10.1080/24694452.2016.1243039.

Carrivick, J.L. et al., 2013: Outburst flood evolution at Russell Glacier, western Greenland: effects of a bedrock channel cascade with intermediary lakes. *Quaternary Sci. Rev.*, **67**, 39–58, doi:10.1016/j.quascirev.2013.01.023.

Carrivick, J.L. and F.S. Tweed, 2016: A global assessment of the societal impacts of glacier outburst floods. *Glob. Planet. Change*, **144**, 1–16, doi:10.1016/j.gloplacha.2016.07.001.

Carroll, B.E., 2012: Worlds in space: American religious pluralism in geographic perspective. *JAAR*, **80**(2), 304–364, doi:10.1093/jaarel/lfs024.

Castebrunet, H. et al., 2014: Projected changes of snow conditions and avalanche activity in a warming climate: the French Alps over the 2020–2050 and 2070–2100 periods. *The Cryosphere*, **8**(5), 1673–1697, doi:10.5194/tc-8-1673-2014.

Cauvy-Fraunié, S. et al., 2016: Ecological responses to experimental glacier-runoff reduction in alpine rivers. *Nat. Commun.*, **7**, 12025, doi:10.1038/ncomms12025.

Ceppi, P., S.C. Scherrer, A.M. Fischer and C. Appenzeller, 2012: Revisiting Swiss temperature trends 1959–2008. *Int. J. Climatol.*, **32** (2), 203–213, doi:10.1002/joc.2260.

Chadburn, S.E. et al., 2017: An observation-based constraint on permafrost loss as a function of global warming. *Nat. Clim. Change*, **7**(5), 340–344, doi:10.1038/nclimate3262.

Chai, M. et al., 2018: Characteristics of asphalt pavement damage in degrading permafrost regions: Case study of the Qinghai–Tibet Highway, China. *J. Cold. Reg. Eng.*, **32**(2), 05018003, doi:10.1061/(asce)cr.1943-5495.0000165.

Chandonnet, A., Z. Mamadalieva and L. Orolbaeva, 2016: *Environment, climate change and migration In the Kyrgyz Republic.* IOM, Kyrgyzstan, 112 pp.

Chen, Y. et al., 2016: Changes in Central Asia's water tower: Past, present and future. *Sci. Rep.*, **6**, 35458, doi:10.1038/srep35458.

Chen, Y. et al., 2014: Comparison of the sensitivity of surface downward longwave radiation to changes in water vapor at two high elevation sites. *Environ. Res. Lett.*, **9**(11), 114015, doi:10.1088/1748-9326/9/11/114015.

Chevallier, P., B. Pouyaud, W. Suarez and T. Condom, 2011: Climate change threats to environment in the tropical Andes: Glaciers and water resources. *Reg. Environ. Change*, **11** (Suppl.1), 179–187, doi:10.1007/s10113-010-0177-6.

Christiansen, H.H. et al., 2010: The thermal state of permafrost in the nordic area during the international polar year 2007–2009. *Permafrost Periglac.*, **21**(2), 156–181, doi:10.1002/ppp.687.

Chudley, T.R., E.S. Miles and I.C. Willis, 2017: Glacier characteristics and retreat between 1991 and 2014 in the Ladakh Range, Jammu and Kashmir. *Remote Sens. Lett.*, **8**(6), 518–527, doi:10.1080/2150704X.2017.1295480.

Clarke, G.K.C. et al., 2015: Projected deglaciation of western Canada in the twenty-first century. *Nat. Geosci.*, **8**(5), 372–377, doi:10.1038/ngeo2407.

Clouse, C., 2016: Frozen landscapes: climate-adaptive design interventions in Ladakh and Zanskar. *Landscape Research*, **41**(8), 821–837, doi:10.1080/01426397.2016.1172559.

Clouse, C., N. Anderson and T. Shippling, 2017: Ladakh's artificial glaciers: climate-adaptive design for water scarcity. *Clim. Dev.*, **9**(5), 428–438, doi:10.1080/17565529.2016.1167664.

Cloutier, C. et al., 2016: Potential impacts of climate change on landslides occurrence in Canada. In: *Slope Safety Preparedness for Impact of Climate Change* [Ho, K., S. Lacasse and L. Picarelli (eds.)]. Taylor & Francis Group, Florida, pp. 71–104. ISBN 978113803230.

Coe, J.A., E.K. Bessette-Kirton and M. Geertsema, 2017: Increasing rock-avalanche size and mobility in Glacier Bay National Park and Preserve, Alaska detected from 1984 to 2016 Landsat imagery. *Landslides*, **15**(3), 393–407, doi:10.1007/s10346-017-0879-7.

Cohen, J., H. Ye and J. Jones, 2015: Trends and variability in rain-on-snow events. *Geophys. Res. Lett.*, **42**(17), 7115–7122, doi:10.1002/2015GL065320.

Cole, J.A., 1985: *The Potosi mita, 1573–1700: Compulsory Indian labor in the Andes.* Stanford University Press, Stanford. 206 pp.

Colombo, N. et al., 2018: Review: Impacts of permafrost degradation on inorganic chemistry of surface fresh water. *Glob. Planet. Change*, **162**, 69–83, doi:10.1016/j.gloplacha.2017.11.017.

Colonia, D. et al., 2017: Compiling an inventory of glacier-bed overdeepenings and potential new lakes in de-glaciating areas of the peruvian andes: Approach, first results, and perspectives for adaptation to climate Change. *Water*, **9**(5), 336, doi:10.3390/w9050336.

Comola, F., B. Schaefli, A. Rinaldo and M. Lehning, 2015: Thermodynamics in the hydrologic response: Travel time formulation and application to Alpine catchments. *Water Resour. Res.*, **51**(3), 1671–1687, doi:10.1002/2014WR016228.

Cook, N. and D. Butz, 2013: The Atta Abad Landslide and Everyday Mobility in Gojal, Northern Pakistan. *Mt. Res. Dev.*, **33**(4), 372–380, doi:10.1659/mrd-journal-d-13-00013.1.

Coppola, E. et al., 2014: Changing hydrological conditions in the Po basin under global warming. *Sci. Total Environ.*, **493**, 1183–1196, doi:10.1016/j.scitotenv.2014.03.003.

Correa-Ibanez, R., G. Keir and N. McIntyre, 2018: Climate-resilient water supply for a mine in the Chilean Andes. In: Proceedings of the Institution of Civil Engineers – Water Management, **171**(4), 203–215. doi.org/10.1680/jwama.16.00129.

Cotto, O. et al., 2017: A dynamic eco-evolutionary model predicts slow response of alpine plants to climate warming. *Nat. Commun.*, **8**, 15399, doi:10.1038/ncomms15399.

Cramer, W. et al., 2014: Detection and attribution of observed impacts. In: Climate Change 2014: Impacts, Adaptation, and Vulnerability. Part A: Global and Sectoral Aspects. Contribution of Working Group II to the Fifth Assessment Report of the Intergovernmental Panel on Climate Change [Field, C.B., V.R. Barros, D.J. Dokken, K.J. Mach, M.D. Mastrandrea, T.E. Bilir, M. Chatterjee, K.L. Ebi, Y.O. Estrada, R.C. Genova, B. Girma, E.S. Kissel, A.N. Levy, S. MacCracken, P.R. Mastrandrea and L.L. White (eds.)]. Cambridge University Press, Cambridge, United Kingdom and New York, NY, USA, 79–1037.

Cuellar, A.D. and D.C. McKinney, 2017: Decision-making methodology for risk management applied to Imja Lake in Nepal. *Water*, **9**(8), 591, doi:10.3390/w9080591.

Cullen, N.J., Sirguey, P., Mölg, T., Kaser, G., Winkler, M., and Fitzsimmons, S.J., 2013: A century of ice retreat on Kilimanjaro: the mapping reloaded. The Cryosphere, **7**, 419–431, doi:10.5194/tc-7-419-2013.

Cunsolo, A. and N.R. Ellis, 2018: Ecological grief as a mental health response to climate change-related loss. *Nat. Clim. Change*, **8**(4), 275–281, doi:10.1038/s41558-018-0092-2.

Curry, A.M., V. Cleasby and P. Zukowskyj, 2006: Paraglacial response of steep, sediment-mantled slopes to post-'Little Ice Age' glacier recession in the central Swiss Alps. *J. Quat. Sci.*, **21**(3), 211–225, doi:10.1002/jqs.954.

Cutter, S. et al., 2012: Managing the risks from climate extremes at the local level. [Field, C.B., V. Barros, T.F. Stocker and Q. Dahe (eds.)]. Cambridge University Press, Cambridge, 291–338.

Cutter, S.L. and D.P. Morath, 2013: The evolution of the social vulnerability index (SoVI). In: Measuring Vulnerability to Natural Hazards. Towards Disaster Resilience Societies [Birkmann, J. (ed.)]. United Nations University Press, New York/Bonn, pp. 304–321.

D'Amico, M.E., M. Freppaz, E. Zanini and E. Bonifacio, 2017: Primary vegetation succession and the serpentine syndrome: the proglacial area of the Verra Grande glacier, North-Western Italian Alps. *Plant Soil*, **415**(1–2), 283--298, doi:10.1007/s11104-016-3165-x.

Da Silva, L. et al., 2019: *Analyse économique des mesures d'adaptation aux changements climatiques appliquée au secteur du ski alpin au Québec.* Ouranos, Montréal, 119 pp.

Dall'Amico, M. et al., 2011: Chapter 4: Local ground movements and effects on infrastructures. In: *Hazards related to permafrost and to permafrost degradation, PermaNET Project Report 6.2*, pp.107–147.

Dame, J. and J.S. Mankelow, 2010: Stongde revisited: Land-use change in central Zangskar. *Erdkunde*, **64**(4), 355–370, doi:10.3112/erdkunde.2010.04.05.

Dame, J. and M. Nüsser, 2011: Food security in high mountain regions: Agricultural production and the impact of food subsidies in Ladakh, Northern India. *Food Security*, **3**(2), 179–194, doi:10.1007/s12571-011-0127-2.

Damm, A., W. Greuell, O. Landgren and F. Prettenthaler, 2017: Impacts of +2°C global warming on winter tourism demand in Europe. *Climate Services, **7***, 31–46, doi:10.1016/j.cliser.2016.07.003.

Dangi, M.B. et al., 2018: Impacts of environmental change on agroecosystems and livelihoods in Annapurna Conservation Area, Nepal. *Environmental Development*, **25**, 59–72, doi:10.1016/j.envdev.2017.10.001.

Darcy, J.L. et al., 2018: Phosphorus, not nitrogen, limits plants and microbial primary producers following glacial retreat. *Sci. Adv.*, **4**(5), doi:10.1126/sciadv.aaq0942.

Darrow, M.M. et al., 2016: Frozen debris lobe morphology and movement: An overview of eight dynamic features, southern Brooks Range, Alaska. *The Cryosphere*, **10**(3), 977–993, doi:10.5194/tc-10-977-2016.

Dawson, J. and D. Scott, 2013: Managing for climate change in the alpine ski sector. *Tourism Management*, **35**, 244–254, doi:10.1016/j.tourman.2012.07.009.

De Boeck, H.J. et al., 2018: Legacy effects of climate extremes in alpine grassland. *Front. Plant Sci.*, **9**(1586), doi:10.3389/fpls.2018.01586.

de Jong, C., 2015: Challenges for mountain hydrology in the third millennium. *Front. Environ. Sci.*, **3**, 38, doi:10.3389/fenvs.2015.00038.

de Kok, R.J., O.A. Tuinenburg, P.N.J. Bonekamp and W.W. Immerzeel, 2018: Irrigation as a Potential Driver for Anomalous Glacier Behavior in High Mountain Asia. *Geophys. Res. Lett.*, **5**(2), 1071–2054, doi:10.1002/2017GL076158.

De Stefano, L. et al., 2010: *Mapping the Resilience of International River Basins to Future Climate Change-Induced Water Variability, Volume 1. Main Report.* World Bank Water Sector Board Discussion Paper No. 15, 56051, The World Bank, Washington, DC., 88 pp.

Deems, J.S. et al., 2013: Combined impacts of current and future dust deposition and regional warming on Colourado River Basin snow dynamics and hydrology. *Hydrol. Earth Syst. Sc.*, **17**(11), 4401–4413, doi:10.5194/hess-17-4401-2013.

Dehecq, A. et al., 2019: Twenty-first century glacier slowdown driven by mass loss in High Mountain Asia. *Nat. Geosci.*, **12**, 22–27, doi:10.1038/s41561-018-0271-9.

Delaloye, R., C. Lambiel and I. Gärtner-Roer, 2010: Overview of rock glacier kinematics research in the Swiss Alps: Seasonal rhythm, interannual variations and trends over several decades. *Geogr. Helv.*, **65**, 135–145, doi:10.5194/gh-65-135-2010.

Deline, P., 2009: Interactions between rock avalanches and glaciers in the Mont Blanc massif during the late Holocene. *Quaternary Sci. Rev.*, **28** (11–12), 1070–1083, doi:10.1016/j.quascirev.2008.09.025.

Deline, P. et al., 2015: Chapter 15: Ice loss and slope stability in High-Mountain Regions. In: *Snow and Ice-Related Hazards, Risks, and Disasters* [Shroder, J.F., W. Haeberli and C. Whiteman (eds.)]. Elsevier, Amsterdam, 521–561.

Deng, M.Z., D.H. Qin and H.G. Zhang, 2012: Public perceptions of climate and cryosphere change in typical arid inland river areas of China: Facts, impacts and selections of adaptation measures. *Quatern. Int.*, **282**, 48–57, doi:10.1016/j.quaint.2012.04.033.

Denning, A., 2014: From Sublime Landscapes to "White Gold": How Skiing Transformed the Alps after 1930. *Environ. Hist.*,, **19**(1), 78–108, doi:10.1093/envhis/emt105.

Dharmadhikary, S., 2008: *Mountains of Concrete: Dam Building in the Himalayas.* eSocialSciences, Working Papers id:1815, [Available at: https://ideas.repec.org/p/ess/wpaper/id1815.html]. Accessed 05/08/2019.

Di Mauro, B. et al., 2015: Mineral dust impact on snow radiative properties in the European Alps combining ground, UAV, and satellite observations. *J. Geophy. Res.*, **120**(12), 6080–6097, doi:10.1002/2015JD023287.

Díaz, S. et al., 2019: *Summary for policymakers of the global assessment report on biodiversity and ecosystem services of the Intergovernmental Science-Policy Platform on Biodiversity and Ecosystem Services (advance unedited version).* [Available at: www.ipbes.net/sites/default/files/downloads/spm_unedited_advance_for_posting_htn.pdf].

Dickerson-Lange, S. et al., 2016: Challenges and successes in engaging citizen scientists to observe snow cover: from public engagement to an educational collaboration. *JCOM*, **15**(1), A01, doi:10.22323/2.15010201.

Dileepkumar, R., K. AchutaRao and T. Arulalan, 2018: Human influence on sub-regional surface air temperature change over India. *Sci. Rep.*, **8**, 8967, doi:10.1038/s41598-018-27185-8.

Dinar, S., D. Katz, L. De Stefano and B. Blankespoor, 2016: *Climate change and water variability: do water treaties contribute to river basin resilience? A review.* Policy Research Working Paper 7855, The World Bank, Washington, D.C. Ding, J. et al., 2016: The permafrost carbon inventory on the Tibetan Plateau: a new evaluation using deep sediment cores. *Glob. Change Biol*, **22**, 2688–2701, doi:10.1111/gcb.13257.

Dixon, E.J., M.E. Callanan, A. Hafner and P.G. Hare, 2014: The emergence of glacial archaeology. *Journal of Glacial Archaeology*, **1**(1), 1–9, doi:10.1558/jga.v1i1.1.

Dixon, E.J., W.F. Manley and C.M. Lee, 2005: The Emerging archaeology of glaciers and ice patches: Examples from Alaska's Wrangell-St. Elias National Park and Preserve. *Am. Antiq.*, **70**(1), 129–143, doi:10.2307/40035272.

Dobrowski, S.Z. and S.A. Parks, 2016: Climate change velocity underestimates climate change exposure in mountainous regions. *Nat. Commun.*, **7**, 1–8, doi:10.1038/ncomms12349.

Dolezal, J. et al., 2016: Vegetation dynamics at the upper elevational limit of vascular plants in Himalaya. *Sci. Rep.*, **6**, 1–13, doi:10.1038/srep24881.

Doré, G., F. Niu and H. Brooks, 2016: Adaptation methods for transportation infrastructure built on degrading permafrost. In: *Permafrost Periglac.*, **27**, 352–364, doi:10.1002/ppp.1919.

Drenkhan, F. et al., 2015: The changing water cycle: climatic and socioeconomic drivers of water-related changes in the Andes of Peru. *WiRes.Water,* **2**(6), 715–733, doi:10.1002/wat2.1105.

Drenkhan, F., C. Huggel, L. Guardamino and W. Haeberli, 2019: Managing risks and future options from new lakes in the deglaciating Andes of Peru: The example of the Vilcanota-Urubamba basin. *Sci. Total Environ.*, **665**, 465–483, doi:10.1016/j.scitotenv.2019.02.070.

Du, J. et al., 2017: Satellite microwave assessment of Northern Hemisphere lake ice phenology from 2002 to 2015. *The Cryosphere*, **11**(1), 47–63, doi:10.5194/tc-11-47-2017.

Duethmann, D. et al., 2015: Attribution of streamflow trends in snow and glacier melt-dominated catchments of the Tarim River, Central Asia. *Water Resour. Res.*, **51**(6), 4727–4750, doi:10.1002/2014wr016716.

Duguay, C.R., M. Bernier, Y. Gauthier and A. Kouraev, 2014: Remote sensing of lake and river ice. In: *Remote Sensing of the Cryosphere* [Tedesco, M. (ed.)]. John Wiley & Sons, Ltd, Chichester, UK, 273–306.

Dunse, T. et al., 2015: Glacier-surge mechanisms promoted by a hydro-thermodynamic feedback to summer melt. *The Cryosphere*, **9**(1), 197–215, doi:10.5194/tc-9-197-2015.

Duntley, M., 2015: Spiritual Tourism and Frontier Esotericism at Mount Shasta, California. *International Journal for the Study of New Religions*, **5**(2), 123–150, doi:10.1558/ijsnr.v5i2.26233.

Duvillard, P.A., L. Ravanel and P. Deline, 2015: Risk assessment of infrastructure destabilisation due to global warming in the high French Alps. *Revue de Géographie Alpine*, **103** (2), doi:10.4000/rga.2896.

Duvillard, P.A., L. Ravanel, M. Marcer and P. Schoeneich, 2019: Recent evolution of damage to infrastructure on permafrost in the French Alps. *Reg. Environ. Change*, **19**(5), 1281–1293, doi:10.1007/s10113-019-01465-z.

Dymov, A.A., E.V. Zhangurov and F. Hagedorn, 2015: Soil organic matter composition along altitudinal gradients in permafrost affected soils of the Subpolar Ural Mountains. *Catena*, **131**, 140–148, doi:10.1016/j.catena.2015.03.020.

Eby, L.A., O. Helmy, L.M. Holsinger and M.K. Young, 2014: Evidence of climate-induced range contractions in bull trout Salvelinus confluentus in a Rocky Mountain watershed, USA. *PLOS ONE*, **9**(6), doi:10.1371/journal.pone.0098812.

Eckert, N. et al., 2013: Temporal trends in avalanche activity in the French Alps and subregions: from occurrences and runout altitudes to unsteady return periods. *J. Glaciol.*, **59**(213), 93–114, doi:10.3189/2013JoG12J091.

Eichel, J., D. Draebing and N. Meyer, 2018: From active to stable: Paraglacial transition of Alpine lateral moraine slopes. *Land Degrad. Dev.*, **29**(11), 4158–4172, doi:10.1002/ldr.3140.

Einarsson, B. and S. Jónsson, 2010: *The effect of climate change on runoff from two watersheds in Iceland*. Icelandic Meteorological Office, Reykjavik. 34 pp.

Eisen, O., W.D. Harrison and C.F. Raymond, 2001: The surges of variegated glacier, Alaska, U.S.A., and their connection to climate and mass balance. *J. Glaciol.*, **47**(158), 351–358, doi:10.3189/172756501781832179.

Elsen, P.R., W.B. Monahan and A.M.Merenlender, 2018: Global patterns of protection of elevational gradients in mountain ranges. *Proc. Natl. Acad. Sci. U.S.A.*, **115**(23), 6004–6009, doi:10.1073/pnas.1720141115.

Elsen, P.R. and M.W. Tingley, 2015: Global mountain topography and the fate of montane species under climate change. *Nat. Clim. Change*, **5**(8), 772–776, doi:10.1038/nclimate2656.

Emmer, A. et al., 2016: 882 lakes of the Cordillera Blanca: An inventory, classification, evolution and assessment of susceptibility to outburst floods. *Catena*, **147**, 269–279, doi:10.1016/j.catena.2016.07.032.

Engelhardt, M. et al., 2017: Melt water runoff in a changing climate (1951–2099) at Chhota Shigri Glacier, Western Himalaya, Northern India. *Ann. Glaciol.*, **58**(75), 47–58, doi:10.1017/aog.2017.13.

Engeset, R.V., T.V. Schuler and M. Jackson, 2005: Analysis of the first jökulhlaup at Blåmannsisen, northern Norway, and implications for future events. *Ann. Glaciol.*, **42**, 35–41, doi:10.3189/172756405781812600.

Eriksen, H. et al., 2018: Recent Acceleration of a Rock Glacier Complex, Ádjet, Norway, Documented by 62 Years of Remote Sensing Observations. *Geophys. Res. Lett.*, **45**(16), 8314–8323, doi:10.1029/2018GL077605.

Erokhin, S.A. et al., 2017: Debris flows triggered from non-stationary glacier lake outbursts: the case of the Teztor Lake complex (Northern Tian Shan, Kyrgyzstan). *Landslides*, **15**(1), 83–98, doi:10.1007/s10346-017-0862-3.

Evangelista, A. et al., 2016: Changes in composition, ecology and structure of high-mountain vegetation: A re-visitation study over 42 years. *AoB Plants*, **8**, 1–11, doi:10.1093/aobpla/plw004.

Evans, D.J.A., M. Ewertowski, S.S.R. Jamieson and C. Orton, 2016: Surficial geology and geomorphology of the Kumtor Gold Mine, Kyrgyzstan: human impacts on mountain glacier landsystems. *J. Maps*, **12**(5), 757–769. doi:10.1080/17445647.2015.1071720.

Evans, S.G. and K.B. Delaney, 2015: Chapter 16: Catastrophic mass flows in the mountain glacial environment. In: *Snow, and Ice-Related Hazards, Risks, and Disasters* [Haeberli, W. and C. Whitemann (eds.)]. Elsevier, Amsterdam, 563–606.

Evans, S.G. et al., 2009: Catastrophic detachment and high-velocity long-runout flow of Kolka Glacier, Caucasus Mountains, Russia in 2002. *Geomorphology*, **105**, 314–321, doi:10.1016/j.geomorph.2008.10.008.

Evette, A., L. Peyras, H. Francois and S. Gaucherand, 2011: Environmental risks and impacts of mountain reservoirs for artificial snow production in a context of climate change. *Journal of Alpine Research | Revue de géographie alpine*, (99–4), doi:10.4000/rga.1471.

Fadrique, B. et al., 2018: Widespread but heterogeneous responses of Andean forests to climate change. *Nature*, **564** (7735), 207–212, doi:10.1038/s41586-018-0715-9.

Faillettaz, J., M. Funk and C. Vincent, 2015: Avalanching glacier instabilities: Review on processes and early warning perspectives. *Rev. Geophys.*, **53**(2), 203–224, doi:10.1002/2014rg000466.

Fairman, T.A., L.T. Bennett, S. Tupper and C.R. Nitschke, 2017: Frequent wildfires erode tree persistence and alter stand structure and initial composition of a fire-tolerant sub-alpine forest. *J. Veg. Sci.*, **28**(6), 1151–1165, doi:10.1111/jvs.12575.

Falk, M., 2016: The stagnation of summer glacier skiing. *Tourism Analysis*, **21**(1), 117–122, doi:10.3727/108354216X14537459509053.

Farbrot, H., K. Isaksen, B. Etzelmüller and K. Gisnås, 2013: Ground thermal regime and permafrost distribution under a changing climate in northern Norway. *Permafrost Periglac.*, **24**(1), 20–38, doi:10.1002/ppp.1763.

Farinotti, D. et al., 2019: A consensus estimate for the ice thickness distribution of all glaciers on Earth. *Nat. Geosci.*, **12**, 168–173, doi:10.1038/s41561-019-0300-3.

Fell, S.C., J.L. Carrivick and L.E. Brown, 2017: The multitrophic effects of climate change and glacier retreat in mountain rivers. *Bioscience*, **67**(10), 897–911, doi:10.1093/biosci/bix107.

Fell, S.C. et al., 2018: Declining glacier cover threatens the biodiversity of alpine river diatom assemblages. *Glob. Change Biol.*, **24**(12), 5828–5840, doi:10.1111/gcb.14454.

Fellman, J.B. et al., 2015: Evidence for the assimilation of ancient glacier organic carbon in a proglacial stream food web. *Limnol, Oceanogr.*, **60**(4), 1118–1128, doi:10.1002/lno.10088.

Fellman, J.B. et al., 2014: Watershed Glacier Coverage Influences Dissolved Organic Matter Biogeochemistry in Coastal Watersheds of Southeast Alaska. *Ecosystems*, **17**(6), 1014–1025, doi:10.1007/s10021-014-9777-1.

Fickert, T. and F. Grüninger, 2018: High-speed colonization of bare ground-permanent plot studies on primary succession of plants in recently deglaciated glacier forelands. *Land Degrad. Dev.*, **29**(8), 2668–2680, doi:10.1002/ldr.3063.

Fickert, T., F. Grüninger and B. Damm, 2016: Klebelsberg revisited: did primary succession of plants in glacier forelands a century ago differ from today? *Alpine Botany*, **127**(1), 17–29, doi:10.1007/s00035-016-0179-1.

Fiddes, J. and S. Gruber, 2012: TopoSUB: a tool for efficient large area numerical modelling in complex topography at sub-grid scales. *Geosci. Model. Dev.* **5**(5), 1245–1257, doi:10.5194/gmd-5-1245-2012.

Figueroa-Armijos, M. and C.B. Valdivia, 2017: Sustainable innovation to cope with climate change and market variability in the Bolivian Highlands. *Innovation and Development*, **7**(1), 17–35, doi:10.1080/2157930X.2017.1281210.

2

Finn, D.S., A.C. Encalada and H. Hampel, 2016: Genetic isolation among mountains but not between stream types in a tropical high-altitude mayfly. *Freshw. Biol.*, **61**(5), 702–714, doi:10.1111/fwb.12740.

Finn, D.S., K. Khamis and A.M. Milner, 2013: Loss of small glaciers will diminish beta diversity in Pyrenean streams at two levels of biological organization. *Global Ecol. Biogeogr.*, **22**(1), 40–51, doi:10.1111/j.1466-8238.2012.00766.x.

Fischer, A., K. Helfricht and M. Stocker-Waldhuber, 2016: Local reduction of decadal glacier thickness loss through mass balance management in ski resorts. *The Cryosphere*, **10**(6), 2941–2952, doi:10.5194/tc-10-2941-2016.

Fischer, L., C. Huggel, A. Kääb and W. Haeberli, 2013: Slope failures and erosion rates on a glacierized high-mountain face under climatic changes. *Earth Surf. Process. Landf.*, **38**(8), 836–846, doi:10.1002/Esp.3355.

Fischer, L. et al., 2012: On the influence of topographic, geological and cryospheric factors on rock avalanches and rockfalls in high-mountain areas. *Nat. Hazard. Earth Sys.*, **12**(1), 241–254, doi:10.5194/nhess-12-241-2012.

Fischlin, A. et al., 2007: Ecosystems, their properties, goods and services. In: Climate change 2007: Impacts, adaptation and vulnerability. Contribution of Working Group II to the Fourth Assessment Report of the Intergovernmental Panel of Climate Change (IPCC) [Parry, M.L., O.F. Canziani, J.P. Palutikof, P.J. van der Linden and C.E. Hanson (eds.)]. Cambridge University Press, Cambridge, UK, 211–272.

Fleming, S.W. and H.E. Dahlke, 2014: Modulation of linear and nonlinear hydroclimatic dynamics by mountain glaciers in Canada and Norway: Results from information-theoretic polynomial selection. *Can. Water. Resour. J.*, **39**(3), 324–341, doi:10.1080/07011784.2014.942164.

Foresta, L. et al., 2016: Surface elevation change and mass balance of Icelandic ice caps derived from swath mode CryoSat-2 altimetry. *Geophys. Res. Lett.*, **43**(23), 12138–12145, doi:10.1002/2016GL071485.

Foresta, L. et al., 2018: Heterogeneous and rapid ice loss over the Patagonian Ice Fields revealed by CryoSat-2 swath radar altimetry. *Remote Sens. Environ.*, **211**, 441–455, doi:10.1016/j.rse.2018.03.041.

Forsythe, N. et al., 2017: Karakoram temperature and glacial melt driven by regional atmospheric circulation variability. *Nat. Clim. Change*, **7** (9), 664–670, doi:10.1038/nclimate3361.

Frans, C. et al., 2015: Predicting glacio-hydrologic change in the headwaters of the Zongo River, Cordillera Real, Bolivia. *Water Resour. Res.*, **51**(11), 9029–9052, doi:10.1002/2014WR016728.

Fraser, B. 2017. Learning from flood-alarm system's fate, EcoAmericas. www.ecoamericas.com/en/story.aspx?id=1776. Accessed on 05/08/2019.

Freeman, B.G., J.A. Lee-Yaw, J.M. Sunday and A.L. Hargreaves, 2018: Expanding, shifting and shrinking: The impact of global warming on species' elevational distributions. *Global Ecol. Biogeogr.*, **27**, 1268–1276, doi:10.1111/geb.12774.

Frei, E.R. et al., 2018: Biotic and abiotic drivers of tree seedling recruitment across an alpine treeline ecotone. *Sci. Rep.*, **8**(1), doi:10.1038/s41598-018-28808-w.

Freudiger, D., I. Kohn, K. Stahl and M. Weiler, 2014: Large-scale analysis of changing frequencies of rain-on-snow events with flood-generation potential. *Hydrol. Earth Syst. Sc.*, **18**(7), 2695–2709, doi:10.5194/hess-18-2695-2014.

Frey, H. et al., 2010: A multi-level strategy for anticipating future glacier lake formation and associated hazard potentials. *Nat. Hazard. Earth Sys.*, **10**(2), 339–352, doi:10.5194/nhess-10-339-2010.

Fuchs, M., P. Kuhry and G. Hugelius, 2015: Low below-ground organic carbon storage in a subarctic Alpine permafrost environment. *The Cryosphere*, **9**(2), 427–438, doi:10.5194/tc-9-427-2015.

Fuhrer, J., P. Smith and A. Gobiet, 2014: Implications of climate change scenarios for agriculture in alpine regions--a case study in the Swiss Rhone catchment. *Sci. Total Environ.*, **493**, 1232–1241, doi:10.1016/j.scitotenv.2013.06.038.

Fujita, K. et al., 2013: Potential flood volume of Himalayan glacial lakes. *Nat. Hazard. Earth Sys.*, **13**(7), 1827–1839, doi:10.5194/nhess-13-1827-2013.

Fürst, J.J. et al., 2018: The ice-free topography of Svalbard. *Geophys. Res. Lett.*, **45** (21), 11,760–11,769, doi:10.1029/2018GL079734.

Furunes, T. and R.J. Mykletun, 2012: Frozen adventure at risk? A 7-year follow-up study of Norwegian glacier tourism. *Scandinavian Journal of Hospitality and Tourism*, **12**(4), 324–348, doi:10.1080/15022250.2012.748507.

Gadek, B. et al., 2017: Snow avalanche activity in Żleb Żandarmerii in a time of climate change (Tatra Mts., Poland). *Catena*, **158**, 201–212, doi:10.1016/j.catena.2017.07.005.

Gagné, K., M.B. Rasmussen and B. Orlove, 2014: Glaciers and society: Attributions, perceptions, and valuations. *WiRes. Clim. Change*, **5**(6), 793–808, doi:10.1002/wcc.315.

Gao, J., 2019: Global population projection grids based on Shared Socioeconomic Pathways (SSPs), downscaled 1-km grids, 2010–2100. Palisades, NY. doi.org/10.7927/H44747X4.

Gao, Y. et al., 2018: Does elevation-dependent warming hold true above 5000 m elevation? Lessons from the Tibetan Plateau. *npj Climate and Atmospheric Science*, **1**(19). doi:10.1038/s41612-018-0030-z.

GAPHAZ, 2017: *Assessment of Glacier and Permafrost Hazards in Mountain Regions – Technical Guidance Document.* [Allen, S., H. Frey, C. Huggel and e. al. (eds.)]. Standing Group on Glacier and Permafrost Hazards in Mountains (GAPHAZ) of the International Association of Cryospheric Sciences (IACS) and the International Permafrost Association (IPA). Zurich, Switzerland / Lima, Peru, 72 pp.

García-Hernández, C. et al., 2017: Reforestation and land use change as drivers for a decrease of avalanche damage in mid-latitude mountains (NW Spain). *Glob. Planet. Change*, **153**, 35–50, doi:10.1016/j.gloplacha.2017.05.001.

Gardelle, J., Y. Arnaud and E. Berthier, 2011: Contrasted evolution of glacial lakes along the Hindu Kush Himalaya mountain range between 1990 and 2009. *Glob. Planet. Change*, **75** (1–2), 47–55, doi:10.1016/j.gloplacha.2010.10.003.

Gardelle, J., E. Berthier and Y. Arnaud, 2012: Slight mass gain of Karakoram glaciers in the early twenty-first century. *Nat. Geosci.*, **5**(5), 322–325, doi:10.1038/ngeo1450.

Gardelle, J., E. Berthier, Y. Arnaud and A. Kääb, 2013: Region-wide glacier mass balances over the Pamir-Karakoram-Himalaya during 1999–2011. *The Cryosphere*, **7**(4), 1263–1286, doi:10.5194/tc-7-1263-2013.

Gardner, A.S. et al., 2013: A reconciled estimate of glacier contributions to sea level rise: 2003 to 2009. *Science*, **340**(6134), 852–857, doi:10.1126/science.1234532.

Gardner, J., J. Sinclair, F. Berkes and R.B. Singh, 2002: Accelerated tourism development and its impacts in Kullu-Manali, H.P., India. *Tourism Recreation Research*, **27**(3), 9–20, doi:10.1080/02508281.2002.11081370.

Gardner, J.S. and J. Dekens, 2006: Mountain hazards and the resilience of social–ecological systems: lessons learned in India and Canada. *Nat. Hazards*, **41**(2), 317–336, doi:10.1007/s11069-006-9038-5.

Gaudard, L., J. Gabbi, A. Bauder and F. Romerio, 2016: Long-term uncertainty of hydropower revenue due to climate change and electricity prices. *Water Resour. Manage.*, **30**(4), 1325–1343, doi:10.1007/s11269-015-1216-3.

Gaudard, L., M. Gilli and F. Romerio, 2013: Climate change impacts on hydropower management. *Water Resour. Manage.*, **27**(15), 5143–5156, doi:10.1007/s11269-013-0458-1.

Gaudard, L. et al., 2014: Climate change impacts on hydropower in the Swiss and Italian Alps. *Sci. Total Environ.*, **493**, 1211–1221, doi:10.1016/j.scitotenv.2013.10.012.

Gebre, S., T. Boissy and K. Alfredsen, 2014: Sensitivity of lake ice regimes to climate change in the Nordic region. *The Cryosphere*, **8** (4), 1589–1605, doi:10.5194/tc-8-1589-2014.

Geertsema, M. and J.J. Clague, 2005: Jokulhlaups at Tulsequah Glacier, northwestern British columbia, Canada. *The Holocene*, **15**(2), 310–316, doi:10.1191/0959683605hl812rr.

Geertsema, M., J.J. Clague, J.W. Schwab and S.G. Evans, 2006: An overview of recent large catastrophic landslides in northern British Columbia, Canada. *Eng. Geol.*, **83**(1–3), 120–143, doi:10.1016/j.enggeo.2005.06.028.

Gentili, R. et al., 2015: Potential warm-stage microrefugia for alpine plants: Feedback between geomorphological and biological processes. *Ecol. Complex.*, **21**, 87–99, doi:10.1016/j.ecocom.2014.11.006.

Gentle, P. and T.N. Maraseni, 2012: Climate change, poverty and livelihoods: adaptation practices by rural mountain communities in Nepal. *Environ. Sci. Policy*, **21**, 24–34, doi:10.1016/j.envsci.2012.03.007.

Gentle, P. and R. Thwaites, 2016: Transhumant pastoralism in the context of socioeconomic and climate change in the mountains of Nepal. *Mt. Res. Dev.*, **36**(2), 173–182, doi:10.1659/mrd-journal-d-15-00011.1.

Giacona, F. et al., 2018: Avalanche activity and socio-environmental changes leave strong footprints in forested landscapes: a case study in the Vosges medium-high mountain range. *Ann. Glaciol.*, **10**(77), 1–23, doi:10.1017/aog.2018.26.

Giersch, J.J. et al., 2017: Climate-induced glacier and snow loss imperils alpine stream insects. *Glob. Change Biol*, **23**(7), 2577–2589, doi:10.1111/gcb.13565.

Giersch, J.J. et al., 2015: Climate-induced range contraction of a rare alpine aquatic invertebrate. *Freshw. Sci.*, **34**(1), 53–65, doi:10.1086/679490.

Giesen, R.H. and J. Oerlemans, 2013: Climate-model induced differences in the 21st century global and regional glacier contributions to sea-level rise. *Clim. Dyn.*, **41**(11–12), 3283–3300, doi:10.1007/s00382-013-1743-7.

Gilbert, A. et al., 2018: Mechanisms leading to the 2016 giant twin glacier collapses, Aru Range, Tibet. *The Cryosphere*, **12**(9), 2883–2900, doi:10.5194/tc-12-2883-2018.

Gilbert, A. et al., 2015: Assessment of thermal change in cold avalanching glaciers in relation to climate warming. *Geophys. Res. Lett.*, **42**(15), 6382–6390, doi:10.1002/2015GL064838.

Gilbert, A. et al., 2012: The influence of snow cover thickness on the thermal regime of Tête Rousse Glacier (Mont Blanc range, 3200 m a.s.l.): Consequences for outburst flood hazards and glacier response to climate change. *J. Geophys. Res-Earth.*, **117**(F4), F04018, doi:10.1029/2011JF002258.

Gilles, J.L., J.L. Thomas, C. Valdivia and E.S. Yucra, 2013: Laggards or Leaders: Conservers of Traditional Agricultural Knowledge in Bolivia. *Rural Sociol.*, **78**(1), 51–74, doi:10.1111/ruso.12001.

Giménez-Benavides, L. et al., 2018: How does climate change affect regeneration of Mediterranean high-mountain plants? An integration and synthesis of current knowledge. *Plant Biology*, **20**, 50–62, doi:10.1111/plb.12643.

Ginot, P. et al., 2014: A 10 year record of black carbon and dust from a Mera Peak ice core (Nepal): variability and potential impact on melting of Himalayan glaciers. *The Cryosphere*, **8**(4), 1479–1496, doi:10.5194/tc-8-1479-2014.

Gitay, H., S. Brown, W. Easterling and B. Jallow, 2001: Ecosystems and their goods and services. In: Climate Change 2001 – Impacts, Adaptation, and Vulnerability. Contribution of Working Group II to the Third Assessment Report of the Intergovernmental Panel of Climate Change (IPCC) [McCarthy, J.J., O.F. Canziani, N.A. Leary, D.J. Dokken and K.S. White (eds.)]. Cambridge University Press, Cambridge, UK, 237–342.

Gonseth, C., 2013: Impact of snow variability on the Swiss winter tourism sector: Implications in an era of climate change. *Clim. Change*, **119**(2), 307–320, doi:10.1007/s10584-013-0718-3.

Gosai, M.A. and L. Sulewski, 2014: Urban attraction: Bhutanese internal rural–urban migration. *Asian Geographer*, **31**(1), 1–16, doi:10.1080/10225706.2013.790830.

Gou, P. et al., 2017: Lake ice phenology of Nam Co, Central Tibetan Plateau, China, derived from multiple MODIS data products. *J. Great. Lakes Res.*, **43**(6), 989–998, doi:10.1016/j.jglr.2017.08.011.

Goulden, M.L. and R.C. Bales, 2014: Mountain runoff vulnerability to increased evapotranspiration with vegetation expansion. *PNAS*, **111**(39), 14071–14075, doi:10.1073/pnas.1319316111.

Graae, B.J. et al., 2018: Stay or go – how topographic complexity influences alpine plant population and community responses to climate change. *Perspect. Plant. Ecol.*, **30**, 41–50, doi:10.1016/J.PPEES.2017.09.008.

Gratzer, G. and W.S. Keeton, 2017: Mountain Forests and Sustainable Development: The Potential for Achieving the United Nations' 2030 Agenda. *Mt. Res. Dev.*, **37**(3), 246–253, doi:10.1659/MRD-JOURNAL-D-17-00093.1.

Grau, H.R. and T.M. Aide, 2007: Are rural–urban migration and sustainable development compatible in mountain systems? *Mt. Res. Dev.*, **27**(2), 119–124, doi:10.1659/mrd.0906.

Groll, M. et al., 2015: Water quality, potential conflicts and solutions-an upstream-downstream analysis of the transnational Zarafshan River (Tajikistan, Uzbekistan). *Environ. Earth Sci.*, **73**(2), 743–763, doi:10.1007/s12665-013-2988-5.

Gruber, S., 2012: Derivation and analysis of a high-resolution estimate of global permafrost zonation. *The Cryosphere*, **6**(1), 221–233, doi:10.5194/tc-6-221-2012.

Gruber, S. et al., 2017: Review article: Inferring permafrost and permafrost thaw in the mountains of the Hindu Kush Himalaya region. *The Cryosphere*, **11**(1), 81–99, doi:10.5194/tc-11-81-2017.

Gruber, S. and W. Haeberli, 2007: Permafrost in steep bedrock slopes and its temperature-related destabilization following climate change. *J. Geophys. Res-Oceans*, **112**(F2), F02S18, doi:10.1029/2006JF000547.

Grünewald, T., F. Wolfsperger and M. Lehning, 2018: Snow farming: conserving snow over the summer season. *The Cryosphere*, **12**(1), 385–400, doi:10.5194/tc-12-385-2018.

Grytnes, J.-A. et al., 2014: Identifying the driving factors behind observed elevational range shifts on European mountains. *Global Ecol. Biogeogr.*, **23**(8), 876–884, doi:10.1111/geb.12170.

Guo, D. and H. Wang, 2016: CMIP5 permafrost degradation projection: A comparison among different regions. *J. Geophys. Res-Atmos.*, **121**(9), 4499–4517, doi:10.1002/2015jd024108.

Guo, D., H. Wang and D. Li, 2012: A projection of permafrost degradation on the Tibetan Plateau during the 21st century. *J. Geophys. Res-Atmos.*, **117**, D05106, doi:10.1029/2011JD016545.

Guo, L. et al., 2016: Displacements of tower foundations in permafrost regions along the Qinghai–Tibet Power Transmission Line. *Cold Reg. Sci. Technol.*, **121**, 187–195, doi:10.1016/j.coldregions.2015.07.012.

Gupta, V. and M.P. Sah, 2008: Impact of the Trans-Himalayan Landslide Lake Outburst Flood (LLOF) in the Satluj catchment, Himachal Pradesh, India. *Nat. Hazards*, **45**(3), 379–390, doi:10.1007/s11069-007-9174-6.

Gutmann, E. et al., 2016: The Intermediate Complexity Atmospheric Research Model (ICAR). *J. Hydrometeorol.*, **17**(3), 957–973, doi:10.1175/jhm-d-15-0155.1.

Haeberli, W. et al., 2016: New lakes in deglaciating high-mountain regions – opportunities and risks. *Clim. Change*, **139**(2), 201–214, doi:10.1007/s10584-016-1771-5.

Haeberli, W., A. Kääb, D.V. Mühll and P. Teysseire, 2001: Prevention of outburst floods from periglacial lakes at Grubengletscher, Valais, Swiss Alps. *J. Glaciol.*, **47** (156), 111–122–122, doi:10.3189/172756501781832575.

Haeberli, W., Y. Schaub and C. Huggel, 2017: Increasing risks related to landslides from degrading permafrost into new lakes in de-glaciating mountain ranges. *Geomorphology*, **293**, 405–417, doi:10.1016/j.geomorph.2016.02.009.

Haeberli, W. and C. Whiteman, 2015: *Snow and Ice-related Hazards, Risks, and Disasters*. Elsevier, Amsterdam. 812 pp. ISBN 9780123948496.

Hagenstad, M., E. Burakowski and R. Hill, 2018: *The economic contributions of winter sports in a changing climate*. Protect our winters. Hagenstad Consulting, Inc., Boulder, USA. 80 pp.

Hall, L.E., A.D. Chalfoun, E.A. Beever and A.E. Loosen, 2016: Microrefuges and the occurrence of thermal specialists: implications for wildlife persistence amidst changing temperatures. *Climate Change Responses*, **3**(1), 8, doi:10.1186/s40665-016-0021-4.

2

Hamududu, B. and A. Killingtveit, 2012: Assessing climate change impacts on global hydropower. *Energies*, **5**(2), 305–322, doi:10.3390/en5020305.

Hänggi, P. and R. Weingartner, 2012: Variations in discharge volumes for hydropower generation in Switzerland. *Water Resour. Manage.*, **26**(5), 1231–1252, doi:10.1007/s11269-011-9956-1.

Hannah, L. et al., 2014: Fine-grain modeling of species' response to climate change: Holdouts, stepping-stones, and microrefugia. *Trends Ecol. Evol.*, **29**, 390–397, doi:10.1016/j.tree.2014.04.006.

Hardy, S.P., D.R. Hardy and K.C. Gil, 2018: Avian nesting and roosting on glaciers at high elevation, Cordillera Vilcanota, Peru. *Wilson J. Ornithol.*, **130**(4), 940--957, doi:10.1676/1559-4491.130.4.940.

Hari, R.E. et al., 2006: Consequences of climatic change for water temperature and brown trout populations in Alpine rivers and streams. *Glob. Change Biol*, **12**(1), 10–26, doi:10.1111/j.1365-2486.2005.001051.x.

Harpold, A.A. and P.D. Brooks, 2018: Humidity determines snowpack ablation under a warming climate. *PNAS*, **115**(6), 1215–1220, doi:10.1073/pnas.1716789115.

Harpold, A.A. and N.P. Molotch, 2015: Sensitivity of soil water availability to changing snowmelt timing in the western U.S. *Geophys. Res. Lett.*, **42**(19), 8011–8020, doi:10.1002/2015GL065855.

Harris, C. et al., 2009: Permafrost and climate in Europe: Monitoring and modelling thermal, geomorphological and geotechnical responses. *Earth-Sci. Rev.*, **92**(3–4), 117–171, doi:10.1016/j.earscirev.2008.12.002.

Harris, C. et al., 2003: Warming permafrost in European mountains. *Glob. Planet. Change*, **39**(3–4), 215–225, doi:10.1016/j.gloplacha.2003.04.001.

Harrison, S. et al., 2018: Climate change and the global pattern of moraine-dammed glacial lake outburst floods. *The Cryosphere*, **12**(4), 1195–1209, doi:10.5194/tc-12-1195-2018.

Harrison, W.D. et al., 2015: Glacier Surges. In: *Snow, and Ice-Related Hazards, Risks, and Disasters* [Haeberli, W. and C. Whitemann (eds.)]. Elsevier, Amsterdam, 437–485.

Hartl, L., A. Fischer, M. Stocker-Waldhuber and J. Abermann, 2016: Recent speed-up of an alpine rock glacier: An updated chronology of the kinematics of outer hochebenkar rock glacier based on geodetic measurements. *Geografiska Annaler. Series A, Physical Geography*, **98**(2), 129–141, doi:10.1111/geoa.12127.

Hartmann, D.L. et al., 2013: Observations: Atmosphere and surface. In: Climate Change 2013: The Physical Science Basis. Contribution of Working Group I to the Fifth Assessment Report of the Intergovernmental Panel on Climate Change [Stocker, T.F., D. Qin, G.-K. Plattner, M. Tignor, S.K. Allen, J. Boschung, A. Nauels, Y. Xia, V. Bex and P.M. Midgley (eds.)]. Cambridge University Press, Cambridge, United Kingdom and New York, NY, USA, 159–254.

Hartmann, H. and L. Andresky, 2013: Flooding in the Indus River basin – A spatiotemporal analysis of precipitation records. *Glob. Planet. Change*, **107**, 25–35, doi:10.1016/j.gloplacha.2013.04.002.

Hasler, A., S. Gruber, M. Font and A. Dubois, 2011: Advective heat transport in frozen rock clefts: Conceptual model, laboratory experiments and numerical simulation. *Permafrost Periglac.*, **22**(4), 378–389, doi:10.1002/ppp.737.

Hauer, C. et al., 2018: State of the art, shortcomings and future challenges for a sustainable sediment management in hydropower: A review. *Renew. Sust. Energ. Rev.*, **98**, 40–55, doi:10.1016/j.rser.2018.08.031.

Hawkings, J. et al., 2016: The Greenland Ice Sheet as a hot spot of phosphorus weathering and export in the Arctic. *Global Biogeochem. Cy.*, **30**(2), 191–210, doi:10.1002/2015GB005237.

He, X., K.S. Burgess, L.M. Gao and D.Z. Li, 2019: Distributional responses to climate change for alpine species of Cyananthus and Primula endemic to the Himalaya-Hengduan Mountains. *Plant Diversity*, **41**, 26–32, doi:10.1016/j.pld.2019.01.004.

Hegglin, E. and C. Huggel, 2008: An integrated assessment of vulnerability to glacial hazards. A case study in the Cordillera Blanca, Peru. *Mt. Res. Dev.*, **28**(3–4), 299–309, doi:10.1659/mrd.0976.

Heikkinen, A., 2017: Climate change in the Peruvian Andes: A case study on small-scale farmers' Vulnerability in the Quillcay River Basin. *Iberoamericana – Nordic Journal of Latin American and Caribbean Studies*, **46**(1), 77–88, doi:10.16993/iberoamericana.211.

Held, I.M. and B.J. Soden, 2006: Robust responses of the hydrological cycle to global warming. *J. Clim.*, **19**(21), 5686–5699, doi:10.1175/JCLI3990.1.

Hermanns, R.L., T. Oppikofer, N.J. Roberts and G. Sandoy, 2014: Catalogue of historical displacement waves and landslide-triggered tsunamis in Norway. In *Engineering Geology for Society and Territory, Vol 4: Marine and Coastal Processes* [Lollino, G., A. Manconi, J. Locat, Y. Huang, and M. Canals Artigas (eds.)]. pp 63–66, doi:10.1007/978-3-319-08660-6_13.

Hewitt, K., 2007: Tributary glacier surges: an exceptional concentration at Panmah Glacier, Karakoram Himalaya. *J. Glaciol.*, **53**(181), 181–188, doi:10.3189/172756507782202829.

Higman, B. et al., 2018: The 2015 landslide and tsunami in Taan Fiord, Alaska. *Sci. Rep.*, **8**, 12993, doi:10.1038/s41598-018-30475-w.

Hilbich, C. et al., 2008: Monitoring mountain permafrost evolution using electrical resistivity tomography: A 7-year study of seasonal, annual, and long-term variations at Schilthorn, Swiss Alps. *J. Geophys. Res-Earth.*, **113**(F1), 1–12, doi:10.1029/2007JF000799.

Hill, A., C. Minbaeva, A. Wilson and R. Satylkanov, 2017: Hydrologic controls and water vulnerabilities in the Naryn River Basin, Kyrgyzstan: A socio-hydro case study of water stressors in Central Asia. *Water*, **9**(5), 325, doi:10.3390/w9050325.

Hill, M., 2013: Adaptive capacity of water governance: Cases from the Alps and the Andes. *Mt. Res. Dev.*, **33**(3), 248–259, doi:10.1659/MRD-JOURNAL-D-12-00106.1.

Hipp, T. et al., 2012: Modelling borehole temperatures in Southern Norway-insights into permafrost dynamics during the 20th and 21st century. *The Cryosphere*, **6**(3), 553–571, doi:10.5194/tc-6-553-2012.

Hirabayashi, Y. et al., 2016: Contributions of natural and anthropogenic radiative forcing to mass loss of Northern Hemisphere mountain glacie Mrs and quantifying their uncertainties. *Sci. Rep.*, **6**, 29723, doi:10.1038/srep29723.

Hirabayashi, Y. et al., 2013: Projection of glacier mass changes under a high-emission climate scenario using the global glacier model HYOGA2. *Hydrol. Res. Lett.*, **7**(1), 6–11, doi:10.3178/hrl.7.6.

Hock, R. et al., 2019: GlacierMIP – A model intercomparison of global-scale glacier mass-balance models and projections. *J. Glaciol.*, **65**(251), 453-467, doi:10.1017/jog.2019.22.

Hodson, A.J., 2014: Understanding the dynamics of black carbon and associated contaminants in glacial systems. *WiRes. Water*, **1**(2), 141–149, doi:10.1002/wat2.1016.

Hoegh-Guldberg, O. et al., 2018: Impacts of 1.5°C global warming on natural and human systems. In: Global Warming of 1.5°C. An IPCC Special Report on the impacts of global warming of 1.5°C above pre-industrial levels and related global greenhouse gas emission pathways, in the context of strengthening the global response to the threat of climate change, sustainable development, and efforts to eradicate poverty [Masson-Delmotte, V., P. Zhai, H.-O. Pörtner, D. Roberts, J. Skea, P.R. Shukla, A. Pirani, W. Moufouma-Okia, C. Péan, R. Pidcock, S. Connors, J.B.R. Matthews, Y. Chen, X. Zhou, M.I. Gomis, E. Lonnoy, T. Maycock, M. Tignor and T. Waterfield (eds.)]. In Press.

Hojesky, H. et al., 2019: *Alpine Climate Target System 2050 – approved by the XV Alpine Conference*. Alpine Climate Board of the Alpine Convention, Innsbruck. www.alpconv.org/en/organization/groups/AlpineClimateBoard/Documents/20190404_ACB_AlpineClimateTargetSystem2050_en.pdf. Accessed on 06/08/2019.

Hood, E. et al., 2015: Storage and release of organic carbon from glaciers and ice sheets. *Nat. Geosci.*, **8**(2), 91–96, doi:10.1038/ngeo2331.

Hood, E. et al., 2009: Glaciers as a source of ancient and labile organic matter to the marine environment. *Nature*, **462**(7276), 1044-U100, doi:10.1038/nature08580.

Hopkins, D. and K. Maclean, 2014: Climate change perceptions and responses in Scotland's ski industry. *Tourism Geographies*, **16**, 400–414, doi:10.1080/14616688.2013.823457.

Hotaling, S. et al., 2018: Demographic modelling reveals a history of divergence with gene flow for a glacially tied stonefly in a changing post-Pleistocene landscape. *J. Biogeogr.*, **45**(2), 304–317, doi:10.1111/jbi.13125.

Hubbard, S.S. et al., 2018: The East River, Colorado, Watershed: A mountainous community testbed for improving predictive understanding of multiscale hydrological–biogeochemical dynamics. *Vadose Zone Journal*, **17**(1), 180061, doi:10.2136/vzj2018.03.0061.

Huggel, C., M. Carey, J.J. Clague and A. Kääb (eds.), 2015a: *The high-mountain cryosphere: Environmental changes and human risks.* Cambridge University Press, Cambridge. 363 pp. ISBN 9781107065840.

Huggel, C. et al., 2019: Loss and Damage in the mountain cryosphere. *Reg. Environ. Change*, **19**(5), 1387–1399, doi:10.1007/s10113-018-1385-8.

Huggel, C. et al., 2015b: A framework for the science contribution in climate adaptation: Experiences from science-policy processes in the Andes. *Environ. Sci. Policy*, **47**, 80–94, doi:10.1016/j.envsci.2014.11.007.

Huggel, C. et al., 2005: The 2002 rock/ice avalanche at Kolka/Karmadon, Russian Caucasus: assessment of extraordinary avalanche formation and mobility, and application of QuickBird satellite imagery. *Nat. Hazard. Earth Sys.*, **5**(2), 173–187, doi:10.5194/nhess-5-173-2005.

Hummel, S., 2017: Relative water scarcity and country relations along cross-boundary rivers: Evidence from the Aral Sea basin. *International Studies Quarterly*, **61**(4), 795–808, doi:10.1093/isq/sqx043.

Huntington, H.P. et al., 2017: How small communities respond to environmental change: patterns from tropical to polar ecosystems. *Ecol. Soc.* **22**(3), 9.

Huss, M. and M. Fischer, 2016: Sensitivity of very small glaciers in the Swiss Alps to future climate change. *Front. Earth Sci.*, **4**, 34, doi:10.3389/feart.2016.00034.

Huss, M. and R. Hock, 2015: A new model for global glacier change and sea-level rise. *Front. Earth Sci.*, **3**, 54, doi:10.3389/feart.2015.00054.

Huss, M. and R. Hock, 2018: Global-scale hydrological response to future glacier mass loss. *Nat. Clim. Change*, **8**(2), 135–140, doi:10.1038/s41558-017-0049-x.

Hussain, A. et al., 2018: Climate change-induced hazards and local adaptations in agriculture: a study from Koshi River Basin, Nepal. *Nat. Hazards*, **91**(3), 1365–1383, doi:10.1007/s11069-018-3187-1.

ICIMOD, 2011: *Glacial Lakes and Glacial Lake Outburst Floods in Nepal.* ICIMOD, Kathmandu. [Available at: http://lib.icimod.org/record/27755]. Accessed 06/08/2019.

IHA, 2018: *Hydropower status report 2018.* International Hydropower Assocation, Sutton, United Kingdom. www.hydropower.org/publications/2018-hydropower-status-report. Accessed 06/08/2019.

IHCAP, 2017: *Mountain and Lowland Linkages: A Climate Change Perspective in the Himalayas.* Indian Himalayas Climate Adaptation Programme (IHCAP). [Available at: http://ihcap.in/?media_dl=872]. Accessed 06/08/2019.

Ikeda, A. and N. Matsuoka, 2002: Degradation of talus-derived rock glaciers in the upper engadin, Swiss alps. *Permafrost Periglac.*, **13**(2), 145–161, doi:10.1002/ppp.413.

Il Jeong, D. and L. Sushama, 2018: Rain-on-snow events over North America based on two Canadian regional climate models. *Clim. Dyn.*, **50**(1–2), 303–316, doi:10.1007/s00382-017-3609-x.

Ilyashuk, B.P. et al., 2018: Rock glaciers in crystalline catchments: Hidden permafrost-related threats to alpine headwater lakes. *Glob. Change Biol*, **24**(4), 1548–1562, doi:10.1111/gcb.13985.

Imperio, S., R. Bionda, R. Viterbi and A. Provenzale, 2013: Climate Change and Human Disturbance Can Lead to Local Extinction of Alpine Rock Ptarmigan: New Insight from the Western Italian Alps. *PLOS ONE*, **8**(11), doi:10.1371/journal.pone.0081598.

Ingty, T., 2017: High mountain communities and climate change: adaptation, traditional ecological knowledge, and institutions. *Clim. Change*, **145**(1–2), 41–55, doi:10.1007/s10584-017-2080-3.

IPCC, 2018: Summary for Policymakers. In: Global Warming of 1.5°C. An IPCC Special Report on the impacts of global warming of 1.5°C above pre-industrial levels and related global greenhouse gas emission pathways, in the context of strengthening the global response to the threat of climate change, sustainable development, and efforts to eradicate poverty [Masson-Delmotte, V., P. Zhai, H. Pörtner, D. Roberts, J. Skea, P. Shukla, A. Pirani, W. Moufouma-Okia, C. Péan, R. Pidcock, S. Connors, J.B.R. Matthews, Y. Chen, X. Zhou, M.I. Gomis, E. Lonnoy, Maycock, M. Tignor and T. Waterfield (eds.)]. World Meteorological Organization, Geneva, Switzerland, 32 pp.

Isaak, D.J. et al., 2016: Slow climate velocities of mountain streams portend their role as refugia for cold-water biodiversity. *PNAS*, **113**(16), 4374–4379, doi:10.1073/pnas.1522429113.

Isaksen, K., P. Holmlund, J.L. Sollid and C. Harris, 2001: Three deep alpine-permafrost boreholes in Svalbard and Scandinavia. *Permafrost Periglac.*, **12**(1), 13–25, doi:10.1002/ppp.380.

Isaksen, K. et al., 2011: Degrading Mountain Permafrost in Southern Norway: Spatial and Temporal Variability of Mean Ground Temperatures, 1999–2009. *Permafrost Periglac.*, **22**, 361–377, doi:10.1002/ppp.728.

Isaksen, K., J.L. Sollid, P. Holmlund and C. Harris, 2007: Recent warming of mountain permafrost in Svalbard and Scandinavia. *J. Geophys. Res-Earth.*, **112**(2), 235, doi:10.1029/2006JF000522.

Islam, S.U., S.J. Déry and A.T. Werner, 2017: Future climate change impacts on snow and water resources of the Fraser River Basin, British Columbia. *J. Hydrometeorol.*, **18**(2), 473–496, doi:10.1175/JHM-D-16-0012.1.

Ives, J.D., R.B. Shrestha and P.K. Mool, 2010: *Formation of glacial lakes in the Hindu Kush-Himalayas and GLOF risk assessment.* ICIMOD, Kathmandu. www.unisdr.org/files/14048_ICIMODGLOF.pdf. Accessed 06/08/2019.

Jackson, M. and G. Ragulina, 2014: *Inventory of glacier-related hazardous events in Norway. Report no. 83 – 2014.* Norwegian Water Resources and Energy Directorate. NVE, Oslo. [Available at: http://asp.bibliotekservice.no/nve/title.aspx?tkey=22514]. Accessed 06/08/2019.

Jacob, D. et al., 2014: EURO-CORDEX: New high-resolution climate change projections for European impact research. *Reg. Environ. Change*, doi:10.1007/s10113-013-0499-2.

Jacobsen, D. et al., 2014: Runoff and the longitudinal distribution of macroinvertebrates in a glacier-fed stream: implications for the effects of global warming. *Freshw. Biol.*, **59**(10), 2038--2050, doi:10.1111/fwb.12405.

Jacobsen, D., A.M. Milner, L.E. Brown and O. Dangles, 2012: Biodiversity under threat in glacier-fed river systems. *Nat. Clim. Change*, **2**(5), 361–364, doi:10.1038/nclimate1435.

Jalilov, S.-M., S.A. Amer and F.A. Ward, 2018: Managing the water-energy-food nexus: Opportunities in Central Asia. *J. Hydrol.*, **557**, 407–425, doi:10.1016/j.jhydrol.2017.12.040.

Jamieson, S.S.R., M.W. Ewertowski and D.J.A. Evans, 2015: Rapid advance of two mountain glaciers in response to mine-related debris loading. *J. Geophys. Res-Earth.*, **120**(7), 1418–1435, doi:10.1002/2015JF003504.

Jamir, O., 2016: Understanding India-Pakistan water politics since the signing of the Indus Water Treaty. *Water Policy*, **18**(5), 1070–1087, doi:10.2166/wp.2016.185.

Jenicek, M., J. Seibert and M. Staudinger, 2018: Modeling of future changes in seasonal snowpack and impacts on summer low flows in Alpine catchments. *Water Resour. Res.*, **54**, 538–556, doi:10.1002/2017WR021648.

Ji, Z.-M., 2016: Modeling black carbon and its potential radiative effects over the Tibetan Plateau. *Adv. Clim. Change Res.*, **7**(3), 139–144, doi:10.1016/J.ACCRE.2016.10.002.

Jin, H. et al., 2009: Changes in frozen ground in the source area of the Yellow River on the Qinghai-Tibet Plateau, China, and their eco-environmental impacts. *Environ. Res. Lett.*, **4**(4), doi:10.1088/1748-9326/4/4/045206.

Johnston, A.N. et al., 2019: Ecological consequences of anomalies in atmospheric moisture and snowpack. *Ecology*, **100**(4), doi:10.1002/ecy.2638.

Johnstone, J.F. et al., 2016: Changing disturbance regimes, ecological memory, and forest resilience. *Front. Ecol. Environ.*, **14**(7), 369–378, doi:10.1002/fee.1311.

Jomelli, V. et al., 2009: Impacts of future climatic change (2070–2099) on the potential occurrence of debris flows: A case study in the Massif des Ecrins (French Alps). *Clim. Change*, **97**(1–2), 171–191, doi:10.1007/s10584-009-9616-0.

Jones, B. and B.C. O'Neill, 2016: Spatially explicit global population scenarios consistent with the Shared Socioeconomic Pathways. *Environ. Res. Lett.*, **11**(2016), 084003, doi:10.1088/1748-9326/11/8/084003.

Jones, D.B., S. Harrison, K. Anderson and R.A. Betts, 2018: Mountain rock glaciers contain globally significant water stores. *Sci. Rep.*, **8**, 2834, doi:10.1038/s41598-018-21244-w.

Jordan, S. et al., 2016: Loss of genetic diversity and increased subdivision in an endemic Alpine stonefly threatened by climate change. *PLOS ONE*, **11**(6), e0157386, doi:10.1371/journal.pone.0157386.

Joshi, S. et al., 2013: Herders' perceptions of and responses to climate change in Northern Pakistan. *Environ. Manage.*, **52**(3), 639–648, doi:10.1007/s00267-013-0062-4.

Jost, G., R. Moore, B. Menounos and R. Wheate, 2012: Quantifying the contribution of glacier runoff to streamflow in the upper Columbia River Basin, Canada. *Hydrol. Earth Syst. Sc.*, **16**(3), 849–860, doi:10.5194/hess-16-849-2012.

Jost, G. and F. Weber, 2013: *Potential Impacts of Climate Change on BC Hydro's Water Resources*. BC Hydro, Canada. www.bchydro.com/content/dam/hydro/medialib/internet/documents/about/climate_change_report_2012.pdf. Accessed on 06/08/2019.

Jurt, C. et al., 2015: Local perceptions in climate change debates: insights from case studies in the Alps and the Andes. *Clim. Change*, **133**(3), 511–523, doi:10.1007/s10584-015-1529-5.

Kääb, A., R. Frauenfelder and I. Roer, 2007: On the response of rockglacier creep to surface temperature increase. *Glob. Planet. Change*, **56**(1), 172–187, doi:10.1016/j.gloplacha.2006.07.005.

Kääb, A. et al., 2018: Massive collapse of two glaciers in western Tibet in 2016 after surge-like instability. *Nat. Geosci.*, **11**(2), 114–120, doi:10.1038/s41561-017-0039-7.

Kaenzig, R., 2015: Can glacial retreat lead to migration? A critical discussion of the impact of glacier shrinkage upon population mobility in the Bolivian Andes. *Popul. Environ.*, **36**(4), 480–496, doi:10.1007/s11111-014-0226-z.

Kaenzig, R., M. Rebetez and G. Serquet, 2016: Climate change adaptation of the tourism sector in the Bolivian Andes. *Tourism Geographies*, **18**(2), 111–128, doi:10.1080/14616688.2016.1144642.

Kainz, M.J., R. Ptacnik, S. Rasconi and H.H. Hager, 2017: Irregular changes in lake surface water temperature and ice cover in subalpine Lake Lunz, Austria. *Inland Waters*, **7**(1), 27–33, doi:10.1080/20442041.2017.1294332.

Kala, C.P., 2014: Deluge, disaster and development in Uttarakhand Himalayan region of India: Challenges and lessons for disaster management. *Int. J. Dis. Risk. Re.*, **8**, 143–152, doi:10.1016/j.ijdrr.2014.03.002.

Kapnick, S. and A. Hall, 2012: Causes of recent changes in western North American snowpack. *Clim. Dyn.*, **38**(9–10), 1885–1899, doi:10.1007/s00382-011-1089-y.

Kapnick, S.B. and T.L. Delworth, 2013: Controls of global snow under a changed climate. *J. Clim.*, **26**(15), 5537–5562, doi:10.1175/JCLI-D-12-00528.1.

Kapnick, S.B. et al., 2014: Snowfall less sensitive to warming in Karakoram than in Himalayas due to a unique seasonal cycle. *Nat. Geosci.*, **7**(11), 834–840, doi:10.1038/ngeo2269.

Kargel, J.S. et al., 2016: Geomorphic and geologic controls of geohazards induced by Nepal's 2015 Gorkha earthquake. *Science*, **351**(6269), aac8353, doi:10.1126/science.aac8353.

Kaspari, S. et al., 2014: Seasonal and elevational variations of black carbon and dust in snow and ice in the Solu-Khumbu, Nepal and estimated radiative forcings. *Atmos. Chem. Phys.*, **14**(15), 8089–8103, doi:10.5194/acp-14-8089-2014.

Katsuyama, Y., M. Inatsu, K. Nakamura and S. Matoba, 2017: Global warming response of snowpack at mountain range in northern Japan estimated using multiple dynamically downscaled data. *Cold Reg. Sci. Technol.*, **136**, 62–71. doi:10.1016/j.coldregions.2017.01.006.

Keiler, M. and S. Fuchs, 2018: Challenges for natural hazard and risk management in mountain regions of Europe. In: *Oxford Research Encyclopedia of Natural Hazard Science*. Oxford University Press, Oxford. doi:10.1093/acrefore/9780199389407.013.322.

Kenner, R. et al., 2017: Factors controlling velocity variations at short-term, seasonal and multiyear time scales, Ritigraben Rock Glacier, western Swiss Alps. *Permafrost Periglac.*, **28**(4), 675–684, doi:10.1002/ppp.1953.

Keuschnig, M. et al., 2015: Permafrost-Related Mass Movements: Implications from a Rock Slide at the Kitzsteinhorn, Austria. In: *Engineering Geology for Society and Territory*, pp. 255–259. doi.org/10.1007/978-3-319-09300-0_48.

Khadim, A.N., 2016: Defending glaciers in Argentina. *Peace Review*, **28**(1), 65–75, doi:10.1080/10402659.2016.1130383.

Khamis, K., L.E. Brown, D.M. Hannah and A.M. Milner, 2015: Experimental evidence that predator range expansion modifies alpine stream community structure. *Freshw. Sci.* **34**(1), 66–80, doi:10.1086/679484.

Khamis, K., L.E. Brown, D.M. Hannah and A.M. Milner, 2016: Glacier-groundwater stress gradients control alpine river biodiversity. *Ecohydrology*, **9**(7), 1263–1275, doi:10.1002/eco.1724.

Khanal, N.R., J.-M. Hu and P. Mool, 2015: Glacial lake outburst flood risk in the Poiqu/Bhote Koshi/Sun Koshi river basin in the Central Himalayas. *Mt. Res. Dev.*, **35**(4), 351–364, doi:10.1659/MRD-JOURNAL-D-15-00009.

Kienholz, C. et al., 2017: Mass balance evolution of black rapids glacier, Alaska, 1980–2100, and its implications for surge recurrence. *Front. Earth Sci.*, **5**, 56, doi:10.3389/feart.2017.00056.

Kirkpatrick, J.B. et al., 2017: Causes and consequences of variation in snow incidence on the high mountains of Tasmania, 1983–2013. *Aust. J. Bot.*, **65**(3), 214–224, doi:10.1071/BT16179.

Kissel, A.M., W.J. Palen, M.E. Ryan and M.J. Adams, 2019: Compounding effects of climate change reduce population viability of a montane amphibian. *Ecol. Appl.*, **29**(2), e01832, doi:10.1002/eap.1832.

Kitzberger, T., D.A. Falk, A.L. Westerling and T.W. Swetnam, 2017: Direct and indirect climate controls predict heterogeneous early-mid 21st century wildfire burned area across western and boreal North America. *PLOS ONE*, **12**(12), e0188486, doi:10.1371/journal.pone.0188486.

Knowles, J.F., N.P. Molotch, E. Trujillo and M.E. Litvak, 2018: Snowmelt-driven trade-offs between early and late season productivity negatively impact forest carbon uptake during drought. *Geophys. Res. Lett.*, **45**(7), 3087–3096, doi:10.1002/2017GL076504.

Konchar, K.M. et al., 2015: Adapting in the shadow of Annapurna: a climate tipping point. *J. Ethnobiol.*, **35**(3), 449–471, doi:10.2993/0278-0771-35.3.449.

Kormann, C., T. Francke, M. Renner and A. Bronstert, 2015: Attribution of high resolution streamflow trends in Western Austria – An approach based on climate and discharge station data. *Hydrol. Earth Syst. Sc.*, **19**(3), 1225–1245, doi:10.5194/hess-19-1225-2015.

Korup, O., T. Gorum and Y. Hayakawa, 2012: Without power? Landslide inventories in the face of climate change. *Earth Surf. Process. Landf.*, **37**(1), 92–99, doi:10.1002/esp.2248.

Kos, A. et al., 2016: Contemporary glacier retreat triggers a rapid landslide response, Great Aletsch Glacier, Switzerland. *Geophys. Res. Lett.*, **43**(24), 12466–12474, doi:10.1002/2016GL071708.

Kraaijenbrink, P.D.A., M.F.P. Bierkens, A.F. Lutz and W.W. Immerzeel, 2017: Impact of a global temperature rise of 1.5 degrees Celsius on Asia's glaciers. *Nature*, **549**(7671), 257–260, doi:10.1038/nature23878.

2

Krautblatter, M., D. Funk and F.K. Guenzel, 2013: Why permafrost rocks become unstable: a rock-ice-mechanical model in time and space. *Earth Surf. Process. Landf.*, **38**(8), 876–887, doi:10.1002/esp.3374.

Kriegel, D. et al., 2013: Changes in glacierisation, climate and runoff in the second half of the 20th century in the Naryn basin, Central Asia. *Glob. Planet. Change*, **110**, 51–61, doi:10.1016/j.gloplacha.2013.05.014.

Kronenberg, J., 2013: Linking ecological economics and political ecology to study mining, glaciers and global warming. *Environmental Policy and Governance*, **23**(2), 75–90, doi:10.1002/eet.1605.

Kropáček, J. et al., 2013: Analysis of ice phenology of lakes on the Tibetan Plateau from MODIS data. *The Cryosphere*, **7**(1), 287–301, doi:10.5194/tc-7-287-2013.

Kuang, X. and J.J. Jiao, 2016: Review on climate change on the Tibetan Plateau during the last half century. *J. Geophys. Res-Atmos.*, **121**(8), 3979–4007, doi:10.1002/2015JD024728.

Kulonen, K., C. Adler, C. Bracher and S. Wymann von Dach, 2019: Spatial context matters for monitoring and reporting on SDGs: Reflections based on research in mountain regions. *GAIA*, **28**(2), 90–94. doi:10.14512/gaia.28.2.5.

Kummert, M., R. Delaloye and L. Braillard, 2017: Erosion and sediment transfer processes at the front of rapidly moving rock glaciers: Systematic observations with automatic cameras in the western Swiss Alps. *Permafrost Periglac.*, **29**(1), 21–33, doi:10.1002/ppp.1960.

Lacelle, D., A. Brooker, R.H. Fraser and S.V. Kokelj, 2015: Distribution and growth of thaw slumps in the Richardson Mountains-Peel Plateau region, northwestern Canada. *Geomorphology*, **235**, 40–51, doi:10.1016/j.geomorph.2015.01.024.

Lafaysse, M. et al., 2014: Internal variability and model uncertainty components in future hydrometeorological projections: The Alpine Durance basin. *Water Resour. Res.*, **50**(4), 3317–3341, doi:10.1002/2013WR014897.

Lamontagne-Hallé, P., J.M. McKenzie, B.L. Kurylyk and S.C. Zipper, 2018: Changing groundwater discharge dynamics in permafrost regions. *Environ. Res. Lett*, **13**(8), 084017, doi:10.1088/1748-9326/aad404.

Lamprecht, A. et al., 2018: Climate change leads to accelerated transformation of high-elevation vegetation in the central Alps. *New Phytologist*, **220**(2), 447–459, doi:10.1111/nph.15290.

Lane, S.N. et al., 2017: Sediment export, transient landscape response and catchment-scale connectivity following rapid climate warming and Alpine glacier recession. *Geomorphology*, **277**, 210–227, doi:10.1016/j.geomorph.2016.02.015.

Langford, H., A.J. Hodson, S. Banwart and C.E. Bøggild, 2010: The microstructure and biogeochemistry of Arctic cryoconite granules. *Ann. Glaciol.*, **51**(56), 87–94, doi:10.3189/172756411795932083.

Larsen, C. et al., 2015: Surface melt dominates Alaska glacier mass balance. *Geophys. Res. Lett.*, **42**(14), 5902–5908, doi:10.1002/2015GL064349.

Lavigne, A., N. Eckert, L. Bel and E. Parent, 2015: Adding expert contributions to the spatiotemporal modelling of avalanche activity under different climatic influences. *J.R. Stat. Soc. C-Appl.*, **64**(4), 651–671, doi:10.1111/rssc.12095.

Lavoie, R.A. et al., 2013: Biomagnification of Mercury in Aquatic Food Webs: A Worldwide Meta-Analysis. *Environ. Sci. Technol.*, **47**(23), 13385–13394. doi:10.1021/es403103t.

Lawrimore, J.H. et al., 2011: An overview of the Global Historical Climatology Network monthly mean temperature data set, version 3. *J. Geopyhs. Res.*, **116**(D19), 1785, doi:10.1029/2011JD016187.

Lazar, B. and M. Williams, 2008: Climate change in western ski areas: Potential changes in the timing of wet avalanches and snow quality for the Aspen ski area in the years 2030 and 2100. *Cold Reg. Sci. Technol.*, **51**(2–3), 219–228. doi:10.1016/j.coldregions.2007.03.015.

Leach, J.A. and R.D. Moore, 2014: Winter stream temperature in the rain-on-snow zone of the Pacific Northwest: influences of hillslope runoff and transient snow cover. *Hydrol. Earth Syst. Sci.*, **18**(2), 819–838, doi:10.5194/hess-18-819-2014.

Lee, C.M., 2012: Withering snow and ice in the mid-latitudes: A new archaeological and paleobiological record for the Rocky Mountain region. *Arctic*, **65**(5), 165–177, doi:10.14430/arctic4191.

Lee, S.-Y., A.F. Hamlet and E.E. Grossman, 2016: Impacts of climate change on regulated streamflow, hydrologic extremes, hydropower production, and sediment discharge in the Skagit river basin. *Northwest Sci.*, **90**(1), 23–43, doi:10.3955/046.090.0104.

Lencioni, V., 2018: Glacial influence and stream macroinvertebrate biodiversity under climate change: Lessons from the Southern Alps. *Sci. Total Environ.*, **622**, 563–575, doi:10.1016/j.scitotenv.2017.11.266.

Lencioni, V., O. Jousson, G. Guella and P. Bernabo, 2015: Cold adaptive potential of chironomids overwintering in a glacial stream. *Physiol. Entomol.*, **40**(1), 43–53, doi:10.1111/phen.12084.

Lesica, P., 2014: Arctic-Alpine plants decline over two decades in Glacier National Park, Montana, U.S.A. *Arct. Antarct. Alp. Res.*, **46**(2), 327–332, doi:10.1657/1938-4246-46.2.327.

Lesica, P. and E.E. Crone, 2016: Arctic and boreal plant species decline at their southern range limits in the Rocky Mountains. *Ecol. Letters*, **20**(2), 166–174, doi:10.1111/ele.12718.

Li, C. et al., 2016: Sources of black carbon to the Himalayan-Tibetan Plateau glaciers. *Nat. Commun.*, **7**(1), 12574, doi:10.1038/ncomms12574.

Li, J. et al., 2017: Evidence for persistent organic pollutants released from melting glacier in the central Tibetan Plateau, China. *Environ. Pollut.*, **220**, 178–185, doi:10.1016/j.envpol.2016.09.037.

Li, X. et al., 2018: Importance of mountain glaciers as a source of dissolved organic carbon. *J. Geopyhs. Res. F- Earth Surface*, **24**(10), GB4033, doi:10.1029/2017JF004333.

Liang, Q. et al., 2018: Shifts in plant distributions in response to climate warming in a biodiversity hotspot, the Hengduan Mountains. *J. Biogeogr.*, **45**, 1334–1344, doi:10.1111/jbi.13229.

Lin, H. et al., 2017: A decreasing glacier mass balance gradient from the edge of the Upper Tarim Basin to the Karakoram during 2000–2014. *Sci. Rep.*, **7**(1), 6712, doi:10.1038/s41598-017-07133-8.

Lin, Z., J. Luo and F. Niu, 2016: Development of a thermokarst lake and its thermal effects on permafrost over nearly 10 yr in the Beiluhe Basin, Qinghai-Tibet Plateau. *Geosphere*, **12**(2), 632–643, doi:10.1130/GES01194.1.

Linsbauer, A. et al., 2016: Modelling glacier-bed overdeepenings and possible future lakes for the glaciers in the Himalaya-Karakoram region. *Ann. Glaciol.*, **57**(71), 119–130, doi:10.3189/2016AoG71A627.

Littell, J.S., 2018: Drought and fire in the western USA: is climate attribution enough? *Curr. Clim. Chang. Rep.*, **4**(4), 396–406, doi:10.1007/s40641-018-0109-y.

Littell, J.S., D. Mckenzie, H.Y. Wan and S.A. Cushman, 2018: Climate change and future wildfire in the western United States: an ecological approach to nonstationarity. *Earth's Future*, **6**(8), 1097–1111, doi:10.1029/2018EF000878.

Liu, G. et al., 2017: Permafrost warming in the context of step-wise climate change in the Tien Shan Mountains, China. *Permafrost Periglac.*, **28**(1), 130–139, doi:10.1002/ppp.1885.

Liu, X., Z. Cheng, L. Yan and Z.-Y. Yin, 2009: Elevation dependency of recent and future minimum surface air temperature trends in the Tibetan Plateau and its surroundings. *Glob. Planet. Change*, **68**(3), 164–174, doi:10.1016/j.gloplacha.2009.03.017.

Liu, Y. et al., 2016: Storage of dissolved organic carbon in Chinese glaciers. *J. Glaciol.*, **62**(232), 402–406, doi:10.1017/jog.2016.47.

Loriaux, T. and G. Casassa, 2013: Evolution of glacial lakes from the Northern Patagonia Icefield and terrestrial water storage in a sea-level rise context. *Glob. Planet. Change*, **102**, 33–40, doi:10.1016/j.gloplacha.2012.12.012.

Lozny, L.R., 2013: *Continuity and Change in Cultural Adaptation to Mountain Environments*. Springer New York Heidelberg Dordrecht London, New York, 410 pp.

Lu, Q., D. Zhao and S. Wu, 2017: Simulated responses of permafrost distribution to climate change on the Qinghai-Tibet Plateau. *Sci. Rep.*, **7**(1), 3845, doi:10.1038/s41598-017-04140-7.

Luethi, R., S. Gruber and L. Ravanel, 2015: Modelling transient ground surface temperatures of past rockfall events: Towards a better understanding of failure mechanisms in changing periglacial environments. *Geografiska Annaler. Series A, Physical Geography*, **97**(4), 753–767, doi:10.1111/geoa.12114.

Luethi, R., M. Phillips and M. Lehning, 2017: Estimating non-conductive heat flow leading to intra-permafrost talik formation at the Ritigraben Rock Glacier (Western Swiss Alps). *Permafrost Periglac.*, **28**(1), 183–194, doi:10.1002/ppp.1911.

Lugon, R. and M. Stoffel, 2010: Rock-glacier dynamics and magnitude-frequency relations of debris flows in a high-elevation watershed: Ritigraben, Swiss Alps. *Glob. Planet. Change*, **73**(3), 202–210, doi:10.1016/j.gloplacha.2010.06.004.

Lutz, A. et al., 2016: Climate change impacts on the upper Indus hydrology: Sources, shifts and extremes. *PLOS ONE*, **11**(11), e0165630, doi:10.1371/journal.pone.0165630.

Lynch, B.D., 2012: Vulnerabilities, competition and rights in a context of climate change toward equitable water governance in Peru's Rio Santa Valley. *Glob. Environ. Change.*, **22**(2), 364–373, doi:10.1016/j.gloenvcha.2012.02.002.

Macfarlane, A., 1976: *Resources and population: A study of the Gurungs of Nepal*. Cambridge University Press, Cambridge, 384 pp. ISBN 101107406862.

Mackintosh, A.N. et al., 2017: Regional cooling caused recent New Zealand glacier advances in a period of global warming. *Nat. Commun.*, **8**, 14202, doi:10.1038/ncomms14202.

Madani, K. and J.R. Lund, 2010: Estimated impacts of climate warming on California's high-elevation hydropower. *Clim. Change*, **102**(3–4), 521–538, doi:10.1007/s10584-009-9750-8.

Magnin, F. et al., 2015: Thermal characteristics of permafrost in the steep alpine rock walls of the Aiguille du Midi (Mont Blanc Massif, 3842 m a.s.l). *The Cryosphere*, **9**(1), 109–121, doi:10.5194/tc-9-109-2015.

Magnin, F. et al., 2017: Modelling rock wall permafrost degradation in the Mont Blanc massif from the LIA to the end of the 21st century. *The Cryosphere*, **11**(4), 1813–1834, doi:10.5194/tc-11-1813-2017.

Mahoney, P.J. et al., 2018: Navigating snowscapes: scale-dependent responses of mountain sheep to snowpack properties. *Ecol. Appl.*, **28**(7), 1715–1729, doi:10.1002/eap.1773.

Mallory, C.D. and M.S. Boyce, 2018: Observed and predicted effects of climate change on Arctic caribou and reindeer. *Environ. Rev.*, **26**, 13–25, doi:10.1139/er-2017-0032.

Mamet, S.D. et al., 2017: Recent increases in permafrost thaw rates and areal loss of palsas in the western Northwest Territories, Canada. *Permafrost Periglac.*, **28**(4), 619–633, doi:10.1002/ppp.1951.

Mankin, J.S. and N.S. Diffenbaugh, 2015: Influence of temperature and precipitation variability on near-term snow trends. *Clim. Dyn.*, **45**(3–4), 1099–1116, doi:10.1007/s00382-014-2357-4.

Mark, B.G. et al., 2010: Climate change and tropical Andean glacier recession: Evaluating hydrologic changes and livelihood vulnerability in the Cordillera Blanca, Peru. *Ann. Am. Assoc. Geogr.*, **100**(4), 794–805, doi:10.1080/00045608.2010.497369.

Mark, B.G. et al., 2017: Glacier loss and hydro-social risks in the Peruvian Andes. *Glob. Planet. Change*, **159**, 61–76, doi:10.1016/j.gloplacha.2017.10.003.

Marke, T. et al., 2015: Scenarios of future snow conditions in Styria (Austrian Alps). *J. Hydrometeorol.*, **16**(1), 261–277, doi:10.1175/JHM-D-14-0035.1.

Marmy, A. et al., 2016: Semi-automated calibration method for modelling of mountain permafrost evolution in Switzerland. *The Cryosphere*, **10**(6), 2693–2719, doi:10.5194/tc-10-2693-2016.

Marmy, A., N. Salzmann, M. Scherler and C. Hauck, 2013: Permafrost model sensitivity to seasonal climatic changes and extreme events in mountainous regions. *Environ. Res. Lett.*, **8**(3), 035048, doi:10.1088/1748-9326/8/3/035048.

Marston, R.A., 2008: Land, life, and environmental change in mountains. *Ann. Am. Assoc. Geogr.*, **98**(3), 507–520, doi:10.1080/00045600802118491.

Marty, C., A.-M. Tilg and T. Jonas, 2017: Recent evidence of large-scale receding snow water equivalents in the European Alps. *J. Hydrometeorol.*, **18**(4), 1021–1031, doi:10.1175/JHM-D-16-0188.1.

Marzeion, B., A.H. Jarosch and J.M. Gregory, 2014: Feedbacks and mechanisms affecting the global sensitivity of glaciers to climate change. *The Cryosphere*, **8**(1), 59–71, doi:10.5194/tc-8-59-2014.

Marzeion, B., A.H. Jarosch and M. Hofer, 2012: Past and future sea-level change from the surface mass balance of glaciers. *The Cryosphere*, **6**(6), 1295–1322, doi:10.5194/tc-6-1295-2012.

Marzeion, B., G. Kaser, F. Maussion and N. Champollion, 2018: Limited influence of climate change mitigation on short-term glacier mass loss. *Nat. Clim. Change*, **8**(4), 305–308, doi:10.1038/s41558-018-0093-1.

Matteodo, M., K. Ammann, E.P. Verrecchia and P. Vittoz, 2016: Snowbeds are more affected than other subalpine–alpine plant communities by climate change in the Swiss Alps. *Ecol. Evol.*, **6**(19), 6969–6982, doi:10.1002/ece3.2354.

Matthews, J.A. and A.E. Vater, 2015: Pioneer zone geo-ecological change: Observations from a chronosequence on the Storbreen glacier foreland, Jotunheimen, southern Norway. *Catena*, **135**, 219–230, doi:10.1016/j.catena.2015.07.016.

Maxwell, J.D., A. Call and S.B. St Clair, 2019: Wildfire and topography impacts on snow accumulation and retention in montane forests. *For. Ecol. Manage.*, **432**, 256–263, doi:10.1016/j.foreco.2018.09.021.

McCabe, G.J. et al., 2007: Rain-on-snow events in the Western United States. *Bull. Am. Meterol. Soc.*, **88**(3), 319–328, doi:10.1175/BAMS-88-3-319.

McClung, D.M., 2013: The effects of El Niño and La Niña on snow and avalanche patterns in British Columbia, Canada, and central Chile. *J. Glaciol.*, **59**(216), 783–792, doi:10.3189/2013JoG12J192.

McColl, S.T., 2012: Paraglacial rock-slope stability. *Geomorphology*, **153–154**, 1–16, doi:10.1016/j.geomorph.2012.02.015.

McDonald, K.I., 1989: Impacts of glacier-related landslides on the settlement at Hopar, Karakoram Himalaya. *Ann. Glaciol.*, **13**, 185–188, doi:10.3189/S0260305500007862.

McDowell, G. et al., 2013: Climate-related hydrological change and human vulnerability in remote mountain regions: a case study from Khumbu, Nepal. *Reg. Environ. Change*, **13**(2), 299–310, doi:10.1007/s10113-012-0333-2.

McDowell, G. et al., 2019: Adaptation action and research in glaciated mountain systems: Are they enough to meet the challenge of climate change? *Glob. Environ. Change*, **54**, 19–30, doi:10.1016/j.gloenvcha.2018.10.012.

McDowell, G. and M.N. Koppes, 2017: Robust adaptation research in high mountains: Integrating the scientific, social, and ecological dimensions of glacio-hydrological change. *Water*, **9**(10), doi:10.3390/w9100739.

McDowell, G., E. Stephenson and J. Ford, 2014: Adaptation to climate change in glaciated mountain regions. *Clim. Change*, **126**(1–2), 77–91, doi:10.1007/s10584-014-1215-z.

Mcdowell, N.G. et al., 2018: Predicting chronic climate-driven disturbances and their mitigation. *Trends Ecol. Evol.*, **33**(1), 15–27, doi:10.1016/j.tree.2017.10.002.

McNabb, R.W. and R. Hock, 2014: Alaska tidewater glacier terminus positions, 1948–2012. *J. Geopyhs. Res.-Earth Surface*, **119**(2), 153–167, doi:10.1002/2013JF002915.

McNabb, R.W., R. Hock and M. Huss, 2015: Variations in Alaska tidewater glacier frontal ablation, 1985–2013. *J. Geophys. Res-Earth.*, **120**(1), 120–136. doi:10.1002/2014JF003276.

McNeeley, S.M., 2017: Sustainable climate change adaptation in Indian Country. *Weather Climate and Society*, **9**(3), 392–403, doi:10.1175/wcas-d-16-0121.1.

Mearns, L. et al., 2017: The NA-CORDEX dataset, version 1.0. NCAR Climate Data Gateway. Boulder, Colourado: doi:10.5065/D6SJ1JCH. Accessed 06/08/2019.

Medwedeff, W.G. and G.H. Roe, 2017: Trends and variability in the global dataset of glacier mass balance. *Clim. Dyn.*, **48**(9–10), 3085–3097, doi:10.1007/s00382-016-3253-x.

Ménégoz, M. et al., 2014: Snow cover sensitivity to black carbon deposition in the Himalayas: from atmospheric and ice core measurements to regional climate simulations. *Atmos. Chem. Phys.*, **14**(8), 4237–4249, doi:10.5194/acp-14-4237-2014.

Menounos, B. et al., 2013: Did rock avalanche deposits modulate the late Holocene advance of Tiedemann Glacier, southern Coast Mountains, British Columbia, Canada? *Earth Planet. Sci. Lett.*, **384**, 154–164, doi:10.1016/j.epsl.2013.10.008.

Menounos, B. et al., 2019: Heterogeneous changes in western north american glaciers linked to decadal variability in zonal wind strength. *Geophys. Res. Lett.*, **46**(1), 200–209, doi:10.1029/2018GL080942.

Merkle, J.A. et al., 2016: Large herbivores surf waves of green-up during spring. *Proc. R. Soc. B-Biol. Sci.*, **283**(1833), doi:10.1098/rspb.2016.0456.

Mernild, S.H. et al., 2013: Global glacier changes: a revised assessment of committed mass losses and sampling uncertainties. *The Cryosphere*, **7**(5), 1565–1577, doi:10.5194/tc-7-1565-2013.

Merrey, D.J. et al., 2018: Evolving high altitude livelihoods and climate change: a study from Rasuwa District, Nepal. *Food Security*, **10**(4), 1055–1071, doi:10.1007/s12571-018-0827-y.

Middleton, A.D. et al., 2018: Green-wave surfing increases fat gain in a migratory ungulate. *Oikos*, **127**(7), 1060–1068, doi:10.1111/oik.05227.

Milan, A. and R. Ho, 2014: Livelihood and migration patterns at different altitudes in the Central Highlands of Peru. *Clim. Dev.*, **6**(1), 69–76, doi:10.1080/17565529.2013.826127.

Millan, R., J. Mouginot and E. Rignot, 2017: Mass budget of the glaciers and ice caps of the Queen Elizabeth Islands, Canada, from 1991 to 2015. *Environ. Res. Lett.*, **12**(2), 024016, doi:10.1088/1748-9326/aa5b04.

Millar, C.I. and N.L. Stephenson, 2015: Temperate forest health in an era of emerging megadisturbance. *Science*, **349** (6250), 823–826, doi:10.1126/science.aaa9933.

Mills, L.S. et al., 2018: Winter colour polymorphisms identify global hot spots for evolutionary rescue from climate change. *Science*, **359**(6379), 1033–1036, doi:10.1126/science.aan8097.

Milner, A.M. et al., 2017: Glacier shrinkage driving global changes in downstream systems. *PNAS*, **114**(37), 9770–9778, doi:10.1073/pnas.1619807114.

Minder, J.R., T.W. Letcher and C. Liu, 2018: The character and causes of elevation-dependent warming in high-resolution simulations of Rocky Mountain climate change. *J. Clim.*, **31**(6), 2093–2113, doi:10.1175/JCLI-D-17-0321.1.

Minville, M., S. Krau, F. Brissette and R. Leconte, 2010: Behaviour and performance of a water resource system in Québec (Canada) under adapted operating policies in a climate change context. *Water Resour. Manage.*, **24**(7), 1333–1352, doi:10.1007/s11269-009-9500-8.

Mishra, A. et al., 2019: Adaptation to climate change in the Hindu Kush Himalaya: Stronger action urgently needed. In: The Hindu Kush Himalaya Assessment: Mountains, Climate Change, Sustainability and People [Wester, P., A. Mishra, A. Mukherji and A.B. Shrestha (eds.)]. Springer International Publishing, Cham, 457–490.

Mock, C.J., K.C. Carter and K.W. Birkeland, 2017: Some Perspectives on Avalanche Climatology. *A. Assoc. Am. Geog.*, **107**(2), 299–308, doi:10.1080/24694452.2016.1203285.

Molden, D.J. et al., 2014: Water infrastructure for the Hindu Kush Himalayas. *Int. J. Water Resour. D.*, **30**(1), 60–77, doi:10.1080/07900627.2013.859044.

Molina, L.T. et al., 2015: Pollution and its Impacts on the South American Cryosphere. *Earth's Future*, **3**, 345–369, doi:10.1002/2015EF000311.

Montana, E., H.P. Diaz and M. Hurlbert, 2016: Development, local livelihoods, and vulnerabilities to global environmental change in the South American Dry Andes. *Reg. Environ. Change*, **16**(8), 2215–2228, doi:10.1007/s10113-015-0888-9.

Moran-Tejéda, E., J.I. López-Moreno, M. Stoffel and M. Beniston, 2016: Rain-on-snow events in Switzerland: recent observations and projections for the 21st century. *Clim. Res.*, **71**(2), 111–125, doi:10.3354/cr01435.

Morueta-Holme, N. et al., 2015: Strong upslope shifts in Chimborazo's vegetation over two centuries since Humboldt. *PNAS*, **112**(41), 12741–12745, doi:10.1073/pnas.1509938112.

Mouginot, J. and E. Rignot, 2015: Ice motion of the Patagonian Icefields of South America: 1984–2014. *Geophys. Res. Lett.*, **42**, 1441–1449, doi:10.1002/2014GL062661.

Mourey, J., M. Marcuzzi, L. Ravanel. and F. Pallandre., 2019: Effects of climate change on high Alpine environments: the evolution of mountaineering routes in the Mont Blanc massif (Western Alps) over half a century. *Arct. Antarct. Alp. Res.* **51**(1), 176–189, doi:10.1080/15230430.2019.1612216.

Mourey, J. and L. Ravanel, 2017: Evolution of access routes to high mountain refuges of the Mer de Glace Basin (Mont Blanc Massif, France). *Revue de Géographie Alpine*, **105**(4), doi:10.4000/rga.3790.

Moyer, A.N., R.D. Moore and M.N. Koppes, 2016: Streamflow response to the rapid retreat of a lake-calving glacier. *Hydrol. Process.*, **30**(20), 3650–3665, doi:10.1002/hyp.10890.

Mu, C. et al., 2017: Relict Mountain Permafrost Area (Loess Plateau, China) Exhibits High Ecosystem Respiration Rates and Accelerating Rates in Response to Warming. *J. Geophys. Res-Biogeo*, **122**(10), 2580–2592, doi:10.1002/2017JG004060.

Mu, C. et al., 2016: Carbon loss and chemical changes from permafrost collapse in the northern Tibetan Plateau. *J. Geophys. Res-Biogeo.*, **121**(7), 1781–1791, doi:10.1002/2015JG003235.

Muellner-Riehl, A.N., 2019: Mountains as evolutionary arenas: Patterns, emerging approaches, paradigm shifts, and their implications for plant phylogeographic research in the Tibeto-Himalayan Region. *Front. Plant. Sci.*, **10**, 1–18, doi:10.3389/fpls.2019.00195.

Mukherji, A. et al., 2019: Contributions of the cryosphere to mountain communities in the Hindu Kush Himalaya: a review. *Reg. Environ. Change*, **42**(2), 228, doi:10.1007/s10113-019-01484-w.

Mukhopadhyay, B. and A. Khan, 2014: Rising river flows and glacial mass balance in central Karakoram. *J. Hydrol.*, **513**, 192–203, doi:10.1016/j.jhydrol.2014.03.042.

Muñoz, R. et al., 2016: Managing glacier related risks disaster in the Chucchún Catchment, Cordillera Blanca, Peru. In: Climate Change Adaption Strategies – An upstream-downstream perspective [Salzmann, N., C. Huggel, S.U. Nussbaumer and G. Ziervogel (eds.)]. Springer International Publishing, Switzerland, 59–78.

Murphy, S.F. et al., 2018: Fire, flood, and drought: extreme climate events alter flow paths and stream chemistry. *J. Geophys. Res.-Biogeosci.*, **123**(8), 2513–2526, doi:10.1029/2017JG004349.

Musselman, K.N. et al., 2018: Projected increases and shifts in rain-on-snow flood risk over western North America. *Nat. Clim. Change*, **8**(9), 808–812, doi:10.1038/s41558-018-0236-4.

Naaim, M., Y. Durand, N. Eckert and G. Chambon, 2013: Dense avalanche friction coefficients: influence of physical properties of snow. *J. Glaciol.*, **59**(216), 771–782, doi:10.3189/2013JoG12J205.

Naaim, M. et al., 2016: Impact of climate warming on avalanche activity in French Alps and increase of proportion of wet snow avalanches. *Houille Blanche*, **59**(6), 12–20, doi:10.1051/lhb/2016055.

Nagorski, S.A. et al., 2014: Spatial distribution of mercury in southeastern Alaskan streams influenced by glaciers, wetlands, and salmon. *Environ. Pollut.*, **184**, 62–72, doi:10.1016/j.envpol.2013.07.040.

Naiman, Z. et al., 2017: Impact of Mountains on Tropical Circulation in Two Earth System Models. *J. Clim.*, **30**(11), 4149–4163, doi:10.1175/JCLI-D-16-0512.1.

Najafi, M.R., F. Zwiers and N. Gillett, 2017: Attribution of the observed spring snowpack decline in British Columbia to anthropogenic climate change. *J. Clim.* **30**, 4113–4130, doi:10.1175/JCLI-D-16-0189.1.

Namgay, K., J.E. Millar, R.S. Black and T. Samdup, 2014: Changes in Transhumant Agro-pastoralism in Bhutan: A Disappearing Livelihood? *Hum. Ecol.*, **42**(5), 779–792, doi:10.1007/s10745-014-9684-2.

Narama, C. et al., 2017: Seasonal drainage of supraglacial lakes on debris-covered glaciers in the Tien Shan Mountains, Central Asia. *Geomorphology*, **286**, 133–142, doi:10.1016/j.geomorph.2017.03.002.

Navarro, F., H. Andrés, F. Acuña and F. José, 2018: Glaciares rocosos en la zona semiárida de Chile: relevancia de un recurso hídrico sin protección normativa. *Cuadernos de Geografía: Revista Colombiana de Geografía*, **27**(2), 338–355, doi:10.15446/rcdg.v27n2.63370.

Nepal, S., 2016: Impacts of climate change on the hydrological regime of the Koshi river basin in the Himalayan region. *Journal of Hydro-Environment Research*, **10**, 76–89, doi:10.1016/j.jher.2015.12.001.

Niedrist, G.H. et al., 2018: Climate warming increases vertical and seasonal water temperature differences and inter-annual variability in a mountain lake. *Clim. Change*, **151**(3–4), 473–490, doi:10.1007/s10584-018-2328-6.

Nilsson, J., L.S. Sørensen, V.R. Barletta and R. Forsberg, 2015: Mass changes in Arctic ice caps and glaciers: implications of regionalizing elevation changes. *The Cryosphere*, **9**, 139–150, doi:10.5194/tc-9-139-2015.

Nitu, R. et al., 2018: *WMO Solid Precipitation Intercomparison Experiment (SPICE) (2012 – 2015)*. Instruments and Observing Methods Report, **131**, World Meteorological Organization, Geneva. www.wmo.int/pages/prog/www/IMOP/publications-IOM-series.html

Niu, F. et al., 2012: Development and thermal regime of a thaw slump in the Qinghai–Tibet plateau. *Cold Reg. Sci. Technol.*, **83–84**, 131–138. doi:10.1016/j.coldregions.2012.07.007.

Noël, B. et al., 2017: A tipping point in refreezing accelerates mass loss of Greenland's glaciers and ice caps. *Nat. Commun.*, **8**, 14730, doi:10.1038/ncomms14730.

Noël, B. et al., 2018: Six decades of glacial mass loss in the Canadian Arctic Archipelago. *J. Geophys. Res-Earth*, **123**(6), 1430–1449, doi:10.1029/2017JF004304.

Noetzli, J. et al., 2018: Permafrost thermal state [in "State of the Climate in 2017"]. *Bull. Am. Meterol. Soc.*

Noetzli, J. and S. Gruber, 2009: Transient thermal effects in Alpine permafrost. *The Cryosphere*, **3**(1), 85–99, doi:10.5194/tc-3-85-2009.

Nothiger, C. and H. Elsasser, 2004: Natural hazards and tourism: New findings on the European Alps. *Mt. Res. Dev.*, **24**(1), 24–27. doi:10.1659/0276-4741(2004)024[0024:NHATNF]2.0.CO;2.

Nüsser, M. and R. Baghel, 2016: Local knowledge and global concerns: Artificial glaciers as a focus of environmental knowledge and development interventions. [Meusburger, P., T. Freytag, T., and L. Suarsana (eds.)]. Ethnic and Cultural Dimensions of Knowledge, 8. Springer, Cham, Switzerland, 191–209. ISBN: 978-3-319-21899-1. doi:10.1007/978-3-319-21900-4.

Nüsser, M. et al., 2018: Socio-hydrology of "artificial glaciers" in Ladakh, India: assessing adaptive strategies in a changing cryosphere. *Reg. Environ. Change*, **48**(2), 1–11, doi:10.1007/s10113-018-1372-0.

Nüsser, M. and S. Schmidt, 2017: Nanga Parbat Revisited: Evolution and Dynamics of Sociohydrological Interactions in the Northwestern Himalaya. *A. Assoc. Am. Geogr.*, **107**(2), 403–415, doi:10.1080/24694452.2016.1235495.

Nüsser, M., S. Schmidt and J. Dame, 2012: Irrigation and development in the upper indus Basin: Characteristics and recent changes of a socio-hydrological system in central Ladakh, India. *Mt. Res. Dev.*, **32**(1), 51–61, doi:10.1659/MRD-JOURNAL-D-11-00091.1.

Nuth, C. et al., 2019: Dynamic vulnerability revealed in the collapse of an Arctic tidewater glacier. *Sci. Rep.*, **9**(1), 5541, doi:10.1038/s41598-019-41117-0.

Nyima, Y. and K.A. Hopping, 2019: Tibetan lake expansion from a pastoral perspective: Local observations and coping strategies for a changing environment. Society & Natural Resources, **32**(9), 965–982, doi:10.1080/08941920.2019.1590667

O'Gorman, P.A., 2014: Contrasting responses of mean and extreme snowfall to climate change. *Nature*, **512**(7515), 416–418, doi:10.1038/nature13625.

Obu, J. et al., 2019: Northern Hemisphere permafrost map based on TTOP modelling for 2000–2016 at 1 km^2 scale. *Earth-Sci. Rev.*, **193**, 299–316, doi:j.earscirev.2019.04.023.

Ohmura, A., 2012: Enhanced temperature variability in high-altitude climate change. *Theor. Appl. Climatol.*, **110**(4), 499–508, doi:10.1007/s00704-012-0687-x.

Oliva, M. and J. Ruiz-Fernández, 2015: Coupling patterns between para-glacial and permafrost degradation responses in Antarctica. *Earth Surf. Process. Landf.*, **40**(9), 1227–1238, doi:10.1002/esp.3716.

Oliver-Smith, A., 1979: Yungay avalanche of 1970 – Anthropological perspectives on disaster and social-change. *Disasters*, **3**(1), 95–101, doi:10.1111/j.1467-7717.1979.tb00205.x.

Oliver-Smith, A., 1996: Anthropological research on hazards and disasters. *Annual Review of Anthropology*, **25**, 303–328, doi:10.1146/annurev.anthro.25.1.303.

Oliver-Smith, A., 2014: Climate Change Adaptation and Disaster Risk Reduction in Highland Peru. [Glavovic, B.C. and G.P. Smith (eds.)]. Adapting to Climate Change: Lessons from Natural Hazards Planning. Springer Netherlands, Dordrecht, 77–100.

Orlove, B. et al., 2019: Framing climate change in frontline communities: anthropological insights on how mountain dwellers in the USA, Peru, and Italy adapt to glacier retreat. *Reg. Environ. Change*, **19**(5), 1295–1309, doi:10.1007/s10113-019-01482-y.

Østby, T.I. et al., 2017: Diagnosing the decline in climatic mass balance of glaciers in Svalbard over 1957–2014. *The Cryosphere*, **11**, 191–215, doi:10.5194/tc-11-191-2017.

Outdoor Industry Association, 2017: *The outdoor recreation economy*. 20 p. [Available at: https://outdoorindustry.org/resource/2017-outdoor-recreation-economy-report/].

Oyler, J.W. et al., 2015: Artificial amplification of warming trends across the mountains of the western United States. *Geophys. Res. Lett.*, **42**(1), 153–161, doi:10.1002/2014GL062803.

Paden, R., L.K. Harmon, C.R. Milling and T.U.o.N.T. Center for Environmental Philosophy, 2013: Philosophical Histories of the Aesthetics of Nature. *Environmental Ethics*, **35**(1), 57–77, doi:10.5840/enviroethics20133516.

Pagán, B.R. et al., 2016: Extreme hydrological changes in the southwestern US drive reductions in water supply to Southern California by mid century. *Environ. Res. Lett.*, **11**, 1–11, doi:10.1088/1748-9326/11/9/094026.

Painter, T.H. et al., 2013: End of the Little Ice Age in the Alps forced by industrial black carbon. *PNAS*, **110**(38), 15216–15221, doi:10.1073/pnas.1302570110.

Painter, T.H. et al., 2018: Variation in rising limb of Colorado River snowmelt runoff hydrograph controlled by dust radiative forcing in snow. *Geophys. Res. Lett.*, **45**(2), 797–808, doi:10.1002/2017GL075826.

Palazzi, E., L. Mortarini, S. Terzago and J. von Hardenberg, 2019: Elevation-dependent warming in global climate model simulations at high spatial resolution. *Clim. Dyn.*, **52**(5–6), 2685–2702, doi:10.1007/s00382-018-4287-z.

Palazzi, E., J. von Hardenberg and A. Provenzale, 2013: Precipitation in the Hindu-Kush Karakoram Himalaya: Observations and future scenarios. *J. Geophys. Res.-Atmos.*, **118**(1), 85–100, doi:10.1029/2012JD018697.

Palazzi, E.L., L. Filippi and J.v. Hardenberg, 2017: Insights into elevation-dependent warming in the Tibetan Plateau-Himalayas from CMIP5 model simulations. *Clim. Dyn.*, **48**(11–12), 3991–4008, doi:10.1007/s00382-016-3316-z.

Palomo, I., 2017: Climate change impacts on ecosystem services in high mountain areas: A literature review. *Mt. Res. Dev.*, **37**(2), 179–187, doi:10.1659/mrd-journal-d-16-00110.1.

Panday, P.K., J. Thibeault and K.E. Frey, 2015: Changing temperature and precipitation extremes in the Hindu Kush-Himalayan region: an analysis of CMIP3 and CMIP5 simulations and projections. *Int. J. Climatol.*, **35**(10), 3058–3077, doi:10.1002/joc.4192.

Panetta, A.M., M.L. Stanton and J. Harte, 2018: Climate warming drives local extinction: Evidence from observation and experimentation. *Science Advances*, **4**(2), eaaq1819, doi:10.1126/sciadv.aaq1819.

Papadaki, C. et al., 2016: Potential impacts of climate change on flow regime and fish habitat in mountain rivers of the south-western Balkans. *Sci. Total Environ.*, **540**, 418–428, doi:10.1016/j.scitotenv.2015.06.134.

Parveen, S., M. Winiger, S. Schmidt and M. Nüsser, 2015: Irrigation in Upper Hunza: Evolution of socio-hydrological interactions in the Karakoram, northern Pakistan. *Erdkunde*, **69**(1), 69–85, doi:10.3112/erdkunde.2015.01.05.

Pathak, T. et al., 2018: Climate change trends and impacts on California agriculture: a detailed review. *Agronomy*, **8**(3), 25, doi:10.3390/agronomy8030025.

Paul, F. and N. Mölg, 2014: Hasty retreat of glaciers in northern Patagonia from 1985 to 2011. *J. Glaciol.*, **60**(224), 1033–1043, doi:10.3189/2014JoG14J104.

Pecl, G.T. et al., 2017: Biodiversity redistribution under climate change: Impacts on ecosystems and human well-being. *Science*, **355**(6332), eaai9214, doi:10.1126/science.aai9214.

Pedersen, S., M. Odden and H.C. Pedersen, 2017: Climate change induced molting mismatch? Mountain hare abundance reduced by duration of snow cover and predator abundance. *Ecosphere*, **8**(3), e01722, doi:10.1002/ecs2.1722.

Pellicciotti, F. et al., 2015: Mass-balance changes of the debris-covered glaciers in the Langtang Himal, Nepal, from 1974 to 1999. *J. Glaciol.*, **61**(226), 373–386, doi:10.3189/2015jog13j237.

Penczykowski, R.M., B.M. Connolly and B.T. Barton, 2017: Winter is changing: Trophic interactions under altered snow regimes. *Food Webs*, **13**, 80–91, doi:10.1016/j.fooweb.2017.02.006.

Peng, H. et al., 2015: Degradation characteristics of permafrost under the effect of climate warming and engineering disturbance along the Qinghai–Tibet Highway. *Nat. Hazards*, **75**(3), 2589–2605, doi:10.1007/s11069-014-1444-5.

Pepin, N. et al., 2015: Elevation-dependent warming in mountain regions of the world. *Nat. Clim. Change*, **5**, 424, doi:10.1038/nclimate2563.

Pepin, N.C. and J.D. Lundquist, 2008: Temperature trends at high elevations: Patterns across the globe. *Geophys. Res. Lett.*, **35**(14), L14701, doi:10.1029/2008GL034026.

PERMOS, 2016: Permafrost in Switzerland 2010/2011 to 2013/2014 [Nötzli, J., R. Luethi and B. Staub (eds.)]. Glaciological Report Permafrost No. 12–15 of the Cryospheric Commission of the Swiss Academy of Sciences, [Available at: https://naturalsciences.ch/service/publications/82035-permafrost-in-switzerland-2010–2011-to-2013–2014]. Accessed on 08/08/2019.

Peter, H. and R. Sommaruga, 2016: Shifts in diversity and function of lake bacterial communities upon glacier retreat. *ISME J.*, **10**(7), 1545–1554, doi:10.1038/ismej.2015.245.

Petrakov, D. et al., 2016: Accelerated glacier shrinkage in the Ak-Shyirak massif, Inner Tien Shan, during 2003–2013. *Sci. Total Environ.*, **562**, 364–378, doi:10.1016/j.scitotenv.2016.03.162.

Philipona, R., 2013: Greenhouse warming and solar brightening in and around the Alps. *Int. J. Climatol.*, **33**(6), 1530–1537, doi:10.1002/joc.3531.

Phillips, M. and S. Margreth, 2008: Effects of ground temperature and slope deformation on the service life of snow-supporting structures in mountain permafrost: Wisse Schijen, Randa, Swiss Alps. In: *Proceedings of the 9th International Conference on Permafrost, Fairbanks, Alaska*, 1990, pp. 1417–1422.

Phillips, M., E.Z. Mutter, M. Kern-Luetschg and M. Lehning, 2009: Rapid degradation of ground ice in a ventilated Talus slope: Flüela Pass, Swiss Alps. *Permafrost Periglac.*, **20**(1), 1–14, doi:10.1002/ppp.638.

Phillips, M., F. Ladner, M. Müller, U. Sambeth, J. Sorg, and P. Teysseire, 2007: Cold Regions Science and Technology, **47** (1–2 Special Issue), 32–42, doi: 10.1016/j.coldregions.2006.08.014.

Phillips, M. et al., 2017: Rock slope failure in a recently deglaciated permafrost rock wall at Piz Kesch (Eastern Swiss Alps), February 2014. *Earth Surf. Process. Landf.*, **42**(3), 426–438, doi:10.1002/esp.3992.

Pickering, C., K. Green, A.A. Barros and S. Venn, 2014: A resurvey of late-lying snowpatches reveals changes in both species and functional composition across snowmelt zones. *Alpine Botany*, **124**(2), 93–103, doi:10.1007/s00035-014-0140-0.

Pielmeier, C., F. Techel, C. Marty and T. Stucki, 2013: Wet snow avalanche activity in the Swiss Alps – Trend analysis for mid-winter season. In: *International Snow Science Workshop Grenoble – Chamonix Mont-Blanc – October 07–11, 2013*, pp. 1240–1246.

Pierce, D.W. et al., 2008: Attribution of declining Western U.S. snowpack to human effects. *J. Clim.*, **21**(23), 6425–6444, doi:10.1175/2008JCLI2405.1.

Pierson, T.C., R.J. Janda, J.C. Thouret and C.A. Borrero, 1990: Perturbation and melting of snow and ice by the 13 November 1985 eruption of Nevado-Del-Ruiz, Colombia, and consequent mobilization, flow and deposition of lLahars. *J. Volcanol. Geoth. Res.*, **41**(1–4), 17–66, doi:10.1016/0377-0273(90)90082-Q. Pintaldi, E. et al., 2017: Sustainable soil management in ski areas: Threats and challenges. *Sustainability*, **9**, 250, doi:10.3390/su9112150.

Plard, F. et al., 2014: Mismatch between birth date and vegetation phenology slows the demography of roe deer. *PLOS Biology*, **12**(4), e1001828, doi:10.1371/journal.pbio.1001828.

Podolskiy, E.A., K. Nishimura, O. Abe and P.A. Chernous, 2010: Earthquake-induced snow avalanches: I. Historical case studies. *J. Glaciol.*, **56**(197), 431–446, doi:10.3189/002214310792447815.

Pogliotti, P. et al., 2015: Warming permafrost and active layer variability at Cime Bianche, Western European Alps. *The Cryosphere*, **9**(2), 647–661, doi:10.5194/tc-9-647-2015.

Polk, M.H. et al., 2017: Exploring hydrologic connections between tropical mountain wetlands and glacier recession in Peru's Cordillera Blanca. *Applied Geography*, **78**, 94–103, doi:10.1016/j.apgeog.2016.11.004.

Pomeroy, J.W., X. Fang and D.G. Marks, 2016: The cold rain-on-snow event of June 2013 in the Canadian Rockies – characteristics and diagnosis. *Hydrol. Process.*, **30**(17), 2899–2914, doi:10.1002/hyp.10905.

Postigo, J.C., 2014: Perception and resilience of Andean populations facing climate change. *J. Ethnobiol.*, **34**(3), 383–400, doi:10.2993/0278-0771-34.3.383.

Prasain, S., 2018: Climate change adaptation measure on agricultural communities of Dhye in Upper Mustang, Nepal. *Clim. Change*, **148**(1–2), 279–291, doi:10.1007/s10584-018-2187-1.

Prasch, M., W. Mauser and M. Weber, 2013: Quantifying present and future glacier melt-water contribution to runoff in a central Himalayan river basin. *The Cryosphere*, **7**(3), 889–904, doi:10.5194/tc-7-889-2013.

Prinz, R. et al., 2016: Climatic controls and climate proxy potential of Lewis Glacier, Mt. Kenya. *The Cryosphere*, **10**(1), 133–148, doi:10.5194/tc-10-133-2016.

Purdie, H., C. Gomez and S. Espiner, 2015: Glacier recession and the changing rockfall hazard: Implications for glacier tourism. *New Zealand Geographer*, **71**(3), 189–202, doi:10.1111/nzg.12091.

Putkonen, J. and G. Roe, 2003: Rain-on-snow events impact soil temperatures and affect ungulate survival. *Geophys. Res. Lett.*, **30**(4), 1188, doi:10.1029/2002GL016326.

Putzer, A. and D. Festi, 2014: Nicht nur Ötzi? – Neufunde aus dem Tisental (Gem. Schnals/Prov. Bozen). *Praehistorische Zeitschrift*, **89**(1), doi:10.1515/pz-2014-0005.

Qian, Y. et al., 2015: Light-absorbing particles in snow and ice: Measurement and modeling of climatic and hydrological impact. *Adv. Atmos. Sci.*, **32**(1), 64–91, doi:10.1007/s00376-014-0010-0.

Qin, J., K. Yang, S. Liang and X. Guo, 2009: The altitudinal dependence of recent rapid warming over the Tibetan Plateau. *Climatic Change.* **97**(1), 321. doi:10.1007/s10584-009-9733-9.

Qixiang, W., M. Wang and X. Fan, 2018: Seasonal patterns of warming amplification of high-elevation stations across the globe. *Int. J. Climatol.*, **38**(8), 3466–3473, doi:10.1002/joc.5509.

Rabatel, A. et al., 2017: Toward an imminent extinction of Colombian glaciers? *Geografiska Annaler. Series A, Physical Geography*, **13**(5), 1–21, doi:10.10 80/04353676.2017.1383015.

Rabatel, A. et al., 2013: Current state of glaciers in the tropical Andes: a multi-century perspective on glacier evolution and climate change. *The Cryosphere*, **7**(1), 81–102, doi:10.5194/tc-7-81-2013.

Radić, V., et al., 2014: Regional and global projections of 21st century glacier mass changes in response to climate scenarios from global climate models. *Clim. Dyn.*, **42**(1-2), 37-58, doi:10.1007/s00382-013-1719-7.

Rajczak, J. and C. Schär, 2017: Projections of future precipitation extremes over europe: a multimodel assessment of climate simulations. *J. Geophys. Res-Atmos.*, **122**(20), 10–773–10–800, doi:10.1002/2017JD027176.

Raman, D., 2018: Damming and Infrastructural Development of the Indus River Basin: Strengthening the Provisions of the Indus Waters Treaty. *Asian Journal of International Law*, **8**(2), 372–402, doi:10.1017/S2044251317000029.

Rangwala, I., E. Sinsky and J.R. Miller, 2013: Amplified warming projections for high altitude regions of the northern hemisphere mid-latitudes from CMIP5 models. *Environ. Res. Lett.*, **8**(2), 024040, doi:10.1088/1748-9326/8/2/024040.

Rasmussen, M.B., 2016: Unsettling Times: Living with the Changing Horizons of the Peruvian Andes. *Latin American Perspectives*, **43**(4), 73–86, doi:10.1177/0094582x16637867.

Rasmussen, M.B., 2019: Rewriting conservation landscapes: protected areas and glacial retreat in the high Andes. *Reg. Environ. Change*, 1–15, doi:10.1007/s10113-018-1376-9.

Rasul, G. and D. Molden, 2019: The global social and economic consequences of mountain cryopsheric change. *Front. Environ. Sci.*, **7**(91), doi:10.3389/fenvs.2019.00091.

Rasul, G., B. Pasakhala, A. Mishra and S. Pant, 2019: Adaptation to mountain cryosphere change: issues and challenges. *Clim. Dev.* doi:10.1080/175655 29.2019.1617099.

Ravanel, L. et al., 2010: Rock falls in the Mont Blanc Massif in 2007 and 2008. *Landslides*, **7**(4), 493–501, doi:10.1007/s10346-010-0206-z.

Ravanel, L. and P. Deline, 2011: Climate influence on rockfalls in high-Alpine steep rockwalls: The north side of the Aiguilles de Chamonix (Mont Blanc massif) since the end of the 'Little Ice Age'. *The Holocene*, **21**(2), 357–365, doi:10.1177/0959683610374887.

Ravanel, L., P. Deline, C. Lambiel and C. Vincent, 2013: Instability of a high alpine rock ridge: the lower Arête Des Cosmiques, Mont Blanc massif, France. *Geogr. Ann. A.*, **95**(1), 51–66, doi:10.1111/geoa.12000.

Ravanel, L., F. Magnin and P. Deline, 2017: Impacts of the 2003 and 2015 summer heatwaves on permafrost-affected rock-walls in the Mont Blanc massif. *Sci. Total Environ.*, **609**, 132–143, doi:10.1016/j.scitotenv.2017.07.055.

Reggiani, P. and T.H.M. Rientjes, 2015: A reflection on the long-term water balance of the Upper Indus Basin. *Hydrol. Res.*, **46**, 446–462, doi:10.2166/nh.2014.060.

Reyer, C.P.O. et al., 2017: Climate change impacts in Central Asia and their implications for development. *Reg. Environ. Change*, **17**(6), 1639–1650, doi:10.1007/s10113-015-0893-z.

Reznichenko, N.V., T.R.H. Davies and D.J. Alexander, 2011: Effects of rock avalanches on glacier behaviour and moraine formation. *Geomorphology*, **132**, 327–338, doi:10.1016/j.geomorph.2011.05.019.

RGI Consortium, 2017: Randolph Glacier Inventory – A dataset of global glacier outlines: Version 6.0: Technical Report, Global Land Ice Measurements from Space, Colorado, USA, Digital Media. doi:10.7265/N5-RGI-60. [Available at: www.glims.org/RGI/randolph60.html].

Rhoades, R.E., X. Zapata Rios and J.A. Ochoa, 2008: Mama Cotacachi: History, local perceptions, and social impacts of climate change and glacier retreat in the Ecuadorian Andes. In: *Darkening Peaks: Glacier Retreat, Science, and Society* [Orlove, B., E. Wiegant and B.H. Luckman (eds.)]. University of California Press, Berkeley, pp. 216–228.

Roberts, N.J., R. McKillop, R.L. Hermanns, J.J. Clague, and T. Oppikofer, 2014: Preliminary global catalogue of displacement waves from subaerial landslides. [Sassa, K., P., Canuti, Y. Yin (eds.)]: Landslide Science for a Safer Geoenvironment. Springer International Publishing. 687–692. ISBN 978-3-319-04996-0.

Rohrer, M., N. Salzmann, M. Stoffel and A.V. Kulkarni, 2013: Missing (in-situ) snow cover data hampers climate change and runoff studies in the Greater Himalayas. *Sci. Total Environ.*, **468–469**, S60–70, doi:10.1016/j.scitotenv.2013.09.056.

Rokaya, P., S. Budhathoki and K.E. Lindenschmidt, 2018: Trends in the Timing and Magnitude of Ice-Jam Floods in Canada. *Sci. Rep.*, **8**, 5834, doi:10.1038/s41598-018-24057-z.

Rosvold, J., 2016: Perennial ice and snow covered land as important ecosystems for birds and mammals. *J. Biogeogr.*, **43**, 3–12, doi:10.1111/jbi.12609.

Round, V. et al., 2017: Surge dynamics and lake outbursts of Kyagar Glacier, Karakoram. *The Cryosphere*, **11**(2), 723–739, doi:10.5194/tc-11-723-2017.

Rumpf, S.B., K. Huelber, N.E. Zimmermann and S. Dullinger, 2019: Elevational rear edges shifted at least as much as leading edges over the last century. *Glob. Ecol. Biogeogr.*, **28**(4), 533–543, doi:10.1111/geb.12865.

Rumpf, S.B. et al., 2018: Range dynamics of mountain plants decrease with elevation. *PNAS*, **115**(8), 1848–1853, doi:10.1073/pnas.1713936115.

Russell, A.J. et al., 2011: A new cycle of jokulhlaups at Russell Glacier, Kangerlussuaq, West Greenland. *J. Glaciol.*, **57**(202), 238–246, doi:10.3189/002214311796405997.

Sæmundsson, Þ. et al., 2018: The triggering factors of the Móafellshyrna debris slide in northern Iceland: Intense precipitation, earthquake activity and thawing of mountain permafrost. *Sci. Total Environ.*, **621**, 1163–1175, doi:10.1016/j.scitotenv.2017.10.111.

Sakai, A. and K. Fujita, 2017: Contrasting glacier responses to recent climate change in high-mountain Asia. *Sci. Rep.*, **7**, 13717, doi:10.1038/s41598-017-14256-5.

Sakakibara, D. and S. Sugiyama, 2014: Ice-front variations and speed changes of calving glaciers in the Southern Patagonia Icefield from 1984 to 2011. *J. Geophys. Res-Earth.*, **119**(11), 2541–2554. doi:10.1002/2014JF003148.

Sanjay, J. et al., 2017: Downscaled climate change projections for the Hindu Kush Himalayan region using CORDEX South Asia regional climate models. *Adv. Clim. Change Res.*, **8**(3), 185–198. doi:10.1016/j.accre.2017.08.003.

Sati, S.P. and V.K. Gahalaut, 2013: The fury of the floods in the north-west Himalayan region: the Kedarnath tragedy. *Geomat. Nat. Haz. Risk*, **4**(3), 193–201, doi:10.1080/19475705.2013.827135.

Schaefli, B. et al., 2019: The role of glacier retreat for Swiss hydropower production. *Renew. Energ.*, **132**, 615–627, doi:10.1016/j.renene.2018.07.104.

Scherrer, D. and C. Körner, 2011: Topographically controlled thermal-habitat differentiation buffers alpine plant diversity against climate warming. *J. Biogeogr.*, **38**(2), 406–416, doi:10.1111/j.1365-2699.2010.02407.x.

Scherrer, S.C., P. Ceppi, M. Croci-Maspoli and C. Appenzeller, 2012: Snow-albedo feedback and Swiss spring temperature trends. *Theor. Appl. Climatol.*, **110**(4), 509–516, doi:10.1007/s00704-012-0712-0.

Schirpke, U., F. Timmermann, U. Tappeiner and E. Tasser, 2016: Cultural ecosystem services of mountain regions: Modelling the aesthetic value. *Ecol. Indic.*, **69**, 78–90, doi:10.1016/j.ecolind.2016.04.001.

Schmidt, K.-H. and D. Morche, 2006: Sediment output and effective discharge in two small high mountain catchments in the Bavarian Alps, Germany. *Geomorphology*, **80**(1–2), 131–145, doi:10.1016/j.geomorph.2005.09.013.

Schmidt, S. and M. Nüsser, 2017: Changes of high altitude glaciers in the Trans-Himalaya of Ladakh over the past five decades (1969–2016). *Geosciences*, **7**(2), 27, doi:10.3390/geosciences7020027.

Schneider, D. et al., 2014: Mapping hazards from glacier lake outburst floods based on modelling of process cascades at Lake 513, Carhuaz, Peru. *Advances in Geosciences*, **35**, 145–155, doi:10.5194/adgeo-35-145-2014.

Schneider, D., C. Huggel, W. Haeberli and R. Kaitna, 2011: Unraveling driving factors for large rock-ice avalanche mobility. *Earth Surf. Process. Landf.*, **36**(14), 1948–1966, doi:10.1002/esp.2218.

Schnorbus, M., A. Werner and K. Bennett, 2014: Impacts of climate change in three hydrologic regimes in British Columbia, Canada. *Hydrol. Process.*, **28**, 1170–1189, doi:10.1002/hyp.9661.

Schoen, E.R. et al., 2017: Future of Pacific salmon in the face of environmental change: Lessons from one of the world's remaining productive salmon regions. *Fisheries*, **42**(10), 538–553, doi:10.1080/03632415.2017.1374251.

Scholz, K., A. Hammerle, E. Hiltbrunner and G. Wohlfahrt, 2018: Analyzing the effects of growing season length on the net ecosystem production of an alpine grassland using model-data fusion. *Ecosystems*, **21**(5), 982–999, doi:10.1007/s10021-017-0201-5.

Schuur, E.A.G. et al., 2015: Climate change and the permafrost carbon feedback. *Nature*, **520**, 171–179, doi:10.1038/nature14338.

Schwanghart, W. et al., 2016: Uncertainty in the Himalayan energy-water nexus: estimating regional exposure to glacial lake outburst floods. *Environ. Res. Lett.*, **11**(7), 074005, doi:10.1088/1748-9326/11/7/074005.

Schweizer, J., J.B. Jamieson and M. Schneebeli, 2003: Snow avalanche formation. *Reviews of Geophysics*, **41**(4), 1016, doi:10.1029/2002RG000123.

Scott, D., R. Steiger, H. Dannevig and C. Aall, 2019: Climate change and the future of the Norwegian alpine ski industry. *Current Issues in Tourism*, doi:10.1080/13683500.2019.1608919.

Scott, D., R. Steiger, M. Rutty and Y. Fang, 2018: The changing geography of the Winter Olympic and Paralympic Games in a warmer world. *Current Issues in Tourism*, **22**(11), 1301–1311, doi:10.1080/13683500.2018.1436161.

SENASA, 2017: Áncash: Vigilancia fitosanitaria en cultivo de rosas. Servicio Nacional de Sanidad Agraria, Ministerio de Agricultura y Riego, Lima [Available at: www.senasa.gob.pe/senasacontigo/ancash-vigilancia-fitosanitaria-en-cultivo-de-rosas/#].

Seneviratne, S.I. et al., 2012: Changes in climate extremes and their impacts on the natural physical environment. [Field, C.B., V. Barros, T.F. Stocker and Q. Dahe (eds.)]. A Special Report of Working Groups I and II of the Intergovernmental Panel on Climate Change (IPCC). Cambridge University Press, Cambridge, UK, and New York, NY, USA, pp. 109–230. Cambridge University Press, Cambridge, 109–230.

Serdeczny, O., 2019: Non-economic loss and damage and the Warsaw International Mechanism. In: *Loss and Damage from Climate Change: Concepts, Methods and Policy Options* [Mechler, R., L.M. Bouwer, T. Schinko, S. Surminski and J. Linnerooth-Bayer (eds.)]. Springer International Publishing, Cham, pp. 205–220.

Serrano, E. et al., 2018: Post-little ice age paraglacial processes and landforms in the high Iberian mountains: A review. *Land Degrad. Dev.*, **29**(11), 4186–4208, doi:10.1002/ldr.3171.

Settele, J. et al., 2014: Terrestrial and inland water systems. In: Climate change 2014: Impacts, Adaptation, and Vulnerability. Part A: Global and Sectoral Aspects. Contribution of Working Group II to the Fifth Assessment Report of the Intergovernmental Panel on Climate Change (IPCC) [Field, C.B., V.R. Barros, D.J. Dokken, K.J. Mach, M.D. Mastrandrea, T.E. Bilir, M. Chatterjee, K.L. Ebi, Y.O. Estrada, R.C. Genova, B. Girma, E.S. Kissel, A.N. Levy, S. MacCracken, P.R. Mastrandrea and L.L. White (eds.)]. Cambridge University Press, Cambridge, UK and New York, NY, USA, 271–359.

Sevestre, H. and D I. Benn, 2015: Climatic and geometric controls on the global distribution of surge-type glaciers: Implications for a unifying model of surging. *J. Glaciol.*, **61**(228), 646–662, doi:10.3189/2015JoG14J136.

Sevestre, H. et al., 2018: Tidewater Glacier Surges Initiated at the Terminus. *J. Geophys. Res-Earth*, **123**(5), 1035–1051, doi:10.1029/2017JF004358.

Seynova, I.B. et al., 2017: Formation of water flow in lahars from active glacier-clad volcanoes. *Earth's Cryosphere*, **21**(6), 103–111, doi:10.21782/EC1560-7496-2017-6(103-111).

Shaheen, F.A., 2016: *The art of glacier grafting: innovative water harvesting techniques in Ladakh*. IWMI-Tata Water Policy Research Highlight, 8. [Available at: https://cgspace.cgiar.org/handle/10568/89600].

Shama, L.N.S. and C.T. Robinson, 2009: Microgeographic life history variation in an alpine caddisfly: plasticity in response to seasonal time constraints. *Freshwater Biol.*, **54**(1), 150–164, doi:10.1111/j.1365-2427.2008.02102.x.

Shangguan, D. et al., 2016: Characterizing the May 2015 Karayaylak Glacier surge in the eastern Pamir Plateau using remote sensing. *J. Glaciol.*, **62**(235), 944–953, doi:10.1017/jog.2016.81.

Shaoliang, Y., M. Ismail and Y. Zhaoli, 2012: Pastoral communities' perspectives on climate change and their adaptation strategies in the Hindukush-Karakoram-Himalaya. [Kreutzmann, H., (ed.)]. Springer Netherlands, Dordrecht, 307–322. doi:10.1007/978-94-007-3846-1, ISBN 978-94-007-3845-4.

Sharma, B.M. et al., 2015: Melting Himalayan glaciers contaminated by legacy atmospheric depositions are important sources of PCBs and high-molecular-weight PAHs for the Ganges floodplain during dry periods. *Environ. Pollut.*, **206**, 588–596, doi:10.1016/j.envpol.2015.08.012.

Sharma, S. et al., 2019: Widespread loss of lake ice around the Northern Hemisphere in a warming world. *Nat. Clim. Change*, **9**(3), 227–231, doi:10.1038/s41558-018-0393-5.

Shen, Y.J. et al., 2018: Trends and variability in streamflow and snowmelt runoff timing in the southern Tianshan Mountains. *J. Hydrol.*, **557**, 173–181, doi:10.1016/j.jhydrol.2017.12.035.

Shrestha, A.B. et al., 2010: Glacial lake outburst flood risk assessment of Sun Koshi basin, Nepal. *Geomat. Nat. Haz. Risk*, **1**(2), 157–169, doi:10.1080/19475701003668968.

Shrestha, N.K., X. Du and J. Wang, 2017: Assessing climate change impacts on fresh water resources of the Athabasca River Basin, Canada. *Sci. Total Environ.*, **601–602**, 425–440, doi:10.1016/j.scitotenv.2017.05.013.

Sietz, D. and G. Feola, 2016: Resilience in the rural Andes: critical dynamics, constraints and emerging opportunities. *Reg. Environ. Change*, **16**(8), 2163–2169, doi:10.1007/s10113-016-1053-9.

Sigl, M. et al., 2018: 19th century glacier retreat in the Alps preceded the emergence of industrial black carbon deposition on high-alpine glaciers. *The Cryosphere*, **12**(10), 3311–3331, doi:10.5194/tc-12-3311-2018.

Sinickas, A., B. Jamieson and M.A. Maes, 2015: Snow avalanches in western Canada: investigating change in occurrence rates and implications for risk assessment and mitigation. *Struct. Infrastruct. E.*, **12**(4), 490–498, doi:10.1080/15732479.2015.1020495.

Skarbø, K. and K. VanderMolen, 2014: Irrigation access and vulnerability to climate-induced hydrological change in the Ecuadorian Andes. *Culture, Agriculture, Food and Environment*, **36**(1), 28–44, doi:10.1111/cuag.12027.

Skiles, S.M. et al., 2018: Radiative forcing by light-absorbing particles in snow. *Nat. Clim. Change*, **8**(11), 965-+, doi:10.1038/s41558-018-0296-5.

Slangen, A.B.A. et al., 2017: A Review of recent updates of sea-level projections at global and regional scales. *Surveys in Geophysics*, **38**(1), 385–406, doi:10.1007/978-3-319-56490-6_17.

Slangen, A.B.A. and R.S.W. Van De Wal, 2011: An assessment of uncertainties in using volume-area modelling for computing the twenty-first century glacier contribution to sea-level change. *The Cryosphere*, **5**(3), doi:10.5194/tc-5-673-2011.

Slater, A.G. and D.M. Lawrence, 2013: Diagnosing present and future permafrost from climate models. *J. Clim.*, **26**(15), 5608–5623, doi:10.1175/jcli-d-12-00341.1.

Slatyer, R.A., M.A. Nash and A.A. Hoffmann, 2017: Measuring the effects of reduced snow cover on Australia's alpine arthropods. *Austral Ecology*, **42**(7), 844–857, doi:10.1111/aec.12507.

Sloat, L.L., A.N. Henderson, C. Lamanna and B.J. Enquist, 2015: The effect of the foresummer drought on carbon exchange in subalpine meadows. *Ecosystems*, **18**(3), 533–545, doi:10.1007/s10021-015-9845-1.

Smadja, J. et al., 2015: Climate change and water resources in the Himalayas: Field study in four geographic units of the Koshi basin, Nepal. *Revue de Géographie Alpine*, **103**(2), doi:10.4000/rga.2910.

Smith, M.W. and D.W. Riseborough, 1996: Permafrost monitoring and detection of climate change. *Permafrost Periglac.*, **7**(4), 301–309, doi:10.1002/(SICI)1099-1530(199610)7:4<301::AID-PPP231>3.0.CO;2-R. Solomina, O.N. et al., 2016: Glacier fluctuations during the past 2000 years. *Quaternary Sci. Rev.*, **149**, 61–90, doi:10.1016/j.quascirev.2016.04.008.

Somers, L.D. et al., 2018: Does hillslope trenching enhance groundwater recharge and baseflow in the Peruvian Andes? *Hydrol. Process.*, **32**(3), 318–331, doi:10.1002/hyp.11423.

Somos-Valenzuela, M.A. et al., 2016: Modeling a glacial lake outburst flood process chain: the case of Lake Palcacocha and Huaraz, Peru. *Hydrol. Earth Syst. Sc.*, **20**(6), 2519–2543, doi:10.5194/hess-20-2519-2016.

Song, C., B. Huang, L. Ke and K.S. Richards, 2014: Remote sensing of alpine lake water environment changes on the Tibetan Plateau and surroundings: A review. *ISPRS J. Photogram.*, **92**, 26–37, doi:10.1016/j.isprsjprs.2014.03.001.

Sørensen, L.S. et al., 2017: The effect of signal leakage and glacial isostatic rebound on GRACE-derived ice mass changes in Iceland. *Geophys. J. Int.*, **209**, 226–233, doi:10.1093/gji/ggx008.

Sorg, A. et al., 2015: Contrasting responses of Central Asian rock glaciers to global warming. *Sci. Rep.*, **5**, 8228, doi:10.1038/srep08228.

Soruco, A. et al., 2015: Contribution of glacier runoff to water resources of La Paz city, Bolivia (16°S). *Ann. Glaciol.*, **56**(70), 147–154, doi:10.3189/2015AoG70A001.

Spandre, P. et al., 2019a: Climate controls on snow reliability in French Alps ski resorts. *Sci. Rep.*, **9**, 8043, doi:10.1038/s41598-019-44068-8.

Spandre, P. et al., 2019b: Winter tourism under climate change in the Pyrenees and the French Alps: relevance of snowmaking as a technical adaptation. *The Cryosphere*, **13**(4), 1325–1347, doi:10.5194/tc-13-1325-2019.

Stäubli, A. et al., 2018: Analysis of Weather – and Climate-Related Disasters in Mountain Regions Using Different Disaster Databases. In: *Climate Change, Extreme Events and Disaster Risk Reduction. Sustainable Development Goals Series* [Mal S., Singh R. and C. Huggel (eds.)]. Springer International Publishing, Cham, 17–41.

Stearns, L.A. et al., 2015: Glaciological and marine geological controls on terminus dynamics of Hubbard Glacier, southeast Alaska. *J. Geophys. Res-Earth*, **120**(6), 1065–1081, doi:10.1002/2014jf003341.

Steiger, R. et al., 2017: A critical review of climate change risk for ski tourism. *Current Issues in Tourism*, **22**(11), 1343–1379, doi:10.1080/13683500.2017.1410110.

Steinbauer, M.J. et al., 2016: Topography-driven isolation, speciation and a global increase of endemism with elevation. *Global Ecol. Biogeogr.*, **25**, 1097–1107, doi:10.1111/geb.12469.

Steinbauer, M.J. et al., 2018: Accelerated increase in plant species richness on mountain summits is linked to warming. *Nature*, **556**(7700), 231–234, doi:10.1038/s41586-018-0005-6.

Steiner, J.F., P.D.A. Kraaijenbrink, S.G. Jiduc and W.W. Immerzeel, 2018: Brief communication: The Khurdopin glacier surge revisited – Extreme flow velocities and formation of a dammed lake in 2017. *The Cryosphere*, **12**(1), 95–101, doi:10.5194/tc-12-95-2018.

Steinkogler, W., B. Sovilla and M. Lehning, 2014: Influence of snow cover properties on avalanche dynamics. *Cold Reg. Sci. Technol.*, **97**, 121–131, doi:10.1016/j.coldregions.2013.10.002.

Stenseth, N.C. et al., 2003: Review article. Studying climate effects on ecology through the use of climate indices: the North Atlantic Oscillation, El Niño Southern Oscillation and beyond. *Proc. Royal Soc. B.*, **270** (1529), 2087–2096, doi:10.1098/rspb.2003.2415.

Stewart, E.J. et al., 2016: Implications of climate change for glacier tourism. *Tourism Geographies*, **18**(4), 377–398, doi:10.1080/14616688.2016.1198416.

Stocker, T.F. et al., 2013: IPCC Technical Summary AR5. In: *Climate Change 2013: The Physical Science Basis. Contribution of Working Group I to the Fifth Assessment Report of the Intergovernmental Panel on Climate Change* [Stocker, T.F., D. Qin, G.-K. Plattner, M. Tignor, S.K. Allen, J. Boschung, A. Nauels, Y. Xia, V. Bex and P.M. Midgley (eds.)]. Cambridge University Press, Cambridge, United Kingdom and New York, NY, USA, 1535 pp.

Stoffel, M. and C. Graf, 2015: Debris-flow activity from high-elevation, periglacial environments. [Huggel, C., M. Carey, J.J. Clague and A. Kääb (eds.)]. Cambridge University Press, Cambridge, 295–314, ISBN 978-1-107-06584-0.

Stucker, D., J. Kazbekov, M. Yakubov and K. Wegerich, 2012: Climate change in a small transboundary tributary of the Syr Darya Calls for effective cooperation and adaptation. *Mt. Res. Dev.*, **32**(3), 275–285, doi:10.1659/MRD-JOURNAL-D-11-00127.1.

Sturm, M., M.A. Goldstein and C. Parr, 2017: Water and life from snow: A trillion dollar science question. *Water Resour. Res.*, **53**(5), 3534–3544, doi:10.1002/2017WR020840.

Suding, K.N. et al., 2015: Vegetation change at high elevation: scale dependence and interactive effects on Niwot Ridge. *Plant Ecol. Divers.*, **8**(5–6), 713–725, doi:10.1080/17550874.2015.1010189.

Sujakhu, N.M. et al., 2016: Farmers' perceptions of and adaptations to changing climate in the Melamchi Valley of Nepal. *Mt. Res. Dev.*, **36**(1), 15–30, doi:10.1659/MRD-JOURNAL-D-15-00032.1.

Sultaire, S.M. et al., 2016: Climate change surpasses land-use change in the contracting range boundary of a winter-adapted mammal. *Proc. R. Soc. B.*, **283**(1831), doi:10.1098/rspb.2016.0899.

Sultana, R. and M. Choi, 2018: Sensitivity of streamflow response in the snow-dominated Sierra Nevada Watershed using projected CMIP5 data. *J. Hydrol. Eng.*, **23**(8), 05018015, doi:10.1061/(ASCE)HE.1943-5584.0001640.

Sun, J. et al., 2018a: Linkages of the dynamics of glaciers and lakes with the climate elements over the Tibetan Plateau. *Earth-Sci. Rev.*, **185**, 308–324, doi:10.1016/j.earscirev.2018.06.012.

Sun, X. et al., 2017: The role of melting alpine glaciers in mercury export and transport: An intensive sampling campaign in the Qugaqie Basin, inland Tibetan Plateau. *Environ. Pollut.*, **220**, 936–945, doi:10.1016/j.envpol.2016.10.079.

Sun, X. et al., 2018b: Mercury speciation and distribution in a glacierized mountain environment and their relevance to environmental risks in the inland Tibetan Plateau. *Sci. Total Environ.*, **631–632**, 270–278, doi:10.1016/j.scitotenv.2018.03.012.

Surfleet, C.G. and D. Tullos, 2013: Variability in effect of climate change on rain-on-snow peak flow events in a temperate climate. *J. Hydrol.*, **479**, 24–34, doi:10.1016/J.JHYDROL.2012.11.021.

Suzuki-Parker, A., Y. Miura, H. Kusaka and M. Kureha, 2018: Assessing the Sustainability of Ski Fields in Southern Japan under Global Warming. *Advances in Meteorology*, **2018**(8529748), 1–10, doi:10.1155/2018/8529748.

Swindles, G.T. et al., 2018: Climatic control on Icelandic volcanic activity during the mid-Holocene. *Geology*, **46**(1), 47–50, doi:10.1130/G39633.1.

Taylor, K.E., R.J. Stouffer and G.A. Meehl, 2012: An overview of CMIP5 and the experiment design. *Bull. Am. Meteorol. Soc.*, **93**, 485–498, doi:10.1175/BAMS-D-11-00094.1.

Teich, M. et al., 2012: Snow and weather conditions associated with avalanche releases in forests: Rare situations with decreasing trends during the last 41 years. *Cold Reg. Sci. Technol.*, **83–84**, 77–88, doi:10.1016/j.coldregions.2012.06.007.

Temme, A.J.A.M., 2015: Using climber's guidebooks to assess rock fall patterns over large spatial and decadal temporal scales: An example from the Swiss Alps. *Geogr. Ann. A.*, **97**(4), 793–807, doi:10.1111/geoa.12116.

Terzago, S., J.v. Hardenberg, E. Palazzi and P. Antonello, 2014: Snowpack changes in the Hindu Kush–Karakoram–Himalaya from CMIP5 Global Climate Models. *J. Hydrometeorol.*, **15**(6), 2293–2313, doi:10.1175/JHM-D-13-0196.1.

Theobald, E.J. et al., 2015: Global change and local solutions: Tapping the unrealized potential of citizen science for biodiversity research. *Biol. Conserv.*, **181**, 236–244, doi:10.1016/j.biocon.2014.10.021.

Thibert, E. et al., 2018: Causes of glacier melt extremes in the Alps since 1949. *Geophys. Res. Lett.*, **45**(2), 817–825, doi:10.1002/2017GL076333.

Thies, H. et al., 2013: Evidence of rock glacier melt impacts on water chemistry and diatoms in high mountain streams. *Cold Reg. Sci. Technol.*, **96**, 77–85, doi:10.1016/j.coldregions.2013.06.006.

Thompson, L.G. et al., 2017: Impacts of Recent Warming and the 2015/2016 El Niño on Tropical Peruvian Ice Fields. *J. Geophys. Res-Earth*, **122**(23), 12,688–12,701, doi:10.1002/2017JD026592.

Tiwari, P.C. and B. Joshi, 2015: Climate Change and Rural Out-migration in Himalaya. *Change and Adaptation in Socio-Ecological Systems*, **2**, 8–25, doi:10.1515/cass-2015-0002.

Torgoev, I. and B. Omorov, 2014: Mass movement in the waste dump of high-altitude Kumtor Goldmine (Kyrgyzstan). [Sassa, K., Canuti, P., Yin, Y. (Eds.)]: Landslide Science for a Safer Geoenvironment. Springer International Publishing. 517–521. ISBN 978-3-319-04996-0.

Toropov, P.A., M.A. Aleshina and A.M. Grachev, 2019: Large-scale climatic factors driving glacier recession in the Greater Caucasus, 20th-21st century. *Int. J. Climatol.*, **39** (12), 4703–4720, doi:10.1002/joc.6101.

Trujillo, E. et al., 2012: Elevation-dependent influence of snow accumulation on forest greening. *Nat. Geosci.*, **5**, 705–709, doi:10.1038/ngeo1571.

Tschakert, P. et al., 2019: One thousand ways to experience loss: A systematic analysis of climate-related intangible harm from around the world. *Glob. Environ. Change*, **55**, 58–72, doi:10.1016/j.gloenvcha.2018.11.006.

Tudoroiu, M. et al., 2016: Negative elevation-dependent warming trend in the Eastern Alps. *Environ. Res. Lett.*, 11(4), doi:10.1088/1748-9326/11/4/044021.

UN, 2015: *Transforming governance for the 2030 agenda for sustainable development.* UN, New York, NY, [Available at: https://sustainabledevelopment.un.org/content/documents/21252030%20Agenda%20for%20Sustainable%20Development%20web.pdf].

UNESCO, 1972: *Convention Concerning the Protection of the World Cultural and Naturral Heritage. Adopted by the General Conference at its seventeenth session Paris, 16 november 1972.* United Nations Educational, Scientific, and Cultural Organisation (UNESCO), Paris, [Available at: https://whc.unesco.org/archive/convention-en.pdf]. Accessed 08/08/2019.

UNESCO, 2012: *Operational Guidelines for the Implementation of the World Heritage Convention.* United Nations Educational, Scientific And Cultural Organisation (UNESCO), Paris. [Available at: http://whc.unesco.org/en/guidelines]. Accessed 08/08/2019.

UNFCCC, 2015: Paris Agreement. United Nations. Climate Change Secretariat, UNEP's Information Unit for Conventions (IUC), Bonn, Germany, 30pp. [Available at: http://unfccc.int/files/essential_background/convention/application/pdf/english_paris_agreement.pdf].

UNFCCC Secretariat, 2014: Subsidiary body for scientific and technological advice. Forty-first session, Lima 1–6 December 2014. Report of the executive committee of the Warsaw international mechanism for loss and damage associated with climate change impacts. UNFCCC, Lima, [Available at: https://unfccc.int/resource/docs/2014/sb/eng/04.pdf]. Accessed 08/08/2019.

UNHRC, 2018: *Resolution adopted by the General Assembly on 26 September 2018: United Nations Declaration on the Rights of Peasants and Other People Working in Rural Areas.* UNHRC 39th Assembly General [Available at: https://undocs.org/A/HRC/39/L.16].

UNISDR, 2015: *Sendai Framework for Disaster Risk Reduction 2015–2030,* Geneva, The United Nations Office for Disaster Risk Reduction. [Available at: www.unisdr.org/files/43291_sendaiframeworkfordrren.pdf].

Uniyal, A., 2013: Lessons from Kedarnath tragedy of Uttarakhand Himalaya, India. *Current Science*, **105**(11), 1472–1474.

Vaidya, R.A. et al., 2019: Disaster Risk Reduction and Building Resilience in the Hindu Kush Himalaya. In: *The Hindu Kush Himalaya Assessment: Mountains, Climate Change, Sustainability and People* [Wester, P., A. Mishra, A. Mukherji and A.B. Shrestha (eds.)]. Springer International Publishing, Cham, pp. 389–419. ISBN 9783319922874.

Valdés-Pineda, R. et al., 2014: Water governance in Chile: Availability, management and climate change. *J. Hydrol.*, **519**, 2538–2567, doi:10.1016/j.jhydrol.2014.04.016.

Vallance, J.W., 2005: Volcanic debris flows. [M. Jakob and O. Hungr (eds.)], Debris-flow Hazards and Related Phenomena. Springer Berlin Heidelberg, Berlin, Heidelberg. 247–274. ISBN 978-3-540-27129-1, doi:10.1007/3-540-27129-5_10.

van der Geest, K. and M. Schindler, 2016: Brief communication: Loss and damage from a catastrophic landslide in Nepal. *Nat. Hazard. Earth Sys.*, **16**(11), 2347–2350, doi:10.5194/nhess-16-2347-2016.

van der Woerd, J. et al., 2004: Giant, ~M8 earthquake-triggered ice avalanches in the eastern Kunlun Shan, northern Tibet: Characteristics, nature and dynamics. *Bull. Geol. Soc. Am.*, **116**(3–4), 394–406, doi:10.1130/B25317.1.

Vanat, L., 2018: 2018 International Report on Snow & Mountain Tourism. Overview of the key industry figures for ski resorts. [Available at: https://vanat.ch/RM-world-report-2018.pdf].

Veh, G., O. Korup, S. Roessner and A. Walz, 2018: Detecting Himalayan glacial lake outburst floods from Landsat time series. *Remote Sens. Environ.*, **207**, 84–97. doi:10.1016/j.rse.2017.12.025.

Veh, G. et al., 2019: Unchanged frequency of moraine-dammed glacial lake outburst floods in the Himalaya. *Nat. Clim. Change*, **9**(5), 379–383, doi:10.1038/s41558-019-0437-5.

Vergara, W. et al., 2007: Economic impacts of rapid glacier retreat in the Andes. *Eos, Trans. AGU*, **88**(25), 261–264, doi:10.1029/2007EO250001.

Vermilyea, A.W. et al., 2017: Continuous proxy measurements reveal large mercury fluxes from glacial and forested watersheds in Alaska. *Sci. Total Environ.*, **599–600**, 145–155, doi:10.1016/j.scitotenv.2017.03.297.

Viazzo, P.P., 1989: *Upland communities: Environment, populations and social structure in the Alps since the sixteenth century.* Cambridge University Press, Cambridge. ISBN: 9780521034166.

Vigano, G. et al., 2016: Effects of Future Climate Change on a River Habitat in an Italian Alpine Catchment. *J. Hydrol. Eng.*, **21**(2), doi:10.1061/(ASCE)HE.1943-5584.0001293.

Vince, G., 2009: Profile: Chewang Norphel. Glacier man. American Association for the Advancement of Science, **326**, 659–661, doi:10.1126/science.326_659.

Viviroli, D. et al., 2011: Climate change and mountain water resources: overview and recommendations for research, management and policy. *Hydrol. Earth Syst. Sc.*, **15**(2), 471–504, doi:10.5194/hess-15-471-2011.

Vors, L.S. and M.S. Boyce, 2009: Global declines of caribou and reindeer. *Glob. Change Biol*, **15**(11), 2626–2633, doi:10.1111/j.1365-2486.2009.01974.x.

Vuille, M., 2013: *Climate change and water resources in the tropical Andes.* Inter-American Development Bank Technical Note 515. [Available at: https://publications.iadb.org/handle/11319/5827].

Vuille, M. et al., 2018: Rapid decline of snow and ice in the tropical Andes – Impacts, uncertainties and challenges ahead. *Earth-Sci. Rev.*, **176**, 195–213, doi:10.1016/j.earscirev.2017.09.019.

Waechter, A., L. Copland and E. Herdes, 2015: Modern glacier velocities across the Icefield Ranges, St Elias Mountains, and variability at selected glaciers from 1959 to 2012. *J. Glaciol.*, **61**(228), 624–634, doi:10.3189/2015JoG14J147.

2

Walter, D., 2017: Percepciones tradicionales del cambio climático en comunidades altoandinas en la Cordillera Blanca, Ancash. *Revista de Glaciares y Ecosistemas de Montaña*, **3**, 9–24.

Wang, L. et al., 2015: Glacier changes in the Sikeshu River basin, Tienshan Mountains. *Quaternary International*, **358**, 153–159, doi:10.1016/j.quaint.2014.12.028.

Wang, Q., X. Fan and M. Wang, 2016a: Evidence of high-elevation amplification versus Arctic amplification. *Sci. Rep.*, **6** (19219), doi:10.1038/srep19219.

Wang, S., Y. He and X. Song, 2010: Impacts of climate warming on Alpine glacier tourism and adaptive measures: A case study of Baishui Glacier No. 1 in Yulong Snow Mountain, Southwestern China. *J. Earth Sci.*, **21**(2), 166–178, doi:10.1007/s12583-010-0015-2.

Wang, S. et al., 2017: Complex responses of spring alpine vegetation phenology to snow cover dynamics over the Tibetan Plateau, China. *Sci. Total Environ.*, **593–594**, 449–461, doi:10.1016/j.scitotenv.2017.03.187.

Wang, X. et al., 2018: Snow cover phenology affects alpine vegetation growth dynamics on the Tibetan Plateau: Satellite observed evidence, impacts of different biomes, and climate drivers. *Agr. Forest Meterol.*, **256–257**, 61–74, doi:10.1016/j.agrformet.2018.03.004.

Wang, X. et al., 2016b: The role of permafrost and soil water in distribution of alpine grassland and its NDVI dynamics on the Qinghai-Tibetan Plateau. *Glob. Planet. Change*, **147**, 40–53, doi:10.1016/J.GLOPLACHA.2016.10.014.

Wangchuk, K. and J. Wangdi, 2018: Signs of climate warming through the eyes of yak herders in northern Bhutan. *Mt. Res. Dev.*, **38**(1), 45–52, doi:10.1659/MRD-JOURNAL-D-17-00094.1.

Warren, F. J. and D.S. Lemmen, 2014: *Canada in a Changing Climate: Sector Perspectives on Impacts and Adaptation*. Government of Canada, Ottawa, ON, 286pp. ISBN: 978-1-100-24142-5. [Available at: www.weadapt.org/sites/weadapt.org/files/2017/february/canadasectorperspectivesfullreport_eng_0.pdf#page=70].

Watson, C.S. and O. King, 2018: Everest's thinning glaciers: implications for tourism and mountaineering. *Geology Today*, doi:10.1111/gto.12215.

Wegner, S.A., 2014: *Lo que el agua se llevó: Consecuencias y lecciones del aluvión de Huaraz de 1941. Notas Técnicas sobre Cambio Climático* 7, Ministerio de Ambiente, Lima, [Available at: https://archive.org/details/NotaTecnica7/page/n1].

Wei, M., N. Fujun, A. Satoshi and A. Dewu, 2006: Slope instability phenomena in permafrost regions of Qinghai-Tibet Plateau, China. *Landslides*, **3**(3), 260–264, doi:10.1007/s10346-006-0045-0.

Welling, J., R. Ólafsdóttir, Þ. Árnason and S. Guðmundsson, 2019: Participatory Planning Under Scenarios of Glacier Retreat and Tourism Growth in Southeast Iceland. *Mt. Res. Dev.*, **39** (2), D1–D13, doi:10.1659/MRD-JOURNAL-D-18-00090.1

Wester, P., A. Mishra, A. Mukherji and A.B. Shrestha (eds.), 2019: The Hindu Kush Himalaya Assessment – Mountains, Climate Change, Sustainability and People. Springer, 627 pp. ISBN 9783319922881.

Westerling, A.L., 2016: Increasing western US forest wildfire activity: sensitivity to changes in the timing of spring. *Philos. Trans. R. Soc. London (Biol).*, **371**(1696), 20150178, doi:10.1098/rstb.2015.0178.

Westermann, S. et al., 2015: A ground temperature map of the North Atlantic permafrost region based on remote sensing and reanalysis data. *The Cryosphere*, **9**(3), 1303–1319, doi:10.5194/tc-9-1303-2015.

White, K.S., D.P. Gregovich and T. Levi, 2017: Projecting the future of an alpine ungulate under climate change scenarios. *Glob. Change Biol*, **24**(3), 113601149, doi:10.1111/gcb.13919.

Wikstrom Jones, K. et al., 2018: Community Snow Observations (CSO): A citizen science campaign to validate snow remote sensing products and hydrological models. In: *International Snow Science Workshop*, Innsbruck, Austria, pp. 420–424.

Williams, C.M., H.A.L. Henry and B.J. Sinclair, 2015: Cold truths: how winter drives responses of terrestrial organisms to climate change. *Biol. Rev.*, **90**(1), 214–235, doi:10.1111/brv.12105.

Williamson, C.J. et al., 2019: Glacier Algae: A Dark Past and a Darker Future. *Front. Microbiol.* **10**, 519, doi:10.3389/fmicb.2019.00524.

Willis, M.J. et al., 2018: Massive destabilization of an Arctic ice cap. *Earth Planet Sc. Lett.*, **502**, 146–155, doi:10.1016/j.epsl.2018.08.049.

Winkler, D.E. et al., 2018: Snowmelt timing regulates community composition, phenology, and physiological performance of alpine plants. *Front. Plant. Sci.* **9**, 1140, doi:10.3389/fpls.2018.01140.

Winkler, D.E., K.J. Chapin and L.M. Kueppers, 2016: Soil moisture mediates alpine life form and community productivity responses to warming. *Ecology*, **97**(6), 1553–1563, doi:10.1890/15-1197.1.

Winski, D. et al., 2017: Industrial-age doubling of snow accumulation in the Alaska Range linked to tropical ocean warming. *Sci. Rep.*, **7**, 17869, doi:10.1038/s41598-017-18022-5.

Winter, K.J.P.M., S. Kotlarski, S.C. Scherrer and C. Schär, 2017: The Alpine snow-albedo feedback in regional climate models. *Clim. Dyn.*, **48**(3–4), 1109–1124, doi:10.1007/s00382-016-3130-7.

Wirz, V., M. Geertsema, S. Gruber and R.S. Purves, 2016: Temporal variability of diverse mountain permafrost slope movements derived from multi-year daily GPS data, Mattertal, Switzerland. *Landslides*, **13** (1), 67–83, doi:10.1007/s10346-014-0544-3.

Wobus, C. et al., 2017: Projected climate change impacts on skiing and snowmobiling: A case study of the United States. *Glob. Environ. Change*, **45**, 1–14 doi:10.1016/j.gloenvcha.2017.04.006.

Wouters, B., A.S. Gardner and G. Moholdt, 2019: Global glacier mass loss during the GRACE satellite mission (2002–2016). *Front. Earth Sci.*, **7**(96), doi:10.3389/feart.2019.00096.

Wrathall, D.J. et al., 2014: Migration Amidst Climate Rigidity Traps: Resource Politics and Social-Ecological Possibilism in Honduras and Peru. *Ann. Am. Assoc. Geogr.*, **104**(2), 292–304, doi:10.1080/00045608.2013.873326.

Wu, Q., Y. Hou, H. Yun and Y. Liu, 2015: Changes in active-layer thickness and near-surface permafrost between 2002 and 2012 in alpine ecosystems, Qinghai-Xizang (Tibet) Plateau, China. *Glob. Planet. Change*, **124**, 149–155, doi:10.1016/j.gloplacha.2014.09.002.

Wu, X. et al., 2018: Uneven winter snow influence on tree growth across temperate China. *Glob. Change Biol*, **25**(1), 144–154, doi:10.1111/gcb.14464.

Würzer, S., T. Jonas, N. Wever and M. Lehning, 2016: Influence of Initial Snowpack Properties on Runoff Formation during Rain-on-Snow Events. *J. Hydrometeorol.*, **17**(6), 1801–1815, doi:10.1175/JHM-D-15-0181.1.

Wymann von Dach, S. et al., 2017: Safer lives and livelihoods in mountains: Making the Sendai framework for disaster risk reduction work for sustainable mountain development. Centre for Development and Environment (CDE), University of Bern, with Bern Open Publishing (BOP), Bern, Switzerland, 82 pp.

Wymann von Dach, S. et al., 2018 *Leaving no one in mountains behind: Localizing the SDGs for resilience of mountain people and ecosystems*. Mountain Research Initiative and Centre for Development and Environment, Bern, Switzerland, [Available at: https://boris.unibe.ch/id/eprint/120130]. Accessed 08/08/2019.

Xenarios, S. et al., 2018: Climate change and adaptation of mountain societies in Central Asia: uncertainties, knowledge gaps, and data constraints. *Reg. Environ. Change*, **31**(3–4), 1113, doi:10.1007/s10113-018-1384-9.

Xiao, C.-D., S.-J. Wang and D.H. Qin, 2015: A preliminary study of cryosphere service function and value evaluation. *Adv. Clim. Change Res.*, **6**(3–4), 181–187, doi:10.1016/j.accre.2015.11.004.

Xie, J. et al., 2018: Relative influence of timing and accumulation of snow on alpine land surface phenology. *J. Geophys. Res.-Biogeosci.*, **123**(2), 561–576, doi:10.1002/2017JG004099.

Xu, Y., V. Ramanathan and W.M. Washington, 2016: Observed high-altitude warming and snow cover retreat over Tibet and the Himalayas enhanced by black carbon aerosols. *Atmos. Chem. Phys.*, **16**(3), 1303–1315, doi:10.5194/acp-16-1303-2016.

Yager, K., 2015: Saiellite Imagery and community perceptions of climate change impacts and landscape change. [Barnes, J. and M. Dove (eds.)], Climate Cultures: Anthropological Perspectives on Climate Change. New Haven, Yale University Press, 146–168.

Yang, K. et al., 2014a: Recent climate changes over the Tibetan Plateau and their impacts on energy and water cycle: A review. *Glob. Planet. Change*, **112**, 79–91, doi:10.1016/J.GLOPLACHA.2013.12.001.

Yang, M. et al., 2010a: Permafrost degradation and its environmental effects on the Tibetan Plateau: A review of recent research. *Earth-Sci. Rev.*, **103** (1–2), 31–44, doi:10.1016/j.earscirev.2010.07.002.

Yang, Y. et al., 2018: Permafrost and drought regulate vulnerability of Tibetan Plateau grasslands to warming. *Ecosphere*, **9**(5), e02233, doi:10.1002/ecs2.2233.

Yang, Y.-C.E. et al., 2014b: Water governance and adaptation to climate change in the Indus River Basin. *J. Hydrol.*, **519**, 2527–2537, doi:10.1016/j.jhydrol.2014.08.055.

Yang, Z.-p. et al., 2010b: Effects of permafrost degradation on ecosystems. *Acta Ecologica Sinica*, **30**(1), 33–39, doi:10.1016/j.chnaes.2009.12.006.

Yao, X. et al., 2016: Spatial-temporal variations of lake ice phenology in the Hoh Xil region from 2000 to 2011. *J. Geogr. Sci.*, **26**(1), 70–82, doi:10.1007/s11442-016-1255-6.

Yasuda, T. and M. Furuya, 2015: Dynamics of surge-type glaciers in West Kunlun Shan, Northwestern Tibet. *J. Geophys. Res-Earth*, **120**(11), 2393–2405, doi:10.1002/2015JF003511.

Yasunari, T.J., R.D. Koster, W.K.M. Lau and K.-M. Kim, 2015: Impact of snow darkening via dust, black carbon, and organic carbon on boreal spring climate in the Earth system. *J. Geophys. Res-Earth*, **120**(11), 5485–5503, doi:10.1002/2014JD022977.

Ye, H., D. Yang and D. Robinson, 2008: Winter rain on snow and its association with air temperature in northern Eurasia. *Hydrol. Process.*, **22**(15), 2728–2736, doi:10.1002/hyp.7094.

You, J. et al., 2018: Response to climate change of montane herbaceous plants in the genus Rhodiola predicted by ecological niche modelling. *Sci. Rep.*, **8**, 1–12, doi:10.1038/s41598-018-24360-9.

You, Q. et al., 2010: Climate warming and associated changes in atmospheric circulation in the eastern and central Tibetan Plateau from a homogenized dataset. *Glob. Planet. Change*, **72**, 11–24, doi:10.1016/j.gloplacha.2010.04.003.

You, Q. et al., 2013: Decadal variation of surface solar radiation in the Tibetan Plateau from observations, reanalysis and model simulations. *Clim. Dyn.*, **40**(7–8), 2073–2086, doi:10.1007/s00382-012-1383-3.

Young, E.F. et al., 2018: Stepping stones to isolation: Impacts of a changing climate on the connectivity of fragmented fish populations. *Evol. Appl.*, **11**(6), 978--994, doi:10.1111/eva.12613.

Young, G. et al., 2010: Vulnerability and adaptation in a dryland community of the Elqui Valley, Chile. *Clim. Change*, **98**(1–2), 245–276, doi:10.1007/s10584-009-9665-4.

Yu, F., J. Qi, X. Yao and Y. Liu, 2013: In-situ monitoring of settlement at different layers under embankments in permafrost regions on the Qinghai-Tibet Plateau. *Eng. Geol.*, **160**, 44–53, doi:10.1016/j.enggeo.2013.04.002.

Yu, W., F. Han, W. Liu and S.A. Harris, 2016: Geohazards and thermal regime analysis of oil pipeline along the Qinghai–Tibet Plateau Engineering Corridor. *Nat. Hazards*, **83**(1), 193–209, doi:10.1007/s11069-016-2308-y.

Yu, Y. et al., 2019: Climate change, water resources and sustainable development in the arid and semi-arid lands of Central Asia in the past 30 years. *J. Arid Land*, **11**(1), 1–14, doi:10.1007/s40333-018-0073-3.

Yucel, I., A. Güventürk and O.L. Sen, 2015: Climate change impacts on snowmelt runoff for mountainous transboundary basins in eastern Turkey. *Int. J. Climatol.*, **35**(2), 215–228, doi:10.1002/joc.3974.

Zaharescu, D.G. et al., 2016: Climate change enhances the mobilisation of naturally occurring metals in high altitude environments. *Sci. Total Environ.*, **560–561**, 73–81, doi:10.1016/j.scitotenv.2016.04.002.

Zazulie, N., 2016: Estudio de los distintos factores que afectaron la evolución de los glaciares en los Andes Centrales del Sur y sus proyecciones ante posibles escenarios de cambio climático. Tesis Doctoral, Facultad de Ciencias Exactas y Naturales. Universidad de Buenos Aires. [Available at: http://hdl.handle.net/20.500.12110/tesis_n5842_Zazulie].

Zazulie, N., M. Rusticucci and G.B. Raga, 2018: Regional climate of the Subtropical Central Andes using high-resolution CMIP5 models. Part II: future projections for the twenty-first century. *Clim. Dyn.*, **51**(7–8), 2913–2925, doi:10.1007/s00382-017-4056-4.

Zdanowicz, C. et al., 2013: Accumulation, storage and release of atmospheric mercury in a glaciated Arctic catchment, Baffin Island, Canada. *Geochimica et Cosmochimica Acta*, **107**, 316–335, doi:10.1016/j.gca.2012.11.028.

Zekollari, H., M. Huss and D. Farinotti, 2019: Modelling the future evolution of glaciers in the European Alps under the EURO-CORDEX RCM ensemble. *The Cryosphere*, **13**(4), 1125–1146, doi:10.5194/tc-13-1125-2019.

Zemp, M. et al., 2015: Historically unprecedented global glacier decline in the early 21st century. *J. Glaciol.*, **61**(228), 745–762, doi:10.3189/2015JoG15J017.

Zemp, M. et al., 2019: Global glacier mass changes and their contributions to sea-level rise from 1961 to 2016 *Nature*, **568**(7752), 382–386 doi:10.1038/s41586-019-1071-0.

Zeng, Z. et al., 2015: Regional air pollution brightening reverses the greenhouse gases induced warming-elevation relationship. *Geophys. Res. Lett.*, **42**(11), 4563–4572, doi:10.1002/2015GL064410.

Zhang, G. et al., 2014: Estimating surface temperature changes of lakes in the Tibetan Plateau using MODIS LST data. *J. Geophys. Res.*, **119**(14), 8552–8567, doi:10.1002/2014JD021615.

Zhang, G. et al., 2015: An inventory of glacial lakes in the Third Pole region and their changes in response to global warming. *Glob. Planet. Change*, **131**, 148–157, doi:10.1016/j.gloplacha.2015.05.013.

Zhang, H.-X. and M.-L. Zhang, 2017: Spatial patterns of species diversity and phylogenetic structure of plant communities in the Tianshan Mountains, arid Central Asia. *Front. Plant Sci.*, **8**, 2134, doi:10.3389/fpls.2017.02134.

Zhang, Q. et al., 2012: Mercury distribution and deposition in glacier snow over western China. *Environ. Sci. Technol.*, **46**(10), 5404–5413. doi:10.1021/es300166x.

Zhang, Y. et al., 2017: Light-absorbing impurities enhance glacier albedo reduction in the southeastern Tibetan plateau. *J. Geophys. Res-Atmos*, **122**(13), 6915–6933, doi:10.1002/2016JD026397.

Zhang, Y. et al., 2018: Black carbon and mineral dust in snow cover on the Tibetan Plateau. *The Cryosphere*, **12**(2), 413–431, doi:10.5194/tc-12-413-2018.

Zhao, L., Q. Wu, S. Marchenko and N. Sharkhuu, 2010: Thermal state of permafrost and active layer in central Asia during the international polar year. *Permafrost Periglac.*, **21**(2), 198–207, doi:10.1002/ppp.688.

Zhou, Y., Z. Li and J. Li, 2017: Slight glacier mass loss in the Karakoram region during the 1970s to 2000 revealed by KH-9 images and SRTM DEM *J. Glaciol.*, **63**(238), 331–342, doi:10.1017/jog.2016.142.

Ziegler, A.D. et al., 2014: Pilgrims, progress, and the political economy of disaster preparedness – the example of the 2013 Uttarakhand flood and Kedarnath disaster. *Hydrol. Process.*, **28**(24), 5985–5990, doi:10.1002/hyp.10349.

Zimmer, A. et al., 2018: Time lag between glacial retreat and upward migration alters tropical alpine communities. *Perspect. Plant Ecol. Evol. Syst.*, **30**, 89–102, doi:10.1016/j.ppees.2017.05.003.

Zimmermann, M. and W. Haeberli, 1992: Climatic change and debris flow activity in high-mountain areas – a case study in the Swiss Alps. *Catena Supplement*, **22**, 59–72.

Zimov, S.A., E.A.G. Schuur and F.S. Chapin, 2006: Permafrost and the global carbon budget. *Science*, **312**(5780), 1612–1613, doi:10.1126/science.1128908.

2

Zimova, M., L.S. Mills and J.J. Nowak, 2016: High fitness costs of climate change-induced camouflage mismatch. *Ecol. Letters*, **19**(3), 299–307, doi:10.1111/ele.12568.

Zimova, M. and Hackländer, K. and Good, J.M. and Melo-Ferreira, J. and Alves, P.C. and Mills, L.S., 2018. Function and underlying mechanisms of seasonal colour moulting in mammals and birds: what keeps them changing in a warming world?. *Biol. Rev.*, **93**(3): 1478–1498. doi:10.1111/brv.12405.

Zou, D. et al., 2017: A new map of permafrost distribution on the Tibetan Plateau. *The Cryosphere*, **11**(6), 2527–2542, doi:10.5194/tc-11-2527-2017.

Zuckerberg, B. and J.N. Pauli, 2018: Conserving and managing the subnivium. *Conserv. Biol.*, **32**(4), 774–781, doi:10.1111/cobi.13091.

Þórhallsdóttir, G. and R. Ólafsson, 2017: A method to analyse seasonality in the distribution of tourists in Iceland. *J. Outdoor Recreat.*, **19**, 17–24, doi:10.1016/j.jort.2017.05.001.

Appendix 2.A: Additional Information on Global and Regional Glacier Mass Change Estimates for 2006–2015

Two global-scale estimates of recent glacier mass changes have been published since AR5 (Wouters et al., 2019; Zemp et al., 2019) that include area-averaged estimates for large-scale glacier regions as defined by the Randolph Glacier Inventory (RGI Consortium, 2017). Zemp et al. (2019) is based on extrapolation of geodetic and glaciological observations, while Wouters et al. (2019) use gravimetric measurements from the Gravity Recovery and Climate Experiment (GRACE). For some regions, additional estimates are available mostly based on remote sensing data (Table 2.A.1).

These estimates were used to derive an average mass change rate for the period 2006–2015 for each glacier region covered in both Chapter 2 and 3. Where several estimates were available for this period or similar periods, these were averaged and uncertainties obtained from standard error propagation assuming the estimates to be independent. The GRACE estimates were only considered in regions with extensive ice cover due to generally large uncertainties in regions with little ice cover (Wouters et al., 2019). The estimates for the polar regions by Box et al. (2018) were not used since they are based on an earlier version of the data by Wouters et al. (2019).

Individual regional estimates for overlapping periods between 2000 and 2017 were recalculated to represent the period 2006–2015, prior to averaging with other existing estimates. For Western Canada and USA the mass change rate by Menounos et al. (2019)

for 2000–2009 was assumed to hold for 2006–2009, and the rate of –12 ± 5 Gt yr^{-1} for 2009–2018 was assumed to be valid for 2010–2015. For Iceland the mass change rate by Björnsson et al. (2013) for 2003–2010 was assumed to hold for 2006–2010, and the rate by Foresta et al. (2016) for 2011–2015 was used for the remaining years. The estimate for Iceland by Nilsson et al. (2015) for the period 2003–2009 is similar to the estimate by Björnsson et al. (2013), but was not used since it is based on spatially relatively scarce remote sensing data compared to Björnsson et al. (2013), which is based on detailed glaciological and geodetic balances. The GRACE estimate for Iceland was not used since it deviates strongly from the estimate by Zemp et al. (2019) which is well-constrained by direct observations in this region, while the GRACE estimate may have been affected by the mass change signal from ice masses in southeast Greenland and processes in the Earth mantle cause by isostatic adjustments since the end of the 19th century (Sørensen et al., 2017). For the Low Latitudes (>99% of glacier area in the Andes) available mass loss estimates differ considerably. Zemp et al. (2019)'s high estimate relies on extrapolation of observations from less than 1% of the glacier area, while the low estimate by Braun et al. (2019) for the Andes may underestimate mass loss due to incomplete coverage and systematic errors in their derived digital elevation models due to radar penetration. In the absence of other estimates for this period the average of both estimates is used. For Arctic Canada and the Southern Andes, the estimates by Zemp et al. (2019) were not considered since they rely on observations from less than 5% of the glacier area. The regional estimates by Gardner et al. (2013) for the period 2003–2009 informed AR5 and are given for comparison but not included in the composite estimate for 2006–2015.

Table 2A.1 | Regional estimates of glacier mass budget in three different units. Only estimates from the studies marked in bold were used to derive the average SROCC estimates. Regional glacier area *A* and volume *V* are taken from the Randolph Glacier Inventory (RGI Consortium, 2017) and Farinotti et al. (2019), respectively. Method *geod.* refers to the geodetic method (using elevation changes) and *gl.* refers to the glaciological method (based on *in situ* mass-balance observations). Results are given for various aggregated areas including among others all regions combined (global), and global excluding the Antarctic (A) and Greenland (G) periphery. All regional estimates (in kg m^{-2} yr^{-1}) are shown in Figures 2.4 and 3.8). SLE is sea level equivalent.

Mass budget		kg m^{-2} yr^{-1}	Gt yr^{-1}	mm SLE yr^{-1}	Reference	Method
Alaska, A=86,725 km^2, V=43.3 ± 11.2 mm SLE	2003–2009	–570 ± 200	–50 ± 17	0.14 ± 0.05	Gardner et al. (2013)	GRACE
	1986–2005	–610 ± 280	–53 ± 24	0.15 ± 0.07	Box et al. (2018)	GRACE, gl.
	1994–2013	–865 ± 130	–75 ± 11	0.21 ± 0.03	Larsen et al. (2015)	geod.
	2006–2015	–710 ± 340	–61 ± 30	0.17 ± 0.08	Box et al. (2018)	GRACE, gl.
	2006–2015	–570 ± 180	–49 ± 16	0.14 ± 0.04	**Wouters et al. (2019)**	GRACE
	2006–2015	–830 ± 190	–71 ± 17	0.20 ± 0.05	**Zemp et al. (2019)**	gl., geod.
	2006–2015	**–700 ± 180**	**–60 ± 16**	**0.17 ± 0.04**	SROCC	
Western Canada and USA, A=14,524 km^2, hV=2. 6± 0.7 mm SLE	2003–2009	–930 ± 230	–14 ± 3	0.04 ± 0.01	Gardner et al. (2013)	gl.
	2000–2009	–200 ± 250	–3 ± 3	0.01 ± 0.01	**Menounos et al. (2019)**	geod.
	2009–2018	–860 ± 320	–12 ± 5	0.03 ± 0.01	**Menounos et al. (2019)**	geod.
	2006–2015	–410 ± 1,500	–6 ± 22	0.02 ± 0.06	**Wouters et al. (2019)**	GRACE
	2006–2015	–800 ± 400	–11 ± 6	0.03 ± 0.02	**Zemp et al. (2019)**	gl., geod.
	2006–2015	**–500 ± 910**	**–8 ± 13**	**0.02 ± 0.04**	SROCC	

Mass budget		kg m⁻² yr⁻¹	Gt yr⁻¹	mm SLE yr⁻¹	Reference	Method
Iceland, **A=11,060 km²,** **V=9.1 ± 2.4 mm SLE***	2003–2009	−910 ± 150	−10 ± 2	0.03 ± 0.01	Gardner et al. (2013)	GRACE, gl.
	1986–2005	−360 ± 630	−4 ± 7	0.01 ± 0.02	Box et al. (2018)	GRACE, gl.
	1995–2010	−860 ± 140	−10 ± 2	0.03 ± 0.00	Björnsson et al. (2013)	gl. geod.
	2003–2010	−995 ± 140	−11 ± 2	0.03 ± 0.00	**Björnsson et al. (2013)**	gl. geod.
	2003–2009	−890 ± 250	−10 ± 3	0.03 ± 0.01	Nilsson et al. (2015)	geod.
	2011–2015	−590 ± 70	−6 ± 1	0.02 ± 0.00	**Foresta et al. (2016)**	geod.
	2006–2015	−910 ± 190	−10 ± 2	0.03 ± 0.01	Wouters et al. (2019)	GRACE
	2006–2015	−620 ± 410	−7 ± 4	0.02 ± 0.01	**Zemp et al. (2019)**	gl., geod.
	2006–2015	**−690 ± 260**	**−7 ± 3**	**0.02 ± 0.01**	SROCC	
Scandinavia, **A=2,949 km²,** **V=0.7 ± 0.2 mm SLE**	2003–2009	−610±140	−2 ± 0	0.01 ± 0.00	Gardner et al. (2013)	gl.
	1986–2005	−120 ± 1,170	−0 ± 3	0.00 ± 0.01	Box et al. (2018)	GRACE, gl.
	2006–2015	230 ± 3,820	1 ± 11	−0.00 ± 0.03	Wouters et al. (2019)	GRACE
	2006–2015	−660 ± 270	−2 ± 1	0.01 ± 0.00	**Zemp et al. (2019)**	gl., geod
	2006–2015	−370 ± 1,220	−1 ± 4	0.00 ± 0.01	Box et al. (2018)	GRACE, gl.
	2006–2015	**−660 ± 270**	**−2 ± 1**	**0.01 ± 0.00**	SROCC	
North Asia, **A=2,410 km²,** **V=0.3 ± 0.1 mm SLE**	2003–2009	−630 ± 310	−2 ± 0	0.01 ± 0.00	Gardner et al. (2013)	gl.
	2006–2015	890 ± 1,850	2 ± 5	−0.01 ± 0.01	Wouters et al. (2019)	GRACE
	2006–2015	−400 ± 310	−1 ± 1	0.00 ± 0.00	**Zemp et al. (2019)**	gl., geod.
	2006–2015	**−400 ± 310**	**−1 ± 1**	**0.00 ± 0.00**	SROCC	
Central Europe, **A=2,092 km²,** **V=0.3 ± 0.1 mm SLE**	2003–2009	−1,060 ± 170	−2 ± 0	0.01 ± 0.00	Gardner et al. (2013)	gl.
	2006–2015	100 ± 510	0 ± 1	−0.00 ± 0.00	Wouters et al. (2019)	GRACE
	2006–2015	−910 ± 70	−2 ± 0	0.01 ± 0.00	**Zemp et al. (2019)**	gl., geod.
	2006–2015	**−910 ± 70**	**−2 ± 0**	**0.01 ± 0.00**	SROCC	
Caucasus and Middle East, **A=1,307 km²,** **V=0.2 ± 0.0 mm SLE**	2003–2009	−900 ± 160	−1 ± 0	0.00 ± 0.00	Gardner et al. (2013)	gl.
	2006–2015	−650 ± 3000	−1 ± 4	0.00 ± 0.01	Wouters et al. (2019)	GRACE
	2006–2015	−880 ± 570	−1 ± 1	0.00 ± 0.00	**Zemp et al. (2019)**	gl., geod.
	2006–2015	**−880 ± 570**	**−1 ± 1**	**0.00 ± 0.00**	SROCC	
High Mountain Asia, **A=97,605 km²,** **V=16.9 ± 2.7 mm SLE**	2003–2009	−220 ± 100	−26 ± 12	−0.07 ± 0.03	Gardner et al. (2013)	GRACE, geod.
	2006–2015	−110 ± 140	−11 ± 14	0.03 ± 0.04	**Wouters et al. (2019)**	GRACE
	2006–2015	−190 ± 70	−18 ± 7	0.05 ± 0.02	**Zemp et al. (2019)**	gl., geod.
	2000–2016	−180 ± 40	−16 ± 4	0.04 ± 0.01	Brun et al. (2017)	geod.
	2006–2015	**−150 ± 110**	**−14 ± 11**	**0.04 ± 0.03**	SROCC	
Low Latitudes, **A=23,41 km²,** **V=0.2 ± 0.1 mm SLE**	2003–2009	−1,080 ± 360	−4 ± 1	0.01 ± 0.00	Gardner et al. (2013)	gl.
	2000–2013	−230 ± 40	−1 ± 0	0.00 ± 0.00	**Braun et al. (2019)**	geod.
	2006–2015	1,560 ± 510	4 ± 1	−0.01 ± 0.00	Wouters et al. (2019)	GRACE
	2006–2015	−940 ± 820	−2 ± 2	0.01 ± 0.00	**Zemp et al. (2019)**	gl., geod
	2006–2015	**−590 ± 580**	**−1 ± 1**	**0.00 ± 0.00**	SROCC	
Southern Andes, **A=29,429 km²,** **V=12.8 ± 3.3 mm SLE**	2003–2009	−990 ± 360	−29 ± 10	0.08 ± 0.03	Gardner et al. (2013)	GRACE
	2006–2015	−1,070 ± 240	−31 ± 7	0.09 ± 0.02	**Wouters et al. (2019)**	GRACE
	2006–2015	−1300 ± 380	−35 ± 11	0.10 ± 0.03	Zemp et al. (2019)	gl., geod
	2000–2015	−640 ± 20	−19 ± 1	0.05 ± 0.00	**Braun et al. (2019)**	geod.
	2011–2017	−1,280 ± 120	−21 ± 2	0.06 ± 0.01	Foresta et al. (2018)**	geod.
	2006–2015	**−860 ± 170**	**−25 ± 4**	**0.07 ± 0.01**	SROCC	

Mass budget		kg m⁻² yr⁻¹	Gt yr⁻¹	mm SLE yr⁻¹	Reference	Method
New Zealand, A=1,162 km², V=0.2 ± 0.0 mm SLE	2003–2009	−320 ± 780	0 ± 1	0.00 ± 0.00	Gardner et al. (2013)	gl.
	2006–2015	110 ± 780	0 ± 1	0.00 ± 0.00	Wouters et al. (2019)	GRACE
	2006–2015	−590 ± 1140	−1 ± 1	0.00 ± 0.00	**Zemp et al. (2019)**	gl., geod
	2006–2015	**−590 ± 1140**	**−1 ± 1**	**0.00 ± 0.00**	SROCC	
Arctic Canada North, A=105,111 km², V=64.8 ± 16.8 mm SLE	2003–2009	−310 ± 40	−33 ± 4	0.09 ± 0.01	Gardner et al. (2013)	GRACE, geod.
	1958–1995	−114 ± 110	−12 ± 12	0.03 ± 0.03	Noël et al. (2018)	Model
	1996–2015	−270 ± 110	−28 ± 12	0.08 ± 0.03	Noël et al. (2018)	Model
	1991–2014	−170 ± 50	−16 ± 2	0.04 ± 0.00	Millan et al. (2017)	Model
	1991–2005	−60 ± 20	−6 ± 1	0.02 ± 0.00	Millan et al. (2017)	
	2005–2014	−340 ± 30	−33 ± 3	0.09 ± 0.01	**Millan et al. (2017)**	
	2003–2009	−260 ± 60	−50 ± 9	0.17 ± 0.02	Nilsson et al. (2015)	geod.
	2006–2015	−400 ± 110	−41 ± 12	0.11 ± 0.03	**Noël et al. (2018)**	Model
	2006–2015	−390 ± 30	−41 ± 4	0.12 ± 0.01	**Wouters et al. (2019)**	GRACE
	2006–2015	−540 ± 800	−56 ± 84	0.15 ± 0.23	Zemp et al. (2019)	gl., geod.
	2006–2015	**−380 ± 80**	**−39 ± 8**	**0.11 ± 0.02**	SROCC	
Arctic Canada South, A=40,888 km², V=20.5 ± 5.3 mm SLE	2003–2009	−660 ± 110	−27 ± 4	0.07 ± 0.01	Gardner et al. (2013)	GRACE, geod.
	1958–1995	−280 ± 100	−12 ± 5	0.03 ± 0.01	Noël et al. (2018)	Model
	1996–2015	−510 ± 100	−22 ± 5	0.06 ± 0.01	Noël et al. (2018)	Model
	2003–2009	−550 ± 130	−23 ± 5	0.06 ± 0.01	Nilsson et al. (2015)	geod.
	2006–2015	−650 ± 100	−28 ± 5	0.08 ± 0.01	**Noël et al. (2018)**	Model
	2006–2015	−940 ± 210	−39 ± 9	0.11 ± 0.02	**Wouters et al. (2019)**	GRACE
	2006–2015	−540 ± 700	−22 ± 28	0.06 ± 0.08	Zemp et al. (2019)	gl., geod.
	2006–2015	**−800 ± 220**	**−33 ± 9**	**0.09 ± 0.03**	SROCC	
Greenland periphery, A=89,717 km², V=33.6 ± 8.7 mm SLE	2003–2009	−420 ± 70	−38 ± 7	0.10 ± 0.02	Gardner et al. (2013)	geod.
	1958–1996	−140 ± 190	−11 ± 16	0.03 ± 0.04	Noël et al. (2017)	Model
	1997–2015	−400 ± 180	−36 ± 16	0.10 ± 0.04	Noël et al. (2017)	Model
	2006–2015	−510 ± 190	−42 ± 16	0.11 ± 0.04	**Noël et al. (2017)**	Model
	2006–2015	−635 ± 200	−53 ± 17	0.15 ± 0.05	**Zemp et al. (2019)**	gl., geod.
	2006–2015	**−570 ± 200**	**−47 ± 16**	**0.13 ± 0.04**	SROCC	
Svalbard, A=33,959 km², V=17.3 ± 4.5 mm SLE***	2003–2009	−130 ± 60	−5 ± 2	0.01 ± 0.01	Gardner et al. (2013)	GRACE, geod.
	1986–2005	−240 ± 120	−8 ± 4	0.02 ± 0.01	Box et al. (2018)	GRACE, gl.
	2003–2009	−120 ± 80	−4 ± 3	0.01 ± 0.01	Nilsson et al. (2015)	geod.
	2003–2013	−260	−9	0.02	**Aas et al. (2016)**	Model
	2004–2013	−210	−7	0.02	Østby et al. (2017)	Model
	2006–2015	−250 ± 160	−8 ± 5	0.02 ± 0.02	Box et al. (2018)	GRACE, gl.
	2006–2015	−200 ± 40	−7 ± 2	0.02 ± 0.00	**Wouters et al. (2019)**	GRACE
	2006–2015	−400 ± 230	−13 ± 7	0.04 ± 0.02	Zemp et al. (2019)	gl., geod
	2006–2015	**−270 ± 170**	**−9 ± 5**	**0.02 ± 0.01**	SROCC	

Mass budget		kg m^{-2} yr^{-1}	Gt yr^{-1}	mm SLE yr^{-1}	Reference	Method
Russian Arctic, **A=51,592 km^2,** **V=32.0 ± 8.3 mm SLE**	2003–2009	−210 ± 80	−11 ± 4	0.03 ± 0.01	Gardner et al. (2013)	GRACE, geod.
	1986–2005	−210 ± 190	−11 ± 10	0.03 ± 0.03	Box et al. (2018)	GRACE, gl.
	2003–2009	−140 ± 50	−7 ± 3	0.02 ± 0.01	Nilsson et al. (2015)	geod.
	2006–2015	−200 ± 250	−11 ± 13	0.03 ± 0.04	Box et al. (2018)	G., gl.
	2006–2015	−220 ± 40	−11 ± 2	0.03 ± 0.01	**Wouters et al. (2019)**	GRACE
	2006–2015	−400 ± 370	−20 ± 16	0.06 ± 0.04	**Zemp et al. (2019)**	gl., geod
	2006–2015	**−300 ± 270**	**−15 ± 12**	**0.04 ± 0.03**	SROCC	
Antarctic periphery, **A=132,867 km^2,** **V=69.4 ± 18 mm SLE**	2003–2009	−50 ± 70	−6 ± 10	0.02 ± 0.03	Gardner et al. (2013)	geod.
	2006–2015	−90 ± 860	−11 ± 108	0.03 ± 0.3	**Zemp et al. (2019)**	gl., geod
	2006–2015	**−90 ± 860**	**−11 ± 108**	**0.03 ± 0.3**	SROCC	
11 Mountain regions covered in Chapter 2, **A=251,604 km^2,** **V=87 ± 15 mm SLE**	**2006–2015**	**−490 ± 100**	**−123 ± 24**	**0.34 ± 0.07**	SROCC	
Arctic regions**** **A =422,000 km^2,** **V =221 ± 25 mm SLE**	**2006–2015**	**−500 ± 70**	**−213 ± 29**	**−0.59 ± 0.08**	SROCC	
Global excl. A+G periphery, **A=483,155 km^2,** **V=221 ± 23 mm SLE**	**2006–2015**	**−460 ± 60**	**−220 ± 30**	**0.61 ± 0.08**	SROCC	
Global, **A =705,739 km^2,** **V =324 ± 84 mm SLE**	**2006–2015**	**−390 ± 160**	**−278 ± 113**	**0.77 ± 0.31**	SROCC	

Notes:

*Björnsson and Pálsson (2008) report a volume of ~9 mm SLE based on radio-echo sounding data.

**only Northern and Southern Patagonian Ice Fields (38% of regional area).

***Fürst et al. (2018) report a volume of 15.3 ± 2.6 mm SLE.

****including Alaska, Iceland, and Scandinavia (covered in Chapter 2), and Arctic Canada, Greenland periphery, Russian Arctic and Svalbard (covered in Chapter 3).

Polar Regions

3

Coordinating Lead Authors:
Michael Meredith (United Kingdom), Martin Sommerkorn (Norway/Germany)

Lead Authors:
Sandra Cassotta (Denmark), Chris Derksen (Canada), Alexey Ekaykin (Russian Federation), Anne Hollowed (USA), Gary Kofinas (USA), Andrew Mackintosh (Australia/New Zealand), Jess Melbourne-Thomas (Australia), Mônica M.C. Muelbert (Brazil), Geir Ottersen (Norway), Hamish Pritchard (United Kingdom), Edward A.G. Schuur (USA)

Contributing Authors:
Nerilie Abram (Australia), Julie Arblaster (Australia), Kevin Arrigo (USA), Kumiko Azetzu-Scott (Canada), David Barber (Canada), Inka Bartsch (Germany), Jeremy Bassis (USA), Dorothea Bauch (Germany), Fikret Berkes (Canada), Philip Boyd (Australia), Angelika Brandt (Germany), Lijing Cheng (China), Steven Chown (Australia), Alison Cook (United Kingdom), Jackie Dawson (Canada), Robert M. DeConto (USA), Thorben Dunse (Norway/Germany), Andrea Dutton (USA), Tamsin Edwards (United Kingdom), Laura Eerkes-Medrano (Canada), Arne Eide (Norway), Howard Epstein (USA), F. Stuart Chapin III (USA), Mark Flanner (USA), Bruce Forbes (Finland), Jeremy Fyke (Canada), Andrey Glazovsky (Russian Federation), Jacqueline Grebmeier (USA), Guido Grosse (Germany), Anne Gunn (Canada), Sherilee Harper (Canada), Jan Hjort (Finland), Will Hobbs (Australia), Eric P. Hoberg (USA), Indi Hodgson-Johnston (Australia), David Holland (USA), Paul Holland (United Kingdom), Russell Hopcroft (USA), George Hunt (USA), Henry Huntington (USA), Adrian Jenkins (United Kingdom), Kit Kovacs (Norway), Gita Ljubicic (Canada), Michael Loranty (USA), Michelle Mack (USA), Andrew Meijers (United Kingdom/Australia), Benoit Meyssignac (France), Hans Meltofte (Denmark), Alexander Milner (United Kingdom), Pedro Monteiro (South Africa), Lawrence Mudryk (Canada), Mark Nuttall (Canada), Jamie Oliver (United Kingdom), James Overland (USA), Keith Reid (United Kingdom), Vladimir Romanovsky (USA/Russian Federation), Don E. Russell (Canada), Christina Schädel (USA/Switzerland), Lars H. Smedsrud (Norway), Julienne Stroeve (Canada/USA), Alessandro Tagliabue (United Kingdom), Mary-Louise Timmermans (USA), Merritt Turetsky (Canada), Michiel van den Broeke (Netherlands), Roderik Van De Wal (Netherlands), Isabella Velicogna (USA/Italy), Jemma Wadham (United Kingdom), Michelle Walvoord (USA), Gongjie Wang (China), Dee Williams (USA), Mark Wipfli (USA), Daqing Yang (Canada)

Review Editors:
Oleg Anisimov (Russian Federation), Gregory Flato (Canada), Cunde Xiao (China)

Chapter Scientist:
Shengping He (Norway/China), Victoria Peck (United Kingdom)

This chapter should be cited as:
Meredith, M., M. Sommerkorn, S. Cassotta, C. Derksen, A. Ekaykin, A. Hollowed, G. Kofinas, A. Mackintosh, J. Melbourne-Thomas, M.M.C. Muelbert, G. Ottersen, H. Pritchard, and E.A.G. Schuur, 2019: Polar Regions. In: *IPCC Special Report on the Ocean and Cryosphere in a Changing Climate* [H.-O. Pörtner, D.C. Roberts, V. Masson-Delmotte, P. Zhai, M. Tignor, E. Poloczanska, K. Mintenbeck, A. Alegría, M. Nicolai, A. Okem, J. Petzold, B. Rama, N.M. Weyer (eds.)]. Cambridge University Press, Cambridge, UK and New York, NY, USA, pp. 203–320. https://doi.org/10.1017/9781009157964.005.

Table of contents

Executive Summary

This chapter assesses the state of physical, biological and social knowledge concerning the Arctic and Antarctic ocean and cryosphere, how they are affected by climate change, and how they will evolve in future. Concurrently, it assesses the local, regional and global consequences and impacts of individual and interacting polar system changes, and it assesses response options to reduce risk and build resilience in the polar regions. Key findings are:

The polar regions are losing ice, and their oceans are changing rapidly. The consequences of this polar transition extend to the whole planet, and are affecting people in multiple ways.

Arctic surface air temperature has *likely*[1] increased by more than double the global average over the last two decades, with feedbacks from loss of sea ice and snow cover contributing to the amplified warming. For each of the five years since the IPCC 5th Asesssment Report (AR5) (2014–2018), Arctic annual surface air temperature exceeded that of any year since 1900. During the winters (January to March) of 2016 and 2018, surface temperatures in the central Arctic were 6°C above the 1981–2010 average, contributing to unprecedented regional sea ice absence. These trends and extremes provide *medium evidence*[2] with *high agreement* of the contemporary coupled atmosphere-cryosphere system moving well outside the 20th century envelope. {Box 3.1; 3.2.1.1}

The Arctic and Southern Oceans are continuing to remove carbon dioxide from the atmosphere and to acidify (*high confidence*). There is *medium confidence* that the amount of CO_2 drawn into the Southern Ocean from the atmosphere has experienced significant decadal variations since the 1980s. Rates of calcification (by which marine organisms form hard skeletons and shells) declined in the Southern Ocean by $3.9 \pm 1.3\%$ between 1998 and 2014. In the Arctic Ocean, the area corrosive to organisms that form shells and skeletons using the mineral aragonite expanded between the 1990s and 2010, with instances of extreme aragonite undersaturation. {3.2.1.2.4}

Both polar oceans have continued to warm in recent years, with the Southern Ocean being disproportionately and increasingly important in global ocean heat increase (*high confidence*). Over large sectors of the seasonally ice-free Arctic, summer upper mixed layer temperatures increased at around 0.5°C per decade during 1982–2017, primarily associated with increased absorbed solar radiation accompanying sea ice loss, and the inflow of ocean heat from lower latitude increased since the 2000s (*high confidence*). During 1970–2017, the Southern Ocean south of 30°S accounted for 35–43% of the global ocean heat gain in the upper

2000 m (*high confidence*), despite occupying ~25% of the global ocean area. In recent years (2005–2017), the Southern Ocean was responsible for an increased proportion of the global ocean heat increase (45–62%) (*high confidence*). {3.2.1.2.1}

Climate-induced changes in seasonal sea ice extent and thickness and ocean stratification are altering marine primary production (*high confidence*), with impacts on ecosystems (*medium confidence*). Changes in the timing, duration and intensity of primary production have occurred in both polar oceans, with marked regional or local variability (*high confidence*). In the Antarctic, such changes have been associated with locally-rapid environmental change, including retreating glaciers and sea ice change (*medium confidence*). In the Arctic, changes in primary production have affected regional species composition, spatial distribution, and abundance of many marine species, impacting ecosystem structure (*medium confidence*). {3.2.1; 3.2.3, 3.2.4}

In both polar regions, climate-induced changes in ocean and sea ice, together with human introduction of non-native species, have expanded the range of temperate species and contracted the range of polar fish and ice-associated species (*high confidence*). Commercially and ecologically important fish stocks like Atlantic cod, haddock and mackerel have expanded their spatial distributions northwards many hundreds of kilometres, and increased their abundance. In some Arctic areas, such expansions have affected the whole fish community, leading to higher competition and predation on smaller sized fish species, while some commercial fisheries have benefited. There has been a southward shift in the distribution of Antarctic krill in the South Atlantic, the main area for the krill fishery (*medium confidence*). These changes are altering biodiversity in polar marine ecosystems (*medium confidence*). {3.2.3; Box 3.4}

Arctic sea ice extent continues to decline in all months of the year (*very high confidence*); the strongest reductions in September (*very likely* $-12.8 \pm 2.3\%$ per decade; 1979–2018) are unprecedented in at least 1000 years (*medium confidence*). Arctic sea ice has thinned, concurrent with a shift to younger ice: since 1979, the areal proportion of thick ice at least 5 years old has declined by approximately 90% (*very high confidence*). Approximately half the observed sea ice loss is attributable to increased atmospheric greenhouse gas concentrations (*medium confidence*). Changes in Arctic sea ice have potential to influence mid-latitude weather on timescales of weeks to months (*low to medium confidence*). {3.2.1.1; Box 3.2}

It is *very likely* that Antarctic sea ice cover exhibits no significant trend over the period of satellite observations

---

[1]   In this Report, the following terms have been used to indicate the assessed likelihood of an outcome or a result: Virtually certain 99–100% probability, Very likely 90–100%, Likely 66–100%, About as likely as not 33–66%, Unlikely 0–33%, Very unlikely 0–10%, and Exceptionally unlikely 0–1%. Additional terms (Extremely likely: 95–100%, More likely than not >50–100%, and Extremely unlikely 0–5%) may also be used when appropriate. Assessed likelihood is typeset in italics, e.g., *very likely* (see Section 1.9.2 and Figure 1.4 for more details). This Report also uses the term '*likely* range' to indicate that the assessed likelihood of an outcome lies within the 17–83% probability range.

[2]   In this Report, the following summary terms are used to describe the available evidence: limited, medium, or robust; and for the degree of agreement: low, medium, or high. A level of confidence is expressed using five qualifiers: very low, low, medium, high, and very high, and typeset in italics, e.g., *medium confidence*. For a given evidence and agreement statement, different confidence levels can be assigned, but increasing levels of evidence and degrees of agreement are correlated with increasing confidence (see Section 1.9.2 and Figure 1.4 for more details).

(1979–2018). While the drivers of historical decadal variability are known with *medium confidence,* there is currently *limited evidence* and *low agreement* concerning causes of the strong recent decrease (2016–2018), and *low confidence* in the ability of current-generation climate models to reproduce and explain the observations. {3.2.1.1}

Shipping activity during the Arctic summer increased over the past two decades in regions for which there is information, concurrent with reductions in sea ice extent (*high confidence*). Transit times across the Northern Sea Route have shortened due to lighter ice conditions, and while long-term, pan-Arctic datasets are incomplete, the distance travelled by ships in Arctic Canada nearly tripled during 1990–2015 (*high confidence*). Greater levels of Arctic ship-based transportation and tourism have socioeconomic and political implications for global trade, northern nations, and economies linked to traditional shipping corridors; they will also exacerbate region specific risks for marine ecosystems and coastal communities if further action to develop and adequately implement regulations does not keep pace with increased shipping (*high confidence*). {3.2.1.1; 3.2.4.2; 3.2.4.3; 3.4.3.3.2; 3.5.2.7}

Permafrost temperatures have increased to record high levels (*very high confidence*), but there is *medium evidence* and *low agreement* that this warming is currently causing northern permafrost regions to release additional methane and carbon dioxide. During 2007–2016, continuous-zone permafrost temperatures in the Arctic and Antarctic increased by $0.39 \pm 0.15°C$ and $0.37 \pm 0.10°C$ respectively. Arctic and boreal permafrost region soils contain 1460–1600 Gt organic carbon (*medium confidence*). Changes in permafrost influence global climate through emissions of carbon dioxide and methane released from the microbial breakdown of organic carbon, or the release of trapped methane. {3.4.1; 3.4.3}

Climate-related changes to Arctic hydrology, wildfire and abrupt thaw are occurring (*high confidence*), with impacts on vegetation and water and food security. Snow and lake ice cover has declined, with June snow extent decreasing $13.4 \pm 5.4\%$ per decade (1967–2018) (*high confidence*). Runoff into the Arctic Ocean increased for Eurasian and North American rivers by $3.3 \pm 1.6\%$ and $2.0 \pm 1.8\%$ respectively (1976–2017; *medium confidence*). Area burned and frequency of fires (including extreme fires) are unprecedented over the last 10,000 years (*high confidence*). There has been an overall greening of the tundra biome, but also browning in some regions of tundra and boreal forest, and changes in the abundance and distribution of animals including reindeer and salmon (*high confidence*). Together, these impact access to (and food availability within) herding, hunting, fishing, forage and gathering areas, affecting the livelihood, health and cultural identity of residents including Indigenous peoples (*high confidence*). {3.4.1; 3.4.3; 3.5.2}

Limited knowledge, financial resources, human capital and organisational capacity are constraining adaptation in many human sectors in the Arctic (*high confidence*). Harvesters of renewable resources are adjusting timing of activities to changes in seasonality and less safe ice travel conditions. Municipalities

and industry are addressing infrastructure failures associated with flooding and thawing permafrost, and coastal communities and cooperating agencies are in some cases planning for relocation (*high confidence*). In spite of these adaptations, many groups are making decisions without adequate knowledge to forecast near- and long-term conditions, and without the funding, skills and institutional support to engage fully in planning processes (*high confidence*). {3.5.2, 3.5.4, Cross-Chapter Box 9}

It is *extremely likely* that the rapid ice loss from the Greenland and Antarctic ice sheets during the early 21st century has increased into the near present day, adding to the ice sheet contribution to global sea level rise. From Greenland, the 2012–2016 ice losses (-247 ± 15 Gt yr^{-1}) were similar to those from 2002 to 2011 (-263 ± 21 Gt yr^{-1}) and *extremely likely* greater than from 1992 to 2001 (-8 ± 82 Gt yr^{-1}). Summer melting of the Greenland Ice Sheet (GIS) has increased since the 1990s (*very high confidence*) to a level unprecedented over at least the last 350 years, and two-to-fivefold the pre-industrial level (*medium confidence*). From Antarctica, the 2012–2016 losses (-199 ± 26 Gt yr^{-1}) were *extremely likely* greater than those from 2002 to 2011 (-82 ± 27 Gt yr^{-1}) and *likely* greater than from 1992 to 2001 (-51 ± 73 Gt yr^{-1}). Antarctic ice loss is dominated by acceleration, retreat and rapid thinning of major West Antarctic Ice Sheet (WAIS) outlet glaciers (*very high confidence*), driven by melting of ice shelves by warm ocean waters (*high confidence*). The combined sea level rise contribution from both ice sheets for 2012–2016 was 1.2 ± 0.1 mm yr^{-1}, a 29% increase on the 2002–2011 contribution and a ~700% increase on the 1992–2001 period. {3.3.1}

Mass loss from Arctic glaciers (-212 ± 29 Gt yr^{-1}) during 2006–2015 contributed to sea level rise at a similar rate (0.6 ± 0.1 mm yr^{-1}) to the GIS (*high confidence*). Over the same period in Antarctic and subantarctic regions, glaciers separate from the ice sheets changed mass by -11 ± 108 Gt yr^{-1} (*low confidence*). {2.2.3, 3.3.2}

There is *limited evidence* and *high agreement* that recent **Antarctic Ice Sheet (AIS) mass losses could be irreversible over decades to millennia.** Rapid mass loss due to glacier flow acceleration in the Amundsen Sea Embayment (ASE) of West Antarctica and in Wilkes Land, East Antarctica, may indicate the beginning of Marine Ice Sheet Instability (MISI), but observational data are not yet sufficient to determine whether these changes mark the beginning of irreversible retreat. {3.3.1; Cross-Chapter Box 8 in Chapter 3; 4.2.3.1.2}

The polar regions will be profoundly different in future compared with today, and the degree and nature of that difference will depend strongly on the rate and magnitude of global climatic change[3]. This will challenge adaptation responses regionally and worldwide.

It is *very likely* that projected Arctic warming will result in continued loss of sea ice and snow on land, and reductions in

---

[3]   Projections for ice sheets and glaciers in the polar regions are summarized in Chapters 4 and 2, respectively.

3

the mass of glaciers. **Important differences in the trajectories of loss emerge from 2050 onwards, depending on mitigation measures taken (*high confidence*).** For stabilised global warming of 1.5°C, an approximately 1% chance of a given September being sea ice free at the end of century is projected; for stabilised warming at a 2°C increase, this rises to 10–35% (*high confidence*). The potential for reduced (further 5–10%) but stabilised Arctic autumn and spring snow extent by mid-century for Representative Concentration Pathway (RCP)2.6 contrasts with continued loss under RCP8.5 (a further 15–25% reduction to end of century) (*high confidence*). Projected mass reductions for polar glaciers between 2015 and 2100 range from 16 ± 7% for RCP2.6 to 33 ± 11% for RCP8.5 (*medium confidence*). {3.2.2; 3.3.2; 3.4.2, Cross-Chapter Box 6 in Chapter 2}

Both polar oceans will be increasingly affected by CO$_2$ uptake, causing conditions corrosive for calcium carbonate shell-producing organisms (*high confidence*), with associated impacts on marine organisms and ecosystems (*medium confidence*). It is *very likely* that both the Southern Ocean and the Arctic Ocean will experience year-round conditions of surface water undersaturation for mineral forms of calcium carbonate by 2100 under RCP8.5; under RCP2.6 the extent of undersaturated waters are reduced markedly. Imperfect representation of local processes and sea ice interaction in global climate models limit the ability to project the response of specific polar areas and the precise timing of undersaturation at seasonal scales. Differences in sensitivity and the scope for adaptation to projected levels of ocean acidification exist across a broad range of marine species groups. {3.2.1; 3.2.2.3; 3.2.3}

Future climate-induced changes in the polar oceans, sea ice, snow and permafrost will drive habitat and biome shifts, with associated changes in the ranges and abundance of ecologically important species (*medium confidence*). Projected shifts will include further habitat contraction and changes in abundance for polar species, including marine mammals, birds, fish, and Antarctic krill (*medium confidence*). Projected range expansion of subarctic marine species will increase pressure for high-Arctic species (*medium confidence*), with regionally variable impacts. Continued loss of Arctic multi-year sea ice will affect ice-related and pelagic primary production (*high confidence*), with impacts for whole ice-associated, seafloor and open ocean ecosystems. On Arctic land, projections indicate a loss of globally unique biodiversity as some high Arctic species will be outcompeted by more temperate species and very limited refugia exist (*medium confidence*). Woody shrubs and trees are projected to expand, covering 24–52% of the current tundra region by 2050. {3.2.2.1; 3.2.3; 3.2.3.1; Box 3.4; 3.4.2; 3.4.3}

The projected effects of climate-induced stressors on polar marine ecosystems present risks for commercial and subsistence fisheries with implications for regional economies, cultures and the global supply of fish, shellfish, and Antarctic krill (*high confidence*). Future impacts for linked human systems depend on the level of mitigation and especially the responsiveness of precautionary management approaches (*medium confidence*). Polar regions support several of the world's largest commercial fisheries. Specific impacts on the stocks and economic value in both regions will depend on future climate

change and on the strategies employed to manage the effects on stocks and ecosystems (*medium confidence*). Under high emission scenarios current management strategies of some high-value stocks may not sustain current catch levels in the future (*low confidence*); this exemplifies the limits to the ability of existing natural resource management frameworks to address ecosystem change. Adaptive management that combines annual measures and within-season provisions informed by assessments of future ecosystem trends reduces the risks of negative climate change impacts on polar fisheries (*medium confidence*). {3.2.4; 3.5.2; 3.5.4}

Widespread disappearance of Arctic near-surface permafrost is projected to occur this century as a result of warming (*very high confidence*), with important consequences for global climate. By 2100, near-surface permafrost area will decrease by 2–66% for RCP2.6 and 30–99% for RCP8.5. This is projected to release 10s to 100s of billions of tons (Gt C), up to as much as 240 Gt C, of permafrost carbon as carbon dioxide and methane to the atmosphere with the potential to accelerate climate change. Methane will contribute a small proportion of these additional carbon emissions, on the order of 0.01–0.06 Gt CH$_4$ yr^{-1}, but could contribute 40–70% of the total permafrost-affected radiative forcing because of its higher warming potential. There is *medium evidence* but with *low agreement* whether the level and timing of increased plant growth and replenishment of soil will compensate these permafrost carbon losses. {3.4.2; 3.4.3}

Projected permafrost thaw and decrease in snow will affect Arctic hydrology and wildfire, with impacts on vegetation and human infrastructure (*medium confidence*). About 20% of Arctic land permafrost is vulnerable to abrupt permafrost thaw and ground subsidence, which is expected to increase small lake area by over 50% by 2100 for RCP8.5 (*medium confidence*). Even as the overall regional water cycle intensifies, including increased precipitation, evapotranspiration, and river discharge to the Arctic Ocean, decreases in snow and permafrost may lead to soil drying (*medium confidence*). Fire is projected to increase for the rest of this century across most tundra and boreal regions, while interactions between climate and shifting vegetation will influence future fire intensity and frequency (*medium confidence*). By 2050, 70% of Arctic infrastructure is located in regions at risk from permafrost thaw and subsidence; adaptation measures taken in advance could reduce costs arising from thaw and other climate change related impacts such as increased flooding, precipitation, and freeze-thaw events by half (*medium confidence*). {3.4.1; 3.4.2; 3.4.3; 3.5.2}

Response options exist that can ameliorate the impacts of polar change, build resilience and allow time for effective mitigation measures. Institutional barriers presently limit their efficacy.

Responding to climate change in polar regions will be more effective if attention to reducing immediate risks (short-term adaptation) is concurrent with long-term planning that builds resilience to address expected and unexpected impacts (*high confidence*). Emphasis on short-term adaptation to specific problems will ultimately not succeed in reducing the risks and vulnerabilities to

3

society given the scale, complexity and uncertainty of climate change. Moving toward a dual focus of short- and long-term adaptation involves knowledge co-production, linking knowledge with decision making and implementing ecosystem-based stewardship, which involves the transformation of many existing institutions (*high confidence*). {3.5.4}

Innovative tools and practices in polar resource management and planning show strong potential in improving society's capacity to respond to climate change (*high confidence*). Networks of protected areas, participatory scenario analysis, decision support systems, community-based ecological monitoring that draws on local and indigenous knowledge, and self assessments of community resilience contribute to strategic plans for sustaining biodiversity and limit risk to human livelihoods and wellbeing. Such practices are most effective when linked closely to the policy process. Experimenting, assessing, and continually refining practices while strengthening the links with decision making has the potential to ready society for the expected and unexpected impacts of climate change (*high confidence*). {3.5.1, 3.5.2, 3.5.4}

Institutional arrangements that provide for strong multiscale linkages with Arctic local communities can benefit from including indigenous knowledge and local knowledge in the formulation of adaptation strategies (*high confidence*). The tightly coupled relationship of northern local communities and their environment provide an opportunity to better understand climate change and its effects, support adaptation and limit unintended consequences. Enabling conditions for the involvement of local communities in climate adaptation planning include investments in human capital, engagement processes for knowledge co-production and systems of adaptive governance. {3.5.3}

The capacity of governance systems in polar regions to respond to climate change has strengthened recently, but the development of these systems is not sufficiently rapid or robust to address the challenges and risks to societies posed by projected changes (*high confidence*). Human responses to climate change in the polar regions occur in a fragmented governance landscape. Climate change, new polar interests from outside the regions, and an increasingly active role played by informal organisations are compelling stronger coordination and integration between different levels and sectors of governance. The governance landscape is currently not sufficiently equipped to address cascading risks and uncertainty in an integrated and precautionary way within existing legal and policy frameworks (*high confidence*). {3.5.3, 3.5.4}

3

3.1 Introduction: Polar Regions, People and the Planet

This chapter provides an integrated assessment of climate change across the physical, biological and human dimensions of the polar regions, based on emerging understanding that assessing these dimensions in isolation is not sufficient or forward-looking. This offers the opportunity, for the first time in a global report, to trace cause and consequence of climate change from polar ocean and cryosphere systems to biological and social impacts, and relate them to responses to reduce risks and enhance adaptation options and resilience. To achieve this, the chapter draws on the body of literature and assessments pertaining to climate-induced dynamics and functioning of the polar regions published since the AR5, which has expanded considerably motivated in large part by growing appreciation of the importance of these regions to planetary systems and to the lives and livelihoods of people across the globe.

As integral parts of the Earth system, the polar regions interact with the rest of the world through shared ocean, atmosphere, ecological and social systems; notably, they are key components of the global climate system. This chapter therefore takes a systems approach that emphasises the interactions of cryosphere and ocean changes and their diverse consequences and impacts to assess key issues of climatic change for the polar regions, the planet and its people (Figure 3.1).

The spatial footprints of the polar regions (Figure 3.2) include a vast share of the world's ocean and cryosphere: they encompass surface areas equalling 20% of the global ocean and more than 90% of the world's continuous and discontinuous permafrost area, 69% of the world's glacier area including both of the world's ice sheets, almost all of the world's sea ice, and land areas with the most persistent winter snow cover.

Important differences in the physical setting of the two polar regions – the Arctic, an ocean surrounded by land, the Antarctic, a continent surrounded by an ocean – structure the nature and magnitude of interactions of cryosphere and ocean systems and their global linkages. The different physical settings have also led to the evolution of unique marine and terrestrial biology in each polar region and shape effects, impacts and adaptation of polar ecosystems.

It is important to recognise the existence of multiple and diverse perspectives of the polar regions, many of them overlapping. These multiple perspectives encompass the polar regions as a source of resources, a key part of the global climate system, a place for preserving intact ecosystems, a place for international cooperation and, importantly, a homeland. While many of these perspectives are equally relevant for both polar regions, only the Arctic has a population for whom the region is a permanent home: approximately four million people reside there, of whom 10% are indigenous. By contrast, the Antarctic population changes seasonally between approximately 1100 and 4400, based predominantly at research stations. When assessing knowledge relating to climate change in the context of adaptation options, limits and enhancing resilience (Cross-Chapter Box 2 in Chapter 1), such differences are important as they are linked to diverse human values, social processes, and use of resources.

Consideration of all peer-reviewed scientific knowledge is a hallmark of the IPCC assessment process. Indigenous knowledge and local knowledge are different and unique sources of knowledge that are increasingly recognised to contribute to observing, understanding, and responding to climate-induced changes (Cross-Chapter Box 4 in Chapter 1). Considering indigenous knowledge and local knowledge facilitates cooperation in the development, identification, and decision making processes for responding to climate change in communities across the Arctic, and better understanding of the challenges facing Indigenous peoples. This chapter incorporates published indigenous knowledge and local knowledge for assessing climate change impacts and responses.

3

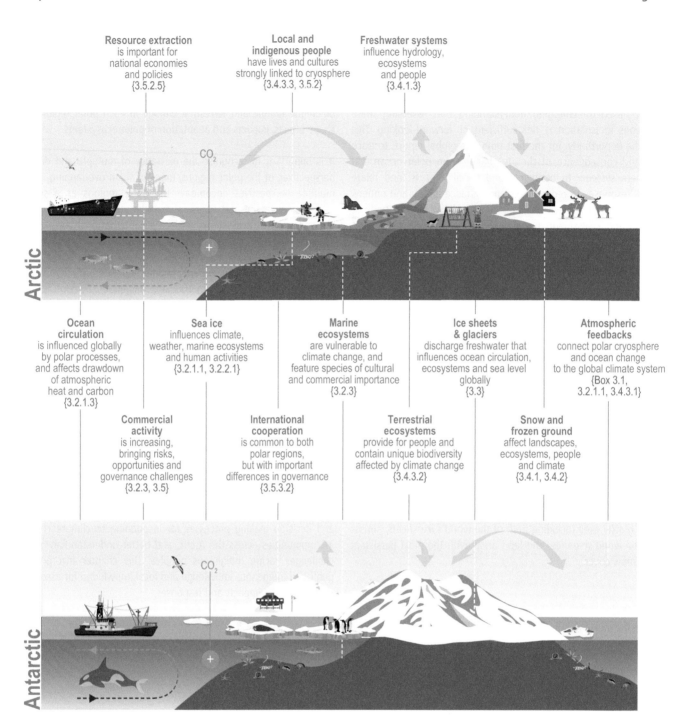

Figure 3.1 | Schematic of some of the key features and mechanisms assessed in this Chapter, and by which the cryosphere and ocean in the polar regions influence climate, ecological and social systems in the regions and across the globe. Specific elements are labelled, and section numbers given for where detailed assessment information can be found.

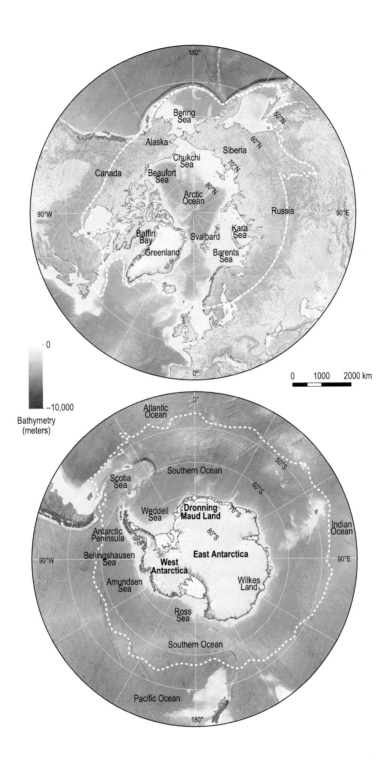

Figure 3.2 | The Arctic (top) and Antarctic (bottom) polar regions. Various place names referred to in the text are marked. Dashed lines denote approximate boundaries for the polar regions; as their spatial footprint varies in relation to particular cryosphere and ocean elements or scientific disciplines, this chapter adopts a purposefully flexible approach to their delineation. The southern polar region encompasses the flow of the Antarctic Circumpolar Current (ACC) at least as far north as the Subantarctic Front and fully encompasses the Convention for the Conservation of Antarctic Marine Living Resources Statistical Areas (CCAMLR, 2017c), the Antarctic continent and Antarctic and subantarctic islands, whilst the marine Arctic includes the areas of the Arctic Large Marine Ecosystems (PAME, 2013). The terrestrial Arctic comprises the areas of the northern continuous and discontinuous permafrost zone, the Arctic biome inclusive of glacial ice, and the parts of the boreal biome that are characterised by cryosphere elements, such as permafrost and persistent winter season snow cover.

Box 3.1 | Polar Region Climate Trends

Over the last two decades, Arctic surface air temperature has increased at more than double the global average (*high confidence*) (Notz and Stroeve, 2016; Richter-Menge et al., 2017). Attribution studies show the important role of anthropogenic increases in greenhouse gases in driving observed Arctic surface temperature increases (Fyfe et al., 2013; Najafi et al., 2015), so there is *high confidence* in projections of further Arctic warming (Overland et al., 2018a). Mechanisms for Arctic amplification are still debated, but include: reduced summer albedo due to sea ice and snow cover loss, the increase of total water vapour content in the Arctic atmosphere, changes in total cloudiness in summer, additional heat generated by newly formed sea ice across more extensive open water areas in the autumn, northward transport of heat and moisture and the lower rate of heat loss to space from the Arctic relative to the subtropics (Serreze and Barry, 2011; Pithan and Mauritsen, 2014; Goosse et al., 2018; Stuecker et al., 2018) (SM3.1.1).

A number of recent events in the Arctic indicate new extremes in the Arctic climate system. Annual Arctic surface temperature for each of the past five years since AR5 (2014–2018; relative to a 1980–2010 base line) exceeded that of any year since 1900 (Overland et al., 2018b). Winter (January to March) near-surface temperature anomalies of +6°C (relative to 1981–2010) were recorded in the central Arctic during both 2016 and 2018, nearly double the previous record anomalies (Overland and Wang, 2018a). These events were caused by a split of the tropospheric polar vortex into two cells, which facilitated the intrusion of subarctic storms (Overland and Wang, 2016). The resulting advection of warm air and moisture from the Pacific and Atlantic Oceans into the central Arctic increased downward longwave radiation, delayed sea ice freeze-up, and contributed to an unprecedented absence of sea ice. Delayed freeze-up of sea ice in subarctic seas (Chukchi, Barents and Kara) acts as a positive feedback allowing warmer temperatures to progress further toward the North Pole (Kim et al., 2017). In addition to dramatic Arctic summer sea ice loss over the past 15 years, all Arctic winter sea ice maxima of the last 4 years were at record low levels relative to 1979–2014 (Overland, 2018). Multi-year, large magnitude extreme positive Arctic temperatures and sea ice minimums (Section 3.2.1.1) since AR5 provide *high agreement* and *medium evidence* of contemporary conditions well outside the envelope of previous experience (1900–2017) (AMAP, 2017d; Walsh et al., 2017).

In contrast to the Arctic, the Antarctic continent has seen less uniform temperature changes over the past 30–50 years, with warming over parts of West Antarctica and no significant overall change over East Antarctica (Nicolas and Bromwich, 2014; Jones et al., 2016; Turner et al., 2016), though there is *low confidence* in these changes given the sparse *in situ* records and large interannual to interdecadal variability. This weaker amplified warming compared to the Arctic is due to deep ocean mixing and ocean heat uptake over the Southern Ocean (Collins et al., 2013). The Southern Annular Mode (SAM), Pacific South American mode (by which tropical Pacific convective heating signals are transmitted to high southern latitudes) and zonal-wave 3 are the dominant large-scale atmospheric circulation drivers of Antarctic surface climate and sea ice changes (SM3.1.3). Over recent decades the SAM has exhibited a positive trend during austral summer, indicating a strengthening of the surface westerly winds around Antarctica. This extended positive phase of the SAM is unprecedented in at least 600 years, according to palaeoclimate reconstructions (Abram et al., 2014; Dätwyler et al., 2017) and is associated with cooler conditions over the continent.

Consistent with AR5, it is *likely* that Antarctic ozone depletion has been the dominant driver of the positive trend in the SAM during austral summer from the late 1970s to the late 1990s (Schneider et al., 2015; Waugh et al., 2015; Karpechko et al., 2018), the period during which ozone depletion was increasing. There is *high confidence* through a growing body of literature that variability of tropical sea surface temperatures can influence Antarctic temperature changes (Li et al., 2014; Turner et al., 2016; Clem et al., 2017; Smith and Polvani, 2017) and the Southern Hemisphere mid-latitude circulation (Li et al., 2015a; Raphael et al., 2016; Turney et al., 2017; Evtushevsky et al., 2018; Yuan et al., 2018). New research suggests a stronger role of tropical sea surface temperatures in driving changes in the SAM since 2000 (Schneider et al., 2015; Clem et al., 2017).

3.2   Sea Ice and Polar Oceans: Changes, Consequences and Impacts

3.2.1   Observed Changes in Sea Ice and Ocean

3.2.1.1   Sea Ice

Sea ice reflects a high proportion of incoming solar radiation back to space, provides thermal insulation between the ocean and atmosphere, influences thermohaline circulation, and provides

habitat for ice-associated species. Sea ice characteristics differ between the Arctic and Antarctic. Expansion of winter sea ice in the Arctic is limited by land, and ice circulates within the central Arctic basin, some of which survives the summer melt season to form multi-year ice. Arctic sea ice variability and impacts on communities includes indigenous knowledge and local knowledge from across the circumpolar Arctic (Cross-Chapter Box 3 in Chapter 1). The Antarctic continent is surrounded by sea ice which interacts with adjacent ice shelves; winter season expansion is limited by the influence of the Antarctic Circumpolar Current (ACC).

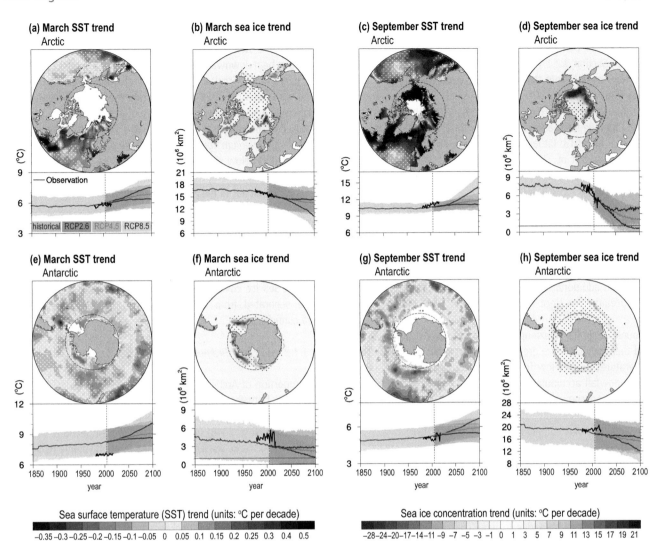

(a) March SST trend Arctic

(b) March sea ice trend Arctic

(c) September SST trend Arctic

(d) September sea ice trend Arctic

(e) March SST trend Antarctic

(f) March sea ice trend Antarctic

(g) September SST trend Antarctic

(h) September sea ice trend Antarctic

Sea surface temperature (SST) trend (units: °C per decade)

−0.35 −0.3 −0.25 −0.2 −0.15 −0.1 −0.05 0 0.05 0.1 0.15 0.2 0.25 0.3 0.4 0.5

Sea ice concentration trend (units: °C per decade)

−28 −24 −20 −17 −14 −11 −9 −7 −5 −3 −1 0 1 3 5 7 9 11 13 15 17 19 21

Figure 3.3 | Maps of linear trends (in °C per decade) of Arctic (a, c) and Antarctic (e, g) sea surface temperature (SST) for 1982–2017 in March (a, e) and September (c, g). (b, d, f, h) same as (a, c, e, g), but for the linear trends of sea ice concentration (in % per decade). Stippled regions indicate the trends that are statistically insignificant. Dashed circles indicate the Arctic/Antarctic Circle. Beneath each map of linear trend shows the time series of SST (area-averaged north of 40°N/south of 40°S) or sea ice extent in the northern/southern hemisphere. Black, green, blue, orange, and red curves indicate observations, Coupled Model Intercomparison Project Phase 5 (CMIP5) historical simulation, Representative Concentration Pathway (RCP)2.6, RCP4.5, and RCP8.5 projections respectively; shading indicates ± standard deviation of multi-models. SST trend was calculated from Hadley Centre Sea Ice and Sea Surface Temperature data set (Version 1, HadISST1; Rayner, 2003). Sea ice concentration trend was calculated from the NOAA/NSIDC Climate Data Record of Passive Microwave Sea Ice Concentration, Version 3 (https://nsidc.org/data/g02202). The time series of observed SST are averages of HadISST1 and NOAA Optimum Interpolation SST dataset (version 2; Reynolds et al., 2002). The time series of observed sea ice extent are the averages of HadISST, the NOAA/NSIDC Climate Data Record of Passive Microwave Sea Ice Concentration, and the Global sea ice concentration reprocessing dataset from EUMETSAT (http://osisaf.met.no/p/ice/ice_conc_reprocessed.html).

3.2.1.1.1 Extent and concentration

The pan-Arctic loss of sea ice cover is a prominent indicator of climate change. Sea ice extent (the total area of the Arctic with at least 15% sea ice concentration) has declined since 1979 in each month of the year (*very high confidence*) (Barber et al., 2017; Comiso et al., 2017b; Stroeve and Notz, 2018) (Figure 3.3). Changes are largest in summer and smallest in winter, with the strongest trends in September (1979–2018; summer month with the lowest sea ice cover) of −83,000 km² yr⁻¹ (−12.8% per decade ± 2.3% relative to 1981–2010 mean), and −41,000 km² yr⁻¹ (−2.7% per decade ± 0.5% relative to 1981–2010 mean) for March (1979–2019; winter month with the greatest sea ice cover) (Onarheim et al., 2018). Regionally, summer ice loss is dominated by reductions in the East Siberian Sea (explains 22% of the September trend), and large declines in the Beaufort,

Chukchi, Laptev and Kara seas (Onarheim et al., 2018). Winter ice loss is dominated by reductions within the Barents Sea, responsible for 27% of the pan-Arctic March sea ice trends (Onarheim and Årthun, 2017). Summer Arctic sea ice loss since 1979 is unprecedented in 150 years based on historical reconstructions (Walsh et al., 2017) and more than 1000 years based on palaeoclimate evidence (Polyak et al., 2010; Kinnard et al., 2011; Halfar et al., 2013) (*medium confidence*).

Approximately half of the observed Arctic summer sea ice loss is driven by increased concentrations of atmospheric greenhouse gases, with the remainder attributed to internal climate variability (Kay et al., 2011; Notz and Marotzke, 2012) (*medium confidence*). The sea ice albedo feedback (increased air temperature reduces sea ice cover, allowing more energy to be absorbed at the surface, fostering more melt) is a key driver of sea ice loss (Perovich and Polashenski, 2012;

Stroeve et al., 2012b; Serreze et al., 2016) and is exacerbated by the transition from perennial to seasonal sea ice (Haine and Martin, 2017; see Section 3.2.1.1.2). Other drivers include increased warm, moist air intrusions into the Arctic during both winter (Box 3.1) and spring (Boisvert et al., 2016; Cullather et al., 2016; Kapsch et al., 2016; Mortin et al., 2016; Graham et al., 2017; Hegyi and Taylor, 2018), radiative feedbacks associated with cloudiness and humidity (Kapsch et al., 2013; Pithan and Mauritsen, 2014; Hegyi and Deng, 2016; Morrison et al., 2018), and increased exchanges of sensible and latent heat flux from the ocean to the atmosphere (Serreze et al., 2012; Taylor et al., 2018). A lack of complete process understanding limits a more definitive differentiation between anthropogenic versus internal drivers of summer Arctic sea ice loss (Serreze et al., 2016; Ding et al., 2017; Meehl et al., 2018). The unabated reduction in Arctic summer sea ice since AR5 means contributions to additional global radiative forcing (Flanner et al., 2011) have continued, with estimates of up to an additional 6.4 ± 0.9 W/m^2 of solar energy input to the Arctic Ocean region since 1979 (Pistone et al., 2014).

Although Arctic ice freeze-up is occurring later (Section 3.2.1.1.3), rapid thermodynamic ice growth occurs over thin ice areas after air temperatures drop below freezing in autumn. Later freeze-up also delays snowfall accumulation on sea ice, leading to a thinner and less insulating snowpack (Section 3.2.1.1.6) (Sturm and Massom, 2016). These two negative feedbacks help to mitigate sudden and irreversible loss of Arctic sea ice (Armour et al., 2011).

Total Antarctic sea ice cover exhibits no significant trend over the period of satellite observations (Figure 3.3; 1979–2018) (*high confidence*) (Ludescher et al., 2018). A significant positive trend in mean annual ice cover between 1979 and 2015 (Comiso et al., 2017a) has not persisted, due to three consecutive years of below average ice cover (2016–2018) driven by atmospheric and oceanic forcing (Turner et al., 2017b; Kusahara et al., 2018; Meehl et al., 2019; Wang et al., 2019). The overall Antarctic sea ice extent trend is composed of near-compensating regional changes, with rapid ice loss in the Amundsen and Bellingshausen seas counteracted by rapid ice gain in the Weddell and Ross seas (Holland, 2014) (Figure 3.3). These regional trends are strongly seasonal in character (Holland, 2014); only the western Ross Sea has a trend that is statistically significant in all seasons, relative to the variance during the period of satellite observations.

Multiple factors contribute to the regionally variable nature of Antarctic sea ice extent trends (Matear et al., 2015; Hobbs et al., 2016b). Sea ice trends are closely related to meridional wind trends (*high confidence*) (Holland and Kwok, 2012; Haumann et al., 2014): poleward wind trends in the Bellingshausen Sea push sea ice closer to the coast (Holland and Kwok, 2012) and advect warm air to the sea ice zone (Kusahara et al., 2017), and the reverse is true over much of the Ross Sea. These meridional wind trends are linked to Pacific variability (Coggins and McDonald, 2015; Meehl et al., 2016; Purich et al., 2016b). Ozone depletion may also affect meridional winds (Fogt and Zbacnik, 2014; England et al., 2016), but there is *low confidence* that this explains observed sea ice trends (Landrum et al., 2017).

Coupled climate models indicate that anthropogenic warming at the surface is delayed by the Southern Ocean circulation, which transports heat downwards into the deep ocean (Armour et al., 2016). This overturning circulation (Cross-Chapter Box 7 in Chapter 3), along with differing cloud and lapse rate feedbacks (Goosse et al., 2018), may explain the weak response of Antarctic sea ice cover to increased atmospheric greenhouse gas concentrations compared to the Arctic (*medium confidence*). Because Antarctic sea ice extent has remained below climatological values since 2016, there is still potential for longer-term changes to emerge in the Antarctic (Meehl et al., 2019), similar to the Arctic.

Historical surface observations (Murphy et al., 2014), reconstructions (Abram et al., 2013b), ship records (de la Mare, 2009; Edinburgh and Day, 2016), early satellite images (Gallaher et al., 2014), and model simulations (Gagné et al., 2015) indicate a decrease in overall Antarctic sea ice cover since the early 1960s which is too modest to be separated from natural variability (Hobbs et al., 2016a) (*high confidence*).

3.2.1.1.2 Age and thickness

The proportion of Arctic sea ice at least 5 years old declined from 30% to 2% between 1979 and 2018; over the same period first-year sea ice proportionally increased from approximately 40% to 60–70% (Stroeve and Notz, 2018) (*very high confidence*) (Sections 3.2.1.1.3 and 3.2.1.1.4). Arctic sea ice has thinned through volume reductions in satellite altimeter retrievals (Laxon et al., 2013; Kwok, 2018), ocean–sea ice reanalyses (Chevallier et al., 2017) and *in situ* measurements (Renner et al., 2014; Haas et al., 2017) (*very high confidence*). Data from multiple satellite altimeter missions show declines in Arctic Basin ice thickness from 2000 to 2012 of -0.58 ± 0.07 m per decade (Lindsay and Schweiger, 2015). Integration of data from submarines, moorings, and earlier satellite radar altimeter missions shows ice thickness declined across the central Arctic by 65%, from 3.59 to 1.25 m between 1975 and 2012 (Lindsay and Schweiger, 2015). There is emerging evidence that this sea ice volume loss may be unprecedented over the past century (Schweiger et al., 2019). New estimates of ice thickness are available for the marginal seas (up to a maximum thickness of ~1 metre) from low-frequency satellite passive microwave measurements (Kaleschke et al., 2016; Ricker et al., 2017) but data are only available since 2010. The shift to thinner seasonal sea ice contributes to further ice extent reductions through enhanced summer season melt via increased energy absorption (Nicolaus et al., 2012), and it is vulnerable to fragmentation from the passage of intense Arctic cyclones in summer and increased ocean swell conditions (Zhang et al., 2013; Thomson and Rogers, 2014).

Surface observations of Antarctic sea ice thickness are extremely sparse (Worby et al., 2008). There are no consistent long-term observations from which trends in ice volume may be derived. Calibrated model simulations suggest that ice thickness trends closely follow those of ice concentration (Massonnet et al., 2013; Holland et al., 2014) (*medium confidence*). Satellite altimeter datasets of Antarctic sea ice thickness are emerging (Paul et al., 2018) but definitive trends are not yet available.

3.2.1.1.3   Seasonality

There is *high confidence* that the Arctic sea ice melt season has extended by 3 days per decade since 1979 due earlier melt onset, and 7 days per decade due to later freeze-up (Stroeve and Notz, 2018). This longer melt season is consistent with the observed loss of sea ice extent and thickness (Sections 3.2.1.1.1; 3.2.1.1.2). While the melt onset trends are smaller, they play a large role in the earlier development of open water (Stroeve et al., 2012b; Serreze et al., 2016) and melt pond development (Perovich and Polashenski, 2012) which enhance the sea ice albedo feedback (Stroeve et al., 2014b; Liu et al., 2015a). Observed reductions in the duration of seasonal sea ice cover are reflected in community-based observations of decreased length of time in which activities can safely take place on sea ice (Laidler et al., 2010; Eisner et al., 2013; Fall et al., 2013; Ignatowski and Rosales, 2013).

Changes in the duration of Antarctic sea ice cover over 1979–2011 largely followed the spatial pattern of sea ice extent trends with reduced ice cover duration in the Amundsen/Bellingshausen Sea region in summer and autumn owing to earlier retreat and later advance, and increases in the Ross Sea due to later ice retreat and earlier advance (Stammerjohn et al., 2012).

3.2.1.1.4   Motion

Winds associated with the climatological Arctic sea level pressure pattern drive the Beaufort Gyre (Dewey et al., 2018; Meneghello et al., 2018) and the Transpolar Drift Stream (Vihma et al., 2012), which retains sea ice within the central Arctic Basin, and exports sea ice out of the Fram Strait, respectively. There is *high confidence* that sea ice drift speeds have increased since 1979, both within the Arctic Basin and through Fram Strait (Rampal et al., 2009; Krumpen et al., 2019), attributed to thinner ice (Spreen et al., 2011) and changes in wind forcing (Olason and Notz, 2014). Fram Strait sea ice area export estimates range between 600,000 to 1 million km² of ice annually, which represents approximately 10% of the ice within the Arctic Basin (*medium confidence*) (Kwok et al., 2013; Krumpen et al., 2016; Smedsrud et al., 2017; Zamani et al., 2019). Sea ice volume flux estimates through Fram Strait are now available from satellite altimeter datasets (Ricker et al., 2018), but they cover too short a time period for robust trend analysis. Observations of extreme Arctic sea ice deformation is attributed to the combination of decreased ice thickness and increased ice motion (Itkin et al., 2017).

Satellite estimates of sea ice drift velocity show significant trends in Antarctic ice drift (Holland and Kwok, 2012). Increased northward drift in the Ross Sea and decreased northward drift in the Bellingshausen and Weddell seas agree with the respective ice extent gains and losses in these regions, but there is only *medium confidence* in these trends due to a small number of ice drift data products derived from temporally inconsistent satellite records (Haumann et al., 2016).

3.2.1.1.5   Landfast ice

Immobile sea ice anchored to land or ice shelves is referred to as 'landfast'. The few long term surface (auger hole) records of Arctic landfast sea ice thickness all exhibit thinning trends in springtime maximum sea ice thickness since the mid-1960s (*high confidence*): declines of 11 cm per decade in the Barents Sea (Gerland et al., 2008), 3.3 cm per decade along the Siberian Coast (Polyakov et al., 2010), and 3.5 cm per decade in the Canadian Arctic Archipelago (Howell et al., 2016). Over a shorter 1976–2007 period, winter season landfast sea ice extent from measurements across the Arctic significantly decreased at a rate of 7% per decade, with the largest decreases in the regions of Svalbard (24% per decade) and the northern coast of the Canadian Arctic Archipelago (20% per decade) (Yu et al., 2013). Svalbard and the Chukchi Sea regions are experiencing the largest declines in landfast sea ice duration (~1 week per decade) since the 1970s (Yu et al., 2013; Mahoney et al., 2014). While most Arctic landfast sea ice melts completely each summer, perennial landfast ice (also termed an 'ice-plug') occurs in Nansen Sound and the Sverdrup Channel in the Canadian Arctic Archipelago. These ice-plugs were in place continuously from the start of observations in the early 1960s, until they disappeared during the anomalously warm summer of 1998, and they have rarely re-formed since 2005 (Pope et al., 2017). The loss of this perennial sea ice is associated with reduced landfast ice duration in the northern Canadian Arctic Archipelago (Galley et al., 2012; Yu et al., 2013) and increased inflow of multi-year ice from the Arctic Ocean into the northern Canadian Arctic Archipelago (Howell et al., 2013).

Arctic landfast ice is important to northern residents as a platform for travel, hunting, and access to offshore regions (Sections 3.4.3.3, 3.5.2.2). Reports of thinning, less stable, and less predictable landfast ice have been documented by residents of coastal communities in Alaska (Eisner et al., 2013; Fall et al., 2013; Huntington et al., 2017), the Canadian Arctic (Laidler et al., 2010), and Chukotka (Inuit Circumpolar Council, 2014). The impact of changing prevailing wind forcing on local ice conditions has been specifically noted (Rosales and Chapman, 2015) including impacts on the landfast ice edge and polynyas (Box 3.3) (Gearheard et al., 2013). Long-term records of Antarctic landfast ice are limited in space and time (Stammerjohn and Maksym, 2016), with a high degree of regional variability in trends (Fraser et al., 2011) (*low confidence*).

3.2.1.1.6   Snow on ice

Snow accumulation on sea ice inhibits sea ice melt through a high albedo, but the insulating properties limit sea ice growth (Sturm and Massom, 2016) and inhibits photosynthetic light (important for in- and under-ice biota) from reaching the bottom of the ice (Mundy et al., 2007). If snow on first-year ice is sufficiently thick, it can depress the ice below the sea level surface, which forms snow-ice due to surface flooding. This process is widespread in the Antarctic (Maksym and Markus, 2008) and the Atlantic Sector of the Arctic (Merkouriadi et al., 2017), and may become more common across the Arctic (with implications for sea ice ecosystems) as the ice regime shifts to thinner seasonal ice (Olsen et al., 2017; Granskog et al., 2018) (*medium confidence*).

Despite the importance of snow on sea ice (Webster et al., 2018), surface or satellite derived observations of snowfall over sea ice, and snow depth on sea ice are lacking (Webster et al., 2014). The primary source of snow depth on Arctic sea ice are based on observations collected decades ago (Warren et al., 1999) the utility of which are impacted by the rapid loss of multi-year ice across the central Arctic (Stroeve and Notz, 2018), and large interannual variability in snow depth on sea ice (Webster et al., 2014). Airborne radar retrievals of snow depth on sea ice provide more recent estimates, but spatial and temporal sampling is highly discontinuous (Kurtz and Farrell, 2011). Multi-source time series provide evidence of declining snow depth on Arctic sea ice (Webster et al., 2014) consistent with estimates of higher fractions of liquid precipitation since 2000 (Boisvert et al.,

2018) but there is *low confidence* because surface measurements for validation are extremely limited and suggest a high degree of regional variability (Haas et al., 2017; Rösel et al., 2018).

Although there are regional estimates of snow depth on Antarctic sea ice from satellite (Kern and Ozsoy-Çiçek, 2016), airborne remote sensing (Kwok and Maksym, 2014), field measurements (Massom et al., 2001) and ship-based observations (Worby et al., 2008), data are not sufficient in time nor space to assess changes in snow accumulation on Antarctic sea ice.

Box 3.2 | Potential for the Polar Cryosphere to Influence Mid-latitude Weather

Since AR5, understanding how observed changes in the Arctic can influence mid-latitude weather has emerged as a societally important topic because hundreds of millions of people can potentially be impacted (Jung et al., 2015). The early to middle part of the Holocene coincided with substantial decreases in net precipitation that may be due to weakening jet stream winds related to Arctic temperatures (Routson et al., 2019). There is only *low* to *medium confidence* in the current nature of Arctic/mid-latitude weather linkages because conclusions of recent analyses are inconsistent (National Research Council, 2014; Barnes and Polvani, 2015; Francis, 2017). The atmosphere interacts with the ocean and cryosphere through radiation, heat, precipitation and wind, but a full understanding of complex interconnected physical processes is lacking. Arctic forcing on the atmosphere from loss of sea ice and terrestrial snow is increasing, but the potential for Arctic/mid-latitude weather linkages varies for different jet stream patterns (Grotjahn et al., 2016; Messori et al., 2016; Overland and Wang, 2018a). Connectivity is reduced by the influence of chaotic internal natural variability and other tropical and oceanic forcing. Part of the scientific disagreement is due to irregular connections in the Arctic to mid-latitude linkage pathways, both within and between years (Overland and Wang, 2018b).

Considerable literature exists on the potential for sea ice loss in the Barents and Kara Seas to drive cold episodes in eastern Asia (Kim et al., 2014; Kretschmer et al., 2016), while sea ice anomalies in the Chukchi Sea and areas west of Greenland are associated with cold events in eastern North America (Kug et al., 2015; Ballinger et al., 2018; Overland and Wang, 2018a). Such connections, however, are only episodic (Cohen et al., 2018). While there is evidence of an increase in the frequency of weak polar vortex events (Screen et al., 2018), studies do not show increases in the number of mid-latitude cold events in observations or model projections (Ayarzaguena and Screen, 2016; Trenary et al., 2016). Potential Arctic/mid-latitude interactions have a more regional tropospheric pathway in November to December (Honda et al., 2009; Chen et al., 2016a; McKenna et al., 2018), whereas January to March has a more hemispheric stratospheric pathway involving migration of the polar vortex off of its usual centred location on the North Pole (Cohen et al., 2012; Nakamura et al., 2016; Zhang et al., 2018b). Overall, changes in the stratospheric polar vortex and Northern Annual Mode are not separable from natural variability, and so cannot be attributed to greenhouse gas forced sea ice loss (Screen et al., 2018).

Only a few studies have focused on the potential impact of Antarctic sea ice changes on the mid-latitude circulation (Kidston et al., 2011; Raphael et al., 2011; Bader et al., 2013; Smith et al., 2017b; England et al., 2018); these find that any impacts on the jet stream are strongly dependent on the season and model examined. England et al. (2018) suggest that the response of the jet stream to future Antarctic sea ice loss may in fact be less seasonal than the response to Arctic sea ice loss.

3.2.1.2   Ocean Properties

The Polar Oceans are amongst the most rapidly changing oceans of the world, with consequences for global-scale storage and cycling of heat, carbon and other climatically and ecologically important properties (SM3.2.1; Figure SM3.2).

3.2.1.2.1   Temperature

Ocean temperatures and associated heat fluxes have a primary influence on sea ice (e.g., Carmack et al., 2015; Steele and Dickinson, 2016). WGI AR5 (their Section 3.2.2) reported that Canada Basin surface waters warmed from 1993 to 2007, and observations over

1950–2010 show the Arctic Ocean water of Atlantic origin (i.e., the Atlantic Water Layer) warming starting in the 1970s. Warming trends have continued: August trends for 1982–2017 reveal summer mixed layer temperatures increasing at about 0.5°C per decade over large sectors of the Arctic basin that are ice-free in summer (Timmermans et al., 2017) (Figure 3.3). This is primarily the result of increased absorption of solar radiation accompanying sea ice loss (Perovich, 2016). Between 1979 and 2011, the decrease in Arctic Ocean albedo corresponded to more solar energy input to the ocean (*virtually certain*) of approximately 6.4 ± 0.9 Wm^{-2} (Pistone et al., 2014), *likely* reducing the growth of sea ice by up to 25% in both Eurasian and Canadian basins (Timmermans, 2015; Ivanov et al., 2016) (Section 3.2.1.1).

Table 3.1 | Ocean heat content trend (0–2000 m depth) during 2005–2017 and 1970–2017 for the global ocean and Southern Ocean. Ordinary Least Square (OLS) method is used; units are 10^{21} J yr^{-1}. Uncertainties denote the 90% confidence interval accounting for the reduction in the degrees of freedom implied by temporal correlations of residuals, as per Section 5.2. Values in curved brackets are percentages of heat gain by the Southern Ocean relative to the global ocean. Data sources are as per Table SM3.1. The mean proportion and its 5–95% confidence interval (1.65 times standard deviation of individual estimates) are in the last column.

OHC Trend (10^{21} J yr^{-1})	Ishii V7.2	IAP	EN4-GR10	IPRC	Scripps	JAMSTEC	Mean [5%, 95%]
Global 2005–17	10.06 ± 1.28	8.45 ± 1.04	10.57 ± 1.17	9.96 ± 1.57	8.38 ± 1.31	9.06 ± 0.67	
South of 30°S 2005–17	5.20 ± 1.03 (52%)	4.55 ± 1.00 (54%)	5.38 ± 1.30 (51%)	6.24 ± 1.80 (63%)	4.22 ± 0.70 (50%)	4.44 ± 0.63 (49%)	53% [45%, 62%]
Global 1970–2017	6.73±0.55	7.02 ± 1.96	5.28 ± 1.01				
South of 30°S 1970–2017	2.42 ± 0.26 (36%)	2.78 ± 0.29 (40%)	2.18 ± 0.36 (41%)				39% [35%, 43%]

While Atlantic Water Layer temperatures appear to show less variability since 2008, total heat content in this layer continues to increase (Polyakov et al., 2017). Recent changes have been dubbed the 'Atlantification' of the Northern Barents Sea and Eurasian Basin (Arthun et al., 2012; Lind et al., 2018), characterised by weaker stratification and enhanced Atlantic Water Layer heat fluxes further northeast (*medium confidence*). Polyakov et al. (2017) estimate 2–4 times larger heat fluxes in 2014–2015 compared with 2007–2008. In the Canadian Basin, the maximum temperature of the Pacific Water Layer increased by ~0.5°C between 2009 and 2013 (Timmermans et al., 2014), with a doubling in integrated heat content over 1987–2017 (Timmermans et al., 2018). Over 2001–2014, heat transport associated with Bering Strait inflow increased by 60%, from around 10 TW in 2001 to 16 TW in 2014, due to increases in both volume flux and temperature (Woodgate et al., 2015; Woodgate, 2018) (*low confidence*).

The Southern Ocean is important for the transfer of heat from the atmosphere to the global ocean, including heat from anthropogenic warming (Frölicher et al., 2015; Shi et al., 2018). The Southern Ocean accounted for ~75% of the global ocean uptake of excess heat during 1870–1995 (Figure SM3.2; Frölicher et al., 2015), of which ~43% resided in the Southern Ocean with the remainder redistributed to lower latitudes. Over 1970–2017, observations show that the upper 2000 m of the ocean south of 30°S was responsible for 35–43% of the increase in global ocean heat content (Table 3.1). Both models and observations show that, relative to its size (Table SM3.1), the Southern Ocean is disproportionately important in the increase in global upper ocean heat content (*high confidence*). Multi-decadal warming of the Southern Ocean has been attributed to anthropogenic factors, especially the role of greenhouse gases but also ozone depletion (Armour et al., 2016; Shi et al., 2018; Swart et al., 2018; Irving et al., 2019) (*medium confidence*).

Surface warming during 1982–2016 was strongest along the northern flank of the ACC, contrasting with cooling further south (Figure 3.3). Interior warming was strongest in the upper 2000 m, peaking around 40°S–50°S (Armour et al., 2016) (SM3.2.1; Figures SM3.2 and SM3.3). There is *high confidence* that this pattern of change is driven by upper-ocean overturning circulation and mixing (Cross-Chapter Box 7 in Chapter 3), whereby heat uptake at the surface by newly upwelled waters is transmitted to the ocean interior in intermediate depth layers (Armour et al., 2016). Whilst temperature trends in the

ACC itself are driven predominantly by air-sea flux changes (Swart et al., 2018), the warming on its northern side appears strongly influenced by wind-forced changes in the thickness and depth of the mode water layer (Desbruyeres et al., 2017; Gao et al., 2018) (*medium confidence*). Below the surface south of the ACC, warming extends close to Antarctica, intruding onto the continental shelf in the Amundsen-Bellingshausen Sea where temperature increases of 0.1°C–0.3°C per decade have been observed over 1983–2012 (Schmidtko et al., 2014) (Section 3.3.1.5). This latter warming may be driven by changes in wind forcing (Spence et al., 2014), and exhibits significant decadal variability (Jenkins et al., 2018).

After around 2005, improved upper ocean heat content estimates became available via Argo profiling floats (Section 1.8.1; Section 5.2). For 2005–2017, multiple datasets show that the heat gained by the Southern Ocean south of 30°S was 45–62% of the global ocean heat gain (Table 3.1) (equivalent figures for other indicative Southern Ocean extents are in Table SM3.2). This accords with Roemmich et al. (2015), who found that during 2006–2013 the ocean south of 20°S accounted for 67–98% of total heat gain in the upper 2000 m of the global ocean. (The smaller proportion for 2005–2017 c.f. 2006–2013 is due to comparatively greater warming in the earlier part of the common period). The recent Southern Ocean heat gain is thus larger than its long-term trend over either the preceding several decades (1970–2004, 30–51%, Table SM3.3) or the full period 1970–2017 (35–43%; Table 3.1 and above). There is *high confidence* that the Southern Ocean has increased its role in global ocean heat content in recent years compared with the past several decades. Attribution of this increased role is currently lacking.

The ocean below 2000 m globally stores ~19% of the excess anthropogenic heat in the Earth system, with a large fraction (6% of global total heat excess) located in the deep Southern Ocean south of 30°S (Frölicher et al., 2015; Talley et al., 2016) (*medium confidence*). The WGI AR5-quantified warming of these waters was recently updated (Desbruyeres et al., 2017) to an equivalent heat uptake of 0.07 ± 0.06 W m^{-2} below 2000 m since the beginning of the century, resulting in an extra 34 ± 14 TW south of 30°S from 1980 to 2012 (Purkey and Johnson, 2013). Antarctic Bottom Water volume is decreasing (Purkey and Johnson, 2012), resulting in a deepening of density surfaces and driving much of the warming on depth surfaces below 2000 m (Desbruyeres et al., 2017). This reduction in bottom water volume is suggestive of a decrease in its production (Purkey and Johnson, 2013). In the Indian and Pacific basins close to Antarctica, bottom water is freshening (Purkey and Johnson, 2013; Menezes et al., 2017) consistent with the uptake of enhanced Antarctic ice shelf and glacial melt (Purkey and Johnson, 2013).

3.2.1.2.2 Salinity

Salinity is the dominant determinant of polar ocean density, and exerts major controls on stratification, circulation and mixing. Salinity changes are induced by freshwater runoff to the ocean (rivers and land ice), net precipitation, sea ice, and advection of mid-latitude waters, with the potential to impact water mass formation and circulation (e.g., Thornalley et al., 2018; see also Section 6.7.1).

Updating WGI AR5 (their Section 3.3.3.3), recent Arctic-wide estimates yield a freshwater increase (relative to salinity of 34.8 on the Practical Salinity Scale, used throughout this chapter) of 600 ± 300 km^3 yr^{-1} over 1992–2012, with about two-thirds concomitant with decreasing salinity, and the remainder with a thickening of the freshwater layer (*medium confidence*) (Rabe et al., 2014; Haine et al., 2015; Carmack et al., 2016). The Beaufort Gyre region has increased its freshwater by ~40% (6600 km^3) over 2003–2017; this, and the Gyre's strengthening, have been attributed to dominance of clockwise wind patterns over the Canadian Basin over 1997–2016 and freshwater accumulation from sea ice-melt (Krishfield et al., 2014; Proshutinsky et al., 2015). Freshwater decreases in the East Siberian, Laptev, Chukchi and Kara seas are estimated to be ~180 km^3 over 2003–2014 (Armitage et al., 2016). During the 2000s, freshwater content in the upper 100 m of the northern Barents Sea declined by about 32%, from a mean of ~2.5 m (relative to a salinity of 35) in 1970–1999, to 1.7 m in 2010–2016 (Lind et al., 2018). An increasing trend of 30 ± 20 km^3 yr^{-1} in freshwater flux through Bering Strait, primarily due to increased volume flux, was measured from 1991 to 2015, with record maximum freshwater influx in 2014 of around 3500 km^3 in that year (Woodgate, 2018). Freshwater fluxes from rivers are also increasing (Section 3.4.1.2.2), and there have been observed increases in discharge of glacial ice from Greenland (Section 3.3.1.3).

Observed Southern Ocean freshening trends are consistent with WGI AR5; subsequent studies have increased confidence in their magnitude and sign, and also attributed them to anthropogenic influences (Swart et al., 2018). Changes over 1950–2010 show persistent surface water freshening over the whole Southern Ocean, with subducted mode/ intermediate waters carrying trends of 0.0002–0.0008 yr^{-1} to below 1500 m (Skliris et al., 2014), whilst de Lavergne et al. (2014) observe a circumpolar freshening south of the ACC of 0.0011 ± 0.0004 yr^{-1} in the upper 100 m since the 1960s (*medium confidence*). This intensifies over the Antarctic continental shelves (except along the Western Antarctic Peninsula), where freshening of up to 0.01 yr^{-1} is observed (Schmidtko et al., 2014). Freshening may be driven by increases in precipitation, but while models (Pauling et al., 2016) and observations suggest an increase may have occurred over the last 60 years, uncertainty is presently too high to quantify its net impact (Skliris et al., 2014). Recently, there has been increased recognition of the importance of sea ice in driving Southern Ocean salinity changes, with Haumann et al. (2016) demonstrating that wind driven

sea ice export has increased by 20 ± 10% from 1982 to 2008, and that this may have driven freshening of 0.002 ± 0.001 yr^{-1} in the surface and intermediate waters. Separately, the central role of sea ice in driving water mass transformations in the Southern Ocean has been highlighted (Abernathey et al., 2016; Pellichero et al., 2018; Swart et al., 2018), hence such changes have the potential to affect overturning circulation (Cross-Chapter Box 7 in Chapter 3). Freshwater input to the ocean from the Antarctic Ice Sheet also has the potential to affect the properties and circulation of Southern Ocean water masses; see Section 3.3.3.

3.2.1.2.3 Stratification

See Supplementary Material (SM3.2.2).

3.2.1.2.4 Carbon and ocean acidification

Various elements of marine biogeochemistry and geochemistry in the polar regions are of global importance. Here we focus on aspects relevant to carbon and ocean acidification; others (e.g., changes in dissolved oxygen) are assessed in Section 5.2.2. Compiled datasets on observed trends in ocean acidification from different observational platforms can be found in Table SM5.3.

About a quarter of carbon dioxide (CO_2) released by human activities is taken up by the ocean (WGI AR5, their Section 3.8). This dissolves in surface water to form carbonic acid, which, upon dissociation, causes a decrease in pH (acidification) and carbonate ion (CO_3^{2-}) concentration. This can affect organisms that form shells and skeletons using calcium carbonate ($CaCO_3$, aragonite and calcite as dominant mineral forms). Since AR5, new observations have demonstrated the spatial and temporal variability of ocean acidification and controlling mechanisms of carbon systems in different regions (Bellerby et al., 2018).

Robbins et al. (2013) showed aragonite undersaturation for about 20% of surface waters in the Canada and Makarov Basins, where substantial sea ice melt occurred. Qi et al. (2017) reported that aragonite undersaturation has expanded northward by at least 5° of latitude, and deepened by ~100 m between the 1990s and 2010 primarily due to increased Pacific Winter Water transport. In the East Siberian Arctic Shelf, extreme aragonite undersaturation was driven by the degradation of terrestrial organic matter and runoff of Arctic river water with elevated CO_2 concentrations, reflecting pH changes in excess of those projected in this region for 2100 (Semiletov et al., 2016) (*high confidence*); this was also observed along the continental margin and traced in the deep Makarov and Canada Basins (Anderson et al., 2017a). The variable buffering capacities of rivers flowing through watersheds with different bedrock geology also influenced the state of ocean acidification in coastal regions (Tank et al., 2012; Azetsu-Scott et al., 2014).

The dissolved inorganic carbon (DIC) concentration increased in subsurface waters (150–1400 m) in the central Arctic between 1991 and 2011 (Ericson et al., 2014). The rate of increase was 0.6–0.9 µmol kg^{-1} yr^{-1} in the Arctic Atlantic Water and 0.4–0.6 µmol kg^{-1} yr^{-1} in the upper Polar Deep Water due to anthropogenic CO_2, while no trend was observed in nutrient concentrations. In waters below 2000 m, no significant trend was observed for DIC and nutrient concentrations. Observation-based estimates (MacGilchrist et al., 2014) revealed a net summertime pan-Arctic export of 231 ± 49 TgC yr^{-1} of DIC across the Arctic Ocean gateways to the North Atlantic; at least 166 ± 60 TgC yr^{-1} of this was sequestered from the atmosphere (*medium confidence*). Similar to other regions (Table SM5.3), observed changes in the carbonate chemistry of the Arctic are indicative of ongoing ocean acidification (*high confidence*).

Studies covering seasonal-to-decadal variability in the Arctic are limited, with most conducted in ice-free or low ice periods during summer to autumn. However, it has been demonstrated that biological processes, respiration and photosynthesis, control the $CaCO_3$ saturation states in Chukchi Sea bottom water (Yamamoto-Kawai et al., 2016). Sea ice formation and melt influence the dynamics of ikaite ($CaCO_3$ precipitation trapped in sea ice during brine rejection), and therefore local carbonate chemistry (Rysgaard et al., 2013; Bates et al., 2014; Geilfus et al., 2016; Fransson et al., 2017). Although the increase of pH and saturation states by biological carbon fixation that consumes DIC in surface water is well documented (Azetsu-Scott et al., 2014; Yamamoto-Kawai et al., 2016) (*high confidence*), it has been shown that long photoperiods in Arctic summers sustain high pH in kelp forests, slowing ocean acidification (Krause-Jensen et al., 2016).

Since AR5, there are new constraints on the seasonal-to-decadal variability in the Southern Ocean CO_2 flux (McNeil and Matear, 2013; Landschützer et al., 2014; Landschützer et al., 2015; Gregor et al., 2017; Ritter et al., 2017; Keppler and Landschutzer, 2019) (Figure SM3.4), with mean annual flux anomalies varying from 0.3 ± 0.1 Pg C yr^{-1} in 2001–2002 to -0.4 Pg C yr^{-1} in 2012 (Landschützer et al., 2015); this can affect the magnitude of the global CO_2 sink (Section 5.2.2). A weakening CO_2 sink during the 1990s (Le Quéré et al., 2007) reversed in the 2000s as part of a decadal cycle (Landschützer et al., 2015; Munro et al., 2015; Williams et al., 2017) (SM3.2.3; Figure SM3.4), with a weakening again since 2011 (Keppler and Landschutzer, 2019). While the weakening sink during the 1990s was explained as a response to changes in the circumpolar winds over the Southern Ocean enhancing the outgassing of natural CO_2, the subsequent changes appear due to a combination of changes in regional winds, temperature and circulation (Landschützer et al., 2015; Gregor et al., 2017; Keppler and Landschutzer, 2019). Data scarcity, especially in winter, remains a challenge (Ritter et al., 2017; Fay et al., 2018; Gruber et al., 2019b); recent data from pH-enabled floats highlighted the potential role for winter outgassing south of the Polar Front (Williams et al., 2017; Gray et al., 2018). Overall, there is *medium confidence* that the Southern Ocean CO_2 sink has experienced significant decadal variations since the 1980s.

Southern Ocean carbon storage is affected by changes in overturning circulation (Cross-Chapter Box 7 in Chapter 3), with the storage of anthropogenic and natural carbon being both variable and out of phase on decadal timescales (DeVries et al., 2017; Tanhua et al., 2017) (Table SM3.4). Mode and intermediate waters are strongly involved in changing storage, also showing high sensitivity to shifts in winds (Swart et al., 2014; Swart et al., 2015a; Tanhua et al., 2017; Gruber et al., 2019a). Zonal basin differences in the uptake and storage of anthropogenic carbon are not well resolved and there is weak agreement between reanalysis products and Coupled Model Intercomparison Project Phase 5 (CMIP5) models (Swart et al., 2014). The presence of subduction hotspots suggest that basin-wide studies may be underestimating the importance of mode water subduction as a principal storage mechanism (Langlais et al., 2017).

Strengthening impacts of Southern Ocean acidification are illustrated by the $3.9 \pm 1.3\%$ decrease in derived calcification rates (1998–2014) (Freeman and Lovenduski, 2015). These have strong regional character, with decreases in the Indian and Pacific sectors (7.5–11.6%) and increases in the Atlantic ($14.3 \pm 5.1\%$). There have also been changes in the seasonality of pCO_2 linked to decreasing buffer capacity (McNeil and Sasse, 2016) (SM3.2.4) or adjustments to primary production (Conrad and Lovenduski, 2015); seasonal changes are discussed further in Section 5.2.2.

3.2.1.3   Ocean Circulation

The major elements of Southern Ocean circulation are assessed in Cross-Chapter Box 7 in Chapter 3; Arctic Ocean circulation is considered here. Arctic processes, such as discharge of freshwater from the Greenland Ice Sheet, have the potential to impact on the formation of the headwaters of the Atlantic Meridional Overturning Circulation (Section 6.7.1), and can impact on the structure and function of the marine ecosystem with implications for commercially-harvested species (Sections 3.2.3, 3.2.4).

Satellite data indicate a general strengthening of the surface geostrophic currents in the Arctic basin (Armitage et al., 2017). Between 2003 and 2014, the strength of some currents in the Beaufort Gyre approximately doubled (Armitage et al., 2017). Over 2001–2014, annual Bering Strait volume transport from the Pacific to the Arctic Ocean increased from 0.7×10^6 m^3s^{-1} to 1.2×10^6 m^3s^{-1} (Woodgate et al., 2015). Mesoscale eddies are characterised by horizontal scales of ~10 km in the Arctic, and are important components of the ocean system. Increased wind power input to the Arctic Ocean system can in principle be compensated by the production of eddy kinetic energy; analysis of observations in the Beaufort Gyre region suggest this is *about as likely as not* (Meneghello et al., 2017). Data of sufficiently high resolution is limited in the boundary regions of the Arctic Ocean, precluding estimates of eddy variability on a basin-wide scale. In the central basin regions, a statistically significant higher concentration of eddies was sampled in the Canadian Basin compared to the Eurasian Basin between 2003 and 2014; further, a medium correspondence was found between eddy activity in the Beaufort Gyre region and intensified gyre flow (Zhao et al., 2014; Zhao et al., 2016).

In contrast to the Southern Ocean (Cross-Chapter Box 7 in Chapter 3), there is comparatively little knowledge on changing Arctic frontal positions and current cores since AR5. An exception is that the Beaufort Gyre expanded to the northwest between 2003 and 2014, contemporaneous with changes in its freshwater accumulation and alterations in wind forcing, resulting in increased proximity to the Chukchi Plateau and Mendeleev Ridge (Armitage et al., 2017; Regan et al., 2019) (Section 3.2.1.2.2).

Cross-Chapter Box 7 | Southern Ocean Circulation: Drivers, Changes and Implications

Authors: Michael P. Meredith (UK), Robert Hallberg (US), Alessandro Tagliabue (UK), Andrew Meijers (UK/Australia), Jamie Oliver (UK), Andrew Hogg (Australia)

Horizontal Circulation and Movement of Fronts

The Southern Ocean is disproportionately important in global climate and ecological systems, being the major connection linking the Atlantic, Pacific and Indian Oceans in the global circulation. The horizontal circulation in the circumpolar Southern Ocean is comprised of an eastward-flowing mean current concentrated in a series of sinuous, braided jets exhibiting strong meandering variability and shedding small-scale transient eddies (Figure CB7.1). The mean flow circumnavigates Antarctica as the world's largest ocean current, the Antarctic Circumpolar Current (ACC), transporting approximately $173.3 \pm 10.7 \times 10^6$ m^3 s^{-1} (Donohue et al., 2016) of water eastward in a geostrophic balance set up by the contrasting properties of waters around Antarctica and those inside the subtropical gyres to the north of ACC. This contrast is maintained by a combination of strong westerly winds and ocean heat loss south of the ACC.

Trends in the atmospheric forcing of the Southern Ocean are dominated by a strengthening of westerly winds in recent decades (Swart et al., 2015a), but there is no evidence that this enhanced wind stress has significantly altered the ACC transport. While the annual mean value of transport is stable in the instrumental period (Chidichimo et al., 2014; Koenig et al., 2014; Donohue et al., 2016) it is difficult to resolve changes in barotropic transport; overall there is *medium confidence* that ACC transport is only weakly sensitive to changes in winds. This is consistent with longer-term analyses that find only minimal changes in ACC transport since the last glaciation (McCave et al., 2013). Theoretical predictions and high-resolution ocean modelling suggest that the weak sensitivity of the ACC to changes in wind stress is a consequence of eddy saturation (Munday et al., 2013), whereby the time-mean state of the ocean remains close to a marginal condition for eddy instability and hence additional energy input from stronger winds cascades rapidly into the smaller-scale eddy field. Satellite measurements of eddy kinetic energy over the last two decades are consistent with this, showing a statistically significant upward trend in eddy energy in the Pacific and Indian Ocean sectors of the Southern Ocean (Hogg et al., 2015) (*medium confidence*). This is supported by eddy-resolving models, which also show a marked regional variability (Patara et al., 2016), and there is evidence that local hotspots in eddy energy, especially downstream of major topographic features including the Drake Passage, Kerguelen Plateau, Campbell Plateau and the East Pacific Rise, may dominate the regional fields (Thompson and Naveira Garabato, 2014).

Working Group I (WGI) of the IPCC's 5th Assessment Report (AR5) assessed that there was *medium confidence* that the mean position of the ACC had moved southwards in response to a contraction of the Southern Ocean circumpolar winds. Such movements can in principle have profound effects on marine ecosystems via, e.g., changing habitat ranges for different species (e.g., Cristofari et al., 2018; Meijers et al., 2019) (Section 3.2.3.2). Since AR5, however, substantial contrary evidence has emerged. While winds have strengthened over the Southern Ocean, reanalysis products show no significant shift in the annual mean latitude of zonal wind jets between 1979–2009 (Swart et al., 2015a). Similarly, a variety of methods applied to satellite data have found no long-term trend and no statistically significant correlation of ACC position with winds (Gille, 2014; Chapman, 2017; Chambers, 2018). The discrepancy between these studies and those assessed in WGI AR5 appears to be caused by issues associated with using a fixed sea surface height contour as a proxy for frontal position in the presence of strongly eddying fields (Chapman, 2014) and large-scale increases in sea surface height consistent with mean global trends in sea level rise (Gille, 2014). The increase in sea surface height is ascribed largely to warming-driven steric expansion in the upper ocean, but the mechanism driving such warming is still uncertain (Gille, 2014). These recent findings do not preclude more local changes in frontal position, but it is now assessed as *unlikely* that there has been a statistically significant net southward movement of the mean ACC position over the past 20 years.

Overturning Circulation and Water Mass Formation

The Southern Ocean is the key region globally for the upwelling of interior ocean waters to the surface, enabling waters that were last ventilated in the pre-industrial era to interact with the industrial-era atmosphere and the cryosphere. New water masses are produced that sink back into the ocean interior. Such export of both extremely cold and dense Antarctic Bottom Water and the lighter mode and intermediate waters (Figure CB7.1) represents important pathways for surface properties to be sequestered from the atmosphere for decades to millennia. This upwelling and sinking constitutes a two-limbed overturning circulation, by which much of the global deep ocean is renewed.

The Southern Ocean overturning circulation plays a strong role in mediating climate change via the transfer of heat and carbon (including that of anthropogenic origin) with the atmosphere (Sections 3.2.1.2; 5.2.2.2); it also has an impact on sea ice extent and concentration, with implications for climate via albedo (Section 3.2.1.1). It acts to oxygenate the ocean interior

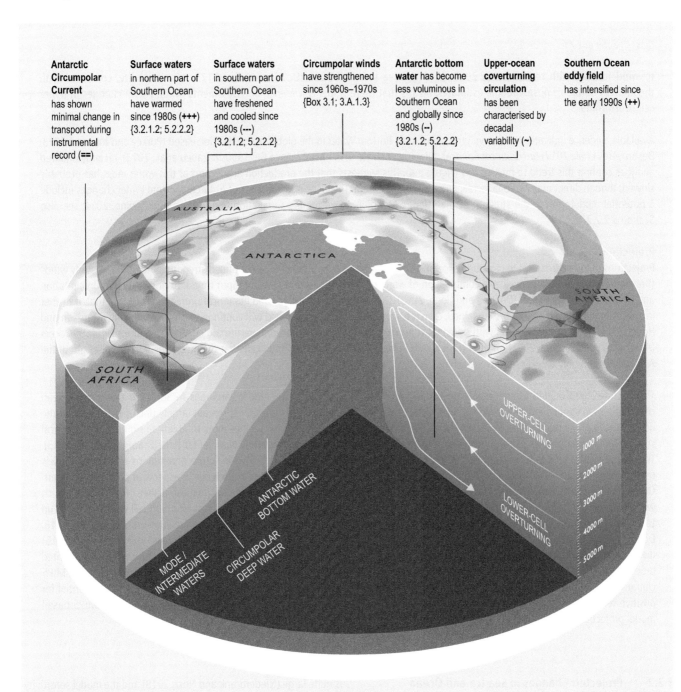

Figure CB7.1 | Schematic of some of the major Southern Ocean changes assessed in this Box and in Chapters 3 and 5. Assessed changes are marked as positive (+), neutral (=), negative (−), or dominated by variability (~). The number of symbols used indicates confidence, from *low* (1) through *medium* (2) to *high* (3). Section numbers indicate the links to further information outside this box.

and sequesters nutrients that ultimately end up supporting a significant fraction of primary production in the rest of the world ocean (Section 5.2.2.2). The upwelling waters in the overturning bring heat to the Antarctic shelf seas, with consequences for ice shelves, marine-terminating glaciers and the stability of the Antarctic Ice Sheet (AIS) (Section 3.3.1). The lower limb of this overturning circulation supplies Antarctic Bottom Water that forms the abyssal layer of much of the world ocean (Section 3.2.1.2; 5.2.2.2).

It is challenging to measure the Southern Ocean overturning directly, and misinterpretation of Waugh et al. (2013) led to AR5 erroneously reporting the upper cell to have slowed (AR5 WGI, Section 3.6.4). However, additional indirect estimates since AR5 provide support for the increase in the upper ocean overturning proposed by Waugh et al. (2013). Waugh (2014) and Ting and Holzer (2017) suggest that over the 1990s–2000s water mass ages changed in a manner consistent with an increase in upwelling and overturning. However, inverse analyses suggest that such overturning experiences significant inter-decadal variability in response

Cross-Chapter Box 7 (continued)

to wind forcing, with reductions in 2000–2010 relative to 1990–2000 (DeVries et al., 2017). This variability, combined with the indirect nature of observational estimates, means that there is *low confidence* in assessments of long-term changes in upper cell overturning.

Available evidence indicates that the volume of Antarctic Bottom Water in the global ocean has decreased (Purkey and Johnson, 2013; Desbruyeres et al., 2017) (*medium confidence*), thinning at a rate of 8.1 m yr^{-1} since the 1950s (Azaneu et al., 2013); recently updated analyses confirm this trend to present day (Figure 5.4). This suggests that the production and export of this water mass has probably slowed, though direct observational evidence is difficult to obtain. The large-scale impacts of Antarctic Bottom Water changes include a potential modulation to the strength of the Atlantic Meridional Overturning Circulation (e.g., Patara and Böning, 2014; see also Section 5.2.2.2.1).

Projections

Projections of future trends in the Southern Ocean are dominated by the potential for a continued strengthening of the westerly winds (Bracegirdle et al., 2013), as well as a combination of warming and increased freshwater input from both increased net precipitation and changes in sea ice export (Downes and Hogg, 2013). Dynamical considerations and numerical simulations indicate that, if further increases in the westerly winds are sustained, then it is *very likely* that the eddy field will continue to grow in intensity (Morrison and Hogg, 2013; Munday et al., 2013), with potential consequences for the upper-ocean overturning circulation and transport of tracers (Abernathey and Ferreira, 2015) (including heat, carbon, oxygen and nutrients), and *likely* that the mean position and strength of the ACC will remain only weakly sensitive to winds.

The considerable Coupled Model Intercomparison Project Phase 5 (CMIP5) inter-model variations in Southern Ocean time-mean circulation projections reported in WGI AR5 (Meijers et al., 2012; Downes and Hogg, 2013) remain largely unchanged. Some of the differences in projected changes have been found to be correlated with biases in the various models' ability to simulate the historical state of the Southern Ocean, such as mixed layer depth (Sallée et al., 2013a) and westerly wind jet latitude (Bracegirdle et al., 2013). This suggests that bias reduction against observed historical metrics (Russell et al., 2018) in future generations of coupled models (e.g., Coupled Model Intercomparison Project Phase 6 (CMIP6)) should lead to improved confidence in aspects of projected Southern Ocean changes. CMIP5 models suggest that the subduction of mode and intermediate water will increase (Sallée et al., 2013b), which will affect oxygen and nutrient transports, and the overall transport of the Southern Ocean upper overturning cell will increase by up to 20% (Downes and Hogg, 2013), but model performance is limited by the inability to explicitly resolve eddy processes (Gent, 2016; Downes et al., 2018). The formation and export of Antarctic Bottom Water is predicted to continue decreasing (Heuzé et al., 2015) due to warming and freshening of surface source waters near the continent. These are, however, some of the most poorly represented processes in global models. Further uncertainty derives from increased meltwater from the AIS not being considered in the CMIP5 climate models, despite its potential for significant impact on Southern Ocean dynamics and the global climate, and its potential for positive feedbacks (Bronselaer et al., 2018). Due to these uncertainties, *low confidence* is therefore ascribed to the CMIP5-based model projections of future Southern Ocean circulation and watermasses.

3.2.2   Projected Changes in Sea Ice and Ocean

3.2.2.1   Sea Ice

The multi-model ensemble of historical simulations from CMIP5 models identify declines in total Arctic sea ice extent and thickness (Sections 3.2.1.1.1; 3.2.1.1.2; Figure 3.3) which agree with observations (Massonnet et al., 2012; Stroeve et al., 2012a; Stroeve et al., 2014a; Stroeve and Notz, 2015). There is a range in the ability of individual models to simulate observed sea ice thickness spatial patterns and sea ice drift rates (Jahn et al., 2012; Stroeve et al., 2014a; Tandon et al., 2018). Reductions in Arctic sea ice extent scale linearly with both global temperatures and cumulative CO_2 emissions in simulations and observations (Notz and Stroeve, 2016), although aerosols influenced historical sea ice trends (Gagné et al., 2017). The uncertainty in sea ice sensitivity (ice extent loss per unit of warming)

is quite large (Niederdrenk and Notz, 2018) and the model sensitivity is too low in most CMIP5 models (Rosenblum and Eisenman, 2017). Emerging evidence suggests, however, that internal variability, including links between the Arctic and lower latitude, strongly influences the ability of models to simulate observed reductions in Arctic sea ice extent (Swart et al., 2015b; Ding et al., 2018).

CMIP5 models project continued declines in Arctic sea ice through the end of the century (Figure 3.3) (Notz and Stroeve, 2016) (*high confidence*). There is a large spread in the timing of when the Arctic may become ice free in the summer, and for how long during the season (Massonnet et al., 2012; Stroeve et al., 2012a; Overland and Wang, 2013) as a result of natural climate variability (Notz, 2015; Swart et al., 2015b; Screen and Deser, 2019), scenario uncertainty (Stroeve et al., 2012a; Liu et al., 2013), and model uncertainties related to sea ice dynamics (Rampal et al., 2011; Tandon et al., 2018) and thermodynamics (Massonnet et al., 2018). Internal climate variability results in an uncertainty of approximately 20 years in the timing of seasonally ice-free conditions (Notz, 2015; Jahn, 2018), but the clear link between summer sea ice extent and cumulative CO_2 emissions provides a basis for when consistent ice-free conditions may be expected (*high confidence*). For stabilised global warming of 1.5°C, sea ice in September is *likely* to be present at end of century with an approximately 1% chance of individual ice-free years (Notz and Stroeve, 2016; Sanderson et al., 2017; Jahn, 2018; Sigmond et al., 2018); after 10 years of stabilised warming at a 2°C increase, more frequent occurrence of an ice-free summer Arctic is expected (around 10–35%) (Mahlstein and Knutti, 2012; Jahn et al., 2016; Notz and Stroeve, 2016). Model simulations show that a temporary temperature overshoot of a warming target has no lasting impact on ice cover (Armour et al., 2011; Ridley et al., 2012; Li et al., 2013).

CMIP5 models show a wide range of mean states and trends in Antarctic sea ice (Turner et al., 2013; Shu et al., 2015). The ensemble mean across multiple models show a decrease in total Antarctic sea ice extent during the satellite era, in contrast to the lack of any observed trend (Figure 3.3; Section 3.2.1.1.1). Interannual sea ice variability in the models is larger than observations (Zunz et al., 2013), which may mask disparity between models and observations. Internal variability (Polvani and Smith, 2013; Zunz et al., 2013), and

model sensitivity to warming (Rosenblum and Eisenman, 2017) are also important sources of uncertainty. During the historical period, regional trends of Antarctic sea ice are not captured by the models, particularly the decrease in the Bellingshausen Sea and the expansion in the Ross Sea (Hobbs et al., 2015). There is a very wide spread of model responses in the Weddell Sea (Hobbs et al., 2015; Ivanova et al., 2016), a region with complex ocean-sea ice interactions that many models do not replicate (de Lavergne et al., 2014).

There is *low confidence* in projections of Antarctic sea ice because there are multiple anthropogenic forcings (ozone and greenhouse gases) and complicated processes involving the ocean, atmosphere, and adjacent ice sheet (Section 3.2.1.1.). Model deficiencies are related to stratification (Sallée et al., 2013a), freshening by ice shelf melt water (Bintanja et al., 2015), atmospheric processes including clouds (Schneider and Reusch, 2015; Hyder et al., 2018), and wind and ocean driven processes (Purich et al., 2016a; Purich et al., 2016b; Schroeter et al., 2017; Purich et al., 2018; Zhang et al., 2018a). Uncertainty in sea ice projections reduces confidence in projections of Antarctic Ice Sheet surface mass balance because sea ice affects Antarctic temperature and precipitation trends (Bracegirdle et al., 2015), and impacts projected changes in the Southern Hemisphere westerly jet (Bracegirdle et al., 2018; England et al., 2018) with implications for the Southern Ocean overturning circulation (Cross-Chapter Box 7 in Chapter 3).

Box 3.3 | Polynyas

Arctic Coastal Polynyas

Arctic polynyas (areas of open water surrounded by sea ice) are important because they ventilate the Arctic Ocean. The polynyas induce bottom reaching convection on shallow shelves (Damm et al., 2018) because the warm and exposed ocean surface creates very high heat fluxes and new sea ice formation during winter, releasing brine and creating dense water (Barber et al., 2012). On the shallow Siberian shelves, the ocean surface waters are dominated by river runoff which is rich with sediments (Damm et al., 2018), which end up both in the dense bottom water and in new sea ice (Bauch et al., 2012; Janout et al., 2015). This process maintains the Arctic Ocean halocline (Bauch et al., 2011), which insulates the sea ice cover from the heat of the underlying Atlantic-derived waters.

Polynyas are projected to change in different ways depending on regional ice conditions and ice formation processes. Further reductions in sea ice are projected for Arctic shelf seas which have already lost ice in recent decades (Barnhart et al., 2015; Onarheim et al., 2018) so polynyas will cease to exist where seasonal sea ice disappears or evolve to become part of the marginal sea ice zone due to changes in ice dynamics (i.e., the North Water polynya and the Circumpolar Flaw Lead); new or enlarged polynyas could result in regions where thinner ice becomes more effectively advected offshore, or where marine terminating glaciers increase land ice fluxes to the marine system (*medium confidence*). The reduced survival rate of sea ice in the Transpolar Drift interrupts the transport of sediment-laden ice produced from Siberian shelf polynyas (Krumpen et al., 2019), with consequences for the associated biogeochemical matter and gas fluxes (Damm et al., 2018) (*medium confidence*).

Projected changes to polynyas are important because the spring phytoplankton bloom starts early as the ocean is often well-ventilated and nutrient rich, so the entire biological range from phytoplankton to seabirds to marine mammals thrive in polynya waters (*high confidence*) (Stirling, 1997; Arrigo and van Dijken, 2004; Karnovsky et al., 2009). Secondary production and upper food web processes are typically adapted to the early availability of energy to the system with arrival of higher trophic species (Asselin et al., 2011). Because of the abundance of marine food resources including seals, whales and fish in and around polynyas, Arctic peoples have hunted regularly in these areas for thousands of years (Barber and Massom, 2007). Recent implementation of Inuit-led marine management areas acknowledge the Inuit knowledge of polynyas, and recognise the potential for development of fisheries and other resources in polynya systems, provided these activities minimise harm on the environment and wildlife. The Inuit Circumpolar Council's Pikialasorsuaq Commission is an example of a proposal to develop an Inuit management area in the North Water Polynya (Cross-Chapter Box 3 in Chapter 1).

Antarctic Coastal Polynyas

The Antarctic continent is surrounded by coastal polynyas, which form from the combined effects of winds and landfast ice in the lee of coastal features that protrude into the westward coastal current (Nihashi and Ohshima, 2015; Tamura et al., 2016). Intense ice growth within these polynyas contributes to the production of Antarctic Bottom Water, the densest and most voluminous water mass in the global ocean (Jacobs, 2004; Nicholls et al., 2008; Orsi and Wiederwohl, 2009; Ohshima et al., 2013). Sea ice production is greatest in Ross and Weddell sea polynyas and around East Antarctica (Drucker et al., 2011; Nihashi and Ohshima, 2015; Tamura et al., 2016) (*high confidence*).

Antarctic coastal polynyas are biological hot-spots that support high rates of primary production (Ainley et al., 2015; Arrigo et al., 2015) due to a combination of both high light (Park et al., 2017) and high nutrient levels, especially iron (Gerringa et al., 2015). Basal ice shelf melt is the primary supplier of iron to coastal polynyas (Arrigo and van Dijken, 2015) although sea ice melt and intrusions of Circumpolar Deep Water are significant in the Ross Sea (McGillicuddy et al., 2015; Hatta et al., 2017). As ice shelves retreat, the polynyas created in their wake also increase local primary production: the new polynyas created after the collapse of the Larsen A and B ice shelves are as productive as other Antarctic shelf regions, *likely* increasing organic matter export and altering marine ecosystem evolution (Cape et al., 2013). The recent calving of Mertz Glacier Tongue in East Antarctica has altered sea ice and ocean stratification (Fogwill et al., 2016) such that polynyas there are now twice as productive (Shadwick et al., 2017).

The productivity associated with these polynyas is a critical food source for some of the most abundant top predators in Antarctic waters, including penguins, albatross and seals (Raymond et al., 2014; Malpress et al., 2017) (Section 3.2.3.2.4). However, only a fraction of the carbon fixed by phytoplankton in coastal polynyas is consumed by upper trophic levels. The rest sinks to the seafloor where it is re-mineralised or sequestered (Shadwick et al., 2017), or is advected off the shelf (Lee et al., 2017b). Given the high amount of residual macronutrients in polynya surface waters, there is evidence that future changes in ice shelf melt rates could increase water column productivity (Gerringa et al., 2015; Rickard and Behrens, 2016; Kaufman et al., 2017), influencing Antarctic coastal ecosystems and increasing the ability of continental shelf waters to sequester atmospheric carbon dioxide (Arrigo and van Dijken, 2015).

The Weddell Polynya

The Weddell Polynya is a large area of open water within the winter ice pack of the Weddell Sea close to the Maud Rise seamount (at approximately 65°S, 3°E), and has importance on a global scale for deep water ventilation. The polynya opens intermittently, and remained open from 1974 to 1976, with an area of 0.2–0.3 million km^2 (Carsey, 1980). A similar polynya appeared in spring 2017, with a smaller area in 2016, but did not occur in 2018 (Campbell et al., 2019; Jena et al., 2019). Based on these recent events, there is *medium confidence* in the drivers of Weddell Polynya formation; it forms over deep water and appears connected to sea ice divergence created by ocean eddies (Holland, 2001) or strong winds (Campbell et al., 2019; Francis et al., 2019; Wilson et al., 2019). Around Maud Rise, the ocean is weakly stratified, and winter sea ice formation causes brine release and the related deepening mixed layer brings warmer deep waters towards the surface. This causes heat loss to the atmosphere above 200 W m^{-2} (Campbell et al., 2019). These polynya formation processes cause deep ocean convection that releases heat from the deep ocean to the atmosphere (Smedsrud, 2005), and may contribute to the uptake of anthropogenic carbon (Bernardello et al., 2014).

In some CMIP5 models, phases of Weddell polynya activity appear for decades or centuries at a time, and then cease for a similar period (Reintges et al., 2017). The observational era is not sufficiently long to rule out this behaviour. Models indicate that under anthropogenic climate change, surface freshening caused by increased precipitation reduces the occurrence of the Weddell polynya (de Lavergne et al., 2014). There are systematic biases in modelled ocean stratification resulting in *low confidence* in future Weddell Polynya projections (Reintges et al., 2017).

3.2.2.2   Physical Oceanography

Consistent with the projected sea ice decline, there is *high confidence* that the Arctic Ocean will warm significantly towards the end of this century at the surface and in the deeper layers. Most CMIP5 models capture the seasonal changes in surface heat and freshwater fluxes for the present day climate, and show that the excess summer solar heating is used to melt sea ice, in a positive ice albedo feedback (Ding et al., 2016). Using RCP8.5, Vavrus et al. (2012) found that the Atlantic layer is projected to warm by 2.5°C at around 400 m depth at the end of the century, but only by 0.5°C in the surface mixed layer.

Consistent results for lower Atlantic Water layer warming were found by Koenigk and Brodeau (2014) for RCP2.5 (0.5°C), RCP4.5 (1.0°C) and RCP8.5 (2.0°C).

Poleward ocean heat transport contributes to Arctic Ocean warming (*medium confidence*). Comparing 20 CMIP5 simulations for RCP8.5, Nummelin et al. (2017) found a 2°C–6°C range in Arctic amplification of surface air temperature north of 70°N, consistent with increased ocean heat transport. Comparing 26 different CMIP5 simulations for RCP4.5, Burgard and Notz (2017) found that ocean heat transport changes explain the Arctic Ocean multi-model mean warming, but that differences between models are compensated by changes in surface fluxes. Increased ocean heat transport into the Barents Sea beyond 2020 appears as a probable mechanism with continued warming (Koenigk and Brodeau, 2014; Årthun et al., 2019). Based on four CMIP5 models, the Barents Sea is projected to become ice-free during winter beyond 2050 under RCP8.5 (Onarheim and Årthun, 2017), to which the main response is an increased ocean-to-atmosphere heat flux and related surface warming (Smedsrud et al., 2013). The ocean heat transport increases in all Arctic gateways, but is dominated by the Barents Sea, and when winter sea ice disappears here the heat loss cannot increase further and the excess ocean heat continues into the Arctic Basin (Koenigk and Brodeau, 2014).

The surface mixed layer of the Arctic Ocean is expected to freshen in future because an intensified hydrological cycle will increase river runoff (Haine et al., 2015) (*medium confidence*). The related increase in stratification has the potential to contribute to the warming of the deep Atlantic Water layer, as upward vertical mixing will be reduced (Nummelin et al., 2016). There are, however, biases in salinity of ~1 across the Arctic Basin for the present day climate (Ilicak et al., 2016) in forced global ice-ocean models with configurations comparable to CMIP5, suggesting limited predictive skill for the Arctic freshwater cycle.

CMIP5 projections (Figure 3.3) indicate that observed Southern Ocean warming trends will continue under RCP4.5 and RCP8.5 scenarios, leading to 1°C–3°C warming by 2100 mostly in the upper ocean (Sallée et al., 2013a). Projections demonstrate a similar distribution of heat storage to historical observations, notably focused in deep pools north of the Subantarctic Front (e.g., Armour et al., 2016). Antarctic Bottom Water becomes coherently warmer by up to 0.3°C by 2100 across the model ensemble under RCP8.5 (Heuzé et al., 2015). The upper ocean also becomes considerably fresher (salinity decrease of approximately 0.1) (Sallée et al., 2013b) with an overall increase in stratification and a shallowing of mixed layers (Sallée et al., 2013a). Although the sign of model changes appear mostly robust, there is *low confidence* in magnitude due to the large inter-model spread in projections and significant warm biases in historical water mass properties (Sallée et al., 2013a) and sea surface temperature, which may be up to 3°C too high in the historical runs (Wang et al., 2014).

Projections of changes in Southern Ocean circulation are discussed in Cross-Chapter Box 7 in Chapter 3.

3.2.2.3   Carbon and Ocean Acidification

The Arctic and Southern Ocean have a systemic vulnerability to aragonite undersaturation (Orr et al., 2005). For the RCP8.5 scenario, the entire Arctic and Southern Ocean surface waters will *very likely* be typified by year-around conditions corrosive for aragonite minerals for 2090–2100 (Figure 3.4) (Hauri et al., 2015; Sasse et al., 2015), whilst under RCP2.6 the extent of undersaturated waters are reduced markedly. At a basin/circumpolar scale, there is *high confidence* in these projections due to our robust understanding of the driving mechanisms. However, there is *medium confidence* for the response of specific locations, due to the need for improved resolution of the local circulation, interactions with sea ice, and other processes that modulate the rate of acidification.

Under RCP8.5, melting ice causes the greatest declining rate of pH and $CaCO_3$ saturation state in the Central Arctic, Canadian Arctic Archipelago and Baffin Bay (Popova et al., 2014). In the Canada Basin, projections using RCP8.5 show reductions in mean surface pH from approximately 8.1 in 1986–2005 to 7.7 by 2066–2085, and aragonite saturation from 1.52–0.74 during the same period (Steiner et al., 2014). A shoaling of the aragonite saturation horizon of approximately 1200 m, a large increase in area extent of undersaturated surface waters, and a pH change in the surface water of –0.19 are projected using the SRES A1B scenario (broadly comparable to RCP6.0) in the Nordic Sea from 2000 to 2065 (Skogen et al., 2014). Under the same scenario, aragonite undersaturation is projected to occur in the bottom waters over the entire Kara Sea shelf by 2040 and over most of the Barents and East Greenland shelves by 2070 due to the accumulation of anthropogenic CO_2 (Wallhead et al., 2017).

Under RCP8.5, the rate of CO_2 uptake by the Southern Ocean is projected to increase from the contemporary 0.91 Pg C yr^{-1} to 2.38 (1.65–2.55) Pg C yr^{-1} by 2100, but the growth in uptake rate will slow and likely stop around 2070 ± 10 corresponding to cumulative CO_2 emissions of 1600 Gt C (Kessler and Tjiputra, 2016;

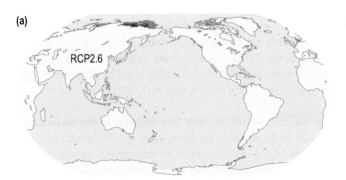

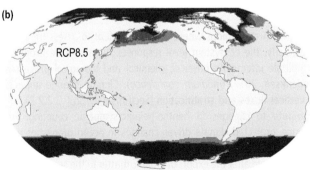

Figure 3.4 | The upper ocean (0–10 m) at end of this century (2081–2100), characterised by year-round undersaturated conditions for aragonite for the Representative Concentration Pathway (RCP)8.5 **(a)** and RCP2.6 **(b)** scenarios in the Coupled Model Intercomparison Project Phase 5 (CMIP5). The medium red shade denotes the area for which the multi-model annual mean of aragonite saturation is <1; the lighter and darker red shades denote the same conditions but calculated using the −90% and +90% confidence intervals for the aragonite saturation state, respectively. Shaded regions are plotted over each other. Blue regions are those without year-round aragonite saturation states <1. Aragonite saturation states are averaged and confidence intervals calculated at each geographic location across the CNRM-CM5, HadGEM2-ES, GFDL-ESM2G, GFDL-ESM2G, IPSL-CM5-LR, IPSL-CM5-MR, MPI-LR, MPI-MR and NCAR-CESM1 models.

Wang et al., 2016b). This halt in the increase in the uptake rate of CO_2 is linked to the combined feedbacks from well-understood reductions in buffering capacity and warming, as well as the increased upwelling rate of carbon-rich Circumpolar Deep Water (Hauck and Volker, 2015) (Cross-Chapter Box 7 in Chapter 3). Although there is *high agreement* amongst models, contemporary biases in the fluxes of CO_2 in CMIP5 models in the Southern Ocean (Mongwe et al., 2018) suggest *medium confidence* levels for these projections.

Alongside the mean state changes, Southern Ocean aragonite saturation is also affected by the seasonal cycle of carbonate as well as by the impact of reduced buffering capacity (SM3.2.4) on the seasonal cycle of CO_2 (Sasse et al., 2015; McNeil and Sasse, 2016). This leads to an amplification of the seasonal variability of pCO_2 (Hauck and Volker, 2015; McNeil and Sasse, 2016; Landschützer et al., 2018) and the hydrogen ion concentration that accelerates the onset of hypercapnia (i.e., high pCO_2 levels; $pCO_2 > 1000$ µatm) to nearly 2 decades (~2085) ahead of anthropogenic CO_2 forcing (McNeil and Sasse, 2016). The seasonal cycles of pH and aragonite saturation will be attenuated (Kwiatkowski and Orr, 2018) (Section 5.2.2.3), however when the mean state changes are combined with the changes in seasonality, the onset of undersaturation is brought forward by 10–20 years (Table SM3.5). Model projections remain uncertain and affected by the resolution of local ocean physics, which leads to overall *medium confidence* in the timing of undersaturation and hypercapnia.

3.2.3 Impacts on Marine Ecosystems

3.2.3.1 Arctic

Climate change has, and is projected to continue to have, significant implications for Arctic marine ecosystems, with consequences at different trophic levels both in the pelagic, benthic, and sympagic (sea ice related) realms (Figure 3.5). Specifically, climate change is projected to alter the distribution and properties of Arctic marine habitats with associated implications for species composition, production and ecosystem structure and function (Frainer et al., 2017; Kaartvedt and Titelman, 2018; Moore et al., 2018). The rate and severity of ecosystem impacts will be spatially heterogeneous and dependent on future emission scenarios.

In the few Arctic regions where data is sufficient to assess trends in biodiversity, the ecosystem level responses appear to be products of multiple interacting physical, chemical and biological processes (Frederiksen, 2017) (*medium confidence*). Climate change impacts on vertical fluxes and stratification (Sections 3.2.1.2.3, 3.2.2.2) will contribute to changes in bentho-pelagic-sympagic coupling. For instance, projected climate driven changes in ocean properties and hydrography (Section 3.2.2.2) and the abundance of pelagic grazers (Box 3.4) could alter the export of organic matter to the sea floor with associated impacts on the benthos in some Arctic shelf ecosystems (Moore and Stabeno, 2015; Stasko et al., 2018) (*low confidence*). Projected future reductions in summer sea ice (Section 3.2.1.1), increased stratification in summer, shifting currents and fronts and increased ocean temperatures (Section 3.2.2.2) and ocean

acidification (Section 3.2.2.3) are all expected to impact the future production and distribution of several marine fish and invertebrates (*high confidence*).

Ocean acidification (Section 3.2.2.3) will affect several key Arctic species (*medium confidence*). The effects of current and projected levels of acidification have been examined for a broad suite of species groups (bivalves, cephalopods, echinoderms, crustaceans, corals and fishes) and these studies reveal species-specific differences in sensitivity, as well as differences in the scope for, and energetic cost of, adaptation (Luckman et al., 2014; Howes et al., 2015; Falkenberg et al., 2018).

3.2.3.1.1 Plankton and primary production

There is evidence that the combination of loss of sea ice, freshening, and regional stratification (Sections 3.2.1.1 and 3.2.1.2) has affected the timing, distribution and production of primary producers (Moore et al., 2018) (*high confidence*). Satellite data show that the decline in ice cover has resulted in a >30% increase in annual net primary production (NPP) in ice-free Arctic waters since 1998 (Arrigo and van Dijken, 2011; Bélanger et al., 2013; Arrigo and van Dijken, 2015; Kahru et al., 2016), a phenomenon corroborated by both *in situ* data (Stanley et al., 2015) and modelling studies (Vancoppenolle et al., 2013; Jin et al., 2016). Ice loss has also resulted in earlier phytoplankton blooms (Kahru et al., 2011) with blooms being dominated by larger-celled phytoplankton (Fujiwara et al., 2016). The longer open water season in the Arctic has also increased the incidence of autumn blooms, a phenomenon previously rarely observed in Arctic waters (Ardyna et al., 2017).

Thinner Arctic sea ice cover has led to the appearance of intense phytoplankton blooms that develop beneath first-year sea ice (*medium confidence*). Blooms of this size (1000s of km²) and intensity (peaks of approximately 30 mg Chla·m⁻³) were previously thought to be restricted to the marginal ice zone and the open ocean where ample light reaches the surface ocean for rapid phytoplankton growth (Arrigo et al., 2012). Evidence shows that these blooms can thrive beneath sea ice in areas of reduced thickness, increased coverage of melt ponds (Arrigo et al., 2014; Zhang et al., 2015; Jin et al., 2016; Horvat et al., 2017), first-year ridges at the snow-ice interface (Fernández-Méndez et al., 2018), and a large number of cracks (high lead fractions) in the ice (Assmy et al., 2017), although the latter has not changed significantly in the last three decades (Wang et al., 2016a). Local features including snow-free or thin snow, hummocks and ridges commonly found on multi-year ice also provide habitat for ice algae (Lange et al., 2017).

The reduction in sea ice area and thickness in the Arctic Ocean appears to be indirectly impacting rates of NPP through increased exposure of the surface ocean to atmospheric forcing (*medium confidence*) and these indirect impacts will possibly increase in the future (*low confidence*). Greater wind stress has been shown to increase upwelling of nutrients at the shelf break both over ice-free waters (Williams and Carmack, 2015) and a partial ice cover (Schulze and Pickart, 2012), leading to more new production (Williams and Carmack, 2015). At the same time, enhanced vertical stratification

(Section 3.2.1.2.2, SM3.2.2) and decreased upwelling of nutrients into surface waters (Capotondi et al., 2012; Nummelin et al., 2016) may reduce Arctic NPP in the future, especially in the central basin (Ardyna et al., 2017). It could also impact phytoplankton community composition and size structure, with small-celled phytoplankton, which require less nutrients, becoming more dominant as nutrient concentrations in surface waters decline (Yun et al., 2015).

In addition to its impact on phytoplankton bloom dynamics, the decline in the proportion of multi-year sea ice and proliferation of a thinner first year sea ice cover may favour growth of microalgae within the ice due to increased light availability (*medium confidence*). Recent studies suggest that the contribution of sea ice algae to total Arctic NPP is higher now than values measured previously (Song et al., 2016), accounting for nearly 10% of total NPP (ice plus water) and as much as 60% in places like the central Arctic (Fernández-Méndez et al., 2015).

Ongoing changes in NPP will impact the biogeochemistry and ecology of large parts of the Arctic Ocean (*high confidence*). In areas of enhanced nutrient availability and greater NPP, dominance by larger-celled microalgae increases vertical export efficiency from the surface downwards in both ice covered (Boetius et al., 2013; Lalande et al., 2014; Mäkelä et al., 2017) and open ocean (Le Moigne et al., 2015) areas. However, because exported biomass production may be increasing in some areas but declining in others, the net impact may be small (Randelhoff and Guthrie, 2016) (Sections 3.2.3.1.2, 5.3.6, SM3.2.6). Phytoplankton may have the capacity to compensate for ocean acidification under a range of temperatures and pH values (Hoppe et al., 2018).

Increased water temperatures (Section 3.2.1) and shifts in the spatial pattern and timing of the ice algal and phytoplankton blooms, have impacted the phenology, magnitude and duration of zooplankton production with associated changes in the zooplankton community composition (*medium confidence*). Negative effects of reductions in ice algae on zooplankton may be partially offset by predicted increases in water column phytoplankton production in the Bering Sea (Wang et al., 2015). Changes in sea ice coverage and thickness may alter the phenology, abundance and distribution of zooplankton in the future. Projected changes will initially have the most pronounced impact on sympagic amphipods, but will subsequently affect food web functioning and carbon dynamics of the pelagic system (Kohlbach et al., 2016).

At the more southern boundaries of the Arctic such as the southeastern Bering Sea, warm conditions have led to reduced production of large copepods and euphausiids (*medium confidence*) (Sigler et al., 2017; Kimmel et al., 2018). On more northern shelves, the increased open water period has led to increases in large copepods over a 60 year period within the Chukchi Sea (Ershova et al., 2015) and in recent years also the Beaufort Sea (Smoot and Hopcroft, 2017), while in the Central Basins zooplankton biomass in general has increased (Hunt et al., 2014; Rutzen and Hopcroft, 2018) (*medium confidence*).

There are inconsistent findings concerning the future development of copepods in the Arctic. Coupled biophysical model results suggest

that sea ice loss will increase primary production and that will primarily be consumed pelagically by zooplankton grazers such as *Calanus hyperboreus*; increasing their abundances in the central Arctic (Kvile et al., 2018). Feng et al. (2018) concluded that *C. glacialis* should continue to benefit from a warmer Arctic Ocean. On the other hand, in the transition zone between Arctic and Atlantic water masses, *C. glacialis* may face increasing competition from the more boreal *C. finmarchicus* (Dalpadado et al., 2016). Renaud et al. (2018) found the lipid content of *Calanus* spp. was related to size and not species. This suggests that climate driven shifts in dominant *Calanus* species may, because of overlap in size spectrum and contrary to earlier assumptions, not negatively impact their consumers in the Barents Sea.

The effects of ocean acidification on Arctic zooplankton and pteropods (small pelagic molluscs) have been examined for only a few species and these studies reveal that the severity of effects is dependent on emission scenarios and the species sensitivity and adaptive capacity. The copepod *C. glacialis* exhibits stage-specific sensitivities to ocean acidification with some stages being relatively insensitive to decreases in pH and other stages exhibiting substantial reductions in scope for growth (Bailey et al., 2017; Thor et al., 2018). Although there is strong evidence that pteropods are sensitive to the effects of ocean acidification (Manno et al., 2017) recent studies indicate they may exhibit some ability to adapt (Peck et al., 2016; Peck et al., 2018). However, the metabolic costs of adaptation may be constraining, especially during periods of low food availability (Lischka and Riebesell, 2016).

3.2.3.1.2 Benthic communities

There is evidence that earlier spring sea ice retreat and later autumn sea ice formation (Section 3.2.1.1) are changing the phenology of primary production with cascading effects on Arctic benthic community biodiversity and production (Link et al., 2013) (*medium confidence*). In the Barents Sea, evidence suggests that factors directly related to climate change (sea ice dynamics, ocean mixing, bottom-water temperature change, ocean acidification, river/glacier freshwater discharge; Sections 3.2.1.1, 3.2.1.2) are impacting the benthic species composition (Birchenough et al., 2015). Other human influenced activities, such as commercial bottom trawling and the introduction of non-native species are also regarded as major drivers of observed and expected changes in benthic community structure (Johannesen et al., 2017), and may interact with climate impacts.

Rapid and extensive structural changes in the rocky-bottom communities of two Arctic fjords in the Svalbard Archipelago during 1980–2010 have been documented and linked to gradually increasing seawater temperature and decreasing sea ice cover (Kortsch et al., 2012; Kortsch et al., 2015). Also, there are indications of declining benthic biomass in the northern Bering Sea (Grebmeier and Cooper, 2016) and southern Chukchi Sea (Grebmeier et al., 2015). It is unclear whether these rapid ecosystem changes will be tipping points for local ecosystems (Chapter 6, Table 6.1; Wassmann and Lenton, 2012). However, biomass of kelps have increased considerably in the intertidal to shallow subtidal in Arctic regions over the last two decades, connected to reduced physical impact by ice scouring and increased light availability as a consequence of warming and concomitant fast-ice retreat (Kortsch et al., 2012; Paar et al., 2016) (*medium confidence*) (see Section 5.3.3 and SM3.2.6 for further information on kelp).

The growth, early survival and production of commercially important crab stocks in the Bering Sea are influenced by time-varying exposure to multiple interacting drivers including bottom temperature, larval advection, predation, competition and fishing (Burgos et al., 2013; Long et al., 2015; Ryer et al., 2016). In Newfoundland and Labrador waters and on the western Scotian Shelf, snow crab (*Chionoecetes opilio*) productivity has declined (Mullowney et al., 2014; Zisserson and Cook, 2017). Contrary to this, snow crabs have expanded their distribution in the Barents Sea and commercial harvesting increased (Hansen, 2016; Lorentzen et al., 2018) (*high confidence*).

Bering sea crabs exhibit species-specific sensitivities to reduced pH (Long et al., 2017; Swiney et al., 2017; Long et al., 2019). However, current pH levels do not appear to have negatively impacted crab production in the Bering or Barents Seas (Mathis et al., 2015; Punt et al., 2016).

3.2.3.1.3 Fish

Since AR5, additional evidence shows climate-induced physical and biogeochemical changes are impacting, and will continue to impact, the distribution and production of marine fish (*medium confidence*). Changes in the spatial distribution and production of Arctic fish are best documented for ecologically and commercially important stocks in the Bering and Barents Seas (Box 3.4; Figure 3.5), while data is severely limited in other Arctic shelf regions and the Central Arctic Ocean (CAO).

Higher temperature and changes in the quality and distribution of prey is already affecting marine fish (Wassmann et al., 2015; Dalpadado et al., 2016; Hunt et al., 2016; Section 3.2.3.1) (*high confidence* for detection, *medium confidence* for attribution). In the northern Barents Sea, Atlantic Sector, higher temperatures (Section 3.2.1.2) have expanded suitable feeding areas for boreal/subarctic species (Box 3.4) and has contributed to increased Atlantic cod (*Gadus morhua*) production (Kjesbu et al., 2014). In contrast, Arctic species like polar cod (*Boreogadus saida*) are expected to be affected negatively by a shortened ice covered season and reduced sea ice extent through loss of spawning habitat and shelter, increased predatory pressure, reduced prey availability (Christiansen, 2017),

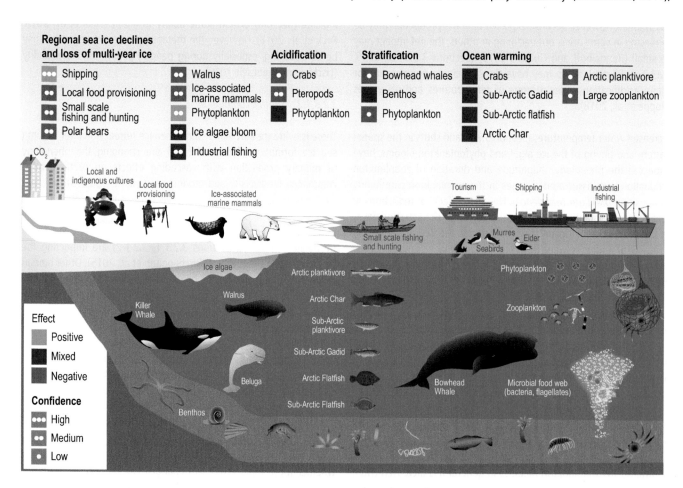

Figure 3.5 | Schematic summary of key drivers that are causing, or are projected to cause, direct effects on Arctic marine ecosystems (Section 3.2.1.2). Effects presented here are described in the main text (Sections 3.2.3.1; 3.2.4.1.1; 3.2.4.2; 3.2.4.3) with associated confidence levels and citations. For mixed effects, no confidence level is given (see main text for details on how multiple drivers cause interacting positive and negative effects). Projected effects are conceptual representations based on high emission scenarios (Section 3.2.1.2). The cross-sectional view of the Arctic ecosystem shows the association of key functional groups (marine mammals, birds, fish, zooplankton, phytoplankton and benthic assemblages) with Arctic marine habitats. Species depicted in the fishing net are not a comprehensive depiction of all target species.

and impaired growth and reproductive success (Nahrgang et al., 2014). These changes may cause structural changes in food webs, with large piscivorous and semipelagic boreal fish species replacing small bodied Arctic benthivores (Box 3.4; Fossheim et al., 2015; Frainer et al., 2017).

Time series on responses of anadromous fish (including salmon) in the high Arctic are limited, although these stocks will also be exposed to a wide range of future stressors (Reist et al., 2016). There is some evidence that environmental variability influences the production of anadromous species such as Arctic char (*Salvelinus alpinus*), brown trout (*Salmo trutta*), and Atlantic salmon (*Salmo salar*) through its influence on growth and winter survival (Jensen et al., 2018).

The scope for adaptation of marine fish to a changing ocean conditions is uncertain, but knowledge is informed by previous biogeographic studies (Chernova, 2011; Lynghammar et al., 2013). The present niche partitioning between subarctic and Arctic pelagic fish species is expected to become more diffuse with potential negative impacts on cold adapted species such as polar cod (Laurel et al., 2017; Alabia et al., 2018; Logerwell et al., 2018) (*low confidence*). Winter ocean conditions in the high Arctic are projected to remain cold in most regions (Section 3.2.3.1), limiting the immigration of subarctic species that spawn in positive temperatures onto the high Arctic shelves (Landa et al., 2014). Projected increases in summer temperature may open gateways to subarctic pelagic foragers in summer, particularly in the inflow regions of the Kara and Chukchi Seas, and the shelf regions of east and west Greenland (Mueter et al., 2017; Joli et al., 2018). For example, the pelagic capelin (*Mallotus villosus*) are capable of entering the CAO, but may be restricted in winter by availability of suitable spawning areas and lack of antifreeze proteins (Hop and Gjøsæter, 2013; Christiansen, 2017).

Regional climate scenarios, derived from down-scaled global climate scenarios, have been used to drive environmentally linked fish population models (Hermann et al., 2016; Holsman et al., 2016; Ianelli et al., 2016; Hermann et al., 2019). Hermann et al. (2019) contrasted future production of copepods and euphausiids in the eastern Bering Sea under scenarios derived from projected downscaled high spatial and temporal resolution ocean habitats under RCP4.5 and RCP8.5. Consistent with AR5, these updated scenarios project future declines in the abundance of large copepods under RCP8.5, a result that has been shown to negatively impact production of walleye pollock, Pacific cod (*Gadus microcephalus*) and arrowtooth flounder (*Atheresthes stomias*) (Sigler et al., 2017; Kimmel et al., 2018) (*medium confidence*). Hedger et al. (2013) predicts increases in Atlantic salmon abundance in northern Norway (river Alta around 70°N) with future warming (*low confidence*). Under end of century RCP8.5 projections, ocean acidification and higher ocean temperatures are expected to reduce production of Barents Sea cod (Stiasny et al., 2016; Koenigstein et al., 2018) (*low confidence*).

3.2.3.1.4   Seabirds and marine mammals

Environmental alterations caused by global warming are resulting in phenological, behavioural, physiological, and distributional changes in Arctic marine mammal and seabird populations (Gilg et al., 2012; Laidre et al., 2015; Gall et al., 2017) (*high confidence*). These changes include altered ecological interactions as well as direct responses to habitat degradation induced especially via loss of sea ice. Population responses to warming have not all been linear, some have been particularly strong and abrupt due to environmental regime shifts, as seen in black-legged kittiwakes (*Rissa tridactyla*). A steep population decline in kittiwake colonies distributed throughout their breeding range coincided with an abrupt warming of sea-surface temperature in the 1990s, while their population dynamics did not seem to be affected during periods of more gradual warming (Descamps et al., 2017).

Seabirds and marine mammals are mobile animals that respond to changes in the distribution of their preferred habitats and prey, by shifting their range, altering the timing or pathways for migration or prey shifting when this is feasible (Post et al., 2013; Hamilton et al., 2019) (*very high confidence*). However, some species display strong site fidelity that can be maladaptive in a changing climate and Arctic endemic marine mammals (all of which are ice-affiliated for breeding) in general have little scope to move northward in response to warming (Kovacs et al., 2012; Hamilton et al., 2015). Changes in the location or availability of polar fronts, polynyas, tidal glacier fronts or ice edges have impacted where Arctic sea birds and marine mammals concentrate because of the influence these physical features have on productivity; traditionally these areas have been key foraging sites for top predators in the Arctic (deHart and Picco, 2015; Hamilton et al., 2017; Hunt et al., 2018).

In some species, shifts in distribution in response to changes in suitable habitat have been associated with increased mortality. Increased mortality rates of walrus (*Odobenus rosmarus*) calves have been observed during on-shore stampedes of unusually large herds, because Pacific walrus females are no longer able to haul out on ice over the shelf in summer due to the retraction of the southern ice edge into the deep Arctic Ocean (Kovacs et al., 2016). Shifts in the temporal and spatial distribution and availability of suitable areas of sea ice for ice-breeding seals have occurred (Bajzak et al., 2011; Øigård et al., 2013) with increases in strandings and pup mortality in years with little ice (Johnston et al., 2012c; Soulen et al., 2013; Stenson and Hammill, 2014).

Climate impacts that reduce the availability of prey resources can negatively impact marine mammals (Asselin et al., 2011; Øigård et al., 2014; Choy et al., 2017) (*very high confidence*). Sea ice changes have increased the foraging effort of ringed seals (*Pusa hispida*) in the marginal ice zone north of Svalbard (Hamilton et al., 2015), also causing diet shifts (Lowther et al., 2017). Ringed seals in Svalbard are using terrestrial haul out sites during summer for the first time in observed history, following major declines in sea ice (Lydersen et al., 2017), an example of an adaptive behavioural response to extreme habitat changes. Sea ice related changes in the export of production to the benthos (Section 3.3.3.1) and associated changes in the benthic community (Section 3.4.1.1.2) may impact marine

3

mammals dependent on benthic prey (e.g., walruses and gray whales, *Eschrichtius robustus*) (Brower et al., 2017; Udevitz et al., 2017; Szpak et al., 2018).

Changes in the timing, distribution and thickness of sea ice and snow (Sections 3.2.1.1, 3.4.1.1) have been linked to phenological shifts, and changes in distribution, denning, foraging behaviour and survival rates of polar bears (*Ursus maritimus*) (Andersen et al., 2012; Hamilton et al., 2017; Escajeda et al., 2018) (*high confidence*). Less ice is also driving polar bears to travel over greater distances and swim more than previously both in offshore and in coastal areas, which can be particularly dangerous for young cubs (Durner et al., 2017; Pilfold et al., 2017; Lone et al., 2018). Cumulatively, changes in sea ice patterns are driving demographic changes in polar bears, including declines in some populations (Lunn et al., 2016; McCall et al., 2016), while others are stable or increasing (Voorhees et al., 2014; Aars et al., 2017). This is because protective management measures have been successful in allowing severely depleted populations to recover or because new food sources, such as carrion, are becoming available to polar bears in some regions (Galicia et al., 2016; Stapleton et al., 2016). Changes in the spatial distribution of polar bears and killer whales can have top-down effects on other marine mammal prey populations (Øigård et al., 2014; Breed et al., 2017; Smith et al., 2017a).

Several studies from different parts of the Arctic show evidence that changing temperatures impact seabird diets (Dorresteijn et al., 2012; Divoky et al., 2015; Vihtakari et al., 2018), reproductive success and body condition (Gaston et al., 2012; Provencher et al., 2012; Gaston and Elliott, 2014) (*high confidence*). Recent studies also show that changes in sea surface temperature and sea ice dynamics have impacts on the distribution and abundance of seabird prey with cascading impacts on seabird community composition (Gall et al., 2017), nutritional stress and decreased reproductive output (Dorresteijn et al., 2012; Divoky et al.; Kokubun et al., 2018) and survival (Renner et al., 2016; Hunt et al., 2018).

3.2.3.2   Southern Ocean

Marine ecosystem dynamics in the Antarctic region are dominated by the ACC and its frontal systems (Cross-Chapter Box 7 in Chapter 3), subpolar gyres, polar seasonality, the annual advance and retreat of sea ice (Section 3.2.1.1) and the supply of limiting micronutrients for productivity (most commonly iron) (Section 5.2.2.5). Antarctic krill (*Euphausia superba*) play a central role in Southern Ocean foodwebs as grazers and as prey items for fish, squid, marine mammals and seabirds (Schmidt and Atkinson, 2016; Trathan and Hill, 2016) (SM3.2.6). This is due in part to the high abundance and circumpolar distribution of Antarctic krill, although the abundance and importance of this species varies between different regions of the Southern Ocean (Larsen et al., 2014; Siegel, 2016; McCormack et al., 2017). Recent work has characterised the nature of habitat change for Southern Ocean biota at regional and circumpolar scales (Constable et al., 2014; Gutt et al., 2015; Constable et al., 2016; Hunt et al., 2016; Gutt et al., 2018), and the direct responses of biota to these changes (Constable et al., 2014) (summarised in Figure 3.6). These findings indicate that overlapping changes in

key ocean and sea ice habitat characteristics (temperature, sea ice cover, iceberg scour, mixed layer depth, aragonite undersaturation; Sections 3.2.1, 3.2.2) will be important in determining future states of Southern Ocean ecosystems (Constable et al., 2014; Gutt et al., 2015) (*medium confidence*). However, there is a need to better characterise the nature and importance of indirect responses to physical change using models and observations. Important advances have also been made since AR5 in (i) identifying key variables to detect and attribute change in Southern Ocean ecosystems, as part of long-term circumpolar modelling designs (Constable et al., 2016), and (ii) refining methods for using sea ice projections from global climate models in ecological studies and ecosystem models for the Southern Ocean (Cavanagh et al., 2017).

3.2.3.2.1   Plankton and pelagic primary production

Changes in column-integrated phytoplankton biomass for the Southern Ocean are coupled with changes in the spatial extent of ice-free waters, suggesting little overall change in biomass per area at the circumpolar scale (Behrenfeld et al., 2016). Arrigo et al. (2008) also report no overall trend in remotely-sensed column-integrated primary production south of 50ºS from 1998 to 2006. At a regional scale, local-scale forcings (e.g., retreating glaciers, topographically steered circulation and sea ice duration) and associated changes in stratification are key determinants of phytoplankton bloom dynamics at coastal stations on the West Antarctic Peninsula (Venables et al., 2013; Schofield et al., 2017; Kim et al., 2018; Schofield et al., 2018) (*medium confidence*). For example, a shallowing trend in mixed layer depth in the southern part of the Peninsula (as opposed to no trend in the north) associated with changes in sea ice duration over a 24-year period (from 1993 to 2017) has been linked to enhanced phytoplankton productivity (Schofield et al., 2018). The phenology of Southern Ocean phytoplankton blooms in this region may also be shifting to earlier in the growth season (Arrigo et al., 2017a). However, the effect of climate change on Southern Ocean pelagic primary production is difficult to determine given that the length of time series data is insufficient (less than 30 years) to enable the climate change signature to be detected and attributed; and that, even when records are of sufficient length, data trends are often reported as being driven by climate change when they are due to a combination of climate change andvariability.

Recent studies on the ecological effects of acidification in coastal waters near the Antarctic continent indicate a detrimental effect of acidification on primary production and changes to the structure and function of microbial communities (Hancock et al., 2017; Deppeler et al., 2018; Westwood et al., 2018) (*medium confidence*). Trimborn et al. (2017) report that Southern Ocean diatoms are more sensitive to ocean acidification and changes in irradiance than the prymnesiophyte *Phaeocystis antarctica*, which may have implications for biogeochemical cycling because diatoms and prymnesiophytes are generally considered key drivers of these cycles. Both laboratory manipulations and *in situ* experiments indicate that sea ice algae are tolerant to acidification (McMinn, 2017) (*medium confidence*). Model projections of trends in primary production in the Southern Ocean due to climate change from Leung et al. (2015) are summarised in Table 3.2.

Table 3.2 | Model projections of trends due to climate change driven alteration of phytoplankton properties under RCP8.5 from 2006 to 2100 across three zones of the Southern Ocean, from Leung et al. (2015). There is *low confidence* in predicted zonal changes in phytoplankton biomass due to *low confidence* regarding future changes in iron supply in the Southern Ocean (Hutchins and Boyd, 2016). Acidification was not reported as an important driver in this modelling experiment.

Zonal Band	Predicted change in phytoplankton biomass	Drivers	Mechanisms
40°S–50°S	⬆	Higher mean underwater irradiance More iron supply	Shallowing of the summertime mixed layer depth Change in iron supply mechanism
50°S–65°S	⬇	Lower mean underwater irradiance	Deeper summertime mixed layer depth Decreased summertime incident radiation (increased cloud fraction)
South of 65°S	⬆	More iron supply Higher mean underwater irradiance Temperature	Melting of sea ice Warming ocean

Previously reported declines in Antarctic krill abundance in the South Atlantic Sector (Atkinson et al., 2004) cited in WGII AR5 (Larsen et al., 2014) may not represent a long-term, climate driven, regional-scale decline (Fielding et al., 2014; Kinzey et al., 2015; Steinberg et al., 2015; Cox et al., 2018) (*medium confidence*) but could reflect a sudden, discontinuous change following an episodic period of anomalous peak abundance for this species (Loeb and Santora, 2015) (*low confidence*). Recent analyses have not detected trends in long-term krill abundance in the South Atlantic Sector in acoustic surveys (Fielding et al., 2014; Kinzey et al., 2015), net-based surveys (Steinberg et al., 2015) or reanalysis of historical data (Cox et al., 2018). Nevertheless, the spatial distribution and size composition of Antarctic krill may already have changed in association with change in the sea ice environment (Atkinson et al., 2019) (*medium confidence*) and may result in different regional trends in numerical krill abundance (Cox et al., 2018; Atkinson et al., 2019) (*medium confidence*).

The distribution of Antarctic krill is expected to change under future climate change because of changes in the location of the optimum conditions for growth and recruitment (Melbourne-Thomas et al., 2016; Piñones and Fedorov, 2016; Meyer et al., 2017; Murphy et al., 2017; Klein et al., 2018). The optimum conditions for krill are predicted to move southwards, with the decreases most apparent in the areas with the most rapid warming (Hill et al., 2013; Piñones and Fedorov, 2016) (Section 3.2.1.2.1) (*medium confidence*). The greatest projected reductions in krill due to the effects of warming and ocean acidification are predicted for the southwest Atlantic/ Weddell Sea region (Kawaguchi et al., 2013; Piñones and Fedorov, 2016) (*low confidence*), which is the area of highest current krill concentrations, contains important foraging grounds for krill predators, and is also the main area of operation of the krill fishery. Modelled effects of warming on krill growth in the Scotia Sea and northern Antarctic Peninsula (AP) region resulted in reductions in total krill biomass under both RCP2.6 and RCP8.5 (Klein et al., 2018). Projections from a food web model for the West Antarctic Peninsula under simple scenarios for change in open water and sea ice-associated primary production from 2010 to 2050 (6, 15, and 41% increases in phytoplankton production with equivalent percentage decreases in ice algal production) indicate a decline in krill biomass with contemporaneous increases in the biomass of gelatinous salps (Suprenand and Ainsworth, 2017).

Current understanding of climate change effects on Southern Ocean zooplankton is largely based on observations and predictions from

the South Atlantic and the West Antarctic Peninsula. Comparison of the mesozooplankton community in the southwestern Atlantic Sector between 1926–1938 and 1996–2013 showed no evidence of change despite surface ocean warming (Tarling et al., 2018). These results suggest that predictions of distributional shifts based on temperature niches may not reflect the actual levels of thermal resilience of key taxa. Sub-decadal cycles of macrozooplankton community composition adjacent to the West Antarctic Peninsula are strongly linked to climate indices, with evidence of increasing abundance for some species over the period from 1993 to 2013 (Steinberg et al., 2015). Pteropods are vulnerable to the effects of acidification, and new evidence indicates that eggs released at high CO_2 concentrations lack resilience to ocean acidification in the Scotia Sea region (Manno et al., 2016) (*medium confidence*).

3.2.3.2.2 Benthic communities

Carbon uptake and storage by Antarctic benthic communities is predicted to increase with sea ice losses, because across-shelf growth gains from longer algal blooms outweigh ice scour mortality in the shallows (Barnes, 2017). Bentho-pelagic coupling and vertical energy flux will also influence Southern Ocean ecosystem responses to climate change (Jansen et al., 2017). Benthic communities in shallow water habitats mostly consist of dark-adapted invertebrates and rely on sea ice to create low-light marine environments. Increases in the amount of light reaching the shallow seabed under climate change may result in ecological regime shifts, in which invertebrate-dominated communities are replaced by macroalgal beds (Clark et al., 2015; Clark et al., 2017) (*low confidence*) (Table 6.1). Griffiths et al. (2017a) modelled distribution changes for 963 benthic invertebrate species in the Southern Ocean under RCP8.5 for 2099. Their results suggest that 79% of Antarctica's endemic species will face a reduction in suitable temperature habitat (an average 12% reduction) over the current century. Predicted reductions in the number of species are most pronounced for the West Antarctic Peninsula and the Scotia Sea region (Griffiths et al., 2017a).

3.2.3.2.3 Fish

Many Antarctic fish have a narrow thermal tolerance as a result of physiological adaptations to cold water (Pörtner et al., 2014; Mintenbeck, 2017), which makes them vulnerable to the effects of increasing temperatures (Mueller et al., 2012; Beers and Jayasundara, 2015). Increasing water temperatures may displace icefish (family *Channichthyidae*) in marginal habitats (e.g., shallow regions around subantarctic islands) as they lack haemoglobin and are unable to adjust blood parameters to an increasing oxygen demand (Mintenbeck et al., 2012) (*low confidence*). Future warming may also reduce the planktonic duration and increase egg and larval mortality for fish species, which is predicted to affect dispersal patterns, with implications for population connectivity and the ability of fish species to adapt to ongoing environmental change (Young et al., 2018). The Antarctic silverfish (*Pleuragramma antarctica*) is an important prey species in some regions of the Southern Ocean, and has an ice-dependent life cycle (Mintenbeck et al., 2012; Vacchi et al., 2012). Documented declines in the abundance of this species in some parts of the West Antarctic Peninsula may have consequences for associated food webs (Parker et al., 2015; Mintenbeck and Torres, 2017) (*low confidence*).

Myctophids and toothfish are important fish groups from both a food web (myctophids) and fishery (toothfish) perspective. Species distribution models for *Electrona antarctica*, a dominant myctophid species in the Southern Ocean, project habitat loss for this species under RCP4.5 (6.2 ± 6.0% loss) and RCP8.5 (13.1 ± 10.2% loss) by 2090, associated with increased sea surface temperature (Freer et al., 2018). There have been no observed effects of climate change on the two species of toothfish that are found in the Southern Ocean: Patagonian and Antarctic toothfish (*Dissostichus eleginoides* and *D. mawsoni*), but recruitment is inversely correlated with sea surface temperature for Patagonian toothfish at South Georgia (Belchier and Collins, 2008). Given differences in temperature tolerances for Patagonian toothfish (with a wide temperature tolerance) and Antarctic toothfish (limited by a low tolerance for water temperatures above 2°C), the latter may be faced with reduced habitat and potential competition with southward-moving Patagonian toothfish under climate change (Mintenbeck, 2017) (*very low confidence*).

3.2.3.2.4 Seabirds and marine mammals

Since AR5, there has been an increasing body of evidence of climate-induced changes in populations of some Antarctic higher predators such as seabirds and marine mammals. These changes vary between different regions of the Southern Ocean and reflect differences in key drivers (Bost et al., 2009; Gutt et al., 2015; Constable et al., 2016; Hunt et al., 2016; Gutt et al., 2018), particularly sea ice extent and food availability (*high confidence*) across regions (Sections 3.2.1.1.1, 5.2.3.1, 5.2.3.2, 5.2.4). The predictability of foraging grounds and ice cover are associated with variations in climate (Dugger et al., 2014; Youngflesh et al., 2017; Abrahms et al., 2018) (Section 3.2.1.1) and are the main drivers of observed population changes of Southern Ocean higher predators (*high confidence*) (Descamps et al., 2015; Jenouvrier et al., 2015; Sydeman et al., 2015; Abadi et al., 2017; Bjorndal et al., 2017; Fluhr et al., 2017; Hinke et al., 2017a; Hinke et al., 2017b; Pardo et al., 2017). The suitability of breeding habitats and the location of environmental features that facilitate the aggregation of prey are also influenced by climate change, and in turn influence the distribution in space and time of marine mammals and birds (Bost et al., 2015; Kavanaugh et al., 2015; Hindell et al., 2016; Santora et al., 2017) (*medium confidence*). Finally, biological parameters (reproductive

success, mortality, fecundity and body condition), life history traits, morphological, physiological and behavioural characteristics of top predators in the Southern Ocean, as well as their patterns of activity (migration, distribution, foraging and reproduction) are also changing as a result of climate change (Braithwaite et al., 2015a; Whitehead et al., 2015; Seyboth et al., 2016; Hinke et al., 2017b) (*high confidence*).

Trends of populations of Antarctic penguins affected by climate change include both increases for gentoo penguins, (*Pygoscelis papua*) (Lynch et al., 2013; Dunn et al., 2016; Hinke et al., 2017a), and decreases for Adélie (*P. adeliae*), chinstrap (*P. antarctica*), king (*Aptenodytes patagonicus*) and Emperor (*A. forsteri*) penguins (Trivelpiece et al., 2011; LaRue et al., 2013; Jenouvrier et al., 2014; Bost et al., 2015; Southwell et al., 2015; Younger et al., 2015; Cimino et al., 2016) (*high confidence*). Yet population shifts in Adélie penguins (Youngflesh et al., 2017) may have resulted from strong interannual environmental variability in good and bad years for prey and breeding habitat rather than climate change (*low confidence*). New evidence suggests that present Emperor penguin population estimates should be evaluated with caution based on the existence of breeding colonies yet to be discovered/confirmed (Ancel et al., 2017) as well as studies that draw conclusions based on trend estimates from single colonies (Kooyman and Ponganis, 2017).

Evidence for climate change impacts on Antarctic flying birds indicates that contraction of sea ice (seasonally and in specific regions), increases in sea surface temperatures, extreme events (snowstorms) and wind regime shifts can reduce breeding success and population growth rates in some species: southern fulmars (*Fulmarus glacialoides*), Antarctic petrels (*Thalassoica antarctica*) and black-browed albatrosses (*Thalassarche melanophris*) (Descamps et al., 2015; Jenouvrier et al., 2015; Pardo et al., 2017) (*low confidence*). Poleward population shifts with increased intensity and frequency of westerly winds affect functional traits, demographic rates, foraging range, rates of travel and flight speeds of flying birds (Weimerskirch et al., 2012; Jenouvrier et al., 2018) but also increase overlap with fisheries activities thus increasing the risk of bycatch and the need for mitigation measures (Krüger et al., 2018) (*medium confidence*).

Changes in local- and regional-scale oceanographic features (Section 3.2.1.2) together with bathymetry control prey aggregation and distribution, and affect the ecological responses and biological traits of higher predators (particularly marine mammals) in the Southern Ocean (Lyver et al., 2014; Bost et al., 2015; Jenouvrier et al., 2015; Whitehead et al., 2015; Cimino et al., 2016; Hinke et al., 2017a; Pardo et al., 2017) (*medium confidence*) and *likely* explain most of the observed population shifts (Kavanaugh et al., 2015; Hindell et al., 2016; Gurarie et al., 2017; Santora et al., 2017). Decadal climate cycles affect access to mesopelagic prey by southern elephant seals (*Mirounga leonina*) in the Indian Sector of the Southern Ocean and breeding females are excluded from highly productive continental shelf waters in years of increased sea ice extent and duration (Hindell et al., 2016) (*medium confidence*). To date there is no unified global estimate of the abundance of Antarctic pack ice seal species (Ross seals (*Ommatophoca rossi*), crabeater seals (*Lobodon carcinophaga*),

3

leopard seals (*Hydrurga leptonyx*) and Weddell seals (*Leptonychotes weddellii*)) as a reference point for understanding climate change impacts on these species (Southwell et al., 2012; Bester et al., 2017), although some regional population estimates for pack ice seals are available (Gurarie et al., 2017 and references therein). Analysis of long-term data suggests a genetic component to adaptation to climate change (*low confidence*) in Antarctic fur seals (*Arctocephalus gazella*, Forcada and Hoffman (2014)) and pigmy blue whales (*Balaenoptera musculus brevicauda*, Attard et al. (2015)).

Population trends of migratory baleen whales have been associated with krill abundance in the Atlantic and Pacific sectors of the Southern Ocean which is reflected in increased reproductive success, body condition and energy allocation (milk availability and transfer) to calves (Braithwaite et al., 2015a; Braithwaite et al., 2015b; Seyboth et al., 2016) (*high confidence*). There have been predictions of negative future impacts of climate change on krill and all whale species, although the magnitude of impacts differs among populations (Tulloch et al., 2019) as for other higher predators (Section 5.2.3). Pacific blue (Tulloch et al., 2019) (*Balaenoptera musculus*), fin (*B. physalus*) and southern right whales (*Eubalaena australis*) are the most at risk but humpback whales (*Megaptera novaeangliae*) are also at risk, as consequence of reduced prey and increasing interspecific competition. Importantly, climate-related risks for whale populations are a product of environmental conditions and connectivity between whale foraging grounds (Southern Ocean) and breeding grounds (lower latitudes) (Section 5.2.3.1).

3.2.3.2.5 Pelagic foodwebs and ecosystem structure

This section assesses the impacts of ocean and sea ice changes on pelagic foodwebs and ecosystem structure. The ecological impacts of loss of ice shelves and retreat of coastal glaciers around Antarctica are assessed in Section 3.3.3.4. Recent syntheses of Southern Ocean ecosystem structure and function recognise the importance of at least two dominant energy pathways in pelagic foodwebs – a short trophic pathway transferring primary production to top predators via krill, and at least one other pathway that moves energy from smaller phytoplankton to top predators via copepods and small mesopelagic fishes – and indicate that the relative importance of these pathways will change under climate change (Murphy et al., 2013; Constable et al., 2016; Constable et al., 2017; McCormack et al., 2017) (*medium confidence*). Using an ecosystem model, Klein et al. (2018) found that the effects of warming on krill growth off the AP and in the Scotia Sea translated to increased risks of declines in krill predator populations, particularly penguins, under both RCP2.6 and RCP8.5. The relative importance of different energy pathways in Southern Ocean foodwebs has important implications for resource management, in particular the management of krill and toothfish fisheries by the Commission for the Conservation of Antarctic Marine Living Resources (CCAMLR) (Constable et al., 2016; Constable et al., 2017) (Sections 3.2.4.1.2, 3.5.3.2.2).

In summary, advances in knowledge regarding the impacts of climate change on Antarctic marine ecosystems since AR5 are consistent with the impacts described in Larsen et al. (2014) (also summarised in Figure 3.6). These advances include further descriptions of local-scale, climate-related influences (sea ice and stratification)

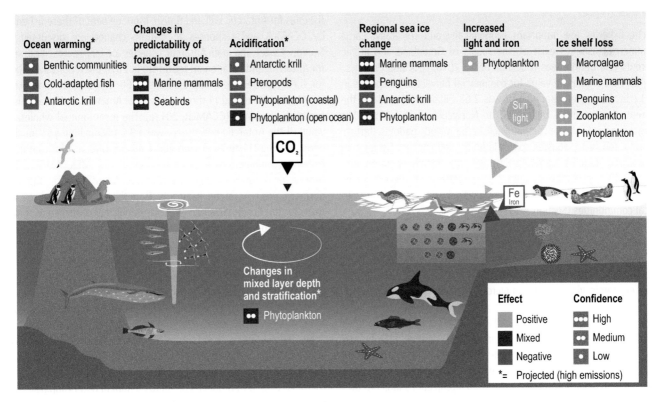

Figure 3.6 | Schematic summary of key drivers that are causing or are projected to cause direct effects on Southern Ocean marine ecosystems. Effects presented here are described in the main text (Sections 3.2.3.2, 3.3.3.4), with associated confidence levels and citations. Projected changes (indicated by an asterisk) are for high emissions scenarios. The cross-sectional view of the Southern Ocean ecosystem shows the association of key functional groups (marine mammals, birds, fish, zooplankton, phytoplankton and benthic assemblages) with Southern Ocean habitats. The configuration of the Southern Ocean foodweb is described in SM3.2.6.

on primary productivity, particularly in the West Antarctic Peninsula region (Section 3.2.3.2.1) (*medium confidence*). At the circumpolar scale, primary production is projected to increase in regions south of 65°S over the period from now to 2100 under RCP8.5 (Leung et al., 2015) (*low confidence*). However, ocean acidification may have a detrimental effect on coastal phytoplankton communities around the Antarctic continent (Section 3.2.3.2.1) (*medium confidence*). Increased information is also available regarding climate-driven changes in Antarctic krill populations in the south Atlantic, including the observed southward shift in the spatial distribution of krill in this region (Atkinson et al., 2019) (*medium confidence*) but evidence of a long-term trend in overall abundance in the region is equivocal (Section 3.2.3.2.1). Further habitat contraction for Antarctic krill is predicted in the future (*medium confidence*) (references detailed in Section 3.2.3.2.1). Under high emissions scenarios the majority of Antarctic seafloor species are projected to be negatively impacted by the end of the century (Griffiths et al., 2017a) (*low confidence*). Observed changes in the geography of ice-associated habitats (sea ice, ice shelves and polynyas) have both positive and negative effects on seabirds and marine mammals, and will interact with ice dependent changes in Antarctic krill populations to compound the impacts on krill dependent predators (Klein et al., 2018) (Sections 3.2.3.2.1, 3.2.3.2.4) (*medium confidence*).

3.2.4　Impacts on Social-Ecological Systems

3.2.4.1　Fisheries

3.2.4.1.1　Arctic

Arctic fisheries are important economically and societally. Large commercial fisheries exist off the coasts of Greenland and in the Barents and Bering Seas (Holsman et al., 2018; Peck and Pinnegar, 2018). First-wholesale value for commercial harvest of all species in 2017 in the Eastern Bering Sea was 2.68 billion USD, and for the Barents Sea around 1 billion USD to Norwegian fishers alone. The target species for these commercial fisheries include gadoids, flatfish, herring, red fish (*Sebastes* sp.), salmonids, and capelin. Fisheries in other Arctic regions are relatively small-scale, locally operated, and target a limited number of species (Reist, 2018). Still, these fisheries are of considerable cultural, economic and subsistence importance to local communities (Section 3.5.2.1).

Climate change will affect the spatial distribution and productivity of some commercially important marine fish and shellfish under most RCPs (Section 3.2.3.1) with associated impacts on the distribution and economic viability of commercial fisheries (*high confidence*). Past performance suggests that high latitude fisheries have been resilient to changing environmental and market drivers. For example, the Norwegian cod fishery has exported dried cod over an unbroken period of more than a thousand years (Barrett et al., 2011), reflecting the resilience of the northern Norwegian cod fisheries to historic climate variability (Eide, 2017). Also, model projections indicate that expansions in suitable habitat for subarctic species and increased production of planktonic prey due to increasing temperatures and ice retreat, will continue to support commercially important fisheries

(Lam et al., 2016; Eide, 2017; Haug et al., 2017; Peck and Pinnegar, 2018) (Section 3.2.3.1.3, Box 3.4) (*medium confidence*).

However, recent studies in the Bering Sea suggest that future fish production will also depend on how climate change and ocean acidification will alter the quality, quantity and availability of suitable prey; the thermal stress and metabolic demands of resident fish; and species interactions (Section 3.2.3.1.3), suggesting that the future of commercial fisheries in Arctic regions is uncertain (Holsman et al., 2018). It is also uncertain whether future autumn and winter ocean conditions will be conducive to the establishment of resident overwintering spawning populations that are large enough to support sustainable commercial fishing operations at higher latitude Arctic shelf regions (Section 3.2.3.1) (*medium confidence*).

Projecting the impacts of climate change on marine fisheries is inextricably intertwined with response scenarios regarding risk tolerance in future management of marine resources, advancements in fish capture technology, and markets drivers (e.g., local and global demand, emerging product lines, competition, processing efficiencies and energy costs) (Groeneveld et al., 2018). Seasonal and interannual variability in ocean conditions influences product quality and costs of fish capture (Haynie and Pfeiffer, 2012) (Table 3.4). Further, past experience suggests that barriers to diversification may limit the portfolio of viable target fisheries available to small-scale fisheries (Ward et al., 2017) (*low confidence*).

3.2.4.1.2　Southern Ocean

This section examines climate change impacts on Southern Ocean fisheries for Antarctic krill and finfish. Management of these fisheries by CCAMLR and responses to climate change are discussed in Section 3.5.2.1. The main Antarctic fisheries are for Antarctic krill, and for Antarctic and Patagonian toothfish; in 2016 the reported catches for these species were approximately 260 thousand tons for krill (CCAMLR, 2017b) and 11 thousand tons for Antarctic and Patagonian toothfish combined (CCAMLR, 2017a). The mean annual wholesale value of the Antarctic krill fishery was 69.5 USD million yr^{-1} for the period from 2011 to 2015, and 206.7 million USD yr^{-1} for toothfish fisheries (combined) over the same period (CCAMLR, 2016b). The fishery for Antarctic krill in the southern Atlantic Sector and the northern West Antarctic Peninsula (together the current area of focus for the fishery) has become increasingly concentrated in space over recent decades, which has raised concern regarding localised impacts on krill predators (Hinke et al., 2017a). The krill fishery has also changed its peak season of operation. In the early years of the fishery, most krill were taken in summer and autumn, with lowest catches being taken in spring. In recent years the lowest catches have occurred over summer, catches have peaked in late autumn, and very little fishing activity has occurred in spring (Nicol and Foster, 2016). Some of these temporal and spatial shifts in the fishery over time have been attributed to reductions in winter sea ice extent in the region (Kawaguchi et al., 2009) (*low confidence*). Recent increases in the use of krill catch to produce krill oil (as a human health supplement) has also led to vessels concentrating on fishing in autumn and winter when krill are richest in lipids (Nicol and Foster, 2016). Available evidence regarding future changes to Antarctic krill populations

(Section 3.2.3.2.1) indicates that the impacts of climate change will be most pronounced in the areas that are currently most important for the Antarctic krill fishery: the Scotia Sea and the northern tip of the AP. Major future changes in the krill fishery itself are expected to be driven by global issues external to the Southern Ocean, including conservation decision making and socioeconomic drivers.

There is limited understanding of the consequences of climate change for Southern Ocean finfish fisheries. Lack of recovery of mackerel icefish (*Champsocephalus gunnari*) after cessation of fishing in 1995 has been related to anomalous water temperatures (~2°C increase related to a strong El Niño) in the subantarctic Indian Ocean and to availability of krill prey in the Atlantic region (Mintenbeck, 2017) (*low confidence*). Differences in temperature tolerance of Patagonian and Antarctic toothfish described in Section 3.2.3.2.3 may have implications for future fisheries of these two species.

3.2.4.2   Tourism

Reductions in sea ice have facilitated an increase in marine and cruise tourism opportunities across the Arctic related to an increase in accessibility (Dawson et al., 2014; Johnston et al., 2017) (*high confidence*). While not exclusively 'polar', Alaska attracts the highest number of cruise passengers annually at just over one million; Svalbard attracts 40,000–50,000; Greenland 20,000–30,000; and Arctic Canada 3,500–5,000 (Johnston et al., 2017). Compared to a decade ago, there are more cruises on offer, ships travel further in a single season, larger vessels with more passenger berths are in operation, more purpose-built polar cruise vessels are being constructed, and private pleasure craft are appearing in the Arctic more frequently (Lasserre and Têtu, 2015; Johnston et al., 2017; Dawson et al., 2018). In Antarctica, almost 37,000 (predominantly shipborne) tourists visited in 2016–2017, with 51,707 during 2017–2018; there were 6700 tourists in 1992–1993 (the first year of record) (ATCM, 2018). Due to accessibility and convenience, these tourism operations are mostly based around the few ice-free areas of Antarctica, concentrated on the AP (Pertierra et al., 2017).

Canada's Northwest Passage (southern route), which only saw occasional cruise ship transits in the early 2000s is now reliably accessible during the summer cruising season, and as a result has experienced a doubling and quadrupling of cruise and pleasure craft activity over the past decade (Johnston et al., 2017; Dawson et al., 2018). There is *high confidence* that demand for Arctic cruise tourism will continue to grow over the coming decade (Johnston et al., 2017). The anticipated implications of future climate change have become a driver for polar tourism. A niche market known as 'last chance tourism' has emerged whereby tourists explicitly seek to experience vanishing landscapes or seascapes, and natural and social heritage in the Arctic and Antarctic, before they disappear (Lemelin et al., 2010; Lamers et al., 2013).

Increases in polar cruise tourism pose risks and opportunities related to development, education, safety (including search and rescue), security within communities and environmental sustainability (Johnston et al., 2012a; Johnston et al., 2012b; Stewart et al., 2013; Dawson et al., 2014; Lasserre and Têtu, 2015; Stewart et al.,

2015). In the Arctic, there are also risks and opportunities related to employment, health and well-being, and the commodification of culture (Stewart et al., 2013; Stewart et al., 2015). There is *high confidence* that biodiversity supported by ice-free areas, particularly those on the AP, are vulnerable to the introduction of terrestrial alien species via tourists and scientists (Chown et al., 2012; Huiskes et al., 2014; Hughes et al., 2015; Duffy et al., 2017; Lee et al., 2017a) (Box 3.3) as well as to the direct impacts of humans (Pertierra et al., 2017). The tourism sector relies on a set of regulations that apply to all types of maritime shipping, yet cruise ships intentionally travel off regular shipping corridors and serve a very different purpose than other vessel types, so there is a need for region-specific governance regimes, specialised infrastructure, and focused policy attention (Dawson et al., 2014; Pashkevich et al., 2015; Pizzolato et al., 2016; Johnston et al., 2017). Private pleasure craft remain almost completely unregulated, and will pose unique risks in the future (Johnston et al., 2017).

3.2.4.3   Transportation

The Arctic is reliant on marine transportation for the import of food, fuel and other goods. At the same time, the global appetite for maritime trade and commerce through the Arctic (including community re-supply, mining and resource development, tourism, fisheries, cargo, research, and military and icebreaking, etc.) is increasing as the region becomes more accessible because of reduced sea ice cover. There are four potential Arctic international trade routes: the Northwest Passage, the Northern Sea Route, the Arctic Bridge and the Transpolar Sea Route. All of these routes offer significant trade benefits because they provide substantial distance savings compared to traditional routes via the Suez or Panama Canals.

There is *high confidence* that shipping activity during the Arctic summer increased over the past two decades in regions for which there is information, concurrent with reductions in Arctic sea ice extent and the shift to predominantly seasonal ice cover (Pizzolato et al., 2014; Eguíluz et al., 2016; Pizzolato et al., 2016). Long term datasets over the pan-Arctic are incomplete, but the distance travelled by ships in Arctic Canada nearly tripled between 1990 and 2015 (from ~365,000 to ~920,000 km) (Dawson et al., 2018). Other non-environmental factors which influence Arctic shipping are natural resource development, regional trade, geopolitics, commodity prices, global economic and social trends, national priorities, tourism demand, ship building technologies and insurance costs (Lasserre and Pelletier, 2011; Têtu et al., 2015; Johnston et al., 2017). Current impacts associated with the observed increase in Arctic shipping include a higher rate of reported accidents per km travelled compared to southern waters (CCA, 2016), increases in vessel noise propagation (Halliday et al., 2017) and air pollution (Marelle et al., 2016). Disruptions to cultural and subsistence hunting activities from increased shipping (Huntington et al., 2015; Olsen et al., 2019) compound climate-related impacts to people (Sections 3.4.3.3.2, 3.4.3.3.3).

It is projected that shipping activity will continue to rise across the Arctic as northern routes become increasingly accessible (Stephenson et al., 2011; Stephenson et al., 2013; Barnhart et al., 2015; Melia et al.,

2016), although mitigating economic and operational factors remain uncertain and could influence future traffic volume (Zhang et al., 2016). The Northern Sea Route is expected to be more viable than other routes because of infrastructure already in place (Milaković et al., 2018); favourable summer ice conditions in recent years have reduced transit times (Aksenov et al., 2017). In comparison, the Northwest Passage and Arctic Bridge presently have limited port and marine transportation infrastructure, incomplete soundings and hydrographic charting, challenging sea ice conditions and limited search and rescue capacity; these compound the risks from shipping activity (Stephenson et al., 2013; Johnston et al., 2017; Andrews et al., 2018).

Future shipping impacts will be regionally diverse considering the unique geographies, sea ice dynamics, infrastructure and service availability and regulatory regimes that exist across different Arctic nations. Considerations include socioeconomic and political implications related to safety (marine and local accidents), security (trafficking, terrorism and local issues), and environmental and cultural sustainability (invasive species, release of biocides, chemicals and other waste, marine mammal strikes, fuel spills, air and underwater noise pollution and impacts to subsistence hunting) (Arctic Council, 2015a; Halliday et al., 2017; Hauser et al., 2018). Black carbon emissions from shipping activity within the Arctic are projected to increase (Arctic Council, 2017) and are more easily deposited at the surface in the region compared with emissions from lower latitudes (Sand et al., 2013). Commercial shipping mainly uses heavy fuel oil, with associated emissions of sulphur, nitrogen, metals, hydrocarbons, organic compounds, black carbon and fly ash to the atmosphere during combustion (Turner et al., 2017a). Mitigation approaches include banning heavy fuel oil as already implemented in Antarctica and the waters around Svalbard, and the use of new technology like scrubbers.

The predominant shipborne activities in Antarctica are fishing, logistic support to land-based stations, and marine research vessels operating for both non-governmental and governmental sectors. Uncertainty in future Antarctic sea ice conditions (Section 3.2.2.1) pose challenges to considering potential impacts on these activities (Chown, 2017).

3.3 Polar Ice Sheets and Glaciers: Changes, Consequences and Impacts

3.3.1 Ice Sheet Changes

Changes in ice sheet mass have been derived repeatedly over the satellite era using complementary methods based on time series of satellite altimetry to measure volume change, ice-flux measurements combined with modelled surface mass balance (SMB) to calculate mass inputs and outputs, and satellite gravimetry to measure regional mass change. Ice sheet changes over earlier periods have also been reconstructed from firn/ice core and geological evidence (SM3.3.1).

3.3.1.1 Antarctic Ice Sheet Mass Change

It is *virtually certain* that the Antarctic Peninsula (AP) and West Antarctic Ice Sheet (WAIS) combined have cumulatively lost mass since widespread measurements began in 1992, and that the rate of loss has increased since around the year 2006 and continued post-AR5 (Martín-Español et al., 2016; Zwally et al., 2017; Bamber et al., 2018; Gardner et al., 2018; The IMBIE Team, 2018; Rignot et al., 2019), extending and reinforcing previous findings (IPCC, 2013) (Figure 3.7, Table 3.3, SM3.3.1.1). From *medium evidence*, there is *high agreement* in the sign and *medium agreement* in the magnitude of both WAIS and AP mass change between the complementary satellite methods (Mémin et al., 2015; The IMBIE Team, 2018).

Table 3.3 | Mass balance (Gt yr⁻¹) of the West Antarctic Ice Sheet (WAIS), Antarctic Peninsula (AP), East Antarctic Ice Sheet (EAIS), the combined Antarctic Ice Sheets (AIS) and the Greenland Ice Sheet (GIS) and the total sea level contribution (mm yr⁻¹).

Ice sheet	1992–1996	1997–2001	2002–2006	2007–2011	2012–2016
WAIS and AP (Bamber et al., 2018)	−55 ± 30	−53 ± 30	−77 ± 17	−197 ± 11	−172 ± 27
WAIS and AP (The IMBIE Team, 2018)	−60 ± 32	−44 ± 31	−85 ± 31	−183 ± 32	−192 ± 31
WAIS only (The IMBIE Team, 2018)	−53 ± 29	−41 ± 28	−65 ± 27	−148 ± 27	−159 ± 26
EAIS (Bamber et al., 2018)	28 ± 76	−50 ± 76	52 ± 37	80 ± 17	−19 ± 20
EAIS (The IMBIE Team, 2018)	11 ± 58	8 ± 56	12 ± 43	23 ± 38	−28 ± 30
GIS (Bamber et al., 2018)	31 ± 83	−47 ± 81	−206 ± 28	−320 ± 10	−247 ± 15
	1992–2006			2007–2016	
WAIS and AP (Bamber et al., 2018; The IMBIE Team, 2018)	−56 ± 20			−185 ± 17	
	1992–2016				
EAIS (Bamber et al., 2018)	18 ± 52				
EAIS (The IMBIE Team, 2018)	15 ± 41				
	1992–2001	2002–2011	2006–2015	2012–2016	
AIS (Bamber et al., 2018; The IMBIE Team, 2018)	−51 ± 73	−82 ± 27	−155 ± 19	−199 ± 26	
GIS (Bamber et al., 2018)	−8 ± 82	−263 ± 21	−278 ± 11	−247 ± 15	
Total sea level contribution (mm yr⁻¹)	0.16 ± 0.3	0.96 ± 0.1	1.20 ± 0.1	1.24 ± 0.1	

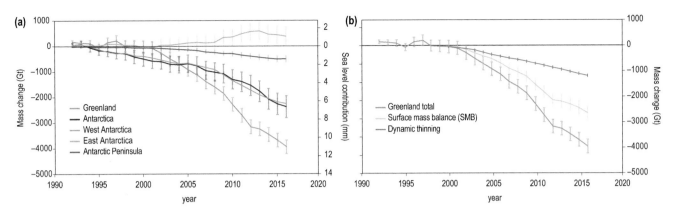

Figure 3.7 | (a) Cumulative Ice Sheet mass change, 1992–2016, (after Bamber et al., 2018; The IMBIE Team, 2018). **(b)** Greenland Ice Sheet (GIS) mass change components from surface mass balance (SMB) (orange) and dynamic thinning (blue) from 2000 to 2016, (after van den Broeke et al., 2016; King et al., 2018). Uncertainties are 1 standard deviation.

WAIS mass loss and recent increases in loss were concentrated in the Amundsen Sea Embayment (ASE) (*high confidence*) with increases particularly in the late 2000s (Mouginot et al., 2014), accounting for most of the -112 ± 10 Gt yr^{-1} WAIS loss from 2003 to 2013 (Martín-Español et al., 2016). The ice sheet margins of nearby Getz Ice Shelf also lost mass rapidly (-67 ± 27 Gt yr^{-1}, 2008–2015) (Gardner et al., 2018). This region also experienced losses during previous warm periods (Cross-Chapter Box 8 in Chapter 3).

On the AP, the Bellingshausen Sea ice sheet margin shifted from close to mass balance in the 2000s to rapid loss from 2009 (-56 ± 8 Gt yr^{-1} from 2010 to 2014) (*high confidence*) (Helm et al., 2014; McMillan et al., 2014b; Wouters et al., 2015; Hogg et al., 2017). This shift accompanied ongoing mass loss (*high confidence*) from the smaller northeastern AP glaciers that fed the former Prince Gustav, Larsen A and B ice shelves, though now at a lower rate than immediately following shelf collapse in 1995 and 2002 (Seehaus et al., 2015; Wuite et al., 2015; Rott et al., 2018). Of 860 marine-terminating AP glaciers, 90% retreated from their 1940s positions (Cook et al., 2014), established in the early to mid-Holocene (Ó Cofaigh et al., 2014) (*medium confidence*). Early 21st century combined AP glacier (Fieber et al., 2018) and ice sheet loss was around -30 Gt yr^{-1} (Table 3.3).

The East Antarctic Ice Sheet (EAIS, covering 85% of the Antarctic Ice Sheet (AIS)) has remained close to balance, with large interannual variability and no clear mass trend over the satellite record (*medium confidence*) (Table 3.3, Figure 3.7, SM3.3.1.2), and relatively large observation uncertainties (SM3.3.1) (Velicogna et al., 2014; Martin-Español et al., 2017; Bamber et al., 2018). SMB trends are particularly ambiguous, leading to disagreement between one altimetry and one flux-based estimate of 136 ± 43 Gt yr^{-1} (spanning 1992–2008) (Zwally et al., 2017), and -41 ± 8 Gt yr^{-1} (1979–2017) (Rignot et al., 2019), respectively. Both differ from the multi-method averages reported here (Table 3.3).

EAIS mass gains on the Siple Coast and Dronning Maud Land (e.g., 63 ± 6 Gt yr^{-1} from 2003 to 2013 (Velicogna et al., 2014)) contrast with Wilkes Land losses e.g., from -17 ± 4 Gt yr^{-1} from the Totten Glacier area, 2003–2013 (Velicogna et al., 2014) that drain a large area of deeply-grounded EAIS with potential for multi-metre sea level contributions (Zwally et al., 2017; Rignot et al., 2019).

Limited palaeo ice sheet evidence suggests that this area has previously lost substantial mass in previous interglacials (*medium confidence*) (Aitken et al., 2016; Wilson et al., 2018).

Overall, 2012–2016 Antarctic Ice Sheet mass losses were *extremely likely* greater than those from 2002 to 2011 and *likely* greater than from 1992 to 2001, and it is *extremely likely* that the negative 2012–2016 Antarctic Ice Sheet mass balance was dominated by losses from WAIS (Table 3.3).

3.3.1.2   Components of Antarctic Ice Sheet Mass Change

Antarctic Ice Sheet mass changes are dominated by changes in snowfall and glacier flow. The WAIS and AP loss trends in recent decades are dominated by glacier flow acceleration (also known as dynamic thinning) (*very high confidence*) (Figure SM3.8). Dynamic thinning losses were -112 ± 12 Gt yr^{-1} for 2003–2013, largely from the Amundsen Sea Embayment (Figure SM3.8) (Martín-Español et al., 2016), which contributed -102 ± 10 Gt yr^{-1} from 2003 to 2011 (Sutterley et al., 2014). Total Amundsen Sea Embayment ice discharge increased by 77% since the 1970s (Mouginot et al., 2014), primarily from acceleration of Pine Island Glacier that began around 1945, Smith, Pope and Kohler glaciers around 1980, and Thwaites Glacier around 2000 (Mouginot et al., 2014; Konrad et al., 2017; Smith et al., 2017c). Dynamic thinning in the Amundsen Sea Embayment and western AP accounted for 88% of the -36 ± 15 Gt yr^{-1} increase in Antarctic Ice Sheet mass loss from 2008 to 2015 (Gardner et al., 2018). Glacier acceleration of up to 25% also affected the Getz Ice Shelf margin from 2007 to 2014 (Chuter et al., 2017).

Reduction or loss of ice shelf buttressing has dominated Antarctic Ice Sheet dynamic thinning (*high confidence*). Ice shelves buttress 90% of Antarctic Ice Sheet outflow (Depoorter et al., 2013; Rignot et al., 2014; Fürst et al., 2016; Reese et al., 2018), and ice shelf thinning increased in WAIS by 70% in the decade to 2012, averaged 8% thickness loss from 1994 to 2012 in the Amundsen Sea Embayment (Paolo et al., 2015), and explains the post-2009 onset of rapid dynamic thinning on the southern-AP Bellingshausen Sea coast (Wouters et al., 2015; Hogg et al., 2017; Martin-Español et al., 2017) (Figure SM3.8). Grounding line retreat, an indicator of thinning, has been observed with *high confidence* (Rignot et al., 2014; Christie et al.,

237

2016; Hogg et al., 2017; Konrad et al., 2018; Roberts et al., 2018). From 2010 to 2016, 22%, 3% and 10% of grounding lines in WAIS, EAIS and the AP respectively retreated at rates faster than 25 m yr^{-1} (the average pace since the Last Glacial Maximum; Konrad et al., 2018), with highest rates along the Amundsen and Bellingshausen Sea coasts, and around Totten Glacier, Wilkes Land, EAIS (Konrad et al., 2018), where dynamic thinning has occurred at least since 1979 (Roberts et al., 2018; Rignot et al., 2019). Ice shelf collapse has driven dynamic thinning in the northern AP over recent decades (*high confidence*) (Seehaus et al., 2015; Wuite et al., 2015; Friedl et al., 2018; Rott et al., 2018).

ASE ice shelf basal melting, grounding line retreat and dynamic thinning have varied with ocean forcing (*medium confidence*) (Dutrieux et al., 2014; Paolo et al., 2015; Christianson et al., 2016; Jenkins et al., 2018) but this variability is superimposed on sustained mass losses compatible with the onset of Marine Ice Sheet Instability (MISI) for several major glaciers (*medium confidence*) (Favier et al., 2014; Joughin et al., 2014; Mouginot et al., 2014; Rignot et al., 2014; Christianson et al., 2016). Whether unstable WAIS retreat has begun or is imminent remains a critical uncertainty (Cross-Chapter Box 8 in Chapter 3).

Mass gains due to increased snowfall have somewhat offset dynamic-thinning losses (*high confidence*). On the AP, snowfall began to increase in the 1930s, accelerated in the 1990s (Thomas et al., 2015; Goodwin et al., 2016), and now offsets sea-level rise by 6.2 ± 1.7 mm per century (Medley and Thomas, 2018). EAIS and WAIS snowfall increases offset 20th century sea-level rise by 7.7 ± 4.0 mm and 2.8 ± 1.7 mm respectively (Medley and Thomas, 2018) (*medium confidence*). AIS snowfall increased by 4 ± 1 then 14 ± 1 Gt per decade over the 19th and 20th centuries, of which EAIS contributed 10% (Thomas et al., 2017b). Longer records suggest either an AIS snowfall decrease over the last 1000 years (Thomas et al., 2017a) or a statistically negligible change over the last 800 years (*low confidence*) (Frezzotti et al., 2013).

Mass balance contributions from ice sheet basal melting were not described in AR5 (IPCC, 2013) and the sensitivity of the AIS subglacial hydrological system to climate change is poorly understood. Around half of the AIS bed melts (Siegert et al., 2017), producing ~65 Gt yr^{-1} of water (Pattyn, 2010) (*low confidence*), some of which refreezes (Bell, 2008) and some accumulates in subglacial lakes with a total volume of 10s of 1000s of km^3 (Popov and Masolov, 2007; Lipenkov et al., 2016; Siegert, 2017). This system contributes fresh water and nutrients to the ocean (Section 3.3.3.3) (Fricker et al., 2007; Siegert et al., 2007; Carter and Fricker, 2012; Horgan et al., 2013; Le Brocq, 2013; Flament et al., 2014; Siegert et al., 2016), and lubricates glacier sliding (e.g., Dow et al., 2018b). Changes in the ice sheet thickness can redistribute subglacial water, affecting drainage pathways and ice flow (Fricker et al., 2016), but hydrological observations are very scarce.

3.3.1.3   Greenland Ice Sheet Mass Change

The Greenland Ice Sheet (GIS) experienced a marked shift to strongly negative mass balance between the early 1990s and mid-2000s (*very high confidence*) (Shepherd et al., 2012; Schrama et al., 2014; Velicogna et al., 2014; van den Broeke et al., 2016; Bamber et al., 2018; King et al., 2018; Sandberg Sørensen et al., 2018; WCRP, 2018).

It is *extremely likely* that the 2002–2011 and 2012–2016 ice losses were greater than in the 1992–2001 period (Bamber et al., 2018) (Table 3.3, Figure 3.7, SM3.3.1.3). Greenland Ice Sheet mass balance is characterised by large interannual variability (e.g., van den Broeke et al., 2017) but from 2005 to 2016, Greenland Ice Sheet was the largest terrestrial contributor to global sea level rise (WCRP, 2018).

A geodetic reconstruction of past ice sheet elevations indicates a Greenland Ice Sheet mass change of −75.1 ± 29.4 Gt yr^{-1} from 1900 to 1983, −73.8 ± 40.5 Gt yr^{-1} from 1983 to 2003, and −186.4 ± 18.9 Gt yr^{-1} from 2003 to 2010, with the losses consistently concentrated along the northwest and southeast coasts, and more locally in the southwest and on the large west coast Jakobshavn Glacier, though intensifying and spreading to the remainder of the coastal ice sheet in the latest period (Kjeldsen et al., 2015). Palaeo evidence also suggests that the GIS has contributed substantially to sea level rise during past warm intervals (Cross-Chapter Box 8 in Chapter 3).

3.3.1.4   Components of Greenland Ice Sheet Mass Change

Ongoing GIS mass loss over recent years has resulted from a combined increase in dynamic thinning and a decrease in SMB. Of these, reduced SMB due to an increase in surface melting and runoff recently came to dominate (*high confidence*) (Andersen et al., 2015; Fettweis et al., 2017; van den Broeke et al., 2017; King et al., 2018), accounting for 42% of losses for 2000–2005, 64% for 2005–2009 and 68% for 2009–2012 (Enderlin et al., 2014) (Figure 3.7).

The GIS was close to balance in the early years of the 1990s (Hanna et al., 2013; Khan et al., 2015), the interior above 2000 m altitude gained mass from 1961 to 1990 (Colgan et al., 2015) and both coastal and ice sheet sites experienced an increasing precipitation trend from 1890 to 2012 and 1890 to 2000 respectively (Mernild et al., 2015), but since the early 1990s multiple observations and modelling studies show strong warming and an increase in runoff (*very high confidence*). High-altitude GIS sites NEEM and Summit warmed by, respectively, 2.7°C ± 0.33°C over the past 30 years (Orsi et al., 2017) and by 2.7°C ± 0.3°C from 1982 to 2011 (McGrath et al., 2013), while satellite thermometry showed statistically significant widespread surface warming over northern GIS from 2000 to 2012 (Hall et al., 2013). The post-1990s period experienced the warmest GIS near-surface summer air temperatures of 1840–2010 (+1.1°C) (statistically highly significant) (Box, 2013), and ice core analysis found the 2000–2010 decade to be the warmest for around 2000 years (Vinther et al., 2009; Masson-Delmotte et al., 2012), and possibly around 7000 years (Lecavalier et al., 2017). This significant summer warming since the early 1990s increased GIS melt event duration (Mernild et al., 2017) and intensity to levels exceptional over at least 350 years (Trusel et al., 2018), and melt frequency to levels unprecedented for at least 470 years (Graeter et al., 2018). GIS melt intensity for 1994–2013 was two to fivefold the pre-industrial intensity (*medium confidence*) (Trusel et al., 2018). In response, GIS meltwater production and runoff increased (Hanna et al., 2012; Box, 2013; Fettweis et al., 2013; Tedstone et al., 2015; van den Broeke et al., 2016; Fettweis et al., 2017), resulting in 1994–2013 runoff being 33% higher the 20th century mean and 50% higher than the 18th century (Trusel et al., 2018), and 80% higher in a western-GIS marginal river catchment in 2003–2014 relative to 1976–2002 (Ahlstrom et al., 2017).

Only around half of the 1960–2014 surface melt ran off, most of the rest being retained in firn and snow (Steger et al., 2017), particularly in recently observed firn aquifers in south and west Greenland (Humphrey et al., 2012; Forster et al., 2013; Munneke et al., 2014; Poinar et al., 2017) that cover up to 5% of GIS (Miège et al., 2016; Steger et al., 2017) and stored around one fifth of the meltwater increase since the late 1990s (Noël et al., 2017) (*medium confidence*). While potential aquifer storage is equivalent to about a quarter of annual GIS melt production (Koenig et al., 2014; van den Broeke et al., 2016) and aquifers have spread to higher altitudes (Steger et al., 2017), their potential to buffer runoff has been reduced by firn densification (Polashenski et al., 2014), diversion of water to the bed via crevasses (Poinar et al., 2017), and the formation of ice layers that prevent drainage and promote surface ponding on the firn (Charalampidis et al., 2016) (*high confidence*). Such ponding lowers the firn albedo, promoting further melting (*high confidence*) (e.g., Charalampidis et al., 2015), but the extent of bare ice is a fivefold stronger control on melt (Ryan et al., 2019). Bare ice produced ~78% of runoff from 1960 to 2014, and its extent is expected to increase non-linearly as snow cover retreats to higher, flatter areas of ice sheet (Steger et al., 2017). This extent is not well reproduced in climate models, however, with biases of –6% to 13% (Ryan et al., 2019).

The remaining ~40% of non-SMB GIS mass loss from 1991 to 2015 has resulted from increased ice discharge due to dynamic thinning (*high confidence*) (Enderlin et al., 2014; van den Broeke et al., 2016; King et al., 2018) (Figure 3.7). From 2000 to 2016, dynamic thinning of 89% of GIS outlet glaciers accounted for –682 ± 31 Gt mass change, of which 92% came from the northwest and southeast GIS (King et al., 2018). Half came from only four glaciers (Jakobshavn Isbræ, Kangerdlugssuaq, Koge Bugt, and Ikertivaq South) (Enderlin et al., 2014). Glacier thinning has decreased glacier discharge, however, reducing the dynamic contribution to GIS mass loss (e.g., from 58% from 2000 to 2005 to 32% between 2009 and 2012; Enderlin et al., 2014). Furthermore, there is now *high confidence* that for most of the GIS, increased surface melt has not led to sustained increases in glacier flux on annual timescales because subglacial drainage networks have evolved to drain away the additional water inputs (e.g., Sole et al., 2013; Tedstone et al., 2015; Stevens et al., 2016; Nienow et al., 2017; King et al.,2018).

3.3.1.5   Drivers of Ice Sheet Mass Change

3.3.1.5.1   Ocean drivers

The reduction of ice shelf buttressing that has dominated AIS mass loss (Section 3.3.1.2) has been driven primarily by increases in sub-ice shelf melting (Khazendar et al., 2013; Pollard et al., 2015; Cook et al., 2016; Rintoul et al., 2016; Walker and Gardner, 2017; Adusumilli et al., 2018; Dow et al., 2018a; Minchew et al., 2018) (*high confidence*). Shoaling of relatively warm Circumpolar Deep Water has controlled recent variability in melting in the Amundsen and Bellingshausen seas, Wilkes Land (Roberts et al., 2018) and the AP (*medium confidence*) (Jacobs et al., 2011; Pritchard et al., 2012;

Depoorter et al., 2013; Rignot et al., 2013; Dutrieux et al., 2014; Paolo et al., 2015; Wouters et al., 2015; Christianson et al., 2016; Cook et al., 2016; Jenkins et al., 2018; Roberts et al., 2018). Changes in winds have driven this shoaling by affecting continental shelf edge undercurrents (Walker et al., 2013; Dutrieux et al., 2014; Kimura et al., 2017) and overturning in coastal polynyas (St-Laurent et al., 2015; Webber et al., 2017) (*medium confidence*). Winds over the Amundsen Sea are highly variable, however, with complex interactions between SAM, El Niño/Southern Oscillation (ENSO), Atlantic Multidecadal Oscillation, and the Amundsen Sea Low (Uotila et al., 2013; Li et al., 2014; Turner et al., 2016) (SM3.1.3).

Through their effects on Antarctic coastal ocean circulation, ENSO or other tropical-ocean variability may have triggered changes to Pine Island Glacier in the 1940s (Smith et al., 2017c) and again in the 1970s and 1990s (Jenkins et al., 2018), and recent ENSO variability is correlated with recent changes in ice shelf thickness (Paolo et al., 2018) (*medium confidence*). Such coupling between wind variability, ocean upwelling, ice shelf melt, buttressing and glacier flow rate has also been observed in EAIS, at Totten Glacier, Wilkes Land (Greene et al., 2017).

Around Greenland, an anomalous inflow of subtropical water driven by wind changes, multi-decadal natural ocean variability (Andresen et al., 2012), and a long-term increase in the North Atlantic's upper ocean heat content since the 1950s (Cheng et al., 2017), all contributed to a warming of the subpolar North Atlantic (Häkkinen et al., 2013) (*medium confidence*). Water temperatures near the grounding zone of GIS outlet glaciers are critically important to their calving rate (O'Leary and Christoffersen, 2013) (*medium confidence*), and warm waters have been observed interacting with major GIS outlet glaciers (*high confidence*) (e.g., Holland et al., 2008; Straneo et al., 2017).

The processes behind warm-water incursions in coastal Greenland that force glacier retreat remain unclear, however (Straneo et al., 2013; Xu et al., 2013b; Bendtsen et al., 2015; Murray et al., 2015; Cowton et al., 2016; Miles et al., 2016), and there is *low confidence* in understanding coastal GIS glacier response to ocean forcing because submarine melt rates, calving rates (Rignot et al., 2010; Todd and Christoffersen, 2014; Benn et al., 2017), bed and fjord geometry and the roles of ice melange and subglacial discharge (Enderlin et al., 2013; Gladish et al., 2015; Slater et al., 2015; Morlighem et al., 2016; Rathmann et al., 2017) are poorly understood, and extrapolation from a small sample of glaciers is impractical (Moon et al., 2012; Carr et al., 2013; Straneo et al., 2016; Cowton et al., 2018).

3.3.1.5.2   Atmospheric drivers

Snow accumulation and surface melt in Antarctica are influenced by the Southern Hemisphere extratropical circulation (SM3.1.3), which has intensified and shifted poleward in austral summer from 1950 to 2012 (Arblaster et al., 2014; Swart et al., 2015a) (*medium confidence*). The austral summer SAM has been in its most positive extended state for the past 600 years (Abram et al., 2014; Dätwyler et al., 2017), and from 1979 to 2013 has contributed to intensified atmospheric circulation and increasing and decreasing snowfall in the western and eastern AP respectively (Marshall et al., 2017) (*medium confidence*). WAIS accumulation trends (Section 3.3.1.2) resulted from a deepening of the Amundsen Sea Low over recent decades (Raphael et al., 2016) (*high confidence*).

During the 1990s, WAIS experienced record surface warmth relative to the past 200 years, though similar conditions occurred for 1% of the preceding 2000 years (Steig et al., 2013), and WAIS surface melting remains limited. In contrast, AP surface melting has intensified since the mid-20th century and the last three decades were unprecedented over 1000 years (Abram et al., 2013a). The northeast AP began warming 600 years ago and past-century rates were unusual over 2000 years (Mulvaney et al., 2012; Stenni et al., 2017). Increased föhn winds due to the more positive SAM (Cape et al., 2015) caused increased surface melting on the Larsen ice shelves (Grosvenor et al., 2014; Luckman et al., 2014; Elvidge et al., 2015) and after 11,000 years intact, the 2002 melt-driven collapse of the Larsen B ice shelf followed strong warming between the mid–1950s and the late 1990s (Domack et al., 2005) (*medium confidence*).

In Greenland, associations between atmospheric pressure indices such as the North Atlantic Oscillation (NAO) and temperature, insolation and snowfall indicate with *high confidence* that, as in Antarctica, variability of large-scale atmospheric circulation is an important driver of SMB changes (Fettweis et al., 2013; Tedesco et al., 2013; Ding et al., 2014; Tedesco et al., 2016b; Ding et al., 2017; Hofer et al., 2017). A post-1990s decrease in summer NAO reflects increased anticyclonic weather (e.g., Tedesco et al., 2013; Hanna et al., 2015) that advected warm air over the GIS, explaining ~70% of summer surface warming from 2003 to 2013 (Fettweis et al., 2013; Tedesco et al., 2013; Mioduszewski et al., 2016), and reduced the cloud cover, increasing shortwave insolation (Tedesco et al., 2013) that, combined with albedo feedbacks (Box et al., 2012; Charalampidis et al., 2015; Tedesco et al., 2016a; Stibal et al., 2017; Ryan et al., 2018) (*high confidence*), explains most of the post-1990s melt increase (Hofer et al., 2017). These drivers culminated in July 2012 in exceptional warmth and surface melt up to the ice sheet summit (Nghiem et al., 2012; Tedesco et al., 2013; Hanna et al., 2014; Hanna et al., 2016; McLeod and Mote, 2016).

3.3.1.6　　Natural and Anthropogenic Forcing

There is *medium agreement* but *limited evidence* of anthropogenic forcing of AIS mass balance through both SMB and glacier dynamics (*low confidence*). Partitioning between natural and human drivers of atmospheric and ocean circulation changes remains very uncertain. Partitioning is challenging because, along with the effects of greenhouse gas increases and stratospheric ozone depletion (Waugh et al., 2015; England et al., 2016; Li et al., 2016a), atmospheric and ocean variability in the areas of greatest AIS mass change are affected by a complex chain of processes (e.g., Fyke et al., 2018; Zhang et al., 2018a) that exhibit considerable natural variability and have multiple interacting links to sea surface conditions in the Pacific (Schneider et al., 2015; England et al., 2016; Raphael et al., 2016; Clem et al., 2017; Steig et al., 2017; Paolo et al., 2018) and Atlantic (Li et al., 2014), with additional local feedbacks (e.g., Stammerjohn et al., 2012; Goosse and Zunz, 2014). Recent AP warming and consequent

ice shelf collapses have evidence of a link to anthropogenic ozone and greenhouse gas forcing via the SAM (e.g., Marshall, 2004; Shindell, 2004; Arblaster and Meehl, 2006; Marshall et al., 2006; Abram et al., 2014) and to anthropogenic Atlantic sea surface warming via the Atlantic Multidecadal Oscillation (e.g., Li et al., 2014). This warming was highly unusual over the last 1000 years but not unprecedented, and along with subsequent cooling is within the bounds of the large natural decadal-scale climate variability in this region (Mulvaney et al., 2012; Turner et al., 2016). More broadly over the AP and coastal WAIS where dynamic mass losses are concentrated, natural variability in atmospheric and ocean forcing appear to dominate observed mass balance (*medium confidence*) (Smith and Polvani, 2017; Jenkins et al., 2018).

Evidence exists for an anthropogenic role in the atmospheric circulation (NAO) changes that have driven GIS mass loss (Section 3.3.1.5.2) (*medium confidence*), although this awaits formal attribution testing (e.g., Easterling et al., 2016). Arctic amplification of anthropogenic warming (e.g., Serreze et al., 2009) affects atmospheric circulation (Francis and Vavrus, 2015; Mann et al., 2017) and has reduced sea ice extent (Section 3.2.1.1.1), feeding back to exacerbate both warming and NAO changes (Screen and Simmonds, 2010) that impact GIS mass balance. Negative-NAO wind patterns increased GIS melt observed in a 40-year runoff signal (Ahlstrom et al., 2017), and an increase in melting beginning in the mid-1800s closely followed the onset of industrial era Arctic warming and emerged beyond the range of natural variability in the last few decades (Graeter et al., 2018; Trusel et al., 2018) (Section 3.3.1.4).

3.3.1.7　　Ice Sheet Projections

Section 4.2 assesses the sea level impacts from observed and projected changes in ice sheets.

3.3.2　　**Polar Glacier Changes**

3.3.2.1　　Observations, Components of Change, and Drivers

Chapter 3 assesses changes in polar glaciers in the Canadian and Russian Arctic, Svalbard, Greenland and Antarctica, independent of the Greenland and Antarctic ice sheets (Figure 3.8). Glaciers in all other regions including Alaska, Scandinavia and Iceland are assessed in Chapter 2.

Changes in the mass of Arctic glaciers for the 'present day' (2006–2015) are assessed using a combination of satellite observations and direct measurements (Figure 3.8; Appendix 2.A, Table 1). During this period, glacier mass loss was largest in the periphery of Greenland (-47 ± 16 Gt yr^{-1}), followed by Arctic Canada North (-39 ± 8 Gt yr^{-1}), Arctic Canada South (-33 ± 9 Gt yr^{-1}), the Russian Arctic (-15 ± 12 Gt yr^{-1}) and Svalbard and Jan Mayen (-9 ± 5 Gt yr^{-1}). When combined with the Arctic regions covered in Chapter 2 (Alaska, the Yukon territory of Canada, Iceland and Scandinavia), Arctic glaciers as a whole lost mass at a rate of -213 ± 29 Gt yr^{-1}, a sea level contribution of 0.59 ± 0.08 mm yr^{-1} (*high confidence*). Overall during this period, Arctic glaciers caused a similar amount of sea

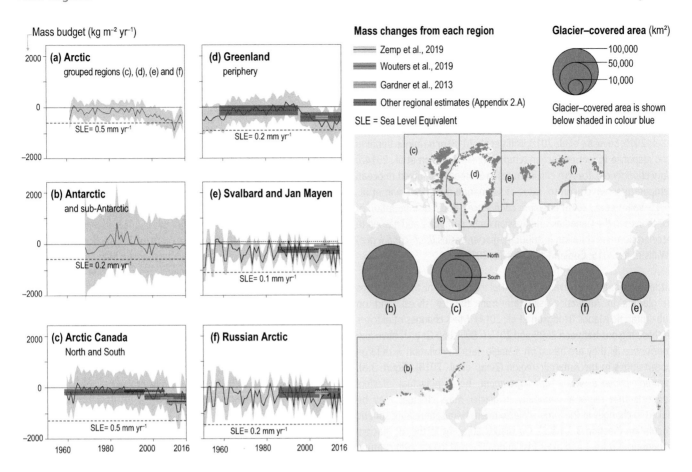

Figure 3.8 | Glacier mass budgets for the six polar regions assessed in Chapter 3. Glacier mass budgets for all other regions (including Iceland, Scandinavia and Alaska) are shown in Chapter 2, Figure 2.4. Regional time series of annual mass change are based on glaciological and geodetic balances (Zemp et al., 2019). Superimposed are multi-year averages by Wouters et al. (2019) and Gardner et al. (2013) from the Gravity Recovery and Climate Experiment (GRACE). Estimates by Gardner et al. (2013) were used in the IPCC 5th Assessment Report (AR5). Additional regional estimates in some regions are listed in Appendix 2.1, Table 1. Annual and time-averaged mass-budget estimates include the errors reported in each study. Glacier outlines and areas are based on RGI Consortium (2017).

level rise to the GIS (Section 3.3.1.3), but their rate of mass loss per unit area was larger (Bolch et al., 2013).

There is *limited evidence* (*high agreement*) that the current rate of glacier mass loss is larger than at any time during the past 4000 years (Fisher et al., 2012; Zdanowicz et al., 2012). Further back in time during the early to mid- Holocene, pre-historic glacial deposits, ice core records, and numerical modelling evidence shows that many Arctic glaciers were at various stages similar to or smaller than present (Gilbert et al., 2017; Zekollari et al., 2017), experienced greater melt rates (Lecavalier et al., 2017), or may have disappeared altogether (Solomina et al., 2015) (*medium confidence*). This evidence, however, does not provide a complete assessment of the rates and magnitudes of past glacier mass loss.

Atmospheric circulation changes (Box et al., 2018) have led to pan-Arctic variability in glacier mass balance (*high confidence*), including different rates of retreat between eastern and western glaciers in Greenland's periphery (Bjørk et al., 2018), and a high rate of surface melt in the Canadian Arctic (Gardner et al., 2013; Van Wychen et al., 2016; Millan et al., 2017) through persistently high summer air temperatures (Bezeau et al., 2014; McLeod and Mote, 2016). Atmospheric circulation anomalies from 2007 to 2012

associated with glacier mass loss are also linked to enhanced GIS melt (Section 3.3.1.4) and Arctic sea ice loss (Section 3.2.1.1), and exceed by a factor of two the interannual variability in daily mean pressure (sea level and 500 hPa) of the Arctic region over the 1871–2014 period (Belleflamme et al., 2015) (Section 3.3.1.6).

Increased surface melt on Arctic glaciers has led to a positive feedback from lowered surface albedo, causing further melt (Box et al., 2012), and in Svalbard, mean glacier albedo has reduced between 1979 and 2015 (Möller and Möller, 2017). Across the Arctic, increased surface melt and subsequent ice-layer formation via refreezing within snow and firn also reduces the ability of glaciers to store meltwater, increasing runoff (Zdanowicz et al., 2012; Gascon et al., 2013a; Gascon et al., 2013b; Noël et al., 2017; Noël et al., 2018).

Between the 1990s and 2017, tidewater glaciers have exhibited regional patterns in glacier dynamics; glaciers in Arctic Canada have largely decelerated, while glaciers in Svalbard and the Russian Arctic have accelerated (Van Wychen et al., 2016; Strozzi et al., 2017). Annual retreat rates of tidewater glaciers in Svalbard and the Russian Arctic for 2000–2010, have increased by a factor 2 and 2.5 respectively, between 1992 and 2000 (Carr et al., 2017). Acceleration due to surging (an internal dynamic instability) of a few key glaciers has dominated dynamic ice discharge on time-scales of years to decades (Van Wychen et al., 2014; Dunse et al., 2015).

The recent acceleration and surge behaviour of polythermal glaciers in Svalbard and the Russian Arctic is caused by destabilisation of the marine termini due to increased surface melt, and changes in basal temperature, lubrication and weakening of subglacial sediments (Dunse et al., 2015; Sevestre et al., 2018; Willis et al., 2018) or terminus thinning and response to warmer ocean temperatures (McMillan et al., 2014a) (*low confidence*). Iceberg calving rates in Svalbard are linked to ocean temperatures which control rates of submarine melt (Luckman et al., 2015; Vallot et al., 2018) (*medium confidence*). Rapid disintegration of ice shelves in the Canadian and Russian Arctic continues and has led to acceleration and thinning in tributary-glacier basins (*high confidence*) (Willis et al., 2015; Copland and Mueller, 2017).

Little information is available on Holocene and historic changes in glaciers in Antarctica (separate from the ice sheet), and on sub-Antarctic islands (Hodgson et al., 2014). Mass changes of glaciers in these regions between 2006 and 2015 (–90 ± 860 Gt yr^{-1}) have *low confidence* as they are based on a single data compilation with large uncertainties in the Antarctic region (Zemp et al., 2019) (Figure 3.8). *Limited evidence* with *high agreement* from individual glaciers suggests that regional variability in glacier mass changes may be linked to changes in the large-scale Southern Hemisphere atmospheric circulation (Section 3.3.1.5.2). On islands adjacent to the AP, glaciers experienced retreat and mass loss during the mid to late 20th century, but since around 2009 there has been a reduction in mass loss rate or a return to slightly positive balance (Navarro et al., 2017; Oliva et al., 2017). Reduced mass loss has been linked to increased winter snow accumulation and decreased summer melt at these locations, associated with recent deepening of the circumpolar pressure trough (Oliva et al., 2017). Conversely, on the sub-Antarctic Kerguelen Islands, increased glacier mass loss (Verfaillie et al., 2015) may be due to reduced snow accumulation rather than increased air temperature as a result of southward migration of storm tracks (Favier et al., 2016).

3.3.2.2   Projections

Projections of all glaciers, including those in polar regions, are covered in Cross-Chapter Box 6 in Chapter 2.

3.3.3   Consequences and Impacts

3.3.3.1   Sea Level

Chapter 4 assesses the sea level impacts from observed and projected changes in ice sheets (Section 3.3.1) and polar glaciers (Section 3.3.2), including uncertainties related to marine ice sheets (Cross-Chapter Box 8 in Chapter 3).

3.3.3.2   Physical Oceanography

The major large-scale impacts of freshwater release from Greenland on ocean circulation relate to the potential modulation/inhibition of the formation of water masses that represent the headwaters of the Atlantic

Meridional Overturning Circulation. The timescales and likelihood of such effects are assessed separately in Chapter 6 (Section 6.7). Freshwater release also affects local circulation within fjords through two principle mechanisms; subglacial release from tidewater glaciers enhances buoyancy driven circulation, whereas runoff from land-terminating glaciers contributes to surface layer freshening and estuarine circulation (Straneo and Cenedese, 2015). There is *limited evidence* that freshening occurred between 2003 and 2015 in North East Greenland fjords and coastal waters (Sejr et al., 2017).

For Antarctica, freshwater input to the ocean from the ice sheet is divided approximately equally between melting of calved icebergs and of ice shelves *in situ* (Depoorter et al., 2013; Rignot et al., 2014). There is *high confidence* that the input of ice shelf meltwater has increased in the Amundsen and Bellingshausen Seas since the 1990s, but *low confidence* in trends in other sectors (Paolo et al., 2015).

Freshwater injected from the AIS affect water mass circulation and transformation, though sea ice dominates upper ocean properties away from the Antarctic ice shelves (Abernathey et al., 2016; Haumann et al., 2016). Over the ice shelf regions, where dense waters sink and flood the global ocean abyss, the role of glacial freshwater input is clearer. From 1980 to 2012, the salinity of Antarctic Bottom Water reduced by an amount equivalent to 73 ± 26 Gt y^{-1} of freshwater added, around half the estimated increase in freshwater input by Antarctic glacial discharge up to that time (Purkey and Johnson, 2013). In some places, notably the Indian-Australian sector, Antarctic Bottom Water freshening may be accelerating (Menezes et al., 2017). There is *medium confidence* in an overall freshening trend and *low confidence* that this is accelerating, given the sparsity of information and significant interannual variability in Antarctic Bottom Water properties at other export locations (Meijers et al., 2016).

For the Southern Ocean, there is *limited evidence* for stratification changes in the post-AR5 period, and *low confidence* in how stratification changes are affecting sea ice and basal ice shelf melt. An increase in stratification caused by release of freshwater from the AIS was invoked as a mechanism to suppress vertical heat flux and permit an increase in sea ice extent (Bintanja et al., 2013; Bronselaer et al., 2018; Purich et al., 2018), though some studies conclude that glacial freshwater input is insufficient to cause a significant sea ice expansion (Swart and Fyfe, 2013; Pauling et al., 2017) (Section 3.2.1.1). In contrast, where warm water intrusions drive melting within ice shelf cavities, a significant entrained heat flux to the surface can exist and increase stratification and potentially reduce sea ice extent (Jourdain et al., 2017; Merino et al., 2018). It has been argued that freshening from glacial melt can enhance basal melting of ice shelves by reducing dense water production and modulating oceanic heat flow into ice shelf cavities (Silvano et al., 2018).

3.3.3.3   Biogeochemistry

Both polar ice sheets have the potential to release dissolved and sediment-bound nutrients and organic carbon directly to the surface ocean via subglacial and surface meltwater, icebergs, melting of the base of ice shelves (Shadwick et al., 2013; Wadham et al., 2013; Hood et al., 2015; Herraiz-Borreguero et al., 2016; Raiswell et al.,

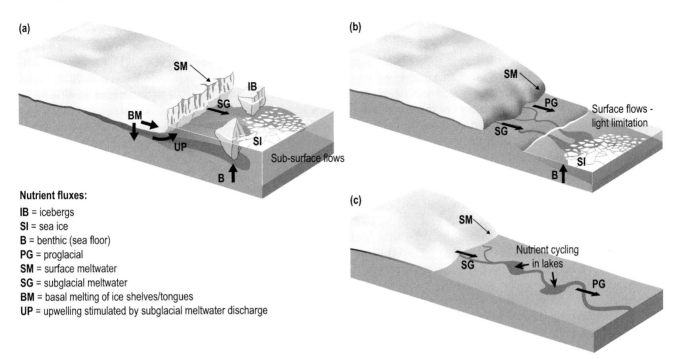

Nutrient fluxes:

IB = icebergs
SI = sea ice
B = benthic (sea floor)
PG = proglacial
SM = surface meltwater
SG = subglacial meltwater
BM = basal melting of ice shelves/tongues
UP = upwelling stimulated by subglacial meltwater discharge

Figure 3.9 | Potential shifts in nutrient fluxes with landward retreat of marine-terminating glaciers **(a)** at different stages (**b** and **c**).

2016; Yager et al., 2016; Hodson et al., 2017), in addition to indirectly stimulating nutrient input via upwelling associated with subglacial meltwater plumes (Meire et al., 2016b; Cape et al., 2018; Hopwood et al., 2018; Kanna et al., 2018) (Figure 3.9). These nutrient additions stimulate primary production in the surrounding ocean waters in some regions (*medium confidence*) (Gerringa et al., 2012; Death et al., 2014; Duprat et al., 2016; Arrigo et al., 2017b). There is also some evidence to support melting ice sheets as source of contaminants (AMAP, 2015).

In Greenland, direct measurements suggest that meltwater is a significant source of bioavailable silica and iron (Bhatia et al., 2013; Hawkings et al., 2014; Meire et al., 2016a; Hawkings et al., 2017) but may be less important for the supply of bioavailable forms of dissolved nitrogen or phosphorous (Hawkings et al., 2016; Wadham et al., 2016), which often limit the integrated primary production during summer in fjords (Meire et al., 2016a; Hopwood et al., 2018). The offshore export of iron, however, has been linked to primary productivity in surface ocean waters in the Labrador Sea (Arrigo et al., 2017b) (*limited evidence, high agreement*).

Subglacial meltwater plumes from tidewater glaciers have emerged recently as an important indirect source of nutrients to fjords, by entraining nutrient-replete seawater (Meire et al., 2016b; Meire et al., 2017; Cape et al., 2018; Hopwood et al., 2018; Kanna et al., 2018) (*medium evidence, high agreement*). There is *medium evidence* with *high agreement* that these upwelled nutrient fluxes enhance primary production in fjords over a distance of up to 100 km along the trajectory of the outflowing plume (Juul-Pedersen et al., 2015; Cape et al., 2018; Kanna et al., 2018).

In Antarctica, there is *medium evidence* with *high agreement* that enhanced input of iron from ice shelves, glacial meltwater and icebergs stimulates primary production in polynyas, coastal regions

and the wider Southern Ocean (Gerringa et al., 2012; Shadwick et al., 2013; Herraiz-Borreguero et al., 2016). Satellite observations and modelling also indicate variable potential for icebergs to fertilise the Southern Ocean beyond the coastal zone (Death et al., 2014; Duprat et al., 2016; Wu and Hou, 2017).

Dissolved nutrient fluxes from ice sheets may be increasing during high melt years (Hawkings et al., 2015). The dominant sediment-bound fraction, however, may not increase with rising melt (Hawkings et al., 2015). Thus, there is *low confidence* overall in the magnitude of the response of direct nutrient fluxes from ice sheets to enhanced melting.

Future predictions of nutrient cycling proximal to ice sheets is made more challenging by the landward progression of marine-terminating glaciers and the collapse of ice shelves (Cook et al., 2016). This has the potential to drive major shifts in nutrient supply to coastal waters (Figure 3.9). The erosion of newly exposed glacial sediments in front of retreating land-terminating glaciers (Monien et al., 2017) and changes in the diffuse nutrient fluxes from newly exposed glacial sediments on the seafloor (Wehrmann et al., 2014) may amplify nutrient supply, whilst other nutrient sources may be cut off (e.g., icebergs, upwelling of marine water; Meire et al., 2017) (*low confidence*).

There is *medium evidence* with *high agreement* that long-term tidewater glacier retreat into shallower water or onto land, a plausible scenario for about 55% of the 243 distinct outlet glaciers in Greenland (Morlighem et al., 2017), will reduce or diminish upwelling a source of nutrients, thereby reducing summer productivity in Greenland fjord ecosystems (Meire et al., 2017; Hopwood et al., 2018).

3.3.3.4   Ecosystems

For Greenland and Svalbard, there is *limited evidence* with *high agreement* that the retreat of marine-terminating glaciers will alter food supply to higher trophic levels of marine food webs (Meire et al., 2017; Milner et al., 2017). The consequences of changes in glacial systems on marine ecosystems are often mediated via the fjordic environments that fringe the edge of the ice sheets, for example changing physical-chemical conditions have affected the benthic ecosystems of Arctic fjords (Bourgeois et al., 2016). The amplification of nutrient fluxes caused by enhanced upwelling at calving fronts (Meire et al., 2017), combined with high carbon/nutrient burial and recycling rates (Wehrmann et al., 2013; Smith et al., 2015), plays an important role in sustaining high productivity of the Arctic fjord ecosystems of Greenland and Svalbard (Lydersen et al., 2014). Glacier retreat, causing glaciers to shift from being marine-terminating to land-terminating, can reduce the productivity in coastal areas off Greenland with potentially large ecological implications, also negatively affecting production of commercially harvested fish (Meire et al., 2017). There is also evidence that marine-terminating glaciers are important feeding areas for marine mammals and seabirds at Greenland (Laidre et al., 2016) and Svalbard (Lydersen et al., 2014).

For Antarctica, there is *high agreement* based on *medium evidence* that ice shelf retreat or collapse is leading to new marine habitats and to biological colonisation (Gutt et al., 2011; Fillinger et al., 2013; Trathan et al., 2013; Hauquier et al., 2016; Ingels et al., 2018). The loss of ice shelves and retreat of coastal glaciers around the AP in the last 50 years has exposed at least 2.4×10^4 km^2 of new open water. These newly-revealed habitats have allowed new phytoplankton blooms to be produced resulting in new marine zooplankton and seabed communities (Gutt et al., 2011; Fillinger et al., 2013; Trathan et al., 2013; Hauquier et al., 2016) (Section 3.2.3.2.1), and have resulted in enhanced carbon uptake by coastal marine ecosystems (*medium confidence*), although quantitative estimates of biological carbon uptake are highly variable (Trathan et al., 2013; Barnes et al., 2018). Newly available habitat on coastlines may also provide breeding or haul out sites for land-based predators such as penguins and seals (Trathan et al., 2013) (*low confidence*). Fjords that have been studied in the subpolar western AP are hotspots of abundance and biodiversity of benthic macro-organisms (Grange and Smith, 2013) and there is evidence that glacier retreat in these environments can impact the structure and function of benthic communities (Moon et al., 2015; Sahade et al., 2015) (*low confidence*).

Cross-Chapter Box 8 | Future Sea Level Changes and Marine Ice Sheet Instability

Authors: Robert M. DeConto (USA), Alexey Ekaykin (Russian Federation), Andrew Mackintosh (Australia), Roderik van de Wal (Netherlands), Jeremy Bassis (USA)

Over the last century, glaciers were the main contributors to increasing ocean water mass (Section 4.2.1.2). However, most terrestrial frozen water is stored in Antarctic and Greenland ice sheets, and future changes in their dynamics and mass balance will cause sea level rise over the 21st century and beyond (Section 4.2.3).

About a third of the Antarctic Ice Sheet (AIS) is 'marine ice sheet', i.e., rests on bedrock below sea level (Figure 4.5), with most of the ice sheet margin terminating directly in the ocean. These features make the overlying ice sheet vulnerable to dynamical instabilities with the potential to cause rapid ice loss; so-called Marine Ice Sheet and Marine Ice Cliff instabilities, as discussed below.

In many places around the AIS margin, the seaward-flowing ice forms floating ice shelves (Figure CB8.1). Ice shelves in contact with bathymetric features on the sea floor or confined within embayments provide back stress (buttressing) that impedes the seaward flow of the upstream ice and thereby stabilises the ice sheet. The ice shelves are thus a key factor controlling AIS dynamics. Almost all Antarctic ice shelves provide substantial buttressing (Fürst et al., 2016) but some are currently thinning at an increasing rate (Khazendar et al., 2016). Today, thinning and retreat of ice shelves is associated primarily with ocean driven basal melt that, in turn, promotes iceberg calving (Section 3.3.1.2).

Accumulation and percolation of surface melt and rain water also impact ice shelves by lowering albedo, deepening surface crevasses, and causing flexural stresses that can lead to hydrofracturing and ice shelf collapse (Macayeal and Sergienko, 2013). In some cases supraglacial (i.e., flowing on the glacier surface) rivers might diminish destabilising impact of surface melt by removing meltwater before it ponds on the ice shelf surface (Bell et al., 2017). In summary, both ocean forcing and surface melt affect ice shelf mechanical stability (*high confidence*), but the precise importance of the different mechanisms remains poorly understood and observed.

The future dynamic response of the AIS to warming will largely be determined by changes in ice shelves, because their thinning or collapse will reduce their buttressing capacity, leading to an acceleration of the grounded ice and to thinning of the ice margin. In turn, this thinning can initiate grounding line retreat (Konrad et al., 2018). If the grounding line is located on bedrock sloping downwards toward the ice sheet interior (retrograde slope), initial retreat can trigger a positive feedback, due to non-linear response of the seaward ice flow to the grounding line thickness change. As a result, progressively more ice will flow into the ocean (Figure CB8.1a).

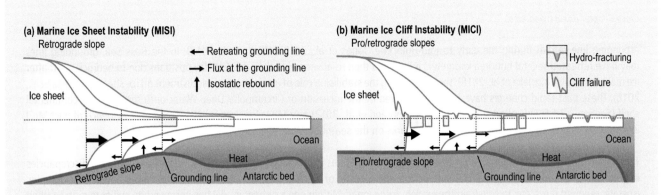

Figure CB8.1 | Schematic representation of Marine Ice Sheet Instability (MISI, a) and Marine Ice Cliff Instability (MICI, b) from Pattyn (2018). **(a)** thinning of the buttressing ice shelf leads to acceleration of the ice sheet flow and thinning of the marine-terminated ice margin. Because bedrock under the ice sheet is sloping towards ice sheet interior, thinning of the ice causes retreat of the grounding line followed by an increase of the seaward ice flux, further thinning of the ice margin, and further retreat of the grounding line. **(b)** disintegration of the ice shelf due to bottom melting and/or hydro-fracturing produces an ice cliff. If the cliff is tall enough (at least ~800 m of total ice thickness, or about 100 m of ice above the water line), the stresses at the cliff face exceed the strength of the ice, and the cliff fails structurally in repeated calving events. Note that MISI requires a retrograde bed slope, while MICI can be realised on a flat or seaward-inclined bed. Like MISI, the persistence of MICI depends on the lack of ice shelf buttressing, which can stop or slow brittle ice failure at the grounding line by providing supportive backstress.

This self-sustaining process is known as Marine Ice Sheet Instability (MISI). The onset and persistence of MISI is dependent on several factors in addition to overall bed slope, including the details of the bed geometry and conditions, ice shelf pinning points, lateral shear from the walls, self-gravitation effects on local sea level and isostatic adjustment. Hence, long-term retreat on every retrograde sloped bed is not necessarily unstoppable (Gomez et al., 2015).

The MISI process might be particularly important in West Antarctica, where most of the ice sheet is grounded on bedrock below sea level (Figure 4.5). Since AR5, there is growing observational and modelling evidence that accelerated retreat may be underway in several major Amundsen Sea outlets, including Thwaites, Pine Island, Smith, and Kohler glaciers (e.g., Rignot et al., 2014) supporting the MISI hypothesis, although observed grounding line retreat on retrograde slope is not definitive proof that MISI isunderway.

It has been shown recently (Barletta et al., 2018) that the Amundsen Sea Embayment (ASE) experiences unexpectedly fast bedrock uplift (up to 41 mm yr^{-1}, due to mantle viscosity much lower than the global average) as an adjustment to reduced ice mass loading, which could help stabilise grounding line retreat.

One of the largest outlets of the East Antarctic Ice Sheet (EAIS), Totten glacier, has also been retreating and thinning in recent decades (Li et al., 2015b). Totten's current behaviour suggests that East Antarctica could become a substantial contributor to future sea level rise, as it has been in the previous warm periods (Aitken et al., 2016). It is not clear, however, if the changes observed recently are a linear response to increased ocean forcing (Section 3.3.1.2), or an indication that MISI has commenced (Roberts et al., 2018).

The disappearance of ice shelves may allow the formation of ice cliffs, which may be inherently unstable if they are tall enough (subaerial cliff height between 100 and 285 m) to generate stresses that exceed the strength of the ice (Parizek et al., 2019). This ice cliff failure can lead to ice sheet retreat via a process called marine ice cliff instability (MICI; Figure CB8.1b), that has been hypothesised to cause partial collapse of the West Antarctic Ice Sheet (WAIS) within a few centuries (Pollard et al., 2015; DeConto and Pollard, 2016).

Limited evidence is available to confirm the importance of MICI. In Antarctica, marine-terminating ice margins with the grounding lines thick enough to produce unstable ice cliffs are currently buttressed by ice shelves, with a possible exception of Crane glacier on the Antarctic Peninsula (Section 4.2.3.1.2).

Overall, there is *low agreement* on the exact MICI mechanism and *limited evidence* of its occurrence in the present or the past. Thus the potential of MICI to impact the future sea level remains very uncertain (Edwards et al., 2019).

Limited evidence from geological records and ice sheet modelling suggests that parts of AIS experienced rapid (i.e., on centennial time-scale) retreat *likely* due to ice sheet instability processes between 20,000 and 9,000 years ago (Golledge et al., 2014; Weber et al., 2014; Small et al., 2019). Both the WAIS (including Pine Island glacier) and EAIS also experienced rapid thinning and

3

Cross-Chapter Box 8 (continued)

grounding line retreat during the early to mid-Holocene (Jones et al., 2015b; Wise et al., 2017). In the Ross Sea, grounding lines may have retreated several hundred kilometers inland and then re-advanced to their present-day positions due to bedrock uplift after ice mass removal (Kingslake et al., 2018), thus supporting the stabilising role of glacial isostatic adjustment on ice sheets (Barletta et al., 2018). These past rapid changes have *likely* been driven by the incursion of Circumpolar Deep Water onto the Antarctic continental shelf (Section 3.3.1.5.1) (Golledge et al., 2014; Hillenbrand et al., 2017) and MISI (Jones et al., 2015b). *Limited evidence* of past MICI in Antarctica is provided by deep iceberg plough marks on the sea-floor (Wise et al., 2017).

The ability of models to simulate the processes controlling MISI has improved since AR5 (Pattyn, 2018), but significant discrepancies in projections remain (Section 4.2.3.2) due to poor understanding of mechanisms and lack of observational data on bed topography, isostatic rebound rates, etc. to constrain the models. Inclusion of MICI in one ice sheet model has improved its ability to match (albeit uncertain) geological sea level targets in the Pliocene (Pollard et al., 2015) and Last Interglacial (DeConto and Pollard, 2016), although the MICI solution may not be unique (Aitken et al., 2016) (Section 4.2.3.1.2).

The Greenland Ice Sheet (GIS) has limited direct access to the ocean through relatively narrow subglacial troughs (Morlighem et al., 2017), and most of the bedrock at the ice sheet margin is above sea level (Figure 4.5). However, since AR5 it has been argued that several Greenland outlet glaciers (Petermann, Kangerdlugssuaq, Jakobshavn Isbræ, Helheim, Zachariæ Isstrøm) and North-East Greenland Ice Stream may contribute more than expected to future sea level rise (Mouginot et al., 2015). It has also been shown that Greenland was nearly ice free for extensive episodic periods during the Pleistocene, suggesting a sensitivity to deglaciation under climates similar to or slightly warmer than present (Schaefer et al., 2016).

A MICI-style behaviour is seen today in Greenland at the termini of Jakobshavn and Helheim glaciers (Parizek et al., 2019), but calving of these narrow outlets is controlled by a combination of ductile and brittle processes, which might not be representative examples of much wider Antarctic outlet glaciers, like Thwaites.

Overall, this assessment finds that unstable retreat and thinning of some Antarctic glaciers, and to a lesser extent Greenland outlet glaciers, may be underway. However, the timescale and future rate of these processes is not well known, casting deep uncertainty on projections of the sea level contributions from the AIS (Cross-Chapter Box 5 in Chapter 1, Section 4.2.3.1).

3.4 Arctic Snow, Freshwater Ice and Permafrost: Changes, Consequences and Impacts

3.4.1 Observations

3.4.1.1 Seasonal Snow Cover

Terrestrial snow cover is a defining characteristic of the Arctic land surface for up to nine months each year, with changes influencing the surface energy budget, ground thermal regime and freshwater budget. Snow cover also interacts with vegetation, influences biogeochemical activity and affects habitats and species, with consequences for ecosystem services. Arctic land areas are almost always completely snow covered in winter, so the transition seasons of autumn and spring are key when characterising variability and change.

3.4.1.1.1 Extent and duration

Dramatic reductions in Arctic (land areas north of 60°N) spring snow cover extent have occurred since satellite charting began in 1967 (Estilow et al., 2015). Declines in May and June of −3.5% (± 1.9%) and −13.4% respectively per decade (± 5.4%) between 1967 and 2018 (relative to the 1981–2010 mean) were determined from

multiple datasets based on the methodology of (Mudryk et al., 2017) (Figure 3.10) (*high confidence*). The loss of spring snow extent is reflected in shorter snow cover duration estimated from surface observations (Bulygina et al., 2011; Brown et al., 2017), satellite data (Wang et al., 2013; Estilow et al., 2015; Anttila et al., 2018), and model-based analyses (Liston and Hiemstra, 2011) (*high confidence*). These trends range between −0.7 and −3.9 days per decade depending on region and time period, but all spring snow cover duration trends from all datasets are negative (Brown et al., 2017). These same multi-source datasets also identify reductions in autumn snow extent and duration (-0.6 to -1.4 days per decade; summarized in Brown et al., 2017) (*high confidence*). There is *low confidence* in positive October and November snow cover extent trends apparent in a single dataset (Hernández-Henríquez et al., 2015) because they are not replicated in other surface, satellite and model datasets (Brown and Derksen, 2013; Mudryk et al., 2017).

3.4.1.1.2 Depth and water equivalent

Weather station observations across the Russian Arctic identify negative trends in the maximum snow depth between 1966 and 2014 (Bulygina et al., 2011; Osokin and Sosnovsky, 2014). There is *medium confidence* in this trend because the pointwise nature of these measurements does not capture prevailing conditions across the landscape. Seasonal maximum snow depth trends over the North American Arctic are mixed and largely statistically insignificant (Vincent et al., 2015; Brown et al., 2017). The timing of maximum snow depth has shifted earlier by 2.7 days per decade for the North American Arctic (Brown et al., 2017); comparable analysis is not available for Eurasia. Gridded products from remote sensing and land surface models identify negative trends in snow water equivalent between 1981 and 2016 for both the Eurasian and North American sectors of the Arctic (Brown et al., 2017). While the snow water equivalent anomaly time series show reasonable consistency between products when averaged at the continental scale, considerable inter-dataset variability in the spatial patterns of change (Liston and Hiemstra, 2011; Park et al., 2012; Brown et al., 2017) mean there is only *medium confidence* in these trends.

3.4.1.1.3   Drivers

Despite uncertainties due to sparse observations (Cowtan and Way, 2014), surface temperature has increased across Arctic land areas in recent decades (Hawkins and Sutton, 2012; Fyfe et al., 2013), driving reductions in Arctic snow extent and duration (*high confidence*). Changes in Arctic snow extent can be directly related to extratropical temperature increases (Brutel-Vuilmet et al., 2013; Thackeray et al., 2016; Mudryk et al., 2017). Based on multiple historical datasets, there is a consistent temperature sensitivity for Arctic snow extent, with approximately 800,000 km² of snow cover lost per degrees Celsius warming in spring (Brown and Derksen, 2013; Brown et al., 2017), and 700,000–800,000 km² lost in autumn (Derksen and Brown, 2012; Brown and Derksen, 2013) (*high confidence*).

There is *high confidence* that darkening of snow through the deposition of black carbon and other light absorbing particles enhances snow melt (Bullard et al., 2016; Skiles et al., 2018; Boy et al., 2019). The global direct radiative forcing for black carbon in seasonal snow and over sea ice is estimated to be 0.04 W m⁻², but the effective forcing can be up to threefold greater at regional scales due to the enhanced albedo feedback triggered by the initial darkening (Bond et al., 2013). Lawrence et al. (2011) estimate the present-day radiative effect of black carbon and dust in land-based snow to be 0.083 W m⁻², only marginally greater than the simulated 1850 effect (0.075 W m⁻²) due to offsetting effects from increased black carbon emissions and reductions in dust darkening (*medium confidence*). Kylling et al. (2018) estimate a surface radiative effect of 0.292 W m⁻² caused by dust deposition (largely transported from Asia) to Arctic snow, approximately half of the black carbon central scenario estimate of Flanner et al. (2007). The forcing from brown carbon deposited in snow (associated with both combustion and secondary organic carbon) is estimated to be 0.09–0.25 W m⁻², with the range due to assumptions of particle absorptivity (Lin et al., 2014) (*low confidence*).

Precipitation remains a sparse and highly uncertain measurement over Arctic land areas: *in situ* datasets remain uncertain (Yang, 2014) and are largely regional (Kononova, 2012; Vincent et al., 2015). Atmospheric reanalyses show increases in Arctic precipitation in recent decades (Lique et al., 2016; Vihma et al., 2016), but there remains *low confidence* in reanalysis-based closure of the Arctic freshwater budget due to a wide spread between available reanalysis derived precipitation estimates (Lindsay et al., 2014). Despite improved process understanding, estimates of sublimation loss during blowing snow events remain a key uncertainty in the mass budget of the Arctic snowpack (Sturm and Stuefer, 2013).

3.4.1.2   Permafrost

3.4.1.2.1   Temperature

Record high temperatures at ~10–20 m depth in the permafrost (near or below the depths affected by intra-annual fluctuation in temperature) have been documented at many long-term monitoring sites in the Northern Hemisphere circumpolar permafrost region (AMAP, 2017d) (Figure 3.10) (*very high confidence*). At some locations, the temperature is 2°C–3°C higher than 30 years ago. During the decade between 2007 and 2016, the rate of increase in permafrost temperatures was 0.39°C ± 0.15°C for colder continuous zone permafrost monitoring sites, 0.20°C ± 0.10°C for warmer discontinuous zone permafrost, giving a global average of 0.29 ± 0.12°C across all polar and mountain permafrost (Biskaborn et al., 2019). Relatively smaller increases in permafrost temperature in warmer sites indicate that permafrost is thawing with heat absorbed by the ice-to-water phase change, and as a result, the active layer may be increasing in thickness. In contrast to temperature, there is only *medium confidence* that active layer thickness across the region has increased. This confidence level is because decadal trends vary across regions and sites (Shiklomanov et al., 2012) and because mechanical probing of the active layer can underestimate the degradation of permafrost in some cases because the surface subsides when ground ice melts and drains (Mekonnen et al., 2016; AMAP, 2017d; Streletskiy et al., 2017). Permafrost in the Southern Hemisphere polar region occurs in ice-free exposed areas (Bockheim et al., 2013), 0.18% of the total land area of Antarctica (Burton-Johnson et al., 2016). This area is three orders of magnitude smaller than the 13–18 × 10⁶ km² area underlain by permafrost in the Northern Hemisphere terrestrial permafrost region (Gruber, 2012). Antarctic permafrost temperatures are generally colder (Noetzli et al., 2017) and increased 0.37°C ± 0.10°C between 2007 and 2016 (Biskaborn et al., 2019).

3.4.1.2.2   Ground ice

Permafrost thaw and loss of ground ice causes the land surface to subside and collapse into the volume previously occupied by ice, resulting in disturbance to overlying ecosystems and human infrastructure (Kanevskiy et al., 2013; Raynolds et al., 2014). Excess ice in permafrost is typical, varying for example from 40% of total volume in some sands up to 80–90% of total volume in fine-grained soil/sediments (Kanevskiy et al., 2013). Ice rich permafrost areas where impacts of thaw could be greatest include the Yedoma deposits in Siberia, Alaska, and the Yukon in Canada, with ice divided between massive wedges interspersed with frozen soil/sediment containing pore ice and smaller ice features (Schirrmeister et al., 2011; Strauss et al., 2017). Other areas including, for example, Northwestern Canada, the Canadian Archipelago, the Yamal and Gydan peninsulas of West Siberia, and smaller portions of Eastern Siberia and Alaska contain buried glacial ice bodies of significant thickness and extent (Lantuit

3

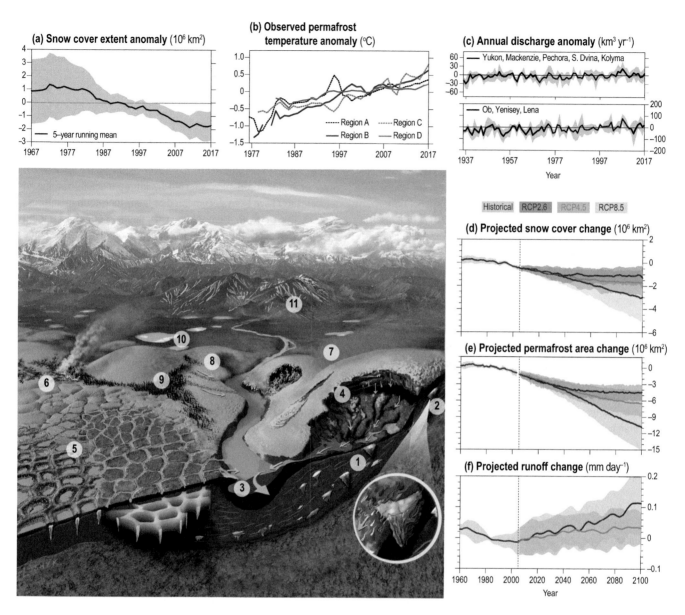

Figure 3.10 | Schematic of important land surface components influenced by the Arctic terrestrial cryosphere: permafrost (1); ground ice (2); river discharge (3); abrupt thaw (4); surface water (5); fire (6); tundra (7); shrubs (8); boreal forest (9); lake ice (10); seasonal snow (11). Time series of snow cover extent anomalies in June (relative to 1981–2010 climatology) from 5 products based on the approach of Mudryk et al. (2017) **(a)**; permafrost temperature change normalised to a baseline period (Romanovsky et al., 2017), Region A: Continuous to discontinuous permafrost in Scandanavia, Svalbard, and Russia/Siberia, Region B: Cold continuous permafrost in northern Alaska, Northwest Territories, and NE Siberia, Region C: Cold continuous permafrost in Eastern and High Arctic Canada, Region D: Discontinuous permafrost in Interior Alaska and Northwest Canada **(b)**, and runoff from northern flowing watersheds normalised to a baseline period (1981–2010) (Holmes et al., 2018), multi-station average (± 1 standard deviation) **(c)**. Coupled Model Intercomparison Project Phase 5 (CMIP5) multi-model average (± 1 standard deviation) projections for different Representative Concentration Pathway (RCP) scenarios for June snow cover extent change (based on Thackeray et al., 2016) **(d)**, area change of near-surface permafrost **(e)**, and runoff change to the Arctic Ocean (based on McGuire et al., 2018) **(f)**.

and Pollard, 2008; Leibman et al., 2011; Kokelj et al., 2017; Coulombe et al., 2019). The location and volume of ground ice integrated across the northern permafrost region (5.63–36.55 × 10³ km³, equivalent to 2–10 cm sea level rise) is known with *medium confidence* and with no recent updates at the circumpolar scale (Zhang et al., 2008).

3.4.1.2.3   Carbon

The permafrost region represents a large, climate sensitive reservoir of organic carbon with the potential for some of this pool to be rapidly decayed and transferred to the atmosphere as CO_2 and methane as

permafrost thaws in a warming climate, thus accelerating the pace of climate change (Schuur et al., 2015). The current best mean estimate of total (surface plus deep) organic soil carbon (terrestrial) in the northern circumpolar permafrost region (17.8 × 10⁶ km² area) is 1460 to 1600 petagrams (*medium confidence*) (Pg; 1 Pg = 1 billion metric tonnes) (Schuur et al., 2018). All permafrost region soils estimated to 3 m in depth (surface) contain 1035 ± 150 Pg C (Tarnocai et al., 2009; Hugelius et al., 2014) (*high confidence*). Of the carbon in the surface, 800–1000 Pg C is perennially frozen, with the remainder contained in seasonally-thawed soils. The northern circumpolar permafrost region occupies only 15% of the total global soil area, but the 1035 Pg C adds another 50% to the rest of the 3 m soil carbon inventory (2050 Pg C for all global biomes excluding tundra and boreal; Jobbágy and Jackson, 2000; Schuur et al., 2015).

Substantial permafrost carbon exists below 3 m depth (*medium confidence*). Deep carbon (>3 m) has been best quantified for the Yedoma region of Siberia and Alaska, characterised by wind- and water-moved permafrost sediments tens of meters thick. The Yedoma region covers a 1.4×10^6 km^2 area that remained ice-free during the last Ice Age (Strauss et al., 2013) and accounts for 327–466 Pg C in deep sediment accumulations below 3 m (Strauss et al., 2017).

The current inventory has also highlighted additional carbon pools that are likely to be present but are so poorly quantified (*low confidence*) that they cannot yet be added into the number reported above. There are deep terrestrial soil/sediment deposits outside of the Yedoma region that may contain about 400 Pg C (Schuur et al., 2015). An additional pool is organic carbon remaining in permafrost but that is now submerged on shallow Arctic sea shelves that were formerly exposed as terrestrial ecosystems during the Last Glacial Maximum ~20,000 years ago (Walter et al., 2007). This permafrost is degrading slowly due to seawater intrusion, and it is not clear what amounts of permafrost and organic carbon still remain in the sediment versus what has already been converted to greenhouse gases. A recent synthesis of permafrost extent for the Beaufort Sea shelf showed that most remaining subsea permafrost in that region exists near shore with much reduced area (*high confidence*) as compared to original subsea permafrost maps that outlined the entire 3×10^6 km^2 shelf area (<120 m below sea level depth) that was formerly exposed as land (Ruppel et al., 2016). These observations are supported by similar studies in the Siberian Arctic Seas (Portnov et al., 2013), and by modelling that suggests that subsea permafrost would be thawed many meters below the seabed under current submerged conditions (Anisimov et al., 2012; AMAP, 2017d; Angelopoulos et al., 2019).

3.4.1.2.4 Drivers

Changes in temperature and precipitation act as gradual 'press' (i.e., continuous) disturbances that directly affect permafrost by modifying the ground thermal regime, as discussed in Section 3.4.1.2.1. Climate change can also modify the occurrence and magnitude of abrupt physical disturbances such as fire, and soil subsidence and erosion resulting from ice rich permafrost thaw (thermokarst). These 'pulse' (i.e., discrete) disturbances (Smith et al., 2009) often are part of the ongoing disturbance and successional cycle in Arctic and boreal ecosystems (Grosse et al., 2011), but changing rates of occurrence alter the landscape distribution of successional ecosystem states, with permafrost characteristics defined by the ecosystem and climate state (Kanevskiy et al., 2013).

Pulse disturbances often rapidly remove the insulating soil organic layer, leading to permafrost degradation (Gibson et al., 2018). Of all pulse disturbance types, wildfire affects the most high-latitude land area annually at the continental scale. In some well-studied regions, there is *high confidence* that area burned, fire frequency and extreme fire years are higher now than the first half of the last century, or even the last 10,000 years (Kasischke and Turetsky, 2006; Flannigan

et al., 2009; Kelly et al., 2013; Hanes et al., 2019). Recent climate warming has been linked to increased wildfire activity in the boreal forest regions in Alaska and western Canada where this has been studied (Gillett, 2004; Veraverbeke et al., 2017). Based on satellite imagery, an estimated 80,000 km^2 of boreal area was burned globally per year from 1997 to 2011 (van der Werf et al., 2010; Giglio et al., 2013). Extreme fire years in northwest Canada during 2014 and Alaska during 2015 doubled the long-term (1997–2011) average area burned annually in this region (Canadian Forest Service, 2017), surpassing Eurasia to contribute 60% of the global boreal area burned (van der Werf et al., 2010; Randerson et al., 2012; Giglio et al., 2013). These extreme North American fire years were balanced by lower-than-average area burned in Eurasian forests, resulting in a 5% overall increase in global boreal area burned. The annual area burned in Arctic tundra is generally small compared to the forested boreal biome. In Alaska – the only region where estimates of burned area exist for both boreal forest and tundra vegetation types – tundra burning averaged approximately 270 km^2 yr^{-1} during the last half century (French et al., 2015), accounting for 7% of the average annual area burned throughout the state (Pastick et al., 2017). There is *high confidence* that changes in the fire regime are degrading permafrost faster than had occurred over the historic successional cycle (Turetsky et al., 2011; Rupp et al., 2016; Pastick et al., 2017), and that the effect of this driver of permafrost change is under-represented in the permafrost temperature observation network.

Abrupt permafrost thaw occurs when changing environmental and ecological conditions interact with geomorphological processes. Melting ground ice causes the ground surface to subside. Pooling or flowing water causes localised permafrost thaw and sometimes mass erosion. Together, these localised feedbacks can thaw through meters of permafrost within a short time, much more rapidly than would be caused by increasing air temperature alone. This process is a pulse disturbance to permafrost that can occur in response to climate, such as an extreme precipitation event (Balser et al., 2014; Kokelj et al., 2015), or coupled with other disturbances such as wildfire that affects the ground thermal regime (Jones et al., 2015a). There is *medium confidence* in the importance of abrupt thaw for driving change in permafrost at the circumpolar scale because it occurs at point locations rather than continuously across the landscape, but the risk for widespread change from this mechanism remains high because of the rapidity of change in these locations (Kokelj et al., 2017; Nitze et al., 2018). New research at the global scale has revealed that 3.6×10^6 km^2, about 20% of the northern permafrost region, appears to be vulnerable to abrupt thaw (Olefeldt et al., 2016).

3.4.1.3 Freshwater Systems

There is increasing awareness of the influence of a changing climate on freshwater systems across the Arctic, and associated impacts on hydrological, biogeophysical and ecological processes (Prowse et al., 2015; Walvoord and Kurylyk, 2016), and northern populations (Takakura, 2018) (Section 3.4.3.3.1). Assessing these impacts requires consideration of complex interconnected processes, many of which are incompletely observed. The increasing imprint of human development, such as flow regulation on major northerly flowing rivers adds complexity to the determination of climate-driven changes.

3.4.1.3.1 Freshwater ice

Long-term *in situ* river ice records indicate that the duration of ice cover in Russian Arctic rivers decreased by 7–20 days between 1955 and 2012 (Shiklomanov and Lammers, 2014) (*high confidence*). This is consistent with historical reductions in Arctic river ice cover derived from models (Park et al., 2015) and regional analysis of satellite data (Cooley and Pavelsky, 2016).

Analysis of satellite imagery between 2000 and 2013 identified a significant trend of earlier spring ice break-up across all regions of the Arctic (Šmejkalová et al., 2016); independent satellite data showed approximately 80% of Arctic lakes experienced declines in ice cover duration during 2002–2015, due to both a later freeze-up and earlier break-up (Du et al., 2017) (*high confidence*). There are indications that lake ice across Alaska has thinned in recent decades (Alexeev et al., 2016), but ice thickness trends are not available at the pan-Arctic scale. Analysis of satellite data over northern Alaska show that approximately one-third of bedfast lakes (the entire water volume freezes by the end of winter) experienced a regime change to floating ice over the 1992–2011 period (Surdu et al., 2014; Arp et al., 2015). This can result in degradation of underlying permafrost (Arp et al., 2016; Bartsch et al., 2017). Lakes of the central and eastern Canadian High Arctic are transitioning from a perennial to seasonal ice regime (Surdu et al., 2016).

3.4.1.3.2 Runoff and surface water

A general trend of increasing discharge has been observed for large Siberian (Troy et al., 2012; Walvoord and Kurylyk, 2016) and Canadian (Ge et al., 2013; Déry et al., 2016) rivers that drain to the Arctic Ocean (*medium confidence*). Between 1976 and 2017, trends are 3.3 ± 1.6% for Eurasian rivers and 2.0 ± 1.8% for North American rivers (Holmes et al., 2018) (Figure 3.10). Extreme regional runoff events have also been identified (Stuefer et al., 2017). An observed increase in baseflow in the North American (Walvoord and Striegl, 2007; St. Jacques and Sauchyn, 2009) and Eurasian Arctic (Smith et al., 2007; Duan et al., 2017) over the last several decades is attributable to permafrost thaw and concomitant enhancement in groundwater discharge. The timing of spring season peak flow is generally earlier (Ge et al., 2013; Holmes et al., 2015). There is consistent evidence of decreasing summer season discharge for the Yenisei, Lena, and Ob watersheds in Siberia (Ye et al., 2003; Yang et al., 2004a; Yang et al., 2004b) and the majority of northern Canadian rivers (Déry et al., 2016). Long-term records indicate water temperature increases (Webb et al., 2008; Yang and Peterson, 2017); attribution to rising air temperatures is complicated by the influence of reservoir regulation over Siberian regions (Liu et al., 2005; Lammers et al., 2007). Increases in discharge and water temperature in the spring season represent notable freshwater and heat fluxes to the Arctic Ocean (Yang et al., 2014).

A large proportion of low-lying Arctic land areas are covered by lakes because permafrost limits surface water drainage and supports ponding even across areas with high moisture deficits (Grosse

et al., 2013). While thaw in continuous permafrost is linked to intensified thermokarst activity and subsequent ponding (resulting in lake/wetland expansion), observations of change in surface water coverage across the Arctic are regionally variable (Nitze et al., 2017; Ulrich et al., 2017; Pastick et al., 2019). In landscapes with degrading ice-wedge polygons, subsidence can reduce inundation, increase runoff, and decrease surface water (Liljedahl et al., 2016; Perreault et al., 2017). In discontinuous permafrost, thaw opens up pathways of subsurface flow, improving the connection among inland water systems which supports the drainage of lakes and overall reduction in surface water cover (Jepsen et al., 2013). Enhanced subsurface connectivity from thaw in discontinuous permafrost serves tempers short-term lake fluctuations (Rey et al., 2019).

3.4.1.3.3 Drivers

There is *high confidence* that environmental drivers of Arctic surface water change are diverse and depend on local and regional factors such as permafrost properties and geomorphology (Nitze et al., 2018). Thermokarst lake expansion has been observed in the continuous permafrost of northern Siberia (Smith et al., 2005; Polishchuk et al., 2015) and Alaska (Jones et al., 2011); surface water area reduction has been observed in discontinuous permafrost of central and southern Siberia (Smith et al., 2005; Sharonov et al., 2012), western Canada (Labrecque et al., 2009; Carroll et al., 2011; Lantz and Turner, 2015) and interior Alaska (Chen et al., 2012; Rover et al., 2012). Increased evaporation from warmer/longer summers, decreased recharge due to reductions in snow melt volume, and dynamic processes such as ice-jam flooding (Chen et al., 2012; Bouchard et al., 2013; Jepsen et al., 2015) are important considerations for understanding observed surface water area change across the Arctic.

Satellite and model-derived estimates of evapotranspiration show increases across the Arctic (Rawlins et al., 2010; Liu et al., 2014; Liu et al., 2015b; Fujiwara et al., 2016; Suzuki et al., 2018) (*medium confidence*). Increases in the seasonal active layer thickness impact temporary water storage and thus runoff regimes in drainage basins. Formation of taliks underneath lakes and rivers may result in reconnection of surface with sub-permafrost ground water aquifers with varying hydrological consequences depending on local geological and hydraulic settings (Wellman et al., 2013).

3.4.2   Projections

3.4.2.1   Seasonal Snow

Historical simulations from CMIP5 models tend to underestimate observed reductions in spring snow cover extent due to uncertainty in the parameterisation of snow processes (Essery, 2013; Thackeray et al., 2014), challenges in simulating snow-albedo feedback (Qu and Hall, 2014; Fletcher et al., 2015; Li et al., 2016b), unrealistic temperature sensitivity (Brutel-Vuilmet et al., 2013; Mudryk et al., 2017), and biases in climatological spring snow cover (Thackeray et al., 2016). The role of precipitation biases is not well understood (Thackeray et al., 2016).

Reductions in Arctic snow cover duration are projected by the CMIP5 multi-model ensemble due to later snow onset in the autumn and earlier snow melt in spring (Brown et al., 2017) driven by increased surface temperature over essentially all Arctic land areas (Hartmann et al., 2013). There is *high confidence* that projected snow cover declines are proportional to the amount of future warming in each model realisation (Thackeray et al., 2016; Mudryk et al., 2017). Projections to mid-century are primarily dependent on natural variability and model dependent uncertainties rather than the choice of forcing scenario (Hodson et al., 2013). By end of century, however, differences between scenarios emerge. Under RCP2.6 and RCP4.5, Arctic snow cover duration stabilises at 5–10% reduction (compared to a 1986–2005 reference period); under RCP8.5, snow cover duration declines reach –15 to –25% (Brown et al., 2017) (Figure 3.10) (*high confidence*).

Positive Arctic snow water equivalent changes emerge across the eastern Eurasian Arctic by mid-century for both RCP4.5 and RCP8.5 (Brown et al., 2017) (*medium confidence*). Projected snow water equivalent increases across the North American Arctic are only modest, emerge later in the century, and only under RCP8.5 (Brown et al., 2017). These projected increases are due to enhanced snowfall (Krasting et al., 2013) from a more moisture-rich Arctic atmosphere coupled with winter season temperatures that remain sufficiently low for precipitation to fall as snow. There is *low confidence* in changes to snow properties such as density and stratigraphy (relevant for understanding the impacts of changes to Arctic snow on ecosystems) which are not resolved directly by climate model simulations, but require detailed snow physics models.

3.4.2.2   Permafrost

Circumpolar- or global-scale models represent permafrost degradation in response to warming scenarios as increases in thaw depth only. The CMIP5 models project with *high confidence* that thaw depth will increase and areal extent of near-surface permafrost will decrease substantially (Koven et al., 2013; Slater and Lawrence, 2013) (Figure 3.10). However, there is only *medium confidence* in the magnitude of these changes due to at least a five-fold range of estimated present day near-surface permafrost area ($<5 - >25 \times 10^6$ km²) by these models. This was caused by a wide range of model sensitivity in permafrost area to air temperature change, resulting in a large range of projected near-surface permafrost loss by 2100: 2–66% for RCP2.6 (24 ± 16%; *likely* range), 15–87% under RCP4.5 and 30–99% (69 ± 20%; *likely* range) under RCP8.5. A more recent analysis of near-surface permafrost trends from a subset of models that self-identified as structurally representing the permafrost region had a significantly smaller range of estimated present day near-surface permafrost area (13.1–19.3 × 10⁶ km²; mean ± SD, 14.1 ± 3.5 × 10⁶ km²) (McGuire et al., 2018). This subset of models also showed large reductions of near-surface permafrost area, averaging a 90% loss (12.7 ± 5.1 × 10⁶ km²) of permafrost area by 2300 for RCP8.5 and 29% loss (4.1 ± 0.6 × 10⁶ km²) for RCP4.5, with much of that long-term loss already occurring by 2100.

Pulse disturbances are not included in the permafrost projections described above, and there is *high confidence* that fire and abrupt thaw will accelerate change in permafrost relative to climate effects alone, if the rates of these disturbances increase. The observed trend of increasing fire is projected to continue for the rest of the century across most of the tundra and boreal region for many climate scenarios, with the boreal region projected to have the greatest increase in total area burned (Balshi et al., 2009; Kloster et al., 2012; Wotton et al., 2017). Due to vegetation-climate interactions, there is only *medium confidence* in projections of future area burned. As fire activity increases, flammable vegetation, such as the black spruce forest that dominates boreal Alaska, is projected to decline as it is replaced by low-flammability deciduous forest (Johnstone et al., 2011; Pastick et al., 2017). In other regions such as western Canada, by contrast, black spruce could be replaced by the even more flammable jack pine, creating regional-scale feedbacks that increase the spread of fire on the landscape (Héon et al., 2014). A regional process-model study of Alaska projected annual median area burned during the 21st century to be 1.3–1.7 times higher compared with the historical average (Pastick et al., 2017). Fire also appears to be expanding as a novel disturbance into tundra and forest-tundra boundary regions previously protected by a cool, moist climate (Jones et al., 2009; Hu et al., 2010; Hu et al., 2015) (*medium confidence*). Annual tundra area burned in Alaska is projected to double under RCP6.0 from a historic rate of 270 km² yr⁻¹ to 500–610 km² yr⁻¹ over the 21st century (Hu et al., 2015). A statistical approach projected a fourfold increase in the 30-year probability of fire occurrence in the forest-tundra boundary by 2100 (Young et al., 2017). In contrast to fire, there has not yet been a comprehensive circumpolar projection of how abrupt thaw rates may change in the future, but one component of abrupt thaw, change in abrupt thaw lake area, has been projected to increase to increase by 53% under RCP8.5 (Walter Anthony et al., 2018) above the 1.4×10^6 km² of small lakes and ponds that currently exist in the permafrost region (Muster et al., 2017). As a result, there is *low confidence* in the ability to assess the magnitude by which abrupt thaw across the entire landscape will affect regional permafrost, even though this mechanism for rapid change appears critically important for projecting future change (Kokelj et al., 2017).

3.4.2.3   Freshwater Systems

Climate model simulations project a warmer and wetter Arctic (Krasting et al., 2013), with increased specific humidity due to enhanced evaporation (Laîné et al., 2014), and moisture flux convergence increases into the Arctic (Skific and Francis, 2013). Increased cold-season precipitation is projected across the Arctic by CMIP5 models (Lique et al., 2016) due to increased moisture flux convergence from outside the Arctic (Zhang et al., 2012) and enhanced moisture availability from reduced sea ice cover (Bintanja and Selten, 2014) (*high confidence*). Increases in precipitation extremes are also projected over northern watersheds (Kharin et al., 2013; Sillmann et al., 2013), while rain on snow events are expected to increase (Hansen et al., 2014). A net increased ratio of precipitation minus evaporation is projected, resulting in increased freshwater flux from the land surface to the Arctic Ocean, projected to be 30% above current values by 2100 under RCP4.5 (Haine et al., 2015) (Figure 3.10). This is consistent with CMIP5 model projections of increased discharge from Arctic watersheds (van Vliet et al., 2013; Gelfan et al., 2016; MacDonald et al., 2018). The water temperature of this increased discharge is projected to be approximately 1°C warmer than current conditions, increasing the heat flux to Arctic Ocean (van Vliet et al., 2013).

Lake ice phenology is sensitive to projected changes in surface temperature (Sharma et al., 2019). Lake ice models project an earlier spring break-up of between 10–25 days by mid-century (compared with 1961–1990), and up to a 15-day delay in the freeze-up for lakes in the North American Arctic, with more extreme reductions for coastal regions (Brown and Duguay, 2011; Dibike et al., 2011; Prowse et al., 2011) (*medium confidence*). Mean maximum ice thickness is projected to decrease by 10–50 cm over the same period (Brown and Duguay, 2011). High-latitude warming is projected to drive earlier river ice break-up in spring due to both decreasing ice strength, and earlier onset of peak discharge (Cooley and Pavelsky, 2016). Complex interplay between hydrology and hydraulics in controlling spring flooding and ice jam events complicate projections of these events (Prowse et al., 2010; Prowse et al., 2011).

3.4.3   Consequences and Impacts

3.4.3.1   Global Climate Feedbacks

3.4.3.1.1   Carbon cycle

Climate warming is expected to change the storage of carbon in vegetation and soils in northern regions, and net carbon transferred to the atmosphere as CO_2 and methane acts as a feedback to accelerate global climate change. There is *high confidence* that the northern region acted as a net carbon sink as carbon accumulated in terrestrial ecosystems over the Holocene (Loisel et al., 2014; Lindgren et al., 2018). There is *medium evidence* with *low agreement* whether changing climate in the modern period has shifted these ecosystems into net carbon sources. Syntheses of ecosystem CO_2 fluxes have alternately showed tundra ecosystems as carbon sinks or neutral averaged across the circumpolar region for the 1990s and 2000s (McGuire et al., 2012), or carbon sources over the same time period (Belshe et al., 2013). Both syntheses agree that the summer growing season is a period of net carbon uptake into terrestrial ecosystems (*high confidence*), and this uptake appears to be increasing as a function of vegetation density/biomass (Ueyama et al., 2013). The discrepancy between these syntheses may be a result of CO_2 release rates during the non-summer season that are now thought to be higher than previously estimated (*high confidence*) (Webb et al., 2016) or the separation of upland and wetland ecosystem types, which was done in one synthesis but not the other. Moisture status is a primary control over ecosystem carbon sink/source strength with wetlands more often than not still acting as annual net carbon sinks even while methane is emitted (Lund et al., 2010). Recent aircraft measurements of atmospheric CO_2 concentrations over Alaska showed that tundra regions of Alaska were a consistent net CO_2 source to the atmosphere, whereas boreal forest regions were either neutral or net CO_2 sinks for the period 2012–2014 (Commane et al., 2017). That study region as a whole was estimated to be a net carbon source of

25 ± 14 Tg CO_2-C yr^{-1} averaged over the land area of both biomes for the entire study period. For comparison to projected global emissions, this would be equivalent to a net source of 0.3 Pg CO_2-C yr^{-1} assuming the Alaska study region (1.6×10^6 km^2) could be scaled to the entire northern circumpolar permafrost region soil area (17.8×10^6 km^2).

The permafrost soil carbon pool is climate sensitive and an order of magnitude larger than carbon stored in plant biomass (Schuur et al., 2018) (*very high confidence*). Initial estimates were converging on a range of cumulative emissions from soils to the atmosphere by 2100, but recent studies have actually widened that range somewhat (Figure 3.11) (*medium confidence*). Expert assessment and laboratory soil incubation studies suggest that substantial quantities of C (tens to hundreds Pg C) could potentially be transferred from the permafrost carbon pool into the atmosphere under RCP8.5 (Schuur et al., 2013; Schädel et al., 2014)[4]. Global dynamical models supported these findings, showing potential carbon release from the permafrost zone ranging from 37–174 Pg C by 2100 under high emission climate warming trajectories, with an average across models of 92 ± 17 Pg C (mean $\pm$ SE) (Zhuang et al., 2006; Koven et al., 2011; Schaefer et al., 2011; MacDougall et al., 2012; Burke et al., 2013; Schaphoff et al., 2013; Schneider von Deimling et al., 2015). This range is generally consistent with several newer data-driven modelling approaches that estimated that soil carbon releases by 2100 (for RCP8.5) will be 57 Pg C (Koven et al., 2015) and 87 Pg C (Schneider von Deimling et al., 2015), as well as an updated estimate of 102 Pg C from one of the previous models (MacDougall and Knutti, 2016). However, the latest model runs performed with either structural enhancements to better represent permafrost carbon dynamics (Burke et al., 2017a), or common environmental input data (McGuire et al., 2016) show similar soil carbon losses, but also indicate the potential for stimulated plant growth (nutrients, temperature/growing season length, CO_2 fertilisation) to offset some (Kleinen and Brovkin, 2018) or all of these losses, at least during this century, by sequestering new carbon into plant biomass and increasing carbon inputs into the surface soil (McGuire et al., 2018). These future carbon emission levels would be a significant fraction of those projected from fossil fuels with implications for allowable carbon budgets that are consistent with limiting global warming, but will also depend on how vegetation responds (*high confidence*). Furthermore, there is *high confidence* that climate scenarios that involve mitigation (e.g., RCP4.5) will help to dampen the response of carbon emissions from the Arctic and boreal regions.

Northern ecosystems contribute significantly to the global methane budget, but there is *low confidence* about the degree to which additional methane from northern lakes, ponds, wetland ecosystems, and the shallow Arctic Ocean shelves is currently contributing to increasing atmospheric concentrations. Analyses of atmospheric concentrations in Alaska concluded that local ecosystems surrounding the observation site have not changed in the exchange of methane from the 1980s until the present, which suggests that either the local wetland ecosystems are responding similarly to other northern wetland ecosystems, or that increasing atmospheric methane concentrations in northern observation sites is derived from methane coming from

---

[4]   For context, total annual anthropogenic CO_2 emissions were 10.8 ± 0.8 GtC yr^{-1} (39.6 ± 2.9 GtCO$_2$ yr^{-1}) on average over the period 2008–2017. Total annual anthropogenic methane emissions were 0.35 ± 0.01 GtCH$_4$ yr^{-1}, on average over the period 2003–2012 (Saunois et al., 2016; Le Quéré et al., 2018).

mid-latitudes (Sweeney et al., 2016). However, this contrasts with indirect integrated estimates of methane emissions from observations of expanding permafrost thaw lakes that suggest a release of an additional 1.6–5 Tg CH_4 yr^{-1} over the last 60 years (Walter Anthony et al., 2014). At the same time, there is *high confidence* that methane fluxes at the ecosystem to regional scale have been under-observed, in part due to the low solubility of methane in water leading to ebullition (bubbling) flux to the atmosphere that is heterogeneous in time and space. Some new quantifications include: cold-season methane emissions that can be >50% of the annual budget of terrestrial ecosystems (Zona et al., 2016); geological methane seeps that may be climate sensitive if permafrost currently serves as a cap preventing atmospheric release (Walter Anthony et al., 2012; Ruppel and Kessler, 2016; Kohnert et al., 2017); estimates of shallow Arctic Ocean shelf methane emissions where the range of estimates based on methane concentrations in air and water has widened with more observations and now ranges from 3 Tg CH_4 yr^{-1} (Thornton et al., 2016) to 17 Tg CH_4 yr^{-1} (Shakhova et al., 2013). Observations such as these underlie the fact that source estimates for methane made from atmospheric observations are typically lower than methane source estimates made from upscaling of ground observations (e.g., Berchet et al., 2016), and this problem has not improved, even at the global scale, over several decades of research (Saunois et al., 2016; Crill and Thornton, 2017).

In many of the dynamical model projections previously discussed, methane release is not explicitly represented because fluxes are small even though higher global warming potential of methane makes these emissions relatively more important than on a mass basis alone. Global models that do include methane show that emissions may already (from 2000 to 2012) be increasing at a rate of 1.2 Tg CH_4 yr^{-1} in the northern region as a direct response to temperature (Riley et al., 2011; Gao et al., 2013; Poulter et al., 2017). A model intercomparison study forecast northern methane emissions to increase from 18 Tg CH_4 yr^{-1} to 42 Tg CH_4 yr^{-1} under RCP8.5 by 2100 largely as a result of an increase in wetland extent (Zhang et al., 2017). However, projected methane emissions are sensitive to changes in surface hydrology (Lawrence et al., 2015) and a suite of models that were thought to perform well in high-latitude ecosystems showed a general soil drying trend even as the overall water cycle intensified (McGuire et al., 2018). Furthermore, most models described above do not include many of the abrupt thaw processes that can result in lake expansion, wetland formation, and massive erosion and exposure to decomposition of previously frozen carbon-rich permafrost, leading to *medium confidence* in future model projections of methane. Recent studies that addressed some of these landscape controls over future emissions projected increases in methane above the current levels on the order 10–60 Tg CH_4 yr^{-1} under RCP8.5 by 2100 (Schuur et al., 2013; Koven et al., 2015; Lawrence et al., 2015; Schneider von Deimling et al., 2015; Walter Anthony et al., 2018). These additional methane fluxes are projected to cause 40–70% of total permafrost-affected radiative forcing in this century even though methane emissions are much less than CO_2 by mass (Schneider von Deimling et al., 2015; Walter Anthony et al., 2018). As with total carbon emissions, there is *high confidence* that mitigation of anthropogenic methane sources could help to dampen the impact of increased methane emissions from the Arctic and boreal regions (Christensen et al., 2019).

3.4.3.1.2 Energy budget

Warming induced reductions in the duration and extent of Arctic spring snow cover (Section 3.4.1.1) lower albedo because snow-free land reflects much less solar radiation than snow. The corresponding increase in net radiation absorption at the surface provides a positive feedback to global temperatures (Flanner et al., 2011; Qu and Hall, 2014; Thackeray and Fletcher, 2016) (*high confidence*). Estimates of increases in global net solar energy flux due to snow cover loss range from 0.10–0.22 W m^{-2} (± 50%; *medium confidence*) depending on dataset and time period (Flanner et al., 2011; Chen et al., 2015; Singh et al., 2015; Chen et al., 2016b). Sources of uncertainty include the range in observed spring snow cover extent trends (Hori et al., 2017) and the influence of clouds on shortwave feedbacks (Sedlar, 2018; Sledd and L'Ecuyer, 2019). Terrestrial snow changes also affect the longwave energy budget via altered surface emissivity (Huang et al., 2018). Climate model simulations show that changes in snow cover dominate land surface related positive feedbacks to atmospheric heating (Euskirchen et al., 2016), but regional variations in surface albedo are also influenced by vegetation (Loranty et al., 2014). There is evidence for positive sensitivity of surface temperatures to increased northern hemisphere boreal and tundra leaf area index, which contributes a positive feedback to warming (Forzieri et al., 2017).

3.4.3.2 Ecosystems and their Services

3.4.3.2.1 Vegetation

Changes in tundra vegetation can have important ecosystem effects, in particular on hydrology, carbon and nutrient cycling and surface energy balance, which together impact permafrost (e.g., Myers-Smith and Hik, 2013; Frost and Epstein, 2014; Nauta et al., 2014). Aside from physical impacts, changing vegetation influences the diversity and abundance of herbivores (e.g., Fauchald et al., 2017b; Horstkotte et al., 2017) in the Arctic. The overall trend for tundra vegetation across the 36–year satellite period (1982–2017) shows increasing above ground biomass (greening) throughout a majority of the circumpolar Arctic (*high confidence*) (Xu et al., 2013a; Ju and Masek, 2016; Bhatt et al., 2017). Increasing greenness has been in some cases linked with shifts in plant species dominance away from graminoids (grasses and sedges) towards shrubs (*high confidence*) (Myers-Smith et al., 2015). Within the overall trend of increases (greening), some tundra areas show declines (browning) (Bhatt et al., 2017).

The spatial variation in greening and browning trends in tundra are also not consistent over time (decadal scale) and can vary across landform/ecosystem types (Lara et al., 2018), suggesting interactions between the changing environment and the biological components of the system that control these trends. There is *high confidence* that increases in summer, spring and winter temperatures lead to tundra greening, as well as increases in growing season length (e.g., Vickers et al., 2016; Myers-Smith and Hik, 2018) that are in part linked to reductions in Arctic Ocean sea ice cover (Bhatt et al., 2017; Macias-Fauria et al., 2017). Other factors that stimulate tundra greening include increases in snow water equivalent and soil moisture (Westergaard-Nielsen et al., 2017), increases in active layer thickness (via nutrient availability or changes in moisture), changes

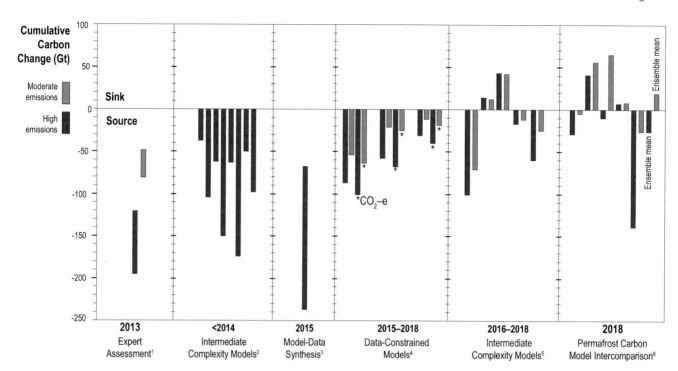

Figure 3.11 | Estimates of cumulative net soil carbon pool change for the northern circumpolar permafrost region by 2100 following medium and high emission scenarios (e.g., Representative Concentration Pathway (RCP)4.5 and RCP8.5 or equivalent). Cumulative carbon amounts are shown in Gigatons C (1 Gt C=1 billion metric tonnes), with source (negative values) indicating net carbon movement from soil to the atmosphere and sink (positive values) indicating the reverse. Some data-constrained models differentiated CO_2 and CH_4; bars show total carbon by weight, paired bars with * indicate CO_2-equivalent, which takes into account the global warming potential of CH_4. Ensemble mean bars refer to the model average for the Permafrost Carbon Model Intercomparison Project [5 models]. Bars that do not start at zero are in part informed by expert assessment and are shown as ranges; all other bars represent model mean estimates. Data are from [1] (Schuur et al., 2013); [2] (Schaefer et al., 2014) [8 models]; [3] (Schuur et al., 2015); [4] (Koven et al., 2015; Schneider von Deimling et al., 2015; Walter Anthony et al., 2018); [5] (MacDougall and Knutti, 2016; Burke et al., 2017a; Kleinen and Brovkin, 2018); [6] (McGuire et al., 2018).

in herbivore activity, and to a lesser degree, human use of the land (e.g., Salmon et al., 2016; Horstkotte et al., 2017; Martin et al., 2017; Yu et al., 2017). Research on tundra browning is more limited but suggests causal mechanisms that include changes in winter climate – specifically reductions in snow cover due to winter warming events that expose tundra to subsequent freezing and desiccation – , insect and pathogen outbreaks, increased herbivore grazing and ground ice melting and subsidence that increases surface water (Phoenix and Bjerke, 2016; Bjerke et al., 2017) (*medium confidence*).

Projections of tundra vegetation distribution across the Arctic by 2050 in response to changing environmental conditions suggest that the areal extent of most tundra types will decrease by at least 50% (Pearson et al., 2013) (*medium confidence*). Woody shrubs and trees are projected to expand to cover 24–52% of the current tundra region by 2050, or 12–33% if tree dispersal is restricted. Adding to this, the expansion of fire into tundra that has not experienced large-scale disturbance for centuries causes large reductions in soil carbon stocks (Mack et al., 2011), shifts in vegetation composition and productivity (Bret-Harte et al., 2013), and can lead to widespread permafrost degradation (Jones et al., 2015a) at faster rates than would occur by changing environmental conditions alone. In tundra regions, graminoid (grasses and sedges) tundra is projected to be replaced by more flammable shrub tundra in future climate scenarios, and tree migration into tundra could further increase fuel loading (Pastick et al., 2017) (*medium confidence*).

Similar to tundra, boreal forest vegetation shows trends of both greening and browning over multiple years in different regions across the satellite record (Beck and Goetz, 2011; Ju and Masek, 2016) (*high confidence*). Here, patterns of changing vegetation are a result of direct responses to changes in climate (temperature, precipitation and seasonality) and other driving factors for vegetation (nutrients, disturbance) similar to what has been reported in tundra. While boreal forest may expand at the northern edge (Pearson et al., 2013), climate projections suggest that it could diminish at the southern edge and be replaced by lower biomass woodland/shrublands (Koven, 2013; Gauthier et al., 2015) (*medium confidence*). Furthermore, changes in fire disturbance are leading to shifts in landscape distribution of early and late successional ecosystem types, which is also a major factor in satellite trends. Fires that burn deeply into the organic soil layer can alter permafrost stability, hydrology and vegetation. Loss of the soil organic layer exposes mineral soil seedbeds (Johnstone et al., 2009), leading to recruitment of deciduous tree and shrub species that do not establish on organic soil (Johnstone et al., 2010). This recruitment has been shown to shift post-fire vegetation to alternate successional trajectories (Johnstone et al., 2010). Model projections suggest that Alaskan boreal forest soon may cross a point where recent increases in fire activity have made deciduous stands as abundant as spruce stands on the landscape (Mann et al., 2012) (*medium confidence*). This projected trend of increasing deciduous forest at the expense of evergreen forest is mirrored in Russian and Chinese boreal forests as well (Shakhova et al., 2013; Shuman et al., 2015; Wu et al., 2017) (*medium confidence*).

3.4.3.2.2 Wildlife

Reindeer and caribou (*Rangifer tarandus*), through their numbers and ecological role as a large-bodied herbivores, are a key driver of Arctic ecology. The seasonal migrations that characterise *Rangifer* link the coastal tundra to the continental boreal forests for some herds, while others live year-round on the tundra. Population estimates and trends exist for most herds, and indicate that pan-Arctic migratory tundra *Rangifer* have declined from about 5 million in the 1990s to about 2 million in 2017 (Gunn, 2016; Fauchald et al., 2017a) (*high confidence*). Numbers have recently increased for two Alaska herds and the Porcupine caribou herd straddling Yukon and Alaska is at a historic high.

There is *low confidence* in understanding the complex drivers of observed *Rangifer* changes. Hunting and predation (the latter exacerbated by modification of the landscape for exploration and resource extraction; Dabros et al., 2018) increase in importance as populations decline. Climate strongly influences productivity: extremes in heat, drought, winter icing and snow depth reduce *Rangifer* survival (Mallory and Boyce, 2017). Changes in the timing of sea ice formation have direct effects on risks during *Rangifer* migration via inter-island movement and connection to the mainland (Poole et al., 2010). Summer warming is changing the composition of tundra plant communities, modifying the relationship between climate, forage and *Rangifer* (Albon et al., 2017), which also impacts other Arctic species such as musk ox (*Ovibos moschatus*) (Schmidt et al., 2015). As polar trophic systems are highly connected (Schmidt et al., 2017), changes will propagate through the ecosystem with effects on other herbivores such as geese and voles, as well as predators such as wolves (Hansen et al., 2013; Klaczek et al., 2016).

In northern Fennoscandia, there are approximately 600,000 semi-domesticated reindeer. Lichen rangelands are key to sustaining reindeer carrying capacity, with variable response to climate change: enhanced summer precipitation increases lichen biomass, while an increase in winter precipitation lowers it (Kumpula et al., 2014). Fire disturbance reduces the amount of pasture available for domestic reindeer and increases predation on herding lands (Lavrillier and Gabyshev, 2017). Later ice formation on waterbodies can impact herding activities (Turunen et al., 2016). Ice formation from rain-on-snow events is associated with population changes including cases of catastrophic mass starvation (Bartsch et al., 2010; Forbes et al., 2016), but there is no evidence of trends in rain-on-snow events (Cohen et al., 2015; Dolant et al., 2017).

Management of keystone species requires an understanding of pathogens and disease in the context of climate warming, but evidence of changing patterns across northern ecosystems (spanning terrestrial, aquatic, and marine environments) is hindered by an incomplete picture of pathogen diversity and distribution (Hoberg, 2013; Jenkins et al., 2013; Cook et al., 2017). Among ungulates, it is *virtually certain* that the emergence of disease attributed to nematode pathogens has accelerated since 2000 in the Canadian Arctic islands and Fennoscandia (Kutz et al., 2013; Hoberg and Brooks, 2015; Laaksonen et al., 2017; Kafle et al., 2018). Discovery of the pathogenic bacterium *Erysipelothrix rhusiopathiae* has been linked to massive and widespread mortality

among muskoxen from the Canadian Arctic Archipelago; loss of >50% of the population since 2010 may be attributable to disease interacting with extreme temperature events, although unequivocal links to climate have not been established (Kutz et al., 2015; Forde et al., 2016a; Forde et al., 2016b). Anthrax is projected to expand northward in response to warming, and resulted in substantial mortality events for reindeer on the Yamal Peninsula of Russia in 2016 with mobilisation of bacteria possibly from a frozen reindeer carcass or melting permafrost (Walsh et al., 2018). In concert with climate forcing, pathogens are *very likely* responsible for increasing mortality in Arctic ungulates (muskox, caribou/reindeer) and alteration of transmission patterns in marine food chains, broadly threatening sustainability of subsistence and commercial hunting and fishing and safety of traditional foods for northern cultures at high latitudes (Jenkins et al., 2013; Kutz et al., 2014; Hoberg et al., 2017).

3.4.3.2.3 Freshwater

Climate-driven changes in seasonal ice and permafrost conditions influence water quality (*high confidence*). Shortened duration of freshwater ice cover (more light absorption, increased nutrient input) is expected to result in higher primary productivity (Hodgson and Smol, 2008; Vincent et al., 2011; Griffiths et al., 2017b) and may also encourage greater methane emissions from Arctic lakes (Greene et al., 2014; Tan and Zhuang, 2015). Thaw slumps, active layer detachments and peat plateau collapse affect surface water connectivity (Connon et al., 2014) and enhance sediment, particulate and solute fluxes in river and stream networks (Kokelj et al., 2013). The transfer of enhanced nutrients from land to water (driven by active layer thickening and thermokarst processes; Abbott et al., 2015; Vonk et al., 2015) has been linked to heightened autotrophic productivity in freshwater ecosystems (Wrona et al., 2016). Still, there is *low confidence* in the influence of permafrost changes on dissolved organic carbon, because of competing mechanisms that influence carbon export. Permafrost thaw could contribute to the mobilisation of previously frozen organic carbon (Abbott et al., 2014; Wickland et al., 2018; Walvoord et al., 2019) thereby enhancing both particulate and dissolved organic carbon export to aquatic systems. Increased delivery of this dissolved carbon from enhanced river discharge to the Arctic Ocean (Section 3.4.3.1.2) can exacerbate regionally extreme aragonite undersaturation of shelf waters (Semiletov et al., 2016) driven by ocean uptake of anthropogenic CO_2 (Section 3.2.1.2.4). Conversely, reduced dissolved organic carbon export could accompany permafrost thaw as (1) water infiltrates deeper and has longer residence times for decomposition (Striegl et al., 2005) and (2) the proportion of groundwater (typically lower in dissolved organic carbon and higher in DIC than runoff) to total streamflow increases (Walvoord and Striegl, 2007). Increased thermokarst also has the potential to impact freshwater cycling of inorganic carbon (Zolkos et al., 2018).

Enhanced subsurface water fluxes resulting from permafrost degradation has consequences for inorganic natural and anthropogenic constituents. Emerging evidence suggests large natural stores of mercury (Schuster et al., 2018; St Pierre et al., 2018) and other trace elements in permafrost (Colombo et al., 2018) may be released upon thaw, thereby having effects (largely unknown at this point) on aquatic ecosystems. In parallel, increased development activity in the Arctic is *likely* to lead to enhanced local sources of anthropogenic chemicals of emerging Arctic concern, including siloxanes, parabens, flame retardants, and per- and polyfluoroalkyl substances (AMAP, 2017c). For legacy pollutants, there is *high confidence* that black carbon and persistent organic pollutants (e.g., hexachlorocyclohexanes, polycyclic aromatic hydrocarbons, and polychlorinated biphenyls) can be transferred downstream and affect water quality (Hodson, 2014). Lakes can become sinks of these contaminants, while floodplains can be contaminated (Sharma et al.,2015).

There is *high confidence* that habitat loss or change due to climate change impact Arctic fishes. Thinning ice on lakes and streams changes the overwintering habitat for aquatic fauna by impacting winter water volumes and dissolved oxygen levels (Leppi et al., 2016). Surface water loss, reduced surface water connectivity among aquatic habitats, and changes to the timing and magnitude of seasonal flows (Section 3.4.1.2) result in a direct loss of spawning, feeding, or rearing habitats (Poesch et al., 2016). Changes to permafrost landscapes have reduced freshwater habitat available for fish and other aquatic biota, including aquatic invertebrates upon which the fish depend for food (Chin et al., 2016). Gullying deepens channels (Rowland et al., 2011; Liljedahl et al., 2016) that otherwise may connect lake habitats occupied by fishes. This can lead to the loss of surface water connectivity, limit fish access to key habitats, and lower fish diversity (Haynes et al., 2014; Laske et al., 2016). Small connecting stream channels, which are vulnerable to drying, provide necessary migratory pathways for fishes, allowing them to access spawning and summer rearing grounds (Heim et al., 2016; McFarland et al., 2017).

Changes to the timing, duration and magnitude of high surface flow events in early and late summer threaten Arctic fish dispersal and migration activities (Heim et al., 2016) (*high confidence*). Timing of important life history events such as spawning can become mismatched with changing stream flows (Lique et al., 2016). There is regional evidence that migration timing has shifted earlier and winter egg incubation temperature has increased for pink salmon (*Oncorhynchus gorbuscha*), directly related to warming (Taylor, 2007). While long-term, pan-Arctic data on run timing of fishes are limited, phenological shifts could create mismatches with food availability or habitat suitability in both marine and freshwater environments for anadromous species, and in freshwater environments for freshwater resident species. Changes to the Arctic growing season (Xu et al., 2013a) increase the risk of drying of surface water habitats and pose a potential mismatch in seasonal availability of food in rearing habitats.

Freshwater systems across the Arctic are relatively shallow, and thus are expected to warm (*high confidence*). This may make some surface waters inhospitably warm for cold water species such as Arctic Grayling (*Thymallus arcticus*) and whitefishes (*Coregonus spp.*), or may increase the risk of *Saprolegnia fungus* that appears to have recently spread rapidly, infecting whitefishes at much higher rates in Arctic Alaska than noted in the past (Sformo et al., 2017). High infection rates may be driven by stress or nutrient enrichment from thawing permafrost, which increases pathogen virulence with fish (Wedekind et al., 2010). Warmer water and longer growing seasons will also affect food abundance because invertebrate life histories and production are temperature and degree-day dependent (Régnière et al., 2012). Increased nutrient export from permafrost loss (Frey et al., 2007), facilitated by warmer temperatures, will *likely* increase food resources for consumers, but the impact on lower trophic levels within food webs is not clearly understood.

Box 3.4 | Impacts and Risks for Polar Biodiversity from Range Shifts and Species Invasions Related to Climate Change

In polar regions climate-induced changes in terrestrial, ocean and sea ice environments, together with human introduction of non-native species, have expanded the range of some temperate species and contracted the range of some polar fish and ice-associated species (Section 3.2.3.2; Duffy et al., 2017) (*high confidence* for detection, *medium confidence* for attribution). In some cases, spatial shifts in distribution have also been influenced by fluctuations in population abundance linked to climate-induced impacts on reproductive success (Section 3.2.3). These changes have the potential to alter biodiversity in polar marine and terrestrial ecosystems (Frenot et al., 2005; Frederiksen, 2017; McCarthy et al., 2019) (*medium confidence*).

Ongoing climate change induced reductions in suitable habitat for Arctic sea ice affiliated endemic marine mammals is an escalating threat (Section 3.2.3.1) (*high confidence*). This is further complicated by the northward expansion of the summer ranges of a variety of temperate whale species, documented recently in both the Pacific and Atlantic sides of the Arctic (Brower et al., 2017; Storrie et al., 2018) and increasing pressure from anthropogenic activities. Also, over the recent decade a northward shift in benthic species, with subsequent changes in community composition has been detected in both the northern Bering Sea (Grebmeier, 2012), off Western Greenland (Renaud et al., 2015) and the Barents Sea (Kortsch et al., 2012) (*medium confidence*). At the same time as these northward expansions or shifts, a number of populations of species as different as polar bear and Arctic char show range contraction or population declines (Winfield et al., 2010; Bromaghin et al., 2015; Laidre et al., 2018).

In the Arctic a number of fish species have changed their spatial distribution substantially over the recent decades (*high confidence*). The most pronounced recent range expansion into the Arctic may be that of the summer feeding distribution of the temperate Atlantic mackerel (*Scomber scombrus*) in the Nordic Seas. From 1997 to 2016 the total area occupied by this large stock expanded from 0.4 to 2.5 million km² and the centre-of-gravity of distribution shifted westward by 1650 km and northward by 400 km (Olafsdottir

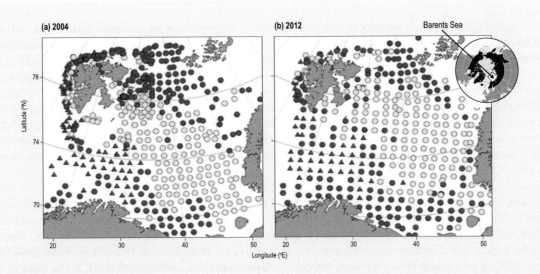

Box 3.4, Figure 1 | Spatial distribution of fish communities identified at bottom trawl stations in the Barents Sea (north of northern Norway and Russia, position indicated by red box in small globe) in **(a)** 2004 and **(b)** 2012. Atlantic (red), Arctic (blue) and Central communities (yellow). Circles: shallow sub-communities, triangles: deep sub-communities. Modified from Fossheim et al. (2015).

et al., 2019), far into Icelandic and Greenland waters and even up to Svalbard (Berge et al., 2015; Jansen et al., 2016; Nøttestad et al., 2016). This range expansion was linked both to a pronounced increase in stock size and warming of the ocean (Berge et al., 2015; Olafsdottir et al., 2019) (*high confidence*). Under RCP4.5 and RCP8.5 further range expansions of mackerel are projected in Greenland waters (Jansen et al., 2016) (*medium confidence*). However, further northwards expansion of planktivorous species may generally be restricted by them not being adapted to lack of primary production during winter (Sundby et al., 2016). Range shifts have also been observed in the Bering Sea since 1993, with warm bottom temperatures being associated with range contractions of Arctic species, and range expansions of sub-arctic species, with responses dependent on species specific vulnerably (Alabia et al., 2018; Stevenson and Lauth, 2018).

In the Barents Sea, major expansions in distribution over the recent years to decades have been well documented for both individual species and whole biological communities (*high confidence*). New information strengthens findings reported in WGII AR5 of ecologically and commercially important fish stocks having extended their habitats markedly to the north and east, concomitant to increased sea temperature and retreating sea ice. This includes capelin (Ingvaldsen and Gjøsæter, 2013), Atlantic cod (Kjesbu et al., 2014) and haddock (Landa et al., 2014). Of even greater importance is novel evidence of distinct distributional changes at the community level (Fossheim et al., 2015; Kortsch et al., 2015; Frainer et al., 2017) (Box 3.4 Figure 1). Until recently, the northern Barents Sea was dominated by small-sized, slow-growing fish species with specialised diets, mostly living in close association with the sea floor. Simultaneous with rising sea temperatures and retreating sea ice, these Arctic fishes are being replaced by boreal, fast-growing, large-bodied generalist fish moving in from the south. These large, migratory predators take advantage of increased production while the Arctic fish species suffer from higher competition and predation and are retracting northwards and eastwards. Consequently, climate change is inducing structural change over large spatial scales, leading to a borealisation ('Atlantification') of the European Arctic biological communities (Fossheim et al., 2015; Kortsch et al., 2015; Frainer et al., 2017) (*medium confidence*).

There is evidence based on population genetics that the ecosystem off Northeast Greenland could also become populated by a larger proportion of boreal species with ocean warming. Andrews et al. (2019) show that Atlantic cod, beaked redfish (*Sebastes mentella*), and deep-sea shrimp (*Pandalus borealis*) recently found on the Northeast Greenland shelf originate from the quite distant Barents Sea, and suggested that pelagic offspring were dispersed via advection across the Fram Strait.

Physical barriers to range expansions into the high Arctic interior shelf systems and the outflow systems of Eurasia and the Canadian Archipelago will continue to govern future expansions of fish populations (*medium confidence*). The limited available information on marine fish from other Arctic shelf regions reveals a latitudinal cline in the abundance of commercially harvestable fish species. For instance, there is evidence of latitudinal partitioning between the four dominant mid-water species (Polar cod, saffron cod (*Eleginus gracilis*), capelin, and Pacific herring (*Clupea pallasii*)) in the Chuckchi and Northern Bering Sea, with Polar cod being most abundant to the north (De Robertis et al., 2017). These latitudinal gradients suggest that future range expansions of fish populations will continue to be governed by a combination of physical factors affecting overwintering success and the availability, quality and quantity of prey (*medium confidence*).

In Antarctic marine systems, there is evidence of recent climate-related range shifts in the southwest Atlantic and West Antarctic Peninsula for penguin species (*Pygoscelis papua* and *P. antarctica*) and for Antarctic krill (*Euphausia superba*), but mesozooplankton communities do not appear to have changed or shifted in response to ocean warming (Section 3.2.3.2). Recent evidence suggests that the ACC and its associated fronts and thermal gradients may be more permeable to biological dispersal than previously thought, with storm-forced surface waves and ocean eddies enhancing oceanographic connectivity for drift particles in surface layers of the Southern Ocean (Fraser et al., 2017; Fraser et al., 2018) (*low confidence*), but it is unclear whether this will be an increasingly important pathway under climate change. Greater ship activity in the Southern Ocean may also present a risk for increasing introduction of non-native marine species, with the potential for these species to become invasive with changing environmental conditions (McCarthy et al., 2019). Current evidence of invasions by shell-crushing crabs on the Antarctic continental slope and shelf remains equivocal (Griffiths et al., 2013; Aronson et al., 2015; Smith et al., 2017d).

On Arctic land, northward range expansions have been recorded in species from all major taxon groups based both on scientific studies and local observations (*high confidence*) (CAFF, 2013a; AMAP, 2017a; AMAP, 2017b; AMAP, 2018). The most recent examples of terrestrial vertebrates expanding northwards include a whole range of mammals in Yakutia, Russia (Safronov, 2016), moose (*Alces alces*) into the Arctic region of both northern continents (Tape et al., 2016) and North American beaver (*Castor canadensis*) in Alaska (Tape et al., 2018). In parallel with these expansions, pathogens and pests are also spreading north (CAFF, 2013a; Taylor et al., 2015; Forde et al., 2016b; Burke et al., 2017b; Kafle et al., 2018). A widespread change is tundra greening, which in some cases is linked to shifting plant dominance within Arctic plant communities, in particular an increase in woody shrub biomass as conditions become more favorable for them (Myers-Smith et al., 2015; Bhatt et al., 2017).

Expansion of subarctic terrestrial species and biological communities into the Arctic and displacing native species is considered a major threat, since unique Arctic species may be less competitive than encroaching subarctic species favoured by changing climatic conditions (CAFF, 2013a). Similar displacements may take place within zones of the Arctic when low- and mid-Arctic species expand northward. Here, the most vulnerable species and communities may be in the species-poor, but unique, northernmost sub-zone of the Arctic because species cannot migrate northward as southern species encroach (CAVM Team, 2003; Walker et al., 2016; AMAP, 2018). This 'Arctic squeeze' is a combined effect of the fact that the area of the globe increasingly shrinks when moving poleward and that there is nowhere further north on land to go for terrestrial biota at the northern coast. The expected overall result of these shifts and limits will be a loss of biodiversity (CAFF, 2013a; CAFF, 2013b; AMAP, 2018) (*medium confidence*). At the southern limit of the Arctic, thermal hotspots may support high biological productivity, but not necessarily high biodiversity (Walker et al., 2015) and may even act as advanced bridgeheads for expansion of subarctic species into the true Arctic (*medium confidence*). At the other end of the Arctic zonal range, a temperature increase of only 1°C–2°C in the northernmost subzone may allow the establishment of woody dwarf shrubs, sedges and other species into bare soil areas that may radically change its appearance and ecological functions (Walker et al., 2015; Myers-Smith et al., 2019) (*medium confidence*).

Range expansions also include the threat from alien species brought in by humans to become invasive and outcompete native species. Relatively few invasive alien species are presently well established in the Arctic, but many are thriving in the subarctic and may expand as a result of climate change (CAFF, 2013a; CAFF, 2013b). Examples of this include: American mink (*Neovison vison*) and Nootka lupin (*Lupinus nootkatensis*) in Arctic western Eurasia, Greenland and Iceland that are already causing severe problems to native fauna and flora (CAFF and PAME, 2017).

Alien species are a major driver of terrestrial biodiversity change also in the Antarctic region (Frenot et al., 2005; Chown et al., 2012; McClelland et al., 2017). The Protocol on Environmental Protection to the Antarctic Treaty restricts the introduction of non-native species to Antarctica as do the management authorities of sub-Antarctic islands (De Villiers et al., 2006). Despite this, alien species and their propagules continue to be introduced to the Antarctic continent and sub-Antarctic islands (Hughes et al., 2015). To date, 14 non-native terrestrial species have colonised the Antarctic Treaty area (excluding sub-Antarctic islands) (Hughes et al., 2015), while the number in the sub-Antarctic is much higher (on the order of 200 species) (Frenot et al., 2005) (*low confidence*). Species distribution models for terrestrial invasive species indicate that climate does not currently constitute a barrier for the establishment of invasive species on all subantarctic islands, and that the AP region will be the most vulnerable location on the Antarctic continent to invasive species establishment under RCP8.5 (Duffy et al., 2017). Thus, for continental Antarctica, existing climatic barriers to alien species establishment will weaken as warming continues across the region (*medium confidence*). An increase in the ice-free area linked to glacier retreat in Antarctica is expected to increase the area available for new terrestrial ecosystems (Lee et al., 2017a). Along with growing number of visitors, this is expected to increase in the establishment probability of terrestrial alien species (Chown et al., 2012; Hughes et al., 2015) (*medium confidence*).

3

3.4.3.3   Impacts on Social-Ecological Systems

The Arctic is home to over four million people, with large regional variation in population distribution and demographics (Heleniak, 2014). 'Connection with nature' is a defining feature of Arctic identity for indigenous communities (Schweitzer et al., 2014) because the lands, waters and ice that surround communities evoke a sense of home, freedom and belonging, and are crucial for culture, life and survival (Cunsolo Willox et al., 2012; Durkalec et al., 2015). Climate-driven environmental changes are affecting local ecosystems and influencing travel, hunting, fishing and gathering practises. This has implications for people's livelihoods, cultural practices, economies and self-determination.

3.4.3.3.1   Food and water security

Impacts of climate change on food and water security in the Arctic can be severe in regions where infrastructure (including ice roads), travel and subsistence practices are reliant on elements of the cryosphere such as snow cover, permafrost and freshwater or sea ice (Cochran et al., 2013; Inuit Circumpolar Council-Alaska (ICC-AK), 2015).

There is *high confidence* in indicators that food insecurity risks are on the rise for Indigenous Arctic peoples. Food is strongly tied to culture, identity, values and ways of life (Donaldson et al., 2010; Cunsolo Willox et al., 2015; Inuit Circumpolar Council-Alaska (ICC-AK), 2015); thus, impacts to food security go beyond access to food and physical health. Food systems in northern communities are intertwined with northern ecosystems because of subsistence hunting, fishing and gathering activities. Environmental changes to animal habitat, population sizes and movement mean that culturally important food species may no longer be found within accessible ranges or familiar areas (Parlee and Furgal, 2012; Rautio et al., 2014; Inuit Circumpolar Council-Alaska (ICC-AK), 2015; Lavrillier et al., 2016) (Section 3.4.3.2.2). This impacts negatively the accessibility of culturally important local food sources (Lavrillier, 2013; Rosol et al., 2016) that make important contributions to a nutritious diet (Donaldson et al., 2010; Hansen et al., 2013; Dudley et al., 2015). Longer open water seasons and poorer ice conditions on lakes impact fishing options (Laidler, 2012) and waterfowl hunting (Goldhar et al., 2014). Permafrost warming and increases in active layer thickness (Section 3.4.1.3) reduce the reliability of permafrost for natural refrigeration. In some cases these changes have reduced access to, and consumption of, locally resourced food and can result in increased incidence of illness (Laidler, 2012; Cochran et al., 2013; Cozzetto et al., 2013; Rautio et al., 2014; Beaumier et al., 2015). These consequences of climate change are intertwined with processes of globalisation, whereby complex social, economic and cultural factors are contributing to a dietary transformation from locally resourced foods to imported market foods across the Arctic (Harder and Wenzel, 2012; Parlee and Furgal, 2012; Nymand and Fondahl, 2014; Beaumier et al., 2015). Limiting exposures to zoonotic, foodborne and waterborne pathogens (Section 3.4.3.2.2) depends on accurate and comprehensive data on species diversity, biology and distribution and pathways for invasion (Hoberg and Brooks, 2015; Kafle et al., 2018).

There is *high confidence* that changes to travel conditions impact food security through access to hunting grounds. Shorter snow cover duration (Section 3.4.1.1), and changes to snow conditions (such as density) make travel more difficult and dangerous (Laidler, 2012; Ford et al., 2019). Changes in dominant wind direction and speed reduce the reliability of traditional navigational indicators such as snow drifts, increasing safety concerns (Ford and Pearce, 2012; Laidler, 2012; Ford et al., 2013; Clark et al., 2016b). Permafrost warming, increased active layer thickness and landscape instability (Section 3.4.1.3), fire disturbance and changes to water levels (Section 3.4.1.2) impact overland navigability in summer (Goldhar et al., 2014; Brinkman et al., 2016; Dodd et al., 2018).

There is *high confidence* that both risks and opportunities arise for coastal communities with changing sea ice and open water conditions. Of particular concern for coastal communities is landfast sea ice (Section 3.3.1.1.5), which creates an extension of the land in winter that facilitates travel (Inuit Circumpolar Council Canada, 2014). The floe edge position, timing and dynamics of freeze-up and break-up, sea ice stability through the winter, and length of the summer open water season are important indicators of changing ice conditions and safe travel (Gearheard et al., 2013; Eicken et al., 2014; Baztan et al., 2017). Warming water temperature, altered salinity profiles, snow properties, changing currents and winds all have consequences for the use of sea ice as a travel or hunting platform (Hansen et al., 2013; Eicken et al., 2014; Clark et al., 2016a). More leads (areas of open water), especially in the spring, can mean more hunting opportunities such as whaling off the coast of Alaska (Hansen et al., 2013; Eicken et al., 2014). In Nunavut, a floe edge closer to shore improves access to marine mammals such as seals or narwhal (Ford et al., 2013). However, these conditions also hamper access to coastal or inland hunting grounds (Hansen et al., 2013; Durkalec et al., 2015), have increased potential for break-off events at the floe edge (Ford et al., 2013), or can result in decreased presence (or total absence) of ice-associated marine mammals with an absence of summer sea ice (Eicken et al., 2014).

Many northern communities rely on ponds, streams and lakes for drinking water (Cochran et al., 2013; Goldhar et al., 2013; Nymand and Fondahl, 2014; Daley et al., 2015; Dudley et al., 2015; Masina et al., 2019), so there is *high confidence* that projected changes in hydrology will impact water supply (Section 3.4.2.2). Surface water is vulnerable to thermokarst disturbance and drainage, as well as bacterial contamination, the risks of which are increased by warming ground and water temperatures (Cozzetto et al., 2013; Goldhar et al., 2013; Dudley et al., 2015; Masina et al., 2019). Icebergs or old multi-year ice are important sources of drinking water for some coastal communities, so reduced accessibility to stable sea ice conditions affects local water security. Small remote communities have limited capacity to respond quickly to water supply threats, which amplifies vulnerabilities to water security (Daley et al., 2015).

3.4.3.3.2   Communities

Culture and knowledge

Spending time on the land is culturally important for indigenous communities (Eicken et al., 2014; Durkalec et al., 2015). There is *high confidence* that daily life is influenced by changes to ice freeze-up and break-up (rivers/lakes/sea ice), snow onset/melt, vegetation phenology, and related wildlife/fish/bird behaviour (Inuit Circumpolar Council-Alaska (ICC-AK), 2015). Inter-generational knowledge transmission of associated values and skills is also influenced by climate change because younger generations do not have the same level of experience or confidence with traditional indicators (Ford, 2012; Parlee and Furgal, 2012; Eicken et al., 2014; Pearce et al., 2015). Climate-driven changes undermine confidence in indigenous knowledge holders in regards to traditional indicators used for safe travel and navigation (Parlee and Furgal, 2012; Golovnev, 2017; Ford et al., 2019).

Economics

The Arctic mixed economy is characterised by a combination of subsistence activities, and employment and cash income. There is *low confidence* about the extent and nature of impact of climate change on local subsistence activities and economic opportunities across the Arctic (e.g., hunting, fishing, resource extraction, tourism and transportation; see Section 3.2.4) because of high variability between communities (Harder and Wenzel, 2012; Cochran et al., 2013; Clark et al., 2016b; Fall, 2016; Ford et al., 2016; Lavrillier et al., 2016). Longer ice-free travel windows in Arctic seas could lower the costs of access and development of northern resources (delivering supplies and shipping resources to markets) and thus, may contribute to increased opportunities for marine shipping, commercial fisheries, tourism and resource development (Sections 3.2.4.2, 3.2.4.3) (Ford et al., 2012; Huskey et al., 2014; Overland et al., 2017). This has important implications for economic development, particularly in relation to local employment opportunities but also raises concerns of detrimental impacts on animals, habitat and subsistence activities (Cochran et al., 2013; Inuit Circumpolar Council-Alaska (ICC-AK),2015).

3.4.3.3.3   Health and wellbeing

For many polar residents, especially Indigenous peoples, the physical environment underpins social determinants of well-being, including physical and mental health. Changes to the environment impact most dimensions of health and well-being (Parlee and Furgal, 2012; Ostapchuk et al., 2015). Climate change consequences in polar regions (Sections 3.3.1.1, 3.4.1.2) have impacted key transportation routes (Gearheard et al., 2006; Laidler, 2006; Ford et al., 2013; Clark et al., 2016a) and pose increased risk of injury and death during travel (Durkalec et al., 2014; Durkalec et al., 2015; Clark et al., 2016b; Driscoll et al., 2016).

Foodborne disease is an emerging concern in the Arctic because warmer waters, loss of sea ice (Section 3.3.1.1) and resultant changes in contaminant pathways can lead to bioaccumulation and biomagnification of contaminants in key food species. While many hypothesised foodborne diseases are not well studied (Parkinson and Berner, 2009), foodborne gastroenteritis is associated with shellfish harvested from warming waters (McLaughlin et al., 2005; Young et al., 2015). Mercury presently stored in permafrost (Schuster et al., 2018) has potential to accumulate in aquatic ecosystems.

Climate change increases the risk of waterborne disease in the Arctic via warming water temperatures and changes to surface hydrology (Section 3.4.1.2) (Parkinson and Berner, 2009; Brubaker et al., 2011;

Dudley et al., 2015). After periods of rapid snowmelt, bacteria can increase in untreated drinking water, with associated increases in acute gastrointestinal illness (Harper et al., 2011). Consumption of untreated drinking water may increase duration and frequency of exposure to local environmental contaminants (Section 3.4.3.2.3) or potential waterborne diseases (Goldhar et al., 2014; Daley et al., 2015). The potential for infectious gastrointestinal disease is not well understood, and there are concerns in relation to the safety of storage containers of raw water in addition to the quality of the source water itself (Goldhar et al., 2014; Wright et al., 2018; Masina et al., 2019).

Climate change has negatively affected place attachment via hunting, fishing, trapping and traveling disruptions, which have important mental health impacts (Cunsolo Willox et al., 2012; Durkalec et al., 2015; Cunsolo and Ellis, 2018). The pathways through which climate change impacts mental wellness in the Arctic varies by gender (Bunce and Ford, 2015; Ostapchuk et al., 2015; Bunce et al., 2016) and age (Petrasek-MacDonald et al., 2013; Ostapchuk et al., 2015). Emotional impacts of climate-related changes in the environment were significantly higher for women compared to men, linked to concern for family members (Ostapchuk et al., 2015). However, men are also vulnerable due to gendered roles in subsistence and cultural activities (Bunce and Ford, 2015). In coastal areas, sea ice means freedom for travel, hunting and fishing, so changes in sea ice affect the experience of and connection with place. In turn, this influences individual and collective mental/emotional health, as well as spiritual and social vitality according to relationships between sea ice use, culture, knowledge and autonomy (Cunsolo Willox et al., 2013a; Cunsolo Willox et al., 2013b; Gearheard et al., 2013; Durkalec et al., 2015; Inuit Circumpolar Council-Alaska (ICC-AK), 2015).

3.4.3.3.4   Infrastructure

Permafrost is undergoing rapid change (Section 3.4.1.2), creating challenges for planners, decision makers and engineers (AMAP, 2017d). The observed changes in the ground thermal regime (Romanovsky et al., 2010; Romanovsky et al., 2017; Biskaborn et al., 2019) threaten the structural stability and functional capacities of infrastructure, in particular that which is located on ice rich frozen ground. Extensive summaries of construction damages along with adaptation and mitigation strategies are available (Larsen et al., 2014; Dore et al., 2016; AMAP, 2017d; Pendakur, 2017; Shiklomanov et al., 2017a; Shiklomanov et al., 2017b; Vincent et al., 2017).

Projections of climate and permafrost suggest that a wide range current infrastructure will be impacted by changing conditions (*medium confidence*). A circumpolar study found that approximately 70% of infrastructure (residential, transportation and industrial facilities), including over 1200 settlements (~40 with population more than 5000) are located in areas where permafrost is projected to thaw by 2050 under RCP4.5 (Hjort et al., 2018). Regions associated with the highest hazard are in the thaw-unstable zone characterised by relatively high ground-ice content and thick deposits of frost susceptible sediments (Shiklomanov et al., 2017b). By 2050, these high hazard environments contain one-third of existing pan-Arctic infrastructure. Onshore hydrocarbon extraction and transportation in the Russian Arctic are at risk: 45% of the oil and natural gas production fields in the Russian Arctic are located in the highest hazard zone.

In a regional study of the state of Alaska, cumulative expenses projected for climate-related damage to public infrastructure totalled USD 5.5 billion between 2015 and 2099 under RCP8.5 (Melvin et al., 2017). The top two causes of damage related costs were projected to be road flooding from increased precipitation, and building damage associated with near-surface permafrost thaw. These costs decreased by 24% to USD 4.2 billion for the same time frame under RCP4.5, indicating that reducing greenhouse gas emissions globally could lessen damages (Figure 3.13). In a related study that included these costs and others, as well as positive gains from climate change in terms of a reduction in heating costs attributable to warmer winter, annual net costs were still USD 340–700 million, or 0.6–1.3% of Alaska's GDP, suggesting that climate change costs will outweigh positive benefits, at least for this region (Berman and Schmidt, 2019).

Winter roads (snow covered ground and frozen lakes) are distinct from the infrastructure considered earlier, but have a strong influence on the reliability and costs of transportation in some remote northern communities and industrial development sites (Parlee and Furgal, 2012; Huskey et al., 2014; Overland et al., 2017). For these communities, changing lake and river levels and the period of safe ice cover all affect the duration of use of overland travel routes and inland waterways, with associated implications for increased travel risks, time, and costs (Laidler, 2012; Ford et al., 2013; Goldhar et al., 2014). There have been recent instances of severely curtailed ice road shipping seasons due to unusually warm conditions in the early winter (Sturm et al., 2017). While the impact of human effort on the maintenance of winter roads is difficult to quantify, a reduction in the operational time window due to winter warming is projected (Mullan et al., 2017).

3.5 Human Responses to Climate Change in Polar Regions

3.5.1 The Polar Context for Responding

Human responses to climate change in the Arctic and Antarctica are shaped by their unique physical, ecological, social, cultural and political conditions. Extreme climatic conditions, remoteness from densely populated regions, limited human mobility, short seasons of biological productivity, high costs in monitoring and research, sovereignty claims to lands and waters by southern-based governments, a rich diversity of indigenous cultures and institutional arrangements that in some cases recognise indigenous rights and support regional and international cooperation in governance are among the many factors that impede and or facilitate adaptation.

The social and cultural differences are an especially noteworthy factor in assessing polar responses. Approximately four million people currently reside in the Arctic with about three quarters residing in urban areas, and approximately 10% being Indigenous (AHDR, 2014). Regions of the Arctic differ widely in population, ranging from 94% of Iceland's population living in urban environments to 68% of Nunavut's population living in rural areas. And while there has been a general movement to greater urbanisation in the Arctic (AHDR, 2014), that trend is not true for all regions (Heleniak, 2014). About 4400 people reside in Antarctic in the summer and about 1100 in the winter, predominantly based at research stations of which approximately 40 are occupied year-round (The World Factbook, 2016).

For most Arctic Indigenous peoples, human responses to climate change are viewed as a matter of cultural survival (Greaves, 2016) (Cross-Chapter Box 3 in Chapter 1). However, Indigenous people are not homogenous in their perspectives. While in some cases Indigenous people are negatively impacted by sectoral activities such as mining and oil and gas development (Nymand and Fondahl, 2014), in other cases they benefit financially (Shadian, 2014), setting up dilemmas and potential internal conflicts (Huskey, 2018; Southcott and Natcher, 2018) (*high confidence*). Geopolitical complexities also confound responses.

Together these conditions make for complexity and uncertainty in human decision making, be it at the household and community levels to the international level. Adding to uncertainty in human choice related to climate change is the interaction of climate with other forces for change, such as globalisation and land and sea-use change. These interactions necessitate that responses to climate change consider cumulative effects as well as context-specific pathways for building resilience (Nymand and Fondahl, 2014; ARR, 2016).

3.5.2 Responses of Human Sectors

The sections below assess human responses to climate change in polar regions by examining various sectors of human-environment activity (i.e., social-ecological subsystems), reviewing their respective systems of governance related to climate change, and considering possible resilience pathways. Table 3.4 summarises the consequences, interacting drivers, responses and assets for responding to climate change by social-ecological subsystems (i.e., sectors) of Arctic and Antarctic regions. An area of response not elaborated in this assessment is geoengineered sea ice remediation to support local-to-regional ecosystem restoration and which may also affect climate via albedo changes. There is an emerging body of literature on this topic (e.g., Berdahl et al., 2014; Desch et al., 2017; Field et al., 2018), which at present is too limited to allow assessing dimensions of feasibility, benefits and risks, and governance.

3.5.2.1 Commercial Fisheries

Responses addressing changes in the abundance and distribution of fish resources (Section 3.2.4.1) differ by region. In some polar regions, strategies of adaptive governance, biodiversity conservation, scenario planning and the precautionary approach are in use (NPFMC, 2018). Further development of coordinated monitoring programs (Cahalan et al., 2014; Ganz et al., 2018), data sharing, social learning and decision support tools that alert managers to climate change impacts on species and ecosystems would allow for appropriate and timely responses including changes in overall fishing capacity, individual stock quotas, shifts between different target species, opening/closure of different geographic areas and balance between different fishing fleets (Busch et al., 2016; NPFMC, 2019; see Section 3.5.4). Scenario planning, adaptive management and similar efforts will contribute to the resilience and conservation of these social-ecological systems (*medium confidence*).

Five Arctic States, known as 'Arctic 5' (Canada, Denmark, Norway, Russia and the United States) have sovereign rights for exploring and exploiting resources within their 200 nautical mile Exclusive Economic Zones (EEZs) in the High Arctic and manage their resources within their own regulatory measures. A review of future harvest in the European Arctic (Haug et al., 2017) points towards high probability of increased northern movement of several commercial fish species (Section 3.3.3.1, Box 3.4), but only to the shelf slope for the demersal species. This shift suggests increased northern fishing activity, but within the EEZs and present management regimes (Haug et al., 2017) (*medium confidence*).

In 2009, a new Marine Resources Act entered into force for Norway's EEZ. This act applies to all wild living marine resources, and states that its purpose is to ensure sustainable and economically profitable management of resources. Conservation of biodiversity is described as an integral part of its sustainable fisheries management and it is mandatory to apply "an ecosystem approach, taking into account habitats and biodiversity" (Gullestad et al., 2017). In addition to national management, the Joint Norwegian-Russian Fisheries Commission provides cooperative management of the most important fish stocks in the Barents and Norwegian Seas. The stipulation of the total quota for the various joint fish stocks is a key element, as is more long-term precautionary harvesting strategies, better allowing for responses to climate change (*medium confidence*). A scenario-based approach to identify management strategies that are effective under changing climate conditions is being explored for the Barents Sea (Planque et al., 2019).

In the US Arctic, an adaptive management approach has been introduced that utilises future ecological scenarios to develop strategies for mitigating the future risks and impacts of climate change (NPFMC, 2018). The fisheries of the southeastern Bering Sea are managed through a complex suite of regulations that includes catch shares (Ono et al., 2017), habitat protections, restrictions on forage fish, bycatch constraints (DiCosimo et al., 2015) and community development quotas. This intricate regulatory framework has inherent risks and benefits to fishers and industry by limiting flexibility (Anderson et al., 2017b). To address these challenges, the North Pacific Fishery Management Council recently adopted a Fishery Ecosystem Plan that includes a multi-model climate change action module (Punt et al., 2016; Holsman et al., 2017; Zador et al., 2017; Holsman et al., 2019). Despite this complex ecosystem-based approach to fisheries management, it may not be possible to prevent projected declines of some high value species at high rates of global warming (Ianelli et al., 2016).

In the US portion of the Chukchi and Beaufort Seas EEZ, fishing is prohibited until sufficient information is obtained to sustainably manage the resource (Wilson and Ormseth, 2009). In the Canadian sector of the Beaufort Sea, commercial fisheries are currently only small-scale and locally operated. However, with decreasing ice cover

and potential interest in expanding fisheries, the Inuvialuit subsistence fishers of the western Canadian Arctic developed a new proactive ecosystem-based Fisheries Management Framework (Ayles et al., 2016). Also in Western Canada, the commercial fishery for Arctic char (*Salvenius alpinus*) in Cambridge Bay is co-managed by local Inuit organisations and Fisheries and Oceans Canada (DFO, 2014).

The high seas region of the CAO is per definition outside of any nation's EEZ. Recent actions of the international community show that a precautionary approach to considerations of CAO fisheries has been adopted (*high confidence*) and that expansion of commercial fisheries into the CAO will be constrained until sufficient information is obtained to manage the fisheries according to an ecosystem approach to fisheries management (*high confidence*). The Arctic 5 officially adopted the precautionary approach to fishing in 2015 by signing the Oslo Declaration concerning the prevention of unregulated fishing in the CAO. The declaration established a moratorium to limit potential expansion of CAO commercial fishing until sufficient information, also on climate change impacts, is available to manage it sustainably. The Arctic 5 and several other nations subsequently agreed to a treaty (the Central Arctic Ocean Fisheries Agreement) that imposed a 16-year moratorium on commercial fishing in the CAO.

CCAMLR is responsible for the conservation of marine resources south of the Antarctic Polar Front (CCAMLR, 1982), and has ecosystem-based fisheries management embedded within its convention (Constable, 2011). This includes the CCAMLR Ecosystem Monitoring Program, which aims to monitor important land-based predators of krill to detect the effects of the krill fishery on the ecosystem. Currently, there is no formal mechanism for choosing which data are needed in a management procedure for krill or how to include such data. However, this information will be important in enabling CCAMLR fisheries management to respond to the effects of climate change on krill and krill predators in the future.

Commercial fisheries management responses to climate change impacts in the Southern Ocean may need to address the displacement of fishing effort due to poleward shifts in species distribution (Pecl et al., 2017) (Box 3.4) (*low confidence*). Fisheries in the Southern Ocean are relatively mobile and are potentially able to respond to range shifts in target species, which is in contrast to small-scale coastal fisheries in other regions. Management responses will also need to adapt to the effects of future changes in sea ice extent and duration on the spatial distribution of fishing operations (ATCM, 2017; Jabour, 2017) (Section 3.2.4).

3.5.2.2   Arctic Subsistence Systems

Subsistence users have responded to climate change by adapting their wildfood production systems and engaging in the climate policy processes at multiple levels of governance. The limitations of many formal institutions, however, suggest that in order to achieve greater resilience of subsistence systems with climate change, transformations in governance are needed to provide greater power sharing, including more resources for engaging in climate change studies and regional-to-national policy making (see Sections 3.2.4.1.1, 3.4.3.2.2, 3.4.3.3.1, 3.4.3.3.2, 3.4.3.3.3, 3.5.3).

Adaptation by subsistence users to climate change falls into several categories. In some cases harvesters are shifting the timing of harvesting and the selection of harvest areas due to changes in seasonality and access to traditional use areas (AMAP, 2017a; AMAP, 2017b; AMAP, 2018). Changes in the navigability of rivers (i.e., shallower) and more open (i.e., dangerous) seas have resulted in harvesters changing harvesting gear, such as shifting from propeller to jet-propelled boats or all-terrain vehicles, and to larger ocean-going vessels for traditional whaling (Brinkman et al., 2016). In many cases, using different gear results in an increase in fuel costs (e.g., jet boats are about 30% less efficient). Unsafe ice conditions have resulted in greater risks of travel on rivers and the ocean in the frozen months. In Savoonga, Alaska, whalers reported limitations in harvesting larger bowhead because of thin ice conditions that do not allow for safe haul outs, and as a result, community residents now anticipate a greater dependence on western Alaska's reindeer as a source of meat in the future (Rosales and Chapman, 2015). Harvesters have also responded with switching of harvested species and in some cases doing without (AMAP, 2018). In many cases, adaption has allowed for continued provisioning of wildfoods in spite of climate change impacts (BurnSilver et al., 2016; AMAP, 2017a; Fauchald et al., 2017b) (*medium confidence*).

The impacts of climate change have also required adaptation to the non-harvesting aspects of wildfood production, such as an abandonment of traditional food storage and drying practices (e.g., ice cellars) and an increased use of household and community freezers (AMAP, 2017a). In several cases there has been an increased emphasis on community self-reliance, such as use of household and community gardens for food production (Loring et al., 2016). In the future, agriculture may be more possible with improved conditions at the southern limit of the Arctic, and could supplement hunting and fishing (AMAP, 2018).

Climate change may in the future bring both new harvestable fish, birds, mammals and berry producing plants to the north, and reduced populations and or access to currently harvested species (AMAP, 2017a; AMAP, 2017b; AMAP, 2018). Adaptive co-management and stronger links of local-to-regional level management with national to international level agreements necessitate consideration for sustainable harvest of new resources, as well as securing sustainable harvest or even full protection of dwindling or otherwise vulnerable populations. In these cases, adaptive co-management could be an efficient tool to achieve consensus on population goals, including international cooperation and agreements regarding migratory species shared between more countries (Kocho-Schellenberg and Berkes, 2014) (Section 3.5.4.3).

While there has been involvement of subsistence users in monitoring and research on climate change (Section 3.5.4.1.1), resource management regimes that regulate harvesting are largely dictated by science-based paradigms that give limited legitimacy to the knowledge and suggested preferences of subsistence users (Section 3.5.4.2, Cross-Chapter Box 4 in Chapter 1).

The social costs and social learning associated with responding to climate change are often related. Involvement in adaptive co-management comes with high transaction costs (e.g., greater demands on overburdened indigenous leaders, added stress of communities living with limited resources) (Forbes et al., 2015). In some cases, co-management has given communities a greater voice in decision making, but when ineffective, these arrangements can perpetuate dominant paradigms of resource management (AMAP, 2018). The perceived risks of climate change can at the same time reinforce cultural identify and motivate greater political involvement, which in turn, gives indigenous leaders experience as agents of change in policy making. Penn et al. (2016) pointed to these conflicting forces, arguing the need for a greater focus on community capacity and cumulative effects.

Greater involvement of indigenous subsistence users in Canada occurs at the national and regional levels through the structures and provisions of indigenous settlement agreements (e.g., 1993 Nunavut Land Claims Agreement, 1984 Inuvialuit Final Agreement), fish and wildlife co-management agreements (e.g., Porcupine Caribou Management Agreement of 1986), and through various boundary organisations (e.g, CircumArctic Rangifer Monitoring and Assessment Network). Home rule in Greenland, established in 1979, gives the Naalakkersuisut (government of Greenland) authority on most domestic matters of governance.

Indigenous leaders are responding to the risks of climate change by engaging in political processes at multiple levels and through different venues. However, indigenous involvement in IPCC assessments remains limited (Ford et al., 2016). At the United Nations Framework Convention on Climate Change (UNFCCC), the discursive space for incorporating perspectives of Indigenous peoples on climate change adaptation has expanded since 2010, which is reflected in texts and engagement with most activity areas (Ford et al., 2015) and by the establishment of the Local Communities and Indigenous Peoples Platform Facilitative Working Group in December 2018. Aleut International Association, Arctic Athabaskan Council, Gwich'in Council International, Inuit Circumpolar Council, Russian Association of Indigenous Peoples of the North, and the Saami Council, which sit as 'Permanent Participants' of the Arctic Council, are involved in many of its working groups and partake also at the political level (Section 3.5.3.2.1).

3.5.2.3   Arctic Reindeer Herding

Herders' responses to climate change have varied by region and respective herding practices, and in some cases are constrained by limited access to pastures (Klokov, 2012; Forbes et al., 2016; Uboni et al., 2016; Mallory and Boyce, 2017). These conditions are exacerbated in some cases by high numbers of predators (Lavrillier and Gabyshev, 2018). In Fennoscandia, husbandry practices of reindeer by some (mostly Sami) include supplemental feeding, which provide a buffer for unfavourable conditions. In Alaska, reindeer herding is primarily free range, where herders must manage herd movements in the event of icing events and the potential loss of reindeer because the movements of caribou herds (wild reindeer), both of which are partially driven by climate. For Nenets of the Yamal, Russia, resilience in herding has been facilitated through herders' own agency and, to some extent, the willingness of the gas industry to observe non-binding guidelines that provide for herders' continued use of traditional migrations routes (Forbes et al., 2015). In response to climate change (i.e., icing events and early spring runoffs blocking migration), the only way of avoiding high deer mortality is to change migration routes or take deer to other pastures. In practice, however, the full set of challenges has meant more Yamal herders opting out of the traditional collective migration partially or entirely to manage their herds privately. The reason to have private herds is one of adaptive advantage; smaller, privately owned herds are nimbler in the face of rapid changes in land cover and the expansion of infrastructure (Forbes, 2013). The same logic has more recently been applied by some herders in the wake of recent rain-on-snow events (Section 3.4.3.2.2) (Forbes et al., 2016). In all these regions, restrictions affecting the movement of reindeer to pastures are expected to negatively interact with the effects of climate and affect the future sustainability of herding systems (*high confidence*).

3.5.2.4   Tourism

The growth of the polar tourism market is, in part, a response to climate change, as travellers seek 'last-chance' opportunities, which, in turn, is creating new challenges in governance (Section 3.2.4.2). Polar-class expedition cruise vessels are now, for the first time, being purposefully built for recreational Arctic sea travel. The anticipated near- and long-term growth of Arctic tourism, especially with small vessels (yachts) (Johnston et al., 2017), points to a deficiency in current regulations and policies to address human safety, environmental risks and cultural impacts. Industry growth also points to the need for operators, governments, destination communities and others to identify and evaluate adaptation strategies, such as disaster relief management plans, updated navigation technologies for vessels, codes of conduct for visitors and improved maps (Pizzolato et al., 2016) and to respond to perceptions of tourism by residents of local destinations (Kaján, 2014; Stokke and Haukeland, 2017). Efforts were initiated with stakeholders in Arctic Canada to identify strategies that would lower risks (Pizzolato et al., 2016); a next step to lower risks and build resilience is to further develop those strategies (AMAP, 2017a; AMAP, 2017b; AMAP, 2018). Opportunities for tourism vessels in the Arctic to contribute to international research activities ('ships of opportunity') may improve sovereignty claims in some regions, contribute to science and enhance education of the public (Stewart et al., 2013; Arctic Council, 2015a; Stewart et al., 2015; de la Barre et al., 2016).

Tourism activities in the Antarctic are conducted in accordance with the Protocol on Environmental Protection to the Antarctic Treaty, which establishes general environmental principles, environmental assessment requirements, a scheme of establishing protected areas and restrictions on waste disposal. Site-specific management tools are in place. While there are varying views amongst Antarctic Treaty Parties on the best management regulations for Antarctic tourism, these Parties continue to work to manage tourism activity, including growth in numbers of visitors. In addition to the Protocol, mandatory measures have been agreed to manage aspects of tourism activity. Industry self-regulation supplements these requirements, coordinated by the International Association of Antarctica Tour Operators, which has worked with Antarctic Treaty Consultative Parties to manage changes in operations and their impact on ice-free areas (ATCM, 2016).

3.5.2.5   Arctic Non-Renewable Extractive Industries

Climate change has resulted a limited response by non-renewable resource extraction industries and agencies in the Arctic to changes in sea ice, thawing permafrost, spring runoffs, and resultant timing of exploration, construction and use of ice roads, and infrastructure design (AHDR, 2014). In some regions, climate change has offered new development opportunities, although off-shore prospects remaining cost prohibitive given current world markets (Petrick et al., 2017). (In the area covered by the Antarctic Treaty, exploitation of mineral resources is prohibited by the Protocol on Environmental Protection to the Antarctic Treaty.)

Climate change in some Arctic regions is facilitating easier access to natural resources (Section 3.5.2.3), which may generate financial capital for Arctic residents and their governments, including Indigenous peoples but also greater exposure to risks such as oil spills and increases in noise. Receding sea ice and glaciers has opened new possibilities for development, such as areas of receding glaciers of eastern Greenland (Smits et al., 2017). As mineral development commenced in Greenland, its home rule government developed environmental impact assessment protocols to provide for improved public participation (Forbes et al., 2015). Indigenous peoples are considered as non-state actors and in many, but not all cases, promote environmental protection in support of the sustainability of their traditional livelihoods. This protection is at times in opposition to the industrial development business sector, which is well-funded and lobbies strongly. Bilateral agreements for resource development in the Arctic are typically state dominated and controlled, and are negotiated with powerful non-state actors, such as state-dominated companies (Young, 2016). Among the non-state actors, new networks and economic forums have been established (Wehrmann, 2016). One example is the Arctic Economic Council, created by the Arctic Council during 2013–2015 as an independent organisation that facilitates Arctic business-to-business activities and supports economic development.

Several regional governments are assessing the long-term viability of ice roads, historically used for accessing mineral development sites, as well as some Arctic human settlements. In Northwest Territories, Canada, several ice roads are being replaced with all-season roads, with other replacements proposed. Assessing future conditions is key for planning and initiating new projects (Hori et al., 2018; Kiani et al., 2018) but is often constrained by uncertainties of available climate models (Mullan et al., 2017).

On the North Slope of Alaska, oil and gas development is now undergoing new expansion, while industry concurrently faces increasing challenges of climate change, such as shorter and warmer winters, the main season for oil exploration and production (Lilly, 2017). The method for building of ice roads on the North Slope has been somewhat modified to account for warmer temperatures during construction. There are also knowledge gaps in understanding implications of seismic studies with climate change on the landscape (Dabros et al., 2018). The issue of cumulative effects also raises questions of current practice of environmental impact assessment to evaluate potential cumulative effects (Kirkfeldt et al., 2016).

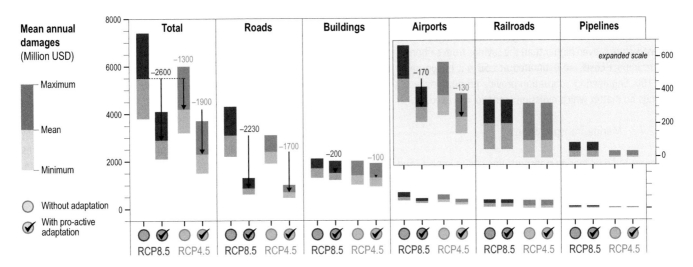

Figure 3.12 | Changes in public infrastructure damage costs in cumulative USD by 2100 in Alaska under different emission scenarios (Representative Concentration Pathways (RCP)). The inset showing airports, railroads, and pipelines has a different in scale than roads, buildings, and the total. Bars over open circles represent climate-related costs of impact with no engineering adaptation measures, whereas bars over check-marked circles represent the costs following savings from engineering adaptation (figure modified from Melvin et al., 2017).

Lilly (2017) reported that optimising Alaska North Slope transportation networks during winter field operations is critical in managing increasing resource development and could potentially provide a better framework for environmentally responsible development. Better understanding of environmental change is also important in ensuring continued oil field operations with protection of natural resources. Improved forecasting of short-term conditions (i.e., snow, soil temperatures, spring runoffs) could allow management agencies to respond to conditions more proactively and give industry more time to plan winter mobilisation, such as construction of ice roads (*low confidence*).

3.5.2.6   Infrastructure

Reducing and avoiding the impacts of climate change on infrastructure will require special attention to engineering, land use planning, maintenance operations, local culture and private and public budgeting (AMAP, 2017a; AMAP, 2017b; AMAP, 2018). In some cases, relocation of human settlements will be required, necessitating more formal methods of assessing relocation needs and identifying sources of funding to support relocations (Cross-Chapter Box 9) (*high confidence*).

A discussion of the relocation of Alaska's coastal villages is found in Cross-Chapter Box 9. Alaskan coastal communities are not the only settlements potentially requiring relocation. Subsidence due to thawing permafrost and river and delta erosion makes other rural communities of Alaska and Russia vulnerable, potentially requiring relocation in the future (Bronen, 2015; Romero Manrique et al., 2018). These situations raise issues of environmental justice and human rights (Bronen, 2017), and illustrate the limits of incremental adaptation when transformation change is needed (Kates et al., 2012). In other cases, cultural resources in the form of historic infrastructure are being threatened and require mitigation (Radosavljevic et al., 2015). Responsibility for funding has been a key issue in the relocation process (Iverson, 2013) as well as the overall role of government and local communities in relocation planning

(Marino, 2012; Romero Manrique et al., 2018). The Alaska Denali Commission, an independent federal agency designed to provide critical utilities, infrastructure and economic support throughout Alaska, is now serving as the lead coordinating organisation for Alaska village relocations and managing federal funding allocations. Several efforts have also been undertaken to provide assessment frameworks and protocols for settlement relocation as an adaptive resource (Bronen, 2015; Ristroph, 2017).

While there has been discussion of future 'climigration' in rural Alaska (Bronen and Chapin, 2013; Matthews and Potts, 2018), a study of Alaska rural villages threated by climate change showed no outmigration response (Hamilton et al., 2016). Several factors explain the lack of outmigration, including an unwillingness to move, attachment to place, people's inability to relocate, the effectiveness of alternative ways of achieving acceptable outcomes and methods of buffering through subsidies (Huntington et al., 2018) (*medium confidence*).

The current pan-Arctic trend of urbanisation (AHDR, 2014), suggests that climate change responses related to infrastructure in towns and cities of the North will require significant adaptation in designs and increases in spending (Streletskiy et al., 2012). These costs do not include costs related to flooding and other stressors associated with warming or additional costs of commercial and industrial operations. Engineers in countries with permafrost are actively working to adapt the design of structures to degrading permafrost conditions (Dore et al., 2016) and the effects of a warming climate, for example the Cold Climate Housing Research Center of Alaska.

An analysis of the costs of total damages from climate change to public infrastructure in Alaska show the financial benefits of proactive adaptation (Melvin et al., 2017) (Figure 3.12). In addition to global carbon emission mitigation, hardening and redesigning of infrastructure can reduce costs of future climate-related impacts. For example, retrofitting and redesigning of infrastructure in order to handle increased precipitation and warmer temperatures can reduce climate-related costs by 50%, from USD 5.5 to 2.9 billion under RCP8.5 by 2100. The cost savings of retrofitting and redesigning infrastructure is even higher than the savings from carbon mitigation, where impact costs are estimated at USD 4.2 billion under RCP4.5 by 2100. Engineering adaptation provide proportionally similar cost savings no matter which emission scenario was used.

3.5.2.7   Marine Transportation

Increases in Arctic marine transportation create impacts and risks for ecosystems and people, such as an increased likelihood of accidents, the introduction of invasive species, oil spills, waste discharges, detrimental impacts on animals, habitat and subsistence activities (Sections 3.2.4.3, 3.4.3.3.2). There has been a rise in geopolitical debate regarding national and international level regulations and policies, and maritime infrastructure to support Arctic shipping development (Heininen and Finger, 2017; AMAP, 2018; Drewniak et al., 2018; Nilsson and Christensen, 2019). Without further action leading to adequate implementation of well-developed management plans and region-specific regulations, anticipated future increases in Arctic shipping will pose a greater risk to people and ecosystems (*high confidence*).

The International Maritime Organization (IMO) has responsibility for the safety and security of shipping and the prevention of marine and atmospheric pollution by ships, including in the Arctic and Antarctic. There are a number of mechanisms standardising regulation and governance, such as the International Convention for the Prevention of Pollution from Ships; the International Convention for the Safety of Life at Sea; the International Convention on Standards of Training, and the Certification and Watchkeeping for Seafarers, and the newly implemented International Code for Ships Operating in Polar Waters, or Polar Code (IMO, 2017).

The Polar Code of 2017 sets new standards for vessels travelling in polar areas to mitigate environmental damage and improve safety (IMO, 2017). The Polar Code, however, currently excludes fishing vessels and vessels on government service, thereby excluding many shipping activities, particularly in the Antarctic region (IMO, 2017). Many ships travelling these waters will therefore continue to pose risks to the environment and to themselves, as they are not regulated under the Polar Code (*high confidence*). The Polar Code does not enhance enforcement capabilities or include environmental protection provisions to address a number of particular polar region-specific risks such as black carbon, ballast water and heavy fuel oil transport and use in the Arctic (Anderson, 2012; Sakhuja, 2014; IMO, 2017). However, both Russian and Canadian legislation provide the possibility for stricter shipping provisions in ice-covered waters. The IMO has prohibited the use of heavy fuel oil in the Antarctic.

States can individually or cooperatively pursue the establishment of Special Areas and Particularly Sensitive Sea Areas at the IMO with a view to protect ecologically unique or vulnerable and economically or culturally important areas in national and international waters from risks and impacts of shipping, including through routing, discharge and equipment measures. Continued, and in some areas, greater, international cooperation on shipping governance can

facilitate addressing emerging climate change issues (Arctic Council, 2015a; ARR, 2016; PEW Charitable Trust, 2016; Chénier et al., 2017; IMO, 2017) (*high confidence*). Cooperation of the member states of the Arctic Council resulted in the 2011 Agreement on Cooperation on Aeronautical and Maritime Search and Rescue in the Arctic and in the 2013 Agreement on Cooperation on Marine Oil Pollution Preparedness and Response in the Arctic. These agreements can, if adequately implemented, reduce risks from increased Arctic shipping (*medium confidence*), however, developing more effective measures is needed as preparedness and response gaps still exist, for example, for the central Arctic Ocean.

Industry has responded to the increase in shipping activity by investing in development of shipping designs for travel in mixed-ice environments (Stephenson et al., 2011; Stephenson et al., 2013). These increases in investments are occurring in spite of the limited total savings when comparing shorter travel to increased CO_2 emissions (Lindstad et al., 2016). In anticipation of spills, research in several regions has explored oil spill response viability and new methods of oil spill response for the Arctic environment (Bullock et al., 2017; Dilliplaine, 2017; Holst-Andersen et al., 2017; Lewis and Prince, 2018) (*medium confidence*). A comparative risk assessment for spills has been developed for the Arctic waters (Robinson et al., 2017) and Statoil has developed and uses risk assessment decision-support tools for environmental management, together with environmental monitoring (Utvik and Jahre-Nilsen, 2016). These tools facilitate the assessment of Arctic oil-spill response capability, ice detection in low visibility, improved management of sea ice and icebergs, and numerical modelling of icing and snow as risk mitigation.

3.5.2.8   Arctic Human Health and Well Being

At present health adaptation to climate change is generally under-represented in policies, planning, and programming (AHDR, 2014). For instance, all initiatives of the Fifth National Communications of Annex I parties to the UNFCCC affect health vulnerability, however, only 15% of initiatives had an explicit human health component described (Lesnikowski et al., 2011). The Arctic is no exception to this global trend. Despite the substantial health risks associated with climate change in the Arctic, health adaptation responses remain sparse and piecemeal (Lesnikowski et al., 2011; Panic and Ford, 2013; Ford et al., 2014b; Loboda, 2014), with the health sector substantially under-represented in adaptation initiatives compared to other sectors (Pearce et al., 2011; Ford et al., 2014b; National Research Council, 2015). Furthermore, the geographic distribution of publicly available documentation on adaptation initiatives is skewed in the Arctic, with more than three-quarters coming from Canada and USA (Ford et al., 2014a; Loboda, 2014).

Many Arctic health adaptation efforts by governments have been groundwork actions, focused increasing awareness of the health impacts of climate change and conducting vulnerability assessments (Lesnikowski et al., 2011; Panic and Ford, 2013; Austin et al., 2015). For instance, in Canada this effort has included training, information resources, frameworks, general outreach and education and dissemination of information to decision makers (Austin et al., 2015). Finland's national adaptation strategy outlines various anticipatory and reactive measures for numerous sectors, including health (Gagnon-Lebrun and Agrawala, 2007). In Alaska, the Arctic Investigations Program responds to infectious disease via advancing molecular diagnostics, integrating data from electronic health records and environmental observing networks, as well as improving access to in-home water and sanitation services. Furthermore, circumpolar efforts are also underway, including a circumpolar working group with experts from public health to assess climate-sensitive infectious diseases, and to identify initiatives that reduce the risks of disease (Parkinson et al., 2014). Importantly, health adaptation is occurring at the local scale in the Arctic (Ford et al., 2014a; Ford et al., 2014b). Adaptation at the local-scale is broad, ranging from community freezers to increase food security, to community-based monitoring programs to detect and respond to climate health events, to Elders mentoring youth in cultural activities to promote mental health when people are 'stuck' in the communities due to unsafe travel conditions (Pearce et al., 2010; Brubaker et al., 2011; Harper et al., 2012;

Brubaker et al., 2013; Douglas et al., 2014; Austin et al., 2015; Bunce et al., 2016; Cunsolo et al., 2017) (*high confidence*). Several regional and national-level initiatives on food security (ICC, 2012), as well as research reporting high levels of household food insecurity (Kofinas et al., 2016; Watts et al., 2017) have prompted greater concerns for climate change (Loring et al., 2013; Beaumier et al., 2015; Islam and Berkes, 2016). A new initiative to operationalise One Health concepts and approaches under the AC's Sustainable Development Working Group has gained momentum since 2015 (Ruscio et al., 2015). One Health approaches seek to link human, animal, and environmental health, using interdisciplinary and participatory methods that can draw on indigenous knowledge and local knowledge (Dudley et al., 2015). Thus far, the initiative has supported new regional-to-international networks, and proposals for its expansion. In the future, the ability to manage, respond, and adapt to climate-related health challenges will be a defining issue for the health sector in the Arctic (Ford et al., 2010; Durkalec et al., 2015) (*medium confidence*).

Table 3.4 | Response of key human sectors /systems to climate change in polar regions. Table 3.4 summarises the consequences, interacting drivers, responses, and assets of climate change responses by select human sectors (i.e., social-ecological systems) of Arctic and Antarctic regions. Also noted are anticipated future conditions and level of certainty and other drivers of change that may interact with climate and affect outcomes. Implications to world demands on natural resources, innovation and development of technologies, population trends and economic growth are likely to affect all systems, as is the Paris Agreement (AMAP, 2017b). In several cases, drivers of change interacting with climate change are regionally specific and not easily captured. In many cases there is limited information on human responses to climate change in the Russian Arctic.

Sector/ System	Consequence of climate change	Documented responses	Key assets and strategies of adaptive and transformative capacity	Anticipated future conditions/level of certainty	Other forces for change that may interact with climate and affect outcomes
Commercial Fisheries	Consequences are multi-dimensional, including impacts to abundance and distribution of different target species differently, by region. Changes in coastal ecosystems affecting fisheries productivity	Implementation of adaptive management practices to assess stocks, change allocations as needed, and address issues of equity	Implementation of adaptive management that is closely linked to monitoring, research, and public participation in decisions	Displacement of fishing effort will impact fishing operations in the eastern Bering Sea and Barents Sea as well as the Convention for the Conservation of Antarctic Marine Living Resources area	Changes in human preference, demand and markets, changes in gear, changes in policies affecting property rights. Changes due to offshore development and transportation
Subsistence (marine and terrestrial)	Changes in species distribution and abundance (not all negative); impediments to access of harvesting areas; safety; changes in seasonality; reduced harvesting success and process of food production (processing, food storage; quality); threats to culture and food security	Change in gear, timing of hunting, species switching; mobilisation to be involved in political action	Systems of adaptive co-management that allow for species switching, changes in harvesting methods and timing, secure harvesting rights	Less access to some areas, more in others. Changes in distribution and abundance of resources. More restrictions with regulations related to species at risk. Adaptation at the individual, household, and community levels may be seriously restricted by conditions where there is poverty (*high confidence*)	Changes in cost of fuel, land use affecting access, food preferences, harvesting rights; international agreements to protect vulnerable species
Reindeer Herding	Rain-on-snow events causing high mortality of herds; shrubification of tundra pasture lowering forage quality	Changes in movement patterns of herders; policies to ensure free-range movements; supplemental feeding	Flexibility in movement to respond to changes in pastures, secure land use rights and adaptive management. Continued economic viability and cultural tradition.	Increased frequency of extreme events and changing forage quality adding to vulnerabilities of reindeer and herders (*medium confidence*)	Change in market value of meat; overgrazing; land use policies affecting access to pasture and migration routes, property rights
Tourism (Arctic and Antarctic)	Warmer conditions, more open water, public perception of 'last chance' opportunities	Increased visitation, (quantity and quality) increase in off-season tourism to polar regions	Policies to ensure safety, cultural integrity, ecological health, adequate quarantine procedures	Increased risk of introduction of alien species and direct effects of tourists on wildlife	Travel costs. Shifting tourism market, more enterprises

Sector / System	Consequence of climate change	Documented responses	Key assets and strategies of adaptive and transformative capacity	Anticipated future conditions / level of certainty	Other forces for change that may interact with climate and affect outcomes
Non-Renewable Resource Extraction (Arctic only)	Reduced sea ice and glaciers offering some new opportunities for development; changes in hydrology (spring runoff), thawing permafrost, and temperature affect production levels, ice roads, flooding events, and infrastructure	Some shifts in practices, greater interest in offshore and on-land development opportunities in some regions	Modification of practices and use of climate change scenario analysis	Increased cost of operations in areas of permafrost thawing; more accessible areas in open waters and receding glaciers	Changes in policies affecting extent of sea and land use area, new extraction technologies (e.g., fracking), changes in markets (e.g., price of barrel of oil)
Infrastructure -urban and rural human settlements, year-round	Thawing permafrost affecting stability of ground; coastal erosion	Damaged and loss of infrastructure, increase in operating costs	Resources for assessments, mitigation, and where needed, relocation	Increasing cost to maintain infrastructure and greater demand for technological solutions to mitigate issues. Shortening windows of operation for use of ice roads; construction of all-season roads	Weak regional and national economies, other disasters that divert resources, disinterest by southern-based law makers
Marine Transportation	Open seas allowing for more vessels; greater constraints in use of ice roads	Increased shipping, tourism, more private vessels. Increased risk of hazardous waste and oil spills and accidents requiring search and rescue	Strong international cooperation leading to agreed-upon and enforced policies that maintain standards for safety; well-developed response plans with readiness by agents in some regions	Continued increases in shipping traffic with increased risks of accidents	Political conflict in other areas that impeded acceptance of policies for safety requirements, timing, and movements. Changing insurance premiums
Human Health	Threats to food security, potential threats to physical and psychological well being	Greater focus on food security research; programs that address fundamental health issues	Human and financial resources to support public programs in hinterland regions; cultural awareness of health issues as related to climate change	Greater likelihood of illnesses, food insecurity, cost of health care	A reduction (of increase) in public resources to support health services to rural community populations, research that links ecological change to human health
Coastal Settlements (see Cross-Chapter Box 9)	Change in extent of sea ice with more storm surges, thawing of permafrost, and coastal erosion	Maintenance of erosion mitigation; relocation planning, some but incomplete allocation for funding	Local leadership and community initiatives to initiate and drive processes, responsive agencies, established processes for assessments and planning, geographic options	Increasing number of communities needing relocation, rising costs for mitigating erosion issues	Limitations of government budgets, other disasters that may take priority for spending, deficiencies in policies for addressing mitigation and relocation

3.5.3   Governance

3.5.3.1   Local to National Governance

Responses to climate change at and across local, regional and national levels occur directly and indirectly through a broad range of governance activities, such as land and sea use planning and regulations, economic development strategies, tax incentives for use of alternative energy technologies, permitting processes, resource management and national security. Increasingly, climate change is considered in environmental assessments and proposals for resource planning of polar regions.

A comprehensive literature review of 157 discrete cases of Arctic adaptation initiatives by Ford et al. (2014b) found that adaptation is primarily local and motivated by reducing risks and their related vulnerabilities (*high confidence*). Several elements for successful climate change adaptation planning at the local level have previously been identified: formal analytical models need to be relevant to the concerns and needs of stakeholders, experts should be made sensitive to community perspectives, information should be packaged and communicated in ways that are accessible to non-experts

and processes of engagement that foster creative problem solving be used. Furthermore, success of local government involvement in adaptation planning has been closely linked to transnational municipal networks that foster social learning and in which local governments assume a role as key players (Sheppard et al., 2011; Fünfgeld, 2015) (*medium confidence*). While transnational networks can be a catalyst for action and promoting innovation, there remain outstanding challenges in measuring the effectiveness of these networks (Fünfgeld, 2015).

Adaptation through formal institutions by Indigenous people is enabled through self-government, land claims and co-management institutions (Baird et al., 2016; Huet et al., 2017). However, organisational capacity is often a limiting factor in involvement (AHDR, 2014; Ford et al., 2014b; Forbes et al., 2015) (*high confidence*). Interactions across scales are also dependent on the extent to which various stakeholders are perceived as legitimate in their perceptions and recommendations, an issue related to the use of local knowledge and indigenous knowledge in governance (Cross-Chapter Box 4 in Chapter 1) (AHDR, 2014; Ford et al., 2014b; Forbes et al., 2015) (*high confidence*).

At a more regional level, Alaska's 'Climate Action for Alaska' was reconstituted in 2017 and is now actively linking local concerns with state-level policies and funding, as well as setting targets for future reductions in the state's carbon emission. The role of cross-scale boundary organisations in climate change adaptation planning, and how central government initiatives can ultimately translate into 'hybrid' forms of adaptation at the local level that allow for actions that are sensitive to local communities has proven important in Norway (Dannevig and Aall, 2015).

At the national level, Norway, Sweden and Finland have engaged in the European Climate Adaptation Platform (Climate-ADAPT), a partnership that aims to support Europe in adapting to climate change by helping users to access and share data and information on expected climate change in Europe, current and future vulnerability of regions and sectors, national and transnational adaptation strategies and actions, adaptation case studies and potential adaptation options, and tools that support adaptation planning. Level of participation by country and the extent to which national level efforts are linked with regional and local adaptation varies. The Canadian government's actions on climate change have been among the most extensive of the Arctic nations, including funding of ArcticNet, a Network of Centres of Excellence, and consideration of climate change by The Northern Contaminants and Nutrition North Canada programs.

3.5.3.2 International Climate Governance and Law: Implications for International Cooperation

The way states and institutions manage international cooperation on environmental governance is changing in response to climate change in the polar regions. Rather than treating regional impacts of climate change and their governance in isolation (i.e., purely with a regional lens), the need to cooperate in a global multi-regulatory fashion across several levels of governance is increasingly realised (Stokke, 2009; Cassotta et al., 2016) (*medium confidence*).

In both polar regions, cooperative approaches to regional governance have been developed to allow for the participation of non-state actors. In some cases, regimes allow for a substantial level of participation by specific groups of the civil society (Jabour, 2017; Keil and Knecht, 2017). For example, in the Antarctic Treaty System, the Antarctic Treaty Parties included the Scientific Committee on Antarctic Research into their Protocol on Environmental Protection to the Antarctic Treaty. In the Arctic, the status of Permanent Participants has enabled the effective participation of Indigenous Peoples in the work of the Council (Pincus and Ali, 2016). Climate change has contributed to modifying the balance between the interests of state and non-state actors, leading to changing forms of cooperation (Young, 2016). While such changes and modifications occur in both the Arctic and Antarctic, the role of states has remained present in all the regimes and sectors of human responses (Young, 2016; Jabour, 2017).

Addressing the risks of climate change impacts in polar regions also requires linking levels of governance and sector governance across local to global scales, considering impacts and human adaptation (Stokke, 2009; Berkman and Vylegzhanin, 2010; Tuori, 2011; Young,

2011; Koivurova, 2013; Prior, 2013; Shibata, 2015; Young, 2016) (*high confidence*). Despite established cooperation in international polar region governance, several authors have come to the conclusion that the current international legal framework is inadequate when applying a precautionary approach at the regional level (*medium confidence*). For example, several studies have shown that the Convention on the Protection of the Marine Environment of the North East Atlantic, which applies only to the North East Atlantic, and that provides a framework for implementation of the United Nations Convention on the Law of the Sea (UNCLOS) and the Convention on Biological Diversity (CBD), are insufficient to deal with risks when applying a precautionary approach (Jakobsen, 2014; Hossain, 2015).

In the Arctic, responses to climate change do not only lead to international governance cooperation but also to competition in access to natural resources, especially hydrocarbons. With ice retreating and thinning, and improved access to natural resources, coastal states are increasingly recurring to the option to invoke Article 76 of the UNCLOS (Art. 76 UNCLOS; Verschuuren, 2013) and seek to demonstrate with scientific data, submitted to the Commission on the Limits of Continental Shelf, and within a set timeline, that their continental shelf is extended. In that case they can enjoy sovereign rights beyond the EEZ. It is *very unlikely* that this new trend from states to refer to Article 76 will lead to future (military) conflicts (Berkman and Vylegzhanin, 2013; Kullerud et al., 2013; Stokke, 2013; Verschuuren, 2013), although the issue cannot be totally dismissed (Kraska, 2011; Åtland, 2013; Huebert, 2013; Cassotta et al., 2015; Barret, 2016; Cassotta et al., 2016).

In the Antarctic, cooperation in general does occur via UNCLOS, the Convention for the Safety of Life at Sea and the Convention for the Prevention of Pollution from Ships and the Polar Code. Global environmental and climate regimes that are implemented and managed through regional regimes (such as the Kyoto Protocol or the Paris Agreement) are also relevant for the Antarctic Treaty and its Protocol on Environmental Protection, which requires, amongst other issues, a minimisation of adverse environmental impacts. Cooperation in the Antarctic also occurs through the the Convention for the Conservation of Antarctic Marine Living Resources. Climate change and its consequences for the marine environment are a central issue for this Convention because it challenges ways to regulate and manage fisheries and designate and manage Marine Protected Areas. Nevertheless, CCAMLR has not agreed to any climate change program and at its most recent meeting, there was again no agreement to do so (Brooks et al. (2018), CCAMLR Report on the 37th Meeting of the Commission, CCAMLR (2018)).

3.5.3.2.1 Formal arrangements: polar conventions and institutions

The Arctic Council

International cooperation on issues related to climate change in the Arctic mainly occurs at the Arctic Council (herein 'the Council'), and consequently in important areas of its mandate: the (marine) environment and scientific research (Morgera and Kulovesi, 2016; Tesar et al., 2016a; Wehrmann, 2016; Young, 2016). The Council is composed of eight Arctic States and six Permanent Participants representing organisations of Arctic Indigenous peoples. Observers status is open to: non-Arctic states, intergovernmental and inter-parliamentary organisations, global and regional non-governmental organisations (NGOs). The Council is an example of cooperation through soft law, a unique institutional body that does not possess a legal personality and is neither an international law nor a completely state-centric institution. However, it is acting state-centric and increasingly operating in a context of the Arctic affected by a changing climate, globalisation and transnationalism (Baker and Yeager, 2015; Cassotta et al., 2015; Pincus and Speth, 2015) (*medium confidence*). In 2013, China, South Korea, Italy, Japan, India and Singapore joined France, Germany, the Netherlands, Poland, Spain and the UK as Observers states to the Arctic Council; Switzerland was granted Observer status in 2017. Non-Arctic States are stimulating the Council towards adopting a new approach for Arctic governance that would leave greater space for their participation.

Despite lacking the role to enact hard law, three binding agreements were negotiated under the auspices of the Council (in its task forces), the latest of which is the Agreement on Enhancing International Arctic Scientific Cooperation, which is an indication that the Council is preparing a regulatory role to respond to climate change in the Arctic using hard-law instruments (Morgera and Kulovesi, 2016; Shapovalova, 2016). Through organising the Task Force on Black Carbon and Methane (Morgera and Kulovesi, 2016), the Council has catalysed action on short-lived climate forcers as the task force was followed by the adoption in 2015 of the Arctic Council Framework for Action on Enhanced Black Carbon and Methane Emission Reductions. In this non-legally binding agreement, Arctic States lay out a common vision for national and collective action to accelerate decline in black carbon and methane emissions (Shapovalova, 2016). The Council thereby moved from merely assessing problems to attempting to solve them (Baker and Yeager, 2015; Young, 2016; Koivurova and Caddell, 2018). While mitigation of global emissions from fossil fuels requires global cooperation, progress with anthropogenic emissions of short-term climate forcers (such as black carbon and methane) may be achieved through smaller groups of countries (Aakre et al., 2018). However, even though the Council has also embraced the Ecosystem Approach, it does not have a mandate to manage climate-related risks and impacts, or apply a precautionary approach, on fisheries issues.

Several studies have shown that the Council has the potential to enhance internal coherence in the current, fragmented landscape of multi-regulatory governance by providing integrated leadership. However, it is *about as likely as not* that the Council could play a strong role in combatting global climate problems and operating successfully within the climate transnational context unless it goes through restructuring and reconfiguration (Stokke, 2013; Baker and Yeager, 2015; Pincus and Speth, 2015; Cassotta et al., 2016; Tesar et al., 2016a; Wehrmann, 2016; Young, 2016; Koivurova and Caddell, 2018).

The future of the governance of the changing Arctic Ocean, including the role of the Council will also depend on the implications of the development for a new agreement on the Conservation and Sustainable use of Marine Biodiversity of Areas beyond National Jurisdictions under the UNCLOS (Baker and Yeager, 2015; De Lucia, 2017; Nengye et al., 2017; Koivurova and Caddell, 2018) (*medium confidence*).

The Antarctic Treaty System

The Antarctic Treaty System (ATS) is the collective term for the Antarctic Treaty and related agreements. The ATS regulates international relations with respect to Antarctica. 54 countries have acceded to the Treaty and 29 of them participate in decision making as Consultative Parties. 27 countries are Party to the CCAMLR, and 40 have ratified the Protocol on Environmental Protection to the Antarctic Treaty. The importance of understanding, mitigating and adapting to the impacts of changes to the Southern Ocean and Antarctic cryosphere has been realised by all of the major bodies responsible for governance in the Antarctic region (south of 60°S). The Antarctic Treaty Consultative Parties agreed that a Climate Change Response Work Programme would address these matters (ATCM, 2016). This led to the establishment of the Subsidiary Group of the Committee for Environmental Protection on Climate Change Response (ATCM, 2017). By contrast, consensus is currently limiting work programme-level responses to climate change by CCAMLR (2017a), while opportunities exist to incorporate climate concerns into mechanisms for implementation and monitoring aimed to conserve ecosystems and the environment (Brooks et al., 2018).

3.5.3.2.2 Informal arrangements

The Antarctic Treaty Consultative Parties, through the Committee for Environmental Protection (CEP) and its Subsidiary Group of the Committee for Environmental Protection on Climate Change Response, continue to work closely with the Scientific Committee on Antarctic Research, the Council of Managers of National Antarctic Programs, the International Association of Antarctica Tour Operators and other NGOs to understand, mitigate and adapt to impacts associated with changes to the Southern Ocean and Antarctic cryosphere. Understanding, mitigating and adapting to climate change are among the key priorities identified for research in the region (Kennicutt et al., 2014a; Kennicutt et al., 2014b) and nationally funded bilateral and multilateral projects are underway.

3.5.3.2.3 Role of informal actors

Several studies show that informal actors of the Arctic can influence decision making process of the Council and shift the Council towards more cooperation with different actors to enhance the co-production of knowledge (Duyck, 2011; Makki, 2012; Keil and Knecht, 2017). Recently, non-state observers at the Council, such as the World Wide Fund for Nature and the Circumpolar Conservation Union have played a role in raising awareness on climate change responses and contributing to the work of the Council's Working Groups and Expert Groups (Keil and Knecht, 2017).

Within the Antarctic Treaty System, several non-state actors play a major role in providing advice and influencing the governance of Antarctica and the Southern Ocean. Among the most prominent actors are formal observers such as the Scientific Committee on Antarctic Research, and invited experts such as the International Association of Antarctica Tour Operators and the Antarctic and Southern Ocean Coalition. At meetings of CCAMLR, the Scientific Committee's 2009 report on Antarctic Climate Change and the Environment (Turner et al., 2009) precipitated an Antarctic Treaty Meeting of Experts on

Table 3.5 | Summary of the assessment of practices, tools and strategies that can contribute to climate resilient pathways. Practices are shown with the potential extent of their contribution to resilience building, considering also seven general strategies (Biggs et al., 2012; Quinlan et al., 2016; Cross-Chapter Box 2 in Chapter 1). Also shown is the current level of their application in polar regions and key conditions facilitating implementation.

Type of resilience-building activity	Practices, tool, or strategy	Potential extent of contribution to resilience building ☐ Large ☐ Moderate ☐ Limited and **Areas of potential contributions to resilience:** DIV = Maintain diversity & redundancy CON = Manage connectivity PAR = Broaden participation LEA = Encourage learning & experimentation SYS = Foster complex system understanding GOV = Enhance polycentric governance SLO = Manage slow variables and feedbacks **Confidence regarding potential contribution to resilience building:** ●●● = high ●● = medium ● =low		Current level of application in polar regions ☐ High ☐ Medium ☐ Low and **Key conditions facilitating implementation:** F = Financial support I = Institutional support T&S = Technical and science support L&I = Local & indigenous capacity and knowledge C = Interdisciplinary and/or cross-cultural cooperation
Knowledge Co-Production and Integration	Community-based monitoring	DIV, PAR, SYS ●●		F, I, T&S, L&I, C
	Understanding regime shifts	LEA, SYS, SLO ●●●		I, T&S, C
	Indicators of resilience and adaptive capacity	PAR, LEA, SYS, SLO ●●		F, L&I, T&S
Linking Knowledge with Decision Making	Participatory scenario analysis and planning	PAR, LEA, SYS ●●		T&S, L&I, C
	Structured decision making	PAR, LEA, SYS ●		I, T&S, C
Resilience-based Ecosystem Stewardship	Adaptive ecosystem governance	DIV, PAR, LEA, SYS, GOV, SLO ●●●		I, T&S, L&I, C
	Spatial planning for biodiversity	DIV, CON, GOV, SLO ●●		I, T&S, L&I, C
	Linking ecosystem services with human livelihoods	DIV, PAR, SYS, GOV, SLO ●●●		I, T&S, L&I, C

Climate Change in 2010 (Antarctic Treaty Meeting of Experts, 2010). The outcomes of the meeting led the Antarctic Treaty's Committee for Environmental Protection to develop a Climate Change Response Work Programme (ATCM, 2017).

3.5.4   Towards Resilient Pathways

This section presents the status of practices, tools and strategies currently employed in the Arctic and or Antarctica that can potentially contribute to climate resilient pathways. Seven general strategies for building resilience have been recognised: i) maintain diversity and redundancy, ii) manage connectivity, iii) manage slow variables and feedbacks, iv) foster an understanding of social-ecological systems as complex adaptive systems, v) encourage learning and experimentation, vi) broaden participation, and vii) promote polycentric governance systems (Biggs et al., 2012; Quinlan et al., 2016) (Cross-Chapter Box 2 in Chapter 1).

The practices listed below are not inclusive of the many resilience-building efforts underway in the polar regions. Those described are well represented in the literature and have shown

sufficient utility to merit further use (ARR, 2016; AMAP, 2017a; AMAP, 2017b; AMAP, 2018) (*high confidence*). Some require more refinement while others are well developed. The following sections assess the extent to which these practices operationalise resilience-building through knowledge co-production, the linking of knowledge with decision making, and implementation of resilience-based ecosystem management, considering also their application level and key facilitating conditions; a summary is presented in Table 3.5.

3.5.4.1   Knowledge Co-production and Integration

The co-production of knowledge and transdisciplinary research are currently contributing to the understanding of polar climate change through the use of a diversity of cultural, geographic, and disciplinary perspectives that provide a holistic framing of problems and possible solutions (Miller and Wyborn, 2018; Robards et al., 2018) (*high confidence*).

Several factors are important in successful knowledge co-production, including use of social-ecological frameworks, engagement of a broad set actors with diverse epistemological orientations, a 'team science' approach to studies, strong leadership, attention to process (vs. only products) and mutual respect for cultural differences (Meadow et al., 2015; National Research Council, 2015; Petrov et al., 2016) (*high confidence*). Knowledge co-production involving Indigenous peoples comes with its own set of challenges (Armitage et al., 2011; Robards et al., 2018). While advancements have been made, the practice of knowledge co-production would benefit from further experimentation and innovation in methodologies and better training of researchers (van der Hel, 2016; Vlasova and Volkov, 2016; Berkes, 2017) (*medium confidence*). Three aspects of knowledge co-production are highlighted below.

3.5.4.1.1 Community-based monitoring

Community-based monitoring (CBM) in the Arctic has emerged as a practice of great interest because of its potential to link western ways of knowing with local knowledge and indigenous knowledge (Retter et al., 2004; Johnson et al., 2015a; Johnson et al., 2015b; Kouril et al., 2016; AMAP, 2017a; Williams et al., 2018). In several CBM programs, innovative approaches using the internet, mobile phones, hand-held information devices, and camera-equipped GPS units are capturing, documenting and communicating local observations of change (Brubaker et al., 2011; Brubaker et al., 2013). The integration of community observations with instrument-based observations and its use in research has proven challenging, with technical and cultural issues (Griffith et al., 2018). Execution of CBM programs in the Arctic has also proven to be labour intensive and difficult to sustain, requiring long-term financial support, agreements specifying data ownership, sufficient human capital, and in some cases, the involvement of boundary organisations that provide technical support (Pulsifer et al., 2012; Eicken et al., 2014) and link CBM with governance (CAFF, 2015b; Robards et al., 2018). As is the case in all knowledge production, power relationships (i.e., who decides what is a legitimate observation, who has access to resources for involvement and who benefits) have been challenging where the legitimacy of local knowledge and indigenous knowledge is questioned (e.g., Pristupa et al., 2018). There is *high agreement* and *limited evidence* that CBM facilitates knowledge co-production and resilience building. More analyses of Arctic communities and their institutional capabilities related to CBM are needed to evaluate the potential of these observation systems, and experimentation and innovation may help determine how CBM can more effectively inform decision making beyond the community (Johnson et al., 2015a; Johnson et al., 2015b) (*medium confidence*).

3.5.4.1.2 Understanding regime shifts

Regime shifts are especially important in polar regions where there are limited data and where rapid directional change suggests the possibility of crossing thresholds that may dramatically alter the flow of ecosystem services (ARR, 2016). Better understanding of the thresholds and dynamics of regime shifts (i.e., SES state changes) is especially important for resilience building (ARR, 2016; Biggs et al., 2018; Rocha et al., 2018) (*high confidence*). While polar regime shifts have been documented (Biggs et al., 2018), most are poorly understood and rarely predictable (Rocha et al., 2018) (*high confidence*). Moreover, the focus on Arctic regime shifts to date has been on almost entirely on biophysical state changes that impact social systems.

A limited number of studies have examined social regime shifts and fewer the feedbacks of social regimes shifts on ecosystems (Gerlach et al., 2017). Future needs for advancing knowledge of regime shifts include: 1) continued and refined updating of details on past regimes shifts, 2) structured comparative analysis of these phenomena to ascertain common patterns and variation, 3) greater investment in research resources on potential large-scale regime shifts, and 4) great attention on how social and economic change may affect ecosystems (ARR, 2016; Biggs et al., 2018).

3.5.4.1.3 Indicators of resilience and adaptive capacity

Well-crafted and effectively communicated indicators of polar geophysical, ecological and human systems have the potential to make complex issues more easily understood by society, including local residents and policy makers seeking to assess the implication of climate change (Petrov et al., 2016; Carson and Sommerkorn, 2017) (*medium confidence*). Having indicators of change is no guarantee they will be used; access to information, awareness of changing conditions, and the motivation to act are also important (e.g., van der Linden et al., 2015).

Indicators of the state of polar geophysical systems, biodiversity, ecosystems and human well-being are monitored as part of polar programs. For example, indicators are reported by the Arctic Council working groups Arctic Monitoring and Assessment Programme and Conservation of Arctic Flora and Fauna (e.g., Odland et al., 2016; CAFF, 2017; Box et al., 2019), the International Arctic Social Science Association (e.g., AHDR, 2014), the CCAMLR Ecosystem Monitoring Programme (e.g., Reid et al., 2005) and the Southern Ocean Observing System (e.g., Meredith et al., 2013).

There is limited development of indicators of social-ecological resilience (Jarvis et al., 2013; Carson and Sommerkorn, 2017). As well, indicators of human adaptive capacity are typically based on qualitative case studies with limited quantitative data, and thus have limited comparability and generalisability (Ford and King, 2013; Petrov et al., 2016; Berman et al., 2017) (*high confidence*). The identification and on-going use of indicators of social-ecological resilience are theoretically best achieved through highly participatory processes that engage stakeholders of a locale, with those processes potentially resulting in self-reflection and actions that improve adaptive capacity (Quinlan et al., 2016; Carson and Sommerkorn, 2017), however, this is untested empirically (*low confidence*).

3.5.4.2 Linking Knowledge with Decision Making

While there is a growing expectation in polar (and other) regions for a more deliberate strategy to link science with social learning and policy making about climate change (and other matters) through iterative interactions of researchers, managers and other stakeholders, meeting that expectation is confounded by several deeply rooted issues (Armitage et al., 2011; ARR, 2016; Tesar et al., 2016b; Baztan et al., 2017; Forbis Jr and Hayhoe, 2018) (*medium confidence*).

In spite of the development of practices like those described above and the establishment of many co-managed arrangements in polar regions, scientists and policy makers often work in separate spheres of influence, tend to maintain different values, interests, concerns, responsibilities and perspectives, and gain limited exposure to the other's knowledge system (see Liu et al., 2008; Armitage et al., 2011). Information exchange flows unequally, as officials struggle with information overload and proliferating institutional voices, and where local residents are mistrusting of scientists (Powledge, 2012). Inherent tensions between science-based assessment and interest-based policy, and many existing institutions often prevent direct connectivity. Further, the longstanding science mandate to remain 'policy neutral' typically leads to norms of constrained interaction (Neff, 2009) (*high confidence*).

Creating pathways towards greater climate resilience will therefore depend, in part, on a redefined 'actionable science' that creates bridges supporting better decisions through more rigorous, accessible, and engaging products, while shaping a narrative that instils public confidence (Beier et al., 2015; Fleming and Pyenson, 2017) (*high confidence*). Stakeholders of polar regions are increasingly using a suite of creative tools and practices for moving from theory to practice in resilience building by informing decision making and fostering long-term planning (Baztan et al., 2017). As noted above, these practices include participatory scenario planning, forecasting for stakeholders, and use structured decision making, solution visualisation tools and decision theatres (e.g., Schartmüller et al., 2015; Kofinas et al., 2016; Garrett et al., 2017; Holst-Andersen et al., 2017; Camus and Smit, 2019). The extent to which these practices can contribute to resilience building in the future will depend, in part, on the willingness of key actors such as scientists, to provide active decision-support services, more often than mere decision-support products (Beier et al., 2015). While progress has been made in linking science with policy, more enhanced data collaboration at every scale, more strategic social engagement, communication that both informs decisions and improves climate literacy and explicit creation of consensus documents that provide interpretive guidance about research implications and alternative choices will be important (*high confidence*).

3.5.4.2.1 *Participatory scenario analysis and planning*

Participatory scenario analysis is a quickly evolving and widely used practice in polar regions, and has proven particularly useful for supporting climate adaptation at multiple scales when it uses a social-ecological perspective (ARR, 2016; AMAP, 2017a; Crépin et al., 2017; Planque et al., 2019) (*medium confidence*). While there are technical dimensions in scenario analysis and planning (e.g., the building of useful simulation models that capture and communicate nuanced social-ecological system dynamics such as long-fuse big bang processes, pathological dynamics, critical thresholds, and unforeseen processes (Crépin et al., 2017), there are also creative aspects, such as the use of art to help in the visualisation of possible future (e.g., Planque et al., 2019).

Participatory scenario analysis has been applied to various problem areas related to climate change responses in the polar regions. Applications demonstrate the utility of the practice for identifying possible local futures that consider climate change or socioeconomic

pathways (e.g., in Alaska, Ernst and van Riemsdijk, 2013; and in Eurasian reindeer-herding systems, van Oort et al., 2015; Nilsson et al., 2017) and interacting drivers of change (e.g., in Antarctica; Liggett et al., 2017). Scenario analysis proved helpful for stakeholders with different expertise and perspectives to jointly develop scenarios to inform ecosystem-based management strategies and adaptation options (e.g., in the Barents region; Nilsson et al., 2017; Planque et al., 2019) and to identify research needs (e.g., in Alaska; Vargas-Moreno et al., 2016), including informing and applying climate downscaling efforts (e.g., in Alaska; Ernst and van Riemsdijk, 2013).

A review of scenario analysis in the Arctic, however, found that while the practice is widespread and many are using best practice methods, less than half scenarios programs incorporated climate projections and that those utilising a backcasting approach had higher local participation than those only using forecasting (Flynn et al., 2018). It noted that integrating different knowledge systems and attention to cultural factors influence program utility and acceptance. Planque et al. (2019) also found that most participating stakeholders had limited experience using scenario analysis, suggesting the importance of process methods for engaging stakeholders when exploring possible, likely, and desirable futures. The long-term utility of this practice in helping stakeholders engage with each other to envision possible futures and be forward thinking in decision making will depend on the science of climate projections, further development of decision support systems to inform decision makers, attention to cultural factors and worldview, as well as refinement of processes that facilitate participants' dialogue (*medium confidence*).

3.5.4.2.2 *Structured decision making*

Structured decision making (SDM) is an emerging practice used with stakeholders to identify alternative actions, evaluate trade-offs, and inform decisions in complex situations (Gregory et al., 2012). Few SDM processes have been undertaken in polar regions, with most as exploratory demonstration projects led by researchers. These have included indigenous residents and researchers identifying trade-offs and actions related to subsistence harvesting in a changing environment (Christie et al., 2018) stakeholder interviews to show how a 'triage method' can link community monitoring with community needs and wildlife management priorities (Wheeler et al., 2018), and the application of multi-criteria decision analysis to address difficult decisions related to mining opportunities in Greenland (Trump et al., 2018). The Decision Theater North at the University of Alaska is also being explored as an innovative method of decision support (Kofinas et al., 2016). SDM may have potential in creating climate resilience pathways in polar regions (*low confidence*), but there is currently limited experience with its application.

3.5.4.3   Resilience-based Ecosystem Stewardship

Renewable resource management and biodiversity conservation that seek to maintain resources in historic levels and reduce uncertainty before taking action remains the dominant paradigm in polar regions (Chapin III et al., 2009; Forbes et al., 2015). The effectiveness of this approach, however, is increasingly challenged as the ranges and populations of species and state of ecosystems are being affected by climate change (Chapin III et al., 2010; Chapin III et al., 2015). Three practices that build and maintain social-ecological resilience in the face of climate change include Adaptive Ecosystem Governance, Spatial Planning for Biodiversity, and Linking Management of Ecosystem Services with Human Livelihoods.

3.5.4.3.1   Adaptive ecosystem governance

'Adaptive Ecosystem Governance' differs from conventional resource management or integrated ecosystem management (Chapin III et al., 2009; Chapin III et al., 2010; Chapin III et al., 2015), with a strong focus on trajectories of change (i.e., emergence), implying that maintaining ecosystems in a state of equilibrium is not possible (Biggs et al., 2012; ARR, 2016). This approach strengthens response options by maintaining or increasing resource diversity (to support human adaptation) and biological diversity (to support ecosystem adaptation) (Biggs et al., 2012; Chapin III et al., 2015; Quinlan et al., 2016) (high confidence). Adaptive ecosystem governance emphasises iterative social learning processes of observing, understanding and acting with collaborative partnerships, such as adaptive co-management arrangements currently used in regions of the Arctic (Armitage et al., 2009; Dale and Armitage, 2011; Chapin III et al., 2015; Arp et al., 2019). This approach is also currently realised through adaptive management of Arctic fisheries in Alaska that combines annual measures and within-season provisions informed by assessments of future ecosystem trends (Section 3.5.2.1), and the use of simulation models with Canadian caribou co-management boards to assess the cumulative effects of proposed land use change with climate change (Gunn et al., 2011; Russell, 2014a; Russell, 2014b). Linking these regional efforts to pan-polar programs can enhance resilience building cross multiple scales (e.g., Gunn et al., 2013) (medium confidence).

3.5.4.3.2   Spatial planning for biodiversity

Shifts in the distribution, abundance and human use of species and populations due to climate-induced cryosphere and ocean change, concurrent with land use changes, increase the risks to ecosystem health and biodiversity (Kaiser et al., 2015). Building resilience in these challenging conditions follows from spatial planning for biodiversity that links multiple scales and considers how impacts to ecosystems may materialise in social-ecological systems elsewhere (Bengtsson et al., 2003; Cumming, 2011; Allen et al., 2016). Developing pathways for spatial resilience in polar regions involves systematic planning and designating networks of protected areas to protect connected tracts of representative habitats, and biologically and ecologically significant features (Ban et al., 2014). Protected area networks that combine both spatially rigid and spatially flexible regimes with climate refugia can support ecological resilience to climate change by maintaining connectivity of populations, foodwebs, and the flow of genes across scales (McLeod et al., 2009). This approach reduces direct pressures on biodiversity, and thus gives biological communities, populations and ecosystems the space to adapt (Nyström and Folke, 2001; Hope et al., 2013; Thomas and Gillingham, 2015) (medium confidence). Networks of protected areas are now being planned (Solovyev et al., 2017) and implemented (Juvonen and Kuhmonen, 2013) in the marine and terrestrial Arctic, respectively;

expanding the terrestrial protected area network in Antarctica is discussed (Coetzee et al., 2017). The planning of protected area networks in polar regions is currently an active topic of international collaboration in both polar regions (Arctic Council, 2015b; CCAMLR, 2016a; Wenzel et al., 2016). Designating marine protected area networks contributes to achieving Sustainable Development Goal 14 and the Aichi Targets of the CBD but is often contested due to competing interests for marine resources.

3.5.4.3.3   Linking eosystem services with human livelihoods

Incorporating measures of ecosystem services into assessments is key in integrating environmental, economic, and social policies that build resilience to climate change in polar regions (CAFF, 2015a; Malinauskaite et al., 2019; Sarkki and Acosta García, 2019) (high confidence). Currently, there is limited recognition of the wide range of benefits people receive from polar ecosystems and a lack of management tools that demonstrate their benefits in decision-making processes (CAFF, 2015a). The concept of ecosystem services is increasingly used in the Arctic, yet there continues to be significant knowledge gaps in mapping, valuation, and the study of the social implications of changes in ecosystem services. There are few Arctic examples of the application of ecosystem services in management (Malinauskaite et al., 2019). A strategy of ecosystem stewardship, therefore, is to maintain a continued flow of ecosystem services, recognising how their benefits provide incentives for preserving biodiversity, while also ensuring options for sustainable development and ecosystem-based adaptation (Chapin III et al., 2015; Guerry et al., 2015; Díaz et al., 2019). Arctic stewardship opportunities at landscape, seascape, and community scales to a great extent lie in supporting culturally engrained (often traditional indigenous) values of respect for land and animals, and reliance on the local environment through the sharing of knowledge and power between local users of renewable resources and agencies responsible for managing resources (Mengerink et al., 2017) (high confidence). In the Antarctic, ecosystem stewardship is dependent on international formally-defined and informally-enacted cooperation, and the recognition of its service to the global community (Section 3.5.3.2).

3.6   Synopsis

This chapter has assessed the consequences of climate change in the polar regions in three sections, focusing on sea ice and the ocean (Section 3.2), glaciers and ice sheets (Section 3.3), and permafrost and snow on land (Section 3.4). A systems approach was taken to assess individual and interacting changes within and between these elements to consider consequences, impacts and risks for marine and terrestrial ecosystems and for people. Mapping on to those observed and projected impacts, Section 3.5 assessed human responses to climate change in the polar regions. This brief synopsis considers the chapter findings across sections, and draws out three key assessment points that inform global responses to polar ocean and cryosphere change.

1. *Climate-induced changes to the polar cryosphere and oceans have global consequences and impacts.*

Loss of Arctic sea ice continues to add to global radiative forcing and has the potential to influence mid-latitude weather on timescales of weeks to months (Section 3.2.1 Box 3.2), the Southern Ocean takes up a disproportionately high amount of atmospheric heat and carbon (Section 3.2.1), melting polar glaciers and the AIS and GIS contribute to observed and projected sea level rise long into the future (Sections 3.3.1, 3.3.2), and projected widespread disappearance of permafrost has the potential to accelerate global warming through the release of carbon dioxide and methane (Sections 3.4.2, 3.4.3).

2. *Across many aspects, the polar regions of the future will appear significantly different from those of today.*

By the end of this century, Arctic sea ice and snow on land are projected to be diminished compared with today, as are the masses of the AIS and GIS and the polar glaciers (Sections 3.2.2, 3.3.2; 3.4.2). Acidification of both polar oceans will have progressed; this, and changing marine habitats associated with ocean warming, are projected to impact marine ecosystems (Sections 3.2.2, 3.2.3). Permafrost thaw and decrease in snow on land are projected to drive habitat and biome shifts, affecting ranges and abundance of ecologically important species, and driving changes in wildfire, vegetation and human infrastructure (Sections 3.4.2, 3.4.3). Collectively, these very different future polar environments pose strong challenges to sustainable use of natural resources, infrastructure, cultures and lives and livelihoods. Although manifested locally, these very different future polar environments have the potential to continue/accelerate the global impacts noted above.

3. *Choices are available that will influence the nature and magnitude of changes, potentially limiting their regional and global impacts and increasing the effectiveness of adaptation actions.*

Compared with high greenhouse gas emissions scenarios, projections using low emissions scenarios result in polar regions that will be significantly less altered. For example, for stabilised global warming of 1.5°C, a sea ice-free Arctic in September is projected to occur significantly less frequently than at 2°C (1% c.f. 10–35%) (Section 3.2.2). The potential for reduced but stabilised Arctic autumn and spring snow extent by mid-century for RCP2.6 contrasts with continued loss under RCP8.5, and the area with near-surface permafrost is projected to decrease less by 2100 under RCP2.6 than under RCP8.5 (Section 3.4.2). Polar glaciers are projected to lose much less mass between 2015 and 2100 under RCP2.6 compared with RCP8.5 (Cross-Chapter Box 6 in Chapter 2). Acidification of the polar oceans will progress more slowly and affect much smaller areas under RCP2.6 compared with RCP8.5 (Section 3.2.2). These differences have strong implications for natural resource management, economic sectors and Arctic cultures (Section 3.5.2). The choices that enable these differences influence the rate and magnitude of polar change, their consequences and impacts at regional and global scales, the effectiveness of adaptation and opportunities for climate resilient pathways (Section 3.5.4).

3.7   Key Knowledge Gaps and Uncertainties

Beyond this report, progress requires that future assessments demonstrate increased confidence in various key aspects; this can be achieved by narrowing numerous gaps in knowledge. Some of the critical ones, which are priorities for future initiatives, are outlined here.

Overturning circulation in the Southern Ocean is a key factor that controls heat and carbon exchanges with the atmosphere, and hence global climate, however there are no direct measures of this and only sparse indirect indicators of how it may be changing. This is a critical weakness in sustained observations of the global ocean.

Snow depth on sea ice is essentially unmeasured, limiting mass balance estimates and ice thickness retrievals. Improved mechanistic understanding of the observed changes and trends in Antarctic sea ice is required, notably the decadal increase and very recent rapid retreat. This has consequences for climate, ecosystems and fisheries; however, lack of understanding and poor model performance translates to very limited predictive skill.

Trends in snow water equivalent over Arctic land are inadequately known, reducing confidence in assessments of snow's role in the water cycle and in insulating the underlying permafrost. Understanding of precipitation in the polar regions is critically limited by sparse observations, and there is a lack of understanding of the processes that drive regional variability in wetting/drying and greening/browning of the Arctic land surface. There is inadequate knowledge concerning carbon dioxide and methane emissions from land and sub-sea permafrost.

There are clear regional gaps in knowledge of polar ecosystems and biodiversity, and insufficient population estimates/trends for many key species. Biodiversity projections are limited by key uncertainties regarding the potential for organisms to adapt to habitat change and the resilience of foodweb structures. Relatedly, knowledge gaps exist concerning how fisheries target levels will change alongside environmental change and how to incorporate this into decision making. Similarly, there are knowledge gaps on the extent to which changes in the availability of resources to subsistence harvesters affects food security of households.

There is a need to better understand the evolution of polar glaciers and ice sheets, and their influences on global sea level. Longer and improved quantifications of their changes are required, especially where mass losses are greatest, and (relatedly) better attribution of natural versus anthropogenic drivers. Better understanding of the sensitivity of Antarctica to marine ice sheet instability is required, and whether recent changes in West Antarctica represent the onset of irreversible change.

There are critical gaps in knowledge concerning interactions between the atmosphere and specific elements of the polar ocean and cryosphere. Detailed assessment of atmospheric processes was outside the remit of this chapter, however such gaps limit understanding of ongoing and future trajectories of the polar regions and their climate systems. Relatedly, there is a paucity of studies analysing differences in the trajectories of polar cryosphere and ocean systems between low and very low greenhouse gas emission scenarios.

There are critical needs to better understand the efficacy and limits of strategies for reducing risk and strengthening resilience for polar ecosystems and people, including the contribution of practices and tools to contribute to climate resilient pathways. Knowledge on how to translate existing theoretical understandings of social-ecological resilience into decision making and governance is limited. There is limited understanding concerning the resources that are needed for successful adaptation responses and about the effectiveness of institutions in supporting adaptation. While the occurrence of regime shifts in polar systems is both documented and anticipated, there is little or no understanding of their preconditions or of indicators that would help pre-empt them.

Frequently Asked Questions

FAQ 3.1 | How do changes in the Polar Regions affect other parts of the world?

Climate change in the Arctic and Antarctic affect people outside of the polar regions in two key ways. First, physical and ecosystem changes in the polar regions have socioeconomic impacts that extend across the globe. Second, physical changes in the Arctic and Antarctic influence processes that are important for global climate and sea level.

Among the risks to societies and economies, aspects of food provision, transport and access to non-renewable resources are of great importance. Fisheries in the polar oceans support regional and global food security and are important for the economies of many countries around the world, but climate change alters Arctic and Antarctic marine habitats, and affects the ability of polar species and ecosystems to withstand or adapt to physical changes. This has consequences for where, when, and how many fish can be captured. Impacts will vary between regions, depending on the degree of climate change and the effectiveness of human responses. While management in some polar fisheries is among the most developed, scientists are exploring modifications to existing precautionary, ecosystem-based management approaches to increase the scope for adaptation to climate change impacts on marine ecosystems and fisheries.

New shipping routes through the Arctic offer cost savings because they are shorter than traditional passages via the Suez or Panama Canals. Ship traffic has already increased and is projected to become more feasible in the coming decades as further reductions in sea ice cover make Arctic routes more accessible. Increased Arctic shipping has significant socioeconomic and political implications for global trade, northern nations and economies strongly linked to traditional shipping corridors, while also increasing environmental risk in the Arctic. Reduced Arctic sea ice cover allows greater access to offshore petroleum resources and ports supporting resource extraction on land.

The polar regions influence the global climate through a number of processes. As spring snow and summer sea ice cover decrease, more heat is absorbed at the surface. There is growing evidence that ongoing changes in the Arctic, primarily sea ice loss, can potentially influence mid-latitude weather. As temperatures increase in the Arctic, permafrost soils in northern regions store less carbon. The release of carbon dioxide and methane from the land to the atmosphere further contributes to global warming.

Melting ice sheets and glaciers in the polar regions cause sea levels to rise, affecting coastal regions and their large populations and economies. At present, the Greenland Ice Sheet (GIS) and polar glaciers are contributing more to sea level rise than the Antarctic Ice Sheet (AIS). However, ice loss from the AIS has continued to accelerate, driven primarily by increased melting of the underside of floating ice shelves, which has caused glaciers to flow faster. Even though it remains difficult to project the amount of ice loss from Antarctica after the second half of the 21st century, it is expected to contribute significantly to future sea level rise.

The Southern Ocean that surrounds Antarctica is the main region globally where waters at depth rise to the surface. Here, they become transformed into cold, dense waters that sink back to the deep ocean, storing significant amounts of human-produced heat and dissolved carbon for decades to centuries or longer, and helping to slow the rate of global warming in the atmosphere. Future changes in the strength of this ocean circulation can so far only be projected with limited certainty.

Acknowledgements

The authors acknowledge the following individuals for their assistance in compiling references containing indigenous knowledge for the Polar Regions Chapter: Claudio Aporta (Canada), David Atkinson (Canada), Todd Brinkmann (USA), Courtney Carothers (USA), Ashlee Cunsolo (Canada), Susan Crate (USA), Bathsheba Demuth (USA), Alexandra Lavrillier (France), Andrey Petrov (USA/Russia), Jon Rosales (USA), Florian Stammler (Finland), Hiroki Takakura (Japan), Wilbert van Rooij (Netherlands), Brent Wolfe (Canada), Torre Jorgenson (USA). Laura Gerrish (UK) is thanked for help with figure preparation.

References

Aakre, S., S. Kallbekken, R. Van Dingenen and D.G. Victor, 2018: Incentives for small clubs of Arctic countries to limit black carbon and methane emissions. *Nature Climate Change*, **8** (1), 85–90, doi:10.1038/s41558-017-0030-8.

Aars, J. et al., 2017: The number and distribution of polar bears in the western Barents Sea. *Polar Research*, **36** (1), 1374125, doi:10.1080/17518369.2017.1374125.

Abadi, F., C. Barbraud and O. Gimenez, 2017: Integrated population modeling reveals the impact of climate on the survival of juvenile emperor penguins. *Global Change Biology*, **23** (3), 1353–1359, doi:10.1111/gcb.13538.

Abbott, B.W. et al., 2015: Patterns and persistence of hydrologic carbon and nutrient export from collapsing upland permafrost. *Biogeosciences*, **12** (12), 3725–3740, doi:10.5194/bg-12-3725-2015.

Abbott, B.W. et al., 2014: Elevated dissolved organic carbon biodegradability from thawing and collapsing permafrost. *Journal of Geophysical Research: Biogeosciences*, **119** (10), 2049–2063, doi:10.1002/2014jg002678.

Abernathey, R. and D. Ferreira, 2015: Southern Ocean isopycnal mixing and ventilation changes driven by winds. *Geophysical Research Letters*, **42** (23), 10,357–10,365, doi:10.1002/2015gl066238.

Abernathey, R.P. et al., 2016: Water-mass transformation by sea ice in the upper branch of the Southern Ocean overturning. *Nature Geoscience*, **9**, 596, doi:10.1038/ngeo2749.

Abrahms, B. et al., 2018: Climate mediates the success of migration strategies in a marine predator. *Ecol Lett*, **21** (1), 63–71, doi:10.1111/ele.12871.

Abram, N.J. et al., 2014: Evolution of the Southern Annular Mode during the past millennium. *Nature Climate Change*, **4** (7), 564–569, doi:10.1038/nclimate2235.

Abram, N.J. et al., 2013a: Acceleration of snow melt in an Antarctic Peninsula ice core during the twentieth century. *Nature Geoscience*, **6** (5), 404–411, doi:10.1038/ngeo1787.

Abram, N.J., E.W. Wolff and M.A.J. Curran, 2013b: A review of sea ice proxy information from polar ice cores. *Quaternary Science Reviews*, **79**, 168–183, doi:10.1016/j.quascirev.2013.01.011.

Adusumilli, S. et al., 2018: Variable Basal Melt Rates of Antarctic Peninsula Ice Shelves, 1994–2016. *Geophysical Research Letters*, **45** (9), 4086–4095, doi:10.1002/2017GL076652.

AHDR, 2014: *Arctic Human Development Report: Regional Processes and Global Linkages*. Nordic Council of Ministers, Larsen, J.N. and G. Fondahl, Copenhagen [Available at: http://norden.diva-portal.org/smash/get/diva2:788965/FULLTEXT03.pdf; Access Date: 10 October 2017].

Ahlstrom, A.P. et al., 2017: Abrupt shift in the observed runoff from the southwestern Greenland ice sheet. *Sci Adv*, **3** (12), e1701169, doi:10.1126/sciadv.1701169.

Ainley, D.G. et al., 2015: Trophic cascades in the western Ross Sea, Antarctica: revisited. *Marine Ecology Progress Series*, **534**, 1–16, doi:10.3354/meps11394.

Aitken, A.R.A. et al., 2016: Repeated large-scale retreat and advance of Totten Glacier indicated by inland bed erosion. *Nature*, **533**, 385, doi:10.1038/nature17447.

Aksenov, Y. et al., 2017: On the future navigability of Arctic sea routes: High-resolution projections of the Arctic Ocean and sea ice. *Marine Policy*, **75**, 300–317, doi:10.1016/j.marpol.2015.12.027.

Alabia, I.D. et al., 2018: Distribution shifts of marine taxa in the Pacific Arctic under contemporary climate changes. *Diversity and Distributions*, **24** (11), 1583–1597, doi:10.1111/ddi.12788.

Albon, S., D. et al., 2017: Contrasting effects of summer and winter warming on body mass explain population dynamics in a food-limited Arctic herbivore. *Global Change Biology*, **23** (4), 1374–1389, doi:10.1111/gcb.13435.

Alexeev, V., A., C.D. Arp, B.M. Jones and L. Cai, 2016: Arctic sea ice decline contributes to thinning lake ice trend in northern Alaska. *Environmental Research Letters*, **11** (7), 074022, doi:10.1088/1748-9326/11/7/074022.

Allen, C.R. et al., 2016: Quantifying spatial resilience. *Journal of Applied Ecology*, **53** (3), 625–635, doi:10.1111/1365-2664.12634.

AMAP, 2015: *AMAP Assessment 2015: Temporal Trends in Persistent Organic Pollutants in the Arctic*. Arctic Monitoring and Assessment Programme (AMAP), Oslo, Norway, vi+71 pp [Available at: www.amap.no/documents/doc/amap-assessment-2015-temporal-trends-in-persistent-organic-pollutants-in-the-arctic/1521; Access Date: 25 October 2018].

AMAP, 2017a: *Adaptation Actions for a Changing Arctic (AACA) – Bering/Chukchi/Beaufort Region Overview report*. Arctic Monitoring and Assessment Programme (AMAP), Oslo, Norway, 24 pp.

AMAP, 2017b: *Adaptation Actions for a Changing Arctic: Perspectives from the Barents Area*. Arctic Monitoring and Assessment Programme (AMAP), xiv + 267 pp.

AMAP, 2017c: *AMAP Assessment 2016: Chemicals of Emerging Arctic Concern*. Arctic Monitoring and Assessment Programme (AMAP), Oslo, Norway, xvi+353 pp [Available at: www.amap.no/documents/doc/AMAP-Assessment-2016-Chemicals-of-Emerging-Arctic-Concern/1624; Access Date: 10 October 2018].

AMAP, 2017d: *Snow, Water, Ice and Permafrost in the Arctic (SWIPA)*. Arctic Council Secretariat, Oslo, Norway, xiv + 269 pp [Available at: www.amap.no/documents/download/2987/inline; Access Date: 10 October 2018].

AMAP, 2018: *Adaptation Actions for a Changing Arctic: Perspectives from the Baffin Bay/Davis Strait Region*. Arctic Monitoring and Assessment Programme (AMAP), Oslo, Norway, xvi+354 pp [Available at: www.amap.no/documents/doc/Adaptation-Actions-for-a-Changing-Arctic-Perspectives-from-the-Baffin-BayDavis-Strait-Region/1630; Access Date: 10 October 2018].

Ancel, A. et al., 2017: Looking for new emperor penguin colonies? Filling the gaps. *Global Ecology and Conservation*, **9**, 171–179, doi:10.1016/j.gecco.2017.01.003.

Andersen, M., A.E. Derocher, Ø. Wiig and J. Aars, 2012: Polar bear (Ursus maritimus) maternity den distribution in Svalbard, Norway. *Polar Biology*, **35** (4), 499–508, doi:10.1007/s00300-011-1094-y.

Andersen, M.L. et al., 2015: Basin-scale partitioning of Greenland ice sheet mass balance components (2007–2011). *Earth and Planetary Science Letters*, **409**, 89–95, doi:10.1016/j.epsl.2014.10.015.

Anderson, E.H., 2012: Polar shipping, the forthcoming polar code implications for the polar environments. *Journal of Maritime Law and Commerce*, **43** (1), 59–64.

Anderson, L.G. et al., 2017a: Export of calcium carbonate corrosive waters from the East Siberian Sea. *Biogeosciences*, **14** (7), 1811–1823, doi:10.5194/bg-14-1811-2017.

Anderson, S.C. et al., 2017b: Benefits and risks of diversification for individual fishers. *Proceedings of the National Academy of Sciences*, **114** (40), 10797–10802, doi:10.1073/pnas.1702506114.

Andresen, C.S. et al., 2012: Rapid response of Helheim Glacier in Greenland to climate variability over the past century. *Nature Geoscience*, **5** (1), 37–41, doi:10.1038/ngeo1349.

Andrews, A.J. et al., 2019: Boreal marine fauna from the Barents Sea disperse to Arctic Northeast Greenland. *Sci Rep*, **9** (1), 5799, doi:10.1038/s41598-019-42097-x.

Andrews, J., D. Babb and D.G. Barber, 2018: Climate change and sea ice: Shipping in Hudson Bay, Hudson Strait, and Foxe Basin (1980–2016). *Elementa-Science of the Anthropocene*, **6** (1), p.19, doi:http://doi.org/10.1525/elementa.281.

Angelopoulos, M. et al., 2019: Heat and Salt Flow in Subsea Permafrost Modeled with CryoGRID2. *Journal of Geophysical Research: Earth Surface*, **124** (4), 920–937, doi:10.1029/2018jf004823.

Anisimov, O.A., I.I. Borzenkova, S.A. Lavrov and J.G. Strelchenko, 2012: Dynamics of sub-aquatic permafrost and methane emission at eastern Arctic sea shelf under past and future climatic changes. *Ice and Snow*, **2**, 97–105, doi:10.1002/lno.10307.

Antarctic Treaty Meeting of Experts, 2010: *Co-Chairs' Report from Antarctic Treaty Meeting of Experts on Implications of Climate Change for Antarctic Management and Governance*. Antarctic Treaty Secretariat, Buenos Aires, Argentina.

Anttila, K. et al., 2018: The role of climate and land use in the changes in surface albedo prior to snow melt and the timing of melt season of seasonal snow in northern land areas of 40°N–80°N during 1982–2015. *Remote Sensing*, **10** (10), 1619, doi:10.3390/rs10101619.

Arblaster, J.M. et al., 2014: Stratospheric ozone changes and climate, Chapter 4 in Scientific Assessment of Ozone Depletion.[World Meteorological Organization, G.S. (ed.)], Geneva, iv + 57 pp; Global Ozone Research and Monitoring Project – Report No. 55.

Arblaster, J.M. and G.A. Meehl, 2006: Contributions of External Forcings to Southern Annular Mode Trends. *Journal of Climate*, **19** (12), 2896–2905, doi:10.1175/jcli3774.1.

Arctic Council, 2015a: *Arctic Marine Tourism Project (AMTP): best practices guidelines*. Protection of the Arctic Marine Environment (PAME), Iceland, 17 pp [Available at: https://oaarchive.arctic-council.org/bitstream/handle/11374/414/AMTP%20Best%20Practice%20Guidelines.pdf?sequence=1&isAllowed=y; Access Date: 10 October 2018].

Arctic Council, 2015b: *Framework for a Pan-Arctic Network of Marine Protected Areas*. Protection of the Arctic Marine Environment (PAME), Iceland, 52 pp [Available at: https://oaarchive.arctic-council.org/handle/11374/417; Access Date: 28 March 2019].

Arctic Council, 2017: *Expert Group on Black Carbon and Methane: Summary of progress and recommendations*. Expert Group on Black Carbon and Methane (EGBCM), 49 pp [Available at: https://oaarchive.arctic-council.org/handle/11374/1936; Access Date: 28 March 2019].

Ardyna, M. et al., 2017: Shelf-basin gradients shape ecological phytoplankton niches and community composition in the coastal Arctic Ocean (Beaufort Sea). *Limnology and Oceanography*, **62** (5), 2113–2132, doi:10.1002/lno.10554.

Armitage, D. et al., 2011: Co-management and the co-production of knowledge: Learning to adapt in Canada's Arctic. *Global Environmental Change*, **21** (3), 995–1004, doi:10.1016/j.gloenvcha.2011.04.006.

Armitage, D.R. et al., 2009: Adaptive co-management for social–ecological complexity. *Frontiers in Ecology and the Environment*, **7** (2), 95–102, doi:10.1890/070089.

Armitage, T.W. et al., 2017: Arctic Ocean surface geostrophic circulation 2003–2014. *The Cryosphere*, **11** (4), 1767, doi:10.5194/tc-11-1767-2017.

Armitage, T.W. et al., 2016: Arctic sea surface height variability and change from satellite radar altimetry and GRACE, 2003–2014. *Journal of Geophysical Research: Oceans*, **121** (6), 4303–4322, doi:10.1002/2015JC011579.

Armour, K.C. et al., 2011: The reversibility of sea ice loss in a state-of-the-art climate model. *Geophysical Research Letters*, **38** (16), L16705, doi:10.1029/2011gl048739.

Armour, K.C. et al., 2016: Southern Ocean warming delayed by circumpolar upwelling and equatorward transport. *Nature Geoscience*, **9** (7), 549, doi:10.1038/Ngeo2731.

Aronson, R.B. et al., 2015: No barrier to emergence of bathyal king crabs on the Antarctic shelf. *Proc Natl Acad Sci U S A*, **112** (42), 12997–13002, doi:10.1073/pnas.1513962112.

Arp, C.D. et al., 2016: Threshold sensitivity of shallow Arctic lakes and sublake permafrost to changing winter climate. *Geophysical Research Letters*, **43** (12), 6358–6365, doi:10.1002/2016gl068506.

Arp, C.D. et al., 2015: Depth, ice thickness, and ice-out timing cause divergent hydrologic responses among Arctic lakes. *Water Resources Research*, **51** (12), 9379–9401, doi:10.1002/2015wr017362.

Arp, C.D. et al., 2019: Ice roads through lake-rich Arctic watersheds: Integrating climate uncertainty and freshwater habitat responses into adaptive management. *Arctic, Antarctic, and Alpine Research*, **51** (1), 9–23, doi:10.1080/15230430.2018.1560839.

ARR, 2016: *Arctic Resilience Report* [Carson, M. and G. Peterson (eds.)]. Arctic Council, Stockholm Environment Institute and Stockholm Resilience Centre, Stockholm.

Arrigo, K.R. et al., 2014: Phytoplankton blooms beneath the sea ice in the Chukchi sea. *Deep Sea Research Part II: Topical Studies in Oceanography*, **105** (Supplement C), 1–16, doi:10.1016/j.dsr2.2014.03.018.

Arrigo, K.R. et al., 2012: Massive phytoplankton blooms under sea ice. *Science*, **336**, 1408, doi:10.1126/science.1215065.

Arrigo, K.R. and G.L. van Dijken, 2004: Annual cycles of sea ice and phytoplankton in Cape Bathurst polynya, southeastern Beaufort Sea, Canadian Arctic. *Geophysical Research Letters*, **31** (8), doi:10.1029/2003gl018978.

Arrigo, K.R. and G.L. van Dijken, 2011: Secular trends in Arctic Ocean net primary production. *J. Geophys. Res.*, **116** (C9), C09011, doi:10.1029/2011jc007151.

Arrigo, K.R. and G.L. van Dijken, 2015: Continued increases in Arctic Ocean primary production. *Progress in Oceanography*, **136**, 60–70, doi:10.1016/j.pocean.2015.05.002.

Arrigo, K.R. et al., 2017a: Early Spring Phytoplankton Dynamics in the Western Antarctic Peninsula. *Journal of Geophysical Research-Oceans*, **122** (12), 9350–9369, doi:10.1002/2017jc013281.

Arrigo, K.R., G.L. van Dijken and S. Bushinsky, 2008: Primary production in the Southern Ocean, 1997–2006. *Journal of Geophysical Research: Oceans*, **113** (C8), doi:10.1029/2007JC004551.

Arrigo, K.R. et al., 2017b: Melting glaciers stimulate large summer phytoplankton blooms in southwest Greenland waters. *Geophysical Research Letters*, **44** (12), 6278–6285, doi:10.1002/2017gl073583.

Arrigo, K.R., G.L. van Dijken and A.L. Strong, 2015: Environmental controls of marine productivity hot spots around Antarctica. *Journal of Geophysical Research: Oceans*, **120** (8), 5545–5565, doi:10.1002/2015jc010888.

Årthun, M., T. Eldevik and L.H. Smedsrud, 2019: The role of Atlantic heat transport in future Arctic winter sea ice loss. *Journal of Climate*, **32** (11), 3327–3341, doi:10.1175/jcli-d-18-0750.1.

Arthun, M. et al., 2012: Quantifying the Influence of Atlantic Heat on Barents Sea Ice Variability and Retreat. *Journal of Climate*, **25** (13), 4736–4743, doi:10.1175/Jcli-D-11-00466.1.

Asselin, N.C. et al., 2011: Beluga (Delphinapterus leucas) habitat selection in the eastern Beaufort Sea in spring, 1975–1979. *Polar Biology*, **34** (12), 1973–1988, doi:10.1007/s00300-011-0990-5.

Assmy, P. et al., 2017: Leads in Arctic pack ice enable early phytoplankton blooms below snow-covered sea ice. *Scientific Reports*, **7**, 40850, doi:10.1038/srep40850.

ATCM, 2016: Final Report of the Thirty-Ninth Antarctic Treaty Consultative Meeting. In: *Antarctic Treaty Consultative Meeting XXXIX*, 23 May – 1 June 2016, Santiago, Chile [Secretariat, A.T. (ed.)], 406 pp.

ATCM, 2017: Final Report of the Fortieth Antarctic Treaty Consultative Meeting. In: *Antarctic Treaty Consultative Meeting XL*, 22 May – 1 June 2017, Beijing, China [Secretariat, A.T. (ed.)], 285 pp.

ATCM, 2018: *IAATO Overview of Antarctic Tourism: 2017–18 Season and Preliminary Estimates for 2018–19 Season*.ATCM, Buenos Aires [Available at: https://iaato.org/documents/10157/2398215/IAATO+overview/bc34db24-e1dc-4eab-997a-4401836b7033].

Atkinson, A. et al., 2019: Krill (Euphausia superba) distribution contracts southward during rapid regional warming. *Nature Climate Change*, **9** (2), 142–147, doi:10.1038/s41558-018-0370-z.

Atkinson, A., V. Siegel, E. Pakhomov and P. Rothery, 2004: Long-term decline in krill stock and increase in salps within the Southern Ocean. *Nature*, **432** (7013), 100–103, doi:10.1038/nature0299.

Åtland, K., 2013: The Security Implications of Climate Change in the Arctic Ocean. In: *Environmental Security in the Arctic Ocean*, Dordrecht, [Berkman, P. and A. Vylegzhanin (eds.)], Springer Netherlands, 205–216.

Attard, C.R. et al., 2015: Low genetic diversity in pygmy blue whales is due to climate-induced diversification rather than anthropogenic impacts. *Biology Letters*, **11** (5), 20141037, doi:10.1098/rsbl.2014.1037.

Austin, S.E. et al., 2015: Public health adaptation to climate change in canadian jurisdictions. *International Journal of Environmental Research and Public Health*, **12** (1), 623–651, doi:10.3390/ijerph120100623.

Ayarzaguena, B. and J.A. Screen, 2016: Future Arctic sea ice loss reduces severity of cold air outbreaks in midlatitudes. *Geophysical Research Letters*, **43** (6), 2801–2809, doi:10.1002/2016gl068092.

Ayles, B., L. Porta and R.M. Clarke, 2016: Development of an integrated fisheries co-management framework for new and emerging commercial fisheries in the Canadian Beaufort Sea. *Marine Policy*, **72**, 246–254, doi:10.1016/j.marpol.2016.04.032.

Azaneu, M., R. Kerr, M.M. Mata and C.A. Garcia, 2013: Trends in the deep Southern Ocean (1958–2010): Implications for Antarctic Bottom Water properties and volume export. *Journal of Geophysical Research: Oceans*, **118** (9), 4213–4227.

Azetsu-Scott, K., M. Starr, Z. Mei and M. Granskog, 2014: Low calcium carbonate saturation state in an Arctic inland sea having large and varying fluvial inputs: The Hudson Bay system. *Journal of Geophysical Research-Oceans*, **119** (9), 6210–6220, doi:10.1002/2014jc009948.

Bader, J. et al., 2013: Atmospheric winter response to a projected future Antarctic sea-ice reduction: a dynamical analysis. *Climate Dynamics*, **40** (11), 2707–2718, doi:10.1007/s00382-012-1507-9.

Bailey, A. et al., 2017: Early life stages of the Arctic copepod Calanus glacialis are unaffected by increased seawater pCO2. *ICES Journal of Marine Science*, **74** (4), 996–1004, doi:10.1093/icesjms/fsw066.

Baird, J., R. Plummer and Ö. Bodin, 2016: Collaborative governance for climate change adaptation in Canada: experimenting with adaptive co-management. *Regional Environmental Change*, **16** (3), 747–758, doi:10.1007/s10113-015-0790-5.

Bajzak, C.E., M.O. Hammill, G.B. Stenson and S. Prinsenberg, 2011: Drifting away: implications of changes in ice conditions for a pack-ice-breeding phocid, the harp seal (Pagophilus groenlandicus). *Canadian Journal of Zoology*, **89** (11), 1050–1062, doi:10.1139/z11-081.

Baker, B. and B. Yeager, 2015: Coordinated Ocean Stewardship in the Arctic: Needs, Challenges and Possible Models for an Arctic Ocean Coordinating Agreement. *Transnational Environmental Law*, **4** (2), 359–394, doi:10.1017/S2047102515000151.

Ballinger, T.J. et al., 2018: Greenland coastal air temperatures linked to Baffin Bay and Greenland Sea ice conditions during autumn through regional blocking patterns. *Climate Dynamics*, **50** (1), 83–100, doi:10.1007/s00382-017-3583-3.

Balser, A.W., J.B. Jones and R. Gens, 2014: Timing of retrogressive thaw slump initiation in the Noatak Basin, northwest Alaska, USA. *Journal of Geophysical Research: Earth Surface*, **119** (5), 1106–1120, doi:10.1002/2013JF002889.

Balshi, M.S. et al., 2009: Vulnerability of carbon storage in North American boreal forests to wildfires during the 21st century. *Global Change Biology*, **15** (6), 1491–1510, doi:10.1111/j.1365-2486.2009.01877.x.

Bamber, J.L., R.M. Westaway, B. Marzeion and B. Wouters, 2018: The land ice contribution to sea level during the satellite era. *Environmental Research Letters*, **13** (6), 063008, doi:10.1088/1748-9326/aac2f0/meta.

Ban, N.C. et al., 2014: Systematic Conservation Planning: A Better Recipe for Managing the High Seas for Biodiversity Conservation and Sustainable Use. *Conservation Letters*, **7** (1), 41–54, doi:10.1111/conl.12010.

Barber, D. et al., 2017: Arctic Sea Ice. In: Snow, Water, Ice and Permafrost in the Arctic (SWIPA). Arctic Monitoring and Assessment Programme, Oslo, 103–136.

Barber, D.G. et al., 2012: Consequences of change and variability in sea ice on marine ecosystem and biogeochemical processes during the 2007–2008 Canadian International Polar Year program. *Climatic Change*, **115** (1), 135–159, doi:10.1007/s10584-012-0482-9.

Barber, D.G. and R.A. Massom, 2007: Chapter 1 The Role of Sea Ice in Arctic and Antarctic Polynyas. In: Elsevier Oceanography Series [Smith, W.O. and D.G. Barber (eds.)]. Elsevier, **74**, 1–54.

Barletta, V.R. et al., 2018: Observed rapid bedrock uplift in Amundsen Sea Embayment promotes ice-sheet stability. *Science*, **360** (6395), 1335–1339, doi:10.1126/science.aao1447.

Barnes, D.K.A., 2017: Polar zoobenthos blue carbon storage increases with sea ice losses, because across-shelf growth gains from longer algal blooms outweigh ice scour mortality in the shallows. *Glob Chang Biol*, **23** (12), 5083–5091, doi:10.1111/gcb.13772.

Barnes, D.K.A. et al., 2018: Icebergs, sea ice, blue carbon and Antarctic climate feedbacks. *Philos Trans A Math Phys Eng Sci*, **376** (2122), doi:10.1098/rsta.2017.0176.

Barnes, E.A. and L.M. Polvani, 2015: CMIP5 Projections of Arctic Amplification, of the North American/North Atlantic Circulation, and of Their Relationship. *Journal of Climate*, **28** (13), 5254–5271, doi:10.1175/jcli-d-14-00589.1.

Barnhart, K.R., C.R. Miller, I. Overeem and J.E. Kay, 2015: Mapping the future expansion of Arctic open water. *Nature Climate Change*, **6**, 280, doi:10.1038/nclimate2848

Barret, J., 2016: Securing the Polar Regions through International Law. In: Security and International Law [Footer, M.E., J. Schmidt, N.D. White and L. Davies (eds.)]. Bright, Oxford and Portland, Oregan, USA. Barrett, J.H. et al., 2011: Interpreting the expansion of sea fishing in medieval Europe using stable isotope analysis of archaeological cod bones. *Journal of Archaeological Science*, **38** (7), 1516–1524, doi:10.1016/j.jas.2011.02.017.

Bartsch, A., T. Kumpula, B.C. Forbes and F. Stammler, 2010: Detection of snow surface thawing and refreezing in the Eurasian Arctic with QuikSCAT: implications for reindeer herding. *Ecological Applications*, **20** (8), 2346–2358, doi:10.1890/09-1927.1.

Bartsch, A. et al., 2017: Circumpolar Mapping of Ground-Fast Lake Ice. *Frontiers in Earth Science*, **5**, 12, doi:10.3389/feart.2017.00012.

Bates, N.R. et al., 2014: Sea-ice melt CO2-carbonate chemistry in the western Arctic Ocean: meltwater contributions to air-sea CO2 gas exchange, mixed-layer properties and rates of net community production under sea ice. *Biogeosciences*, **11** (23), 6769–6789, doi:10.5194/bg-11-6769-2014.

Bauch, D. et al., 2012: Impact of Siberian coastal polynyas on shelf-derived Arctic Ocean halocline waters. *Journal of Geophysical Research: Oceans*, **117** (C9), doi:10.1029/2011JC007282.

3

Bauch, D. et al., 2011: Origin of freshwater and polynya water in the Arctic Ocean halocline in summer 2007. *Progress in Oceanography*, **91** (4), 482–495, doi:10.1016/j.pocean.2011.07.017.

Baztan, J. et al., 2017: Life on thin ice: Insights from Uummannaq, Greenland for connecting climate science with Arctic communities. *Polar Science*, **13**, 100–108, doi:10.1016/j.polar.2017.05.002.

Beaumier, M.C., J.D. Ford and S. Tagalik, 2015: The food security of Inuit women in Arviat, Nunavut: the role of socio-economic factors and climate change. *Polar Record*, **51** (5), 550–559, doi:10.1017/s0032247414000618.

Beck, P.S.A. and S.J. Goetz, 2011: Satellite observations of high northern latitude vegetation productivity changes between 1982 and 2008: ecological variability and regional differences. *Environmental Research Letters*, **6** (4), 049501.

Beers, J.M. and N. Jayasundara, 2015: Antarctic notothenioid fish: what are the future consequences of 'losses' and 'gains' acquired during long-term evolution at cold and stable temperatures? *J Exp Biol*, **218** (Pt 12), 1834–45, doi:10.1242/jeb.116129.

Behrenfeld, M.J. et al., 2016: Annual boom–bust cycles of polar phytoplankton biomass revealed by space-based lidar. *Nature Geoscience*, **10**, 118, doi:10.1038/ngeo2861.

Beier, P. et al., 2015: *Guiding Principles and Recommended Practices for Co-Producing Actionable Science*. A How-to Guide for DOI Climate Science Centers and the National Climate Change and Wildlife Science Center. Report to the Secretary of the Interior Advisory Committee on Climate Change and Natural Resource Science, Washington, DC. Bélanger, S., M. Babin and J.-É. Tremblay, 2013: Increasing cloudiness in Arctic damps the increase in phytoplankton primary production due to sea ice receding. *Biogeosciences*, **10** (6), 4087, doi:10.5194/bg-10-4087-2013.

Belchier, M. and M.A. Collins, 2008: Recruitment and body size in relation to temperature in juvenile Patagonian toothfish (Dissostichus eleginoides) at South Georgia. *Marine Biology*, **155** (5), 493, doi:10.1007/s00227-008-1047-3.

Bell, R.E., 2008: The role of subglacial water in ice-sheet mass balance. *Nature Geoscience*, **1**, 297, doi:10.1038/ngeo186.

Bell, R.E. et al., 2017: Antarctic ice shelf potentially stabilized by export of meltwater in surface river. *Nature*, **544** (7650), 344–348, doi:10.1038/nature22048.

Belleflamme, A., X. Fettweis and M. Erpicum, 2015: Recent summer Arctic atmospheric circulation anomalies in a historical perspective. *The Cryosphere*, **9** (1), 53–64, doi:10.5194/tc-9-53-2015.

Bellerby, R. et al., 2018: *Arctic Ocean acidification: an update*. AMAP Assessment 2018 Arctic Monitoring and Assessment Programme (AMAP), AMAP, Tromsø, Norway., vi+187 pp.

Belshe, E.F., E.A.G. Schuur and B.M. Bolker, 2013: Tundra ecosystems observed to be CO₂ sources due to differential amplification of the carbon cycle. *Ecology Letters*, **16** (10), 1307–1315, doi:10.1111/ele.12164.

Bendtsen, J., J. Mortensen, K. Lennert and S. Rysgaard, 2015: Heat sources for glacial ice melt in a west Greenland tidewater outlet glacier fjord: The role of subglacial freshwater discharge. *Geophysical Research Letters*, **42** (10), 4089–4095, doi:10.1002/2015GL063846.

Bengtsson, J. et al., 2003: Reserves, resilience and dynamic landscapes. *AMBIO*, **32** (6), 389–96, doi:10.1579/0044-7447-32.6.389.

Benn, D.I. et al., 2017: Melt-under-cutting and buoyancy-driven calving from tidewater glaciers: new insights from discrete element and continuum model simulations. *Journal of Glaciology*, **63** (240), 691–702, doi:10.1017/jog.2017.41.

Berchet, A. et al., 2016: Atmospheric constraints on the methane emissions from the East Siberian Shelf. *Atmos. Chem. Phys.*, **16** (6), 4147–4157, doi:10.5194/acp-16-4147-2016.

Berdahl, M. et al., 2014: Arctic cryosphere response in the Geoengineering Model Intercomparison Project G3 and G4 scenarios. *Journal of Geophysical Research: Atmospheres*, **119** (3), 1308–1321, doi:10.1002/2013jd020627.

Berge, J. et al., 2015: First Records of Atlantic Mackerel (Scomber scombrus) from the Svalbard Archipelago, Norway, with Possible Explanations for the Extension of Its Distribution. *Arctic*, **68** (1), 54–61, doi:10.14430/arctic4455.

Berkes, F., 2017: Environmental Governance for the Anthropocene? Social-Ecological Systems, Resilience, and Collaborative Learning. *Sustainability*, **9** (7), 1232, doi:10.3390/su9071232.

Berkman, A. and A. Vylegzhanin, 2013: *Environmental Security in the Arctic Ocean*. NATO Science for Peace and Security Series -C: Environmental Security., Springer.

Berkman, P.A. and A.N. Vylegzhanin, 2010: Environmental Security in the Arctic Ocean. In: *Environmental Security in the Arctic Ocean*, Dordrecht, [Berkman, P.A. and A.N. Vylegzhanin (eds.)], Springer Netherlands.

Berman, M., G. Kofinas and S. BurnSilver, 2017: Measuring Community Adaptive and Transformative Capacity in the Arctic Context. In: Northern Sustainabilities: Understanding and Addressing Change in the Circumpolar World [Fondahl, G. and G.N. Wilson. (eds.)]. Springer International Publishing, Inc., Cham, Switzerland, 59–75.

Berman, M. and J.I. Schmidt, 2019: Economic Effects of Climate Change in Alaska. *Weather, Climate, and Society*, **11** (2), 245–258, doi:10.1175/wcas-d-18-0056.1.

Bernardello, R. et al., 2014: Impact of Weddell Sea deep convection on natural and anthropogenic carbon in a climate model. *Geophysical Research Letters*, **41** (20), 7262–7269, doi:10.1002/2014gl061313.

Bester, M.N., H. Bornemann and T. McIntyre, 2017: *Sea Ice, Third Edition* [Thomas, D.N. (ed.)]. Antarctic marine mammals and sea ice, John Wiley & Sons, Ltd.

Bezeau, P., M. Sharp and G. Gascon, 2014: Variability in summer anticyclonic circulation over the Canadian Arctic Archipelago and west Greenland in the late 20th/early 21st centuries and its effect on glacier mass balance. *International Journal of Climatology*, **35** (4), 540–557, doi:10.1002/joc.4000.

Bhatia, M.P. et al., 2013: Erratum: Greenland meltwater as a significant and potentially bioavailable source of iron to the ocean. *Nature Geoscience*, **6** (6), 503–503, doi:10.1038/ngeo1833.

Bhatt, U. et al., 2017: Changing seasonality of panarctic tundra vegetation in relationship to climatic variables. *Environmental Research Letters*, **12** (5), doi:10.1088/1748-9326/aa6b0b.

Biggs, R., G.D. Peterson and J.C. Rocha, 2018: The Regime Shifts Database: a framework for analyzing regime shifts in social-ecological systems. *Ecology and Society*, **23** (3), doi:10.5751/es-10264-230309.

Biggs, R. et al., 2012: Toward Principles for Enhancing the Resilience of Ecosystem Services. *Annual Review of Environment and Resources, Vol 37*, **37** (1), 421–448, doi:10.1146/annurev-environ-051211-123836.

Bintanja, R. and F.M. Selten, 2014: Future increases in Arctic precipitation linked to local evaporation and sea-ice retreat. *Nature*, **509**, 479, doi:10.1038/nature13259.

Bintanja, R. et al., 2013: Important role for ocean warming and increased ice-shelf melt in Antarctic sea-ice expansion. *Nature Geoscience*, **6** (5), 376–379, doi:10.1038/Ngeo1767.

Bintanja, R., G.J. Van Oldenborgh and C.A. Katsman, 2015: The effect of increased fresh water from Antarctic ice shelves on future trends in Antarctic sea ice. *Annals of Glaciology*, **56** (69), 120–126, doi:10.3189/2015AoG69A001.

Birchenough, S.N.R. et al., 2015: Climate change and marine benthos: a review of existing research and future directions in the North Atlantic. *Wiley Interdisciplinary Reviews: Climate Change*, **6** (2), 203–223, doi:10.1002/wcc.330.

Biskaborn, B.K. et al., 2019: Permafrost is warming at a global scale. *Nat Commun*, **10** (1), 264, doi:10.1038/s41467-018-08240-4.

Bjerke, J. et al., 2017: Understanding the drivers of extensive plant damage in boreal and Arctic ecosystems: Insights from field surveys in the

3

aftermath of damage. *Science of the Total Environment*, **599**, 1965–1976, doi:10.1016/j.scitotenv.2017.05.050.

Bjørk, A.A. et al., 2018: Changes in Greenland's peripheral glaciers linked to the North Atlantic Oscillation. *Nature Climate Change*, **8** (1), 48–52, doi:10.1038/s41558-017-0029-1.

Bjorndal, K.A. et al., 2017: Ecological regime shift drives declining growth rates of sea turtles throughout the West Atlantic. *Global Change Biology*, **23** (11), 4556–4568, doi:10.1111/gcb.13712.

Bockheim, J. et al., 2013: Climate warming and permafrost dynamics in the Antarctic Peninsula region. *Global and Planetary Change*, **100**, 215–223, doi:10.1016/j.gloplacha.2012.10.018.

Boetius, A. et al., 2013: Export of Algal Biomass from the Melting Arctic Sea Ice. *Science*, **339** (6126), 1430–1432, doi:10.1126/science.1231346.

Boisvert, L.N., A.A. Petty and J.C. Stroeve, 2016: The Impact of the Extreme Winter 2015/16 Arctic Cyclone on the Barents–Kara Seas. *Monthly Weather Review*, **144** (11), 4279–4287, doi:10.1175/mwr-d-16-0234.1.

Boisvert, L.N. et al., 2018: Intercomparison of Precipitation Estimates over the Arctic Ocean and Its Peripheral Seas from Reanalyses. *Journal of Climate*, **31** (20), 8441–8462, doi:10.1175/jcli-d-18-0125.1.

Bolch, T. et al., 2013: Mass loss of Greenland's glaciers and ice caps 2003–2008 revealed from ICESat laser altimetry data. *Geophysical Research Letters*, **40** (5), 875–881, doi:10.1002/grl.50270.

Bond, T.C. et al., 2013: Bounding the role of black carbon in the climate system: A scientific assessment. *Journal of Geophysical Research: Atmospheres*, **118** (11), 5380–5552, doi:10.1002/jgrd.50171.

Bost, C.A. et al., 2009: The importance of oceanographic fronts to marine birds and mammals of the southern oceans. *Journal of Marine Systems*, **78** (3), 363–376, doi:10.1016/j.jmarsys.2008.11.022.

Bost, C.A. et al., 2015: Large-scale climatic anomalies affect marine predator foraging behaviour and demography. *Nat Commun*, **6**, 8220, doi:10.1038/ncomms9220.

Bouchard, F. et al., 2013: Vulnerability of shallow subarctic lakes to evaporate and desiccate when snowmelt runoff is low. *Geophysical Research Letters*, **40** (23), 6112–6117, doi:10.1002/2013gl058635.

Bourgeois, S. et al., 2016: Glacier inputs influence organic matter composition and prokaryotic distribution in a high Arctic fjord (Kongsfjorden, Svalbard). *Journal of Marine Systems*, **164**, 112–127, doi:10.1016/j.jmarsys.2016.08.009.

Box, J.E., 2013: Greenland ice sheet mass balance reconstruction. Part II: Surface mass balance (1840–2010). *Journal of Climate*, **26** (18), 6974–6989, doi:10.1175/JCLI-D-12-00518.1.

Box, J.E. et al., 2019: Key indicators of Arctic climate change: 1971–2017. *Environmental Research Letters*, **14** (4), 045010, doi:10.1088/1748-9326/aafc1b.

Box, J.E. et al., 2018: Global sea-level contribution from Arctic land ice: 1971–2017. *Environmental Research Letters*, **13** (12), 125012, doi:10.1088/1748-9326/aaf2ed.

Box, J.E. et al., 2012: Greenland ice sheet al.edo feedback: thermodynamics and atmospheric drivers. *The Cryosphere*, **6** (4), 821–839, doi:10.5194/tc-6-821-2012.

Boy, M. et al., 2019: Interactions between the atmosphere, cryosphere, and ecosystems at northern high latitudes. *Atmospheric Chemistry and Physics*, **19** (3), 2015–2061, doi:10.5194/acp-19-2015-2019.

Bracegirdle, T.J., P. Hyder and C.R. Holmes, 2018: CMIP5 Diversity in Southern Westerly Jet Projections Related to Historical Sea Ice Area: Strong Link to Strengthening and Weak Link to Shift. *Journal of Climate*, **31** (1), 195–211, doi:10.1175/jcli-d-17-0320.1.

Bracegirdle, T.J. et al., 2013: Assessment of surface winds over the Atlantic, Indian, and Pacific Ocean sectors of the Southern Ocean in CMIP5 models: historical bias, forcing response, and state dependence. *Journal of Geophysical Research: Atmospheres*, **118** (2), 547–562, doi:10.1002/jgrd.50153.

Bracegirdle, T.J., D.B. Stephenson, J. Turner and T. Phillips, 2015: The importance of sea ice area biases in 21st century multimodel projections of Antarctic temperature and precipitation. *Geophysical Research Letters*, **42** (24), 10832–10839, doi:10.1002/2015gl067055.

Braithwaite, J.E., J.J. Meeuwig and M.R. Hipsey, 2015a: Optimal migration energetics of humpback whales and the implications of disturbance. *Conserv Physiol*, **3** (1), doi:10.1093/conphys/cov001.

Braithwaite, J.E. et al., 2015b: From sea ice to blubber: linking whale condition to krill abundance using historical whaling records. *Polar Biology*, **38** (8), 1195–1202, doi:10.1007/s00300-015-1685-0.

Breed, G.A. et al., 2017: Sustained disruption of narwhal habitat use and behavior in the presence of Arctic killer whales. *Proceedings of the National Academy of Sciences*, **114** (10), 2628–2633, doi:10.1073/pnas.1611707114.

Bret-Harte, M.S. et al., 2013: The response of Arctic vegetation and soils following an unusually severe tundra fire. *Philosophical Transactions of the Royal Society B: Biological Sciences*, **368** (1624), doi:10.1098/rstb.2012.0490.

Brinkman, T.J. et al., 2016: Arctic communities perceive climate impacts on access as a critical challenge to availability of subsistence resources. *Climatic Change*, **139** (3), 413–427, doi:10.1007/s10584-016-1819-6.

Bromaghin, J.F. et al., 2015: Polar bear population dynamics in the southern Beaufort Sea during a period of sea ice decline. *Ecological Applications*, **25** (3), 634–651, doi:10.1890/14-1129.1.

Bronen, R., 2015: Climate-induced community relocations: using integrated social-ecological assessments to foster adaptation and resilience. *Ecology and Society*, **20** (3), doi:10.5751/es-07801-200336.

Bronen, R., 2017: The Human Rights of Climate-Induced community relocation. In: Climate change, migration, and human rights [Mano, D., A. Baldwin, D. Cubie, A. Mihr and T. Thorp (eds.)]. Routledge Press., Abingdon UK, 129–148.

Bronen, R. and F.S. Chapin, 3rd, 2013: Adaptive governance and institutional strategies for climate-induced community relocations in Alaska. *Proc Natl Acad Sci U S A*, **110** (23), 9320–5, doi:10.1073/pnas.1210508110.

Bronselaer, B. et al., 2018: Change in future climate due to Antarctic meltwater. *Nature*, **564** (7734), 53–58, doi:10.1038/s41586-018-0712-z.

Brooks, C.M. et al., 2018: Antarctic fisheries: factor climate change into their management. *Nature*, **558** (7709), 177–180, doi:10.1038/d41586-018-05372-x.

Brower, A.A. et al., 2017: Gray whale distribution relative to benthic invertebrate biomass and abundance: Northeastern Chukchi Sea 2009–2012. *Deep Sea Research Part II: Topical Studies in Oceanography*, **144**, 156–174, doi:10.1016/j.dsr2.2016.12.007.

Brown, L.C. and C.R. Duguay, 2011: The fate of lake ice in the North American Arctic. *The Cryosphere*, **5** (4), 869–892, doi:10.5194/tc-5-869-2011.

Brown, R. and C. Derksen, 2013: Is Eurasian October snow cover extent increasing? *Environmental Research Letters*, **8**, doi:10.1088/1748-9326/8/2/024006.

Brown, R. et al., 2017: Arctic terrestrial snow cover. In: Snow, Water, Ice and Permafrost in the Arctic (SWIPA) 2017 Assessment. Arctic Monitoring and Assessment Programme, Oslo, Norway.

Brubaker, M., J. Berner, R. Chavan and J. Warren, 2011: Climate change and health effects in Northwest Alaska. *Global Health Action*, **4** (1), 8445, doi:10.3402/gha.v4i0.8445.

Brubaker, M., J. Berner and M. Tcheripanoff, 2013: LEO, the Local Environmental Observer Network: a community-based system for surveillance of climate, environment, and health events. *International Journal of Circumpolar Health*, **72**, 513–514.

Brutel-Vuilmet, C., M. Ménégoz and G. Krinner, 2013: An analysis of present and future seasonal Northern Hemisphere land snow cover simulated by CMIP5 coupled climate models. *The Cryosphere*, **7** (1), 67–80, doi:10.5194/tc-7-67-2013.

Bullard, J.E. et al., 2016: High-latitude dust in the Earth system. *Reviews of Geophysics*, **54** (2), 447–485, doi:10.1002/2016rg000518.

Bullock, R., S. Aggarwal, R.A. Perkins and W. Schnabel, 2017: *Scale-up considerations for surface collecting agent assisted in-situ burn crude oil spill response experiments in the Arctic: Laboratory to field-scale investigations*.

Bulygina, O.N., P.Y. Groisman, V.N. Razuvaev and N.N. Korshunova, 2011: Changes in snow cover characteristics over Northern Eurasia since 1966. *Environmental Research Letters*, **6** (4), 045204, doi:10.1088/1748-9326/6/4/045204.

Bunce, A. and J. Ford, 2015: How is adaptation, resilience, and vulnerability research engaging with gender? *Environmental Research Letters*, **10** (12), 123003, doi:10.1088/1748-9326/10/12/123003/meta.

Bunce, A. et al., 2016: Vulnerability and adaptive capacity of Inuit women to climate change: a case study from Iqaluit, Nunavut. *Natural Hazards*, **83** (3), 1419–1441, doi:10.1007/s11069-016-2398-6.

Burgard, C. and D. Notz, 2017: Drivers of Arctic Ocean warming in CMIP5 models. *Geophysical Research Letters*, **44** (9), 4263–4271, doi:10.1002/2016GL072342.

Burgos, J., B. Ernst, D. Armstrong and J. Orensanz, 2013: Fluctuations in Range and Abundance of Snow Crab (Chionoecetes Opilio) from the Eastern Bering Sea: What Role for Pacific Cod (Gadus Macrocephalus) Predation? *Bulletin of Marine Science*, **89** (1), 57–81, doi:10.5343/bms.2011.1137.

Burke, E.J. et al., 2017a: Quantifying uncertainties of permafrost carbon–climate feedbacks. *Biogeosciences*, **14** (12), 3051–3066, doi:10.5194/bg-14-3051-2017.

Burke, E.J., C.D. Jones and C.D. Koven, 2013: Estimating the permafrost-carbon climate response in the CMIP5 climate models using a simplified approach. *Journal of Climate*, **26** (14), 4897–4909, doi:10.1175/jcli-d-12-00550.1.

Burke, J.L., J. Bohlmann and A.L. Carroll, 2017b: Consequences of distributional asymmetry in a warming environment: invasion of novel forests by the mountain pine beetle. *Ecosphere*, **8** (4), e01778, doi:10.1002/ecs2.1778.

BurnSilver, S. et al., 2016: Are Mixed Economies Persistent or Transitional? Evidence Using Social Networks from Arctic Alaska. *American Anthropologist*, **118** (1), 121–129, doi:10.1111/aman.12447.

Burton-Johnson, A., M. Black, P.T. Fretwell and J. Kaluza-Gilbert, 2016: An automated methodology for differentiating rock from snow, clouds and sea in Antarctica from Landsat 8 imagery: a new rock outcrop map and area estimation for the entire Antarctic continent. *The Cryosphere*, **10** (4), 1665–1677, doi:10.5194/tc-10-1665-2016.

Busch, D.S. et al., 2016: Climate science strategy of the US National Marine Fisheries Service. *Marine Policy*, **74**, 58–67, doi:10.1016/j.marpol.2016.09.001.

CAFF, 2013a: *Arctic Biodiversity Assessment. Status and trends in Arctic biodiversity*. Conservation of Arctic Flora and Fauna (CAFF), Akureyri, Iceland, 28 pp [Available at: http://arcticlcc.org/assets/resources/ABA2013Science.pdf; Accecss Date: 10 October 2018].

CAFF, 2013b: *Arctic Biodiversity Assessment: Report for Policy Makers*. Conservation of Arctic Flora and Fauna (CAFF), Akureyri, Iceland, 678 pp.

CAFF, 2015a: *The Economics of Ecosystems and Biodiversity (TEEB) Scoping Study for the Arctic*. Conservation of Arctic Flora and Fauna (CAFF), Akureyri, Iceland, 168 pp.

CAFF, 2015b: *Traditional Knowledge & Community Based Monitoring Progress report 2015*. Conservation of Arctic Flora and Fauna (CAFF), Akureyri, Iceland, 4 pp [Available at: https://oaarchive.arctic-council.org/handle/11374/397; Access Date: 13 April 2019].

CAFF, 2017: *State of the Arctic Marine Biodiversity Report*. Conservation of Arctic Flora and Fauna International Secretariat, Akureyri, Iceland, 200 pp.

CAFF and PAME, 2017: *Arctic Invasive Alien Species: Strategy and Action Plan*. Conservation of Arctic Flora and Fauna and Protection of the Arctic Marine Environment Akureyri, Iceland, 20 pp.

Cahalan, J., J. Gasper and J. Mondragon, 2014: *Catch sampling and estimation in the federal groundfish fisheries off Alaska, 2015edition*. U.S. Dep.

Commer., NOAA Tech. Memo., NMFS-AFSC-286, 46 p [Available at: www.afsc.noaa.gov/Publications/AFSC-TM/NOAA-TM-AFSC-286.pdf].

Campbell, E.C. et al., 2019: Antarctic offshore polynyas linked to Southern Hemisphere climate anomalies. *Nature*, **in press**, doi:10.1038/s41586-019-1294-0.

Camus, L. and M.G.D. Smit, 2019: Environmental effects of Arctic oil spills and spill response technologies, introduction to a 5 year joint industry effort. *Mar Environ Res*, **144**, 250–254, doi:10.1016/j.marenvres.2017.12.008.

Canadian Forest Service, 2017: National Fire Database; Agency Fire Data., Edmonton, Alberta.

Cape, M., R., M. Vernet, M. Kahru and G. Spreen, 2013: Polynya dynamics drive primary production in the Larsen A and B embayments following ice shelf collapse. *Journal of Geophysical Research: Oceans*, **119** (1), 572–594, doi:10.1002/2013jc009441.

Cape, M.R. et al., 2018: Nutrient release to oceans from buoyancy-driven upwelling at Greenland tidewater glaciers. *Nature Geoscience*, **12** (1), 34–39, doi:10.1038/s41561-018-0268-4.

Cape, M.R. et al., 2015: Foehn winds link climate-driven warming to ice shelf evolution in Antarctica. *Journal of Geophysical Research-Atmospheres*, **120** (21), 11037–11057, doi:10.1002/2015jd023465.

Capotondi, A. et al., 2012: Enhanced upper ocean stratification with climate change in the CMIP3 models. *Journal of Geophysical Research: Oceans*, **117** (C4), C04031, doi:10.1029/2011JC007409.

Carmack, E. et al., 2015: Toward Quantifying the Increasing Role of Oceanic Heat in Sea Ice Loss in the New Arctic. *Bulletin of the American Meteorological Society*, **96** (12), 2079–2105, doi:10.1175/Bams-D-13-00177.1.

Carmack, E.C. et al., 2016: Freshwater and its role in the Arctic Marine System: Sources, disposition, storage, export, and physical and biogeochemical consequences in the Arctic and global oceans. *Journal of Geophysical Research: Biogeosciences*, **121** (3), 675–717, doi:10.1002/2015JG003140.

Carr, J.R., C.R. Stokes and A. Vieli, 2013: Recent progress in understanding marine-terminating Arctic outlet glacier response to climatic and oceanic forcing. *Progress in Physical Geography*, **37** (4), 436–467, doi:10.1177/0309133313483163.

Carr, J.R. et al., 2017: Basal topographic controls on rapid retreat of Humboldt Glacier, northern Greenland. *Journal of Glaciology*, **61** (225), 137–150, doi:10.3189/2015JoG14J128.

Carroll, M.L. et al., 2011: Shrinking lakes of the Arctic: Spatial relationships and trajectory of change. *Geophysical Research Letters*, **38** (20), doi:10.1029/2011GL049427.

Carsey, F.D., 1980: Microwave Observation of the Weddell Polynya. *Monthly Weather Review*, **108** (12), 2032–2044, doi:10.1175/1520-0493(1980)108.

Carson, M. and M. Sommerkorn, 2017: *A resilience approach to adaptation actions*. Chapter 8 of Adaptation Actions for a Changing Arctic: Perspectives from the Barents Area, Arctic Monitoring and Assessment Programme (AMAP), Oslo, Norway, 195–218 [Available at: www.amap.no/documents/doc/adaptation-actions-for-a-changing-arctic-perspectives-from-the-barents-area/1604].

Carter, S.P. and H.A. Fricker, 2012: The supply of subglacial meltwater to the grounding line of the Siple Coast, West Antarctica. *Ann. Glaciol.*, **53**, 267–280, doi:10.3189/2012AoG60A119.

Cassotta, S., K. Hossain, J. Ren and M.E. Goodsite, 2015: Climate Change and China as a Global Emerging Regulatory Sea Power in the Arctic Ocean: Is China a Threat for Arctic Ocean Security? *Beijing Law Review*, **6** (3), 119–207, doi:10.4236/blr.2015.63020.

Cassotta, S., K. Hossain, J. Ren and M.E. Goodsite, 2016: Climate Change and Human Security in a Regulatory Multilevel and Multidisciplinary Dimension: The Case of the Arctic Environmental Ocean. In: Climate Change Adaptation, Resilience and Hazards [Leal Filho, W., H. Musa, G. Cavan, P. O'Hare and J. Seixas (eds.)]. Springer International Publishing, Cham, 71–91.

Cavanagh, R.D. et al., 2017: A Synergistic Approach for Evaluating Climate Model Output for Ecological Applications. *Frontiers in Marine Science*, **4**, 1–12, doi:https://doi.org/10.3389/fmars.2017.00308.

3

CAVM Team, 2003: Circumpolar Arctic Vegetation Map. Scale 1:7,500,000. Conservation of Arctic Flora and Fauna (CAFF) Map No. 1., Anchorage, Alaska.

CCA, 2016: *Commercial Marine Shipping Accidents: Understanding the Risks in Canada*. Council of Canadian Academies (CCA), Ottawa (ON); Workshop Report, pp. 84 [Available at: https://clearseas.org/en/research_project/characterization-of-risk-of-marine-shipping-in-canadian-waters/].

CCAMLR, 1982: *Convention on the Conservation of Antarctic Marine Living Resources, opened for signature 20 May 1980, 1329 UNTS 47 (entered into force 7 April 1982) ('CCAMLR')*. Canberra [Available at: https://treaties.un.org/pages/showDetails.aspx?objid=08000002800dc364; Access Date: 05 December 2018].

CCAMLR, 2016a: *Commission for the Conservation of Antarctic Marine Living Resources: Report of the Thirty-fifth Meeting of the Commission*. Report of the meeting of the Commission, 35, Hobart, Tasmania, Australia, 222.

CCAMLR, 2016b: *The value of marine resources harvested in the CCAMLR Convention Area – an assessment of GVP*. Commission for the Conservation of Antarctic Marine Living Resources [Available at: www.ccamlr.org/en/ccamlr-xxxv/10; Access Date: 05 December, 2018].

CCAMLR, 2017a: *Fishery Reports 2016*. Commission for the Conservation of Antarctic Marine Living Resources, Hobart, Tasmania [Available at: www.ccamlr.org/en/publications/fishery-reports-2016; Access Date: 05 December, 2018].

CCAMLR, 2017b: *Krill Fishery Report 2016*. Commission for the Conservation of Antarctic Marine Living Resources, Hobart, Tasmania. [Available at: www.ccamlr.org/en/document/publications/krill-fishery-report-2016; Access Date: 05 December, 2018].

CCAMLR, 2017c: *Map of the CAMLR Convention Area*. **2017**, www.ccamlr.org/node/86816 [Available at: www.ccamlr.org/node/86816; Access Date: 05 December, 2018].

CCAMLR, 2018: *Report of the thirty-seventh meeting of the commission*. Commission for the Conservation of Antarctic Marine Living Resources, Hobart, Australia [Available at: www.ccamlr.org/en/system/files/e-cc-xxxvii.pdf; Access Date: 05 December, 2018].

Chambers, D.P., 2018: Using kinetic energy measurements from altimetry to detect shifts in the positions of fronts in the Southern Ocean. *Ocean Science*, **14** (1), 105, doi:10.5194/os-14-105-2018.

Chapin III, F.S., G.P. Kofinas and C. Folke, Eds., 2009: Principles of Ecosystem Stewardship: Resilience-Based Natural Resource Management in a Changing World. Springer-Verlag, New York, USA., 401 pp.

Chapin III, F.S. et al., 2010: Resilience of Alaska's boreal forest to climatic change. In: The Dynamics of Change in Alaska's Boreal Forests: Resilience and Vulnerability in Response to Climate Warming. Canadian Journal of Forest Research. NRC Research Press, **40**, 1360–1370.

Chapin III, F.S., M. Sommerkorn, M.D. Robards and K. Hillmer-Pegram, 2015: Ecosystem stewardship: A resilience framework for arctic conservation. *Global Environmental Change*, **34**, 207–217, doi:doi.org/10.1016/j.gloenvcha.2015.07.003.

Chapman, C.C., 2014: Southern Ocean jets and how to find them: Improving and comparing common jet detection methods. *Journal of Geophysical Research: Oceans*, **119** (7), 4318–4339, doi:10.1002/2014JC009810.

Chapman, C.C., 2017: New perspectives on frontal variability in the Southern Ocean. *Journal of Physical Oceanography*, **47** (5), 1151–1168, doi:10.1175/JPO-D-16-0222.1.

Charalampidis, C. et al., 2015: Changing surface–atmosphere energy exchange and refreezing capacity of the lower accumulation area, West Greenland. *The Cryosphere*, **9** (6), 2163–2181, doi:10.5194/tc-9-2163-2015.

Charalampidis, C. et al., 2016: Thermal tracing of retained meltwater in the lower accumulation area of the Southwestern Greenland ice sheet. *Annals of Glaciology*, **57** (72), 1–10, doi:10.1017/aog.2016.2.

Chen, H.W., F. Zhang and R.B. Alley, 2016a: The Robustness of Midlatitude Weather Pattern Changes due to Arctic Sea Ice Loss. *Journal of Climate*, **29** (21), 7831–7849, doi:10.1175/jcli-d-16-0167.1.

Chen, M. et al., 2012: Temporal and spatial pattern of thermokarst lake area changes at Yukon Flats, Alaska. *Hydrological Processes*, **28** (3), 837–852, doi:10.1002/hyp.9642.

Chen, X. et al., 2015: Observed contrast changes in snow cover phenology in northern middle and high latitudes from 2001–2014. *Scientific Reports*, **5**, 16820, doi:10.1038/srep16820.

Chen, X.N., S.L. Liang and Y.F. Cao, 2016b: Satellite observed changes in the Northern Hemisphere snow cover phenology and the associated radiative forcing and feedback between 1982 and 2013. *Environmental Research Letters*, **11** (8), 084002, doi:10.1088/1748-9326/11/8/084002.

Cheng, L. et al., 2017: Improved estimates of ocean heat content from 1960 to 2015. *Science Advances*, **3** (3), doi:10.1126/sciadv.1601545

Chénier, R., L. Abado, O. Sabourin and L. Tardif, 2017: Northern marine transportation corridors: Creation and analysis of northern marine traffic routes in Canadian waters. *Transactions in GIS*, **21** (6), 1085–1097, doi:10.1111/tgis.12295.

Chernova, N.V., 2011: Distribution patterns and chorological analysis of fish fauna of the Arctic Region. *Journal of Ichthyology*, **51** (10), 825–924, doi:10.1134/S0032945211100043.

Chevallier, M. et al., 2017: Intercomparison of the Arctic sea ice cover in global ocean–sea ice reanalyses from the ORA-IP project. *Climate Dynamics*, **49** (3), 1107–1136, doi:10.1007/s00382-016-2985-y.

Chidichimo, M.P., K.A. Donohue, D.R. Watts and K.L. Tracey, 2014: Baroclinic transport time series of the Antarctic Circumpolar Current measured in Drake Passage. *Journal of Physical Oceanography*, **44** (7), 1829–1853, doi:10.1175/JPO-D-13-071.1.

Chin, K.S. et al., 2016: Permafrost thaw and intense thermokarst activity decreases abundance of stream benthic macroinvertebrates. *Glob Chang Biol*, **22** (8), 2715–28, doi:10.1111/gcb.13225.

Chown, S.L., 2017: Antarctic environmental challenges in a global context. In: The Handbook on the Politics of Antarctica [Dodds, K. and A. Hemmings (eds.)]. Edward Elgar Publishing, Cheltenham, 523–539.

Chown, S.L. et al., 2012: Continent-wide risk assessment for the establishment of nonindigenous species in Antarctica. *Proceedings of the National Academy of Sciences*, **109** (13), 4938, doi:10.1073/pnas.1119787109.

Choy, E.S., B. Rosenberg, J.D. Roth and L.L. Loseto, 2017: Inter-annual variation in environmental factors affect the prey and body condition of beluga whales in the eastern Beaufort Sea. *Marine Ecology Progress Series*, **579**, 213–225, doi:10.3354/meps12256.

Christensen, T.R. et al., 2019: Tracing the climate signal: mitigation of anthropogenic methane emissions can outweigh a large Arctic natural emission increase. *Sci Rep*, **9** (1), 1146, doi:10.1038/s41598-018-37719-9.

Christiansen, J.S., 2017: No future for Euro-Arctic ocean fishes? *Marine Ecology Progress Series*, **575**, 217–227, doi:10.3354/meps12192.

Christianson, K. et al., 2016: Sensitivity of Pine Island Glacier to observed ocean forcing. *Geophysical Research Letters*, **43** (20), 10817–10825, doi:10.1002/2016gl070500.

Christie, F.D. et al., 2016: Four-decade record of pervasive grounding line retreat along the Bellingshausen margin of West Antarctica. *Geophysical Research Letters*, **43** (11), 5741–5749.

Christie, K.S., T.E. Hollmen, H.P. Huntington and J.R. Lovvorn, 2018: Structured decision analysis informed by traditional ecological knowledge as a tool to strengthen subsistence systems in a changing Arctic. *Ecology and Society*, **23** (4), 42, doi:https://doi.org/10.5751/ES-10596-230442.

Chuter, S.J., A. Martín-Español, B. Wouters and J.L. Bamber, 2017: Mass balance reassessment of glaciers draining into the Abbot and Getz Ice Shelves of West Antarctica. *Geophysical Research Letters*, **44** (14), 7328–7337, doi:10.1002/2017GL073087.

Cimino, M.A., H.J. Lynch, V.S. Saba and M.J. Oliver, 2016: Projected asymmetric response of Adelie penguins to Antarctic climate change. *Sci Rep*, **6**, 28785, doi:10.1038/srep28785.

Clark, D.G. et al., 2016a: The role of environmental factors in search and rescue incidents in Nunavut, Canada. *Public Health*, **137** (Supplement C), 44–49, doi:10.1016/j.puhe.2016.06.003.

Clark, D.G., J.D. Ford, T. Pearce and L. Berrang-Ford, 2016b: Vulnerability to unintentional injuries associated with land-use activities and search and rescue in Nunavut, Canada. *Social Science and Medicine*, **169**, 18–26, doi:10.1016/j.socscimed.2016.09.026.

Clark, G.F. et al., 2015: Vulnerability of Antarctic shallow invertebrate-dominated ecosystems. *Austral Ecology*, **40** (4), 482–491, doi:10.1111/aec.12237.

Clark, G.F. et al., 2017: The Roles of Sea-Ice, Light and Sedimentation in Structuring Shallow Antarctic Benthic Communities. *Plos One*, **12** (1), e0168391, doi:10.1371/journal.pone.0168391.

Clem, K.R., J.A. Renwick and J. McGregor, 2017: Relationship between eastern tropical Pacific cooling and recent trends in the Southern Hemisphere zonal-mean circulation. *Climate Dynamics*, **49** (1–2), 113–129, doi:10.1007/s00382-016-3329-7.

Cochran, P. et al., 2013: Indigenous frameworks for observing and responding to climate change in Alaska. *Climatic Change*, **120** (3), 557–567, doi:10.1007/s10584-013-0735-2.

Coetzee, B.W.T., P. Convey and S.L. Chown, 2017: Expanding the Protected Area Network in Antarctica is Urgent and Readily Achievable. *Conservation Letters*, **10** (6), 670–680, doi:10.1111/conl.12342.

Coggins, J.H.J. and A.J. McDonald, 2015: The influence of the Amundsen Sea Low on the winds in the Ross Sea and surroundings: Insights from a synoptic climatology. *Journal of Geophysical Research-Atmospheres*, **120** (6), 2167–2189, doi:10.1002/2014JD022830.

Cohen, J., K. Pfeiffer and J.A. Francis, 2018: Warm Arctic episodes linked with increased frequency of extreme winter weather in the United States. *Nature Communications*, **9** (1), 869, doi:10.1038/s41467-018-02992-9.

Cohen, J., H. Ye and J. Jones, 2015: Trends and variability in rain-on-snow events. *Geophysical Research Letters*, **42** (17), 7115–7122, doi:10.1002/2015gl065320.

Cohen, J.L. et al., 2012: Arctic warming, increasing snow cover and widespread boreal winter cooling. *Environmental Research Letters*, **7** (1), 014007, doi:10.1088/1748-9326/7/1/014007.

Colgan, W. et al., 2015: Greenland high-elevation mass balance: inference and implication of reference period (1961–90) imbalance. *Annals of Glaciology*, **56** (70), 105–117, doi:10.3189/2015AoG70A967.

Collins, M. et al., 2013: *Climate Change 2013: The Physical Science Basis. Contribution of Working Group I to the Fifth Assessment Report of the Intergovernmental Panel on Climate Change*. Long-term climate change: Projections, commitments and irreversibility, Cambridge University Press, Cambridge, UK and New York, USA. Colombo, N. et al., 2018: Review: Impacts of permafrost degradation on inorganic chemistry of surface fresh water. *Global and Planetary Change*, **162**, 69–83, doi:10.1016/j.gloplacha.2017.11.017.

Comiso, J.C. et al., 2017a: Positive Trend in the Antarctic Sea Ice Cover and Associated Changes in Surface Temperature. *Journal of Climate*, **30** (6), 2251–2267, doi:10.1175/Jcli-D-16-0408.1.

Comiso, J.C., W.N. Meier and R. Gersten, 2017b: Variability and trends in the Arctic Sea ice cover: Results from different techniques. *Journal of Geophysical Research: Oceans*, **122** (8), 6883–6900, doi:10.1002/2017jc012768.

Commane, R. et al., 2017: Carbon dioxide sources from Alaska driven by increasing early winter respiration from Arctic tundra. *Proceedings of the National Academy of Sciences USA*, **114** (21), 5361–5366, doi:10.1073/pnas.1618567114.

Connon, R.F., W.L. Quinton, J.R. Craig and M. Hayashi, 2014: Changing hydrologic connectivity due to permafrost thaw in the lower Liard River valley, NWT, Canada. *Hydrological Processes*, **28** (14), 4163–4178, doi:10.1002/hyp.10206.

Conrad, C.J. and N.S. Lovenduski, 2015: Climate-Driven Variability in the Southern Ocean Carbonate System. *Journal of Climate*, **28** (13), 5335–5350, doi:papers2://publication/doi/10.1175/JCLI-D-14-00481.1.

Constable, A., J. Melbourne-Thomas, R. Trebilco and A.J. Press, 2017: *ACE CRC Position Analysis: Managing change in Southern Ocean ecosystems*. Centre, A.C.E.C.R., Hobart, 1–40 [Available at: http://acecrc.org.au/wp-content/uploads/2017/10/2017-ACECRC-Position-Analysis-Southern-Ocean-Ecosystems.pdf; Access Date: 05 December 2018].

Constable, A.J., 2011: Lessons from CCAMLR on the implementation of the ecosystem approach to managing fisheries. *Fish and Fisheries*, **12** (2), 138–151, doi:10.1111/j.1467-2979.2011.00410.x.

Constable, A.J. et al., 2016: Developing priority variables ("ecosystem Essential Ocean Variables" – eEOVs) for observing dynamics and change in Southern Ocean ecosystems. *Journal of Marine Systems*, **161**, 26–41, doi:10.1016/j.jmarsys.2016.05.003.

Constable, A.J. et al., 2014: Climate change and Southern Ocean ecosystems I: how changes in physical habitats directly affect marine biota. *Global Change Biology*, **20** (10), 3004–3025, doi:10.1111/gcb.12623.

Cook, A.J. et al., 2016: Ocean forcing of glacier retreat in the western Antarctic Peninsula. *Science*, **353** (6296), 283, doi:10.1126/science.aae0017.

Cook, A.J., D.G. Vaughan, A.J. Luckman and T. Murray, 2014: A new Antarctic Peninsula glacier basin inventory and observed area changes since the 1940s. *Antarctic Science*, **26** (6), 614–624, doi:10.1017/S0954102014000200.

Cook, J.A. et al., 2017: The Beringian Coevolution Project: holistic collections of mammals and associated parasites reveal novel perspectives on evolutionary and environmental change in the North. *Arctic Science*, **3** (3), 585–617, doi:10.1139/as-2016-0042.

Cooley, S.W. and T.M. Pavelsky, 2016: Spatial and temporal patterns in Arctic river ice breakup revealed by automated ice detection from MODIS imagery. *Remote Sensing of Environment*, **175** (Supplement C), 310–322, doi:10.1016/j.rse.2016.01.004.

Copland, L. and D. Mueller, 2017: *Arctic Ice Shelves and Ice Islands*. Springer Polar Sciences, Springer, Dordrecht.

Coulombe, S. et al., 2019: Origin, burial and preservation of late Pleistocene-age glacier ice in Arctic permafrost (Bylot Island, NU, Canada). *The Cryosphere*, **13** (1), 97–111, doi:10.5194/tc-13-97-2019.

Cowtan, K. and R.G. Way, 2014: Coverage bias in the HadCRUT4 temperature series and its impact on recent temperature trends. *Quarterly Journal of the Royal Meteorological Society*, **140** (683), 1935–1944, doi:10.1002/qj.2297.

Cowton, T. et al., 2016: Controls on the transport of oceanic heat to Kangerdlugssuaq Glacier, East Greenland. *Journal of Glaciology*, **62** (236), 1167–1180, doi:10.1017/jog.2016.117.

Cowton, T.R. et al., 2018: Linear response of east Greenland's tidewater glaciers to ocean/atmosphere warming. *Proceedings of the National Academy of Sciences*, **115** (31), 7907, doi:10.1073/pnas.1801769115.

Cox, M.J. et al., 2018: No evidence for a decline in the density of Antarctic krill Euphausia superba Dana, 1850, in the Southwest Atlantic sector between 1976 and 2016. *Journal of Crustacean Biology*, **38** (6), 656–661, doi:10.1093/jcbiol/ruy072.

Cozzetto, K. et al., 2013: Climate change impacts on the water resources of American Indians and Alaska Natives in the U.S. *Climatic Change*, **120** (3), 569–584, doi:10.1007/s10584-013-0852-y.

Crépin, A.-S., Å. Gren, G. Engström and D. Ospina, 2017: Operationalising a social–ecological system perspective on the Arctic Ocean. *AMBIO*, **46** (3), 475–485, doi:10.1007/s13280-017-0960-4.

Crill, P.M. and B.F. Thornton, 2017: Whither methane in the IPCC process? *Nature Climate Change*, **7**, 678, doi:10.1038/nclimate3403.

Cristofari, R. et al., 2018: Climate-driven range shifts of the king penguin in a fragmented ecosystem. *Nature Climate Change*, **8** (3), 245–251, doi:10.1038/s41558-018-0084-2.

Cullather, R.I. et al., 2016: Analysis of the warmest Arctic winter, 2015–2016. *Geophysical Research Letters*, **43** (20), 10,808–10,816, doi:10.1002/2016gl071228.

Cumming, G.S., 2011: Spatial resilience: integrating landscape ecology, resilience, and sustainability. *Landscape Ecology*, **26** (7), 899–909, doi:10.1007/s10980-011-9623-1.

Cunsolo, A. and N.R. Ellis, 2018: Ecological grief as a mental health response to climate change-related loss. *Nature Climate Change*, **8** (4), 275–281, doi:10.1038/s41558-018-0092-2.

Cunsolo, A., I. Shiwak and M. Wood, 2017: "You Need to Be a Well-Rounded Cultural Person": Youth Mentorship Programs for Cultural Preservation, Promotion, and Sustainability in the Nunatsiavut Region of Labrador. In: Northern Sustainabilities: Understanding and Addressing Change in the Circumpolar World [Fondahl, G. and G. Wilson (eds.)]. Springer Polar Sciences, Springer, 285–303.

Cunsolo Willox, A. et al., 2013a: The land enriches the soul: On climatic and environmental change, affect, and emotional health and well-being in Rigolet, Nunatsiavut, Canada. *Emotion, Space and Society*, **6**, 14–24.

Cunsolo Willox, A. et al., 2013b: Climate change and mental health: an exploratory case study from Rigolet, Nunatsiavut, Canada. *Climatic Change*, **121** (2), 255–270, doi:10.1007/s10584-013-0875-4.

Cunsolo Willox, A. et al., 2012: From this place and of this place:" Climate change, sense of place, and health in Nunatsiavut, Canada. *Social Science & Medicine*, **75** (3), 538–547, doi:10.1016/j.socscimed.2012.03.043.

Cunsolo Willox, A. et al., 2015: Examining relationships between climate change and mental health in the Circumpolar North. *Regional Environmental Change*, **15** (1), 169–182, doi:10.1007/s10113-014-0630-z.

Dabros, A., M. Pyper and G. Castilla, 2018: Seismic lines in the boreal and arctic ecosystems of North America: environmental impacts, challenges, and opportunities. *Environmental Reviews*, **26** (2), 214–229, doi:10.1139/er-2017-0080.

Dale, A. and D. Armitage, 2011: Marine mammal co-management in Canada's Arctic: Knowledge co-production for learning and adaptive capacity. *Marine Policy*, **35** (4), 440–449, doi:10.1016/j.marpol.2010.10.019.

Daley, K. et al., 2015: Water systems, sanitation, and public health risks in remote communities: Inuit resident perspectives from the Canadian Arctic. *Social Science & Medicine*, **135** (Supplement C), 124–132, doi:10.1016/j.socscimed.2015.04.017.

Dalpadado, P. et al., 2016: Distribution and abundance of euphausiids and pelagic amphipods in Kongsfjorden, Isfjorden and Rijpfjorden (Svalbard) and changes in their relative importance as key prey in a warming marine ecosystem. *Polar Biology*, **39** (10), 1765–1784, doi:10.1007/s00300-015-1874-x.

Damm, E. et al., 2018: The Transpolar Drift conveys methane from the Siberian Shelf to the central Arctic Ocean. *Sci Rep*, **8** (1), 4515, doi:10.1038/s41598-018-22801-z.

Dannevig, H. and C. Aall, 2015: The regional level as boundary organization? An analysis of climate change adaptation governance in Norway. *Environmental Science & Policy*, **54**, 168–175, doi:10.1016/j.envsci.2015.07.001.

Dätwyler, C. et al., 2017: Teleconnection stationarity, variability and trends of the Southern Annular Mode (SAM) during the last millennium. *Climate Dynamics*, **51** (5–6), 2321–2339, doi:10.1007/s00382-017-4015-0.

Dawson, J., M.E. Johnston and E.J. Stewart, 2014: Governance of Arctic expedition cruise ships in a time of rapid environmental and economic change. *Ocean and Coastal Management*, **89**, 88–99, doi:10.1016/j.ocecoaman.2013.12.005.

Dawson, J. et al., 2018: Temporal and Spatial Patterns of Ship Traffic in the Canadian Arctic from 1990 to 2015. *Arctic*, **71** (7), 15–26, doi:10.14430/arctic4698.

de la Barre, S. et al., 2016: Tourism and Arctic Observation Systems: exploring the relationships. *Polar Research*, **35** (1), 24980, doi:10.3402/polar.v35.24980.

de la Mare, W.K., 2009: Changes in Antarctic sea-ice extent from direct historical observations and whaling records. *Climatic Change*, **92** (3), 461–493, doi:10.1007/s10584-008-9473-2.

de Lavergne, C. et al., 2014: Cessation of deep convection in the open Southern Ocean under anthropogenic climate change. *Nature Climate Change*, **4**, 278–282, doi:10.1038/nclimate2132.

De Lucia, V., 2017: The Arctic environment and the BBNJ negotiations. Special rules for special circumstances? *Marine Policy*, **86**, 234–240, doi:10.1016/j.marpol.2017.09.011.

De Robertis, A., K. Taylor, C.D. Wilson and E.V. Farley, 2017: Abundance and distribution of Arctic cod (Boreogadus saida) and other pelagic fishes over the U.S. Continental Shelf of the Northern Bering and Chukchi Seas. *Deep Sea Research Part II: Topical Studies in Oceanography*, **135**, 51–65, doi:10.1016/j.dsr2.2016.03.002.

De Villiers, M.S. et al., 2006: Conservation management at Southern Ocean islands: towards the development of best-practice guidelines. *Polarforschung*, **75** (2/3), 113–131.

Death, R. et al., 2014: Antarctic ice sheet fertilises the Southern Ocean. *Biogeosciences*, **11** (10), 2635–2643, doi:10.5194/bg-11-2635-2014.

DeConto, R.M. and D. Pollard, 2016: Contribution of Antarctica to past and future sea-level rise. *Nature*, **531** (7596), 591–597, doi:10.1038/nature17145.

deHart, P.A.P. and C.M. Picco, 2015: Stable oxygen and hydrogen isotope analyses of bowhead whale baleen as biochemical recorders of migration and arctic environmental change. *Polar Science*, **9** (2), 235–248, doi:https://doi.org/10.1016/j.polar.2015.03.002.

Depoorter, M.A. et al., 2013: Calving fluxes and basal melt rates of Antarctic ice shelves. *Nature*, **502**, 89, doi:10.1038/nature12567.

Deppeler, S. et al., 2018: Ocean acidification of a coastal Antarctic marine microbial community reveals a critical threshold for CO2 tolerance in phytoplankton productivity. *Biogeosciences*, **15** (1), 209–231, doi:10.5194/bg-15-209-2018.

Derksen, C. and R. Brown, 2012: Spring snow cover extent reductions in the 2008–2012 period exceeding climate model projections. *Geophysical Research Letters*, **39** (19), L19504, doi:10.1029/2012gl053387.

Déry, S.J., T.A. Stadnyk, M.K. MacDonald and B. Gauli-Sharma, 2016: Recent trends and variability in river discharge across northern Canada. *Hydrol. Earth Syst. Sci.*, **20** (12), 4801–4818, doi:10.5194/hess-20-4801-2016.

Desbruyeres, D., E.L. McDonagh, B.A. King and V. Thierry, 2017: Global and Full-Depth Ocean Temperature Trends during the Early Twenty-First Century from Argo and Repeat Hydrography. *Journal of Climate*, **30** (6), 1985–1997, doi:10.1175/JCLI-D-16-0396.1.

Descamps, S. et al., 2017: Circumpolar dynamics of a marine top-predator track ocean warming rates. *Global Change Biology*, **23** (9), 3770–3780, doi:10.1111/gcb.13715.

Descamps, S. et al., 2015: Demographic effects of extreme weather events: snow storms, breeding success, and population growth rate in a long-lived Antarctic seabird. *Ecology and Evolution*, **5** (2), 314–325, doi:10.1002/ece3.1357.

Desch, S.J. et al., 2017: Arctic ice management. *Earth's Future*, **5** (1), 107–127, doi:10.1002/2016ef000410.

DeVries, T., M. Holzer and F. Primeau, 2017: Recent increase in oceanic carbon uptake driven by weaker upper-ocean overturning. *Nature*, **542** (7640), 215, doi:10.1038/nature21068.

Dewey, S. et al., 2018: Arctic Ice-Ocean Coupling and Gyre Equilibration Observed With Remote Sensing. *Geophysical Research Letters*, **45** (3), 1499–1508, doi:10.1002/2017gl076229.

DFO, 2014: *Integrated Fisheries Management Plan Cambridge Bay Arctic Char Commercial Fishery, Nunavut Settlement Area Effective 2014. Arctic Char (Salvelinus alpinus)*. Fisheries and Oceans Canada, Central and Arctic Region, Resource Management and Aboriginal Affairs, 501 University Crescent, Winnipeg, MB, R3T 2N6, 38p.

Díaz, S. et al., 2019: *Summary for policymakers of the global assessment report on biodiversity and ecosystem services – unedited advance version* [Manuela Carneiro da Cunha, Georgina Mace and Harold Mooney (eds.)]. The Intergovernmental Science-Policy Platform on Biodiversity and

Ecosystem Services (IPBES) [Available at: www.ipbes.net/system/tdf/spm_global_unedited_advance.pdf?file=1&type=node&id=35245].

Dibike, Y., T. Prowse, T. Saloranta and R. Ahmed, 2011: Response of Northern Hemisphere lake-ice cover and lake-water thermal structure patterns to a changing climate. *Hydrological Processes*, **25** (19), 2942–2953, doi:10.1002/hyp.8068.

DiCosimo, J., S. Cunningham and D. Brannan, 2015: Pacific Halibut Bycatch Management in Gulf of Alaska Groundfish Trawl Fisheries. In: Fisheries Bycatch: Global Issues and Creative Solutions [Kruse, G.H., H.C. An, J. DiCosimo, C.A. Eischens, G.S. Gislason, D.N. McBride, C.S. Rose and C.E. Siddo (eds.)]. Alaska Sea Grant, University of Alaska Fairbanks, Fairbanks, Alaska, 19.

Dilliplaine, K.B., 2017: The effect of under ice crude oil spills on sympagic biota of the Arctic: a mesocosm approach. University of Alaska Fairbanks, Fairbanks, Alaska.

Ding, Q. et al., 2017: Influence of high-latitude atmospheric circulation changes on summertime Arctic sea ice. *Nature Climate Change*, **7**, 289, doi:10.1038/nclimate3241

Ding, Q. et al., 2018: Fingerprints of internal drivers of Arctic sea ice loss in observations and model simulations. *Nature Geoscience*, **12** (1), 28–33, doi:10.1038/s41561-018-0256-8.

Ding, Q. et al., 2014: Tropical forcing of the recent rapid Arctic warming in northeastern Canada and Greenland. *Nature*, **509** (7499), 209–12, doi:10.1038/nature13260.

Ding, Y.N. et al., 2016: Seasonal heat and freshwater cycles in the Arctic Ocean in CMIP5 coupled models. *Journal of Geophysical Research-Oceans*, **121** (4), 2043–2057, doi:10.1002/2015jc011124.

Divoky, G.J., P.M. Lukacs and M.L. Druckenmiller, 2015: Effects of recent decreases in arctic sea ice on an ice-associated marine bird. *Progress in Oceanography*, **136**, 151–161, doi:10.1016/j.pocean.2015.05.010.

Dodd, W. et al., 2018: Lived experience of a record wildfire season in the Northwest Territories, Canada. *Can J Public Health*, **109** (3), 327–337, doi:10.17269/s41997-018-0070-5.

Dolant, C. et al., 2017: Meteorological inventory of rain-on-snow events in the Canadian Arctic Archipelago and satellite detection assessment using passive microwave data. *Physical Geography*, **39** (5), 428–444, doi:10.1080/02723646.2017.1400339.

Domack, E. et al., 2005: Stability of the Larsen B ice shelf on the Antarctic Peninsula during the Holocene epoch. *Nature*, **436**, 681, doi:10.1038/nature03908.

Donaldson, S.G. et al., 2010: Environmental contaminants and human health in the Canadian Arctic. *Science of the Total Environment*, **408** (22), 5165–5234, doi:10.1016/j.scitotenv.2010.04.059.

Donohue, K. et al., 2016: Mean Antarctic Circumpolar Current transport measured in Drake Passage. *Geophysical Research Letters*, **43** (22).

Dore, G., F.J. Niu and H. Brooks, 2016: Adaptation Methods for Transportation Infrastructure Built on Degrading Permafrost. *Permafrost and Periglacial Processes*, **27** (4), 352–364, doi:10.1002/ppp.1919.

Dorresteijn, I. et al., 2012: Climate affects food availability to planktivorous least auklets Aethia pusilla through physical processes in the southeastern Bering Sea. *Marine Ecology Progress Series*, **454**, 207–220, doi:10.3354/meps09372.

Douglas, V. et al., 2014: Reconciling traditional knowledge, food security, and climate change: experience from Old Crow, YT, Canada. *Progress in community health partnerships: research, education, and action*, **8** (1), 21–7, doi:10.1353/cpr.2014.0007.

Dow, C.F. et al., 2018a: Basal channels drive active surface hydrology and transverse ice shelf fracture. *Sci Adv*, **4** (6), eaao7212, doi:10.1126/sciadv.aao7212.

Dow, C.F. et al., 2018b: Dynamics of Active Subglacial Lakes in Recovery Ice Stream. *Journal of Geophysical Research: Earth Surface*, **123** (4), 837–850, doi:10.1002/2017jf004409.

Downes, S., P. Spence and A. Hogg, 2018: Understanding variability of the Southern Ocean overturning circulation in CORE-II models. *Ocean Modelling*, **123**, 98–109, doi:10.1016/j.ocemod.2018.01.005.

Downes, S.M. and A.M. Hogg, 2013: Southern Ocean circulation and eddy compensation in CMIP5 models. *Journal of Climate*, **26** (18), 7198–7220, doi:10.1175/JCLI-D-12-00504.1.

Drewniak, M. et al., 2018: Geopolitics of Arctic shipping: the state of icebreakers and future needs. *Polar Geography*, **41** (2), 107–125, doi:10.1080/1088937x.2018.1455756.

Driscoll, D.L. et al., 2016: Assessing the health effects of climate change in Alaska with community-based surveillance. *Climatic Change*, **137** (3), 455–466, doi:10.1007/s10584-016-1687-0.

Drucker, R., S. Martin and R. Kwok, 2011: Sea ice production and export from coastal polynyas in the Weddell and Ross Seas. *Geophysical Research Letters*, **38**, L17502, doi:10.1029/2011gl048668.

Du, J. et al., 2017: Satellite microwave assessment of Northern Hemisphere lake ice phenology from 2002 to 2015. *The Cryosphere*, **11** (1), 47–63, doi:10.5194/tc-11-47-2017.

Duan, L., X. Man, B. Kurylyk and T. Cai, 2017: Increasing Winter Baseflow in Response to Permafrost Thaw and Precipitation Regime Shifts in Northeastern China. *Water*, **9** (1), 25, doi:10.3390/w9010025.

Dudley, J.P., E.P. Hoberg, E.J. Jenkins and A.J. Parkinson, 2015: Climate Change in the North American Arctic: A One Health Perspective. *EcoHealth*, **12** (4), 713–725, doi:10.1007/s10393-015-1036-1.

Duffy, G.A. et al., 2017: Barriers to globally invasive species are weakening across the Antarctic. *Diversity and Distributions*, **23** (9), 982–996, doi:10.1111/ddi.12593.

Dugger, K.M. et al., 2014: AdÃ©lie penguins coping with environmental change: results from a natural experiment at the edge of their breeding range. *Frontiers in Ecology and Evolution*, **2**, 1–12, doi:10.3389/fevo.2014.00068.

Dunn, M.J. et al., 2016: Population Size and Decadal Trends of Three Penguin Species Nesting at Signy Island, South Orkney Islands. *Plos One*, **11** (10), e0164025, doi:10.1371/journal.pone.0164025.

Dunse, T. et al., 2015: Glacier-surge mechanisms promoted by a hydro-thermodynamic feedback to summer melt. *The Cryosphere*, **9** (1), 197–215, doi:10.5194/tc-9-197-2015.

Duprat, L.P.A.M., G.R. Bigg and D.J. Wilton, 2016: Enhanced Southern Ocean marine productivity due to fertilization by giant icebergs. *Nature Geoscience*, **9**, 219, doi:10.1038/ngeo2633.

Durkalec, A., C. Furgal, M.W. Skinner and T. Sheldon, 2014: Investigating environmental determinants of injury and trauma in the Canadian north. *Int J Environ Res Public Health*, **11** (2), 1536–1548, doi:10.3390/ijerph110201536.

Durkalec, A., C. Furgal, M.W. Skinner and T. Sheldon, 2015: Climate change influences on environment as a determinant of Indigenous health: Relationships to place, sea ice, and health in an Inuit community. *Social Science and Medicine*, **136–137**, 17–26, doi:10.1016/j.socscimed.2015.04.026.

Durner, G.M. et al., 2017: Increased Arctic sea ice drift alters adult female polar bear movements and energetics. *Global Change Biology*, **23** (9), 3460–3473, doi:10.1111/gcb.13746.

Dutrieux, P. et al., 2014: Strong Sensitivity of Pine Island Ice-Shelf Melting to Climatic Variability. *Science*, **343** (6167), 174–178, doi:10.1126/science.1244341.

Duyck, S., 2011: Participation of Non-State Actors in Arctic Environmental Governance *Nordia Geographical Publications*, **40** (4), 99–110.

Easterling, D.R., K.E. Kunkel, M.F. Wehner and L. Sun, 2016: Detection and attribution of climate extremes in the observed record. *Weather and Climate Extremes*, **11**, 17–27, doi:10.1016/j.wace.2016.01.001.

Edinburgh, T. and J.J. Day, 2016: Estimating the extent of Antarctic summer sea ice during the Heroic Age of Antarctic Exploration. *The Cryosphere*, **10** (6), 2721–2730, doi:10.5194/tc-10-2721-2016.

Edwards, T.L. et al., 2019: Revisiting Antarctic ice loss due to marine ice-cliff instability. *Nature*, **566** (7742), 58–64, doi:10.1038/s41586-019-0901-4.

Eguíluz, V.M., J. Fernández-Gracia, X. Irigoien and C.M. Duarte, 2016: A quantitative assessment of Arctic shipping in 2010–2014. *Scientific Reports*, **6**, 30682, doi:10.1038/srep30682.

Eicken, H. et al., 2014: A framework and database for community sea ice observations in a changing Arctic: an Alaskan prototype for multiple users. *Polar Geography*, **37** (1), 5–27, doi:10.1080/1088937x.2013.873090.

Eide, A., 2017: Climate change, fisheries management and fishing aptitude affecting spatial and temporal distributions of the Barents Sea cod fishery. *AMBIO*, **46** (Suppl 3), 387–399, doi:10.1007/s13280-017-0955-1.

Eisner, W.R., K.M. Hinkel, C.J. Cuomo and R.A. Beck, 2013: Environmental, cultural, and social change in Arctic Alaska as observed by Iñupiat elders over their lifetimes: a GIS synthesis. *Polar Geography*, **36** (3), 221–231, doi:10.1080/1088937x.2012.724463.

Elvidge, A.D. et al., 2015: Foehn jets over the Larsen C Ice Shelf, Antarctica. *Quarterly Journal of the Royal Meteorological Society*, **141** (688), 698–713, doi:10.1002/qj.2382.

Enderlin, E.M. et al., 2014: An improved mass budget for the Greenland ice sheet. *Geophysical Research Letters*, **41** (3), doi:10.1002/2013GL059010.

Enderlin, E.M., I.M. Howat and A. Vieli, 2013: High sensitivity of tidewater outlet glacier dynamics to shape. *Cryosphere*, **7** (3), 1007–1015, doi:10.5194/tc-7-1007-2013.

England, M., L. Polvani and L. Sun, 2018: Contrasting the Antarctic and Arctic Atmospheric Responses to Projected Sea Ice Loss in the Late Twenty-First Century. *Journal of Climate*, **31** (16), 6353–6370, doi:10.1175/jcli-d-17-0666.1.

England, M.R. et al., 2016: Robust response of the Amundsen Sea Low to stratospheric ozone depletion. *Geophysical Research Letters*, **43** (15), 8207–8213, doi:10.1002/2016gl070055.

Ericson, Y. et al., 2014: Increasing carbon inventory of the intermediate layers of the Arctic Ocean. *Journal of Geophysical Research-Oceans*, **119** (4), 2312–2326, doi:10.1002/2013jc009514.

Ernst, K.M. and M. van Riemsdijk, 2013: Climate change scenario planning in Alaska's National Parks: Stakeholder involvement in the decision-making process. *Applied Geography*, **45**, 22–28, doi:10.1016/j.apgeog.2013.08.004.

Ershova, E.A., R.R. Hopcroft and K.N. Kosobokova, 2015: Inter-annual variability of summer mesozooplankton communities of the western Chukchi Sea: 2004–2012. *Polar Biology*, **38** (9), 1461–1481, doi:10.1007/s00300-015-1709-9.

Escajeda, E. et al., 2018: Identifying shifts in maternity den phenology and habitat characteristics of polar bears (Ursus maritimus) in Baffin Bay and Kane Basin. *Polar Biology*, **41** (1), 87–100, doi:10.1007/s00300-017-2172-6.

Essery, R., 2013: Large-scale simulations of snow albedo masking by forests. *Geophysical Research Letters*, **40** (20), 5521–5525, doi:10.1002/grl.51008.

Estilow, T.W., A.H. Young and D.A. Robinson, 2015: A long-term Northern Hemisphere snow cover extent data record for climate studies and monitoring. *Earth System Science Data*, **7**, 137–142, doi:10.5194/essd-7-137-2015.

Euskirchen, E.S. et al., 2016: Consequences of changes in vegetation and snow cover for climate feedbacks in Alaska and northwest Canada. *Environmental Research Letters*, **11** (10), 105003, doi:10.1088/1748-9326/11/10/105003.

Evtushevsky, O.M., A.V. Grytsai and G.P. Milinevsky, 2018: Decadal changes in the central tropical Pacific teleconnection to the Southern Hemisphere extratropics. *Climate Dynamics*, **52** (7–8), 4027–4055, doi:10.1007/s00382-018-4354-5.

Falkenberg, J. et al., 2018: *AMAP Assessment 2018: Arctic Ocean Acidification*. Biological responses to ocean acidification, Arctic Monitoring and Assessment Programme (AMAP), Tromsø, Norway, 187 [Available at: www.amap.no/documents/doc/AMAP-Assessment-2018-Arctic-Ocean-Acidification/1659].

Fall, J., 2016: Regional Patterns of Fish and Wildlife Harvests in Contemporary Alaska. *Arctic*, **69** (1), 47–54, doi:10.14430/arctic4547.

Fall, J.A. et al., 2013: Continuity and change in subsistence harvests in five Bering Sea communities: Akutan, Emmonak, Savoonga, St. Paul, and Togiak. *Deep-Sea Research Part II*, **94**, 274–291, doi:10.1016/j.dsr2.2013.03.010.

Fauchald, P., V.H. Hausner, J.I. Schmidt and D.A. Clark, 2017a: Transitions of social-ecological subsistence systems in the Arctic. *International Journal of the Commons*, **11** (1), 275–329, doi:10.18352/ijc.698.

Fauchald, P. et al., 2017b: Arctic greening from warming promotes declines in caribou populations. *Sci Adv*, **3** (4), e1601365, doi:10.1126/sciadv.1601365.

Favier, L. et al., 2014: Retreat of Pine Island Glacier controlled by marine ice-sheet instability. *Nature Climate Change*, **4** (2), 117–121, doi:10.1038/Nclimate2094.

Favier, V. et al., 2016: Atmospheric drying as the main driver of dramatic glacier wastage in the southern Indian Ocean. *Scientific Reports*, **6**, 32396, doi:10.1038/srep32396.

Fay, A.R. et al., 2018: Utilizing the Drake Passage Time-series to understand variability and change in subpolar Southern Ocean pCO2. *Biogeosciences*, **15** (12), 3841–3855, doi:10.5194/bg-15-3841-2018.

Feng, Z. et al., 2018: Biogeographic responses of the copepod Calanus glacialis to a changing Arctic marine environment. *Glob Chang Biol*, **24** (1), e159-e170, doi:10.1111/gcb.13890.

Fernández-Méndez, M. et al., 2015: Photosynthetic production in the central Arctic Ocean during the record sea-ice minimum in 2012. *Biogeosciences*, **12** (11), 3525–3549, doi:10.5194/bg-12-3525-2015.

Fernández-Méndez, M. et al., 2018: Algal Hot Spots in a Changing Arctic Ocean: Sea-Ice Ridges and the Snow-Ice Interface. *Frontiers in Marine Science*, **5** (75), doi:10.3389/fmars.2018.00075.

Fettweis, X. et al., 2017: Reconstructions of the 1900–2015 Greenland ice sheet surface mass balance using the regional climate MAR model. *The Cryosphere*, **11** (2), 1015–1033, doi:10.5194/tc-11-1015-2017.

Fettweis, X. et al., 2013: Brief communication Important role of the mid-tropospheric atmospheric circulation in the recent surface melt increase over the Greenland ice sheet. *Cryosphere*, **7** (1), 241–248, doi:10.5194/tc-7-241-2013.

Fieber, K.D. et al., 2018: Rigorous 3D change determination in Antarctic Peninsula glaciers from stereo WorldView-2 and archival aerial imagery. *Remote Sensing of Environment*, **205**, 18–31, doi:10.1016/j.rse.2017.10.042.

Field, L. et al., 2018: Increasing Arctic Sea Ice Albedo Using Localized Reversible Geoengineering. *Earths Future*, **6** (6), 882–901, doi:10.1029/2018ef000820.

Fielding, S. et al., 2014: Interannual variability in Antarctic krill (Euphausia superba) density at South Georgia, Southern Ocean: 1997–2013. *ICES Journal of Marine Science*, **71** (9), 2578–2588, doi:10.1093/icesjms/fsu104.

Fillinger, L., D. Janussen, T. Lundälv and C. Richter, 2013: Rapid Glass Sponge Expansion after Climate-Induced Antarctic Ice Shelf Collapse. *Current Biology*, **23** (14), 1330–1334, doi:10.1016/j.cub.2013.05.051.

Fisher, D. et al., 2012: Recent melt rates of Canadian arctic ice caps are the highest in four millennia. *Global and Planetary Change*, **84–85**, 3–7, doi:10.1016/j.gloplacha.2011.06.005.

Flament, T., E. Berthier and F. Rémy, 2014: Cascading water underneath Wilkes Land, East Antarctic ice sheet, observed using altimetry and digital elevation models. *The Cryosphere*, **8** (2), 673–687, doi:10.5194/tc-8-673-2014.

Flanner, M.G. et al., 2011: Radiative forcing and albedo feedback from the Northern Hemisphere cryosphere between 1979 and 2008. *Nature Geoscience*, **4** (3), 151–155, doi:10.1038/ngeo1062.

Flanner, M.G., C.S. Zender, J.T. Randerson and P.J. Rasch, 2007: Present-day climate forcing and response from black carbon in snow. *Journal of Geophysical Research*, **112** (D11), D11202, doi:10.1029/2006jd008003.

Flannigan, M., B. Stocks, M. Turetsky and M. Wotton, 2009: Impacts of climate change on fire activity and fire management in the circumboreal forest. *Global Change Biology*, **15** (3), 549–560, doi:10.1111/j.1365-2486.2008.01660.x.

3

Fleming, A.H. and N.D. Pyenson, 2017: How to Produce Translational Research to Guide Arctic Policy. *Bioscience*, **67** (6), 490–493, doi:10.1093/biosci/bix002.

Fletcher, C.G., C.W. Thackeray and T.M. Burgers, 2015: Evaluating biases in simulated snow albedo feedback in two generations of climate models. *Journal of Geophysical Research: Atmospheres*, **120** (1), 12–26, doi:10.1002/2014jd022546.

Fluhr, J. et al., 2017: Weakening of the subpolar gyre as a key driver of North Atlantic seabird demography: a case study with Brunnich's guillemots in Svalbard. *Marine Ecology Progress Series*, **563**, 1–11, doi:10.3354/meps11982.

Flynn, M., J.D. Ford, T. Pearce and S.L. Harper, 2018: Participatory scenario planning and climate change impacts, adaptation and vulnerability research in the Arctic. *Environmental Science & Policy*, **79**, 45–53, doi:10.1016/j.envsci.2017.10.012.

Fogt, R.L. and E.A. Zbacnik, 2014: Sensitivity of the Amundsen Sea Low to Stratospheric Ozone Depletion. *Journal of Climate*, **27** (24), 9383–9400, doi:10.1175/JCLI-D-13-00657.1.

Fogwill, C.J. et al., 2016: Brief communication: Impacts of a developing polynya off Commonwealth Bay, East Antarctica, triggered by grounding of iceberg B09B. *The Cryosphere*, **10** (6), 2603–2609, doi:10.5194/tc-10-2603-2016.

Forbes, B.C., 2013: Cultural Resilience of Social-ecological Systems in the Nenets and Yamal-Nenets Autonomous Okrugs, Russia: A Focus on Reindeer Nomads of the Tundra. *Ecology and Society*, **18** (4), doi:10.5751/ES-05791-180436.

Forbes, B.C. et al., 2015: *Arctic Human Development Report II* [Larsen, J.N. and G. Fondal (eds.)]. Chapter 7 Resource Governanc, Denmark, 253–289.

Forbes, B.C. et al., 2016: Sea ice, rain-on-snow and tundra reindeer nomadism in Arctic Russia. *Biology Letters*, **12** (11), doi:10.1098/rsbl.2016.0466.

Forbis Jr, R. and K. Hayhoe, 2018: Does Arctic governance hold the key to achieving climate policy targets? *Environmental Research Letters*, **13** (2), 020201, doi:10.1088/1748-9326/aaa359.

Forcada, J. and J.I. Hoffman, 2014: Climate change selects for heterozygosity in a declining fur seal population. *Nature*, **511** (7510), 462–5, doi:10.1038/nature13542.

Ford, J.D., 2012: Indigenous health and climate change. *American Journal of Public Health*, **102** (7), 1260–1266, doi:10.2105/AJPH.2012.300752.

Ford, J.D., L. Berrang-Ford, M. King and C. Furgal, 2010: Vulnerability of Aboriginal health systems in Canada to climate change. *Global Environmental Change*, **20** (4), 668–680, doi:10.1016/j.gloenvcha.2010.05.003.

Ford, J.D. et al., 2012: Mapping Human Dimensions of Climate Change Research in the Canadian Arctic. *AMBIO*, **41** (8), 808–822, doi:10.1007/s13280-012-0336-8.

Ford, J.D. et al., 2016: Including indigenous knowledge and experience in IPCC assessment reports. *Nature Climate Change*, **6**, 349, doi:10.1038/nclimate2954

Ford, J.D. et al., 2019: Changing access to ice, land and water in Arctic communities. *Nature Climate Change*, **9** (4), 335–339, doi:10.1038/s41558-019-0435-7.

Ford, J.D. and D. King, 2013: A framework for examining adaptation readiness. *Mitigation and Adaptation Strategies for Global Change*, **20** (4), 505–526, doi:10.1007/s11027-013-9505-8.

Ford, J.D., G. McDowell and J. Jones, 2014a: The state of climate change adaptation in the Arctic. *Environmental Research Letters*, **9** (10), 104005, doi:10.1088/1748-9326/9/10/104005.

Ford, J.D., G. McDowell and T. Pearce, 2015: The adaptation challenge in the Arctic. *Nature Climate Change*, **5**, 1046, doi:10.1038/nclimate2723.

Ford, J.D. and T. Pearce, 2012: Climate change vulnerability and adaptation research focusing on the Inuit subsistence sector in Canada: Directions for future research. *The Canadian Geographer / Le Géographe canadien*, **56** (2), 275–287, doi:10.1111/j.1541-0064.2012.00418.x.

Ford, J.D. et al., 2014b: Adapting to the Effects of Climate Change on Inuit Health. *American Journal of Public Health*, **104** (S3), e9-e17, doi:10.2105/AJPH.2013.301724.

Ford, J.D. a. et al., 2013: The dynamic multiscale nature of climate change vulnerability: An Inuit harvesting example. *Annals of the Association of American Geographers*, **103** (5), 1193–1211, doi:10.1080/00045608.2013.776880.

Forde, T. et al., 2016a: Genomic analysis of the multi-host pathogen Erysipelothrix rhusiopathiae reveals extensive recombination as well as the existence of three generalist clades with wide geographic distribution. *BMC Genomics*, **17**, 461, doi:10.1186/s12864-016-2643-0.

Forde, T.L. et al., 2016b: Bacterial Genomics Reveal the Complex Epidemiology of an Emerging Pathogen in Arctic and Boreal Ungulates. *Frontiers in Microbiology*, **7**, 1759, doi:10.3389/fmicb.2016.01759.

Forster, R.R. et al., 2013: Extensive liquid meltwater storage in firn within the Greenland ice sheet. *Nature Geoscience*, **7** (2), 95–98, doi:10.1038/ngeo2043.

Forzieri, G., R. Alkama, D.G. Miralles and A. Cescatti, 2017: Satellites reveal contrasting responses of regional climate to the widespread greening of Earth. *Science*, **356** (6343), 1180–1184, doi:10.1126/science.aal1727.

Fossheim, M. et al., 2015: Recent warming leads to a rapid borealization of fish communities in the Arctic. *Nature Climate Change*, **5**, 673, doi:10.1038/nclimate2647.

Frainer, A. et al., 2017: Climate-driven changes in functional biogeography of Arctic marine fish communities. *Proc Natl Acad Sci U S A*, **114** (46), 12202–12207, doi:10.1073/pnas.1706080114.

Francis, D., C. Eayrs, J. Cuesta and D. Holland, 2019: Polar Cyclones at the Origin of the Reoccurrence of the Maud Rise Polynya in Austral Winter 2017. *Journal of Geophysical Research-Atmospheres*, **124** (10), 5251–5267, doi:10.1029/2019jd030618.

Francis, J.A., 2017: Why Are Arctic Linkages to Extreme Weather Still up in the Air? *Bulletin of the American Meteorological Society*, **98** (12), 2551–2557, doi:10.1175/bams-d-17-0006.1.

Francis, J.A. and S.J. Vavrus, 2015: Evidence for a wavier jet stream in response to rapid Arctic warming. *Environmental Research Letters*, **10** (1), 014005, doi:10.1088/1748-9326/10/1/014005.

Fransson, A. et al., 2017: Effects of sea-ice and biogeochemical processes and storms on under-ice water fCO2 during the winter-spring transition in the high Arctic Ocean: Implications for sea-air CO2 fluxes. *Journal of Geophysical Research-Oceans*, **122** (7), 5566–5587, doi:10.1002/2016jc012478.

Fraser, A.D. et al., 2011: East Antarctic Landfast Sea Ice Distribution and Variability, 2000–08. *Journal of Climate*, **25** (4), 1137–1156, doi:10.1175/jcli-d-10-05032.1.

Fraser, C.I., G.M. Kay, M. d. Plessis and P.G. Ryan, 2017: Breaking down the barrier: dispersal across the Antarctic Polar Front. *Ecography*, **40** (1), 235–237, doi:10.1111/ecog.02449.

Fraser, C.I. et al., 2018: Antarctica's ecological isolation will be broken by storm-driven dispersal and warming. *Nature Climate Change*, **8**, 704–708, doi:10.1038/s41558-018-0209-7.

Frederiksen, M., 2017: Synthesis: Status and trends of Arctic marine biodiversity and monitoring. In: CAFF State of the Arctic Marine Biodiversity Report. Conservation of Arctic Flora and Fauna International Secretariat Akureyri, Iceland, 175–195.

Freeman, N.M. and N.S. Lovenduski, 2015: Decreased calcification in the Southern Ocean over the satellite record. *Nature Geoscience*, **42** (6), 1834–1840, doi:papers2://publication/uuid/E879F895-C356-42B6-B1DE-7E33D4676735.

Freer, J.J. et al., 2018: Predicting ecological responses in a changing ocean: the effects of future climate uncertainty. *Mar Biol*, **165** (1), 7, doi:10.1007/s00227-017-3239-1.

French, N.H. et al., 2015: Fire in Arctic tundra of Alaska: past fire activity, future fire potential, and significance for land management and ecology.

International Journal of Wildland Fire, **24** (8), 1045–1061, doi:10.1071/WF14167.

Frenot, Y. et al., 2005: Biological invasions in the Antarctic: extent, impacts and implications. *Biological Reviews*, **80** (1), 45–72, doi:10.1017/s1464793104006542.

Frey, K.E., J.W. McClelland, R.M. Holmes and L.C. Smith, 2007: Impacts of climate warming and permafrost thaw on the riverine transport of nitrogen and phosphorus to the Kara Sea. *Journal of Geophysical Research: Biogeosciences*, **112** (G4), G04S5, doi:10.1029/2006jg000369.

Frezzotti, M. et al., 2013: A synthesis of the Antarctic surface mass balance during the last 800 yr. *Cryosphere*, **7** (1), 303–319, doi:10.5194/tc-7-303-2013.

Fricker, H.A., T. Scambos, R. Bindschadler and L. Padman, 2007: An active subglacial water system in West Antarctica mapped from space. *Science*, **315** (5818), 1544–1548, doi:10.1126/science.1136897

Fricker, H.A., M.R. Siegfried, S.P. Carter and T.A. Scambos, 2016: A decade of progress in observing and modelling Antarctic subglacial water systems. *Philosophical Transactions of the Royal Society A: Mathematical, Physical and Engineering Sciences*, **374** (2059), 20140294, doi:10.1098/rsta.2014.0294.

Friedl, P. et al., 2018: Recent dynamic changes on Fleming Glacier after the disintegration of Wordie Ice Shelf, Antarctic Peninsula. *The Cryosphere*, **12** (4), 1347–1365, doi:10.5194/tc-12-1347-2018.

Frölicher, T.L. et al., 2015: Dominance of the Southern Ocean in anthropogenic carbon and heat uptake in CMIP5 models. *Journal of Climate*, **28** (2), 862–886, doi:10.1175/JCLI-D-14-00117.1.

Frost, G.V. and H.E. Epstein, 2014: Tall shrub and tree expansion in Siberian tundra ecotones since the 1960s. *Global Change Biology*, **20** (4), 1264–1277, doi:10.1111/gcb.12406.

Fujiwara, A. et al., 2016: Influence of timing of sea ice retreat on phytoplankton size during marginal ice zone bloom period on the Chukchi and Bering shelves. *Biogeosciences*, **13** (1115–131), doi:10.5194/bg-13-115-2016.

Fünfgeld, H., 2015: Facilitating local climate change adaptation through transnational municipal networks. *Current Opinion in Environmental Sustainability*, **12**, 67–73, doi:10.1016/j.cosust.2014.10.011.

Fürst, J.J. et al., 2016: The safety band of Antarctic ice shelves. *Nature Climate Change*, **6**, 479, doi:10.1038/nclimate2912.

Fyfe, J.C. et al., 2013: One hundred years of Arctic surface temperature variation due to anthropogenic influence. *Scientific Reports*, **3**, 2645, doi:10.1038/srep02645

Fyke, J. et al., 2018: An Overview of Interactions and Feedbacks Between Ice Sheets and the Earth System. *Reviews of Geophysics*, **56** (2), 361–408, doi:10.1029/2018rg000600.

Gagné, M.-È. et al., 2017: Aerosol-driven increase in Arctic sea ice over the middle of the twentieth century. *Geophysical Research Letters*, **44** (14), 7338–7346, doi:10.1002/2016gl071941.

Gagné, M.È., N.P. Gillett and J.C. Fyfe, 2015: Observed and simulated changes in Antarctic sea ice extent over the past 50 years. *Geophysical Research Letters*, **42** (1), 90–95, doi:10.1002/2014gl062231.

Gagnon-Lebrun, F. and S. Agrawala, 2007: Implementing adaptation in developed countries: an analysis of progress and trends Implementing adaptation in developed countries: an analysis of progress and trends. *Climate Policy*, **7** (5), 37–41, doi:10.1080/14693062.2007.9685664.

Galicia, M.P. et al., 2016: Dietary habits of polar bears in Foxe Basin, Canada: possible evidence of a trophic regime shift mediated by a new top predator. *Ecology and Evolution*, **6** (16), 6005–6018, doi:10.1002/ece3.2173.

Gall, A.E., T.C. Morgan, R.H. Day and K.J. Kuletz, 2017: Ecological shift from piscivorous to planktivorous seabirds in the Chukchi Sea, 1975–2012. *Polar Biology*, **40** (1), 61–78, doi:10.1007/s00300-016-1924-z.

Gallaher, D.W., G.G. Campbell and W.N. Meier, 2014: Anomalous Variability in Antarctic Sea Ice Extents During the 1960s With the Use of Nimbus Data. *IEEE Journal of Selected Topics in Applied Earth Observations and Remote Sensing*, **7**, 881–887, doi:10.1109/jstars.2013.2264391.

Galley, R. et al., 2012: Landfast sea ice conditions in the Canadian Arctic, 1983–2009. *Arctic*, **65** (2), 133–144, doi:10.14430/arctic4195.

Ganz, P. et al., 2018: *Deployment performance review of the 2017 North Pacific Observer Program*. U.S. Dep. Commer., NOAA Tech. Memo., NMFS-AFSC-379, 68p [Available at: www.afsc.noaa.gov/publications/AFSC-TM/NOAA-TM-AFSC-379.pdf].

Gao, L., S.R. Rintoul and W. Yu, 2018: Recent wind-driven change in Subantarctic Mode Water and its impact on ocean heat storage. *Nature Climate Change*, **8** (1), 58, doi:10.1038/s41558-017-0022-8.

Gao, X. et al., 2013: Permafrost degradation and methane: low risk of biogeochemical climate-warming feedback. *Environmental Research Letters*, **8** (3), 035014, doi:10.1088/1748-9326/8/3/035014.

Gardner, A.S. et al., 2013: A reconciled estimate of glacier contributions to sea level rise: 2003 to 2009. *Science*, **340** (6134), 852–857, doi:10.1126/science.1234532.

Gardner, A.S. et al., 2018: Increased West Antarctic and unchanged East Antarctic ice discharge over the last 7 years. *The Cryosphere*, **12** (2), 521–547, doi:10.5194/tc-12-521-2018.

Garrett, R.A., T.C. Sharkey, M. Grabowski and W.A. Wallace, 2017: Dynamic resource allocation to support oil spill response planning for energy exploration in the Arctic. *European Journal of Operational Research*, **257** (1), 272–286, doi:10.1016/j.ejor.2016.07.023.

Gascon, G. et al., 2013a: Changes in accumulation-area firn stratigraphy and meltwater flow during a period of climate warming: Devon Ice Cap, Nunavut, Canada. *Journal of Geophysical Research: Earth Surface*, **118** (4), 2380–2391, doi:10.1002/2013JF002838.

Gascon, G., M. Sharp and A. Bush, 2013b: Changes in melt season characteristics on Devon Ice Cap, Canada, and their association with the Arctic atmospheric circulation. *Annals of Glaciology*, **54** (63), 101–110, doi:10.3189/2013AoG63A601.

Gaston, A.J. and K.H. Elliott, 2014: Seabird diet changes in northern Hudson Bay, 1981–2013, reflect the availability of schooling prey. *Marine Ecology Progress Series*, **513**, 211–223, doi:10.3354/meps10945.

Gaston, A.J., P.A. Smith and J.F. Provencher, 2012: Discontinuous change in ice cover in Hudson Bay in the 1990s and some consequences for marine birds and their prey. *ICES Journal of Marine Science*, **69** (7), 1218–1225, doi:10.1093/icesjms/fss040.

Gauthier, S. et al., 2015: Boreal forest health and global change. *Science*, **349** (6250), 819–22, doi:10.1126/science.aaa9092.

Ge, S., D. Yang and D.L. Kane, 2013: Yukon River Basin long-term (1977–2006) hydrologic and climatic analysis. *Hydrological Processes*, **27**, 2475–2484, doi:10.1002/hyp.9282.

Gearheard, S. et al., 2006: "It's Not that Simple": A Collaborative Comparison of Sea Ice Environments, Their Uses, Observed Changes, and Adaptations in Barrow, Alaska, USA, and Clyde River, Nunavut, Canada. *AMBIO: A Journal of the Human Environment*, **35** (4), 203–211, doi:10.1579/0044-7447(2006)35[203:intsac]2.0.co;2.

Gearheard, S.F. et al., 2013: *The meaning of ice: People and sea ice in three Arctic communities*. International Polar Institute, Montreal.

Geilfus, N.X. et al., 2016: Estimates of ikaite export from sea ice to the underlying seawater in a sea ice-seawater mesocosm. *Cryosphere*, **10** (5), 2173–2189, doi:10.5194/tc-10-2173-2016.

Gelfan, A. et al., 2016: Climate change impact on the water regime of two great Arctic rivers: modeling and uncertainty issues. *Climatic Change*, **141** (3), 499–515, doi:10.1007/s10584-016-1710-5.

Gent, P.R., 2016: Effects of Southern Hemisphere wind changes on the meridional overturning circulation in ocean models. *Annual Review of Marine Science*, **8**, 79–94, doi:10.1146/annurev-marine-122414-033929.

Gerlach, C., P.A. Loring, G. Kofinas and H. Penn, 2017: *Resilience to rapid change in Bering, Chukchi, and Beaufort communities*. Chapter 6 of Adaptation Actions for a Changing Arctic: Perspectives from the Bering-Chukchi-Beaufort Region, Arctic Monitoring and Assessment Programme (AMAP), Oslo, Norway, 155–176 [Available at: www.amap.no/documents/doc/

Adaptation-Actions-for-a-Changing-Arctic-Perspectives-from-the-Bering-Chukchi-Beaufort-Region/1615].

Gerland, S. et al., 2008: Decrease of sea ice thickness at Hopen, Barents Sea, during 1966–2007. *Geophysical Research Letters*, **35** (6), doi:10.1029/2007gl032716.

Gerringa, L.J.A. et al., 2012: Iron from melting glaciers fuels the phytoplankton blooms in Amundsen Sea (Southern Ocean): Iron biogeochemistry. *Deep Sea Research Part II: Topical Studies in Oceanography*, **71–76**, 16–31, doi:10.1016/j.dsr2.2012.03.007.

Gerringa, L.J.A. et al., 2015: Sources of iron in the Ross Sea Polynya in early summer. *Marine Chemistry*, **177**, 447–459, doi:10.1016/j.marchem.2015.06.002.

Gibson, C.M. et al., 2018: Wildfire as a major driver of recent permafrost thaw in boreal peatlands. *Nat Commun*, **9** (1), 3041, doi:10.1038/s41467-018-05457-1.

Giglio, L., J.T. Randerson and G.R. van der Werf, 2013: Analysis of daily, monthly, and annual burned area using the fourth-generation global fire emissions database (GFED4). *Journal of Geophysical Research: Biogeosciences*, **118** (1), 317–328, doi:10.1002/jgrg.20042.

Gilbert, A. et al., 2017: The projected demise of Barnes Ice Cap: Evidence of an unusually warm 21st century Arctic. *Geophysical Research Letters*, **44** (6), 2810–2816, doi:10.1002/2016gl072394.

Gilg, O. et al., 2012: Climate change and the ecology and evolution of Arctic vertebrates. *Annals of the New York Academy of Sciences*, **1249** (1), 166–190, doi:10.1111/j.1749-6632.2011.06412.x.

Gille, S.T., 2014: Meridional displacement of the Antarctic Circumpolar Current. *Phil. Trans. R. Soc. A*, **372** (2019), 20130273, doi:10.1098/rsta.2013.0273.

Gillett, N.P., 2004: Detecting the effect of climate change on Canadian forest fires. *Geophysical Research Letters*, **31** (18), L18211, doi:10.1029/2004gl020876.

Gladish, C.V., D.M. Holland and C.M. Lee, 2015: Oceanic Boundary Conditions for Jakobshavn Glacier. Part I: Variability and Renewal of Ilulissat Icefjord Waters, 2001–14*. *Journal of Physical Oceanography*, **45** (2003), 33–63, doi:10.1175/JPO-D-14-0045.1.

Goldhar, C., T. Bell and J. Wolf, 2013: Rethinking Existing Approaches to Water Security in Remote Communities: An Analysis of Two Drinking Water Systems in Nunatsiavut, Labrador, Canada. *Water Alternatives*, **6** (3), 462–486.

Goldhar, C., T. Bell and J. Wolf, 2014: Vulnerability to Freshwater Changes in the Inuit Settlement Region of Nunatsiavut, Labrador: A Case Study from Rigolet. *Arctic*, **67** (1), 71–83, doi:10.14430/arctic4365.

Golledge, N.R. et al., 2014: Antarctic contribution to meltwater pulse 1A from reduced Southern Ocean overturning. *Nature Communications*, **5** (5107), 1–10, doi:10.1038/ncomms6107.

Golovnev, A., 2017: Challenges to Arctic Nomadism: Yamal Nenets Facing Climate Change Era Calamities. *Arctic Anthropology*, **54**, 40–51.

Gomez, N., D. Pollard and D. Holland, 2015: Sea-level feedback lowers projections of future Antarctic Ice-Sheet mass loss. *Nature Communications*, **6**, 8798, doi:10.1038/ncomms9798.

Goodwin, B.P. et al., 2016: Accumulation variability in the Antarctic Peninsula: The role of large-scale atmospheric oscillations and their interactions. *Journal of Climate*, **29** (7), 2579–2596, doi:10.1175/JCLI-D-15-0354.1.

Goosse, H. et al., 2018: Quantifying climate feedbacks in polar regions. *Nature Communications*, **9** (1), 1919, doi:10.1038/s41467-018-04173-0.

Goosse, H. and V. Zunz, 2014: Decadal trends in the Antarctic sea ice extent ultimately controlled by ice-ocean feedback. *The Cryosphere*, **8**, 453–470, doi:10.5194/tc-8-435-2014.

Graeter, K.A. et al., 2018: Ice Core Records of West Greenland Melt and Climate Forcing. *Geophysical Research Letters*, **45** (7), 3164–3172, doi:10.1002/2017gl076641.

Graham, R.M. et al., 2017: Increasing frequency and duration of Arctic winter warming events. *Geophysical Research Letters*, **44** (13), 6974–6983, doi:10.1002/2017gl073395.

Grange, L.J. and C.R. Smith, 2013: Megafaunal communities in rapidly warming fjords along the West Antarctic Peninsula: hotspots of abundance and beta diversity. *Plos One*, **8** (12), e77917, doi:10.1371/journal.pone.0077917.

Granskog, M.A., I. Fer, A. Rinke and H. Steen, 2018: Atmosphere-Ice-Ocean-Ecosystem Processes in a Thinner Arctic Sea Ice Regime: The Norwegian Young Sea ICE (N-ICE2015) Expedition. *Journal of Geophysical Research: Oceans*, **123** (3), 1586–1594, doi:10.1002/2017jc013328.

Gray, A.R. et al., 2018: Autonomous Biogeochemical Floats Detect Significant Carbon Dioxide Outgassing in the High-Latitude Southern Ocean. *Geophysical Research Letters*, **45** (17), 9049–9057, doi:10.1029/2018gl078013.

Greaves, W., 2016: Arctic (in)security and Indigenous peoples: Comparing Inuit in Canada and Sámi in Norway. *Security Dialogue*, **47** (6), 461–480, doi:10.1177/0967010616665957.

Grebmeier, J.M., 2012: Shifting patterns of life in the Pacific Arctic and sub-Arctic Seas. *Annual Reviews in Marine Science*, **2012** (4), 63–78, doi:10.1146/annurev-marine-120710-100926.

Grebmeier, J.M. et al., 2015: Ecosystem characteristics and processes facilitating persistent macrobenthic biomass hotspots and associated benthivory in the Pacific Arctic. *Progress in Oceanography*, **136** (Supplement C), 92–114, doi:10.1016/j.pocean.2015.05.006.

Grebmeier, J.M. and L.W. Cooper, 2016: The Saint Lawrence Island Polynya: A 25-Year Evaluation of an Analogue for Climate Change in Polar Regions. In: Aquatic Microbial Ecology and Biogeochemistry: A Dual Perspective [Glibert, P.M. and T.M. Kana (eds.)]. Springer International Publishing, Cham, 171–183.

Greene, C.A. et al., 2017: Wind causes Totten Ice Shelf melt and acceleration. *Sci Adv*, **3** (11), e1701681, doi:10.1126/sciadv.1701681.

Greene, S. et al., 2014: Modeling the impediment of methane ebullition bubbles by seasonal lake ice. *Biogeosciences*, **11** (23), 6791–6811, doi:10.5194/bg-11-6791-2014.

Gregor, L., S. Kok and P.M.S. Monteiro, 2017: Empirical methods for the estimation of Southern Ocean CO2: support vector and random forest regression. *Biogeosciences*, **14** (23), 5551–5569, doi:10.1002/2016GB005541.

Gregory, R. et al., 2012: *Structured Decision Making*. John Wiley & Sons Ltd, Chichester, United Kingdom.

Griffith, D.L., L. Alessa and A. Kliskey, 2018: Community-based observing for social-ecological science: lessons from the Arctic. *Frontiers in Ecology and the Environment*, **16** (S1), S44-S51, doi:10.1002/fee.1798.

Griffiths, H.J., A.J.S. Meijers and T.J. Bracegirdle, 2017a: More losers than winners in a century of future Southern Ocean seafloor warming. *Nature Climate Change*, **7** (10), 749–754, doi:10.1038/nclimate3377.

Griffiths, H.J. et al., 2013: Antarctic Crabs: Invasion or Endurance? *Plos One*, **8** (7), e66981, doi:10.1371/journal.pone.0066981.

Griffiths, K. et al., 2017b: Ice-cover is the principal driver of ecological change in High Arctic lakes and ponds. *Plos One*, **12** (3), e0172989, doi:10.1371/journal.pone.0172989.

Groeneveld, R.A. et al., 2018: Defining scenarios of future vectors of change in marine life and associated economic sectors. *Estuarine, Coastal and Shelf Science*, **201**, 164–171, doi:10.1016/j.ecss.2015.10.020.

Grosse, G. et al., 2011: Vulnerability of high-latitude soil organic carbon in North America to disturbance. *Journal of Geophysical Research*, **116** (G4), G00K06, doi:10.1029/2010jg001507.

Grosse, G., B. Jones and C. Arp, 2013: Thermokarst lakes, drainage, and drained basins. *Treatise on Geomorphology*, **8**, 325–353, doi:10.1016/b978-0-12-374739-6.00216-5.

Grosvenor, D.P., J.C. King, T.W. Choularton and T. Lachlan-Cope, 2014: Downslope föhn winds over the antarctic peninsula and their effect on the larsen ice shelves. *Atmospheric Chemistry and Physics*, **14** (18), 9481–9509, doi:10.5194/acp-14-9481-2014.

Grotjahn, R. et al., 2016: North American extreme temperature events and related large scale meteorological patterns: a review of statistical methods, dynamics, modeling, and trends. *Climate Dynamics*, **46** (3–4), 1151–1184, doi:10.1007/s00382-015-2638-6.

Gruber, N. et al., 2019a: The oceanic sink for anthropogenic CO2 from 1994 to 2007. *Science*, **363** (6432), 1193–1199, doi:10.1126/science.aau5153.

Gruber, N., P. Landschutzer and N.S. Lovenduski, 2019b: The Variable Southern Ocean Carbon Sink. *Ann Rev Mar Sci*, **11**, 159–186, doi:10.1146/annurev-marine-121916-063407.

Gruber, S., 2012: Derivation and analysis of a high-resolution estimate of global permafrost zonation. *The Cryosphere*, **6** (1), 221–233, doi:10.5194/tc-6-221-2012.

Guerry, A.D. et al., 2015: Natural capital and ecosystem services informing decisions: From promise to practice. *Proceedings of the National Academy of Sciences*, **112** (24), 7348, doi:10.1073/pnas.1503751112.

Gullestad, P. et al., 2017: Towards ecosystem-based fisheries management in Norway – Practical tools for keeping track of relevant issues and prioritising management efforts. *Marine Policy*, **77**, 104–110, doi:10.1016/j.marpol.2016.11.032.

Gunn, A., 2016: Rangifer tarandus. The IUCN Red List of Threatened Species 2016:e.T29742A22167140;www.iucnredlist.org/species/29742/22167140.

Gunn, A. et al., 2011: Understanding the cumulative effects of human activities on barren-ground caribou. In: Cumulative Effects in Wildlife Management: Impact Mitigation [Krausman, P.R. and L.K. Harris (eds.)]. CRC Press, Boca Raton, 113–134.

Gunn, A. et al., 2013: CARMA's approach for the collaborative and inter-disciplinary assessment of cumulative effects. *Rangifer*, **33** (2), 161–166, doi:10.7557/2.33.2.2540.

Gurarie, E. et al., 2017: Distribution, density and abundance of Antarctic ice seals off Queen Maud Land and the eastern Weddell Sea. *Polar Biology*, **40** (5), 1149–1165, doi:10.1007/s00300-016-2029-4.

Gutt, J. et al., 2011: Biodiversity change after climate-induced ice-shelf collapse in the Antarctic. *Deep Sea Research Part II: Topical Studies in Oceanography*, **58** (1), 74–83, doi:10.1016/j.dsr2.2010.05.024.

Gutt, J. et al., 2015: The Southern Ocean ecosystem under multiple climate change stresses--an integrated circumpolar assessment. *Glob Chang Biol*, **21** (4), 1434–53, doi:10.1111/gcb.12794.

Gutt, J. et al., 2018: Cross-disciplinarity in the advance of Antarctic ecosystem research. *Mar Genomics*, **37**, 1–17, doi:10.1016/j.margen.2017.09.006.

Haas, C. et al., 2017: Ice and Snow Thickness Variability and Change in the High Arctic Ocean Observed by In Situ Measurements. *Geophysical Research Letters*, **44** (20), 10,462–10,469, doi:10.1002/2017gl075434.

Haine, T.W.N. et al., 2015: Arctic freshwater export: Status, mechanisms, and prospects. *Global and Planetary Change*, **125** (Supplement C), 13–35, doi:10.1016/j.gloplacha.2014.11.013.

Haine, T.W.N. and T. Martin, 2017: The Arctic-Subarctic sea ice system is entering a seasonal regime: Implications for future Arctic amplification. *Sci Rep*, **7** (1), 4618, doi:10.1038/s41598-017-04573-0.

Häkkinen, S., P.B. Rhines and D.L. Worthen, 2013: Northern North Atlantic sea surface height and ocean heat content variability. *Journal of Geophysical Research: Oceans*, **118** (7), 3670–3678, doi:10.1002/jgrc.20268.

Halfar, J. et al., 2013: Arctic sea-ice decline archived by multicentury annual-resolution record from crustose coralline algal proxy. *Proc Natl Acad Sci U S A*, **110** (49), 19737–41, doi:10.1073/pnas.1313775110.

Hall, D.K. et al., 2013: Variability in the surface temperature and melt extent of the Greenland ice sheet from MODIS. *Geophysical Research Letters*, **40** (10), 2114–2120, doi:10.1002/grl.50240.

Halliday, W.D. et al., 2017: Potential impacts of shipping noise on marine mammals in the western Canadian Arctic. *Marine Pollution Bulletin*, **123** (1), 73–82, doi:10.1016/j.marpolbul.2017.09.027.

Hamilton, C.D. et al., 2017: An Arctic predator–prey system in flux: climate change impacts on coastal space use by polar bears and ringed seals. *Journal of Animal Ecology*, **86** (5), 1054–1064, doi:10.1111/1365-2656.12685.

Hamilton, C.D., C. Lydersen, R.A. Ims and K.M. Kovacs, 2015: Predictions replaced by facts: a keystone species' behavioural responses to declining arctic sea-ice. *Biology Letters*, **11** (11), doi:10.1098/rsbl.2015.0803.

Hamilton, C.D. et al., 2019: Contrasting changes in space use induced by climate change in two Arctic marine mammal species. *Biol Lett*, **15** (3), 20180834, doi:10.1098/rsbl.2018.0834.

Hamilton, L.C. et al., 2016: Climigration? Population and climate change in Arctic Alaska. *Population and Environment*, **38** (2), 115–133, doi:10.1007/s11111-016-0259-6.

Hancock, A.M. et al., 2017: Ocean acidification changes the structure of an Antarctic coastal protistan community. *Biogeosciences Discussions*, **2017**, 1–32, doi:10.5194/bg-2017-224.

Hanes, C.C. et al., 2019: Fire-regime changes in Canada over the last half century. *Canadian Journal of Forest Research*, **49** (3), 256–269, doi:10.1139/cjfr-2018-0293.

Hanna, E., T.E. Cropper, R.J. Hall and J. Cappelen, 2016: Greenland Blocking Index 1851–2015: a regional climate change signal. *International Journal of Climatology*, **36** (15), 4847–4861, doi:10.1002/joc.4673.

Hanna, E. et al., 2015: Recent seasonal asymmetric changes in the NAO (a marked summer decline and increased winter variability) and associated changes in the AO and Greenland Blocking Index. *International Journal of Climatology*, **35** (9), 2540–2554, doi:10.1002/joc.4157.

Hanna, E. et al., 2014: Atmospheric and oceanic climate forcing of the exceptional Greenland ice sheet surface melt in summer 2012. *International Journal of Climatology*, **34** (4), 1022–1037, doi:10.1002/joc.3743.

Hanna, E., S.H. Mernild, J. Cappelen and K. Steffen, 2012: Recent warming in Greenland in a long-term instrumental (1881–2012) climatic context: I. Evaluation of surface air temperature records. *Environmental Research Letters*, **7** (4), 045404–045404, doi:10.1088/1748-9326/7/4/045404.

Hanna, E. et al., 2013: Ice-sheet mass balance and climate change. *Nature*, **498** (7452), 51–59, doi:10.1038/nature12238.

Hansen, B., B. et al., 2014: Warmer and wetter winters: characteristics and implications of an extreme weather event in the High Arctic. *Environmental Research Letters*, **9** (11), 114021, doi:10.1088/1748-9326/9/11/114021.

Hansen, B.B. et al., 2013: Climate Events Synchronize the Dynamics of a Resident Vertebrate Community in the High Arctic. *Science*, **339** (6117), 313, doi:10.1126/science.1226766.

Hansen, H.S.B., 2016: Three major challenges in managing non-native sedentary Barents Sea snow crab (Chionoecetes opilio). *Marine Policy*, **71**, 38–43, doi:10.1016/j.marpol.2016.05.013.

Harder, M.T. and G.W. Wenzel, 2012: Inuit subsistence, social economy and food security in Clyde River, Nunavut. *Arctic*, **65** (3), 305–318.

Harper, S.L., V.L. Edge, A. Cunsolo Willox and G. Rigolet Inuit Community, 2012: 'Changing climate, changing health, changing stories' profile: Using an EcoHealth approach to explore impacts of climate change on inuit health. *EcoHealth*, **9** (1), 89–101, doi:10.1007/s10393-012-0762-x.

Harper, S.L. et al., 2011: Weather, Water Quality and Infectious Gastrointestinal Illness in Two Inuit Communities in Nunatsiavut, Canada: Potential Implications for Climate Change. *EcoHealth*, **8** (1), 93–108, doi:10.1007/s10393-011-0690-1.

Hartmann, D.L. et al., 2013: Observations: atmosphere and surface. In: Climate Change 2013: The Physical Science Basis. Contribution of Working Group I to the Fifth Assessment Report of the Intergovernmental Panel on Climate Change. [Stocker, T.F., D. Qin, G.K. Plattner, M. Tignor, S.K. Allen, J. Boschung, A. Nauels, Y. Xia, V. Bex and P.M. Midgley (eds.)]. Cambridge University Press.

Hatta, M. et al., 2017: The relative roles of modified circumpolar deep water and benthic sources in supplying iron to the recurrent phytoplankton blooms above Pennell and Mawson Banks, Ross Sea, Antarctica. *Journal of Marine Systems*, **166**, 61–72, doi:10.1016/j.jmarsys.2016.07.009.

3

Hauck, J. and C. Volker, 2015: Rising atmospheric CO2 leads to large impact of biology on Southern Ocean CO2 uptake via changes of the Revelle factor. *Geophys Res Lett*, **42** (5), 1459–1464, doi:10.1002/2015GL063070.

Haug, T. et al., 2017: Future harvest of living resources in the Arctic Ocean north of the Nordic and Barents Seas: A review of possibilities and constraints. *Fisheries Research*, **188**, 38–57, doi:10.1016/j.fishres.2016.12.002.

Haumann, F.A. et al., 2016: Sea-ice transport driving Southern Ocean salinity and its recent trends. *Nature*, **537** (7618), 89–92, doi:10.1038/nature19101.

Haumann, F.A., D. Notz and H. Schmidt, 2014: Anthropogenic influence on recent circulation-driven Antarctic sea ice changes. *Geophysical Research Letters*, **41** (23), 8429–8437, doi:10.1002/2014gl061659.

Hauquier, F., L. Ballesteros-Redondo, J. Gutt and A. Vanreusel, 2016: Community dynamics of nematodes after Larsen ice-shelf collapse in the eastern Antarctic Peninsula. *Ecol Evol*, **6** (1), 305–17, doi:10.1002/ece3.1869.

Hauri, C., T. Friedrich and A. Timmermann, 2015: Abrupt onset and prolongation of aragonite undersaturation events in the Southern Ocean. *Nature Climate Change*, **6** (2), 172–176, doi:10.1038/nclimate2844.

Hauser, D.D.W., K.L. Laidre and H.L. Stern, 2018: Vulnerability of Arctic marine mammals to vessel traffic in the increasingly ice-free Northwest Passage and Northern Sea Route. *Proceedings of the National Academy of Sciences*, **115** (29), 7617, doi:10.1073/pnas.1803543115.

Hawkings, J. et al., 2016: The Greenland Ice Sheet as a hot spot of phosphorus weathering and export in the Arctic. *Global Biogeochemical Cycles*, **30** (2), 191–210, doi:10.1002/2015GB005237.

Hawkings, J.R. et al., 2017: Ice sheets as a missing source of silica to the polar oceans. *Nat Commun*, **8**, 14198, doi:10.1038/ncomms14198.

Hawkings, J.R. et al., 2015: The effect of warming climate on nutrient and solute export from the Greenland Ice Sheet. *Geochemical Perspectives Letters*, **1** (0), 94–104, doi:10.7185/geochemlet.1510.

Hawkings, J.R. et al., 2014: Ice sheets as a significant source of highly reactive nanoparticulate iron to the oceans. *Nat Commun*, **5**, 3929, doi:10.1038/ncomms4929.

Hawkins, E. and R. Sutton, 2012: Time of emergence of climate signals. *Geophysical Research Letters*, **39** (1), doi:10.1029/2011gl050087.

Haynes, T.B. et al., 2014: Patterns of lake occupancy by fish indicate different adaptations to life in a harsh Arctic environment. *Freshwater Biology*, **59** (9), 1884–1896, doi:10.1111/fwb.12391.

Haynie, A.C. and L. Pfeiffer, 2012: Why economics matters for understanding the effects of climate change on fisheries. *ICES Journal of Marine Science*, **69** (7), 1160–1167, doi:10.1093/icesjms/fss021.

Hedger, R.D. et al., 2013: Predicting climate change effects on subarctic-Arctic populations of Atlantic salmon (Salmo salar). *Canadian Journal of Fisheries and Aquatic Sciences*, **70** (2), 159–168, doi:10.1139/cjfas-2012-0205.

Hegyi, B.M. and Y. Deng, 2016: Dynamical and Thermodynamical Impacts of High- and Low-Frequency Atmospheric Eddies on the Initial Melt of Arctic Sea Ice. *Journal of Climate*, **30** (3), 865–883, doi:10.1175/jcli-d-15-0366.1.

Hegyi, B.M. and P.C. Taylor, 2018: The Unprecedented 2016–2017 Arctic Sea Ice Growth Season: The Crucial Role of Atmospheric Rivers and Longwave Fluxes. *Geophysical Research Letters*, **45** (10), 5204–5212, doi:10.1029/2017gl076717.

Heim, K.C. et al., 2016: Seasonal cues of Arctic grayling movement in a small Arctic stream: the importance of surface water connectivity. *Environmental Biology of Fishes*, **99** (1), 49–65, doi:10.1007/s10641-015-0453-x.

Heininen, L. and M. Finger, 2017: The "Global Arctic" as a New Geopolitical Context and Method. *Journal of Borderlands Studies*, **33** (2), 199–202, doi:10.1080/08865655.2017.1315605.

Heleniak, T., 2014: Arctic Populations and Migration. In: Arctic Human Development Report: Regional Processes and Global Linkages [Nymand Larsen, J. and G. Fondhal (eds.)]. Nordic Council of Ministers, Copenhagen, 53–104.

Helm, V., A. Humbert and H. Miller, 2014: Elevation and elevation change of Greenland and Antarctica derived from CryoSat-2. *The Cryosphere*, **8** (4), 1539–1559, doi:10.5194/tc-8-1539-2014.

Héon, J., D. Arseneault and M.A. Parisien, 2014: Resistance of the boreal forest to high burn rates. *Proc Natl Acad Sci U S A*, **111** (38), 13888–93, doi:10.1073/pnas.1409316111.

Hermann, A.J. et al., 2016: Projected future biophysical states of the Bering Sea. *Deep Sea Research Part II: Topical Studies in Oceanography*, **134** (Supplement C), 30–47, doi:10.1016/j.dsr2.2015.11.001.

Hermann, A.J. et al., 2019: Projected biophysical conditions of the Bering Sea to 2100 under multiple emission scenarios. *ICES Journal of Marine Science*, **76** (5), 1280–1304, doi:10.1093/icesjms/fsz043.

Hernández-Henríquez, M., S. Déry and C. Derksen, 2015: Polar amplification and elevation-dependence in trends of Northern Hemisphere snow cover extent. *Environmental Research Letters*, **10**, 044010, doi:doi:10.1088/1748-9326/10/4/044010.

Herraiz-Borreguero, L. et al., 2016: Large flux of iron from the Amery Ice Shelf marine ice to Prydz Bay, East Antarctica. *Journal of Geophysical Research: Oceans*, **121** (8), 6009–6020, doi:10.1002/2016jc011687.

Heuzé, C., K.J. Heywood, D.P. Stevens and J.K. Ridley, 2015: Changes in global ocean bottom properties and volume transports in CMIP5 models under climate change scenarios. *Journal of Climate*, **28** (8), 2917–2944, doi:10.1175/JCLI-D-14-00381.1.

Hill, S.L., T. Phillips and A. Atkinson, 2013: Potential Climate Change Effects on the Habitat of Antarctic Krill in the Weddell Quadrant of the Southern Ocean. *Plos One*, **8** (8), e72246, doi:10.1371/journal.pone.0072246.

Hillenbrand, C.-D. et al., 2017: West Antarctic Ice Sheet retreat driven by Holocene warm water incursions. *Nature*, **547**, 43, doi:10.1038/nature22995.

Hindell, M.A. et al., 2016: Circumpolar habitat use in the southern elephant seal: implications for foraging success and population trajectories. *Ecosphere*, **7** (5), e01213, doi:10.1002/ecs2.1213.

Hinke, J.T. et al., 2017a: Identifying Risk: Concurrent Overlap of the Antarctic Krill Fishery with Krill-Dependent Predators in the Scotia Sea. *Plos One*, **12** (1), e0170132, doi:10.1371/journal.pone.0170132.

Hinke, J.T., S.G. Trivelpiece and W.Z. Trivelpiece, 2017b: Variable vital rates and the risk of population declines in Adelie penguins from the Antarctic Peninsula region. *Ecosphere*, **8** (1), 1–13, doi:10.1002/ecs2.1666.

Hjort, J. et al., 2018: Degrading permafrost puts Arctic infrastructure at risk by mid-century. *Nat Commun*, **9** (1), 5147, doi:10.1038/s41467-018-07557-4.

Hobbs, W., M. Curran, N. Abram and E.R. Thomas, 2016a: Century-scale perspectives on observed and simulated Southern Ocean sea ice trends from proxy reconstructions. *Journal of Geophysical Research: Oceans*, **121** (10), 7804–7818, doi:10.1002/2016jc012111.

Hobbs, W.R., N.L. Bindoff and M.N. Raphael, 2015: New Perspectives on Observed and Simulated Antarctic Sea Ice Extent Trends Using Optimal Fingerprinting Techniques. *Journal of Climate*, **28** (4), 1543–1560, doi:10.1175/Jcli-D-14-00367.1.

Hobbs, W.R. et al., 2016b: A review of recent changes in Southern Ocean sea ice, their drivers and forcings. *Global and Planetary Change*, **143**, 228–250, doi:10.1016/j.gloplacha.2016.06.008.

Hoberg, e. a. Arctic Biodiversity Assessment 2013: Chapter 15, Parasites. [Available at: www.caff.is/assessment-series/arctic-biodiversity-assessment/220-arctic-biodiversity-assessment-2013-chapter-15-parasites]

Hoberg, E.P. and D.R. Brooks, 2015: Evolution in action: climate change, biodiversity dynamics and emerging infectious disease. *Philos Trans R Soc Lond B Biol Sci*, **370** (1665), 20130553, doi:10.1098/rstb.2013.0553.

Hoberg, E.P. et al., 2017: Arctic systems in the Quaternary: ecological collision, faunal mosaics and the consequences of a wobbling climate. *J Helminthol*, **91** (4), 409–421, doi:10.1017/S0022149X17000347.

Hodgson, D.A. et al., 2014: Terrestrial and submarine evidence for the extent and timing of the Last Glacial Maximum and the onset of deglaciation on the maritime-Antarctic and sub-Antarctic islands. *Quaternary Science Reviews*, **100**, 137–158, doi:10.1016/j.quascirev.2013.12.001.

Hodgson, D.A. and J.P. Smol, 2008: High latitude paleolimnology. In: In Polar lakes and rivers – Limnology of Arctic and Antarctic aquatic ecosystems [Vincent, W.F. and J. Laybourn-Parry (eds.)]. Oxford University Press, Oxford, 43–64.

Hodson, A. et al., 2017: Climatically sensitive transfer of iron to maritime Antarctic ecosystems by surface runoff. *Nat Commun*, **8**, 14499, doi:10.1038/ncomms14499.

Hodson, A.J., 2014: Understanding the dynamics of black carbon and associated contaminants in glacial systems. *Wiley Interdisciplinary Reviews: Water*, **1** (2), 141–149, doi:10.1002/wat2.1016.

Hodson, D.L.R. et al., 2013: Identifying uncertainties in Arctic climate change projections. *Climate Dynamics*, **40** (11), 2849–2865, doi:10.1007/s00382-012-1512-z.

Hofer, S., A.J. Tedstone, X. Fettweis and J.L. Bamber, 2017: Decreasing cloud cover drives the recent mass loss on the Greenland Ice Sheet. *Sci Adv*, **3** (6), e1700584, doi:10.1126/sciadv.1700584.

Hogg, A.E. et al., 2017: Increased ice flow in Western Palmer Land linked to ocean melting. *Geophysical Research Letters*, **44** (9), 4159–4167, doi:10.1002/2016GL072110.

Hogg, A.M. et al., 2015: Recent trends in the Southern Ocean eddy field. *Journal of Geophysical Research: Oceans*, **120** (1), 257–267, doi:10.1002/2014JC010470.

Holland, D.M., 2001: Explaining the Weddell Polynya – a large ocean eddy shed at Maud Rise. *Science*, **292** (5522), 1697–1700, doi:10.1126/science.1059322.

Holland, D.M. et al., 2008: Acceleration of Jakobshavn Isbræ triggered by warm subsurface ocean waters. *Nature Geoscience*, **1**, 659, doi:10.1038/ngeo316.

Holland, P.R., 2014: The seasonality of Antarctic sea ice trends. *Geophysical Research Letters*, **41**, 4230–4237, doi:10.1002/2014GL060172.

Holland, P.R. et al., 2014: Modeled trends in Antarctic sea ice thickness. *Journal of Climate*, **27**, 3784–3801, doi:10.1175/JCLI-D-13-00301.1.

Holland, P.R. and R. Kwok, 2012: Wind-driven trends in Antarctic sea-ice drift. *Nature Geoscience*, **5** (12), 872–875, doi:Doi 10.1038/Ngeo1627.

Holmes, R.M. et al., 2018: *River Discharge* [Jeffries, M.O., J. Richter-Menge and J.E. Overland (eds.)]. Arctic Report Card, Update for 2018 [Available at: www.arctic.noaa.gov/Report-Card/Report-Card-2018/ArtMID/7878/ArticleID/786/River-Discharge].

Holmes, R.M. et al., 2015: *River Discharge* [Jeffries, M.O., J. Richter-Menge and J.E. Overland (eds.)]. Arctic Report Card, 2015, 60–65 [Available at: https://arctic.noaa.gov/Report-Card/Report-Card-2015/ArtMID/5037/ArticleID/227/River-Discharge].

Holsman, K. et al., 2018: Chapter 6: Climate change impacts, vulnerabilities and adaptations: North Pacific and Pacific Arctic marine fisheries. In: Impacts of climate change on fisheries and aquaculture: synthesis of current knowledge, adaptation and mitigation options. FAO Fisheries and Aquaculture Technical Paper No. 627. [Barange, M., T. Bahri, M.C.M. Beveridge, K.L. Cochrane, S. Funge-Smith and F. Poulain (eds.)], Rome.

Holsman, K. et al., 2017: An ecosystem-based approach to marine risk assessment. *Ecosystem Health and Sustainability*, **3** (1), e01256, doi:10.1002/ehs2.1256.

Holsman, K.K. et al., 2019: Towards climate resiliency in fisheries management. *ICES Journal of Marine Science*, **76** (5), 1368–1378, doi:10.1093/icesjms/fsz031.

Holsman, K.K. et al., 2016: A comparison of fisheries biological reference points estimated from temperature-specific multi-species and single-species climate-enhanced stock assessment models. *Deep Sea Research Part II: Topical Studies in Oceanography*, **134** (Supplement C), 360–378, doi:10.1016/j.dsr2.2015.08.001.

Holst-Andersen, J.P. et al., 2017: Impact of Arctic Met ocean Conditions on Oil Spill Response: Results of the First Circumpolar Response Viability Analysis. *International Oil Spill Conference Proceedings*, **2017** (1), 2017172, doi:10.7901/2169-3358-2017.1.000172.

Honda, M., J. Inoue and S. Yamane, 2009: Influence of low Arctic sea-ice minima on anomalously cold Eurasian winters. *Geophysical Research Letters*, **36** (8), L08707, doi:10.1029/2008gl037079.

Hood, E. et al., 2015: Storage and release of organic carbon from glaciers and ice sheets. *Nature Geosci*, **8** (2), 91–96, doi:10.1038/ngeo2331.

Hop, H. and H. Gjøsæter, 2013: Polar cod (Boreogadus saida) and capelin (Mallotus villosus) as key species in marine food webs of the Arctic and the Barents Sea. *Marine Biology Research*, **9** (9), 878–894, doi:10.1080/17451000.2013.775458.

Hope, A.G. et al., 2013: Future distribution of tundra refugia in northern Alaska. *Nature Climate Change*, **3**, 931, doi:10.1038/nclimate1926.

Hoppe, C.J.M. et al., 2018: Compensation of ocean acidification effects in Arctic phytoplankton assemblages. *Nature Climate Change*, **8** (6), 529–533, doi:10.1038/s41558-018-0142-9.

Hopwood, M.J. et al., 2018: Non-linear response of summertime marine productivity to increased meltwater discharge around Greenland. *Nat Commun*, **9** (1), 3256, doi:10.1038/s41467-018-05488-8.

Horgan, H.J. et al., 2013: Estuaries beneath ice sheets. *Geology*, **41** (11), 1159–1162, doi:10.1130/G34654.1.

Hori, M. et al., 2017: A 38-year (1978–2015) Northern Hemisphere daily snow cover extent product derived using consistent objective criteria from satellite-borne optical sensors. *Remote Sensing of Environment*, **191**, 402–418, doi:doi:10.1016/j.rse.2017.01.023.

Hori, Y. et al., 2018: Implications of projected climate change on winter road systems in Ontario's Far North, Canada. *Climatic Change*, **148** (1–2), 109–122, doi:10.1007/s10584-018-2178-2.

Horstkotte, T. et al., 2017: Human–animal agency in reindeer management: Sámi herders' perspectives on vegetation dynamics under climate change. *Ecosphere*, **8** (9), e01931, doi:10.1002/ecs2.1931.

Horvat, C. et al., 2017: The frequency and extent of sub-ice phytoplankton blooms in the Arctic Ocean. *Science Advances*, **3** (3), e1601191, doi:10.1126/sciadv.1601191.

Hossain, K., 2015: Governance of Arctic Ocean Marine Resources. In: Climate Change Impacts on Ocean and Coastal Law: U.S. and International Perspectives [Abate, R.S. (ed.)]. Oxford Scholarship Online, New York.

Howell, S.E.L. et al., 2016: Landfast ice thickness in the Canadian Arctic Archipelago from observations and models. *The Cryosphere*, **10** (4), 1463–1475, doi:10.5194/tc-10-1463-2016.

Howell, S.E.L. et al., 2013: Recent changes in the exchange of sea ice between the Arctic Ocean and the Canadian Arctic Archipelago. *Journal of Geophysical Research: Oceans*, **118** (7), 3595–3607, doi:10.1002/jgrc.20265.

Howes, E., F. Joos, M. Eakin and J.-P. Gattuso, 2015: An updated synthesis of the observed and projected impacts of climate change on the chemical, physical and biological processes in the oceans. *Frontiers in Marine Science*, **2** (36), doi:10.3389/fmars.2015.00036.

Hu, F.S. et al., 2015: Arctic tundra fires: natural variability and responses to climate change. *Frontiers in Ecology and the Environment*, **13** (7), 369–377, doi:10.1890/150063.

Hu, F.S. et al., 2010: Tundra burning in Alaska: Linkages to climatic change and sea ice retreat. *Journal of Geophysical Research: Biogeosciences*, **115** (G4), G04002, doi:10.1029/2009JG001270.

Huang, X. et al., 2018: Improved Representation of Surface Spectral Emissivity in a Global Climate Model and Its Impact on Simulated Climate. *Journal of Climate*, **31** (9), 3711–3727, doi:10.1175/jcli-d-17-0125.1.

Huebert, R., 2013: Cooperation or Conflict in the New Arctic? Too Simple of a Dichotomy! In: *Environmental Security in the Arctic Ocean*, Dordrecht, [P., B. and V.A. (eds.)], Springer Netherlands, 195–203.

Huet, C. et al., 2017: Food insecurity and food consumption by season in households with children in an Arctic city: a cross-sectional study. *BMC Public Health*, **17** (1), 578, doi:10.1186/s12889-017-4393-6.

Hugelius, G. et al., 2014: Estimated stocks of circumpolar permafrost carbon with quantified uncertainty ranges and identified data gaps. *Biogeosciences*, **11** (23), 6573–6593, doi:10.5194/bg-11-6573-2014.

Hughes, K.A., L.R. Pertierra, M.A. Molina-Montenegro and P. Convey, 2015: Biological invasions in terrestrial Antarctica: what is the current status and can we respond? *Biodiversity and Conservation*, **24** (5), 1031–1055, doi:10.1007/s10531-015-0896-6.

Huiskes, A.H.L. et al., 2014: Aliens in Antarctica: Assessing transfer of plant propagules by human visitors to reduce invasion risk. *Biological Conservation*, **171**, 278–284, doi:10.1016/j.biocon.2014.01.038.

Humphrey, N.F., J.T. Harper and W.T. Pfeffer, 2012: Thermal tracking of meltwater retention in Greenland's accumulation area. *Journal of Geophysical Research: Earth Surface*, **117** (1), 1–11, doi:10.1029/2011JF002083.

Hunt, B.P.V. et al., 2014: Zooplankton community structure and dynamics in the Arctic Canada Basin during a period of intense environmental change (2004–2009). *Journal of Geophysical Research: Oceans*, **119** (4), 2518–2538, doi:10.1002/2013JC009156.

Hunt, G.L. et al., 2016: Advection in polar and sub-polar environments: Impacts on high latitude marine ecosystems. *Progress in Oceanography*, **149**, 40–81, doi:10.1016/j.pocean.2016.10.004.

Hunt, G.L. et al., 2018: Timing of sea-ice retreat affects the distribution of seabirds and their prey in the southeastern Bering Sea. *Marine Ecology Progress Series*, **593**, 209–230, doi:10.3354/meps12383.

Huntington, H.P. et al., 2015: Vessels, risks, and rules: Planning for safe shipping in Bering Strait. *Marine Policy*, **51**, 119–127, doi:10.1016/j.marpol.2014.07.027.

Huntington, H.P. et al., 2018: Staying in place during times of change in Arctic Alaska: the implications of attachment, alternatives, and buffering. *Regional Environmental Change*, **18** (2), 489–499, doi:10.1007/s10113-017-1221-6.

Huntington, H.P., L.T. Quakenbush and M. Nelson, 2017: Evaluating the Effects of Climate Change on Indigenous Marine Mammal Hunting in Northern and Western Alaska Using Traditional Knowledge. *Frontiers in Marine Science*, **4**, 319, doi:10.3389/fmars.2017.00319.

Huskey, L., 2018: An Arctic development strategy? The North Slope Inupiat and the resource curse. *Canadian Journal of Development Studies / Revue canadienne d'études du développement*, **39** (1), 89–100, doi:10.1080/02255189.2017.1391067.

Huskey, L., I. Maenpaa and A. Pelyasov, 2014: Economic Systems. In: Arctic Human Development Report [Nymand Larson, J. and G. Fondahl (eds.)]. Nordic Council of Ministers, Copenhagen, 151–184.

Hutchins, D.A. and P.W. Boyd, 2016: Marine phytoplankton and the changing ocean iron cycle. *Nature Climate Change*, **6** (12), 1072, doi:10.1038/nclimate3147.

Hyder, P. et al., 2018: Critical Southern Ocean climate model biases traced to atmospheric model cloud errors. *Nat Commun*, **9** (1), 3625, doi:10.1038/s41467-018-05634-2.

Ianelli, J., K.K. Holsman, A.E. Punt and K. Aydin, 2016: Multi-model inference for incorporating trophic and climate uncertainty into stock assessments. *Deep Sea Research Part II: Topical Studies in Oceanography*, **134** (Supplement C), 379–389, doi:10.1016/j.dsr2.2015.04.002.

ICC, 2012: *Food Security across the Arctic, Background paper of the Steering Committee of the Circumpolar Inuit Health Strategy Inuit Circumpolar Council*. Canada [Available at: www.inuitcircumpolar.com/uploads/3/0/5/4/30542564/icc_food_security_across_the_arctic_may_2012.pdf].

Ignatowski, J. and J. Rosales, 2013: Identifying the exposure of two subsistence villages in Alaska to climate change using traditional ecological knowledge. *Climatic Change*, **121** (2), 285–299, doi:10.1007/s10584-013-0883-4.

Ilicak, M. et al., 2016: An assessment of the Arctic Ocean in a suite of interannual CORE-II simulations. Part III: Hydrography and fluxes. *Ocean Modelling*, **100**, 141–161, doi:10.1016/j.ocemod.2016.02.004.

IMO, 2017: *International Maritime Organisation, International Code for Ships Operating in Polar Waters ('Polar Code')*.

Ingels, J., R.B. Aronson and C.R. Smith, 2018: The scientific response to Antarctic ice-shelf loss. *Nature Climate Change*, **8** (10), 848–851, doi:10.1038/s41558-018-0290-y.

Ingvaldsen, R.B. and H. Gjøsæter, 2013: Responses in spatial distribution of Barents Sea capelin to changes in stock size, ocean temperature and ice cover. *Marine Biology Research*, **9** (9), 867–877, doi:10.1080/17451000.2013.775450.

Inuit Circumpolar Council-Alaska (ICC-AK), 2015: *Alaskan Inuit Food Security Conceptual Framework: How to Assess the Arctic from an Inuit Perspective -- Summary and Recommendations Report. Report created as part of 2015 Alaskan Inuit Food Security Conceptual Framework Technical Report*. [Available at: https://tribalclimateguide.uoregon.edu/literature/inuit-circumpolar-council-alaska-icc-ak-2015-alaskan-inuit-food-security-conceptual].

Inuit Circumpolar Council, 2014: *The Sea Ice Never Stops. Circumpolar Inuit Reflections on Sea Ice Use and Shipping in Inuit Nunaat*. (ICC), I.C.C., Canada [Available at: http://hdl.handle.net/11374/1478].

Inuit Circumpolar Council Canada, 2014: *The Sea Ice Never Stops: Reflections on sea ice use and shipping in Inuit Nunaat*. Inuit Circumpolar Council – Canada, Ottawa, Canada [Available at: https://oaarchive.arctic-council.org/handle/11374/410].

IPCC, 2013: *Climate Change 2013: The Physical Science Basis. Contribution of Working Group I to the Fifth Assessment Report of the Intergovernmental Panel on Climate Change* [Stocker, T.F., D. Qin, G.-K. Plattner, M. Tignor, S.K. Allen, J. Boschung, A. Nauels, Y. Xia, V. Bex and P.M. Midgley (eds.)]. Cambridge University Press, Cambridge, United Kingdom and New York, NY, USA, 1535 pp [Available at: www.climatechange2013.org/report/full-report/].

Irving, D.B., S. Wijffels and J.A. Church, 2019: Anthropogenic Aerosols, Greenhouse Gases, and the Uptake, transport, and Storage of Excess Heat in the Climate System. *Geophysical Research Letters*, **46** (9), 4894–4903, doi:10.1029/2019gl082015.

Islam, D. and F. Berkes, 2016: Indigenous peoples' fisheries and food security: a case from northern Canada. *Food Security*, **8** (4), 815–826, doi:10.1007/s12571-016-0594-6.

Itkin, P. et al., 2017: Thin ice and storms: Sea ice deformation from buoy arrays deployed during N-ICE2015. *Journal of Geophysical Research: Oceans*, **122** (6), 4661–4674, doi:10.1002/2016jc012403.

Ivanov, V. et al., 2016: Arctic Ocean heat impact on regional ice decay: A suggested positive feedback. *Journal of Physical Oceanography*, **46** (5), 1437–1456, doi:10.1175/JPO-D-15-0144.1.

Ivanova, D.P. et al., 2016: Moving beyond the Total Sea Ice Extent in Gauging Model Biases. *Journal of Climate*, **29** (24), 8965–8987, doi:10.1175/Jcli-D-16-0026.1.

Iverson, J., 2013: Funding Alaska Village Relocation Caused by Climate Change and Preserving Cultural Values During Relocation. *Seattle Journal for Social Justice*, **12** (2), Article 12.

Jabour, J., 2017: 25. Southern Ocean search and rescue: platforms and procedures. *Handbook on the Politics of Antarctica*, 392.

Jacobs, S.S., 2004: Bottom water production and its links with the thermohaline circulation. *Antarctic Science*, **16** (4), 427–437, doi:10.1017/S095410200400224x.

Jacobs, S.S., A. Jenkins, C.F. Giulivi and P. Dutrieux, 2011: Stronger ocean circulation and increased melting under Pine Island Glacier ice shelf. *Nature Geoscience*, **4** (8), 519, doi:10.1038/ngeo1188.

Jahn, A., 2018: Reduced probability of ice-free summers for 1.5°C compared to 2°C warming. *Nature Climate Change*, **8** (5), 409–413, doi:10.1038/s41558-018-0127-8.

3

Jahn, A., J.E. Kay, M.M. Holland and D.M. Hall, 2016: How predictable is the timing of a summer ice-free Arctic? *Geophysical Research Letters*, **43** (17), 9113–9120, doi:10.1002/2016gl070067.

Jahn, A. et al., 2012: Late-Twentieth-Century Simulation of Arctic Sea Ice and Ocean Properties in the CCSM4. *Journal of Climate*, **25** (5), 1431–1452, doi:10.1175/jcli-d-11-00201.1.

Jakobsen, I., 2014: Extractive Industries in Arctic: The International Legal Framework for the Protection of the Environment. *Nordic Environmental Law Journal I*, **1**, 39–52.

Janout, M. et al., 2015: Episodic warming of near-bottom waters under the Arctic sea ice on the central Laptev Sea shelf. *Geophysical Research Letters*, **43** (1), 264–272, doi:10.1002/2015GL066565.

Jansen, J. et al., 2017: Abundance and richness of key Antarctic seafloor fauna correlates with modelled food availability. *Nature Ecology & Evolution*, 1–13, doi:10.1038/s41559-017-0392-3.

Jansen, T. et al., 2016: Ocean warming expands habitat of a rich natural resource and benefits a national economy. *Ecological Applications*, **26** (7), 2021–2032, doi:10.1002/eap.1384.

Jarvis, D. et al., 2013: *Review of the evidence on Indicators, metrics and monitoring systems*. Department for International Development, UK Government [Available at: http://r4d.dfid.gov.uk/Output/192446/Default.aspx].

Jena, B., M. Ravichandran and J. Turner, 2019: Recent Reoccurrence of Large Open-Ocean Polynya on the Maud Rise Seamount. *Geophysical Research Letters*, **46** (8), 4320–4329, doi:10.1029/2018gl081482.

Jenkins, A. et al., 2018: West Antarctic Ice Sheet retreat in the Amundsen Sea driven by decadal oceanic variability. *Nature Geoscience*, **11** (10), 733–738, doi:10.1038/s41561-018-0207-4.

Jenkins, E.J. et al., 2013: Tradition and transition: parasitic zoonoses of people and animals in Alaska, northern Canada, and Greenland. *Adv Parasitol*, **82**, 33–204, doi:10.1016/B978-0-12-407706-5.00002-2.

Jenouvrier, S. et al., 2018: Climate change and functional traits affect population dynamics of a long-lived seabird. *Journal of Animal Ecology*, **87** (4), 906–920, doi:10.1111/1365-2656.12827.

Jenouvrier, S. et al., 2014: Projected continent-wide declines of the emperor penguin under climate change. *Nature Climate Change*, **4** (8), 715–718, doi:10.1038/nclimate2280.

Jenouvrier, S., C. Péron and H. Weimerskirch, 2015: Extreme climate events and individual heterogeneity shape life-history traits and population dynamics. *Ecological Monographs*, **85** (4), 605–624, doi:10.1890/14-1834.1.

Jensen, A.J., B. Finstad and P. Fiske, 2018: Evidence for the linkage of survival of anadromous Arctic char and brown trout during winter to marine growth during the previous summer. *Canadian Journal of Fisheries and Aquatic Sciences*, **75** (5), 663–672, doi:10.1139/cjfas-2017-0077.

Jepsen, S.M. et al., 2013: Linkages between lake shrinkage/expansion and sublacustrine permafrost distribution determined from remote sensing of interior Alaska, USA. *Geophysical Research Letters*, **40** (5), 882–887, doi:doi:10.1002/grl.50187.

Jepsen, S.M., M.A. Walvoord, C.I. Voss and J. Rover, 2015: Effect of permafrost thaw on the dynamics of lakes recharged by ice-jam floods: case study of Yukon Flats, Alaska. *Hydrological Processes*, **30** (11), 1782–1795, doi:doi:10.1002/hyp.10756.

Jin, M. et al., 2016: Ecosystem model intercomparison of under-ice and total primary production in the Arctic Ocean. *Journal of Geophysical Research: Oceans*, **121** (1), 934–948, doi:10.1002/2015JC011183.

Jobbágy, E.G. and R.B. Jackson, 2000: The vertical distribution of soil organic carbon and its relation to climate and vegetation. *Ecological Applications*, **10** (2), 423–436, doi:10.1890/1051-0761.

Johannesen, E. et al., 2017: Large-scale patterns in community structure of benthos and fish in the Barents Sea. *Polar Biology*, **40** (2), 237–246, doi:10.1007/s00300-016-1946-6.

Johnson, N. et al., 2015a: The Contributions of Community-Based Monitoring and Traditional Knowledge to Arctic Observing Networks: Reflections on the State of the Field. *Arctic*, **68** (28–40), doi:10.14430/arctic4447.

Johnson, N. et al., 2015b: *Community-Based Monitoring and Indigenous Knowledge in a Changing Arctic: A Review for the Sustaining Arctic Observing Networks*. Brown University's Voss Interdisciplinary Postdoctoral Fellowship, Brown University.

Johnston, A., M.E. Johnston, J. Dawson and E.J. Stewart, 2012a: Challenges of changes in Arctic cruise tourism: perspectives of federal government stakeholders. *Journal of Maritime Law and Commerce*, **43** (3), 335–347.

Johnston, A. et al., 2012b: Perspectives of decision makers and regulators on climate change and adaptation in expedition cruise ship tourism in Nunavut. *Northern Review*, **35**, 69–85.

Johnston, D.W., M.T. Bowers, A.S. Friedlaender and D.M. Lavigne, 2012c: The Effects of Climate Change on Harp Seals (Pagophilus groenlandicus). *Plos One*, **7** (1), e29158, doi:10.1371/journal.pone.0029158.

Johnston, M., J. Dawson, E. De Souza and E.J. Stewart, 2017: Management challenges for the fastest growing marine shipping sector in Arctic Canada: pleasure crafts. *Polar Record*, **53** (1), 67–78, doi:10.1017/s0032247416000565.

Johnstone, J. et al., 2009: Postfire seed rain of black spruce, a semiserotinous conifer, in forests of interior Alaska. *Canadian Journal of Forest Research*, **39** (8), 1575–1588, doi:10.1139/X09-068.

Johnstone, J.F. et al., 2010: Fire, climate change, and forest resilience in interior AlaskaThis article is one of a selection of papers from The Dynamics of Change in Alaska's Boreal Forests: Resilience and Vulnerability in Response to Climate Warming. *Canadian Journal of Forest Research*, **40** (7), 1302–1312, doi:10.1139/X10-061.

Johnstone, J.F., T.S. Rupp, M. Olson and D. Verbyla, 2011: Modeling impacts of fire severity on successional trajectories and future fire behavior in Alaskan boreal forests. *Landscape Ecology*, **26** (4), 487–500, doi:10.1007/s10980-011-9574-6.

Joli, N. et al., 2018: Need for focus on microbial species following ice melt and changing freshwater regimes in a Janus Arctic Gateway. *Scientific Reports*, **8** (1), 9405, doi:10.1038/s41598-018-27705-6.

Jones, B. et al., 2009: Fire behavior, weather, and burn severity of the 2007 Anaktuvuk river tundra fire, North Slope, Alaska. *Arctic, Antarctic, and Alpine Research*, **41** (3), 309–316, doi:10.1657/1938-4246-41.3.309.

Jones, B.M. et al., 2011: Modern thermokarst lake dynamics in the continuous permafrost zone, northern Seward Peninsula, Alaska. *Journal of Geophysical Research: Biogeosciences*, **116** (G2), G00M03, doi:10.1029/2011JG001666.

Jones, B.M. et al., 2015a: Recent Arctic tundra fire initiates widespread thermokarst development. *Scientific Reports*, **5**, 15865, doi:10.1038/srep15865.

Jones, J.M. et al., 2016: Assessing recent trends in high-latitude Southern Hemisphere surface climate. *Nature Climate Change*, **6**, 917, doi:10.1038/nclimate3103.

Jones, R.S. et al., 2015b: Rapid Holocene thinning of an East Antarctic outlet glacier driven by marine ice sheet instability. *Nature Communications*, **6**, 8910, doi:10.1038/ncomms9910.

Joughin, I., B.E. Smith and B. Medley, 2014: Marine Ice Sheet Collapse Potentially Under Way for the Thwaites Glacier Basin, West Antarctica. *Science*, **344** (6185), 735–738, doi:10.1126/science.1249055.

Jourdain, N.C. et al., 2017: Ocean circulation and sea-ice thinning induced by melting ice shelves in the Amundsen Sea. *Journal of Geophysical Research: Oceans*, **122** (3), 2550–2573, doi:10.1002/2016JC012509.

Ju, J. and J.G. Masek, 2016: The vegetation greenness trend in Canada and US Alaska from 1984–2012 Landsat data. *Remote Sensing of Environment*, **176**, 1–16, doi:10.1016/j.rse.2016.01.001.

Jung, T. et al., 2015: Polar Lower-Latitude Linkages and Their Role in Weather and Climate Prediction. *Bulletin of the American Meteorological Society*, **96** (11), Es197-Es200, doi:10.1175/Bams-D-15-00121.1.

Juul-Pedersen, T. et al., 2015: Seasonal and interannual phytoplankton production in a sub-Arctic tidewater outlet glacier fjord, SW Greenland. *Marine Ecology Progress Series*, **524**, 27–38, doi:10.3354/meps11174.

3

Juvonen, S.-K. and A.E. Kuhmonen, 2013: Evaluation of the Protected Area Network in the Barents Region Using the Programme of Work on Protected Areas of the Convention on Biological Diversity as a Tool. In: Reports of the Finnish Environment Institute, Helsinki, **37**.

Kaartvedt, S. and J. Titelman, 2018: Planktivorous fish in a future Arctic Ocean of changing ice and unchanged photoperiod. *ICES Journal of Marine Science*, **75** (7), 2312–2318, doi:10.1093/icesjms/fsx248.

Kafle, P. et al., 2018: Temperature-dependent development and freezing survival of protostrongylid nematodes of Arctic ungulates: implications for transmission. *Parasites & Vectors*, **11** (1), 400, doi:10.1186/s13071-018-2946-x.

Kahru, M., V. Brotas, M. Manzano-Sarabia and B.G. Mitchell, 2011: Are phytoplankton blooms occurring earlier in the Arctic? *Global Change Biology*, **17** (4), 1733–1739, doi:10.1111/j.1365-2486.2010.02312.x.

Kahru, M., Z. Lee, B.G. Mitchell and C.D. Nevison, 2016: Effects of sea ice cover on satellite-detected primary production in the Arctic Ocean. *Biology Letters*, **12** (11), doi:10.1098/rsbl.2016.0223.

Kaiser, B.A. et al., 2015: Spatial issues in Arctic marine resource governance workshop summary and comment. *Marine Policy*, **58**, 1–5, doi:10.1016/j.marpol.2015.03.033.

Kaján, E., 2014: Arctic Tourism and Sustainable Adaptation: Community Perspectives to Vulnerability and Climate Change. *Scandinavian Journal of Hospitality and Tourism*, **14** (1), 60–79, doi:10.1080/15022250.2014.886097.

Kaleschke, L. et al., 2016: SMOS sea ice product: Operational application and validation in the Barents Sea marginal ice zone. *Remote Sensing of Environment*, **180**, 264–273, doi:10.1016/j.rse.2016.03.009.

Kanevskiy, M. et al., 2013: Ground ice in the upper permafrost of the Beaufort Sea coast of Alaska. *Cold Regions Science and Technology*, **85**, 56–70, doi:10.1016/j.coldregions.2012.08.002.

Kanna, N. et al., 2018: Upwelling of Macronutrients and Dissolved Inorganic Carbon by a Subglacial Freshwater Driven Plume in Bowdoin Fjord, Northwestern Greenland. *Journal of Geophysical Research: Biogeosciences*, **123** (5), 1666–1682, doi:10.1029/2017JG004248.

Kapsch, M.-L., R.G. Graversen and M. Tjernström, 2013: Springtime atmospheric energy transport and the control of Arctic summer sea-ice extent. *Nature Climate Change*, **3**, 744, doi:10.1038/nclimate1884.

Kapsch, M.-L., R.G. Graversen, M. Tjernström and R. Bintanja, 2016: The Effect of Downwelling Longwave and Shortwave Radiation on Arctic Summer Sea Ice. *Journal of Climate*, **29** (3), 1143–1159, doi:10.1175/jcli-d-15-0238.1.

Karnovsky, N.J., K.A. Hobson, Z.W. Brown and G.L. Hunt, 2009: Distribution and diet of Ivory Gulls (Pagophila eburnea) in the North Water Polynya. *Arctic*, **62**, 65–74, doi:10.14430/arctic113

Karpechko, A. et al., 2018: *Scientific Assessment of Ozone Depletion: 2018* [Cagnazzo, C. and L. Polvani (eds.)]. Stratospheric Ozone and Climate, Chapter 5, World Meteorological Organization, G.S., Geneva, Switzerland [Available at: www.esrl.noaa.gov/csd/assessments/ozone/2018/report/Chapter5_2018OzoneAssessment.pdf].

Kasischke, E.S. and M.R. Turetsky, 2006: Recent changes in the fire regime across the North American boreal region – Spatial and temporal patterns of burning across Canada and Alaska. *Geophysical Research Letters*, **33** (9), L09703, doi:10.1029/2006gl025677.

Kates, R.W., W.R. Travis and T.J. Wilbanks, 2012: Transformational adaptation when incremental adaptations to climate change are insufficient. *Proceedings of the National Academy of Sciences of the United States of America*, **109** (19), 7156–7161, doi:10.1073/pnas.1115521109.

Kaufman, D.E. et al., 2017: Climate change impacts on southern Ross Sea phytoplankton composition, productivity, and export. *Journal of Geophysical Research: Oceans*, **122** (3), 2339–2359, doi:10.1002/2016jc012514.

Kavanaugh, M.T. et al., 2015: Effect of continental shelf canyons on phytoplankton biomass and community composition along the western Antarctic Peninsula. *Marine Ecology Progress Series*, **524**, 11–26, doi:10.3354/meps11189.

Kawaguchi, S. et al., 2013: Risk maps for Antarctic krill under projected Southern Ocean acidification. *Nature Climate Change*, **3**, 843, doi:10.1038/nclimate1937.

Kawaguchi, S., S. Nicol and A.J. Press, 2009: Direct effects of climate change on the Antarctic krill fishery. *Fisheries Management and Ecology*, **16** (5), 424–427, doi:10.1111/j.1365-2400.2009.00686.x.

Kay, J.E., M.M. Holland and A. Jahn, 2011: Inter-annual to multi-decadal Arctic sea ice extent trends in a warming world. *Geophysical Research Letters*, **38** (15), L15708, doi:10.1029/2011gl048008.

Keil, K. and S.E. Knecht, 2017: *Governing Arctic Change: Global Perspectives*. Palgrave Macmillan, Basingstoke, UK. Kelly, R. et al., 2013: Recent burning of boreal forests exceeds fire regime limits of the past 10,000 years. *Proceedings of the National Academy of Sciences USA*, **110** (32), 13055–13060, doi:10.1073/pnas.1305069110.

Kennicutt, M.C. et al., 2014a: Polar research: Six priorities for Antarctic science. *Nature*, (512), 23–25, doi:10.1038/512023a.

Kennicutt, M.C. et al., 2014b: A roadmap for Antarctic and Southern Ocean science for the next two decades and beyond. *Antarctic Science*, **27** (1), 3–18, doi:10.1017/S0954102014000674.

Keppler, L. and P. Landschutzer, 2019: Regional Wind Variability Modulates the Southern Ocean Carbon Sink. *Sci Rep*, **9** (1), 7384, doi:10.1038/s41598-019-43826-y.

Kern, S. and B. Ozsoy-Çiçek, 2016: Satellite Remote Sensing of Snow Depth on Antarctic Sea Ice: An Inter-Comparison of Two Empirical Approaches. *Remote Sensing*, **8** (6), 450, doi:10.3390/rs8060450.

Kessler, A. and J. Tjiputra, 2016: The Southern Ocean as a constraint to reduce uncertainty in future ocean carbon sinks. *Earth System Dynamics*, **7** (2), 295–312, doi:papers2://publication/doi/10.5194/esd-7-295-2016.

Khan, S.A. et al., 2015: Greenland ice sheet mass balance: a review. *Reports on Progress in Physics*, **046801**, 1–26, doi:10.1088/0034-4885/78/4/046801.

Kharin, V.V., F.W. Zwiers, X. Zhang and M. Wehner, 2013: Changes in temperature and precipitation extremes in the CMIP5 ensemble. *Climatic Change*, **119** (2), 345–357, doi:10.1007/s10584-013-0705-8.

Khazendar, A. et al., 2016: Rapid submarine ice melting in the grounding zones of ice shelves in West Antarctica. *Nature Communications*, **7**, 13243, doi:10.1038/ncomms13243.

Khazendar, A. et al., 2013: Observed thinning of Totten Glacier is linked to coastal polynya variability. *Nature Communications*, **4**, 2857, doi:10.1038/ncomms3857.

Kiani, S. et al., 2018: Effects of recent temperature variability and warming on the Oulu-Hailuoto ice road season in the northern Baltic Sea. *Cold Regions Science and Technology*, **151**, 1–8, doi:10.1016/j.coldregions.2018.02.010.

Kidston, J., A.S. Taschetto, D.W.J. Thompson and M.H. England, 2011: The influence of Southern Hemisphere sea-ice extent on the latitude of the mid-latitude jet stream. *Geophysical Research Letters*, **38**, 5, doi:10.1029/2011gl048056.

Kim, B.-M. et al., 2017: Major cause of unprecedented Arctic warming in January 2016: Critical role of an Atlantic windstorm. *Scientific Reports*, **7**, 40051, doi:10.1038/srep40051.

Kim, B.M. et al., 2014: Weakening of the stratospheric polar vortex by Arctic sea-ice loss. *Nat Commun*, **5**, 4646, doi:10.1038/ncomms5646.

Kim, H. et al., 2018: Inter-decadal variability of phytoplankton biomass along the coastal West Antarctic Peninsula. *Philos Trans A Math Phys Eng Sci*, **376** (2122), 1–21, doi:10.1098/rsta.2017.0174.

Kimmel, D.G., L.B. Eisner, M.T. Wilson and J.T. Duffy-Anderson, 2018: Copepod dynamics across warm and cold periods in the eastern Bering Sea: Implications for walleye pollock (Gadus chalcogrammus) and the Oscillating Control Hypothesis. *Fisheries Oceanography*, **27** (2), 143–158, doi:10.1111/fog.12241.

Kimura, S. et al., 2017: Oceanographic Controls on the Variability of Ice-Shelf Basal Melting and Circulation of Glacial Meltwater in the Amundsen Sea Embayment, Antarctica. *Journal of Geophysical Research: Oceans*, **122** (12), 10131–10155, doi:10.1002/2017JC012926.

King, M.D. et al., 2018: Seasonal to decadal variability in ice discharge from the Greenland Ice Sheet. *The Cryosphere*, **12** (12), 3813–3825, doi:10.5194/tc-12-3813-2018.

Kingslake, J. et al., 2018: Extensive retreat and re-advance of the West Antarctic Ice Sheet during the Holocene. *Nature*, **558** (7710), 430–434, doi:10.1038/s41586-018-0208-x.

Kinnard, C. et al., 2011: Reconstructed changes in Arctic sea ice over the past 1,450 years. *Nature*, **479** (7374), 509–12, doi:10.1038/nature10581.

Kinzey, D., G.M. Watters and C.S. Reiss, 2015: Selectivity and two biomass measures in an age-based assessment of Antarctic krill (Euphausia superba). *Fisheries Research*, **168**, 72–84, doi:10.1016/j.fishres.2015.03.023.

Kirkfeldt, T.S. et al., 2016: Why cumulative impacts assessments of hydrocarbon activities in the Arctic fail to meet their purpose. *Regional Environmental Change*, **17** (3), 725–737, doi:10.1007/s10113-016-1059-3.

Kjeldsen, K.K. et al., 2015: Spatial and temporal distribution of mass loss from the Greenland Ice Sheet since AD 1900. *Nature*, **528**, 396, doi:10.1038/nature16183.

Kjesbu, O.S. et al., 2014: Synergies between climate and management for Atlantic cod fisheries at high latitudes. *Proceedings of the National Academy of Sciences*, **111** (9), 3478–3483, doi:10.1073/pnas.1316342111.

Klaczek, M.R., C.J. Johnson and H.D. Cluff, 2016: Wolf–caribou dynamics within the central Canadian Arctic. *The Journal of Wildlife Management*, **80** (5), 837–849, doi:10.1002/jwmg.1070.

Klein, E.S. et al., 2018: Impacts of rising sea temperature on krill increase risks for predators in the Scotia Sea. *Plos One*, **13** (1), e0191011, doi:10.1371/journal.pone.0191011.

Kleinen, T. and V. Brovkin, 2018: Pathway-dependent fate of permafrost region carbon. *Environmental Research Letters*, **13** (9), 094001, doi:10.1088/1748-9326/aad824.

Klokov, K., 2012: Changes in reindeer population numbers in Russia: an effect of the political context or climate? *Rangifer*, **32** (1), 19, doi:10.7557/2.32.1.2234.

Kloster, S., N.M. Mahowald, J.T. Randerson and P.J. Lawrence, 2012: The impacts of climate, land use, and demography on fires during the 21st century simulated by CLM-CN. *Biogeosciences*, **9** (1), 509–525, doi:10.5194/bg-9-509-2012.

Kocho-Schellenberg, J.-E. and F. Berkes, 2014: Tracking the development of co-management: using network analysis in a case from the Canadian Arctic. *Polar Record*, **51** (4), 422–431, doi:10.1017/S0032247414000436.

Koenig, L.S., C. Miège, R.R. Forster and L. Brucker, 2014: Initial in situ measurements of perennial meltwater storage in the Greenland firn aquifer. *Geophysical Research Letters*, **41** (1), 81–85, doi:10.1002/2013GL058083.

Koenigk, T. and L. Brodeau, 2014: Ocean heat transport into the Arctic in the twentieth and twenty-first century in EC-Earth. *Climate Dynamics*, **42** (11–12), 3101–3120, doi:10.1007/s00382-013-1821-x.

Koenigstein, S. et al., 2018: Forecasting future recruitment success for Atlantic cod in the warming and acidifying Barents Sea. *Glob Chang Biol*, **24** (1), 526–535, doi:10.1111/gcb.13848.

Kofinas, G. et al., 2016: Building resilience in the Arctic: From theory to practice. In: Arctic Resilience Report [Council, A. (ed.)][Carson, M. and G. Peterson (eds.)]. Stockholm Environment Institute and Stockholm Resilience Centre, Stockholm, 180–208.

Kohlbach, D. et al., 2016: The importance of ice algae-produced carbon in the central Arctic Ocean ecosystem: Food web relationships revealed by lipid and stable isotope analyses. *Limnology and Oceanography*, **61** (6), 2027–2044, doi:10.1002/lno.10351.

Kohnert, K. et al., 2017: Strong geologic methane emissions from discontinuous terrestrial permafrost in the Mackenzie Delta, Canada. *Scientific Reports*, **7** (1), 5828, doi:10.1038/s41598-017-05783-2.

Koivurova, T., 2013: Gaps in International Regulatory Frameworks for the Arctic Ocean. In: *Environmental Security in the Arctic Ocean*, 2013//, Dordrecht, [Berkman, P.A. and A.N. Vylegzhanin (eds.)], Springer Netherlands, 139–155.

Koivurova, T. and R. Caddell, 2018: Managing Biodiversity Beyond National Jurisdiction in the Changing Arctic. *The American Society of International Law*, **112**, 134–138, doi:10.1017/aju.2018.44.

Kokelj, S.V. et al., 2013: Thawing of massive ground ice in mega slumps drives increases in stream sediment and solute flux across a range of watershed scales. *Journal of Geophysical Research: Earth Surface*, **118** (2), 681–692, doi:10.1002/jgrf.20063.

Kokelj, S.V. et al., 2017: Climate-driven thaw of permafrost preserved glacial landscapes, northwestern Canada. *Geology*, **45** (4), 371–374, doi:10.1130/G38626.1.

Kokelj, S.V. et al., 2015: Increased precipitation drives mega slump development and destabilization of ice-rich permafrost terrain, northwestern Canada. *Global and Planetary Change*, **129**, 56–68, doi:10.1016/j.gloplacha.2015.02.008.

Kokubun, N. et al., 2018: Inter-annual climate variability affects foraging behavior and nutritional state of thick-billed murres breeding in the southeastern Bering Sea. *Marine Ecology Progress Series*, **593**, 195–208, doi:10.3354/meps12365.

Kononova, N.K., 2012: The influence of atmospheric circulation on the formation of snow cover on the north eastern Siberia. *Ice and Snow*, **1**, 38/53, doi:10.15356/2076-6734-2012-1-38-53.

Konrad, H. et al., 2017: Uneven onset and pace of ice-dynamical imbalance in the Amundsen Sea Embayment, West Antarctica. *Geophysical Research Letters*, **44** (2), 910–918, doi:10.1002/2016GL070733.

Konrad, H. et al., 2018: Net retreat of Antarctic glacier grounding line. *Nature Geosciences*, **11**, 258–262, doi:10.1038/s41561-018-0082-z.

Kooyman, G.L. and P.J. Ponganis, 2017: Rise and fall of Ross Sea emperor penguin colony populations: 2000 to 2012. *Antarctic Science*, **29** (3), 201–208, doi:10.1017/S0954102016000559.

Kortsch, S. et al., 2012: Climate-driven regime shifts in Arctic marine benthos. *Proceedings of the National Academy of Sciences of the United States of America*, **109** (35), 14052–14057, doi:10.1073/pnas.1207509109.

Kortsch, S. et al., 2015: Climate change alters the structure of arctic marine food webs due to poleward shifts of boreal generalists. *Proceedings of the Royal Society B: Biological Sciences*, **282** (1814), 20151546, doi:10.1098/rspb.2015.1546.

Kouril, D., C. Furgal and T. Whillans, 2016: Trends and key elements in community-based monitoring: a systematic review of the literature with an emphasis on Arctic and Subarctic regions. *Environmental Reviews*, **24** (2), 151–163, doi:10.1139/er-2015-0041.

Kovacs, K.M. et al., 2012: Global threats to pinnipeds. *Marine Mammal Science*, **28** (2), 414–436, doi:10.1111/j.1748-7692.2011.00479.x.

Kovacs, K.M., P. Lemons, J.G. MacCracken and C. Lydersen, 2016: *Walruses in a time of climate change*. Arctic Report Card: Update for 2015 [Available at: www.arctic.noaa.gov/Report-Card].

Koven, C.D., 2013: Boreal carbon loss due to poleward shift in low-carbon ecosystems. *Nature Geoscience*, **6** (6), 452–456, doi:10.1038/ngeo1801.

Koven, C.D., W.J. Riley and A. Stern, 2013: Analysis of Permafrost Thermal Dynamics and Response to Climate Change in the CMIP5 Earth System Models. *Journal of Climate*, **26** (6), 1877–1900, doi:10.1175/jcli-d-12-00228.1.

Koven, C.D. et al., 2011: Permafrost carbon-climate feedbacks accelerate global warming. *Proceedings of the National Academy of Sciences USA*, **108** (36), 14769–14774, doi:10.1073/pnas.1103910108.

Koven, C.D. et al., 2015: A simplified, data-constrained approach to estimate the permafrost carbon–climate feedback. *Philosophical Transactions of the Royal Society A: Mathematical, Physical and Engineering Sciences*, **373** (2054), 20140423, doi:10.1098/rsta.2014.0423.

Kraska, J., 2011: *Arctic Security in an Age of Climate Change* [Kraska, J. (ed.)]. Cambridge University Press, Cambridge.

Krasting, J.P., A.J. Broccoli, K.W. Dixon and J.R. Lanzante, 2013: Future Changes in Northern Hemisphere Snowfall. *Journal of Climate*, **26** (20), 7813–7828, doi:10.1175/jcli-d-12-00832.1.

3

Krause-Jensen, D. et al., 2016: Long photoperiods sustain high pH in Arctic kelp forests. *Science Advances*, **2** (12), 8, doi:10.1126/sciadv.1501938.

Kretschmer, M., D. Coumou, J.F. Donges and J. Runge, 2016: Using Causal Effect Networks to Analyze Different Arctic Drivers of Midlatitude Winter Circulation. *Journal of Climate*, **29** (11), 4069–4081, doi:10.1175/jcli-d-15-0654.1.

Krishfield, R.A. et al., 2014: Deterioration of perennial sea ice in the Beaufort Gyre from 2003 to 2012 and its impact on the oceanic freshwater cycle. *Journal of Geophysical Research: Oceans*, **119** (2), 1271–1305, doi:10.1002/2013JC008999.

Krüger, L. et al., 2018: Projected distributions of Southern Ocean albatrosses, petrels and fisheries as a consequence of climatic change. *Ecography*, **41** (1), 195–208, doi:10.1111/ecog.02590.

Krumpen, T. et al., 2019: Arctic warming interrupts the Transpolar Drift and affects long-range transport of sea ice and ice-rafted matter. *Sci Rep*, **9** (1), 5459, doi:10.1038/s41598-019-41456-y.

Krumpen, T. et al., 2016: Recent summer sea ice thickness surveys in Fram Strait and associated ice volume fluxes. *The Cryosphere*, **10** (2), 523–534, doi:10.5194/tc-10-523-2016.

Kug, J.S. et al., 2015: Two distinct influences of Arctic warming on cold winters over North America and East Asia. *Nature Geoscience*, **8** (10), 759, doi:10.1038/ngeo2517.

Kullerud, L. et al., 2013: The Arctic Ocean and UNCLOS Article 76: Are There Any Commons? In: *Environmental Security in the Arctic Ocean*, Dordrecht, [Berkman, P. and A. Vylegzhanin (eds.)], Springer Netherlands, 185–194.

Kumpula, J., M. Kurkilahti, T. Helle and A. Colpaert, 2014: Both reindeer management and several other land use factors explain the reduction in ground lichens (Cladonia spp.) in pastures grazed by semi-domesticated reindeer in Finland. *Regional Environmental Change*, **14** (2), 541–559, doi:10.1007/s10113-013-0508-5.

Kurtz, N.T. and S.L. Farrell, 2011: Large-scale surveys of snow depth on Arctic sea ice from Operation IceBridge. *Geophysical Research Letters*, **38** (20), L20505, doi:10.1029/2011gl049216.

Kusahara, K. et al., 2018: An ocean-sea ice model study of the unprecedented Antarctic sea ice minimum in 2016. *Environmental Research Letters*, **13** (8), 084020, doi:10.1088/1748-9326/aad624.

Kusahara, K. et al., 2017: Roles of wind stress and thermodynamic forcing in recent trends in Antarctic sea ice and Southern Ocean SST: An ocean-sea ice model study. *Global and Planetary Change*, **158**, 103–118, doi:10.1016/j.gloplacha.2017.09.012.

Kutz, S. et al., 2015: Erysipelothrix rhusiopathiae associated with recent widespread muskox mortalities in the Canadian Arctic. *Can Vet J*, **56** (6), 560–563.

Kutz, S.J. et al., 2013: Invasion, establishment, and range expansion of two parasitic nematodes in the Canadian Arctic. *Glob Chang Biol*, **19** (11), 3254–62, doi:10.1111/gcb.12315.

Kutz, S.J. et al., 2014: A walk on the tundra: Host-parasite interactions in an extreme environment. *Int J Parasitol Parasites Wildl*, **3** (2), 198–208, doi:10.1016/j.ijppaw.2014.01.002.

Kvile, K.O. et al., 2018: Pushing the limit: Resilience of an Arctic copepod to environmental fluctuations. *Glob Chang Biol*, **24** (11), 5426–5439, doi:10.1111/gcb.14419.

Kwiatkowski, L. and J.C. Orr, 2018: Diverging seasonal extremes for ocean acidification during the twenty-first century. *Nature Climate Change*, **8** (2), 141–145, doi:10.1038/s41558-017-0054-0.

Kwok, R., 2018: Arctic sea ice thickness, volume, and multiyear ice coverage: losses and coupled variability (1958–2018). *Environmental Research Letters*, **13** (10), 105005, doi:10.1088/1748-9326/aae3ec.

Kwok, R. and T. Maksym, 2014: Snow depth of the Weddell and Bellingshausen sea ice covers from IceBridge surveys in 2010 and 2011: An examination. *Journal of Geophysical Research: Oceans*, **119** (7), 4141–4167, doi:10.1002/2014jc009943.

Kwok, R., G. Spreen and S. Pang, 2013: Arctic sea ice circulation and drift speed: Decadal trends and ocean currents. *Journal of Geophysical Research: Oceans*, **118** (5), 2408–2425, doi:10.1002/jgrc.20191.

Kylling, A., C.D. Groot Zwaaftink and A. Stohl, 2018: Mineral Dust Instantaneous Radiative Forcing in the Arctic. *Geophysical Research Letters*, **45** (9), 4290–4298, doi:10.1029/2018gl077346.

Laaksonen, S. et al., 2017: Filarioid nematodes, threat to arctic food safety and security. In: Food safety and security [Paulsen, P., A. Bauer and F.J.M. Smulders (eds.)]. Wageningen Academic Publishers, 101–120.

Labrecque, S. et al., 2009: Contemporary (1951–2001) evolution of lakes in the Old Crow Basin, Northern Yukon, Canada: Remote Sensing, numerical modeling, and stable isotope analysis. *Arctic*, **62**, 225–238, doi:10.14430/arctic13.

Laidler, G., 2012: Societal Aspects of Changing Cold Environments. In: Changing Cold Environments: A Canadian Perspective [French, H. and O. Slaymaker (eds.)]. Wiley-Blackwell, Oxford, 267–300.

Laidler, G.J., 2006: Inuit and Scientific Perspectives on the Relationship Between Sea Ice and Climate Change: The Ideal Complement? *Climatic Change*, **78** (2), 407, doi:10.1007/s10584-006-9064-z.

Laidler, G.J. et al., 2010: *Mapping Sea-Ice Knowledge, Use, and Change in Nunavut, Canada (Cape Dorset, Igloolik, Pangnirtung)*. SIKU: Knowing Our Ice, Documenting Inuit Sea-Ice Knowledge and Use, Springer, Dordrecht.

Laidre, K.L. et al., 2018: Range contraction and increasing isolation of a polar bear subpopulation in an era of sea-ice loss. *Ecology and Evolution*, **8** (4), 2062–2075, doi:10.1002/ece3.3809.

Laidre, K.L. et al., 2016: Use of glacial fronts by narwhals (Monodon monoceros) in West Greenland. *Biol Lett*, **12** (10), doi:10.1098/rsbl.2016.0457.

Laidre, K.L. et al., 2015: Arctic marine mammal population status, sea ice habitat loss, and conservation recommendations for the 21st century. *Conservation Biology*, **29** (3), 724–737, doi:10.1111/cobi.12474.

Laîné, A., H. Nakamura, K. Nishii and T. Miyasaka, 2014: A diagnostic study of future evaporation changes projected in CMIP5 climate models. *Climate Dynamics*, **42** (9), 2745–2761, doi:10.1007/s00382-014-2087-7.

Lalande, C. et al., 2014: Variability in under-ice export fluxes of biogenic matter in the Arctic Ocean. *Global Biogeochemical Cycles*, **28** (5), 571–583, doi:10.1002/2013GB004735.

Lam, V.W.Y., W.W.L. Cheung and U.R. Sumaila, 2016: Marine capture fisheries in the Arctic: winners or losers under climate change and ocean acidification? *Fish and Fisheries*, **17** (2), 335–357, doi:10.1111/faf.12106.

Lamers, M.A.J., E. Eijgelaar and B. Amelung, 2013: Last chance tourism in Antarctica Cruising for change? In: Last chance tourism: Adapting tourism opportunities in a changing world [Lemelin, R.H., J. Dawson and E.J. Stewart (eds.)]. Routledge, London, 24–41.

Lammers, R.B., J.W. Pundsack and A.I. Shiklomanov, 2007: Variability in river temperature, discharge, and energy flux from the Russian pan-Arctic landmass. *Journal of Geophysical Research: Biogeosciences*, **112** (G4), G04S59, doi:10.1029/2006jg000370.

Landa, C.S. et al., 2014: Recruitment, distribution boundary and habitat temperature of an arcto-boreal gadoid in a climatically changing environment: a case study on Northeast Arctic haddock (Melanogrammus aeglefinus). *Fisheries Oceanography*, **23** (6), 506–520, doi:10.1111/fog.12085.

Landrum, L.L., M.M. Holland, M.N. Raphael and L.M. Polvani, 2017: Stratospheric Ozone Depletion: An Unlikely Driver of the Regional Trends in Antarctic Sea Ice in Austral Fall in the Late Twentieth Century. *Geophysical Research Letters*, **44** (21), 11062–11070, doi:10.1002/2017gl075618.

Landschützer, P., N. Gruber, D.C.E. Bakker and U. Schuster, 2014: Recent variability of the global ocean carbon sink. *Global Biogeochemical Cycles*, **28** (9), 927–949, doi:10.1002/2014GB004853.

Landschützer, P. et al., 2018: Strengthening seasonal marine CO_2 variations due to increasing atmospheric CO_2. *Nature Climate Change*, **8** (2), 146–150, doi:10.1038/s41558-017-0057-x.

Landschützer, P. et al., 2015: The reinvigoration of the Southern Ocean carbon sink. *Science*, **349** (6253), 1221–1224, doi:10.1126/science.aab2620

Lange, B.A. et al., 2017: Pan-Arctic sea ice-algal chl a biomass and suitable habitat are largely underestimated for multiyear ice. *Glob Chang Biol*, **23** (11), 4581–4597, doi:10.1111/gcb.13742.

Langlais, C.E. et al., 2017: Stationary Rossby waves dominate subduction of anthropogenic carbon in the Southern Ocean. *Sci Rep*, **7** (1), 17076, doi:10.1038/s41598-017-17292-3.

Lantuit, H. and W. Pollard, 2008: Fifty years of coastal erosion and retrogressive thaw slump activity on Herschel Island, southern Beaufort Sea, Yukon Territory, Canada. *Geomorphology*, **95** (1–2), 84–102, doi:10.1016/j.geomorph.2006.07.040.

Lantz, T.C. and K.W. Turner, 2015: Changes in lake area in response to thermokarst processes and climate in Old Crow Flats, Yukon. *Journal of Geophysical Research: Biogeosciences*, **120** (3), 513–524, doi:10.1002/2014jg002744.

Lara, M.J. et al., 2018: Reduced arctic tundra productivity linked with landform and climate change interactions. *Sci Rep*, **8** (1), 2345, doi:10.1038/s41598-018-20692-8.

Larsen, J.N. et al., 2014: Polar regions. In: Climate Change 2014: Impacts, Adaptation, and Vulnerability. Part B: Regional Aspects. Contribution of Working Group II to the Fifth Assessment Report of the Intergovernmental Panel of Climate Change [Barros, V.R., C.B. Field, D.J. Dokken, M.D. Mastrandrea, K.J. Mach, T.E. Bilir, M. Chatterjee, K.L. Ebi, Y.O. Estrada, R.C. Genova, B. Girma, E.S. Kissel, A.N. Levy, S. MacCracken, P.R. Mastrandrea and L.L. White (eds.)]. Cambridge University Press, Cambridge, United Kingdom and New York, NY, USA, 1567–1612.

LaRue, M.A. et al., 2013: Climate change winners: receding ice fields facilitate colony expansion and altered dynamics in an Adelie penguin metapopulation. *Plos One*, **8** (4), e60568, doi:10.1371/journal.pone.0060568.

Laske, S.M. et al., 2016: Surface water connectivity drives richness and composition of Arctic lake fish assemblages. *Freshwater Biology*, **61** (7), 1090–1104, doi:10.1111/fwb.12769.

Lasserre, F. and S. Pelletier, 2011: Polar super seaways? Maritime transport in the Arctic: an analysis of shipowners' intentions. *Journal of Transport Geography*, **19** (6), 1465–1473, doi:10.1016/j.jtrangeo.2011.08.006.

Lasserre, F. and P.-L. Têtu, 2015: The cruise tourism industry in the Canadian Arctic: analysis of activities and perceptions of cruise ship operators. *Polar Record*, **51** (1), 24–38, doi:10.1017/s0032247413000508.

Laurel, B.J., L.A. Copeman, M. Spencer and P. Iseri, 2017: Temperature-dependent growth as a function of size and age in juvenile Arctic cod (Boreogadus saida). *ICES Journal of Marine Science*, **74** (6), 1614–1621, doi:10.1093/icesjms/fsx028.

Lavrillier, A., 2013: Climate change among nomadic and settled Tungus of Siberia: continuity and changes in economic and ritual relationships with the natural environment. *Polar Record*, **49** (3), 260–271, doi:10.1017/s0032247413000284.

Lavrillier, A. and S. Gabyshev, 2017: *An Arctic Indigenous Knowledge System of Landscape, Climate, and Human Interactions. Evenki Reindeer Herders and Hunters*. Fürstenberg/Havel: Verlag der Kulturstiftung Sibirien SEC Publications, Published online by Cambridge University Press, 467 pp.

Lavrillier, A. and S. Gabyshev, 2018: An emic science of climate. Reindeer Evenki environmental knowledge and the notion of an "extreme process", Études mongoles et sibériennes, centrasiatiques et tibétaines. 49.

Lavrillier, A., S. Gabyshev and M. Rojo, 2016: The Sable for Evenk Reindeer Herders in Southeastern Siberia: Interplaying Drivers of Changes on Biodiversity and Ecosystem Services. In: Indigenous and Local Knowledge of Biodiversity and Ecosystems Services in Europe and Central Asia: Contributions to an IPBES regional assessment [Roué, M. and Z. Molnar (eds.)]. UNESCO, Paris, 111–128.

Lawrence, D.M. et al., 2015: Permafrost thaw and resulting soil moisture changes regulate projected high-latitude CO 2 and CH 4 emissions. *Environmental Research Letters*, **10** (9), 094011.

Lawrence, D.M. et al., 2011: The CCSM4 Land Simulation, 1850–2005: Assessment of Surface Climate and New Capabilities. *Journal of Climate*, **25** (7), 2240–2260, doi:10.1175/jcli-d-11-00103.1.

Laxon, S.W. et al., 2013: CryoSat-2 estimates of Arctic sea ice thickness and volume. *Geophysical Research Letters*, **40** (4), 732–737, doi:10.1002/grl.50193.

Le Brocq, A.M., 2013: Evidence from ice shelves for channelizing meltwater flow beneath the Antarctic ice sheet. *Nat. Geosci.*, **6**, 945–948, doi:10.1038/ngeo1977.

Le Moigne, F.A.C. et al., 2015: Carbon export efficiency and phytoplankton community composition in the Atlantic sector of the Arctic Ocean. *Journal of Geophysical Research: Oceans*, **120** (6), 3896–3912, doi:10.1002/2015JC010700.

Le Quéré, C. et al., 2018: Global Carbon Budget 2018. *Earth System Science Data*, **10** (4), 2141–2194, doi:10.5194/essd-10-2141-2018.

Le Quéré, C. et al., 2007: Saturation of the Southern Ocean CO2 Sink Due to Recent Climate Change. *Science*, **316** (5832), 1735–1738, doi:10.1126/science.1136188

Lecavalier, B.S. et al., 2017: High Arctic Holocene temperature record from the Agassiz ice cap and Greenland ice sheet evolution. *Proc Natl Acad Sci U S A*, **114** (23), 5952–5957, doi:10.1073/pnas.1616287114.

Lee, J.R. et al., 2017a: Climate change drives expansion of Antarctic ice-free habitat. *Nature*, **547** (7661), 49, doi:10.1038/nature22996.

Lee, S. et al., 2017b: Evidence of minimal carbon sequestration in the productive Amundsen Sea polynya. *Geophysical Research Letters*, **44** (15), 7892–7899, doi:10.1002/2017gl074646.

Leibman, M. et al., 2011: Sulfur and carbon isotopes within atmospheric, surface and ground water, snow and ice as indicators of the origin of tabular ground ice in the Russian Arctic. *Permafrost and Periglacial Processes*, **22** (1), 39–48, doi:10.1002/ppp.716.

Lemelin, H. et al., 2010: Last-chance tourism: the boom, doom, and gloom of visiting vanishing destinations. *Current Issues in Tourism*, **13** (5), 477–493, doi:10.1080/13683500903406367.

Leppi, J.C., C.D. Arp and M.S. Whitman, 2016: Predicting Late Winter Dissolved Oxygen Levels in Arctic Lakes Using Morphology and Landscape Metrics. *Environmental Management*, **57** (2), 463–473, doi:10.1007/s00267-015-0622-x.

Lesnikowski, A.C. et al., 2011: Adapting to health impacts of climate change: a study of UNFCCC Annex I parties. *Environmental Research Letters*, **6** (4), 044009, doi:10.1088/1748-9326/6/4/044009.

Leung, S., A. Cabré and I. Marinov, 2015: A latitudinally banded phytoplankton response to 21st century climate change in the Southern Ocean across the CMIP5 model suite. *Biogeosciences*, **12** (19), 5715–5734, doi:10.5194/bg-12-5715-2015.

Lewis, A. and R.C. Prince, 2018: Integrating Dispersants in Oil Spill Response in Arctic and Other Icy Environments. *Environ Sci Technol*, **52** (11), 6098–6112, doi:10.1021/acs.est.7b06463.

Li, C., D. Notz, S. Tietsche and J. Marotzke, 2013: The Transient versus the Equilibrium Response of Sea Ice to Global Warming. *Journal of Climate*, **26** (15), 5624–5636, doi:10.1175/jcli-d-12-00492.1.

Li, F. et al., 2016a: Impacts of Interactive Stratospheric Chemistry on Antarctic and Southern Ocean Climate Change in the Goddard Earth Observing System, Version 5 (GEOS-5). *Journal of Climate*, **29** (9), 3199–3218, doi:10.1175/jcli-d-15-0572.1.

Li, X., D.M. Holland, E.P. Gerber and C. Yoo, 2014: Impacts of the north and tropical Atlantic Ocean on the Antarctic Peninsula and sea ice. *Nature*, **505**, 538, doi:10.1038/nature12945.

Li, X., D.M. Holland, E.P. Gerber and C. Yoo, 2015a: Rossby Waves Mediate Impacts of Tropical Oceans on West Antarctic Atmospheric Circulation in Austral Winter. *Journal of Climate*, **28** (20), 8151–8164, doi:10.1175/jcli-d-15-0113.1.

Li, X. et al., 2015b: Grounding line retreat of Totten Glacier, East Antarctica, 1996 to 2013. *Geophysical Research Letters*, **42** (19), 8049–8056, doi:10.1002/2015GL065701.

Li, Y. et al., 2016b: Evaluating biases in simulated land surface albedo from CMIP5 global climate models. *Journal of Geophysical Research: Atmospheres*, **121** (11), 6178–6190, doi:10.1002/2016jd024774.

Liggett, D., B. Frame, N. Gilbert and F. Morgan, 2017: Is it all going south? Four future scenarios for Antarctica. *Polar Record*, **53** (5), 459–478, doi:10.1017/S0032247417000390.

Liljedahl, A.K. et al., 2016: Pan-Arctic ice-wedge degradation in warming permafrost and its influence on tundra hydrology. *Nature Geoscience*, **9** (4), 312–318, doi:10.1038/ngeo2674.

Lilly, M.R., 2017: *Alaskan North Slope Oil & Gas Transportation Support*. United States [Available at: www.osti.gov/scitech/servlets/purl/1350972].

Lin, G. et al., 2014: Radiative forcing of organic aerosol in the atmosphere and on snow: Effects of SOA and brown carbon. *Journal of Geophysical Research: Atmospheres*, **119** (12), 7453–7476, doi:10.1002/2013jd021186.

Lind, S., R.B. Ingvaldsen and T. Furevik, 2018: Arctic warming hotspot in the northern Barents Sea linked to declining sea-ice import. *Nature Climate Change*, **8** (7), 634–639, doi:10.1038/s41558-018-0205-y.

Lindgren, A., G. Hugelius and P. Kuhry, 2018: Extensive loss of past permafrost carbon but a net accumulation into present-day soils. *Nature*, **560** (7717), 219–222, doi:10.1038/s41586-018-0371-0.

Lindsay, R. and A. Schweiger, 2015: Arctic sea ice thickness loss determined using subsurface, aircraft, and satellite observations. *The Cryosphere*, **9** (1), 269–283, doi:10.5194/tc-9-269-2015.

Lindsay, R., M. Wensnahan, A. Schweiger and J. Zhang, 2014: Evaluation of Seven Different Atmospheric Reanalysis Products in the Arctic. *Journal of Climate*, **27** (7), 2588–2606, doi:10.1175/jcli-d-13-00014.1.

Lindstad, H., R.M. Bright and A.H. Strømman, 2016: Economic savings linked to future Arctic shipping trade are at odds with climate change mitigation. *Transport Policy*, **45**, 24–30, doi:10.1016/j.tranpol.2015.09.002.

Link, H., D. Piepenburg and P. Archambault, 2013: Are Hotspots Always Hotspots? The Relationship between Diversity, Resource and Ecosystem Functions in the Arctic. *Plos One*, **8** (9), e74077, doi:10.1371/journal.pone.0074077.

Lipenkov, V., A.A. Ekaykin, E.V. Polyakova and D. Raynaud, 2016: Characterization of subglacial Lake Vostok as seen from physical and isotope properties of accreted ice. *Phil. Trans. R. Soc. A*, **374** (2059), 20140303, doi:10.1098/rsta.2014.0303.

Lique, C. et al., 2016: Modeling the Arctic freshwater system and its integration in the global system: Lessons learned and future challenges. *Journal of Geophysical Research: Biogeosciences*, **121** (3), 540–566, doi:10.1002/2015jg003120.

Lischka, S. and U. Riebesell, 2016: Metabolic response of Arctic pteropods to ocean acidification and warming during the polar night/twilight phase in Kongsfjord (Spitsbergen). *Polar Biology*, **40** (6), 1211–1227, doi:10.1007/s00300-016-2044-5.

Liston, G. and C. Hiemstra, 2011: The changing cryosphere: pan-Arctic snow trends (1979–2009). *Journal of Climate*, **24**, 5691–5712, doi:10.1175/jcli-d-11-00081.1.

Liu, B., D. Yang, B. Ye and S. Berezovskaya, 2005: Long-term open-water season stream temperature variations and changes over Lena River Basin in Siberia. *Global and Planetary Change*, **48** (1), 96–111, doi:10.1016/j.gloplacha.2004.12.007.

Liu, J., M. Song, R.M. Horton and Y. Hu, 2013: Reducing spread in climate model projections of a September ice-free Arctic. *Proceedings of the National Academy of Sciences*, **110** (31), 12571–12576, doi:10.1073/pnas.1219716110.

Liu, J., M. Song, M.H. Radley and Y. Hu, 2015a: Revisiting the potential of melt pond fraction as a predictor for the seasonal Arctic sea ice extent minimum.

Environmental Research Letters, **10** (5), 054017, doi:10.1088/1748-9326/10/5/054017/meta.

Liu, Y., H. Gupta, E. Springer and T. Wagener, 2008: Linking science with environmental decision making: Experiences from an integrated modeling approach to supporting sustainable water resources management. *Environmental Modelling & Software*, **23** (7), 846–858, doi:10.1016/j.envsoft.2007.10.007.

Liu, Y. et al., 2015b: Evapotranspiration in Northern Eurasia: Impact of forcing uncertainties on terrestrial ecosystem model estimates. *Journal of Geophysical Research: Atmospheres*, **120** (7), 2647–2660, doi:10.1002/2014jd022531.

Liu, Y. et al., 2014: Response of evapotranspiration and water availability to the changing climate in Northern Eurasia. *Climatic Change*, **126** (3), 413–427, doi:10.1007/s10584-014-1234-9.

Loboda, T.V., 2014: Adaptation strategies to climate change in the Arctic: a global patchwork of reactive community-scale initiatives. *Environmental Research Letters*, **9** (11), 111006, doi:10.1088/1748-9326/9/11/111006.

Loeb, V.J. and J.A. Santora, 2015: Climate variability and spatiotemporal dynamics of five Southern Ocean krill species. *Progress in Oceanography*, **134**, 93–122, doi:10.1016/j.pocean.2015.01.002.

Logerwell, E., K. Rand, S. Danielson and L. Sousa, 2018: Environmental drivers of benthic fish distribution in and around Barrow Canyon in the northeastern Chukchi Sea and western Beaufort Sea. *Deep Sea Research Part II: Topical Studies in Oceanography*, **152**, 170–181, doi:10.1016/j.dsr2.2017.04.012.

Loisel, J. et al., 2014: A database and synthesis of northern peatland soil properties and Holocene carbon and nitrogen accumulation. *The Holocene*, **24** (9), 1028–1042, doi:10.1177/0959683614538073.

Lone, K. et al., 2018: Aquatic behaviour of polar bears (Ursus maritimus) in an increasingly ice-free Arctic. *Scientific Reports*, **8** (1), 9677, doi:10.1038/s41598-018-27947-4.

Long, W.C., P. Pruisner, K.M. Swiney and R. Foy, 2019: Effects of ocean acidification on the respiration and feeding of juvenile red and blue king crabs. *ICES Journal of Marine Science*, **76** (5), 1335–1343, doi:https://doi.org/10.1093/icesjms/fsz090.

Long, W.C., S.B. Van Sant and J.A. Haaga, 2015: Habitat, predation, growth, and coexistence: Could interactions between juvenile red and blue king crabs limit blue king crab productivity? *Journal of Experimental Marine Biology and Ecology*, **464**, 58–67, doi:10.1016/j.jembe.2014.12.011.

Long, W.C. et al., 2017: Survival, growth, and morphology of blue king crabs: effect of ocean acidification decreases with exposure time. *ICES Journal of Marine Science*, **74** (4), 1033–1041, doi:10.1093/icesjms/fsw197.

Loranty, M.M. et al., 2014: Vegetation controls on northern high latitude snow-albedo feedback: observations and CMIP5 model simulations. *Global Change Biology*, **20** (2), 594–606, doi:10.1111/gcb.12391.

Lorentzen, G. et al., 2018: Current Status of the Red King Crab (Paralithodes camtchaticus) and Snow Crab (Chionoecetes opilio) Industries in Norway. *Reviews in Fisheries Science & Aquaculture*, **26** (1), 42–54, doi:10.1080/23308249.2017.1335284.

Loring, P., S. Gerlach and H. Harrison, 2013: Seafood as Local Food: Food Security and Locally Caught Seafood on Alaska's Kenai Peninsula. *Journal of Agriculture, Food Systems, and Community Development*, **3** (3), 13–30, doi:10.5304/jafscd.2013.033.006.

Loring, P.A., S.C. Gerlach and H.J. Penn, 2016: "Community Work" in a Climate of Adaptation: Responding to Change in Rural Alaska. *Human Ecology*, **44** (1), 119–128, doi:10.1007/s10745-015-9800-y.

Lowther, A.D., A. Fisk, K.M. Kovacs and C. Lydersen, 2017: Interdecadal changes in the marine food web along the west Spitsbergen coast detected in the stable isotope composition of ringed seal (Pusa hispida) whiskers. *Polar Biology*, **40** (10), 2027–2033, doi:10.1007/s00300-017-2122-3.

Luckman, A. et al., 2015: Calving rates at tidewater glaciers vary strongly with ocean temperature. *Nature Communications*, **6**, 8566, doi:10.1038/ncomms9566.

Luckman, A. et al., 2014: Surface melt and ponding on Larsen C Ice Shelf and the impact of föhn winds. *Antarctic Science*, **26** (06), 625–635, doi:10.1017/S0954102014000339.

Ludescher, J., N. Yuan and A. Bunde, 2018: Detecting the statistical significance of the trends in the Antarctic sea ice extent: an indication for a turning point. *Climate Dynamics*, **53** (1–2), 237–244, doi:10.1007/s00382-018-4579-3.

Lund, M. et al., 2010: Variability in exchange of CO2 across 12 northern peatland and tundra sites. *Global Change Biology*, **16** (9), 2436–2448, doi:10.1111/j.1365-2486.2009.02104.x.

Lunn, N.J. et al., 2016: Demography of an apex predator at the edge of its range: impacts of changing sea ice on polar bears in Hudson Bay. *Ecological Applications*, **26** (5), 1302–1320, doi:doi:10.1890/15-1256.

Lydersen, C. et al., 2014: The importance of tidewater glaciers for marine mammals and seabirds in Svalbard, Norway. *Journal of Marine Systems*, **129**, 452–471, doi:10.1016/j.jmarsys.2013.09.006.

Lydersen, C. et al., 2017: Novel terrestrial haul-out behaviour by ringed seals (Pusa hispida) in Svalbard, in association with harbour seals (Phoca vitulina). *Polar Research*, **36** (1), 1374124, doi:10.1080/17518369.2017.1374124.

Lynch, H.J., R. Naveen and P. Casanovas, 2013: Antarctic Site Inventory breeding bird survey data, 1994–2013. *Ecology*, **94** (11), 2653–2653.

Lynghammar, A. et al., 2013: Species richness and distribution of chondrichthyan fishes in the Arctic Ocean and adjacent seas. *Biodiversity*, **14** (1), 57–66, doi:10.1080/14888386.2012.706198.

Lyver, P.O. et al., 2014: Trends in the Breeding Population of Adelie Penguins in the Ross Sea, 1981–2012: A Coincidence of Climate and Resource Extraction Effects. *Plos One*, **9** (3), 10, doi:10.1371/journal.pone.0091188.

Macayeal, D.R. and O.V. Sergienko, 2013: The flexural dynamics of melting ice shelves. *Annals of Glaciology*, **54** (63), 1–10, doi:10.3189/2013AoG63A256.

MacDonald, M.K. et al., 2018: Impacts of 1.5 and 2.0°C Warming on Pan-Arctic River Discharge Into the Hudson Bay Complex Through 2070. *Geophysical Research Letters*, **45** (15), 7561–7570, doi:10.1029/2018gl079147.

MacDougall, A.H., C.A. Avis and A.J. Weaver, 2012: Significant contribution to climate warming from the permafrost carbon feedback. *Nature Geoscience*, **5** (10), 719–721, doi:10.1038/ngeo1573.

MacDougall, A.H. and R. Knutti, 2016: Projecting the release of carbon from permafrost soils using a perturbed parameter ensemble modelling approach. *Biogeosciences*, **13** (7), 2123–2136, doi:10.5194/bg-13-2123-2016.

MacGilchrist, G.A. et al., 2014: The Arctic Ocean carbon sink. *Deep-Sea Research Part I-Oceanographic Research Papers*, **86**, 39–55, doi:10.1016/j.dsr.2014.01.002.

Macias-Fauria, M., S. Karlsen and B. Forbes, 2017: Disentangling the coupling between sea ice and tundra productivity in Svalbard. *Scientific Reports*, **7** (1), 8586, doi:10.1038/s41598-017-06218-8.

Mack, M.C. et al., 2011: Carbon loss from an unprecedented Arctic tundra wildfire. *Nature*, **475** (7357), 489–492, doi:10.1038/nature10283.

Mahlstein, I. and R. Knutti, 2012: September Arctic sea ice predicted to disappear near 2°C global warming above present. *Journal of Geophysical Research: Atmospheres*, **117** (D6), D06104, doi:10.1029/2011jd016709.

Mahoney, A.R., H. Eicken, A.G. Gaylord and R. Gens, 2014: Landfast sea ice extent in the Chukchi and Beaufort Seas: The annual cycle and decadal variability. *Cold Regions Science and Technology*, **103** (Supplement C), 41–56, doi:10.1016/j.coldregions.2014.03.003.

Mäkelä, A., U. Witte and P. Archambault, 2017: Ice algae versus phytoplankton: resource utilization by Arctic deep sea macroinfauna revealed through isotope labelling experiments. *Marine Ecology Progressive Series*, **572**, 1–18, doi:10.3354/meps12157.

Makki, M., 2012: Evaluating arctic dialogue: A case study of stakeholder relations for sustainable oil and gas development. *Journal of Sustainable Development*, **5** (3), 34–45, doi:10.5539/jsd.v5n3p34.

Maksym, T. and T. Markus, 2008: Antarctic sea ice thickness and snow-to-ice conversion from atmospheric reanalysis and passive microwave snow depth. *Journal of Geophysical Research: Oceans*, **113** (C2), C02S12, doi:10.1029/2006jc004085.

Malinauskaite, L. et al., 2019: Ecosystem services in the Arctic: a thematic review. *Ecosystem Services*, **36**, 100898, doi:10.1016/j.ecoser.2019.100898.

Mallory, C.D. and M.S. Boyce, 2017: Observed and predicted effects of climate change on Arctic caribou and reindeer. *Environmental Reviews*, **26** (1), 13–25, doi:10.1139/er-2017-0032.

Malpress, V. et al., 2017: Bio-physical characterisation of polynyas as a key foraging habitat for juvenile male southern elephant seals (Mirounga leonina) in Prydz Bay, East Antarctica. *Plos One*, **12** (9), e0184536, doi:10.1371/journal.pone.0184536.

Mann, D.H., T. Scott Rupp, M.A. Olson and P.A. Duffy, 2012: Is Alaska's Boreal Forest Now Crossing a Major Ecological Threshold? *Arctic, Antarctic, and Alpine Research*, **44** (3), 319–331, doi:10.1657/1938-4246-44.3.319.

Mann, M.E. et al., 2017: Influence of Anthropogenic Climate Change on Planetary Wave Resonance and Extreme Weather Events. *Scientific Reports*, **7** (1), 45242, doi:10.1038/srep45242.

Manno, C. et al., 2017: Shelled pteropods in peril: Assessing vulnerability in a high CO 2 ocean. *Earth-Science Reviews*, **169**, 132–145, doi:10.1016/j.earscirev.2017.04.005.

Manno, C., V.L. Peck and G.A. Tarling, 2016: Pteropod eggs released at high pCO2 lack resilience to ocean acidification. *Sci Rep*, **6**, 25752, doi:10.1038/srep25752.

Marelle, L. et al., 2016: Air quality and radiative impacts of Arctic shipping emissions in the summertime in northern Norway: from the local to the regional scale. *Atmospheric Chemistry and Physics*, **16** (4), 2359–2379, doi:10.5194/acp-16-2359-2016.

Marino, E., 2012: The long history of environmental migration: Assessing vulnerability construction and obstacles to successful relocation in Shishmaref, Alaska. *Global Environmental Change*, **22** (2), 374–381, doi:10.1016/j.gloenvcha.2011.09.016.

Marshall, G.J., 2004: Causes of exceptional atmospheric circulation changes in the Southern Hemisphere. *Geophysical Research Letters*, **31** (14), L1420, doi:10.1029/2004gl019952.

Marshall, G.J., A. Orr, N.P.M. van Lipzig and J.C. King, 2006: The Impact of a Changing Southern Hemisphere Annular Mode on Antarctic Peninsula Summer Temperatures. *Journal of Climate*, **19** (20), 5388–5404, doi:10.1175/jcli3844.1.

Marshall, G.J., D.W.J. Thompson and M.R. Van den Broeke, 2017: The signature of Southern Hemisphere atmospheric circulation patterns in Antarctic precipitation. *Geophysical Research Letters*, **44** (22), 11580–11589, doi:10.1002/2017GL075998.

Martin-Español, A., J.L. Bamber and A. Zammit-Mangion, 2017: Constraining the mass balance of East Antarctica. *Geophysical Research Letters*, **44** (9), 4168–4175, doi:10.1002/2017GL072937.

Martín-Español, A. et al., 2016: Spatial and temporal Antarctic Ice Sheet mass trends, glacio-isostatic adjustment, and surface processes from a joint inversion of satellite altimeter, gravity, and GPS data. *Journal of Geophysical Research: Earth Surface*, **121** (2), 182–200, doi:10.1002/2015JF003550.

Martin, A. et al., 2017: Shrub growth and expansion in the Arctic tundra: an assessment of controlling factors using an evidence-based approach. *Environmental Research Letters*, **12** (8), 085007, doi:10.1088/1748-9326/aa7989.

Masina, S. et al., 2019: Weather, environmental conditions, and waterborne Giardia and Cryptosporidium in Iqaluit, Nunavut. *J Water Health*, **17** (1), 84–97, doi:10.2166/wh.2018.323.

Massom, R.A. et al., 2001: Snow on Antarctic sea ice. *Reviews of Geophysics*, **39** (3), 413–445, doi:10.1029/2000rg000085.

Masson-Delmotte, V. et al., 2012: Greenland climate change: from the past to the future. *Wiley Interdisciplinary Reviews: Climate Change*, **3** (5), 427–449, doi:10.1002/wcc.186.

Massonnet, F. et al., 2012: Constraining projections of summer Arctic sea ice. *The Cryosphere*, **6** (6), 1383–1394, doi:10.5194/tc-6-1383-2012.

3

Massonnet, F. et al., 2013: A model reconstruction of the Antarctic sea ice thickness and volume changes over 1980–2008 using data assimilation. *Ocean Modelling*, **64**, 67–75, doi:DOI 10.1016/j.ocemod.2013.01.003.

Massonnet, F. et al., 2018: Arctic sea-ice change tied to its mean state through thermodynamic processes. *Nature Climate Change*, **8** (7), 599–603, doi:10.1038/s41558-018-0204-z.

Matear, R.J., T.J. O'Kane, J.S. Risbey and M. Chamberlain, 2015: Sources of heterogeneous variability and trends in Antarctic sea-ice. *Nature Communications*, **6**, 8656, doi:10.1038/Ncomms9656.

Mathis, J.T. et al., 2015: Ocean acidification risk assessment for Alaska's fishery sector. *Progress in Oceanography*, **136**, 71–91, doi:10.1016/j.pocean.2014.07.001.

Matthews, T. and R. Potts, 2018: Planning for climigration: a framework for effective action. *Climatic Change*, **148** (4), 607–621, doi:10.1007/s10584-018-2205-3.

McCall, A.G., N.W. Pilfold, A.E. Derocher and N.J. Lunn, 2016: Seasonal habitat selection by adult female polar bears in western Hudson Bay. *Population Ecology*, **58** (3), 407–419, doi:10.1007/s10144-016-0549-y.

McCarthy, A.H., L.S. Peck, K.A. Hughes and D.C. Aldridge, 2019: Antarctica: The final frontier for marine biological invasions. *Glob Chang Biol*, **25** (7), 2221–2241, doi:10.1111/gcb.14600.

McCave, I.N. et al., 2013: Minimal change in Antarctic Circumpolar Current flow speed between the last glacial and Holocene. *Nature Geoscience*, **7**, 113, doi:10.1038/ngeo2037.

McClelland, G.T.W. et al., 2017: Climate change leads to increasing population density and impacts of a key island invader. *Ecological Applications*, **28** (1), 212–224, doi:10.1002/eap.1642.

McCormack, S.A. et al., 2017: Simplification of complex ecological networks – species aggregation in Antarctic food web models. In: *MODSIM2017, 22nd International Congress on Modelling and Simulation*, [Syme, G., D. Hatton MacDonald, E. Fulton and J. Piantadosi (eds.)], Modelling and Simulation Society of Australia and New Zealand, 264–270.

McFarland, J.J., M.S. Wipfli and M.S. Whitman, 2017: Trophic pathways supporting Arctic grayling in a small stream on the Arctic Coastal Plain, Alaska. *Ecology of Freshwater Fish*, **27** (1), 184–197, doi:10.1111/eff.12336.

McGillicuddy, D.J. et al., 2015: Iron supply and demand in an Antarctic shelf ecosystem. *Geophysical Research Letters*, **42** (19), 8088–8097, doi:10.1002/2015gl065727.

McGrath, D. et al., 2013: Recent warming at Summit, Greenland: Global context and implications. *Geophysical Research Letters*, **40** (10), 2091–2096, doi:10.1002/grl.50456.

McGuire, A.D. et al., 2012: An assessment of the carbon balance of Arctic tundra: comparisons among observations, process models, and atmospheric inversions. *Biogeosciences*, **9** (8), 3185–3204, doi:10.5194/bg-9-3185-2012.

McGuire, A.D. et al., 2016: Variability in the sensitivity among model simulations of permafrost and carbon dynamics in the permafrost region between 1960 and 2009. *Global Biogeochemical Cycles*, **30** (7), 1015–1037, doi:10.1002/2016gb005405.

McGuire, A.D. et al., 2018: Dependence of the evolution of carbon dynamics in the northern permafrost region on the trajectory of climate change. *Proc Natl Acad Sci U S A*, **115** (15), 3882–3887, doi:10.1073/pnas.1719903115.

McKenna, C.M. et al., 2018: Arctic Sea Ice Loss in Different Regions Leads to Contrasting Northern Hemisphere Impacts. *Geophysical Research Letters*, **45** (2), 945–954, doi:10.1002/2017gl076433.

McLaughlin, J.B. et al., 2005: Outbreak of Vibrio parahaemolyticus Gastroenteritis Associated with Alaskan Oysters. *New England Journal of Medicine*, **353** (14), 1463–1470, doi:10.1056/NEJMoa051594.

McLeod, E., R. Salm, A. Green and J. Almany, 2009: Designing marine protected area networks to address the impacts of climate change. *Frontiers in Ecology and the Environment*, **7** (7), 362–370, doi:10.1890/070211.

McLeod, J.T. and T.L. Mote, 2016: Linking interannual variability in extreme Greenland blocking episodes to the recent increase in summer melting across the Greenland ice sheet. *International Journal of Climatology*, **36** (3), 1484–1499, doi:10.1002/joc.4440.

McMillan, M. et al., 2014a: Rapid dynamic activation of a marine-based Arctic ice cap. *Geophysical Research Letters*, **41** (24), 8902–8909, doi:10.1002/2014GL062255.

McMillan, M. et al., 2014b: Increased ice losses from Antarctica detected by CryoSat-2. *Geophysical Research Letters*, **41** (11), 3899–3905, doi:10.1002/2014gl060111.

McMinn, A., 2017: Reviews and syntheses: Ice acidification, the effects of ocean acidification on sea ice microbial communities. *Biogeosciences*, **14** (17), 3927–3935, doi:10.5194/bg-14-3927-2017.

McNeil, B.I. and R.J. Matear, 2013: The non-steady state oceanic CO_2 signal: its importance, magnitude and a novel way to detect it. *Biogeosciences*, **10** (4), 2219–2228, doi:10.5194/bg-10-2219-2013.

McNeil, B.I. and T.P. Sasse, 2016: Future ocean hypercapnia driven by anthropogenic amplification of the natural CO_2 cycle. *Nature*, **529** (7586), 383–6, doi:10.1038/nature16156.

Meadow, A.M. et al., 2015: Moving toward the Deliberate Coproduction of Climate Science Knowledge. *Weather, Climate, and Society*, **7** (2), 179–191, doi:10.1175/WCAS-D-14-00050.1.

Medley, B. and E.R. Thomas, 2018: Increased snowfall over the Antarctic Ice Sheet mitigated twentieth-century sea-level rise. *Nature Climate Change*, **9** (1), 34–39, doi:10.1038/s41558-018-0356-x.

Meehl, G.A. et al., 2016: Antarctic sea-ice expansion between 2000 and 2014 driven by tropical Pacific decadal climate variability. *Nature Geoscience*, **9** (8), 590–595, doi:10.1038/Ngeo2751.

Meehl, G.A. et al., 2019: Sustained ocean changes contributed to sudden Antarctic sea ice retreat in late 2016. *Nat Commun*, **10** (1), 14, doi:10.1038/s41467-018-07865-9.

Meehl, G.A. et al., 2018: Tropical Decadal Variability and the Rate of Arctic Sea Ice Decrease. *Geophysical Research Letters*, **45** (20), 11,326–11,333, doi:10.1029/2018gl079989.

Meijers, A. et al., 2016: Wind-driven export of Weddell Sea slope water. *Journal of Geophysical Research: Oceans*, **121** (10), 7530–7546, doi:10.1002/2016JC011757.

Meijers, A.J.S. et al., 2019: The role of ocean dynamics in king penguin range estimation. *Nature Climate Change*, **9** (2), 120–121, doi:10.1038/s41558-018-0388-2.

Meijers, A.J.S. et al., 2012: Representation of the Antarctic Circumpolar Current in the CMIP5 climate models and future changes under warming scenarios. *Journal of Geophysical Research: Oceans*, **117** (C12), 19 pp, doi:10.1029/2012JC008412.

Meire, L. et al., 2016a: High export of dissolved silica from the Greenland Ice Sheet. *Geophysical Research Letters*, **43** (17), 9173–9182, doi:10.1002/2016GL070191.

Meire, L. et al., 2017: Marine-terminating glaciers sustain high productivity in Greenland fjords. *Glob Chang Biol*, **23** (12), 5344–5357, doi:10.1111/gcb.13801.

Meire, L. et al., 2016b: Spring bloom dynamics in a subarctic fjord influenced by tidewater outlet glaciers (Godthabsfjord, SW Greenland). *Journal of Geophysical Research-Biogeosciences*, **121** (6), 1581–1592, doi:10.1002/2015jg003240.

Mekonnen, A., J.A. Renwick and A. Sánchez-Lugo, 2016: Regional climates [in "State of the Climate in 2015"]. *Bulletin of the American Meteorological Society*, **97** (8), S173–S226.

Melbourne-Thomas, J. et al., 2016: Under ice habitats for Antarctic krill larvae: Could less mean more under climate warming? *Geophysical Research Letters*, **43** (19), 10,322–10,327, doi:10.1002/2016gl070846.

Melia, N., K. Haines and E. Hawkins, 2016: Sea ice decline and 21st century trans-Arctic shipping routes. *Geophysical Research Letters*, **43** (18), 9720–9728, doi:10.1002/2016gl069315.

3

Melvin, A.M. et al., 2017: Climate change damages to Alaska public infrastructure and the economics of proactive adaptation. *Proceedings of the National Academy of Sciences*, **114** (2), E122, doi:10.1073/pnas.1611056113.

Mémin, A. et al., 2015: Interannual variation of the Antarctic Ice Sheet from a combined analysis of satellite gravimetry and altimetry data. *Earth and Planetary Science Letters*, **422**, 150–156, doi:10.1016/j.epsl.2015.03.045.

Meneghello, G. et al., 2018: The Ice-Ocean Governor: Ice-Ocean Stress Feedback Limits Beaufort Gyre Spin-Up. *Geophysical Research Letters*, **45** (20), 11,293–11,299, doi:10.1029/2018gl080171.

Meneghello, G., J. Marshall, S.T. Cole and M.-L. Timmermans, 2017: Observational Inferences of Lateral Eddy Diffusivity in the Halocline of the Beaufort Gyre. *Geophysical Research Letters*, **44** (24), 12,331–12,338, doi:10.1002/2017gl075126.

Menezes, V.V., A.M. Macdonald and C. Schatzman, 2017: Accelerated freshening of Antarctic Bottom Water over the last decade in the Southern Indian Ocean. *Science Advances*, **3** (1), e1601426, doi:10.1126/sciadv.1601426.

Mengerink, K., D. Roche and G. Swanson, 2017: Understanding Arctic Co-Management: The U.S. Marine Mammal Approach. *The Yearbook of Polar Law Online*, **8** (1), 76–102, doi:10.1163/22116427_008010007.

Meredith, M.P. et al., 2013: The vision for a Southern Ocean Observing System. *Current Opinion in Environmental Sustainability*, **5** (3–4), 306–313, doi:10.1016/j.cosust.2013.03.002.

Merino, N. et al., 2018: Impact of increasing antarctic glacial freshwater release on regional sea-ice cover in the Southern Ocean. *Ocean Modelling*, **121**, 76–89, doi:10.1016/j.ocemod.2017.11.009.

Merkouriadi, I. et al., 2017: Critical Role of Snow on Sea Ice Growth in the Atlantic Sector of the Arctic Ocean. *Geophysical Research Letters*, **44** (20), 10,479–10,485, doi:10.1002/2017gl075494.

Mernild, S.H. et al., 2015: Greenland precipitation trends in a long-term instrumental climate context (1890–2012): Evaluation of coastal and ice core records. *International Journal of Climatology*, **35** (2), 303–320, doi:10.1002/joc.3986.

Mernild, S.H., T.L. Mote and G.E. Liston, 2017: Greenland ice sheet surface melt extent and trends: 1960–2010. *Journal of Glaciology*, **57** (204), 621–628, doi:10.3189/002214311797409712.

Messori, G., R. Caballero and M. Gaetani, 2016: On cold spells in North America and storminess in western Europe. *Geophysical Research Letters*, **43** (12), 6620–6628, doi:10.1002/2016gl069392.

Meyer, B. et al., 2017: The winter pack-ice zone provides a sheltered but food-poor habitat for larval Antarctic krill. *Nat Ecol Evol*, **1** (12), 1853–1861, doi:10.1038/s41559-017-0368-3.

Miège, C. et al., 2016: Spatial extent and temporal variability of Greenland firn aquifers detected by ground and airborne radars. *Journal of Geophysical Research: Earth Surface*, **121** (12), 2381–2398, doi:10.1002/2016JF003869.

Milaković, A.-S. et al., 2018: Current status and future operational models for transit shipping along the Northern Sea Route. *Marine Policy*, **94**, 53–60, doi:10.1016/j.marpol.2018.04.027.

Miles, V.V., M.W. Miles and O.M. Johannessen, 2016: Satellite archives reveal abrupt changes in behavior of Helheim Glacier, southeast Greenland. *Journal of Glaciology*, **62** (231), 137–146, doi:10.1017/jog.2016.24.

Millan, R., J. Mouginot and E. Rignot, 2017: Mass budget of the glaciers and ice caps of the Queen Elizabeth Islands, Canada, from 1991 to 2015. *Environmental Research Letters*, **12** (2), 024016.

Miller, C.A. and C. Wyborn, 2018: Co-production in global sustainability: Histories and theories. *Environmental Science & Policy*, doi:10.1016/j.envsci.2018.01.016.

Milner, A.M. et al., 2017: Glacier shrinkage driving global changes in downstream systems. *Proceedings of the National Academy of Sciences*, **114** (37), 9770–9778.

Minchew, B.M. et al., 2018: Modeling the dynamic response of outlet glaciers to observed ice-shelf thinning in the Bellingshausen Sea Sector, West Antarctica. *Journal of Glaciology*, **64** (244), 333–342, doi:10.1017/jog.2018.24.

Mintenbeck, K., 2017: Impacts of Climate Change on the Southern Ocean. In: Climate Change Impacts on Fisheries and Aquaculture [Phillips, B.F. and M. Pérez-Ramírez (eds.)]. John Wiley & Sons New Jersey, 663–701.

Mintenbeck, K. et al., 2012: Impact of Climate Change on Fishes in Complex Antarctic Ecosystems. *Advances in Ecological Research, Vol 46: Global Change in Multispecies Systems, Pt 1*, **46**, 351–426, doi:10.1016/B978-0-12-396992-7.00006-X. Mintenbeck, K. and J.J. Torres, 2017: Impact of Climate Change on the Antarctic Silverfish and Its Consequences for the Antarctic Ecosystem. In: The Antarctic Silverfish: a Keystone Species in a Changing Ecosystem [Vacchi, M., E. Pisano and L. Ghiglotti (eds.)]. Springer Nature, New York, 253–286.

Mioduszewski, J.R. et al., 2016: Atmospheric drivers of Greenland surface melt revealed by self-organizing maps. *Journal of Geophysical Research: Atmospheres*, **121** (10), 5095–5114, doi:10.1002/2015JD024550.

Möller, M. and R. Möller, 2017: Modeling glacier-surface albedo across Svalbard for the 1979–2015 period: The HiRSvaC500-🔲 data set. *Journal of Advances in Modeling Earth Systems*, **9** (1), 404–422, doi:10.1002/2016MS000752.

Mongwe, N.P., M. Vichi and P.M.S. Monteiro, 2018: The seasonal cycle of pCO2 and CO2 fluxes in the Southern Ocean: diagnosing anomalies in CMIP5 Earth system models. *Biogeosciences*, **15** (9), 2851–2872, doi:10.5194/bg-15-2851-2018.

Monien, D. et al., 2017: Meltwater as a source of potentially bioavailable iron to Antarctica waters. *Antarctic Science*, **29** (3), 277–291, doi:10.1017/S095410201600064X. Moon, H.W., W.M.R.W. Hussin, H.C. Kim and I.Y. Ahn, 2015: The impacts of climate change on Antarctic nearshore mega-epifaunal benthic assemblages in a glacial fjord on King George Island: Responses and implications. *Ecological Indicators*, **57**, 280–292, doi:10.1016/j.ecolind.2015.04.031.

Moon, T., I. Joughin, B. Smith and I. Howat, 2012: 21st-Century Evolution of Greenland Outlet Glacier Velocities. *Science*, **336** (6081), 576, doi:10.1126/science.1219985.

Moore, S.E. and P.J. Stabeno, 2015: Synthesis of Arctic Research (SOAR) in marine ecosystems of the Pacific Arctic. *Progress in Oceanography*, **136**, 1–11, doi:10.1016/j.pocean.2015.05.017.

Moore, S.E., P.J. Stabeno, J.M. Grebmeier and S.R. Okkonen, 2018: The Arctic Marine Pulses Model: linking annual oceanographic processes to contiguous ecological domains in the Pacific Arctic. *Deep Sea Research Part II: Topical Studies in Oceanography*, **152**, 8–21, doi:10.1016/j.dsr2.2016.10.011.

Morgera, E. and K. Kulovesi, 2016: Research Handbook on International Law and Natural Resources.[Morgera, E. and K. Kulovesi (eds.)]. Research Handbooks in International Law, 349–365.

Morlighem, M. et al., 2016: Modeling of Store Gletscher's calving dynamics, West Greenland, in response to ocean thermal forcing. *Geophysical Research Letters*, **43** (6), 2659–2666, doi:10.1002/2016gl067695.

Morlighem, M. et al., 2017: BedMachine v3: Complete Bed Topography and Ocean Bathymetry Mapping of Greenland From Multibeam Echo Sounding Combined With Mass Conservation. *Geophysical Research Letters*, **44** (21), 11,051–11,061, doi:10.1002/2017GL074954.

Morrison, A.K. and A. McC. Hogg, 2013: On the Relationship between Southern Ocean Overturning and ACC Transport. *Journal of Physical Oceanography*, **43** (1), 140–148, doi:10.1175/jpo-d-12-057.1.

Morrison, A.L. et al., 2018: Isolating the Liquid Cloud Response to Recent Arctic Sea Ice Variability Using Spaceborne Lidar Observations. *Journal of Geophysical Research: Atmospheres*, **123** (1), 473–490, doi:10.1002/2017jd027248.

Mortin, J. et al., 2016: Melt onset over Arctic sea ice controlled by atmospheric moisture transport. *Geophysical Research Letters*, **43** (12), 6636–6642, doi:10.1002/2016gl069330.

Mouginot, J., E. Rignot and B. Scheuchl, 2014: Sustained increase in ice discharge from the Amundsen Sea Embayment, West Antarctica, from

1973 to 2013. *Geophysical Research Letters*, **41** (5), 1576–1584, doi:10.1002/2013gl059069.

Mouginot, J. et al., 2015: Fast retreat of Zachariæ Isstrøm, northeast Greenland. *Science*, 350 (6266), 1357–1361, doi:10.1126/science.aac7111.

Mudryk, L., P. Kushner, C. Derksen and C. Thackeray, 2017: Snow cover response to temperature in observational and climate model ensembles. *Geophysical Research Letters*, **44** (2), 919–926, doi:doi:10.1002/2016GL071789.

Mueller, I.A. et al., 2012: Exposure to critical thermal maxima increases oxidative stress in hearts of white- but not red-blooded Antarctic notothenioid fishes. *The Journal of Experimental Biology*, **215** (20), 3655, doi:10.1242/jeb.071811.

Mueter, F.J., J. Weems, E.V. Farley and M.F. Sigler, 2017: Arctic Ecosystem Integrated Survey (Arctic Eis): Marine ecosystem dynamics in the rapidly changing Pacific Arctic Gateway. *Deep Sea Research Part II: Topical Studies in Oceanography*, **135** (Supplement C), 1–6, doi:10.1016/j.dsr2.2016.11.005.

Mullan, D. et al., 2017: Climate change and the long-term viability of the World's busiest heavy haul ice road. *Theoretical and Applied Climatology*, **129** (3), 1089–1108, doi:10.1007/s00704-016-1830-x.

Mullowney, D.R.J., E.G. Dawe, E.B. Colbourne and G.A. Rose, 2014: A review of factors contributing to the decline of Newfoundland and Labrador snow crab (Chionoecetes opilio). *Reviews in Fish Biology and Fisheries*, **24** (2), 639–657, doi:10.1007/s11160-014-9349-7.

Mulvaney, R. et al., 2012: Recent Antarctic Peninsula warming relative to Holocene climate and ice-shelf history. *Nature*, **489** (7414), 141–4, doi:10.1038/nature11391.

Munday, D.R., H.L. Johnson and D.P. Marshall, 2013: Eddy Saturation of Equilibrated Circumpolar Currents. *Journal of Physical Oceanography*, **43** (3), 507–532, doi:10.1175/JPO-D-12-095.1.

Mundy, C.J., J.K. Ehn, D.G. Barber and C. Michel, 2007: Influence of snow cover and algae on the spectral dependence of transmitted irradiance through Arctic landfast first-year sea ice. *Journal of Geophysical Research: Oceans*, **112** (C3), C03007, doi:10.1029/2006jc003683.

Munneke, P.K. et al., 2014: Explaining the presence of perennial liquid water bodies in the firn of the Greenland Ice Sheet. *Geophysical Research Letters*, **41** (2), 476–483, doi:10.1002/2013gl058389.

Munro, D.R. et al., 2015: Recent evidence for a strengthening CO_2 sink in the Southern Ocean from carbonate system measurements in the Drake Passage (2002–2015). *Geophysical Research Letters*, **42** (18), 7623–7630, doi:10.1002/2015GL065194.

Murphy, E.J., A. Clarke, N.J. Abram and J. Turner, 2014: Variability of sea-ice in the northern Weddell Sea during the 20th century. *Journal of Geophysical Research: Oceans*, **119** (7), 4549–4572, doi:10.1002/2013jc009511.

Murphy, E.J. et al., 2013: Comparison of the structure and function of Southern Ocean regional ecosystems: The Antarctic Peninsula and South Georgia. *Journal of Marine Systems*, **109**, 22–42, doi:10.1016/j.jmarsys.2012.03.011.

Murphy, E.J. et al., 2017: Restricted regions of enhanced growth of Antarctic krill in the circumpolar Southern Ocean. *Sci Rep*, **7** (1), 6963, doi:10.1038/s41598-017-07205-9.

Murray, T. et al., 2015: Extensive retreat of Greenland tidewater glaciers, 2000–2010. *Arctic Antarctic and Alpine Research*, **47** (3), 427–447, doi:10.1657/Aaar0014-049.

Muster, S. et al., 2017: PeRL: a circum-Arctic Permafrost Region Pond and Lake database. *Earth System Science Data*, **9** (1), 317–348, doi:10.5194/essd-9-317-2017.

Myers-Smith, I. and D. Hik, 2013: Shrub canopies influence soil temperatures but not nutrient dynamics: An experimental test of tundra snow-shrub interactions. *Ecology and Evolution*, **3** (11), 3683–3700, doi:10.1002/ece3.710.

Myers-Smith, I. and D. Hik, 2018: Climate warming as a driver of tundra shrubline advance. *Journal of Ecology*, **106** (2), 547–560, doi:10.1111/1365-2745.12817.

Myers-Smith, I.H. et al., 2015: Climate sensitivity of shrub growth across the tundra biome. *Nature Clim. Change*, **5** (9), 887–891, doi:10.1038/nclimate2697.

Myers-Smith, I.H. et al., 2019: Eighteen years of ecological monitoring reveals multiple lines of evidence for tundra vegetation change. *Ecological Monographs*, **89** (2), e01351, doi:10.1002/ecm.1351.

Nahrgang, J. et al., 2014: Gender Specific Reproductive Strategies of an Arctic Key Species (Boreogadus saida) and Implications of Climate Change. *Plos One*, **9** (5), e98452, doi:10.1371/journal.pone.0098452.

Najafi, M.R., F.W. Zwiers and N.P. Gillett, 2015: Attribution of Arctic temperature change to greenhouse-gas and aerosol influences. *Nature Climate Change*, **5** (3), 246–249, doi:10.1038/nclimate2524.

Nakamura, T. et al., 2016: The stratospheric pathway for Arctic impacts on midlatitude climate. *Geophysical Research Letters*, **43** (7), 3494–3501, doi:10.1002/2016gl068330.

National Research Council, 2014: *Linkages Between Arctic Warming and Mid-Latitude Weather Patterns: Summary of a Workshop*. The National Academies Press, Washington, DC, 85 pp.

National Research Council, 2015: *Enhancing the Effectiveness of Team Science*. The National Academies Press, Washington, DC, 280 pp.

Nauta, A.L. et al., 2014: Permafrost collapse after shrub removal shifts tundra ecosystem to a methane source. *Nature Climate Change*, **5**, 67, doi:10.1038/nclimate2446

Navarro, F.J., U.Y. Jonsell, M.I. Corcuera and A. Martín-Español, 2017: Decelerated mass loss of Hurd and Johnsons Glaciers, Livingston Island, Antarctic Peninsula. *Journal of Glaciology*, **59** (213), 115–128, doi:10.3189/2013JoG12J144.

Neff, T., 2009: Connecting Science and Policy to Combat Climate Change. Scientific American; www.scientificamerican.com/article/connecting-science-and-po/.

Nengye, L., A.K. Elizabeth and H. Tore, 2017: *The European Union and the Arctic*. Brill, Leiden, The Netherlands.

Nghiem, S.V. et al., 2012: The extreme melt across the Greenland ice sheet in 2012. *Geophysical Research Letters*, **39** (20), 6–11, doi:10.1029/2012GL053611.

Nicholls, K.W., L. Boehme, M. Biuw and M.A. Fedak, 2008: Wintertime ocean conditions over the southern Weddell Sea continental shelf, Antarctica. *Geophysical Research Letters*, **35** (21), L21605, doi:10.1029/2008gl035742.

Nicol, S. and J. Foster, 2016: The Fishery for Antarctic Krill: Its Current Status and Management Regime. In: Biology and Ecology of Antarctic Krill [Siegel, V. (ed.)]. Springer, New York, 387–421.

Nicolas, J.P. and D.H. Bromwich, 2014: New reconstruction of antarctic near-surface temperatures: Multidecadal trends and reliability of global reanalyses. *Journal of Climate*, **27** (21), 8070–8093, doi:10.1175/JCLI-D-13-00733.1.

Nicolaus, M., C. Katlein, J. Maslanik and S. Hendricks, 2012: Changes in Arctic sea ice result in increasing light transmittance and absorption. *Geophysical Research Letters*, **39** (24), L24501, doi:10.1029/2012gl053738.

Niederdrenk, A.L. and D. Notz, 2018: Arctic Sea Ice in a 1.5°C Warmer World. *Geophysical Research Letters*, **45** (4), 1963–1971, doi:10.1002/2017gl076159.

Nienow, P.W., A.J. Sole, D.A. Slater and T.R. Cowton, 2017: Recent Advances in Our Understanding of the Role of Meltwater in the Greenland Ice Sheet System. *Current Climate Change Reports*, **3** (4), 330–344, doi:10.1007/s40641-017-0083-9.

Nihashi, S. and K.I. Ohshima, 2015: Circumpolar mapping of Antarctic coastal polynyas and landfast sea ice: relationship and variability. *Journal of Climate*, **28**, 3650–3670, doi:10.1175/JCLI-D-14-00369.1.

Nilsson, A.E. et al., 2017: Towards extended shared socioeconomic pathways: A combined participatory bottom-up and top-down methodology with results from the Barents region. *Global Environmental Change-Human and Policy Dimensions*, **45**, 124–132, doi:10.1016/j.gloenvcha.2017.06.001.

3

Nilsson, A.E. and M. Christensen, 2019: *Arctic Geopolitics, Media and Power*. Taylor & Francis Group, London.

Nitze, I. et al., 2018: Remote sensing quantifies widespread abundance of permafrost region disturbances across the Arctic and Subarctic. *Nat Commun*, **9** (1), 5423, doi:10.1038/s41467-018-07663-3.

Nitze, I. et al., 2017: Landsat-Based Trend Analysis of Lake Dynamics across Northern Permafrost Regions. *Remote Sensing*, **9** (7), 640, doi:10.3390/rs9070640.

Noël, B. et al., 2017: A tipping point in refreezing accelerates mass loss of Greenland's glaciers and ice caps. *Nature Communications*, **8**, 14730, doi:10.1038/ncomms14730.

Noël, B. et al., 2018: Six Decades of Glacial Mass Loss in the Canadian Arctic Archipelago. *Journal of Geophysical Research: Earth Surface*, **123** (6), 1430–1449, doi:10.1029/2017jf004304.

Noetzli, J. et al., 2017: *Permafrost thermal state* [Blunden, J. and D.S. Arndt (eds.)]. State of the Climate in 2017, **99**, Bull. Amer. Meteor. Soc., Si–S332.

Nøttestad, L. et al., 2016: Quantifying changes in abundance, biomass, and spatial distribution of Northeast Atlantic mackerel (Scomber scombrus) in the Nordic seas from 2007 to 2014. *ICES Journal of Marine Science*, **73** (2), 359–373, doi:10.1093/icesjms/fsv218.

Notz, D., 2015: How well must climate models agree with observations? *Philosophical Transactions of the Royal Society A: Mathematical, Physical and Engineering Sciences*, **373** (2052), 20140164, doi:10.1098/rsta.2014.0164.

Notz, D. and J. Marotzke, 2012: Observations reveal external driver for Arctic sea-ice retreat. *Geophysical Research Letters*, **39** (8), L08502, doi:10.1029/2012gl051094.

Notz, D. and J. Stroeve, 2016: Observed Arctic sea-ice loss directly follows anthropogenic CO2 emission. *Science*, **354** (6313), 747–750, doi:10.1126/science.aag2345.

NPFMC, 2018: Draft Bering Sea Fishery Ecosystem Plan. North Pacific Fishery Management Council, 605 West 4th, Site 306, Anchorage, Alaska.

NPFMC, 2019: Bering Sea Fishery Ecosystem Plan. North Pacific Fishery Management Council, 605 West 4th, Suite 306. Anchorage, Alaska 99501.

Nummelin, A., M. Ilicak, C. Li and L.H. Smedsrud, 2016: Consequences of future increased Arctic runoff on Arctic Ocean stratification, circulation, and sea ice cover. *Journal of Geophysical Research: Oceans*, **121** (1), 617–637, doi:10.1002/2015JC011156.

Nummelin, A., C. Li and P.J. Hezel, 2017: Connecting ocean heat transport changes from the midlatitudes to the Arctic Ocean. *Geophysical Research Letters*, **44** (4), 1899–1908, doi:10.1002/2016GL071333.

Nymand, L.J. and G. Fondahl, 2014: Major Findings and Emerging Trends in Arctic Human Development. In: Arctic Human Development Report: Regional Processes and Global Linkages [Nymand Larsen, J. and G. Fondhal (eds.)]. Nordic Council of Ministers, Copenhagen, 479–502.

Nyström, M. and C. Folke, 2001: Spatial Resilience of Coral Reefs. *Ecosystems*, **4** (5), 406–417, doi:10.1007/s10021-001-0019-y.

O'Leary, M. and P. Christoffersen, 2013: Calving on tidewater glaciers amplified by submarine frontal melting. *Cryosphere*, **7** (1), 119–128, doi:10.5194/tc-7-119-2013.

Ó Cofaigh, C. et al., 2014: Reconstruction of ice-sheet changes in the Antarctic Peninsula since the Last Glacial Maximum. *Quaternary Science Reviews*, **100**, 87–110, doi:10.1016/j.quascirev.2014.06.023.

Odland, J.O., S. Donaldson, A. Dudarev and A. Carlsen, 2016: AMAP assessment 2015: human health in the Arctic. *Int J Circumpolar Health*, **75**, 33949, doi:10.3402/ijch.v75.33949.

Ohshima, K.I. et al., 2013: Antarctic Bottom Water production by intense sea-ice formation in the Cape Darnley polynya. *Nature Geoscience*, **6**, 235, doi:10.1038/ngeo1738.

Øigård, T.A., T. Haug and K.T. Nilssen, 2014: Current status of hooded seals in the Greenland Sea. Victims of climate change and predation? *Biological Conservation*, **172**, 29–36, doi:10.1016/j.biocon.2014.02.007.

Øigård, T.A. et al., 2013: Functional relationship between harp seal body condition and available prey in the Barents Sea. *Marine Ecology Progress Series*, **484**, 287–301, doi:10.3354/meps10272.

Olafsdottir, A.H. et al., 2019: Geographical expansion of Northeast Atlantic mackerel (Scomber scombrus) in the Nordic Seas from 2007 to 2016 was primarily driven by stock size and constrained by low temperatures. *Deep Sea Research Part II: Topical Studies in Oceanography*, **159**, 152–168, doi:10.1016/j.dsr2.2018.05.023.

Olason, E. and D. Notz, 2014: Drivers of variability in Arctic sea-ice drift speed. *Journal of Geophysical Research: Oceans*, **119** (9), 5755–5775, doi:10.1002/2014jc009897.

Olefeldt, D. et al., 2016: Circumpolar distribution and carbon storage of thermokarst landscapes. *Nature Communications*, **7**, 13043, doi:10.1038/ncomms13043.

Oliva, M. et al., 2017: Recent regional climate cooling on the Antarctic Peninsula and associated impacts on the cryosphere. *Sci Total Environ*, **580**, 210–223, doi:10.1016/j.scitotenv.2016.12.030.

Olsen, J., N.A. Carter, J. Dawson and W. Coetzee, 2019: Community perspectives on the environmental impacts of Arctic shipping: Case studies from Russia, Norway and Canada. *Cogent Social Sciences*, **5** (1), 1609189, doi:10.1080/23311886.2019.1609189.

Olsen, L.M. et al., 2017: The seeding of ice algal blooms in Arctic pack ice: The multiyear ice seed repository hypothesis. *Journal of Geophysical Research: Biogeosciences*, **122** (7), 1529–1548, doi:10.1002/2016jg003668.

Onarheim, I.H. and M. Årthun, 2017: Toward an ice-free Barents Sea. *Geophysical Research Letters*, **44** (16), 8387–8395, doi:10.1002/2017GL074304.

Onarheim, I.H., T. Eldevik, L.H. Smedsrud and J.C. Stroeve, 2018: Seasonal and regional manifestation of Arctic sea ice loss. *Journal of Climate*, **31** (12), 4917–4932, doi:10.1175/jcli-d-17-0427.1.

Ono, K. et al., 2017: Management strategy analysis for multispecies fisheries including technical interactions and human behavior in modeling management decisions and fishing. *Canadian Journal of Fisheries and Aquatic Sciences*, **75** (8), 1185–1202, doi:10.1139/cjfas-2017-0135.

Orr, J.C. et al., 2005: Anthropogenic ocean acidification over the twenty-first century and its impact on calcifying organisms. *Nature*, **437** (7059), 681–6, doi:10.1038/nature04095.

Orsi, A.H. and C.L. Wiederwohl, 2009: A recount of Ross Sea waters. *Deep-Sea Research Part Ii-Topical Studies in Oceanography*, **56** (13–14), 778–795, doi:10.1016/j.dsr2.2008.10.033.

Orsi, A.J. et al., 2017: The recent warming trend in North Greenland. *Geophysical Research Letters*, **44** (12), 6235–6243, doi:10.1002/2016gl072212.

Osokin, N.I. and A.V. Sosnovsky, 2014: Spatial and temporal variability of depth and density of the snow cover in Russia. *Ice and Snow*, **4**, 72–80, doi:10.15356/2076-6734-2014-4-72-80.

Ostapchuk, J. et al., 2015: Exploring Elders' and Seniors' Perceptions of How Climate Change is Impacting Health and Well-being in Rigolet, Nunatsiavut. *Journal of Aboriginal Health*, **9**, 6–24, doi:10.18357/ijih92201214358.

Overland, J. et al., 2018a: The urgency of Arctic change. *Polar Science*, doi:10.1016/j.polar.2018.11.008.

Overland, J. et al., 2017: *Synthesis: summary and implications of findings*. Snow, Water, Ice and Permafrost in the Arctic (SWIPA), Arctic Monitoring and Assessment Programme, Oslo, Norway, 269p.

Overland, J.E., 2018: Sea ice Index. Boulder, Colorado USA. NASA National Snow and Ice Data Center Distributed Active Archive Center.

Overland, J.E. et al., 2018b: *Surface air temperature*. [in Arctic Report Card 2018] [Available at: https://arctic.noaa.gov/Report-Card/Report-Card-2018/ArtMID/7878/ArticleID/783/Surface-Air-Temperature].

Overland, J.E. and M.Y. Wang, 2013: When will the summer Arctic be nearly sea ice free? *Geophysical Research Letters*, **40** (10), 2097–2101, doi:10.1002/grl.50316.

Overland, J.E. and M.Y. Wang, 2016: Recent Extreme Arctic Temperatures are due to a Split Polar Vortex. *Journal of Climate*, **29** (15), 5609–5616, doi:10.1175/jcli-d-16-0320.1.

Overland, J.E. and M.Y. Wang, 2018a: Arctic-midlatitude weather linkages in North America. *Polar Science*, (16), 1–9, doi:10.1016/j.polar.2018.02.001.

Overland, J.E. and M.Y. Wang, 2018b: Resolving Future Arctic/ Midlatitude Weather Connections. *Earth's Future*, 6 (8), 1146–1152, doi:10.1029/2018ef000901.

Paar, M. et al., 2016: Temporal shift in biomass and production of macrozoobenthos in the macroalgal belt at Hansneset, Kongsfjorden, after 15 years. *Polar Biology*, 39 (11), 2065–2076, doi:10.1007/s00300-015-1760-6.

PAME, 2013: Large Marine Ecosystems (LMEs) of the Arctic area. Revision of the Arctic LME map. PAME Secretariat, Akureyri, Iceland.

Panic, M. and J.D. Ford, 2013: A review of national-level adaptation planning with regards to the risks posed by climate change on infectious diseases in 14 OECD nations. *International Journal of Environmental Research and Public Health*, 10 (12), 7083–7109, doi:10.3390/ijerph10127083.

Paolo, F.S., H.A. Fricker and L. Padman, 2015: Volume loss from Antarctic ice shelves is accelerating. *Science*, 348 (6232), 327–331, doi:10.1126/science.aaa0940.

Paolo, F.S. et al., 2018: Response of Pacific-sector Antarctic ice shelves to the El Niño/Southern Oscillation. *Nature Geoscience*, 11 (2), 121–126, doi:10.1038/s41561-017-0033-0.

Pardo, D., S. Jenouvrier, H. Weimerskirch and C. Barbraud, 2017: Effect of extreme sea surface temperature events on the demography of an age-structured albatross population. *Philos Trans R Soc Lond B Biol Sci*, 372 (1723), 1–10, doi:10.1098/rstb.2016.0143.

Parizek, B.R. et al., 2019: Ice-cliff failure via retrogressive slumping. *Geology*, 47 (5), 449–452, doi:10.1130/g45880.1.

Park, H., H. Yabuki and T. Ohata, 2012: Analysis of satellite and model datasets for variability and trends in Arctic snow extent and depth, 1948–2006. *Polar Science*, 6, 23–37, doi:10.1016/j.polar.2011.11.002.

Park, H. et al., 2015: Quantification of Warming Climate-Induced Changes in Terrestrial Arctic River Ice Thickness and Phenology. *Journal of Climate*, 29 (5), 1733–1754, doi:10.1175/jcli-d-15-0569.1.

Park, J. et al., 2017: Light availability rather than Fe controls the magnitude of massive phytoplankton bloom in the Amundsen Sea polynyas, Antarctica. *Limnology and Oceanography*, 62 (5), 2260–2276, doi:10.1002/lno.10565.

Parker, M.L. et al., 2015: Assemblages of micronektonic fishes and invertebrates in a gradient of regional warming along the Western Antarctic Peninsula. *Journal of Marine Systems*, 152, 18–41, doi:10.1016/j.jmarsys.2015.07.005.

Parkinson, A.J. and J. Berner, 2009: Climate change and impacts on human health in the Arctic: an international workshop on emerging threats and the response of Arctic communities to climate change. *International Journal of Circumpolar Health*, 68 (1), 84–91, doi:10.3402/ijch.v68i1.18295.

Parkinson, A.J. et al., 2014: Climate change and infectious diseases in the Arctic: establishment of a circumpolar working group. *International Journal of Circumpolar Health*, 73 (1), 25163, doi:10.3402/ijch.v73.25163.

Parlee, B. and C. Furgal, 2012: Well-being and environmental change in the arctic: a synthesis of selected research from Canada's International Polar Year program. *Climatic Change*, 115 (1), 13–34, doi:10.1007/s10584-012-0588-0.

Pashkevich, A., J. Dawson and E.J. Stewart, 2015: Governance of expedition cruise ship tourism in the Arctic: A comparison of the Canadian and Russian Arctic. *Tourism in Marine Environments*, 10 (3/4), 225–240, doi:10.1016/j.ocecoaman.2013.12.005.

Pastick, N.J. et al., 2017: Historical and projected trends in landscape drivers affecting carbon dynamics in Alaska. *Ecol Appl*, 27 (5), 1383–1402, doi:10.1002/eap.1538.

Pastick, N.J. et al., 2019: Spatiotemporal remote sensing of ecosystem change and causation across Alaska. *Glob Chang Biol*, 25 (3), 1171–1189, doi:10.1111/gcb.14279.

Patara, L. and C.W. Böning, 2014: Abyssal ocean warming around Antarctica strengthens the Atlantic overturning circulation. *Geophysical Research Letters*, 41 (11), 3972–3978, doi:10.1002/2014GL059923.

Patara, L., C.W. Böning and A. Biastoch, 2016: Variability and trends in Southern Ocean eddy activity in 1/12 ocean model simulations. *Geophysical Research Letters*, 43 (9), 4517–4523, doi:10.1002/2016GL069026.

Pattyn, F., 2010: Antarctic subglacial conditions inferred from a hybrid ice sheet/ice stream model. *Earth Planet. Sci. Lett.*, 295, 451–461, doi:10.1016/j.epsl.2010.04.025.

Pattyn, F., 2018: The paradigm shift in Antarctic ice sheet modelling. *Nature Communications*, 9 (2728), 1–3, doi:10.1038/s41467-018-05003-z.

Paul, S. et al., 2018: Empirical parametrization of Envisat freeboard retrieval of Arctic and Antarctic sea ice based on CryoSat-2: progress in the ESA Climate Change Initiative. *The Cryosphere*, 12 (7), 2437–2460, doi:10.5194/tc-12-2437-2018.

Pauling, A.G., C.M. Bitz, I.J. Smith and P.J. Langhorne, 2016: The Response of the Southern Ocean and Antarctic Sea Ice to Freshwater from Ice Shelves in an Earth System Model. *Journal of Climate*, 29 (5), 1655–1672, doi:10.1175/jcli-d-15-0501.1.

Pauling, A.G., I.J. Smith, P.J. Langhorne and C.M. Bitz, 2017: Time-dependent freshwater input from ice shelves: Impacts on Antarctic sea ice and the Southern Ocean in an Earth System Model. *Geophysical Research Letters*, 44 (20), 10454–10461, doi:10.1002/2017GL075017.

Pearce, T., J. Ford, A.C. Willox and B. Smit, 2015: Inuit Traditional Ecological Knowledge (TEK), Subsistence Hunting and Adaptation to Climate Change in the Canadian Arctic. *Arctic*, 68 (2), 233–245, doi:10.14430/arctic4475.

Pearce, T. et al., 2011: Advancing adaptation planning for climate change in the Inuvialuit Settlement Region (ISR): a review and critique. *Regional Environmental Change*, 11 (1), 1–17, doi:10.1007/s10113-010-0126-4.

Pearce, T. et al., 2010: Inuit vulnerability and adaptive capacity to climate change in Ulukhaktok, Northwest Territories, Canada. *Polar Record*, 46 (2), 157–177, doi:10.1017/S0032247409008602.

Pearson, R.G. et al., 2013: Shifts in Arctic vegetation and associated feedbacks under climate change. *Nature Climate Change*, 3 (7), 673–677, doi:10.1038/nclimate1858.

Peck, M. and J.K. Pinnegar, 2018: Chapter 5: Climate change impacts, vulnerabilities and adaptations: North Atlantic and Atlantic Arctic marine fisheries. In: Impacts of climate change on fisheries and aquaculture: synthesis of current knowledge, adaptation and mitigation options. FAO Fisheries and Aquaculture Technical Paper No. 627. [Barange, M., T. Bahri, M.C.M. Beveridge, K.L. Cochrane, S. Funge-Smith and F. Poulain (eds.)], Rome.

Peck, V.L. et al., 2018: Pteropods counter mechanical damage and dissolution through extensive shell repair. *Nature Communications*, 9 (1), 1–7, doi:10.1038/s41467-017-02692-w.

Peck, V.L. et al., 2016: Outer organic layer and internal repair mechanism protects pteropod Limacina helicina from ocean acidification. *Deep Sea Research Part II: Topical Studies in Oceanography*, 127, 41–52, doi:10.1016/j.dsr2.2015.12.005.

Pecl, G.T. et al., 2017: Biodiversity redistribution under climate change: Impacts on ecosystems and human well-being. *Science*, 355 (6332), eaai9214, doi:10.1126/science.aai9214

Pellichero, V., J.-B. Sallée, C.C. Chapman and S.M. Downes, 2018: The southern ocean meridional overturning in the sea-ice sector is driven by freshwater fluxes. *Nature Communications*, 9 (1), 1789, doi:10.1038/s41467-018-04101-2.

Pendakur, K., 2017: Northern Territories. In: Climate risks and adaptation practices for the Canadian transportation sector 2016 [Palko, K. and D.S. Lemmen (eds.)]. Government of Canada, Ottawa, 27–64.

Penn, H.J.F., S.C. Gerlach and P.A. Loring, 2016: Seasons of Stress: Understanding the Dynamic Nature of People's Ability to Respond to Change and Surprise. *Weather, Climate, and Society*, 8 (4), 435–446, doi:10.1175/WCAS-D-15-0061.1.

Perovich, D.K., 2016: Sea ice and sunlight. In: Sea Ice [Thomas, D.N. (ed.)]. Wiley Online Library, 110–137.

Perovich, D.K. and C. Polashenski, 2012: Albedo evolution of seasonal Arctic sea ice. *Geophysical Research Letters*, **39** (8), L08501, doi:10.1029/2012gl051432.

Perreault, N. et al., 2017: Remote sensing evaluation of High Arctic wetland depletion following permafrost disturbance by thermo-erosion gullying processes. *Arctic Science*, **3** (2), 237–253, doi:10.1139/as-2016-0047.

Pertierra, L.R., K.A. Hughes, G.C. Vega and M.A. Olalla-Tarraga, 2017: High Resolution Spatial Mapping of Human Footprint across Antarctica and Its Implications for the Strategic Conservation of Avifauna. *Plos One*, **12** (1), e0168280, doi:10.1371/journal.pone.0168280.

Petrasek-MacDonald, J., J.D. Ford, A.C. Willox and N.A. Ross, 2013: A review of protective factors and causal mechanisms that enhance the mental health of Indigenous Circumpolar youth. *International Journal of Circumpolar Health*, **72** (1), 21775, doi:10.3402/ijch.v72i0.21775.

Petrick, S. et al., 2017: Climate change, future Arctic Sea ice, and the competitiveness of European Arctic offshore oil and gas production on world markets. *AMBIO*, **46** (Suppl 3), 410–422, doi:10.1007/s13280-017-0957-z.

Petrov, A.N. et al., 2016: Arctic sustainability research: toward a new agenda. *Polar Geography*, **39** (3), 165–178, doi:10.1080/1088937x.2016.1217095.

PEW Charitable Trust, 2016: *The integrated arctic corridors framework: planning for responsible shipping in Canadian arctic waters*. [Available at: www.pewtrusts.org/en/research-and-analysis/reports/2016/04/the-integrated-arctic-corridors-framework].

Phoenix, G.K. and J.W. Bjerke, 2016: Arctic browning: extreme events and trends reversing Arctic greening. *Global Change Biology*, **22** (9), 2960–2962, doi:10.1111/gcb.13261.

Pilfold, N., W. et al., 2017: Migratory response of polar bears to sea ice loss: to swim or not to swim. *Ecography*, **40** (1), 189–199, doi:10.1111/ecog.02109.

Pincus, R. and S.H. Ali, 2016: Have you been to 'The Arctic'? Frame theory and the role of media coverage in shaping Arctic discourse. *Polar Geography*, **39** (2), 83–97, doi:10.1080/1088937X.2016.1184722.

Pincus, R. and J.G. Speth, 2015: Security in the Arctic

A Receding Wall. In: Diplomacy on Ice [Pincus, R. and S.H. Ali (eds.)]. Yale University Press, 161–168.

Piñones, A. and A.V. Fedorov, 2016: Projected changes of Antarctic krill habitat by the end of the 21st century. *Geophysical Research Letters*, **43** (16), 8580–8589, doi:10.1002/2016gl069656.

Pistone, K., I. Eisenman and V. Ramanathan, 2014: Observational determination of albedo decrease caused by vanishing Arctic sea ice. *Proceedings of the National Academy of Sciences*, **111** (9), 3322–3326.

Pithan, F. and T. Mauritsen, 2014: Arctic amplification dominated by temperature feedbacks in contemporary climate models. *Nature Geoscience*, **7**, 181, doi:10.1038/ngeo2071.

Pizzolato, L. et al., 2016: The influence of declining sea ice on shipping activity in the Canadian Arctic. *Geophysical Research Letters*, **43** (23), 12,146–12,154, doi:10.1002/2016gl071489.

Pizzolato, L. et al., 2014: Changing sea ice conditions and marine transportation activity in Canadian Arctic waters between 1990 and 2012. *Climatic Change*, **123** (2), 161–173, doi:10.1007/s10584-013-1038-3.

Planque, B. et al., 2019: A participatory scenario method to explore the future of marine social-ecological systems. *Fish and Fisheries*, **20** (3), 434–451, doi:10.1111/faf.12356.

Poesch, M., S. et al., 2016: Climate Change Impacts on Freshwater Fishes: A Canadian Perspective. *Fisheries*, **41** (7), 385–391, doi:10.1080/036324 15.2016.1180285.

Poinar, K. et al., 2017: Drainage of Southeast Greenland Firn Aquifer Water through Crevasses to the Bed. *Frontiers in Earth Science*, **5**, 1–15, doi:10.3389/feart.2017.00005.

Polashenski, C. et al., 2014: Observations of pronounced Greenland ice sheet firn warming and implications for runoff production. *Geophysical Research Letters*, **41** (12), 4238–4246, doi:10.1002/2014GL059806.

Polishchuk, Y.M., N.A. Bryksina and V.Y. Polishchuk, 2015: Remote analysis of changes in the number of small thermokarst lakes and their distribution with respect to their sizes in the cryolithozone of Western Siberia, 2015. *Izvestiya, Atmospheric and Oceanic Physics*, **51** (9), 999–1006, doi:10.1134/s0001433815090145.

Pollard, D., R.M. DeConto and R.B. Alley, 2015: Potential Antarctic Ice Sheet retreat driven by hydrofracturing and ice cliff failure. *Earth and Planetary Science Letters*, **412**, 112–121, doi:10.1016/j.epsl.2014.12.035.

Polvani, L.M. and K.L. Smith, 2013: Can natural variability explain observed Antarctic sea ice trends? New modeling evidence from CMIP5. *Geophysical Research Letters*, **40** (12), 3195–3199, doi:10.1002/grl.50578.

Polyak, L. et al., 2010: History of sea ice in the Arctic. *Quaternary Science Reviews*, **29** (15–16), 1757–1778, doi:10.1016/j.quascirev.2010.02.010.

Polyakov, I.V. et al., 2017: Greater role for Atlantic inflows on sea-ice loss in the Eurasian Basin of the Arctic Ocean. *Science*, **356** (6335), 285–291, doi:10.1126/science.aai8204

Polyakov, I.V. et al., 2010: Arctic Ocean Warming Contributes to Reduced Polar Ice Cap. *Journal of Physical Oceanography*, **40** (12), 2743–2756, doi:10.1175/2010jpo4339.1.

Poole, K.G., A. Gunn, B.R. Patterson and M. Dumond, 2010: Sea Ice and Migration of the Dolphin and Union Caribou Herd in the Canadian Arctic: An Uncertain Future. *Arctic*, **63** (4), 414–428, doi:10.14430/arctic3331.

Pope, S., L. Copland and B. Alt, 2017: Recent changes in sea ice plugs along the northern Canadian Arctic Archipelago. In: Arctic Ice Shelves and Ice Islands [Copland, L. and D. Mueller (eds.)]. Springer Nature, Dordrecht, 317–342.

Popov, S.V. and V.N. Masolov, 2007: Forty-seven new subglacial lakes in the 0–110*E sector of East Antarctica. *J. Glaciol.*, **53** (181), 289–297, doi:10.3189/172756507782202856.

Popova, E.E. et al., 2014: Regional variability of acidification in the Arctic: a sea of contrasts. *Biogeosciences*, **11** (2), 293–308, doi:10.5194/bg-11-293-2014.

Pörtner, H.-O. et al., 2014: *Ocean systems* [Field, C.B., V.R. Barros, D.J. Dokken, K.J. Mach, M.D. Mastrandrea, T.E. Bilir, M. Chatterjee, K.L. Ebi, Y.O. Estrada, R.C. Genova, B. Girma, E.S. Kissel, A.N. Levy, S. MacCracken, P.R. Mastrandrea and L.L. White (eds.)]. Climate Change 2014: Impacts, Adaptation, and Vulnerability. Part A: Global and Sectoral Aspects. Contribution of Working Group II to the Fifth Assessment Report of the Intergovernmental Panel on Climate Change, Press, C.U., Cambridge, United Kingdom and New York, NY, USA, 411–484.

Portnov, A. et al., 2013: Offshore permafrost decay and massive seabed methane escape in water depths >20 m at the South Kara Sea shelf. *Geophysical Research Letters*, **40** (15), 3962–3967, doi:10.1002/grl.50735.

Post, E. et al., 2013: Ecological Consequences of Sea-Ice Decline. *Science*, **341** (6145), 519–524, doi:10.1126/science.1235225.

Poulter, B. et al., 2017: Global wetland contribution to 2000–2012 atmospheric methane growth rate dynamics. *Environmental Research Letters*, **12** (9), 094013, doi:10.1088/1748-9326/aa8391.

Powledge, F., 2012: Scientists, Policymakers, and a Climate of UncertaintyCan research gain a foothold in the politics of climate change? *Bioscience*, **62** (1), 8–13, doi:10.1525/bio.2012.62.1.3.

Prior, T.L., 2013: Breaking the Wall of Monocentric Governance: Polycentricity in the Governance of Persistent Organic Pollutants in the Arctic. *The Yearbook of Polar Law Online*, **5** (1), 185–232, doi:10.1163/22116427-91000123.

Pristupa, A.O., M. Lamers, M. Tysiachniouk and B. Amelung, 2018: Reindeer Herders Without Reindeer. The Challenges of Joint Knowledge Production on Kolguev Island in the Russian Arctic. *Society & Natural Resources*, **32** (3), 338–356, doi:10.1080/08941920.2018.1505012.

Pritchard, H.D. et al., 2012: Antarctic ice-sheet loss driven by basal melting of ice shelves. *Nature*, **484** (7395), 502, doi:10.1038/nature10968.

3

Proshutinsky, A. et al., 2015: Arctic circulation regimes. *Phil. Trans. R. Soc. A,* **373** (2052), 20140160, doi:10.1098/rsta.2014.0160.

Provencher, J.F., A.J. Gaston, P.D. O'Hara and H.G. Gilchrist, 2012: Seabird diet indicates changing Arctic marine communities in eastern Canada. *Marine Ecology Progress Series,* **454**, 171–182, doi:10.3354/meps09299.

Prowse, T. et al., 2011: Effects of Changes in Arctic Lake and River Ice. *AMBIO,* **40** (1), 63–74, doi:10.1007/s13280-011-0217-6.

Prowse, T. et al., 2015: Arctic Freshwater Synthesis: Summary of key emerging issues. *Journal of Geophysical Research: Biogeosciences,* **120** (10), 1887–1893, doi:10.1002/2015JG003128.

Prowse, T., R. Shrestha, B. Bonsal and Y. Dibike, 2010: Changing spring air-temperature gradients along large northern rivers: Implications for severity of river-ice floods. *Geophysical Research Letters,* **37** (19), L19706, doi:10.1029/2010gl044878.

Pulsifer, P. et al., 2012: The role of data management in engaging communities in Arctic research: overview of the Exchange for Local Observations and Knowledge of the Arctic (ELOKA). *Polar Geography,* **35** (3–4), 271–290, doi:10.1080/1088937X.2012.708364.

Punt, A.E. et al., 2016: Effects of long-term exposure to ocean acidification conditions on future southern Tanner crab (Chionoecetes bairdi) fisheries management. *ICES Journal of Marine Science: Journal du Conseil,* **73** (3), 849–864, doi:10.1093/icesjms/fsv205.

Purich, A., W.J. Cai, M.H. England and T. Cowan, 2016a: Evidence for link between modelled trends in Antarctic sea ice and underestimated westerly wind changes. *Nature Communications,* **7**, 10409, doi:10.1038/ncomms10409.

Purich, A. et al., 2016b: Tropical Pacific SST Drivers of Recent Antarctic Sea Ice Trends. *Journal of Climate,* **29** (24), 8931–8948, doi:10.1175/jcli-d-16-0440.1.

Purich, A. et al., 2018: Impacts of Broad-Scale Surface Freshening of the Southern Ocean in a Coupled Climate Model. *Journal of Climate,* **31** (7), 2613–2632, doi:10.1175/jcli-d-17-0092.1.

Purkey, S.G. and G.C. Johnson, 2012: Global Contraction of Antarctic Bottom Water between the 1980s and 2000s*. *Journal of Climate,* **25** (17), 5830–5844, doi:10.1175/jcli-d-11-00612.1.

Purkey, S.G. and G.C. Johnson, 2013: Antarctic Bottom Water warming and freshening: Contributions to sea level rise, ocean freshwater budgets, and global heat gain. *Journal of Climate,* **26** (16), 6105–6122, doi:10.1175/JCLI-D-12-00834.1.

Qi, D. et al., 2017: Increase in acidifying water in the western Arctic Ocean. *Nature Climate Change,* **7** (3), 195, doi:10.1038/nclimate3228.

Qu, X. and A. Hall, 2014: On the persistent spread in snow-albedo feedback. *Climate Dynamics,* **42** (1), 69–81, doi:10.1007/s00382-013-1774-0.

Quinlan, A.E. et al., 2016: Measuring and assessing resilience: broadening understanding through multiple disciplinary perspectives. *Journal of Applied Ecology,* **53** (3), 677–687, doi:10.1111/1365-2664.12550.

Rabe, B. et al., 2014: Arctic Ocean basin liquid freshwater storage trend 1992–2012. *Geophysical Research Letters,* **41** (3), 961–968, doi:10.1002/2013GL058121.

Radosavljevic, B. et al., 2015: Erosion and Flooding – Threats to Coastal Infrastructure in the Arctic: A Case Study from Herschel Island, Yukon Territory, Canada. *Estuaries and Coasts,* **39** (4), 900–915, doi:10.1007/s12237-015-0046-0.

Raiswell, R. et al., 2016: Potentially Bioavailable Iron Delivery by Iceberg hosted Sediments and Atmospheric Dust to the Polar Oceans. *Biogeosciences Discussions,* **13** (13), 3887–3900, doi:doi:10.5194/bg-2016-20.

Rampal, P., J. Weiss, C. Dubois and J.-M. Campin, 2011: IPCC climate models do not capture Arctic sea ice drift acceleration: Consequences in terms of projected sea ice thinning and decline. *Journal of Geophysical Research: Oceans,* **116** (C8), C00D07, doi:10.1029/2011JC007110.

Rampal, P., J. Weiss and D. Marsan, 2009: Positive trend in the mean speed and deformation rate of Arctic sea ice, 1979–2007. *Journal of Geophysical Research: Oceans,* **114** (C5), C05013, doi:10.1029/2008jc005066.

Randelhoff, A. and J.D. Guthrie, 2016: Regional patterns in current and future export production in the central Arctic Ocean quantified from nitrate fluxes. *Geophysical Research Letters,* **43** (16), 8600–8608, doi:10.1002/2016GL070252.

Randerson, J.T. et al., 2012: Global burned area and biomass burning emissions from small fires. *Journal of Geophysical Research: Biogeosciences,* **117** (G4), G04012, doi:10.1029/2012JG002128.

Raphael, M.N., W. Hobbs and I. Wainer, 2011: The effect of Antarctic sea ice on the Southern Hemisphere atmosphere during the southern summer. *Climate Dynamics,* **36** (7–8), 1403–1417, doi:10.1007/s00382-010-0892-1.

Raphael, M.N. et al., 2016: The Amundsen Sea Low Variability, Change, and Impact on Antarctic Climate. *Bulletin of the American Meteorological Society,* **97** (1), 111–121, doi:10.1175/bams-d-14-00018.1.

Rathmann, N.M. et al., 2017: Highly temporally resolved response to seasonal surface melt of the Zachariae and 79N outlet glaciers in northeast Greenland. *Geophysical Research Letters,* **44** (19), 9805–9814, doi:10.1002/2017gl074368.

Rautio, A., B. Poppel and K. Young, 2014: Human Health and Well-Being. In: Arctic Human Development Report [Nymand Larson, J. and G. Fondahl (eds.)]. Nordic Council of Ministers, Copenhagen, 299–348.

Rawlins, M.A. et al., 2010: Analysis of the Arctic System for Freshwater Cycle Intensification: Observations and Expectations. *Journal of Climate,* **23** (21), 5715–5737, doi:10.1175/2010jcli3421.1.

Raymond, B. et al., 2014: Important marine habitat off east Antarctica revealed by two decades of multi-species predator tracking. *Ecography,* **38** (2), 121–129, doi:10.1111/ecog.01021.

Rayner, N.A., 2003: Global analyses of sea surface temperature, sea ice, and night marine air temperature since the late nineteenth century. *Journal of Geophysical Research,* **108** (D14), 4407, doi:10.1029/2002jd002670.

Raynolds, M.K. et al., 2014: Cumulative geoecological effects of 62 years of infrastructure and climate change in ice-rich permafrost landscapes, Prudhoe Bay Oilfield, Alaska. *Global Change Biology,* **20** (4), 1211–1224, doi:10.1111/gcb.12500.

Reese, R., G.H. Gudmundsson, A. Levermann and R. Winkelmann, 2018: The far reach of ice-shelf thinning in Antarctica. *Nature Climate Change,* **8** (1), 53–57, doi:10.1038/s41558-017-0020-x.

Regan, H.C., C. Lique and T.W.K. Armitage, 2019: The Beaufort Gyre Extent, Shape, and Location Between 2003 and 2014 From Satellite Observations. *Journal of Geophysical Research: Oceans,* **124** (2), 844–862, doi:10.1029/2018jc014379.

Régnière, J., J. Powell, B. Bentz and V. Nealis, 2012: Effects of temperature on development, survival and reproduction of insects: Experimental design, data analysis and modeling. *Journal of Insect Physiology,* **58** (5), 634–647, doi:10.1016/j.jinsphys.2012.01.010.

Reid, K., J. Croxall, D. Briggs and E. Murphy, 2005: Antarctic ecosystem monitoring: quantifying the response of ecosystem indicators to variability in Antarctic krill. *ICES Journal of Marine Science,* **62** (3), 366–373, doi:10.1016/j.icesjms.2004.11.003.

Reintges, A., T. Martin, M. Latif and N.S. Keenlyside, 2017: Uncertainty in twenty-first century projections of the Atlantic Meridional Overturning Circulation in CMIP3 and CMIP5 models. *Climate Dynamics,* **49** (5–6), 1495–1511, doi:10.1007/s00382-016-3180-x.

Reist, J., 2018: Fisheries. In: Marine Fisheries of Arctic Canada [Coad, B.W. and J. Reist (eds.)]. Canadian Museum of Nature and University of Toronto Press, Toronto, 618p.

Reist, J.D., C.D. Sawatzky and L. Johnson, 2016: The Arctic 'Great' Lakes of Canada and their fish faunas – An overview in the context of Arctic change. *Journal of Great Lakes Research,* **42** (2), 173–192, doi:10.1016/j.jglr.2015.10.008.

Renaud, P.E. et al., 2018: Pelagic food-webs in a changing Arctic: a trait-based perspective suggests a mode of resilience. *ICES Journal of Marine Science,* **75** (6), 1871–1881, doi:10.1093/icesjms/fsy063.

3

Renaud, P.E. et al., 2015: The future of Arctic benthos: Expansion, invasion, and biodiversity. *Progress in Oceanography*, **139**, 244–257, doi:10.1016/j.pocean.2015.07.007.

Renner, A.H.H. et al., 2014: Evidence of Arctic sea ice thinning from direct observations. *Geophysical Research Letters*, **41** (14), 5029–5036, doi:10.1002/2014gl060369.

Renner, M. et al., 2016: Timing of ice retreat alters seabird abundances and distributions in the southeast Bering Sea. *Biology Letters*, **12** (9), 20160276, doi:10.1098/rsbl.2016.0276.

Retter, G.-B. et al., 2004: *Community-based Monitoring Discussion paper*. Supporting publication to the CAFF Circumpolar Biodiversity Monitoring Program – Framework Document, CAFF CBMP Report No. 9, CAFF International Secretariat, Council, A., Akureyri, Iceland, 21 pp [Available at: https://oaarchive.arctic-council.org/handle/11374/178].

Rey, D.M. et al., 2019: Investigating lake-area dynamics across a permafrost-thaw spectrum using airborne electromagnetic surveys and remote sensing time-series data in Yukon Flats, Alaska. *Environmental Research Letters*, **14** (2), 025001, doi:10.1088/1748-9326/aaf06f.

Reynolds, R.W. et al., 2002: An Improved In Situ and Satellite SST Analysis for Climate. *Journal of Climate*, **15** (13), 1609–1625, doi:10.1175/1520-0442.

RGI Consortium, 2017: Randolph Glacier Inventory – A Dataset of Global Glacier Outlines: Version 6.0: Technical Report, Global Land Ice Measurements from Space. Digital Media, Colorado, USA, doi: https://doi.org/10.7265/N5-RGI-60.

Richter-Menge, J., J.E. Overland, J.T. Mathis and E.E. Osborne, 2017: *Arctic Report Card 2017*. [Available at: www.arctic.noaa.gov/Report-Card].

Rickard, G. and E. Behrens, 2016: CMIP5 Earth System Models with biogeochemistry: a Ross Sea assessment. *Antarctic Science*, **28** (5), 327–346, doi:10.1017/s0954102016000122.

Ricker, R., F. Girard-Ardhuin, T. Krumpen and C. Lique, 2018: Satellite-derived sea ice export and its impact on Arctic ice mass balance. *The Cryosphere*, **12** (9), 3017–3032, doi:10.5194/tc-12-3017-2018.

Ricker, R. et al., 2017: A weekly Arctic sea-ice thickness data record from merged CryoSat-2 and SMOS satellite data. *The Cryosphere*, **11** (4), 1607–1623, doi:10.5194/tc-11-1607-2017.

Ridley, J.K., J.A. Lowe and H.T. Hewitt, 2012: How reversible is sea ice loss? *The Cryosphere*, **6** (1), 193–198, doi:10.5194/tc-6-193-2012.

Rignot, E., S. Jacobs, J. Mouginot and B. Scheuchl, 2013: Ice-shelf melting around Antarctica. *Science*, **341** (6143), 266–70, doi:10.1126/science.1235798.

Rignot, E., M. Koppes and I. Velicogna, 2010: Rapid submarine melting of the calving faces of West Greenland glaciers. *Nature Geoscience*, **3** (3), 187–191, doi:10.1038/Ngeo765.

Rignot, E. et al., 2014: Widespread, rapid grounding line retreat of Pine Island, Thwaites, Smith, and Kohler glaciers, West Antarctica, from 1992 to 2011. *Geophysical Research Letters*, **41** (10), 3502–3509, doi:10.1002/2014gl060140.

Rignot, E. et al., 2019: Four decades of Antarctic Ice Sheet mass balance from 1979–2017. *Proc Natl Acad Sci U S A*, **116** (4), 1095–1103, doi:10.1073/pnas.1812883116.

Riley, W.J. et al., 2011: Barriers to predicting changes in global terrestrial methane fluxes: analyses using CLM4Me, a methane biogeochemistry model integrated in CESM. *Biogeosciences*, **8** (7), 1925–1953, doi:10.5194/bg-8-1925-2011.

Rintoul, S.R. et al., 2016: Ocean heat drives rapid basal melt of the Totten Ice Shelf. *Science Advances*, **2** (12), doi:10.1126/sciadv.1601610.

Ristroph, E.B., 2017: Presenting a Picture of Alaska Native Village Adaptation: A Method of Analysis. *Sociology and Anthropology*, **5** (9), 762–775, doi:10.13189/sa.2017.050908.

Ritter, R. et al., 2017: Observation-Based Trends of the Southern Ocean Carbon Sink. *Geophysical Research Letters*, **44** (24), 12,339–12,348, doi:papers2://publication/doi/10.1175/JTECH-D-13-00137.1.

Robards, M.D. et al., 2018: Understanding and adapting to observed changes in the Alaskan Arctic: Actionable knowledge co-production with Alaska Native communities. *Deep Sea Research Part II: Topical Studies in Oceanography*, doi:10.1016/j.dsr2.2018.02.008.

Robbins, L.L. et al., 2013: Baseline Monitoring of the Western Arctic Ocean Estimates 20% of Canadian Basin Surface Waters Are Undersaturated with Respect to Aragonite. *Plos One*, **8** (9), 15, doi:10.1371/journal.pone.0073796.

Roberts, J. et al., 2018: Ocean forced variability of Totten Glacier mass loss. *Geological Society, London, Special Publications*, **461** (1), 175–186, doi:10.1144/sp461.6.

Robinson, H., W. Gardiner, R.J. Wenning and M.A. Rempel-Hester, 2017: Spill Impact Mitigation Assessment Framework for Oil Spill Response Planning in the Arctic Environment. *International Oil Spill Conference Proceedings*, **2017** (1), 1325–1344, doi:10.7901/2169-3358-2017.1.1325.

Rocha, J.C., G. Peterson, O. Bodin and S. Levin, 2018: Cascading regime shifts within and across scales. *Science*, **362** (6421), 1379–1383, doi:10.1126/science.aat7850.

Roemmich, D. et al., 2015: Unabated planetary warming and its ocean structure since 2006. *Nature Climate Change*, **5** (3), 240, doi:10.1038/nclimate2513.

Romanovsky, V. et al., 2017: Changing permafrost and its impacts. In: Snow, Water, Ice and Permafrost in the Arctic (SWIPA). Arctic Monitoring and Assessment Programme (AMAP), Oslo, Norway, 65–102.

Romanovsky, V.E., S.L. Smith and H.H. Christiansen, 2010: Permafrost thermal state in the polar Northern Hemisphere during the international polar year 2007–2009: a synthesis. *Permafrost and Periglacial Processes*, **21** (2), 106–116, doi:10.1002/ppp.689.

Romero Manrique, D., S. Corral and Â. Guimarães Pereira, 2018: Climate-related displacements of coastal communities in the Arctic: Engaging traditional knowledge in adaptation strategies and policies. *Environmental Science & Policy*, **85**, 90–100, doi:10.1016/j.envsci.2018.04.007.

Rosales, J. and L.J. Chapman, 2015: Perceptions of Obvious and Disruptive Climate Change: Community-Based Risk Assessment for Two Native Villages in Alaska. *Climate*, **3** (4), 812–832, doi:10.3390/cli3040812.

Rösel, A. et al., 2018: Thin Sea Ice, Thick Snow, and Widespread Negative Freeboard Observed During N-ICE2015 North of Svalbard. *Journal of Geophysical Research: Oceans*, **123** (2), 1156–1176, doi:10.1002/2017jc012865.

Rosenblum, E. and I. Eisenman, 2017: Sea Ice Trends in Climate Models Only Accurate in Runs with Biased Global Warming. *Journal of Climate*, **30** (16), 6265–6278, doi:10.1175/jcli-d-16-0455.1.

Rosol, R., S. Powell-Hellyer and L.H.M. Chan, 2016: Impacts of decline harvest of country food on nutrient intake among Inuit in Arctic Canada: Impact of climate change and possible adaptation plan. *International Journal of Circumpolar Health*, **75**, 31127, doi:10.3402/ijch.v75.31127.

Rott, H. et al., 2018: Changing pattern of ice flow and mass balance for glaciers discharging into the Larsen A and B embayments, Antarctic Peninsula, 2011 to 2016. *The Cryosphere*, **12** (4), 1273–1291, doi:10.5194/tc-12-1273-2018.

Routson, C.C. et al., 2019: Mid-latitude net precipitation decreased with Arctic warming during the Holocene. *Nature*, **568** (7750), 83–87, doi:10.1038/s41586-019-1060-3.

Rover, J., L. Ji, B.K. Wylie and L.L. Tieszen, 2012: Establishing water body areal extent trends in interior Alaska from multi-temporal Landsat data. *Remote Sensing Letters*, **3** (7), 595–604, doi:10.1080/01431161.2011.643507.

Rowland, J.C. et al., 2011: Arctic Landscapes in Transition: Responses to Thawing Permafrost. *Eos, Transactions American Geophysical Union*, **91** (26), 229–230, doi:10.1029/2010eo260001.

Rupp, T.S. et al., 2016: Climate Scenarios, Land Cover, and Wildland Fire. In: *Baseline and Projected Future Carbon Storage and Greenhouse-Gas Fluxes in Ecosystems of Alaska* [Zhu, Z. and A.D. McGuire (eds.)], USGS Professional Paper 1826, 196.

Ruppel, C.D., B.M. Herman, L.L. Brothers and P.E. Hart, 2016: Subsea ice-bearing permafrost on the U.S. Beaufort Margin: 2. Borehole constraints. *Geochemistry, Geophysics, Geosystems*, **17** (11), 4333–4353, doi:10.1002/2016GC006582.

Ruppel, C.D. and J.D. Kessler, 2016: The interaction of climate change and methane hydrates. *Reviews of Geophysics*, **55** (1), 126–168, doi:10.1002/2016RG000534.

Ruscio, B.A. et al., 2015: One Health – a strategy for resilience in a changing arctic. *International Journal of Circumpolar Health*, **74** (1), 27913, doi:10.3402/ijch.v74.27913.

Russell, D., 2014a: *Energy-protein modeling of North Baffin Island caribou in relation to the Mary River Project: a reassessment from Russell (2012)*. Prepared for EDI Environmental Dynamics Inc., Whitehorse YT and Baffinland Iron Mines Corporation, Oakville Ontario.

Russell, D., 2014b: *Kiggavik Project Effects: Energy-Protein and Population Modeling of the Qamanirjuaq Caribou Herd*. Prepared for EDI Environmental Dynamics Inc., Whitehorse YT and AREVA Resources Canada.

Russell, J.L. et al., 2018: Metrics for the Evaluation of the Southern Ocean in Coupled Climate Models and Earth System Models. *Journal of Geophysical Research: Oceans*, **123** (5), 3120–3143, doi:10.1002/2017JC013461.

Rutzen, I. and R.R. Hopcroft, 2018: Abundance, biomass and community structure of epipelagic zooplankton in the Canada Basin. *Journal of Plankton Research*, **40** (4), 486–499, doi:10.1093/plankt/fby028.

Ryan, J.C. et al., 2018: Dark zone of the Greenland Ice Sheet controlled by distributed biologically-active impurities. *Nature Communications*, **9** (1), 1065, doi:10.1038/s41467-018-03353-2.

Ryan, J.C. et al., 2019: Greenland Ice Sheet surface melt amplified by snowline migration and bare ice exposure. *Sci Adv*, **5** (3), eaav3738, doi:10.1126/sciadv.aav3738.

Ryer, C.H. et al., 2016: Temperature-Dependent Growth of Early Juvenile Southern Tanner Crab Chionoecetes bairdi: Implications for Cold Pool Effects and Climate Change in the Southeastern Bering Sea. *Journal of Shellfish Research*, **35** (1), 259–267, doi:10.2983/035.035.0128.

Rysgaard, S. et al., 2013: Ikaite crystal distribution in winter sea ice and implications for CO2 system dynamics. *Cryosphere*, **7** (2), 707–718, doi:10.5194/tc-7-707-2013.

Safronov, V.M., 2016: Climate change and mammals of Yakutia. *Biology Bulletin*, **43** (9), 1256–1270, doi:10.1134/S1062359016110121.

Sahade, R. et al., 2015: Climate change and glacier retreat drive shifts in an Antarctic benthic ecosystem. *Science Advances*, **1**, doi:10.1126/sciadv.1500050

Sakhuja, V., 2014: The Polar Code and Arctic Navigation. *Strategic Analysis*, **38** (6), 803–811, doi:10.1080/09700161.2014.952943.

Sallée, J.-B. et al., 2013a: Assessment of Southern Ocean mixed-layer depths in CMIP5 models: Historical bias and forcing response. *Journal of Geophysical Research: Oceans*, **118** (4), 1845–1862, doi:10.1002/jgrc.20157.

Sallée, J.-B. et al., 2013b: Assessment of Southern Ocean water mass circulation and characteristics in CMIP5 models: Historical bias and forcing response. *Journal of Geophysical Research: Oceans*, **118** (4), 1830–1844, doi:10.1002/jgrc.20135.

Salmon, V.G. et al., 2016: Nitrogen availability increases in a tundra ecosystem during five years of experimental permafrost thaw. *Global Change Biology*, **22** (5), 1927–1941, doi:10.1111/gcb.13204.

Sand, M., T.K. Berntsen, Ø. Seland and J.E. Kristjánsson, 2013: Arctic surface temperature change to emissions of black carbon within Arctic or midlatitudes. *Journal of Geophysical Research: Atmospheres*, **118** (14), 7788–7798, doi:10.1002/jgrd.50613.

Sandberg Sørensen, L. et al., 2018: 25 years of elevation changes of the Greenland Ice Sheet from ERS, Envisat, and CryoSat-2 radar altimetry. *Earth and Planetary Science Letters*, **495**, 234–241, doi:10.1016/j.epsl.2018.05.015.

Sanderson, B.M. et al., 2017: Community climate simulations to assess avoided impacts in 1.5 and 2°C futures. *Earth Syst. Dynam.*, **8** (3), 827–847, doi:10.5194/esd-8-827-2017.

Santora, J.A. et al., 2017: Impacts of ocean climate variability on biodiversity of pelagic forage species in an upwelling ecosystem. *Marine Ecology Progress Series*, **580**, 205–220, doi:10.3354/meps12278.

Sarkki, S. and N. Acosta García, 2019: Merging social equity and conservation goals in IPBES. *Conservation Biology*, **33** (5), 1214–1218, doi:10.1111/cobi.13297.

Sasse, T.P., B.I. McNeil, R.J. Matear and A. Lenton, 2015: Quantifying the influence of CO2 seasonality on future aragonite undersaturation onset. *Biogeosciences*, **12** (20), 6017–6031, doi:10.5194/bg-12-6017-2015.

Saunois, M. et al., 2016: The global methane budget 2000–2012. *Earth System Science Data*, **8** (2), 697–751, doi:10.5194/essd-8-697-2016.

Schädel, C. et al., 2014: Circumpolar assessment of permafrost C quality and its vulnerability over time using long-term incubation data. *Global Change Biology*, **20** (2), 641–652, doi:10.1111/gcb.12417.

Schaefer, J.M. et al., 2016: Greenland was nearly ice-free for extended periods during the Pleistocene. *Nature*, **540** (7632), 252–255, doi:10.1038/nature20146.

Schaefer, K. et al., 2014: The impact of the permafrost carbon feedback on global climate. *Environmental Research Letters*, **9** (8), 085003, doi:10.1088/1748-9326/9/8/085003.

Schaefer, K., T. Zhang, L. Bruhwiler and A.P. Barrett, 2011: Amount and timing of permafrost carbon release in response to climate warming. *Tellus B: Chemical and Physical Meteorology*, **63** (2), 165–180, doi:10.1111/j.1600-0889.2011.00527.x.

Schaphoff, S. et al., 2013: Contribution of permafrost soils to the global carbon budget. *Environmental Research Letters*, **8** (1), 014026, doi:10.1088/1748-9326/8/1/014026.

Schartmüller, B., A.-S. Milaković, M. Bergström and S. Ehlers, 2015: A Simulation-Based Decision Support Tool for Arctic Transit Transport. (56567), V008T07A006, doi:10.1115/OMAE2015-41375.

Schirrmeister, L. et al., 2011: Fossil organic matter characteristics in permafrost deposits of the northeast Siberian Arctic. *Journal of Geophysical Research: Biogeosciences*, **116** (G2), G00M02, doi:10.1029/2011jg001647.

Schmidt, K. and A. Atkinson, 2016: Feeding and Food Processing in Antarctic Krill (*Euphausia superba* Dana). In: Biology and Ecology of Antarctic Krill [Siegel, V. (ed.)]. Springer, New York, 175–224.

Schmidt, N.M. et al., 2017: Interaction webs in arctic ecosystems: Determinants of arctic change? *AMBIO*, **46** (1), 12–25, doi:10.1007/s13280-016-0862-x.

Schmidt, N.M., S.H. Pedersen, J.B. Mosbacher and L.H. Hansen, 2015: Long-term patterns of muskox (Ovibos moschatus) demographics in high arctic Greenland. *Polar Biology*, **38** (10), 1667–1675, doi:10.1007/s00300-015-1733-9.

Schmidtko, S., K.J. Heywood, A.F. Thompson and S. Aoki, 2014: Multidecadal warming of Antarctic waters. *Science*, **346** (6214), 1227–1231.

Schneider, D.P., C. Deser and T. Fan, 2015: Comparing the Impacts of Tropical SST Variability and Polar Stratospheric Ozone Loss on the Southern Ocean Westerly Winds. *Journal of Climate*, **28** (23), 9350–9372, doi:10.1175/jcli-d-15-0090.1.

Schneider, D.P. and D.B. Reusch, 2015: Antarctic and Southern Ocean Surface Temperatures in CMIP5 Models in the Context of the Surface Energy Budget. *Journal of Climate*, **29** (5), 1689–1716, doi:10.1175/JCLI-D-15-0429.1.

Schneider von Deimling, T. et al., 2015: Observation-based modelling of permafrost carbon fluxes with accounting for deep carbon deposits and thermokarst activity. *Biogeosciences*, **12** (11), 3469–3488, doi:10.5194/bg-12-3469-2015.

Schofield, O. et al., 2018: Changes in the upper ocean mixed layer and phytoplankton productivity along the West Antarctic Peninsula. *Philos Trans A Math Phys Eng Sci*, **376** (2122), doi:10.1098/rsta.2017.0173.

Schofield, O. et al., 2017: Decadal variability in coastal phytoplankton community composition in a changing West Antarctic Peninsula. *Deep*

Sea Research Part I: Oceanographic Research Papers, **124**, 42–54, doi:10.1016/j.dsr.2017.04.014.

Schrama, E.J.O., B. Wouters and R. Rietbroek, 2014: A mascon approach to assess ice sheet and glacier mass balances and their uncertainties from GRACE data. *Journal of Geophysical Research: Solid Earth*, **119** (7), 6048–6066, doi:10.1002/2013jb010923.

Schroeter, S., W. Hobbs and N.L. Bindoff, 2017: Interactions between Antarctic sea ice and large-scale atmospheric modes in CMIP5 models. *Cryosphere*, **11** (2), 789–803, doi:10.5194/tc-11-789-2017.

Schulze, L.M. and R.S. Pickart, 2012: Seasonal variation of upwelling in the Alaskan Beaufort Sea: Impact of sea ice cover. *Journal of Geophysical Research: Oceans*, **117** (C6), C06022, doi:10.1029/2012JC007985.

Schuster, P.F. et al., 2018: Permafrost Stores a Globally Significant Amount of Mercury. *Geophysical Research Letters*, **45** (3), 1463–1471, doi:10.1002/2017GL075571.

Schuur, E.A.G. et al., 2013: Expert assessment of vulnerability of permafrost carbon to climate change. *Climatic Change*, **119** (2), 359–374, doi:10.1007/s10584-013-0730-7.

Schuur, E.A.G. et al., 2018: Chapter 11: Arctic and boreal carbon. In: Second State of the Carbon Cycle Report (SOCCR2): A Sustained Assessment Report [Cavallaro, N., G. Shrestha, R. Birdsey, M.A. Mayes, R.G. Najjar, S.C. Reed, P. Romero-Lankao and Z. Zhu (eds.)]. U.S. Global Change Research Program, Washington, DC, USA, 428–468.

Schuur, E.A.G. et al., 2015: Climate change and the permafrost carbon feedback. *Nature*, **520** (7546), 171–179, doi:10.1038/nature14338.

Schweiger, A.J., K.R. Wood and J. Zhang, 2019: Arctic sea ice volume variability over 1901–2010: A model-based reconstruction. *Journal of Climate*, **32** (15), 4731–4752, doi:10.1175/jcli-d-19-0008.1.

Schweitzer, P., P. Sköld and O. Ulturgasheva, 2014: Cultures and Identities. In: Arctic Human Development Report: Regional Processes and Global Linkages [Nymand Larsen, J. and G. Fondhal (eds.)]. Nordic Council of Ministers, Copenhagen, 105–150.

Screen, J.A., T.J. Bracegirdle and I. Simmonds, 2018: Polar Climate Change as Manifest in Atmospheric Circulation. *Curr Clim Change Rep*, **4** (4), 383–395, doi:10.1007/s40641-018-0111-4.

Screen, J.A. and C. Deser, 2019: Pacific Ocean Variability Influences the Time of Emergence of a Seasonally Ice-Free Arctic Ocean. *Geophysical Research Letters*, **46** (4), 2222–2231, doi:10.1029/2018gl081393.

Screen, J.A. and I. Simmonds, 2010: The central role of diminishing sea ice in recent Arctic temperature amplification. *Nature*, **464** (7293), 1334–7, doi:10.1038/nature09051.

Sedlar, J., 2018: Spring Arctic Atmospheric Preconditioning: Do Not Rule Out Shortwave Radiation Just Yet. *Journal of Climate*, **31** (11), 4225–4240, doi:10.1175/jcli-d-17-0710.1.

Seehaus, T. et al., 2015: Changes in ice dynamics, elevation and mass discharge of Dinsmoor–Bombardier–Edgeworth glacier system, Antarctic Peninsula. *Earth and Planetary Science Letters*, **427**, 125–135, doi:10.1016/j.epsl.2015.06.047.

Sejr, M.K. et al., 2017: Evidence of local and regional freshening of Northeast Greenland coastal waters. *Sci Rep*, **7** (1), 13183, doi:10.1038/s41598-017-10610-9.

Semiletov, I. et al., 2016: Acidification of East Siberian Arctic Shelf waters through addition of freshwater and terrestrial carbon (vol 9, pg 361, 2016). *Nature Geoscience*, **9** (9), 1, doi:10.1038/ngeo2799.

Serreze, M.C., A.P. Barrett and J. Stroeve, 2012: Recent changes in tropospheric water vapor over the Arctic as assessed from radiosondes and atmospheric reanalyses. *Journal of Geophysical Research: Atmospheres*, **117** (D10), n/a-n/a, doi:10.1029/2011jd017421.

Serreze, M.C. et al., 2009: The emergence of surface-based Arctic amplification. *The Cryosphere*, **3** (1), 11–19, doi:10.5194/tc-3-11-2009.

Serreze, M.C. and R.G. Barry, 2011: Processes and impacts of Arctic amplification: A research synthesis. *Global and Planetary Change*, **77** (1–2), 85–96, doi:10.1016/j.gloplacha.2011.03.004.

Serreze, M.C., J. Stroeve, A.P. Barrett and L.N. Boisvert, 2016: Summer atmospheric circulation anomalies over the Arctic Ocean and their influences on September sea ice extent: A cautionary tale. *Journal of Geophysical Research: Atmospheres*, **121** (19), 11,463–11,485, doi:10.1002/2016jd025161.

Sevestre, H. et al., 2018: Tidewater Glacier Surges Initiated at the Terminus. *Journal of Geophysical Research: Earth Surface*, **123** (5), 1035–1051, doi:10.1029/2017JF004358.

Seyboth, E. et al., 2016: Southern Right Whale (Eubalaena australis) Reproductive Success is Influenced by Krill (Euphausia superba) Density and Climate. *Sci Rep*, **6**, 28205, doi:10.1038/srep28205.

Sformo, T.L. et al., 2017: Observations and first reports of saprolegniosis in Aanaakłiq, broad whitefish (Coregonus nasus), from the Colville River near Nuiqsut, Alaska. *Polar Science*, **14**, 78–82, doi:10.1016/j.polar.2017.07.002.

Shadian, J.M., 2014: *The Politics of Arctic Sovereignty: oil, ice and Inuit goverance*. Routledge Press, London.

Shadwick, E.H. et al., 2013: Glacier tongue calving reduced dense water formation and enhanced carbon uptake. *Geophysical Research Letters*, **40** (5), 904–909, doi:10.1002/grl.50178.

Shadwick, E.H., B. Tilbrook and K.I. Currie, 2017: Late-summer biogeochemistry in the Mertz Polynya: East Antarctica. *Journal of Geophysical Research: Oceans*, **122** (9), 7380–7394, doi:10.1002/2017jc013015.

Shakhova, N. et al., 2013: Ebullition and storm-induced methane release from the East Siberian Arctic Shelf. *Nature Geoscience*, **7**, 64, doi:10.1038/ngeo2007.

Shapovalova, D., 2016: The Effectiveness of the Regulatory Regime for Black Carbon Mitigation in the Arctic. *Arctic Review on Law and Politics*, **7** (2), 136–151, doi:10.17585/arctic.v7.427.

Sharma, B.M. et al., 2015: Melting Himalayan glaciers contaminated by legacy atmospheric depositions are important sources of PCBs and high-molecular-weight PAHs for the Ganges floodplain during dry periods. *Environmental Pollution*, **206** (Supplement C), 588–596, doi:10.1016/j.envpol.2015.08.012.

Sharma, S. et al., 2019: Widespread loss of lake ice around the Northern Hemisphere in a warming world. *Nature Climate Change*, **9** (3), 227–231, doi:10.1038/s41558-018-0393-5.

Sharonov, D.S., N.A. Bryksina, V.Y. Polishuk and Y.M. Polishuk, 2012: Comparative analysis of thermokarst dynamics in permafrost territory of Western Siberia and Gorny Altai on the basis of space images. *Current problems in remote sensing of the earth from space*, **9** (1), 313–319.

Shepherd, A. et al., 2012: A reconciled estimate of ice-sheet mass balance. *Science*, **338** (6111), 1183–9, doi:10.1126/science.1228102.

Sheppard, S.R.J. et al., 2011: Future visioning of local climate change: A framework for community engagement and planning with scenarios and visualisation. *Futures*, **43** (4), 400–412, doi:10.1016/j.futures.2011.01.009.

Shi, J.-R., S.-P. Xie and L.D. Talley, 2018: Evolving Relative Importance of the Southern Ocean and North Atlantic in Anthropogenic Ocean Heat Uptake. *Journal of Climate*, **31** (18), 7459–7479, doi:10.1175/jcli-d-18-0170.1.

Shibata, A., 2015: Japan and 100 Years of Antarctic Legal Order: Any Lessons for the Arctic? *The Yearbook of Polar Law Online*, **7** (1), 1–54, doi:10.1163/2211-6427_002.

Shiklomanov, A.I. and R.B. Lammers, 2014: River ice responses to a warming Arctic — recent evidence from Russian rivers. *Environmental Research Letters*, **9** (3), 035008, doi:10.1088/1748-9326/9/3/035008.

Shiklomanov, N.I., D.A. Streletskiy, V.I. Grebenets and L. Suter, 2017a: Conquering the permafrost: urban infrastructure development in Norilsk, Russia. *Polar Geography*, **40** (4), 273–290, doi:10.1080/1088937X.2017.1329237.

Shiklomanov, N.I., D.A. Streletskiy and F.E. Nelson, 2012: Northern Hemisphere component of the global Circumpolar Active Layer Monitoring (CALM) program. In: *10th International Conference on Permafrost*, Salekhard, Russia, 377–382.

Shiklomanov, N.I., D.A. Streletskiy, T.B. Swales and V.A. Kokorev, 2017b: Climate Change and Stability of Urban Infrastructure in Russian Permafrost Regions: Prognostic Assessment based on GCM Climate Projections. *Geographical Review*, **107** (1), 125–142, doi:10.1111/gere.12214.

Shindell, D.T., 2004: Southern Hemisphere climate response to ozone changes and greenhouse gas increases. *Geophysical Research Letters*, **31** (18), doi:10.1029/2004gl020724.

Shu, Q., Z. Song and F. Qiao, 2015: Assessment of sea ice simulations in the CMIP5 models. *Cryosphere*, **9** (1), 399–409, doi:10.5194/tc-9-399-2015.

Shuman, J.K. et al., 2015: Forest forecasting with vegetation models across Russia. *Canadian Journal of Forest Research*, **45** (2), 175–184, doi:10.1139/cjfr-2014-0138.

Siegel, V., Eds., 2016: Biology and Ecology of Antarctic Krill. Springer, 1–458 pp.

Siegert, M.J., 2017: A 60-year international hystory of Antarctic subglacial lake exploration. In: Exploration of subsurface Antarctica: Uncovering past changes and modern processes [Siegert, M.J., S.S.R. Jamieson and D.A. White (eds.)]. Geological Society, London, **461**, 7–21.

Siegert, M.J. et al., 2017: Antarctic glacial groundwater: a concept paper on its measurement and potential influence on ice flow. In: Exploration of subsurface Antarctica: uncovering past changes and modern processes [Siegert, M.J., S.S.R. Jamieson and D.A. White (eds.)]. The Geological Society, London, **461**, 197–213.

Siegert, M.J., A.M. Le Brocq and A.J. Payne, 2007: Hydrological connections between Antarctic subglacial lakes, the flow of water beneath the East Antarctic ice sheet and implications for sedimentary processes. In: Glacial sedimentary processes and products [Hambrey, M.J., P. Christoffersen, N.F. Glasser and B. Hubbard (eds.)]. International Association of Sedimentologists, 3–22.

Siegert, M.J., N. Ross and A.M. Le Brocq, 2016: Recent advances in understanding Antarctic subglacial lakes and hydrology. *Phil. Trans. R. Soc. A*, **374** (20140306), doi:10.1098/rsta.2014.0306.

Sigler, M.F. et al., 2017: Late summer zoogeography of the northern Bering and Chukchi seas. *Deep Sea Research Part II: Topical Studies in Oceanography*, **135** (Supplement C), 168–189, doi:10.1016/j.dsr2.2016.03.005.

Sigmond, M., J.C. Fyfe and N.C. Swart, 2018: Ice-free Arctic projections under the Paris Agreement. *Nature Climate Change*, **8** (5), 404–408, doi:10.1038/s41558-018-0124-y.

Sillmann, J. et al., 2013: Climate extremes indices in the CMIP5 multimodel ensemble: Part 2. Future climate projections. *Journal of Geophysical Research: Atmospheres*, **118** (6), 2473–2493, doi:10.1002/jgrd.50188.

Silvano, A. et al., 2018: Freshening by glacial meltwater enhances melting of ice shelves and reduces formation of Antarctic Bottom Water. *Sci Adv*, **4** (4), eaap9467, doi:10.1126/sciadv.aap9467.

Singh, D., M.G. Flanner and J. Perket, 2015: The global land shortwave cryosphere radiative effect during the MODIS era. *The Cryosphere*, **9** (6), 2057–2070, doi:10.5194/tc-9-2057-2015.

Skific, N. and J.A. Francis, 2013: Drivers of projected change in arctic moist static energy transport. *Journal of Geophysical Research: Atmospheres*, **118** (7), 2748–2761, doi:10.1002/jgrd.50292.

Skiles, S.M. et al., 2018: Radiative forcing by light-absorbing particles in snow. *Nature Climate Change*, **8** (11), 964–971, doi:10.1038/s41558-018-0296-5.

Skliris, N. et al., 2014: Salinity changes in the World Ocean since 1950 in relation to changing surface freshwater fluxes. *Climate Dynamics*, **43** (3–4), 709–736, doi:10.1007/s00382-014-2131-7.

Skogen, M.D. et al., 2014: Modelling ocean acidification in the Nordic and Barents Seas in present and future climate. *Journal of Marine Systems*, **131**, 10–20, doi:10.1016/j.jmarsys.2013.10.005.

Slater, A.G. and D.M. Lawrence, 2013: Diagnosing Present and Future Permafrost from Climate Models. *Journal of Climate*, **26** (15), 5608–5623, doi:10.1175/jcli-d-12-00341.1.

Slater, D.A. et al., 2015: Effect of near-terminus subglacial hydrology on tidewater glacier submarine melt rates. *Geophysical Research Letters*, **42** (8), 2861–2868, doi:10.1002/2014GL062494.

Sledd, A. and T. L'Ecuyer, 2019: How much do clouds mask the impacts of Arctic sea ice and snow cover variations? Different perspectives from observations and reanalyses. *Atmosphere*, **10** (1), 12, doi:10.3390/atmos10010012.

Small, D. et al., 2019: Antarctic ice sheet palaeo-thinning rates from vertical transects of cosmogenic exposure ages. *Quaternary Science Reviews*, **206**, 65–80, doi:10.1016/j.quascirev.2018.12.024.

Smedsrud, L.H., 2005: Warming of the deep water in the Weddell Sea along the Greenwich meridian: 1977–2001. *Deep-Sea Research Part I-Oceanographic Research Papers*, **52** (2), 241–258, doi:10.1016/j.dsr.2004.10.004.

Smedsrud, L.H. et al., 2013: The Role of the Barents Sea in the Arctic Climate System. *Reviews of Geophysics*, **51** (3), 415–449, doi:10.1002/rog.20017.

Smedsrud, L.H. et al., 2017: Fram Strait sea ice export variability and September Arctic sea ice extent over the last 80 years. *The Cryosphere*, **11** (1), 65–79, doi:10.5194/tc-11-65-2017.

Šmejkalová, T., M.E. Edwards and J. Dash, 2016: Arctic lakes show strong decadal trend in earlier spring ice-out. *Scientific Reports*, **6**, 38449, doi:10.1038/srep38449.

Smith, A.J. et al., 2017a: Beluga whale summer habitat associations in the Nelson River estuary, western Hudson Bay, Canada. *Plos One*, **12** (8), e0181045, doi:10.1371/journal.pone.0181045.

Smith, D.M. et al., 2017b: Atmospheric Response to Arctic and Antarctic Sea Ice: The Importance of Ocean–Atmosphere Coupling and the Background State. *Journal of Climate*, **30** (12), 4547–4565, doi:10.1175/JCLI-D-16-0564.1.

Smith, J.A. et al., 2017c: Sub-ice-shelf sediments record history of twentieth-century retreat of Pine Island Glacier. *Nature*, **541**, 77, doi:10.1038/nature20136.

Smith, K.E. et al., 2017d: Climate change and the threat of novel marine predators in Antarctica. *Ecosphere*, **8** (11), 1–3, doi:10.1002/ecs2.2017.

Smith, K.L. and L.M. Polvani, 2017: Spatial patterns of recent Antarctic surface temperature trends and the importance of natural variability: lessons from multiple reconstructions and the CMIP5 models. *Climate Dynamics*, **48** (7–8), 2653–2670, doi:10.1007/s00382-016-3230-4.

Smith, L.C. et al., 2007: Rising minimum daily flows in northern Eurasian rivers: A growing influence of groundwater in the high-latitude hydrologic cycle. *Journal of Geophysical Research*, **112** (G04S47), doi:doi:10.1029/2006JG000327.

Smith, L.C., Y. Sheng, G.M. MacDonald and L.D. Hinzman, 2005: Disappearing Arctic Lakes. *Science*, **308** (5727), 1429, doi:10.1126/science.1108142.

Smith, M.D., A.K. Knapp and S.L. Collins, 2009: A framework for assessing ecosystem dynamics in response to chronic resource alterations induced by global change. *Ecology*, **90** (12), 3279–3289, doi:10.1890/08-1815.1.

Smith, R.W. et al., 2015: High rates of organic carbon burial in fjord sediments globally. *Nature Geosci*, **8** (6), 450–453, doi:10.1038/ngeo2421.

Smits, C.C.A., J. van Leeuwen and J.P.M. van Tatenhove, 2017: Oil and gas development in Greenland: A social license to operate, trust and legitimacy in environmental governance. *Resources Policy*, **53**, 109–116, doi:10.1016/j.resourpol.2017.06.004.

Smoot, C.A. and R.R. Hopcroft, 2017: Depth-stratified community structure of Beaufort Sea slope zooplankton and its relations to water masses. *Journal of Plankton Research*, **39** (1), 79–91, doi:10.1093/plankt/fbw087.

Sole, A. et al., 2013: Winter motion mediates dynamic response of the Greenland Ice Sheet to warmer summers. *Geophysical Research Letters*, **40** (15), 3940–3944, doi:10.1002/grl.50764.

Solomina, O.N. et al., 2015: Holocene glacier fluctuations. *Quaternary Science Reviews*, **111**, 9–34, doi:10.1016/j.quascirev.2014.11.018.

Solovyev, B. et al., 2017: Identifying a network of priority areas for conservation in the Arctic seas: Practical lessons from Russia. *Aquatic Conservation: Marine and Freshwater Ecosystems*, **27** (S1), 30–51, doi:10.1002/aqc.2806.

Song, H.J. et al., 2016: In-situ measured primary productivity of ice algae in Arctic sea ice floes using a new incubation method. *Ocean Science Journal*, **51** (3), 387–396, doi:10.1007/s12601-016-0035-7.

3

Soulen, B.K., K. Cammen, T.F. Schultz and D.W. Johnston, 2013: Factors Affecting Harp Seal (Pagophilus groenlandicus) Strandings in the Northwest Atlantic. *Plos One*, **8** (7), e68779, doi:10.1371/journal.pone.0068779.

Southcott, C. and D. Natcher, 2018: Extractive industries and Indigenous subsistence economies: a complex and unresolved relationship. *Canadian Journal of Development Studies / Revue canadienne d'études du développement*, **39** (1), 137–154, doi:10.1080/02255189.2017.1400955.

Southwell, C. et al., 2012: A review of data on abundance, trends in abundance, habitat utilisation and diet for Southern Ocean ice-breeding seals. *CCAMLR Science*, **19**, 1–49.

Southwell, C. et al., 2015: Spatially Extensive Standardized Surveys Reveal Widespread, Multi-Decadal Increase in East Antarctic Adelie Penguin Populations. *Plos One*, **10** (10), 18, doi:10.1371/journal.pone.0139877.

Spence, P. et al., 2014: Rapid subsurface warming and circulation changes of Antarctic coastal waters by poleward shifting winds. *Geophysical Research Letters*, **41** (13), 4601–4610, doi:10.1002/2014gl060613.

Spreen, G., R. Kwok and D. Menemenlis, 2011: Trends in Arctic sea ice drift and role of wind forcing: 1992–2009. *Geophysical Research Letters*, **38** (19), L19501, doi:10.1029/2011gl048970.

St-Laurent, P., J.M. Klinck and M.S. Dinniman, 2015: Impact of local winter cooling on the melt of Pine Island Glacier, Antarctica. *Journal of Geophysical Research: Oceans*, **120** (10), 6718–6732, doi:doi:10.1002/2015JC010709.

St Pierre, K.A. et al., 2018: Unprecedented Increases in Total and Methyl Mercury Concentrations Downstream of Retrogressive Thaw Slumps in the Western Canadian Arctic. *Environ Sci Technol*, **52** (24), 14099–14109, doi:10.1021/acs.est.8b05348.

St. Jacques, J.-M. and D.J. Sauchyn, 2009: Increasing winter baseflow and mean annual streamflow from possible permafrost thawing in the Northwest Territories, Canada. *Geophysical Research Letters*, **36** (1), L01401, doi:10.1029/2008gl035822.

Stammerjohn, S. and T. Maksym, 2016: Gaining (and losing) Antarctic sea ice: variability, trends and mechanisms. In: Sea Ice. John Wiley & Sons, Ltd, 261–289.

Stammerjohn, S., R. Massom, D. Rind and D. Martinson, 2012: Regions of rapid sea ice change: An inter-hemispheric seasonal comparison. *Geophysical Research Letters*, **39** (6), L06501, doi:10.1029/2012GL050874.

Stanley, R.H.R., Z.O. Sandwith and W.J. Williams, 2015: Rates of summertime biological productivity in the Beaufort Gyre: A comparison between the low and record-low ice conditions of August 2011 and 2012. *Journal of Marine Systems*, **147** (Supplement C), 29–44, doi:10.1016/j.jmarsys.2014.04.006.

Stapleton, S., E. Peacock and D. Garshelis, 2016: Aerial surveys suggest long-term stability in the seasonally ice-free Foxe Basin (Nunavut) polar bear population. *Marine Mammal Science*, **32** (1), 181–201, doi:10.1111/mms.12251.

Stasko, A.D. et al., 2018: Benthic-pelagic trophic coupling in an Arctic marine food web along vertical water mass and organic matter gradients. *Marine Ecology Progress Series*, **594**, 1–19, doi:10.3354/meps12582.

Steele, M. and S. Dickinson, 2016: The phenology of Arctic Ocean surface warming. *J Geophys Res Oceans*, **121** (9), 6847–6861, doi:10.1002/2016JC012089.

Steger, C.R. et al., 2017: Firn Meltwater Retention on the Greenland Ice Sheet: A Model Comparison. *Frontiers in Earth Science*, **5**, doi:10.3389/feart.2017.00003.

Steig, E.J., Q. Ding, D.S. Battisti and A. Jenkins, 2017: Tropical forcing of Circumpolar Deep Water Inflow and outlet glacier thinning in the Amundsen Sea Embayment, West Antarctica. *Annals of Glaciology*, **53** (60), 19–28, doi:10.3189/2012AoG60A110.

Steig, E.J. et al., 2013: Recent climate and ice-sheet changes in West Antarctica compared with the past 2,000 years. *Nature Geoscience*, **6** (5), 372–375, doi:10.1038/ngeo1778.

Steinberg, D.K. et al., 2015: Long-term (1993–2013) changes in macrozooplankton off the Western Antarctic Peninsula. *Deep Sea Research Part I: Oceanographic Research Papers*, **101**, 54–70, doi:10.1016/j.dsr.2015.02.009.

Steiner, N.S. et al., 2014: Future ocean acidification in the Canada Basin and surrounding Arctic Ocean from CMIP5 earth system models. *Journal of Geophysical Research-Oceans*, **119** (1), 332–347, doi:10.1002/2013jc009069.

Stenni, B. et al., 2017: Antarctic climate variability on regional and continental scales over the last 2000 years. *Climate of the Past*, **13** (11), 1609–1634, doi:10.5194/cp-13-1609-2017.

Stenson, G.B. and M.O. Hammill, 2014: Can ice breeding seals adapt to habitat loss in a time of climate change? *ICES Journal of Marine Science*, **71** (7), 1977–1986, doi:10.1093/icesjms/fsu074.

Stephenson, S.R., L.C. Smith and J.A. Agnew, 2011: Divergent long-term trajectories of human access to the Arctic. *Nature Climate Change*, **1**, 156, doi:10.1038/nclimate1120.

Stephenson, S.R., L.C. Smith, L.W. Brigham and J.A. Agnew, 2013: Projected 21st-century changes to Arctic marine access. *Climatic Change*, **118** (3), 885–899, doi:10.1007/s10584-012-0685-0.

Stevens, L.A. et al., 2016: Greenland Ice Sheet flow response to runoff variability. *Geophysical Research Letters*, **43** (21), 11,295–11,303, doi:10.1002/2016GL070414.

Stevenson, D.E. and R.R. Lauth, 2018: Bottom trawl surveys in the northern Bering Sea indicate recent shifts in the distribution of marine species. *Polar Biology*, **42** (2), 407–421, doi:10.1007/s00300-018-2431-1.

Stewart, E., J. Dawson and M. Johnston, 2015: Risks and opportunities associated with change in the cruise tourism sector: community perspectives from Arctic Canada. *The Polar Journal*, **5** (2), 403–427, doi:10.1080/2154896x.2015.1082283.

Stewart, E.J. et al., 2013: Local-level responses to sea ice change and cruise tourism in Arctic Canada's Northwest Passage. *Polar Geography*, **36** (1–2), 142–162, doi:10.1080/1088937x.2012.705352.

Stiasny, M.H. et al., 2016: Ocean Acidification Effects on Atlantic Cod Larval Survival and Recruitment to the Fished Population. *Plos One*, **11** (8), e0155448, doi:10.1371/journal.pone.0155448.

Stibal, M. et al., 2017: Algae Drive Enhanced Darkening of Bare Ice on the Greenland Ice Sheet. *Geophysical Research Letters*, **44** (22), 11, 463–11,471, doi:10.1002/2017GL075958.

Stirling, I., 1997: The importance of polynyas, ice edges, and leads to marine mammals and birds. *Journal of Marine Systems*, **10** (1), 9–21, doi:10.1016/S0924-7963(96)00054-1.

Stokke, K.B. and J.V. Haukeland, 2017: Balancing tourism development and nature protection across national park borders – a case study of a coastal protected area in Norway. *Journal of Environmental Planning and Management*, **61** (12), 2151–2165, doi:10.1080/09640568.2017.1388772.

Stokke, O.S., 2009: Protecting the Arctic Environment: The Interplay of Global and Regional Regimes. *The Yearbook of Polar Law Online*, **1** (1), 349–369, doi:10.1163/22116427-91000018.

Stokke, O.S., 2013: Political Stability and Multi-level Governance in the Arctic. In: *Environmental Security in the Arctic Ocean*, Dordrecht, [Berkman, P. and A. Vylegzhanin (eds.)], Springer Netherlands, 297–311.

Storrie, L. et al., 2018: Determining the species assemblage and habitat use of cetaceans in the Svalbard Archipelago, based on observations from 2002 to 2014. *Polar Research*, **37** (1), 1463065, doi:10.1080/17518369.2018.1463065.

Straneo, F. and C. Cenedese, 2015: The Dynamics of Greenland's Glacial Fjords and Their Role in Climate. *Ann Rev Mar Sci*, **7**, 89–112, doi:10.1146/annurev-marine-010213-135133.

Straneo, F., G.S. Hamilton, L.A. Stearns and D.A. Sutherland, 2016: Connecting the Greenland Ice Sheet and the ocean: A case study of Helheim Glacier and Semilik Fjord. *Oceanography*, **29** (4), 34–45, doi:10.5670/oceanog.2016.97.

Straneo, F. et al., 2013: Challenges to understanding the dynamic response of Greenland's marine terminating glaciers to oc eanic and atmospheric forcing. *Bulletin of the American Meteorological Society*, **94** (8), 1131–1144, doi:10.1175/BAMS-D-12-00100.1.

Straneo, F. et al., 2017: Characteristics of ocean waters reaching Greenland's glaciers. *Annals of Glaciology*, **53** (60), 202–210, doi:10.3189/2012AoG60A059.

Strauss, J. et al., 2017: Deep Yedoma permafrost: A synthesis of depositional characteristics and carbon vulnerability. *Earth-Science Reviews*, **172**, 75–86, doi:10.1016/j.earscirev.2017.07.007.

Strauss, J. et al., 2013: The Deep Permafrost Carbon Pool of the Yedoma Region in Siberia and Alaska. *Geophysical Research Letters*, **40**, 6165–6170, doi:10.1002/2013gl058088.

Streletskiy, D., N. Shiklomanov and E. Hatleberg, 2012: Infrastructure and a Changing Climate in the Russian Arctic: A Geographic Impact Assessment. In: *Proceedings of the 10th International Conference on Permafrost*, June 25 – 29, 2012, Salekhard, Russia, **1**, 407–412.

Streletskiy, D.A. et al., 2017: Thaw Subsidence in Undisturbed Tundra Landscapes, Barrow, Alaska, 1962–2015. *Permafrost and Periglacial Processes*, **28** (3), 566–572, doi:10.1002/ppp.1918.

Striegl, R.G. et al., 2005: A decrease in discharge-normalized DOC export by the Yukon River during summer through autumn. *Geophysical Research Letters*, **32** (21), doi:10.1029/2005gl024413.

Stroeve, J., A. Barrett, M. Serreze and A. Schweiger, 2014a: Using records from submarine, aircraft and satellites to evaluate climate model simulations of Arctic sea ice thickness. *The Cryosphere*, **8** (5), 1839–1854, doi:10.5194/tc-8-1839-2014.

Stroeve, J. and D. Notz, 2015: Insights on past and future sea-ice evolution from combining observations and models. *Global and Planetary Change*, **135** (Supplement C), 119–132, doi:10.1016/j.gloplacha.2015.10.011.

Stroeve, J. and D. Notz, 2018: Changing state of Arctic sea ice across all seasons. *Environmental Research Letters*, **13** (10), 103001, doi:10.1088/1748-9326/aade56.

Stroeve, J.C. et al., 2012a: Trends in Arctic sea ice extent from CMIP5, CMIP3 and observations. *Geophysical Research Letters*, **39** (16), L16502, doi:10.1029/2012gl052676.

Stroeve, J.C. et al., 2014b: Changes in Arctic melt season and implications for sea ice loss. *Geophysical Research Letters*, **41** (4), 1216–1225, doi:10.1002/2013gl058951.

Stroeve, J.C. et al., 2012b: The Arctic's rapidly shrinking sea ice cover: a research synthesis. *Climatic Change*, **110** (3), 1005–1027, doi:10.1007/s10584-011-0101-1.

Strozzi, T. et al., 2017: Circum-Arctic Changes in the Flow of Glaciers and Ice Caps from Satellite SAR Data between the 1990s and 2017. *Remote Sensing*, **9** (9), 947, doi:10.3390/rs9090947.

Stuecker, M.F. et al., 2018: Polar amplification dominated by local forcing and feedbacks. *Nature Climate Change*, **8** (12), 1076–1081 doi:10.1038/s41558-018-0339-y.

Stuefer, S.L., C.D. Arp, D.L. Kane and A.K. Liljedahl, 2017: Recent Extreme Runoff Observations From Coastal Arctic Watersheds in Alaska. *Water Resources Research*, **53** (11), 9145–9163, doi:doi:10.1002/2017WR020567.

Sturm, M., M.A. Goldstein, H. Huntington and T.A. Douglas, 2017: Using an option pricing approach to evaluate strategic decisions in a rapidly changing climate: Black–Scholes and climate change. *Climatic Change*, **140** (3), 437–449, doi:10.1007/s10584-016-1860-5.

Sturm, M. and R.A. Massom, 2016: Snow in the sea ice system: friend or foe? In: Sea Ice [Thomas, D.N. (ed.)]. Wiley-Blackwell, 652.

Sturm, M. and S. Stuefer, 2013: Wind-blown flux rates derived from drifts at arctic snow fences. *Journal of Glaciology*, **59** (213), 21–34, doi:10.3189/2013JoG12J110.

Sundby, S., K.F. Drinkwater and O.S. Kjesbu, 2016: The North Atlantic Spring-Bloom System – Where the Changing Climate Meets the Winter Dark. *Frontiers in Marine Science*, **3**, doi:10.3389/fmars.2016.00028.

Suprenand, P.M. and C.H. Ainsworth, 2017: Trophodynamic effects of climate change-induced alterations to primary production along the western Antarctic Peninsula. *Marine Ecology Progress Series*, **569**, 37–54, doi:10.3354/meps12100.

Surdu, C.M., C.R. Duguay, L.C. Brown and D. Fernández Prieto, 2014: Response of ice cover on shallow lakes of the North Slope of Alaska to contemporary climate conditions (1950–2011): radar remote-sensing and numerical modeling data analysis. *The Cryosphere*, **8** (1), 167–180, doi:10.5194/tc-8-167-2014.

Surdu, C.M., C.R. Duguay and D. Fernández Prieto, 2016: Evidence of recent changes in the ice regime of lakes in the Canadian High Arctic from spaceborne satellite observations. *The Cryosphere*, **10** (3), 941–960, doi:10.5194/tc-10-941-2016.

Sutterley, T.C. et al., 2014: Mass loss of the Amundsen Sea Embayment of West Antarctica from four independent techniques. *Geophysical Research Letters*, **41** (23), 8421–8428, doi:10.1002/2014GL061940.

Suzuki, K. et al., 2018: Hydrological Variability and Changes in the Arctic Circumpolar Tundra and the Three Largest Pan-Arctic River Basins from 2002 to 2016. *Remote Sensing*, **10** (3), doi:10.3390/rs10030402.

Swart, N.C. and J.C. Fyfe, 2013: The influence of recent Antarctic ice sheet retreat on simulated sea ice area trends. *Geophysical Research Letters*, **40**, 4328–4332, doi:10.1002/grl.50820.

Swart, N.C., J.C. Fyfe, N. Gillett and G.J. Marshall, 2015a: Comparing Trends in the Southern Annular Mode and Surface Westerly Jet. *Journal of Climate*, **28** (22), 8840–8859, doi:10.1175/JCLI-D-14-00716.s1.

Swart, N.C. et al., 2015b: Influence of internal variability on Arctic sea-ice trends. *Nature Climate Change*, **5**, 86, doi:10.1038/nclimate2483.

Swart, N.C., J.C. Fyfe, O.A. Saenko and M. Eby, 2014: Wind-driven changes in the ocean carbon sink. *Biogeosciences*, **11** (21), 6107–6117, doi:10.5194/bg-11-6107-2014.

Swart, N.C., S.T. Gille, J.C. Fyfe and N.P. Gillett, 2018: Recent Southern Ocean warming and freshening driven by greenhouse gas emissions and ozone depletion. *Nature Geoscience*, **11** (11), 836–841, doi:10.1038/s41561-018-0226-1.

Sweeney, C. et al., 2016: No significant increase in long-term CH_4 emissions on North Slope of Alaska despite significant increase in air temperature. *Geophysical Research Letters*, **43** (12), 6604–6611, doi:10.1002/2016GL069292.

Swiney, K.M., W.C. Long and R.J. Foy, 2017: Decreased pH and increased temperatures affect young-of-the-year red king crab (Paralithodes camtschaticus). *ICES Journal of Marine Science*, **74** (4), 1191–1200, doi:10.1093/icesjms/fsw251.

Sydeman, W.J., E. Poloczanska, T.E. Reed and S.A. Thompson, 2015: Climate change and marine vertebrates. *Science*, **350** (6262), 772–777, doi:10.1126/science.aac9874.

Szpak, P., M. Buckley, C.M. Darwent and M.P. Richards, 2018: Long-term ecological changes in marine mammals driven by recent warming in northwestern Alaska. *Global Change Biology*, **24** (1), 490–503, doi:10.1111/gcb.13880.

Takakura, H., 2018: Local Perception of River Thaw and Spring Flooding of the Lena River. In: Global Warming and Human-Nature Dimension in Northern Eurasia [Hiyama, T. and H. Takakura (eds.)]. Springer, Singapore, 29–51.

Talley, L.D. et al., 2016: Changes in Ocean Heat, Carbon Content, and Ventilation: A Review of the First Decade of GO-SHIP Global Repeat Hydrography. *Ann Rev Mar Sci*, **8**, 185–215, doi:10.1146/annurev-marine-052915-100829.

Tamura, T., K.I. Ohshima, A.D. Fraser and G.D. Williams, 2016: Sea ice production variability in Antarctic coastal polynyas. *Journal of Geophysical Research: Oceans*, **121** (5), 2967–2979, doi:10.1002/2015jc011537.

Tan, Z. and Q. Zhuang, 2015: Arctic lakes are continuous methane sources to the atmosphere under warming conditions. *Environmental Research Letters*, **10** (5), 054016, doi:10.1088/1748-9326/10/5/054016.

Tandon, N.F. et al., 2018: Reassessing Sea Ice Drift and Its Relationship to Long-Term Arctic Sea Ice Loss in Coupled Climate Models. *Journal of Geophysical Research: Oceans*, **123** (6), 4338–4359, doi:10.1029/2017jc013697.

Tanhua, T. et al., 2017: Temporal changes in ventilation and the carbonate system in the Atlantic sector of the Southern Ocean. *Deep-Sea Research Part II*, **138**, 26–38, doi:10.1016/j.dsr2.2016.10.004.

Tank, S.E. et al., 2012: A land-to-ocean perspective on the magnitude, source and implication of DIC flux from major Arctic rivers to the Arctic Ocean. *Global Biogeochemical Cycles*, **26** (4), GB4018, doi:10.1029/2011gb004192.

Tape, K.D. et al., 2016: Range Expansion of Moose in Arctic Alaska Linked to Warming and Increased Shrub Habitat. *Plos One*, **11** (4), e0152636, doi:10.1371/journal.pone.0152636.

Tape, K.D. et al., 2018: Tundra be dammed: Beaver colonization of the Arctic. *Global Change Biology*, **24** (10), 4478–4488, doi:10.1111/gcb.14332.

Tarling, G.A., P. Ward and S.E. Thorpe, 2018: Spatial distributions of Southern Ocean mesozooplankton communities have been resilient to long-term surface warming. *Glob Chang Biol*, **24** (1), 132–142, doi:10.1111/gcb.13834.

Tarnocai, C. et al., 2009: Soil organic carbon pools in the northern circumpolar permafrost region. *Global Biogeochemical Cycles*, **23**, GB2023, doi:10.1029/2008gb003327.

Taylor, D.J., M.J. Ballinger, A.S. Medeiros and A.A. Kotov, 2015: Climate-associated tundra thaw pond formation and range expansion of boreal zooplankton predators. *Ecography*, **39** (1), 43–53, doi:10.1111/ecog.01514.

Taylor, P., B. Hegyi, R. Boeke and L. Boisvert, 2018: On the Increasing Importance of Air-Sea Exchanges in a Thawing Arctic: A Review. *Atmosphere*, **9** (2), doi:10.3390/atmos9020041.

Taylor, S.G., 2007: Climate warming causes phenological shift in Pink Salmon, Oncorhynchus gorbuscha, behavior at Auke Creek, Alaska. *Global Change Biology*, **14** (2), 229–235, doi:10.1111/j.1365-2486.2007.01494.x.

Tedesco, M. et al., 2016a: The darkening of the Greenland ice sheet: trends, drivers, and projections (1981–2100). *The Cryosphere*, **10** (2), 477–496, doi:10.5194/tc-10-477-2016.

Tedesco, M. et al., 2013: Evidence and analysis of 2012 Greenland records from spaceborne observations, a regional climate model and reanalysis data. *The Cryosphere*, **7** (2), 615–630, doi:10.5194/tc-7-615-2013.

Tedesco, M. et al., 2016b: Arctic cut-off high drives the poleward shift of a new Greenland melting record. *Nat Commun*, **7**, 11723, doi:10.1038/ncomms11723.

Tedstone, A.J. et al., 2015: Decadal slowdown of a land-terminating sector of the Greenland Ice Sheet despite warming. *Nature*, **526**, 692, doi:10.1038/nature15722.

Tesar, C., M.-A. Dubois, M. Sommerkorn and A. Shestakov, 2016a: Warming to the subject: the Arctic Council and climate change. *The Polar Journal*, **6** (2), 417–429, doi:10.1080/2154896X.2016.1247025.

Tesar, C., M.A. Dubois and A. Shestakov, 2016b: Toward strategic, coherent, policy-relevant arctic science. *Science*, **353** (6306), 1368–1370, doi:10.1126/science.aai8198.

Têtu, P.-L., J.-F. Pelletier and F. Lasserre, 2015: The mining industry in Canada north of the 55th parallel: a maritime traffic generator? *Polar Geography*, **38** (2), 107–122, doi:10.1080/1088937x.2015.1028576.

Thackeray, C.W. and C.G. Fletcher, 2016: Snow albedo feedback: Current knowledge, importance, outstanding issues and future directions. *Progress in Physical Geography*, **40** (3), 392–408, doi:10.1177/0309133315620999.

Thackeray, C.W., C.G. Fletcher and C. Derksen, 2014: The influence of canopy snow parameterizations on snow albedo feedback in boreal forest regions. *Journal of Geophysical Research: Atmospheres*, **119** (16), 9810–9821, doi:10.1002/2014jd021858.

Thackeray, C.W., C.G. Fletcher, L.R. Mudryk and C. Derksen, 2016: Quantifying the Uncertainty in Historical and Future Simulations of Northern Hemisphere Spring Snow Cover. *Journal of Climate*, **29** (23), 8647–8663, doi:10.1175/jcli-d-16-0341.1.

The IMBIE Team, 2018: Mass balance of the Antarctic Ice Sheet from 1992 to 2017. *Nature*, **558** (7709), 219–222, doi:10.1038/s41586-018-0179-y.

The World Factbook, 2016: *The World Factbook 2016–17*. Central Intelligence Agency, Washington, DC [Available at: www.cia.gov/library/publications/the-world-factbook/index.html].

Thomas, C.D. and P.K. Gillingham, 2015: The performance of protected areas for biodiversity under climate change. *Biological Journal of the Linnean Society*, **115** (3), 718–730, doi:10.1111/bij.12510.

Thomas, E.R. et al., 2015: Twentieth century increase in snowfall in coastal West Antarctica. *Geophysical Research Letters*, **42** (21), 9387–9393, doi:10.1002/2015GL065750.

Thomas, E.R. et al., 2017a: Regional Antarctic snow accumulation over the past 1000 years. *Climate of the Past*, **13** (11), 1491–1513, doi:10.5194/cp-13-1491-2017.

Thomas, E.R. et al., 2017b: Review of regional Antarctic snow accumulation over the past 1000 years. *Climate of the Past Discussions*, 1–42, doi:10.5194/cp-2017-18.

Thompson, A.F. and A.C. Naveira Garabato, 2014: Equilibration of the Antarctic Circumpolar Current by Standing Meanders. *Journal of Physical Oceanography*, **44** (7), 1811–1828, doi:10.1175/JPO-D-13-0163.1.

Thomson, J. and W.E. Rogers, 2014: Swell and sea in the emerging Arctic Ocean. *Geophysical Research Letters*, **41** (9), 3136–3140, doi:10.1002/2014gl059983.

Thor, P. et al., 2018: Contrasting physiological responses to future ocean acidification among Arctic copepod populations. *Glob Chang Biol*, **24** (1), e365-e377, doi:10.1111/gcb.13870.

Thornalley, D.J.R. et al., 2018: Anomalously weak Labrador Sea convection and Atlantic overturning during the past 150 years. *Nature*, **556** (7700), 227–230, doi:10.1038/s41586-018-0007-4.

Thornton, B.F. et al., 2016: Methane fluxes from the sea to the atmosphere across the Siberian shelf seas. *Geophysical Research Letters*, **43** (11), 5869–5877, doi:10.1002/2016GL068977.

Timmermans, M.-L., C. Ladd and K. Wood, 2017: *Sea surface temperature* [NOAA (ed.)]. Arctic Report Card, NOAA, https://arctic.noaa.gov/Report-Card/Report-Card-2017/ArtMID/7798/ArticleID/698/Sea-Surface-Temperature).

Timmermans, M.-L., J. Toole and R. Krishfield, 2018: Warming of the interior Arctic Ocean linked to sea ice losses at the basin margins. *Science Advances*, **4** (8), doi:10.1126/sciadv.aat6773.

Timmermans, M.L., 2015: The impact of stored solar heat on Arctic sea ice growth. *Geophysical Research Letters*, **42** (15), 6399–6406, doi:10.1002/2015GL064541.

Timmermans, M.L. et al., 2014: Mechanisms of Pacific summer water variability in the Arctic's Central Canada basin. *Journal of Geophysical Research: Oceans*, **119** (11), 7523–7548, doi:10.1002/2014JC010273.

Ting, Y.-H. and M. Holzer, 2017: Decadal changes in Southern Ocean ventilation inferred from deconvolutions of repeat hydrographies. *Geophysical Research Letters*, **44** (11), 5655–5664, doi:10.1002/2017gl073788.

Todd, J. and P. Christoffersen, 2014: Are seasonal calving dynamics forced by buttressing from ice melange or undercutting by melting? Outcomes from full-Stokes simulations of Store Glacier, West Greenland. *Cryosphere*, **8** (6), 2353–2365, doi:10.5194/tc-8-2353-2014.

Trathan, P.N., S.M. Grant, V. Siegel and K.H. Kock, 2013: Precautionary spatial protection to facilitate the scientific study of habitats and communities under ice shelves in the context of recent, rapid, regional climate change. *CCAMLR Science*, **20**, 139–151.

Trathan, P.N. and S.L. Hill, 2016: The Importance of Krill Predation in the Southern Ocean. In: Biology and Ecology of Antarctic Krill [Siegel, V. (ed.)]. Springer, 321–350.

Trenary, L., T. DelSole, M.K. Tippett and B. Doty, 2016: Extreme eastern U.S. winter of 2015 not symptomatic of climate change [in "Explaining Extreme Events of 2015 from a Climate Perspective"].. **97**, S31-S35, doi:10.1175/BAMS-D-16-0156.1.

Trimborn, S. et al., 2017: Two Southern Ocean diatoms are more sensitive to ocean acidification and changes in irradiance than the prymnesiophyte

Phaeocystis antarctica. *Physiol Plant*, **160** (2), 155–170, doi:10.1111/ppl.12539.

Trivelpiece, W.Z. et al., 2011: Variability in krill biomass links harvesting and climate warming to penguin population changes in Antarctica. *Proc Natl Acad Sci U S A*, **108** (18), 7625–8, doi:10.1073/pnas.1016560108.

Troy, T.J., J. Sheffield and E.F. Wood, 2012: The role of winter precipitation and temperature on northern Eurasian streamflow trends. *Journal of Geophysical Research: Atmospheres*, **117** (D5), D05131, doi:10.1029/2011jd016208.

Trump, B.D., M. Kadenic and I. Linkov, 2018: A sustainable Arctic: Making hard decisions. *Arctic, Antarctic, and Alpine Research*, **50** (1), doi:10.1080/15230430.2018.1438345.

Trusel, L.D. et al., 2018: Nonlinear rise in Greenland runoff in response to post-industrial Arctic warming. *Nature*, **564** (7734), 104–108, doi:10.1038/s41586-018-0752-4.

Tulloch, V.J.D. et al., 2019: Future recovery of baleen whales is imperiled by climate change. *Glob Chang Biol*, doi:10.1111/gcb.14573.

Tuori, K., 2011: The Disputed Roots of Legal Pluralism. *Law, Culture and the Humanities*, **9** (2), 330–351, doi:10.1177/1743872111412718.

Turetsky, M.R. et al., 2011: Recent acceleration of biomass burning and carbon losses in Alaskan forests and peatlands. *Nature Geoscience*, **4** (1), 27–31, doi:https://doi.org/10.1038/ngeo1027.

Turner, D.R., I.-M. Hassellöv, E. Ytreberg and A. Rutgersson, 2017a: Shipping and the environment: Smokestack emissions, scrubbers and unregulated oceanic consequences. *Elem Sci Anth*, **5**, 45, doi:http://doi.org/10.1525/elementa.167.

Turner, J. et al., 2009: *Antarctic Climate Change and the Environment*. Scientific Committee on Antarctic Research Scott Polar Research Institute, Cambridge, UK. Turner, J. et al., 2013: An Initial Assessment of Antarctic Sea Ice Extent in the CMIP5 Models. *Journal of Climate*, **26** (5), 1473–1484, doi:10.1175/jcli-d-12-00068.1.

Turner, J. et al., 2016: Absence of 21st century warming on Antarctic Peninsula consistent with natural variability. *Nature*, **535** (7612), 411–415, doi:10.1038/nature18645.

Turner, J. et al., 2017b: Unprecedented springtime retreat of Antarctic sea ice in 2016. *Geophysical Research Letters*, **44** (13), 6868–6875, doi:10.1002/2017gl073656.

Turney, C.S.M. et al., 2017: Tropical forcing of increased Southern Ocean climate variability revealed by a 140-year subantarctic temperature reconstruction. *Clim. Past*, **13** (3), 231–248, doi:10.5194/cp-13-231-2017.

Turunen, M.T. et al., 2016: Coping with difficult weather and snow conditions: Reindeer herders' views on climate change impacts and coping strategies. *Climate Risk Management*, **11**, 15–36, doi:10.1016/j.crm.2016.01.002.

Uboni, A. et al., 2016: Long-term trends and role of climate in the population dynamics of eurasian reindeer. *Plos One*, **11** (6), 1–20, doi:10.1371/journal.pone.0158359.

Udevitz, M.S. et al., 2017: Forecasting consequences of changing sea ice availability for Pacific walruses. *Ecosphere*, **8** (11), e02014, doi:10.1002/ecs2.2014.

Ueyama, M. et al., 2013: Growing season and spatial variations of carbon fluxes of Arctic and boreal ecosystems in Alaska (USA). *Ecological Applications*, **23** (8), 1798–1816, doi:10.1890/11-0875.1.

Ulrich, M. et al., 2017: Differences in behavior and distribution of permafrost-related lakes in Central Yakutia and their response to climatic drivers. *Water Resources Research*, **53** (2), 1167–1188, doi:10.1002/2016wr019267.

Uotila, P., T. Vihma and M. Tsukernik, 2013: Close interactions between the Antarctic cyclone budget and large-scale atmospheric circulation. *Geophysical Research Letters*, **40** (12), 3237–3241, doi:10.1002/grl.50560.

Utvik, T.I.R. and C. Jahre-Nilsen, 2016: The Importance of Early Identification of Safety and Sustainability Related Risks in Arctic Oil and Gas Operations. In: *SPE International Conference and Exhibition on Health, Safety, Security, Environment, and Social Responsibility*, 2016/4/11/, Stavanger, Norway, Society of Petroleum Engineers, SPE, doi:10.2118/179325-MS.

Vacchi, M. et al., 2012: A nursery area for the Antarctic silverfish Pleuragramma antarcticum at Terra Nova Bay (Ross Sea): first estimate of distribution and abundance of eggs and larvae under the seasonal sea-ice. *Polar Biology*, **35** (10), 1573–1585, doi:10.1007/s00300-012-1199-y.

Vallot, D. et al., 2018: Effects of undercutting and sliding on calving: a global approach applied to Kronebreen, Svalbard. *The Cryosphere*, **12** (2), 609–625, doi:10.5194/tc-12-609-2018.

van den Broeke, M. et al., 2017: Greenland Ice Sheet Surface Mass Loss: Recent Developments in Observation and Modeling. *Current Climate Change Reports*, **3** (4), 345–356, doi:10.1007/s40641-017-0084-8.

van den Broeke, M.R. et al., 2016: On the recent contribution of the Greenland ice sheet to sea level change. *The Cryosphere*, **10** (5), 1933–1946, doi:10.5194/tc-10-1933-2016.

van der Hel, S., 2016: New science for global sustainability? The institutionalisation of knowledge co-production in Future Earth. *Environmental Science & Policy*, **61**, 165–175, doi:10.1016/j.envsci.2016.03.012.

van der Linden, S., E. Maibach and A. Leiserowitz, 2015: Improving Public Engagement With Climate Change: Five "Best Practice" Insights From Psychological Science. *Perspectives on Psychological Science*, **10** (6), 758–763, doi:10.1177/1745691615598516.

van der Werf, G.R. et al., 2010: Global fire emissions and the contribution of deforestation, savanna, forest, agricultural, and peat fires (1997–2009). *Atmospheric Chemistry and Physics*, **10** (23), 11707–11735, doi:10.5194/acp-10-11707-2010.

van Oort, B., M. Bjørkan and E.M. Klyuchnikova, 2015: *Future narratives for two locations in the Barents region. CICERO Report*. CICERO Report, Research, C.C. f. I.C. a. E., Oslo [Available at: http://hdl.handle.net/11250/2367371].

van Vliet, M.T.H. et al., 2013: Global river discharge and water temperature under climate change. *Global Environmental Change*, **23** (2), 450–464, doi:10.1016/j.gloenvcha.2012.11.002.

Van Wychen, W. et al., 2014: Glacier velocities and dynamic ice discharge from the Queen Elizabeth Islands, Nunavut, Canada. *Geophysical Research Letters*, **41** (2), 484–490, doi:10.1002/2013gl058558.

Van Wychen, W. et al., 2016: Characterizing interannual variability of glacier dynamics and dynamic discharge (1999–2015) for the ice masses of Ellesmere and Axel Heiberg Islands, Nunavut, Canada. *Journal of Geophysical Research: Earth Surface*, **121** (1), 39–63, doi:10.1002/2015jf003708.

Vancoppenolle, M. et al., 2013: Future arctic ocean primary productivity from CMIP5 simulations: Uncertain outcome, but consistent mechanisms. *Global Biogeochemical Cycles*, **27** (3), 605–619, doi:10.1002/gbc.20055.

Vargas-Moreno, J.C., B. Fradkin, S. Emperador and O.L. (eds), 2016: *Project Summary: Prioritizing Science Needs Through Participatory Scenarios for Energy and Resource Development on the North Slope and Adjacent Seas*. GeoAdaptive, LLC, Boston, Massachusetts. [Available at: http://northslope.org/scenarios/].

Vavrus, S.J. et al., 2012: Twenty-First-Century Arctic Climate Change in CCSM4. *Journal of Climate*, **25** (8), 2696–2710, doi:10.1175/jcli-d-11-00220.1.

Velicogna, I., T.C. Sutterley and M.R. Van Den Broeke, 2014: Regional acceleration in ice mass loss from Greenland and Antarctica using GRACE time-variable gravity data. *Geophysical Research Letters*, **41** (22), 8130–8137, doi:10.1002/2014GL061052.

Venables, H.J., A. Clarke and M.P. Meredith, 2013: Wintertime controls on summer stratification and productivity at the western Antarctic Peninsula. *Limnology and Oceanography*, **58** (3), 1035–1047, doi:10.4319/lo.2013.58.3.1035.

Veraverbeke, S. et al., 2017: Lightning as a major driver of recent large fire years in North American boreal forests. *Nature Climate Change*, **7** (7), 529–534, doi:10.1038/nclimate3329.

Verfaillie, D. et al., 2015: Recent glacier decline in the Kerguelen Islands (49°S, 69°E) derived from modeling, field observations, and satellite

3

data. *Journal of Geophysical Research: Earth Surface*, **120** (3), 637–654, doi:10.1002/2014JF003329.

Verschuuren, J., 2013: Legal Aspects of Climate Change Adaptation In: Climate Change and the Law Part III: Comparative Perspectives on Law and Justice. [Hollo, E., K. Kulovesi and M. Mehling (eds.)]. Spinger, Dordrecht, **21**.

Vickers, H. et al., 2016: Changes in greening in the high Arctic: insights from a 30 year AVHRR max NDVI dataset for Svalbard. *Environmental Research Letters*, **11** (10), doi:10.1088/1748-9326/11/10/105004.

Vihma, T. et al., 2016: The atmospheric role in the Arctic water cycle: A review on processes, past and future changes, and their impacts. *Journal of Geophysical Research: Biogeosciences*, **121** (3), 586–620, doi:10.1002/2015jg003132.

Vihma, T., P. Tisler and P. Uotila, 2012: Atmospheric forcing on the drift of Arctic sea ice in 1989–2009. *Geophysical Research Letters*, **39** (2), doi:10.1029/2011gl050118.

Vihtakari, M. et al., 2018: Black-legged kittiwakes as messengers of Atlantification in the Arctic. *Scientific Reports*, **8** (1), 1178, doi:10.1038/s41598-017-19118-8.

Vincent et al., 2015: Observed trends in Canada's climate and influence of low-frequency variability modes. *Journal of Climate*, **28**, 4545–4560, doi:10.1175/jcli-d-14-00697.1.

Vincent, W.F. et al., 2011: Ecological Implications of Changes in the Arctic Cryosphere. *AMBIO*, **40** (1), 87–99, doi:10.1007/s13280-011-0218-5.

Vincent, W.F., M. Lemay and M. Allard, 2017: Arctic permafrost landscapes in transition: towards an integrated Earth system approach. *Arctic Science*, **3** (2), 39–64, doi:10.1139/as-2016-0027.

Vinther, B.M. et al., 2009: Holocene thinning of the Greenland ice sheet. *Nature*, **461** (7262), 385–8, doi:10.1038/nature08355.

Vlasova, T. and S. Volkov, 2016: Towards transdisciplinarity in Arctic sustainability knowledge co-production: Socially-Oriented Observations as a participatory integrated activity. *Polar Science*, **10** (3), 425–432, doi:10.1016/j.polar.2016.06.002.

Vonk, J.E. et al., 2015: Reviews and syntheses: Effects of permafrost thaw on Arctic aquatic ecosystems. *Biogeosciences*, **12** (23), 7129–7167, doi:10.5194/bg-12-7129-2015.

Voorhees, H., R. Sparks, H.P. Huntington and K.D. Rode, 2014: Traditional Knowledge about Polar Bears (Ursus maritimus) in Northwestern Alaska. *Arctic*, **67** (4), 523–536.

Wadham, J.L. et al., 2013: The potential role of the Antarctic Ice Sheet in global biogeochemical cycles. *Earth and Environmental Science Transactions of the Royal Society of Edinburgh*, **104** (1), 55–67, doi:10.1017/S1755691013000108.

Wadham, J.L. et al., 2016: Sources, cycling and export of nitrogen on the Greenland Ice Sheet. *Biogeosciences*, **13** (22), 6339–6352, doi:10.5194/bg-13-6339-2016.

Walker, C.C. and A.S. Gardner, 2017: Rapid drawdown of Antarctica's Wordie Ice Shelf glaciers in response to ENSO/Southern Annular Mode-driven warming in the Southern Ocean. *Earth and Planetary Science Letters*, **476**, 100–110, doi:10.1016/j.epsl.2017.08.005.

Walker, D.A. et al., 2015: A hierarchic approach for examining panarctic vegetation with a focus on the linkages between remote-sensing and plot-based studies: A prototype example from Toolik Lake, Alaska. In: *AGU Fall meeting 14–19 Dec 2015*, San Francisco, USA, AGU Fall Meeting.

Walker, D.A. et al., 2016: Circumpolar Arctic vegetation: a hierarchic review and roadmap toward an internationally consistent approach to survey, archive and classify tundra plot data. *Environmental Research Letters*, **11** (5), 055005, doi:10.1088/1748-9326/11/5/055005.

Walker, D.P. et al., 2013: Oceanographic observations at the shelf break of the Amundsen Sea, Antarctica. *Journal of Geophysical Research: Oceans*, **118** (6), 2906–2918, doi:10.1002/jgrc.20212.

Wallhead, P.J. et al., 2017: Bottom Water Acidification and Warming on the Western Eurasian Arctic Shelves: Dynamical Downscaling Projections.

Journal of Geophysical Research: Oceans, **122** (10), 8126–8144, doi:10.1002/2017jc013231.

Walsh, J.E., F. Fetterer, J. Scott Stewart and W.L. Chapman, 2017: A database for depicting Arctic sea ice variations back to 1850. *Geographical Review*, **107** (1), 89–107, doi:10.1111/j.1931-0846.2016.12195.x.

Walsh, M.G., A.W. de Smalen and S.M. Mor, 2018: Climatic influence on anthrax suitability in warming northern latitudes. *Sci Rep*, **8** (1), 9269, doi:10.1038/s41598-018-27604-w.

Walter Anthony, K. et al., 2018: 21st-century modeled permafrost carbon emissions accelerated by abrupt thaw beneath lakes. *Nat Commun*, **9** (1), 3262, doi:10.1038/s41467-018-05738-9.

Walter Anthony, K.M., P. Anthony, G. Grosse and J. Chanton, 2012: Geologic methane seeps along boundaries of Arctic permafrost thaw and melting glaciers. *Nature Geoscience*, **5** (6), 419–426, doi:10.1038/ngeo1480.

Walter Anthony, K.M. et al., 2014: A shift of thermokarst lakes from carbon sources to sinks during the Holocene epoch. *Nature*, **511** (7510), 452–456, doi:10.1038/nature13560.

Walter, K.M. et al., 2007: Thermokarst Lakes as a Source of Atmospheric CH_4 During the Last Deglaciation. *Science*, **318** (5850), 633–636, doi:10.1126/science.1142924.

Walvoord, M.A. and B.L. Kurylyk, 2016: Hydrologic Impacts of Thawing Permafrost – A Review. *Vadose Zone Journal*, **15** (6), doi:10.2136/vzj2016.01.0010.

Walvoord, M.A. and R.G. Striegl, 2007: Increased groundwater to stream discharge from permafrost thawing in the Yukon River basin: Potential impacts on lateral export of carbon and nitrogen. *Geophysical Research Letters*, **34** (L12402), doi:doi:10.1029/2007GL030216.

Walvoord, M.A., C.I. Voss, B.A. Ebel and B.J. Minsley, 2019: Development of perennial thaw zones in boreal hillslopes enhances potential mobilization of permafrost carbon. *Environmental Research Letters*, **14** (1), doi:10.1088/1748-9326/aaf0cc.

Wang, C. et al., 2014: A global perspective on CMIP5 climate model biases. *Nature Climate Change*, **4** (3), 201, doi:10.1038/nclimate2118.

Wang, G. et al., 2019: Compounding tropical and stratospheric forcing of the record low Antarctic sea-ice in 2016. *Nat Commun*, **10** (1), 13, doi:10.1038/s41467-018-07689-7.

Wang, L., C. Derksen and R. Brown, 2013: Recent changes in pan-Arctic melt onset from satellite passive microwave measurements. *Geophysical Research Letters*, **40**, 522–528, doi:10.1002/grl.50098.

Wang, Q. et al., 2016a: Sea ice leads in the Arctic Ocean: Model assessment, interannual variability and trends. *Geophysical Research Letters*, **43** (13), 7019–7027, doi:10.1002/2016GL068696.

Wang, S.W. et al., 2015: Importance of sympagic production to Bering Sea zooplankton as revealed from fatty acid-carbon stable isotope analyses. *Marine Ecology Progress Series*, **518**, 31–50, doi:10.3354/meps11076.

Wang, Y. et al., 2016b: A Comparison of Antarctic Ice Sheet Surface Mass Balance from Atmospheric Climate Models and In Situ Observations. *Journal of Climate*, **29** (14), 5317–5337, doi:10.1175/JCLI-D-15-0642.1.

Ward, E.J. et al., 2017: Effects of increased specialization on revenue of Alaskan salmon fishers over four decades. *Journal of Applied Ecology*, **55** (3), 1082–1091, doi:10.1111/1365-2664.13058.

Warren, S.G. et al., 1999: Snow Depth on Arctic Sea Ice. *Journal of Climate*, **12** (6), 1814–1829, doi:10.1175/1520-0442(1999)012<1814:sdoasi>2.0.co;2.

Wassmann, P. et al., 2015: The contiguous domains of Arctic Ocean advection: Trails of life and death. *Progress in Oceanography*, **139**, 42–65, doi:10.1016/j.pocean.2015.06.011.

Wassmann, P. and T.M. Lenton, 2012: Arctic Tipping Points in an Earth System Perspective. *AMBIO*, **41** (1), 1–9, doi:10.1007/s13280-011-0230-9.

Watts, P., K. Koutouki, S. Booth and S. Blum, 2017: Inuit food security in canada: arctic marine ethnoecology. *Food Security*, **9** (3), 421–440, doi:10.1007/s12571-017-0668-0.

Waugh, D.W., 2014: Changes in the ventilation of the southern oceans. *Phil. Trans. R. Soc. A*, **372** (2019), 20130269, doi:10.1126/science.1225411.

Waugh, D.W., C.I. Garfinkel and L.M. Polvani, 2015: Drivers of the Recent Tropical Expansion in the Southern Hemisphere: Changing SSTs or Ozone Depletion? *Journal of Climate*, **28** (16), 6581–6586, doi:10.1175/jcli-d-15-0138.1.

Waugh, D.W., F. Primeau, T. DeVries and M. Holzer, 2013: Recent Changes in the Ventilation of the Southern Oceans. *Science*, **339** (6119), 568–570, doi:10.1029/2008JC004864.

WCRP, 2018: Global sea-level budget 1993–present. *Earth System Science Data*, **10** (3), 1551–1590, doi:10.5194/essd-10-1551-2018.

Webb, B.W. et al., 2008: Recent advances in stream and river temperature research. *Hydrological Processes*, **22** (7), 902–918, doi:10.1002/hyp.6994.

Webb, E.E. et al., 2016: Increased wintertime CO2loss as a result of sustained tundra warming. *Journal of Geophysical Research: Biogeosciences*, **121** (2), 249–265, doi:10.1002/2014jg002795.

Webber, B.G.M. et al., 2017: Mechanisms driving variability in the ocean forcing of Pine Island Glacier. *Nature Communications*, **8**, 14507, doi:10.1038/ncomms14507.

Weber, M.E. et al., 2014: Millennial-scale variability in Antarctic ice-sheet discharge during the last deglaciation. *Nature*, **510**, 134, doi:10.1038/nature13397.

Webster, M. et al., 2018: Snow in the changing sea-ice systems. *Nature Climate Change*, **8** (11), 946–953, doi:10.1038/s41558-018-0286-7.

Webster, M.A. et al., 2014: Interdecadal changes in snow depth on Arctic sea ice. *Journal of Geophysical Research: Oceans*, **119** (8), 5395–5406, doi:10.1002/2014jc009985.

Wedekind, C. et al., 2010: Elevated resource availability sufficient to turn opportunistic into virulent fish pathogens. *Ecology*, **91** (5), 1251–1256, doi:10.1890/09-1067.1.

Wehrmann, D., 2016: The Polar Regions as "barometers" in the Anthropocene: towards a new significance of non-state actors in international cooperation? *The Polar Journal*, **6** (2), 379–397, doi:10.1080/2154896X.2016.1241483.

Wehrmann, L.M. et al., 2013: The evolution of early diagenetic signals in Bering Sea subseafloor sediments in response to varying organic carbon deposition over the last 4.3Ma. *Geochimica et Cosmochimica Acta*, **109**, 175–196, doi:10.1016/j.gca.2013.01.025.

Wehrmann, L.M. et al., 2014: Iron and manganese speciation and cycling in glacially influenced high-latitude fjord sediments (West Spitsbergen, Svalbard): Evidence for a benthic recycling-transport mechanism. *Geochimica et Cosmochimica Acta*, **141**, 628–655, doi:10.1016/j.gca.2014.06.007.

Weimerskirch, H., M. Louzao, S. de Grissac and K. Delord, 2012: Changes in wind pattern alter albatross distribution and life-history traits. *Science*, **335** (6065), 211–4, doi:10.1126/science.1210270.

Wellman, T.P., C.I. Voss and M.A. Walvoord, 2013: Impacts of climate, lake size, and supra- and sub-permafrost groundwater flow on lake-talik evolution, Yukon Flats, Alaska (USA). *Hydrogeology Journal*, **21** (1), 281–298, doi:10.1007/s10040-012-0941-4.

Wenzel, L. et al., 2016: Polar opposites? Marine conservation tools and experiences in the changing Arctic and Antarctic. *Aquatic Conservation: Marine and Freshwater Ecosystems*, **26**, 61–84, doi:10.1002/aqc.2649.

Westergaard-Nielsen, A. et al., 2017: Transitions in high-Arctic vegetation growth patterns and ecosystem productivity tracked with automated cameras from 2000 to 2013. *AMBIO*, **46** (1), 39–52, doi:10.1007/s13280-016-0864-8.

Westwood, K.J. et al., 2018: Ocean acidification impacts primary and bacterial production in Antarctic coastal waters during austral summer. *Journal of Experimental Marine Biology and Ecology*, **498**, 46–60, doi:10.1016/j.jembe.2017.11.003.

Wheeler, H.C. et al., 2018: Identifying key needs for the integration of social-ecological outcomes in arctic wildlife monitoring. *Conserv Biol*, doi:10.1111/cobi.13257.

Whitehead, A.L. et al., 2015: Factors driving Adélie penguin chick size, mass and condition at colonies of different sizes in the Southern Ross Sea. *Marine Ecology Progress Series*, **523**, 199–213, doi:10.3354/meps11130.

Wickland, K.P. et al., 2018: Dissolved organic carbon and nitrogen release from boreal Holocene permafrost and seasonally frozen soils of Alaska. *Environmental Research Letters*, **13** (6), 065011, doi:10.1088/1748-9326/aac4ad.

Williams, N.L. et al., 2017: Calculating surface ocean pCO2 from biogeochemical Argo floats equipped with pH: An uncertainty analysis. *Global Biogeochemical Cycles*, **31** (3), 591–604, doi:10.1002/2016GB005541.

Williams, P. et al., 2018: Community-based observing networks and systems in the Arctic: Human perceptions of environmental change and instrument-derived data. *Regional Environmental Change*, **18** (2), 547–559, doi:10.1007/s10113-017-1220-7.

Williams, W.J. and E.C. Carmack, 2015: The 'interior' shelves of the Arctic Ocean: Physical oceanographic setting, climatology and effects of sea-ice retreat on cross-shelf exchange. *Progress in Oceanography*, **139**, 24–41, doi:10.1016/j.pocean.2015.07.008.

Willis, M.J., A.K. Melkonian and M.E. Pritchard, 2015: Outlet glacier response to the 2012 collapse of the Matusevich Ice Shelf, Severnaya Zemlya, Russian Arctic. *Journal of Geophysical Research: Earth Surface*, **120** (10), 2040–2055, doi:10.1002/2015JF003544.

Willis, M.J. et al., 2018: Massive destabilization of an Arctic ice cap. *Earth and Planetary Science Letters*, **502**, 146–155, doi:10.1016/j.epsl.2018.08.049.

Wilson, D.J. et al., 2018: Ice loss from the East Antarctic Ice Sheet during late Pleistocene interglacials. *Nature*, **561** (7723), 383–386, doi:10.1038/s41586-018-0501-8.

Wilson, E.A., S.C. Riser, E.C. Campbell and A.P.S. Wong, 2019: Winter Upper-Ocean Stability and Ice–Ocean Feedbacks in the Sea Ice–Covered Southern Ocean. *Journal of Physical Oceanography*, **49** (4), 1099–1117, doi:10.1175/jpo-d-18-0184.1.

Wilson, W.J. and O.A. Ormseth, 2009: A new management plan for Arctic waters of the United States. *Fisheries*, **34** (11), 555–558.

Winfield, I.J. et al., 2010: Population trends of Arctic charr (Salvelinus alpinus) in the UK: assessing the evidence for a widespread decline in response to climate change. *Hydrobiologia*, **650** (1), 55–65, doi:10.1007/s10750-009-0078-1.

Wise, M.G., J.A. Dowdeswell, M. Jakobsson and R.D. Larter, 2017: Evidence of marine ice-cliff instability in Pine Island Bay from iceberg-keel plough marks. *Nature*, **550** (7677), 506–510, doi:10.1038/nature24458.

Woodgate, R.A., 2018: Increases in the Pacific inflow to the Arctic from 1990 to 2015, and insights into seasonal trends and driving mechanisms from year-round Bering Strait mooring data. *Progress in Oceanography*, **160**, 124–154, doi:10.1016/j.pocean.2017.12.007.

Woodgate, R.A., K.M. Stafford and F.G. Prahl, 2015: A synthesis of year-round interdisciplinary mooring measurements in the Bering Strait (1990–2014) and the RUSALCA years (2004–2011). *Oceanography*, **28** (3), 46–67, doi:10.5670/oceanog.2015.57.

Worby, A.P. et al., 2008: Thickness distribution of Antarctic sea ice. *Journal of Geophysical Research-Oceans*, **113** (C5), C05s92, doi:10.1029/2007jc004254.

Wotton, B.M., M.D. Flannigan and G.A. Marshall, 2017: Potential climate change impacts on fire intensity and key wildfire suppression thresholds in Canada. *Environmental Research Letters*, **12** (9), 095003, doi:10.1088/1748-9326/aa7e6e.

Wouters, B., A.S. Gardner and G. Moholdt, 2019: Global Glacier Mass Loss During the GRACE Satellite Mission (2002–2016). *Frontiers in Earth Science*, **7**, doi:10.3389/feart.2019.00096.

Wouters, B. et al., 2015: Dynamic thinning of glaciers on the Southern Antarctic Peninsula. *Science*, **348** (6237), 899–903, doi:10.1126/science.aaa5727.

Wright, C.J. et al., 2018: Water quality and health in northern Canada: stored drinking water and acute gastrointestinal illness in Labrador Inuit. *Environ Sci Pollut Res Int*, **25** (33), 32975–32987, doi:10.1007/s11356-017-9695-9.

Wrona, F.J. et al., 2016: Transitions in Arctic ecosystems: Ecological implications of a changing hydrological regime. *Journal of Geophysical Research: Biogeosciences*, **121** (3), 650–674, doi:10.1002/2015jg003133.

Wu, C. et al., 2017: Present-day and future contribution of climate and fires to vegetation composition in the boreal forest of China. *Ecosphere*, **8** (8), e01917, doi:10.1002/ecs2.1917.

Wu, S.-Y. and S. Hou, 2017: Impact of icebergs on net primary productivity in the Southern Ocean. *The Cryosphere*, **11** (2), 707–722, doi:10.5194/tc-11-707-2017.

Wuite, J. et al., 2015: Evolution of surface velocities and ice discharge of Larsen B outlet glaciers from 1995 to 2013. *The Cryosphere*, **9** (3), 957–969, doi:10.5194/tc-9-957-2015.

Xu, L. et al., 2013a: Temperature and vegetation seasonality diminishment over northern lands. *Nature Climate Change*, **3**, 581, doi:10.1038/nclimate1836

Xu, Y. et al., 2013b: Subaqueous melting of Store Glacier, west Greenland from three-dimensional, high-resolution numerical modeling and ocean observations. *Geophysical Research Letters*, **40** (17), 4648–4653, doi:10.1002/grl.50825.

Yager, P.L. et al., 2016: A carbon budget for the Amundsen Sea Polynya, Antarctica: Estimating net community production and export in a highly productive polar ecosystem. *Elementa-Science of the Anthropocene*, **4**, 000140, doi:http://doi.org/10.12952/journal.elementa.000140.

Yamamoto-Kawai, M., T. Mifune, T. Kikuchi and S. Nishino, 2016: Seasonal variation of CaCO3 saturation state in bottom water of a biological hotspot in the Chukchi Sea, Arctic Ocean. *Biogeosciences*, **13** (22), 6155–6169, doi:10.5194/bg-13-6155-2016.

Yang, D., 2014: Double Fence Intercomparison Reference (DFIR) vs. Bush Gauge for "true" snowfall measurement. *Journal of Hydrology*, **509** (Supplement C), 94–100, doi:10.1016/j.jhydrol.2013.08.052.

Yang, D., P. Marsh and S. Ge, 2014: Heat flux calculations for Mackenzie and Yukon Rivers. *Polar Science*, **8** (3), 232–241, doi:10.1016/j.polar.2014.05.001.

Yang, D. and A. Peterson, 2017: River Water Temperature in Relation to Local Air Temperature in the Mackenzie and Yukon Basins. *Arctic*, **70** (1), 47–58, doi:10.14430/arctic4627.

Yang, D., B. Ye and D.L. Kane, 2004a: Streamflow changes over Siberian Yenisei River Basin. *Journal of Hydrology*, **296** (1), 59–80, doi:10.1016/j.jhydrol.2004.03.017.

Yang, D., B. Ye and A. Shiklomanov, 2004b: Discharge Characteristics and Changes over the Ob River Watershed in Siberia. *Journal of Hydrometeorology*, **5** (4), 595–610, doi:10.1175/1525-7541.

Ye, B., D. Yang and D.L. Kane, 2003: Changes in Lena River streamflow hydrology: Human impacts versus natural variations. *Water Resources Research*, **39** (7), 1200, doi:10.1029/2003wr001991.

Young, A.M., P.E. Higuera, P.A. Duffy and F.S. Hu, 2017: Climatic thresholds shape northern high-latitude fire regimes and imply vulnerability to future climate change. *Ecography*, **40** (5), 606–617, doi:10.1111/ecog.02205.

Young, E.F. et al., 2018: Stepping stones to isolation: Impacts of a changing climate on the connectivity of fragmented fish populations. *Evol Appl*, **11** (6), 978–994, doi:10.1111/eva.12613.

Young, I., K. Gropp, A. Fazil and B.A. Smith, 2015: Knowledge synthesis to support risk assessment of climate change impacts on food and water safety: A case study of the effects of water temperature and salinity on Vibrio parahaemolyticus in raw oysters and harvest waters. *Food Research International*, **68**, 86–93, doi:10.1016/j.foodres.2014.06.035.

Young, O.R., 2011: If an Arctic Ocean treaty is not the solution, what is the alternative? *Polar Record*, **47** (4), 327–334, doi:10.1017/S0032247410000677.

Young, O.R., 2016: The shifting landscape of Arctic politics: implications for international cooperation. *The Polar Journal*, **6** (2), 209–223, doi:10.1080/2154896X.2016.1253823.

Younger, J.L. et al., 2015: Too much of a good thing: sea ice extent may have forced emperor penguins into refugia during the last glacial maximum. *Glob Chang Biol*, **21** (6), 2215–26, doi:10.1111/gcb.12882.

Youngflesh, C. et al., 2017: Circumpolar analysis of the Adelie Penguin reveals the importance of environmental variability in phenological mismatch. *Ecology*, **98** (4), 940–951, doi:10.1002/ecy.1749.

Yu, Q., H. Epstein, R. Engstrom and D. Walker, 2017: Circumpolar arctic tundra biomass and productivity dynamics in response to projected climate change and herbivory. *Global Change Biology*, **23** (9), 3895–3907, doi:10.1111/gcb.13632.

Yu, Y. et al., 2013: Interannual Variability of Arctic Landfast Ice between 1976 and 2007. *Journal of Climate*, **27** (1), 227–243, doi:10.1175/jcli-d-13-00178.1.

Yuan, X., M.R. Kaplan and M.A. Cane, 2018: The Interconnected Global Climate System – A Review of Tropical–Polar Teleconnections. *Journal of Climate*, **31** (15), 5765–5792, doi:10.1175/jcli-d-16-0637.1.

Yun, M.S. et al., 2015: Regional productivity of phytoplankton in the Western Arctic Ocean during summer in 2010. *Deep Sea Research Part II: Topical Studies in Oceanography*, **120** (Supplement C), 61–71, doi:10.1016/j.dsr2.2014.11.023.

Zador, S.G. et al., 2017: Ecosystem considerations in Alaska: the value of qualitative assessments. *ICES Journal of Marine Science*, **74** (1), 421–430, doi:10.1093/icesjms/fsw144.

Zamani, B., T. Krumpen, L.H. Smedsrud and R. Gerdes, 2019: Fram Strait sea ice export affected by thinning: comparing high-resolution simulations and observations. *Climate Dynamics*, **53** (5–6), 3257–3270, doi:10.1007/s00382-019-04699-z.

Zdanowicz, C. et al., 2012: Summer melt rates on Penny Ice Cap, Baffin Island: Past and recent trends and implications for regional climate. *Journal of Geophysical Research: Earth Surface*, **117** (F2), doi:10.1029/2011JF002248.

Zekollari, H., B.S. Lecavalier and P. Huybrechts, 2017: Holocene evolution of Hans Tausen Iskappe (Greenland) and implications for the palaeoclimatic evolution of the high Arctic. *Quaternary Science Reviews*, **168**, 182–193, doi:10.1016/j.quascirev.2017.05.010.

Zemp, M. et al., 2019: Global glacier mass changes and their contributions to sea-level rise from 1961 to 2016. *Nature*, **568** (7752), 382–386, doi:10.1038/s41586-019-1071-0.

Zhang, J. et al., 2015: The influence of sea ice and snow cover and nutrient availability on the formation of massive under-ice phytoplankton blooms in the Chukchi Sea. *Deep Sea Research Part II: Topical Studies in Oceanography*, **118** (Part A), 122–135, doi:10.1016/j.dsr2.2015.02.008.

Zhang, J., R. Lindsay, A. Schweiger and M. Steele, 2013: The impact of an intense summer cyclone on 2012 Arctic sea ice retreat. *Geophysical Research Letters*, **40** (4), 720–726, doi:10.1002/grl.50190.

Zhang, L., T.L. Delworth, W. Cooke and X. Yang, 2018a: Natural variability of Southern Ocean convection as a driver of observed climate trends. *Nature Climate Change*, **9** (1), 59–65, doi:10.1038/s41558-018-0350-3.

Zhang, P. et al., 2018b: A stratospheric pathway linking a colder Siberia to Barents-Kara Sea sea ice loss. *Sci Adv*, **4** (7), eaat6025, doi:10.1126/sciadv.aat6025.

Zhang, T. et al., 2008: Statistics and characteristics of permafrost and ground-ice distribution in the Northern Hemisphere. *Polar Geography*, **31** (1–2), 47–68, doi:10.1080/10889370802175895.

Zhang, X. et al., 2012: Enhanced poleward moisture transport and amplified northern high-latitude wetting trend. *Nature Climate Change*, **3** (1), 47–51, doi:10.1038/nclimate1631.

Zhang, Y., Q. Meng and S.H. Ng, 2016: Shipping efficiency comparison between Northern Sea Route and the conventional Asia-Europe shipping route via Suez Canal. *Journal of Transport Geography*, **57**, 241–249, doi:10.1016/j.jtrangeo.2016.09.008.

3

Zhang, Z. et al., 2017: Emerging role of wetland methane emissions in driving 21st century climate change. *Proceedings of the National Academy of Sciences*, **114** (36), 9647.

Zhao, M. et al., 2014: Characterizing the eddy field in the Arctic Ocean halocline. *Journal of Geophysical Research: Oceans*, **119** (12), 8800–8817.

Zhao, M. et al., 2016: Evolution of the eddy field in the Arctic Ocean's Canada Basin, 2005–2015. *Geophysical Research Letters*, **43** (15), 8106–8114, doi:10.1002/2016GL069671.

Zhuang, Q. et al., 2006: CO_2 and CH_4 exchanges between land ecosystems and the atmosphere in northern high latitudes over the 21st century. *Geophysical Research Letters*, **33** (17), L17403, doi:10.1029/2006gl026972.

Zisserson, B. and A. Cook, 2017: Impact of bottom water temperature change on the southernmost snow crab fishery in the Atlantic Ocean. *Fisheries Research*, **195** (Supplement C), 12–18, doi:10.1016/j.fishres.2017.06.009.

Zolkos, S., S.E. Tank and S.V. Kokelj, 2018: Mineral Weathering and the Permafrost Carbon-Climate Feedback. *Geophysical Research Letters*, **45** (18), 9623–9632, doi:10.1029/2018gl078748.

Zona, D. et al., 2016: Cold season emissions dominate the Arctic tundra methane budget. *Proceedings of the National Academy of Sciences USA*, **113** (1), 40–45, doi:10.1073/pnas.1516017113.

Zunz, V., H. Goosse and F. Massonnet, 2013: How does internal variability influence the ability of CMIP5 models to reproduce the recent trend in Southern Ocean sea ice extent? *The Cryosphere*, **7** (2), 451–468, doi:10.5194/tc-7-451-2013.

Zwally, H.J. et al., 2017: Mass gains of the Antarctic ice sheet exceed losses. *Journal of Glaciology*, **61** (230), 1019–1036, doi:10.3189/2015JoG15J071.

3

Sea Level Rise and Implications for Low-Lying Islands, Coasts and Communities

4

Coordinating Lead Authors:
Michael Oppenheimer (USA), Bruce C. Glavovic (New Zealand/South Africa)

Lead Authors:
Jochen Hinkel (Germany), Roderik van de Wal (Netherlands), Alexandre K. Magnan (France), Amro Abd-Elgawad (Egypt), Rongshuo Cai (China), Miguel Cifuentes-Jara (Costa Rica), Robert M. DeConto (USA), Tuhin Ghosh (India), John Hay (Cook Islands), Federico Isla (Argentina), Ben Marzeion (Germany), Benoit Meyssignac (France), Zita Sebesvari (Hungary/Germany)

Contributing Authors:
Robbert Biesbroek (Netherlands), Maya K. Buchanan (USA), Ricardo Safra de Campos (UK), Gonéri Le Cozannet (France), Catia Domingues (Australia), Sönke Dangendorf (Germany), Petra Döll (Germany), Virginie K.E. Duvat (France), Tamsin Edwards (UK), Alexey Ekaykin (Russian Federation), Donald Forbes (Canada), James Ford (UK), Miguel D. Fortes (Philippines), Thomas Frederikse (Netherlands), Jean-Pierre Gattuso (France), Robert Kopp (USA), Erwin Lambert (Netherlands), Judy Lawrence (New Zealand), Andrew Mackintosh (New Zealand), Angélique Melet (France), Elizabeth McLeod (USA), Mark Merrifield (USA), Siddharth Narayan (US), Robert J. Nicholls (UK), Fabrice Renaud (UK), Jonathan Simm (UK), AJ Smit (South Africa), Catherine Sutherland (South Africa), Nguyen Minh Tu (Vietnam), Jon Woodruff (USA), Poh Poh Wong (Singapore), Siyuan Xian (USA)

Review Editors:
Ayako Abe-Ouchi (Japan), Kapil Gupta (India), Joy Pereira (Malaysia)

Chapter Scientist:
Maya K. Buchanan (USA)

This chapter should be cited as:
Oppenheimer, M., B.C. Glavovic , J. Hinkel, R. van de Wal, A.K. Magnan, A. Abd-Elgawad, R. Cai, M. Cifuentes-Jara, R.M. DeConto, T. Ghosh, J. Hay, F. Isla, B. Marzeion, B. Meyssignac, and Z. Sebesvari, 2019: Sea Level Rise and Implications for Low-Lying Islands, Coasts and Communities. In: *IPCC Special Report on the Ocean and Cryosphere in a Changing Climate* [H.-O. Pörtner, D.C. Roberts, V. Masson-Delmotte, P. Zhai, M. Tignor, E. Poloczanska, K. Mintenbeck, A. Alegría, M. Nicolai, A. Okem, J. Petzold, B. Rama, N.M. Weyer (eds.)]. Cambridge University Press, Cambridge, UK and New York, NY, USA, pp. 321–445. https://doi.org/10.1017/9781009157964.006.

Table of contents

4

Executive Summary

This chapter assesses past and future contributions to global, regional and extreme sea level changes, associated risk to low-lying islands, coasts, cities, and settlements, and response options and pathways to resilience and sustainable development along the coast.

Observations

Global mean sea level (GMSL) is rising (*virtually certain*[1]) and accelerating (*high confidence*[2]). The sum of glacier and ice sheet contributions is now the dominant source of GMSL rise (*very high confidence*). GMSL from tide gauges and altimetry observations increased from 1.4 mm yr^{-1} over the period 1901–1990 to 2.1 mm yr^{-1} over the period 1970–2015 to 3.2 mm yr^{-1} over the period 1993–2015 to 3.6 mm yr^{-1} over the period 2006–2015 (*high confidence*). The dominant cause of GMSL rise since 1970 is anthropogenic forcing (*high confidence*). {4.2.2.1.1, 4.2.2.2}

GMSL was considerably higher than today during past climate states that were warmer than pre-industrial, including the Last Interglacial (LIG; 129–116 ka), when global mean surface temperature was 0.5°C–1.0°C warmer, and the mid-Pliocene Warm Period (mPWP; ~3.3 to 3.0 million years ago), 2°C–4°C warmer. Despite the modest global warmth of the Last Interglacial, GMSL was *likely* 6–9 m higher, mainly due to contributions from the Greenland and Antarctic ice sheets (GIS and AIS, respectively), and *unlikely* more than 10m higher (*medium confidence*). Based on new understanding about geological constraints since the IPCC 5th Assessment Report (AR5), 25 m is a plausible upper bound on GMSL during the mPWP (*low confidence*). Ongoing uncertainties in palaeo sea level reconstructions and modelling hamper conclusions regarding the total magnitudes and rates of past sea level rise (SLR). Furthermore, the long (multi-millennial) time scales of these past climate and sea level changes, and regional climate influences from changes in Earth's orbital configuration and climate system feedbacks, lead to *low confidence* in direct comparisons with near-term future changes. {Cross-Chapter Box 5 in Chapter 1, 4.2.2, 4.2.2.1, 4.2.2.5, SM 4.1}

Non-climatic anthropogenic drivers, including recent and historical demographic and settlement trends and anthropogenic subsidence, have played an important role in increasing low-lying coastal communities' exposure and vulnerability to SLR and extreme sea level (ESL) events (*very high confidence*). In coastal deltas, for example, these drivers have altered freshwater and sediment availability (*high confidence*). In low-lying coastal areas more broadly, human-induced changes can be rapid and modify coastlines over short periods of time, outpacing the effects of SLR (*high confidence*). Adaptation can be undertaken in the short- to medium-term by targeting local drivers of exposure and vulnerability, notwithstanding uncertainty about local SLR impacts in coming decades and beyond (*high confidence*). {4.2.2.4, 4.3.1, 4.3.2.2, 4.3.2.3}

Coastal ecosystems are already impacted by the combination of SLR, other climate-related ocean changes, and adverse effects from human activities on ocean and land (*high confidence*). Attributing such impacts to SLR, however, remains challenging due to the influence of other climate-related and non-climatic drivers such as infrastructure development and human-induced habitat degradation (*high confidence*). Coastal ecosystems, including saltmarshes, mangroves, vegetated dunes and sandy beaches, can build vertically and expand laterally in response to SLR, though this capacity varies across sites (*high confidence*). These ecosystems provide important services that include coastal protection and habitat for diverse biota. However, as a consequence of human actions that fragment wetland habitats and restrict landward migration, coastal ecosystems progressively lose their ability to adapt to climate-induced changes and provide ecosystem services, including acting as protective barriers (*high confidence*). {4.3.2.3}

Coastal risk is dynamic and increased by widely observed changes in coastal infrastructure, community livelihoods, agriculture and habitability (*high confidence*). As with coastal ecosystems, attribution of observed changes and associated risk to SLR remains challenging. Drivers and processes inhibiting attribution include demographic, resource and land use changes and anthropogenic subsidence. {4.3.3, 4.3.4}

A diversity of adaptation responses to coastal impacts and risks have been implemented around the world, but mostly as a reaction to current coastal risk or experienced disasters (*high confidence*). Hard coastal protection measures (dikes, embankments, sea walls and surge barriers) are widespread, providing predictable levels of safety in northwest Europe, East Asia, and around many coastal cities and deltas. Ecosystem-based adaptation (EbA) is continuing to gain traction worldwide, providing multiple co-benefits, but there is still *low agreement* on its cost and long-term effectiveness. Advance, which refers to the creation of new land by building into the sea (e.g., land reclamation), has a long history in most areas where there are dense coastal populations. Accommodation measures, such as early warning systems (EWS) for ESL events, are widespread. Retreat is observed but largely restricted

---

[1] Each finding is grounded in an evaluation of underlying evidence and agreement. A level of confidence is expressed using five qualifiers: very low, low, medium, high and very high, and typeset in italics, e.g., *medium confidence*. The following terms have been used to indicate the assessed likelihood of an outcome or a result: virtually certain 99–100% probability, very likely 90–100%, likely 66–100%, about as likely as not 33–66%, unlikely 0–33%, very unlikely 0–10%, exceptionally unlikely 0–1%. Assessed likelihood is typeset in italics, e.g., *very likely*. This is consistent with AR5 and the other AR6 Special Reports. Additional terms (extremely likely 95–100%, more likely than not >50–100%, more unlikely than likely 0–<50%, extremely unlikely 0–5%) are used when appropriate. This Report also uses the term '*likely* range' or '*very likely* range' to indicate that the assessed likelihood of an outcome lies within the 17–83% or 5–95% probability range. For more details see {1.9.2, Figure 1.4}.

[2] Statements about uncertainty in Section 4.2 are contingent upon the RCP or other emissions assumptions that accompany them. In Section 4.4, the entirety of information facing a decision maker is taken into consideration, including the unknown path of future emissions, in assessing uncertainty. Depending on which perspective is chosen, uncertainty may or may not be characterised as 'deep'.

to small communities or carried out for the purpose of creating new wetland habitat. {4.4.2.3, 4.4.2.4, 4.4.2.5}

Projections

Future rise in GMSL caused by thermal expansion, melting of glaciers and ice sheets and land water storage changes, is strongly dependent on which Representative Concentration Pathway (RCP) emission scenario is followed. SLR at the end of the century is projected to be faster under all scenarios, including those compatible with achieving the long-term temperature goal set out in the Paris Agreement. GMSL will rise between 0.43 m (0.29–0.59 m, *likely* range; RCP2.6) and 0.84 m (0.61–1.10 m, *likely* range; RCP8.5) by 2100 (*medium confidence*) relative to 1986–2005. Beyond 2100, sea level will continue to rise for centuries due to continuing deep ocean heat uptake and mass loss of the GIS and AIS and will remain elevated for thousands of years (*high confidence*). Under RCP8.5, estimates for 2100 are higher and the uncertainty range larger than in AR5. Antarctica could contribute up to 28 cm of SLR (RCP8.5, upper end of *likely* range) by the end of the century (*medium confidence*). Estimates of SLR higher than the *likely* range are also provided here for decision makers with low risk tolerance. {SR1.5, 4.1, 4.2.3.2, 4.2.3.5}

Under RCP8.5, the rate of SLR will be 15 mm yr^{-1} (10–20 mm yr^{-1}, *likely* range) in 2100, and could exceed several cm yr^{-1} in the 22nd century. These high rates challenge the implementation of adaptation measures that involve a long lead time, but this has not yet been studied in detail. {4.2.3.2, 4.4.2.2.3}

Processes controlling the timing of future ice shelf loss and the spatial extent of ice sheet instabilities could increase Antarctica's contribution to SLR to values higher than the *likely* range on century and longer time scales (*low confidence*). Evolution of the AIS beyond the end of the 21st century is characterized by deep uncertainty as ice sheet models lack realistic representations of some of the underlying physical processes. The few model studies available addressing time scales of centuries to millennia indicate multi-metre (2.3–5.4 m) rise in sea level for RCP8.5 (*low confidence*). There is *low confidence* in threshold temperatures for ice sheet instabilities and the rates of GMSL rise they can produce. {Cross-Chapter Box 5 in Chapter 1, Cross-Chapter Box 8 in Chapter 3, and Sections 4.1, 4.2.3.1.1, 4.2.3.1.2, 4.2.3.6}

Sea level rise is not globally uniform and varies regionally. Thermal expansion, ocean dynamics and land ice loss contributions will generate regional departures of about ±30% around the GMSL rise. Differences from the global mean can be greater than ±30% in areas of rapid vertical land movements, including those caused by local anthropogenic factors such as groundwater extraction (*high confidence*). Subsidence caused by human activities is currently the most important cause of relative sea level rise (RSL) change in many delta regions. While the comparative importance of climate-driven RSL rise will increase over time, these findings on anthropogenic subsidence imply that a consideration of

local processes is critical for projections of sea level impacts at local scales (*high confidence*). {4.2.1.6, 4.2.2.4}

Due to projected GMSL rise, ESLs that are historically rare (for example, today's hundred-year event) will become common by 2100 under all RCPs (*high confidence*). Many low-lying cities and small islands at most latitudes will experience such events annually by 2050. Greenhouse gas (GHG) mitigation envisioned in low-emission scenarios (e.g., RCP2.6) is expected to sharply reduce but not eliminate risk to low-lying coasts and islands from SLR and ESL events. Low-emission scenarios lead to slower rates of SLR and allow for a wider range of adaptation options. For the first half of the 21st century differences in ESL events among the scenarios are small, facilitating adaptation planning. {4.2.2.5, 4.2.3.4}

Non-climatic anthropogenic drivers will continue to increase the exposure and vulnerability of coastal communities to future SLR and ESL events in the absence of major adaptation efforts compared to today (*high confidence*). {4.3.4, Cross-Chapter Box 9}

The expected impacts of SLR on coastal ecosystems over the course of the century include habitat contraction, loss of functionality and biodiversity, and lateral and inland migration. Impacts will be exacerbated in cases of land reclamation and where anthropogenic barriers prevent inland migration of marshes and mangroves and limit the availability and relocation of sediment (*high confidence*). Under favourable conditions, marshes and mangroves have been found to keep pace with fast rates of SLR (e.g., >10 mm yr^{-1}), but this capacity varies significantly depending on factors such as wave exposure of the location, tidal range, sediment trapping, overall sediment availability and coastal squeeze (*high confidence*). {4.3.3.5.1}

In the absence of adaptation, more intense and frequent ESL events, together with trends in coastal development will increase expected annual flood damages by 2–3 orders of magnitude by 2100 (*high confidence*). However, well designed coastal protection is very effective in reducing expected damages and cost efficient for urban and densely populated regions, but generally unaffordable for rural and poorer areas (*high confidence*). Effective protection requires investments on the order of tens to several hundreds of billions of USD yr^{-1} globally (*high confidence*). While investments are generally cost efficient for densely populated and urban areas (*high confidence*), rural and poorer areas will be challenged to afford such investments with relative annual costs for some small island states amounting to several percent of GDP (*high confidence*). Even with well-designed hard protection, the risk of possibly disastrous consequences in the event of failure of defences remains. {4.3.4, 4.4.2.2, 4.4.3.2, Cross-Chapter Box 9}

Risk related to SLR (including erosion, flooding and salinisation) is expected to significantly increase by the end of this century along all low-lying coasts in the absence of major additional adaptation efforts (*very high confidence*). While only urban atoll islands and some Arctic communities are expected to experience moderate to high risk relative to today in a low emission pathway, almost high to very high risks are expected in all low-lying coastal settings at the upper end of the *likely* range for high emission pathways (*medium confidence*). However, the transition from moderate to high and from high to very high risk will vary from one coastal setting to another (*high confidence*). While a slower rate of SLR enables greater opportunities for adapting, adaptation benefits are also expected to vary between coastal settings. Although ambitious adaptation will not necessarily eradicate end-century SLR risk (*medium confidence*), it will help to buy time in many locations and therefore help to lay a robust foundation for adaptation beyond 2100. {4.1.3, 4.3.4, Box 4.1, SM4.2}

Choosing and Implementing Responses

All types of responses to SLR, including protection, accommodation, EbA, advance and retreat, have important and synergistic roles to play in an integrated and sequenced response to SLR (*high confidence*). Hard protection and advance (building into the sea) are economically efficient in most urban contexts facing land scarcity (*high confidence*), but can lead to increased exposure in the long term. Where sufficient space is available, EbA can both reduce coastal risks and provide multiple other benefits (*medium confidence*). Accommodation such as flood proofing buildings and EWS for ESL events are often both low-cost and highly cost-efficient in all contexts (*high confidence*). Where coastal risks are already high, and population size and density are low, or in the aftermath of a coastal disaster, retreat may be especially effective, albeit socially, culturally and politically challenging. {4.4.2.2, 4.4.2.3, 4.4.2.4, 4.4.2.5, 4.4.2.6, 4.4.3}

Technical limits to hard protection are expected to be reached under high emission scenarios (RCP8.5) beyond 2100 (*high confidence*) and biophysical limits to EbA may arise during the 21st century, but economic and social barriers arise well before the end of the century (*medium confidence*). Economic challenges to hard protection increase with higher sea levels and will make adaptation unaffordable before technical limits are reached (*high confidence*). Drivers other than SLR are expected to contribute more to biophysical limits of EbA. For corals, limits may be reached during this century, due to ocean acidification and ocean warming, and for tidal wetlands due to pollution and infrastructure limiting their inland migration. Limits to accommodation are expected to occur well before limits to protection occur. Limits to retreat are uncertain, reflecting research gaps. Social barriers (including governance challenges) to adaptation are already encountered. {4.4.2.2, 4.4.2.3., 4.4.2.3.2, 4.4.2.5, 4.4.2.6, 4.4.3, Cross-Chapter Box 9}

Choosing and implementing responses to SLR presents society with profound governance challenges and difficult social choices, which are inherently political and value laden (*high confidence*). The large uncertainties about post 2050 SLR, and the substantial impact expected, challenge established planning and decision making practises and introduce the need for coordination within and between governance levels and policy domains. SLR responses also raise equity concerns about marginalising those most vulnerable and could potentially spark or compound social conflict (*high confidence*). Choosing and implementing responses is further challenged through a lack of resources, vexing trade-offs between safety, conservation and economic development, multiple ways of framing the 'sea level rise problem', power relations, and various coastal stakeholders having conflicting interests in the future development of heavily used coastal zones (*high confidence*). {4.4.2, 4.4.3}

Despite the large uncertainties about post 2050 SLR, adaptation decisions can be made now, facilitated by using decision analysis methods specifically designed to address uncertainty (*high confidence*). These methods favour flexible responses (i.e., those that can be adapted over time) and periodically adjusted decisions (i.e., adaptive decision making). They use robustness criteria (i.e., effectiveness across a range of circumstances) for evaluating alternative responses instead of standard expected utility criteria (*high confidence*). One example is adaptation pathway analysis, which has emerged as a low-cost tool to assess long-term coastal responses as sequences of adaptive decisions in the face of dynamic coastal risk characterised by deep uncertainty (*medium evidence, high agreement*). The range of SLR to be considered in decisions depends on the risk tolerance of stakeholders, with stakeholders whose risk tolerance is low also considering SLR higher than the *likely* range. {4.1, 4.4.4.3}

Adaptation experience to date demonstrates that using a locally appropriate combination of decision analysis, land use planning, public participation and conflict resolution approaches can help to address the governance challenges faced in responding to SLR (*high confidence*). Effective SLR responses depend, first, on taking a long-term perspective when making short-term decisions, explicitly accounting for uncertainty of locality-specific risks beyond 2050 (*high confidence*), and building governance capabilities to tackle the complexity of SLR risk (*medium evidence, high agreement*). Second, improved coordination of SLR responses across scales, sectors and policy domains can help to address SLR impacts and risk (*high confidence*). Third, prioritising consideration of social vulnerability and equity underpins efforts to promote fair and just climate resilience and sustainable development (*high confidence*) and can be helped by creating safe community arenas for meaningful public deliberation and conflict resolution (*medium evidence, high agreement*). Finally, public awareness and understanding about SLR risks and responses can be improved by drawing on local, indigenous and scientific knowledge systems, together with social learning about locality-specific SLR risk and response potential (*high confidence*). {4.4.4.2, 4.4.5, Table 4.9}

Achieving the United Nations Sustainable Development Goals (SDGs) and charting Climate Resilient Development Pathways depends in part on ambitious and sustained mitigation efforts to contain SLR coupled with effective adaptation actions to reduce SLR impacts and risk (*medium evidence, high agreement*).

4

4.1 Synthesis

4.1.1 Purpose, Scope, and Structure of this Chapter

This chapter assesses the literature published since the AR5 on past and future contributions to global, regional and ESL changes, associated risk to low-lying islands, coasts, cities and settlements, and response options and pathways to resilience and sustainable development along the coast. The chapter follows the risk framework of AR5, in which risk is assessed in terms of hazard, exposure and vulnerability (Cross-Chapter Box 1 Chapter 1; Box 4.1), and is structured as follows (Figure 4.1):

- Section 4.1 (this section) presents a high-level synthesis of our assessment and provides entry points to more specific content found in the other sections.
- Section 4.2 assesses the current understanding of processes contributing to mean and extreme SLR globally, regionally and locally, with an emphasis on new insights about the AIS contribution.
- Section 4.3 assesses how mean and extreme sea level changes translate into coastal hazards (e.g., flooding, erosion and salinity intrusion), how these interact with socioeconomic drivers of coastal exposure and vulnerability, and how this interaction translates into observed impacts and projected risks for ecosystems, natural resources and human systems.
- Section 4.4 assesses the cost, effectiveness, co-benefits, efficiency, and technical limits of different types of SLR responses and identifies governance challenges (also called barriers) associated

with choosing and implementing responses. Next, planning, public participation, conflict resolution and decision analysis methods for addressing the identified governance challenges are assessed, as well as practical lessons learned in local cases.

4.1.2 Future Sea level Rise and Implications for Responses

For understanding responses to climate-change induced SLR, two aspects of sea level are important to note initially:

1. Climate-change induced GMSL rise is caused by thermal expansion of ocean water and ocean mass gain, the latter primarily due to a decrease in land-ice mass. However, responses to SLR are local and hence always based on RSL experienced at a particular location. GMSL is modified regionally by climate processes and locally by a variety of factors, some driven or influenced by human activity. Of particular relevance for responding to SLR is anthropogenic subsidence, which can lead to rates of RSL rise that exceed those of climate-induced SLR by an order of magnitude, specifically in delta regions and near cities (4.2.2.4). In these subsiding regions, one available response to prepare for future climate-induced SLR is to manage and reduce anthropogenic subsidence (4.4.2).

2. The combination of gradual change of mean sea level with ESL events such as tides, surges and waves causes coastal impacts (4.2.3). ESL events at the coast that are rare today will become more frequent in the future, which means that for many locations, the main starting point for coastal planning

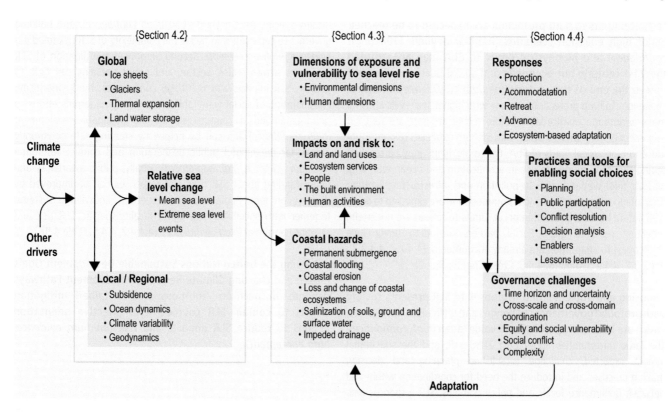

Figure 4.1 | Schematic illustration of the interconnection of Chapter 4 themes, including drivers of sea level rise (SLR) and (extreme) sea level hazards (Section 4.2), exposure, vulnerability, impacts and risk related to SLR (Section 4.3), and responses, associated governance challenges and practises and tools for enabling social choices and addressing governance challenges (Section 4.4).

and decision making is information on current and future ESL events. One important response for preparing for future SLR is to improve observational systems (tide gauges, wave buoys and remote sensing techniques), because in many places around the world current frequencies and intensities of ESL events are not well understood due to a lack of observational data (4.2.3.4.1).

After an increase of sea level from 1–2 mm yr^{-1} in most regions over the past century, rates of 3–4 mm yr^{-1} are now being experienced that will further increase to 4–9 mm yr^{-1} under RCP2.6 and 10–20 mm yr^{-1} at the end of the century under RCP8.5. Nevertheless, up to 2050, uncertainty in climate change-driven future sea level is relatively small, which provides a robust basis for short-term (£30 years) adaptation planning. GMSL will rise between 0.24 m (0.17–0.32 m, *likely* range) under RCP2.6 and 0.32 m (0.23–0.40 m, *likely* range) under RCP8.5 (*medium confidence*; 4.2.3). The combined effect of mean and extreme sea levels results in events which are rare in the historical context (return period of 100 years or larger; probability <0.01 yr^{-1}) occurring yearly at some locations by the middle of this century under all emission scenarios (4.2.3.4.1; *high confidence*). This includes, for instance, those parts of the intertropical low-lying coasts that are currently exposed to storm surges only infrequently. Hence, additional adaptation is needed irrespective of the uncertainties in future global GHG emissions and the Antarctic contribution to SLR.

Beyond 2050, uncertainty in climate change induced SLR increases substantially due to uncertainties in emission scenarios and the associated climate changes, and the response of the AIS in a warmer world. Combining process-model based studies in which there is *medium confidence*, it is found that GMSL is projected to rise between 0.43 m (0.29–0.59 m, *likely* range) under RCP 2.6 and 0.84 m (0.61–1.10 m,

likely range) under RCP 8.5 by 2100 (Figure 4.3). The range that needs to be considered for planning and implementing coastal responses depends on the risk tolerance of stakeholders (i.e., those deciding and those affected by a decision; 4.4.4.3.2). Stakeholders that are risk tolerant (e.g., those planning for investments that can be very easily adapted to unforeseen conditions) may prefer to use the *likely* ranges of RCP2.6 and RCP8.5 for long-term adaptation planning. Stakeholders with a low risk tolerance (e.g., those planning for coastal safety in cities and long term investment in critical infrastructure) may also consider SLR above this range, because there is a 17% chance that GMSL will exceed 0.59 m under RCP2.6 and 1.10 m under RCP8.5 in 2100. Process-model based studies cannot yet provide this information, but expert elicitation studies show that a GMSL of 2 m in 2100 cannot be ruled out (4.2.3).

Despite the large uncertainty in late 21st century SLR, progress in adaptation planning and implementation is feasible today and may be economically beneficial. Many coastal decisions with time horizons of decades to over a century are made today (e.g., critical infrastructure, coastal protection works, city planning, etc.) and accounting for relative SLR can improve these decisions. Decision-analysis methods specifically targeting situations of large uncertainty are available and, combined with suitable planning, public participation and conflict resolution processes, can improve outcomes (*high confidence*; 4.4.4.2, 4.4.4.3). For example, adaptation pathway analysis recognises and enables sequenced long-term decision making in the face of dynamic coastal risk characterised by deep uncertainty (*medium evidence, high agreement*; 4.4.4.3.4). The use of these decision-analysis tools can be integrated into statutory land use or spatial planning provisions to formalise these decisions and enable effective implementation by relevant governing authorities (4.4.4.2).

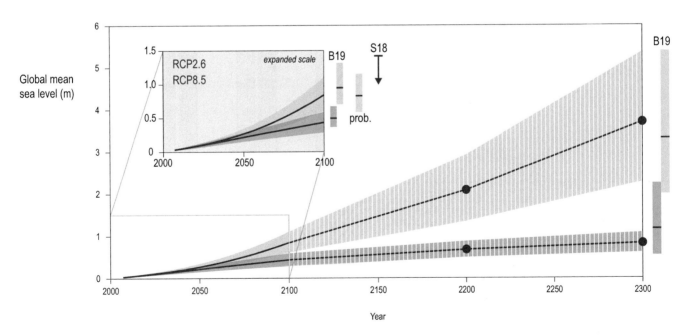

Figure 4.2 | Projected sea level rise (SLR) until 2300. The inset shows an assessment of the *likely* range of the projections for RCP2.6 and RCP8.5 up to 2100 (*medium confidence*). Projections for longer time scales are highly uncertain but a range is provided (4.2.3.6; *low confidence*). For context, results are shown from other estimation approaches in 2100 and 2300. The two sets of two bars labelled B19 are from an expert elicitation for the Antarctic component (Bamber et al., 2019), and reflect the *likely* range for a 2°C and 5°C temperature warming (*low confidence*; details section 4.2.3.3.1). The bar labelled "prob." indicates the *likely* range of a set of probabilistic projections (4.2.3.2). The arrow indicated by S18 shows the result of an extensive sensitivity experiment with a numerical model for the Antarctic Ice Sheet (AIS) combined, like the results from B19 and "prob.", with results from Church et al. (2013) for the other components of SLR. S18 also shows the *likely* range.

Beyond 2100, sea level will continue to rise for centuries and will remain elevated for thousands of years (*high confidence;* 4.2.3.5). Only a few modelling studies are available for SLR beyond 2100. However, all studies agree that the difference in GMSL between RCP2.6 and RCP8.5 increases substantially on multi-centennial and millennial time scales (*very high confidence*). On a millennial time scale, this difference is about 10 metres in some model simulations, whereas it is only several decimetres at the end of 21st century. The larger the emissions the larger the risks associated with SLR as already assessed in SR1.5. Under RCP8.5 the few available studies indicate a *likely* range of 2.3–5.4 m (*low confidence*) in 2300. With strong mitigation efforts (RCP2.6), SLR will be kept to a *likely* range of 0.6–1.1 m (Figure 4.2). Regardless, ambitious and sustained adaptation efforts are needed to reduce risks.

4.1.3　Sea Level Rise Impacts and Implications for Responses

Rising mean and increasingly extreme sea level threaten coastal zones through a range of coastal hazards including (i) the permanent submergence of land by higher mean sea levels or mean high tides; (ii) more frequent or intense coastal flooding; (iii) enhanced coastal erosion; (iv) loss and change of coastal ecosystems; (v) salinisation of soils, ground and surface water; and (vi) impeded drainage. At the century scale and without adaptation, the vast majority of low-lying islands, coasts and communities face substantial risk from these coastal hazards, whether they are urban or rural, continental or island, at any latitude, and irrespective of their level of development (Section 4.3.4; Figure 4.3; *high confidence*). In the absence of an ambitious increase in adaptation efforts compared to those currently underway, high to very high risks are expected in many coastal geographies at the upper end of the RCP8.5 *likely* range. These include resource-rich coastal cities, urban atoll islands, densely populated deltas, and Arctic communities (Chapter 4 Box 4; Figure 4.3 and Section 4.3.4). At the same time coastal protection is very effective and cost-efficient for cities but not for less densely populated rural areas. Some geographies, such as urban atoll islands and Arctic communities face moderate to high risk even under RCP2.6 (*medium confidence*).

In many places, however, non SLR-related, local environmental and human dimensions of exposure and vulnerability play a critical role in increasing exposure and vulnerability to coastal hazards (Section 4.3.2.5). For example, the ability of morphological and ecological systems (Sections 4.3.3.3 and 4.3.3.5) to protect human settlements and infrastructure by attenuating ESL events and stabilising shorelines is progressively being lost due to coastal squeeze, pollution, habitat degradation and fragmentation (Section 4.3.3.5.4; *high confidence*). Hence, an important near term response to RSL rise is to reduce these adverse environmental and human dimensions of exposure and vulnerability. In addition, the drivers of exposure and vulnerability vary across different coastal contexts ranging from resource-rich cities to small islands (Sections 4.3.3, 4.3.4). Accordingly, effective responses need to be context-specific, and address the locality-specific drivers of risk.

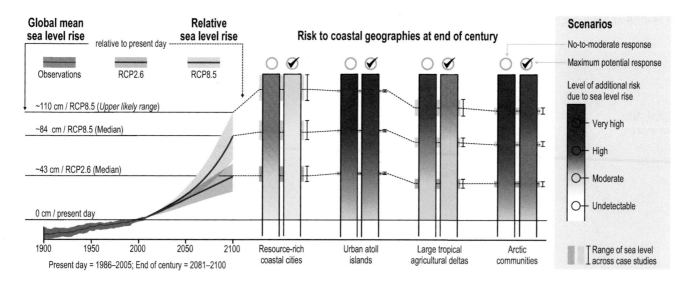

Figure 4.3 | Additional risk related to sea level rise (SLR) for low-lying coastal areas by the end of the 21st century. Section 4.3.4 provides a synthesis of the assessment methodology and the findings, while SM4.3 provides details. Left-hand panel describes global mean sea level (GMSL) rise observations for the Present-Day (1986–2005) and projections under RCP2.6 and RCP8.5 by 2100 relative to the Present-Day according to advances in this chapter. Relative sea level (RSL) changes at specific locations are represented by the coloured blocs (range of the real-world case studies used) and coloured dotted lines (mean) at the background of the middle panel, which describes risk to illustrative geographies as assessed in this chapter. Each illustrative geography is supported by real-world case studies described in the literature (Box 4.1, 4.3.4.1 and Table SM4.2.5): three for resource-rich coastal cities, three for urban atoll islands, two for large tropical agricultural deltas, five for Arctic communities. N.B (1): Only Arctic communities remote from regions of rapid glacial-isostatic adjustment have been selected for this assessment. N.B (2): according to the specific scope of the chapter, this assessment focuses on the additional risks due to SLR and does not account for changes in extreme event climatology (Sections 4.2.3.4.1 to 4.2.3.4.3, 6.3.1.1 to 6.3.1.3), which in some cases would imply a different level of risk than assessed here. The middle panel also distinguishes between two adaptation scenarios. (A) 'No-to-moderate response' represents a business-as-usual scenario where no major additional adaptation efforts compared to today's level of effort are implemented (i.e., neither further significant intensification of action nor new types of actions). (B) 'Maximum potential response' represents the opposite situation, that is, an ambitious combination of both incremental and transformational adaptation that leads to significant additional efforts compared to today and to (A). Here, the authors assume adaptation implemented at its full potential, that is, the extent of adaptation that is technologically possible, with minimal financial, social and political barriers.

4.1.4   Response Options, Governance Challenges and Ways Forward

Responding to SLR refers to reducing hazards, exposure and vulnerability of low-lying coastal areas. It can be approached in fundamentally different ways and five major categories are described in this chapter (Box 4.3): Protection reduces coastal risk and impacts by blocking the inland propagation and other effects of mean or extreme sea levels hazards (e.g., through dikes, seawalls, storm surge barriers, breakwaters, beach-dune systems, etc.). Advance creates new land by building seawards (e.g., reclamation of new land above sea levels or planting vegetation with the specific intention to support natural accretion of land). Ecosystem-based adaptation (EbA) provides a combination of the benefits of protect and advance strategies based on the conservation and restoration of ecosystems such as reefs and coastal vegetation. Accommodation includes a diverse set of biophysical and institutional responses to reduce vulnerability of coastal residents, human activities, ecosystems and the built environment (e.g., raising buildings, planting salt tolerant crops, insurance and EWS for ESL events). Retreat reduces exposure to coastal hazards by moving people, assets and human activities out of the exposed coastal area.

Each type of response has particular advantages and disadvantages, and may play a synergistic role in an integrated and sequenced response to SLR. For example, hard protection needs less space and its effectiveness is more predictable than for EbA (*high confidence*; 4.4.2.2.4, 4.4.2.3.4). EbA has advantages of contributing to conservation goals and providing additional ecosystem services such as carbon sequestration and improved water quality (4.4.2.2.5). EbA can become more effective over time, because coastal ecosystems can migrate inland with rising sea levels, provided this is not restricted by infrastructure (4.4.2.2.4). In practise, hard, sediment-based and ecosystem-based protection responses are often combined and there is *high agreement* that such hybrid approaches are a promising way forward (4.4.2.3.1). Advance is an option widely practised when land is scarce and offers the opportunity to finance adaptation through land sale revenues, but can also increase exposure and destroy coastal wetlands and their protective function (4.4.2.4). Accommodation measures such as flood proofing buildings, flood forecasting, early warning and emergency planning have high benefit-cost ratios, which means that implementing them is much cheaper than doing nothing (4.4.2.5.6). Retreat, and avoidance of development in some locations, are the only types of responses that eliminate residual risks, assuming there is sufficiently safe alternative land to retreat to or develop (4.4.2.6, Cross-Chapter Box 9).

Given diverse geographies and contexts (4.1.3), and the pros and cons of different responses, there is no silver bullet for responding to SLR. Rather, each coastal locality requires a tailor-made response that uses an appropriate mix of measures, sequenced over time as sea level rises. Possible integrated response strategies are illustrated for two contrasting types of settlements: densely populated urban and sparsely populated rural coasts.

For densely populated urban low elevation areas, including continental and island cities and megacities, hard protection has played and will continue to play the central role in response strategies (4.4.2.2, Box 4.1). In general, it is technologically feasible and economically efficient to protect large parts of cities against 21st century SLR (*high confidence*; 4.4.2.2.4, 4.4.2.7). However, questions of affordability remain for poorer and developing regions (4.3.3.4, 4.4.2.2.3). In cities, advance can offer a way to finance coastal protection through revenues generated from newly created land (4.4.2.4), but raises equity concerns with regard to the distribution of costs and access to the new land (4.4.2.4.6). Where space is available, EbA can supplement hard protection (4.4.2.3), except in situations where other human interventions, like infrastructure and pollution, interfere with EbA, especially for RCP8.5 (Cross-Chapter Box 9). Retreat may currently be favoured over rebuilding in the aftermath of major flooding disasters, but in densely populated areas protected by hard infrastructure, general retreat need not be considered until later in the century once it is known whether or not SLR will reach the higher end of the projections (1.1 m or more by 2100; 4.4.2.6).

Along sparsely populated rural coasts, safeguarding communities by conserving coastal ecosystems and natural morphodynamic processes, and restoring those already degraded, is the central element of an integrated strategy. Intact coastal ecosystems can protect settlements and, in some contexts, natural sedimentation processes and avoiding sand mining can help to raise exposed land (4.4.2.2). Hard coastal protection can lead to flooding or erosion elsewhere (4.4.2.2.5), and the destruction of ecosystems and the coastal protection they offer (4.3.3.5). Ecosystem health can be further maintained by reducing non-climatic drivers such as those that interrupt sediment flows in deltas and estuaries (4.3.2.3). Hard protection may be appropriate for areas containing high value assets (e.g., settlements and cultural sites). Retreat is worth considering now where coastal population size and density is low, risks are already high, and the economic, cultural and sociopolitical impacts of retreat and resettlement are carefully considered and addressed by at-risk communities and their governing authorities.

Designing and implementing an appropriate mix of responses is not only a technical task but also an inherently political and value-laden social choice that involves trade-offs between multiple values, goals and interests (Section 4.4.3). Specifically, distinctive features of SLR together with this complex nature of social choices give rise to five overarching governance challenges (Section 4.4.3.3):

1. *Time horizon and uncertainty* associated with SLR beyond 2050 challenge standard planning and decision making practises (*high confidence*).
2. *Cross-scale and cross-domain coordination* linking differing jurisdictional levels, sectors and policy domains is often needed for effective responses (*medium confidence*).
3. *Equity and social vulnerability* are often negatively affected by SLR and also responses to SLR, which can undermine societal aspirations such as achieving the SDGs (*high confidence*).
4. *Social conflict* (i.e., nonviolent struggle between groups, organisations and communities over values, interests, resources, and influence or power) caused or exacerbated by SLR could escalate over time and become very difficult to resolve (*high confidence*).

5. **Complexity**, reinforced by the combination of the above challenges, makes it difficult to understand and address SLR (*high confidence*).

These governance challenges can be addressed through an integrated combination of well-established and emerging planning, public participation and conflict resolution practices (Section 4.4.4.2), decision analysis methods (Section 4.4.4.3) and enabling conditions (Section 4.4.5). For example, iterative planning and flexible, adaptive and robust decision making (RDM) can help coastal communities to plan for the future and account for SLR uncertainty. Planning can also enable thinking and action across spatial, temporal and governance scales and thus help to coordinate roles and responsibilities across multiple governance levels. Public participation approaches can be designed to account for divergent perspectives in making difficult social choices, enhancing social learning, experimentation and innovation in developing locally appropriate SLR responses. Conflict resolution approaches have considerable potential to improve adaptation prospects by harnessing the productive potential of nonviolent conflict.

4.2 Physical Basis for Sea Level Change and Associated Hazards

As a consequence of natural and anthropogenic changes in the climate system, sea level changes are occurring on temporal and spatial scales that threaten coastal communities, cities, and low-lying islands. Sea level in this context means the time average height of the sea surface, thus eliminating short duration fluctuations like waves, surges and tides. GMSL rise refers to an increase in the volume of ocean water caused by warmer water having a lower density, and by the increase in mass caused by loss of land ice or a net loss in terrestrial water reservoirs. Spatial variations in volume changes are related to spatial changes in the climate. In addition, mass changes due to the redistribution of water on the Earth's surface and deformation of the lithosphere leads to a change in the Earth's rotation and gravitational field, producing distinct spatial patterns in regional sea level change. In addition to the regional changes associated with contemporary ice and water redistribution, the solid Earth may cause sea level changes due to tectonics, mantle dynamics or glacial isostatic adjustment (see Section 4.2.1.5). These processes cause vertical land motion (VLM) and sea surface height changes at coastlines. Hence, RSL change is defined as the change in the difference in elevation between the land and the sea surface at a specific time and location (Farrell and Clark, 1976). Here, regional sea level refers to spatial scales of around 100 km, while local sea level refers to spatial scales smaller than 10 km.

In most places around the world, current annual mean rates of RSL changes are typically on the order of a few mm yr^{-1} (see Figure 4.6). Risk associated with changing sea level also is related to individual events that have a limited duration, superimposed on the background of these gradual changes. As a result, the gradual changes in time and space have to be assessed together with processes that lead to flooding and erosion events. These processes include storm surges, waves and tides or a combination of these processes and

lead to ESL events (see Figure 4.4). In this section, newly emerging understanding of these different episodic and gradual aspects of sea level change are assessed, within a context of sea level changes measured directly over the last century, and those inferred for longer geological time scales. This longer-term perspective is important for contextualising future projections of sea level and providing guidance for process-based models of the individual components of SLR, in particular the ice sheets. In addition, anthropogenic subsidence may affect local sea level substantially in many locations but this process is not taken into account in values reported here for projected SLR unless specifically noted.

4.2.1   Processes of Sea Level Change

Sea level changes have been discussed throughout the various IPCC assessment reports as SLR is a key feature of climate change. Complex interactions between the oceans and ice sheets only recently have been recognised as important drivers of processes that can lead to rapid dynamical changes in the ice sheets. Understanding of basal melt below the ice shelves, ice calving processes and glacial hydrological processes was also limited. Projections of future sea level in the IPCC 4th Assessment Report (AR4; Lemke et al., 2007) were presented with the caveat that dynamical ice sheet processes were not accounted for, as our physical understanding of these processes was too rudimentary and no literature could be assessed (Bindoff et al., 2007). In AR5 (Church et al., 2013), a first attempt was made to quantify the dynamic contribution of the ice sheets, although still with modeling based on limited physcis, relying mainly on an extrapolation of existing observations (Little et al., 2013) and a single process based case study (Bindschadler et al., 2013). Here the focus is on sea level changes around coastlines and low-lying islands, updating the GMSL rise by including a new estimate of the dynamic contribution of Antarctica. The mechanism driving past and contemporary sea level changes and episodic extremes of sea level is explained, and confidence in regional projections of future sea level over the 21st century and beyond is assessed.

4.2.1.1   Ice Sheets and Ice Shelves

The ice sheets on Greenland and Antarctica contain most of the fresh water on the Earth's surface. As a consequence, they have the greatest potential to cause changes in sea level. Figure 4.4 illustrates the size of land ice reservoirs and the most important processes that drive mass changes of ice sheets.

Ice sheets change sea level through the loss or gain of ice above flotation, defined as the ice thickness in exceedance of the smallest thickness that would remain in contact with the sea floor at hydrostatic equilibrium. The GIS is currently losing mass at roughly twice the pace of the AIS (Table 4.1). However, Antarctica contains eight times more ice above flotation than Greenland. Furthermore, a substantial fraction of the AIS rests on bedrock below sea level, making the ice sheet responsive to changes in ocean-driven melt and possibly vulnerable to marine ice sheet instabilities (Cross-Chapter Box 8 in Chapter 3) that can drive rapid mass loss.

Ice sheets gain or lose mass through changes in surface mass balance (SMB), the sum of accumulation and ablation controlled by atmospheric processes, the loss of ice to the ocean though melting of ice shelves, and by calving (breaking off of ice bergs) at marine-terminating ice fronts (see Chapter 3). Ice shelves, the floating extensions of grounded ice flowing into the ocean (Figure 4.4) do not directly contribute to sea level, but they play an important role in ice sheet dynamics by providing resistance to the seaward flow of the grounded ice upstream (Fürst et al., 2016; Reese et al., 2018b). Ice shelves gain mass through the inflow of ice from the ice sheet, precipitation, and accretion at the ice-ocean interface. They lose mass through a combination of calving and by melting from below, especially where basal ice is in contact with warm water (Paolo et al., 2015, Khazendar et al., 2016). Calving rates at the terminus of marine terminating ice fronts are governed by complex ice-mechanical processes, the internal strength of the ice, and interaction with ocean waves and tides (Benn et al., 2007; Bassis, 2011; Massom et al., 2018). Sub-ice shelf melts rates are controlled by ice-ocean interactions involving the large-scale circulation, more localised heat and fresh water fluxes, and micro (mm)-scale processes in the ice-ocean boundary layer (Gayen et al., 2015; Dinniman et al., 2016; Schodlok et al., 2016). Ice shelves are also impacted by surface processes. Where surface melt rates are high, ice shelves not only lose mass, they can collapse (hydrofracture) from flexural stresses caused by the movement of the meltwater and the deepening of water-filled crevasses (Banwell et al., 2013; Macayeal and Sergienko, 2013; Kuipers Munneke et al., 2014). These complex ice-ocean interactions, calving and hydrofracture processes remain difficult to model, particularly at the scale of ice sheets.

Our understanding of ice sheets has progressed substantially since AR5, although deep uncertainty (Cross-Chapter Box 5 in Chapter 1) remains with regard to their potential contribution to future SLR on time scales longer than a century under any given emissions scenario. This is particularly true for Antarctica.

4.2.1.2   Glaciers

Glaciers outside of the GIS and AIS are important contributors to sea level change (Figure 4.4). Because of their specific accumulation and ablation rates, which are often high compared to those of the ice sheets, they are sensitive indicators of climate change and respond quickly to changes in climate. Over the past century, glaciers have added more mass to the ocean than the GIS and AIS combined (Gregory et al., 2013). However, the mass of glaciers is small by comparison, equivalent to only 0.32 ± 0.08 m mean SLR if only the fraction of ice above sea level is considered (Farinotti et al., 2019). Sections 2.2.3, 3.3.2 and Cross-Chapter Box 6 in Chapter 2 provide a detailed discussion of glacier response to climate change.

4.2.1.3   Ocean Processes

In general, increasing temperatures lead to a lower density ('thermal expansion') and therefore the larger its volume per unit of mass. Thus, warming leads to a higher sea level even when the ocean mass remains constant. Over at least the last 1500 years changes in sea level were related to global mean temperatures (Kopp et al., 2016), partly

because of ice mass loss, and partly because of thermal expansion. Models and observations indicate that over recent decades, more than 90% of the increase in energy in the climate system has been stored in the ocean. Hence, thermal expansion provides insight into climate sensitivity (Church et al., 2013). Findings from sea level studies and the energy budget are consistent (Otto et al., 2013). As thermal expansion per degree is dependent on the temperature itself, heat uptake by a warm region has a larger impact on SLR than heat uptake by a cold region. This contributes to regional changes in sea level, which are also caused by the water temperature and salinity variations (e.g., Lowe and Gregory, 2006; Suzuki and Ishii, 2011; Bouttes et al., 2014; Saenko et al., 2015). Regional patterns in sea level change are also modified from the global average by oceanic and atmospheric (fluid) dynamics (Griffies and Greatbatch, 2012), including trends in ocean currents, redistribution of temperature and salinity (sea water density), buoyancy, and atmospheric pressure. An analysis of these trends in Coupled Model Intercomparison Project Phase 5 (CMIP5) General Circulation Models (GCMs; Yin, 2012) demonstrates the potential for >15 cm of SLR by 2100 and >30 cm by 2300 (RCP8.5) along the east coast of the USA and Canada from fluid dynamical processes alone. However, Coupled Model Intercomparison Project Phase 6 (CMIP6) GCM simulations are not yet available for an updated analysis of these processes in SROCC.

4.2.1.4   Terrestrial Reservoirs

Global sea level changes are also affected by changes in terrestrial reservoirs of liquid water. Withdrawal of groundwater and storage of fresh water through dam construction (Chao et al., 2008; Fiedler and Conrad, 2010) in the earlier parts of the 20th century dominated, leading to sea level fall, but in recent decades, land water depletion due to domestic, agricultural and industrial usage has begun to contribute to sea level change (Wada et al., 2017). Changes in terrestrial reservoirs may also be related to climate variability: in particular, the El Niño Southern Oscillation (ENSO) has a strong impact on precipitation distribution and temporary storage of water on continents (Boening et al., 2012; Cazenave et al., 2012; Fasullo et al., 2013).

4.2.1.5   Geodynamic Processes

Changing distributions of water mass between land, ice and ocean reservoirs cause nearly instantaneous changes in the Earth's gravity field and rotation, and elastic deformation of the solid Earth. These processes combine to produce spatially varying patterns of sea level change (Mitrovica et al., 2001; Mitrovica et al., 2011). For example, adjacent to an ice sheet losing mass, reduced gravitational attraction between the ice and nearby ocean causes RSL to fall, despite the rise in GMSL from the input of melt water to the ocean. The opposite effect is found far from the ice sheet, where RSL rise can be enhanced as much as 30% relative to the global average.

On time scales longer than the elastic Earth response, redistributions of water and ice cause time-dependent, visco-elastic deformation. This is observed in regions previously covered by ice during the Last Glacial Maximum (LGM), including much of Scandinavia and parts of North America (Lambeck et al., 1998; Peltier, 2004), where glacio-isostatic adjustment (GIA) is causing uplift and a lowering of RSL that continues today. In other locations proximal to the previous ice load, and where a glacial forebulge once existed, the relaxing forebulge can contribute to a relative SLR, as currently being experienced along the coastline of the northeast United States. Water being syphoned to high latitudes as the peripheral bulges collapse leads to a widespread RSL fall in equatorial regions, while the overall loading of ocean crust by melt water can cause uplift of land areas near continental margins, far from the location of previous ice loading (Mitrovica and Milne, 2003; Milne and Mitrovica, 2008). Rates of modern VLM associated with these post-glacial processes are generally on the order of a few mm yr^{-1} or less, but can exceed 1 cm yr^{-1} in some places. Because these gravity, rotation, and deformation (GRD) processes control spatial patterns of SLR from melting land ice, they need to be accounted for in regional-to-local sea level assessments. GRD processes are also important for marine-based ice sheets themselves, because they locally reduce RSL at retreating grounding lines which can slow and reduce retreat (Gomez et al., 2015; see 4.3.3.1.2 and Cross-chapter Box 8 in Chapter 3; Larour et al., 2019).

VLM from tectonics and dynamic topography associated with viscous mantle processes also affect spatial patterns of relative sea level change. These geological processes are important for reconstructing ancient sea levels based on geological indicators (Austermann and Mitrovica, 2015; see SM4.1). Along with other natural and anthropogenic processes including volcanism, compaction, and anthropogenic subsidence from ground water extraction (Section 4.2.2.4) these geodynamic processes can be locally important, producing rates of VLM comparable to or greater than recent climate-driven rates of GMSL change (Wöppelmann and Marcos, 2016). In this chapter, GIA

and anthropogenic subsidence are used, and other components of VLM are ignored unless explicitly stated.

4.2.1.6   Extreme Sea Level Events

Superimposed on gradual changes in RSL, as described in the previous sections, tides, storm surges, waves and other high-frequency processes (Figure 4.4) can be important. Understanding the localised impact of such processes requires detailed knowledge of bathymetry, erosion and sedimentation, as well as a good description of the temporal variability of the wind fields generating waves and storm surges. The potential for compounding effects, like storm surge and high SLR, are of particular concern as they can contribute significantly to flooding risks and extreme events (Little et al., 2015a). These processes can be captured by hydrodynamical models (see Section 4.2.3.4).

4.2.2   Observed Changes in Sea Level (Past and Present)

Sea level changes in the distant geologic past provide information on the size of the ice sheets in climate states different from today. Past intervals with temperatures comparable to or warmer than today are of particular interest, and since AR5 (Masson-Delmotte et al., 2013) they have been increasingly used to test and calibrate process-based ice sheet models used in future projections (DeConto and Pollard, 2016; Edwards et al., 2019; Golledge et al., 2019). These intervals include the mPWP around 3.3–3.0 Ma, when atmospheric CO_2 concentrations were similar to today (~300–450 ppmv; Badger et al., 2013; Martínez-Botí et al., 2015; Stap et al., 2016) and global

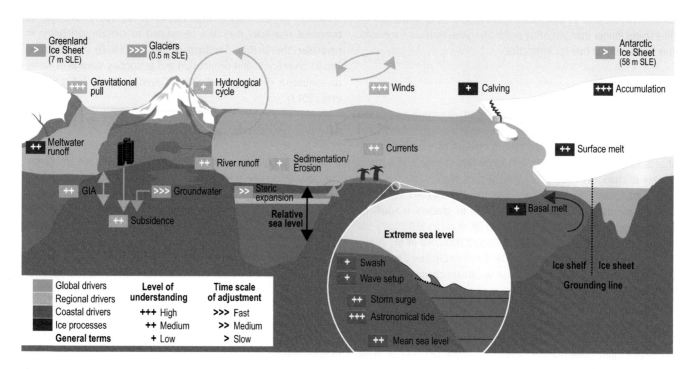

Figure 4.4 | A schematic illustration of the climate and non-climate driven processes that can influence global, regional (green colours), relative and extreme sea level (ESL) events (red colours) along coasts. Major ice processes are shown in purple and general terms in black. SLE stands for Sea Level Equivalent and reflects the increase in GMSL if the mentioned ice mass is melted completely and added to the ocean.

mean temperature was 2°C–4°C warmer than pre-industrial (Dutton et al., 2015a; Haywood et al., 2016) and the LIG around 129–116 ka, when global mean temperature was 0.5°C–1.0°C warmer (Capron et al., 2014; Dutton et al., 2015a; Fischer et al., 2018) and sea surface temperatures were similar to today (Hoffman et al., 2017). Updated reconstructions of GMSL (Dutton et al., 2015a) based on ancient shoreline elevations corrected to account for geodynamic processes (4.2.1.5), and geochemical records extracted from marine sediment cores, indicate sea levels were >5 m higher than today during these past warm periods (*medium confidence*).

Most estimates of peak GMSL during the mPWP range between 6 and 30 m higher than today (Miller et al., 2012; Rovere et al., 2014; Dutton et al., 2015a) but with deep uncertainty (Cross-Chapter Box 5 in Chapter 1) and few constraints on the high end of the range. The large uncertainty is contributed by uncertain GIA corrections applied to palaeo shoreline indicators (Raymo et al., 2011; Rovere et al., 2014), dynamic topography, the vertical land surface motion associated with Earth's mantle flow (Rowley et al., 2013), and possible biases in geochemical records of ice volume derived from marine sediments (Raymo et al., 2018). Estimates of GMSL >10 m higher than today require a meltwater contribution from the East Antarctic Ice Sheet in addition to the GIS and West Antarctic Ice Sheets (WAIS; Miller et al., 2012; Dutton et al., 2015a). Pliocene modelling studies appearing since AR5 (Masson-Delmotte et al., 2013) demonstrate the potential for substantial retreat of East Antarctic ice into deep submarine basins (Austermann and Mitrovica, 2015; Pollard et al., 2015; Aitken et al., 2016; DeConto and Pollard, 2016; Gasson et al., 2016; Golledge et al., 2019), as does emerging geological evidence from marine sediment cores recovered from the East Antarctic margin (Cook et al., 2013; Patterson et al., 2014; Bertram et al., 2018). However, the range of maximum Pliocene GMSL contributions from Antarctic modelling (Austermann and Mitrovica, 2015; Pollard et al., 2015; Yamane et al., 2015; DeConto and Pollard, 2016; Gasson et al., 2016) remains large (5.4–17.8 m), providing little additional constraint on the geological estimates. Land surface exposure measurements on sediment sourced from East Antarctica (Shakun et al., 2018) suggests Pliocene ice loss was limited to marine-based ice, where the bedrock is below sea level and possibly prone to marine ice sheet instabilities (Cross-Chapter box 8 in Chapter 3). The total potential contribution to GMSL rise from marine-based ice in Antarctica is ~22.5 m (Fretwell et al., 2013). Combined with the complete loss of the GIS, this could conceivably produce ~30 m of GMSL rise. However, this would require maximum retreat of GIS and AIS to be synchronous, which is not probable due to orbitally paced, inter-hemispheric asymmetries in Greenland and Antarctic climate (de Boer et al., 2017). As such, 25 m is found to be a reasonable upper bound on GMSL during the mPWP, but with *low confidence*.

An updated estimate of maximum GMSL during the more geologically recent LIG ranges between 6–9 m higher than today (Dutton et al., 2015a). This is close to the values reported by a probabilistic analysis of globally distributed sea level indicators (Kopp et al., 2009), but slightly higher than AR5's central estimate of 6 m. Like the mid-Pliocene, the LIG estimates also suffer from uncertainties in GIA corrections and dynamic topography. Düsterhus et al. (2016) applied data assimilation techniques including GIA corrections to the same LIG dataset used

by Kopp et al. (2009) and found good agreement (7.5 ± 1.1 m likely range) with Kopp et al. (2009) and Dutton et al. (2015a), but the upper range remains poorly constrained. Their estimates of peak LIG sea level are sensitive to the assumed ice history before and after the LIG, consistent with the results of other studies (Lambeck et al., 2012; Dendy et al., 2017). Austermann et al. (2017) compared a compilation of LIG shoreline indicators with dynamic topography simulations. They found that vertical surface motions driven by mantle convection can produce several metres of uncertainty in LIG sea level estimates, but their mean and most probable estimates of 6.7 m and 6.4 m are broadly in line with other estimates.

Greenland and Antarctic climate change on these time scales is influenced by inter-hemispheric differences in polar amplification (Stap et al., 2018), changes in Earth's orbit, and long-term climate system processes. This complicates relationships between global mean temperature and ice sheet response. On Greenland, the magnitude of LIG summer warming and changes in ice sheet volume continue to be contested. Extreme summer warming of 6°C or more, reconstructed from ice cores (Dahl-Jensen et al., 2013; Landais et al., 2016; Yau et al., 2016) and lake archives (McFarlin et al., 2018) is in apparent conflict with a persistent, spatially extensive GIS reconstructed from ice cores and radar imaging (Dahl-Jensen et al., 2013). Maximum retreat of the GIS during the LIG varies widely among modelling studies, ranging from ~1 m to ~6 m (Helsen et al., 2013; Quiquet et al., 2013; Dutton et al., 2015a; Goelzer et al., 2016; Yau et al., 2016); however, the models consistently indicate a small Greenland contribution to GMSL early in the interglacial, implying Antarctica was the dominant contributor to the early interglacial highstand of 6 ± 1.5 m, beginning around 129 ka (Dutton et al., 2015b). An early LIG loss of Antarctic ice is consistent with recent ice sheet modelling (DeConto and Pollard, 2016; Goelzer et al., 2016). Due to its bedrock configuration and susceptibility to marine ice sheet instabilities (Cross-Chapter Box 8 in Chapter 3), the WAIS would have been especially vulnerable to subsurface ocean warming during the LIG (Sutter et al., 2016). However, most evidence of WAIS retreat during the LIG remains indirect (Steig et al., 2015) and firm geological evidence has yet to be uncovered. A recent analysis of East Antarctic sediments provides evidence of some ice retreat in Wilkes subglacial basin during the LIG (Wilson and Forsyth, 2018), but the volume of ice loss is not quantified.

GMSL during the LIG was at times higher than today (*virtually certain*), with a *likely range* between 6–9 m, and not expected to be more than 10 m (*medium confidence*). Due to ongoing uncertainties in the evolution of atmospheric and oceanic warming over and around the ice sheets, and low confidence in the relative contributions of Antarctic versus Greenland meltwater to GMSL change, the LIG is not used here to directly assess the sensitivity of the ice sheets under current or future climate conditions. There is *low confidence* in the utility of changes in either mPWP or LIG sea level changes to quantitatively inform near-term future rates of GMSL rise.

An expanded summary of recent advances and ongoing difficulties in reconstructing these time periods in terms of climate, sea level, and implications for the future evolution of ice sheets and sea level is provided in SM4.1.

4.2.2.1   Global Mean Sea Level Changes During the Instrumental Period

Observational estimates of the sea level variations over past millennia rely essentially on proxy-based regional relative sea level reconstructions corrected for GIA. Since AR5, the increasing availability of regional proxy-based reconstructions enables the estimation of GMSL change over the last ~3 kyr. The first statistical integration of the available reconstructions shows that the GMSL experienced variations of ±9 [±7 to ±11] cm (5–95% uncertainty range; Kopp et al., 2016) over the 2400 years preceding the 20th century (*medium confidence*). This is more tightly bound than the AR5 assessment which indicated a variability in GMSL that was <±25 cm over the same period. This progress since AR5 confirms that it is *virtually certain* that the mean rate of GMSL has increased during the last two centuries from relatively low rates of change during the late Holocene (order tenths of mm yr^{-1}) to modern rates (order mm yr^{-1}; Woodruff et al., 2013).

Over the last two centuries, sea level observations have mostly relied on tide gauge measurements. These records, beginning around 1700 in some locations (Holgate et al., 2012; PSMSL, 2019), provide insight into historic sea level trends. Since 1992, the emergence of precise satellite altimetry has advanced our knowledge on GMSL and regional sea level changes considerably through a combination of near global ocean coverage and high spatial resolution. It has also enabled more detailed monitoring of land ice loss. Since 2002, high precision gravity measurements provided by the Gravity Recovery and Climate Experiment (GRACE) and GRACE Follow-On missions show the loss of land ice in Greenland and Antarctica, and confirm independent assessments of ice sheet mass changes based on satellite altimetry (Shepherd et al., 2012; The Imbie team, 2018) and InSAR measurements combined with ice sheet SMB estimates (Noël et al., 2018; Rignot et al., 2019). Since 2006, when the array of Argo profiling floats reached near-global coverage, it has been possible to get an accurate estimate of the ocean thermal expansion (down to 2000 m depth) and test the closure of the sea level budget. The combined analysis of the different observing systems that are available has improved significantly the understanding of the magnitude and relative contributions of the different processes causing sea level change. In particular, important progress has been achieved since AR5 on estimating and understanding the increasing contribution of the ice sheets to SLR.

4.2.2.1.1   Tide gauge records

The number of tide gauges has increased over time from only a few in northern Europe in the 18th century to more than 2000 today along the world's coastlines. Because of their location and limited number, tide gauges sample the ocean sparsely and non-uniformly with a bias towards the Northern Hemisphere. Most tide gauge records are short and have significant gaps. In addition, tide gauges are anchored on land and are affected by the vertical motion of Earth's crust caused by both natural processes (e.g., GIA, tectonics and sediment compaction; Wöppelmann and Marcos, 2016; Pfeffer et al., 2017) and anthropogenic activities (e.g., groundwater depletion, dam building

or settling of landfill in urban areas; Raucoules et al., 2013; Pfeffer et al., 2017). When estimating the GMSL due to the ocean thermal expansion and land ice melt, tide gauges must be corrected for this VLM, where VLM = GIA + anthropogenic subsidence + (tectonics, natural subsidence). This is possible with stations of the Global Positioning System (GPS) network when they are co-located with tide gauges (Santamaría-Gómez et al., 2017; Kleinherenbrink et al., 2018). However, this approach provides information on the VLM over the past two to three decades and has limited value over longer time scales for places where the VLM has varied significantly through the last century (Riva et al., 2017).

AR5 assessed the different strategies to estimate the 20th century GMSL changes. These strategies only accounted for the inhomogeneous space and time coverage of tide gauge data and for the VLM induced by GIA (Figure 4.5). Since AR5 two new approaches have been developed. The first one uses a Kalman smoother which combines tide gauge records with the spatial patterns associated with ocean dynamic change, change in land ice and GIA. It enables accounting for the inhomogeneous distribution of tide gauges and the VLM associated with both GIA and current land ice loss (Hay et al., 2015; Figure 4.5). The second approach uses ad hoc corrections to tide gauge records with an additional spatial pattern associated with changes in terrestrial water storage to account for the inhomogeneous distribution in tide gauges. It also accounts for the total VLM (Dangendorf et al., 2017; Figure 4.5). Both methods yield significantly lower GMSL changes over the period 1950–1970 than previous estimates, leading to long-term trends since 1900 that are smaller than previous estimates by 0.4 mm yr^{-1} (Figure 4.5). Different arguments including biases in the tide gauge datasets (Hamlington and Thompson, 2015), biases in the averaging technique and absence of VLM correction (Dangendorf et al., 2017), or in the spatial patterns associated with the sea level contributions (Hamlington et al., 2018) have been proposed to explain these smaller GMSL rates. There is no agreement yet on which is the primary reason for the differences and it is not clear whether all the reasons invoked can actually explain all the differences across reconstructions. As there is no clear evidence to discard any reconstruction, this assessment considers the ensemble of AR5 sea level reconstructions augmented by the two recent reconstructions from Hay et al. (2015) and Dangendorf et al. (2017) to evaluate the GMSL changes over the 20th century. On this basis, it is estimated that it is *very likely* that the long-term trend in GMSL estimated from tide gauge records is 1.5 (1.1–1.9) mm yr^{-1} between 1902 and 2010 for a total SLR of 0.16 (0.12–0.21) m (see also Table 4.1) over this period. This estimate is consistent with the AR5 assessment (but with an increased uncertainty range) and confirms that it is *virtually certain* that GMSL rates over the 20th century are several times as large as GMSL rates during the late Holocene (see 4.2.2.1). Over the 20th century the GMSL record also shows an acceleration (*high confidence*) as now four out of five reconstructions extending back to at least 1902 show a robust acceleration (Jevrejeva et al., 2008; Church and White, 2011; Ray and Douglas, 2011; Haigh et al., 2014b; Hay et al., 2015; Watson, 2016; Dangendorf et al., 2017). The estimates of the acceleration ranges between -0.002–0.019 mm yr^{-1} over 1902–2010 are consistent with AR5.

4.2.2.1.2   Satellite altimetry

High precision satellite altimetry started in October 1992 with the launch of the TOPEX/Poseidon and Jason series of spacecraft. Since then, 11 satellite altimeters have been launched providing nearly global sea level measurements (up to ±82°latitude) over more than 25 years. Six groups (AVISO/CNES, SL_cci/ESA, University of Colorado, CSIRO, NASA/GSFC, NOAA; Nerem et al., 2010; Henry et al., 2014; Leuliette, 2015; Watson et al., 2015; Beckley et al., 2017; Legeais et al., 2018) provide altimetry-based GMSL time series. Since AR5, several studies using two independent approaches based on tide gauge records (Watson et al., 2015) and the sea level budget closure (Chen et al., 2017; Dieng et al., 2017) identified a drift of 1.5 (0.4–3.4) mm yr^{-1} in TOPEX A from January 1993 to February 1999. Accounting for this drift leads to a revised GMSL rate from satellite altimetry of 3.16 (2.79–3.53) for 1993–2015 (WCRP Global Sea Level Budget Group, 2018; see Table 4.1) compared to 3.3 mm yr^{-1} (2.7–3.9) for 1993–2010 in AR5. Compared to AR5, the revised satellite altimetry GMSL estimates now show with *high confidence* an acceleration of 0.084 (0.059–0.090) mm yr^{-1} over 1993–2015 (5–95% uncertainty range; Watson et al., 2015; Nerem et al., 2018). This acceleration is due to an increase in Greenland mass loss since the 2000s (Chen et al., 2017; Dieng et al., 2017) and a slight increase in all other contributions probably partly due to the recovery from the Pinatubo volcanic eruption in 1991 (Fasullo et al., 2016) and partly due to increased GHG concentrations e.g., (Slangen et al., 2016; *high confidence*). The current sea level rise is 3.6 ± 0.3 mm yr^{-1} over 2006–2015 (90% confidence level). This is the highest rate measured by satellite altimetry (Ablain et al., 2019; *medium confidence*). Before the satellite altimetry era, the highest rate of sea level rise recorded was reached during the period 1935–1944. It amounted 2.5 ± 0.7 mm yr^{-1} (estimate at the 90% confidence level from sea level reconstructions; Church and White, 2011; Ray and Douglas, 2011; Jevrejeva et al., 2008; Hay et al., 2015; Dangendorf et al., 2017). This is expected to be smaller than the current rate of sea level rise, making the current sea level rise the highest on instrumental record (*medium confidence*).

4.2.2.2   Contributions to Global Mean Sea Level Change During the Instrumental Period

The different contributions to the GMSL rise are independently observed over various time scales. They are compared with simulated estimates from climate model experiments of CMIP5 (Taylor et al., 2012) when available (see Table 4.1). The observations are compared with experiments beginning in the mid-19th century, forced with past time-dependent anthropogenic changes in atmospheric composition, natural forcings due to volcanic aerosols and variations in solar irradiance (Taylor et al., 2012). The objective is first, to assess understanding of the causes of observed sea level changes and second, to evaluate the ability of coupled climate models to simulate these causes. It enables the evaluation of the confidence level there is in current coupled climate models that form the basis of future sea level projections.

4.2.2.2.1   Thermal expansion contribution

The ocean thermal expansion is caused by excess heat being absorbed by the ocean, as the climate warms. Thermal expansion is estimated from *in situ* ocean observations and ocean heat content reanalyses that rely on assimilation of data into numerical models (Storto et al., 2017; Sections 1.8.1.1 and 1.8.1.4; WCRP Global Sea Level Budget Group, 2018). Full-depth, high-quality and unbiased ocean temperature profile data with adequate metadata and spatio-temporal coverage are required to estimate thermal expansion and to understand drivers of variability and long-term change (Pfeffer et al., 2018; Section 5.2.2.2.2).

Historically, however, observational gaps exist and some ocean regions remain under-sampled to date (Sections 1.8.1.1 and 5.2.2.2.2; Figure 1.3; Appendix 1.A, Figure 1.1). Other factors also introduce uncertainty in estimates of thermal expansion like changes in instrumentation, systematic instrumental errors, changes in the quality control of the data and the mapping method used to produce regular grids (Section 5.2.2.2.2; Palmer et al., 2010). In the upper 700 m, the largest sources of uncertainty for estimates of global mean thermal expansion from 1970 to 2004 are the choice of mapping methods (Boyer et al., 2016), followed by the choice of bias correction for the bathythermographic observations (Cheng et al., 2016; Section 5.2.2.2.2). From 2006 onwards, the uncertainty is considerably reduced (Roemmich et al., 2015; von Schuckmann et al., 2016; Wijffels et al., 2016), because the Argo array reached its targeted near-global (up to ±60°latitude) coverage for the upper 2000 m in November 2007 (Riser et al., 2016; Section 5.2.2.2.2).

Since AR5, in a community effort, the (WCRP Global Sea Level Budget Group, 2018) revisited the global mean thermal expansion estimates based on observations only. On the basis of a full-depth 13-member ensemble of global mean thermal expansion time series developed with the latest data and corrections available, they estimated that the global thermal expansion was 1.40 (1.08–1.72) mm yr^{-1} for 2006–2015, 1.36 (0.96–1.76) mm yr^{-1} for 1993–2015 (see Table 4.1). While the relative contribution of the upper 300 m did not change (~70%) between 2006–2015 and 1993–2015, the 700–2000 m contribution increased around 10% over the Argo decade (2006–2015), when observations for that depth interval soared (Figure 1.3; Appendix 1.A, Figure 1.1). This suggests that observed changes for 700–2000 m may have been underestimated for 1993–2005. Before 1993, estimates are based on a smaller ensemble of 4 datasets in which no thermal expansion is assumed below 2000 m because of lack of data (see Section 5.2.2.2.2 for more details). This ensemble shows a thermal expansion linear rate of 0.89 (0.84–0.94) mm yr^{-1} for 1970–2015 (see Table 4.1).

Coupled climate models simulate the historical thermal expansion (see Table 4.1). However, for models that omit the volcanic forcing in their control experiment, the imposition of the historical volcanic forcing during the 20th century results in a spurious time mean negative forcing and a spurious persistent ocean cooling related to the control climate (Gregory, 2010; Gregory et al., 2013). Since AR5, the magnitude of this effect has been estimated from historical

Table 4.1 | Global mean sea level (GMSL) budget over different periods from observations and from climate model base contributions. All values are in mm yr^{-1}. Values in brackets in 4.2 are uncertainties ranging from 5–95%. The climate model historical simulations end in 2005; projections for Representative Concentration Pathway (RCP)8.5 are used for 2006–2015. The modelled thermal expansion, glacier and ice sheet surface mass balance (SMB) contributions are computed from the Coupled Model Intercomparison Project Phase 5 (CMIP5) models as in Slangen et al. (2017b). For the model contributions, uncertainties are estimated from the spread of the ensemble of model simulations following Slangen et al. (2017b), see the footnotes for the details on the uncertainty propagation. GIS is Greenland Ice Sheet.

Source	1901–1990	1970–2015	1993–2015	2006–2015
Observed contribution to GMSL rise				
Thermal expansion		0.89 (0.84–0.94)[a]	1.36 (0.96–1.76)[a]	1.40 (1.08–1.72)[a]
Glaciers except in Greenland and Antarctica	0.49 (0.34–0.64)[b]	0.46 (0.21–0.72)[o]	0.56 (0.34–0.78)[p]	0.61 (0.53–0.69)[n]
GIS including peripheral glaciers	0.40 (0.23–0.57)[c]		0.46 (0.21–0.71)[d]	0.77 (0.72–0.82)[d]
Antarctica ice sheet including peripheral glaciers			0.29 (0.11–0.47)[e]	0.43 (0.34–0.52)[e]
Land water storage	–0.12[f]	–0.07[f]	0.09[f]	–0.21 (–0.36–0.06)[g]
Ocean mass				2.23 (2.07–2.39)[h]
Total contributions			**2.76 (2.21–3.31)[i]**	**3.00 (2.62–3.38)[i]**
Observed GMSL rise from tide gauges and altimetry	**1.38 (0.81–1.95)**	**2.06 (1.77–2.34)[j]**	**3.16 (2.79–3.53)[k]**	**3.58 (3.10–4.06)[k]**
Modelled contributions to GMSL rise				
Thermal expansion	0.32 (0.04–0.60)	0.97 (0.45–1.48)	1.48 (0.86–2.11)	1.52 (0.96–2.09)
Glaciers	0.53 (0.38–0.68)	0.73 (0.50–0.95)	0.99 (0.60–1.38)	1.10 (0.64–1.56)
Greenland SMB	–0.02 (–0.05–0.02)	0.03 (–0.01–0.07)	0.08 (–0.01–0.16)	0.12 (–0.02–0.26)
Total including land water storage and ice discharge[l]	0.71 (0.39–1.03)	1.88 (1.31–2.45)	3.13 (2.38–3.88)	3.54 (2.79–4.29)
Residual with respect to observed GMSL rise[m]	**0.67 (0.02–1.32)**	**0.18 (–0.46–0.82)**	**0.03 (–0.81–0.87)**	**0.04 (–0.85–0.93)**

Notes:

(a) The number is built from WCRP Global Sea Level Budget Group (2018) estimate of the 0–700 m depth thermal expansion, assuming no trend below 2000 m depth before 1992 and the mean value from Purkey and Johnson (2010), and Desbruyères et al. (2017) afterwards.

(b) The number is calculated as the mean between the estimate from a reconstruction of glacier mass balance based on glacier length (update of Leclercq et al. (2011)) and the estimate from a mass balance model forced with atmospheric observations (Marzeion et al., 2015). The uncertainty is assumed to be a gaussian with a standard deviation of half the difference between the two estimates.

(c) The number is calculated as the sum of the Greenland Ice Sheet (GIS) contribution from Kjeldsen et al. (2015) and the peripheral glaciers' contribution. The peripheral glaciers' contribution and the associated uncertainty are computed from a mass balance model forced with atmospheric observations (Marzeion et al., 2015). The total uncertainty is computed assuming that both uncertainties from the GIS contribution and from the peripheral glaciers' contribution are independent.

(d) Numbers from Bamber et al. (2018). See Section 3.3.1 for more details.

(e) These numbers are the weighted average of the numbers from Bamber et al. (2018) and from The Imbie team (2018). The weights in the average are based on the uncertainty associated to each estimate. See Section 3.3.1 for more details.

(f) Only direct anthropogenic contribution, from Wada et al. (2016).

(g) Land water storage estimated from Gravity Recovery and Climate Experiment (GRACE) excluding glaciers, from WCRP Global Sea Level Budget Group (2018).

(h) Direct estimate of ocean mass from GRACE from WCRP Global Sea Level Budget Group (2018).

(i) Sum of the thermal expansion and the contributions from glaciers, GIS, Antarctica Ice Sheet (AIS) and land water storage. Uncertainties in the different contributions are assumed as independent.

(j) Sea level reconstructions that end before 2015 have been extended to 2015 with the satellite altimetry record from Legeais et al. (2018). The uncertainty is derived from the uncertainty of individual sea level reconstructions over the longest period available that start in 1970. The uncertainty from different sea level reconstructions are assumed as independent.

(k) The mean estimate is from the satellite altimetry estimate in WCRP Global Sea Level Budget Group (2018) corrected for GIA and for the elastic response of the ocean crust to present day mass redistribution (Frederikse et al., 2017; Lickley et al., 2018). The uncertainty is computed using the updated error budget of Ablain et al. (2015).

(l) Land water storage is estimated from Wada et al. (2016) and ice discharge is deduced from Shepherd et al. (2012). The ice discharge contribution is assumed to be zero before 1992. The uncertainties in the different contributions from coupled climate models are assumed independent.

(m) The uncertainties in the observed GMSL and the coupled climate models' estimate of GMSL are assumed independent for the computation of the uncertainties in the residuals.

(n) Numbers taken from Appendix 2.A.

(o) Numbers taken from Zemp et al. (2019), see Sections 2.2.3 and 3.3.2 for more details.

(p) The Number is calculated as the mean of the estimates of Zemp et al. (2019) and Bamber et al. (2018). The uncertainties of the two estimates are assumed to be independent of each other to obtain the uncertainty estimate of the mean.

simulations forced by only natural radiative forcing. Then it has been used to correct the historical simulations forced with the full 20th century forcing (Slangen et al., 2016; Slangen et al., 2017b). The resulting ensemble mean of simulated thermal expansion provides a good fit to the observations within the uncertainty ranges of both models and observations (Slangen et al., 2017b; Cheng et al., 2019; Table 4.1). The spread, which is essentially due to uncertainty in radiative forcing and uncertainty in the modelled climate sensitivity and ocean heat uptake efficiency (Melet and Meyssignac, 2015), is still larger than the observational uncertainties (Gleckler et al., 2016; Cheng et al., 2017; Table 4.1). Compared to AR5, the availability of improved observed and modelled estimates of thermal expansion and the good agreement between both confirm the *high confidence* level in the simulated thermal expansion using climate models and the *high confidence* level in their ability to project future thermal expansion.

4.2.2.2.2   Ocean mass observations from GRACE and GRACE Follow-On

The ocean mass changes correspond to the sum of land ice and terrestrial water storage changes. Since 2002, the GRACE and GRACE follow-on missions provide direct estimates of the ocean mass changes and thus they provide an independent estimate of the sum of land ice and terrestrial water storage contributions to sea level. Since AR5, GRACE-based estimates of the ocean mass rates are increasingly consistent (WCRP Global Sea Level Budget Group, 2018) because of the extended length of GRACE missions' observations (over 15 years), the improved understanding of data and methods for addressing GRACE limitations (e.g., noise filtering, leakage correction and low-degree spherical harmonics estimates), and the improved knowledge of geophysical corrections applied to GRACE data (e.g., GIA). The most recent estimates (Dieng et al., 2015b; Reager et al., 2016; Rietbroek et al., 2016; Chambers et al., 2017; Blazquez et al., 2018; Uebbing et al., 2019) report a global ocean mass increase of 1.7 (1.4–2.0) mm yr^{-1} over 2003–2015 (see also Table 4.1). The uncertainty arises essentially from differences in the inversion method to compute the ocean mass (Chen et al., 2013; Jensen et al., 2013; Johnson and Chambers, 2013; Rietbroek et al., 2016), uncertainties in the geocentre motion and uncertainty in the GIA correction (Blazquez et al., 2018; Uebbing et al., 2019). The consistency between estimates of the global mean ocean mass on a monthly time scale has also increased since AR5.

4.2.2.2.3   Glaciers

To assess the mass contribution of glaciers to sea level change, global estimates are required. Recent updates and temporal extensions of estimates obtained by different methods continue to provide *very high confidence* in continuing glacier mass loss on the global scale during the past decade (Bamber et al., 2018; Wouters et al., 2019; Zemp et al., 2019, see Section 2.2.3 and Appendix 2.A for a detailed discussion also on regional scales). Updates of the reconstructions of Cogley (2009), Leclercq et al. (2011) and Marzeion et al. (2012), presented and compared in Marzeion et al. (2015), show increased agreement on rates of mass loss during the entire 20th century (Marzeion et al., 2015), compared to earlier estimates reported by AR5. The contribution of glaciers that may be missing in inventories or have already melted during the 20th century is hard to constrain (Parkes and Marzeion, 2018), and there is *low confidence* in their estimated contribution. These glaciers are thus neglected in the assessment of the sea level budget (Table 4.1).

While the agreement between the observational estimates of glacier mass changes and the modelled estimates from glacier models forced with climate model simulations has increased since AR5 (Slangen et al., 2017b), there is only *medium confidence* in the use of glacier models to reconstruct sea level change because of the limited number of well-observed glaciers available to evaluate models on long time scales, and because of the small number of model-based global glacier reconstructions.

4.2.2.2.4   Greenland and Antarctic ice sheets

Frequent observations of ice sheet mass changes have only been available since the advent of space observations (see Section 3.3.1). In the pre-satellite era, mass balance was geodetically reconstructed only for the GIS (Kjeldsen et al., 2015). These geodetic reconstructions empirically constrain the contribution of the GIS to SLR between 1900 and 1983 to 17.2 (10.7–23.2; Kjeldsen et al., 2015). During the satellite era, three approaches have been developed to estimate ice sheet mass balance: 1) Mass loss is estimated by direct measurements of ice sheet height changes with satellite laser or radar altimetry in combination with climatological/glaciological models for firn density and compaction, 2) the input–output method combines measurements of ice flow velocities estimated from satellite (synthetic aperture radar or optical imagery) across key outlets with estimates of net surface balance derived from ice thickness data, 3) space gravimetry data yields direct estimate of the mass changes by inversion of the anomalies in the gravity field (see Section 3.3.1 for more details). AR5 concluded that the three space-based methods give consistent results. They agree in showing that the rate of SLR due to the GIS and AIS' contributions has increased since the early 1990s. Since AR5, up-to-date observations confirm this statement with increased confidence for both ice sheets (Rignot et al., 2019; see Section 3.3.1). The assessment of the literature since AR5 made in Section 3.3.1 shows that the contribution from Greenland to SLR over 2012–2016 (0.68 (0.64–0.72) mm yr^{-1}) was similar to the contribution over 2002–2011 (0.73 (0.67–0.79) mm yr^{-1}) and *extremely likely* greater than over 1992–2001 (0.02 (0.21–0.25) mm yr^{-1}). The contribution from Antarctica over 2012–2016 (0.55 (0.48–0.62) mm yr^{-1}) was *extremely likely* greater than over the 2002–2011 period (0.23 (0.16–0.30) mm yr^{-1}) and *likely* greater than over the period 1992–2001 (0.14 (0.12–0.16); see Section 3.3.1 for more details).

Here, the approach of Section 3.3.1 is followed, using the two multi-method assessments from Bamber et al. (2018) and the IMBIE team (2018) to evaluate the contribution of ice sheet mass loss to SLR over 1993–2015 and 2006–2015 (see Table 4.1). These two studies agree with results from the WCRP Global Sea Level Budget Group (2018). For the estimation of the AIS contribution, Bamber et al. (2018) and The IMBIE team (2018) use similar but not identical data sources and processing. Both studies find consistent results within uncertainties over both periods. In Table 4.1, the results of these two studies were averaged, and weighted the average on the basis of their uncertainties, because there is no apparent reason to discount either study. For the estimation of the GIS contribution only the Bamber et al. (2018) estimate is used, as there is no other multi-method assessment available.

4.2.2.2.5   Contributions from water storage on land

Water is stored on land not only in the form of ice but snow, surface water, soil moisture and groundwater. Temporal changes in land water storage, defined as all forms of water stored on land excluding land ice, contribute to observed changes in ocean mass and thus sea level on annual to centennial time scales (Döll et al., 2016; Reager et al., 2016; Hamlington et al., 2017; Wada et al., 2017). They are caused by both climate variability and direct human interventions, at the multi-decadal to centennial time scales. Over the past century, the main cause for land water storage changes are the groundwater depletion and impoundment of water behind dams in reservoirs (Döll et al., 2016; Wada et al., 2016). While the rate of groundwater depletion and thus its contribution to SLR increased during the 20th century and up to today (Wada et al., 2016), its effect on sea level was more than balanced by the increase in land water storage due to dam construction between 1950 and 2000 (Wada et al., 2016). Since about 2000, based on hydrological models, the combined effect of both processes is a positive contribution to SLR (Wada et al., 2016). Decreased water storage in lakes, wetlands and soils due to human activities are less important for ocean mass changes (Wada et al., 2016). Overall, the integrated effects of the direct human intervention on land hydrology have reduced land water storage during the last decade, increasing the rate of SLR by 0.15–0.24 mm yr^{-1} (Wada et al., 2016; Wada et al., 2017; Scanlon et al., 2018; WCRP Global Sea Level Budget Group, 2018). Over periods of a few decades, land water storage was affected significantly by climate variability (Dieng et al., 2015a; Reager et al., 2016; Dieng et al., 2017). Net land water storage change driven by both climate and direct human interventions can be determined based on GRACE observations and global hydrological modelling. They indicate different estimates of the rate of SLR. Over the period 2002–2014 GRACE-based estimates of the net land water storage (i.e., not including glaciers) show a negative contribution to sea level (e.g., Scanlon et al., 2018) resulting in the negative value after 2006 in Table 1 while hydrological models determined a slightly positive one. The reasons for this difference between estimates are not elucidated. There is scientific consensus that uncertainties of both net land water storage contribution to sea level and its individual contributions remain high (WCRP Global Sea Level Budget Group, 2018). The differences in estimates and the lack of multiple consistent studies give *low confidence* in the net land water storage contribution to current SLR.

4.2.2.2.6   Budget of global mean sea level change

Drawing on previous sections, the budget of GMSL rise (Table 4.1, Figure 4.5) is assessed with observations over 4 periods: 1901–1990 (which corresponds to the period in the 20th century that is prior to the increase in ice sheet contributions to GMSL rise), 1970–2015 (when ocean observations are sufficiently accurate to estimate the global ocean thermal expansion and when glacier mass balance reconstructions start), 1993–2015 (when precise satellite altimetry is available) and 2006–2015 (when GRACE data is available in addition to satellite altimetry and when the Argo network reaches a near-global coverage). The budget of GMSL rise is also assessed with sea level contributions simulated by climate models over the same periods (Table 4.1, Figure 4.5). The periods 1993–2015 and 2006–2015 are only 23 and 10 years long respectively, short enough so that they can be affected by internal climate variability. Therefore, it is not expected that observations over these periods will be precisely reproduced by climate model historical experiments. For the contribution from land water storage, the estimated effect of direct human intervention was used, neglecting climate-related variations until 2002 (Ngo-Duc et al., 2005). From 2002 to 2015, total land water storage estimated with GRACE was used. In general, historical simulations of climate models end in 2005. Historical simulations were extended here to 2015 using the RCP8.5 scenario. This choice of RCP scenario is not critical for the simulated sea level, as the different scenarios only start to diverge significantly after the year 2030 (Church et al., 2013).

For 1993–2015 and 2006–2015, the observed GMSL rise is consistent within uncertainties with the sum of the estimated observed contributions (Table 4.1). Over the period 1993–2015 the two largest terms are the ocean thermal expansion (accounting for 43% of the observed GMSL rise) and the glacier mass loss (accounting for a further 20%). Compared to AR5, the extended observations corrected for the TOPEX-A drift (see Section 4.2.2.1.2) allow us now to identify an acceleration in the observed SLR over 1993–2015 and to attribute this acceleration mainly to Greenland ice loss along with an acceleration in Antarctic ice loss (Velicogna et al., 2014; Harig and Simons, 2015; Chen et al., 2017; Dieng et al., 2017; Yi et al., 2017; see also Sections 4.2.2.2.2, 4.2.2.3.4, 3.3.1). Since 2006, land ice, collectively from glaciers and the ice sheets has become the most important contributor to GMSL rise over the thermal expansion with mountain glaciers contributing 20% and ice sheets 33% (see Table 4.1). Over the periods 1993–2015, the sum of the observed sea level contributions is consistent with the total observed sea level within uncertainties at monthly-scales (not shown, e.g., Dieng et al., 2017). This is also true for the period 2006–2015, when uncertainties are significantly smaller. This agreement at monthly time scales represents a significant advance since the AR5 in physical understanding of the causes of past GMSL change. It provides an improved basis for the evaluation of models. Given these elements there is *high confidence* that the current observing system is capable of resolving decadal to multidecadal changes in GMSL and its components (with an uncertainty of <0.7 mm yr^{-1} at decadal and longer time scales, see Table 4.1 and for example, WCRP Global Sea Level Budget Group, 2018). However, despite this advance since AR5 there are still no comprehensive observations of ocean thermal expansion below 2000 m, in regions covered by sea ice and in marginal seas. The understanding of glacier mass loss can be improved at regional scale and the understanding of the land water storage contribution is still limited. Thus, for smaller changes in sea level of the order of a few tenths of a mm yr^{-1} at decadal time scales and shorter time scales there is *medium confidence* in the capability of the current observing system to resolve them (e.g., WCRP Global Sea Level Budget Group, 2018).

Before 1992, observations are not sufficient to confidently estimate the ice sheet mass balance and before 1970, the space and time sampling of ocean observations are not sufficient to estimate the global ocean thermal expansion. For these reasons, it is difficult to assess the closure of the GMSL rise budget over 1901–1990 and 1970–2015 (Church et al., 2013; Gregory et al., 2013; Jevrejeva et al., 2017; Meyssignac et al., 2017c; Slangen et al., 2017b; Parkes and Marzeion, 2018). For the period 1970–2015, the thermal expansion of the ocean represents 43% of the observed GMSL rise while the glaciers' contribution represents 22% (see Table 4.1). This result indicates a slightly smaller contribution from glaciers than reported by AR5. If the GIS contribution and the Antarctic SMB is added, then the sum of the contributors to sea level is in agreement with the low end observed SLR estimates over 1970–2015 (Frederikse et al., 2018). This result suggests that the contribution of Antarctica ice sheet dynamics to SLR has been small, if any, before the 1990s.

Since AR5, extended simulations along with recent findings in observations and improved model estimates allow for a new more robust, consistent and comprehensive comparison between sea level estimates based on observations and climate model simulations (e.g., Meyssignac et al., 2017c; Slangen et al., 2017b; Parkes and Marzeion, 2018). Compared to AR5, the simulated thermal expansion from climate models has improved with a new correction for the volcanic activity (see Section 4.2.2.2.1). The glacier contribution from glacier models forced with inputs from climate models is updated with a new glacier inventory and improvements to the glacier mass balance model (Marzeion et al., 2015). The simulated Greenland SMB is estimated with a new regional SMB-component downscaling

technique, which accounts for the regional variations in components of the Greenland SMB (Noël et al., 2015; Meyssignac et al., 2017a). In addition, an updated groundwater extraction contribution from Döll et al. (2014) is now used for the land water storage contribution.

For the periods 1970–2015, 1993–2015 and 2006–2015 the simulated contributions from thermal expansion, glaciers mass loss and Greenland SMB explain respectively 84%, 81% and 77% of the observed GMSL (see Table 4.1). For all these periods the residual is consistent within uncertainty with the sum of the contribution from land water storage and ice discharge from Greenland and Antarctica. For each period the consistency is improved compared to AR5 (see Table 4.1) although the uncertainty on the residual is slightly larger because of a larger uncertainty in simulated Glaciers and Greenland SMB contributions.

For the period 1901–1990 the simulated contributions from thermal expansion, glaciers mass loss and Greenland SMB explain only 60% of the observed GMSL and the residual is too large to be explained by the sum of the contribution from land water storage and ice discharge from Greenland and Antarctica. The gap can be explained by a bias in the simulated Greenland SMB and glacier ice loss around Greenland in the early 20th century (Slangen et al., 2017b). When the glacier model and the Greenland SMB downscaling technique are forced with observed climate from atmospheric reanalyses, rather than the simulated climate from coupled climate models, simulated SLR becomes consistent with the observed SLR (see the dashed blue

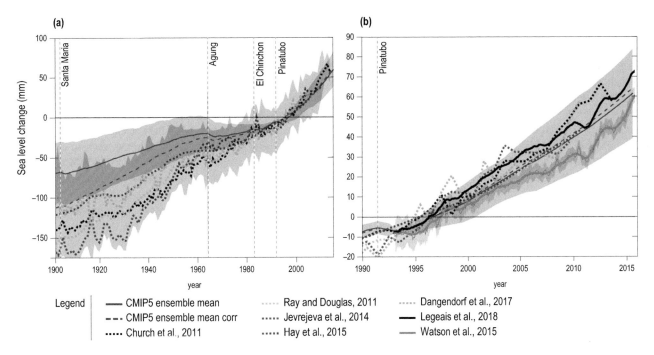

Figure 4.5 | Comparison of simulated (by coupled climate models as in Section 4.4.2.6) and observed global mean sea level change (GMSL) since 1901 (a) and since 1993 (b). The average estimate of 12 Coupled Model Intercomparison Project Phase 5 (CMIP5) climate model simulations is shown in blue with the 5–95% uncertainty range shaded in blue and calculated according to the procedures in Church et al. (2013). The average of the 12 model estimates corrected for the bias in glaciers mass loss and Greenland surface mass balance (SMB) over 1900–1940 (see Section 4.2.2.2.6) is shown in dashed blue. The estimates from tide gauge reconstructions is shown in other colours in panel a), with the 5–95% uncertainty range shaded in grey. The satellite altimetry observations from Legeais et al. (2018) is shown in black in panel b). GMSL from altimetry corrected for the TOPEX-A drift (Watson et al., 2015) in orange as well as the tide gauge reconstruction. The 5–95% uncertainty range is shaded in orange (Ablain et al., 2015). All curves in (a) represent anomalies in sea level with respect to the period 1986–2005 (i.e., with zero time-mean over the period 1986–2005) in order to be consistent with sea level projections in Section 4.2.3. Vertical lines indicate the occurrence of major volcanic eruptions, which cause temporary drops in GMSL. Updated from Slangen et al. (2017b).

line on Figure 4.5). This is because atmospheric reanalyses show an increase in air temperatures in and around Greenland over the period 1900–1940, which lead to increased melt in Greenland (Bjørk et al., 2012; Fettweis et al., 2017) and surrounding glaciers in the first half of the 20th century. This increase in air temperature over 1900–1940 is not reproduced by climate models (Slangen et al., 2017b). It may be because this increase in air temperature was due to internal climate variability on temporal and spatial scales that cannot be precisely reproduced by climate models. It may also be due to a bias in atmospheric circulation in climate models (Fettweis et al., 2017), or an issue with the spatial pattern of the historical aerosol forcing.

In summary, the agreement between climate model simulations and observations of the global thermal expansion, glacier mass loss and Greenland SMB has improved compared to AR5 for periods starting after 1970. However, for periods prior to 1970, significant discrepancies between climate models and observations arise from the inability of climate models to reproduce some observed regional changes in glacier and GIS SMB around the southern tip of Greenland. It is not clear whether this bias in climate models is due to the internal variability of the climate system or deficiencies in climate models. For this reason, there is still *medium confidence* in the ability of climate models to simulate past and future changes in glaciers mass loss and Greenland SMB.

4.2.2.3   Regional Sea Level Changes During the Instrumental Period

Sea level does not rise uniformly. Observations from tide gauges and satellite altimetry (Figure 4.6) indicate that sea level shows substantial regional variability at decadal to multi-decadal time scales (e.g., Carson et al., 2017; Hamlington et al., 2018). These regional changes are essentially due to changing winds, air-sea heat and freshwater fluxes, atmospheric pressure loading and the addition of melting ice into the ocean, which alters the ocean circulation (Stammer et al., 2013; Forget and Ponte, 2015; Meyssignac et al., 2017b). The addition of water into the ocean also change the geoid, alter the rotation of the Earth and deform the ocean floor which in turn change sea level (e.g., Tamisiea, 2011; Stammer et al., 2013).

Sea level is rising in all ocean basins (*virtually certain*; Legeais et al. 2018). Part of this regional sea level rise is due to global sea level rise of which a majority is attributable to anthropogenic greenhouse gas emissions (*high confidence*; Slangen et al. 2016). The remaining part of the regional sea-level rise in ocean basins is a combination of the response to anthropogenic GHG emissions and internal variability (e.g., Stammer et al. 2013; *medium confidence*).

In the open ocean, the spatial variability and trends in sea level observed during the recent altimetry era or reconstructed over the previous decades are dominated by the thermal expansion of the ocean. In shallow shelf seas and at high latitudes (>60°N and <55°S), the effect of dynamic mass redistribution becomes important. At local scale, salinity changes can also generate sizeable changes in the ocean density similar to thermal expansion and lead to significant variability in sea level (Forget and Ponte, 2015; Meyssignac et al., 2017b). On global average, the heat and freshwater fluxes from the

atmosphere into the ocean are responsible for the total heat that enters the ocean and for the associated GMSL rise. At regional scale and local scale, both the ocean transport divergences caused by wind stress anomalies and the spatial variability in atmospheric heat fluxes are responsible for the spatial variability in thermal expansion and thus for most of the regional sea level departures around the GMSL rise (e.g., Stammer et al., 2013; Forget and Ponte, 2015).

Over the Pacific, the surface wind anomalies responsible for the sea level spatio-temporal variability are associated with the ENSO, Pacific Decadal Oscillation (PDO) and North Pacific Gyre Oscillation modes (Hamlington et al., 2013; Moon et al., 2013; Palanisamy et al., 2015; Han et al., 2017). In the Indian Ocean they are associated with the ENSO and Indian Ocean Dipole (IOD) modes (Nidheesh et al., 2013; Han et al., 2014; Thompson et al., 2016; Han et al., 2017). In particular, the PDO is responsible for most of the intensified SLR that has been observed in the western tropical Pacific Ocean since the 1990s (Moon et al., 2013; Han et al., 2014; Thompson and Mitchum, 2014). Several studies suggested that in addition to the PDO signal, warming of the tropical Indian and Atlantic Oceans enhanced surface easterly trade winds and thus also contributes to the intensified SLR in the western tropical Pacific (England et al., 2014; Hamlington et al., 2014; McGregor et al., 2014).

Over the Atlantic, the regional sea level variability at interannual to multi-decadal time scales, is generated by surface wind anomalies and heat fluxes associated with the North Atlantic Oscillation (NAO; Han et al., 2017) and also by ocean heat transport due to changes in the Atlantic Meridional Overturning Circulation (AMOC; McCarthy et al., 2015). Both mechanisms are not independent as heat fluxes and wind stress anomalies associated with NAO can induce changes in the AMOC (Schloesser et al., 2014; Yeager and Danabasoglu, 2014). In the Southern Ocean, the sea level variability is dominated by the SAM influence in particular in the Indian and Pacific sectors. The Southern Annular Mode (SAM) influence becomes weaker equator-wards in these sectors while the influence of PDO, ENSO and IOD increases (Frankcombe et al., 2015). In the southern ocean, the zonal asymmetry in westerly winds associated to the SAM, generates convergent and divergent transport in the Antarctic Circumpolar Current which may have contributed to the regional asymmetry of decadal sea level variations during most of the twentieth century (Thompson and Mitchum, 2014).

As for GMSL, net regional sea level changes can be estimated from a combination of the various contributions to sea level change. The contributions from dynamic sea level, atmospheric loading, glacier mass changes and ice sheet SMB can be derived from CMIP5 climate model outputs either directly or through downscaling techniques (Perrette et al., 2013; Kopp et al., 2014; Slangen et al., 2014a; Bilbao et al., 2015; Carson et al., 2016; Meyssignac et al., 2017a). The contributions from groundwater depletion, reservoir storage and dynamic ice sheet mass changes are not simulated by coupled climate models over the 20th century and have to be estimated from observations. The sum of all contributions, including the GIA contribution, provides a modelled estimate of the 20th century net regional sea level changes that can be compared with observations from satellite altimetry and tide-gauge records (see Figure 4.6).

In terms of interannual to multi-decadal variability, there is a general agreement between the simulated regional sea level and tide gauge records, over the period 1900–2015 (see inset figures in Figure 4.6). The relatively large, short-term oscillations in observed sea level (black lines in insets in Figure 4.6), which are due to the natural internal climate variability, are included in general within the modelled internal variability of the climate system represented by the blue shaded area (5–95% uncertainty). But, as for GMSL, climate models tend to systematically underestimate the observed sea level trends from tide gauge records, particularly in the first half of the 20th century. This underestimation is explained by a bias identified in modelled Greenland SMB, and glacier ice loss around Greenland in the early 20th century (see Section 4.2.2.2.6; Slangen et al., 2017b). The correction of this bias improves the agreement between the spatial variability in sea level trends from observations and from climate models (see Figure 4.6). Climate models indicate that the spatial variability in sea level trends observed by tide-gauge records over the 20th century is dominated by the GIA contribution and the thermal expansion contribution over 1900–2015. Locally all contributions to sea level changes are important as any contribution can cause

significant local deviations. Around India for example, groundwater depletion is responsible for the low 20th century SLR (because the removal of groundwater mass generated a local decrease in geoid that made local SLR slower; Meyssignac et al., 2017c).

These results show the ability of models to reproduce the major 20th century regional sea level changes due to GIA, thermal expansion, glacier mass loss and ice sheet SMB. This is tangible progress since AR5. But some doubts remain regarding the ability of climate models to reproduce local variations such as the glaciers and the Greenland SMB contributions to sea level in the region around the southern tip of Greenland (Slangen et al., 2017b) or such as the thermal expansion in some eddy active regions (Sérazin et al., 2016). Because of these doubts there is still *medium confidence* in climate models to project future regional sea level changes associated with thermal expansion, glacier mass loss and ice sheet SMB. Coupled climate models have not simulated the other contributions to 20th century sea level, including the growing ice sheet dynamical contribution and land water storage changes.

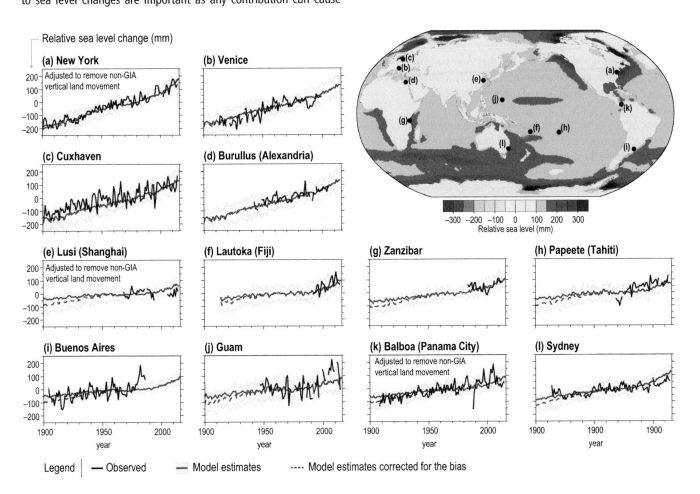

Figure 4.6 | 20th century simulated regional sea level changes by coupled climate models and comparison with a selection of local tide gauge time series. In the upper left corner: map of changes in simulated relative sea level (RSL) for the period 1901–1920 to 1996–2015 estimated from climate model outputs. Insets: Observed RSL changes (black lines) from selected tide gauge stations for the period 1900–2015. For comparison, the estimate of the simulated RSL change at the tide gauge station is also shown (blue plain line for the model estimates and blue dashed line for the model estimates corrected for the bias in glaciers mass loss and Greenland surface mass balance (SMB) over 1900–1940, see Section 4.2.2.2.6). The relatively large, short-term oscillations in observed local sea level (black lines) are due to the natural internal climate variability. For Mediterranean tide gauges, that is, Venice and Alexandria, the local simulated sea level has been computed with the simulated sea level in the Atlantic ocean at the entrance of the strait of Gibraltar following (Adloff et al., 2018). Tide gauge records have been corrected for vertical land motion (VLM) not associated with GIA where available, that is, for New York, Balboa and Lusi. Updated from Meyssignac et al. (2017b) to mimic RSL as good as possible.

4.2.2.4   Local Coastal Sea Level

Since the local coastal sea level (scale ~10 km) is affected by global, regional (scale ~100 km) and coastal scale features and processes like anthropogenic subsidence, it may differ substantially from the regional sea level. At the coast, the sea level change is additionally affected by wave run up, tidal level, wind forcing, sea level pressure (SLP), the dominant modes of climate variability, seasonal climatic periodicities, mesoscale eddies, changes in river flow, as well as anthropogenic subsidence (see also Box 4.1). These local contributions, combined with sea level events generated by storm surges and tides result in anomalous conditions (ESL) which last for a short time in contrast to the gradual increase over time from for instance ice mass loss. Flood risk due to ESL is exacerbated due to its interaction with RSL and hence physical vulnerability assessments combine uncertainties around ESL and RSL, both in terms of contemporary assessments and future projections (Little et al., 2015b; Vousdoukas, 2016; Vousdoukas et al., 2016; Wahl et al., 2017). Changes in mean sea level have been dealt with in previous sections (e.g., Section 4.2.2.2.6). Here the focus is on some of the components of ESL that have been assessed in combination with changes in RSL. Church et al. (2013) concluded that change in sea level extremes is *very likely* to be caused by a RSL increase, and that storminess and surges will contribute towards these extremes; however, it was noted that there was *low confidence* in region-specific projections as there was only a limited number of studies with a poor geographical coverage available.

Recent advances in statistical and dynamical modelling of wave effects at the coast, storm surges and inundation risk have reduced the uncertainties around the inundation risks at the coast (Vousdoukas et al., 2016; Vitousek et al., 2017; Melet et al., 2018; Vousdoukas et al., 2018c) and assessments of the resulting highly resolved coastal sea levels are now emerging (Cid et al., 2017; Muis et al., 2017; Wahl et al., 2017). This progress was facilitated due to the availability of, for example, the Global Extreme Sea Level Analysis (GESLA-2; Woodworth et al., 2016) high-frequency (hourly) datasets, advances in the Coordinated Ocean Wave Climate Project (COWCLIP; Hemer et al., 2013), coastal altimetry datasets (Cipollini et al., 2017), and the Global Tide and Surge Reanalysis (GTSR; Muis et al., 2016), while new analyses of datasets that have been available since before the publication of AR5 have continued (e.g., PSML; Holgate et al., 2012).

Although ESL is experienced episodically by definition, Marcos et al. (2015) examined the long-term behaviour of storm surge models and detected decadal and multidecadal variations in storm surge that are not related to changes in RSL. They found that, although 82% of their observed time series showed synchronous patterns at regional scales, the pattern tended to be non-linear, implying that it would be difficult to infer future behaviour unless the physical basis for the responses was understood. An analysis of the relative contributions of SLR and ESL due to storminess showed that in the US Pacific northwest since the early 1980s, increases in wave height and period have had a larger effect on coastal flooding and erosion than RSL (Ruggiero, 2012) since the early 1980s. This is also true in other regions distributed over the entire globe (Melet et al., 2016; Melet et al., 2018). Changes since 1990 in the sea level harmonics and seasonal phases and amplitudes of the wave period and significant wave height were found for the Gulf of Mexico coast and along the US east coast (Wahl et al., 2014; Wahl and Plant, 2015). These authors found that high waters have increased twice as much as one would expect from long-term SLR alone, because of additional changes in the seasonal cycle, yielding a 30% increase in risk of flooding. Such effects are *likely* to be highly dependent on the local conditions. For example, using WAVEWATCH III, TOPEX/Poseidon altimetry tide model data and atmospheric forcing physically downscaled using Delft3D-WAVE and Delft3D-FLOW in what they call the Coastal Storm Modeling System (CoSMoS), Vitousek et al. (2017) were able to detect local inundation hazards (at a scale of hundreds of metres) across regions along the Californian coast. Similarly, Castrucci and Tahvildari (2018) simulated the impact of SLR along the Mid-Atlantic region in the USA. A study for the Maldives shows that the contribution of wave setup is essential to estimate flood risks (Wadey et al., 2017).

In deltas, the local sea level can be dominated by anthropogenic subsidence more than by the processes outlined above. It is often a primary driver of elevated local SLR and increased flood hazards in those regions. This is particularly true for deltaic systems, where fertile soils, low-relief topography, freshwater access, and strategic ports have encouraged the development of many of the world's most densely populated coastlines and urban centres. For example, globally, one in fourteen humans resides in mid-to-low latitude deltas (Day et al., 2016). Although in these areas RSL is dominated by anthropogenic subsidence, climate effects need to be included for estimating risks associated with RSL (Syvitski et al., 2009).

Deltas are formed by the accumulation of unconsolidated river born sediments and porous organic material, both of which are particularly prone to compaction. It is the compaction which causes a drop in land elevation that increases the rate of local SLR above what would be observed along a static coastline or one where only climatological forced processes control the RSL. Under stable deltaic conditions, the accumulation of fluvially-sourced surficial sediment and organic matter offsets this natural subsidence (Syvitski and Saito, 2007); however, in many cases this natural process of delta construction has been disturbed by reductions in fluvial sediment supply via upstream dams and fluvial channelisation (Vörösmarty et al., 2003; Syvitski and Saito, 2007; Syvitski et al., 2009; Luo et al., 2017). Further, the extraction of fluids and gas that fill the pore space of deltaic sediments and provide support for overlying material has significantly increased the rate of compaction and resultant anthropogenic subsidence along many populated deltas (Higgins, 2016). In addition, Nicholls (2011) pointed to anthropogenic subsidence by the weight of buildings in megacities in South-East Asia.

Average natural and anthropogenic subsidence rates of 6–9 mm yr^{-1} are reported for the highly populated areas of Ganges-Brahmaputra-Meghna delta in the urban centres of Kolkata and Dhaka (Brown and Nicholls, 2015). A fraction of these subsidence rates might be caused by long-term processes of increased sediment loading during the Holocene resulting from changes in the monsoon system (Karpytchev et al., 2018). Subsidence rates are expected to decrease in the Ganges-Brahmaputra-Meghna delta in the near future due to planned dam projects and an estimated 21% drop in resulting sediment supply (Tessler et al., 2018). Observations of enhanced natural and anthropogenic subsidence on the Ganges-Brahmaputra-Meghna are common to most heavily populated deltaic systems. Coastal mega-cities that have been particularly prone to human-enhanced subsidence include Bangkok, Ho Chi Minh city (Vachaud et al., 2018), Jakarta, Manila, New Orleans, West Netherlands and Shanghai (Yin et al., 2013; Cheng et al., 2018). On a global scale, observed rates of modern deltaic anthropogenic subsidence range from 6–100 mm yr^{-1} (Bucx et al., 2015; Higgins, 2016). Rates of recent deltaic subsidence over the last few decades have been at least twice the 3 mm yr^{-1} rate of GMSL rise observed over this same interval (Higgins, 2016; Tessler et al., 2018). Numerical models that have reproduced these observed rates of anthropogenic deltaic subsidence by considering human-induced compaction and reduced sediment supply, support anthropogenic causes for elevated rates of subsidence (Tessler et al., 2018).

In summary, ESL interacts with RSL rise including anthropogenic subsidence in many vulnerable areas (see Box 4.1). Therefore, it is concluded with *high confidence* that the inclusion of local processes (wave effects, storm surges, tides, erosion, sedimentation and compaction) is essential to estimate local, relative and changes in ESL events. Although the effect of anthropogenic subsidence may be very large locally, it is not accounted for in the projection sections of this chapter as no global data sets are available which are consistent with RCP scenarios, and because the scale at which these processes take place is often smaller than the spatial scale used in climate models.

4.2.2.5 Attribution of Sea Level Change to Anthropogenic Forcing

Bindoff et al. (2013) concluded that it is *very likely* that there has been a substantial contribution to ocean heat content from anthropogenic forcing (i.e., anthropogenic greenhouse gases, anthropogenic aerosols and land use change) since the 1970s, that it is *likely* that loss of land ice is partly caused by anthropogenic forcing, and that as a result, it is *very likely* that there is an anthropogenic contribution to the observed trend in GMSL rise since 1970. However, these conclusions were based on the understanding of the responsible physical processes, since formal attribution studies dedicated to quantifying the effect of individual external forcings were not available for GMSLR. Since AR5, such formal studies have attributed changes in individual components of sea level change (i.e., thermosteric sea level change and glacier mass loss), and in the total GMSL, to anthropogenic forcing.

4.2.2.5.1 Attribution of individual components of sea level change to anthropogenic forcing

Marcos and Amores (2014) found that during the period 1970–2005, 87% (95% confidence interval: 72–100%) of the observed thermosteric SLR in the upper 700 m of the ocean was anthropogenic. Slangen et al. (2014b) included the full ocean depth in their analysis. They concluded that a combination of anthropogenic and natural forcing is necessary to explain the temporal evolution of observed global mean thermosteric sea level change during the period 1957–2005. Anthropogenic forcing was responsible for the amplitude of observed thermosteric sea level change, while natural forcing caused the forced variability of observations. Observations could best be reproduced by scaling the patterns from 'natural-only'

forcing experiments by using a factor of 0.70 ± 0.30 (2 standard deviations of the CMIP5 ensemble subset used), indicating a potential overestimation of forced variability in the CMIP5 ensemble. Patterns from the 'anthropogenic-only' forcing experiments needed to be scaled by a factor of 1.08 ± 0.13 (2 standard deviations of the CMIP5 ensemble subset used), indicating a realistic response of the CMIP5 ensemble to anthropogenic forcing.

For the glacier contribution to GMSL, Marzeion et al. (2014) concluded that while natural climate forcing and long-term adjustment of the glaciers to the end of the preceding Little Ice Age lead to continuous glacier mass loss throughout the simulation period of 1851–2010, the observed rates of glacier mass loss since 1990 can only be explained by including anthropogenic forcing. During the period 1851–2010, only 25 ± 35% of global glacier mass loss can be attributed to anthropogenic forcing, but 69 ± 24% during the period 1991–2010 (see Section 2.2.3 for a more detailed discussion of attribution of glacier mass change on regional scales).

There is *medium confidence* in evidence linking GIS mass loss to anthropogenic climate change, and *low confidence* in the evidence that AIS mass balance can be attributed to anthropogenic forcing (see Section 3.3.1.6 for a detailed discussion). The effects of groundwater depletion and reservoir impoundment on sea level change are anthropogenic by definition (e.g., Wada et al., 2012).

4.2.2.5.2 Attribution of global mean sea level change to anthropogenic forcing

By estimating a probabilistic upper range of long-term persistent natural sea level variability, Dangendorf et al. (2015) detected a fraction of observed sea level change that is unexplained by natural variability and concluded by inference that it is *virtually certain* that at least 45% of the observed increase in GMSL since 1900 is attributable to anthropogenic forcing. Similarly, Becker et al. (2014) provided statistical evidence that the observed sea level trend, both in the global mean and at selected tide gauge locations, is not consistent with unforced, internal variability. They inferred that more than half of the observed GMSL trend during the 20th century is attributable to anthropogenic forcing.

Slangen et al. (2016) reconstructed GMSL from 1900 to 2005 based on CMIP5 model simulations separating individual components of radiative climate forcing and combining the contributions of thermosteric sea level change with glacier and ice sheet mass loss. They found that the naturally caused sea level change, including the long-term adjustment of sea level to climate change preceding 1900, caused 67 ± 23% of observed change from 1900 to 1950, but only 9 ± 18% between 1970 and 2005. Anthropogenic forcing was found to have caused 15 ± 55% of observed sea level change during 1900–1950, but 69 ± 31% during 1970–2005. The sum of all contributions explains only 74 ± 22% of observed GMSL change during the period 1900–2005 considering the mean of the reconstructions of Church and White (2011), Ray and Douglas (2011), Jevrejeva et al. (2014b) and Hay et al. (2015). However, the budget could be closed taking into contribution of glaciers that are missing from the global glacier

4

inventory or have already melted (Parkes and Marzeion, 2018) which were not considered in Slangen et al. (2016).

Based on these multiple lines of evidence, there is *high confidence* that anthropogenic forcing *very likely* is the dominant cause of observed GMSL rise since 1970.

4.2.3   Projections of Sea Level Change

As a consequence of climate change, the global and regional mean sea level will change. Coupled climate models are used to make projections of the climate changes and the associated SLR. Results from the CMIP5 model archive used for AR5 provide information on expected changes in the oceans and on the evolution of climate, glaciers and ice sheets. New estimates from CMIP6 are not yet available and will be discussed in the IPCC 6th Assessment Report (AR6), hence only a partly updated projection can be presented here.

Coupled climate models can be applied on century time scales, to provide estimates of the steric (temperature and salinity effects on sea water density) and ocean dynamical (ocean circulation) components of sea level change, both globally and regionally. However, the glacier and ice sheet component are calculated off-line based on temperature and precipitation changes. In the AR5 report, changes in the SMB of glaciers and ice sheet were calculated from the global surface air temperature. In addition, GCMs also resolve climate variability related to changes in precipitation and evaporation. These changes are used to calculate short duration sea level changes (Cazenave and Cozannet, 2014; Hamlington et al., 2017). With various degrees of success those models capture ENSO, PDO and other modes of variability (e.g., Yin et al., 2009; Zhang and Church, 2012), which affect sea level through redistributions of energy and salt in the ocean on slightly longer time scales. Off-line temperature and precipitation fields can be dynamically or statistically downscaled to match the high spatial resolution required for ice sheets and glaciers, but serious limitations remain. This deficiency limits adequate representation of potentially important feedbacks between changes in ice sheet geometry and climate, for example through fresh water and iceberg production that impact on ocean circulation and sea ice, which can have global consequences (Lenaerts et al., 2016; Donat-Magnin et al., 2017). Another limitation is the lack of coupling with the solid Earth which controls the ice sheet evolution (Whitehouse et al., 2019). Dynamics of the interaction of ice streams with bedrock and till at the ice base remain difficult to model due to lack of direct observations. Nevertheless, several new ice sheet models have been generated over the last few years, particularly for Antarctica (Section 4.2.3.1) focusing on the dynamic contribution of the ice sheet to sea level change, which remains the key uncertainty in future projections (Church et al., 2013), particularly beyond 2050 (Kopp et al., 2014; Nauels et al., 2017b; Slangen et al., 2017a; Horton et al., 2018).

Information beyond that provided by climate models is needed to describe local and RSL changes. Geodynamic models are used to calculate RSL changes due to changes in ice mass in the past and future. This includes solid Earth deformation, gravitational and rotational changes, as ice and water are redistributed around the globe. Input for those models is provided by the mass changes following from the off-line land ice models, time series of terrestrial water mass changes which typically require climate input, and reconstruction of past ice sheet changes over the last glacial cycle provided by coupled ice-Earth models (de Boer et al., 2017). Combining these different models leads to projections of RSL (Section 4.2.3.2).

At the local spatial scales of specific cities, islands and stretches of coastlines, hydrodynamical models (Section 4.2.3.3) and knowledge about anthropogenic subsidence are necessary to analyse the impacts of highly variable processes leading to ESL, such as tropical cyclone-driven storm surges. These hydrodynamical models are capable of providing statistics on the variability or the change in variability of the water level required for flood risk calculations at specific locations and at spatial scales of less than 1 km. The models also rely on input from climate models, like temperature, precipitation, wind regime, and storm tracks (Colbert et al., 2013; Garner et al., 2017).

In summary, climate models play an important role at the various stages of projections in providing, together with emission scenarios, geodynamic, ice-dynamic and hydrodynamic models, the required information for hazard estimation for coasts and low-lying islands. This report relies on results of the CMIP5 model runs.

4.2.3.1   Contribution of Ice Sheets to GMSL

4.2.3.1.1   Greenland

The GIS is currently losing mass at roughly twice the pace of the AIS (see Chapter 3 and Table 4.1). About 60% of the mass loss between 1991 and 2015 has been attributed to increasingly negative SMB from surface melt and runoff on the lower elevations of the ice sheet margin. Ice dynamical changes and increased discharge of marine-terminating glaciers account for the remaining 40% of mass loss (Csatho et al., 2014; Enderlin, 2014; van den Broeke et al., 2016). The ability of firn on Greenland to retain meltwater until it refreezes has diminished markedly since the late 1990s, especially in lower elevations and on peripheral ice caps (Noël et al., 2017). Patterns of surface melt on Greenland are highly dependent on regional atmospheric patterns (Bevis et al., 2019), adding uncertainty to future projections of SMB. Melt-albedo feedbacks associated with darkening of the ice surface from ponded water, changes in snow and firn properties, and accumulation of impurities are also important, because they can strongly enhance surface melt (Tedesco et al., 2016; Ryan et al., 2018; Trusel et al., 2018; Ryan et al., 2019). These processes are not fully captured by most Greenland-scale models which is an important deficiency, because surface processes tend to dominate uncertainty in future GIS model projections (e.g., Edwards et al., 2014; Aschwanden et al., 2019). Increases in meltwater and changes in the basal hydrologic regime, once thought to have a possible destabilising effect on the ice sheet (Zwally et al., 2002), have been linked with recent reductions in ice velocity in western Greenland. On decadal time scales the effect of meltwater on ice dynamics are now assessed to be small (van de Wal et al., 2015; Flowers, 2018), which is supported by ice sheet model experiments (Shannon et al., 2013).

In sum, uncertain climate projections (Edwards et al., 2014), albedo evolution, uncertainties around meltwater buffering by firn, complex processes linking surface, englacial and basal hydrology with ice dynamics (Goelzer et al., 2013; Stevens et al., 2016; Noël et al., 2017; Hempelmann et al., 2018) and meltwater induced melting at marine-terminating ice fronts (Chauché et al., 2014), and coarse spatial model resolution (Pattyn et al., 2018), all continue to provide substantial challenges for ice sheet and SMB models.

Greenland-scale ice sheet modelling since AR5 (Edwards et al., 2014; Fürst et al., 2015; Vizcaino et al., 2015; Calov et al., 2018; Golledge et al., 2019; Aschwanden et al., 2019) has built upon earlier work by coupling the ice models with regional climate models and using multiple climate and ice sheet models within single studies (Edwards et al., 2014). Recent modelling studies use higher-order representations of ice flow (Fürst et al., 2015), include more explicit representations of ice sheet processes including subglacial hydrology (Calov et al., 2018), run the models at higher resolution and with updated boundary conditions (Aschwanden et al., 2019), and account for two-way coupling between the ice sheet and the global ocean (Vizcaino et al., 2015; Golledge et al., 2019). Among these studies, Fürst et al. (2015), Vizcaino et al. (2015), and Aschwanden et al. (2019) provide projections following RCP2.6, RCP4.5, and RCP8.5 emissions scenarios. Calov et al. (2018) and Golledge et al. (2019) did not consider RCP2.6. Edwards et al. (2014) used the Special Report on Emissions Scenarios (SRES) A1B scenario which isn't directly comparable to the other studies assessed here, but they do provide a rigorous analysis of uncertainty contributed by different climate forcings, varying simplifications of ice flow equations and height-SMB feedbacks.

Fürst et al. (2015) used ten different CMIP5 Atmosphere-Ocean General Circulation Model (AOGCM) simulations to provide offline SMB and ocean forcing for their Greenland-wide ice sheet model, accounting for influences of warming subsurface ocean temperatures and basal lubrication on ice dynamics. In their RCP8.5 ensemble, they found a GIS contribution to GMSL in 2100 of 10.15 cm ± 3.24 cm. Similarly, Calov et al. (2018) found a range of GMSL contributions between 4.6–13 cm, depending on which CMIP5 GCM is used to force their regional climate model to produce SMB forcing. The wide range of RCP8.5 results in these studies highlights the substantial

climate-driven uncertainty in 21st century projections of the GIS as emphasised by Edwards et al. (2014). It was found that central estimates and ranges for RCP8.5 simulated by Fürst et al. (2015), Calov et al. (2018), and Golledge et al. (2019) are in reasonable agreement with previous multi-model results (Bindschadler et al., 2013) and the assessment of AR5 (Church et al., 2013), which reported a *likely* RCP8.5 range of Greenland's contribution to GMSL between 7–21 cm by 2100 (Table 4.2.). The GIS simulations provided by Vizcaino et al. (2015), using a relatively course-resolution ice model (10 km) with SMB forcing provided by a single GCM, estimate much less ice loss than other recent studies. Their GMSL projections (Vizcaino et al., 2015) also fall below the *likely* range of AR5 estimates. In contrast, the study by Aschwanden et al. (2019) shows a significantly higher contribution to GMSL than the other studies, especially under RCP8.5 and beyond 2100 (see 4.2.3.5). This may be due to their SMB forcing, which is based on spatially uniform warming derived from future CMIP5 GCM climatologies averaged over the entire Greenland region. As noted by earlier work (e.g., Van de Wal and Wild, 2001; Gregory and Huybrechts, 2006), this approach can overestimate melt rates in the ablation zone, which could account for their higher projected ice loss. It is noted that the process-based estimates of future GMSL rise from Greenland found in Aschwanden et al. (2019) are closest to those from an updated, structured judgement of glaciological and modelling experts (Bamber et al., 2019). Calculations from the expert elicitation (Bamber et al., 2019) result in higher estimates of Greenland ice loss than any of the process-based studies, with a mean and standard deviation of 33 ± 30 cm and a 17–83% range of 10–60 cm by 2100, following a climate scenario comparable to RCP8.5. The combination of the new process-based studies produces central estimates (Table 4.2) consistent with the *likely* ranges for Greenland's contribution to GMSL in 2100 assessed by AR5.

Complimentary to the ice sheet scale simulations discussed above, Nick et al. (2013) used detailed flowline models of four Greenland outlet glaciers (Petermann, Kangerdlugssuaq, Jakobshavn Isbræ, and Helheim) to estimate a dynamical contribution to sea level in an RCP8.5 scenario of 11.3–17.5 mm by 2100, and 29–49 mm, by 2200. This demonstrates the limited potential of Greenland outlet glaciers alone to drive GMSL rise. Greenland-wide modelling studies (Table 4.2) consistently find a dominant role of runoff relative to dynamic discharge of ice loss, and a long-term reduction in the rate of

Table 4.2 | Estimates of the Greenland Ice Sheet (GIS) contribution to Global Mean Sea Level (GMSL; cm) in 2100 reported by process-based modelling studies including the effects of both surface mass balance (SMB) and ice dynamics published since the IPCC 5th Assessment Report (AR5). Only model results including elevation-SMB feedback are shown. All values are reported as the contribution to GMSL in 2100 relative to 2000, with the exception of Aschwanden et al. (2019) who report values relative to 2008. The median estimate for comparison with AR5 is based on the average of the three simulations in Calov et al. (2018) using different General Climate Models (GCMs), combined with the central estimates from the other studies. RMSD (Fürst et al., 2015) is the Root Mean Squared Deviation from their ensemble median. The range reported by Aschwanden et al. (2019) refers to the 16–84% interval of a 500 member ensemble with varying model physical parameters. RCP is Representative Concentration Pathway.

Study	RCP2.6	RCP4.5	RCP8.5	Reported uncertainty
Aschwanden et al. (2019)	5–19	8–23	14–33	16–84% range
Calov et al. (2018)		1.9–5.6	4.6–13.0	Range of three GCMs
Fürst et al. (2015)	4.2 ± 1.8	5.5 ± 1.86	10.2 ± 3.24	RMSD from ensemble median
Golledge et al. (2019)		10.9	11.2	
Vizcaino et al. (2015)	2.7	3.4	5.8	
Process based median	6.3	7.8	11.9	
IPCC AR5 Table 13.5	4–10	4–13	7–21	*likely* range

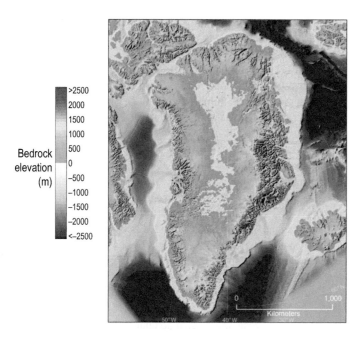

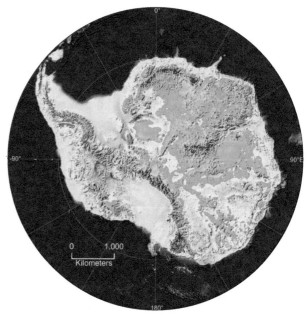

Figure 4.7 | Bedrock topography below the existing ice sheets in Greenland (Morlighem et al., 2017) and Antarctica (right) (Fretwell et al., 2013). Horizontal scales are not the same in both panels. Note the deep subglacial basins in West Antarctica and the East Antarctic margin. The ice above floatation in these areas is equivalent to >20 m of Global Mean Sea Level (GMSL).

dynamic ice discharge to the ocean as the ice sheet margin thins and the termini of outlet glaciers retreat from the coast (Goelzer et al., 2013; Lipscomb et al., 2013). Greenland's bedrock geography and the limited, direct access of thick interior ice to the ocean ultimately limits the potential pace of GMSL rise from the GIS. Figure 4.7 illustrates a fundamental difference between Greenland and Antarctica. In Greenland, most of the bedrock at the ice sheet margin is above sea level (land terminating), with relatively narrow (generally <10 km wide) outlet glaciers reaching the ocean. In contrast, Antarctica has extensive areas with subglacial bedrock below sea level, and thick marine-terminating ice in direct contact with the open ocean. Recent subglacial mapping and mass conservation calculations since AR5 (Morlighem et al., 2014; Morlighem et al., 2017) revise earlier bathymetric maps under and around the ice sheet, and reveal deeper and more extensive valley networks extending into the GIS interior than previously known. Accurate subglacial topography is important for modelling individual Greenland outlet glaciers (Aschwanden et al., 2016; Morlighem et al., 2016); however, the importance of these revised bedrock boundary conditions for the broader ice sheet has yet to be fully tested. Based on the limited cross sectional area of subglacial valleys and outlet glaciers on Greenland (Figure 4.7) and the results of Nick et al. (2013), the effects of uncertain bathymetric boundary conditions are assessed to be small relative to the uncertainties in future SMB forcing (*medium confidence*).

In summary, new modelling since AR5 is consistent with previous studies suggesting future Greenland ice loss over the 21st century will be dominated by surface processes, rather than dynamic ice discharge to the ocean, regardless of which emissions scenario is followed (*high confidence*). Based on these modelling studies, the GIS is not expected to contribute more than 20 cm of GMSL rise by 2100 in a RCP8.5 scenario, similar to the upper end of the *likely* range reported by AR5 (Church et al., 2013). GIS simulations

are most sensitive to uncertainties in the applied climate forcing, especially over this century (Edwards et al., 2014), but updated climate projections since AR5 are not yet available. Because of the consistency of recent modelling with the assessment of Church et al. (2013), Greenland's contribution to future sea level reported in AR5 was used in our projections of GMSL.

4.2.3.1.2   Antarctica

Unlike Greenland, most of the AIS margin terminates in the ocean. The AIS also contains almost eight times more glacial ice above flotation than Greenland, and nearly half of this ice is marine-based, that is, grounded on bedrock hundreds of metres (or more) below sea level (Figure 4.7; Fretwell et al., 2013). In places where the subglacial bedrock slopes downward away from the coast (reverse-sloped), the marine-based glacial ice is susceptible to dynamical instabilities (Weertman, 1974; Schoof, 2007b; Pollard et al., 2015) that can contribute rapid ice loss (Cross-Chapter Box 8 in Chapter 3). The instabilities can be triggered by the loss or thinning of ice shelves through changes in the surrounding ocean and increased sub-ice melt rates and changes in the overlying atmosphere affecting SMB and surface meltwater production. Much progress has been made since AR5 in the understanding of these processes, but their representation in continental-scale models continue to be heavily parameterised in most cases. Complex interactions between the ice sheet, ocean, atmosphere and underlying bedrock also remain difficult to simulate collectively.

In contrast to Greenland, Antarctica's recent contribution to SLR has been dominated by ice-dynamical processes rather than changes in SMB (Mouginot et al., 2014; Rignot et al., 2014; Scheuchl et al., 2016; Shen et al., 2018; The IMBIE team, 2018). Since AR5, it has become increasingly evident that this ice loss is being driven by sub-ice oceanic melt (thinning) of ice shelves (Paolo et al., 2015; Wouters et al., 2015) and the resulting loss of back stress (buttressing) that impedes the seaward flow of grounded ice upstream. Elevated melt rates are generally associated with the increased presence of warm Circumpolar Deep Water (CDW) on the continental shelf (Khazendar et al., 2016). Dynamic ice loss driven by ocean changes have also been observed on the East Antarctic margin (Li et al., 2016; Shen et al., 2018). This is an important development, because East Antarctica contains much more ice than West Antarctica, so even minor changes there could make major contributions to sea level in the future.

Several of West Antarctica's major outlet glaciers, including Pine Island Glacier, and Thwaites Glacier in the Amundsen Sea (Figure 4.8) have grounding lines currently retreating on retrograde bedrock (Rignot et al., 2014). Thwaites Glacier is particularly important (Figure 4.8), because it extends into the interior of the WAIS, where the bed is >2000 m below sea level in places. By itself, the Thwaites drainage area contains the equivalent of ~0.4 m GMSL (Holt et al., 2006; Millan et al., 2017), but loss of the glacier could have a destabilising impact on the entire WAIS (Feldmann and Levermann, 2015). The WAIS contains enough ice to raise GMSL by ~3.4 m (Fretwell et al., 2013). Since AR5, a number of ice sheet modelling studies have focussed on limited fractions of Antarctica and so are not included in estimating the SROCC Antarctic contribution to GMSL (see Section 4.2.3.2). However, these studies do allow an assessment of the potential for persistent and increasing ice loss, and the role of the marine ice sheet instability (MISI, see Cross-Chapter Box 8 in Chapter 3).

Joughin et al. (2014) modelled the response of the Thwaites Glacier to a combination of elevated sub-ice melt rates and increased precipitation and found persistent future retreat, despite either the partial compensation of increased accumulation or a future reduction in melt. Sub-ice melt rates sustained at current levels were found to generate >1 mm yr^{-1} equivalent GMSL rise within a millennium. Higher melt rates and an assumed weak ice shelf triggered rapid retreat within a few centuries. Similarly, Waibel et al. (2018) used the BISICLES ice sheet model (Cornford et al., 2015) to investigate the potential for self-sustained retreat of Thwaites Glacier, by incrementally increasing sub-ice melt rates until retreat is triggered, and then returning to pre-retreat melt rates. Consistent with Joughin et al. (2014), they found self-sustained retreat of Thwaites Glacier through MISI. Most uncertainty in their future WAIS simulations arises from uncertainties in the long-term response of Thwaites Glacier (Figure 4.8). Nias et al. (2016) demonstrated model sensitivity of Thwaites Glacier to poorly resolved bedrock boundary conditions (small scale topography), pointing to the need for better geophysical information to reduce model uncertainty (Schlegel et al., 2018). Arthern and Williams (2017) used adaptive mesh techniques, but with a different formulation than Cornford et al., (2015), to simulate the future response of Amundsen Sea outlet glaciers. They demonstrate a sustained, but slow future retreat when sub-ice melt is maintained at current rates, and a direct relationship between the strength of ocean forcing and the pace of MISI-driven ice loss. Yu et al. (2018) simulate future Thwaites retreat using a range of model formulations with varying approximations of ice stress balance, different ocean melt schemes, and different basal friction laws. Like Arthern and Williams (2017) they find model-specific dependencies in the rate

of ice loss, but all of their simulations demonstrate sustained ice loss and a bathymetrically controlled future acceleration.

Like Thwaites, the neighbouring Pine Island Glacier (PIG) has also been thinning and retreating at an accelerating rate in recent decades, in response to incursions of warm CDW in the waters underlying the glacier's ice shelf. These incursions of CDW are controlled in part by sea floor bathymetry and climatic variability (Dutrieux et al., 2014). Favier et al. (2014) used three models with differing formulations to simulate PIG's response to elevated sub-ice melt. Consistent with modelling of Thwaites Glacier (Joughin et al., 2014), all three models demonstrate sustained future retreat at an increasing rate, as the glacier backs onto its retrograde bed. Only one of the three models used by Favier et al. (2014) demonstrates the possibility that the glacier can recover if sub-ice melt rates are reduced enough to allow the ice shelf to thicken and pin on bathymetric features to provide buttressing. These results highlight the long-term commitment to marine-based ice loss.

While limited to 50 year simulations, Seroussi et al. (2017) provide the first interactively coupled ice-ocean model simulations of Thwaites Glacier at a high spatial resolution. Their model demonstrates MISI-like grounding line retreat at a rate of ~1 km yr^{-1}, comparable to observations between 1992 and 2011 (Rignot et al., 2014). The retreat is interrupted when the main trunk of the glacier stabilises on a bathymetric ridge, ~20 km upstream of the present-day grounding line (Figure 4.8), but due to the short duration of the simulation, the long-term potential for additional retreat into the interior of the ice sheet is not captured.

Despite the use of independent model formulations, forcings, and different geographic settings, the overall agreement among these highly-resolved regional modelling studies and their ability to capture current rates of retreat, increases confidence since AR5 that observed retreat of Amundsen Sea outlet glaciers is driven by processes consistent with MISI theory (*medium confidence*), will continue (*medium confidence*), and could accelerate (*medium confidence*).

Observations of rapid bedrock uplift in the Amundsen Sea, low viscosity of the underlying mantle, and short GIA response times to glacial unloading suggest ice-Earth interactions could be important there (Barletta et al., 2018). Bedrock uplift and reduced gravitational attraction between the ice sheet and ocean as an ice margin loses mass reduces RSL at the grounding line, promoting stability and providing a negative feedback on retreat (Adhikari et al., 2014; Gomez et al., 2015). Using a high-resolution ice sheet-Earth model, Larour et al. (2019) showed that long-term future retreat of Amundsen Sea grounding lines are slowed by these processes, but the effect is found to be minimal until after ~2250. This agrees with other recent modelling accounting for ice-Earth interactions, including the viscoelastic Earth response to changing ice loads and self-gravitation (Gomez et al., 2015; Konrad et al., 2015; Pollard et al., 2017). These studies also showed a small negative feedback on future retreat over the next several centuries, particularly under strong climate forcing. However, the viscosity structure of the Earth under the AIS is not well resolved, and lateral variations in Earth structure could impact these results (Hay et al., 2017). Based on these consistent model

4

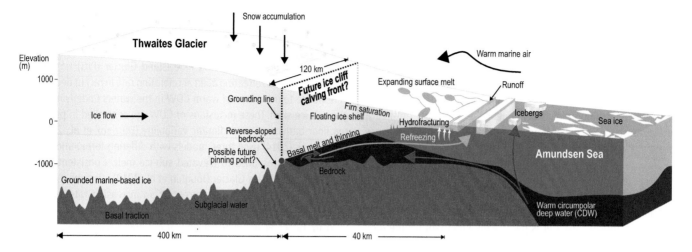

Figure 4.8 | Processes affecting the Thwaites Glacier in the Amundsen Sea sector of Antarctica (adapted from Scambos et al., 2017). The grounding line is currently retreating on reverse-sloped bedrock at a water depth of ~600 m (Joughin et al., 2014; Mouginot et al., 2014). The glacier terminus is ~120 km wide, widens upstream, and is minimally buttressed by a laterally discontinuous ~40 km long ice shelf. The remaining shelf is thinning in response to warm, sub-shelf incursions of circumpolar deep water (CDW), with melt rates up 200 m yr⁻¹ near the groundling line in some places (Milillo et al., 2019). The bathymetry upstream of the grounding zone is complex, but it generally slopes downward into a deep basin, up to 2000 m below sea level under the centre of the West Antarctic Ice Sheet (WAIS) (far left), making the glacier vulnerable to marine ice sheet instabilities (Cross-Chapter Box 8 in Chapter 3).

results, and new observational evidence that PIG has been retreating on reverse-sloped bedrock for a half-century or more (Smith et al., 2017), ice-Earth interactions are not expected to substantially slow GMSL rise from marine-based ice in Antarctica over the 21st century (*medium confidence*). However, these processes could become important for GMSL rise on multi-century and longer time scales.

Atmospheric forcing is also becoming increasingly recognised to be an important factor for the future of the AIS. A sustained (15 days) melt event over the Ross Sea sector of the WAIS in 2016 illustrated both the connectivity of Antarctica to the tropics and El Niño, and the possibility that future meltwater production on ice shelf surfaces could change in the near future (Nicolas et al., 2017). This was highlighted by Trusel et al. (2015), who evaluated the future expansion of surface meltwater using the snow component in the RACMO2 regional atmospheric model (Kuipers Munneke et al., 2012) and output from CMIP5 GCMs. Under RCP8.5, they found a substantial expansion of surface meltwater production on ice shelves late in the 21st century that exceed melt rates observed before the 2002 collapse of the Larsen B Ice Shelf. Surface meltwater is important for both ice dynamics and SMB due to its potential to reduce albedo, saturate the firn layer, deepen surface crevasses, and to cause flexural stresses that can contribute to ice shelf break-up (hydrofracturing) (Banwell et al., 2013; Kuipers Munneke et al., 2014). The presence of surface meltwater does not necessarily lead to immediate ice shelf collapse (Bell et al., 2017b; Kingslake et al., 2017), although surface meltwater was a precursor on ice shelves which have collapsed (Scambos et al., 2004; Banwell et al., 2013). This dichotomy illustrates the uncertain role of meltwater and the need for additional study. When and if melt rates will be sufficiently high in future warming scenarios to trigger widespread hydrofracturing is a key question, because the loss of ice shelves is associated with the onset of marine ice sheet instabilities (Cross-chapter Box 8 in Chapter 3). Based on the single modelling study by Trusel et al. (2015), it is not expected that widespread ice shelf loss will occur before the end of the 21st century, but due to

limited observations and modelling to date, there is *low confidence* in this assessment.

Continental-scale ice sheet simulations are ultimately required to provide projections of future GMSL rise from Antarctica. At this spatial scale, most models rely on simplifying approximations of the equations representing three-dimensional ice flow, and in some cases they parameterise ice flow at the grounding line (Schoof, 2007b) to improve computational efficiency. Such simplifications are necessary to allow long simulations that can be validated against geological information, in addition to modern observations (Briggs et al., 2013; Pollard et al., 2016), however processes related to MISI are best represented at high spatial resolution and without simplifications of the underlying physics (Pattyn et al., 2013; Reese et al., 2018c).

Various ice sheet model formulations, including the choice of grounding line parameterisations and basal sliding schemes can strongly affect model response to a given forcing (Brondex et al., 2017; Pattyn, 2017), although sophisticated statistical methodologies have been increasingly used since AR5 to quantitatively gauge model uncertainty (Bulthuis et al., 2019; Edwards et al., 2019). Accurate atmospheric forcing (SMB) and sub-ice melt are also prerequisite to resolving the time-evolving dynamics of the system, with sub-ice melt rates being particularly important (Schlegel et al., 2018). An important ongoing deficiency is the lack of ice-ocean coupling in most continental-scale studies, which remains too computationally expensive to simulate the ocean at the spatial scales necessary to capture circulation in ice shelf cavities and time-evolving ice-ocean interactions (Donat-Magnin et al., 2017; Hellmer et al., 2017). Instead, melt rates are often parameterised as a depth dependent function of nearby ocean temperature derived from offline ocean models, but the lack of ice-ocean interaction can seriously overestimate melt rates in some settings (de Rydt et al., 2015; Seroussi et al., 2017). Approaches that link offline ocean temperatures with efficient box models of the circulation in ice shelf cavities have been developed (Lazeroms et al., 2018; Reese et al., 2018a) and used in long-term

4

future simulations (Bulthuis et al., 2019), although they still require uncoupled ocean models to provide time-evolving ocean conditions outside the cavities.

Ritz et al. (2015) used a hybrid physical-statistical modelling approach, whereby the timing of MISI onset is determined statistically rather than physically. They estimated probabilities of MISI onset in eleven different sectors around the ice sheet margin based on observations of continent-wide retreat and thinning over the last few decades, and expected future climate change following an IPCC SRES A1B emission scenario only. In places where MISI is projected to begin, the persistence and rate of grounding-line retreat is parameterised as a function of the local bedrock topography (slope), ice thickness at grounding lines following Schoof (2007b), and basal friction. This study represents a statistically rigorous approach in which model parameters are based on a synthesis of observations and projected surface and sub-shelf forcing, rather than coming directly from climate and ocean models. However, the model calibrations rely on recent observations, which may not provide adequate guidance under warmer future conditions.

Levermann et al. (2014) use simplified emulations of temperature increase in order to estimate both SMB and sub-ice melt (including a parameterised delay for ocean warming) to determine the linearised response of five ice sheet models calibrated against recent rates of retreat. Substantial uncertainty arises from the different model treatments of grounding line dynamics and ice shelves. However, they conclude that the single greatest source of uncertainty stems from the external forcing.

Golledge et al. (2015) used PISM (Parallel Ice Sheet Model; Winkelmann et al., 2011) to simulate the future response of the AIS to RCP emission scenarios. PISM links grounded, streaming, and shelf flow, and has freely evolving grounding lines required to capture MISI. PISM's parameterised treatment of sub-ice melt applies melt under partially grounded grid cells (Feldmann and Levermann, 2015), making the model sensitive to subsurface ocean warming, although the validity of this approach is contested (Arthern and Williams, 2017; Seroussi and Morlighem, 2018; Yu et al., 2018). While providing alternative outcomes with the two basal melt rate parameterisations, the model is not calibrated to observations and doesn't provide a probability distribution. In a subsequent study Golledge et al. (2019) used PISM, but with updated RCP climate forcing based on CMIP5 GCMs, and with sub-ice ocean melt calibrated to observations. An offline, intermediate-complexity climate model was used to capture global ice-climate feedbacks ignored in most other studies, but the simulations only include RCP4.5 and RCP8.5 and do not extend beyond 2100. Accounting for the climatic effects of meltwater input from Greenland and Antarctica nearly doubled their estimates of Antarctic's contribution to GMSL in 2100 from 2.4 cm to 4.6 cm in RCP4.5, and from 7.7 cm to 14 cm in RCP8.5. The increase is caused by a combination of SMB decrease over the WAIS, combined with subsurface ocean warming that increases sub-ice melt. However, the climate model used to diagnose the spatial patterns of the atmospheric and oceanic response to the meltwater input is simplistic. Bronselaer et al. (2018) tested the global climatic response to future meltwater input from Antarctica using an ensemble of GCM simulations, but

without an interactive ice sheet. They simulated an RCP8.5 scenario with and without a massive input of meltwater into the Southern Ocean and demonstrate that the addition of Antarctic meltwater expands sea ice in the Southern Ocean, delays the trajectory of global warming, and moderates atmospheric warming around the Antarctic coastline. Consistent with Golledge et al. (2019), they found meltwater-induced stratification around Antarctica warms subsurface ocean temperatures, indicating the potential for a positive meltwater feedback on ice shelf melt. These studies reinforce the need for continental-scale studies to consider two-way ice-climate coupling, but with limited published studies to draw from and no simulations run beyond 2100, firm conclusions regarding the net importance of atmospheric versus ocean melt feedbacks on the long-term future of Antarctica can not be made.

Bulthuis et al. (2019) used a different continental-scale ice sheet model (Pattyn, 2017) with the same simplified atmospheric and ocean forcing used by Golledge et al. (2015) to simulate RCP2.6, RCP4.5, and RCP8.5 scenarios. Simulations with varying model parameters were used to quantify uncertainties related to the atmospheric forcing, various ice-model physics, and bedrock response to changing ice loads. A key finding was that irrespective of model parametric uncertainty, the strongly mitigated RCP2.6 scenario prevents catastrophic WAIS collapse over the coming centuries. The probabilistic projections of Antarctic GMSL contributions (Bulthuis et al., 2019) represent a rigorous blending of physical ice sheet modelling and uncertainty quantification (UQ) techniques, albeit with a simplistic representation of future climate and using a relatively coarse-resolution ice sheet model. These results are well-supported by Schlegel et al. (2018), who blend UQ with a higher resolution ice sheet model than used by Bulthuis et al. (2019), but using an idealised climate forcing scheme not directly linked to time-evolving future climate trajectories. Their 800 simulations, run to 2100, provide not only probabilistic constraints on future GMSL-rise from Antarctica, but an assessment of key drivers of uncertainty, including uniform and regional dependencies on model physical parameters, climate forcing, and boundary conditions. Sub-ice shelf melt rates provide the greatest source of uncertainty in their projections, although the source region dominating the GMSL contribution is found to be dependent on the climate forcing applied, and different from those found by Golledge et al. (2015).

DeConto and Pollard (2016) used an ice sheet model with a formulation similar to that used by Golledge et al. (2015) and Bulthuis et al. (2019) but they include glaciological processes not accounted for in other continental-scale models: 1) surface melt and rain water influence on hydrofracturing of ice shelves; and 2) brittle failure of thick, marine-terminating ice fronts that have lost their buttressing ice shelves. Where the ice fronts are thick enough to form tall ice cliffs above the waterline, they can produce stresses exceeding the strength of the ice, causing calving (Bassis and Walker, 2012). Once initiated, ice-cliff calving has been hypothesised to produce a self-sustaining Marine Ice Cliff Instability (MICI; Cross-chapter Box 8, Chapter 3). The validity of MICI remains unproven (Edwards et al., 2019) and is considered to be characterised by 'deep uncertainty', but it has the potential to raise GMSL faster than MISI. DeConto and Pollard (2016) represent hydrofracturing and ice-cliff calving with simple parameterisations, but the glaciological processes themselves are supported by more detailed modelling and observations (Scambos et al., 2009; Banwell et al., 2013; Ma et al., 2017; Wise et al., 2017; Parizek et al., 2019). DeConto and Pollard (2016) provide four ensembles for RCP2.6, RCP4.5, and RCP8.5 scenarios, representing two alternative ocean model treatments and two alternative palaeo sea level targets used to tune their model physical parameters. However, their ensembles do not explore the full range of model parameter space or provide a probabilistic assessment (Kopp et al., 2017; Edwards et al., 2019). Under RCP2.6, DeConto and Pollard (2016) find very little GMSL rise from Antarctica by 2100 (0.02–0.16 m), consistent with the findings of Golledge et al. (2015) and Bulthuis et al. (2019). In contrast, their four ensemble means range between 0.26–0.58 m for RCP4.5, and 0.64–1.14 m for RCP8.5. In RCP8.5, rates of GMSL rise from Antarctica exceed 5 cm yr^{-1} in the 22nd century and contribute as much as 15 m of GMSL rise by 2500, largely due to the ice cliff calving process. The climate forcing used by DeConto and Pollard (2016) simulates the appearance of extensive surface meltwater several decades earlier than indicated by other CMIP5 climate simulations (Trusel et al., 2015). Because their model physics are sensitive to melt water through hydrofracturing, this makes the timing and magnitude of their simulated ice loss too uncertain to include in SROCC sea level projections. However, their results do demonstrate the potential for brittle ice sheet processes not considered by AR5 to exert a strong influence on future rates of GMSL rise and the possibility that GMSL beyond 2100 could be considerably higher than the *likely* range projected by models that do not include these processes.

4.2.3.2   Global and Regional Projections of Sea Level Rise

In addition to the model including MICI from DeConto and Pollard (2016), only a subset of studies (Levermann et al., 2014; Golledge et al., 2015; Ritz et al., 2015; Bulthuis et al., 2019; Golledge et al., 2019), and statistical emulation of DeConto and Pollard (2016) by Edwards et al. (2019) provide continental-scale estimates of future Antarctic ice loss, under a range of GHG emissions scenarios. They all provide probabilistic information, but vary considerably, both in their physical approaches and their resulting projections of Antarctica's future contribution to GMSL. Such variations facilitate the first quantitative uncertainty assessment of the full dynamical contribution of Antarctica, which could not be made by Church et al. (2013) in AR5. The assessment by Church et al. (2013), based on a single statistical-physical model, reported median values (and *likely* ranges) of 0.05 m (-0.04–0.13) and 0.04 m (-0.06–0.12), for RCP4.5 and RCP8.5, respectively, for the total Antarctic contribution in 2081–2100 relative to 1986–2005, and added the following: 'Based on current understanding, only the collapse of marine-based sectors of the AIS, if initiated, could cause GMSL to rise substantially above the *likely* range during the 21st century. This potential additional contribution cannot be precisely quantified but there is *medium confidence* that it would not exceed several tenths of a metre of SLR during the 21st century (Church et al., 2013). Given the above-mentioned publications after AR5, Antarctica's contribution to sea level change was reassessed and now include the possibility of MISI allowing for a more complete assessment of the *likely* range of the projections for three RCP scenarios. Our assessment is based

on process-based numerical models of the AIS, driven by diverse climate scenarios. Results are discussed in the context of an expert elicitation study (Bamber et al., 2019), probabilistic studies (Perrette et al., 2013; Slangen et al., 2014a; Grinsted et al., 2015; Jackson and Jevrejeva, 2016) and a sensitivity study (Schlegel et al., 2018) assessing the uncertainty in snow accumulation, ocean-induced melting, ice viscosity, basal friction, bedrock elevation and the effect of ice shelves on ice mass loss in 2100, Figure 4.4.

Ritz et al. (2015) is difficult to contextualise as they only provided estimates for the A1B scenario and not for the RCP scenarios. Despite this limitation their results, which are close to the other studies, are included as if they represent RCP8.5 and as such supports the assessment. The results by DeConto and Pollard (2016) indicate significantly higher mass loss even for RCP4.5, potentially related to their high surface melt rates on the ice shelves as contested by Trusel et al. (2015). This early onset of high surface melt rates in DeConto and Pollard (2016) leads to extensive hydrofracturing of ice shelves before the end of the 21st century and therefore to rapid ice mass loss. For this reason, their results and probabilistic (e.g., Kopp et al., 2017; Le Bars et al., 2017) and statistical emulation estimates that build on them (Edwards et al., 2019), are not used in SROCC sea level projections. Consequently, the process-based studies by Golledge et al. (2015), Ritz et al. (2015), Levermann et al. (2014), Golledge et al. (2019), and Bulthuis et al. (2019) are used to assess the Antarctic contribution for the different RCP scenarios. The study by Schlegel et al. (2018) does not provide RCP based scenarios, but is considered as an extensive sensitivity estimate providing a high-end estimate based on physical process understanding of the Antarctic contribution.

Each study expresses an uncertainty in the Antarctic contribution to GMSL rise which is, in part, dependent on a common driver, namely regional warming. The uncertainties were therefore interpreted as being dependent and propagate the total uncertainty accordingly. As a result, the total uncertainty exceeds that of the individual studies, which reflects that the individual studies only sample a fraction of the total uncertainty. The uncertainty estimates of Levermann et al. (2014) concentrate on the oceanic basal melt rates including a time delay between atmosphere and ocean temperature, but do not consider other sources of uncertainty. Ritz et al. (2015) is constrained by observations and provides an asymmetric distribution of the rate of mass loss. The ice sheet simulations by Golledge et al. (2015) and Golledge et al. (2019) only provide two alternative subgrid parameterisations for sub-ice melt, rather than a statistical estimate of the uncertainty. The more sensitive of these two parameterisations which induces more ice loss is challenged by Seroussi and Morlighem (2018). In order to assess a realistic uncertainty for the total Antarctic contribution, it was first assumed that Golledge et al. (2015) and Golledge et al. (2019) are dependent, because they use similar parameterisations. For each study, a probabilistic distribution is used, assuming a normal distribution with a *likely* range bounded by the high and low estimate from those studies. Levermann et al. (2014) also provides two alternatives, one with and one without a time delay between oceanic temperatures below the Antarctic ice shelves and global mean atmospheric temperature. As it is unclear which version best matches the updated record of ice loss presented by The IMBIE team, (2018), results are combined assuming full probabilistic dependence as for the two Golledge studies. Bulthuis et al. (2019) uses a simplified ice sheet model to study the uncertainty caused by the atmospheric forcing, ice dynamics, ice and bed rheology, calving and sub-shelf melting. Finally, the studies by Ritz et al. (2015), Bulthuis et al. (2019) and the averages for Golledge and Levermann are combined to identify a best estimate for the Antarctic contribution under RCP8.5. This results in a median contribution of 16 cm in 2100 under RCP8.5. A Monte Carlo technique is used to combine the uncertainties in the aforementioned studies, assuming mutual dependence. The resulting 5–95 percentile range, 2–37 cm in 2100 under RCP8.5, is assessed as the *likely* range. This assessment is used in order to reflect ongoing limited understanding of the physics and the fact that the individual studies only reflect part of the total uncertainty. The distribution is slightly skewed to higher values, because of an underlying skewness in the studies of Levermann et al. (2014) and Ritz et al. (2015). This skewed distribution is supported by an expert elicitation study (Bamber et al., 2009). The expert elicitation approach (Bamber et al., 2018), which applied elicitation to both ice sheets, suggests considerably higher values for total SLR for RCP2.6, RCP4.5 and RCP8.5 than provided in Table 4.3.

As the importance of MISI and MICI is difficult to assess on longer time scales, there remains deep uncertainty for the Antarctic contribution to GMSL after 2100 (Cross-Chapter Box 4 in Chapter 1). Results on these long-time scales are discussed in 4.2.3.5.

There is limited evidence for major changes since AR5 in the non-Antarctic components. Recent projections of the glacier contribution are nearly identical to AR5 results used here (see Cross-Chapter Box 6 in Chapter 2). Greenland, thermal expansion and land water storage are also not updated, mainly due to a lack of updated CMIP simulations. Hence, our revised projections replace only the AR5 estimate for Antarctica by a new assessment as outlined in the previous paragraph based on post-AR5 literature and maintaining identical contributions for the non-Antarctic components. As no general dependence between the Antarctic contribution and the non-Antarctic components can be derived from the four studies, independent uncertainties are assumed, which is close to the uncertainty propagation by Church et al. (2013).

Time series for the different RCP scenarios are shown in Figure 4.9 indicating a divergence in median and upper *likely* range for RCP8.5 during the second half of the century between this report and the AR5 projections (Church et al., 2013). The value of the Antarctic contribution in 2081–2100 under RCP8.5 is the individual component with the largest uncertainty. As a consequence, the uncertainty in the GMSL projections is slightly increased compared to Church et al. (2013). Nevertheless, results can also be considered to be consistent with Church et al. (2013). In AR5, the potential additional contribution by ice dynamics, was estimated to be not more than several tenths of a metre but excluded from projections; here this

Table 4.3 | An overview of different studies estimating the future Antarctic contribution to sea level rise (SLR), listed here are median values. Estimates from Golledge et al. (2015) are based on the average contribution to Global Mean Sea Level (GMSL) over the full 21st century, based on two alternative ensembles using different sub-ice melt schemes. This average is not explicitly reported in the original paper where the individual values of 0.1 and 0.39 m are reported. SMB is the surface mass balance, BMB the basal melt balance, LIG is Last Interglacial, MICI is marine ice cliff instability, GCM is General Circulation Model, PDD is positive-degree day.

	Levermann et al. (2014)	Ritz et al. (2015)	Golledge et al. (2015)	Golledge et al. (2019)	DeConto and Pollard (2016)	Bulthuis et al. (2019)
	RCP2.6/ RCP4.5/ A1B/ RCP8.5	RCP2.6/ RCP4.5/ A1B/ RCP8.5	RCP2.6/ RCP4.5/ A1B/ RCP8.5	RCP2.6/ RCP4.5/ A1B/ RCP8.5	RCP2.6/ RCP4.5/ A1B/ RCP8.5	RCP2.6/ RCP4.5/ A1B/ RCP8.5
Antarctica 2050 (m)	0.03/0.03/-/-0.03	-/-/0.03/-	0.00/0.01/-/-0.02	-/0.0/-/0.02	0.02/0.03/-/0.04	0.01/0.01/-/-0.03
Antarctica 2100 (m)	0.07/0.09/-/-0.11	-/-/0.12/-	0.02/0.05/-/0.18	-/0.04/-/0.11	0.14/0.41/-/0.79	0.03/0.05/-/-0.11
Antarctica 2200 (m)	0.16/0.25/-/-0.54	-/-/0.41/-	0.10/0.32/-/1.15	-/-/-/-	0.35/1.67/-/5.39/	0.08/0.15/-/0.45
Uncertainties	Ensembles	Quantiles	High-average	High-average	Ensemble selections	Stochastic sensitivities
Tuning targets	variable	Present-Day rates from observations	None	None	LIG and Pliocene	Present-Day rates from observations
Grounding Line	Poor	Conditional on bed slope and Schoof flux	Subgrid parameterisation	Sub-grid parameterisation	Pollard and DeConto (2012)	Schoof (2007a); Tsai et al. (2015)
Dynamics	Traditional	Several basal friction laws	Hybrid, 10–20 km grid Till friction angle	Hybrid, 10–20 km grid Till friction angle	Hybrid, 10 km grid	Hybrid, 20 km grid
Hydrofracturing	No	No	No	No	Yes	No
MICI	No	No	No	No	Yes	No
Initialisation	variable	Observed rates	Focus on long time scales	Focus on long time scales	1950	Close to steady state
SMB	No	Parameterised	PDD scheme	PDD scheme	Regional Climate Model	Van Wessem et al. (2014)
BMB	Linear perturbation	Parameterised	Slab Ocean GCM	Slab Ocean GCM	NCAR CCSM4	Reese et al. (2018a)
Driving mechanism for retreat	Ocean only	Observations, statistics	Ocean (2/3)	Intermediate complexity	Atmospheric forcing dominates	Atmospheric and ocean forcing

4

Table 4.4 | Median values and *likely* ranges for projections of global mean sea level (GMSL) rise in metres in 2081–2100 relative to 1986–2005 for three scenarios. In addition, values of GMSL rise are given for 2046-2065 and 2100, and the rate of GMSL rise is given for 2100. Values between parentheses reflect the *likely* range. SMB is surface mass balance, DYN is dynamical contribution, LWS is land water storage. Total AR5 minus Antarctica AR5 is the GMSL rise contribution in Church et al. (2013) without the Antarctic contribution of Church et al. (2013). The newly derived Antarctic contribution is added to this to arrive at the GMSL rise.

	RCP2.6	RCP4.5	RCP8.5	Comments
Thermal expansion	0.14 (0.10–0.18)	0.19 (0.14–0.23)	0.27 (0.21–0.33)	AR5
Glaciers	0.10 (0.04–0.16)	0.12 (0.06–0.18)	0.16 (0.09–0.23)	AR5
Greenland SMB	0.03 (0.01–0.07)	0.04 (0.02–0.09)	0.07 (0.03–0.17)	AR5
Greenland DYN	0.04 (0.01–0.06)	0.04 (0.01–0.06)	0.05 (0.02–0.07)	AR5
LWS	0.04 (−0.01–0.09)	0.04 (−0.01–0.09)	0.04 (−0.01–0.09)	AR5
Total AR5 – Antarctica AR5*; 2081–2100	0.35 (0.23–0.48)	0.43 (0.30–0.57)	0.60 (0.43–0.78)	SROCC implicit in AR5
Total AR5 – Antarctica AR5; 2046–2065	0.22 (0.15–0.29)	0.24 (0.17–0.31)	0.28 (0.20–0.36)	SROCC implicit in AR5
Antarctica 2031–2050	0.01 (0.00–0.03)	0.01 (0.00–0.03)	0.02 (0.00–0.05)	SROCC
Antarctica 2046–2065	0.02 (0.00–0.05)	0.02 (0.01–0.05)	0.03 (0.00–0.08)	SROCC
Antarctica 2081–2100	0.04 (0.01–0.10)	0.05 (0.01–0.13)	0.10 (0.02–0.23)	SROCC
Antarctica 2100	0.04 (0.01–0.11)	0.06 (0.01–0.15)	0.12 (0.03–0.28)	SROCC
GMSL 2031–2050	0.17 (0.12–0.22)	0.18 (0.13–0.23)	0.20 (0.15–0.26)	SROCC
GMSL 2046–2065	0.24 (0.17–0.32)	0.26 (0.19–0.34)	0.32 (0.23–0.40)	SROCC
GMSL 2081–2100	0.39 (0.26–0.53)	0.49 (0.34–0.64)	0.71 (0.51–0.92)	SROCC
GMSL in 2100	0.43 (0.29–0.59)	0.55 (0.39–0.72)	0.84 (0.61–1.10)	SROCC
Rate (mm yr^{-1})	4(2–6)	7(4–9)	15(10–20)	SROCC

Notes:

*The uncertainty in this value is calculated as in Church et al. (2013).

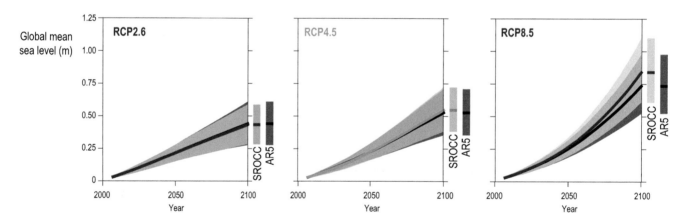

Figure 4.9 | Time series of Global Mean Sea Level (GMSL) for Representative Concentration Pathway (RCP)2.6, RCP4.5 and RCP8.5 as used in this report and, for reference the IPCC 5th Assessment Report (AR5) results (Church et al., 2013). Results are based on AR5 results for all components except the Antarctic contribution. Results for the Antarctic contribution in 2081–2100 are provided in Table 4.4. The shaded region is considered to be the *likely range*.

value was assessed to be 16 cm (5–95 percentile; 2–37 cm) and include it in the projections. As the projections build on the CMIP5 work presented in AR5, and also given the limited exploration of uncertainty in estimates from each individual study, the results of the 5–95 percentile are interpreted to represent the *likely* range, that is, the 17–83 percentile, as assessed by Church et al. (2013) and as assessed in AR5 for other CMIP5-derived results.

Projections as presented in Table 4.4 are used to calculate the regional RSL projections as outlined in AR5 by including gravitational and rotational patterns as shown in Figure 4.10 and subsequently used in 4.2.3.4 to calculate ESL projections. Including the updated results in terms of magnitude and uncertainty for the Antarctic component also changes the regional patterns in sea level projections. Results of the regional patterns in Figure 4.10 show an increased SLR with respect to the results presented in AR5 nearly everywhere for RCP8.5 because of the increased Antarctic contribution.

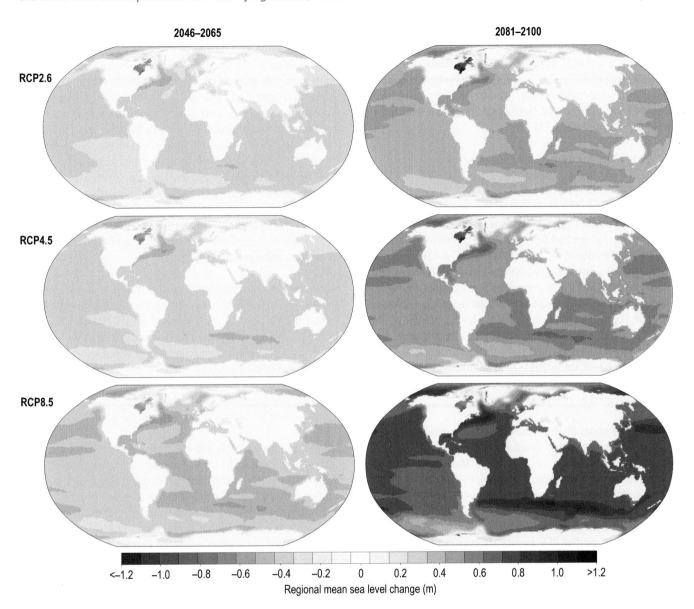

Figure 4.10 | Regional sea level change for RCP2.6, RCP4.5 and RCP8.5 in metres as used in this report for extreme sea level (ESL) events. Results are median values based on the values in Table 4.4 for Antarctica including GIA and the gravitational and rotational effects, and results by Church et al. (2013) for glaciers, land water storage (LWS) and Greenland. The left column is for the time slice 2046–2065 and the right column for 2081–2100.

4.2.3.3 Probabilistic Sea Level Projections

Since AR5, several studies have produced SLR projections in coherent frameworks that link together global-mean and RSL rise projections. The approaches are generally similar to those adopted by AR5 for its global-mean sea level projections: a bottom-up accounting of different contributing processes (e.g., land-ice mass loss, thermal expansion, dynamic sea level), of which many are 'probabilistic', in that they attempt to describe more comprehensive probability distributions of sea level change than the *likely* ranges presented by Church et al. (2013). An overview of probabilistic approaches is presented in Garner et al. (2017), indicating higher values for post AR5 studies mainly reflecting increased uncertainty based on a single contested study for the Antarctic contribution (DeConto and Pollard, 2016). As such many of these probabilistic studies present full

probability density function conditional not only on an RCP scenario, but with additional and equally important a priori assumptions concerning for instance the Antarctic contribution over which a consensus has yet to solidify. An example is the study by Le Bars et al. (2017) who expand the projection by Church et al. (2013) in a probabilistic way with the Antarctic projections by DeConto and Pollard (2016) to obtain a full probability density function for SLR for RCP8.5. Other probabilistic approaches are provided by Kopp et al. (2014) and Jackson and Jevrejeva (2016) using different ice sheet representations drawing on expert elicitation (Bamber and Aspinall, 2013). Probabilistic estimates are useful for a quantitative risk management perspective (see Section 4.3.3). An even more general approach than the probabilistic estimates has been taken by Le Cozannet et al. (2017) who frame a 'possibilistic' framework of SLR including existing probabilistic estimates and combining them.

This section first briefly reviews key sources of information for probabilistic projections (Section 4.2.3.3.1), with a focus on new results since AR5, then summarises the different global and regional projections (Section 4.2.3.3.2). Eventually, bottom-up projections were distinguished which explicitly describe the different components of SLR (Section 4.2.3.3.3) from semi-empirical projections (Section 4.2.3.3.4).

4.2.3.3.1   Components of probabilistic global mean sea level projections

Thermal expansion: Global mean thermal expansion projections rely on coupled climate models projections (Kopp et al., 2014; Slangen et al., 2014a; Jackson and Jevrejeva, 2016) or simple climate model projections (Perrette et al., 2013; Nauels et al., 2017b; Wong et al., 2017), and are substantively unchanged since AR5. For those studies relying on the CMIP5 GCM ensemble, interpretations of the model output differ mainly with regard to how the range is understood. For example, Kopp et al. (2014), interprets the 5–95 percentile of CMIP5 values as a *likely* range of thermal expansion. The differences among the studies yield discrepancies smaller than 10 cm, e.g., Slangen et al. (2014a) use 20–36 cm in 2081–2100 with respect to 1986–2005, while (Kopp et al., 2014) project a *likely* range of 28–46 cm in 2081–2099 with respect to 1991–2009.

Glaciers: Projections of glacier mass change rely either on models of glacier SMB and geometry, forced by temperature and precipitation fields (Slangen and Van de Wal, 2011; Marzeion et al., 2012; Hirabayashi et al., 2013; Radić et al., 2014; Huss and Hock, 2015), or simple scaling relationships with global mean temperature (Perrette et al., 2013; Bakker et al., 2017; Nauels et al., 2017a). Glacier mass change projections published since AR5, based on newly developed glacier models, confirm the overall assessment of AR5 (see also Section 4.2.3.2).

Land water storage: Projections of the GMSL rise contributions due to dam impoundment and groundwater withdrawal are generally either calibrated to hydrological models (e.g., Wada et al., 2012) or neglected. Recent coupled climate-hydrological modelling suggests that a significant minority of pumped groundwater remains on land, which may reduce total GMSL rise relative to studies assuming full drainage to the ocean (Wada et al., 2016). Kopp et al. (2014) estimated land water storage based on population projections. However, there are no substantive updates to projections of the future land-water storage contribution to GMSL rise since AR5.

Ice sheets: GMSL projections in previous IPCC assessments were based on results from physical models of varying degree of complexity interpreted using expert judgment of the assessment authors (Meehl et al., 2007; Church et al., 2013). AR5 (Church et al., 2013) used this approach and is partly based on the assessment of statistical-physical modelling of the Antarctic contribution (Little et al., 2013). As an alternative to the model-based approach, several studies have applied structured expert elicitation to the GMSL contribution of ice sheets. This approach is based on a more formal expert elicitation protocol (Cooke, 1991; Bamber and Aspinall, 2013; Bamber et al., 2019) instead of physically based models. Combining the Antarctic contribution from

the expert elicitation with the non-Antarctic components from AR5 as done for Table 4.4 leads to an estimated SLR of 0.95 m (median) for the high scenario and an upper *likely* range of 1.32 m (Figure 4.2), which is slightly higher than the process-based results. Results by Bamber and Aspinall (2013) were criticised because of their procedure for post-processing the expert data of individual ice sheets to a total sea level contribution from the ice sheets (de Vries and van de Wal, 2015; Bamber et al., 2016; de Vries and van de Wal, 2016). Bamber et al. (2019) avoids this issue by eliciting expert judgments about ice sheet dependence. Alternatively, Horton et al. (2014) used a simpler elicitation protocol focusing on the total SLR rather than the ice sheet contribution alone. Finally, several probabilistic studies (e.g., Bakker et al., 2017; Kopp et al., 2017; Le Bars et al., 2017) used the results of a single ice sheet model study from DeConto and Pollard (2016) as the Antarctic contribution to GMSL.

Beside the total contribution of ice sheets several studies address the individual contribution of either Greenland or Antarctica (see Section 4.2.3.1.1 and 4.2.3.1.2) based on ice dynamical studies. Critical for GMSL projections is the low confidence in the dynamic contribution of the AIS beyond 2050 in previous assessments, as discussed in Section 4.2.3.1.2.

4.2.3.3.2   From probabilistic global mean sea level projections to regional relative sea level change

Differences between GMSL and RSL change are driven by three main factors: (1) changes in the ocean, for instance, the thermal expansion component and the circulation driven changes, (2) gravitational and rotational effects caused by redistribution of mass within cryosphere and hydrosphere, leading to spatial patterns, and (3) long term processes caused by GIA that lead to horizontal and VLM. Finally, the inverse barometer effect caused by changes in the atmospheric pressure, sometimes neglected in projections, can also make a small contribution, particularly on shorter time scales. For the 21st century as a whole, estimates of the latter are smaller than 5 cm at local scales (Church et al., 2013; Carson et al., 2016).

Ocean Dynamic sea level: Projections of dynamic sea level change are necessarily derived through interpretations of coupled climate model projections. As with thermal expansion projections, interpretations of the CMIP5 ensemble differ with regard to how the model range is understood and the manner of drift correction, if any (Jackson and Jevrejeva, 2016). However, relative to tide-gauge observations, coupled climate models tend to overestimate the memory in dynamic sea level; thus, they may underestimate the emergence of the externally forced signal of DSL change above scenario uncertainty (Becker et al., 2016). ODSL from coupled climate models does not include the changes resulting from ice melt because ice melt is calculated off-line.

Gravitational-rotational and deformational effects (GRD; Gregory et al., 2019): All projections of RSL change include spatial patterns in sea level for cryospheric changes, which however may differ in the details with which these are represented. Some studies also include a spatial pattern for land-water storage change (Slangen et al., 2014a), anthropogenic subsidence is not included. Recent work indicates that, for some regions with low mantle viscosity, spatial patterns cannot be treated as fixed on multi-century time scales (Hay et al., 2017). This effect has not yet been incorporated into comprehensive RSL projections, but is probably only of relevance near ice sheets. For adaptation purposes, Larour et al. (2017) developed a mapping method to indicate which areas of ice mass loss are important for which major port city. There is *high confidence* in the patterns caused by GRD, as in AR5.

Vertical land motion (VLM): These processes can be an important driver of RSL change, particularly in the near- to intermediate-field of the large ice sheets of the LGM (e.g., North America and northern Europe). This process is incorporated either by physical modelling (Slangen et al., 2014a) or by estimation of a long-term trend from tide-gauge data (e.g., Kopp et al., 2014), which is then spatially extrapolated. In the former case, only the long-term GIA process is included in the projections, but it excludes other important local factors contributing to VLM (e.g., tectonic uplift/subsidence and groundwater/hydrocarbon withdrawal); by using only tide gauge measurements, projections may assume that these other processes proceed at a steady rate and thus do not allow for management changes that affect groundwater extraction.

4.2.3.3.3 Semi-empirical projections

Semi-empirical models provide an alternative approach to process-based models aiming to close the budget between the observed SLR and the sum of the different components contributing to SLR. In general, motivated by a mechanistic understanding, semi-empirical models use statistical correlations from time series analysis of observations to generate projections (Rahmstorf, 2007; Vermeer and Rahmstorf, 2009; Grinsted et al., 2010; Kemp et al., 2011; Kopp et al., 2016). They implicitly assume that the processes driving the observations and feedback mechanisms remain similar over the past and future. In the past, differences between semi-empirical projections and process-based models were significant but for

Table 4.5 | Sources of Information Underlying Probabilistic Projections of Sea level Rise (SLR) Projections. CMIP5 is Coupled Model Intercomparison Project Phase 5, GRD is gravitational, rotational and deformation effects, SMB is surface mass balance, AR4 is IPCC 4th Assessment Report, VLM is vertical land motion, GIA is glacio-isostatic adjustment.

Study	Thermal expansion	Glaciers	Land water storage	Ice Sheets	Dynamic sea level	GRD	VLM
Perrette et al. (2013)	CMIP5	Global SMB sensitivity and exponent from AR4; total glacier volume from Radić and Hock (2010)	Not included	Greenland's SMB from AR4; semi-empirical model using historical observations.	CMIP5	Bamber et al. (2009)	Not included
Grinsted et al. (2015)	CMIP5	Church et al. (2013)	Wada et al. (2012)	Church et al. (2013); Expert elicitation from Bamber and Aspinall (2013)	CMIP5	Bamber et al. (2009)	GIA projections from Hill et al. (2010) using observations
Slangen et al. (2014a)	CMIP5	CMIP5; glacier area inventory Radić and Hock (2010) in a glacier mass loss model	Wada et al. (2012)	SMB Meehl et al. (2007), ice dynamics Meehl et al. (2007) and Katsman et al. (2011)	CMIP5	Slangen et al. (2014a)	GIA resulting of ice sheet melt from glacier mass loss model
Kopp et al. (2014)	CMIP5	CMIP5; Marzeion et al. (2012)	Chambers et al. (2017); Konikow (2011)	Church et al. (2013); Expert elicitation from Bamber and Aspinall (2013)	CMIP5	Mitrovica et al. (2011)	GIA, tectonics, and subsidence from Kopp et al. (2013)
Kopp et al. (2017)	CMIP5	CMIP5; Marzeion et al. (2012)	Chambers et al. (2017); Konikow (2011)	DeConto and Pollard (2016)	CMIP5	Mitrovica et al. (2011)	GIA, tectonics, and subsidence from Kopp et al. (2013)
Le Bars et al. (2017)	CMIP5	Four glacier models: Giesen and Oerlemans (2013) Marzeion et al. (2012), Radić et al. (2014) Slangen and Van de Wal (2011)	Wada et al. (2012)	DeConto and Pollard (2016); Fettweis et al. (2013) Church et al. (2013)	CMIP5	–	–
Jackson and Jevrejeva (2016)	CMIP5	Marzeion et al. (2012)	Wada et al. (2012)	Church et al. (2013); Expert elicitation from Bamber and Aspinall (2013)	CMIP5	Bamber et al. (2009)	GIA resulting of ice sheet melt from glacier mass loss model Peltier et al. (2015)
de Winter et al. (2017)	CMIP5	CMIP5; glacier area inventory Radić and Hock (2010) in a glacier mass loss model	Wada et al. (2012)	Church et al. (2013); Expert elicitation de Vries and van de Wal (2015); Ritz et al. (2015)	CMIP5	Mitrovica et al. (2001)	GIA resulting of ice sheet melt from glacier mass loss model

more recent studies the differences are vanishingly small. Ongoing advances in closing the sea level budget and in the process understanding of the dynamics of ice have reduced the salience of estimates from semi-empirical models. Moreover, the results from semi-empirical models (Kopp et al., 2016; Mengel et al., 2016) are in general agreement with Church et al. (2013), except when those results reflect the combined hydrofracturing and ice cliff instability mechanism as presented by DeConto and Pollard (2016). At the same time, semi-empirical models based on past observations capture poorly or miss altogether the recent observed changes in Antarctica. MISI may lend a very different character to ice sheet evolution in the near future than in the recent past and hydrofracturing remains impossible to quantify from observational records only. For this reason, a new generation of semi-empirical models and emulators has been developed that estimate individual components of SLR, which the former models do not (Mengel et al., 2018). These newer models aim to emulate the response of more complex models providing more detailed information for different climate scenarios or probability estimates than process-based models (Bakker et al., 2017; Nauels et al., 2017a; Wong et al., 2017; Edwards et al., 2019).

4.2.3.3.4 Recent probabilistic and semi-empirical projections

A wide range of probabilistic sea level projections exist, ranging from simple scaling relations to partly process-based components combined with scaling relations. Table 4.5 illustrates the overlap between many of the studies, a complete overview is presented by Garner et al. (2017), and differences between different classes of models are discussed in Horton et al. (2018). Many studies rely on CMIP simulations for an important part of their sea level components. The largest difference can be found in the treatment of the ice dynamics, particularly for Antarctica, which are usually not CMIP5 based. Instead, each derives from one of several estimates of the Antarctic contribution. These results are useful for the purposes of elucidating sensitivities of process-based studies and effects of changing components to the total projection. This report relies on the Antarctic component from Section 4.2.3.2 for calculating the *likely* range of RSL. Hence the values in Table 4.5 are not used for the final assessment of RSL including the SROCC specific Antarctic contribution presented in Section 4.2.3.2. Comparing the probabilistic projections (Table 4.6) is difficult because of the subtle differences between their assumptions. Nevertheless, values range much more for 2100 than for 2050.

Table 4.6 | Median and *likely* Global Mean Sea Level (GMSL) rise projections (m). Values between brackets are *likely* range, if no values are given the *likely* range is not available. The table shows result from the probabilistic and semi-empirical results. A is 2000 as base line year up to 2100; B is the average of 1986–2005 as base line for the projection up to 2081–2100, C 1980–1999 as baseline up to 2090–2099.

	Period	2050			2100		
		RCP2.6	RCP4.5	RCP8.5	RCP2.6	RCP4.5	RCP8.5
Perrette et al. (2013)	C		0.28 (0.23–0.32)	0.28 (0.23–0.34)		0.86 (0.66–1.11)	1.06 (0.78–1.43)
Grinsted et al. (2015)	A						0.8 (0.58–1.20)
Slangen et al. (2014a)	B AB B					0.54 (0.35–0.73)	0.71 (0.43–0.99)
Kopp et al. (2014)	A	0.25 (0.21–0.29)	0.26 (0.21–0.31)	0.29 (0.24–0.34)	0.50 (0.37–0.65)	0.59 (0.45–0.77)	0.79 (0.62–1.00)
Kopp et al. (2017)	A	0.23 (0.16–0.33)	0.26 (0.18–0.36)	0.31 (0.22–0.40)	0.56 (0.37–0.78)	0.91 (0.66–1.25)	1.46 (1.09–2.09)
de Winter et al. (2017)	B						0.68/0.86
Jackson and Jevrejeva (2016)	B					0.54 (0.36–0.72)	0.75 (0.54–0.98)
Le Bars et al. (2017)	B					1.06 (0.65–1.47)	1.84 (1.24–2.46)
Nauels et al. (2017b)	B	0.24 (0.19–0.30)	0.25 (0.21–0.30)	0.27 (0.23–0.33)	0.45 (0.35–0.56)	0.55 (0.45–0.67)	0.79 (0.65–0.97)
Bakker et al. (2017)	A	0.20	0.23	0.25	0.53	0.72	1.16
Wong et al. (2017)	A	0.26	0.28	0.30	0.55	0.77	1.50
Jevrejeva et al. (2014a)	A						0.80 (0.6–1.2)
Schaeffer et al. (2012)	A					0.90	1.02
Mengel et al. (2016)	B	0.18	0.18	0.21	0.39	0.53	0.85

4.2.3.4 Changes in Extreme Sea Level events

ESL events are water level heights that consist of contributions from mean sea level, storm surges and tides. Compound effects of surges and tides are drivers of the ESL events. Section 4.2.3.4.1 discusses the combination of mean sea level change with a characterisation of the ESL events derived from tide gauges over the historical period and the sections 4.2.3.4.2 and 4.2.3.4.3 evaluate possible changes in these characteristics caused by cyclones and waves. This section discusses the importance of ESL and different modelling strategies to improve our understanding of ESL projections.

Even a small increase in mean sea level can significantly augment the frequency and intensity of flooding. This is because SLR elevates the platform for storm surges, tides, and waves, and because there is a log-linear relationship between a flood's height and its occurrence interval. Changes are most pronounced in shelf seas. Roughly 1.3% of the global population is exposed to a 1 in 100-year^{-1} flood (Muis et al., 2016). This exposure to ESL and resulting damage could increase significantly with SLR, potentially amounting to 10% of the global GDP by the end of the century in the absence of adaptation (Hinkel et al., 2014).

The frequency and intensity of ESL events can be estimated with statistical models or hydrodynamical models constrained by observations. Hydrodynamic models simulate a series of ESL events over time, which can then be fitted by extreme value distributions to estimate the frequency and intensity (e.g. the return level of an event occurring with a period of 100 years or frequency of 0.01 yr^{-1}, also called the 100-year event). A tide model is sometimes included and sometimes added offline to estimate the ESL events. Statistical models fit tide gauge observations to extreme value distributions to directly estimate ESL events or combine probabilistic RSL scenarios with storm surge modelling. This can be done on global scale or local scale. For example, Lin et al. (2016) and Garner et al. (2017) estimate the increase in flood frequency along the US east coast. Both of these modelling approaches can account for projections of SLR. Rasmussen et al. (2018) used a combination of a global network of tide gauges and a probabilistic localised SLR to estimate expected ESL events showing inundation reductions for different temperature stabilisation targets as shown in the SR15 report.

An advantage of the use of hydrodynamic models is that they can quantify interactions between the different components of ESL (Arns et al., 2013). Hydrodynamical models can be executed over the entire ocean with flexible grids at a high resolution (up to 1/20°or ~5 km) where necessary, appropriate for local estimates (Kernkamp et al., 2011). Input for these models are wind speed and direction, and atmospheric pressure. Results of those models show that the Root Mean Squared Error between modelled and observed sea level is less than 0.2 m for 80% of a data set of 472 stations covering the global coastline (Muis et al., 2016) at 10-minute temporal resolution over a reference period from 1980–2011. This implies that for most locations it can be used to describe the variability in ESL. However, the areas where ESL is dominated by tropical storms are problematic for hydrodynamical models. Another difficulty arises when these models are forced with climate models: they inherit the limitations

(resolution, precision and accuracy) of wind and pressure in climate projections, which is often insufficient to describe the role of waves.

Statistical models have shown that the estimation of ESL is highly sensitive to the characterisation of SLR and flood frequency distributions (Buchanan et al., 2017). This is confirmed by Wahl et al. (2017) who estimate that the 5–95 percentile uncertainty range, attained through the application of different statistical extreme value methods and record lengths, of the current 100-year event is on average 40 cm, whereas the corresponding range in projected GMSL of AR5 under RCP8.5 is 37 cm. For ESL events with a higher return period, differences will be larger. Capturing changes in the ESL return periods in the future is even more complicated because both the changing variability over time and the uncertainty in the mean projection must be combined. A statistical framework to combine RSL and ESL, based on historical tide gauge data was applied to the US coastlines (Buchanan et al., 2016). Hunter (2012) and the AR5 (Church et al., 2013) projected changes in flood frequency worldwide; however, these analyses used the Gumbel distribution for high water return periods, which implies that the frequency of all ESLs (e.g., whether the 1in 10-year or 1 in 500-year) will change by the same magnitude for a given RSL, an approximation that can underestimate or overestimate ESL (Buchanan et al., 2017). Hence, the amplification factors of future storm return frequency in AR5 WGI Figure 13.25 may underestimate flood hazards in some areas, while overestimating them in others. By using the Gumbel distribution, Muis et al. (2016) may also inadequately estimate flood frequencies.

4.2.3.4.1   Relative sea level and extreme sea level events based on tide gauge records

Changes in ESL are presented here, based on the projections as presented in 4.2.3.2 at the tide gauge locations in the GESLA2 database (Woodworth et al., 2016). Results include GIA effects, but anthropogenic subsidence is not prescribed. These calculations serve as a signal to guide adaption to SLR (Stephens et al., 2018). Return periods are calculated as a combination of regional RSL projections and a probabilistic characterisation of the variability in sea level as derived from the GESLA2 data set which contains a quasi-global set of tide gauges. By doing so, it is assumed that the variability in the tide gauge record does not change over time. Models are not accurate enough to address whether this is correct or not.

To quantify the average return period of ESL events, a peak-over-threshold method is applied following Arns et al. (2013) and Wahl et al. (2017). Tide gauge records are detrended by subtracting a running mean of one year. Peaks above the 99th percentile of hourly water levels are extracted and declustered by applying a minimum time between peaks of 72 hours. This threshold of 99% was recommended by Wahl et al. (2017) for global applications. Using a maximum likelihood estimator, a Generalized Pareto Distribution (GPD) is fitted to these peaks, allowing for an extrapolation to return periods beyond the available period of observations. Changes in ESL events due to regional mean SLR are quantified following Hunter (2010). Uncertainties in the GPD parameters and projections are propagated using a Monte Carlo approach, from which a best estimate is derived (see SM4.2). Only tide gauge records of 20 years

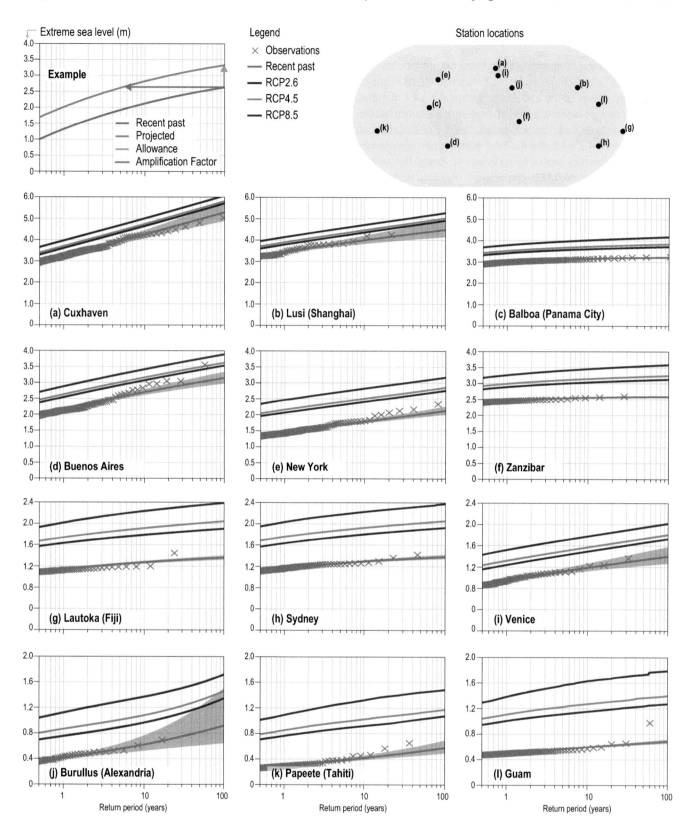

Figure 4.11 | The relation between expected extreme sea level (ESL) events and return period at a set of characteristic tide gauge locations (see upper left for their location), referenced to recent past mean sea level, based on observations in the GESLA-2 data base (grey lines) and 2081–2100 conditions for three different RCP scenarios as presented in Section 4.2.3.2. The grey bands represent the 5–95% uncertainty range in the fit of the extreme value distribution to observations. The upper right hand panel provides an example illustrating the relationship between ESL events and return period for historical and future conditions; the blue line in this panel shows the best estimate ESL event above the 1986–2005 reference mean sea level. The coloured lines for the different locations show this expected ESL events for different RCP scenarios. The horizontal line denoting the amplification factor expresses the increase in frequency of events which historically have a return period of once every 100 years. In the example, a water level of 2.5 m above mean sea level, recurring in the recent climate approximately every 100 years in recent past climate, will occur every 2 to 3 years under future climate conditions. The allowance expresses the increase in ESL for events that historically have a return period of 100 years.

of longer, which are at least 70% complete, are used. However, as can be seen for Guam (Fig 4.9), this does not ensure a good fit of the GPD to all peaks, as rare events may have been captured in this relatively short record.

Projected changes in ESL events are shown for 12 selected tide gauges in Figure 4.11. The magnitude of these changes depends on the relation between ESL events and the associated return periods, as well as regional sea level projections, and the uncertainty therein (see inset Figure 4.11). The change in ESL events is commonly expressed in terms of the amplification factor and the allowance. The amplification factor denotes the amplification in the average occurrence frequency of a certain extreme event, often referenced to the water level with a 100-year return period during the historic period. The allowance denotes the increased height of the water level with a given return period. This

allowance equals the regional projection of SLR with an additional height related to the uncertainty in the projection (Hunter, 2012).

Amplification factors are strongly determined by the local variability in ESL events. Locations where this variability is large due to large storm surges and astronomical tides (e.g., Cuxhaven, see Figure 4.9) will experience a relatively moderate amplification of the occurrence frequency of extremes. In comparison, locations with small variability in ESL events (e.g., Lautoka and Papeete) will experience large amplifications even for a moderate rise in mean sea level (Vitousek et al., 2017). Globally, this contrast between regions with large and small amplification factors becomes clear for projections by mid-century (Fig 4.11, left panels). Although regional differences in projected mean SLR are small for the coming centuries, regional contrasts in amplification factors are considerable. In particular,

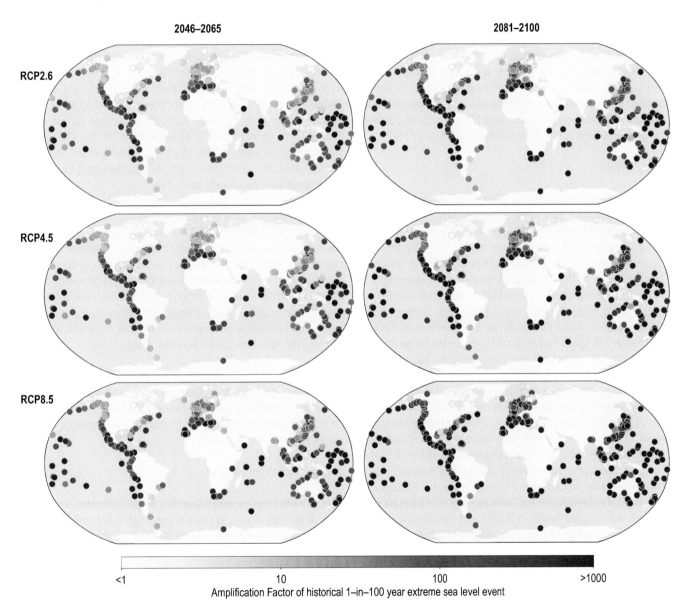

Figure 4.12 | The colours of the dots express the factor by which the frequency of extreme sea level (ESL) events increase in the future for events which historically have a return period of 100 years. Hence a value of 50 means that what is currently 1-in-100 year event will happen every 2 years due to a rise in mean sea level. Results are shown for three RCP scenarios and two future time slices as median values. Results are shown for tide gauges in the GESLA2 database. The accompanying confidence interval can be found in SM4.2 as well as a list of all locations. The data underlying the graph are identical to those presented in Figure 4.11. The amplification factor is schematically explained in the upper right panel of figure 4.11. Storm climatology is constant in these projections.

many coastal areas in the lower latitudes may expect amplification factors of 100 or larger by mid-century, regardless of the scenario as also shown in SR15 and Rasmussen et al. (2018). This indicates that, at these locations, water levels with return periods of 100 years during recent past will become annual or more frequent events by mid-century. By end-century and in particular under RCP8.5, such amplification factors are widespread along the global coastlines (Vousdoukas et al., 2018a).

In summary, ESL events estimates as presented in this subsection, clearly show that as a consequence of SLR, events which are currently rare (e.g., with an average return period of 100 years), will occur annually or more frequently at most available locations for RCP8.5 by the end of the century (*high confidence*). For some locations, this change will occur as soon as mid-century for RCP8.5 and by 2100 for all emission scenarios. The affected locations are particularly located in low-latitude regions, away from the tropical cyclone (TC) tracks. In these locations, historical sea level variability due to tides and storm surges is small compared to projected mean SLR. Therefore, even limited changes in mean sea level will have a noticeable effect on ESLs, and for some locations, even RCP2.6 will lead to the annual occurrence of historically rare events by mid-century. Results should be treated with caution in regions where TCs are important as they are underrepresented in the observations (Haigh et al., 2014a).

4.2.3.4.2   Waves

A warming climate is expected to affect wind patterns and storm characteristics, which in turn will impact wind waves that contribute to high coastal water levels. Wind-wave projections are commonly based on dynamical and statistical wave models forced by projected surface winds from GCMs, notably those participating in the CMIP. In the framework of the Coordinated Ocean Wave Climate Project (COWCLIP), an ensemble of Coupled Model Intercomparison Project Phase 3 (CMIP3)-based global wave projections (Hemer et al., 2013) was produced and the results were summarised in the AR5 (WGI Chapter 13). Casas-Prat et al. (2018) expanded the geographic domain to include the Arctic Ocean, highlighting the vulnerability of high-latitude coastlines to wave action as ice retreats. Reduced sea ice allows larger waves and stronger cyclones in the Arctic Ocean, which can further disrupt and break up sea ice (e.g., Thomson and Rogers, 2014; Day and Hodges, 2018). A review and consensus-based analysis of regional and global scale wave projections, including CMIP5-based projections, has been provided by Morim et al. (2018) as part of COWCLIP. Projections of annual and seasonal mean significant wave height changes agree on an increase in the Southern Ocean, tropical eastern Pacific and Baltic Sea; and on a decrease over the North Atlantic, northwestern Pacific and Mediterranean Sea. Projections of mean significant wave height lack consensus over the eastern North Pacific and southern Indian and Atlantic Oceans. Projections of future extreme significant wave height are consistent in projecting an increase over the Southern Ocean and a decrease over the northeastern Atlantic and Mediterranean Sea. Regional projections of wind-waves have mostly been applied to Europe so far, while highly vulnerable regions have been largely overlooked. This is the case for low-lying islands where impacts of SLR and wave-induced

flooding are expected to be severe and adaptive capacity is reduced (Hoeke et al., 2013; Albert et al., 2016).

A number of studies have included waves, in addition to tides and sea level anomalies, to assess coastal vulnerability to SLR using dynamical and statistical approaches. CoSMoS (Barnard et al., 2014) includes a series of embedded wave models to estimate high resolution projections of total water levels along the Southern California coast for different extreme scenarios (O'Neill et al., 2017). Arns et al. (2017) find that an increase in sea level may reduce the depth-limitation of waves, thereby resulting in waves with greater energy approaching the coast. Including wave effects is crucial for coastal adaptation and planning (e.g., Isobe, 2013). For example Arns et al. (2017) report that coastal protection design heights need to be increased by 48–56% in the German Bight region relative to a design height based on the effect of SLR on ESL only. Combining SLR with extreme value theory applied to past observations of tides, storm surges and waves, Vitousek et al. (2017) found that a 10–20 cm SLR could result in a doubling of coastal flooding frequency in the tropics. For the southern North Sea region, Weisse et al. (2012) argue that increasing storm activity also increases hazards from ESL events. Global-scale projections of ESL event changes including wave setup indicate a *very likely* increase of the global average 100-year ESL of 58–172 cm under RCP8.5 (Vousdoukas et al., 2018c). Changes in storm surges and waves enhance the effects of relative SLR along the majority of northern European coasts, with contributions up to 40% in the North Sea (Vousdoukas et al., 2017).

A stationarity of the wave climate is often assumed for projections of ESL events (Vitousek et al., 2017). Yet, wave contributions to coastal sea level changes (setup and swash) depend on several factors that can vary in response to internal climate variability and climate change, including deep-water wave field, water-depth, and geomorphology. Melet et al. (2018) reported that over recent decades, wave setup and swash interannual-to-decadal changes induced by deep-water wave height and period changes alone were sizeable compared to steric and land-ice mass loss coastal sea level changes.

Comprehensive broad-scale projections of sea level at the coast including regional sea level changes, tides, waves, storm surges, interactions between these processes and accounting for changes in period and height of waves and frequency and intensity of storm surges are yet to be performed.

4.2.3.4.3   Effects of cyclones

Tropical and extratropical cyclones (TCs and ETCs) tend to determine ESL events, such as coastal storm surges, high water events, coastal floods, and their associated impacts on coastal communities around the world. The projected potential future changes in TCs and ETCs frequency, track and intensity is therefore of great importance. After AR5, it was realised that the modelled global frequency of TCs is underestimated and that the geographical pattern is poorly resolved in the case of TC tracks, very intense TCs (i.e., Category 4/5) and TC formation by using low resolution climate models (Camargo, 2013). Over recent years, multiple methods including downscaling

CMIP5 climate models (Knutson et al., 2015; Yamada et al., 2017), high-resolution simulations (Camargo, 2013; Yamada et al., 2017), TC–ocean interaction (Knutson et al., 2015; Yamada et al., 2017), statistical models (Ellingwood and Lee, 2016) and statistical-deterministic models (Emanuel et al., 2008) have been developed, and the ability to simulate TCs has been substantially improved. Most models still project a decrease or constant global frequency of TCs, but a robust increase in the lifetimes, precipitation, landfalls and ratio of intense TCs under global warming. This is consistent with IPCC AR5 and many additional studies (Emanuel et al., 2008; Holland et al., 2008; Knutson et al., 2015; Kanada et al., 2017; Nakamura et al., 2017; Scoccimarro et al., 2017; Zheng et al., 2017). It is expected that these projected increases are intensified by favourable marine environmental conditions, expansion of the tropical belt, or ocean warming in the northwest Pacific and north Atlantic, and increasing water vapour in the atmosphere (Kossin et al., 2014; Moon et al., 2015; Cai et al., 2016; Mei and Xie, 2016; Cai et al., 2017; Kossin, 2017; Scoccimarro et al., 2017; Kossin, 2018). However, it is noted that, in contrast to most models, some models do predict an increase in global TC frequency during the 21st century (Emanuel, 2013; Bhatia et al., 2018).

Previous extensive studies indicated the important role of warming oceans in the TC activity (Emanuel, 2005; Mann and Emanuel, 2006; Trenberth and Fasullo, 2007; Trenberth and Fasullo, 2008; Villarini and Vecchi, 2011; Trenberth et al., 2018) and also revealed TCs stir the ocean and mix the subsurface cold water to the surface (Shay et al., 1992; Lin et al., 2009). The resulting increased thermal stratification of the upper ocean under global warming will reduce the projected intensification of TCs (Emanuel, 2015; Huang et al., 2015; Tuleya et al., 2016). A recent study suggests a strengthening effect of ocean freshening in TC intensification, opposing the thermal effect (Balaguru et al., 2016). It is concluded that it is *likely* that the intensity of severe TCs will increase in a warmer climate, but there is still *low confidence* in the frequency change of TCs in the future.

Recent projection studies indicate that trends in regional ETCs vary from region to region, for example, a projected increase in the frequency of ETCs in the South and the northeast North Atlantic, the South Indian Ocean, and the Pacific (Colle et al., 2013; Zappa et al., 2013; Cheng et al., 2017; Michaelis et al., 2017) and a decrease in the numbers of ETCs in the North Atlantic basin and the Mediterranean (Zappa et al., 2013; Michaelis et al., 2017). Note that the projected frequency in ETCs still remains uncertain due to different definitions of cyclone, model biases or climate variability (Chang, 2014; Cheng et al., 2016). Considering these processes implies that changes in TC and ETC characteristics will vary locally and therefore there is *low confidence* in the regional storm changes, which is in agreement with AR5 WGI Chapter 14 (Christensen et al., 2013).

Observed damages from ETCs/TCs to coastal regions has increased over the past 30 years and will continue in the future (Ranson et al., 2014). The global population exposed to ETCs/TCs hazards is expected to continue to increase in a warming climate (Peduzzi et al., 2012; Blöschl et al., 2017; Emanuel, 2017a; Michaelis et al., 2017). The probabilities of sea level extreme events induced by TC

storm surge are *very likely* to increase significantly over the 21st century. Risk from TCs increases in highly vulnerable coastal regions (Hallegatte et al., 2013), e.g., on coasts of China (Feng and Tsimplis, 2014), west Florida, north of Queensland, the Persian Gulf, and even in well protected area such as the Greater Tokyo area (Tebaldi et al., 2012; Lin and Emanuel, 2015; Ellingwood and Lee, 2016; Hoshino et al., 2016; Dinan, 2017; Emanuel, 2017b; Lin and Shullman, 2017). The ESL return period has greatly decreased over recent decades and is also expected to decrease greatly in the near future, for example, in NYC (by 2030–2045; Garner et al., 2017). It is *very likely* that the ESL return period in low-lying areas such as coastal megacities decreased over the 20th century and frequencies of still unusual ESL events are expected to increase in frequency in the future. In addition, the compound effects of SLR, storm surge and waves on ESL events and the associated flood hazard are assessed in Chapter 6 (Section 6.3.3.3 and 6.3.4).

4.2.3.5 Long-Term Scenarios, Beyond 2100

Sea level at the end of the century will be higher than present day and continuing to rise in all cases even if the Paris Agreement is followed (Nicholls, 2018). The reasons for this are mainly related to the slow response of glacier melt, thermal expansion and ice sheet mass loss (Solomon et al., 2009). These processes operate on long time scales, implying that even if the rise in global temperature slows or the trend reverses, sea level will continue to rise (SR1.5 report, AR5). A study by Levermann et al. (2013) based on palaeo-evidence and physical models formed the basis of the assessment by Church et al. (2013) indicating that committed SLR is approximately 2.3 m per degree warming for the next 2000 years with respect to pre-industrial temperatures. This rate is based on a relation between ocean warming and basal melt as used by Levermann et al. (2013), without accounting for surface melt, hydrofracturing of ice shelves and subsequent ice cliff failure, suggested to be a dominant long term mechanism for ice mass loss (DeConto and Pollard, 2016). Deep uncertainty (Cross-Chapter Box 5 in Chapter 1) remains on the ice dynamical contribution from Antarctica after 2100.

Beyond the 21st century, the relative importance of the long-term contributions of the various components of SLR changes markedly. For glaciers, the long-term is of limited importance, because the sea level equivalent of all glaciers is restricted to 0.32 ± 0.08 m when taking account of ice mass above present day sea level (Farinotti et al., 2019). Hence, there is *high confidence* that the contribution of glaciers to SLR expressed as a rate will decrease over the 22nd century under RCP8.5 (Marzeion et al., 2012). For thermal expansion the gradual rate of heat absorption in the ocean will lead to a further SLR for several centuries (Zickfeld et al., 2017). By far, the most important uncertainty on long time scales arises from the contribution of the major ice sheets. The time scale of response of ice sheets is thousands of years. Hence, if ice sheets contribute significantly to sea level in 2100, they will necessarily also contribute to sea level in the centuries to follow. Only for low emission scenarios, like RCP2.6, can substantial ice loss be prevented, according to ice dynamical models (Levermann et al., 2014; Golledge et al., 2015; DeConto and Pollard, 2016; Bulthuis et al., 2019). For Greenland, surface warming may lead

4

to ablation becoming larger than accumulation, and the associated surface lowering increases ablation further (positive feedback). As a consequence, the ice sheet will significantly retreat. Church et al. (2013) concluded that the threshold for perpetual negative mass balance based on modelling studies lies between 1°C (Robinson et al., 2012; *low confidence*) and 4°C (*medium confidence*) above pre-industrial temperatures. Pattyn et al. (2018) demonstrated that with more than 2.0°C of summer warming, it becomes *more likely than not* that the GIS crosses a tipping point, and the ice sheet will enter a long-term state of decline with the potential loss of most or all of the ice sheet over thousands of years. If the warming is sustained, ice loss could become irreversible due to the initiation of positive feedbacks associated with elevation-SMB feedback (reinforced surface melt as the ice sheet surface lowers into warmer elevations), and albedo-melt feedback associated with darkening of the ice surface due to the presence of liquid water, loss of snow, changes in firn and biological processes (Tedesco et al., 2016; Ryan et al., 2018). The precise temperature threshold and duration of warming required to trigger such irreversible retreat remains very uncertain, and more research is still needed.

The mechanisms for decay of the AIS are related to ice shelf melt by the ocean, followed by accelerated loss of grounded ice and MISI, possibly exacerbated by hydrofracturing of the ice shelves and ice cliff failure (Cross-Chapter Box 8 in Chapter 3). The latter processes have the potential to drive faster rates of ice mass loss than the SMB processes that are *likely* to dominate the future loss of ice on Greenland. Furthermore, the loss of marine-based Antarctic ice represents a long-term (millennial) commitment to elevated SLR, due to the long thermal memory of the ocean. Once marine based Antarctic ice is lost, local ocean temperatures will have to cool sufficiently for buttressing ice shelves to reform, allowing retreated grounding lines to re-advance (DeConto and Pollard, 2016). A minimum time scale, whereby the majority of West Antarctica decays, was derived from a schematic experiment with an ice flow model by Golledge et al. (2017), where ice shelves were removed instantaneously and prohibited from re-growing. Results of this experiment indicate that most of West Antarctica's ice is lost in about a century.

Gradual melt of ice shelves accompanied by partial retreat of East Antarctic ice would yield greater ice melt but on a time scale of millennial or longer (Cross-Chapter Box 8 in Chapter 3). Prescribing a uniform warming of 2°C–3°C in the Southern Ocean triggers an accelerated decay of West Antarctica in a coarse resolution model with a temperature-driven basal melt formulation yielding 1–2 m SLR by the year 3000 and up to 4 m by the year 5000 (Sutter et al., 2016). Formulating an ice sheet model with Coulomb friction in the grounding line zone yields a SLR of 2 m after 500 year for a sub-ice shelf melt of 20 m a^{-1} (Pattyn, 2017). On decadal to millennial time scales the interaction between ice and the solid Earth indicates the possibility of a negative feedback slowing retreat by viscoelastic uplift and gravitational effects that reduce the water depth at the grounding line (Gomez et al., 2010; de Boer et al., 2014; Gomez et al., 2015; Konrad et al., 2015; Pollard et al., 2017; Barletta et al., 2018; Section 4.2.3.1.2).

A blended statistical and physical model, calibrated by observed recent ice loss in a few basins (Ritz et al., 2015) projects an Antarctic contribution to sea level of 30 cm by 2100 and 72 cm by 2200, following the SRES A1B scenario, roughly comparable to RCP6.0. The projected contribution of WAIS was found to be limited to 48 cm in 2200 following the A1B scenario. The key uncertainty in these calculations comes from the dependency on the relation between the sliding velocity and the friction at the ice-bedrock interface. Several parameterisations are used to describe this process. Golledge et al. (2015) present values between 0.6–3 m by 2300 for the RCP8.5 scenario. In contrast to the previous studies, Cornford et al. (2015) used an adaptive grid model, which can describe more accurately grounding line migration (Cross-Chapter Box 8 in Chapter 3). Due to the computational complexity of their model, simulations are limited to West Antarctica. Starting from present-day observations, they find that the results are critically dependent on initial conditions, sub ice shelf melt rates, and grid resolution. The glacier with the most uncertain vulnerability is the 120 km-wide Thwaites Glacier, in the Amundsen Sea sector of West Antarctica. Thwaites Glacier is currently retreating in a reverse-sloped trough extending into the central WAIS (Figure 4.8), where the bed is up to 2 km below sea level. In addition to Thwaites, several smaller outlet glaciers and ice streams may contribute to sea level on long time scales, but in the study by Cornford et al. (2015), a full West Antarctic retreat does not occur with limited oceanic heating under the two major ice shelves (Filchner-Ronne and Ross) keeping ice streams flowing into the Ross and Weddell Seas in place. However, the representation of these processes remains simplistic at the continental ice sheet scale (Cross-Chapter Box 8 in Chapter 3).

Nonetheless, recent studies using independently developed Antarctic ice dynamical models (Golledge et al., 2015; DeConto and Pollard, 2016; Bulthuis et al., 2019) agree that low emission scenarios, are required to prevent substantial future ice loss (*medium confidence*). However, observations (Rignot et al., 2014) and modelling of the Thwaites Glacier in West Antarctica (Joughin et al., 2014), suggest grounding line retreat on the glacier's reverse sloped bedrock is already underway and possibly capable of driving major WAIS retreat on century time scales. Whether the retreat is driven by ocean changes driven by climate change or by climate variability (Jenkins et al., 2018) is still under debate. Hence it is not possible to determine whether a low emission scenario would prevent substantial future ice loss (*medium confidence*). This is a further elaboration on the SR15 assertion that the chance for passing a threshold is larger for 2°C warming than for 1.5°C warming.

A study by Clark et al. (2016) addresses the evolution of the ice sheets over the next 10,000 years and concludes that given a climate model with an equilibrium climate sensitivity of 3.5°C, the estimated combined loss of Greenland and Antarctica ranges from 25–52 m of equivalent sea level, depending on the emission scenario considered, with rates of GMSL as high as 2–4 m per century. A worst-case scenario was explored with an intermediate complexity climate model coupled to a dynamical ice model (Winkelmann et al., 2015), in which all readily available fossil fuels are combusted at present-day rates until they are exhausted. The associated climate warming leads to the disappearance of the entire AIS with rates of SLR up to around

3 m per century. A follow up study by Clark et al. (2018) addressing the long-term commitment of SLR based on cumulative carbon dioxide emissions points to SLR as an additional measure for setting emission targets. It shows that a 2°C scenario would result in 0.9 m in 2300 and around 7.4 m in the year 9000 CE.

Similar to the strategy for the 21st century, the long-term projections of sea level were assessed. Since no new CMIP runs are available there are no major new insights in the thermal expansion and glacier component which deviate from the AR5 assessment for the long-term contribution of these components. Some studies updated the contribution of the GIS on long time scales. Vizcaino et al. (2015) used a GCM coupled to an ice sheet model to calculate the Greenland contribution which is within the range of estimates presented by Church et al. (2013). This is also true for the ice sheet simulations by Calov et al. (2018) which are based on off line simulations with a regional climate model forced by RCP4.5 and RCP8.5 scenarios of three different CMIP5 models. On the other hand, Aschwanden et al. (2019) used temperatures to calculate SMB which was used to force an ice sheet model to arrive at much higher values for SLR. However, they used a spatially uniform temperature forcing, which is in conflict with earlier work and overestimated temperatures in the ablation zone (e.g., Van de Wal and Wild, 2001; Gregory and Huybrechts, 2006). Given this limited and contrasting evidence for Greenland, the assessed values presented in Table 13.8 of Church et al. (2013) were also used, but again replacing the Antarctic component by the assessed value from the process and climate scenario-based studies published after 2013. The low scenario in Table 13.8 of Church et al. (2013) without the Antarctic contribution was combined with the RCP2.6 estimates for Antarctica simulated by Golledge et al. (2015), the mean of the RCP2.6 simulations with and without time delay between global mean atmosphere and ocean temperature around Antarctica of Levermann et al. (2014), and the model results of Bulthuis et al. (2019). The medium scenario from Church et al. (2013) is combined with RCP4.5 results and the high scenario with RCP8.5. Results are shown in Figure 4.2, Section 4.1 and show a strong divergence of RSL rise over time, whereby the estimates in 2300 range from about 1–2 m under RCP2.6 up to 2–5.5 m for RCP8.5.

The specific trajectories that will be followed may depend critically on if and when certain tipping points are reached. Most critical in that respect are presumably the tipping points corresponding (1) to the threshold where the ablation in Greenland becomes larger than the accumulation, causing an irreversible and nearly full retreat of the ice sheet; and (2) the thresholds for ice shelf stability in West Antarctica, which depend on surface melt and sub-ice melt, combined with uncertainties surrounding MISI and/or MICI. There is deep uncertainty about whether and when a tipping point will be passed. For RCP8.5, the chance of passing a tipping-point are considered to be substantially higher than for RCP2.6.

In summary, there is *high confidence* in continued thermal expansion and the loss of ice from both the GIS and AIS sheets beyond 2100. A complete loss of Greenland ice contributing about 7 m to sea level over a millennium or more would occur for sustained GMST between 1°C (*low confidence*) and 4°C (*medium confidence*) above pre-industrial levels. Due to deep uncertainties regarding the dominant processes that could trigger a major retreat, there is *low confidence* in the estimates of the contribution of the AIS beyond 2100, but our estimates (2.3–5.4 m in 2300) for RCP8.5 are considerably higher than presented in AR5. High-emission scenarios or exhaustion of fossil fuels over a multi-century period lead to rates of SLR as high as several metres per century in the long term (*low confidence*). Low-emission scenarios lead to a limited contribution over multi-century time scales (*high confidence*). Discriminating between 1.5°C and 2°C scenarios in terms of long-term sea level change is not possible with the limited evidence. Hence, it is concluded that the SLR on millennial time scales is strongly dependent on the emission scenario followed. This, combined with the lack in predictability of the tipping points, indicates the importance of emissions mitigation for minimising the risk to low-lying coastlines and islands (*high confidence*).

Box 4.1 | Case Studies of Coastal Hazard and Response

This box illustrates current coastal flood risk management and adaptation practices through four case studies from around the world, showing how current approaches could be refined using the new seal level rise (SLR) projections of this report, as well as findings on adaptation options, decision making approaches and governance (called Practice Consistent with SROCC Assessment in Tables 1–3). In an effort to illustrate some of the diverse social-ecological settings in this report, the locations are Nadi in Fiji, the Nile delta in Egypt, New York and Shanghai. The latter two studies are framed as a comparison. For each case, Current Practice reflects understanding, policy planning, and implementation that existed prior to SROCC. Recent improvements in understanding documented in this chapter suggest that significant, beneficial changes in the basis for design and planning are feasible in each case for addressing future risk.

Responding to Coastal Flooding and Inundation, Nadi, Fiji

Hazards that contribute to riverine flooding and coastal inundation for Nadi Town and the wider Nadi Basin are heavy rainfall, elevated sea levels and subsidence of the delta. People and built assets in the Nadi River floodplain are already being affected by climate change. Observed sea level shows an increase of 4 mm yr^{-1} over the period 1992–2018. Over the past 75 years, extreme rainfall events have become more frequent. Of the 84 floods which occurred in the Nadi River Basin since 1870, 54 were post 1980, with 26 major floods since 1991 (Hay, 2017). In January 2009, large areas of Fiji were inundated by devastating floods which claimed at least 11 lives, left 12,000 people temporarily homeless and caused 54 million USD in damage. Worst hit was the Nadi area, with total damage estimated at 39 million USD (Hay, 2017). The increased frequency of flooding is not all attributable to increases in sea level and extreme rainfall events. River channels have become filled with sediment over time, largely owing to deforestation of the

Box 4.1 (continued)

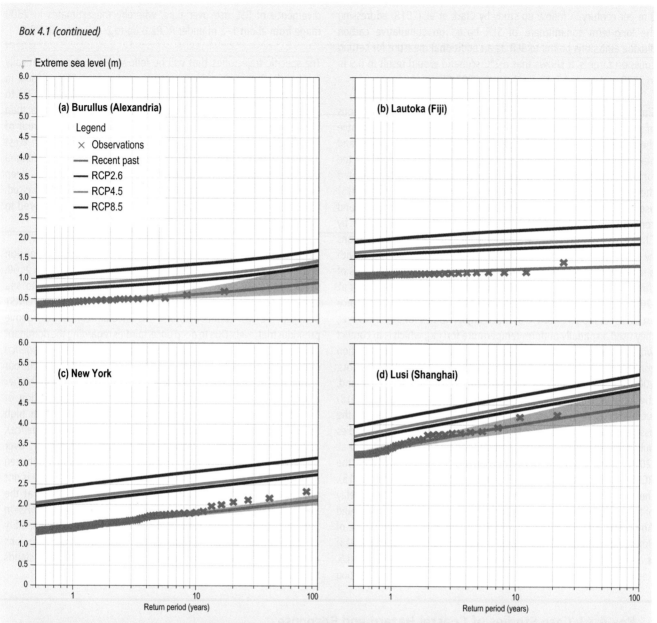

Box 4.1, Figure 1 | Historical and projected extreme sea level (ESL) events at four stations discussed in this box. The heights of ESL events are shown as a function of their return period. Observations (crosses) are derived from tide-gauge records. The historical return height (grey) is the best fit through these observations, and the 5–95% confidence intervals (grey band) are shown. Note that the confidence interval for Lautoka is too narrow to be visible. Future ESL events represent the effect of regional sea level change for the period 2081–2100 for scenarios RCP2.6 (blue) and RCP8.5 (red). The increased height of the 100-year event for scenario RCP2.6 and RCP8.5 is 0.43 m and 0.80 m respectively for Burullus; 0.55 m and 1.03 m for Lautoka; 0.63 m and 1.04 m for New York; and 0.44 m and 0.79 m for Lusi. The increased frequency of the historical 100-year event for scenario RCP2.6 and RCP8.5 is a factor of 15 and 777 for Burullus; >1200 and >1200 for Lautoka; 67 and 541 for New York; and 6 and 26 for Lusi. The notation >1200 indicates that the methodology allows for estimation of only a lower bound on the increased frequency.

hinterland. Much of the mangrove fringe has been sacrificed for development of various kinds. The Nadi River Delta is subsiding, exacerbating the effects of SLR (Chandra and Gaganis, 2016). Various initiatives to help alleviate flooding and inundation in the Nadi Basin have been proposed. These include both hard protection and engineering (e.g., ring dikes, river widening, bridge rebuilding, retarding basins, shortcutting tributaries, dams and diversion channels) and accommodation (e.g., early flood warnings and improved land management practices in upper basin) measures (Box 4.1, Table 1).

A Comparison of New York City and Shanghai Coastal Flood Adaptation Measures

Hurricane Sandy (2012) and Typhoon Winnie (1997) are considered the largest recorded historical flood events for New York City and Shanghai, respectively (Xian et al., 2018). Hurricane Sandy killed 55 people in New York City and neighbouring states and caused over 19 billion USD losses to New York City (Rosenzweig and Solecki, 2014). Typhoon Winnie killed more than 310 people and

Box 4.1 (continued)

Box 4.1, Table 1 | Current coastal flood risk management and adaptation practices in the Nadi Basin and possible refinements using the new SLR and ESL projections of this report, as well as findings on adaptation options, decision making approaches and governance (Practice Consistent with SROCC Assessment). See Hay (2017) for background on current practice and practice consistent with SROCC assessment; see Box 4.1, Figure 1 for ESL event values. VLM is vertical land motion.

	Current practise	Practise consistent with SROCC Assessment
Hazards	Design storm tide and design two-day rainfall used for existing flood control works: 1.2 m and 436 mm, respectively. Subsidence: Not considered.	Design storm tide in 2081–2100 (100-yr ESL): 1.9 m and 2.4 m (RCP2.6 and RCP8.5); Two-day design rainfall, (100-yr event): 670 mm (RCP8.5); Subsidence and other VLM considered.
Exposure and Vulnerability	Exposure and vulnerability assessed for present day only; thus static, with no reference to drivers.	Exposure and vulnerability assessed for the present day and future time periods, with the projections taking into account both biophysical and human drivers, drawing on community-based assessments (Sections 4.3, 4.4.4.2).
Levels of Risk	Reflect current levels of risk, with no allowance for changes in climatic, biophysical or socioeconomic conditions.	Use risk scenarios reflecting the full suite of biophysical and socioeconomic changes over the life of the planned investment project, including their interactions (Section 4.4.4.3).
Response Options	Interventions based entirely on reducing current levels of risk, with the primary focus on hard protection measures to reduce flood hazard.	Consider locally appropriate, sequenced mix of hard, soft and ecosystem-based measures to reduce risks expected to occur over the lifetime of the investment (Sections 4.4.2, 4.4.4.3).
Planning and Decision Making	Responses take a narrow 'flood control' approach aimed at 'controlling' single hazards, rather than managing the multiple and interacting risks in their broader contexts.	Take into account vulnerability and equity implications, and use community-based approaches (Section 4.4.5). Responses reflect a risk-based, flexible design approach that addresses the tension between a fixed design standard on the one hand and, on the other hand, increasing flood risk over time due to further floodplain development, climate change leading to higher peak flows and inundation, and river channel bed aggradation.
Governance and Institutional Dimensions	Nadi Town Drainage Plan completed in August 2000. Aimed to address drainage problems and reduce flood damage. However, the Plan was developed without distinction between inland water and river water, without hydraulic analysis and verification, and without a scientific basis. Retention dams were constructed, but other initiatives in the plan have not seen substantive implementation.	A comprehensive 'Flood Management Strategy and Plan' for the Nadi River Basin would be one way to achieve the desired complementarity between hard, soft and accommodation measures. Strengthening of policies on water and land management, as well as addressing related legal and institutional issues, could be undertaken at the national, basin and local levels. Government, civil society, indigenous iTaukei Fijians, and the private sector, have valuable contributions to make in preparing an integrated approach to flood and coastal hazard risk reduction (Section 4.4.5).

caused damage exceeding 3.2 billion USD in China. Many dikes and floodwalls along coastal Shanghai and Zhejiang were breached by surge-driven floodwaters (Gu, 2005).

Past updates of the flood defences in Shanghai occurred after extreme flood events (i.e., typhoons in 1962, 1974 and 1981; Xian et al., 2018). In contrast, the flood tide of Hurricane Sandy stands out in the record at the Battery tide gauge. Unlike the numerous episodes of severe inundation experienced by Shanghai, NYC suffered relatively moderate consequences from individual events before Hurricane Sandy. This and other factors led to higher-standard flood protection measures in Shanghai, such as sea walls designed to protect its coastlines and critical infrastructure in developed areas against a 200-year (0.005 annual chance) flood level, and flood walls with 1000-year (0.001 annual chance) riverine flood return level along the Huangpu River. New York City, on the other hand, has relatively low protection, consisting of sandy dunes (e.g., on Staten Island), vegetation (e.g., in Queens) and low-rise sea walls or bulkheads in lower Manhattan. Another reason why Shanghai has repeatedly updated its flood protection while New York City failed to do so lies in their differing governance structures (Wei and Leung, 2005; Yin et al., 2011; Rosenzweig and Solecki, 2014), as well as rapid economic growth in China, providing funding for large-scale infrastructure (Zhang, 2003).

Since Hurricane Sandy in 2012, implementation of the 'Big U' project, a coastal protection system for lower Manhattan, has begun, and a variety of measures are planned, and some undertaken, to protect the subway system to a flood level of 4.3 m above street level (Jacob Balter, 2017; MTA, 2017). Newly built critical facilities will be located outside the flood zone and siting guidelines for publicly financed projects in the current and future flood zones have been tightened (New York City Mayor's Office of Recovery and Resiliency, 2019). The degree to which these protection projects will be completed and the guidelines enforced remains uncertain. Home buyouts have enabled some permanent relocation away from hazardous areas, but relocation impacts on social networks and place-based ties hamper long-term recovery (Binder et al., 2019; Buchanan et al., 2019).

Box 4.1 (continued)

Box 4.1, Table 2 | Current coastal flood risk management and adaptation practices in New York City and Shanghai and possible refinements using the new sea level rise (SLR) and extreme sea level (ESL) projections of this report, as well as findings on adaptation options, decision making approaches and governance (Practice Consistent with SROCC Assessment). See Xian et al. (2018) for background on current practice and practice consistent with SROCC assessment. See Box 4.1, Figure 1 for ESL event values. EbA is ecosystem-based adaptation.

	Current Practice	**Practice Consistent with SROCC Assessment**
Hazards	Design storm tide: 4.6 m (200-yr ESL) Shanghai vs 2.1m (100-yr event) NYC; Subsidence + GIA ~5–7 mm yr^{-1} Shanghai vs 1–1.5 mm yr^{-1} NYC	Design storm tide in 2081–2100: 5.1 and 5.5 m (200-yr ESL) Shanghai vs 2.8 and 3.2 m (100-yr ESL) NYC (RCP2.6 and RCP8.5); Subsidence could significantly increase these values for Shanghai.
Exposure and Vulnerability	Current exposure and vulnerability considers topographic elevation and human drivers.	Exposure and vulnerability for current and future time periods would take into account projections that consider both biophysical and human drivers (Section 4.3.3).
Levels of Risk	Reflect current hazard plus various freeboard for NYC; 0.5 m for Shanghai, with no consideration of changes to date in storm characteristics and socioeconomic factors.	Use risk scenarios to reflect the full suite of climate change and socio-economic changes and their interdependency for the planned coastal projects (Section 4.4.4.3); use freeboard that accounts for ESL uncertainty, such as ESL with probability >5% during planning horizon.
Response Options	Protection measures: fixed-height sea wall for both NYC and Shanghai; for NYC: building retrofit, ecosystem-based measures such as dune enhancement. Accommodation measures for NYC: insurance, building codes.	Consider locally appropriate, sequenced mix of hard protection, EbA, accommodation and retreat measures to minimise combined total costs; more flexible design that reflects the dynamic risk (Section 4.4.2).
Planning and Decision Making	Funding and governance issues slow down long-term planning and implementation process, especially for NYC.	Flexible adaptation responses to uncertain long-term risk facilitate SLR planning and implementation (Section 4.4.4.3). Take into account vulnerability and equity implications, and continuously involve stakeholders in locally appropriate ways.
Governance and Institutional Dimensions	Shanghai has relatively high autonomy in China context and alignment of central and local Governments' objectives; NYC's decision making occurs in context of city-to-national multi-level governance.	More effective coordination of planning, finance, and transparency among local, regional, and national government agencies, and between government, civil society, and the private sector, enables response to increasing risks (Section 4.4.5).

Climate Change Adaptation in Nile Delta Regions of Egypt

Coastal hazard for the Nile delta arises because large portions lie only ±1.5 m above sea level (Shaltout et al., 2015) and while the Delta includes only 2% of Egypt's total area, it held 41% of its population as of 2006 (Hereher, 2010; World Bank, 2017) and is key to Egypt's economy (Bucx et al., 2010). The delta is an important resource for Egypt's fish farms (Hereher, 2009; El-Sayed, 2016) and contains more than 63% of Egypt's cultivated lands (Hereher, 2010). The Nile Delta's coastal lagoons are internationally renowned for abundant bird life, account for one fourth of Mediterranean wetlands and 60% of Egypt's fish catch (Government of Egypt, 2016). Coastal flooding and salinisation of freshwater lagoons would negatively affect fisheries and biodiversity (UNDP, 2017).

The Nile Delta's low elevation translates into high exposure to SLR (Shaltout et al., 2015), and the level of protection varies greatly from place to place (Frihy et al., 2010). Box 4.1, Figure 1 indicates that episodic flooding will increase substantially without effective adaptation measures. An estimated 2660 km^2 in the northern delta will be inundated by 2100 for GMSL of 0.44 m (Gebremichael et al., 2018; *low confidence*), which is comparable to the RCP2.6 emission scenario. In addition, subsidence due to sediment diversion by the Aswan High Dam, water and natural gas extraction (Gebremichael et al., 2018) and some other critical natural aspects (Frihy et al., 2010) heightens vulnerability to coastal flooding (Box 4.1, Table 3) and reduces fresh water supply to the delta. Subsidence rates range from 0.4 mm yr^{-1} in the west delta to 1.1 mm yr^{-1} in the mid-delta and 3.4 mm yr^{-1} in the east delta (Elshinnawy et al., 2010), although rates as high as 10 mm yr^{-1} near natural gas extraction operations are also reported (Gebremichael et al., 2018). While there is *low confidence* in reported values, these indicate that subsidence makes locally a substantial contribution to RSL. Future construction of Ethiopia's Grand Renaissance Dam (Stanley and Clemente, 2017) may heighten problems of fresh water availability and reduce hydropower production.

The low-lying northern coast and Nile Delta region are a high priority for adaptation to climate change (UNDP, 2017). The Egyptian government has committed 200 million USD to hard coastal protection at Alexandria and adopted integrated coastal zone management for the northern coast. Recent activities include integrating SLR risks within adaptation planning for social-ecological systems, with special focus on coastal urban areas, agriculture, migration and other human security dimensions (Government of Egypt, 2016; UNDP, 2017).

Box 4.1 (continued)

Box 4.1, Table 3 | Current coastal flood risk management and adaptation practices in the Nile Delta and possible refinements using the new sea level rise (SLR) and extreme sea level (ESL) projections of this report, as well as findings on adaptation options, decision making approaches and governance (Practice Consistent with SROCC Assessment). See Elshinnawy et al. (2010) and text above for data sources for current practice. See Box 4.1, Figure 1 for ESL event values.

	Current Practice	Practice Consistent with SROCC Assessment
Hazards	Design storm tide unclear; 0.9m corresponds to current 100-yr ESL. Tide gauge trends: 1.6–5.3 mm yr^{-1}. Subsidence: 0.4–10 mm yr^{-1}.	Design storm tide in 2081–2100: 1.4 and 1.7 m (100-yr ESL; RCP2.6 and RCP8.5); Subsidence: 0.4–10 mm yr^{-1}
Exposure and Vulnerability	High coastal flood risk, especially affecting coastal cities, fisheries, farming and ecosystems (UNDP, 2017). Frameworks under development to assess future risk.	Full assessment of current and future exposure and vulnerability scenarios to account for interactions of changing demographic, industrial, and ecological characteristics of the Delta, especially with respect to agricultural land, fisheries, and coastal cities (UNDP, 2017; Section 4.3).
Levels of Risk	Risk is high and has increased due to regional SLR and subsidence. (UNDP, 2017). Highly variable protection against current risk.	Use plausible risk scenarios, including regional sea level trends and possibly subsidence that could lead to higher flood levels (UNDP, 2017) for potentially exposed and vulnerable communities.
Response Options	Unconnected and small incremental steps toward increasing management capability in Egypt to confront coastal flood risks associated with SLR, including enhancing planning paradigms, interventions that account for climate change threats, community-based measures, and adaptation and improving resilience (UNDP, 2017).	Consideration and deployment of a sequenced mix of tailor-made options, including hard protection, ecosystem-based measures, advance, and retreat. Comprehensive monitoring and evaluation of effectiveness of measures. Timing of suitable options a key aspect (Section 4.4.4.3).
Planning and Decision Making	"One of the most prominent obstacles to integrated coastal zone management in Egypt is the complex and sometimes unclear institutional framework for addressing development activities; the limited and often ad-hoc approach between different agencies" (UNDP, 2017).	An adaptation based coastal planning approach able to manage the range of climatic risks on both natural and built environments that will cover hard protection (Ghoneim et al., 2015), accommodation and ecosystem-based measures (UNDP, 2017), involving potentially transformational approaches that take dynamic long term risk into account (UNDP, 2017; Sections 4.4.2, 4.4.4).
Governance and Institutional Dimensions	Some recent projects and proposals aim to "integrate the management of SLR risks into the development of Egypt's Low Elevation Coastal Zone (LECZ) in the Nile Delta" (UNDP, 2017).	Mainstream long-term SLR risk into all elements of decision making that directly or indirectly affect the Nile Delta (4.4.5). Strengthen coordination of SLR-relevant provisions across scales, sectors and policy domains, including arrangements to involve government, civil society, science and the private sector (Section 4.4.5).

4.3 Exposure, Vulnerability, Impacts and Risk Related to Sea Level Rise

4.3.1 Introduction

Section 4.2 demonstrates that sea level is rising and accelerating over time, and that it will continue to rise throughout the 21st century and for centuries beyond. It also shows that ESL events that are historically rare, will become common by 2100 under all emission scenarios, leading to severe flooding in the absence of ambitious adaptation efforts (*high confidence*). In both RCP2.6 and RCP8.5 emission scenarios, many low-lying coastal areas at all latitudes will experience such events annually by 2050. In this context, Section 4.3 updates knowledge on the recent methodological advances in exposure and vulnerability assessments (Box 4.2), dimensions of exposure and vulnerability (Section 4.3.2) and observed and projected impacts (Section 4.3.3). It concludes with a synthesis on future risks to illustrative low-lying coastal geographies (resource-rich coastal cities, urban atoll island, large tropical agricultural deltas and Arctic communities), and according to various adaptation scenarios (Section 4.3.4).

4.3.2 Dimensions of Exposure and Vulnerability to Sea Level Rise

4.3.2.1 Point of Departure

4.3.2.1.1 Environmental dimension of exposure and vulnerability

Exposure of coastal natural ecosystems to RSL and related coastal hazards change by two means: alterations in the spatial coverage and distribution of ecosystems within the potentially exposed area; and changes in the size of the exposed area caused by relative SLR. The vulnerability of coastal ecosystems to SLR and related coastal hazards differs strongly across ecosystem types and depends on human interventions (e.g., land use change and fragmentation, coastal squeeze and anthropogenic subsidence) and degradation (e.g., pollution), as well as climate change, including changes in temperature and precipitation patterns. SLR and its physical impacts, such as flooding or salinisation, also increase ecosystems' vulnerability and decrease the ecosystems' ability to support livelihoods and provide ecosystem services such as coastal protection. Healthy, diverse, connected coastal ecosystems support

Box 4.2 | Methodological Advances in Exposure and Vulnerability Assessments

This box highlights recent advances in methodologies in assessing exposure and vulnerability to sea level rise (SLR) and its physical impacts, such as coastal flooding since the IPCC 5th Assessment Report (AR5). In few cases it also leverages methodological advances, which have not been yet applied in the coastal context but have great potential to inform coastal assessments.

Improved spatial-temporal exposure assessments

Exposure assessment is frequently based on census data, which is available at coarse resolutions. However, new technologies (e.g., drones and mobile phone data) and more available satellite products provide new tools for exposure analysis. Exposure assessment is increasingly based on the combination of high resolution satellite imagery and spatio-temporal population modelling as well as improved quality of digital elevation models (DEM; Kulp and Strauss, 2017). This is used to understand better exposure to coastal flooding (Kulp and Strauss, 2017), diurnal differences in flood risk exposure (Smith et al., 2016), dynamic gridded population information for daily and seasonal differences in exposure (Renner et al., 2017), a combination of remotely-sensed and geospatial data with modelling for a gridded prediction of population density at ~100 m spatial resolution (Stevens et al., 2015), or open building data using building locations, footprint areas and heights (Figueiredo and Martina, 2016). In addition, methods based on mobile phone data (Deville et al., 2014; Ahas et al., 2015), and social media-based participation are increasingly available for population distribution mapping (Steiger et al., 2015). Some of these methodologies have been already applied in coastal assessments (Smith et al., 2016). Integrating daily and seasonal changes with the distribution of population improves population exposure information for risk assessments especially in areas with highly dynamic population distributions, as shown in high tourism areas in mountain regions (e.g., Renner et al., 2017), which would have advantages at touristic coastal areas as well.

Projections of future exposure

Recent studies assess exposure considering not only projected sea levels but also expected changes in population size (Jongman et al., 2012; Hauer et al., 2016). It involves different socioeconomic scenarios together with changing growth rates for coastal areas and the hinterland (Neumann et al., 2015) and using spatially explicit simulation models for urban, residential and rural areas (Sleeter et al., 2017). Migration-based changes in population distribution (Merkens et al., 2016; Hauer, 2017) are also considered, as well as simulated future land use (specifically urban growth) to investigate future exposure to SLR (Song et al., 2017). Other studies assess future exposure trends by accounting for the role of varying patterns of topography and development projections leading to different rates of anticipated future exposure (Kulp and Strauss, 2017), which influence how effectively coastal communities can adapt. Recent studies aim to account for the sociodemographic characteristics of potentially exposed future populations (Shepherd and Binita, 2015), and anticipate future risk by projecting the evolution of the exposure of vulnerable populations and groups (Hardy and Hauer, 2018). Using social heterogeneity modelling when developing future exposure scenarios enhances the quality of risk assessments in coastal areas (Rao et al., 2017; Hardy and Hauer, 2018). Subnational population dynamics combined with an extended coastal narrative-based version of the five shared socioeconomic pathways (SSP) for global coastal population distribution was used for assessing global climate impacts at the coast, highlighting regions where high coastal population growth is expected and which therefore face increased exposure to coastal flooding (Merkens et al., 2016). SSPs have also been used to estimate future population in regional coastal-hazard risk exposure studies (Vousdoukas et al., 2018b).

Advances in vulnerability assessment

Since the IPCC Special Report on Managing the Risks of Extreme Events and Disasters (SREX) report, vulnerability has been more consistently considered in climate risk assessments (*medium confidence*). It is recognised that climate risk is not only hazard-driven, but also a sociopolitical and economic phenomenon that evolves with changing societal and institutional conditions (*high confidence*). Many studies related to climate risk and adaptation include vulnerability assessments, most of them considering vulnerability as a pre-existing condition while some interpret vulnerability as an outcome (Jurgilevich et al., 2017).

Increasing importance of dynamic assessments

The dynamic nature of vulnerability, and the need to align climate forecasts with socioeconomic scenarios, was a key message of IPCC SREX. Challenges in methodology and data availability, particularly of future socioeconomic data is overcome by extrapolating empirical information of past trends in vulnerability to flooding (Jongman et al., 2015; Mechler and Bouwer, 2015; Kreibich et al., 2017), downscaling global scenarios, for example, the SSPs (Van Ruijven et al., 2014; Viguié et al., 2014; Absar and Preston, 2015), or by using participatory methods, surveys and interviews to develop future scenarios (Ordóñez and Duinker, 2015; Tellman et al., 2016). The uncertainty of the downscaled projections needs to be considered along with the limitation that, even if population data projections are available, the future level of education, poverty, etc. is hard to predict (Jurgilevich et al., 2017). Suggestions to overcome these shortcomings entail the use of a combination of different data sources for triangulation and inclusion of uncertainties (Hewitson et al., 2014), or the meaningful involvement of stakeholders to project plausible future socioeconomic conditions through co-production (Jurgilevich et al., 2017). Recent innovations in (flood) risk assessment include the integration of behaviour into risk assessments (Aerts et al., 2018b) as well as vulnerabilities related to cascading events (Serre and Heinzlef, 2018).

Social-ecological vulnerability assessments
Especially in rural, natural resource-dependent settings, where the population directly rely on the services provided by ecosystems, the vulnerability of the ecosystems (e.g., fragmented, degraded ecosystems with low biodiversity) directly influence that of the population. Since AR5, several methods have been developed and piloted to assess and map social-ecological vulnerability. Examples include the use of i) the sustainable livelihood approach and resource dependence metrics for Australian coastal communities (Metcalf et al., 2015), ii) integration of local climate forecasts for coral reef fisheries in Papua New Guinea (Maina et al., 2016), iii) ecosystem supply-demand model for an integrated vulnerability assessment in Rostock, Germany (Beichler, 2015), iv) participatory indicator development for multiple hazards in river deltas (Hagenlocher et al., 2018), and v) human-nature dependencies and ecosystem services for small-scale fisheries in French Polynesia (Thiault et al., 2018). Areas, where social vulnerability prevail may be, but are not necessarily associated with hotspots of ecosystem vulnerability, highlighting the need to specifically adapt management interventions to local social-ecological settings and to adaptation goals (Hagenlocher et al., 2018; Thiault et al., 2018). The number of assessments considering both the social and the ecological part of the system are increasingly used (Sebesvari et al., 2016).

Assessment of vulnerability to multiple hazards simultaneously
Increasingly, multi-hazard risk assessments are undertaken at the coast (e.g., flooding and inundation of coastal lands in India; Kunte et al., 2014), to understand the inter-relationships between hazards (e.g., Gill and Malamud, 2014), and by focusing on hazard interactions where one hazard triggers another or increases the probability of others occurring. Liu et al. (2016a) provide a systematic hazard interaction classification based on the geophysical environment that allows for the consideration of all possible interactions (independent, mutex, parallel and series) between different hazards, and for the calculation of the probability and magnitude of multiple interacting natural hazards occurring together. Advances have been reported since AR5 by using, for example, modular sets of vulnerability indicators, flexibly adapting to the hazard situation (Hagenlocher et al., 2018).

Using vulnerability functions, thresholds, innovative ways of aggregation in indicator-based assessment, improved data sources
The use of vulnerability functions has been shown to be helpful in assessing the damage response of buildings to tsunamis (Tarbotton et al., 2015), to coastal surge and wave hazards (Hatzikyriakou and Lin, 2017) and accounting for non-linear relationships between mortality and temperature above a 'comfort temperature' (El-Zein and Tonmoy, 2017). Acknowledging the non-compensatory nature of different vulnerability indicators (e.g., proximity to the sea cannot always be fully compensated by being wealthy), the concepts of preference, indifference and dominance thresholds have been applied as a form of data aggregation (Tonmoy and El-Zein, 2018). Similar to advances in exposure assessments, freely available data and mobile technologies hold promise for enabling better input data for vulnerability assessments. Examples include using a combination of mobile phone and satellite data to determine and monitor vulnerability indicators such as poverty (Steele et al., 2017), and using data on subnational dependency ratios and high resolution gridded age/sex group datasets (Pezzulo et al., 2017).

local adaptation to SLR and its consequences (*high confidence*). This section explores new knowledge since AR5 regarding changes in ecosystem's exposure and vulnerability as well as processes affecting the ability of ecosystems to adapt to SLR, and associated impacts, such as flooding or salinisation, on coastal social-ecological systems and coast-dependent livelihoods.

Changes in the exposure of coastal ecosystems
The effects of coastal habitat loss on ecosystem exposure are well documented (Lavery et al., 2013; Serrano et al., 2014; Short et al., 2014; Yaakub et al., 2014; Cullen-Unsworth and Unsworth, 2016; Breininger et al., 2017), and depend on the type of ecosystem, its conservation status, and interactions with SLR and human interventions such as coastal squeeze, which prevents inland

migration (Kirwan and Megonigal, 2013; Schile et al., 2014; Hopper and Meixler, 2016). For instance, coastal habitat loss due to human growth and encroachment due to development, and human structures that restrict tides and, thus, interrupt mass flow processes (water, nutrients and sediments) impact tidal ecosystems depending on the type of restriction, its severity and the geomorphology of the system (Burdick and Roman, 2012). Coastal dunes, for example, although threatened, are well maintained by protected areas in some localities, like Italy, but climate change could cause a drastic drop in their protection (Prisco et al., 2013). In addition, seagrass and other benthic ecosystems, for example, are declining across their range at unprecedented rates (Telesca et al., 2015; Unsworth et al., 2015; Samper-Villarreal et al., 2016; Balestri et al., 2017), due to degrading water quality (i.e., increased nutrient and sediment or dissolved organic carbon loads) from upland-based activities, which include deforestation, agriculture, aquaculture, fishing, and urbanisation, port development, channel deepening, dredging and anchoring of boats (Saunders et al., 2013; Ray et al., 2014; Deudero et al., 2015; Abrams et al., 2016; Benham et al., 2016; Mayer-Pinto et al., 2016; Thorhaug et al., 2017). The exact magnitude of area loss is still uncertain, especially at smaller scales (Yaakub et al., 2014; Telesca et al., 2015) and the implications of habitat shifts for ecosystem attributes and processes, and the services they deliver, remain poorly understood (Ray et al., 2014; Tuya et al., 2014).

Changes in the vulnerability of coastal ecosystems

Global and local-scale processes influence the stability of coastal ecosystems and can interact to restrict ecosystem responses to SLR, and thus increase their vulnerability. At the global scale, changes in precipitation and air temperature represent a potentially significant risk that increases the vulnerability of ecosystems to SLR and related hazards (Garner et al., 2015; Osland et al., 2017). Maximum temperature and mean precipitation change over the last 100 years are main drivers of global ecosystem instability (Mantyka-Pringle et al., 2013), with marked regional and local variations. In addition, seawater warming may affect marine communities and ecosystems but research remains sparse and results are contradictory (Crespo et al., 2017; Hernán et al., 2017). The synergistic effects between climate change and habitat loss due to human impact and urban development are increasingly well-documented but the effects are still not well-known at larger spatial and temporal scales (Kaniewski et al., 2014; Sherwood and Greening, 2014). Although evidence is limited, recurrent disturbances may lead to losses in ecosystem adaptive capacity (Villnäs et al., 2013).

At smaller scales, conversion of coastal areas to urban, agricultural and industrial uses exacerbates pressure on ecosystems, increases their vulnerability to natural hazards, including SLR, and decreases their ability to support coastal livelihoods and deliver ecosystem services, such as coastal protection, fisheries, wildlife habitat, recreational use and tourism (*high confidence*; Foster et al., 2017). The vulnerability of exposed ecosystems is highly variable, as shown in intertidal rocky reef habitats in Australia (*high confidence*; Thorner et al., 2014). Even without SLR, the transition zone between two coastal ecosystems and adjacent uplands responds dynamically and rapidly to interannual changes in inundation, with local factors, such as management of water control structures, outweighing regional ones (Wasson et al., 2013). The resulting interaction of these variables and dynamics with fragmentation, land use planning and management (Richards and Friess, 2017) has only recently been investigated.

Research to date has focused on identifying synergisms among stressors (Campbell and Fourqurean, 2014; Lefcheck et al., 2017; Moftakhari et al., 2017; Noto and Shurin, 2017), but antagonisms and other feedbacks may be just as common (Brown et al., 2013; Conlisk et al., 2013; Maxwell et al., 2015; Crotty et al., 2017), and are seldom investigated, as is also true for thresholds and tipping points in coastal ecosystem stability and vulnerability (Connell et al., 2017; O'Meara et al., 2017; Wu et al., 2017). This precludes complete understanding of their complex responses, which may be greater than additive responses alone (Crotty et al., 2017), their adequate

management, or restoration regimes (Maxwell et al., 2015; Unsworth et al., 2015). Furthermore, although local management efforts cannot prevent severe climate change impacts on ecosystems, they can attempt to slow down adverse impacts, and allow the degree of evolutionary adaptation that is feasible given the trajectory of global GHG emission reductions (Brown et al., 2013).

In contrast, ecosystems with strong physical influences controlling elevation (sediment accretion and subsidence), even where mangrove replacement of salt marsh is expected, do not show changes in their vulnerability to SLR (McKee and Vervaeke, 2018), suggesting strong resilience of some coastal ecosystems. In areas such as South Florida, the wider Caribbean and the India-Pacific mangrove region (Lovelock et al., 2015), however, mangroves cannot outpace current SLR rates, and are at risk of disappearing. These regional and local effects are highly variable (even contradictory between studies; e.g., Smoak et al., 2013; Koch et al., 2015) and are related to local conditions shaping vulnerability such as topography and controls over salinity from freshwater and inputs (Flower et al., 2017), but further research on the mass and surface energy balance is needed (Barr et al., 2013). In addition, the responses and behaviour of private landowners, who could impede landward migration of ecosystems, is incipient, but needs to be taken into account in assessing the ability of coastal ecosystems to respond to climate change (Field et al., 2017). Overall, the long-term resilience of some coastal vegetation communities, and their ability to respond to rapid changes in sea level, is not well developed (Foster et al., 2017).

In summary, coastal ecosystems' with responses to SLR around the globe are complex and variable, with many specific responses at the ecosystem level or from keystone (foundation) species remaining poorly understood (Thompson et al., 2015). Moreover, responses are studied independently when holistic approaches may be required to understand how multiple threats affect ecosystem components, structure and functions (Giakoumi et al., 2015), and how human behaviour enables or constrains ecosystem responses to climate change (Field et al., 2017). In addition to the intrinsic coastal ecosystem values at stake, increasing exposure and vulnerability of these ecosystems contributes to increasing human exposure and vulnerability to SLR (*medium evidence, high agreement*; Arkema et al., 2013).

4.3.2.1.2   Point of departure on the human dimensions

The 2012 Special Report on Managing the Risks of Extreme Events and Disasters (SREX) acknowledged that patterns of human development create and compound exposure and vulnerability to climate-related hazards, including SLR (*high confidence*). The recent IPBES report also discusses the role of anthropogenic drivers in biodiversity loss (Díaz et al., 2019). Climate change-focussed studies have progressively moved from the analysis of various parameters' influence taken individually (education, poverty, etc.) to a more systemic approach that describes combinations of parameters, e.g., coastal urbanisation and settlement patterns (see Section 4.3.2.2) resulting from urban-rural discrepancies and trends in sociopolitical and economic inequalities. The AR5 also started differentiating between contemporary and historically-rooted drivers (e.g., trends in social systems over recent decades; Marino, 2012; Duvat et al., 2017; Fawcett et al., 2017), and reported some progress in the development of context-specific studies, especially on coastal megacities, major deltas and small islands (Cross-Chapter Box 9, Box 4.1).

AR5 also concluded with *very high confidence* that both RSL rise and related impacts are influenced by a variety of local social and/ or environmental processes unrelated to climate (e.g., anthropogenic subsidence, glacial isostatic adjustment, sediment supply and coastal squeeze). Some of these processes are partly attributable as anthropogenic drivers, and although they may or may not be directly related to RSL rise, they do cause changes in coastal ecosystem habitat connectivity and ecosystem health conditions, for instance, and consequently influence the ability of coastal social-ecological systems as a whole to cope with and adapt to SLR and its impacts.

However, the scientific literature still barely deals with the exposure and vulnerability of social-ecological systems to SLR specifically. Papers predominantly analyse the immediate and delayed consequences of extreme events such as TCs, storms and distant swells (see Section 6.3.3), for instance, and the resulting exposure and vulnerability 'in the context of SLR' (Woodruff et al., 2013). One reason for this touches on the difficulty for society to fully comprehend and for science to fully analyse long-term gradual changes like SLR (Fincher et al., 2014; Oppenheimer and Alley, 2016; Elrick-Barr et al., 2017). Consequently, Sections 4.3.2.2 to 4.3.2.5 concentrate on highlighting the anthropogenic or systemic drivers that have the potential to influence exposure and vulnerability to slow-onset sea level related hazards.

4.3.2.2   Settlement Trends

Major changes in coastal settlement patterns have occurred in the course of the 20th century, and are continuing to take place due to various complex interacting processes (Moser et al., 2012; Bennett et al., 2016) that together configure and concentrate exposure and vulnerability to climate change and SLR along the coast (Newton et al., 2012; Bennett et al., 2016). These processes include population growth and demographic changes (Smith, 2011; Neumann et al., 2015), urbanisation and a rural exodus, tourism development, and displacement or (re)settlement of some indigenous communities (Ford et al., 2015). This has resulted in a growing number of people living in the Low Elevation Coastal Zone (LECZ, coastal areas below 10 m of elevation; around 11% of the world's population in 2010; Neumann et al., 2015; Jones and O'Neill, 2016; Merkens et al., 2016) and in significant infrastructure and assets being located in risk-prone areas (*high confidence*). High density coastal urban development is commonplace in both developed and developing countries, as documented in recent case studies, for example in Canada (Fawcett et al., 2017), China (Yin et al., 2015; Lilai et al., 2016; Yan et al., 2016), Fiji (Hay, 2017), France (Genovese and Przyluski, 2013; Chadenas et al., 2014; Magnan and Duvat, 2018), Israel (Felsenstein and Lichter, 2014), Kiribati (Storey and Hunter, 2010; Duvat et al., 2013), New Zealand (Hart, 2011) and the USA (Heberger, 2012; Grifman et al., 2013; Liu et al., 2016b). This has implications for levels of SLR risk at regional and local scales (*medium evidence, high agreement*). In

Latin America and the Caribbean, for example, it is estimated that 6–8% of the population live in areas that are at high or very high risk of being affected by coastal hazards (Reguero et al., 2015; Calil et al., 2017; Villamizar et al., 2017), with higher percentages in Caribbean islands (Mycoo, 2018). In the Pacific, ~57% of Pacific Island countries' built infrastructure are located in risk-prone coastal areas (Kumar and Taylor, 2015). In Kiribati, due to the flow of outer, rural populations to limited, low-elevated capital islands, together with constraints inherent in the sociocultural land tenure system, the built area located <20 m from the shoreline quadrupled between 1969 and 2007–2008 (Duvat et al., 2013). Other examples of rural exodus are reported in the recent literature, for example in the Maldives (Speelman et al., 2017).

Population densification also affects rural areas' exposure and vulnerability, and interacts with other factors shaping settlement patterns, such as the fact that 'indigenous peoples in multiple geographical contexts have been pushed into marginalised territories that are more sensitive to climate impacts, in turn limiting their access to food, cultural resources, traditional livelihoods and place-based knowledge (…) [and therefore undermining] aspects of social-cultural resilience' (Ford et al., 2016b, p. 350). In the Pacific, for example, 'while traditional settlements on high islands (…) were often located inland, the move to coastal locations was encouraged by colonial and religious authorities and more recently through the development of tourism' (Ballu et al., 2011; Nurse et al., 2014, p. 1623; Duvat et al., 2017). Although these population movements are orders of magnitude smaller than the global trends described above, they play a critical role at the very local scale in explaining the emergence of, or changes in exposure and vulnerability. In atoll contexts, for example, the growing pressure on freshwater resources together with a loss in local knowledge (e.g., how to collect water from palm trees), result in increased exposure of communities to brackish, polluted groundwater, inducing water insecurity and health problems (Storey and Hunter, 2010; Lazrus, 2015).

4.3.2.3   Terrestrial Processes Shaping Coastal Exposure and Vulnerability

Coastal areas, including deltas, are highly dynamic as they are affected by natural and/or human-induced processes locally or originating from both the land and the sea. Changes within the catchment can therefore have severe consequences for coastal areas in terms of sediment supply, pollution, and/or land subsidence. Sediment supply reaching the coast is a critical factor for delta sustainability (Tessler et al., 2018) and has declined drastically in the last few decades due to dam construction, land use changes and sand mining (Ouillon, 2018; *high confidence*). For instance, Anthony et al. (2015) reported large-scale erosion affecting over 50% of the delta shoreline in the Mekong delta between 2003 and 2012, which was attributed in part to a reduction in surface-suspended sediments in the Mekong river potentially linked to dam construction within the river basin, sand mining in the river channels, and land subsidence linked to groundwater over-abstraction locally. Schmitt et al. (2017) demonstrated that these and other drivers in sediment budget changes can have severe effects on the very physical existence of the Mekong delta by the end of this century, with the most important single driver leading to inundation of large portions of the delta being ground-water pumping induced land subsidence. Thi Ha et al. (2018) estimated the decline in sediment supply to the Mekong delta to be around 75% between the 1970s and the period 2009–2016. In the Red River, the construction of the Hoa Binh Dam in the 1980s led to a 65% drop in sediment supply to the sea (Vinh et al., 2014). Based on projections of historical and 21st century sediment delivery to the Ganges-Brahmaputra-Meghna, Mahanadi and Volta deltas, Dunn et al. (2018) showed that these deltas fall short in sediment and may not be able to maintain their current elevation relative to sea level, suggesting increasing salinisation, erosion, flood hazards and adaptation demands.

Another rarely considered factor is the shift in TC climatology which also plays a critical role in explaining changes in fluvial suspended sediment loads to deltas as demonstrated by Darby et al. (2016), again for the Mekong delta. More generally, most conventional engineering strategies that are commonly employed to reduce flood risk (including levees, sea walls, and dams) disrupt a delta's natural mechanisms for building land. These approaches are rather short-term solutions which overall reduce the long-term resilience of deltas (Tessler et al., 2015; Welch et al., 2017). Systems particularly prone to flood risk due to anthropogenic activities include North America's Mississippi River delta, Europe's Rhine River delta, and deltas in East Asia (Renaud et al., 2013; Day et al., 2016). In regions where suspended sediments are still available in relatively large quantities, rates of sedimentation can vary depending on multiple factors, including the type of infrastructure present locally, as was shown by Rogers and Overeem (2017) for the Ganges-Brahmaputra-Meghna (Bengal) delta in Bangladesh as well as seasonal differences in sediment supply and place of deposition. For example, in meso-tidal and macro-tidal estuaries, during floods most of the sediments are depositing in the coastal zones and a large part of these sediments are brought back to the estuary during the low flow season by tidal pumping. This can lead to significantly higher deposition rates in the dry season as shown by Lefebvre et al. (2012) in the lower Red River estuary and by Gugliotta et al. (2018) in the Mekong delta. Enhanced sedimentation further upstream in estuaries and a silting-up of estuarine navigation channels can have high economic consequences for cities with a large estuarine harbour. In Haiphong city, in North Vietnam, the authorities decided to build a new harbour further downstream, for a cost estimated at 2 billion USD (Duy Vinh et al., 2018).

Overall, reduced freshwater and sediment inputs from the river basins are critical factors determining delta sustainability (Renaud et al., 2013; Day et al., 2016). In some contexts, this can be addressed through basin-scale management which allow more natural flows of water and sediments through the system, including methods for long-term flood mitigation such as improved river-floodplain connectivity, the controlled redirection of a river (i.e., avulsions) during times of elevated sediment loads, the removal of levees, and the redirection of future development to lands less prone to extreme flooding (Renaud et al., 2013; Day et al., 2016; Brakenridge et al., 2017). These actions could potentially increase the persistence of coastal landforms in the context of SLR. Next to decreasing sediment inputs to the coast, river bed and beach sand mining has been shown

to contribute to shoreline erosion, for example, for shorelines of Crete (Foteinis and Synolakis, 2015), and several sub-Saharan countries such Kenya, Madagascar, Mozambique, South Africa and Tanzania (UNEP, 2015). At the global scale, 24% of the world's sandy beaches are eroding at rates exceeding 0.5 m yr^{-1}, while 28% are accreting for the period 1984–2016. The largest and longest eroding sandy coastal stretches are in North America (Texas; Luijendijk et al., 2018).

Shoreline erosion leads to coastal squeeze if the eroding coastline approaches fixed and hard built or natural structures as noted in AR5 (Pontee, 2013; Wong et al., 2014), a process to which SLR also contributes (Doody, 2013; Pontee, 2013). The AR5 further noted that coastal squeeze is expected to accelerate due to rising sea levels (Wong et al., 2014). Doody (2013) characterised coastal squeeze as coastal habitats being pushed landward through the effects of SLR and other coastal processes on the one hand and, on the other hand, the presence of static natural or artificial barriers effectively blocking this migration, thereby squeezing habitats into an ever narrowing space. Distinctions are made between coastal squeeze being limited to (1) the consequences of SLR vs. other environmental changes on the coastline and (2) the presence of only coastal defence structures vs. natural sloping land or other artificial infrastructure (Pontee, 2013). Recent publications have emphasised coastal squeeze related to SLR, although inland infrastructure blocking habitat migration is not necessarily limited to defence structures (Torio and Chmura, 2015; McDougall, 2017). Coastal ecosystem degradation by human activities leading to coastal erosion is also an important consideration (McDougall, 2017). Taking into consideration the current challenges to attribute coastal impacts to SLR (Section 4.3.3.1), it can be hypothesised here that as long as SLR impacts remain moderate, the dominant driving factor of coastal squeeze will be anthropogenic land-based development (e.g., Section 4.3.2.2). With higher SLR scenarios and in the case of no further development at the coast, SLR may become the dominant driver before the end of this century.

Preserved coastal habitats can play important roles in reducing risks related to some coastal hazards and initiatives are being put in place to reduce coastal squeeze, such as managed realignment (Sections 4.1, 4.4.3.1) which includes removing inland barriers (Doody, 2013). Coastal squeeze can lead to degradation of coastal ecosystems and species (Martínez et al., 2014), but if inland migration is unencumbered, observation data and modelling have shown that the net area of coastal ecosystems could increase under various scenarios of SLR, depending on the ecosystems considered (Torio and Chmura, 2015; Kirwan et al., 2016; Mills et al., 2016). However, recent modelling research has shown that rapid SLR in a context of coastal squeeze could be detrimental to the areal extent and functionality of coastal ecosystems (Mills et al., 2016) and, for marshes, could lead to a reduction of habitat complexity and loss of connectivity, thus affecting both aquatic and terrestrial organisms (Torio and Chmura, 2015). Contraction of marsh extent is also identified by Kirwan et al. (2016) when artificial barriers to landward migration are in place. Adaptation to SLR therefore needs to account for both development and conservation objectives so that trade-offs between protection and realignment that satisfy both objectives can be identified (Mills et al., 2016).

In summary, catchment-scale changes have very direct impacts on the coastline, particularly in terms of water and sediment budgets (*high confidence*). The changes can be rapid and modify coastlines over short periods of time, outpacing the effects of SLR and leading to increased exposure and vulnerability of social-ecological systems (*high confidence*). Without losing sight of this fact, management of catchment-level processes contribute to limiting rapid increases in exposure and vulnerability. Further to hinterland influences, coastal squeeze increases coastal exposure as well as vulnerability by the loss of a buffer zone between the sea and infrastructure behind the habitat undergoing coastal squeeze. The clear implication is that coastal ecosystems progressively lose their ability to provide regulating services with respect to coastal hazards, including as a defence against SLR driven inundation and salinisation (*high confidence*). Vulnerability is also increased if freshwater resources become salinised, particularly if these resources are already scarce. The exposure and vulnerability of human communities is exacerbated by the loss of other provisioning, supporting and cultural services generated by coastal ecosystems, which is especially problematic for coast-dependent communities (*high confidence*).

4.3.2.4 Other Human Dimensions

The development of local scale case studies from a social science perspective, for example, in the Arctic (Ford et al., 2012; Ford et al., 2014), small islands (Petzold, 2016; Duvat et al., 2017) and within cities (Rosenzweig and Solecki, 2014; Paterson et al., 2017; Texier-Teixeira and Edelblutte, 2017) or at the household level (Koerth et al., 2014) support a better understanding of the anthropogenic drivers of exposure and vulnerability. Four examples of drivers that were only emerging at the time of the AR5 are discussed below. Very importantly, another major emerging dimension that is not discussed here but rather in Section 4.4.4, relates to power asymmetries, politics, and the prevailing political economy, which are important drivers of exposure and vulnerability to SLR-related coastal hazards, and consequently adaptation prospects (Eriksen et al., 2015; Dolšak and Prakash, 2018). Recent literature provides examples in coastal megacities like Jakarta, Indonesia (Shatkin, 2019) as well as in smaller cities, like Maputo, Mozambique (Broto et al., 2015) and Surat, India (Chu, 2016a; Chu, 2016b), and many other coastal cities and settlements around the world (*high confidence*; Jones et al., 2015; Allen et al., 2018; Hughes et al., 2018; Sovacool, 2018).

4.3.2.4.1 Gender inequality

Gender inequality came to prominence only recently in climate change studies (~15 years ago; see Pearse, 2017). In light of sea-related hazards and SLR specifically, the issue is still mainly investigated in the context of developing countries, although growing attention is paid to the issue in developed countries (e.g., Lee et al., 2015; Pearse, 2017). Recent studies in southern coastal Bangladesh, for example, show that women get less access than men to climate- and disaster-related information (both emergency information and training programmes), decision making processes at the household and community levels, economic resources including financial means such as micro-credit, land ownership, and mobility within and outside the villages (Rahman, 2013; Alam and Rahman, 2014; Garai, 2016). Gender inequity may

be inherent in unfavourable background conditions (higher illiteracy rates, deficiencies in food and calories intake and poorer health conditions) as a result of, among other things, traditions, social norms and patriarchy. Together, these barriers disadvantage women more than men in developing effective responses to anticipate gradual environmental changes such as persistent coastal erosion, flooding and soil salinisation (*medium evidence, high agreement*). Such conclusions are in line with the literature on gender inequality and climate change at large (Alston, 2013; Pearse, 2017), thus suggesting no major SLR-inherent specificities.

4.3.2.4.2 Loss of indigenous knowledge and local knowledge

Despite the identification of this issue in AR4, its treatment in AR5 remained limited. Recent literature partly focussing on SLR reaffirms that indigenous knowledge and local knowledge (IK and LK; Cross-Chapter Box 4 in Chapter 1 and Glossary) are key to determining how people recognise and respond to environmental risk (Bridges and McClatchey, 2009; Lefale, 2010; Leonard et al., 2013; Lazrus, 2015), and therefore to increasing adaptive capacity and reducing long-term vulnerability (Ignatowski and Rosales, 2013; McMillen et al., 2014; Hesed and Paolisso, 2015; Janif et al., 2016; Morrison, 2017).

IK and LK contribute both as a foundation for and an outcome of customary resource management systems aimed at regulating resource use and securing critical ecosystem protection (examples in Indonesia; Hiwasaki et al., 2015), structuring the relationship between people and authorities, and framing and maintaining a strong sense of place in the community (examples in Timor Leste; Hiwasaki et al., 2015). In turn, this allows local communities to predict and prepare for both sudden shock events that have historical precedent and, when IK and LK are embedded in day-to-day rituals and decision making processes, to also anticipate the consequences of gradual changes, as in sea level (examples in Indonesia; Hiwasaki et al., 2015). Customary resource management systems based on IK and elders' leadership – for instance, Rahui in French Polynesia (Gharasian, 2016), or Mo in the Marshall Islands (Bridges and McClatchey, 2009) – also allow communities to diversify access to marine and terrestrial resources using seasonal calendars, to ensure collective food and water security, and to maintain ecological integrity (McMillen et al., 2014). In rural Pacific atolls, traditional food preservation and storage (e.g., storing germinated coconuts or drying fish) still play a role in anticipating disruptions in natural resource availability (Campbell, 2015; Lazrus, 2015). Such practices have enabled the survival of isolated communities from the Arctic to tropical islands in constraining sea environments for centuries to millennia (McMillen et al., 2014; Nunn et al., 2017a). Morrison (2017) argues that IK and LK can also play a role in supporting internal migration in response to SLR, by avoiding social and cultural uprooting (Cross-Chapter Box 4 in Chapter 1).

In some specific contexts, climate change will also imply no-analogue changes, such as rapid ice-melt and changing conditions in the Arctic that have no precedent in the modern era, and could thus limit the relevance of IK and LK in efforts to address significantly different circumstances. Except in these specific situations, the literature

4

suggests that the loss of IK and LK, and related social norms and mechanisms, will increase populations' exposure and vulnerability to SLR impacts (Nakashima et al., 2012). The literature notably points out that modern, externally-driven socioeconomic dynamics, such as the introduction of imported food (noodles, rice, canned meat and fish, etc.), diminish the cultural importance of IK-based practices and diets locally, together with introducing dependency on monetisation and external markets (Hay, 2013; Campbell, 2015).

As a result, the loss of IK and LK may increase long-term vulnerability to SLR (*medium evidence, high agreement*). Given that IK and LK are largely based on observing and 'making sense' of the surrounding environment (moon, waves, winds, animal behaviours, topography, etc.), such a loss reflects a more general concern about the weakening of environmental connectedness in contemporary societies, which is not limited to remote, rural and developing communities (*medium confidence*). In developed contexts too, the loss of LK has played a critical role in recent coastal disasters (e.g., Katrina in 2005 in the USA, Kates et al., 2006) and increasing vulnerability to SLR (e.g., Newton and Weichselgartner, 2014; Wong et al., 2014).

4.3.2.4.3   Social capital

Coastal communities draw on social structures and capabilities that can reduce risk and increase adaptive capacity in the face of coastal hazards (Aldrich, 2017; Petzold, 2018). Although the term is subject to debate (Meyer, 2018), social capital – that is, the level of cohesion between individuals, between groups of individuals, and between people and institutions, within and between communities – is considered to be a key enabler for collective action to reduce risk and build adaptive capacity (Adger, 2010; Aldrich and Meyer, 2015; Petzold and Ratter, 2015). Levels of social capital can be influenced by underlying social processes, such as socioeconomic (in)equalities, gender issues, health, social networks and social media. It applies to both developing and developed countries, for example in densely populated deltas (Jordan, 2015), European coasts (Jones and Clark, 2014; Petzold, 2016), Asian urban or semi-urban coastal areas (Lo et al., 2015; Triyanti et al., 2017) and Pacific islands (Neef et al., 2018). Social capital framed as an enabler for reducing vulnerability has been studied in the context of extreme events (risk prevention mechanisms, emergency responses and post-crisis actions) and collective environmental management (e.g., replanting mangroves, beach cleaning, etc.). Social capital also enables adaptation prospects. For example, its role has been explored in public acceptability of long-term coastal adaptation policies in the UK (Jones and Clark, 2014; Jones et al., 2015). The role of social capital in building resilience to climate stress in coastal Bangladesh was explored by Jordan (2015), who found complex and even contradictory interactions between social capital and resilience to climate stress. Among others, Jordan (2015) also advises caution about uncritical importation of such Westernised concepts in seeking to understand and address coastal vulnerability in developing countries.

4.3.2.4.4   Risk perception

Risk perception, which is context-specific and varies from one individual to another, may influence communities' exposure and vulnerability as it shapes authorities' and people's attitudes towards sudden and slow onset hazards, as shown by Terpstra (2011), Lazrus (2015), Elrick-Barr et al. (2017) and O'Neill et al. (2016) in the Netherlands, Tuvalu, Australia and Ireland, respectively. The progressive discounting of coastal hazard risks and subsequent loss of risk memory also played a role in coastal disasters such as Hurricane Katrina in 2005 in the USA (Burby, 2006; Kates et al., 2006) and Storm Xynthia in 2010 in France (Vinet et al., 2012; Genovese and Przyluski, 2013; Chadenas et al., 2014).

Risk perceptions stem from intertwined predictors such as 'gender, political party identification, cause-knowledge, impact-knowledge, response-knowledge, holistic affect, personal experience with extreme weather events, [social norms] and biospheric value orientations' (Kellens et al., 2011; Carlton and Jacobson, 2013; Lujala et al., 2015; van der Linden, 2015, p. 112; Weber, 2016; Elrick-Barr et al., 2017; Goeldner-Gianella et al., 2019). In general, there is a lack of education, training and thus knowledge and literacy on recent and projected trends in sea level, which compromises ownership of science facts and projections at all levels, from individuals and institutions to society at large.

While some studies have begun to highlight the influence of the distance from the sea on risk perceptions (Milfont et al., 2014; Lujala et al., 2015; O'Neill et al., 2016), there is still little knowledge about how risk perceptions vary across different geographical and social contexts, and how this influences exposure and vulnerability to coastal hazards (e.g., Terpstra, 2011; van der Linden, 2015). There is a critical lack of studies specifically addressing SLR. Some recent works conducted in coastal Australia suggest that while people are confident about their ability to cope with an already experienced event, when it comes to SLR, the dominant narrative is articulated around the barriers related to the 'uncertainty in the nature and scale of the impacts as well as the response options available' (Elrick-Barr et al., 2017, p. 1147). Similar conclusions have been highlighted in the Caribbean islands of St. Vincent (Smith, 2018) and the Bahamas (Thomas and Benjamin, 2018). SLR is rarely addressed separately from sea-related extreme events, which masks a crucial difference between already-observed and delayed impacts. Climate change is considered a "distant psychological risk" (Spence et al., 2012), making it and SLR per se 'markedly different from the way that our ancestors have traditionally perceived threats in their local environment' (Milfont et al., 2014; Lujala et al., 2015; van der Linden, 2015, p. 112; O'Neill et al., 2016).

4.3.2.5   Towards a Synthetic Understanding of the Drivers of Exposure and Vulnerability

Recent literature confirms that anthropogenic drivers played an important role, over the last century, in increasing exposure and vulnerability worldwide, and indicates that they will continue to do so in the absence of adaptation (*medium evidence, high agreement*). Some scholars argue that 'even with pervasive and extensive environmental change associated with ~2°C warming, it is non-climatic factors that primarily determine impacts, response options and barriers to adapting' (Ford et al., 2015, p. 1046). Although it is the interaction of climate and non-climate factors that eventually determine the level of impacts, acknowledging the role of a range of purely anthropogenic drivers has important implications for action. It suggests that major action can be taken now to enhance long-term adaptation prospects, notwithstanding uncertainty about local RSL rise and resultant impacts in the distant future (*medium evidence, high agreement*; Magnan et al., 2016). Acting on the human-driven drivers and root causes of vulnerability could yield co-benefits, for example by improving the state and condition of coastal ecosystems – and hence the capacity to cope with or adapt to SLR impacts – or, in deltaic regions, lowering the rates of anthropogenic subsidence and, in turn, minimising changes in sea level.

In addition, coastal ecosystem degradation is acknowledged as another major non-climatic driver of exposure and vulnerability (*high confidence*). The ability of coastal ecosystems to serve as a buffer zone between the sea and human assets (settlements and infrastructure), and to provide regulating services with respect to SLR-related coastal hazards (including inundation and salinisation), is progressively being lost due to coastal squeeze, pollution, and habitat and land degradation mainly due to land-use conversion.

We now better understand the diversity and interactions of the climate and non-climate drivers of exposure and vulnerability, as well as their dynamics over time (Bennett et al., 2016; Duvat et al., 2017). As a result, it is now realised how many context-specificities interact (including geography, economic development, social inequity, power and politics, and risk perceptions) and play a critical role in shaping the direction and influence of individual drivers and of their possible combinations on the ground (*medium evidence, high agreement*; Eriksen et al., 2015; Hesed and Paolisso, 2015; McCubbin et al., 2015). This also provides a stronger foundation to identify the range of possible responses (Sections 1.6.1, 1.6.2 and 4.4.3) to observed impacts and projected risks, as well as critical areas of action to enhance adaptation pathways (Section 4.4.4).

Recent studies (e.g., cited in Sections 4.3.2.1.1, 4.3.2.2, 4.3.2.4.2 and 4.3.2.4.4) also confirm AR5 conclusions that both developing and developed countries are exposed and vulnerable to SLR (*high confidence*).

4.3.3 Observed Impacts, and Current and Future Risk of Sea Level Rise

SLR leads to hazards and impacts that are also partly inherent in other processes such as starvation of sediments provided by rivers (Kondolf et al., 2014); permafrost thaw and ice retreat; or the disruption of natural dynamics by land reclamation or sediment mining. Six main concerns for low-lying coasts (Figure 4.13) are: (i) permanent submergence of land by mean sea levels or mean high tides; (ii) more frequent or intense flooding; (iii) enhanced erosion; (iv) loss and change of ecosystems; (v) salinisation of soils, ground

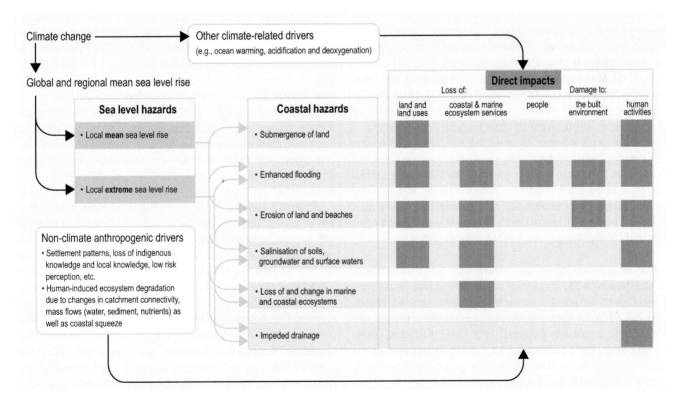

Figure 4.13 | Overview of the main cascading effects of sea level rise (SLR). Colours of lines (light green and light orange) and boxes are used only for the readability of the figure. Sea level hazards are discussed in Section 4.2. The various impacts listed in this figure are discussed in the sections below: Submergence of land and enhanced flooding (4.3.3.2); Erosion of land and beaches (4.3.3.3); Salinisation (4.3.3.4); Loss of and changes in ecosystems (4.3.3.5); Loss of land and land uses (4.3.3.2); Loss of ecosystems services (4.3.3.5); Damage to people and to the built environment (4.3.3.2, 4.3.3.3, 4.3.3.4 and 4.3.3.6); Damage to human activities (4.3.3.6). Non-climate anthropogenic drivers are discussed in Section 4.3.2 and other climate-related drivers are notably discussed in Section 5.2.1 and 5.2.2.

and surface water; and (vi) impeded drainage. This section discusses some of these hazards (flooding, erosion, salinisation) as well as observed and projected impacts on some critical marine ecosystems (marshes, mangroves, lagoons, coral reefs and seagrasses), ecosystem services (coastal protection) and human societies (people, assets, infrastructures, economic and subsistence activities, inequity and well-being, etc.). In many cases, the Chapter 4 assessment of impacts and responses uses results from literature based on values of SLR and ESL events prior to SROCC. However, the general findings reported here also carry forward with the new SROCC SLR and ESL values. Except in the case of submergence and flooding of coastal areas (Section 4.3.3.2), this section assumes no major additional adaptation efforts compared to today (i.e., neither significant intensification of ongoing action nor new types of action), thus reflecting the state of knowledge in the literature.

4.3.3.1   Attribution of Observed Physical Changes to Sea Level Rise

The AR5 concludes that attribution of coastal changes to SLR is difficult because 'the coastal sea level change signal is often small when compared to other processes' (Wong et al., 2014: 375). New literature, however, shows that extreme water levels at the coast are rising due to mean SLR (4.2.2.4 for observations, and 4.3.5 for projections), with observable impacts on chronic flooding in some regions (Sweet and Park, 2014; Strauss et al., 2016).

On coastal morphological changes for example, contemporary SLR currently acts as a 'background driver', with extreme events, changes in wave patterns, tides and human intervention often described as the prevailing drivers of observed changes (Grady et al., 2013; Albert et al., 2016). Morphological changes are also interacting with other impacts of SLR, such as coastal flooding (Pollard et al., 2018). Despite the complexity of the attribution issue (Romine et al., 2013; Le Cozannet et al., 2014), recent literature suggests possibly emerging signs of the direct influence of recent SLR on shoreline behaviour, for example on small highly-sensitive reef islands in New Caledonia (Garcin et al., 2016) and in the Solomon Islands (Albert et al., 2016). Early signs of the direct influence of SLR on estuaries' water salinity are also emerging, for example, in the Delaware, USA, where Ross et al. (2015) estimate a rate of salinity increase by as much as 4.4 psu (Practical Salinity Unit) per metre of SLR since the 1950s.

Overall, while the literature suggests that it is still too early to attribute coastal impacts to SLR in most of the world's coastal areas, there is *very high confidence* that as sea level continues to rise (Sections 4.2.3.2, 4.2.3.3), the frequency, severity and duration of hazards and related impacts increases (Woodruff et al., 2013; Lilai et al., 2016; Vitousek et al., 2017; Sections 4.2.3.4, 6.3.1.3). Detectable impacts and attributable impacts on shoreline behaviour are expected as soon as the second half of the 21st century (Nicholls and Cazenave, 2010; Storlazzi et al., 2018).

4.3.3.2   Submergence and Flooding of Coastal Areas

Since AR5, a number of continental and global scale coastal exposure studies have accounted for sub-national human dynamics such as coastward migration or coastal urbanisation. These studies project a population increase in the LECZ (coastal areas below 10 m of elevation) by 2100 of 85 to 239 million people as compared to only considering national dynamics (Merkens et al., 2016; Section 4.3.2). Under the five SSPs and without SLR, the population living in the LECZ increases from 640–700 million in 2000 to over one billion in 2050 under all SSPs, and then declines to 500–900 million in 2100 under all SSPs, except for SSP3 (i.e., a world in which countries will increasingly focus on domestic issues, or at best regional ones), for which the coastal population reaches 1.1–1.2 billion (Jones and O'Neill, 2016; Merkens et al., 2016).

The population exposed to mean and ESL events will grow significantly during the 21st century (*high confidence*) with socioeconomic development and SLR contributing roughly equally (*medium confidence*). Considering an average relative SLR of 0.7–0.9 m but no population growth, the number of people living below the hundred-year ESL in Latin America and the Caribbean will increase from 7.5 million in 2011 to 9 million by the end of the century (Reguero et al., 2015). Considering population growth and urbanisation, only 21 cm of global mean SLR by 2060 would increase the global population living below the hundred-year ESL from about 189 million in 2000 to 316–411 million in 2060, with the largest absolute changes in South and Southeast Asia and the largest relative changes in Africa (Neumann et al., 2015). Considering population growth, Hauer et al. (2016) estimate that 4.3 and 13.1 million people in the USA would live below the levels of 0.9 and 1.8 m SLR by 2100.

New coastal flood risk studies conducted since AR4 at global, continental and city scale, reinforce AR5 findings that if coastal societies do not adapt, flood risks will increase by 2–3 orders of magnitude reaching catastrophic levels by the end of the century, even under the lower end SLR expected under RCP2.6 (*high confidence*; Hinkel et al., 2014; Abadie et al., 2016; Diaz, 2016; Hunter et al., 2017; Lincke and Hinkel, 2018; Abadie, 2018; Brown et al., 2018a; Nicholls, 2018). In combination, these studies take into account a SLR scenario range wider than the *likely* range of AR5 but consistent with the range of projections assessed in this report (Section 4.2.3.2). For example, considering 25–123 cm of SLR in 2100, all SSPs and no adaptation, Hinkel et al. (2014) find that 0.2–4.6% of global population is expected to be flooded annually in 2100, with expected annual damages (EAD) amounting to 0.3–9.3% of global GDP. Assessing 120 cities globally, Abadie (2018) find that under a weighted combination of the probabilistic scenarios, New Orleans and Guangzhou Guangdong rank highest with EAD above 1 trillion USD (not discounted) in each city. For Europe, EAD are expected to rise from 1.25 billion EUR today to 93–960 billion EUR by the end of the century (Vousdoukas et al., 2018b). Already today, many small islands face large flood damages relative to their GDP specifically through TCs (Cashman and Nagdee, 2017) and under SLR EAD can reach up to several percent of GDP in 2100, as highlighted in

AR5 (Wong et al., 2014). Similar to the exposure studies, estimates of future flood risk without considering adaptation, as presented in this paragraph, do not provide a meaningful characterisation of coastal flood risks, because adaptation and specifically hard protection is expected to be widespread during the 21st century in urban areas and cities (*high confidence*; Section 4.4.3.2.2). Rather, these estimates need to be seen as illustrations of the scale of adaptation needed to offset risk.

Flood risk studies that have included adaptation find that hard coastal protection is generally very effective in reducing flood risks during the 21st century even under high SLR scenarios (*high confidence*; Hinkel et al., 2014; Diaz, 2016; Brown et al., 2018a; Hinkel et al., 2018; Lincke and Hinkel, 2018; Tamura et al., 2019) (Section 4.4.2.2.2). For example, Hinkel et al. (2014) find that under 25–123 cm of SLR in 2100 and all SSPs, hard coastal protection reduces the annual number of people affected by coastal floods and EAD by 2–3 orders of magnitude. Under high-end SLR and beyond the 21st century, effectiveness of coastal adaptation is expected to decline rapidly, but there is a lack of studies addressing this issue. Furthermore, there is a lack of studies taking into account responses beyond hard protection such as ecosystem-based adaptation, accommodation, advance and retreat (Sections 4.4.2).

Studies also confirm AR5 findings that the relative costs and benefits of coastal adaptation are distributed unequally across countries and regions (*high confidence*; Wong et al., 2014; Diaz, 2016; Lincke and Hinkel, 2018; Tamura et al., 2019). For example, while the median cost of protection and retreat under RCP8.5 in 2050 has been estimated to be under 0.09% of national GDP, large relative costs are found for small island states such as the Marshall Islands (7.6%), the Maldives (7.5%), Tuvalu (4.6%) and Kiribati (4.1%; Diaz, 2016). Furthermore, on a global average and for urban and densely populated regions, hard protection is highly cost efficient with benefit-cost ratios up to 10^4, but for poorer and less densely populated areas benefit-cost ratios are generally smaller than one (Lincke and Hinkel, 2018). Hence, without substantial transfer payments supporting poor areas, coastal flood risks will evolve unequally during this century, with richer and densely populated areas well protected behind hard structures and poorer less densely populated areas suffering losses and damages, and eventually retreating from the coast.

While continental to global scale flood exposure and risk studies have also explored a wider range of uncertainty as compared to AR5, much remains to be done. All of these studies rely on global elevation data, but few studies have explored the underlying bias. For example, for the Po delta in Italy, it was found that elevation data based on the widely used Shuttle Radar Topography Mission (SRTM), Reuter et al. (2007) overestimates the 100-year floodplain by about 50% as compared to local Lidar data (Wolff et al., 2016), while in the Ria Formosa region in Portugal SRTM underestimates EAD by up to 50% depending on the resampled resolution of the Lidar data (Vousdoukas et al., 2018a). For the USA, SRTM data systemically underestimates population exposure below 3 m by more than 60% as compared to coastal Lidar data (Kulp and Strauss, 2016). A global scale comparison of major contributors to flood risk uncertainty finds that uncertainty in digital elevation data is roughly at equal

footing with uncertainties in socioeconomic development, emission scenarios, and SLR in determining the magnitude of flood risks in the 21st century (Hinkel et al., 2014). At a European level, the number of people living in the 100-year coastal floodplain can vary between 20–70% depending on the different inundation models used and the inclusion or exclusion of wave set up (Vousdoukas, 2016). Comparing damage functions attained in different studies for European cities, Prahl et al. (2018) find up to four-fold differences in damages for floods above 3 m. Another major source of uncertainty relates to uncertainties in present-day ESL events due to the application of different extreme value methods (Wahl et al., 2017; Section 4.2.3.4). While all of the uncertainties reported above affected the actual size of exposure and flood risk figures, they do not affect the overall conclusions drawn here.

4.3.3.3 Coastal Erosion and Projected Global Impacts of Enhanced Erosion on Human Systems

Recent global assessments of coastal erosion indicate that land losses currently dominate over land gains and that human interventions are a major driver of shoreline changes (Cazenave and Cozannet, 2014; Luijendijk et al., 2018; Mentaschi et al., 2018). Luijendijk et al. (2018) estimate that over the 1984–2016 period, about a quarter of the world's sandy beaches eroded at rates exceeding 0.5 m yr^{-1} while about 28% accreted. While such global results can be challenged due to the relatively large detection threshold used (±0.5 m yr^{-1}), there is growing literature indicating that coastal erosion is occurring or increasing, e.g. in the Arctic (Barnhart et al., 2014a; Farquharson et al., 2018; Irrgang et al., 2019), Brazil (Amaro et al., 2015), China (Yang et al., 2017), Colombia (Rangel-Buitrago et al., 2015), India (Kankara et al., 2018), and along a large number of deltaic systems worldwide (e.g., Section 4.2.2.4).

Since AR5, however, there is growing appreciation and understanding of the ability of coastal systems to respond dynamically to SLR (Passeri et al., 2015; Lentz et al., 2016; Deng et al., 2017). Most low-lying coastal systems exhibit important feedbacks between biological and physical processes (e.g., Wright and Nichols, 2018), that have allowed them to maintain a relatively stable morphology under moderate rates of SLR (<0.3 cm yr^{-1}) over the past few millennia (Woodruff et al., 2013; Cross-Chapter Box 5 in Chapter 1). In a global review on multi-decadal changes in the land area of 709 atoll islands, Duvat (2019) shows that in a context of more rapid SLR than the global mean (Becker et al., 2012; Palanisamy et al., 2014), 73.1% of islands were stable in area, while respectively 15.5% and 11.4% increased and decreased in size. While anthropogenic drivers played a major role, especially in urban islands (e.g., shoreline stabilisation by coastal defences, increase in island size as a result of reclamation works), this study and others (e.g., McLean and Kench, 2015) suggest that these islands have had the capacity to maintain their land area by naturally adjusting to SLR over the past decades (*high confidence*). However, it has been argued that this capacity could be reduced in the coming decades, due to the combination of higher rates of SLR, increased wave energy (Albert et al., 2016), changes in run-up (Shope et al., 2017) and storm wave direction (Harley et al., 2017), effects of ocean warming and acidification on critical ecosystems such as coral reefs (Section 4.3.3.5.2), and a continued increase in anthropogenic pressure.

From a global scale perspective, based on AR4 SLR scenarios and without considering the potential benefits of adaptation, Hinkel et al. (2013b) estimate that about 6000 to 17,000 km^2 of land is expected to be lost during the 21st century due to enhanced coastal erosion associated with SLR, in combination with other drivers. This could lead to a displacement of 1.6–5.3 million people and associated cumulative costs of 300 to 1000 billion USD (Section 4.4.3.5). Importantly, these global figures mask the wide diversity of local situations; and some literature is emerging on the non-physical and non-quantifiable impacts of coastal erosion, for example, on the loss of recreational grounds and the induced risks to the associated social dimensions (i.e., how local communities experience coastal erosion impacts; Karlsson et al., 2015).

4.3.3.4   Salinisation

With rising sea levels, saline water intrusion into coastal aquifers and surface waters and soils is expected to be more frequent and enter farther landwards. Salinisation of groundwater, surface water and soil resources also increases with land-based drought events, decreasing river discharges in combination with water extraction and SLR (*high confidence*).

4.3.3.4.1   Coastal aquifers and groundwater lenses

Groundwater volumes will primarily be affected by variations in precipitation patterns (Taylor et al., 2013; Jiménez Cisneros et al., 2014), which are expected to increase water stress in small islands (Holding et al., 2016). While SLR will mostly impact groundwater quality (Bailey et al., 2016) and in turn exacerbate salinisation induced by marine flooding events (Gingerich et al., 2017), it will also affect the watertable height (Rotzoll and Fletcher, 2013; Jiménez Cisneros et al., 2014; Masterson et al., 2014; Werner et al., 2017). In addition, the natural migration of groundwater lenses inland in response to SLR can also be severely constrained by urbanisation, for example, in semi-arid South Texas, USA (Uddameri et al., 2014).

These changes will affect both freshwater availability (for drinking water supply and agriculture) and vegetation dynamics. At many locations, however, direct anthropogenic influences, such as groundwater pumping for agricultural or urban uses, already impact salinisation of coastal aquifers more strongly than what is expected from SLR in the 21st century (Ferguson and Gleeson, 2012; Jiménez Cisneros et al., 2014; Uddameri et al., 2014), with trade-offs in terms of groundwater depletion that may contribute to anthropogenic subsidence and thus increase coastal flood risk. Recent studies also suggest that the influence of land-surface inundation on seawater intrusion and resulting salinisation of groundwater lenses on small islands has been underestimated until now (Ataie-Ashtiani et al., 2013; Ketabchi et al., 2014). Such impacts will potentially also combine with a projected drying of most of the tropical-to-temperate islands by mid-century (Karnauskas et al., 2016).

4.3.3.4.2   Surface waters

The quality of surface water resources (in estuaries, rivers, reservoirs, etc.) can be affected by the intrusion of saline water, both in a direct (increased salinity) and indirect way (altered environmental conditions which change the behaviour of pollutants and microbes). In terms of direct impacts, statistical models and long-term (1950 to present) records of salinity show significant upward trends in salinity and a positive correlation between rising sea levels and increasing residual salinity, for example in the Delaware Estuary, USA (Ross et al., 2015). Higher salinity levels, further inland, have also been reported in the Gorai river basin, southwestern Bangladesh (Bhuiyan and Dutta, 2012), and in the Mekong Delta, Vietnam. In the Mekong Delta for instance, salinity intrusion extends around 15 km inland during the rainy season and typically around 50 km during dry season (Gugliotta et al., 2017). Importantly, salinity intrusion in these deltas is caused by a variety of factors such as changes in discharge and water abstraction along with relative SLR. More broadly, the impact of salinity intrusion can be significant in river deltas or low-lying wetlands, especially during low-flow periods such as in the dry season (Dessu et al., 2018). In Bangladesh, for instance, some freshwater fish species are expected to lose their habitat with increasing salinity, with profound consequences on fish-dependent communities (Dasgupta et al., 2017). In the Florida Coastal Everglades, sea level increasingly exceeds ground surface elevation at the most downstream freshwater sites, affecting marine-to-freshwater hydrologic connectivity and transport of salinity and phosphorous upstream from the Gulf of Mexico. The impact of SLR is higher in the dry season when there is practically no freshwater inflow (Dessu et al., 2018). Salinity intrusion was shown to cause shifts in the diatom assemblages, with expected cascading effects through the food web (Mazzei and Gaiser, 2018). Salinisation of surface water may lead to limitations in drinking water supply (Wilbers et al., 2014), as well as to future fresh water shortage in reservoirs, for example in Shanghai (Li et al., 2015). Salinity changes the partitioning and mobility of some metals, and hence their concentration or speciation in the water bodies (Noh et al., 2013; Wong et al., 2015; de Souza Machado et al., 2018). Varying levels of salinity also influence the abundance and toxicity of *Vibrio cholerae* in the Ganges Delta (Batabyal et al., 2016).

4.3.3.4.3   Soils

Salinisation is one of the major drivers of soil degradation, with sea water intrusion being one of the common causes (Daliakopoulos et al., 2016). In a study in the Ebro Delta, Spain, for instance, soil salinity was shown to be directly related to distances to the river, to the delta inner border, and to the old river mouth (Genua-Olmedo et al., 2016). Land elevation was the most important variable in explaining soil salinity.

SLR was also shown to decrease organic carbon (C_{org}) concentrations and stocks in sediments of salt marshes as reworked marine particles contribute with a lower amount of C_{org} than terrigenous sediments. C_{org} accumulation in tropical salt marshes can be as high as in mangroves and the reduction of C_{org} stocks by ongoing SLR might cause high CO_2 releases (Ruiz-Fernández et al., 2018). In many cases attribution to SLR is missing, but, independent from clear attribution, sea water intrusion leads to a salinisation of exposed soils with changes in carbon dynamics (Ruiz-Fernández et al., 2018) and microbial communities (Sánchez-Rodríguez et al., 2017), soil enzyme activity and metal toxicity (Zheng et al., 2017). Water salinity levels in the pores of coastal marsh soils can become significantly elevated in just one week of flooding by sea water, which can potentially negatively impact associated microbial communities for significantly longer time periods (McKee et al., 2016). SLR will also alter the frequency and magnitude of wet/dry periods and salinity levels in coastal ecosystems, with consequences for the formation of climate relevant GHGs (Liu et al., 2017b) and therefore feedbacks to the climate.

Soil salinisation affects agriculture directly with impacts on plant germination (Sánchez-García et al., 2017), plant biomass (rice and cotton) production (Yao et al., 2015), and yield (Genua-Olmedo et al., 2016). Impact on agriculture is especially relevant in low-lying coastal areas where agricultural production is a major land use, such as in river deltas.

4.3.3.5 Ecosystems and Ecosystem Services

4.3.3.5.1 Tidal wetlands

Global coastal wetlands have been reduced by a half since the pre-industrial period due to the impacts of both climatic and non-climatic drivers such as flooding, coastal urbanisation, alterations in drainage and sediment supply. (Sections 4.3.2.3, 5.3.2). Potentially one of the most important of the eco-morphodynamic feedbacks allowing for relatively stable morphology under SLR is the ability of marsh and mangrove systems to enhance the trapping of sediment, which in turn allows tidal wetlands to grow and increase the production and accumulation of organic material (Kirwan and Megonigal, 2013). When ecosystem health is maintained and sufficient sediment exists to support their growth, this particular feedback has generally allowed marshes and mangrove systems to build vertically at rates equal to or greater than SLR up to the present day (Kirwan et al., 2016; Woodroffe et al., 2016).

While recent reviews suggest that mangroves' surface accretion rate will keep pace with a high SLR scenario (RCP8.5) up to years 2055 and 2070 in fringe and basin mangrove settings, respectively (Sasmito et al., 2016), process-based models of vertical marsh growth that incorporate biological and physical feedbacks support survival under rates of SLR as high as 1–5 cm yr^{-1} before drowning (Kirwan et al., 2016). Threshold rates of SLR before marsh drowning however vary significantly from site-to-site and can be substantially lower than 1 cm yr^{-1} in micro-tidal regions where the tidal trapping of sediment is reduced and/or in areas with low sediment availability (Lovelock et al., 2015; Ganju et al., 2017; Jankowski et al., 2017; Watson et al., 2017). Global environmental change may also to lead to changes in growth rates, productivity and geographic distribution of different mangrove and marsh species, including the replacement of environmentally sensitive species by those possessing greater climatic tolerance (Krauss et al., 2014; Reef and Lovelock, 2014; Coldren et al., 2019). Processes impacting lateral changes at the marsh boundary including wave erosion are just as important, if not more, than vertical accretion rates in determining coastal wetland survival (e.g., Mariotti and Carr, 2014). For most low-lying coastlines, a seaward loss of wetland area due to marsh retreat could be offset by a similar landward migration of coastal wetlands (Kirwan and Megonigal, 2013; Schile et al., 2014), this landward migration having

the potential to maintain and even increase the extent of coastal wetlands globally (Morris et al., 2012; Kirwan et al., 2016; Schuerch et al., 2018). This natural process will however be constrained in areas with steep topography or hard engineering structures (i.e., coastal squeeze, Section 4.3.2.4). Seawalls, levees and dams can also prevent the fluvial and marine transport of sediment to wetland areas and reduce their resilience further (Giosan, 2014; Tessler et al., 2015; Day et al., 2016; Spencer et al., 2016).

4.3.3.5.2 Coral reefs

Coral reefs are considered to be the marine ecosystem most threatened by climate-related ocean change, especially ocean warming and acidification, even under an RCP2.6 scenario (Gattuso et al., 2015; Albright et al., 2018; Hoegh-Guldberg et al., 2018; Díaz et al., 2019; Section 5.3.4). AR5 concluded that 'a number of coral reefs could [...] keep up with the maximum rate of SLR of 15.1 mm yr^{-1} projected for the end of the century [...] (*medium confidence*) [but a future net accretion rate lower] than during the Holocene (Perry et al., 2013) and increased turbidity (Storlazzi et al., 2011) will weaken this capability (*very high confidence*)' (Wong et al., 2014: 379). Subsequently, some studies suggested that SLR may have negligible impacts on coral reefs' vertical growth because the projected rate and magnitude of SLR by 2100 are within the potential accretion rates of most coral reefs (van Woesik et al., 2015). Other studies, however, stressed that the overall net vertical accretion of reefs may decrease after the first 30 years of rise in a 1.2 m SLR scenario (Hamylton et al., 2014), and that most reefs will not be able to keep up with SLR under RCP4.5 and beyond (Perry et al., 2018). The SR1.5 also concludes that coral reefs 'are projected to decline by a further 70–90% at 1.5°C (*high confidence*) with larger losses (>99%) at 2°C (*very high confidence*)' (Hoegh-Guldberg et al., 2018: 10). A key point is that SLR will not act in isolation of other drivers. Cumulative impacts, including anthropogenic drivers, are estimated to reduce the ability of coral reefs to keep pace with future SLR (Hughes et al., 2017; Yates et al., 2017) and thereby reduce the capacity of reefs to provide sediments and protection to coastal areas. For example, the combination of reef erosion due to acidification and human-induced mechanical destruction is altering seafloor topography, increasing risks from SLR in carbonate sediment dominated regions (Yates et al., 2017). Both ocean acidification (Albright et al., 2018; Eyre et al., 2018) and ocean warming (Perry and Morgan, 2017) have been considered to slow future growth rates and reef accretion (Section 5.3.4). Recent literature also shows that alterations of coral reef 3D structure from changes in growth, breakage, disease or acidification can profoundly affect their ability to buffer waves impacts (through wave breaking and wave energy damping), and therefore keep-up with SLR (Yates et al., 2017; Harris et al., 2018). Such prospects contribute to raise concerns about the future ability of atoll islands to adjust naturally to SLR and persist (Section 4.3.3.3, Cross-Chapter Box 9). Another concern is that locally, even minimal SLR can increase turbidity on fringing reefs, reducing light and, therefore, photosynthesis and calcification. SLR-induced turbidity can be caused by increased coastal erosion and the transfer of sediment to nearby reefs and enhanced sediment resuspension (Field et al., 2011).

4

4.3.3.5.3　Seagrasses

Due to their natural capacity to enhance accretion and in the absence of mechanical or chemical destruction by human activities, seagrasses are not expected to be severely affected by SLR, except indirectly through the increase of the impacts of extreme weather events and waves on coastal morphology (i.e., erosion) as well as through changes in light levels and through effects on adjacent ecosystems (Saunders et al., 2013). Extreme flooding events have also been shown to cause large-scale losses of seagrass habitats (Bandeira and Gell, 2003), for example seagrasses in Queensland, Australia, were lost in a disastrous flooding event (Campbell and McKenzie, 2004). Changes in ocean currents can have either positive or negative effects on seagrasses, creating new space for seagrasses to grow or eroding seagrass beds (Bjork et al., 2008). But overall, seagrass will primarily be negatively affected by the direct effects of increased sea temperature on growth rates and the occurrence of disease (Marba and Duarte, 2010; Burge et al., 2013; Koch et al., 2013; Thompson et al., 2015; Chefaoui et al., 2018; Gattuso et al., 2018; Section 5.3.2) as well as by heavy rains that may dilute the seawater to a lower salinity. Such impacts will be exacerbated by major causes of seagrass decline including coastal eutrophication, siltation and coastal development (Waycott et al., 2009). Noteworthy is that some positive impacts are expected, as ocean acidification is expected to benefit photosynthesis and growth rates of seagrass (Repolho et al., 2017).

4.3.3.5.4　Coastal protection by coastal and marine ecosystems

Major 'protection' benefits derived from the above-mentioned coastal ecosystems include wave attenuation and shoreline stabilisation, for example, by coral reefs (Elliff and Silva, 2017; Siegle and Costa, 2017), mangroves (Zhang et al., 2012; Barbier, 2016; Menéndez et al., 2018) or salt marshes (Möller et al., 2014; Hu et al., 2015). Recently, a global meta-analysis of 69 studies demonstrated that, on average, these ecosystems together reduced wave heights between 35–71% at the limited locations considered (Narayan et al., 2016), with coral reefs, salt marshes, mangroves and seagrass/kelp beds reducing wave heights by 54–81%, 62–79%, 25–37% and 25–45% respectively (see Narayan et al., 2016 for map of locations considered). Additional studies suggest greater wave attenuation in mangrove systems (Horstman et al., 2014), and highlight broader complexities in wave attenuation related to total tidal wetland extent, water depth, and species. Global analyses show that natural and artificial seagrasses can attenuate wave height and energy by as much as 40% and 50%, respectively (Fonseca and Cahalan, 1992; John et al., 2015), while coral reefs have been observed to reduce total wave energy by 94–98% (n = 13; Ferrario et al., 2014) and wave driven flooding volume by 72% (Beetham et al., 2017). In addition, storm surge attenuation based on a recent literature review by Stark et al. (2015) range from -2–25 cm km^{-1} length of marsh, where the negative value denotes actual amplification. Other ecosystems provide coastal protection, including macroalgae, oyster and mussel beds, and also beaches, dunes and barrier islands, but there is less understanding of the level of protection conferred by these other organisms and habitats (Spalding et al., 2014).

While there is little literature on the extent to which SLR specifically will affect coastal protection by coastal and marine ecosystems, it is estimated that SLR may reduce this ecosystem service (*limited evidence, high agreement*) through the above-described impacts on the ecosystems themselves, and in combination with the impacts of other climate-related changes to the ocean (e.g., ocean warming and acidification; Sections 5.3.1 to 5.3.6, 5.4.1). Wave attenuation by coral reefs, for example, is estimated to be negatively affected in the near future by changes in coral reefs' structural complexity more than by SLR (Harris et al., 2018); changes in mean and ESL events will rather add a layer of stress. Beck et al. (2018) estimate that under RCP8.5 by 2100, a 1 m loss in coral reefs' height will increase the global area flooded under a 100-year storm event by 116% compared to today, against +66% with no reef loss.

4.3.3.6　Human Activities

4.3.3.6.1　Coastal agriculture

SLR will affect agriculture mainly through land submergence, soil and fresh groundwater resources salinisation, and land loss due to permanent coastal erosion, with consequences on production, livelihood diversification and food security, especially in heavily coastal agriculture-dependent countries such as Bangladesh (Khanom, 2016). Recent literature confirms that salinisation is already a major problem for traditional agriculture in deltas (Wong et al., 2014; Khai et al., 2018) and low-lying island nations where some edible cultivated plants such as taro patches are threatened (Nunn et al., 2017b). Taking the case of rice cultivation, recent works emphasise the prevailing role of combined surface elevation and soil salinity, such as in the Mekong delta (Vietnam; Smajgl et al., 2015) and in the Ebro delta (Spain; Genua-Olmedo et al., 2016), estimating for the latter a decrease in the rice production index from 61.2% in 2010 to 33.8% by 2100 in a 1.8 m SLR scenario. For seven wetland species occurring in coastal freshwater marshes in central Veracruz on the Gulf of Mexico, an increase in salinity was shown to affect the germination process under wetland salt intrusion (Sánchez-García et al., 2017). In coastal Bangladesh, oilseed, sugarcane and jute cultivation was reported to be already discontinued due to challenges to cope with current salinity levels (Khanom, 2016), and salinity is projected to have an unambiguously negative influence on all dry-season crops over the next 15–45 years (especially in the southwest; Clarke et al., 2018; Kabir et al., 2018). Salinity intrusion and salinisation can trigger land use changes towards brackish or saline aquaculture such as shrimp or rice-shrimp systems with impacts on environment, livelihoods and income stability (Renaud et al., 2015). However, increasing salinity is only one of the land use change drivers along with, for example, policy changes and market prices at the household level (Renaud et al., 2015).

4.3.3.6.2　Coastal tourism and recreation

SLR may significantly affect tourism and recreation through impacts on landscapes (e.g., beaches), cultural features (e.g., Marzeion and Levermann, 2014; Fang et al., 2016), and critical transportation infrastructures such as harbours and airports (Monioudi et al., 2018). Coastal areas' future tourism and recreation attractiveness will

4

however also depend on changes in air temperature, seasonality and sea surface temperature (including induced effects such as invasive species, e.g., jellyfishes, and disease spreading; Burge et al., 2014; Weatherdon et al., 2016; Hoegh-Guldberg et al., 2018; Section 5.4.2). Future changes in climatic conditions in tourists' areas of origin will also play a role in reshaping tourism flows (Bujosa and Rosselló, 2013; Amelung and Nicholls, 2014), in addition to mitigation policies on air transportation, non-climatic features (e.g., accommodation and travel prices) and tourists' and tourism developers' perceptions of climate-related changes (Shakeela and Becken, 2015). Since AR5, forecasting the consequences of climate change effects on global-to-local tourism flows has remained challenging (Rosselló-Nadal, 2014; Wong et al., 2014; Hoegh-Guldberg et al., 2018). There are also concerns about the effect of SLR on tourism facilities, for example hotels in Ghana (Sagoe-Addy and Addo, 2013), in a context where tourism infrastructure often contributes to the degradation of natural buffering environments through, for example, coastal squeeze (e.g., Section 4.3.2.4) and human-driven coastal erosion. Again, forecasting is constrained by the lack of scientific studies on tourism stakeholders' long-term strategies and adaptive capacity (Hoogendoorn and Fitchett, 2018).

4.3.3.6.3   Coastal fisheries and aquaculture

Recent studies support the AR5 conclusion that ocean warming and acidification are considered more influential drivers of change in fisheries and aquaculture than SLR (Larsen et al., 2014; Nurse et al., 2014; Wong et al., 2014). The negative effects of SLR on fisheries and aquaculture are indirect, through adverse impacts on habitats (e.g., coral reef degradation, reduced water quality in deltas and estuarine environments, soil salinisation, etc.), as well as on facilities (e.g., damage to small and large harbours). This makes future projections on SLR implications for coastal and marine fisheries and aquaculture an understudied field of research. Conclusions only state that future impacts will be highly context-specific due to local manifestations of SLR and local fishery-dependent communities' ability to adapt to alterations in fish and aquaculture conditions and productivity (Hollowed et al., 2013; Weatherdon et al., 2016). Salinity intrusion also contributes to conversion of land or freshwater ponds to brackish or saline aquaculture in many low-lying coastal areas of Southeast Asia such as in the Mekong Delta in Vietnam (Renaud et al., 2015).

4.3.3.6.4   Social values

Social values refer to what people consider of critical importance about the places in which they live, and range from material to immaterial things (assets, beliefs, etc.; Hurlimann et al., 2014; Rouse et al., 2017). Consideration of social values offers an opportunity to address a wider perspective on impacts on human systems, for example, complementary to quantitative assessments of health impacts (e.g., loss of source of calories and food insecurity; Keim, 2010). This also encompasses immaterial dimensions, such as threats to cultural heritage (Marzeion and Levermann, 2014; Fatorić and Seekamp, 2017a), socialising activities (Karlsson et al., 2015), integration of marginalised groups (Maldonado, 2015) and cultural ecosystem services (Fish et al., 2016), and provides an opportunity to

better reflect context-specificities in valuing the physical/ecological/human/cultural impacts' importance for and distribution within a given society (Fatorić and Seekamp, 2017b). This field of research (no detailed mention found in AR5) is just emerging due to the transdisciplinary and qualitative nature of the topic. Graham et al. (2013) advance a 5-category framing of social values specifically at risk from SLR: health (i.e., the social determinants of survival such as environmental and housing quality and healthy lifestyles), feeling of safety (e.g., financial and job security), belongingness (i.e., attachment to places and people), self-esteem (e.g., social status or pride that can be affected by coastal retreat), and self-actualisation (i.e., people's efforts to define their own identity). Another emerging issue relates to social values at risk due to land submergence in low-lying islands (Yamamoto and Esteban, 2014) and parts of countries and individual properties (Marino, 2012; Maldonado et al., 2013; Aerts, 2017; Allgood and McNamara, 2017). Recent studies also highlight the potential additional risks to social values in areas where displaced people relocate (Davis et al., 2018).

4.3.4   Conclusion on Coastal Risk: Reasons for Concern and Future Risks

SLR projections for the 21st century, together with other ocean related changes (e.g., acidification and warming) and the possible increase in human-driven pressures at the coast (e.g., demographic and settlement patterns), make low-lying islands, coasts and communities relevant illustrations of some of the five Reasons for Concern (RFCs) developed by the IPCC since the Third Assessment Report (McCarthy et al., 2001; Smith et al., 2001) to assess risks from a global perspective. The AR5 Synthesis Report (IPCC, 2014) as well as the more recent SR1.5 (Hoegh-Guldberg et al., 2018) refined the RFC approach. The AR5 Synthesis Report (IPCC, 2014) developed two additional RFCs related to the coasts, subsequently updated along with the other RFCs (O'Neill et al., 2017). One refers to risks to marine species arising from ocean acidification, and the other one refers to risks to human and natural systems from SLR. Despite the difficulty in attributing observed impacts to SLR per se (Section 4.3.3.1), O'Neill et al. (2017) estimate that risks related to SLR are already detectable globally and will increase rapidly, so that high risk may occur before a 1m rise level is reached. O'Neill et al. (2017) also suggest that limits to coastal protection and EbA by 2100 could occur in a 1 m SLR rise scenario. Previous assessments however left gaps, including quantifying the benefits from adaptation in terms of risk reduction.

4.3.4.1   Methodological Advances

Rather than revisiting the AR5 and O'Neill et al. (2017) assessments from the particular perspective of risk related to SLR and for the global scale, this section provides a complementary perspective by assessing risks for specific geographies (resource-rich coastal cities, urban atoll islands, large tropical agricultural deltas and selected Arctic communities), based on the methodological advances below.

Scale of analysis and geographical scope – To date, the RFCs and associated burning embers have been developed at a global scale (Oppenheimer et al., 2014; Gattuso et al., 2015; O'Neill et al., 2017) and do not address the spatial variability of risk highlighted in this report (Sections 4.3.2.7, 4.3.4, 5.3.7, Cross-Chapter Box 9, Box 4.1). In addition, assessments usually identify risks either for global human dimensions (e.g., to people, livelihood, breakdown of infrastructures, biodiversity, global economy, etc.; IPCC, 2014; Oppenheimer et al., 2014; O'Neill et al., 2017) or for ecosystems and ecosystem services (Gattuso et al., 2015; Hoegh-Guldberg et al., 2018) (Section 5.3.7). This section moves the focus from the global to more local scales by considering four generic categories of low-lying coastal areas (Figure 4.3, Panel B): selected Arctic communities remote from regions of rapid GIA, large tropical agricultural deltas, urban atoll islands, and resource-rich coastal cities. Each of these categories is informed by several real-world case studies.

Risks considered – In line with the AR5 (IPCC, 2014), current and future risks result from the interaction of SLR-related hazards with the vulnerability of exposed ecosystems and societies. According to the specific scope of the chapter, this assessment focusses on the additional risks due to SLR and does not account for changes in extreme event climatology. Hazards considered are coastal flooding (Section 4.3.4.2), erosion (Section 4.3.4.3) and salinisation (Section 4.3.4.4). The proxies used to describe exposure and vulnerability are the density of assets at the coast (Section 4.3.2.2) and the level of degradation of natural buffering by marine and terrestrial ecosystems (Sections 4.3.2.3, 4.3.3.5.4, and 5.3.2 to 5.3.4). The assessment especially addresses risks to human assets at the coast, including populations, infrastructures and livelihoods. Specific metrics were developed (see SM4.3 for details), and their contribution to present-day observed impacts and to end-century risk have been assessed based on the authors' expert judgment and a methodological grid presented in SM4.3 (SM4.3.1 to SM4.3.6). The author's expert judgment draws on Sections 4.3.3.2 to 4.3.3.5 as well as additional literature for local scale perspectives (SM4.3.9).

Sea level rise scenarios – Based on the updates for ranges and mean values developed in this chapter (Section 4.2, Table 4.3), this assessment considers the end-century GMSL (2100) relative to 1986–2005 levels for two scenarios, SROCC RCP2.6 and SROCC RCP8.5. Both mean values and the SROCC RCP8.5 upper end of the *likely* range are used to assess risk transitions (Figure 4.3, Panel A). For the sake of readability, the following values were used: 43 cm (mean SROCC RCP2.6), 84 cm (mean SROCC RCP8.5) and 110 cm (SROCC RCP8.5 upper end of *likely* range). While GMSL serves as a representation of different possible climate change scenarios (see Panel A in Figure 4.3, Section 4.1.2), the assessment of additional risks due to SLR on specific geographies is developed against end-century relative SLR (RSL) in order to allow a geographically accurate approach (Panel B, Figure 4.3). Accordingly, risk was assessed to illustrative geographies based on RSLs for each of the two SROCC RCP scenarios and each of the real-world case studies to (SM4.3.6 and Table SM4.3.2; see dotted lines in Panel B of Figure 4.3). RSL observations include some or all of the following VLMs: both uplift (e.g., due to tectonics) and subsidence due to natural (e.g., tectonics, sediment compaction) and human (e.g., oil/gas/water extraction, mining activities) factors, as well as to GIA. However, in SROCC, numerical RSL projections only include GIA and the regional gravitational, rotational, and deformational responses (GRD, see Section 4.2.1.5) to ice mass loss.

The main reason is the difficulty of projecting the influence on some factors such as human interventions to the end of the century.

Adaptation scenarios – Risk will also depend on the effectiveness of coastal societies' responses to both extreme events and slow onset changes. To capture the response dimension, four metrics have been considered that refer to the implementation of adequately calibrated hard, engineered coastal defences (Section 4.4.2.2), the restoration of the degraded ecosystems or the creation of new natural buffers areas (Section 4.4.2.2 and 4.4.2.3), planned and local-scale relocation (Section 4.4.2.6), and measures to limit human-induced subsidence (Sections 4.4.2.2, 4.4.2.5). On these bases, two contrasting adaptation scenarios were considered. The first one is called 'No-to-moderate response' (see (A) bars in Panel B, Figure 4.3) and represents a business-as-usual scenario where no major additional adaptation efforts compared to today are implemented. That is, neither substantial intensification of current actions nor new types of actions, e.g., only moderate raising of existing protections in high-density areas or sporadic episodes of relocation or beach nourishment where largescale efforts are not already underway. The second one, called 'Maximum potential response' (bars (B) in Figure 4.3), refers to an ambitious combination of both incremental and transformational adaptation (i.e., significantly upscaled effort); for example, relocation of entire districts or raised protections in some cities, or creation/ restoration at a significant scale of beach-dune systems including indigenous vegetation.

4.3.4.2 Key Findings on Future Risks and Adaptation Benefits

4.3.4.2.1 Future risks

The findings suggest that risks from SLR are already detectable for all of the geographies considered (Panel B in Figure 4.3), and that risk is expected to increase over this century in virtually all low-lying coastal areas whatever their context-specificities or nature (island/continental, developed/developing county) (Cross-Chapter Box 9). In the absence of high adaptation (bars (A)), risk is expected to significantly increase in urban atoll islands and the selected Arctic coastal communities even in a SROCC RCP2.6 scenario, and all geographies are expected to experience almost high to very high risks at the upper *likely* range of SROCC RCP8.5. These results allow refining AR5 conclusions by showing, first, that high risk can indeed occur before the 1m rise benchmark (Oppenheimer et al., 2014; O'Neill et al., 2017) and, second, that risk as a function of SLR is highly variable from one geography to another. Some rationale is provided below for our assessment of illustrative geographies, summarising the more detailed description provided in SM4.3 (SM4.3.6 to SM4.3.8). Note however that the text below is not intended to be fully comprehensive and does not necessarily include all elements for which there is a substantive body of literature, nor does it necessarily include all elements which are of particular interest to decision makers.

Resource-rich coastal cities (SM4.3.8.1, Panel B in Figure 4.3) – Resource-rich coastal cities considered in this analysis are Shanghai, New York (see Box 4.1 for further details and references on Shanghai and NYC), and Rotterdam (Brinke et al., 2010; Hinkel et al., 2018). High, and in many cases, growing population density and total population, and high exposure of people and infrastructure to GMSL rise and ESL events characterise coastal megacities (Hanson et al., 2011). These are high concentrations of income and wealth in geographic terms but within relatively small area exhibit large distributional differences of both with important implications for emergency response and adaptation. Concentration translates into high exposure of monetary value to coastal hazards and the cities noted here have both historical and recent experience with damaging ESL events, such as Typhoon Winnie which struck Shanghai in 1997 (Xian et al., 2018), Hurricane Sandy in New York in 2012 (Rosenzweig and Solecki, 2014), and the North Sea storm of 1953 which impacted the Rotterdam area (Gerritsen, 2005; Jonkman et al., 2008). However, high density, limited space and high cost of land leads to development of below-ground space for transportation (e.g., subways, road tunnels; MTA, 2017) and storage, and even habitation, creating vulnerabilities not seen in low-density areas. Natural ecosystems within the megacity boundaries and nearby have been exploited for centuries and in some cases decimated or even extirpated (Hartig et al., 2002). Accordingly, they provide limited benefits in terms of coastal protection for the densest part of these cities but can be critically important for protection of lower-density areas, for example, wetlands and sandy beaches in the Jamaica Bay/Rockaway sector of New York that protect nearby residential communities (Hartig et al., 2002). Space limitations also constrain the potential benefits of EbA measures. Instead, resource-rich coastal cities depend largely on hard defences like sea walls and surge barriers for coastal protection (Section 4.4.2.2). Such defences are costly but generally cost effective due to the aforementioned concentration of population and value. However, barriers to planning and implementing adaptation include governance challenges (Section 4.4.2) such as limited control over finances and the intermittent nature of ESLs which inhibit focused attention over the long time scales needed to plan and implement hard defences (Section 4.4.2.2). As a result, coastal adaptation for resource-rich cities is uneven and the three presented here were selected with a view toward exhibiting a range of current and potential future effectiveness.

Urban atoll islands (SM4.3.8.2, Panel B in Figure 4.3) – The capital islands (or groups of islands) of three atoll nations in the Pacific and Indian Oceans are considered here: Fongafale (Funafuti Atoll, Tuvalu), the South Tarawa Urban District (Tarawa Atoll, Kiribati) and Male' (North Kaafu Atoll, Maldives). Urban atoll islands have low elevation (<4 m above mean sea level; in South Tarawa, e.g., lagoon sides where settlement concentrates are <1.80 m in elevation) (Duvat, 2013) and are mainly composed of reef-derived unconsolidated material. Their future is of nation-wide importance as they concentrate populations, economic activities and critical infrastructure (airports, main harbours). They illustrate the prominence of anthropogenic-driven disturbances to marine and terrestrial ecosystems (e.g., mangrove clearing in South Tarawa or human-induced coral reef degradation through land reclamation in Male'; Duvat et al., 2013; Naylor, 2015) and therefore to services such as coastal protection delivered by the coral reef (i.e., wave energy attenuation that reduces flooding and erosion, and sediment provision that contributes to island persistence over time) (McLean and Kench, 2015; Quataert et al., 2015; Elliff and Silva, 2017; Storlazzi et al., 2018).

The controlling factors of urban atoll islands' future habitability are the density of assets exposed to marine flooding and coastal erosion (SM4.3.8.2), future trends in these hazards, and ecosystem response to both ocean-climate related pressures and human activities. Urban atoll islands already experience coastal flooding, for example, in Male' (Wadey et al., 2017) and Funafuti (Yamano et al., 2007; McCubbin et al., 2015). Coastal erosion is also a major concern along non-armoured shoreline in South Tarawa (Duvat et al., 2013) and Fongafale (Onaka et al., 2017), but not in Male' where surrounding fortifications have extended along almost the entire shoreline from several decades (Naylor, 2015). Salinisation already affects groundwater lenses, but its contribution to risk varies from one case to another, from low in Male' (relying on desalinised seawater) to important for human consumption and agriculture in South Tarawa (Bailey et al., 2014; Post et al., 2018).

Together, high population densities (from ~3,200 people per km^2 in South Tarawa to ~65,700 people per km^2 in Male') (Government of the Maldives, 2014; McIver et al., 2015) and the concentration of critical infrastructure and settlements in naturally low-lying flood-prone areas already substantially contribute to coastal risk (Duvat et al., 2013; Field et al., 2017). Even stabilised densities in the future would translate into a substantial increase of risk under a 43cm GMSL rise. Risk will also be exacerbated by the negative effects of ocean warming and acidification, especially on coral reef and mangrove capacity to cope with SLR (Pendleton et al., 2016; Van Hooidonk et al., 2016; Perry and Morgan, 2017; Perry et al., 2018) (Sections 4.3.3.5, 5.3). In addition, even small values of SLR will significantly increase risk to atoll islands' aquifers (Bailey et al., 2016; Storlazzi et al., 2018). Finally, land scarcity in atoll environments will exacerbate the importance of SLR induced damages (on housing, agriculture and infrastructure especially) and cascading impacts (on livelihoods, for example, as a result of groundwater and soil salinisation).

Large tropical agricultural deltas (SM4.3.8.3, Panel B in Figure 4.3) – River deltas considered in this analysis are the Mekong Delta and the Ganges-Brahmaputra-Meghna Delta. Both deltas are large, low-lying and dominated by agricultural production. The risk assessment to SLR considered the entire delta area (not only the coastal fringe; see SM4.3.6 for explanation). High population densities (1280 people per km^2 and 433 people per km^2 in the Ganges-Brahmaputra-Meghna and Mekong deltas, respectively) (Ericson et al., 2006; Government of the Maldives, 2014) and the removal of natural vegetation buffers contribute to high exposure rates to coastal flooding, erosion, and salinisation. Agricultural production contributes to GDP strongly (Smajgl et al., 2015; Hossain et al., 2018), making agricultural fields important assets. In both deltas, mangroves are partially degraded (Ghosh et al., 2018; Veettil et al., 2018) as well as other wetlands at the coast and further inland (Quan et al., 2018a; Rahman et al., 2018). Currently, riverine flooding dominates in both deltas (Auerbach et al., 2015; Rahman and Rahman, 2015; Ngan et al., 2018). However, high tides and cyclones can generate large coastal flooding events, especially in the Ganges-Brahmaputra-Meghna Delta (Auerbach et al., 2015; Rahman and Rahman, 2015). Human-induced subsidence increases the likelihood of flooding in both deltas (Brown et al., 2018b). Coastal and river bank erosion is already a problem in both delta

4

(Anthony et al., 2015; Brown and Nicholls, 2015; Li et al., 2017) as well as salinity intrusion, which is impacting coastal aquifers, soils and surface waters (Anthony et al., 2015; Brown and Nicholls, 2015; Li et al., 2017). Salinisation of water and soil resources remains a coastal phenomenon (Smajgl et al., 2015), but salinity intrusion can reach far inland in some extreme years and significantly contribute to risk at the delta scale (Section 4.3.3.4.2). Both deltas are partly protected with hard engineered defences such as dikes and sluice gates to prevent riverine flooding, and polders and dikes in some coastal stretches to prevent salinity intrusion and storm surges (Smajgl et al., 2015; Rogers and Overeem, 2017; Warner et al., 2018a). Today, in both deltas, the measures implemented to restore natural buffers are still limited to mangroves ecosystems (Quan et al., 2018a; Rahman et al., 2018), and the measures aiming at reducing subsidence are underdeveloped (Schmidt, 2015; Schmitt et al., 2017). Assuming stable population densities in the future, coastal flooding will contribute increasingly to risk at the delta level (Brown and Nicholls, 2015; Brown et al., 2018a; Dang et al., 2018). Coastal erosion will increase (Anthony et al., 2015; Liu et al., 2017a; Uddin et al., 2019) and salinisation of coastal waters and soils will be more significant (Tran Anh et al., 2018; Vu et al., 2018; Rakib et al., 2019) and will strongly impact agriculture and water supply for the entire delta (Jiang et al., 2018; Timsina et al., 2018; Nhung et al., 2019). Without increased adaptation, coastal ecosystems will be largely destroyed at 110 cm of SLR (Schmitt et al., 2017; Mehvar et al., 2019; Mukul et al., 2019). Given the size of these deltas, it is only under high emission scenarios, that flooding, erosion and salinisation lead to high risk at the entire delta scale.

Arctic communities (SM4.3.8.4, Panel B in Figure 4.3) – Five small indigenous settlements located on the Arctic Coastal Plain are considered in this analysis: Bykovsky (Lena Delta, Russian Federation), Shishmaref and Kivalina (Alaska, USA), and Shingle Point and Tuktoyaktuk (Mackenzie Delta, Canada). They lie on exposed coasts composed of unlithified ice-rich sediments in permafrost, in areas with seasonal sea ice and slow to moderate SLR. These communities have populations ranging from 380 to 900 (fewer and seasonal at Shingle Point) that are heavily dependent on marine subsistence resources (Forbes, 2011; Ford et al., 2016a). Shishmaref and Kivalina are located on low-lying barrier islands highly susceptible to rising sea level (Marino, 2012; Bronen and Chapin, 2013; Fang et al., 2018; Rolph et al., 2018). Shingle Point is situated on an active gravel spit; Tuktoyaktuk is built on low ground with high concentrations of massive ice; and Bykovsky is mostly situated on an ice-rich eroding terrace about 20 m above sea level. All the selected communities are remote from regions of rapid positive GIA; many other areas in the Arctic experience rapid GIA uplift (James et al., 2015; Forbes et al., 2018) and have very low sensitivity to SLR, which may in fact help to reduce shoaling.

Especially in the Arctic, anthropogenic drivers in recent decades resulted in the induced settlement of indigenous peoples in marginalised climate-sensitive communities (Ford et al., 2016b) and the construction of infrastructure in nearshore areas, with the assumption of stable coastlines. This resulted in increased exposure to coastal hazards. Coastal erosion is already a major problem in all of the case studies, where space for building is usually limited.

Accelerating permafrost thaw is promoting rapid erosion of ice-rich sediments, e.g., at Bykovsky (Myers, 2005; Lantuit et al., 2011; Vanderlinden et al., 2018) and Tuktoyaktuk (Lamoureux et al., 2015; Ford et al., 2016a). Related to this, Kivalina, Shishmaref, Shingle Point, Tuktoyaktuk, and parts of the Lena delta (less so for Bykovsky) are already facing high risk of flooding. Shishmaref, for example, experienced 10 flooding events between 1973 and 2015 that resulted in emergency declarations (Bronen and Chapin, 2013; Lamoureux et al., 2015; Irrgang et al., 2019). There is however no evidence of salinisation in the selected communities, but brackish water flooding of the outer Mackenzie Delta caused by a 1999 storm surge (a rare event due to upwelling ahead of the storm) led to widespread die-off of vegetation with negative ecosystem impacts (Pisaric et al., 2011; Kokelj et al., 2012).

Permafrost thaw is already accelerating due to increasing ground temperatures that weaken the mechanical stability of frozen ground (Section 3.4.2.2). Arctic SLR and sea surface warming have the potential to substantially contribute to this thawing (Forbes, 2011; Barnhart et al., 2014b; Lamoureux et al., 2015; Fritz et al., 2017). An additional factor unique to the polar regions is the decrease in seasonal sea ice extent in the Arctic (Sections 3.2.1 and 3.2.2), which together with a lengthening open water season, provides less protection from storm impacts, particularly later in the year when storms are prevalent (Forbes, 2011; Lantuit et al., 2011; Barnhart et al., 2014a; Melvin et al., 2017; Fang et al., 2018; Forbes, 2019) and therefore reduces the physical protection of the land (Section 6.3.1.3).

4.3.4.2.2 Adaptation benefits

The assessment also shows that benefits in terms of risk reduction over this century are to be expected from ambitious adaptation efforts (bars (B), Sections 4.4.2, 4.4.3 and 4.4.3). In the case of resource-rich coastal cities especially, adequately engineered coastal defences can play a decisive role in reducing risk (Section 4.4.2.2, Box 4.1), for example from high to moderate at the SROCC RCP8.5 upper *likely* range. In other contexts, such as atoll islands for example, while engineered protection structures will reduce risk of flooding, they will not necessarily prevent seawater infiltration due to the permeable nature of the island substratum. So even adequate coastal protection would not eliminate risk (SM4.3.8.3). In urban atoll islands, large tropical agricultural deltas and the selected Arctic communities, ambitious adaptation efforts mixing adequate coastal defences, the restoration and creation of buffering ecosystems (e.g., coral reefs), and a moderate amount of relocation are expected to reduce risk. For resource-rich coastal cities, adequately engineered hard protection can virtually eliminate risk of flooding up to 84 cm except for residual risk of structural failure (Sections 4.4.2 to 4.4.5). Benefits are relatively important in a 84 cm SLR scenario, as they reduce risk from high-to-very-high to moderate-to-high (atolls, Arctic) and from moderate-to-high to moderate (deltas). These benefits become more modest when approaching the upper *likely* range of SROCC RCP8.5, and risk tends to return to high-to-very-high (atolls, Arctic) levels once the 110 cm rise in sea level is reached. Noteworthy in urban atoll islands, intensified proactive coastal relocation (e.g., relocation of buildings and infrastructures that are very close to the shoreline) is expected to play a substantial role in risk reduction under all SLR scenarios. Proactive relocation can indeed compensate for the increasing extent of coastal flooding and associated damages (SM4.3.8.3). When taken to the extreme, relocation could lead to the elimination of risk in situ, for example in the case of the relocation of the full population of urban atoll islands either elsewhere in the country (e.g., on another island) or abroad (i.e., international migration). This is an extreme situation where it is hard to distinguish whether the measure is an impact of SLR (and ocean change more broadly), for example, displacement, or an adaptation solution. In addition, relocation of people displaces pressure to destination areas, with a potential increase of risk for the latter. In other words, the broader 'coastal retreat' category (Section 4.4.2.6) raises the issue of the 'limits to adaptation', which is not represented in Figure 4.3.

These conclusions must be nuanced, first, by the fact that our assessment does not consider either financial or social aspects that can act as limiting factors to the development of adaptation options (Sections 4.4.3 and 4.4.5), for instance, hard engineering coastal defences (Hurlimann et al., 2014; Jones et al., 2014; Elrick-Barr et al., 2017; Hinkel et al., 2018). However, from a general perspective, these findings suggest that although ambitious adaptation will not necessarily eradicate end-century risk from SLR across all low-lying coastal areas around the world, it will help to buy time in many locations and therefore contribute to developing a robust foundation for adaptation beyond 2100. Second, the future of other climate-related drivers of risk (such as ESL, waves and cyclones; Sections 4.2.3.4.1 to 4.2.3.4.3, 6.3.1.1 to 6.3.1.3) is not fully and systematically included in each risk assessment above, so that much larger risks than assessed here are to be expected.

4.4 Responding to Sea Level Rise

4.4.1 Introduction

SLR responses refer to legislation, plans and actions undertaken to reduce risk and build resilience in the face of SLR (see Cross-Chapter Box 3 in Chapter 1). These responses range from protecting the coast, accommodating SLR impacts, retreating from the coast, advancing into the ocean by building seawards and EbA (Box 4.3). Identifying the most appropriate way to respond to SLR is not straightforward and is politically and socially contested with a range of governance challenges (also called barriers) arising. This section first assesses the post-AR5 literature on the different types of SLR responses (i.e., protection, accommodation, advance, retreat and EbA) in terms of their effectiveness, technical limits, costs, benefits, co-benefits, drawbacks, economic efficiency and barriers, and the specific governance challenges associated with each type of response (Section 4.4.2). It then identifies a set of overarching governance challenges that arise from the nature of SLR, such as its long-term commitment and uncertainty, and the associated politically and socially contested choices that need to be made (Section 4.4.3). Next, planning, public participation, conflict resolution and decision analysis approaches and tools are assessed that, when applied in combination, can help to address the governance challenges identified, facilitating social choices about SLR responses (Section 4.4.4). Finally, enablers and lessons learned from practical efforts to implement SLR responses are assessed (Section 4.4.5), concluding with a synthesis emphasising the utility of climate resilient development pathways (Section 4.4.6).

Box 4.3 | Responses to Sea Level Rise

Protection reduces coastal risk and impacts by blocking the inland propagation and other effects of mean or extreme sea levels (ESL). This includes: i) **hard protection** such as dikes, seawalls, breakwaters, barriers and barrages to protect against flooding, erosion and salt water intrusion (Nicholls, 2018), ii) **sediment-based protection** such as beach and shore nourishment, dunes (also referred to as soft structures), and iii) ecosystem-based adaptation (EbA) (see below). The three subcategories are often applied in combination as so-called hybrid measures. Examples are a marsh green-belt in front of a sea wall, or a sea wall especially designed to include niches for habitat formation (Coombes et al., 2015).

Accommodation includes diverse biophysical and institutional responses that mitigate coastal risk and impacts by reducing the vulnerability of coastal residents, human activities, ecosystems and the built environment, thus enabling the habitability of coastal zones despite increasing levels of hazard occurrence. Accommodation measures for erosion and flooding include building codes, raising house elevation (e.g., on stilts), lifting valuables to higher floors and floating houses and gardens (Trang, 2016). Accommodation measures for salinity intrusion include changes in land use (e.g., rice to brackish/salt shrimp aquaculture) or changes to salt tolerant crop varieties. Institutional accommodation responses include EWS, emergency planning, insurance schemes and setback zones (Nurse et al., 2014; Wong et al., 2014).

Advance creates new land by building seaward, reducing coastal risks for the hinterland and the newly elevated land. This includes land reclamation above sea levels by land filling with pumped sand or other fill material, planting vegetation with the specific intention to support natural accretion of land and surrounding low areas with dikes, termed polderisation, which also requires drainage and often pumping systems (Wang et al., 2014; Donchyts et al., 2016).

Retreat reduces coastal risk by moving exposed people, assets and human activities out of the coastal hazard zone. This includes the following three forms: i) **Migration,** which is the voluntary permanent or semi-permanent movement by a person at least for one year (Adger et al., 2014). ii) **Displacement,** which refers to the involuntary and unforeseen movement of people due to environment-related

4

Box 4.3 (continued)

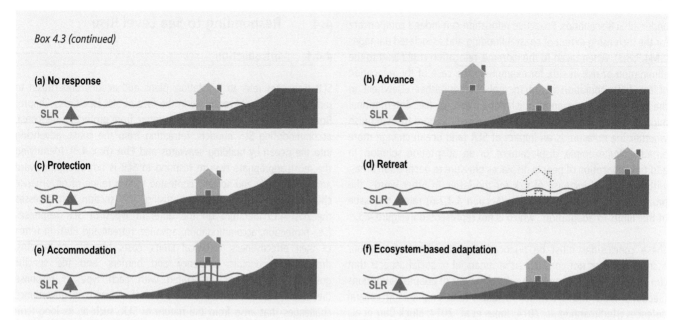

Box 4.3, Figure 1 | Different types of responses to coastal risk and sea level rise (SLR).

impacts or political or military unrest (Black et al., 2013; Islam and Khan, 2018; McLeman, 2018; Mortreux et al., 2018). iii) **Relocation**, also termed resettlement, managed retreat or managed realignment, which is typically initiated, supervised and implemented by governments from national to local levels and usually involves small sites and/or communities (Wong et al., 2014; Hino et al., 2017; Mortreux et al., 2018). Managed realignment may also be conducted for the purpose of creating new habitat. These three sub-categories are not neatly separable– any household's decision to retreat may be 'voluntary' in theory, but in practice, may result from very limited choices. Displacement certainly occurs in response to extreme events but some of those retreating may have other options. Relocation programs may rely on incentives such as land buyouts that households adopt voluntarily. The need for retreat and other response measures can be reduced by avoiding new development commitments in areas prone to severe SLR hazards (Section 4.4.4.2).

Ecosystem-based adaptation (EbA) responses provide a combination of protect and advance benefits based on the sustainable management, conservation and restoration of ecosystems (Van Wesenbeeck et al., 2017). Examples include the conservation or restoration of coastal ecosystems such as wetlands and reefs. EbA measures protect the coastline by (i) attenuating waves, and, in the case of wetlands storm surge flows, by acting as obstacles and providing retention space (Krauss et al., 2009; Zhang et al., 2012; Vuik et al., 2015; Rupprecht et al., 2017); and (ii) by raising elevation and reducing rates of erosion through trapping and stabilising coastal sediments (Shepard et al., 2011), as well as building-up of organic matter and detritus (Shepard et al., 2011; McIvor et al., 2012a; McIvor et al., 2012b; Cheong et al., 2013; McIvor et al., 2013; Spalding et al., 2014). EbA is also referred to by various other names, including Natural and Nature-based Features, Nature-based Solutions, Ecological Engineering, Ecosystem-based Disaster Risk Reduction or Green Infrastructure (Bridges, 2015; Pontee et al., 2016).

4.4.2   Observed and Projected Responses, their Costs, Benefits, Co-benefits, Drawbacks, Efficiency and Governance

4.4.2.1   Types of Responses and Framework for Assessment

Following earlier IPCC Reports Protection, Retreat and Accommodation responses to SLR and its impacts are distinguished between (Nicholls et al., 2007; Wong et al., 2014), and Advance is added as a fourth type of response that consists in building seaward and upward (Box 4.3). Advance had not received much attention in the climate change literature but plays an important role in coastal development across the world (e.g., Institution of Civil Engineers, 2010; Lee, 2014; Donchyts et al., 2016). The broader term response is used here instead of adaptation, because some responses such as

retreat may or may not be meaningfully considered to be adaptation (Hinkel et al., 2018). Responses that address the causes of climate change, such as mitigating GHGs or geoengineering temperature and sea level responses to emissions fall beyond the scope of this chapter, and are addressed in SR1.5 (Hoegh-Guldberg et al., 2018). In coastal areas where anthropogenic subsidence contributes to relative SLR, another important type of response is the management of subsidence by, for instance, restricting ground fluid abstraction. Although this type of measure is considered in the risk assessment developed in Section 4.3.4, it is not assessed here due to a lack of space.

Observed coastal responses are rarely responses to climate-change induced SLR only, but also to relative SLR caused by land subsidence as well as current coastal risks and many socioeconomic factors and related hazards. As a consequence, coastal responses have been practised for centuries, and there are many experiences specifically in places that have subsided up to several metres due to earthquakes or anthropogenic ground fluid abstraction in the last century that responding to climate-change induced SLR can draw upon (Esteban et al., 2019). Finally, in practise, many responses are hybrid, applying combinations of protection, accommodation, retreat, advance and EbA.

Since AR5, the literature on SLR responses has grown significantly. It is assessed in this section for the five above-described broad types of responses in terms of the following six criteria:

- **Observed responses** across geographies, describing where the different types of responses have been implemented.
- **Projected responses**, which refers to the potential extent of responses in the future, as assessed in the literature through modelling or in a more qualitative way.
- **Cost of responses**, which refers to the costs of implementing and maintaining responses. Other costs that arise due to negative side-effects of implementing a response are captured under the criterion 'co-benefits and drawbacks'.
- **Effectiveness of responses** in terms of reducing SLR risks and impacts. This includes biophysical and technical limits beyond which responses cease to be effective.
- **Co-benefits and drawbacks** of responses that occur next to the intended benefits of reducing SLR risks and impacts.
- **Governance challenges (or barriers)**, which refers to institutional and organisational factors that have been found to hinder the effective, efficient and equitable implementation of responses (see also Section 4.4.3).
- **Economic efficiency** of responses, which refers to the overall monetised balance of costs, benefits (in terms of the effectiveness of responses), co-benefits and drawbacks. Economic barriers arise if responses have a negative net benefit or a benefit-cost ratio smaller than one. While it would be desirable to have information on the economic efficiency of integrated responses combining different response types, an assessment cannot be provided here due to the lack of literature.

4.4.2.2 Hard and Sediment-Based Protection

4.4.2.2.1 *Observed hard and sediment-based protection across geographies*

Coastal protection through hard measures is widespread around the world, although it is difficult to provide estimates on how many people benefit from them. Currently, at least 20 million people living below normal high tides are protected by hard structures (and drainage) in countries such as Belgium, Canada, China, Germany, Italy, Japan, the Netherlands, Poland, Thailand, the UK, and the USA (Nicholls, 2010). Many more people living above high tides are also protected against ESL by hard structures in major cities around the world. There is a concentration of these measures in northwest Europe and East Asia, although extensive defences are also found in and around many coastal cities and deltas. For example, large scale coastal protection exists in Vancouver (Canada), Alexandria (Egypt) and Keta (Ghana; Nairn et al., 1999) and 6000 km of polder dikes

in coastal Bangladesh. Gittman et al. (2015) estimate that 14% of the total US coastline has been armoured, with New Orleans being an example of an area below sea level dependent on extensive engineered protection (Kates et al., 2006; Rosenzweig and Solecki, 2014; Cooper et al., 2016). Defences built and raised for tsunami protection, such as post-2011 in Japan (Raby et al., 2015), also provide protection against SLR.

The application of sediment-based protection measures also has a long history, offering multiple benefits in terms of enhancing safety, recreation and natural systems (JSCE, 2000; Dean, 2002; Hanson et al., 2002; Cooke et al., 2012). About 24% of the world's sandy beaches are currently eroding by rates faster than 0.5 m yr^{-1} (Luijendijk et al., 2018). In the USA, Europe and Australia, these responses are often driven by the recreational value of beaches and the high economic benefits associated with beach tourism. More recently, sediment-based measures are implemented as effective and yet flexible measures to address SLR (Kabat et al., 2009) and experiments are being conducted with innovative decadal scale application of sediments such as the sand engine in the Netherlands (Stive et al., 2013).

There is *high confidence* that most major upgrades in defences happen after coastal disasters (Box 4.1). Dikes were raised and reienforced after the devastating coastal flood of 1953 in the Netherlands and the UK, and in 1962 in Germany. In New Orleans, investments in the order of 15 billion USD, including a major storm surge barrier, followed Hurricane Katrina in 2005 (Fischetti, 2015), and in New York the Federal Government made available 16 billion USD for disaster recovery and adaptation after Superstorm Sandy in 2012 (NYC, 2015). Examples in which SLR has been considered proactively in the planning process include SLR safety margins in, for example, the UK, Germany and France, upgrading defences according to cost-benefit analysis in the Netherlands, and SLR guidance in the USA (USACE, 2011).

4.4.2.2.2 *Projected hard and sediment-based protection*

There is *high confidence* that hard coastal protection will continue to be a widespread response to SLR in densely populated and urban areas during the 21st century, because this response is widely practised (Section 4.4.2.2.2), effective in reducing current (Section 4.4.2.2.2) and future flood risk (Section 4.3.3.2) and highly cost efficient in urban and densely populated areas (Section 4.4.2.7). There is, however, *low agreement* on the level of hard coastal protections to expect, with projections being based on different assumptions. A model assuming that coastal societies upgrade hard protection following scenario-based cost-benefit analysis finds that 22% of the global coastline will be protected under various SSPs and 1 m of 21st century global mean SLR (Nicholls et al., 2019). Another model assuming that only areas for which benefit-cost ratios are above 1 under SLR scenarios up to 2 m, all SSPs and discount rates up to 6%, finds that this would lead to protecting 13% of the global coastline (Lincke and Hinkel, 2018; Figure 4.14).

Table 4.7 | Capital and maintenance costs of hard protection measures.

Measure	Capital cost (in million USD unless stated otherwise)	Annual Maintenance Cost (% of capital cost)
Sea Wall	0.4–27.5 per km length and metre height (Linham et al., 2010)	1–2% per annum (Jonkman et al., 2013)
Sea Dike	0.9–69.9 per km length and metre height (Jonkman et al., 2013; Nicholls et al., 2019; Tamura et al., 2019)	1–2% per annum (Jonkman et al., 2013)
Breakwater	2.5–10.0 per km length (Narayan et al., 2016)	1% per annum (Jonkman et al., 2013)
Storm Surge Barrier	0.9–2.7 (Jonkman et al., 2013) or 2.2 (Mooyaart and Jonkman, 2017) million EUR per metre width	1% per annum (Mooyaart and Jonkman, 2017) or 5–10% per annum (Nicholls et al., 2007)
Saltwater Intrusion Barriers	Limited knowledge	Limited knowledge

4.4.2.2.3   Cost of hard and sediment-based protection

There is *medium evidence* and *medium agreement* on the costs of hard protection. Data on the costs of hard defences is only available for few countries and unit costs estimated from this data vary substantially depending on building/fill material used, labour cost, urban versus rural settings, hydraulic loads, etc. (Jonkman et al., 2013; Lenk et al., 2017; Aerts, 2018; Nicholls et al., 2019). In general, there has been limited systematic data collection across sites, although useful national guidance does exist in some cases (Environment Agency, 2015). Defences depend on good maintenance to remain effective. For some types of infrastructure such as surge barriers, maintenance costs are poorly described and hence more uncertain (Nicholls et al., 2007). Protection-based adaptation to saltwater intrusion is more complex than adaptation to flooding and erosion, and there is less experience to draw upon.

Based on these unit cost estimates, and different assumptions on future protection, global annual protection costs have been estimated to be 12–71 billion USD considering coastal dikes only (Hinkel et al., 2014) and about 40–170 billion USD yr^{-1} considering coastal dikes, river dikes and storm surge barriers, under RCP2.6, and about 25–200 billion USD yr^{-1} considering coastal dikes only (Tamura et al. 2019) under RCP8.5. If protection is widely practised through the 21st century, the bulk of the costs will be maintenance rather than capital costs (Nicholls et al., 2019).

Sediment-based measures are generally costed as the unit cost of sand (or gravel) delivery multiplied by the volumetric demand. Unit costs range from 3–21 USD m^{-3} sand, with some high outlier costs in, for example, the UK, South Africa and New Zealand (Linham et al., 2010; Aerts, 2018). Costs are small where sources of sand are plentiful and close to the sites of demand. Costs are further reduced by shoreface nourishment approaches. The Netherlands maintains its entire open coast with large-scale shore nourishment (Mulder et al., 2011) and the innovative sand engine has been implemented as a full-scale decadal experiment (Stive et al., 2013). The capital costs for dunes are similar to beach nourishment, although placement and planting vegetation may raise costs. Maintenance costs vary from almost nothing to several million USD km^{-1}, although costs are usually at the lower end of this range (Environment Agency, 2015).

4.4.2.2.4   Effectiveness of hard and sediment-based protection

There is *high confidence* that well designed and maintained hard and sediment-based protection is very effective in reducing risk to the impacts of SLR and ESL (Horikawa, 1978; USACE, 2002; CIRIA, 2007). This includes situations in which coastal megacities in river deltas have experienced, and adapted to, relative SLR of several metres caused by land subsidence during the 20th century (Kaneko and Toyota, 2011; Esteban et al., 2019; Box 4.1). In principle, there are no technological limits to protect the coast during the 21st century even under high-end SLR of 2 m (Section 4.3.3.2), but technological challenges can make protection very expensive and hence unaffordable in some areas (Hinkel et al., 2018). Examples include southeast Florida, because protected areas can be flooded by rising groundwater through underlying porous limestone (Bloetscher et al., 2011). Gradually rising water tables behind defences is also an issue, which can be managed by increasing pumping and drainage (Aerts, 2018). Maintaining this effectiveness over time requires regular monitoring and maintenance, accounting for changing conditions such as SLR and widespread erosional trends in front of the defences. There will always be residual risks, which can be reduced, but never eliminated, by engineering protection infrastructure to very high standards, such as so-called 'unbreakable dikes' (de Bruijn et al., 2013).

It is difficult to assess at what point in time and for which amount of SLR technical limits for coastal protection will be reached. Parts of Tokyo have been protected against five metres of relative SLR during the 21st century (Kaneko and Toyota, 2011) and it has been argued that it is possible to preserve territorial integrity of the Netherlands even under 5 m SLR, using current engineering technology (Aerts et al., 2008; Olsthoorn et al., 2008). This suggests that under RCP2.6, technical limits to adaptation will be rare even under longer-term SLR. Protecting against high-end SLR will be increasingly technically challenging as we move beyond the 21st century. This is not only due to the absolute amount of SLR, but also due to the very high rates of annual SLR (e.g., 10–20 mm yr^{-1} *likely* range under RCP8.5 in 2100), which challenge the planning and implementation of hard protection because major protection infrastructure requires decades to plan and implement (Gilbert et al., 1984; Burcharth et al., 2014). In summary, the higher and faster SLR, the more challenging coastal protection will be, but quantifying this is difficult. In any case, before technical limits are reached, economic and social limits will be reached because societies are neither economically able nor socially willing to invest in coastal protection (Sections 4.4.2.2 and 4.3.3.2; Hinkel et al., 2018; Esteban et al., 2019).

4.4.2.2.5 Co-benefits and drawbacks of hard and sediment-based protection

When space is limited (e.g., in an urban setting), co-benefits can be generated through multi-functional hard flood defences, which combine flood protection with other urban functions, such as car parks, buildings, roads or recreational spaces into one multifunctional structure (Stalenberg, 2013; van Loon-Steensma and Vellinga, 2014). An important co-benefit of sediment-based protection, such as beach nourishment and dune management, is that it preserves beach and associated environments, as well as tourism (Everard et al., 2010; Hinkel et al., 2013a; Stive et al., 2013).

Drawbacks of hard protection include the alteration of hydrodynamic and morphodynamic patterns, which in turn may export flooding and erosion problems downdrift (Masselink and Gehrels, 2015; Nicholls et al., 2015). For example, protection of existing shoreline in estuaries and tidal creeks may increase tidal amplification in the upper parts (Lee et al., 2017). Hard protection also hinders or prohibits the onshore migration of geomorphic features and ecosystems (called coastal squeeze; Pontee, 2013; Gittman et al., 2016), leading to both a loss of habitat as well as of the protection function of ecosystems (see Sections 4.3.2.4 and 4.4.2.2). Another drawback of raising hard structures, also emphasised in AR5, is the risk of lock-in to a development pathway in which development intensifies behind higher and higher defences, with escalating severe consequences in the event of protection failure (Wong et al., 2014; Welch et al., 2017), as experienced in Hurricane Katrina impacted New Orleans (Burby, 2006; Freudenburg et al., 2009). This lock-in results from protection attracting further economic development in the flood zone within defenses, which then leads to further raising defences with SLR, and the growing value of exposed assets.

Seabed dredging of sand and gravel can have negative impacts on marine ecosystems such as seagrass meadows and corals (Erftemeijer and Lewis III, 2006; Erftemeijer et al., 2012). Nourishment practices on sandy beaches have also been shown to have drawbacks for local ecosystems if local habitat factors are not taken into consideration when planning and implementing nourishment and maintenance (Speybroeck et al., 2006). A further emerging issue is beach material scarcity mainly driven by demand of sand and gravel for construction, but also for beach and shore nourishment (Peduzzi, 2014; Torres et al., 2017), which makes sourcing the increasing volumes of beach materials required to sustain beaches in the face of SLR more expensive and challenging (Roelvink, 2015).

4.4.2.2.6 Governance of hard and sediment-based protection

Reviews and comparative case studies confirm findings of AR5 that governance challenges are amongst the most common hindrance to implementing coastal measures (Ekstrom and Moser, 2014; Hinkel et al., 2018). One main issue to resolve is conflicting stakeholder interests. This includes conflicts between those favouring protection and those being negatively affected by adaptation measures. In Catalonia, for example, the tourism sector welcomes beach nourishment because it provides direct benefits, whereas those dependent upon natural resources (e.g., fishermen) are increasingly in opposition because they fear that sand mining destroys coastal habitat and livelihood prospects (González-Correa et al., 2008).

There is also conflict related to the distribution of public money between communities receiving public support for adaptation and non-coastal communities who pay for this support through taxes (Elrick-Barr et al., 2015). Generally, access to financial resources for adaptation, including from public sources, development and climate finance or capital markets, frequently constrain adaptation (Ekstrom and Moser, 2014; Hinkel et al., 2018). For example, homeowners are often not willing to pay taxes or levies for public protection or sediment-base measures even if they directly benefit, as found, for example in communities on the US east coast where beach nourishment is used to maintain recreational and tourism amenities (Mullin et al., 2019). In many parts of the world, coastal adaptation governance is further complicated by existing conflicts over resources. For example, illegal coastal sand mining is currently a major driver of coastal erosion in many parts of the developing world (Peduzzi, 2014). Examples of this can be found in Ghana (Addo, 2015) and the Comoros (Betzold and Mohamed, 2017).

An associated governance challenge is ensuring the effective maintenance of coastal protection. Ineffective maintenance has contributed to many coastal disasters in the past, such as in New Orleans (Andersen, 2007). AR5 highlighted that effective maintenance is challenging in a small island context due to a lack of adequate funds, policies and technical skills (Nurse et al., 2014). In some countries in which coastal defence systems have a long history, effective governance arrangements for maintenance, such as the Water Boards in the Netherlands, have emerged. In Bangladesh, where Dutch-like polders were introduced in the 1960s, maintenance has been a challenge due to shifts in multi-level governance structures associated with independence, national policy priorities and donor involvement (Dewan et al., 2015).

4.4.2.2.7 Economic efficiency of hard and sediment-based protection

At global scales, new economic assessments of responses have mostly focused on the direct costs of hard protection and the benefits of reducing coastal extreme event flood risks. These studies confirm AR5 findings that the benefits of reducing coastal flood risk through hard protection exceed the costs of protection, on a global average, and for cities and densely populated areas, during the 21st century even under high-end SLR (*medium evidence, high agreement*; Hallegatte et al., 2013; Wong et al., 2014; Diaz, 2016; Lincke and Hinkel, 2018). For example, Lincke and Hinkel (2018) find that, during the 21st century, it is economically efficient to protect 13% of the global coastline, which corresponds to 90% of global floodplain population, under SLR scenarios from 0.3–2.0 m, five SSPs and discount rates up to 6% (Figure 4.14). While the above two studies have not considered the effects of hard protection in reducing the area of coastal wetlands, it is expected that coastal hard protection in densely populated areas and conserving wetlands in sparsely populated areas can go hand in hand. Protecting less than 42% of the global coastline would leave coastal wetlands sufficient accommodation space to even grow in areas under rising sea levels during the 21st century (Schuerch et al., 2018). Diaz (2016), who includes the cost of wetland loss, using a simpler wetland model, finds that both protection and retreat reduce the global net present costs of SLR by a factor of seven as compared to no adaptation (applying a discount rate of 4%) under 21st century SLR of 0.3–1.3 m and SSP2. There is no global study that has considered social costs and benefits of responses (e.g., health, beach amenity, etc.) or looked at the economics of accommodate, retreat and advance responses.

At local scales, a large number of economic assessments of response options are available but mostly in the grey literature and again with a focus on hard and sediment-based protection. Similar to the global studies, hard protection is generally found to be economically efficient for urban and densely populated areas such as New York, USA (Aerts et al., 2014) and Ho Chi Minh City, Vietnam (Scussolini et al., 2017). Both global and local studies show that sediment-based protection, such as beach nourishment is economically efficient in areas of intensive tourism development due to the large revenues generated within this sector (Rigall-I-Torrent et al., 2011; Hinkel et al., 2013a).

4.4.2.3   Ecosystem-based Adaptation

4.4.2.3.1   Observed ecosystem-based adaptation across geographies

Relative to hard adaptation measures whose global distribution is not known in detail (Scussolini et al., 2015), the current global distribution of coastal ecosystems is well-studied (e.g., for saltmarshes and mangroves, respectively; Giri et al., 2011; Mcowen

et al., 2017). EbA, by definition, can only exist and function where the environmental conditions are appropriate for a given ecosystem. Mangroves, salt marshes and reefs occur along about 40–50% of the world's coastlines (Wessel and Smith, 1996; Burke, 2011; Giri et al., 2011; Mcowen et al., 2017). However, there is no clear estimate on the global length of coastline covered by ecosystems relevant for EbA in the face of SLR in part because of a mismatch between the spatial resolutions of different estimates available. Mangroves occur on tropical and subtropical coasts, and cover 138,000–152,000 km^2 across about 120 countries (Spalding et al., 2010; Giri et al., 2011). At least 150,000 km of coastline in over 100 countries benefit from the presence of coral reefs (Burke, 2011) and these are estimated to protect over 100 million people from wave-induced flooding globally (Ferrario et al., 2014). The extent of other coastal habitats is less well known: salt marshes are estimated to occur in 99 countries, especially in temperature to high latitude locations, with nearly 5,500,000 ha mapped across 43 countries (Mcowen et al., 2017).

Since AR5 there has been growing recognition of the value of conserving existing coastal ecosystems, and where possible restoring them, for the flood protection and multiple other benefits they provide (Temmerman et al., 2013; Arkema et al., 2015). In parallel, EbA measures are increasingly being incorporated and required within national plans, strategies and targets (Lo, 2016), international adaptation funding mechanisms, such as the Adaptation Fund (AF; e.g., in Sri Lanka and India; Epple et al., 2016), and national natural capital valuations (Beck and Lange, 2016). Given their relative novelty, there is widespread interest in building and collecting knowledge of EbA implementation case-studies and examples

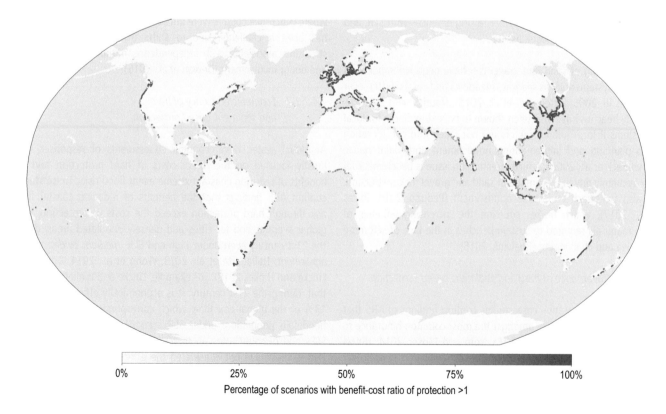

0%   25%   50%   75%   100%

Percentage of scenarios with benefit-cost ratio of protection >1

Figure 4.14 | Economic robustness of coastal protection under sea level rise (SLR) scenarios from 0.3–2.0 m, the five Shared Socioeconomic Pathways (SSPs) and discount rates of up to 6%. Coastlines are coloured according to the percentage of scenarios under which the benefit-cost ratio of protection (reduced flood risk divided by the cost of protection) are above 1. Source: Lincke and Hinkel (2018).

(Table 4.7). Meanwhile, coastal communities around the globe are already implementing EbA responses at local scales, with emphasis on community participation and ownership and local priorities, needs and capacities (Reid, 2016; see Section 4.4.4.4).

EbA has been used as an integral part of some retreat, advance and accommodation responses. For example, on coastlines where high-risk properties are relocated inland, space can be made for ecosystem restoration to enhance natural biodiversity and provide coastal protection (French, 2006; Coastal Protection and Restoration Authority of Louisiana, 2017). There are also examples of ecosystem restoration to advance coastlines and build land elevation (Chung, 2006). EbA can also be an element of accommodation responses by, for example, restoring or creating marshes to provide space for flood water (Temmerman et al., 2013).

4.4.2.3.2   Projected ecosystem-based adaptation

While there are projections available of ecosystem responses to climate change and SLR (Section 4.3.3), to date, there are no large-scale projections available on the future extent of EbA. However, several coastal nations, particularly Small Island Developing States (SIDS) explicitly advocate EbA measures as a means to address future coastal hazard and SLR concerns. Based on Nationally Determined Contributions (NDCs) submitted to the United Nations Framework Convention on Climate Change (UNFCCC), more than 30 SIDS cite EbA as a preferred SLR response, with mangrove planting being the most common measure (Wong, 2018).

4.4.2.3.3   Cost of ecosystem-based adaptation

There is *limited evidence and low agreement* on the costs of ecosystem-based measures to make generally valid estimations of the unit costs across large spatial scales. The total cost of an ecosystem-based measure includes capital costs, maintenance costs, the cost of land and, in some situations, permitting costs (Bilkovic, 2017). The costs of restoring and maintaining coastal habitats depend on coastal setting, habitat type and project conditions. In general, unit restoration costs are lowest for mangroves, higher for salt marshes and oyster reefs, and highest for seagrass beds and coral reefs (Table 4.8).

The conservation of coral reefs and other coastal habitats may also entail substantial opportunity costs because alternative uses of this land, such as through agricultural production, industry and settlements, are generally of high economic value (Stewart et al., 2003; Balmford et al., 2004; Adams et al., 2011; Hunt, 2013). The high value of these alternative uses are the reason why globally, coastal ecosystems are amongst the ecosystems that face the highest rates of anthropogenic destruction, with estimated annual losses of 1–3% of mangroves area, 2–5% seagrass area and 4–9% corals (Duarte et al., 2013). Conserving these areas means reversing these trends.

Under the right conditions, and to some extent, EbA measures are free of maintenance costs, because they respond and adapt to changes in their coastal environment. However, maintenance can become important in the aftermath of damage by storms or human action, for example, when wetlands and reefs can be damaged by high winds, waves and surges, or affected by dredging operations (Smith III et al., 2009; Puotinen et al., 2016). At present, there is limited evidence about the conditions under which EbA measures can self-adapt and when they require human intervention to recover.

4.4.2.3.4   Effectiveness of ecosystem-based adaptation

While EbA has been able to reduce the impacts of sea level related hazards, there is still *little agreement* on the size of the effect (Gedan et al., 2011; Doswald et al., 2012; Lo, 2016; Renaud et al., 2016). Dozens of independent field, experimental and numerical studies have observed and measured the wave attenuation and flood reduction benefits provided by natural habitats, such as marsh and mangrove wetlands (Barbier and Enchelmeyer, 2014; Möller et al., 2014; Rupprecht et al., 2017), coral reefs (Ferrario et al., 2014; Storlazzi et al., 2017), oyster reefs (Scyphers et al., 2011) and submerged seagrass beds (Infantes et al., 2012). Local and global numerical studies indicate that marshes and mangroves can reduce present-day surge-related flood damages by >15% annually, and the loss of a metre of living coral reef can double annual wave-related flood damages (Narayan et al., 2017; Beck et al., 2018). Artificial reef restoration along tens of metres of coastline using Reef Ball™ and other structures has been shown to reduce wave heights and stabilise beach widths (Reguero et al., 2018a; Torres-Freyermuth et al., 2018).

Table 4.8 | Costs of ecosystem-based adaptation (EbA). MPA is marine protected area.

Type of measure	Capital Costs	Maintenance Costs
Wetland Conservation (Marshes/ Mangroves, Maritime Forests)	No data available	Thinning, clearing debris after storms, etc.: Mangrove: 5000 USD ha^{-1} yr^{-1} in Florida (Lewis, 2001) to 11,000 ha^{-1} yr^{-1} (Aerts, 2018). For mangroves globally, 7–85 USD ha^{-1} yr^{-1} (Aerts et al., 2018a); For marshes in the Wadden Sea, 25 USD m^{-1} yr^{-1} (Vuik et al., 2019).
Wetland Restoration (Marshes/ Mangroves, Maritime Forests)	Wetlands: 85,000–230,000 USD ha^{-1} (Aerts et al., 2018a); Mangroves: USD 9000 ha^{-1} (median; Bayraktarov et al., 2016); 2000–13,000 USD ha^{-1} in American Samoa (Gilman and Ellison, 2007); Salt Marshes: 67,000 USD ha^{-1} (Bayraktarov et al., 2016); Brushwood dams for marsh restoration 150 m^{-1} (Vuik et al., 2019).	Similar to maintenance costs for Wetland Conservation.
Reef Conservation (Coral/ Oyster)	For example, start-up costs for Reef MPAs: 96–40,000 USD km^{-2} (McCrea-Strub et al., 2011).	For MPAs, 12 million USD yr^{-1} for the Great Barrier Reef (Balmford et al., 2004).
Reef Restoration (Coral/ Oyster)	165,600 USD ha^{-1} (median; Bayraktarov et al., 2016); Oyster Reefs: 66,800 USD ha^{-1} (median; Bayraktarov et al., 2016); Artificial Reefs in the UK 30,000–90,000 USD 100 m^{-1} (Aerts et al., 2018a).	Similar to maintenance costs for Reef Conservation.

The effectiveness of EbA measures, however, varies considerably depending on storm, wetland, reef and landscape parameters (Koch et al., 2009; Loder et al., 2009; Wamsley et al., 2010; Pinsky et al., 2013; Quataert et al., 2015), which makes it difficult to extrapolate the physical and economic benefits across geographies. Depending on these parameters, rates of surge attenuation can vary between 5–70 cm km^{-1} (Krauss et al., 2009; Vuik et al., 2015).

Critical gaps remain in our understanding about those parameters that together affect the success of ecosystem-based measures including choice of species and restoration techniques, lead time, natural variability and residual risk, temperature, salinity, wave energy and tidal range (Smith, 2006; Stiles Jr, 2006). Among reasons commonly cited for the failure of mangrove restoration projects are poor choice of mangrove species, planting in the wrong tidal zones and in areas of excessive wave energy (Primavera and Esteban, 2008; Bayraktarov et al., 2016; Kodikara et al., 2017).

The effectiveness of ecosystem-based measures also exhibits high seasonal, annual and longer-term variability. For example, marsh and seagrass wetlands typically have lower densities in winter which reduces their coastal protection capacity (Möller and Spencer, 2002; Paul and Amos, 2011; Schoutens et al., 2019). In the long-term, there is *limited evidence* and *low agreement* on how changes in sea level, sediment inputs, ocean temperature and ocean acidity will influence the extent, distribution and health of marsh and mangrove wetlands, coral reefs and oyster reefs (Hoegh-Guldberg et al., 2007; Lovelock et al., 2015; Crosby et al., 2016; Albert et al., 2017). EbA measures may have differential lead times before they are effective. For example, newly planted mangroves provide less wave attenuation until they mature (~3–5 years; Mazda et al., 1997). In contrast, a reef restoration project that uses submerged concrete structures performs as a breakwater as soon as the sub-structure is in place (Reguero et al., 2018a).

4.4.2.3.5   Co-benefits and drawbacks of ecosystem-based adaptation

There is high confidence that ecosystem-based measures provide multiple co-benefits such as sequestering carbon (Siikamäki et al., 2012; Hamilton and Friess, 2018), income from tourism (Carr and Mendelsohn, 2003; Spalding et al., 2017), enhancing coastal fishery productivity (Carrasquilla-Henao and Juanes, 2017; Taylor et al., 2018), improving water quality (Coen et al., 2007; Lamb et al., 2017), providing raw material for food, medicine, fuel and construction (Hussain and Badola, 2010; Uddin et al., 2013), and a range of intangible and cultural benefits (Scyphers et al., 2015) that help improve the resilience of communities vulnerable to sea level hazards (Sutton-Grier et al., 2015).

In comparison to hard structures like seawalls, EbA measures, particularly coastal wetlands, require more land (The Royal Society Science Policy Centre, 2014), and competition for land is often why the ecosystems have declined in the first place (4.4.2.3.1). On developed coasts, this land is often not available. In such cases, hybrid measures that either combine EbA measures with structural measures like mangrove forests in front of dikes (Dasgupta et al., 2019), or build ecological enhancements into engineered structures can provide an

effective solution. Like any other feature that interacts with coastal processes, natural wetlands and reefs can increase flooding in some instances, for example, due to the redistribution or acceleration of flows in channels within a wetland system (Marsooli et al., 2016), or an increase in infragravity wave (i.e., surface gravity waves with frequencies lower than wind waves) energy behind a reef (Roeber and Bricker, 2015).

4.4.2.3.6   Governance of ecosystem-based adaptation

The coastal protection benefits of natural ecosystems are increasingly being recognised within international discourse and national coastal adaptation, resilience and sustainable development plans and strategies (Section 4.4.2.3.1). In general, obtaining permits for EbA remains more difficult compared to established hard protection measures, in places like the USA (Bilkovic, 2017). However, there are examples of instruments specifically tailored to retain the protective function of EbA (Borges et al., 2009; Government of India, 2018). The Living Shorelines Regulations of the state government of Maryland in the USA (Maryland DEP, 2013), for instance, requires that private properties must include marsh creation or other non-structural measures when stabilising their shorelines, unless a waiver is obtained.

There are an increasing number of public and private financial mechanisms and policy instruments to encourage the use and implementation of EbA measures (Colgan et al., 2017; Sutton-Grier et al., 2018). For example, a regulation by the Federal Emergency Management Agency (FEMA) of the USA, allows proponents of hazard mitigation projects, such as state, territorial and local governments, to take into account the co-benefits of EbA when assessing benefit-cost ratios of FEMA-funded recovery projects (FEMA, 2015). International guidelines are being developed for designing and implementing EbA measures, with the intention to support wider implementation of these responses (Hardaway Jr and Duhring, 2010; Van Slobbe et al., 2013; Van Wesenbeeck et al., 2017; Bridges et al., 2018).

4.4.2.3.7   Economic efficiency of ecosystem-based adaptation

There is *limited evidence* regarding the economic efficiency of EbA, mainly due to the *low agreement* about EbA effectiveness (Section 4.4.2.3.2) and costs (Section 4.4.2.3.2). A study of coastal protection measures on the Gulf of Mexico coastline, USA, estimated that EbA measures have average benefit-cost ratios above 3.5 for 2030 flood risk conditions, assuming a discount rate of 2% (Reguero et al., 2018b; see Section 4.4.2.3.2). This study also finds that EbA are nearly four times more cost-efficient along developed coastlines as compared to conservation-priority areas because protection benefits are higher in the former case due to the level of asset exposure.

4.4.2.4   Advance

4.4.2.4.1   Observed advance across geographies

Advance has a long history in most areas where there are dense coastal populations and a shortage of land (*very high confidence*). This includes land reclamation through polders around the southern North Sea (Germany, the Netherlands, Belgium and England) and

China (Wang et al., 2014), which coincides with regions where there is extensive hard protection in place (Section 4.4.2.4). Land reclamation has also taken place in all major coastal cities to some degree, even if only for the creation of port and harbour areas by raising coastal flats above normal tidal levels through sediment infill. On some steep coasts, where there is little flat land, such as the Hong Kong Special Administrative Region of China, material from elevated areas has been excavated to create fill material to build land out into the sea.

Globally, it is estimated that about 33,700 km^2 of land has been gained from the sea during the last 30 years (about 50% more than has been lost), with the biggest gains being due to land reclamation in places like Dubai, Singapore and China (Wang et al., 2014; Donchyts et al., 2016). In Shanghai alone, 590 km^2 land has been reclaimed during the same period (Sengupta et al., 2018). In Lagos, 25 km^2 of new land is currently being reclaimed (www.ekoatlantic. com). Land reclamation is also popular in some small island settings. The Maldives has recently increased the land area of their capital region by constructing a new island called Hulhumalé, which has been built 60 cm higher than the normal island elevation of 1.5 m, in order to take into account future SLR (Hinkel et al., 2018).

4.4.2.4.2 Projected advance

Advance was not primarily a response to SLR in the past, but due to a range of drivers, including land scarcity, population pressure and extreme events, future advance measures are expected to become more integrated with coastal adaptation and might even be seen as an opportunity to support and fund adaptation in some cases (Linham and Nicholls, 2010; RIBA and ICE, 2010; Nicholls, 2018). While there is no literature on this topic, significant further advance measures can be expected in land scarce situations, such as found in China, Japan and Singapore, in coming decades.

4.4.2.4.3 Costs of advance

Contrary to protection measures, little systematic monetary information is available about costs of advance measures, specifically not in the peer reviewed literature. The costs of land reclamation are extremely variable and depend on the unit cost of fill versus the volumetric requirement to raise the land. Hence, filling shallow areas is preferred on a cost basis.

4.4.2.4.4 Effectiveness of advance

Similar to hard protection, land reclamation is mature and effective technology and can provide predictable levels of safety. If the entire land area is raised above the height of ESLs, residual risks are lower as compared to hard protection as there is no risk of catastrophic defence failure.

4.4.2.4.5 Co-benefits and drawbacks of advance

The major co-benefit of advance is the creation of new land. The major drawbacks include groundwater salinisation, enhanced erosion and loss of coastal ecosystems and habitat, and the growth

of the coastal floodplain (Li et al., 2014; Nadzir et al., 2014; Wang et al., 2014; Chee et al., 2017). In China, for example, about 50% of coastal ecosystems have been lost due to land reclamation, leading to a range of impacts such as loss of biodiversity, decline of bird species and fisheries resources, reduced water purification, and more frequent harmful algal blooms (Wang et al., 2014). For example, the reclamation of about 29,000 ha of land in Saemangeum, Republic of Korea, in 2006, has led to a decrease in shorebird numbers by over 30% in two years, probably caused by mortality (Moores et al., 2016). Inadvertently, historic land reclamation through polderisation may have enhanced exposure and risk to coastal flooding by creating new populated floodplains, but this has not been evaluated.

4.4.2.4.6 Governance of advance

Land reclamation raises equity issues with regards to access and distribution of the new land created, specifically due to the political economy associated with high coastal land values, and the involvement of private capital and interests (Bisaro and Hinkel, 2018), but this has hardly been explored in the literature.

4.4.2.4.7 Economic efficiency of advance

There is *limited evidence* on the efficiency of advance responses in the scientific literature. Benefit-cost ratios of land reclamation can be very high in urban areas due to high land and real estate prices (Bisaro and Hinkel, 2018).

4.4.2.5 Accommodation

4.4.2.5.1 Observed accommodation across geographies

There is a *high agreement* that accommodation is a core element of adaptation, and it is taking place on various scales based on measures such as flood proofing and raising buildings, implementing drainage systems, land use changes as well as EWS, emergency planning, setback zones and insurance schemes. However, no literature is available that summarises observed accommodation worldwide. There is *low evidence* of accommodation occurring directly as a consequence of SLR but *high evidence* of accommodation measures being implemented in response to coastal hazards such as coastal flooding, salinisation and other sea-borne hazards such as cyclones.

Flood proofing may include the use of building designs and materials which make structures less vulnerable to flood damages and/or prevent floodwaters from entering structures. Examples include floating houses in Asia, such as in Vietnam (Trang, 2016), raising the floor of houses in the lower Niger delta (Musa et al., 2016), construction of verandas with sandbags and shelves in houses to elevate goods during floods in coastal communities in Cameroon (Munji et al., 2013). In Semarang City, Indonesia, residents adapted to coastal flooding by elevation of their houses by 50–400 cm or by moving their goods to safer places, without making structural changes (Buchori et al., 2018). Residents of Can Tho City of the Mekong Delta, Vietnam elevated houses in response to tidal flooding (Garschagen, 2015). In urban areas extensive drainage systems contribute to accommodation such as in Hong Kong and Singapore, which rely on urban drainage systems to handle large volumes of surface runoff generated during storm events (Chan et al., 2018). Farming practices have been adapted to frequent flooding in the lower Niger delta: farmers raise crops above floodwaters by planting on mounds of soil and apply ridging and terracing on farmlands to form barriers (Musa et al., 2016). In the floodplains of Bangladesh, floating gardens help to maintain food production even if the area is submerged (Irfanullah et al., 2011). Here, the traditional way to build homesteads is on a raised mound, built with earth from the excavation of canals and ponds (ADPC, 2005). Coastal infrastructure, such as ports, having a functional need to be at the coast, accommodate SLR with elevated piers and critical infrastructure. One example is Los Angeles, where PierS was raised to an elevation of 6 m (Aerts, 2018).

Communities in the Netherlands are experimenting with floating/amphibious houses capable of adapting to different water levels, and similar considerations are also discussed in other geographies, such as in Bangkok (Nilubon et al., 2016). Flood proofing is widely applied in the USA, where wet and dry flood proofing measures are recognised: wet flood proofing reduces damage from flooding while dry flood proofing makes a building watertight or substantially impermeable to floodwaters up to the expected flood height (FEMA, 2014). In that sense, dry flood proofing could also be interpreted as a protection measure on the level of individual structures.

Physical accommodation to salinisation and saline water intrusion is more poorly documented. It mainly entails agricultural adaptation to soil salinity, and saline surface and ground water, as described for the land use changes aimed at alternating rice-shrimp systems and shrimp aquaculture in the Mekong Delta (Renaud et al., 2015) or using methods which decrease soil salinity, such as flushing rice fields with fresh water to wash out salinity (Renaud et al., 2015), or applying maize straw in wheat fields (Xie et al., 2017). Coastal communities are also experimenting with the use of salt tolerant varieties as a result of breeding programmes, for example, in Indonesia (Rumanti et al., 2018), or saline irrigation water in conjunction with fresh water, such as for maize in coastal Bangladesh (Murad et al., 2018).

Adaptation planning for SLR has been incorporated into land use planning in several states in the USA (Butler et al., 2016b). In the Yangtze River Delta, landscape planning designs floodplain zones to accept floodwaters (Seavitt, 2013). In the Mekong Delta, different land use options, including shifting from freshwater agriculture to brackish and saline agriculture, were proposed as seawater intrudes farther inland (Smajgl et al., 2015).

EWS are frequently incorporated into overall risk reduction strategies and are applied for various coastal hazards such as tsunamis in coastal areas of Indonesia (Lauterjung et al., 2017) and hydro-meteorological coastal hazards in Bangladesh and Uruguay (Leal Filho et al., 2018). They fall under 'accommodation' as they allow people to remain in the hazard-prone area but provide advance warning or evacuation in the face of imminent danger. In contrast to hard protection measures, EWS have shorter installation time and lower impact on the environment (Sättele et al., 2015). They can work effectively to reduce risk arising from predictable hazardous events but are less

well-suited to accommodate slow onset change (i.e., events or processes that happen with high certainty under different climate change scenarios).

Climate risk insurance schemes have been recently developed to address sudden, and in rare cases, slow onset hazards at the coast, and to increase overall resilience. For coastal risks, insurance is mainly applicable for sudden onset hazards, including storm surges and coastal flooding, to buffer against the financial impacts of loss events. For slow onset hazards, insurance schemes are not the first-best tool, whereas resilience building and prevention of loss and damage in such instances may be more cost-effective ways to address these risks (Warner et al., 2013). In this context, index based insurance products are increasingly offered, particularly in low-income countries and have also been included in a number of countries in their NDCs and in some cases in their National Adaptation Plans (NAPs; Kreft et al., 2017). Countries with existing climate risk insurance schemes include, for example, Haiti, Maldives, Seychelles and Vietnam. The InsuResilience Global Partnership for Climate and Disaster Risk Finance and Insurance Solutions was launched at the 2017 UN Climate Conference (COP 23) in Bonn. InsuResilience aims to enable more timely response after a disaster and helps to better prepare for climate and disaster risk through the use of climate and disaster risk finance and insurance solutions. So far, climate risk insurance was used mainly in the context of agriculture, where it has showed great efficacy in boosting investments for increasing productivity (Fernandez and Schäfer, 2018). However, on the global scale, the uptake of index insurance is still low (Yuzva et al., 2018).

4.4.2.5.2 Projected accommodation

While there is no literature on projected accommodation, current trends suggest further uptake of accommodation approaches in coming decades, especially where protection approaches are not economically viable. Flood proofing of houses and establishment of new building codes to accommodate coastal hazards is also expected to become more common in coming decades. Similarly, accommodation measures for salinity are under further development, such as rice breeding programs to improve salt tolerance (Linh et al., 2012; Quan et al., 2018b). However, the achievements to improve salinity tolerance in rice are rather modest so far (Hoang et al., 2016) although efforts are expected to continue or even intensify. Given that index based insurance products have been included in NDCs and NAPs in a number of countries (Kreft et al., 2017), uptake is expected to grow. Ports can continue elevating hazard-prone facilities and the critical parts of port infrastructure can be protected by flood walls. Alternatively, ports can use advance measures to develop port facilities seaward (Aerts, 2018).

In summary, due to the large variety of different measures implemented in ad hoc ways worldwide, there is *low confidence* in quantitative projections of accommodation measures in response to SLR. However, there is *high confidence* that accommodation measures will continue to be a widespread adaptation option especially in combination with protection and retreat measures.

4.4.2.5.3 Cost of accommodation

The cost of accommodation varies widely with the measures taken as well as the expected flood height. For flood proofing of buildings in New York City for instance, Aerts et al. (2014) provided an economic rationale for the implementation of improved building codes, such as elevating new buildings and protecting critical infrastructure (see also Box 4.1). Flood proofing can also be undertaken by individuals and even small, inexpensive flood proofing efforts can result in reductions in flood damage (Zhu et al., 2010). In general, costs for flood proofing increase as the flood protection elevation increases. Other costs include those for maintenance and, if applicable, insurance premiums. For example, deciding to elevate a building in the USA will increase the project's cost; however, the additional elevation may lead to significant savings on flood insurance premiums (FEMA, 2014).

4.4.2.5.4 Effectiveness of accommodation

Accommodation measures can be very effective for current conditions and small amounts of SLR, also buying time to prepare for future SLR. Success stories include the case of Bangladesh where improved early warnings, the construction of shelters, and development of evacuation plans, helped to reduce fatalities as a result of flooding and cyclones (Haque et al., 2012). Illiteracy, lack of awareness and poor communication are, however, still hampering the effectiveness of early warnings (Haque et al., 2012). If well designed, and if the premiums reflect individual risks, insurance can effectively discourage further investments in risky areas as insurance cost provides information on the nature of locality-specific risks and can incentivise investment in risk reduction by requiring that certain minimum standards are met before granting insurance coverage (Kunreuther, 2015). Limits to such accommodation occur much earlier compared to protect, advance and retreat measures. While dikes can be raised to 10 m, and retreat can be implemented to the 10 m contour or higher, accommodating SLR has practical and economic limits, and ultimately retreat or protection will be required.

4.4.2.5.5 Co-benefits and drawbacks of accommodation

The major co-benefit of accommodation is improved resilience of *in situ* communities without retreat or the use of land and resources for the construction of protection measures. Flood proofing, for example, helps prevent demolition or relocation of structures and it is often an affordable and cost effective approach to reducing flood risk (Zhu et al., 2010). Specific accommodation measures have different co-benefits such as that stilt houses not only protect from flooding but also from wild animals (Biswas et al., 2015). Accommodation – depending on the measure implemented – has the potential to maintain landscape connectivity allowing access to the ocean as well as landward migration of ecosystems, at least to some degree. It also retains flood dynamics and with that the benefits of flooding such as sediment re-distribution. Stilt houses leave space for the floodwater while wet-flood proofing maintains a low hydrostatic pressure on the buildings so that structures are less prone to failure during flooding (FEMA, 2014).

The major drawback of accommodation is that it actually does not prevent flooding or salinisation, which might have consequences not addressed by the accommodation measure itself. Examples include inundation of an area where houses are flood proofed but schooling of children and business operations are nevertheless disrupted. Significant clean up may also be needed after flood water enters buildings, including the removal of sediment, debris or chemical residues (FEMA, 2014). Also, flood proofing measures require the current risk of flooding to be known and communicated to and understood by the public through flood hazard mapping studies and flood warning information (Zhu et al., 2010). Small businesses in particular may face difficulties to recover from flooding due to lack of forward planning (Hoggart et al., 2014).

Co-benefits of insurance include the possibility that sovereign level insurance may improve the credit ratings of vulnerable countries, reducing the cost of capital and allowing them to borrow to invest in resilient infrastructure (Buhr et al., 2018). Major natural disasters can weaken sovereign ratings, especially if there is no insurance in place (Moritz Kramer, 2015). One much discussed drawback of insurance is the moral hazard that may result: since someone else bears the costs of a loss, those insured may be less inclined to take precautionary measures or may act recklessly (Duus-Otterström and Jagers, 2011).

4.4.2.5.6 Governance of accommodation

While accommodation measures to coastal hazards are often taking place at the local level, and are decided by individual homeowners, farmers or communities, from a governance perspective it is important to provide guidance on how and to what extent owners can retrofit their homes to reduce the risk to coastal flooding. In New York City, for instance, changes to building codes, require elevating, or flood proofing of existing and new buildings in the 100-year floodplain, and prevent construction of critical infrastructure like hospitals in the flood zone (NYC, 2014; see also Box 4.1).

Effective coastal risk management efforts rely on good governance that includes understanding the probability and consequences of hazard impacts like flooding and salinisation, and implementing mechanisms to prevent or manage all possible events (EEA, 2013). The effectiveness of accommodation measures based on institutional measures, such as EWS and evacuation plans, largely depends on the governance capabilities they are embedded in.

4.4.2.5.7 Economic-efficiency of accommodation

There is *high confidence* that many accommodation measures are very cost-efficient. Flood EWS coupled with precautionary measures have been shown to produce significant economic benefits (Parker, 2017). Elevating areas at high risk and retrofitting buildings in Ho Chi Minh City, for example, have benefit-cost ratios of 15 under SLR of 180 cm and a discount rate of 5% during the 21st century (Scussolini et al., 2017). In the context of the National Flood Insurance Program in the USA, it has been estimated that elevating new houses by 60 cm might raise mortgage payments by 240 USD yr^{-1}, but reduce flood insurance by 1000–2000 USD yr^{-1} depending on the flood zone (FEMA, 2018), although this only addresses present extremes and ignores future SLR (Zhu et al., 2010). In Europe, the benefits of installing a cross-border continental-scale flood EWS are estimated at 400 EUR per EUR invested (Pappenberger et al., 2015).

4.4.2.6 Retreat

4.4.2.6.1 Observed retreat across geographies

There is *limited evidence* of migration occurring directly as a consequence of impacts associated with environmental change generally and SLR specifically. Research examining the linkages between migration and environmental change has been conducted in the Pacific (Connell, 2012; Janif et al., 2016; Perumal, 2018), South Asia (Szabo et al., 2016; Call et al., 2017; Stojanov et al., 2017), Latin America (Nawrotzki and DeWaard, 2016; Nawrotzki et al., 2017), Alaska, in North America (Marino and Lazrus, 2015; Hamilton et al., 2016) and Africa (Gray and Wise, 2016). While some limited evidence was found on population movement inland associated with shoreline encroachment in Louisiana, USA (Hauer et al., 2018), this research emphasises that the relationship between climate change impacts including SLR and migration is more nuanced than suggested by simplified cause-and-effect models (Adger et al., 2015). Migration is driven by a large number of individual, social, economic, political, demographic and environmental push and pull factors (Black et al., 2011; Koubi et al., 2016), interwoven with mega-trends such as urbanisation, land use change and globalisation, and is influenced by development and political practices and discourses (Bettini and Gioli, 2016; Cross-Chapter Box 7). For example, asset endowed individuals and households are more able to migrate out from flood-prone areas (Milan and Ruano, 2014; Logan et al., 2016), while the poorest households are significantly susceptible to material and human losses following an extreme event or disruptive environmental change (Call et al., 2017). Individual and social drivers include perceptions of environmental change (Koubi et al., 2016), formed by both direct experience of change and indirect information from social networks, mass media and governmental agencies. Environmental factors include the longer term impacts of climate variability and change, which can erode the capacity of ecosystems to provide essential services such as availability of freshwater, soil fertility and energy production acting as a threat multiplier for other drivers of migration (Hunter et al., 2015; McLeman, 2018).

There is *robust evidence* of disasters displacing people worldwide, but *limited evidence* that climate change or SLR is the direct cause. In 2017, 18.8 million people were displaced by disasters, of which 18 million were displaced by weather-related events including 8.6 million people displaced by floods and 7.5 million by storms, with hundreds of millions more at risk (IDMC, 2017; Islam and Khan, 2018). The majority of resultant population movements tend to occur within the borders of affected countries (Warner and Afifi, 2014; Hunter et al., 2015; Nawrotzki et al., 2017).

We find *robust evidence* of planned relocation taking place worldwide in low-lying zones exposed to the impacts of coastal hazards (Hino et al., 2017; Mortreux et al., 2018). While relocation plans are usually discussed after an extreme event occurs, they generally target the reduction of long-term environmental risks, including those of

SLR (McAdam and Ferris, 2015; Hino et al., 2017; Morrison, 2017). For example, in the aftermath of Hurricane Katrina, the Louisiana Comprehensive Master Plan for a Sustainable Coast recommended the relocation of several communities in the next 50 years due to expected RSL rise, and relocation of inhabitants from Isle de Jean Charles is already taking place (Barbier, 2015; Coastal Protection and Restoration Authority of Louisiana, 2017). In Shismaref, an Iñupiat community in Alaska, increased shoreline erosion triggered government-led relocation (Bronen and Chapin, 2013; Maldonado et al., 2013). In the Pacific, current coastal risks aggravated by rising sea level are driving the government led relocation of the inhabitants of Taro, the provincial capital of Choiseul Province in the Solomon Islands (Albert et al., 2018). In 2014, the government of Kiribati purchased land on Vanua Levu, the second largest island of Fiji, with the purpose of economic development and food security, but many i-Kiribati associated the acquisition with future relocation to Fiji (Hermann and Kempf, 2017). In southeast Asia, the government of Vietnam assists and manages rural populations' relocation from disaster prone areas exposed to coastal risks in the Mekong Delta to large industrial areas with high labour demand, such as Ho Chi Minh City and Can Tho City (Collins et al., 2017). Managed realignment carried out for the purposes of habitat creation, improved flood risk management and more affordable coastal protection, is increasingly popular in Europe, but usually involves small-scale projects and few people if any (Esteves, 2013). Most of the managed realignment projects in the UK and Germany have been carried out for habitat creation and to reduce spending on coastal defences (Hino et al., 2017).

4.4.2.6.2 Projected retreat

There is *high agreement* that climate change has the potential to drastically alter the size and direction of migration flows (Connell, 2012; Gray and Wise, 2016; Janif et al., 2016; Nawrotzki and DeWaard, 2016; Szabo et al., 2016; Call et al., 2017; Nawrotzki et al., 2017), but there is *low confidence* in quantitative projections of migration in response to SLR and extremes of sea level. The number of modelling studies of migration in response to environmental drivers has increased rapidly over the past decade (Kumari et al. 2018), but only a small portion of these model studies address migration in response to SLR and sea level extremes. Amongst these, a variety of different modelling approaches have been applied, but no model currently accounts for all push and pull factors influencing migration decisions (see Section 4.4.2.6.1).

A model projecting future US county-level populations exposed to permanent inundation was combined with an empirical model of potential migration destinations to produce the first sea level/migration analysis of migrant destinations (Hauer, 2017). Assuming that households with incomes above 100,000 USD yr^{-1} would have resources to stay and adapt, it was found that 1.8 m SLR by 2100 would displace over two million people in south Florida. Projected population gains due to SLR reach several hundred thousand for some inland urban areas. A gravity model modified to account for both distance to destinations and their attractiveness (deriving from such factors as economic opportunity and environmental amenities) projects a net migration into and out of the East African coastal zone, ranging from out-migration of 750,000 people between 2020 and 2050 to a small in-migration (Kumari et al., 2018). However, this range includes migration stimulated by freshwater availability as well as SLR and episodic flooding. A generalised radiation or diffusion model predicts 0.9 million people will migrate due to SLR in Bangladesh by 2050 and 2.1 million by 2100, largely internally, with substantial implications for nutrition, shelter and employment in destination areas (Davis et al., 2018).

A global dynamic general equilibrium framework (Desmet et al., 2018) provides a more comprehensive approach to accounting for economic factors including changes to trade, innovation, and agglomeration, and political factors, such as policy barriers to mobility, all of which influence the migration response to environmental change. Agent-based models attempt to simulate decisions by individuals who face a variety of socioeconomic and environmental changes (Kniveton et al., 2012). However, neither general equilibrium nor agent-based frameworks have been applied yet to migration responses to SLR. Econometric models, common in climate/migration studies (Millock, 2015), likewise have yet to be applied to the SLR context, except for a single case study where an econometric model was used to interpret the outcome of a discrete choice experiment (Buchanan et al., 2019). For example, an interesting distinction between migration responses to long term temperature and precipitation trends in contrast to extreme events like flooding has been noted (Bohra-Mishra et al., 2014; Mueller et al., 2014), but similar econometric studies have yet to be done comparing responses to gradual land loss versus flooding during ESL events.

4.4.2.6.3 Cost of retreat

We have *limited evidence* of estimates on the cost of retreat. There are few cost estimates in the literature and these are based on stylised assumptions as little empirical data is available.

The cost of managed relocation, including land acquisition, building of roads and infrastructure and other subsidies, was found to vary from 10,000–270,000 GBP per home in United Kingdom Coastal Change Pathfinder projects (Regeneris Consulting, 2011), and between 10,000 USD in Fiji and 100,000 USD per person in Alaska and in the Isle of Jean Charles in the USA (Hino et al., 2017). For people involved in planned relocation in Shaanxi Province, Northwest China, households receive subsidies ranging from 1200–5100 USD (Lei et al., 2017). The Louisiana's National Disaster Resilience Competition, Phase II Application states that the proposed relocation of 40 households in the Isle de Jean Charles in Louisiana is estimated to cost 48,379,249 USD, including the cost for land acquisition, infrastructure and construction of new dwellings (State of Louisiana, 2015). Generally, maintenance costs do not arise if people are moved completely out of the hazard zone (Suppasri et al., 2015; Hino et al., 2017). In cases in which people are only moved so that short-term but not long-term risk is reduced, follow up costs for further responses will occur.

The individual costs associated with displacement after an environmental disaster are difficult to obtain. In the literature, there are limited estimates of the social costs to residents of Guadeloupe, Saint Croix, St. Thomas, Puerto Rico, and the southeast USA displaced after Hurricanes Hugo (1989) and Katrina (2005). A survey conducted

across 18 parishes (i.e. counties) in Louisiana in 2006 revealed that non-displaced households had an average income of 36,000 USD compared to an average income of 30,000 USD recorded for displaced households (Hori and Schafer, 2010).

4.4.2.6.4 Effectiveness of retreat

There is *very high confidence* that retreat is effective in reducing the risks and impacts of SLR as retreat directly reduces exposure of human settlements and activities (Gioli et al., 2016; Shayegh et al., 2016; Hauer, 2017; Morrison, 2017).

4.4.2.6.5 Co-benefits and drawbacks of retreat

The other outcomes of retreat responses, beyond the one of effectively reducing SLR risks and impacts, are complex and affect both origin and destination. Generally, retreat impacts social networks, access to services and economic and social opportunities, and several well-being indicators (Jones and Clark, 2014; Adams, 2016; Herath et al., 2017; Kura et al., 2017; McNamara et al., 2018). The socioeconomic benefits of migration to individuals and households may include improved access to health and education services, as well as labour markets (Wrathall and Suckall, 2016). Destination areas may gain economically as populations and capital relocate and provide a new source of labour, capital and innovation to inland areas (see Section 4.4.2.6.2; de Haas, 2010). Income inequality may be reduced, but only through migration to areas with growing economies. Remittances can provide flexibility in livelihood options, supply capital for investment and spread risk (Scheffran et al., 2012).

Drawbacks of migration and displacement at the destination can be increased competition for resources and within labour markets, pressure on frontline services and on social cohesion as a result of heightened cultural or ethnic tension (Werz and Hoffman, 2015), as well as cultural, social and psychological losses related to disruptions to sense of place and identity, self-efficacy, and rights to ancestral land and culture (McNamara et al., 2018). The unplanned and unassisted voluntary relocation of the inhabitants of Nuatambu and Nusa Hope in the Solomon Islands to areas further from the coast poses a series of practical challenges with sanitation, access to drinking water and transport (Albert et al., 2018).

The success of planned relocation in terms of the balance of co-benefits and drawbacks varies across relocation schemes (Hino et al., 2017) and outcomes are highly uneven (Genovese and Przyluski, 2013; Ford et al., 2015; Nordstrom et al., 2015; Bukvic and Owen, 2017; Hino et al., 2017; Jamero et al., 2017). On the one hand, well designed and carefully implemented programmes, such as the ongoing resettlement of indigenous communities in Alaska, can improve housing standards and reduce vulnerability (Suppasri et al., 2015; Albert et al., 2018). On the other hand, relocated communities have often become further impoverished (Wilmsen and Webber, 2015), because they are removed from cultural and material resources on which they rely, compounded by poor implementation processes that may fail to ensure fairness, social and environmental

4

justice and well-being (Herath et al., 2017; Mortreux et al., 2018; Nygren and Wayessa, 2018).

4.4.2.6.6　Governance of retreat

Environmentally driven migration and displacement gained major attention over the last decade in the international policy community (Goodwin-Gill and McAdam, 2017). Worldwide programmes, such as the Nansen Initiative, signed by 110 countries to address the serious legal gap around the protection of cross-border migrants impacted by natural disasters, have been implemented (Gemenne and Brücker, 2015). In 2016, the Platform on Disaster Displacement was established to follow up on the work conducted by the Nansen Initiative with the objective of implementing the recommendations of the Protection Agenda (McAdam and Ferris, 2015). Governments are further encouraged by civil society to relocate people at risk and displaced populations out of disaster-prone areas to avoid potential casualties (Lei et al., 2017; Mortreux et al., 2018). There have been discussions among Pacific Island countries and territories and other nations in the Pacific Rim around new policy mechanisms that would facilitate adaptive migration in the region in response to natural hazards including SLR (Burson and Bedford, 2015). There have been cases presented at the Immigration and Protection Tribunal of New Zealand testing refugee claims associated with climate change from Tuvaluan and i-Kiribati applicants, both citing environmental change on their home islands as grounds for remaining in New Zealand. One applicant was successful in the quest to remain in New Zealand on humanitarian grounds, but not on the grounds of refugee status (Farbotko et al., 2016).

The is *high agreement* that outcomes can be improved by upholding the principle of procedural justice and respecting the autonomy of individuals and their decisions about where and how they live (Warner et al., 2013; Schade et al., 2015; McNamara et al., 2018). However, there are cases where logistical and political stances constrain the application of such approach, such as when the government of Sri Lanka prohibited rebuilding along the coastline of the country after the 2004 tsunami (Hino et al., 2017). Proactive planning, including participation and consultation with those in peril, has the potential to improve outcomes (*medium confidence*; de Sherbinin et al., 2011; Gemenne and Blocher, 2017). Governments can assist migrants through policy reforms to enable relocation to fast growing economic regions in the country. An example of this approach was adopted in Vietnam by both the National Target Program to Respond to Climate Change and the National Strategy for Natural Disaster Prevention, Response and Mitigation targeted at locations within the Mekong Delta exposed to the impacts of SLR (Nguyen et al., 2015; Collins et al., 2017). Outcomes of retreat for both community of origin and destination can also be improved by building the human capital of migrants (skills, health and education), reducing costs of migration and remittance transfer, and provision of improved safety nets for migrants at their destinations (*high agreement*) (Gemenne and Blocher, 2017).

4.4.2.6.7　Economic efficiency of retreat

There is *limited evidence* on the efficiency of retreat responses in the scientific literature.

4.4.3　Governance Challenges in Responding to Sea Level Rise

4.4.3.1　Introduction

Governance is pivotal to shaping SLR responses. The assessment of SLR responses above has shown that each type of response raises specific governance challenges associated with the distribution of costs, benefits and negative consequences of responses across societal actors. Hence, SLR responses require governance efforts if social conflicts are to be resolved and mutual opportunities amongst all actors realised. Generally, responses involve the interaction of diverse public and private actors at different levels of decision making with divergent values, interests and goals on coastal activities, lifestyles, livelihoods, risks, resilience and sustainability (*high confidence*; Dovers and Hezri, 2010; Foerster et al., 2015; Giddens, 2015; Mills et al., 2016; Dolšak and Prakash, 2018; Hinkel et al., 2018; Hoegh-Guldberg et al., 2018; AR5). This leads to a number of overarching governance challenges that arise from the nature of SLR, which will be assessed in this section.

While there is a substantial literature on coastal governance, little attention has been focused explicitly on SLR governance, as was also the case in AR5 (Wong et al., 2014). Furthermore, much of the adaptation governance literature has focused on putting forward normative prescriptions on how governance arrangements ought to be (e.g., transformative governance; Chaffin et al., 2016), but with limited empirical evidence on the actual effectiveness of these prescriptions (Klostermann et al., 2018; Runhaar et al., 2018). Hence, understanding the social mechanisms leading to the emergence of particular governance arrangements, and how effective they are in addressing climate change and SLR, is limited (Wong et al., 2014; Bisaro and Hinkel, 2016; Oberlack, 2017; Bisaro et al., 2018; Roggero et al., 2018a; Roggero et al., 2018b). An important post-AR5 development has thus been to move beyond descriptions and normative prescriptions about 'good governance' to explore which factors help (called enablers) or hinder (called barriers) how social choices are made and implemented on complex issues like climate change and SLR, as elaborated in the next subsection.

4.4.3.2　Understanding Barriers to Adaptation as Governance Challenges

AR5 stated that there are many reasons why adaptation governance is complex (Klein et al., 2014). The first generation of studies that investigated this question empirically identified many (lists of) barriers that people have experienced in adaptation governance in specific case contexts, including political, institutional, social-cognitive, economic, financial, biophysical and technical barriers (Klein et al., 2014). Although insightful for these specific cases, including SLR (Hinkel et al., 2018), accumulation of empirical

4

findings in building theory proved to be limited, and it did not result in more evidence-informed advice to policy makers on how to deal with barriers (Biesbroek et al., 2013; Eisenack et al., 2014).

In response, in a second generation of studies, several frameworks have been proposed and tested to advance scholarship on barriers to adaptation (Eisenack and Stecker, 2012; Barnett et al., 2015; Lehmann et al., 2015; Bisaro and Hinkel, 2016). A frequently used framework was developed by Moser and Ekstrom (2010) who identified and linked key barriers to certain stages of the policy process: understanding, planning and management stages. Moser and Ekstrom (2010) argue that conditions, such as the scope and scale of adaptation, have significant implications for which barriers are activated in the policy process, and how persistent and difficult they are to overcome. This and other frameworks have been applied in a diversity of contexts, providing valuable insights about the governance challenges involved in adapting to climate change and suggestions for improvement (Ekstrom and Moser, 2014; Rosendo et al., 2018; Thaler et al., 2019).

A recent generation of studies takes more theory based approaches and includes contextual factors to analyse the key social mechanisms that explain why adaptation processes are often complex, result in deadlocks, delays or even failure (Biesbroek et al., 2014; Eisenack et al., 2014; Wellstead et al., 2014; Biesbroek et al., 2015; Bisaro and Hinkel, 2016; Oberlack and Eisenack, 2018; Sieber et al., 2018; Wellstead et al.). Such insights are critical as they can be used by practitioners for policy design (e.g., to prevent certain deadlocks from emerging by (re)designing contextual conditions), or provide insights on strategic interventions in ongoing processes to revitalise deadlocked adaptation governance (Biesbroek et al., 2017).

4.4.3.3 Governance Challenges in the Face of Sea Level Rise

There is a wide diversity of governance challenges and opportunities for tackling SLR, with marked differences within and between coastal communities in developed and developing counties. Five salient overarching governance challenges that arise due to distinctive features of SLR were highlighted. This typology is then used to assess how planning, participation and conflict resolution (Section 4.4.4.2), decision analysis methods (Section 4.4.4.3), and enabling conditions (Section 4.5) can help to address these five challenges.

Time horizon and uncertainty: The long-term commitment to SLR (Section 4.2.3.5) and the large and deep uncertainty about the magnitude and timing of SLR beyond 2050 (Section 4.4.4.3.2), challenge standard planning and decision making practises for several reasons (*high confidence*; Peters et al., 2017; Pot et al., 2018; Hall et al., 2019; Hinkel et al., 2019). The time horizon of SLR extends beyond usual political, electoral and budget cycles. Furthermore, many planning and decision making practices strive for predictability and certainty, which is at odds with the dynamic risk and deep uncertainty characterising SLR (Hall et al., 2019). Tensions can arise between established risk-based planning that seeks to measure risk, and adaptation responses that embrace uncertainty and complexity (Kuklicke and Demeritt, 2016; Carlsson Kanyama et al., 2019). For example, tensions arise because of the mismatch between the

relative inflexibility of existing law and institutions and the evolving nature of SLR risk and impacts (Cosens et al., 2017; Craig et al., 2017; DeCaro et al., 2017). Possible limits of in situ responses to ongoing SLR (e.g., protection and accommodation), bring into question prevailing legal approaches to property rights and land use regulation (Byrne, 2012). In addition, because uncertainty about SLR makes it difficult to decide when to wait and when to act, public actors fear being held accountable for misjudgments (Kuklicke and Demeritt, 2016). The long time horizon and uncertainty of SLR make it difficult to mobilise political will and the leadership required to take visionary action (Cuevas et al., 2016; Gibbs, 2016; Yusuf et al., 2016; Yusuf et al., 2018b).

Cross-scale and cross-domain coordination: SLR creates new coordination problems across jurisdictional levels and domains, because impacts cut across scales, sectors and policy domains and responding often exceeds the capacities of local governments and communities (*medium confidence*; Araos et al., 2017; Termeer et al., 2017; Pinto et al., 2018; Clar, 2019; Clar and Steurer, 2019; Sections 4.3.2 and 4.4.2). Local responses are generally nested within a hierarchy of local, regional, national and international governance arrangements and cut across sectors (Cuevas, 2018; Chhetri et al., 2019; Clar, 2019). Furthermore responding to SLR is only one administrative priority amongst many, and the choice of SLR response is influenced by multiple co-existing functional responsibilities and perspectives (e.g., planning, emergency management, asset management and community development) that compete for legitimacy, further complicating the coordination challenge (Klein et al., 2016; Vij et al., 2017; Jones et al., 2019).

Equity and social vulnerability: SLR and responses may affect communities and society in ways that are not evenly distributed, which can compound vulnerability and inequity, and undermine societal aspirations, such as achieving SDGs (*high confidence*; Section 4.3.3.2; Eriksen et al., 2015; Foerster et al., 2015; Sovacool et al., 2015; Clark et al., 2016; Gorddard et al., 2016; Adger et al., 2017; Holland, 2017; Dolšak and Prakash, 2018; Lidström, 2018; Matin et al., 2018; Paprocki and Huq, 2018; Sovacool, 2018; Warner et al., 2018a). Costs and benefits of action and inaction are distributed unevenly, with some coastal nations, particularly small island states, being confronted with adaptation costs amounting to several percent of GDP in the 21st century (Section 4.3.3.2). Land use planning for climate adaptation can exacerbate sociospatial inequalities at the local level, as illustrated in a study of eight cities, namely Boston (USA), New Orleans (USA), Medellin (Colombia), Santiago (Chile), Metro Manila (Philippines), Jakarta (Indonesia), Surat (India), and Dhaka (Bangladesh; Anguelovski et al., 2016). Private responses may also exacerbate inequalities as, for example, in Miami, USA, where purchase of homes in areas at higher elevation has increased property prices displacing poorer communities from these areas (Keenan et al., 2018). In Bangladesh, some adaptation practices have enabled land capture by elites, public servants, the military and roving gangs, and resulted in various forms of marginalisation that compound vulnerability and risk (Sovacool, 2018); a reality also faced by many other coastal communities around the world (Sovacool et al., 2015).

4

Social conflict: Ongoing SLR could become a catalyst for possibly intractable social conflict by impacting human activities, infrastructure and development along low-lying shorelines (*high confidence*). Social conflict refers here to the non-violent struggle between groups, organisations and communities over values, interests, resources, and influence or power, whereby parties seek to achieve their own goals, and may seek to prevent others from realising their goals and possibly harm rivals (Coser, 1967; Oberschall, 1978; Pruitt et al., 2003). SLR impacts that could contribute to conflict include: disruptions to critical infrastructure, cultural ties to the coast, livelihoods, coastal economies, public health, well-being, security, identity and the sovereignty of some low-lying island nations (Sections 4.3.2.4, 4.3.3.2, 4.3.3.6; Mills et al., 2016; Yusuf et al., 2016; Nursey-Bray, 2017; Hinkel et al., 2018). SLR responses inevitably raise difficult trade-offs between private and public interests, short- and long-term concerns, and security and conservation goals, which are difficult to reconcile due to divergent problem framing, interests, values and ethical positions (Eriksen et al., 2015; Foerster et al., 2015; Mills et al., 2016; Termeer et al., 2017; Sovacool, 2018). To some countries, SLR presents a security risk due to the scale of potential displacement and migration of people (Section 4.4.2.6). Climate change, and rising seas in particular, could compound sociopolitical stressors (Sovacool et al., 2015), challenge the efficacy of prevailing legal processes (Byrne, 2012; Busch, 2018; Setzer and Vanhala, 2019), and spark or escalate conflict (Lusthaus, 2010; Nursey-Bray, 2017).

Complexity: SLR introduces novel and complex problems that are difficult to understand and address (*high confidence*; Moser et al., 2012; Alford and Head, 2017; Wright and Nichols, 2018; Hall et al., 2019). As a result of the preceding features of the SLR problem, and the complexity of the nonlinear interactions between biogeophysical and human systems, SLR challenges may be difficult to frame, understand and respond to. Often, disciplinary science is not sufficient for understanding complex problems like SLR and traditional technical problem solving may not be well suited for crafting enduring SLR responses (Lawrence et al., 2015; Termeer et al., 2015). SLR poses a challenge for bridging gaps between science, policy and practice (Hall et al., 2019). The complexity and rapid pace of SLR in some localities is also challenging traditional community decision making practices, for example, in some Pacific Island communities (Nunn et al., 2014).

4.4.4 Planning, Engagement and Decision Tools for Choosing Responses

4.4.4.1 Introduction

A range of established and emerging planning, public participation and conflict resolution practices can help in making social choices and addressing governance challenges in responding to SLR (Section 4.4.4.2). Social choice and decision making processes may also involve the application of formal decision analysis methods for appraising and choosing responses (Section 4.4.4.3). Making social choices involves deciding which combination of decision methods, modes of participation, conflict resolution strategies, and planning processes to use when and how, and then implementing the resultant decisions, monitoring and reviewing progress over time, and making

appropriate adjustments in the light of experience, change and emerging knowledge and needs. The legal literature on roles played by law (Byrne, 2012; Vidas, 2014; Reiblich et al., 2019; Setzer and Vanhala, 2019), or informal approaches such as indigenous decision making practices (Carter, 2018) were not assessed.

4.4.4.2 Planning, Public Participation and Conflict Resolution in the Face of SLR

Land use or spatial planning has the potential to help communities prepare for the future and decide how to manage coastal activities and land use taking into account the uncertainty, complexity and contestation that characterise SLR (*high confidence*; Hurlimann and March, 2012; Hurlimann et al., 2014; Berke and Stevens, 2016; King et al., 2016; Reiblich et al., 2017) Planners work with governing authorities, the private sector, and local communities to integrate and apply tailor-made decision analysis, public participation and conflict resolution approaches that can be institutionalised in statutory provisions, and aligned with informal institutional structures and processes carried out at various scales (Hurlimann and March, 2012; Smith and Glavovic, 2014; Berke and Stevens, 2016).

Planning can play an important role in crafting SLR responses, addressing several of the governance challenges identified above (Section 4.4.3). Planning is future focused and can assist communities to develop and pursue a shared vision, and understand and address SLR concerns in locality-specific ways (Hurlimann and March, 2012; Berke and Stevens, 2016). Planning can help articulate and clarify roles and responsibilities through statutory planning provisions, complemented by non-statutory processes (Vella et al., 2016). It can build social and administrative networks that mobilise cross-scale SLR responses, and facilitate integration of diverse mitigation and adaptation goals alongside other public aspirations and policy imperatives (Hurlimann and March, 2012; Vella et al., 2016). Planning can also facilitate the establishment of collaborative regional forums that cross jurisdictional boundaries and assist local governments and other stakeholders to pool resources and coordinate roles and responsibilities across multiple governance levels, such as the Southeast Florida Regional Climate Change Compact, USA (Shi et al., 2015; Vella et al., 2016). Regulatory planning can be used by governing authorities to steer future infrastructure, housing, industry and related development away from areas exposed to SLR (Hurlimann and March, 2012; Hurlimann et al., 2014; Smith and Glavovic, 2014; Berke and Stevens, 2016).

The extent to which planning is effective in reducing coastal risk, however, varies widely between and within coastal nations (Glavovic and Smith, 2014; Shi et al., 2015; Cuevas et al., 2016; King et al., 2016; Woodruff and Stults, 2016). Planning can fail to prevent development in at-risk locations, and may even accelerate such development, as experienced in settings as diverse as Java, Indonesia (Suroso and Firman, 2018), the Philippines (Cuevas, 2018), Australia (Hurlimann et al., 2014), and the USA (Vella et al., 2016; Woodruff and Stults, 2016). Planning has exacerbated sociospatial inequalities in cities like Boston, USA, Santiago, Chile, and Jakarta, Indonesia (Anguelovski et al., 2016). A study of vulnerability dynamics in Houston, New Orleans and Tampa, USA shows that vulnerability can be reinforced or ameliorated through adaptation planning and decision making processes (Kashem et al., 2016). Regulatory planning may be non-existent in some settings, such as informal settlements, or when used can paradoxically entrench vulnerability and compound risk (Berquist et al., 2015; Amoako, 2016; Ziervogel et al., 2016b). Planning practice is thus both a contributor to and an outcome of local politics and power. Recognising and navigating these challenges is key to realising the promise of planning for reducing SLR risk, and participatory planning processes that reconcile divergent interests are central to this endeavour (Forester, 2006; Smith and Glavovic, 2014; Anguelovski et al., 2016; Cuevas et al., 2016).

Public participation refers to directly involving citizens in decision making processes rather than only indirectly via voting. Citizen participation is commonplace in public decision making that addresses important societal concerns like SLR (Sarzynski, 2015; Berke and Stevens, 2016; Gorddard et al., 2016; Baker and Chapin III, 2018; Yusuf et al., 2018b). Practices sit along a continuum from manipulation to minimal involvement and more empowering and self-determining practices (Arnstein, 1969; International Association for Public Participation, 2018). Public participation draws on a wide variety of tailored engagement processes and practices, from 'serious games' (Wu and Lee, 2015) to role-play simulations (Rumore et al., 2016), and deliberative-analytical engagement (Webler et al., 2016).

There has been a proliferation of public engagement approaches and practices applied to adaptation in recent decades (Webler et al., 2016; Kirshen et al., 2018; Mehring et al., 2018; Nkoana et al., 2018; Yusuf et al., 2018a; Uittenbroek et al., 2019). Increasing citizen participation in adaptation and other public decision making processes shifts the role of government from a chiefly steering and regulating role towards more responsive and enabling roles, sometimes referred to as co-design, co-production, and co-delivery of adaptation responses (Ziervogel et al., 2016a; Mees et al., 2019). Engagement strategies grounded in community deliberation can help to improve understanding about SLR and response options, reducing the polarising effect of alternative political allegiances and worldviews (Akerlof et al., 2016; Uittenbroek et al., 2019). Public participation has also the potential to successfully include vulnerable groups in multi-level adaptation processes (Kirshen et al., 2018), promote justice and enable transformative change (Broto et al., 2015; Schlosberg et al., 2017).

It is widely recognised that authentic and meaningful public participation is important and can help in crafting effective and enduring adaptation responses, but is invariably difficult to achieve in practice (Barton et al., 2015; Cloutier et al., 2015; Sarzynski, 2015; Serrao-Neumann et al., 2015; Berke and Stevens, 2016; Chu et al., 2016; Schlosberg et al., 2017; Baker and Chapin III, 2018; Kirshen et al., 2018; Lawrence et al., 2018; Mehring et al., 2018; Lawrence et al., 2019; Uittenbroek et al., 2019). There is limited empirical evidence that public participation per se improves environmental outcomes (Callahan, 2007; Reed, 2008; Newig and Fritsch, 2009). Major factors determining outcomes are tacit, including trust, environmental preferences, power relationships and the true motivations of sponsor and participants (Reed, 2008; Newig and Fritsch, 2009). Difficulties in realising the anticipated benefits of public participation have been

shown in coastal settings including Queensland, Australia (Burton and Mustelin, 2013), Germany's Baltic Sea (Schernewski et al., 2018), England (Mehring et al., 2018), Sweden (Brink and Wamsler, 2019), and South Africa (Ziervogel, 2019). Research by Uittenbroek et al. (2019) in the Netherlands, for example, shows that public participation objectives are more probable if participation objectives and process design principles and practices are co-produced by community and government stakeholders. In some cities in the Global South, experience shows that a focus on building effective multi-sector governance institutions can facilitate ongoing public involvement in adaptation planning and implementation, and enhance long-term adaptation prospects (Chu, 2016b).

Conflict resolution refers to formal and informal processes that enable parties to create peaceful solutions for their disputes (Bercovitch et al., 2008). They range from litigation and adjudication to more collaborative processes based on facilitation, mediation and negotiation (Susskind et al., 1999; Bercovitch et al., 2008). Such processes can be used in the public domain to make difficult social choices. Whilst it may be impossible to eliminate controversy and disputes due to SLR, conflict resolution can be foundational for achieving effective, fair and just outcomes for coastal communities (Susskind et al., 2015; Nursey-Bray, 2017). Whereas some responses to social conflict (see definition in Section 4.4.3.3) can be destructive (e.g., resorting to violence), constructive approaches to conflict resolution (e.g., negotiation and mediation) can help disputants satisfy their interests and even have transformational adaptation potential (Laws et al., 2014; Nursey-Bray, 2017). Laws et al. (2014), for example, use the term 'hot adaptation' to describe adaptation efforts that harness the energy and engagement that conflict provokes; and create opportunities for public deliberation and social learning about complex problems like SLR. Such an approach has particular relevance in settings most at risk to SLR. Realising this potential is, however, challenging in the face of local politics and the differential power and influence of disputants. These realities have been accounted for in public conflict resolution scholarship and practice for many decades (Forester, 1987; Dukes, 1993; Forester, 2006), and lessons learned are beginning to be applied to adaptation (Laws et al., 2014; Nursey-Bray, 2017; Sultana and Thompson, 2017) and SLR response planning (Susskind et al., 2015). Conflict was turned into cooperation in some villages in floodplains in Bangladesh, for example, by facilitated dialogue and incentivised cooperation between local communities and government, with external facilitator assistance, leading to improved water security in a climate stressed environment (Sultana and Thompson, 2017). At a larger scale, the Mekong River Commission, with its water diplomacy framework, provides an institutional structure and processes, with technical support, and legal and strategic mechanisms, that help to negotiate solutions for complex delta problems and, in so doing, help avert widespread destruction of livelihoods and conflict (Kittikhoun and Staubli, 2018).

Many of the techniques used in planning, public participation and conflict resolution, at times together with decision analysis and support tools, are being applied in combination. In New Zealand, for example, a participatory approach was used to combine dynamic adaptive pathways planning with multi-criteria and real options analysis (Section 4.4.4.3.4) to develop a 100-year strategy to manage coastal hazard risk (Lawrence et al., 2019; see Box 4.1). Public participation

4

thereby helped to shift communities towards a longer-term view and towards considering a wider range of adaptation options and pathways. Such combined approaches are also sometimes referred to as Community Based Adaptation, which involve local people directly in understanding and addressing the climate change risks they face (Box 4.4). These processes and practices are used in many settings, from small, isolated indigenous communities to large-scale coastal infrastructure projects in both the Global North and South. See Table 4.9 in Section 4.4.5 for illustrative examples.

4.4.4.3　　Decision Analysis Methods

4.4.4.3.1　Introduction

Decision analysis methods are formal methods that help to identify alternatives that perform best or well with regard to given objectives. An alternative (also called response option or, as a sequence of options over time: adaptation pathway) is a specific combination of SLR responses (See Section 4.4.3). Each alternative is characterised for each possible future state-of-the world (e.g., levels of SLR or socioeconomic development) by one or several attributes, which may measure any relevant social, ecological, or economic effect associated with choosing and implementing the alternative (Kleindorfer et al., 1993). Attributes commonly used include cost of adaptation alternatives, monetary and non-monetary benefits of the SLR impacts avoided, or net present value (NPV), which is the difference between discounted monetised benefits over time and discounted costs over time. Formal decision analysis is one way to support social choices that is generally suggested for decision support if decisions are complex and involve large investments, as is frequently the case in coastal contexts in the face of SLR.

In order to be effective, decision analysis needs to be embedded in a governance process that accounts for societal needs and objectives (Sections 4.4.4.2 and 4.4.5). This is because decision analysis entails a number of normative choices about the objectives chosen, the criteria used, the specific methods and data applied, the set of alternatives considered, and the attributes used to characterise alternatives. These choices need to reflect the diversity of values, preference and goals of all stakeholders involved in and affected by a decision. Furthermore, decision analysis needs to consider all available knowledge, including all major uncertainties in both climate and non-climate factors, ambiguities in expert opinions, and differences in approaches, because a partial consideration of uncertainty and ambiguity could misguide the choice of adaptation alternatives (*high confidence*; Renn, 2008; Jones et al., 2014; Hinkel and Bisaro, 2016).

Since AR5, the literature on coastal decision analysis has advanced significantly, specifically addressing the large uncertainty about post-2050 SLR through i) using robust decision approaches instead of expected utility, ii) iterating or adapting decisions over time, and iii) increasing flexibility of responses. Each advance is elaborated below. Furthermore, the coastal decision analysis literature also stresses the consideration of multiple criteria or attributes, because adaptation often involves stakeholders with differing objectives and ways of valuing alternatives (Oddo et al., 2017). Many decision making methods combine each of the three advances highlighted here (Marchau et al., 2019). The suitability of each method depends strongly on the specific context, including available resources, technical capabilities, policy objectives, stakeholder preferences and available information.

4.4.4.3.2　Using robustness criteria instead of expected utility

A growing literature on decision analysis of coastal adaptation advocates the use of RDM approaches instead of maximising expected utility approaches (Hallegatte et al., 2012; Haasnoot et al., 2013; Lempert et al., 2013; Wong et al., 2017). The core criterion to be considered for choosing between the two types of approaches is whether one is confronted with a situation of shallow or deep uncertainty (*high confidence*) (Lempert and Schlesinger, 2001; Kwakkel et al., 2010; Kwakkel et al., 2016b; Hinkel et al., 2019). Uncertainty is shallow when a single unambiguous objective or subjective probability distribution can be attached to states-of-the-world. Uncertainty is deep, when this is not possible, either because there is no unambiguous method for deriving objective probabilities or the subjective probability judgements of parties involved differ (Cross-Chapter Box 4 in Chapter 1; Type 2).

Expected utility approaches can only be applied for identifying an optimal response in situations of shallow uncertainty. This is because these approaches require a probability distribution over states of the world in order to identify the optimal alternative which leads to the highest expected utility (i.e., the probability weighted sum of the utilities of all outcomes under a given alternative and all states-of-the-world; Simpson et al., 2016). A prominent example of this approach is cost-benefit analysis under risk, which assesses expected outcomes across states of the world in terms of NPV (the discounted stream of net benefits). Cost-benefit analysis has several well-known limitations, such as its sensitivity to discount rates and the difficulty to monetise ecological, cultural and other intangible benefits (Section 1.1.4) that have been widely discussed in the climate change literature (Chambwera et al., 2014; Kunreuther et al., 2014; Dennig, 2018).

In the context of coastal adaptation, uncertainty is only shallow if projected SLR does not significantly differ between low end (e.g., RCP2.6) and high end (e.g., RCP8.5) scenarios (Hinkel et al., 2019). The point in time when this is the case (i.e., time of scenario divergence) depends on what difference in expected utility matters to the particular stakeholders involved in a decision. The time of scenario divergence also differs across locations. In locations where the internal sea level variability is large as compared to relative SLR, it takes longer before the differences in sea levels under low end and high end scenarios become apparent. Figure 4.15 illustrates this effect for the ESL projections of this report (Sections 4.2.3.2 and 4.2.3.4), following the approach of Hinkel et al. (2019). Under the assumption that a 10% statistical distance between the distributions of RCP2.6 and RCP8.5 is decision relevant, scenario divergence occurs before 2050 for approximately two thirds of coastal sites with sufficient observational data, but for 7% of locations this occurs later than 2070.

In principle, a single unambiguous probability distribution on future sea levels could also be attained beyond the time of scenario divergence by attributing subjective probabilities to emission scenarios, but individuals may significantly disagree in their subjective probabilities, which again results in deep uncertainty (Lempert and Schlesinger, 2001; Stirling, 2010). For this reason, very few studies that assign subjective probabilities to emission scenarios are found in the literature (Woodward et al., 2014; Abadie, 2018). But even before the year of scenario divergence is reached, uncertainty about relative SLR can be deep, because of deep uncertainties about non-climatic contributors to relative sea level change, such as VLM during or after earthquakes and human-induced subsidence (Cross-Chapter 5 in Chapter 1; Section 4.2.2.4; Hinkel et al., 2019).

Under situations of deep uncertainty, RDM approaches aim to identify alternatives that perform reasonably well (i.e., 'are robust') under a wide range of states-of-the-world or scenarios and hence do not require probability assessments. These approaches include minimax or minimax regret (Savage, 1951), info gap theory (Ben-Haim, 2006), robust optimisation (Ben-Tal et al., 2009) and exploratory modelling methods that create a large ensemble of plausible future scenarios for each alternative, and then use search and visualisation techniques to extract robust alternatives (Lempert and Schlesinger, 2000). SLR examples of RDM include Brekelmans (2012) who minimise the average and maximum regret across a range of SLR scenarios for investments in dike rings in the Netherlands and Lempert et al. (2013) who apply RDM in Hoh-Chi-Minh City.

But even if SLR uncertainty is shallow, RDM are more suitable than expected utility approaches if parties involved or affected by a decision have a low uncertainty tolerance, because the goal of the uncertainty intolerant decision maker is to avoid major damages under most or all circumstances (Hinkel et al., 2019).

An adaptation strategy developed based on the maximisation of expected utility may not meet this goal, because worst case damages occurring can exceed expected damages by orders of magnitude.

The uncertainty tolerance of stakeholders is also determining how large of a SLR range needs to be considered in RDM. Stakeholders (i.e., those deciding and those affected by a decision) that have a high uncertainty tolerance (e.g., those planning for investments that can be very easily adapted) can use the combined *likely* range of RCP2.6 and RCP8.5 (0.29–1.10 m by 2100) for long-term adaptation decisions. For stakeholders with a low uncertainty tolerance (e.g. those planning for coastal safety in cities and long term investment in critical infrastructure) it is meaningful to also consider SLR above this range, because a 17% chance of GMSL exceeding this range under RCP8.5 is too high to be tolerated from this point of view (Ranger et al., 2013; Hinkel et al., 2015; Hinkel et al., 2019).

Independent of the debate about whether to apply expected utility or robust decision making approaches, there is an extensive literature that applies scenario-based cost-benefit analysis. For example, this approach has been applied for setting the safety standards of Dutch dike rings (Kind, 2014; Eijgenraam et al., 2016), exploring future protection alternatives for New York (Aerts et al., 2014), Ho Chi Minh City (Scussolini et al., 2017), and for many other locations. Scenario-based cost-benefit analysis differs from cost-benefit analysis under risk discussed above in that scenario-based cost-benefit analysis is not applied to rank alternatives across scenarios, but a 'separate' cost-benefit analysis is applied within each emission or SLR scenario considered. While this identifies the optimal alternative under each scenario, it does not formally address the problem faced by a coastal decision maker, namely to decide across scenarios (Lincke and Hinkel, 2018). Nevertheless, the results of scenario-based cost-benefit analysis (i.e., NPV of each alternative under each scenario) provide guidance for decision makers and can also be used as inputs (i.e., as attributes) to robust and flexible decision making approaches.

4.4.4.3.3 Adapting decisions over time

Irrespective of whether expected utility or robustness criteria are applied, there is *high confidence* that an effective way of dealing with large uncertainties is adaptive decision making (also called iterative decision making, adaptive planning or adaptive management), which maintains that decision and decision analysis should be conducted within an iterative policy cycle. This approach includes monitoring of sea level variables and evaluation of alternatives in this light in order to learn from past decisions and collect information to inform future decisions (Haasnoot et al., 2013; Barnett et al., 2014; Burch et al., 2014; Jones et al., 2014; Wise et al., 2014; Kelly, 2015; Lawrence and Haasnoot, 2017). Such a staged approach is especially suitable for coastal adaptation due to the long lead and lifetimes of many coastal adaptation measures and the deep uncertainties in future sea levels (Hallegatte, 2009; Kelly, 2015). Prominent representatives of methods that entail this idea are Dynamic Adaptive Policy Pathways (Haasnoot et al., 2013) and Dynamic Adaptation Planning (Walker et al., 2001). An important prerequisite for any adaptive decision-making approach is a monitoring system that can detect sea level signals sufficiently early to enable the required responses (Hermans et al., 2017; Haasnoot et al., 2018; Stephens et al., 2018).

In recent years, many different frameworks for adaptive decision making have been put forward, including Adaptive Policy Making (Walker et al., 2001), Dynamic Adaptive Policy Pathways (Haasnoot et al., 2013), Dynamic Adaptive Planning (Walker et al., 2013), Iterative risk management (Jones et al., 2014) and Engineering Options Analysis (de Neufville and Smet, 2019). Each frameworks emphasises particular aspects of adaptive decision making and has merits in specific situations depending on the preferences, goals, uncertainties and information at stake (Marchau et al., 2019). Nevertheless, all of these frameworks share the following generic and iterative steps

1. Set the stage: Identify current situation, objectives, options (alternatives) and uncertainties.
2. Develop a dynamic plan, which consists of a basic plan plus contingency actions to be carried out based on observed triggers.
3. Implement basic plan and monitor system for triggers.
4. Monitor and act upon triggers.

4

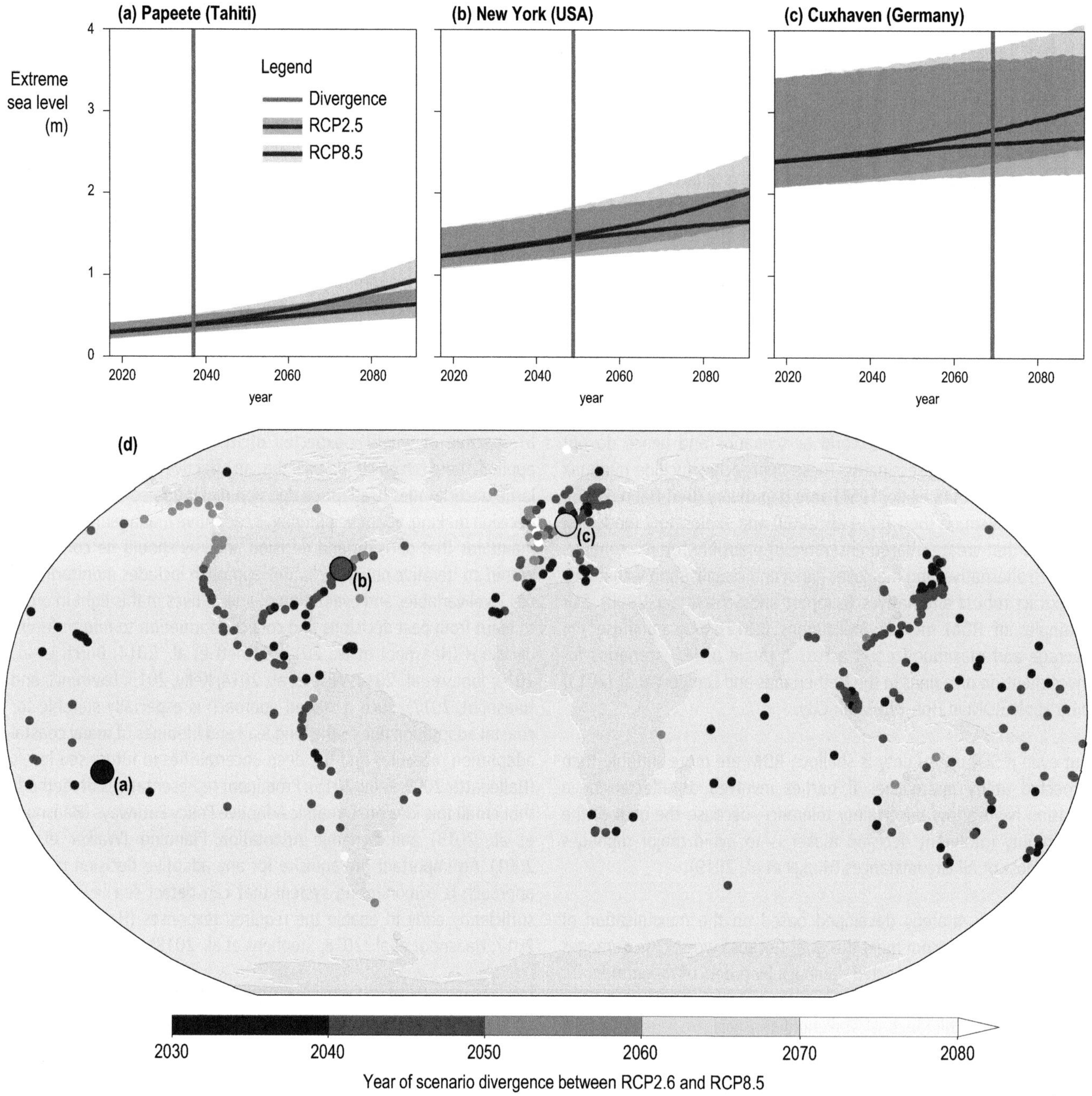

Figure 4.15 | Year of scenario divergence between extreme sea level projections for Representative Concentration Pathway (RCP)2.6 and RCP8.5 for all tide-gauge locations with sufficient observational data relative to a 1986–2005 baseline (bottom panel). Time of divergence is defined using a 10% threshold in the statistical distance between the two distributions, which can be graphically interpreted as the first year in which at least 10% of the area under the probability distribution function (PDF) of RCP8.5 lies outside of the area under the upper half (i.e., above the 50th percentile) of the PDF of RCP2.6. Upper panels indicate the median and 5–95% range of future extreme sea level (ESL) relative to the 1986–2005 baseline for three tide gauge locations with low variability (Papeete), medium variability (New York) and high variability (Cuxhaven). Locations with low variability have a relatively early scenario divergence.

4.4.4.3.4   Increasing flexibility of responses

An idea closely related to adaptive decision making is to keep future alternatives open by favouring flexible alternatives over non-flexible ones. An alternative is said to be 'flexible' if it allows switching to other alternatives once the implemented alternative is no longer effective. For example, a flexible protection approach would be to build small dikes on foundations designed for higher dikes, in order to be able to raise dikes in the future should SLR necessitate this.

A prominent and straightforward method that addresses the objective of flexibility is adaptation pathways analysis (Haasnoot et al., 2011; Haasnoot et al., 2012), which is one component of Dynamic Adaptive Policy Pathways. The method graphically represents alternative combinations of measures over time together with information on the conditions under which alternatives cease to be effective in meeting agreed objectives, as well as possible alternatives that will then be available. As time and SLR progress, monitoring may trigger a decision to switch to another alternative. Adaptation pathway analysis has been widely applied both in the scientific literature as well as in practical cases. Applications after AR5 include Indonesia (Butler et al., 2014), New York City (Rosenzweig and Solecki, 2014), Singapore (Buurman and Babovic, 2016) and Australia (Lin and Shullman, 2017). In New Zealand, the method has been included in national guidance for coastal hazard and climate change decision making (Lawrence et al., 2018). There is *high confidence* that the method is useful in interaction with decision makers and other stakeholders, helping to identify possible alternative sequences of measures over time, avoiding lock-in, and showing decision makers that there are several possible pathways leading to the same desired future (Haasnoot et al., 2012; Haasnoot et al., 2013; Brown et al., 2014; Werners et al., 2015).

Alternatives can also be characterised through multiple attributes such as costs, effectiveness, co-benefits, social acceptability, etc., which in turn can be used in multi-attribute decision making methods (Haasnoot et al., 2013). An important attribute is transfer cost, which is the cost of course correction (switching from one alternative to another), reflecting the potential for path dependency (Haasnoot et al., 2019). Delaying decisions and opting for flexible measures introduces extra costs, such as transfer costs. Also, flexible measures are often more expensive than inflexible ones, and damages may occur whilst delaying the decision. An important question therefore is whether it is cheaper to implement a flexible measure now or to wait and implement a less flexible (i.e., cheaper) measure later in time when more information is at hand.

Technically more demanding methods such as real-options analysis (Dixit et al., 1994), and decision tree analysis (Conrad, 1980), can also find pathways that are economically efficient in terms of flexibility and timing of adaptation. There is little application of these approaches in the SLR literature. For example, Woodward et al. (2014) applied real-options analysis to determine flood defences around the Thames Estuary, London, England; Buurman and Babovic (2016) for climate-proofing drainage networks in Singapore; Dawson et al. (2018) for coastal rail infrastructure in southern England; and Kim et al. (2018) for assessing flood defences in southern England. A requirement for applying real-options analysis and decision tree analysis is to quantify today how much will have been learned at a given point in time in the future. The few applications of these methods to SLR-related decisions in the literature have generally used ad-hoc assumptions. For example, Woodward et al. (2011) assumed either perfect learning (i.e., in 2040, which SLR trajectory is occurring will be known) or no learning (i.e., uncertainty ranges and confidence in these remains as today). Others have derived learning rates from comparing past progress in SLR projections and then applied these to the future. An example is given by Dawson et al. (2018) who derive learning rates from the 2002 and 2009 SLR projections of the UK Climate Impacts Programme and apply these in real-options analysis.

4.4.4.3.5   Research needs

Four general gaps can be identified in the literature. First, the generation of SLR information is insufficiently coupled to the use of this information in decision analysis. This constitutes a limitation, as different coastal decision contexts require different decision analysis methods, which in turn require different SLR information. Specifically, applications of decision analysis methods generally convert existing sea level information to fit their method, often misinterpreting the information, making arbitrary assumptions or losing essential information in the process (Hinkel et al., 2015; Bakker et al., 2017; Van der Pol and Hinkel, 2018). Second, with the exception of adaptation pathway analysis, methods of robust and flexible decision making are under-represented in the literature despite their suitability (Van der Pol and Hinkel, 2018). Third, research is necessary to compare the various methods, to identity which methods are most suitable in which context and to develop consistent categorisations of methods (Hallegatte et al., 2012; Haasnoot et al., 2013; Hinkel et al., 2015; Watkiss et al., 2015; Dittrich et al., 2016; Suckall et al., 2018). Fourth, future research needs to address how to embed decision analysis better in real world planning and decision making processes, recognising that adaptation to SLR is a multi-stakeholder process often characterised by conflicting interests and interdependence between stakeholders (Section 4.4.3). Addressing these gaps requires closer cooperation between SLR sciences, decision science, and planning and governance scholars. An underlying challenge is to design and integrate relevant formal decision making approaches into the heterogeneous reality of local planning and decision making cultures, institutions, processes and practices, often with community-specific needs and requirements (see Box 4.4).

Box 4.4 | Community Based Experiences: Canadian Arctic and Hawkes Bay, New Zealand

Climate-Change Adaptation on the Canadian Arctic Coast

Communities of the Inuvialuit Settlement Region (ISR), established under the Inuvialuit Final Agreement (Government of Canada, 1984), include the delta communities of Aklavik and Inuvik (the regional hub) and the coastal hamlets of Tuktoyaktuk, Paulatuk, Sachs Harbour, and Ulukhaktok. The Inuvialuit Regional Corporation (IRC) administers Inuvialuit lands, a portfolio of businesses, and social and cultural services, including co-management of food harvest resources. Other social, education, health, and infrastructure services are managed by the Government of the Northwest Territories and municipal Councils. Community Corporations and Hunters and Trappers Committees handle other aspects of governance and socioeconomic development. Very high ground-ice content renders the coast and coastal infrastructure in this region sensitive to rising temperatures and largely precludes conventional hard shore protection. Higher temperatures (>3°C rise since 1948), combined with rising sea level and a lengthening open-water season, contribute to accelerating coastal erosion, threatening infrastructure, cultural resources, and the long-term viability of Tuktoyaktuk Harbour, while impacting winter travel on ice, access to subsistence resources, food security, safety and well-being (Lamoureux et al., 2015). Despite ongoing shore recession, there is strong attachment to the most vulnerable sites and a reluctance to relocate. Adaptation challenges include technical issues (e.g., the ice-rich substrate, sea ice impacts), high transportation costs (until recent completion of an all-weather road to Tuktoyaktuk, heavy or bulky material had to come in by sea or ice road), availability of experienced labour, and, crucially, financial resources. Other inhibitors of adaptation include access to knowledge in suitable forms for uptake, gaps in understanding, research readiness, and institutional barriers related to multiple levels of decision making (Ford et al., 2016a). The IRC, as the indigenous leadership organisation in the ISR, is moving to play a more proactive role in driving adaptation at the regional level (IRC, 2016), as indigenous leaders across the country are demanding more control of the northern research agenda for adaptation action (Bell, 2016; ITK, 2018). For a number of years, IRC has promoted community-based monitoring, incorporating Inuvialuit knowledge in partnership with trusted research collaborators. Recently IRC is exploring partnership with the community-based ice awareness service and social enterprise, SmartICE Inc. Despite the inherent adaptability of Inuit culture, concentration in locality bound communities dependent on physical infrastructure has increased vulnerability, as changing climate has raised exposure. Various government and academic initiatives and tools over many years to promote resilience and adaptation strategies have had limited impact. Current engagement supporting locally driven knowledge acquisition and management capacity, combined with IRC institutional leadership with government support, are expected to enable a more effective co-designed and co-delivered adaptation agenda.

Clifton to Tangoio Coastal Hazards Strategy 2120 Hawkes Bay, New Zealand:
A community-based and science-informed decision making process

New Zealand is applying tools for decision making under deep uncertainty (Lawrence and Haasnoot, 2017; Lawrence et al., 2019) to address the changing risk and uncertainties related to SLR impacts on coastal settlements. An opportunity arose in 2017 for the Resilience to Nature's Challenges National Science Challenge, "The Living Edge" research project (https://resiliencechallenge. nz/edge), to co-develop a Coastal Hazards Strategy for the Clifton to Tangoio coast in Hawkes Bay, New Zealand. The Strategy, a joint council/ community/ iwi Māori initiative (Kench et al., 2018), planned to use Multi-Criteria Decision Analysis (MCDA) within their decision framework – a static tool in time and space, unsuited for decision making where changing risk and uncertainties exist over long timeframes. The Dynamic Adaptive Policy Pathways (Haasnoot et al., 2013) approach and a modified Real Options Analysis were proposed by the 'Edge Team' and integrated with MCDA (Lawrence et al., 2019) in a process comprising a Technical Advisory Group of councils and two community panels of directly affected communities, infrastructure agencies, business, conservation interests, and iwi Māori. Many adaptation options and pathways were assessed for their ability to reduce risk exposure and maintain flexibility for switching pathways over a 100-year timeframe, which is the planning timeframe mandated by the New Zealand Coastal Policy Statement (Minister of Conservation, 2010) with the force of law under the Resource Management Act, 1991. The recently revised New Zealand national coastal hazards and climate change guidance for local government (Bell et al., 2017a) provided context. This novel assessment, engagement and planning approach to the formulation of a Coastal Hazards Strategy was undertaken through a non-statutory planning process. The agreed options and pathways have yet to be implemented through statutory processes that will test the risk tolerance of the wider community. This example illustrates how tailor-made assessment that addresses SLR uncertainty and change by keeping options open and reducing path-dependency, and engagement and planning processes can be initiated with leadership across councils, sectors and stakeholders, before being implemented, thus reducing contestation. Lessons learned include the central role of: leadership, governance and iwi Māori; Local Authority collaboration; taking time to build trust; independent knowledge brokers for credibility; nuanced project leadership and facilitation, enabling a community-based process. The preferred intervention options and pathways have been agreed for implementation. The remaining challenges are to cost the range of actions, decide funding formulae, develop physical and socioeconomic signals and triggers for monitoring changing risk, embed the strategy in statutory plans and practices, and socialise the strategy with the wider public in the context of competing priorities.

4.4.5   Enabling Conditions and Lessons Learned From 'Practice'

In addition to the literature on planning, public participation, conflict resolution and decision making assessed in the last Section, much is being learned from practical experiences gained in adapting to climate change and SLR at the coast. Some salient enabling conditions and lessons learnt are illustrated in Table 4.9 through case studies or examples of real-world experience in diverse coastal communities around the world, structured according to the five overarching SLR governance challenges identified in Section 4.4.3. In these cases, the following stands out as being foundational for enabling the implementation of SLR responses and addressing the governance challenges that arise. First, effective SLR responses take a long-term perspective (e.g., 100 years and beyond) and explicitly account for the uncertainty of locality-specific risks beyond 2050. Second, given the locality-specific but cross-cutting nature of SLR impacts, improving cross-scale and cross-domain coordination of SLR responses may be beneficial. Third, prioritising social vulnerability and equity in SLR responses may be essential because SLR impacts and risks are spread unevenly across society, and within and between coastal communities. Fourth, safe community arenas for working together constructively can help to resolve social conflict arising from SLR. Fifth, a sharp increase may be needed in governance capabilities to tackle the complex problems caused by SLR. There is, however, no one-size-fits-all solution to SLR, and responses need to be tailored to the environmental, social, economic, political, technological, and cultural context in which they are to be implemented. Enablers that work in one context might not be effective in another case. As sea level rises, more experience in addressing SLR governance challenges will be gained, which can in turn be evaluated in order to obtain a better contextual understanding of enabling conditions and effective SLR governance.

Table 4.9 | Enablers and lessons learned to overcome governance challenges arising from sea level rise (SLR).

Governance challenges	Enablers and lessons learned	Illustrative examples
Time horizon and uncertainty	Take action now with the long-term in mind, keeping options open so that new responses can be developed over time (*high confidence*) (Section 4.4.2) (Haasnoot et al., 2013; Hurlimann et al., 2014; Dewulf and Termeer, 2015; Termeer et al., 2015; Stephens et al., 2018; OECD, 2019)	**Participatory scenario planning has been used widely** including in Lagos, Nigeria (Ajibade et al., 2016), Dhaka, Bangladesh (Ahmed et al., 2018), Rotterdam, Netherlands, Hong Kong and Guangzhou, China (Francesch-Huidobro et al., 2017), Maputo, Mozambique (Broto et al., 2015), Santos, Brazil (Marengo et al., 2019), Arctic (Flynn et al., 2018), Indonesia (Butler et al., 2016a), Dutch delta (Dewulf and Termeer, 2015; Termeer et al., 2015; Bloemen et al., 2019) and Bangladesh (Paprocki and Huq, 2018). Lessons include: – Develop shared coastal visions (Tuts et al., 2015; Brown et al., 2016; OECD, 2019) – Use participatory planning processes that respect and reconcile different values, belief systems and cultures (Flynn et al., 2018) – Address power imbalances and human development imperatives (Broto et al., 2015; Butler et al., 2016a) **Long-term adaptation pathways have been developed in New Zealand** using 'serious games' (Flood et al., 2018) and hybrid processes to integrate decision analysis methods (Section 4.4.2) with public participation and planning (Section 4.4.3) (Cradock-Henry et al., 2018; Lawrence et al., 2019). Lessons include: – Develop enabling national guidance, policy and legislation that requires a long-term focus (e.g., 100 years) and prioritises measures to minimise risk escalation – Secure buy-in from key governance actors – Involve coastal stakeholders in adaptation planning – Draw on local, indigenous and scientific knowledges
	Avoid new development commitments in high-risk locations (Section 4.4.3) (*medium evidence, high agreement*) (Hurlimann and March, 2012; Glavovic and Smith, 2014; Hurlimann et al., 2014; Tuts et al., 2015; Berke and Stevens, 2016; Butler et al., 2016b; OECD, 2019)	**Spatial planning to regulate development at risk from SLR is** underway in many locations, including Victoria, Australia (Hurlimann et al., 2014) and Florida, USA (Butler et al., 2016a; Vella et al., 2016). Limiting future development in high risk areas is much easier than dealing with existing assets at risk (Tuts et al., 2015; OECD, 2019) **Proactive managed retreat through flexible, tailor-made provisions that address distinctive local circumstances is under way** in, for example, USA and Australia, revealing the importance of understanding risks politicians face from local opposition, and distributional impacts (Dyckman et al., 2014; Gibbs, 2016; Siders, 2019). Post Hurricane Sandy managed retreat from Staten Island, New York City, USA, was enabled by community receptivity to buyouts and political expedience (Koslov, 2019; Box 4.1). Lessons include: – Limit new development commitments in high risk areas – Facilitate property abandonment as inundation occurs – Leverage the window of opportunity coastal disasters create (Kousky, 2014)

Governance challenges	Enablers and lessons learned	Illustrative examples
Cross-scale and cross-domain coordination	Build vertical and horizontal governance networks and linkages across policy domains and sectors to legitimise decisions, build trust and improve coordination (*high confidence*). (Glavovic and Smith, 2014; Colenbrander and Sowman, 2015; Dutra et al., 2015; Sowman et al., 2016; Van Putten et al., 2016; Forino et al., 2018; Lund, 2018; Pinto et al., 2018; Clar, 2019; Pittman and Armitage, 2019)	In the Lesser Antilles multiple state and non-state actors are working together, building trust, and coordinating activities through decentralisation and self-organisation (Pittman and Armitage, 2019). Lessons include: – Participation in collaborative projects – Multilateral agreements between states – Boundary spanning organisations connecting governance actors, citizens and states – Extreme events can be a catalyst for raising awareness and political salience – Leadership by central actors and capacitated teams – Mobilise capabilities of communities and non-state actors In South Devon, UK, decentralisation, privatisation and fragmentation impacts adaptation (Den Uyl and Russel, 2018). Lessons include: – Identify policy inconsistencies and clarify problem-ownership, responsibility and accountability – Explore ways to leverage national funding to support local action – Establish networks to facilitate interaction, dialogue and coordination
	Social learning, experimentation and innovation inform technical solutions, build shared understanding and develop locally appropriate SLR responses (*high confidence*). (Dyckman et al., 2014; Glavovic and Smith, 2014; Dutra et al., 2015; Ensor and Harvey, 2015; Chu et al., 2018; McFadgen and Huitema, 2018; Mazeka, 2019; Wolfram et al., 2019)	Innovation is underway to enhance social learning, reflexivity and coalition building (Chu et al., 2018; Bellinson and Chu, 2019; Wolfram et al., 2019), e.g., Surat, India (Chu, 2016a; Chu, 2016b), Santos, Brazil (Marengo et al., 2019), Portland, USA (Fink, 2019), and port cities in Europe and East Asia (Blok and Tschötschel, 2016), In Surat, for example, adaptation experiments created valuable arenas for engaging governance actors and stakeholders, understanding climate and development co-benefits, and testing new ideas (Chu, 2016b). Lessons include: – Design experiments to account for how local political economic factors shape adaptation, for example, understanding local history and politics reveals how adaptation trade-offs are made in city decision making – Ensure experiments generate socioeconomic benefits and climate-development co-benefits Accelerate social learning and governance innovations through transnational municipal networks together with local efforts (Hughes et al., 2018), with processes developed and institutionalised through political negotiation, e.g., Rotterdam, Netherlands, and Berkeley, USA (Bellinson and Chu, 2019)
Equity and social vulnerability	Recognise the political nature of adaptation and explicitly address vulnerability and equity implications to achieve enduring, enabling impact of responses (*high confidence*). (Eriksen et al., 2015; Sovacool et al., 2015; Tuts et al., 2015; Adger et al., 2017; Hardy et al., 2017; Holland, 2017; Dolšak and Prakash, 2018; Finkbeiner et al., 2018; Sovacool, 2018; Warner et al., 2018b; OECD, 2019)	Rights-based approach to participatory adaptation planning in Maputo, Mozambique, fosters a more inclusive and potentially fairer city (Broto et al., 2015). Lessons include: – Expose drivers of structural inequity and vulnerability – Link adaptation and human development imperatives – Raise awareness and public support for adaptation with equity Race-aware adaptation planning can reveal racial inequalities and overcome passive indifference as shown in, for example, Sapelo Island, Georgia, USA (Hardy et al., 2017). Lessons include: – Develop an understanding of historical racial drivers of coastal land ownership, development and risk – Address barriers African Americans face in participating in adaptation planning
	Focus on enabling community capabilities for responding to SLR, where necessary complementing community knowledge, skills and resources, and political influence and problem solving abilities, with external assistance and government support (*high confidence*). (Schlosberg, 2012; Musa et al., 2016; Vedeld et al., 2016; Elrick-Barr et al., 2017; Warrick et al., 2017; Dolšak and Prakash, 2018)	Various professionals can play valuable support roles in **leveraging and building adaptive capacity and resilience of small island communities,** recognising diverse needs and capabilities (Robinson, 2017; Weir et al., 2017; Kelman, 2018; Petzold and Magnan, 2019). For example, in poor Caribbean communities, social workers are helping strengthen social capital, enabling individuals to understand and integrate risk, resilience and sustainability principles into day-to-day decision making, and promoting socially and environmentally just adaptation (Joseph, 2017). In the Solomon Islands, Pacific, **community-based approaches enhance community capacity** to work with external organisations to plan together, obtain resources, and respond to SLR on their own terms (Warrick et al., 2017). The value of integrating traditional community responses with local government efforts has been demonstrated in Micronesian islands (Nunn et al., 2017b). **Local collective action in Monkey River, Belize, helped to overcome power asymmetries and to obtain support otherwise unavailable to vulnerable community members.** Working with journalists, researchers and local NGOs, was key for villagers to have concerns heard and a solution found for coastal erosion (Karlsson and Hovelsrud, 2015). **Rural coastal community resilience boosted in Albemarle Pamlico Peninsula of North Carolina, USA,** by focused attention on local needs through capacity building, and ensuring local voices heard in adaptation planning (Jurjonas and Seekamp, 2018).

4

Governance challenges	Enablers and lessons learned	Illustrative examples
Social conflict	Social conflict can be reduced by tailor-made design and facilitation of participation processes, and involving stakeholders early and consistently throughout decision making and implementation of SLR responses (*medium evidence, high agreement*). (Burton and Mustelin, 2013; Berke and Stevens, 2016; Gorddard et al., 2016; Webler et al., 2016; Schlosberg et al., 2017; Kirshen et al., 2018; Lawrence et al., 2018; Mehring et al., 2018; Nkoana et al., 2018; Schernewski et al., 2018; Yusuf et al., 2018b; Uittenbroek et al., 2019)	**Public participation has been foundational for South Africa's coastal management, risk reduction and adaptation efforts since 1994** (Celliers et al., 2013; Daron and Colenbrander, 2015; Desportes and Colenbrander, 2016; Glavovic et al., 2018; Colenbrander, 2019). Lessons include: – Create opportunities to understand and address technical, sociopolitical and economic realities in an integrated way (Colenbrander and Sowman, 2015; Daron and Colenbrander, 2015) – Incorporate conflict resolution mechanisms into engagement processes (Daron and Colenbrander, 2015; Colenbrander et al., 2016; Colenbrander and Bavinck, 2017) – Align informal engagement processes with formal statutory provisions (Colenbrander and Bavinck, 2017), taking into account visible formal procedures and 'invisible' and informal ways in which knowledge is shared and shapes government decision making (Leck and Roberts, 2015) – Independent facilitators can play a crucial role bringing contending parties together; local government officials can work as bureaucratic activists to create more inclusive, iterative and reflexive participation (Desportes and Colenbrander, 2016) – Sustain engagement, sequence participatory interventions with political and bureaucratic cycles (Pasquini et al., 2013) and secure enabling resources, including channelling adaptation finance to local level (Colenbrander, 2019) – Use practical ways to involve historically disadvantaged and socially vulnerable groups and communities, for example, by choosing accessible locations, language(s) and culturally appropriate meeting protocols (Sowman and Gawith, 1994; Ziervogel et al., 2016b) – Dedicated environmental champions within local political leadership play a key role in mainstreaming adaptation into local decision making (Pasquini et al., 2015)
	Social conflict can be managed by creating safe arenas for inclusive, informed and meaningful deliberation, negotiation and collaborative problem-solving (*medium evidence, high agreement*). (Susskind et al., 1999; Laws et al., 2014; Susskind et al., 2015; Glavovic, 2016; Nursey-Bray, 2017; Sultana and Thompson, 2017)	**Turning conflict into cooperation in Baragaon village, northeast Bangladesh, and eight villages in Narial district, southwest Bangladesh:** A flexible and enabling process, founded on local institutions judged robust and fair, prompted government investment in communities beyond their traditional focus on water infrastructure, paid attention to local social dynamics and reduced elite domination and local conflict (Sultana & Thompson 2017). Lessons include: – Use local knowledge to inform adaptation actions – Encourage institutional improvisation to address local concerns, for example, shifting government investment from water infrastructure to community development – Use external facilitation – Incentivise participation by disadvantaged groups **Innovative collective coastal risk management process, New England (USA) Climate Adaptation Project,** developed by university researchers with partner communities to address coastal conflict by: – Building community risk literacy, optimism and collaborative problem-solving capacity to take action – Joint fact-finding, scenario planning, negotiating trade-offs, facilitated public dialogue, and securing support for collaborative adaptation – Establishing forums for ongoing public deliberation and social learning; and committing to continual adjustments in face of change (Rumore, 2014; Susskind et al., 2015)
Complexity	Drawing upon multiple knowledge systems to co-design and co-produce SLR responses results in more acceptable and implementable responses (*high confidence*). (Dannevig and Aall, 2015; Dutra et al., 2015; Sovacool et al., 2015; Desportes and Colenbrander, 2016; Adger et al., 2017; Betzold and Mohamed, 2017; Onat et al., 2018; Warner et al., 2018b; St. John III and Yusuf, 2019)	**The merits of drawing on scientific and local and indigenous knowledges is recognised in diverse settings** such as Australia (Dutra et al., 2015), Comoros (Betzold and Mohamed, 2017), Arctic (Flynn et al., 2018; Huntington et al., 2019), Canada (Chouinard et al., 2015; Chouinard et al., 2017), Portugal (Costas et al., 2015) and Brazil (Marengo et al., 2019). **Storytelling can build shared knowledge and understanding** because stories are engaging, help people visualise problems, see things from different positions, and recognise shared goals (Dutra et al., 2015; Elrick-Barr et al., 2017). Māori, indigenous people of New Zealand, use oral history and storytelling to describe their relationship to the coast, which informs how New Zealand responds to SLR (Carter, 2018; Lawrence et al., 2018). **Gaps between SLR science, policy and practice can be bridged by adaptation policy experiments with support of actors and organisations who work across organisational boundaries to bring parties together** (Dannevig and Aall, 2015; St. John III and Yusuf, 2019).

4

Governance challenges	Enablers and lessons learned	Illustrative examples
Complexity	**Build governance capabilities to tackle complex problems** (*medium evidence, high agreement*) (Moser et al., 2012; Head, 2014; Dewulf and Termeer, 2015; Head and Alford, 2015; Termeer et al., 2015; Kwakkel et al., 2016a; Termeer et al., 2016; Alford and Head, 2017; Daviter, 2017; Head, 2018; McConnell, 2018)	**The Dutch Delta Programme aims to future-proof the Netherlands against SLR** (Bloemen et al., 2019). Lessons learned in building governance capabilities to deal with associated complex problems include (Dewulf and Termeer, 2015; Bloemen et al., 2018; Bloemen et al., 2019): – Committing to long-term policy implementation at Cabinet level – Allocate necessary dedicated budget and build capacity of government agencies to tackle complex problems, for example, Senate resolution and programme uniting government and knowledge institutes on adaptation – Flexible and robust governance approaches and solutions build resilience, for example, independent programme alongside traditional administrative structures is more agile – Adaptation pathways help overcome the temporal mismatch between short-term decisions and long-term goals, explicitly accounting for uncertainty – Enabling provisions for fit-for-purpose local-level policy and practice are key to translating national programme goals into local action, for example, liaison officers can bridge local, regional and national decision making arenas – Institutionalise monitoring and lesson learning (e.g., annual reporting to parliament, forums for politicians to share experiences) to track progress, deal with multiple legitimate perspectives and tackle emergent problems – Responsive governance arrangements address competing demands legitimately and timeously, for example, steering groups, workshops and social media reveal stakeholder concerns – Policy deadlocks or lock-in due to vested interests or short-term priorities can be tackled by taking a long-term perspective, exploring alternative scenarios and incentivising novel solutions

4.4.6   Towards Climate Resilient Development Pathways

Our assessment shows that failure to mitigate GHG emissions or to adapt to SLR will cause major disruptions to many low-lying coastal communities and jeopardise achievement of all UN SDGs and other societal aspirations. Immediate and ambitious GHG emissions reduction is necessary (Hoegh-Guldberg et al., 2018) to contain the rate and magnitude of SLR, and consequently adaptation prospects. Under unmitigated emissions (RCP8.5), coastal societies, especially poorer, rural and small islands societies, will struggle to maintain their livelihoods and settlements during the 21st century (Sections 4.3.4; 4.4.2). Without mitigation, sea levels will continue to rise for centuries, reaching 2.3–5.4 m by 2300 (*likely* range) and much more beyond (Section 4.2.3.5), making adaptation extremely challenging, if not impossible, for all low-lying coasts, including more intensively developed urbanised coasts. But even with ambitious mitigation (RCP2.6), sea levels will continue to rise, reaching 0.6–1.1 m by 2300 (*likely* range; Figure 4.2 Panel B). Hence, adaptation will continue to be imperative irrespective of the uncertainties about future GHG emissions and key physical processes such as those determining the Antarctic contribution to SLR. Our assessment also shows that all types of responses, from hard protection to EbA, advance and retreat, have important and synergistic roles to play in an integrated and sequenced response to SLR. The merits of a particular type of response, at a particular point in time, critically depends on the biophysical, cultural, economic, technical, institutional and political context.

In this context, AR5 put forward the vision of Climate Resilient Development Pathways, which is "a continuing process for managing changes in the climate and other driving forces affecting development, combining flexibility, innovativeness, and participative problem solving with effectiveness in mitigating and adapting to climate change" (Denton et al., 2014: 1106). Charting Climate Resilient Development Pathways in the face of rising sea level depends on how well mitigation, adaption and other sustainable development efforts are combined, and the governance challenges introduced by SLR are resolved. There are no panaceas for solving these complex issues. However, the wise application of the planning, public participation, conflict resolution, and decision analysis methods assessed above can help coastal communities, cities and settlements develop locally relevant, enabling and adaptive SLR responses. Difficult social choices will nonetheless need to be made as sea levels continue to rise. Given the SLR projections outlined here, it is concluded that global resilience and sustainability prospects depend, to a large extent, on how effectively coastal communities develop and implement ambitious, forward-looking adaptation plans in synchrony with drastic mitigation of GHG emissions.

FAQ 4.1 | What challenges does the inevitability of sea level rise present to coastal communities and how can communities adapt?

As the global climate changes, rising sea levels, combined with high tides, storms and flooding, put coastal and island communities increasingly at risk. Protection can be achieved by building dikes or seawalls and by maintaining natural features like mangroves or coral reefs. Communities can also adjust by reclaiming land from the sea and adapting buildings to cope with floods. However, all measures have their limits, and once these are reached people may ultimately have to retreat. Choices made today influence how coastal ecosystems and communities can respond to sea level rise (SLR) in the future. Reducing greenhouse gas (GHG) emissions would not just reduce risks, but also open up more adaptation options.

Global Mean Sea Level (GMSL) is rising and it will continue to do so for centuries. Sustainable development aspirations are at risk because many people, assets and vital resources are concentrated along low-lying coasts around the world. Many coastal communities have started to consider the implications of SLR. Measures are being taken to address coastal hazards exacerbated by rising sea level, such as coastal flooding due to extreme events (e.g. storm surges, tropical cyclones, coastal erosion and salinisation). However, many coastal communities are still not sufficiently adapted to today's ESLs.

Scientific evidence about SLR is clear: GMSL rose by 1.5 mm yr^{-1} during the period 1901–1990, accelerating to 3.6 mm yr^{-1} during the period 2005–2015. It is likely to rise 0.61–1.10 m by 2100 if global GHG emissions are not mitigated (RCP8.5). However, a rise of two or more metres cannot be ruled out. It could rise to more than 3 m by 2300, depending on the level of GHG emissions and the response of the AIS, which are both highly uncertain. Even if efforts to mitigate emissions are very effective, ESL events that were rare over the last century will become common before 2100, and even by 2050 in many locations. Without ambitious adaptation, the combined impact of hazards like coastal storms and very high tides will drastically increase the frequency and severity of flooding on low-lying coasts.

SLR, as well as the context for adaptation, will vary regionally and locally, thus action to reduce risks related to SLR takes different forms depending on the local circumstances. 'Hard protection', like dikes and seawalls, can effectively reduce risk under two or more metres of SLR but it is inevitable that limits will be reached. Such protection produces benefits that exceed its costs in low-lying coastal areas that are densely populated, as is the case for many coastal cities and some small islands, but in general, poorer regions will not be able to afford hard protection. Maintaining healthy coastal ecosystems, like mangroves, seagrass beds or coral reefs, can provide 'soft protection' and other benefits. SLR can also be 'accommodated' by raising buildings on the shoreline, for example. Land can be reclaimed from the sea by building outwards and upwards. In coastal locations where the risk is very high and cannot be effectively reduced, 'retreat' from the shoreline is the only way to eliminate such risk. Avoiding new development commitments in areas exposed to coastal hazards and SLR also avoids additional risk.

For those unable to afford protection, accommodation or advance measures, or when such measures are no longer viable or effective, retreat becomes inevitable. Millions of people living on low-lying islands face this prospect, including inhabitants of Small Island Developing States (SIDS), of some densely populated but less intensively developed deltas, of rural coastal villages and towns, and of Arctic communities who already face melting sea ice and unprecedented changes in weather. The resultant impacts on distinctive cultures and ways of life could be devastating. Difficult trade-offs are therefore inevitable when making social choices about rising sea level. Institutionalising processes that lead to fair and just outcomes is challenging, but vitally important.

Choices being made now about how to respond to SLR profoundly influence the trajectory of future exposure and vulnerability to SLR. If concerted emissions mitigation is delayed, risks will progressively increase as SLR accelerates. Prospects for global climate-resilience and sustainable development therefore depend in large part on coastal nations, cities and communities taking urgent and sustained locally-appropriate action to mitigate GHG emissions and adapt to SLR.

References

Abadie, L.M., 2018: Sea level damage risk with probabilistic weighting of IPCC scenarios: An application to major coastal cities. *J. Clean. Prod.*, **175**, 582–598, doi:10.1016/j.jclepro.2017.11.069.

Abadie, L.M., E. Sainz de Murieta and I. Galarraga, 2016: Climate risk assessment under uncertainty: an application to main European coastal cities. *Front. Mar. Sci.*, **3**, 265.

Ablain, M. et al., 2015: Improved sea level record over the satellite altimetry era (1993–2010) from the Climate Change Initiative project. *Ocean Sci.*, **11**(1), 67–82.

Ablain, Michaël, et al., 2019: Uncertainty in satellite estimates of global mean sea-level changes, trend and acceleration. *Earth System Science Data* **11**(3), 1189–1202.

Abrams, J.F. et al., 2016: The impact of Indonesian peatland degradation on downstream marine ecosystems and the global carbon cycle. *Global Change Biol.*, **22**(1), 325–337, doi:10.1111/gcb.13108.

Absar, S.M. and B.L. Preston, 2015: Extending the Shared Socioeconomic Pathways for sub-national impacts, adaptation, and vulnerability studies. *Global Environ. Chang.*, **33**, 83–96.

Adams, H., 2016: Why populations persist: mobility, place attachment and climate change. *Popul. Environ.*, **37**(4), 429–448.

Adams, V.M., M. Mills, S. D. Jupiter and R. L. Pressey, 2011: Improving social acceptability of marine protected area networks: A method for estimating opportunity costs to multiple gear types in both fished and currently unfished areas. *Biol. Conserv.*, **144**, 350–361.

Addo, K.A., 2015: Monitoring sea level rise-induced hazards along the coast of Accra in Ghana. *Nat. Hazards*, **78**(2), 1293–1307.

Adger, W.N., 2010: Social capital, collective action, and adaptation to climate change. In: *Der klimawandel* [Voss M. (eds.)]. VS Verlag für Sozialwissenschaften, pp. 327–345.

Adger, W.N. et al., 2015: Focus on environmental risks and migration: causes and consequences. *Environ. Res. Lett.*, **10**(6), 1–6.

Adger, W.N., C. Butler and K. Walker-Springett, 2017: Moral reasoning in adaptation to climate change. *Environ. Politics*, **26**(3), 371–390.

Adger, W.N. et al., 2014: Human security. In: Climate Change 2014: Impacts, Adaptation, and Vulnerability. Part A: Global and Sectoral Aspects. Contribution of Working Group II to the Fifth Assessment Report of the Intergovernmental Panel of Climate Change. [Field, C.B., V.R. Barros, D.J. Dokken, K.J. Mach, M.D. Mastrandrea, T.E. Bilir, M. Chatterjee, K.L. Ebi, Y.O. Estrada, R.C. Genova, B. Girma, E.S. Kissel, A.N. Levy, S. MacCracken, P.R. Mastrandrea and L L. White (eds.)]. Cambridge University Press, Cambridge, United Kingdom and New York, NY, USA.

Adhikari, S. et al., 2014: Future Antarctic bed topography and its implications for ice sheet dynamics. *Solid Earth*, **5**, 569–584, doi:10.5194/se-5-569-2014.

Adloff, F. et al., 2018: Improving sea level simulation in Mediterranean regional climate models. *Clim. Dyn.*, **51**(3), 1167–1178.

ADPC, 2005: *Handbook on Design and Construction of Housing for Flood-prone Rural Areas of Bangladesh*. Asian Disaster Preparedness Center [Available at: www.adpc.net/igo/category/ID189/doc/2013-p74Wob-ADPC-handbook_complete-b.pdf]. Accessed: 2019/09/20.

Aerts, J., 2018: A review of cost estimates for flood adaptation. *Water*, **10**(11), 1646.

Aerts, J. et al., 2008: Aandacht voor veiligheid, leven met water, klimaat voor ruimte, dg water. VU, Amsterdam, Amsterdam.

Aerts, J.C. et al., 2018a: Pathways to resilience: adapting to sea level rise in Los Angeles. *Ann. NY. Acad. Sci.*, **1427**(1), 1–90.

Aerts, J.C.J.H., 2017: Climate-induced migration: Impacts beyond the coast. *Nat. Clim. Change.* **7**(5), 315–316.

Aerts, J.C.J.H. et al., 2018b: Integrating human behaviour dynamics into flood disaster risk assessment. *Nat. Clim. Change*, **8**(3), 193–199.

Aerts, J.C.J.H. et al., 2014: Evaluating Flood Resilience Strategies for Coastal Megacities. *Science*, **344**, 473–475, doi:10.1126/science.1248222.

Ahas, R. et al., 2015: Everyday space–time geographies: using mobile phone-based sensor data to monitor urban activity in Harbin, Paris, and Tallinn. *Int. J. Geogr. Info. Sci.*, **29**(11), 2017–2039.

Ahmed, F. et al., 2018: Tipping points in adaptation to urban flooding under climate change and urban growth: The case of the Dhaka megacity. *Land Use Policy*, **79**, 496–506.

Aitken, A.R.A. et al., 2016: Repeated large-scale retreat and advance of Totten Glacier indicated by inland bed erosion. *Nature*, **533**, 385, doi:10.1038/nature17447.

Ajibade, I., M. Pelling, J. Agboola and M. Garschagen, 2016: Sustainability transitions: exploring risk management and the future of adaptation in the megacity of Lagos. *Journal of Extreme Events*, **3**(03), 1650009.

Akerlof, K.L. et al., 2016: Risky business: Engaging the public on sea level rise and inundation. *Environ. Sci. Policy*, **66**, 314–323, doi:10.1016/j.envsci.2016.07.002.

Alam, K. and M. H. Rahman, 2014: Women in natural disasters: a case study from southern coastal region of Bangladesh. *Int. J. Disast. Risk Reduc.*, **8**, 68–82.

Albert, S. et al., 2018: Heading for the hills: climate-driven community relocations in the Solomon Islands and Alaska provide insight for a 1.5 C future. *Reg. Environ. Change*, **18**(8), 2261–2272.

Albert, S. et al., 2016: Interactions between sea level rise and wave exposure on reef island dynamics in the Solomon Islands. *Environ. Res. Lett.*, **11**(5), 054011.

Albert, S. et al., 2017: Winners and losers as mangrove, coral and seagrass ecosystems respond to sea level rise in Solomon Islands. *Environ. Res. Lett.*, **12**(9), 2–11.

Albright, R. et al., 2018: Carbon dioxide addition to coral reef waters suppresses net community calcification. *Nature*, **555**(7697), 516–519, doi:10.1038/nature25968.

Aldrich, D.P., 2017: The importance of social capital in building community resilience. In: *Rethinking Resilience, Adaptation and Transformation in a Time of Change*. Springer, Cham, 357–364.

Aldrich, D.P. and M. A. Meyer, 2015: Social capital and community resilience. *Am. Behav. Sci.*, **59**(2), 254–269.

Alford, J. and B.W. Head, 2017: Wicked and less wicked problems: a typology and a contingency framework. *Policy and Society*, **36**(3), 397–413.

Allen, J.S., S.B. Longo and T.E. Shriver, 2018: Politics, the state, and sea level rise: the treadmill of production and structural selectivity in North Carolina's coastal resource commission. *Sociol. Q.*, **59**(2), 320–337.

Allgood, L. and K.E. McNamara, 2017: Climate-induced migration: Exploring local perspectives in Kiribati. *Singapore Journal of Tropical Geography*, **38**(3), 370–385.

Alston, M., 2013: Women and adaptation. *WiRes. Clim. Change*, **4**(5), 351–358.

Amaro, V.E. et al., 2015: Multitemporal analysis of coastal erosion based on multisource satellite images, Ponta Negra Beach, Natal City, Northeastern Brazil. *Marine Geodesy*, **38**(1), 1–25.

Amelung, B. and S. Nicholls, 2014: Implications of climate change for tourism in Australia. *Tourism Manage.*, **41**, 228–244.

Amoako, C., 2016: Brutal presence or convenient absence: The role of the state in the politics of flooding in informal Accra, Ghana. *Geoforum*, **77**, 5–16.

Andersen, C.F., 2007: *The New Orleans hurricane protection system: what went wrong and why: A report*. [American Society of Civil Engineers (eds.)]. Reston, Virginia, ISBN-13: 978-0-7844-0893-3.

Anguelovski, I. et al., 2016: Equity impacts of urban land use planning for climate adaptation: Critical perspectives from the global north and south. *J. Plan. Educ. Res.*, **36**(3), 333–348.

Anthony, E. J. et al., 2015: Linking rapid erosion of the Mekong River delta to human activities. *Sci. Rep.*, **5**, 14745, doi:10.1038/srep14745.

Araos, M. et al., 2017: Climate change adaptation planning for Global South megacities: the case of Dhaka. *J. Environ. Pol. Plan.*, **19** 6), 682–696.

Arkema, K.K. et al., 2013: Coastal habitats shield people and property from sea level rise and storms. *Nat. Clim. Change*, **3**(10), 913.

Arkema, K.K. et al., 2015: Embedding ecosystem services in coastal planning leads to better outcomes for people and nature. *PNAS*, **112**(24), 7390–7395.

Arns, A. et al., 2017: Sea level rise induced amplification of coastal protection design heights. *Sci. Rep.*, **7**, 40171, doi:10.1038/srep40171.

Arns, A. et al., 2013: Estimating extreme water level probabilities: A comparison of the direct methods and recommendations for best practise. *Coast. Eng. J.*, **81**, 51–66.

Arnstein, S. R., 1969: A ladder of citizen participation. *Journal of the American Institute of planners*, **35**(4), 216–224.

Arthern, R. J. and C. R. Williams, 2017: The sensitivity of West Antarctica to the submarine melting feedback. *Geophys. Res. Lett.*, **44**(5), 2352–2359.

Aschwanden, A. et al., 2019: Contribution of the Greenland Ice Sheet to sea level over the next millennium. *Sci. Adv.*, **5**, 1–11.

Aschwanden, A., M.A. Fahnestock and M. Truffer, 2016: Complex Greenland outlet glacier flow captured. *Nat. Commun.*, **7**, 10524, doi:10.1038/ncomms10524.

Ataie-Ashtiani, B. et al., 2013: How important is the impact of land-surface inundation on seawater intrusion caused by sea level rise? *Hydrogeol. J.*, **21**(7), 1673–1677.

Auerbach, L. et al., 2015: Flood risk of natural and embanked landscapes on the Ganges–Brahmaputra tidal delta plain. *Nat. Clim. Change*, **5**(2), 153.

Austermann, J. and J.X. Mitrovica, 2015: Calculating gravitationally self-consistent sea level changes driven by dynamic topography. *Geophys. J. Int.*, **203**(3), 1909–1922, doi:10.1093/gji/ggv371.

Austermann, J., J.X. Mitrovica, P. Huybers and A. Rovere, 2017: Detection of a dynamic topography signal in last interglacial sea level records. *Sci. Adv.*, **3**(7), e1700457.

Badger, M.P.S., D.N. Schmidt, A. Mackensen and R. D. Pancost, 2013: High-resolution alkenone palaeobarometry indicates relatively stable CO^2 during the Pliocene (3.3–2.8 Ma). *Philos. Trans. Roy. Soc. A.*, **371**(2001), 1–16, doi:10.1098/rsta.2013.0094.

Bailey, R.T., K. Barnes and C.D. Wallace, 2016: Predicting Future Groundwater Resources of Coral Atoll Islands. *Hydrol. Process.*, **30**(13), 2092–2105.

Bailey, R.T., A. Khalil and V. Chatikavanij, 2014: Estimating transient freshwater lens dynamics for atoll islands of the Maldives. *J. Hydrol.*, **515**, 247–256.

Baker, S. and F.S. Chapin III, 2018: Going beyond "it depends:" the role of context in shaping participation in natural resource management. *Ecol. Soc.*, **23**(1), 1–11.

Bakker, A.M., T.E. Wong, K.L. Ruckert and K. Keller, 2017: Sea level projections representing the deeply uncertain contribution of the West Antarctic ice sheet. *Sci. Rep.*, **7**(1), 3880, doi:10.1038/s41598-017-04134-5.

Balaguru, K., G.R. Foltz, L.R. Leung and K.A. Emanuel, 2016: Global warming-induced upper-ocean freshening and the intensification of super typhoons. *Nat. Commun.*, **7**, 13670, doi:10.1038/ncomms13670.

Balestri, E., F. Vallerini and C. Lardicci, 2017: Recruitment and Patch Establishment by Seed in the Seagrass Posidonia oceanica: Importance and Conservation Implications. *Front. Plant Sci.*, **8**, 1067.

Ballu, V. et al., 2011: Comparing the role of absolute sea level rise and vertical tectonic motions in coastal flooding, Torres Islands (Vanuatu). *PNAS*, **108**(32), 13019–13022.

Balmford, A. et al., 2004: The worldwide costs of marine protected areas. *PNAS*, **101**(26), 9694–9697.

Bamber, J. and W. Aspinall, 2013: An expert judgement assessment of future sea level rise from the ice sheets. *Nat. Clim. Change*, **3**(4), 424–427.

Bamber, J., W. Aspinall and R. Cooke, 2016: A commentary on "how to interpret expert judgment assessments of twenty-first century sea level rise" by Hylke de Vries and Roderik SW van de Wal. *Clim. Change*, **137**(3–4), 321–328.

Bamber, J.L. et al., 2019: Ice sheet contributions to future sea level rise from structured expert judgment. *PNAS*, **116**(23), 11195–11200.

Bamber, J.L., R.E. Riva, B.L.A. Vermeersen and A.M. LeBrocq, 2009: Reassessment of the potential sea level rise from a collapse of the West Antarctic Ice Sheet. *Science*, **324**(5929), 901–903, doi:10.1126/science.1169335.

Bamber, J.L., R.M. Westaway, B. Marzeion and B. Wouters, 2018: The land ice contribution to sea level during the satellite era. *Environ. Res. Lett.* **13**(6), 1–21.

Bandeira, S. and F. Gell, 2003: The seagrasses of Mozambique and southeastern Africa. In: *World Atlas of Seagrasses* [Green, E. P. (ed.)], 93–100, ISBN 0520240472.

Banwell, A.F., D.R. MacAyeal and O.V. Sergienko, 2013: Breakup of the Larsen B Ice Shelf triggered by chain reaction drainage of supraglacial lakes. *Geophys. Res. Lett.*, **40**(22), 5872–5876, doi:10.1002/2013GL057694.

Barbier, E.B., 2015: Hurricane Katrina's lessons for the world. *Nature*, **524**(7565), 285.

Barbier, E.B., 2016: The protective service of mangrove ecosystems: a review of valuation methods. *Mar. Pollut. Bull.*, **109**(2), 676–681.

Barbier, E.B. and B.S. Enchelmeyer, 2014: Valuing the storm surge protection service of US Gulf Coast wetlands. *J. Environ. Econ. Policy*, **3**, 167–185.

Barletta, V.R. et al., 2018: Observed rapid bedrock uplift in Amundsen Sea Embayment promotes ice sheet stability. *Science*, **360**(6395), 1335–1339, doi:10.1126/science.aao1447.

Barnard, P.L. et al., 2014: Development of the Coastal Storm Modeling System (CoSMoS) for predicting the impact of storms on high-energy, active-margin coasts. *Nat. Hazards*, **74**(2), 1095–1125.

Barnett, J. et al., 2015: From barriers to limits to climate change adaptation: path dependency and the speed of change. *Ecol. Soc.*, **20**(3), 1–11.

Barnett, J. et al., 2014: A local coastal adaptation pathway. *Nat. Clim. Change*, **4**(12), 1103–1108.

Barnhart, K., I. Overeem and R. Anderson, 2014a: The effect of changing sea ice on the physical vulnerability of Arctic coasts. *The Cryosphere*, **8**(5), 1777–1799.

Barnhart, K.R. et al., 2014b: Modeling erosion of ice-rich permafrost bluffs along the Alaskan Beaufort Sea coast. *J. Geophys. Res-Earth*, **119**(5), 1155–1179.

Barr, J.G. et al., 2013: Summertime influences of tidal energy advection on the surface energy balance in a mangrove forest. *Biogeosciences*, **10**, 501–511, doi:10.5194/bg-10-501-2013.

Barton, J.R., K. Krellenberg and J.M. Harris, 2015: Collaborative governance and the challenges of participatory climate change adaptation planning in Santiago de Chile. *Clim. Dev.*, **7**(2), 175–184.

Bassis, J. and C. Walker, 2012: Upper and lower limits on the stability of calving glaciers from the yield strength envelope of ice. *Proc. Royal Soc. Lond, A.*, **468**(2140), 913–931.

Bassis, J.N., 2011: The statistical physics of iceberg calving and the emergence of universal calving laws. *J. Glaciol.*, **57**(201), 3–16, doi:10.3189/002214311795306745.

Batabyal, P. et al., 2016: Environmental drivers on seasonal abundance of riverine-estuarine V. cholerae in the Indian Sundarban mangrove. *Ecol. Indic.*, **69**, 59–65.

Bayraktarov, E. et al., 2016: The cost and feasibility of marine coastal restoration. *Ecol. Appl.*, **26**, 1055–1074.

Beck, M.W. and G.M. Lange, 2016: *Managing Coasts with Natural Solutions: Guidelines for Measuring and Valuing the Coastal Protection Services of Mangroves and Coral Reefs*, The World Bank, Washington, DC, 166-pp.

Beck, M.W. et al., 2018: The global flood protection savings provided by coral reefs. *Nat. Commun.*, **9**(1), 2186.

Becker, M., M. Karpytchev and S. Lennartz-Sassinek, 2014: Long-term sea level trends: Natural or anthropogenic? *Geophys. Res. Lett.*, **41**(15), 5571–5580.

Becker, M. et al., 2016: Do climate models reproduce complexity of observed sea level changes? *Geophys. Res. Lett.*, **43**(10), 5176–5184.

Becker, M. et al., 2012: Sea level variations at tropical Pacific islands since 1950. *Global Planet. Change*, **80**, 85–98.

Beckley, B. et al., 2017: On the "Cal-Mode" Correction to TOPEX Satellite Altimetry and Its Effect on the Global Mean Sea Level Time Series. *Journal of Geophysical Research: Oceans*, **122**(11), 8371–8384.

Beetham, E., P.S. Kench and S. Popinet, 2017: Future Reef Growth Can Mitigate Physical Impacts of Sea-Level Rise on Atoll Islands. *Earth's Future*, **5**(10), 1002–1014.

Beichler, S.A., 2015: Exploring the link between supply and demand of cultural ecosystem services–towards an integrated vulnerability assessment. *International Journal of Biodiversity Science, Ecosystem Services & Management*, **11**(3), 250–263.

Bell, R. et al., 2017a: *Coastal Hazards and Climate Change: Guidance for Local Government*. Ministry for the Environment, New Zealand [Available at: www.mfe.govt.nz/sites/default/files/media/Climate%20Change/coastal-hazards-guide-final.pdf]. Accessed: 2019/09/20.

Bell, R.E. et al., 2017b: Antarctic ice shelf potentially stabilized by export of meltwater in surface river. *Nature*, **544** (7650), 344–348.

Bell, T., 2016: Turning Research Inside Out: Labrador Inuit focus on research priorities that strengthen community sustainability and well-being. *Newfoundland Quarterly*, **109**(1), 37–41.

Bellinson, R. and E. Chu, 2019: Learning pathways and the governance of innovations in urban climate change resilience and adaptation. *J. Environ. Pol. Plan.*, **21** (1), 76–89.

Ben-Haim, Y., 2006: *Info-gap decision theory: decisions under severe uncertainty*. Academic Press, Oxford, ISBN-13 978-0-12-373552-2, 361-pp.

Ben-Tal, A., L. El Ghaoui and A. Nemirovski, 2009: *Robust optimization*. Princeton University Press, **28**, ISBN 978-0-691-14368-2.

Benham, C.F., S.G. Beavis, R.A. Hendry and E.L. Jackson, 2016: Growth effects of shading and sedimentation in two tropical seagrass species: Implications for port management and impact assessment. *Mar. Pollut. Bull.*, **109**(1), 461–470, doi:10.1016/J.MARPOLBUL.2016.05.027.

Benn, D.I., C.R. Warren and R.H. Mottram, 2007: Calving processes and the dynamics of calving glaciers. *Earth-Sci. Rev.*, **82**(3), 143–179, doi:10.1016/j.earscirev.2007.02.002.

Bennett, N.J., J. Blythe, S. Tyler and N.C. Ban, 2016: Communities and change in the anthropocene: understanding social-ecological vulnerability and planning adaptations to multiple interacting exposures. *Reg. Environ. Change*, **16**(4), 907–926.

Bercovitch, J., V. Kremenyuk and I. W. Zartman, 2008: *The SAGE handbook of conflict resolution*. Sage, Great Britain, Cromwell Press Ltd, ISBN 978-1-4129-2192-3.

Berke, P.R. and M.R. Stevens, 2016: Land use planning for climate adaptation: Theory and practice. *J. Plan. Educ. Res.*, **36**(3), 283–289.

Berquist, M., A. Daniere and L. Drummond, 2015: Planning for global environmental change in Bangkok's informal settlements. *J. Environ. Plan. Manage.*, **58**(10), 1711–1730.

Bertram, R.A. et al., 2018: Pliocene deglacial event timelines and the biogeochemical response offshore Wilkes Subglacial Basin, East Antarctica. *Earth Planet. Sci. Lett.*, **494**, 109–116, doi:10.1016/j.epsl.2018.04.054.

Bettini, G. and G. Gioli, 2016: Waltz with development: insights on the developmentalization of climate-induced migration. *Migration and Development*, **5**(2), 171–189.

Betzold, C. and I. Mohamed, 2017: Seawalls as a response to coastal erosion and flooding: a case study from Grande Comore, Comoros (West Indian Ocean). *Reg. Environ. Change*, **17**(4), 1077–1087.

Bevis, M. et al., 2019: Accelerating changes in ice mass within Greenland, and the ice sheet's sensitivity to atmospheric forcing. *PNAS*, **116**(6), 1934–1939, doi:10.1073/pnas.1806562116.

Bhatia, K. et al., 2018: Projected response of tropical cyclone intensity and intensification in a global climate model. *J. Clim.*, **31**(20), 8281–8303.

Bhuiyan, M.J.A.N. and D. Dutta, 2012: Assessing impacts of sea level rise on river salinity in the Gorai river network, Bangladesh. *Estuar. Coast. Shelf Sci.*, **96**, 219–227.

Biesbroek, G.R., J.E. Klostermann, C.J. Termeer and P. Kabat, 2013: On the nature of barriers to climate change adaptation. *Reg. Environ. Change*, **13**(5), 1119–1129.

Biesbroek, G.R., C.J. Termeer, J.E. Klostermann and P. Kabat, 2014: Rethinking barriers to adaptation: Mechanism-based explanation of impasses in the governance of an innovative adaptation measure. *Global Environ. Chang.*, **26**, 108–118.

Biesbroek, R. et al., 2015: Opening up the black box of adaptation decision-making. *Nat. Clim. Change*, **5**(6), 493.

Biesbroek, R., J. Dupuis and A. Wellstead, 2017: Explaining through causal mechanisms: resilience and governance of social–ecological systems. *Curr. Opin. Environ. Sustain.*, **28**, 64–70.

Bilbao, R.A., J.M. Gregory and N. Bouttes, 2015: Analysis of the regional pattern of sea level change due to ocean dynamics and density change for 1993–2099 in observations and CMIP5 AOGCMs. *Clim. Dyn.*, **45** (9–10), 2647–2666.

Bilkovic, D., Mitchell, M., Peyre, M. La, Toft, J., 2017: *Living shorelines: the science and management of nature-based coastal protection*. CRC Press, Boca Raton, FL, USA, 1–498, ISBN 13: 978-1-4987-4002-9.

Binder, S.B., J.P. Barile, C.K. Baker and B. Kulp, 2019: Home buyouts and household recovery: neighborhood differences three years after Hurricane Sandy. *Environ. Hazards*, **18**(2), 127–145.

Bindoff, N.L. et al., 2013: Detection and Attribution of Climate Change: from Global to Regional. In: Climate Change 2013: The Physical Science Basis. Contribution of Working Group I to the Fifth Assessment Report of the Intergovernmental Panel on Climate Change. [Stocker, T.F., D. Qin, G.K. Plattner, M. Tignor, S.K. Allen, J. Boschung, A. Nauels, Y. Xia, V. Bex and P.M. Midgley (eds.)]. Cambridge University Press, Cambridge, United Kingdom and New York, NY, USA, 867–952.

Bindoff, N.L. et al., 2007: Observations: Oceanic Climate Change and Sea Level. In: Climate Change 2007: The Physical Science Basis. Contribution of Working Group I to the Fourth Assessment Report of the Intergovernmental Panel on Climate Change. [Solomon, S., D. Qin, M. Manning, Z. Chen, M. Marquis, K.B. Averyt, M. Tignor and H.L. Miller (eds.)]. Cambridge University Press, Cambridge, United Kingdom and New York, NY, USA, 385–432.

Bindschadler, R.A. et al., 2013: Ice sheet model sensitivities to environmental forcing and their use in projecting future sea level (the SeaRISE project). *J. Glaciol.*, **59**(214), 195–224.

Bisaro, A. and J. Hinkel, 2016: Governance of social dilemmas in climate change adaptation. *Nat. Clim. Change*, **6**(4), 354–359.

Bisaro, A. and J. Hinkel, 2018: Mobilizing private finance for coastal adaptation: A literature review. *WiRes. Clim. Change*, **9**(3), e514.

Bisaro, A., M. Roggero and S. Villamayor-Tomas, 2018: Institutional analysis in climate change adaptation research: A systematic literature review. *Ecol. Econ.*, **151**, 34–43.

Biswas, S., M.A. Hasan and M.S. Inslam, 2015: Stilt Housing Technology for Flood Disaster Reduction in the Rural Areas of Bangladesh. *Int. J. Res. Civ. Eng.*, **3**, 1–6.

Bjørk, A.A. et al., 2012: An aerial view of 80 years of climate-related glacier fluctuations in southeast Greenland. *Nat. Geosci.*, **5**(6), 427–432.

Bjork, M., F.T. Short, E. Mcleod and S. Beer, 2008: *Managing Seagrasses for Resilience to Climate Change*. IUCN Global Marine Programme, Gland, Switzerland, 1–55, ISBN 978-2-8317-1089-1.

Black, R. et al., 2011: The effect of environmental change on human migration. *Global Environ. Chang.*, **21**, S3–S11.

Black, R. et al., 2013: Migration, immobility and displacement outcomes following extreme events. *Environ. Sci. Policy*, **27**, S32–S43.

4

Blazquez, A. et al., 2018: Exploring the uncertainty in GRACE estimates of the mass redistributions at the Earth surface: implications for the global water and sea level budgets. *Geophys. J. Int.*, **215**(1), 415–430.

Bloemen, P. et al., 2018: Lessons learned from applying adaptation pathways in flood risk management and challenges for the further development of this approach. *Mitig. Adapt. Strat. Gl.*, **23**(7), 1083–1108.

Bloemen, P., M. Van Der Steen and Z. Van Der Wal, 2019: Designing a century ahead: Climate change adaptation in the Dutch Delta. *Policy and Society*, **38** (1), 58–76.

Bloetscher, F., B. Heimlich and D. E. Meeroff, 2011: Development of an adaptation toolbox to protect southeast Florida water supplies from climate change. *Environ. Rev.*, **19**, 397–417.

Blok, A. and R. Tschötschel, 2016: World port cities as cosmopolitan risk community: Mapping urban climate policy experiments in Europe and East Asia. *Environ. Plan. C.*, **34**(4), 717–736.

Blöschl, G. et al., 2017: Changing climate shifts timing of European floods. *Science*, **357**(6351), 588–590.

Boening, C. et al., 2012: The 2011 La Niña: So strong, the oceans fell. *Geophys. Res. Lett.*, **39** (19), 1–5, doi:10.1029/2012GL053055.

Bohra-Mishra, P., M. Oppenheimer and S.M. Hsiang, 2014: Nonlinear permanent migration response to climatic variations but minimal response to disasters. *PNAS*, **111**(27), 9780–9785.

Borges, A. et al., 2009: Convenient solutions to an inconvenient truth: Ecosystem-based approaches to climate change.

Bouttes, N., J. Gregory, T. Kuhlbrodt and R.J.C.D. Smith, 2014: The drivers of projected North Atlantic sea level change. **43** (5–6), 1531–1544.

Boyer, T. et al., 2016: Sensitivity of global upper-ocean heat content estimates to mapping methods, XBT bias corrections, and baseline climatologies. *J. Clim.*, **29**(13), 4817–4842.

Brakenridge, G. et al., 2017: Design with nature: Causation and avoidance of catastrophic flooding, Myanmar. *Earth-Sci. Rev.*, **165**, 81–109.

Breininger, D.R., R.D. Breininger and C.R. Hall, 2017: Effects of surrounding land use and water depth on seagrass dynamics relative to a catastrophic algal bloom. *Conserv. Biol.*, **31**(1), 67–75.

Brekelmans, R., Hertog, D. den, Roos, K., Eijgenraam, C., 2012: Safe Dike Heights at Minimal Costs: The Nonhomogeneous Case. *Oper. Res.*, **60**, 1342–1355, doi:10.1287/opre.1110.1028.

Bridges, K.W. and W.C. McClatchey, 2009: Living on the margin: ethnoecological insights from Marshall Islanders at Rongelap atoll. *Global Environ. Chang.*, **19**(2), 140–146.

Bridges, T., J. Simm, N. Pontee and J. Guy, 2018: International guidance on use of natural and nature-based features in flood and coastal management. [Available at:https://ewn.el.erdc.dren.mil/nnbf-guidelines.html.] Accessed: 2019/09/20.

Bridges, T., Wagner, P.W., Burks-Copes, K.A., Bates, M.E., Collier, Z.A., Fischenich, C.J., Gailani, J.Z., Leuck, L.D., Piercy, C.D., Rosati, J.D., Russo, E.J., Shafer, D.J., Suedel, B.C., Vuxton, E.A., Wamsley, T. V., 2015: *Use of Natural and Nature-Based Features (NNBF) for Coastal Resilience, North Atlantic Coast Comprehensive Study: Resilient Adaptation to Increasing Risk*. US Army Corps of Engineers: Engineer Research and Development Center, Vicksburg, MS, USA, 412-pp.´

Briggs, R., D. Pollard and L. Tarasov, 2013: A glacial systems model configured for large ensemble analysis of Antarctic deglaciation. *The Cryosphere*, **7**(6), 1949–1970.

Brink, E. and C. Wamsler, 2019: Citizen engagement in climate adaptation surveyed: The role of values, worldviews, gender and place. *J. Clean. Prod.*, **209**, 1342–1353.

Brinke, T. et al., 2010: Contingency planning for large-scale floods in the Netherlands. *J. Contingencies Crisis Manage.*, **18** (1), 55–69.

Brondex, J., O. Gagliardini, F. Gillet-Chaulet and G. Durand, 2017: Sensitivity of grounding line dynamics to the choice of the friction law. *J. Glaciol.*, **63**(241), 854–866, doi:10.1017/jog.2017.51.

Bronen, R. and F. S. Chapin, 2013: Adaptive governance and institutional strategies for climate-induced community relocations in Alaska. *PNAS*, **110**(23), 9320–9325.

Bronselaer, B. et al., 2018: Change in future climate due to Antarctic meltwater. *Nature*, **564**(7734), 53–58, doi:10.1038/s41586-018-0712-z.

Broto, V.C., E. Boyd and J. Ensor, 2015: Participatory urban planning for climate change adaptation in coastal cities: lessons from a pilot experience in Maputo, Mozambique. *Curr. Opin. Environ. Sustain.*, **13**, 11–18.

Brown, C.J., M.I. Saunders, H.P. Possingham and A.J. Richardson, 2013: Managing for Interactions between Local and Global Stressors of Ecosystems. *PLoS One*, **8**(6), 1–10, doi:10.1371/journal.pone.0065765.

Brown, I., J. Martin-Ortega, K. Waylen and K. Blackstock, 2016: Participatory scenario planning for developing innovation in community adaptation responses: three contrasting examples from Latin America. *Reg. Environ. Change*, **16**(6), 1685–1700.

Brown, S. and R. Nicholls, 2015: Subsidence and human influences in mega deltas: the case of the Ganges–Brahmaputra–Meghna. *Sci. Total Environ.*, **527**, 362–374.

Brown, S. et al., 2018a: Quantifying Land and People Exposed to Sea-Level Rise with No Mitigation and 1.5 and 2.0°C Rise in Global Temperatures to Year 2300. *Earth's Future*, **6**(3), 583–600.

Brown, S. et al., 2014: Shifting perspectives on coastal impacts and adaptation. *Nat. Clim. Change*, **4**(9), 752–755.

Brown, S. et al., 2018b: What are the implications of sea level rise for a 1.5, 2 and 3°C rise in global mean temperatures in the Ganges-Brahmaputra-Meghna and other vulnerable deltas? *Reg. Environ. Change*, **18**(6), 1829–1842.

Buchanan, M.K., R.E. Kopp, M. Oppenheimer and C. Tebaldi, 2016: Allowances for evolving coastal flood risk under uncertain local sea level rise. *Clim. Change*, **137**(3–4), 347–362.

Buchanan, M.K., M. Oppenheimer and R.E. Kopp, 2017: Amplification of flood frequencies with local sea level rise and emerging flood regimes. *Environ. Res. Lett.*, **12**(6), 064009.

Buchanan, M.K., M. Oppenheimer and A. Parris, 2019: Values, bias, and stressors affect intentions to adapt to coastal flood risk: a case study from NYC. *Weather, Climate, and Society*, **11**(4), 809–821.

Buchori, I. et al., 2018: Adaptation to coastal flooding and inundation: Mitigations and migration pattern in Semarang City, Indonesia. *Ocean Coast. Manage.*, **163**, 445–455.

Bucx, T., M. Marchand, B. Makaske and C. van de Guchte, 2010: *Comparative assessment of the vulnerability and resilience of 10 deltas: work document (No.1)*, 177, Deltares, Netherlands, ISBN 9789490070397.

Bucx, T., C. van Ruiten, G. Erkens and G. de Lange, 2015: An integrated assessment framework for land subsidence in delta cities. *Proc. Int. Assoc. Hydrology. Sci.*, **372**, 485.

Buhr, B. et al., 2018: *Climate Change and the Cost of Capital in Developing Countries*. UN Environment, UNEP Inquiry, Imperial College Business School Center for Climate Finance and Investment and SOAS [Available at: http://unepinquiry.org/wp-content/uploads/2018/07/Climate_Change_and_the_Cost_of_Capital_in_Developing_Countries.pdf]. Accessed: 2019/09/20.

Bujosa, A. and J. Rosselló, 2013: Climate change and summer mass tourism: the case of Spanish domestic tourism. *Clim. Change*, **117**(1–2), 363–375.

Bukvic, A. and G. Owen, 2017: Attitudes towards relocation following Hurricane Sandy: should we stay or should we go? *Disasters*, **41**(1), 101–123.

Bulthuis, K., M. Arnst, S. Sun and F. Pattyn, 2019: Uncertainty quantification of the multi-centennial response of the Antarctic ice sheet to climate change. *The Cryosphere*, **13**, 1349–1380.

Burby, R.J., 2006: Hurricane Katrina and the paradoxes of government disaster policy: Bringing about wise governmental decisions for hazardous areas. *Ann. Am. Acad. Political. Soc. Sci.*, **604**(1), 171–191.

Burch, S., A. Shaw, A. Dale and J. Robinson, 2014: Triggering transformative change: a development path approach to climate change response in communities. *Clim. Policy*, **14**(4), 467–487.

4

Burcharth, H.F., T.L. Andersen and J.L. Lara, 2014: Upgrade of coastal defence structures against increased loadings caused by climate change: A first methodological approach. *Coast. Eng. J.*, **87**, 112–121.

Burdick, D. M. and C. T. Roman (eds.), 2012: Salt marsh responses to tidal restriction and restoration. In: *Tidal marsh restoration*. Springer, pp. 373–382.

Burge, C.A. et al., 2014: Climate change influences on marine infectious diseases: implications for management and society, *Annual review of marine science*, **6**, 249–277.

Burge, C.A., C.J. Kim, J.M. Lyles and C.D. Harvell, 2013: Special issue oceans and humans health: the ecology of marine opportunists. *Microb. Ecol.*, **65**(4), 869–879.

Burke, L., K. Reytar, M. Spalding, A. Perry, 2011: *Reefs at Risk Revisited*. World Resources Institute, Washington, DC, 130-pp.

Burson, B. and R. Bedford, 2015: Facilitating voluntary adaptive migration in the Pacific. *Forced Migration Review*, **(49)**, 54.

Burton, P. and J. Mustelin, 2013: Planning for climate change: is greater public participation the key to success? *Urban Policy Res.*, **31**(4), 399–415.

Busch, S.V., 2018: Sea Level Rise and Shifting Maritime Limits: Stable Baselines as a Response to Unstable Coastlines. *Arctic Rev.*, **9**, 174–194.

Butler, J. et al., 2016a: Scenario planning to leap-frog the Sustainable Development Goals: an adaptation pathways approach. *Clim. Risk Manage.*, **12**, 83–99.

Butler, J. et al., 2014: Framing the application of adaptation pathways for rural livelihoods and global change in eastern Indonesian islands. *Global Environ. Chang.*, **28**, 368–382.

Butler, W.H., R.E. Deyle and C. Mutnansky, 2016b: Low-regrets incrementalism: Land use planning adaptation to accelerating sea level rise in Florida's Coastal Communities. *J. Plan. Educ. Res.*, **36**(3), 319–332.

Buurman, J. and V. Babovic, 2016: Adaptation Pathways and Real Options Analysis: An approach to deep uncertainty in climate change adaptation policies. *Policy and Society*, **35**(2), 137150.

Byrne, J.P., 2012: The cathedral engulfed: Sea level rise, property rights, and time. *La. L. Rev.*, **73**, 69.

Cai, R., H. Tan and H. Kontoyiannis, 2017: Robust Surface Warming in Offshore China Seas and Its Relationship to the East Asian Monsoon Wind Field and Ocean Forcing on Interdecadal Time Scales. *J. Clim.*, **30**(22), 8987–9005.

Cai, R., H. Tan and Q. Qi, 2016: Impacts of and adaptation to inter-decadal marine climate change in coastal China seas. *Int. J. Climatol.*, **36**(11), 3770–3780, doi:10.1002/joc.4591.

Calil, J. et al., 2017: Comparative Coastal Risk Index (CCRI): A multidisciplinary risk index for Latin America and the Caribbean. *PLoS One*, **12**(11), e0187011.

Call, M.A., C. Gray, M. Yunus and M. Emch, 2017: Disruption, not displacement: environmental variability and temporary migration in Bangladesh. *Global Environ. Chang.*, **46**, 157–165.

Callahan, K., 2007: Citizen participation: Models and methods. *Int. J. Public Admin.*, **30**(11), 1179–1196.

Calov, R. et al., 2018: Simulation of the future sea level contribution of Greenland with a new glacial system model. *The Cryosphere*, **12**, 3097–3121.

Camargo, S.J., 2013: Global and regional aspects of tropical cyclone activity in the CMIP5 models. *J. Clim.*, **26**(24), 9880–9902, doi:10.1175/JCLI-D-12-00549.1.

Campbell, J.E. and J.W. Fourqurean, 2014: Ocean acidification outweighs nutrient effects in structuring seagrass epiphyte communities. *J. Ecol.*, **102**(3), 730–737, doi:10.1111/1365-2745.12233.

Campbell, J.R., 2015: Development, global change and traditional food security in Pacific Island countries. *Reg. Environ. Change*, **15**(7), 1313–1324.

Campbell, S.J. and L.J. McKenzie, 2004: Flood related loss and recovery of intertidal seagrass meadows in southern Queensland, Australia. *Estuar. Coast. Shelf Sci.*, **60**(3), 477–490.

Capron, E. et al., 2014: Temporal and spatial structure of multi-millennial temperature changes at high latitudes during the Last Interglacial. *Quat. Sci. Rev.*, **103**, 116–133.

Carlsson Kanyama, A., P. Wikman-Svahn and K. Mossberg Sonnek, 2019: "We want to know where the line is": comparing current planning for future sea level rise with three core principles of robust decision support approaches. *J. Environ. Plan. Manage.*, **62**(8), 1339–1358.

Carlton, S.J. and S.K. Jacobson, 2013: Climate change and coastal environmental risk perceptions in Florida. *J. Environ. Manage.*, **130**, 32–39.

Carr, L. and R. Mendelsohn, 2003: Valuing Coral Reefs: A Travel Cost Analysis of the Great Barrier Reef. *AMBIO*, **32**, 353–357.

Carrasquilla-Henao, M. and F. Juanes, 2017: Mangroves enhance local fisheries catches: a global meta-analysis. *Fish Fish.*, **18**, 79–93.

Carson, M. et al., 2017: Regional Sea Level Variability and Trends, 1960–2007: A Comparison of Sea Level Reconstructions and Ocean Syntheses. *J. Geophys. Res-Oceans*, **122**(11), 9068–9091.

Carson, M. et al., 2016: Coastal sea level changes, observed and projected during the 20th and 21st century. *Clim. Change*, **134** (1–2), 269–281.

Carter, L., 2018: *Indigenous Pacific Approaches to Climate Change: Aotearoa/New Zealand*, 1–103, Springer, Cham, Switzerland, ISBN 978-3-319-96438-6.

Cashman, A. and M.R. Nagdee, 2017: Impacts of climate change on settlements and infrastructure in the coastal and marine environments of Caribbean small island developing states (SIDS). *Sci. Rev.*, **2017**, 155–173.

Castrucci, L. and N. J. M. T. S. J. Tahvildari, 2018: Modeling the Impacts of Sea Level Rise on Storm Surge Inundation in Flood-Prone Urban Areas of Hampton Roads, Virginia. **52**(2), 92–105.

Cazenave, A. and G.L. Cozannet, 2014: Sea level rise and its coastal impacts. *Earth's Future*, **2**(2), 15–34.

Cazenave, A. et al., 2012: Estimating ENSO influence on the global mean sea level, 1993–2010. *Marine Geodesy*, **35** (sup1), 82–97.

Celliers, L., S. Rosendo, I. Coetzee and G. Daniels, 2013: Pathways of integrated coastal management from national policy to local implementation: Enabling climate change adaptation. *Mar. Policy*, **39**, 72–86.

Chadenas, C., A. Creach and D. Mercier, 2014: The impact of storm Xynthia in 2010 on coastal flood prevention policy in France. *J. Coast. Conserv.*, **18**(5), 529–538.

Chaffin, B. C. et al., 2016: Transformative environmental governance. *Annu. Rev. Environ. Resourc.*, **41**, 399–423.

Chambers, D. P. et al., 2017: Evaluation of the global mean sea level budget between 1993 and 2014. *Surv. Geophys.*, **38**(1), 309–327.

Chambwera, M. et al., 2014: Economics of adaptation. In: Climate Change 2014: Impacts, Adaptation, and Vulnerability. Part A: Global and Sectoral Aspects. Contribution of Working Group II to the Fifth Assessment Report of the Intergovernmental Panel of Climate Change. [Field, C.B., V.R. Barros, D.J. Dokken, K.J. Mach, M.D. Mastrandrea, T.E. Bilir, M. Chatterjee, K.L. Ebi, Y.O. Estrada, R.C. Genova, B. Girma, E.S. Kissel, A.N. Levy, S. MacCracken, P.R. Mastrandrea and L.L. White (eds.)]. Cambridge University Press, Cambridge, United Kingdom and New York, NY, USA.

Chan, F. et al., 2018: Towards resilient flood risk management for Asian coastal cities: Lessons learned from Hong Kong and Singapore. *J. Clean. Prod.*, **187**, 576–589.

Chandra, A. and P. Gaganis, 2016: Deconstructing vulnerability and adaptation in a coastal river basin ecosystem: a participatory analysis of flood risk in Nadi, Fiji Islands. *Clim. Dev.*, **8**(3), 256–269.

Chang, E.K.M., 2014: Impacts of background field removal on CMIP5 projected changes in Pacific winter cyclone activity. *J. Geophys. Res-Atmos.*, **119**(8), 4626–4639.

Chao, B.F., Y. Wu and Y. Li, 2008: Impact of artificial reservoir water impoundment on global sea level. *Science*, **320**(5873), 212–214.

Chauché, N. et al., 2014: Ice–ocean interaction and calving front morphology at two west Greenland tidewater outlet glaciers. *The Cryosphere*, **8**(4), 145–1468, doi:10.5194/tc-8-1457-2014.

Chee, S.Y. et al., 2017: Land reclamation and artificial islands: Walking the tightrope between development and conservation. *Global Ecol. Conserv.*, **12**, 80–95, doi:10.1016/j.gecco.2017.08.005.

Chefaoui, R.M., C.M. Duarte and E.A. Serrão, 2018: Dramatic loss of seagrass habitat under projected climate change in the Mediterranean Sea. *Global Change Biol.*, **24**(10), 4919–4928.

Chen, J., C. Wilson and B. Tapley, 2013: Contribution of ice sheet and mountain glacier melt to recent sea level rise. *Nat. Geosci.*, **6**(7), 549–552.

Chen, X. et al., 2017: The increasing rate of global mean sea level rise during 1993–2014. *Nat. Clim. Change*, **7**(7), 492–495.

Cheng, H. et al., 2018: Mapping Sea Level Rise Behavior in an Estuarine Delta System: A Case Study along the Shanghai Coast. *Engineering*, **4**(1), 156–163.

Cheng, L., J. Abraham, Z. Hausfather and K.E.J.S. Trenberth, 2019: How fast are the oceans warming? **363**(6423), 128–129.

Cheng, L. et al., 2017: Improved estimates of ocean heat content from 1960 to 2015. *Sci. Adv.*, **3**(3), 1–10, doi:10.1126/sciadv.1601545.

Cheng, L. et al., 2016: Observed and simulated full-depth ocean heat-content changes for 1970–2005. *Ocean Sci*, **12**, 925–935.

Cheong, S.-M. et al., 2013: Coastal adaptation with ecological engineering. *Nature Clim. Change*, **3**, 787–791.

Chhetri, N., M. Stuhlmacher and A. Ishtiaque, 2019: Nested pathways to adaptation. *Environ. Res. Commun.*, **1**(1), 015001.

Chouinard, O. et al., 2017: The Participative Action Research Approach to Climate Change Adaptation in Atlantic.

Canadian Coastal Communities. In: *Climate Change Adaptation in North America.* Springer, 67–87, ISBN 978-3-319-53741-2.

Chouinard, O., S. Weissenberger and D. Lane, 2015: L'adaptation au changement climatique en zone côtière selon l'approche communautaire: études de cas de projets de recherche-action participative au Nouveau-Brunswick (Canada). *VertigO-la revue électronique en sciences de l'environnement*, (Hors-série 23).

Christensen, J.H. et al., 2013: Climate Phenomena and their Relevance for Future Regional Climate Change. In: Climate Change 2013: The Physical Science Basis. Contribution of Working Group I to the Fifth Assessment Report of the Intergovernmental Panel on Climate Change [Stocker, T.F., D. Qin, G.K. Plattner, M. Tignor, S.K. Allen, J. Boschung, A. Nauels, Y. Xia, V. Bex and P.M. Midgley (eds.)], Cambridge, United Kingdom and New York, NY, USA.

Chu, E., 2016a: The political economy of urban climate adaptation and development planning in Surat, India. *Environ. Plan. C.*, **34**(2), 281–298.

Chu, E., I. Anguelovski and J. Carmin, 2016: Inclusive approaches to urban climate adaptation planning and implementation in the Global South. *Clim. Policy*, **16**(3), 372–392.

Chu, E.K., 2016b: The governance of climate change adaptation through urban policy experiments. *Environ. Policy Governance*, **26**(6), 439–451.

Chu, E.K., S. Hughes and S.G. Mason (eds.), 2018: Conclusion: Multilevel Governance and Climate Change Innovations in Cities. In: *Climate Change in Cities.* Springer, Cham, pp. 361–378, ISBN 978-3-319-65003-6.

Chung, C.-H., 2006: Forty years of ecological engineering with Spartina plantations in China. *Ecol. Eng.*, **27**, 49–57.

Church, J.A. et al., 2013: Sea Level Change. In: Climate Change 2013: The Physical Science Basis. Contribution of Working Group I to the Fifth Assessment Report of the Intergovernmental Panel on Climate Change. [Stocker, T.F., D. Qin, G.K. Plattner, M. Tignor, S.K. Allen, J. Boschung, A. Nauels, Y. Xia, V. Bex and P.M. Midgley (eds.)]. Cambridge University Press, Cambridge, United Kingdom and New York, NY, USA.

Church, J.A. and N.J. White, 2011: Sea level rise from the late 19th to the early 21st century. *Surv. Geophys.*, **32**(4–5), 585–602.

Cid, A. et al., 2017: Global reconstructed daily surge levels from the 20th Century Reanalysis (1871–2010). *Global Planet. Change*, **148**, 9–21.

Cipollini, P. et al., 2017: Monitoring sea level in the coastal zone with satellite altimetry and tide gauges. *Surv. Geophys.*, **38**(1), 33–57.

CIRIA, 2007: *The Rock Manual: The Use of Rock for Hydraulic Engineering.* C683, CIRIA, CUR and CETMEF, London.

Clar, C., 2019: Coordinating climate change adaptation across levels of government: the gap between theory and practice of integrated adaptation strategy processes. *J. Environ. Plan. Manage.*, **61**(11), 1–20.

Clar, C. and R. Steurer, 2019: Climate change adaptation at different levels of government: Characteristics and conditions of policy change. *Natural Resources Forum*, **43**(2), 121–131.

Clark, P.U. et al., 2018: Sea level commitment as a gauge for climate policy. *Nat. Clim. Change*, **8**(8), 653–655.

Clark, P.U. et al., 2016: Consequences of twenty-first-century policy for multi-millennial climate and sea level change. *Nat. Clim. Change*, **6**(4), 360–369, doi:10.1038/nclimate2923.

Clarke, D., A.N. Lázár, A.F.M. Saleh and M. Jahiruddin, 2018: Prospects for Agriculture Under Climate Change and Soil Salinisation. In: *Ecosystem Services for Well-Being in Deltas.* Springer, pp. 447–467.

Cloutier, G. et al., 2015: Planning adaptation based on local actors' knowledge and participation: A climate governance experiment. *Clim. Policy*, **15**(4), 458–474.

Coastal Protection and Restoration Authority of Louisiana, 2017: *Louisiana's Comprehensive Master Plan for a Sustainable Coast.* Coastal Protection and Restoration Authority of Louisiana, Louisiana, USA.

Coen, L.D. et al., 2007: Ecosystem services related to oyster restoration. *Mar. Ecol. Prog. Ser.*, **341**, 303–307.

Cogley, J.G., 2009: Geodetic and direct mass-balance measurements: comparison and joint analysis. *Ann. Glaciol.*, **50**(50), 96–100.

Colbert, A.J., B.J. Soden, G.A. Vecchi and B.P. Kirtman, 2013: The impact of anthropogenic climate change on North Atlantic tropical cyclone tracks. *J. Clim.*, **26**(12), 4088–4095.

Coldren, G.A., J.A. Langley, I.C. Feller and S.K. Chapman, 2019: Warming accelerates mangrove expansion and surface elevation gain in a subtropical wetland. *J. Ecol.*, **107**(1), 79–90.

Colenbrander, D., 2019: Dissonant discourses: revealing South Africa's policy-to-praxis challenges in the governance of coastal risk and vulnerability. *J. Environ. Plan. Manage.*, **62**(10), 1–20.

Colenbrander, D. and M. Bavinck, 2017: Exploring the role of bureaucracy in the production of coastal risks, City of Cape Town, South Africa. *Ocean Coast. Manage*, **150**, 35–50.

Colenbrander, D., A. Cartwright and A. Taylor, 2016: Drawing a line in the sand: managing coastal risks in the City Of Cape Town, South African Geographical Journal 97(1), 1–17. *South African Geographical Journal*, **98**(1), 104.

Colenbrander, D.R. and M.R. Sowman, 2015: Merging Socioeconomic Imperatives with Geospatial Data: A Non-Negotiable for Coastal Risk Management in South Africa. *Coastal Manage.*, **43**(3), 270–300.

Colgan, C.S., M.W. Beck and S. Narayan, 2017: *Financing Natural Infrastructure for Coastal Flood Damage Reduction*. Lloyd's Tercentenary Research Foundation, London.

Colle, B.A. et al., 2013: Historical evaluation and future prediction of eastern North American and western Atlantic extratropical cyclones in the CMIP5 models during the cool season. *J. Clim.*, **26**(18), 6882–6903.

Collins, N., S. Jones, T.H. Nguyen and P. Stanton, 2017: The contribution of human capital to a holistic response to climate change: learning from and for the Mekong Delta, Vietnam. *Asia Pacific Business Review*, **23**(2), 230–242.

Conlisk, E. et al., 2013: Uncertainty in assessing the impacts of global change with coupled dynamic species distribution and population models. *Global Change Biol.*, **19**(3), 858–869, doi:10.1111/gcb.12090.

Connell, J., 2012: Population Resettlement in the Pacific: lessons from a hazardous history? *Austral. Geogr.*, **43**(2), 127–142.

Connell, S.D. et al., 2017: Testing for thresholds of ecosystem collapse in seagrass meadows. *Conserv. Biol.*, **31**(5), 1196–1201, doi:10.1111/cobi.12951.

Conrad, J.M., 1980: Quasi-option value and the expected value of information. *Quart. J. Econ.*, **94**(4), 813–820.

Cook, C.P. et al., 2013: Dynamic behaviour of the East Antarctic ice sheet during Pliocene warmth. *Nat. Geosci.*, **6**(9), 765–769, doi:10.1038/ngeo1889.

Cooke, B.C., A.R. Jones, I.D. Goodwin and M.J. Bishop, 2012: Nourishment practices on Australian sandy beaches: a review. *J. Environ. Manage.*, **113**, 319–327.

Cooke, R., 1991: *Experts in uncertainty: opinion and subjective probability in science*. Oxford University Press on Demand, New York, NY, USA, 319-pp., ISBN 0-19-506465-8.

Coombes, M.A., E.C. La Marca, L.A. Naylor and R.C. Thompson, 2015: Getting into the groove: opportunities to enhance the ecological value of hard coastal infrastructure using fine-scale surface textures. *Ecol. Eng.*, **77**, 314–323.

Cooper, J., M. O'Connor and S. McIvor, 2016: Coastal defences versus coastal ecosystems: a regional appraisal. *Mar. Policy*.

Cornford, S.L. et al., 2015: Century-scale simulations of the response of the West Antarctic Ice Sheet to a warming climate. *The Cryosphere*, **9**(4), 1579–1600, doi:10.5194/tc-9-1579-2015.

Cosens, B.A. et al., 2017: The role of law in adaptive governance. *Ecol. Soc.* **22**(1), 1–30.

Coser, L.A., 1967: Continuities in the study of social conflict. Free Press, New York, NY, USA.

Costas, S., O. Ferreira and G. Martinez, 2015: Why do we decide to live with risk at the coast? *Ocean Coast. Manage*, **118**, 1–11.

Cradock-Henry, N.A. et al., 2018: Dynamic adaptive pathways in downscaled climate change scenarios. *Clim. Change*, **150**(3–4), 333–341.

Craig, R.K. et al., 2017: Balancing stability and flexibility in adaptive governance: an analysis of tools available in US environmental law. *Ecol. Soc.* **22**(2), 1–3.

Crespo, D. et al., 2017: New climatic targets against global warming: Will the maximum 2°C temperature rise affect estuarine benthic communities? *Sci. Rep.*, **7**(1), 1–14, doi:10.1038/s41598-017-04309-0.

Crosby, S.C. et al., 2016: Salt marsh persistence is threatened by predicted sea level rise. *Estuar. Coast. Shelf Sci.*, **181**, 93–99.

Crotty, S.M., C. Angelini and M.D. Bertness, 2017: Multiple stressors and the potential for synergistic loss of New England salt marshes. *PLoS One*, **12**(8), 1–13, doi:10.1371/journal.pone.0183058.

Csatho, B.M. et al., 2014: Laser altimetry reveals complex pattern of Greenland Ice Sheet dynamics. *PNAS*, **111**(52), 18478–18483.

Cuevas, S.C., 2018: Institutional dimensions of climate change adaptation: insights from the Philippines. *Clim. Policy*, **18**(4), 499–511.

Cuevas, S.C., A. Peterson, C. Robinson and T.H. Morrison, 2016: Institutional capacity for long-term climate change adaptation: evidence from land use planning in Albay, Philippines. *Reg. Environ. Change*, **16**(7), 2045–2058.

Cullen-Unsworth, L.C. and R.K.F. Unsworth, 2016: Strategies to enhance the resilience of the world's seagrass meadows. *J. Appl. Ecol.*, **53**(4), 967–972, doi:10.1111/1365-2664.12637.

Dahl-Jensen, D. et al., 2013: Eemian interglacial reconstructed from a Greenland folded ice core. *Nature*, **493**(7433), 489–494, doi:10.1038/nature11789.

Daliakopoulos, I. et al., 2016: The threat of soil salinity: A European scale review. *Sci. Total Environ.*, **573**, 727–739.

Dang, T.D., T.A. Cochrane and M.E. Arias, 2018: Future hydrological alterations in the Mekong Delta under the impact of water resources development, land subsidence and sea level rise. *J. Hydrol. Reg. Stud.*, **15**, 119–133.

Dangendorf, S. et al., 2015: Detecting anthropogenic footprints in sea level rise. *Nat. Commun.*, **6**, 1–9.

Dangendorf, S. et al., 2017: Reassessment of 20th century global mean sea level rise. *PNAS*, **114**(23), 5946–5951.

Dannevig, H. and C. Aall, 2015: The regional level as boundary organization? An analysis of climate change adaptation governance in Norway. *Environ. Sci. Policy*, **54**, 168–175.

Darby, S.E. et al., 2016: Fluvial sediment supply to a mega-delta reduced by shifting tropical-cyclone activity. *Nature*, **539**(7628), 276.

Daron, J.D. and D.R. Colenbrander, 2015: A critical investigation of evaluation matrices to inform coastal adaptation and planning decisions at the local scale. *J. Environ. Plan. Manage.*, **58**(12), 2250–2270.

Dasgupta, S. et al., 2017: The impact of aquatic salinization on fish habitats and poor communities in a changing climate: evidence from southwest coastal Bangladesh. *Ecol. Econ.*, **139**, 128–139.

Dasgupta, S. et al., 2019: Quantifying the protective capacity of mangroves from storm surges in coastal Bangladesh. *PLoS One*, **14**(3), e0214079.

Davis, K.F., A. Bhattachan, P. D'Odorico and S. Suweis, 2018: A universal model for predicting human migration under climate change: examining future sea level rise in Bangladesh. *Environ. Res. Lett.*, **13**(6), 064030.

Daviter, F., 2017: Coping, taming or solving: alternative approaches to the governance of wicked problems. *Policy Studies*, **38**(6), 571–588.

Dawson, D.A., A. Hunt, J. Shaw and W.R. Gehrels, 2018: The Economic Value of Climate Information in Adaptation Decisions: Learning in the Sea level Rise and Coastal Infrastructure Context. *Ecol. Econ.*, **150**, 1–10.

Day, J.J. and K.I. Hodges, 2018: Growing land-sea temperature contrast and the intensification of Arctic cyclones. *Geophys. Res. Lett.*, **45**(8), 3673–3681.

Day, J.W. et al., 2016: Approaches to defining deltaic sustainability in the 21st century. *Estuar. Coast. Shelf Sci.*, **183**, 275–291.

de Boer, B., P. Stocchi and R. Van De Wal, 2014: A fully coupled 3-D ice sheet-sea level model: algorithm and applications. *Geosci. Model Dev.*, **7**(5), 2141–2156.

de Boer, B., P. Stocchi, P.L. Whitehouse and R. van de Wal, 2017: Current state and future perspectives on coupled ice sheet–sea level modelling. *Quaternary Science Reviews*, **169**, 13–28.

de Bruijn, K., F. Klijn and J. Knoeff, 2013: Unbreachable embankments? In pursuit of the most effective stretches for reducing fatality risk. In: *Comprehensive flood risk management. Research for policy and practice. Proceedings of the 2nd European Conference on Flood Risk Management*, FLOODrisk2012, Rotterdam, the Netherlands, pp. 19–23.

de Haas, H., 2010: Migration and development: A theoretical perspective. *Int. Migr. Rev.*, **44**(1), 227–264.

de Neufville, R. and K. Smet, 2019: Engineering Options Analysis (EOA). In: *Decision Making under Deep Uncertainty: From Theory to Practice* [Marchau, V.A.W.J., W.E. Walker, P.J.T.M. Bloemen and S.W. Popper (eds.)]. Springer International Publishing, Cham, pp. 117–132.

de Rydt, J., G. Gudmundsson, H. Rott and J. Bamber, 2015: Modeling the instantaneous response of glaciers after the collapse of the Larsen B Ice Shelf. *Geophys. Res. Lett.*, **42**(13), 5355–5363.

de Sherbinin, A. et al., 2011: Climate change. Preparing for resettlement associated with climate change. *Science*, **334**(6055), 456–457, doi:10.1126/science.1208821.

de Souza Machado, A.A., K.L. Spencer, C. Zarfl and F.T. O'Shea, 2018: Unravelling metal mobility under complex contaminant signatures. *Sci. Total Environ.*, **622**, 373–384.

de Vries, H. and R.S.W. van de Wal, 2015: How to interpret expert judgment assessments of 21st century sea level rise. *Clim. Change*, **130**(2), 87–100, doi:10.1007/s10584-015-1346-x.

de Vries, H. and R.S.W. van de Wal, 2016: Response to commentary by J. L. Bamber, W. P. Aspinall and R. M. Cooke (2016). *Clim. Change*, **137**(3–4), 329–332, doi:10.1007/s10584-016-1712-3.

de Winter, R. et al., 2017: Impact of asymmetric uncertainties in ice sheet dynamics on regional sea level projections. *Nat. Hazards Earth Syst. Sci.*, **17**(12), 2125–2141.

Dean, R.G., 2002: *Beach Nourishment: Theory and Practice*. World Scientific, Singapore, 420-pp.

DeCaro, D.A. et al., 2017: Legal and institutional foundations of adaptive environmental governance. *Ecol. Soc.* **22**(1), 1–32.

DeConto, R.M. and D. Pollard, 2016: Contribution of Antarctica to past and future sea level rise. *Nature*, **531**(7596), 591–597, doi:10.1038/nature17145.

Den Uyl, R.M. and D.J. Russel, 2018: Climate adaptation in fragmented governance settings: the consequences of reform in public administration. *Environ. Politics*, **27**(2), 341–361.

4

Dendy, S., J. Austermann, J. Creveling and J.J.Q.S.R. Mitrovica, 2017: Sensitivity of Last Interglacial sea level high stands to ice sheet configuration during Marine Isotope Stage 6. *Quaternary Science Reviews*, **171**, 234–244.

Deng, J., C.D. Woodroffe, K. Rogers and J. Harff, 2017: Morphogenetic modelling of coastal and estuarine evolution. *Earth-Sci. Rev.* **171**, 254–271.

Dennig, F., 2018: Climate change and the re-evaluation of cost-benefit analysis. *Clim. Change*, **151**(1), 43–54.

Denton, F. et al., 2014: Climate-resilient pathways: adaptation, mitigation, and sustainable development. In: Climate Change 2014: Working II Group Contribution to the 5th Assessment Report of the Intergovernmental Panel on Climate Change. e [Field, C.B., V.R. Barros, D.J. Dokken, K.J. Mach, M.D. Mastrandrea, T.E. Bilir, M. Chatterjee, K.L. Ebi, Y.O. Estrada, R.C. Genova, B. Girma, E.S. Kissel, A.N. Levy, S. MacCracken, P.R. Mastrandrea, and L.L.White (eds.)]. Cambridge University Press, Cambridge, United Kingdom and New York, NY, USA, pp. 1101–1131.

Desbruyères, D., E. L. McDonagh, B. A. King and V. Thierry, 2017: Global and Full-Depth Ocean Temperature Trends during the Early Twenty-First Century from Argo and Repeat Hydrography. *J. Clim.*, **30**(6), 1985–1997.

Desmet, K., D.K. Nagy and E. Rossi-Hansberg, 2018: The geography of development. *J. Politic. Econ.*, **126** (3), 903–983.

Desportes, I. and D.R. Colenbrander, 2016: Navigating interests, navigating knowledge: Towards an inclusive set-back delineation along Cape Town's coastline. *Habitat Int.*, **54**, 124–135.

Dessu, S.B., R.M. Price, T.G. Troxler and J.S. Kominoski, 2018: Effects of sea level rise and freshwater management on long-term water levels and water quality in the Florida Coastal Everglades. *J. Environ. Manage.*, **211**, 164–176.

Deudero, S., M. Vázquez-Luis and E. Álvarez, 2015: Human stressors are driving coastal benthic long-lived sessile fan mussel Pinna nobilis population structure more than environmental stressors. *PLoS One*, **10**(7), e0134530.

Deville, P. et al., 2014: Dynamic population mapping using mobile phone data. *PNAS*, **111**(45), 15888–15893.

Dewan, C., A. Mukherji and M.-C. Buisson, 2015: Evolution of water management in coastal Bangladesh: from temporary earthen embankments to depoliticized community-managed polders. *Water Int.*, **40**(3), 401–416.

Dewulf, A. and C. Termeer, 2015: Governing the future? The potential of adaptive delta management to contribute to governance capabilities for dealing with the wicked problem of climate change adaptation. *J. Water Clim. Change*, **6**(4), 759–771.

Diaz, D.B., 2016: Estimating global damages from sea level rise with the Coastal Impact and Adaptation Model (CIAM). *Clim. Change*, **137** (1–2), 143–156.

Díaz, S. et al., 2019: *Summary for policymakers of the global assessment report on biodiversity and ecosystem services of the Intergovernmental Science-Policy Platform on Biodiversity and Ecosystem Services* [Cunha, M.C.d., G. Mace and H. Mooney (eds.)]. Intergovernmental Science-Policy Platform on Biodiversity and Ecosystem Services, IPBES Secretariat, Bonn, Germany, 45-pp.

Dieng, H., A. Cazenave, B. Meyssignac and M. Ablain, 2017: New estimate of the current rate of sea level rise from a sea level budget approach. *Geophys. Res. Lett.*, **44**(8), 3744–3751.

Dieng, H.B. et al., 2015a: Total land water storage change over 2003–2013 estimated from a global mass budget approach. *Environ. Res. Lett.*, **10**(12), 124010.

Dieng, H.B. et al., 2015b: The sea level budget since 2003: inference on the deep ocean heat content. *Surv. Geophys.*, **36**(2), 209–229.

Dinan, T., 2017: Projected Increases in Hurricane Damage in the United States: The Role of Climate Change and Coastal Development. *Ecol. Econ.*, **138**, 186–198.

Dinniman, M.S. et al., 2016: Modeling ice shelf/ocean interaction in Antarctica: A review. *Oceanography*, **29**(4), 144–153.

Dittrich, R., A. Wreford and D. Moran, 2016: A survey of decision-making approaches for climate change adaptation: Are robust methods the way forward? *Ecol. Econ.*, **122**, 79–89, doi:10.1016/j.ecolecon.2015.12.006.

Dixit, A.K., R.K. Dixit, R.S. Pindyck and R. Pindyck, 1994: *Investment under uncertainty*. Princeton University Press, Princeton, NJ, USA, 445-pp., ISBN 978-0-691-03410-2.

Döll, P. et al., 2016: Modelling freshwater resources at the global scale: Challenges and prospects. *Surv. Geophys.*, **37**(2), 195–221.

Döll, P. et al., 2014: Global-scale assessment of groundwater depletion and related groundwater abstractions: Combining hydrological modeling with information from well observations and GRACE satellites. *Water Resourc. Res.*, **50**(7), 5698–5720.

Dolšak, N. and A. Prakash, 2018: The politics of climate change adaptation. *Annu. Rev. Environ. Resourc.*, **43**, 317–341.

Donat-Magnin, M. et al., 2017: Ice shelf Melt Response to Changing Winds and Glacier Dynamics in the Amundsen Sea Sector, Antarctica. *J. Geophys. Res-Oceans*, **122**(12), 10206–10224, doi:10.1002/2017JC013059.

Donchyts, G. et al., 2016: Earth's surface water change over the past 30 years. *Nature Clim. Change*, **6**(9), 810–813, doi:10.1038/nclimate3111.

Doody, J.P., 2013: Coastal squeeze and managed realignment in southeast England, does it tell us anything about the future? *Ocean Coast. Manage*, **79**, 34–41.

Doswald, N. et al., 2012: Review of the evidence base for ecosystem-based approaches for adaptation to climate change. *Environ. Evid.*, **1**, 1–11.

Dovers, S.R. and A.A. Hezri, 2010: Institutions and policy processes: the means to the ends of adaptation. *WiRes. Clim. Change*, **1**(2), 212–231.

Duarte, C.M. et al., 2013: The role of coastal plant communities for climate change mitigation and adaptation. *Nat. Clim. Change*, **3**(11), 961.

Dukes, F., 1993: Public conflict resolution: A transformative approach. *Negot. J.*, **9**(1), 45–57.

Dunn, F.E. et al., 2018: Projections of historical and 21st century fluvial sediment delivery to the Ganges-Brahmaputra-Meghna, Mahanadi, and Volta deltas. *Sci. Total Environ.*, **642**, 105–116.

Düsterhus, A., M.E. Tamisiea and S. Jevrejeva, 2016: Estimating the sea level highstand during the last interglacial: a probabilistic massive ensemble approach. *Geophys. J. Int.*, **206**(2), 900–920, doi:10.1093/gji/ggw174.

Dutra, L.X. et al., 2015: Organizational drivers that strengthen adaptive capacity in the coastal zone of Australia. *Ocean Coast. Manage*, **109**, 64–76.

Dutrieux, P. et al., 2014: Strong sensitivity of Pine Island ice shelf melting to climatic variability. *Science*, **343**, 174–178.

Dutton, A. et al., 2015a: Sea level rise due to polar ice sheet mass loss during past warm periods. *Science*, **349** (6244).

Dutton, A., J.M. Webster, D. Zwartz and K. Lambeck, 2015b: Tropical tales of polar ice: evidence of Last Interglacial polar ice sheet retreat recorded by fossil reefs of the granitic Seychelles islands. *Quat. Sci. Rev.*, **107**, 182–196.

Duus-Otterström, G. and S.C. Jagers, 2011: Why (most) climate insurance schemes are a bad idea. *Environ. Politics*, **20**(3), 322–339.

Duvat, V., 2013: Coastal protection structures in Tarawa atoll, Republic of Kiribati. *Sustain. Sci.*, **8**(3), 363–379.

Duvat, V., A. Magnan and F. Pouget, 2013: Exposure of atoll population to coastal erosion and flooding: a South Tarawa assessment, Kiribati. *Sustain. Sci.*, **8** (3), 423–440.

Duvat, V. et al., 2017: Trajectories of exposure and vulnerability of small islands to climate change. *WiRes. Clim. Change*, 8(6), 1–14.

Duvat, V.K., 2019: A global assessment of atoll island planform changes over the past decades. *WiRes. Clim. Change*, **10**(1), e557.

Duy Vinh, V., S. Ouillon and D. Van Uu, 2018: Estuarine Turbidity Maxima and variations of aggregate parameters in the Cam-Nam Trieu estuary, North Vietnam, in early wet season. *Water*, **10**(1), 68.

Dyckman, C.S., C.S. John and J.B. London, 2014: Realizing managed retreat and innovation in state-level coastal management planning. *Ocean Coast. Manage*, **102**, 212–223.

Edwards, T. et al., 2014: Effect of uncertainty in surface mass balance-elevation feedback on projections of the future sea level contribution of the Greenland ice sheet. *The Cryosphere*, **8**(1), 195.

4

Edwards, T.L. et al., 2019: Revisiting Antarctic ice loss due to marine ice-cliff instability. *Nature*, **566**(7742), 58–64, doi:10.1038/s41586-019-0901-4.

EEA, 2013: *Late lessons from early warnings: science, precaution, innovation*. European Environment Agency [Available at: www.eea.europa.eu/publications/late-lessons-2]. Accessed: 2019/09/20.

Eijgenraam, C., R. Brekelmans, D. den Hertog and K. Roos, 2016: Optimal Strategies for Flood Prevention. *Manag. Sci.*, **63**, 1644–1656, doi:10.1287/mnsc.2015.2395.

Eisenack, K. et al., 2014: Explaining and overcoming barriers to climate change adaptation. *Nat. Clim. Change*, **4**(10), 867–872.

Eisenack, K. and R. Stecker, 2012: A framework for analyzing climate change adaptations as actions. *Mitig. Adapt. Strat. Gl.*, **17**(3), 243–260.

Ekstrom, J.A. and S C. Moser, 2014: Identifying and overcoming barriers in urban climate adaptation: case study findings from the San Francisco Bay Area, California, USA. *Urban Clim.*, **9**, 54–74.

El-Sayed, A.-F.M., 2016: Fish and fisheries in the Nile Delta. In: *The Nile Delta: the handbook of environmental chemistry*, Springer, Cham, pp. 495–516, ISBN: 978-3-319-56124-0.

El-Zein, A. and F.N. Tonmoy, 2017: Nonlinearity, fuzziness and incommensurability in indicator-based assessments of vulnerability to climate change: A new mathematical framework. *Ecol. Indic.*, **82**, 82–93.

Elliff, C.I. and I.R. Silva, 2017: Coral reefs as the first line of defense: Shoreline protection in face of climate change. *Mar. Environ. Res.*, **127**, 148–154.

Ellingwood, B.R. and J.Y. Lee, 2016: Managing risks to civil infrastructure due to Nat. Hazards: communicating long-term risks due to climate change. In: *Risk Analysis of Natural Hazards*. Gardoni P., Murphy C., Rowell A. (eds.) Springer, pp. 97–112, ISBN: 987-3-319-22126-7.

Elrick-Barr, C., B.C. Glavovic and R. Kay, 2015: A tale of two atoll nations: A comparison of risk, resilience, and adaptive response of Kiribati and the Maldives. In: *Climate Change and the Coast: Building Resilient Communities*. Bruce Glavovic; Michael Kelly; Robert Kay; Ailbhe Travers (eds). CRC Press, pp. 313–336, ISBN: 9780415464871.

Elrick-Barr, C.E., D.C. Thomsen, B.L. Preston and T.F. Smith, 2017: Perceptions matter: household adaptive capacity and capability in two Australian coastal communities. *Reg. Environ. Change*, **17**(4), 1141–1151.

Elshinnawy, I. et al., 2010: *Climate Change Risks to Coastal Development and Adaptation Options in the Nile Delta*. United Nations Development Programme; Stockholm Environment Institute; Coastal Research Institute, Fund, M.A. [Available at: www.nile-delta-adapt.org/index.php?view=DownLoadAct&id=4]. Accessed: 2019/09/20.

Emanuel, K., 2005: Increasing destructiveness of tropical cyclones over the past 30 years. *Nature*, **436**, 686, doi:10.1038/nature03906.

Emanuel, K., 2013: Downscaling CMIP5 climate models shows increased tropical cyclone activity over the 21st century. *PNAS*, **110**(30), 12219–12224, doi:10.1073/pnas.1301293110.

Emanuel, K., 2015: Effect of Upper-Ocean Evolution on Projected Trends in Tropical Cyclone Activity. *J. Clim.*, **28**(20), 8165–8170, doi:10.1175/JCLI-D-15-0401.1.

Emanuel, K., 2017a: Assessing the present and future probability of Hurricane Harvey's rainfall. *PNAS*, **114**(48), 12681–12684.

Emanuel, K., 2017b: Will Global Warming Make Hurricane Forecasting More Difficult? *Bull. Am. Meteorol. Soc.*, **98**(3), 495–501, doi:10.1175/BAMS-D-16-0134.1.

Emanuel, K., R. Sundararajan and J. Williams, 2008: Hurricanes and global warming: Results from downscaling IPCC AR4 simulations. *Bull. Am. Meteorol. Soc.*, **89**(3), 347–367, doi:10.1175/bams-89-3-347.

Enderlin, E.M., I.M. Howat, S. Jeong, M.-J. Noh, J.H. van Angelen, and M.R. van den Broeke, 2014: An improved mass budget for the Greenland ice sheet. *Geophys. Res. Lett*, **41**, 866–872, doi:10.1002/2013GL059010.

England, M.H. et al., 2014: Recent intensification of wind-driven circulation in the Pacific and the ongoing warming hiatus. *Nat. Clim. Change*, **4**(3), 222.

Ensor, J. and B. Harvey, 2015: Social learning and climate change adaptation: evidence for international development practice. *WiRes. Clim. Change*, **6**(5), 509–522.

Environment Agency, 2015: *Cost estimation for coastal protection – summary of evidence*. Environment Agency, Bristol, UK, Report SC080039/R7.

Epple, C. et al., 2016: Shared goals – joined-up approaches? Why action under the Paris Agreement, the Sustainable Development Goals and the Strategic Plan for Biodiversity 2011–2020 needs to come together at the landscape level. In: *CBD COP 13*, UNEP-WCMC and IUCN, IIED code: G04113. [Available at: https://pubs.iied.org/G04113/]. Accessed: 2019/09/20.

Erftemeijer, P.L. and R.R.R. Lewis III, 2006: Environmental impacts of dredging on seagrasses: a review. *Mar. Pollut. Bull.*, **52**(12), 1553–1572.

Erftemeijer, P.L., B. Riegl, B.W. Hoeksema and P.A. Todd, 2012: Environmental impacts of dredging and other sediment disturbances on corals: a review. *Mar. Pollut. Bull.*, **64**(9), 1737–1765.

Ericson, J.P. et al., 2006: Effective sea level rise and deltas: causes of change and human dimension implications. *Global Planet. Change*, **50**(1–2), 63–82.

Eriksen, S.H., A.J. Nightingale and H. Eakin, 2015: Reframing adaptation: The political nature of climate change adaptation. *Global Environ. Chang.*, **35**, 523–533.

Esteban, M. et al., 2019: Adaptation to sea level rise on low coral islands: Lessons from recent events. *Ocean Coast. Manage*, **168**, 35–40.

Esteves, L.S., 2013: Is managed realignment a sustainable long-term coastal management approach? *J. Coast. Res.*, **65**(sp1), 933–939.

Everard, M., L. Jones and B. Watts, 2010: Have we neglected the societal importance of sand dunes? An ecosystem services perspective. *Aquat. Conserv.: Mar. Freshw. Ecosyst.*, **20**(4), 476–487.

Eyre, B.D. et al., 2018: Coral reefs will transition to net dissolving before end of century. *Science*, **359**(6378), 908–911.

Fang, Y., J. Yin and B. Wu, 2016: Flooding risk assessment of coastal tourist attractions affected by sea level rise and storm surge: a case study in Zhejiang Province, China. *Nat. Hazards*, **84**(1), 611–624.

Fang, Z., P.T. Freeman, C.B. Field and K.J. Mach, 2018: Reduced sea ice protection period increases storm exposure in Kivalina, Alaska. *Arctic Science*, **4**(4), 525–537.

Farbotko, C., E. Stratford and H. Lazrus, 2016: Climate migrants and new identities? The geopolitics of embracing or rejecting mobility. *Social & Cultural Geography*, **17**(4), 533–552.

Farinotti, D. et al., 2019: A consensus estimate for the ice thickness distribution of all glaciers on Earth. *Nature Geoscience*, **12**(3), 168–173.

Farquharson, L. et al., 2018: Temporal and spatial variability in coastline response to declining sea-ice in northwest Alaska. *Mar. Geol.*, **404**, 71–83.

Farrell, W.E. and J.A. Clark, 1976: On Postglacial Sea Level. *Geophysical Journal of the Royal Astronomical Society*, **46**(3), 647–667, doi:10.1111/j.1365-246X.1976.tb01252.x.

Fasullo, J., R. Nerem and B. Hamlington, 2016: Is the detection of accelerated sea level rise imminent? *Sci. Rep.*, **6**, 31245.

Fasullo, J.T., C. Boening, F.W. Landerer and R.S. Nerem, 2013: Australia's unique influence on global sea level in 2010–2011. *Geophys. Res. Lett.*, **40**(16), 4368–4373, doi:10.1002/grl.50834.

Fatorić, S. and E. Seekamp, 2017a: Are cultural heritage and resources threatened by climate change? A systematic literature review. *Clim. Change*, **142**(1–2), 227–254.

Fatorić, S. and E. Seekamp, 2017b: Evaluating a decision analytic approach to climate change adaptation of cultural resources along the Atlantic Coast of the United States. *Land Use Policy*, **68**, 254–263.

Favier, L. et al., 2014: Retreat of Pine Island Glacier controlled by marine ice sheet instability. *Nat. Clim. Change*, **4**(2), 117–121.

Fawcett, D., T. Pearce, J.D. Ford and L. Archer, 2017: Operationalizing longitudinal approaches to climate change vulnerability assessment. *Global Environ. Chang.*, **45**, 79–88.

Feldmann, J. and A. Levermann, 2015: Collapse of the West Antarctic Ice Sheet after local destabilization of the Amundsen Basin. *PNAS*, **112**(46), 14191–14196.

Felsenstein, D. and M. Lichter, 2014: Social and economic vulnerability of coastal communities to sea level rise and extreme flooding. *Nat. Hazards*, **71**(1), 463–491.

FEMA, 2014: *Homeowner's Guide to Retrofitting P-312*. Federal Emergency Management Agency, Washington, D.C., USA. [Available at: www.fema.gov/media-library-data/1404148604102-f210b5e43aba0fb393443fe7ae9cd953/FEMA_P-312.pdf]. Accessed: 2019/09/20.

FEMA, 2015: *Hazard Mitigation Assistance Guidance*. Hazard Mitigation Grant Program, Pre-Disaster Mitigation Program, and Flood Mitigation Assistance Program, Federal Emergency Management Agency, Washington, D.C., USA [Available at: www.fema.gov/media-library-data/1424983165449–38f5dfc69c0bd4ea8a161e8bb7b79553/HMA_Guidance_022715_508.pdf]. Accessed: 2019/09/20.

FEMA, 2018: *2017 Hurricane Season: FEMA After-Action Report*. Federal Emergency Management Agency, Washington, D.C., USA [Available at: www.fema.gov/media-library-data/1531743865541-d16794d43d308254 4435e1471da07880/2017FEMAHurricaneAAR.pdf]. Accessed: 2019/09/20.

Feng, X. and M.N. Tsimplis, 2014: Sea level extremes at the coasts of China. *J. Geophys. Res-Oceans*, **119**(3), 1593–1608, doi:10.1002/2013JC009607.

Ferguson, G. and T. Gleeson, 2012: Vulnerability of coastal aquifers to groundwater use and climate change. *Nat. Clim. Change*, **2**(5), 342.

Fernandez, R. and L. Schäfer, 2018: *Impact evaluation of climate risk insurance approaches. Status quo and way forward*. [Series, D.P. (ed.)]. United Nations University-Institute for Environment and Human Security, Bonn, Germany [Available at: https://collections.unu.edu/eserv/UNU:6699/Discussion_paper_MCII_Final_June_11_vs.pdf]. Accessed: 2019/09/20.

Ferrario, F. et al., 2014: The effectiveness of coral reefs for coastal hazard risk reduction and adaptation. *Nat. Commun.*, **5**, 1–9.

Fettweis, X. et al., 2017: Reconstructions of the 1900–2015 Greenland ice sheet surface mass balance using the regional climate MAR model. *The Cryosphere*, **11**(2), 1015.

Fettweis, X. et al., 2013: Estimating the Greenland ice sheet surface mass balance contribution to future sea level rise using the regional atmospheric climate model MAR. *The Cryosphere*, **7**, 469–489.

Fiedler, J.W. and C.P. Conrad, 2010: Spatial variability of sea level rise due to water impoundment behind dams. *Geophys. Res. Lett.*, **37**(12).

Field, C.R., A.A. Dayer and C.S. Elphick, 2017: Landowner behavior can determine the success of conservation strategies for ecosystem migration under sea level rise. *PNAS*, **114**(34), 9134–9139, doi:10.1073/pnas.1620319114.

Field, M.E., A.S. Ogston and C.D. Storlazzi, 2011: Rising sea level may cause decline of fringing coral reefs. *Eos*, **92**(33), 273–274.

Figueiredo, R. and M. Martina, 2016: Using open building data in the development of exposure data sets for catastrophe risk modelling. *Nat. Hazards Earth Syst. Sci.*, **16**(2), 417.

Fincher, R., J. Barnett, S. Graham and A. Hurlimann, 2014: Time stories: Making sense of futures in anticipation of sea level rise. *Geoforum*, **56**, 201–210.

Fink, J.H., 2019: Contrasting governance learning processes of climate-leading and -lagging cities: Portland, Oregon, and Phoenix, Arizona, USA. *J. Environ. Pol. Plan.*, **21**(1), 16–29.

Finkbeiner, E.M. et al., 2018: Exploring trade-offs in climate change response in the context of Pacific Island fisheries. *Mar. Policy*, **88**, 359–364.

Fischer, H. et al., 2018: Palaeoclimate constraints on the impact of 2°C anthropogenic warming and beyond, *Nature Geoscience*, **11**(7). 474–485.

Fischetti, M., 2015: Is New Orleans Safer Today Than When Katrina Hit 10 Years Ago? *Sci. American*, [Available at: www.scientificamerican.com/article/is-new-orleans-safer-today-than-when-katrina-hit-10-years-ago/]. Accessed: 2019/09/20.

Fish, R., A. Church and M. Winter, 2016: Conceptualising cultural ecosystem services: a novel framework for research and critical engagement. *Ecosyst. Serv.*, **21**, 208–217.

Flood, S., N.A. Cradock-Henry, P. Blackett and P. Edwards, 2018: Adaptive and interactive climate futures: Systematic review of 'serious games' for engagement and decision-making. *Environ. Res. Lett.*, **13**(6), 063005.

Flower, H., M. Rains and C. Fitz, 2017: Visioning the Future: Scenarios Modeling of the Florida Coastal Everglades. *Environ. Manage.*, **60**(5), 989–1009, doi:10.1007/s00267-017-0916-2.

Flowers, G.E., 2018: Hydrology and the future of the Greenland Ice Sheet. *Nat. Commun.*, **9**(1), 2729, doi:10.1038/s41467-018-05002-0.

Flynn, M. et al., 2018: Participatory scenario planning and climate change impacts, adaptation and vulnerability research in the Arctic. *Environ. Sci. Policy*, **79**, 45–53.

Foerster, A., A. Macintosh and J. McDonald, 2015: Trade-Offs in Adaptation Planning: Protecting Public Interest Environmental Values. *J. Environ. Law*, **27**(3), 459–487.

Fonseca, M.S. and J.A. Cahalan, 1992: A preliminary evaluation of wave attenuation by four species of seagrass. *Estuar. Coast. Shelf Sci.*, **35**(6), 565–576.

Forbes, D.L., 2011: *State of the Arctic coast 2010: scientific review and outlook*. Land-Ocean Interactions in the Coastal Zone, Institute of Coastal Research.

Forbes, D.L., 2019: Arctic Deltas and Estuaries: A Canadian Perspective. *Coasts and Estuaries*. [Wolanski, E., Day, J., Elliott, M. and Ramachandran, R. (eds.)]. Elsevier, 123–147, ISBN 9780128140031.

Forbes, D.L. et al., 2018: Coastal environments and drivers. In: *From Science to Policy in the Eastern Canadian Arctic: An Integrated Regional Impact Study (IRIS) of Climate Change and Moderization* [Bell, T. and T.M. Brown (eds.)]. ArcticNet, Quebec, pp. 210–249.

Ford, J., T. Bell and N. Couture, 2016a: Perspectives of Canada's North Coast region. *Climate Change Impacts and Adaptation Assessment of Canada's Marine Coasts*, [D.S. Lemmen, F.J. Warren, T.S. James and C.S.L. Mercer Clarke; Government of Canada (eds.)], Ottawa, Canada, pp. 153–206.

Ford, J.D. et al., 2012: Mapping human dimensions of climate change research in the Canadian Arctic. *Ambio*, **41**(8), 808–822.

Ford, J.D. et al., 2016b: Including indigenous knowledge and experience in IPCC assessment reports. *Nat. Clim. Change*, **6**(4), 349–353.

Ford, J.D., G. McDowell and J. Jones, 2014: The state of climate change adaptation in the Arctic. *Environ. Res. Lett.*, **9**(10), 104005.

Ford, J.D., G. McDowell and T. Pearce, 2015: The adaptation challenge in the Arctic. *Nat. Clim. Change*, **5**(12), 1046–1053.

Forester, J., 1987: Planning in the face of conflict: Negotiation and mediation strategies in local land use regulation. *J. Am. Plan. Assoc.*, **53**(3), 303–314.

Forester, J., 2006: Making participation work when interests conflict: Moving from facilitating dialogue and moderating debate to mediating negotiations. *J. Am. Plan. Assoc.*, **72**(4), 447–456.

Forget, G. and R.M. Ponte, 2015: The partition of regional sea level variability. *Progr. Oceanogr.*, **137**, 173–195.

Forino, G., J. Von Meding and G.J. Brewer, 2018: Challenges and opportunities for Australian local governments in governing climate change adaptation and disaster risk reduction integration. *Int. J. Disaster Resilience Built Environ.* **9**(3), 258–272.

Foster, T.E. et al., 2017: Modeling vegetation community responses to sea level rise on Barrier Island systems: A case study on the Cape Canaveral Barrier Island complex, Florida, USA. *PLoS One*, **12**(8), e0182605.

Foteinis, S. and C.E. Synolakis, 2015: Beach erosion threatens Minoan beaches: a case study of coastal retreat in Crete. *Shore Beach*, **83**(1), 53–62.

Francesch-Huidobro, M. et al., 2017: Governance challenges of flood-prone delta cities: Integrating flood risk management and climate change in spatial planning. *Progress in Planning*, **114**, 1–27.

Frankcombe, L.M., S. McGregor and M.H. England, 2015: Robustness of the modes of Indo-Pacific sea level variability. *Clim. Dyn.*, **45**(5–6), 1281–1298.

Frederikse, T., S. Jevrejeva, R.E. Riva and S. Dangendorf, 2018: A consistent sea level reconstruction and its budget on basin and global scales over 1958–2014. *J. Clim.*, **31**(3), 1267–1280.

Frederikse, T., R.E. Riva and M.A.J.G.R.L. King, 2017: Ocean bottom deformation due to present-day mass redistribution and its impact on sea level observations. *Geophysical Research Letters,* **44**(24), 306–314.

French, P.W., 2006: Managed realignment–the developing story of a comparatively new approach to soft engineering. *Estuar. Coast. Shelf Sci.,* **67**(3), 409–423.

Fretwell, P. et al., 2013: Bedmap2: improved ice bed, surface and thickness datasets for Antarctica. *The Cryosphere,* **7**(1), 375–393.

Freudenburg, W.R., R.B. Gramling, S. Laska and K. Erikson, 2009: *Catastrophe in the making: the engineering of Katrina and the disasters of tomorrow.* Island Press, Washington, D.C., USA, 201-pp., ISBN 978-1-59726-682-6.

Frihy, O.E.S., E.A. Deabes, S.M. Shereet and F.A. Abdalla, 2010: Alexandria-Nile Delta coast, Egypt: update and future projection of relative sea level rise. *Environ. Earth Sci.,* **61**(2), 253–273.

Fritz, M., J. E. Vonk and H. Lantuit, 2017: Collapsing arctic coastlines. *Nat. Clim. Change,* **7**(1), 6.

Fürst, J., H. Goelzer and P. Huybrechts, 2015: Ice-dynamic projections of the Greenland ice sheet in response to atmospheric and oceanic warming. *The Cryosphere,* **9**(3), 1039–1062.

Fürst, J.J. et al., 2016: The safety band of Antarctic ice shelves. *Nat. Clim. Change,* **6**, 479, doi:10.1038/nclimate2912.

Ganju, N.K. et al., 2017: Spatially integrative metrics reveal hidden vulnerability of microtidal salt marshes. *Nat. Commun.,* **8**, 1–7.

Garai, J., 2016: Gender Specific Vulnerability in Climate Change and Possible Sustainable Livelihoods of Coastal People. A Case from Bangladesh. *Revista de Gestão Costeira Integrada-Journal of Integrated Coastal Zone Management,* **16**(1), 79–88.

Garcin, M. et al., 2016: Lagoon islets as indicators of recent environmental changes in the South Pacific–The New Caledonian example. *Cont. Shelf Res.,* **122**, 120–140.

Garner, A.J. et al., 2017: Impact of climate change on NYC's coastal flood hazard: Increasing flood heights from the preindustrial to 2300 CE. *PNAS,* **114**(45), 11861–11866.

Garner, K.L. et al., 2015: Impacts of sea level rise and climate change on coastal plant species in the central California coast. *PeerJ,* **3**, e958–e958, doi:10.7717/peerj.958.

Garschagen, M., 2015: Risky change? Vietnam's urban flood risk governance between climate dynamics and transformation. *Pac. Aff.,* **88**(3), 599–621.

Gasson, E., R.M. DeConto and D. Pollard, 2016: Modeling the oxygen isotope composition of the Antarctic ice sheet and its significance to Pliocene sea level. *Geology,* **44**(10), 827–830.

Gattuso, J.-P. et al., 2015: Contrasting futures for ocean and society from different anthropogenic CO_2 emissions scenarios. *Science,* **349**(6243), aac4722.

Gattuso, J.-P. et al., 2018: Ocean solutions to address climate change and its effects on marine ecosystems. *Front. Mar. Sci.,* **5**, 337.

Gayen, B., R.W. Griffiths and R.C. Kerr, 2015: Melting Driven Convection at the Ice-seawater Interface. *Procedia IUTAM,* **15**, 78–85, doi:10.1016/j.piutam.2015.04.012.

Gebremichael, E. et al., 2018: Assessing Land Deformation and Sea Encroachment in the Nile Delta: A Radar Interferometric and Inundation Modeling Approach. *J. Geophys. Res-Earth,* **123**(4), 3208–3224.

Gedan, K.B. et al., 2011: The present and future role of coastal wetland vegetation in protecting shorelines: answering recent challenges to the paradigm. *Clim. Change,* **106**(1), 7–29.

Gemenne, F. and J. Blocher, 2017: How can migration serve adaptation to climate change? Challenges to fleshing out a policy ideal. *Geogr. J.,* **183**(4), 336–347, doi:10.1111/geoj.12205.

Gemenne, F. and P. Brücker, 2015: From the guiding principles on internal displacement to the Nansen initiative: What the governance of environmental migration can learn from the governance of internal displacement. *Int. J. Refug. Law,* **27**(2), 245–263.

Genovese, E. and V. Przyluski, 2013: Storm surge disaster risk management: the Xynthia case study in France. *J. Risk Res.,* **16**(7), 825–841.

Genua-Olmedo, A., C. Alcaraz, N. Caiola and C. Ibáñez, 2016: Sea level rise impacts on rice production: The Ebro Delta as an example. *Sci. Total Environ.,* **571**, 1200–1210.

Gerritsen, H., 2005: What happened in 1953? The Big Flood in the Netherlands in retrospect. *Philos. Trans. Roy. Soc. A.,* **363**(1831), 1271–1291.

Gharasian, C., 2016: Protection of natural resources through a sacred prohibition: The rahui on Rapa iti. In: *The Rahui: legal pluralism in Polynesian traditional management of resources and territories* [Bambridge, T. (ed.)]. Australian National University Press, Acton, Australian, pp. 139–153, ISBN: 9781925022919.

Ghoneim, E. et al., 2015: Nile Delta exhibited a spatial reversal in the rates of shoreline retreat on the Rosetta promontory comparing pre-and post-beach protection. *Geomorphology,* **228**, 1–14.

Ghosh, M.K., L. Kumar and P.K. Langat, 2018: Mapping tidal channel dynamics in the Sundarbans, Bangladesh, between 1974 and 2017, and implications for the sustainability of the Sundarbans mangrove forest. *Environ. Monit. Assess.,* **190**(10), 582.

Giakoumi, S. et al., 2015: Towards a framework for assessment and management of cumulative human impacts on marine food webs. *Conserv. Biol.,* **29**(4), 1228–1234, doi:10.1111/cobi.12468.

Gibbs, M.T., 2016: Why is coastal retreat so hard to implement? Understanding the political risk of coastal adaptation pathways. *Ocean Coast. Manage,* **130**, 107–114.

Giddens, A., 2015: The politics of climate change. *Policy & Politics,* **43**(2), 155–162.

Giesen, R.H. and J. Oerlemans, 2013: Climate-model induced differences in the 21st century global and regional glacier contributions to sea level rise. *Clim. Dyn.,* **41**(11–12), 3283–3300.

Gilbert, S., R. Horner, S. Gilbert and R. Horner, 1984: *The Thames Barrier.* Society of Civil Engineers, London, ISBN 978-0727701824.

Gill, J.C. and B.D. Malamud, 2014: Reviewing and visualizing the interactions of Nat. Hazards. *Rev. Geophys.,* **52**(4), 680–722.

Gilman, E. and J. Ellison, 2007: Efficacy of alternative low-cost approaches to mangrove restoration, American Samoa. *Estuaries Coasts,* **30**(4), 641–651.

Gingerich, S.B., C.I. Voss and A.G. Johnson, 2017: Seawater-flooding events and impact on freshwater lenses of low-lying islands: Controlling factors, basic management and mitigation. *J. Hydrol.,* **551**, 676–688.

Gioli, G., G. Hugo, M.M. Costa and J. Scheffran, 2016: Human mobility, climate adaptation, and development. *Migration and Development,* **5**(2), 165–170.

Giosan, L., 2014: Protect the world's deltas. *Nature,* **516**(7529), 31.

Giri, C. et al., 2011: Status and distribution of mangrove forests of the world using earth observation satellite data. *Glob. Ecol. Biogeogr,* **20**, 154–159.

Gittman, R.K. et al., 2015: Engineering away our natural defenses: an analysis of shoreline hardening in the US. *Front. Ecol. Environ.,* **13**(6), 301–307.

Gittman, R.K. et al., 2016: Ecological consequences of shoreline hardening: a meta-analysis. *BioScience,* **66**(9), 763–773.

Glavovic, B.C., 2016: Towards deliberative coastal governance: insights from South Africa and the Mississippi Delta. *Reg. Environ. Change,* **16**(2), 353–365.

Glavovic, B.C., C. Cullinan and M. Groenink, 2018: The Coast. In: *Fuggle and Rabie's Environmental Management in South Africa, 3rd Edition* [King, N.D., H.A. Strydom and F.P. Retief (eds.)]. Juta and Co., Cape Town, pp. 653–733.

Glavovic, B.C. and G.P. Smith, 2014: *Adapting to climate change: Lessons from Natural Hazards planning.* Springer, New York, NY, USA, ISBN 978-94-017-8630-0, 451-pp.

Gleckler, P. et al., 2016: Industrial-era global ocean heat uptake doubles in recent decades. *Nat. Clim. Change,* **6**(4), 394–398.

Goeldner-Gianella, L. et al., 2019: The perception of climate-related coastal risks and environmental changes on the Rangiroa and Tikehau atolls,

French Polynesia: The role of sensitive and intellectual drivers. *Ocean Coast. Manage.*, **172**, 14–29.

Goelzer, H. et al., 2013: Sensitivity of Greenland ice sheet projections to model formulations. *J. Glaciol.*, **59**(216), 733–749.

Goelzer, H., P. Huybrechts, M.-F. Loutre and T. Fichefet, 2016: Last Interglacial climate and sea level evolution from a coupled ice sheet–climate model. *Clim. Past*, **12**, 2195–2213, doi:10.5194/cp-12-2195-2016.

Golledge, N.R. et al., 2019: Global environmental consequences of twenty-first-century ice sheet melt. *Nature*, **566**(7742), 65–72, doi:10.1038/s41586-019-0889-9.

Golledge, N.R. et al., 2015: The multi-millennial Antarctic commitment to future sea level rise. *Nature*, **526**(7573), 421–425.

Golledge, N.R., R.H. Levy, R.M. McKay and T.R. Naish, 2017: East Antarctic ice sheet most vulnerable to Weddell Sea warming. *Geophys. Res. Lett.*, **44**(5), 2343–2351, doi:10.1002/2016GL072422.

Gomez, N., J.X. Mitrovica, P. Huybers and P.U. Clark, 2010: Sea level as a stabilizing factor for marine-ice sheet grounding lines. *Nat. Geosci.*, **3**(12), 850.

Gomez, N., D. Pollard and D. Holland, 2015: Sea level feedback lowers projections of future Antarctic Ice sheet mass loss. *Nat. Commun.*, **6**, 8798.

González-Correa, J.M., Y.F. Torquemada and J.L.S. Lizaso, 2008: Long-term effect of beach replenishment on natural recovery of shallow Posidonia oceanica meadows. *Estuar. Coast. Shelf Sci.*, **76**(4), 834–844.

Goodwin-Gill, G. S. and J. McAdam, 2017: *Climate Change Disasters and Displacement*. UNHCR [Available at: www.unhcr.org/596f25467.pdf]. Accessed: 2019/09/20.

Gorddard, R. et al., 2016: Values, rules and knowledge: adaptation as change in the decision context. *Environ. Sci. Policy*, **57**, 60–69.

Government of Canada, 1984: Western Arctic (Inuvialuit) Claims Settlement Act, S.C. 1984, c. 24 (amended 2003). Minister of Justice, Ottawa.

Government of Egypt, 2016: *Egypt Third National Communication: Under the United Nations Framework Convention on Climate Change*. Ministry of State for Environmental Affairs Egyptian Environmental Affairs Agency; United Nations Development Program; Global Environment Facility, Agency, E.E.A., Cairo, Egypt [Available at: https://unfccc.int/files/national_reports/non-annex_i_parties/biennial_update_reports/application/pdf/tnc_report.pdf].

Government of India, 2018: *Draft Coastal Regulation Zone Notification, 2018*. New Delhi, India. [Available at: www.indiaenvironmentportal.org.in/files/file/Draft%20Coastal%20Regulation%20Zone%20Notification,%202018.pdf]. Accessed 2019/09/20.

Government of the Maldives. Census 2014. [Available at: http://statistics maldives.gov.mv/census-2014/]. Accessed 2019/09/20.

Grady, A. et al., 2013: The influence of sea level rise and changes in fringing reef morphology on gradients in alongshore sediment transport. *Geophys. Res. Lett.*, **40**(12), 3096–3101.

Graham, S. et al., 2013: The social values at risk from sea level rise. *Environ. Impact Assess. Rev.*, **41**, 45–52.

Gray, C. and E. Wise, 2016: Country-specific effects of climate variability on human migration. *Clim. Change*, **135**(3–4), 555–568.

Gregory, J. et al., 2019: Concepts and terminology for sea level – mean, variability and change, both local and global. *Surveys of Geophysics*, 39-pp.

Gregory, J. and P. Huybrechts, 2006: Ice sheet contributions to future sea level change. *Philos. Trans. Roy. Soc. A.*, **364**(1844), 1709–1732.

Gregory, J.M., 2010: Long-term effect of volcanic forcing on ocean heat content. *Geophys. Res. Lett.*, **37**(22).

Gregory, J.M. et al., 2013: Twentieth-century global-mean sea level rise: Is the whole greater than the sum of the parts? *J. Clim.*, **26**(13), 4476–4499, doi:10.1175/JCLI-D-12-00319.1.

Griffies, S.M. and R.J. Greatbatch, 2012: Physical processes that impact the evolution of global mean sea level in ocean climate models. *Ocean Modelling*, **51**, 37–72, doi:10.1016/j.ocemod.2012.04.003.

Grifman, P. et al., 2013: Sea Level Rise Vulnerability Study for the City of Los Angeles. *University of Southern California*, USCSG-TR-05–2013 [Available at: https://dornsife.usc.edu/assets/sites/291/docs/pdfs/City_of_LA_SLR_Vulnerability_Study_FINAL_Summary_Report_Online_Hyperlinks.pdf]. Accessed 2019/09/20.

Grinsted, A., S. Jevrejeva, R.E.M. Riva and D. Dahl-Jensen, 2015: Sea level rise projections for Northern Europe under RCP8.5. *Clim. Res.*, **64**(1), 15–23, doi:10.3354/cr01309.

Grinsted, A., J.C. Moore and S. Jevrejeva, 2010: Reconstructing sea level from palaeo and projected temperatures 200 to 2100 AD. *Clim. Dyn.*, **34**(4), 461–472.

Gu, X., 2005: Retrospect and prospect of 50 years construction of Huangpu River flood control wall in Shanghai. *Water*, **21**(2), 15–25 (in Chinese).

Gugliotta, M. et al., 2018: Sediment distribution and depositional processes along the fluvial to marine transition zone of the Mekong River delta, Vietnam. *Sedimentology*, **66**(1), 146–164.

Gugliotta, M. et al., 2017: Process regime, salinity, morphological, and sedimentary trends along the fluvial to marine transition zone of the mixed-energy Mekong River delta, Vietnam. *Cont. Shelf Res.*, **147**, 7–26.

Haasnoot, M., J.H. Kwakkel, W.E. Walker and J. ter Maat, 2013: Dynamic adaptive policy pathways: A method for crafting robust decisions for a deeply uncertain world. *Global Environ. Chang.*, **23**, 485–498, doi:10.1016/j.gloenvcha.2012.12.006.

Haasnoot, M. et al., 2012: Exploring pathways for sustainable water management in river deltas in a changing environment. *Clim. Change*, **115**, 795–819, doi:10.1007/s10584-012-0444-2.

Haasnoot, M., H. Middelkoop, E. van Beek and W.P.A. van Deursen, 2011: A method to develop sustainable water management strategies for an uncertain future. *Sustain. Dev.*, **19**, 369–381, doi:10.1002/sd.438.

Haasnoot, M. et al., 2019: Investments under non-stationarity: economic evaluation of adaptation pathways. *Clim. Change*, 1–13.

Haasnoot, M., S. van't Klooster and J. van Alphen, 2018: Designing a monitoring system to detect signals to adapt to uncertain climate change. *Global Environ. Chang.*, **52**, 273–285.

Hagenlocher, M., F.G. Renaud, S. Haas and Z. Sebesvari, 2018: Vulnerability and risk of deltaic social-ecological systems exposed to multiple hazards. *Sci. Total Environ.*, **631**, 71–80.

Haigh, I.D. et al., 2014a: Estimating present day extreme water level exceedance probabilities around the coastline of Australia: tropical cyclone-induced storm surges. *Clim. Dyn.*, **42**(1–2), 139–157.

Haigh, I.D. et al., 2014b: Timescales for detecting a significant acceleration in sea level rise. *Nat. Commun.*, **5**, 3635.

Hall, J.A. et al., 2019: Rising sea levels: Helping decision-makers confront the inevitable. *Coastal Manage.*, **47**(2), 1–24.

Hallegatte, S., 2009: Strategies to adapt to an uncertain climate change. *Global Environ. Chang.*, **19**(2), 240–247.

Hallegatte, S., C. Green, R.J. Nicholls and J. Corfee-Morlot, 2013: Future flood losses in major coastal cities. *Nat. Clim. Change*, **3**(9), 802.

Hallegatte, S. et al., 2012: *Investment Decision Making under Deep Uncertainty – Application to Climate Change*. Policy Research Working Papers, The World Bank [Available at: http://hdl.handle.net/10986/12028]. Accessed: 2019/09/20.

Hamilton, L.C. et al., 2016: Climigration? Population and climate change in Arctic Alaska. *Popul. Environ.*, **38**(2), 115–133.

Hamilton, S.E. and D.A. Friess, 2018: Global carbon stocks and potential emissions due to mangrove deforestation from 2000 to 2012. *Nature Clim. Change*, 8(3), 240–244.

Hamlington, B. et al., 2018: Observation-Driven Estimation of the Spatial Variability of 20th Century Sea Level Rise. *J. Geophys. Res-Oceans*, **123**(3), 2129–2140.

Hamlington, B. et al., 2013: Contribution of the Pacific Decadal Oscillation to global mean sea level trends. *Geophys. Res. Lett.*, **40**(19), 5171–5175.

4

Hamlington, B. et al., 2017: Separating decadal global water cycle variability from sea level rise. *Scientific Reports*, **7**(995), 1–7.

Hamlington, B. et al., 2014: Uncovering an anthropogenic sea level rise signal in the Pacific Ocean. *Nat. Clim. Change*, **4**(9), 782.

Hamlington, B. and P. Thompson, 2015: Considerations for estimating the 20th century trend in global mean sea level. *Geophys. Res. Lett.*, **42**(10), 4102–4109.

Hamylton, S., J.X. Leon, M.I. Saunders and C. Woodroffe, 2014: Simulating reef response to sea level rise at Lizard Island: A geospatial approach. *Geomorphology*, **222**, 151–161.

Han, W. et al., 2014: Intensification of decadal and multi-decadal sea level variability in the western tropical Pacific during recent decades. *Clim. Dyn.*, **43**(5–6), 1357–1379.

Han, W. et al., 2017: Spatial patterns of sea level variability associated with natural internal climate modes. *Surv. Geophys.*, **38**(1), 217–250.

Hanson, H. et al., 2002: Beach nourishment projects, practices, and objectives – a European overview. *Coast. Eng. J.*, **47**(2), 81–111.

Hanson, S. et al., 2011: A global ranking of port cities with high exposure to climate extremes. *Clim. Change*, **104**(1), 89–111.

Haque, U. et al., 2012: Reduced death rates from cyclones in Bangladesh: what more needs to be done? *Bull. World Health Organ.*, **90**, 150–156.

Hardaway Jr, C.S. and K. Duhring, 2010: *Living Shoreline Design Guidelines for Shore Protection in Virginia's Estuarine Environments Verson 1.2.* Virginia Institute of Marine Science, College of William and Mary, Gloucester Point, Virginia [Available at: www.vims.edu/research/departments/physical/programs/ssp/_docs/living_shorelines_guidelines.pdf]. Accessed: 2019/09/20.

Hardy, R.D. and M.E. Hauer, 2018: Social vulnerability projections improve sea level rise risk assessments. *Appl. Geogr.*, **91**, 10–20.

Hardy, R.D., R.A. Milligan and N. Heynen, 2017: Racial coastal formation: The environmental injustice of colorblind adaptation planning for sea level rise. *Geoforum*, **87**, 62–72.

Harig, C. and F.J. Simons, 2015: Accelerated West Antarctic ice mass loss continues to outpace East Antarctic gains. *Earth Planet. Sci. Lett.*, **415**, 134–141.

Harley, M.D. et al., 2017: Extreme coastal erosion enhanced by anomalous extratropical storm wave direction. *Sci. Rep.*, **7**(1), 6033.

Harris, D.L. et al., 2018: Coral reef structural complexity provides important coastal protection from waves under rising sea levels. *Sci. Adv.*, **4**(2).

Hart, G., 2011: *Vulnerability and adaptation to sea level rise in Auckland, New Zealand.* New Zealand Climate Change Research Institute, Victoria University of Wellington, Wellington, New Zealand. [Available at: www.victoria.ac.nz/sgees/research-centres/documents/vulnerability-and-adaptation-to-sea-level-rise-in-auckland-new-zealand.pdf]. Accessed: 2019/09/20.

Hartig, E.K. et al., 2002: Anthropogenic and climate-change impacts on salt marshes of Jamaica Bay, NYC. *Wetlands*, **22**(1), 71–89.

Hatzikyriakou, A. and N. Lin, 2017: Simulating storm surge waves for structural vulnerability estimation and flood hazard mapping. *Nat. Hazards*, **89**(2), 939–962.

Hauer, M.E., 2017: Migration induced by sea level rise could reshape the US population landscape. *Nat. Clim. Change*, **7**(5), 321.

Hauer, M.E., J.M. Evans and D.R. Mishra, 2016: Millions projected to be at risk from sea–level rise in the continental United States. *Nat. Clim. Change*, **6**(7), 691.

Hauer, M.E., R.D. Hardy, D.R. Mishra and J.S. Pippin, 2018: No landward movement: examining 80 years of population migration and shoreline change in Louisiana. *Popul. Environ.*, **40**(4), 1–19.

Hay, C.C. et al., 2017: Sea Level Fingerprints in a Region of Complex Earth Structure: The Case of WAIS. *Journal of Climate*, **30**(6), 1881–1892.

Hay, C.C., E. Morrow, R.E. Kopp and J.X. Mitrovica, 2015: Probabilistic reanalysis of twentieth-century sea level rise. *Nature*, **517**(7535), 481–484.

Hay, J.E., 2013: Small island developing states: coastal systems, global change and sustainability. *Sustain. Sci.*, **8**(3), 309–326.

Hay, J.E., 2017: *Nadi flood control project. Climate risk and vulnerability assessment.* 52.

Haywood, A.M., H.J. Dowsett and A.M. Dolan, 2016: Integrating geological archives and climate models for the mid-Pliocene warm period. *Nat. Commun.*, **7**, 1–14 doi:10.1038/ncomms10646.

Head, B.W., 2014: Evidence, uncertainty, and wicked problems in climate change decision making in Australia. *Environ. Plan. C.*, **32**(4), 663–679.

Head, B.W., 2018: Forty years of wicked problems literature: forging closer links to policy studies. *Policy and Society*, **38**(2), 1–18.

Head, B.W. and J. Alford, 2015: Wicked problems: Implications for public policy and management. *Adm. Soc.*, **47**(6), 711–739.

Heberger, M., 2012: *The impacts of sea level rise on the San Francisco Bay.* California Energy Commission. [Available at: https://ww2.energy.ca.gov/2012publications/CEC-500-2012-014/CEC-500-2012-014.pdf]. Accessed 2019/09/20.

Hellmer, H.H., F. Kauker, R. Timmermann and T. Hattermann, 2017: The Fate of the Southern Weddell Sea Continental Shelf in a Warming Climate. *J. Clim.*, **30**(12), 4337–4350, doi:10.1175/jcli-d-16-0420.1.

Helsen, M.M. et al., 2013: Coupled regional climate–ice sheet simulation shows limited Greenland ice loss during the Eemian. *Clim. Past*, **9**(4), 1773–1788, doi:10.5194/cp-9-1773-2013.

Hemer, M.A. et al., 2013: Projected changes in wave climate from a multi-model ensemble. *Nat. Clim. Change*, **3**(5), 471.

Hempelmann, N. et al., 2018: Web processing service for climate impact and extreme weather event analyses. Flyingpigeon (Version 1.0). *Comput. Geosci.*, **110**, 65–72.

Henry, O. et al., 2014: Effect of the processing methodology on satellite altimetry-based global mean sea level rise over the Jason-1 operating period. *Journal of Geodesy*, **88**(4), 351–361.

Herath, D., R. Lakshman and A. Ekanayake, 2017: Urban Resettlement in Colombo from a Wellbeing Perspective: Does Development-Forced Resettlement Lead to Improved Wellbeing? *J. Refug. Stud.*, **30**(4), 554–579.

Hereher, M., 2009: Inventory of agricultural land area of Egypt using MODIS data. *Egypt J. Remote Sens. Space Sci.*, **12**, 179–184.

Hereher, M.E., 2010: Vulnerability of the Nile Delta to sea level rise: an assessment using remote sensing. *Geomat. Nat. Hazards Risk*, **1**(4), 315–321.

Hermann, E. and W. Kempf, 2017: Climate change and the imagining of Migration: Emerging discourses on Kiribati's land purchase in Fiji. *Contemp. Pac.*, **29**(2), 231–263.

Hermans, L.M., M. Haasnoot, J. ter Maat and J.H. Kwakkel, 2017: Designing monitoring arrangements for collaborative learning about adaptation pathways. *Environ. Sci. Policy*, **69**, 29–38.

Hernán, G. et al., 2017: Future warmer seas: increased stress and susceptibility to grazing in seedlings of a marine habitat-forming species. *Global Change Biol.*, **23**(11), 4530–4543, doi:10.1111/gcb.13768.

Hesed, C.D.M. and M. Paolisso, 2015: Cultural knowledge and local vulnerability in African American communities. *Nat. Clim. Change*, **5**(7), 683–687.

Hewitson, B. et al., 2014: Regional context. In: Climate Change 2014: Impacts, Adaptation, and Vulnerability. Part B: Regional Aspects. Contribution of Working Group II to the Fifth Assessment Report of the Intergovernmental Panel on Climate Change. [VR Barros et al. (eds.)] Cambridge University Press, Cambridge, United Kingdom and New York, NY, USA, 1133–1197.

Higgins, S.A., 2016: Advances in delta-subsidence research using satellite methods. *Hydrogeol. J.*, **24**(3), 587–600.

Hill, E.M., J.L. Davis, M.E. Tamisiea and M. Lidberg, 2010: Combination of geodetic observations and models for glacial isostatic adjustment fields in Fennoscandia. *J. Geophys. Res-Earth*, **115**(B07403), 1–12.

Hinkel, J. et al., 2018: The ability of societies to adapt to twenty-first-century sea level rise. *Nat. Clim. Change*, **8**(7), 570–578, doi:10.1038/s41558-018-0176-z.

Hinkel, J. and A. Bisaro, 2016: Methodological choices in solution-oriented adaptation research: a diagnostic framework. *Reg. Environ. Change*, **16**, 7–20, doi:10.1007/s10113-014-0682-0.

Hinkel, J. et al., 2019: Meeting user needs for sea-level rise information: a decision analysis perspective. *Earth's Future*, **7**(3), 320–337.

Hinkel, J. et al., 2015: Sea level rise scenarios and coastal risk management. *Nat. Clim. Change*, **5**(3), 188–190.

Hinkel, J. et al., 2014: Coastal flood damage and adaptation costs under 21st century sea level rise. *PNAS*, **111**(9), 3292–3297.

Hinkel, J. et al., 2013a: A global analysis of erosion of sandy beaches and sea level rise: An application of DIVA. *Global Planet. Change*, **111**, 150–158.

Hinkel, J., D. Vuuren, R. Nicholls and R.T. Klein, 2013b: The effects of adaptation and mitigation on coastal flood impacts during the 21st century. An application of the DIVA and IMAGE models. *Clim. Change*, **117**, 783–794.

Hino, M., C.B. Field and K.J. Mach, 2017: Managed retreat as a response to natural hazard risk. *Nat. Clim. Change*.

Hirabayashi, Y. et al., 2013: Projection of glacier mass changes under a high-emission climate scenario using the global glacier model HYOGA2. *Hydrol. Res. Lett.*, **7**(1), 6–11.

Hiwasaki, L., E. Luna and J.A. Marçal, 2015: Local and indigenous knowledge on climate-related hazards of coastal and small island communities in Southeast Asia. *Clim. Change*, **128**(1–2), 35–56.

Hoang, T.M.L. et al., 2016: Improvement of salinity stress tolerance in rice: challenges and opportunities. *Agronomy*, **6**(4), 54.

Hoegh-Guldberg, O. et al., 2018: Impacts of 1.5°C global warming on natural and human systems. In: Global Warming of 1.5°C. An IPCC Special Report on the impacts of global warming of 1.5°C above pre-industrial levels and related global greenhouse gas emission pathways, in the context of strengthening the global response to the threat of climate change, sustainable development, and efforts to eradicate poverty [Masson-Delmotte, V., P. Zhai, H.-O. Pörtner, D. Roberts, J. Skea, P.R. Shukla, A. Pirani, W. Moufouma-Okia, C. Péan, R. Pidcock, S. Connors, J.B.R. Matthews, Y. Chen, X. Zhou, M.I. Gomis, E. Lonnoy, T. Maycock, M. Tignor and T. Waterfield (eds.)], 175–311.

Hoegh-Guldberg, O. et al., 2007: Coral Reefs Under Rapid Climate Change and Ocean Acidification. *Science*, **80**(318).

Hoeke, R.K. et al., 2013: Widespread inundation of Pacific islands triggered by distant-source wind-waves. *Global Planet. Change*, **108**, 128–138.

Hoffman, J.S., P.U. Clark, A.C. Parnell and F. He, 2017: Regional and global sea-surface temperatures during the last interglaciation. *Science*, **355**(6322), 276–279.

Hoggart, S. et al., 2014: The consequences of doing nothing: the effects of seawater flooding on coastal zones. *Coast. Eng. J.*, **87**, 169–182.

Holgate, S.J. et al., 2012: New data systems and products at the permanent service for mean sea level. *J. Coast. Res.*, **29**(3), 493–504.

Holland, B., 2017: Procedural justice in local climate adaptation: political capabilities and transformational change. *Environ. Politics*, **26**(3), 391–412.

Holland, P.R., A. Jenkins and D.M. Holland, 2008: The response of ice shelf basal melting to variations in ocean temperature. *J. Clim.*, **21**(11), 2558–2572.

Hollowed, A.B. et al., 2013: Projected impacts of climate change on marine fish and fisheries. *ICES J. Mar. Sci.*, **70**(5), 1023–1037.

Holt, J.W. et al., 2006: New boundary conditions for the West Antarctic Ice Sheet: Subglacial topography of the Thwaites and Smith glacier catchments. *Geophys. Res. Lett.*, **33**(9).

Hoogendoorn, G. and J.M. Fitchett, 2018: Tourism and climate change: A review of threats and adaptation strategies for Africa. *Current Issues in Tourism*, **21**(7), 742–759.

Hopper, T. and M.S. Meixler, 2016: Modeling Coastal Vulnerability through Space and Time. *PLoS One*, **11**(10), e0163495–e0163495, doi:10.1371/journal.pone.0163495.

Hori, M. and M.J. Schafer, 2010: Social costs of displacement in Louisiana after Hurricanes Katrina and Rita. *Popul. Environ.*, **31**(1–3), 64–86.

Horikawa, K., 1978: *Coastal Engineering*. University of Tokyo Press, 402-pp.

Horstman, E. et al., 2014: Wave attenuation in mangroves: A quantitative approach to field observations. *Coast. Eng. J.*, **94**, 47–62.

Horton, B.P. et al., 2018: Mapping sea level change in time, space, and probability. *Annu. Rev. Environ. Resourc.*, **43**, 481–521.

Horton, B.P., S. Rahmstorf, S.E. Engelhart and A.C. Kemp, 2014: Expert assessment of sea level rise by AD 2100 and AD 2300. *Quat. Sci. Rev.*, **84**, 1–6.

Hoshino, S. et al., 2016: Estimation of increase in storm surge damage due to climate change and sea level rise in the Greater Tokyo area. *Nat. Hazards*, **80**(1), 539–565.

Hossain, M., M. Ahmed, E. Ojea and J.A. Fernandes, 2018: Impacts and responses to environmental change in coastal livelihoods of south-west Bangladesh. *Sci. Total Environ.*, **637**, 954–970.

Hu, K., Q. Chen and H. Wang, 2015: A numerical study of vegetation impact on reducing storm surge by wetlands in a semi-enclosed estuary. *Coast. Eng. J.*, **95**, 66–76.

Huang, P., I.I. Lin, C. Chou and R.-H. Huang, 2015: Change in ocean subsurface environment to suppress tropical cyclone intensification under global warming. *Nat. Commun.*, **6**, 7188, doi:10.1038/ncomms8188.

Hughes, S., E.K. Chu and S.G. Mason, 2018: Climate change in cities. *Innovations in Multi-Level Governance.* Cham: Springer International Publishing (The Urban Book Series), ISBN: 978-3-319-65003-6.

Hughes, T.P. et al., 2017: Coral reefs in the Anthropocene. *Nature*, **546**(7656), 82.

Hunt, C., 2013: Benefits and opportunity costs of Australia's Coral Sea marine protected area: A precautionary tale. *Mar. Policy*, **39**(352–260).

Hunter, J., 2010: Estimating sea level extremes under conditions of uncertain sea level rise. *Clim. Change*, **99**(3–4), 331–350.

Hunter, J., 2012: A simple technique for estimating an allowance for uncertain sea level rise. *Clim. Change*, **113**(2), 239–252, doi:10.1007/s10584-011-0332-1.

Hunter, J., P. Woodworth, T. Wahl and R. Nicholls, 2017: Using global tide gauge data to validate and improve the representation of extreme sea levels in flood impact studies. *Global Planet. Change*, **156**, 34–45.

Hunter, L.M., J.K. Luna and R.M. Norton, 2015: Environmental dimensions of migration. *Annu. Rev. Sociol.*, **41**, 377–397.

Huntington, H.P. et al., 2019: Climate change in context: putting people first in the Arctic. *Reg. Environ. Change*, **19**(4), 1217–1223.

Hurlimann, A. et al., 2014: Urban planning and sustainable adaptation to sea level rise. *Landscape and Urban Planning*, **126**, 84–93.

Hurlimann, A.C. and A.P. March, 2012: The role of spatial planning in adapting to climate change. *WiRes. Clim. Change*, **3**(5), 477–488.

Huss, M. and R. Hock, 2015: A new model for global glacier change and sea level rise. *Front. Earth Sci.*, **3**, 54.

Hussain, S.A. and R. Badola, 2010: Valuing mangrove benefits: contribution of mangrove forests to local livelihoods in Bhitarkanika Conservation Area, East Coast of India. *Wetl. Ecol. Manag.*, **18**, 321–331.

IDMC, 2017: *Global Report on International Displacement*. International Displacement Monitoring Centre [Available at: www.internal-displacement.org/global-report/grid2017/#download]. Accessed: 2019/09/20.

Ignatowski, J.A. and J. Rosales, 2013: Identifying the exposure of two subsistence villages in Alaska to climate change using traditional ecological knowledge. *Clim. Change*, **121**(2), 285–299.

Infantes, E. et al., 2012: Effect of a seagrass (Posidonia oceanica) meadow on wave propagation. *Mar. Ecol. Prog. Ser.*, **456**, 63–72.

Institution of Civil Engineers, 2010: *Facing-up to rising sea levels. Retreat? Defend? Attack?*, RIBA Royal Institute of British Architects, London [Available at: www.ice.org.uk/getattachment/news-and-insight/policy/facing-up-to-rising-sea-levels/Facing-Up-to-Rising-Sea-Levels-Document-Final.pdf.aspx]. Accessed 2019/09/20.

International Association for Public Participation, 2018: IAP2 Spectrum of Public Participation IAP2 International Federation, Australasia. [Available at: https://cdn.ymaws.com/www.iap2.org/resource/resmgr/pillars/Spectrum_8.5x11_Print.pdf]. Accessed 2019/09/20.

4

IPCC, 2014: *Climate Change 2014: Synthesis Report. Contribution of Working Groups I, II and III to the Fifth Assessment Report of the Intergovernmental Panel on Climate Change*. IPCC, Geneva, Switzerland. 151-pp.

IRC, 2016: *Inuvialuit on the Frontline of Climate Change: Development of a Regional Climate Change Adaptation Strategy*. Inuvialuit Regional Corporation, Inuvik, 181-pp. (Available at: www.irc.inuvialuit.com/system/files/Inuvialuit%20on%20the%20Frontline%20of%20Climate%20Change-Final-Feb2018%20%28SMALL%29.pdf). Accessed: 2019/09/20.

Irfanullah, H., A.K. Azad, Kamruzzaman and A. Wahed, 2011: Floating Gardening in Bangladesh: a means to rebuild lives after devastating flood. *Indian J. Tradit. Know.*, **10**(1), 31–38.

Irrgang, A.M. et al., 2019: Impacts of past and future coastal changes on the Yukon coast – threats for cultural sites, infrastructure, and travel routes. *Arctic Science*, **5**(2), 107–126.

Islam, M.R. and N.A. Khan, 2018: Threats, vulnerability, resilience and displacement among the climate change and natural disaster-affected people in South-East Asia: an overview. *Journal of the Asia Pacific Economy*, 23(2), 297–323.

Isobe, M., 2013: Impact of global warming on coastal structures in shallow water. *Ocean Engineering*, **71**, 51–57.

ITK, 2018: National Inuit Strategy on Research. Inuit Tapiriit Kanatami, 43-pp., ISBN: 978-0-9699774-2-1.

Jackson, L.P. and S. Jevrejeva, 2016: A probabilistic approach to 21st century regional sea level projections using RCP and high-end scenarios. *Global Planet. Change*, **146**, 179–189.

Jacob Balter, R.B., et al., 2017: *MTA Climate Adaptation Task Force Resiliency Report* [Projjal Dutta, M.J., Madeline Smith, Nelson Smith, Susan Yoon (ed.)]. MTA, MTA, New York (Available at: http://web.mta.info/sustainability/pdf/ResiliencyReport.pdf). Accessed: 2019/09/20.

Jamero, M.L. et al., 2017: Small-island communities in the Philippines prefer local measures to relocation in response to sea level rise. *Nat. Clim. Change*, **7**(8), 581–586.

James, T. et al., 2015: Tabulated values of relative sea level projections in Canada and the adjacent mainland United States. Geological Survey of Canada, Open File 7942. doi:10.4095/297048.

Janif, S. et al., 2016: Value of traditional oral narratives in building climate–change resilience: insights from rural communities in Fiji. *Ecol. Soc.*, **21**(2), 1–10.

Jankowski, K.L., T.E. Törnqvist and A.M. Fernandes, 2017: Vulnerability of Louisiana's coastal wetlands to present-day rates of relative sea level rise. *Nat. Commun.*, **8**, 14792.

Jenkins, A. et al., 2018: West Antarctic Ice Sheet retreat in the Amundsen Sea driven by decadal oceanic variability. *Nat. Geosci.*, 11(10), 733–741.

Jensen, L., R. Rietbroek and J. Kusche, 2013: Land water contribution to sea level from GRACE and Jason-1measurements. *J. Geophys. Res-Oceans*, **118**(1), 212–226.

Jevrejeva, S., A. Grinsted and J.C. Moore, 2014a: Upper limit for sea level projections by 2100. *Environ. Res. Lett.*, **9**(10), 104008.

Jevrejeva, S., A. Matthews and A. Slangen, 2017: The twentieth-century sea level budget: recent progress and challenges. In: *Integrative Study of the Mean Sea Level and Its Components*. Springer, Cham, pp. 301–313, ISBN: 978-3-319-56490-6.

Jevrejeva, S. et al., 2014b: Trends and acceleration in global and regional sea levels since 1807. *Global Planet. Change*, **113**, 11–22.

Jevrejeva, S., J. Moore, A. Grinsted and P. Woodworth, 2008: Recent global sea level acceleration started over 200 years ago? *Geophys. Res. Lett.*, **35**(8), 1–4.

Jiang, Z. et al., 2018: Future changes in rice yields over the Mekong River Delta due to climate change – Alarming or alerting? *Theor. Appl. Climatol.*, **127**(1–2), 545–555.

Jiménez Cisneros, B.E. et al., 2014: Freshwater resources. In Climate Change 2014: Impacts, Adaptation, and Vulnerability. Part A: Global and Sectoral Aspects. Contribution of Working Group II to the Fifth Assessment Report of the Intergovernmental Panel on Climate Change. Cambridge University Press, Cambridge, United Kingdom and New York, NY, USA.

John, B.M., K.G. Shirlal and S. Rao, 2015: Effect of Artificial Sea Grass on Wave Attenuation-An Experimental Investigation. *Aquatic Procedia*, **4**, 221–226.

Johnson, G.C. and D.P. Chambers, 2013: Ocean bottom pressure seasonal cycles and decadal trends from GRACE Release-05: Ocean circulation implications. *J. Geophys. Res-Oceans*, **118**(9), 4228–4240.

Jones, B. and B. O'Neill, 2016: Spatially explicit global population scenarios consistent with the Shared Socioeconomic Pathways. *Environ. Res. Lett.*, **11**(8), 1–10.

Jones, N. and J. Clark, 2014: Social capital and the public acceptability of climate change adaptation policies: a case study in Romney Marsh, UK. *Clim. Change*, **123**(2), 133–145.

Jones, N., J.R. Clark and C. Malesios, 2015: Social capital and willingness-to-pay for coastal defences in south-east England. *Ecol. Econ.*, **119**, 74–82.

Jones, R. et al., 2014: Foundations for decision making. In: Climate Change 2014: Impacts, Adaptation, and Vulnerability. Part A: Global and Sectoral Aspects. Contribution of Working Group II to the Fifth Assessment Report of the Intergovernmental Panel on Climate Change. [Field, C.B., V.R. Barros, D.J. Dokken, K.J. Mach, M.D. Mastrandrea, T.E. Bilir, M. Chatterjee, K.L. Ebi, Y.O. Estrada, R.C. Genova, B. Girma, E.S. Kissel, A.N. Levy, S. MacCracken, P.R. Mastrandrea and L.L. White (eds.)]. Cambridge University Press, Cambridge, United Kingdom and New York, NY, USA.

Jones, S. et al., 2019: Roads to Nowhere in Four States: State and Local Governments in the Atlantic Southeast Facing Sea level Rise. *Colum. J. Envtl. L.*, **44**, 67.

Jongman, B., P.J. Ward and J.C. Aerts, 2012: Global exposure to river and coastal flooding: Long term trends and changes. *Global Environ. Chang.*, **22**(4), 823–835.

Jongman, B. et al., 2015: Declining vulnerability to river floods and the global benefits of adaptation. *PNAS*, **112**(18), E2271-E2280.

Jonkman, S., J. Vrijling and A. Vrouwenvelder, 2008: Methods for the estimation of loss of life due to floods: a literature review and a proposal for a new method. *Nat. Hazards*, **46**(3), 353–389.

Jonkman, S.N. et al., 2013: Costs of adapting coastal defences to sea level rise – New estimates and their implications. *J. Coast. Res.*, **290**, 1212–1226.

Jordan, J.C., 2015: Swimming alone? The role of social capital in enhancing local resilience to climate stress: a case study from Bangladesh. *Clim. Dev.*, **7**(2), 110–123.

Joseph, D.D., 2017: Social work models for climate adaptation: the case of small islands in the Caribbean. *Reg. Environ. Change*, **17**(4), 1117–1126.

Joughin, I., B.E. Smith and B. Medley, 2014: Marine ice sheet collapse potentially under way for the thwaites glacier basin, West Antarctica. *Science*, **344**(6185), 735–738, doi:10.1126/science.1249055.

JSCE, 2000: Design Manual for Coastal Facilities. Japanese Society of Civil Engineers, Toyko, 577-pp.

Jurgilevich, A., A. Räsänen, F. Groundstroem and S. Juhola, 2017: A systematic review of dynamics in climate risk and vulnerability assessments. *Environ. Res. Lett.*, **12**(1), 013002.

Jurjonas, M. and E. Seekamp, 2018: Rural coastal community resilience: Assessing a framework in eastern North Carolina. *Ocean Coast. Manage*, **162**, 137–150.

Kabat, P. et al., 2009: Dutch coasts in transition. *Nat. Geosci.*, **2**(7), 450.

Kabir, M.J., D.S. Gaydon, R. Cramb and C.H. Roth, 2018: Bio-economic evaluation of cropping systems for saline coastal Bangladesh: I. Biophysical simulation in historical and future environments. *Agric. Syst.*, **162**, 107–122.

Kanada, S. et al., 2017: A Multimodel Intercomparison of an Intense Typhoon in Future, Warmer Climates by Four 5-km-Mesh Models. *J. Clim.*, **30**(15), 6017–6036.

Kaneko, S. and T. Toyota, 2011: Long-term urbanization and land subsidence in Asian Megacities: an indicators system approach. In: *Groundwater and Subsurface Environments* [Taniguchi, M. (ed.)]. Springer, Tokyo, pp. 249–270, ISBN: 978-4-431-53904-9.

Kaniewski, D. et al., 2014: Vulnerability of mediterranean ecosystems to long-term changes along the coast of Israel. *PLoS One*, **9**(7), 1–9, doi:10.1371/journal.pone.0102090.

Kankara, R.S., M.V. Ramana Murthy and M. Rajeevan, 2018: *National Assessment of Shoreline changes along Indian Coast: Status report for 26 years (1990–2016)*. Ministry of Earth Sciences, National Centre for Coastal Research, Chennai [Available at: www.indiaspend.com/wp-content/uploads/2018/11/National-Assessment-of-Shoreline-Changes-NCCR-report.pdf]. Accessed: 2019/09/20.

Karlsson, M. and G.K. Hovelsrud, 2015: Local collective action: Adaptation to coastal erosion in the Monkey River Village, Belize. *Global Environ. Chang.*, **32**, 96–107.

Karlsson, M., B. van Oort and B. Romstad, 2015: What we have lost and cannot become: societal outcomes of coastal erosion in southern Belize. *Ecol. Soc.*, **20**(1), 1–13.

Karnauskas, K.B., J.P. Donnelly and K.J. Anchukaitis, 2016: Future freshwater stress for island populations. *Nat. Clim. Change*, **6**(7), 720–725.

Karpytchev, M. et al., 2018: Contributions of a Strengthened Early Holocene Monsoon and Sediment Loading to Present-Day Subsidence of the Ganges-Brahmaputra Delta. *Geophys. Res. Lett.*, **45**(3), 1433–1442.

Kashem, S.B., B. Wilson and S. Van Zandt, 2016: Planning for climate adaptation: Evaluating the changing patterns of social vulnerability and adaptation challenges in three coastal cities. *J. Plan. Educ. Res.*, **36**(3), 304–318.

Kates, R.W., C.E. Colten, S. Laska and S.P. Leatherman, 2006: Reconstruction of New Orleans after Hurricane Katrina: a research perspective. *PNAS*, **103**(40), 14653–14660.

Katsman, C.A. et al., 2011: Exploring high-end scenarios for local sea level rise to develop flood protection strategies for a low-lying delta – the Netherlands as an example. *Clim. Change*, **109**(3–4), 617–645.

Keenan, J.M., T. Hill and A. Gumber, 2018: Climate gentrification: from theory to empiricism in Miami-Dade County, Florida. *Environ. Res. Lett.*, **13**(5), 054001.

Keim, M.E., 2010: Sea level-rise disaster in Micronesia: sentinel event for climate change? *Disaster Med. Public Health Prep.*, **4**(1), 81–87.

Kellens, W. et al., 2011: An analysis of the public perception of flood risk on the Belgian coast. *Risk Anal.*, **31**(7), 1055–1068.

Kelly, P.M., 2015: Climate drivers in coastal zone. In: *Climate Change and the Coast: Building Resilient Communities* [Glavovic, B., M. Kelly, R. Kay and A. Travers (eds.)]. CRC Press, Boca Raton; London; New York, pp. 29–49, ISBN 978-1-4822-8858-2.

Kelman, I., 2018: Islandness within climate change narratives of small island developing states (SIDS). *Isl. Stud. J.*, **13**(1), 149–166.

Kemp, A.C. et al., 2011: Climate related sea level variations over the past two millennia. *PNAS*, **108**(27), 11017–11022.

Kench, P.S. et al., 2018: Co-creating resilience solutions to coastal hazards through an interdisciplinary research project in New Zealand. *J. Coast. Res.*, **85**(sp1), 1496–1500.

Kernkamp, H.W.J., A. Van Dam, G.S. Stelling and E.D. De Goede, 2011: Efficient scheme for the shallow water equations on unstructured grids with application to the Continental Shelf. *Ocean Dynam.*, **61**(8), 1175–1188, doi:10.1007/s10236-011-0423-6.

Ketabchi, H. et al., 2014: Sea-level rise impact on fresh groundwater lenses in two-layer small islands. *Hydrol. Process.*, **28**(24), 5938–5953.

Khai, H.V., N.H. Dang and M. Yabe, 2018: Impact of Salinity Intrusion on Rice Productivity in the Vietnamese Mekong Delta. 九州大学大学院農学研究院紀要, **63**(1), 143–148.

Khanom, T., 2016: Effect of salinity on food security in the context of interior coast of Bangladesh. *Ocean Coast. Manage*, **130**, 205–212.

Khazendar, A. et al., 2016: Rapid submarine ice melting in the grounding zones of ice shelves in West Antarctica. *Nat. Commun.*, **7**, 13243.

Kim, M.J., R.J. Nicholls, J.M. Preston and G.A. de Almeida, 2018: An assessment of the optimum timing of coastal flood adaptation given sea-level rise using real options analysis. *J. Flood Risk Manage.*, e12494, 1–17.

Kind, J.M., 2014: Economically efficient flood protection standards for the Netherlands: Efficient flood protection standards for the Netherlands. *J. Flood Risk Manage.*, **7**, 103–117, doi:10.1111/jfr3.12026.

King, D. et al., 2016: Land use planning for disaster risk reduction and climate change adaptation: Operationalizing policy and legislation at local levels. *Int. J. Disaster Resilience Built Environ.* **7**(2), 158–172.

Kingslake, J., J.C. Ely, I. Das and R.E. Bell, 2017: Widespread movement of meltwater onto and across Antarctic ice shelves. *Nature*, **544**(7650), 349–352.

Kirshen, P. et al., 2018: Engaging Vulnerable Populations in Multi-Level Stakeholder Collaborative Urban Adaptation Planning for Extreme Events and Climate Risks – A Case Study of East Boston USA. *Journal of Extreme Events*, **5**(02n03), 1850013.

Kirwan, M.L. and J.P. Megonigal, 2013: Tidal wetland stability in the face of human impacts and sea level rise. *Nature*, **504**(7478), 53–60.

Kirwan, M.L. et al., 2016: Overestimation of marsh vulnerability to sea level rise. *Nat. Clim. Change*, **6**(3), 253.

Kittikhoun, A. and D.M. Staubli, 2018: Water diplomacy and conflict management in the Mekong: From rivalries to cooperation. *J. Hydrol.*, **567**, 654–667.

Kjeldsen, K.K. et al., 2015: Spatial and temporal distribution of mass loss from the Greenland Ice Sheet since AD 1900. *Nature*, **528**(7582), 396.

Klein, J., R. Mäntysalo and S. Juhola, 2016: Legitimacy of urban climate change adaptation: a case in Helsinki. *Reg. Environ. Change*, **16**(3), 815–826.

Klein, R.J. et al., 2014: Adaptation opportunities, constraints, and limits. *Constraints*, **16**, 4.

Kleindorfer, P.R., H.G. Kunreuther and P.J. Schoemaker, 1993: *Decision sciences: an integrative perspective*. Cambridge University Press, Cambridge, UK, pp 55, ISBN 0-521-32867-5.

Kleinherenbrink, M., R. Riva and T.J.O.S. Frederikse, 2018: A comparison of methods to estimate vertical land motion trends from GNSS and altimetry at tide gauge stations. *Ocean Sci.*, **14**(2), 187–204.

Klostermann, J. et al., 2018: Towards a framework to assess, compare and develop monitoring and evaluation of climate change adaptation in Europe. *Mitig. Adapt. Strat. Gl.*, **23**(2), 187–209.

Kniveton, D.R., C.D. Smith and R. Black, 2012: Emerging migration flows in a changing climate in dryland Africa. *Nat. Clim. Change*, **2**(6), 444.

Knutson, T.R. et al., 2015: Global projections of intense tropical cyclone activity for the late twenty-first century from dynamical downscaling of CMIP5/RCP4.5 scenarios. *J. Clim.*, **28**(18), 7203–7224, doi:10.1175/JCLI-D-15-0129.1.

Koch, E.W. et al., 2009: Non-linearity in ecosystem services: temporal and spatial variability in coastal protection. *Front. Ecol. Environ.*, **7**, 29–37.

Koch, M., G. Bowes, C. Ross and X. H. Zhang, 2013: Climate change and ocean acidification effects on seagrasses and marine macroalgae. *Global Change Biol.*, **19**(1), 103–132.

Koch, M.S. et al., 2015: Climate Change Projected Effects on Coastal Foundation Communities of the Greater Everglades Using a 2060 Scenario: Need for a New Management Paradigm. *Environ. Manage.*, **55**(4), 857–875, doi:10.1007/s00267-014-0375-y.

Kodikara, K.A S. et al., 2017: Have mangrove restoration projects worked? An in-depth study in Sri Lanka. *Restor. Ecol.*, **25**(5), 705–716.

Koerth, J. et al., 2014: A typology of household-level adaptation to coastal flooding and its spatio-temporal patterns. *SpringerPlus*, **3**(1), 466.

Kokelj, S.V. et al., 2012: Using multiple sources of knowledge to investigate northern environmental change: regional ecological impacts of a storm surge in the outer Mackenzie Delta, NWT. *Arctic*, 257–272.

Kondolf, G.M. et al., 2014: Sustainable sediment management in reservoirs and regulated rivers: Experiences from five continents. *Earth's Future*, **2**(5), 256–280.

Konikow, L F., 2011: Contribution of global groundwater depletion since 1900 to sea-level rise. *Geophys. Res. Lett.*, **38**(17), 1–5.

Konrad, H., I. Sasgen, D. Pollard and V. Klemann, 2015: Potential of the solid-Earth response for limiting long-term West Antarctic Ice Sheet retreat in a warming climate. *Earth Planet. Sci. Lett.*, **432**, 254–264.

Kopp, R.E. et al., 2017: Evolving Understanding of Antarctic Ice sheet Physics and Ambiguity in Probabilistic Sea level Projections. *Earth's Future*, **5**(12), 1217–1233, doi:10.1002/2017ef000663.

Kopp, R.E. et al., 2014: Probabilistic 21st and 22nd century sea-level projections at a global network of tide-gauge sites. *Earth's Future*, **2**(8), 383–406.

Kopp, R.E. et al., 2016: Temperature-driven global sea level variability in the Common Era. *PNAS*, **113**(11), E1434–E1441.

Kopp, R.E. et al., 2009: Probabilistic assessment of sea level during the last interglacial stage. *Nature*, **462**(7275), 863–867, doi:10.1038/nature08686.

Kopp, R.E. et al., 2013: A probabilistic assessment of sea level variations within the last interglacial stage. *Geophys. J. Int.*, **193**(2), 711–716.

Koslov, L., 2019: Avoiding Climate Change: "Agnostic Adaptation" and the Politics of Public Silence. *Ann. Am. Assoc. Geogr.*, **109**(2), 568–580.

Kossin, J.P., 2017: Hurricane intensification along United States coast suppressed during active hurricane periods. *Nature*, **541**(7637), 390.

Kossin, J.P., 2018: A global slowdown of tropical–cyclone translation speed. *Nature*, **558**(7708), 104.

Kossin, J.P., K.A. Emanuel and G.A. Vecchi, 2014: The poleward migration of the location of tropical cyclone maximum intensity. *Nature*, **509**(7500), 349–352, doi:10.1038/nature13278.

Koubi, V., G. Spilker, L. Schaffer and T. Böhmelt, 2016: The role of environmental perceptions in migration decision-making: evidence from both migrants and non-migrants in five developing countries. *Popul. Environ.*, **38**(2), 134–163.

Kousky, C., 2014: Managing shoreline retreat: a US perspective. *Clim. Change*, **124**(1–2), 9–20.

Krauss, K.W. et al., 2009: Water level observations in mangrove swamps during two hurricanes in Florida. *Wetlands*, **29**, 142–149.

Krauss, K.W. et al., 2014: How mangrove forests adjust to rising sea level. *New Phytol.*, **202**(1), 19–34, doi:10.1111/nph.12605.

Kreft, S., L. Schaefer, E. Behre and D. Matias, 2017: *Climate Risk Insurance for Resilience: Assessing Countries Implementation Plans.*, United Nations University Institute for Environment and Human Security (UNU-EHS), Bonn, Germany [Available at: https://collections.unu.edu/eserv/UNU:6321/MCII_DIE_171107_meta.pdf]. Accessed: 2019/09/20.

Kreibich, H. et al., 2017: Adaptation to flood risk: Results of international paired flood event studies. *Earth's Future*, **5**(10), 953–965.

Kuipers Munneke, P., S.R.M. Ligtenberg, M.R. Van Den Broeke and D.G. Vaughan, 2014: Firn air depletion as a precursor of Antarctic ice shelf collapse. *J. Glaciol.*, **60**(220), 205–214, doi:10.3189/2014JoG13J183.

Kuipers Munneke, P. et al., 2012: Insignificant change in Antarctic snowmelt volume since 1979. *Geophys. Res. Lett.*, **39**(1), 1–5.

Kuklicke, C. and D. Demeritt, 2016: Adaptive and risk-based approaches to climate change and the management of uncertainty and institutional risk: The case of future flooding in England. *Global Environ. Chang.*, **37**, 56–68.

Kulp, S. and B.H. Strauss, 2016: Global DEM errors underpredict coastal vulnerability to sea level rise and flooding. *Front. Earth Sci.*, **4**, 36.

Kulp, S. and B.H. Strauss, 2017: Rapid escalation of coastal flood exposure in US municipalities from sea level rise. *Clim. Change*, **142**(3–4), 477–489.

Kumar, L. and S. Taylor, 2015: Exposure of coastal built assets in the South Pacific to climate risks. *Nat. Clim. Change*, **5**(11), 992–996.

Kumari, R. et al., 2018: *Groundswell: Preparing for Internal Climate Migration. Chapter 2.*, The World Bank, Washington, DC [Available at: www.worldbank.org/en/news/infographic/2018/03/19/groundswell--preparing-for-internal-climate-migration]. Accessed 2019/09/20.

Kunreuther, H., 2015: The role of insurance in reducing losses from extreme events: The need for public–private partnerships. *The Geneva Papers on Risk and Insurance-Issues and Practice*, **40**(4), 741–762.

Kunreuther, H. et al., 2014: Integrated Risk and Uncertainty Assessment of Climate Change Response Policies. In: Climate Change 2014: Mitigation of Climate Change. Contribution of Working Group III to the Fifth Assessment Report of the Intergovernmental Panel on Climate Change. [Edenhofer, O., R. Pichs-Madruga, Y. Sokona, E. Farahani, S. Kadner, K. Seyboth, A. Adler, I. Baum, S. Brunner, P. Eickemeier, B. Kriemann, J. Savolainen, S. Schlömer, C.v. Stechow, T. Zwickel and J.C. Minx (eds.)]. Cambridge University Press, Cambridge, United Kingdom and New York, NY, USA.

Kunte, P.D. et al., 2014: Multi-hazards coastal vulnerability assessment of Goa, India, using geospatial techniques. *Ocean Coast. Manage*, **95**, 264–281.

Kura, Y., O. Joffre, B. Laplante and B. Sengvilaykham, 2017: Coping with resettlement: A livelihood adaptation analysis in the Mekong River basin. *Land Use Policy*, **60**, 139–149.

Kwakkel, J.H., M. Haasnoot and W.E. Walker, 2016a: Comparing robust decision-making and dynamic adaptive policy pathways for model-based decision support under deep uncertainty. *Environ. Modell. Softw.*, **86**, 168–183.

Kwakkel, J.H., W.E. Walker and M. Haasnoot, 2016b: Coping with the wickedness of public policy problems: approaches for decision making under deep uncertainty. *J. Water Res. Plan. Man.*, **142**(3), 1–5.

Kwakkel, J.H., W.E. Walker and V.A. Marchau, 2010: Classifying and communicating uncertainties in model-based policy analysis. *Int. J. Technol. Policy Manage.*, **10**(4), 299–315.

Lamb, J.B. et al., 2017: Seagrass ecosystems reduce exposure to bacterial pathogens of humans, fishes, and invertebrates. *Science*, **355**, 731–733.

Lambeck, K., A. Purcell and A. Dutton, 2012: The anatomy of interglacial sea levels: The relationship between sea levels and ice volumes during the Last Interglacial. *Earth Planet. Sci. Lett.*, **315–316**, 4–11, doi:10.1016/j.epsl.2011.08.026.

Lambeck, K., C. Smither and P. Johnston, 1998: Sea level change, glacial rebound and mantle viscosity fornorthern Europe. *Geophys. J. Int.*, **134**(1), 102–144, doi:10.1046/j.1365-246x.1998.00541.x.

Lamoureux, S. et al., 2015: The impact of climate change on infrastructure in the western and central Canadian Arctic. In: *From Science to Policy in the Western and Central Canadian Arctic: an Integrated Regional Impact Study (IRIS) of Climate Change and Modernization* [Stern, G.A. and A. Gaden (eds.)]. ArcticNet, Quebec, pp. 300–341.

Landais, A. et al., 2016: How warm was Greenland during the last interglacial period? *Clim. Past*, **12**(9), 1933–1948, doi:10.5194/cp-12-1933-2016.

Lantuit, H. et al., 2011: Coastal erosion dynamics on the permafrost-dominated Bykovsky Peninsula, north Siberia, 1951–2006. *Polar Res.*, **30**(1), 7341.

Larour, E., E.R. Ivins and S. Adhikari, 2017: Should coastal planners have concern over where land ice is melting? *Sci. Adv.*, **3**(11), e1700537.

Larour, E. et al., 2019: Slowdown in Antarctic mass loss from solid Earth and sea level feedbacks. *Science*, **364**(6444), doi:10.1126/science.aav7908.

Larsen, J.N. et al., 2014: Polar regions. In *Climate Change 2014: Impacts, Adaptation, and Vulnerability Part B: Regional Aspects Contribution of Working Group II to the Fifth Assessment Report of the Intergovernmental Panel on Climate Change* [Barros, V.R. et al. (eds.)], Cambridge University Press, Cambridge, United Kingdom and New York, NY, USA, pp. 1567–1612.

Lauterjung, J. et al., 2017: *10 Years Indonesian Tsunami Early Warning System: Experiences, Lessons Learned and Outlook* [Lauterjung, J. and H. Letz (eds.)]. GFZ German Research Centre for Geosciences Potsdam, Germany, 68-pp. [Available at: http://gfzpublic.gfz-potsdam.de/pubman/item/escidoc:2431901:13/component/escidoc:2469889/10_years_InaTEWS_2431901.pdf]. Accessed: 2019/09/201.

Lavery, P.S., M.Á. Mateo, O. Serrano and M. Rozaimi, 2013: Variability in the Carbon Storage of Seagrass Habitats and Its Implications for Global Estimates of Blue Carbon Ecosystem Service. *PLoS One*, **8**(9), 1–12, doi:10.1371/journal.pone.0073748.

Lawrence, J. et al., 2018: National guidance for adapting to coastal hazards and sea level rise: Anticipating change, when and how to change pathway. *Environ. Sci. Policy*, **82**, 100–107.

Lawrence, J., R. Bell and A. Stroombergen, 2019: A Hybrid Process to Address Uncertainty and Changing Climate Risk in Coastal Areas Using Dynamic

Adaptive Pathways Planning, Multi-Criteria Decision Analysis & Real Options Analysis: A New Zealand Application. *Sustainability*, **11**(2), 406.

Lawrence, J. and M. Haasnoot, 2017: What it took to catalyse uptake of dynamic adaptive pathways planning to address climate change uncertainty. *Environ. Sci. Policy*, **68**, 47–57.

Lawrence, J. et al., 2015: Adapting to changing climate risk by local government in New Zealand: institutional practice barriers and enablers. *Local Environ.*, **20**(3), 298–320.

Laws, D., D. Hogendoorn and H. Karl, 2014: Hot adaptation: what conflict can contribute to collaborative natural resource management. *Ecol. Soc.*, **19**(2), 1–9.

Lazeroms, W.M., A. Jenkins, G.H. Gudmundsson and R.S. van de Wal, 2018: Modelling present-day basal melt rates for Antarctic ice shelves using a parametrization of buoyant meltwater plumes. *The Cryosphere*, **12**(1), 49.

Lazrus, H., 2015: Risk perception and climate adaptation in Tuvalu: a combined cultural theory and traditional knowledge approach. *Hum. Organ.*, **74**(1), 52–61.

Le Bars, D., S. Drijfhout and H. De Vries, 2017: A high-end sea level rise probabilistic projection including rapid Antarctic ice sheet mass loss. *Environ. Res. Lett.*, **12**(4), doi:10.1088/1748-9326/aa6512.

Le Cozannet, G. et al., 2014: Approaches to evaluate the recent impacts of sea level rise on shoreline changes. *Earth-Sci. Rev.*, **138**, 47–60.

Le Cozannet, G., J.-C. Manceau and J. Rohmer, 2017: Bounding probabilistic sea level projections within the framework of the possibility theory. *Environ. Res. Lett.*, **12**(1), 014012.

Leal Filho, W. et al., 2018: Fostering coastal resilience to climate change vulnerability in Bangladesh, Brazil, Cameroon and Uruguay: a cross-country comparison. *Mitig. Adapt. Strat. Gl.*, **23**(4), 579–602.

Leck, H. and D. Roberts, 2015: What lies beneath: understanding the invisible aspects of municipal climate change governance. *Curr. Opin. Environ. Sustain.*, **13**, 61–67.

Leclercq, P.W., J. Oerlemans and J.G. Cogley, 2011: Estimating the glacier contribution to sea level rise for the period 1800–2005. *Surv. Geophys.*, **32**(4–5), 519.

Lee, S.B., M. Li and F. Zhang, 2017: Impact of sea level rise on tidal range in Chesapeake and Delaware Bays. *J. Geophys. Res-Oceans*, **122**(5), 3917–3938.

Lee, T.M. et al., 2015: Predictors of public climate change awareness and risk perception around the world. *Nat. Clim. Change*, **5**(11), 1014–1020.

Lee, Y., 2014: Coastal Planning Strategies for Adaptation to Sea Level Rise: A Case Study of Mokpo, Korea. *J. Building Constr. Plan. Res.*, **2**(01), 74–81.

Lefale, P.F., 2010: Ua 'afa le Aso Stormy weather today: traditional ecological knowledge of weather and climate. The Samoa experience. *Clim. Change*, **100**(2), 317–335.

Lefcheck, J.S. et al., 2017: Multiple stressors threaten the imperiled coastal foundation species eelgrass (Zostera marina) in Chesapeake Bay, USA. *Global Change Biol.*, **23**(9), 3474–3483.

Lefebvre, J.-P. et al., 2012: Seasonal variability of cohesive sediment aggregation in the Bach Dang–Cam Estuary, Haiphong (Vietnam). *Geo-Marine Letters*, **32**(2), 103–121.

Legeais, J.-F. et al., 2018: An improved and homogeneous altimeter sea level record from the ESA Climate Change Initiative. *Earth Syst. Sci. Data*, **10**, 281–301.

Lehmann, P. et al., 2015: Barriers and opportunities for urban adaptation planning: analytical framework and evidence from cities in Latin America and Germany. *Mitig. Adapt. Strat. Gl.*, **20**(1), 75–97.

Lei, Y. et al., 2017: Using Government Resettlement Projects as a Sustainable Adaptation Strategy for Climate Change. *Sustainability*, **9**(8), 1373.

Lemke, P. et al., 2007: Observations: Changes in Snow, Ice and Frozen Ground. In *Climate Change 2007: The Physical Science Basis. Contribution of Working Group I to the Fourth Assessment Report of the Intergovernmental Panel on Climate Change*. [Solomon S. et al. (eds.)], Cambridge University Press Cambridge, UK, pp. 337–383.

Lempert, R. et al., 2013: *Ensuring Robust Flood Risk Management in Ho Chi Minh City*. The World Bank Sustainable Development Network. [Available at: www.researchgate.net/publication/255698092_Ensuring_Robust_Flood_Risk_Management_in_Ho_Chi_Minh_City]. Accessed: 2019/09/20.

Lempert, R. and M. Schlesinger, 2001: Climate-change strategy needs to be robust. *Nature*, **412**(6845), 375–375, doi:10.1038/35086617.

Lempert, R. and M.E. Schlesinger, 2000: Robust strategies for abating climate change. *Clim. Change*, **45**, 387–401.

Lenaerts, J. et al., 2016: Meltwater produced by wind–albedo interaction stored in an East Antarctic ice shelf. *Nat. Clim. Change*, **7**(1), 58.

Lenk, S. et al., 2017: Costs of sea dikes–regressions and uncertainty estimates. *Nat. Hazards Earth Syst. Sci.*, **17**(5), 765–779.

Lentz, E.E. et al., 2016: Evaluation of dynamic coastal response to sea level rise modifies inundation likelihood. *Nat. Clim. Change*, **6**(7), 696–700.

Leonard, S., M. Parsons, K. Olawsky and F. Kofod, 2013: The role of culture and traditional knowledge in climate change adaptation: Insights from East Kimberley, Australia. *Global Environ. Chang.*, **23**(3), 623–632.

Leuliette, E.W., 2015: The balancing of the sea level budget. *Curr. Clim.*, **1**(3), 185–191.

Levermann, A. et al., 2013: The multimillennial sea level commitment of global warming. *PNAS*, **110**(34), 13745–13750, doi:10.1073/pnas.1219414110.

Levermann, A. et al., 2014: Projecting Antarctic ice discharge using response functions from SeaRISE ice sheet models. *Earth Syst. Dyn.*, **5**(2), 271.

Lewis, R.R., 2001: Mangrove restoration-Costs and benefits of successful ecological restoration. In: *Proceedings of the Mangrove Valuation Workshop, Universiti Sains Malaysia, Penang*, Penang, Malaysia, Beijer International Institute of Ecological Economics, pp. 4–8.

Li, X., J.P. Liu, Y. Saito and V.L. Nguyen, 2017: Recent evolution of the Mekong Delta and the impacts of dams. *Earth-Sci. Rev.*, **175**, 1–17.

Li, X. et al., 2015: Grounding line retreat of Totten Glacier, East Antarctica, 1996 to 2013. *Geophys. Res. Lett.*, **42**(19), 8049–8056.

Li, X., E. Rignot, J. Mouginot and B. Scheuchl, 2016: Ice flow dynamics and mass loss of Totten Glacier, East Antarctica, from 1989 to 2015. *Geophys. Res. Lett.*, **43**(12), 6366–6373.

Li, Y. et al., 2014: Coastal wetland loss and environmental change due to rapid urban expansion in Lianyungang, Jiangsu, China. *Reg. Environ. Change*, **14**(3), 1175–1188.

Lickley, M.J., C.C. Hay, M.E. Tamisiea and J.X. Mitrovica, 2018: Bias in estimates of global mean sea level change inferred from satellite altimetry. *J. Clim.*, **31**(13), 5263–5271.

Lidström, S., 2018: Sea level rise in public science writing: history, science and reductionism. *Environ. Commun.*, **12**(1), 15–27.

Lilai, X., H. Yuanrong and H. Wei, 2016: A multi-dimensional integrated approach to assess flood risks on a coastal city, induced by sea level rise and storm tides. *Environ. Res. Lett.*, **11**(1), 014001.

Lin, I.I. et al., 2009: Warm ocean anomaly, air sea fluxes, and the rapid intensification of tropical cyclone Nargis (2008). *Geophys. Res. Lett.*, **36**(3), 1–5, doi:10.1029/2008GL035815.

Lin, N. and K. Emanuel, 2015: Grey swan tropical cyclones. *Nat. Clim. Change*, **6**(1), 106–111, doi:10.1038/nclimate2777.

Lin, N., R.E. Kopp, B.P. Horton and J.P. Donnelly, 2016: Hurricane Sandy's flood frequency increasing from year 1800 to 2100. *PNAS*, **113**(43), 12071–12075.

Lin, N. and E. Shullman, 2017: Dealing with hurricane surge flooding in a changing environment: part I. Risk assessment considering storm climatology change, sea level rise, and coastal development. *Stoch. Env. Res. Risk A.*, **31**(9), 2379–2400, doi:10.1007/s00477-016-1377-5.

Lincke, D. and J. Hinkel, 2018: Economically robust protection against 21st century sea level rise. *Global Environ. Chang.*, **51**, 67–73.

Linh, L.H. et al., 2012: Molecular breeding to improve salt tolerance of rice (Oryza sativa L.) in the Red River Delta of Vietnam. *Int. J. Plant Genom*, **2012**, 1–9.

Linham, M., C. Green and R. Nicholls, 2010: *AVOID Report on the Costs of adaptation to the effects of climate change in the world's large port cities*. AV/WS2. Available at: www.avoid.uk.net/2010/07/avoid-1-costs-of-adaptation-to-the-effects-of-climate-change-in-the-worlds-large-port-cities/

Linham, M. and R. J. Nicholls, 2010: *Technologies for Climate Change Adaptation: Coastal Erosion and Flooding* [Zhu, X. (ed.)]. TNA Guidebook Series, UNEP Risø Centre on Energy, Climate and Sustainable Development Publishing, M.C., Roskilde, Denmark [Available at: www.researchgate. net/publication/216584246_Technologies_for_Climate_Change_ Adaptation_-_Coastal_Erosion_and_Flooding]. Accessed: 2019/09/20.

Lipscomb, W.H. et al., 2013: Implementation and initial evaluation of the glimmer community ice sheet model in the community earth system model. *J. Clim.*, **26**(19), 7352–7371.

Little, C.M. et al., 2015a: Joint projections of US East Coast sea level and storm surge. *Nat. Clim. Change*, **5**(12), 1114–1120.

Little, C.M. et al., 2015b: Uncertainty in twenty-first-century CMIP5 sea level projections. *J. Clim.*, **28**(2), 838–852.

Little, C.M., N.M. Urban and M. Oppenheimer, 2013: Probabilistic framework for assessing the ice sheet contribution to sea level change. *PNAS*, **110**(9), 3264–3269.

Liu, B., Y.L. Siu and G. Mitchell, 2016a: Hazard interaction analysis for multi-hazard risk assessment: a systematic classification based on hazard-forming environment. *Nat. Hazards Earth Syst. Sci.*, **16**(2), 629.

Liu, H., J.G. Behr and R. Diaz, 2016b: Population vulnerability to storm surge flooding in coastal Virginia, USA. *Integr. Environ. Assess.*, **12**(3), 500–509.

Liu, J.P. et al., 2017a: Stratigraphic formation of the Mekong River Delta and its recent shoreline changes. *Oceanography*, **30**(3), 72–83.

Liu, X. et al., 2017b: Effects of salinity and wet–dry treatments on C and N dynamics in coastal-forested wetland soils: Implications of sea level rise. *Soil Biol. Biochem.*, **112**, 56–67.

Lo, A.Y., B. Xu, F.K. Chan and R. Su, 2015: Social capital and community preparation for urban flooding in China. *Appl. Geogr.*, **64**, 1–11.

Lo, V., 2016: *Synthesis report on experiences with ecosystem-based approaches to climate change adaptation and disaster risk reduction, Technical Series no. 85,* Secretariat of the Conservation on Biological Diversity, Montreal, Canada, 1–110, ISBN: 9789292256432.

Loder, N., J.L. Irish, M. Cialone and T. Wamsley, 2009: Sensitivity of hurricane surge to morphological parameters of coastal wetlands. *Estuar. Coast. Shelf Sci.*, **84**(4), 625–636.

Logan, J.R., S. Issar and Z. Xu, 2016: Trapped in Place? Segmented Resilience to Hurricanes in the Gulf Coast, 1970–2005. *Demography*, **53**(5), 1511–1534.

Lovelock, C.E. et al., 2015: The vulnerability of Indo-Pacific mangrove forests to sea level rise. *Nature*, **526**(7574), 559–563.

Lowe, J.A. and J.M. Gregory, 2006: Understanding projections of sea level rise in a Hadley Centre coupled climate model. *J. Geophys. Res-Oceans*, **111**(C11).

Luijendijk, A. et al., 2018: The State of the World's Beaches. *Nature Sci. Rep.*, **8**, 1–11.

Lujala, P., H. Lein and J.K. Rød, 2015: Climate change, Nat. Hazards, and risk perception: the role of proximity and personal experience. *Local Environ.*, **20**(4), 489–509.

Lund, D.H., 2018: Governance innovations for climate change adaptation in urban Denmark. *J. Environ. Pol. Plan.*, **20**(5), 632–644.

Luo, X. et al., 2017: New evidence of Yangtze delta recession after closing of the Three Gorges Dam. *Sci. Rep.*, **7**, 41735.

Lusthaus, J., 2010: Shifting sands: sea level rise, maritime boundaries and inter-state conflict. *Politics*, **30**(2), 113–118.

Ma, Y., C.S. Tripathy and J.N. Bassis, 2017: Bounds on the calving cliff height of marine terminating glaciers. *Geophys. Res. Lett.*, **44**(3), 1369–1375.

Macayeal, D.R. and O.V. Sergienko, 2013: The flexural dynamics of melting ice shelves. *Ann. Glaciol.*, **54**(63), 1–10, doi:10.3189/2013AoG63A256.

Magnan, A. et al., 2016: Addressing the risk of maladaptation to climate change. *WiRes. Clim. Change*, **7**(5), 646–665.

Magnan, A.K. and V.K.E. Duvat, 2018: Unavoidable solutions for coastal adaptation in Reunion Island (Indian Ocean). *Environ. Sci. Policy*, **89**, 393–400, doi:10.1016/j.envsci.2018.09.002.

Maina, J. et al., 2016: Integrating social–ecological vulnerability assessments with climate forecasts to improve local climate adaptation planning for coral reef fisheries in Papua New Guinea. *Reg. Environ. Change*, **16**(3), 881–891.

Maldonado, J.K., 2015: Everyday practices and symbolic forms of resistance: adapting to environmental change in coastal Louisiana. In: *Hazards, Risks and Disasters in Society*. Elsevier, pp. 199–216, ISBN: 978-0-12-396451-9.

Maldonado, J.K. et al., 2013: The impact of climate change on tribal communities in the US: displacement, relocation, and human rights. *Clim. Change*, **120**(3), 601–614.

Mann, M.E. and K.A. Emanuel, 2006: Atlantic hurricane trends linked to climate change. *Eos*, **87**(24), 233–241, doi:10.1029/2006EO240001.

Mantyka-Pringle, C.S., T.G. Martin and J.R. Rhodes, 2013: Interactions between climate and habitat loss effects on biodiversity: a systematic review and meta-analysis. *Global Change Biol.*, **19**(5), 1642–1644, doi:10.1111/gcb.12148.

Marba, N. and C.M. Duarte, 2010: Mediterranean warming triggers seagrass (Posidonia oceanica) shoot mortality. *Global Change Biol.*, **16**(8), 2366–2375.

Marchau, V.A.W.J., W.E. Walker, P.J.T.M. Bloemen and S. Popper, (eds)., 2019: *Decision Making under Deep Uncertainty*. Springer International Publishing, Cham, 1–401, ISBN: 9783030052522.

Marcos, M. and A. Amores, 2014: Quantifying anthropogenic and natural contributions to thermosteric sea level rise. *Geophys. Res. Lett.*, **41**(7), 2502–2507, doi:10.1002/2014GL059766.

Marcos, M., F.M. Calafat, Á. Berihuete and S. Dangendorf, 2015: Long-term variations in global sea level extremes. *J. Geophys. Res-Oceans*, **120**(12), 8115–8134.

Marengo, J.A., F. Muller-Karger, M. Pelling and C.J. Reynolds, 2019: The METROPOLE Project–An Integrated Framework to Analyse Local Decision Making and Adaptive Capacity to Large-Scale Environmental Change: Decision Making and Adaptation to Sea Level Rise in Santos, Brazil. In: *Climate Change in Santos Brazil: Projections, Impacts and Adaptation Options* [Hidalgo Nunes, L., R. Greco and J.A. Marengo (eds.)]. Springer, pp. 3–15, ISBN: 978-3-319-96534-5, 285-pp.

Marino, E., 2012: The long history of environmental migration: Assessing vulnerability construction and obstacles to successful relocation in Shishmaref, Alaska. *Global Environ. Chang.*, **22**(2), 374–381.

Marino, E. and H. Lazrus, 2015: Migration or forced displacement?: the complex choices of climate change and disaster migrants in Shishmaref, Alaska and Nanumea, Tuvalu. *Hum. Organ.*, (2015), 341–350.

Mariotti, G. and J. Carr, 2014: Dual role of salt marsh retreat: Long-term loss and short-term resilience. *Water Resourc. Res.*, **50**(4), 2963–2974.

Marsooli, R., P.M. Orton, N. Georgas and A.F. Blumberg, 2016: Three-dimensional hydrodynamic modeling of coastal flood mitigation by wetlands. *Coast. Eng. J.*, **111**, 83–94.

Martínez, M.L., G. Mendoza-González, R. Silva-Casarín and E. Mendoza-Baldwin, 2014: Land use changes and sea level rise may induce a "coastal squeeze" on the coasts of Veracruz, Mexico. *Global Environ. Chang.*, **29**, 180–188.

Martínez-Botí, M.A. et al., 2015: Plio-Pleistocene climate sensitivity evaluated using high-resolution CO2 records. *Nature*, **518**(7537), 49.

Maryland DEP, 2013: New Tidal Wetland Regulations for Living Shorelines. 1–10. Available at: https://mde.state.md.us/programs/Water/Wetlandsand Waterways/Documents/www.mde.state.md.us/assets/document/ wetlandswaterways/Living%20Shoreline%20Regulations.Final.Effective% 2002–04–13.pdf.

Marzeion, B., J.G. Cogley, K. Richter and D. Parkes, 2014: Attribution of global glacier mass loss to anthropogenic and natural causes. *Science*, **345**(6199), 919–921.

Marzeion, B., A. Jarosch and M. Hofer, 2012: Past and future sea level change from the surface mass balance of glaciers. *The Cryosphere*, **6**(6), 1295.

Marzeion, B., P. Leclercq, J. Cogley and A. Jarosch, 2015: Brief Communication: Global reconstructions of glacier mass change during the 20th century are consistent. *The Cryosphere*, **9**(6), 2399–2404.

4

Marzeion, B. and A. Levermann, 2014: Loss of cultural world heritage and currently inhabited places to sea level rise. *Environ. Res. Lett.*, **9**(3), 034001.

Masselink, G. and R. Gehrels, 2015: *Coastal Environments and Global Change*, Hoboken, NJ, USA, pp. 432, ISBN: 9781119117261.

Massom, R.A. et al., 2018: Antarctic ice shelf disintegration triggered by sea ice loss and ocean swell. *Nature*, **558**(7710), 383–389, doi:10.1038/s41586-018-0212-1.

Masson-Delmotte, V. et al., 2013: Information from Palaeoclimate Archives. In: Climate Change 2013: The Physical Science Basis: Contribution of Working Group I to the Fifth Assessment Report of the Intergovernmental Panel on Climate Change. [Stocker, T.F., D. Qin, G.K. Plattner, M. Tignor, S.K. Allen, J. Boschung, A. Nauels, Y. Xia, V. Bex and P.M. Midgley (eds.)]. Cambridge University Press, Cambridge, United Kingdom and New York, NY, USA.

Masterson, J.P. et al., 2014: Effects of sea-level rise on barrier island groundwater system dynamics–ecohydrological implications. *Ecohydrology*, **7**(3), 1064–1071.

Matin, N., J. Forrester and J. Ensor, 2018: What is equitable resilience? *World Dev.*, **109**, 197–205.

Maxwell, P.S. et al., 2015: Identifying habitats at risk: simple models can reveal complex ecosystem dynamics. *Ecol. Appl.*, **25**(2), 573–587, doi:10.1890/14-0395.1.

Mayer-Pinto, M., M.G. Matias and R.A. Coleman, 2016: The interplay between habitat structure and chemical contaminants on biotic responses of benthic organisms. *PeerJ*, **4**, e1985-e1985, doi:10.7717/peerj.1985.

Mazda, Y., M. Magi, M. Kogo and P.N. Hong, 1997: Mangroves as a coastal protection from waves in the Tong King delta, Vietnam. *Mangroves and Salt Marshes*, **1**(2), 127–135.

Mazeka, B., Sutherland, C., Buthelezi, S., Khumalo, D., 2019: Community-Based Mapping Methodology for Climate Change Adaptation: A Case Study of Quarry Road West Informal Settlement, Durban, South Africa. In: *The Geography of Climate Change Adaptation in Urban Africa* [Cobbinah, P.B. and M. Addaney (eds.)]. Palgrave McMillan, Cham, pp. 57–88, ISSN: 9783030048730.

Mazzei, V. and E. Gaiser, 2018: Diatoms as tools for inferring ecotone boundaries in a coastal freshwater wetland threatened by saltwater intrusion. *Ecol. Indic.*, **88**, 190–204, doi:10.1016/j.ecolind.2018.01.003.

McAdam, J. and E. Ferris, 2015: Planned relocations in the context of climate change: unpacking the legal and conceptual issues. *Cambridge J. Intl & Comp. L.*, **4**, 137.

McCarthy, G.D. et al., 2015: Ocean impact on decadal Atlantic climate variability revealed by sea level observations. *Nature*, **521**(7553), 508.

McCarthy, J.J. et al., (eds.), 2001: Climate change 2001: impacts, adaptation, and vulnerability: contribution of Working Group II to the third assessment report of the Intergovernmental Panel on Climate Change. Cambridge University Press, Cambridge, UK.

McConnell, A., 2018: Rethinking wicked problems as political problems and policy problems. *Policy & Politics*, **46**(1), 165–180.

McCrea-Strub, A. et al., 2011: Understanding the cost of establishing marine protected areas. *Mar. Policy*, **35**(1), 1–9.

McCubbin, S., B. Smit and T. Pearce, 2015: Where does climate fit? Vulnerability to climate change in the context of multiple stressors in Funafuti, Tuvalu. *Global Environ. Chang.*, **30**, 43–55.

McDougall, C., 2017: Erosion and the beaches of Negril. *Ocean Coast. Manage*, **148**, 204–213.

McFadgen, B. and D. Huitema, 2018: Experimentation at the interface of science and policy: a multi-case analysis of how policy experiments influence political decision-makers. *Policy Sciences*, **51**(2), 161–187.

McFarlin, J.M. et al., 2018: Pronounced summer warming in northwest Greenland during the Holocene and Last Interglacial. *PNAS*, **115**(25), 6357–6362, doi:10.1073/pnas.1720420115.

McGregor, S. et al., 2014: Recent Walker circulation strengthening and Pacific cooling amplified by Atlantic warming. *Nat. Clim. Change*, **4**(10), 888.

McIver, L. et al., 2015: Climate change, overcrowding and non-communicable diseases: The 'triple whammy' of tuberculosis transmission risk in Pacific atoll countries. *Annals of the ACTM: An International Journal of Tropical and Travel Medicine*, **16**(3), 57.

McIvor, A., I. Möller, T. Spencer and M. Spalding, 2012a: Reduction of wind and swell waves by mangroves (Natural Coastal Protection Series: Report 1. Cambridge Coastal Research Unit Working Paper 40). The Nature Conservancy and Wetlands International, Cambridge, UK, ISSN: 2050–7941, pp. 27.

McIvor, A., T. Spencer, I. Möller and M. Spalding, 2012b: *Storm surge reduction by mangroves*. Natural Coastal Protection Series, Cambridge Coastal Research Unit, University of Cambridge, Conservancy, T.N. and W. International [Available at: www.conservationgateway.org/ConservationPractices/Marine/crr/library/Documents/storm-surge-reduction-by-mangroves-report.pdf]. Accessed: 2019/09/20.

McIvor, A., T. Spencer, I. Möller and M. Spalding, 2013: *The response of mangrove soil surface elevation to sea level rise*. Natural Coastal Protection Series, Cambridge Coastal Research Unit, University of Cambridge, Conservancy, T. N. and W. International [Available at: www.conservationgateway.org/ConservationPractices/Marine/crr/library/Documents/mangrove-surface-elevation-and-sea level-rise.pdf]. Accessed: 2019/09/20.

McKee, K.L. and W.C. Vervaeke, 2018: Will fluctuations in salt marsh-mangrove dominance alter vulnerability of a subtropical wetland to sea level rise? *Global Change Biol.*, **24**(3), 1224–1238, doi:10.1111/gcb.13945.

McKee, M., J. White and L. Putnam-Duhon, 2016: Simulated storm surge effects on freshwater coastal wetland soil porewater salinity and extractable ammonium levels: Implications for marsh recovery after storm surge. *Estuar. Coast. Shelf Sci.*, **181**, 338–344.

McLean, R. and P. Kench, 2015: Destruction or persistence of coral atoll islands in the face of 20th and 21st century sea-level rise? *WiRes. Clim. Change*, **6**(5), 445–463.

McLeman, R., 2018: Thresholds in climate migration. *Popul. Environ.*, **39**(4), 319–338.

McMillen, H. et al., 2014: Small islands, valuable insights: systems of customary resource use and resilience to climate change in the Pacific. *Ecol. Soc.*, **19**(4), 1–17.

McNamara, K.E., R. Bronen, N. Fernando and S. Klepp, 2018: The complex decision-making of climate-induced relocation: adaptation and loss and damage. *Clim. Policy*, **18**(1), 111–117.

Mcowen, C.J. et al., 2017: A global map of saltmarshes. *Biodiversity data journal*, **5**(e11764), 1–13.

Mechler, R. and L.M. Bouwer, 2015: Understanding trends and projections of disaster losses and climate change: is vulnerability the missing link? *Clim. Change*, **133**(1), 23–35.

Meehl, G.A. et al., 2007: Global climate projections. In: Climate change 2007: the physical science basis. Contribution of Working Group 1 to the Fourth Assessment Report of the Intergovernmental Panel on Climate Change, [Solomon S., Q.D. Manning, M.Z. Chen, M. Marquis, K.B. Averyt, M. Tignor, H.L. Miller (eds.)]. Cambridge University Press, Cambridge, United Kingdom and New York, NY, USA.

Mees, H.L., C.J. Uittenbroek, D.L. Hegger and P.P. Driessen, 2019: From citizen participation to government participation: A n exploration of the roles of local governments in community initiatives for climate change adaptation in the N etherlands. *Environ. Policy Governance*, **29**, 198–208.

Mehring, P., H. Geoghegan, H.L. Cloke and J. Clark, 2018: What is going wrong with community engagement? How flood communities and flood authorities construct engagement and partnership working. *Environ. Sci. Policy*, **89**, 109–115.

Mehvar, S. et al., 2019: Climate change-driven losses in ecosystem services of coastal wetlands: A case study in the West coast of Bangladesh. *Ocean Coast. Manage*, **169**, 273–283.

4

Mei, W. and S.-P. Xie, 2016: Intensification of landfalling typhoons over the northwest Pacific since the late 1970s. *Nat. Geosci.*, **9**(10), 753–757, doi:10.1038/ngeo2792.

Melet, A., S. Legg and R. Hallberg, 2016: Climatic impacts of parameterized local and remote tidal mixing. *J. Clim.*, **29**(10), 3473–3500.

Melet, A. and B. Meyssignac, 2015: Explaining the spread in global mean thermosteric sea level rise in CMIP5 climate models. *J. Clim.*, **28**(24), 9918–9940.

Melet, A., B. Meyssignac, R. Almar and G. Le Cozannet, 2018: Under-estimated wave contribution to coastal sea level rise. *Nat. Clim. Change*, 1.

Melvin, A.M. et al., 2017: Climate change damages to Alaska public infrastructure and the economics of proactive adaptation. *PNAS*, **114**(2), E122–E131.

Menéndez, P. et al., 2018: Valuing the protection services of mangroves at national scale: The Philippines. *Ecosyst. Serv.*, **34**, 24–36.

Mengel, M. et al., 2016: Future sea level rise constrained by observations and long-term commitment. *PNAS*, **113**(10), 2597–2602, doi:10.1073/pnas.1500515113.

Mengel, M., A. Nauels, J. Rogelj and C.-F. Schleussner, 2018: Committed sea level rise under the Paris Agreement and the legacy of delayed mitigation action. *Nat. Commun.*, **9**(1), 601.

Mentaschi, L. et al., 2018: Global long-term observations of coastal erosion and accretion. *Sci. Rep.*, **8**(1), 12876.

Merkens, J.-L., L. Reimann, J. Hinkel and A.T. Vafeidis, 2016: Gridded population projections for the coastal zone under the Shared Socioeconomic Pathways. *Global Planet. Change*, **145**, 57–66.

Metcalf, S.J. et al., 2015: Measuring the vulnerability of marine social-ecological systems: a prerequisite for the identification of climate change adaptations. *Ecol. Soc.*, **20**(2), 1–21.

Meyer, M.A., 2018: Social capital in disaster research. In: *Handbook of disaster research* [Rodríguez, H., W. Donner and J.E. Trainor (eds.)]. Springer, Cham, Switzerland, pp. 263–286, ISBN: 978-3-319-63253-7.

Meyssignac, B., X. Fettweis, R. Chevrier and G. Spada, 2017a: Regional Sea Level Changes for the Twentieth and the Twenty-First Centuries Induced by the Regional Variability in Greenland Ice Sheet Surface Mass Loss. *J. Clim.*, **30**(6), 2011–2028, doi:10.1175/jcli-d-16-0337.1.

Meyssignac, B. et al., 2017b: Causes of the regional variability in observed sea level, sea surface temperature and ocean colour over the period 1993–2011. *Surv. Geophys.*, **38**, 187–215.

Meyssignac, B. et al., 2017c: Evaluating model simulations of twentieth-century sea level rise. Part II: Regional sea level changes. *J. Clim.*, **30**(21), 8565–8593.

Michaelis, A.C., J. Willison, G.M. Lackmann and W.A. Robinson, 2017: Changes in winter North Atlantic extratropical cyclones in high-resolution regional pseudo-global warming simulations. *J. Clim.*, **30**(17), 6905–6925, doi:10.1175/JCLI-D-16-0697.1.

Milan, A. and S. Ruano, 2014: Rainfall variability, food insecurity and migration in Cabricán, Guatemala. *Clim. Dev.*, **6**(1), 61–68.

Milfont, T.L. et al., 2014: Proximity to coast is linked to climate change belief. *PLoS One*, **9**(7), e103180.

Milillo, P. et al., 2019: Heterogeneous retreat and ice melt of Thwaites Glacier, West Antarctica. *Sci. Adv.*, **5**(1), doi:10.1126/sciadv.aau3433.

Millan, R. et al., 2017: Bathymetry of the Amundsen Sea Embayment sector of West Antarctica from Operation IceBridge gravity and other data. *Geophys. Res. Lett.*, **44**(3), 1360–1368.

Miller, K.G. et al., 2012: High tide of the warm Pliocene: Implications of global sea level for Antarctic deglaciation. *Geology*, doi:10.1130/g32869.1.

Millock, K., 2015: Migration and environment. *Annu. Rev. Resour. Econ.*, **7**, 35–60.

Mills, M. et al., 2016: Reconciling Development and Conservation under Coastal Squeeze from Rising Sea Level. *Conserv. Lett.*, **9**(5), 361–368.

Milne, G.A. and J.X. Mitrovica, 2008: Searching for eustasy in deglacial sea level histories. *Quat. Sci. Rev.*, **27**(25–26), 2292–2302.

Minister of Conservation, 2010: *New Zealand Coastal Policy Statement*. New Zealand Department of Conservation, Team, D.o.C.P., Wellington, New Zealand. 30-pp. [Available at: www.doc.govt.nz/Documents/conservation/marine-and-coastal/coastal-management/nz-coastal-policy-statement-2010.pdf]. Accessed: 2019/09/20.

Mitrovica, J. et al., 2011: On the robustness of predictions of sea level fingerprints. *Geophys. J. Int.*, **187**(2), 729–742.

Mitrovica, J., M.E. Tamisiea, J.L. Davis and G.A. Milne, 2001: Recent mass balance of polar ice sheets inferred from patterns of global sea level change. *Nature*, **409**, 1026–1029, doi:10.1038/35059054.

Mitrovica, J.X. and G.A. Milne, 2003: On post-glacial sea level: I. General theory. *Geophys. J. Int.*, **154**(2), 253–267.

Moftakhari, H.R. et al., 2017: Compounding effects of sea level rise and fluvial flooding. *PNAS*, **114**(37), 9785–9790.

Möller, I. et al., 2014: Wave attenuation over coastal salt marshes under storm surge conditions. *Nat. Geosci.*, **7**(10), 727.

Möller, I. and T. Spencer, 2002: Wave dissipation over macro-tidal saltmarshes: Effects of marsh edge typology and vegetation change. *J. Coast. Res.*, **36**(sp1), 506–521.

Monioudi, I.N. et al., 2018: Climate change impacts on critical international transportation assets of Caribbean Small Island Developing States (SIDS): the case of Jamaica and Saint Lucia. *Reg. Environ. Change*, **18**(8), 1–15.

Moon, I.-J., S.-H. Kim, P. Klotzbach and J.C.L. Chan, 2015: Roles of interbasin frequency changes in the poleward shifts of the maximum intensity location of tropical cyclones. *Environ. Res. Lett.*, **10**(10), 1–9, doi:10.1088/1748-9326/10/10/104004.

Moon, J.H., Y.T. Song, P.D. Bromirski and A.J. Miller, 2013: Multidecadal regional sea level shifts in the Pacific over 1958–2008. *J. Geophys. Res-Oceans*, **118**(12), 7024–7035.

Moores, N., D.I. Rogers, K. Rogers and P.M. Hansbro, 2016: Reclamation of tidal flats and shorebird declines in Saemangeum and elsewhere in the Republic of Korea. *Emu.*, **116**(2), 136–146.

Mooyaart, L. and S. Jonkman, 2017: Overview and Design Considerations of Storm Surge Barriers. *J. Waterway Port Coast.*, **143**(4), 06017001.

Morim, J. et al., 2018: On the concordance of 21st century wind-wave climate projections. *Global Planet. Change*, **167**, 160–171.

Moritz Kramer, M.M., A. Petrov, B. Glass, 2015: *Storm Alert: Natural Disasters Can Damage Sovereign Creditworthiness*. Standard and Poor's Financial Services LLC [Available at: https://unepfi.org/pdc/wp-content/uploads/StormAlert.pdf]. Accessed: 2019/09/20.

Morlighem, M. et al., 2016: Modeling of Store Gletscher's calving dynamics, West Greenland, in response to ocean thermal forcing. *Geophys. Res. Lett.*, **43**(6), 2659–2666, doi:10.1002/2016gl067695.

Morlighem, M. et al., 2014: Deeply incised submarine glacial valleys beneath the Greenland ice sheet. *Nat. Geosci.*, **7**(6), 418–422.

Morlighem, M. et al., 2017: BedMachine v3: Complete bed topography and ocean bathymetry mapping of Greenland from multibeam echo sounding combined with mass conservation. *Geophys. Res. Lett.*, **44**(21), 11051–11061, doi:10.1002/2017GL074954.

Morris, J.T., J. Edwards, S. Crooks and E. Reyes, 2012: Assessment of carbon sequestration potential in coastal wetlands. In: *Recarbonization of the Biosphere*. Springer, pp. 517–531.

Morrison, K., 2017: The Role of Traditional Knowledge to Frame Understanding of Migration as Adaptation to the "Slow Disaster" of Sea Level Rise in the South Pacific. In: *Identifying Emerging Issues in Disaster Risk Reduction, Migration, Climate Change and Sustainable Development*. Springer, pp. 249–266.

Mortreux, C. et al., 2018: Political economy of planned relocation: A model of action and inaction in government responses. *Global Environ. Chang.*, **50**, 123–132.

Moser, S.C. and J.A. Ekstrom, 2010: A framework to diagnose barriers to climate change adaptation. *PNAS*, **107**(51), 22026–22031.

Moser, S.C., S. Jeffress Williams and D.F. Boesch, 2012: Wicked challenges at land's end: Managing coastal vulnerability under climate change. *Annu. Rev. Environ. Resourc.*, **37**, 51–78.

Mouginot, J., E. Rignot and B. Scheuchl, 2014: Sustained increase in ice discharge from the Amundsen Sea Embayment, West Antarctica, from 1973 to 2013. *Geophys. Res. Lett.*, **41**(5), 1576–1584.

MTA, 2017: MTA Announces Superstorm Sandy Recovery and Resiliency Progress 5 Years After Storm. *Metropolitan Transportation Authority*, NY, New York, USA. Available at: www.mta.info/news-mta-new-york-city-transit-bridges-tunnels/2017/10/30/mta-announces-superstorm-sandy-recovery.

Mueller, V., C. Gray and K. Kosec, 2014: Heat stress increases long-term human migration in rural Pakistan. *Nat. Clim. Change*, **4**(3), 182.

Muis, S. et al., 2017: A comparison of two global datasets of extreme sea levels and resulting flood exposure. *Earth's Future*, **5**(4), 379–392.

Muis, S. et al., 2016: A global reanalysis of storm surges and extreme sea levels. *Nat. Commun.*, **7**, doi:10.1038/ncomms11969.

Mukul, S.A. et al., 2019: Combined effects of climate change and sea level rise project dramatic habitat loss of the globally endangered Bengal tiger in the Bangladesh Sundarbans. *Sci. Total Environ.*, **663**, 830–840.

Mulder, J.P., S. Hommes and E.M. Horstman, 2011: Implementation of coastal erosion management in the Netherlands. *Ocean Coast. Manage*, **54**(12), 888–897.

Mullin, M., M.D. Smith and D.E. McNamara, 2019: Paying to save the beach: effects of local finance decisions on coastal management. *Clim. Change*, **152**(2), 275–289.

Munji, C.A. et al., 2013: Vulnerability to coastal flooding and response strategies: the case of settlements in Cameroon mangrove forests. *Environ. Dev.*, **5**, 54–72.

Murad, K.F.I. et al., 2018: Conjunctive use of saline and fresh water increases the productivity of maize in saline coastal region of Bangladesh. *Agric. Water Manage.*, **204**, 262–270.

Musa, Z.N., I. Popescu and A. Mynett, 2016: Assessing the sustainability of local resilience practices against sea level rise impacts on the lower Niger delta. *Ocean Coast. Manage*, **130**, 221–228.

Mycoo, M.A., 2018: Beyond 1.5 C: vulnerabilities and adaptation strategies for Caribbean Small Island developing states. *Reg. Environ. Change*, **18**(8), 2341–2353.

Myers, S.L., Revkin, A.C., Romero, S., Krauss, C., 2005: Old ways of life are fading as the Arctic thaws. *New York Times*, Available at: www.nytimes.com/2005/10/20/science/earth/old-ways-of-life-are-fading-as-the-arctic-thaws.html

Nadzir, N.M., M. Ibrahim and M. Mansor, 2014: Impacts of coastal reclamation to the quality of life: Tanjung Tokong community, Penang. *Procedia-Social and Behavioral Sciences*, **153**, 159–168.

Nairn, R. et al., 1999: Coastal Erosion at Keta Lagoon, Ghana–Large Scale Solution to a Large Scale Problem. In: Coastal Engineering 1998 [Edge, B. L. (ed.)]. American Society of Civil Engineers, 3192–3205.

Nakamura, J. et al., 2017: Western North Pacific Tropical Cyclone Model Tracks in Present and Future Climates. *J. Geophys. Res-Atmos.*, **122**(18), 9721–9744, doi:10.1002/2017JD027007.

Nakashima, D.J. et al., 2012: Weathering uncertainty: traditional knowledge for climate change assessment and adaptation. UNESCO and UNU, Paris and Darwin, 120-pp, ISBN: 978-92-3-001068-3.

Narayan, S. et al., 2016: The effectiveness, costs and coastal protection benefits of natural and nature-based defences. *PLoS One*, **11**(5), e0154735.

Narayan, S. et al., 2017: The Value of Coastal Wetlands for Flood Damage Reduction in the Northeastern USA. *Sci. Rep.*, **7**(1), 9463.

Nauels, A. et al., 2017a: Synthesizing long-term sea level rise projections–the MAGICC sea level model v2. 0. *Geosci. Model Dev.*, **10**(6), 2495.

Nauels, A. et al., 2017b: Linking sea level rise and socioeconomic indicators under the Shared Socioeconomic Pathways. *Environ. Res. Lett.*, **12**(11), 114002.

Nawrotzki, R.J. and J. DeWaard, 2016: Climate shocks and the timing of migration from Mexico. *Popul. Environ.*, **38**(1), 72–100.

Nawrotzki, R.J., J. DeWaard, M. Bakhtsiyarava and J.T. Ha, 2017: Climate shocks and rural-urban migration in Mexico: exploring nonlinearities and thresholds. *Clim. Change*, **140**(2), 243–258.

Naylor, A. K., 2015: Island morphology, reef resources, and development paths in the Maldives. *Progr. Phys. Geogr.*, **39**(6), 728–749.

Neef, A. et al., 2018: Climate adaptation strategies in Fiji: The role of social norms and cultural values. *World Dev.*, **107**, 125–137.

Nerem, R.S. et al., 2018: Climate-change–driven accelerated sea level rise detected in the altimeter era. *PNAS*, **115**(9), 2022–2025, doi:10.1073/pnas.1717312115.

Nerem, R.S., D.P. Chambers, C. Choe and G.T. Mitchum, 2010: Estimating mean sea level change from the TOPEX and Jason altimeter missions. *Mar. Geod.*, **33**(S1), 435–446.

Neumann, B., A.T. Vafeidis, J. Zimmermann and R.J. Nicholls, 2015: Future coastal population growth and exposure to sea level rise and coastal flooding-a global assessment. *PLoS One*, **10**(3), e0118571.

Newig, J. and O. Fritsch, 2009: Environmental governance: participatory, multi-level–and effective? *Environ. Policy Governance*, **19**(3), 197–214.

Newton, A., T. J. Carruthers and J. Icely, 2012: The coastal syndromes and hotspots on the coast. *Estuar. Coast. Shelf Sci.*, **96**, 39–47.

Newton, A. and J. Weichselgartner, 2014: Hotspots of coastal vulnerability: a DPSIR analysis to find societal pathways and responses. *Estuar. Coast. Shelf Sci.*, **140**, 123–133.

Ngan, L.T. et al., 2018: Interplay between land use dynamics and changes in hydrological regime in the Vietnamese Mekong Delta. *Land use policy*, **73**, 269–280.

Ngo-Duc, T. et al., 2005: Effects of land water storage on global mean sea level over the past half century. *Geophys. Res. Lett.*, **32**(9).

Nguyen, L.D., K. Raabe and U. Grote, 2015: Rural–urban migration, household vulnerability, and welfare in Vietnam. *World Dev.*, **71**, 79–93.

Nhung, T.T., P. Le Vo, V. Van Nghi and H.Q. Bang, 2019: Salt intrusion adaptation measures for sustainable agricultural development under climate change effects: A case of Ca Mau Peninsula, Vietnam. *Clim. Risk Manage.*, **23**, 88–100.

Nias, I.J., S.L. Cornford and A.J. Payne, 2016: Contrasting the modelled sensitivity of the Amundsen Sea Embayment ice streams. *J. Glaciol.*, **62**(233), 552–562.

Nicholls, R.J., 2010: Impacts of and responses to sea-level rise. In: *Understanding sea level rise and variability*, 17–51, [Church et al. (eds.)], Blackwell Publishing Ltd, West Sussex, UK, ISBN: 9781444323276.

Nicholls, R.J., 2011: Planning for the Impacts of Sea Level Rise. *Oceanography*, **24**(2), 144–157, doi:10.5670/oceanog.2011.34.

Nicholls, R.J., 2018: Adapting to Sea level Rise. In: *Resilience: The Science of Adaptation to Climate Change* [Zommers, Z. and K. Alverson (eds.)]. Elsevier, Oxford, UK, pp. 13–29, ISBN: 978-0-12-811891-7.

Nicholls, R.J. and A. Cazenave, 2010: Sea level rise and its impact on coastal zones. *Science*, **328** (5985), 1517–1520.

Nicholls, R.J., N.J. Cooper and I.H. Townend, 2007: The management of coastal flooding and erosion. In: *Future flooding and coastal erosion risks* [Thorne, C. (ed.)]. Thomas Telford, London, pp. 392–413, ISBN: 978-0-7277-3449-5.

Nicholls, R.J., R.J. Dawson and S.A. Day, (eds.), 2015: Broad Scale Coastal Simulation. Springer, London, UK, 395-pp., ISBN: 978-94-007-5257-3.

Nicholls, R.J., D. Lincke, J. Hinkel and T. van der Pol, 2019: *Global Investment Costs for Coastal Defence Through the 21st Century*. World Bank, Group, W.B.G.S.D.P. [Available at: http://documents.worldbank.org/curated/en/433981550240622188/pdf/WPS8745.pdf]. Accessed: 2019/09/20.

Nick, F.M. et al., 2013: Future sea level rise from Greenland's main outlet glaciers in a warming climate. *Nature*, **497**(7448), 235–238, doi:10.1038/nature12068.

Nicolas, J.P. et al., 2017: January 2016 extensive summer melt in West Antarctica favoured by strong El Niño. *Nat. Commun.*, **8**, ncomms15799.

4

Nidheesh, A. et al., 2013: Decadal and long-term sea level variability in the tropical Indo-Pacific Ocean. *Clim. Dyn.*, **41**(2), 381–402.

Nilubon, P., W. Veerbeek and C. Zevenbergen, 2016: Amphibious Architecture and Design: A Catalyst of Opportunistic Adaptation?–Case Study Bangkok. *Procedia-Social and Behavioral Sciences*, **216**, 470–480.

Nkoana, E., A. Verbruggen and J. Hugé, 2018: Climate change adaptation tools at the community level: An integrated literature review. *Sustainability*, **10**(3), 796.

Noël, B. et al., 2015: Evaluation of the updated regional climate model RACMO2. 3: summer snowfall impact on the Greenland Ice Sheet. *The Cryosphere*, **9**(5), 1831–1844.

Noël, B. et al., 2017: A tipping point in refreezing accelerates mass loss of Greenland's glaciers and ice caps. *Nat. Commun.*, **8**, 14730, doi:10.1038/ncomms14730.

Noël, B. et al., 2018: Modelling the climate and surface mass balance of polar ice sheets using RACMO2-Part 1: Greenland (1958–2016). *The Cryosphere*, **12**(3), 811–831.

Noh, S. et al., 2013: Influence of salinity intrusion on the speciation and partitioning of mercury in the Mekong River Delta. *Geochimica et Cosmochimica Acta*, **106**, 379–390.

Nordstrom, K.F., C. Armaroli, N.L. Jackson and P. Ciavola, 2015: Opportunities and constraints for managed retreat on exposed sandy shores: Examples from Emilia-Romagna, Italy. *Ocean Coast. Manage*, **104**, 11–21.

Noto, A.E. and J.B. Shurin, 2017: Early Stages of Sea level Rise Lead To Decreased Salt Marsh Plant Diversity through Stronger Competition in Mediterranean- Climate Marshes. *PLoS One*, **12**(1), 1–11, doi:10.1371/journal.pone.0169056.

Nunn, P.D., W. Aalbersberg, S. Lata and M. Gwilliam, 2014: Beyond the core: community governance for climate-change adaptation in peripheral parts of Pacific Island Countries. *Reg. Environ. Change*, **14**(1), 221–235.

Nunn, P.D., A. Kohler and R. Kumar, 2017a: Identifying and assessing evidence for recent shoreline change attributable to uncommonly rapid sea level rise in Pohnpei, Federated States of Micronesia, Northwest Pacific Ocean. *J. Coast. Conserv.* **21**(6), 719–730.

Nunn, P.D., J. Runman, M. Falanruw and R. Kumar, 2017b: Culturally grounded responses to coastal change on islands in the Federated States of Micronesia, northwest Pacific Ocean. *Reg. Environ. Change*, **17**(4), 959–971.

Nurse, L.A. et al., 2014: Small islands. In: Climate Change 2014: Impacts, Adaptation, and Vulnerability. Part B: Regional Aspects. Contribution of Working Group II to the Fifth Assessment Report of the Intergovernmental Panel of Climate Change. [Barros, V.R., C.B. Field, D.J. Dokken, M.D. Mastrandrea, K.J. Mach, T.E. Bilir, M. Chatterjee, K.L. Ebi, Y.O. Estrada, R.C. Genova, B. Girma, E.S. Kissel, A.N. Levy, S. MacCracken, P.R. Mastrandrea and L. L. White (eds.)]. Cambridge University Press, Cambridge, United Kingdom and New York, NY, USA.

Nursey-Bray, M., 2017: Towards socially just adaptive climate governance: the transformative potential of conflict. *Local Environ.*, **22**(2), 156–171.

NYC, 2014: *Retrofitting Buildings for Flood Risk*. NYC Department of City Planning, New York. [Available at: www1.nyc.gov/assets/planning/download/pdf/plans-studies/retrofitting-buildings/retrofitting_complete.pdf]. Accessed: 2019/09/20.

NYC, 2015: The City of New York CDBG-DR Action Plan Incorporating Amendments 1–10. *NYC Community Development Block Grant Disaster Recovery*. Available at: www1.nyc.gov/site/cdbgdr/action-plan/action-plan.page.

NYC Mayor's Office of Recovery and Resiliency, 2019: *Climate Resiliency Design Guidelines*. 3.0, Resiliency. New York Mayor's Office, New York, 66-pp. [Available at: www1.nyc.gov/assets/orr/pdf/NYC_Climate_Resiliency_Design_Guidelines_v3–0.pdf]. Accessed: 2019/09/20.

Nygren, A. and G. Wayessa, 2018: At the intersections of multiple marginalisations: displacements and environmental justice in Mexico and Ethiopia. *Environmental Sociology*, **4**(1), 148–161.

O'Meara, T.A., J.R. Hillman and S.F. Thrush, 2017: Rising tides, cumulative impacts and cascading changes to estuarine ecosystem functions. *Sci. Rep.*, **7**, 1–7, doi:10.1038/s41598-017-11058-7.

O'Neill, B.C. et al., 2017: IPCC reasons for concern regarding climate change risks. *Nat. Clim. Change*, **7**(1), 28.

O'Neill, E., F. Brereton, H. Shahumyan and J.P. Clinch, 2016: The Impact of Perceived Flood Exposure on Flood-Risk Perception: The Role of Distance. *Risk Anal.*, **36**(11), 2158–2186.

Oberlack, C., 2017: Diagnosing institutional barriers and opportunities for adaptation to climate change. *Mitig. Adapt. Strat. Gl.*, **22**(5), 805–838.

Oberlack, C. and K. Eisenack, 2018: Archetypical barriers to adapting water governance in river basins to climate change. *Journal of Institutional Economics*, **14**(3), 527–555.

Oberschall, A., 1978: Theories of social conflict. *Annu. Rev. Sociol.*, **4**(1), 291–315.

Oddo, P.C. et al., 2017: Deep Uncertainties in Sea level Rise and Storm Surge Projections: Implications for Coastal Flood Risk Management: Deep Uncertainties in Coastal Flood Risk Management. *Risk Anal.*, 1–16, doi:10.1111/risa.12888.

OECD, 2019: *Responding to Rising Seas*. OECD Publishing, Paris, 176-pp.

Olsthoorn, X., P. van der Werff, L.M. Bouwer and D. Huitema, 2008: Neo-Atlantis: The Netherlands under a 5-m sea level rise. *Clim. Change*, **91**(1–2), 103–122.

Onaka, S. et al., 2017: Effectiveness of gravel beach nourishment on Pacific Island. In: Asian And Pacific Coast 2017: Proceedings Of The 9th International Conference On Apac 2017 [Suh, K.-D., E.C. Cruz and Y. Tajima (eds.)]. 651–662, World Scientific, London, UK.

Onat, Y., O.P. Francis and K. Kim, 2018: Vulnerability assessment and adaptation to sea level rise in high-wave environments: A case study on O'ahu, Hawai'i. *Ocean Coast. Manage*, **157**, 147–159.

Oppenheimer, M. and R.B. Alley, 2016: How high will the seas rise? *Science*, **354**(6318), 1375–1377.

Oppenheimer, M. et al., 2014: Emergent risks and key vulnerabilities. In: *Climate Chagne 2014 Impacts, Adaptation and Vulnerability*, 1039–1099, Cambridge University Press, Cambridge, UK, ISBN: 9781107415379.

Ordóñez, C. and P. Duinker, 2015: Climate change vulnerability assessment of the urban forest in three Canadian cities. *Clim. Change*, **131**(4), 531–543.

Osland, M.J. et al., 2017: Assessing coastal wetland vulnerability to sea level rise along the northern Gulf of Mexico coast: Gaps and opportunities for developing a coordinated regional sampling network. *PLoS One*, **12**(9), 1–23, doi:10.1371/journal.pone.0183431.

Otto, A. et al., 2013: Energy budget constraints on climate response. *Nat. Geosci.*, **6**(6), 415–416, doi:10.1038/ngeo1836.

Ouillon, S., 2018: Why and How Do We Study Sediment Transport? Focus on Coastal Zones and Ongoing Methods. *Water*, **10**(4), 1–34.

Palanisamy, H., A. Cazenave, T. Delcroix and B. Meyssignac, 2015: Spatial trend patterns in the Pacific Ocean sea level during the altimetry era: the contribution of thermocline depth change and internal climate variability. *Ocean Dynam.*, **65**(3), 341–356.

Palanisamy, H. et al., 2014: Regional sea level variability, total relative sea level rise and its impacts on islands and coastal zones of Indian Ocean over the last sixty years. *Global Planet. Change*, **116**, 54–67.

Palmer, M. et al., 2010: Future observations for monitoring global ocean heat content. In: *Proceedings of OceanObs'09: Sustained Ocean Observations and Information for Society*, Venice, Italy, 1–13-pp.

Paolo, F.S., H.A. Fricker and L. Padman, 2015: Volume loss from Antarctic ice shelves is accelerating. *Science*, **348**(6232), 327–331.

Pappenberger, F. et al., 2015: The monetary benefit of early flood warnings in Europe. *Environ. Sci. Policy*, **51**, 278–291.

Paprocki, K. and S. Huq, 2018: Shrimp and coastal adaptation: on the politics of climate justice. *Clim. Dev.*, **10**(1), 1–3.

Parizek, B.R. et al., 2019: Ice-cliff failure via retrogressive slumping. *Geology*, **47**(5), 1–4, doi:10.1130 /G45880.1.

4

Parker, D.J., 2017: Risk Communication and Warnings, Risk Management, Response, Floods, Preparedness. *Natural Hazard Science*, doi: 10.1093/acrefore/9780199389407.013.84.

Parkes, D. and B. Marzeion, 2018: Twentieth-century contribution to sea level rise from uncharted glaciers. *Nature*, **563**(7732), 551.

Pasquini, L., R. Cowling and G. Ziervogel, 2013: Facing the heat: Barriers to mainstreaming climate change adaptation in local government in the Western Cape Province, South Africa. *Habitat Int.*, **40**, 225–232.

Pasquini, L., G. Ziervogel, R.M. Cowling and C. Shearing, 2015: What enables local governments to mainstream climate change adaptation? Lessons learned from two municipal case studies in the Western Cape, South Africa. *Clim. Dev.*, **7**(1), 60–70.

Passeri, D.L. et al., 2015: The dynamic effects of sea level rise on low-gradient coastal landscapes: A review. *Earth's Future*, **3**(6), 159–181.

Paterson, S.K. et al., 2017: Size does matter: City scale and the asymmetries of climate change adaptation in three coastal towns. *Geoforum*, **81**, 109–119.

Patterson, M.O. et al., 2014: Orbital forcing of the East Antarctic ice sheet during the Pliocene and Early Pleistocene. *Nat. Geosci.*, **7**, 841–847, doi:10.1038/ngeo2273.

Pattyn, F., 2017: Sea level response to melting of Antarctic ice shelves on multi-centennial timescales with the fast Elementary Thermomechanical Ice Sheet model (f. ETISh v1. 0). *The Cryosphere*, **11**(4), 1851–1878.

Pattyn, F. et al., 2013: Grounding-line migration in plan-view marine ice sheet models: results of the ice2sea MISMIP3d intercomparison. *J. Glaciol.*, **59**(215), 410–422.

Pattyn, F. et al., 2018: The Greenland and Antarctic ice sheets under 1.5°C global warming. *Nat. Clim. Change*, **8**(12), 1053–1061, doi:10.1038/s41558-018-0305-8.

Paul, M. and C. Amos, 2011: Spatial and seasonal variation in wave attenuation over Zostera noltii. *J. Geophys. Res-Oceans*, **116**(C8).

Pearse, R., 2017: Gender and climate change. *WiRes. Clim. Change*, **8**(2), 1–16.

Peduzzi, P., 2014: Sand, rarer than one thinks. *Environ. Dev.*, **11**, 208–218.

Peduzzi, P. et al., 2012: Global trends in tropical cyclone risk. *Nat. Clim. Change*, **2**(4), 289–294, doi:10.1038/nclimate1410.

Peltier, W., 2004: Global glacial isostasy and the surface of the ice-age earth: the ICE-5G(VM2) model and GRACE. *Annu. Rev. Earth Planet. Sci.*, **32**, 111–149, doi:10.1146/annurev.earth.32.082503.144359.

Peltier, W., D. Argus and R. Drummond, 2015: Space geodesy constrains ice age terminal deglaciation: The global ICE-6G_C (VM5a) model. *J. Geophys. Res-Earth*, **120**(1), 450–487.

Pendleton, L. et al., 2016: Coral reefs and people in a high-CO2 world: Where can science make a difference to people? *PLoS One*, **11**(11), e0164699.

Perrette, M. et al., 2013: A scaling approach to project regional sea level rise and its uncertainties. *Earth Syst. Dyn.*, **4**(1), 11–29, doi:10.5194/esd-4-11-2013.

Perry, C. and K. Morgan, 2017: Bleaching drives collapse in reef carbonate budgets and reef growth potential on southern Maldives reefs. *Sci. Rep.*, **7**, 40581.

Perry, C.T. et al., 2018: Loss of coral reef growth capacity to track future increases in sea level. *Nature*, **558**(7710), 396.

Perry, C.T. et al., 2013: Caribbean-wide decline in carbonate production threatens coral reef growth. *Nat. Commun.*, **4**, 1402.

Perumal, N., 2018: "The place where I live is where I belong": community perspectives on climate change and climate-related migration in the Pacific island nation of Vanuatu. *Isl. Stud. J.*, **13**(1), 45–64.

Peters, B.G., A. Jordan and J. Tosun, 2017: Over-reaction and under-reaction in climate policy: An institutional analysis. *J. Environ. Pol. Plan.*, **19**(6), 612–624.

Petzold, J., 2016: Limitations and opportunities of social capital for adaptation to climate change: a case study on the Isles of Scilly. *Geogr. J.*, **182**(2), 123–134.

Petzold, J., 2018: Social adaptability in ecotones: sea level rise and climate change adaptation in Flushing and the Isles of Scilly, UK. *Isl. Stud. J.*, **13**(1), 101–118.

Petzold, J. and A.K. Magnan, 2019: Climate change: thinking small islands beyond Small Island Developing States (SIDS). *Clim. Change*, **152**(1), 145–165.

Petzold, J. and B.M. Ratter, 2015: Climate change adaptation under a social capital approach–An analytical framework for small islands. *Ocean Coast. Manage*, **112**, 36–43.

Pezzulo, C. et al., 2017: Sub-national mapping of population pyramids and dependency ratios in Africa and Asia. *Sci. Data*, **4**, 170089.

Pfeffer, J. et al., 2017: Decoding the origins of vertical land motions observed today at coasts. *Geophys. J. Int.*, **210**(1), 148–165.

Pfeffer, J., P. Tregoning, A. Purcell and M. Sambridge, 2018: Multitechnique Assessment of the Interannual to Multidecadal Variability in Steric Sea Levels: A Comparative Analysis of Climate Mode Fingerprints. *J. Clim.*, **31**(18), 7583–7597.

Pinsky, M.L., G. Guannel and K.K. Arkema, 2013: Quantifying wave attenuation to inform coastal habitat conservation. *Ecosphere*, **4**(8), 1–16.

Pinto, P.J., G.M. Kondolf and P.L.R. Wong, 2018: Adapting to sea level rise: Emerging governance issues in the San Francisco Bay Region. *Environ. Sci. Policy*, **90**, 28–37.

Pisaric, M.F. et al., 2011: Impacts of a recent storm surge on an Arctic delta ecosystem examined in the context of the last millennium. *PNAS*, **108**(22), 8960–8965.

Pittman, J. and D. Armitage, 2019: Network Governance of Land-Sea Social-Ecological Systems in the Lesser Antilles. *Ecol. Econ.*, **157**, 61–70.

Pollard, D. and R. DeConto, 2012: Description of a hybrid ice sheet-shelf model, and application to Antarctica. *Geosci. Model Dev.*, **5**(5), 1273.

Pollard, D., R.M. DeConto and R.B. Alley, 2015: Potential Antarctic Ice Sheet retreat driven by hydrofracturing and ice cliff failure. *Earth Planet. Sci. Lett.*, **412**, 112–121.

Pollard, D., N. Gomez and R.M. Deconto, 2017: Variations of the Antarctic Ice Sheet in a Coupled Ice Sheet-Earth-Sea Level Model: Sensitivity to Viscoelastic Earth Properties. *J. Geophys. Res-Earth*, **122**(11), 2124–2138.

Pollard, J., T. Spencer and S. Brooks, 2018: The interactive relationship between coastal erosion and flood risk. *Progress in Physical Geography: Earth and Environment*, **43**(4), 574–585.

Pontee, N., 2013: Defining coastal squeeze: A discussion. *Ocean Coast. Manage*, **84**, 204–207.

Pontee, N., S. Narayan, M.W. Beck and A.H. Hosking, 2016: Nature-based solutions: lessons from around the world. *Maritime Engineering*, **169**(1), 29–36.

Post, V.E. et al., 2018: On the resilience of small-island freshwater lenses: Evidence of the long-term impacts of groundwater abstraction on Bonriki Island, Kiribati. *J. Hydrol.*, **564**, 133–148.

Pot, W. et al., 2018: What makes long-term investment decisions forward looking: A framework applied to the case of Amsterdam's new sea lock. *Technol. Forecast. Soc. Change*, **132**, 174–190.

Prahl, B.F. et al., 2018: Damage and protection cost curves for coastal floods within the 600 largest European cities. *Sci. Data*, **5**, 180034.

Primavera, J.H. and J.M.A. Esteban, 2008: A review of mangrove rehabilitation in the Philippines: successes, failures and future prospects. *Wetl. Ecol. Manag.*, **16**(5), 345–358.

Prisco, I., M. Carboni and A.T.R. Acosta, 2013: The Fate of Threatened Coastal Dune Habitats in Italy under Climate Change Scenarios. *PLoS One*, **8**(7), 1–14, doi:10.1371/journal.pone.0068850.

Pruitt, D., J. Rubin and S.H. Kim, 2003: *Social Conflict: Escalation, Stalemate, and Settlement (Third edition)*. McGraw-Hill Education, 316-pp, McGraw-Hill, Boston, USA.

PSMSL, 2019: Tide Gauge Data. Permanent Service for Mean Sea Level. Available at: www.psmsl.org/

4

Puotinen, M. et al., 2016: A robust operational model for predicting where tropical cyclone waves damage coral reefs. *Sci. Rep.*, **6**, 26009.

Purkey, S.G. and G.C. Johnson, 2010: Warming of global abyssal and deep Southern Ocean waters between the 1990s and 2000s: Contributions to global heat and sea level rise budgets. *J. Clim.*, **23**(23), 6336–6351.

Quan, N.H. et al., 2018a: Conservation of the Mekong Delta wetlands through hydrological management. *Ecol. Res.*, **33**(1), 87–103.

Quan, R. et al., 2018b: Improvement of Salt Tolerance Using Wild Rice Genes. *Front. Plant Sci.*, **8**, 2269.

Quataert, E. et al., 2015: The influence of coral reefs and climate change on wave-driven flooding of tropical coastlines. *Geophys. Res. Lett.*, **42**(15), 6407–6415.

Quiquet, A., C. Ritz, H.J. Punge and D. Salas y Mélia, 2013: Greenland ice sheet contribution to sea level rise during the last interglacial period: a modelling study driven and constrained by ice core data. *Clim. Past*, **9**(1), 353–366, doi:10.5194/cp-9-353-2013.

Raby, A. et al., 2015: Implications of the 2011 Great East Japan Tsunami on sea defence design. *Int. J. Disast. Risk Reduc.*, **14**, 332–346.

Radić, V. et al., 2014: Regional and global projections of twenty-first century glacier mass changes in response to climate scenarios from global climate models. *Clim. Dyn.*, **42**(1–2), 37–58.

Radić, V. and R. Hock, 2010: Regional and global volumes of glaciers derived from statistical upscaling of glacier inventory data. *J. Geophys. Res-Earth*, **115**(F1), 1–10.

Rahman, M.A. and S. Rahman, 2015: Natural and traditional defense mechanisms to reduce climate risks in coastal zones of Bangladesh. *Weather and Climate Extremes*, **7**, 84–95.

Rahman, M.M., Y. Jiang and K. Irvine, 2018: Assessing wetland services for improved development decision-making: a case study of mangroves in coastal Bangladesh. *Wetl. Ecol. Manag.*, **26**(4), 563–580.

Rahman, M.S., 2013: Climate change, disaster and gender vulnerability: A study on two divisions of Bangladesh. *Am. J. Hum. Ecol.*, **2**(2), 72–82.

Rahmstorf, S., 2007: A semi-empirical approach to projecting future sea level rise. *Science*, **315**(5810), 368–370.

Rakib, M., J. Sasaki, H. Matsuda and M. Fukunaga, 2019: Severe salinity contamination in drinking water and associated human health hazards increase migration risk in the southwestern coastal part of Bangladesh. *J. Environ. Manage.*, **240**, 238–248.

Rangel-Buitrago, N.G., G. Anfuso and A.T. Williams, 2015: Coastal erosion along the Caribbean coast of Colombia: magnitudes, causes and management. *Ocean Coast. Manage*, **114**, 129–144.

Ranger, N., T. Reeder and J. Lowe, 2013: Addressing 'deep' uncertainty over long-term climate in major infrastructure projects: four innovations of the Thames Estuary 2100 Project. *EURO Journal on Decision Processes*, **1**(3–4), 233f262.

Ranson, M. et al., 2014: Tropical and extratropical cyclone damages under climate change. *Clim. Change*, **127**(2), 227f241, doi:10.1007/s10584-014-1255-4.

Rao, N.D., B.J. van Ruijven, K. Riahi and V. Bosetti, 2017: Improving poverty and inequality modelling in climate research. *Nat. Clim. Change*, **7**(12), 857.

Rasmussen, D. et al., 2018: Extreme sea level implications of 1.5°C, 2.0°C, and 2.5°C temperature stabilization targets in the 21st and 22nd centuries. *Environ. Res. Lett.*, **13**(3), 034040.

Raucoules, D. et al., 2013: High nonlinear urban ground motion in Manila (Philippines) from 1993 to 2010 observed by DInSAR: implications for sea level measurement. *Remote Sens. Environ.*, **139**, 386–397.

Ray, B.R., M.W. Johnson, K. Cammarata and D.L. Smee, 2014: Changes in Seagrass Species Composition in Northwestern Gulf of Mexico Estuaries: Effects on Associated Seagrass Fauna. *PLoS One*, **9**(9), e107751–e107751, doi:10.1371/journal.pone.0107751.

Ray, R.D. and B.C. Douglas, 2011: Experiments in reconstructing twentieth-century sea levels. *Progr. Oceanogr.*, **91**(4), 496–515.

Raymo, M.E. et al., 2018: The accuracy of mid-Pliocene δ18O-based ice volume and sea level reconstructions. *Earth-Sci. Rev.*, **177**, 291–302, doi:10.1016/j.earscirev.2017.11.022.

Raymo, M.E. et al., 2011: Departures from eustasy in Pliocene sea level records. *Nat. Geosci.*, **4**(5), 328.

Reager, J.T. et al., 2016: A decade of sea level rise slowed by climate-driven hydrology. *Science*, **351**(6274), 699–703, doi:10.1126/science.aad8386.

Reed, M.S., 2008: Stakeholder participation for environmental management: a literature review. *Biol. Conserv.*, **141**(10), 2417–2431.

Reef, R. and C.E. Lovelock, 2014: Historical analysis of mangrove leaf traits throughout the 19th and 20th centuries reveals differential responses to increases in atmospheric CO_2. *Global Ecol. Biogeogr.*, **23**(11), 1209–1214.

Reese, R. et al., 2018a: Antarctic sub-shelf melt rates via PICO. *The Cryosphere*, **12**(6), 1969–1985, doi:10.5194/tc-12-1969-2018.

Reese, R., G.H. Gudmundsson, A. Levermann and R. Winkelmann, 2018b: The far reach of ice shelf thinning in Antarctica. *Nat. Clim. Change*, **8**(1), 53.

Reese, R., R. Winkelmann and G.H. Gudmundsson, 2018c: Grounding-line flux formula applied as a flux condition in numerical simulations fails for buttressed Antarctic ice streams. *The Cryosphere*, **12**(10), 3229–3242, doi:10.5194/tc-12-3229-2018.

Regeneris Consulting, 2011: *Coastal Pathfinder Evaluation: An Assessment of the Five Largest Pathfinder Projects* Department for Environment Food and Rural Affairs, Affairs, London. [Available at: https://assets.publishing.service.gov.uk/government/uploads/system/uploads/attachment_data/file/69509/pb13721-coastal-pathfinder-evaluation.pdf]. Accessed: 2019/09/20.

Reguero, B. et al., 2018a: Coral reefs for coastal protection: A new methodological approach and engineering case study in Grenada. *J. Environ. Manage.*, **210**, 146–161.

Reguero, B.G. et al., 2018b: Comparing the cost effectiveness of nature-based and coastal adaptation: A case study from the Gulf Coast of the United States. *PLoS One*, **13**(4), e0192132.

Reguero, B.G. et al., 2015: Effects of climate change on exposure to coastal flooding in Latin America and the Caribbean. *PLoS One*, **10**(7), e0133409.

Reiblich, J. et al., 2019: Bridging climate science, law, and policy to advance coastal adaptation planning. *Mar. Policy*, **104**, 125–134.

Reiblich, J., L. Wedding and E. Hartge, 2017: Enabling and Limiting Conditions of Coastal Adaptation: Local Governments, Land Uses, and Legal Challenges. *Ocean Coast. Law J.*, **22**(2), 156–194.

Reid, H., 2016: Ecosystem-and community-based adaptation: learning from community-based natural resource management. *Clim. Dev.*, **8**(1), 4–9.

Renaud, F.G. et al., 2015: Resilience and shifts in agro-ecosystems facing increasing sea level rise and salinity intrusion in Ben Tre Province, Mekong Delta. *Clim. Change*, **133**(1), 69–84.

Renaud, F.G., K. Sudmeier-Rieux, M. Estrella and U. Nehren, 2016: *Ecosystem-based disaster risk reduction and adaptation in practice*. Springer, Cham, Switzerland, 593-pp., IBSN: 978–3-319–43631–9.

Renaud, F.G. et al., 2013: Tipping from the Holocene to the Anthropocene: How threatened are major world deltas? *Curr. Opin. Environ. Sustain.*, **5**(6), 644–654.

Renn, O., 2008: *Risk governance: coping with uncertainty in a complex world*. Earthscan, Sterling, VA, USA, 445 pp., ISBN: 978-1-84407-291-0.

Renner, K. et al., 2017: Spatio-temporal population modelling as improved exposure information for risk assessments tested in the Autonomous Province of Bolzano. *Int. J. Disast. Risk Reduc.*, **27**, 470–479.

Repolho, T. et al., 2017: Seagrass ecophysiological performance under ocean warming and acidification. *Sci. Rep.*, **7**, 41443.

Reuter, H.I., A. Nelson and A. Jarvis, 2007: An evaluation of void-filling interpolation methods for SRTM data. *Int. J. Geogr. Info. Sci.*, **21**(9), 983–1008.

RIBA and ICE, 2010: *Facing-up to rising sea levels. Retreat? Defend? Attack?*, Royal Institute of British Architects; Institution of Civil Engineers, London [Available at: https://tamug-ir.tdl.org/bitstream/handle/1969.3/29265/Facing_Up_To_Rising_Sea_Levels.pdf?sequence=1&isAllowed=y].

Richards, D.R. and D.A. Friess, 2017: Characterizing Coastal Ecosystem Service Trade-offs with Future Urban Development in a Tropical City. *Environ. Manage.*, **60**(5), 961–973.

Rietbroek, R. et al., 2016: Revisiting the contemporary sea level budget on global and regional scales. *PNAS*, **113**(6), 1504–1509.

Rigall-I-Torrent, R. et al., 2011: The effects of beach characteristics and location with respect to hotel prices. *Tourism Manage.*, **32**(5), 1150–1158.

Rignot, E. et al., 2014: Widespread, rapid grounding line retreat of Pine Island, Thwaites, Smith, and Kohler glaciers, West Antarctica, from 1992 to 2011. *Geophys. Res. Lett.*, **41**(10), 3502–3509.

Rignot, E. et al., 2019: Four decades of Antarctic Ice Sheet mass balance from 1979–2017. *Proceeding of the National Academy of Sciences*, **116**(4), 1095–1103, doi:10.1073/pnas.1812883116.

Riser, S.C. et al., 2016: Fifteen years of ocean observations with the global Argo array. *Nat. Clim. Change*, **6**(2), 145.

Ritz, C. et al., 2015: Potential sea level rise from Antarctic ice sheet instability constrained by observations. *Nature*, **528**(7580), 115–118.

Riva, R.E. et al., 2017: Brief communication: The global signature of post-1900 land ice wastage on vertical land motion. *The Cryosphere*, **11**(3), 1327–1332.

Robinson, A., R. Calov and A. Ganopolski, 2012: Multistability and critical thresholds of the Greenland ice sheet. *Nat. Clim. Change*, **2**, 429–432, doi:10.1038/nclimate1449 www.nature.com/articles/nclimate1449#supplementary-information.

Robinson, S.-A., 2017: Climate change adaptation trends in small island developing states. *Mitig. Adapt. Strat. Gl.*, **22**(4), 669–691.

Roeber, V. and J.D. Bricker, 2015: Destructive tsunami-like wave generated by surf beat over a coral reef during Typhoon Haiyan. *Nat. Commun.*, **6**, 7854.

Roelvink, D., 2015: Addressing local and global sediment imbalances: coastal sediments as rare minerals. In: *The Proceedings of the Coastal Sediments 2015* [Wang, P., J.D. Rosati and J. Cheng (eds.)]. World Scientific, London, UK, ISBN: 978-981-4689-97-9.

Roemmich, D. et al., 2015: Unabated planetary warming and its ocean structure since 2006. *Nat. Clim. Change*, **5**(3), 240–245.

Rogers, K.G. and I. Overeem, 2017: Doomed to drown? Sediment dynamics in the human-controlled floodplains of the active Bengal Delta. *Elementa-Sci. Anthrop.*, **5**(66), 1–15, doi:10.1525/elementa.250.

Roggero, M., A. Bisaro and S. Villamayor-Tomas, 2018a: Institutions in the climate adaptation literature: a systematic literature review through the lens of the Institutional Analysis and Development framework. *Journal of Institutional Economics*, **14**(3), 423–448.

Roggero, M. et al., 2018b: Introduction to the special issue on adapting institutions to climate change. *Journal of institutional economics*, **14**(3), 409–422.

Rolph, R.J., A.R. Mahoney, J. Walsh and P.A. Loring, 2018: Impacts of a lengthening open water season on Alaskan coastal communities: deriving locally relevant indices from large-scale datasets and community observations. *Cryosphere*, **12**(5), 1779–1790.

Romine, B.M. et al., 2013: Are beach erosion rates and sea level rise related in Hawaii? *Global Planet. Change*, **108**, 149–157.

Rosendo, S., L. Celliers and M. Mechisso, 2018: Doing more with the same: A reality-check on the ability of local government to implement Integrated Coastal Management for climate change adaptation. *Mar. Policy*, **87**, 29–39.

Rosenzweig, C. and W. Solecki, 2014: Hurricane Sandy and adaptation pathways in New York: lessons from a first-responder city. *Global Environ. Chang.*, **28**, 395–408.

Ross, A.C. et al., 2015: Sea level rise and other influences on decadal-scale salinity variability in a coastal plain estuary. *Estuar. Coast. Shelf Sci.*, **157**, 79–92.

Rosselló-Nadal, J., 2014: How to evaluate the effects of climate change on tourism. *Tourism Manage.*, **42**, 334–340.

Rotzoll, K. and C.H. Fletcher, 2013: Assessment of groundwater inundation as a consequence of sea level rise. *Nat. Clim. Change*, **3**(5), 477.

Rouse, H. et al., 2017: Coastal adaptation to climate change in Aotearoa-New Zealand. *New Zeal. J. Mar. Fresh.*, **51**(2), 183–222.

Rovere, A. et al., 2014: The Mid-Pliocene sea level conundrum: Glacial isostasy, eustasy and dynamic topography. *Earth Planet. Sci. Lett.*, **387**, 27–33, doi:10.1016/j.epsl.2013.10.030.

Rowley, D.B. et al., 2013: Dynamic topography change of the eastern United States since 3 million years ago. *Science*, **340**(6140), 1560–1563.

Ruggiero, P., 2012: Is the intensifying wave climate of the US Pacific Northwest increasing flooding and erosion risk faster than sea level rise? *J. Waterway Port Coast.*, **139**(2), 88–97.

Ruiz-Fernández, A. et al., 2018: Carbon burial and storage in tropical salt marshes under the influence of sea level rise. *Sci. Total Environ.*, **630**, 1628–1640.

Rumanti, I.A. et al., 2018: Development of tolerant rice varieties for stress-prone ecosystems in the coastal deltas of Indonesia. *Field Crops Research*, **223**, 75–82.

Rumore, D., 2014: Building the Capacity of Coastal Communities to Adapt to Climate Change through Participatory Action Research. *Carolina Planning*, **39**, 16–12.

Rumore, D., T. Schenk and L. Susskind, 2016: Role-play simulations for climate change adaptation education and engagement. *Nat. Clim. Change*, **6**(8), 745.

Runhaar, H. et al., 2018: Mainstreaming climate adaptation: taking stock about "what works" from empirical research worldwide. *Reg. Environ. Change*, **18**(4), 1201–1210.

Rupprecht, F. et al., 2017: Vegetation-wave interactions in salt marshes under storm surge conditions. *Ecol. Eng.*, **100**, 301–315.

Ryan, J.C. et al., 2018: Dark zone of the Greenland Ice Sheet controlled by distributed biologically-active impurities. *Nat. Commun.*, **9**(1), 1065, doi:10.1038/s41467-018-03353-2.

Ryan, J.C. et al., 2019: Greenland Ice Sheet surface melt amplified by snowline migration and bare ice exposure. *Sci. Adv.*, **5**(3), eaav3738, doi:10.1126/sciadv.aav3738.

Saenko, O.A. et al., 2015: Separating the influence of projected changes in air temperature and wind on patterns of sea level change and ocean heat content. *J. Geophys. Res-Oceans*, **120**(8), 5749–5765.

Sagoe-Addy, K. and K.A. Addo, 2013: Effect of predicted sea level rise on tourism facilities along Ghana's Accra coast. *J. Coast. Conserv.*, **17**(1), 155–166.

Samper-Villarreal, J. et al., 2016: Organic carbon in seagrass sediments is influenced by seagrass canopy complexity, turbidity, wave height, and water depth. *Limnol. Oceanogr.*, **61**(3), 938–952, doi:10.1002/lno.10262.

Sánchez-García, E.A., K. Rodríguez-Medina and P. Moreno-Casasola, 2017: Effects of soil saturation and salinity on seed germination in seven freshwater marsh species from the tropical coast of the Gulf of Mexico. *Aquat. Bot.*, **140**, 4–12.

Sánchez-Rodríguez, A.R. et al., 2017: Comparative effects of prolonged freshwater and saline flooding on nitrogen cycling in an agricultural soil. *Appl. Soil Ecol*, **125**, 56–70.

Santamaría-Gómez, A. et al., 2017: Uncertainty of the 20th century sea level rise due to vertical land motion errors. *Earth Planet. Sci. Lett.*, **473**, 24–32.

Sarzynski, A., 2015: Public participation, civic capacity, and climate change adaptation in cities. *Urban Climate*, **14**, 52–67, doi:10.1016/j.uclim.2015.08.002.

Sasmito, S.D., D. Murdiyarso, D.A. Friess and S. Kurnianto, 2016: Can mangroves keep pace with contemporary sea level rise? A global data review. *Wetl. Ecol. Manag.*, **24**(2), 263–278.

Sättele, M., M. Bründl and D. Straub, 2015: Reliability and effectiveness of early warning systems for Nat. Hazards: Concept and application to debris flow warning. *Reliability Engineering & System Safety*, **142**, 192–202.

Saunders, M.I. et al., 2013: Coastal retreat and improved water quality mitigate losses of seagrass from sea level rise. *Global Change Biol.*, **19**(8), 2569–2583, doi:10.1111/gcb.12218.

Savage, L.J., 1951: The Theory of Statistical Decision. *J. Am. Stat. Assoc.*, **46**, 55–67, doi:10.1080/01621459.1951.10500768.

Scambos, T.A., J. Bohlander, C.U. Shuman and P. Skvarca, 2004: Glacier acceleration and thinning after ice shelf collapse in the Larsen B embayment, Antarctica. *Geophys. Res. Lett.*, **31**(18).

Scambos, T.A. et al., 2009: Ice shelf disintegration by plate bending and hydro-fracture: Satellite observations and model results of the 2008 Wilkins ice shelf break-ups. *Earth Planet. Sci. Lett.*, **280**(1–4), 51–60.

Scanlon, B.R. et al., 2018: Global models underestimate large decadal declining and rising water storage trends relative to GRACE satellite data. *PNAS*, **115**(6), 1080–1089.

Schade, J. et al., 2015: Climate change and climate policy induced relocation: a challenge for social justice. University Bielefeld, Centre on Migration, Citizenship and Development. Available at: www.ssoar.info/ssoar/handle/document/50741.

Schaeffer, M., W. Hare, S. Rahmstorf and M. Vermeer, 2012: Long-term sea level rise implied by 1.5 C and 2°C warming levels. *Nat. Clim. Change*, **2**(12), 867–870.

Scheffran, J., E. Marmer and P. Sow, 2012: Migration as a contribution to resilience and innovation in climate adaptation: Social networks and co-development in Northwest Africa. *Appl. Geogr.*, **33**, 119–127.

Schernewski, G., J. Schumacher, E. Weisner and L. Donges, 2018: A combined coastal protection, realignment and wetland restoration scheme in the southern Baltic: planning process, public information and participation. *J. Coast. Conserv.*, **22**(3), 533–547.

Scheuchl, B. et al., 2016: Grounding line retreat of Pope, Smith, and Kohler Glaciers, West Antarctica, measured with Sentinel-1a radar interferometry data. *Geophys. Res. Lett.*, **43**(16), 8572–8579.

Schile, L.M. et al., 2014: Modeling tidal marsh distribution with sea level rise: Evaluating the role of vegetation, sediment, and upland habitat in marsh resiliency. *PLoS One*, **9**(2), e88760.

Schlegel, N.J. et al., 2018: Exploration of Antarctic Ice Sheet 100-year contribution to sea level rise and associated model uncertainties using the ISSM framework. *The Cryosphere*, **12**(11), 3511–3534, doi:10.5194/tc-12-3511-2018.

Schloesser, F., R. Furue, J. McCreary and A. Timmermann, 2014: Dynamics of the Atlantic meridional overturning circulation. Part 2: Forcing by winds and buoyancy. *Progr. Oceanogr.*, **120**, 154–176.

Schlosberg, D., 2012: Climate justice and capabilities: a framework for adaptation policy. *Ethics & International Affairs*, **26**(4), 445–461.

Schlosberg, D., L.B. Collins and S. Niemeyer, 2017: Adaptation policy and community discourse: risk, vulnerability, and just transformation. *Environ. Politics*, **26**(3), 413–437.

Schmidt, C.W., 2015: Delta subsidence: an imminent threat to coastal populations. *Environ Health Perspect*, **123**(8), A204-A209.

Schmitt, R., Z. Rubin and G. Kondolf, 2017: Losing ground-scenarios of land loss as consequence of shifting sediment budgets in the Mekong Delta. *Geomorphology*, **294**, 58–69.

Schodlok, M.P., D. Menemenlis and E.J. Rignot, 2016: Ice shelf basal melt rates around Antarctica from simulations and observations. *J. Geophys. Res-Oceans*, **121**(2), 1085–1109, doi:10.1002/2015JC011117.

Schoof, C., 2007a: Ice sheet grounding line dynamics: Steady states, stability, and hysteresis. *J. Geophys. Res-Earth*, **112**(F3), 1–19, doi:10.1029/2006JF000664.

Schoof, C., 2007b: Marine ice sheet dynamics. Part 1. the case of rapid sliding. *J. Fluid Mech.*, **573**, 27–55, doi:10.1017/S0022112006003570.

Schoutens, K. et al., 2019: How effective are tidal marshes as nature-based shoreline protection throughout seasons? *Limnol. Oceanogr*, **64**, 1750–1762.

Schuerch, M. et al., 2018: Future response of global coastal wetlands to sea level rise. *Nature*, **561**(7722), 231.

Scoccimarro, E. et al., 2017: Tropical Cyclone Rainfall Changes in a Warmer Climate. In: *Hurricanes and Climate Change* [Collins, J.M. and K. Walsh (eds.)]. Springer International Publishing, Cham, **3**, 243–255.

Scussolini, P. et al., 2015: FLOPROS: an evolving global database of flood protection standards. *Nat. Hazards Earth Syst. Sci. Discuss*, **3**, 7275–309.

Scussolini, P. et al., 2017: Adaptation to Sea Level Rise: A Multidisciplinary Analysis for Ho Chi Minh City, Vietnam. *Water Resourc. Res.*, **53**, 10841–10857, doi:10.1002/2017WR021344.

Scyphers, S.B., J.S. Picou and S.P. Powers, 2015: Participatory conservation of coastal habitats: the importance of understanding homeowner decision making to mitigate cascading shoreline degradation. *Conserv. Lett.*, **8**(1), 41–49.

Scyphers, S.B., S.P. Powers, K.L. Heck Jr and D. Byron, 2011: Oyster reefs as natural breakwaters mitigate shoreline loss and facilitate fisheries. *PLoS One*, **6**(8), e22396.

Seavitt, C., 2013: Yangtze river delta project. *Scenario 03: Rethinking Infrastructure*. Available at: https://scenariojournal.com/article/yangtze-river-delta-project/.

Sebesvari, Z. et al., 2016: A review of vulnerability indicators for deltaic social–ecological systems. *Sustain. Sci.*, **11**(4), 575–590.

Sengupta, D., R. Chen and M.E. Meadows, 2018: Building beyond land: An overview of coastal land reclamation in 16 global megacities. *Appl. Geogr.*, **90**, 229–238.

Sérazin, G. et al., 2016: Quantifying uncertainties on regional sea level change induced by multidecadal intrinsic oceanic variability. *Geophys. Res. Lett.*, **43**(15), 8151–8159.

Seroussi, H. and M. Morlighem, 2018: Representation of basal melting at the grounding line in ice flow models. *The Cryosphere*, **12**(10), 3085–3096, doi:10.5194/tc-12-3085-2018.

Seroussi, H. et al., 2017: Continued retreat of Thwaites Glacier, West Antarctica, controlled by bed topography and ocean circulation. *Geophys. Res. Lett.*, **44**(12), 6191–6199.

Serrano, O., P.S. Lavery, M. Rozaimi and M.Á. Mateo, 2014: Influence of water depth on the carbon sequestration capacity of seagrasses. *Glob. Biogeochem. Cy.*, **28**(9), 950–961.

Serrao-Neumann, S. et al., 2015: Maximising synergies between disaster risk reduction and climate change adaptation: Potential enablers for improved planning outcomes. *Environ. Sci. Policy*, **50**, 46–61.

Serre, D. and C. Heinzlef, 2018: Assessing and mapping urban resilience to floods with respect to cascading effects through critical infrastructure networks. *Int. J. Disast. Risk Reduc.*, **30**, 235–243.

Setzer, J. and L. Vanhala, 2019: Climate change litigation: a review of research on courts and litigants in climate government. *WiRes. Clim. Change*. **10**(3), 1–19.

Shakeela, A. and S. Becken, 2015: Understanding tourism leaders' perceptions of risks from climate change: An assessment of policy-making processes in the Maldives using the social amplification of risk framework (SARF). *J. Sustain. Tourism*, **23**(1), 65–84.

Shakun, J.D. et al., 2018: Minimal East Antarctic Ice Sheet retreat onto land during the past eight million years. *Nature*, **558**(7709), 284–287, doi:10.1038/s41586-018-0155-6.

Shaltout, M., K. Tonbol and A. Omstedt, 2015: Sea level change and projected future flooding along the Egyptian Mediterranean coast. *Oceanologia*, **57**(4), 293–307.

Shannon, S.R. et al., 2013: Enhanced basal lubrication and the contribution of the Greenland ice sheet to future sea level rise. *PNAS*, **110**(35), 14156–14161.

Shatkin, G., 2019: Futures of Crisis, Futures of Urban Political Theory: Flooding in Asian Coastal Megacities. *Int. J. Urban Reg. Res.*, **43**(2), 207–226.

Shay, L.K. et al., 1992: Upper ocean response to Hurricane Gilbert. *J. Geophys. Res-Oceans*, **97**(C12), 20227–20248, doi:10.1029/92JC01586.

Shayegh, S., J. Moreno-Cruz and K. Caldeira, 2016: Adapting to rates versus amounts of climate change: a case of adaptation to sea level rise. *Environ. Res. Lett.*, **11**(10), 104007.

Shen, Q. et al., 2018: Recent high-resolution Antarctic ice velocity maps reveal increased mass loss in Wilkes Land, East Antarctica. *Sci. Rep.*, **8**(1), 4477, doi:10.1038/s41598-018-22765-0.

4

Shepard, C.C., C.M. Crain and M.W. Beck, 2011: The protective role of coastal marshes: a systematic review and meta-analysis. *PLoS One*, **6**(11), e27374.

Shepherd, A. et al., 2012: A Reconciled Estimate of Ice sheet Mass Balance. *Science*, **338**(6111), 1183–1189, doi:10.1126/science.1228102.

Shepherd, M. and K.C. Binita, 2015: Climate change and African Americans in the USA. *Geogr. Compass*, **9**(11), 579–591, doi:10.1111/gec3.12244.

Sherwood, E.T. and H.S. Greening, 2014: Potential Impacts and Management Implications of Climate Change on Tampa Bay Estuary Critical Coastal Habitats. *Environ. Manage.*, **53**(2), 401–415, doi:10.1007/s00267-013-0179-5.

Shi, L., E. Chu and J. Debats, 2015: Explaining progress in climate adaptation planning across 156 US municipalities. *J. Am. Plan. Assoc.*, **81**(3), 191–202.

Shope, J.B., C.D. Storlazzi and R.K. Hoeke, 2017: Projected atoll shoreline and run-up changes in response to sea level rise and varying large wave conditions at Wake and Midway Atolls, Northwestern Hawaiian Islands. *Geomorphology*, **295**, 537–550.

Short, F.T. et al., 2014: Monitoring in the Western Pacific region shows evidence of seagrass decline in line with global trends. *Mar. Pollut. Bull.*, **83**(2), 408–416, doi:10.1016/J.MARPOLBUL.2014.03.036.

Siders, A., 2019: Social justice implications of US managed retreat buyout programs. *Clim. Change*, **152**(2), 239–257.

Sieber, I.M., R. Biesbroek and D. de Block, 2018: Mechanism-based explanations of impasses in the governance of ecosystem-based adaptation. *Reg. Environ. Change*, **18**(8), 2379–2390.

Siegle, E. and M.B. Costa, 2017: Nearshore Wave Power Increase on Reef-Shaped Coasts Due to Sea-Level Rise. *Earth's Future*, **5**(10), 1054–1065.

Siikamäki, J., J.N. Sanchirico and S.L. Jardine, 2012: Global economic potential for reducing carbon dioxide emissions from mangrove loss. *PNAS*, **109**(36), 14369–14374.

Simpson, M. et al., 2016: Decision analysis for management of Nat. Hazards. *Annu. Rev. Environ. Resourc.*, **41**, 489–516.

Slangen, A. et al., 2014a: Projecting twenty-first century regional sea level changes. *Clim. Change*, **124**(1–2), 317–332, doi:10.1007/s10584-014-1080-9.

Slangen, A. et al., 2016: Anthropogenic forcing dominates global mean sea level rise since 1970. *Nat. Clim. Change*, **6**(7), 701–705.

Slangen, A., J.A. Church, X. Zhang and D. Monselesan, 2014b: Detection and attribution of global mean thermosteric sea level change. *Geophys. Res. Lett.*, **41**(16), 5951–5959.

Slangen, A. and R. Van de Wal, 2011: An assessment of uncertainties in using volume-area modelling for computing the twenty-first century glacier contribution to sea level change. *The Cryosphere*, **5**(3), 673.

Slangen, A. et al., 2017a: The impact of uncertainties in ice sheet dynamics on sea level allowances at tide gauge locations. *J. Mar. Sci Eng.*, **5**(2), 1–20, doi:10.3390/jmse5020021.

Slangen, A.B. et al., 2017b: Evaluating model simulations of twentieth-century sea level rise. Part I: Global mean sea level change. *J. Clim.*, **30**(21), 8539–8563.

Sleeter, B.M., N.J. Wood, C.E. Soulard and T.S. Wilson, 2017: Projecting community changes in hazard exposure to support long-term risk reduction: a case study of tsunami hazards in the US Pacific Northwest. *Int. J. Disast. Risk Reduc.*, **22**, 10–22.

Smajgl, A. et al., 2015: Responding to rising sea levels in the Mekong Delta. *Nat. Clim. Change*, **5**(2), 167–174.

Smith, A., D. Martin and S. Cockings, 2016: Spatio-temporal population modelling for enhanced assessment of urban exposure to flood risk. *Appl. Spat. Anal. Policy*, **9**(2), 145–163.

Smith, G.P. and B.C. Glavovic, 2014: Conclusions: integrating Nat. Hazards risk management and climate change
adaptation through Nat. Hazards planning. In: *Adapting to Climate Change*. Springer, pp. 405–450, NY, New York, USA, ISBN 978-94-017-8630-0.

Smith III, T.J. et al., 2009: Cumulative impacts of hurricanes on Florida mangrove ecosystems: sediment deposition, storm surges and vegetation. *Wetlands*, **29**(1), 24–34.

Smith, J.B. et al., 2001: Vulnerability to climate change and reasons for concern: a synthesis. *Clim. Change*, 913–967.

Smith, K., 2011: We are seven billion. *Nat. Clim. Change*, **1**, 331–335.

Smith, K.M., 2006: Integrating habitat and shoreline dynamics into living shoreline applications. In: *Management, Policy, Science, and Engineering of Nonstructural Erosion Control in the Chesapeake Bay*, 9–11, Proceedings of the 2006 Living Shoreline Summit, Edgewater, MD, USA.

Smith, L.C. et al., 2017: Direct measurements of meltwater runoff on the Greenland ice sheet surface. *PNAS*, **114**(50), E10622–E10631, doi:10.1073/pnas.1707743114.

Smith, R.-A., 2018: Risk perception and adaptive responses to climate change and climatic variability in northeastern St. Vincent. *J. Environ. Stud. Sci.*, **8**(1), 73–85.

Smoak, J.M., J.L. Breithaupt, T.J. Smith and C.J. Sanders, 2013: Sediment accretion and organic carbon burial relative to sea level rise and storm events in two mangrove forests in Everglades National Park. *CATENA*, **104**, 58–66, doi:10.1016/J.CATENA.2012.10.009.

Solomon, S., G.-K. Plattner, R. Knutti and P. Friedlingstein, 2009: Irreversible climate change due to carbon dioxide emissions. *PNAS*, **106**(6), 1704–1709.

Song, J. et al., 2017: An examination of land use impacts of flooding induced by sea level rise. *Nat. Hazards Earth Syst. Sci.*, **17**(3), 315.

Sovacool, B.K., 2018: Bamboo beating bandits: Conflict, inequality, and vulnerability in the political ecology of climate change adaptation in Bangladesh. *World Dev.*, **102**, 183–194.

Sovacool, B.K., B.-O. Linnér and M.E. Goodsite, 2015: The political economy of climate adaptation. *Nat. Clim. Change*, **5**(7), 616.

Sowman, M. and M. Gawith, 1994: Participation of disadvantaged communities in project planning and decision-making: A case-study of Hout Bay. *Development Southern Africa*, **11**(4), 557–571.

Sowman, M., D. Scott and C. Sutherland, 2016: *Governance and Social Justice Position Paper: Milnerton Beach*. ERMD, City of Cape Town.

Spalding, M. et al., 2017: Mapping the global value and distribution of coral reef tourism. *Mar. Policy*, **82**, 104–113.

Spalding, M., M. Kainuma and L. Collins, 2010: *World Atlas of Mangroves*. Earthscan, London, UK, 336-pp., ISBN: 978-1-84407–657–4.

Spalding, M. et al., 2014: Coastal ecosystems: a critical element of risk reduction. *Conserv. Lett.*, **7**(3), 293–301.

Speelman, L.H., R.J. Nicholls and J. Dyke, 2017: Contemporary migration intentions in the Maldives: the role of environmental and other factors. *Sustain. Sci.*, **12**(3), 433–451.

Spence, A., W. Poortinga and N. Pidgeon, 2012: The psychological distance of climate change. *Risk Anal.*, **32**(6), 957–972.

Spencer, T. et al., 2016: Global coastal wetland change under sea level rise and related stresses: The DIVA Wetland Change Model. *Global Planet. Change*, **139**, 15–30.

Speybroeck, J. et al., 2006: Beach nourishment: an ecologically sound coastal defence alternative? A review. *Aquat. Conserv.: Mar. Freshw. Ecosyst.*, **16**(4), 419–435.

St. John III, B. and J.-E. Yusuf, 2019: Perspectives of the Expert and Experienced on Challenges to Regional Adaptation for Sea Level Rise: Implications for Multisectoral Readiness and Boundary Spanning. *Coastal Manage.*, **47**(2) 1–18.

Stalenberg, B., 2013: Innovative flood defences in highly urbanised water cities. In: *Climate Adaptation and Flood Risk in Coastal Cities*, [Aerts et al. (eds.)], 145–164. Earthscan, Oxon, UK, ISBN: 978-1-84971-346-7.

Stammer, D., A. Cazenave, R.M. Ponte and M.E. Tamisiea, 2013: Causes for contemporary regional sea level changes. *Annu. Rev. Mar. Sci.*, **5**, 21–46.

Stanley, J.-D. and P. L. Clemente, 2017: Increased land subsidence and sea level rise are submerging Egypt's Nile Delta coastal margin. *GSA Today*, **27**(5), 4–11.

Stap, L.B. et al., 2016: CO2 over the past 5 million years: Continuous simulation and new δ11B-based proxy data. *Earth Planet. Sci. Lett.*, **439**, 1–10, doi:10.1016/j.epsl.2016.01.022.

Stap, L.B. et al., 2018: Modeled Influence of Land Ice and CO2 on Polar Amplification and Palaeoclimate Sensitivity During the Past 5 Million Years. *Paleoceanogr. Paleocl.*, **33**(4), 381–394, doi:10.1002/2017pa003313.

4

Stark, J., T. Van Oyen, P. Meire and S. Temmerman, 2015: Observations of tidal and storm surge attenuation in a large tidal marsh. *Limnol. Oceanogr.*, **60**(4), 1371–1381.

State of Louisiana, 2015: *National Disaster Resilience Competition, Phase II Application*. State of Louisiana, Louisiana.

Steele, J.E. et al., 2017: Mapping poverty using mobile phone and satellite data. *J. R. Soc. Interface*, **14**(127), 20160690.

Steig, E.J. et al., 2015: Influence of West Antarctic Ice Sheet collapse on Antarctic surface climate. *Geophys. Res. Lett.*, **42**(12), 4862–4868, doi:10.1002/2015GL063861.

Steiger, E., R. Westerholt, B. Resch and A. Zipf, 2015: Twitter as an indicator for whereabouts of people? Correlating Twitter with UK census data. *Comput. Environ. Urban*, **54**, 255–265.

Stephens, S.A., R.G. Bell and J. Lawrence, 2018: Developing signals to trigger adaptation to sea level rise. *Environ. Res. Lett.*, **13**(10), 104004.

Stevens, F.R., A.E. Gaughan, C. Linard and A.J. Tatem, 2015: Disaggregating census data for population mapping using random forests with remotely-sensed and ancillary data. *PLoS One*, **10**(2), e0107042.

Stevens, L.A. et al., 2016: Greenland Ice Sheet flow response to runoff variability. *Geophys. Res. Lett.*, **43**(21), 11295–11303, doi:10.1002/2016GL070414.

Stewart, R., T. Noyce and H. Possingham, 2003: Opportunity cost of ad hoc marine reserve design decisions: an example from South Australia. *Mar. Ecol. Prog. Ser.*, **253**, 25–38.

Stiles Jr, W.A., 2006: Living Shorelines: A Strategic Approach to Making it Work on the Ground in Virginia in Management, Policy, Science, and Engineering of Nonstructural Erosion Control in the Chesapeake Bay. In: *Living Shoreline Summit*, [Erdle, S.Y., J.L.D. Davis and K.G. Sellner (eds.)], CRC Press, Edgewater, MD, USA, pp. 99–105.

Stirling, A., 2010: Keep it complex. *Nature*, **468**, 1029–1031, doi:10.1038/4681029a.

Stive, M.J. et al., 2013: A new alternative to saving our beaches from sea level rise: The sand engine. *J. Coast. Res.*, **29**(5), 1001–1008.

Stojanov, R. et al., 2017: Local perceptions of climate change impacts and migration patterns in Malé, Maldives. *Geogr. J.*, **183**(4), 370–385.

Storey, D. and S. Hunter, 2010: Kiribati: an environmental 'perfect storm'. *Austral. Geogr.*, **41**(2), 167–181.

Storlazzi, C., E. Elias, M. Field and M. Presto, 2011: Numerical modeling of the impact of sea level rise on fringing coral reef hydrodynamics and sediment transport. *Coral Reefs*, **30**(1), 83–96.

Storlazzi, C. et al., 2018: Most atolls will be uninhabitable by the mid-21st century due to sea level rise exacerbating wave-driven flooding. *Sci. Adv.*, **4**(4), 1–9.

Storlazzi, C. et al., 2017: Rigorously valuing the role of coral reefs in coastal protection: An example from Maui, Hawaii, USA. In: *Coastal Dynamics 2017*, Helsingør, Denmark, [Aagaard, T., R. Deigaard and D. Fuhrman (eds.)], pp. 665–674.

Storto, A. et al., 2017: Steric sea level variability (1993–2010) in an ensemble of ocean reanalyses and objective analyses. *Clim. Dyn.*, **49**(3), 709–729.

Strauss, B.H., R.E. Kopp, W.V. Sweet and K. Bittermann, 2016: *Unnatural coastal floods: Sea level rise and the human fingerprint on US floods since 1950*. Climate Central, Princeton, NJ, USA, 1–16, Available at: http://sealevel. climatecentral.org/uploads/research/Unnatural-Coastal-Floods-2016.pdf

Suckall, N. et al., 2018: A framework for identifying and selecting long term adaptation policy directions for deltas. *Sci. Total Environ.*, **633**, 946–957.

Sultana, P. and P.M. Thompson, 2017: Adaptation or conflict? Responses to climate change in water management in Bangladesh. *Environ. Sci. Policy*, **78**, 149–156.

Suppasri, A. et al., 2015: A decade after the 2004 Indian Ocean Tsunami: The progress in disaster preparedness and future challenges in Indonesia, Sri Lanka, Thailand and the Maldives. *Pure Appl. Geophys.*, **172**(12), 3313–3341.

Suroso, D. S. A. and T. Firman, 2018: The role of spatial planning in reducing exposure towards impacts of global sea level rise case study: Northern coast of Java, Indonesia. *Ocean Coast. Manage.*, **153**, 84–97.

Susskind, L., D. Rumore, C. Hulet and P. Field, 2015: *Managing climate risks in coastal communities: Strategies for engagement, readiness and adaptation*. Anthem Environment and Sustainability, Anthem Press.

Susskind, L.E., S. McKearnen and J. Thomas-Lamar, (eds.), 1999: *The consensus building handbook: A comprehensive guide to reaching agreement*. SAGE Publications.

Sutter, J. et al., 2016: Ocean temperature thresholds for Last Interglacial West Antarctic Ice Sheet collapse. *Geophys. Res. Lett.*, **43**(6), 2675–2682, doi:10.1002/2016GL067818.

Sutton-Grier, A.E. et al., 2018: Investing in Natural and Nature-Based Infrastructure: Building Better Along Our Coasts. *Sustainability*, **10**(2), 523.

Sutton-Grier, A.E., K. Wowk and H. Bamford, 2015: Future of our coasts: the potential for natural and hybrid infrastructure to enhance the resilience of our coastal communities, economies and ecosystems. *Environ. Sci. Policy*, **51**, 137–148.

Suzuki, T. and M. Ishii, 2011: Regional distribution of sea level changes resulting from enhanced greenhouse warming in the Model for Interdisciplinary Research on Climate version 3.2. *Geophys. Res. Lett.*, **38**(2).

Sweet, W.V. and J. Park, 2014: From the extreme to the mean: Acceleration and tipping points of coastal inundation from sea level rise. *Earth's Future*, **2**(12), 579–600.

Syvitski, J.P. et al., 2009: Sinking deltas due to human activities. *Nat. Geosci.*, **2**(10), 681–686.

Syvitski, J.P. and Y. Saito, 2007: Morphodynamics of deltas under the influence of humans. *Global Planet. Change*, **57**(3–4), 261–282.

Szabo, S. et al., 2016: Soil salinity, household wealth and food insecurity in tropical deltas: evidence from south-west coast of Bangladesh. *Sustain. Sci.*, **11**(3), 411–421.

Tamisiea, M.E., 2011: Ongoing glacial isostatic contributions to observations of sea level change. *Geophys. J. Int.*, **186**(3), 1036–1044.

Tamura, M., N. Kumano, M. Yotsukuri and H. Yokoki, 2019: Global assessment of the effectiveness of adaptation in coastal areas based on RCP/SSP scenarios. *Clim. Change*, **152**(3–4), 1–15.

Tarbotton, C., F. Dall'Osso, D. Dominey-Howes and J. Goff, 2015: The use of empirical vulnerability functions to assess the response of buildings to tsunami impact: comparative review and summary of best practice. *Earth-Sci. Rev.*, **142**, 120–134.

Taylor, K.E., R.J. Stouffer and G.A. Meehl, 2012: An overview of CMIP5 and the experiment design. *Bull. Am. Meteorol. Soc.*, **93**(4), 485–498.

Taylor, M.D., T.F. Gaston and V. Raoult, 2018: The economic value of fisheries harvest supported by saltmarsh and mangrove productivity in two Australian estuaries. *Ecol. Indic.*, **84**, 701–709.

Taylor, R.G. et al., 2013: Ground water and climate change. *Nat. Clim. Change*, **3**(4), 322–329.

Tebaldi, C., B.H. Strauss and C.E. Zervas, 2012: Modelling sea level rise impacts on storm surges along US coasts. *Environ. Res. Lett.*, **7**(1), 014032.

Tedesco, M. et al., 2016: The darkening of the Greenland ice sheet: trends, drivers, and projections (1981–2100). *The Cryosphere*, **10**(2), 477–496, doi:10.5194/tc-10-477-2016.

Telesca, L. et al., 2015: Seagrass meadows (Posidonia oceanica) distribution and trajectories of change. *Sci. Rep.*, **5**, 12505–12505, doi:10.1038/srep12505.

Tellman, B., J.E. Saiers and O.A.R. Cruz, 2016: Quantifying the impacts of land use change on flooding in data-poor watersheds in El Salvador with community-based model calibration. *Reg. Environ. Change*, **16**(4), 1183–1196.

Temmerman, S. et al., 2013: Ecosystem-based coastal defence in the face of global change. *Nature*, **504**(7478), 79.

Termeer, C. et al., 2016: Coping with the wicked problem of climate adaptation across scales: The Five R Governance Capabilities. *Landscape Urban Plan.*, **154**, 11–19.

Termeer, C.J., A. Dewulf and G.R. Biesbroek, 2017: Transformational change: governance interventions for climate change adaptation from a continuous change perspective. *J. Environ. Plan. Manage.*, **60**(4), 558–576.

Termeer, C.J., A. Dewulf, G. Breeman and S. J. Stiller, 2015: Governance capabilities for dealing wisely with wicked problems. *Adm. Soc.*, **47**(6), 680–710.

Terpstra, T., 2011: Emotions, trust, and perceived risk: Affective and cognitive routes to flood preparedness behavior. *Risk Anal.*, **31**(10), 1658–1675.

Tessler, Z. et al., 2015: Profiling risk and sustainability in coastal deltas of the world. *Science*, **349**(6248), 638–643.

Tessler, Z.D., C.J. Vörösmarty, I. Overeem and J.P. Syvitski, 2018: A model of water and sediment balance as determinants of relative sea level rise in contemporary and future deltas. *Geomorphology*, **305**, 209–220.

Texier-Teixeira, P. and E. Edelblutte, 2017: Jakarta: Mumbai – Two Megacities Facing Floods Engaged in a Marginalization Process of Slum Areas. In: *Identifying Emerging Issues in Disaster Risk Reduction, Migration, Climate Change and Sustainable Development.* Springer, pp. 81–99.

Thaler, T. et al., 2019: Drivers and barriers of adaptation initiatives–How societal transformation affects natural hazard management and risk mitigation in Europe. *Sci. Total Environ.*, **650**, 1073–1082.

The Imbie team, 2018: Mass balance of the Antarctic Ice Sheet from 1992 to 2017. *Nature*, **558**(7709), 219–222, doi:10.1038/s41586-018-0179-y.

The Royal Society Science Policy Centre, 2014: *Resilience to Extreme Weather*. The Royal Society, London, 124-pp. [Available at: https://royalsociety.org/-/media/policy/projects/resilience-climate-change/resilience-full-report.pdf]. Accessed: 2019/09/20.

Thi Ha, D., S. Ouillon and G. Van Vinh, 2018: Water and Suspended Sediment Budgets in the Lower Mekong from High-Frequency Measurements (2009–2016). *Water*, **10**(7), 846.

Thiault, L. et al., 2018: Space and time matter in social-ecological vulnerability assessments. *Mar. Policy*, **88**, 213–221.

Thomas, A. and L. Benjamin, 2018: Perceptions of climate change risk in The Bahamas. *J. Environ. Stud. Sci.*, **8**(1), 63–72.

Thompson, E.L. et al., 2015: Differential proteomic responses of selectively bred and wild-type Sydney rock oyster populations exposed to elevated CO_2. *Mol. Ecol.*, **24**(6), 1248–1262.

Thompson, P. and G. Mitchum, 2014: Coherent sea level variability on the North Atlantic western boundary. *J. Geophys. Res-Oceans*, **119**(9), 5676–5689.

Thompson, P. et al., 2016: Forcing of recent decadal variability in the Equatorial and North Indian Ocean. *J. Geophys. Res-Oceans*, **121**(9), 6762–6778.

Thomson, J. and W.E. Rogers, 2014: Swell and sea in the emerging Arctic Ocean. *Geophys. Res. Lett.*, **41**(9), 3136–3140.

Thorhaug, A. et al., 2017: Seagrass blue carbon dynamics in the Gulf of Mexico: Stocks, losses from anthropogenic disturbance, and gains through seagrass restoration. *Sci. Total Environ.*, **605–606**, 626–636, doi:10.1016/J.SCITOTENV.2017.06.189.

Thorner, J., L. Kumar and S.D.A. Smith, 2014: Impacts of climate-change-driven sea level rise on intertidal rocky reef habitats will be variable and site specific. *PLoS One*, **9**(1), 1–7, doi:10.1371/journal.pone.0086130.

Timsina, J. et al., 2018: Can Bangladesh produce enough cereals to meet future demand? *Agric. Syst.*, **163**, 36–44.

Tonmoy, F. N. and A. El-Zein, 2018: Vulnerability to sea level rise: A novel local-scale indicator-based assessment methodology and application to eight beaches in Shoalhaven, Australia. *Ecol. Indic.*, **85**, 295–307.

Torio, D. D. and G. L. Chmura, 2015: Impacts of sea level rise on marsh as fish habitat. *Estuaries Coasts*, **38**(4), 1288–1303.

Torres, A., J. Brandt, K. Lear and J. Liu, 2017: A looming tragedy of the sand commons. *Science*, **357**(6355), 970–971.

Torres-Freyermuth, A. et al., 2018: On the assesment of detached breakwaters on a sea-breeze dominated beach. *Coast. Eng. Proc.*, **1**(36), 36.

Tran Anh, D., L. Hoang, M. Bui and P. Rutschmann, 2018: Simulating future flows and salinity intrusion using combined one-and two-dimensional hydrodynamic modelling – the case of Hau River, Vietnamese Mekong delta. *Water*, **10**(7), 897.

Trang, N.T.T., 2016: Architectural Approaches to a Sustainable Community with Floating Housing Units Adapting to Climate Change and Sea Level Rise in Vietnam. *World Academy of Science, Engineering and Technology, International Journal of Civil, Environmental, Structural, Construction and Architectural Engineering*, **10**(2), 168–179.

Trenberth, K.E. et al., 2018: Hurricane Harvey links to Ocean Heat Content and Climate Change Adaptation. *Earth's Future*, **6**(5), 730–744.

Trenberth, K.E. and J. Fasullo, 2007: Water and energy budgets of hurricanes and implications for climate change. *J. Geophys. Res-Atmos.*, **112**(D23), 1–10, doi:10.1029/2006JD008304.

Trenberth, K. E. and J. Fasullo, 2008: Energy budgets of Atlantic hurricanes and changes from 1970. *Geochemistry, Geophysics, Geosystems*, **9**(9), 1–12, doi:10.1029/2007GC001847.

Triyanti, A., M. Bavinck, J. Gupta and M.A. Marfai, 2017: Social capital, interactive governance and coastal protection: The effectiveness of mangrove ecosystem-based strategies in promoting inclusive development in Demak, Indonesia. *Ocean Coast. Manage*, **150**, 3–11.

Trusel, L.D. et al., 2018: Nonlinear rise in Greenland runoff in response to post-industrial Arctic warming. *Nature*, **564**(7734), 104–108, doi:10.1038/s41586-018-0752-4.

Trusel, L.D. et al., 2015: Divergent trajectories of Antarctic surface melt under two twenty-first-century climate scenarios. *Nat. Geosci.*, **8**, 927, doi:10.1038/ngeo2563.

Tsai, V.C., A.L. Stewart and A.F. Thompson, 2015: Marine ice sheet profiles and stability under Coulomb basal conditions. *J. Glaciol.*, **61**(226), 205–215.

Tuleya, R.E. et al., 2016: Impact of Upper-Tropospheric Temperature Anomalies and Vertical Wind Shear on Tropical Cyclone Evolution Using an Idealized Version of the Operational GFDL Hurricane Model. *J. Atmos. Sci.*, **73**(10), 3803–3820, doi:10.1175/JAS-D-16-0045.1.

Tuts, R. et al., 2015: *Guiding principles for city climate action planning* [Osanjo, T. (ed.)]. UN-Habitat, Nairobi, Kenya [Available at: http://e-lib.iclei.org/wp-content/uploads/2016/02/Guiding-Principles-for-City-Climate-Action-Planning.pdf]. Accessed: 2019/09/20.

Tuya, F. et al., 2014: Ecological structure and function differs between habitats dominated by seagrasses and green seaweeds. *Mar. Environ. Res.*, **98**, 1–13, doi:10.1016/j.marenvres.2014.03.015.

Uddameri, V., S. Singaraju and E.A. Hernandez, 2014: Impacts of sea level rise and urbanization on groundwater availability and sustainability of coastal communities in semi-arid South Texas. *Environ. Earth Sci.*, **71**(6), 2503–2515.

Uddin, M.N. et al., 2019: Mapping of climate vulnerability of the coastal region of Bangladesh using principal component analysis. *Appl. Geogr.*, **102**, 47–57.

Uddin, M.S., E.d.R. van Steveninck, M. Stuip and M.A.R. Shah, 2013: Economic valuation of provisioning and cultural services of a protected mangrove ecosystem: a case study on Sundarbans Reserve Forest, Bangladesh. *Ecosyst. Serv.*, **5**, 88–93.

Uebbing, B., J. Kusche, R. Rietbroek and F.W. Landerer, 2019: Processing choices affect ocean mass estimates from GRACE. *J. Geophys. Res-Oceans*, **124**(2), 1029–1044.

Uittenbroek, C.J., H.L. Mees, D.L. Hegger and P.P. Driessen, 2019: The design of public participation: who participates, when and how? Insights in climate adaptation planning from the Netherlands. *J. Environ. Plan. Manage.*, 1–19.

UNDP, 2017: *Enhancing Climate Change Adaptation in the North Coast and Nile Delta Regions in Egypt (Funding proposal FP053)*. United Nations Development Programme, Egypt. [Available at: www.greenclimate.fund/documents/20182/574760/Funding_proposal_-_FP053_-_UNDP_-_Egypt.pdf/6f006804-3009-43dc-9c49-4ad01c5d7e05]. Accessed: 2019/09/20.

UNEP, 2015: Regional State of the Coast Report: Western Indian Ocean. United Nations Environment Programme, Nairobi, Kenya, 546.

Unsworth, R.K. et al., 2015: A framework for the resilience of seagrass ecosystems. *Mar. Pollut. Bull.*, **100**(1), 34–46, doi:10.1016/J.MARPOLBUL.2015.08.016.

USACE, 2002: *Coastal Engineering Manual*. Department of the Army, U.S. Army Corps of Engineers, Engineers, Vicksburg, Mississippi, USA [Available at:

www.plainwater.com/pubs/em/em-1110–2-1100-coastal-engineering-manual/]. Accessed: 2019/09/20.

USACE, 2011: *Sea level Change Considerations for Civil Works Programs*. Department of the Army, U.S. Army Corps of Engineers, Engineers, Washington, DC [Available at: https://web.archive.org/web/20160519022621/www.corpsclimate.us/docs/EC_1165–2-212%20-Final_10_Nov_2011.pdf].

Vachaud, G. et al., 2018: Flood-related risks in Ho Chi Minh City and ways of mitigation. *J. Hydrol.*, **573**, 1021–1027.

Van de Wal, R. and M. Wild, 2001: Modelling the response of glaciers to climate change by applying volume-area scaling in combination with a high resolution GCM. *Clim. Dyn.*, **18**(3–4), 359–366.

van de Wal, R.W. et al., 2015: Self-regulation of ice flow varies across the ablation area in south-west Greenland. *The Cryosphere*, **9**(2), 603–611, doi:10.5194/tc-9-603-2015.

van den Broeke, M.R. et al., 2016: On the recent contribution of the Greenland ice sheet to sea level change. *The Cryosphere*, **10**, 1933–1946, doi:10.5194/tc-10-1933-2016.

van der Linden, S., 2015: The social-psychological determinants of climate change risk perceptions: Towards a comprehensive model. *J. Environ. Psychol.*, **41**, 112–124.

Van der Pol, T. and J. Hinkel, 2018: Uncertainty Representations of Mean Sea level Change: A Telephone Game? *Clim. Change*, **152**(3–4), 393–411.

Van Hooidonk, R. et al., 2016: Local-scale projections of coral reef futures and implications of the Paris Agreement. *Sci. Rep.*, **6**, 39666.

van Loon-Steensma, J. and P. Vellinga, 2014: Robust, multifunctional flood defenses in the Dutch rural riverine area. *Nat. Hazards Earth. Syst. Sci.*, **14**(5), 1085–1098.

Van Putten, I.E. et al., 2016: Objectives for management of socio-ecological systems in the Great Barrier Reef region, Australia. *Reg. Environ. Change*, **16**(5), 1417–1431.

Van Ruijven, B.J. et al., 2014: Enhancing the relevance of Shared Socioeconomic Pathways for climate change impacts, adaptation and vulnerability research. *Clim. Change*, **122**(3), 481–494.

Van Slobbe, E. et al., 2013: Building with Nature: in search of resilient storm surge protection strategies. *Nat. Hazards*, **66**(3), 1461–1480.

Van Wesenbeeck, B. et al., 2017: *Implementing nature based flood protection: principles and implementation guidance*. World Bank Group, Washington, D.C. [Available at: http://documents.worldbank.org/curated/en/739421509427698706/pdf/120735-REVISED-PUBLIC-Brochure-Implementing-nature-based-flood-protection-web.pdf]. Acccessed: 2019/09/201.

Van Wessem, J. et al., 2014: Improved representation of East Antarctic surface mass balance in a regional atmospheric climate model. *J. Glaciol.*, **60**(222), 761–770.

van Woesik, R., Y. Golbuu and G. Roff, 2015: Keep up or drown: adjustment of western Pacific coral reefs to sea level rise in the 21st century. *R. Soc. Open Sci.*, **2**(7), 150181.

Vanderlinden, J.-P. et al., 2018: Scoping the risks associated with accelerated coastal permafrost thaw: lessons from Bykovsky (Sakha Republic, Russian Federation) and Tuktoyaktuk (Northwest Territories, Canada). In: *European Geosciences Union General Assembly*, Vienna, Austria, Geophysical Research Abstracts 20.

Vedeld, T., A. Coly, N.M. Ndour and S. Hellevik, 2016: Climate adaptation at what scale? Multi-level governance, resilience, and coproduction in Saint Louis, Senegal. *Nat. Hazards*, **82**(2), 173–199.

Veettil, B.K. et al., 2018: Mangroves of Vietnam: Historical development, current state of research and future threats. *Estuar. Coast. Shelf Sci.*, **218**, 212–236.

Velicogna, I., T. Sutterley and M. Van Den Broeke, 2014: Regional acceleration in ice mass loss from Greenland and Antarctica using GRACE time-variable gravity data. *Geophys. Res. Lett.*, **41**(22), 8130–8137.

Vella, K. et al., 2016: Voluntary collaboration for adaptive governance: the southeast Florida regional climate change compact. *J. Plan. Educ. Res.*, **36**(3), 363–376.

Vermeer, M. and S. Rahmstorf, 2009: Global sea level linked to global temperature. *PNAS*, **106**(51), 21527–21532.

Vidas, D., 2014: Sea level rise and international law: At the convergence of two epochs. *Clim. Law*, **4**(1–2), 70–84.

Viguié, V., S. Hallegatte and J. Rozenberg, 2014: Downscaling long term socioeconomic scenarios at city scale: A case study on Paris. *Technol. Forecast. Soc. Change*, **87**, 305–324.

Vij, S. et al., 2017: Climate adaptation approaches and key policy characteristics: Cases from South Asia. *Environ. Sci. Policy*, **78**, 58–65.

Villamizar, A. et al., 2017: Climate adaptation in South America with emphasis in coastal areas: the state-of-the-art and case studies from Venezuela and Uruguay. *Clim. Dev.*, **9**(4), 364–382.

Villarini, G. and G.A. Vecchi, 2011: North Atlantic Power Dissipation Index (PDI) and Accumulated Cyclone Energy (ACE): Statistical Modeling and Sensitivity to Sea Surface Temperature Changes. *J. Clim.*, **25**(2), 625–637, doi:10.1175/JCLI-D-11-00146.1.

Villnäs, A. et al., 2013: The role of recurrent disturbances for ecosystem multifunctionality. *Ecology*, **94**(10), 2275–2287.

Vinet, F., D. Lumbroso, S. Defossez and L. Boissier, 2012: A comparative analysis of the loss of life during two recent floods in France: the sea surge caused by the storm Xynthia and the flash flood in Var. *Nat. Hazards*, **61**(3), 1179–1201.

Vinh, V.D., S. Ouillon, T.D. Thanh and L. Chu, 2014: Impact of the Hoa Binh dam (Vietnam) on water and sediment budgets in the Red River basin and delta. *Hydrol. Earth Syst. Sci.*, **18**(10), 3987–4005.

Vitousek, S. et al., 2017: Doubling of coastal flooding frequency within decades due to sea level rise. *Sci. Rep.*, **7**(1), 1399.

Vizcaino, M. et al., 2015: Coupled simulations of Greenland Ice Sheet and climate change up to AD 2300. *Geophys. Res. Lett.*, **42**(10), 3927–3935.

von Schuckmann, K. et al., 2016: An imperative to monitor Earth's energy imbalance. *Nat. Clim. Change*, **6**, 138–144, doi:10.1038/nclimate2876.

Vörösmarty, C.J. et al., 2003: Anthropogenic sediment retention: major global impact from registered river impoundments. *Global Planet. Change*, **39**(1–2), 169–190.

Vousdoukas, M.I., 2016: Developments in large-scale coastal flood hazard mapping. *Nat. Hazards Earth Syst. Sci.*, **16**(8), 1841.

Vousdoukas, M.I. et al., 2018a: Understanding epistemic uncertainty in large-scale coastal flood risk assessment for present and future climates. *Nat. Hazards Earth Syst. Sci.*, **18**(8), 2127–2142.

Vousdoukas, M.I. et al., 2018b: Climatic and socioeconomic controls of future coastal flood risk in Europe. *Nat. Clim. Change*, **8**(9), 776.

Vousdoukas, M.I. et al., 2017: Extreme sea levels on the rise along Europe's coasts. *Earth's Future*, **5**(3), 304–323.

Vousdoukas, M.I. et al., 2018c: Global probabilistic projections of extreme sea levels show intensification of coastal flood hazard. *Nat. Commun.*, **9**(1), 2360.

Vousdoukas, M.I. et al., 2016: Projections of extreme storm surge levels along Europe. *Clim. Dyn.*, **47**(9–10), 3171–3190.

Vu, D., T. Yamada and H. Ishidaira, 2018: Assessing the impact of sea level rise due to climate change on seawater intrusion in Mekong Delta, Vietnam. *Water Sci. Technol.*, **77**(6), 1632–1639.

Vuik, V., B.W. Borsje, P.W. Willemsen and S.N. Jonkman, 2019: Salt marshes for flood risk reduction: Quantifying long-term effectiveness and life-cycle costs. *Ocean Coast. Manage*, **171**, 96–110.

Vuik, V. et al., 2015: Nature-based flood protection: the efficiency of vegetated foreshores in reducing wave run-up. In: *Proceedings of the 36th IAHR World Congr*, **36**, 1–7.

Wada, Y. et al., 2012: Past and future contribution of global groundwater depletion to sea-level rise. *Geophys. Res. Lett.*, **39**(9), 1–6.

4

Wada, Y. et al., 2016: Fate of water pumped from underground and contributions to sea level rise. *Nat. Clim. Change*, **6**(8), 777–780, doi:10.1038/nclimate3001.

Wada, Y. et al., 2017: Recent Changes in Land Water Storage and its Contribution to Sea Level Variations. *Surv. Geophys.*, **38**(1), 131–152.

Wadey, M., S. Brown, R.J. Nicholls and I. Haigh, 2017: Coastal flooding in the Maldives: an assessment of historic events and their implications. *Nat. Hazards*, **89**(1), 131–159.

Wahl, T., F.M. Calafat and M.E. Luther, 2014: Rapid changes in the seasonal sea level cycle along the US Gulf coast from the late 20th century. *Geophys. Res. Lett.*, **41**(2), 491–498.

Wahl, T. et al., 2017: Understanding extreme sea levels for broad-scale coastal impact and adaptation analysis. *Nat. Commun.*, **8**, 1–12, doi:10.1038/ncomms16075.

Wahl, T. and N. G. Plant, 2015: Changes in erosion and flooding risk due to long-term and cyclic oceanographic trends. *Geophys. Res. Lett.*, **42**(8), 2943–2950.

Waibel, M.S., C.L. Hulbe, C.S. Jackson and D.F. Martin, 2018: Rate of mass loss across the instability threshold for Thwaites Glacier determines rate of mass loss for entire basin. *Geophys. Res. Lett.*, **45**(2), 809–816, doi:10.1002/2017GL076470.

Walker, W.E., V.A.W.J. Marchau and J.H. Kwakkel, 2013: Uncertainty in the Framework of Policy Analysis. In: *Public Policy Analysis: New Developments* [Thissen, W.A.H. and W.E. Walker (eds.)]. Springer US, Boston, pp. 215–261.

Walker, W.E., S.A. Rahman and J. Cave, 2001: Adaptive policies, policy analysis, and policy-making. *European J. Oper. Res. Soc.*, **128**(2), 282–289, doi:10.1016/0377-2217(00)00071-0.

Wamsley, T.V. et al., 2010: The potential of wetlands in reducing storm surge. *Ocean Engineering*, **37**(1), 59–68.

Wang, W., H. Liu, Y. Li and J. Su, 2014: Development and management of land reclamation in China. *Ocean Coast. Manage*, **102**, 415–425.

Warner, J.F., M.F. van Staveren and J. van Tatenhove, 2018a: Cutting dikes, cutting ties? Reintroducing flood dynamics in coastal polders in Bangladesh and the netherlands. *Int. J. Disast. Risk Reduc.*, **32**, 106–112.

Warner, J.F., A.J. Wesselink and G.D. Geldof, 2018b: The politics of adaptive climate management: Scientific recipes and lived reality. *WiRes. Clim. Change*, **9**(3), e515.

Warner, K. and T. Afifi, 2014: Where the rain falls: Evidence from 8 countries on how vulnerable households use migration to manage the risk of rainfall variability and food insecurity. *Clim. Dev.*, **6**(1), 1–17.

Warner, K. et al., 2013: *Changing climate, moving people: framing migration, displacement and planned relocation*. UNU-EHS, Bonn. (Available at: https://i.unu.edu/media/migration.unu.edu/publication/229/Policybrief_8_web.pdf). Accessed: 2019/09/20.

Warrick, O. et al., 2017: The 'Pacific adaptive capacity analysis framework': guiding the assessment of adaptive capacity in Pacific Island communities. *Reg. Environ. Change*, **17**(4), 1039–1051.

Wasson, K., A. Woolfolk and C. Fresquez, 2013: Ecotones as Indicators of Changing Environmental Conditions: Rapid Migration of Salt Marsh-Upland Boundaries. *Estuaries Coasts*, **36**(3), 654–664, doi:10.1007/s12237-013-9601-8.

Watkiss, P., A. Hunt, W. Blyth and J. Dyszynski, 2015: The use of new economic decision support tools for adaptation assessment: A review of methods and applications, towards guidance on applicability. *Clim. Change*, **132**, 401–416, doi:10.1007/s10584-014-1250-9.

Watson, C.S. et al., 2015: Unabated global mean sea level rise over the satellite altimeter era. *Nat. Clim. Change*, **5**(6), 565–568, doi:10.1038/nclimate2635.

Watson, E. et al., 2017: Anthropocene Survival of Southern New England's Salt Marshes. *Estuaries Coasts*, **40**(3), 617–625.

Watson, P.J., 2016: A new perspective on global mean sea level (GMSL) acceleration. *Geophys. Res. Lett.*, **43**(12), 6478–6484.

Waycott, M. et al., 2009: Accelerating loss of seagrasses across the globe threatens coastal ecosystems. *PNAS*, **106**(30), 12377–12381.

WCRP Global Sea Level Budget Group, 2018: Global sea level budget 1993-present. *Earth Syst. Sci. Data*, **10**(3), 1551–1590.

Weatherdon, L.V. et al., 2016: Observed and projected impacts of climate change on marine fisheries, aquaculture, coastal tourism, and human health: an update. *Front. Mar. Sci.*, **3**, 48.

Weber, E.U., 2016: What shapes perceptions of climate change? New research since 2010. *WiRes. Clim. Change*, **7**(1), 125–134.

Webler, T. et al., 2016: Design and evaluation of a local analytic-deliberative process for climate adaptation planning. *Local Environ.*, **21**(2), 166–188.

Weertman, J., 1974: Stability of the junction of an ice sheet and an ice shelf. *J. Glaciol.*, **13**(67), 3–11.

Wei, Y.D. and C.K. Leung, 2005: Development zones, foreign investment, and global city formation in Shanghai. *Growth Change*, **36**(1), 16–40.

Weir, T., L. Dovey and D. Orcherton, 2017: Social and cultural issues raised by climate change in Pacific Island countries: an overview. *Reg. Environ. Change*, **17**(4), 1017–1028.

Weisse, R., H. von Storch, H.D. Niemeyer and H. Knaack, 2012: Changing North Sea storm surge climate: An increasing hazard? *Ocean Coast. Manage*, **68**, 58–68.

Welch, A., R. Nicholls and A. Lázár, 2017: Evolving deltas: Coevolution with engineered interventions. *Elem Sci Anth*, **5**(49), 1–18.

Wellstead, A. et al., 2018: Overcoming the 'Barriers' Orthodoxy: A New Approach to Understanding Climate Change Adaptation and Mitigation Governance Challenges in the Canadian Forest Sector. *Canadian Journal of Forest Research* **48**(10), 1241–1245.

Wellstead, A., J. Rayner and M. Howlett, 2014: Beyond the black box: forest sector vulnerability assessments and adaptation to climate change in North America. *Environ. Sci. Policy*, **35**, 109–116.

Werner, A.D. et al., 2017: Hydrogeology and management of freshwater lenses on atoll islands: Review of current knowledge and research needs. *J. Hydrol.*, **551**, 819–844.

Werners, S. et al., 2015: Turning points in climate change adaptation. *Ecol. Soc.*, **20**(4).

Werz, M. and M. Hoffman, 2015: Climate change, migration, and the demand for greater resources: challenges and responses. *SAIS Rev. Int. Aff.*, **35**(1), 99–108.

Wessel, P. and W.H. Smith, 1996: A global, self-consistent, hierarchical, high-resolution shoreline database. *J. Geophys. Res-Earth*, **101**(B4), 8741–8743.

Whitehouse, P.L., N. Gomez, M.A. King and D.A. Wiens, 2019: Solid Earth change and the evolution of the Antarctic Ice Sheet. *Nat. Commun.*, **10**(1), 503.

Wijffels, S. et al., 2016: Ocean temperatures chronicle the ongoing warming of Earth. *Nat. Clim. Change*, **6**(2), 116–118.

Wilbers, G.-J., M. Becker, Z. Sebesvari and F. G. Renaud, 2014: Spatial and temporal variability of surface water pollution in the Mekong Delta, Vietnam. *Sci. Total Environ.*, **485**, 653–665.

Wilmsen, B. and M. Webber, 2015: What can we learn from the practice of development-forced displacement and resettlement for organised resettlements in response to climate change? *Geoforum*, **58**, 76–85.

Wilson, A.M.W. and C. Forsyth, 2018: Restoring near-shore marine ecosystems to enhance climate security for island ocean states: Aligning international processes and local practices. *Mar. Policy.*, **93**, 284–294.

Winkelmann, R., A. Levermann, A. Ridgwell and K. Caldeira, 2015: Combustion of available fossil fuel resources sufficient to eliminate the Antarctic Ice Sheet. *Sci. Adv.*, **1**(8), e1500589.

Winkelmann, R. et al., 2011: The Potsdam parallel ice sheet model (PISM-PIK)-Part 1: Model description. *The Cryosphere*, **5**(3), 715.

Wise, M.G., J.A. Dowdeswell, M. Jakobsson and R.D. Larter, 2017: Evidence of marine ice-cliff instability in Pine Island Bay from iceberg-keel plough marks. *Nature*, **550**(7677), 506–510, doi:10.1038/nature24458.

Wise, R. et al., 2014: Reconceptualising adaptation to climate change as part of pathways of change and response. *Global Environ. Chang.*, **28**, 325–336.

Wolff, C. et al., 2016: Effects of scale and input data on assessing the future impacts of coastal flooding: an application of DIVA for the Emilia-Romagna coast. *Front. Mar. Sci.*, **3**, 41.

Wolfram, M., J. van der Heijden, S. Juhola and J. Patterson, 2019: Learning in urban climate governance: concepts, key issues and challenges. *J. Environ. Pol. Plan.*, **21**(1), 1–15.

Wong, P.P., 2018: Coastal Protection Measures–Case of Small Island Developing States to Address Sea level Rise. *Asian Journal of Environment & Ecology*, **6**(3), 1–14.

Wong, P.P. et al., 2014: Coastal systems and low-lying areas. In: Climate Change 2014: Impacts, Adaptation, and Vulnerability. Part A: Global and Sectoral Aspects. Contribution of Working Group II to the Fifth Assessment Report of the Intergovernmental Panel on Climate Change [Barros, V.R., C.B. Field, D.J. Dokken, M.D. Mastrandrea, K.J. Mach, T.E. Bilir, M. Chatterjee, K.L. Ebi, Y.O. Estrada, R.C. Genova, B. Girma, E.S. Kissel, A.N. Levy, S. MacCracken, P.R. Mastrandrea, L.L. White, R.J. Nicholls and F. Santos (eds.)]. Cambridge University Press, Cambridge, United Kingdom and New York, NY, USA. 361–409.

Wong, T.E., A.M. Bakker and K. Keller, 2017: Impacts of Antarctic fast dynamics on sea level projections and coastal flood defense. *Clim. Change*, **144**(2), 347–364.

Wong, V.N. et al., 2015: Seawater inundation of coastal floodplain sediments: short-term changes in surface water and sediment geochemistry. *Chem. Geol.*, **398**, 32–45.

Woodroffe, C.D. et al., 2016: Mangrove sedimentation and response to relative sea level rise. *Annu. Rev. Mar. Sci.*, **8**, 243–266.

Woodruff, J.D., J.L. Irish and S.J. Camargo, 2013: Coastal flooding by tropical cyclones and sea level rise. *Nature*, **504**(7478), 44–52.

Woodruff, S.C. and M. Stults, 2016: Numerous strategies but limited implementation guidance in US local adaptation plans. *Nat. Clim. Change*, **6**(8), 796.

Woodward, M. et al., 2011: R eal O ptions in flood risk management decision making. *J. Flood Risk Manage.*, **4**(4), 339–349.

Woodward, M., Z. Kapelan and B. Gouldby, 2014: Adaptive Flood Risk Management Under Climate Change Uncertainty Using Real Options and Optimization: Adaptive Flood Risk Management. *Risk Anal.*, **34**, 75–92, doi:10.1111/risa.12088.

Woodworth, P. et al., 2016: Towards a global higher-frequency sea level dataset. *Geosci. Data J.*, **3**(2), 50–59.

Wöppelmann, G. and M. Marcos, 2016: Vertical land motion as a key to understanding sea level change and variability. *Rev. Geophys.*, **54**(1), 64–92.

World Bank, 2017: Population of Egypt. The World Bank.

Wouters, B., A.S. Gardner and G. Moholdt, 2019: Global glacier mass loss during the GRACE satellite mission (2002–2016). *Front. Earth Sci.*, **7**, 96.

Wouters, B. et al., 2015: Dynamic thinning of glaciers on the Southern Antarctic Peninsula. *Science*, **348**(6237), 899–903.

Wrathall, D.J. and N. Suckall, 2016: Labour migration amidst ecological change. *Migration and Development*, **5**(2), 314–329.

Wright, L.D. and C.R. Nichols, (eds.), 2018: *Tomorrow's Coasts: Complex and Impermanent*. Springer, Heidelberg, Germany.

Wu, J.S. and J.J. Lee, 2015: Climate change games as tools for education and engagement. *Nat. Clim. Change*, **5**(5), 413.

Wu, W., P. Biber and M. Bethel, 2017: Thresholds of sea level rise rate and sea level rise acceleration rate in a vulnerable coastal wetland. *Ecol. Evol.*, **7**(24), 10890–10903, doi:10.1002/ece3.3550.

Xian, S., J. Yin, N. Lin and M. Oppenheimer, 2018: Influence of risk factors and past events on flood resilience in coastal megacities: comparative analysis of NYC and Shanghai. *Sci. Total Environ.*, **610**, 1251–1261.

Xie, W. et al., 2017: Effects of straw application on coastal saline topsoil salinity and wheat yield trend. *Soil Tillage Res.*, **169**, 1–6.

Yaakub, S.M. et al., 2014: Courage under fire: Seagrass persistence adjacent to a highly urbanised city–state. *Mar. Pollut. Bull.*, **83**(2), 417–424, doi:10.1016/J.MARPOLBUL.2014.01.012.

Yamada, Y. et al., 2017: Response of Tropical Cyclone Activity and Structure to Global Warming in a High-Resolution Global Nonhydrostatic Model. *J. Clim.*, **30**(23), 9703–9724, doi:10.1175/jcli-d-17-0068.1.

Yamamoto, L. and M. Esteban, 2014: *Atoll Island States and International Law Climate Change Displacement and Sovereignty*. Springer, Berlin.

Yamane, M. et al., 2015: Exposure age and ice sheet model constraints on Pliocene East Antarctic ice sheet dynamics. *Nat. Commun.*, **6**, doi:10.1038/ncomms8016.

Yamano, H. et al., 2007: Atoll island vulnerability to flooding and inundation revealed by historical reconstruction: Fongafale Islet, Funafuti Atoll, Tuvalu. *Global Planet. Change*, **57**(3–4), 407–416.

Yan, B. et al., 2016: Socioeconomic vulnerability of the megacity of Shanghai (China) to sea level rise and associated storm surges. *Reg. Environ. Change*, **16**(5), 1443–1456.

Yang, H. et al., 2017: Erosion potential of the Yangtze Delta under sediment starvation and climate change. *Sci. Rep.*, **7**(1), 10535.

Yao, R.-J. et al., 2015: Determining soil salinity and plant biomass response for a farmed coastal cropland using the electromagnetic induction method. *Comput. Electron. Agr.*, **119**, 241–253.

Yates, K.K., D.G. Zawada, N.A. Smiley and G. Tiling-Range, 2017: Divergence of seafloor elevation and sea level rise in coral reef ecosystems. *Biogeosciences*, **14**(6), 1739.

Yau, A.M., M. Bender, A. Robinson and E. Brook, 2016: Reconstructing the last interglacial at Summit, Greenland: Insights from GISP2. *PNAS*, **113**(35), 9710–9715, doi:10.1073/pnas.1524766113.

Yeager, S. and G. Danabasoglu, 2014: The origins of late-twentieth-century variations in the large-scale North Atlantic circulation. *J. Clim.*, **27**(9), 3222–3247.

Yi, S., K. Heki and A. Qian, 2017: Acceleration in the global mean sea level rise: 2005–2015. *Geophys. Res. Lett.*, **44**(23), 11,905–11,913.

Yin, J., 2012: Century to multi-century sea level rise projections from CMIP5 models. *Geophys. Res. Lett.*, **39**(17), 1–7, doi:10.1029/2012GL052947.

Yin, J., M.E. Schlesinger and R.J. Stouffer, 2009: Model projections of rapid sea level rise on the northeast coast of the United States. *Nat. Geosci.*, **2**(4), 262.

Yin, J., M. Ye, Z. Yin and S. Xu, 2015: A review of advances in urban flood risk analysis over China. *Stoch. Env. Res. Risk A.*, **29**(3), 1063–1070.

Yin, J. et al., 2011: Monitoring urban expansion and land use/land cover changes of Shanghai metropolitan area during the transitional economy (1979–2009) in China. *Environ. Monit. Assess.*, **177**(1–4), 609–621.

Yin, J. et al., 2013: Modelling the combined impacts of sea level rise and land subsidence on storm tides induced flooding of the Huangpu River in Shanghai, China. *Clim. Change*, **119**(3–4), 919–932.

Yu, H., E. Rignot, H. Seroussi and M. Morlighem, 2018: Retreat of Thwaites Glacier, West Antarctica, over the next 100 years using various ice flow models, ice shelf melt scenarios and basal friction laws. *The Cryosphere Discussions*, **12**(12), 3861–3876, doi:10.5194/tc-2018-104.

Yusuf, J.-E. et al., 2016: The sea is rising… but not onto the policy agenda: A multiple streams approach to understanding sea level rise policies. *Environ. Plan. C.*, **34**(2), 228–243.

Yusuf, J.-E. W. et al., 2018a: Participatory GIS as a Tool for Stakeholder Engagement in Building Resilience to Sea Level Rise: A Demonstration Project. *Mar. Technol. Soc. J.*, **52**(2), 45–55.

Yusuf, J.E., B. St. John III, M. Covi and J.G. Nicula, 2018b: Engaging Stakeholders in Planning for Sea Level Rise and Resilience. *J. Contemp. Water Res. Educ.*, **164**(1), 112–123.

Yuzva, K., W. W. Botzen, J. Aerts and R. Brouwer, 2018: A global review of the impact of basis risk on the functioning of and demand for index insurance. *Int. J. Disast. Risk Reduc.*, **28**, 845–853.

Zappa, G., L.C. Shaffrey and K.I. Hodges, 2013: The ability of CMIP5 models to simulate North Atlantic extratropical cyclones. *J. Clim.*, **26**(15), 5379–5396.

Zemp, M. et al., 2019: Global glacier mass changes and their contributions to sea level rise from 1961 to 2016. *Nature*, **568**(7752), 382–386, doi:10.1038/s41586-019-1071-0.

Zhang, K. et al., 2012: The role of mangroves in attenuating storm surges. *Estuar. Coast. Shelf Sci.*, **102**, 11–23.

Zhang, L.-Y., 2003: Economic development in Shanghai and the role of the state. *Urban Studies*, **40**(8), 1549–1572.

Zhang, X. and J. A. Church, 2012: Sea level trends, interannual and decadal variability in the Pacific Ocean. *Geophys. Res. Lett.*, **39**(21).

Zheng, L. et al., 2017: Impact of salinity and Pb on enzyme activities of a saline soil from the Yellow River delta: A microcosm study. *Phys. Chem. Earth*, **97**, 77–87.

Zhu, X., M. M. Linham and R. J. Nicholls, 2010: *Technologies for Climate Change Adaptation: Coastal Erosion and Flooding*. TNA Guidebook Series, Roskilde: Danmarks Tekniske Universitet, Risø Nationallaboratoriet for Bæredygtig Energi [Available at: http://orbit.dtu.dk/files/5699563/Technologies%20for%20Climate%20Change%20Adaptation-Coastal%20Erosion%20and%20Flooding.pdf]. Accessed: 2019/09/20.

Zickfeld, K., S. Solomon and D.M. Gilford, 2017: Centuries of thermal sea level rise due to anthropogenic emissions of short-lived greenhouse gases. *PNAS*, **114**(4), 657–662, doi:10.1073/pnas.1612066114.

Ziervogel, G., 2019: Building transformative capacity for adaptation planning and implementation that works for the urban poor: Insights from South Africa. *Ambio*, **48**(5), 1–13.

Ziervogel, G., E. Archer van Garderen and P. Price, 2016a: Strengthening the knowledge–policy interface through co-production of a climate adaptation plan: leveraging opportunities in Bergrivier Municipality, South Africa. *Environ. Urban.*, **28**(2), 455–474.

Ziervogel, G., J. Waddell, W. Smit and A. Taylor, 2016b: Flooding in Cape Town's informal settlements: barriers to collaborative urban risk governance. *South African Geographical Journal*, **98**(1), 1–20.

Zwally, H.J. et al., 2002: Surface Melt-Induced Acceleration of Greenland Ice sheet Flow. *Science*, **297**(5579), 218–222, doi:10.1126/science.1072708.

4

5

Changing Ocean, Marine Ecosystems, and Dependent Communities

Coordinating Lead Authors
Nathaniel L. Bindoff (Australia), William W. L. Cheung (Canada), James G. Kairo (Kenya)

Lead Authors
Javier Arístegui (Spain), Valeria A. Guinder (Argentina), Robert Hallberg (USA), Nathalie Hilmi (Monaco/France), Nianzhi Jiao (China), Md saiful Karim (Australia), Lisa Levin (USA), Sean O'Donoghue (South Africa), Sara R. Purca Cuicapusa (Peru), Baruch Rinkevich (Israel), Toshio Suga (Japan), Alessandro Tagliabue (United Kingdom), Phillip Williamson (United Kingdom)

Contributing Authors
Sevil Acar (Turkey), Juan Jose Alava (Ecuador/Canada), Eddie Allison (United Kingdom), Brian Arbic (USA), Tamatoa Bambridge (French Polynesia), Inka Bartsch (Germany), Laurent Bopp (France), Philip W. Boyd (Australia/ United Kingdom), Thomas Browning (Germany/United Kingdom), Jorn Bruggeman (Netherlands), Momme Butenschön (Germany), Francisco P. Chávez (USA), Lijing Cheng (China), Mine Cinar (USA), Daniel Costa (USA), Omar Defeo (Uruguay), Salpie Djoundourian (Lebanon), Catia Domingues (Australia), Tyler Eddy (Canada), Sonja Endres (Germany), Alan Fox (UK), Christopher Free (USA), Thomas Frölicher (Switzerland), Jean-Pierre Gattuso (France), Gemma Gerber (South Africa), Charles Greene (USA), Nicolas Gruber (Switzerland), Gustaav Hallegraef (Australia), Matthew Harrison (USA), Sebastian Hennige (UK), Mark Hindell (Australia), Andrew Hogg (Australia), Taka Ito (USA), Tiff-Annie Kenny (Canada), Kristy Kroeker (USA), Lester Kwiatkowski (France/UK), Vicky W. Y. Lam (China/Canada), Charlotte Laüfkotter (Switzerland/German), Philippe LeBillon (Canada), Nadine Le Bris (France), Heike Lotze (Canada), Jennifer MacKinnon (USA), Annick de Marffy-Mantuano (Monaco), Patrick Martel (South Africa), Nadine Marshall (Australia), Kathleen McInnes (Australia), Jorge García Molinos (Japan/Spain), Serena Moseman-Valtierra (USA), Andries Motau (South Africa), Sandor Mulsow (Brazil), Kana Mutombo (South Africa), Andreas Oschlies (Germany), Muhammed Oyinlola (Nigeria), Elvira S. Poloczanska (Australia), Nicolas Pascal (France), Maxime Philip (France), Sarah Purkey (USA), Saurabh Rathore (India), Xavier Rebelo (South Africa), Gabriel Reygondeau (France), Jake Rice (Canada), Anthony Richardson (Australia), Ulf Riebesell (Germany), Christopher Roach (France/Australia), Joacim Rocklöv (Sweden), Murray Roberts (United Kingdom), Alain Safa (France), Sunke Schmidtko (Germany), Gerald Singh (Canada), Bernadette Sloyan (Australia), Karinna von Schuckmann (France), Manal Shehabi (England), Matthew Smith (USA), Amy Shurety (South Africa), Fernando Tuya (Spain), Cristian Vargas (Chile), Colette Wabnitz (France), Caitlin Whalen (USA)

Review Editors
Manuel Barange (South Africa), Brad Seibel (USA)

Chapter Scientist
Axel Durand (Australia)

This chapter should be cited as:
Bindoff, N.L., W.W.L. Cheung, J.G. Kairo, J. Arístegui, V.A. Guinder, R. Hallberg, N. Hilmi, N. Jiao, M.S. Karim, L. Levin, S. O'Donoghue, S.R. Purca Cuicapusa, B. Rinkevich, T. Suga, A. Tagliabue, and P. Williamson, 2019: Changing Ocean, Marine Ecosystems, and Dependent Communities. In: *IPCC Special Report on the Ocean and Cryosphere in a Changing Climate* [H.-O. Pörtner, D.C. Roberts, V. Masson-Delmotte, P. Zhai, M. Tignor, E. Poloczanska, K. Mintenbeck, A. Alegría, M. Nicolai, A. Okem, J. Petzold, B. Rama, N.M. Weyer (eds.)]. Cambridge University Press, Cambridge, UK and New York, NY, USA, pp. 447–587. https://doi.org/10.1017/9781009157964.007.

Table of contents

5

Executive Summary

The ocean is essential for all aspects of human well-being and livelihood. It provides key services like climate regulation, through the energy budget, carbon cycle and nutrient cycle. The ocean is the home of biodiversity ranging from microbes to marine mammals that form a wide variety of ecosystems in open pelagic and coastal ocean.

Observations: Climate-related trends, impacts, adaptation

Carbon emissions from human activities are causing ocean warming, acidification and oxygen loss with some evidence of changes in nutrient cycling and primary production. The warming ocean is affecting marine organisms at multiple trophic levels, impacting fisheries with implications for food production and human communities. Concerns regarding the effectiveness of existing ocean and fisheries governance have already been reported, highlighting the need for timely mitigation and adaptation responses.

The ocean has warmed unabated since 2005, continuing the clear multi-decadal ocean warming trends documented in the IPCC Fifth Assessment Report (AR5). The warming trend is further confirmed by the improved ocean temperature measurements over the last decade. The 0–700 m and 700–2000 m layers of the ocean have warmed at rates of 5.31 ± 0.48 and 4.02 ± 0.97 ZJ yr^{-1} from 2005 to 2017. The long-term trend for 0–700 m and 700–2000 m layers have warmed 4.35 ± 0.8 and 2.25 ± 0.64 ZJ yr^{-1} from between the averages of 1971–1990 and 1998–2017 and is attributed to anthropogenic influences. It is *likely*[1] the ocean warming has continued in the abyssal and deep ocean below 2000 m (southern hemisphere and Southern Ocean). {1.8.1, 1.2, 5.2.2}

It is *likely* that the rate of ocean warming has increased since 1993. The 0–700 m and 700–2000 m layers of the ocean have warmed by 3.22 ± 1.61 ZJ and 0.97 ± 0.64 ZJ from 1969 to 1993, and 6.28 ± 0.48 ZJ and 3.86 ± 2.09 ZJ from 1993 to 2017. This represents at least a two-fold increase in heat uptake. {Table 5.1, 5.2.2}

The upper ocean is *very likely* to have been stratifying since 1970. Observed warming and high-latitude freshening are making the surface ocean less dense over time relative to the deeper ocean (*high confidence*[2]) and inhibiting the exchange between surface and deep waters. The upper 200 m stratification increase is in the *very likely* range of between 2.18–2.42% from 1970 to 2017. {5.2.2}

Multiple datasets and models show that the rate of ocean uptake of atmospheric CO$_2$ has continued to strengthen in the recent two decades in response to the increasing concentration of CO$_2$ in the atmosphere. The *very likely* range for ocean uptake is between 20–30% of total anthropogenic emissions in the recent two decades. Evidence is growing that the ocean carbon sink is dynamic on decadal timescales, especially in the Southern Ocean, which has affected the total global ocean carbon sink (*medium confidence*). {5.2.2.3}

The ocean is continuing to acidify in response to ongoing ocean carbon uptake. The open ocean surface water pH is observed to be declining (*virtually certain*) by a *very likely* range of 0.017–0.027 pH units per decade since the late 1980s across individual time series observations longer than 15 years. The anthropogenic pH signal is *very likely* to have emerged for three-quarters of the near-surface open ocean prior to 1950 and it is *very likely* that over 95% of the near surface open ocean has already been affected. These changes in pH have reduced the stability of mineral forms of calcium carbonate due to a lowering of carbonate ion concentrations, most notably in the upwelling and high-latitude regions of the ocean. {5.2.2.3, Box 5.1}

There is a growing consensus that the open ocean is losing oxygen overall with a *very likely* loss of 0.5–3.3% between 1970–2010 from the ocean surface to 1000 m (*medium confidence*). Globally, the oxygen loss due to warming is reinforced by other processes associated with ocean physics and biogeochemistry, which cause the majority of the observed oxygen decline (*high confidence*). The oxygen minimum zones (OMZs) are expanding by a *very likely* range of 3–8%, most notably in the tropical oceans, but there is substantial decadal variability that affects the attribution of the overall oxygen declines to human activity in tropical regions (*high confidence*). {5.2.2.4}

In response to ocean warming and increased stratification, open ocean nutrient cycles are being perturbed and there is *high confidence* that this is having a regionally variable impact on primary producers. There is currently *low confidence* in appraising past open ocean productivity trends, including those determined by satellites, due to newly identified region-specific drivers of microbial growth and the lack of corroborating *in situ* time series datasets. {5.2.2.5, 5.2.2.6}

Ocean warming has contributed to observed changes in biogeography of organisms ranging from phytoplankton to marine mammals (*high confidence*), consequently changing community composition (*high confidence*), and in some cases, altering interactions between organisms (*medium confidence*). Observed rate of range shifts since the 1950s and *its*

---

[1]   In this Report, the following terms have been used to indicate the assessed likelihood of an outcome or a result: Virtually certain 99–100% probability, Very likely 90–100%, Likely 66–100%, About as likely as not 33–66%, Unlikely 0–33%, Very unlikely 0–10%, and Exceptionally unlikely 0–1%. Additional terms (Extremely likely: 95–100%, More likely than not >50–100%, and Extremely unlikely 0–5%) may also be used when appropriate. Assessed likelihood is typeset in italics, e.g., *very likely* (see Section 1.9.2 and Figure 1.4 for more details). This Report also uses the term '*likely* range' to indicate that the assessed likelihood of an outcome lies within the 17–83% probability range.

[2]   In this Report, the following summary terms are used to describe the available evidence: limited, medium, or robust; and for the degree of agreement: low, medium, or high. A level of confidence is expressed using five qualifiers: very low, low, medium, high, and very high, and typeset in italics, e.g., *medium confidence*. For a given evidence and agreement statement, different confidence levels can be assigned, but increasing levels of evidence and degrees of agreement are correlated with increasing confidence (see Section 1.9.2 and Figure 1.4 for more details).

very likely range are estimated to be 51.5 ± 33.3 km per decade and 29.0 ± 15.5 km per decade for organisms in the epipelagic and seafloor ecosystems, respectively. The direction of the majority of the shifts of epipelagic organisms are consistent with a response to warming (*high confidence*). {5.2.3, 5.3}

Warming-induced range expansion of tropical species to higher latitudes has led to increased grazing on some coral reefs, rocky reefs, seagrass meadows and epipelagic ecosystems, leading to altered ecosystem structure (*medium confidence*). Warming, sea level rise (SLR) and enhanced loads of nutrients and sediments in deltas have contributed to salinisation and deoxygenation in estuaries (*high confidence*), and have caused upstream redistribution of benthic and pelagic species according to their tolerance limits (*medium confidence*). {5.3.4, 5.3.5, 5.3.6, 5.2.3}

Fisheries catches and their composition in many regions are already impacted by the effects of warming and changing primary production on growth, reproduction and survival of fish stocks (*high confidence*). Ocean warming and changes in primary production since the 20th century are related to changes in productivity of many fish stocks (*high confidence*), with an average decrease of approximately 3% per decade in population replenishment and 4.1% (*very likely range* of 9.0% decline to 0.3% increase) in maximum catch potential (*robust evidence, low agreement* between fish stocks, *medium confidence*). Species composition of fisheries catches since the 1970s in many shelf seas ecosystems of the world is increasingly dominated by warm water species (*medium confidence*). {5.2.3, 5.4.1}

Warming-induced changes in spatial distribution and abundance of fish stocks have already challenged the management of some important fisheries and their economic benefits (*high confidence*). For existing international and national ocean and fisheries governance, there are concerns about the reduced effectiveness to achieve mandated ecological, economic, and social objectives because of observed climate impacts on fisheries resources (*high confidence*). {5.4.2, 5.5.2}

Coastal ecosystems are observed to be under stress from ocean warming and SLR that are exacerbated by non-climatic pressures from human activities on ocean and land (*high confidence*). Global wetland area has declined by nearly 50% relative to pre-industrial level as a result of warming, SLR, extreme climate events and other human impacts (*medium confidence*). Warming related mangrove encroachment into subtropical salt marshes has been observed in the past 50 years (*high confidence*). Distributions of seagrass meadows and kelp forests are contracting at low-latitudes that is attributable to warming (*high confidence*), and in some areas a loss of 36–43% following heat waves (*medium confidence*). Inundation, coastline erosion and salinisation are causing inland shifts in plant species distributions, which has been accelerating in the last decades (*medium confidence*). Warming has increased the frequency of large-scale coral bleaching events, causing worldwide reef degradation since 1997–1998 with cases of shifts to algal-dominated reefs (*high confidence*). Sessile calcified organisms (e.g., barnacles and mussels) in intertidal rocky shores are highly sensitive to extreme temperature events and acidification (*high*

confidence), a reduction in their biodiversity and abundance have been observed in naturally-acidified rocky reef ecosystems (*medium confidence*). Increased nutrient and organic matter loads in estuaries since the 1970s have exacerbated the effects of warming on bacterial respiration and eutrophication, leading to expansion of hypoxic areas (*high confidence*). {5.3.1, 5.3.2, 5.3.4, 5.3.6}

Coastal and near-shore ecosystems including salt marshes, mangrove forests and vegetated dunes in sandy beaches have a varying capacity to build vertically and expand laterally in response to SLR. These ecosystems provide important services including coastal protection, carbon sequestration and habitat for diverse biota (*high confidence*). The carbon emission associated with the loss of vegetated coastal ecosystems is estimated to be 0.04–1.46 Gt C yr^{-1} (*high confidence*). The natural capacity of ecosystems to adapt to climate impacts may be limited by human activities that fragment wetland habitats and restrict landward migration (*high confidence*). {5.3.2, 5.3.3, 5.4.1, 5.5.1}

Three out of the four major Eastern Boundary Upwelling Systems (EBUS) have shown large-scale wind intensification in the past 60 years (*high confidence*). However, the interaction of coastal warming and local winds may have affected upwelling strength, with the direction of changes varies between and within EBUS (*low confidence*). Increasing trends in ocean acidification in the California Current EBUS and deoxygenation in California Current and Humboldt Current EBUS are observed in the last few decades (*high confidence*), although there is *low confidence* to distinguish anthropogenic forcing from internal climate variability. The expanding California EBUS OMZ has altered ecosystem structure and fisheries catches (*medium confidence*). {Box 5.3}

Since the early 1980s, the occurrence of harmful algal blooms (HABs) and pathogenic organisms (e.g., *Vibrio*) has increased in coastal areas in response to warming, deoxygenation and eutrophication, with negative impacts on food provisioning, tourism, the economy and human health (*high confidence*). These impacts depend on species-specific responses to the interactive effects of climate change and other human drivers (e.g., pollution). Human communities in poorly monitored areas are among the most vulnerable to these biological hazards (*medium confidence*). {Box 5.4, 5.4.2}

Many frameworks for climate resilient coastal adaptation have been developed since AR5, with substantial variations in approach between and within countries, and across development status (*high confidence*). Few studies have assessed the success of implementing these frameworks due to the time-lag between implementation, monitoring, evaluation and reporting (*medium confidence*). {5.5.2}

Projections: scenarios and time horizons

Climate models project significant changes in the ocean state over the coming century. Under the high emissions scenario (Representative Concentration Pathway (RCP)8.5) the impacts by 2090 are substantially larger and more widespread than for

the low emissions scenario (RCP2.6) throughout the surface and deep ocean, including: warming (*virtually certain*); ocean acidification (*virtually certain*); decreased stability of mineral forms of calcite (*virtually certain*); oxygen loss (*very likely*); reduced near-surface nutrients (*likely as not*); decreased net primary productivity (*high confidence*); reduced fish production (*likely*) and loss of key ecosystems services (*medium confidence*) that are important for human well-being and sustainable development. {5.2.2, Box 5.1, 5.2.3, 5.2.4, 5.4}

By 2100 the ocean is *very likely* to warm by 2 to 4 times as much for low emissions (RCP2.6) and 5 to 7 times as much for the high emissions scenario (RCP8.5) compared with the observed changes since 1970. The 0–2000 m layer of the ocean is projected to warm by a further 2150 ZJ (*very likely* range 1710–2790 ZJ) between 2017 and 2100 for the RCP8.5 scenario. The 0–2000 m layer is projected to warm by 900 ZJ (*very likely* range 650–1340 ZJ) by 2100 for the RCP2.6 scenario, and the overall warming of the ocean will continue this century even after radiative forcing and mean surface temperatures stabilise (*high confidence*). {5.2.2.2}

The upper ocean will continue to stratify. By the end of the century the annual mean stratification of the top 200 m (averaged between 60°S–60°N relative to the 1986–2005 period) is projected to increase in the *very likely* range of 1–9% and 12–30% for RCP2.6 and RCP8.5 respectively. {5.2.2.2}

It is *very likely* that the majority of coastal regions will experience statistically significant changes in tidal amplitudes over the course of the 21st century. The sign and amplitude of local changes to tides are *very likely* to be impacted by both human coastal adaptation measures and climate drivers. {5.2.2.2.3}

It is *virtually certain* that surface ocean pH will decline, by 0.036–0.042 or 0.287–0.29 pH units by 2081–2100, relative to 2006–2015, for the RCP2.6 or RCP8.5 scenarios, respectively. These pH changes are *very likely* to cause the Arctic and Southern Oceans, as well as the North Pacific and Northwestern Atlantic Oceans to become corrosive for the major mineral forms of calcium carbonate under RCP8.5, but these changes are *virtually certain* to be avoided under the RCP2.6 scenario. There is increasing evidence of an increase in the seasonal exposure to acidified conditions in the future (*high confidence*), with a *very likely* increase in the amplitude of seasonal cycle of hydrogen iron concentrations of 71–90% by 2100, relative to 2000 for the RCP8.5 scenario, especially at high latitudes. {5.2.2.3}

Oxygen is projected to decline further. Globally, the oxygen content of the ocean is *very likely* to decline by 3.2–3.7% by 2081–2100, relative to 2006–2015, for the RCP8.5 scenario or by 1.6–2.0% for the RCP2.6 scenario. The volume of the oceans OMZ is projected to grow by a *very likely* range of 7.0 ± 5.6% by 2100 during the RCP8.5 scenario, relative to 1850–1900. The climate signal of oxygen loss will *very likely* emerge from the historical climate by 2050 with a *very likely* range of 59–80% of ocean area being affected by 2031–2050 and rising with a *very likely* range of 79–91% by 2081–2100 (RCP8.5). The emergence of oxygen

loss is *very likely* smaller in area for the RCP2.6 scenario in the 21st century and by 2090 the emerged area is declining. {5.2.2.4, Box 5.1 Figure 1}

Overall, nitrate concentrations in the upper 100 m are *very likely* to decline by 9–14% across CMIP5 models by 2081–2100, relative to 2006–2015, in response to increased stratification for RCP8.5, with *medium confidence* in these projections due to the *limited evidence* of past changes that can be robustly understood and reproduced by models. There is *low confidence* regarding projected increases in surface ocean iron levels due to systemic uncertainties in these models. {5.2.2.5}

Climate models project that net primary productivity will *very likely* decline by 4–11% for RCP8.5 by 2081–2100, relative to 2006–2015. The decline is due to the combined effects of warming, stratification, light, nutrients and predation and will show regional variations between low and high latitudes (*low confidence*). The tropical ocean NPP will *very likely* decline by 7–16% for RCP8.5, with *medium confidence* as there are improved constraints from historical variability in this region. Globally, the sinking flux of organic matter from the upper ocean into the ocean interior is *very likely* to decrease by 9–16% for RCP8.5 in response to increased stratification and reduced nutrient supply, especially in tropical regions (*medium confidence*), which will reduce organic carbon supply to deep sea ecosystems (*high confidence*). The reduction in food supply to the deep sea is projected to lead to a 5–6% reduction in biomass of benthic biota over more than 97% of the abyssal seafloor by 2100 (*medium confidence*). {5.2.2.6, 5.2.4.2}

New ocean states for a broad suite of climate indices will progressively emerge over a substantial fractions of the ocean in the coming century (relative to past internal ocean variability), with Earth System Models (ESMs) showing an ordered emergence of first pH, followed by sea surface temperature (SST), interior oxygen, upper ocean nutrient levels and finally net primary production (NPP). The anthropogenic pH signal has *very likely* emerged for three quarters of the ocean prior to 1950, with little difference between scenarios. Oxygen changes will *very likely* emerge over 59–80% of the ocean area by 2031–2050 and rises to 79–91% by 2081–2100 (RCP8.5). The projected time of emergence for five primary drivers of marine ecosystem change (surface warming and acidification, oxygen loss, nitrate content and net primary production change) are all prior to 2100 for over 60% of the ocean area under RCP8.5 and over 30% under RCP2.6 (*very likely*). {Box 5.1, Box 5.1 Figure 1}

Simulated ocean warming and changes in NPP during the 21st century are projected to alter community structure of marine organisms (*high confidence*), reduce global marine animal biomass (*medium confidence*) and the maximum potential catches of fish stocks (*medium confidence*) with regional differences in the direction and magnitude of changes (*high confidence*). The global biomass of marine animals, including those that contribute to fisheries, is projected to decrease with a *very likely* range under RCP2.6 and RCP8.5 of 4.3 ± 2.0% and 15.0 ± 5.9%, respectively, by 2080–2099 relative to 1986–2005. The maximum catch potential is projected to decrease by 3.4% to 6.4% (RCP2.6) and 20.5% to 24.1% (RCP8.5) in the 21st century. {5.4.1}

Projected decreases in global marine animal biomass and fish catch potential could elevate the risk of impacts on income, livelihood and food security of the dependent human communities (*medium confidence*). Projected climate change impacts on fisheries also increase the risk of potential conflicts among fishery area users and authorities or among different communities within the same country (*medium confidence*), exacerbated through competing resource exploitation from international actors and mal-adapted policies (*low confidence*). {5.2.3, 5.4, 5.5.3}

Projected decrease in upper ocean export of organic carbon to the deep seafloor is expected to result in a loss of animal biomass on the deep seafloor by 5.2–17.6% by 2090–2100 compared to the present (2006–2015) under RCP8.5 with regional variations (*medium confidence*). Some increases are projected in the polar regions, due to enhanced stratification in the surface ocean, reduced primary production and shifts towards small phytoplankton (*medium confidence*). The projected impacts on biomass in the abyssal seafloor are larger under RCP8.5 than RCP4.5 (*very likely*). The increase in climatic hazards beyond thresholds of tolerance of deep sea organisms will increase the risk of loss of biodiversity and impacts on functioning of deep water column and seafloor that is important to support ecosystem services, such as carbon sequestration (*medium confidence*). {5.2.4}

Structure and functions of all types of coastal ecosystems will continue to be at moderate to high risk under the RCP2.6 scenario (*medium confidence*) and will face high to very high risk under the RCP8.5 scenario (*high confidence*) by 2100. Seagrass meadows (*high confidence*) and kelp forests (*high confidence*) will face moderate to high risk at temperature above 1.5°C global sea surface warming. Coral reefs will face very high risk at temperatures 1.5°C of global sea surface warming (*very high confidence*). Intertidal rocky shores are also expected to be at very high risk (transition above 3°C) under the RCP8.5 scenario (*medium confidence*). These ecosystems have low to moderate adaptive capacity, as they are highly sensitive to ocean temperatures and acidification. The ecosystems with moderate to high risk (transition above 1.8°C) under future emissions scenarios are mangrove forests, sandy beaches, estuaries and salt marshes (*medium confidence*). Estuaries and sandy beaches are subject to highly dynamic hydrological and geomorphological processes, giving them more natural adaptive capacity to climate hazards. In these systems, sediment relocation, soil accretion and landward expansion of vegetation may initially mitigate against flooding and habitat loss, but salt marshes in particular will be at very high risk in the context of SLR and extreme climate-driven erosion under RCP8.5. {5.3, Figure 5.16}

Expected coastal ecosystem responses over the 21st century are habitat contraction, migration and loss of biodiversity and functionality. Pervasive human coastal disturbances will limit natural ecosystem adaptation to climate hazards (*high confidence*). Global coastal wetlands will lose between 20–90% of their area

depending on emissions scenario with impacts on their contributions to carbon sequestration and coastal protection (*high confidence*). Kelp forests at low-latitudes and temperate seagrass meadows will continue to retreat as a result of intensified extreme temperatures, and their low dispersal ability will elevate the risk of local extinction under RCP8.5 (*high confidence*). Intertidal rocky shores will continue to be affected by ocean acidification, warming, and extreme heat exposure during low tide emersion, causing reduction of calcareous species and loss of ecosystem biodiversity and complexity shifting towards algae dominated habitats (*high confidence*). Salinisation and expansion of hypoxic conditions will intensify in eutrophic estuaries, especially in mid and high latitudes with microtidal regimes (*high confidence*). Sandy beach ecosystems will increasingly be at risk of eroding, reducing the habitable area for dependent organisms (*high confidence*). {5.3, 5.4.1}

Almost all coral reefs will degrade from their current state, even if global warming remains below 2°C (*very high confidence*), and the remaining shallow coral reef communities will differ in species composition and diversity from present reefs (*very high confidence*). These declines in coral reef health will greatly diminish the services they provide to society, such as food provision (*high confidence*), coastal protection (*high confidence*) and tourism (*medium confidence*). {5.3.4, 5.4.1}

Multiple hazards of warming, deoxygenation, aragonite under-saturation and decrease in flux of organic carbon from the surface ocean will decrease calcification and exacerbate the bioerosion and dissolution of the non-living component of cold water coral. Habitat-forming, cold water corals will be vulnerable where temperature and oxygen exceed the species' thresholds (*medium confidence*). Reduced particulate food supply is projected to be experienced by 95% of cold water coral ecosystems by 2100 under RCP8.5 relative to the present, leading to a *very likely* range of 8.6 ± 2% biomass loss (*medium confidence*). {5.2.4, Box 5.2}

Anthropogenic changes in EBUS will emerge primarily in the second half of the 21st century (*medium confidence*). EBUS will be impacted by climate change in different ways, with strong regional variability with consequences for fisheries, recreation and climate regulation (*medium confidence*). The Pacific EBUS are projected to have calcium carbonate undersaturation in surface waters within a few decades under RCP8.5 (*high confidence*); combined with warming and decreasing oxygen levels, this will increase the impacts on shellfish larvae, benthic invertebrates and demersal fishes (*high confidence*) and related fisheries and aquaculture (*medium confidence*). The inherent natural variability of EBUS, together with uncertainties in present and future trends in the intensity and seasonality of upwelling, coastal warming and stratification, primary production and biogeochemistry of source waters poses large challenges in projecting the response of EBUS to climate change and to the adaptation of governance of biodiversity conservation and living marine resources in EBUS (*high confidence*). {Box 5.3}

Climate change impacts on ecosystems and their goods and services threatens key cultural dimensions of lives and livelihoods. These threats include erosion of Indigenous and non-indigenous culture, their knowledge about the ocean and knowledge transmission, reduced access to traditional food, loss of opportunities for aesthetic and spiritual appreciation of the ecosystems, and marine recreational activities (*medium confidence*). Ultimately, these can lead to the loss of part of people's cultural identity and values beyond the rate at which identify and values can be adjusted or substituted (*medium confidence*). {5.4.2}

Climate change increases the exposure and bioaccumulation of contaminants such as persistent organic pollutants and mercury (*medium confidence*), and their risk of impacts on marine ecosystems and seafood safety (*high agreement, medium evidence, medium confidence*). Such risks are particularly large for top predators and for human communities that have high consumption on these organisms, including coastal Indigenous communities (*medium confidence*). {5.4.2}

Shifting distributions of fish stocks between governance jurisdictions will increase the risk of potential conflicts among fishery area users and authorities or different communities within the same country (*medium confidence*). These fishery governance related risks are widespread under high emissions scenarios with regional hotspots (*medium confidence*), and highlight the limits of existing natural resource management frameworks for addressing ecosystem change (*high confidence*). {5.2.5, 5.4.2.1.3, 5.5, 5.5.2}

Response options to enhance resilience

There is clear evidence for observed climate change impacts throughout the ocean with consequences for human communities and require options to reduce risks and impacts. Coastal blue carbon can contribute to mitigation for many nations but its global scope is modest (offset of <2% of current emissions) (*likely*). Some ocean indices are expected to emerge earlier than others (e.g., warming, acidification and effects on fish stocks) and could therefore be used to prioritise planning and building resilience. The survival of some keystone ecosystems (e.g., coral reefs) are at risk, while governance structures are not well-matched to the spatial and temporal scale of climate change impacts on ocean systems. Ecosystem restoration may be able to locally reduce climate risks (*medium confidence*) but at relatively high cost and effectiveness limited to low emissions scenarios and to less sensitive systems (*high confidence*). {5.2, 5.3, 5.4, 5.5}

Coastal blue carbon ecosystems, such as mangroves, salt marshes and seagrasses, can help reduce the risks and impacts of climate change, with multiple co-benefits. Some 151 countries around the world contain at least one of these coastal blue carbon ecosystems and 71 countries contain all three. Below-ground carbon storage in vegetated marine habitats can be up to 1000 tC ha^{-1}, much higher than most terrestrial ecosystems (*high confidence*). Successful implementation of measures to maintain and promote carbon storage in such coastal ecosystems could assist several countries in achieving a balance between emissions and removals of greenhouse gases (*medium confidence*). Conservation of these habitats would also sustain the wide range of ecosystem services they provide and assist with climate adaptation through improving critical habitats for biodiversity, enhancing local fisheries production, and protecting coastal communities from SLR and storm events (*high confidence*). The climate mitigation effectiveness of other natural carbon removal processes in coastal waters, such as seaweed ecosystems and proposed non-biological marine CO_2 removal methods, are smaller or currently have higher associated uncertainties. Seaweed aquaculture warrants further research attention. {5.5.1.1, 5.5.1.1, 5.5.1, 5.5.2, 5.5.1.1.3, 5.5.1.1.4}

The potential climatic benefits of blue carbon ecosystems can only be a very modest addition to, and not a replacement for, the very rapid reduction of greenhouse gas emissions. The maximum global mitigation benefits of cost-effective coastal wetland restoration is *unlikely* to be more than 2% of current total emissions from all sources. Nevertheless, the protection and enhancement of coastal blue carbon can be an important contribution to both mitigation and adaptation at the national scale. The feasibility of climate mitigation by open ocean fertilisation of productivity is limited to negligible, due to the likely decadal-scale return to the atmosphere of nearly all the extra carbon removed, associated difficulties in carbon accounting, risks of unintended side effects and low acceptability. Other human interventions to enhance marine carbon uptake, for example, ocean alkalinisation (enhanced weathering), would also have governance challenges, with the increased risk of undesirable ecological consequences (*high confidence*). {5.5.1.2}

Socioinstitutional adaptation responses are more frequently reported in the literature than ecosystem-based and built infrastructure approaches. Hard engineering responses are more effective when supported by ecosystem-based adaptation approaches (*high agreement*), and both approaches are enhanced by combining with socioinstitutional approaches for adaptation (*high confidence*). Stakeholder engagement is necessary (*robust evidence, high agreement*). {5.5.2}

Ecosystem-based adaptation is a cost-effective coastal protection tool that can have many co-benefits, including supporting livelihoods, contributing to carbon sequestration and the provision of a range of other valuable ecosystem services (*high confidence*). Such adaptation does, however, assume that the climate can be stabilised. Under changing climatic conditions there are limits to the effectiveness of ecosystem-based adaptation, and these limits are currently difficult to determine. {5.5.2.1}

Socioinstitutional adaptation responses, including community-based adaptation, capacity-building, participatory processes, institutional support for adaptation planning and support mechanisms for communities are important tools to address climate change impacts (*high confidence*). For fisheries management, improving coordination of integrated coastal management and marine protected areas (MPAs) have emerged in the literature as important adaptation governance responses (*robust evidence, medium agreement*). {5.5.2.2, 5.5.2.6}

Observed widespread decline in warm water corals has led to the consideration of alternative restoration approaches to enhance climate resilience. Approaches, such as 'coral reef gardening' have been tested, and ecological engineering and other approaches such as assisted evolution, colonisation and chimerism are being researched for reef restoration. However, the effectiveness of these approaches to increase resilience to climate stressors and their large-scale implementation for reef restoration will be limited unless warming and ocean acidification are rapidly controlled (*high confidence*). {Box 5.5, 5.5.2}

Existing ocean governance structures are already facing multi-dimensional, scale-related challenges because of climate change. This trend of increasing complexity will continue (*high confidence*). The mechanisms for the governance of marine Areas Beyond National Jurisdiction (ABNJ), such as ocean acidification, would benefit from further development (*high confidence*). There is also scope to increase the overall effectiveness of international and national ocean governance regimes by increasing cooperation, integration and widening participation (*medium confidence*). Diverse adaptations of ocean related governance are being tried, and some are producing promising results. However, rigorous evaluation is needed of the effectiveness of these adaptations in achieving their goals. {5.5.3}

There are a broad range of identified barriers and limits for adaptation to climate change in ecosystems and human systems (*high confidence*). Limitations include the space that ecosystems require, non-climatic drivers and human impacts that need to be addressed as part of the adaptation response, the lowering of adaptive capacity of ecosystems because of climate change, and the slower ecosystem recovery rates relative to the recurrence of climate impacts, availability of technology, knowledge and financial support and existing governance structures (*medium confidence*). {5.5.2}

5.1   Introduction

The ocean is a key component of the Earth system (Chapter 1) as it provides essential life supporting services (Inniss et al. 2017). For example, it stores heat trapped in the atmosphere caused by increasing concentrations of greenhouse gases, it masks and slows surface warming, it stores excess carbon dioxide and is an important component of global biogeochemical cycles. The ocean is the home to the largest continuous ecosystem, provides habitats for rich marine biodiversity, is an essential source of food and contributes to human health, livelihood and security. The ocean also supports other services to humans, for example, transport and trade, tourism, renewable energy, and cultural services such as aesthetic appeal, local and traditional knowledge and religious practices. Governance of the ocean has a unique set of challenges and opportunities compared with land systems and requiring different treatment under a changing climate.

AR5 from Working Group I (WGI) showed that there are ongoing changes to the physical and chemical state of the ocean. The AR5 WGI report (IPCC, 2013) concluded that (1) 'ocean warming dominates the increased energy stored in the climate system with more than 90% of the energy accumulated since 1971'; (2) 'the ocean has absorbed about 30% of the emitted anthropogenic carbon causing ocean acidification' since pre-industrial times; and (3) it is '*extremely likely* that human influence has been the dominant cause of warming since mid 20th century'.

The IPCC AR5 Working Group II (WGII) concluded that changes in the ocean such as warming, acidification and deoxygenation are affecting marine life from molecular processes to organisms and ecosystems, with major impacts on the use of marine systems by human societies (Pörtner, 2012). IPCC Special Report on the Impacts of Global Warming of 1.5°C above pre-industrial levels and related global greenhouse gas emission pathways (SR15) also concluded that reducing these risks by 'limiting warming to 1.5°C above pre-industrial levels would require transformative systemic change, integrated with sustainable development' and that 'adaptation needs will be lower in a 1.5°C world compared to a 2°C world.' (de Coninck et al. 2018; Hoegh-Guldberg et al. 2018).

This report updates earlier assessments, evaluating new research and knowledge regarding changing ocean climate and ecosystems, risks to ecosystem services, and vulnerability of the dependent communities including governance. It also delves into changes the ocean that were beyond the scope of the previous reports. Radiation management techniques (also known as sunlight reflection methods) are excluded here. Such geo-engineering approaches are addressed in the SR15. However, natural carbon uptake and stores in the marine environment are included (Section 5.5.1).

The chapter is structured around three guiding questions:

- What are the key changes in the physical and biogeochemical properties of the ocean? (Section 5.2.2)
- How have these changes impacted key ecosystems, risks to ecosystems services and human well-being? (Section 5.2.3, 5.2.4, 5.3, 5.4)

- Are there effective pathways for adaptation and nature-based solutions to risk reduction for marine dependent communities? (Section 5.5)

This chapter covers both regional and global scales and across natural and human systems. Chapter 3 covers the polar regions, including their oceans, Chapter 4 covers sea level rise and its implications, and Chapter 6 covers extremes and abrupt events. This chapter uses IPCC calibrated language around scientific uncertainty, as described in Section 1.8.3. Two emissions scenarios, RCP2.6 and RCP8.5, are used for projections of climate change (see Cross-Chapter Box 1 in Chapter 1).

5.2   Changing Ocean and Biodiversity

5.2.1   Introduction

This section assesses changes in the ocean. It includes the physical and chemical properties (Section 5.2.2), their impacts on the pelagic ecosystem (Section 5.2.3) and deep seafloor system (Section 5.2.4). In this assessment, the open ocean and deep seafloor includes areas where the water column is deeper than 200 m; it is the main subject of Section 5.2. Coastal and shelf seas are primarily discussed in Section 5.3.

5.2.2   Changes in Physical and Biogeochemical Properties

5.2.2.1   Introduction to Changing Open Ocean

The ocean is getting progressively warmer, with parallel changes in ocean chemistry such as acidification and oxygen loss, as documented in the AR5 (Rhein et al. 2013). The global scale warming and acidification trends are readily detectable in oceanic observations, well understood scientifically, and consistently projected by ESMs. Each of these has been directly attributed to anthropogenic forcing from changing concentrations of greenhouse gases and aerosols (Bindoff et al. 2013). These trends in the global average ocean temperature will continue for centuries after the anthropogenic forcing is stabilised (Collins et al. 2013).

The impacts on ocean ecosystems and human societies are primarily driven by regional trends and by the local manifestation of the global-scale changes. At these smaller scales, the temperature, acidification, salinity, nutrient and oxygen concentrations in the ocean are also expected to exhibit basin and local-scale changes. However, the ocean also has significant natural variability at basin and local-scales with time scales from minutes to decades and longer (Rhein et al. 2013), which can mask the underlying observed and projected trends (see Box 5.1). The impact of multiple stressors on marine ecosystems is one of the main subjects of this chapter (Section 5.2.3, 5.2.4, 5.3), including new evidence and understanding since the last assessment report (e.g., Gunderson et al. 2016). The most severe impacts of a changing climate will typically be experienced when conditions are driven outside the range of previous experience at rates that are faster than human or ecological systems can adapt (Pörtner et al. 2014; Box 5.1).

This section summarises our emerging understanding of the primary changes to the ocean, along with an assessment of several key areas of scientific uncertainty about these changes. Because many of these long-term trends have already been extensively discussed in previous assessments (IPCC, 2013), much of this summary of the physical changes is brief except where there are significant new findings.

5.2.2.2   Changing Temperature, Salinity, Circulation

Historically, scientific research expeditions starting in the 19th century have provided occasional sections measuring deep ocean properties (Roemmich et al. 2012). Greater spatial and temporal coverage of temperatures down to about 700 m was obtained using expendable bathythermographs along commercial shipping tracks starting in the 1970s (Abraham et al. 2013). Since the early 2000s, thousands of autonomous profiling floats (Argo floats) have provided high-quality temperature and salinity profiles of the upper 2000 m in ice-free regions of the ocean (Abraham et al. 2013; Riser et al. 2016). Further advances in autonomous floats have been developed that now allow these floats to operate in seasonally ice covered oceans (Wong and Riser, 2011; Wong and Riser, 2013), and more recently to profile the entire depth of the water column down to 4000 or 6000 m (Johnson et al. 2015; Zilberman, 2017) and to include biogeochemical properties (Johnson et al. 2017). Autonomous floats have revolutionised our sampling and accuracy of the global ocean temperature and salinity records and increased certainty and confidence in global estimates of the earth heat (temperature) budget, particularly since 2004 (Von Schuckmann et al. 2014; Roemmich et al. 2015; Riser et al. 2016), as demonstrated by the convergence of observational estimates of the changes in the heat budget of the upper 2000 m (Figure 5.1). New findings using data collected from such observing platforms mark significant progress since AR5.

To understand the recent and future climate, we use ensembles of coupled ocean-atmosphere-cryosphere-ecosystem models (ESMs) with the full-time history of atmospheric forcing (greenhouse gases, aerosols, solar radiation and volcanic eruptions) for the historical period and projections of the concentrations or emissions of these forcings to 2100. For these projections the RCPs of atmospheric emissions scenarios are used as specified by the Coupled Model Intercomparison Project, Phase 5 (CMIP5) (see Section 1.8.2.3, Cross-Chapter Box 1, and also IPCC AR5)[3]. This chapter focuses on the low and high emissions scenarios RCP2.6 and RCP8.5, respectively. When these scenarios are used to drive ESMs, it is possible to simulate the recent and future patterns of changes in the ocean temperature, salinity and circulation (and other oceanic properties such as ocean oxygen concentration and acidification, Section 5.2.2.3 and 5.2.2.4). Finally, the projections of ocean changes also informs the detection,

attribution and projection of risk and impacts on ecosystems (Sections 5.2.3, 5.2.4 and 5.3), ecosystem services (Section 5.4.1) and human well-being (Section 5.4.2) under climate change.

5.2.2.2.1   Observed and projected global ocean heat uptake

As AR5 concluded, the ocean is warming as a direct result of anthropogenic changes to the radiative properties of the atmosphere and the heat budget of the Earth (*very likely*) (Bindoff et al. 2013). Over the past few decades our ocean observing system has measured an increase in ocean temperature (Figure 5.1). This temperature increase corresponds to an uptake of over 90% of the excess heat accumulated in the Earth system over this period (Bindoff et al. 2013; Rhein et al. 2013). This heat in the ocean also causes it to expand and has contributed about 43% of the observed global mean SLR from 1970–2015 (Section 4.2.2.3.6).

Since AR5, there have been further improvements in our ability to understand and correct instrumental errors and new estimates also attempt to minimise biases in estimating temperature changes arising from traditional data-void filling strategies (Abraham et al. 2013; Durack, 2015; Cheng and Chen, 2017; Cheng et al. 2017). New estimates from ocean observations of ocean heat uptake in the top 2000 m between 1993 and 2017 *very likely* range from 9.2 ± 2.3 ZJ yr^{-1} to 12.1 ± 3.1 ZJ yr^{-1} (Johnson et al. 2018)[4]. Three recent independent estimates do a better job of accounting for instrumental biases and the sparseness of historical ocean temperature measurements than the older studies assessed in AR5, and provide larger and more consistent estimates of heat uptake rates for the 0–2000 m layer of 5.8 ± 1.0 ZJ yr^{-1} (Cheng and Chen, 2017; Cheng et al. 2017; Ishii et al. 2017), 6.0 ± 0.8 ZJ yr^{-1} (updated from Domingues et al. (2008)) and 6.3 ± 1.8 ZJ yr^{-1} (Cheng and Chen, 2017; Cheng et al. 2017; Ishii et al. 2017) for the 1971–2010 period assessed by AR5.

Based on these new published methods and revised atlases we update the estimates for ocean heat uptake (Table 5.1, and SM5.1). For all of the periods assessed in Table 5.1, it is *virtually certain* that the upper ocean (0–700 m) has warmed. These results are consistent with earlier research into the duration of record needed to detect a significant signal in global ocean heat content (Gleckler et al. 2012). Critically, the *high confidence* and *high agreement* in the ocean temperature data means we can detect discernable rates of increase in ocean heat uptake (Gleckler et al. 2012; Cheng et al. 2019). The rate of heat uptake in the upper ocean (0–700 m) is *very likely* higher in the 1993–2017 (or 2005–2017) period compared with the 1969–1993 period (see Table 5.1). The deeper layer (700–2000 m) heat uptake rate is *likely* to be higher in the 1993–2017 period compared with the 1969–1993 period.

---

[3]   The 30 CMIP5 ESMs used in here in various contexts were selected based on the availability of ocean data from the historical period, RCP2.6 and RCP8.5 projections, and corresponding control runs to correct for model drift. The models used include: ACCESS1.0, ACCESS1.3, BNU-ESM, BCC-CSM1-1, CCSM4, CESM1, CMCC-CESM, CMCC-CMS, CNRM-CM5, CSIRO-Mk3, CanESM2, FGOALS-S2.0, GFDL-CM3, GFDL-ESM2G, GFDL-ESM2M, GISS-E2-H, GISS-E2-R, HadGEM2-AO, HadGEM2-CC, HadGEM2-ES, INM-CM4, IPSL-CM5A-LR, IPSL-CM5A-MR, IPSL-CM5B-LR, MIROC-ESM, MIROC5, MPI-ESM-LR, MPI-ESM-MR, MRI-CGCM3, and NorESM1-M. Up to 3 ensemble members or variants were included per model, and all changes are relative to a control run with an identical initial condition but with preindustrial forcing. A table with a description and citations for each of these models, along with more detailed discussion of the use of ESM output, can be found in Flato et al. (2013).

[4]   ZJ is Zettajoule and is equal to 10^{21} Joules. Warming the entire ocean by 1°C requires about 5500 ZJ; 144 ZJ would warm the top 100 m by about 1°C.

Table 5.1 | The assessed rate of increase in ocean heat content in the two depth layers 0–700 m and 700–2000 m and their *very likely* ranges. Fluxes in W m⁻² are averaged over the Earth's entire surface area. The four periods cover earlier and more recent trends; the 2005–2017 period has the most complete interior ocean data coverage and the greatest consistency between estimates, while longer trends are better for distinguishing between forced changes and internal variability. These observationally-estimated rates come from an assessment of the recent research (see SM5.1), while the Coupled Model Intercomparison Project Phase 5 (CMIP5) Earth System Models (ESM) estimates are based on a combined 28-member ensemble of historical, Representative Concentration Pathway (RCP)2.6 and RCP8.5 simulations.

	Ocean Heat Uptake Rate, ZJ yr⁻¹				Ocean Heat Uptake as Average Fluxes, W m⁻²			
Period	1969–1993	1993–2017	1970–2017	2005–2017	1969–1993	1993–2017	1970–2017	2005–2017
Observationally Based Ocean Heat Uptake Estimates:								
0–700 m	3.22 ± 1.61	6.28 ± 0.48	4.35 ± 0.80	5.31 ± 0.48	0.20 ± 0.10	0.39 ± 0.03	0.27 ± 0.05	0.33 ± 0.03
700–2000 m	0.97 ± 0.64	3.86 ± 2.09	2.25 ± 0.64	4.02 ± 0.97	0.06 ± 0.04	0.24 ± 0.13	0.14 ± 0.04	0.25 ± 0.06
CMIP5 ESM Ensemble-mean Ocean Heat Uptake with 90% Certainty Range from Ensemble Spread:								
0–700 m	3.60 ± 1.92	7.37 ± 2.09	5.64 ± 1.90	7.85 ± 2.71	0.22 ± 0.12	0.46 ± 0.13	0.35 ± 0.12	0.49 ± 0.17
700–2000 m	1.32 ± 1.49	2.72 ± 1.41	1.99 ± 1.51	3.33 ± 1.75	0.08 ± 0.09	0.17 ± 0.09	0.12 ± 0.09	0.21 ± 0.11

The direct comparison of the observed changes in ocean heat content and the simulated historical changes is undertaken to detect climate change, to attribute the causes of climate change to the forcings in the system, and to evaluate the performance of ESMs. Attribution studies also reject competing hypotheses to explain the global ocean changes such as natural forcing from solar variability or volcanic eruptions (see Section 1.3) (Bindoff et al. 2013). Detection and attribution studies have since been used to detect changes in the rate of ocean heat uptake and to attribute these changes to human activity (Gleckler et al. 2016).

Updated observationally-based estimates of ocean heat uptake are consistent with simulations of equivalent time-periods from an ensemble of CMIP5 ESMs (Table 5.1 and the inset panel in Figure 5.1) (*high confidence*), once the limitations of the historical ocean observing network and the internally generated variability with a single realisation of the real world are taken into account (see Section 5.2.2.2). Following the CMIP5 protocol, the ESMs are radiatively forced with observationally derived estimates of greenhouse gas concentrations and aerosols, including natural forcing variations from volcanic eruptions and solar forcing, through 2005; after 2006 each of the ESMs uses either the RCP2.6 or RCP8.5 emissions scenarios.

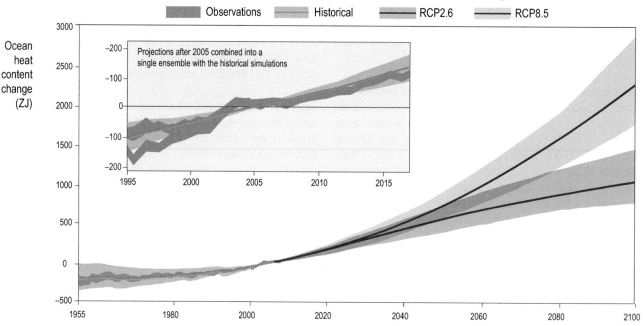

Simulated and observed global 0–200 m ocean heat content change

Figure 5.1 | Time series of globally integrated upper 2000 m ocean heat content changes in ZJ, relative to the 2000–2010 period average, as inferred from observations (magenta) and as simulated for historical (tan), Representative Concentration Pathway (RCP)2.6 (blue) and RCP8.5 (red) forcing by a 25-member ensemble of Coupled Model Intercomparison Project Phase 5 (CMIP5) Earth System Models (ESMs) (Cheng et al. 2019). The shaded magenta in the outer panel is the *very likely* range determined by combining data from 4 long-term estimates (Palmer et al. 2007; Levitus et al. 2012; Lyman and Johnson, 2014; Cheng and Chen, 2017; Cheng et al. 2017; Ishii et al. 2017) processed as in Johnson et al. (2018). The tan, blue and red lines are the ESM ensemble means, while shading shows each ensemble's 5th to 95th percentile range. In the inset subpanel, the four different shaded magenta areas are the reported *very likely* range of heat content changes as inferred from observations by four independent groups (Magenta shading; Palmer et al. 2007; Lyman and Johnson, 2014; Cheng and Chen, 2017; Cheng et al. 2017; Ishii et al. 2017) processed as in Johnson et al. (2018). In the inset subpanel the RCP2.6 and RCP8.5 projections after 2005 are combined into a single ensemble with the historical simulations.

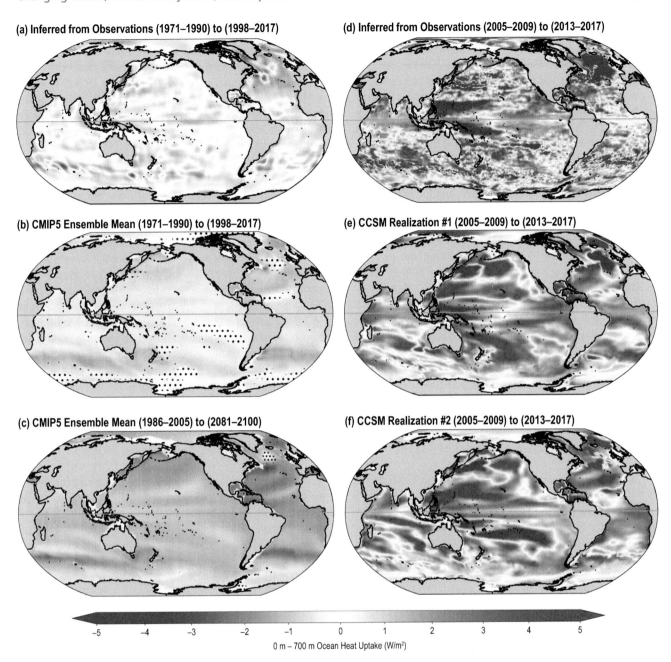

(a) Inferred from Observations (1971–1990) to (1998–2017)

(b) CMIP5 Ensemble Mean (1971–1990) to (1998–2017)

(c) CMIP5 Ensemble Mean (1986–2005) to (2081–2100)

(d) Inferred from Observations (2005–2009) to (2013–2017)

(e) CCSM Realization #1 (2005–2009) to (2013–2017)

(f) CCSM Realization #2 (2005–2009) to (2013–2017)

-5 -4 -3 -2 -1 0 1 2 3 4 5

0 m – 700 m Ocean Heat Uptake (W/m²)

Figure 5.2 | Heat uptake by the top 700 m of the ocean, as determined by differences between the averages over two 5- or 20-year intervals converted to a heat flux into the ocean (W m⁻²), either from observationally-based analyses or a 38-member ensemble of Coupled Model Intercomparison Project Phase 5 (CMIP5) Earth System Models (ESMs). **(a)** Change between (1971–1990) and (1998–2017) as inferred from observations (Good et al. 2013); **(b)** The ensemble mean change in CMIP5 ESMs for the same time periods as in (a); **(c)** Projected ensemble mean change in CMIP5 ESMs between (1986–2005) and (2081–2100) for the RCP8.5 forcing scenario. In panels (b) and (c), stippling indicates regions where the ensemble mean change is not significantly different from 0 at the 95% confidence level based on the models' temporal variability. **(d)** Change between (2004–2008) and (2013–2017) as inferred from observations by the SODA 3.4.2 reanalysis product (Carton et al. 2018); **(e)** and **(f)** Estimates of change in heat uptake as in (d) but from two individual realisations of the CCSM ESM (Table SM5.2). These two realisations are identical apart from their initial conditions, which leads to different timing in their internal modes of variability; they were selected from the full CMIP5 ensemble as examples where one is reminiscent of the recent observed changes while the other has regional changes that have dissimilar timing.

The *very likely* ranges of the observed trends of heat uptake for the four periods and two layers all fall within the *very likely* range of simulated heat uptake from the ESM ensemble (Table 5.1). The difference between observations and average of the simulations in the upper ocean is an overestimate of heat uptake by about 20% and for the deeper layer there an underestimate by a similar amount, but this difference is still well within the *very likely* range from the ensemble of simulations. The overall consistency between observationally-based estimates and

ESM simulations of the historical period gives greater confidence in the projections; it is *very likely* that historical simulations agree with observations of the global ocean heat uptake (Table 5.1).

While the collection of the worlds' ESMs have been criticised for having an ensemble mean that does not exhibit the observed 'hiatus' or 'slowdown' of global mean surface temperature increase in the early 21st century (Meehl et al. 2011; Trenberth et al. 2016), it is increasingly clear that this is at least in part due to the redistribution of heat within the climate system from the surface into the interior ocean and between ocean basins. Individual realisations of ESMs do show decades with slow increases in mean surface temperature change comparable to what was observed, even though these cases exhibit continued interior ocean heat uptake, and every ensemble member exhibits surface warming closer to the ensemble-mean over multi-decadal timescales (Meehl et al. 2011; England et al. 2015; Knutson et al. 2016).

The ocean will continue to take up heat in the coming decades for all plausible scenarios. As depicted in Figure 5.1, the ensemble of CMIP5 ESMs used by Cheng et al. (2019) project that under RCP2.6, the top 2000 m of the ocean will take up 935 ZJ of heat between 2015 and 2100 (with a *very likely* range of 650–1340 ZJ based on the 5th and 95th percentiles of the 25 ESMs used here that have available data from the historical, scenario and control runs for RCP2.6). Under RCP8.5 this ensemble projects heat uptake of 2180 ZJ (with a *very likely* range of 1710–2790 ZJ, based on 35 ESMs) between 2015 and 2100. By 2100 the ocean is *very likely* to warm by 2 to 4 times as much for low emissions (RCP2.6) and 5 to 7 times as much for the high emissions scenario (RCP8.5) compared with the observed changes since 1970. With the RCP8.5 scenario, the ocean is *very likely* to take up about twice as much heat as RCP2.6 (Figure. 5.1). Even under RCP2.6 the ocean will continue to warm for several centuries to come (Collins et al. 2013). It is *virtually certain* that the ocean will continue to take up heat throughout the 21st century, and the rate of uptake will depend upon on the emissions scenario we collectively choose to follow.

5.2.2.2.2   Structure of anthropogenic climate changes in the ocean

The ensemble average of the CMIP5 ESMs projects widespread ocean warming over the coming century, concentrated in the upper ocean (Figures 5.2c and 5.3) (Kuhlbrodt and Gregory, 2012). The anthropogenic heat will penetrate into the ocean following well-established circulation pathways (Jones et al. 2016a). The greatest vertically integrated heat uptake occurs where there is already the formation of interior waters, such as Antarctic Intermediate Water along the Antarctic Circumpolar Current (Frölicher et al. 2015) or NADW precursors in the Nordic Seas (Figure 5.2c), but all water-masses[5] that are subducted over decades are expected to experience significant warming (see Figure 5.3). The warming in the subtropical gyres penetrates deeper into the ocean than other gyres (roughly 15°N–45°N and 15°S–45°S in Figure 5.3), following the wind-driven bowing down of the density surfaces (the solid lines in Figure 5.3) in these gyres (Terada and Minobe, 2018). The greater warming at 700–2000 m in the Atlantic than the Pacific or Indian Oceans (Figure 5.3) reflects the strong southward transport of recently formed NADW at these depths by the AMOC. Two areas that commonly exhibit substantially reduced near-surface warming over the course of the 21st century are the northern north Atlantic, where a slowing AMOC (see Section 6.7.1.1) reduces the northward heat transport and brings the surface temperatures closer to what is found in other ocean basins at these latitudes (Collins et al. 2013), and the southern side of the Southern Ocean, where water upwells

that has been submerged for so long that it has not yet experienced significant anthropogenic climate change (Armour et al. 2016). Most of these projected warming patterns are broadly consistent across the current and previous generations of climate models (Mitchell et al. 1995; Collins et al. 2014) as well as observations and theoretical understanding. These multiple lines of evidence give *high confidence* that the projections describe the changes in the real world (*high agreement, robust evidence*).

The near surface salinity of the ocean is both observed and projected to evolve in ways that reflect the increased intensity of the Earth's hydrologic cycle (Durack, 2015) and the increasing near-surface ocean stratification (Zika et al. 2018). As described in WGI AR5, the ocean surface in areas that currently have net evaporation are expected to become saltier, while areas with net precipitation are expected to get fresher (Rhein et al. 2013), as the patterns of precipitation and evaporation are generally expected to be amplified (Held and Soden, 2006). At longer time-scales of decades, the larger scale changes in the ocean circulation and basin integrated freshwater imbalances emerge in the near-surface salinity changes, as shown in Figure 5.3b, with an increasingly salty tropical and subtropical Atlantic and Mediterranean contrasting with a freshening Pacific and polar Arctic emerging as robust signals across the suite of ESMs (Collins et al. 2013). The freshening of the high latitudes in the north Atlantic and Arctic basin is consistent with the widely expected weakening of the AMOC (also discussed in Section 6.7), hydrological cycle changes and a decline in the volume of sea ice (discussed in Section 3.2.2).

Projected salinity changes in the subsurface ocean reflect changes in the rates of formation of water masses or their newly formed properties (Purich et al. 2018). Thus, projected freshening of the Southern Ocean surface leads to a freshening of the Antarctic Intermediate Water that is subducted there, flowing northward from the Southern Ocean as a relatively fresh water-mass at depths of 500–1500 m (Figure 5.3b). Increased surface salinity in the Atlantic subtropical gyres are pumped into the interior by the winds, leading to an increased salinity of the interior subtropical gyres, along with contributions from increasingly salty Mediterranean water (Jordà et al. 2017). Conversely, freshwater capping of the northwestern north Atlantic is projected to inhibit deep convection in the Labrador Sea and the consequent production of Labrador Sea Water in some models (Collins et al. 2013), and contributes to the increased salinity of the north Atlantic between 1000–2000 m depths (Figure 5.3b).

Identifying the specific patterns of anthropogenic climate changes in oceanic observations is complicated by the presence of basin-scale natural variability with timescales ranging from tidal to multi-decadal, and due to the difficulties associated with maintaining high-precision observing systems spanning the ocean basins and limited observational coverage of the extratropical Southern Hemisphere before 2006 (Rhein et al. 2013). Inferences based on oceanographic observations from the 1970s onward show wide-spread warming of the upper 700 m (Figure 5.2a), in broad agreement with the

---

[5]   Following common oceanographic practice dating back to Helland-Hansen (1916) and discussed in detail by Sverdrup et al. (1942), an ocean water-mass is defined as a large volume of seawater with a characteristic range of temperature and salinity properties, typically falling along a line in temperature-salinity space, often with common formation processes and locations.

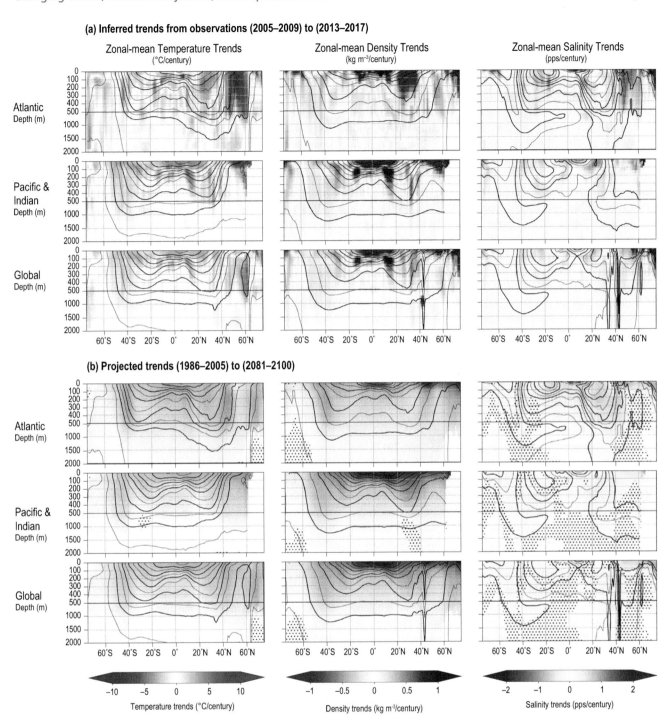

(a) Inferred trends from observations (2005–2009) to (2013–2017)

Zonal-mean Temperature Trends (°C/century)

Zonal-mean Density Trends (kg m⁻³/century)

Zonal-mean Salinity Trends (pps/century)

(b) Projected trends (1986–2005) to (2081–2100)

Temperature trends (°C/century)

Density trends (kg m⁻³/century)

Salinity trends (pps/century)

Figure 5.3 | Side-view basin-averaged zonal-mean trends (change per century) in water-mass properties in the top 2000 m by basin **(a)** as inferred from observations (average of 2013–2017 minus average of 2005–2009) and **(b)** Coupled Model Intercomparison Project Phase 5 (CMIP5) model projections with Representative Concentration Pathway (RCP)8.5 forcing (average of 2081–2100 minus average of 1981–2000) trends in water-mass changes forcing. Subpanels within each group: top-to-bottom (Atlantic, combined Pacific and Indian, Global); left-to-right (Temperature, *in situ* Density, Salinity). Shaded areas show where the projected changes are not statistically significant at the 95% level. This figure uses the same observationally-derived reanalysis datasets and ensemble of Earth System Models (ESMs) as in Figure 5.2c and 5.2d. Solid lines show present contours of these fields; the notable structure in the northern hemisphere of the global-zonal mean contours of density and salinity are due to the relatively salty Mediterranean and fresh Black seas.

ensemble of historical CMIP5 ESM simulations (Figure 5.2b). These ESMs indicate that anthropogenic regional warming over the past half-century should be discernable at the 95% confidence level in much of the upper oceans (un-stippled areas in Figure 5.2b). Most of the areas where observational analyses (Figure 5.2a) exhibit long-term cooling are either regions where the internally generated

variability is large enough to mask the trends (e.g., the Eastern Tropical Pacific, Northwest Atlantic, and Kurushio extension east of Japan, which are stippled in Figure 5.2b), or where the observational coverage early in the record is limited and different analyses can disagree about trends (e.g., the Southern Ocean and extratropical South Pacific). When internal variability is taken into account, the broad consistency in the magnitude and regional distribution of observed and simulated 50-year trends gives confidence to the ESM projections of longer-term oceanic changes described previously.

Detailed regional patterns of trends in temperature and heat content at depths of 0–2000 m during the early 21st century are consistent in various analysis, owing to the improved observing network (Roemmich et al. 2015; Desbruyères et al. 2016a) (Figure 5.2d). At depths of 700–2000 m, observations in all of the ocean basins show broadly warming trends in the well-observed Argo era (2006 to present), with particularly significant warming patterns in the Southern Hemisphere extratropics around 40°S and the subpolar north Atlantic (Figure 5.3a). These observed changes support the notion that deep ocean heat content has been continuously increasing. As a result, regional climate change signatures emerge from confounding natural variability sooner in the 700–2000 m depth range than in upper 700 m of the ocean, where interannual modes of variability have a larger influence on the circulation (for a more complete discussion see Johnson et al. (2018)). Despite regional patches of cooling water in the upper 700 m (Figure 5.2d), every one of the world's ocean basins volume averaged over depths of 0–2000 m has experienced significant warming over the last decade (Figure 5.3, and also Desbruyères et al. (2016a)). The greatest warming of the top 2000 m has been in the Southern Ocean (Roemmich et al. 2015; Trenberth et al. 2016), the tropical and subtropical Pacific Ocean (Roemmich et al. 2015), and the tropical and subtropical Atlantic Ocean (Cheng and Chen, 2017). The Southern Hemisphere extratropical oceans accounted for 67–98% of the total ocean heat increase in the uppermost 2000 m for the period of 2006–2013 (Roemmich et al. 2015). Shi et al. (2018) suggest that the dominant ocean heat uptake by the Southern Hemisphere in the early 21st century is expected to become more balanced between the hemispheres as the asymmetric cooling by aerosols decreases.

Large-scale patterns of natural variability at interannual to decadal time scales can mask the long-term warming trend in the upper 700 m, particularly in the tropical Pacific and Indian Oceans (England et al. 2014; Liu et al. 2016) and in the north Atlantic (Buckley and Marshall, 2015). The most significant upper 700 m warming between five-year averages centered on 2007–2015 occurred in a large extratropical band of the Southern Hemisphere between 30°S–60°S, and in the tropical Indian Ocean, the eastern North Pacific and western subtropical north Atlantic (Figure 5.2d). Warming of the southern hemisphere subtropical gyres is driven, in part, by an intensification of Southern Ocean winds in recent decades, facilitating the penetration of heat to deeper depths (Gao et al. 2018). Marginal seas, such as the Mediterranean and Red seas have also exhibited notable warming. Conversely, over this timeframe there were also regions of cooling in the upper 700 m, notably in the north Atlantic around 40°N–60°N and in the western tropical Pacific (Figure 5.2d). Recent relatively cold and fresh surface and subsurface conditions in the north Atlantic have been attributed to anomalous atmospheric forcing (Josey et al. 2018) or weakened transport by the north Atlantic Current and AMOC (Smeed et al. 2018), and in turn may have contributed to an intensification of deep convection in the Labrador Sea since 2012 (Yashayaev and Loder, 2017). All these observed decadal changes can be related to internal decadal variability (Robson et al. 2014;

Yeager et al. 2015) even though they resemble expected longer-term anthropogenically forced trends. Substantial decadal-scale warming and cooling trends in the tropical Pacific and Indian oceans can arise from natural El Niño-Southern Oscillation (ENSO) and Indian Ocean Dipole variability (Han et al. 2014). Large ensembles of freely running CMIP5 ESM simulations also show that internal variability can dominate the regional manifestation of the anthropogenic climate signal on decadal timescales (Kay et al. 2014). This is illustrated by the differing warming trends in Figure 5.2e and 5.2f from two identical ESMs that differ only in the weather in their 1850 initial conditions, averaged over the whole 21st century, by contrast, the ensemble of CMIP5 models project statistically significant anthropogenic regional upper 700 m heat content trends almost everywhere (Figure 5.2c).

There are well documented changes in observed ocean temperatures and salinities (Abraham et al. 2013; Ishii et al. 2017). However, attributing these changes in the state of the ocean to anthropogenic causes can be challenging due to the presence of internally generated variability, which can swamp the underlying climate change signal in short records and on regional scales. As can be seen in Figure 5.2, the observed long-term trends (Figure 5.2a) exhibit a striking similarity to the CMIP5 ensemble mean in areas where the models suggest that anthropogenic changes should be statistically significant (Figure 5.2b). However, the trends in the shorter well-observed period covering 2005–2017 (Figure 5.2d) exhibits strong trends from internal variability, as illustrated by the differences of two ensemble members of the same ESM with the same forcing but initialised with different weather (Figure 5.2e and 5.2f). Detection and Attribution studies take the internal variability into account and separate the underlying climate signals with the same spatio-temporal sampling as the observations, and apply a range of statistical tests to determine the coherence of the observations with the co-sampled observations (Bindoff et al. 2013; AR5 WG1 Box 10.1).

Since AR5, the use of different and updated oceanographic data sets and increase in the number of ensembles of the CMIP5 simulations (Kay et al. 2014) has improved the overall detection and attribution of human influence. Together these measures increase the coherence of the simulations and reduce noise. For example, an isotherm approach used to reduce the noise from the displacement of isotherms in the upper water column allowing detection in each of the mid-latitude ocean basins was achieved on 60-year time series (Weller et al. 2016). Using all the available ocean temperature and salinity profiles from the Southern Ocean, Swart et al. (2018) show that the warming and freshening patterns were consistent primarily with increased human induced greenhouse gases and secondarily from ozone depletion in the stratosphere, but inconsistent with internal variability. Together the evidence from the AR5, and the discussion above with the new evidence on regional scales across the global oceans, we conclude that the observed long-term upper ocean temperature changes are *very likely* to have a substantial contribution from anthropogenic forcing.

The wind-driven ocean circulation at the end of the 21st century is expected to be qualitatively similar to that in the present day, even as important buoyancy-loss driven overturning circulations are expected to weaken. ESM projections suggest that some major ocean current transports will exhibit a modest increase (such as the Kuroshio

Extension (Terada and Minobe, 2018) or a small decrease such as for the Indonesian Throughflow (Sen Gupta et al. 2016); many predominantly wind-driven current-system transports are expected to exhibit smaller than 20% changes by 2100 with RCP8.5 forcing. Climate-change induced changes of the circulation in other mid-latitude basins may be difficult to detect or reliably project because of significant natural variability at inter-annual (e.g., El Niño) to decadal (e.g., the Pacific Decadal Oscillation) timescales. The Antarctic Circumpolar Current is projected to be subject to strengthening westerly winds and substantially reduced rates of Antarctic Bottom Water (AABW) formation, as assessed in the Cross-Chapter Box 7 in Chapter 3. The heat transported by the buoyancy-loss driven AMOC, in particular, contributes to the relatively clement climate of northern Europe and the north Atlantic Basin as a whole, although the wind-driven ocean gyres also contribute to the meridional ocean heat transport (see the review by Buckley and Marshall (2015)). As a result, there is a concern that significant changes in ocean circulation could lead to localised climate changes that are much larger than the global mean. Projected and observed changes in the AMOC and the rates of formation of deep water-masses in the north Atlantic are discussed in Chapter 6.7.1, along with the possibility of abrupt or enduring changes resulting from forcing by Greenlandic meltwater. A significant reduction in AMOC would, in turn, modestly weaken the Gulf Stream transport, which also has a substantial wind driven component (Frajka-Williams et al. 2016). Most aspects of the large-scale wind-driven ocean circulation are *very likely* to be qualitatively similar to the circulation in the present day, with only modest changes in transports and current location.

The global ocean below 2000 m has warmed significantly between the 1980s and 2010s (Figure 5.4), contributing to ocean heat uptake and through thermal expansion to SLR (Purkey and Johnson, 2010; Desbruyères et al. 2016b). The observed deep warming rate varies regionally and by depth reflecting differences in the waters influencing particular regions. The deep and abyssal north Atlantic, fed by North Atlantic Deep Water (NADW), has reversed from warming to cooling over the past decade, possibly associated with the North Atlantic

Oscillation (NAO) (e.g., Yashayaev, 2007; Desbruyères et al. 2014) or longer-term weakening in north Atlantic overturning circulation (Caesar et al. 2018; Thornalley et al. 2018). The strongest warming is observed in regions of the deep ocean AABW (Purkey et al. 2014). Regions of the ocean fed by AABW from the Weddell Sea have exhibited a possible slowdown in local AABW warming rates (Lyman and Johnson, 2014), while the Pacific, fed by AABW from the shelves along the Ross and Adelie Coast, has continued to warm at an accelerating rate between 1990 and 2018 (Desbruyères et al. 2016b).

To date, assessment of deep ocean (below 2000 m) heat content has mostly been from ship-based data collected along decadal repeats of oceanographic transects (Figure 5.4b) (Talley et al. 2016). While relatively sparse in space and time compared to the upper ocean, these transects were positioned to optimise sampling of most deep ocean basins and provide the highest quality of salinity, temperature and pressure data. Argo floats capable of sampling to 6000 m have just started to populate select deep ocean basins; this Deep Argo data has just started providing regional deep ocean warming estimates (Johnson et al. 2019). Decadal monitoring by the full global Deep Argo array (Johnson et al. 2015), complemented by indirect estimates from space (Llovel et al. 2014; Von Schuckmann et al. 2014), will strongly reduce the currently large uncertainties of deep ocean heat content change estimates in the future.

The spatial and temporal sparseness of observations below 4000 m, along with significant differences between various ESMs, limits our understanding of the exact mechanisms driving the abyssal ocean variability. However, ESMs consistently predict an anthropogenic climate-change induced long-term abyssal warming trend originating in the Southern Ocean due to a reduction in the formation rates of cold AABW (Heuzé et al. 2015). Although the abyssal modes of natural variability are not as pronounced as closer to the surface, deep ocean heat content can vary on relatively short time scales through the communication of topographic and planetary waves driven by changes in the rate of deep water formation at high latitudes (Kawase, 1987;

(a)

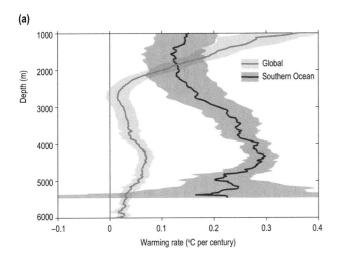

(b)

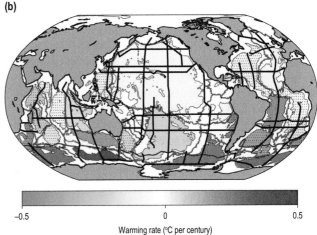

Figure 5.4 | Observed rates of warming from 1981 to 2019 (a) as a function of depth globally (orange) and south of the Sub-Antarctic Front (the purple line in (b) at about 55°S) (purple) with 90% confidence intervals and (b) average warming rate (colours) in the abyss (below 4000 m) over various ocean basins (whose boundaries are shown in grey lines), with stippling indicating basins with no significant changes. The black lines show the repeat hydrographic sections used to make these estimates. These figures use updated GoShip data and the techniques of Purkey and Johnson (2010).

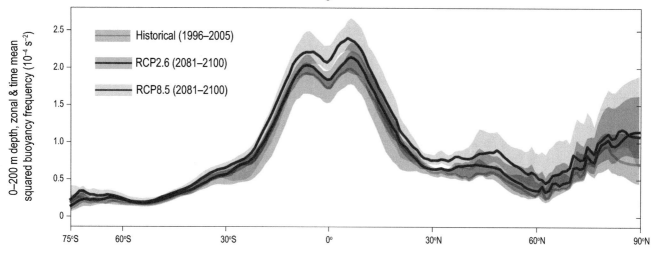

Zonal average 0–200 m stratification

Historical (1996–2005)

RCP2.6 (2081–2100)

RCP8.5 (2081–2100)

Figure 5.5 | Zonal and 20-year mean stratification averaged over the top 200 m of the ocean for the Coupled Model Intercomparison Project Phase 5 (CMIP5) ensemble of simulations at the end of the historical runs (green), and for the end of the 21st century for Representative Concentration Pathway (RCP)2.6 (blue) and RCP8.5 (red) scenarios. The values between the 5th and 95th percentiles of the ensembles are shaded, while the lines are the ensemble mean. These model results are not adjusted by the control-run, so the spread in the various estimates primarily reflect model formulation differences. The average squared buoyancy frequency shown here is nearly linearly proportional to the density difference between the surface and 200 m, and is a measure of the density stratification of the upper ocean.

Masuda et al. 2010; Spence et al. 2017). AABW has shown variability in properties and production rates over the past half century (Purkey and Johnson, 2013; Menezes et al. 2017). A slowdown in AABW formation rates may arise from freshening of shelf waters, changes in local winds driving cross shelf mixing, or larger scale dynamics controlling the spin up or down of Southern Ocean gyres influencing the density of outflowing waters over deep sills. Large-scale circulation changes can also alter the properties of the ambient water that is entrained as dense water descends along the Antarctic continental slopes (Spence et al. 2017). Evolving AABW properties may also reflect changes in deep Southern Ocean convection. The Weddell Polynya is a large opening in the wintertime ice of the Weddell Sea that is kept ice-free despite intense heat loss to the atmosphere by convective mixing bringing up warm and salty water from the deep ocean. (See Box 3.2 for a more extensive discussion of polynyas and the Weddell Polynya in particular). The Weddell Polynya was present in three of the first years of infrared satellite observations of wintertime sea ice concentrations in the mid-1970s, but it has been closed since 1976, only to reopen in 2016 and 2017. The prominent Weddell Polynya in the mid-1970s greatly increased the volume of the coldest waters in the deep Weddell Sea. Weddell Polynyas are documented to drive abyssal cold and salty signals and can spread thermal signals as waves further and faster than could be explained by slow advective signals (Martin et al. 2015; Zanowski and Hallberg, 2017); these waves do not directly heat individual water parcels, but instead warm the ocean where they cause the coldest deep layers to spread laterally and thin. However, recovery from the large Weddell polynya of the early 1970s can only explain about 20% of the observed abyssal warming trend (Zanowski et al. 2015).

The ocean's properties are changing most rapidly in the near surface waters that are more immediately exposed to atmospheric forcing. As a result of the surface-intensified warming, the upper few hundred meters of the ocean are becoming more stably stratified (Helm et al. 2011; Talley et al. 2016). The combination of surface

intensified warming and near-surface freshening at high latitudes leading to a projection of more intense near-surface stratification (the downward-increasing vertical gradient of density) across all ocean basins (Figures 5.3 and 5.5) is a robust result with a *high agreement* across successive generations of coupled climate models (Capotondi et al. 2012; Bopp et al. 2013). Based on the projected changes from individual models between 1986–2005 and 2081–2100, the mean stratification of the upper 200 m averaged between 60°S–60°N, normalised by the ensemble mean value from 1986–2005 will *very likely* increase by between 1.0–9.3% (with 95% confidence and a CMIP5 median change of 2.6%) for RCP2.6, and by between 12.2–30.0% (median value 21.2%) for RCP8.5. Inferences from oceanic observations (Good et al. 2013) suggest that the 20-year mean stratification averaged between 60°S–60°N and over the top 200 m *very likely* increased by between 2.18–2.42% from 1971–1990 to 1998–2017. By contrast, the bottom intensified warming in the abyss (see Figure 5.4) which is consistent with a slowing in the rate of AABW formation, is also associated with a reduction in the abyssal stratification of the ocean (Lyman and Johnson, 2014; Desbruyères et al. 2016b). Both of these changes have consequences for the evolving turbulence and ocean water-mass structure. Based on observational evidence, theoretical understanding and robust ESM projections, it is *very likely* that stratification in the upper few hundred meters of the ocean below the mixed layer will increase significantly in the 21st century over most ocean basins as a result of climate change, and abyssal stratification will *likely* decrease.

Many dynamical consequences of increased stratification are understood with *very high confidence* (see, for instance, Gill (1982) and Vallis (2017)). For the same turbulent kinetic energy dissipation, locally increased stratification reduces the turbulent vertical diffusivity of heat, salinity, oxygen and nutrients (see Section 5.2.2.2.4). Increased stratification in the tropics and subtropical gyres will *likely* lead to a net reduction in the vertical diffusivities of nutrients and other gases within the main thermocline, reducing the flux of nutrients into the euphotic zone and increasing the gradient in oxygen concentrations between the near surface ocean and the interior. Increasing upper ocean stratification (Figure 5.5) acts to restrict the depth of the ocean's surface mixed layer. Increasing stratification increases the buoyancy frequency and the lateral propagation speed of internal gravity waves and boundary waves by about half the percentage change of the stratification itself. Increasing stratification increases both the length of the internal deformation radius (a typical length scale in baroclinic eddy dynamics) and the horizontal scales of internal tides (see Section 5.2.2.2.3) proportionately with the changes in the internal gravity wave speeds. An increase in stratification will increase the lateral propagation of internal Rossby waves (which set up the basin-scale ocean density structure) proportionately. For the same forcing, increasing stratification reduces the geostrophically balanced slope of density surfaces, and hence the vertical extent of basin-scale wind-driven gyres or coastal upwelling circulations. The flattening of density surfaces by increased stratification inhibits advective exchange between the surface and interior ocean (Wang et al. 2015a), with consequences for the uptake of anthropogenic carbon (Section 5.2.2.3), the evolving oxygen distribution (Section 5.2.2.4) and the supply of nutrients to support primary production (Section 5.2.2.5).

5.2.2.2.3   Tides and coastal physical changes in a changing climate

Coastal systems are subject to the same large-scale warming trends as the open ocean, but the local response may be dominated by a complex of localised changes in factors such as circulation, mixing, river plumes or the seasonal upwelling of cold water. Using ESMs to project how these factors will interact often requires much finer resolution than is currently affordable in global models, however regional high-resolution models can be effective, especially in marginal seas like the Mediterranean with restricted interactions with the open ocean and that respond primarily to local forcing (Adloff et al. 2015). High resolution regional models have also been used to project robust localised ocean climate changes in wide shelf seas with more extensive interactions with the open ocean, like those in northwestern Europe (Tinker et al. 2016). The technical difficulties of using nested regional models are much greater in coasts adjacent to energetic large-scale currents like the Gulf Stream, Kuroshio, and Agulhas, and projecting detailed coastal climate change such places may require the use of expensive high resolution global models (Saba et al. 2016). These physical coastal changes have consequences that cascade through ecosystems to people, as is illustrated in detail for eastern boundary upwelling systems in Box 5.2.

Both human structures and ecological systems in the coastal zone are directly impacted by tidal amplitudes, which contribute to high-water levels and the tidal flushing rates of estuaries, embayments, marshes and mangroves. The tides are the response of a forced-damped-resonance system (Arbic et al. 2009). The M_2 tide is the dominant tidal constituent in most places, with a period of half a lunar day, or 12 hours, 25 minutes; the M_2 tides are created by the differential motion of the solid Earth and oceans in response to the gravitational attraction of the moon (Newton, 1687; Laplace, 1799). The astronomical forcing evolves only slowly, however the tidal damping and basin resonance at tidal frequencies can change in response to changes in sea level, stratification and coastal conditions

(Müller, 2012; Schindelegger et al. 2018). Several recent studies have analysed historical coastal tide gauge data and found amplitude trends of order 1–4% per century (Ray, 2009; Woodworth, 2010; Müller et al. 2011). In some locations, the changes in the tides have been of comparable importance to changes in mean sea level for explaining changes in high water levels (Jay, 2009). For many individual tide gauges, the trends in tidal amplitude are strongly positively or negatively correlated with local time-mean sea level trends (Devlin et al. 2017). Another source of secular tidal changes, changes in oceanic stratification, modifies the rate of energy conversion from the barotropic tides to the internal tides (Jayne and St. Laurent, 2001), the vertical profile of turbulent viscosity on shelves (Müller, 2012), and the propagation speed of the internal tides (Zhao, 2016). For example, Colosi and Munk (2006) found an increase in the amplitude of the principal lunar semidiurnal tide M_2 in Honolulu of about 1 cm over the past 100 years, which they attributed primarily to changes in oceanic stratification bringing about local changes in relative phases of the internal and external M_2 tides, increasing constructive interference. Both sea level and stratification are expected to exhibit robust secular positive trends in the coming century due to climate change, at rates that are significantly larger than historical trends, and people may choose to replace natural beaches and marshes with sea-walls in response to rising sea levels. As a result, it is *very likely* that the majority of coastal regions will experience statistically significant changes in tidal amplitudes over the course of the 21st century.

Because coastal tides are near resonance in many locations, small changes in sea level and bay shape can change the local tides significantly. For example, the insertion of tidal power plants can have a significant impact on the local tides (Ward et al. 2012). Various observational and modeling studies demonstrate that SLR has spatial heterogeneous impacts on the tides, with some locations experiencing decreased tidal amplitudes and others experiencing increased tidal amplitudes (Pickering et al. 2012; Devlin et al. 2017; Pickering et al. 2017). Projections of tidal changes indicate that the patterns and even the sign of changes in tidal amplitudes depend on whether the coastlines are allowed to recede with rising sea levels or are held in place (Pickering et al. 2017; Schindelegger et al. 2018). Pelling et al. (2013) and Hwang et al. (2014) demonstrate that the rapid coastline changes in China's Bohai Sea have already altered the tides in that region and throughout the Yellow Sea (Hwang et al. 2014). Pelling and Green (2014) examine the impact of flood defenses as well as SLR on tides on the European Shelf. Such tidal changes have implications for designing flood defenses, for tidal renewable energy, for tidal flushing timescales of estuaries and embayments, and for navigational dredging requirements (Pickering et al. 2012) (Section 5.4.2). The sign and amplitude of local changes to tides are *very likely* to be impacted by both human coastal adaptation measures and climate drivers (listed above).

5.2.2.2.4   Systematic sources of uncertainty in projections of ocean physical changes

ESMs are able to capture the dynamics of the climate system, but all numerical models have approximations and biases. The most commonly used type of ocean component in ESMs is known to exhibit numerically induced vertical mixing that can be a significant fraction

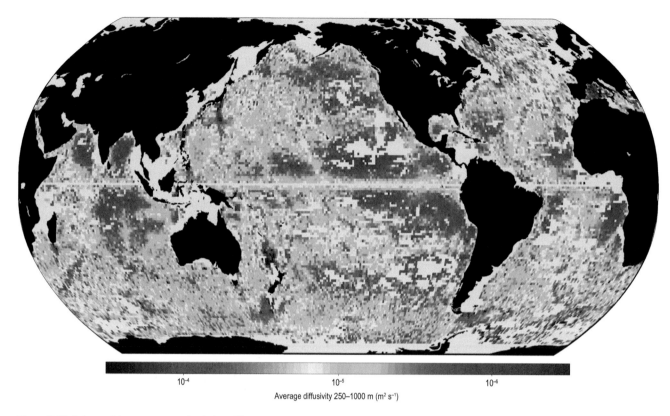

Average diffusivity 250–1000 m (m² s⁻¹)

Figure 5.6 | Estimate of the average vertical turbulent diffusivity between 250–1000 m calculated by applying fine structure techniques to Argo float data from below the well-mixed near-surface boundary layer. Only bins with at least three estimates are plotted and regions with insufficient data are coloured grey. This figure was created using updated data through April, 2018 with the techniques from Whalen et al. (2012).

of the physical mixing (Ilıcak et al. 2012; Megann, 2018). Because so many ocean models exhibit the same sign of bias, there is a systematic warming of the lower-main thermocline that is not cancelled out when taking the average over the ensemble of all the models in CMIP5. These biases are widely known within the ocean modelling community, and various groups are working to reduce these biases in future ESMs with better ocean model numerics and parameterisations. To correct for model biases, ESM projections are always taken as the difference from a control run without the anomalous forcing. However, some aspects of the ocean response to climate change are nonlinear, and model biases can introduce uncertainties into climate projections. In the case of heat uptake, this is of the order of 10% uncertainty, while for the rate of steric SLR (which depends on the nonlinear equation of state of seawater) the uncertainty in CMIP5 models is of the order of 20% (Hallberg et al. 2012).

Mesoscale eddies (geostrophic rotating vortices with spatial scales of 10–100 km that penetrate deeply into the water column, and are often described as the ocean's weather) play an important role in regulating the changes to the larger scale ocean circulation, especially in the Antarctic Circumpolar current, as is discussed in Cross Chapter Box 7. In addition, sub-mesoscale eddies (rotationally influenced motions with smaller horizontal scales of hundreds of metres to about 10 km and intrinsic timescales of a few days that especially arise in association with fronts in the ocean's surface properties) are known to be particularly important in the dynamics of the near-surface ocean boundary layer (see the review by Mahadevan (2016)). Sub-mesoscale instabilities are associated with

re-stratifying overturning circulations that can limit the thickness of the well-mixed ocean surface boundary layer near fronts (Bachman et al. 2017). Moreover, sub-mesoscale motions generate strong vertical velocities that drive fluxes of nutrients from the interior ocean into the euphotic zone or create pockets of reduced mixing with increased phytoplankton residency time within the euphotic zone (Lévy et al. 2012). Intense mesoscale eddies are known to create favourable conditions for sub-mesoscale instabilities as shown in both observational (Bachman et al. 2017) and numerical studies (Brannigan et al. 2017). Intensifying Southern Ocean eddy fields will have a significant local impact on biological productivity, ecosystem structure, and carbon uptake, both directly and via sub-mesoscale processes. At typical CMIP5 ESM resolutions, it is only in the tropics that mesoscale eddies are adequately resolved to explicitly model their effects (Hallberg, 2013), while sub-mesoscale eddies are not resolved anywhere, so eddy effects need to be parameterised in ESMs. Despite great progress over the past 30 years in parameterising eddy effects, uncertainties in these parameterisations and how eddies will respond to novel conditions continue to contribute to uncertainties in projections of oceanic climate change (*medium confidence*).

Ocean turbulent mixing is a key process regulating the ocean circulation and climate. Turbulent mixing is important for the uptake and redistribution of heat, carbon, nutrients, oxygen and other tracers (properties that are carried along with the flow of water) in the ocean (Schmittner et al. 2009; MacKinnon et al. 2017). Both observations and theory indicate that turbulent mixing in the ocean is not constant in space or time. Global estimates of both the turbulent kinetic energy dissipation rate and the vertical diffusivity, two measures of ocean turbulence, vary over several orders of magnitude throughout the ocean (Figure 5.6) (Polzin et al. 1997; Waterman et al. 2012; Whalen et al. 2012; Alford et al. 2013; Hummels et al. 2013; Sheen et al. 2013; Waterhouse et al. 2014; Kunze, 2017). For a given energy dissipation rate, the turbulent diffusivities of heat, salinity, nutrients and other tracers tend to be smaller with stronger stratification. This dependency on stratification helps explain why the observationally inferred diffusivity in the heavily stratified main thermocline (250–1000 m depth) is of similar magnitude to those deeper in the water column, while the turbulent energy density and dissipation rate are much stronger at the shallower depths (Whalen et al. 2012). Oceanic turbulence also fluctuates in time, is modulated by tidal cycles (Klymak et al. 2008), the mesoscale eddy field and seasonal changes (Whalen et al. 2018). In the mixed layer and directly below, turbulence changes according to local conditions, such as the winds, heating rates and local stratification (Sloyan et al. 2010; Moum et al. 2013; D'Asaro, 2014; Tanaka et al. 2015) at diurnal to seasonal and longer timescales. These variations in near-surface turbulence need to be taken into account for ESMs to reproduce more accurately the observed seasonal cycle of surface properties and spatial structure of the depth of the thermally well-mixed near surface layer of the ocean. The spatial and temporal patterns of ocean turbulence help shape ocean tracer distributions (heat, dissolved greenhouse gases and nutrients) and how they will evolve in a changing climate (*high confidence*).

Ocean turbulent mixing requires energy sources, many of which are expected to change with a changing climate. Surface wind and buoyancy forcing, the mean and eddying larger-scale ocean circulation itself, and the barotropic tides are all thought to be significant sources of the energy that drives mixing (Wunsch and Ferrari, 2004). Often this energy first passes through the ocean's pervasive field of internal gravity waves that propagate and refract through the varying ocean circulation, often breaking into turbulent mixing far from their sources (Eden and Olbers, 2014; Alford et al. 2016; Melet et al. 2016; Meyer et al. 2016; Zhao et al. 2016b). The energy contributing to the internal waves from the winds and the subsequent turbulence will be altered by changes in tropical storm activity or sea ice coverage. For example, the increasing extent of ice-free Arctic Ocean has already been observed to lead to increased wind-driven internal waves (Dosser and Rainville, 2016). The Southern Annular Mode is expected to intensify as a result of climate change (Young et al. 2011; Jones et al. 2016b), bringing with it stronger winds, and more wind-energy input over most of the Southern Ocean and a more intense mesoscale eddy field (Hogg et al. 2015). Changes in the near-bottom stratification will alter the rate that the barotropic tides generate internal waves, thereby altering the strength and distribution of the tidally generated mixing. Some of the parameterisations of interior ocean mixing used in CMIP5 ESMs take some changing turbulent energy sources into account (Jayne and St. Laurent, 2001), and more comprehensive mixing treatments are being developed for use in future generations of ESMs (Eden and Olbers, 2014). However, not all of the physical processes leading to the rich structure of mixing shown in Figure 5.6 are well understood or included in ESMs; the prospect of significant changes in the patterns and intensity of ocean turbulent mixing is a potential source of uncertainty (probably at the 10% level) in projections of physical and ecological changes in

the ocean, including heat uptake, stratification changes, steric SLR, deoxygenisation and nutrient fluxes (*medium confidence*).

5.2.2.3   Changes in Ocean Carbon

Since AR5, new global-scale data synthesis products, novel methods for their analyses, as well as progress in modeling have substantially increased our quantitative understanding of the role of the ocean in absorbing and storing CO_2 from the atmosphere. The most important progress concerns the data-based quantification of the temporal variability of the ocean carbon sink. While AR5 assessed primarily the climatological mean processes governing the ocean carbon cycle, the most recent work now permits us to assess how these processes have changed in recent decades in response to climate variability and change. Here we focus specifically on the open ocean carbon cycle.

5.2.2.3.1   Ocean carbon fluxes and inventories

The analyses of the steadily growing number of surface ocean CO_2 observations (now more than 20 million observations, SOCATv6 (www.socat.info/index.php/2018/06/19/v6-release) demonstrate that the net ocean uptake of CO_2 from the atmosphere has increased from around 1.2 ± 0.5 Pg C yr^{-1} in the early 1980s to 2.0 ± 0.5 Pg C yr^{-1} in the years 2010–2015 (Rödenbeck et al. 2014; Landschützer et al. 2016). Once new estimates of the outgassing flux stemming from river derived carbon of 0.8 Pg C yr^{-1} (Resplandy et al. 2018) are accounted for, these new observations imply that the rate of global ocean uptake of anthropogenic CO_2 increased from 2.0 ± 0.5 Pg C yr^{-1} to 2.8 ± 0.5 Pg C yr^{-1} between the early 1980s and 2010–2015 (Rödenbeck et al. 2014; Landschützer et al. 2016; Le Quéré et al. 2018). This increase is supported by the current generation of ocean carbon cycle models (Le Quéré et al. 2018), and commensurate with the increase in atmospheric CO_2.

The continuing efforts to re-measure dissolved inorganic carbon (DIC) along many of the repeat hydrographic lines that were occupied during the 1980s and 1990 (Talley et al. 2016), alongside the preparation of a global quality controlled database of ocean interior observations (Olsen et al. 2016a), have led to progress since AR5 regarding to the oceanic interior storage of anthropogenic CO_2. Several studies analysed the changes in the amount of anthropogenic CO_2 that have accumulated between different occupations in the different ocean basins (Wanninkhof et al. 2010; Pérez et al. 2013; Woosley et al. 2016; Carter et al. 2017), confirming that the anthropogenic CO_2 taken up from the atmosphere is transported to depth, where most of it is stored. Using a newly developed reconstruction method, Gruber et al. (2019) extended these results to the globe. They find that between 1994 and 2007, across two standard deviations, that the global ocean has accumulated an additional 30–38 Pg C of anthropogenic CO_2, which is equivalent to an air-sea CO_2 flux of between 2.3–2.9 Pg C yr^{-1} (coherent with surface ocean CO_2 observations), bringing the total inventory for the year 2007 to 150 ± 20 Pg C. Extrapolating this estimate to the year 2010 gives an inventory of 158 ± 18 Pg C, which is statistically indistinguishable from the 'best' estimate provided by Khatiwala et al. (2013) of 155 ± 31 Pg C and more recently also found from a steady-state ocean model (DeVries, 2014) for this reference year. If the inventory-based estimates are

5

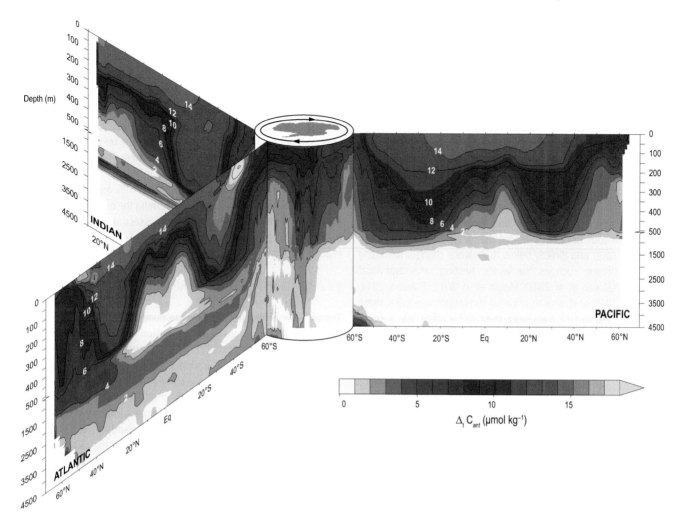

Figure 5.7 | Vertical sections of the change in anthropogenic CO_2 from 1994 to 2007 represented by the zonal mean sections in each ocean basin, organised around the Southern Ocean in the centre. The upper 500 m are expanded. Contour intervals of anthropogenic CO_2 are 2 μmol kg^{-1} (Gruber, 2019).

adjusted for the loss of natural carbon, a *very likely* total increase in storage between 1994 and 2007 of 24–34 Pg C, or around 25% of total emissions, is found (Gruber, 2019).

Thus, there is *very high confidence* from surface ocean and ocean interior carbon data that the strength of the ocean sink for anthropogenic carbon has increased in the last two decades in response to the growth of atmospheric CO_2. Multiple lines of evidence indicate that it is *very likely* that the ocean has taken up 20–30% of the global emissions of CO_2 from the burning of fossil fuels, cement production, and land-use change since the mid 1980s. The consistency between independent surface ocean observations and the ocean interior data-based reconstructions supports the assessment of *very high confidence* and provides *robust evidence* that fraction of emissions taken up by the ocean has not changed in a statistically significant manner in the last few decades and remains consistent with AR5.

Alongside a globally integrated perspective, these new surface ocean observations also reveal a substantial degree of variability at interannual and decadal scales (Rödenbeck et al. 2015; Landschützer et al. 2016; Le Quéré et al. 2018). Most notable are the air-sea CO_2 flux variations in the tropics linked to ENSO variations (Rödenbeck et al. 2015;

Landschützer et al. 2016), as well as the strong decadal variations in the high latitudes, especially the Southern Ocean (Landschützer et al. 2015; Munro et al. 2015; Ritter et al. 2017), discussed further in Chapter 3 (Section 3.2.1.2.4). Fluctuations in the Southern Ocean CO_2 flux are important as they impart a substantial imprint also on the global uptake fluxes. For instance, reduced Southern Ocean uptake in the 1990–2000 period coincided with an exceptionally weak global net uptake of only about 0.8 ± 0.5 Pg C yr^{-1}.

Thus, there is growing evidence from multiple datasets that the ocean carbon sink exhibits decadal variability at regional scales that significantly alter the globally integrated sink (*medium confidence*).

Detailed analyses of the spatial structure of the change in storage of anthropogenic CO_2 confirm the variable nature of the ocean carbon sink suggested by the surface observations (Pérez et al. 2013), which are most likely a consequence of changes in ocean circulation (DeVries and Weber, 2017). The increase in anthropogenic CO_2 between 1994 and 2007 occurs throughout the upper 1000 m, but with very different penetration depths, reflecting largely differences in the efficiency, with which the anthropogenic CO_2 is transported from the surface to depth (Gruber et al. 2019) (Figure 5.7). This spatial distribution of how the amount of anthropogenic CO_2 has changed between 1994 and 2007 is similar to the distribution of anthropogenic CO_2 reconstructed for 1994 (Sabine et al. 2004), although the imprint of regional variations in ocean circulation and transport are discernible (Gruber, 2019).

5.2.2.3.2 Ocean carbon chemistry

Analyses of direct measurements of ocean chemistry from time series stations and merged shipboard studies show consistent decreases in surface-ocean pH over the past few decades. Reductions range between 0.013–0.03 pH units decade^{-1} over records that span up to 25 years (Table SM5.3). Focusing on the individual time series locations with records longer than 15 years, there is an overall decline of 0.017–0.027 (across 99% confidence intervals). Trends calculated from repeat measurements on ocean surveys show a consistent value of around –0.02 pH units decade^{-1} for diverse oceanic regions (Table SM5.3), with greater subsurface than surface trends reported in the subtropical oceans (Dore et al. 2009). At larger spatial scales, surface-ocean pH trends are assessed using shipboard observations of the fugacity of CO_2 and estimates of ocean alkalinity (Takahashi et al. 2014; Lauvset et al. 2015). Between 1991–2011, mean surface-ocean pH has declined by 0.018 ± 0.004 units decade^{-1} in 70% of ocean biomes, with the largest declines in the Indian Ocean (–0.027 units decade^{-1}), eastern Equatorial Pacific (–0.026 units decade^{-1}) and the South Pacific subtropical (–0.022 units decade^{-1}) biomes (Lauvset et al. 2015). Due to the close link between carbonate ion concentrations and pH, mean trends in the stability of mineral forms of aragonite and calcite (known as the 'saturation state') that are important for organisms such as coccolithophorids, pteropods and corals follow those of pH, with high-latitude regions most vulnerable to under-saturation due to naturally lower mean values.

It is *virtually certain* that ocean pH is declining, and the *very likely* range of this decline is 0.017–0.027 pH units per decade for the 8 locations where individual time series observations longer than 15 years exist. This trend is lowering the chemical stability of mineral forms of calcium carbonate and can be attributed to rising atmospheric CO_2 levels.

CMIP5 models are in good agreement with historical observations of declining surface-ocean pH (Figure 5.8a). Models project global surface-ocean declines between 2006–2015 and 2081–2100 of 0.287–0.291 and 0.036–0.042 pH units (both across 99% confidence intervals) for the RCP2.6 and RCP8.5 scenarios, respectively, with higher reductions in the subsurface of subtropical oceans (Bopp et al. 2013; Gattuso et al. 2015). These changes in pH will be greatest in the Arctic Ocean and the high latitudes of the Atlantic and Pacific Oceans due to their lower buffer capacity and are lowest in contemporary upwelling systems (Figure 5.8b) and will also reduce the stability of calcite minerals (Bopp et al. 2013; Gattuso et al. 2015). The area of the surface ocean (0–10 m) characterised by undersaturated conditions in CMIP5 models by 2081–2100 reduces from a *very likely* range of 6.4–9.5 × 10^{12} m^2 or 5.5–7.3 × 10^{13} m^2 under RCP8.5 (as much as 16–20% of ocean surface area for aragonite), to just 0.01–0.2 × 10^{12} m^2 or 0.01–0.13 × 10^{13} m^2 under RCP2.6 for either calcite or aragonite minerals, respectively.

Under RCP8.5, hotspots for undersaturated waters for calcite remain restricted to the Arctic Ocean, while for aragonite, much of the Southern Ocean and the North Pacific and Northwestern Atlantic Oceans are also projected to become undersaturated (Orr et al. 2005; Hauri et al. 2015; Sasse et al. 2015). These results arise from the very well understood reductions in carbonate ion concentrations at lower pH, the vulnerability of regions with naturally low mean values, and the greater overall sensitivity of aragonite solubility. Regional models, with higher resolution that ESMs, also project year-round corrosive conditions for aragonite in some eastern boundary upwelling systems (Franco et al. 2018a). In the ocean interior, the decline in pH and calcium carbonate saturation state is more uncertain across models (Steiner et al. 2014) as it is modulated by changes to ocean overturning and water mass subduction (Resplandy et al. 2013; Chen et al. 2017). Projected benthic changes in pH over the next century are highly localised and are linked to transport of surface anomalies to depth, with over 20% of the north Atlantic sea floor deeper than 500 m projected to experience pH reductions greater than 0.2 units by 2100 under the RCP8.5 scenario (Gehlen et al. 2014). Changes in pH in the abyssal ocean (>3000 m deep) are greatest in the Atlantic and Arctic Oceans, with lesser impact in the Southern and Pacific Oceans by 2100, mainly due to the circulation timescales (Sweetman et al. 2017).

Overall, it is *virtually certain* that the future surface open ocean will experience pH drops of either 0.036–0.042 (RCP2.6) or 0.287–0.291 (RCP8.5) pH units by 2081–2100, relative to 2006–2105. These pH changes are *very likely* to cause 16–20% of the surface ocean, specifically the Arctic and Southern Oceans, as well as the northern Pacific and northwestern Atlantic Oceans, to experience year-round corrosive conditions for aragonite by 2081–2100. It is *virtually certain* these impacts will be avoided under the RCP2.6 scenario. There is *medium confidence*, due to the potential for parallel changes in ocean circulation, that the Arctic and north Atlantic seafloors will experience the largest pH changes over the next century.

Although ocean acidification results in long-term trends in mean ocean chemistry, it can also influence seasonal cycles. Observation-based products indicate that the seasonal cycle of global surface-ocean pCO$_2$ increased in amplitude by 2.2 ± 0.4 µatm between 1982 and 2014 (Landschützer et al. 2018). CMIP5 models and data-based products similarly project consistent future increases in the seasonal cycle of surface-ocean pCO$_2$ under the RCP8.5 emissions scenario, with enhanced amplification in high-latitude waters (McNeil and Sasse, 2016). The amplitude of the seasonal cycle of global surface-ocean free acidity ([H$^+$]) is projected to increase by 71–91% (across 90% confidence intervals) over the 21st century under RCP8.5, also with greater amplification in the high-latitudes (Kwiatkowski and Orr, 2018). Conversely, models project a 12–20% reduction (across 90% confidence intervals) in the seasonal amplitude of surface-ocean pH, as changes in pH represent relative changes in [H$^+$] due to their logarithmic relationship, and there are typically greater projected increases in annual mean state [H$^+$] than the seasonal amplitude of [H$^+$]. Models also project a 4–14% (across 90% confidence intervals) reduction in the seasonal amplitude of global mean surface-ocean aragonite saturation state under RCP8.5, with a slight amplification in the subtropics being outweighed by dampening elsewhere.

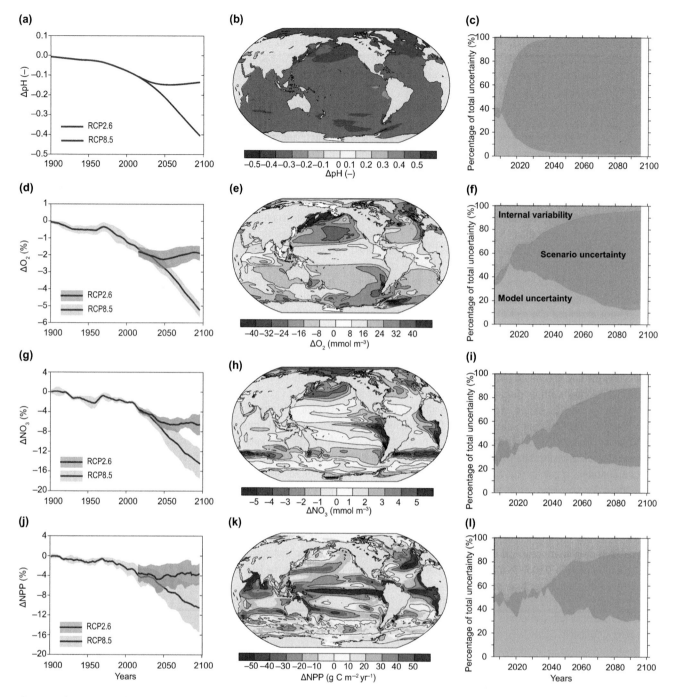

Figure 5.8 | Panels **(a)**, **(d)**, **(g)** and **(j)** display simulated global changes over the period of 1900–2100 (with solid lines representing the multi-model mean and the envelope representing 90% confidence intervals for RCP8.5 and RCP2.6), for surface pH, O_2 concentration averaged over 100–600 m depth, upper 100 m nitrate concentrations and NPP integrated over the top 100 m. Differences are calculated relative to the 1850–1900 period. Panels **(b)**, **(e)**, **(h)** and **(k)** show spatial patterns of simulated change in surface pH, upper 100 m nitrate concentrations, O_2 concentration averaged over 100 to 600 m depth, and NPP integrated over the top 100 m averaged over 2081–2100, relative to 1850–1900 for RCP8.5. Panels **(c)**, **(f)**, **(i)** and **(l)** display time series of the percentage of total uncertainty ascribed to internal variability uncertainty, model uncertainty, and scenario uncertainty in projections of global annual mean changes. Figure adapted after (Frölicher et al. 2016). Please note that confidence intervals can be affected by the different number of models available for the RCP8.5 and RCP2.6 scenarios and for different variables. See also Table SM5.4.

The contrasting changes in the seasonal amplitudes of ocean carbonate chemistry variables derive from different sensitivities to atmospheric CO_2 and climate change and to diverging trends in the seasonal cycles of DIC, alkalinity and temperature. Model skill at simulating the seasonal cycles of carbonate chemistry is moderate, with persistent biases in the Southern Ocean, particularly for pCO_2, [H^+] and pH (Kwiatkowski and Orr, 2018; Mongwe et al. 2018).

Overall, we assess that alongside the strong mean state changes, it is *very likely* that the amplitude of the seasonal cycle in free acidity will increase by 71–91%, while it is *very likely* that the seasonal cycles of pH and aragonite saturation will decrease by 12–20% and 4–14%, respectively.

5.2.2.4   Changing Ocean Oxygen

Ocean oxygen (O_2) levels at the surface are controlled by the balance between oxygen production during photosynthesis, temperature-controlled solubility and air-sea exchange. Deeper in the water column, consumption of oxygen during respiration and redistribution by ocean circulation and mixing are dominant processes. In theory, a warmer more stratified ocean would have a reduced oxygen content, due to the combined influence of lowered gas solubility and a greater interior respiration of organic matter due to enhanced physical isolation of subsurface waters. In accord, global changes in ocean oxygen assessed from three different analyses of compiled global oxygen datasets going back to the 1960s agree that there is a net loss of oxygen from the ocean over all depths (see Table 5.2). For the 0–1000 m depth stratum that contains the most data and is common to all three analyses, oxygen is assessed to have declined by a *very likely* range of 0.5–3.3% between 1970 and 2010. For the surface ocean (0–100 m) and the thermocline later of 100–600 m the *very likely* range of oxygen declines are 0.2–2.1% and 0.7–3.5%, respectively (Table 5.2). Across two studies, global oxygen is assessed to have declined by a *very likely* range of 0.3–2.0%, with a similar range of decline for waters deeper than 600 m (Table 5.2). The regions of lowest oxygen, known as OMZs, with oxygen levels lower than 80 mmol L^{-1}), are observed to be expanding by a *very likely* range of 3.0–8.3% across the three studies.

Regionally, all studies agree that the north Pacific and Southern Oceans have shown the largest overall oxygen declines (Figure 5.9), but there is some disagreement regarding the magnitude of the oxygen change in the tropical ocean, with some studies suggesting significant declines (Schmidtko et al. 2017) and other reporting more modest reductions (Helm et al. 2011; Ito et al. 2017) and data coverage is still limited for some regions and deeper than 1000 m. Based on the available data, the strongest declines in deep ocean oxygen have occurred in the Equatorial Pacific, North Pacific, Southern Ocean and South Atlantic, with intermediate declines in the Arctic, South Pacific and Equatorial Atlantic, while the north Atlantic has experienced a moderate oxygen increase below 1200 m (Figure 5.9). A particular difference between parallel oxygen analyses concerns the means of integrating and mapping sparse data across the ocean, both horizontally and vertically, with different studies making specific decisions about averaging grids and integration methods. Moreover, data remains sparse for some ocean regions, depths and periods. Taken together, the challenges of data sparsity, regional differences and the relatively large uncertainties on the oxygen changes across different studies, but also recognising that oxygen declines are significantly different to zero, leads to *medium confidence* in the observed oxygen decline.

Syntheses of datasets from local time series tend to document stronger trends, with oxygen declines of over 20% at sites in the northeastern Pacific between 1956–2006 (Whitney et al. 2007), the Northwestern Pacific between 1954–2014 (Sasano et al. 2015) and the California Current between 1984–2011 (Bograd et al. 2015). Despite holding the highest inventory of oxygen in the ocean, oxygen levels in Southern Ocean contributed 25% to the global decline between 1970–1992 (Helm et al. 2011) and have

fallen by over 150 Tmol per decade from the 1960s to present (Schmidtko et al. 2017). Observations along ocean cruises as part of the CLIVAR programme have also documented broad thermocline oxygen declines in the northern hemisphere oceans, accompanied by well understood oxygen increases in subtropical and southern hemispheres (Talley et al. 2016).

Overall there is *medium confidence* that the oxygen content of the upper 1000 m has declined with a *very likely* loss of 0.5–3.3% between 1970–2010. OMZ are expanding in volume, by a *very likely* range of 3.0–8.3%. There is *medium confidence* that the largest regional changes have occurred in the Southern Ocean, equatorial regions, North Pacific and South Atlantic due to *medium agreement* among studies.

The role of ocean warming alone in driving the oxygen changes can be appraised using solubility estimates, which vary between around 15–50% for the upper 1000 m oxygen trend between studies (Helm et al. 2011; Ito et al. 2017; Schmidtko et al. 2017). The role of other processes, linked to changing ocean ventilation and respiration are challenging to appraise directly, but tend to reinforce the impacts from warming and are probably predominant overall (Oschlies et al. 2018). Indeed, that the observed oxygen decline is negatively correlated with ocean heat content changes (Ito et al. 2017) reflects the overriding role of changing ocean ventilation and associated processes (see also Section 5.2.2). That the ratio of the associated oxygen to heat changes is larger than would be expected from thermal processes alone also highlights the role played by other processes (Oschlies et al. 2018). Local oxygen trends have emphasised the role of changes to ocean physics in western Northern Pacific (Whitney et al. 2013); Sasano et al. (2015), the southern California Current region (Goericke et al. 2015), and the Santa Barbara Basin (Goericke et al. 2015). In regions of high mesoscale activity, such as the tropical north Atlantic, low oxygen eddies can have a significant impact on oxygen dynamics (Karstensen et al. 2015; Grundle et al. 2017). Oxygen fluctuations in the deep ocean have been linked to changes in large scale ocean circulation (Watanabe et al. 2003; Stendardo and Gruber, 2012) and at the global scale, the observed oxygen decline is negatively correlated with ocean heat content changes (Ito et al. 2017). Changes to respiration rates, either due to temperature enhancement or in the amount/quality of organic material can also be important and the enhanced respiratory demand associated with an intensified monsoon has been invoked as a driver of the expansion of the Arabian Sea OMZ (Lachkar et al. 2018).

Ocean oxygen changes are also affected by climate variability on interannual and decadal timescales, especially for the tropical ocean OMZs (Deutsch et al. 2011). ENSO variability in particular affects the thermocline structure, which then alongside changes in circulation modulates oxygen solubility and respiratory demand in this region (Ito and Deutsch, 2013; Eddebbar et al. 2017). These drivers may then be combined with modifications to overturning and ventilation of OMZs by lateral jets and equatorial current intensity (Duteil et al. 2014). Centennial scale studies based on isotope proxies for low oxygen regions have demonstrated fluctuations in OMZ extent linked to decadal changes in tropical trade winds that affects interior ocean respiratory oxygen demand, which implies that it will be difficult

Table 5.2 | Observed oxygen changes for the period 1970–2010 for 6 different layers within the ocean. The changes are shown as percentage change of global averages. The layers are depths 0–100, 100–600, 0–1000, and 600–bottom are in metres. The oxygen minimum zone (OMZ) is defined as the ocean volume change that is less than 80 mmol L^{-1}. The estimates and confidence intervals are based published papers (Schmidtko et al. 2018, Ito et al. 2017 and Helm et al. 2011). The assessed change is the average of the available estimates and the 90% Confidence Interval (CI) combines the confidence as their standard deviation with two degrees of freedom.

		Schmidtko		Ito		Helm		Assessed Change	
Layer	Period	Change	90 CI	Change	90 CI	Change	90 CI	Change	90 CI
0–100	1970–2010	–0.38%	±1.06%	–1.65%	±0.63%	–1.30%	±0.54%	–1.11%	±0.95%
100–600	1970–2010	–1.06%	±1.36%	–3.17%	±1.34%	–2.04%	±0.60%	–2.09%	±1.42%
0–1000	1970–2010	–1.35%	±1.38%	–2.70%	±1.30%	–1.74%	±0.54%	–1.93%	±1.39%
600–bottom	1970–2010	–1.51%	±0.62%	n.a.	n.a.	–0.81%	±0.57%	–1.16%	±0.84%
OMZ	1970–2010	6.33%	±2.52%	6.10%	1.2%	4.49%	±2.25%	5.64%	±2.66%
Global	1970–2010	–1.43%	±0.70%	n.a.	n.a.	–0.87%	±0.53%	–1.15%	±0.88%

to attribute recent changes in the Pacific OMZ to anthropogenic forcing alone (Deutsch et al. 2015). Parallel work based on oxygen observations (Llanillo et al. 2013), as well as modelling (Duteil et al. 2018) supports the importance of decadal scale variability in the eastern tropical Pacific OMZ. There is some evidence for the potential of a modulating impact on tropical Pacific oxygen at interannual timescales from atmospheric deposition of nitrogen and iron (Ito et al. 2016; Yang and Gruber, 2016).

At the global scale, there is *high confidence* that the impact of a warmer ocean on oxygen levels is reinforced by other processes associated with ocean physics and biogeochemistry, which cause the majority of the observed oxygen decline. For the tropical Pacific OMZ, there is *medium confidence* arising from *medium agreement* from *medium evidence* that low frequency decadal changes in ocean physics have controlled past fluctuations in OMZ extent.

Future changes in oxygen can be appraised from ESMs that account for the combined effects of ocean physics and biogeochemistry. Globally, these models project that it is *very likely* oxygen will decline by 3.2–3.7% or 1.6–2.0% (both across 90% confidence limits) for RCP8.5 or RCP2.6, respectively, relative to 2000 (Bopp et al. 2013). Focussing on the 100–600 m depth stratum, O_2 changes by –4 to –3.1% for the RCP8.5 or by –0.5–0.1% for the RCP2.6 scenario (relative to 2006–2015, Figure 5.8d). It should be noted that ESMs appear to be underestimating the rate of oxygen change from available datasets from the historical period (Oschlies et al. 2018). Increased tropical ocean stratification reduces interior ocean oxygen by diminishing pathways of ventilation in the subtropical gyres and by inhibiting turbulent mixing with the oxygen-rich surface ocean (see Section 5.2.2.2.4). This relatively robust global modelled trend (Figure 5.8d) however masks important uncertainties in the projection of regional trends (Figure 5.8e), particularly in the tropical ocean OMZs (Bopp et al. 2013; Cocco et al. 2013; Cabré et al. 2015). The uncertainty in the trends in tropical ocean OMZs arises due to the fact that oxygen depletion due to warming induced reductions in oxygen saturation are opposed by oxygen enrichment due to reduced oxygen consumption during respiration in response to predicted declines in marine export production, as well as biases due to model

resolution in the tropics and the length of the model spin up (Bopp et al. 2017). The 80 mmol L^{-1} threshold that may be used to define the volume of the oxygen minimum is projected to grow by a *very likely* range of 7.0 ± 5.6% by 2100 during the RCP8.5 scenario or show virtually no change during the RCP2.6 scenario, relative to

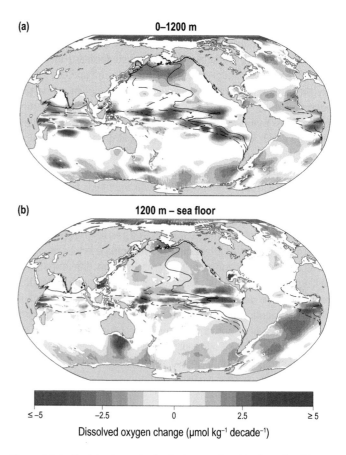

Figure 5.9 | Absolute change in dissolved oxygen (umol kg^{-1} per decade) between water depths of **(a)** 0 and 1200 m, and **(b)** 1200 m and the sea floor over the period 1960–2010. Lines indicate boundaries of OMZs with less than 80 µ mol kg^{-1} oxygen anywhere within the water column (dashed/dotted), less than 40 µ mol kg^{-1} (dashed) and less than 20 µ mol kg^{-1} (solid). Redrawn from Oschlies et al. (2018).

5

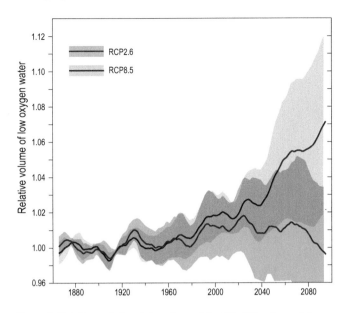

Figure 5.10 | The evolution of the volume of the 100–600 m layer of the ocean with oxygen concentrations less than 80 mmol L^{-1} for the RCP8.5 (red line) and the RCP2.6 (blue line), normalised to the volume in 1850–1900. Dashed lines indicated the *very likely* range (90% confidence intervals) across the CMIP5 models (CNRM-CM5, GFDL-ESM2M, GFDL-ESM2G, IPSL-CM5A-LR, IPSL-CM5A-MR, MPI-ESM-LR, MPI-ESM-MR and the NCAR-CESM1 models). Models are corrected for drift in O$_2$ using their control simulations.

a 1850–1900 reference period (Figure 5.10). At the seafloor, between 200–3000 m depth strata, the north Pacific, north Atlantic, Arctic and Southern Oceans may see oxygen declines by 0.3–3.7% by 2100 (relative to 2005), with abyssal ocean changes being lower and more localised around regions in the north Atlantic and Southern Ocean (Sweetman et al. 2017), but will be modulated by any future changes in overturning strength. There is *high confidence* that the largest changes in deep sea systems will occur after 2100 (Battaglia and Joos, 2018).

Simulations extended to 2300 suggest that by 2150 the trend of declining tropical ocean oxygen (both in terms of concentrations and volume of low oxygen waters) may reverse itself, mainly due to the effect of strong declines in primary production and organic matter fluxes to the ocean interior (Fu et al. 2018) or due to enhanced Antarctic ventilation (Yamamoto et al. 2015), but with *low confidence* due to *limited evidence*. At the global scale, 10,000 year intermediate complexity model simulations find that overall ocean oxygen loss shows near linear relationships to equilibrium temperature, itself linearly related to cumulative emissions, and any climate mitigation scenario will reduce peak oxygen loss by 4.4% per degree Celsius of avoided warming (Battaglia and Joos, 2018).

In summary, the total oxygen content of the ocean is *very likely* to decline by 3.2–3.7% by 2100, relative to 2000, for RCP8.5 or by between 1.6–2.0% for RCP2.6 with *medium confidence*. There is *medium confidence* that sea floor changes will be more localised in the north Atlantic and Southern Oceans by 2100, but *high confidence* that the largest deep sea floor changes in oxygen will occur after 2100.

5.2.2.5   Changing Ocean Nutrients

Changes to ocean nutrient cycling are driven by modifications to ocean mixing and transport (Section 5.2.2.2.2), internal biogeochemical cycling and fluctuations in external supply, particularly from rivers and the atmosphere. This assessment will focus on the main nutrients important for driving microbial growth (Section 5.2.2.6), namely nitrogen, phosphorus and iron.

Diverse studies (including shipboard experiments and use of protein biomarkers) have highlighted nitrogen and phosphorus limitation in the stratified tropical ocean regions accompanied by widespread iron limitation at high latitudes and in upwelling regions that typically have elevated levels of productivity (Figure 5.11) (Moore et al. 2013; Saito et al. 2014; Browning et al. 2017; Tagliabue et al. 2017). Moreover, more extensive experimental work has demonstrated overlapping nitrogen-iron co-limitation at the boundaries between gyre and upwelling regimes (Browning et al. 2017). There is *high confidence* arising from *robust evidence* and *high agreement* across different types of studies that the main limiting nutrient is either iron (in most major upwelling regions and the Southern, north Atlantic and sub-Arctic Pacific Oceans) or nitrogen and phosphorus (in the low productivity tropical ocean gyres).

There is *limited evidence* on contemporary trends in nutrient levels, either from time series sites or broader meta-analyses. Increasing inputs of anthropogenic nitrogen from the atmosphere are perturbing ocean nutrient levels (Jickells et al. 2017). In the North Pacific in particular, additional atmospheric nitrogen input has raised the nitrogen to phosphorus ratio between 1988–2011 and induced a progressive shift towards phosphorus limitation in this region (Kim et al. 2011; Kim et al. 2014; Ren et al. 2017). This tendency is supported by modelling experiments that find enhanced atmospheric nitrogen input only has a small influence on productivity due to expanded phosphorus limitation (Yang and Gruber, 2016) and other nitrogen cycle feedbacks (Somes et al. 2016; Landolfi et al. 2017).

In general, future increases in stratification (Dave and Lozier, 2013; Talley et al. 2016; Kwiatkowski et al. 2017; and see also Section 5.2.2.2) will trap nutrients in the ocean interior and reduce upper ocean nutrient levels, alongside an additional local impact from changes to atmospheric delivery. However, no CMIP5 models accounted for changes in nutrient delivery from dust and anthropogenic aerosols during their experiments, which could be an important component of regional change (Wang et al. 2015b; Somes et al. 2016; Yang and Gruber, 2016). ESMs project a decline in the nitrate content of the upper 100 m of 9–14% or 1.5–6% (across 90% confidence intervals) for the RCP8.5 or RCP2.6 scenario, respectively, by 2081–2100 relative to 2006–2015 (Figure 5.8g). The largest absolute declines in nitrate content is projected in the present day upwelling zones (Figure 5.8h). Projected changes to upper 100 m nitrate concentrations are significantly different to zero for both RCP8.5 and RCP2.6 at the 90% confidence level, but are overall lower for the RCP2.6. Scenario, internal variability and inter-model variability contribute roughly equally to the overall projection uncertainty in 2100 (Figure 5.8i) and there is no clear separation of nitrate trends between RCP8.5 and RCP2.6 outside the model uncertainty (Figure 5.8h).

5

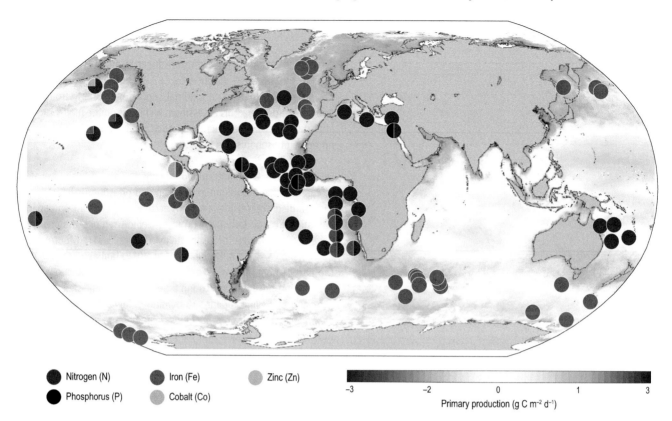

Figure 5.11 | Map of the dominant limiting resource (Moore et al. 2013), updated to include new experiments from the north Pacific, tropical Atlantic and south east Atlantic (Browning et al. 2017; Shilova et al. 2017). The background is depth integrated primary productivity using the Vertically Generalized Production Model algorithm. Colouring of the circles indicates the primary limiting nutrients inferred from chlorophyll and/or primary productivity increases following artificial amendment of: N (blue), P (black), Fe (red), Co (yellow) and Zn (cyan). Divided circles indicate potentially co-limiting nutrients, for example, a red-blue divided circle indicates Fe-N co-limitation.

Iron concentrations are projected to increase in the future from ESM simulations, due to enhanced lateral transport into high-latitude oceans and reduced biological consumption in regions of declining nitrate (Misumi et al. 2013). Other modelling efforts also suggest greater levels of the more biologically available Fe(II) species in a warmer and more acidic ocean (Tagliabue and Völker, 2011). These modelling studies tend to indicate greater ocean iron availability in the future overall, but the very limited skill of contemporary global ocean iron models in reproducing observations available from the new basin scale datasets from the international GEOTRACES program and neglect for parallel dust supply changes lower the confidence in the models' projected changes (Tagliabue et al. 2016).

Overall, nitrate concentrations in the upper 100 m are *very likely* to decline by 9–14% by 2081–2100, relative to 2006–2015 for RCP8.5 or 1.5–6% for RCP2.6, in response to increased stratification, with *medium confidence* in these projections due to the *limited evidence* of past changes that can be robustly understood and reproduced by models. Surface ocean iron levels is projected to increase in the 21st century with *low confidence* due to systemic uncertainties in these models.

5.2.2.6   Changing Ocean Primary and Export Production

Ocean primary productivity is a key process in the ocean carbon cycle (see Section 5.2.2.3), as well as for supporting pelagic ocean ecosystems (see Section 5.2.3). NPP is the product of phytoplankton

growth rate and standing stock. Phytoplankton growth is controlled by the combination of temperature, light and nutrients, while the phytoplankton standing stock is modified by both gains from growth and losses due to grazing by zooplankton (Figure 5.12). Export production is here defined as the sinking flux of particulate organic carbon (produced by NPP) across a specified depth horizon. Otherwise known as the biological pump, export production is also a key component of the global carbon cycle (see Section 5.2.2.3) and an essential food supply to benthic organisms (see Section 5.2.3.2). Export production is regulated by the level of primary production and the transfer efficiency with depth, itself controlled by the type of sinking organic carbon, which is affected by the upper ocean food web structure (Boyd et al. 2019).

Satellite datasets that use mathematical algorithms to convert ocean colour, often alongside other remotely sensed information, into chlorophyll or other indexes of phytoplankton biomass and NPP provide the potential to deliver a global meta-analysis of changes in NPP. Since AR5, a variety of studies have reported relatively insignificant changes in overall open ocean chlorophyll levels of $<\pm1\%$ yr^{-1} for individual time periods (Boyce et al. 2014; Gregg and Rousseaux, 2014; Boyce and Worm, 2015; Hammond et al. 2017). Regionally, trends of ±4% between 2002–2015 for different regions are found when different satellite products are merged, with increases at high latitudes and moderate decreases at low latitudes (Mélin et al. 2017). While some studies report good comparability of merged products (Mélin et al. 2017), others highlight significant mismatches regarding absolute values and decadal trends in NPP between NPP algorithms (Gómez-Letona et al. 2017). Satellite derived NPP shows significant mismatches when compared to *in situ* data and reducing uncertainties in derived NPP is a high priority for the community (Lee et al. 2015), although there is a reasonable correlation in higher biomass coastal regions (Kahru et al. 2009). Importantly, satellite records are not yet long enough to unambiguously isolate long term climate related trends from natural variability (Beaulieu et al. 2013). Overall, there is *low confidence* in satellite-based trends in global ocean NPP due to the time series length and lack of corroborating *in situ* measurements or other validation time series. This is especially true at regional scales where distinct sets of poorly understood processes dominate.

Future changes in NPP will result from the changing influence from temperature, light, nutrients and grazing (Figure 5.12). Across CMIP models, NPP is predicted to broadly decline or remain constant by 2081–2100, with mean changes by 2100 of –3.8 to –10.6% and –1.1–0.8% across 90% confidence intervals for the RCP85 and RCP26 scenario, respectively (all relative to 2006–2015), with a strong degree of regional symmetry (Figure 5.8k). As seen for nitrate, changes are most marked in low-latitude upwelling regions, which are projected to show the largest absolute declines. As for nitrate, projected NPP changes are lower for the RCP26 scenario (Figure 5.8j), but the overall uncertainty is dominated by internal and inter-model variability in 2100 (Figure 5.8l) which results in no clear separation of NPP trends between the RCP85 and RCP26 (Figure 5.8j). Tropical ocean NPP is projected to show a large decline, but is underpinned by substantial intermodal uncertainty, with mean changes of 11 ± 24% across the suite of CMIP5 models by 2100, relative to 2000 under RCP8.5 (Laufkötter et al. 2015). However, if emergent constraints from the historical record that link the variability of tropical productivity to temperature anomalies then a four-fold decline in inter-model uncertainty results. This leads to a projected tropical ocean decline of 11 ± 6%, or from 6.8–16.2% across 90% confidence limits, depending on which historical constraint is used (Kwiatkowski et al. 2017). NPP is projected to increases for higher latitude regions, such as the Arctic and Southern Oceans.

Detailed analyses of the interplay between different drivers of NPP, including temperature, light, nutrient levels and grazing from a subset of CMIP5 models, reveals a complex interplay with a strong latitudinal dependence (Laufkötter et al. 2015) summarised in Figure 5.12. Warming acts to enhance growth, most notably at lower latitudes, while light conditions are also predicted to improve, mostly at the poles. Nutrient limitation shows a much more complex response across models, but tends to increase in the tropics and northern high latitudes, with little change in the Southern Ocean. Taken together there is a tendency for reduced growth rates across the entire ocean, but there is a large amount of inter-model variability. The changes in growth are allied to a consistent increase in the grazing loss of biomass to upper trophic levels. Since AR5, we have an increasing body of literature concerning role of biological feedbacks, especially due to interactions between organisms, specific physiological responses and from upper trophic levels on nutrient concentrations, linked to variable food quality (Kwiatkowski et al. 2018), resource recycling (Boyd et al. 2015a; Tagliabue et al. 2017)

and interactions between organisms (Lima-Mendez et al. 2015), but their role in shaping the response of NPP to climate change remains a major unknown. Lastly, modelling work suggests that the increasing deposition of anthropogenic aerosols (supplying N and Fe) stimulates biological activity (Wang et al. 2015b) and may compensate for warming driven reductions in primary productivity (Wang et al. 2015b), but these effects do not form part of the CMIP5 projections assessed here.

CMIP5 models show a strong negative relationship between changes in stratification that reduces net nutrient supply and integrated export production (Fu et al. 2016). Export production is projected to decline by 8.9–15.8% or 1.6–4.9% (across 90% confidence intervals) by 2100, relative to 2000 for the RCP8.5 or RCP2.6 scenario, respectively (Bopp et al. 2013; Fu et al. 2016; Laufkötter et al. 2016). The projected changes in export production can be larger than global primary production because they are affected by both the NPP changes, but also how shifts in food web structure modulates the 'transfer efficiency' of particulate organic material (Guidi et al. 2016; Tréguer et al. 2018), which then affects the sinking speed and lability of exported particles through the ocean interior to the sea floor (Bopp et al. 2013; Fu et al. 2016; Laufkötter et al. 2016). Declines in export production over much of the ocean mean that the flux arriving at the sea floor is also predicted to decline, while increases in export production are projected in the polar regions that see enhanced NPP (Sweetman et al. 2017).

The realism in model projections can be appraised via their ability to accurately simulate the limiting nutrient in specific ocean regions (Figure 5.11), with high model skill in reproducing surface distributions of nitrate and phosphate (Laufkötter et al. 2015), raising confidence in projections in nitrogen and phosphorus limited systems, but poor skill in reproducing iron distributions (Tagliabue et al. 2016) lowering confidence in iron limited regions (Figure 5.11). In addition to concentrations of specific nutrients, the response of NPP to environmental change is strongly controlled by accurate representation of the ratio of resources (Moreno et al. 2017). Overall CMIP5 models skill in reproducing patterns of NPP and export production from limited satellite derived estimates range from poor to average (correlation coefficients of 0.1–0.6 across different models (Laufkötter et al. 2016; Moreno et al. 2017)), but it should be noted that complete comprehensive observational datasets do not exist for these metrics with very few *in situ* observations. As export production is a much better understood net integral of changing net nutrient supply (Sarmiento and Gruber, 2002) and can be constrained by interior ocean nutrient and oxygen levels, there is *medium confidence* in these projections for global changes. Improving the ability of models to reproduce historical NPP is crucial for more accurate projections as model biases in simulating contemporary ocean biogeochemistry play a key role in driving future projections (Fu et al. 2016).

Overall, these assessments balance the range of projections across models alongside the strength of different kinds of observational constraints available, as well as our theoretical or experimental understanding of the impact of a warmer, more stratified ocean on NPP and export production. As for AR5, net primary productivity is *very likely* to decline by 4–11% by 2081–2100, relative to 1850–1900, across CMIP5 models for RCP8.5, but there is *low confidence* for this estimate due to the *medium agreement* among models and the *limited evidence* from observations. It is *very likely* that tropical NPP will decline by 7–16% by 2100 for RCP8.5 with *medium confidence*, as there are improved constraints from historical variability in this region. Globally, the increased stratification in the future is *very likely* to reduce export production by 9–16% in response to reduced nutrient supply, especially in tropical regions (*medium confidence*).

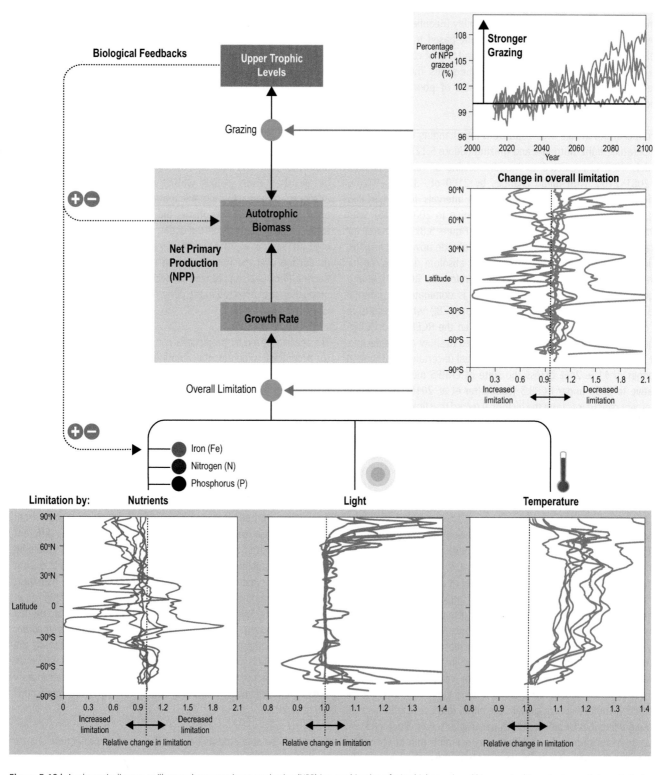

Figure 5.12 | A schematic diagram to illustrate how net primary production (NPP) is a combination of microbial growth and biomass. In this context, growth is controlled by three limiting factors (nutrients, light and temperature), while biomass is affected by grazing. The grey lines in the plots represent results from different Coupled Model Intercomparison Project Phase 5 (CMIP5) models as reported by Laufkötter et al. (2015). Poorly understood feedbacks from upper trophic levels on autotroph biomass and nutrients are represented by dashed arrows.

Box 5.1 | Time of Emergence and Exposure to Climate Hazards

The concept of time of emergence (ToE) is defined as the time at which the *signal* of climate change in a given variable emerges from a measure of the background variability or *noise* (SROCC Glossary). In associating a calendar date with the detection, attribution and projection of climate trends, the concept of a ToE has proved useful for policy and planning particularly through informing important climatic thresholds and the uncertainties associated with past and future climate change (Hawkins and Sutton, 2012). However, there is not a single agreed metric and the ToE for a given variable thus depends on choices regarding the space and time scale, the threshold at which emergence is defined and the reference period (IPCC 5th Asseessment Report (AR5) Working Group I (WGI) Section 11.3.2.1). Recently, the ToE concept has been expanded to consider variables related to climatic hazards to marine organisms and ecosystems such as pH, carbonate ion concentrations, aragonite and calcite saturation states, nutrient levels and marine primary productivity (Box 5.1, Figure 1) (Ilyina et al. 2009; Friedrich et al. 2012; Keller et al. 2014b; Lovenduski et al. 2015; Rodgers et al. 2015). ToE assessments for the ocean typically quantify the internal variability using the standard deviation of the detrended data over a given time period (Keller et al. 2014b; Rodgers et al. 2015; Henson et al. 2016; Henson et al. 2017), the scenario and model uncertainty associated with different climate scenarios and across available ESMs (Frölicher et al. 2016), and in some cases the autocorrelation of noise (Weatherhead et al. 1998). As more components of 'noise' are accounted for, the ToE lengthens and the ToE is also affected by whether a control simulation or historical variability is used to determine the noise (Hameau et al. 2019).

This assessment considers the ToE of hazards exposed to by marine organisms and ecosystems. These biological components of the ocean respond to climate hazards that emerge locally, rather than to the global and basin-scale averages reported in WGI AR5 (Stocker et al. 2013). Overall, ESMs show that there is an ordered emergence of the climate variables, with pH emerging rapidly across the entire open ocean, followed by sea surface temperature (SST), interior oxygen, upper ocean nutrient levels and finally NPP under both Representative Concentration Pathway (RCP)2.6 and RCP8.5 relative to the 1861–1900 reference period (Box 5.1, Figure 1). Anthropogenic signals remain detectable for over large parts of the ocean even for the RCP2.6 scenario for pH and SST, but are *likely* lowered for nutrients and NPP in the 21st century. For example, for the open ocean, the anthropogenic pH signal in Earth System Models (ESM) historical simulations is *very likely* to have emerged for three-quarters of the ocean prior to 1950 and it is *very likely* over 95% of the ocean has already been affected, with little discernable difference between scenarios. The climate signal of oxygen loss will *very likely* emerge from the historical climate by 2050 with a *very likely* range of 59–80% by 2031–2050 and rising with a *very likely* range of 79–91% of the ocean area by 2081–2100 (RCP8.5 emissions scenario). The emergence of oxygen loss is smaller in area under RCP2.6 scenario in the 21st century and by 2090 the emerged area is declining (Henson et al. 2017) (Box 5.1 Figure 1). It has also been shown that changes to oxygen solubility or utilisation may emerge earlier than bulk oxygen levels (Hameau et al. 2019).

It must be noted that variability will be greater in the coastal ocean than for the open ocean, which will be important for both hazard exposure for coastal species and the detection of trends. For example, although signals of anthropogenic influences have already emerged from internal variability in the late 20th century for global and basin-scale averaged ocean surface and sub-surface temperature (*very likely*) (AR5 WGI Summary for Policymakers), their ToE and level of confidence vary greatly at local scales and in coastal seas (Frölicher et al. 2016). Pelagic organisms with small range size may thus be more (or less) at risk to warming with earlier (or later) ToE at the scale of the area that they inhabit. From an observational standpoint, analyses that account for autocorrelation of noise suggest time series of around a decade are sufficient to detect a trend in pH or SST, whereas datasets spanning 30 years or longer are typically needed for detection of emergence at local scales for oxygen, nitrate and primary productivity (Henson et al. 2016).

The rapidity of change and its geographic scope, encompassed in the ToE, can be linked to concepts of exposure to hazard and vulnerability of biota. As organisms have evolved to be adaptable to natural variations in the environmental conditions of their habitats, changes to their habitat conditions larger than that typically experienced or specific biological thresholds such as upper temperature or oxygen tolerance may become hazardous (Mora et al. 2013). This would then move from the statistical nature of the 'detection and attribution' nature of the ToE discussed above towards timescales of impacts on organisms useful for ecosystem projections. In doing so, it will be important to think about the differences in habitat suitability between different organisms, including their specific thresholds for specific drivers, for example, temperature, oxygen or calcium carbonate stability. Further, thresholds vary depending on habitat, for example, warming thresholds for coral bleaching (Pendleton et al. 2016) may differ from the temperature and oxygen thresholds for fishes such as Atlantic cod and tunas (Deutsch et al. 2015). Moreover, species with fast generation times relative to the ToE of key habitat conditions (e.g., phytoplankton) may evolve more quickly to environmental change and be less vulnerable to climate change than longer-lived, slower generation time species (e.g., large sharks) (Jones and Cheung, 2018). However, evidence on evolutionary adaptation to expected climate change is limited, thus while shorter generation time may facilitate adaptation to environmental change, it does not necessarily result in successful adaptation of organisms (Section 5.2.3.1).

5

Box 5.1 (continued)

Earlier ToE and their subsequent biological impacts on organisms and ecosystems increase the urgency of policy responses through both climate mitigation and adaptation (Sections 5.5). However, the rapid emergence of hazards at the local scale in the near-term (already past or in this decade) such as warming and ocean acidification and the resulting impacts on some of the more sensitivity or less adaptive biodiversity and ecosystem services may post challenges for international and regional policies as their often require multiple decades to designate and implement (Box 5.6). In contrast, scope for adaptation for national and local ocean governance can be more responsive to rapid changes (Sections 5.5.2, 5.5.3). This highlights the opportunities for multi-level adaptation that allows for reducing climate risks that are expected to emergence of stressors and impacts at different time frame (Mackenzie et al. 2014).

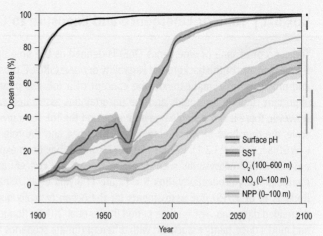

Box 5.1, Figure 1 | Time of emergence of key ocean condition variables: sea surface temperature (SST), surface pH, 100–600 m oxygen (O_2), 0–100 m nitrate (NO_3), and 0–100 m integrated net primary production (NPP). The year of emergence represents the year when the mean change relative to the reference period of 1861–1900 is above the standard deviation of each variable over the historical period (Frölicher et al. 2016) and is expressed here in terms of the rate at which different climate signals emerge as a proportion of total ocean area for the Representative Concentration Pathway (RCP)8.5 scenario. The final area (and standard deviation) by 2100 under the RCP2.6 scenario is indicated by vertical lines at 2100.

5.2.3   Impacts on Pelagic Ecosystems

Marine pelagic ecosystems (the water column extending from the surface ocean down to the deep sea floor) face increasing climate related hazards from the changing environmental conditions (see Section 5.2.2). WGII AR5 (Pörtner et al. 2014) concluded, as also confirmed in Section 5.2.2, that long time series of more than three or four decades in length are necessary for determining biological trends in the ocean. However, long-term biological observations of pelagic ecosystems are rare and biased toward mid to high-latitude systems in the Northern Hemisphere (Edwards et al. 2013; Poloczanska et al. 2013; Poloczanska et al. 2016). This assessment, therefore, combines multiple lines of evidence ranging from experiments, field observations to model simulations to detect and attribute drivers of biological changes in the past, project future climate impacts and risks of pelagic ecosystems. In this section the pelagic ecosystem is subdivided into the surface, epipelagic ocean (<200 m, the uppermost part of the ocean that receives enough sunlight to allow photosynthesis) (Section 5.2.3.1) and the deep pelagic ocean, comprising the twilight, mesopelagic zone (200–1000 m) and the dark, bathypelagic zone (>1000 m deep) (Section 5.2.3.2). Although the WGII AR5 Chapter 30 defined the deep sea as below 1000 m (Hoegh-Guldberg et al. 2014), the absence of photosynthetically useful light and ensuing critical ecological, biogeochemical transformations, and altered human interactions that occur on much of the sea floor below 200 m have led both pelagic and benthic biologists to include the ocean waters and seafloor below 200 m within the definition of the deep sea (Herring and Dixon, 1998; Gage, 2003).

5.2.3.1   The Epipelagic Ocean

This section synthesises new evidence since AR5 to assess observed changes in relation to the effects of and the interactions between multiple climate and non-climate hazards, and to project future risks of impacts from these hazards on the epipelagic organisms, communities and food web interactions, and their scope and limitation to adapt.

5.2.3.1.1   Detection and attribution of biological changes in the epipelagic ocean

Temperature-driven shifts in distribution and phenology
WGII AR5 concluded that the vulnerability of most organisms to warming is set by their physiology, which defines their limited temperature ranges and thermal sensitivity (Pörtner et al. 2014). Although different hypotheses have been proposed since AR5 to explain the mechanism linking temperature sensitivity of marine organisms and their physiological tolerances (Schulte, 2015; Pörtner et al. 2017; Somero et al. 2017), evidence from physiological experiments and observations from paleo- and contemporary periods continue to support the conclusion from AR5 on the impacts of temperature change beyond thermal tolerance ranges on biological functions such as metabolism, growth and reproduction (Payne et al. 2016; Pörtner and Gutt, 2016; Gunderson et al. 2017), contributing to changes in biogeography and community structure (Beaugrand et al. 2015; Stuart-Smith et al. 2015) (*high agreement, high confidence*). Comparison of biota across land and ocean suggests that marine species are generally inhabiting environment that is closer to their upper temperature limits, explaining the substantially higher rate of

local extirpation related to warming relative to those on land (Pinsky et al. 2019). Hypoxia and acidification can also limit the temperature ranges of organisms and exacerbate their sensitivity to warming (Mackenzie et al. 2014; Rosas-Navarro et al. 2016; Pörtner et al. 2017), although interactions vary strongly between species and biological processes (Gobler and Baumann, 2016; Lefevre, 2016).

Shifts in distribution of marine species from phytoplankton to marine mammals continued to be observed since AR5 across all ocean regions (Poloczanska et al. 2016). Recent evidence continues to support that a large proportion of records of observed range shifts in the epipelagic ecosystem (Poloczanska et al. 2016) are correlated with ocean temperature, with an estimated average shift in distribution (including range centroids, northward and southward boundaries) from these records of 51.5 ± 33.3 km per decade since the 1950s (Figure 5.13). Such rate of shift is significantly faster than those records for organisms in the seafloor; the latter has an average rate of distribution shift of 29.0 ± 15.5 km per decade (44% of the records for seafloor species with range shifts that are consistent with expectation from the observed temperature changes) (*very likely*) (Figure 5.13). Comparison of global seafloor-derived planktonic foraminifera from pre-industrial age with recent (from year 1978) communities show that the recent assemblages differ from their pre-industrial with increasing dominance of warmer or cooler species that are mostly consistent with temperature changes (Jonkers et al. 2019). Rate of observed responses also varies between and within animal groups among ocean regions, with zooplankton and fishes having faster recorded range shifts (Pinsky et al. 2013; Asch, 2015; Jones and Cheung, 2015; Poloczanska et al. 2016). For example, analysis of the Continuous Plankton Recorder (CPR) data-series from the north Atlantic in the last decades shows that the range of dinoflagellates tended to closely track the velocity of climate change (the rate of isotherm movement). In contrast, the distribution range of diatoms shifted much more slowly (Chivers et al. 2017) and its distribution seems to be primary influenced by multi-decadal variability rather than from secular temperature trends. The CPR surveys have also provided evidence that some calanoid copepods are expanding poleward in the Northeast Atlantic, at a rate up to 232 km per decade (Beaugrand, 2009; Chivers et al. 2017), although different calanoid species respond differently in the rate and direction of shifts (Philippart et al. 2003; Edwards and Richardson, 2004; Asch, 2015; Crespo et al. 2017). Overall, the observed changes in biogeography are consistent with expected responses to changes in ocean temperature for the majority of marine biota (*high confidence*). This is also consistent with theories and experimental evidence that scale from individual organisms' physiological responses to community level effects (*high confidence*). Sensitivity of organisms' biogeography varies between taxonomic groups (*high confidence*).

The rate and direction of observed range shifts are shaped by the interaction between climatic and non-climatic factors (Poloczanska et al. 2013; Sydeman et al. 2015; Poloczanska et al. 2016), such as local temperature and oxygen gradients in the habitat across depth (Cheung et al. 2013; Deutsch et al. 2015), latitude and longitude (Burrows et al. 2014; Barton et al. 2016), ocean currents (Sunday et al. 2015; Barton et al. 2016; García Molinos et al. 2017),

bathymetry in all or part of their life stages (for organisms living on or close to the seafloor) (Pinsky et al. 2013; Kleisner et al. 2015), geographical barriers (Pinsky et al. 2013; Burrows et al. 2014), availability of food and critical habitat (Sydeman et al. 2015), fishing and other non-climatic human impacts (Engelhard et al. 2014; Hoegh-Guldberg et al. 2014). Moreover, observed range shifts in respond to climate change in some regions such as the north Atlantic are strongly influenced by warming due to multi-decadal variability (Edwards et al. 2013; Harris et al. 2014), suggesting that there is a longer time-of-emergence of range shifts from natural variability and a need for longer biological time series for robust attribution. The rate of shifts in biogeography of organism is influenced by multiple climatic and non-climatic factors (*high confidence*) that can result in non-synchronous shifts in community composition (*high confidence*). There is general under-representation of biogeographical records in low latitudes (Dornelas et al. 2018), rendering detection and attribution of shifts in biogeography in these regions having *medium confidence*. The variation in responses of marine biota to range shifts can cause spatial restructuring of the pelagic ecosystem with consequences for organisms at higher trophic levels (Chivers et al. 2017; Pecl et al. 2017) (*high confidence*). Marine ectotherms have demonstrated some capacity for physiological adjustment and evolutionary adaptation that lowers their sensitivity to warming and decrease in oxygen (Pörtner et al. 2014; Cavallo et al. 2015) (*low confidence*). However, historical responses in abundance and ranges of marine species to ocean warming suggest that adaptation not always suffices to mitigate projected impacts (WGII AR5 Chapter 6) (*high confidence*).

Marine reptiles, seabirds and mammals breathe air, instead of obtaining oxygen from water, and many of them spend some of their life cycle on land, being their abundance and distribution still affected by temperature (Pörtner et al. 2014). Long term population changes and shifts in distribution associated with climate change have been observed for temperate species of seabirds and marine mammals (Henderson et al. 2014; Hiscock and Chilvers, 2014; Ramp et al. 2015) (*high confidence*). For example, Laysan, *Phoebastria immutabilis,* and Wandering, *Diomedea exulans,* albatross have responded positively to climate change as they have been able to take advantage of the increased intensity of winds. This has allowed them to forage farther and faster, making their foraging trips shorter, increasing their foraging efficiency and breeding success (Descamps et al. 2015; Thorne et al. 2016). For reptiles, like sea turtles and snakes, temperature directly affects important life history traits including hatchling size, sex, viability and performance (*high confidence*) (Hays et al. 2003; Pike, 2014; Dudley et al. 2016; Santos et al. 2017). This is particularly important for marine turtles as changing temperatures will affect the hatchling sex ratio because sex is determined by nest site temperature (*high confidence*) (Hatfield et al. 2012; Santidrián Tomillo et al. 2014; Patricio et al. 2017). Loss of breeding substrate, including mostly coastal habitats such as sandy beaches (Section 5.3.3), can reduce the available nesting or pupping habitat for land breeding marine turtles, lizards, seabirds and pinnipeds (Fish et al. 2005; Fuentes et al. 2010; Funayama et al. 2013; Reece et al. 2013; Katselidis et al. 2014; Patino-Martinez et al. 2014; Pike et al. 2015; Reynolds et al.

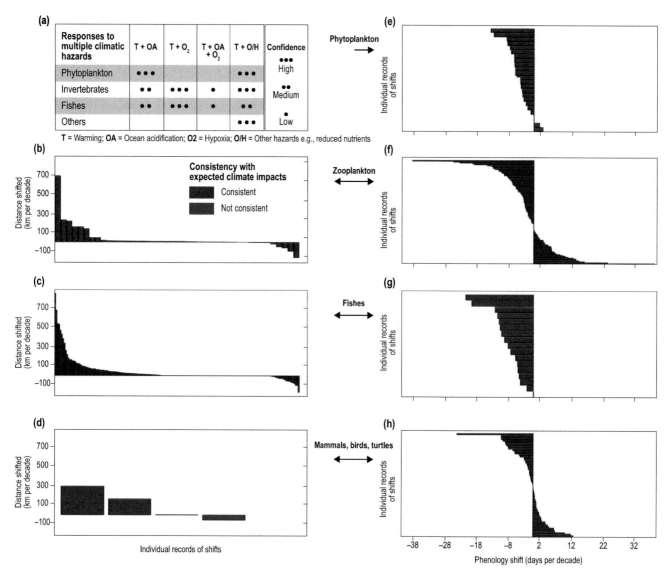

Figure 5.13 | Evidence of climate change responses of marine organisms to changes in ocean conditions under climate change. (a) evidence of interactive effects (including synergistic and antagonistic) of multiple climatic hazards (based on Przeslawski et al. (2015); Lefevre (2016); Section 5.2.2, 5.2.3, 5.2.4, 5.3). 'Others' mainly include mammals, seabirds and marine reptiles). The lighter-coloured cell represents insufficient information to draw conclusion; (b–d) observations on changes in latitudinal range and (e–h) phenology (based on Poloczanska et al. 2013). For b–h, each bar represents one record.

2015; Marshall et al. 2017) (*high confidence*). Climatic hazards such as SLR contributes to the loss of these coastal habitats (see Section 5.3 and Chapter 3). Changes in ocean temperature will also indirectly impact marine mammals, seabirds and reptiles by changing the abundance and distribution of their prey (Polovina, 2005; Polovina et al. 2011; Doney et al. 2012; Sydeman et al. 2015; Briscoe et al. 2017; Woodworth-Jefcoats et al. 2017) (*high confidence*). The distributions of some of these large animals is determined by the occurrence and persistence of oceanic bridges and barriers that are related to climate driven processes (Ascani et al. 2016; McKeon et al. 2016). For example, the decline of Arctic sea ice is affecting the range and migration patterns of some species and is allowing the exchange of species previously restricted to either the Pacific or Atlantic oceans (Alter et al. 2015; George et al. 2015; Laidre et al. 2015; MacIntyre et al. 2015; McKeon et al. 2016; Breed et al. 2017; Hauser et al. 2017) (Chapter 3). Also, the range expansion of some of these predatory megafauna can affect

species endemic to the habitat; for example, while the decrease in summer sea ice in the Arctic may favour the expansion of killer whales (*Orcinus orca*), their occurrence can result in narwhale (*Monodon monoceros*) to avoid the use of key habitats to reduce the risk of killer whales' predation (Bost et al. 2009; Sydeman et al. 2015; Breed et al. 2017) (see Chapter 3; section 3.2.1.4). In addition, marine mammals, seabirds and sea turtles present habitat requirements associated with bathymetric and mesoscale features that facilitate the aggregation of their prey (Bost et al. 2015; Kavanaugh et al. 2015; Hindell et al. 2016; Hunt et al. 2016; Santora et al. 2017). The persistence and location of these features are linked to variations in climate (Crocker et al. 2006; Baez et al. 2011; Dugger et al. 2014; Abrahms et al. 2017; Youngflesh et al. 2017) and to foraging success, juvenile recruitment, breeding phenology, growth rates and population stability (Costa et al. 2010; Ancona and Drummond, 2013; Ducklow et al. 2013; Chambers et al. 2014; Descamps et al. 2015; Abadi et al. 2017; Bjorndal et al. 2017; Fluhr et al. 2017; Youngflesh et al. 2017) (*high confidence*). Overall, recent evidence further support that impacts of climate change on some marine reptiles, mammals and birds have been observed in recent decades (*high confidence*) and that the direction of impacts vary between species, population and geographic locations (Trivelpiece et al. 2011; Hazen et al. 2013; Clucas et al. 2014; Constable et al. 2014; George et al. 2015) (*high confidence*).

Warming has contributed also to observed changes in phenology (timing of repeated seasonal activities) of marine organisms (Gittings et al. 2018), although observations are biased towards the northeast Atlantic (Poloczanska et al. 2016; Thackeray et al. 2016). Shifts in the timing of interacting species have occurred in the last decades, eventually leading to uncoupling between prey and predators, with cascading community and ecosystem consequences (Kharouba et al. 2018; Neuheimer et al. 2018). Timing of spring phenology of marine organisms is shifting to earlier in the year under warming, at an average rate of 4.4 ± 1.1 days per decade (Poloczanska et al. 2013), although it is variable among taxonomic groups and among ocean regions (Lindley and Kirby, 2010). This is consistent with the expectations based on the close relationship between temperature and these biological events, supporting evidence from AR5 (Bruge et al. 2016; Poloczanska et al. 2016). Thus, the growing amount of literature and new studies since AR5 WGII and SR15 further support that phenology of marine ectotherms in the epipelagic systems are related to ocean warming (*high confidence*) and that the timing of biological events has shifted earlier (*high confidence*).

Observed impacts of multiple climatic hazards

WGII AR5 concludes that multiple climatic hazards from ocean acidification, hypoxia and decrease in nutrient and food supplies pose risks to marine ecosystems, and the risk can be elevated when combined with warming (Riebesell and Gattuso, 2014; Gattuso et al. 2015). In a recent meta-analysis of 632 published experiments, primary production by temperate non-calcifying plankton increases with elevated temperature and CO_2, whereas tropical plankton decreases productivity because of acidification (Nagelkerken and Connell, 2015). Also, temperature increases consumption and metabolic rates of herbivores but not secondary production; the latter decreases with acidification in calcifying and non-calcifying species. These effects together create a mismatch with carnivores whose metabolic and foraging costs increase with temperature (Nagelkerken and Connell, 2015). Warming may also exacerbate the effects of ocean acidification on the rate of photosynthesis in phytoplankton (Lefevre, 2016). There is some, but limited, reports of observed impacts on calcified pelagic organisms that are attributed to secular trend in ocean acidification and warming (Harvey et al. 2013; Kroeker et al. 2013; Nagelkerken et al. 2015; Boyd et al. 2016). For example, Rivero-Calle et al. (2015) reported, using CPR archives, that stocks of coccolithophores (a group of phytoplankton that forms calcium carbonate plateles) have increased by 2% to over 20% in the north Atlantic over the last five decades, and that this increase is linked to synergistic effects of increasing anthropogenic CO_2 and rising temperatures, as supported by their statistical analysis and a number of experimental studies. Most of the available evidence

supports that ocean acidification and hypoxia can act additively or synergistically between each other and with temperature across different groups of biota (Figure 5.13). Limitation of nutrient and food availability and predation pressures can further increase the sensitivity of organismal groups to climate change in specific ecosystems (Riebesell et al. 2017). Climate change also affects organisms indirectly through the impacts on competitiveness between organisms that favour those that are more adaptive to the changing environmental conditions (Alguero-Muniz et al. 2017) and changes in trophic interactions (Seebacher et al. 2014). Overall, direct *in situ* observations and laboratory experiments show that there are significant responses to the multiple stressors of warming, ocean acidification and low oxygen on phytoplankton, zooplankton and fishes and that these responses can be additive or synergistic (*high confidence*, Figure 5.13).

5.2.3.1.2 Future changes in the epipelagic ocean

WGII AR5 and SR15 conclude that projected ocean warming will continue to cause poleward shifts in the distribution and biomass of pelagic species, paralleled by altered seasonal timing of their activities, species abundance, migration pattern and reduction in body size in the 21st century under scenarios of increasing greenhouse gas emission (Pörtner et al. 2014; Hoegh-Guldberg et al. 2018). Simultaneously, projected expansion of OMZ and ocean acidification could lead to shifts in community composition toward hypoxia-tolerant and non-calcified organisms, respectively. However, these projected biological changes in the ocean raise questions about how individuals, communities and food webs will respond to the multiple impacts from climatic and non-climatic stressors in the future, and the feedbacks of the effects of their ecological impacts on modifying the physical and biogeochemical conditions of the ocean (Schaum et al. 2013; Boyd et al. 2016; O'Brien et al. 2016; Moore, 2018). This section focuses on addressing these questions in order to assess the future risk of impacts of climate change on the epipelagic ecosystem.

Future projections on phytoplankton distribution, community structure and biomass

While analysis of outputs from CMIP5 ESMs project that global average NPP and biomass of phytoplankton community will decrease in the 21st century under RCP2.6 and RCP8.5 (see Section 5.2.2.6). However, the future risk of impacts of epipelagic ecosystem can also depend on changes in community structure of phytoplankton species. Barton et al. (2016) projected the biogeography of 87 taxa of phytoplankton (diatoms and dinoflagellates) in the north Atlantic to 2051–2100 relative to the past (1951–2000) with scenarios of changes in temperature and other ocean conditions such as salinity, density and nutrients under RCP8.5. The study found that 74% of the studied taxa exhibit a poleward shift at a median rate of 12.9 km per decade, but 90% of the taxa shift eastward at a median rate of 42.7 km per decade. Such changes may affect food webs and biogeochemical cycles, and with consequence to the productivity of living marine resources (Stock et al. 2014; Barton et al. 2016).

5

Outputs from CMIP5 ESMs suggest that projected warming and reduction in nutrient availability in low latitudes, as a result of increasing stratification of the ocean under climate change, will increase the dominance of small-sized phytoplankton, growing more efficiently than larger taxa at low nutrient levels (Dutkiewicz et al. 2013b). Dominant groups in subtropical oceans, like the picoplanktonic cyanobacteria *Synechococcus* and *Prochlorococcus*, are projected to expand their range of distribution towards higher latitudes and increase their abundances by 14–29%, respectively, under a future warmer ocean (Flombaum et al. 2013), although synergistic effects of warming and CO_2 on photosynthetic rates could lead to a dominance of *Synechococcus* over *Prochlorococcus* (Fu et al. 2007) (*low confidence*). Similarly, temperature-driven range shifts towards higher latitudes are also likely for tropical diazotrophic (N_2-fixing) cyanobacteria, although they could disappear from parts of their current tropical ranges where future warming may exceed their maximum thermal tolerance limits (Hutchins and Fu, 2017) (*low confidence*). Modelling experiments show that the effects of warming on phytoplankton community will be exacerbated by ocean acidification at levels expected in the 21st century for RCP8.5, leading to increasing growth rate responses of some phytoplankton groups, such as diazotrophs and *Synechococcus*, with predicted increases in biomass up to 10% in tropical and subtropical waters (Dutkiewicz et al. 2015) (*low confidence*). Furthermore, warming is projected to interact with decreasing oxygen levels and increases in iron in the nutrient-impoverished subtropical waters, favoring the dominance of the diazotrophic colonial cyanobacteria *Trichodesmium* (Sohm et al. 2011; Boyd et al. 2013; Ward et al. 2013; Hutchins and Fu, 2017) (*medium confidence*).

Regional differences in the changes in phytoplankton community and their impacts on epipelagic ecosystem are however complex and depends on multiple interactions of co-varying climate change stressors at regional level (Boyd and Hutchins, 2012). Based on global ocean model simulations, Boyd et al. (2015b) show that the interaction between warming, increased CO_2 and a decline in phosphate and silicate would benefit coccolithophores against diatoms in the northern north Atlantic, despite decreasing rates of calcification. Evidence, based on long-term experiments of acclimation or adaptation to increasing temperatures in combination with elevated CO_2, show that individual growth and carbon fixation rates of coccolithophores at high CO_2 are modulated by temperature, light, nutrients and UV radiation, and could increase calcification while the responses are also species-specific (Lohbeck et al. 2012; Khanna et al. 2013). Calcification of planktonic foraminifera will be however negatively affected by acidification (Roy et al. 2015), and their populations are predicted to experience the greatest decrease in diversity and abundance in sub-polar and tropical areas, under RCP8.5 (Brussaard et al. 2013), however environmental controls of calcite production by foraminifera are still poorly understood (*low confidence*). Boyd et al. (2015b) analysis indicate also that diatoms would benefit from the synergistic effects of increased warming and iron supply in the northern Southern Ocean, as supported by laboratory experiments and field studies with polar diatoms (Rose et al. 2009) (*low confidence*). At low-latitude provinces, projected concurrent increases of CO_2 and iron, and decreases in both nitrate and phosphate supply, may favour nitrogen fixers, but with ocean regional variability, since iron is thought to limit N_2 fixation in the eastern Pacific and phosphorus in the Atlantic Ocean (Gruber, 2019; Wang et al. 2019). However, recent experimental work with the diazotrophic colonial *Trichodesmium* and the unicellular *Crocosphaera* have shown a broad range of responses from rising CO_2, with either increases or decreases in N_2 fixation rates, and with mixed evidence on co-limiting processes (Eichner et al. 2014; Garcia et al. 2014; Gradoville et al. 2014; Walworth et al. 2016; Hong et al. 2017; Luo et al. 2019) (*low confidence*).

Overall, the response of phytoplankton to the interactive effects of multiple drivers is complex, and presently ESMs do not resolve the full complexity of their physiological responses (Breitberg et al. 2015; Hutchins and Boyd, 2016; O'Brien et al. 2016), precluding a clear assessment of the effects of these regional distinctive multi-stressor patterns (*high confidence*).

Future projections on zooplankton distribution and biomass

An ensemble of 12 CMIP5 ESMs project average declines of 6.4 ± 0.79% (95% confident limits) and 13.6 ± 1.70%) in zooplankton biomass in the 21st century relative to 1990–1999 historical values under RCP2.6 and RCP8.5 (Kwiatkowski et al. 2019). Also, production of mesozooplankton is projected from a single ESM to decrease by 7.9% between 1951–2000 and 2051–2100 under RCP8.5 (Stock et al. 2014). Such projected decreases in zooplankton biomass and production are partly contributed by climate-induced reduction in phytoplankton production and trophic transfer efficiency particularly in low-latitude ecosystems (Stock et al. 2014) (5.2.2.6). The impacts may be larger than these projections if changes in the relative abundance of carbon, nitrogen and phosphorus are considered by the models (Kwiatkowski et al. 2019). The overall projected decrease in zooplankton biomass is characterised by a strong latitudinal differences, with the largest decrease in tropical regions and increase in the polar regions, particularly the Arctic Ocean (Chust et al. 2014; Stock et al. 2014; Kwiatkowski et al. 2019) (Chapter 3) (*high agreement*). However, the projected increase in zooplankton biomass in the polar region may be affected by the seasonality of light cycle at high latitudes that may limit the bloom season at high latitude (Sundby et al. 2016). The projected decrease in zooplankton abundance, particularly in tropical regions, can impact marine organisms higher in the foodweb, including fish populations that are important to fisheries (Woodworth-Jefcoats et al. 2017). Therefore, there is *high agreement* in model projections that global zooplankton biomass will *very likely* reduce in the 21st century, with projected decline under RCP8.5 almost doubled that of RCP2.6 (*very likely*). However, the strong dependence of the projected declines on phytoplankton production (*low confidence*, 5.2.2.6) and simplification in representation of the zooplankton communities and foodweb render their projections having *low confidence*.

Future responses of zooplankton species and communities to climate change are however affected by interactions between multiple climatic drivers. Experiments in laboratory show that acidification could partly counteract some observed effects of increased temperature on zooplankton, although the level and direction of the biological responses vary largely between species (Mayor et al. 2015;

5

Garzke et al. 2016), with results ranging from no effects (Weydmann et al. 2012; McConville et al. 2013; Cripps et al. 2014; Alguero-Muniz et al. 2016; Bailey et al. 2016), to negative effects (Lischka et al. 2011; Cripps et al. 2014; Alguero-Muniz et al. 2017) or positive effects (Alguero-Muniz et al. 2017; Taucher et al. 2017). These differences in response can affect trophic interactions between zooplankton species; for example, some predatory non-calcifying zooplankton may perform better under warmer and lower pH conditions, leading to increased predation on other zooplankton species (Caron and Hutchins, 2012; Winder et al. 2017). Therefore, the large variation in sensitivity between zooplankton to future conditions of warming and ocean acidification suggests elevated risk on community structure and inter-specific interactions of zooplankton in the 21st century (*medium confidence*). Consideration of these species-specific responses may further modify the projected changes in zooplankton biomass by ESMs (Boyd et al. 2015a).

Future projections on fish distribution, size and biomass

Recent model projections since AR5 and SR15 continue to support global-scale range shifts of marine fishes at rates of tens to hundreds of km per decade in the 21st century, with rate of shifts being substantially higher under RCP8.5 than RCP2.6 (Jones and Cheung, 2015; Robinson et al. 2015; Morley et al. 2018). Globally, the general direction of range shifts of epipelagic fishes is poleward (Jones and Cheung, 2015; Robinson et al. 2015), while the projected directions of regional and local range shifts generally follow temperature gradients (Morley et al. 2018). Polewards range shifts are projected to result in decreases in species richness in tropical oceans, and increases in mid to high-latitude regions leading to global-scale species turnover (sum of species local extinction and expansion) (Ben Rais Lasram et al. 2010; Jones and Cheung, 2015; Cheung and Pauly, 2016; Molinos et al. 2016) (*medium confidence* on trends, *low confidence* on magnitude because of model uncertainties and limited number of published model simulations). For example, species turnover relative to their present day richness in the tropical oceans (30°N–30°S) is projected to be 14–21% and 37–39% by 2031–2050 and 2081–2100 under RCP8.5 (ranges of mean projections from two sets of simulation for marine fish distributions) (Jones and Cheung, 2015; Molinos et al. 2016). In contrast, high-latitude regions (>60°N–60°S) is projected to have higher rate of species turnover than the tropics (an average of 48% between the two data sets for region >60°N). The high species turnover in the Arctic is explained by species' range expansion from lower-latitude and the relatively lower present day fish species

richness in the Arctic. The projected intensity of species turnover is lower under lower emission scenarios (Jones and Cheung, 2015; Molinos et al. 2016) (see also Section 5.4.1) (*high confidence*). Projections from multiple fish species distribution models show hotspots of decrease in species richness in the Indo-Pacific region, and semi-enclosed seas such as the Red Sea and Persian Gulf (Cheung et al. 2013; Burrows et al. 2014; García Molinos et al. 2015; Jones and Cheung, 2015; Wabnitz et al. 2018) (*medium evidence, high agreement*). In addition, geographic barriers such as land boundaries in the poleward species range edge in semi-enclosed seas or lower oxygen water in deeper waters are projected to limit range shifts, resulting in larger relative decrease in species richness (*medium confidence*) (Cheung et al. 2013; Burrows et al. 2014; García Molinos et al. 2015; Jones and Cheung, 2015; Rutterford et al. 2015).

Warming and decrease in oxygen content is projected to impact growth of fishes, leading to reduction in body size and contraction of suitable environmental conditions (Deutsch et al. 2015; Pauly and Cheung, 2017), with the intensity of impacts being directly related to the level of climate change. The projected reduction in abundance of larger-bodied fishes could reduce predation and exacerbate the increase in dominance of smaller-bodied fishes in the epipelagic ecosystem (Lefort et al. 2015). Fishes exposed to ocean acidification level expected under RCP8.5 showed impairments of sensory ability and alteration of behaviour including olfaction, hearing, vision, homing and predator avoidance (Kroeker et al. 2013; Heuer and Grosell, 2014; Nagelkerken et al. 2015). The combined effects of warming, ocean deoxygenation and acidification in the 21st century are projected to exacerbate the impacts on the body size, growth, reproduction and mortality of fishes, and consequently increases their risk of population decline (*medium evidence, high agreement, high confidence*).

An ensemble of global-scale marine ecosystem and fisheries models that are part of the Fisheries and Marine Ecosystems Impact Models Intercomparison Project (FISHMIP) undertook coordinated simulation experiments and projected future changes in marine animals (mainly invertebrate and fish) globally under climate change (Lotze et al. 2018). These models represent marine biota and ecosystems differently, ranging from population-based to functional traits- and size-based structure and their responses are driven primarily by temperature and NPP, although oxygen, salinity and ocean advection are considered in a subset of models

Table 5.3 | Projected changes in total animal biomass by the mid- and end- of the 21st century under Representative Concentration Pathway (RCP)2.6 and RCP8.5. Total animal biomass is based on 10 sets of projections for each RCP under the Fisheries and Marine Ecosystems Impact Model Intercomparison Project (FISMIP) (Lotze et al. 2018). The *very likely* ranges of the projections (95% confidence intervals) are provided. Reference period is 1986–2005.

	Total animal biomass (%)			
	RCP2.6		RCP8.5	
Region	2031–2050	2081–2100	2031–2050	2081–2100
>60°N	8.4 ± 9.3	8.5 ± 13.7	7 ± 9.2	−1.1 ± 20.2
30°N–50°N	−8.1 ± 4	−4.5 ± 3.6	−10.1 ± 4.7	−21.3 ± 9.4
30°N–30°S	−7.2 ± 2.7	−7.3 ± 3.1	−9 ± 3.6	−23.2 ± 9.5
30°S–50°S	−3.3 ± 2.1	−3.5 ± 2.5	−4.2 ± 2.9	−9 ± 9.8
<60°S	1.7 ± 4.5	−0.9 ± 2.9	0.7 ± 3.9	12.4 ±11.9

and play a secondary role in affecting the projected changes in biomass (Blanchard et al. 2012; Fernandes et al. 2013; Carozza et al. 2016; Cheung et al. 2016a). Overall, potential total marine animal biomass is projected to decrease by 4.3 ± 2.0% (95% confident intervals) and 15.0 ± 5.9% under RCP2.6 and RCP8.5, respectively, by 2080–2099 relative to 1986–2005, while the decrease is around 4.9% by 2031–2050 across all RCP2.6 and RCP8.5 (*very likely*) (Figure 5.14). Accounting for the removal of biomass by fishing exacerbates the decrease in biomass for large-bodied animals which are particularly sensitive to fishing (*likely* for the direction of changes). Regionally, total animal biomass decreases largely in tropical and mid-latitude oceans (*very likely*) (Table 5.3, Figure 5.14) (Bryndum-Buchholz et al. 2019). The high uncertainty and the *low confidence* in the projection in the Arctic Ocean (Chapter 3) is because of the large variations in simulation results for this region between the ESMs and between the FISHMIP models, as well as the insufficient understanding of the oceanographic changes and their biological implications in the Arctic Ocean. In the Southern Ocean, the decrease in consumer biomass is mainly in the southern Indian Ocean while other parts of the Southern Ocean are projected to have an increase in animal biomass by 2100 under RCP8.5, reflecting mainly the projected pattern of changes in NPP from the ESMs (see Section 5.2.2.6).

Future projections on epipelagic components of the biological pump

A wide range of studies, from laboratory experiments, mesocosm enclosures, synthesis of observations to modeling experiments, provide insights into how the multi-faceted components of the 'biological pump' (the physical and biologically mediated processes responsible for transporting organic carbon from the upper ocean to depth) are projected to be altered in the coming decades. A synthesis of the individual components reported to both influence the performance of the biological pump, and which are sensitive to changing ocean conditions, is presented in Table 5.4. The table lists the putative controlling of each environmental factor, such as warming, that influences the biological pump, and the reported modification (where available) of each individual factor by changing ocean conditions for both the epipelagic ocean and the deep ocean. Analyses of long-term trends in primary production and particle export production, as well as model simulations, reveal that increasing temperatures, leading to enhanced stratification and nutrient limitation, will have the greatest influence on decreasing the flux of particulate organic carbon (POC) to the deep ocean (Bopp et al. 2013; Boyd et al. 2015a; Fu et al. 2016; Laufkötter et al. 2016). However, different lines of evidence (including observation, modeling and experimental studies) provide *low confidence* on the mechanistic understanding of how climatic drivers affect different components of the biological pump in the epipelagic ocean, as well as changes in the efficiency and magnitude of carbon export in the deep ocean (see section below and Table 5.4); this renders the projection of future contribution of the biological carbon pump to the export of POC to the deep ocean having *low confidence*.

(a) RCP2.6

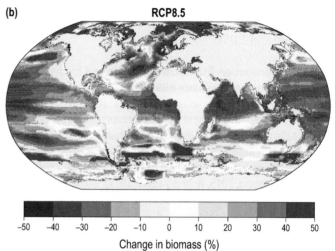

(b) RCP8.5

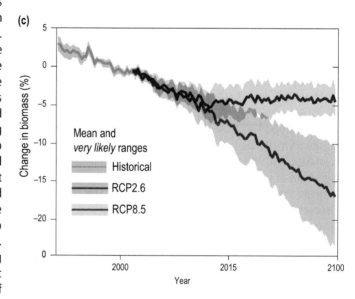

Change in biomass (%)

Figure 5.14 | Projected changes in total animal biomass (including fishes and invertebrates) based on outputs from 10 sets of projections for each Representative Concentration Pathway (RCP) from the Fisheries and Marine Ecosystems Impact Model Intercomparison Project (FISMIP, www.isimip.org/gettingstarted/marine-ecosystems-fisheries) (Lotze et al. 2018); (a, b) multi-model mean change (%) in un-fished total marine animal biomass in 2085–2099 relative to 1986–2005 under RCP2.6 and RCP8.5, respectively. Dotted area represents 8 out of 10 sets of model projections agree in the direction of change (c) projected change in global total animal biomass from 1970 to 2099 under RCP2.6 (red) and RCP8.5 (blue). Variability among different ecosystem and Earth-system model combinations (n=10) expressed as the *very likely* range (95% confidence interval).

Table 5.4 | Projected future changes to the ocean biological pump (adapted from Boyd et al. (2015a)). Environmental controls on individual factors that influence downward POC flux are based on published reports from experiments (denoted by **E**), modelling simulations (**M**) and observations (**O**). In some cases, due to the paucity, and regional specificity, of published reports it has been indicated the sign of the projected change on export (in italics), as opposed to the magnitude. NPP: Net Primary Production; POC: Particulate Organic Carbon; DOC: Dissolved Organic Carbon; TEP: Transparent Exopolymer Particles; OA: Ocean Acidification. Climate change denotes multiple controls such as nutrients, temperature and irradiance, as parameterised in coupled ocean atmosphere models. *denotes observation for low latitudes only. ** represents major uncertainty over environmental modulation of this component of the biological pump. ***denotes joint influence of temperature and acidification.

Pump component	Oceanic driver	Projected change (by year 2100)	Confidence	References & Lines of evidence
Epipelagic Ocean				
Phytoplankton growth	Temperature (warming)	~10% Faster (nutrient-replete) no change (nutrient-deplete)	High	(Boyd et al. 2013) **E**; (Maranon et al. 2014) **O***
NPP	Climate change (temperature, nutrients, CO_2)	10–20% decrease (low latitudes); 10–20% increase (high latitudes)	Medium	(Bopp et al. 2013) **M**
Partitioning of NPP (POC, TEP, DOC)	OA	~20% increase in TEP production	Medium	(Engel et al. 2014) **E**; (Riebesell et al. 2007) **E**; (Seebah et al. 2014) **E**
Food web retention of NPP	OA	Enhanced transfer of organic matter to higher trophic levels, reduced N and P sedimentation by 10%	Low	(Boxhammer et al. 2018) **E**
Floristic shifts	Climate change (warming, salinity, OA, iron)	Shift to smaller or larger cells (*less export vs more export; inconclusive*)	Low	(Morán et al. 2010) **O**; (Li et al. 2009) **O**; (Dutkiewicz et al. 2013a) **M**; (Tréguer et al. 2018) **O**; (Sett et al. 2014) **E**
Differential susceptibility	Temperature (warming)	Growth-rate of grazers more temperature dependent than prey (*less export*)	Low	(Rose and Caron, 2007) **O**
Bacterial hydrolytic effects	Warming, OA	Increase under warming and low pH (variable response in different plankton communities)	Low	(Burrell et al. 2017) **E**
Grazer physiological responses	Warming	Copepods had faster respiration and ingestion rates, but higher mortality (*inconclusive*)	Low	(Isla et al. 2008) **E**
Faunistic shifts	Temperate and subpolar zooplankton species shifts	Temperature (*inconclusive*)	Low	(Edwards et al. 2013) **O**
Food web amplification	Warming	Zooplankton negatively amplify the climate change signal that propagates up from phytoplankton in tropical regions, and positively amplify in polar regions	Low	(Chust et al. 2014) **M**; (Stock et al. 2014) **M**
Deep Ocean				
Bacterial hydrolytic enzyme activity	Temperature	20% increase (resource-replete) to no change (resource-deplete)	Low	(Wohlers-Zöllner et al. 2011) **E**; (Endres et al. 2014) **E**; (Bendtsen et al. 2015) **E**; (Piontek et al. 2015) **E****
Particle sinking rates (viscosity)	Warming	5% faster sinking/°C warming	Low	(Taucher et al. 2014) **M**
Mesozooplankton community composition	Temperature**	Shifts which increase/decrease particle transformations (*less/more export, respectively*)	Low	(Burd and Jackson, 2002) **M**; (Ikeda et al. 2001) **O**
Vertical migrators	Climate change (irradiance, temperature)	(*more export*)	Low	(Almén et al. 2014) **O**; (Berge et al. 2014) **O**
Deoxygenation	Climate change	(*more export*)	Low	(Rykaczewski and Dunne, 2010) **M**; (Cocco et al. 2013) **O**; (Hofmann and Schellnhuber, 2009) **M**

5.2.3.2 The Deep Pelagic Ocean

5.2.3.2.1 Detection and attribution of biological changes in the deep ocean

The pelagic realm of the deep ocean represents a key site for remineralisation of organic matter and long-term biological carbon storage and burial in the biosphere (Arístegui et al. 2009), but the observed effects of climate change on deep sea organisms, communities and biological processes are largely unknown (*high confidence*). Observational and model-based methods provide

limited evidence that the transfer efficiency of organic carbon to the sea floor is partly controlled by temperature and oxygen in the mesopelagic zone, affecting microbial metabolism and zooplankton community structure, with highest efficiencies for high-latitude and OMZ) (see Section 5.2.2.4 for more detail on OMZs), while below 1000 m organic carbon transfer is controlled by particle sinking speed (Boyd et al. 2015a; Marsay et al. 2015; DeVries and Weber, 2017). However, there are contrasting results and *low confidence* on whether transfer efficiencies are highest at low or high latitudes (Boyd et al. 2015a; Marsay et al. 2015; Guidi et al. 2016; DeVries and Weber, 2017; Sweetman et al. 2017). There is also *low confidence* on the effects of increasing temperatures on POC remineralisation to CO_2 versus POC solubilisation to dissolved organic carbon (DOC) by microbial communities and its storage as refractory DOC (i.e., with life times of >16,000 years) (Legendre et al. 2015).

5.2.3.2.2   *Future changes in the deep ocean*

The global magnitude of the biological pump and how this will be affected by climate change is also uncertain. Model-based studies agree in projecting a global decline in particle gravitational flux to the deep sea floor, but with regional variability in both the total particle export flux and transfer efficiency (DeVries and Weber, 2017; Sweetman et al. 2017) (see Sections 5.2.2 and 5.2.4). However, recent evidence suggest that other physical and biological processes may contribute nearly as much as the gravitational flux to the carbon transport from the surface to the deep ocean (Boyd et al. 2019), with *low confidence* on the future rate of change in magnitude and direction of these processes. In particular, the 'active flux' of organic carbon due to vertical migration of zooplankton and fishes has been reported to account from 10 to 40% of the gravitational sinking flux (Bianchi et al. 2013; Davison et al. 2013; Hudson et al. 2014; Jónasdóttir et al. 2015; Aumont et al. 2018; Gorgues et al. 2019). Predictions based on model studies suggest that mesopelagic zooplankton and fish communities living at deep scattering layers (DSLs) will increase their biomass by 2100, enhancing their trophic efficiency, because of deep-ocean warming (Section 5.2.2.1; Figures 5.2 and 5.3) and shallowing of DSL (Proud et al. 2017) (*low confidence*). Expansion of OMZs (see Section 5.2.2.4) will also widen the DSL and increase the exposure of mesopelagic organisms to shallower depths (Gilly et al. 2013; Netburn and Anthony Koslow, 2015). In the California Current, the abundance of mesopelagic fishes is closely tied to variations in the OMZ, whose dynamic is linked to the Pacific Decadal Oscillation and ENSO cycles (Koslow et al. 2015). Some large predators, like the Humboldt squid, could indirectly benefit from expanding OMZs due to the aggregation of their primary food source, myctophid fishes (Stewart et al. 2014). However, many non-adapted fish and invertebrates (like diurnal vertical migrators) will have their depth distributions compressed, affecting the carbon transport and trophic efficiency of food webs in the mesopelagic (Stramma et al. 2011; Brown and Thatje, 2014; Rogers, 2015) (*low confidence*). In OMZ waters, where zooplankton is almost absent, like in the Eastern Tropical North Pacific, the microbial remineralisation efficiency of sinking particles would be reduced, eventually increasing the transfer efficiency of organic matter to the deep ocean and thus biological carbon storage (Cavan et al. 2017) (*low confidence;* Table 1). However, increases in ocean temperature may also lead to shallower remineralisation of POC in warm tropical regions, countering the storage of carbon in the dark ocean (Marsay et al. 2015). Overall, the direct impacts of climate change on the biological pump are not well understood for the deep pelagic organisms and ecosystems (Pörtner et al. 2014), and there is *low confidence* on the effect of climate change drivers on biological processes in the deep ocean (Table 5.1).

5.2.4   Impacts on Deep Seafloor Systems

5.2.4.1   Changes on the Deep Seafloor

The deep seafloor is assessed here as the vast area of the ocean bottom >200 m deep, beyond most continental shelves (Levin and Sibuet, 2012; Boyd et al. 2019) (Figure 5.15). Below 200 m changes in light, food supply and the physical environment lead to altered benthic (seafloor) animal taxonomic composition, morphologies, lifestyles and body sizes collectively understood to represent the deep sea (Tyler, 2003).

Most deep seafloor ecosystems globally are experiencing rising temperatures, declining oxygen levels, and elevated CO_2, leading to lower pH and carbonate undersaturation (WGII AR5 30.5.7; Section 5.2.2.3). Small changes in exposure to these hazards by deep seafloor ecosystem have been confirmed by observation over the past 50 years. However, analysis using direct seafloor observations of these hazards over the past 15–29 years suggest that the environmental conditions are highly variable over time because of the strong and variable influences by ocean conditions from the sea surface (Frigstad et al. 2015; Thomsen et al. 2017). Such high environmental variability makes it difficult to attribute observed trends to anthropogenic drivers using existing datasets (Smith et al. 2013; Hartman et al. 2015; Soltwedel et al. 2016; Thomsen et al. 2017) (*high confidence*). Projections from global ESMs suggest large changes for temperature by 2100 and beyond under RCP8.5 (relative to present day variation) (Mora et al. 2013; Sweetman et al. 2017; FAO, 2019). The magnitude of the projected changes is lower under RCP2.6, and in some cases the direction of projected change to 2100 varies regionally under either scenario (FAO, 2019) (*high confidence*).

5.2.4.2   Open Ocean Seafloor – Abyssal Plains (3000–6000 m)

Abyssal communities (3000–6000 m) cover over 50% of the ocean's surface and are considered to be extremely food limited (Gage and Tyler, 1992; Smith et al. 2018). There is a strong positive relationship between surface primary production, export flux, and organic matter supply to the abyssal seafloor (Smith et al. 2008), with pulses of surface production reflected as carbon input on the deep seafloor in days to months (Thomsen et al. 2017). Both vertical and horizontal transport contribute organic matter to the sea floor (Frischknecht et al. 2018). Food supply to the seafloor regulates faunal biomass, explaining the strong positive relationships documented between surface production and seafloor faunal biomass in the Pacific Ocean (Smith et al. 2013), Gulf of Mexico (Wei et al. 2011) and north Atlantic Ocean (Hartman et al. 2015). Extended time series and broad spatial coverage reveal strong positive relationship between annual POC flux and abyssal sediment community oxygen consumption (Rowe et al. 2008; Smith et al. 2016a). Observed reduction in in POC flux at the abyssal seafloor enhances the relative importance of the microbial loop and reduces the importance of benthic invertebrates in carbon transfer (Dunlop et al. 2016) (single study, *limited evidence*). However, changes in the overlying mesopelagic and bathypelagic communities (see Section 5.2.3.2) will also affect food flux to the deep seafloor, as nekton and zooplankton transfer energy to depth through diel (daily day-night) vertical migrations, ontogenetic (life staged-based) migrations and falls of dead carcasses (Gage, 2003).

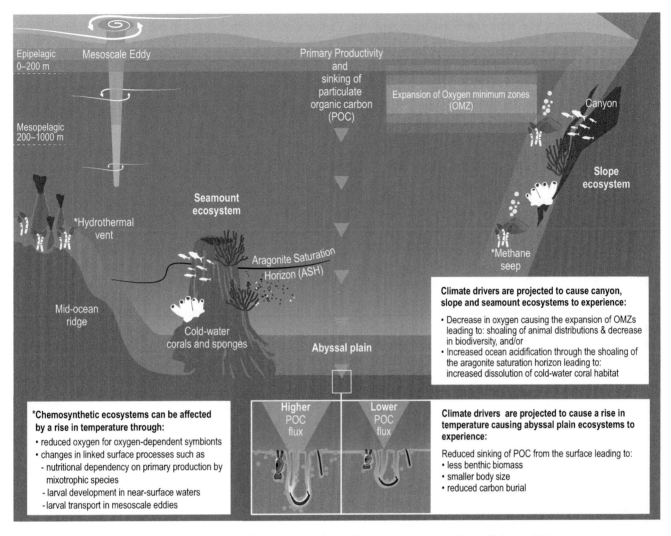

Figure 5.15 | A conceptual diagram illustrating how climate drivers are projected to modify deep sea ecosystems as discussed in Section 5.2.4.

Therefore, climate change impacts on organic carbon export from the epipelagic (Section 5.2.3.1) and deeper pelagic systems (Section 5.2.3.2) can affect the energy available to support the abyssal seafloor ecosystems (*medium confidence*). However, because observations on historical changes in POC flux in abyssal seafloor ecosystems are limited to a few locations, long-term records show high variability, and mechanistic understanding of factors affecting the biological carbon pump is incomplete, there is *limited evidence* that the abyssal seafloor ecosystem has already been affected by changes in POC flux as a result of climate change. The metabolic rate of deep seafloor ectotherms, and consequently their demand for food, increases with temperature. Thus, observed warming in deep sea ecosystems (Hoegh-Guldberg et al. 2014) (Section 5.2.2.2.1) is expected to increase the sensitivity of deep seafloor biota to decrease in food supplies associated with a change in POC flux (*high confidence*). However, there is *limited evidence* of observed changes in abyssal biota. Small deep sea biota demonstrate increased efficiency (effective use of food energy for growth and metabolism with minimal loss) at low food inputs (due to small size and dominance by prokaryotic taxa) (Gambi et al. 2017). Adaptation to low food availability in abyssal ecosystems may confer higher capacity to adjust to reduced food availability than for shallow biota

(*limited evidence*). Overall, the risk of impacts of climate change on abyssal ecosystems through reduction in food supplies from declining POC flux in the present day is low with *low confidence*.

The globally integrated export flux of carbon is projected to decrease in the open ocean in the 21st century under RCP2.6 (by 1.6–4.9%) and RCP8.5 (by 8.9–15.8%) relative to 2000 (*medium confidence*) (Section 5.2.2.6). This change in export flux of carbon is projected to yield declines in POC flux at the abyssal seafloor (representing food supply to benthos) of up to –27% in the Atlantic and up to –31 to –40% in the Pacific and Indian Oceans, with some increases in polar regions (Sweetman et al. 2017). In some models, additional dissolution of calcium carbonate due to ocean acidification further lowers POC flux, causing the projected export production declines to be up to 38% at the northeast Atlantic seafloor (Jones et al. 2014). Lower POC fluxes to the abyss reduce food supply and have been projected to cause a size-shift towards smaller organisms (Jones et al. 2014), resulting in rising respiration rates, lower biomass production efficiency, and lesser energy transfer to higher trophic levels (Brown et al. 2004) (*medium confidence*). Changes are projected to be largest for macrofauna and lesser and similar for megafauna and meiofauna (Jones et al. 2014) (*limited evidence, low confidence*).

Projections using outputs from seven CMIP5 models suggest that 97.8 ± 0.6% (95% CI) of the abyssal seafloor area will experience a biomass decline by 2091–2100 relative to 2006–2015 under RCP8.5. The projected decreases in overall POC flux to the abyssal seafloor are projected to cause a 5.2–17.6% reduction in seafloor biomass in 2090–2100, relative to 2006–2015 under RCP8.5 (Jones et al. 2014). The projected impacts on abyssal seafloor biomass are significantly larger under RCP8.5 than RCP4.5 (Jones et al. 2014). However, existing estimates are based on total POC flux changes and do not account for changes in the type or quality of the sinking material, to which macrofaunal and meiofaunal invertebrates are highly sensitive (Smith et al. 2008; Smith et al. 2009; Tittensor et al. 2011). The projections also do not account for direct faunal responses to changes in temperature, oxygen or the carbonate system, all of which will influence benthic responses to changing food availability (AR5 Chapter 30.5.7), reducing to *medium confidence* the risk assessment that is based on these projections (Figure 5.16).

Regionally, while reductions in POC flux are projected at low and mid latitudes in the Pacific, Indian and Atlantic Oceans, increases are projected at high latitudes associated in part with reduction in sea ice cover (Yool et al. 2013; Rogers, 2015; Sweetman et al. 2017; Yool et al. 2017; FAO 2019) (see Chapter 3) (*medium confidence*). Notably, Arctic and Southern Ocean POC fluxes at the abyssal seafloor are projected to increase by up to 38% and 21%, respectively by 2100 under RCP8.5 (Sweetman et al. 2017). While an increase in food supply may yield higher benthic biomass at high latitudes, warmer temperatures and reduced pH projected for the polar regions (Chapter 3) would elevate faunal metabolic demands, likely diminishing the benefit of elevated food supply to an unknown extent (Sweetman et al. 2017). Overall, given the limited food availability for fauna in the abyssal plains and the projected warming (Section 5.2.2.2.2) that increases the demand for food to support the elevated metabolic rates, the projected decrease in influx of organic matter and seafloor biomass will result in high risks of impacts to abyssal ecosystems by the end of the 21st century under RCP8.5 (*medium confidence*) (Figure 5.16). The risk of impacts is projected to be substantially lower under RCP4.5 or RCP2.6 (*high confidence*). The impacts on abyssal seafloor ecosystems affect functions that are important to support ecosystem services (see Section 5.4.1). For example, smaller-sized organisms exhibit reduced bioturbation intensity and depth of mixing causing reduced carbon sequestration (Smith et al. 2008) (Figure 5.15).

5.2.4.3   Bathyal Ecosystems (200–3000 m)

Bathyal ecosystems consist of numerous geomorphic features with steep topography (Figure 5.15). These include continental slopes covering 5.2% of the seafloor, over 9400 steep-sided canyons, and >9000 conical seamounts (submarine volcanos which are mainly inactive), as well as guyots and ridges which together cover ~6% of the seafloor (Harris et al. 2014). Seamounts and canyons support high animal densities and biomass including cold water coral, sponge and bryozoan reefs, exhibit high secondary production supported by locally enhanced primary production and intensified water flow, function as diversity hotspots and serve as stepping stones for larval dispersal (Rowden et al. 2010). Canyons transport particulate organic matter, migrating plankton and coarse material from the shelf, and

are sites where intensified mixing and advection of water masses occurs (De Leo et al. 2010; Levin and Sibuet, 2012; Fernandez-Arcaya et al. 2017). Slopes, canyons and seamounts exhibit strong vertical temperature, oxygen and pH gradients generating sharp ecological zonation (Levin and Sibuet, 2012), thus changes in exposures are expected to alter the distributions of their communities (Figure 5.15, 5.16) (*medium confidence*).

In some regions, observational records document changing conditions in bathyal ecosystems (Levin, 2018; Section 5.2.2.4). In the Northeast Pacific continental slopes associated with the California Current ecosystem, observations over the past 25 years show high variability but an overall trend of decreasing ocean oxygen and pH levels with oxygen declines of up to 40% and pH declines of 0.08 units in California. (Goericke et al. 2015) (high agreement, robust evidence, high confidence). Large oxygen declines are linked to past warming events on continental margins, over multiple time scales from 1–100 ky (Dickson et al. 2012; Moffitt et al. 2015). Studies across modern oxygen gradients on slopes reveal that suboxic (5–10 µMol kg^{-1} O$_2$) values lead to loss of biodiversity of fish (Gallo and Levin, 2016), invertebrates (Levin, 2003; Gallo and Levin, 2016; Sperling et al. 2016), and protozoans (Bernhard and Reimers, 1991; Gooday et al. 2000; Moffitt et al. 2014) (high confidence). Shoaling oxyclines on continental slopes have altered depth distributions of multiple co-occurring echinoid species over the past 25 years (Sato et al. 2017) and can reduce the growth rate, and change the skeletal structure and biochemical composition of a common sea urchin (Sato et al. 2018). In central Pacific oceanic canyons, fish abundance and diversity are reduced at 4 to 5 times higher oxygen concentrations than on continental slopes (<31 µMol kg^{-1} O$_2$) (De Leo et al. 2012). Low oxygen on continental slopes causes reductions in faunal body size and bioturbation (Diaz and Rosenberg, 1995; Levin, 2003; Middelburg and Levin, 2009; Sturdivant et al. 2012), simplification of trophic structure reducing energy flow to upper trophic levels (Sperling et al. 2013), shifts in carbon processing pathways from metazoans to protozoans (Woulds et al. 2009), and reduced colonisation potential (Levin et al. 2013). These changes are expected to lead to altered ecosystem structure and function, with lower carbon burial (Smith et al. 2000; Levin and Dayton, 2009) (medium confidence). Both carbon sequestration and nitrogen recycling are highly sensitive to small changes in oxygenation within the suboxic zone (Deutsch et al. 2011).

Bathyal species adapted to OMZs where CO$_2$ levels are characteristically high, appear less vulnerable to the negative impacts of ocean acidification (Taylor et al. 2014). Benthic foraminifera, which are often the numerically dominant deep sea taxon, show no significant effect of short-term exposure to ocean acidification on survival of multiple species (Dissard et al. 2010; Haynert et al. 2011; Keul et al. 2013; McIntyre-Wressnig et al. 2014; Wit et al. 2016) and in fact hypoxia in combination with elevated pCO$_2$ favors survival of some foraminifera (Wit et al. 2016). However, lower pH exacerbates shallow foraminiferal sensitivity to warming (Webster et al. 2016). Limited evidence suggests that combined declines in pH and oxygen may lead to increase in some agglutinating taxa and a decrease in carbonate-producing foraminifera, including those using carbonate cement (van Dijk et al. 2017). Exposure to acidification (0.4 unit pH decrease) reduces fecundity and embryo development rate in a bathyal polychaete. Where both oxygen and CO_2 stress occur together on bathyal slopes, oxygen can be the primary driver of change (Taylor et al. 2014; Sato et al. 2018). Nematodes are sensitive to changes in temperature (Danovaro et al. 2001; Danovaro et al. 2004; Yodnarasri et al. 2008) and elevated CO_2 (Barry et al. 2004; Fleeger et al. 2006; Fleeger et al. 2010). There is low agreement about the direction of meiofaunal responses among studies, reflecting opposing responses in different regions. However, there is high agreement that meiofauna are sensitive to change in environment and food supply (*medium confidence*). Additional research is needed across all taxa on how hypoxia and pH interact (Gobler and Baumann, 2016).

Continental slopes, seamounts and canyons (200–2500 m) are projected to experience significant warming, pH decline, oxygen loss and decline in POC flux by 2081–2100 (compared to 1951–2000) under RCP8.5 (Table 5.5). In contrast, the average changes are projected to be 30–50% less under RCP2.6 (Table 5.5) by 2081–2100. Most ocean regions at bathyal depths (200–2500 m) except the Southern and Arctic Oceans are predicted to experience on average declining export POC flux under RCP8.5 by 2081–2100 (Yool et al. 2017; FAO, 2019) with the largest declines of 0.7–8.1 mg C m^{-2} d^{-1} in the Northeast Atlantic (FAO, 2019). There is a strong macroecological relationship between depth, export POC flux, biomass and zonation of macrobenthos on continental slopes (Wei et al. 2011), such that lower POC fluxes will alter seafloor community biomass and structure (*medium confidence*) (See also Section 5.2.4.1). This is modified on the local scale by near-bottom currents, which alter sediment grain size, food availability, and larval dispersal (Wei et al. 2011).

Declines in faunal biomass (6.1 ± 1.6% 95% C.I) are predicted for 96.6 ± 1.2% of seamounts under RCP8.5 by 2091–2100 relative to 2006–2015, driven by a projected 13.8 ± 3.3% drop in POC flux (Jones et al. 2014). The majority (85%) of mapped canyons are projected to

experience comparable benthic biomass declines (Jones et al. 2014). By 2100 under RCP8.5, pH reductions exceeding -0.2 pH units are projected in ~23% of north Atlantic deep sea canyons and 8% of seamounts (Gehlen et al. 2014), with potential negative consequences for their cold water coral habitats (See Box 5.2).

Mean temperature (warming) signals are projected to emerge from background variability before 2040 in canyons of the Antarctic, northwest Atlantic, and South Pacific (FAO, 2019). Enhanced stratification and change in the intensity and frequency of downwelling processes under atmospheric forcing (including storms and density-driven cascading events would alter organic matter transported through canyons (Allen and Durrieu de Madron, 2009) (*low confidence*). Changes in the quantity and quality of transferred particulate organic matter, as well as physical disturbance during extreme events cause a complex combination of positive and negative impacts at different depths along the canyon floor (Canals et al. 2006; Pusceddu et al. 2010). Canyons and slopes are recognised as hosting many methane seeps and other chemosynthetic habitats (e.g., whale and wood falls) supported by massive transport of terrestrial organic matter (Pruski et al. 2017); their climate vulnerabilities are discussed below.

Seamounts have been proposed to serve as refugia for cold water corals facing shoaling aragonite saturation horizons (Tittensor et al. 2011), but could become too warm for deep-water corals in some regions (e.g., projections off Australia) (Thresher et al. 2015) (*one study, low confidence*). Seamounts are major spawning grounds for fishes; reproduction on seamounts may be disrupted by warming (Henry et al. 2016) (*one study, low confidence*). In the north Atlantic, models suggest seamounts are an important source of cold water coral larvae that maintain resilience under shifting NAO conditions (Fox et al. 2016), thus loss of suitable seamount habitat may have far-reaching consequences (Gehlen et al. 2014) (*low confidence*) (also see Box 5.2).

Table 5.5 | Projected climate changes from the present to 2081–2100 given as mean (min, max) at the deep seafloor for continental slopes, canyons, seamounts and cold water corals mapped from 200–2500 m under RCP8.5 and RCP2.6 Projections are based on three 3D, fully coupled earth system models (ESMs) (as part of CMIP5): the Geophysical Fluid Dynamics Laboratory's ESM 2G (GFDL-ESM-2G); the Institut Pierre Simon Laplace's CM6-MR (IPSL-CM5A-MR); and (iii) the Max Planck Institute's ESM-MR (MPI-ESM-MR). Export flux at 100 m was converted to export POC flux at the seafloor (*epc*) using the Martin curve following the equation: $epc = epc100 \, (depth/export\ depth)^{-0.858}$. Projections were made onto the (i) slope from a global ocean basin mask from World Ocean Atlas 2013 V2 (NOAA, 2013), (ii) global distribution of submarine canyons with canyon heads shallower than 1500 m (Harris and Whiteway, 2011); (iii) global distribution of seamounts with summits between 200–2500 m (Kim et al. 2011); and (iv) global occurrence of cold water corals between 200–2500 m (Freiwald et al. 2017).

	Temperature (°C)	pH	DO (µMol kg^{-1})	POC flux (mgC m^{-2} d^{-1})
	RCP2.6	RCP2.6	RCP2.6	RCP2.6
Continental slopes	+0.30 (−0.44, + 2.30)	−0.06 (−0.19, −0.02)	−3.1 (−49.3, +61.7)	−0.39 (−16.0, +3.9)
Canyons	+0.31 (−0.27, +1.76)	-0.05 (-0.13, +0.01)	−3.5 (−44.7, +29.3)	−0.33 (−10.53, +3.53)
Seamounts	+0.13 (+0.01, +0.67)	-0.02 (-0.11, +0.005)	−3.46 (−18.9, +4.1)	−0.15 (−2.20, +1.33)
Cold water corals	+4.3 (−0.29, +1.85)	-0.07 (-0.13, 0.0)	−3.5 (−25.6, +24.7)	−0.7 (−10.5, +3.4)
	RCP8.5	RCP8.5	RCP8.5	RCP8.5
Continental slopes	+0.75 (−8.4, +4.4)	−0.14 (−0.44, −0.02)	−10.2 (−67.8, +53.8)	−0.66 (−33.33, +10.3)
Canyons	+0.19 (−0.03, +1.14)	−0.11 (−0.35, +0.02)	−0.8 (−28.8, +10.1)	−0.80 (−28.76, +10.07)
Seamounts	+0.66 (−0.75, +3.19)	−0.03 (−0.19, +0.001)	−0.50 (−7.2, +3.0)	−0.50 (−7.18, +2.98)
Cold water corals	+0.96 (−0.42, +3.84)	−0.15 (−0.39, +0.001)	−10.6 (−59.2, +11.1)	−1.69 (−20.1, +4.6)

5

5.2.4.4   Chemosynthetic Ecosystems

Despite having nutrition derived largely from chemosynthetic sources fueled by fluids from the earth's interior, hydrothermal vent and methane seep ecosystems are linked to surface ocean environments and water-column processes in many ways that can expose them to aspects of climate change (*medium confidence*). The reliance of vent and seep mussels on surface-derived photosynthetic production to supplement chemosynthetic food sources (Riou et al. 2010; Riekenberg et al. 2016; Demopoulos et al. 2019), and in some cases as a cue for synchronised gametogenesis (sperm and egg production) (Dixon et al. 2006; Tyler et al. 2007) can make them vulnerable to changing amounts or timing of POC flux to the deep seabed in most areas except high latitudes, or to changes in timing of surface production (see Section 5.2.2.5) (*limited evidence*) Most of the large, habitat-forming (foundation) species at vents and seeps such as mussels, tubeworms, and clams require oxygen to serve as electron acceptor for aerobic hydrogen-, sulfide- and methane oxidation (Dubilier et al. 2008) and appear unable to grow under dysoxic conditions (<5–10 µmol kg^{-1} O$_2$) (Sweetman et al. 2017) (*medium confidence*). The distributions of these taxa at seeps could be constrained by climate-driven expansion of midwater oxygen minima (Stramma et al. 2008; Schmidtko et al. 2017), which is occurring at water depths where seep ecosystems typically occur on continental margins (200–1000 m). Rising bottom temperatures or shifting of warm currents on continental margins could increase dissociation of buried gas hydrates on margins (Phrampus and Hornbach, 2012) (*low confidence*) potentially intensifying

anaerobic methane oxidation (which produces hydrogen sulfide) (Boetius and Wenzhoefer, 2013) and expanding cover of methane seep communities (*limited* evidence). Larvae of vent species such as bathymodiolin mussels, alvinocarid shrimp, and some limpets that develop in or near surface waters (Herring and Dixon, 1998; Arellano et al. 2014), are likely to be exposed to warming waters, decreasing pH and carbonate saturation states, and in some places, reduced phytoplankton availability (Section 5.2.2), causing reduced calcification and growth rates (as in shallow water mussel larvae, Frieder et al. (2014)) (*limited evidence, low confidence*). Larvae originating at vents or seeps beneath upwelling regions may also be impaired by effects of hypoxia associated with expanding OMZ (Stramma et al. 2008) during migration to the surface (*limited evidence*). Warming and its effects on climate cycles have the potential to alter patterns of larval transport and population connectivity through changes in circulation (Fox et al. 2016) or surface generated mesoscale eddies (Adams et al. 2011) (*limited evidence; low confidence*). Climate-induced changes in the distribution and cover of vent and seep foundation species may involve alteration of attachment substrate, food and refuge for the many habitat-endemic species that rely on them (Cordes et al. 2010) and for the surrounding deep sea ecosystems which interact through transport of nutrients and microbes, movement of vagrant predators and scavengers, and plankton interactions (Levin et al. 2016) (*limited evidence; low confidence*). There is, however, insufficient analysis of faunal symbiont and nutritional requirements, life histories, larval transport and cross-system interaction to quantify the extent of the consequences described above under future climate conditions.

Box 5.2 | Cold Water Corals and Sponges

Cold water corals and sponges form large reefs at the deep seafloor mostly between 200–1500 m, creating complex 3D habitat that supports high biodiversity; they are found at the highest densities on hard substrates of continental slopes, canyons, and seamounts (Buhl-Mortensen et al. 2010). The meta-analysis reported in AR5 Chapter 6 Table 6–3 (Pörtner et al. 2014), identifies 10 studies involving 6 species of cold water corals that suggest low vulnerability to CO$_2$ changes at RCP6.0 and medium vulnerability at RCP8.5, with negative effects starting at pCO$_2$ of 445 µatm.

Scleractinian corals have the capacity to acclimate to high CO$_2$ conditions due to their capacity to upregulate the pH at the calcification site (Form and Riebesell, 2011; Rodolfo-Metalpa et al. 2015; Gori et al. 2016). The most widely distributed, habitat-forming species in deep water (e.g., *Lophelia pertusa* [renamed *Desmophyllum pertusum*) (Addamo et al. 2016) can continue to calcify at aragonite undersaturation and high CO$_2$ levels projected for 2100 (750–1100 uatm) based on experiments (Georgian et al. 2016; Kurman et al. 2017) and observations along the natural gradient of carbon chemistry in their distributions (Fillinger and Richter, 2013; Movilla et al. 2014; Baco et al. 2017) (Appendix 1) (*robust evidence, medium agreement, medium confidence*) and thus appear to be able to acclimate to rising CO$_2$ levels (Hennige et al. 2015). However, net calcification rates (difference between calcification and dissolution) of *L. pertusa* exposed to aragonite-undersaturated conditions ($\Omega_{arag} < 1$, where Ω_{arag} = aragonite saturation state) often decreases to close to zero or even becomes negative (Lunden et al. 2014; Hennige et al. 2015; Büscher et al. 2017), with genetic variability underpinning ability to calcify at low aragonite saturation states (Kurman et al. 2017). Additionally, skeletons become longer, thinner and weaker (Hennige et al. 2015), and bioerosion is enhanced (e.g., by bacteria, fungi, annelids and sponges) (Schönberg et al. 2017), exacerbating effects of dissolution of the skeleton. *L. pertusa* can calcify when exposed to multiple environmental stresses in the laboratory (Hennige et al. 2015; Büscher et al. 2017), but cannot survive with warming above water temperatures of 14°C–15°C or oxygen concentrations below 1.6 ml l^{-1} in the Gulf of Mexico, 3.3 ml l^{-1} in the north Atlantic, 2 ml l^{-1} in the Mediterranean, and 0.5–1.5 ml l^{-1} in the SE Atlantic (Brooke et al. 2013; Lunden et al. 2014; Hanz et al. 2019), highlighting the existence of critical thresholds for cold water coral populations living at the edge of their tolerance. The role of temporal dynamics, species-specific thermal tolerances, and food availability in mediating the response to combinations of stressors is recognised but is still poorly studied under *in situ* conditions (Lartaud et al. 2014; Naumann et al. 2014; Baco et al. 2017).

Box 5.2 (continued)

Sponges also form critical habitat in the deep ocean but are much less well studied than cold-water corals with respect to climate change. The geologic record, modern distributions and evolutionary and metabolic pathways suggest that sponges are more tolerant to warm temperatures, high CO_2 and low oxygen than are cold-water corals (Schulz et al. 2013). One habitat forming, deep sea sponge along with its microbiome (microbial inhabitants) has been shown in laboratory experiments to tolerate a 5°C increase in temperature, albeit with evidence of stress (Strand et al. 2017), while ocean acidification (pH 7.5) reduces the feeding of two deep sea demosponge taxa (Robertson et al. 2017).

Generally, the deep sea areas where cold water corals may be found are projected to be exposed to multiple climate hazards in the 21st century because of the projected ocean warming, oxygen loss, and decrease in POC flux (Table 5.5) under scenarios of greenhouse gas emissions. The average changes in these climate hazards for coral-water corals are projected to be almost halved under RCP2.6 relative to RCP8.5 (Table 5.5). Under RCP8.5, 95 ± 2% (95% CI) of cold-water coral habitats are projected to experience animal biomass decline (–8.6 ± 2.0%) globally by 2091–2100 relative to 2006–2015, driven by a projected 21 ± 9% drop in POC flux (*medium confidence*) (Jones et al. 2014). However, nutritional co-reliance of cold-water corals on zooplankton (Höfer et al. 2018) and carbon fixation by symbiotic microbes (Middelburg et al. 2015), is not incorporated into the models, adding uncertainty to these estimates. Regionally, suitable habitat for coral-water corals in the NE Atlantic is projected to decrease with multiple climatic hazards (warming, acidification, decreases in oxygen and POC flux) under RCP8.5 for 2081–2100 (FAO, 2019), with up to 98% loss of suitable habitat by 2099 due to shoaling aragonite saturation horizons. In the Southern hemisphere, a tolerance threshold of 7°C and decline of aragonite saturation below that required for survival (Ω_{arag} <0.84) can cause large loss of cold water corals habitat (*Solenosmilia variabilis*) on seamounts off Australia and New Zealand under future projections of warming and acidification to 2099 at RCP4.5 and nearly complete loss under RCP8.5 (Thresher et al. 2015).

Overall, cold water corals can survive conditions of aragonite-undersaturation associated with ocean acidification but sensitivity varies among species and skeletons will be weakened (*medium confidence*). The largest impacts on calcification and growth will occur when aragonite saturation is accompanied by warming and/or decrease in oxygen concentration beyond the tolerance limits of these corals (*medium confidence*). Given present day occurrence of 95% of cold water corals above the aragonite saturation horizon (Guinotte et al. 2006) and that no adaptation has been detected with regard to increased dissolution of exposed aragonite (Eyre et al. 2014), there is limited scope for the non-living components of cold water corals and for the large, non-living reef framework that comprises deep water reefs to avoid dissolution under RCP8.5 in the 21st century (*high confidence*). Multiple climatic hazards of warming, deoxygenation, aragonite under-saturation and decrease in POC flux are projected to negatively affect cold water corals worldwide from the present day by 2100 (*high confidence*). Uncertainty remains in the adaptive capacity of living cold water corals to cope with these changes and in the influence of altered regional current patterns on connectivity (Fox et al. 2016; Roberts et al. 2017). Sponges and the habitat they form may be less vulnerable than cold water corals to warming, acidification and deoxygenation that will occur under RCP8.5 in 2100 (*low confidence*).

5.2.5   Risk Assessment of Open Ocean Ecosystems

This section synthesises the assessment of climate impacts on open ocean and deep seafloor ecosystem structure and functioning and the levels of risk under future conditions of global warming (see SM5.2). The format for Figure 5.16 matches that of Figure 19.4 of AR5 (Pörtner et al. 2014) and Figure 3.20 of SR15 (Hoegh-Guldberg et al. 2018), indicating the levels of additional risk as colours (white, yellow, red and purple). Each column in Figure 5.16 indicates how risks increase with ocean warming, acidification (OA), deoxygenation, and POC flux with a focus on present day conditions (2000s) and future conditions by the year 2100 under low (RCP2.6) and high (RCP8.5) CO_2 emission scenarios. The transition between the levels of risk to each type of ecosystem is estimated from key evidence assessed in earlier parts of this chapter (Sections 5.2.2, 5.2.3, 5.2.4). SST is chosen to provide an indication of the changes in all these variables because it is closely related to cumulative carbon emission (Gattuso et al. 2015) which is the main climatic driver of the hazards. SST scales with Global

Mean Surface Temperature (GMST) by a factor of 1.44 according to changes in an ensemble of RCP8.5 simulations; with an uncertainty of about 4% in this scaling factor based on differences between the RCP2.6 and RCP8.5 scenarios. The transition values may have an error of ±0.3°C depending on the consensus of expert judgment. The deep seafloor embers are generated based on earth system model projection of climate variables to the seafloor under RCP2.6 and RCP8.5 scenarios, and then translated to RCP associated change in SST. The assessed confidence in assigning the levels of risk at present day and future scenarios are *low, medium, high* and *very high* levels of confidence. A detailed account of the procedures involved in the ember for each type of ecosystem, such as their exposure to climate hazards, sensitivity of key biotic and abiotic components, natural adaptive capacity, observed impacts and projected risks, and regional hotspots of vulnerability is provided in the SM5.2 and Table 5.5. The risk assessment for cold water corals is in agreement with the conclusions in AR5 Chapter 6.3.1.4.1, although more recent literature is assessed in Box 5.2 and Table SM5.5.

Overall, the upper ocean (0–700 m) and 700–2000 m layers have both warmed from 2004 to 2016 (*virtually certain*) and the abyssal ocean continues to warm in the Southern Hemisphere (*high confidence*). The ocean is stratifying; observed warming and high-latitude freshening are both surface intensified trends making the surface ocean lighter at a faster rate than deeper in the ocean (*high confidence*) (Section 5.2.2.2). It is *very likely* that stratification in the upper few hundred meters of the ocean will increase significantly in the 21st century. It is *virtually certain* that ocean pH is declining by ~0.02 pH units per decade where time series observations exist (Section 5.2.2.3). The anthropogenic pH signal has already emerged over the entire surface ocean (*high confidence*) and emission scenarios are the most important control of surface ocean pH relative to internal variability for most of the 21st century at both global and local scale (*virtually certain*). The oxygen content of the global ocean has declined by about 0.5–3.3% in 0–1000 m layer (Section 5.2.2.4). Over the next century oxygen declines of 3.5% by 2100 are predicted by CMIP5 models globally (*medium confidence*), with *low confidence* at regional scales, especially in the tropics. The largest changes in the deep sea will occur after 2100 (Section 5.2.2.3). CMIP5 models project a decrease in global NPP (*medium confidence*) with increases in high-latitude (*low confidence*) and decreases in low latitude (*medium confidence*) (Section 5.2.2.6) in response to changes in ocean nutrient supply (Section 5.2.2.5). These models also project reductions by 8.9–15.8% in the globally integrated POC flux for RCP8.5, with decreases in tropical regions and increases at higher latitudes (*medium confidence*), affecting the organic carbon supply to the deep sea floor ecosystems (*high confidence*) (Section 5.2.2.6). However, there is *low confidence* on the mechanistic understanding of how climatic drivers will affect the different components of the biological pump in the epipelagic ocean (Table 5.4). Therefore, the exposure to hazard for epipelagic ecosystems ranges from moderate (RCP2.6) to high (RCP8.5), with uncertain effects and tolerance of planktonic organisms, fishes and large vertebrates to interactive climate stressors. Major risks are predicted for declining productivity and fish biomass in tropical and subtropical waters (RCP8.5) (SM5.2).

The climatic hazards for pelagic organisms from plankton to mammals are driving changes in eco-physiology, biogeography and ecology and biodiversity (*high confidence*) (Section 5.2.3.1). Observed and projected population declines in the equator-ward range boundary (*medium confidence*), expansion in the poleward boundary (*high confidence*), earlier timing of biological events (*high confidence*), overall shift species composition (*high confidence*) and decreases in animal biomass (*medium confidence*), are consistent with expected responses to climate change (Section 5.2.3; Figure 5.13). It is *likely* that increased OA has not yet caused sufficient reduction in fitness to decrease abundances of calcifying phytoplankton and zooplankton, but *is very likely* (*high confidence*) that calcifying planktonic organisms will experience great decreases in abundance and diversity under high emission scenarios by the end of the century. Therefore, impacts to the epipelagic ecosystems are already observed in the present day (Figure 5.16). Based on simulation modelling and experimental findings, the combined effects of warming, ocean deoxygenation, OA and changes in NPP in the 21st century are projected to exacerbate

the impacts on the growth, reproduction and mortality of fishes, and consequently increase the risk of population decline (*high confidence*) (Section 5.2.3.1). There may be some capacity for adjustment and evolutionary adaptation that lowers their sensitivity to warming and decrease in oxygen (*low confidence*). However, historical responses in abundance and ranges of marine fishes to ocean warming and decrease in oxygen in the past suggest that adaptation is not always sufficient to mitigate the observed impacts (*medium confidence*) (Section 5.2.3) (SM5.2).

Despite its remoteness, most of the deep seafloor ecosystems already have or are projected to experience rising temperatures and declining oxygen, pH and POC flux beyond natural variability within the next half century (See Section 5.2.4). On slopes, seamounts and canyons these changes are projected to be much larger under RCP8.5 than under RCP2.6 (*high confidence*), with greatest effects on seafloor community diversity and function from expansion of low oxygen zones and aragonite undersaturation (*medium confidence*). As critical thresholds of temperature, oxygen and CO_2 are exceeded, coral species will alter their depth distributions, non-living carbonate will experience dissolution and bioerosion, and stress will be exacerbated by lower food supply. These changes are projected to cause loss of cold water coral habitat with highest climate hazard in the Arctic and north Atlantic Ocean (*medium confidence*), while sponges may be more tolerant (Box 5.2) (*low confidence*). Projected changes in food supply to the seafloor at abyssal depths combined with warmer temperatures are anticipated to cause reductions in biomass and body size (*medium confidence*) that could affect the carbon cycle in this century under RCP8.5 (*low confidence*). Even at hydrothermal vents and methane seeps, some dominant species such as mussels may be vulnerable to reduced photosynthetically-based food supply or have planktonic larvae or oxidising symbionts that are negatively affected by warming, acidification and oxygen loss (*low confidence*).

Widespread attributes of deep seafloor fauna (e.g., great longevity, high levels of habitat specialisation including well-defined physiological tolerances and thresholds, dependence on environmental triggers for reproduction, and highly developed mutualistic interactions) can increase the vulnerability of selected taxa to changing conditions (FAO, 2019) (*medium confidence*). However, some deep sea taxa (e.g., foraminifera and nematodes) may be more resilient to environmental change than their shallow-water counterparts (*low confidence*). Observations, experiments and model projections indicate that impacts of climate change have or are expected to take place in this century, indicating a transition from undetectable risk to moderate risk at <1.5°C warming of sea surface temperature for continental slope, canyon and seamount habitats, and for cold water corals (Figure 5.16). Emergence of risk is expected to occur later at around the mid-21st century under RCP8.5 for abyssal plain and chemosynthetic ecosystems (vents and seeps) (Figure 5.16). All deep seafloor ecosystems are expected to be subject to at least moderate risk under RCP8.5 by the end of the 21st century, with cold water corals experiencing a transition from moderate to high risk below 3°C (SM5.2).

5.3 Changing Coastal Ecosystems and Biodiversity

The world's shelf seas and coastal waters (hereafter 'coastal seas') extend from the coastline to the 200 m water depth contour. They encompass diverse ecosystems, including estuaries, sandy beaches, kelp forests, mangroves and coral reefs. Although they occupy a small part of the global ocean (7.6%), coastal seas provide up to 30% of global marine primary production and about 50% of the organic carbon supplied to the deep ocean (Chen, 2003; Bauer et al. 2013) (Sections 5.2.4.1 and 5.4.1.1). Coastal seas include several frontal and upwelling areas (Box 5.3) that support high fisheries yields (Scales et al. 2014), and productive coastal ecosystems, such as wetlands (McLeod et al. 2011). Mangrove forests, seagrass meadows and kelp forests form important habitats supporting high biodiversity while offering opportunities for climate change mitigation and adaptation (Section 5.5.1.2) (Duarte et al. 2013), with mangrove forests providing physical protection against extreme events such as storms and floods (Kelleway et al. 2017a) (Sections 5.4.1.2 and 4.3.3.5.4). The regional characteristics and habitat heterogeneity of many coastal seas support endemic fauna and flora (e.g., seagrass meadows in the Mediterranean), which makes them particularly vulnerable to climate change impacts with high risk of diversity loss and alterations in ecosystem structure and functioning (Rilov, 2016; Chefaoui et al. 2018).

Near-shore coastal ecosystems are classified by their geomorphological structure (e.g., estuaries, sandy beaches and rocky shores) or foundation species (e.g., salt marshes, seagrass meadows, mangrove forests, coral reefs and kelp forests). All these coastal ecosystems are threatened to a varying degree by SLR, warming, acidification, deoxygenation and extreme weather events (Sections 5.3.1 to 5.3.7). Unlike the open ocean where detection and attribution of climate driven-physical and chemical changes are robust (Section 5.2.2), coastal ecosystems display regional complexity that can render the conclusive detection and attribution of climate effects uncertain. The hydrological complexity of coastal ecosystems that affects their biota is driven by the interactions between the land (e.g., river and groundwater discharges), the sea (e.g., circulation, tides) (Section 5.2.2.2.3) and seabed structures and substrates (Sharples et al. 2017; Chen et al. 2018; Laurent et al. 2018; Zahid et al. 2018).

Additionally, the high density of human populations on coastal land causes most of the adjacent marine ecosystems to be impacted by local anthropogenic disturbances such as eutrophication, coastline modifications, pollution and overfishing (Levin et al. 2015; Diop and Scheren, 2016; Maavara et al. 2017; Dunn et al. 2018) (Section 4.3.2.2, Cross-Chapter Box 9). Climate driven impacts interact with such human disturbances and pose a serious risk to ecosystems structure and functioning (Gattuso et al. 2015). Projections of the ecological impacts of climate change in coastal ecosystems must therefore deal with many emerging complexities such as the differentiation between the long-term climate trends (e.g., progressive ocean acidification) and the short-term natural fluctuations (Boyd et al. 2018), ranging from the seasons to interannual climate oscillations like El Niño. The 'time of emergence' for specific climate drivers to exceed background variability varies between ecosystems and is strongly sensitive to projected emission scenarios (Hammond et al. 2017; Reusch et al. 2018) (Box 5.1).

This section summarises our updated understanding of ecological and functional changes that coastal ecosystems are experiencing due to multiple climate and non-climatic human drivers, and their synergies. Additional experimental and long-term observational evidence since AR5 WGII (Wong et al. 2014a) and SR15 (Hoegh-Guldberg et al. 2018) improves the attribution of impacts on all the types of coastal ecosystems assessed here to climate trends (Sections 5.3.1 to 5.3.6). Moreover, the emergent impacts detected in the present strengthen the projection of risk of each ecosystem under future emission scenarios by 2100, depending on their exposure to different climate hazards (Section 5.3.7).

5.3.1 Estuaries

Estuarine ecosystems are defined by the river-sea interface that provides high habitat heterogeneity and supports high biodiversity across freshwater and subtidal zones (Basset et al. 2013). AR5 WGII (Wong et al. 2014a) and SR15 (Hoegh-Guldberg et al. 2018) concluded that estuarine ecosystems have been impacted by SLR and human influences that drive salinisation, resulting in increased flooding, land degradation and erosion of coastal areas around estuaries.

Observations since AR5 provide further evidence that SLR increases seawater intrusions and raises salinity in estuaries. Salinisation of estuaries can be exacerbated by droughts and modifications of drainage area by human activities (Ross et al. 2015; Cardoso-Mohedano et al. 2018; Hallett et al. 2018; Zahid et al. 2018). The changing salinity gradients in estuaries have been linked to the observed upstream expansion of brackish and marine benthic and pelagic communities, and a reduction in the diversity and richness of freshwater fauna (Robins et al. 2016; Raimonet and Cloern, 2017; Hallett et al. 2018; Addino et al. 2019) (*medium confidence*). However, because the distribution of benthic species in estuaries is strongly determined by sediment properties like grain size, the gradient of sediment types in estuaries can be a barrier to upstream shifts of brackish and marine benthic biota, leading to a reduction in species richness in mid- to upper-estuarine areas and altering food webs (Little et al. 2017; Hudson et al. 2018; Addino et al. 2019). Similarly, estuarine wetlands (Section 5.3.2) and tidal flats (Murray et al. 2019) have reduced their extent and productivity in response to increased salinity, inundation and wave exposure, especially in areas with limited capacity for soil accretion or inland migration due to coastal squeezing (Sections 4.3.2.3, 5.3.2) (*high confidence*). Poleward migration of tropical and sub-tropical biota between estuaries has been observed in response to warming (Hallett et al. 2018) (*medium confidence*), in agreement with the global trend of biogeographic shifts of marine organisms (Sections 5.2.3.1.1; 5.3.2–5.3.6).

Intensive human activities around estuaries and river deltas worldwide has substantially increased nutrient and organic matter inputs into such systems since the 1970s (Maavara et al. 2017). Increased organic matter accumulation has been shown to interact with warming, resulting in intensification of bacterial degradation and eutrophication (Maavara et al. 2017; Chen et al. 2018; Fennel and Testa, 2019), contributing to an increase in the frequency and extent of hypoxic zones (Breitberg et al. 2015; Gobler and Baumann, 2016).

5

The interaction between warming, increased nutrient loading, and hypoxia has shown to be related to the increased occurrences of HABs (Anderson et al. 2015; Paerl et al. 2018) (Box 5.4) (*high confidence*), pathogenic bacteria such as *Vibrio* species (Baker-Austin et al. 2017; Kopprio et al. 2017) (Section 5.4.2) (*low confidence*), and mortalities of invertebrates and fish communities (Jeppesen et al. 2018; Warwick et al. 2018) (*medium confidence*).

Fluctuations in estuarine salinity, turbidity and nutrient gradients are influenced by changes in precipitation and wind-stress caused by large-scale climatic variations such as the ENSO, the NAO and the South Atlantic Meridional Overturning Circulation (SAMOC) which have shown persistent anomalies associated with climate change since the 1970s (Wang and Cai, 2013; Delworth and Zeng, 2016; García-Moreiras et al. 2018). Similarly, storm surges and heat waves have increased nutrients and sediment loads in estuaries (Tweedley et al. 2016; Arias-Ortiz et al. 2018; Chen et al. 2018). Sustained long-term observations (15–40 years) provide evidence that large-scale climatic variations and extreme events affect plankton phenology and composition in estuaries worldwide with regional differences in the characteristics of the responses (Thompson et al. 2015; Abreu et al. 2017; Marques et al. 2017; Arias-Ortiz et al. 2018; López-Abbate et al. 2019) (*high confidence*). Although these changes in ecosystem components may be attributed to climate variability (Box 5.1), they demonstrate the sensitivity of estuarine ecosystems to climate change. Also, these large-scale climate events are *likely* to be intensified in the 21st century (Stocker, 2014) (Section 6.5.1).

Salinisation in estuaries is projected to continue in response to SLR, warming and droughts under global warming greater than 1.5°C (*high confidence*), and will pose further risks to ecosystems biodiversity and functioning (Zhou et al. 2017; Hallett et al. 2018; Zahid et al. 2018; Elliott et al. 2019) (Section 4.3.3.4, Cross-Chapter Box 7) (*medium confidence*). Estuarine wetlands are resilient to modest rates of SLR due to their sediment relocation capacity, but such adaptation is not expected to keep pace with projected rates of SLR under the RCP8.5 climate scenario (Section 5.3.2) (*high confidence*). Moreover, human activities that inhibit sediment movement and deposition in coastal deltas increase the likelihood of their shrinking as a result of SLR (Brown et al. 2018b; Schuerch et al. 2018) (*medium confidence*).

Oxygen-depleted dead zones in coastal areas are already a problem; they are projected to increase under the co-occurrence and intensification of climate threats and eutrophication (Breitburg et al. 2018; Laurent et al. 2018) (Section 5.2.2.4). While warming is the primary climate driver of deoxygenation in the open ocean, eutrophication is projected to increase in estuaries due to human activities and intensified precipitation increasing riverine nitrogen loads under both RCP2.6 and RCP8.5 scenarios, both mid-century (2031–2060) and later (2071–2100) (Sinha et al. 2017). Moreover, enhanced stratification in estuaries in response to warming is also expected to increase the risk of hypoxia through reduced vertical mixing (Du et al. 2018; Hallett et al. 2018; Warwick et al. 2018). The effects of warming will be more pronounced on high-latitude and temperate shallow estuaries with limited exchange with the open ocean (e.g., Río de La Plata Estuary, Baltic Sea and Chesapeake Bay) and seasonality that already leads to dead zone development when summertime temperatures reach critical values (e.g., Black Sea) (Altieri and Gedan, 2015) (*medium confidence*). The coastal acidification related to this expansion of hypoxic zones (Zhang and Gao, 2016; Cai et al. 2017; Laurent et al. 2017) imposes risk for sensitive organisms (Beck et al. 2011; Duarte et al. 2013; Feely et al. 2016; Carstensen et al. 2018).

The interaction of SLR and changes in precipitation will have a more severe impact on shallow estuaries (<10 m) than on deep basin estuaries (>10 m) (Hallett et al. 2018; Elliott et al. 2019) (*medium confidence*). For a projected SLR of 1 m, climate-related risks for shallow estuaries ecosystems are estimated to increase through increased tidal current amplitudes (by 5% on average), energy dissipation, vertical mixing and salinity intrusion (Prandle and Lane, 2015). Estuaries with high tidal exchanges and associated well-developed sediment areas are more resilient to global climate changes than estuaries with low tidal exchanges and sediment supply, since the latter are more vulnerable to SLR and changes in river flow (Brown et al. 2018b; Warwick et al. 2018) (*medium confidence*).

Overall, this assessment concludes that there is evidence of upstream redistribution of marine biotic communities in estuaries driven by increased sea water intrusion (*medium confidence*). Such distribution shifts are limited by physical barriers such as the availability of benthic substrates leading to reduction of suitable habitats for estuarine communities (*medium confidence*). Warming has led to poleward range shifts of biota between estuaries (*medium confidence*). Increased nutrient inputs from intensive human development in deltas increases bacterial respiration, which in turn is exacerbated by warming, leading to an expansion of suboxic and anoxic areas (*high confidence*). These changes reduce the survival of estuarine animals (*medium confidence*), and increase the occurrence of HABs and pathogenic microbes (*medium confidence*). Projected warming, SLR and tidal changes in the 21st century will continue to expand salinisation and hypoxia in estuaries (*medium confidence*). These impacts will be more pronounced under higher emission scenarios, and in temperate and high-latitude estuaries that are eutrophic, shallow and that naturally have low sediment supply.

5.3.2 Coastal Wetlands (Salt Marshes, Seagrass Meadows and Mangrove Forests)

Coastal vegetated wetlands include salt marshes, mangrove forests and subtidal seagrass meadows ecosystems, considered to be the main 'blue carbon' habitats (Sections 5.4.1 and 5.5.1.1) (McLeod et al. 2011). AR5 WGII and SR15 concluded that wetland salinisation is occurring at a large geographic scale (*high confidence*); that rising water temperatures has led to shifts in plant species distribution (*medium confidence*) (Wong et al. 2014b); and that SLR and storms are causing wetland erosion and habitat loss, enhanced by human disturbances (*high confidence*) (Section 4.3.3.5.1) (Wong et al. 2014b). This section assesses new evidence since AR5 and SR15 of observed climate impacts and future risks of these vegetated wetlands in terms of their role in supporting biodiversity and key ecosystem functions. The recent literature confirms and strengthens the SR15 conclusions (Section 5.3.7 and Figure 5.16).

Nearly 50% of the pre-industrial, natural extent of global coastal wetlands have been lost since the 19th century (Li et al. 2018a). Such a reduction in wetlands is primarily caused by non-climatic drivers such as alteration of drainage, agriculture development, coastal settlement, hydrological alterations and reductions in sediment supply (Adam, 2002; Wang et al. 2014; Kroeger et al. 2017; Thomas et al. 2017; Li et al. 2018a). However, large-scale mortality events of mangroves from 'natural causes' has also occurred globally since the 1960s; ~70% of this loss has resulted from low frequency, high intensity weather events, such as tropical cyclones (45%) and climatic extremes such as droughts, SLR variations and heat waves (Sippo et al. 2018) (*high confidence*). In Australia, the mangrove loss due to heat waves accounted for 22% of global mangrove forests (Sippo et al. 2018), with negative impacts on ecosystem biodiversity and the provisioning of services (Carugati et al. 2018; Saintilan et al. 2018) (Section 5.4). In coastal areas with sufficient sediment supply across the Indo-Pacific region, inland expansion of mangroves is occurring as a result of vertical accretion and root growth, allowing them to keep pace with current SLR (Lovelock et al. 2015). In seagrass meadows, temperature is the main limiting range factor, and over the past decades there have been several global die-off events (Hoegh-Guldberg et al. 2018). The vulnerability of seagrasses to warming varies locally depending on soil accretion and herbivory (El-Hacen et al. 2018; Marbà et al. 2018; Vergés et al. 2018) and on the population assemblages (e.g., expansion at high latitudes) (Beca-Carretero et al. 2018; Duarte et al. 2018). The compounding effects of heat waves, hypersaline conditions and increased turbidity and nutrient levels associated with floods have been shown to cause negative changes in the composition and biomass of co-occurring seagrass species (Nowicki et al. 2017; Arias-Ortiz et al. 2018; Lin et al. 2018) (*high confidence*). For example, in Shark Bay, Western Australia, a marine heat wave in austral summer 2010/2011 caused widespread losses (36% of area) of seagrass meadows, with negative implications for carbon storage (Arias-Ortiz et al. 2018). The poleward expansion of tropical mangroves into subtropical salt marshes as a result of increase in temperature has been also observed over the past half century on five continents (Saintilan et al. 2014; Saintilan et al. 2018) (*high confidence*); for example, in the Texas Gulf Coast (Armitage et al. 2015). The loss of open areas with herbaceous plants (salt marshes) reduces food and habitat availability for resident and migratory animals (Kelleway et al. 2017a; Lin et al. 2018) (Section 5.4.1.2).

The ability of salt marshes to increase their elevation and withstand erosion under SLR depends on the development of new soil by the external supply of mineral sediments and organic accretion by local biota (Section 5.4.1, Figure 5.19) (Bouma et al. 2016). In some places, critical organic accretion rates are declining due to reduced plant productivity from stress by more frequent inundation, and increased plant and microbial respiration rates as a result of warming; consequently, the elevation of marshes from soil accretion is slower than the rate of rising sea level, resulting in reduction of salt marsh area (Carey et al. 2017; Watson et al. 2017b). Vegetation loss rates were significantly negatively correlated with marsh elevation, suggesting inundation due to SLR since 1970 as the main driver, enhanced by storms and increased tidal range in back barrier marshes (Watson et al. 2017b). Plant species that are more sensitive to higher temperatures and increases in saltwater intrusion were found to be less abundant and in some cases replaced by salinity-tolerant species (Janousek et al. 2017; Piovan et al. 2019). Plant community restructuring has resulted in biodiversity loss (Pratolongo et al. 2013; Raposa et al. 2017) and reduced above- and below-ground productivity (McLeod et al. 2011; Watson et al. 2017b). As a result of tidal flooding, salt marsh soils do not dry out and high levels of carbon can accumulate under anaerobic conditions. This is coupled with generally low rates of methane emission which is strongly limited in saline marshes (Poffenbarger et al. 2011; Martin and Moseman-Valtierra, 2015; Kroeger et al. 2017; Tong et al. 2018) (*high confidence*).

Non-climatic human pressures on wetland ecosystems, including overfishing (Crotty et al. 2017), eutrophication (Legault II et al. 2018), and invasive species (Zhang et al. 2016), interact with climate change drivers and affect wetlands composition and structure, with the impacts varying between regions and species (Tomas et al. 2015; O'Brien et al. 2017; Pagès et al. 2017; York et al. 2017). The intensity of herbivory on seagrasses is expected to increase with global warming, particularly in temperate areas, because of the migration of tropical herbivores into temperate seagrass meadows (Hyndes et al. 2016; Vergés et al. 2018) (*medium confidence*, Section 5.2.3.1.1). Warming also reduces the fitness of seedlings by increasing necrosis and susceptibility to consumers and pathogenic pressure while reducing establishment potential and nutritional (Olsen et al. 2016b; Hernán et al. 2017). Because herbivores play a key role in modulating the biomass of plant communities, their more intense activity affects the provision of services in these ecosystems (Scott et al. 2018) (Section 5.4).

Globally, between 20–90% of existing coastal wetland area is projected to be lost by 2100 (Blankespoor et al. 2014; Crosby et al. 2016; Spencer et al. 2016), depending on different SLR projections under future emission scenarios. These projected changes vary regionally and between different types of wetlands. Gaining area may be possible, at least locally, if vertical sediment accretion occurs together with lateral re-accommodation (Brown et al. 2018b; Schuerch et al. 2018) (Section 4.3.3.5.1). Local losses may also be higher; for example, in New England, where regional rates of SLR have been as much as 50% greater than the global average (from 1–5.83 mm yr^{-1}; 1979–2015) (Watson et al. 2017a) and where projections suggest that 40–95% of salt marshes will be submerged by the end of this century (Valiela et al. 2018). In some species of seagrasses, enhanced temperature-driven flowering (Ruiz-Frau et al. 2017) and greater biomass production in response to elevated CO_2 (Campbell and Fourqurean, 2018) may increase resilience to warming. Nevertheless, severe habitat loss (70%) of endemic species such as *Posidonia oceanica* is projected by 2050 with the potential for functional extinction by 2100 under RCP8.5 climate scenario. For *Cymodosea nodosa*, the species with the highest thermal optima (Savva et al. 2018), warming is expected to lead to significant reduction of meadows (46% under RCP8.5) in the Mediterranean, although potentially compensated in part by future expansion into the Atlantic (Chefaoui et al. 2018).

The mangrove habitats of small islands, with lack of rivers, steep topography, sediment-starved areas, groundwater extraction and coastal development, are particularly vulnerable to SLR. Although mangrove ecosystems may survive the increased storm intensity and sea levels projected until 2100 under RCP2.6 (Ward et al. 2016), for RCP8.5 they are only resilient up to 2050 conditions (Sasmito et al. 2016). Negative climate impacts will be exacerbated in cases where anthropogenic barriers cause further 'coastal squeeze' that prevents inland movement of plants and limits relocation of sediment (*medium confidence*) (Enwright et al. 2016; Borchert et al. 2018).

In conclusion, substantial evidence supports with *high confidence* that warming and salinisation of wetlands caused by SLR are causing shifts in the distribution of plant species inland and poleward, such as mangrove encroachment into subtropical salt marshes (*high confidence*) or seagrass meadows contraction at low latitudes (*high confidence*). Plants with low tolerance to flooding and extreme temperatures are particularly vulnerable and may be locally extirpated (*medium confidence*). The flooded area of salt marshes can become a mudflat or be colonised by more tolerant, invasive species, whose expansion is favoured by combined effects of warming, rising CO_2 and nutrient enrichment (*medium confidence*). The loss of vegetated coastal ecosystems causes a reduction in carbon storage with positive feedbacks to the climate system (*high confidence*) (Section 5.4.1.2). SLR and warming are expected to continue to reduce the area of coastal wetlands, with a projected global loss of 20–90% by the end of the century depending on emission scenarios. High risk of total local loss is projected under the RCP8.5 emission scenario by 2100 (*medium confidence*), especially if landward migration and sediment supply is constrained by human modification of shorelines and river flows (*medium confidence*).

5.3.3   Sandy Beaches

Sandy beaches represent 31% of the world's ice-free shoreline (Luijendijk et al. 2018). They provide habitat for dune vegetation, benthic fauna and sea birds, nesting areas for marine turtles (Defeo et al. 2009), and several key ecosystem services (Drius et al. 2019) (Section 5.4.1.2). Sandy beach ecosystems are physically dynamic, where sediment movement is a key driver of benthic flora and fauna zonation (Schlacher and Thompson, 2013; van Puijenbroek et al. 2017). In AR5 WGII (Wong et al. 2014b) and SR15 (Hoegh-Guldberg et al. 2018), climate impacts on sandy beach ecosystems were not assessed individually but together with other coastal systems that included beaches, barriers, sand dunes, rocky coasts, aquifers and lagoons. Those assessments concluded with *high confidence* that SLR, storminess, wave energy and weathering regimes will continue to erode coastal shorelines and affect the soil accretion and land-based ecosystems, with highly site-specific effects (*high confidence*). Infrastructure and geological constraints reduce shoreline movement and cause coastal squeeze (*high confidence*). Assessment in Section 4.3.3.3 supports the conclusions in AR5 and SR15 regarding the erosion of sandy coastlines. This section specifically assesses the combined climate and non-climatic impacts on sandy beach biodiversity, ecosystem structure and functioning.

Worldwide, sandy beaches show vegetation transformations caused by erosion following locally severe wave events (Castelle et al. 2017; Delgado-Fernandez et al. 2019; Zinnert et al. 2019) (Table SM5.7). The original dense vegetation is replaced by sparser vegetation (Zinnert et al. 2019) and has a generally slow recovery (multiple years to decades) (Castelle et al. 2017). In some instances, the changes persist over decades, resulting in a regime shift in the beach morphology (Kuriyama and Yanagishima, 2018). Such changes in vegetation and beach morphology in response to local disturbances were also related to shifts in the associated fauna composition (Carcedo et al. 2017; Delgado-Fernandez et al. 2019). Direct attribution of these observed events to climate change is not available despite early evidence (since the 1970s) and an emerging literature (Section 4.3.3.1, Table SM5.7).

Sandy beaches show similar patterns of biogeographical shifts following warming, with increased dominance of species more tolerant to higher temperatures, as observed in other ocean ecosystems (Section 5.2.3.1.1, Table SM5.7). Examples of these observed shifts in abundance and distribution of benthic fauna in sandy beaches are found in the Pacific and Atlantic coasts of North and South America, and in Australia, including increased mortality of clam populations close to their upper temperature limits with low population recovery (Orlando et al. 2019), and poleward expansion of crabs since the 1980s that were related to warming (Schoeman et al. 2015) (Table SM5.7). Also, mass mortalities of beach clams have occurred during warm phases of El Niño events (Orlando et al. 2019) (Table SM5.7), parasite infestations on dense populations (Vázquez et al. 2016) and high wave exposure (Turra et al. 2016).

Human disturbances have caused coastal squeeze and morphological changes in sandy beaches (Martínez et al. 2017; Rêgo et al. 2018; Delgado-Fernandez et al. 2019). Along with SLR and climate-driven intensification of waves and offshore winds, these hazards have increased erosion rates suggesting a reduced resilience due to insufficient sediment supply and accretion capacity (Castelle et al. 2017; Houser et al. 2018; Kuriyama and Yanagishima, 2018). Narrow sandy beaches such as those in south California (Vitousek et al. 2017) or central Chile (Martínez et al. 2017) are particularly vulnerable to climate hazards when combined with human disturbances and where landward retreat of beach profile and benthic organisms is constrained due to increasing urbanisation (Hubbard et al. 2014) (Section 4.3.2.3).

Notwithstanding the uncertainty in projecting future interactions of SLR with other natural and human impacts on sandy shorelines (Le Cozannet et al. 2019; Orlando et al. 2019), they are expected to continue to reduce their area and change their topography due to SLR and increased extreme climatic erosive events. This will be especially important in low-lying coastal areas with high population and building densities (*medium confidence*, SM 4.2). Megafauna that use sandy beaches during vulnerable parts of their life cycles could be particularly impacted (Laloë et al. 2017). For example, the modelled incubation temperatures of green turtles have increased by 1°C since the mid-1970s, resulting in an average 20% increase in the proportion of female hatchlings over this period (Patrício et al. 2019). By 2100, global temperatures will approach lethal levels for incubation in existing nesting sites, and hatchling success is expected to drop to 32% under RCP8.5 scenario, with 93% of the hatchlings expected to be female (76% under RCP4.5). A possible microhabitat adaptation such as shadowed vegetated areas, however, could allow for continued male production throughout the 21st century (Patrício et al. 2019). In addition, a projected global mean SLR of ~1.2 m under the upper likely range of RCP8.5 by 2100 implies a loss of 59% and 67% in the present nesting area of the green turtle and the loggerhead respectively in the Mediterranean (Varela et al. 2019), and a loss of 43% in the nesting area of green turtles in West Africa (Patrício et al. 2019). Moreover, benthic crustaceans of sandy beaches, including isopods, crabs and amphipods, generally follow the temperature-body size gradient in which body size decreases towards warmer lower-latitude regions (Jaramillo et al. 2017). Assuming that the physiological underpinning of the relationship between body size and temperature can be applied to warming (see Section 5.2.2, *medium confidence*), the body size of sandy beach crustaceans is expected to decrease under warming (*low evidence, medium agreement*).

Overall, changes in sandy beach morphology have been observed from climate related events, such as storm surges, intensified offshore winds, and from coastal degradation caused by humans (*high confidence*), with impacts on beach habitats (e.g., benthic megafauna) (*medium confidence*). The direct influence of contemporary SLR on shoreline behaviour is emerging, but attribution of such changes to SLR remains difficult (Section 4.3.3.1). Projected changes in mean and extreme sea levels (Section 4.2.3) and warming (Section 5.2.1) under RCP8.5 are expected to result in high risk of impacts on sandy beach ecosystems by the end of the 21st century (*medium confidence*, Figure 5.16), taking account of the slow recovery rate of sandy beach vegetation, the direct loss of habitats and the high climatic sensitivity of some fauna. Under RCP2.6, the risk of impacts on sandy beaches is expected to be only slightly higher than the present day level (*low confidence*, Figure 5.16). However, pervasive coastal urbanisation lowers the buffering capacity and recovery potential of sandy beach ecosystems to impacts from SLR and warming and thus is expected to limit their resilience to climate change (*high confidence*).

5.3.4   Coral Reefs

Human activities and warming have already led to major impacts on shallow water tropical coral reefs caused by species replacement, bleaching and decreased coral cover while warming, ocean acidification and climate hazards will put warm water corals at very high risk even if global warming can be limited to 1.5°C above pre-industrial level (Hoegh-Guldberg et al. 2018; Kubicek et al. 2019; Sully et al. 2019). While providing new evidence to support these previous assessments (Kleypas, 2019), this assessment focuses on evaluating the variations in sensitivities and responses of coral reefs and their associated biota to highlight comparative risks and resiliences.

New evidence since AR5 and SR15 confirms the impacts of ocean warming (Kao et al. 2018; Jury and Toonen, 2019) and acidification (Jiang et al. 2018; Mollica et al. 2018; Bove et al. 2019) on coral reefs (*high confidence*), enhancing reef dissolution and bioerosion (*high confidence*), affecting coral species distribution, and leading

to community changes (Agostini et al. 2018) (*high confidence*). The rate of SLR (primarily noticed in small reef islands) may outpace the growth of reefs to keep up although there is *low agreement* in the literature (Brown et al. 2011; Perry et al. 2018) (*low confidence*). Reefs are further exposed to other increased impacts, such as enhanced storm intensity (Lavender et al. 2018), turbidity and increased runoff from the land (Kleypas, 2019) (*high confidence*). Recovery of coral reefs resulting from repeated disturbance events is slow (Hughes et al. 2019a; Ingeman et al. 2019) (*high confidence*). Only few coral reef areas show some resilience to global change drivers (Fine et al. 2019) (*low confidence*).

Globally, coral reefs and their associated communities are projected to change their species composition and biodiversity as a result of future interactions of multiple climatic and non-climatic hazards (Kleypas, 2019; Kubicek et al. 2019; Rinkevich, 2019) (*high evidence, very high agreement, very high confidence*). Multiple stressors act together to increase the risk of population declines or local extinction of reef-associated species through impacts of warming and ocean acidification on physiology and behaviours (Gunderson et al. 2017) (*high confidence*). Alteration of composition of coral reef-associated biota is exacerbated by changes in habitat conditions through increased sedimentation and nutrient concentrations from human coastal activities (Fabricius, 2005) (*high confidence*). Coral ecosystems in tropical small islands are also at high risk of being affected by extreme events, including storms, with their impacts exacerbated by SLR (Duvat et al. 2017; Harborne et al. 2017) (*high confidence*). Such risks on coral reef associated communities are substantially elevated when the level of these climatic and non-climatic hazards are above thresholds that may cause phase shifts in reef communities (McCook, 1999; Hughes et al. 2010; Graham et al. 2013; Hughes et al. 2018) (*high confidence*). A phase shift is characterised by an abrupt decrease in coral abundance or cover, with concurrent increase in the dominance of non-reef building organisms, such as algae and soft corals (Kleypas, 2019). Such phase shifts have already been observed in many coral reefs worldwide (Wernberg et al. 2016; Kleypas, 2019).

Notwithstanding the conclusion that coral reefs globally are projected to greatly decline at 2°C warming relative to pre-industrial level (Cacciapaglia and van Woesik, 2018; Dietz et al. 2018; Hoegh-Guldberg et al. 2018), climate impacts can be affected by variations in the sensitivity and adaptive capacity across coral species and coral reef ecosystems. Laboratory experiments show that some warm water corals possess the cellular, physiological or molecular machineries that could help them acclimatise or adapt to the effects of global change (*medium confidence*) (DeBiasse and Kelly, 2016; Gibbin et al. 2017; Wall et al. 2017; Camp et al. 2018; Donelson et al. 2018; Drake et al. 2018; Veilleux and Donelson, 2018; Hughes et al. 2019b). For example, there are species or genotypes that show less impacts by either ocean acidification or increased temperatures (Cornwall et al. 2018; Gintert et al. 2018). Some corals and their symbionts might be able to use epigenetic (heritable phenotype changes that do not involve alterations in the DNA sequences) mechanisms to reduce their sensitivity to temperature changes in their environment and to pass such traits to their offspring (Liew et al. 2017; Torda et al. 2017; Li et al. 2018b; Liew et al. 2018). The variations in sensitivity and adaptive capacity of coral species to warming and ocean acidification contribute to changes in species composition of coral reefs as they are exposed to climatic and non-climatic hazards (Ingeman et al. 2019; Kleypas, 2019; Kubicek et al. 2019) (*high confidence*). However, it has not yet been established whether coral and coral associated biota adaptation may hold beyond 1.5°C warming. The onset of coral bleaching in the last decade has occurred at higher SSTs (~0.5°C) than in the previous decade, suggesting that coral populations that remain after preceding bleaching events may have a higher thermal threshold (Sully et al. 2019) (*medium confidence*), potentially as a result of the increased dominance of species with lower sensitivity or higher adaptive capacity (Schulz et al. 2013; McClanahan et al. 2014; Mumby and van Woesik, 2014; Pandolfi, 2015; Folkersen, 2018) (*medium confidence*).

Coral reefs in deeper or mesophotic waters (found in tropical/subtropical regions at 30–150 m depth) may serve as refuges and sources for larval supply to those reefs exposed to disturbances (e.g., bleaching, storms, floods from land, sedimentation and tourism impacts) (Bridge et al. 2013; Thomas et al. 2015; Lindfield et al. 2016; Smith et al. 2016b; Bongaerts et al. 2017). Reefs exposed to local oceanographic characteristics that reduce warming, such as upwelling, may similarly provide refuges and larval sources (Tkachenko and Soong, 2017). However, recent evidence suggests that mesophotic coral reefs are at higher risk than previously indicated (Rocha et al. 2018). Monitoring of coral reefs worldwide shows that some areas in the eastern tropical Pacific Ocean (Smith et al. 2017), the Caribbean (Chollett and Mumby, 2013), the Red Sea (Fine et al. 2013; Osman et al. 2017), the Persian Gulf (Coles and Riegl, 2013) and the Great Barrier Reef, Australia (Hughes et al. 2010; Morgan et al. 2017) have recovered more rapidly after bleaching than the larger-scale average (*medium confidence*). There are regional differences in reef vulnerability when considering scales larger than 100 km or over latitudinal gradients (van Hooidonk et al. 2013; Heron et al. 2016; Langlais et al. 2017; McClenachan et al. 2017) (*high confidence*).

Based on findings from simulation modelling, SR15 concluded that "coral reefs are projected to decline by a further 70–90% at 1.5°C (*very high confidence*) with larger losses (>99%) at 2°C (*very high confidence*)". The variations in exposure, sensitivity and adaptive capacity between coral populations and regions are further projected to cause large changes in the composition and structure of the remaining coral reefs, with large regional differences (van Hooidonk et al. 2016; Hoegh-Guldberg et al. 2018; Kleypas, 2019; Kubicek et al. 2019; Sully et al. 2019).

5.3.5   Rocky Shores

Rocky shore ecosystems span the intertidal and shallow subtidal zones of the world's temperate coasts and are typically dominated by calcareous mussels or seaweeds (macroalgae). Other organisms that inhabit rocky shores are coralline algae (i.e., maerl beds), polychaetes, molluscs, bryozoans and sponges. Intertidal habitats are characterised by strong environmental gradients, and are exposed to marine and atmospheric climate regimes (Hawkins et al. 2016).

IPCC AR5 (Wong et al. 2014a) concluded that rocky shores are among the better-understood coastal ecosystems in terms of potential impacts of climate variability and change. The high sensitivity of sessile organisms (e.g., barnacles, mussels) to extreme temperature events (e.g., mass mortality and drastic biodiversity loss of mussels beds), and to acidification (widely observed in manipulative experiments) gives *high confidence* that rocky shore species are at high risk of changes in distribution and abundance from these two drivers. SR15 (Hoegh-Guldberg et al. 2018) concluded that rocky coasts are already experiencing large-scale changes, and critical thresholds are expected to be reached at warming of 1.5°C and above (*high confidence*).

More observational and empirical evidence since AR5 and SR15 confirms that climate change poses high risk to rocky shore ecosystems' biodiversity, structure and functioning through warming, acidification, SLR and extreme events (Agostini et al. 2018; Duarte and Krause-Jensen, 2018; Ullah et al. 2018; Milazzo et al. 2019). Immobile intertidal organisms are especially vulnerable to warming, due to the potential for extreme heat exposure during low tide emersion and prolonged desiccation events (Hawkins et al. 2016; Zamir et al. 2018) (*high confidence*). This effect is expected to lower the upper vertical limit of intertidal communities (Hawkins et al. 2016), reducing their suitable habitat (Harley, 2011), and accompanied by temperature-induced increases in predation by consumers (Sanford, 1999). While previous studies have documented a poleward shift in species distributions of rocky intertidal and reef algae (Duarte et al. 2013; Nicastro et al. 2013) and faunal species (Barry et al. 1995; Mieszkowska et al. 2006; Lima et al. 2007), local extinctions at the equatorial or warm edge of species ranges are increasingly being attributed to climate change (Yeruham et al. 2015; Sorte et al. 2017) (*high confidence*). Extreme heat waves are expected to cause mortality among rocky shore species (Gazeau et al. 2014; Jurgens et al. 2015) and subsequent declines or losses in important species can have cascading effects on the whole intertidal community and the services it provides (Gatti et al. 2017; Sorte et al. 2017; Sunday et al. 2017). Coralline fauna adapted to narrow environmental conditions seem especially vulnerable to heat waves, with observed mass mortalities in the Adriatic Sea in response to extreme summer temperatures (Kružić et al. 2016). The loss of thermal refugia associated with continued warming could exacerbate the impacts of heat stress on rocky intertidal communities (Lima et al. 2016). Nevertheless, experimental data indicate that some coralline algae that are well adapted to highly variable transitional environments can tolerate the warming projected for 2100 under RCP8.5; for these species, ocean acidification will constitute the main hazard (Nannini et al. 2015).

Ocean acidification is expected to decrease the net calcification (*high confidence*) and abundance (*medium confidence*) of rocky intertidal and reef-associated species (Kroeker et al. 2013), and the dissolution of calcareous species has already been documented in tide-pool communities (Kwiatkowski et al. 2016; Duarte and Krause-Jensen, 2018). Recent experimental and field studies, however, have demonstrated the importance of food resources in mediating the effects of ocean acidification on vulnerable rocky shores species (Ciais et al. 2013; Ramajo et al. 2016), suggesting that species' vulnerability to ocean acidification may be most pronounced in areas of high heat stress and low food availability (*medium confidence*) (Kroeker et al. 2017). There is increasing evidence that the interactions between multiple climate drivers will determine species vulnerability and the ecosystem impacts of climate change (Hewitt et al. 2016).

Studies on naturally acidified rocky reef ecosystems suggest ocean acidification will simplify rocky shore ecosystems, due to an overgrowth by macroalgae, a reduction in biodiversity and a reduction in the abundance of calcareous species (*medium confidence*) (Kroeker et al. 2013; Linares et al. 2015). These shifts in community structure and function have been observed in CO_2 seep communities (Hall-Spencer et al. 2008), already exposed to levels of pCO_2 expected to generally occur by the end of the century (Agostini et al. 2018). Reductions in the abundance of calcareous herbivores that can create space for rarer species by grazing the dominant algae, are expected to contribute to the overgrowth of fleshy macroalgae on rocky shores (Baggini et al. 2015). This shift towards macroalgae is associated with a simplification of the food web at lower trophic levels (Kroeker et al. 2011).

At the local scale, warming and ocean acidification are expected to change energy flows within rocky shores ecosystems (*medium confidence*). Experiments indicate that both climate drivers may boost primary productivity in some cases (Goldenberg et al. 2017); however, increased metabolic demands and greater consumption by predators under warmer temperature increase the strength of top-down control (predation mortalities of herbivores) and thus counteracts the effects of increased bottom-up productivity (Goldenberg et al. 2017; Kordas et al. 2017). Ocean acidification could also increase species energetic costs and the grazing rate of herbivores, affecting ecosystem responses to increased primary productivity (Ghedini et al. 2015). Although these increasingly complex experiments have highlighted the potential for species interactions to mediate the effects of climate change, our understanding of the effects on intact, functioning ecosystems is limited. Despite predictions for increased production and herbivory with warming and acidification, an experimental study of a more complex food web revealed an overall reduction in the energy flow to higher trophic levels and a shift towards detritus-based food webs (Ullah et al. 2018).

Overall, intertidal rocky shores ecosystems are highly sensitive to ocean warming, acidification and extreme heat exposure during low tide emersion (*high confidence*). More field and experimental evidence shows that these ecosystems are at a moderate risk at present and this level is expected to rise to very high under the RCP8.5 scenario by the end of the century (see Section 5.3.7). Benthic species will continue to relocate in the intertidal zones and experience mass mortality events due to warming (*high confidence*). Interactive effects between acidification and warming will exacerbate the negative impacts on rocky shore communities, causing a shift towards a less diverse ecosystem in terms of species richness and complexity, increasingly dominated by macroalgae (*high confidence*).

5.3.6   Kelp Forests

Kelp forests are three-dimensional, highly productive coastal ecosystems with a reported global NPP between 1.02–1.96 GtC yr^{-1} (Krause-Jensen and Duarte, 2016). They cover about 25% of the world's coastline (Filbee-Dexter et al. 2016), mostly temperate and polar (Steneck et al. 2003). Canopy-forming macroalgae provide habitat for many associated invertebrates and fish communities (Pessarrodona et al. 2019). This assessment synthesises new evidence since SR15 on climate risks and impacts, and their interactions with non-climatic drivers on ecosystem biodiversity, structure and functioning.

Observational and experimental evidence since SR15 (Hoegh-Guldberg et al. 2018) supports its conclusions that kelp forests are already experiencing large-scale changes, and that critical thresholds occur for some forests at 1.5°C of global warming (*high confidence*). Due to their low capacity to relocate and high sensitivity to warming, kelp forests are projected to experience higher frequency of mass mortality events as the exposure to extreme temperature rises (*very high confidence*). Moreover, changes in ocean currents have facilitated the entry of tropical herbivorous fish into temperate kelp forests decreasing their distribution and abundance (*medium confidence*). More evidence from model projections in the 21st century supports this observed range contraction of kelp forests at the warm end of their distributional margins and expansion at the poleward end with the rate being faster for high emission scenarios (*high confidence*).

New global estimates show that the abundance of kelp forests has decreased at a rate of ~2% per year over the past half century (Wernberg et al. 2019), mainly due to ocean warming and marine heat waves (e.g., in western Australia a mean loss of 43% in area followed a marine heat weave in summer 2010–2011 (Wernberg et al. 2016), Section 6.4.2.1), as well as from other human stressors (*high confidence*) (Filbee-Dexter and Wernberg, 2018). At some localities, human-driven environmental changes such as coastal eutrophication and pollution is causing severe deterioration of kelp forests adding to the loss of these ecosystems from warming, storms and heat weaves (Andersen et al. 2013; Filbee-Dexter and Wernberg, 2018).

Two global datasets and one dataset covering European coastlines (Araujo et al. 2016; Krumhans et al. 2016; Poloczanska et al. 2016) identify large local and regional variations in kelp abundance over the past half century with 38% of these ecoregions showing a decline, 27% an increase and 35% no change (Krumhans et al. 2016). These data reflect the high spatio-temporal variability and resilience of kelp forests (Reed et al. 2016; Wernberg et al. 2018). For example, a 34 year dataset of kelp canopy biomass along the California coastline does not yet show a significant response to global warming because this ecosystem responds to low frequency marine climate oscillations (Bell et al. 2018c). However, between 1950–2010 regional warming caused consistent negative responses in abundance, phenology, demography and calcification of macroalgae for the northeast Atlantic and southeast Indian Ocean (Poloczanska et al. 2016). Declines in kelp forest abundance attributed to climate change and not related to sea urchin overgrazing (which is a major driver of decline and regime shift; Ling et al. (2014)) have been documented since the 1970s and evidence has increased within the last two decades (Filbee-Dexter and Wernberg, 2018). Despite a lack of data from some regions such as South America (Pérez-Matus et al. 2017), observational evidence since SR15 supports with *very high confidence* that warming is driving a contraction of kelp forests at low latitudes (Franco et al. 2018b; Casado-Amezúa et al. 2019; Pessarrodona et al. 2019) and expansion in polar regions (*medium confidence*) (Section 3.2.3.1.2) (Bartsch et al. 2016; Paar et al. 2016).

In many areas worldwide where the distribution range of kelp has contracted due to climatic and non-climatic drivers, it has been replaced by a less diverse and less complex turf-dominated ecosystem (Filbee-Dexter and Wernberg, 2018) (*high confidence*). Kelp supports other ecosystem components by providing food, substrate for spawning and habitat that mediate trophic interactions (O'Brien et al. 2018); its degradation therefore reduces species richness, biomass production and dependent flora and fauna species (Teagle and Smale, 2018; Pessarrodona et al. 2019). In the northeast Atlantic, the warm water species *Laminaria ochroleuca* is expanding poleward into regions previously dominated by the cold water species *L. hyperborea* which is retreating at its southern edge. These two kelp species are similar in morphology, but the cold water *L. hyperborea* hosts sessile communities of algae and invertebrates 12 times more diverse and richer in biomass than the warm water kelp species (Teagle and Smale, 2018). Climate-driven shifts in the species composition also affect carbon cycling, because warm-temperate kelps produce larger pools of organic matter than cold-temperate species, and their detritus is degraded faster (Pessarrodona et al. 2019).

New empirical eco-physiological studies in combination with field surveys support the evidence for climate change causing kelp forest degradation and range shifts (Franco et al. 2018b; Wernberg et al. 2018). For example, interactive effects of ocean warming and acidification cause kelp degradation and disease-like symptoms, with detrimental effects on photosynthetic efficiency (Qiu et al. 2019). Enhanced herbivory due to warming and the establishment of herbivorous fish species in temperate kelp forest has been observed to enhance ecosystem degradation (Vergés et al. 2016). However, invader seaweed species driven by warming can create more complex trophic interactions, reducing the consumption by herbivorous gastropods (Miranda et al. 2019). Increased physical stress by storm events also alters the kelps community, affecting the recruitment time of kelp species. The resulting dominance of younger stages favors species with a year-round spore production or an opportunistic life strategy, reducing the kelp canopy (Pereira et al. 2017).

Projections of future distribution of kelp species based on their physiological thresholds show major species-specific range shifts under different emission scenarios. For example, under RCP2.6, laminaria and other canopy-forming seaweed species in the Northwest Atlantic are projected to show northward range shifts at their southern (warm) edge of ≤40 km, with some equatorial range expansion from 2050 to 2100. That northward range shift increases to 406 km under RCP8.5 (at 13–19 km per decade, including contractions of their warmer edges) (Wilson et al. 2019). Whilst no changes in species richness are projected under RCP2.6, more than 50% richness loss is projected under RCP8.5 in some

areas (Wilson et al. 2019). Overall, model projections show that worldwide range contractions of kelps can be expected to continue at the warm end of distributional margins and range expansions at their poleward end (*high confidence*) (Raybaud et al. 2013; Assis et al. 2016; Assis et al. 2018; Wilson et al. 2019).

In summary, kelp forests have experienced large-scale habitat loss and degradation of ecosystem structure and functioning over the past half century, implying a moderate to high level of risk at present conditions of global warming (*high confidence*) (Section 5.3.7). The loss of kelp forests is followed by the colonisation of turfs, which contributes to the reduction in habitat complexity, carbon storage and diversity (*high confidence*). Kelp ecosystems are expected to continue to decline in temperate regions driven by ocean warming and intensification of extreme climate events (*high confidence*). The level of risk for the ecosystem is projected to rise to very high under RCP8.5 scenario by 2100 (*high confidence*).

5.3.7   Risk Assessment for Coastal Ecosystems

This section synthesises the assessment of climate impacts on coastal ecosystems' biodiversity, structure and functioning and the levels of risk under contrasting future conditions of global warming. As described in Section 5.2.5, the format for Figure 5.16 matches that of Figure 19.4 of AR5 (Oppenheimer et al. 2015) and Figure 3.20 of SR15 (Hoegh-Guldberg et al. 2018), indicating the levels of additional risk as colours (white, yellow, red and purple). The elements or burning embers for coastal ecosystems (Figure 5.16) indicate how risks increase with ocean warming, acidification, deoxygenation, SLR and extreme events with a comparison between present day conditions (2000s) and future conditions by the year 2100 under low (RCP2.6) and high (RCP8.5) CO_2 emission scenarios. The transition between the levels of risk for each type of coastal ecosystem is estimated from key evidence assessed in Sections 5.3.1 to 5.3.6. The embers are based on SST and the transition-values may have an error of ±0.3°C depending on the consensus of expert judgment. The assessed confidence in assigning the levels of risk at present day and future scenarios are *low*, *medium*, *high* and *very high* levels of confidence. A detailed account of the procedures involved in developing the ember for each type of coastal ecosystem is given in the supplementary material (SM5.3). This includes the description of climate hazards, sensitivity of key biotic and abiotic components, natural adaptive capacity, and observed impacts and projected risks. The burning embers for seagrass meadows, warm water corals and mangrove forests are in agreement with the conclusions in SR15 (Hoegh-Guldberg et al. 2018). The more recent literature assessed here strengthens the overall confidence in the assignment of transition and the level of risk for each ecosystem.

Detection and attribution studies show that climate change impacts began over the past 50 years in coastal ecosystems, indicating a transition from undetectable risk (white areas in Figure 5.16) to moderate risk below recent sea surface temperatures for some ecosystems (*high confidence*). This transition occurs at lower global levels of sea surface warming for coral reefs (0.2°C–0.4°C) (*high confidence*), seagrass meadows (0.5°C–0.8°C) (*very high confidence*) and kelp forests (0.6°C–1.0°C) (*high confidence*), with coral reefs already at high risk (0.4°C–0.6°C) for the present day (*very high confidence*). Global common responses include large-scale coral bleaching events (Section 5.3.4) and contraction of seagrass meadows (Section 5.3.2) and kelp forests (Section 5.3.6) at low-latitudes (*high confidence*), in response to warming and marine heat waves. Degraded coral reefs and kelp forests have shifted to algal and turf-dominated ecosystem at several regions worldwide, causing loss of habitat complexity and biodiversity.

The transition from undetectable to moderate risk in salt marshes (Section 5.3.2) and rocky shores (Section 5.3.5) takes place between 0.7°C–1.2°C of global sea surface warming (*medium/high confidence*), and between 0.9°C–1.8°C (*medium confidence*) in sandy beaches (Section 5.3.3), estuaries (Section 5.3.1) and mangrove forests (Section 5.3.2) (Figure 5.16). In all these coastal ecosystems, the detection and attribution of changes in biodiversity, structure and functioning

are not as robust as in coral, seagrass and kelp ecosystems that have been extensively studied over the past decades and are highly sensitive to extreme climate events. Estuaries and sandy beaches are highly dynamic in terms of hydrological and geomorphological processes, giving them more natural adaptive capacity to climate impacts. In these systems, sediment relocation, soil accretion and landward expansion of vegetation may mitigate against flooding and habitat loss in the context of SLR and extreme climate-driven erosion. Common global responses observed since 1970 include poleward expansion of mangrove forests due to warming; transformation of salt marshes into mudflats; shifts in species composition in response to flooding and salinisation; upstream migration of estuarine biota; and redistribution of macrobenthic communities in sandy beaches. Calcified organisms in intertidal rocky shores are highly sensitive to ocean warming and acidification, marine heat waves and heat exposure during low tide, with observed mass mortality events and reduced calcification.

(a) Open ocean

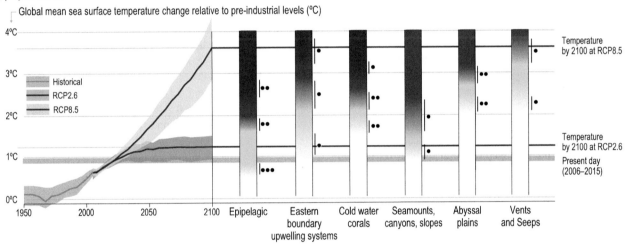

(b) Coastal ecosystems

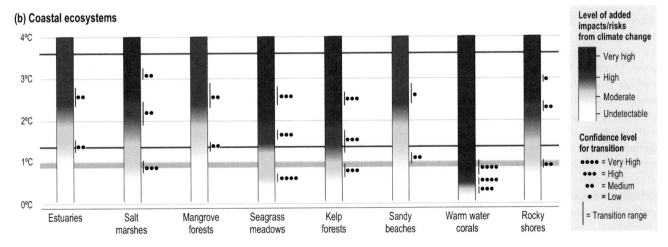

Figure 5.16 | Risk scenarios for open ocean (upper panel) and coastal (lower panel) ecosystems based on observed and projected climate impacts. 'Present day' corresponds to the 2000s, whereas the different greenhouse emissions scenarios: Representative Concentration Pathway (RCP)2.6 and RCP8.5 correspond to year 2100. Multiple climatic hazards are considered, including ocean warming, deoxygenation, acidification, changes in nutrients, particulate organic carbon flux and sea level rise (SLR) (see sections 5.2 and 5.3). The projected changes in sea surface temperature (SST) from an ensemble of general circulation models (left panels) indicate the level of ocean changes under RCP2.6 and RCP8.5 (see Cross Chapter Box 1 Table CB1 for the projected global average changes in average air temperature, SST and other selected ocean variables). Global average impacts/risks are represented. Regional variations of risks/impacts are described in Section 5.2.5, 5.3.7, SM5.2 and SM5.5. Impact/risk levels do not consider human risk reduction strategies such as societal adaptation, or future changes in non-climatic hazards. The grey vertical bars indicate the transition between the levels of risks, with their confidence level based on expert judgment. Note: The figure depicts climate change impacts and risks on warm water corals taken from SR15, based on global models. Observed impacts on coral reefs ecosystems outlined in Section 5.3.4 and Box 5.5 reveal a more complex situation that may result in regional differences in confidence levels.

In all coastal ecosystems, multiple climate hazards will emerge from historical variability in the 21st century under RCP8.5 (Box 5.1), while the time of emergence will be later and with less climate hazard under RCP2.6. Non-climatic human impacts such as eutrophication add to, and in some cases, exacerbate these large-scale slow climate drivers beyond biological thresholds at local scale (e.g., deoxygenation).

All coastal ecosystems will experience high to very high risk under RCP8.5 by the end of the 21st century. The ecosystems expected to be at very high risk under the high emission scenario are coral reefs (transition from high to very high risk 0.6°C–1.2°C) (*very high confidence*), seagrasses meadows (2.2°C–3.0°C) (*high confidence*), kelp forests (2.2°C–2.8°C) (*high confidence*) and rocky shores (2.9°C–3.4°C) (*medium confidence*). These ecosystems have low to moderate adaptive capacity, as they are highly sensitive to ocean warming, marine heat waves and acidification. For example, kelp forests at low-latitudes and temperate seagrass meadows with endemic species will continue to retreat with more frequent extreme temperatures, and their low dispersal ability will elevate the risk of local extinction. Biogenic shallow reefs with calcified organisms (e.g., corals, mussels, calcified algae) are particularly sensitive to ocean acidification and compound effects with rising temperatures, deoxygenation, SLR and increasing extreme events, making these ecosystems highly vulnerable (with low resilience) to future emission scenarios. Furthermore, almost all coral reefs will greatly decline from their current levels, even if global warming remains below 2°C (*very high confidence*). Any coral reefs that do survive to the end of the century will not be the same because of irreversible changes in habitat structure and functioning, including species extinctions and food web disruptions; these changes are already taking place (e.g., the Caribbean reefs). The transition to new ecosystem states driven by unpredictable pulses of disturbance and progressive climate hazards will have negative impacts on ecosystem services (Section 5.4).

The ecosystems at moderate to high risk under future emission scenarios (Figure 5.16) are mangrove forests (transition from moderate to high risk at 2.5°C–2.7°C of global sea surface warming), estuaries and sandy beaches (2.3°C–3.0°C) and salt marshes (transition from moderate to high risk at 1.8°C–2.7°C and from high to very high risk at 3.0°C–3.4°C) (*medium confidence*). Mangrove forests and salt marshes can initially cope with SLR by plant biomass accumulation, soil accretion and sediment relocation, but the evidence shows they are unlikely to withstand the SLR projected under RCP8.5. Moreover, pervasive coastal squeeze and human-driven habitat deterioration will reduce the natural capacity of these ecosystems to adapt to climate impacts (*high confidence*). Projected warming and SLR by the end of the century will continue to expand salinisation and hypoxia in estuaries with high risk of impacts for benthic and pelagic biota. These impacts will be more pronounced under RCP8.5 in more vulnerable eutrophic, shallow and microtidal estuaries in temperate and high latitudes. Erosion in sandy beach ecosystems will continue with global warming, rising sea level and more intense and frequent storm surges and marine heat waves. The risk of losing habitats for flora and fauna is expected to rise to high level under the high emission scenario by the end of the 21st century (*medium confidence*, Figure 5.16). By contrast, the risk of impacts is expected to be only slightly higher than present for a low emission scenario than today (*medium confidence*, Figure 5.16).

All types of ecosystems that have been assessed in the open ocean (Sections 5.2.3 and 5.2.4) and coastal areas (Sections 5.3.1 to 5.3.6) show increased risk under both the low and the high emission scenarios (RCP2.6 and RCP8.5) compared with the present level of change (Figure 5.16). In all assessed cases with all of the factors considered (climate drivers and physiological understanding), RCP2.6 has a lower level of risk than RCP8.5 (*very high confidence*).

5.4 Changing Marine Ecosystem Services and Human Well-being

Ecosystem services are the environmental processes and functions that have monetary or intrinsic value to human society; they render benefits to people and support human well-being (Tallis et al. 2010; Costanza et al. 2014). Marine ecosystem services are generated throughout the ocean, from shallow water to the deep sea (Armstrong et al. 2012; Thurber et al. 2014). Although all ecosystem services are interconnected (Leadley et al. 2014), they can be broadly divided into provisioning services, regulating services, supporting services and cultural services (Sandifer and Sutton-Grier, 2014), as considered below (Section 5.4.1) together with their implications for human well-being (Section 5.4.2). Ecosystem services have also been described as 'nature's contribution to people' (Díaz et al. 2018).

5.4.1 Changes in Key Ecosystem Services

AR5 WGII concluded that climate change increases the risk of impacts on the goods and services derived from marine biodiversity and ecosystems (Pörtner et al. 2014). SR15 concluded that current ecosystem services from the ocean are expected to be reduced at 1.5°C of global warming, with losses being even greater at 2°C of global warming. These reductions in services are driven by decreasing ocean productivity, biogeographic shifts, damage to ecosystems, loss of fisheries productivity and changes to ocean chemistry (*high confidence*) (Hoegh-Guldberg et al. 2018). Building on these previous assessments, this section assesses new evidence on observed impacts and future risk of climate change on ecosystem goods and services from the open ocean (Section 5.2) and coastal ecosystems (Section 5.3). Chapter 3 assesses ecosystem services in polar oceans.

5.4.1.1 Provisioning Services

Fisheries are an important provisioning service from marine ecosystems, providing food, nutrition, income and livelihoods for many millions of people around the world (FAO, 2018). Globally, total fish catches amount to 80–105 Mt annually in the 2000s (FAO, 2016; FAO 2018; Pauly and Zeller, 2016), directly generating over 80 billion USD of revenue (Sumaila et al. 2015). Most global fisheries are considered to be fully- to over-exploited (FAO, 2018). Over 80% of the global fish catch is estimated to be from coastal and shelf seas with less than 20% from the high seas (Sumaila et al. 2015) (Figure 5.17).

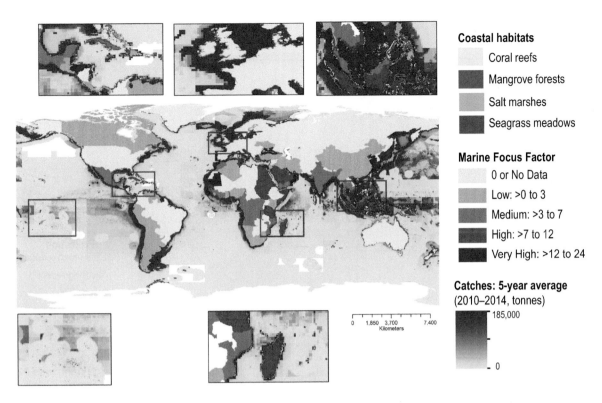

Figure 5.17 | Global distribution of fish catches (average 2010–2014, based on (Pauly and Zeller, 2016)), coastal habitats including seagrasses (UNEP-WCMC and FT, 2017) salt marshes (Mcowen et al. 2017), mangrove forests (Spalding, 2010), coral reefs (UNEP-WCMC and WRI, 2010) and an index (called Marine Focus Factor) for the inclusion of the ocean in the Nationally Determined Contributions (NDCs) published by each country (Gallo et al. 2017). The higher the Marine Focus Factor, the more frequent use of ocean in the country's NDCs.

Observed fish catches have been related to NPP and water temperature, with the direction and magnitude of the relationship varying between regions and fish stocks (Cheung et al. 2008; McOwen et al. 2015; Britten et al. 2016; Stock et al. 2017). The maximum catch potential of large marine ecosystems generally increases with their NPP and energy transfer efficiency, but the relative importance of total NPP to fisheries production is lower in nutrient-poor systems with microbially-dominated foodwebs (Section 5.2) and empirical relationships between NPP and fisheries production over-estimate potential catches in polar regions (Stock et al. 2017) (Chapter 3). Here, potential fish catch or maximum catch potential refers to the potential of the fish stocks to provide long-term fish catches; it is considered a proxy of maximum sustainable yield (MSY). However, the actual catches realised by fisheries will depend strongly on past and present fishing effort and the exploitation status of the resources (Cheung et al. 2018a; Barange, 2019). Observed variations between regions suggest that changes in temperature and NPP in the past (Section 5.2.2, 5.2.3) may have also affected maximum catch potential (*medium evidence, high agreement, medium confidence*).

Changes in fish catches from 1998 to 2006 in 47 large marine ecosystems around the world were found to be significantly related to: changes in estimated cholorophyll *a* (a proxy for phytoplankton biomass) in 18 of these ecosystems (mostly tropical and eastern boundary upwelling systems); changes in SST in 12 of these ecosystems (mostly mid-latitude); and changes in fishing intensity in 16 of these ecosystems (widely spread) (McOwen et al. 2015). Analysis of population data since the 1950s for 262 fish stocks across

39 large marine ecosystems and the high seas suggest that average recruitment to the stocks has declined by around 3% of the historical maximum per decade with variations between regions and stocks (Britten et al. 2016). The declines (69% of the studied stocks, 31 of the 39 assessed large marine ecosystems) are significantly related to estimated chlorophyll *a* concentration and the intensity of fishing, with the North Atlantic showing the steepest declines (Britten et al. 2016). In addition, recent meta-analysis of population data from 235 fish stocks worldwide from 1930 to 2010 suggest that the maximum catch potential from these populations decreased by 4.1% (95% confidence span 9.0% decline to 0.3% increase) during this period with variations between fish stocks and regions (Free et al. 2019). Specifically, temperature is a significant factor explaining changes in catch potential of 12% of the fish stocks, with East Asian regions having the largest stock declines related to warming. In intermediate latitudes across the Atlantic, Indian and Pacific Oceans, catches of tropical tunas, including skipjack and yellowfin tuna, are significantly and positively related to increases in SST, although the overall catches across latitudinal zones do not show significant change (Monllor-Hurtado et al. 2017). Observational evidence from spatial and temporal linkages between catches and oceanographic variables therefore supports the conclusions from AR5 WGII and SR15 that potential fisheries catches have already been impacted by the effects of warming and changing primary production on growth, reproduction and survival of fish stocks (*robust evidence, high agreement, high confidence*).

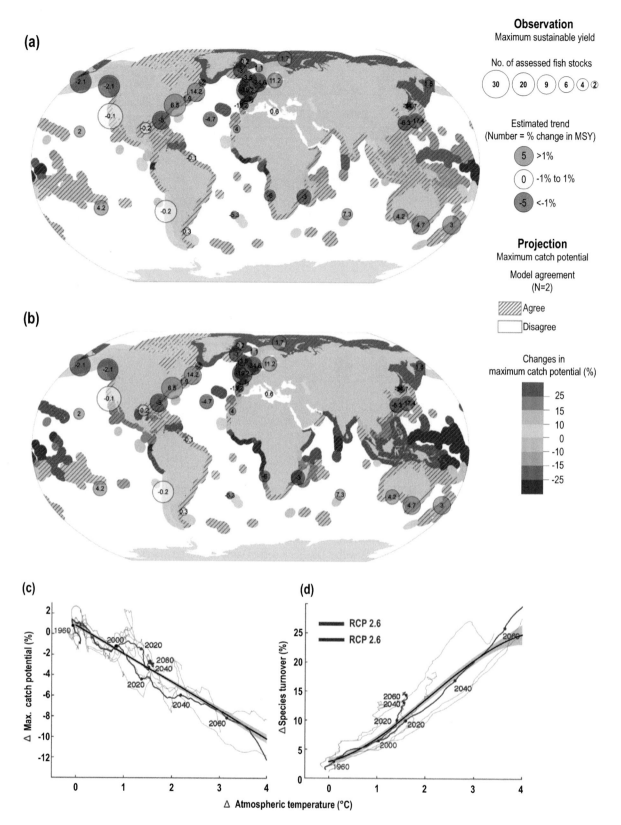

Figure 5.18 | Historical and projected maximum sustainable yield (MSY) and maximum fish catch potential by region. Historical trends in MSY is based on time series of fish stock assessment data (Free et al. 2019) represented as circles in panels (a) and (b). The size of the circle represents the number of assessed fish stocks while the number in the circle represents the estimated percent change in MSY since the 1930s. Projected changes in maximum catch potential by 2050 (average between 2041–2060) relative to 2000 (1991–2010) under (a) RCP2.6 and (b) RCP8.5 scenarios from two models: Dynamic Bioclimate Envelope Model and dynamic size-spectrum foodweb model with the colour in each ocean region representing the projected level of change and the shading representing where both models agree in the direction of change (Cheung et al. 2018a). Also presented is the scaling between projected global atmospheric warming (relative to 1950–1961) and (c) changes in maximum fish catch potential and (d) species turnover using the Dynamic Bioclimate Envelope Model and outputs from three Coupled Model Intercomparison Project Phase 5 (CMIP5) Earth System Models (ESMs) (Cheung and Pauly, 2016). Model projections (a,b) are provided by Plymouth Marine Laboratory, Euro-Meditteranean Centre for Climate Change and Fisheries and Marine Ecosystem Impact Model Intercomparison Project (FISHMIP) of coastal ecosystems.

There are substantial variations in the direction of changes and the attribution of climatic drivers between regions and fish stocks, and the availability of datasets is biased towards mid-latitude areas and epipelagic and coastal ecosystems. As a result, quantitative attribution of climate impacts on the productivity of specific fish stocks has *low confidence*. Changes in catch potential for fish stocks and regions worldwide that were considered overfished were most sensitive to warming (Essington et al. 2015; Britten et al. 2016; Free et al. 2019). This suggest that climatic drivers and overfishing have interacted synergistically in impacting some fish stocks and their catches (*high confidence*). In addition, analysis of historical catch records since AR5 show further warming related changes in species composition, with an increased dominance of warm water species in coastal and shelf seas since the 1970s (Cheung et al. 2013; Keskin and Pauly, 2014; Tsikliras et al. 2014; Maharaj et al. 2018). Many marine ecosystems worldwide have shown an increased dominance of warm water species following increases in sea water temperature (Section 5.2.3, 5.3), with parallel changes in the species composition of fish catches since the 1970s in many of the studied shelf seas (*high confidence*).

Based on CMIP5 ESM projections of changes in temperature, net primary production, oxygen, salinity and sea ice extent, two marine ecosystem and fisheries models project a decrease in maximum catch potential under RCP 2.6 of 3.9–8.5% by 2041–2060 and 3.4–6.4% by 2081–2100 relative to 1986–2005 (based on model projections described in Barange et al. 2018). Under RCP 8.5, the projected decrease was larger: 8.6–14.2% and 20.5–24.1% by the mid- and end- of the 21st century (Figure 5.18). The trends agree with the projected changes in total marine animal biomass for the 21st century (Blanchard et al. 2017; Lotze et al. 2018) (Section 5.2.3). A single fisheries model with atmospheric warming projected a potential catch loss of 3.4 million tonnes and decreases of 6.4% of catch potential of the exploited species per degree Celsius atmospheric warming relative to 1951–1960 level (Cheung et al. 2016b) (Figure 5.18). Interactions between temperature, net primary production and transfer efficiency of energy across the foodweb are projected to amplify these trends, with projected decreases greater than 50% in some regions by 2100 under high emissions scenarios (Stock et al. 2017). Thus, there is high model agreement that ocean warming and changes in NPP in the 21st century will reduce the global maximum catch potential, particularly in tropical oceans (*high confidence*) and alter the distribution and composition of exploited species (*high confidence*). The projected risk of these fisheries impacts increases with increasing greenhouse gas emissions (*high confidence*). However, given the uncertainties of projected changes in ocean conditions from ESMs (Section 5.2.2), and that most global scale fisheries models are largely driven by changes in temperature and primary production while other changes in ocean biogeochemical changes are not explicitly considered (Tittensor et al. 2018), the quantitative magnitude of the projected changes in maximum catch potential is considered to have *medium* and *low* confidence at global and regional scales, respectively. Given the significant interactions between catch potential and level of fisheries exploitation, the realised catches in the 21st century would depend on future scenarios of fishing and fisheries governance (Section 5.4.2, 5.5). As a result, projections of realised catches have *low confidence*.

Tropical oceans are projected to experience much larger impacts (three times or more decrease in catch potential) than the global average, particularly the western central Pacific Ocean, eastern central Atlantic Ocean and the western Indian Ocean, by the end of the 21st century under RCP8.5 (Blanchard et al. 2017). For example, around the exclusive economic zones of the Pacific Islands states, more than 50% of exploited fishes and invertebrates are projected to become locally extinct in many regions by 2100 relative to the recent past under RCP8.5 (Asch et al. 2018). These factors cause 74% of the area to experience a projected loss in catch potential of more than 50%. Under RCP2.6, the area of large projected catch loss is projected to be halved (Asch et al. 2018). However, while temperate commercially-important tunas species such as albacore, Atlantic and southern bluefin) are projected to shift poleward and decrease in abundance in the tropics, some tropical species such as skipjack tuna are projected to remain abundant, but with changes in distribution patterns in low-latitude regions by the mid-21st century, with some models projecting subsequent decrease under RCP8.5 (Lehodey et al. 2013; Dueri et al. 2014; Erauskin-Extramiana et al. 2019). Recent evidence therefore supports the conclusion from previous assessments (Pörtner et al. 2014; Hoegh-Guldberg et al. 2018) that low-latitude fish catch potential are projected to have a high risk of climate impacts, which will be exacerbated by higher greenhouse gas emissions (*medium evidence, high agreement, high confidence*). Tropical fish catch potential of some species resilient to the changing environment may have lower climate risk in the near-term although their risk increases substantially further into the 21st century under RCP8.5 (*medium confidence*). In contrast, the catch potential in the Arctic is projected to increase, although with high inter-model variability (*medium evidence, low agreement, low confidence*) (Cheung and Pauly, 2016; Blanchard et al. 2017) (Chapter 3).

Although demersal fisheries in the deep ocean represent a small proportion of global fisheries catches, they are economically valuable for some countries, and there is increasing commercial interest in mesopelagic (deep pelagic ocean) fisheries (St. John et al. 2016). Commercially-exploited fish and shellfish from deep sea ecosystems will be exposed to climate risks from physical and chemical changes in ocean conditions including warming, decreased oxygen, reduced aragonite saturation state, and decreased supply of particulate organic matter from the upper ocean (Section 5.2.3, 5.2.4) (FAO, 2019). These biogeochemical changes may reduce the growth, reproduction and survivorship of deep-ocean fish stocks, which will alter their distributions, in similar ways to those in the surface ocean, impacting their fish catch potential (FAO, 2019). For example, in the eastern Pacific near-bottom oxygen concentration is positively correlated with biomass of commercially harvested species (Keller et al. 2010) and catch per unit effort (Banse, 1968; Rosenberg et al. 1983; Keller et al. 2015); some commercially harvested species only appear during oxygenation events associated with El Niño (Arntz et al. 2006). In the mesopelagic zone, expansion of the OMZ results in habitat compression that can increase catchability of fish stocks such as tunas (Prince et al. 2010; Stramma et al. 2011). Also, as OMZ expands, the potential may exist for increased availability and harvest of hypoxia-tolerant species such as Humboldt squid (*Dosidicus gigas*), thornyheads (*Sebastolobus* spp.) or dover sole (*Microstomus pacificus*) (Gilly et al. 2013; Gallo and Levin, 2016). However, any expansion of the OMZ will interact with other climatic hazards such as warming, which then adds to the overall risk of impacts on fish stocks and their catches (Breitburg et al. 2018). Overall, the abundance of fisheries resources and potential catches from the deep sea will be at high risk of impacts in the 21st century under RCP8.5 (*low confidence*), with reduced risk under RCP2.6 (*medium confidence*).

In addition to capture fisheries, mariculture (marine aquaculture) is also an important marine ecosystem provisioning service, contributing about 27.7 million tonnes of seafood in 2016 (FAO, 2018). Recent projections of climate change impacts on mariculture, based on thermal tolerance and the effects of changing temperature, primary production and ocean acidification, suggest an overall decline in mariculture potential

by 2100 under RCP8.5 with large regional variations (Froehlich et al. 2018). Modelling analyses for farmed Atlantic salmon, cobia and seabream also suggest that climate change would reduce their growth potential in ocean areas where temperature is projected to increase to levels outside the thermal tolerance ranges of these species (Klinger et al. 2017). This decrease in growth could therefore translate into a decrease in the general productivity of the sector (*limited evidence, low confidence*); however, new potential areas and the use of more climate resilient strains or species for mariculture may emerge that could reduce the risk of impacts on potential mariculture production (*limited evidence, low confidence*).

Box 5.3 | Responses of Coupled Human-Natural Eastern Boundary Upwelling Systems to Climate Change

Eastern Boundary Upwelling Systems (EBUS) are among the world's most productive ocean ecosystems (Kämpf and Chapman, 2016). They directly support livelihoods in coastal communities and provide many wider benefits to human society (García-Reyes et al. 2015; Levin and Le Bris, 2015). The high productivity of EBUS is supported by the upwelling of cold and nutrient-rich waters, itself driven by equator-ward alongshore winds that cause the displacement of surface waters offshore and their replacement by deeper waters. Total annual fish catches from the four main EBUS (California Current, Humboldt Current, Canary Current and Benguela Current) were 16–24 tonnes yr^{-1} in the 2000s, providing around 17% of the global catch (Pauly and Zeller, 2016). These catches are consumed locally, as well as being processed and exported as seafood, fish meals and oils to support aquaculture and livestock production. Upwelling of cold deeper water also increases the condensation of humid air in coastal areas, benefitting coastal vegetation and agriculture and suppressing forest fires (Black et al. 2014). The high concentration of marine mammals attracted by the productive upwelling ecosystem support lucrative eco-tourism, such as whale watching in the California Current (Kämpf and Chapman, 2016). The total economic value of the goods and services provided by the Humboldt Current alone is estimated to be 19.45 billion USD per year (Gutiérrez et al.). Thus, although their area is small compared to other pelagic ecosystems, climate change impacts on EBUS will have disproportionately large consequences for human society (*very high confidence*).

The coupled human-natural EBUS are vulnerable to the multiple effects of climate change with large regional variation (Blasiak et al. 2017). Observations and modelling analyses suggest that winds have intensified in most EBUS (except the Canary Current) during the last 60 years, with several hypotheses proposed to explain the mechanisms (Sydeman et al. 2014; García-Reyes et al. 2015; Rykaczewski et al. 2015; Varela et al. 2015). ESMs predict reduction of wind and upwelling intensity in EBUS at low latitudes and enhancement at high latitudes for Representative Concentration Pathway (RCP)8.5, with an overall reduction in either upwelling intensity or extension (Belmadani et al. 2014; Rykaczewski et al. 2015; Sousa et al. 2017). However, coastal warming and wind intensification may lead to variable countervailing responses to upwelling intensification at local scales (García-Reyes et al. 2015; Wang et al. 2015a; Oyarzún and Brierley, 2018; Xiu et al. 2018). Local winds and mesoscale oceanographic features (not resolved in most global Earth System Models (ESMs)) are thought to have a greater impact on regional productivity than large-scale wind patterns (Renault et al. 2016; Xiu et al. 2018).

There is conflicting evidence in sea surface temperature (SST) trends in recent decades, even among the same EBUS, due to varying spatio-temporal resolution of SST data and the superimposed effects of interannual to multi-decadal variability (García-Reyes et al. 2015). Some EBUS are close to important thresholds in terms of oxygenation and ocean acidification (Gruber et al. 2012; Franco et al. 2018a; Levin, 2018). Large-scale coastal and offshore data for the California Current indicate that there have been decadal decreases in pH and dissolved oxygen affecting organisms and ecosystems (Alin et al. 2012; Bednaršek et al. 2014; Breitburg et al. 2018; Levin, 2018). Model projections for 2100 suggest strong effects of deoxygenation and reduced pH in the Humboldt Current and the California Current under RCP8.5 (Gruber et al. 2012; García-Reyes et al. 2015), affecting seafloor habitats and invertebrate fisheries (Marshall et al. 2017; Hodgson et al. 2018). For instance, the Humboldt Current is projected to experience widespread aragonite undersaturation within a few decades (Franco et al. 2018a), with strong impacts on calcified organisms. Such ocean acidification could be worsened by synergistic effects of ocean warming and deoxygenation (Lachkar, 2014).

The climate change impacts on ecosystem services from EBUS vary according to the biophysical and the socioeconomic characteristics of the upwelling systems (García-Reyes et al. 2015) (SM5.4). The fisheries are not only highly sensitive to upwelling conditions but also by fishing effects on the exploited populations. For example, the anchoveta population collapsed in the Humboldt Current after an El Niño in the 1970s (Gutiérrez et al. 2017). Because small pelagic fisheries from upwelling regions are the main source of the global fishmeal market, decreases in their catches increase the international fishmeal price, increasing the price of other food commodities (like aquaculture derived fish) that rely on fishmeal for their production (Merino et al. 2010; Carlson et al. 2017).

Box 5.3 (continued)

Any decrease in fish catches in EBUS will affect regional food security. For example, coastal fisheries in the Canary Current are an important source of micronutrients to nearby West African countries (Golden et al. 2016) that have particularly high susceptibility to climate change impacts and low adaptive capacity, because of their strong dependence on the fisheries resources, a rapidly growing population and regional conflicts. Decreased small pelagic fish stocks also increase the mortality and reduce reproduction of larger vertebrates such as hake (Guevara-Carrasco and Lleonart, 2008), whales and seabirds (Essington et al. 2015). Impacts on these organisms affect other non-fishing sectors that are dependent on EBUS, such as whale watching in the California Current, and generally degrade their intrinsic value.

Overall, EBUS have been changing with intensification of winds that drives the upwelling, leading to changes in water temperature and other ocean biogeochemistry (*medium confidence*). Three out of the four major EBUS have shown upwelling intensification in the past 60 years, with strongly increasing trends in ocean acidification and deoxygenation in the two Pacific EBUS in the last few decades (*high confidence*). The expanding oxygen minimum zone in the California EBUS has altered ecosystem structure and and fisheries catches (*medium confidence*). However, the direction and magnitude of observed changes vary among and within EBUS, with uncertainties regarding the driving mechanisms behind this variability. Moreover, the high natural variability of EBUS and their insufficient representation by global ESMs gives *low confidence* that these observed changes can be attributed to anthropogenic causes, which are predicted to emerge primarily in the second half of the 21st century (*medium confidence*) (Brady et al. 2017). Given the high sensitivity of the coupled human-natural EBUS to oceanographic changes, the future sustainable delivery of key ecosystem services from EBUS is at risk under climate change; those that are most at risk in the 21st century include fisheries (*high confidence*), aquaculture (*medium confidence*), coastal tourism (*low confidence*) and climate regulation (*low confidence*). For vulnerable human communities with a strong dependence on EBUS services and low adaptive capacity, such as those along the Canary Current system (Belhabib et al. 2016; Blasiak et al. 2017), unmitigated climate change effects on EBUS (complicated by other non-climatic stresses such as social unrest) have a high risk of altering their development pathways (*high confidence*).

5.4.1.2 Regulating Services

Regulating services are those ecosystem functions, like climate regulation, that allow the environment to be in conditions conducive to human well-being and development (Costanza et al. 2017). AR5 WGII concluded that climate change will alter biological, chemical and physical processes in the ocean that provide feedback on the climate system through their effects on atmospheric composition (*high confidence*) (Pörtner et al. 2014). Sections 5.2 and 5.3 consider new evidence since AR5 regarding climate impacts on marine ecosystems and associated risks; their implications for regulating services are examined here.

A major regulating service provided by marine ecosystems is carbon sequestration. The observed net carbon uptake from the atmosphere to the global ocean varied between 1.0–2.5 GtC yr^{-1} between 2000 and 2012, with a *very likely* uptake of 30–38 Gt of anthropogenic C over the period 1994–2007 (Section 5.2.2.3, Gruber et al. 2019). Estimates of carbon sequestered in the deep ocean range from 0.4 GtC yr^{-1} (Rogers, 2015) to 1.6 GtC yr^{-1} (Armstrong et al. 2010) with the annual burial rate (permanent removal to sediment) around 0.2 GtC yr^{-1} (Armstrong et al. 2010).

Deep sea ecosystems also contribute to the removal of methane released from the beneath the seabed through microbial anaerobic oxidation and the sequestration of methane-derived carbon in carbonate (Marlow et al. 2014; Thurber et al. 2014). In coastal ecosystems, carbon is biologically sequestered in coastal sediments, commonly known as 'blue carbon' (Section 5.5.1). Tidal wetlands play disproportionately important roles in coastal carbon budgets,

forming critical linkages between rivers, estuaries, and oceans (Najjar et al. 2018). Mean carbon storage in the top meter of soil is estimated at 280 MgC ha^{-1} for mangroves, 250 MgC ha^{-1} for salt marshes, and 140 MgC ha^{-1} for seagrass meadows, with long-term rates of carbon accumulation in sediments of salt marshes, mangroves, and seagrasses ranging from 18–1713 gC m^{-2} yr^{-1} (Pendleton et al. 2012). These values are, however, highly variable (Section 5.5.1.2). The large space and time scales mean that there is a long time-lag between seafloor change and detectable changes in carbon sequestration. These large lags, in turn render assessment of climate impacts on regulatory services in the deep ocean having *low confidence*.

Under RCP2.6, CMIP5 ESMs project a reduced net ocean carbon uptake by 2080, to around 1.0 GtC yr^{-1}. Under RCP8.5, net ocean carbon uptake increases to a net sink of around 5.5 GtC yr^{-1}, but with variability between models (Lovenduski et al. 2016). Although the open ocean biological pump contributes only part of current carbon uptake (Boyd et al. 2019), the downward carbon flux at 1000 m is projected to decrease by 9–16% globally under RCP8.5 by 2100. A projected decrease in carbon sequestration in the North Atlantic by 27–41% has been estimated to represent a loss of 170–3000 billion USD in abatement (mitigation) costs and 23–401 billion USD in social costs (Barange et al. 2017). Others have highlighted the declining value of open ocean carbon sequestration in the eastern tropical Pacific (Martin et al. 2016b) and the Mediterranean (Melaku Canu et al. 2015). The open ocean therefore seems *very likely* to reduce its carbon uptake by the end of the 21st century, with the reduction *very likely* being greater under RCP8.5 than for RCP2.6; however, specific projections only have *medium confidence* due to uncertainties associated with the structure of the models and with the future behaviour of the biological carbon pump (Section 5.2.2.3.1, 5.2.3).

Coastal blue carbon ecosystems provide climate regulatory services through their carbon removal and storage (Section 5.3.3). The current rates of loss of blue carbon ecosystems, partly due to climate change (Section 5.3) results in release of their stored CO_2 to the atmosphere (Section 5.5.1.2.2). However, increases in carbon sequestration are also possible; for example, temperature-driven displacement of salt marsh plants by mangrove trees may increase carbon uptake in coastal wetlands (Megonigal et al. 2016). Different rates of SLR may have opposite effects, with potential increases in net carbon uptake for slowly rising sea levels (assuming inland habitat migration is possible), but net carbon release for more rapid SLR (Figure 5.19). Such contrasting feedbacks between scenarios arise from the different responses of plant biomass, sediment accretion and inundation that control the overall response of vegetated coastal ecosystems to rising sea level (Gonneea et al. 2019). Thus, under high emission scenarios, SLR and warming are expected to reduce carbon sequestration by vegetated coastal ecosystems (*medium confidence*); however, under conditions of slow SLR, there may be net increase in carbon uptake by some coastal wetlands (*medium confidence*).

Coastal vegetation-rich ecosystems such as mangrove forests, coral reefs and salt marshes reduce storm impacts, protect the coastline from erosion, and help buffer the impacts of SLR, wave action and even moderate-sized tsunamis (Orth et al. 2006; Ferrario et al. 2014; Rao et al. 2015) (Section 5.5.2.2). Their loss or degradation under climate change (Sections 5.3) would therefore reduce the benefits of these regulatory services to coastal human communities (Perry et al. 2018), increasing the risk of damage and mortality from natural disasters (Rao et al. 2015) (*high confidence*). In some locations where climate-induced range expansion of coastal wetlands occurs, regulatory services such as storm protection and nutrient storage may be enhanced; however, the replacement of an existing ecosystem by others (e.g., salt marshes replaced by mangroves) may reduce habitat availability for fauna requiring specific vegetation structure and consequently other types of ecosystem services (Kelleway et al. 2017b; Sheng and Zou, 2017).

5.4.1.3   Supporting Services

Supporting ecosystem services are structures and processes, such as habitats, biodiversity and productivity, that maintain the ecosystem functions that deliver other services (Costanza et al. 2017). Marine supporting services include: primary and secondary production; habitat provision for feeding, spawning or nursery grounds, and refugia; and biodiversity. All these provide essential support for provisioning, regulating or cultural services (Haines-Young and

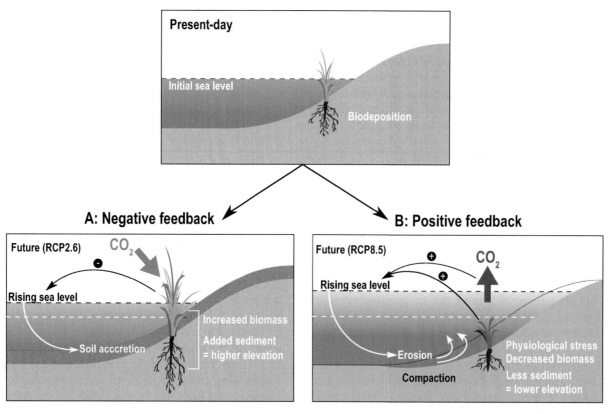

Figure 5.19 | Biogeomorphic climate feedbacks involving plant biomass, sediment accretion and inundation that control the response of vegetated coastal ecosystems to rising sea levels. (A) Under high rates of soil formation plants are able to offset gradual sea level rise (SLR) and may produce a negative feedback by increasing the uptake of atmospheric CO_2. In addition, below ground root production contributes to the formation of new soils and consolidates the seabed substrates. (B) Under low rate of soil formation, and when SLRs exceed critical thresholds, plants become severely stressed by inundation leading to less organic accretion and below ground subsidence and decay, producing a positive feedback by net CO_2 outgassing. This figure does not consider landward movements, controlled by topography and human land-use.

Potschin, 2013; Bopp et al. 2017). Therefore, climate change impacts on supporting services provided by marine ecosystems are directly dependent on the risks and impacts on their biodiversity and ecosystem functions, which are assessed in Sections 5.2.3, 5.2.4 and 5.3. Previously, AR5 highlighted the importance of the potential loss or degradation of habitat forming calcifying algae and corals, and the projected changes in waterways for Arctic shipping (Pörtner et al. 2014). The latter topic is considered in Chapter 3 and Section 5.4.2.4.

Publications since AR5 provide further evidence that coastal habitats are at risk from SLR, warming and other climate-related hazards (see Section 5.3). All these changes to supporting services have implications for other ecosystem services (Costanza et al. 2014), such as altering fish catches and their composition (Pratchett et al. 2014; Carrasquilla-Henao and Juanes, 2017; Maharaj et al. 2018) (Section 5.4.1.1) and carbon sequestration (Section 5.4.1.2). In the epipelagic ocean, climate change affects the pattern and magnitude of global NPP (Section 5.2.2.6) and the export of organic matter; both of these processes support ecosystem services in the deep ocean (Section 5.2.4) and elsewhere. Projected ocean acidification and oxygen loss will also affect deep ocean biodiversity and habitats that are linked to provisioning services in the deep ocean (Section 5.2.3.2, 5.2.4). Overall, there is *high confidence* that marine habitat loss and degradation have already impacted supporting services from many marine ecosystems worldwide. The confidence on the attribution of those impacts to climate change depends on the assessment of the ocean and coastal ecosystems (Section 5.2.3, 5.2.4, 5.3). Projected climate-driven alterations of marine habitats will increase the future risks of impacts on supporting services (*high confidence*).

5.4.1.4   Cultural Services

Cultural ecosystem services include recreation, tourism, aesthetic, cultural identity and spiritual experiences. These services are a product of humans experiencing nature and the availability of nature to provide the experiences (Chan et al. 2012). There is increasing evidence to support the conclusion in WGII AR5 that the intrinsic values and cultural importance of marine ecosystems, such as indigenous culture, recreational fishing and tourism, that are dependent on biodiversity and other ecosystem functions, are at risk from climate change. Since marine cultural services are inherently integrated with human well-being, their assessment is provided in Section 5.4.2.

5.4.2   Climate Risk, Vulnerability and Exposure of Human Communities and their Well-being

Human communities heavily depend on the ocean through the goods and services provided by marine ecosystems (Section 5.4.1) (Hilmi et al. 2015). The values of ocean-based economic activities are estimated to be trillions of USD, generating hundreds of millions of jobs (Hoegh-Guldberg, 2015; Spalding, 2016). As climate change is impacting marine biodiversity and ecosystem services (Section 5.3.1), human communities and their well-being will also be affected. This section is based on diverse types of information, from quantitative modelling to qualitative studies, using expert opinion, local knowledge and Indigenous knowledge (Cross-Chapter Box 4 in Chapter 1). Projection and assessment of risk and vulnerabilities not only depend on climate change scenarios but are also strongly dependent on scenarios of future social-economic development (Cross-Chapter Box 1 in Chapter 1).

This assessment divides the linkages between ecosystem services and human communities and their well-being into the three pillars of sustainable development, as used by the World Commission on Environment and Development. The three pillars are social and cultural, economic and environmental. Table 5.6 lists the specific dimensions under these pillars that are assessed in this section. Synthesis of risks and opportunities of climate change on human communities and well-being is at the end of this section through the lens of ocean economy and the United Nations' Sustainable Development Goals (SDGs).

Table 5.6 | The social, cultural and economic dimensions assessed in Section 5.4.2.

Dimensions	Sections under 5.4.2
Human and environmental health	Water-borne diseases (5.4.2.1.1) Harmful algal blooms (HABs) (Box 5.4) Interactions with contaminants (5.4.2.1.2) Food security (5.4.2.1.3)
Culture and other social dimensions	Cultural and aesthetic values (5.4.2.2.1) Potential conflicts in resource utilisation (5.4.2.2.2)
Monetary and material wealth	Fisheries (5.4.2.3.1) Coastal and marine tourism (5.4.2.3.2) Property values and coastal infrastructure (5.4.2.3.3)

5.4.2.1   Human Health and Environmental Health

5.4.2.1.1   *Water-borne diseases*

SR15 concluded that climate change will result in an aerial expansion and increased risk of water-borne disease with regional differences (*high to very high confidence*) (Hoegh-Guldberg et al. 2018). AR5 concluded that warming, excessive nutrient and seawater inundation due to SLR are projected to exacerbate the expansion and threat of cholera (Pörtner et al. 2014) (*medium confidence*). This assessment focuses on health risks caused by *Vibrio* bacteria and HABs. *Vibrio cholerae* (causing cholera) is estimated to be responsible for around 760,000 and 650,000 cases of human illness and death respectively in the world in 2010 (Kirk et al. 2015). An assessment of HABs is given in Box 5.4.

Vibrio species naturally occur in warm, nutrient-rich and low salinity coastal waters. Since AR5, analysis of the the Continuous Plankton Recorder dataset (Section 5.2.3) has shown a significant increase in *Vibrio* abundance in the North Sea over the period 1958–2011 related to sea surface warming (Vezzulli et al. 2016). Other time series data have confirmed a poleward expansion of *Vibrio* pathogens in mid- to high-latitude regions, ascribed at least partly to climate change (Baker-Austin et al. 2013; Baker-Austin et al. 2017). Extreme weather events such as flooding and tropical cyclones are also linked to increased incidences of *Vibrio*-related disease, suggested to be caused by the increased exposure of human populations to the pathogens during these extreme events (Baker-Austin et al. 2017). New evidence since AR5 therefore increases support for the linkages between warming, extreme weather events and increased risk of diseases caused by *Vibrio* bacteria (*very high confidence*).

Extrapolating from the observed relationship between environmental conditions and current *Vibrio* distributions, coastal areas that experience future warming, changes in precipitation and increases in nutrient inputs can be expected to see an increase in prevalence of *Vibrio* pathogens. These effects have been simulated in a global-scale model that relates occurrences of *Vibrio* with SST, pH, dissolved oxygen and chlorophyll *a* concentration under the SRES B1 scenario (Escobar et al. 2015). In the Baltic Sea, a nearly two-fold increase in the area suitable for *Vibrio* is projected between 2015 and 2050 for both RCP4.5 and RCP8.5 scenarios (relating to projected SST increase of 4°C–5°C), resulting in an elevated risk of *Vibrio* infections (Semenza et al. 2017). Projected conditions of increased coastal flooding from storm surges and SLR (Section 5.2.2) will also increase exposure to waterborne disease (Ashbolt, 2019), such as *Vibrio* (*medium confidence*). However, uncertainty in the socioeconomic factors affecting the future vulnerabilities of human populations render quantitative projections of the magnitude of health impacts uncertain (Lloyd et al. 2016).

Box 5.4 | Harmful Algal Blooms and Climate Change

Harmful Algal Blooms (HABs) are proliferations of phytoplankton (mostly dinoflagellates, diatoms and cyanobacteria) and macroaglae that have negative effects on marine environments and associated biota. Impacts include water discolouration and foam accumulation, anoxia, contamination of seafood with toxins, disruption of food webs and massive large-scale mortality of marine biota (Hallegraeff, 2010; Quillien et al. 2015; Amaya et al. 2018; García-Mendoza et al. 2018; Álvarez et al. 2019). The IPCC 5th Assessment Report (AR5) concluded that harmful algal outbreaks had increased in frequency and intensity, caused partly by warming, nutrient fluctuations in upwelling areas, and coastal eutrophication (*medium confidence*); however, there was *limited evidence* and *low confidence* for future climate change effects on HABs (AR5 Chapters 5, 6) (Pörtner et al. 2014; Wong et al. 2014b). Since AR5, HABs have increasingly affected human society, with negative impacts on food provisioning, tourism, the economy and human health (Anderson et al. 2015; Berdalet et al. 2017). For example, HABs caused an estimated loss of 42 million USD for the tuna industry in Baja California, Mexico (García-Mendoza et al. 2018) and mortality of more than 40,000 tonnes of cultivated salmon in Chile (Díaz et al. 2019). This additional observational and experimental evidence has improved detection and attribution of HABs to climate change, demonstrating that shifts in biogeography, increased abundance and increased toxicity of HABs in recent years have been partly or wholly caused by warming and by other, more direct human drivers.

New studies since AR5 show range expansion of warm water HAB species, such as *Gambierdiscus* that causes ciguatera fish poisoning (Kohli et al. 2014; Bravo et al. 2015; Sparrow et al. 2017); contraction of cold water species (Tester et al. 2010; Rodríguez et al. 2017); the detection of novel phycotoxins and toxic species (Akselman et al. 2015; Guinder et al. 2018; Paredes et al. 2019; Tillmann et al. 2019); and regional increases in the occurrence and intensity of toxic phytoplankton blooms (McKibben et al. 2017; Díaz et al. 2019) in relation to ocean warming. For example, growth of the toxic dinoflagellates *Alexandrium* and *Dinophysis*, producers of paralytic shellfish poisoning and okadaic acid, respectively, is enhanced by warmer conditions in the North Atlantic and North Pacific (Gobler et al. 2017), whilst environmental conditions linked with warm phases of El Niño Southern Oscillation ENSO are associated with blooms of toxic *Pseudo-nitzschia* species in the Northern California Current (McKibben et al. 2017), with devastating effects on coastal ecosystems (McCabe et al. 2016; Ritzman et al. 2018). Regional variations of trends in HAB occurrences can be explained by spatial differences in climate drivers (temperature, water column stratification, ocean acidification, precipitation and extreme weather events), as well as non-climatic drivers, such as eutrophication and pollution (Hallegraeff, 2010; Hallegraeff, 2016; Glibert et al. 2018; Paerl et al. 2018).

Experimental studies have provided additional evidence for the role of environmental drivers in inducing HABs and their degree of impact. These studies include those showing that toxin production can be affected by grazers (Tammilehto et al. 2015; Xu and Kiørboe, 2018) and changing nutrient levels (Van de Waal et al. 2013; Brunson et al. 2018). The biosynthesis of domoic acid by some *Pseudo-nitschia* species is induced by combined phosphate limitation and high CO_2 conditions (Brunson et al. 2018), with their growth and toxicity enhanced by warming in incubation experiments (Zhu et al. 2017). Recent mesocosm experiments using natural subtropical planktonic communities found that simulated CO_2 emission scenarios (between Representative Concentration Pathway (RCP)2.6 and RCP8.5 by 2100) improved the competitive fitness of the toxic microalgae *Vicicitus globosus* for CO_2 treatments above 600 µatm, and induced blooms above 800 µatm, with severe negative impacts for other components of the planktonic food web (Riebesell et al. 2018). Experiments with the toxic dinoflagellate *Akashiwo sanguinea* (hemolytic activity) have also shown that a combination of high CO_2 levels, warming and high irradiance stimulate the growth and toxicity of this HAB species (Ou et al. 2017).

Box 5.4 (continued)

Given the worldwide distribution of the key toxic species of *Alexandrium*, *Pseudo-nitzschia* and *Dinophysis*, if the current relationship between warming and the occurrences of HABs associated with these species persists in the future (Gobler et al. 2017; Townhill et al. 2018) (*medium confidence*), the projected changes in ocean conditions can be expected to intensify HAB-related risks for coastal biodiversity and ecosystems services (*high confidence*). The greatest risk is expected for estuarine organisms (Section 5.3.1) because HABs occurrences are stimulated by riverine nutrient loads, and exacerbated by warming and the lower dissolved oxygen and pH in estuarine environments (Gobler and Baumann, 2016; Paredes-Banda et al. 2018).

Local scale sustained monitoring programmes and early warning systems for HABs can alert resource managers and stakeholders of their potential occurrences so that they can take actions (e.g., toxic seafood alerts or relocation of activities) to reduce the impacts of HABs (Anderson et al. 2015; Wells et al. 2015) (*high confidence*). There is *limited evidence* in determining the degree to which reduction of non-climatic anthropogenic stressors can reduce risk of HABs (Section 5.5.2), although this approach may be effective in some areas (*low confidence*); for example, controlling nutrient inputs from human sources may reduce the risk of occurrence of HABs in the Baltic Sea. Other techniques such as active chemical and biological interventions are at experimental stage.

Overall, the occurrence of HABs, their toxicity and risk on natural and human systems are projected to continue to increase with warming and rising CO_2 in the 21st century (Glibert et al. 2014; Martín-García et al. 2014; McCabe et al. 2016; Paerl et al. 2016; Gobler et al. 2017; McKibben et al. 2017; Rodríguez et al. 2017; Paerl et al. 2018; Riebesell et al. 2018) (*high confidence*). Moreover, poleward distributional shifts of HAB species are expected to continue as a result of warming (Townhill et al. 2018). The increasing likelihood of occurrences of HABs under climate change also elevates their risks on ecosystem services such as fisheries, aquaculture and tourism as well as public health (Section 5.4.2, *high confidence*). Such risks will be greatest in poorly monitored areas (Borbor-Córdova et al. 2018; Cuellar-Martinez et al. 2018).

5.4.2.1.2   Interactions between climate change and contaminants

Climate change–contaminant interactions can alter the bioaccumulation and amplify biomagnification of several contaminant classes (Boxall et al. 2009; Alava et al. 2018). This section assesses two types of contaminants that are of concern to environmental and human health as examples of other contaminants with similar properties (Alava et al. 2017). These two types of contaminants are the toxic and fat-soluble persistent organic pollutants (POPs), such as polychlorinated biphenyls (PCBs), as well as the neurotoxic and protein-binding organic form of mercury, methylmercury (MeHg) (Alava et al. 2017). POPs and MeHg are bioaccumulated by marine organisms and biomagnified in food webs, reaching exposure concentrations that become harmful and toxic to populations of apex predators such as marine mammals (Desforges et al. 2017; Desforges et al. 2018) (Figure 5.20). Human exposure to POPs and MeHg can lead to serious health effects (Ishikawa and Ikegaki, 1980; UNEP, 2013; Fort et al. 2015; Scheuhammer et al. 2015).

Inorganic forms of mercury are more soluble in low pH water, while higher temperature increases mercury uptake and the metabolic activity of bacteria, thereby increasing mercury methylation, uptake by organisms and bioaccumulation rates (Scheuhammer, 1991; Celo et al. 2006; López et al. 2010; Macdonald and Loseto, 2010; Riget et al. 2010; Corbitt et al. 2011; Krabbenhoft and Sunderland, 2013; Roberts et al. 2013; de Orte et al. 2014; McKinney et al. 2015), although there is *limited evidence* on the extent of exacerbation by ocean acidification expected in the 21st century. Increased melting of snow and ice from alpine ecosystems and mountains (Chapter 2) can also increase the release of POPs and MeHg from land-based sources into coastal ecosystems (Morrissey et al. 2005). Modelling projections for the Faroe Islands region suggest increased bioaccumulation of

methyl mercury under climate change, with an average increases in MeHg concentrations in marine species of 1.6–1.8% and 4.1–4.7% under ocean warming scenarios of 0.8°C and 2.0°C, respectively, with an associated increase in potential human intake of mercury beyond levels recommended by the World Health Organization (Booth and Zeller, 2005). Foodweb modeling for the northeastern Pacific projects that concentrations of MeHg and PCBs in top predators could increase by 8% and 3%, respectively, by 2100 under RCP8.5 relative to current levels (Alava et al. 2018). Climate-related pollution risks are of particular concern in Arctic ecosystems and their associated indigenous communities because of the bioaccumulation of POPs and MeHg, causing long-term contamination of traditional seafoods (Marques et al. 2010; Tirado et al. 2010; Alava et al. 2017) of high dietary importance (Cisneros-Montemayor et al. 2016).

Overall, climate change can increase the exposure and bioaccumulation of contaminants and thus the risk of impacts of POPs and MeHg on marine ecosystems and their dependent human communities as suggested by indirect evidence and model simulations (Marques et al. 2010; Tirado et al. 2010; Alava et al. 2017) (*high agreement*). However, there is *limited evidence* on observed increase in POPs and MeHg due to climate change. Apex predators and human communities that consume them, including Arctic communities and other coastal indigenous populations, are thus vulnerable to increase in exposure to these contaminants and the resulting health effects (*medium evidence, medium agreement*).

The risk of microplastics has become a major concern for the ocean as they are highly persistent and have accumulated in many different marine environments, including the deep sea (Woodall et al. 2014; GESAMP, 2015; van Sebille et al. 2015; Waller et al. 2017;

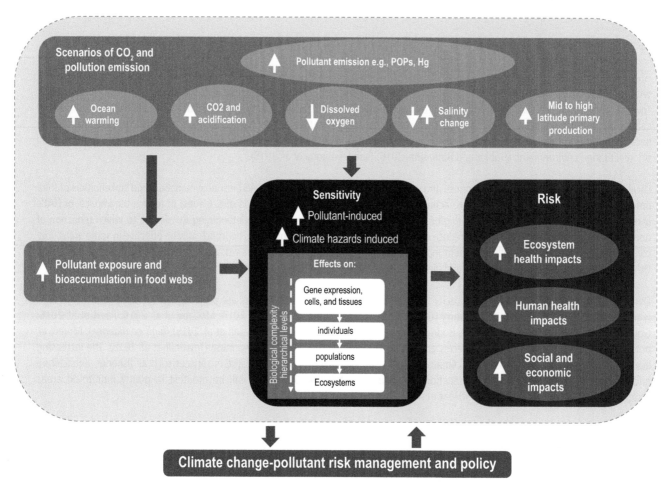

Figure 5.20 | The pathways through which scenario of climatic and pollutant hazards (orange boxes) and their interactions can lead to increases in exposure to hazards by the biota, ecosystems and people, their sensitivity (blue box) and the risk of impacts to ecosystem and human health and societies (red box). Such risks will interact with climate-pollutant risk management and policy. The synthesis is based on literature review presented in Alava et al. (2017). Figure adapted from Alava et al. (2017).

de Sá et al. 2018; Everaert et al. 2018; Botterell et al. 2019). There is *limited evidence* at present to assess their risk to marine ecosystems, wildlife and potentially humans through human consumption of seafood under climate change.

5.4.2.1.3　Food security

Seafood provides protein, fatty acids, vitamins and other micronutrients essential for human health such as iodine and selenium (Golden et al. 2016). Over 4.5 billion people in the world obtain more than 15% of their protein intake from seafood, including algae and marine mammals as well as fish and shellfish (Béné et al. 2015; FAO, 2017). Around 1.39 billion people obtain at least 20% of their supply of essential micronutrients from fish (Golden et al. 2016). SR15 concluded that global warming poses large risks to food security globally and regionally, especially in low-latitude areas, including fisheries (*medium confidence*) (Hoegh-Guldberg et al. 2018). This section builds on the assessment on observed and projected climate impacts on fish catches (Section 5.4.1.1) and further assess how such impacts interact with other climatic and non-climatic drivers in affecting food security through fisheries.

Many populations that are already facing challenges in food insecurity reside in low-latitude regions such as in the Pacific Islands

and West Africa where maximum fisheries catch potential is projected to decrease under climate change security (Golden et al. 2016; Hilmi et al. 2017) (Section 5.4.1; Figure 5.21) and where land-based food production is also at risk (Blanchard et al. 2017) (*medium confidence*). Populations in these regions are also estimated to have the highest proportion of their micronutrient intake relative to the total animal sourced food (Golden et al. 2016) (ASF; Figure 5.21). This highlights their strong dependence on seafood as a source of nutrition that further elevates their vulnerability to food security from climate change impacts on seafood supply (*high confidence*). Modeling of seafood trade networks suggests that Central and West African nations are particularly vulnerable to shocks from decrease in seafood supply from international imports; thus their climate risks of seafood insecurity could be exacerbated by climate impacts on catches and seafood supply elsewhere (Gephart et al. 2016). In addition, experimental studies suggest that warming and ocean acidification reduce the nutritional quality of some seafood by reducing levels of protein, lipid and omega-3 fatty acids (Tate et al. 2017; Ab Lah et al. 2018; Lemasson et al. 2019).

Non-climatic factors may exacerbate climate effects on seafood security. Over-exploitation of fish stocks reduces fish catches (Section 5.4.1.1) (Golden et al. 2016), whilst strong cultural dependence on seafood in many coastal communities may pose constraints in their adaptive

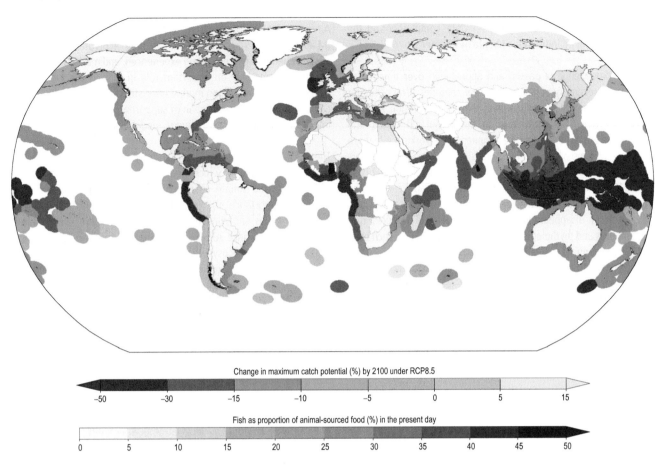

Change in maximum catch potential (%) by 2100 under RCP8.5

-50   -30   -15   -10   -5   0   5   15

Fish as proportion of animal-sourced food (%) in the present day

0   5   10   15   20   25   30   35   40   45   50

Figure 5.21 | Over the ocean the projected changes in catch potential (Section 5.4.1.1), and on land, each countries current proportion of fish micronutrient intake relative to the total animal sourced food (ASF) (Golden et al. 2016). The colour scale on land is the proportion of fish micronutrient intake relative to the total ASF; the scale on the ocean is projected change in maximum catch potential under Representative Concentration Pathway (RCP)8.5 by 2100 relative to the 2000s.

capacity to changing fish availability (Marushka et al. 2019). The shift from traditional nutritious wild caught seafood-based diets of coastal indigenous communities, towards increased consumption of processed energy dense foods high in fat, refined sugar and sodium, due to social and economic changes (Kuhnlein and Receveur, 1996; Shannon, 2002; Charlton et al. 2016; Batal et al. 2017), has important consequences on diet quality and nutritional status (Thaman, 1982; Quinn et al. 2012; Luick et al. 2014). This has led to an increased prevalence of obesity, diabetes, and other diet-related chronic diseases (Gracey, 2007; Sheikh et al. 2011) as well as the related decrease in access to culturally or religiously significant food items. The risk of climate change on coastal communities through the ocean could therefore be increased by non-climatic factors such as economic development, trade, effectiveness of resource governance and cultural changes (*high confidence*).

In summary, the food security of many coastal communities, particularly in low-latitude developing regions, is vulnerable to decreases in seafood supply (*medium confidence*) because of their strong dependence on seafood to meet their basic nutritional requirements (*medium confidence*), limited alternative sources of some of the essential nutrients obtained from seafood (*medium confidence*), and exposure to multiple hazards on their food security (*high confidence*). Although direct evidence from attribution analysis is not available, climate change may have already contributed to malnutrition by decreasing seafood supply in these vulnerable

communities (*low confidence*) and reduce coastal Indigenous communities' reliance on seafood-based diets (*low confidence*). Projected decreases in potential fish catches in tropical areas (*high confidence*) and a possible decrease in the nutritional content of seafood (*low confidence*) will further increase the risk of impacts on food security in low-latitude developing regions, with that risk being greater under high emission scenarios (*medium confidence*).

5.4.2.2   Cultural and Other Social Dimensions

5.4.2.2.1   Cultural and aesthetic values

Climate change threatens key cultural dimensions of lives and livelihoods (Adger et al. 2012), because people develop strong cultural ties and associate distinctive meanings with many natural places and biota in the form of traditions, customs and ways of life (Marshall et al. 2018). These impacts have been felt both by indigenous and non-indigenous peoples. Recent estimates suggest that there are more than 1900 indigenous groups along the coastline with around 27 million people across 87 countries (Cisneros-Montemayor et al. 2016). AR5 concluded that climate change will affect the harvests of marine species with spiritual and aesthetic importance to indigenous cultures (Pörtner et al. 2014). This section further assesses the effects of climate change on Indigenous knowledge and local knowledge and their transmission

5

and the implication for well-being of people, complementing the assessment for Arctic indigenous people in Chapter 3.

Indigenous knowledge is passed and appreciated over timeframes ranging from several generations to a few centuries (Cross-Chapter Box 4 in Chapter 1). The adjustment of the transmission and the network of Indigenous knowledge on the ocean and coasts, and related perceptions and practice, implies a reworking of these knowledge systems where the individuals and the groups are actors in a narrative and historical construction (Roué, 2012; Alderson-Day et al. 2015). SLR is already transforming the seascape, such as the shape of shores in many low-lying islands in the Pacific, leading to modification or disappearance of geomorphological features that represent gods and mythological ancestors (Camus, 2017; Kench et al. 2018). These changing seascape also affects the mobility of people and residence patterns, and consequently, the structure and transmission of Indigenous knowledge (Camus, 2017). The fear of SLR and climate change encourage security measures and the grouping of local people to the safest places, contributing to the erosion of indigenous culture and their knowledge about the ocean (Bambridge and Le Meur, 2018), and impairment of opportunity for social elevation for some Pacific indigenous communities (Borthwick, 2016).

Climate change is also projected to shift the biogeography and potential catches of fishes and invertebrates (5.2.3.1, 5.3, 5.4.1.1) that form an integral part of the culture, economy and diet of many indigenous communities, such as those situated along the Pacific Coast of North America (Lynn et al. 2013). Indigenous fishing communities that depend on traditional marine resources for food and economic security are particularly vulnerable to climate change through reduced capacity to conduct traditional harvests because of reduced access to, or availability of, resources (Larsen et al. 2014; Weatherdon et al. 2016). Overall, the transmission of Indigenous culture and knowledge is at risk because of SLR affecting sea- and land-scapes, the availability and access to culturally important marine species, and communities' reliance on the ocean for their livelihood and their cultural beliefs (*medium confidence*). Strong attachment to traditional marine-based livelihoods has also been reported for non-indigenous communities in Canada (Davis, 2015), the USA (Paolisso et al. 2012), Spain (Ruiz et al. 2012) and Australia (Metcalf et al. 2015). Reduction in populations of fish species that have supported livelihoods for generations, and deteriorations of iconic elements of seascapes are putting the well-being of these communities at risk (*high confidence*).

Other cultural values supported by the ocean are diverse. They include education, based on knowledge of marine environments. Such education can increase knowledge and awareness of climate change impacts and the efficacy of their mitigation (Meadows, 2011); it can also influence the extent to which stewardship activities are adopted (von Heland et al. 2014; Wynveen and Sutton, 2015; Bennett et al. 2018), and can help develop new networks between coastal people and environmental managers for the purposes of planning and implementing new adaptation strategies (Wynveen and Sutton, 2015). A critical element in reducing vulnerability to climate change is to educate people that they are an integral part of the Earth system and have a huge influence on the balance of the system.

An important marine ecosystem service is to support such education (Malone, 2016). Thus, education can play a pivotal role in how climate change is perceived and experienced, and marine biodiversity and ecosystems play an important role in this. At the same time, climate change impacts on marine ecosystems (Sections 5.2.3, 5.2.4) can affect the role of the ocean in supporting such public education (*medium evidence, high agreement, medium confidence*).

The aesthetic appreciation of natural places is one of the fundamental ways in which people relate to their environment. AR5 noted that climate change may impact marine species with aesthetic importance that affect local and indigenous cultures, local economies and challenge cultural preservation (Pörtner et al. 2014). Evidence since AR5 confirms that aesthetically appreciated aspects of marine ecosystems are important for supporting local and international economies (especially through tourism), human well-being, and stewardship. For example, Marshall et al. (2018) found that aesthetic values are a critically important cultural value for all cultural groups, and are important for maintaining sense of place, pride, identity and opportunities for inspiration, spirituality, recreation and well-being. However, climate change induced degradation and loss of biodiversity and habitats (Section 5.2.3, 5.2.4, 5.3) can also negatively impact the ecosystem features that are currently appreciated by human communities, such as coral reefs, mangroves, charismatic species (such as some marine mammals and seabirds) and geomorphological features (e.g., sandy beaches). There are also aesthetic and inspirational values of marine biodiversity and ecosystems that are important to the psychological and spiritual well-being of people, including film, literature, art and recreation (Pescaroli and Magni, 2015). Other cultural dimensions that are becoming more widely acknowledged as potentially disturbed by climate change include the appreciation of scientific, artistic, spiritual, and health opportunities, as well as appreciation of biodiversity, lifestyle and aesthetics (Marshall et al. 2018). Thus, climate change may also affect the way in which marine ecosystems support human well-being through cultural dimensions. However, the difficulties in evaluating the importance of aesthetic aspects of marine ecosystems, and in detecting and attributing of climate change impacts, result in such assessment having *low confidence*.

Climate change affects human cultures and well-being differently. For example, Marshall et al. (2018) assessed the importance of identity, pride, place, aesthetics, biodiversity, lifestyle, scientific value and well-being within the Great Barrier Reef region by 8,300 people across multiple cultural groups. These groups included indigenous and non-indigenous local residents, Australians (non-local), international and domestic tourists, tourism operators, and commercial fishers. They found that all groups highly rated all (listed) cultural values, suggesting that these values are critically associated with iconic ecosystems. Climate change impacts upon the Great Barrier Reef, through increased temperatures, cyclones and SLR that cumulatively degrade the quality of the Reef, are therefore liable to result in cultural impacts for all groups. However, survey that assess the emotional responses to degradation of the Great Barrier Reef by similar stakeholder groups reported different levels of impacts among these groups (Marshall et al. 2019). Therefore, many ocean and coastal dependent communities value marine ecosystems highly and climate impacts can affect their well-being, although the sensitivity to such impacts can vary among stakeholder groups (Marshall et al. 2019) (*low confidence*).

Climate change may alter the environment too rapidly for cultural adaptation to keep pace. This is because the culture that forms around a natural environment can be so integral to people's lives that disassociation from that environment can induce a sense of disorientation and disempowerment (Fisher and Brown, 2015). The adaptive capacity of people to moderate or influence cultural impacts, and thereby reduce vulnerability to such impacts, is also culturally determined (Cinner et al. 2018). For example, when a resource user such as a fisher, farmer, or forester is suddenly faced with the prospect that their resource-based occupation is no longer viable, they lose not only a means of earning an income but also an important part of their identity (Marshall et al. 2012; Tidball, 2012). Loss of identity can, in turn, have severe economic, psychological and cultural impacts (Turner et al. 2008). Climate change can quickly alter the quality of, or access to, a natural resource through degradation or coastal inundation, so that livelihoods and lifestyles are no longer able to be supported by that resource. When people are displaced from places that they value, there is strong evidence that their cultures are diminished, and in many cases endangered. There are no effective substitutions for, or adequate compensation for, lost sites of significance (Adger et al. 2012). As sensitive marine ecosystems such as coral reefs and kelp forest are impacted by climate change at rapid rate (Section 5.3), these can lead to the loss of part of people's cultural identity and values beyond the rate at which identify and values can be adjusted or substituted (*medium confidence*).

5.4.2.2.2 Potential conflicts in resource utilisation

Redistribution of marine species in response to direct and indirect effects of climate change may also disrupt existing marine resource sharing and associated governance (Miller and Russ, 2014; Pinsky et al. 2018). These effects have contributed to disputes in international fisheries management for North Atlantic mackerel (Spijkers and Boonstra, 2017) and Pacific salmon (Miller and Russ, 2014). These disagreements have stressed diplomatic relations in some cases (Pinsky et al. 2018). Decreases and fluctuations in fish stock abundance and fish catches have also contributed to past disputes (Belhabib et al. 2016; Pomeroy et al. 2016; Blasiak et al. 2017). Under climate change, shifts in abundance and distribution of fish stocks are projected to intensify in the 21st century (Sections 5.2.3, 5.3, 5.4.1.1). Stocks may locally increase and decrease elsewhere. New or increased fishing opportunities may be created when exploited fish stocks shift their distribution into a country's waters where their abundance was previously too low to support viable fisheries (Pinsky et al. 2018). The number of new transboundary stocks occurring in exclusive economic zones worldwide was projected to be around 46 and 60 under RCP2.6 and RCP8.5, respectively, by 2060 relative to 1950–2014 (Pinsky et al. 2018). However, such alteration of the sharing of resources between countries would challenge existing international fisheries governance regimes and, without sufficient adaptation responses, increase the potential for disputes in resource allocation and management (Belhabib et al. 2018; Pinsky et al. 2018). Overall, projected climate change impacts on fisheries in the

21st century increase the risk of potential conflicts among fishery area users and authorities or between two different communities within the same country (Ndhlovu et al. 2017; Shaffril et al. 2017; Spijkers and Boonstra, 2017) (*medium confidence*), exacerbated through competing resource exploitation from international actors and mal-adapted policies (*low confidence*). Such risks can be reduced by appropriate fisheries governance responses that are discussed in Sections 5.5.2 and 5.5.3.

5.4.2.3 Monetary and Material Wealth

5.4.2.3.1 Wealth generated from fisheries

Global gross revenues from marine fisheries were around 150 billion in 2010 USD (Swartz et al. 2013; Tai et al. 2017). Capture fisheries provide full-time and part-time jobs for an estimated 260 ± 6 million people in the 2000s period, of whom 22 ± 0.45 million are small-scale fishers (Teh and Sumaila, 2013). Small-scale fisheries are important for the livelihood and viability of coastal communities worldwide (Chuenpagdee, 2011). AR5 concluded with *low confidence* that climate change will lead to a global decrease in revenue with regional differences that are driven by spatial variations of climate impacts on and the flexibility and capacities of food production systems (Pörtner et al. 2014). AR5 also highlighted the high vulnerability of mollusc aquaculture to ocean acidification. For example, the oyster industry in the Pacific has lost nearly 110 million USD in annual revenue due to ocean acidification (Ekstrom et al. 2015). This section examines the rapidly growing literature assessing the risks of climate change on fisheries and aquaculture sectors, and the potential interaction between climatic and non-climatic drivers on the economics of fisheries. However, new evidence on observed economic impacts of climate change on fisheries since AR5 is limited.

Since AR5, projections on climate change impacts on the economics of marine fisheries have incorporated a broader range of social-economic considerations. Driven by shifts in species distributions and maximum catch potential of fish stocks (Section 5.4.1), if the ex-vessel price of catches remains the same, marine fisheries maximum revenue potential are projected to be negatively impacted in 89% of the world's fishing countries under the RCP8.5 scenario by the 2050s relative to the current status, with projected global decreases of $10.4 \pm 4.2\%$ and $7.1 \pm 3.5\%$ under RCP8.5 and RCP2.6, respectively, by 2050 relative to 2000 (Lam et al. 2016). While the projected changes in revenues are sensitive to price scenarios (Lam et al. 2016), future maximum revenue potential is reduced under high emission scenarios (Sumaila et al. 2019). For example, when the elasticity of seafood price in relation to their supply was modelled explicitly, fisheries maximum revenue potential under a 1.5°C atmospheric warming scenario was projected to be higher than for 3.5°C warming by 7.4% (13.1 billion USD) $\pm 2.3\%$, across projections from three CMIP5 models (Sumaila et al. 2019). Accounting for the subsequent impacts on the dependent communities and relative to the 1.5°C warming scenario, that study also projected a decrease in seafood workers' incomes of 7.8% (3.7 billion USD) $\pm 2.3\%$ and an increase in households' seafood expenditure by the global population of 3.2% (6.3 billion USD) $\pm 3.9\%$ annually under a 3.5°C warming scenario (Sumaila et al. 2019).

5

Fisheries management strategies and fishing effort affect the realised catch and economic benefits of fishing (Barange, 2019). Modelling analysis of fish stocks with available data worldwide showed that for RCP6.0, adaptation of fisheries by accommodating shifts in species distribution and abundance, as well as rebuilding existing overexploited or depleted fish stocks, is projected to lead to substantially higher global profits (154%), harvest (34%), and biomass (60%) in the future, relative to a no adaptation scenario. However, the total profit, harvest and biomass are negatively affected even with the full adaptation scenario under RCP8.5 (Gaines et al. 2018). Overall, climate change impacts on the abundance, distribution and potential catches of fish stocks (see Section 5.3.1) are expected to reduce the maximum potential revenues of global fisheries (*high agreement, medium evidence, medium confidence*). These impacts on fisheries will increase the risk of impacts on the income and livelihoods of people working in these economic sectors by 2050 under high greenhouse gas emission scenarios relative to low emission scenario (*high confidence*). Rebuilding overexploited or depleted fisheries can help improve economic efficiency and reduce climate risk, provided that emissions are greatly reduced (*medium confidence*).

The economic implications of climate change on fisheries vary between regions and countries because of the differences in exposure to revenue changes and the sensitivity and adaptive capacity of the fishing communities to these changes (Hilmi et al. 2015). Regions where the maximum potential revenue is projected to decrease coincide with areas where indicators such as human development index suggest high economic vulnerability to climate change (Barbier, 2015; Lam et al. 2016). Many coastal communities in these regions rely heavily on fish and fisheries as a major source of animal proteins, nutritional needs, income and job opportunities (FAO, 2019). Negative impacts on the catch and total fisheries revenues for these countries are expected to have greater implications for jobs, economies, food and nutritional security than the impacts on regions with high Human Development Index (Allison et al. 2009; Srinivasan et al. 2010; Golden et al. 2016; Blasiak et al. 2017). Climate change impacts to coral reefs and other fish habitats, as well as to targeted fish and invertebrate species themselves are expected to reduce harvests from small-scale, coastal fisheries by up to 20% by 2050, and by up to 50% by 2100, under RCP8.5 (Bell et al. 2018a). Therefore, climate risk to communities that are strongly dependent on fisheries associated with ecosystems that are particularly sensitive to climate change such as coral reefs will have be particularly high (Cinner et al. 2016) (*high confidence*).

Climate change may also worsen by non-climate related socioeconomic shocks and stresses, and hence is an obstacle to economic developments (Hallegatte et al. 2015). Climate risk on the economics of fishing is projected to be higher for tropical developing countries where existing adaptive capacity to the risk is lower, thereby challenging their sustainable economic development (*high confidence*). However, observed impacts are not yet well documented (Lacoue-Labarthe et al. 2016), and there are many uncertainties relating to how climate change would affect the dynamics of fishing costs, with consequent adjustment of fishing effort that might intensify or lessen the overcapacity issue. Studies have attempted to project how fishers may respond to changes in fish distribution and

abundance by incorporating different management systems (Haynie and Pfeiffer, 2012; Galbraith et al. 2017). However, the impacts of climate change on management effectiveness and trade practices is still inadequately understood (Galbraith et al. 2017).

5.4.2.3.2 Wealth generated from coastal and marine tourism sector

Tourism is one of the largest sectors in the global economy. Between 1995–1998 and 2011–2014, the average total contribution of tourism to global GDP increased from 69 billion USD (6.8%) to 166 billion USD (8.5%) respectively, and generated more than 21 million jobs between 2011–2014 (UNCTAD, 2018). Coastal tourism and other marine-related recreational activities contributes substantially to the tourism sector (Cisneros-Montemayor et al. 2013; O'Malley et al. 2013; Spalding et al. 2017; Giorgio et al. 2018; UNWTO, 2018). For example, it is estimated that around 121 million people a year participated in marine-based recreational activities, generating 47 billion in 2003 USD in expenditures and supporting one million jobs (Cisneros-Montemayor and Sumaila, 2010). Tourism is one of the main industries that provides opportunities for social and economic development (Jiang and DeLacy, 2014), and marine tourism is particularly important for many coastal developing countries and Small Island Developing States (SIDS). AR5 identified the tourism sector in the Caribbean region as particularly vulnerable to climate change effects, due to hurricanes, whilst SR15 concluded that warming will directly affect climate-dependent tourism markets on a worldwide basis (*medium confidence*) (Hoegh-Guldberg et al. 2018). This assessment provides updates since AR5 and SR15.

Empirical modelling of future risks to tourism is based on projected climate impacts (Section 5.3) for relevant coastal ecosystems, including degradation or loss of beach and coral reef assets (Weatherdon et al. 2016) (Section 4.3.3.6.2). These projections are developed from the relationship between the economic benefits generated from coral reef related tourism with observed characteristics of coral reefs, the characteristics of tourism activities. Based on scenarios of projected future warming and decreases in coral reef coverage, a global loss of tourism and recreation value in the near-future (2031–2050) of 2.57–2.95 billion yr^{-1} in 2000 USD is projected under RCP2.6, and of 3.88–5.80 billion yr^{-1} in 2000 USD under RCP8.5 (Chen et al. 2015). Opinion surveys in four countries suggest that if severe coral bleaching persists in the Great Barrier Reef, tourism in adjacent areas could greatly decline, from 2.8 million to around 1.7 million visitors per year, equivalent to more than 1 billion AUS (~0.69 billion USD using exchange rate in 2019), that is, in tourism expenditure and with potential loss of around 10,000 jobs (Swann and Campbell, 2016).

Many coastal tourism destinations are exposed to risks of flooding, SLR and coastal squeeze on coastal ecosystems (Lithgow et al. 2019) (Section 5.3); there are also other climate related-risks. Droughts, which are projected to be more frequent, will also impact the tourism industry (and local food security) through water and food shortages (Pearce et al. 2018). If climate change and ocean acidification reduce the seafood supply, the attractiveness of coastal regions for tourists will also decrease (Wabnitz et al. 2017). North Atlantic hurricanes and tropical storms have increased in intensity over the last 30 years, with climate projections indicating an increasing trend in hurricane intensity (Chapter 6). Three major Caribbean storms, Harvey, Irma and Maria, occurred in 2017, with loss and damage to the tourism industries of Dominica, the British Virgin Islands, and Antigua and Barbuda estimated at 2.2 billion USD, and environmental recovery costs estimated at 6.8 million USD (UNDP, 2017). Pacific tourist destinations, which tend to focus on nature-based and marine activities, are also at high risk of extreme events and other climate change impacts (Klint et al. 2015). However, global tourism has a high carbon footprint (flights, cruises, etc.) (Lenzen et al. 2018), so any reduction in the intensity of this sector would help mitigate climate change.

Evidence from recent studies on projected climate risks on recreational fishing is equivocal, with the direction of impacts depending on the location, species targeted and societal context. For example: poleward range shifts of marine fish (Section 5.2.3) could yield new opportunities for recreational fishing in mid- to high-latitude regions (DiSegni and Shechter, 2013); projected increases in air temperature may enable longer fishing days in some area (Dundas and von Haefen, 2015); and extreme events may alter the composition of recreational fishing catches (Santos et al. 2016). Since climate risks to recreational fishing vary largely depending on the responses of the targeted species to climate-related pressures, there is *low confidence* in the overall risk to the activity.

Overall, evidence since AR5 and SR15 confirms that climate impacts to coastal ecosystems would increase risks to coastal tourism, particularly under high emission scenarios (*medium confidence*). Economic impacts will be greatest for those developing countries where tourism is the main source of foreign revenue (*medium to high evidence*).

5.4.2.3.3 Property values

The integrity of ecosystems and their services can affect the value of human assets, particularly coastal properties and infrastructure (Hoegh-Guldberg et al. 2018). Climate change is expected to have negative impacts on coastal properties and their value through the loss and damage caused by SLR, increased storm intensity (hurricanes and cyclones), heat waves, floods, droughts and other extreme events, particularly in tropical SIDS (Chapter 4). Natural disasters already cost Pacific Island Countries and Territories between 0.5–6.6% of GDP yr^{-1} (World Bank, 2017), with localised damages and losses from individual storms far exceeding these estimates (e.g., 64% of Vanuatu's GDP for Cyclone Pam in 2015). The impacts of natural disasters on Jamaica's coastal transport infrastructure are currently estimated to be a significant proportion of their GDP, and such costs are projected to increase substantially in the next few decades under climate change (UNCTAD, 2017; Monioudi et al. 2018). In 2015, tropical storm Erika devastated Dominica causing 483 million USD in damages and losses (mostly related to transport, housing and agriculture), equivalent to 90% of Dominica's GDP (World Bank, 2017). For the USA, Ackerman and Stanton (2007) forecast that annual real estate losses due to climate change could increase from 0.17% of GDP in 2025 to 0.36% in 2100, with Atlantic and Gulf Coast states being the most vulnerable. Other North

American studies have shown that informed coastal property owners are willing to initially invest in infrastructure to counter climate change impacts (McNamara and Keeler, 2013); however, they would avoid further investment if adaptation costs increase substantially and there are greater risks of long-term impacts (Putra et al. 2015).

The impacts of changing marine ecosystems and ecosystem services on the value of human assets need to consider the risk perception, future development and adaptation responses of human communities (Section 5.5.2, Chapter 4) (Bunten and Kahn, 2014). For example, the potential for climate impacts on the value of coastal real estate will depend on the changing insurance market or the cost of adaptation measures, which in turn depend on the willingness to pay by asset holders and wider society, including local and national governments. Further research is needed to discount valuations for potential losses that may occur in the future but with uncertain occurrence, and to improve real estate loss estimates over local to regional scales.

Marine ecosystem services contribute to climate moderation and coastal defenses (Section 5.4.1.2). However, while the above studies in this section acknowledge the contribution of many climate impacts on real estate and infrastructure through ecosystem losses and degradation, often they are not accounted for in quantitative economic impact assessments. Overall, there is *high confidence* that SLR, increases in storm intensity and other extreme events will impact the values of coastal real estates and infrastructure, particularly in tropical SIDS, through the risk and impacts of direct physical damages. However, there is *low confidence* that impacts due to underlying loss and damage of ecosystems and their services are being similarly accounted for.

5.4.2.4 Risk and Opportunities for Ocean Economy

The 'ocean economy' refers to the sustainable use of ocean resources for economic growth, improved livelihoods and jobs, and ocean ecosystem health (World Bank, 2017). In SR15 (Hoegh-Guldberg et al. 2018) and elsewhere here (Chapters 3 and 5), the risks and opportunities of specific sectors that contribute to the ocean economy under climate change are assessed. The fishing industry is particularly important in this context. As previously noted, warming has already directly impacted coastal and open ocean fishing activities in some regions (Section 5.4.1.1, 5.4.2.3.1); the risk of fishery impacts is exacerbated by the observed climate-driven changes to coral reefs and other coastal ecosystems that contribute to the productivity of exploited fish species (Section 5.4.1.3, 5.4.2.3.1); and there are challenges to sustainable management of transboundary fisheries resources caused by species' range shifts and associated governance challenges (Section 5.4.2.2.2).

Fisheries-related national and local economies of many tropical developing countries are exposed high climate risks (Section 5.4.2.3.1) (Blasiak et al. 2017), as a result of the projected large decrease in maximum catch and revenue potential under RCP8.5 in the 21st century (Section 5.4.1.1). Historical examples from fishery over-exploitation indicate that a large decrease in catches for specific fish stocks have had substantial negative effects for dependent economies and communities (Brierley and Kingsford, 2009; Davis, 2015).

Moreover, coastal economies that are dependent on marine tourism and recreational activities are also exposed to elevated risks from impacts on biota that are important for these sectors (Section 5.4.2.3.2). Nevertheless, new opportunities for coastal tourism may occur in future for some regions as a result of species' biogeographic shifts (Section 5.4.2.3.2) and increased accessibility, such as in the Arctic (Chapter 3).

Decrease in sea ice in the Arctic is opening up economic opportunities for the oil and gas exploration, mining industries and shipping that are currently important economic sectors in the ocean (Pelletier and Guy, 2012; George, 2013) (Section 3.4.3; 3.5.3). Although the Arctic region has oil and gas reserves estimated to account for one-tenth of world oil and a quarter of global gas (*U.S. Geological Survey released on 24 July 2008*), offshore oil and gas exploration with poor regulation or as a result of accidents poses additional risk of impacts on species, populations, assemblages, to ecosystems by modifying a variety of ecological parameters (e.g., biodiversity, biomass, and productivity) (Cordes et al. 2016) threatening the sensitive Arctic ecosystems and the livelihood of dependent communities (Section 3.5.3.3).

Similarly, global warming and changing weather patterns may have a substantial impact on global trade and transport pathways (Koetse and Rietveld, 2009); for example, the reduction in sea ice in the Arctic Ocean during summer opens up the possibility for sea transport on the Northwest or Northeast Passage for several months per year (Ng et al. 2018) (Section 3.5.3.2). Both routes may provide opportunities for more efficient transport between North America, Europe, Russia and China for fleets with established Arctic equipment, and may open up access to known natural resources which have so far been covered by ice (Guy and Lasserre, 2016). However, whether the Arctic shipping routes will be a realistic alternative depends not only on regulatory frameworks and economic aspects (such as infrastructure and reliability of the routes) but also on societal trends and values, demographics and tourism demand (Prowse et al. 2009; Wassmann et al. 2010; Pelletier and Guy, 2012; George, 2013; Hodgson et al. 2016; Pizzolato et al. 2016; Dawson, 2017) (Section 3.2.4.2, 3.4.3.3). Simultaneously, shipping routes through the Arctic pose additional risk from human impact such as pollution, introduction of invasive species and collision with marine mammals, and emission of short-lived climate forcers that can amplify warming in the region and accelerate localised warming (Wan et al. 2016) (Section 3.5.3.2).

Existing governance may not be sufficient to limit the elevated risk on Arctic ecosystems and their dependent economies from increased shipping activities (Section 3.4.3, 3.5.3). Climate change may bring new economic opportunities, particularly for polar oil and gas development (*medium confidence*), shipping (*medium confidence*) and tourism (*low confidence*) although realisation of these opportunities will pose uncertain ecological risks to sensitive ecosystems and biota, and the dependent human communities in the region (*high confidence*).

Ocean renewable energy provides an emerging alternative to fossil fuels and comprises energy extraction from offshore winds, tides, waves, ocean thermal gradients, currents and salinity gradients (Harrison and Wallace, 2005; Koetse and Rietveld, 2009; Bae et al.

2010; Jaroszweski et al. 2010; O Rourke et al. 2010; Hooper and Austen, 2013; Kempener and Neumann, 2014b; Kempener and Neumann, 2014a; Abanades et al. 2015; Astariz et al. 2015; Borthwick, 2016; Foteinis and Tsoutsos, 2017; Manasseh et al. 2017; Becker et al. 2018; Gattuso et al. 2018; Hemer et al. 2018; Dinh and McKeogh, 2019b; Dinh and McKeogh, 2019a). Other potential sources of marine renewable energy include algal biofuels (Greene et al. 2010; Greene et al. 2016). While such approaches offers a way to mitigate climate change, changes in climatic conditions (such as waves and winds) may impact marine renewable energy installations and their effectiveness (Harrison and Wallace, 2005). A more comprehensive assessment of these issues is expected to be provided by IPCC WGIII in the AR6 full report.

Overall, some major existing ocean economy sectors such as fishing, coastal tourism and recreation are already at risk by climate change (*medium confidence*), and all sectors are expected to have elevated risks with high future emission scenarios (*high confidence*). The emerging demand for alternative energy sources is expected to generate economic opportunities for the ocean renewable energy section (*high confidence*), although their potential may also be affected by climate change (*low confidence*).

5.4.2.5 Impacts of Changing Ocean on Sustainable Development Goals

Climate change impacts will have consequences for the ability of human society to achieve sustainable development. SR15 concludes that "Limiting global warming to 1.5°C rather than 2°C would make it markedly easier to achieve many aspects of sustainable development, with greater potential to eradicate poverty and reduce inequalities (*medium evidence, high agreement*)". This assessment focuses on how climate change impacts on marine ecosystems would challenge sustainable development, using the United Nations SDGs as a framework to discuss the linkages between those issues.

Climate impacts on marine ecosystems affect their ability to provide seafood and raw materials, and to support biodiversity, habitats and other regulating processes (Section 5.4.1), and these impacts on the ocean affect people directly and indirectly (Sections 5.4.2.1, 5.4.2.2, 5.4.2.3). SDG 14 is the goal that is most directly relevant: "Life below water: including indicators for marine pollution, habitat restoration and protected areas, ocean acidification, fisheries, and coastal development."

Climate impacts in the ocean to other SDGs are mediated through social and economic factors when the SDG targets are affected (Singh et al. 2019). For example, climate impacts on marine ecosystem services related to primary industries that provide food, income and livelihood to people have direct implications for a range of SDGs. These SDGs include 'no poverty' (SDG 1), 'zero hunger' (SDG 2), 'decent work and economic growth' (SDG 8), 'reduced inequalities' (SDG 10) and 'responsible consumption and production' (SGD 12) (Singh et al. 2019, Figure 5.22). These impacts relate to changing ocean under climate change that affect the pathways to build sustainable economies and eliminate poverty (Sections 5.4.2.4), eliminate hunger and achieve food security (Section 5.4.2.1.3), reduce inequalities (Sections 5.4.2.2) and

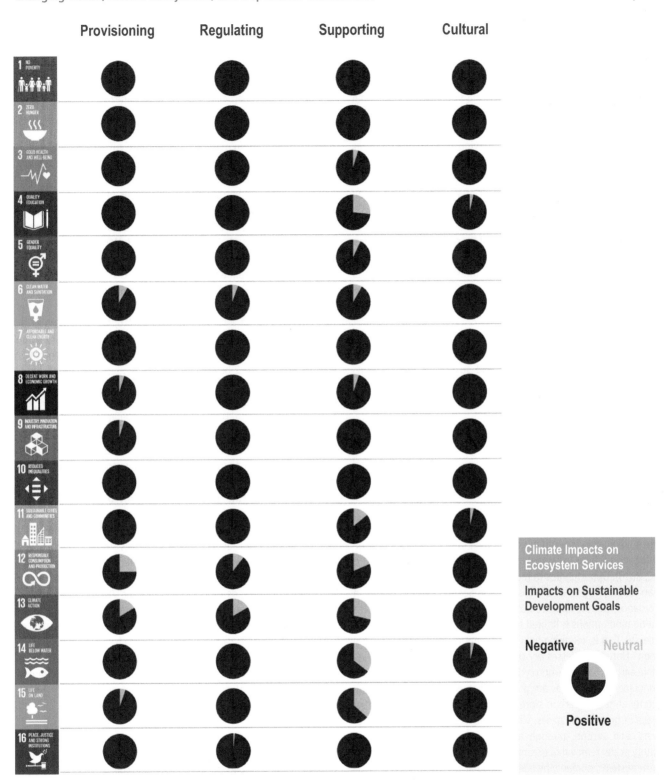

Figure 5.22 | Summary of the types of relationships (negative, neutral and positive) between impacted marine ecosystem services (Provisioning, Regulating, Supporting and Cultural) and the Sustainable Development Goals (SDGs) based on literature review and expert-based analysis (Singh et al. 2019). Pie charts represent the proportion of targets within SDGs that a particular ocean SDG target contributes to according to the literature reviewed and expert-based analysis presented in Singh et al. (2019).

achieve responsible consumption and production (Sections 5.4.2.3.1) (Carvalho et al. 2017; Castells-Quintana et al. 2017). Climate change is also creating living conditions in coastal areas that are less suitable to human settlement and changing distributions of marine disease vectors (Section 5.4.2.1.1, 5.4.2.3.3), reducing our chances of

achieving the goal for good health and well-being (SDG 3) (Pearse, 2017; Wouters et al. 2017). Women are often engaged in jobs and livelihood sources that are more exposed to climate change impacts from the ocean such as impacts on fisheries (Section 5.4.2.3.1) and impacts of SLR on coastal regions (Chapter 4). For example, in Senegal, women disproportionately engage in rice crop cultivation in coastal flood plain (Linares, 2009), and are thus exposed to the risks on their livelihood from rising sea levels and resulting salinisation (Dennis et al. 1995). Flooding in Bangladesh has increased the vulnerability of women to harassment and abuse as the flooding upends normal life and increases crime rates (Azad et al. 2013). As such, climate change may negatively affect our ability to achieve "gender equality" (SDG 5) (Salehyan, 2008). Impacts on living conditions as well as changing recreational, aesthetic, and spiritual experiences also affect our ability to achieve 'sustainable cities and communities' (SDG 11) (Section 5.4.2.2.1). The consequences of climate change in the ocean to achieving the remaining SDGs are less clear. However, the SDGs are interlinked, and achieving SDG 14, and especially the targets of increasing economic benefits to SIDS and Least Developed Countries, as well as eliminating illegal fishing and overfishing, will benefits all other SDGs (Singh et al. 2017). The interlinkages among SDGS mean climate change impact on the ocean will affect all other SDGs beside SDG14 in various ways, some possible direct and many indirect (*low confidence*).

Overall, climate change impacts on the ocean will negatively affect the chance of achieving the SDGs and sustaining their benefits (*medium confidence*).

5.5   Risk-reduction Responses and their Governance

5.5.1   Ocean-based Mitigation

5.5.1.1   Context for Blue Carbon and Overview Assessment

There is political and scientific agreement on the need for a wide range of mitigation actions to avoid dangerous climate change (UNEP, 2017; IPCC, 2018). Opportunities to reduce emissions by the greater use of ocean renewable energy are identified in Section 5.4.2.3.2. Here, in accordance with the approved scoping of this report, the assessment of mitigation options is limited to the management of natural ocean processes, that is, requiring policy intervention, with a focus on 'blue carbon'. Natural processes *per se*, although important to the climate system and the global carbon cycle, are not a mitigation response. Two management approaches are possible: first, actions to maintain the integrity of natural carbon stores, thereby decreasing their potential release of greenhouse gases, whether caused by human or climate-drivers; and second, through actions that enhance the longterm (century-scale) removal of greenhouse gases from the atmosphere by marine systems, primarily by biological means.

These mitigation approaches match those proposed using terrestrial natural processes (Griscom et al. 2017), with extensive afforestation and reforestation included in all climate models that limit future warming to 1.5°C (de Coninck et al. 2018). As on land, reliable carbon accounting is a critical consideration (Grassi et al. 2017), together with confidence in the longterm security of carbon storage. The feasibility of climatically-significant (and societally acceptable) mitigation using marine natural processes therefore depends on a robust quantitative understanding of how human actions can affect the uptake and

release of greenhouse gases from different marine environments, interacting with natural biological, physical and chemical processes. Whilst CO_2 is the most important greenhouse gas, marine fluxes of methane and nitrous oxide can also be important, for both coastal regions and the open ocean (Arévalo-Martínez et al. 2015; Borges et al. 2016; Hamdan and Wickland, 2016).

The term 'blue carbon' was originally used to cover biological carbon in all marine ecosystems (Nellemann et al. 2009). Subsequent use of the term has focused on carbon-accumulating coastal habitats structured by rooted plants, such as mangroves, tidal salt marshes and seagrass meadows, that are relatively amenable to management (McLeod et al. 2011; Pendleton et al. 2012; Thomas, 2014; Macreadie et al. 2017a; Alongi, 2018; Windham-Myers et al. 2019; Lovelock and Duarte, 2019). Comparisons across the full range of freshwater and saline wetland types are assisted by standardised approaches (Nahlik and Fennessy, 2016; Vázquez-González et al. 2017). Seaweeds (macroalgae) can also be considered as coastal blue carbon (Krause-Jensen and Duarte, 2016; Krause-Jensen et al. 2018; Raven, 2018), however, because of differences in their carbon processing, their climate mitigation potential is assessed separately within Section 5.5.1.2 below.

In the open ocean, the biological carbon pump is driven by the combination of photosynthesis by phytoplankton and downward transfer of particulate carbon by a variety of processes (Henson et al. 2010; DeVries et al. 2017); it results in large-scale transfer of around 10 GtC yr^{-1} carbon from near-surface waters to the ocean interior (Boyd et al. 2019). Most of this carbon is respired in the mesopelagic and contributes to the 37,000 GtC inventory of DIC, with around ~0.1 GtC yr^{-1} eventually being permanently removed in deep sea sediments (Cartapanis et al. 2018). In addition, the microbial carbon pump (Jiao et al. 2010) produces refractory dissolved organic molecules throughout the water column at a rate of around 0.4 GtC yr^{-1} (Jiao et al. 2014b), which due to their residence time of hundreds to thousands of years maintain the 700 GtC inventory of dissolved organic carbon in the ocean (Jiao et al. 2010; Jiao et al. 2014a; Legendre et al. 2015; Jiao et al. 2018a). The natural removal of carbon by the various carbon pumps is closely balanced by upwelling and outgassing, with the ocean a moderate source of CO_2 under pre-industrial conditions (Ciais et al. 2013). The mitigation potential of managing natural processes in the open ocean is only briefly assessed here (Section 5.5.1.3).

Gattuso et al. (2018) provide an overview assessment of the environmental, technical and societal feasibilities of using a range of ocean management actions to reduce climate change and its impacts. Their results for nine actions based on natural processes are summarised in Figure 5.23, also including marine renewable energy (wind, wave and tidal) for comparison. Eight semi-quantitative criteria were used to assess each action: maximum potential effectiveness by 2100 in reducing climatic drivers (ocean warming, ocean acidification and SLR), assuming full theoretical implementation; technological readiness and lead time to full potential effectiveness (subsequently combined as technical feasibility); duration of benefits; co-benefits; trade-offs (originally described as dis-benefits); cost-effectiveness; and governability (capability of implementation, and management

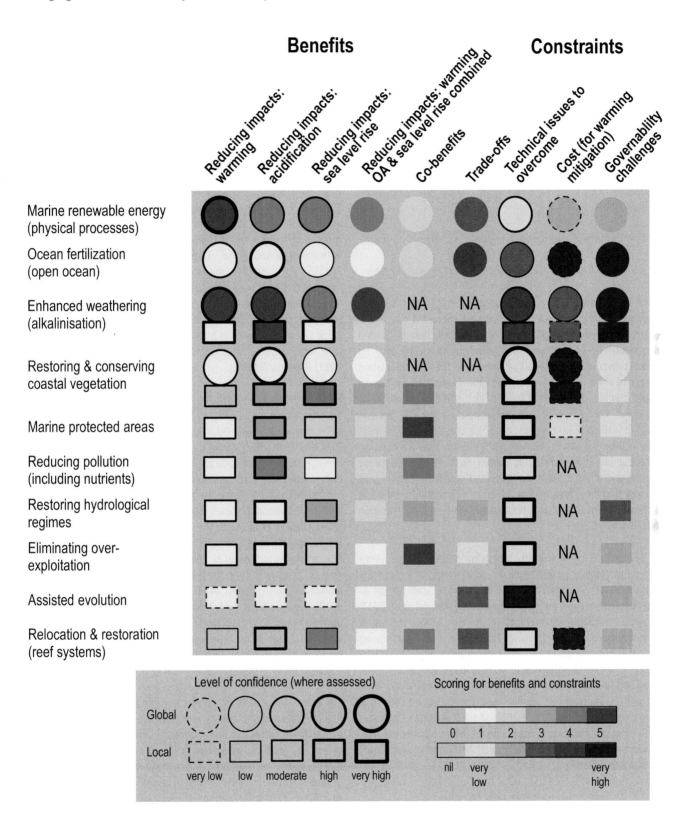

Figure 5.23 | Summary of potential benefits and constraints of ocean-based risk-reduction options using natural processes, from literature-based expert assessments by Gattuso et al. (2018). Mitigation effectiveness was quantified relative to Representative Concentration Pathway (RCP)8.5, assuming maximum theoretical implementation, with reduction of climate-related drivers considered at either global or local (<100 km²) scale, shown as circles or rectangles respectively. Impact reduction, co-benefits and trade-offs are in the context of eight sensitive marine ecosystems and ecosystem services. 'Technical issues to overcome' is based on scores for technological readiness, lead time for full implementation and duration of effects. Cost is based on USD per tonne of CO₂ either not released or removed from the atmosphere (for global measures) or per hectare of coastal area with action implemented (for local measures). 'Governance challenges' shows the potential difficulty of implementation by the international community. NA, not assessed. Additional information on scoring methods is given in SM5.4, Tables SM5.9a and SM5.9b.

of any associated conflicts). Here, governability is considered as a constraint (governability challenges) reversing the scoring scale used by Gattuso et al. (2018).

Global measures (circles in Figure 5.23) can be regarded as mitigation, reducing drivers; local measures (rectangles), are primarily ecosystem-based adaptation (EbA), reducing impacts (Section 5.5.2), although they may also contribute to mitigation; two actions were considered at both scales. Gattuso et al. (2018) did not consider the effects of actions on ocean oxygenation, notwithstanding the importance of deoxygenation as a component of climate change. Additional detail is given in SM5.4.

5.5.1.2 Climate Mitigation in the Coastal Ocean

5.5.1.2.1 Opportunities and challenges relating to coastal carbon

Estuaries, shelf seas and a wide range of other intertidal and shallow-water habitats (Section 5.3) play an important role in the global carbon cycle through their primary production by rooted plants, seaweeds (macroalgae) and phytoplankton, and also by processing riverine organic carbon. However, the natural carbon dynamics of these systems have been greatly changed by human activities (Regnier et al. 2013; Cloern et al. 2016; Day and Rybczyk, 2019) (*high confidence*). Direct anthropogenic impacts include coastal land-use change (Ramesh et al. 2015; Li et al. 2018a); indirect effects include increased nutrient delivery and other changes in river catchments (Jiao et al. 2011; Regnier et al. 2013), and marine resource exploitation in shelf seas (Bauer et al. 2013). There is *high confidence* that these human-driven changes will continue, reflecting coastal settlement trends and global population growth (Barragán and de Andrés, 2015).

Policy recognition of the mitigation benefits of coastal ecosystems requires quantitative information on their actual and potential carbon uptake and storage at the local and national scale, within an international framework for carbon accounting (Crooks et al. 2011; Hejnowicz et al. 2015). Such methods are being developed for coastal habitats structured by rooted plants (Needelman et al. 2018; Troxler et al. 2018; Needelman et al. 2019), considered here as 'coastal vegetation', linked to protocols for verification of longterm carbon removal and financial incentives (Crooks et al. 2011; Hejnowicz et al. 2015) and building on techniques used for managing terrestrial carbon sinks (Ahmed and Glaser, 2016b; Aziz et al. 2016). Proposals to apply carbon accounting to seaweeds, the water column and shelf sea sediments (Krause-Jensen and Duarte, 2016; Zhang et al. 2017) are less well-developed.

5.5.1.2.2 Coastal vegetation: mangrove, salt marsh and seagrass ecosystems

Mangrove, salt marsh and seagrass habitats are widely recognised as blue carbon ecosystems with mitigation potential (Chmura et al. 2003; Duarte et al. 2005; Kennedy et al. 2010; McLeod et al. 2011). Although covering only ~0.1% of the Earth's surface, these three ecosystems together have been estimated to support 1–10% of global marine primary production (Duarte et al. 2017). More than

150 countries contain at least one of these ecosystems; 71 countries contain all three (Herr and Landis, 2016), and 74 countries mention such coastal wetlands (five specifically as blue carbon) in their Nationally Determined Contributions (NDCs) to the Paris Agreement (Martin et al. 2016a; Gallo et al. 2017).

These three vegetated coastal habitats are characterised by high, yet variable, organic carbon storage in their soils and sediments on a per unit area basis (*high confidence*). In the humid tropics, mangrove below-ground organic carbon is typically 500–1000 tC ha^{-1} (Donato et al. 2011; Alongi and Mukhopadhyay, 2015; Howard et al. 2017), although only ~50 tC ha^{-1} in arid regions (Almahasheer et al. 2017). Australian salt marshes show particularly wide variation in organic carbon storage, ranging from 15–1000 tC ha^{-1} (top 1 m) with mean of 165 tC ha^{-1} (Kelleway et al. 2016; Macreadie et al. 2017b). For seagrass meadows, storage values are typically 400–1600 tC ha^{-1} but can exceed 2000 tC ha^{-1} (Serrano et al. 2014). These accumulations have occurred over decadal to millennial time scales (McKee et al. 2007; Lo Iacono et al. 2008). Such blue carbon stock values are similar to freshwater wetlands and peat, but higher than for most forest soils (Laffoley and Grimsditch, 2009; Pan et al. 2011) (*high confidence*).

When vegetated coastal ecosystems are disturbed, a proportion of their stored carbon is released back to the atmosphere, along with other greenhouse gases (Marba and Duarte, 2009; Duarte et al. 2010; Pendleton et al. 2012; Lovelock et al. 2017). Globally, around 25–50% of vegetated coastal habitats have already been lost or degraded due to coastal agricultural developments, urbanisation and other human disturbance during the past 100 years (McLeod et al. 2011). The highest historical losses (60–90%) have occurred in Europe and China (Jickells et al. 2015; Gu et al. 2018; Li et al. 2018a). Current losses are estimated at 0.2–3.0% yr^{-1}, depending on vegetation type and location (FAO et al. 2014; Alongi and Mukhopadhyay, 2015; Atwood et al. 2017) (*medium confidence*). Associated global carbon emissions are estimated at 0.04–0.28 GtC yr^{-1} (Pendleton et al. 2012); 0.06–0.61 GtC yr^{-1} (Howard et al. 2017); 0.10–1.46 GtC yr^{-1} (Lovelock et al. 2017); and 0.007 GtC yr^{-1} (mangroves only) (Taillardat et al. 2018). This range of values reflects uncertainties regarding the global rate of habitat loss, and the proportion of carbon remineralised to CO_2.

Mitigation through emission reduction can therefore be achieved by habitat protection, to greatly reduce or end the human-driven loss of mangrove, salt marsh and seagrass ecosystems. Such action could potentially produce nationally-significant mitigation (>1% of fossil fuel emissions) for several countries (Taillardat et al. 2018). However, there are still many uncertainties in quantifying carbon release due to habitat degradation and loss (Lovelock et al. 2017), and hence in determining emission reductions. Furthermore, this mitigation option is not available to those countries where habitat loss is not currently occurring, for example, in Bangladesh (Taillardat et al. 2018). Since legal structures already exist in many countries to protect coastal wetlands, the main policy need may be the enforcement of national regulation and site-specific MPAs (Miteva et al. 2015; Herr et al. 2017; Howard et al. 2017).

The alternative mitigation approach using coastal blue carbon ecosystems is to enhance the natural carbon uptake of such habitats, not only by increasing their spatial coverage through habitat restoration and new habitat creation, but also by taking management measures to maximise the carbon uptake and storage for existing coastal ecosystems. Such measures include reducing anthropogenic nutrient inputs and other pollutants; restoring hydrology, by removing barriers to tidal flow and sediment delivery; and reinstating predators (to reduce carbon loss caused by some bioturbators) (Macreadie et al. 2017a). Per unit area of habitat created, restored or rehabilitated, such actions may offer high rates of carbon removal: widely-quoted values are 226 ± 39 gC m^{-1} yr^{-1} for mangroves, 218 ± 24 gC m^{-1} yr^{-1} for salt marsh and 138 ± 38 gC m^{-1} yr^{-1} for seagrass ecosystems (McLeod et al. 2011; Isensee et al. 2019).

Around 90 restoration and rehabilitation projects for mangroves have been documented (López-Portillo et al. 2017), with associated development of a range of restoration evaluation methods (Zhao et al. 2016a). Salt marsh restoration is reviewed by Adam (2019) and seagrass restoration by van Katwijk et al. (2016). Consistent conclusions, supported by other studies (Bayraktarov et al. 2016; Wylie et al. 2016) are that: natural regeneration increases the likelihood of longterm survival; higher success rates are achieved with strong stakeholder engagement; and it is critical that the (human) factors causing original loss and degradation have been properly addressed (*high confidence*).

Quantification of the climatic benefits of such actions is, however, not straightforward. Measurements of carbon burial rates show high site-specific variability, being strongly affected by a wide range of environmental factors for mangroves (Adame et al. 2017; Schile et al. 2017), seagrasses (Lavery et al. 2013) and salt marshes (Kelleway et al. 2017b). The reliable determination of sediment accumulation rates is a key consideration, with associated uncertainties not fully reflected in the McLeod et al. (2011) estimates given above. In particular, geochemical-based studies have indicated that seagrass carbon burial may have been greatly overestimated (Johannessen and Macdonald, 2016). These issues are contentious (Johannessen and Macdonald, 2018a; Johannessen and Macdonald, 2018b; Macreadie et al. 2018; Oreska et al. 2018); their scientific resolution is highly desirable. Additional complexities relating to the mitigation role of coastal blue carbon ecosystems include the following:

- Emissions of other greenhouse gases also need to be taken into account (Keller, 2019b). Methane release from mangrove habitats can reduce the scale of their climatic benefits by 18–22% (Adams et al. 2012; Chen and Ganapin, 2016; Chmura et al. 2016; Rosentreter et al. 2018; Cameron et al. 2019) and nitrous oxide and methane together may offset salt marsh CO_2 uptake by 24–31% (Adams et al. 2012). Nitrous oxide emissions are strongly affected by nutrient loading (Chmura et al. 2016); under pristine conditions, mangroves can provide a sink rather than a source (Maher et al. 2016). Note that values of the 'offset' depend on the metrics used for determining CO_2 equivalents.
- Carbonate formation, releasing CO_2, may also reduce the benefits of carbon storage by similar proportions (Howard et al. 2017; Macreadie et al. 2017a; Kennedy et al. 2018; Saderne et al. 2019).

- Lateral transfers are not well-quantified. Whilst some of the carbon stored in coastal marine sediments may be recalcitrant carbon from terrestrial or atmospheric sources (and should therefore be excluded) (Chew and Gallagher, 2018), export of dissolved organic carbon, inorganic carbon and alkalinity may be considered as additional sequestration (Maher et al. 2018; Santos et al. 2019).
- The permanence of vegetated coastal systems, even if well-protected, cannot be assumed under future temperature regimes (Ward et al. 2016; Duke et al. 2017; Jennerjahn et al. 2017; Nowicki et al. 2017).
- Responses to future SLR are also uncertain and complex (Kirwan and Megonigal, 2013; Spencer et al. 2016). However, impacts are not necessarily negative: carbon sequestration capacity may increase where totally new habitats are created (Barnes, 2017), or if mangroves replace salt marshes (Kelleway et al. 2016).

In summary, a combination of both conservation and restoration of mangrove, salt marsh and seagrass habitats can contribute to national mitigation effort for those countries with relatively large coastlines where such ecosystems naturally occur (Murdiyarso et al. 2015; Atwood et al. 2017). However, the associated current uncertainties in quantifying relevant carbon storage and flows are expected to be problematic for reliable measurement, reporting and verification (*high confidence*).

At the global scale, synthesis studies have estimated the potential additional sequestration achieved by cost effective coastal blue carbon restoration as ~0.05 GtC yr^{-1} (Griscom et al. 2017) and 0.04 GtC yr^{-1} (National Academies of Sciences, Engineering, and Medicine, 2019), assuming that a relatively high proportion of vegetated ecosystems can be re-instated to their 1980–1990 extents. These values compare to current net anthropogenic emissions from all sources of 10.0 GtC yr^{-1} (Le Quéré et al. 2018), and are consistent with the 'very low' scores by (Gattuso et al. 2018) for the climate mitigation benefits of conserving and restoring coastal vegetation (Figure 5.23). Coastal ecosystem restoration could theoretically achieve higher sequestration, around ~0.2 GtC yr^{-1} (Griscom et al. 2017), but would be challenging, because of the semi-permanent and on-going nature of most coastal land-use change, such as human settlement, conversion to agriculture and aquaculture, shoreline hardening and port development (Gittman et al. 2015; Li et al. 2018a).

Restoration costs could also be an important constraint for large-scale application. Based on published data from 246 observations, Bayraktarov et al. (2016) estimated median total costs for restoration of one hectare of mangrove, salt marsh and seagrass habitat to be ~2,508, 151,129 and 383,672 respectively, in 2010 USD. For each ecosystem, there was high variability in costs according to the economy of the country where the restoration projects were carried out, and the restoration technique applied. Assessment of coastal conservation and restoration costs is also given in Section 4.4.2.3, in Box 5.5 (in the context of coral reef restoration costs) and Section 5.5.2.5.

Measures to protect and restore coastal blue carbon habitats provide many other societal benefits in addition to climate regulation (Section 5.4.1). In particular, there is *high confidence* that coastal

5

wetlands benefit local fisheries, enhance biodiversity, give storm protection, reduce coastal erosion, improve water quality and support local livelihoods (Costanza et al. 2008; Spalding et al. 2014). Coastal ecosystems may keep pace with sufficiently gradual SLR, and may be more cost-effective in flood protection than hard infrastructure like seawalls (Temmerman et al. 2013; Möller, 2019). Coastal blue carbon can therefore be considered as a 'no regrets' mitigation option at the national level in many countries, in addition to (not a replacement for) more effective mitigation measures. Additional research is needed over the full range of environmental conditions to improve knowledge and understanding of the complex carbon dynamics of coastal vegetation and associated systems, to enable well-quantified and cost-effective carbon sequestration enhancement (Vázquez-González et al. 2017; Windham-Myers et al. 2019).

5.5.1.2.3 Seaweeds (macroalgae)

Seaweeds do not directly transfer carbon to marine sediments, unlike the rooted coastal vegetation considered above (Howard et al. 2017). Nevertheless, seaweed detritus can deliver carbon to sedimentary sites (Hill et al. 2015) and may provide a source of refractory dissolved organic (Krause-Jensen and Duarte, 2016). Recent studies indicate that globally important amounts of carbon may be involved in these processes (Krause-Jensen and Duarte, 2016; Krause-Jensen et al. 2018; Smale et al. 2018). There is, however, currently *low confidence* that enhancement of natural seaweed production can provide a significant mitigation response, due to large uncertainties relating to sequestration duration and effectiveness. Such considerations relate to transport pathways, the fate of material transported to deeper water, and the timescales of its subsequent return to the atmosphere over decadal to century timescales.

Seaweed aquaculture is inherently more manageable as a mitigation response (N'Yeurt et al. 2012; Chung et al. 2013; Chung et al. 2017; Duarte et al. 2017). If linked to biofuel or biogas production (N'Yeurt and Iese, 2014; Moreira and Pires, 2016; Sondak et al. 2017), there would be potential to reduce emissions (as an alternative to fossil fuels); if also linked to carbon capture and storage (Hughes et al. 2012), it may be possible to achieve negative emissions (net CO_2 removal from the atmosphere). Full life cycle analyses are needed to assess the energy efficiency of such approaches, and the viability of scaling them up to climatically-important levels, taking account of associated environmental and socioeconomic implications.

A different mitigation option using seaweeds relates to their use as a dietary supplement for ruminants to suppress methane production. In vitro studies have given promising results (Dubois et al. 2013; Machado et al. 2016; Machado et al. 2018). However, because the potential scale of real-world benefits have yet to be quantified, there is *low confidence* in this approach as a mitigation option.

5.5.1.2.4 Land-sea integrated eco-engineering

Land-based nutrient management could, in theory, be used to enhance carbon storage in coastal seas and deeper waters, by increasing the amount of refractory dissolved organic carbon (Jiao et al. 2011; Jiao et al. 2014b; Jiao et al. 2018b). This idea is supported by a statistical

analysis of the relationship between organic carbon and nitrate in various natural environments (Taylor and Townsend, 2010) as well as by experimental results in estuarine and offshore waters (Yuan et al. 2010; Jiao et al. 2011; Jiao et al. 2014b). Delivery of nutrients from agricultural fertilisers and sewage discharge to coastal waters may currently promote the microbial breakdown of river-derived terrestrial dissolved organic carbon, reducing carbon storage (Liu et al. 2014). Thus reducing nutrient inputs in the future may expand carbon storage by favouring the microbial carbon pump, in addition to the multiple co-benefits of reduced nutrient loads related to HABs, oxygenation and ocean acidification (Miranda et al. 2013; Jiao et al. 2018a; Zhang et al. 2018). Although there is some evidence for the impact of dissolved organic carbon variations on global scale climate (Rothman et al. 2003) the benefits of this approach have yet to be determined quantitatively and uncertainties remain regarding the longevity of removal and associated carbon accounting (measurement, reporting and verification). Until such issues are better resolved, there is *low confidence* that stimulation of refractory dissolved organic carbon production could provide an operational long-term mitigation measure.

5.5.1.2.5 Control of sediment disturbance, enhanced weathering and other geochemical approaches

Anthropogenic sediment disturbance, through fishing, dredging and the installation of offshore structures, affects the security of carbon storage in shelf sea sediments (Hale et al. 2017). Management of such activities might therefore increase carbon retention, over relatively large areas of shelf seas (Avelar et al. 2017; Luisetti et al. 2019). However, there is a lack of data and understanding of the complex processes that affect carbon storage in the potentially mobile fraction of marine sediments (van de Velde et al. 2018); exceptions are provided by Hu et al. (2016) and Diesing et al. (2017). Due to these uncertainties, there is currently *low confidence* that control of sediment disturbance can be used for climate mitigation.

There is theoretically greater potential for carbon removal by 'enhanced weathering' using mineral additions to coastal waters (and the open ocean) (Rau, 2011; Renforth and Henderson, 2017). These approaches are based on increasing the naturally-occurring uptake of CO_2 by carbonates (e.g., calcite and dolomite) or silicate minerals (such as olivine). Such rock-weathering currently sequesters ~0.25 GtC yr^{-1}, on land and at sea (Taylor et al. 2015) and provides the longterm control of atmospheric CO_2 concentrations. It could be enhanced by adding ground minerals to beaches (Montserrat et al. 2017) or the sea surface. Other geochemical approaches for adding alkalinity that are less directly based on natural processes (Rau et al. 2012; GESAMP, 2019) are not considered here.

Enhanced weathering methods might be used to reduce local impacts, for example, for coral reefs (Albright et al. 2016b; Feng et al. 2016), as well as contributing to wider mitigation of climate change. However, their climatic benefits would be difficult to quantify, with other constraints on their development and deployment relating to the governance, cost and uncertain environmental impacts of large-scale application (Gattuso et al. 2018). The combination of these factors results in *low confidence* that enhanced weathering can provide a viable and acceptable climate mitigation approach.

5.5.1.3 Climate Mitigation in the Open Ocean

Recent reviews of the scope for using natural processes in the open ocean for climate mitigation are provided by Keller (2019a) and GESAMP (2019). The summary assessment given here is limited to direct and indirect biologically-based approaches, consistent with the scoping of this report and the major governance constraints on the large-scale application of open ocean interventions.

Current NPP by marine phytoplankton is estimated to be 58 ± 7 GtC yr^{-1} (Legendre et al. 2015), similar to terrestrial primary production and around 6 times greater than anthropogenic emissions (Le Quere et al. (2016). However, over 99% of the biologically-fixed carbon returns to the atmosphere over a range of timescales (Cartapanis et al. 2018).

The direct method of increasing marine productivity involves adding land-derived nutrients that may currently limit primary production, particularly iron. This approach has been investigated experimentally, by modelling and by observations of natural system behaviour (Keller et al. 2014a; Bowie et al. 2015; Tagliabue et al. 2017). The 13 experimental studies to date (seven in the Southern Ocean, five in the Pacific, and one in the sub-tropical Atlantic) have shown that primary production can be, but is not always, enhanced by the addition of iron (Boyd et al. 2007; Yoon et al. 2016; GESAMP, 2019).

The difficulties arise in demonstrating the time-scale of additional carbon removal, and in obtaining information on the consequences of the fertilisation for other marine ecosystem components, including ocean acidification and other potential side-effects (Williamson and Turley, 2012). Modelling studies (Aumont and Bopp, 2006) indicate that the climatic benefits could be relatively short-lived. Furthermore, public and political acceptability for ocean fertilisation is low (Williamson et al. 2012; Boyd and Bressac, 2016; Williamson and Bodle, 2016; Fuentes-George, 2017; McGee et al. 2018). Ocean iron fertilisation is regulated by the London Protocol, with amendments prohibiting such action unless constituting legitimate scientific research authorised under permit (see Section 5.5.4.1). There are additional governance constraints for the Southern Ocean where ocean iron fertilisation is theoretically considered to be most effective (Robinson et al. 2014).

Open ocean fertilisation by macro-nutrients (e.g., nitrate) has also been proposed, with modelled potential for gigaton-scale carbon removal (Harrison, 2017). Similar technical and governance considerations apply with regard to the quantification of mitigation benefits, the monitoring of potential adverse impacts, and the political acceptability of large-scale deployment. This approach would also involve higher costs, because of the much greater quantities of nutrients required (Williamson and Turley, 2012).

The indirect method of enhancing marine productivity uses physical devices to increase upwelling, thereby increasing the supply of a wide range of naturally-occurring nutrients from deeper water. This technique risks releasing additional CO_2 to the atmosphere, reducing its potential for climate mitigation (Bauman et al. 2014). There may also be other undesirable climatic consequences, including

disruption of regional weather patterns and long-term warming rather than cooling, if enhanced upwelling is deployed at large scale (Kwiatkowski et al. 2015).

Because of the many technical, environmental and governance issues relating to marine productivity enhancement, by either direct fertilisation or upwelling, there is *low confidence* that such open ocean manipulations provide a viable mitigation measure.

5.5.2 Ocean-based Adaptation

The AR5 concluded, with *high agreement* but *limited evidence*, that climate change impacts on coastal human settlements and communities could be reduced through coastal adaptation activities (Wong et al. 2014a). The limited evidence of the context-specific application of adaptation principles to support the assessment was highlighted as a knowledge gap for future research. This assessment reports progress made with developing such evidence and assesses human adaptation response to climate change in ecosystems, coastal communities and marine environments.

Components of human adaptation responses include risk assessment, risk reduction, and pathways towards resilience (Cross-Chapter Box 2; Chapter 1.6). Residual risk remains where hazard, vulnerability and exposure intersect, subsequent to an adaptation pathway response. Here we focus on adaptation responses within ecosystems and in human systems, as framed in Chapter 1, and defined by:

- *Nature-based* or *ecosystem-based adaptation* (5.5.2.1). The use of biodiversity and ecosystem services as part of an overall adaptation strategy to help people to adapt to the adverse effects of climate change. EbA uses the range of opportunities for the sustainable management, conservation, and restoration of ecosystems to provide services that enable people to adapt to the impacts of climate change (Narayan et al. 2016; Moosavi, 2017).
- *Human systems – Built environment adaptation* (5.5.2.3.1) Adaptation solutions pertaining to coastal built infrastructure and the systems that support such infrastructure (Mutombo and Ölçer, 2016; Forzieri et al. 2018).
- *Human systems – Socioinstitutional adaptation* (5.5.2.3) Adaptation responses within human social, governance and economic systems and sectors (Oswald Beiler et al. 2016; Thorne et al. 2017). This includes, but is not limited to *community-based adaptation* by coastal communities (5.5.2.3.2) based on empowering and promoting the adaptive capacity of communities, through appropriate use of context, culture, knowledge, agency, and community preferences (Archer et al. 2014; Shaffiril et al. 2017).

To avoid duplication, detailed consideration of adaptation responses to SLR and extreme events (including heat waves, and compound and cascading events) are avoided here, as they are covered by Chapter 4 and Chapter 6, respectively. Tables 5.7 and 5.8 provide a summary assessment of climate change impacts, human adaptation response and benefits in ecosystems and human systems respectively. Details of the assessed literature are in SM Table 5.7. Climate drivers and

5

Table 5.7 | Summary of reported Adaptation responses (A), the Impacts (I) they aimed to address, and the expected Benefits (B) in coastal ecosystems within Physical, Ecological, Social, Governance, Economic and Knowledge categories. For further details of impacts on ecosystems see Section 5.3. Legend: a + sign indicates *robust evidence*, a triangle indicates *medium evidence* and an underline indicates *limited evidence*. Dark blue cells indicate *high agreement*, blue indicates *medium agreement* and light blue indicates either *low agreement* (denoted by presence of a sign) if sufficient papers were reviewed for an assessment or no assessment (if less than three papers were assessed per cell). The papers used for this assessment can be found in SM5.5.

Cat.	Impacts (I)	Adaptation responses (A)	Coral reefs I	A	B	Mangroves I	A	B	Salt marshes/wetlands I	A	B	Estuaries I	A	B	Sandy beaches/dunes I	A	B	Multiple ecosystems I	A	B	Benefits (B)
Physical	Coastal physical processes disrupted	Supporting physical processes	Δ		Δ	Δ	Δ	+	+	—	Δ	+	—	Δ	+		+				Physical processes supported
	Catchment physical processes disrupted	Hard engineering responses										—									Coastal infrastructure resilience increased
	Coastal infrastructure damage	Soft engineering responses and buffers										—					Δ				Improved infrastructure functionality
	Disruption of urban systems	Integrated hard and soft engineering															Δ				Increased structural heterogeneity
	Land subsidence	Managed retreat and coastal realignment																			
Ecological	Ecosystem degradation and loss	Ecosystem restoration and protection	+	Δ	+	+	+	+	Δ			—	—	Δ	Δ		—				Ecosystem/ecological resilience supported
	Biodiversity and genetic diversity loss	Bioengineering							Δ	—											Physical processes supported
	Habitat range shifts	Assisted evolution and relocation		Δ									Δ								Coastal infrastructure resilience increased
	Sub-lethal species impacts	Nature based solutions					Δ														Increased biodiversity
	Invasive alien species																				Habitat range shifts accommodated
						Δ															Improved organismal fitness
																					Genetic heterogeneity supported
																					Strengthened socio-ecological system
Social	Decreased access to ecosystem services	Improving access to/storage of natural resources		Δ					+			Δ									Access to sustainable ecosystem services
	Local decline in agriculture and fisheries	Improving agricultural or fisheries practices																			Improved access to community services
	Increasing living costs	Supporting nature-based industries																			Increasing resilience in human systems
	Livelihoods impacts	Sustainable resource use															—				Improved socio-economic services
	Increased food insecurity	Maintaining or switching livelihoods									—										Improved employment and livelihoods
	Public health risks increased	Community participatory programmes						—													Improved health
	Cultural and traditional knowledge impacts	Developing adaptive networks																			Improved community participation
	Gender-related impacts	Sustainable household management																			Better informed communities
	Increased social vulnerability	Improving access to community services																			Improved integration of knowledge systems
	Decreased access to local government services	Empowering communities and addressing inequality																			Empowering women and children
	Socio-economic entrapment and decline	Building socio-ecological resilience																			Increased adaptive capacity
	Global declines in foodstocks																				Improved disaster preparedness

	Impacts (I)	Adaptation responses (A)	Coral reefs I	A	B	Mangroves I	A	B	Salt marshes/wetlands I	A	B	Estuaries I	A	B	Sandy beaches/dunes I	A	B	Multiple ecosystems I	A	B	Benefits (B)
Social (cont.)	Public areas access restrictions																				Empowered communities
	Decline in perceived value of human systems																				Improved community cohesion
	Conflict and migration																				Reduced inequality
Governance	Capacity challenges	Adopting/mainstreaming sustainability policies																			Political and institutional capacity developed
	Increased geopolitical tensions	Improving disaster response programmes																			Strengthened participatory governance
	Growing inequalities	Improving implementation and coordination of policies																			Better planning processes supported
		Developmental controls																			Improved coordination and decision making
		Evidence-based implementation																			Improved implementation and policies
		Improving ICM/MPAs																			Better communication
		Horizontal/vertical integration of governance																			Improved transparency and trust
		Developing partnerships and building capacity																			Climate justice advanced
		Improving access to community services																			Reduced conflict
		Pursuing climate justice																			Improved security
																					Improved adaptive management
																					Development supported
Economic	Increased business and living costs	Improving financial resources availability																			Increased revenue/income
	Business disruptions and losses	Improving access to insurance products																			Increased financial resources available
	Decreased value of assets/products	Economic diversification																			Reduced operational and capital costs
		Improving access to international funding programmes																			Investment strengthened
Knowledge	Uncertainty for decision makers	Better monitoring and modelling				–															Informed decision making tools
		Improving planning processes																			Improved co-production of knowledge
		Improving forecasting and early warning systems																			Improved relevance of products
		Improving decision support frameworks																			Improved education and outreach
		Improving participatory processes																			Improved awareness
		Coordinating top down and bottom up approaches																			
		Integrating knowledge systems																	Δ			
		Improving location and context specific knowledge																			
		Improving scientific communication																			
		Stakeholder identification, outreach and education				–															

impacts reported in the adaptation literature are consistent with those reported in Sections 5.2 and 5.3. Physical impacts include the disruption of physical coastal processes, like sediment dynamics, leading to, for example, erosion, flooding and coastal infrastructure damage (see Tables 5.7 and 5.8). Ecological impacts include the loss of ecosystems and biodiversity (Sections 5.2.3, 5.2.4, 5.3), which affected provision of ecosystem services, like coastal protection or food provision. The most commonly reported non-climate human drivers are growing human coastal populations (Elliff and Silva, 2017; van Oppen et al. 2017a; Gattuso et al. 2018) with poorly planned or managed urban development (Barbier, 2015; Wigand et al. 2017), land use change (Robins et al. 2016), loss of ecosystems (Runting et al. 2017), socioeconomic vulnerability (Broto et al. 2015; Bennett et al. 2016) of many coastal communities, ineffective governance and knowledge gaps for implementation.

5.5.2.1   Ecosystem-based Adaptation

This section assesses adaptation response in coastal ecosystems, beginning with biological adaptation in species, and followed by a summary assessment of EbA as a response to climate change.

5.5.2.1.1   Biological adaptation

There are many studies on biological climate change adaptation responses (Crozier and Hutchings, 2014; Miller et al. 2017; Diamond, 2018). Sections 5.2.3 and 5.3.3 discuss three main types of biological adaptation, broadly defined: evolutionary (genetic) adaptation through natural selection; phenotypic plasticity (acclimatisation), within an organism's lifetime; and individual or population mobility towards more favourable conditions. There are, however, expected to be limits to such natural adaptation, and large variations between species and populations (Gienapp and Merilä, 2018).

An accurate understanding of climate change impacts upon species, their sensitivity and adaptive capacity and consequent ecological effects (considering both indirect as well as direct impacts) is used to estimate extinction risk, so that an appropriate management response can be developed (Butt et al. 2016). EbA takes these complex interactions into account (Hobday et al. 2015), including the disruptive impacts of alien invasive species (Ondiviela et al. 2014; Wigand et al. 2017). Effective adaptation action, therefore, contains a broader consideration than historical conservation practices (*medium evidence, high agreement*), including the development of international collaborations and databases to improve ocean-scale understanding of climate change impacts (Okey et al. 2014; Young et al. 2015). A key knowledge gap relates to the critical thresholds for irreversible change for species (Powell et al. 2017).

5.5.2.1.2   Adaptation in coral reefs

Coral reefs are currently threatened by the continuous global degradation of warm water coral reef ecosystems and the failure of traditional conservation actions to revive most of the degrading reefs (Rinkevich, 2008; Miller and Russ, 2014). Interventions to rehabilitate degraded coral reef ecosystems can be categorised as preventive

('passive' restoration) or adaptive ('active' restoration) (Miller and Russ, 2014; Linden and Rinkevich, 2017) (see Box 5.5).

Inspired by silviculture (forestation) approaches to terrestrial ecosystem restoration, studies (Rinkevich, 1995; Rinkevich, 2005; Rinkevich, 2006; Rinkevich, 2008; Bongiorni et al. 2011) have proposed a two step restoration strategy for warm water coral reefs termed gardening of denuded coral reefs. In the first step, a large pool of coral colonies (derived from coral nubbins and fragments, and from sexually derived spat) are farmed in underwater nurseries, preferably on mid-water floating devices installed in sheltered zones, in which coral material can be cultured for up to several years. In the second step, nursery-grown coral colonies, together with recruited associated biota, are transplanted to degraded reef sites (Shafir and Rinkevich, 2008; Mbije et al. 2010; Shaish et al. 2010b; Shaish et al. 2010a; Bongiorni et al. 2011; Horoszowski-Fridman et al. 2011; Linden and Rinkevich, 2011; Mbije et al. 2013; Cruz et al. 2014; Chavanich et al. 2015; Horoszowski-Fridman et al. 2015; Lirman and Schopmeyer, 2016; Montoya Maya et al. 2016; Ng et al. 2016; Lohr and Patterson, 2017; Rachmilovitz and Rinkevich, 2017). Active restoration of coral reefs, while still in its infancy and facing a variety of challenges (Rinkevich, 2015b; Hein et al. 2017), has been suggested to potentially improve the ecological status of degraded coral reefs and the socioeconomic benefits that the reefs provide (Rinkevich, 2014; Rinkevich, 2015b; Linden and Rinkevich, 2017).

Ecological engineering approaches may promote coral reef adaptation (Rinkevich, 2014; Forsman et al. 2015; Coelho et al. 2017; Horoszowski-Fridman and Rinkevich, 2017; Linden and Rinkevich, 2017; Rachmilovitz and Rinkevich, 2017). They also include: augmenting functional diversity, including that of the microbiome (Casey et al. 2015; Horoszowski-Fridman and Rinkevich, 2017; Shaver and Silliman, 2017); transplantating whole habitats (Shaish et al. 2010b; Gómez et al. 2014); and enhancing genetic diversity (Iwao et al. 2014; Drury et al. 2016; Horoszowski-Fridman and Rinkevich, 2017). Active restoration can contribute to reef rehabilitation in all major reef regions (Rinkevich, 2014; Rinkevich, 2015b). However, there is *limited evidence* on how resistant these manipulated corals are to global change drivers (Shaish et al. 2010b; Shaish et al. 2010a) or how the nursery time affects biological traits like reproduction in coral transplants (Horoszowski-Fridman et al. 2011). Coral epigenetics may also be used as an adaptive management tool for reef rehabilitation (*low confidence*), as suggested by studies on coral adaptation (Brown et al. 2002; Horoszowski-Fridman et al. 2011; Palumbi et al. 2014; Putnam and Gates, 2015; Putnam et al. 2016).

Research on active coral reef restoration (Box 5.5) suggests the potential to help rehabilitate degraded coral reefs, provided that the underlying drivers of the impacts are mitigated (*high confidence*). Ongoing and new research in active coral reef restoration may further improve active reef restoration outcomes (Box 5.5) (*low confidence*). However, these coral reef restoration options may be ineffectual if global warming exceeds 1.5°C relative to pre-industrial levels (Hoegh-Guldberg et al. 2018; IPCC, 2018).

Box 5.5 | Coral Reef Restoration as Ocean-based Adaptation

Anthropogenic global change is impacting all warm water corals and the reef structures (Section 5.2.2.3.3; IPCC 5th Assessment Report (AR5)). These impacts are rapidly increasing in scale and intensity, exposing coral reefs to enhanced degradation rates and diminishing capacities to maintain ecological resilience, to absorb disturbances, and to adapt to the changes (Box 5.1) (Graham et al. 2014; Rinkevich, 2015a; Harborne et al. 2017). With the growing awareness that traditional reef conservation measures are insufficient to address climate change impacts on coral reefs (Section 5.2.2.1), adaptation interventions to enhance the resilience of coral reefs are being called for (Rinkevich, 1995; Rinkevich, 2000; Barton et al. 2017). Intervention strategies that are still at the 'proof-of-concept' stage, include: 'assisted colonisation' – actively moving species that are confined to disappearing habitats (Hoegh-Guldberg et al. 2008; Chauvenet et al. 2013); 'assisted evolution' – developing corals resistant to climate change via accelerated natural evolution processes (van Oppen et al. 2015); assisted coral chimerism (Rinkevich, 2019); novel coral symbiont associations (McIlroy and Coffroth, 2017); and coral microbiome manipulation (Bourne et al. 2016; Sweet and Bulling, 2017; van Oppen et al. 2017b). In contrast, the 'coral gardening' approachs – coral farmed in nurseries and transplanted using a range of tactics to increase survivability, growth rates and reproduction (Rinkevich, 2006; Rinkevich, 2014) – is already in use. Other interventions that have already been implemented in some coral reefs, such as the use of artificial reefs (Ng et al. 2017) are limited in impacts, and all are also revealing considerable challenges (Riegl et al. 2011; Coles and Riegl, 2013; Ferrario et al. 2014).

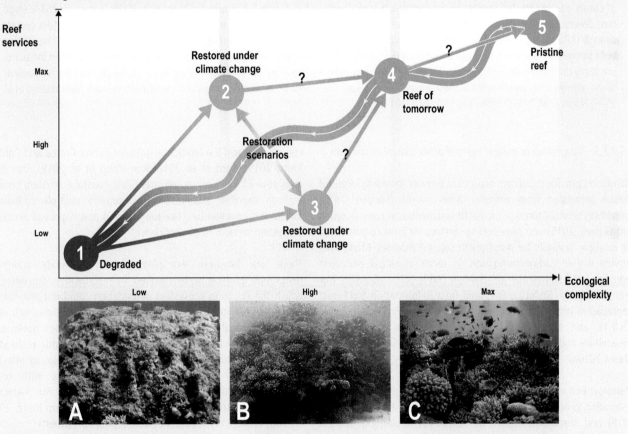

Box 5.5, Figure 1 | Coral reef restoration as an ocean-based adaptation tool to climate change. The squiggly line represents non-linear ecological statuses along a trajectory and five reef states (circles 1–5; in varying ecological complexity [x-axis] and service levels [y-axis]) including two extreme statuses (a pristine versus a highly degraded state, circles 5 and 1, respectively). Two 'restored reef-state' scenarios (circles 2, 3), lead to the state of the restored 'reef of tomorrow' (circle 4). The route from the state of the 'reef of tomorrow' (circle 4) to a pristine state (circle 5) is doubtful (the question mark) and is still at a theoretical level. The routes from the two 'restored reef-state' scenarios to the 'reef of tomorrow' are under investigations (the question marks). Based on Rinkevich (2014) (Figure 1). A–C represent different reef statuses. A = a denuded knoll at the Dekel Beach, Eilat, Israel before reef transplantation (November 2005; Photo: Y. Horoszowski-Fridman); B = the same knoll, restored (June 2016; photo by Shai Shafir). More than 300 nursery-grown colonies of 7 coral species were transplanted during three successive transplantations (years 2005, 2007, 2009). In 2016 the knoll was surrounded by reef inhabiting schools of fish. C = a pristine reef, not existing under current and anticipated reef conditions. Restoration scenarios are developed along paths from a degraded reef (low ecological complexity, minimal reef services) toward a healthy 'reef of tomorrow', passing through two restored reef states that are impacted by climate change (Shaish et al. 2010a; Schopmeyer et al. 2012; Hernández-Delgado et al. 2014; Rinkevich, 2015a). The employment of ecological engineering approaches may help in moving the ecological states from either restored reef to the 'reef of tomorrow' status (*medium confidence*).

Box 5.5 (continued)

Many of the alternative interventions that aim to increase the climate resilience of coral reefs involve culturing, selectively breeding and transplanting corals to enhance the adaptability of reef organisms to climate change, for example, by supporting the natural poleward range expansion of corals (West et al. 2017; Vergés et al. 2019). Advances in reef restoration techniques have been made in the last two decades (Rinkevich, 2014; Lirman and Schopmeyer, 2016), but assessments of the effectiveness of these techniques have mostly focused on the short-term feasibility of the technique (Frias-Torres and van de Geer, 2015; Lirman and Schopmeyer, 2016; Montoya Maya et al. 2016; Jacob et al. 2017; Rachmilovitz and Rinkevich, 2017), while longer-term evaluation in the context of all the pillars of sustainable development (Section 5.4.2) is limited (Rinkevich, 2015b; Barton et al. 2017; Flores et al. 2017; Hein et al. 2017). These alternative interventions, primarily the coral gardening approach, face two challenges. The first is scaling up; currently, these interventions have been tested at scales of hundreds of meters, while application at larger scale is lacking (Rinkevich, 2014). The second challenge (Box 5.5, Figure 1) is the effectiveness of active reef restoration to mitigate or rehabilitate global change impacts (Shaish et al. 2010a; Schopmeyer et al. 2012; Coles and Riegl, 2013; Hernández-Delgado et al. 2014; Rinkevich, 2015a; Wilson and Forsyth, 2018) and whether it can keep up with rising sea levels (Perry et al. 2018), especially in low-lying ocean states.

Altogether, coral reefs of the future will not resemble those of today because of the projected decline and changes in the composition of corals and associated species in the remaining reefs (Section 5.3.4, Box 5.5 Figure 1) (Rinkevich, 2008; Ban et al. 2014) (*high confidence*). The very high vulnerability of coral reefs to warming, ocean acidification, increasing storm intensity and SLR under climate change (AR5 WG2), including enhanced bioerosion (Schönberg et al. 2017) (*high confidence*) point to the importance of considering both mitigation (Section 5.5.1) and adaptation (Section 5.3.3.6) for coral reefs. Extensive research has explored adaptation measures involving the cultivation and transplantation of corals; however, the literature contains *limited evidence* on the comprehensive analysis of the relative costs and benefits of these interventions across the economic, ecological, social and cultural dimensions (Bayraktarov et al. 2016; Flores et al. 2017; Linden and Rinkevich, 2017).

5.5.2.1.3   Adaptation in mangroves and other coastal ecosystems

Mangroves provide significant ecosystem services, including localised coastal protection from extreme storm events (Section 5.4.1), supporting services through increased sedimentation rates (Hayden and Granek, 2015) and provisioning services for local communities, for example, habitats for nurseries to support fisheries. Mangroves provide limited carbon mitigation, in terms of global emissions reduction, and substantial job creation (Table 5.7) co-benefits (for example through Reducing Emissions from Deforestation and Forest Degradation programmes) when managed properly (Section 5.4.1, 5.5.1.1), and there is evidence of their value in supporting aquaculture and fishery initiatives (Huxham et al. 2015; Ahmed and Glaser, 2016a).

Mangrove EbA responses most commonly reported included ecosystem restoration (Sierra-Correa and Cantera Kintz, 2015; Romañach et al. 2018) and management such as mangroves re-planting through community participation programmes (Nanlohy et al. 2015; Nguyen et al. 2017; Triyanti et al. 2017). Mangrove EbA has been reported to provide multiple co-benefits in terms of improvement in support for coastal physical processes, including: shoreline stabilisation (Hayden and Granek, 2015; Nanlohy et al. 2015); ecological functioning (Sierra-Correa and Cantera Kintz, 2015; Miller et al. 2017) with improved ecosystem services (Alongi, 2015; Nanlohy et al. 2015; Palacios and Cantera, 2017); carbon mitigation (5.5.1.1); supporting livelihoods (Nanlohy et al. 2015; Nguyen et al. 2017); and reductions in coastal infrastructure damage and community vulnerability to climate change impacts. Managed retreat to counter coastal squeeze (Section 5.3) through improved governance, creation of finance and land use planning can allow mangroves to move up the shoreline

contour or down the latitudinal gradient (Sierra-Correa and Cantera Kintz, 2015; Ward et al. 2016; Romañach et al. 2018). Therefore, mangrove EbA responses can strengthen coastal ecosystem services through shoreline stabilisation and provide multiple co-benefits for coastal communities, like job creation and improved access to ecosystem services (*high confidence*).

There are, however, examples where community mangrove restoration projects have resulted in maladaptive outcomes, in which the resulting ecosystem degradation could not provide the ecosystem services required (Nguyen et al. 2017; Romañach et al. 2018). Such maladaptation can be a result of poor governance processes or a lack of community compliance with restoration plans. These examples emphasise the value of designing effective governance to implement adaptation responses with broad community participation to improve the climate risk reduction outcomes and co-benefits (Sierra-Correa and Cantera Kintz, 2015; Nguyen et al. 2017) (*medium evidence, high agreement*).

Mangrove and other coastal ecosystems restoration and management can be applied through reducing non-climatic hazards (Gilman et al. 2008; Ataur Rahman and Rahman, 2015; Sierra-Correa and Cantera Kintz, 2015; Ahmed and Glaser, 2016a; Nguyen et al. 2017; Romañach et al. 2018). Coastal and catchment development, including wetland transformation and degradation (Miloshis and Fairfield, 2015; Schaeffer-Novelli et al. 2016; Watson et al. 2017a; Schuerch et al. 2018), the disruption of physical processes impacting sedimentation rates (Watson et al. 2017a) and coastal squeeze compound coastal climate change impacts like erosion, flooding and saltwater intrusion (Ondiviela et al. 2014; Miloshis and Fairfield, 2015; Schaeffer-Novelli et al. 2016; Wigand et al. 2017) (Section 5.3). This reduces the ability

5

of these ecosystems to provide protection from wave and storm impacts, whilst positive feedbacks may occur that cause a net release of carbon into the atmosphere, for example, in salt marshes (Wong et al. 2014a) (Section 5.4.1). In some cases, effective interventions requires management at a broad spatial scale that includes a variety of ecosystems, for example, including ecosystems like mussel beds on the seaward side of seagrass beds to reduce wave energy and erosion (Ondiviela et al. 2014). Where sediment accretion matches the SLR rate, wetlands and salt marshes provide effective coastal protection and other important ecosystem services (*high confidence*).

Coastal dune systems are widely transformed globally. Human disturbance and the limited stabilising ability of dune vegetation are key causes of degradation (Onaka et al. 2015; Ranasinghe, 2016; MacDonald et al. 2017; Pranzini, 2017; Salgado and Martinez, 2017; Vikolainen et al. 2017; Gracia et al. 2018), while restoration efforts can be supported by both hard (Sutton-Grier et al. 2015; Pranzini, 2017) and soft (Sutton-Grier et al. 2015; Vikolainen et al. 2017) engineering responses. Reduced coastal erosion (Sánchez-Arcilla et al. 2016; Goreau and Prong, 2017; Vikolainen et al. 2017; Carro, 2018; Gracia et al. 2018) and flood risk (Onaka et al. 2015; MacDonald et al. 2017; Nehren et al. 2017) through maintaining dunes as natural buffers against wave energy (Nehren et al. 2017) can increase resilience to climate change impacts (Sutton-Grier et al. 2015; Magnan and Duvat, 2018). Engineered responses and sand replenishment are considered complementary approaches (Onaka et al. 2015; Martínez et al. 2017). Section 4.4.4.1 provides an overview of sediment-based adaptation response measures, including cost estimates for beach nourishment and dune maintenance, a discussion of co-benefits and drawbacks of combining hard and soft infrastructure measures, and challenges with sourcing sediment for beach replenishment. In some cases dune restoration and sand replenishment projects have not been successful, due to fire damage (Shumack and Hesse, 2017) or the rapid loss of sand within replenishment schemes due to coastal processes and stakeholder rejection of adaptation activities (Pranzini, 2017). Coastal dune restoration and beach replenishment are effective responses against coastal erosion and flooding, where sufficient materials and space to implement are available (*medium confidence*).

5.5.2.1.4 Ecosystem-based adaptation

There is a growing body of literature regarding the effectiveness and economics of EbA. In addition to building resilience to climate change, EbA is expected to bring a wide range of co-benefits that include increasing ecological complexity, with multiple ecosystem services, and other economic co-benefits (Perkins et al. 2015; Perry, 2015; Moosavi, 2017; Scarano, 2017). The cost-effectiveness of EbA approaches varies between marine ecosystem types; for example, coral reefs (Perkins et al. 2015; Beetham et al. 2017; Elliff and Silva, 2017; Beck et al. 2018; Comte and Pendleton, 2018) and salt-marshes (Ondiviela et al. 2014; Miloshis and Fairfield, 2015; Schaeffer-Novelli et al. 2016; Wigand et al. 2017) performed best at reducing wave heights, whilst salt marshes and mangroves were two to five times cheaper than submerged breakwaters for wave heights of less than half a meter. Although low regrets, win-win approaches like EbA are supported in the literature (Watkiss et al. 2014; Barange et al. 2018), syntheses of experience from context-specific practical

implementation of EbA and assessment of their cost-effectiveness are limited (Narayan et al. 2016). Therefore, EbA can be a cost-effective approach for securing climate change-related ecosystem services with multiple co-benefits (*medium evidence, high agreement*).

The application of EbA approaches can be more effective when incorporating local knowledge and Indigenous knowledge and cultural practices into adaptation responses (Ataur Rahman and Rahman, 2015; Perkins et al. 2015; Sutton-Grier et al. 2015; Sánchez-Arcilla et al. 2016; van der Nat et al. 2016). The application of synergistic combinations of adaptation responses in multiple ecosystems can provide a range of co-benefits, and this approach is strengthened when combined with socioinstitutional approaches (Kochnower et al. 2015; MacDonald et al. 2017). Research to improve and refine EbA approaches and increase their specificity to local context is important for their effectiveness in reducing climate risks and generating co-benefits (Sutton-Grier et al. 2015). Conversely, a lack of inclusion of local communities and economic undervaluation of specific coastal and marine ecosystems, compounded by gaps in scientific data, can undermine the potential effectiveness of EbA approaches (Perkins et al. 2015; Hernández-González et al.; Narayan et al. 2016; Roberts et al. 2017).

Despite the abundance of EbA examples in the literature, knowledge gaps pertaining to their implementation and limitations remain. Developing this literature could help with understanding context specific application of EbA and improve their effectiveness (*medium confidence*).

5.5.2.2 Human Systems

Many of the world's great cities lie within the coastal region, and climate change impacts put these cities, their inhabitants and their economic activities at risk. Section 5.5.2.2 assesses the impacts of climate change, adaptation response and benefits upon human systems, including coastal communities, built infrastructure, fisheries and aquaculture, coastal tourism, government and health systems. Table 5.8 provides a summary of the assessment, with citations provided in the Supplementary Material Table 5.7.

Poorly planned (Ataur Rahman and Rahman, 2015), located (Abedin et al. 2014; Betzold and Mohamed, 2017; Linkon, 2018) and managed urban settlements or human systems, driven by growing human coastal populations (Perkins et al. 2015; Moosavi, 2017; Carter, 2018) and compounded by the disruption of coastal and catchment physical processes (Nagy et al. 2014; Broto et al. 2015; Marfai et al. 2015; Kabisch et al. 2017) and pollution (Zikra et al. 2015; Peng et al. 2017) are major human drivers of change compounding the impacts of climate change.

Coastal communities, built infrastructure and fisheries and aquaculture (Table 5.8) are likely to be significantly affected through the disruption of coastal physical processes (DasGupta and Shaw, 2015; Betzold and Mohamed, 2017; Hagedoorn et al. 2019) leading to coastal erosion, flooding, salt water intrusion and built infrastructure damage (Dhar and Khirfan, 2016; Hobday et al. 2016a; Jurjonas and Seekamp, 2018) (*robust evidence, high agreement*).

5

Table 5.8 | Summary of reported Adaptation responses (A), the Impacts (I) they aimed to address, and the expected Benefits (B) in human systems within Physical, Ecological, Social, Governance, Economic and Knowledge categories. Legend: a + sign indicates *robust evidence*, a triangle indicates *medium evidence* and an underline indicates *limited evidence*. Dark blue cells indicate *high agreement*, blue indicates *medium agreement* and light blue indicates either *low agreement* (denoted by presence of a sign) if sufficient papers were reviewed for an assessment or no assessment (if less than three papers were assessed per cell). Papers used for this assessment can be found in SM5.6.

	Impacts (I)	Adaptation responses (A)	Coastal communities I	A	B	Built infrastructure I	A	B	Fisheries and aquaculture I	A	B	Coastal tourism I	A	B	Government I	A	B	Health I	A	B	Benefits (B)
Physical	Coastal physical processes disrupted	Supporting physical processes	+			+			+												Physical processes supported
	Catchment physical processes disrupted	Hard engineering responses		+			+	+													Coastal infrastructure resilience increased
	Coastal infrastructure damage	Soft engineering responses and buffers	+	−		+				−											Improved infrastructure functionality
	Disruption of urban systems	Integrated hard and soft engineering	△				−	+													Increased structural heterogeneity
	Land subsidence	Managed retreat and coastal realignment							−												
Ecological	Ecosystem degradation and loss	Ecosystem restoration and protection	+	+					+	+	−	−					−				Ecosystem/ecological resilience supported
	Biodiversity and genetic diversity loss	Bioengineering	△			△			+												Physical processes supported
	Habitat range shifts	Assisted evolution and relocation																			Coastal infrastructure resilience increased
	Sub-lethal species impacts	Nature based solutions	−			△	−		+												Increased biodiversity
	Invasive alien species																				Habitat range shifts accommodated
																					Improved organismal fitness
																					Genetic heterogeneity supported
																					Strengthened socio-ecological system
Social	Decreased access to ecosystem services	Improving access to/storage of natural resources	△	△					△	+											Access to sustainable ecosystem services
	Local decline in agriculture and fisheries	Improving agricultural or fisheries practices	+	−					+	+											Improved access to community services
	Increasing living costs	Supporting nature-based industries											−								Increasing resilience in human systems
	Livelihoods impacts	Sustainable resource use	+	−	−				+	△											Improved socio-economic services
	Increased food insecurity	Maintaining or switching livelihoods	−		−				+	−	+										Improved employment and livelihoods
	Public health risks increased	Community participatory programmes	−	+					+	+							−				Improved health
	Cultural and traditional knowledge impacts	Developing adaptive networks																			Improved community participation
	Gender-related impacts	Sustainable household management		−	−																Better informed communities
	Increased social vulnerability	Improving access to community services	+						+												Improved integration of knowledge systems
	Decreased access to local government services	Empowering communities and addressing inequality									△										Empowering women and children
	Socio-economic entrapment and decline	Building socio-ecological resilience									△										Increased adaptive capacity
	Global declines in foodstocks																				Improved disaster preparedness

Category	Impacts (I)	Adaptation responses (A)	CC I	A	B	BI I	A	B	FA I	A	B	CT I	A	B	GOV I	A	B	H I	A	B	Benefits (B)
Social (cont.)	Public areas access restrictions									–											Empowered communities
	Decline in perceived value of human systems						–														Improved community cohesion
	Conflict and migration																				Reduced inequality
Governance	Capacity challenges	Adopting/mainstreaming sustainability policies					Δ			–						–					Political and institutional capacity developed
	Increased geopolitical tensions	Improving disaster response programmes								–											Strengthened participatory governance
	Growing inequalities	Improving implementation and coordination of policies								+	–					+	–				Better planning processes supported
		Developmental controls																			Improved coordination and decision making
		Evidence-based implementation															–				Improved implementation and policies
		Improving ICM/MPAs								+						+					Better communication
		Horizontal/vertical integration of governance																			Improved transparency and trust
		Developing partnerships and building capacity		–						–						Δ					Climate justice advanced
		Improving access to community services								–											Reduced conflict
		Pursuing climate justice																			Improved security
										+											Improved adaptive management
																					Development supported
Economic	Increased business and living costs	Improving financial resources availability							–												Increased revenue/income
	Business disruptions and losses	Improving access to insurance products	+						+												Increased financial resources available
	Decreased value of assets/products	Economic diversification																			Reduced operational and capital costs
		Improving access to international funding programmes																			Investment strengthened
Knowledge	Uncertainty for decision makers	Better monitoring and modelling					+	+	+	Δ						Δ					Informed decision making tools
		Improving planning processes			Δ	Δ															Improved co-production of knowledge
		Improving forecasting and early warning systems								Δ											Improved relevance of products
		Improving decision support frameworks				Δ				+			–			–					Improved education and outreach
		Improving participatory processes				Δ				+						–					Improved awareness
		Coordinating top down and bottom up approaches																			
		Integrating knowledge systems	+							Δ											
		Improving location and context specific knowledge	Δ																		
		Improving scientific communication								–											
		Stakeholder identification, outreach and education	–							Δ						–					

Ecosystem degradation and biodiversity loss will further compound impacts in coastal communities and fisheries and aquaculture (Ataur Rahman and Rahman, 2015; Petzold and Ratter, 2015; Dhar and Khirfan, 2016), with sub-lethal species impacts like changes in the productivity and distribution of fisheries target species reported for the latter (Gourlie et al. 2018; Nursey-Bray et al. 2018; Pinsky et al. 2018) (*high confidence*). This is likely to result in decreased access to ecosystem services (Asch et al. 2018; Cheung et al. 2018b; Finkbeiner et al. 2018) (*medium evidence, high agreement*), local declines in agriculture and fisheries (Cvitanovic et al. 2016; Faraco et al. 2016) (*high confidence*) and livelihood impacts (Harkes et al. 2015; Busch et al. 2016; Valmonte-Santos et al. 2016) (*high confidence*) in coastal communities and fisheries and aquaculture, particularly increased food insecurity and health risk in the latter (*high confidence*). These livelihood impacts are likely to increase social vulnerability (*high confidence*). Businesses within coastal communities are likely to experience disruptions and losses (*robust evidence, high agreement*).

5.5.2.2.1   Coastal communities

This section describes a range of adaptation responses reported at the level of the individual or community. Hard engineering responses included small scale hard infrastructure coastal defenses (Betzold and Mohamed, 2017; Jamero et al. 2018), design responses at the household level (Ataur Rahman and Rahman, 2015; Linkon, 2018) and retreat (Marfai et al. 2015). Ecosystem restoration and protection, particularly in mangroves (Ataur Rahman and Rahman, 2015; Bennett et al. 2016; Jamero et al. 2018; Hagedoorn et al. 2019) through community participation programmes (Barbier, 2015; Petzold and Ratter, 2015; Bennett et al. 2016; Dhar and Khirfan, 2016; Jamero et al. 2018) was strongly supported in the literature as a means to improve access to or storage of natural resources (*medium evidence, high agreement*).

Social responses include increasing climate change awareness, improving participatory decision making through bottom-up approaches, community organisation for action and engagements with local management authorities (Dutra et al. 2015; Tapsuwan and Rongrongmuang, 2015; Galappaththi et al. 2017; Ray et al. 2017; Cinner et al. 2018; Hagedoorn et al. 2019). In coastal communities, and indeed in most other sectors, despite consensus on the importance of cooperation in tackling climate change (Elrick-Barr et al. 2016), adaptation progress may be hampered by competing economic interests and worldviews (Hamilton and Safford, 2015), which can be compounded by limited climate change knowledge (Nanlohy et al. 2015). Factors like home ownership and a general future planning ability support resilience (Elrick-Barr et al. 2016). Climate change adaptation capacity is shaped by historical path dependencies, local context and international linkages, while action is shaped by science, research partnerships and citizen participation (Hernández-Delgado, 2015; Sheller and León, 2016). Locally context-specific data to guide appropriate adaptation response remains a knowledge gap (Abedin and Shaw, 2015; Hobday et al. 2015; Lirman and Schopmeyer, 2016; Williams et al. 2016).

Coastal and oceanic adaptation responses are greatly complicated by the presence of competing interests (either between user groups, communities or nations), where considerations other than climate change need to be incorporated into cooperation agreements and policy (Wong et al. 2014a). The deployment of either built or natural protection systems, or adopting a 'wait and see' approach, is subject to the social acceptance of these approaches in communities (Poumadère et al. 2015; Sherren et al. 2016; Torabi et al. 2018). Similarly, the willingness to move away from climate change impacted zones is dependent upon a range of other socioeconomic factors like age, access to resources and crime (Bukvic et al.; Rulleau and Rey-Valette, 2017). Adaptation to climate change includes a range of non-climatic and social variables that complicate implementation of adaptation plans (*robust evidence, high agreement*).

Improving community participation and integrating knowledge systems (local, traditional and scientific) supports coastal community adaptation responses (*high confidence*), providing improved co-production of knowledge (*medium evidence, high agreement*), improved community awareness (*medium evidence, medium agreement*) and better-informed, more cohesive coastal communities (*limited evidence, medium agreement*).

5.5.2.2.2   Built infrastructure

Built infrastructure impacts are most frequently addressed through hard engineering approaches including: construction of groins, seawalls, revetments, gabions and breakwaters (Friedrich and Kretzinger, 2012; Vikolainen et al. 2017); improving drainage and raising the height of roadways and other fixed-location infrastructure (Perkins et al. 2015; Becker et al. 2016; Colin et al. 2016; Asadabadi and Miller-Hooks, 2017; Brown et al. 2018a); erosion control systems (Jeong et al. 2014); and the relocation of infrastructure (Friedrich and Kretzinger, 2012; Colin et al. 2016). Nature-based responses are increasingly being reported as complementary and supporting tools (van der Nat et al. 2016; Kabisch et al. 2017; Gracia et al. 2018) using ecological engineering (Perkins et al. 2015; van der Nat et al. 2016; Moosavi, 2017) combined with innovative construction strategies (Moosavi, 2017).

When implemented together, hard and soft engineering responses provide social (Gracia et al. 2018; Martínez et al. 2018; Woodruff, 2018) and ecological (Perkins et al. 2015; van der Nat et al. 2016; Gracia et al. 2018) co-benefits with reduced damage costs (Jeong et al. 2014). Constraints on implementation include the space and extra cost required by ecological infrastructure, sub-optimal performance when impacted by natural physical processes that are disrupted (Gracia et al. 2018) or restrictions associated with governance (Vikolainen et al. 2017). Adaptation planning including local communities can improve implementation and help fill knowledge gaps (Kaja and Mellic, 2017; Moosavi, 2017; Martínez et al. 2018; Mikellidou et al. 2018). Benefits include increased resilience in coastal infrastructure and better informed decision making tools (*medium confidence*).

5.5.2.2.3   Adaptation in fisheries and aquaculture

Sixty percent of assessed species are projected to be at high risk from both overfishing and climate change by 2050 (RCP8.5), particularly tropical and subtropical species (Cheung et al. 2018b). Overfishing is one of the most important non-climatic drivers affecting the

5

sustainability of fisheries (Islam et al. 2013; Heenan et al. 2015; Faraco et al. 2016; Dasgupta et al. 2017; Cheung et al. 2018b; Harvey et al. 2018). Pursuing sustainable fisheries practices under a low emissions scenario would decrease risk by 63%. This highlights the importance of effective fisheries management (Gaines et al. 2018). Eliminating overfishing would, however, require reducing current levels of fishing effort, with a potential short-term reduction in catches impacting livelihoods and the food security of coastal communities (Hobday et al. 2015; Dey et al. 2016; Rosegrant et al. 2016; Campbell, 2017; Finkbeiner et al. 2018). Despite consensus on the effectiveness of eliminating overfishing in supporting climate change adaptation in fisheries (*robust evidence, high agreement*), successful adaptation outcomes remain aspirational.

Range shifts under ocean warming (Section 5.2.3) will alter the distribution of fish stocks across political boundaries, thus demand for transboundary fisheries management will increase. Redistribution of transboundary fish stocks between countries (Ho et al. 2016; Gourlie et al. 2017; Asch et al. 2018) could destabilise existing international fisheries agreements and increase the risk of international conflicts (Section 5.4.2). Adaptation to reduce risks in international fisheries management could involve improving planning for cooperative management between countries informed by reliable predictions (Payne et al. 2017) and projections (Pinsky et al. 2018) of species shifts and associated uncertainties. Cooperative international fisheries arrangements, such as flexible fishing effort allocation and adaptive frameworks (Colburn et al. 2016; Cvitanovic et al. 2016; Faraco et al. 2016) may also improve the robustness of fisheries management (Miller et al. 2013). Thus, although range shifts pose significant challenges to transboundary fisheries management, proactive planning and adjustment of fisheries management arrangements, informed by scientific projections, could help improve adaptive capacity (*medium confidence*). The effectiveness of incorporating MPAs as an adaptation strategy to climate change can be improved by considering climate impacts in the design of MPAs (*medium, high agreement*).

Improving integrated coastal management and better planning for MPAs by incorporating projected shifting biological communities, abundance and life history changes (Álvarez-Romero et al. 2018) due to climate change could contribute towards improved fisheries adaptive management by, for example, increasing resilience of habitats, providing refugia for species with shifting distributions and by conserving biodiversity (Faraco et al. 2016; Valmonte-Santos et al. 2016; Dasgupta et al. 2017; Le Cornu et al. 2017; Roberts et al. 2017; Asch et al. 2018; Cheung et al. 2018b; Harvey et al. 2018; Jones et al. 2018; O'Leary and Roberts, 2018) (Sections 5.2.3, 5.3, 5.4.1), but MPAs may also reduce access to subsistence fishers, increasing their vulnerability to food insecurity (Bennett et al. 2016; Faraco et al. 2016). The global area of MPAs is rapidly increasing towards the United Nations' target of 10% of the global ocean. While this is encouraging, it is estimated that only 2% of the ocean is well enough managed, as described in (Edgar et al. 2014), to meet conservation goals (Sala et al. 2018). Improving the implementation and coordination of policies, and improving integrated coastal management and MPAs have emerged in the literature as important adaptation governance responses (*robust evidence, medium agreement*).

Governance responses to support adaptation in fisheries communities include conducting vulnerability assessments, improving monitoring of ecosystem indicators and evaluating management strategies (Himes-Cornell and Kasperski, 2015b; Busch et al. 2016). Socioeconomic factors like access to alternative income, mobility, gender and religion collectively shape a community's adaptation response (Arroyo Mina et al. 2016). In West Africa, the industrial fishery response to climate change induced reductions in landings was the expansion of fishing grounds, which increased operational costs (Belhabib et al. 2016). This response is not available to artisanal and local fishing communities, who are considered highly vulnerable (Kais and Islam, 2017). Access to finance to support these communities or their governments could help them reach novel fishing grounds, and, therefore, potentially reduce their vulnerability. Food security linked to fisheries depends on stock recovery, but also on access to and distribution of the harvest, as well as gender considerations (Béné et al. 2015). Hence, granting preferential access to dependent coastal communities should be considered in examining policy options. Other adaptation responses include improved fishing gear and technology, use of fish aggregating devices and uptake of insurance products (Zougmoré et al. 2016) [see Barange et al. (2018) for a summary of possible adaptation responses]. Community response as a part of climate change adaptation for local fisheries is an important element in assessing adaptive capacity (*medium evidence, good agreement*).

Fisheries management strategies depend heavily upon data collection and monitoring systems. These include the accuracy of data collected in respect of predicting environmental conditions, over time scales from months to decades (Dunstan et al. 2018), effective monitoring and evaluative mechanisms (Le Cornu et al. 2017; Gourlie et al. 2018), controlling for aspects of fish population dynamics like recruitment success and fish movement (Mace, 2001). Seasonal to decadal climate prediction systems allow for skillful predictions of climate variables relevant to fisheries management strategies (Hobday et al. 2016b; Payne et al. 2017). Effective fisheries adaptation responses will require knowledge development including better monitoring, modelling and improving decision support frameworks (*medium evidence, high agreement*) and improving forecasting and early warning systems (*medium evidence, medium agreement*).

In considering a participatory decision making approach for fisheries management that responds to climate change, Heenan et al. (2015) provided a number of key elements that contribute towards a successful outcome. These include expert knowledge of climate change threats to fish habitats, stocks and landings, the necessity of transdisciplinary collaboration and stakeholder participation, broadening the range and scope of fisheries systems and increased commitment of resources and capacity. This was considered in the context of the ability of developing countries to sustainably exploit fisheries resources and related ecosystems. More research is required on socio-ecological responses to climate change impacts on fishery communities, including such aspect as like risk reduction, adaptive capacity through knowledge attainment and social networks, developing alternative skills and participatory approaches to decision making (Dubey et al. 2017; Shaffril et al. 2017; Finkbeiner et al. 2018). Important fisheries adaptation responses in relation to knowledge management include improving participatory processes (*robust evidence, high agreement*), integrating knowledge systems (*medium evidence, high agreement*), and stakeholder identification, outreach and education (*medium evidence, medium agreement*). Ecosystem-based adaptation, community participatory programmes, and improving agricultural and fisheries practices are very strongly supported in the literature (*high confidence*).

Less still is known about how climate change will affect the deep oceans and its fisheries (Section 5.2.3 and 5.2.4), the vulnerability of its habitats to fishing disturbance and future effects on resources not currently harvested (FAO, 2019). Johnson et al. (2019) concluded that in a 20- to 50-year timeframe, the effectiveness of virtually all north Atlantic deep water and open ocean area-based management tools can be expected to be affected. They concluded that more precise and detailed oceanographic data are needed to determine possible refugia, and more research on adaptation and resilience in the deep sea is needed to predict ecosystem response times.

As with fisheries, community- and ecosystem-based adaptation responses, an integrated coastal management framework is considered useful for planning for anticipated challenges for aquaculture (Ahmed and Diana, 2015b; Barange et al. 2018). Where *in situ* adaptation is not possible, translocation and polyculture (Ahmed and Diana, 2015a; Bunting et al. 2017) have been suggested as appropriate responses, but this would suit commercial rather than subsistence interests. Policy, economic, knowledge and other types of support are required to build socio-ecological resilience of vulnerable coastal aquaculture communities (Harkes et al. 2015; Bunting et al. 2017; Rodríguez-Rodríguez and Bande Ramudo, 2017), which requires a deep understanding of the nature of stressors and a commitment for collective action (Galappaththi et al. 2017). Climate resilient pathway development (see Cross-Chapter Box 2) is considered a useful framework for Sri Lankan shrimp aquaculture (Harkes et al. 2015). Another example of successful aquaculture adaptation is the employment of near real time monitoring technology to track the carbonate chemistry in water to reduce bio-erosion in shellfish from acidification (Barton et al. 2015; Cooley et al. 2016). Numerous adaptation responses are available for aquaculture, but some options, like translocation and technological responses may not be available to subsistence-based communities (*medium evidence*).

An example of eco-engineering-based adaptation option in seaweed aquaculture under climate change is artificial upwelling, as shown by experiments and observations. Artificial upwelling powered by green energy (solar, wind, wave or tidal energy) to seaweeds (Jiao et al. 2014b; Zhang et al. 2015; Pan and Schimel, 2016) can moderate the amount of deep water upwelled to the euphotic zone to just meet the demands of nutrients and DIC by the seaweed for photosynthesis, while avoiding the acidification and hypoxia that often occur in natural upwelling systems (Jiao et al. 2018a; Jiao et al. 2018b) (*high confidence*). Such artificial upwelling based eco-engineering may also gradually release the 'bomb' of rich nutrients and hypoxia in the bottom water, which could otherwise breakout following storms (Daneri et al. 2012) (*high confidence*).

5.5.2.3.4   Coastal tourism

The coastal tourism economic sector is highly sensitive to climate change. Tourism response, in terms of mitigating carbon emissions and adapting to climate change impacts, are assessed here. Coastal tourism is likely to be impacted by ecosystem degradation and loss (*limited evidence, medium agreement*), which underscores the importance of nature-based tourism. An example of coastal erosion in Latin America illustrates this, whereby SLR interacting with non-climate change impacts including sand mining, inappropriate development and habitat destruction (e.g., mangroves), resulted in declines in tourism (Rangel-Buitrago et al. 2015). The management recommendation was appropriate legislation with a marine spatial planning emphasis, enforcement, sustainable funding mechanisms and support networks for decision making.

Climate change impacts upon tourism are nuanced and not restricted to just physical impacts on tourism establishments (Biggs et al. 2015). Understanding the drivers of tourist choices could help support adaptation in the industry through marine spatial planning processes (Papageorgiou, 2016). For example, in an survey ranking mitigation and adaptation responses in Greece, tourists prioritised rational energy use, energy efficiency and water saving measures (Michailidou et al. 2016b). Location specific information of tourist choices could help shape local industries. In one example from the Thailand dive industry, climate change adaptation responses of participants were reported to be based on misconceptions about climate change and personal observations (Tapsuwan and Rongrongmuang, 2015). To improve community-based adaptation, efforts aimed at broadening the level of awareness about climate change could improve decision making processes (Tapsuwan and Rongrongmuang, 2015). Tourist behaviour is shaped by changing ocean physical processes and degrading ecosystems at tourist destinations, which drive destination changes, economic flows and market share adjustments. (Bujosa et al. 2015; De Urioste-Stone et al. 2016).

It is very likely that climate change will have direct and nuanced impacts upon coastal tourism. Improving decision support frameworks (*low evidence, medium agreement*) for better-informed decision making tools could contribute towards increasing resilience in coastal tourism (*low evidence, limited agreement*).

5.5.2.2.5   Government responses

Government responses included adopting and mainstreaming sustainability policies, including investments and policies for climate change (Aylett, 2015; Buurman and Babovic, 2016) and applying the precautionary principle in the absence of precise scientific guidance (Johnson et al. 2018). Developing adequate governance and management systems (Johnson et al. 2018), strengthening capacity (Gallo et al. 2017; Paterson et al. 2017), increasing cooperation (Nunn et al. 2014; Gormley et al. 2015) and aligning policies of local authorities (Porter et al. 2015; Gallo et al. 2017; Rosendo et al. 2018) could help to improve implementation (Sano et al. 2015; Elsharouny, 2016). This includes planning for MPAs and improving integrated coastal management (Abelshausen et al. 2015; Roberts et al. 2017; Rosendo et al. 2018) by incorporating climate science (Hopkins et al. 2016; Johnson et al. 2018) to optimise priority marine habitats (Gormley et al. 2015; Jones et al. 2018). An advantage of integrated coastal management is that it helps manage the interactions between multiple climate and non-climatic drivers of coastal ecosystems and sectors. Incorporating stakeholder participation with local knowledge and Indigenous knowledge could help to reduce the risk of maladaptation, and increase buy-in for implementation (Serrao-Neumann et al. 2013). Improving participatory processes strengthens governance decision making and flexible risk management processes (Gerkensmeier and Ratter, 2018; Rosendo et al. 2018), while stimulating bi-directional knowledge flow and improving social learning (Abelshausen et al. 2015).

Technology for environmental monitoring, for example using drones (Clark, 2017), web-based coastal information systems (Mayerle et al. 2016; Newell and Canessa, 2017), the Internet of Things and machine learning solutions promise to improve the local scale knowledge base, which should improve climate adaptation planning and resilience effort and environmental management decisions (Conde et al. 2015). Where such knowledge gaps persist, the implementation of climate change adaptation measures could proceed on the basis of a set of general principals of best practice (Sheaves et al. 2016; Thorne et al. 2017).

Benefits of effective government adaptation response includes the promotion of sustainable use, development and protection of coastal ecosystems (Rosendo et al. 2018) and the protection of biodiversity through setting appropriate conservation priorities (Gormley et al. 2015). Improved governance includes consideration of social processes in risk management (Gerkensmeier and Ratter, 2018; Rosendo et al. 2018) and improved systematic conservation planning (Johnson et al. 2018). At a local level, this translates into sustained service delivery (Aylett, 2015), improved rationality and effective policy making (Serrao-Neumann et al. 2013; Rosendo et al. 2018).

Improving the implementation and coordination of policies and improving integrated coastal management are both considered important climate change adaptation governance responses (*robust evidence, high agreement*), as are developing partnerships and building capacity (*medium evidence, high agreement*) and adopting or mainstreaming sustainability policies (*limited evidence, medium agreement*). Benefits include improved ecosystem resilience, better planning processes, implementation and policies (all *limited evidence, medium agreement*).

5.5.2.3   *Ocean-based Climate Change Adaptation Frameworks*

Adaptation action in pursuit of a climate resilient development pathway is likely to have a deeper transformative outcome than stepwise or ad hoc responses (Cross-Chapter Box 2 in Chapter 1). Recent literature highlighting the effectiveness of components of adaptation planning includes quantitative assessments of vulnerability in ecosystems (Kuhfuss et al. 2016), species (Cheung et al. 2015; Cushing et al. 2018), and communities (Islam et al. 2013; Himes-Cornell and Kasperski, 2015b), and integrated assessments of all of the above (Peirson et al. 2015; Kaplan-Hallam et al. 2017; McNeeley et al. 2017; Ramm et al. 2017; Mavromatidi et al. 2018).

Seasonal and decadal forecasting tools have improved rapidly since AR5, especially in supporting management of living marine resources (Payne et al. 2017) and modelling to support decision making processes (Čerkasova et al. 2016; Chapman and Darby, 2016; Jiang et al. 2016; Justic et al. 2016; Joyce et al. 2017; Mitchell et al. 2017). Decision making processes are supported by economic evaluations (Bujosa et al. 2015; Jones et al. 2015), evaluations of ecosystem services (MacDonald et al. 2017; Micallef et al. 2018), participatory processes (Byrne et al. 2015) and social learning outcomes, the development of adaptation pathways, frameworks and decision making (Buurman and Babovic, 2016; Dittrich et al. 2016; Michailidou et al. 2016a; Osorio-Cano et al. 2017; Cumiskey et al. 2018), and indicators to support evaluation of adaptation actions (Carapuço et al. 2016; Nguyen et al. 2016) through monitoring frameworks (Huxham et al. 2015). Climate change adaptation responses are more effective when developed within institutional frameworks that include effective planning and cross-sector integration.

Evidence-based decision making for climate adaptation is strongly supported in the literature (Endo et al. 2017; Thorne et al. 2017) through better understanding of coastal ecosystems and human adaptation responses (Dutra et al. 2015; Cvitanovic et al. 2016), as well as consideration of non-climate change related factors. Relevant research includes the topics of: multiple-stakeholder participatory planning (Archer et al. 2014; Abedin and Shaw, 2015); trans-boundary ocean management (Gormley et al. 2015; Williams et al. 2016); ecosystem-based adaptation (Hobday et al. 2015; Dalyander et al. 2016; McNeeley et al. 2017; Osorio-Cano et al. 2017); and community-based adaptation with socioeconomic outcomes (Merkens et al. 2016). Research on applying 'big data' and high end computational capabilities could also help develop a comprehensive understanding of climate and non-climate variables in planning for coastal adaptation (Rumson et al. 2017). New knowledge from these research areas could substantially improve planning, implementation and monitoring of climate adaptation responses for marine systems, if research processes are participatory and inclusive (*medium confidence*).

Despite such interest, evaluations of the planning, implementation and monitoring of adaptation actions remain scarce (Miller et al. 2017). In a global analysis of 401 local governments, only 15% reported on adaptation actions (mostly large cities in high income countries), and 18% reported on planning towards adaptation policy (Araos et al. 2016). Thus, integrated adaptation planning with non-climate change related impacts remains an under-achieved ambition, especially in developing countries (Finkbeiner et al. 2018). Challenges reported for adaptation planning include uncoordinated, top-down approaches, a lack of political will, insufficient resources (Elias and Omojola, 2015; Porter et al. 2015), and access to information (Thorne et al. 2017).

Characteristics of successful adaptation frameworks include: a robust but flexible approach, accounting for deep uncertainty through well-coordinated participatory processes (Dutra et al. 2015; Jiao et al. 2015; Buurman and Babovic, 2016; Dittrich et al. 2016); well-developed monitoring systems (Barrett et al. 2015; Bell et al. 2018b); and taking a whole systems approach (Sheaves et al. 2016), with the identification of co-benefits for human development and the environment (Wise et al. 2016). The coastal adaptation framework

5

literature is dominated by Australian, North American and European cities, with fewer studies from African and Caribbean sites, least developed countries and SIDS (Kuruppu and Willie, 2015; Torresan et al. 2016).

In contrast with the many examples of proposed frameworks for climate resilient coastal adaptation, few studies have assessed their success, possibly due to the time-lag between implementation, monitoring, evaluation and reporting. Nevertheless, there is substantial support for 'no regrets' approaches addressing both proximate and systematic underlying drivers of vulnerability (Sánchez-Arcilla et al. 2016; Pentz and Klenk, 2017; Zandvoort et al. 2017) with leadership, adaptive management, capacity and the monitoring and evaluation of actions considered useful in governance responses (Dutra et al. 2015; Doherty et al. 2016). More extensive learning processes could help build decision makers' capacity to tackle systemic drivers, guide pursuance climate change appropriate policies (Barange et al. 2018) and to scrutinise potentially maladaptive infrastructural investments (Wise et al. 2016). More effective coordination across a range of stakeholders, within and between organisations, especially in developing countries, would strengthen the global coastal adaptation response (*medium confidence*).

5.5.2.4   The Role of Education and Local Knowledge in Adapting to Climate Change

Education can help improve understanding of issues related to climate change and increase adaptive capacity (Fauville et al. 2011; Marshall et al. 2013; von Heland et al. 2014; Pescaroli and Magni, 2015; Tapsuwan and Rongrongmuang, 2015; Wynveen and Sutton, 2015). Participatory processes can facilitate the development of networks between coastal communities and environmental managers for the purposes of developing and implementing adaptation strategies (Wynveen and Sutton, 2015). Education, combined with other forms of institutional support empowers fisheries and aquaculture communities (Table 5.8) to make informed adaptation decisions and take action (*medium evidence, medium agreement*).

Local knowledge and Indigenous knowledge systems can complement scientific knowledge by, for example, improving community ability to understand their local environment (Andrachuk and Armitage, 2015), forecast extreme events (Audefroy and Sánchez, 2017) and help to increase community resilience (Leon et al. 2015; Sakakibara, 2017; Cinner et al. 2018; Panikkar et al. 2018). Committing resources could strengthen local level adaptation planning (Alam et al. 2016; Novak Colwell et al. 2017) through the inclusion of cultural practices (Audefroy and Sánchez, 2017; Fatorić and Seekamp, 2017) and Indigenous knowledge systems (Kuruppu and Willie, 2015; von Storch et al. 2015). Local knowledge can, however, act as a barrier to adaptation where there is a strong dependency upon such knowledge for immediate survival, to the detriment of long-term adaptation planning (Marshall et al. 2013; Metcalf et al. 2015). There is evidence, however, to suggest that vulnerability in fisheries communities and coastal tourism operators with high levels of Local knowledge is reduced where they have a correspondingly high level of adaptive capacity (Marshall et al. 2013). Resource users with high levels of local knowledge may also be able to identify signals of

change within their environment, and recognise the need to adapt. In these instances, fishers with higher local knowledge are expected to demonstrate a higher adaptive capacity than fishers with lower local knowledge, and can be expected to progress towards developing new strategies to combat the impacts of climate change (Kittinger et al. 2012). In these instances, local knowledge acts to promote adaptation (*medium confidence*).

Localised, individual-scale behaviors can aggregate rapidly and contribute to the global adaptation response. This can be supported by clear messaging that clarifies the role of individuals, households and local businesses in addressing climate change. Coastal communities can improve the co-production of climate change knowledge (*medium evidence, good agreement*) through the integration of knowledge systems (Table 5.8). In fisheries and aquaculture, better-informed decision making tools (*medium evidence, medium agreement*) are supported by improved participatory processes (*high confidence*), integrating knowledge systems (*medium evidence, good agreement*) and improving decision support frameworks (*medium evidence, medium agreement*).

5.5.2.5   Costs and Limits for Coastal Climate Change Adaptation

Challenges persist in conducting economic assessments for built infrastructure adaptation due to complicated uncertainties such as the accuracy of climate projections and limited information regarding paths for future economic growth and adaptation technologies. Annual investment and maintenance costs of protecting coasts were projected to be 12–71 billion USD (Hinkel et al. 2014), which was considered significantly less than damage costs in the absence of such action. In an analysis of twelve Pacific island countries, 57% of assessed built infrastructure was located within 500 m of coastlines, requiring a replacement value of 21.9 billion USD. Substantial coastal adaptation costs (and international financing) are likely to be required in these countries (*medium confidence*).

In West African fisheries, loss of coastal ecosystems and productivity are estimated to require 5–10% of countries' GDP in adaptation costs (Zougmoré et al. 2016). Similarly, for Pacific Islands and Coastal Territories, fisheries adaptation will require significant investment from local governments and the private sector (Rosegrant et al. 2016), with adaptation costs considered beyond the means of most of these countries (Campbell, 2017). In SIDS, tourism could provide the funding for climate change adaptation, but concerns with creating investment barriers, assumptions around cost-effectiveness and consumer driven demand remain barriers (Hess and Kelman, 2017). MPAs with multiple co-benefits, are considered a cost-effective strategy (Byrne et al. 2015). In 2004, the annual cost of managing 20–30% of global seas as MPAs was estimated at between 5–19 billion USD, with the creation of approximately one million jobs (Balmford et al. 2004).

Estimating adaptation costs is challenging because of wide ranging regional responses and uncertainty (Dittrich et al. 2016). Despite these challenges, the protection from flooding and frequent storms that coral reefs provide has been quantified by (Beck et al. 2018), who estimated that without reefs, damage from flooding and costs from frequent storms would double and triple respectively, while countries from Southeast Asia, East Asia and Central America could each save in excess of 400 million USD through good reef management. Although quantifying global adaptation costs remains challenging because of a wide range of regional responses and contexts, it is likely that managing ecosystems will contribute towards reducing costs associated with climate change associated coastal storms (*medium confidence*). Further research evaluating natural infrastructure is required (Roberts et al. 2017) to better understand costs and benefits of EBA.

There is a broad range of reported barriers and limits to climate change adaptation for both ecosystems and human systems. Coastal ecosystem-based adaptation can be physically constrained by space requirements and coastal squeeze (Sutton-Grier et al. 2015; Robins et al. 2016; Sánchez-Arcilla et al. 2016; Ahmed et al. 2017; Peña-Alonso et al. 2017; Salgado and Martinez, 2017; Triyanti et al. 2017; Schuerch et al. 2018), while the pace of climate change may exceed the adaptive capacity of ecosystems, for example, SLR may outpace the vertical reef accretion rate (Beetham et al. 2017; Elliff and Silva, 2017; Joyce et al. 2017). One technical limit for coral reef adaptation is that tools have not yet been developed for large-scale implementation (van Oppen et al. 2017a). Ecosystems may also have physiological and ecological constraints which are exceeded by climate change impacts (Miller et al. 2017; Wigand et al. 2017), and the recovery periods of natural systems (Gracia et al. 2018) and for ecological succession (Salgado and Martinez, 2017) may be outpaced by climate change impacts. The performance of ecosystems in EBA projects may be inhibited by the poor condition of the ecosystem (Nehren et al. 2017), highlighting the importance of effective implementation (Salgado and Martinez, 2017).

Social and cultural norms with conflicting and competing values (Miller et al. 2017), public lack of knowledge on climate change and distrust of information sources (Wynveen and Sutton, 2015), as well as populations increasingly distanced from, and unconcerned about nature (Romañach et al. 2018), may constrain ecosystem-based adaptation response. Examples of governance adaptation constraints include: inadequate policy, governance and institutional structures (Sánchez-Arcilla et al. 2016; Miller et al. 2017; Wigand et al. 2017), limited capacity (Sutton-Grier et al. 2015; Thorne et al. 2017), ineffective implementation (Nguyen et al. 2017; Comte and Pendleton, 2018), and poor enforcement (Nguyen et al. 2017). Governance constraints are compounded by lack of finances (Miller et al. 2017), financial costs of design and implementation (Gallagher et al. 2015) and the high cost of coastal land (Gracia et al. 2018), although ecosystem-based adaptation is considered cheaper than human-made structures (Nehren et al. 2017; Salgado and Martinez, 2017; Vikolainen et al. 2017; Gracia et al. 2018).

Knowledge limitations can include a lack of data (Sutton-Grier et al. 2015; Wigand et al. 2017; Romañach et al. 2018), for example, when an absence of baseline data may undermine coastline management (Perkins et al. 2015). Scale-relevant information may be required for local decision making (Robins et al. 2016; Thorne et al. 2017) and to comply with localised design requirements (Vikolainen et al. 2017). Other knowledge barriers include inherent uncertainties in models (Schaeffer-Novelli et al. 2016) and complexity of coastal

systems (Wigand et al. 2017). A more nuanced knowledge barrier is the disconnect between scientific, community and decision making processes (Romañach et al. 2018).

Substantial knowledge gaps are reported for ecosystem-based adaptation, including restoration of coral reef systems as an adaptation tool (Comte and Pendleton, 2018), managing mangrove and human response to climate change (Ward et al. 2016), advancing coastal EBA science by quantifying ecosystem services (Hernández-González et al.), and evaluating natural infrastructure (Roberts et al. 2017). Few syntheses of the context-specific application and cost-effectiveness of EBA approaches are to be found in the literature (Narayan et al. 2016).

Human systems have similar limitations. Improved understanding of limitations in built infrastructure, beach nourishment and nature-based adaptation responses, especially with respect to cost effectiveness and resilience, would substantially aid shoreline stabilisation attempts (Mackey and Ware, 2018). For artisanal fisheries, a range of physical and socioinstitutional limits and barriers to adaptation have been reported, including increasing occurrence and severity of storms limiting fishing time, technologically poor boats and fishing equipment and lack of access to credit and markets, among others (Islam et al. 2013). Conflicting interests and values of stakeholders (Evans et al. 2016), the path-dependent nature of organisations and resistance to change (Evans et al. 2016) and inadequate collaboration and public awareness (Oulahen et al. 2018) have been reported as socioinstitutional barriers. A knowledge gap persists in understanding how such limits and barriers interact to suppress adaptation response.

In some communities, climate change may not be prioritised in the face of chronic, daily challenges to secure livelihoods (Esteban et al. 2017; Fischer, 2018) or risk severity may be underestimated due to a high frequency of exposure in the recent past (Esteban et al. 2017). In a world with competing risks and urgent priorities, some local inhabitants appear to be unable to avoid, or are willing to carry, the risk associated with a climate impact in order to meet other, more pressing needs. This example reflects the reality of many poor, informal settlement dwellers in coastal areas around the world (*medium confidence*). Other human system barriers to effective adaptation action include insufficient climate change knowledge, inappropriate coping strategies, high dependency upon natural resources, level of exposure to hazards and weak community networks (Islam et al. 2013; Nanlohy et al. 2015; Lohmann, 2016; Koya et al. 2017; Senapati and Gupta, 2017; Cumiskey et al. 2018).

In summary, it is concluded that the broad range of reported barriers and limits to climate change adaptation for ecosystem and human system adaptation responses (*high confidence*). Limitations include the space that ecosystems require, non-climatic drivers and human impacts that need to be addressed as part of the adaptation response, the lowering of adaptive capacity of ecosystems because of climate change, and slower ecosystem recovery rates relative to the recurrence of climate impacts, availability of technology, knowledge and financial support and existing governance structures (*medium confidence*). (5.5.2.5)

5

5.5.2.6   Summary

There has been a substantial amount of literature focused on coastal and oceanic adaptation since AR5. Socio-institutional adaptation responses are the more numerous of the three types of adaptation responses assessed in this chapter. There is broad agreement that hard engineering responses are optimally supported by ecosystem-based adaptation approaches, and both approaches should be augmented by socioinstitutional approaches for adaptation (*high confidence*) (Nicholls et al. 2015; Peirson et al. 2015; Sánchez-Arcilla et al. 2016; van der Nat et al. 2016; Francesch-Huidobro et al. 2017; Khamis et al. 2017). In planning adaptation responses, awareness-raising and stakeholder engagement processes are important for buy-in and ownership of responses (*robust evidence, high agreement*) as is institutional capacity within local government organisations, whose importance in coastal adaptation initiatives has been emphasised in the recent literature (*robust evidence, high agreement*). With all three types of adaptation, basic good governance and effective implementation of service delivery processes are prerequisites for successful adaptation planning and response.

5.5.3   Governance Across All Scales

There are many global, regional, national and local governance structures with interests in climate-driven ocean warming, acidification, deoxygenation and SLR, and their impacts on marine ecosystems and dependent communities (Galland et al. 2012; Stephens, 2015; Fennel and VanderZwaag, 2016; Diamond, 2018). The legal, policy and institutional response is therefore shared by many institutions developed for a number of distinct but inter-related fields, including governance regimes for ocean systems, climate change, marine environment, fisheries and the environment generally. A changing ocean poses several scale-related challenges for these governance institutions and processes, arising from:

- The global and transboundary scales of the major changes to ocean properties (temperature, circulation, oxygen loss, acidification, etc.), with variability in their local expression;
- The regional scales of changes in ecosystem services following from the changes in ocean properties (including services provided to humans living far from the coasts);
- The global scales of land-based drivers of those changes (both greenhouse gas emissions and changes in ecosystems services), which often motivate policy responses (primarily at the national level) and behavioural responses (primarily at the community level);
- The scale dependent need for coordinated responses by the different governance structures, to ensure their overall effectiveness (see also Chapter 1).

For all of these challenges, the scales of the climate-related issues may be poorly matched to the scales of most governance institutions and processes, making effective responses or proactive initiatives difficult. Sections 5.2 to 5.4 provide evidence, through case histories and thematic overviews, that illustrates these four types of challenges. In some cases, more than one type of challenge is illustrated in a single example, such as when a change in an amount or availability of an ecosystem service is discussed in the context of factors influencing the vulnerability of socio-ecological systems to climate change (Sections 5.2., 5.3 and 5.4).

Existing ocean governance structures for the ocean already face multi-dimensional challenges because of climate change, and this trend of increasing complexity will continue (Galaz et al. 2012). Current international governance regimes and structures for fisheries and the ocean environment do not yet adequately address the issues of ocean warming, acidification and deoxygenation (Oral, 2018; Box 5.6). At the time of the initial development and adoption of these legal and governance regimes, minimal attention was given to climate change and the effects of carbon dioxide emissions on the ocean, with associated impacts on the interacting physical, chemical, biological properties of the ecosystems, and the resulting risks and vulnerabilities of dependent communities and economic sectors. In particular, the governance of ocean ABNJ is a major challenge (Levin and Le Bris, 2015); the collaborative structures and mechanisms for environmental assessment in ABNJ need further development (Warner, 2018) (*high confidence*). Negotiations are currently ongoing regarding a new international agreement for marine biodiversity of ABNJ (UNEP, 2016).

The following changes in governance may improve the ability of governance institutions and processes to address the challenges identified above:

- Cooperation on regional and global scales through various types of agreements of varying degrees of formality for States and other participants in governance;
- Increasing the voice and role in decision making for non-governmental participants such as Indigenous peoples, social and labour organisations;
- Increasing the horizontal integration of decision making across industry and societal sectors, under processes such as 'integrated management' and 'marine spatial planning';
- Increasing resource mobilisation at the community scale to enable communities to experiment and innovate to address the challenges, and then to share their experiences with other communities and build cooperative approaches to promote strategies with successful outcomes.

These governance innovation strategies have the potential to increase the ability of the governance institutions and processes to successfully respond to all four types of scale-related challenges listed earlier. However, any of them also have the potential to fail to address their intended concerns effectively if implemented inappropriately, or to create new challenges as the initial priorities are addressed. In some countries, lack of capacity of the existing governance institutions, lack of access to basic facilities, insufficient income diversification and illiteracy are major hindrance for ocean governance in a changing climate (Bennett et al. 2014; Salik et al. 2015; Weng et al. 2015; Karim and Uddin, 2019; Sarkodie and Strezov, 2019) (*high confidence*).

Table 5.9 | Ocean Governance and Climate Change: Major Issues.

Area of Governance	Major Legal Instruments	Major Issues and Actions
Marine Environment Generally	UNCLOS, CBD, CITES, WHC, MARPOL and other IMO legal instruments, regional seas conventions and other legal instruments	UNCLOS imposes obligations on state parties to take action to combat the main sources of ocean pollution. Tools and techniques in UNCLOS may need adjustment in response to the emerging challenges created by ocean climate change (Redgwell, 2012). However, success of the umbrella regulatory framework of UNCLOS depends heavily on the further development, modification and implementation of detailed regulations by relevant international, regional and national institutions (Karim, 2015). The London Protocol to the London Convention was amended in 2006 to address the issue of carbon dioxide storage processes for sequestration. Two subsequent amendments concern sharing transboundary sub-seabed geological formations for sequestration projects, and ocean fertilisation and other marine geoengineering. One of these new amendments prohibits ocean fertilisation except for research purposes (Dixon et al. 2014). The issue of ocean acidification has been considered within the framework of the OSPAR Convention, the CCAMLR Convention (Herr et al. 2014), and the CBD (CBD,2014) ; this issue is discussed further in Box 5.6. The CBD has also considered regulatory issues relating to ocean fertilisation and other (marine) geoengineering (Williamson and Bodle, 2016). In 2018, the CBD adopted *Voluntary Guidelines for the Design and Effective Implementation of Ecosystem-Based Approaches to Climate Change Adaptation and Disaster Risk Reduction*. However, even if Parties to the Convention choose to adopt the voluntary guidelines, there is no mechanism to implement them beyond their exclusive economic zones in the water column and their extended continental shelves (if recognised) in the seabed. Most of the 29 world heritage listed coral reefs are facing severe heat stress (Heron, 2017) and the WHC may play a role for coral reef protection.
Climate Change	UNFCCC, Paris Agreement, MARPOL Convention and other legal instruments	Existing international legal instruments do not adequately address climate change challenges for the open ocean and coastal seas (Galland et al. 2012; Redgwell, 2012; Herr et al. 2014; Magnan et al. 2016; Gallo et al. 2017; Heron, 2017). Nevertheless, ocean and coastal areas will benefit from the overall UNFCCC goal for preventing dangerous interference with the climate system. A study of the 161 national pledges for climate change mitigation and adaptation (NDCs) identified 'gaps between scientific [understanding] and government attention, including on ocean deoxygenation, which is barely mentioned' (Gallo et al. 2017). In 2011, the MARPOL convention was amended to include technical and operational measures for the reduction of greenhouse gas emissions from ships. However, the effectiveness of these provisions depends on the national implementation by flag, port and coastal states, with no international enforcement authority (Karim, 2015).
Fisheries	UNCLOS, UN Fish Stocks Agreement, FAO Compliance Agreement, FAO PSMA, Regional Fisheries Agreements and other legal instruments	The impact of climate change on marine fisheries is expected to be very significant (Sections 5.3, 5.4) (Barange et al. 2018; FAO, 2019), with adverse impacts on food security, livelihood and national development in many coastal countries; least developed countries seem particularly vulnerable (Blasiak et al. 2017). Regional fisheries management systems need to address these emerging challenges (Brooks et al. 2013). The ecological and socio-ecological criteria and standards for performance can be set at regional levels where Regional Fisheries Management Organizations have been established, but their effectiveness is variable depending on the characteristics of regulatory instruments and other factors (Ojea et al. 2017). The current international regulatory framework for fisheries management has a responsiveness gap, since it does not fully incorporate issues related to the fluctuating and changing distribution of fisheries (Pentz and Klenk, 2017; Pinsky et al. 2018). However, some regional fisheries management organisations (RFMOs) have initiated processes to improve the equity of sharing fishery resources affected by climate change (Aqorau et al. 2018). A climate-informed ecosystem-based fisheries governance approach has been suggested for enhancing climate change resilience of marine fisheries in the developing world (Heenan et al. 2015), but robust and effective management, policy, legislation and planning based on flexibility and scientific understanding will be required for coastal fisheries (Gourlie et al. 2017). The existing failing condition of many stocks, coupled with maladaptive responses to climate change, may create serious challenges for the sustainability of global fisheries; improved fisheries governance can offset some of these challenges (Gaines et al. 2018). The fisheries agreements and the provisions in UNCLOS have helped RFMOs to increase the sustainability of fisheries on stocks in or migrating through international waters, and equity of access to them. Because the distribution of many stocks changes with changes in physical oceanic conditions (particularly temperature and current regimes), many of the measures and access arrangements negotiated and adopted by the RFMOs have reduced effectiveness in a changing climate. New arrangements have been difficult to negotiate, in part because of concerns that the distributions and productivities will continue to change as climate change continues to drive changes on ocean conditions (Blasiak et al. 2017; Ojea et al. 2017; Pentz and Klenk, 2017; Aqorau et al. 2018; Pinsky et al. 2018).

Acronyms and organisations: CBD, Convention on Biological Diversity; CCAMLR, Convention on the Conservation of Antarctic Marine Living Resources; CITES, Convention on International Trade in Endangered Species of Wild Fauna and Flora; IMO: International Maritime Organization; London Convention: Convention on the Prevention of Marine Pollution by Dumping of Wastes and Other Matter; London Protocol: 1996 Protocol to the Convention on the Prevention of Marine Pollution by Dumping of Wastes and Other Matter; MARPOL Convention: International Convention for the Prevention of Pollution from Ships; NDCs: Nationally Determined Contributions; OSPAR Convention: Convention for the Protection of the Marine Environment of the North-East Atlantic; UN Fish Stocks Agreement: The Agreement for the Implementation of the Provisions of the United Nations Convention on the Law of the Sea of 10 December 1982 relating to the Conservation and Management of Straddling Fish Stocks and Highly Migratory Fish Stocks; UNCLOS: United Nations Convention on the Law of the Sea; UNFCCC: United Nations Framework Convention on Climate Change; WHC, World Heritage Convention: Convention Concerning the Protection of the World Cultural and Natural Heritage; FAO Compliance Agreement: The Agreement to Promote Compliance with International Conservation and Management Measures by Fishing Vessels on the High Seas; FAO PSMA: The Agreement on Port State Measures.

5

Additional considerations identified by recent studies of ocean related mitigation and adaptation include the need for: early warning and precautionary management; multi-level and multi-sectoral governance responses; holistic, integrated and flexible management systems; integration of scientific and local knowledge as well as natural, social and economic investigation; identification and incorporation of a set of social indicators and checklists; adaptive governance; and incorporation of climate change effects in marine spatial planning (Hiwasaki et al. 2014; Kettle et al. 2014; Hernández-Delgado, 2015; Himes-Cornell and Kasperski, 2015a; Pittman et al.

2015; Colburn et al. 2016; Creighton et al. 2016; Hobday et al. 2016a; Audefroy and Sánchez, 2017; Gissi et al. 2019; Tuda et al. 2019). Diverse adaptations of governance are being tried, and some are producing promising results (Sections 5.2, 5.3 and 5.4). However, rigorous further evaluation is needed regarding the effectiveness of these adaptations in achieving their goals in addressing specific governance challenges. Robust conclusions on the effectiveness of specific types of governance adaptations in various socio-ecological contexts would require a targeted assessment of ocean (and terrestrial) governance in a changing climate, possible as a key part of AR6.

Box 5.6 | Policy Responses to Ocean Acidification: Is there an International Governance Gap?

Ocean acidification is not specifically mentioned in the Paris Agreement on climate change (UNFCCC, 2015) and has only been given limited attention to date in other UNFCCC discussions. Nevertheless, ocean acidification is widely considered to be part of the climate system: it is one of seven state-of-the-climate indicators used by the World Meteorological Organization (WMO, 2019); it featured strongly in AR5, being covered by both WGI and WGII; its impacts are assessed in many sections of this Chapter; and concerns regarding ocean acidification have been raised through many international governance structures, including the United Nations Convention on the Law of the Sea (UNCLOS), the Convention on Biological Diversity (CBD), the United Nations Environment Programme (UNEP), and the Intergovernmental Oceanographic Commission of the United Nations Educational, Scientific and Cultural Organization (IOC-UNESCO).

Although many bodies have interests in ocean acidification, no unifying treaty or single instrument has been developed (Herr et al. 2014; Harrould-Kolieb and Hoegh-Guldberg, 2019) and there has been only limited governance action that is specific to the problem (Fennel and VanderZwaag, 2016; Jagers et al. 2018). Exceptions to this generalisation are the development of coordinated monitoring through the Global Ocean Acidification Observing Network (Newton et al. 2015), with associated scientific support through the International Atomic Energy Agency (IAEA) (Osborn et al. 2017; Watson-Wright and Valdés, 2018); and SDG14.3, with its non-binding, and relatively general, commitment to 'minimise and address the impacts of ocean acidification, including through enhanced scientific cooperation at all levels'.

One possible response to the fragmented responsibilities for ocean acidification governance would be the development of a new UN mechanism specifically to address ocean acidification (Kim, 2012). This option would take time and political will, and has not been widely supported (Harrould-Kolieb and Herr, 2012). One pragmatic approach could be enhancing the involvement of UNFCCC with acidification governance (Herr et al. 2014) together with increased use of multilateral environment agreements (Harrould-Kolieb and Herr, 2012) (medium confidence).

UNFCCC action to stabilise the climate by reducing CO_2 emissions also necessarily addresses the problem of ocean acidification, which is primarily caused by anthropogenic CO_2 dissolving in seawater and lowering pH. Nevertheless, there are also distinct ocean acidification mitigation and adaptation issues, including:

- Climate mitigation measures that might be focused on greenhouse gases other than CO_2;
- pH-associated thresholds or tipping-points (Hughes et al. 2013; Good et al. 2018) that have implications for scenario-modelling of emission reductions (Steinacher et al. 2013);
- The large-scale use of bioenergy with carbon capture and storage (BECCS) as a mitigation option, if this involved sub-seafloor CO_2 storage, with risk of leakage and hence ocean acidification impacts (Blackford et al. 2014);
- The use of other CO_2 removal techniques (negative emissions) such as ocean fertilisation (Section 5.5.1.3), or solar radiation management, without CO_2 emission reductions; both approaches would worsen ocean acidification (Williamson and Turley, 2012; Keller et al. 2014a).

Adaptation to climate change could also include a more integrated approach to reduce ocean acidification impacts (Section 5.5.2). Proposed adaptation actions for ocean acidification (Kelly et al. 2011; Billé et al. 2013; Strong et al. 2014; Albright et al. 2016a) include reduction of pollution and other stressors (thereby strengthening resilience); water treatment (e.g., for high value aquaculture); and the use of seaweed cultivation and seagrass restoration to slow longterm pH changes (although short-term variability may be increased) (Sabine, 2018). These measures are generally applicable to relatively limited spatial scales; whilst they may succeed in 'buying time', their future effectiveness will decrease unless underlying global drivers are also addressed (high confidence).

5.6   Synthesis

This chapter has documented an extraordinary array of observed changes in the open ocean, deep sea and coasts. It draws on evidence from thousands of references from the literature, millions of observations and hundreds of simulations of the past and future scenarios. The ocean climate and its state, ecosystems and human systems have changed (Section 4.2.2.6, 5.2, 5.3, 5.4, Figure 5.24) and are projected to change further. The ocean is a highly connected environment allowing water and living organisms to move freely. Change is observed across physical conditions that pose hazards to ecosystems in all regions from the surface to the deepest parts (Figure 5.24). All types of human and managed systems that have been covered in this chapter have evidence of mostly negative impacts but also some positive, some very significantly, some less so (Figure 5.24). Overall the multiple lines of evidence from the literature and the assessment in this chapter's Executive Summary point to profound and pervasive changes on regional and global scales (Figure 5.24).

The level of knowledge and confidence of the changes in the marine environment that are particularly relevant to ecosystems and human systems ranges from *virtually certain* to *low confidence* (see Figure 5.24). Many of the observed changes in some variables can be directly attributed to human influence from rising greenhouse gases and other anthropogenic forcings (Section 5.2.2 and 5.2.3). For other variables and in some systems the evidence is less direct, but the cascading of risks from changing ocean, marine ecosystems and dependent communities remains robust when considered as a whole. The observed and projected changes in the ocean systems that are covered in this chapter are consistent with our understanding of ocean chemistry and circulation, and our knowledge of the ecosystems responses. In many cases, the assessments of risk level of ecosystems for the recent past and long-term future are based on multiple lines of evidence, combining ecological and physiological knowledge (from experiments, direct observations and model projections) with the major climate drivers (e.g., Sections 5.2.5 and 5.3.4). Globally, all the marine ecosystems assessed here have elevated risk for biodiversity, ecosystem function, structure and services with increasing greenhouse gas emissions (Figure 5.16) (*high confidence*). These risks result from ocean warming, stratification, acidification, deoxygenation, SLR and associated changes as well as interactions with non-climatic human drivers. Most importantly, all the coastal ecosystems that were assessed, where linkages between natural systems and human

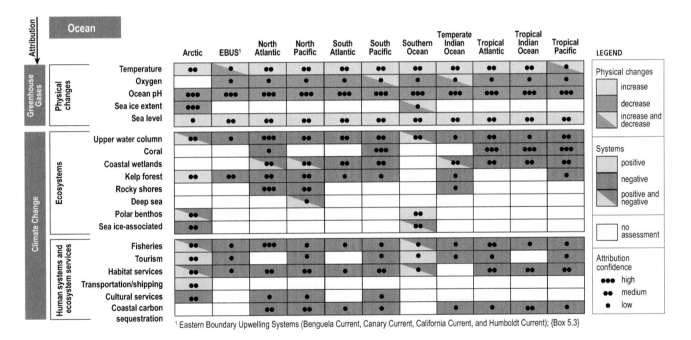

Figure 5.24 | Synthesis of observed regional hazards and impacts in the ocean assessed in SROCC. For the ocean, physical changes, impacts on key ecosystems, and impacts on human systems and ecosystem services are shown. For physical changes, yellow/green refers to an increase/decrease, respectively, in amount or frequency of the measured variable. For impacts on ecosystems, human systems and ecosystems services blue or red depicts whether an observed impact is positive (beneficial) or negative (adverse), respectively, to the given system or service. Cells assigned 'increase and decrease' indicate that within that region, both increases and decreases of physical changes are found, but are not necessarily equal; the same holds for cells showing 'positive and negative' attributable impacts. For ocean regions, the confidence level refers to the confidence in attributing observed changes to changes in greenhouse gas forcing for physical changes and to climate change for ecosystem, human systems, and ecosystem services. No assessment means: not applicable, not assessed at regional scale, or the evidence is insufficient for assessment. The physical changes in the ocean are defined as: Temperature change in 0–700 m layer of the ocean except for Southern Ocean (0–2000 m) and Arctic Ocean (upper mixed layer and major inflowing branches); Oxygen in the 0–1200 m layer or oxygen minimum layer; Ocean pH as surface pH (decreasing pH corresponds to increasing ocean acidification). Ecosystems in the ocean: Coral refers to warm-water coral reefs and cold-water corals. The 'upper water column' category refers to epipelagic zone for all ocean regions except Polar Regions, where the impacts on some pelagic organisms in open water deeper than the upper 200 m were included. Coastal wetland includes salt marshes, mangroves and seagrasses. Kelp forests are habitats of a specific group of macroalgae. Rocky shores are coastal habitats dominated by immobile calcified organisms such as mussels and barnacles. Deep sea is seafloor ecosystems that are 3000–6000 m deep. Sea-ice associated includes ecosystems in, on and below sea ice. Habitat services refer to supporting structures and services (e.g., habitat, biodiversity, primary production). Coastal Carbon Sequestration refers to the uptake and storage of carbon by coastal blue carbon ecosystems. Impacts on tourism refer to the operating conditions for the tourism sector. Cultural services include cultural identity, sense of home, and spiritual and intrinsic and aesthetic values. The underlying information for ocean regions in Tables SM5.10, SM5.11, SM3.8, SM3.9, and SM3.10. {3.2.1; 3.2.3; 3.2.4; 3.3.3; 3.4.1; 3.4.3; 3.5.2; Box 3.4, 4.2.2, 5.2.2, 5.2.3, 5.3.3, 5.4, 5.6, Figure 5.24, Box 5.3}

communities are the strongest, had increased risk, and none saw a risk reduction from a warming climate (*high confidence*).

The observed and projected changes in the open ocean and coastal seas have consequences on human communities and affect all aspects of well-being and have social, economic and environmental costs (Section 5.4, *high confidence*). The range and diversity of impacts is striking, with varying consequences for the wider community when analysed across the key marine ecosystems services. These consequences clearly affect the capacity for human society to achieve the SDGs (e.g., Figure 5.22). The evidence of climate change in the ocean is a pervasive thread through all types of coupled human-natural systems and projections amplify these observed impacts with the least impact from lower emission scenarios.

Risk-reduction responses and their governance through adaptation at the local scale are the most common responses to climate change from ocean systems (Section 5.5.2). It is clear that there are many choices for reducing risk of climate change. Many of the actions have benefits and relatively few dis-benefits, while others have large dis-benefits and marginal effectiveness (Section 5.5.1, Figure 5.23, Table 5.7 and Table 5.8). Many of the risk reduction approaches are limited in their capacity to reduce the risks of climate change, or are at best temporary solutions, which is a significant challenge to adapting to climate change (*high confidence*). In particular, the effectiveness of the assessed risk reduction measures are minimal under high greenhouse gas emission scenarios, highlighting the critical importance of mitigation. The assessment points to the increased effectiveness and importance of a portfolio of different types of mitigation and adaptation options. Governance is also a critical element in the portfolio of options and occurs at local, national and international scales. Such responses can be more effective with the support of scientific information, Local knowledge and Indigenous knowledge, and the consideration of local context and the inclusion of stakeholders.

5.7   Key Uncertainties and Gaps

This chapter was designed around three guiding questions (Section 5.1). These guiding questions mean that the report covers both regional and global scales of the ocean and many aspects of human systems, including governance and institutions, and adaptation pathways for dependent communities. This assessment is new linking together a broad and complex set of ocean disciplines and therefore also provides a unique perspective on key uncertainties and gaps in these systems. These gaps limit the extent of the assessments that were possible in this report. Notable outstanding uncertainties and gaps from this assessment include the following:

Physical and biogeochemical processes: While the Earth system is better monitored and the relevant data are more accessible than the other areas of assessment there is considerable room to improve these capabilities. For example, gaps remain in predictive modelling of climate change in coastal areas, deep ocean temperature and salinity measurements for sea level and closure of the energy budget, and oxygen and carbon measurements dense enough to measure

deoxygenation of the world ocean and track the mechanisms driving the ocean carbon cycle. Our capacity to understand and model net primary productivity and the rates of carbon burial in coastal sediments are also significant weaknesses. Projections of future changes in the Earth system depend on the use of ESMs, in which there are uncertainties arising from physical or ecological processes that are either omitted or incompletely understood. Most ESMs still rely on relatively simple representations of ocean biogeochemical cycling and the linkages to ocean ecosystem structure and function (Section 5.2.3). Other examples of under-assessed biogeochemical process in the ocean that may have implications for the Earth system under climate change include the fate of methane in the deep ocean (Section 5.2.4). Open ocean primary productivity and its projections requires critical corroborating measurements and improved understanding of its drivers to project changes in ocean productivity with higher confidence (Sections 5.2.2 and 5.2.3).

Biological processes and monitoring: There are a number of marine environments (e.g., on the deep sea floor) and ecosystem components (e.g., viruses and protists) where insufficient scientific understanding limits the assessments of risks to *low confidence or no assessment*. Examples of gaps include the narrow range of climate and non-climatic hazards and their interactions in simulation models, the linkages between single organisms to communities of organisms, knowledge of climate feedbacks in biological systems (Section 5.3.4, Section 5.2.4), and the capacity and limits of biological adaptation for many ecosystems (Section 5.2.2, 5.2.3, 5.2.4, 5.3). Increasing observational capacity can help provide the data to improve understanding and modelling of these important biophysical responses to climate change.

Variance in human systems and effectiveness of responses: The wide range of contributing factors (physical, social and economic) that interact with localised climate projections make projecting site-specific costs of impacts and benefits of adaptation difficult. There were few examples in the literature evaluating implemented adaptation actions, and there was *low confidence* in their reliability and provenance, thus largely precluding any assessments of their cost effectiveness. This lack of evidence on costs and benefits particularly affected assessments in Section 5.4 and 5.5. Adaptation responses to climate change have been undertaken by communities, industry and governments. However, their effectiveness for mitigating the risks of climate change (e.g., different types of adaptation response on the coasts, Section 5.5.2) is largely unassessed here, and consequently precludes a global understanding of the capacity in the world to address the risks of climate change in coastal seas, open ocean and the deep sea. A partial solution would be establishing an appropriate ocean and coasts database, including costs-benefits, for these types of studies.

FAQ 5.1 | How is life in the sea affected by climate change?

Climate change poses a serious threat to life in our seas, including coral reefs and fisheries, with impacts on marine ecosystems, economies and societies, especially those most dependent upon natural resources. The risk posed by climate change can be reduced by limiting global warming to no more than 1.5°C.

Life in most of the global ocean, from pole to pole and from sea surface to the abyssal depths, is already experiencing higher temperatures due to human-driven climate change. In many places, that increase may be barely measurable. In others, particularly in near-surface waters, warming has already had dramatic impacts on marine animals, plants and microbes. Due to closely linked changes in seawater chemistry, less oxygen remains available (in a process called ocean deoxygenation). Seawater contains more dissolved carbon dioxide, causing ocean acidification. Non-climatic effects of human activities are also ubiquitous, including over-fishing and pollution. Whilst these stressors and their combined effects are likely to be harmful to almost all marine organisms, food-webs and ecosystems, some are at greater risk (FAQ5.1, Figure 1). The consequences for human society can be serious unless sufficient action is taken to constrain future climate change.

Warm water coral reefs host a wide variety of marine life and are very important for tropical fisheries and other marine and human systems. They are particularly vulnerable, since they can suffer high mortalities when water temperatures persist above a threshold of between 1°C–2°C above the normal range. Such conditions occurred in many tropical seas between 2015 and 2017 and resulted in extensive coral bleaching, when the coral animal hosts ejected the algal partners upon which they depend. After mass coral mortalities due to bleaching, reef recovery typically takes at least 10–15 years. Other impacts of climate change include SLR, acidification and reef erosion. Whilst some coral species are more resilient than others, and impacts vary between regions, further reef degradation due to future climate change now seems inevitable, with serious consequences for other marine and coastal ecosystems, like loss of coastal protection for many islands and low-lying areas and loss of the high biodiversity these reefs host. Coral habitats can also occur in deeper waters and cooler seas, and more research is needed to understand impacts in these reefs. Although these cold water corals are not at risk from bleaching, due to their cooler environment, they may weaken or dissolve under ocean acidification, and other ocean changes.

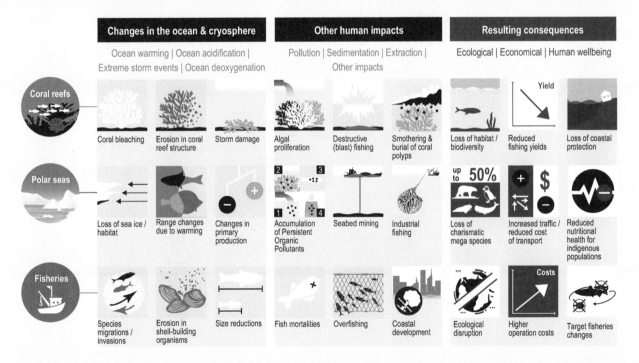

FAQ5.1, Figure 1 | Summary schematic of the impacts and resulting consequences of climate change (warming, acidification, storminess and deoxygenation) and other human impacts, on coral reefs, polar seas and fisheries, discussed in this FAQ.

Mobile species, such as fish, may respond to climate change by moving to more favorable regions, with populations shifting poleward or to deeper water, to find their preferred range of water temperatures or oxygen levels. As a result, projections of total future fishery yields under different climate change scenarios only show a moderate decrease of around 4% (~3.4 million tonnes) per degree Celsius warming. However, there are dramatic regional variations. With high levels of climate change, fisheries in tropical regions could lose up to half of their current catch levels by the end of this century. Polar catch levels may increase slightly, although the extent of such gains is uncertain, because fish populations that are currently depleted by overfishing and subject to other stressors may not be capable of migrating to polar regions, as assumed in models.

In polar seas, species adapted to life on or under sea ice are directly threatened by habitat loss due to climate change. The Arctic and Southern Oceans are home to a rich diversity of life, from tiny plankton to fish, krill and seafloor invertebrates to whales, seals, polar bears or penguins. Their complex interactions may be altered if new warmer-water species extend their ranges as sea temperatures rise. The effects of acidification on shelled organisms, as well as increased human activities (e.g., shipping) in ice-free waters, can amplify these disruptions.

Whilst some climate change impacts (like possible increased catch levels in polar regions) may benefit humans, most will be disruptive for ecosystems, economies and societies, especially those that are highly dependent upon natural resources. However, the impacts of climate change can be much reduced if the world as a whole, through inter-governmental interventions, manages to limit global warming to no more than 1.5°C.

References

Ab Lah, R., B.P. Kelaher, D. Bucher and K. Benkendorff, 2018: Ocean warming and acidification affect the nutritional quality of the commercially-harvested turbinid snail Turbo militaris. *Mar. Environ. Res.*, **141**, 100–108, doi:10.1016/j.marenvres.2018.08.009.

Abadi, F., C. Barbraud and O. Gimenez, 2017: Integrated population modeling reveals the impact of climate on the survival of juvenile emperor penguins. *Global Change Biol.*, **23**(3), 1353–1359, doi:10.1111/gcb.13538.

Abanades, J., D. Greaves and G. Iglesias, 2015: Coastal defence using wave farms: The role of farm-to-coast distance. *Renew. Energ.*, **75**, 572–582, doi:10.1016/j.renene.2014.10.048.

Abedin, M.A., U. Habiba and R. Shaw, 2014: Community Perception and Adaptation to Safe Drinking Water Scarcity: Salinity, Arsenic, and Drought Risks in Coastal Bangladesh. *Int. J. Disast. Risk Sci.*, **5**(2), 110–124, doi:10.1007/s13753-014-0021-6.

Abedin, M.A. and R. Shaw, 2015: The role of university networks in disaster risk reduction: Perspective from coastal Bangladesh. *Int. J. Disast. Risk Reduc.*, **13**, 381–389, doi:10.1016/j.ijdrr.2015.08.001.

Abelshausen, B., T. Vanwing and W. Jacquet, 2015: Participatory integrated coastal zone management in Vietnam: Theory versus practice case study: Thua Thien Hue province. *Journal of Marine and Island Cultures*, **4**(1), 42–53, doi:10.1016/j.imic.2015.06.004.

Abraham, J.P. et al., 2013: A review of global ocean temperature observations: Implications for ocean heat content estimates and climate change. *Rev. Geophys.*, **51**(3), 450–483, doi:10.1002/rog.20022.

Abrahms, B. et al., 2017: Climate mediates the success of migration strategies in a marine predator. *Ecol. Lett.*, **14**, 21: 63–71, doi:10.1111/ele.12871.

Abreu, P.C., J. Marangoni and C. Odebrecht, 2017: So close, so far: differences in long-term chlorophyll a variability in three nearby estuarine-coastal stations. *Mar. Biol. Res.*, **13**(1), 9–21, doi:10.1080/17451000.2016.1189081.

Ackerman, F. and E.A. Stanton, 2007: *The cost of climate change: what we'll pay if global warming continues unchecked.*, Natural Resources Defense Council, New York. 33 pp. https://www.nrdc.org/sites/default/files/cost.pdf.

Adam, P., 2002: Saltmarshes in a time of change. *Environ. Conserv.*, **29**(1), 39–61.

Adam, P., 2019: Salt marsh restoration. In: *Coastal Wetlands*. [G.M.E. Perillo, E. Wolanski, D.R. Cahoon, C.S. Hopkinson, eds.]Elsevier, Amsterdam, Netherlands, pp. 817–861. ISBN: 978-0-444-63893-9.

Adame, M.F., S. Cherian, R. Reef and B. Stewart-Koster, 2017: Mangrove root biomass and the uncertainty of belowground carbon estimations. *Forest Ecol. Manag.*, **403**, 52–60, doi:10.1016/j.foreco.2017.08.016.

Adams, C.A., J.E. Andrews and T. Jickells, 2012: Nitrous oxide and methane fluxes vs. carbon, nitrogen and phosphorous burial in new intertidal and saltmarsh sediments. *Sci. Total Environ.*, **434**, 240–251, doi:10.1016/j.scitotenv.2011.11.058.

Adams, D.K. et al., 2011: Surface-generated mesoscale eddies transport deep sea products from hydrothermal vents. *Science*, **332**(6029), 580–583, doi:10.1126/science.1201066.

Addamo, A.M. et al., 2016: Merging scleractinian genera: the overwhelming genetic similarity between solitary *Desmophyllum* and colonial *Lophelia*. *BMC Evol. Biol.*, **16**(1), 108.

Addino, M.S. et al., 2019: Growth changes of the stout razor clam Tagelus plebeius (Lightfoot, 1786) under different salinities in SW Atlantic estuaries. *J. Sea Res.*, **146**, 14–23, doi:10.1016/j.seares.2019.01.005.

Adger, W.N. et al., 2012: Cultural dimensions of climate change impacts and adaptation. *Nat. Clim. Change*, **3**, 112, doi:10.1038/nclimate1666.

Adloff, F. et al., 2015: Mediterranean Sea response to climate change in an ensemble of twenty first century scenarios. *Clim. Dyn.*, **45**(9), 2775–2802, doi:10.1007/s00382-015-2507-3.

Agostini, S. et al., 2018: Ocean acidification drives community shifts towards simplified non-calcified habitats in a subtropical–temperate transition zone. *Sci. Rep.*, **8**(1), 11354, doi:10.1038/s41598-018-29251-7.

Ahmed, N., W.W.L. Cheung, S. Thompson and M. Glaser, 2017: Solutions to blue carbon emissions: Shrimp cultivation, mangrove deforestation and climate change in coastal Bangladesh. *Mar. Policy*, **82**, 68–75, doi:10.1016/j.marpol.2017.05.007.

Ahmed, N. and J.S. Diana, 2015a: Coastal to inland: Expansion of prawn farming for adaptation to climate change in Bangladesh. *Aquacult. Rep.*, **2**, 67–76, doi:10.1016/j.aqrep.2015.08.001.

Ahmed, N. and J.S. Diana, 2015b: Threatening "white gold": Impacts of climate change on shrimp farming in coastal Bangladesh. *Ocean Coast. Manage.*, **114**, 42–52, doi:10.1016/j.ocecoaman.2015.06.008.

Ahmed, N. and M. Glaser, 2016a: Can "Integrated Multi-Trophic Aquaculture (IMTA)" adapt to climate change in coastal Bangladesh? *Ocean Coast. Manage.*, **132**, 120–131, doi:10.1016/j.ocecoaman.2016.08.017.

Ahmed, N. and M. Glaser, 2016b: Coastal aquaculture, mangrove deforestation and blue carbon emissions: Is REDD+ a solution? *Mar. Policy*, **66**, 58–66, doi:10.1016/j.marpol.2016.01.011.

Akselman, R. et al., 2015: Protoceratium reticulatum (Dinophyceae) in the austral Southwestern Atlantic and the first report on YTX-production in shelf waters of Argentina. *Harmful Algae*, **45**, 40–52.

Alam, G.M.M., K. Alam and S. Mushtaq, 2016: Influence of institutional access and social capital on adaptation decision: Empirical evidence from hazard-prone rural households in Bangladesh. *Ecol. Econ.*, **130**, 243–251, doi:10.1016/j.ecolecon.2016.07.012.

Alava, J.J., W.W.L. Cheung, P.S. Ross and U.R. Sumaila, 2017: Climate change-contaminant interactions in marine food webs: Toward a conceptual framework. *Global Change Biol.*, **23**(10), 3984–4001, doi:10.1111/gcb.13667.

Alava, J.J., A.M. Cisneros-Montemayor, U.R. Sumaila and W.W.L. Cheung, 2018: Projected amplification of food web bioaccumulation of MeHg and PCBs under climate change in the Northeastern Pacific. *Sci. Rep.*, **8**(1), 13460, doi:10.1038/s41598-018-31824-5.

Albright, R. et al., 2016a: Ocean acidification: Linking science to management solutions using the Great Barrier Reef as a case study. *J. Environ. Manage.*, **182**, 641–650, doi:10.1016/j.jenvman.2016.07.038.

Albright, R. et al., 2016b: Reversal of ocean acidification enhances net coral reef calcification. *Nature*, **531**, 362, doi:10.1038/nature17155.

Alderson-Day, B., S. McCarthy-Jones and C. Fernyhough, 2015: Hearing voices in the resting brain: A review of intrinsic functional connectivity research on auditory verbal hallucinations. *Neurosci. Biobehav. Rev.*, **55**(Supplement C), 78–87, doi:10.1016/j.neubiorev.2015.04.016.

Alford, M.H. et al., 2013: Turbulent mixing and hydraulic control of abyssal water in the Samoan Passage. *Geophys. Res. Lett.*, **40**(17), 4668–4674, doi:10.1002/grl.50684.

Alford, M.H., J.A. MacKinnon, H.L. Simmons and J.D. Nash, 2016: Near-Inertial Internal Gravity Waves in the Ocean. *Annu. Rev. Mar. Sci.*, **8**(1), 95–123, doi:10.1146/annurev-marine-010814-015746.

Alguero-Muniz, M. et al., 2017: Ocean acidification effects on mesozooplankton community development: Results from a long-term mesocosm experiment. *PLoS One*, **12**(4), doi:10.1371/journal.pone.0175851.

Alguero-Muniz, M. et al., 2016: Withstanding multiple stressors: ephyrae of the moon jellyfish (Aurelia aurita, Scyphozoa) in a high-temperature, high-CO_2 and low-oxygen environment. *Mar. Biol.*, **163**(9), doi:10.1007/s00227-016-2958-z.

Alin, S.R. et al., 2012: Robust empirical relationships for estimating the carbonate system in the southern California Current System and application to CalCOFI hydrographic cruise data (2005–2011). *J. Geophys. Res-Oceans*, **117**(C5), doi:10.1029/2011JC007511.

Allen, S.E. and X. Durrieu de Madron, 2009: A review of the role of submarine canyons in deep-ocean exchange with the shelf. *Ocean Sci.*, **5**(4), 607–620, doi:10.5194/os-5-607-2009.

5

Allison, E.H. et al., 2009: Vulnerability of national economies to the impacts of climate change on fisheries. *Fish Fish.*, **10**(2), 173–196, doi:10.1111/j.1467-2979.2008.00310.x.

Almahasheer, H. et al., 2017: Low Carbon sink capacity of Red Sea mangroves. *Sci. Rep.*, **7**(1), 9700, doi:10.1038/s41598-017-10424-9.

Almén, A.-K., A. Vehmaa, A. Brutemark and J. Engström-Öst, 2014: Coping with climate change? Copepods experience drastic variations in their physicochemical environment on a diurnal basis. *J. Exp. Mar. Biol. Ecol.*, **460**, 120–128, doi:10.1016/j.jembe.2014.07.001.

Alongi, D.M., 2015: The impact of climate change on mangrove forests. *Curr. Clim. Change Rep.*, **1**(1), 30–39.

Alongi, D.M., 2018: *Blue Carbon: Coastal Sequestration for Climate Change Mitigation*. Springer, Cham, Switzerland. ISBN: 978-3-319-91697-2.

Alongi, D.M. and S.K. Mukhopadhyay, 2015: Contribution of mangroves to coastal carbon cycling in low latitude seas. *Agric. For. Meteorol.*, **213**, 266–272, doi:10.1016/j.agrformet.2014.10.005.

Alter, S.E. et al., 2015: Climate impacts on transocean dispersal and habitat in gray whales from the Pleistocene to 2100. *Mol. Ecol.*, **24**(7), 1510–1522, doi:10.1111/mec.13121.

Altieri, A.H. and K.B. Gedan, 2015: Climate change and dead zones. *Global Change Biol.*, **21**(4), 1395–1406.

Álvarez, G. et al., 2019: Paralytic Shellfish Toxins in Surf Clams Mesodesma donacium during a Large Bloom of Alexandrium catenella Dinoflagellates Associated to an Intense Shellfish Mass Mortality. *Toxins*, **11**(4), 188. doi:10.3390/toxins11040188.

Álvarez-Romero, J.G. et al., 2018: Designing connected marine reserves in the face of global warming. *Global Change Biol.*, **24**(2), e671–e691, doi:10.1111/gcb.13989.

Amaya, O. et al., 2018: Large-Scale sea turtle mortality events in El Salvador attributed to paralytic shellfish toxin-producing algae blooms. *Front. Mar. Sci.*, **5**(411), doi:10.3389/fmars.2018.00411.

Ancona, S. and H. Drummond, 2013: Life History Plasticity of a Tropical Seabird in Response to El Nino Anomalies during Early Life. *PLoS One*, **8**(9), doi:10.1371/journal.pone.0072665.

Andersen, G.S., M.F. Pedersen and S.L. Nielsen, 2013: Temperature Acclimation and Heat Tolerance of Photosynthesis in Norwegian Saccharina Latissima (Laminariales, Phaeophyceae). *J. Phycol.*, **49**(4), 689–700, doi:10.1111/jpy.12077.

Anderson, C.R. et al., 2015: Living with harmful algal blooms in a changing world: strategies for modeling and mitigating their effects in coastal marine ecosystems. In Castal and Marine Hazards, Risks, and Disasters [J.F. Shroder, J.T. Ellis, D.J. Sherman eds.] Elsevier BV, Amsterdam, pp. 495–561. ISBN: 978-0-12-396483-0.

Andrachuk, M. and D. Armitage, 2015: Understanding social-ecological change and transformation through community perceptions of system identity. *Ecol. Soc.*, **20**(4). 26. http://dx.doi.org/10.5751/ES-07759-200426

Aqorau, T., J. Bell and J.N. Kittinger, 2018: Good governance for migratory species. *Science*, **361**(6408), 1208, doi:10.1126/science.aav2051.

Araos, M. et al., 2016: Climate change adaptation planning in large cities: A systematic global assessment. *Environ. Sci. Policy*, **66**, 375–382, doi:10.1016/j.envsci.2016.06.009.

Araujo, R.M. et al., 2016: Status, trends and drivers of kelp forests in Europe: an expert assessment. *Biodivers. Conserv.*, **25**(7), 1319–1348, doi:10.1007/s10531-016-1141-7.

Arbic, B.K., R.H. Karsten and C. Garrett, 2009: On tidal resonance in the global ocean and the back-effect of coastal tides upon open-ocean tides. *Atmos. Ocean*, **47**(4), 239–266, doi:10.3137/OC311.2009.

Archer, D. et al., 2014: Moving towards inclusive urban adaptation: approaches to integrating community-based adaptation to climate change at city and national scale. *Clim. Dev.*, **6**(4), 345–356, doi:10.1080/17565529.2014.918868.

Arellano, S.M. et al., 2014: Larvae from deep sea methane seeps disperse in surface waters. *Proc. Roy. Soc. B. Biol.*, **281**(1786), 20133276, doi:10.1098/rspb.2013.3276.

Arévalo-Martínez, D.L. et al., 2015: Massive nitrous oxide emissions from the tropical South Pacific Ocean. *Nat. Geosci.*, **8**, 530, doi:10.1038/ngeo2469.

Arias-Ortiz, A. et al., 2018: A marine heatwave drives massive losses from the world's largest seagrass carbon stocks. *Nat. Clim. Change*, **8**, 338–344.

Arístegui, J., M. Gasol Josep, M. Duarte Carlos and J. Herndld Gerhard, 2009: Microbial oceanography of the dark ocean's pelagic realm. *Limnol. Oceanogr.*, **54**(5), 1501–1529, doi:10.4319/lo.2009.54.5.1501.

Armitage, A.R., W.E. Highfield, S.D. Brody and P. Louchouarn, 2015: The Contribution of Mangrove Expansion to Salt Marsh Loss on the Texas Gulf Coast. *PLoS One*, **10**(5), e0125404, doi:10.1371/journal.pone.0125404.

Armour, K.C. et al., 2016: Southern Ocean warming delayed by circumpolar upwelling and equatorward transport. *Nat. Geosci.*, **9**(7), 549.

Armstrong, C.W., N. Foley, R. Tinch and S. van den Hove, 2010: Ecosystem goods and services of the deep sea. *Deliverable D6*, Universititet i Tromsø, Tromsø, 68 pp. https://www.pik-potsdam.de/news/public-events/archiv/alter-net/former-ss/2010/13.09.2010/van_den_hove/d6-2-final.pdf.

Armstrong, C.W., N.S. Foley, R. Tinch and S. van den Hove, 2012: Services from the deep: Steps towards valuation of deep sea goods and services. *Ecosyst. Serv.*, **2**, 2–13, doi:10.1016/j.ecoser.2012.07.001.

Arntz, W.E. et al., 2006: El Niño and similar perturbation effects on the benthos of the Humboldt, California, and Benguela Current upwelling ecosystems. *Adv. Geosci.*, **6**, 243–265, doi:10.5194/adgeo-6-243-2006.

Arroyo Mina, J.S., D.A. Revollo Fernandez, A. Aguilar Ibarra and N. Georgantzis, 2016: Economic behavior of fishers under climate-related uncertainty: Results from field experiments in Mexico and Colombia. *Fish. Res.*, **183**, 304–317, doi:10.1016/j.fishres.2016.05.020.

Asadabadi, A. and E. Miller-Hooks, 2017: Assessing strategies for protecting transportation infrastructure from an uncertain climate future. *Transport. Res. A-Pol.*, **105**, 27–41.

Ascani, F. et al., 2016: Juvenile recruitment in loggerhead sea turtles linked to decadal changes in ocean circulation. *Global Change Biol.*, **22**(11), 3529–3538, doi:10.1111/gcb.13331.

Asch, R.G., 2015: Climate change and decadal shifts in the phenology of larval fishes in the California Current ecosystem. *PNAS*, **112**(30), E4065–E4074, doi:10.1073/pnas.1421946112.

Asch, R.G., W.W.L. Cheung and G. Reygondeau, 2018: Future marine ecosystem drivers, biodiversity, and fisheries maximum catch potential in Pacific Island countries and territories under climate change. *Mar. Policy*, **88**, 285–294, doi:10.1016/j.marpol.2017.08.015.

Ashbolt, N.J., 2019: Flood and Infectious Disease Risk Assessment. In: *Health in Ecological Perspectives in the Anthropocene* [t. Watanabe, C. Watanabe eds]. Springer, Singapore. pp. 145–159. ISBN: 978-981-13-2525-0.

Assis, J., M.B. Araújo and E.A. Serrão, 2018: Projected climate changes threaten ancient refugia of kelp forests in the North Atlantic. *Global Change Biol.*, **24**(1), e55–e66, doi:10.1111/gcb.13818.

Assis, J., A.V. Lucas, I. Barbara and E.A. Serrao, 2016: Future climate change is predicted to shift long-term persistence zones in the cold-temperate kelp Laminaria hyperborea. *Mar. Environ. Res.*, **113**, 174–182, doi:10.1016/j.marenvres.2015.11.005.

Astariz, S., C. Perez-Collazo, J. Abanades and G. Iglesias, 2015: Towards the optimal design of a co-located wind-wave farm. *Energy*, **84**, 15–24, doi:10.1016/j.energy.2015.01.114.

Ataur Rahman, M. and S. Rahman, 2015: Natural and traditional defense mechanisms to reduce climate risks in coastal zones of Bangladesh. *Weather and Climate Extremes*, **7**, 84–95, doi:10.1016/j.wace.2014.12.004.

Atwood, T.B. et al., 2017: Global patterns in mangrove soil carbon stocks and losses. *Nat. Clim. Change*, **7**, 523, doi:10.1038/nclimate3326.

Audefroy, J.F. and B.N.C. Sánchez, 2017: Integrating local knowledge for climate change adaptation in Yucatán, Mexico. *Int. J. Sustain. Built Environ.*, **6**(1), 228–237, doi:10.1016/j.ijsbe.2017.03.007.

5

Aumont, O. and L. Bopp, 2006: Globalizing results from ocean in situ iron fertilization studies. *Global Biogeochem. Cy.*, **20**(2), doi:10.1029/2005GB002591.

Aumont, O., O. Maury, S. Lefort and L. Bopp, 2018: Evaluating the Potential Impacts of the Diurnal Vertical Migration by Marine Organisms on Marine Biogeochemistry. *Global Biogeochem. Cy.*, **32**(11), 1622–1643, doi:10.1029/2018GB005886.

Avelar, S., T.S. van der Voort and T.I. Eglinton, 2017: Relevance of carbon stocks of marine sediments for national greenhouse gas inventories of maritime nations. *Carbon Bal. Manage.*, **12**(1), 10, doi:10.1186/s13021-017-0077-x.

Aylett, A., 2015: Institutionalizing the urban governance of climate change adaptation: Results of an international survey. *Urban Clim.*, **14**, 4–16, doi:10.1016/j.uclim.2015.06.005.

Azad, A.K., K.M. Hossain and M. Nasreen, 2013: Flood-induced vulnerabilities and problems encountered by women in northern Bangladesh. *Int. J. Disast. Risk Sci.*, **4**(4), 190–199, doi:10.1007/s13753-013-0020-z.

Aziz, A.A., S. Thomas, P. Dargusch and S. Phinn, 2016: Assessing the potential of REDD+ in a production mangrove forest in Malaysia using stakeholder analysis and ecosystem services mapping. *Mar. Policy*, **74**, 6–17, doi:10.1016/j.marpol.2016.09.013.

Bachman, S.D., J.R. Taylor, K.A. Adams and P.J. Hosegood, 2017: Mesoscale and Submesoscale Effects on Mixed Layer Depth in the Southern Ocean. *J. Phys. Oceanogr.*, **47**(9), 2173–2188, doi:10.1175/JPO-D-17-0034.1.

Baco, A.R. et al., 2017: Defying dissolution: discovery of deep sea scleractinian coral reefs in the North Pacific. *Sci. Rep.*, **7**(1), 5436.

Bae, Y.H., K.O. Kim and B.H. Choi, 2010: Lake Sihwa tidal power plant project. *Ocean Eng.*, **37**(5), 454–463, doi:10.1016/j.oceaneng.2010.01.015.

Baez, J.C. et al., 2011: The North Atlantic Oscillation and sea surface temperature affect loggerhead abundance around the Strait of Gibraltar. *Sci. Mar.*, **75**(3), 571–575, doi:10.3989/scimar.2011.75n3571.

Baggini, C., Y. Issaris, M. Salomidi and J. Hall-Spencer, 2015: Herbivore diversity improves benthic community resilience to ocean acidification. *J. Exp. Mar. Biol. Ecol.*, **469**, 98–104, doi:10.1016/j.jembe.2015.04.019.

Bailey, A. et al., 2016: Early life stages of the Arctic copepod Calanus glacialisare unaffected by increased seawater pCO₂. *ICES J. Mar. Sci.*, **74**(4), 996–1004 doi:10.1093/icesjms/fsw066.

Baker-Austin, C., J. Trinanes, N. Gonzalez-Escalona and J. Martinez-Urtaza, 2017: Non-Cholera Vibrios: The Microbial Barometer of Climate Change. *Trends Microbiol.*, **25**(1), 76–84, doi:10.1016/j.tim.2016.09.008.

Baker-Austin, C. et al., 2013: Emerging Vibrio risk at high latitudes in response to ocean warming. *Nat. Clim. Change*, **3**(1), 73–77, doi:10.1038/NCLIMATE1628.

Balmford, A. et al., 2004: The worldwide costs of marine protected areas. *PNAS*, **101**(26), 9694–9697, doi:10.1073/pnas.0403239101.

Bambridge, T. and P.Y. Le Meur, 2018: Savoirs locaux et biodiversité aux îles Marquises: don, pouvoir et perte. *Revue d'anthropologie et des connaissances*, **12**(1), 29–55.

Ban, S.S., N.A.J. Graham and S.R. Connolly, 2014: Evidence for multiple stressor interactions and effects on coral reefs. *Global Change Biol.*, **20**(3), 681–697, doi:10.1111/gcb.12453.

Banse, K., 1968: Hydrography of the Arabian Sea Shelf of India and Pakistan and effects on demersal fishes. *Deep Sea Res. Pt. I*, **15**(1), 45–79, doi:10.1016/0011-7471(68)90028-4.

Barange, M., 2019: Avoiding misinterpretation of climate change projections of fish catches. *ICES J. Mar. Sci.*, doi:10.1093/icesjms/fsz061.

Barange, M. et al., 2017: The Cost of Reducing the North Atlantic Ocean Biological Carbon Pump. *Front. Mar. Sci.*, **3**, 290.

Barbier, E.B., 2015: Climate change impacts on rural poverty in low-elevation coastal zones. *Estuar. Coast. Shelf Sci.*, **165**, A1–A13, doi:10.1016/j.ecss.2015.05.035.

Barnes, R.S.K., 2017: Are seaward pneumatophore fringes transitional between mangrove and lower-shore system compartments? *Mar. Environ. Res.*, **125**, 99–109, doi:10.1016/j.marenvres.2017.01.008.

Barragán, J.M. and M. de Andrés, 2015: Analysis and trends of the world's coastal cities and agglomerations. *Ocean Coast. Manage.*, **114**, 11–20, doi:10.1016/j.ocecoaman.2015.06.004.

Barrett, J. et al., 2015: Development of an estuarine climate change monitoring program. *Ecol. Indic.*, **53**, 182–186, doi:10.1016/j.ecolind.2015.01.039.

Barry, J.P., C.H. Baxter, R.D. Sagarin and S.E. Gilman, 1995: Climate-related, long-term faunal changes in a california rocky intertidal community. *Science*, **267**(5198), 672–675, doi:10.1126/science.267.5198.672.

Barry, J.P. et al., 2004: Effects of Direct Ocean CO₂ Injection on Deep sea Meiofauna. *J. Oceanogr.*, **60**(4), 759–766, doi:10.1007/s10872-004-5768-8.

Barton, A. et al., 2015: Impacts of coastal acidification on the Pacific Northwest shellfish industry and adaptation strategies implemented in response. *Oceanography*, **28**(2), 146–159.

Barton, A.D., A.J. Irwin, Z.V. Finkel and C.A. Stock, 2016: Anthropogenic climate change drives shift and shuffle in North Atlantic phytoplankton communities. *PNAS*, **113**(11), 2964–2969, doi:10.1073/pnas.1519080113.

Barton, J.A., B.L. Willis and K.S. Hutson, 2017: Coral propagation: a review of techniques for ornamental trade and reef restoration. *Rev. Aquacult.*, **9**(3), 238–256, doi:10.1111/raq.12135.

Bartsch, I. et al., 2016: Changes in kelp forest biomass and depth distribution in Kongsfjorden, Svalbard, between 1996–1998 and 2012–2014 reflect Arctic warming. *Polar Biol.*, **39**(11), 2021–2036, doi:10.1007/s00300-015-1870-1.

Basset, A., M. Elliott, R.J. West and J.G. Wilson, 2013: Estuarine and lagoon biodiversity and their natural goods and services. *Estuar. Coast. Shelf Sci.*, **132**, 1–4, doi:10.1016/j.ecss.2013.05.018.

Batal, M. et al., 2017: Quantifying associations of the dietary share of ultra-processed foods with overall diet quality in First Nations peoples in the Canadian provinces of British Columbia, Alberta, Manitoba and Ontario. *Public Health Nutr.*, **21**(1), 103–113. doi:10.1017/S1368980017001677.

Battaglia, G. and F. Joos, 2018: Hazards of decreasing marine oxygen: the near-term and millennial-scale benefits of meeting the Paris climate targets. *Earth Syst. Dyn.*, **9**(2), 797.

Bauer, J.E. et al., 2013: The changing carbon cycle of the coastal ocean. *Nature*, **504**(7478), 61–70, doi:10.1038/nature12857.

Bauman, S.J. et al., 2014: Augmenting the biological pump: The shortcomings of geoengineered upwelling. *Oceanography*, **27**(3), 17–23.

Bayraktarov, E. et al., 2016: The cost and feasibility of marine coastal restoration. *Ecol. Appl.*, **26**(4), 1055–1074, doi:10.1890/15-1077.

Beaugrand, G., 2009: Decadal changes in climate and ecosystems in the North Atlantic Ocean and adjacent seas. *Deep Sea Res. Pt. II*, **56**(8–10), 656–673, doi:10.1016/j.dsr2.2008.12.022.

Beaugrand, G. et al., 2015: Future vulnerability of marine biodiversity compared with contemporary and past changes. *Nat. Clim. Change*, **5**(7), 695–701, doi:10.1038/nclimate2650.

Beaulieu, C. et al., 2013: Factors challenging our ability to detect long-term trends in ocean chlorophyll. *Biogeosciences*, **10**(4), 2711–2724, doi:10.5194/bg-10-2711-2013.

Beca-Carretero, P., B. Olesen, N. Marbà and D. Krause-Jensen, 2018: Response to experimental warming in northern eelgrass populations: comparison across a range of temperature adaptations. *Mar. Ecol. Prog. Ser.*, **589**, 59–72.

Beck, M.W. et al., 2011: Oyster Reefs at Risk and Recommendations for Conservation, Restoration, and Management. *BioScience*, **61**(2), 107–116, doi:10.1525/bio.2011.61.2.5.

Beck, M.W. et al., 2018: The global flood protection savings provided by coral reefs. *Nat. Commun.*, **9**(1), 2186, doi:10.1038/s41467-018-04568-z.

Becker, A., K.Y. Ng Adolf, D. McEvoy and J. Mullett, 2018: Implications of climate change for shipping: Ports and supply chains. *WiRes. Clim. Change*, **9**(2), e508, doi:10.1002/wcc.508.

Becker, A.H. et al., 2016: A method to estimate climate-critical construction materials applied to seaport protection. *Global Environ. Change*, **40**, 125–136.

5

Bednaršek, N., R.A. Feely, J.C.P. Reum, B. Peterson, J. Menkel, S.R. Alin, and B. Hales. "Limacina helicina shell dissolution as an indicator of declining habitat suitability owing to ocean acidification in the California Current Ecosystem." Proceedings of the Royal Society B: Biological Sciences 281, no. 1785 (2014): 20140123.

Beetham, E., P.S. Kench and S. Popinet, 2017: Future Reef Growth Can Mitigate Physical Impacts of Sea level Rise on Atoll Islands. *Earth's Future*, 5(10), 1002–1014, doi:10.1002/2017ef000589.

Belhabib, D. et al., 2018: Impacts of anthropogenic and natural "extreme events" on global fisheries. *Fish Fish.*, doi:10.1111/faf.12314.

Belhabib, D., V.W.Y. Lam and W.W.L. Cheung, 2016: Overview of West African fisheries under climate change: Impacts, vulnerabilities and adaptive responses of the artisanal and industrial sectors. *Mar. Policy*, 71(Supplement C), 15–28, doi:10.1016/j.marpol.2016.05.009.

Bell, J. et al., 2018a: Climate change impacts, vulnerabilities and adaptations: Western and Central Pacific Ocean marine fisheries. In: *Impacts of climate change on fisheries and aquaculture* [Barange, M., Bahri, T., Beveridge, M.C.M., Cochrane, K.L., Funge-Smith, S. & Poulain, F. (eds.)]. FAO Fisheries and Aquaculture Technical Paper T, FAO, Rome, Italy. 305–324. ISBN: 978-92-5-130607-9.

Bell, J.D. et al., 2018b: Adaptations to maintain the contributions of small-scale fisheries to food security in the Pacific Islands. *Mar. Policy*, 88, 303–314, doi:10.1016/j.marpol.2017.05.019.

Bell, T.W., J.G. Allen, K.C. Cavanaugh and D.A. Siegel, 2018c: Three decades of variability in California's giant kelp forests from the Landsat satellites. *Remote Sens. Environ.*, doi:10.1016/j.rse.2018.06.039.

Belmadani, A. et al., 2014: What dynamics drive future wind scenarios for coastal upwelling off Peru and Chile? *Clim. Dyn.*, 43(7), 1893–1914, doi:10.1007/s00382-013-2015-2.

Ben Rais Lasram, F. et al., 2010: The Mediterranean Sea as a 'cul-de-sac' for endemic fishes facing climate change. *Global Change Biol.*, 16(12), 3233–3245, doi:10.1111/j.1365-2486.2010.02224.x.

Bendtsen, J., J. Mortensen, K. Lennert and S. Rysgaard, 2015: Heat sources for glacial ice melt in a west Greenland tidewater outlet glacier fjord: The role of subglacial freshwater discharge. *Geophys. Res. Lett.*, 42(10), 4089–4095, doi:10.1002/2015GL063846.

Béné, C. et al., 2015: Feeding 9 billion by 2050 – Putting fish back on the menu. *Food Secur.*, 7(2), 261–274, doi:10.1007/s12571-015-0427-z.

Bennett, N.J., P. Dearden, G. Murray and A. Kadfak, 2014: The capacity to adapt?: communities in a changing climate, environment, and economy on the northern Andaman coast of Thailand. *Ecol. Soc.*, 19(2), doi:10.5751/ES-06315-190205.

Bennett, N.J., A. Kadfak and P. Dearden, 2016: Community-based scenario planning: a process for vulnerability analysis and adaptation planning to social–ecological change in coastal communities. *Environ. Dev. Sustain.*, 18(6), 1771–1799, doi:10.1007/s10668-015-9707-1.

Bennett, N.J. et al., 2018: Environmental Stewardship: A Conceptual Review and Analytical Framework. *Environ. Manage.*, 61(4), 597–614, doi:10.1007/s00267-017-0993-2.

Berdalet, E. et al., 2017: GlobalHAB: a new program to promote international research, observations, and modeling of harmful algal blooms in aquatic systems. *Oceanography*, 30(1), 70–81.

Berge, J. et al., 2014: Arctic complexity: a case study on diel vertical migration of zooplankton. *J. Plankton Res.*, 36(5), 1279–1297, doi:10.1093/plankt/fbu059.

Bernhard, J.M. and C.E. Reimers, 1991: Benthic foraminiferal population fluctuations related to anoxia: Santa Barbara Basin. *Biogeochemistry*, 15(2), 127–149, doi:10.1007/BF00003221.

Betzold, C. and I. Mohamed, 2017: Seawalls as a response to coastal erosion and flooding: a case study from Grande Comore, Comoros (West Indian Ocean). *Reg. Environ. Change*, 17(4), 1077–1087, doi:10.1007/s10113-016-1044-x.

Bianchi, D., C. Stock, E.D. Galbraith and J.L. Sarmiento, 2013: Diel vertical migration: Ecological controls and impacts on the biological pump in a one-dimensional ocean model. *Global Biogeochem. Cy.*, 27(2), 478–491, doi:10.1002/gbc.20031.

Biggs, D., C.C. Hicks, J.E. Cinner and C.M. Hall, 2015: Marine tourism in the face of global change: The resilience of enterprises to crises in Thailand and Australia. *Ocean Coast. Manage.*, 105, 65–74, doi:10.1016/j.ocecoaman.2014.12.019.

Billé, R. et al., 2013: Taking Action Against Ocean Acidification: A Review of Management and Policy Options. *Environ. Manage.*, 52(4), 761–779, doi:10.1007/s00267-013-0132-7.

Bindoff, N.L. et al., 2013: Detection and Attribution of Climate Change: from Global to Regional. In: Climate Change 2013: The Physical Science Basis. Contribution of Working Group I to the Fifth Assessment Report of the Intergovernmental Panel on Climate Change [Stocker, T.F., D. Qin, G.K. Plattner, M. Tignor, S.K. Allen, J. Boschung, A. Nauels, Y. Xia, V. Bex and P.M. Midgley (eds.)]. Cambridge University Press, Cambridge, United Kingdom and New York, NY, USA, 867–952.

Bjorndal, K.A. et al., 2017: Ecological regime shift drives declining growth rates of sea turtles throughout the West Atlantic. *Global Change Biol.*, 23(11), 4556–4568, doi:10.1111/gcb.13712.

Black, B.A. et al., 2014: Six centuries of variability and extremes in a coupled marine-terrestrial ecosystem. *Science*, 345(6203), 1498.

Blackford, J. et al., 2014: Detection and impacts of leakage from sub-seafloor deep geological carbon dioxide storage. *Nat. Clim. Change*, 4, 1011, doi:10.1038/nclimate2381.

Blanchard, J.L. et al., 2012: Potential consequences of climate change for primary production and fish production in large marine ecosystems. *Philos. Trans. Roy. Soc. B.*, 367(1605), 2979–2989.

Blanchard, J.L. et al., 2017: Linked sustainability challenges and trade-offs among fisheries, aquaculture and agriculture. *Nat. Ecol. Evol.*, 1(9), 1240–1249, doi:10.1038/s41559-017-0258-8.

Blankespoor, B., S. Dasgupta and B. Laplante, 2014: Sea level rise and coastal wetlands. *Ambio*, 43(8), 996–1005, doi:10.1007/s13280-014-0500-4.

Blasiak, R. et al., 2017: Climate change and marine fisheries: Least developed countries top global index of vulnerability. *PLoS One*, 12(6), e0179632, doi:10.1371/journal.pone.0179632.

Boetius, A. and F. Wenzhoefer, 2013: Seafloor oxygen consumption fuelled by methane from cold seeps. *Nat. Geosci.*, 6(9), 725–734, doi:10.1038/NGEO1926.

Bograd, S.J. et al., 2015: Changes in source waters to the Southern California Bight. *Deep Sea Res. Pt. II*, 112, 42–52, doi:10.1016/j.dsr2.2014.04.009.

Bongaerts, P. et al., 2017: Deep reefs are not universal refuges: Reseeding potential varies among coral species. *Sci. Adv.*, 3(2), e1602373, doi:10.1126/sciadv.1602373.

Bongiorni, L. et al., 2011: First step in the restoration of a highly degraded coral reef (Singapore) by in situ coral intensive farming. *Aquaculture*, 322–323(Supplement C), 191–200.

Booth, S. and D. Zeller, 2005: Mercury, food webs, and marine mammals: Implications of diet and climate change for human health. *Environ. Health. Perspect.*, 113(5), 521–526, doi:10.1289/ehp.7603.

Bopp, L. et al., 2013: Multiple stressors of ocean ecosystems in the 21st century: projections with CMIP5 models. *Biogeosciences*, 10(10), 6225–6245, doi:10.5194/bg-10-6225-2013.

Bopp, L. et al., 2017: Ocean (de)oxygenation from the Last Glacial Maximum to the twenty-first century: insights from Earth System models. *Philos. Trans. Roy. Soc. A.*, 375(2102), 20160323, doi:10.1098/rsta.2016.0323.

Borbor-Córdova, M.J. et al., 2018: Risk Perception of Coastal Communities and Authorities on Harmful Algal Blooms in Ecuador. *Front. Mar. Sci.*, 5(365), doi:10.3389/fmars.2018.00365.

5

Borchert, S.M., M.J. Osland, N.M. Enwright and K. Griffith, 2018: Coastal wetland adaptation to sea level rise: Quantifying potential for landward migration and coastal squeeze. *J. Appl. Ecol.*, **55**(6), 2876–2887, doi:10.1111/1365-2664.13169.

Borges, A.V. et al., 2016: Massive marine methane emissions from near-shore shallow coastal areas. *Sci. Rep.*, **6**, 27908, doi:10.1038/srep27908.

Borthwick, A.G.L., 2016: Marine Renewable Energy Seascape. *Engineering*, **2**(1), 69–78, doi:10.1016/J.ENG.2016.01.011.

Bost, C.A. et al., 2009: The importance of oceanographic fronts to marine birds and mammals of the southern oceans. *J. Mar. Syst.*, **78**(3), 363–376, doi:10.1016/j.jmarsys.2008.11.022.

Bost, C.A. et al., 2015: Large-scale climatic anomalies affect marine predator foraging behaviour and demography. *Nat. Commun.*, **6**, 8220, doi:10.1038/ncomms9220.

Botterell, Z.L.R. et al., 2019: Bioavailability and effects of microplastics on marine zooplankton: A review. *Environ. Pollut.*, **245**, 98–110, doi:10.1016/j.envpol.2018.10.065.

Bouma, T.J. et al., 2016: Short-term mudflat dynamics drive long-term cyclic salt marsh dynamics. *Limnol. Oceanogr.*, **61**(6), 2261–2275, doi:10.1002/lno.10374.

Bourne, D.G., K.M. Morrow and N.S. Webster, 2016: Insights into the Coral Microbiome: Underpinning the Health and Resilience of Reef Ecosystems. *Annu. Rev. Microbiol.*, **70**(1), 317–340, doi:10.1146/annurev-micro-102215-095440.

Bove, C.B. et al., 2019: Common Caribbean corals exhibit highly variable responses to future acidification and warming. *Proc. Roy. Soc. B.*, **286**(1900), 20182840.

Bowie, A.R. et al., 2015: Iron budgets for three distinct biogeochemical sites around the Kerguelen Archipelago (Southern Ocean) during the natural fertilisation study, KEOPS-2. *Biogeosciences*, **12**(14), 4421–4445.

Boxall, A.B.A. et al., 2009: Impacts of climate change on indirect human exposure to pathogens and chemicals from agriculture. *Environ. Health. Perspect.*, **117**(4), 508–514, doi:10.1289/ehp.0800084.

Boxhammer, T. et al., 2018: Enhanced transfer of organic matter to higher trophic levels caused by ocean acidification and its implications for export production: A mass balance approach. *PLoS One*, **13**(5), e0197502, doi:10.1371/journal.pone.0197502.

Boyce, D.G., M. Dowd, M.R. Lewis and B. Worm, 2014: Estimating global chlorophyll changes over the past century. *Progr. Oceanogr.*, **122**, 163–173, doi:10.1016/j.pocean.2014.01.004.

Boyce, D.G. and B. Worm, 2015: Patterns and ecological implications of historical marine phytoplankton change. *Mar. Ecol. Prog. Ser.*, **534**, 251–272, doi:10.3354/meps11411.

Boyd, P.W. and M. Bressac, 2016: Developing a test-bed for robust research governance of geoengineering: the contribution of ocean iron biogeochemistry. *Philos. Trans. Roy. Soc. A.*, **374**(2081).

Boyd, P.W. et al., 2019: Multi-faceted particle pumps drive carbon sequestration in the ocean. *Nature*, **568**(7752), 327–335, doi:10.1038/s41586-019-1098-2.

Boyd, P.W. et al., 2018: Experimental strategies to assess the biological ramifications of multiple drivers of global ocean change – A review. *Global Change Biol.*, **24**(6), 2239–2261, doi:10.1111/gcb.14102.

Boyd, P.W. et al., 2016: Biological responses to environmental heterogeneity under future ocean conditions. *Global Change Biol.*, **22**(8), 2633–2650, doi:10.1111/gcb.13287.

Boyd, P.W. et al., 2015a: Physiological responses of a Southern Ocean diatom to complex future ocean conditions. *Nat. Clim. Change*, **6**(2), 207–213, doi:10.1038/nclimate2811.

Boyd, P.W. and D.A. Hutchins, 2012: Understanding the responses of ocean biota to a complex matrix of cumulative anthropogenic change. *Mar. Ecol. Prog. Ser.*, **470**, 125–135.

Boyd, P.W. et al., 2007: Mesoscale Iron Enrichment Experiments 1993–2005: Synthesis and Future Directions. *Science*, **315**(5812), 612, doi:10.1126/science.1131669.

Boyd, P.W. et al., 2013: Marine Phytoplankton Temperature versus Growth Responses from Polar to Tropical Waters – Outcome of a Scientific Community-Wide Study. *PLoS One*, **8**(5), e63091, doi:10.1371/journal.pone.0063091.

Boyd, P.W. et al., 2015b: Why are biotic iron pools uniform across high- and low-iron pelagic ecosystems? *Global Biogeochem. Cy.*, **29**(7), 1028–1043, doi:10.1002/2014GB005014.

Brady, R.X., M.A. Alexander, N.S. Lovenduski and R.R. Rykaczewski, 2017: Emergent anthropogenic trends in California Current upwelling. *Geophys. Res. Lett.*, **44**(10), 5044–5052, doi:10.1002/2017GL072945.

Brannigan, L. et al., 2017: Submesoscale Instabilities in Mesoscale Eddies. *J. Phys. Oceanogr.*, **47**(12), 3061–3085, doi:10.1175/JPO-D-16-0178.1.

Bravo, J., F. Suárez, A. Ramírez and F. Acosta, 2015: Ciguatera, an emerging human poisoning in Europe. *J. Aquac. Mar. Biol*, **3**, 00053.

Breed, G.A. et al., 2017: Sustained disruption of narwhal habitat use and behavior in the presence of Arctic killer whales. *PNAS*, **114**(10), 2628–2633, doi:10.1073/pnas.1611707114.

Breitberg, D. et al., 2015: And on Top of All That… Coping with Ocean Acidification in the Midst of Many Stressors. *Oceanography*, **25**(2), 48–61, doi:10.5670/oceanog.2015.31.

Breitburg, D. et al., 2018: Declining oxygen in the global ocean and coastal waters. *Science*, **359**(6371).

Bridge, T.C.L. et al., 2013: Depth-dependent mortality of reef corals following a severe bleaching event: implications for thermal refuges and population recovery. *F1000Research*, **2**, 187, doi:10.12688/f1000research.2-187.v3.

Brierley, A.S. and M.J. Kingsford, 2009: Impacts of Climate Change on Marine Organisms and Ecosystems. *Curr. Biol.*, **19**(14), R602–R614, doi:10.1016/j.cub.2009.05.046.

Briscoe, D.K. et al., 2017: Ecological bridges and barriers in pelagic ecosystems. *Deep sea Res. Pt. II*, **140**, 182–192, doi:10.1016/j.dsr2.2016.11.004.

Britten, G.L., M. Dowd and B. Worm, 2016: Changing recruitment capacity in global fish stocks. *PNAS*, **113**(1), 134–139, doi:10.1073/pnas.1504709112.

Brooke, S. et al., 2013: Temperature tolerance of the deep sea coral Lophelia pertusa from the southeastern United States. *Deep sea Res. Pt. II*, **92**, 240–248.

Brooks, C.M., J.B. Weller, K. Gjerde and U.R. Sumaila, 2013: Challenging the right to fish in a fast-changing ocean. *Stan. Envtl. LJ*, **33**, 289.

Broto, V.C., E. Boyd and J. Ensor, 2015: Participatory urban planning for climate change adaptation in coastal cities: lessons from a pilot experience in Maputo, Mozambique. *Curr. Opin. Environ. Sustain.*, **13**, 11–18, doi:10.1016/j.cosust.2014.12.005.

Brown, A. and S. Thatje, 2014: The effects of changing climate on faunal depth distributions determine winners and losers. *Global Change Biol.*, **21**(1), 173–180, doi:10.1111/gcb.12680.

Brown, B., R. Dunne, M. Goodson and A. Douglas, 2002: Experience shapes the susceptibility of a reef coral to bleaching. *Coral Reefs*, **21**(2), 119–126.

Brown, B.E., R.P. Dunne, N. Phongsuwan and P.J. Somerfield, 2011: Increased sea level promotes coral cover on shallow reef flats in the Andaman Sea, eastern Indian Ocean. *Coral Reefs*, **30**(4), 867, doi:10.1007/s00338-011-0804-9.

Brown, J.H. et al., 2004: Toward a Metabolic Theory of Ecology. *Ecology*, **85**(7), 1771–1789, doi:10.1890/03-9000.

Brown, J.M. et al., 2018a: A coastal vulnerability assessment for planning climate resilient infrastructure. *Ocean Coast. Manage.*, **163**, 101–112.

Brown, S. et al., 2018b: What are the implications of sea level rise for a 1.5, 2 and 3°C rise in global mean temperatures in the Ganges-Brahmaputra-Meghna and other vulnerable deltas? *Reg. Environ. Change*, **18**(6), 1829–1842, doi:10.1007/s10113-018-1311-0.

Browning, T.J. et al., 2017: Nutrient co-limitation at the boundary of an oceanic gyre. *Nature*, **551**(7679), 242–246, doi:10.1038/nature24063.

5

Bruge, A. et al., 2016: Thermal Niche Tracking and Future Distribution of Atlantic Mackerel Spawning in Response to Ocean Warming. *Front. Mar. Sci.*, **3**(86), doi:10.3389/fmars.2016.00086.

Brunson, J.K. et al., 2018: Biosynthesis of the neurotoxin domoic acid in a bloom-forming diatom. *Science*, **361**(6409), 1356–1358.

Brussaard, C. et al., 2013: Arctic microbial community dynamics influenced by elevated CO_2 levels. *Biogeosciences*, **10**(2), 719–731.

Bryndum-Buchholz, A. et al., 2019: Twenty-first-century climate change impacts on marine animal biomass and ecosystem structure across ocean basins. *Global Change Biol.*, **25**(2), 459–472, doi:10.1111/gcb.14512.

Buckley, M.W. and J. Marshall, 2015: Observations, inferences, and mechanisms of the Atlantic Meridional Overturning Circulation: A review. *Rev. Geophys.*, **54**(1), 5–63, doi:10.1002/2015RG000493.

Buhl-Mortensen, L. et al., 2010: Biological structures as a source of habitat heterogeneity and biodiversity on the deep ocean margins. *Mar. Ecol-Evol. Persp.*, **31**(1), 21–50, doi:10.1111/j.1439-0485.2010.00359.x.

Bujosa, A., A. Riera and C.M. Torres, 2015: Valuing tourism demand attributes to guide climate change adaptation measures efficiently: The case of the Spanish domestic travel market. *Tourism Manage.*, **47**, 233–239, doi:10.1016/j.tourman.2014.09.023.

Bukvic, A., A. Smith and A. Zhang, 2015: Evaluating drivers of coastal relocation in Hurricane Sandy affected communities. *Int. J. Disast. Risk Reduc.*, **13**, 215–228, doi:10.1016/j.ijdrr.2015.06.008.

Bunten, D. and M. Kahn, 2014: *The Impact of Emerging Climate Risks on Urban Real Estate Price Dynamics*. National Bureau of Economic Research, Cambridge, MA [Available at: www.nber.org/papers/w20018.pdf]. Accessed: 2019/09/30.

Bunting, S.W., N. Kundu and N. Ahmed, 2017: Evaluating the contribution of diversified shrimp-rice agroecosystems in Bangladesh and West Bengal, India to social-ecological resilience. *Ocean Coast. Manage.*, **148**, 63–74, doi:10.1016/j.ocecoaman.2017.07.010.

Burd, A.B. and G.A. Jackson, 2002: Modeling steady-state particle size spectra. *Environ. Sci. Technol.*, **36**(3), 323–327.

Burrell, T.J., E.W. Maas, D.A. Hulston and C.S. Law, 2017: Variable response to warming and ocean acidification by bacterial processes in different plankton communities. *Aqut. Microb. Ecol.*, **79**(1), 49–62.

Burrows, M.T. et al., 2014: Geographical limits to species-range shifts are suggested by climate velocity. *Nature*, **507**(7493), 492–495, doi:10.1038/nature12976.

Busch, D.S. et al., 2016: Climate science strategy of the US National Marine Fisheries Service. *Mar. Policy*, **74**, 58–67, doi:10.1016/j.marpol.2016.09.001.

Büscher, J.V., A.U. Form and U. Riebesell, 2017: Interactive Effects of Ocean Acidification and Warming on Growth, Fitness and Survival of the Cold water Coral Lophelia pertusa Under Different Food Availabilities. *Front. Mar. Sci.*, **4**, 119, doi:10.3389/fmars.2017.00101.

Butt, N. et al., 2016: Challenges in assessing the vulnerability of species to climate change to inform conservation actions. *Biol. Conserv.*, **199**, 10–15, doi:10.1016/j.biocon.2016.04.020.

Buurman, J. and V. Babovic, 2016: Adaptation Pathways and Real Options Analysis: An approach to deep uncertainty in climate change adaptation policies. *Policy Soc.*, **35**(2), 137–150, doi:10.1016/j.polsoc.2016.05.002.

Byrne, J.A., A.Y. Lo and Y. Jianjun, 2015: Residents' understanding of the role of green infrastructure for climate change adaptation in Hangzhou, China. *Landscape Urban Plan.*, **138**, 132–143, doi:10.1016/j.landurbplan.2015.02.013.

Cabré, A., I. Marinov, R. Bernardello and D. Bianchi, 2015: Oxygen minimum zones in the tropical Pacific across CMIP5 models: mean state differences and climate change trends. *Biogeosciences*, **12**(18), 5429–5454, doi:10.5194/bg-12-5429-2015.

Cacciapaglia, C. and R. van Woesik, 2018: Marine species distribution modelling and the effects of genetic isolation under climate change. *J. Biogeogr.*, **45**(1), 154–163, doi:10.1111/jbi.13115.

Caesar, L. et al., 2018: Observed fingerprint of a weakening Atlantic Ocean overturning circulation. *Nature*, **556**(7700), 191–196, doi:10.1038/s41586-018-0006-5.

Cai, W.-J. et al., 2017: Redox reactions and weak buffering capacity lead to acidification in the Chesapeake Bay. *Nat. Commun.*, **8**(1), 369, doi:10.1038/s41467-017-00417-7.

Cameron, C., L.B. Hutley and D.A. Friess, 2019: Estimating the full greenhouse gas emissions offset potential and profile between rehabilitating and established mangroves. *Sci. Total Environ.*, **665**, 419–431, doi:10.1016/j.scitotenv.2019.02.104.

Camp, E.F. et al., 2018: The Future of Coral Reefs Subject to Rapid Climate Change: Lessons from Natural Extreme Environments. *Front. Mar. Sci.*, **5**, 4.

Campbell, J.E. and J.W. Fourqurean, 2018: Does Nutrient Availability Regulate Seagrass Response to Elevated CO_2? *Ecosystems*, **21**(7), 1269–1282.

Campbell, J.R., 2017: *Climate Change Impacts on Atolls and Island Nations in the South Pacific.* Encyclopedia of the Anthropocene: Volume 2. [Dellasala, D.A., Goldstein, M.I. (eds.)] Elsevier, New York. pp. 227–232. ISBN: 978-0-12-813576-1.

Camus, V.G., 2017: *Le cas de l'atoll de Tabiteuea, république de Kiribati.* In: Les atolls du Pacifique face au changement climatique. Une comparaison Tuamotu-Kiribati, Karthala [T. Bambridge and J.-P. Latouche (eds.)], Karthala, Chavannes de Bogis, Switzerland,122 pp. ISBN: 978-2811117399.

Canals, M. et al., 2006: Flushing submarine canyons. *Nature*, **444**(7117), 354–357, doi:10.1038/nature05271.

Capotondi, A. et al., 2012: Enhanced upper ocean stratification with climate change in the CMIP3 models. *J. Geophys. Res-Oceans*, **117**(C4), doi:10.1029/2011JC007409.

Carapuço, M.M. et al., 2016: Coastal geoindicators: Towards the establishment of a common framework for sandy coastal environments. *Earth-Sci. Rev.*, **154**, 183–190, doi:10.1016/j.earscirev.2016.01.002.

Carcedo, M.C., S.M. Fiori and C.S. Bremec, 2017: Zonation of macrobenthos across a mesotidal sandy beach: Variability based on physical factors. *J. Sea Res.*, **121**, 1–10.

Cardoso-Mohedano, J.-G. et al., 2018: Sub-tropical coastal lagoon salinization associated to shrimp ponds effluents. *Estuar. Coast. Shelf Sci.*, **203**, 72–79, doi:10.1016/j.ecss.2018.01.022.

Carey, J. et al., 2017: The declining role of organic matter in New England salt marshes. *Estuar. Coast.*, **40**(3), 626–639.

Carlson, A.K., W.W. Taylor, J. Liu and I. Orlic, 2017: The telecoupling framework: an integrative tool for enhancing fisheries management. *Fisheries*, **42**(8), 395–397.

Caron, D.A. and D.A. Hutchins, 2012: The effects of changing climate on microzooplankton grazing and community structure: drivers, predictions and knowledge gaps. *J. Plankton Res.*, **35**(2), 235–252, doi:10.1093/plankt/fbs091.

Carozza, D.A., D. Bianchi and E.D. Galbraith, 2016: The ecological module of BOATS-1.0: a bioenergetically constrained model of marine upper trophic levels suitable for studies of fisheries and ocean biogeochemistry. *Geosci. Model Dev.*, **9**(4), 1545–1565, doi:10.5194/gmd-9-1545-2016.

Carrasquilla-Henao, M. and F. Juanes, 2017: Mangroves enhance local fisheries catches: a global meta-analysis. *Fish Fish.*, **18**(1), 79–93, doi:10.1111/faf.12168.

Carro, I., Seijo, L., Nagy, G.J., Lagos, X. and Gutiérrez, O., 2018: Building capacity on ecosystem-based adaption strategy to cope with extreme events and sea level rise on the Uruguayan coast. *Int. J. Clim. Change Strategies Manage.*, **10**(4), 504–522, doi:10.1108/IJCCSM-07-2017-0149.

Carstensen, J., M. Chierici, B. G. Gustafsson and E. Gustafsson, 2018: Long-Term and Seasonal Trends in Estuarine and Coastal Carbonate Systems. *Global Biogeochem. Cy.*, **32**(3), 497–513, doi:10.1002/2017GB005781.

Cartapanis, O., E.D. Galbraith, D. Bianchi and S. L. Jaccard, 2018: Carbon burial in deep sea sediment and implications for oceanic inventories of carbon and alkalinity over the last glacial cycle. *Clim. Past*, **14**(11), 1819–1850, doi:10.5194/cp-14-1819-2018.

5

Carter, B. et al., 2017: Two decades of Pacific anthropogenic carbon storage and ocean acidification along Global Ocean Ship-based Hydrographic Investigations Program sections P16 and P02. *Global Biogeochem. Cy.*, **31**(2), 306–327.

Carter, J.G., 2018: Urban climate change adaptation: Exploring the implications of future land cover scenarios. *Cities*, **77**, 73–80.

Carton, J.A., G.A. Chepurin and L. Chen, 2018: SODA3: A New Ocean Climate Reanalysis. *J. Clim.*, **31**(17), 6967–6983, doi:10.1175/JCLI-D-18-0149.1.

Carugati, L. et al., 2018: Impact of mangrove forests degradation on biodiversity and ecosystem functioning. *Sci. Rep.*, **8**(1), 13298, doi:10.1038/s41598-018-31683-0.

Carvalho, B., E. Rangel and M. Vale, 2017: Evaluation of the impacts of climate change on disease vectors through ecological niche modelling. *Bulletin of entomological research*, **107**(4), 419–430.

Casado-Amezúa, P. et al., 2019: Distributional shifts of canopy-forming seaweeds from the Atlantic coast of Southern Europe. *Biodivers. Conserv.*, **28**(5), 1151–1172, doi:10.1007/s10531-019-01716-9.

Casey, J.M., S.R. Connolly and T.D. Ainsworth, 2015: Coral transplantation triggers shift in microbiome and promotion of coral disease associated potential pathogens. *Sci. Rep.*, **5**(1), 833, doi:10.1038/srep11903.

Castelle, B., S. Bujan, S. Ferreira and G. Dodet, 2017: Foredune morphological changes and beach recovery from the extreme 2013/2014 winter at a high-energy sandy coast. *Mar. Geol.*, **385**, 41–55, doi:10.1016/j.margeo.2016.12.006.

Castells-Quintana, D., M.d.P. Lopez-Uribe and T.K. McDermott, 2017: Geography, institutions and development: a review of the long-run impacts of climate change. *Clim. Dev.*, **9**(5), 452–470.

Cavallo, C. et al., 2015: Predicting climate warming effects on green turtle hatchling viability and dispersal performance. *Funct. Ecol.*, **29**(6), 768–778, doi:10.1111/1365-2435.12389.

Cavan, E.L., M. Trimmer, F. Shelley and R. Sanders, 2017: Remineralization of particulate organic carbon in an ocean oxygen minimum zone. *Nat. Commun.*, **8**, 14847, doi:10.1038/ncomms14847.

Celo, V., D.R.S. Lean and S.L. Scott, 2006: Abiotic methylation of mercury in the aquatic environment. *Sci. Total Environ.*, **368**(1), 126–137, doi:10.1016/j.scitotenv.2005.09.043.

Čerkasova, N. et al., 2016: Curonian Lagoon drainage basin modelling and assessment of climate change impact. *Oceanologia*, **58**(2), 90–102, doi:10.1016/j.oceano.2016.01.003.

Chambers, L.E., P. Dann, B. Cannell and E.J. Woehler, 2014: Climate as a driver of phenological change in southern seabirds. *International Journal of Biometeorology*, **58**(4), 603–612, doi:10.1007/s00484-013-0711-6.

Chan, K.M., T. Satterfield and J. Goldstein, 2012: Rethinking ecosystem services to better address and navigate cultural values. *Ecol. Econ.*, **74**, 8–18.

Chapman, A. and S. Darby, 2016: Evaluating sustainable adaptation strategies for vulnerable mega-deltas using system dynamics modelling: Rice agriculture in the Mekong Delta's An Giang Province, Vietnam. *Sci. Total Environ.*, **559**, 326–338, doi:10.1016/j.scitotenv.2016.02.162.

Charlton, K.E. et al., 2016: Fish, food security and health in Pacific Island countries and territories: a systematic literature review. *BMC Public Health*, **16**(1), 285, doi:10.1186/s12889-016-2953-9.

Chauvenet, A.L.M. et al., 2013: Maximizing the success of assisted colonizations. *Animal Conserv.*, **16**(2), 161–169, doi:10.1111/j.1469-1795.2012.00589.x.

Chavanich, S. et al., 2015: Conservation, management, and restoration of coral reefs. *Animal evolution: early emerging animals matter*, **118**(2), 132–134.

Chefaoui, R. M., C. M. Duarte and E. A. Serrão, 2018: Dramatic loss of seagrass habitat under projected climate change in the Mediterranean Sea. *Global Change Biol.*, **24(10)**, 4919–4928, doi:10.1111/gcb.14401.

Chen, C.-T. A., 2003: New vs. export production on the continental shelf. *Deep Sea Res. Pt. II*, **50**(6), 1327–1333, doi:10.1016/S0967-0645(03)00026-2.

Chen, C.-T. A. et al., 2017: Deep oceans may acidify faster than anticipated due to global warming. *Nat. Clim. Change*, **7**(12), 890–894, doi:10.1038/s41558-017-0003-y.

Chen, N. et al., 2018: Storm induced estuarine turbidity maxima and controls on nutrient fluxes across river-estuary-coast continuum. *Sci. Total Environ.*, **628**, 1108–1120.

Chen, P.-Y., C.-C. Chen, L. Chu and B. McCarl, 2015: Evaluating the economic damage of climate change on global coral reefs. *Global Environ. Change*, **30**(Supplement C), 12–20, doi:10.1016/j.gloenvcha.2014.10.011.

Chen, S. and D. Ganapin, 2016: Polycentric coastal and ocean management in the Caribbean Sea Large Marine Ecosystem: harnessing community-based actions to implement regional frameworks. *Environ. Dev.*, **17**, 264–276, doi:10.1016/j.envdev.2015.07.010.

Cheng, H.-Q. and J.-Y. Chen, 2017: Adapting cities to sea level rise: A perspective from Chinese deltas. *Advances in Climate Change Research*, **8**(2), 130–136, doi:10.1016/j.accre.2017.05.006.

Cheng, L., J. Abraham, Z. Hausfather and K.E. Trenberth, 2019: How fast are the oceans warming? *Science*, **363**(6423), 128, doi:10.1126/science.aav7619.

Cheng, L. et al., 2017: Improved estimates of ocean heat content from 1960 to 2015. *Sci. Adv.*, **3**(3), e1601545, doi:10.1126/sciadv.1601545.

Cheung, W.W.L., R.D. Brodeur, T.A. Okey and D. Pauly, 2015: Projecting future changes in distributions of pelagic fish species of Northeast Pacific shelf seas. *Progr. Oceanogr.*, **130**, 19–31, doi:10.1016/j.pocean.2014.09.003.

Cheung, W.W.L., J. Bruggeman and M. Butenschön, 2018a: Projected changes in global and national potential marine fisheries catch under climate change scenarios in the twenty-first century. In: Impacts of climate change on fisheries and aquaculture [Barange, M., Bahri, T., Beveridge, M.C.M., Cochrane, K.L., Funge-Smith, S. & Poulain, F. (eds.)]. FAO Fisheries and Aquaculture Technical Paper T, FAO, Rome, Italy. 63–86. ISBN: 978-92-5-130607-9.

Cheung, W.W.L. et al., 2008: Application of macroecological theory to predict effects of climate change on global fisheries potential. *Mar. Ecol. Prog. Ser.*, **365**, 187–197.

Cheung, W.W.L. et al., 2016a: Transform high seas management to build climate resilience in marine seafood supply. *Fish Fish.*, **18**(2), 254–263, doi:10.1111/faf.12177.

Cheung, W.W.L., M.C. Jones, G. Reygondeau and T.L. Frölicher, 2018b: Opportunities for climate-risk reduction through effective fisheries management. *Global Change Biol.*, **2108**, 1–15, doi:doi:10.1111/gcb.14390.

Cheung, W.W.L. and D. Pauly, 2016: Impacts and Effects of Ocean Warming on Marine Fishes. 239–253.

Cheung, W.W.L., G. Reygondeau and T.L. Frölicher, 2016b: Large benefits to marine fisheries of meeting the 1.5°C global warming target. *Science*, **354**(6319), 1591–1594, doi:10.1126/science.aag2331.

Cheung, W.W.L., R. Watson and D. Pauly, 2013: Signature of ocean warming in global fisheries catch. *Nature*, **497**, 365, doi:10.1038/nature121.

Chew, S.T. and J.B. Gallagher, 2018: Accounting for black carbon lowers estimates of blue carbon storage services. *Sci. Rep.*, **8**(1), 2553, doi:10.1038/s41598-018-20644-2.

Chivers, W.J., A.W. Walne and G.C. Hays, 2017: Mismatch between marine plankton range movements and the velocity of climate change. *Nat. Commun.*, **8**, doi:10.1038/ncomms14434.

Chmura, G.L., S.C. Anisfeld, D.R. Cahoon and J.C. Lynch, 2003: Global carbon sequestration in tidal, saline wetland soils. *Global Biogeochem. Cy.*, **17**(4), doi:10.1029/2002GB001917.

Chmura, G.L., L. Kellman, L. van Ardenne and G.R. Guntenspergen, 2016: Greenhouse Gas Fluxes from Salt Marshes Exposed to Chronic Nutrient Enrichment. *PLoS One*, **11**(2), e0149937, doi:10.1371/journal.pone.0149937.

Chollett, I. and P.J. Mumby, 2013: Reefs of last resort: Locating and assessing thermal refugia in the wider Caribbean. *Biol. Conserv.*, **167**, 179–186, doi:10.1016/j.biocon.2013.08.010.

Chuenpagdee, R., 2011: *World small-scale fisheries: contemporary visions*. EburonAcademic Publishers, Delft, The Netherlands. 400 pp. ISBN: 978-90-5972-539-3.

Chung, I.K. et al., 2013: Installing kelp forests/seaweed beds for mitigation and adaptation against global warming: Korean Project Overview. *ICES J. Mar. Sci.*, **70**(5), 1038–1044, doi:10.1093/icesjms/fss206.

5

Chung, I.K., C.F.A. Sondak and J. Beardall, 2017: The future of seaweed aquaculture in a rapidly changing world. *European J. Phycol.*, **52**(4), 495–505, doi:10.1080/09670262.2017.1359678.

Chust, G. et al., 2014: Biomass changes and trophic amplification of plankton in a warmer ocean. *Global Change Biol.*, **20**(7), 2124–2139, doi:10.1111/gcb.12562.

Ciais, P. et al., 2013: Carbon and Other Biogeochemical Cycles. In: Climate Change 2013: The Physical Science Basis. Contribution of Working Group I to the Fifth Assessment Report of the Intergovernmental Panel on Climate Change [Stocker, T.F., D. Qin, G.-K. Plattner, M. Tignor, S.K. Allen, J. Boschung, A. Nauels, Y. Xia, V. Bex and P.M. Midgley(eds.)]. Cambridge University Press, Cambridge, United Kingdom and New York, NY, USA, 465–570.

Cinner, J.E. et al., 2018: Building adaptive capacity to climate change in tropical coastal communities. *Nat. Clim. Change*, **8**(2), 117–123, doi:10.1038/s41558-017-0065-x.

Cinner, J.E. et al., 2016: A framework for understanding climate change impacts on coral reef social–ecological systems. *Reg. Environ. Change*, **16**(4), 1133–1146, doi:10.1007/s10113-015-0832-z.

Cisneros-Montemayor, A.M. et al., 2013: Global economic value of shark ecotourism: implications for conservation. *Oryx*, **47**(3), 381–388.

Cisneros-Montemayor, A.M., D. Pauly, L.V. Weatherdon and Y. Ota, 2016: A Global Estimate of Seafood Consumption by Coastal Indigenous Peoples. *PLoS One*, **11**(12), doi:10.1371/journal.pone.0166681.

Cisneros-Montemayor, A.M. and U.R. Sumaila, 2010: A global estimate of benefits from ecosystem-based marine recreation: potential impacts and implications for management. *Journal of Bioeconomics*, **12**(3), 245–268.

Clark, A., 2017: Small unmanned aerial systems comparative analysis for the application to coastal erosion monitoring. *Geo. Res. J.*, **13**, 175–185, doi:10.1016/j.grj.2017.05.001.

Cloern, J.E. et al., 2016: Human activities and climate variability drive fast-paced change across the world's estuarine–coastal ecosystems. *Global Change Biol.*, **22**(2), 513–529.

Clucas, G.V. et al., 2014: A reversal of fortunes: climate change 'winners' and 'losers'; in Antarctic Peninsula penguins. *Sci. Rep.*, **4**, 5024, doi:10.1038/srep05024.

Cocco, V. et al., 2013: Oxygen and indicators of stress for marine life in multi-model global warming projections. *Biogeosciences*, **10**(3), 1849–1868, doi:10.5194/bg-10-1849-2013.

Coelho, V.R. et al., 2017: Shading as a mitigation tool for coral bleaching in three common Indo-Pacific species. *J. Exp. Mar. Biol. Ecol.*, **497**(Supplement C), 152–163.

Colburn, L.L. et al., 2016: Indicators of climate change and social vulnerability in fishing dependent communities along the Eastern and Gulf Coasts of the United States. *Mar. Policy*, **74**, 323–333, doi:10.1016/j.marpol.2016.04.030.

Coles, S.L. and B.M. Riegl, 2013: Thermal tolerances of reef corals in the Gulf: A review of the potential for increasing coral survival and adaptation to climate change through assisted translocation. *Mar. Pollut. Bull.*, **72**(2), 323–332, doi:10.1016/j.marpolbul.2012.09.006.

Colin, M., F. Palhol and A. Leuxe, 2016: Adaptation of transport infrastructures and networks to climate change. *Transp. Res. Proc.*, **14**, 86–95.

Collins, M. et al., 2013: Long-term Climate Change: Projections, Commitments and Irreversibility. In: Climate Change 2013: The Physical Science Basis. Contribution of Working Group I to the Fifth Assessment Report of the Intergovernmental Panel on Climate Change [Stocker, T.F., D. Qin, G.-K. Plattner, M. Tignor, S.K. Allen, J. Boschung, A. Nauels, Y. Xia, V. Bex and P.M. Midgley (eds.)]. Cambridge University Press, Cambridge, United Kingdom and New York, NY, USA, 1029–1136.

Collins, S., B. Rost and T.A. Rynearson, 2014: Evolutionary potential of marine phytoplankton under ocean acidification. *Evol. Appl.*, **7**(1), 140–155, doi:doi:10.1111/eva.12120.

Colosi, J.A. and W. Munk, 2006: Tales of the Venerable Honolulu Tide Gauge. *J. Phys. Oceanogr.*, **36**(6), 967–996, doi:10.1175/JPO2876.1.

Comte, A. and L.H. Pendleton, 2018: Management strategies for coral reefs and people under Global Environ. Change: 25 years of scientific research. *J. Environ. Manage.*, **209**, 462–474, doi:10.1016/j.jenvman.2017.12.051.

Conde, D. et al., 2015: Solutions for Sustainable Coastal Lagoon Management. In: Coastal Zones – Solutions for the 21st Century. [Baztan, J., Chouinard, O., Jorgensen, B., Tett, P., Vanderlinden, J-P., Vasseur, L. (eds)]. Elsevier, New York. pp. 217–250. ISBN: 978-0-12-802748-6.

Constable, A.J. et al., 2014: Climate change and Southern Ocean ecosystems I: how changes in physical habitats directly affect marine biota. *Global Change Biol.*, **20**(10), 3004–3025, doi:10.1111/gcb.12623.

Cooley, S.R., C.R. Ono, S. Melcer and J. Roberson, 2016: Community-Level Actions that Can Address Ocean Acidification. *Front. Mar. Sci.*, **2**(128), 1–12.

Corbitt, E.S. et al., 2011: Global Source–Receptor Relationships for Mercury Deposition Under Present-Day and 2050 Emissions Scenarios. *Environ. Sci. Technol.*, **45**(24), 10477–10484, doi:10.1021/es202496y.

Cordes, E.E. et al., 2010: The influence of geological, geochemical, and biogenic habitat heterogeneity on seep biodiversity. *Mar. Ecol-Evol. Persp.*, **31**(1), 51–65, doi:10.1111/j.1439-0485.2009.00334.x.

Cordes, E.E. et al., 2016: Environmental Impacts of the Deep-Water Oil and Gas Industry: A Review to Guide Management Strategies. *Front. Environ. Sci.*, **4**, 58.

Cornwall, C.E. et al., 2018: Resistance of corals and coralline algae to ocean acidification: physiological control of calcification under natural pH variability. *Proc. Roy. Soc. B. Biol.*, **285**(1884). https://doi.org/10.1098/rspb.2018.1168.

Costa, D.P. et al., 2010: Approaches to studying climatic change and its role on the habitat selection of antarctic pinnipeds. *Integr. Comp. Biol.*, **50**(6), 1018–1030, doi:10.1093/icb/icq054.

Costanza, R. et al., 2017: Twenty years of ecosystem services: How far have we come and how far do we still need to go? *Ecosyst. Serv.*, **28**, 1–16.

Costanza, R. et al., 2014: Changes in the global value of ecosystem services. *Global Environ. Change*, **26**, 152–158, doi:10.1016/j.gloenvcha.2014.04.002.

Costanza, R. et al., 2008: The value of coastal wetlands for hurricane protection. *Ambio*, **37**(4), 241–248.

Creighton, C., A.J. Hobday, M. Lockwood and G.T. Pecl, 2016: Adapting Management of Marine Environments to a Changing Climate: A Checklist to Guide Reform and Assess Progress. *Ecosystems*, **19**(2), 187–219, doi:10.1007/s10021-015-9925-2.

Crespo, O., D. Guillermo and C. Daniel, 2017: A review of the impacts of fisheries on open-ocean ecosystems. *ICES J. Mar. Sci.*, **74**(9), 2283–2297, doi:10.1093/icesjms/fsx084.

Cripps, G., P. Lindeque and K.J. Flynn, 2014: Have we been underestimating the effects of ocean acidification in zooplankton? *Global Change Biol.*, **20**(11), 3377–3385, doi:10.1111/gcb.12582.

Crocker, D.E. et al., 2006: Impact of El Niño on the foraging behavior of female northern elephant seals. *Mar. Ecol. Prog. Ser.*, **309**(1), 1–10, doi:10.3354/meps309001.

Crooks, S. et al., 2011: Mitigating climate change through restoration and management of coastal wetlands and near-shore marine ecosystems: challenges and opportunities. Environment Department Paper 121, World Bank, Washington, D.C. pp 59.

Crosby, S.C. et al., 2016: Salt marsh persistence is threatened by predicted sea level rise. *Estuar. Coast. Shelf Sci.*, **181**, 93–99, doi:10.1016/j.ecss.2016.08.018.

Crotty, S.M., C. Angelini and M.D. Bertness, 2017: Multiple stressors and the potential for synergistic loss of New England salt marshes. *PLOS ONE*, **12**(8), e0183058.

Crozier, L.G. and J.A. Hutchings, 2014: Plastic and evolutionary responses to climate change in fish. *Evol. Appl.*, **7**(1), 68–87, doi:doi:10.1111/eva.12135.

Cruz, D.W.d., R.D. Villanueva and M.V.B. Baria, 2014: Community-based, low-tech method of restoring a lost thicket of Acropora corals. *ICES J. Mar. Sci.*, **71**(7), 1866–1875, doi:10.1093/icesjms/fst228.

5

Cuellar-Martinez, T. et al., 2018: Addressing the Problem of Harmful Algal Blooms in Latin America and the Caribbean – A Regional Network for Early Warning and Response. *Front. Mar. Sci.*, **5**(409), doi:10.3389/fmars.2018.00409.

Cumiskey, L. et al., 2018: A framework to include the (inter)dependencies of Disaster Risk Reduction measures in coastal risk assessment. *Coast. Eng.*, **134**, 81–92, doi:10.1016/j.coastaleng.2017.08.009.

Cushing, D. A., D. D. Roby and D. B. Irons, 2018: Patterns of distribution, abundance, and change over time in a subarctic marine bird community. *Deep Sea Res. Pt. II*, **147**, 148–163, doi:10.1016/j.dsr2.2017.07.012.

Cvitanovic, C. et al., 2016: Linking adaptation science to action to build food secure Pacific Island communities. *Clim. Risk Manage.*, **11**, 53–62, doi:10.1016/j.crm.2016.01.003.

D'Asaro, E.A., 2014: Turbulence in the Upper-Ocean Mixed Layer. *Annu. Rev. Mar. Sci.*, **6**(1), 101–115, doi:10.1146/annurev-marine-010213-135138.

Dalyander, P.S. et al., 2016: Use of structured decision-making to explicitly incorporate environmental process understanding in management of coastal restoration projects: Case study on barrier islands of the northern Gulf of Mexico. *J. Environ. Manage.*, **183**(3), 497–509, doi:10.1016/j.jenvman.2016.08.078.

Daneri, G. et al., 2012: Wind forcing and short-term variability of phytoplankton and heterotrophic bacterioplankton in the coastal zone of the Concepción upwelling system (Central Chile). *Progr. Oceanogr.*, **92–95**(Supplement C), 92–96.

Danovaro, R., A. Dell'Anno and A. Pusceddu, 2004: Biodiversity response to climate change in a warm deep sea. *Ecol. Lett.*, **7**(9), 821–828, doi:10.1111/j.1461-0248.2004.00634.x.

Danovaro, R. et al., 2001: Deep sea ecosystem response to climate changes: the eastern Mediterranean case study. *Trends Ecol. Evol.*, **16**(9), 505–510, doi:10.1016/S0169-5347(01)02215-7.

DasGupta, R. and R. Shaw, 2015: An indicator based approach to assess coastal communities' resilience against climate related disasters in Indian Sundarbans. *J. Coast. Conserv.*, **19**(1), 85–101, doi:10.1007/s11852-014-0369-1.

Dasgupta, S. et al., 2017: The Impact of Aquatic Salinization on Fish Habitats and Poor Communities in a Changing Climate: Evidence from Southwest Coastal Bangladesh. *Ecol. Econ.*, **139**, 128–139, doi:10.1016/j.ecolecon.2017.04.009.

Dave, A.C. and M.S. Lozier, 2013: Examining the global record of interannual variability in stratification and marine productivity in the low-latitude and mid-latitude ocean. *J. Geophys. Res-Oceans*, **118**(6), 3114–3127, doi:10.1002/jgrc.20224.

Davis, R., 2015: 'All in': Snow crab, capitalization, and the future of small-scale fisheries in Newfoundland. *Mar. Policy*, **61**, 323–330, doi:10.1016/j.marpol.2015.04.008.

Davison, P.C., D.M. Checkley, J.A. Koslow and J. Barlow, 2013: Carbon export mediated by mesopelagic fishes in the northeast Pacific Ocean. *Progr. Oceanogr.*, **116**, 14–30, doi:10.1016/j.pocean.2013.05.013.

Dawson, J., 2017: Climate Change Adaptation Strategies and Policy Options for Arctic Shipping. Transport Canada. Ottawa, Canada. 154 pp. http://hdl.handle.net/10393/36016.

Day, J.W. and J.M. Rybczyk, 2019: Global Change Impacts on the Future of Coastal Systems: Perverse Interactions Among Climate Change, Ecosystem Degradation, Energy Scarcity, and Population. In: *Coasts and Estuaries*. Elsevier, 621–639.

de Coninck, H. et al., 2018: Chapter 4: Strengthening and implementing the global response. In: An IPCC Special Report on the impacts of global warming of 1.5°C above pre-industrial levels and related global greenhouse gas emission pathways, in the context of strengthening the global response to the threat of climate change, sustainable development, and efforts to eradicate poverty [Masson-Delmotte, V.P., Zhai, P., Pörtner, H.-O., Roberts, D., Skea, J., Shukla, P.R., Pirani, A., Moufouma-Okia, W., Péan, C., Pidcock, R., Connors, S., Matthews, J.B.R., Chen, Y., Zhou, X., Gomis, M.I., Lonnoy, E., Maycock, T., Tignor, M., Waterfield, T. (eds.)]. In Press.

De Leo, F.C. et al., 2012: The effects of submarine canyons and the oxygen minimum zone on deep sea fish assemblages off Hawai'i. *Deep sea Res. Pt. I*, **64**, 54–70, doi:10.1016/j.dsr.2012.01.014.

De Leo, F.C. et al., 2010: Submarine canyons: hotspots of benthic biomass and productivity in the deep sea. *Proc. Biol. Sci.*, **277**(1695), 2783–2792, doi:10.1098/rspb.2010.0462.

de Orte, M.R. et al., 2014: Effects on the mobility of metals from acidification caused by possible CO_2 leakage from sub-seabed geological formations. *Sci. Total Environ.*, **470–471**(Supplement C), 356–363, doi:10.1016/j.scitotenv.2013.09.095.

de Sá, L.C. et al., 2018: Studies of the effects of microplastics on aquatic organisms: What do we know and where should we focus our efforts in the future? *Sci. Total Environ.*, **645**, 1029–1039, doi:10.1016/j.scitotenv.2018.07.207.

De Urioste-Stone, S.M., L. Le, M.D. Scaccia and E. Wilkins, 2016: Nature-based tourism and climate change risk: Visitors' perceptions in mount desert island, Maine. *J. Outdoor Recreat. Tour.*, **13**, 57–65, doi:10.1016/j.jort.2016.01.003.

DeBiasse, M.B. and M.W. Kelly, 2016: Plastic and Evolved Responses to Global Change: What Can We Learn from Comparative Transcriptomics? *J. Hered.*, **107**(1), 71–81, doi:10.1093/jhered/esv073.

Defeo, O. et al., 2009: Threats to sandy beach ecosystems: A review. *Estuar. Coast. Shelf Sci.*, **81**(1), 1–12, doi:10.1016/j.ecss.2008.09.022.

Delgado-Fernandez, I., N. O'Keeffe and R.G.D. Davidson-Arnott, 2019: Natural and human controls on dune vegetation cover and disturbance. *Sci. Total Environ.*, **672**, 643–656, doi:10.1016/j.scitotenv.2019.03.494.

Delworth, T.L. and F. Zeng, 2016: The impact of the North Atlantic Oscillation on climate through its influence on the Atlantic Meridional Overturning Circulation. *J. Clim.*, **29**(3), 941–962, doi:10.1175/JCLI-D-15-0396.1.

Demopoulos, A.W.J. et al., 2019: Examination of Bathymodiolus childressi nutritional sources, isotopic niches, and food-web linkages at two seeps in the US Atlantic margin using stable isotope analysis and mixing models. *Deep sea Res. Pt. I*, doi:10.1016/j.dsr.2019.04.002.

Dennis, K.C., I. Niang-Diop and R.J. Nicholls, 1995: Sea level Rise and Senegal: Potential Impacts and Consequences. *J. Coast. Res.*, 243–261.

Desbruyères, D., E. L. McDonagh, B.A. King and V. Thierry, 2016a: Global and Full-Depth Ocean Temperature Trends during the Early Twenty-First Century from Argo and Repeat Hydrography. *J. Clim.*, **30**(6), 1985–1997, doi:10.1175/JCLI-D-16-0396.1.

Desbruyères, D.G. et al., 2014: Full-depth temperature trends in the northeastern Atlantic through the early 21st century. *Geophys. Res. Lett.*, **41**(22), 7971–7979, doi:10.1002/2014GL061844.

Desbruyères, D.G. et al., 2016b: Deep and abyssal ocean warming from 35 years of repeat hydrography. *Geophys. Res. Lett.*, **43**(19), 10,356–10,365, doi:10.1002/2016gl070413.

Descamps, S. et al., 2015: Demographic effects of extreme weather events: snow storms, breeding success, and population growth rate in a long-lived Antarctic seabird. *Ecol. Evol.*, **5**(2), 314–325, doi:10.1002/ece3.1357.

Desforges, J.-P. et al., 2018: Predicting global killer whale population collapse from PCB pollution. *Science*, **361**(6409), 1373, doi:10.1126/science.aat1953.

Desforges, J.-P. et al., 2017: Effects of Polar Bear and Killer Whale Derived Contaminant Cocktails on Marine Mammal Immunity. *Environ. Sci. Technol.*, **51**(19), 11431–11439, doi:10.1021/acs.est.7b03532.

Deutsch, C. et al., 2011: Climate-Forced Variability of Ocean Hypoxia. *Science*, **333**(6040), 336.

Deutsch, C. et al., 2015: Climate change tightens a metabolic constraint on marine habitats. *Science*, **348**(6239), 1132.

Devlin, A.T. et al., 2017: Tidal Variability Related to Sea Level Variability in the Pacific Ocean. *J. Geophys. Res-Ocean*, **122(11)**, 8445–8463 doi:10.1002/2017JC013165.

DeVries, T., 2014: The oceanic anthropogenic CO_2 sink: Storage, air-sea fluxes, and transports over the industrial era. *Global Biogeochem. Cy.*, **28**(7), 631–647, doi:10.1002/2013GB004739.

5

DeVries, T., M. Holzer and F. Primeau, 2017: Recent increase in oceanic carbon uptake driven by weaker upper-ocean overturning. *Nature*, **542**(7640), 215.

DeVries, T. and T. Weber, 2017: The export and fate of organic matter in the ocean: New constraints from combining satellite and oceanographic tracer observations. *Global Biogeochem. Cy.*, **31**(3), 535–555, doi:10.1002/2016GB005551.

Dey, M.M. et al., 2016: Analysis of the economic impact of climate change and climate change adaptation strategies for fisheries sector in Pacific coral triangle countries: Model, estimation strategy, and baseline results. *Mar. Policy*, **67**, 156–163, doi:10.1016/j.marpol.2015.12.011.

Dhar, T.K. and L. Khirfan, 2016: Community-based adaptation through ecological design: lessons from Negril, Jamaica. *J. Urban Des.*, **21**(2), 234–255, doi:10.1080/13574809.2015.1133224.

Diamond, S.E., 2018: Contemporary climate-driven range shifts: Putting evolution back on the table. *Funct. Ecol.*, **32**(7), 1652–1665, doi:10.1111/1365-2435.13095.

Díaz, P.A. et al., 2019: Impacts of harmful algal blooms on the aquaculture industry: Chile as a case study. *Perspect. Phycol.*, **6**, 1–2. doi: 10.1127/pip/2019/0081.

Diaz, R.J. and R. Rosenberg, 1995: Marine benthic hypoxia: a review of its ecological effects and the behavioural responses of benthic macrofauna. *Oceanogr. Mar. Biol.*, **33**, 245–03.

Díaz, S. et al., 2018: Assessing nature's contributions to people. *Science*, **359**(6373), 270.

Dickson, A.J., A.S. Cohen and A.L. Coe, 2012: Seawater oxygenation during the Paleocene-Eocene Thermal Maximum. *Geology*, **40**(7), 639–642, doi:10.1130/G32977.1.

Diesing, M. et al., 2017: Predicting the standing stock of organic carbon in surface sediments of the North–West European continental shelf. *Biogeochemistry*, **135**(1), 183–200, doi:10.1007/s10533-017-0310-4.

Dietz, S. et al., 2018: The Economics of 1.5°C Climate Change. *Annu. Rev. Environ. Resourc.*, **43**(1), 455–480, doi:10.1146/annurev-environ-102017-025817.

Dinh, V.N. and E. McKeogh, 2019a: Offshore Wind Energy: Technology Opportunities and Challenges. In: *Proceedings of the 1st Vietnam Symposium on Advances in Offshore Engineering*, [Randolph, M.F., D.H. Doan, A.M. Tang, M. Bui and V.N. Dinh (eds.)], Springer Singapore, pp. 3–22.

Dinh, V.N. and E. McKeogh, 2019b: Offshore Wind Energy: Technology Opportunities and Challenges. In: Proceedings of the 1st Vietnam Symposium on Advances in Offshore Engineering. Energy and Geotechnics [Marco di Prisco, S.-H. C., Giovanni Solari, Ioannis Vayas (ed.)] [Randolph, M.F., D.H. Doan, A.M. Tang, M. Bui and V.N. Dinh (eds.)]. Springer, Singapore, 3–22.

Diop, S. and P. Scheren, 2016: Sustainable oceans and coasts: Lessons learnt from Eastern and Western Africa. *Estuar. Coast. Shelf Sci.*, **183**, 327–339.

DiSegni, D.M. and M. Shechter, 2013: Socioeconomic Aspects: Human Migrations, Tourism and Fisheries. In: *The Mediterranean Sea*. Springer Netherlands, Dordrecht, pp. 571–575.

Dissard, D., G. Nehrke, G. J. Reichart and J. Bijma, 2010: Impact of seawater CO_2 on calcification and Mg/Ca and Sr/Ca ratios in benthic foraminifera calcite: results from culturing experiments with Ammonia tepida. *Biogeosciences*, **7**(1), 81–93, doi:10.5194/bg-7-81-2010.

Dittrich, R., A. Wreford and D. Moran, 2016: A survey of decision-making approaches for climate change adaptation: Are robust methods the way forward? *Ecol. Econ.*, **122**, 79–89, doi:10.1016/j.ecolecon.2015.12.006.

Dixon, D.R. et al., 2006: Evidence of seasonal reproduction in the Atlantic vent mussel Bathymodiolus azoricus, and an apparent link with the timing of photosynthetic primary production. *J. Mar. Biol. Assoc. U.K.*, **86**(6), 1363–1371, doi:10.1017/S0025315406014391.

Dixon, T., J. Garrett and E. Kleverlaan, 2014: Update on the London Protocol – Developments on Transboundary CCS and on Geoengineering. *12th International Conference on Greenhouse Gas Control Technologies, GHGT-12*, **63**(Supplement C), 6623–6628, doi:10.1016/j.egypro.2014.11.698.

Doherty, M., K. Klima and J.J. Hellmann, 2016: Climate change in the urban environment: Advancing, measuring and achieving resiliency. *Environ. Sci. Policy*, **66**, 310–313, doi:10.1016/j.envsci.2016.09.001.

Domingues, C.M. et al., 2008: Improved estimates of upper-ocean warming and multi-decadal sea level rise. *Nature*, **453**, 1090, doi:10.1038/nature07080.

Donato, D.C. et al., 2011: Mangroves among the most carbon-rich forests in the tropics. *Nat. Geosci.*, **4**, 293, doi:10.1038/ngeo1123.

Donelson, J.M., S. Salinas, P.L. Munday and L.N.S. Shama, 2018: Transgenerational plasticity and climate change experiments: Where do we go from here? *Glob Chang Biol*, **24**(1), 13–34, doi:10.1111/gcb.13903.

Doney, S.C. et al., 2012: Climate change impacts on marine ecosystems. *Annu. Rev. Mar. Sci.*, **4**(1), 11–37, doi:10.1146/annurev-marine-041911-111611.

Dore, J.E. et al., 2009: Physical and biogeochemical modulation of ocean acidification in the central North Pacific. *PNAS*, **106**(30), 12235, doi:10.1073/pnas.0906044106.

Dornelas, M. et al., 2018: BioTIME: A database of biodiversity time series for the Anthropocene. *Global Ecol. Biogeogr.*, **27**(7), 760–786, doi:10.1111/geb.12729.

Dosser, H.V. and L. Rainville, 2016: Dynamics of the Changing Near-Inertial Internal Wave Field in the Arctic Ocean. *J. Phys. Oceanogr.*, **46**(2), 395–415, doi:10.1175/jpo-d-15-0056.1.

Drake, J.L. et al., 2018: Molecular and geochemical perspectives on the influence of CO_2 on calcification in coral cell cultures. *Limnol. Oceanogr.*, **63**(1), 107–121, doi:10.1002/lno.10617.

Drius, M. et al., 2019: Not just a sandy beach. The multi-service value of Mediterranean coastal dunes. *Sci. Total Environ.*, **668**, 1139–1155, doi:10.1016/j.scitotenv.2019.02.364.

Drury, C. et al., 2016: Genomic variation among populations of threatened coral: Acropora cervicornis. *BMC Genomics*, **17**(1), 958, doi:10.1186/s12864-016-2583-8.

Du, J. et al., 2018: Worsened physical condition due to climate change contributes to the increasing hypoxia in Chesapeake Bay. *Sci. Total Environ.*, **630**, 707–717, doi:10.1016/j.scitotenv.2018.02.265.

Duarte, B. et al., 2018: Climate Change Impacts on Seagrass Meadows and Macroalgal Forests: An Integrative Perspective on Acclimation and Adaptation Potential. *Front. Mar. Sci.*, **5**, 190.

Duarte, C.M. et al., 2013: Is ocean acidification an open-ocean syndrome? Understanding anthropogenic impacts on seawater pH. *Estuar. Coast.*, **36**(2), 221–236.

Duarte, C.M. and D. Krause-Jensen, 2018: Greenland Tidal Pools as Hot Spots for Ecosystem Metabolism and Calcification. *Estuar. Coast.*, **41**(5), 1314–1321, doi:10.1007/s12237-018-0368-9.

Duarte, C.M. et al., 2010: Seagrass community metabolism: Assessing the carbon sink capacity of seagrass meadows. *Global Biogeochem. Cy.*, **24**(4), doi:10.1029/2010GB003793.

Duarte, C.M., J.J. Middelburg and N. Caraco, 2005: Major role of marine vegetation on the oceanic carbon cycle. *Biogeosciences*, **2**(1), 1–8, doi:10.5194/bg-2-1-2005.

Duarte, C.M. et al., 2017: Can Seaweed Farming Play a Role in Climate Change Mitigation and Adaptation? *Front. Mar. Sci.*, **4**, 100.

Dubey, S.K. et al., 2017: Farmers' perceptions of climate change, impacts on freshwater aquaculture and adaptation strategies in climatic change hotspots: A case of the Indian Sundarban delta. *Environ. Dev.*, **21**, 38–51, doi:10.1016/j.envdev.2016.12.002.

Dubilier, N., C. Bergin and C. Lott, 2008: Symbiotic diversity in marine animals: the art of harnessing chemosynthesis. *Nat. Rev. Microbiol.*, **6**, 725, doi:10.1038/nrmicro1992.

Dubois, B. et al., 2013: Effect of Tropical Algae as Additives on Rumen in Vitro Gas Production and Fermentation Characteristics. *Am. J. Plant Sci.*, **04**(12), 34–43, doi:10.4236/ajps.2013.412A2005.

Ducklow, H.W. et al., 2013: West Antarctic Peninsula: An Ice-Dependent Coastal Marine Ecosystem in Transition. *Oceanography*, **26**(3), 190–203, doi:10.5670/oceanog.2013.62.

Dudley, P.N., R. Bonazza and W. P. Porter, 2016: Climate change impacts on nesting and internesting leatherback sea turtles using 3D animated computational fluid dynamics and finite volume heat transfer. *Ecol. Modell.*, **320**(Supplement C), 231–240, doi:10.1016/j.ecolmodel.2015.10.012.

Dueri, S., L. Bopp and O. Maury, 2014: Projecting the impacts of climate change on skipjack tuna abundance and spatial distribution. *Global Change Biol.*, **20**(3), 742–753, doi:10.1111/gcb.12460.

Dugger, K.M. et al., 2014: Adelie penguins coping with environmental change: results from a natural experiment at the edge of their breeding range. *Frontiers in Ecol. Evol.*, **2**, doi:10.3389/fevo.2014.00068.

Duke, N.C. et al., 2017: Large-scale dieback of mangroves in Australia's Gulf of Carpentaria: a severe ecosystem response, coincidental with an unusually extreme weather event. *Mar. Freshw. Res.*, **68**(10), 1816–1829.

Dundas, S.J. and R.H. von Haefen, 2015: *Weather effects on the demand for coastal recreational fishing: Implications for a changing climate.* CEnREP Working Paper No. 15–015, 63 pp. doi: 10.22004/ag.econ.264980.

Dunlop, K.M. et al., 2016: Carbon cycling in the deep eastern North Pacific benthic food web: Investigating the effect of organic carbon input. *Limnol. Oceanogr.*, **61**(6), 1956–1968, doi:10.1002/lno.10345.

Dunn, F.E. et al., 2018: Projections of historical and 21st century fluvial sediment delivery to the Ganges-Brahmaputra-Meghna, Mahanadi, and Volta deltas. *Sci. Total Environ.*, **642**, 105–116.

Dunstan, P.K. et al., 2018: How can climate predictions improve sustainability of coastal fisheries in Pacific Small-Island Developing States? *Mar. Policy*, 88, 295–302. doi:10.1016/j.marpol.2017.09.033.

Durack, P.J., 2015: Ocean Salinity and the Global Water Cycle. *Oceanography*, **28**(1), 20–31.

Duteil, O., A. Oschlies and C.W. Böning, 2018: Pacific Decadal Oscillation and recent oxygen decline in the eastern tropical Pacific Ocean. *Biogeosciences*, **15**, 7111–7126.

Duteil, O., F.U. Schwarzkopf, C.W. Böning and A. Oschlies, 2014: Major role of the equatorial current system in setting oxygen levels in the eastern tropical Atlantic Ocean: A high-resolution model study. *Geophys. Res. Lett.*, **41**(6), 2033–2040, doi:10.1002/2013GL058888.

Dutkiewicz, S. et al., 2015: Impact of ocean acidification on the structure of future phytoplankton communities. *Nat. Clim. Change*, **5**, 1002, doi:10.1038/nclimate2722.

Dutkiewicz, S., J.R. Scott and M. Follows, 2013a: Winners and losers: Ecological and biogeochemical changes in a warming ocean. *Global Biogeochem. Cy.*, **27**(2), 463–477.

Dutkiewicz, S., J.R. Scott and M.J. Follows, 2013b: Winners and losers: Ecological and biogeochemical changes in a warming ocean. *Global Biogeochem. Cy.*, **27**(2), 463–477, doi:10.1002/gbc.20042.

Dutra, L.X.C. et al., 2015: Organizational drivers that strengthen adaptive capacity in the coastal zone of Australia. *Ocean Coast. Manage.*, **109**, 64–76, doi:10.1016/j.landusepol.2015.09.003.

Duvat, V.K.E. et al., 2017: Trajectories of exposure and vulnerability of small islands to climate change. *WiRes. Clim. Change*, **8**(6), e478 doi:10.1002/wcc.478.

Eddebbar, Y.A. et al., 2017: Impacts of ENSO on air-sea oxygen exchange: Observations and mechanisms. *Global Biogeochem. Cy.*, **31**(5), 901–921, doi:doi:10.1002/2017GB005630.

Eden, C. and D. Olbers, 2014: An Energy Compartment Model for Propagation, Nonlinear Interaction, and Dissipation of Internal Gravity Waves. *J. Phys. Oceanogr.*, **44**(8), 2093–2106, doi:10.1175/JPO-D-13-0224.1.

Edgar, G.J. et al., 2014: Global conservation outcomes depend on marine protected areas with five key features. *Nature*, **506**, 216, doi:10.1038/nature13022.

Edwards, M. et al., 2013: Marine Ecosystem Response to the Atlantic Multidecadal Oscillation. *PLoS One*, **8**(2), doi:10.1371/journal.pone.0057212.

Edwards, M. and A.J. Richardson, 2004: Impact of climate change on marine pelagic phenology and trophic mismatch. *Nature*, **430**(7002), nature02808-884, doi:10.1038/nature02808.

Eichner, M., B. Rost and S.A. Kranz, 2014: Diversity of ocean acidification effects on marine N2 fixers. *J. Exp. Mar. Biol. Ecol.*, **457**, 199–207, doi:10.1016/j.jembe.2014.04.015.

Ekstrom, J.A. et al., 2015: Vulnerability and adaptation of US shellfisheries to ocean acidification. *Nat. Clim. Change*, **5**, 207, doi:10.1038/nclimate2508.

El-Hacen, E.-H. M. et al., 2018: Evidence for 'critical slowing down' in seagrass: a stress gradient experiment at the southern limit of its range. *Sci. Rep.*, **8**(1), 17263, doi:10.1038/s41598-018-34977-5.

Elias, P. and A. Omojola, 2015: Case study: The challenges of climate change for Lagos, Nigeria. *Curr. Opin. Environ. Sustain.*, **13**, 74–78, doi:10.1016/j.cosust.2015.02.008.

Elliff, C.I. and I.R. Silva, 2017: Coral reefs as the first line of defense: Shoreline protection in face of climate change. *Mar. Environ. Res.*, **127**, 148–154, doi:10.1016/j.marenvres.2017.03.007.

Elliott, M., J.W. Day, R. Ramachandran and E. Wolanski, 2019: Chapter 1 – A Synthesis: What Is the Future for Coasts, Estuaries, Deltas and Other Transitional Habitats in 2050 and Beyond? In: *Coasts and Estuaries* [Wolanski, E., J.W. Day, M. Elliott and R. Ramachandran (eds.)]. Elsevier, pp. 1–28. ISBN: 9780128140031.

Elrick-Barr, C.E. et al., 2016: How are coastal households responding to climate change? *Environ. Sci. Policy*, **63**, 177–186, doi:10.1016/j.envsci.2016.05.013.

Elsharouny, M.R.M.M., 2016: Planning Coastal Areas and Waterfronts for Adaptation to Climate Change in Developing Countries. *Procedia Environ. Sci.*, **34**, 348–359, doi:10.1016/j.proenv.2016.04.031.

Endo, H., K. Suehiro, X. Gao and Y. Agatsuma, 2017: Interactive effects of elevated summer temperature, nutrient availability, and irradiance on growth and chemical compositions of juvenile kelp, Eisenia bicyclis. *Phycol. Res.*, **65**(2), 118–126, doi:10.1111/pre.12170.

Endres, S. et al., 2014: Stimulated Bacterial Growth under Elevated pCO2: Results from an Off-Shore Mesocosm Study. *PLoS One*, **9**(6), e99228, doi:10.1371/journal.pone.0099228.

Engel, J. et al., 2014: Towards the Disease Biomarker in an Individual Patient Using Statistical Health Monitoring. *PLoS One*, **9**(4), e92452, doi:10.1371/journal.pone.0092452.

Engelhard, G.H., D.A. Righton and J.K. Pinnegar, 2014: Climate change and fishing: a century of shifting distribution in North Sea cod. *Global Change Biol.*, **20**(8), 2473–2483, doi:10.1111/gcb.12513.

England, M.H., J.B. Kajtar and N. Maher, 2015: Robust warming projections despite the recent hiatus. *Nat. Clim. Change*, **5**, 394, doi:10.1038/nclimate2575.

England, M.H. et al., 2014: Recent intensification of wind-driven circulation in the Pacific and the ongoing warming hiatus. *Nat. Clim. Change*, **4**, 222, doi:10.1038/nclimate2106.

Enwright, N.M., K.T. Griffith and M.J. Osland, 2016: Barriers to and opportunities for landward migration of coastal wetlands with sea level rise. *Front. Ecol. Environ.*, **14**(6), 307–316, doi:10.1002/fee.1282.

Erauskin-Extramiana, M. et al., 2019: Large-scale distribution of tuna species in a warming ocean. *Global Change Biol.*, **25**(6), 2043–2060, doi:10.1111/gcb.14630.

Escobar, L.E. et al., 2015: A global map of suitability for coastal Vibrio cholerae under current and future climate conditions. *Acta Trop.*, **149**(Supplement C), 202–211, doi:10.1016/j.actatropica.2015.05.028.

Essington, T.E. et al., 2015: Fishing amplifies forage fish population collapses. *PNAS*, **112**(21), 6648.

Esteban, M. et al., 2017: Awareness of coastal floods in impoverished subsiding coastal communities in Jakarta: Tsunamis, typhoon storm surges and dyke-induced tsunamis. *Int. J. Disast. Risk Reduc.*, **23**, 70–79, doi:10.1016/j.ijdrr.2017.04.007.

Evans, L.S. et al., 2016: Structural and Psycho-Social Limits to Climate Change Adaptation in the Great Barrier Reef Region. *PLoS One*, **11**(3), e0150575.

5

Everaert, G. et al., 2018: Risk assessment of microplastics in the ocean: Modelling approach and first conclusions. *Environ. Pollut.*, **242**, 1930–1938, doi:10.1016/j.envpol.2018.07.069.

Eyre, B.D., A.J. Andersson and T. Cyronak, 2014: Benthic coral reef calcium carbonate dissolution in an acidifying ocean. *Nat. Clim. Change*, **4**, 969 EP-, doi:10.1038/nclimate2380.

Fabricius, K.E., 2005: Effects of terrestrial runoff on the ecology of corals and coral reefs: review and synthesis. *Mar. Pollut. Bull.*, **50**(2), 125–146, doi:10.1016/j.marpolbul.2004.11.028.

FAO, 2016: *The State of World Fisheries and Aquaculture* 2016. Contributing to food security and nutrition for all, FAO, Rome, 200 pp. ISBN 978-92-5-109185-2.

FAO, IFAD, UNICEF, WFP, and WHO, 2017: *The State of Food Security and Nutrition In The World, Building Resilience for Peace and Food Security. FAO, Rome*, 132 pp. ISBN: 978-92-5-109888-2.

FAO, 2018: *The State of World Fisheries and Aquaculture 2018 – Meeting the sustainable development goals*. FAO, Rome, pp 1–227. ISBN: 978-92-5-1305562-1.

FAO, 2019: *Deep-ocean climate change impacts on habitat, fish and fisheries* [Levin, L.A., M. Baker and A. Thompson (eds.)]. 638, FAO, Rome, 186 pp.

Faraco, L.F.D. et al., 2016: Vulnerability Among Fishers in Southern Brazil and its Relation to Marine Protected Areas in a Scenario of Declining Fisheries. *Desenvolvimento e Meio Ambiente*, **38**(1), 51–76, doi:10.5380/dma.v38i0.45850.

Fatorić, S. and E. Seekamp, 2017: Securing the Future of Cultural Heritage by Identifying Barriers to and Strategizing Solutions for Preservation under Changing Climate Conditions. *Sustainability*, **9**(11), 1–20, doi:10.3390/su9112143.

Fauville, G. et al., 2011: *Virtual Ocean Acidification Laboratory as an Efficient Educational Tool to Address Climate Change Issues in The Economic, Social and Political Elements of Climate Change* [W. Filho Leal ed.]. Springer Berlin Heidelberg, pp. 825–836. ISBN: 978-3-642-14776-0.

Feely, R.A. et al., 2016: Chemical and biological impacts of ocean acidification along the west coast of North America. *Estuar. Coast. Shelf Sci.*, **183**, 260–270, doi:10.1016/j.ecss.2016.08.043.

Feng, E.Y., P.K. David, K. Wolfgang and O. Andreas, 2016: Could artificial ocean alkalinization protect tropical coral ecosystems from ocean acidification? *Environ. Res. Lett.*, **11**(7), 074008.

Fennel, K. and J.M. Testa, 2019: Biogeochemical Controls on Coastal Hypoxia. *Annu. Rev. Mar. Sci.*, **11**(1), 105–130, doi:10.1146/annurev-marine-010318-095138.

Fennel, K. and D.L. VanderZwaag, 2016: Ocean acidification: Scientific surges, lagging law and policy responses in *Routledge handbook of maritime regulation and enforcement* [R. Warner and S. Kaye eds]. Routledge 1st eddition, pp. 342–362. London, UK.

Fernandes, J.A. et al., 2013: Modelling the effects of climate change on the distribution and production of marine fishes: accounting for trophic interactions in a dynamic bioclimate envelope model. *Global Change Biol.*, **19**(8), 2596–2607, doi:10.1111/gcb.12231.

Fernandez-Arcaya, U. et al., 2017: Ecological Role of Submarine Canyons and Need for Canyon Conservation: A Review. *Front. Mar. Sci.*, **4**, 69, doi:10.3389/fmars.2017.00005.

Ferrario, F. et al., 2014: The effectiveness of coral reefs for coastal hazard risk reduction and adaptation. *Nat. Commun.*, **5**, 3794, doi:10.1038/ncomms4794.

Filbee-Dexter, K., C.J. Feehan and R.E. Scheibling, 2016: Large-scale degradation of a kelp ecosystem in an ocean warming hotspot. *Mar. Ecol. Prog. Ser.*, **543**, 141–152, doi:10.3354/meps11554.

Filbee-Dexter, K. and T. Wernberg, 2018: Rise of turfs: A new battlefront for globally declining kelp forests. *BioScience*, **68**(2), 64–76.

Fillinger, L. and C. Richter, 2013: Vertical and horizontal distribution of *Desmophyllum dianthus* in Comau Fjord, Chile: a cold water coral thriving at low pH. *Peerj*, **1**, e194.

Fine, M. et al., 2019: Coral reefs of the Red Sea – Challenges and potential solutions. *Reg. Stud. Mar. Sci.*, **25**, 100498, doi:10.1016/j.rsma.2018.100498.

Fine, M., H. Gildor and A. Genin, 2013: A coral reef refuge in the Red Sea. *Global Change Biol.*, **19**(12), 3640–3647, doi:10.1111/gcb.12356.

Finkbeiner, E.M. et al., 2018: Exploring trade-offs in climate change response in the context of Pacific Island fisheries. *Mar. Policy*, **88**, 359–364, doi:10.1016/j.marpol.2017.09.032.

Fischer, A.P., 2018: Pathways of adaptation to external stressors in coastal natural-resource-dependent communities: Implications for climate change. *World Dev.*, **108**, 235–248, doi:10.1016/j.worlddev.2017.12.007.

Fish, M.R. et al., 2005: Predicting the impact of sea level rise on Caribbean sea turtle nesting habitat. *Conserv. Biol.*, **19**(2), 482–491, doi:10.1111/j.1523-1739.2005.00146.x.

Fisher, J.A. and K. Brown, 2015: Reprint of" Ecosystem services concepts and approaches in conservation: Just a rhetorical tool?". *Ecol. Econ.*, **117**, 261–269.

Flato, G. et al., 2013: Evaluation of Climate Models. In: Climate Change 2013: The Physical Science Basis. Contribution of Working Group I to the Fifth Assessment Report of the Intergovernmental Panel on Climate Change [Stocker, T.F., D. Qin, G.-K. Plattner, M. Tignor, S.K. Allen, J. Boschung, A. Nauels, Y. Xia, V. Bex and P.M. Midgley (eds.)]. Cambridge University Press, Cambridge, United Kingdom and New York, NY, USA, 741–866.

Fleeger, J.W. et al., 2006: Simulated sequestration of anthropogenic carbon dioxide at a deep sea site: Effects on nematode abundance and biovolume. *Deep sea Res. Pt. I*, **53**(7), 1135–1147, doi:10.1016/j.dsr.2006.05.007.

Fleeger, J.W. et al., 2010: The response of nematodes to deep sea CO_2 sequestration: A quantile regression approach. *Deep sea Res. Pt. I*, **57**(5), 696–707, doi:10.1016/j.dsr.2010.03.003.

Flombaum, P. et al., 2013: Present and future global distributions of the marine Cyanobacteria, Prochlorococcus and Synechococcus. *PNAS*, **110**(24), 9824, doi:10.1073/pnas.1307701110.

Flores, R. et al., 2017: Application of Transplantation Technology to Improve Coral Reef Resources for Sustainable Fisheries and Underwater Tourism. *Int. J. Environ. Sci. Dev.*, **8**(1), 44.

Fluhr, J. et al., 2017: Weakening of the subpolar gyre as a key driver of North Atlantic seabird demography: a case study with Brunnich's guillemots in Svalbard. *Mar. Ecol. Prog. Ser.*, **563**, 1–11, doi:10.3354/meps11982.

Folkersen, M.V., 2018: Ecosystem valuation: Changing discourse in a time of climate change. *Ecosyst. Serv.*, **29**, 1–12, doi:10.1016/j.ecoser.2017.11.008.

Form, A.U. and U. Riebesell, 2011: Acclimation to ocean acidification during long-term CO_2 exposure in the cold water coral Lophelia pertusa. *Global Change Biol.*, **18**(3), 843–853, doi:10.1111/j.1365-2486.2011.02583.x.

Forsman, Z.H., C.A. Page, R.J. Toonen and D. Vaughan, 2015: Growing coral larger and faster: micro-colony-fusion as a strategy for accelerating coral cover. *Peerj*, **3**(1), e1313, doi:10.7717/peerj.1313.

Fort, J. et al., 2015: Mercury in wintering seabirds, an aggravating factor to winter wrecks? *Sci. Total Environ.*, **527–528**(Supplement C), 448–454, doi:10.1016/j.scitotenv.2015.05.018.

Forzieri, G. et al., 2018: Escalating impacts of climate extremes on critical infrastructures in Europe. *Global Environ. Change*, **48**, 97–107, doi:10.1016/j.gloenvcha.2017.11.007.

Foteinis, S. and T. Tsoutsos, 2017: Strategies to improve sustainability and offset the initial high capital expenditure of wave energy converters (WECs). *Renew. Sustain. Energ. Rev.*, **70**, 775–785, doi:10.1016/j.rser.2016.11.258.

Fox, A.D., L.-A. Henry, D.W. Corne and J.M. Roberts, 2016: Sensitivity of marine protected area network connectivity to atmospheric variability. *R. Soc. Open Sci.*, **3**(11), 160494.

Frajka-Williams, E. et al., 2016: Compensation between meridional flow components of the Atlantic MOC at 26N. *Ocean Sci.*, **12**(2), 481–493.

Francesch-Huidobro, M. et al., 2017: Governance challenges of flood-prone delta cities: Integrating flood risk management and climate change in spatial planning. *Progr. Plan.*, **114**, 1–27, doi:10.1016/j.progress.2015.11.001.

5

Franco, A.C., N. Gruber, T.L. Frölicher and L. Kropuenske Artman, 2018a: Contrasting Impact of Future CO_2 Emission Scenarios on the Extent of CaCO3 Mineral Undersaturation in the Humboldt Current System. *J. Geophys. Res-Oceans*, **123**(3), 2018–2036, doi:10.1002/2018JC013857.

Franco, J.N. et al., 2018b: The 'golden kelp' Laminaria ochroleuca under global change: Integrating multiple eco-physiological responses with species distribution models. *J. Ecol.*, **106**(1), 47–58, doi:10.1111/1365-2745.12810.

Free, C.M. et al., 2019: Impacts of historical warming on marine fisheries production. *Science*, **363**(6430), 979, doi:10.1126/science.aau1758.

Freiwald A., Rogers A., Hall-Spencer J., Guinotte J.M., Davies A.J., Yesson C., Martin C.S. & Weatherdon L.V. 2017. Global distribution of cold-water corals (version 3.0) [online[. Second update to the dataset in Freiwald et al. (2004) by UNEP-WCMC, in collaboration with Andre Freiwald and John Guinotte. Cambridge (UK). UNEP World Conservation Monitoring Centre. http://data.unep-wcmc.org/datasets/3.

Frias-Torres, S. and C. van de Geer, 2015: Testing animal-assisted cleaning prior to transplantation in coral reef restoration. *Peerj*, **3**, e1287, doi:10.7717/peerj.1287.

Frieder, C.A. et al., 2014: Evaluating ocean acidification consequences under natural oxygen and periodicity regimes: Mussel development on upwelling margins. *Global Change Biol.*, (20), 754–764.

Friedrich, E. and D. Kretzinger, 2012: Vulnerability of wastewater infrastructure of coastal cities to sea level rise: A South African case study. *Water SA*, **38**(5), 755–764.

Friedrich, T. et al., 2012: Detecting regional anthropogenic trends in ocean acidification against natural variability. *Nat. Clim. Change*, **2**(3), 167–171, doi:10.1038/nclimate1372.

Frigstad, H. et al., 2015: Links between surface productivity and deep ocean particle flux at the Porcupine Abyssal Plain sustained observatory. *Biogeosciences*, **12**(19), 5885–5897, doi:10.5194/bg-12-5885-2015.

Frischknecht, M., M. Münnich and N. Gruber, 2018: Origin, Transformation, and Fate: The Three-Dimensional Biological Pump in the California Current System. *J. Geophys. Res-Oceans*, **123**(11), 7939–7962, doi:10.1029/2018JC013934.

Froehlich, H.E. et al., 2018: Comparative terrestrial feed and land use of an aquaculture-dominant world. *PNAS*, **115**(20), 5295, doi:10.1073/pnas.1801692115.

Frölicher, T.L., K.B. Rodgers, C.A. Stock and W.W.L. Cheung, 2016: Sources of uncertainties in 21st century projections of potential ocean ecosystem stressors. *Global Biogeochem. Cy.*, **30**(8), 1224–1243, doi:10.1002/2015gb005338.

Frölicher, T.L. et al., 2015: Dominance of the Southern Ocean in Anthropogenic Carbon and Heat Uptake in CMIP5 Models. *J. Clim.*, **28**(2), 862–886, doi:10.1175/jcli-d-14-00117.1.

Fu, F.-X. et al., 2007: Effects of increased temperature and CO_2 on photosynthesis, growth and elemental ratios in marine Synechococcus and Prochlorococcus (Cyanobacteria). *J. Phycol.*, **43**(3), 485–496, doi:10.1111/j.1529-8817.2007.00355.x.

Fu, W. et al., 2018: Reversal of Increasing Tropical Ocean Hypoxia Trends with Sustained Climate Warming. *Global Biogeochem. Cy.*, **32**(4), 551–564, doi:10.1002/2017gb005788.

Fu, W., J.T. Randerson and J.K. Moore, 2016: Climate change impacts on net primary production (NPP) and export production (EP) regulated by increasing stratification and phytoplankton community structure in the CMIP5 models. *Biogeosciences*, **13**(18), 5151–5170, doi:10.5194/bg-13-5151-2016.

Fuentes, M., C.J. Limpus, M. Hamann and J. Dawson, 2010: Potential impacts of projected sea level rise on sea turtle rookeries. *Aquat. Conserv. Mar. Frewshw. Ecosyst.*, **20**(2), 132–139, doi:10.1002/aqc.1088.

Fuentes-George, K., 2017: Consensus, Certainty, and Catastrophe: Discourse, Governance, and Ocean Iron Fertilization. *Global Environmental Politics*, **17**(2), 125–143, doi:10.1162/GLEP_a_00404.

Funayama, K., E. Hines, J. Davis and S. Allen, 2013: Effects of sea level rise on northern elephant seal breeding habitat at Point Reyes Peninsula, California. *Aquat. Conserv. Mar. Frewshw. Ecosyst.*, **23**(2), 233–245, doi:10.1002/aqc.2318.

Gage, J.D., 2003: Food inputs, utilization, carbon flow and energetics. In: *Ecosystems of the Deep Sea* [Tyler, P.A. (ed.)]. Elsevier, Amsterdam, Volume 28, 1st eddition, pp. 313–380. ISBN: 9780080494654.

Gage, J.D. and P.A. Tyler, 1993: *Deep sea biology: a natural history of organisms at the deep sea floor* [Gage, J.D. and P.A. Tyler eds.]. Cambridge University Press, Paperbackk eddition, 504 pp, *Journal of the Marine Biological Association of the United Kingdom* 73, no. 1, doi: 10.1017/S0025315400070156.

Gaines, S.D. et al., 2018: Improved fisheries management could offset many negative effects of climate change. *Sci. Adv.*, **4**(8), eaao1378, doi:10.1126/sciadv.aao1378.

Galappaththi, I.M., E.K. Galappaththi and S.S. Kodithuwakku, 2017: Can start-up motives influence social-ecological resilience in community-based entrepreneurship setting? Case of coastal shrimp farmers in Sri Lanka. *Mar. Policy*, **86**, 156–163, doi:10.1016/j.marpol.2017.09.024.

Galaz, V. et al., 2012: Polycentric systems and interacting planetary boundaries – Emerging governance of climate change-ocean acidification-marine biodiversity. *Ecol. Econ.*, **81**, 21–32, doi:10.1016/j.ecolecon.2011.11.012.

Galbraith, E.D., D.A. Carozza and D. Bianchi, 2017: A coupled human-Earth model perspective on long-term trends in the global marine fishery. *Nat. Commun.*, **8**, 14884 EP -, doi:10.1038/ncomms14884.

Gallagher, R.V., R.O. Makinson, P.M. Hogbin and N. Hancock, 2015: Assisted colonization as a climate change adaptation tool. *Austral Ecol.*, **40**(1), 12–20, doi:10.1111/aec.12163.

Galland, G., E. Harrould-Kolieb and D. Herr, 2012: The ocean and climate change policy. *Clim. Policy*, **12**(6), 764–771, doi:10.1080/14693062.2012.692207.

Gallo, N.D. and L.A. Levin, 2016: Fish Ecol. Evol. in the World's Oxygen Minimum Zones and Implications of Ocean Deoxygenation. *Adv. Mar. Biol.*, **74**, 117–198, doi:10.1016/bs.amb.2016.04.001.

Gallo, N.D., D.G. Victor and L.A. Levin, 2017: Ocean commitments under the Paris Agreement. *Nat. Clim. Change*, **7**(11), 833–838, doi:10.1038/NCLIMATE3422.

Gambi, C. et al., 2017: Functional response to food limitation can reduce the impact of global change in the deep-sea benthos. *Global Ecol. Biogeogr.*, **26**(9), 1008–1021, doi:10.1111/geb.12608.

Gao, L., S.R. Rintoul and W. Yu, 2018: Recent wind-driven change in Subantarctic Mode Water and its impact on ocean heat storage. *Nat. Clim. Change*, **8**(1), 58–63, doi:10.1038/s41558-017-0022-8.

García Molinos, J., M.T. Burrows and E.S. Poloczanska, 2017: Ocean currents modify the coupling between climate change and biogeographical shifts. *Sci. Rep.*, **7**(1), 1332, doi:10.1038/s41598-017-01309-y.

García Molinos, J. et al., 2015: Climate velocity and the future global redistribution of marine biodiversity. *Nat. Clim. Change*, **6**, 83, doi:10.1038/nclimate2769.

Garcia, N.S., F. Fu, P.N. Sedwick and D.A. Hutchins, 2014: Iron deficiency increases growth and nitrogen-fixation rates of phosphorus-deficient marine cyanobacteria. *The Isme Journal*, **9**, 238, doi:10.1038/ismej.2014.104.

García-Mendoza, E. et al., 2018: Mass Mortality of Cultivated Northern Bluefin Tuna Thunnus thynnus orientalis Associated With Chattonella Species in Baja California, Mexico. *Front. Mar. Sci.*, **5**(454), doi:10.3389/fmars.2018.00454.

García-Moreiras, I., V. Pospelova, S. García-Gil and C. Muñoz Sobrino, 2018: Climatic and anthropogenic impacts on the Ría de Vigo (NW Iberia) over the last two centuries: A high-resolution dinoflagellate cyst sedimentary record. *Palaeogeogr. Palaeoclim. Palaeoecol.*, **504**, 201–218, doi:10.1016/j.palaeo.2018.05.032.

García-Reyes, M. et al., 2015: Under Pressure: Climate Change, Upwelling, and Eastern Boundary Upwelling Ecosystems. *Front. Mar. Sci.*, **2**, 109.

5

Garzke, J., T. Hansen, S.M.H. Ismar and U. Sommer, 2016: Combined Effects of Ocean Warming and Acidification on Copepod Abundance, Body Size and Fatty Acid Content. *PLoS One*, **11**(5), e0155952, doi:10.1371/journal.pone.0155952.

Gatti, G. et al., 2017: Observational information on a temperate reef community helps understanding the marine climate and ecosystem shift of the 1980–90s. *Mar. Pollut. Bull.*, **114**(1), 528–538.

Gattuso, J.-P. et al., 2015: OCEANOGRAPHY. Contrasting futures for ocean and society from different anthropogenic CO_2 emissions scenarios. *Science*, **349**(6243), 1–10, doi:10.1126/science.aac4722.

Gattuso, J.-P. et al., 2018: Ocean Solutions to Address Climate Change and Its Effects on Marine Ecosystems. *Front. Mar. Sci.*, **5**(337), doi:10.3389/fmars.2018.00337.

Gazeau, F. et al., 2014: Impact of ocean acidification and warming on the Mediterranean mussel (Mytilus galloprovincialis). *Front. Mar. Sci.*, **1**, 62, doi:10.3389/fmars.2014.00062.

Gehlen, M. et al., 2014: Projected pH reductions by 2100 might put deep North Atlantic biodiversity at risk. *Biogeosciences*, **11**(23), 6955–6967, doi:10.5194/bg-11-6955-2014.

George, J.C. et al., 2015: Bowhead whale body condition and links to summer sea ice and upwelling in the Beaufort Sea. *Progr. Oceanogr.*, **136**, 250–262, doi:10.1016/j.pocean.2015.05.001.

George, R., 2013: *Ninety percent of everything: inside shipping, the invisible industry that puts clothes on your back, gas in your car, and food on your plate.* Macmillan-Picador, 1st eddition, 304 pp, USA, ISBN: 9781250058294.

Georgian, S.E. et al., 2016: Biogeographic variability in the physiological response of the cold water coral *Lophelia pertusa* to ocean acidification. *Mar. Ecol.*, **37**(6), 1345–1359, doi:10.1111/maec.12373.

Gephart, J.A. et al., 2016: Vulnerability to shocks in the global seafood trade network. *Environ. Res. Lett.*, **11**(3), 035008.

Gerkensmeier, B. and B.M.W. Ratter, 2018: Governing coastal risks as a social process – Facilitating integrative risk management by enhanced multi-stakeholder collaboration. *Environ. Sci. Policy*, **80**, 144–151, doi:10.1016/j.envsci.2017.11.011.

GESAMP, 2015: *Sources, fate and effects of microplastics in the marine environment: a global assessment.* [Kershaw, P.J. (ed.)]. International Maritime Organization, 96 pp., London, UK, ISSN: 1020-4873.

GESAMP, 2019: *High Level Review of a Wide Range of Proposed Marine Geoengineering Techniques* [Boyd, P.W. and C.M.G. Vivian (eds.)]. IMO/FAO/UNESCO-IOC/UNIDO/WMO/IAEA/UN/UN Environment/UNDP/ISA Joint Group of Experts on the Scientific Aspects of Marine Environmental Protection, GESAMP, International Maritime Organization, No. 98, 144 pp. London, UK, ISSN: 1020-4973.

Ghedini, G., B.D. Russell and S.D. Connell, 2015: Trophic compensation reinforces resistance: herbivory absorbs the increasing effects of multiple disturbances. *Ecol. Lett.*, **18**(2), 182–187.

Gibbin, E.M. et al., 2017: The evolution of phenotypic plasticity under global change. *Sci. Rep.*, **7**(1), 17253, doi:10.1038/s41598-017-17554-0.

Gienapp, P. and J. Merilä, 2018: Evolutionary Responses to Climate Change. In: *Encyclopedia of the Anthropocene* [Dellasala, D.A. and M.I. Goldstein (eds.)]. Elsevier, Oxford, pp. 51–59., ISBN: 9780128096659.

Gill, A.E., 1982: *Atmosphere – ocean dynamics* [W.L Donn ed.] International Geophysics Series, **30**, 662 pp., Academic Press, San Diego, Claifornia, USA, ISBN: 0-12-283520-4.

Gilly, W.F., J.M. Beman, S.Y. Litvin and B.H. Robison, 2013: Oceanographic and Biological Effects of Shoaling of the Oxygen Minimum Zone. *Annu. Rev. Mar. Sci.*, **5**(1), 393–420, doi:10.1146/annurev-marine-120710-100849.

Gilman, E.L., J. Ellison, N.C. Duke and C. Field, 2008: Threats to mangroves from climate change and adaptation options: a review. *Aquat. Bot.*, **89**(2), 237–250.

Gintert, B.E. et al., 2018: Marked annual coral bleaching resilience of an inshore patch reef in the Florida Keys: A nugget of hope, aberrance, or last man standing? *Coral Reefs*, **37**(2), 533–547, doi:10.1007/s00338-018-1678-x.

Giorgio, A. et al., 2018: Coastal Tourism Importance and Beach Users' Preferences: The "Big Fives" Criterions and Related Management Aspects. *J. Tourism Hospit.*, **7**(347), 2167–0269.1000347.

Gissi, E., S. Fraschetti and F. Micheli, 2019: Incorporating change in marine spatial planning: A review. *Environ. Sci. Policy*, **92**, 191–200, doi:10.1016/j.envsci.2018.12.002.

Gittings, J.A., D.E. Raitsos, G. Krokos and I. Hoteit, 2018: Impacts of warming on phytoplankton abundance and phenology in a typical tropical marine ecosystem. *Sci. Rep.*, **8**(1), 2240, doi:10.1038/s41598-018-20560-5.

Gittman, R.K. et al., 2015: Engineering away our natural defenses: an analysis of shoreline hardening in the US. *Front. Ecol. Environ.*, **13**(6), 301–307, doi:10.1890/150065.

Gleckler, P.J. et al., 2016: Industrial-era global ocean heat uptake doubles in recent decades. *Nat. Clim. Change*, **6**, 394, doi:10.1038/nclimate2915.

Gleckler, P.J. et al., 2012: Human-induced global ocean warming on multidecadal timescales. *Nat. Clim. Change*, **2**, 524, doi:10.1038/nclimate1553.

Glibert, P.M. et al., 2018: Key Questions and Recent Research Advances on Harmful Algal Blooms in Relation to Nutrients and Eutrophication. In: *Global Ecology and Oceanography of Harmful Algal Blooms* [Glibert, P.M., E. Berdalet, M.A. Burford, G.C. Pitcher and M. Zhou (eds.)]. Springer International Publishing, Cham, pp. 229–259, ISBN: 978-3-319-70069-4.

Glibert, P.M. et al., 2014: Vulnerability of coastal ecosystems to changes in harmful algal bloom distribution in response to climate change: projections based on model analysis. *Global Change Biol.*, **20**(12), 3845–3858.

Gobler, C.J. and H. Baumann, 2016: Hypoxia and acidification in ocean ecosystems: coupled dynamics and effects on marine life. *Biol. Lett.*, **12**(5), 20150976, doi:10.1098/rsbl.2015.0976.

Gobler, C.J. et al., 2017: Ocean warming since 1982 has expanded the niche of toxic algal blooms in the North Atlantic and North Pacific oceans. *PNAS*, **114**(19): 4975–4980, doi: 10.1073/pnas.1619575114.

Goericke, R., S.J. Bograd and D.S. Grundle, 2015: Denitrification and flushing of the Santa Barbara Basin bottom waters. *Deep Sea Res. Pt. II*, **112**, 53–60, doi:10.1016/j.dsr2.2014.07.012.

Golden, C.D. et al., 2016: Nutrition: Fall in fish catch threatens human health. *Nature*, **534**(7607), 317–320, doi:10.1038/534317a.

Goldenberg, S.U. et al., 2017: Boosted food web productivity through ocean acidification collapses under warming. *Global Change Biol.*, **23**(10), 4177–4184.

Gómez, C.E.G. et al., 2014: Responses of the tropical gorgonian coral Eunicea fusca to ocean acidification conditions. *Coral Reefs*, **34**, 451–460.

Gómez-Letona, M., A.G. Ramos, J. Coca and J. Arístegui, 2017: Trends in Primary Production in the Canary Current Upwelling System – A Regional Perspective Comparing Remote Sensing Models. *Front. Mar. Sci.*, **4**, 1–18, doi:10.3389/fmars.2017.00370.

Gonneea, M.E. et al., 2019: Salt marsh ecosystem restructuring enhances elevation resilience and carbon storage during accelerating relative sea level rise. *Estuar. Coast. Shelf Sci.*, **217**, 56–68, doi:10.1016/j.ecss.2018.11.003.

Good, P. et al., 2018: Recent progress in understanding climate thresholds: Ice sheets, the Atlantic meridional overturning circulation, tropical forests and responses to ocean acidification. *Prog. Phys. Geog.*, **42**(1), 24–60, doi:10.1177/0309133317751843.

Good, S.A., M.J. Martin and N.A. Rayner, 2013: EN4: Quality controlled ocean temperature and salinity profiles and monthly objective analyses with uncertainty estimates. *J. Geophys. Res-Oceans*, **118**(12), 6704–6716, doi:10.1002/2013JC009067.

Gooday, A.J., J.M. Bernhard, L.A. Levin and S.B. Suhr, 2000: Foraminifera in the Arabian Sea oxygen minimum zone and other oxygen-deficient settings: taxonomic composition, diversity, and relation to metazoan faunas. *Deep Sea Res. Pt. II*, **47**(1), 25–54, doi:10.1016/S0967-0645(99)00099-5.

Goreau, T.J.F. and P. Prong, 2017: Biorock Electric Reefs Grow Back Severely Eroded Beaches in Months. *J. Mar. Sci. Eng.*, **5**(4), 48.

5

Gorgues, T., O. Aumont and L. Memery, 2019: Simulated Changes in the Particulate Carbon Export Efficiency due to Diel Vertical Migration of Zooplankton in the North Atlantic. *Geophys. Res. Lett.*, **46**(10); 5387–5395, doi:10.1029/2018GL081748.

Gori, A. et al., 2016: Physiological response of the cold water coral *Desmophyllum dianthusto* thermal stress and ocean acidification. *Peerj*, **4**, e1606, doi:10.7717/peerj.1606.

Gormley, K.S.G. et al., 2015: Adaptive management, international co-operation and planning for marine conservation hotspots in a changing climate. *Mar. Policy*, **53**, 54–66, doi:10.1016/j.marpol.2014.11.017.

Gourlie, D. et al., 2017: Performing "A New Song": Suggested Considerations for Drafting Effective Coastal Fisheries Legislation Under Climate Change. *Mar. Policy*, **88**; 342–349, doi:10.1016/j.marpol.2017.06.012.

Gourlie, D. et al., 2018: Performing "A New Song": Suggested Considerations for Drafting Effective Coastal Fisheries Legislation Under Climate Change. *Mar. Policy*, **88**, 342–349, doi:10.1016/j.marpol.2017.06.012.

Gracey, M.S., 2007: Nutrition-related disorders in Indigenous Australians: how things have changed. *Med, J. Aust.*, **186**(1), 15.

Gracia, A., N. Rangel-Buitrago, J.A. Oakley and A.T. Williams, 2018: Use of ecosystems in coastal erosion management. *Ocean Coast. Manage.*, **156**, 277–289, doi:10.1016/j.ocecoaman.2017.07.009.

Gradoville, M.R. et al., 2014: Diversity trumps acidification: Lack of evidence for carbon dioxide enhancement of Trichodesmium community nitrogen or carbon fixation at Station ALOHA. *Limnol. Oceanogr.*, **59**(3), 645–659, doi:10.4319/lo.2014.59.3.0645.

Graham, N.A.J. et al., 2013: Managing resilience to reverse phase shifts in coral reefs. *Front. Ecol. Environ.*, **11**(10), 541–548, doi:10.1890/120305.

Graham, N.A.J., J.E. Cinner, A.V. Norstrom and M. Nystrom, 2014: Coral reefs as novel ecosystems: embracing new futures. *Curr. Opin. Environ. Sustain.*, **7**, 9–14, doi:10.1016/j.cosust.2013.11.023.

Grassi, G. et al., 2017: The key role of forests in meeting climate targets requires science for credible mitigation. *Nat. Clim. Change*, **7**, 220, doi:10.1038/nclimate3227.

Greene, C., B. Monger and M. Huntley, 2010: Geoengineering: The inescapable truth of getting to 350. *Solutions*, **1**(5), 57–66.

Greene, C.H. et al., 2016: Marine microalgae: Climate, energy, and food security from the sea. *Oceanography*, **29**(4), 10–15.

Gregg, W.W. and C.S. Rousseaux, 2014: Decadal trends in global pelagic ocean chlorophyll: A new assessment integrating multiple satellites, in situ data, and models. *J. Geophys. Res-Oceans*, **119**(9), 5921–5933, doi:10.1002/2014JC010158.

Griscom, B.W. et al., 2017: Natural climate solutions. *PNAS*, **114**(44), 11645.

Gruber, N., 2019: A diagnosis for marine nitrogen fixation. *Nature*, **566**(7743), 191–193.

Gruber, N. et al., 2019: The oceanic sink for anthropogenic CO_2 from 1994 to 2007. *Science*, **363**(6432), 1193, doi:10.1126/science.aau5153.

Gruber, N. et al., 2012: Rapid Progression of Ocean Acidification in the California Current System. *Science*, **337**(6091), 220.

Grundle, D.S. et al., 2017: Low oxygen eddies in the eastern tropical North Atlantic: Implications for N2O cycling. *Sci. Rep.*, **7**(1), 4806, doi:10.1038/s41598-017-04745-y.

Gu, J. et al., 2018: Losses of salt marsh in China: Trends, threats and management. *Estuar. Coast. Shelf Sci.*, **214**, 98–109, doi:10.1016/j.ecss.2018.09.015.

Guevara-Carrasco, R. and J. Lleonart, 2008: Dynamics and fishery of the Peruvian hake: Between nature and man. *J. Mar. Syst.*, **71**(3), 249–259, doi:10.1016/j.jmarsys.2007.02.030.

Guidi, L. et al., 2016: Plankton networks driving carbon export in the oligotrophic ocean. *Nature*, **532**, 465, doi:10.1038/nature16942.

Guinder, V.A. et al., 2018: Plankton multiproxy analyses in the Northern Patagonian Shelf, Argentina: community structure, phycotoxins and characterization of Alexandrium strains. *Front. Mar. Sci.*, **5**, 394.

Guinotte, J.M. et al., 2006: Will human-induced changes in seawater chemistry alter the distribution of deep sea scleractinian corals? *Front. Ecol. Environ.*, **4**(3), 141–146, doi:10.1890/1540-9295(2006)004[0141:WHCISC]2.0.CO;2.

Gunderson, A.R., E.J. Armstrong and J.H. Stillman, 2016: Multiple Stressors in a Changing World: The Need for an Improved Perspective on Physiological Responses to the Dynamic Marine Environment. *Annu. Rev. Mar. Sci.*, **8**(1), 357–378, doi:10.1146/annurev-marine-122414-033953.

Gunderson, A.R., B. Tsukimura and J.H. Stillman, 2017: Indirect Effects of Global Change: From Physiological and Behavioral Mechanisms to Ecological Consequences. *Integr. Comp. Biol.*, **57**(1), 48–54, doi:10.1093/icb/icx056.

Gutiérrez, M.T., P. Jorge Castillo, B. Laura Naranjo and M.J. Akester, 2017: Current state of goods, services and governance of the Humboldt Current Large Marine Ecosystem in the context of climate change. *Environ. Dev.*, **22**, 175–190, doi:10.1016/j.envdev.2017.02.006.

Guy, E. and F. Lasserre, 2016: Commercial shipping in the Arctic: new perspectives, challenges and regulations. *Polar Record*, **52**(3), 294–304, doi:10.1017/S0032247415001011.

Hagedoorn, L.C. et al., 2019: Community-based adaptation to climate change in small island developing states: an analysis of the role of social capital. *Clim. Dev.*, 1–12, doi:10.1080/17565529.2018.1562869.

Haines-Young, R. and M. Potschin, 2013: *Common International Classification of Eco- system Services (CICES): Consultation on Version 4, August–December 2012*, pp 34.

Hale, R. et al., 2017: Mediation of macronutrients and carbon by post-disturbance shelf sea sediment communities. *Biogeochemistry*, **135**(1), 121–133, doi:10.1007/s10533-017-0350-9.

Hall-Spencer, J.M. et al., 2008: Volcanic carbon dioxide vents show ecosystem effects of ocean acidification. *Nature*, **454**, 96, doi:10.1038/nature07051.

Hallberg, R., 2013: Using a resolution function to regulate parameterizations of oceanic mesoscale eddy effects. *Ocean Model.*, **72**, 92–103, doi:10.1016/j.ocemod.2013.08.007.

Hallberg, R. et al., 2012: Sensitivity of Twenty-First-Century Global-Mean Steric Sea Level Rise to Ocean Model Formulation. *J. Clim.*, **26**(9), 2947–2956, doi:10.1175/JCLI-D-12-00506.1.

Hallegatte, S. et al., 2015: *Shock waves: managing the impacts of climate change on poverty*. The World Bank.

Hallegraeff, G.M., 2010: Ocean climate change, phytoplankton community responses, and harmful algal blooms: a formidable predictive challenge1. *J. Phycol.*, **46**(2), 220–235.

Hallegraeff, G.M., 2016: Impacts and effects of ocean warming on marine phytoplankton and harmful algal blooms. in *Explaining ocean warming: Causes, scale, effects and consequences. Full Report.* [D. Laffoley and J.M. Baxter eds.], 456 pp, IUCN, Gland, Switzerland, ISBN: 978-8317-1806-4.

Hallett, C.S. et al., 2018: Observed and predicted impacts of climate change on the estuaries of south-western Australia, a Mediterranean climate region. *Reg. Environ. Change*, **18**(5), 1357–1373, doi:10.1007/s10113-017-1264-8.

Hamdan, L.J. and K.P. Wickland, 2016: Methane emissions from oceans, coasts, and freshwater habitats: New perspectives and feedbacks on climate. *Limnol. Oceanogr.*, **61**(S1), S3–S12, doi:10.1002/lno.10449.

Hameau, A., J. Mignot and F. Joos, 2019: Assessment of time of emergence of anthropogenic deoxygenation and warming: insights from a CESM simulation from 850 to 2100 CE. *Biogeosciences*, **16**(8), 1755–1780, doi:10.5194/bg-16-1755-2019.

Hamilton, L.C. and T.G. Safford, 2015: Environmental Views from the Coast: Public Concern about Local to Global Marine Issues. *Society & Natural Resources*, **28**(1), 57–74, doi:10.1080/08941920.2014.933926.

Hammond, M.L., C. Beaulieu, S.K. Sahu and S.A. Henson, 2017: Assessing trends and uncertainties in satellite-era ocean chlorophyll using space-time modeling. *Global Biogeochem. Cy.*, **31**(7), 1103–1117, doi:10.1002/2016gb005600.

Han, W. et al., 2014: Indian Ocean Decadal Variability: A Review. *Bull. Am. Meteorol. Soc.*, **95**(11), 1679–1703, doi:10.1175/BAMS-D-13-00028.1.

5

Hanz, U. et al., 2019: Environmental factors influencing cold water coral ecosystems in the oxygen minimum zones on the Angolan and Namibian margins. *Biogeosciences*, (*In review*) 1–37.

Harborne, A.R. et al., 2017: Multiple Stressors and the Functioning of Coral Reefs. *Annu. Rev. Mar. Sci., Vol 8*, **9**(1), 445–468, doi:10.1146/annurev-marine-010816-060551.

Harkes, I.H.T. et al., 2015: Shrimp aquaculture as a vehicle for Climate Compatible Development in Sri Lanka. The case of Puttalam Lagoon. *Mar. Policy*, **61**, 273–283, doi:10.1016/j.marpol.2015.08.003.

Harley, C.D.G., 2011: Climate change, keystone predation, and biodiversity loss. *Science*, **334**(6059), 1124–1127, doi:10.1126/science.1210199.

Harris, P.T., M. Macmillan-Lawler, J. Rupp and E.K. Baker, 2014: Geomorphology of the oceans. *Mar. Geol.*, **352**(Supplement C), 4–24.

Harris, P.T. and T. Whiteway, 2011: Global distribution of large submarine canyons: Geomorphic differences between active and passive continental margins. *Mar. Geol.*, **285**(1), 69–86, doi:10.1016/j.margeo.2011.05.008.

Harrison, D.P., 2017: Global negative emissions capacity of ocean macronutrient fertilization. *Environ. Res. Lett.*, **12**(3), 035001.

Harrison, G.P. and A.R. Wallace, 2005: Climate sensitivity of marine energy. *Renew. Energ.*, **30**(12), 1801–1817, doi:10.1016/j.renene.2004.12.006.

Harrould-Kolieb, E.R. and D. Herr, 2012: Ocean acidification and climate change: synergies and challenges of addressing both under the UNFCCC. *Clim. Policy*, **12**(3), 378–389, doi:10.1080/14693062.2012.620788.

Harrould-Kolieb, E.R. and O. Hoegh-Guldberg, 2019: A governing framework for international ocean acidification policy. *Mar. Policy*, **102**, 10–20, doi:10.1016/j.marpol.2019.02.004.

Hartman, S.E. et al., 2015: Biogeochemical variations at the Porcupine Abyssal Plain sustained Observatory in the northeast Atlantic Ocean, from weekly to inter-annual timescales. *Biogeosciences*, **12**(3), 845–853, doi:10.5194/bg-12-845-2015.

Harvey, B.J., K.L. Nash, J.L. Blanchard and D.P. Edwards, 2018: Ecosystem-based management of coral reefs under climate change. *Ecol. Evol.*, **8**(12), 6354–6368, doi:10.1002/ece3.4146.

Harvey, B.P., D. Gwynn-Jones and P.J. Moore, 2013: Meta-analysis reveals complex marine biological responses to the interactive effects of ocean acidification and warming. *Ecol. Evol.*, **3**(4), 1016–1030, doi:10.1002/ece3.516.

Hatfield, J.S., M.H. Reynolds, N.E. Seavy and C.M. Krause, 2012: Population dynamics of Hawaiian seabird colonies vulnerable to sea level rise. *Conserv. Biol.*, **26**(4), 667–78, doi:10.1111/j.1523-1739.2012.01853.x.

Hauri, C., T. Friedrich and A. Timmermann, 2015: Abrupt onset and prolongation of aragonite undersaturation events in the Southern Ocean. *Nat. Clim. Change*, **6**, 172, doi:10.1038/nclimate2844.

Hauser, D.D.W. et al., 2017: Decadal shifts in autumn migration timing by Pacific Arctic beluga whales are related to delayed annual sea ice formation. *Global Change Biol.*, **23**(6), 2206–2217, doi:10.1111/gcb.13564.

Hawkins, E. and R. Sutton, 2012: Time of emergence of climate signals. *Geophys. Res. Lett.*, **39**(1); 1–6, doi:10.1029/2011gl050087.

Hawkins, S. et al., 2016: *Impacts and effects of ocean warming on intertidal rocky habitats* in *Explaining ocean warming: Cause, scale, effects and consequences. Full report.* [D. Laffoley and J.M. Baxter eds.] IUCN, 147–176, Gland, CH, ISBN: 978-2-8317-1806-4.

Hayden, H.L. and E.F. Granek, 2015: Coastal sediment elevation change following anthropogenic mangrove clearing. *Estuar. Coast. Shelf Sci.*, **165**, 70–74, doi:10.1016/j.ecss.2015.09.004.

Haynert, K. et al., 2011: Biometry and dissolution features of the benthic foraminifer Ammonia aomoriensis at high pCO$_2$. *Mar. Ecol. Prog. Ser.*, **432**, 53–67.

Haynie, A.C. and L. Pfeiffer, 2012: Why economics matters for understanding the effects of climate change on fisheries. *ICES J. Mar. Sci.*, **69**(7), 1160–1167, doi:10.1093/icesjms/fss021.

Hays, G.C., A.C. Broderick, F. Glen and B.J. Godley, 2003: Climate change and sea turtles: a 150-year reconstruction of incubation temperatures at a major marine turtle rookery. *Global Change Biol.*, **9**(4), 642–646, doi:10.1046/j.1365-2486.2003.00606.x.

Hazen, E.L. et al., 2013: Predicted habitat shifts of Pacific top predators in a changing climate. *Nat. Clim. Change*, **3**(3), 234–238, doi:10.1038/nclimate1686.

Heenan, A. et al., 2015: A climate-informed, ecosystem approach to fisheries management. *Mar. Policy*, **57**, 182–192, doi:10.1016/j.marpol.2015.03.018.

Hein, M.Y., B.L. Willis, R. Beeden and A. Birtles, 2017: The need for broader ecological and socioeconomic tools to evaluate the effectiveness of coral restoration programs. *Restor. Ecol.*, **25**(6), 873–883, doi:10.1111/rec.12580.

Hejnowicz, A.P., H. Kennedy, M.A. Rudd and M.R. Huxham, 2015: Harnessing the climate mitigation, conservation and poverty alleviation potential of seagrasses: prospects for developing blue carbon initiatives and payment for ecosystem service programmes. *Front. Mar. Sci.*, **2**, 32.

Held, I.M. and B.J. Soden, 2006: Robust Responses of the Hydrological Cycle to Global Warming. *J. Clim.*, **19**(21), 5686–5699, doi:10.1175/JCLI3990.1.

Helland-Hansen, B., 1916: *Nogen hydrografiske metoder*. Scand. Naturforsker Mote, Kristiana, Oslo.

Helm, K.P., N.L. Bindoff and J.A. Church, 2011: Observed decreases in oxygen content of the global ocean. *Geophys. Res. Lett.*, **38**(23), doi:10.1029/2011GL049513.

Hemer, M.A. et al., 2018: Perspectives on a way forward for ocean renewable energy in Australia. *Renew. Energ.*, **127**, 733–745, doi:10.1016/j.renene.2018.05.036.

Henderson, E.E. et al., 2014: Effects of fluctuations in sea-surface temperature on the occurrence of small cetaceans off Southern California. *Fish-B NOAA*, **112**(2–3), 159–177, doi:10.7755/fb.112.2-3.5.

Hennige, S.J. et al., 2015: Hidden impacts of ocean acidification to live and dead coral framework. *Proc. Biol. Sci.*, **282**(1813), doi:10.1098/rspb.2015.0990.

Henry, L.A. et al., 2016: Seamount egg-laying grounds of the deep-water skate Bathyraja richardsoni. *J. Fish Biol.*, **89**(2), 1473–1481, doi:10.1111/jfb.13041.

Henson, S.A. et al., 2017: Rapid emergence of climate change in environmental drivers of marine ecosystems. *Nat. Commun.*, **8**, 14682, doi:10.1038/ncomms14682.

Henson, S.A., C. Beaulieu and R. Lampitt, 2016: Observing climate change trends in ocean biogeochemistry: when and where. *Global Change Biol.*, **22**(4), 1561–1571, doi:10.1111/gcb.13152.

Henson, S.A. et al., 2010: Detection of anthropogenic climate change in satellite records of ocean chlorophyll and productivity. *Biogeosciences*, **7**(2), 621–640, doi:10.5194/bg-7-621-2010.

Hernán, G. et al., 2017: Future warmer seas: increased stress and susceptibility to grazing in seedlings of a marine habitat-forming species. *Global Change Biol.*, **23**(11), 4530–4543, doi:10.1111/gcb.13768.

Hernández-Delgado, E.A., 2015: The emerging threats of climate change on tropical coastal ecosystem services, public health, local economies and livelihood sustainability of small islands: Cumulative impacts and synergies. *Mar. Pollut. Bull.*, **101**(1), 5–28, doi:10.1016/j.marpolbul.2015.09.018.

Hernández-Delgado, E.A. et al., 2014: Community-Based Coral Reef Rehabilitation in a Changing Climate: Lessons Learned from Hurricanes, Extreme Rainfall, and Changing Land Use Impacts. *Open Ecol. J.*, **04**(14), 918–944, doi:10.4236/oje.2014.414077.

Hernández-González, Y. et al., 2016: *Perspectives on contentions about climate change adaptation in the Canary Islands: A case study for Tenerife*. Publication Office of the European Union, Luxembourg, 74 pp, ISBN: 978-92-79-64595-2.

Heron, S.F., 2017: *Impacts of Climate Change on World Heritage Coral Reefs: A First Global Scientic Assessment*. UNESCO World Heritage Centre, Paris, 16 pp.

Heron, S.F., J.A. Maynard, R. van Hooidonk and C.M. Eakin, 2016: Warming Trends and Bleaching Stress of the World's Coral Reefs 1985–2012. *Sci. Rep.*, **6**(1), doi:10.1038/srep38402.

Herr, D., K. Isensee, E. Harrould-Kolieb and C. Turley, 2014: *Ocean Acidification, iv*. IUCN, Gland, Switzerland, 52 pp.

Herr, D. and E. Landis, 2016: *Coastal blue carbon ecosystems. Opportunities for nationally determined contributions.* Policy Brief. Gland, Switzerland: IUCN and Washington, DC, USA: TNC.

Herr, D., M. Unger, D. Laffoley and A. McGivern, 2017: Pathways for implementation of blue carbon initiatives. *Aquat. Conserv. Mar. Freshw. Ecosyst.*, **27**(S1), 116–129, doi:10.1002/aqc.2793.

Herring, P.J. and D.R. Dixon, 1998: Extensive deep sea dispersal of postlarval shrimp from a hydrothermal vent. *Deep sea Res. Pt. I*, **45**(12), 2105–2118, doi:10.1016/S0967-0637(98)00050-8.

Hess, J. and I. Kelman, 2017: Tourism Industry Financing of Climate Change Adaptation: Exploring the Potential in Small Island Developing States. *Clim. Disast. Dev. J.*, **2**(2), 33–45.

Heuer, R.M. and M. Grosell, 2014: Physiological impacts of elevated carbon dioxide and ocean acidification on fish. *Am. J. Physiol. Regul. Integr. Comp. Physiol.*, **307**(9), R1061–R1084, doi:10.1152/ajpregu.00064.2014.

Heuzé, C., K.J. Heywood, D.P. Stevens and J.K. Ridley, 2015: Changes in global ocean bottom properties and volume transports in CMIP5 models under climate change scenarios. *J. Clim.*, **28**(8), 2917–2944.

Hewitt, J.E., J.I. Ellis and S.F. Thrush, 2016: Multiple stressors, nonlinear effects and the implications of climate change impacts on marine coastal ecosystems. *Global Change Biol.*, **22**(8), 2665–2675.

Hill, R. et al., 2015: Can macroalgae contribute to blue carbon? An Australian perspective. *Limnol. Oceanogr.*, **60**(5), 1689–1706, doi:10.1002/lno.10128.

Hilmi, N. et al., 2017: Ocean acidification in the Middle East and North African region. *Region et Developpement*, **46**, 43–57 pp, LEADm Universite du Sud - Toulon Var.

Hilmi, N. et al., 2015: *Bridging the gap between ocean acidification impacts and economic valuation: regional impacts of ocean acidification on fisheries and aquaculture.* Brochure of The Third International Monaco Workshop on Economics of Ocean Acidification, Monaco.

Himes-Cornell, A. and S. Kasperski, 2015a: Assessing climate change vulnerability in Alaska's fishing communities. *Fish. Res.*, **162**, 1–11, doi:10.1016/j.fishres.2014.09.010.

Himes-Cornell, A. and S. Kasperski, 2015b: Assessing climate change vulnerability in Alaska's fishing communities. *Fish. Res.*, **162**, 1–11, doi:10.1016/j.fishres.2014.09.010.

Hindell, M.A. et al., 2016: Circumpolar habitat use in the southern elephant seal: implications for foraging success and population trajectories. *Ecosphere*, **7**(5), e01213, doi:10.1002/ecs2.1213.

Hinkel, J. et al., 2014: Coastal flood damage and adaptation costs under 21st century sea level rise. *PNAS*, **(9)** 3292–3297, doi:10.1073/pnas.1222469111.

Hiscock, J.A. and B.L. Chilvers, 2014: Declining eastern rockhopper (Eudyptes filholi) and erect-crested (E-sclateri) penguins on the Antipodes Islands, New Zealand. *New Zeal. J. Ecol.*, **38**(1), 124–131.

Hiwasaki, L., E. Luna, Syamsidik and R. Shaw, 2014: Process for integrating local and indigenous knowledge with science for hydro-meteorological disaster risk reduction and climate change adaptation in coastal and small island communities. *Int. J. Disast. Risk Reduc.*, **10**, 15–27, doi:10.1016/j.ijdrr.2014.07.007.

Ho, C.-H. et al., 2016: Mitigating uncertainty and enhancing resilience to climate change in the fisheries sector in Taiwan: Policy implications for food security. *Ocean Coast. Manage.*, **130**, 355–372, doi:10.1016/j.ocecoaman.2016.06.020.

Hobday, A.J. et al., 2015: Reconciling conflicts in pelagic fisheries under climate change. *Deep Sea Res. Pt. II*, **113**, 291–300, doi:10.1016/j.dsr2.2014.10.024.

Hobday, A.J. et al., 2016a: Planning adaptation to climate change in fast-warming marine regions with seafood-dependent coastal communities. *Rev. Fish Biol. Fisher.*, **26**(2), 249–264, doi:10.1007/s11160-016-9419-0.

Hobday, A.J., C.M. Spillman, J. Paige Eveson and J.R. Hartog, 2016b: Seasonal forecasting for decision support in marine fisheries and aquaculture. *Fish. Oceanogr.*, **25**(S1), 45–56, doi:doi:10.1111/fog.12083.

Hodgson, E.E. et al., 2018: Consequences of spatially variable ocean acidification in the California Current: Lower pH drives strongest declines in benthic species in southern regions while greatest economic impacts occur in northern regions. *Ecol. Modell.*, **383**, 106–117, doi:10.1016/j.ecolmodel.2018.05.018.

Hodgson, J., W. Russell and M. Megannety, 2016: Exploring plausible futures for marine transportation in the Canadian arctic, a scenarios based approach. Prepared for Transport Canada. Hodgson and Associates, Vancouver, Canada, 120 pp.

Hoegh-Guldberg, O., 2015: *Reviving the Ocean Economy: the case for action-2015.* WWF International. Gland, Switzerland, Geneva.

Hoegh-Guldberg, O. et al., 2014: The Ocean. In: Climate Change 2014: Impacts, Adaptation, and Vulnerability. Part B: Regional Aspects. Contribution of Working Group II to the Fifth Assessment Report of the Intergovernmental Panel of Climate Change [Barros, V.R., C.B. Field, D.J. Dokken, M.D. Mastrandrea, K.J. Mach, T.E. Bilir, M. Chatterjee, K.L. Ebi, Y.O. Estrada, R.C. Genova, B. Girma, E.S. Kissel, A.N. Levy, S. MacCracken, P.R. Mastrandrea and L.L. White (eds.)]. Cambridge University Press, Cambridge, United Kingdom and New York, NY, USA, 1655–1731 pp., ISBN: 978-1-107-05807-1.

Hoegh-Guldberg, O. et al., 2008: Assisted Colonization and Rapid Climate Change. *Science*, **321**(5887), 345.

Hoegh-Guldberg, O. et al., 2018: Impacts of 1.5°C Global Warming on Natural and Human Systems. In: Global Warming of 1.5°C. An IPCC Special Report on the impacts of global warming of 1.5°C above pre-industrial levels and related global greenhouse gas emission pathways, in the context of strengthening the global response to the threat of climate change, sustainable development, and efforts to eradicate poverty [Masson-Delmotte, V., P. Zhai, H.O. Pörtner, D. Roberts, J. Skea, P.R. Shukla, A. Pirani, W. Moufouma-Okia, C. Péan, R. Pidcock, S. Connors, J.B.R. Matthews, Y. Chen, X. Zhou, M.I. Gomis, E. Lonnoy, T. Maycock, M. Tignor and T. Waterfield (eds.)]. 630 pp., In Press.

Höfer, J. et al., 2018: All you can eat: the functional response of the cold water coral *Desmophyllum dianthus* feeding on krill and copepods. *Peerj*, **6**, e5872.

Hofmann, M. and H.-J. Schellnhuber, 2009: Oceanic acidification affects marine carbon pump and triggers extended marine oxygen holes. *PNAS*, **106**(9), 3017.

Hogg, A.M. et al., 2015: Recent trends in the Southern Ocean eddy field. *J. Geophys. Res-Oceans*, **120**(1), 257–267, doi:10.1002/2014JC010470.

Hong, H. et al., 2017: The complex effects of ocean acidification on the prominent N$_2$-fixing cyanobacterium Trichodesmium. *Science*, **356**(6337), 527, doi:10.1126/science.aal2981.

Hooper, T. and M. Austen, 2013: Tidal barrages in the UK: Ecological and social impacts, potential mitigation, and tools to support barrage planning. *Renew. Sustain. Energ. Rev.*, **23**, 289–298, doi:10.1016/j.rser.2013.03.001.

Hopkins, C.R., D.M. Bailey and T. Potts, 2016: Perceptions of practitioners: Managing marine protected areas for climate change resilience. *Ocean Coast. Manage.*, **128**, 18–28, doi:10.1016/j.ocecoaman.2016.04.014.

Horoszowski-Fridman, Y.B., J.-C. Brêthes, N. Rahmani and B. Rinkevich, 2015: Marine silviculture: Incorporating ecosystem engineering properties into reef restoration acts. *Ecol. Eng.*, **82**(Supplement C), 201–213.

Horoszowski-Fridman, Y.B., I. Izhaki and B. Rinkevich, 2011: Engineering of coral reef larval supply through transplantation of nursery-farmed gravid colonies. *J. Exp. Mar. Biol. Ecol.*, **399**(2), 162–166, doi:10.1016/j.jembe.2011.01.005.

Horoszowski-Fridman, Y.B. and B. Rinkevich, 2017: Restoration of the Animal Forests: Harnessing Silviculture Biodiversity Concepts for Coral

5

Transplantation in Marine Animal Forests.[Rossi, S., L. Bramanti, A. Gori and C. Orejas (eds.)]. Springer International Publishing, Cham, pp. 1–2, ISBN: 978-3-319-21011-7.

Houser, C., P. Wernette and B.A. Weymer, 2018: Scale-dependent behavior of the foredune: Implications for barrier island response to storms and sea level rise. *Geomorphology*, **303**, 362–374, doi:10.1016/j.geomorph.2017.12.011.

Howard, J. et al., 2017: Clarifying the role of coastal and marine systems in climate mitigation. *Front. Ecol. Environ.*, **15**(1), 42–50, doi:10.1002/fee.1451.

Hu, L. et al., 2016: Recent organic carbon sequestration in the shelf sediments of the Bohai Sea and Yellow Sea, China. *J. Mar. Syst.*, **155**, 50–58, doi:10.1016/j.jmarsys.2015.10.018.

Hubbard, D., J. Dugan, N. Schooler and S. Viola, 2014: Local extirpations and regional declines of endemic upper beach invertebrates in southern California. *Estuar. Coast. Shelf Sci.*, **150**, 67–75.

Hudson, D.M. et al., 2018: Physiological and behavioral response of the Asian shore crab, *Hemigrapsus sanguineus*, to salinity: implications for estuarine distribution and invasion. *Peerj*, **6**, e5446, doi:10.7717/peerj.5446.

Hudson, J.M. et al., 2014: Myctophid feeding ecology and carbon transport along the northern Mid-Atlantic Ridge. *Deep sea Res. Pt. I*, **93**, 104–116, doi:10.1016/j.dsr.2014.07.002.

Hughes, A.D. et al., 2012: Does seaweed offer a solution for bioenergy with biological carbon capture and storage? *Greenh, Gases:*, **2**(6), 402–407, doi:10.1002/ghg.1319.

Hughes, T.P. et al., 2018: Spatial and temporal patterns of mass bleaching of corals in the Anthropocene. *Science*, **359**(6371), 80, doi:10.1126/science.aan8048.

Hughes, T.P. et al., 2010: Rising to the challenge of sustaining coral reef resilience. *Trends Ecol. Evol.*, **25**(11), 633–642, doi:10.1016/j.tree.2010.07.011.

Hughes, T.P. et al., 2019a: Global warming impairs stock–recruitment dynamics of corals. *Nature*, **568**(7752), 387–390, doi:10.1038/s41586-019-1081-y.

Hughes, T.P. et al., 2019b: Ecological memory modifies the cumulative impact of recurrent climate extremes. *Nat. Clim. Change*, **9**(1), 40–43, doi:10.1038/s41558-018-0351-2.

Hughes, T.P. et al., 2013: Living dangerously on borrowed time during slow, unrecognized regime shifts. *Trends Ecol. Evol.*, **28**(3), 149–155, doi:10.1016/j.tree.2012.08.022.

Hummels, R., M. Dengler and B. Bourlès, 2013: Seasonal and regional variability of upper ocean diapycnal heat flux in the Atlantic cold tongue. *Progr. Oceanogr.*, **111**, 52–74, doi:10.1016/j.pocean.2012.11.001.

Hunt, G.L. et al., 2016: Advection in polar and sub-polar environments: Impacts on high latitude marine ecosystems. *Progr. Oceanogr.*, **149**(40), 40–81, doi:10.1016/j.pocean.2016.10.004.

Hutchins, D.A. and P.W. Boyd, 2016: Marine phytoplankton and the changing ocean iron cycle. *Nat. Clim. Change*, **6**(12), 1072–1079, doi:10.1038/NCLIMATE3147.

Hutchins, D.A. and F. Fu, 2017: Microorganisms and ocean global change. *Nature Microbiol.*, **2**, 17058, doi:10.1038/nmicrobiol.2017.58.

Huxham, M. et al., 2015: Applying Climate Compatible Development and economic valuation to coastal management: A case study of Kenya's mangrove forests. *J. Environ. Manage.*, **157**, 168–181, doi:10.1016/j.jenvman.2015.04.018.

Hwang, J.H. et al., 2014: The physical processes in the Yellow Sea. *Ocean Coast. Manage.*, **102**, 449–457.

Hyndes, G.A. et al., 2016: Accelerating Tropicalization and the Transformation of Temperate Seagrass Meadows. *BioScience*, **66**(11), 938–948, doi:10.1093/biosci/biw111.

Ikeda, T., Y. Kanno, K. Ozaki and A. Shinada, 2001: Metabolic rates of epipelagic marine copepods as a function of body mass and temperature. *Mar. Biol.*, **139**(3), 587–596, doi:10.1007/s002270100608.

Ilıcak, M., A.J. Adcroft, S.M. Griffies and R.W. Hallberg, 2012: Spurious dianeutral mixing and the role of momentum closure. *Ocean Model.*, **45–46**(Supplement C), 37–58, doi:10.1016/j.ocemod.2011.10.003.

Ilyina, T., R.E. Zeebe, E. Maier-Reimer and C. Heinze, 2009: Early detection of ocean acidification effects on marine calcification. *Global Biogeochem. Cy.*, **23**(1); 1–11, doi:10.1029/2008gb003278.

Ingeman, K.E., J.F. Samhouri and A.C. Stier, 2019: Ocean recoveries for tomorrow's Earth: Hitting a moving target. *Science*, **363**(6425), eaav1004, doi:10.1126/science.aav1004.

Inniss, L. et al., 2017: *The First Global Integrated Marine Assessment: World Ocean Assessment I.* United Nations, New York, 1752 pp.

IPCC, 2013: Climate Change 2013: The Physical Science Basis. Contribution of Working Group I to the Fifth Assessment Report of the Intergovernmental Panel on Climate Change [Stocker, T.F., D. Qin, G.-K. Plattner, M. Tignor, S.K. Allen, J. Boschung, A. Nauels, Y. Xia, V. Bex and P.M. Midgley (eds.)]. Cambridge University Press, Cambridge, United Kingdom and New York, NY, USA, 1535 pp.

IPCC, 2014: *2013 Supplement to the 2006 IPCC Guidelines for National Greenhouse Gas Inventories: Wetlands* [Hiraishi, T., Krug, T., Tanabe, K., Srivastava, N., Baasansuren, J., Fukuda, M. and Troxler, T.G. eds.].IPCC, Switzerland, 354 pp., ISBN: 978-92-9169-139-5.

IPCC, 2018: Global Warming of 1.5°C. An IPCC Special Report on the impacts of global warming of 1.5°C above pre-industrial levels and related global greenhouse gas emission pathways, in the context of strengthening the global response to the threat of climate change, sustainable development, and efforts to eradicate poverty [Masson-Delmotte, V., P. Zhai, H.-O. Pörtner, D. Roberts, J. Skea, P.R. Shukla, A. Pirani, W. Moufouma-Okia, C. Péan, R. Pidcock, S. Connors, J.B.R. Matthews, Y. Chen, X. Zhou, M.I. Gomis, E. Lonnoy, T. Maycock, M. Tignor and T. Waterfield (eds.)]. Cambridge University Press, Cambridge, United Kingdom and New York, NY, USA., 630.

Isensee, K., J. Howard, E. Pidgeon and J. Ramos, 2019: Coastal blue carbon. In: *WMO Statement on the State of the Global Climate in 2018.* WMO, Geneva, pp. 10–11, ISBN: 978-92-63-11233-0.

Ishii, M. et al., 2017: Accuracy of Global Upper Ocean Heat Content Estimation Expected from Present Observational Data Sets. *SOLA*, **13**, 163–167.

Ishikawa, T. and Y. Ikegaki, 1980: Control of Mercury Pollution in Japan and the Minamata Bay Cleanup. *J. Water Pollut. Contro Fed.*, **52**(5), 1013–1018.

Isla, J.A., K. Lengfellner and U. Sommer, 2008: Physiological response of the copepod Pseudocalanus sp in the Baltic Sea at different thermal scenarios. *Global Change Biol.*, **14**(4), 895–906, doi:10.1111/j.1365-2486.2008.01531.x.

Islam, M.M., S. Sallu, K. Hubacek and J. Paavola, 2013: Vulnerability of fishery-based livelihoods to the impacts of climate variability and change: insights from coastal Bangladesh. *Reg. Environ. Change*, **14**(1), 281–294, doi:10.1007/s10113-013-0487-6.

Ito, T. and C. Deutsch, 2013: Variability of the oxygen minimum zone in the tropical North Pacific during the late twentieth century. *Global Biogeochem. Cy.*, **27**(4), 1119–1128, doi:10.1002/2013gb004567.

Ito, T., S. Minobe, M.C. Long and C. Deutsch, 2017: Upper ocean O2 trends: 1958–2015. *Geophys. Res. Lett.*, **44**(9), 4214–4223, doi:10.1002/2017GL073613.

Ito, T. et al., 2016: Acceleration of oxygen decline in the tropical Pacific over the past decades by aerosol pollutants. *Nat. Geosci.*, **9**, 443, doi:10.1038/ngeo2717.

Iwao, K., N. Wada, A. Ohdera and M. Omori, 2014: How many donor colonies should be cross-fertilized for nursery farming of sexually propagated corals? *Natural Resources*, **05**(10), 521–526, doi:10.4236/nr.2014.510047.

Jacob, C., A. Buffard, S. Pioch and S. Thorin, 2017: Marine ecosystem restoration and biodiversity offset. *Ecol. Eng.*, doi:10.1016/j.ecoleng.2017.09.007.

Jagers, S.C. et al., 2018: Societal causes of, and responses to, ocean acidification. *Ambio*, **(48)**8, 816–830, doi:10.1007/s13280-018-1103-2.

Jamero, M.L., M. Onuki, M. Esteban and N. Tan, 2018: Community-based adaptation in low-lying islands in the Philippines: challenges and lessons learned. *Reg. Environ. Change*, **18**(8), 2249–2260, doi:10.1007/s10113-018-1332-8.

Janousek, C.N. et al., 2017: Inundation, vegetation, and sediment effects on litter decomposition in Pacific Coast tidal marshes. *Ecosystems*, **20**(7), 1296–1310.

5

Jaramillo, E. et al., 2017: Macroscale patterns in body size of intertidal crustaceans provide insights on climate change effects. *PLoS One*, **12**(5), e0177116, doi:10.1371/journal.pone.0177116.

Jaroszweski, D., L. Chapman and J. Petts, 2010: Assessing the potential impact of climate change on transportation: the need for an interdisciplinary approach. *J. Transport. Geogr.*, **18**(2), 331–335, doi:10.1016/j.jtrangeo.2009.07.005.

Jay, D.A., 2009: Evolution of tidal amplitudes in the eastern Pacific Ocean. *Geophys. Res. Lett.*, **36**(4), n/a–n/a, doi:10.1029/2008GL036185.

Jayne, S.R. and L.C. St. Laurent, 2001: Parameterizing tidal dissipation over rough topography. *Geophys. Res. Lett.*, **28**(5), 811–814.

Jennerjahn, T.C. et al., 2017: Mangrove Ecosystems under Climate Change. In: *Mangrove Ecosystems: A Global Biogeographic Perspective* [V.H. Rivera-Monroy, S.Y. Lee, E. Kristensen, and R.R. Twilley eds.]. Springer, pp. 211–244, ISBN: 978-3-319-62206-4.

Jeong, H., H. Lee, H. Kim and H. Kim, 2014: Algorithm for economic assessment of infrastructure adaptation to climate change. In *ISARC Proceedings of the International Symposium on Automation and Robotics in Construction*. IAARC Publications, Australia, **31**, 1.ISBN: 978-0-64-659711-9.

Jeppesen, R. et al., 2018: Effects of Hypoxia on Fish Survival and Oyster Growth in a Highly Eutrophic Estuary. *Estuar. Coast.*, **41**(1), 89–98, doi:10.1007/s12237-016-0169-y.

Jiang, J. et al., 2016: Defining the next generation modeling of coastal ecotone dynamics in response to global change. *Ecol. Model.*, **326**, 168–176, doi:10.1016/j.ecolmodel.2015.04.013.

Jiang, L. et al., 2018: Increased temperature mitigates the effects of ocean acidification on the calcification of juvenile Pocillopora damicornis, but at a cost. *Coral Reefs*, **37**(1), 71–79.

Jiang, M. and T. DeLacy, 2014: 14 A climate change adaptation framework for Pacific Island tourism. In T. DeLacy, M. Jiang, G. Lipman and S. Vorster (Eds), *Green Growth and Travelism: Concept, Policy and Practice for Sustainable Tourism*, Routledge, 225.

Jiao, N. et al., 2018a: Unveiling the enigma of refractory carbon in the ocean. *Natl. Sci. Rev.*, **5**(4), 459–463. doi:10.1093/nsr/nwy020.

Jiao, N. et al., 2010: Microbial production of recalcitrant dissolved organic matter: long-term carbon storage in the global ocean. *Nat. Rev. Microbiol.*, **8**(8), 593–599, doi:10.1038/Nrmicro2386.

Jiao, N. et al., 2014a: Presence of Prochlorococcus in the aphotic waters of the western Pacific Ocean. *Biogeosciences*, **11**(8), 2391–2400, doi:10.5194/bg-11-2391-2014.

Jiao, N. et al., 2014b: Mechanisms of microbial carbon sequestration in the ocean – future research directions. *Biogeosciences*, **11**(19), 5285–5306, doi:10.5194/bg-11-5285-2014.

Jiao, N., K. Tang, H. Cai and Y. Mao, 2011: Increasing the microbial carbon sink in the sea by reducing chemical fertilization on the land. *Nat. Rev. Microbiol.*, **9**(1), doi:10.1038/nrmicro2386-c2.

Jiao, N., H. Wang, G. Xu and S. Aricò, 2018b: Blue Carbon on the Rise:Challenges and Opportunities. *Natl. Sci. Rev.*, **5**(4), 464–468 doi:10.1093/nsr/nwy030.

Jiao, N.-Z. et al., 2015: Climate change and anthropogenic impacts on marine ecosystems and countermeasures in China. *Advances in Climate Change Research*, **6**(2), 118–125, doi:10.1016/j.accre.2015.09.010.

Jickells, T.D., J.E. Andrews and D.J. Parkes, 2015: Direct and Indirect Effects of Estuarine Reclamation on Nutrient and Metal Fluxes in the Global Coastal Zone. *Aquat. Geochem.*, **22**(4), 337–348, doi:10.1007/s10498-015-9278-7.

Jickells, T.D. et al., 2017: A reevaluation of the magnitude and impacts of anthropogenic atmospheric nitrogen inputs on the ocean. *Global Biogeochem. Cy.*, **31**(2), 289–305, doi:10.1002/2016gb005586.

Johannessen, S.C. and R.W. Macdonald, 2018a: Reply to Oreska et al 'Comment on Geoengineering with seagrasses: is credit due where credit is given?'. *Environ. Res. Lett.*, **13**(3), 038002.

Johannessen, S.C. and R.W. Macdonald, 2016: Geoengineering with seagrasses: is credit due where credit is given? *Environ. Res. Lett.*, **11**(11), 113001.

Johannessen, S.C. and R.W. Macdonald, 2018b: Reply to Macreadie et al Comment on 'Geoengineering with seagrasses: is credit due where credit is given?'. *Environ. Res. Lett.*, **13**(2), 028001.

Johnson, D., M. Adelaide Ferreira and E. Kenchington, 2018: Climate change is likely to severely limit the effectiveness of deep sea ABMTs in the North Atlantic. *Mar. Policy*, **87**, 111–122, doi:10.1016/j.marpol.2017.09.034.

Johnson, G.C., J.M. Lyman and S.G. Purkey, 2015: Informing Deep Argo Array Design Using Argo and Full-Depth Hydrographic Section Data. *J. Atmos. Ocean. Tech.*, **32**(11), 2187–2198, doi:10.1175/JTECH-D-15-0139.1.

Johnson, G.C., S.G. Purkey, N.V. Zilberman and D. Roemmich, 2019: Deep Argo Quantifies Bottom Water Warming Rates in the Southwest Pacific Basin. *Geophys. Res. Lett.*, **46**(5), 2662–2669, doi:10.1029/2018GL081685.

Johnson, K.S. et al., 2017: Biogeochemical sensor performance in the SOCCOM profiling float array. *J. Geophys. Res-Oceans*, **122**(8), 6416–6436, doi:10.1002/2017JC012838.

Jónasdóttir, S.H., A.W. Visser, K. Richardson and M.R. Heath, 2015: Seasonal copepod lipid pump promotes carbon sequestration in the deep North Atlantic. *PNAS*, **112**(39), 12122.

Jones, D.C. et al., 2016a: How does subantarctic mode water ventilate the Southern Hemisphere subtropics? *J. Geophys. Res-Oceans*, **121**(9), 6558–6582.

Jones, D.O. et al., 2014: Global reductions in seafloor biomass in response to climate change. *Glob Chang Biol*, **20**(6), 1861–72, doi:10.1111/gcb.12480.

Jones, J.M. et al., 2016b: Assessing recent trends in high-latitude Southern Hemisphere surface climate. *Nat. Clim. Change*, **6**, 917, doi:10.1038/nclimate3103.

Jones, K.R. et al., 2018: The Location and Protection Status of Earth's Diminishing Marine Wilderness. *Curr. Biol.*, **28**(15), 2506–2512.e3, doi:10.1016/j.cub.2018.06.010.

Jones, M.C. and W.W.L. Cheung, 2015: Multi-model ensemble projections of climate change effects on global marine biodiversity. *ICES J. Mar. Sci.*, **72**(3), 741–752, doi:10.1093/icesjms/fsu172.

Jones, M.C. and W.W.L. Cheung, 2018: Using fuzzy logic to determine the vulnerability of marine species to climate change. *Glob Chang Biol*, **24**(2), e719–e731, doi:10.1111/gcb.13869.

Jones, N., J.R.A. Clark and C. Malesios, 2015: Social capital and willingness-to-pay for coastal defences in south-east England. *Ecol. Econ.*, **119**, 74–82, doi:10.1016/j.ecolecon.2015.07.023.

Jonkers, L., H. Hillebrand and M. Kucera, 2019: Global change drives modern plankton communities away from the pre-industrial state. *Nature*, **570**, 372–375, doi:10.1038/s41586-019-1230-3.

Jordà, G. et al., 2017: The Mediterranean Sea heat and mass budgets: Estimates, uncertainties and perspectives. *Progr. Oceanogr.*, **156**, 174–208, doi:10.1016/j.pocean.2017.07.001.

Josey, S.A. et al., 2018: The Recent Atlantic Cold Anomaly: Causes, Consequences, and Related Phenomena. *Annu. Rev. Mar. Sci.*, **10**(1), 475–501, doi:10.1146/annurev-marine-121916-063102.

Joyce, J. et al., 2017: Developing a multi-scale modeling system for resilience assessment of green-grey drainage infrastructures under climate change and sea level rise impact. *Environ. Modell. Softw.*, **90**, 1–26, doi:10.1016/j.envsoft.2016.11.026.

Jurgens, L.J. et al., 2015: Patterns of mass mortality among rocky shore invertebrates across 100 km of northeastern Pacific coastline. *PLoS One*, **10**(6), e0126280.

Jurjonas, M. and E. Seekamp, 2018: Rural coastal community resilience: Assessing a framework in eastern North Carolina. *Ocean Coast. Manage.*, **162**, 137–150, doi:10.1016/j.ocecoaman.2017.10.010.

Jury, C.P. and R.J. Toonen, 2019: Adaptive responses and local stressor mitigation drive coral resilience in warmer, more acidic oceans. *Proc. Roy. Soc. B.*, **286**(1902), 20190614.

Justic, D. et al., 2016: Chapter 11 – Coastal Ecosystem Modeling in the Context of Climate Change: An Overview With Case Studies. In: *Developments*

5

in Environmental Modelling, Volume 28 [Sven Erik, J. (ed.)]. Elsevier, Netherlands, pp. 227–260. ISSN: 0167-8892.

Kabisch, N., H. Korn, J. Stadler and A. Bonn, 2017: *Nature-based Solutions to Climate Change Adaptation in Urban Areas. Linkages between Science, Policy and Practice. Theory and Practice of Urban Sustainability Transitions,* Springer Open, 337 pp.

Kahru, M., R. Kudela, M. Manzano-Sarabia and B.G. Mitchell, 2009: Trends in primary production in the California Current detected with satellite data. *J. Geophys. Res-Oceans,* **114**(C2).

Kais, S.M. and M.S. Islam, 2017: Impacts of and resilience to climate change at the bottom of the shrimp commodity chain in Bangladesh: A preliminary investigation. *Aquaculture,* doi:10.1016/j.aquaculture.2017.05.024.

Kaja, N. and M. Mellic, 2017: Climate change: Issues of Built Heritage Structures in the coastal region. *Journal of Scientific Research,* **13,** 54–60.

Kämpf, J. and P. Chapman, 2016: Upwelling Systems of the World.Springer International Publishing Switzerland. ISBN 978-3-319-42522-1.

Kao, K.-W. et al., 2018: Repeated and Prolonged Temperature Anomalies Negate Symbiodiniaceae Genera Shuffling in the Coral Platygyra verweyi (Scleractinia; Merulinidae). *Zool. Stud.,* **57**(55).

Kaplan-Hallam, M., N.J. Bennett and T. Satterfield, 2017: Catching sea cucumber fever in coastal communities: Conceptualizing the impacts of shocks versus trends on social-ecological systems. *Global Environ. Change,* **45,** 89–98, doi:10.1016/j.gloenvcha.2017.05.003.

Karim, M.S., 2015: *Prevention of Pollution of the Marine Environment from Vessels.* Springer International Publishing, Cham.ISBN 978-3-319-10608-3.

Karim, M.S. and M.M. Uddin, 2019: Swatch-of-no-ground marine protected area for sharks, dolphins, porpoises and whales: Legal and institutional challenges. *Mar. Pollut. Bull.,* **139,** 275–281, doi:10.1016/j.marpolbul.2018.12.037.

Karl, T.R. et al., 2015: Possible artifacts of data biases in the recent global surface warming hiatus. *Science,* **348**(6242), 1469, doi:10.1126/science.aaa5632.

Karstensen, J. et al., 2015: Open ocean dead zones in the tropical North Atlantic Ocean. *Biogeosciences,* **12**(8), 2597–2605, doi:10.5194/bg-12-2597-2015.

Katselidis, K.A. et al., 2014: Employing sea level rise scenarios to strategically select sea turtle nesting habitat important for long-term management at a temperate breeding area. *J. Exp. Mar. Biol. Ecol.,* **450,** 47–54, doi:10.1016/j.jembe.2013.10.017.

Kavanaugh, M.T. et al., 2015: Effect of continental shelf canyons on phytoplankton biomass and community composition along the western Antarctic Peninsula. *Mar. Ecol. Prog. Ser.,* **524,** 11–26, doi:10.3354/meps11189.

Kawase, M., 1987: Establishment of Deep Ocean Circulation Driven by Deep-Water Production. *J. Phys. Oceanogr.,* **17**(12), 2294–2317, doi:10.1175/1520-0485(1987)017<2294:EODOCD>2.0.CO;2.

Kay, J.E. et al., 2014: The Community Earth System Model (CESM) Large Ensemble Project: A Community Resource for Studying Climate Change in the Presence of Internal Climate Variability. *Bull. Am. Meteorol. Soc.,* **96**(8), 1333–1349, doi:10.1175/BAMS-D-13-00255.1.

Keller, A.A. et al., 2015: Occurrence of demersal fishes in relation to near-bottom oxygen levels within the California Current large marine ecosystem. *Fish. Oceanogr.,* **24**(2), 162–176, doi:10.1111/fog.12100.

Keller, A.A. et al., 2010: Demersal fish and invertebrate biomass in relation to an offshore hypoxic zone along the US West Coast. *Fish. Oceanogr.,* **19**(1), 76–87, doi:10.1111/j.1365-2419.2009.00529.x.

Keller, D.P., 2019a: Marine climate engineering. In: *Handbook on Marine Environment Protection: Science, Impacts and Sustainable Management* [Salomon, M. and T. Markus (eds.)]. Springer, Switzerland. ISBN: 978-3-319-60154-0.

Keller, D.P., E.Y. Feng and A. Oschlies, 2014a: Potential climate engineering effectiveness and side effects during a high carbon dioxide-emission scenario. *Nat. Commun.,* **5,** 3304, doi:10.1038/ncomms4304.

Keller, J.K., 2019b: Greenhouse Gases. In: A Blue Carbon Primer, The State of Coastal Wetland Carbon Science, Practice and Policy [Windham-Myers, L., S. Crooks and T.G. Troxler (eds.)]. Taylor and Francis Group, United States. ISBN: 978-1-4987-6909-9.

Keller, K.M., F. Joos and C.C. Raible, 2014b: Time of emergence of trends in ocean biogeochemistry. *Biogeosciences,* **11**(13), 3647–3659, doi:10.5194/bg-11-3647-2014.

Kelleway, J.J. et al., 2017a: Review of the ecosystem service implications of mangrove encroachment into salt marshes. *Global Change Biol.* **23**(10), 3967–3983.

Kelleway, J.J. et al., 2017b: Geochemical analyses reveal the importance of environmental history for blue carbon sequestration. *J. Geophys. Res-Biogeo.,* **122**(7), 1789–1805, doi:10.1002/2017JG003775.

Kelleway, J.J. et al., 2016: Seventy years of continuous encroachment substantially increases 'blue carbon' capacity as mangroves replace intertidal salt marshes. *Glob Chang Biol,* **22**(3), 1097–109, doi:10.1111/gcb.13158.

Kelly, R.P. et al., 2011: Mitigating Local Causes of Ocean Acidification with Existing Laws. *Science,* **332**(6033), 1036.

Kempener, R. and F. Neumann, 2014a: Salinity gradient energy – technology brief. *IRENA Ocean Energy Technology,* **4.**

Kempener, R. and F. Neumann, 2014b: *Tidal Energy: Technology Brief.* International Renewable Energy Agency (IRENA). Abu Dhabi.

Kench, P.S., M.R. Ford and S.D. Owen, 2018: Patterns of island change and persistence offer alternate adaptation pathways for atoll nations. *Nat. Commun.,* **9**(1), 605, doi:10.1038/s41467-018-02954-1.

Kennedy, H. et al., 2010: Seagrass sediments as a global carbon sink: Isotopic constraints. *Global Biogeochem. Cy.,* **24**(4), n/a–n/a, doi:10.1029/2010GB003848.

Kennedy, H., J.W. Fourqurean and S. Papadimitriou, 2018: The Calcium Carbonate Cycle in Seagrass Ecosystems. In: *A Blue Carbon Primer.* CRC Press, pp. 107–119.

Keskin, C. and D. Pauly, 2014: Changes in the 'Mean Temperature of the Catch': application of a new concept to the North-eastern Aegean Sea. *Acta Adriatica: international journal of Marine Sciences,* **55**(2), 213–218.

Kettle, N.P. et al., 2014: Integrating scientific and local knowledge to inform risk-based management approaches for climate adaptation. *Clim. Risk Manage.,* **4–5,** 17–31, doi:10.1016/j.crm.2014.07.001.

Keul, N., G. Langer, L.J. de Nooijer and J. Bijma, 2013: Effect of ocean acidification on the benthic foraminifera Ammonia sp. is caused by a decrease in carbonate ion concentration. *Biogeosciences,* **10**(10), 6185–6198, doi:10.5194/bg-10-6185-2013.

Khamis, Z.A., R. Kalliola and N. Käyhkö, 2017: Geographical characterization of the Zanzibar coastal zone and its management perspectives. *Ocean Coast. Manage.,* **149,** 116–134, doi:10.1016/j.ocecoaman.2017.10.003.

Khanna, N., J.A. Godbold, W.E.N. Austin and D.M. Paterson, 2013: The Impact of Ocean Acidification on the Functional Morphology of Foraminifera. *PLoS One,* **8**(12), e83118, doi:10.1371/journal.pone.0083118.

Kharouba, H.M. et al., 2018: Global shifts in the phenological synchrony of species interactions over recent decades. *PNAS,* **115**(20), 5211, doi:10.1073/pnas.1714511115.

Khatiwala, S. et al., 2013: Global ocean storage of anthropogenic carbon. *Biogeosciences,* **10**(4), 2169–2191.

Kim, I.-N. et al., 2014: Increasing anthropogenic nitrogen in the North Pacific Ocean. *Science,* **346**(6213), 1102, doi:10.1126/science.1258396.

Kim, R.E., 2012: Is a New Multilateral Environmental Agreement on Ocean Acidification Necessary? *Review of European Community & International Environmental Law,* **21**(3), 243–258, doi:10.1111/reel.12000.x.

Kim, T.-W. et al., 2011: Increasing N Abundance in the Northwestern Pacific Ocean Due to Atmospheric Nitrogen Deposition. *Science,* **334**(6055), 505, doi:10.1126/science.1206583.

Kirk, M.D. et al., 2015: World Health Organization estimates of the global and regional disease burden of 22 foodborne bacterial, protozoal, and viral diseases, 2010: a data synthesis. *PLoS medicine,* **12**(12), e1001921.

Kirwan, M.L. and J.P. Megonigal, 2013: Tidal wetland stability in the face of human impacts and sea level rise. *Nature,* **504,** 53, doi:10.1038/nature12856.

Kittinger, J.N., E.M. Finkbeiner, E.W. Glazier and L.B. Crowder, 2012: Human dimensions of coral reef social-ecological systems. *Ecol. Soc.,* **17**(4).

5

Kleisner, K.M. et al., 2015: Evaluating changes in marine communities that provide ecosystem services through comparative assessments of community indicators. *Ecosyst. Serv.*, **16**(Supplement C), 413–429, doi:10.1016/j. ecoser.2015.02.002.

Kleypas, J.A.K.A., 2019: Climate change and tropical marine ecosystems: A review with an emphasis on coral reefs. *UNED Research Journal*, **11**(1), 24–35.

Klinger, D.H., S.A. Levin and J.R. Watson, 2017: The growth of finfish in global open-ocean aquaculture under climate change. *Proc. Roy. Soc. B. Biol.*, **284**(1864).

Klint, L., T. DeLacy and S. Filep, 2015: A Focus on the South Pacific. In: *Small Islands and Tourism: Current Issues and Future Challenges. Tourism in Pacific Islands: Current Issues and Future Challenges.* [Pratt, S., D. Harrison. (ed.)]. Routledge, London. ISBN: 978-1-315-77382-7.

Klymak, J.M., R. Pinkel and L. Rainville, 2008: Direct Breaking of the Internal Tide near Topography: Kaena Ridge, Hawaii. *J. Phys. Oceanogr.*, **38**(2), 380–399, doi:10.1175/2007JPO3728.1.

Knutson, T.R., R. Zhang and L.W. Horowitz, 2016: Prospects for a prolonged slowdown in global warming in the early 21st century. *Nat. Commun.*, **7**, 13676, doi:10.1038/ncomms13676.

Kochnower, D., S.M.W. Reddy and R.E. Flick, 2015: Factors influencing local decisions to use habitats to protect coastal communities from hazards. *Ocean Coast. Manage.*, **116**, 277–290, doi:10.1016/j.ocecoaman.2015.07.021.

Koetse, M.J. and P. Rietveld, 2009: The impact of climate change and weather on transport: An overview of empirical findings. *Transport. Res. D. Tr. E.*, **14**(3), 205–221, doi:10.1016/j.trd.2008.12.004.

Kohli, G.S. et al., 2014: High abundance of the potentially maitotoxic dinoflagellate Gambierdiscus carpenteri in temperate waters of New South Wales, Australia. *Harmful Algae*, **39**, 134–145, doi:10.1016/j.hal.2014.07.007.

Kopprio, G.A. et al., 2017: Biogeochemical and hydrological drivers of the dynamics of Vibrio species in two Patagonian estuaries. *Sci. Total Environ.*, **579**, 646–656, doi:10.1016/j.scitotenv.2016.11.045.

Kordas, R.L., I. Donohue and C.D. Harley, 2017: Herbivory enables marine communities to resist warming. *Sci. Adv.*, **3**(10), e1701349.

Koslow, J.A., E.F. Miller and J.A. McGowan, 2015: Dramatic declines in coastal and oceanic fish communities off California. *Mar. Ecol. Prog. Ser.*, **538**, 221–227.

Koya, M. et al., 2017: Vulnerability of coastal fisher households to climate change: a case study fom Gujarat, India. *Turkish Journal of Fisheries and Aquatic Sciences*, **17**, 193–203, doi:10.4194/1303-2712-v17_1_21.

Krabbenhoft, D.P. and E.M. Sunderland, 2013: Global Change and Mercury. *Science*, **341**(6153), 1457.

Krause-Jensen, D. and C.M. Duarte, 2016: Substantial role of macroalgae in marine carbon sequestration. *Nat. Geosci.*, **9**(10), 737–742, doi:10.1038/ngeo2790.

Krause-Jensen, D. et al., 2018: Sequestration of macroalgal carbon: the elephant in the Blue Carbon room. *Biol. Lett.*, **14**(6), 20180236, doi:10.1098/rsbl.2018.0236.

Kroeger, K.D., S. Crooks, S. Moseman-Valtierra and J. Tang, 2017: Restoring tides to reduce methane emissions in impounded wetlands: A new and potent Blue Carbon climate change intervention. *Sci. Rep.*, **7**(1), 11914.

Kroeker, K.J. et al., 2013: Impacts of ocean acidification on marine organisms: quantifying sensitivities and interaction with warming. *Global Change Biol.*, **19**(6), 1884–1896.

Kroeker, K.J., R.L. Kordas and C.D. Harley, 2017: Embracing interactions in ocean acidification research: confronting multiple stressor scenarios and context dependence. *Biol. Lett.*, **13**(3), 20160802.

Kroeker, K.J., F. Micheli, M.C. Gambi and T.R. Martz, 2011: Divergent ecosystem responses within a benthic marine community to ocean acidification. *PNAS*, **108**(35), 14515–14520, doi:10.1073/pnas.1107789108.

Krumhans, K.A. et al., 2016: Global patterns of kelp forest change over the past half-century. *PNAS*, **113**(48), 13785–13790, doi:10.1073/pnas.1606102113.

Kružić, P., P. Rodić, A. Popijač and M. Sertić, 2016: Impacts of temperature anomalies on mortality of benthic organisms in the Adriatic Sea. *Mar. Ecol.*, **37**(6), 1190–1209, doi:10.1111/maec.12293.

Kubicek, A., B. Breckling, O. Hoegh-Guldberg and H. Reuter, 2019: Climate change drives trait-shifts in coral reef communities. *Sci. Rep.*, **9**(1), 3721, doi:10.1038/s41598-019-38962-4.

Kuhfuss, L. et al., 2016: Evaluating the impacts of sea level rise on coastal wetlands in Languedoc-Roussillon, France. *Environ. Sci. Policy*, **59**, 26–34, doi:10.1016/j.envsci.2016.02.002.

Kuhlbrodt, T. and J. Gregory, 2012: Ocean heat uptake and its consequences for the magnitude of sea level rise and climate change. *Geophys. Res. Lett.*, **39**(18).

Kuhnlein, H.V. and O. Receveur, 1996: Dietary Change and Traditional Food Systems of Indigenous Peoples. *Annu. Rev. Nutr.*, **16**(1), 417–442, doi:10.1146/annurev.nu.16.070196.002221.

Kunze, E., 2017: Internal-Wave-Driven Mixing: Global Geography and Budgets. *J. Phys. Oceanogr.*, **47**(6), 1325–1345, doi:10.1175/JPO-D-16-0141.1.

Kuriyama, Y. and S. Yanagishima, 2018: Regime shifts in the multi-annual evolution of a sandy beach profile. *Earth Surface Proc. Landf.*, **43**(15), 3133–3141, doi:10.1002/esp.4475.

Kurman, M.D. et al., 2017: Intra-Specific Variation Reveals Potential for Adaptation to Ocean Acidification in a Cold water Coral from the Gulf of Mexico. *Front. Mar. Sci.*, **4**, 111.

Kuruppu, N. and R. Willie, 2015: Barriers to reducing climate enhanced disaster risks in Least Developed Country-Small Islands through anticipatory adaptation. *Weather and Climate Extremes*, **7**, 72–83, doi:10.1016/j.wace.2014.06.001.

Kwiatkowski, L., O. Aumont and L. Bopp, 2019: Consistent trophic amplification of marine biomass declines under climate change. *Global Change Biol.*, **25**(1), 218–229, doi:10.1111/gcb.14468.

Kwiatkowski, L. and J.C. Orr, 2018: Diverging seasonal extremes for ocean acidification during the twenty45 first century. *Nature Climate Change*, **8**(2), 141–145, doi:10.1038/s41558-017-0054-0.

Kwiatkowski, L., O. Aumont, L. Bopp and P. Ciais, 2018: The Impact of Variable Phytoplankton Stoichiometry on Projections of Primary Production, Food Quality, and Carbon Uptake in the Global Ocean. *Global Biogeochem. Cy.*, **32**(4), 516–528, doi:10.1002/2017GB005799.

Kwiatkowski, L. et al., 2017: Emergent constraints on projections of declining primary production in the tropical oceans. *Nat. Clim. Change*, **7**(5), 355–358, doi:10.1038/nclimate3265.

Kwiatkowski, L. et al., 2016: Nighttime dissolution in a temperate coastal ocean ecosystem under acidification. *Sci. Rep.*, **6**(1), 22984, doi:10.1038/srep22984.

Kwiatkowski, L., K.L. Ricke and K. Caldeira, 2015: Atmospheric consequences of disruption of the ocean thermocline. *Environ. Res. Lett.*, **10**(3) 034016, doi:10.1088/1748-9326/10/3/034016.

Lachkar, Z., 2014: Effects of upwelling increase on ocean acidification in the California and Canary Current systems. *Geophys. Res. Lett.*, **41**(1), 90–95, doi:10.1002/2013GL058726.

Lachkar, Z., M. Lévy and S. Smith, 2018: Intensification and deepening of the Arabian Sea oxygen minimum zone in response to increase in Indian monsoon wind intensity. *Biogeosciences*, **15**(1),159–186.

Lacoue-Labarthe, T. et al., 2016: Impacts of ocean acidification in a warming Mediterranean Sea: An overview. *Reg. Stud. Mar. Sci.*, **5**, 1–11, doi:10.1016/j.rsma.2015.12.005.

Laffoley, D. and G.D. Grimsditch, 2009: *The management of natural coastal carbon sinks*. IUCN, Gland, Switzerland. 53.: ISBN: 978-2-8317-1205-5.

Laidre, K.L. et al., 2015: Arctic marine mammal population status, sea ice habitat loss, and conservation recommendations for the 21st century. *Conserv. Biol.*, **29**(3), 724–737, doi:10.1111/cobi.12474.

Laloë, J.O. et al., 2017: Climate change and temperature-linked hatchling mortality at a globally important sea turtle nesting site. *Global Change Biol.*, **23**(11), 4922–4931.

5

Lam, V.W.Y., W.W.L. Cheung, G. Reygondeau and U.R. Sumaila, 2016: Projected change in global fisheries revenues under climate change. *Sci. Rep.*, **6**, 32607 EP –, doi:10.1038/srep32607.

Landolfi, A. et al., 2017: Oceanic nitrogen cycling and N2O flux perturbations in the Anthropocene. *Global Biogeochem. Cy.*, **31**(8), 1236–1255, doi:10.1002/2017GB005633.

Landschützer, P., N. Gruber and D.C.E. Bakker, 2016: Decadal variations and trends of the global ocean carbon sink. *Global Biogeochem. Cy.*, **30**(10), 1396–1417, doi:10.1002/2015gb005359.

Landschützer, P. et al., 2018: Strengthening seasonal marine CO_2 variations due to increasing atmospheric CO_2. *Nat. Clim. Change*, **8**(2), 146–150, doi:10.1038/s41558-017-0057-x.

Landschützer, P. et al., 2015: The reinvigoration of the Southern Ocean carbon sink. *Science*, **349**(6253), 1221–1224.

Langlais, C.E. et al., 2017: Coral bleaching pathways under the control of regional temperature variability. *Nat. Clim. Change*, **7**(11), nclimate3399-844, doi:10.1038/nclimate3399.

Laplace, P.S., 1799: *Traité de Mécanique Céleste, Vol. 1*. Duprat, Paris.

Larsen, J.N. et al., 2014: Polar regions. In: Climate Change 2014: Impacts, Adaptation, and Vulnerability. Part B: Regional Aspects. Contribution of Working Group II to the Fifth Assessment Report of the Intergovernmental Panel on Climate Change [Barros, V.R., C.B. Field and D.J. Dokken (eds.)]. World Meteorological Organization, Geneva, Switzerland, 1567–1612.

Lartaud, F. et al., 2014: Temporal changes in the growth of two Mediterranean cold water coral species, in situ and in aquaria. *Deep Sea Res. Pt. II*, **99**, 64–70, doi:10.1016/j.dsr2.2013.06.024.

Laufkötter, C. et al., 2015: Drivers and uncertainties of future global marine primary production in marine ecosystem models. *Biogeosciences*, **12**(23), 6955–6984, doi:10.5194/bg-12-6955-2015.

Laufkötter, C. et al., 2016: Projected decreases in future marine export production: the role of the carbon flux through the upper ocean ecosystem. *Biogeosciences*, **13**(13), 4023–4047, doi:10.5194/bg-13-4023-2016.

Laurent, A. et al., 2017: Eutrophication-induced acidification of coastal waters in the northern Gulf of Mexico: Insights into origin and processes from a coupled physical-biogeochemical model. *Geophys. Res. Lett.*, **44**(2), 946–956.

Laurent, A., K. Fennel, D.S. Ko and J. Lehrter, 2018: Climate change projected to exacerbate impacts of coastal eutrophication in the northern Gulf of Mexico. *J. Geophys. Res-Oceans.***123**(5), 3408–3426.

Lauvset, S.K. et al., 2015: Trends and drivers in global surface ocean pH over the past 3 decades. *Biogeosciences*, **12**(5), 1285–1298, doi:10.5194/bg-12-1285-2015.

Lavender, S.L., R.K. Hoeke and D.J. Abbs, 2018: The influence of sea surface temperature on the intensity and associated storm surge of tropical cyclone Yasi: a sensitivity study. *Nat. Hazards Earth Syst. Sci.*, **18**(3), 795–805, doi:10.5194/nhess-18-795-2018.

Lavery, P.S., M.-Á. Mateo, O. Serrano and M. Rozaimi, 2013: Variability in the Carbon Storage of Seagrass Habitats and Its Implications for Global Estimates of Blue Carbon Ecosystem Service. *PLoS One*, **8**(9), e73748, doi:10.1371/journal.pone.0073748.

Le Cornu, E. et al., 2017: Spatial management in small-scale fisheries: A potential approach for climate change adaptation in Pacific Islands. *Mar. Policy*,**88**, 350–358. doi:10.1016/j.marpol.2017.09.030.

Le Cozannet, G. et al., 2019: Quantifying uncertainties of sandy shoreline change projections as sea level rises. *Sci. Rep.*, **9**(1), 42, doi:10.1038/s41598-018-37017-4.

Le Quere, C. et al., 2016: Global Carbon Budget 2016. *Earth Syst. Sci. Data*, **8**(2), 605–649, doi:10.5194/essd-8-605-2016.

Le Quéré, C. et al., 2018: Global Carbon Budget 2017. *Earth Syst. Sci. Data*, **10**(1), 405–448, doi:10.5194/essd-10-405-2018.

Leadley, P. et al., 2014: Interacting regional-scale regime shifts for biodiversity and ecosystem services. *BioScience*, biu093.

Lee, Z., J. Marra, M.J. Perry and M. Kahru, 2015: Estimating oceanic primary productivity from ocean color remote sensing: A strategic assessment. *J. Mar. Syst.*, **149**, 50–59, doi:10.1016/j.jmarsys.2014.11.015.

Lefevre, S., 2016: Are global warming and ocean acidification conspiring against marine ectotherms? A meta-analysis of the respiratory effects of elevated temperature, high CO_2 and their interaction. *Conserv. Physiol.*, **4**(1), cow009–cow009, doi:10.1093/conphys/cow009.

Lefort, S. et al., 2015: Spatial and body-size dependent response of marine pelagic communities to projected global climate change. *Global Change Biol.*, **21**(1), 154–164, doi:10.1111/gcb.12679.

Legault II, R., G.P. Zogg and S.E. Travis, 2018: Competitive interactions between native Spartina alterniflora and non-native Phragmites australis depend on nutrient loading and temperature. *PLoS One*, **13**(2), e0192234.

Legendre, L. et al., 2015: The microbial carbon pump concept: Potential biogeochemical significance in the globally changing ocean. *Progr. Oceanogr.*, **134**, 432–450, doi:10.1016/j.pocean.2015.01.008.

Lehodey, P. et al., 2013: Modelling the impact of climate change on Pacific skipjack tuna population and fisheries. *Clim. Change*, **119**(1), 95–109, doi:10.1007/s10584-012-0595-1.

Lemasson, A.J., J.M. Hall-Spencer, V. Kuri and A.M. Knights, 2019: Changes in the biochemical and nutrient composition of seafood due to ocean acidification and warming. *Mar. Environ. Res.*, **143**, 82–92, doi:10.1016/j.marenvres.2018.11.006.

Lenzen, M. et al., 2018: The carbon footprint of global tourism. *Nat. Clim. Change*, **8**(6), 522–528, doi:10.1038/s41558-018-0141-x.

Leon, J.X. et al., 2015: Supporting Local and Traditional Knowledge with Science for Adaptation to Climate Change: Lessons Learned from Participatory Three-Dimensional Modeling in BoeBoe, Solomon Islands. *Coast. Manage.*, **43**(4), 424–438, doi:10.1080/08920753.2015.1046808.

Levin, L.A., 2003: Oxygen minimum zone benthos: Adaptation and community response to hypoxia. *Oceanogr. Mar. Biol.*, **41**, 1–45.

Levin, L.A., 2018: Manifestation, Drivers, and Emergence of Open Ocean Deoxygenation. *Annu. Rev. Mar. Sci.*, **10**(1), 229–260, doi:10.1146/annurev-marine-121916-063359.

Levin, L.A. et al., 2016: Hydrothermal Vents and Methane Seeps: Rethinking the Sphere of Influence. *Front. Mar. Sci.*, **3**, 72.

Levin, L.A. and P.K. Dayton, 2009: Ecological theory and continental margins: where shallow meets deep. *Trends Ecol. Evol.*, **24**(11), 606–617, doi:10.1016/j.tree.2009.04.012.

Levin, L.A. and N. Le Bris, 2015: The deep ocean under climate change. *Science*, **350**(6262), 766–768, doi:10.1126/science.aad0126.

Levin, L.A. et al., 2015: Comparative biogeochemistry–ecosystem–human interactions on dynamic continental margins. *J. Mar. Syst.*, **141**, 3–17, doi:10.1016/j.jmarsys.2014.04.016.

Levin, L.A. et al., 2013: Macrofaunal colonization across the Indian margin oxygen minimum zone. *Biogeosciences*, **10**(11), 7161–7177, doi:10.5194/bg-10-7161-2013.

Levin, L.A. and M. Sibuet, 2012: Understanding Continental Margin Biodiversity: A New Imperative. *Annu. Rev. Mar. Sci.*, **4**(1), 79–112, doi:10.1146/annurev-marine-120709-142714.

Levitus, S. et al., 2012: World ocean heat content and thermosteric sea level change (0–2000 m), 1955–2010. *Geophys. Res. Lett.*, **39**(10), doi:10.1029/2012GL051106.

Lévy, M. et al., 2012: Bringing physics to life at the submesoscale. *Geophys. Res. Lett.*, **39**(14), doi:10.1029/2012GL052756.

Li, H. et al., 2009: The Sequence Alignment/Map format and SAMtools. *Bioinformatics*, **25**(16), 2078–2079, doi:10.1093/bioinformatics/btp352.

Li, X., R. Bellerby, C. Craft and S.E. Widney, 2018a: Coastal wetland loss, consequences, and challenges for restoration. *Anthropocene Coasts*, **1**(0), 1–15.

Li, Y. et al., 2018b: DNA methylation regulates transcriptional homeostasis of algal endosymbiosis in the coral model Aiptasia. *Sci. Adv.*, **4**(8), eaat2142, doi:10.1126/sciadv.aat2142.

Liew, Y.J. et al., 2017: Condition-specific RNA editing in the coral symbiont Symbiodinium microadriaticum. *PLOS Genetics*, **13**(2), e1006619, doi:10.1371/journal.pgen.1006619.

Liew, Y.J. et al., 2018: Epigenome-associated phenotypic acclimatization to ocean acidification in a reef-building coral. *Sci. Adv.*, **4**(6), eaar8028, doi:10.1126/sciadv.aar8028.

Lima, F.P. et al., 2016: Loss of thermal refugia near equatorial range limits. *Global Change Biol.*, **22**(1), 254–263.

Lima, F.P. et al., 2007: Do distributional shifts of northern and southern species of algae match the warming pattern? *Global Change Biol.*, **13**(12), 2592–2604, doi:10.1111/j.1365-2486.2007.01451.x.

Lima-Mendez, G. et al., 2015: Determinants of community structure in the global plankton interactome. *Science*, **348**(6237), 1262073, doi:10.1126/science.1262073.

Lin, H.-J. et al., 2018: The effects of El Niño-Southern Oscillation events on intertidal seagrass beds over a long-term timescale. *Global Change Biol.*, **0**(0), doi:10.1111/gcb.14404.

Linares, C. et al., 2015: Persistent natural acidification drives major distribution shifts in marine benthic ecosystems. *Proc. Roy. Soc. B. Biol.*, **282**(1818) 20150587.

Linares, O.F., 2009: From past to future agricultural expertise in Africa: Jola women of Senegal expand market-gardening. *PNAS*, **106**(50), 21074.

Linden, B. and B. Rinkevich, 2011: Creating stocks of young colonies from brooding coral larvae, amenable to active reef restoration. *J. Exp. Mar. Biol. Ecol.*, **398**(1), 40–46.

Linden, B. and B. Rinkevich, 2017: Elaborating an eco-engineering approach for stock enhanced sexually derived coral colonies. *J. Exp. Mar. Biol. Ecol.*, **486**(Supplement C), 314–321.

Lindfield, S.J., E.S. Harvey, A.R. Halford and J.L. McIlwain, 2016: Mesophotic depths as refuge areas for fishery-targeted species on coral reefs. *Coral Reefs*, **35**(1), 125–137, doi:10.1007/s00338-015-1386-8.

Lindley, J.A. and R.R. Kirby, 2010: Climate-induced changes in the North Sea Decapoda over the last 60 years. *Clim. Res.*, **42**(3), 257–264.

Ling, S.D. et al., 2014: Global regime shift dynamics of catastrophic sea urchin overgrazing. *Philos. Trans. Roy. Soc. B. Biol.*, **370**(1659), 20130269–20130269, doi:10.1098/rstb.2013.0269.

Linkon, S.B., 2018: *Autonomy in Building Process to Adapt the Climate Change Impacts: A Study of the Coastal Settlements in Bangladesh*. International Journal of Environment and Sustainability [IJES], **6**(2), 19–39 ISSN 1927-9566.

Lirman, D. and S. Schopmeyer, 2016: Ecological solutions to reef degradation: optimizing coral reef restoration in the Caribbean and Western Atlantic. *Peerj*, **4**, e2597, doi:10.7717/peerj.2597.

Lischka, S., J. Büdenbender, T. Boxhammer and U. Riebesell, 2011: Impact of ocean acidification and elevated temperatures on early juveniles of the polar shelled pteropod Limacina helicina: mortality, shell degradation, and shell growth. *Biogeosciences*, **8**(4), 919–932, doi:10.5194/bg-8-919-2011.

Lithgow, D. et al., 2019: Exploring the co-occurrence between coastal squeeze and coastal tourism in a changing climate and its consequences. *Tourism Manage.*, **74**, 43–54, doi:10.1016/j.tourman.2019.02.005.

Little, S., P.J. Wood and M. Elliott, 2017: Quantifying salinity-induced changes on estuarine benthic fauna: The potential implications of climate change. *Estuar. Coast. Shelf Sci.*, **198**, 610–625, doi:10.1016/j.ecss.2016.07.020.

Liu, J., N. Jiao and K. Tang, 2014: An experimental study on the effects of nutrient enrichment on organic carbon persistence in the western Pacific oligotrophic gyre. *Biogeosciences*, **11**(18), 5115–5122, doi:10.5194/bg-11-5115-2014.

Liu, W., S.-P. Xie and J. Lu, 2016: Tracking ocean heat uptake during the surface warming hiatus. *Nat. Commun.*, **7**, 10926, doi:10.1038/ncomms10926.

Llanillo, P.J., J. Karstensen, J.L. Pelegrí and L. Stramma, 2013: Physical and biogeochemical forcing of oxygen and nitrate changes during El Niño/El Viejo and La Niña/La Vieja upper-ocean phases in the tropical eastern South Pacific along 86° W. *Biogeosciences*, **10**(10), 6339–6355, doi:10.5194/bg-10-6339-2013.

Llovel, W., J.K. Willis, F.W. Landerer and I. Fukumori, 2014: Deep-ocean contribution to sea level and energy budget not detectable over the past decade. *Nat. Clim. Change*, **4**, 1031, doi:10.1038/nclimate2387.

Lloyd, S.J. et al., 2016: Modelling the influences of climate change-associated sea level rise and socioeconomic development on future storm surge mortality. *Clim. Change*, **134**(3), 441–455, doi:10.1007/s10584-015-1376-4.

Lo Iacono, C. et al., 2008: Very high-resolution seismo-acoustic imaging of seagrass meadows (Mediterranean Sea): Implications for carbon sink estimates. *Geophys. Res. Lett.*, **35**(18), n/a–n/a, doi:10.1029/2008GL034773.

Lohbeck, K.T., U. Riebesell and T.B.H. Reusch, 2012: Adaptive evolution of a key phytoplankton species to ocean acidification. *Nat. Geosci.*, **5**, 346, doi:10.1038/ngeo1441.

Lohmann, H., 2016: Comparing vulnerability and adaptive capacity to climate change in individuals of coastal Dominican Republic. *Ocean Coast. Manage.*, **132**, 111–119, doi:10.1016/j.ocecoaman.2016.08.009.

Lohr, K.E. and J.T. Patterson, 2017: Intraspecific variation in phenotype among nursery-reared staghorn coral Acropora cervicornis (Lamarck, 1816). *J. Exp. Mar. Biol. Ecol.*, **486**(Supplement C), 87–92.

López, I.R., J. Kalman, C. Vale and J. Blasco, 2010: Influence of sediment acidification on the bioaccumulation of metals in Ruditapes philippinarum. *Environ. Sci. Pollut. Res.*, **17**(9), 1519–1528, doi:10.1007/s11356-010-0338-7.

López-Abbate, M.C. et al., 2019: Long-term changes on estuarine ciliates linked with modifications on wind patterns and water turbidity. *Mar. Environ. Res.*, **144**, 46–55, doi:10.1016/j.marenvres.2018.12.001.

López-Portillo, J., A.L. Lara-Domínguez, G. Vázquez and J.A. Aké-Castillo, 2017: Water Quality and Mangrove-Derived Tannins in Four Coastal Lagoons from the Gulf of Mexico with Variable Hydrologic Dynamics. *In*: Martinez, M.L.; Taramelli, A., and Silva, R. (eds.), *Coastal Resilience: Exploring the Many Challenges from Different Viewpoints. Journal of Coastal Research*, Special Issue No. 77, pp. 28–38. Coconut Creek (Florida), ISSN 0749-0208 doi:10.2112/SI77-004.1.

Lotze, H.K. et al., 2018: Ensemble projections of global ocean animal biomass with climate change. *bioRxiv*, 467175, doi:10.1101/467175.

Lovelock, C.E. et al., 2015: The vulnerability of Indo-Pacific mangrove forests to sea level rise. *Nature*, **526**, 559, doi:10.1038/nature15538.

Lovelock, C.E. and C.M. Duarte, 2019: Dimensions of Blue Carbon and emerging perspectives. *Biol. Lett.*, **15**(3), 20180781.

Lovelock, C.E., J.W. Fourqurean and J.T. Morris, 2017: Modeled CO_2 Emissions from Coastal Wetland Transitions to Other Land Uses: Tidal Marshes, Mangrove Forests, and Seagrass Beds. *Front. Mar. Sci.*, **4**, 143.

Lovenduski, N.S., M.C. Long and K. Lindsay, 2015: Natural variability in the surface ocean carbonate ion concentration. *Biogeosciences*, **12**(21), 6321–6335, doi:10.5194/bg-12-6321-2015.

Lovenduski, N.S. et al., 2016: Partitioning uncertainty in ocean carbon uptake projections: Internal variability, emission scenario, and model structure. *Global Biogeochem. Cy.*, **30**(9), 1276–1287, doi:10.1002/2016gb005426.

Luick, B., A. Bersamin and J.S. Stern, 2014: Locally harvested foods support serum 25-hydroxyvitamin D sufficiency in an indigenous population of Western Alaska. *Int. J. Circumpolar Health*, **73**(1), 22732, doi:10.3402/ijch.v73.22732.

Luijendijk, A. et al., 2018: The State of the World's Beaches. *Sci. Rep.*, **8**(1), 6641, doi:10.1038/s41598-018-24630-6.

Luisetti, T. et al., 2019: Quantifying and valuing carbon flows and stores in coastal and shelf ecosystems in the UK. *Ecosyst. Serv.*, **35**, 67–76, doi:10.1016/j.ecoser.2018.10.013.

Lunden, J.J. et al., 2014: Acute survivorship of the deep sea coral *Lophelia pertusa* from the Gulf of Mexico under acidification, warming, and deoxygenation. *Front. Mar. Sci.*, **1**, 419, doi:10.3389/fmars.2014.00078.

Luo, Y.-W. et al., 2019: Reduced nitrogenase efficiency dominates response of the globally important nitrogen fixer Trichodesmium to ocean acidification. *Nat. Commun.*, **10**(1), 1521, doi:10.1038/s41467-019-09554-7.

5

Lyman, J.M. and G.C. Johnson, 2014: Estimating global ocean heat content changes in the upper 1800 m since 1950 and the influence of climatology choice. *J. Clim.*, **27**(5), 1945–1957.

Lynn, K. et al., 2013: The impacts of climate change on tribal traditional foods. *Clim. Change*, **120**(3), 545–556, doi:10.1007/s10584-013-0736-1.

Maavara, T., R. Lauerwald, P. Regnier and P. Van Cappellen, 2017: Global perturbation of organic carbon cycling by river damming. *Nat. Commun.*, **8**, 15347.

MacDonald, M.A. et al., 2017: Benefits of coastal managed realignment for society: Evidence from ecosystem service assessments in two UK regions. *Estuar. Coast. Shelf Sci.*, doi:10.1016/j.ecss.2017.09.007.

Macdonald, R.W. and L.L. Loseto, 2010: Are Arctic Ocean ecosystems exceptionally vulnerable to global emissions of mercury? A call for emphasised research on methylation and the consequences of climate change. *Environ. Chem.*, **7**(2), 133–138.

Mace, 2001: A new role for MSY in single-species and ecosystem approaches to fisheries stock assessment and management. *Fish Fish.*, **2**(1), 2–32, doi:10.1046/j.1467-2979.2001.00033.x.

Machado, L. et al., 2016: Dose-response effects of Asparagopsis taxiformis and Oedogonium sp. on in vitro fermentation and methane production. *J. App. Phycol.*, **28**(2), 1443–1452, doi:10.1007/s10811-015-0639-9.

Machado, L. et al., 2018: In Vitro Response of Rumen Microbiota to the Antimethanogenic Red Macroalga Asparagopsis taxiformis. *Microb. Ecol.*, **75**(3), 811–818, doi:10.1007/s00248-017-1086-8.

MacIntyre, K.Q. et al., 2015: The relationship between sea ice concentration and the spatio-temporal distribution of vocalizing bearded seals (Erignathus barbatus) in the Bering, Chukchi, and Beaufort Seas from 2008 to 2011. *Progr. Oceanogr.*, **136**, 241–249, doi:10.1016/j.pocean.2015.05.008.

Mackenzie, C.L. et al., 2014: Ocean Warming, More than Acidification, Reduces Shell Strength in a Commercial Shellfish Species during Food Limitation. *PLoS One*, **9**(1), e86764, doi:10.1371/journal.pone.0086764.

Mackey, B. and D. Ware, 2018: Limits to Capital Works Adaptation in the Coastal Zones and Islands: Lessons for the Pacific. In: *Limits to Climate Change Adaptation* [Leal Filho, W. and J. Nalau (eds.)]. Springer International Publishing, Cham, pp. 301–323. ISBN: 978-3-319-64599-5.

MacKinnon, J.A. et al., 2017: Climate Process Team on Internal-Wave Driven Ocean Mixing. *Bull. Am. Meteorol. Soc.*, **98**(11), 2429–2454. doi:10.1175/BAMS-D-16-0030.1.

Macreadie, P.I. et al., 2018: Comment on 'Geoengineering with seagrasses: is credit due where credit is given?'. *Environ. Res. Lett.*, **13**(2), 028002.

Macreadie, P.I. et al., 2017a: Can we manage coastal ecosystems to sequester more blue carbon? *Front. Ecol. Environ.*, **15**(4), 206–213, doi:10.1002/fee.1484.

Macreadie, P.I. et al., 2017b: Carbon sequestration by Australian tidal marshes. *Sci. Rep.*, **7**, 44071, doi:10.1038/srep44071.

Magnan, A.K. et al., 2016: Implications of the Paris Agreement for the ocean. *Nat. Clim. Change*, **6**(8), 732–735.

Magnan, A.K. and V.K.E. Duvat, 2018: Unavoidable solutions for coastal adaptation in Reunion Island (Indian Ocean). *Environ. Sci. Policy*, **89**, 393–400, doi:10.1016/j.envsci.2018.09.002.

Mahadevan, A., 2016: The Impact of Submesoscale Physics on Primary Productivity of Plankton. *Annu. Rev. Mar. Sci.*, **8**(1), 161–184, doi:10.1146/annurev-marine-010814-015912.

Maharaj, R.R., V.W.Y. Lam, D. Pauly and W.W.L. Cheung, 2018: Regional variability in the sensitivity of Caribbean reef fish assemblages to ocean warming. *Mar. Ecol. Prog. Ser.*, **590**, 201–209.

Maher, D.T., M. Call, I.R. Santos and C.J. Sanders, 2018: Beyond burial: lateral exchange is a significant atmospheric carbon sink in mangrove forests. *Biol. Lett.*, **14**(7), 20180200.

Maher, D.T. et al., 2016: Pristine mangrove creek waters are a sink of nitrous oxide. *Sci. Rep.*, **6**, 25701, doi:10.1038/srep25701.

Malone, K., 2016: Reconsidering children's encounters with nature and place using posthumanism. *Aust. J. Environ. Educ.*, **32**(1), 42–56.

Manasseh, R. et al., 2017: Integration of wave energy and other marine renewable energy sources with the needs of coastal societies. *The International Journal of Ocean and Climate Systems*, **8**(1), 19–36, doi:10.1177/1759313116683962.

Maranon, E. et al., 2014: Resource Supply Overrides Temperature as a Controlling Factor of Marine Phytoplankton Growth. *PLoS One*, **9**(6), doi:10.1371/journal.pone.0099312.

Marba, N. and C.M. Duarte, 2009: Mediterranean warming triggers seagrass (Posidonia oceanica) shoot mortality. *Global Change Biol.*, **16**(8), 2366–2375, doi:10.1111/j.1365-2486.2009.02130.x.

Marbà, N., D. Krause-Jensen, P. Masqué and C.M. Duarte, 2018: Expanding Greenland seagrass meadows contribute new sediment carbon sinks. *Sci. Rep.*, **8**(1), 14024, doi:10.1038/s41598-018-32249-w.

Marfai, M.A., A. Sekaranom and P. Ward, 2015: Community responses and adaptation strategies toward flood hazard in Jakarta, Indonesia. *Natural Hazards: Journal of the International Society for the Prevention and Mitigation of Natural Hazards*, **75**, 1127–1144.

Marlow, J.J. et al., 2014: Carbonate-hosted methanotrophy represents an unrecognized methane sink in the deep sea. *Nat. Commun.*, **5**, 5094, doi:10.1038/ncomms6094.

Marques, A., M.L. Nunes, S.K. Moore and M.S. Strom, 2010: Climate change and seafood safety: Human health implications. *Food Res. Int.*, **43**(7), 1766–1779, doi:10.1016/j.foodres.2010.02.010.

Marques, S.C. et al., 2017: Evidence for Changes in Estuarine Zooplankton Fostered by Increased Climate Variance. *Ecosystems*, **21**(1), 56–67, doi:10.1007/s10021-017-0134-z.

Marsay, C.M. et al., 2015: Attenuation of sinking particulate organic carbon flux through the mesopelagic ocean. *PNAS*, **112**(4), 1089.

Marshall, K.N. et al., 2017: Risks of ocean acidification in the California Current food web and fisheries: ecosystem model projections. *Global Change Biol.*, **23**(4), 1525–1539, doi:10.1111/gcb.13594.

Marshall, N. et al., 2019: Reef Grief: investigating the relationship between place meanings and place change on the Great Barrier Reef, Australia. *Sustain. Sci.*, **14**(3), 579–587, doi:10.1007/s11625-019-00666-z.

Marshall, N.A. et al., 2018: Measuring What Matters in the Great Barrier Reef. *Front. Ecol. Environ.*, **16**(5), 271–27.

Marshall, N.A. et al., 2012: Transformational capacity and the influence of place and identity. *Environ. Res. Lett.*, **7**(3), 034022.

Marshall, N.A. et al., 2013: Social Vulnerability of Marine Resource Users to Extreme Weather Events. *Ecosystems*, **16**(5), 797–809, doi:10.1007/s10021-013-9651-6.

Martin, A. et al., 2016a: *Blue Carbon – Nationally Determined Contributions Inventory. Appendix to: Coastal blue carbon ecosystems. Opportunities for Nationally Determined Contributions*. GRID-Arendal, Norway [Available at: http://bluecsolutions.org/dev/wp-content/uploads/Blue-Carbon-NDC-Appendix.pdf]. Accessed: 2019/09/30.

Martin, R.M. and S. Moseman-Valtierra, 2015: Greenhouse gas fluxes vary between Phragmites australis and native vegetation zones in coastal wetlands along a salinity gradient. *Wetlands*, **35**(6), 1021–1031.

Martin, S.L., L.T. Ballance and T. Groves, 2016b: An Ecosystem Services Perspective for the Oceanic Eastern Tropical Pacific: Commercial Fisheries, Carbon Storage, Recreational Fishing, and Biodiversity. *Front. Mar. Sci.*, **3**, 50.

Martin, T., W. Park and M. Latif, 2015: Southern Ocean forcing of the North Atlantic at multi-centennial time scales in the Kiel Climate Model. *Deep Sea Res. Pt. II*, **114**(Supplement C), 39–48, doi:10.1016/j.dsr2.2014.01.018.

Martín-García, L. et al., 2014: Predicting the potential habitat of the harmful cyanobacteria Lyngbya majuscula in the Canary Islands (Spain). *Harmful Algae*, **34**, 76–86.

Martínez, C. et al., 2017: Coastal erosion in central Chile: A new hazard? *Ocean Coast. Manage.*, **156**, 141–155. doi:10.1016/j.ocecoaman.2017.07.011.

Martínez, C.I.P., W.H.A. Piña and S.F. Moreno, 2018: Prevention, mitigation and adaptation to climate change from perspectives of urban population in an emerging economy. *J. Clean. Prod.*, **178**, 314–324.

Marushka, L. et al., 2019: Potential impacts of climate-related decline of seafood harvest on nutritional status of coastal First Nations in British Columbia, Canada. *PLoS One*, **14**(2), e0211473, doi:10.1371/journal.pone.0211473.

Masuda, S. et al., 2010: Simulated Rapid Warming of Abyssal North Pacific Waters. *Science*, **329**(5989), 319.

Mavromatidi, A., E. Briche and C. Claeys, 2018: Mapping and analyzing socioenvironmental vulnerability to coastal hazards induced by climate change: An application to coastal Mediterranean cities in France. *Cities*, **72, Part A**, 189–200, doi:10.1016/j.cities.2017.08.007.

Mayerle, R. et al., 2016: Development of a coastal information system for the management of Jeddah coastal waters in Saudi Arabia. *Comput. Geosci-UK*, **89**, 71–78, doi:10.1016/j.cageo.2015.12.006.

Mayor, D. J., U. Sommer, K.B. Cook and M.R. Viant, 2015: The metabolic response of marine copepods to environmental warming and ocean acidification in the absence of food. *Sci. Rep.*, **5**, 13690, doi:10.1038/srep13690.

Mbije, N.E., E. Spanier and B. Rinkevich, 2013: A first endeavour in restoring denuded, post-bleached reefs in Tanzania. *Estuar. Coast. Shelf Sci.*, **128**(Supplement C), 41–51.

Mbije, N.E.J., E. Spanier and B. Rinkevich, 2010: Testing the first phase of the 'gardening concept' as an applicable tool in restoring denuded reefs in Tanzania. *Ecol. Eng.*, **36**(5), 713–721, doi:10.1016/j.ecoleng.2009.12.018.

McCabe, R.M. et al., 2016: An unprecedented coastwide toxic algal bloom linked to anomalous ocean conditions. *Geophys. Res. Lett.*, **43**(19).

McClanahan, T.R., N.A.J. Graham and E.S. Darling, 2014: Coral reefs in a crystal ball: predicting the future from the vulnerability of corals and reef fishes to multiple stressors. *Curr. Opin. Environ. Sustain.*, **7**, 59–64, doi:10.1016/j.cosust.2013.11.028.

McClenachan, L. et al., 2017: Ghost reefs: Nautical charts document large spatial scale of coral reef loss over 240 years. *Sci. Adv.*, **3**(9) e1603155, doi:10.1126/sciadv.1603155.

McConville, K. et al., 2013: Effects of elevated CO_2 on the reproduction of two calanoid copepods. *Mar. Pollut. Bull.*, **73**(2), 428–434, doi:10.1016/j.marpolbul.2013.02.010.

McCook, L.J., 1999: Macroalgae, nutrients and phase shifts on coral reefs: scientific issues and management consequences for the Great Barrier Reef. *Coral Reefs*, **18**(4), 357–367, doi:10.1007/s003380050213.

McGee, J., K. Brent and W. Burns, 2018: Geoengineering the oceans: an emerging frontier in international climate change governance. *Australian Journal of Maritime & Ocean Affairs*, **10**(1), 67–80, doi:10.1080/18366503.2017.1400899.

McIlroy, S.E. and M.A. Coffroth, 2017: Coral ontogeny affects early symbiont acquisition in laboratory-reared recruits. *Coral Reefs*, **36**(3), 927–932, doi:10.1007/s00338-017-1584-7.

McIntyre-Wressnig, A., J.M. Bernhard, J.C. Wit and D.C. McCorkle, 2014: Ocean acidification not likely to affect the survival and fitness of two temperate benthic foraminiferal species: results from culture experiments. *J. Foramin. Res.*, **44**(4), 341–351.

McKee, K.L., D.R. Cahoon and I.C. Feller, 2007: Caribbean mangroves adjust to rising sea level through biotic controls on change in soil elevation. *Global Ecol. Biogeogr.*, **16**(5), 545–556, doi:10.1111/j.1466-8238.2007.00317.x.

McKeon, C.S. et al., 2016: Melting barriers to faunal exchange across ocean basins. *Global Change Biol.*, **22**(2), 465–473, doi:10.1111/gcb.13116.

McKibben, S.M. et al., 2017: Climatic regulation of the neurotoxin domoic acid. *PNAS*, **114**(2), 239–244.

McKinney, M.A. et al., 2015: A review of ecological impacts of global climate change on persistent organic pollutant and mercury pathways and exposures in arctic marine ecosystems. *Curr. Zool.*, **61**(4), 617–628, doi:10.1093/czoolo/61.4.617.

McLeod, E. et al., 2011: A blueprint for blue carbon: toward an improved understanding of the role of vegetated coastal habitats in sequestering CO_2. *Front. Ecol. Environ.*, **9**(10), 552–560, doi:10.1890/110004.

McNamara, D.E. and A. Keeler, 2013: A coupled physical and economic model of the response of coastal real estate to climate risk. *Nat. Clim. Change*, **3**(6), 559–562, doi:10.1038/nclimate1826.

McNeeley, S.M. et al., 2017: Expanding vulnerability assessment for public lands: The social complement to ecological approaches. *Clim. Risk Manage.*, **16**, 106–119, doi:10.1016/j.crm.2017.01.005.

McNeil, B.I. and T.P. Sasse, 2016: Future ocean hypercapnia driven by anthropogenic amplification of the natural CO_2 cycle. *Nature*, **529**, 383, doi:10.1038/nature16156.

Mcowen, C.J. et al., 2015: Is fisheries production within Large Marine Ecosystems determined by bottom-up or top-down forcing? *Fish Fish.*, **16**(4), 623–632, doi:10.1111/faf.12082.

Mcowen, C.J. et al., 2017: A global map of saltmarshes. *Biodiversity data journal*,(5), e11764.

Meadows, P.S., 2011: Ecosystem Sustainability, Climate Change, and Rural Communities. *J. Anim. Plant Sci.*, **21**, 317–332.

Meehl, G.A. et al., 2011: Model-based evidence of deep-ocean heat uptake during surface-temperature hiatus periods. *Nat. Clim. Change*, **1**, 360, doi:10.1038/nclimate1229.

Megann, A., 2018: Estimating the numerical diapycnal mixing in an eddy-permitting ocean model. *Ocean Model.*, **121**, 19–33, doi:10.1016/j.ocemod.2017.11.001.

Megonigal, J.P. et al., 2016: 3.4 Impacts and effects of ocean warming on tidal marsh and tidal freshwater forest ecosystems. In: Laffoley, D., & Baxter, J.M. (editors). 2016. Explaining ocean warming: Causes, scale, effects and consequences. Full report. Gland, Switzerland: IUCN,105-210. ISBN: 978-2-1806-4.

Melaku Canu, D. et al., 2015: Estimating the value of carbon sequestration ecosystem services in the Mediterranean Sea: An Ecol. Econ. approach. *Global Environ. Change*, **32**(Supplement C), 87–95, doi:10.1016/j.gloenvcha.2015.02.008.

Melet, A., S. Legg and R. Hallberg, 2016: Climatic Impacts of Parameterized Local and Remote Tidal Mixing. *J. Clim.*, **29**(10), 3473–3500, doi:10.1175/jcli-d-15-0153.1.

Mélin, F. et al., 2017: Assessing the fitness-for-purpose of satellite multi-mission ocean color climate data records: A protocol applied to OC-CCI chlorophyll-a data. *Remote Sens. Environ.*, **203**, 139–151, doi:10.1016/j.rse.2017.03.039.

Menezes, V.V., A.M. Macdonald and C. Schatzman, 2017: Accelerated freshening of Antarctic Bottom Water over the last decade in the Southern Indian Ocean. *Sci. Adv.*, **3**(1), e1601426, doi:10.1126/sciadv.1601426.

Merino, G., M. Barange and C. Mullon, 2010: Climate variability and change scenarios for a marine commodity: Modelling small pelagic fish, fisheries and fishmeal in a globalized market. *J. Mar. Syst.*, **81**(1), 196–205, doi:10.1016/j.jmarsys.2009.12.010.

Merkens, J.-L., L. Reimann, J. Hinkel and A.T. Vafeidis, 2016: Gridded population projections for the coastal zone under the Shared Socioeconomic Pathways. *Global Planet. Change*, **145**, 57–66, doi:10.1016/j.gloplacha.2016.08.009.

Metcalf, S.J. et al., 2015: Measuring the vulnerability of marine social-ecological systems: a prerequisite for the identification of climate change adaptations. *Ecol. Soc.*, **20**(2): 35, doi:10.5751/ES-07509-200235.

Meyer, A., K L. Polzin, B.M. Sloyan and H.E. Phillips, 2016: Internal Waves and Mixing near the Kerguelen Plateau. *J. Phys. Oceanogr.*, **46**(2), 417–437, doi:10.1175/jpo-d-15-0055.1.

Micallef, S., A. Micallef and C. Galdies, 2018: Application of the Coastal Hazard Wheel to assess erosion on the Maltese coast. *Ocean Coast. Manage.*, **156**, 209–222, doi:10.1016/j.ocecoaman.2017.06.005.

Michailidou, A.V., C. Vlachokostas and N. Moussiopoulos, 2016a: Interactions between climate change and the tourism sector: Multiple-criteria decision analysis to assess mitigation and adaptation options in tourism areas. *Tourism Manage.*, **55**, 1–12, doi:10.1016/j.tourman.2016.01.010.

5

Michailidou, A.V., C. Vlachokostas and N. Moussiopoulos, 2016b: Interactions between climate change and the tourism sector: Multiple-criteria decision analysis to assess mitigation and adaptation options in tourism areas. *Tourism Manage.*, **55**(Supplement C), 1–12.

Middelburg, J.J. and L.A. Levin, 2009: Coastal hypoxia and sediment biogeochemistry. *Biogeosciences*, **6**(7), 1273–1293, doi:10.5194/bg-6-1273-2009.

Middelburg, J.J. et al., 2015: Discovery of symbiotic nitrogen fixation and chemoautotrophy in cold water corals. *Sci. Rep.*, **5**, 17962.

Mieszkowska, N. et al., 2006: Changes in the range of some common rocky shore species in Britain – a response to climate change? *Hydrobiologia*, **555**, 241–251.

Mikellidou, C.V., L.M. Shakou, G. Boustras and C. Dimopoulos, 2018: Energy critical infrastructures at risk from climate change: A state of the art review. *Saf. Sci.*, **110**, 110–120.

Milazzo, M. et al., 2019: Biogenic habitat shifts under long-term ocean acidification show nonlinear community responses and unbalanced functions of associated invertebrates. *Sci. Total Environ.*, **667**, 41–48, doi:10.1016/j.scitotenv.2019.02.391.

Miller, D.D. et al., 2017: Adaptation strategies to climate change in marine systems. *Global Change Biol.*, **24**, e1–e14.

Miller, K.A., G.R. Munro, U.R. Sumaila and W.W. Cheung, 2013: Governing marine fisheries in a changing climate: A game-theoretic perspective. *Can. J. gr. Econ.*, **61**(2), 309–334.

Miller, K.I. and G.R. Russ, 2014: Studies of no-take marine reserves: Methods for differentiating reserve and habitat effects. *Ocean Coast. Manage.*, **96**(Supplement C), 51–60.

Miloshis, M. and C.A. Fairfield, 2015: Coastal wetland management: A rating system for potential engineering interventions. *Ecol. Eng.*, **75**, 195–198.

Miranda, P.M.A., J.M.R. Alves and N. Serra, 2013: Climate change and upwelling: response of Iberian upwelling to atmospheric forcing in a regional climate scenario. *Clim. Dyn.*, **40**(11–12), 2813–2824, doi:10.1007/s00382-012-1442-9.

Miranda, R.J. et al., 2019: Invasion-mediated effects on marine trophic interactions in a changing climate: positive feedbacks favour kelp persistence. *Proc. Roy. Soc. B.*, **286**(1899), 20182866.

Misumi, K. et al., 2013: The iron budget in ocean surface waters in the 20th and 21st centuries: projections by the Community Earth System Model version 1. *Biogeosciences*, **10**(5), 8505–8559.

Mitchell, J.F., T. Johns, J.M. Gregory and S. Tett, 1995: Climate response to increasing levels of greenhouse gases and sulphate aerosols. *Nature*, **376**(6540), 501.

Mitchell, S., I. Boateng and F. Couceiro, 2017: Influence of flushing and other characteristics of coastal lagoons using data from Ghana. *Ocean Coast. Manage.*, **143**, 26–37, doi:10.1016/j.ocecoaman.2016.10.002.

Miteva, D.A., B.C. Murray and S.K. Pattanayak, 2015: Do protected areas reduce blue carbon emissions? A quasi-experimental evaluation of mangroves in Indonesia. *Ecol. Econ.*, **119**, 127–135.

Moffitt, S.E. et al., 2014: Vertical oxygen minimum zone oscillations since 20 ka in Santa Barbara Basin: A benthic foraminiferal community perspective. *Paleoceanography*, **29**(1), 44–57, doi:10.1002/2013pa002483.

Moffitt, S.E., T.M. Hill, P.D. Roopnarine and J. P. Kennett, 2015: Response of seafloor ecosystems to abrupt global climate change. *PNAS*, **112**(15), 4684–4689, doi:10.1073/pnas.1417130112.

Molinos, J.G. et al., 2016: Climate velocity and the future global redistribution of marine biodiversity. *Nat. Clim. Change*, **6**(1), 83–88, doi:10.1038/NCLIMATE2769.

Möller, I., 2019: Applying Uncertain Science to Nature-Based Coastal Protection: Lessons From Shallow Wetland-Dominated Shores. *Front. Environ. Sci.*, **7**(49), doi:10.3389/fenvs.2019.00049.

Mollica, N.R. et al., 2018: Ocean acidification affects coral growth by reducing skeletal density. *PNAS*, **115**(8), 1754, doi:10.1073/pnas.1712806115.

Mongwe, N.P., M. Vichi and P.M.S. Monteiro, 2018: The seasonal cycle of pCO_2 and CO_2 fluxes in the Southern Ocean: diagnosing anomalies in CMIP5 Earth system models. *Biogeosciences*, **15**(9), 2851.

Monioudi, I.N. et al., 2018: Climate change impacts on critical international transportation assets of Caribbean Small Island Developing States (SIDS): the case of Jamaica and Saint Lucia. *Reg. Environ. Change*, **18**(8), 2211–2225.

Monllor-Hurtado, A., M.G. Pennino and J.L. Sanchez-Lizaso, 2017: Shift in tuna catches due to ocean warming. *PLoS One*, **12**(6), e0178196, doi:10.1371/journal.pone.0178196.

Montoya Maya, P.H., K.P. Smit, A.J. Burt and S. Frias-Torres, 2016: Large-scale coral reef restoration could assist natural recovery in Seychelles, Indian Ocean. *Nat. Conserv.*, **16**(3), 1–17, doi:10.3897/natureconservation.16.8604.

Montserrat, F. et al., 2017: Olivine Dissolution in Seawater: Implications for CO_2 Sequestration through Enhanced Weathering in Coastal Environments. *Environ. Sci. Technol.*, **51**(7), 3960–3972, doi:10.1021/acs.est.6b05942.

Moore, C.M. et al., 2013: Processes and patterns of oceanic nutrient limitation. *Nat. Geosci.*, **6**(9), ngeo1765, doi:10.1038/ngeo1765.

Moore, J.C., 2018: Predicting tipping points in complex environmental systems. *PNAS*, **115**(4), 635, doi:10.1073/pnas.1721206115.

Moosavi, S., 2017: Ecological Coastal Protection: Pathways to Living Shorelines. *Procedia Eng.*, **196**, 930–938, doi:10.1016/j.proeng.2017.08.027.

Mora, C. et al., 2013: The projected timing of climate departure from recent variability. *Nature*, **502**(7470), 183–7, doi:10.1038/nature12540.

Moràn, X.A.G., Á. Lòpez-Urrutia, A. Calvo-DÍAz and W.K.W. Li, 2010: Increasing importance of small phytoplankton in a warmer ocean. *Global Change Biol.*, **16**(3), 1137–1144, doi:10.1111/j.1365-2486.2009.01960.x.

Moreira, D. and J.C.M. Pires, 2016: Atmospheric CO_2 capture by algae: Negative carbon dioxide emission path. *Bioresour. Technol.* **215**, 371–379, doi:10.1016/j.biortech.2016.03.060.

Moreno, A.R. et al., 2017: Marine Phytoplankton Stoichiometry Mediates Nonlinear Interactions Between Nutrient Supply, Temperature, and Atmospheric CO_2. *Biogeosciences*, 1–28, doi:10.5194/bg-2017-367.

Morgan, K.M., C.T. Perry, J.A. Johnson and S.G. Smithers, 2017: Nearshore Turbid-Zone Corals Exhibit High Bleaching Tolerance on the Great Barrier Reef Following the 2016 Ocean Warming Event. *Front. Mar. Sci.*, **4**, 224, doi:10.3389/fmars.2017.00224.

Morley, J.W. et al., 2018: Projecting shifts in thermal habitat for 686 species on the North American continental shelf. *PLoS One*, **13**(5), e0196127, doi:10.1371/journal.pone.0196127.

Morrissey, C.A., L.I. Bendell-Young and J.E. Elliott, 2005: Identifying Sources and Biomagnification of Persistent Organic Contaminants in Biota from Mountain Streams of Southwestern British Columbia, Canada. *Environ. Sci. Technol.*, **39**(20), 8090–8098, doi:10.1021/es050431n.

Moum, J.N., A. Perlin, J.D. Nash and M.J. McPhaden, 2013: Seasonal sea surface cooling in the equatorial Pacific cold tongue controlled by ocean mixing. *Nature*, **500**, 64, doi:10.1038/nature12363.

Movilla, J. et al., 2014: Resistance of two Mediterranean cold water coral species to low-pH conditions. *Water*, **6**(1), 59–67.

Müller, M., 2012: The influence of changing stratification conditions on barotropic tidal transport and its implications for seasonal and secular changes of tides. *Cont. Shelf Res.*, **47**(Supplement C), 107–118, doi:10.1016/j.csr.2012.07.003.

Müller, M., B.K. Arbic and J.X. Mitrovica, 2011: Secular trends in ocean tides: Observations and model results. *J. Geophys. Res-Oceans*, **116**(C5), n/a–n/a, doi:10.1029/2010JC006387.

Mumby, P.J. and R. van Woesik, 2014: Consequences of Ecological, Evolutionary and Biogeochemical Uncertainty for Coral Reef Responses to Climatic Stress. *Curr. Biol.*, **24**(10), R413–R423, doi:10.1016/j.cub.2014.04.029.

Munro, D.R. et al., 2015: Recent evidence for a strengthening CO_2 sink in the Southern Ocean from carbonate system measurements in the Drake Passage (2002–2015). *Geophys. Res. Lett.*, **42**(18), 7623–7630.

5

Murdiyarso, D. et al., 2015: The potential of Indonesian mangrove forests for global climate change mitigation. *Nat. Clim. Change*, **5**, 1089, doi:10.1038/nclimate2734.

Mutombo, K. and A. Ölçer, 2016: Towards Port Infrastructure: A Global Port Climate Risk Analysis. *WMU Journal of Maritime Affairs*, **16**, 161, doi:10.1007/s13437-016-0113-9.

N'Yeurt, A.R. et al., 2012: Negative carbon via Ocean Afforestation. *Process Saf. Environ.*, **90**(6), 467–474, doi:10.1016/j.psep.2012.10.008.

N'Yeurt, A.d.R. and V. Iese, 2014: The proliferating brown alga Sargassum polycystum in Tuvalu, South Pacific: assessment of the bloom and applications to local agriculture and sustainable energy. *J. App. Phycol.*, **27**(5), 2037–2045, doi:10.1007/s10811-014-0435-y.

Nagelkerken, I. and S.D. Connell, 2015: Global alteration of ocean ecosystem functioning due to increasing human CO_2 emissions. *PNAS*, **112**(43), 13272–13277, doi:10.1073/pnas.1510856112.

Nagelkerken, I., M. Sheaves, R. Baker and R.M. Connolly, 2015: The seascape nursery: a novel spatial approach to identify and manage nurseries for coastal marine fauna. *Fish Fish.*, **16**(2), 362–371, doi:10.1111/faf.12057.

Nagy, G.J., L. Seijo, J.E. Verocai and M. Bidegain, 2014: Stakeholders' climate perception and adaptation in coastal Uruguay. *International Journal of Climate Change Strategies and Management*, **6**(1), 63–84, doi:doi:10.1108/IJCCSM-03-2013-0035.

Nahlik, A.M. and M.S. Fennessy, 2016: Carbon storage in US wetlands. *Nat. Commun.*, **7**, 13835, doi:10.1038/ncomms13835.

Najjar, R. et al., 2018: Carbon budget of tidal wetlands, estuaries, and shelf waters of Eastern North America. *Global Biogeochem. Cy.*, **32**(3), 389–416.

Nanlohy, H., A.N. Bambang, Ambariyanto and S. Hutabarat, 2015: Coastal Communities Knowledge Level on Climate Change as a Consideration in Mangrove Ecosystems Management in the Kotania Bay, West Seram Regency. *Procedia Environ. Sci.*, **23**, 157–163, doi:10.1016/j.proenv.2015.01.024.

Nannini, M., L. De Marchi, C. Lombardi and F. Ragazzola, 2015: Effects of thermal stress on the growth of an intertidal population of Ellisolandia elongata (Rhodophyta) from N–W Mediterranean Sea. *Mar. Environ. Res.*, **112**, 11–19, doi:10.1016/j.marenvres.2015.05.005.

Narayan, S. et al., 2016: The Effectiveness, Costs and Coastal Protection Benefits of Natural and Nature-Based Defences. *PLoS One*, **11**(5), e0154735, doi:10.1371/journal.pone.0154735.

Naumann, M.S., C. Orejas and C. Ferrier-Pagès, 2014: Species-specific physiological response by the cold water corals Lophelia pertusa and Madrepora oculata to variations within their natural temperature range. *Deep Sea Res. Pt. II*, **99**, 36–41.

Ndhlovu, N., O. Saito, R. Djalante and N. Yagi, 2017: Assessing the Sensitivity of Small-Scale Fishery Groups to Climate Change in Lake Kariba, Zimbabwe. *Sustainability*, **9**(12), 2209.

Needelman, B.A. et al., 2018: The Science and Policy of the Verified Carbon Standard Methodology for Tidal Wetland and Seagrass Restoration. *Estuar. Coast.*, **41**(8), 2159–2171, doi:10.1007/s12237-018-0429-0.

Needelman, B.A., I.M. Emmer, M.P. Oreska and J.P. Megonigal, 2019: Blue carbon accounting for carbon markets. In: *A Blue Carbon Primer*. [Windham-Myers, L., Crooks, S. and Troxler, T. G. (eds.)]. CRC Press, Boca Raton, FL, pp. 283–292, ISBN: 978-1-4987-6909-9.

Nehren, U. et al., 2017: Sand Dunes and Mangroves for Disaster Risk Reduction and Climate Change Adaptation in the Coastal Zone of Quang Nam Province, Vietnam. In: *Land Use and Climate Change Interactions in Central Vietnam: LUCCi* [Nauditt, A. and L. Ribbe (eds.)]. Springer Singapore, Singapore, pp. 201–222. ISBN: 978-981-10-2624-9.

Nellemann, C. et al., 2009: *Blue carbon: the role of healthy oceans in binding carbon: a rapid response assessment*. UNEP/Earthprint, Arendal, Norway, 78 p. ISBN: 978-82-7701-060-1.

Netburn, A.N. and J. Anthony Koslow, 2015: Dissolved oxygen as a constraint on daytime deep scattering layer depth in the southern California current ecosystem. *Deep-Sea Res. Pt. I*, **104**, 149–158, doi:10.1016/j.dsr.2015.06.006.

Neuheimer, A.B., B.R. MacKenzie and M.R. Payne, 2018: Temperature-dependent adaptation allows fish to meet their food across their species' range. *Sci. Adv.*, **4**(7), eaar4349, doi:10.1126/sciadv.aar4349.

Newell, R. and R. Canessa, 2017: Picturing a place by the sea: Geovisualizations as place-based tools for collaborative coastal management. *Ocean Coast. Manage.*, **141**, 29–42, doi:10.1016/j.ocecoaman.2017.03.002.

Newton, I., 1687: *Philosophiæ Naturalis Principia Mathematica*. London.

Newton, J. et al., 2015: Global ocean acidification observing network: requirements and governance plan. *Global Ocean Acidification Observing Network: Requirements and Governance Plan*, pp 57.

Ng, A.K.Y. et al., 2018: Implications of climate change for shipping: Opening the Arctic seas. *WiRes. Clim. Change*, **9**(2), e507, doi:10.1002/wcc.507.

Ng, C.S.L., T.C. Toh and L.M. Chou, 2016: Coral restoration in Singapore's sediment-challenged sea. *Regional Studies in Marine Science*, **8**(3), 422–429.

Ng, C.S.L., T.C. Toh and L.M. Chou, 2017: Artificial reefs as a reef restoration strategy in sediment-affected environments: Insights from long-term monitoring. *Aquat. Conserv. Mar. Freshw. Ecosyst.*, **27**(5), 976–985, doi:10.1002/aqc.2755.

Nguyen, T.P., T.T. Luom and K.E. Parnell, 2017: Mangrove allocation for coastal protection and livelihood improvement in Kien Giang province, Vietnam: Constraints and recommendations. *Land Use Policy*, **63**, 401–407, doi:10.1016/j.landusepol.2017.01.048.

Nguyen, T.T.X., J. Bonetti, K. Rogers and C.D. Woodroffe, 2016: Indicator-based assessment of climate-change impacts on coasts: A review of concepts, methodological approaches and vulnerability indices. *Ocean Coast. Manage.*, **123**, 18–43, doi:10.1016/j.ocecoaman.2015.11.022.

Nicastro, K.R. et al., 2013: Shift happens: trailing edge contraction associated with recent warming trends threatens a distinct genetic lineage in the marine macroalga Fucus vesiculosus. *BMC Biology*, **11**(1), 6, doi:10.1186/1741-7007-11-6.

Nicholls, R. et al., 2015: Chapter 2 – Developing a Holistic Approach to Assessing and Managing Coastal Flood Risk. In: *Coastal Risk Management in a Changing Climate*. [Zanuttigh, B., Nicholls, R.J., Vanderlinden, J-P, Burcharth, H.F. and Thompson, R.C. (eds.)]. Butterworth-Heinemann, Boston, pp. 9–53. ISBN: 978-0-12-397310-8.

NOAA, 2013: *World Ocean Atlas 2013 version 2*. [Available at: www.nodc.noaa.gov/OC5/woa13]. Accessed: 2019/09/30.

Novak Colwell, J.M., M. Axelrod, S.S. Salim and S. Velvizhi, 2017: A Gendered Analysis of Fisherfolk's Livelihood Adaptation and Coping Responses in the Face of a Seasonal Fishing Ban in Tamil Nadu & Puducherry, India. *World Dev.*, **98**, 325–337, doi:10.1016/j.worlddev.2017.04.033.

Nowicki, R.J. et al., 2017: Predicting seagrass recovery times and their implications following an extreme climate event. *Mar. Ecol. Prog. Ser.*, **567**, 79–93.

Nunn, P.D., W. Aalbersberg, S. Lata and M. Gwilliam, 2014: Beyond the core: community governance for climate-change adaptation in peripheral parts of Pacific Island Countries. *Reg. Environ. Change*, **14**(1), 221–235, doi:10.1007/s10113-013-0486-7.

Nursey-Bray, M., P. Fidelman and M. Owusu, 2018: Does co-management facilitate adaptive capacity in times of environmental change? Insights from fisheries in Australia. *Mar. Policy*, **96**, 72–80, doi:10.1016/j.marpol.2018.07.016.

O Rourke, F., F. Boyle and A. Reynolds, 2010: Tidal energy update 2009. *Appl. Energy*, **87**(2), 398–409, doi:10.1016/j.apenergy.2009.08.014.

O'Brien, B.S., K. Mello, A. Litterer and J.A. Dijkstra, 2018: Seaweed structure shapes trophic interactions: A case study using a mid-trophic level fish species. *J. Exp. Mar. Biol. Ecol.*, **506**, 1–8, doi:10.1016/j.jembe.2018.05.003.

O'Brien, K.R. et al., 2017: Seagrass ecosystem trajectory depends on the relative timescales of resistance, recovery and disturbance. *Mar. Pollut. Bull.*, **134**, 166–176. doi:10.1016/j.marpolbul.2017.09.006.

O'Leary, B.C. and C.M. Roberts, 2018: Ecological connectivity across ocean depths: Implications for protected area design. *Global Ecol. Conserv.*, **15**, p.e00431.

O'Brien, P.A., K.M. Morrow, B.L. Willis and D.G. Bourne, 2016: Implications of Ocean Acidification for Marine Microorganisms from the Free-Living

5

to the Host-Associated. *Front. Mar. Sci.*, **3**(fiv142), 1029, doi:10.3389/fmars.2016.00047.

O'Malley, M.P., K. Lee-Brooks and H.B. Medd, 2013: The global economic impact of manta ray watching tourism. *PLoS One*, **8**(5), e65051.

Ojea, E., I. Pearlman, S.D. Gaines and S.E. Lester, 2017: Fisheries regulatory regimes and resilience to climate change. *Ambio*, **46**(4), 399–412.

Okey, T.A., H.M. Alidina, V. Lo and S. Jessen, 2014: Effects of climate change on Canada's Pacific marine ecosystems: a summary of scientific knowledge. *Rev. Fish Biol. Fisher.*, **24**(2), 519–559, doi:10.1007/s11160-014-9342-1.

Olsen, A. et al., 2016a: The Global Ocean Data Analysis Project version 2 (GLODAPv2)–an internally consistent data product for the world ocean. *Earth Syst. Sci. Data (Online)*, **8**(2), 297–323.

Olsen, J.L. et al., 2016b: The genome of the seagrass Zostera marina reveals angiosperm adaptation to the sea. *Nature*, **530**, 331, doi:10.1038/nature16548.

Onaka, S., H. Hashimoto, S.R. Nashreen Banu Soogun and A. Jheengut, 2015: Chapter 26 – Coastal Erosion and Demonstration Project as Coastal Adaptation Measures in Mauritius. In: *Handbook of Coastal Disaster Mitigation for Engineers and Planners*. [Esteban, M., H. Takagi and T. Shibayama (eds.)]. Butterworth-Heinemann, Boston, pp. 561–577. ISBN: 978-0-12-801060-0.

Ondiviela, B. et al., 2014: The role of seagrasses in coastal protection in a changing climate. *Coast. Eng.*, **87**(Supplement C), 158–168, doi:10.1016/j.coastaleng.2013.11.005.

Oppenheimer, M. et al., 2015: Emergent risks and key vulnerabilities. In: *Climate Change 2014 Impacts, Adaptation and Vulnerability: Part A: Global and Sectoral Aspects*. [Field, C.B., V.R. Barros, D.J. Dokken, K.J. Mach, M.D. Mastrandrea, T.E. Bilir, M. Chatterjee, K.L. Ebi, Y.O. Estrada, R.C. Genova, B. Girma, E.S. Kissel, A.N. Levy, S. MacCracken, P.R. Mastrandrea, and L.L. White (eds.)]. Cambridge University Press, Cambridge, United Kingdom and New York, NY, USA, pp. 1039–1100. ISBN: ISBN 978-1-107-05807-1.

Oral, N., 2018: Ocean Acidification: Falling Between the Legal Cracks of UNCLOS and the UNFCCC. *Ecology Law Quarterly*, **45**(1), 9.

Oreska, M.P. et al., 2018: Comment on Geoengineering with seagrasses: is credit due where credit is given? *Environ. Res. Lett.*, **13**(3), 038001.

Orlando, L., L. Ortega and O. Defeo, 2019: Multi-decadal variability in sandy beach area and the role of climate forcing. *Estuar. Coast. Shelf Sci.*, **218**, 197–203, doi:10.1016/j.ecss.2018.12.015.

Orr, J.C. et al., 2005: Anthropogenic ocean acidification over the twenty-first century and its impact on calcifying organisms. *Nature*, **437**(7059), 681–686, doi:10.1038/nature04095.

Orth, R.J. et al., 2006: A Global Crisis for Seagrass Ecosystems. *BioScience*, **56**(12), 987–996, doi:10.1641/0006-3568(2006)56[987:AGCFSE]2.0.CO;2.

Osborn, D., S. Dupont, L. Hansson and M. Metian, 2017: Ocean acidification: Impacts and governance. In: *Handbook on the Economics and Management of Sustainable Oceans* [Nunes, P.A.L.D., L.E. Svensson and A. Marikandya (eds.)], Cheltenham, UK, pp. 396–415. ISBN: 978-1-78643-071-7.

Oschlies, A., P. Brandt, L. Stramma and S. Schmidtko, 2018: Drivers and mechanisms of ocean deoxygenation. *Nat. Geosci.*, **11**(7), 467–473, doi:10.1038/s41561-018-0152-2.

Osman, E.O. et al., 2017: Thermal refugia against coral bleaching throughout the northern Red Sea. *Global Change Biol.*, **52**, 716, doi:10.1111/gcb.13895.

Osorio-Cano, J.D., A.F. Osorio and D.S. Peláez-Zapata, 2017: Ecosystem management tools to study natural habitats as wave damping structures and coastal protection mechanisms. *Ecol. Eng.*, **130**, 282–295, doi:10.1016/j.ecoleng.2017.07.015.

Oswald Beiler, M., L. Marroquin and S. McNeil, 2016: State-of-the-practice assessment of climate change adaptation practices across metropolitan planning organizations pre- and post-Hurricane Sandy. *Transport. Res. A-Pol.*, **88**, 163–174, doi:10.1016/j.tra.2016.04.003.

Ou, G., H. Wang, R. Si and W. Guan, 2017: The dinoflagellate Akashiwo sanguinea will benefit from future climate change: The interactive effects of ocean acidification, warming and high irradiance on photophysiology and hemolytic activity. *Harmful Algae*, **68**, 118–127, doi:10.1016/j.hal.2017.08.003.

Oulahen, G. et al., 2018: Barriers and Drivers of Planning for Climate Change Adaptation across Three Levels of Government in Canada. *Planning Theory & Practice*, **19**(3), 405–421, doi:10.1080/14649357.2018.1481993.

Oyarzún, D. and C.M. Brierley, 2018: The future of coastal upwelling in the Humboldt current from model projections. *Clim. Dyn.*, **52**, 599–615. doi:10.1007/s00382-018-4158-7.

Paar, M. et al., 2016: Temporal shift in biomass and production of macrozoobenthos in the macroalgal belt at Hansneset, Kongsfjorden, after 15 years. *Polar Biol.*, **39**(11), 2065–2076, doi:10.1007/s00300-015-1760-6.

Paerl, H.W. et al., 2016: Mitigating cyanobacterial harmful algal blooms in aquatic ecosystems impacted by climate change and anthropogenic nutrients. *Harmful Algae*, **54**, 213–222, doi:10.1016/j.hal.2015.09.009.

Paerl, H.W., T.G. Otten and R. Kudela, 2018: Mitigating the Expansion of Harmful Algal Blooms Across the Freshwater-to-Marine Continuum. *Environ. Sci. Technol.*, **52**(10), 5519–5529, doi:10.1021/acs.est.7b05950.

Pagès, J.F. et al., 2017: Contrasting effects of ocean warming on different components of plant-herbivore interactions. *Mar. Pollut. Bull.*, **134**, 55–65. doi:10.1016/j.marpolbul.2017.10.036.

Palacios, M.L. and J.R. Cantera, 2017: Mangrove timber use as an ecosystem service in the Colombian Pacific. *Hydrobiologia*, **803**(1), 345–358.

Palmer, M., K. Haines, S. Tett and T. Ansell, 2007: Isolating the signal of ocean global warming. *Geophys. Res. Lett.*, **34**(23), 1–6.

Palumbi, S.R., D.J. Barshis, N. Traylor-Knowles and R.A. Bay, 2014: Mechanisms of reef coral resistance to future climate change. *Science*, **344**(6186), 895–898, doi:10.1126/science.1251336.

Pan, Y. et al., 2011: A Large and Persistent Carbon Sink in the World's Forests. *Science*, **333**(6045), 988.

Pan, Y. and D. Schimel, 2016: Synergy of a warm spring and dry summer. *Nature*, **534**, 483, doi:10.1038/nature18450.

Pandolfi, J.M., 2015: Incorporating Uncertainty in Predicting the Future Response of Coral Reefs to Climate Change. *Annu. Rev. Ecol. Evol. Syst.*, **46**(1), 281–303, doi:10.1146/annurev-ecolsys-120213-091811.

Panikkar, B., B. Lemmond, B. Else and M. Murray, 2018: Ice over troubled waters: navigating the Northwest Passage using Inuit knowledge and scientific information. *Clim. Res.*, **75**(1), 81–94.

Paolisso, M. et al., 2012: Climate Change, Justice, and Adaptation among African American Communities in the Chesapeake Bay Region. *Weather, Clim. Soc.*, **4**(1), 34–47, doi:10.1175/WCAS-D-11-00039.1.

Papageorgiou, M., 2016: Coastal and marine tourism: A challenging factor in Marine Spatial Planning. *Ocean Coast. Manage.*, **129**, 44–48, doi:10.1016/j.ocecoaman.2016.05.006.

Paredes, J. et al., 2019: Population Genetic Structure at the Northern Edge of the Distribution of Alexandrium catenella in the Patagonian Fjords and Its Expansion Along the Open Pacific Ocean Coast. *Front. Mar. Sci.*, **5**(532), 1–13. doi:10.3389/fmars.2018.00532.

Paredes-Banda, P. et al., 2018: Association of the Toxigenic Dinoflagellate Alexandrium ostenfeldii With Spirolide Accumulation in Cultured Mussels (Mytilus galloprovincialis) From Northwest Mexico. *Front. Mar. Sci.*, **5**(491), 1–13. doi:10.3389/fmars.2018.00491.

Paterson, S.K. et al., 2017: Size does matter: City scale and the asymmetries of climate change adaptation in three coastal towns. *Geoforum*, **81**, 109–119, doi:10.1016/j.geoforum.2017.02.014.

Patino-Martinez, J., A. Marco, L. Quinones and L.A. Hawkes, 2014: The potential future influence of sea level rise on leatherback turtle nests. *J. Exp. Mar. Biol. Ecol.*, **461**, 116–123, doi:10.1016/j.jembe.2014.07.021.

Patricio, A.R. et al., 2017: Balanced primary sex ratios and resilience to climate change in a major sea turtle population. *Mar. Ecol. Prog. Ser.*, **577**, 189–203, doi:10.3354/meps12242.

Patrício, A.R. et al., 2019: Climate change resilience of a globally important sea turtle nesting population. *Global Change Biol.*, **25**(2), 522–535, doi:10.1111/gcb.14520.

Pauly, D. and W.W.L. Cheung, 2017: Sound physiological knowledge and principles in modeling shrinking of fishes under climate change. *Global Change Biol.*, **25**(2), n/a–n/a, doi:10.1111/gcb.13831.

Pauly, D. and D. Zeller, 2016: Catch reconstructions reveal that global marine fisheries catches are higher than reported and declining. *Nat. Commun.*, **7**, 10244 EP 1–9, doi:10.1038/ncomms10244.

Payne, M.R. et al., 2017: Lessons from the First Generation of Marine Ecological Forecast Products. *Front. Mar. Sci.*, **4**(289), doi:10.3389/fmars.2017.00289.

Payne, N.L. et al., 2016: Temperature dependence of fish performance in the wild: links with species biogeography and physiological thermal tolerance. *Funct. Ecol.*, **30**(6), 903–912, doi:10.1111/1365-2435.12618.

Pearce, T., R. Currenti, A. Mateiwai and B. Doran, 2018: Adaptation to climate change and freshwater resources in Vusama village, Viti Levu, Fiji. *Reg. Environ. Change*, **18**(2), 501–510, doi:10.1007/s10113-017-1222-5.

Pearse, R., 2017: Gender and climate change. *WiRes. Clim. Change*, **8**(2), 1–16, e451. doi: 10.1002/wcc.451.

Pecl, G.T. et al., 2017: Biodiversity redistribution under climate change: Impacts on ecosystems and human well–being. *Science*, **355**(6332), eaai9214, doi:10.1126/science.aai9214.

Peirson, W. et al., 2015: Opportunistic management of estuaries under climate change: A new adaptive decision-making framework and its practical application. *J. Environ. Manage.*, **163**, 214–223, doi:10.1016/j.jenvman.2015.08.021.

Pelletier, J.F. and E. Guy, 2012: Évaluation des activités de transport maritime en arctique canadien. *Cahiers Scientifiques Du Transport*, (61), 3–33.

Pelling, H.E. and J.A.M. Green, 2014: Impact of flood defences and sea level rise on the European Shelf tidal regime. *Cont. Shelf Res.*, **85**(Supplement C), 96–105, doi:10.1016/j.csr.2014.04.011.

Pelling, H.E., K. Uehara and J.A.M. Green, 2013: The impact of rapid coastline changes and sea level rise on the tides in the Bohai Sea, China. *J. Geophys. Res-Oceans*, **118**(7), 3462–3472, doi:10.1002/jgrc.20258.

Peña-Alonso, C., L. Hernández-Calvento, E. Pérez-Chacón and E. Ariza-Solé, 2017: The relationship between heritage, recreational quality and geomorphological vulnerability in the coastal zone: A case study of beach systems in the Canary Islands. *Ecol. Indic.*, **82**, 420–432, doi:10.1016/j.ecolind.2017.07.014.

Pendleton, L. et al., 2012: Estimating Global "Blue Carbon" Emissions from Conversion and Degradation of Vegetated Coastal Ecosystems. *PLoS One*, **7**(9), e43542, doi:10.1371/journal.pone.0043542.

Pendleton, L.H., O. Thébaud, R.C. Mongruel and H. Levrel, 2016: Has the value of global marine and coastal ecosystem services changed? *Mar. Policy*, **64**(Supplement C), 156–158, doi:10.1016/j.marpol.2015.11.018.

Peng, L., M.G. Stewart and R.E. Melchers, 2017: Corrosion and capacity prediction of marine steel infrastructure under a changing environment. *Struct. Infrastruct. E.*, **13**(8), 988–1001.

Pentz, B. and N. Klenk, 2017: The 'responsiveness gap' in RFMOs: The critical role of decision-making policies in the fisheries management response to climate change. *Ocean Coast. Manage.*, **145**, 44–51, doi:10.1016/j.ocecoaman.2017.05.007.

Pereira, T.R. et al., 2017: Population dynamics of temperate kelp forests near their low-latitude limit. *Aquat. Bot.*, **139**, 8–18, doi:10.1016/j.aquabot.2017.02.006.

Pérez, F.F. et al., 2013: Atlantic Ocean CO 2 uptake reduced by weakening of the meridional overturning circulation. *Nat. Geosci.*, **6**(2), 146.

Pérez-Matus, A. et al., 2017: Exploring the effects of fishing pressure and upwelling intensity over subtidal kelp forest communities in Central Chile. *Ecosphere*, **8**(5), e01808, doi:10.1002/ecs2.1808.

Perkins, M.J. et al., 2015: Conserving intertidal habitats: What is the potential of ecological engineering to mitigate impacts of coastal structures? *Estuar. Coast. Shelf Sci.*, **167**, 504–515, doi:10.1016/j.ecss.2015.10.033.

Perry, C.T. et al., 2018: Loss of coral reef growth capacity to track future increases in sea level. *Nature*, **558**(7710), 396–400, doi:10.1038/s41586-018-0194-z.

Perry, J., 2015: Climate change adaptation in the world's best places: A wicked problem in need of immediate attention. *Landscape Urban Plan.*, **133**, 1–11, doi:10.1016/j.landurbplan.2014.08.013.

Pescaroli, G. and M. Magni, 2015: Flood warnings in coastal areas: how do experience and information influence responses to alert services? *Nat. Hazards Earth Syst. Sci.*, **15**(4), 703–714, doi:10.5194/nhess-15-703-2015.

Pessarrodona, A., A. Foggo and D.A. Smale, 2019: Can ecosystem functioning be maintained despite climate-driven shifts in species composition? Insights from novel marine forests. *J. Ecol.*, **107**(1), 91–104, doi:10.1111/1365-2745.13053.

Petzold, J. and B.M.W. Ratter, 2015: Climate change adaptation under a social capital approach – An analytical framework for small islands. *Ocean Coast. Manage.*, **112**, 36–43, doi:10.1016/j.ocecoaman.2015.05.003.

Philippart, C.J.M. et al., 2003: Climate-related changes in recruitment of the bivalve Macoma balthica. *Limnol. Oceanogr.*, **48**(6), 2171–2185, doi:10.4319/lo.2003.48.6.2171.

Phrampus, B.J. and M.J. Hornbach, 2012: Recent changes to the Gulf Stream causing widespread gas hydrate destabilization. *Nature*, **490**(7421), 527–+, doi:10.1038/nature11528.

Pickering, M.D. et al., 2017: The impact of future sea level rise on the global tides. *Cont. Shelf Res.*, **142**, 50–68, doi:10.1016/j.csr.2017.02.004.

Pickering, M.D., N.C. Wells, K.J. Horsburgh and J.A.M. Green, 2012: The impact of future sea level rise on the European Shelf tides. *Cont. Shelf Res.*, **35**(Supplement C), 1–15, doi:10.1016/j.csr.2011.11.011.

Pike, D.A., 2014: Forecasting the viability of sea turtle eggs in a warming world. *Global Change Biol.*, **20**(1), 7–15, doi:10.1111/gcb.12397.

Pike, D.A., E.A. Roznik and I. Bell, 2015: Nest inundation from sea level rise threatens sea turtle population viability. *R. Soc. Open Sci.*, **2**(7), 150127, doi:10.1098/rsos.150127.

Pinsky, M.L. et al., 2019: Greater vulnerability to warming of marine versus terrestrial ectotherms. *Nature*, **569**(7754), 108–111, doi:10.1038/s41586-019-1132-4.

Pinsky, M.L. et al., 2018: Preparing ocean governance for species on the move. *Science*, **360**(6394), 1189.

Pinsky, M.L. et al., 2013: Marine Taxa Track Local Climate Velocities. *Science*, **341**(6151), 1239–1242, doi:10.1126/science.1239352.

Piontek, J., M. Sperling, E.-M. Noethig and A. Engel, 2015: Multiple environmental changes induce interactive effects on bacterial degradation activity in the Arctic Ocean. *Limnol. Oceanogr.*, **60**(4), 1392–1410, doi:10.1002/lno.10112.

Piovan, M.J. et al., 2019: Germination Response to Osmotic Potential, Osmotic Agents, and Temperature of Five Halophytes Occurring along a Salinity Gradient. *Int. J. Plant Sci.*, **180**(4), 345–355, doi:10.1086/702663.

Pittman, J. et al., 2015: Governance fit for climate change in a Caribbean coastal-marine context. *Mar. Policy*, **51**, 486–498, doi:10.1016/j.marpol.2014.08.009.

Pizzolato, L. et al., 2016: The influence of declining sea ice on shipping activity in the Canadian Arctic. *Geophys. Res. Lett.*, **43**(23).

Poffenbarger, H.J., B.A. Needelman and J.P. Megonigal, 2011: Salinity influence on methane emissions from tidal marshes. *Wetlands*, **31**(5), 831–842.

Poloczanska, E.S. et al., 2013: Global imprint of climate change on marine life. *Nat. Clim. Change*, **3**(10), 919–925, doi:10.1038/NCLIMATE1958.

Poloczanska, E.S. et al., 2016: Responses of Marine Organisms to Climate Change across Oceans. *Front. Mar. Sci.*, **3**(28), 515, doi:10.3389/fmars.2016.00062.

Polovina, J.J., 2005: Climate variation, regime shifts, and implications for sustainable fisheries. *Bulletin of Marine Science*, **76**(2), 233–244.

Polovina, J.J., J.P. Dunne, P.A. Woodworth and E.A. Howell, 2011: Projected expansion of the subtropical biome and contraction of the temperate and equatorial upwelling biomes in the North Pacific under global warming. *ICES J. Mar. Sci.*, **68**(6), 986–995, doi:10.1093/icesjms/fsq198.

Polzin, K.L., J.M. Toole, J.R. Ledwell and R.W. Schmitt, 1997: Spatial Variability of Turbulent Mixing in the Abyssal Ocean. *Science*, **276**(5309), 93.

Pomeroy, R., J. Parks, K.L. Mrakovcich and C. LaMonica, 2016: Drivers and impacts of fisheries scarcity, competition, and conflict on maritime security. *Mar. Policy*, **67**(Supplement C), 94–104, doi:10.1016/j.marpol.2016.01.005.

5

Porter, J.J., D. Demeritt and S. Dessai, 2015: The right stuff? informing adaptation to climate change in British Local Government. *Global Environ. Change*, **35**, 411–422, doi:10.1016/j.gloenvcha.2015.10.004.

Pörtner, H.-O., C. Bock and F.C. Mark, 2017: Oxygen- and capacity-limited thermal tolerance: bridging ecology and physiology. *J. Exp. Biol.*, **220**(15), 2685.

Pörtner, H.O., 2012: Integrating climate-related stressor effects on marine organisms: unifying principles linking molecule to ecosystem-level changes. *Mar. Ecol. Prog. Ser.*, **470**, 273–290, doi:10.3354/meps10123.

Pörtner, H.O. and J. Gutt, 2016: Impacts of Climate Variability and Change on (Marine) Animals: Physiological Underpinnings and Evolutionary Consequences. *Integr. Comp. Biol.*, **56**(1), 31–44, doi:10.1093/icb/icw019.

Pörtner, H.O. et al., 2014: Ocean systems. In: Climate Change 2014: Impacts, Adaptation, and Vulnerability. Part A: Global and Sectoral Aspects. Contribution of Working Group II to the Fifth Assessment Report of the Intergovernmental Panel of Climate Change [Field, C.B., V.R. Barros, D.J. Dokken, K.J. Mach, M.D. Mastrandrea, T.E. Bilir, M. Chatterjee, K.L. Ebi, Y.O. Estrada, R.C. Genova, B. Girma, E.S. Kissel, A.N. Levy, S. MacCracken, P.R. Mastrandrea and L.L. White (eds.)]. Cambridge University Press, Cambridge, United Kingdom and New York, NY, USA, 411–484.

Poumadère, M. et al., 2015: Coastal vulnerabilities under the deliberation of stakeholders: The case of two French sandy beaches. *Ocean Coast. Manage.*, **105**, 166–176, doi:10.1016/j.ocecoaman.2014.12.024.

Powell, E.J. et al., 2017: A synthesis of thresholds for focal species along the U.S. Atlantic and Gulf Coasts: A review of research and applications. *Ocean Coast. Manage.*, **148**, 75–88, doi:10.1016/j.ocecoaman.2017.07.012.

Prandle, D. and A. Lane, 2015: Sensitivity of estuaries to sea level rise: Vulnerability indices. *Estuar. Coast. Shelf Sci.*, **160**, 60–68, doi:10.1016/j.ecss.2015.04.001.

Pranzini, E., 2017: Shore protection in Italy: From hard to soft engineering… and back. *Ocean Coast. Manage.*, **156**, 43–57. doi:10.1016/j.ocecoaman.2017.04.018.

Pratchett, M.S., A.S. Hoey and S.K. Wilson, 2014: Reef degradation and the loss of critical ecosystem goods and services provided by coral reef fishes. *Curr. Opin. Environ. Sustain.*, **7**(Supplement C), 37–43, doi:10.1016/j.cosust.2013.11.022.

Pratolongo, P. et al., 2013: Land cover changes in tidal salt marshes of the Bahía Blanca estuary (Argentina) during the past 40 years. *Estuar. Coast. Shelf Sci.*, **133**, 23–31.

Prince, E.D. et al., 2010: Ocean scale hypoxia-based habitat compression of Atlantic istiophorid billfishes. *Fish. Oceanogr.*, **19**(6), 448–462, doi:10.1111/j.1365-2419.2010.00556.x.

Proud, R., M.J. Cox and A.S. Brierley, 2017: Biogeography of the Global Ocean's Mesopelagic Zone. *Curr Biol*, **27**(1), 113–119, doi:10.1016/j.cub.2016.11.003.

Prowse, T.D. et al., 2009: Implications of Climate Change for Economic Development in Northern Canada: Energy, Resource, and Transportation Sectors. *Ambio*, **38**(5), 272–281, doi:10.1579/0044-7447-38.5.272.

Pruski, A.M. et al., 2017: Energy transfer in the Congo deep sea fan: From terrestrially-derived organic matter to chemosynthetic food webs. *Deep Sea Res. Pt. II*, **142**, 197–218, doi:10.1016/j.dsr2.2017.05.011.

Przeslawski, R., M. Byrne and C. Mellin, 2015: A review and meta-analysis of the effects of multiple abiotic stressors on marine embryos and larvae. *Global Change Biol.*, **21**(6), 2122–2140, doi:10.1111/gcb.12833.

Purich, A. et al., 2018: Impacts of broad-scale surface freshening of the Southern Ocean in a coupled climate model. *J. Clim.*, **31**(7), 2613–2632.

Purkey, S.G. and G.C. Johnson, 2010: Warming of Global Abyssal and Deep Southern Ocean Waters between the 1990s and 2000s: Contributions to Global Heat and Sea Level Rise Budgets. *J. Clim.*, **23**(23), 6336–6351, doi:10.1175/2010JCLI3682.1.

Purkey, S.G. and G.C. Johnson, 2013: Antarctic Bottom Water Warming and Freshening: Contributions to Sea Level Rise, Ocean Freshwater Budgets, and Global Heat Gain. *J. Clim.*, **26**(16), 6105–6122, doi:10.1175/JCLI-D-12-00834.1.

Purkey, S.G., G.C. Johnson and P. Chambers Don, 2014: Relative contributions of ocean mass and deep steric changes to sea level rise between 1993 and 2013. *J. Geophys. Res-Oceans*, **119**(11), 7509–7522, doi:10.1002/2014JC010180.

Pusceddu, A. et al., 2010: Ecosystem effects of dense water formation on deep Mediterranean Sea ecosystems: an overview. *Adv. Oceanogr. Limnol.*, **1**(1), 67–83, doi:10.1080/19475721003735765.

Putnam, H.M., J.M. Davidson and R.D. Gates, 2016: Ocean acidification influences host DNA methylation and phenotypic plasticity in environmentally susceptible corals. *Evol. Appl.*, **9**(9), 1165–1178, doi:10.1111/eva.12408.

Putnam, H.M. and R.D. Gates, 2015: Preconditioning in the reef-building coral Pocillopora damicornis and the potential for trans-generational acclimatization in coral larvae under future climate change conditions. *J. Exp. Biol.*, **218**(15), 2365–2372, doi:10.1242/jeb.123018.

Putra, H.C., H. Zhang and C. Andrews, 2015: Modeling Real Estate Market Responses to Climate Change in the Coastal Zone. *JASSS J. Artific. Soc. S.*, **18**(2), doi:10.18564/jasss.2577.

Qiu, Z. et al., 2019: Future climate change is predicted to affect the microbiome and condition of habitat-forming kelp. *Proc. Roy. Soc. B.*, **286**(1896), 20181887.

Quillien, N. et al., 2015: Effects of macroalgal accumulations on the variability in zoobenthos of high-energy macrotidal sandy beaches. *Mar. Ecol. Prog. Ser.*, **522**, 97–114.

Quinn, R.W., G.M. Spreitzer and C.F. Lam, 2012: Building a Sustainable Model of Human Energy in Organizations: Exploring the Critical Role of Resources. *Acad. Manag.*, **6**(1), 337–396, doi:10.1080/19416520.2012.676762.

Rachmilovitz, E.N. and B. Rinkevich, 2017: Tiling the reef – Exploring the first step of an ecological engineering tool that may promote phase-shift reversals in coral reefs. *Ecol. Eng.*, **105**(Supplement C), 150–161.

Raimonet, M. and J.E. Cloern, 2017: Estuary-ocean connectivity: fast physics, slow biology. *Global Change Biol.*, **23**(6), 2345–2357, doi:10.1111/gcb.13546.

Ramajo, L. et al., 2016: Food supply confers calcifiers resistance to ocean acidification. *Sci. Rep.*, **6**(1), 19374. doi:10.1038/srep19374.

Ramesh, R. et al., 2015: Land–Ocean Interactions in the Coastal Zone: Past, present & future. *Anthropocene*, **12**, 85–98, doi:10.1016/j.ancene.2016.01.005.

Ramm, T.D., C.J. White, A.H.C. Chan and C.S. Watson, 2017: A review of methodologies applied in Australian practice to evaluate long-term coastal adaptation options. *Clim. Risk Manage.*, **17**, 35–51, doi:10.1016/j.crm.2017.06.005.

Ramp, C. et al., 2015: Adapting to a Warmer Ocean-Seasonal Shift of Baleen Whale Movements over Three Decades. *PLoS One*, **10**(3), e0121374, doi:10.1371/journal.pone.0121374.

Ranasinghe, R., 2016: Assessing climate change impacts on open sandy coasts: A review. *Earth-Sci. Rev.*, **160**, 320–332, doi:10.1016/j.earscirev.2016.07.011.

Rangel-Buitrago, N.G., G. Anfuso and A.T. Williams, 2015: Coastal erosion along the Caribbean coast of Colombia: Magnitudes, causes and management. *Ocean Coast. Manage.*, **114**, 129–144, doi:10.1016/j.ocecoaman.2015.06.024.

Rao, N.S., A. Ghermandi, R. Portela and X. Wang, 2015: Global values of coastal ecosystem services: A spatial economic analysis of shoreline protection values. *Ecosyst. Serv.*, **11**, 95–105, doi:10.1016/j.ecoser.2014.11.011.

Raposa, K.B., R.L. Weber, M.C. Ekberg and W. Ferguson, 2017: Vegetation dynamics in Rhode Island salt marshes during a period of accelerating sea level rise and extreme sea level events. *Estuar. Coast.*, **40**(3), 640–650.

Rau, G.H., 2011: CO_2 Mitigation via Capture and Chemical Conversion in Seawater. *Environ. Sci. Technol.*, **45**(3), 1088–1092, doi:10.1021/es102671x.

Rau, G.H., E.L. McLeod and O. Hoegh-Guldberg, 2012: The need for new ocean conservation strategies in a high-carbon dioxide world. *Nat. Clim. Change*, **2**, 720, doi:10.1038/nclimate1555.

Raven, J., 2018: Blue carbon: past, present and future, with emphasis on macroalgae. *Biol. Lett.*, **14**(10), 20180336.

Ray, A., L. Hughes, D.M. Konisky and C. Kaylor, 2017: Extreme weather exposure and support for climate change adaptation. *Global Environ. Change*, **46**, 104–113, doi:10.1016/j.gloenvcha.2017.07.002.

5

Ray, R.D., 2009: Secular changes in the solar semidiurnal tide of the western North Atlantic Ocean. *Geophys. Res. Lett.*, **36**(19), 1–5, doi:10.1029/2009GL040217.

Raybaud, V. et al., 2013: Decline in Kelp in West Europe and Climate. *PLoS One*, **8**(6), e66044, doi:10.1371/journal.pone.0066044.

Redgwell, C., 2012: UNCLOS and Climate Change. *Proceedings of the Annual Meeting (American Society of International Law)*, **106**, 406, doi:10.5305/procannmeetasil.106.0406.

Reece, J.S. et al., 2013: Sea level rise, land use, and climate change influence the distribution of loggerhead turtle nests at the largest USA rookery (Melbourne Beach, Florida). *Mar. Ecol. Prog. Ser.*, **493**, 259–274, doi:10.3354/meps10531.

Reed, D. et al., 2016: Extreme warming challenges sentinel status of kelp forests as indicators of climate change. *Nat. Commun.*, **7**, 13757, doi:10.1038/ncomms13757.

Regnier, P. et al., 2013: Anthropogenic perturbation of the carbon fluxes from land to ocean. *Nat. Geosci.*, **6**(8), 597–607, doi:10.1038/ngeo1830.

Rêgo, J.C.L., A. Soares-Gomes and F. S. da Silva, 2018: Loss of vegetation cover in a tropical island of the Amazon coastal zone (Maranhão Island, Brazil). *Land Use Policy*, **71**, 593–601, doi:10.1016/j.landusepol.2017.10.055.

Ren, H. et al., 2017: 21st-century rise in anthropogenic nitrogen deposition on a remote coral reef. *Science*, **356**(6339), 749, doi:10.1126/science.aal3869.

Renault, L. et al., 2016: Partial decoupling of primary productivity from upwelling in the California Current system. *Nat. Geosci.*, **9**(7), 505–508, doi:10.1038/ngeo2722.

Renforth, P. and G. Henderson, 2017: Assessing ocean alkalinity for carbon sequestration. *Rev. Geophys.*, **55**(3), 636–674, doi:10.1002/2016RG000533.

Resplandy, L., L. Bopp, J.C. Orr and J.P. Dunne, 2013: Role of mode and intermediate waters in future ocean acidification: Analysis of CMIP5 models. *Geophys. Res. Lett.*, **40**(12), 3091–3095, doi:10.1002/grl.50414.

Resplandy, L. et al., 2018: Revision of global carbon fluxes based on a reassessment of oceanic and riverine carbon transport. *Nat. Geosci.*, **11**(7), 504–508.

Reusch, T.B.H. et al., 2018: The Baltic Sea as a time machine for the future coastal ocean. *Sci. Adv.*, **4**(5), eaar8195, doi:10.1126/sciadv.aar8195.

Reynolds, M.H. et al., 2015: Will the Effects of Sea level Rise Create Ecological Traps for Pacific Island Seabirds? *PLoS One*, **10**(9), e0136773, doi:10.1371/journal.pone.0136773.

Rhein, M. et al., 2013: Observations: Ocean. In: Climate Change 2013: The Physical Science Basis. Contribution of Working Group I to the Fifth Assessment Report of the Intergovernmental Panel on Climate Change [Stocker, T.F., D. Qin, G.-K. Plattner, M. Tignor, S.K. Allen, J. Boschung, A. Nauels, Y. Xia, V. Bex and P.M. Midgley (eds.)]. Cambridge University Press, Cambridge, United Kingdom and New York, NY, USA, 255–316.

Riebesell, U. et al., 2018: Toxic algal bloom induced by ocean acidification disrupts the pelagic food web. *Nat. Clim. Change*, **8**(12), 1082–1086, doi:10.1038/s41558-018-0344-1.

Riebesell, U. et al., 2017: Ocean acidification impairs competitive fitness of a predominant pelagic calcifier, Nat. Geosci., **10**, 19–24.

Riebesell, U. and J.-P. Gattuso, 2014: Lessons learned from ocean acidification research. *Nat. Clim. Change*, **5**(1), 12–14, doi:10.1038/nclimate2456.

Riebesell, U. et al., 2007: Enhanced biological carbon consumption in a high CO_2 ocean. *Nature*, **450**, 545, doi:10.1038/nature06267.

Riegl, B.M. et al., 2011: Present Limits to Heat-Adaptability in Corals and Population-Level Responses to Climate Extremes. *PLoS One*, **6**(9), e24802, doi:10.1371/journal.pone.0024802.

Riekenberg, P.M., R. Carney and B. Fry, 2016: Trophic plasticity of the methanotrophic mussel Bathymodiolus childressi in the Gulf of Mexico. *Mar. Ecol. Prog. Ser.*, **547**, 91–106.

Riget, F., K. Vorkamp and D. Muir, 2010: Temporal trends of contaminants in Arctic char (Salvelinus alpinus) from a small lake, southwest Greenland during a warming climate. *J. Environ. Monit.*, **12**(12), 2252–2258, doi:10.1039/C0EM00154F.

Rilov, G., 2016: Multi-species collapses at the warm edge of a warming sea. *Sci. Rep.*, **6**, 36897, doi:10.1038/srep36897.

Rinkevich, B., 1995: Restoration Strategies for Coral Reefs Damaged by Recreational Activities: The Use of Sexual and Asexual Recruits. *Restor. Ecol.*, **3**(4), 241–251, doi:10.1111/j.1526-100X.1995.tb00091.x.

Rinkevich, B., 2000: Steps towards the evaluation of coral reef restoration by using small branch fragments. *Mar. Biol.*, **136**(5), 807–812, doi:10.1007/s002270000293.

Rinkevich, B., 2005: What do we know about Eilat (Red Sea) reef degradation? A critical examination of the published literature. *J. Exp. Mar. Biol. Ecol.*, **327**(2), 183–200.

Rinkevich, B., 2006: The coral gardening concept and the use of underwater nurseries: lessons learned from silvics and silviculture. In: *Coral Reef Restoration Handbook* [Precht, W.F. (ed.)]. CRS/Taylor; Francis Boca Raton, pp. 291–302. ISBN: 9780429117886.

Rinkevich, B., 2008: Management of coral reefs: We have gone wrong when neglecting active reef restoration. *Mar. Pollut. Bull.*, **56**(11), 1821–1824, doi:10.1016/j.marpolbul.2008.08.014.

Rinkevich, B., 2014: Rebuilding coral reefs: does active reef restoration lead to sustainable reefs? *Curr. Opin. Environ. Sustain.*, **7**(Supplement C), 28–36, doi:10.1016/j.cosust.2013.11.018.

Rinkevich, B., 2015a: Climate Change and Active Reef Restoration – Ways of Constructing the "Reefs of Tomorrow". *J. Mar. Sci. Eng.*, **3**(1), 111–127, doi:10.3390/jmse3010111.

Rinkevich, B., 2015b: Novel tradable instruments in the conservation of coral reefs, based on the coral gardening concept for reef restoration. *J. Environ. Manage.*, **162**(Supplement C), 199–205.

Rinkevich, B., 2019: Coral chimerism as an evolutionary rescue mechanism to mitigate global climate change impacts. *Global Change Biol.*, **25**(4), 1198–1206, doi:10.1111/gcb.14576.

Riou, V. et al., 2010: Mixotrophy in the deep sea: a dual endosymbiotic hydrothermal mytilid assimilates dissolved and particulate organic matter. *Mar. Ecol. Prog. Ser.*, **405**, 187–201.

Riser, S.C. et al., 2016: Fifteen years of ocean observations with the global Argo array. *Nat. Clim. Change*, **6**, 145, doi:10.1038/nclimate2872.

Ritter, R. et al., 2017: Observation-Based Trends of the Southern Ocean Carbon Sink. *Geophys. Res. Lett.*, **44**(24), 12,339–12,348.

Ritzman, J. et al., 2018: Economic and sociocultural impacts of fisheries closures in two fishing-dependent communities following the massive 2015 U.S. West Coast harmful algal bloom. *Harmful Algae*, **80**, 35–45, doi:10.1016/j.hal.2018.09.002.

Rivero-Calle, S. et al., 2015: Multidecadal increase in North Atlantic coccolithophores and the potential role of rising CO_2. *Science*, **350**(6267), 1533.

Roberts, C.M. et al., 2017: Marine reserves can mitigate and promote adaptation to climate change. *PNAS*, **114**(24), 6167–6175.

Roberts, D.A. et al., 2013: Ocean acidification increases the toxicity of contaminated sediments. *Global Change Biol.*, **19**(2), 340–351, doi:10.1111/gcb.12048.

Roberts, J.J. et al., 2016: Habitat-based cetacean density models for the U.S. Atlantic and Gulf of Mexico. *Sci. Rep.*, **6**, 22615, doi:10.1038/srep22615.

Robertson, L.M., J.-F. Hamel and A. Mercier, 2017: Feeding in deep sea demosponges: Influence of abiotic and biotic factors. *Deep sea Res. Pt. I*, **127**(Supplement C), 49–56.

Robins, P.E. et al., 2016: Impact of climate change on UK estuaries: A review of past trends and potential projections. *Estuar. Coast. Shelf Sci.*, **169**, 119–135, doi:10.1016/j.ecss.2015.12.016.

Robinson, J. et al., 2014: How deep is deep enough? Ocean iron fertilization and carbon sequestration in the Southern Ocean. *Geophys. Res. Lett.*, **41**(7), 2489–2495, doi:10.1002/2013GL058799.

Robinson, L.M. et al., 2015: Rapid assessment of an ocean warming hotspot reveals "high" confidence in potential species' range extensions. *Global Environ. Change*, **31**, 28–37, doi:10.1016/j.gloenvcha.2014.12.003.

5

Robson, J., R. Sutton and D. Smith, 2014: Decadal predictions of the cooling and freshening of the North Atlantic in the 1960s and the role of ocean circulation. *Clim. Dyn.*, **42**(9), 2353–2365, doi:10.1007/s00382-014-2115-7.

Rocha, L.A. et al., 2018: Mesophotic coral ecosystems are threatened and ecologically distinct from shallow water reefs. *Science*, **361**(6399), 281, doi:10.1126/science.aaq1614.

Rödenbeck, C. et al., 2015: Data-based estimates of the ocean carbon sink variability–first results of the Surface Ocean pCO$_2$ Mapping intercomparison (SOCOM). *Biogeosciences*, **12**, 7251–7278.

Rödenbeck, C. et al., 2014: Interannual sea-air CO$_2$ flux variability from an observation-driven ocean mixed-layer scheme. *Biogeosciences*, **11**, 3167–3207.

Rodgers, K.B., J. Lin and T.L. Frölicher, 2015: Emergence of multiple ocean ecosystem drivers in a large ensemble suite with an Earth system model. *Biogeosciences*, **12**(11), 3301–3320, doi:10.5194/bg-12-3301-2015.

Rodolfo-Metalpa, R. et al., 2015: Calcification is not the Achilles' heel of cold water corals in an acidifying ocean. *Global Change Biol.*, **21**(6), 2238–2248, doi:10.1111/gcb.12867.

Rodríguez, F. et al., 2017: Canary Islands (NE Atlantic) as a biodiversity 'hotspot' of Gambierdiscus: Implications for future trends of ciguatera in the area. *Harmful Algae*, **67**, 131–143.

Rodríguez-Rodríguez, G. and R. Bande Ramudo, 2017: Market driven management of climate change impacts in the Spanish mussel sector. *Mar. Policy*, **83**, 230–235, doi:10.1016/j.marpol.2017.06.014.

Roemmich, D. et al., 2015: Unabated planetary warming and its ocean structure since 2006. *Nat. Clim. Change*, **5**, 240, doi:10.1038/nclimate2513.

Roemmich, D., W. John Gould and J. Gilson, 2012: 135 years of global ocean warming between the Challenger expedition and the Argo Programme. *Nat. Clim. Change*, **2**, 425, doi:10.1038/nclimate1461.

Rogers, A.D., 2015: Environmental Change in the Deep Ocean. *Annu. Rev. Environ. Resourc.*, *Vol 41*, **40**(1), 1–38, doi:10.1146/annurev-environ-102014-021415.

Romañach, S.S. et al., 2018: Conservation and restoration of mangroves: Global status, perspectives, and prognosis. *Ocean Coast. Manage.*, **154**, 72–82.

Rosas-Navarro, A., G. Langer and P. Ziveri, 2016: Temperature affects the morphology and calcification of Emiliania huxleyi strains. *Biogeosciences*, **13**(10), 2913–2926, doi:10.5194/bg-13-2913-2016.

Rose, J.M. and D.A. Caron, 2007: Does low temperature constrain the growth rates of heterotrophic protists? Evidence and implications for algal blooms in cold waters. *Limnol. Oceanogr.*, **52**(2), 886–895.

Rose, J.M. et al., 2009: Synergistic effects of iron and temperature on Antarctic phytoplankton and microzooplankton assemblages. *Biogeosciences*, **6**(12), 3131–3147, doi:10.5194/bg-6-3131-2009.

Rosegrant, M.W., M.M. Dey, R. Valmonte-Santos and O.L. Chen, 2016: Economic impacts of climate change and climate change adaptation strategies in Vanuatu and Timor-Leste. *Mar. Policy*, **67**, 179–188, doi:10.1016/j.marpol.2015.12.010.

Rosenberg, R. et al., 1983: Benthos biomass and oxygen deficiency in the upwelling system off Peru. *J. Mar. Res.*, **41**(2), 263–279, doi:10.1357/002224083788520153.

Rosendo, S., L. Celliers and M. Mechisso, 2018: Doing more with the same: A reality-check on the ability of local government to implement Integrated Coastal Management for climate change adaptation. *Mar. Policy*, **87**, 29–39, doi:10.1016/j.marpol.2017.10.001.

Rosentreter, J.A. et al., 2018: Methane emissions partially offset "blue carbon" burial in mangroves. *Sci. Adv.*, **4**(6), eaao4985.

Ross, A.C. et al., 2015: Sea level rise and other influences on decadal-scale salinity variability in a coastal plain estuary. *Estuar. Coast. Shelf Sci.*, **157**, 79–92, doi:10.1016/j.ecss.2015.01.022.

Rothman, D.H., J.M. Hayes and R.E. Summons, 2003: Dynamics of the Neoproterozoic carbon cycle. *PNAS*, **100**(14), 8124, doi:10.1073/pnas.0832439100.

Roué, M., 2012: History and Epistemology of Local and Indigenous Knowledge: from Tradition to Trend. *Revue d'ethnoécologie*, (1), doi:10.4000/ethnoecologie.813.

Rowden, A.A. et al., 2010: Paradigms in seamount ecology: fact, fiction and future. *Mar. Ecol.*, **31**(s1), 226–241, doi:10.1111/j.1439-0485.2010.00400.x.

Rowe, G.T., J. Morse, C. Nunnally and G.S. Boland, 2008: Sediment community oxygen consumption in the deep Gulf of Mexico. *Deep-Sea Res. Pt. II*, **55**(24), 2686–2691, doi:10.1016/j.dsr2.2008.07.018.

Roy, T., F. Lombard, L. Bopp and M. Gehlen, 2015: Projected impacts of climate change and ocean acidification on the global biogeography of planktonic Foraminifera. *Biogeosciences*, **12**(10), 2873–2889, doi:10.5194/bg-12-2873-2015.

Ruiz, J., L. Prieto and D. Astorga, 2012: A model for temperature control of jellyfish (Cotylorhiza tuberculata) outbreaks: A causal analysis in a Mediterranean coastal lagoon. *Ecol. Model.*, **233**, 59–69, doi:10.1016/j.ecolmodel.2012.03.019.

Ruiz-Frau, A. et al., 2017: Current state of seagrass ecosystem services: Research and policy integration. *Ocean Coast. Manage.*, **149**, 107–115, doi:10.1016/j.ocecoaman.2017.10.004.

Rulleau, B. and H. Rey-Valette, 2017: Forward planning to maintain the attractiveness of coastal areas: Choosing between seawalls and managed retreat. *Environ. Sci. Policy*, **72**, 12–19, doi:10.1016/j.envsci.2017.01.009.

Rumson, A.G., S.H. Hallett and T.R. Brewer, 2017: Coastal risk adaptation: the potential role of accessible geospatial Big Data. *Mar. Policy*, **83**, 100–110, doi:10.1016/j.marpol.2017.05.032.

Runting, R.K., C.E. Lovelock, H.L. Beyer and J.R. Rhodes, 2017: Costs and Opportunities for Preserving Coastal Wetlands under Sea Level Rise. *Conserv. Lett.*, **10**(1), 49–57, doi:doi:10.1111/conl.12239.

Rutterford, L.A. et al., 2015: Future fish distributions constrained by depth in warming seas. *Nat. Clim. Change*, **5**, 569, doi:10.1038/nclimate2607.

Rykaczewski, R.R. and J.P. Dunne, 2010: Enhanced nutrient supply to the California Current Ecosystem with global warming and increased stratification in an earth system model. *Geophys. Res. Lett.*, **37**(21), L21606. doi:10.1029/2010GL045019.

Rykaczewski, R.R. et al., 2015: Poleward displacement of coastal upwelling-favorable winds in the ocean's eastern boundary currents through the 21st century. *Geophys. Res. Lett.*, **42**(15), 6424–6431, doi:10.1002/2015GL064694.

Saba, V.S. et al., 2016: Enhanced warming of the Northwest Atlantic Ocean under climate change. *J. Geophys. Res-Oceans*, **121**(1), 118–132, doi:10.1002/2015JC011346.

Sabine, C.L., 2018: Good news and bad news of blue carbon. *PNAS*, **115**(15), 3745–3746.

Sabine, C.L. et al., 2004: The oceanic sink for anthropogenic CO$_2$. *Science*, **305**(5682), 367–71, doi:10.1126/science.1097403.

Saderne, V. et al., 2019: Role of carbonate burial in Blue Carbon budgets. *Nat. Commun.*, **10**(1), 1106.

Saintilan, N. et al., 2018: Climate Change Impacts on the Coastal Wetlands of Australia. *Wetlands*, doi:10.1007/s13157-018-1016-7.

Saintilan, N. et al., 2014: Mangrove expansion and salt marsh decline at mangrove poleward limits. *Global Change Biol.*, **20**(1), 147–157, doi:10.1111/gcb.12341.

Saito, M.A. et al., 2014: Multiple nutrient stresses at intersecting Pacific Ocean biomes detected by protein biomarkers. *Science*, **345**(6201), 1173–1177, doi:10.1126/science.1256450.

Sakakibara, C., 2017: People of the Whales: Climate Change and Cultural Resilience Among Iñupiat of Arctic Alaska. *Geogr. Rev.*, **107**(1), 159–184.

Sala, E. et al., 2018: Assessing real progress towards effective ocean protection. *Mar. Policy*, **91**, 11–13, doi:10.1016/j.marpol.2018.02.004.

Salehyan, I., 2008: From climate change to conflict? No consensus yet. *J. Peace Res.*, **45**(3), 315–326.

Salgado, K. and M.L. Martinez, 2017: Is ecosystem-based coastal defense a realistic alternative? Exploring the evidence. *J. Coast. Conserv.*, **21**(6), 837–848, doi:10.1007/s11852-017-0545-1.

Salik, K.M., S. Jahangir, W.u.Z. Zahdi and S.u. Hasson, 2015: Climate change vulnerability and adaptation options for the coastal communities of Pakistan. *Ocean Coast. Manage.*, **112**, 61–73, doi:10.1016/j.ocecoaman.2015.05.006.

Sánchez-Arcilla, A. et al., 2016: Managing coastal environments under climate change: Pathways to adaptation. *Sci. Total Environ.*, **572**, 1336–1352, doi:10.1016/j.scitotenv.2016.01.124.

Sandifer, P.A. and A.E. Sutton-Grier, 2014: Connecting stressors, ocean ecosystem services, and human health. *Natural Resources Forum*, **38**(3), 157–167, doi:10.1111/1477-8947.12047.

Sanford, E., 1999: Regulation of keystone predation by small changes in ocean temperature. *Science*, **283**(5410), 2095–2097, doi:10.1126/science.283.5410.2095.

Sano, M. et al., 2015: Coastal vulnerability and progress in climate change adaptation: An Australian case study. *Reg. Stud. Mar. Sci.*, **2**, 113–123, doi:10.1016/j.rsma.2015.08.015.

Santidrián Tomillo, P. et al., 2014: High beach temperatures increased female-biased primary sex ratios but reduced output of female hatchlings in the leatherback turtle. *Biol. Conserv.*, **176**(Supplement C), 71–79, doi:10.1016/j.biocon.2014.05.011.

Santora, J.A. et al., 2017: Impacts of ocean climate variability on biodiversity of pelagic forage species in an upwelling ecosystem. *Mar. Ecol. Prog. Ser.*, **580**, 205–220, doi:10.3354/meps12278.

Santos, I.R. et al., 2019: Carbon outwelling and outgassing vs. burial in an estuarine tidal creek surrounded by mangrove and saltmarsh wetlands. *Limnol. Oceanogr.*, **64**(3), 996–1013, doi:10.1002/lno.11090.

Santos, K.C., M. Livesey, M. Fish and A.C. Lorences, 2017: Climate change implications for the nest site selection process and subsequent hatching success of a green turtle population. *Mitig. Adapt. Strat. Gl.*, **22**(1), 121–135, doi:10.1007/s11027-015-9668-6.

Santos, R., J.S. Rehage, R. Boucek and J. Osborne, 2016: Shift in recreational fishing catches as a function of an extreme cold event. *Ecosphere*, **7**(6), e01335.

Sarkodie, S.A. and V. Strezov, 2019: Economic, social and governance adaptation readiness for mitigation of climate change vulnerability: Evidence from 192 countries. *Sci. Total Environ.*, **656**, 150–164, doi:10.1016/j.scitotenv.2018.11.349.

Sarmiento, J.L. and N. Gruber, 2002: Sinks for Anthropogenic Carbon. *Physics Today*, **55**(8), 30–36, doi:10.1063/1.1510279.

Sasano, D. et al., 2015: Multidecadal trends of oxygen and their controlling factors in the western North Pacific. *Global Biogeochem. Cy.*, **29**(7), 935–956, doi:10.1002/2014gb005065.

Sasmito, S.D., D. Murdiyarso, D.A. Friess and S. Kurnianto, 2016: Can mangroves keep pace with contemporary sea level rise? A global data review. *Wetlands Ecol. Manage.*, **24**(2), 263–278, doi:10.1007/s11273-015-9466-7.

Sasse, T.P., B.I. McNeil, R.J. Matear and A. Lenton, 2015: Quantifying the influence of CO_2 seasonality on future aragonite undersaturation onset. *Biogeosciences*, **12**(20), 6017–6031, doi:10.5194/bg-12-6017-2015.

Sato, K.N., L.A. Levin and K. Schiff, 2017: Habitat compression and expansion of sea urchins in response to changing climate conditions on the California continental shelf and slope (1994–2013). *Deep Sea Res. Pt. II*, **137**, 377–389, doi:10.1016/j.dsr2.2016.08.012.

Sato KN, Andersson AJ, Day JMD, Taylor JRA, Frank MB, Jung J-Y, McKittrick J and Levin LA (2018) Response of Sea Urchin Fitness Traits to Environmental Gradients Across the Southern California Oxygen Minimum Zone. Front. Mar. Sci. 5:258. doi: 10.3389/fmars.2018.00258.

Savva, I. et al., 2018: Thermal tolerance of Mediterranean marine macrophytes: Vulnerability to global warming. *Ecol. Evol.*, **8**(23), 12032–12043, doi:10.1002/ece3.4663.

Scales, K.L. et al., 2014: Review: On the Front Line: frontal zones as priority at-sea conservation areas for mobile marine vertebrates. *J. Appl. Ecol.*, **51**(6), 1575–1583, doi:10.1111/1365-2664.12330.

Scarano, F.R., 2017: Ecosystem-based adaptation to climate change: concept, scalability and a role for conservation science. *Perspect. Ecol. Conserv.*, **15**(2), 65–73, doi:10.1016/j.pecon.2017.05.003.

Schaeffer-Novelli, Y. et al., 2016: Climate changes in mangrove forests and salt marshes. *Brazilian J. Oceanogr.*, **64**((spe2)), 37–52.

Schaum, E., B. Rost, A.J. Millar and S. Collins, 2013: Variation in plastic responses of a globally distributed picoplankton species to ocean acidification. *Nat. Clim. Change*, **3**(3), 298–302, doi:10.1038/NCLIMATE1774.

Scheuhammer, A. et al., 2015: Recent progress on our understanding of the biological effects of mercury in fish and wildlife in the Canadian Arctic. *Sci. Total Environ.*, **509**, 91–103.

Scheuhammer, A.M., 1991: Effects of acidification on the availability of toxic metals and calcium to wild birds and mammals. *Environ. Pollut.*, **71**(2), 329–375, doi:10.1016/0269-7491(91)90036-V.

Schile, L.M., J.C. Callaway, K.N. Suding and N.M. Kelly, 2017: Can community structure track sea level rise? Stress and competitive controls in tidal wetlands. *Ecol. Evol.*, **7**(4), 1276–1285, doi:10.1002/ece3.2758.

Schindelegger, M., J.A.M. Green, S.B. Wilmes and I.D. Haigh, 2018: Can We Model the Effect of Observed Sea Level Rise on Tides? *J. Geophys. Res-Oceans*, **123**(7), 4593–4609, doi:10.1029/2018JC013959.

Schlacher, T.A. and L. Thompson, 2013: Spatial structure on ocean-exposed sandy beaches: faunal zonation metrics and their variability. *Mar. Ecol. Prog. Ser.*, **478**, 43–55.

Schmidtko, S., L. Stramma and M. Visbeck, 2017: Decline in global oceanic oxygen content during the past five decades. *Nature*, **542**(7641), 335–339, doi:10.1038/nature21399.

Schmittner, A., M. Urban Nathan, K. Keller and D. Matthews, 2009: Using tracer observations to reduce the uncertainty of ocean diapycnal mixing and climate–carbon cycle projections. *Global Biogeochem. Cy.*, **23**(4), doi:10.1029/2008GB003421.

Schoeman, D.S. et al., 2015: Edging along a warming coast: a range extension for a common sandy beach crab. *PLoS One*, **10**(11), e0141976.

Schönberg, C.H.L. et al., 2017: Bioerosion: the other ocean acidification problem. *ICES J. Mar. Sci.*, **74**(4), 895–925, doi:10.1093/icesjms/fsw254.

Schopmeyer, S., A. et al., 2012: In Situ Coral Nurseries Serve as Genetic Repositories for Coral Reef Restoration after an Extreme Cold-Water Event. *Restor. Ecol.*, **20**(6), 696–703, doi:10.1111/j.1526-100X.2011.00836.x.

Schuerch, M. et al., 2018: Future response of global coastal wetlands to sea level rise. *Nature*, **561**(7722), 231–234, doi:10.1038/s41586-018-0476-5.

Schulte, P.M., 2015: The effects of temperature on aerobic metabolism: towards a mechanistic understanding of the responses of ectotherms to a changing environment. *J. Exp. Biol.*, **218**(12), 1856, doi:10.1242/jeb.118851.

Schulz, K.G. et al., 2013: Temporal biomass dynamics of an Arctic plankton bloom in response to increasing levels of atmospheric carbon dioxide. *Biogeosciences (BG)*, **10**, 161–180.

Scott, A.L. et al., 2018: The Role of Herbivory in Structuring Tropical Seagrass Ecosystem Service Delivery. *Front. Plant Sci,* **9**(127), doi:10.3389/fpls.2018.00127.

Seebacher, F., C.R. White and C.E. Franklin, 2014: Physiological plasticity increases resilience of ectothermic animals to climate change. *Nat. Clim. Change*, **5**, 61, doi:10.1038/nclimate2457.

Seebah, S., C. Fairfield, M.S. Ullrich and U. Passow, 2014: Aggregation and Sedimentation of Thalassiosira weissflogii (diatom) in a Warmer and More Acidified Future Ocean. *PLoS One*, **9**(11), e112379, doi:10.1371/journal.pone.0112379.

Semenza, J.C. et al., 2017: Environmental Suitability of Vibrio Infections in a Warming Climate: An Early Warning System. *Environ. Health. Perspect.*, **125**(10), 107004. doi:papers3://publication/doi/10.1289/EHP2198.

5

Sen Gupta, A. et al., 2016: Future changes to the Indonesian Throughflow and Pacific circulation: The differing role of wind and deep circulation changes. *Geophys. Res. Lett.*, **43**(4), 1669–1678, doi:10.1002/2016GL067757.

Senapati, S. and V. Gupta, 2017: Socioeconomic vulnerability due to climate change: Deriving indicators for fishing communities in Mumbai. *Mar. Policy*, **76**, 90–97, doi:10.1016/j.marpol.2016.11.023.

Serrano, O., P.S. Lavery, M. Rozaimi and M.Á. Mateo, 2014: Influence of water depth on the carbon sequestration capacity of seagrasses. *Global Biogeochem. Cy.*, **28**(9), 950–961, doi:10.1002/2014GB004872.

Serrao-Neumann, S. et al., 2013: Improving cross-sectoral climate change adaptation for coastal settlements: insights from South East Queensland, Australia. *Reg. Environ. Change*, **14**(2), 489–500, doi:10.1007/s10113-013-0442-6.

Sett, S. et al., 2014: Temperature Modulates Coccolithophorid Sensitivity of Growth, Photosynthesis and Calcification to Increasing Seawater pCO(2). *PLoS One*, **9**(2), doi:10.1371/journal.pone.0088308.

Shaffiril, H.A.M., A.A. Samah and J. Lawrence, 2017: Adapting towards climate change impacts: Strategies for small-scale fishermen in Malaysia. *Mar. Policy*, **81**, 196–201.

Shaffril, H.A.M., A. Abu Samah and J.L. D'Silva, 2017: Climate change: Social adaptation strategies for fishermen. *Mar. Policy*, **81**, 256–261, doi:10.1016/j.marpol.2017.03.031.

Shafir, S. and B. Rinkevich, 2008: Chapter 9 – The underwater silviculture approach for reef restoration: an emergent aquaculture theme. In: *Aquaculture Research Trends* [Schwartz, S.H.]. *Nova Science Publications, New York*, pp. 279–295. ISBN: 9781604562170.

Shaish, L., G. Levy, G. Katzir and B. Rinkevich, 2010a: Coral Reef Restoration (Bolinao, Philippines) in the Face of Frequent Natural Catastrophes. *Restor. Ecol.*, **18**(3), 285–299, doi:10.1111/j.1526-100X.2009.00647.x.

Shaish, L., G. Levy, G. Katzir and B. Rinkevich, 2010b: Employing a highly fragmented, weedy coral species in reef restoration. *Ecol. Eng.*, **36**(10), 1424–1432, doi:10.1016/j.ecoleng.2010.06.022.

Shannon, C., 2002: Acculturation: Aboriginal and Torres Strait Islander nutrition. *Asia Pacific Journal of Clinical Nutrition*, **11**, S576–S578, doi:10.1046/j.0964-7058.2002.00352.x.

Sharples, J., J.J. Middelburg, K. Fennel and T.D. Jickells, 2017: What proportion of riverine nutrients reaches the open ocean? *Global Biogeochem. Cy.*, **31**(1), 39–58.

Shaver, E.C. and B.R. Silliman, 2017: Time to cash in on positive interactions for coral restoration. *Peerj*, **5**, e3499, doi:10.7717/peerj.3499.

Sheaves, M. et al., 2016: Principles for operationalizing climate change adaptation strategies to support the resilience of estuarine and coastal ecosystems: An Australian perspective. *Mar. Policy*, **68**, 229–240, doi:10.1016/j.marpol.2016.03.014.

Sheen, K.L. et al., 2013: Rates and mechanisms of turbulent dissipation and mixing in the Southern Ocean: Results from the Diapycnal and Isopycnal Mixing Experiment in the Southern Ocean (DIMES). *J. Geophys. Res-Oceans*, **118**(6), 2774–2792, doi:10.1002/jgrc.20217.

Sheikh, N., G.M. Egeland, L. Johnson-Down and H.V. Kuhnlein, 2011: Changing dietary patterns and body mass index over time in Canadian Inuit communities. *Int. J. Circumpolar Health*, **70**(5), 511–519, doi:10.3402/ijch.v70i5.17863.

Sheller, M. and Y.M. León, 2016: Uneven socioecologies of Hispaniola: Asymmetric capabilities for climate adaptation in Haiti and the Dominican Republic. *Geoforum*, **73**, 32–46, doi:10.1016/j.geoforum.2015.07.026.

Sheng, Y.P. and R. Zou, 2017: Assessing the role of mangrove forest in reducing coastal inundation during major hurricanes. *Hydrobiologia*, **803**(1), 87–103, doi:10.1007/s10750-017-3201-8.

Sherren, K., L. Loik and J.A. Debner, 2016: Climate adaptation in 'new world' cultural landscapes: The case of Bay of Fundy agricultural dykelands (Nova Scotia, Canada). *Land Use Policy*, **51**, 267–280, doi:10.1016/j.landusepol.2015.11.018.

Shi, J.-R., S.-P. Xie and L.D. Talley, 2018: Evolving Relative Importance of the Southern Ocean and North Atlantic in Anthropogenic Ocean Heat Uptake. *J. Clim.*, **31**(18), 7459–7479, doi:10.1175/jcli-d-18-0170.1.

Shilova, I.N. et al., 2017: Differential effects of nitrate, ammonium, and urea as N sources for microbial communities in the North Pacific Ocean. *Limnol. Oceanogr.*, **62**(2), 2550–2574. doi:10.1002/lno.10590.

Shumack, S. and P. Hesse, 2017: Assessing the geomorphic disturbance from fires on coastal dunes near Esperance, Western Australia: Implications for dune de-stabilisation. *Aeolian Research*, 31, 29–49. doi:10.1016/j.aeolia.2017.08.005.

Sierra-Correa, P.C. and J.R. Cantera Kintz, 2015: Ecosystem-based adaptation for improving coastal planning for sea level rise: A systematic review for mangrove coasts. *Mar. Policy*, **51**, 385–393, doi:10.1016/j.marpol.2014.09.013.

Singh, G.G. et al., 2017: A rapid assessment of co-benefits and trade-offs among Sustainable Development Goals. *Mar. Policy*, **93**, 223–231.

Singh, G.G. et al., 2019: Climate impacts on the ocean are making the Sustainable Development Goals a moving target travelling away from us. *People and Nature*, **1**(3), 317–330. doi:10.1002/pan3.26.

Sinha, P.R. et al., 2017: Evaluation of ground-based black carbon measurements by filter-based photometers at two Arctic sites. *J. Geophys. Res-Atmos.*, **122**(6), 3544–3572, doi:10.1002/2016JD025843.

Sippo, J.Z. et al., 2018: Mangrove mortality in a changing climate: An overview. *Estuar. Coast. Shelf Sci.*, **215**, 241–249, doi:10.1016/j.ecss.2018.10.011.

Sloyan, B.M. et al., 2010: Antarctic Intermediate Water and Subantarctic Mode Water Formation in the Southeast Pacific: The Role of Turbulent Mixing. *J. Phys. Oceanogr.*, **40**(7), 1558–1574, doi:10.1175/2010JPO4114.1.

Smale, D.A. et al., 2018: Appreciating interconnectivity between habitats is key to blue carbon management. *Front. Ecol. Environ.*, **16**(2), 71–73, doi:10.1002/fee.1765.

Smeed, D.A. et al., 2018: The North Atlantic Ocean Is in a State of Reduced Overturning. *Geophys. Res. Lett.*, **45**(3), 1527–1533, doi:10.1002/2017gl076350.

Smith, B., I. Burton, R. Klein and J. Wandel, 2000: An anatomy of adaptation to climate change and variability. *Clim. Change*, **45**(1), 223–251, doi:10.1023/A:1005661622966.

Smith, C.R. et al., 2008: Abyssal food limitation, ecosystem structure and climate change. *Trends Ecol. Evol.*, **23**(9), 518–528, doi:10.1016/j.tree.2008.05.002.

Smith, K.L., C.L. Huffard, A.D. Sherman and H.A. Ruhl, 2016a: Decadal Change in Sediment Community Oxygen Consumption in the Abyssal Northeast Pacific. *Aquat. Geochem.*, **22**(5), 401–417, doi:10.1007/s10498-016-9293-3.

Smith, K.L. et al., 2018: Episodic organic carbon fluxes from surface ocean to abyssal depths during long-term monitoring in NE Pacific. *PNAS*, **115**(48), 12235, doi:10.1073/pnas.1814559115.

Smith, K.L.J. et al., 2009: Climate, carbon cycling, and deep-ocean ecosystems. *PNAS*, **106**(46), 19211–19218, doi:10.1073/pnas.0908322106.

Smith, K.L.J. et al., 2013: Deep ocean communities impacted by changing climate over 24 y in the abyssal northeast Pacific Ocean. *PNAS*, **110**(49), 19838–19841, doi:10.1073/pnas.1315447110.

Smith, T.B. et al., 2016b: Caribbean mesophotic coral ecosystems are unlikely climate change refugia. *Global Change Biol.*, **22**(8), 2756–2765, doi:10.1111/gcb.13175.

Smith, T.B., J.L. Maté and J. Gyory, 2017: Thermal Refuges and Refugia for Stony Corals in the Eastern Tropical Pacific. In: *Coral Reefs of the Eastern Tropical Pacific* [Glynn, P.W., D.P. Manzello and I.C. Enochs (eds.)]. Springer Netherlands, Dordrecht, pp. 501–515. ISBN 978-94-017-7498-7.

Sohm, J.A., E.A. Webb and D.G. Capone, 2011: Emerging patterns of marine nitrogen fixation. *Nat. Rev. Microbiol.*, **9**, 499, doi:10.1038/nrmicro2594.

Soltwedel, T. et al., 2016: Natural variability or anthropogenically-induced variation? Insights from 15 years of multidisciplinary observations at the arctic marine LTER site HAUSGARTEN. *Ecol. Indic.*, **65**, 89–102, doi:10.1016/j.ecolind.2015.10.001.

Somero, G., B.L. Lockwood and L. Tomanek, 2017: *Biochemical adaptation: response to environmental challenges, from life's origins to the Anthropocene*. Sinauer Associates, Incorporated Publishers, Oxford University Press, Sunderland, Massachusetts, p. 572. ISBN: 9781605355641.

Somes, C.J., A. Landolfi, W. Koeve and A. Oschlies, 2016: Limited impact of atmospheric nitrogen deposition on marine productivity due to biogeochemical feedbacks in a global ocean model. *Geophys. Res. Lett.*, **43**(9), 4500–4509, doi:10.1002/2016GL068335.

Sondak, C.F.A. et al., 2017: Carbon dioxide mitigation potential of seaweed aquaculture beds (SABs). *J. App. Phycol.*, **29**(5), 2363–2373, doi:10.1007/s10811-016-1022-1.

Sorte, C.J. et al., 2017: Long-term declines in an intertidal foundation species parallel shifts in community composition. *Global Change Biol.*, **23**(1), 341–352.

Sousa, M.C. et al., 2017: Why coastal upwelling is expected to increase along the western Iberian Peninsula over the next century? *Sci. Total Environ.*, **592**, 243–251, doi:10.1016/j.scitotenv.2017.03.046.

Spalding, M., 2010: *World atlas of mangroves*. Routledge. Earthscan, London, UK. p. 319.ISBN: 978-1844076574.

Spalding, M. et al., 2017: Mapping the global value and distribution of coral reef tourism. *Mar. Policy*, **82**(Supplement C), 104–113, doi:10.1016/j.marpol.2017.05.014.

Spalding, M.D. et al., 2014: The role of ecosystems in coastal protection: Adapting to climate change and coastal hazards. *Ocean Coast. Manage.*, **90**, 50–57, doi:10.1016/j.ocecoaman.2013.09.007.

Spalding, M.J., 2016: The new blue economy: the future of sustainability. *J. Ocean Coast. Econ.*, **2**(2), 8.

Sparrow, L., P. Momigliano, G.R. Russ and K. Heimann, 2017: Effects of temperature, salinity and composition of the dinoflagellate assemblage on the growth of Gambierdiscus carpenteri isolated from the Great Barrier Reef. *Harmful Algae*, **65**, 52–60, doi:10.1016/j.hal.2017.04.006.

Spence, P. et al., 2017: Localized rapid warming of West Antarctic subsurface waters by remote winds. *Nat. Clim. Change*, **7**, 595, doi:10.1038/nclimate3335.

Spencer, T. et al., 2016: Global coastal wetland change under sea level rise and related stresses: The DIVA Wetland Change Model. *Global Planet. Change*, **139**, 15–30, doi:10.1016/j.gloplacha.2015.12.018.

Sperling, E.A., C.A. Frieder and L.A. Levin, 2016: Biodiversity response to natural gradients of multiple stressors on continental margins. *Proc. Biol. Sci.*, **283**(1829), doi:10.1098/rspb.2016.0637.

Sperling, E.A. et al., 2013: Oxygen, ecology, and the Cambrian radiation of animals. *PNAS*, **110**(33), 13446–13451, doi:10.1073/pnas.1312778110.

Spijkers, J. and W.J. Boonstra, 2017: Environmental change and social conflict: the northeast Atlantic mackerel dispute. *Reg. Environ. Change*, **17**(6), 1835–1851, doi:10.1007/s10113-017-1150-4.

Srinivasan, U.T., W.W.L. Cheung, R. Watson and U. R. Sumaila, 2010: Food security implications of global marine catch losses due to overfishing. *Journal of Bioeconomics*, **12**(3), 183–200.

St. John, M.A. et al., 2016: A Dark Hole in Our Understanding of Marine Ecosystems and Their Services: Perspectives from the Mesopelagic Community. *Front. Mar. Sci.*, **3**, 31.

Steinacher, M., F. Joos and T.F. Stocker, 2013: Allowable carbon emissions lowered by multiple climate targets. *Nature*, **499**, 197, doi:10.1038/nature12269.

Steiner, N.S. et al., 2014: Future ocean acidification in the Canada Basin and surrounding Arctic Ocean from CMIP5 earth system models. *J. Geophys. Res-Oceans*, **119**(1), 332–347, doi:10.1002/2013JC009069.

Stendardo, I. and N. Gruber, 2012: Oxygen trends over five decades in the North Atlantic. *J. Geophys. Res-Oceans*, **117**(C11), doi:10.1029/2012JC007909.

Steneck, R.S. et al., 2003: Kelp forest ecosystems: biodiversity, stability, resilience and future. *Environ. Conserv.*, **29**(04), 436–459, doi:10.1017/S0376892902000322.

Stephens, T., 2015: Ocean acidification.[Rayfuse, R. (ed.)]. Edward Elgar Publishing, **106**, 406.

Stewart, J.S. et al., 2014: Combined climate- and prey-mediated range expansion of Humboldt squid (Dosidicus gigas), a large marine predator in the California Current System. *Global Change Biol.*, **20**(6), 1832–1843, doi:10.1111/gcb.12502.

Stock, C.A., J.P. Dunne and J.G. John, 2014: Drivers of trophic amplification of ocean productivity trends in a changing climate. *Biogeosciences*, **11**(24), 7125.

Stock, C.A. et al., 2017: Reconciling fisheries catch and ocean productivity. *PNAS*, **114**(8), E1441–E1449, doi:10.1073/pnas.1610238114.

Stocker, T.F., D. Qin, G.-K. Plattner, LV. Alexander, S.K. Allen, N.L. Bindoff, F.-M. Bréon, J.A. Church, U. Cubasch, S. Emori, P. Forster, P. Friedlingstein, N. Gillett, J.M. Gregory, D.L. Hartmann, E. Jansen, B. Kirtman, R. Knutti, K. Krishna Kumar, P. Lemke, J. Marotzke, V. Masson-Delmotte, G.A. Meehl, I.I. Mokhov, S. Piao, V. Ramaswamy, D. Randall, M. Rhein, M. Rojas, C. Sabine, D. Shindell, L.D. Talley, D.G. Vaughan, 2014: Technical Summary. In: Climate Change 2013 – The Physical Science Basis: Working Group I Contribution to the Fifth Assessment Report of the Intergovernmental Panel on Climate Change [Intergovernmental Panel on Climate, C. (ed.)]. Cambridge University Press, Cambridge, 31–116.

Stocker, T. F. et al., 2013: Technical Summary. In: Climate Change 2013: The Physical Science Basis. Contribution of Working Group I to the Fifth Assessment Report of the Intergovernmental Panel on Climate Change [Stocker, T.F., D. Qin, G.-K. Plattner, M. Tignor, S.K. Allen, J. Boschung, A. Nauels, Y. Xia, V. Bex and P.M. Midgley (eds.)]. Cambridge University Press, Cambridge, United Kingdom and New York, NY, USA, 33–115.

Stramma, L., G.C. Johnson, J. Sprintall and V. Mohrholz, 2008: Expanding Oxygen-Minimum Zones in the Tropical Oceans. *Science*, **320**(5876), 655.

Stramma, L. et al., 2011: Expansion of oxygen minimum zones may reduce available habitat for tropical pelagic fishes. *Nat. Clim. Change*, **2**, 33, doi:10.1038/nclimate1304.

Strand, R. et al., 2017: The response of a boreal deep sea sponge holobiont to acute thermal stress. *Sci. Rep.*, **7**(1), 1660, doi:10.1038/s41598-017-01091-x.

Strong, A.L. et al., 2014: Ocean Acidification 2.0: Managing our Changing Coastal Ocean Chemistry. *BioScience*, **64**(7), 581–592, doi:10.1093/biosci/biu072.

Stuart-Smith, R.D. et al., 2015: Thermal biases and vulnerability to warming in the world's marine fauna. *Nature*, **528**, 88, doi:10.1038/nature16144.

Sturdivant, S.K., R.J. Díaz and G.R. Cutter, 2012: Bioturbation in a Declining Oxygen Environment, in situ Observations from Wormcam. *PLoS One*, **7**(4), e34539, doi:10.1371/journal.pone.0034539.

Sully, S. et al., 2019: A global analysis of coral bleaching over the past two decades. *Nat. Commun.*, **10**(1), 1264, doi:10.1038/s41467-019-09238-2.

Sumaila, U.R. et al., 2015: Winners and losers in a world where the high seas is closed to fishing. *Sci. Rep.*, **5**, 8481, doi:10.1038/srep08481.

Sumaila, U.R. et al., 2019: Benefits of the Paris Agreement to ocean life, economies, and people. *Sci. Adv.*, **5**(2), eaau3855, doi:10.1126/sciadv.aau3855.

Sunday, J.M. et al., 2017: Ocean acidification can mediate biodiversity shifts by changing biogenic habitat. *Nat. Clim. Change*, **7**(1), 81.

Sunday, J.M. et al., 2015: Species traits and climate velocity explain geographic range shifts in an ocean-warming hotspot. *Ecol. Lett.*, **18**(9), 944–953, doi:10.1111/ele.12474.

Sundby, S., K.F. Drinkwater and O.S. Kjesbu, 2016: The North Atlantic Spring-Bloom System – Where the Changing Climate Meets the Winter Dark. *Front. Mar. Sci.*, **3**(28), doi:10.3389/fmars.2016.00028.

Sutton-Grier, A.E., K. Wowk and H. Bamford, 2015: Future of our coasts: The potential for natural and hybrid infrastructure to enhance the resilience of our coastal communities, economies and ecosystems. *Environ. Sci. Policy*, **51**, 137–148, doi:10.1016/j.envsci.2015.04.006.

Sverdrup, H.U., M.W. Johnson and R.H. Fleming, 1942: *The Oceans: Their physics, chemistry, and general biology*. Prentice-Hall, New York.

Swann, T. and R. Campbell, 2016: *Great Barrier Bleached: Coral bleaching, the Great Barrier Reef and potential impacts on tourism*. Australia Institute, Canberra, 41.

5

Swart, N.C., S.T. Gille, J.C. Fyfe and N.P. Gillett, 2018: Recent Southern Ocean warming and freshening driven by greenhouse gas emissions and ozone depletion. *Nat. Geosci.*, **11**, 836–841, doi:10.1038/s41561-018-0226-1.

Swartz, W., R. Sumaila and R. Watson, 2013: Global Ex-vessel Fish Price Database Revisited: A New Approach for Estimating 'Missing' Prices. *Environ. Resour. Econ.*, **56**(4), 467–480.

Sweet, M.J. and M.T. Bulling, 2017: On the Importance of the Microbiome and Pathobiome in Coral Health and Disease. *Front. Mar. Sci.*, **4**(9), doi:10.3389/fmars.2017.00009.

Sweetman, A.K. et al., 2017: Major impacts of climate change on deep sea benthic ecosystems. *Elementa Science Anthropocene*, **5**(0), 4, doi:10.1525/elementa.203.

Sydeman, W.J. et al., 2014: Climate change and wind intensification in coastal upwelling ecosystems. *Science*, **345**(6192), 77–80, doi:10.1126/science.1251635.

Sydeman, W.J., E. Poloczanska, T.E. Reed and S.A. Thompson, 2015: Climate change and marine vertebrates. *Science*, **350**(6262), 772–777, doi:10.1126/science.aac9874.

Tagliabue, A. et al., 2016: How well do global ocean biogeochemistry models simulate dissolved iron distributions? *Global Biogeochem. Cy.*, **30**(2), 149–174, doi:10.1002/2015gb005289.

Tagliabue, A. et al., 2017: The integral role of iron in ocean biogeochemistry. *Nature*, **543**(7643), 51–59, doi:10.1038/nature21058.

Tagliabue, A. and C. Völker, 2011: Towards accounting for dissolved iron speciation in global ocean models. *Biogeosciences*, **8**(10), 3025–3039, doi:10.5194/bg-8-3025-2011.

Tai, T.C. et al., 2017: Ex-vessel Fish Price Database: Disaggregating Prices for Low-Priced Species from Reduction Fisheries. *Front. Mar. Sci.*, **4**(363), doi:10.3389/fmars.2017.00363.

Taillardat, P., D.A. Friess and M. Lupascu, 2018: Mangrove blue carbon strategies for climate change mitigation are most effective at the national scale. *Biol. Lett.*, **14**(10), 20180251.

Takahashi, T. et al., 2014: Climatological distributions of pH, pCO$_2$, total CO$_2$, alkalinity, and CaCO3 saturation in the global surface ocean, and temporal changes at selected locations. *Mar. Chem.*, **164**, 95–125, doi:10.1016/j.marchem.2014.06.004.

Talley, L.D. et al., 2016: Changes in Ocean Heat, Carbon Content, and Ventilation: A Review of the First Decade of GO-SHIP Global Repeat Hydrography. *Annu. Rev. Mar. Sci.*, **8**(1), 185–215, doi:10.1146/annurev-marine-052915-100829.

Tallis, H. et al., 2010: The many faces of ecosystem-based management: Making the process work today in real places. *Mar. Policy*, **34**(2), 340–348.

Tammilehto, A. et al., 2015: Induction of domoic acid production in the toxic diatom Pseudo-nitzschia seriata by calanoid copepods. *Aquatic Toxicology*, **159**, 52–61.

Tanaka, Y., T. Hibiya and H. Sasaki, 2015: Downward lee wave radiation from tropical instability waves in the central equatorial Pacific Ocean: A possible energy pathway to turbulent mixing. *J. Geophys. Res-Oceans*, **120**(11), 7137–7149, doi:10.1002/2015JC011017.

Tapsuwan, S. and W. Rongrongmuang, 2015: Climate change perception of the dive tourism industry in Koh Tao island, Thailand. *J. Outdoor Recreat. Tour.*, **11**, 58–63, doi:10.1016/j.jort.2015.06.005.

Tate, R.D., K. Benkendorff, R. Ab Lah and B.P. Kelaher, 2017: Ocean acidification and warming impacts the nutritional properties of the predatory whelk, Dicathais orbita. *J. Exp. Mar. Biol. Ecol.*, **493**, 7–13, doi:10.1016/j.jembe.2017.03.006.

Taucher, J., L.T. Bach, U. Riebesell and A. Oschlies, 2014: The viscosity effect on marine particle flux: A climate relevant feedback mechanism. *Global Biogeochem. Cy.*, **28**(4), 415–422, doi:10.1002/2013GB004728.

Taucher, J. et al., 2017: Influence of ocean acidification on plankton community structure during a winter-to-summer succession: An imaging approach indicates that copepods can benefit from elevated CO$_2$ via indirect food web effects. *PLoS One*, **12**(2), e0169737, doi:10.1371/journal.pone.0169737.

Taylor, J.R. et al., 2014: Physiological effects of environmental acidification in the deep sea urchin *Strongylocentrotus fragilis*. *Biogeosciences*, **11**(5), 1413–1423, doi:10.5194/bg-11-1413-2014.

Taylor, L.L. et al., 2015: Enhanced weathering strategies for stabilizing climate and averting ocean acidification. *Nat. Clim. Change*, **6**, 402, doi:10.1038/nclimate2882.

Taylor, P.G. and A.R. Townsend, 2010: Stoichiometric control of organic carbon-nitrate relationships from soils to the sea. *Nature*, **464**(7292), 1178–1181, doi:10.1038/nature08985.

Teagle, H. and D.A. Smale, 2018: Climate-driven substitution of habitat-forming species leads to reduced biodiversity within a temperate marine community. *Divers. Distrib.*, **24**(10), 1367–1380, doi:10.1111/ddi.12775.

Teh, L.C.L. and U.R. Sumaila, 2013: Contribution of marine fisheries to worldwide employment. *Fish Fish.*, **14**(1), 77–88, doi:10.1111/j.1467-2979.2011.00450.x.

Temmerman, S. et al., 2013: Ecosystem-based coastal defence in the face of global change. *Nature*, **504**, 79, doi:10.1038/nature12859.

Terada, M. and S. Minobe, 2018: Projected sea level rise, gyre circulation and water mass formation in the western North Pacific: CMIP5 inter-model analysis. *Clim. Dyn.*, **50**(11), 4767–4782, doi:10.1007/s00382-017-3902-8.

Tester, P.A. et al., 2010: Ciguatera fish poisoning and sea surface temperatures in the Caribbean Sea and the West Indies. *Toxicon*, **56**(5), 698–710.

Thackeray, S.J. et al., 2016: Phenological sensitivity to climate across taxa and trophic levels. *Nature*, **535**(7611), 241–245, doi:10.1038/nature18608.

Thaman, R.R., 1982: Deterioration of traditional food systems, increasing malnutrition and food dependency in the Pacific Islands. *J. Food. Nutr.*, **39**(3) 109–121.

Thomas, C.J. et al., 2015: Connectivity between submerged and near-sea-surface coral reefs: can submerged reef populations act as refuges? *Divers. Distrib.*, **21**(10), 1254–1266, doi:10.1111/ddi.12360.

Thomas, N. et al., 2017: Distribution and drivers of global mangrove forest change, 1996–2010. *PLoS One*, **12**(6), e0179302, doi:10.1371/journal.pone.0179302.

Thomas, S., 2014: Blue carbon: Knowledge gaps, critical issues, and novel approaches. *Ecol. Econ.*, **107**(Supplement C), 22–38.

Thompson, P.A. et al., 2015: Climate variability drives plankton community composition changes: the 2010–2011 El Niño to La Niña transition around Australia. *J. Plankton Res.*, **37**(5), 966–984, doi:10.1093/plankt/fbv069.

Thomsen, L. et al., 2017: The Oceanic Biological Pump: Rapid carbon transfer to depth at Continental Margins during Winter. *Sci. Rep.*, **7**(1), 10763, doi:10.1038/s41598-017-11075-6.

Thornalley, D.J.R. et al., 2018: Anomalously weak Labrador Sea convection and Atlantic overturning during the past 150 years. *Nature*, **556**(7700), 227–230, doi:10.1038/s41586-018-0007-4.

Thorne, K.M. et al., 2017: Are coastal managers ready for climate change? A case study from estuaries along the Pacific coast of the United States. *Ocean Coast. Manage.*, **143**, 38–50, doi:10.1016/j.ocecoaman.2017.02.010.

Thorne, L.H. et al., 2016: Effects of El Niño-driven changes in wind patterns on North Pacific albatrosses. *J R Soc Interface*, **13**(119), 20160196, doi:10.1098/rsif.2016.0196.

Thresher, R.E., J.M. Guinotte, R.J. Matear and A.J. Hobday, 2015: Options for managing impacts of climate change on a deep sea community. *Nat. Clim. Change*, **5**(7), 635–639, doi:10.1038/nclimate2611.

Thurber, A.R. et al., 2014: Ecosystem function and services provided by the deep sea. *Biogeosciences*, **11**(14), 3941–3963, doi:10.5194/bg-11-3941-2014.

Tidball, K., 2012: Urgent biophilia: human-nature interactions and biological attractions in disaster resilience. *Ecol. Soc.*, 17(2):5, http://dx.doi.org/10.5751/ES-04596-170205.

Tillmann, U. et al., 2019: High abundance of Amphidomataceae (Dinophyceae) during the 2015 spring bloom of the Argentinean Shelf and a new, non-toxigenic ribotype of Azadinium spinosum. *Harmful Algae*, **84**, 244–260, doi:10.1016/j.hal.2019.01.008.

5

Tinker, J. et al., 2016: Uncertainty in climate projections for the 21st century northwest European shelf seas. *Progr. Oceanogr.*, **148**, 56–73, doi:10.1016/j.pocean.2016.09.003.

Tirado, M.C. et al., 2010: Climate change and food safety: A review. *Food Res. Int.*, **43**(7), 1745–1765, doi:10.1016/j.foodres.2010.07.003.

Tittensor, D.P. et al., 2018: A protocol for the intercomparison of marine fishery and ecosystem models: Fish-MIP v1.0. *Geosci. Model Dev.*, **11**(4), 1421–1442, doi:10.5194/gmd-11-1421-2018.

Tittensor, D.P. et al., 2011: Species-energy relationships in deep sea molluscs. *Biol. Lett.*, **7**(5), 718–722, doi:10.1098/rsbl.2010.1174.

Tkachenko, K.S. and K. Soong, 2017: Dongsha Atoll: A potential thermal refuge for reef-building corals in the South China Sea. *Mar. Environ. Res.*, **127**, 112–125, doi:10.1016/j.marenvres.2017.04.003.

Tomas, F., B. Martínez-Crego, G. Hernán and R. Santos, 2015: Responses of seagrass to anthropogenic and natural disturbances do not equally translate to its consumers. *Global Change Biol.*, **21**(11), 4021–4030, doi:10.1111/gcb.13024.

Tong, C. et al., 2018: Changes in pore-water chemistry and methane emission following the invasion of Spartina alterniflora into an oliogohaline marsh. *Limnol. Oceanogr.*, **63**(1), 384–396, doi:10.1002/lno.10637.

Torabi, E., A. Dedekorkut-Howes and M. Howes, 2018: Adapting or maladapting: Building resilience to climate-related disasters in coastal cities. *Cities*, **72**, 295–309, doi:10.1016/j.cities.2017.09.008.

Torda, G. et al., 2017: Rapid adaptive responses to climate change in corals. *Nat. Clim. Change*, **7**, 627, doi:10.1038/nclimate3374.

Torresan, S. et al., 2016: DESYCO: A decision support system for the regional risk assessment of climate change impacts in coastal zones. *Ocean Coast. Manage.*, **120**, 49–63, doi:10.1016/j.ocecoaman.2015.11.003.

Townhill, B.L. et al., 2018: Harmful algal blooms and climate change: exploring future distribution changes. *ICES J. Mar. Sci.*, **75**(6), 1882–1893, doi:10.1093/icesjms/fsy113.

Tréguer, P. et al., 2018: Influence of diatom diversity on the ocean biological carbon pump. *Nat. Geosci.*, **11**(1), 27–37, doi:10.1038/s41561-017-0028-x.

Trenberth, K.E., M. Marquis and S. Zebiak, 2016: The vital need for a climate information system. *Nat. Clim. Change*, **6**, 1057, doi:10.1038/nclimate3170.

Trivelpiece, W.Z. et al., 2011: Variability in krill biomass links harvesting and climate warming to penguin population changes in Antarctica. *PNAS*, **108**(18), 7625–7628, doi:10.1073/pnas.1016560108.

Triyanti, A., M. Bavinck, J. Gupta and M.A. Marfai, 2017: Social capital, interactive governance and coastal protection: The effectiveness of mangrove ecosystem-based strategies in promoting inclusive development in Demak, Indonesia. *Ocean Coast. Manage.*, **150**, 3–11, doi:10.1016/j.ocecoaman.2017.10.017.

Troxler, T.G., H.A. Kennedy, S. Crooks and A.E. Sutton-Grier, 2018: Introduction of Coastal Wetlands into the IPCC Greenhouse Gas Inventory Methodological Guidance. Editors: Windham-Myers, Crooks, Troxler, In: *A Blue Carbon Primer*. CRC Press, Boca Raton, pp. 217–234, eBook ISBN9780429435362, https://doi.org/10.1201/9780429435362.

Tsikliras, A.C. et al., 2014: Shift in trophic level of Mediterranean mariculture species. *Conserv Biol*, **28**(4), 1124–8, doi:10.1111/cobi.12276.

Tuda, A.O., S. Kark and A. Newton, 2019: Exploring the prospects for adaptive governance in marine transboundary conservation in East Africa. *Mar. Policy*, **104**, 75–84, doi:10.1016/j.marpol.2019.02.051.

Turner, N. et al., 2008: From Invisibility to Transparency: Identifying the Implications. *Ecol. Soc.*, **13**(2): 7. [online] URL: http://www.ecologyandsociety.org/vol13/iss2/art7/.

Turra, A. et al., 2016: Frequency, magnitude, and possible causes of stranding and mass-mortality events of the beach clam Tivela mactroides (Bivalvia: Veneridae). *PLoS One*, **11**(1), e0146323.

Tweedley, J.R. et al., 2016: The hypoxia that developed in a microtidal estuary following an extreme storm produced dramatic changes in the benthos. *Mar. Freshw. Res.*, **67**(3), 327–341.

Tyler, P. et al., 2007: Gametogenic periodicity in the chemosynthetic cold-seep mussel "Bathymodiolus" childressi. *Mar. Biol.*, **150**(5), 829–840, doi:10.1007/s00227-006-0362-9.

Tyler, P.A et al., (eds.), 2003: Ecosystems of the Deep Ocean. Elsevier Science, Amsterdam, 582 pp, eBook ISBN: 9780080494654.

Ullah, H., I. Nagelkerken, S.U. Goldenberg and D.A. Fordham, 2018: Climate change could drive marine food web collapse through altered trophic flows and cyanobacterial proliferation. *PLoS Biology*, **16**(1), e2003446.

UNCTAD, 2017: *Climate change impacts on coastal transport infrastructure in the Caribbean: enhancing the adaptive capacity of Small Island Developing States (SIDS), JAMAICA: A case study.*, UNDA project 14150.

UNCTAD, 2018: *Economic Development in Africa Report 2018*. United Nations, UNCTAD/ALDC/AFRICA/2018 ISBN: 978-92-1-112924-3.

UNDP, 2017: *Regional overview: Impact of hurricanes Irma and Maria. Conference supporting document. Report prepared with support of ACAPS, OCHOA and UNDP.* 39pp.

UNEP, 2013: *Minamata convention on Mercury.* [Available at: www.mercuryconvention.org/Convention]. Accessed: 2019/09/30.

UNEP, 2016: *Regional Seas Programmes and other UNEP Activities Relevant to Marine Biodiversity in Areas beyond National Jurisdiction.* Development of an international legally-binding instrument on the conservation and sustainable use of marine biodiversity of areas beyond national jurisdiction under the United Nations Convention on Law of the Sea, 8 pp, https://www.un.org/depts/los/biodiversity/prepcom_files/UNEP_and_BBNJ_PrepCom2.pdf.

UNEP, 2017: *The Emissions Gap Report.* United Natoins Environment Programme, Nairobi [Available at: www.worldcat.org/title/emissions-gap-report-2017-a-un-environment-synthesis-report/oclc/1009432397]. Accessed: 2019/09/30.

UNEP-WCMC and S. FT, 2017: Global distribution of seagrasses (version 5.0). Fourth update to the data layer used in Green and Short (2003). Cambridge (UK): UNEP World Conservation Monitoring Centre [Available at http://data.unep-wcmc.org/datasets/7]. Accessed: 2019/09/30.

UNEP-WCMC, W. C. and T. WRI, 2010: Global distributin of warm water coral reefs, compiled from multiple sources including the Millennium Coral Reef Mapping Project. Version 1.3.

UNFCCC, 2015: *Adoption of the Paris Agreement.* United Nations Framework Convention on Climate Change, Twenty-first Session of Conference of the Parties, https://unfccc.int/resource/docs/2015/cop21/eng/l09r01.pdf.

UNWTO, 2018: *Tourism in Small Island Developing States.* World Tourism Organization, Madrid, Spain. 5p. http://cf.cdn.unwto.org/sites/all/files/docpdf/tourisminsids.pdf.

Valiela, I. et al., 2018: Transient coastal landscapes: Rising sea level threatens salt marshes. *Sci. Total Environ.*, **640–641**, 1148–1156, doi:10.1016/j.scitotenv.2018.05.235.

Vallis, G.K., 2017: *Atmospheric and oceanic fluid dynamics.* Cambridge University Press. 946 pp. ISBN: 978-1-107-06550-5

Valmonte-Santos, R., M. W. Rosegrant and M.M. Dey, 2016: Fisheries sector under climate change in the coral triangle countries of Pacific Islands: Current status and policy issues. *Mar. Policy*, **67**, 148–155, doi:10.1016/j.marpol.2015.12.022.

van de Velde, S. et al., 2018: Anthropogenic disturbance keeps the coastal seafloor biogeochemistry in a transient state. *Sci. Rep.*, **8**(1), 5582, doi:10.1038/s41598-018-23925-y.

Van de Waal, D.B. et al., 2013: Nutrient pulse induces dynamic changes in cellular C: N: P, amino acids, and paralytic shellfish poisoning toxins in Alexandrium tamarense. *Mar. Ecol. Prog. Ser.*, **493**, 57–69.

van der Nat, A., P. Vellinga, R. Leemans and E. van Slobbe, 2016: Ranking coastal flood protection designs from engineered to nature-based. *Ecol. Eng.*, **87**, 80–90, doi:10.1016/j.ecoleng.2015.11.007.

van Dijk, I. et al., 2017: Combined Impacts of Ocean Acidification and Dysoxia On Survival and Growth of Four Agglutinating Foraminifera. *J. Foramin. Res.*, **47**(3), 294–303.

5

van Hooidonk, R. et al., 2016: Local-scale projections of coral reef futures and implications of the Paris Agreement. *Sci. Rep.*, **6**, 39666, doi:10.1038/srep39666.

van Hooidonk, R., J.A. Maynard and S. Planes, 2013: Temporary refugia for coral reefs in a warming world. *Nat. Clim. Change*, **3**(5), 508–511, doi:10.1038/NCLIMATE1829.

van Katwijk, M.M. et al., 2016: Global analysis of seagrass restoration: the importance of large-scale planting. *J. Appl. Ecol.*, **53**(2), 567–578, doi:10.1111/1365-2664.12562.

van Oppen, M.J.H. et al., 2017a: Shifting paradigms in restoration of the world's coral reefs. *Global Change Biol.*, **23**(9), 3437–3448, doi:10.1111/gcb.13647.

van Oppen, M.J.H. et al., 2017b: Shifting paradigms in restoration of the world's coral reefs. *Global Change Biol.*, **23**(9), 3437–3448, doi:10.1111/gcb.13647.

van Oppen, M.J.H., J.K. Oliver, H.M. Putnam and R.D. Gates, 2015: Building coral reef resilience through assisted evolution. *PNAS*, **112**(8), 2307.

van Puijenbroek, M.E.B. et al., 2017: Exploring the contributions of vegetation and dune size to early dune development using unmanned aerial vehicle (UAV) imaging. *Biogeosciences*, **14**(23), 5533–5549, doi:10.5194/bg-14-5533-2017.

van Sebille, E. et al., 2015: A global inventory of small floating plastic debris. *Environ. Res. Lett.*, **10**(12), 124006, doi:10.1088/1748-9326/10/12/124006.

Varela, M.R. et al., 2019: Assessing climate change associated sea level rise impacts on sea turtle nesting beaches using drones, photogrammetry and a novel GPS system. *Global Change Biol.*, **25**(2), 753–762, doi:10.1111/gcb.14526.

Varela, R. et al., 2015: Has upwelling strengthened along worldwide coasts over 1982–2010? *Sci. Rep.*, **5**, 10016, doi:10.1038/srep10016.

Vázquez, N.G. et al., 2016: Mass Mortalities Affecting Populations of the Yellow Clam *Amarilladesma mactroide* Along Its Geographic Range. *J. of Shellfish Research*, **35**(4), 739–745.

Vázquez-González, C. et al., 2017: Mangrove and Freshwater Wetland Conservation Through Carbon Offsets: A Cost-Benefit Analysis for Establishing Environmental Policies. *Environ. Manage.*, **59**(2), 274–290, doi:10.1007/s00267-016-0790-3.

Veilleux, H.D. and J.M. Donelson, 2018: Reproductive gene expression in a coral reef fish exposed to increasing temperature across generations. *Conserv. Physiol.*, **6**(1), cox077–cox077, doi:10.1093/conphys/cox077.

Vergés, A. et al., 2018: Latitudinal variation in seagrass herbivory: Global patterns and explanatory mechanisms. *Global Ecol. Biogeogr.*, **27**(9), 1068–1079, doi:10.1111/geb.12767.

Vergés, A. et al., 2016: Long-term empirical evidence of ocean warming leading to tropicalization of fish communities, increased herbivory, and loss of kelp. *PNAS*, **113**(48), 13791, doi:10.1073/pnas.1610725113.

Vergés, A. et al., 2019: Tropicalisation of temperate reefs: Implications for ecosystem functions and management actions. *Funct. Ecol.*, **33**(6), 1000-1013. doi:10.1111/1365-2435.13310.

Vezzulli, L. et al., 2016: Climate influence on *Vibrio* and associated human diseases during the past half-century in the coastal North Atlantic. *PNAS*, **113**(34), E5062-E5071.

Vikolainen, V., J. Flikweert, H. Bressers and K. Lulofs, 2017: Governance context for coastal innovations in England: The case of Sandscaping in North Norfolk. *Ocean Coast. Manage.*, **145**, 82–93, doi:10.1016/j.ocecoaman.2017.05.012.

Vitousek, S. et al., 2017: A model integrating longshore and cross-shore processes for predicting long-term shoreline response to climate change. *J. Geophys. Res-Earth*, **122**(4), 782–806.

von Heland, F., J. Clifton and P. Olsson, 2014: Improving Stewardship of Marine Resources: Linking Strategy to Opportunity. *Sustainability*, **6**(7), 4470–4496. doi:10.3390/su6074470.

Von Schuckmann, K. et al., 2014: Consistency of the current global ocean observing systems from an Argo perspective. *Ocean Sci.*, **10**(3), 547–557.

von Storch, H. et al., 2015: Making coastal research useful – cases from practice. *Oceanologia*, **57**(1), 3–16, doi:10.1016/j.oceano.2014.09.001.

Wabnitz, C.C.C., A.M. Cisneros-Montemayor, Q. Hanich and Y. Ota, 2017: Ecotourism, climate change and reef fish consumption in Palau: Benefits, trade-offs and adaptation strategies. *Mar. Policy*, **88**, 323–332. doi:10.1016/j.marpol.2017.07.022.

Wabnitz, C.C.C. et al., 2018: Climate change impacts on marine biodiversity, fisheries and society in the Arabian Gulf. *PLoS One*, **13**(5), e0194537, doi:10.1371/journal.pone.0194537.

Wall, C.B. et al., 2017: Elevated pCO(2) affects tissue biomass composition, but not calcification, in a reef coral under two light regimes. *R. Soc. Open Sci.*, **4**(11), 170683, doi:10.1098/rsos.170683.

Waller, C.L. et al., 2017: Microplastics in the Antarctic marine system: An emerging area of research. *Sci. Total Environ.*, **598**, 220–227, doi:10.1016/j.scitotenv.2017.03.283.

Walworth, N.G. et al., 2016: Mechanisms of increased Trichodesmium fitness under iron and phosphorus co-limitation in the present and future ocean. *Nat. Commun.*, **7**, 12081–12081, doi:10.1038/ncomms12081.

Wan, Z., M. Zhu, S. Chen and D. Sperling, 2016: Pollution: Three steps to a green shipping industry.

Wang, D., T.C. Gouhier, B.A. Menge and A.R. Ganguly, 2015a: Intensification and spatial homogenization of coastal upwelling under climate change. *Nature*, **518**(7539), 390–394, doi:10.1038/nature14235.

Wang, G. and W. Cai, 2013: Climate-change impact on the 20th-century relationship between the Southern Annular Mode and global mean temperature. *Sci. Rep.*, **3**(1), 2039, doi:10.1038/srep02039.

Wang, R. et al., 2015b: Influence of anthropogenic aerosol deposition on the relationship between oceanic productivity and warming. *Geophys. Res. Lett.*, **42**(24), 10745–10754, doi:10.1002/2015GL066753.

Wang, W., H. Liu, Y. Li and J. Su, 2014: Development and management of land reclamation in China. *Ocean Coast. Manage.*, **102**, 415–425.

Wang, W.-L., J. K. Moore, A.C. Martiny and F.W. Primeau, 2019: Convergent estimates of marine nitrogen fixation. *Nature*, **566**(7743), 205–211, doi:10.1038/s41586-019-0911-2.

Wanninkhof, R. et al., 2010: Detecting anthropogenic CO_2 changes in the interior Atlantic Ocean between 1989 and 2005. *J. Geophys. Res-Oceans*, **115**(C11). https://doi.org/10.1029/2010JC006251.

Ward, B.A., S. Dutkiewicz, C.M. Moore and M.J. Follows, 2013: Iron, phosphorus, and nitrogen supply ratios define the biogeography of nitrogen fixation. *Limnol. Oceanogr.*, **58**(6), 2059–2075, doi:10.4319/lo.2013.58.6.2059.

Ward, R.D., D.A. Friess, R.H. Day and R.A. MacKenzie, 2016: Impacts of climate change on mangrove ecosystems: a region by region overview. *Ecosyst. Health Sustain.*, **2**(4), e01211, doi:10.1002/ehs2.1211.

Ward, S.L., J.A.M. Green and H.E. Pelling, 2012: Tides, sea level rise and tidal power extraction on the European shelf. *Ocean Dyn.*, **62**(8), 1153–1167, doi:10.1007/s10236-012-0552-6.

Warner, R.M., 2018: Oceans in Transition: Incorporating Climate-Change Impacts into Environmental Impact Assessment for Marine Areas Beyond National Jurisdiction. Ecology Law Quarterly, **45**(1), https://doi.org/10.15779/Z38M61BQ0J.

Warwick, R.M., J.R. Tweedley and I.C. Potter, 2018: Microtidal estuaries warrant special management measures that recognise their critical vulnerability to pollution and climate change. *Mar. Pollut. Bull.*, **135**, 41–46, doi:10.1016/j.marpolbul.2018.06.062.

Wassmann, P., M. Duarte Carlos, S. Agustí and K. Sejr Mikael, 2010: Footprints of climate change in the Arctic marine ecosystem. *Global Change Biol.*, **17**(2), 1235–1249, doi:10.1111/j.1365-2486.2010.02311.x.

Watanabe, Y.W. et al., 2003: Synchronous bidecadal periodic changes of oxygen, phosphate and temperature between the Japan Sea deep water and the North Pacific intermediate water. *Geophys. Res. Lett.*, **30**(24), doi:10.1029/2003GL018338.

5

Waterhouse, A.F. et al., 2014: Global Patterns of Diapycnal Mixing from Measurements of the Turbulent Dissipation Rate. *J. Phys. Oceanogr.*, **44**(7), 1854–1872, doi:10.1175/JPO-D-13-0104.1.

Waterman, S., A.C. Naveira Garabato and K.L. Polzin, 2012: Internal Waves and Turbulence in the Antarctic Circumpolar Current. *J. Phys. Oceanogr.*, **43**(2), 259–282, doi:10.1175/JPO-D-11-0194.1.

Watkiss, P., A. Hunt and M. Savaga, 2014: *Early Value-for-Money Adaptation: Delivering VfM Adaptation using Iterative Frameworks and LowRegret Options*, Global Climate Adaptation Partnership, UK Department for International Development, London, 53 pp. DOI:http://dx.doi.org/10.12774/eod_cr.july2014.watkisspetal.

Watson, E.B. et al., 2017a: Anthropocene Survival of Southern New England's Salt Marshes. *Estuar. Coast.*, **40**(3), 617–625, doi:10.1007/s12237-016-0166-1.

Watson, E.B. et al., 2017b: Wetland Loss Patterns and Inundation-Productivity Relationships Prognosticate Widespread Salt Marsh Loss for Southern New England. *Estuar. Coast.*, **40**(3), 662–681, doi:10.1007/s12237-016-0069-1.

Watson-Wright, W. and J.L. Valdés, 2018: Fragmented governance of our one global ocean. In: *The Future of Ocean Governance and Capacity Development* [Institute, I.O. (ed.)]. Brill, Leiden, Netherland. 562 pp. ISBN: 978-90-04-38027-1.

Weatherdon, L.V. et al., 2016: Observed and Projected Impacts of Climate Change on Marine Fisheries, Aquaculture, Coastal Tourism, and Human Health: An Update. *Front. Mar. Sci.*, **3**(36), 473, doi:10.3389/fmars.2016.00048.

Weatherhead, E.C. et al., 1998: Factors affecting the detection of trends: Statistical considerations and applications to environmental data. *J. Geophys. Res-Atmos.*, **103**(D14), 17149–17161, doi:10.1029/98jd00995.

Webster, N.S. et al., 2016: Host-associated coral reef microbes respond to the cumulative pressures of ocean warming and ocean acidification. *Sci. Rep.*, **6**(1), doi:10.1038/srep19324.

Wei, C.-L. et al., 2011: Global Patterns and Predictions of Seafloor Biomass Using Random Forests. *PLoS One*, **5**(12), e15323, doi:10.1371/journal.pone.0015323.

Weller, E. et al., 2016: Multi-model attribution of upper-ocean temperature changes using an isothermal approach. *Sci. Rep.*, **6**, 26926, doi:10.1038/srep26926.

Wells, M.L. et al., 2015: Harmful algal blooms and climate change: Learning from the past and present to forecast the future. *Harmful Algae*, **49**, 68–93, doi:10.1016/j.hal.2015.07.009.

Weng, K.C., E. Glazier, S.J. Nicol and A.J. Hobday, 2015: Fishery management, development and food security in the Western and Central Pacific in the context of climate change. *Deep Sea Res. Pt. II*, **113**, 301–311, doi:10.1016/j.dsr2.2014.10.025.

Wernberg, T. et al., 2016: Climate-driven regime shift of a temperate marine ecosystem. *Science*, **353**(6295), 169, doi:10.1126/science.aad8745.

Wernberg, T. et al., 2018: Genetic diversity and kelp forest vulnerability to climatic stress. *Sci. Rep.*, **8**(1), 1851, doi:10.1038/s41598-018-20009-9.

Wernberg, T., K. Krumhansl, K. Filbee-Dexter and M.F. Pedersen, 2019: Status and trends for the world's kelp forests. In: *World Seas: An Environmental Evaluation*. [Sheppard, C. (ed.)]. Elsevier, New York. pp. 57–78. ISBN: 978-0-12-805052-1.

West, J.M. et al., 2017: Climate-Smart Design for Ecosystem Management: A Test Application for Coral Reefs. *Environ. Manage.*, **59**(1), 102–117, doi:10.1007/s00267-016-0774-3.

Weydmann, A., J.E. Søreide, S. Kwasniewski and S. Widdicombe, 2012: Influence of CO_2-induced acidification on the reproduction of a key Arctic copepod Calanus glacialis. *J. Exp. Mar. Biol. Ecol.*, **428**, 39–42, doi:10.1016/j.jembe.2012.06.002.

Whalen, C.B., J.A. MacKinnon and L.D. Talley, 2018: Large-scale impacts of the mesoscale environment on mixing from wind-driven internal waves. *Nat. Geosci.*, **11**(11), 842–847, doi:10.1038/s41561-018-0213-6.

Whalen, C.B., L.D. Talley and J.A. MacKinnon, 2012: Spatial and temporal variability of global ocean mixing inferred from Argo profiles. *Geophys. Res. Lett.*, **39**(18), doi:10.1029/2012GL053196.

Whitney, F.A., S.J. Bograd and T. Ono, 2013: Nutrient enrichment of the subarctic Pacific Ocean pycnocline. *Geophys. Res. Lett.*, **40**(10), 2200–2205, doi:10.1002/grl.50439.

Whitney, F.A., H.J. Freeland and M. Robert, 2007: Persistently declining oxygen levels in the interior waters of the eastern subarctic Pacific. *Progr. Oceanogr.*, **75**(2), 179–199, doi:10.1016/j.pocean.2007.08.007.

Wigand, C. et al., 2017: A climate change adaptation strategy for management of coastal marsh systems. *Estuar. Coast.*, **40**(3), 682–693.

Williams, G.A. et al., 2016: Meeting the climate change challenge: Pressing issues in southern China and SE Asian coastal ecosystems. *Reg. Stud. Mar. Sci.*, **8**, 373–381, doi:10.1016/j.rsma.2016.07.002.

Williamson, P. and R. Bodle, 2016: Update on Climate Geoengineering in Relation to the Convention on Biological Diversity: Potential Impacts and Regulatory Framework. CBD *Technical Series* No. 85. Secretariat of the Convention on Biological Diversity, Montreal, 158 pp. ISBN: 9789292256425.

Williamson, P. and C. Turley, 2012: Ocean acidification in a geoengineering context. *Philos. Trans. Roy. Soc. A.*, **370**(1974), 4317.

Williamson, P. et al., 2012: Ocean fertilization for geoengineering: A review of effectiveness, environmental impacts and emerging governance. *Process Saf. Environ.*, **90**(6), 475–488, doi:10.1016/j.psep.2012.10.007.

Wilson, A.M.W. and C. Forsyth, 2018: Restoring near-shore marine ecosystems to enhance climate security for island ocean states: Aligning international processes and local practices. *Mar. Policy*, **93**, 284–294 doi:10.1016/j.marpol.2018.01.018.

Wilson, K.L., M.A. Skinner and H.K. Lotze, 2019: Projected 21st-century distribution of canopy-forming seaweeds in the Northwest Atlantic with climate change. *Divers. Distrib.*, **25**(4), 582–602, doi:10.1111/ddi.12897.

Winder, M. et al., 2017: The land–sea interface: A source of high-quality phytoplankton to support secondary production. *Limnol. Oceanogr.*, **62**(S1), S258-S271.

Windham-Myers, L., S. Crooks and T.G. Troxler, 2019: *A blue carbon primer: the state of coastal wetland carbon science, practice and policy*. CRC Press, Boca Raton, Florida. 481 pp. ISBN: 978-1-4987-6909-9.

Wise, R.M. et al., 2016: How climate compatible are livelihood adaptation strategies and development programs in rural Indonesia? *Clim. Risk Manage.*, **12**, 100–114, doi:10.1016/j.crm.2015.11.001.

Wit, J.C., M.M. Davis, D.C. McCorkle and J.M. Bernhard, 2016: A short-term survival experiment assessing impacts of ocean acidification and hypoxia on the benthic foraminifer Globobulimina turgida. *J. Foramin. Res.*, **46**(1), 25–33.

WMO, 2019: *WMO Statement on the State of the Global Climate in 2018*. World Meterological Organization, Geneva. 39 pp. ISBN 978-92-63-11233-0.

Wohlers-Zöllner, J. et al., 2011: Temperature and nutrient stoichiometry interactively modulate organic matter cycling in a pelagic algal–bacterial community. *Limnol. Oceanogr.*, **56**(2), 599–610, doi:10.4319/lo.2011.56.2.0599.

Wong, A.P.S. and S.C. Riser, 2011: Profiling Float Observations of the Upper Ocean under Sea Ice off the Wilkes Land Coast of Antarctica. *J. Phys. Oceanogr.*, **41**(6), 1102–1115, doi:10.1175/2011JPO4516.1.

Wong, A.P.S. and S.C. Riser, 2013: Modified shelf water on the continental slope north of Mac Robertson Land, East Antarctica. *Geophys. Res. Lett.*, **40**(23), 6186–6190, doi:10.1002/2013gl058125.

Wong, P.P., et al. 2014a: *Coastal systems and low-lying areas*. In: Climate Change 2014: Impacts, Adaptation and Vulnerability. Part A: Global and Sectoral Aspects. Contribution of Working Group II to the Fifth Assessment Report of the Intergovernmental Panel on Climate Change. [Field, C.B., et al. (eds.)]. Cambridge University Press, Cambridge, United Kingdom and New York, USA, pp. 361–409. ISBN: 978-1-107-05807-1.

Wong, P.P. et al., 2014b: Coastal systems and low-lying areas. *Clim. Change*, **2104**, 361–409.

5

Woodall, L.C. et al., 2014: The deep sea is a major sink for microplastic debris. *R. Soc. Open Sci.*, **1**(4), 140317, doi:doi:10.1098/rsos.140317.

Woodruff, S.C., 2018: City membership in climate change adaptation networks. *Environ. Sci. Policy*, **84**, 60–68.

Woodworth, P L., 2010: A survey of recent changes in the main components of the ocean tide. *Cont. Shelf Res.*, **30**(15), 1680–1691, doi:10.1016/j.csr.2010.07.002.

Woodworth-Jefcoats, P.A., J.J. Polovina and J.C. Drazen, 2017: Climate change is projected to reduce carrying capacity and redistribute species richness in North Pacific pelagic marine ecosystems. *Global Change Biol.*, **23**(3), 1000–1008, doi:10.1111/gcb.13471.

Woosley, R.J., F.J. Millero and R. Wanninkhof, 2016: Rapid anthropogenic changes in CO_2 and pH in the Atlantic Ocean: 2003–2014. *Global Biogeochem. Cy.*, **30**(1), 70–90.

World Bank, 2017: *Pacific Possible : long-term economic opportunities and challenges for Pacific Island Countries (English). Pacific possible series.* The World Bank, Washington, DC. 130 p. http://documents.worldbank.org/curated/en/168951503668157320/pdf/ACS22308-PUBLIC-P154324-ADD-SERIES-PPFullReportFINALscreen.pdf.

Woulds, C. et al., 2009: The short-term fate of organic carbon in marine sediments: Comparing the Pakistan margin to other regions. *Deep Sea Res. Pt. II*, **56**(6–7), 393–402, doi:10.1016/j.dsr2.2008.10.008.

Wouters, H. et al., 2017: Heat stress increase under climate change twice as large in cities as in rural areas: A study for a densely populated midlatitude maritime region. *Geophys. Res. Lett.*, **44**(17), 8997–9007.

Wunsch, C. and R. Ferrari, 2004: Vertical mixing, energy, and the general circulation of the oceans. *Annu. Rev. Fluid Mech.*, **36**(1), 281–314, doi:10.1146/annurev.fluid.36.050802.122121.

Wylie, L., A.E. Sutton-Grier and A. Moore, 2016: Keys to successful blue carbon projects: Lessons learned from global case studies. *Mar. Policy*, **65**, 76–84, doi:10.1016/j.marpol.2015.12.020.

Wynveen, C.J. and S.G. Sutton, 2015: Engaging the public in climate change-related pro-environmental behaviors to protect coral reefs: The role of public trust in the management agency. *Mar. Policy*, **53**, 131–140, doi:10.1016/j.marpol.2014.10.030.

Xiu, P., F. Chai, E.N. Curchitser and F.S. Castruccio, 2018: Future changes in coastal upwelling ecosystems with global warming: The case of the California Current System. *Sci. Rep.*, **8**(1), 2866, doi:10.1038/s41598-018-21247-7.

Xu, J. and T. Kiørboe, 2018: Toxic dinoflagellates produce true grazer deterrents. *Ecology*.

Yamamoto, A. et al., 2015: Global deep ocean oxygenation by enhanced ventilation in the Southern Ocean under long-term global warming. *Global Biogeochem. Cy.*, **29**(10), 1801–1815, doi:10.1002/2015GB005181.

Yang, S. and N. Gruber, 2016: The anthropogenic perturbation of the marine nitrogen cycle by atmospheric deposition: Nitrogen cycle feedbacks and the 15N Haber-Bosch effect. *Global Biogeochem. Cy.*, **30**(10), 1418–1440, doi:10.1002/2016GB005421.

Yashayaev, I., 2007: Hydrographic changes in the Labrador Sea, 1960–2005. *Progr. Oceanogr.*, **73**(3), 242–276, doi:10.1016/j.pocean.2007.04.015.

Yashayaev, I. and J.W. Loder, 2017: Further intensification of deep convection in the Labrador Sea in 2016. *Geophys. Res. Lett.*, **44**(3), 1429–1438, doi:10.1002/2016gl071668.

Yeager, S.G., A.R. Karspeck and G. Danabasoglu, 2015: Predicted slowdown in the rate of Atlantic sea ice loss. *Geophys. Res. Lett.*, **42**(24), 10,704–10,713, doi:10.1002/2015gl065364.

Yeruham, E., G. Rilov, M. Shpigel and A. Abelson, 2015: Collapse of the echinoid Paracentrotus lividus populations in the Eastern Mediterranean – result of climate change? *Sci. Rep.*, **5**, 13479.

Yodnarasri, S. et al., 2008: Is there any seasonal variation in marine nematodes within the sediments of the intertidal zone? *Mar. Pollut. Bull.*, **57**(1), 149–154, doi:10.1016/j.marpolbul.2008.04.016.

Yool, A. et al., 2017: Big in the benthos: Future change of seafloor community biomass in a global, body size-resolved model. *Global Change Biol.*, **23**(9), 3554–3566, doi:10.1111/gcb.13680.

Yool, A. et al., 2013: Climate change and ocean acidification impacts on lower trophic levels and the export of organic carbon to the deep ocean. *Biogeosciences*, **10**(9), 5831–5854, doi:10.5194/bg-10-5831-2013.

Yoon, J.E. et al., 2016: Ocean iron fertilization experiments: Past-Present-Future with introduction to Korean Iron Fertilization Experiment in the Southern Ocean (KIFES) project. *Biogeosciences Discuss.*, **2016**, 1–41, doi:10.5194/bg-2016-472.

York, P.H. et al., 2017: Identifying knowledge gaps in seagrass research and management: An Australian perspective. *Mar. Environ. Res.*, **127**, 163–172, doi:10.1016/j.marenvres.2016.06.006.

Young, I.R., S. Zieger and A.V. Babanin, 2011: Global Trends in Wind Speed and Wave Height. *Science*, **332**(6028), 451.

Young, J.W. et al., 2015: The trophodynamics of marine top predators: Current knowledge, recent advances and challenges. *Deep Sea Res. Pt. II*, **113**, 170–187, doi:10.1016/j.dsr2.2014.05.015.

Youngflesh, C. et al., 2017: Circumpolar analysis of the Adelie Penguin reveals the importance of environmental variability in phenological mismatch. *Ecology*, **98**(4), 940–951, doi:10.1002/ecy.1749.

Yuan, X. et al., 2010: Bacterial production and respiration in subtropical Hong Kong waters: influence of the Pearl River discharge and sewage effluent. *Aqut. Microb. Ecol.*, **58**(2), 167–179, doi:10.3354/ame03146.

Zahid, A. et al., 2018: Model Impact of Climate Change on the Groundwater Flow and Salinity Encroachment in the Coastal Areas of Bangladesh. In: Groundwater of South Asia. Springer, pp. 545–568.

Zamir, R., P. Alpert and G. Rilov, 2018: Increase in Weather Patterns Generating Extreme Desiccation Events: Implications for Mediterranean Rocky Shore Ecosystems. *Estuar. Coast.*, **41**(7), 1868–1884, doi:10.1007/s12237-018-0408-5.

Zandvoort, M. et al., 2017: Adaptation pathways in planning for uncertain climate change: Applications in Portugal, the Czech Republic and the Netherlands. *Environ. Sci. Policy*, **78**, 18–26, doi:10.1016/j.envsci.2017.08.017.

Zanowski, H. and R. Hallberg, 2017: Weddell Polynya Transport Mechanisms in the Abyssal Ocean. *J. Phys. Oceanogr.*, doi:10.1175/JPO-D-17-0091.1.

Zanowski, H., R. Hallberg and J.L. Sarmiento, 2015: Abyssal Ocean Warming and Salinification after Weddell Polynyas in the GFDL CM2G Coupled Climate Model. *J. Phys. Oceanogr.*, **45**(11), 2755–2772, doi:10.1175/JPO-D-15-0109.1.

Zhang, D. et al., 2016: Reviews of power supply and environmental energy conversions for artificial upwelling. *Renew. Sustain. Energy Rev.*, **56**, 659–668, doi:10.1016/j.rser.2015.11.041.

Zhang, J. and X. Gao, 2016: Nutrient distribution and structure affect the acidification of eutrophic ocean margins: A case study in southwestern coast of the Laizhou Bay, China. *Mar. Pollut. Bull.*, **111**(1–2), 295–304.

Zhang, J. et al., 2015: Phylogeographic data revealed shallow genetic structure in the kelp Saccharina japonica (Laminariales, Phaeophyta). *BMC Evol. Biol.*, **15**, 237, doi:10.1186/s12862-015-0517-8.

Zhang, S. et al., 2018: Phosphorus release from cyanobacterial blooms during their decline period in eutrophic Dianchi Lake, China. *Environ. Sci. Pollut. Res.*, doi:10.1007/s11356-018-1517-1.

Zhang, Y. et al., 2017: Carbon sequestration processes and mechanisms in coastal mariculture environments in China. *Science China Earth Sciences*, **60**(12), 2097–2107, doi:10.1007/s11430-017-9148-7.

Zhao, Q. et al., 2016a: A review of methodologies and success indicators for coastal wetland restoration. *Ecol. Indic.*, **60**, 442–452, doi:10.1016/j.ecolind.2015.07.003.

Zhao, Z., 2016: Internal tide oceanic tomography. *Geophys. Res. Lett.*, **43**(17), 9157–9164, doi:10.1002/2016GL070567.

Zhao, Z. et al., 2016b: Global Observations of Open-Ocean Mode-1 M2 Internal Tides. *J. Phys. Oceanogr.*, **46**(6), 1657–1684, doi:10.1175/JPO-D-15-0105.1.

Zhou, X. et al., 2017: Prospective scenarios of the saltwater intrusion in an estuary under climate change context using Bayesian neural networks. *Stochastic Environmental Research and Risk Assessment*, **31**(4), 981–991.

Zhu, Z. et al., 2017: Understanding the blob bloom: Warming increases toxicity and abundance of the harmful bloom diatom Pseudo-nitzschia in California coastal waters. *Harmful Algae*, **67**, 36–43, doi:10.1016/j. hal.2017.06.004.

Zika, J.D. et al., 2018: Improved estimates of water cycle change from ocean salinity: the key role of ocean warming. *Environ. Res. Lett.*, **13**(7), 074036, doi:10.1088/1748-9326/aace42.

Zikra, M., S. Suntoyo and L. Lukijanto, 2015: Climate Change Impacts on Indonesian Coastal Areas. *Procedia Earth Planet. Sci.*, **14**, 57–63.

Zilberman, N., 2017: Deep Argo – Sampling the total ocean volume. *Bull. Am. Meteorol. Soc., State of the Climate in 2016 report*, **8**(98), 73–74.

Zinnert, J.C. et al., 2019: Connectivity in coastal systems: barrier island vegetation influences upland migration in a changing climate. *Global Change Biol.*, **25**(7), 2419-2430. doi:10.1111/gcb.14635.

Zougmoré, R. et al., 2016: Toward climate-smart agriculture in West Africa: a review of climate change impacts, adaptation strategies and policy developments for the livestock, fishery and crop production sectors. *Agriculture & Food Security*, **5**(1), 26, doi:10.1186/s40066-016-0075-3.

5

6

Extremes, Abrupt Changes and Managing Risks

Coordinating Lead Authors
Matthew Collins (UK), Michael Sutherland (Trinidad and Tobago)

Lead Authors
Laurens Bouwer (Netherlands), So-Min Cheong (Republic of Korea), Thomas Frölicher (Switzerland), Hélène Jacot Des Combes (Fiji), Mathew Koll Roxy (India), Iñigo Losada (Spain), Kathleen McInnes (Australia), Beate Ratter (Germany), Evelia Rivera-Arriaga (Mexico), Raden Dwi Susanto (Indonesia), Didier Swingedouw (France), Lourdes Tibig (Philippines)

Contributing Authors
Pepijn Bakker (Netherlands), C. Mark Eakin (USA), Kerry Emanuel (USA), Michael Grose (Australia), Mark Hemer (Australia), Laura Jackson (UK), Andreas Kääb (Norway), Jules Kajtar (UK), Thomas Knutson (USA), Charlotte Laufkötter (Switzerland), Ilan Noy (New Zealand), Mark Payne (Denmark), Roshanka Ranasinghe (Netherlands), Giovanni Sgubin (Italy), Mary-Louise Timmermans (USA)

Review Editors
Amjad Abdulla (Maldives), Marcelino Hernádez González (Cuba), Carol Turley (UK)

Chapter Scientist
Jules Kajtar (UK)

This chapter should be cited as:
Collins M., M. Sutherland, L. Bouwer, S.-M. Cheong, T. Frölicher, H. Jacot Des Combes, M. Koll Roxy, I. Losada, K. McInnes, B. Ratter, E. Rivera-Arriaga, R.D. Susanto, D. Swingedouw, and L. Tibig, 2019: Extremes, Abrupt Changes and Managing Risk. In: *IPCC Special Report on the Ocean and Cryosphere in a Changing Climate* [H.-O. Pörtner, D.C. Roberts, V. Masson-Delmotte, P. Zhai, M. Tignor, E. Poloczanska, K. Mintenbeck, A. Alegría, M. Nicolai, A. Okem, J. Petzold, B. Rama, N.M. Weyer (eds.)]. Cambridge University Press, Cambridge, UK and New York, NY, USA, pp. 589–655. https://doi.org/10.1017/9781009157964.008.

Table of contents

Executive Summary

This chapter assesses extremes and abrupt or irreversible changes in the ocean and cryosphere in a changing climate, to identify regional hot spots, cascading effects, their impacts on human and natural systems, and sustainable and resilient risk management strategies. It is not comprehensive in terms of the systems assessed and some information on extremes, abrupt and irreversible changes, in particular for the cryosphere, may be found in other chapters.

Ongoing and Emerging Changes in the Ocean and Cryosphere, and their Impacts on Ecosystems and Human Societies

Anthropogenic climate change has increased observed precipitation (*medium confidence*), winds (*low confidence*), and extreme sea level events (*high confidence*) associated with some tropical cyclones, which has increased intensity of multiple extreme events and associated cascading impacts (*high confidence*). Anthropogenic climate change may have contributed to a poleward migration of maximum tropical cyclone intensity in the western North Pacific in recent decades related to anthropogenically-forced tropical expansion (*low confidence*). There is emerging evidence for an increase in the annual global proportion of Category 4 or 5 tropical cyclones in recent decades (*low confidence*). {6.3, Table 6.2, Figure 6.2, Box 6.1}

Changes in Arctic sea ice have the potential to influence mid-latitude weather (*medium confidence*), but there is *low confidence* in the detection of this influence for specific weather types. {6.3}

Extreme wave heights, which contribute to extreme sea level events, coastal erosion and flooding, have increased in the Southern and North Atlantic Oceans by around 1.0 cm yr^{-1} and 0.8 cm yr^{-1} over the period 1985–2018 (*medium confidence*). Sea ice loss in the Arctic has also increased wave heights over the period 1992–2014 (*medium confidence*). {6.3}

Marine heatwaves (MHWs), periods of extremely high ocean temperatures, have negatively impacted marine organisms and ecosystems in all ocean basins over the last two decades, including critical foundation species such as corals, seagrasses and kelps (*very high confidence*). Globally, marine heat related events have increased; marine heatwaves, defined when the daily sea surface temperature exceeds the local 99th percentile over the period 1982 to 2016, have doubled in frequency and have become longer-lasting, more intense and more extensive (*very likely*). It is *very likely* that between 84–90% of marine heatwaves that occurred between 2006 and 2015 are attributable to the anthropogenic temperature increase. {6.4, Figures 6.3, 6.4}

Both palaeoclimate and modern observations suggest that the strongest El Niño and La Niña events since the pre-industrial period have occurred during the last fifty years (*medium confidence*). There have been three occurrences of extreme El Niño events during the modern observational period (1982–1983, 1997–1998, 2015–2016), all characterised by pronounced rainfall in the normally dry equatorial East Pacific. There have been two occurrences of extreme La Niña (1988–1989, 1998–1999). El Niño and La Niña variability during the last 50 years is unusually high compared with average variability during the last millennium. {6.5, Figure 6.5}

The equatorial Pacific trade wind system experienced an unprecedented intensification during 2001–2014, resulting in enhanced ocean heat transport from the Pacific to the Indian Ocean, influencing the rate of global temperature change (*medium confidence*). In the last two decades, total water transport from the Pacific to the Indian Ocean by the Indonesian Throughflow (ITF), and the Indian Ocean to Atlantic Ocean has increased (*high confidence*). Increased ITF has been linked to Pacific cooling trends and basin-wide warming trends in the Indian Ocean. Pacific sea surface temperature (SST) cooling trends and strengthened trade winds have been linked to an anomalously warm tropical Atlantic. {6.6, Figure 6.7}

Observations, both in situ (2004–2017) and based on sea surface temperature reconstructions, indicate that the Atlantic Meridional Overturning Circulation (AMOC) has weakened relative to 1850–1900 (*medium confidence*). There is insufficient data to quantify the magnitude of the weakening, or to properly attribute it to anthropogenic forcing due to the limited length of the observational record. Although attribution is currently not possible, CMIP5 model simulations of the period 1850–2015, on average, exhibit a weakening AMOC when driven by anthropogenic forcing. {6.7, Figure 6.8}

Climate change is modifying multiple types of climate-related events or hazards in terms of occurrence, intensity and periodicity. It increases the likelihood of compound hazards that comprise simultaneously or sequentially occurring events to cause extreme impacts in natural and human systems. Compound events in turn trigger cascading impacts (*high confidence*). Three case studies are presented in the chapter, (i) Tasmania's Summer of 2015–2016, (ii) The Coral Triangle and (ii) Hurricanes of 2017. {6.8, Box 6.1}

6

Projections of Ocean and Cryosphere Change and Hazards to Ecosystems and Human Society Under Low and High Emission Futures

The average intensity of tropical cyclones, the proportion of Category 4 and 5 tropical cyclones and the associated average precipitation rates are projected to increase for a 2°C global temperature rise above any baseline period (*medium confidence*). Rising mean sea levels will contribute to higher extreme sea levels associated with tropical cyclones (*very high confidence*). Coastal hazards will be exacerbated by an increase in the average intensity, magnitude of storm surge and precipitation rates of tropical cyclones. There are greater increases projected under RCP8.5 than under RCP2.6 from around mid-century to 2100 (*medium confidence*). There is *low confidence* in changes in the future frequency of tropical cyclones at the global scale. {6.3.1}

Significant wave heights (the average height from trough to crest of the highest one-third of waves) are projected to increase across the Southern Ocean and tropical eastern Pacific (*high confidence*) and Baltic Sea (*medium confidence*) and decrease over the North Atlantic and Mediterranean Sea under RCP8.5 (*high confidence*). Coastal tidal amplitudes and patterns are projected to change due to sea level rise and coastal adaptation measures (*very likely*). Projected changes in waves arising from changes in weather patterns, and changes in tides due to sea level rise, can locally enhance or ameliorate coastal hazards (*medium confidence*). {6.3.1, 5.2.2}

Marine heatwaves are projected to further increase in frequency, duration, spatial extent and intensity (maximum temperature) (*very high confidence*). Climate models project increases in the frequency of marine heatwaves by 2081–2100, relative to 1850–1900, by approximately 50 times under RCP8.5 and 20 times under RCP2.6 (*medium confidence*). The largest increases in frequency are projected for the Arctic and the tropical oceans (*medium confidence*). The intensity of marine heatwaves is projected to increase about 10-fold under RCP8.5 by 2081–2100, relative to 1850–1900 (*medium confidence*). {6.4}

Extreme El Niño and La Niña events are projected to *likely* increase in frequency in the 21st century and to *likely* intensify existing hazards, with drier or wetter responses in several regions across the globe. Extreme El Niño events are projected to occur about as twice as often under both RCP2.6 and RCP8.5 in the 21st century when compared to the 20th century (*medium confidence*). Projections indicate that extreme Indian Ocean Dipole events also increase in frequency (*low confidence*). {6.5; Figures 6.5, 6.6}

Lack of long-term sustained Indian and Pacific Ocean observations, and inadequacies in the ability of climate models to simulate the magnitude of trade wind decadal variability and the inter-ocean link, mean there is *low confidence* in future projections of the trade wind system. {6.6, Figure 6.7}

The AMOC will *very likely* weaken over the 21st century (*high confidence*), although a collapse is *very unlikely* (*medium confidence*). Nevertheless, a substantial weakening of the AMOC remains a physically plausible scenario. Such a weakening would strongly impact natural and human systems, leading to a decrease in marine productivity in the North Atlantic, more winter storms in Europe, a reduction in Sahelian and South Asian summer rainfall, a decrease in the number of TCs in the Atlantic, and an increase in regional sea level around the Atlantic especially along the northeast coast of North America (*medium confidence*). Such impacts would be superimposed on the global warming signal. {6.7, Figure 6.8}

Impacts from further changes in TCs and ETCs, MHWs, extreme El Niño and La Niña events and other extremes will exceed the limits of resilience and adaptation of ecosystems and people, leading to unavoidable loss and damage (*medium confidence*). {6.9.2}

Strengthening the Global Responses in the Context of Sustainable Development Goals (SDGs) and Charting Climate Resilient Development Pathways for Oceans and Cryosphere

There is *medium confidence* that including extremes and abrupt changes, such as AMOC weakening, ice sheet collapse (West Antarctic Ice Sheet (WAIS) and Greenland Ice Sheet (GIS)), leads to a several-fold increase in the cost of carbon emissions (*medium confidence*). If carbon emissions decline, the risk of extremes and abrupt changes are reduced, creating co-benefits. {6.8.6}

For TCs and ETCs, investment in disaster risk reduction, flood management (ecosystem and engineered) and early warning systems decreases economic loss (*medium confidence*), but such investments may be hindered by limited local capacities, such as increased losses and mortality from extreme winds and storm surges in less developed countries despite adaptation efforts. There is emerging evidence of increasing risks for locations impacted by unprecedented storm trajectories (*low confidence*). Managing the risk from such changing storm trajectories and intensity proves challenging because of the difficulties of early warning and its receptivity by the affected population (*high confidence*). {6.3, 6.9}

Limiting global warming would reduce the risk of impacts of MHWs, but critical thresholds for some ecosystems (e.g., kelp forests, coral reefs) will be reached at relatively low levels of future global warming (*high confidence*). Early warning systems, producing skillful forecasts of MHWs, can further help to reduce the vulnerability in the areas of fisheries, tourism and conservation, but are yet unproven at large scale (*medium confidence*). {6.4}

Sustained long-term monitoring and improved forecasts can be used in managing the risks of extreme El Niño and La Niña events associated with human health, agriculture, fisheries, coral reefs, aquaculture, wildfire, drought and flood management (*high confidence*). {6.5}

Extreme change in the trade wind system and its impacts on global variability, biogeochemistry, ecosystems and society have not been adequately understood and represent significant knowledge gaps. {6.6}

By 2300, an AMOC collapse is *as likely as not* for high emission pathways and *very unlikely* for lower ones, highlighting that an AMOC collapse can be avoided in the long term by CO_2 mitigation (*medium confidence*). Nevertheless, the human impact of these physical changes have not been sufficiently quantified and there are considerable knowledge gaps in adaptation responses to a substantial AMOC weakening. {6.7}

The ratio between risk reduction investment and reduction of damages of extreme events varies. Investing in preparation and prevention against the impacts from extreme events is *very likely* less than the cost of impacts and recovery (*medium confidence*). Coupling insurance mechanisms with risk reduction measures can enhance the cost-effectiveness of adapting to climate change (*medium confidence*). {6.9}

Climate change adaptation and disaster risk reduction require capacity building and an integrated approach to ensure trade-offs between short- and long-term gains in dealing with the uncertainty of increasing extreme events, abrupt changes and cascading impacts at different geographic scales (*high confidence*). {6.9}

Limiting the risk from the impact of extreme events and abrupt changes leads to successful adaptation to climate change with the presence of well-coordinated climate-affected sectors and disaster management relevant agencies (*high confidence*). Transformative governance inclusive of successful integration of disaster risk management (DRM) and climate change adaptation, empowerment of vulnerable groups, and accountability of governmental decisions promotes climate-resilient development pathways (*high confidence*). {6.9}

6.1   Introduction

This chapter assesses extremes and abrupt or irreversible changes in the ocean and cryosphere in a changing climate, to identify regional hot spots, cascading effects, their impacts on human and natural systems, and sustainable and resilient risk management strategies. While not comprehensive in terms of discussing all such phenomena, it addresses a number of issues that are prominent in both the policy area and in the scientific literature. Further information may also be found in Chapters 2 to 4 for other aspects of the ocean and cryosphere.

Building on the Special Report on Managing the Risks of Extreme Events and Disasters to Advance Climate Change Adaptation (SREX; IPCC, 2012), IPCC 5th Assessment Report (AR5; IPCC, 2013; IPCC, 2014) assessments and the Special Report on Global Warming of 1.5°C (SR15; IPCC, 2018), for each of the topics addressed, we provide an assessment of:

- Key processes and feedbacks, observations, detection and attribution, projections;
- Impacts on human and natural systems;
- Monitoring and early warning systems;
- Risk management and adaptation, sustainable and resilient pathways.

The chapter is organised in terms of the space- and time-scales of different phenomena. We move from small-scale TCs, which last for days to weeks, to the global-scale AMOC, which has time scales of decades to centuries. A common risk framework is adopted, based on that used in AR5 and introduced in Chapter 1, Section 1.5 and Cross-Chapter Box 1 in Chapter 1 (Figure 6.1).

While much of what is discussed within the chapter concerns the ocean, we also summarise abrupt events in the cryosphere in Section 6.2, drawing information from Chapters 2 to 4, where the main assessment of those phenomena may be found.

6.1.1   Definitions of Principal Terms

In discussing concepts such as abrupt changes, irreversibility, tipping points and extreme events it is important to define precisely what is meant by those terms. The following definitions are therefore adopted (based on either AR5, Special Report on Global Warming of 1.5°C (SR15) or Special Report on Climate Change and Land (SRCCL) Glossaries):

Abrupt climate change: A large-scale change in the climate system that takes place over a few decades or less, persists (or is anticipated to persist) for at least a few decades, and causes substantial disruptions in human and natural systems.

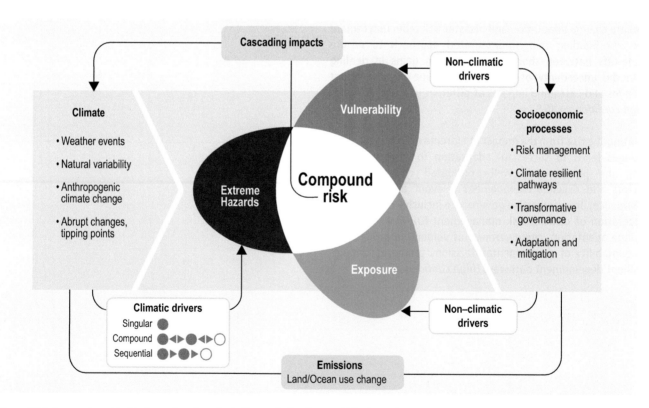

Figure 6.1 | Framework used in this chapter (see discussion in Chapter 1). Singular or multiple climate drivers can lead to extreme hazards and associated cascading impacts, which combined with non-climatic drivers affect exposure and vulnerability, leading to compound risks. Extremes discussed are tropical cyclones (TCs) and extratropical cyclones (ETCs) and associated sea surface dynamics (Section 6.3); marine heatwaves (MHWs) (Section 6.4), extreme El Niño and La Niña events (Section 6.5); and extreme oceanic decadal variability (Section 6.6). Examples of abrupt events, irreversibility and tipping points discussed are the Atlantic Meridional Overturning Circulation (AMOC) and subpolar gyre (SPG) system (Section 6.7). Section 6.2 also collects examples of such events from the rest of the Special Report on the Oceans and Cryosphere in a Changing Climate (SROCC) and compiles examples of events whose occurrence or severity has been linked to climate change. Cascading impacts and compound events are discussed in Section 6.8 and three examples are given in Box 6.1. Section 6.9 discusses risk management, climate resilience pathways, transformative governance adaptation and mitigation required to address societal and environmental risks.

Extreme weather/climate event: An extreme event is an event that is rare at a particular place and time of year. Definitions of 'rare' vary, but an extreme event would normally be as rare as or rarer than the 10th or 90th percentile of a probability density function estimated from observations. By definition, the characteristics of what is called an extreme event may vary from place to place in an absolute sense. When a pattern of extreme weather persists for some time, such as a season, it may be classed as an extreme climate event, especially if it yields an average or total that is itself extreme (e.g., high temperature, drought, or total rainfall over a season).

Irreversibility: A perturbed state of a dynamical system is defined as irreversible on a given timescale, if the recovery timescale from this state due to natural processes is significantly longer than the time it takes for the system to reach this perturbed state. In the context of this report, the recovery time scale of interest is hundreds to thousands of years.

Tipping point: A level of change in system properties beyond which a system reorganises, often in a nonlinear manner, and does not return to the initial state even if the drivers of the change are abated. For the climate system, the term refers to a critical threshold when global or regional climate changes from one stable state to another stable state. Tipping points are also used when referring to impact; the term can imply that an impact tipping point is (about to be) reached in a natural or human system.

These above four terms generally refer to aspects of the physical climate system. Here we extend their definitions to natural and human systems. For example, there may be gradual physical climate change which causes an irreversible change in an ecosystem. An adaptation tipping point could be reached when an adaptation option no longer remains effective. There may be a tipping point within a governance structure.

We also introduce two new key terms relevant for discussing risk-related concepts:

Compound events refer to the combination of multiple drivers and/or hazards that contribute to societal or environmental risks.

Cascading impacts from extreme weather/climate events occur when an extreme hazard generates a sequence of secondary events in natural and human systems that result in physical, natural, social or economic disruption, whereby the resulting impact is significantly larger than the initial impact. Cascading impacts are complex and multi-dimensional, and are associated more with the magnitude of vulnerability than with that of the hazard.

6.2   Climate Change influences on Abrupt Changes, Irreversibility, Tipping Points and Extreme Events

6.2.1   Introduction

Some potentially abrupt or irreversible events are assessed in other chapters, hence Table 6.1 presents a cross-chapter summary of those. Subsection numbers indicate where detailed information may be found.

Table 6.1 | Cross-Chapter assessment of abrupt and irreversible phenomena related to the ocean and cryosphere. The column on the far right of the table indicates the likelihood of an abrupt/irreversible change based on the assessed literature which, in general, assesses Representative Concentration Pathway (RCP) scenarios. Assessments of likelihood and confidence are made according to IPCC guidance on uncertainties.

Change in system component	Potentially abrupt	Irreversibility if forcing reversed (time scales indicated)	Impacts on natural and human systems; global vs. regional vs. local	Projected likelihood and/or confidence level in 21st century under scenarios considered
Ocean				
Atlantic Meridional Overturning Circulation (AMOC) collapse (Section 6.7)	Yes	Unknown	Widespread; increased winter storms in Europe, reduced Sahelian rainfall and agricultural capacity, variations in tropical storms, increased sea levels on Atlantic coasts	*Very unlikely*, but physically plausible
Subpolar gyre (SPG) cooling (Section 6.7)	Yes	Irreversible within decades	Similar to AMOC impacts but considerably smaller	*Medium confidence*
Marine heatwave (MHW) increase (Section 6.4)	Yes	Reversible within decades to centuries	Coral bleaching, loss of biodiversity and ecosystem services, harmful algal blooms, species redistribution	*Very likely* (*very high confidence*) for physical change *High confidence* for impacts
Arctic sea ice retreat (Section 3.3)	Yes	Reversible within decades to centuries	Coastal erosion in Arctic (may take longer to reverse), impact on mid-latitude storms (*low confidence*); rise in Arctic surface temperatures (*high confidence*)	*High confidence*
Ocean deoxygenation and hypoxic events (Section 5.2)	Yes	Reversible at surface, but irreversible for centuries to millennia at depth	Major changes in ocean productivity, biodiversity and biogeochemical cycles	*Medium confidence*
Ocean acidification (Section 5.2)	Yes	Reversible at surface, but irreversible for centuries to millennia at depth	Changes in growth, development, calcification, survival and abundance of species, for example, from algae to fish	*Virtually certain* (*very high confidence*)

6

Change in system component	Potentially abrupt	Irreversibility if forcing reversed (time scales indicated)	Impacts on natural and human systems; global *vs.* regional *vs.* local	Projected likelihood and/or confidence level in 21st century under scenarios considered
Cryosphere				
Methane release from permafrost (Section 3.4)	Yes	Reversible due to short lifetime of methane in the atmosphere	Further increased global temperatures through climate feedback	*Medium confidence*
CO_2 release from permafrost (Section 3.4)	Yes	Irreversible for millennia due to long lifetime of CO_2 in the atmosphere	Further increased global temperatures through climate feedback	*Low confidence*
Partial West Antarctic Ice Sheet (WAIS) collapse (Cross Chapter Box 2 in Chapter 1, Section 4.2)	Yes (late 21st century, under RCP8.5 only)	Irreversible for decades to millennia	Significant contribution to sea level rise (SLR) and local decrease in ocean salinity	*Low confidence*
Greenland Ice sheet (GIS) decay (Cross Chapter Box 8, Section 4.2)	No	Irreversible for millennia	Significant contribution to SLR, shipping (icebergs)	*High confidence* for decay contributing 10s of cm of SLR
Ice-shelf collapses (Cross Chapter Box 8, Sections 3.3, 4.2)	Yes	Possibly irreversible for centuries	May lead to SLR from contributing glaciers; some shelves more prone than others	*Low confidence*
Glacier avalanches, surges, and collapses (Section 2.3)	Yes	Variable	Local hazard; may accelerate SLR; local iceberg production; local ecosystems	*Medium confidence* for occurrence *Low confidence* for increase in frequency/magnitude
Strong shrinkage or disappearance of individual glaciers (Sections 2.2, 3.3)	Yes	Reversible within decades to centuries	Regional impact on water resources, tourism, ecosystems and global sea level	*Medium confidence*
Landslides related to glaciers and permafrost, glacier lake outbursts (Section 2.3)	Yes	Irreversible for rock slopes; reversible within decades to centuries for glaciers, debris and lakes	Local direct impact on humans, land use, infrastructure (hazard), and ecosystems	*Medium confidence for increase in frequency*
Change in biodiversity in high mountain areas (impact – Section 2.4)	Yes	In many cases irreversible (e.g., extinction of species)	Local impacts on ecosystems and ecosystem services	*Medium confidence*

6.2.2 Recent Anomalous Extreme Climate Events and their Causes

The attribution of changes in the observed statistics of extremes are generally addressed using well-established detection-attribution methods. In contrast, record-breaking weather and climate events are by definition unique, and can be expected to occur with or without climate change as the observed record lengthens. Therefore, event attribution begins with the premise that the climate is changing, the goal being to determine statistically how much climate change has contributed to the severity of the event (Trenberth et al. 2015; Shepherd, 2016). Annual reports dedicated to extreme event attribution (Peterson et al. 2012; Peterson et al. 2013; Herring et al. 2014; Herring et al. 2015; Herring et al. 2018) have helped stimulate studies that adopt recognised methods for extreme event attribution. The increasing pool of studies allows different approaches

to be contrasted and builds consensus on the role of climate change when individual climate events are studied by multiple teams using different methods. A number of these events are summarised in Table 6.2 and Figure 6.2. Collectively, these studies show that the role of climate change in the ocean and cryosphere extreme events is increasingly driving extreme climate and weather events across the globe including compound events (*high confidence*). Some regions including Africa and the Pacific have had relatively fewer event attribution studies undertaken, possibly reflecting the lack of capacity by regional and national technical institutions. A caveat of this approach is that there is a potential for 'null results', that is, cases where attribution is not possible, to be reported. Nevertheless, there is no evidence that this is the case, and the number of recent studies and wide range of phenomena addressed suggests increasing influence of climate change on extreme events.

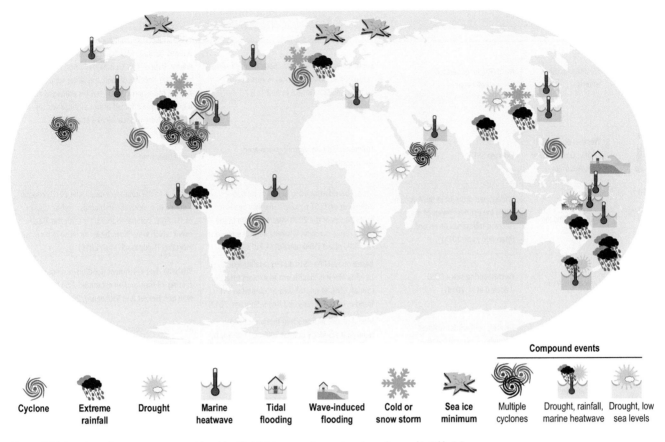

Cyclone | Extreme rainfall | Drought | Marine heatwave | Tidal flooding | Wave-induced flooding | Cold or snow storm | Sea ice minimum | **Compound events** Multiple cyclones | Drought, rainfall, marine heatwave | Drought, low sea levels

Figure 6.2 | Locations where extreme events with an identified link to ocean changes have been discussed in Table 6.2.

Table 6.2 | A selection of extreme events with links to oceans and cryosphere. In many of these studies the method of event attribution has been used to estimate the role of climate change using either a probabilistic approach (using ensembles of climate models to assess how much more likely the event has become with anthropogenic climate change compared to a world without) or a storyline approach which examines the components of the climate system that contribute to the events and how changes in the climate system affect them (Shepherd, 2016).

Year/type of hazard	Region	Severe hazard	Attribution to anthropogenic climate change	Impact, costs
1998	Western equatorial Pacific, Great Barrier Reef, Australia	Extreme sea surface temperatures (SSTs)	Unknown if global warming has increased the probability	Coral bleaching
2003	Mediterranean Sea	June to August with sea water temperatures 1°C–3°C above climatological mean (Olita et al. 2007; Garrabou et al. 2009; Galli et al. 2017)	Increase in air temperature and a reduction of wind stress and air-sea exchanges (Olita et al. 2007). Unknown if global warming has increased the probability	Mass mortality of macro-invertebrate species; amplified heatwave over central Europe in 2003
2004	South Atlantic	First hurricane in the South Atlantic since 1970	Increasing trend to positive Southern Annular Mode (SAM) could favour the synoptic conditions for such events in the future (Pezza and Simmonds, 2005)	Three deaths, 425 million USD damage (McTaggart-Cowan et al. 2006)
2005	North Atlantic	Record number of tropical storms, hurricanes and Category 5 hurricanes since 1970	Trend in SST due to global warming contributed to half of the total SST anomaly. Atlantic Multi-decadal Variability (AMV) and the after-effects of the 2004–2005 El Niño also played a role (Trenberth and Shea, 2006)	Costliest US natural disaster; 1,836 deaths and 30 billion USD in direct economic costs in Louisiana due to Hurricane Katrina (Link, 2010)
2007	Arabian Sea	Strongest tropical cyclone (TC) (Gonu) attaining sustained winds of 270 kph and gustiness of 315 kph	No attribution study done, although it was noted that this Category 5 TC had followed an unusual path (Dibajnia et al. 2010)	Caused around 4 billion USD in damages (Fritz et al. 2010; Coles et al. 2015)

6

Year/type of hazard	Region	Severe hazard	Attribution to anthropogenic climate change	Impact, costs
2008	Western Pacific Islands	North Pacific generated wave-swell event	Event shown to have been made more extreme compared to other historical events due to La Niña and SLR (Hoeke et al. 2013)	Wave-induced inundation in islands of six Pacific nations (Kiribati, Marshall Islands, Micronesia, Nauru, Papua New Guinea, Solomon Islands), salt water flooding of food and water supplies in Kosrae, Micronesia, 1,408 houses damaged and 63,000 people affected across eight provinces in Papua New Guinea (Hoeke et al. 2013)
2010	Western equatorial Pacific, Great Barrier Reef, Australia	Extreme SST	Unknown if global warming increased the probability	Coral bleaching
2010	Southern Amazon	Widespread drought in the Amazon led to lowest river levels of major Amazon tributaries on record (Marengo et al. 2011)	Model-based attribution indicates human influences and SST natural variability increased probabilities of the 2010 severe drought in the South Amazon region whereas aerosol emissions had little effect (Shiogama et al. 2013)	Relative to the long-term mean, the 2010 drought resulted in a reduction in biomass carbon uptake of 1.1 Pg C, compared to 1.6 Pg C for the 2005 event which was driven by an increase in biomass mortality (Feldpausch et al. 2016)
2010–2011	Eastern Australia	Wettest spring since 1900 (Leonard et al. 2014)	Based on La Niña SSTs during satellite era, La Niña alone is insufficient to explain total rainfall. 25% of rainfall was attributed to SST trend in region (Evans and Boyer-Souchet, 2012)	Brisbane river catchment flooding in January 2011, costing 23 lives and an estimated 2.55 billion USD (van den Honert and McAneney, 2011)
2010–2011	UK	Severely cold winter (coldest December since 1910 and second coldest since 1659)	Model results indicate that human influence reduced the odds by at least 20% and possibly by as much as 4 times with a best estimate that the odds have been halved (Christidis and Stott, 2012)	Many adverse consequences of the extreme temperatures, including closed schools and airports (Christidis and Stott, 2012)
2011	Western North Pacific	Tropical Storm Washi (also known as TS Sendong) was world's deadliest storm in 2011	No attribution done; disaster was the outcome of interplay of climatic, environmental and social factors (Espinueva et al. 2012)	Fatalities: >1,250; injured: 2,002; missing: 1,049 (Rasquinho et al. 2013) Socioeconomic costs: 63.3 million USD (Espinueva et al. 2012)
2011	Western Australia	Most extreme warming event in the region in the last 140 years during which sea temperature anomalies of 2°C–4°C persisted for more than 10 weeks along >2,000 km of coastline from Ningaloo (22°S) to Cape Leeuwin (34°S); up to 5°C warmer SSTs than normal (Feng et al. 2013; Pearce and Feng, 2013; Benthuysen et al. 2014; Caputi et al. 2016; Perkins-Kirkpatrick et al. 2016)	Warming of poleward-flowing Leeuwin Current in Austral summer forced by oceanic and atmospheric teleconnections associated with the 2010–2011 La Niña (Feng et al. 2013). Conditions increased since 1970's by negative Interdecadal Pacific Oscillation (IPO) and anthropogenic global warming (Feng et al. 2015). Shift of temperate marine ecosystem was climate-driven	Widespread coral bleaching and fish kills. Biodiversity patterns of temperate seaweeds, sessile invertebrates and demersal fish were altered leading to reduced abundance of habitat-forming seaweeds (Wernberg et al. 2013)
2011	Golden Bay, New Zealand	In December, Extreme two day total rainfall was experienced (one in 500-year event)	Model based attribution indicated total moisture available for precipitation in Golden Bay, New Zealand was 1–5% higher due to anthropogenic emissions (Dean et al. 2013)	In town of Takaka, 453 mm was recorded in 24 hours and 674 mm in 48 hours (Dean et al. 2013)
2012	Arctic	Arctic sea ice minimum	Model-based attribution indicated the exceptional 2012 sea ice loss was due to sea ice memory and positive feedback of warm atmospheric conditions, both contributing approximately equally (Guemas et al. 2013) and extremely unlikely to have occurred due to internal climate variability alone based on observations and model-based attribution (Zhang and Knutson, 2013)	Up to 60% higher contribution of sea ice algae in the central Arctic (Fernández-Méndez et al. 2015; see also chapter 3.2.3)
2012	US East coast	Hurricane Sandy	Relative SLR shown to have increased probabilities of exceeding peak impact elevations since the mid-20th century (Sweet et al. 2013; Lackmann, 2015)	Repair and mitigation expenditures funded at 60.2 billion USD. Losses of fishing vessels estimated at 52 million USD (Sainsbury et al. 2018)
2012	Northwest Atlantic	First half of 2012, record-breaking SSTs (1°C–3°C above normal) from the Gulf of Maine to Cape Hatteras (Mills et al. 2013; Chen et al. 2014; Pershing et al. 2015; Zhou et al. 2015)	Local warming from the atmosphere due to anomalous atmospheric jet stream position (Chen et al. 2014). Unknown if global warming increased the probability	Northward movement of warm water species and local migrations of lobsters earlier in the season (Mills et al. 2013; Pershing et al. 2015)

Year/type of hazard	Region	Severe hazard	Attribution to anthropogenic climate change	Impact, costs
2013	UK	Extreme winter rainfall	Some evidence for a human-induced increase in extreme winter rainfall in the UK for events with time scales of 10 days (Christidis and Stott, 2015)	Tidal surges, widespread floodplain inundation, and pronounced river flows leading to damages in transport infrastructure, business and residential properties and a cost of 560 million GBP in recovery schemes (Department for Communities and Local Government, 2014; Huntingford et al. 2014). Unprecedented deaths of over 4,400 Puffins found on UK and Scottish coasts linked to cold and strong winds during this event (Harris and Elkins, 2013)
2013	Western North Pacific	Strongest and fastest Super Typhoon Haiyan (Category 5) in the region	Occurred in a season with remarkably warm SSTs, (David et al. 2013; Takagi and Esteban, 2016). Ocean heat content and sea levels had increased since 1998 due to the negative Pacific Decadal Oscillation (PDO) phase but impacts were worsened by thermodynamic effects on SSTs, SLR and storm surges due to climate change (Trenberth et al. 2015)	Deadliest and most expensive natural disaster in the Philippines (Fatalities: 6,245; Injured: 28,626; Missing: 1,039). Damage to mangroves was still apparent 18 months after the storm (Sainsbury et al. 2018)
2013–2015	Northeast Pacific Ocean	Largest heatwave ever recorded (often called 'The Blob'; Bond et al. 2015), with maximum SST anomalies of 6°C off Southern California (Jacox et al. 2016; Gentemann et al. 2017; Rudnick et al. 2017) and subsurface warm anomalies in the deep British Columbia Fjord that persisted through the beginning of 2018 (Jackson and Wood, 2018)	Emerged in 2013 in response to teleconnections between North Pacific and the weak El Niño that drove strong positive sea level pressure anomalies across the northeast Pacific inducing smaller heat loss (Bond et al. 2015; Di Lorenzo and Mantua, 2016). Global warming increased the probability of occurrence for regional parts of the MHW (Weller et al. 2015; Jacox et al. 2018; Newman et al. 2018)	Major impacts on entire marine food web. Caused a major outbreak of a toxic algal bloom along the US West Coast leading to impacts on fisheries (McCabe et al. 2016) . Increased mortality of sea birds (Jones et al. 2018). Contributed to drought conditions across the US West Coast
2014	Hawaiian hurricane season	Extremely active hurricane season in the eastern and central Pacific Ocean, particularly around Hawaii	Anthropogenic forcing could have contributed to the unusually large number of hurricanes in Hawaii in 2015, in combination with the moderately favourable El Niño event conditions (Murakami et al. 2015)	Acute disturbance of coral along Wai'ōpae coastline (southeastern tip of Hawai'i Island) due to passages of Hurricanes Iselle, Julio and Ana that caused high waves, increased runoff and elevated SSTs associated with the 2014–2015 El Niño (Burns et al. 2016)
2014	Arabian Sea	Cyclone Nilofar was the first severe TC to be recorded in the Arabian Sea in post-monsoon cyclone season (Murakami et al. 2017)	Anthropogenic global warming has been shown to have increased the probability of post-monsoon TCs over the Arabian Sea (Murakami et al. 2017)	Cyclone did not make landfall but produced heavy rainfall on western Indian coasts (Bhutto et al. 2017)
2014	Northland New Zealand	Extreme five day rainfall in Northland	Extreme five day rainfall over Northland, New Zealand was influenced by human-induced climate change (Rosier et al. 2015)	18.8 million NZD in insurance claims (Rosier et al. 2015)
2014–2017	Western equatorial Pacific, Great Barrier Reef, Australia	Extreme SSTs	Global warming increased probability of occurrence for regional parts of the MHW (Weller et al. 2015; Oliver et al. 2018b)	Anthropogenic greenhouse gas (GHG) emission increased the risk of coral bleaching through anomalously high SSTs and accumulation of heat stress (Lewis and Mallela, 2018)
2015	North America	Anomalously low temperatures with intense snowstorms	Reduced Arctic sea ice and anomalous SSTs may have contributed to establishing and sustaining the anomalous meander of the jet stream, and could enhance the probability of such extreme cold spells over North America (Bellprat et al. 2016)	Several intense snowstorms resulting in power outages and large economic losses (Munich RE, 2016)
2015	Arctic	Record low Northern Hemisphere (NH) sea ice extent in March 2015	Record low in NH sea ice maximum could not have been reached without human-induced change in climate, with the surface atmospheric conditions, on average, contributing 54% to the change (Fuckar et al. 2016)	March NH sea ice content reached the lowest winter maximum in 2015. Emerging evidence of increased snow fall over regions outside the Arctic (see 3.4.1.1) due to sea ice reduction as well as changes in the timing, duration and intensity of primary production, which affect secondary production (3.2.3.1)
2015	Florida	Sixth largest flood in Virginia Key, Florida since 1994, with the fifth highest in response to hurricanes	The probability of a 0.57 m flood has increased by 500% (Sweet et al. 2016)	Flooding in several Miami-region communities with 0.57 m of ocean water on a sunny day

6

Year/type of hazard	Region	Severe hazard	Attribution to anthropogenic climate change	Impact, costs
2015–2016	Ethiopia and Southern Africa	One of the worst droughts in 50 years, also intensified flash droughts characterised by severe heatwaves	Anthropogenic warming contributed substantially to the very warm 2015–2016 El Niño SSTs, land local air temperatures thereby reducing Northern Ethiopia and Southern Africa rainfall and runoff (Funk et al. 2018; Yuan et al. 2018)	A 9 million tonne cereal deficit resulted in more than 28 million people in need of humanitarian aid (Funk et al. 2018)
2015	Eastern North Pacific	TC Patricia, the most intense and rapidly intensifying storm in the Western Hemisphere (estimated mean sea level (MSL) pressure of 872 hPa (Rogers et al. 2017), intensified rapidly into a Category 5 TC (Diamond and Schreck, 2016)	A near-record El Niño combined with a positive Pacific Meridional Mode provided extreme record SSTs and low vertical wind shear that fuelled the 2015 eastern North Pacific hurricane season to near-record levels (Collins et al. 2016)	Approximately 9,000 homes and agricultural croplands, including banana crops, were damaged by wind and rain from Patricia that made landfall near Jalisco, Mexico (Diamond and Schreck, 2016)
2015	Arabian Sea, Somalia and Yemen	Cyclones Chapala and Megh occurred within a week of each other and both tracked westward across Socotra Island and Yemen. Rainfall from Chapala was seven times the annual average	Anthropogenic global warming has been shown to have increased the probability of post-monsoon TCs over the Arabian Sea (Murakami et al. 2017)	Death toll in Yemen from Chapala and Megh was 8 and 20 respectively. Thousands of houses and businesses damaged or destroyed by both cyclones and fishing disrupted. The coastal town of Al Mukalla experienced a 10 m storm surge that destroyed the seafront (Kruk, 2016). Flooding in Somalia led to thousands of livestock killed and damage to infrastructure (IFRC, 2016)
2015–2016	Northern Australia (Gulf of Carpentaria)	High temperatures, low rainfall, extended drought period and low sea levels	Attributed to anomalously high temperatures and low rainfall and low sea levels associated with El Niño (Duke et al. 2017)	1,000 km of mangrove tidal wetland dieback (>74,000 ha). with potential flow-on consequences to Gulf of Carpentaria fishing industry worth 30 million AUS yr^{-1} due to loss of recruitment habitat
2015–2016	Tasman Sea	MHW lasted for 251 days with maximum SSTs of 2.9°C above the 1982–2005 average (Oliver et al. 2017)	Enhanced southward transport in the East Australian current driven by increased wind stress (Oliver et al. 2017). The intensity and duration of the MHW were unprecedented and both had a clear human signature (Oliver et al. 2017)	Disease outbreaks in farmed shellfish, mortality in wild shellfish and species found further south than previously recorded. Drought followed by severe rainfall caused severe bushfires and flooding in northeast Tasmania (see Box 6.1)
2016	Arctic	Record high air temperatures and record low sea ice were observed in the Arctic winter/spring of 2016 (Petty et al. 2017)	Would not have been possible without anthropogenic forcing (Kam et al. 2018), however the relative role of preconditioning, seasonal atmospheric/ocean forcing and storm activity in determining the evolution of the Arctic sea ice cover is still highly uncertain (Petty et al. 2018)	Impacts on Arctic ecosystems (e.g., Post et al. 2013; Meier et al. 2014), potential changes to mid-latitude weather (e.g., Cohen et al. 2014; Francis and Skific, 2015; Screen et al. 2015) and human activities in the Arctic
2016	Bering Sea/Gulf of Alaska	Record-setting warming with peak SSTs of 6°C above the 1981–2010 climatology (Walsh et al. 2017; Walsh et al. 2018)	Nearly fully attributed to human-induced climate change (Oliver et al. 2018b; Walsh et al. 2018)	Impacts on marine ecosystems in Alaska, included favouring some phytoplankton species, but resulted in one of the largest harmful algal blooms on record which reached the Alaska coast in 2015 (Peterson et al. 2017), uncommon paralytic shellfish poisoning events in Kachemak Bay and oyster farm closures in 2015 and 2016, dramatic mortality events in seabird species such as common murres in 2015–2016 (Walsh et al. 2018)
2016	East China Sea	MHW	Warming predominantly attributable to combined effects of oceanic advection (-0.18°C, 24%) and net heat flux (-0.44°C, 58%; Tan and Cai, 2018)	Impacts on marine organisms (Kim and Han, 2017)
2016	Eastern China	Super cold surge	This cold surge would have been stronger if there was no anthropogenic warming (Qian et al. 2018; Sun and Miao, 2018)	Extreme weather brought by the cold surge caused significant impacts on >1 billion people in China in terms of transportation and electricity transmission systems, agriculture and human health (Qian et al. 2018)
2016	Antarctic	Antarctic sea ice extent decreased at a record rate 46% faster than the mean rate and 18% faster than any spring rate in the satellite era producing a record minimum for the satellite period (1979–2016) (Turner et al. 2017)	Largely attributable to thermodynamic surface forcing (53%), while wind stress and the sea ice and oceanic conditions from the previous summer (January 2016) explain the remaining 34% and 13%, respectively (Kusahara et al. 2018) linked with a shift to positive phase of PDO and negative SAM in late 2016 (Meehl et al. 2019; see also 3.2.1.1)	Potential impacts on ecosystems and fisheries are poorly known (chapter 3.6)

6

Year/type of hazard	Region	Severe hazard	Attribution to anthropogenic climate change	Impact, costs
2017	Yellow Sea/ East China Sea	SSTs 2°C–7°C higher than normal (Kim and Han, 2017; Tan and Cai, 2018)	Unknown if global warming has increased the probability	Impacts on marine organisms
2017	Western North Atlantic	Hurricanes Harvey, Irma and Maria	Rainfall intensity in Harvey attributed to climate change and winds for Irma and Maria attributed to climate change. (Emanuel, 2017; Risser and Wehner, 2017; van Oldenborgh et al. 2017; see Box 6.1)	Extensive impacts (see Box 6.1)
2017	Europe	Storm Ophelia	In agreement with projections of increase of cyclones of tropical origin hitting European coasts (Haarsma et al. 2013)	Largest ever recorded hurricane in East Atlantic; extreme winds and coastal erosion in Ireland
2017	Persian Gulf	Severe warming in the Gulf with reef bottom temperatures resulting in 5.5°C-weeks of thermal stress as degree heating weeks (Burt et al. 2019)	Mortality of corals shown to have been caused by increases in sea-bottom temperatures (Burt et al. 2019)	94.3% of corals bleached in the Gulf
2017	East Africa	Drought (across Tanzania, Ethiopia, Kenya and Somalia)	Extremely warm 'Western V' (stretching poleward and eastward from a point near the Maritime Continent) SST doubled the probability of drought (Funk et al. 2018)	Contributed to extreme food insecurity (Funk et al. 2018) approaching near-famine conditions (FEWS NET and FSNAU, 2017; WFP et al. 2017)
2017	Peru	Extremely wet rainy season	Human influence is estimated to make such events at least 1.5 times more likely (Christidis et al. 2018a)	Widespread flooding and landslides 1.7 million people, a death toll of 177 and an estimated damages of 3.1 billion USD (Christidis et al. 2018a)
2017	Bangladesh	Pre-monsoon extreme six day rainfall event	The likelihood of this 2017 pre-monsoon extreme rainfall is nearly doubled by anthropogenic climate change; although this contribution is sensitive to the climatological period used (Rimi et al. 2018)	Triggered flash floods affecting 850,000 households and 220,000 hectares of harvestable crops leading to a 30% rice price hike (FAO, 2017)
2017	Uruguay, South America	April-May heavy precipitation	The risk of the extreme rainfall in the Uruguay River increased two-fold by anthropogenic climate change	Triggered wide-spread overbank flooding along the Uruguay River causing economic loss of 102 million USD (FAMURS, 2017) and displacement of 3,500 people (de Abreu et al. 2019)
2017	Northeast China	Persistent summer-spring hot and dry extremes	Risk of persistent spring-summer hot and dry extremes is increased by 5–55% and 37–113%, respectively, by anthropogenic climate change (Wang et al. 2018)	Affected more than 7.4 million km² of crops and herbage and direct economic loss of about 10 billion USD (Zhang et al. 2017c)
2017	Coastal Peru	Strong shallow ocean warming of up to 10°C off the northern coast of Peru	Unknown if global warming increased the probability	Caused heavy rainfall and flooding (ENFEN, 2017; Garreaud, 2018). Affected anchovies (decreased fat content and early spawning as a reproductive strategy; IMARPE, 2017)
2017	Southwestern Atlantic	SSTs were 1.7°C higher than previous maximum from February to March 2017 between 32°S–38°S (Manta et al. 2018)	High air temperature and low wind speed led to MHW. Unknown if global warming increased the probability	Fish species mass mortalities

6.3 Changes in Tracks, Intensity, and Frequency of Tropical and Extratropical Cyclones and Associated Sea Surface Dynamics

This section addresses new literature on TCs and ETCs and their effects on the ocean in the context of understanding how the changing nature of extreme events can cause compound hazards, risk and cascading impacts (discussed in Section 6.8). These topics are also discussed in Chapter 4 in the context of changes to ESLs (see Section 4.2.3.4).

6.3.1 Changes in Storms and Associated Sea Surface Dynamics

6.3.1.1 Tropical Cyclones

IPCC AR5 concluded that there was *low confidence* in any long-term increases in TC activity globally and in attribution of global changes to any particular cause (Bindoff et al. 2013; Hartmann et al. 2013). Based on process understanding and agreement in 21st century projections, it is *likely* that the global TC frequency will either decrease or remain essentially unchanged, while global mean

6

601

TC maximum wind speed and precipitation rates will *likely* increase although there is *low confidence* in region-specific projections of frequency and intensity (Christensen et al. 2013). The AR5 concluded that circulation features have moved poleward since the 1970s, associated with a widening of the tropical belt, a poleward shift of storm tracks and jet streams, and contractions of the northern polar vortex and the Southern Ocean westerly wind belts. However it is noted that natural modes of variability on interannual to decadal time scales prevent the detection of a clear climate change signal (Hartmann et al. 2013).

Since the AR5 and Knutson et al. (2010), palaeoclimatic surveys of coastal overwash sediments and stalagmites have provided further evidence of historical TC variability over the past several millennia. Patterns of storm activity across TC basins show variations through time that appear to be correlated with El Niño-Southern Oscillation (ENSO), North Atlantic Oscillation (NAO), and changes in atmospheric dynamics related to changes in precession of the sun (Toomey et al. 2013; Denommee et al. 2014; Denniston et al. 2015).

Further studies have investigated the dynamics of TCs. A modelling study investigated a series of low-frequency increases and decreases in TC activity over the North Atlantic over the 20th century (Dunstone et al. 2013). These variations, culminating in a recent rise in activity, are thought to be due in part to atmospheric aerosol forcing variations (aerosol forcing), which exerts a cooling effect (Booth et al. 2012; Dunstone et al. 2013). However, the relative importance of internal variability vs. radiative forcing for multidecadal variability in the Atlantic basin, including TC variability, remains uncertain (Weinkle et al. 2012; Zhang et al. 2013; Vecchi et al. 2017; Yan et al., 2017). Although the aerosol cooling effect has largely cancelled the increases in potential intensity over the observational period, according to Coupled Model Intercomparison Project Phase 5 (CMIP5) model historical runs, further anthropogenic warming in the future is expected to dominate the aerosol cooling effect leading to increasing TC intensities (Sobel et al. 2016).

TCs amplify wave heights along the tracks of rapidly moving cyclones (e.g., Moon et al. 2015a) and can therefore increase mixing to the surface of cooler subsurface water. Several studies found that TCs reduce the projected thermal stratification of the upper ocean in CMIP5 models under global warming, thereby slightly offsetting the simulated TC-intensity increases under climate warming conditions (Emanuel, 2015; Huang et al. 2015b; Tuleya et al. 2016). On the other hand, freshening of the upper ocean by TC rainfall enhances density stratification by reducing near-surface salinity and this reduces the ability of TC's to cool the upper ocean, thereby having an influence opposite to the thermal stratification effect (Balaguru et al. 2015). In the late 21st century, increased salinity stratification was found to offset about 50% of the suppressive effects that TC mixing has on temperature stratification (Balaguru et al. 2015). Coupled ocean-atmosphere models still robustly project an increase of TC intensity with climate warming, and particularly for new TC-permitting coupled climate model simulations that compute internally consistent estimates of thermal stratification change (e.g., Kim et al. 2014a; Bhatia et al. 2018). Higher TC intensities in turn may further aggravate the impacts of SLR on TC-related coastal inundation extremes (Timmermans et al. 2017).

Kossin et al. (2014) identified a poleward expansion of the latitudes of maximum TC intensity in recent decades, which has been linked to an anthropogenically-forced tropical expansion (Sharmila and Walsh, 2018) and a continued poleward shift of cyclones projected over the western North Pacific in a warmer climate (Kossin et al. 2016). A 10% slowdown in translation speed of TCs over the 1949–2016 period has been linked to the weakening of the tropical summertime circulation associated with tropical expansion and a more pronounced slowdown in the range 16–22% was found over land areas affected by TCs in the western North Pacific, North Atlantic and Australian regions (Kossin, 2018). Slow-moving TCs together with higher moisture carrying capacity can cause significantly greater flood hazards (Emanuel, 2017; Risser and Wehner, 2017; van Oldenborgh et al. 2017; see also Table 6.2 and Box 6.1).

Trends in TCs over decades to a century or more have been investigated in several new studies. Key findings include: i) decreasing frequency of severe TCs that make landfall in eastern Australia since the late 1800s (Callaghan and Power, 2011); ii) increase in frequency of moderately large US storm surge events since 1923 (Grinsted et al. 2012); iii) recent increase of extremely severe cyclonic storms over the Arabian Sea in the post-monsoon season (Murakami et al. 2017); iv) intense TCs that make landfall in East and Southeast Asia in recent decades (Mei and Xie, 2016; Li et al. 2017); and v) an increase in annual global proportion of hurricanes reaching Category 4 or 5 intensity in recent decades (Holland and Bruyère, 2014).

Rapid intensification of tropical cyclones (RITCs) poses forecast challenges and increased risks for coastal communities (Emanuel, 2017). Warming of the upper ocean in the central and eastern tropical Atlantic associated with the positive phase of the Atlantic Multidecadal Oscillation (AMO) (Balaguru et al. 2018) and in the western North Pacific in recent decades due to a La Niña-like pattern (Zhao et al. 2018) has favoured RITCs in these regions. One new modelling study suggests there has been a detectable increase in RITC occurrence in the Atlantic basin in recent decades, with a positive contribution from anthropogenic forcing (Bhatia et al. 2019). Nonetheless, the background conditions that favour RITC's across the Atlantic basin as a whole tend to be associated with less favourable conditions for TC occurrence along the US east coast (Kossin, 2017).

New studies have used event attribution to explore attribution of certain individual TC events or anomalous seasonal cyclone activity events to anthropogenic forcing (Lackmann, 2015; Murakami et al. 2015; Takayabu et al. 2015; Zhang et al. 2016; Emanuel, 2017; see also Table 6.2 and Box 6.1). Risser and Wehner (2017) and van Oldenborgh et al. (2017) concluded that for the Hurricane Harvey event, there is a detectable human influence on extreme precipitation in the Houston area, although their detection analysis is for extreme precipitation in general and not specifically for TC-related precipitation.

There have been more TC dynamical or statistical/dynamical downscaling studies and higher resolution General Circulation Model (GCM) experiments (e.g., Emanuel, 2013; Manganello et al. 2014; Knutson et al. 2015; Murakami et al. 2015; Roberts et al. 2015; Wehner et al. 2015; Yamada et al. 2017). The findings of these studies generally support the AR5 projections of a general increase in intensity of the most intense TCs and a decline in TC frequency overall. However, the projected increase in global TC frequency by Emanuel (2013) and Bhatia et al. (2018) differed from most other TC frequency projections and previous assessments. For studies into future track changes of TCs under climate warming scenarios (Li et al. 2010; Kim and Cai, 2014; Manganello et al. 2014; Knutson et al. 2015; Murakami et al. 2015; Roberts et al. 2015; Wehner et al. 2015; Nakamura et al. 2017; Park et al. 2017; Sugi et al. 2017; Yamada et al. 2017; Yoshida et al. 2017; Zhang et al. 2017a), it is difficult to identify a robust consensus of projected change in TC tracks, although several of the studies found either poleward or eastward expansion of TC occurrence over the North Pacific region resulting in greater storm occurrence in the central North Pacific. There have been new studies on storm size (Kim et al. 2014a; Knutson et al. 2015; Yamada et al. 2017) under climate warming scenarios. These project TC size changes of up to ±10% between basins and studies and provide preliminary findings on this issue that future studies will continue to investigate. Several studies of TC storm surge (e.g., Lin et al. 2012; Garner et al. 2017) suggest that SLR will dominate the increased height of storm surge due to TCs under climate change.

Taking the above into account, the following is a summary assessment of TC detection and attribution. The observed poleward migration of the latitude of maximum TC intensity in the western North Pacific appears to be unusual compared to expected natural variability and therefore there is *low to medium confidence* that this change represents a detectable climate change, though with only *low confidence* that the observed shift has a discernible positive contribution from anthropogenic forcing. Anthropogenic forcing is believed to be producing some poleward expansion of the tropical circulation with climate warming. Additional studies of observed long-term TC changes such as: an increase in annual global proportion of Category 4 or 5 TCs in recent decades, severe TCs occurring in the Arabian Sea, TCs making landfall in East and Southeast Asia, the increasing frequency of moderately large US storm surge events since 1923 and the decreasing frequency of severe TCs that make landfall in eastern Australia since the late 1800s, may each represent emerging anthropogenic signals, but still with *low confidence* (*limited evidence*). The lack of confident climate change detection for most TC metrics continues to limit confidence in both future projections and in the attribution of past changes and TC events, since TC event attribution in most published studies is generally being inferred without support from a confident climate change detection of a long-term trend in TC activity.

TCs projections for the late 21st century are summarised as follows: 1) there is *medium confidence* that the proportion of TCs that reach Category 4–5 levels will increase, that the average intensity of TCs will increase (by roughly 1–10%, assuming a 2°C global temperature rise), and that average TCs precipitation rates (for a given storm) will increase by at least 7% per degree Celsius SST warming, owing to higher atmospheric water vapour content, 2) there is *low confidence* (*low agreement, medium evidence*) in how global TC frequency will change, although most modelling studies project some decrease in global TC frequency and 3) SLR will lead to higher storm surge levels for the TCs that do occur, assuming all other factors are unchanged (*very high confidence*).

6.3.1.2 Extratropical Cyclones and Blocking

ETCs form in the mid-latitudes of the North Atlantic, North Pacific and Southern Oceans, and the Mediterranean Sea. The storm track regions are characterised by large surface equator-to-pole temperature gradients and baroclinic instability, and jet streams influence the direction and speed of movement of ETCs in this region. The thermodynamic response of the atmosphere to CO_2 tends to have opposing influences on storm tracks; surface shortwave cloud radiative changes increase the equator-to-pole temperature gradient whereas longwave cloud radiative changes reduce it (Shaw et al. 2016). AR5 concluded that the global number of ETCs is not expected to decrease by more than a few percent due to anthropogenic change. The Southern Hemisphere (SH) storm track is projected to have a small poleward shift, but the magnitude is model dependent (Christensen et al. 2013). AR5 also found a *low confidence* in the magnitude of regional storm track changes and the impact of such changes on regional surface climate (Christensen et al. 2013).

A 'blocking' event is an extratropical weather system in which the anticyclone (region of high pressure) becomes quasi-stationary and interrupts the usual westerly flow and/or storm tracks for up to a week or more (Woollings et al. 2018). Recent attention has focused on whether Arctic warming is linked to increased blocking and mid-latitude weather extremes (Barnes and Screen, 2015; Francis and Skific, 2015; Francis and Vavrus, 2015; Kretschmer et al. 2016), such as drought in California due to sea ice changes that cause a reorganisation of tropical convection (Cvijanovic et al. 2017), cold and snowy winters over Europe and North America (Liu et al. 2012; Cohen et al. 2018), extreme summer weather (Tang et al. 2013; Coumou et al. 2014) and Balkan flooding (Stadtherr et al. 2016). Studies suggest how blocking may influence arctic sea ice extent (Gong and Luo, 2017) and various pathways whereby Arctic warming could influence extreme weather (Barnes and Screen, 2015) such as reducing the equator to pole temperature gradient, slowing the jet stream thereby increasing its meandering behaviour (Röthlisberger et al. 2016; Mann et al. 2017) or causing it to split (Coumou et al. 2014), changing local dynamics in the vicinity of the sea ice edge (Screen and Simmonds, 2013) or weakening the stratospheric polar vortex (Cohen et al. 2014). However, sensitivity to choice of methodology (Screen and Simmonds, 2013) and large internal atmospheric variability masks the detection of such links in past records, and climate change can lead to opposing effects on the mid-latitude jet stream response leading to large uncertainty in future changes (Barnes and Polvani, 2015; Barnes and Screen, 2015).

New studies of future storm track behaviour in the NH, include Harvey et al. (2014) who find that the future changes to upper and lower tropospheric equator-to-pole temperature differences by the end of the century in a CMIP5 multi-model RCP8.5 ensemble are not well correlated and the lower temperature gradient dominates the summer storm track response whereas both upper and lower temperature gradients play a role in winter. In the northern North Atlantic storm track region, projected changes are found to be more strongly associated with changes in the lower rather than upper tropospheric equator-to-pole temperature difference (Harvey et al. 2015). In the SH, Harvey et al. (2014) find equator-to-pole

6

temperature differences in the upper and lower troposphere in the future climate across a multi-model ensemble are well correlated with a general strengthening of the storm track. The total number of ETCs in a CMIP5 GCM multi-model ensemble decreased in the future climate, whereas the number of strong ETCs increased in most models and in the ensemble mean (Grieger et al. 2014). This was associated with a general poleward shift related to both tropical upper tropospheric warming and shifting meridional SST gradients in the Southern Ocean. The poleward movement of baroclinic instability and associated storm formation over the observational period due to external radiative forcing, is projected to continue, with associated declining rainfall trends in the mid-latitudes and positive trends further polewards (Frederiksen et al. 2017).

A number of new studies have found links between Arctic amplification, blocking events and various types of weather extremes in NH mid-latitudes in recent decades. However, the sensitivity of results to analysis technique and the generally short record with respect to internal variability means that at this stage there is *low confidence* in these connections. Consistent with the AR5, projected changes to NH storm tracks exhibit large differences between responses, causal mechanisms and ocean basins and so there remains *low confidence* in future changes in blocking and storm tracks in the NH. The storm track projections for the SH remain consistent with previous studies in indicating an observed poleward contraction and a continued strengthening and southward contraction of storm tracks in the future (*medium confidence*).

6.3.1.3   Waves and Extreme Sea Levels

AR5 also concluded that there is *medium confidence* that mean significant wave height has increased in the North Atlantic north of 45°N based on ship observations and reanalysis-forced wave model hindcasts. ESL events have increased since 1970, mainly due to a rise in mean sea levels (MSLs) over this period (Rhein et al. 2013). There is *medium confidence* that mid-latitude jets will move 1–2 degrees further poleward by the end of the 21st century under RCP8.5 in both hemispheres with weaker shifts in the NH. In the SH during austral summer, the poleward movement of the mid-latitude westerlies under climate change is projected to be partially offset by stratospheric ozone recovery. There is *low confidence* in projections of NH storm tracks particularly in the North Atlantic. Tropical expansion is *likely* to continue causing wider tropical regions and poleward movement of the subtropical dry zones (Collins et al. 2013). In the SH, it is *likely* that enhanced wind speeds will cause an increase in annual mean significant wave heights. Wave swells generated in the Southern Ocean may also affect wave heights, periods and directions in adjacent ocean basins. The projected reduction in sea ice extent in the Arctic Ocean (Holland et al. 2006) will increase wave heights and wave season length (Church et al. 2013).

Since AR5, new studies have shown observed changes in wave climate. Satellite observations from 1985–2018, showed small increases in significant wave height (+0.3 cm/year) and larger increases in extreme wave heights (90th percentiles), especially in the Southern (+1 cm/year) and North Atlantic (+0.8 cm/year) Oceans (Young and Ribal, 2019) as well as positive trends in wave height in the Arctic

over 1992–2014 due to sea ice loss (Stopa et al. 2016; Thomson et al. 2016). Based on a wave reanalysis and satellite observations, Reguero et al. (2019) found that the global wave power, which represents the transport of the energy transferred from the wind into the sea surface motion, therefore including wave height, period and direction, has increased globally at a rate of 0.41% yr⁻¹ between 1948 and 2008, with large variations across oceans. Long-term correlations are found between the increase in wave power and SSTs, particularly between the tropical Atlantic temperatures and the wave power in high southern latitudes, the most energetic region globally.

The results of several new global wave climate projection studies are consistent with those presented in IPCC AR5. Mentaschi et al. (2017) find up to a 30% increase in 100-year return level wave energy flux (the rate of transfer of wave energy) for the majority of coastal areas in the southern temperate zone, and a projected decrease in wave energy flux for most NH coastal areas at the end of the century in wave model simulations forced by six CMIP5 RCP8.5 simulations. The most significant long-term trends in extreme wave energy flux are explained by their relationship to modelled climate indices (Arctic Oscillation, ENSO and NAO). Wang et al. (2014b) assessed the climate change signal and uncertainty in a 20-member ensemble of wave height simulations, and found model uncertainty (inter-model variability) is significant globally, being about 10 times as large as the variability between RCP4.5 and RCP8.5 scenarios. In a study focussing on the western north Pacific wave climate, Shimura et al. (2015) associate projected regions of future change in wave climate with spatial variation of SSTs in the tropical Pacific Ocean. A review of 91 published global and regional scale wind-wave climate projection studies found a consensus on a projected increase in significant wave height over the Southern Ocean, tropical eastern Pacific (*high confidence*) and Baltic Sea (*medium confidence*), and decrease over the North Atlantic and Mediterranean Sea. They found little agreement between studies of projected changes over the Atlantic Ocean, southern Indian and eastern North Pacific Ocean and no regional agreement of projected changes to extreme wave height. It was noted that few studies focussed on wave direction change, which is important for shoreline response (Morim et al. 2018).

Significant developments have taken place since the AR5 to model storm surges and tides at the global scale. An unstructured global hydrodynamic modelling system has been developed with maximum coastal resolution of 5 km (Verlaan et al. 2015) and used to develop a global climatology of ESLs due to the combination of storm surge and tide (Muis et al. 2016). A global modelling study finds that under SLR of 0.5–10 m, changes to astronomical tidal mean high water exceed the imposed SLR by 10% or more at around 10% of coastal cities when coastlines are held fixed. When coastal recession is permitted a reduction in tidal range occurs due to changes in the period of oscillation of the basin under the changed coastline morphology (Pickering et al. 2017). A recent study on global probabilistic projections of ESLs considering MSL, tides, wind-waves and storm surges shows that under RCP4.5 and RCP8.5, the global average 100-year ESL is *very likely* to increase by 34–76 cm and 58–172 cm, respectively between 2000–2100 (Vousdoukas et al. 2018). Despite the advancements in global tide and surge modelling, using CMIP GCM multi-model ensembles to examine the effects of future weather and circulation changes on storm surges in a globally consistent way is still a challenge because of the *low confidence* in GCMs being able to represent small scale weather systems such as TCs. To date only a small number of higher resolution GCMs are able to produce credible cyclone climatologies (e.g., Murakami et al. 2012) although this will probably improve with further GCM development and increases to GCM resolution (Walsh et al. 2016).

The role of austral winter swell waves on ESL have been investigated in the Gulf of Guinea (Melet et al. 2016) and the Maldives (Wadey et al. 2017). Multivariate statistical analysis and probabilistic modelling is used to show that flood risk in the northern Gulf of Mexico is higher than determined from short observational records (Wahl et al. 2016). In Australia, changes in ESLs were modelled using four CMIP5 RCP8.5 simulations (Colberg et al. 2019). On the southern mainland coast, the southward movement of the subtropical ridge in the climate models led to small reductions (up to 0.4 m) in the modelled 20-year (5% probability of occurring in a year) storm surge. Over the Gulf of Carpentaria in the north, changes were largest and positive during austral summer in two out of the four models in response to a possible eastward shift in the northwest monsoon. Synthetic cyclone modelling was used to evaluate probabilities, interannual variability and future changes of extreme water levels from tides and TC-induced storm surge (storm tide) along the coastlines of Fiji (McInnes et al. 2014) and Samoa (McInnes et al. 2016). Higher resolution modelling for Apia, Samoa incorporating waves highlights that although SLR reduces wave setup and wind setup by 10–20%, during storm surges it increases wave energy reaching the shore by up to 200% (Hoeke et al. 2015).

In the German Bight, Arns et al. (2015) show that under SLR, increases in extreme water levels occur due to a change in phase of tidal propagation; which more than compensates for a reduction in storm surge due to deeper coastal sea levels. Vousdoukas et al. (2017) develop ESL projections for Europe that account for changes in waves and storm surge. In 2100, increases of up to 0.35 m relative to the SLR projections occur towards the end of the century under RCP8.5 along the North Sea coasts of northern Germany and Denmark and the Baltic Sea coast, whereas little to negative change is found for the southern European coasts.

In the USA, Garner et al. (2017) combine downscaled TCs, storm surge models, and probabilistic SLR projections to assess flood hazard associated with changing storm characteristics and SLR in New York City from the pre-industrial era to 2300. Increased storm intensity was found to compensate for offshore shifts in storm tracks leading to minimal change in modelled storm surge heights through 2300. However, projected SLR leads to large increases in future overall flood heights associated with TCs in New York City. Consequently, flood height return periods that were ~500y (0.2% probability of occurring in a given year) during the pre-industrial era have fallen to ~25y (4% probability of occurring annually) at present and are projected to fall to ~5y (20% probability of occurring annually) within the next three decades.

In summary, new studies on observed wave climate change from 1985–2018 showed small increases in significant wave height of +0.3 cm/year and larger increases in 90th percentile wave heights

of +1 cm/year in the Southern Ocean and +0.8 cm/year in the North Atlantic ocean (*medium confidence*). Sea ice loss in the Arctic has also increased wave heights over the period 1992–2014 (*medium confidence*). Global wave power has increased over the last six decades with differences across oceans related to long-term correlations with SST (*low confidence*). Future projections indicate an increase of the mean significant wave height across the Southern Ocean and tropical eastern Pacific (*high confidence*) and Baltic Sea (*medium confidence*) and decrease over the North Atlantic and Mediterranean Sea under RCP8.5 (*high confidence*). Extreme waves are projected to increase in the Southern Ocean and decrease in the North Atlantic and Mediterranean Sea under RCP4.5 and RCP8.5 (*high confidence*). There is still limited knowledge on projected wave period and direction. For coastal ESLs, new studies at the regional to global scale have generally had a greater focus on multiple contributing factors such as waves, tides, storm surges and SLR. At the global scale, probabilistic projections of extreme sea levels considering these factors projects the global average 100-year ESL is *very likely* to increase by 34–76 cm and 58–172 cm, under RCP4.5 and RCP8.5, respectively between 2000–2100.

6.3.2   Impacts

As shown in previous assessments, increasing exposure is a major driver of increased cyclone risk (wind damages), as well as flood risk associated with cyclone rainfall and surge, besides possible changes in hazard intensities from anthropogenic climate change (Handmer et al. 2012; Arent et al. 2014). Changes in TC trajectories are potentially a major source of increased risk, as the degree of vulnerability is typically much higher in locations that were previously not exposed to the hazard (Noy, 2016). Typhoon Haiyan's move to the south of the usual trajectories of TCs in the western North Pacific basin (Yonson et al. 2018) made the evacuation more difficult as people were less willing to heed storm surge warnings they received.

Abrupt changes in impacts therefore are not only determined by changes in cyclone hazard, but also by the sensitivity or tipping points that are crossed in terms of flooding for instance, that can be driven by SLR but also by changes in local exposure. The frequency of nuisance flooding along the US east coast is expected to accelerate further in the future (Sweet and Park, 2014). The loss of coral reef cover and mangrove forests have also been shown to increase damages from storm surge events (e.g., Beck et al. 2018). Cyclones also affect marine life, habitats and fishing. There is some evidence that fish may evacuate storm areas or be redistributed by storm waves and currents (FAO, 2018; Sainsbury et al. 2018). Other examples of damage to fisheries from cyclones and storm surges can be found in FAO (2018: Chapter V, Table 1).

With regard to property losses, according to most projections, increasing losses from more intense cyclones are not offset by a possible reduction in frequency (Handmer et al. 2012). While the relation between aggregate damages and frequency may be linear, the relationship between intensity and damages is most probably highly nonlinear; with research suggesting a 10% increase in wind speed associated with a 30–40% increase in damages (e.g., Strobl, 2012). Although it is clear that direct damages from cyclones could

6

increase, investigations into the economic impact of past cyclone events is less common, as these are much more difficult to identify. Examples of such work include Strobl (2012) on hurricane impacts in the Caribbean, Haque and Jahan (2016) on TC Sidr in Bangladesh, Jakobsen (2012) on Hurricane Mitch in Nicaragua, and Taupo and Noy (2017) on TC Pam in Tuvalu. The relation between changes in TCs and property losses is complex, and there are indications that wind shear changes may have larger impact than changes in global temperatures (Wang and Toumi, 2016). With regard to loss of life, total fatalities and mortality from cyclone-related coastal flooding is globally declining, probably as a result of improved forecasting and evacuation, although in some low-income countries mortality is still high (Paul, 2009; Lumbroso et al. 2017; Bouwer and Jonkman, 2018). A global analysis finds that despite adaptation efforts, further SLR could increase storm surge mortality in many parts of the developing world (Lloyd et al. 2016).

An assessment of future changes in coastal impacts based on direct downscaling of indicators of flooding such as total water level and number of hours per year with breakwater overtopping over a given threshold for port operability is provided by Camus et al. (2017). These indicators are multivariable and include the combined effect of SLR, storm surge, astronomical tide and waves. Regional projected wave climate is downscaled from global multi-model projections from 30 CMIP5 model realisations. For example, projections by 2100 under the RCP8.5 scenario show a spatial variability along the coast of Chile with port operability loss between 600–800 h yr^{-1} and around 200 h yr^{-1} relative to present (1979–2005) conditions. Although wave changes are included in projected overtopping distributions, future changes of operability are mainly due to the SLR contribution.

6.3.3   Risk Management and Adaptation

The most effective risk management strategy in the last few decades has been the development of early warning systems for cyclones (Hallegatte, 2013). Generally, however, a lack of familiarity with the changed nature of storms prevails. Powerful storms often generate record storm surges (Needham et al. 2015), such as in the cases of Cyclone Nargis and Typhoon Haiyan but surge warnings had been less well understood and followed because they had tended to be new or rare to the locality (Lagmay et al. 2015). A US study on storm surge warnings highlights the issue of the right timing to warn, as well as the difficulty in delivering accurate surge maps (Morrow et al. 2015). Previous experience with warnings that were not followed by hazard events show the 'crying wolf' problem leading many to ignore future warnings (Bostrom et al. 2018).

There is scant literature on the management of storms that follow less common trajectories. The most recent and relatively well-studied ones are Superstorm Sandy in 2012 in the USA and Typhoon Haiyan in 2013 in the Philippines. These two storms were unexpected and having underestimated the levels of impact, people ignored warnings and evacuation directives. In the case of Typhoon Haiyan, the dissemination of warnings via scripted text messages were ineffective without an explanation of the difference between Haiyan's accompanying storm surge and that of other 'normal'

storms to which people were used to (Lejano et al. 2016). Negative experiences of previous evacuations also lead to the reluctance of authorities to issue mandatory evacuation orders, for example, during Superstorm Sandy (Kulkarni et al. 2017), and contributes to a preventable high number of casualties (Dalisay and De Guzman, 2016). These examples also show that saving lives and assets through warning and evacuation is limited. Providing biophysical protection measures as well as improving self-reliance during such events can complement warning and evacuation.

After the storms, retreat or rebuild options exist. Rebuilding options can depend on whether insurance is still affordable after the event. Buyout programs, a form of 'managed retreat' whereby government agencies pay people affected by extreme weather events to relocate to safer areas, gained traction in recent years as a potential solution to reduce exposure to changing storm surge and flood risk. The decision to retreat or rebuild *in situ* depends, at least partially, on how communities have recovered in the past and therefore on the perceived success of a future recovery (Binder, 2014). However, political and jurisdictional conflicts between local, regional, and national government over land management responsibilities, lack of coordinated nation-wide adaptation plans, and clashes between individual and community needs have led to some unpopular buyout programs after Hurricane Sandy (Boet-Whitaker, 2017). Relocation (i.e., managed retreat) is often very controversial, can incur significant political risk even when it is in principle voluntary (Gibbs et al. 2016), and is rarely implemented with much success at larger scales (Beine and Parsons, 2015; Hino et al. 2017). In addition, managed retreats are often fraught with legal, distributional and human rights issues, as seen in the case of resettlements after Typhoon Haiyan (Thomas, 2015; see also Cross-Chapter Box 5 in Chapter 1), and extend to loss of cultural heritage and indigenous qualities in the case of small island states.

If rebuilding *in situ* is pursued after catastrophic events and without decreased exposure, it is often accompanied by actions that aim to reduce vulnerability in order to adapt to the increasing risk (Harman et al. 2013). In many cases, resilient designs and sustainable urban plans integrating climate change concerns, that are inclusive of vegetation barriers as coastal defences and hybrid designs, are considered (Cheong et al. 2013; Saleh and Weinstein, 2016). However, often more physical structures that are known to be less sustainable in the long-term, but potentially more protective in the short-term, are constructed (Knowlton and Rotkin-Ellman, 2014; Rosenzweig and Solecki, 2014). Anticipatory planning approaches are under way to warn and enable decision making in time (Bloemen et al. 2018; Lawrence et al. 2018).

6.4   Marine Heatwaves and their Implications

AR5 concluded that it is *virtually certain* that the global ocean temperature in the upper few hundred meters has increased from 1971–2010 (Rhein et al. 2013), and that the temperature is projected to further increase during the 21st century (Collins et al. 2013). For an update on observed and projected long-term changes in ocean temperature and heat, see Chapter 5.

6

Superimposed onto the long-term ocean warming trend are short-term extreme warming events, called MHWs, during which ocean temperatures are extremely high. Whereas the response of marine organisms and ecosystems to gradual trends in temperature has been assessed in AR5 (e.g., Hoegh-Guldberg et al. 2014; Pörtner et al. 2014), research on the response of the natural, physical and socioeconomic systems to MHWs has newly emerged since AR5. Notable exceptions are studies on the effect of MHWs on intertidal systems and tropical coral reef ecosystems, which have been already assessed in AR5 (Gattuso et al. 2014; Pörtner et al. 2014).

MHWs are periods of extremely high ocean temperatures that persist for days to months, can extend up to thousands of kilometres and can penetrate multiple hundreds of metres into the deep ocean (see SROCC Glossary; Hobday et al. 2016a; Scannell et al. 2016; Benthuysen et al. 2018). A MHW is an event at a particular place and time of the year that is rare and predominately, but not exclusively, defined with a relative threshold; that is, an event rarer than 90th or 99th percentile of a probability density function. By definition, the characteristics of what is called a MHW may therefore vary from place to place in an absolute sense. Different metrics are used to quantify changes in MHW characteristics, such as frequency, duration, intensity, spatial extent and severity. To monitor and predict coral bleaching risk, the metric degree heating week (DHW; e.g., Eakin et al. 2010) is often used, which combines the effect of duration and magnitude of the heatwave.

6.4.1 Observations and Key Processes, Detection and Attribution, Projections

6.4.1.1 Recent Documented MHWs and Key Driving Mechanisms

MHWs have been observed and documented in all ocean basins over the last two decades (Figure 6.3a, Figure 6.2, Table 6.2). Prominent examples include the Northeast Pacific 2013–2015 MHW (often called 'The Blob'; Bond et al. 2015), the Yellow Sea/East China Sea 2016 MHW (KMA, 2016; KMA, 2017; KMA, 2018), the Western Australia 2011 MHW (Pearce and Feng, 2013; Kataoka et al. 2014), and the Northwest Atlantic 2012 MHW (Mills et al. 2013).

The dominant ocean and/or atmospheric processes leading to the buildup, persistence and decay of MHWs vary greatly among the individual MHWs and depend on the location and time of occurrence. One of the most important global driver of MHWs are El Niño events (Oliver et al. 2018a). During El Niño events, the SST, in particular of the central and eastern equatorial Pacific and the Indian Ocean, are anomalously warm (see Section 6.5). MHWs may also be associated with other large-scale modes of climate variability, such as the Pacific Decadal Oscillation (PDO), AMO, Indian Ocean Dipole (IOD), North Pacific Oscillation and NAO, which modulate ocean temperatures at the regional scale (Benthuysen et al. 2014; Bond et al. 2015; Chen et al. 2015b; Di Lorenzo and Mantua, 2016). These modes can change the strength, direction and location of ocean currents that build up areas of extreme warm waters, or they can change the air-sea heat flux, leading to a warming of the ocean surface from the atmosphere.

For example, predominant La Niña conditions in 2010 and 2011 strengthened and shifted the Leeuwin Current southward along the west coast of Australia leading to the Western Australia 2011 MHW (Pearce and Feng, 2013; Kataoka et al. 2014). Another example is The Blob, which emerged in 2013 in response to teleconnections between the North Pacific and the weak El Niño that drove strong positive sea level pressure anomalies across the northeast Pacific inducing a smaller heat loss from the ocean (Bond et al. 2015; Di Lorenzo and Mantua, 2016). Low sea ice concentrations in the Arctic, however, may have also played a role (Lee et al. 2015a).

The buildup and decay of extreme warm SSTs may also be caused by small-scale atmospheric and oceanic processes, such as ocean mesoscale eddies or local atmospheric weather patterns (Carrigan and Puotinen, 2014; Schlegel et al. 2017a; Schlegel et al. 2017b). For example, the Tasman Sea 2015–2016 MHW was caused by enhanced southward transport in the East Australian current driven by increased wind stress curl across the mid-latitude South Pacific (Oliver and Holbrook, 2014; Oliver et al. 2017) with local downwelling-favourable winds also having played a role in the subsurface intensification of the MHW (Schaeffer and Roughan, 2017). In addition, the 2016 MHW in the southern part of the Great Barrier Reef was mitigated by the ETC Winston that passed over Fiji on February 20th. The cyclone caused strong winds, cloud cover and rain, which lowered SST and prevented corals from bleaching (Hughes et al. 2017b).

6.4.1.2 Detection and Attribution of MHW Events

The upper ocean temperature has significantly increased in most regions over the last few decades, with anthropogenic forcing *very likely* being the main driver (Bindoff et al. 2013). Concurrent with the long-term increase in upper ocean temperatures, MHWs have become more frequent, extensive and intense (Frölicher and Laufkötter, 2018; Oliver et al. 2018a; Smale et al. 2019). Analysis of satellite daily SST data reveal that the number of MHW days exceeding the 99th percentile, calculated over the 1982–2016 period, has doubled globally between 1982 and 2016, from about 2.5 heatwave days yr^{-1} to 5 heatwave days yr^{-1} (Frölicher et al. 2018; Oliver et al. 2018a). At the same time, the maximum intensity of MHWs has increased by 0.15°C and the spatial extent by 66% (Frölicher et al. 2018). Using a classification system to separate MHWs into categories (I-IV, depending on the level to which SSTs exceed local averages), Hobday et al. (2018) show that the occurrence of MHWs has increased for all categories over the past 35 years with the largest increase (24%) in strong (Category II) MHW events. In 2016, about a quarter of the surface ocean experienced either the longest or most intense MHW (Hobday et al. 2016a; Figure 6.3b).

The observed trend towards more frequent, intense and extensive MHWs, defined relative to a fixed baseline period, is *very likely* due to the long-term anthropogenic increase in mean ocean temperatures, and cannot be explained by natural climate variability (Frölicher et al. 2018; Oliver et al. 2018a; Oliver, 2019). As climate models project a long-term increase in ocean temperatures over the 21st century (Collins et al. 2013), a further increase in the probability of MHWs under continued global warming can be expected (see Section 6.4.1.3). Extending the analysis to the pre-satellite period (before 1982) by using a combination of daily *in situ* measurements and gridded monthly *in situ* based data

6

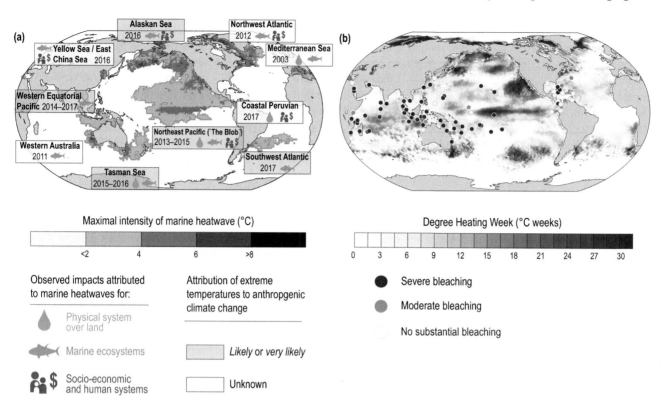

Figure 6.3 | Examples of recent marine heatwaves (MHWs) and their observed impacts. **(a)** Examples of documented MHWs over the last two decades and their impacts on natural, physical and socioeconomic systems. The colour map shows the maximum sea surface temperature (SST) anomaly during the MHW using the National Oceanic and Atmospheric Administration's (NOAA) daily Optimum Interpolation SST dataset (Reynolds et al. 2007; Banzon et al. 2016). A MHW is defined here as a set of spatially and temporally coherent grid points exceeding the 99th percentile. The 99th percentile is calculated over the 1982–2011 reference period after de-seasonalising the data. Red shading of the boxes indicates if the likelihood of MHW occurrence has increased due to anthropogenic climate change, and symbols denote observed impacts on physical systems over land, marine ecosystems, and socioeconomic and human systems. Figure is updated from Frölicher and Laufkötter (2018) and is not a complete compilation of all documented MHWs. **(b)** The record warming years 2015 and 2016 and the global extent of mass bleaching of corals during these years. The colour map shows the Degree Heating Week (DHW) annual maximum over 2015 and 2016 from NOAA's Coral Reef Watch Daily Global 5 km Satellite Coral Bleaching Heat Stress Monitoring Product Suite v.3.1 (Liu et al. 2014a). The DHW describes how much heat has accumulated in an area over the past twelve weeks by adding up any temperatures that exceed 1°C above the maximum summertime mean (e.g., Eakin et al. 2010). Symbols show reef locations that are assessed in Hughes et al. (2018a) and indicate where severe bleaching affected more than 30% of corals (purple circles), moderate bleaching affected less than 30% of corals (blue circles), and no substantial bleaching was recorded (light blue circles).

sets, Oliver et al. (2018a) show that the global frequency and duration of MHWs have increased since 1925. At regional scale, MHWs have become more common in 38% of the world's coastal ocean over the last few decades (Lima and Wethey, 2012). In tropical reef systems, the interval between recurrent MHWs and associated coral bleaching events has diminished steadily since 1980, from once every 25 to 30 years in early 1980s to once every 6 years in 2016 (Hughes et al. 2018a). Due to the scarcity of below surface temperature data with high temporal and spatial resolution, it is currently unknown if and how MHWs at depth have changed over the past decades.

Several attribution studies (summarised in Table 6.2) have investigated if the likelihood of individual MHW events has changed due to anthropogenic warming. On a global scale and at present day (2006–2015), climate models suggest that 84–90% (*very likely* range) of all globally occurring MHWs are attributable to the temperature increase since 1850–1900 (Fischer and Knutti, 2015; Frölicher et al. 2018). Attribution studies on individual MHW events show that the intensity of the western tropical Pacific MHW in 2014 (Weller et al. 2015), the intensity of the Alaskan Sea 2016 MHW (Oliver et al. 2018b; Walsh et al. 2018) and the extreme SSTs in the central equatorial Pacific in 2015–2016 can be fully attributed to anthropogenic warming.

In other words, the aforementioned studies show that such events could not have occurred without the temperature increase since 1850–1900. In addition, extreme SSTs in the northeast Pacific in 2014 have become about five times more likely with human-induced global warming (Wang et al. 2014a; Kam et al. 2015; Weller et al. 2015). The Tasman Sea 2015–2016 MHW was 330 times (for duration) and 6.8 times (for intensity) more likely with anthropogenic climate change than without (Oliver et al. 2017), and the northern Australia 2016 MHW was up to fifty times more likely due to anthropogenic climate change (Weller et al. 2015; King et al. 2017; Lewis and Mallela, 2018; Newman et al. 2018; Oliver et al. 2018b). Also the risk of the Great Barrier Reef bleaching event in 2016 was increased due to anthropogenic climate change (King et al. 2017; Lewis and Mallela, 2018). Even though natural variability is still needed for the events to occur, these studies show that most of the individual MHW events analysed so far have a clear human-induced signal. However, such attribution studies have not been undertaken for all major individual MHW events yet (e.g., five out of ten MHWs indicated in Figure 6.3a have not been assessed), and it is therefore still unknown for some of the observed individual MHW events if they have an anthropogenic signal or not (labelled as 'unknown' in Figure 6.3a).

6

We conclude that it is *very likely* that MHWs have increased in frequency, duration and intensity since pre-industrial (1850–1900), and that between 2006–2015 most MHWs (84–90%; *very likely* range) are attributable to the temperature increase since 1850–1900. Only few studies on the attribution of individual MHW events exist, but they all point to human influence on recent MHW events.

6.4.1.3   Future Changes

MHWs will increase in frequency, duration, spatial extent and intensity throughout the ocean under future global warming (Oliver et al. 2017; Ramírez and Briones, 2017; Alexander et al. 2018; Frölicher et al. 2018; Frölicher and Laufkötter, 2018; Darmaraki et al. 2019). Projections based on 12 CMIP5 Earth system models suggest that, on global scale, the probability of MHWs exceeding the pre-industrial (1850–1900) 99th percentile will *very likely* increase by a factor of 20–27 by 2031–2050 and *very likely* by a factor of 46–55 by 2081–2100 under the RCP8.5 greenhouse gas (GHG) scenario (Figure 6.4a; Frölicher et al. 2018). In other words, a one-in-hundred-day event at pre-industrial levels is projected to become a one-in-four-day event by 2031–2050 and a one-in-two-day event by

2081–2100. The duration of MHW is projected to *very likely* increase from 8–10 days at 1850–1900, to 126–152 days in 2081–2100 under the RCP8.5 scenario (Frölicher et al. 2018). The maximum intensity (maximum exceedance of the 1850–1900 99th percentile) will *very likely* increase from 0.3°C–0.4°C in 1850–1900, to 3.1°C–3.8°C in 2081–2100 under the RCP8.5 scenario. Under the RCP2.6 scenario, the magnitude of changes in the different MHW metrics would be substantially reduced (Frölicher et al. 2018). For example, the probability ratio would *very likely* increase by a factor of 16–24 by 2081–2100 for RCP2.6; less than half of that is projected for the RCP8.5. The magnitude of changes in the probability ratio scales with global mean atmospheric surface temperature and is independent of the warming path (Figure 6.4b), that is, it does not depend on whether a particular warming level is reached sooner (RCP8.5) or later (RCP2.6).

The projected changes in MHWs will not be globally uniform. CMIP5 models project that the largest increases in the probability of MHWs will occur in the tropical ocean, especially in the western tropical Pacific, and the Arctic Ocean (Figure 6.4c,d), and that most of the large marine ecosystems will also experience large increases in the

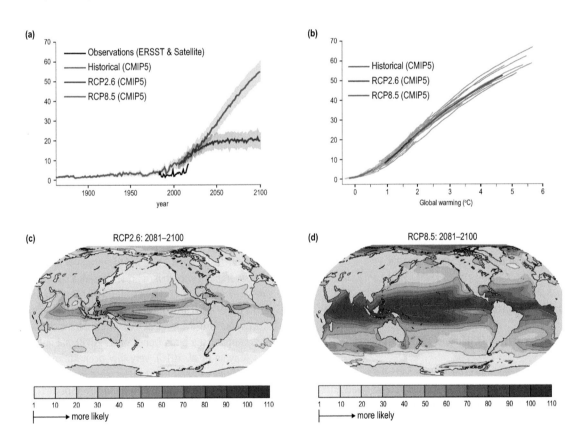

Figure 6.4 | Global and regional changes in the probability ratio of marine heatwaves (MHWs). The probability ratio is the fraction by which the number of MHW days yr⁻¹ has changed since 1850–1900. **(a)** Changes in the annual mean probability ratio of MHWs exceeding the 99th percentile of pre-industrial local daily sea surface temperature (SST) averaged over the ocean. The thick lines represent the multi-model averages of 12 climate models that participated in the Coupled Model Intercomparison Project Phase 5 (CMIP5) covering the 1861–2100 period for the Representative Concentration Pathway (RCP) 8.5 and RCP2.6 scenarios, respectively. The shaded bands indicate the 90% confidence interval of the standard error of the mean. The black line shows an observational-based estimate. As daily SST data are available only for the 1982–2016 period, we assume that the observed mean temperature change is the main cause of the change in frequency of extremes (Frölicher et al. 2018; Oliver, 2019). We therefore subtracted first the differences between 1854–1900 and 1982–2016 obtained from the extended reconstructed SST Version 4 dataset (ERSSTv4; Huang et al. 2015a) from the daily satellite data before calculating the 99th percentile for the observations. **(b)** Same as (a), but the probability ratio is plotted for different levels of global surface atmospheric warming and for the individual models. The simulated time series in (b) are smoothed with a 10-year running mean. **(c,d)** Simulated regional changes in the multi-model mean probability ratio of MHWs exceeding the preindustrial 99th percentile in 2081–2100 for the (c) RCP2.6 scenario and the (d) RCP8.5 scenario. The grey contours in (c,d) highlight the spatial pattern. Figure is modified from Frölicher et al. (2018).

number of MHW days (Alexander et al. 2018; Frölicher et al. 2018). Smallest increases are projected for the Southern Ocean. In addition, MHW events in the Great Barrier Reef, such as the one associated with the bleaching in 2016, are projected to be at least twice as frequent under 2°C global warming than they are today (King et al. 2017). The magnitude of projected changes at the local scale is uncertain, partly due to issues of horizontal and vertical resolution of CMIP5-type Earth system models. Only a few studies have used higher resolution oceanic models (eddy-resolving) to assess the local-to-regional changes in MHW characteristics. For example, regional high-resolution coupled climate model simulations suggest that the Mediterranean Sea will experience at least one long lasting MHW every year by the end of the 21st century under the RCP8.5 scenario (Darmaraki et al. 2019), and eddy-resolving ocean model simulations project a further increase in the likelihood of extreme temperature events in the Tasman Sea (Oliver et al. 2014; Oliver et al. 2015; Oliver et al. 2017).

Most of the global changes in the probability of MHWs, when defined relative to a fixed temperature climatology and using coarse resolution CMIP5-type climate models, are driven by the global-scale shift in the mean ocean temperature (Alexander et al. 2018; Frölicher et al. 2018). However, previously ice-covered regions, such as the Arctic Ocean, will exhibit larger SST variability under future global warming. This is because of an enhanced SST increase in summer due to sea ice retreat, but SST remaining near the freezing point in winter (Carton et al. 2015; Alexander et al. 2018). When contrasting the changes in the probability of MHWs with land-based heatwaves (Fischer and Knutti, 2015), it is evident that MHWs are projected to occur more frequently (Frölicher et al. 2018; Frölicher and Laufkötter, 2018). This is because the temperature variability is much smaller in ocean surface waters than in the atmosphere (Frölicher and Laufkötter, 2018).

We conclude that there is *very high confidence* that MHWs will increase in frequency, duration, spatial extent and intensity in all ocean basins under future global warming, mainly because of an increase in mean ocean temperature. However, higher resolution models are needed to make robust projections at the local-to-regional scale.

6.4.2   Impacts on Natural, Physical and Human Systems

6.4.2.1   Impacts on Marine Organisms and Ecosystems

Temperature plays an essential role in the biology and ecology of marine organisms (e.g., Pörtner, 2002; Pörtner and Knust, 2007; Poloczanska et al. 2013; Hoegh-Guldberg et al. 2014), and therefore extreme high ocean temperature can have large impacts on marine ecosystems. Recent studies show that MHWs have strongly impacted marine organisms and ecosystem services in all ocean basins (Smale et al. 2019) over the last two decades. Impacts include coral bleaching and mortality (Hughes et al. 2017b; Hughes et al. 2018a; Hughes et al. 2018b), loss of seagrass and kelp forests (Smale et al. 2019), shifts in species range (Smale and Wernberg, 2013), and local (Wernberg et al. 2013; Wernberg et al. 2016) and potentially global extinctions of coral species (Brainard et al. 2011).

A growing number of studies have reported that MHWs negatively affect corals and coral reefs through bleaching, disease, and mortality (see Chapter 5 for an extensive discussion on coral reefs and coral bleaching). The recent (2014–2017) high ocean temperatures in the tropics and subtropics triggered a pan-tropical episode of unprecedented mass bleaching of corals (100s of km²), the third global-scale event after 1997–1998 and 2010 (Heron et al. 2016; Eakin et al. 2017; Hughes et al. 2017b; Eakin et al. 2018; Hughes et al. 2018a). The heat stress during this event was sufficient to cause bleaching at 75% of global reefs (Hughes et al. 2018a; Figure 6.3b) and mortality at 30% (Eakin et al. 2017), much more than any previously documented global bleaching event. In some locations, many reefs bleached extensively for the first time on record, and over half of the reefs bleached multiple times during the three year event. However, there were distinct geographical variations in bleaching, mainly determined by the spatial pattern and magnitude of the MHW (Figure 6.3b). For example, bleaching was extensive and severe in the northern regions of the Great Barrier Reef, with 93% of the northern Australian Great Barrier Reef coral suffering bleaching in 2016, but impacts were moderate at the southern coral reefs of the Great Barrier Reef (Brainard et al. 2018; Stuart-Smith et al. 2018).

Apart from strong impacts on corals, recent MHWs have demonstrated their potential impacts on other marine ecosystems and ecosystems services (Ummenhofer and Meehl, 2017; Smale et al. 2019). Two of the best studied MHWs with extensive ecological implications are the Western Australia 2011 MHW and the Northeast Pacific 2013–2015 MHW. The Western Australia 2011 MHW resulted in a regime shift of the temperate reef ecosystem (Wernberg et al. 2013; Wernberg et al. 2016). The abundance of the dominant habitat-forming seaweeds *Scytohalia dorycara* and *Ecklonia radiata* became significantly reduced and *Ecklonia* kelp forest was replaced by small turf-forming algae with wide ranging impacts on associated sessile invertebrates and demersal fish. The sea grass *Amphibolis antarctica* in Shark Bay underwent defoliation after the MHW (Fraser et al. 2014), and together with the loss of other sea grass species, these lead to releases of 2–9 Tg CO_2 to the atmosphere during the subsequent three years after the MHW (Arias-Ortiz et al. 2018). In addition, coral bleaching and adverse impacts on invertebrate fisheries were documented (Depczynski et al. 2013; Caputi et al. 2016). The Northeast Pacific 2013–2015 MHW also caused extensive alterations to open ocean and coastal ecosystems (Cavole et al. 2016). Impacts included increased mortality events of sea birds (Jones et al. 2018), salmon and marine mammals (Cavole et al. 2016), very low ocean primary productivity (Whitney, 2015; Jacox et al. 2016), an increase in warm water copepod species (Di Lorenzo and Mantua, 2016) and novel species compositions (Peterson et al. 2017). In addition, a coast wide bloom of the toxigenic diatom *Pseudo-nitzschia* resulted in the largest ever recorded outbreak of domoic acid along the North American west coast (McCabe et al. 2016). Domoic acid was detected in many marine mammals, such as whales, dolphins, porpoises, seals and sea lions. The elevated toxins in commercially harvested fish and invertebrates resulted in prolonged and geographically extensive closure of razor clam and crab fisheries.

Other MHWs also demonstrated the vulnerability of marine organisms and ecosystems to extremely high ocean temperatures. The Northwest Atlantic 2012 MHW strongly impacted coastal ecosystems by causing a northward movement of warm water species and local migrations of some species (e.g., lobsters) earlier in the season (Mills et al. 2013; Pershing et al. 2015). The Mediterranean Sea 2003 MHW lead to mass mortalities of macro-invertebrate species (Garrabou et al. 2009) and the Tasman Sea 2015–2016 MHW had impacts on sessile, sedentary and cultured species in the shallow, near-shore environment including outbreaks of disease in commercially viable species (Oliver et al. 2017). *Vibrio* outbreaks were also observed in the Baltic Sea in response to elevated SSTs (Baker-Austin et al. 2013). The Alaskan Sea 2016 MHW favoured some phytoplankton species, leading to harmful algal blooms, shellfish poisoning events and mortality events in seabirds (Walsh et al. 2018; see chapter 3 for more details). Also, lower than average size of multiple groundfish species were observed including Pollock, Pacific cod, and Chinook salmon (Zador and Siddon, 2016). The Yellow Sea/East China Sea 2016 MHW killed a large number of different marine organisms in coastal and bay areas around South Korea (Kim and Han, 2017) and the Southwest Atlantic 2017 MHW lead to toxic algal blooms (Manta et al. 2018). The Coastal Peruvian 2017 MHW affected anchovies, which showed decreased fat content and early spawning as a reproductive strategy (IMPARPE, 2017), a behaviour usually seen during warm El Niño conditions (Ñiquen and Bouchon, 2004).

Based on the examples described above we conclude with *very high confidence* that a range of organisms and ecosystems have been impacted by MHWs across all ocean basins over the last two decades. Given that MHWs will *very likely* increase in intensity and frequency with further climate warming, we conclude with *high confidence* that this will push some marine organisms, fisheries and ecosystem beyond the limits of their resilience. These impacts will occur on top of those expected from a progressive shift in global mean ocean temperatures.

6.4.2.2   Impacts on the Physical System

MHWs can impact weather patterns over land via teleconnections causing drought, heavy precipitation or heat wave events. For example, the Northeast Pacific 2013–2015 MHW and the associated persistent atmospheric high-pressure ridge prevented normal winter storms from reaching the West Coast of the US and may have contributed to the drought conditions across the entire West Coast (Seager et al. 2015; Di Lorenzo and Mantua, 2016). The Tasman Sea 2015–2016 MHW has increased the intensity of rainfall that caused flooding in northeast Tasmania in January 2016 (see Box 6.1) and the Coastal Peruvian 2017 MHW caused heavy rainfall and flooding on the west coast of tropical South America (ENFEN, 2017; Echevin et al. 2018; Garreaud, 2018; Takahashi et al. 2018). Similarly, MHWs in the Mediterranean Sea may have amplified heatwaves (Feudale and Shukla, 2007; García-Herrera et al. 2010) and heavy precipitation events over central Europe (Messmer et al. 2017), as well as trigger intense ETCs over the Mediterranean Sea (González-Alemán et al. 2019). Such physical changes induced by MHWs may then also affect ecosystems and human systems on land (Reimer et al. 2015).

It should be noted that past and future impacts of MHWs on weather patterns over land depend not only on the duration and intensity of MHWs, but also on a wide range of different additional processes in the climate system such as the large-scale circulation of the atmosphere and oceans, and changes in the mean climate. Therefore, we conclude that there is currently *low confidence* in how MHWs impact the weather systems over land.

6.4.2.3   Impacts on the Human System

MHWs can also lead to significant socioeconomic ramifications when affecting aquaculture or important fishery species, or when triggering heavy rain or drought events on land. The Northwest Atlantic 2012 MHW, for example, had major economic impacts on the US lobster industry in 2015 (Mills et al. 2013). The MHWs lead to changes in lobster fishing practices and harvest patterns, because the lobsters moved from the deep offshore waters into shallower coastal areas much earlier in the season than usual causing a rapid rise in lobster catch rates. Together with a supply chain bottleneck, the record catch outstripped market demand and contributed to a collapse in lobster prices (Mills et al. 2013). Even though high catch volumes were reported, the price collapse threatened the economic viability of many US and Canadian lobster fisheries. Economic impacts through changes in fisheries were also reported during the Northeast Pacific 2013–2015 MHW and the Alaskan Sea 2016 MHW. The Northeast Pacific 2013–2015 MHW led to closing of both commercial and recreational fisheries resulting in millions of USD in losses among fishing industries (Cavole et al. 2016). In addition, the toxin produced by the harmful algal blooms can be transferred through the marine food web and humans who eat contaminated fish, shellfish or crustaceans (Berdalet et al. 2016; Du et al. 2016; McCabe et al. 2016). The ingestion of such contaminated seafood products, the inhalation of aerosolised toxins or the skin contact with toxin-contaminated water may cause toxicity in humans. Symptoms in human associated with the ingestion of the contaminated seafood range from mild gastrointestinal distress to seizures, coma, permanent short-term memory loss and death (Perl et al. 1990). The ecological changes associated with the Alaskan Sea 2016 MHW impacted subsistence and commercial activities. For example, ice-based harvesting of seals, crabs and fish in western Alaska was delayed due to the lack of winter sea ice. MHWs can also impact the socioeconomic and human system through changes to weather patterns. For example, heavy rain associated with the Coastal Peruvian 2017 MHW triggered numerous landslides and flooding, which resulted in a death toll of several hundred, and widespread damage to infrastructure and civil works (United Nations, 2017).

Studies on the impact of MHWs on human systems are still relatively scarce, even though many show negative impacts on human health and economy. We therefore conclude with *medium confidence* that MHWs can negatively impact human health and economy.

6.4.3   Risk Management and Adaptation, Monitoring and Early Warning Systems

Risk management strategies to respond to MHWs include early warning systems as well as seasonal (weeks to several months) and multi-annual predictions systems. Since 1997, the National Oceanic and Atmospheric Administration's (NOAA) Coral Reef Watch has used satellite SST data to provide near real-time warning of coral bleaching (Liu et al. 2014a). These satellite-based products, along with NOAA Coral Reef Watch's four month coral bleaching outlook based on operational climate forecast models (Liu et al. 2018), and coral disease outbreak risk (Heron et al. 2010) provide critical guidance to coral reef managers, scientists, and other stakeholders (Tommasi et al. 2017b; Eakin et al. 2018). These products are also used to implement proactive bleaching response plans (Rosinski et al. 2017), brief stakeholders, and allocate monitoring resources in advance of bleaching events, such as the 2014–2017 global coral bleaching event (Eakin et al. 2017). For example, Thailand closed ten reefs for diving in advance of the bleaching peak in 2016, while Hawaii immediately began preparation of resources both to monitor the 2015 bleaching and to place specimens of rare corals in climate controlled, onshore nurseries in response to these forecast systems (Tommasi et al. 2017b). New measurement techniques, such as Argo and deep Argo floats, may help to further develop prediction systems for subsurface MHWs, but such systems are not yet in place.

SST forecasts ranging from seasonal to decadal (5–10 years) have also been used or are planned to be used as early warning systems for multiple other ecosystems and fisheries in addition to coral reefs, including aquaculture, lobster, sardine, and tuna fisheries (Hobday et al. 2016b; Payne et al. 2017; Tommasi et al. 2017b). For example, seasonal forecasts of SST around Tasmania may help farm managers of salmon aquaculture to prepare and respond to upcoming MHWs by changing stocking densities, varying feed mixes, transferring fish to different locations in the farming region and implementing disease management (Spillman and Hobday, 2014; Hobday et al. 2016b). Skilful multi-annual to decadal SST predictions may also inform and improve decisions about spatial and industrial planning, as well as the management of various extractive sectors such as the adjustments to quotas for internationally shared fish stocks (Tommasi et al. 2017a). It has been shown that global climate forecasts have significant skill in predicting the occurrence of above average warm or cold SST events at decadal timescales in coastal areas (Tommasi et al. 2017a), but barriers to their widespread usage in fishery and aquaculture industry still exist (Tommasi et al. 2017b).

Even with a monitoring and prediction system in place, MHWs have developed without warning and had catastrophic effects (Payne et al. 2017). For example, governmental agencies, socioeconomic sectors, public health officials and citizens were not forewarned of the Coastal Peruvian 2017 MHW, despite a basin-wide monitoring system across the Pacific. The reason was partly due to a coastal El Niño definition problem and a new government (in Nicaragua) that may have hindered actions (Ramírez and Briones, 2017). Therefore, early warning systems should not only provide predictions of physical changes, but should also connect different institutions to assist decision makers in performing time-adaptive measures (Chang et al. 2013).

Monitoring and prediction systems are important and can be advanced by the use of common metrics to describe MHWs. So far, MHWs are often defined differently in the literature, and it is only recently that a categorising scheme (Categories I to IV; based on the degree to which temperatures exceed the local climatology), similar to what is used for hurricanes, has been developed (Hobday et al. 2018). Such a categorising scheme, can easily be applied to real data and

predictions, and may facilitate comparison, public communication and familiarity with MHWs. Similar metrics (e.g., DHW) have been successfully developed and used to identify ocean regions where conditions conducive to coral bleaching are developing.

6.5 Extreme ENSO Events and Other Modes of Interannual Climate Variability

6.5.1 Key Processes and Feedbacks, Observations, Detection and Attribution, Projections

6.5.1.1 Extreme El Niño, La Niña

AR5 (Christensen et al. 2013) and SREX do not provide a definition for an extreme El Niño but mention such events, especially in the context of the 1997–1998 El Niño and its impacts. AR5 and SREX concluded that confidence in any specific change in ENSO variability in the 21st century is low. However, they did note that due to increased moisture availability, precipitation variability associated with ENSO is likely to intensify. Since AR5 and SREX, there is now a limited body of literature that examines the impact of climate change on ENSO over the historical period.

Palaeo-ENSO studies suggest that ENSO was highly variable throughout the Holocene, with no evidence for a systematic trend in ENSO variance (Cobb et al. 2013) but with some indication that the ENSO variance over 1979–2009 has been much larger than that over 1590–1880 (McGregor et al. 2013). Palaeo-ENSO reconstruction for the past eight centuries suggests that central Pacific ENSO activity has increased between the last two decades (1980–2015; Liu et al. 2017b), with an increasing number of central Pacific El Niño events compared to east Pacific El Niño events (Freund et al. 2019). Further proxy evidence exists for changes in the mean state of the equatorial Pacific in the last 2000 years (Rustic et al. 2015; Henke et al. 2017). Simulations using an Earth System Model indicate significantly higher ENSO variance during 1645–1715 than during the 21st century warm period, though it is unclear whether these simulated changes are realistic (Keller et al. 2015). For the 20th century, the frequency and intensity of El Niño events were high during 1951–2000, in comparison with the 1901–1950 period (Lee and McPhaden, 2010; Kim et al. 2014b; Roxy et al. 2014). Current instrumental observational records are not long enough and the quality of data before 1950 is limited, to assert these changes with *high confidence* (Wittenberg, 2009; Stevenson et al. 2010) though the palaeo records mentioned here signal the emergence of a statistically significant increase in ENSO variance in recent decades.

Since SREX and AR5, an extreme El Niño event occurred in 2015–2016. This has resulted in significant new literature regarding physical processes and impacts but there are no firm conclusions regarding the impact of climate change on the event. The SST anomaly peaked toward the central equatorial Pacific causing floods in many regions of the world such as those in the west coasts of the USA and other parts of North America, some parts of South America close to Argentina and Uruguay, the UK and China (Ward et al. 2014; Ward et al. 2016; Zhai et al. 2016; Scaife et al. 2017; Whan and Zwiers, 2017; Sun and Miao, 2018; Yuan et al. 2018).

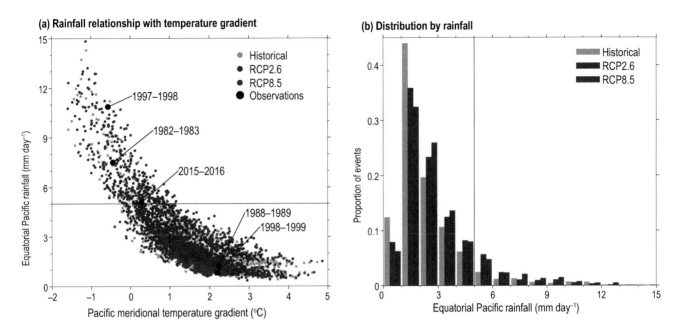

(a) Rainfall relationship with temperature gradient

(b) Distribution by rainfall

Figure 6.5 | Frequency of extreme El Niño Southern Oscillation (ENSO) events, adapted from Cai et al. (2014a). **(a)** December to February mean meridional sea surface temperature (SST) gradient (x-axis: 5°N–10°N, 210°E–270°E minus 2.5°S–2.5°N, 210°E–270°E) versus equatorial Pacific anomalous rainfall (y-axis: 5°S–5°N, 210°E–270°E). Data from only those Coupled Model Intercomparison Project Phase 5 (CMIP5) models that capture the observed relationship between Pacific SST and rainfall are shown. Black dots are from observations with extreme El Niño and extreme La Niña years indicated. The horizontal line denotes the threshold of 5 mm day^{-1} for an extreme event. **(b)** Histogram showing the relative frequency of rainfall rates. The vertical line denotes the 5 mm day^{-1} threshold. Higher counts of extreme events under the Representative Concentration Pathway (RCP)8.5 scenario suggest an increase in the frequency of extreme El Niño under global warming.

The main new body of literature concerns future projections of the frequency of occurrence and variability of extreme ENSO events with improved confidence (Cai et al. 2014a; Cai et al. 2018). These studies define extreme El Niño events as those El Niño events which are characterised by a pronounced eastward extension of the west Pacific warm pool and development of atmospheric convection, and hence a rainfall increase of greater than 5 mm day^{-1} during December to February (above 90th percentile), in the usually cold and dry equatorial eastern Pacific (Niño 3 region, 150°W–90°W, 5°S–5°N; Cai et al. 2014a), such as the 1982–1983, 1997–1998 and 2015–2016 El Niños (Santoso et al. 2017; Figure 6.5).

The background long-term warming puts the 2015–2016 El Niño among the three warmest in the instrumental records (24 El Niño events occurred during 1900–2018; Huang et al. 2016; Santoso et al. 2017). The 2015–2016 event can be viewed as the first emergence of an extreme El Niño in the 21st century – one which satisfies the rainfall threshold definition, but not characterised by the eastward extension of the west Pacific warm pool (L'Heureux et al. 2017; Santoso et al. 2017).

Based on the precipitation threshold, extreme El Niño frequency is projected to increase with the global mean temperatures (*medium confidence*) with a doubling in the 21st century under 1.5°C of global warming, from about one event every 20 years during 1891–1990, to one every 10 years (Cai et al. 2014a; Figure 6.5). The increase in frequency continues for up to a century even after global mean temperature has stabilised at 1.5°C, thereby challenging the limits to adaptation, and hence indicates high risk even at the 1.5°C threshold (Wang et al. 2017; Hoegh-Guldberg et al. 2018). Meanwhile, the La Niña events also tend to increase in frequency and double under

RCP8.5 (Cai et al. 2015), but indicate no further significant changes after global mean temperatures have stabilised (Wang et al. 2017). Particularly concerning is that swings from extreme El Niño to extreme La Niña (opposite of extreme El Niño) have been projected to occur more frequently under greenhouse warming (Cai et al. 2015). The increasing ratio of Central Pacific El Niño events to East Pacific El Niño events is projected to continue, under increasing emissions (Freund et al. 2019). Further, CMIP5 models indicate that the risk of major rainfall disruptions has already increased for countries where the rainfall variability is linked to ENSO variability. This risk will remain elevated for the entire 21st century, even if substantial reductions in global GHG emissions are made (*medium confidence*). The increase in disruption risk is caused by anthropogenic warming that drives an increase in the frequency and magnitude of ENSO events and also by changes in background SST patterns (Power et al. 2013; Chung et al. 2014; Huang and Xie, 2015). While many of these studies have adopted the precipitation view of an extreme El Nino, studies also indicate an increase in SST variability for events with their main SST anomalies in the east Pacific (Cai et al. 2018). Also, a role of cross-equatorial winds has been identified (Hu and Fedorov, 2018).

6.5.1.2 Indian Ocean Basin-wide Warming and Changes in Indian Ocean Dipole (IOD) Events

The Indian Ocean has experienced consistent warming from the surface to 2,000 m during 1960–2015, with most of the warming occurring in the upper 300 m (Cheng et al. 2015; Nieves et al. 2015; Cheng et al. 2017; Gnanaseelan et al. 2017). New historical ocean heat content (OHC) estimates show an abrupt increase in the Indian Ocean upper 700 m OHC after 1998, contributing to more than 21% of the global ocean heat gain, despite representing only about 12%

6

of the global ocean area (Cheng et al. 2017; Makarim et al. 2019). The tropical Indian Ocean SST has warmed by 1.04°C during 1950–2015, while the tropical SST warming is 0.83°C and the global SST warning is 0.65°C. More than 90% of the surface warming in the Indian Ocean has been attributed to changes in GHG emissions (Dong et al. 2014), with the heat redistributed in the basin via local ocean and atmospheric dynamics (Liu et al. 2015b), the ITF (Section 6.6.1; Susanto et al. 2012; Sprintall and Revelard, 2014; Lee et al. 2015b; Susanto and Song, 2015; Zhang et al. 2018) and the Walker circulation (Roxy et al. 2014; Abish et al. 2018).

The dynamic processes related to the projected changes in IOD under global warming have a large inter-model spread (Cai et al. 2013). The frequency of extreme positive IOD events are projected to increase by almost a factor of three, from a one-in-seventeen-year event in the 20th century to a one-in-six-year event in the 21st century (*low confidence*). The bias in the CMIP5 models and internal variability could enlarge the projected increase in the extreme positive IOD events (Li et al. 2016a; Hui and Zheng, 2018). The increase in IOD events is not linked to the change in the frequency of El Niño events but instead to mean state change—with weakening of both equatorial westerly winds and eastward oceanic currents in association with a faster warming in the western than the eastern equatorial Indian Ocean (Cai et al. 2014b). A combination of extreme ENSO and IOD events has led to a northward shift in the Intertropical Convergence Zone (ITCZ) during 1979–2015, which is expected to increase further in the future (Freitas et al. 2017).

6.5.2 Impacts on Human and Natural Systems

Increasing frequency of extreme ENSO and IOD events have the potential to have widespread impacts on natural and human systems in many parts of the globe. Though the occurrence of the extreme 2015–2016 El Niño has produced a large body of literature, it is still not clear how climate change may have altered such an impact, nor how such impacts might change in the future with increasing frequency of extreme ENSO events. We highlight here some studies that have attempted to assess the joint impact of mean change and variability. In addition to observed high variability of rainfall, severe weather events and impacts on TCs activity (Yonekura and Hall, 2014; Zhang and Guan, 2014; Wang and Liu, 2016; Zhan, 2017), extreme El Nino events have substantial impacts on natural systems which include those on marine ecosystems (Sanseverino et al. 2016; Mogollon and Calil, 2017; Ohman et al. 2017), such as severe and repeated bleaching of corals (Hughes et al. 2017a; Hughes et al. 2017b; Eakin et al. 2018), and glacial growth and retreat (Thompson et al. 2017). On the other hand, impacts on human, including managed systems are: increased incidences of forest fires (Christidis et al. 2018b; Tett et al. 2018), degraded air quality (Koplitz et al. 2015; Chang et al. 2016; Zhai et al. 2016) such as the dense haze over most parts of Indonesia and the neighboring countries in Southeast Asia as a result of prolonged Indonesian wildfires, thus imposing adverse impacts on public health in the affected areas (Koplitz et al. 2015; WMO, 2016), decreased agricultural yields in many parts of the globe (e.g., in most of the Pacific Islands countries, Thailand, eastern and southern Africa and others which resulted food insecurity, particularly in eastern and

southern Africa (UNSCAP, 2015; WMO, 2016; Christidis et al. 2018b; Funk et al. 2018), and regional uptick in the number of reported cases of plague and hantavirus in Colorado and New Mexico, cholera in Tanzania, dengue in Brazil and Southeast Asia (Anyamba et al. 2019) and Zika virus in South America (Caminade et al. 2017), including increases in heat stroke cases (Christidis et al. 2018b). Substantial economic losses had resulted from droughts and floods across various parts of the globe due to teleconnections. For instance, direct losses of 10 billion USD (Sun and Miao, 2018; Yuan et al. 2018) and 6.5 billion USD (Christidis et al. 2018b) were estimated to have been incurred from severe urban inundation in cities along the Yangtze River in China and the extreme drought in Thailand, respectively.

ENSO events affect TCs activity through variations in the low-level wind anomalies, vertical wind shear, mid-level relative humidity, steering flow, the monsoon trough and the western Pacific subtropical high in Asia (Yonekura and Hall, 2014; Zhang and Guan, 2014). The subsurface heat discharge due to El Niño can intensify TCs in the eastern Pacific (Jin et al. 2014; Moon et al. 2015b). TCs are projected to become more frequent (~20–40%) during future-climate El Niño events compared with present climate El Niño events (*medium confidence*), and less frequent during future-climate La Niña events, around a group of small island nations (for example, Fiji, Vanuatu, Marshall Islands and Hawaii) in the Pacific (Chand et al. 2017). The Indian Ocean basin-wide warming has led to an increase in TC heat potential in the Indian Ocean over the last 30 years, however the link to the changes in the frequency of TCs is not robust (Rajeevan et al. 2013).

During the early stages of an extreme El Niño event (2015–2016 El Niño), there is an initial decrease in atmospheric CO_2 concentrations over the tropical Pacific Ocean, due to suppression of equatorial upwelling, reducing the supply of CO_2 to the surface (Chatterjee et al. 2017), followed by a rise in atmospheric CO_2 concentrations due reduced terrestrial CO_2 uptake and increased fire emissions (Bastos et al. 2018). It is not clear how a future increase in the frequency extreme events would modulate the carbon cycle on longer decadal time scales.

Studies on projections of changes in ENSO impacts or teleconnections are rather limited. Nevertheless, Power and Delage (2018) provide a multi-model assessment of CMIP5 models and their simulated changes in the precipitation response to El Niño in the future (Figure 6.6). They identify different combinations of changes that might further impact natural and human systems. El Niño causes either positive or negative precipitation anomalies in diverse regions of the globe. Dry El Niño teleconnection anomalies may be further strengthened by, either mean climate drying in the region (Amazon, Central America and Australia in June to August (JJA)), or a strengthening of the El Niño dry teleconnection, or both. Conversely, wet El Niño teleconnections can be further strengthened by either increases in mean precipitation (East Africa and southeastern South America in December to February (DJF)) or a strengthening of the El Niño wet teleconnection (southeastern South America in JJA), or both (Tibetan Plateau, DJF). However, a present day dry El Niño response may be dampened by a wet mean response (South, East and Southeast Asia in JJA) or a wet present day El Niño response may be weakened by a dry mean change (Southern Europe/Mediterranean

(a) June-August

(b) December-February

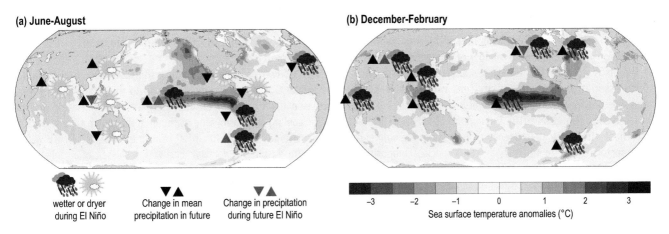

wetter or dryer
during El Niño

Change in mean
precipitation in future

Change in precipitation
during future El Niño

Sea surface temperature anomalies (°C)

Figure 6.6 | Schematic figure indicating future changes in El Niño teleconnections based on the study of Power and Delage (2018). The background pattern of sea surface temperature (SST) anomalies (°C) are averaged from June 2015 to August 2015 (panel a) and December 2015 to February 2016 (panel b), during the most recent extreme El Niño event (anomalies computed with respect to 1986–2005). Symbols indicate present day teleconnections for El Niño events. Black arrows indicate if there is a model consensus on change in mean rainfall in the region. Red arrows indicate if there is a model consensus on change in the rainfall anomaly under a future El Niño event. Direction of the arrow indicates whether the response in precipitation is increasing (up) or decreasing (down). Significance is determined when two-thirds or more of the models agree on the sign.

and West Coast South America in JJA). Finally, changes in the mean and El Niño response may be in the opposite direction (Southeast Asia, JJA and Central North America, DJF). Such changes could have an impact on phenomena such as wildfires (Fasullo et al. 2018). However, in many other regions that are currently impacted by El Niño, e.g., regions of South America, studies have found no significant changes in the ENSO-precipitation relationship (Tedeschi and Collins, 2017) and agreement between models for many regions suggests *low confidence* in projections of teleconnection changes (Yeh et al. 2018).

Along with extreme El Niño events, abrupt warming in the Indian Ocean and extreme IOD events have largely altered the Asian and African monsoon, impacting the food and water security over these regions. As a response to rising global SSTs and partially due to extreme El Niño events, the NH summer monsoon showed substantial intensification during 1979–2011, with an increase in rainfall by 9.5% per degree Celsius of global warming (Wang et al. 2013). However, the Indian summer monsoon circulation and rainfall exhibits a statistically significant weakening since the 1950s. This weakening has been hypothesised to be a response to the Indian Ocean basin-wide warming (Mishra et al. 2012; Roxy et al. 2015) and also to increased aerosol emissions (Guo et al. 2016) and changes in land use (Paul et al. 2016). Warming in the north Indian Ocean has resulted in increasing fluctuations in the southwest monsoon winds and a three-fold increase in extreme rainfall events across central India (Roxy et al. 2017). The frequency and duration of heatwaves have increased over the Indian subcontinent, and these events are associated with the Indian Ocean basin-wide warming and frequent El Niños (Rohini et al. 2016). In April 2016, as a response to the extreme El Niño, Southeast Asia experienced surface air temperatures that surpassed national records, increased energy consumption, disrupted agriculture and resulted in severe human discomfort (Thirumalai et al. 2017). A strong negative IOD event in 2016 led to large climate impact on East African rainfall, with some regions recording below 50% of normal rainfall, leading to devastating drought, food insecurity and unsafe drinking water for over 15 million people in Somalia, Ethiopia and Kenya.

6.5.3   Risk Management and Adaptation

Risk management of ENSO events has focussed on two main aspects: better prediction and early warning systems, and better mechanisms for reducing risks to agriculture, infrastructure, fisheries and aquaculture, wildfire and flood management. Extreme ENSO events are rare, with three such events since 1950 and they are difficult to predict due to the different drivers influencing them (Puy et al. 2017). The impacts of ENSO events also vary between events and between the different regions affected (Murphy et al. 2014; Fasullo et al. 2018; Power and Delage, 2018) however, there is limited literature on the change in the impacts of extreme ENSO compared to other ENSO events. In addition, there are also no specific risk management and adaptation strategies for human and natural systems for more extreme events other than what is in place for ENSO events (see also Chapter 4, Section 4.4 for the response to sea level change, an observed impact of ENSO). A first step in risk management and adaptation is thus to better understand the impacts these events have and to identify conditions that herald such extreme events that could be used to better predict extreme ENSO events.

Monitoring and forecasting are the most developed ways to manage extreme ENSOs. Several systems are already in place for monitoring and predicting seasonal climate variability and ENSO occurrence. However, the sustainability of the observing system is challenging and currently the Tropical Pacific Observing System 2020 (TPOS 2020) has the task of redesigning such a system, with ENSO prediction as one of its main objectives. These systems could be further elaborated to include extreme ENSO events. Westerly wind events in the Western Tropical Pacific, (Lengaigne et al. 2004; Chen et al. 2015a; Fedorov et al. 2015) strong easterly wind events in the tropical Pacific (Hu and Fedorov, 2016; Puy et al. 2017), nonlinear interaction between air-sea fluxes and atmospheric deep convection (Bellenger et al. 2014; Takahashi and Dewitte, 2016) and advection of mean temperature by anomalous eastward zonal currents (Kim and Cai, 2014) are some of the factors that play an important role in the evolution of extreme ENSO events, which can be considered while improving the monitoring and forecasting system.

6

Despite the specificity of each extreme El Niño event, their forecasting is expected to improve through monitoring of recently identified precursory signals that peak in a window of two years before the event (Varotsos et al. 2016). An early warning system for coral bleaching associated, among other stressors, with extreme ENSO heat stress is provided by the NOAA Coral Reef Watch service with a 5 km resolution (Liu et al. 2018). The impacts of ENSO-associated extreme heat stress are heterogeneous, indicating the influence of other factors either biotic such as coral species composition, local adaptation by coral taxa reef depth or abiotic such as local upwelling or thermal anomalies (Claar et al. 2018). When identified and quantified, these factors can be used for risk analysis and risk management for these ecosystems.

In principle, it is easier to transfer the financial risk associated with extreme ENSO events through, for example, insurance products or other risk transfer instruments such as Catastrophe Bonds, than for more moderate events. An accurate prediction system is not required, but the measurement of these events, and quantification of likely impacts is required. As in other types of insurance systems, this can be done through, for example, calculations of average annual losses associated with extreme ENSO, and the design of appropriate financial instruments. Examples of research that can support the design of risk transfer instruments include Anderson et al. (2018) and Gelcer et al. (2018) for specific crops yields, and Aguilera et al. (2018) and Broad et al. (2002) for specific fisheries. Several risk transfer instruments have been implemented to deal with ENSO impacts, including parametric insurance based on SSTs for heavy rainfall damages, and another scheme for agricultural damages, both in Peru. Other examples include forecast-based financial aid (Red Cross Climate Centre, 2016). More broadly, other forms of risk management and governance can be designed with better information about the likely impacts of extreme ENSO events (e.g., Vignola et al. 2018).

6.6   Inter-Ocean Exchanges and Global Change

Section 3.6.5.1 in AR5 briefly described the Indonesian Throughflow (ITF) but did not explain its variability and impacts. Palaeoclimate record, observations, and climate model studies suggest that ITF plays an integral role in global ocean circulation, directly impacting mass, heat and freshwater budgets of the Pacific and Indian Oceans (*high confidence*). ITF is influenced by equatorial Pacific trade wind system which experienced an unprecedented intensification during 2001–2014, resulting in enhanced ocean heat transport from the Pacific to the Indian Ocean and influencing the rate of global temperature change (*medium confidence*). Yet, numerical models are not able to simulate the magnitude of decadal variability and the inter-ocean link, which means there is *low confidence* in future projections of the trade wind system.

6.6.1   Key Processes and Feedbacks, Observations, Detection and Attribution, Projections

In the last two decades, total water transport from the Pacific to the Indian Ocean and the Indian Ocean to the Atlantic Ocean has increased (*high confidence*). Increased ITF has been attributed to Pacific cooling and basin-wide warming in the Indian Ocean. The ITF annual average is 15×10^6 m^3 s^{-1} (Susanto et al. 2012). ITF varies from intraseasonal to decadal time scales. On seasonal time scale, South China Sea Throughflow controls freshwater flux and modulates the main ITF (Fang et al. 2010; Susanto et al. 2013; Lee et al. 2019; Wang et al. 2019; Wei et al. 2019). During the extreme El Niño of 1997–1998, the ITF transport was reduced to 9.2×10^6 m^3 s^{-1}. Based on observations and proxy records from satellite altimetry and gravimetry, in the last two decades, 1992–2012, ITF has been stronger (Sprintall and Revelard, 2014; Liu et al. 2015a; Susanto and Song, 2015), which translates to an increase in ocean heat-flux into the Indian Ocean (Lee et al. 2015b). Exchanges of heat and fresh water between ocean basins are important at the global scale (Flato et al. 2013). ITF may have played a key role in the slowdown of the Pacific SST warming during 1998–2013, and the rapid warming in the surface and subsurface Indian Ocean during this period (Section 6.5.1.2; Makarim et al. 2019), by transferring warm water from the western Pacific into the Indian Ocean (Lee et al. 2015b; Dong and McPhaden, 2018).

Under 1.5°C warming both El Niño and La Niña frequencies may increase (see Section 6.5) and hence ITF variability may also increase. ITF is also influenced by the IOD events, with an increase in transport during a positive IOD and vice-versa during a negative IOD event (Potemra and Schneider, 2007; Pujiana et al. 2019). Positive IODs are projected to increase threefold in the 21st century as a response to changes in the mean state rather than changes in the El Niño frequency (Section 6.5.1.2; Cai et al. 2014b) and this may have an impact on the ITF, additional to the changes due to increasing extreme ENSO events. In response to greenhouse warming, climate models predict that on interannual time scale, it is *likely* that the mean ITF may decrease due to wind variability (Sen Gupta et al. 2016), but recent observation trend tends to strengthen which has led to speculations about the fidelity of the current climate models (Chung et al. 2019). On multidecadal and centennial timescales, it is *likely* that mean ITF decreases which is not associated with wind variability but due to reduction of net deep ocean upwelling in the tropical South Pacific (Sen Gupta et al. 2016; Feng et al. 2017; Feng et al. 2018). Due to a lack of long-term sustained ITF observations, their impacts on Indo-Pacific climate varibility, biogeochemisty, ecosystem as well as society are not fully understood.

Pacific SST cooling trends and strengthened the equatorial Pacific trade winds have been linked to anomalously warm tropical Indian and Atlantic oceans. The period following the mid-1990s saw a marked strengthening of both the easterly trade winds in the central equatorial Pacific (Figure 6.7) and the Walker circulation (L'Heureux et al. 2013; England et al. 2014). Both the magnitude and duration of this trend are large when compared with past variability reconstructed using atmosphere reanalyses. (The 1886–1905 extreme weakening trend is poorly constrained by observations and we note the disparity

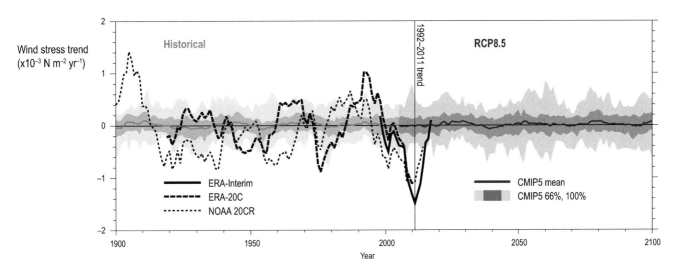

Figure 6.7 | Running twenty-year trends of zonal wind stress over the central Pacific (area-averaged over 8°S–8°N and 160°E–150°W) in Coupled Model Intercomparison Project Phase 5 (CMIP5) models and three reanalyses: European Centre for Medium-Range Weather Forecasts (ECMWF) Interim re-analysis, ERA-Interim (Dee et al. 2011), ECMWF 20th century reanalysis, ERA-20C (Poli et al. 2016), and the National Oceanic and Atmospheric Administration's (NOAA) 20th century reanalysis, NOAA 20CR v2c (Compo et al. 2011). The 66% and 100% ranges of all available CMIP5 historical simulations with Representative Concentration Pathway (RCP)8.5 extension are shown.

between reanalysis products going back in time.) Moreover, it is very unusual when model simulations are used as an estimate of internal climate variability (Figure 6.7; England et al. 2014; Kociuba and Power, 2015). The slowdown in global surface warming is dominated by the cooling in the Pacific SSTs, which is associated with a strengthening of the Pacific trade winds (Kosaka and Xie, 2013). This pattern leads to cooling over land and possibly to additional heat uptake by the ocean, although recent studies suggest that ocean heat uptake may even slow down during surface warming slowdown periods (Xie et al. 2016; von Känel et al. 2017). The intensification of the Pacific trade winds has been related to inter-ocean basin SST trends, with rapid warming in the Indian (see section 6.5.1.2) and Atlantic Oceans both hypothesised as drivers (Kucharski et al. 2011; Luo et al. 2012; McGregor et al. 2014; Zhang and Karnauskas, 2017). While the extreme event of strengthening trade winds are potentially a result of natural internal variability, a role of anthropogenic contribution has not been ruled out. Nevertheless, the CMIP5 models indicate no general change in trends into the future (Figure 6.7), giving more weight to natural internal variability as an explanation.

Among the number of potential causes of this decadal variability in surface global temperature, a prolonged negative phase of the Pacific Decadal Oscillation/Interdecadal Pacific Oscillation (PDO/IPO) was suggested as a contributor. Because of the magnitude and duration of this Pacific-centred variability (Figure 6.7), it is identified as an extreme decadal climate event. One line of research has explored the role of the warm tropical Atlantic decadal variability in forcing the trade wind trends and associated cooling Pacific SST trends (Kucharski et al. 2011; McGregor et al. 2014; Li et al. 2016b). It appears that climate models may misrepresent this link due to tropical Atlantic biases (Kajtar et al. 2018; McGregor et al. 2018) and thus potentially underestimate global mean temperature decadal variability. Nevertheless, there is no indication that such an underestimation of global temperature variability is evident in the models (Flato et al. 2013; Marotzke and Forster, 2015). The impact of modes of natural variability on global mean temperature decadal variability remains an active area of research.

In the Indian Ocean, water exits the Indonesian Seas mostly flowing westward along with the South Equatorial Current, and some supplying the Leeuwin Current. The South Equatorial Current feeds the heat and biogeochemical signatures from the Indian Ocean into the Agulhas Current, which transports it further into the Atlantic Ocean. Observations of Mozambique Channel inflow from 2003–2012 measured a mean transport of 16.7×10^6 m^3 s^{-1} with a maximum in austral winter, and IOD related interannual variability of 8.9×10^6 m^3 s^{-1} (Ridderinkhof et al. 2010). A multidecadal proxy, from three years of mooring data and satellite altimetry, suggests that the Agulhas Current has been broadening since the early 1990s due to an increase in eddy kinetic energy (Beal and Elipot, 2016). Numerical model experiments suggest an intensification of the Agulhas leakage since the 1960s, which has contributed to the warming in the upper 300 m of the tropical Atlantic Ocean (Lübbecke et al. 2015). Agulhas leakage is found to covary with the AMOC on decadal and multi-decadal timescales and has *likely* contributed to the AMOC slowdown (Biastoch et al. 2015; Kelly et al. 2016). Meanwhile, climate projections indicate that Agulhas leakage is *likely* to strengthen and may partially compensate the AMOC slowdown projected by coarse-resolution climate models (Loveday et al. 2015).

6.6.2 Impacts on Natural and Human Systems

Interannual to decadal variability of Indo-Pacific SST variability is *likely* to affect extreme hydroclimate in East Africa (Ummenhofer et al. 2018). The Pacific cooling pattern is often synonymous with predominance of La Niña events in 1998 and 2012 is linked to megadroughts in the USA (Baek et al. 2019). On decadal to multidecadal time scales, PDO/IPO and Atlantic variability may have impacts on megadroughts in North America (Coats et al. 2016; Diodato et al. 2019) and Australia (Vance et al. 2015) as well as Indian subcontinent (Bao et al. 2015; Joshi and Rai, 2015). It is *likely* that occurrence of megadroughts in North America and Australia increased (Kiem et al. 2016; Baek et al. 2019). PDO and North Pacific Gyre Oscillation may also

6

influence the decadal variability of North Pacific nutrient, chlorophyll and zooplankton taxa (Di Lorenzo et al. 2013).

The Pacific cooling pattern may have significant impacts on terrestrial carbon uptake via teleconnections. The reduced ecosystem respiration due to the smaller warming over land has significantly accelerated the net biome productivity and therefore increased the terrestrial carbon sink (Ballantyne et al. 2017) and paused the growth rate of atmospheric CO_2 despite increasing anthropogenic carbon emissions (Keenan et al. 2016). During the 2000s, the global ocean carbon sink has also strengthened (Fay and McKinley, 2013; Landschützer et al. 2014; Majkut et al. 2014; Landschützer et al. 2015; Munro et al. 2015), reversing a trend of stagnant or declining carbon uptake during the 1990s. It has been suggested that the upper ocean overturning circulation has weakened during the 2000s thereby decreasing the outgassing of natural CO_2, especially in the Southern Ocean (Landschützer et al. 2015), and enhanced the global ocean CO_2 sink (DeVries et al. 2017). How this is connected to the global warming slowdown is currently unclear.

6.7   Risks of Abrupt Change in Ocean Circulation and Potential Consequences

6.7.1   Key Processes and Feedbacks, Observations, Detection and Attribution, Projections

6.7.1.1   Observational and Model Understanding of Atlantic Ocean Circulation Changes

Palaeo-reconstructions indicate that the North Atlantic is a region where rapid climatic variations can occur (IPCC, 2013). Deep waters formed in the northern North Atlantic induces a large-scale AMOC which transports large amounts of heat northward across the hemispheres, explaining part of the difference in temperature between the two hemispheres, as well as the northward location of the ITCZ (e.g., Buckley and Marshall, 2016). This circulation system is believed to be a key tipping point of the Earth's climate system (IPCC, 2013).

Considerable effort has been dedicated in the last decades to improve the observation system of the large-scale ocean circulation (e.g., Argo and its array of about 3,800 free-drifting profiling floats), including the AMOC through dedicated large-scale observing arrays (at 16°N (Send et al. 2011) and 26°N (McCarthy et al. 2015b), in the subpolar gyre (SPG) (Lozier et al. 2017), between Portugal and the tip of Greenland (Mercier et al. 2015), at 34.5°S (Meinen et al. 2013), among others). The strength of the AMOC at 26°N has been continuously estimated since 2004 with an annual mean estimate of $17 \pm 1.9 \times 10^6$ m^3 s^{-1} over the 2004–2017 period (Smeed et al. 2018). The AMOC at 26°N has been 2.7×10^6 m^3 s^{-1} weaker in 2008–2017 than in the first four years of measurement (Smeed et al. 2018). However, the record is not yet long enough to determine if there is a long-term decline of the AMOC. McCarthy et al. (2012) reported a 30% reduction in the AMOC in 2009–2010, followed by a weaker minimum a year later. Analysis of forced ocean models suggests such events may occur once every two or three decades

(Blaker et al. 2015). At 34.5°S, the mean AMOC is estimated as $14.7 \pm 8.3 \times 10^6$ m^3 s^{-1} over the period 2009–2017 (Meinen et al. 2018) also with large interannual variability, while no trend has been identified at this latitude. Estimates based on ocean reanalyses show considerable diversity in their AMOC mean state, and its evolution over the last 50 years (Karspeck et al. 2017; Menary and Hermanson, 2018), because only very few deep ocean observations before the Argo era, starting around 2004, are available. During the Argo era, the reanalyses agree better with each other (Jackson et al. 2016).

During the last interglacial warm period, palaeo-data suggest that the AMOC may have been weaker (Govin et al. 2012) and also show proxy record evidences of instabilities (Galaasen et al. 2014). Based on an AMOC reconstruction using SST fingerprints, it has been suggested that the AMOC may have experienced around $3 \pm 1 \times 10^6$ m^3 s^{-1} of weakening (about 15% decrease) since the mid-20th century (Caesar et al. 2018). Such a trend in AMOC was also suspected in a former study using Principal Component Analysis of SST (Dima and Lohmann, 2010). Palaeo-proxies also highlight that the historical era may exhibit an unprecedented low AMOC over the last 1,600 years (Sherwood et al. 2011; Rahmstorf et al. 2015; Thibodeau et al. 2018; Thornalley et al. 2018). Nevertheless, these proxy records are indirect measurements of the AMOC so that considerable uncertainty remains concerning these results. Moreover, the exact mechanisms to explain such a long-term weakening are not fully understood and some reconstructions show a weakening starting very early in the historical era, when the level of anthropogenic perturbation and warming was very low. Climate model simulations (Figure 6.8) do show a weakening over the historical era, but this weakening is mainly occurring over the recent decades. Climate projections exhibit a weakening of around $1.4 \pm 1.4 \times 10^6$ m^3 s^{-1} for present day (2006–2015) minus pre-industrial (1850–1900), highlighting that anthropogenic warming may have already forced an AMOC weakening. Nevertheless, no proper detection and attribution of the on-going changes has been led so far due to still limited observational evidences. Thus, we conclude that there is *medium confidence* that the AMOC has weakened over the historical era but there is insufficient evidence to quantify a *likely* range of the magnitude of the change.

Examination of 14 models from the CMIP5 archive, which do not take into account the melting (either from runoff, basal melting or icebergs) from the GIS (cf. Section 6.7.1.2), led to the assessment that the AMOC is *very unlikely* to collapse in the 21st century in response to increasing GHG concentrations (IPCC, 2013). Nonetheless, the CMIP5 models agree that a weakening of the AMOC into the 21st century will lead to localised cooling (relative to the global mean) centred in the North Atlantic SPG (Menary and Wood, 2018), although the precise location as well as the extension of this cooling patch, notably towards Europe, remains uncertain (Sgubin et al. 2017; Menary and Wood, 2018).

Abrupt variations in SST or sea ice cover have been found in 19 out of the 40 models of the CMIP5 archive (Drijfhout et al. 2015). Large cooling trends, which can occur in a decade, are found in the subpolar North Atlantic in 9 out of 40 models. Results show that the heat transport in the AMOC plays a role in explaining such a rapid cooling, but other processes are also key for setting the rapid (decadal-scale) timeframe of SPG cooling, notably vertical heat transport in the ocean and interactions with sea ice and the atmosphere (Sgubin et al. 2017). Using the representation of stratification as an emergent constraint, rapid changes in subpolar convection and associated cooling are occurring in the 21st century in 5 of the 11 best models (Sgubin et al. 2017). The poor representation of ocean deep convection in most CMIP5 models has been confirmed in Heuze (2017), which can notably limit a key feedback mechanism related with warm summer in the North Atlantic and its impact on oceanic convection in winter (Oltmanns et al. 2018). Thus, there is *low confidence* in the projections of SPG fate. Increasing the horizontal resolution of the ocean in next generation climate models might be a way to increase confidence in ocean convection future changes.

The SPG dynamical system has been identified as a tipping element of the climate system (Mengel et al. 2012; Born et al. 2013). If this element reaches its tipping point, the SPG circulation can change very abruptly between different stable steady states, due to positive feedback between convective activity and salinity transport within the gyre (Born et al. 2016). It has been argued that a transition between two SPG stable states can explain the onset of the Little Ice Age that may have occurred around the 14–15th century (Lehner et al. 2013; Schleussner et al. 2015; Moreno-Chamarro et al. 2017) possibly triggered by large volcanic eruption (Schleussner and Feulner, 2013). Furthermore a few CMIP5 climate models also

showed a rapid cooling in the SPG within the 1970s cooling events, as a nonlinear response to aerosols (Bellucci et al. 2017). The SPG therefore appears as a tipping element in the climate system, with a faster (decade) response than the AMOC (century), but with lower induced SST cooling. Thus, the SPG system can cross a threshold in climate projections when surface water in the subpolar becomes lighter due to increase in temperature and decrease in salinity related with changes in radiative forcing (Sgubin et al. 2017).

Evaluation of AMOC variations in the CMIP5 database has been further analysed in this report (Figure 6.8) using almost twice as many models as in AR5 (IPCC, 2013). The AR5 assessment of a *very unlikely* AMOC collapse has been confirmed, although one model (FGOALS-s2) does show such a collapse (e.g., decrease larger than 80% relative to present day) before the end of the century for RCP8.5 scenario (Figure 6.8). Now based on up to 27 model simulations, the decrease of the AMOC is assessed to be of $-2.1 \pm 2.6 \times 10^6$ m^3 s^{-1} ($-11 \pm 14\%$, *likely* range) in 2081–2100 relative to present day (2006–2015) for RCP2.6 scenario and $-5.5 \pm 2.7 \times 10^6$ m^3 s^{-1} ($-32 \pm 14\%$) for RCP8.5 scenario, in line with a process-based probabilistic assessment (Schleussner et al. 2014). Furthermore, the uncertainty in AMOC changes has been shown to be mainly related to the spread in model responses rather than scenarios (RCP4.5 and RCP8.5) or internal variability uncertainty (Reintges et al. 2017). This behaviour is very different from the uncertainty in global SST

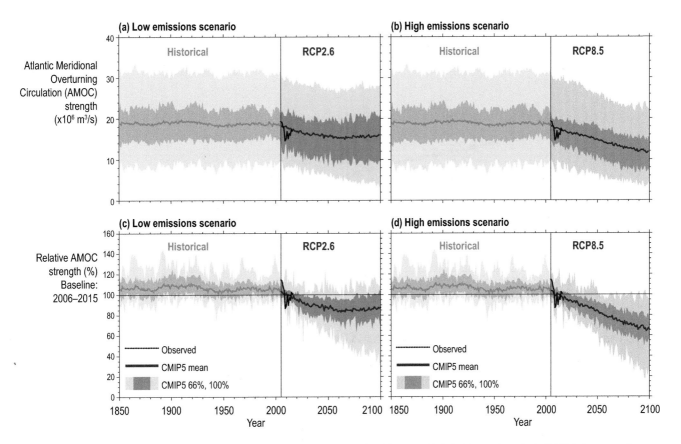

Figure 6.8 | Atlantic Meridional Overturning Circulation (AMOC) changes at 26°N as simulated by 27 models (only 14 were shown in the IPCC 5th Assessment Report (AR5); IPCC, 2013). The dotted line shows the observation-based estimate at 26°N (McCarthy et al. 2015b) and the thick grey/blue/red lines the multi-model ensemble mean. Values of AMOC maximum at 26°N (in units 10^6 m^3 s^{-1}) are shown in historical simulations (most of the time 1850–2005) followed for 2006–2100 by **a)** Representative Concentration Pathway (RCP)2.6 simulations and **b)** RCP8.5 simulations. In c) and d), the time series show the AMOC strength relative to the value during 2006–2015, a period over which observations are available. **c)** shows historical followed by RCP2.6 simulations and **d)** shows historical followed by RCP8.5 simulations. The 66% and 100% ranges of all-available CMIP5 simulations are shown in grey for historical, blue for RCP2.6 scenario and red for RCP8.5 scenario.

changes, which is mainly driven by emission scenario after a few decades (Frölicher et al. 2016). To explain the AMOC decline, a new mechanism has been proposed on top of the classical changes in heat and freshwater forcing (Gregory et al. 2016). A potential role for sea ice decrease has been highlighted (Sevellec et al. 2017), due to large heat uptake increase in the Arctic leading to a strong warming of the North Atlantic, increasing the vertical stability of the upper ocean, as already observed in the Greenland and Iceland seas (Moore et al. 2015). It has also been showed that convection sites may move northward in future projections, following the sea ice edge (Lique and Thomas, 2018).

6.7.1.2   Role of GIS Melting and their Freshwater Release Sources

Satellite data indicate accelerated mass loss from the GIS beginning around 1996, and freshwater contributions to the subpolar North Atlantic from Greenland, Canadian Arctic Archipelago glaciers and sea ice melt totalling around 60,000 m³ s⁻¹ in 2013, a 50% increase since the mid-1990s (Yang et al. 2016b), in line with more recent estimates (Bamber et al. 2018). This increase in GIS melting is unprecedented over the last 350 years (Trusel et al. 2018). Since the mid-1990s, there has been about a 50% decrease in the thickness of the dense water mass formed in the Labrador Sea, suggesting a possible relationship

between enhanced freshwater fluxes and suppressed formation of North Atlantic Deep Water (Yang et al. 2016b). This hypothesis has been further supported by high-resolution ocean-only simulations showing that GIS melting may have affected the Labrador Sea convection since 2010, which may imply an emerging on-going impact of this melting on the SPG but a still non-detectable impact on the AMOC (Boning et al. 2016). Thus, while some studies argue that this melting may have affected the evolution of the AMOC over the 20th century (Rahmstorf et al. 2015; Yang et al. 2016b), considerable variability and limitation in ocean models restrain the full validation of this hypothesis, which remains model dependent (Proshutinsky et al. 2015; Dukhovskoy et al. 2016). Furthermore, some deep convection events resumed since 2014 (Yashayaev and Loder, 2017).

The impact of GIS melting is neglected in AR5 projections (Swingedouw et al. 2013) but has been considered in a recent multi-model study (Bakker et al. 2016; Figure 6.9). The decrease of the AMOC in projections including this melting term is depicted in Figure 6.9. GIS melting estimates added in those simulations were based on the Lenaerts et al. (2015) approach, using a regional atmosphere model to estimate GIS mass balance. Results from eight climate models and an extrapolation by an emulator calibrated on these models showed that GIS melting has an impact on the AMOC, potentially adding up to around 5–10% more AMOC weakening in

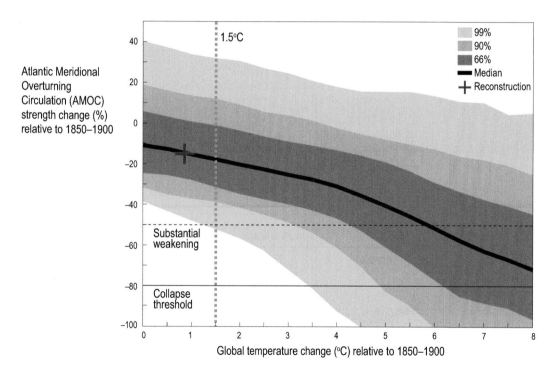

Figure 6.9 | The changes in the Atlantic Meridional Overturning Circulation (AMOC) strength as a function of transient changes in global mean temperature for projections from RCP4.5 and RCP8.5 scenario. This probabilistic assessment of annual mean AMOC strength changes (%) at 26°N (below 500 m and relative to 1850–1900) as a function of global temperature change (degrees Celsius; relative to 1850–1900) results from 10,000 RCP4.5 and 10,000 RCP8.5 experiments over the period 2006–2300, which are derived from an AMOC emulator calibrated with simulations from eight climate models including the Greenland Ice Sheet (GIS) melting (Bakker et al. 2016). The annual mean AMOC strength changes are taken from transient simulations and are therefore not equilibrium values *per se*. Moreover, it should be stressed that the results stem from future runs, not past or historical runs. Thus, due to internal variability both in the global mean temperature and AMOC in this transient simulation, large weakening can be found even at 0°C global warming. The ranges (66%, 90% and 99%) correspond to the amount of simulations that are within each envelope. The thick black line corresponds to the ensemble mean, while the different colours stand for different probability quantiles. The horizontal black thick line corresponds to the value of 80% of AMOC decrease, which can be seen as an almost total collapse of the AMOC. The horizontal black dashed thick line corresponds to a reduction of 50% of the AMOC, which can be considered as a substantial weakening. The vertical dashed green line stands for the 1.5°C of global warming threshold (relative to 1850–1900). The violet cross stands for the observation-based reduction estimate from Caesar et al. (2018). The size of the cross represents the uncertainty in this estimate.

2100 under RCP8.5. Based on Figure 6.8 and 6.9, the risk of collapse before the end of the century is *very unlikely*, although biases in present-day climate models only provide *medium confidence* in this assessment. By 2290–2300, Bakker et al. (2016; Figure 6.9) estimated at 44% the likelihood of an AMOC collapse in RCP8.5 scenario, while the AMOC weakening stabilises in RCP4.5 (37% reduction, (15–65%) *very likely* range). This result suggests that an AMOC collapse can be avoided in the long term by mitigation.

Concerning the question of the reversibility of the AMOC, a few ramp-up/ramp-down simulations have been performed to evaluate it for transient time scales (a few centuries, while millennia will be necessary for a full steady state). Results usually show a reversibility of the AMOC (Jackson et al. 2014; Sgubin et al. 2015) although the timing and amplitude is highly model dependent (Palter et al. 2018). A hysteresis behaviour of the AMOC in response to freshwater release has been found in a few climate models (Hawkins et al. 2011; Jackson et al. 2017) even at the eddy resolving resolution (Mecking et al. 2016; Jackson and Wood, 2018). This is in line with the possibility of tipping point in the AMOC system. The biases of present-day models in representing the transport at 30°S (Deshayes et al. 2013; Liu et al. 2017a; Mecking et al. 2017) or the salinity in the tropical era (Liu et al. 2014b) may considerably affect the sensitivity of the models to freshwater release, but more on the multi-centennial time scale.

Regarding the near-term changes of the AMOC, decadal prediction systems are now in place. They indicate a clear impact of the AMOC on the climate predictability horizon (Robson et al. 2012; Persechino et al. 2013; Robson et al. 2013; Wouters et al. 2013; Msadek et al. 2014; Robson et al. 2018), and a possible weakening of the AMOC in the coming decade (Smith et al. 2013; Hermanson et al. 2014; Yeager et al. 2015; Robson et al. 2016), although not true in all decadal prediction systems (Yeager et al. 2018). All these prediction systems do not account for future melting of the GIS yet.

6.7.2   Impacts on Climate, Natural and Human Systems

Even though the AMOC is *very unlikely* to collapse over the 21st century, its weakening may be substantial, which may therefore induce strong and large-scale climatic impacts with potential far-reaching impacts on natural and human systems (e.g., Good et al. 2018). Furthermore, the SPG subsystem has been shown to potentially shift, in the future, into a cold state over a decadal time scale, with significant climatic implications for the North Atlantic bordering regions (Sgubin et al. 2017). There have been far more studies analysing impacts on climate of an AMOC weakening than SPG collapse. We will thus in the following mainly depict impacts of an AMOC substantial weakening.

The AR5 report concludes that based on palaeoclimate data, large changes in the Atlantic Ocean circulation can cause worldwide climatic impacts (Masson-Delmotte et al. 2013), with notably, for an AMOC weakening, a cooling of the North Atlantic, a warming of the South Atlantic, less evaporation and therefore precipitation over the North Atlantic, and a shift of the ITCZ. Impacts of AMOC

or SPG changes and their teleconnections in the atmosphere and ocean are supported by a large amount of palaeo-evidence (Lynch-Stieglitz, 2017). Such impacts and teleconnections have been further evaluated over the last few years both using new palaeo-data and higher resolution models. Furthermore, multi-decadal variations in SST observed over the last century, the so-called Atlantic Multidecadal Variability (AMV) or Atlantic Multidecadal Oscillation (AMO), also provide observational evidence of potential impacts of changes in ocean circulation. Nevertheless, due to a lack of long-term direct measurements of the Atlantic Ocean circulation, the exact link between SST and circulation remains controversial (Clement et al. 2015; Zhang, 2017).

The different potential impacts of large changes in the Atlantic Ocean circulation are summarised in Figure 6.10. Based on variability analysis, it has been shown that a decrease in the AMOC strength has impacts on storm track position and intensity in the North Atlantic (Gastineau et al. 2016), with a potential increase in the number of winter storms hitting Europe (Woollings et al. 2012; Jackson et al. 2015), although some uncertainty remains with respect to the models considered (Peings et al. 2016). The influence on the Arctic sea ice cover has also been evidenced at the decadal scale, with a lower AMOC limiting the retreat of Arctic sea ice (Yeager et al. 2015; Delworth and Zeng, 2016). The climatic impacts could be substantial over Europe (Jackson et al. 2015), where an AMOC weakening can lead to high pressure over the British Isles in summer (Haarsma et al. 2015), reminiscent of a negative summer NAO, inducing an increase in precipitation in Northern Europe and a decrease in Southern Europe. In winter, the response of atmospheric circulation may help to reduce the cooling signature over Europe (Yamamoto and Palter, 2016), notably through an enhancement of warming maritime effect due to a stronger storm track (Jackson et al. 2015), driving more powerful storms in the North Atlantic (Hansen et al. 2016). The observed extreme low AMOC in 2009–2010, which was followed by a reduction in ocean heat content to the north (Cunningham et al. 2013), has been possibly implicated in cold European weather events in winter 2009–2010 and December 2010 (Buchan et al. 2014) although a robust attribution is missing. In summer, cold anomalies in the SPG, like the one occurring during the so-called cold blob in 2015 (Josey et al. 2018), have been suspected to potentially enhance the probability of heatwaves over Europe in summer (Duchez et al. 2016). Nevertheless, considerable uncertainties remain with regard to this aspect due to the lack of historical observations before 2004 and due to poor model resolution of small-scale processes related to frontal dynamics around the Gulf Stream region (Vanniere et al. 2017). In addition, oceanic changes in the Gulf Stream region may occur in line with AMOC weakening (Saba et al. 2016) with potential rapid warming due to a northward shift of the Gulf Stream. However, these changes are largely underestimated in coarse resolution models (Saba et al. 2016) . In North America, a negative phase of the AMV, reminiscent of a weakening of the AMOC, lowers agricultural production in a few Mexican coastal states (Azuz-Adeath et al. 2019).

Changes in ocean circulation can also strongly impact sea level in the regions bordering the North Atlantic (McCarthy et al. 2015a; Palter et al. 2018). A collapse of the AMOC or of the SPG could induce substantial increase of sea level up to a few tens

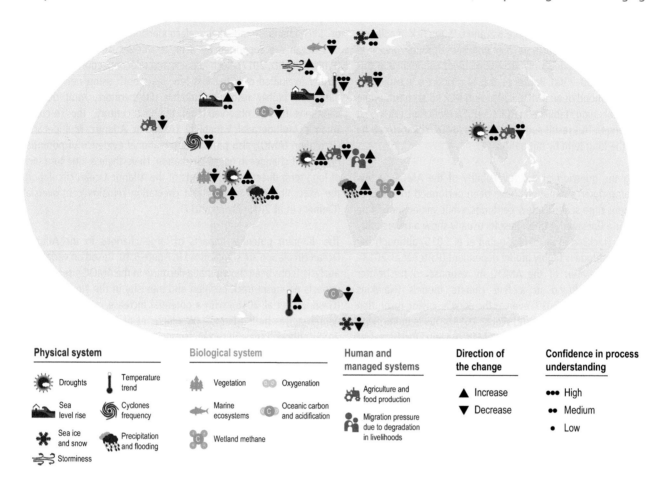

Physical system

- Droughts
- Sea level rise
- Sea ice and snow
- Storminess
- Temperature trend
- Cyclones frequency
- Precipitation and flooding

Biological system

- Vegetation
- Marine ecosystems
- Wetland methane
- Oxygenation
- Oceanic carbon and acidification

Human and managed systems

- Agriculture and food production
- Migration pressure due to degradation in livelihoods

Direction of the change

- ▲ Increase
- ▼ Decrease

Confidence in process understanding

- ●●● High
- ●● Medium
- ● Low

Figure 6.10 | Infographic on teleconnections and impacts due to Atlantic Meridional Overturning Circulation (AMOC) collapse or substantial weakening. Changes in circulation have multiple impacts around the Atlantic Basin, but also include remote impacts in Asia and Antarctica. Reductions in AMOC lead to an excess of heat in the South Atlantic, leading to increased flooding, methane emissions and drought, and a concomitant negative impact on food production and human systems. In the North Atlantic region hurricane frequency is decreased on the western side of the basin, but storminess increases in the east. Marine and terrestrial ecosystems, including food production, are impacted while sea level rise (SLR) is seen on both sides of the Atlantic. The arrows indicate the direction of the change associated with each icon and is put on its right. An assessment of the confidence level in the understanding of the processes at play is indicated below each arrow.

of centimetres along the western boundary of the North Atlantic (Ezer et al. 2013; Little et al. 2017; cf. Chapter 5). For instance, such a link may explain 30% of the extreme observed SLR event (a short-lived increase of 12 mm during 2 years) in northeast America in 2009–2010 (Ezer, 2015; Goddard et al. 2015). This illustrates that monitoring changes in AMOC may have practical implications for coastal protection.

The AMOC teleconnections are widespread and notably strongly affect the tropical area, as evidenced in palaeo-data for the Sahel region (Collins et al. 2017; Mulitza et al. 2017) and in model simulations (Jackson et al. 2015; Delworth and Zeng, 2016). These teleconnections may affect vulnerable populations. For instance, Defrance et al. (2017) found that a substantial decrease in the AMOC, at the very upper end of potential changes, may strongly diminish precipitation in the Sahelian region, decreasing the millet and sorghum emblematic crop production, which may impact subsistence of tens of millions of people, increasing their potential for migration. Smaller amplitude variations in Sahelian rainfall, driven by North Atlantic SST, has been found to be predictable up to a decade ahead (Gaetani and Mohino, 2013; Mohino et al. 2016; Sheen et al. 2017), potentially providing mitigation and adaptation

opportunities. The number of tropical storms in the North Atlantic has been found to be very sensitive to the AMOC (Delworth and Zeng, 2016; Yan et al. 2017) as well as to the SPG (Hermanson et al. 2014) variations, so that a large weakening of the AMOC or cooling of the SPG may decrease the number of Atlantic tropical storms. The Asian monsoon may also potentially weaken in the case of large changes in the AMOC (Marzin et al. 2013; Jackson et al. 2015; Zhou et al. 2016; Monerie et al. 2019) implying substantial adverse impacts on populations. The interactions of the Atlantic basin with the Pacific has also been largely discussed over the last few years, with the supposed influence of a cool North Atlantic inducing a warm tropical Pacific (McGregor et al. 2014; Chafik et al. 2016; Li et al. 2016b), although not found in all models (Swingedouw et al. 2017), which may induce stronger amplitudes of El Niño (Dekker et al. 2018).

The AMOC plays an important function in transporting excess heat and anthropogenic carbon from the surface to the deep ocean (Kostov et al. 2014; Romanou et al. 2017), and therefore in setting the pace of global warming (Marshall et al. 2014). A large potential decline in the AMOC strength reduces global surface warming. This is due to changes in the location of ocean heat uptake and

6

associated expansion of the cryosphere around the North Atlantic, which increases surface albedo (Rugenstein et al. 2013; Winton et al. 2013), as well as cloud cover variations and modifications in water vapour content (Trossman et al. 2016). As the uptake of excess heat occurs preferentially in regions with delayed warming (Winton et al. 2013; Frölicher et al. 2015; Armour et al. 2016), a potential large reduction of the AMOC may shift the uptake of excess heat from the low to the high latitudes (Rugenstein et al. 2013; Winton et al. 2013), where the atmosphere is more sensitive to external forcing (Winton et al. 2010; Rose et al. 2014; Rose and Rayborn, 2016; Rugenstein et al. 2016). A decrease in AMOC may also decrease the subduction of anthropogenic carbon to deeper waters (Zickfeld et al. 2008; Winton et al. 2013; Randerson et al. 2015; Rhein et al. 2017). A potential impact of methane emissions has also been highlighted for past Heinrich events during which massive icebergs discharge in the North Atlantic may have led to large AMOC disruptions. Large increases (>100 ppb) in methane production have been associated with these events (Rhodes et al. 2015) potentially due to increased wetland production in the SH, related to teleconnections of the North Atlantic with tropical area (Ringeval et al. 2013; Zurcher et al. 2013). All these different effects indicate a potentially positive feedback of the AMOC on the carbon cycle (Parsons et al. 2014), although other elements from the terrestrial biosphere may limit its strength or even reverse its sign (Bozbiyik et al. 2011).

Changes in Atlantic Ocean circulation can also strongly impact marine life and can be seen at all levels of different ecosystems. For instance, changes in the abundance and distribution of species in response to circulation changes in the SPG have been documented amongst plankton (Hátún et al. 2009), fish (Payne et al. 2012; Miesner and Payne, 2018), seabirds (Descamps et al. 2013) and top predators such as tuna, billfish and pilot whales (Hátún et al. 2009; MacKenzie et al. 2014). Nutrient concentrations in the northeast Atlantic have also been shown to be limited by the recent weakening of the SPG, with concomitant ecosystem impacts (Johnson et al. 2013; Hátún et al. 2016). The influence of SPG circulation also extends to ecosystems beyond from the immediate area, and has a clear impact on the productivity of cod (*Gadus morhua*) in the Barents Sea, for example (Årthun et al. 2017; Årthun et al. 2018). On a broader scale, changes in the AMOC are an important driver of AMV, which has also been linked to substantial changes in marine ecosystems on both sides of the North Atlantic (Alheit et al. 2014; Nye et al. 2014). Recent AMOC weakening is also suspected to explain large marine deoxygenation in the northwest coastal Atlantic (Claret et al. 2018). In addition, a recent study using a marine productivity proxy from Greenland ice cores suggest that net primary productivity has decreased by 10 ± 7% in the subarctic Atlantic over the past two centuries possibly related to changes in AMOC (Osman et al. 2019). Finally, a model study investigated the impact of mitigation by reversing the forcing from a RCP8.5 scenario from 2100 and found that global marine net productivity may recover very rapidly and even overshoot contemporary values at the end of the reversal, highlighting the potential benefit of mitigation (John et al. 2015).

Following all these potential impacts, it has been suggested that a collapse of the AMOC may have the potential to induce a cascade of abrupt events, related to the crossing of thresholds from different tipping points, itself potentially driven by GIS rapid melting. For example, a collapse of the AMOC may induce causal interactions like changes in ENSO characteristics (Rocha et al. 2018), dieback of the Amazon rainforest and shrinking of the WAIS due to seesaw effect, ITCZ southern migration and large warming of the Southern Ocean (Cai et al. 2016). However, such a worst case scenario remains very poorly constrained quantitatively due to the large uncertainty in GIS and AMOC response to global warming.

The potential impacts of such rapid changes in ocean circulation on agriculture, economy and human health remain poorly evaluated up to now with very few studies on the topic (Kopits et al. 2014). The available impact literature on AMOC weakening has focussed on impacts from temperature change only (reduced warming), globally leading to economic benefits (e.g., Anthoff et al. 2016), and local losses can amount to a few percent of gross domestic product (GDP), however under a complete shutdown (Link and Tol, 2011). Declines in Barents Sea fish species could lead to economic losses (Link and Tol, 2009), but more comprehensive economic studies are lacking.

6.7.3   Risk Management and Adaptation

The numerous potential impacts of AMOC weakening (see Section 6.7.2) require adaptation responses. A specific adaptation action is a monitoring and early warning system using observation and prediction systems, which can help to respond in time to effects of an AMOC decline. Although it is difficult to warn very early for large changes in AMOC to come, notably due to large natural decadal variability of the AMOC (Boulton et al. 2014), the observation arrays that are in place may allow the development of such an early warning system. Nevertheless, the prospects for its operational use for early warnings have not yet been fully developed. In this respect, developing early warning systems that do not depend on statistical timeseries analysis of long observational record might be seen as an important research goal in the future.

Decadal prediction systems can help fill this gap. Skilful prediction of AMOC variation has been demonstrated on the multi-annual scale (Matei et al. 2012) and retrospective prediction experiments have demonstrated that the large changes in the SPG seen in the mid-1990s could have been foreseen several years in advance (Wouters et al. 2013; Msadek et al. 2014). The World Climate Research Programme's grand challenge of launching decadal predictions every year (Kushnir et al. 2019) is an important step towards anticipating rapid changes in the near term and can drive decadal-scale climate services. For example, a few studies have already shown that small variations anticipated by decadal predictions (e.g., Sheen et al. 2017) can be useful for the development of climate services, notably for agriculture in south and east Africa (Nyamwanza et al. 2017). Decadal predictions also match the decision making time horizons of many users of the ocean (Tommasi et al. 2017b) and are expected to play an increasingly important role in this sector in the future (Payne et al. 2017).

6.8   Compound Events and Cascading Impacts

6.8.1   Concepts

Compound events refer to events that are characterised by multiple failures that can amplify overall risk and/or cause cascading impacts (Helbing, 2013; Gallina et al. 2016; Figure 6.1). These impacts may be triggered by multiple hazards that occur coincidently or sequentially and can lead to substantial disruption of natural or human systems (Leonard et al. 2014; Oppenheimer et al. 2014; Gallina et al. 2016; Zscheischler et al. 2018). These concepts are illustrated in a series of recent case studies that show how compound events interact with multiple elements of the ecosystem and society to create compound risk and cascading impacts (Box 6.1).

Compound events and cascading impacts are examples of deep uncertainty because data deficiency often prevents the assessment of probabilities and consequences of the risks from compound events. Furthermore, climate drivers that contribute to compound events could cross tipping points in the future (e.g., Cai et al. 2016; Cross-Chapter Box 4 in Chapter 1). Concepts and methods for addressing compound events and cascading impacts have a solid foundation in disaster risk reduction frameworks (Scolobig, 2017) where they may be assessed with scenarios, risk mapping, and participatory governance (Marzocchi et al. 2012; Komendantova et al. 2014). However, these approaches have tended to not consider the effects of climate change, rather considering hazards and vulnerability as stationary entities (Gallina et al. 2016). Trends in geophysical and meteorological extreme events and their interaction with more complex social, economic and environmental vulnerabilities overwhelm existing governance and institutional capacities (Shimizu and Clark, 2015) because of the aggregated cascading impacts.

6.8.2   Multiple Hazards

Understanding regions where changes in the climate system could increase the likelihood or severity of multiple hazards is relevant to understanding compound events (Figure 6.1). Several recent studies have highlighted coastal regions that are becoming more susceptible to multiple hazards from changes in regional climate. Warming and poleward expansion of the warm western boundary current regions (WBCs; Yang et al. 2016a) together with intensified cyclogenesis in these WBC regions; the Gulf Stream (Booth et al. 2012), the Kuroshio (Hirata et al. 2016) and the East Australian Current (EAC; Pepler et al. 2016a) can increase the likelihood of multiple hazards. These include increased rates of SLR (Brunnabend et al. 2017; Zhang et al. 2017b) together with increases in severe rainfall, storm surges and associated flooding (Thompson et al. 2013; Oey and Chou, 2016; Pepler et al. 2016a). WBCs have undergone an intensification and poleward expansion in all but the Gulf Stream where the weakening of the AMOC cancelled this effect (Seager and Simpson, 2016; Yang et al. 2016a).

Acknowledging the dual role of regional SLR and TCs frequency and intensity changes for future flood risk, Little et al. (2015) developed a flood index that takes account of local projected SLR along with TC frequency and intensity changes in a CMIP5 multi-model ensemble. They find that relative to 1986–2005, the Flood Index is 4–75 times higher by 2080–2099 for RCP2.6 (10–90th percentile range) and 35–350 times higher for RCP8.5. In the vicinity of the East Australian Current, Pepler et al. (2016b) found warmer SSTs boost the intensification of weak to moderate ETC's. Neglecting the compounding effects of flood and extreme sea level drivers can cause significant underestimation of flood risk and projected failure probability (Wahl et al. 2016; Moftakhari et al. 2017).

Over the last decade, several efforts have been made to address long-term shoreline change driven by the cascading impact of SLR, waves and MSL. Ranasinghe et al. (2012) presented the Probabilistic Coastline Recession model, which provides probabilistic estimates of coastline recession in response to both storms and SLR in the 21st century. Dune recession is estimated for each storm considering the recovery between storms, which is obtained empirically. More recently, Toimil et al. (2017) developed a methodology to address shoreline change over this century due to the action of waves, storm surges, astronomical tides in combination with SLR. The methodology considers the generation of thousands of multi-variate hourly time series of waves and storm surges to reconstruct future shoreline evolution probabilistically, which enables estimates of extreme recessions and long-term coastline change to be obtained. The model proposed by Vitousek et al. (2017) integrates longshore and cross-shore transport induced by GCM-projected waves and SLR, which allows it to be applied to both long and pocket sandy beaches. The analysis provides only one instance of what coastline change over the 21st century may be.

To summarise, new studies highlight regions such as coasts including those adjacent to WBCs, that are experiencing larger changes to multiple phenomena simultaneously such as SLR and cyclone intensity linked to higher SST increases (*medium confidence*), which increases the likelihood of extremes from multiple hazards occurring (*medium confidence*). Failing to account for the multiple factors responsible for extreme events will lead to an underestimation of the probabilities of occurrence (*high confidence*).

6.8.3   Cascading Impacts on Ecosystems

Damage and loss of ecosystems (mangrove, coral reefs, polar deserts, wetlands and salt marshes); or regime shifts in ecosystem communities lead to reduced resilience of all the ecosystems and possible flow-on effects to human systems. For example, recent studies showed that living corals and reef structures have experienced significant losses from human-related drivers such as coastal development; sand and coral mining; overfishing, acidification, and climate-related storms and bleaching events (Smith, 2011; Nielsen et al. 2012; Hilmi et al. 2013; Graham et al. 2015; Lenoir and Svenning, 2015; Hughes et al. 2017b). As a consequence, reef flattening is taking place globally due the loss of corals and from the bio-erosion and dissolution of the underlying reef carbonate structures (Alvarez-Filip et al. 2009). Reef mortality and flattening due to non-climate and climate-related drivers trigger cascading impacts and risks due to the loss of the protection services provided to coastal areas. High emission scenarios are expected to lead to almost the complete loss of coral cover by 2100, although policies aiming to lower the combined aerosol-radiation interaction and aerosol-cloud interaction (e.g., IPCC RCP 6.0) may partially limit the impacts on coral reefs and the associated habitat loss, thereby preserving an estimated 14 to 20 billion USD in consumer surplus 2100 (2014 USD, 3% discount; Speers et al. 2016). Moreover, projected SLR will increase flooding risks, and these risks will be even greater if reefs that now help protect coasts from waves are lost due to bleaching-induced mortality.

6.8.4   Cascading Impacts on Social Systems

Impacts of compound events also have significant multi-effects in the societal system. Cascading impacts are particularly driven by the loss or (temporary) disruption of critical infrastructure (Pescaroli and Alexander, 2018), such as communications, transport, and power supply, on housing, dams and flood protection; as well as health provision. Repeated extreme and compound events are leading to critical transitions in social systems (Kopp et al. 2016) which may cause the disruption of (local) communities, creating cascading impacts consisting of short-term impacts as well as long-lasting economic effects, and in some cases migration. When the responses of the economic sector to short term weather variations are applied to long term-climate projections, risks associated with climate change on different sectors are projected to result in an average 1.2% of decrease of US GDP per degree Celsius of warming. Furthermore, broad geographical discrepancies generate a large transfer of value northwards and westwards with the expected consequence of increased economic inequality (Hsiang et al. 2017). The severity and intensity of the cascading impacts also depend on the affected societies' vulnerability and resilience. For example, the intensity and influence of compound events are dependent on the size and scale of the affected society and the percentage of economy or GDP impacted (Handmer et al. 2012 in IPCC SREX). Smaller countries and especially small islands face the challenge of being unable to 'hedge' the risk through geographical redistribution (see Cross-Chapter Box 9).

Impacts from the natural system can descend into a cascade of disasters. For example, in 2005, Hurricane Katrina led to heavy flooding in the coastal area, dike breaches, emergency response failures, chaos in evacuation (traffic jams) and social disruption. Flooding in Thailand in 2011 led to the closure of many factories which not only impacted on the country's economy but impaired the global automobile and electronic industry (Kreibich et al. 2014). Female-owned establishments are more challenged with failures than businesses owned by men due to less experience, shorter duration and smaller size of businesses (Haynes et al. 2011; Marshall et al. 2015). The impact of compound events on ecosystems can also, in the long run, have devastating impacts on societal systems, for example, impacts from tropical storms can lead to coral degradation, which leads to increased wave impact and subsequent accelerated coastal erosion and impacts on fishing resources. This

subsequently can have an impact on local economies, potentially leading to social disruption and migration (Saha, 2017). Impacts on marine ecosystems and habitats will also affect subsistence and commercial fisheries and, as a result, food security (Barrow et al. 2018). Climate-induced community relocations in Alaska stem from repeated extreme weather events coupled with climate change-induced coastal erosion and these impact the habitability of the whole community (Bronen, 2011; Durrer and Adams, 2011; Marino, 2011; Marino, 2012; Bronen and Chapin, 2013; see also Cross-Chapter Boxes 2 and 5 in Chapter 1).

6.8.5   Risk Management and Adaptation, Sustainable and Resilient Pathways

The management of compound events and cascading impacts in the context of governance poses challenges, partly because it is place dependent and heavily influenced by local parameters such as hazard experience and cultural values. Moreover, in some cases, people perceive that their community or country is less affected than others, leading to a 'spatial optimism bias' that delays or reduces the scope of actions (Nunn et al. 2016). In other cases it is unclear who will take responsibility when compound events and cascading impacts occur (Scolobig, 2017), although for some compound risks (e.g., na-tech disasters – when natural hazards trigger technological disasters), the private sector cooperates with governments to manage and respond to risks (Krausmann et al. 2017). Considerable variations exist among and inside countries. The level of engagement depends on the process of cascading impacts and the role of governance arrangement at the country level (Lawrence et al. 2018), countries' capacity to develop integrated risk and disaster frameworks and regulations, viable multi-stakeholder and public-private partnership in the case of multiple technological and natural hazards (Gerkensmeier and Ratter, 2018), the initiatives of local governments to exercise compound risk operations, and experience in interagency cooperation (Scolobig, 2017). The importance of local knowledge and traditional practices in disaster risk prevention and reduction is widely recognised (Hiwasaki et al. 2014; Hilhorst et al. 2015; Audefroy and Sánchez, 2017) (*high confidence*). The need to strengthen DRM is evident and can be improved and communicated effectively by integrating local knowledge such as Inuit's indigenous knowledge and local knowledge in Alaska (Pearce et al. 2015; Cross-Chapter Box 3 in Chapter 1) since it is easier for communities to accept than pure science-based DRM (Ikeda et al. 2016).

Despite difficulties of governance and decision making, many researchers and policy makers have recognised the need to study combined climatic and other hazards and their impacts. Several methods are now being employed to assess climatic hazards and compound events simultaneously, and also in combination (Klerk et al. 2015; van den Hurk et al. 2015; Wahl et al. 2015; Zscheischler and Seneviratne, 2017; Wu et al. 2018; Zscheischler et al. 2018). Policy makers can also begin to plan for disaster risk reduction and adaptation, based on these analyses of compound events and risks. Addressing limitations in understanding the compound hazards, as well as adequate mechanisms of the cascading impacts is needed. Finally, there are limits to resources to study these complex interactions in sufficient detail, as well as limits to data and information on past events that would allow the simulation of these effects, including economic impacts.

6.8.6   Global Impact of Tipping Points

A small number of studies (Lontzek et al. 2015; Cai et al. 2016; Lemoine and Traeger, 2016) use different versions of the Dynamic Integrated Climate-Economy assessment model (Nordhaus, 1992; Nordhaus, 2017) to assess the impact of diverse sets of tipping points and causal interactions between them on the socially optimal reduction of gas emissions and the present social cost of carbon, representing the economic cost caused by an additional ton of CO_2 emissions or its equivalent.

Cai et al. (2016) consider five interacting, stochastic, potential climate tipping points: reorganisation of the AMOC; disintegration of the GIS; collapse of the WAIS; dieback of the Amazon Rain Forest; and shift to a more persistent El Niño regime. The deep uncertainties associated with the likelihood of each of these tipping points and the dependence of them on the state of the others is addressed through expert elicitation. There *is limited evidence*, but *high agreement* that present costs of carbon are clearly underestimated. Double (Lemoine and Traeger, 2016), triple (Ceronsky et al. 2011), to eightfold (Cai et al. 2016) increase of the carbon price are suggested, depending on the working hypothesis. Cai et al. (2016) indicate that with the prospect of multiple interacting tipping points, the present social cost of carbon increases from 15 to 116 USD per tonne of CO_2, and conclude that stringent efforts are needed to reduce CO_2 emission if these impacts are to be avoided.

Box 6.1 | Multiple Hazards, Compound Risk and Cascading Impacts

The following case studies illustrate that anthropogenic climate change including ocean changes is increasingly having a discernible influence on elements of the climate system by exacerbating extreme events and causing multiple hazards, often with compound or sequential characteristics. In turn these elements are interacting with vulnerability and exposure to trigger compound events and cascading impacts.

Case Study 1: Tasmania's Summer of 2015–2016
Tasmania in southeast Australia experienced multiple extreme climate events in 2015–2016, driven by the combined effects of natural modes of climate variability and anthropogenic climate change, with impacts on the energy sector, fisheries and emergency services. The driest warm season on record (October to April), together with the warmest summer on record, brought agricultural and hydrological droughts to Tasmania and preconditioned the sensitive highland environment for major fires during the summer. Thousands of lightning strikes during the first two months of the year led to more than 165 separate vegetation fires, which burned more than 120,000 hectares including highland zones and the World Heritage Area and incurred costs to the state of more than 50 million AUD (Press, 2016).

In late January an intense cutoff low-pressure system brought heavy rainfall and floods, so that emergency services were simultaneously dealing with highland fires and floods in the east and north. The floods were followed by an extended wet period for Tasmania, with the wettest wet season (April to November) on record in 2016. Meanwhile, an intense marine heatwave (MHW) off the east coast persisted for 251 days from spring 2015 through to autumn 2016 (Oliver et al. 2017).

The driest October on record was influenced by both the El Niño and anthropogenic forcing (Karoly et al. 2016). Warmer sea surface temperatures (SSTs) due to anthropogenic warming may have increased the intensity of rainfall during the floods in January (e.g., Pepler et al. 2016a). The intensity and the duration of the MHW was unprecedented and both aspects had a clear human signature (Oliver et al. 2017).

Tasmania primarily relies on hydro-electric power generation and the trading of power over an undersea cable to mainland Australia, 'Basslink', for its energy needs. Lake levels in hydro-electric dams were at relatively low levels in early spring 2015, and the extended dry period led to further reductions and significantly reduced capacity to generate power (Hydro Tasmania, 2016). An unanticipated failure of the Basslink cable subsequently necessitated the use of emergency diesel generators (Hydro Tasmania, 2016).

The compound events caused many impacts on natural systems, agriculture, infrastructure and communities. Additional emergency services from outside the state were needed to deal with the fires. The MHW caused disease outbreaks in farmed shellfish, mortality in wild shellfish and species found further south than previously recorded. The energy sector experienced a severe cascade of impacts due to climate stressors and system inter-dependencies. The combination of drought, fires, floods and MHW reduced output from the agriculture, forestry, fishing and energy sectors and reduced the State of Tasmania gross state product (GSP) to 1.3%, well below the anticipated growth of 2.5%. To address the energy shortages, Tasmania's four largest industrial energy users, responsible for 60% of Tasmania's electricity usage, agreed to a series of voluntary load reductions of up to 100 MW on a sustained basis, contributing to a 1.7% reduction in the output of the manufacturing sector (Eslake, 2016). The total cost of the fires and floods was assessed at

Box 6.1 (continued)

300 million USD. In response funding has been increased to government agencies responsible for managing floods and bushfires, and multiple independent reviews have recommended major policy reforms that are now under consideration (Blake, 2017; Tasmanian Climate Change Office, 2017).

This case illustrates the concepts presented in Figure 6.1. Anthropogenic climate change *likely* contributed to the severity of multiple hazards; including coincident and sequential events (droughts and bushfires, followed by extreme rainfall and floods). Compound risks, including risks for the safety of residents affected by floods and fires, the natural environment affected by MHWs and fires and the economy in the food and energy sectors arose from these climate events with cascading impacts on the industrial sector more broadly as it responded to the shortfall in energy supply.

Extremes experienced in 2015–2016 in Tasmania are projected to become more frequent or more intense due to climate change, including dry springs and summers (Bureau of Meteorology and Australian CSIRO, 2007), intense lows bringing extreme rains and floods in summer (Grose et al. 2012), and MHWs on the east coast associated with convergence of heat linked to the East Australia Current (Oliver et al. 2017) indicating that climate change by increasing the frequency or intensity of multiple climate events will *likely* increase compound risk and cascading impacts (*high confidence*).

Case Study 2: The Coral Triangle
The Coral Triangle is under the combined threats of mean warming, ocean acidification, temperature and sea level variability (often associated with both El Niño and La Niña), coastal development and overfishing, leading to reduced ecosystem services and loss of biodiversity. The Coral Triangle covers 4 million square miles of ocean and coastal waters in Southeast Asia and the Pacific, in the area surrounding Indonesia, Malaysia, Papua New Guinea, the Philippines, Timor Leste and the Solomon Islands. It is the centre of the highest coastal marine biodiversity in the world due to its geological setting, physical environment, and an array of ecological and evolutionary processes which makes it a conservation priority. Together with mangroves and seagrass beds, the 605 species of corals including 15 regional endemics (Veron et al. 2011) provide ecosystem services to over 100 million people from diverse and rich cultures, in particular for food, building materials and coastal protection.

The riches of the ecosystems in the Coral Triangle led to expanding human activities, such as coastal development to accommodate a booming tourism sector and overfishing. There is agreement that these activities, including coastal deforestation, coastal reclamation, destructive fishing methods and over-exploitation of marine life generate important pressures on the ecosystem (Pomeroy et al. 2015; Ferrigno et al. 2016; Huang and Coelho, 2017). As a result, the coastal ecosystems of the Coral Triangle have already lost 40% of their coral reefs and mangroves over the past 40 years (Hoegh-Guldberg et al. 2009).

Risks from compound events include increase in sea surface temperature (SST), SLR and increased human activities. The increasing trend in SSTs was estimated to be 0.1°C per decade between 1960 and 2007 (Kleypas et al. 2015) but increased to 0.2°C per decade from 1985 to 2006 (Penaflor et al. 2009), an estimation comparable with that in the South China Sea (Zuo et al. 2015). However, waters in the northern and eastern parts are warming faster than the rest of the region, and this variability is increased by local parameters linked to the complex bathymetry and oceanography of the region (Kleypas et al. 2015). Areas in the eastern part have experienced more thermal stress events, and these appear to be more likely during La Niña events, which generate heat pulses in the region, leading to bleaching events, some of them already triggered by El Niño Southern Oscillation (ENSO) events. In the Coral Triangle, El Niño events have a relative cooling effect, while La Niña events are accompanied by warming (Penaflor et al. 2009). The 1997–1998 El Niño was followed by a strong La Niña so that degree heating weeks (DHW) values in many parts of the region were greater than four, which caused widespread coral bleaching (DHW values greater than zero indicate there is thermal stress, while DHW values of 4 and greater indicate the existence of sufficient thermal stress to produce significant levels of coral bleaching; Kayanne, 2017). However, in Indonesia, the 2015–2016 El Niño event had impacted shallow water reefs well before high SSTs could trigger any coral bleaching (Ampou et al. 2017). Sea level in Indonesia had been at its lowest in the past 12 years following this El Niño event and this had affected corals living in shallow waters. Substantial mortality was likely caused by higher daily aerial exposure during low tides and warmer SST associated with shallow waters. Another climate change-associated impact in the Coral Triangle is ocean acidification. Although less exposed than other reefs at higher latitudes (van Hooidonk et al. 2013), changes in pH are expected to affect coral calcification (DeCarlo et al. 2017), with an impact on coral reef fisheries (Speers et al. 2016).

At present, different approaches are used to manage the different risks to coral ecosystems in the Coral Triangle such as fisheries management (White et al. 2014) and different conservation initiatives (Beger et al. 2015), including coral larval replenishment (dela Cruz and Harrison, 2017) and the establishment of a region-wide marine protected area system (e.g., Christie et al. 2016).

Box 6.1 (continued)

There is *high confidence* that reefs with high species diversity are more resilient to stress, including bleaching (e.g., Ferrigno et al. 2016; Mellin et al. 2016; Mori, 2016). Sustainable Management of coastal resources, such as marine protected areas is thus a commonly used management approach (White et al. 2014; Christie et al. 2016), supported in some cases by ecosystem modelling projections (Weijerman et al. 2015; Weijerman et al. 2016). Evaluations of these management approaches led to the development of guiding frameworks and supporting tools for coastal area managers (Anthony et al. 2015); however biological and ecological factors are still expected to limit the adaptive capacity of these ecosystems to changes (Mora et al. 2016).

Case Study 3: Severe Atlantic Hurricanes of 2017
The above-average hurricane activity of the 2017 season led to the sequential occurrence of Hurricanes Harvey, Irma and Maria on the Caribbean and southern US coasts (Klotzbach and Bell, 2017), collectively causing 265 billion USD damage and making 2017 the costliest hurricane season on record (Blake et al. 2011; Blake and Zelinsky, 2018).

The role of climate change in contributing to the severity of these recent hurricanes has been much discussed in the public and media. It has not been possible to identify robust long-term trends in either hurricane frequency or strength given the large natural variability, which makes trend detection challenging especially given the opposing influences of greenhouse gases (GHGs) and aerosols on past changes. However, observational data shows a warming of the surface waters of the Gulf of Mexico, and indeed most of the world's oceans, over the past century as human activities have had an increasing impact on our climate (Sobel et al. 2016).

Hurricane Harvey brought unprecedented rainfall to Texas and produced a storm surge that exceeded 2 m in some regions (Shuckburgh et al. 2017). Climate change increased the rainfall intensity associated with Harvey by at least 8% (8–19%; Risser and Wehner, 2017; van Oldenborgh et al. 2017) (*high confidence*). Emanuel (2017) estimated that the annual probability of 500 mm of area-averaged rainfall had increased from 1% in the period 1981–2000 to 6% in 2017. Furthermore, if society were to follow RCP8.5, the probability would increase to 18% over the period 2081–2100.

The event attribution method of Emanuel (2017) indicates that for TC Irma, which impacted the Caribbean islands of Barbuda and Cuba, the annual probability of encountering Irma's peak wind of 160 knots within 300 km of Barbuda increased from 0.13% in the period 1981–2000 to 0.43% by 2017, and will further increase to 1.3% by 2081–2100 assuming RCP8.5. TC Maria followed Irma, and made landfalls on the island of Dominica, Puerto Rico, and Turks and Caicos Islands. The annual probability of encountering Maria's peak wind of 150 knots within 150 km of 17°N, 64°W increased from 0.5% during 1981–2000 to 1.7% in 2017, and will increase to 5% by 2081–2100 assuming RCP8.5.

At least 68 people died from the direct effects of Harvey in Houston (Blake and Zelinsky, 2018). The Houston metropolitan area was devastated with the release of about 4.6 million pounds of contaminants from petrochemical plants and refineries. Irma caused 44 direct deaths (Cangialosi et al. 2018) and wiped out housing, schools, fisheries and livestock in Barbuda, Antigua, St. Martin and the British Virgin Islands (ACAPS et al. 2017). Maria caused 31 direct deaths in Dominica, two in Guadeloupe and around 65 in Puerto Rico (Pasch et al. 2018), and completely vacated Barbuda. Maria destroyed almost all power lines, buildings and 80% of crops in Puerto Rico (Rexach et al. 2017; Rosselló, 2017), and damaged pharmaceutical industries that provided 33% of Puerto Rico's gross domestic product (GDP) causing shortages of some medical supplies in the USA (Sacks et al. 2018). The effects of Maria are expected to increase the poverty rate by 14% because of unemployment in tourism and agriculture sectors for more than a year in Dominica (The Government of the Commonwealth of Dominica, 2017), and resulted in outmigration to neighboring countries or the USA (ACAPS et al. 2017; Rosselló, 2017). These economic and social consequences are indicative of the cascading impact of the 2017 hurricanes. The post-disaster reconstruction plan is to renovate telecommunications, develop climate resilient building plans and emergency coordination (Rosselló, 2017; The Government of the Commonwealth of Dominica, 2017).

Collectively, these case studies indicate that climate change has played a role in multiple coincident or sequential extreme events that have led to cascading impacts (*high confidence*). Climate change is projected to increase the frequency or intensity of multiple climate events in the future and this will *likely* increase risks of compound events and cascading impacts (*high confidence*).

6.9   Governance and Policy Options, Risk Management, Including Disaster Risk Reduction and Enhancing Resilience

6.9.1   Decision Making for Abrupt Change and Extreme Events

As outlined earlier in this report, several approaches exist for adaptive responses towards climate change impacts. Other sections that deal with adaptation responses to extremes include Section 1.5.2, Section 4.4 (SLR and coastal flooding), Cross Chapter Box 4 in Chapter 1 and Section 5.5.2.5 in Chapter 5 (adaptation limits for coastal infrastructure and ecosystems). Here, we address adaptation responses especially to abrupt and extreme changes (for responses to special abrupt changes (e.g., AMOC; see also Section 6.7).

Since AR5, growing discussions have advocated for transformative adaptation, implying that they support fundamental societal shift towards sustainability and climate-resilient development pathways (Moloney et al. 2017; IPCC, 2018; Morchain, 2018). Successful adaptation to abrupt change and extreme events incorporates climate change concerns and the impact of climate extremes on vulnerable populations taking into account community participation and local knowledge (Tozier de la Poterie and Baudoin, 2015). These interventions reduce risk and enhance resilience, and contribute to the SDGs and social justice (Mal et al. 2018). Temporal scales denote before and after abrupt changes and extreme events (prevention and post-event response), long- and short-term adaptation measures, and the lag time between forecast, warning and event (Field et al. 2012; IPCC, 2012). Spatial dimensions include local risk management and adaptation as well as regional and international coordination to prepare for unexpected extremes tackling the impacts at multiple geographic scales (Devine-Wright, 2013; Barnett et al. 2014; Lyth et al. 2016; Barange et al. 2018).

Decision making about abrupt change or extreme events is not autonomous; it is constrained by formal and informal institutional processes such as regulatory structures, property rights, as well as culture, traditions and social norms (Field et al. 2012; IPCC, 2012). Efforts in various countries and large cities to improve resilience and adaptation are growing, and these efforts are linked to a global network of research, information and best practices (e.g., Aerts et al. 2014). In both northern and southern high latitudes, extreme climatic conditions and remoteness from densely populated regions constrain human choices. The question is whether responses to extremes and abrupt changes require approaches that are different from the anticipatory management of adaptation to changes in climate and weather extremes. While there are several impact studies on extreme events and abrupt change, very few focus on the necessity of dedicated individual, governmental or business adaptive responses (Tol et al. 2006; Anthoff et al. 2010; Anthoff et al. 2016).

Making appropriate decisions to manage abrupt change and extreme events given deep uncertainty is challenging (Weaver et al. 2013; see Cross-Chapter Boxes 4 and 5 in Chapter 1). This requires the construction of new models integrating different uncertainties under extreme or abrupt scenarios and evaluation of value for money

(Weaver et al. 2013). Examples include the inclusion of rapid SLR for assessing coastal impacts and adaptation options (Ranger et al. 2013; Haasnoot et al. 2018; see Sections 6.4 and 6.7). Decision analysis frameworks such as 'Robust Decision Making', 'Decision Scaling', 'Assess Risk of Policy', 'Info-gap', 'Dynamic Adaptation Policy Pathways', 'Dynamic Adaptive Pathways Planning', 'Multi-Criteria Decision Analysis', 'Real Options Analysis' and 'Context-First' accommodate a wide range of uncertainties with subsequent socio-ecological impact (Weaver et al. 2013). The central question remains, however, how one can overcome path dependencies which may cause technical lock-ins in the current system. Monitoring systems of climatic and derived variables, in order to predict necessary shifts in adaptation policies are in development (Haasnoot et al. 2015). However, these frameworks have so far been mostly applied to more gradual shifts of climate change, rather than extreme events and abrupt changes.

Request for the use of 'actionable' information and communication based on climate science and modelling will increase (McNie, 2007; Moser and Boykoff, 2013). Such information can only be effective when it is perceived as 'credible, salient, and legitimate' (Paton, 2007; Paton, 2008; Dilling et al. 2015). Since SREX (IPCC, 2012), there is *medium confidence* that trust in the information and the institution (Hardin, 2002; Townley and Garfield, 2013) that governs extreme events and abrupt change (Malka et al. 2009; Birkmann et al. 2011; Schoenefeld and McCauley, 2016) is important. Trust in expert and scientific knowledge helps people make sense of climate change impact and engage with adaptation measures (Moser and Boykoff, 2013; Yeh, 2016). Without such knowledge, people have little recourse to believe and evaluate relevant information (Bråten et al. 2011). Individuals who trust their government can be complacent and do not prepare for the consequences of extremes (Simpson, 2012; Edmondson and Levy, 2019), and shift the responsibility to the government (Edmondson and Levy, 2019). Familiarity with and information about hazards, community characteristics, as well as the relationship between people and government agencies influence the level of trust (Paton, 2007).

Recent literature shows that there are crucial differences between the ethical challenges of mitigation and those of adaptation (Wallimann-Helmer, 2015; Wallimann-Helmer, 2016) in their dealings with Loss and Damage (L&D); and the ongoing analysis disputes how to distribute responsibilities between mitigation and adaptation based on climate justice criteria (Wallimann-Helmer et al. 2019). The Warsaw International Mechanism on L&D under the United Nations Framework Convention on Climate Change (UNFCCC) addresses irreversible changes and limits to adaptation at the global scale (see also Cross-Chapter Box 1 in Chapter 1). This is in contrast to national and local policies, addressing impacts and adaptation. Within the SROCC report, several of the documented and projected irreversible or unavoidable and thus residual impacts beyond adaptation would potentially fall under this category (e.g., Warner and van der Geest, 2013; Huggel et al. 2019; Mechler et al. 2019), including impacts from SLR, land erosion and reduced freshwater resources on small islands, changes in high mountains and cryosphere changes, as well as changes in ocean species and resources. Apart from climate hazards, risks for L&D are also determined by increasing exposure and vulnerability

(Birkmann and Welle, 2015). Such impacts can be assessed using conventional frameworks, but the debate on the precise scope of such impacts remains, including those from anthropogenic climate change impacts as well as natural climate variability and extremes (e.g., James et al. 2014). More work is required to explore the range of activities available for responding to L&D resulting from slow onset processes in the scope of the SROCC report such as ocean acidification (Harrould-Kolieb and Hoegh-Guldberg, 2019) and mountain cryosphere changes (Huggel et al. 2019).

Under the same L&D mechanism, risk transfer mechanisms and insurance have been suggested as a specific adaptation policy option. Several forms of 'climate change' insurance have been proposed recently, but their potential for adaptation has met with criticism, importantly because of the costs of formal insurance and other risk transfer options, as well as issues with sustainability given the lack of loss prevention and adaptation (Surminski et al. 2016; Linnerooth-Bayer et al. 2019). A compensation mechanism for low-lying small islands inclusive of L&D proposal is in progress (Adelman, 2016). Insurance (see also Section 4.4.4) can help absorb extreme shocks for both individuals, using traditional insurance and parametric insurance. Sovereign insurance mechanisms can help governments absorb large losses (Linnerooth-Bayer et al. 2019), but eventually they need to be coupled with other incentives for adaptation and risk reduction measures to be cost-effective (Botzen, 2013) (*medium confidence*).

There is a consensus that investing in disaster risk reduction has economic benefits, although there is *medium evidence* about the range of the estimated benefits which varies from a global estimate of two to four dollars saved for each dollar invested (Kull et al. 2013; Mechler, 2016) to about 400 EUR per invested 1 EUR in the case of flood early warning systems in Europe (Pappenberger et al. 2015). The US Federal Emergency Management Agency indicated that a 1% increase in annual investment in flood management decreases flood damage by 2.1% (Davlasheridze et al. 2017). Conserving ecosystems that provide services for risk reduction also has monetary benefits. Wetlands have been observed to reduce damages during storms. Wetlands and floodplains in Otter Creek (Vermont, USA) reduced damages caused by storms by 54–78% and 84–95%, respectively, for Tropical Storm Irene (Watson et al. 2016). For the whole of the USA, wetlands provide 23.2 billion USD yr^{-1} in storm protection services and the loss of 1 hectare of wetland is estimated to correspond to an average 33,000 USD increase in storm damage from specific storms (Costanza et al. 2008). Engineered structures are also expected to reduce risks. In Europe, to maintain the coastal flood loss constant relative to the size of the economy, flood defence structures need to be able to protect coastal areas for a projected increase of sea level between 0.5–2.5 m. Without these risk reduction actions, the expected damages from coastal floods could increase by two or three degrees of magnitude compared to the present (Vousdoukas et al. 2018). Although risk reduction actions are generally considered an effective way to reduce the damages by shifting the loss-exceedance curve, cost-benefit analysis of disaster risk reduction actions faces several challenges, including its limited role in informing decisions, spatial and temporal uncertainty scales, and discounting and choice of discount rate that affect cost-benefit analysis results heavily (Mechler, 2016).

6.9.2 Transformative Governance and Integrating Disaster Risk Reduction and Climate Change Adaptation

Governance for effective adaptation defined as changes in practice, process and structure (Smit et al. 2001) considers equity, legitimacy and co-benefits (Patterson et al. 2018) appropriate to the issue (Young, 2002). Countries, sectors and localities place different values and perspectives on these categories, and they can change over time (Plummer et al. 2017; see Cross-Chapter Boxes 1 and 2 in Chapter 1). Transformative governance embraces a wider application of climate change-induced mitigation and adaptation strategies to generate fundamental change. It is society-wide and goes beyond the goals of climate change policies and measures (IPCC, 2013; Patterson et al. 2018). It is distinguished from conventional strategies and solutions, as it includes both natural and human systems and intertwines with the SDGs (Fleurbaey et al. 2014; Tàbara et al. 2019). Transformational adaptation is also needed when incremental adaptation to extreme events and abrupt changes is insufficient (Kates et al. 2012). Planned retreat from SLR and climate refugees illustrate the need for transformative governance as the current coastal and risk management regimes do not have the capacity to handle these issues adequately. Inclusion of bilateral and regional agreements related to climate-induced migration (McAdam, 2011), land use planning frameworks to respond to policy, institutional and cultural implications of migration (Matthews and Potts, 2018), and identification of beneficiaries of managed retreat (Hino et al. 2017) along with positive opportunities for migrants to diversify income and avoid being in harm's way (Gemenne, 2015) are steps towards transformative governance. Retreat and migration entail local responses that include indigenous and local knowledges and perspectives that can be applied to solve these issues (Farbotko and Lazrus, 2012; Hilhorst et al. 2015; Tharakan, 2015; Iloka, 2016; Nunn et al. 2016; see also Cross-Chapter Boxes 2 and 5 in Chapter 1). Another example is the Polar region which has started to pursue transformative governance given the potential for increased tourism and cooperation that require changed governance structure (see Sections 3.5.2; 3.5.5 and Table 3.7 in Chapter 3). Accountability for transformations and transitions has been identified as a crucial factor to support responsible action and strengthen climate governance (Edmondson and Levy, 2019).

Though discourse abounds related to transformative governance, it falls short of its ideal in climate change action plans as it is unclear whether communities have the capacity to engage in substantive change to build low-carbon and resilient communities (Burch et al. 2014). The results of a study on the USA by Tang and Dessai (2012) indicate that climate adaptation and mitigation plans' treatment of extreme climate conditions and disaster preparedness is limited. Moreover, risk communication with the public is part of an integrated disaster warning system, but behavioural response to disaster warnings are often governed by personal beliefs about the nature of the hazard; and ultimately swaying individual decisions to comply with or ignore the warning message (Mayhorn and McLaughlin, 2014). New approaches such as the 'first mile' of early warning systems, built on the specific needs from beneficiary communities instead of on technological progress, are being implemented (Zommers et al. 2017); but they have not yet been assessed.

Coupling disaster risk reduction and management with climate change adaptation effort – following the set targets of UNFCCC and the Sendai Framework – has shown progress since SREX and AR5 (e.g., Lawrence and Saunders, 2017). Substantial literature exists on the topic, but there is little assessment of practices on the ground in the implementation of integrated disaster management and climate change adaptation (Nalau et al. 2016) including health (Banwell et al. 2018). Mainstreaming disaster risk reduction and climate change adaptation within and across sectors is considered essential to ensure administrative coordination and coherence across sectoral plans and policies (Shimizu and Clark, 2015) (*medium confidence*). Financial and technological support and capacity building especially related to public works, savings or loans enable households to build assets and improve livelihoods (Ulrichs et al. 2019). No assessment is available so far of the efficiency and effectiveness of mainstreaming especially related to the integration of climate change adaptation and disaster risk reduction, let alone for abrupt and extreme impacts.

Case studies of integration note major problems, for example, weak coordination among government agencies (Seidler et al. 2018); lack of data and user-friendly information to guide decision making at the local level (Jones et al. 2017) and the need for the central governmental support for data availability (Putra et al. 2018); fragmentation due to competing local objectives (Forino et al. 2017); dependence on regional and international frameworks in the absence of a national framework (Rivera and Wamsler, 2014); limited availability of formal training in integration (Hemstock et al. 2017); and turf wars between responsible government agencies (Nemakonde and Van Niekerk, 2017). The case of Pacific islands such as Vanuatu is indicative of these problems. Though they have coupled disaster risk reduction with climate change adaptation, problems manifest in relationships, responsibilities, capacity and expectations between government agencies and other actors (e.g., international donors and non-governmental organisations), as analysed by Vanuatu's response to the Category 5 TC Pam (Nalau et al. 2017). Some solutions are proposed such as getting all the actors on the same page and focusing on reducing vulnerability to longer-term environmental hazards (Schipper et al. 2016); focussing on specific goals, objectives and strategies (Organization of American States, 2014); assigning a single department to handle integration (APEC, 2016); and citing real-life decision examples in national guidelines (Bell et al. 2017). Place-based responses also entail the inclusion and the acknowledgement of indigenous and local knowledge for an enhanced resilience pathway (Hilhorst et al. 2015; Tharakan, 2015; Iloka, 2016; Nunn et al. 2016).

Given the significance of disaster risk reduction to enhance climate change adaptation regardless of the integration of the two, the Sendai Framework for Disaster Risk Reduction 2015–2030 focuses on seven targets and four priorities that foster participation beyond information sharing and include partnerships and collaborations within society (UNISDR, 2015). Inclusion of, and coordination between, different stakeholders is a key component for managing risks of extreme events, including in a changing climate (*medium confidence*). In the Wadden Sea coastal area, for example, crucial parts of coordinating disaster risk reduction, include (i) responsibility-sharing among authorities, sectors and stakeholders, (ii) all-of-society engagement and partnership with empowerment and inclusive participation, and (iii) development of international, regional, subregional and transboundary cooperation schemes (González-Riancho et al. 2017). In India, a change in the coordination structure was pivotal in reducing fatalities from over 10,000 to 45 between cyclones Orissa (Odisha) in 1999 and Phailin in 2013. In this case, the Disaster Management Act of 2005 established a comprehensive policy and command and control system during disaster response that empowered the most qualified government officials regardless of their rank. This system provides authority to and holds accountability for those in charge of ground operations. Though this rigid system may sometimes be questioned, a unified and top-down command structure works better when there is a lack of mature disaster management system (Pal et al. 2017).

In sum, limiting the risk from the impact of extreme events and abrupt changes leads to successful adaptation to climate change if climate-affected sectors and disaster management relevant agencies coordinate well (*high confidence*). Transformative governance, including successful integration of disaster risk management and climate change adaptation, empowerment of vulnerable groups, accountability of governmental decisions, and longer-term planning promotes climate-resilient development pathways (*high confidence*). An enhanced understanding of the institutional capacity as well as the legal framework addressing abrupt changes and extreme events is especially important (*medium confidence*).

Knowledge gaps limit the identification of the most relevant actions to achieve and pursue climate-resilient development pathways. Since SREX and AR5, there is little research on indirect impacts of climatic extremes on ecosystems and consequences on poverty and livelihoods critical to the SDGs. For example, adaptation solutions and limitations, including governance challenges, for the ocean do not include extreme events (Sections 5.5.2 and 5.5.3 in Chapter 5). Further, there is only scant literature on L&D, including non-economic impacts, resulting from well-documented processes such as MHWs (Section 6.4), SLR impacts on low-lying coasts (Section 4.3), and cryosphere changes (Section 2.3; Chapter 3) (*high confidence*). Limited information is available concerning the cost-benefit and effectiveness of risk-reduction measures. Coupling risk transfer and insurance mechanisms with risk reduction measures, for example, can enhance the cost-effectiveness of adapting to climate change (*medium confidence*).

6

FAQ 6.1 | How can risks of abrupt changes in the ocean and cryosphere related to climate change be addressed?

Reducing greenhouse gas (GHG) emissions will reduce the occurrence of extreme events and the likelihood of abrupt changes. Abrupt changes can be irreversible on human time scales and, as tipping points, bring natural systems to novel conditions. To reduce risks that emerge from these impacts of climate change, communities can protect themselves or accommodate to the new environment. In the last resort, they may retreat from exposed areas. Governance that builds on diverse expertise and considers a variety of actions is best equipped to manage remaining risks.

Climate change is projected to influence extreme events and to potentially cause abrupt changes in the ocean and the cryosphere. Both these phenomena can add to the other, slow-onset impacts of climate change, such as a global warming or sea level rise (SLR). In addition, abrupt changes can be tipping points, bringing the ocean, cryosphere, as well as their ecosystems, or the whole climate system, to new conditions instead or going back to the ones prevailing before the abrupt change.

In the ocean, a possible abrupt change is associated with an interruption of the Atlantic Meridional Overturning Circulation (AMOC), an important component of global ocean circulation. A slowdown of the AMOC could have consequences around the world: rainfall in the Sahel region could reduce, hampering crop production; the summer monsoon in Asia could weaken; regional SLR could increase around the Atlantic, and there might be more winter storms in Europe. The collapse of the West Antarctic Ice Sheet (WAIS) is considered to be one of the tipping points for the global climate. Such an event can be triggered when ice shelves break and ice flows towards the ocean. While, in general, it is difficult to assess the probability of occurrence of abrupt climate events they are physically plausible events that could cause large impacts on ecosystems and societies and may be irreversible.

Reducing GHG emissions is the main action to limit global warming to acceptable levels and reduce the occurrence of extreme events and abrupt changes. However, in addition to mitigation, a variety of measures and risk management strategies supports adaptation to future risks. Future risks linked to abrupt changes are strongly influenced by local conditions and different characteristics of the events themselves and evolve differently depending on the circumstances. One major factor for adaptation is whether the extreme events will simply amplify the known impacts or whether they will cause completely new conditions, which may be related to a tipping point. Another essential factor is whether an extreme event or abrupt change will happen in isolation or in conjunction with other events, in a chain of cascading impacts or as part of a compound risk where several events happen at the same time so that impacts can multiply each other. Also, impacts are heavily aggravated by increasing exposure and changes in vulnerability, for example reducing the availability of food, water and energy supply, and not just the occurrence of extremes themselves.

Successful management of extreme events and abrupt changes in the ocean and cryosphere involves all available resources and governance approaches, including among others land-use and spatial planning, indigenous knowledge and local knowledge. The management of the risks to ecosystems include their preservation, the sustainable use of resources and the recognition of the value of ecosystem services. There are three general approaches that, alone or in combination, can enable communities to adapt to these events: retreat from the area, accommodation to new conditions and protection. All have advantages and limitations and their success will depend on the specific circumstances and the community's level of adaptability. But only transformative governance that integrates a variety of strategies and benefits from institutional change helps to address larger risks posed by compound events. Integrating risk-reduction approaches into institutional practices and inclusive decision making that builds on the respective competences of different government agencies and other stakeholders can support management of these extremes. A change of lifestyles and livelihoods might further support the adaptation to new conditions.

6

6.10   Knowledge Gaps

A comprehensive, detailed list of all the knowledge gaps that have been identified during the assessment performed in this chapter is not possible, hence we focus here on gaps that are relevant for multiple phenomena.

Detection, attribution and projection of physical aspects of climate change at regional and local scales are generally limited by uncertainties in the response of climate models to changes in GHGs and other forcing agents. Additionally, regionally-based attribution studies for extreme events may be lacking in some areas, possibly reflecting the lack of capacity or imperative by regional and national technical institutions to undertake such studies. Thermodynamic aspects of change may be more robust than those involving changes in dynamics e.g., the tracks of TCs or ocean dynamical components of MHW formation. Increasing resolution and improvements in climate models may help to reduce uncertainty. However, because extreme events and highly nonlinear changes (e.g., AMOC collapse) are, by definition, found in the 'tails' of distributions, ensembles or long climate model runs may be required.

While it may not be possible to quantify the likelihood of very rare events or irreversible phenomena, it may be possible to quantify their impacts on natural and human systems. Such information may be more useful to policy makers (Sutton, 2018). Impacts on natural systems (e.g., marine ecosystems) are in general better quantified than impacts on human systems, but there are still many gaps in the literature for the phenomena assessed here (e.g., future impacts of extreme El Niño and La Niña events). The body of literature on compound risks and cascading impacts is growing but is still rather small. One area where there seems to be a serious lack of literature is in the assessment of the economic impacts of extreme and abrupt/irreversible events.

Literature on managing risks and adaptation strategies for abrupt and irreversible events is sparse, as is the literature on the combined impacts of climate-driven events and societal development or maladaptation. The same is true for compound risks and cascading impacts. Theory on transformative governance is emerging but practical demonstrations are few.

Finally, research is still often 'siloed' in physical modelling, ecosystem modelling, social sciences etc. Researchers who can cross boundaries between these disciplines will help accelerate research in the areas covered by this chapter.

6

References

Abish, B., A. Cherchi and S.B. Ratna, 2018: ENSO and the recent warming of the Indian Ocean. *Int. J. Climatol.*, **38**(1), 203–214, doi:10.1002/joc.5170.

ACAPS, OCHA and UNDP, 2017: *Regional Overview: Impact of Hurricanes Irma and Maria*. Conference Supporting Document, UN Development Programme. New York, New York, USA, 39pp. Available at: https://reliefweb.int/sites/reliefweb.int/files/resources/UNDP%20%20Regional%20Overview%20Impact%20of%20Hurricanes%20Irma%20and%20Maria.pdf Accessed 2018/04/19.

Adelman, S., 2016: Climate justice, loss and damage and compensation for small island developing states. *Journal of Human Rights and the Environment*, **7**(1), 32–53, doi:10.4337/jhre.2016.01.02.

Aerts, J.C.J.H. et al., 2014: Evaluating flood resilience strategies for coastal megacities. *Science*, **344**(6183), 473–475, doi:10.1126/science.1248222.

Aguilera, S.E., K. Broad and C. Pomeroy, 2018: Adaptive Capacity of the Monterey Bay Wetfish Fisheries: Proactive Responses to the 2015–16 El Niño Event. *Soc. Nat. Resour.*, **31**(12), 1338–1357, doi:10.1080/08941920.2018.1471176.

Alexander, M.A. et al., 2018: Projected sea surface temperatures over the 21st century: Changes in the mean, variability and extremes for large marine ecosystem regions of Northern Oceans. *Elementa-Sci Anthrop.*, **6**(1), 9, doi:10.1525/elementa.191.

Alheit, J. et al., 2014: Atlantic Multidecadal Oscillation (AMO) modulates dynamics of small pelagic fishes and ecosystem regime shifts in the eastern North and Central Atlantic. *J. Marine Syst.*, **131**, 21–35, doi:10.1016/j.jmarsys.2013.11.002.

Altman, J., O.N. Ukhvatkin, A.M. Omelko, M. Macek, T. Plener, V. Pejcha, T. Cerny, P. Petrik, M. Srutek, J.-S. Song, A.A. Zhmerenetsky, A.S. Vozmishcheva, P.V. Krestov, T.Y. Petrenko, K. Treydte, and J. Dolezal, 2018: Poleward migration of the destructive effects of tropical cyclones during the 20th century. *PNAS*, **115**(45), 11543–11548, doi:10.1073/pnas.1808979115.

Alvarez-Filip, L. et al., 2009: Flattening of Caribbean coral reefs: region-wide declines in architectural complexity. *Proc. Royal Soc. B.*, **276**(1669), 3019–3025, doi:10.1098/rspb.2009.0339.

Ampou, E.E. et al., 2017: Coral mortality induced by the 2015-2016 El Nino in Indonesia: the effect of rapid sea level fall. *Biogeosciences*, **14**, 817–826, doi:10.5194/bg-14-817-2017.

Anderson, W., R. Seager, W. Baethgen and M. Cane, 2018: Trans-Pacific ENSO teleconnections pose a correlated risk to agriculture. *Agr. Forest Meteorol.*, **262**, 298–309, doi:10.1016/j.agrformet.2018.07.023.

Anthoff, D., F. Estrada and R.S.J. Tol, 2016: Shutting Down the Thermohaline Circulation. *Am. Econ. Rev.*, **106**(5), 602–606, doi:10.1257/aer.p20161102.

Anthoff, D., R.J. Nicholls and R.S. Tol, 2010: The economic impact of substantial sea level rise. *Mit. Adapt. Strat. Gl.*, **15**(4), 321–335, doi:10.1007/s11027-010-9220-7.

Anthony, K.R.N. et al., 2015: Operationalizing resilience for adaptive coral reef management under global environmental change. *Global Change Biol.*, **21**, 48–61, doi:10.1111/gcb.12700.

Anyamba, A. et al., 2019: Global Disease outbreaks Associated with the 2015–2016 El Niño event. *Sci. Rep.*, **9**(1), 1930, doi:10.1038/s41598-018-38034-z.

APEC, 2016: *Disaster Risk Reduction Action Plan*. 10th Senior Disaster Management Officials Forum, Iquitos, Peru, Asia-Pacific Economic Cooperation. Singapore, 15pp. Available at: http://www.apec-epwg.org/public/uploadfile/act/d20829852d84ae1cb0aba86b475e8f82.pdf Accessed 2018/10/13.

Arent, D.J. et al., 2014: Key economic sectors and services. In: Climate Change 2014: Impacts, Adaptation, and Vulnerability. Part A: Global and Sectoral Aspects. Contribution of Working Group II to the Fifth Assessment Report of the Intergovernmental Panel of Climate Change [Field, C.B., V.R. Barros, D.J. Dokken, K.J. Mach, M.D. Mastrandrea, T.E. Bilir, M. Chatterjee, K.L. Ebi, Y.O. Estrada, R.C. Genova, B. Girma, E.S. Kissel, A.N. Levy, S. MacCracken, P.R. Mastrandrea and L.L. White (eds.)]. Cambridge University Press, Cambridge, United Kingdom and New York, NY, USA, 659–708.

Arias-Ortiz, A. et al., 2018: A marine heatwave drives massive losses from the world's largest seagrass carbon stocks. *Nat. Clim. Change*, **8**(4), 338–344, doi:10.1038/s41558-018-0096-y.

Armour, K.C. et al., 2016: Southern Ocean warming delayed by circumpolar upwelling and equatorward transport. *Nat. Geosci.*, **9**, 549, doi:10.1038/ngeo2731.

Arns, A., T. Wahl, I.D. Haigh and J. Jensen, 2015: Determining return water levels at ungauged coastal sites: a case study for northern Germany. *Ocean Dynam.*, **65**(4), 539–554, doi:10.1007/s10236-015-0814-1.

Årthun, M. et al., 2018: Climate based multi-year predictions of the Barents Sea cod stock. *PLoS ONE*, **13**(10), e0206319–e0206319, doi:10.1371/journal.pone.0206319.

Årthun, M. et al., 2017: Skillful prediction of northern climate provided by the ocean. *Nat. Commun.*, **8**(May), 15875–15875, doi:10.1038/ncomms15875.

Audefroy, J.F. and B.N.C. Sánchez, 2017: Integrating local knowledge for climate change adaptation in Yucatán, Mexico. *Int. J. Sustain. Built Environ.*, **6**(1), 228–237, doi:10.1016/j.ijsbe.2017.03.007.

Azuz-Adeath, I., C. Gonzalez-Campos and A. Cuevas-Corona, 2019: Predicting the Temporal Structure of the Atlantic Multidecadal Oscillation (AMO) for Agriculture Management in Mexico's Coastal Zone. *J. Coastal Res.*, **35**(1), 210–226, doi:10.2112/jcoastres-d-18-00030.1.

Baek, S.H. et al., 2019: Pacific Ocean Forcing and Atmospheric Variability Are the Dominant Causes of Spatially Widespread Droughts in the Contiguous United States. *J. Geophys. Res.-Atmos.*, **124**(5), 2507–2524, doi:10.1029/2018JD029219.

Baker-Austin, C. et al., 2013: Emerging Vibrio risk at high latitudes in response to ocean warming. *Nat. Clim. Change*, **3**(1), 73–77, doi:10.1038/nclimate1628.

Bakker, P. et al., 2016: Fate of the Atlantic Meridional Overturning Circulation: Strong decline under continued warming and Greenland melting. *Geophys. Res. Lett.*, **43**(23), 12,252–12,260, doi:10.1002/2016gl070457.

Balaguru, K., G.R. Foltz and L.R. Leung, 2018: Increasing Magnitude of Hurricane Rapid Intensification in the Central and Eastern Tropical Atlantic. *Geophys. Res. Lett.*, **45**(9), 423–4247, doi:10.1029/2018gl077597.

Balaguru, K. et al., 2015: Dynamic Potential Intensity: An improved representation of the ocean's impact on tropical cyclones. *Geophys. Res. Lett.*, **42**(16), 6739–6746, doi:10.1002/2015GL064822.

Ballantyne, A. et al., 2017: Accelerating net terrestrial carbon uptake during the warming hiatus due to reduced respiration. *Nat. Clim. Change*, **7**, 148–152, doi:10.1038/nclimate3204.

Bamber, J.L. et al., 2018: Land ice freshwater budget of the Arctic and North Atlantic Oceans: 1. Data, methods, and results. *J. Geophys. Res.-Oceans*, **123**(3), 1827-–1837, doi:10.1002/2017JC013605.

Banwell, N., S. Rutherford, B. Mackey and C. Chu, 2018: Towards Improved Linkage of Disaster Risk Reduction and Climate Change Adaptation in Health: A Review. *Int. J. Environ. Res. Public Health*, **15**(4), 793, doi:10.3390/ijerph15040793.

Banzon, V. et al., 2016: A long-term record of blended satellite and in situ sea-surface temperature for climate monitoring, modeling and environmental studies. *Earth Syst. Sci. Data*, **8**(1), 165–176, doi:10.5194/essd-8-165-2016.

Bao, G., Y. Liu, N. Liu and H.W. Linderholm, 2015: Drought variability in eastern Mongolian Plateau and its linkages to the large-scale climate forcing. *Clim. Dynam.*, **44**(3–4), 717–733, doi:10.1007/s00382-014-2273-7.

Barange, M. et al., 2018: *Impacts of Climate Change on fisheries and aquaculture: Synthesis of current knowledge, adaptation and mitigation options*. **627**, FAO Fisheries Technical Paper, 654 pp. http://www.fao.org/3/I9705EN/i9705en.pdf. Accessed 2019/08/20.

Barnes, E.A. and L.M. Polvani, 2015: CMIP5 Projections of Arctic Amplification, of the North American/North Atlantic Circulation, and of Their Relationship. *J. Clim.*, **28**(13), 5254–5271, doi:10.1175/jcli-d-14-00589.1.

Barnes, E.A. and J.A. Screen, 2015: The impact of Arctic warming on the midlatitude jet-stream: Can it? Has it? Will it? *WiRes. Clim. Change*, **6**(3), 277–286, doi:10.1002/wcc.337.

Barnett, J. et al., 2014: A local coastal adaptation pathway. *Nat. Clim. Change*, **4**(12), 1103–1108, doi:10.1038/Nclimate2383.

Barrow, J., J. Ford, R. Day and J. Morrongiello, 2018: Environmental drivers of growth and predicted effects of climate change on a commercially important fish, Platycephalus laevigatus. *Mar. Ecol. Prog. Ser.*, **598**, 201–212, doi:10.3354/meps12234.

Bastos, A. et al., 2018: Impact of the 2015/2016 El Niño on the terrestrial carbon cycle constrained by bottom-up and top-down approaches. *Philos. Trans. R. Soc. London. B.*, **373**(1760), 20170304, doi:10.1098/rstb.2017.0304.

Beal, L.M. and S. Elipot, 2016: Broadening not strengthening of the Agulhas Current since the early 1990s. *Nature*, **540** (7634), 570–573, doi:10.1038/nature19853.

Beck, M.W. et al., 2018: The global flood protection savings provided by coral reefs. *Nat. Commun.*, **9**(1), 2186, doi:10.1038/s41467-018-04568-z.

Beger, M. et al., 2015: Integrating regional conservation priorities for multiple objectives into national policy. *Nat. Commun.*, **6**, 8208, doi:10.1038/ncomms9208.

Beine, M. and C. Parsons, 2015: Climatic factors as determinants of international migration. *Scand. J. Econ.*, **117**(2), 723–767, doi:10.1111/sjoe.12098.

Bell, R. et al., 2017: *Coastal hazards and climate change: Guidance for local government*. Wellington, New Zealand. Published in December 2017 by the Ministry for the Environment, Manatū Mō Te Taiao, PO Box 10362, Wellington 6143, New Zealand ISBN: 978-1-98-852535-8

Bellenger, H. et al., 2014: ENSO representation in climate models: From CMIP3 to CMIP5. *Clim. Dynam.*, **42**(7–8), 1999–2018, doi:10.1007/s00382-013-1783-z.

Bellprat, O. et al., 2016: The Role of Arctic Sea Ice and Sea Surface Temperatures on the Cold 2015 February Over North America. *Bull. Am. Meterol. Soc.*, **97**(12), S36–S41, doi:10.1175/bams-d-16-0159.1.

Bellucci, A., A. Mariotti and S. Gualdi, 2017: The Role of Forcings in the Twentieth-Century North Atlantic Multidecadal Variability: The 1940-75 North Atlantic Cooling Case Study. *J. Clim.*, **30**(18), 7317–7337, doi:10.1175/jcli-d-16-0301.1.

Benthuysen, J.A., M. Feng and L. Zhong, 2014: Spatial patterns of warming off Western Australia during the 2011 Ningaloo Niño: Quantifying impacts of remote and local forcing. *Cont. Shelf Res.*, **91**(Supplement C), 232–246, doi:10.1016/j.csr.2014.09.014.

Benthuysen, J.A., E.C.J. Oliver, M. Feng and A.G. Marshall, 2018: Extreme Marine Warming Across Tropical Australia During Austral Summer 2015–2016. *J. Geophys. Res.-Oceans*, **123**(2), 1301–1326, doi:10.1002/2017JC013326.

Berdalet, E. et al., 2016: Marine harmful algal blooms, human health and wellbeing: challenges and opportunities in the 21st century. *J. Mar. Biol. Assoc. U.K.*, **96**(1), 61–91, doi:10.1017/S0025315415001733.

Bhatia, K. et al., 2018: Projected response of tropical cyclone intensity and intensification in a global climate model. *J. Clim.*, **31**(20), 8281–8303, doi:10.1175/JCLI-D-17-0898.1.

Bhatia, K.T. et al., 2019: Recent increases in tropical cyclone intensification rates. *Nat. Commun.*, **10**(1), 635, doi:10.1038/s41467-019-08471-z.

Bhutto, A.Q., M.J. Iqbal and M.J. Baig, 2017: Abrupt Intensification and Dissipation of Tropical Cyclones in Indian Ocean: A Case Study of Tropical Cyclone Nilofar – 2014. *Journal of Basic & Applied Sciences*, **13**, 566–576, doi:10.6000/1927-5129.2017.13.92.

Biastoch, A. et al., 2015: Atlantic multi-decadal oscillation covaries with Agulhas leakage. *Nat. Commun.*, **6**, 10082, doi:10.1038/ncomms10082.

Binder, S.B., 2014: *Resilience and postdisaster relocation: A study of New York's home buyout plan in the wake of Hurricane Sandy*. University of Hawai'i at Manoa. [Doctoral dissertation, Honolulu, University of Hawaii at Manoa, May 2014]

Bindoff, N.L. et al., 2013: Detection and Attribution of Climate Change: from Global to Regional. In: Climate Change 2013: The Physical Science Basis. Contribution of Working Group I to the Fifth Assessment Report of the Intergovernmental Panel on Climate Change [Stocker, T.F., D. Qin, G.-K. Plattner, M. Tignor, S.K. Allen, J. Boschung, A. Nauels, Y. Xia, V. Bex and P.M. Midgley (eds.)]. Cambridge University Press, Cambridge, United Kingdom and New York, NY, USA, 867–952.

Birkmann, J., D.C. Seng and D.-C. Suarez, 2011: *Adaptive Disaster Risk Reduction: Enhancing Methods and Tools of Disaster Risk Reduction in the light of Climate Change*. DKKV, 52 pp. www.dkkv.org/fileadmin/user_upload/Veroeffentlichungen/Publikationen/DKKV_43_Adaptive_Disaster_Risk_Reduction.pdf. Accessed 2019/08/20.

Birkmann, J. and T. Welle, 2015: Assessing the risk of loss and damage: exposure, vulnerability and risk to climate-related hazards for different country classifications. *Int. J. Global Warm.*, **8**(2), 191–212, doi:10.1504/ljgw.2015.071963.

Blake, E.S., C.W. Landsea and E.J. Gibney, 2011: *The deadliest, costliest, and most intense United States tropical cyclones from 1851 to 2010 (and other frequently requested hurricane facts)*. NOAA/National Weather Service, National Centers for Environmental Prediction, National Hurricane Center Miami, Florida, 47 pp. www.census.gov/history/pdf/nws-nhc-6.pdf. Accessed 2019/08/20.

Blake, E.S. and D.A. Zelinsky, 2018: Hurricane Harvey. Tropical Cyclone Report, National Hurricane Center, Miami, Florida, 76 pp. www.nhc.noaa.gov/data/tcr/AL092017_Harvey.pdf. Accessed 2019/08/20.

Blake, M., 2017: Report of the Independent Review into the Tasmanian Floods of June and July 2016. 138 pp. www.dpac.tas.gov.au/__data/assets/pdf_file/0015/332610/floodreview.pdf. Accessed 2019/08/20.

Blaker, A.T. et al., 2015: Historical analogues of the recent extreme minima observed in the Atlantic meridional overturning circulation at 26A degrees N. *Clim. Dynam.*, **44**(1-2), 457–473, doi:10.1007/s00382-014-2274-6.

Bloemen, P., M. Van Der Steen and Z. Van Der Wal, 2018: Designing a century ahead: climate change adaptation in the Dutch Delta. *Policy and Society*, **38**, 58–76, doi:10.1080/14494035.2018.1513731.

Boet-Whitaker, S.K., 2017: Buyouts as resiliency planning in New York City after Hurricane Sandy. Masters dissertation, Cambridge, Massachusetts Institute of Technology, June 2017.

Bond, N.A., M.F. Cronin, H. Freeland and N. Mantua, 2015: Causes and impacts of the 2014 warm anomaly in the NE Pacific. *Geophys. Res. Lett.*, **42**(9), 3414–3420, doi:10.1002/2015gl063306.

Boning, C.W. et al., 2016: Emerging impact of Greenland meltwater on deepwater formation in the North Atlantic Ocean. *Nat. Geosci.*, **9**(7), 523–527, doi:10.1038/ngeo2740.

Booth, J.F., L. Thompson, J. Patoux and K.A. Kelly, 2012: Sensitivity of midlatitude storm intensification to perturbations in the sea surface temperature near the Gulf Stream. *Mon. Weather Rev.*, **140**(4), 1241–1256, doi:10.1175/mwr-d-11-00195.1.

Born, A., T.F. Stocker, C.C. Raible and A. Levermann, 2013: Is the Atlantic subpolar gyre bistable in comprehensive coupled climate models? *Clim. Dynam.*, **40**(11–12), 2993–3007, doi:10.1007/s00382-012-1525-7.

Born, A., T.F. Stocker and A.B. Sando, 2016: Transport of salt and freshwater in the Atlantic Subpolar Gyre. *Ocean Dynam.*, **66**(9), 1051–1064, doi:10.1007/s10236-016-0970-y.

Bostrom, A. et al., 2018: Eyeing the storm: How residents of coastal Florida see hurricane forecasts and warnings. *Int. J. Disast. Risk Re.*, **30**, 105–119, doi:10.1016/j.ijdrr.2018.02.027.

Botzen, W.J.W., 2013: *Managing extreme climate change risks through insurance*. Cambridge University Press, Cambridge CB2 8RU, UK ISBN: 978-1-107-03327-6

Boulton, C.A., L.C. Allison and T.M. Lenton, 2014: Early warning signals of Atlantic Meridional Overturning Circulation collapse in a fully coupled climate model. *Nat. Commun.*, **5**, 5752, doi:10.1038/ncomms6752.

Bouwer, L.M. and S.N. Jonkman, 2018: Global mortality from storm surges is decreasing. *Environ. Res. Lett.*, **13**, 014008, doi:10.1088/1748-9326/aa98a3.

Bozbiyik, A. et al., 2011: Fingerprints of changes in the terrestrial carbon cycle in response to large reorganizations in ocean circulation. *Clim. Past*, **7**(1), 319–338, doi:10.5194/cp-7-319-2011.

6

Brainard, R.E. et al., 2011: *Status Review Report of 82 Candidate Coral Species Petitioned Under the U.S. Endangered Species Act*. NOAA Technical Memorandum, Pacific Islands Fisheries Science Center, 530 pp. http://citeseerx.ist.psu.edu/viewdoc/download?doi=10.1.1.365.1819&rep=rep1&type=pdf. Accessed 2018/04/19.

Brainard, R.E. et al., 2018: Ecological Impacts of the 2015/16 El Niño in the Central Equatorial Pacific. *Bull. Am. Meterol. Soc.*, **99**(1), S21–S26, doi:10.1175/Bams-D-17-0128.1.

Bråten, I., H.I. Strømsø and L. Salmerón, 2011: Trust and mistrust when students read multiple information sources about climate change. *Learning and Instruction*, **21**(2), 180–192, doi:10.1016/j.learninstruc.2010.02.002.

Broad, K., A.S. Pfaff and M.H. Glantz, 2002: Effective and equitable dissemination of seasonal-to-interannual climate forecasts: policy implications from the Peruvian fishery during El Nino 1997–98. *Clim. Change*, **54**(4), 415–438, doi:10.1023/A:1016164706290.

Bronen, R., 2011: Case 1: Network, the first village in Alaska to relocate due to climate change. In: *North by 2020: perspectives on Alaska's changing social-ecological systems* [Lovecraft, A. L. and H. Eicken (eds.)]. University of Alaska Press, pp. 257-260. ISBN: 978-1-602-23142-9.

Bronen, R. and F.S. Chapin, 2013: Adaptive governance and institutional strategies for climate-induced community relocations in Alaska. *PNAS*, **110**(23), 9320–9325, doi:10.1073/pnas.1210508110.

Brunnabend, S.-E. et al., 2017: Changes in extreme regional sea level under global warming. *Ocean Sci.*, **13**(1), 47–60, doi:10.5194/os-13-47-2017.

Buchan, J., J.J. M. Hirschi, A.T. Blaker and B. Sinha, 2014: North Atlantic SST Anomalies and the Cold North European Weather Events of Winter 2009/10 and December 2010. *Mon. Weather Rev.*, **142**(2), 922–932, doi:10.1175/mwr-d-13-00104.1.

Buckley, M.W. and J. Marshall, 2016: Observations, inferences, and mechanisms of the Atlantic Meridional Overturning Circulation: A review. *Rev. Geophys.*, **54**(1), 5–63, doi:10.1002/2015rg000493.

Burch, S., A. Shaw, A. Dale and J. Robinson, 2014: Triggering transformative change: a development path approach to climate change response in communities. *Climate Policy*, **14**(4), 467–487, doi:10.1080/14693062.2014.876342.

Bureau of Meteorology and Australian CSIRO, 2007: *Climate change in Australia: technical report 2007*. CSIRO Marine and Atmospheric Research. Aspendale, Victoria, 148 pp. ISBN: 978-1-921-23293-0.

Burns, J. et al., 2016: Assessing the impact of acute disturbances on the structure and composition of a coral community using innovative 3D reconstruction techniques. *Methods in Oceanography*, **15-16**, 49–59, doi:10.1016/j.mio.2016.04.001.

Burt, J.A. et al., 2019: Causes and consequences of the 2017 coral bleaching event in the southern Persian/Arabian Gulf. *Coral Reefs*, **38**(4), 567–589, doi:10.1007/s00338-019-01767-y.

Caesar, L. et al., 2018: Observed fingerprint of a weakening Atlantic Ocean overturning circulation. *Nature*, **556** (7700), 191–196, doi:10.1038/s41586-018-0006-5.

Cai, W. et al., 2014a: Increasing frequency of extreme El Nino events due to greenhouse warming. *Nat. Clim. Change*, **4**(2), 111–116, doi:10.1038/NCLIMATE2100.

Cai, W. et al., 2014b: Increased frequency of extreme Indian Ocean Dipole events due to greenhouse warming. *Nature*, **510**(7504), 254–258, doi:10.1038/nature13327.

Cai, W. et al., 2018: Increased variability of eastern Pacific El Niño under greenhouse warming. *Nature*, **564**(7735), 201–206, doi:10.1038/s41586-018-0776-9.

Cai, W. et al., 2015: Increased frequency of extreme La Niña events under greenhouse warming. *Nat. Clim. Change*, **5**(2), 132–137, doi:10.1038/Nclimate2492.

Cai, W. et al., 2013: Projected response of the Indian Ocean Dipole to greenhouse warming. *Nat. Geosci.*, **6**(12), 999–1007, doi:10.1038/ngeo2009.

Cai, Y.Y., T.M. Lenton and T.S. Lontzek, 2016: Risk of multiple interacting tipping points should encourage rapid CO_2 emission reduction. *Nat. Clim. Change*, **6**(5), 520–525, doi:10.1038/NCLIMATE2964.

Callaghan, J. and S.B. Power, 2011: Variability and decline in the number of severe tropical cyclones making land-fall over eastern Australia since the late nineteenth century. *Clim. Dynam.*, **37**(3), 647–662, doi:10.1007/s00382-010-0883-2.

Caminade, C. et al., 2017: Global risk model for vector-borne transmission of Zika virus reveals the role of El Niño 2015. *PNAS*, **114**(1), 119–124, doi:10.1075/pnas.1614303114.

Camus, P. et al., 2017: Statistical wave climate projections for coastal impact assessments. *Earth's Future*, **5**(9), 918–933, doi:10.1002/2017EF000609.

Cangialosi, J.P., A.S. Latto and R. Berg, 2018: *Hurricane Irma*. Tropical Cyclone Report, National Hurricane Center, Center, N. H., Miami, FL, 111 pp. www.nhc.noaa.gov/data/tcr/AL112017_Irma.pdf. Accessed: 2019/20/08.

Caputi, N. et al., 2016: Management adaptation of invertebrate fisheries to an extreme marine heat wave event at a global warming hot spot. *Ecol. Evol.*, **6**(11), 3583–3593, doi:10.1002/ece3.2137.

Carrigan, A.D. and M. Puotinen, 2014: Tropical cyclone cooling combats region-wide coral bleaching. *Global Change Biol.*, **20**(5), 1604–1613, doi:10.1111/gcb.12541.

Carton, J.A., Y. Ding and K.R. Arrigo, 2015: The seasonal cycle of the Arctic Ocean under climate change. *Geophys. Res. Lett.*, **42**(18), 7681–7686, doi:10.1002/2015gl064514.

Cavole, L. et al., 2016: Biological Impacts of the 2013–2015 Warm-Water Anomaly in the Northeast Pacific: Winners, Losers, and the Future. *Oceanography*, **29**(2), 273–285, doi:10.5670/oceanog.2016.32.

Ceronsky, M., D. Anthoff, C. Hepburn and R.S. Tol, 2011: *Checking the price tag on catastrophe: the social cost of carbon under non-linear climate response*. ESRI working paper, Economic and Social Research Institute (ESRI), Dublin, Ireland, 34 pp. www.econstor.eu/bitstream/10419/50174/1/663372984.pdf. Accessed 2018/10/12.

Chafik, L. et al., 2016: Global linkages originating from decadal oceanic variability in the subpolar North Atlantic. *Geophys. Res. Lett.*, **43**(20), 10909–10919, doi:10.1002/2016GL071134.

Chand, S.S., K.J. Tory, H. Ye and K.J. Walsh, 2017: Projected increase in El Niño-driven tropical cyclone frequency in the Pacific. *Nat. Clim. Change*, **7**, 123–127, doi:10.1038/nclimate3181.

Chang, L.Y., J.M. Xu, X.X. Tie and J.B. Wu, 2016: Impact of the 2015 El Nino event on winter air quality in China. *Sci. Rep.*, **6**, 34275, doi:10.1038/srep34275.

Chang, Y., M.-A. Lee, K.-T. Lee and K.-T. Shao, 2013: Adaptation of fisheries and mariculture management to extreme oceanic environmental changes and climate variability in Taiwan. *Mar. Policy*, **38**, 476–482, doi:10.1016/j.marpol.2012.08.002.

Chatterjee, A. et al., 2017: Influence of El Niño on atmospheric CO_2 over the tropical Pacific Ocean: Findings from NASA's OCO-2 mission. *Science*, **358** (6360), eaam5776, doi:10.1126/science.aam5776.

Chen, D.K. et al., 2015a: Strong influence of westerly wind bursts on El Nino diversity. *Nat. Geosci.*, **8**(5), 339–345, doi:10.1038/NGEO2399.

Chen, K., G. Gawarkiewicz, Y.-O. Kwon and W.G. Zhang, 2015b: The role of atmospheric forcing versus ocean advection during the extreme warming of the Northeast U.S. continental shelf in 2012. *J. Geophys. Res.-Oceans*, **120**(6), 4324-4339, doi:10.1002/2014JC010547.

Chen, K., G.G. Gawarkiewicz, S.J. Lentz and J.M. Bane, 2014: Diagnosing the warming of the Northeastern U.S. Coastal Ocean in 2012: A linkage between the atmospheric jet stream variability and ocean response. *J. Geophys. Res.-Oceans*, **119**(1), 218–227, doi:10.1002/2013jc009393.

Cheng, L. et al., 2017: Improved estimates of ocean heat content from 1960 to 2015. *Sci. Adv.*, **3**(3), e1601545, doi:10.1126/sciadv.1601545.

Cheng, L., F. Zheng and J. Zhu, 2015: Distinctive ocean interior changes during the recent warming slowdown. *Sci. Rep.*, **5**, 14346, doi:10.1038/srep14346.

Cheong, S.-M. et al., 2013: Coastal adaptation with ecological engineering. *Nat. Clim. Change*, **3**(9), 787, doi:10.1038/Nclimate1854.

Christensen, J.H. et al., 2013: Climate Phenomena and their Relevance for Future Regional Climate Change. In: Climate Change 2013: The Physical Science Basis. Contribution of Working Group I to the Fifth Assessment Report of the Intergovernmental Panel on Climate Change [Stocker, T.F., D. Qin, G.-K. Plattner, M. Tignor, S.K. Allen, J. Boschung, A. Nauels, Y. Xia, V. Bex and P.M. Midgley (eds.)]. Cambridge University Press, Cambridge, United Kingdom and New York, NY, USA, 1217–1308.

Christidis, N., R.A. Betts and P.A. Stott, 2018a: The Extremely Wet March of 2017 in Peru [in "Explaining Extremes of 2017 from a Climate Perspective"]. *Bull. Am. Meteorol. Soc.*, **100**(1), S31–S37, doi:10.1175/BAMS-ExplainingExtremeEvents2017.1.

Christidis, N., K. Manomaiphiboon, A. Ciavarella and P.A. Stott, 2018b: The hot and dry April of 2016 in Thailand [in "Explaining Extreme Events of 2016 from a Climate Perspective"]. *Bull. Am. Meteorol. Soc.*, **99**(1), S128–S132, doi:10.1175/BAMS-ExplainingExtremeEvents2016.1.

Christidis, N. and P.A. Stott, 2012: Lengthened odds of the cold UK winter of 2010/11 attributable to human influence. *Bull. Am. Meteorol. Soc.*, **93**(7), 1060–1062, doi:10.1175/BAMS-D-12-00021.1.

Christidis, N. and P.A. Stott, 2015: Extreme rainfall in the United Kingdom during winter 2013/14: the role of atmospheric circulation and climate change. *Bull. Am. Meteorol. Soc.*, **96**(12), S46–S50, doi:10.1175/BAMS-D-15-00094.1.

Christie, P. et al., 2016: Improving human and environmental conditions through the Coral Triangle Initiative: progress and challenges. *Curr. Opin. Env. Sust.*, **19**, 169–181, doi:10.1016/j.cosust.2016.03.002.

Chung, C.T.Y. et al., 2014: Nonlinear precipitation response to El Nino and global warming in the Indo-Pacific. *Clim. Dynam.*, **42**(7–8), 1837-1856, doi:10.1007/s00382-013-1892-8.

Chung, E.-S. et al., 2019: Reconciling opposing Walker circulation trends in observations and model projections. *Nat. Clim. Change*, **9**(5), 405–412, doi:10.1038/s41558-019-0446-4.

Church, J.A. et al., 2013: Sea Level Change. In: Climate Change 2013: The Physical Science Basis. Contribution of Working Group I to the Fifth Assessment Report of the Intergovernmental Panel on Climate Change [Stocker, T.F., D. Qin, G.-K. Plattner, M. Tignor, S.K. Allen, J. Boschung, A. Nauels, Y. Xia, V. Bex and P.M. Midgley (eds.)]. Cambridge University Press, Cambridge, United Kingdom and New York, NY, USA, 1137-1216.

Claar, D. et al., 2018: Global patterns and impacts of El Nino events on coral reefs: A meta-analysis. *PLoS ONE*, **13**(2), e0190967, doi:10.1371/journal.pone.0190957.

Claret, M. et al., 2018: Rapid coastal deoxygenation due to ocean circulation shift in the northwest Atlantic. *Nat. Clim. Change*, **8**(10), 868–872, doi:10.1038/s41558-018-0263-1.

Clement, A. et al., 2015: The Atlantic Multidecadal Oscillation without a role for ocean circulation. *Science*, **350**(6258), 320–324, doi:10.1126/science.aab3980.

Coats, S. et al., 2016: Internal ocean-atmosphere variability drives megadroughts in Western North America. *Geophys. Res. Lett.*, **43**(18), 9886–9894, doi:10.1002/ 2016GL070105.

Cobb, K.M. et al., 2013: Highly variable El Niño–Southern Oscillation throughout the Holocene. *Science*, **339** (6115), 67–70, doi:10.1126/science.1228246.

Cohen, J., K. Pfeiffer and J.A. Francis, 2018: Warm Arctic episodes linked with increased frequency of extreme winter weather in the United States. *Nat. Commun.*, **9**(1), 869, doi:10.1038/s41467-018-02992-9.

Cohen, J. et al., 2014: Recent Arctic amplification and extreme mid-latitude weather. *Nat. Geosci.*, **7**, 627, doi:10.1038/ngeo2234.

Colberg, F., K.L. McInnes, J. O'Grady and R. Hoeke, 2019: Atmospheric circulation changes and their impact on extreme sea levels around Australia. *Nat. Hazard. Earth Sys.*, **19**, 1–20, doi:10.5194/nhess-2018-64.

Coles, S.L., E. Looker and J. A. Burt, 2015: Twenty-year changes in coral near Muscat, Oman estimated from manta board tow observations. *Mar. Environ. Res.*, **103**, 66–73, doi:10.1016/j.marenvres.2014.11.006.

Collins, J.A. et al., 2017: Rapid termination of the African Humid Period triggered by northern high-latitude cooling. *Nat. Commun.*, **8**, 1372, doi:10.1038/s41467-017-01454-y.

Collins, J.M. et al., 2016: The record-breaking 2015 hurricane season in the eastern North Pacific: An analysis of environmental conditions. *Geophys. Res. Lett.*, **43**(17), 9217–9224, doi:10.1002/2016GL070597.

Collins, M. et al., 2013: Long-term Climate Change: Projections, Commitments and Irreversibility. In: Climate Change 2013: The Physical Science Basis. Contribution of Working Group I to the Fifth Assessment Report of the Intergovernmental Panel on Climate Change [Stocker, T.F., D. Qin, G.-K. Plattner, M. Tignor, S.K. Allen, J. Boschung, A. Nauels, Y. Xia, V. Bex and P.M. Midgley (eds.)]. Cambridge University Press, Cambridge, United Kingdom and New York, NY, USA, 1029-1136.

Compo, G.P. et al., 2011: The Twentieth Century Reanalysis Project. *Q. R. Roy. Meteorol. Soc.*, **137**(654), 1–28, doi:10.1002/qj.776.

Costanza, R. et al., 2008: The value of coastal wetlands for hurricane protection. *AMBIO*, **37**(4), 241–248, doi:10.1579/0044-7447(2008)37[241:TVOCWF]2.0.CO;2.

Coumou, D. et al., 2014: Quasi-resonant circulation regimes and hemispheric synchronization of extreme weather in boreal summer. *PNAS*, **111**(34), 12331, doi:10.1073/pnas.1412797111.

Cunningham, S. A. et al., 2013: Atlantic Meridional Overturning Circulation slowdown cooled the subtropical ocean. *Geophys. Res. Lett.*, **40**(23), 6202–6207, doi:10.1002/2013gl058464.

Cvijanovic, I. et al., 2017: Future loss of Arctic sea-ice cover could drive a substantial decrease in California's rainfall. *Nat. Commun.*, **8**(1), 1947, doi:10.1038/s41467-017-01907-4.

Dalisay, S.N. and M.T. De Guzman, 2016: Risk and culture: the case of typhoon Haiyan in the Philippines. *Disaster Prevention and Management: An International Journal*, **25**(5), 701–714, doi:10.1108/Dpm-05-2016-0097.

Daloz, A.S. and S. J. Camargo, 2018: Is the poleward migration of tropical cyclone maximum intensity associated with a poleward migration of tropical cyclone genesis? *Clim. Dynam.*, **50**, 705–715, doi:10.1007/s00382-017-3636-7.

Darmaraki, S. et al., 2019: Future evolution of Marine Heat Waves in the Mediterranean Sea. *Clim. Dynam.*, **53**(3-4), 1371–1392, doi:10.1007/s00382-019-04661-z.

David, C.P., B.A.B. Racoma, J. Gonzales and M.V. Clutario, 2013: A manifestation of climate change? A look at Typhoon Yolanda in relation to the historical tropical cyclone archive. *Science Diliman*, **25**(2), 78–86.

Davlasheridze, M., K. Fisher-Vanden and H. A. Klaiber, 2017: The effects of adaptation measures on hurricane induced property losses: Which FEMA investments have the highest returns? *Journal of Environmental Economics and Management*, **81**, 93–114, doi: 10.1016/j.jeem.2016.09.005.

de Abreu, R.C. et al., 2019: Contribution of Anthropogenic Climate Change to April–May 2017 Heavy Precipitation over the Uruguay River Basin [in "Explaining Extremes of 2017 from a Climate Perspective"]. *Bull. Am. Meteorol. Soc.*, **100**(1), S37–S41, doi:10.1175/BAMS-D-18-0102.1.

Dean, S.M., S. Rosier, T. Carey-Smith and P.A. Stott, 2013: The role of climate change in the two-day extreme rainfall in Golden Bay, New Zealand, December 2011 [in "Explaining Extreme Events of 2012 from a Climate Perspective"]. *Bull. Am. Meteorol. Soc.*, **94**(9), S61–S63, doi:10.1175/BAMS-D-13-00085.1.

DeCarlo, T.M. et al., 2017: Community production modulates coral reef pH and the sensitivity of ecosystem calcification to ocean acidification. *J. Geophys. Res.-Oceans*, **122**(1), 745–761, doi:10.1002/2016jc012326.

Dee, D.P. et al., 2011: The ERA-Interim reanalysis: Configuration and performance of the data assimilation system. *Q. R. Roy. Meteorol. Soc.*, **137**(656), 553–597, doi:10.1002/qj.828.

Defrance, D. et al., 2017: Consequences of rapid ice sheet melting on the Sahelian population vulnerability. *PNAS*, **114**(25), 6533–6538, doi:10.1073/pnas.1619358114.

6

Dekker, M.M., A.S. von der Heydt and H.A. Dijkstra, 2018: Cascading transitions in the climate system. *Earth Syst. Dynam.*, 9(4), 1243–1260, doi:10.5194/esd-9-1243-2018.

dela Cruz, D.W. and P.L. Harrison, 2017: Enhanced larval supply and recruitment can replenish reef corals on degraded reefs. *Sci. Rep.*, 7(1), 13985, doi:10.1038/s41598-017-14546-y.

Delworth, T.L. and F.R. Zeng, 2016: The impact of the North Atlantic Oscillation on climate through its influence on the Atlantic Meridional Overturning Circulation. *J. Clim.*, 29(3), 941–962, doi:10.1175/jcli-d-15-0396.1.

Denniston, R. F. et al., 2015: Extreme rainfall activity in the Australian tropics reflects changes in the El Niño/Southern Oscillation over the last two millennia. *PNAS*, 112 (15), 4576, doi:10.1073/pnas.1422270112.

Denommee, K.C., S.J. Bentley and A.W. Droxler, 2014: Climatic controls on hurricane patterns: a 1200-y near-annual record from Lighthouse Reef, Belize. *Sci. Rep.*, 4, 3876, doi:10.1038/srep03876.

Department for Communities and Local Government, 2014: *Winter 2013/14 severe weather recovery progress report*. Department for Communities and Local Government, Fry Building, 2 Marsham Street, London, SW1P 4DF, UK, 38pp. Available at: https://assets.publishing.service.gov.uk/government/uploads/system/uploads/attachment_data/file/380573/Winter_2013-14_severe_weather_recovery_progress_report.pdf.

Depczynski, M. et al., 2013: Bleaching, coral mortality and subsequent survivorship on a West Australian fringing reef. *Coral Reefs*, 32(1), 233–238, doi:10.1007/s00338-012-0974-0.

Descamps, S., H. Strøm and H. Steen, 2013: Decline of an arctic top predator: synchrony in colony size fluctuations, risk of extinction and the subpolar gyre. *Oecologia*, 173(4), 1271–1282, doi:10.1007/s00442-013-2701-0.

Deshayes, J. et al., 2013: Oceanic hindcast simulations at high resolution suggest that the Atlantic MOC is bistable. *Geophys. Res. Lett.*, 40(12), 3069–3073, doi:10.1002/grl.50534.

Devine-Wright, P., 2013: Think global, act local? The relevance of place attachments and place identities in a climate changed world. *Global Environ. Change*, 23(1), 61–69, doi:10.1016/j.gloenvcha.2012.08.003.

DeVries, T., M. Holzer and F. Primeau, 2017: Recent increase in oceanic carbon uptake driven by weaker upper-ocean overturning. *Nature*, 542, 215, doi:10.1038/nature21068.

Di Lorenzo, E. et al., 2013: Synthesis of Pacific Ocean climate and ecosystem dynamics. *Oceanography*, 26(4), 68–81, doi:10.5670/oceanog.2013.76.

Di Lorenzo, E. and N. Mantua, 2016: Multi-year persistence of the 2014/15 North Pacific marine heatwave. *Nat. Clim. Change*, 6(11), 1042–1047, doi:10.1038/nclimate3082.

Diamond, H. J. and C. J. Schreck, 2016: Tropical Cyclones [in "State of the Climate in 2015"]. *Bull. Am. Meteorol. Soc.*, 97(8), S104–S130, doi:10.1175/2016BAMSStateoftheClimate.1.

Dibajnia, M., M. Soltanpour, R. Nairn and M. Allahyar, 2010: Cyclone Gonu: the most intense tropical cyclone on record in the Arabian Sea. In: *Indian ocean tropical cyclones and climate change*. EYassine Charabi (Ed.), Springer Netherlands Springer, 149–157, ISBN: 978-90-481-3108-2.

Dilling, L. et al., 2015: The dynamics of vulnerability: why adapting to climate variability will not always prepare us for climate change. *WiRes. Clim. Change*, 6(4), 413–425, doi:10.1002/wcc.341.

Dima, M. and G. Lohmann, 2010: Evidence for two distinct modes of large-scale ocean circulation changes over the last century. *J. Clim.*, 23(1), 5–16, doi:10.1175/2009JCLI2867.1.

Diodato, N., L. de Guenni, M. Garcia and G. Bellocchi, 2019: Decadal Oscillation in the Predictability of Palmer Drought Severity Index in California. *Climate*, 7(1), 6, doi:10.3390/cli7010006.

Dong, L. and M.J. McPhaden, 2018: Unusually warm Indian Ocean sea surface temperatures help to arrest development of El Niño in 2014. *Sci. Rep.*, 8(1), 2249, doi:10.1038/s41598-018-20294-4.

Dong, L., T. Zhou and B. Wu, 2014: Indian Ocean warming during 1958–2004 simulated by a climate system model and its mechanism. *Clim. Dynam.*, 42(1–2), 203–217, doi:10.1007/s00382-013-1722-z.

Drijfhout, S. et al., 2015: Catalogue of abrupt shifts in Intergovernmental Panel on Climate Change climate models. *PNAS*, 112(43), E5777-E5786, doi:10.1073/pnas.1511451112.

Du, X. et al., 2016: Initiation and development of a toxic and persistent Pseudo-nitzschia bloom off the Oregon coast in spring/summer 2015. *PLoS ONE*, 11(10), e0163977, doi:10.1371/journal.pone.0163977.

Duchez, A. et al., 2016: Potential for seasonal prediction of Atlantic sea surface temperatures using the RAPID array at 26N. *Clim. Dynam.*, 46 (9–10), 3351–3370, doi:10.1007/s00382-015-2918-1.

Duke, N.C. et al., 2017: Large-scale dieback of mangroves in Australia's Gulf of Carpentaria: a severe ecosystem response, coincidental with an unusually extreme weather event. *Mar. Freshwater Res.*, 68(10), 1816–1829, doi:10.1071/MF16322.

Dukhovskoy, D.S. et al., 2016: Greenland freshwater pathways in the sub-Arctic Seas from model experiments with passive tracers. *J. Geophys. Res.-Oceans*, 121(1), 877–907, doi:10.1002/2015jc011290.

Dunstone, N. et al., 2013: Anthropogenic aerosol forcing of Atlantic tropical storms. *Nat. Geosci.*, 6(7), 534–539, doi:10.1038/NGEO1854.

Durrer, P. and E. Adams, 2011: Case 3: Finding Ways to Move: The Challenges of Relocation in Kivalina, Northwest Alaska. In: *North by 2020: perspectives on Alaska's changing social-ecological systems* [Lovecraft, A. L. and H. Eicken (eds.)]. University of Alaska Press, 265-268. ISBN: 9781602231429.

Eakin, C.M. et al., 2017: Ding, dong, the witch is dead (?) – three years of global coral bleaching 2014-2017. *Reef Encounter*, 45(32), 33–38.

Eakin, C.M. et al., 2018: Unprecedented three years of global coral bleaching 2014-2017 [Sidebar 3.1, in "State of the Climate in 2017"]. *Bull. Am. Meterol. Soc.*, 99(8), S74–S75, doi:10.1175/2018BAMSStateoftheClimate.1.

Eakin, C.M. et al., 2010: Monitoring coral reefs from space. *Oceanography*, 23 (4), 118–133, doi:10.2307/24860867.

Echevin, V.M. et al., 2018: Forcings and evolution of the 2017 coastal El Niño off Northern Peru and Ecuador. *Front. Mar. Sci.*, 5, 367, doi:10.3389/fmars.2018.00367.

Edmondson, B. and S. Levy, 2019: *Transformative Climates and Accountable Governance*. Palgrave Macmillan, Cham. ISBN: 978-3-319-97399-9.

Emanuel, K.A., 2013: Downscaling CMIP5 climate models shows increased tropical cyclone activity over the 21st century. *PNAS*, 110(30), 12219–12224, doi:10.1073/pnas.1301293110.

Emanuel, K. A., 2015: Effect of upper-ocean evolution on projected trends in tropical cyclone activity. *J. Clim.*, 28(20), 8165–8170, doi:10.1175/Jcli-D-15-0401.1.

Emanuel, K.A., 2017: Assessing the present and future probability of Hurricane Harvey's rainfall. *PNAS*, 114(48), 12681–12684, doi:10.1073/pnas.1716222114.

ENFEN, 2017: *El Niño Costero 2017*. Informe Técnico Extraordinario, N°001-2017/ENFEN, Estudio Nacional del Fenómeno "El Niño", Estudio Nacional Del Fenómeno "El Niño" (ENFEN). Peru. 31 pp. www.imarpe.pe/imarpe/archivos/informes/imarpe_inftco_informe__tecnico_extraordinario_001_2017.pdf. Accessed 2018/09/28.

England, M.H. et al., 2014: Recent intensification of wind-driven circulation in the Pacific and the ongoing warming hiatus. *Nat. Clim. Change*, 4(3), 222–227, doi:10.1038/NCLIMATE2106.

Eslake, S., 2016: *Tasmania Report 2016*. Tasmania Chamber of Commerce and Industry, Tasmanian Chamber of Commerce and Industry. Hobart, Tasmania, Australia. 92pp. http://www.tcci.com.au/getattachment/Events/Tasmania-Report-2016/Tasmania-Report-2016-FINAL.pdf.aspx, Accessed 2018/04/01.

Espinueva, S.R., E.O. Cayanan and N.C. Nievares, 2012: A retrospective on the devastating impacts of Tropical Storm Washi. *Tropical Cyclone Research and Review*, 1(2), 163–176, doi:10.6057/2012TCRR02.11.

Evans, J.P. and I. Boyer-Souchet, 2012: Local sea surface temperatures add to extreme precipitation in northeast Australia during La Niña. *Geophys. Res. Lett.*, 39(10), L10803, doi:10.1029/2012GL052014.

Ezer, T., 2015: Detecting changes in the transport of the Gulf Stream and the Atlantic overturning circulation from coastal sea level data: The extreme decline in 2009–2010 and estimated variations for 1935–2012. *Glob. Planet. Change*, **129**, 23–36, doi:10.1016/j.gloplacha.2015.03.002.

Ezer, T., L.P. Atkinson, W.B. Corlett and J.L. Blanco, 2013: Gulf Stream's induced sea level rise and variability along the U.S. mid-Atlantic coast. *J. Geophys. Res.-Oceans*, **118**(2), 685–697, doi:10.1002/jgrc.20091.

FAMURS, 2017: *Sobe para R$ 339 milhões o valor dos prejuízos com o temporal no RS*. Federação das Associações de Municípios do Rio Grande do Sul (FAMURS). Porto Allegre, Brazil. Available at: http://www.famurs.com.br/noticias/sobe-para-r-339-milhoes-o-valor-dos-prejuizos-com-o-temporal-no-rs/. Accessed 2019/05/20.]

Fang, G. et al., 2010: Volume, heat, and freshwater transports from the South China Sea to Indonesian seas in the boreal winter of 2007–2008. *J. Geophys. Res.-Oceans*, **115**, C12020, doi:10.1029/2010jc006225.

FAO, 2017: *Bangladesh: Severe floods in 2017 affected large numbers of people and caused damage to the agriculture sector*. GIEWS Update, Food and Agriculture Organization of the United Nations, Rome, Italy, 6 pp. www.fao.org/3/a-i7876e.pdf. Accessed: 2019/20/08.

FAO, 2018: *The impact of disasters and crises on agriculture and food security 2017*. Food and Agriculture Organization of the United Nations, Rome, Italy, 168 pp. www.fao.org/3/I8656EN/i8656en.pdf. Accessed: 2019/20/08.

Farbotko, C. and H. Lazrus, 2012: The first climate refugees? Contesting global narratives of climate change in Tuvalu. *Global Environ. Change*, **22** (2), 382–390, doi:10.1016/j.gloenvcha.2011.11.014.

Fasullo, J., B. Otto-Bliesner and S. Stevenson, 2018: ENSO's Changing Influence on Temperature, Precipitation, and Wildfire In a Warming Climate. *Geophys. Res. Lett.*, **45** (17), 9216–9225, doi:10.1029/2018gl079022.

Fay, A.R. and G.A. McKinley, 2013: Global trends in surface ocean pCO2 from in situ data. *Global Biogeochem. Cy.*, **27**(2), 541–557, doi:10.1002/gbc.20051.

Fedorov, A., S. Hu, M. Lengaigne and E. Guilyardi, 2015: The impact of westerly wind bursts and ocean initial state on the development, and diversity of El Nino events. *Clim. Dynam.*, **44**(5-6), 1381–1401, doi:10.1007/s00382-014-2126-4.

Feldpausch, T.R. et al., 2016: Amazon forest response to repeated droughts. *Global Biogeochem. Cy.*, **30**(7), 964–982, doi:10.1002/2015GB005133.

Feng, M. et al., 2015: Decadal increase in Ningaloo Niño since the late 1990s. *Geophys. Res. Lett.*, **42**(1), 104–112, doi:10.1002/2014GL062509.

Feng, M., M.J. McPhaden, S.-P. Xie and J. Hafner, 2013: La Niña forces unprecedented Leeuwin Current warming in 2011. *Sci. Rep.*, **3**, 1277, doi:10.1038/srep01277.

Feng, M., N. Zhang, Q. Liu and S. Wijffels, 2018: The Indonesian throughflow, its variability and centennial change. *Geoscience Letters*, **5**(1), 3, doi:10.1186/s40562-018-0102-2.

Feng, M., X. Zhang, B. Sloyan and M. Chamberlain, 2017: Contribution of the deep ocean to the centennial changes of the Indonesian Throughflow. *Geophys. Res. Lett.*, **44**(6), 2859–2867, doi:10.1002/2017GL072577.

Fernández-Méndez, M. et al., 2015: Photosynthetic production in the central Arctic Ocean during the record sea-ice minimum in 2012. *Biogeosciences*, **12**(11), 3525–3549, doi:10.5194/bg-12-3525-2015.

Ferrigno, F. et al., 2016: Corals in high diversity reefs resist human impact. *Ecol. Indic.*, **70**, 106–113, doi:10.1016/j.ecolind.2016.05.050.

Feudale, L. and J. Shukla, 2007: Role of Mediterranean SST in enhancing the European heat wave of summer 2003. *Geophys. Res. Lett.*, **34**(3), L03811, doi:10.1029/2006GL027991.

FEWS NET and FSNAU, 2017: *Risk of Famine (IPC Phase 5) persists in Somalia*. Somalia Food Security Outlook, Famine Early Warning Systems Network; Food Security and Nutrition Analysis Unit - Somalia, 16 pp. www.fsnau.org/downloads/FEWSNET-FSNAU-Joint-Somalia-Food-Security-Outlook-February-to-September-2017_0.pdf. Accessed 2019/20/08.

Field, C.B., V. Barros, T.F. Stocker and Q. Dahe, 2012: Managing the risks of extreme events and disasters to advance climate change adaptation: special report of the intergovernmental panel on climate change. Cambridge University Press. ISBN 978-1-107-02506-6.

Fischer, E.M. and R. Knutti, 2015: Anthropogenic contribution to global occurrence of heavy-precipitation and high-temperature extremes. *Nat. Clim. Change*, **5**(6), 560–564, doi:10.1038/nclimate2617.

Flato, G. et al., 2013: Evaluation of Climate Models. In: Climate Change 2013: The Physical Science Basis. Contribution of Working Group I to the Fifth Assessment Report of the Intergovernmental Panel on Climate Change [Stocker, T.F., D. Qin, G.-K. Plattner, M. Tignor, S.K. Allen, J. Boschung, A. Nauels, Y. Xia, V. Bex and P.M. Midgley (eds.)]. Cambridge University Press, Cambridge, United Kingdom and New York, NY, USA, 741-866.

Fleurbaey, M. et al., 2014: Sustainable development and equity. In: Climate Change 2014: Mitigation of climate change. Contribution of Working Group III to the Fifth Assessment Report of the Intergovernmental Panel on Climate Change [Edenhofer, O., R. Pichs-Madruga, Y. Sokona, E. Farahani, S. Kadner, K. Seyboth, A. Adler, I. Baum, S. Brunner, P. Eickemeier, B. Kriemann, J. Savolainen, S. Schlömer, C. von Stechow, T. Zwickel and J.C. Minx (eds.)]. Cambridge University Press, Cambridge, United Kingdom and New York, NY, USA, 283–350.

Forino, G., J. von Meding, G. Brewer and D. van Niekerk, 2017: Climate Change Adaptation and Disaster Risk reduction integration: Strategies, Policies, and Plans in three Australian Local Governments. *Int. J. Disast. Risk Re.*, **24**, 100–108, doi:10.1016/j.ijdrr.2017.05.021.

Francis, J. and N. Skific, 2015: Evidence linking rapid Arctic warming to mid-latitude weather patterns. *Philos. Trans. Royal Soc. A.*, **373**(2045), 20140170, doi:10.1098/rsta.2014.0170.

Francis, J.A. and S.J. Vavrus, 2015: Evidence for a wavier jet stream in response to rapid Arctic warming. *Environ. Res. Lett.*, **10**(1), 014005, doi:10.1088/1748-9326/10/1/014005.

Fraser, M.W. et al., 2014: Extreme climate events lower resilience of foundation seagrass at edge of biogeographical range. *J. Ecol.*, **102**(6), 1528 doi:10.1098/rsta.2014.0170.1536, doi:10.1111/1365-2745.12300.

Frederiksen, C.S., J.S. Frederiksen, J.M. Sisson and S.L. Osbrough, 2017: Trends and projections of Southern Hemisphere baroclinicity: the role of external forcing and impact on Australian rainfall. *Clim. Dynam.*, **48**(9), 3261–3282, doi:10.1007/s00382-016-3263-8.

Freitas, A.C.V., L. Aímola, T. Ambrizzi and C.P. de Oliveira, 2017: Extreme Intertropical Convergence Zone shifts over Southern Maritime Continent. *Atmos. Sci. Lett.*, **18** (1), 2–10, doi:10.1002/asl.716.

Freund, M.B. et al., 2019: Higher frequency of Central Pacific El Niño events in recent decades relative to past centuries. *Nat. Geosci.*, **12**, 450–455, doi:10.1038/s41561-019-0353-3.

Fritz, H.M., C.D. Blount, F.B. Albusaidi and A.H.M. Al-Harthy, 2010: Cyclone Gonu storm surge in Oman. *Estuar. Coast. Shelf.*, **86**(1), 102–106, doi:10.1016/j.ecss.2009.10.019.

Frölicher, T.L., E.M. Fischer and N. Gruber, 2018: Marine heat waves under global warming. *Nature*, **560**(7718), 360–-364, doi:10.1038/s41586-018-0383-9.

Frölicher, T.L. and C. Laufkötter, 2018: Emerging risks from marine heat waves. *Nat. Commun.*, **9**(1), 650, doi:10.1038/s41467-018-03163-6.

Frölicher, T.L., K.B. Rodgers, C.A. Stock and W.W.L. Cheung, 2016: Sources of uncertainties in 21st century projections of potential ocean ecosystem stressors. *Global Biogeochem. Cy.*, **30**(8), 1224–1243, doi:10.1002/2015gb005338.

Frölicher, T.L. et al., 2015: Dominance of the Southern Ocean in Anthropogenic Carbon and Heat Uptake in CMIP5 Models. *J. Clim.*, **28**(2), 862–886, doi:10.1175/jcli-d-14-00117.1.

Fuckar, N.S. et al., 2016: Record low northern hemisphere sea ice extent in March 2015. *Bull. Am. Meterol. Soc.*, **97**(12), S136–S143, doi:10.1175/Bams-D-16-0153.1.

6

Funk, C. et al., 2018: Anthropogenic enhancement of moderate-to-strong El Nino events likely contributed to drought and poor harvests in Southern Africa during 2016 [in "Explaining Extreme Events of 2016 from a Climate Perspective"]. *Bull. Am. Meteorol. Soc.*, **99**(1), S91–S101, doi:10.1175/BAMS-ExplainingExtremeEvents2016.1.

Gaetani, M. and E. Mohino, 2013: Decadal Prediction of the Sahelian Precipitation in CMIP5 Simulations. *J. Clim.*, **26**(19), 7708–7719, doi:10.1175/JCLI-D-12-00635.1.

Galaasen, E.V. et al., 2014: Rapid reductions in North Atlantic deep water during the peak of the last interglacial period. *Science*, **343**(6175), 1129–1132, doi:10.1126/science.1248667.

Galli, G., C. Solidoro and T. Lovato, 2017: Marine Heat Waves Hazard 3D Maps and the Risk for Low Motility Organisms in a Warming Mediterranean Sea. *Front. Mar. Sci.*, **4**, 136, doi:10.3389/fmars.2017.00136.

Gallina, V. et al., 2016: A review of multi-risk methodologies for natural hazards: Consequences and challenges for a climate change impact assessment. *J. Environ. Manage.* **168**, 123–132, doi:10.1016/j.jenvman.2015.11.011.

García-Herrera, R. et al., 2010: A Review of the European Summer Heat Wave of 2003. *Crit. Rev. Env. Sci. Tec.*, **40**(4), 267–306, doi:10.1080/10643380802238137.

Garner, A.J. et al., 2017: Impact of climate change on New York City's coastal flood hazard: Increasing flood heights from the preindustrial to 2300 CE. *PNAS*, **114**(45), 11861–-11866, doi:10.1073/pnas.1703568114.

Garrabou, J. et al., 2009: Mass mortality in Northwestern Mediterranean rocky benthic communities: effects of the 2003 heat wave. *Global Change Biol.*, **15**(5), 1090–1103, doi:10.1111/j.1365-2486.2008.01823.x.

Garreaud, R.D., 2018: A plausible atmospheric trigger for the 2017 coastal El Niño. *Int. J. Climatol.*, **38**, e1296–-e1302, doi:10.1002/joc.5426.

Gastineau, G., B. L'Heveder, F. Codron and C. Frankignoul, 2016: Mechanisms determining the winter atmospheric response to the Atlantic Overturning Circulation. *J. Clim.*, **29**(10), 3767–3785, doi:10.1175/jcli-d-15-0326.1.

Gattuso, J.P., O. Hoegh-Guldberg and H.O. Pörtner, 2014: Cross-chapter box on coral reefs. In: Climate Change 2014: Impacts, Adaptation, and Vulnerability. Part A: Global and Sectoral Aspects. Contribution of Working Group II to the Fifth Assessment Report of the Intergovernmental Panel of Climate Change [Field, C.B., V.R. Barros, D.J. Dokken, K.J. Mach, M.D. Mastrandrea, T.E. Bilir, M. Chatterjee, K.L. Ebi, Y.O. Estrada, R.C. Genova, B. Girma, E.S. Kissel, A.N. Levy, S. MacCracken, P.R. Mastrandrea and L.L. White (eds.)]. Cambridge University Press, Cambridge, United Kingdom and New York, NY, USA, 97–100.

Gelcer, E. et al., 2018: Influence of El Niño-Southern oscillation (ENSO) on agroclimatic zoning for tomato in Mozambique. *Agr. Forest Meteorol.*, **248**, 316–328, doi:10.1016/j.agrformet.2017.10.002.

Gemenne, F., 2015: One good reason to speak of 'climate refugees'. *Forced Migration Review*, **49**, 70–71.

Gentemann, C.L., M.R. Fewings and M. García-Reyes, 2017: Satellite sea surface temperatures along the West Coast of the United States during the 2014–2016 northeast Pacific marine heat wave. *Geophys. Res. Lett.*, **44**(1), 312–319, doi:10.1002/2016gl071039.

Gerkensmeier, B. and B.M.W. Ratter, 2018: Multi-risk, multi-scale and multi-stakeholder–the contribution of a bow-tie analysis for risk management in the trilateral Wadden Sea Region. *J. Coast. Conserv.*, **22**(1), 145–156, doi:10.1007/s11852-016-0454-8.

Gibbs, L., H.C. Gallagher, K. Block and E. Baker, 2016: Post-bushfire relocation decision-making and personal wellbeing: a case study from Victoria, Australia. In: *Planning for Community-based Disaster Resilience Worldwide*. Routledge, UK, pp. 355–378. ISBN: 9781472468154.

Gnanaseelan, C., M.K. Roxy and A. Deshpande, 2017: Variability and Trends of Sea Surface Temperature and Circulation in the Indian Ocean. In: *Observed climate variability and change over the Indian region* [Rajeevan, M.N. and S. Nayak (eds.)]. Springer Singapore, Singapore, pp. 165–179. ISBN: 978-981-10-2530-3.

Goddard, P.B., J.J. Yin, S.M. Griffies and S.Q. Zhang, 2015: An extreme event of sea level rise along the Northeast coast of North America in 2009–2010. *Nat. Commun.*, **6**, 6346, doi:10.1038/ncomms7346.

Gong, T. and D. Luo, 2017: Ural Blocking as an Amplifier of the Arctic Sea Ice Decline in Winter. *J. Clim.*, **30**(7), 2639–2654, doi:10.1175/jcli-d-16-0548.1.

González-Riancho, P., B. Gerkensmeier and B.M. Ratter, 2017: Storm surge resilience and the Sendai Framework: Risk perception, intention to prepare and enhanced collaboration along the German North Sea coast. *Ocean. Coast. Manage.*, **141**, 118–131, doi:10.1016/j.ocecoaman.2017.03.006.

González-Alemán, J.J. et al., 2019: Potential increase in hazard from Mediterranean hurricane activity with global warming. *Geophys. Res. Lett.*, **46**(3), 1754–1764, doi:10.1029/2018GL081253.

Good, P. et al., 2018: Recent progress in understanding climate thresholds: Ice sheets, the Atlantic meridional overturning circulation, tropical forests and responses to ocean acidification. *Prog. Phys. Geo.*, **42**(1), 24–60, doi:10.1177/0309133317751843.

Govin, A. et al., 2012: Persistent influence of ice sheet melting on high northern latitude climate during the early Last Interglacial. *Clim. Past*, **8**(2), 483–507, doi:10.5194/cp-8-483-2012.

Graham, N.A. et al., 2015: Predicting climate-driven regime shifts versus rebound potential in coral reefs. *Nature*, **518** (7537), 94–97, doi:10.1038/nature14140.

Gregory, J.M. et al., 2016: The Flux-Anomaly-Forced Model Intercomparison Project (FAFMIP) contribution to CMIP6: investigation of sea level and ocean climate change in response to CO_2 forcing. *Geosci. Model. Dev.*, **9**(11), 3993–4017, doi:10.5194/gmd-9-3993-2016.

Grieger, J. et al., 2014: Southern Hemisphere winter cyclone activity under recent and future climate conditions in multi-model AOGCM simulations. *Int. J. Climatol.*, **34**(12), 3400–3416, doi:10.1002/joc.3917.

Grinsted, A., J.C. Moore and S. Jevrejeva, 2012: Homogeneous record of Atlantic hurricane surge threat since 1923. *PNAS*, **109**(48), 19601–19605, doi:10.1073/pnas.1209542109.

Grose, M.R. et al., 2012: The simulation of cutoff lows in a regional climate model: reliability and future trends. *Clim. Dynam.*, **39**(1–2), 445–459, doi:10.1007/s00382-012-1368-2.

Guemas, V. et al., 2013: September 2012 Arctic sea ice minimum: discriminating between sea ice memory, the August 2012 extreme storm, and prevailing warm conditions. In: *"Explaining Extremes of 2012 from a Climate Perspective"*. *Bull. Am. Meteorol. Soc.*, **94**(9), S20–S22, doi:10.1175/BAMS-D-13-00085.1.

Guo, L., A.G. Turner and E.J. Highwood, 2016: Local and remote impacts of aerosol species on Indian summer monsoon rainfall in a GCM. *J. Clim.*, **29**(19), 6937–6955, doi:10.1175/Jcli-D-15-0728.1.

Haarsma, R.J. et al., 2013: More hurricanes to hit western Europe due to global warming. *Geophys. Res. Lett.*, **40**(9), 1783–1788, doi:10.1002/grl.50360.

Haarsma, R.J., F.M. Selten and S.S. Drijfhout, 2015: Decelerating Atlantic meridional overturning circulation main cause of future west European summer atmospheric circulation changes. *Environ. Res. Lett.*, **10**(9), 094007, doi:10.1088/1748-9326/10/9/094007.

Haasnoot, M. et al., 2018: *Mogelijke gevolgen van versnelde zeespiegelstijging voor het Deltaprogramma: een verkenning*, Detares. Delft, Netherlands, 43 pp. www.deltacommissaris.nl/binaries/deltacommissaris/documenten/publicaties/2018/09/18/dp2019-b-rapport-deltares/DP2019+B+Rapport+Deltares.pdf. Accessed 2018/10/05.

Haasnoot, M. et al., 2015: Transient scenarios for robust climate change adaptation illustrated for water management in The Netherlands. *Environ. Res. Lett.*, **10**(10), 105008, doi:10.1088/1748-9326/10/10/105008.

Hallegatte, S., 2013: A Cost Effective Solution to Reduce Disaster Losses in Developing Countries: Hydro-Meteorological Services, Early Warning, and Evacuation. In: *Global Problems, Smart Solutions: Costs and Benefits* [Lomborg, B. (ed.)]. Cambridge University Press, pp. 481–499. ISBN: 9781107039599.

6

Handmer, J. et al., 2012: Changes in Impacts of Climate Extremes: Human Systems and Ecosystems. In: Managing the Risks of Extreme Events and Disasters to Advance Climate Change Adaptation. A Special Report of Working Groups I and II of the Intergovernmental Panel on Climate Change (IPCC) [Field, C.B., V. Barros, T.F. Stocker, D. Qin, D.J. Dokken, K.L. Ebi, M.D. Mastrandrea, K.J. Mach, G.K. Plattner, S.K. Allen, M. Tignor and P.M. Midgley (eds.)]. Cambridge University Press, Cambridge, United Kingdom and New York, NY, USA, 231-290.

Hansen, J. et al., 2016: Ice melt, sea level rise and superstorms: evidence from paleoclimate data, climate modeling, and modern observations that 2 A degrees C global warming could be dangerous. *Atmos. Chem. Phys.*, **16**(6), 3761–3812, doi:10.5194/acp-16-3761-2016.

Haque, A. and S. Jahan, 2016: Regional impact of cyclone sidr in Bangladesh: A multi-sector analysis. *International Journal of Disaster Risk Science*, **7**(3), 312–327, doi:10.1007/s13753-016-0100-y.

Hardin, R., 2002: *Trust and trustworthiness*. Russell Sage Foundation, New York, NY. ISBN: 978-1-61044-271-8.

Harman, B.P., S. Heyenga, B.M. Taylor and C.S. Fletcher, 2013: Global lessons for adapting coastal communities to protect against storm surge inundation. *J. Coastal Res.*, **31**(4), 790–801, doi:10.2112/JCOASTRES-D-13-00095.1.

Harris, M.P. and N. Elkins, 2013: An unprecedented wreck of Puffins in eastern Scotland in March and April 2013. *Scottish Birds*, **33**(2), 157–159.

Harrould-Kolieb, E.R. and O. Hoegh-Guldberg, 2019: A governing framework for international ocean acidification policy. *Mar. Policy*, **102**, 10–20, doi:10.1016/j.marpol.2019.02.004.

Hartmann, D. L. et al., 2013: Observations: Atmosphere and Surface. In: Climate Change 2013: The Physical Science Basis. Contribution of Working Group I to the Fifth Assessment Report of the Intergovernmental Panel on Climate Change [Stocker, T.F., D. Qin, G.-K. Plattner, M. Tignor, S.K. Allen, J. Boschung, A. Nauels, Y. Xia, V. Bex and P.M. Midgley (eds.)]. Cambridge University Press, Cambridge, United Kingdom and New York, NY, USA, 159–254.

Harvey, B.J., L.C. Shaffrey and T.J. Woollings, 2014: Equator-to-pole temperature differences and the extra-tropical storm track responses of the CMIP5 climate models. *Clim. Dynam.*, **43**(5), 1171–1182, doi:10.1007/s00382-013-1883-9.

Harvey, B.J., L.C. Shaffrey and T.J. Woollings, 2015: Deconstructing the climate change response of the Northern Hemisphere wintertime storm tracks. *Clim. Dynam.*, **45**(9), 2847–2860, doi:10.1007/s00382-015-2510-8.

Hátún, H. et al., 2016: An inflated subpolar gyre blows life toward the northeastern Atlantic. *Progress in Oceanography*, **147**, 49–66, doi:10.1016/j.pocean.2016.07.009.

Hátún, H. et al., 2009: Large bio-geographical shifts in the north-eastern Atlantic Ocean: From the subpolar gyre, via plankton, to blue whiting and pilot whales. *Progress in Oceanography*, **80**(3–4), 149–162, doi:10.1016/j.pocean.2009.03.001.

Hawkins, E. et al., 2011: Bistability of the Atlantic overturning circulation in a global climate model and links to ocean freshwater transport. *Geophys. Res. Lett.*, **38**, L10605, doi:10.1029/2011GL048997.

Haynes, G.W., S.M. Danes and K. Stafford, 2011: Influence of federal disaster assistance on family business survival and success. *J. Conting. Crisis Man.*, **19**(2), 86–98, doi:10.1111/j.1468-5973.2011.00637.x.

Helbing, D., 2013: Globally networked risks and how to respond. *Nature*, **497**(7447), 51, doi:10.1038/nature12047.

Hemstock, S.L. et al., 2017: A case for formal education in the Technical, Vocational Education and Training (TVET) sector for climate change adaptation and disaster risk reduction in the Pacific Islands region. In: *Climate Change Adaptation in Pacific Countries.* Springer Nature. Gewerbestrasse 11, 6330 Cham, Switzerland, 309-324. ISBN: 978-3-319-50093-5.

Henke, L.M.K., F.H. Lambert and D.J. Charman, 2017: Was the Little Ice Age more or less El Niño-like than the Medieval Climate Anomaly? Evidence from hydrological and temperature proxy data. *Clim. Past*, **13**(3), 267–301, doi:10.5194/cp-13-267-2017.

Hermanson, L. et al., 2014: Forecast cooling of the Atlantic subpolar gyre and associated impacts. *Geophys. Res. Lett.*, **41**(14), 5167–5174, doi:10.1002/2014GL060420.

Heron, S.F., J.A. Maynard, R. van Hooidonk and C. . Eakin, 2016: Warming Trends and Bleaching Stress of the World's Coral Reefs 1985–2012. *Sci. Rep.*, **6**, 38402, doi:10.1038/srep38402.

Heron, S.F. et al., 2010: Summer Hot Snaps and Winter Conditions: Modelling White Syndrome Outbreaks on Great Barrier Reef Corals. *PLoS ONE*, **5**(8), e12210, doi:10.1371/journal.pone.0012210.

Herring, S.C. et al., 2018: Introduction to explaining extreme events of 2016 from a climate perspective. In *"Explaining Extreme Events of 2016 from a Climate Perspective"*. *Bull. Am. Meterol. Soc.*, **99**(1), S54–S59, doi:10.1175/BAMS-D-17-0118.1.

Herring, S.C., M.P. Hoerling, T.C. Peterson and P.A. Stott, 2014: Explaining extreme events of 2013 from a climate perspective. *Bull. Am. Meterol. Soc.*, **95**(9), S1–S96, doi:10.1175/1520-0477-95.9.S1.1.

Herring, S.C. et al., 2015: Explaining extreme events of 2014 from a climate perspective. *Bull. Am. Meterol. Soc.*, **96**, S1–S172, doi:10.1175/BAMS-ExplainingExtremeEvents2014.1.

Heuze, C., 2017: North Atlantic deep water formation and AMOC in CMIP5 models. *Ocean Sci.*, **13**(4), 609–622, doi:10.5194/os-13-609-2017.

Hilhorst, D., J. Baart, G. van der Haar and F.M. Leeftink, 2015: Is disaster "normal" for indigenous people? Indigenous knowledge and coping practices. *Disaster Prevention and Management: An International Journal*, **24**(4), 506–522, doi:10.1108/DPM-02-2015-0027.

Hilmi, N. et al., 2013: Towards improved socio-economic assessments of ocean acidification's impacts. *Mar. Biol.*, **160**(8), 1773–1787, doi:10.1007/s00227-012-2031-5.

Hino, M., C.B. Field and K.J. Mach, 2017: Managed retreat as a response to natural hazard risk. *Nat. Clim. Change*, **7**(5), 364–370, doi:10.1038/Nclimate3252.

Hirata, H., R. Kawamura, M. Kato and T. Shinoda, 2016: Response of rapidly developing extratropical cyclones to sea surface temperature variations over the western Kuroshio–Oyashio confluence region. *J. Geophys. Res.-Atmos.*, **121**(8), 3843–3858, doi:10.1002/2015JD024391.

Hiwasaki, L., E. Luna and R. Shaw, 2014: Process for integrating local and indigenous knowledge with science for hydro-meteorological disaster risk reduction and climate change adaptation in coastal and small island communities. *Int. J. Disast. Risk Re.*, **10**, 15–27, doi:10.1016/j.ijdrr.2014.07.007.

Hobday, A.J. et al., 2016a: A hierarchical approach to defining marine heatwaves. *Progress in Oceanography*, **141**, 227–238, doi:10.1016/j.pocean.2015.12.014.

Hobday, A.J. et al., 2018: Categorizing and naming marine heatwaves. *Oceanography*, **31**(2), doi:10.5670/oceanog.2018.205.

Hobday, A.J., C.M. Spillman, J.P. Eveson and J.R. Hartog, 2016b: Seasonal forecasting for decision support in marine fisheries and aquaculture. *Fish. Oceanogr.*, **25**(S1), 45–56, doi:10.1111/fog.12083.

Hoegh-Guldberg, O. et al., 2014: The Ocean. In: Climate Change 2014: Impacts, Adaptation, and Vulnerability. Part B: Regional Aspects. Contribution of Working Group II to the Fifth Assessment Report of the Intergovernmental Panel of Climate Change [Barros, V.R., C.B. Field, D.J. Dokken, M.D. Mastrandrea, K.J. Mach, T.E. Bilir, M. Chatterjee, K.L. Ebi, Y.O. Estrada, R.C. Genova, B. Girma, E.S. Kissel, A.N. Levy, S. MacCracken, P.R. Mastrandrea and L.L. White (eds.)]. Cambridge University Press, Cambridge, United Kingdom and New York, NY, USA, 1655-1731.

Hoegh-Guldberg, O. et al., 2009: *The Coral Triangle and Climate Change Ecosystems, People and Societies at Risk*. WWF Australia, Brisbane, 276 pp. ISBN: 978-1-921031-35-9.

6

Hoegh-Guldberg, O. et al., 2018: Impacts of 1.5 °C global warming on natural and human systems. In: Global Warming of 1.5°C. An IPCC Special Report on the impacts of global warming of 1.5°C above pre-industrial levels and related global greenhouse gas emission pathways, in the context of strengthening the global response to the threat of climate change, sustainable development, and efforts to eradicate poverty [Masson-Delmotte, V., P. Zhai, H.-O. Pörtner, D. Roberts, J. Skea, P.R. Shukla, A. Pirani, W. Moufouma-Okia, C. Péan, R. Pidcock, S. Connors, J.B.R. Matthews, Y. Chen, X. Zhou, M.I. Gomis, E. Lonnoy, T. Maycock, M. Tignor, and T. Waterfield (eds.)], in press

Hoeke, R.K., K. McInnes and J. O'Grady, 2015: Wind and Wave Setup Contributions to Extreme Sea Levels at a Tropical High Island: A Stochastic Cyclone Simulation Study for Apia, Samoa. *J. Mar. Sci. Tech.*, **3**(3), 1117, doi:10.3390/jmse3031117.

Hoeke, R.K. et al., 2013: Widespread inundation of Pacific islands triggered by distant-source wind-waves. *Glob. Planet. Change*, **108**, 128–138, doi:10.1016/j.gloplacha.2013.06.006.

Holland, G. and C.L. Bruyère, 2014: Recent intense hurricane response to global climate change. *Clim. Dynam.*, **42**(3), 617–627, doi:10.1007/s00382-013-1713-0.

Holland, M. M., C. M. Bitz and B. Tremblay, 2006: Future abrupt reductions in the summer Arctic sea ice. *Geophys. Res. Lett.*, **33**(23), L23503, doi:10.1029/2006GL028024.

Hsiang, S. et al., 2017: Estimating economic damage from climate change in the United States. *Science*, **356**(6345), 1362–1369, doi:10.1126/science.aal4369.

Hu, S. and A.V. Fedorov, 2018: Cross-equatorial winds control El Niño diversity and change. *Nat. Clim. Change*, **8**(9), 798–802, doi:10.1038/s41558-018-0248-0.

Hu, S.N. and A.V. Fedorov, 2016: Exceptionally strong easterly wind burst stalling El Nino of 2014. *PNAS*, **113**(8), 2005–2010, doi:10.1073/pnas.1514182113.

Huang, B. et al., 2015a: Extended reconstructed sea surface temperature version 4 (ERSST. v4). Part I: upgrades and intercomparisons. *J. Clim.*, **28**(3), 911–930, doi:10.1175/JCLI-D-14-00006.1.

Huang, B., M. L'Heureux, Z.Z. Hu and H.M. Zhang, 2016: Ranking the strongest ENSO events while incorporating SST uncertainty. *Geophys. Res. Lett.*, **43**(17), 9165–9172, doi:10.1002/2016gl070888.

Huang, P., I.I. Lin, C. Chou and R.-H. Huang, 2015b: Change in ocean subsurface environment to suppress tropical cyclone intensification under global warming. *Nat. Commun.*, **6**, 7188, doi:10.1038/ncomms8188.

Huang, P. and S.P. Xie, 2015: Mechanisms of change in ENSO-induced tropical Pacific rainfall variability in a warming climate. *Nat. Geosci.*, **8**(12), 922–926, doi:10.1038/NGEO2571.

Huang, Y. and V.R. Coelho, 2017: Sustainability performance assessment focusing on coral reef protection by the tourism industry in the Coral Triangle region. *Tourism Manage.*, **59**, 510–527, doi:10.1016/j.tourman.2016.09.008.

Huggel, C. et al., 2019: Loss and Damage in the mountain cryosphere. *Reg. Environ. Change*, **19**(5), 1387–1399, doi:10.1007/s10113-018-1385-8.

Hughes, T.P. et al., 2018a: Spatial and temporal patterns of mass bleaching of corals in the Anthropocene. *Science*, **359** (6371), 80-83, doi:10.1126/science.aan8048.

Hughes, T.P. et al., 2017a: Coral reefs in the Anthropocene. *Nature*, **546**(7656), 82–90, doi:10.1038/nature22901.

Hughes, T.P. et al., 2017b: Global warming and recurrent mass bleaching of corals. *Nature*, **543**(7645), 373–377, doi:10.1038/nature21707.

Hughes, T.P. et al., 2018b: Global warming transforms coral reef assemblages. *Nature*, **556**, 492–496, doi:10.1038/s41586-018-0041-2.

Hui, C. and X.-T. Zheng, 2018: Uncertainty in Indian Ocean Dipole response to global warming: the role of internal variability. *Clim. Dynam.*, **51**(9), 3597–3611, doi:10.1007/s00382-018-4098-2.

Huntingford, C. et al., 2014: Potential influences on the United Kingdom's floods of winter 2013/14. *Nat. Clim. Change*, **4**(9), 769, doi:10.1038/nclimate2314.

Hydro Tasmania, 2016: *Annual Report*. Hydro Tasmania, 128pp. URL: https://www.hydro.com.au/docs/default-source/about-us/our-governance/annual-reports/hydro-tasmania-annual-report-2016.pdf?sfvrsn=1c551328_2 Publisher: Hydro-Electric Corporation. 4 Elizabeth Street, Hobart, Tasmania, 7000, Australia.

IFRC, 2016: *Somalia: Tropical Cyclone Chapala*. Emergency Plan of Action Final Report, International Federation of Red Cross and Red Crescent Societies, 11 pp. https://reliefweb.int/sites/reliefweb.int/files/resources/MDRSO004FR.pdf. Accessed 2019/20/08.

Ikeda, N., C. Narama and S. Gyalson, 2016: Knowledge sharing for disaster risk reduction: Insights from a glacier lake workshop in the Ladakh region, Indian Himalayas. *Mt. Res. Dev.*, **36**(1), 31–40, doi:10.1659/Mrd-Journal-D-15-00035.1.

Iloka, N.G., 2016: Indigenous knowledge for disaster risk reduction: An African perspective. *Jàmbá: Journal of Disaster Risk Studies*, **8**(1), 1–7, doi:10.4102/jamba.v8i1.272.

IMARPE, 2017: *Informe integrado de la operación EUREKA LXIX* Instituto del Mar del Perú, 23 pp [Available at: http://www.imarpe.pe/imarpe/archivos/informes/Informe_Operacion_EUREKA_LXIX%20_21_23feb_2017.pdf]. [publisher information: Instituto del Mar del Perú (IMARPE). Peru. Accessed 2018/09/28.]

IPCC, 2012: *Managing the Risks of Extreme Events and Disasters to Advance Climate Change Adaptation. A Special Report of Working Groups I and II of the Intergovernmental Panel on Climate Change*. Cambridge University Press, Cambridge, UK, and New York, NY, USA, 582 pp. [Field, C.B., V. Barros, T.F. Stocker, D. Qin, D.J. Dokken, K.L. Ebi, M.D. Mastrandrea, K.J. Mach, G.-K. Plattner, S.K. Allen, M. Tignor, and P.M. Midgley (Eds.)]

IPCC, 2013: *Climate Change 2013: The Physical Science Basis. Contribution of Working Group I to the Fifth Assessment Report of the Intergovernmental Panel on Climate Change* [Stocker, T.F., D. Qin, G.-K. Plattner, M. Tignor, S.K. Allen, J. Boschung, A. Nauels, Y. Xia, V. Bex and P.M. Midgley (eds.)]. Cambridge University Press, Cambridge, United Kingdom and New York, NY, USA, 1535 pp.

IPCC, 2014: *Climate Change 2014: Impacts, Adaptation, and Vulnerability. Part A: Global and Sectoral Aspects. Contribution of Working Group II to the Fifth Assessment Report of the Intergovernmental Panel on Climate Change* [Field, C.B., V.R. Barros, D.J. Dokken, K.J. Mach, M.D. Mastrandrea, T.E. Bilir, M. Chatterjee, K.L. Ebi, Y. O. Estrada, R.C. Genova, B. Girma, E.S. Kissel, A.N. Levy, S. MacCracken, P.R. Mastrandrea and L.L. White (eds.)]. Cambridge University Press, Cambridge, United Kingdom and New York, NY, USA.

IPCC, 2018: Global Warming of 1.5° C: An IPCC Special Report on the Impacts of Global Warming of 1.5° C Above Pre-industrial Levels and Related Global Greenhouse Gas Emission Pathways, in the Context of Strengthening the Global Response to the Threat of Climate Change, Sustainable Development, and Efforts to Eradicate Poverty [Masson-Delmotte, V., P. Zhai, H.-O. Pörtner, D. Roberts, J. Skea, P.R. Shukla, A. Pirani, W. Moufouma-Okia, C. Péan, R. Pidcock, S. Connors, J.B.R. Matthews, Y. Chen, X. Zhou, M.I. Gomis, E. Lonnoy, T. Maycock, M. Tignor, and T. Waterfield (eds.)], in press.

Jackson, L.C. et al., 2015: Global and European climate impacts of a slowdown of the AMOC in a high resolution GCM. *Clim. Dynam.*, **45**(11–12), 3299–3316, doi:10.1007/s00382-015-2540-2.

Jackson, L.C., K.A. Peterson, C.D. Roberts and R.A. Wood, 2016: Recent slowing of Atlantic overturning circulation as a recovery from earlier strengthening. *Nat. Geosci.*, **9**(7), 518–522, doi:10.1038/ngeo2715.

Jackson, L. C. et al., 2014: Response of the Atlantic meridional overturning circulation to a reversal of greenhouse gas increases. *Clim. Dynam.*, **42**(11–12), 3323–3336, doi:10.1007/s00382-013-1842-5.

Jackson, L.C., R.S. Smith and R.A. Wood, 2017: Ocean and atmosphere feedbacks affecting AMOC hysteresis in a GCM. *Clim. Dynam.*, **49**(1–2), 173–191, doi:10.1007/s00382-016-3336-8.

Jackson, L.C. and R.A. Wood, 2018: Hysteresis and Resilience of the AMOC in an Eddy-Permitting GCM. *Geophys. Res. Lett.*, **45**(16), 8547–8556, doi:10.1029/2018gl078104.

Jacox, M.G. et al., 2018: Forcing of Multiyear Extreme Ocean Temperatures that Impacted California Current Living Marine Resources in 2016. *Bull. Am. Meterol. Soc.*, **99**(1), S27–S33, doi:10.1175/bams-d-17-0119.1.

Jacox, M.G. et al., 2016: Impacts of the 2015–2016 El Niño on the California Current System: Early assessment and comparison to past events. *Geophys. Res. Lett.*, **43**(13), 7072–7080, doi:10.1002/2016gl069716.

Jakobsen, K.T., 2012: In the eye of the storm – The welfare impacts of a hurricane. *World Dev.*, **40**(12), 2578–2589, doi:10.1016/j.worlddev.2012.05.013.

James, R. et al., 2014: Characterizing loss and damage from climate change. *Nat. Clim. Change*, **4**(11), 938–939, doi:10.1038/nclimate2411.

Jin, F.-F., J. Boucharel and I.-I. Lin, 2014: Eastern Pacific tropical cyclones intensified by El Niño delivery of subsurface ocean heat. *Nature*, **516**(7529), 82–85, doi:10.1038/nature13958.

John, J.G., C.A. Stock and J.P. Dunne, 2015: A more productive, but different, ocean after mitigation. *Geophys. Res. Lett.*, **42**(22), 9836–9845, doi:10.1002/2015gl066160.

Johnson, C., M. Inall and S. Hakkinen, 2013: Declining nutrient concentrations in the northeast Atlantic as a result of a weakening Subpolar Gyre. *Deep-Sea Res. Pt. I.*, **82**, 95–107, doi:10.1016/j.dsr.2013.08.007.

Jones, L. et al., 2017: Constraining and enabling factors to using long-term climate information in decision-making. *Climate Policy*, **17**(5), 551–572, doi:10.1080/14693062.2016.1191008.

Jones, T. et al., 2018: Massive Mortality of a Planktivorous Seabird in Response to a Marine Heatwave. *Geophys. Res. Lett.*, **45**, 3193–3202, doi:10.1002/2017GL076164.

Josey, S.A. et al., 2018: The Recent Atlantic Cold Anomaly: Causes, Consequences, and Related Phenomena. *Annu. Rev. Mar. Sci.*, **10**(1), 475–501, doi:10.1146/annurev-marine-121916-063102.

Joshi, M.K. and A. Rai, 2015: Combined interplay of the Atlantic multidecadal oscillation and the interdecadal Pacific oscillation on rainfall and its extremes over Indian subcontinent. *Clim. Dynam.*, **44**(11–12), 3339–3359, doi:10.1007/s00382-014-2333-z.

Kajtar, J.B. et al., 2018: Model under-representation of decadal Pacific trade wind trends and its link to tropical Atlantic bias. *Clim. Dynam.*, **50**(3–4), 1471–1484, doi:10.1007/s00382-017-3699-5.

Kam, J., T.R. Knutson, F. Zeng and A.T. Wittenberg, 2015: Record Annual Mean Warmth Over Europe, the Northeast Pacific, and the Northwest Atlantic During 2014: Assessment of Anthropogenic Influence. *Bull. Am. Meteorol. Soc.*, **96**(12), S61–S65, doi:10.1175/Bams-D-15-00101.1.

Kam, J., T.R. Knutson, F. Zeng and A.T. Wittenberg, 2018: CMIP5 model-based assessment of anthropoegenic influence on highly anomalous Arctic Warmth during November-December 2016. In *"Explaining Extreme Events of 2016 from a Climate Perspective"*. *Bull. Am. Meteorol. Soc.*, **99**(1), S34–S38, doi:10.1175/BAMS-ExplainingExtremeEvents2016.1.

Karoly, D.J., M. Black, M.R. Grose and A.D. King, 2016: The roles of climate change and El Niño in the record low rainfall in October 2015 in Tasmania, Australia. *Bull. Am. Meteorol. Soc.*, **97**(12), S18, doi:10.1175/Bams-D-16-0139.1.

Karspeck, A.R. et al., 2017: Comparison of the Atlantic meridional overturning circulation between 1960 and 2007 in six ocean reanalysis products. *Clim. Dynam.*, **49**(3), 957–982, doi:10.1007/s00382-015-2787-7.

Kataoka, T., T. Tozuka, S. Behera and T. Yamagata, 2014: On the Ningaloo Niño/Niña. *Clim. Dynam.*, **43**(5), 1463–1482, doi:10.1007/s00382-013-1961-z.

Kates, R.W., W.R. Travis and T.J. Wilbanks, 2012: Transformational adaptation when incremental adaptations to climate change are insufficient. *PNAS*, **109**(19), 7156–7161, doi:10.1073/pnas.1115521109.

Kayanne, H., 2017: Validation of degree heating weeks as a coral bleaching index in the northwestern Pacific. *Coral Reefs*, **36**(1), 63–70, doi:10.1007/s00338-016-1524-y.

Keenan, T.F. et al., 2016: Recent pause in the growth rate of atmospheric CO_2 due to enhanced terrestrial carbon uptake. *Nat. Commun.*, **7**, 13428, doi:10.1038/ncomms13428.

Keller, K.M., F. Joos, F. Lehner and C.C. Raible, 2015: Detecting changes in marine responses to ENSO from 850 to 2100 CE: Insights from the ocean carbon cycle. *Geophys. Res. Lett.*, **42**(2), 518–525, doi:10.1002/2014gl062398.

Kelly, K.A. et al., 2016: Impact of slowdown of Atlantic overturning circulation on heat and freshwater transports. *Geophys. Res. Lett.*, **43**(14), 7625–7631, doi:10.1002/2016gl069789.

Kiem, A.S. et al., 2016: Natural hazards in Australia: droughts. *Clim. Change*, **139**(1), 37–54, doi:10.1007/s10584-016-1798-7.

Kim, H.S. et al., 2014a: Tropical Cyclone Simulation and Response to CO_2 Doubling in the GFDL CM2.5 High-Resolution Coupled Climate Model. *J. Clim.*, **27**(21), 8034–8054, doi:10.1175/jcli-d-13-00475.1.

Kim, J.-Y. and I.-S. Han, 2017: Sea Surface Temperature Time Lag Due to the Extreme Heat Wave of August 2016. *J. Korean Soc. Mar. Environ. Saf.*, **23**(6), 677–683, doi:10.7837/kosomes.2017.23.6.677.

Kim, S.T. et al., 2014b: Response of El Niño sea surface temperature variability to greenhouse warming. *Nat. Clim. Change*, **4**(9), 786–790, doi:10.1038/nclimate2326.

Kim, W. and W. Cai, 2014: The importance of the eastward zonal current for generating extreme El Niño. *Clim. Dynam.*, **42**(11–12), 3005–3014, doi:10.1007/s00382-013-1792-y.

King, A.D., D.J. Karoly and B.J. Henley, 2017: Australian climate extremes at 1.5°C and 2°C of global warming. *Nat. Clim. Change*, **7**, 412, doi:10.1038/nclimate3296.

Klerk, W.-J. et al., 2015: The co-incidence of storm surges and extreme discharges within the Rhine–Meuse Delta. *Environ. Res. Lett.*, **10**(3), 035005, doi:10.1088/1748-9326/10/3/035005.

Kleypas, J.A., F.S. Castruccio, E.N. Curchitser and E. Mcleod, 2015: The impact of ENSO on coral heat stress in the western equatorial Pacific. *Global Change Biol.*, **21**(7), 2525–2539, doi:10.1111/gcb.12881.

Klotzbach, P.J. and M.M. Bell, 2017: *Summary of 2017 Atlantic Tropical Cyclone Activity and Verification of Authors' Seasonal and Two-Week Forecasts*. Colorado State University, Department of Atmospheric Science, Fort Collins, USA, 37 pp. https://tropical.colostate.edu/media/sites/111/2017/11/2017-11.pdf. Accessed 2019/20/08.

KMA, 2016: *Abnormal Climate Report 2016*. Korea Meteorological Administration, 61 16-Gil Yeouidaebang-ro, Dongjak-gu, Seoul 07062, Republic of Korea, 192pp.

KMA, 2017: *Abnormal Climate Report 2017*. Korea Meteorological Administration, 61 16-Gil Yeouidaebang-ro, Dongjak-gu, Seoul 07062, Republic of Korea, 218pp.

KMA, 2018: *Abnormal Climate Report 2018*. Korea Meteorological Administration, 61 16-Gil Yeouidaebang-ro, Dongjak-gu, Seoul 07062, Republic of Korea, 200pp.

Knowlton, K. and M. Rotkin-Ellman, 2014: *Preparing for Climate Change: Lessons for Coastal Cities from Hurricane Sandy*. Natural Resources Defense Council Report, New York City, NY, 25 pp. www.nrdc.org/sites/default/files/hurricane-sandy-coastal-flooding-report.pdf. Accessed 2019/20/08.

Knutson, T.R. et al., 2010: Tropical cyclones and climate change. *Nat. Geosci.*, **3**(3), 157–163, doi:10.1038/Ngeo779.

Knutson, T.R. et al., 2015: Global Projections of Intense Tropical Cyclone Activity for the Late Twenty-First Century from Dynamical Downscaling of CMIP5/RCP4.5 Scenarios. *J. Clim.*, **28**(18), 7203–7224, doi:10.1175/JCLI-D-15-0129.1.

Kociuba, G. and S.B. Power, 2015: Inability of CMIP5 Models to Simulate Recent Strengthening of the Walker Circulation: Implications for Projections. *J. Clim.*, **28**(1), 20–35, doi:10.1175/JCLI-D-13-00752.1.

Komendantova, N. et al., 2014: Multi-hazard and multi-risk decision-support tools as a part of participatory risk governance: Feedback from civil protection stakeholders. *Int. J. Disast. Risk Re.*, **8**, 50–67, doi:10.1016/j.ijdrr.2013.12.006.

Kopits, E., A. Marten and A. Wolverton, 2014: Incorporating 'catastrophic' climate change into policy analysis. *Climate Policy*, **14**(5), 637–664, doi:10.1080/14693062.2014.864947.

Koplitz, S.N. et al., 2015: Public health impacts of the severe haze in Equatorial Asia in September–October 2015: demonstration of a new framework for informing fire management strategies to reduce downwind smoke exposure. *Environ. Res. Lett.*, **11**, 094023, doi:10.1088/1748-9326/11/9/094023.

Kopp, R.E., R.L. Shwom, G. Wagner and J. Yuan, 2016: Tipping elements and climate–economic shocks: Pathways toward integrated assessment. *Earth's Future*, **4**(8), 346–372, doi:10.1002/2016ef000362.

Korea Meteorological Administration, 2016: *Annual Report 2016*. 48pp [Available at: http://www.kma.go.kr/download_01/Annual_Report_2016.pdf]. [Publisher: Korea Meteorological Administration. 61 16-Gil Yeouidaebang-ro, Dongjak-gu, Seoul 07062, Republic of Korea.]

Kosaka, Y. and S.P. Xie, 2013: Recent global-warming hiatus tied to equatorial Pacific surface cooling. *Nature*, **501**(7467), 403–407, doi:10.1038/nature12534.

Kossin, J.P., 2017: Hurricane intensification along United States coast suppressed during active hurricane periods. *Nature*, **541**, 390, doi:10.1038/nature20783.

Kossin, J.P., 2018: A global slowdown of tropical-cyclone translation speed. *Nature*, **558**(7708), 104–107, doi:10.1038/s41586-018-0158-3.

Kossin, J.P., K.A. Emanuel and S.J. Camargo, 2016: Past and Projected Changes in Western North Pacific Tropical Cyclone Exposure. *J. Clim.*, **29**(16), 5725–5739, doi:10.1175/Jcli-D-16-0076.1.

Kossin, J.P., K.A. Emanuel and G.A. Vecchi, 2014: The poleward migration of the location of tropical cyclone maximum intensity. *Nature*, **509**(7500), 349–352, doi:10.1038/nature13278.

Kostov, Y., K.C. Armour and J. Marshall, 2014: Impact of the Atlantic meridional overturning circulation on ocean heat storage and transient climate change. *Geophys. Res. Lett.*, **41**(6), 2108–2116, doi:10.1002/2013gl058998.

Krausmann, E., A.M. Cruz and E. Salzano, 2017: Reducing Natech Risk: Organizational Measures. In: *Natech Risk Assessment and Management: Reducing the Risk of Natural-Hazard Impact on Hazardous Installations* [Krausmann E., A.M. Cruz, E. Salzano (eds.)]. Elsevier, Amsterdam, The Netherlands, 243-252. ISBN: 978-0-12-803807-9]

Kreibich, H. et al., 2014: Costing natural hazards. *Nat. Clim. Change*, **4**, 303, doi:10.1038/nclimate2182.

Kretschmer, M., D. Coumou, J.F. Donges and J. Runge, 2016: Using Causal Effect Networks to Analyze Different Arctic Drivers of Midlatitude Winter Circulation. *J. Clim.*, **29**(11), 4069–4081, doi:10.1175/jcli-d-15-0654.1.

Kruk, M.C., 2016: Tropical cyclones: North Indian Ocean. In *"State of the Climate in 2015"*. *Bull. Am. Meteor. Soc.*, **97**(8), S114–S115 doi:10.1175/2016BAMSStateoftheClimate.1.

Kucharski, F., I. Kang, R. Farneti and L. Feudale, 2011: Tropical Pacific response to 20th century Atlantic warming. *Geophys. Res. Lett.*, **38**, L03702, doi:10.1029/2010GL046248.

Kulkarni, P.A. et al., 2017: Evacuations as a Result of Hurricane Sandy: Analysis of the 2014 New Jersey Behavioral Risk Factor Survey. *Disaster Med. Public.*, **11**(6), 720–728, doi:10.1017/dmp.2017.21.

Kull, D., R. Mechler and S. Hochrainer-Stigler, 2013: Probabilistic cost-benefit analysis of disaster risk management in a development context. *Disasters*, **37**(3), 374–400, doi:10.1111/disa.12002.

Kusahara, K. et al., 2018: An ocean-sea ice model study of the unprecedented Antarctic sea ice minimum in 2016. *Environ. Res. Lett.*, **13**(8), 084020, doi:10.1088/1748-9326/aad624.

Kushnir, Y. et al., 2019: Towards operational predictions of the near-term climate. *Nat. Clim. Change*, **9**(February), 94–101, doi:10.1038/s41558-018-0359-7.

L'Heureux, M.L., S. Lee and B. Lyon, 2013: Recent multidecadal strengthening of the Walker circulation across the tropical Pacific. *Nat. Clim. Change*, **3**(6), 571–576, doi:10.1038/NCLIMATE1840.

L'Heureux, M.L. et al., 2017: Observing and predicting the 2015/16 El Niño. *Bull. Am. Meteor. Soc.*, **98**(7), 1363–1382, doi:10.1175/BAMS-D-16-0009.1.

Lackmann, G.M., 2015: Hurricane Sandy before 1900 and after 2100. *Bull. Am. Meteor. Soc.*, **96**(4), 547–560, doi:10.1175/Bams-D-14-00123.1.

Lagmay, A.M.F. et al., 2015: Devastating storm surges of Typhoon Haiyan. *Int. J. Disast. Risk Re.*, **11**, 1–12, doi:10.1016/j.ijdrr.2014.10.006.

Landschützer, P., N. Gruber, D.C.E. Bakker and U. Schuster, 2014: Recent variability of the global ocean carbon sink. *Global Biogeochem. Cy.*, **28**(9), 927–949, doi:10.1002/2014gb004853.

Landschützer, P. et al., 2015: The reinvigoration of the Southern Ocean carbon sink. *Science*, **349**(6253), 1221–1224, doi:10.1126/science.aab2620.

Lawrence, J. et al., 2018: National guidance for adapting to coastal hazards and sea level rise: Anticipating change, when and how to change pathway. *Environ. Sci. Policy*, **82**, 100–107, doi:10.1016/j.envsci.2018.01.012.

Lawrence, J. and W. Saunders, 2017: The Planning Nexus Between Disaster Risk Reduction and Climate Change Adaptation. In: *The Routledge Handbook of Disaster Risk Reduction Including Climate Change Adaptation. Routledge* [Kelman I., J. Mercer, J.C. Gaillard. (eds)], pp. 418–428. ISBN: 9781138924567.

Lee, M.Y., C.C. Hong and H.H. Hsu, 2015a: Compounding effects of warm sea surface temperature and reduced sea ice on the extreme circulation over the extratropical North Pacific and North America during the 2013–2014 boreal winter. *Geophys. Res. Lett.*, **42**(5), 1612–1618, doi:10.1002/2014GL062956.

Lee, S.-K. et al., 2015b: Pacific origin of the abrupt increase in Indian Ocean heat content during the warming hiatus. *Nat. Geosci.*, **8**(6), 445–449, doi:10.1038/ngeo2438.

Lee, T., S. Fournier, A.L. Gordon and J. Sprintall, 2019: Maritime Continent water cycle regulates low-latitude chokepoint of global ocean circulation. *Nat. Commun.*, **10**(1), 2103, doi:10.1038/s41467-019-10109-z.

Lee, T. and M.J. McPhaden, 2010: Increasing intensity of El Niño in the central-equatorial Pacific. *Geophys. Res. Lett.*, **37**(14), L14603, doi:10.1029/2010GL044007.

Lehner, F., A. Born, C.C. Raible and T.F. Stocker, 2013: Amplified Inception of European Little Ice Age by Sea Ice-Ocean-Atmosphere Feedbacks. *J. Clim.*, **26**(19), 7586–7602, doi:10.1175/jcli-d-12-00690.1.

Lejano, R.P., J.M. Tan and A.M.W. Wilson, 2016: A Textual Processing Model of Risk Communication: Lessons from Typhoon Haiyan. *Weather Clim. Soc.*, **8**(4), 447–463, doi:10.1175/WCAS-D-16-0023.1.

Lemoine, D. and C.P. Traeger, 2016: Economics of tipping the climate dominoes. *Nat. Clim. Change*, **6**(5), 514, doi:10.1038/Nclimate2902.

Lenaerts, J.T.M. et al., 2015: Representing Greenland ice sheet freshwater fluxes in climate models. *Geophys. Res. Lett.*, **42**(15), 6373–6381, doi:10.1002/2015gl064738.

Lengaigne, M. et al., 2004: Westerly Wind Events in the Tropical Pacific and their Influence on the Coupled Ocean-Atmosphere System: A Review. In: *Earth's Climate: The Ocean-Atmosphere Interaction* [Wang, C., S. Xie and J. Carton (eds.)], American Geophysical Union. Washington DC, USA, 49-69. ISBN: 9780875904122.

Lenoir, J. and J.C. Svenning, 2015: Climate-related range shifts–a global multidimensional synthesis and new research directions. *Ecography*, **38**(1), 15–28, doi:10.1111/ecog.00967.

Leonard, M. et al., 2014: A compound event framework for understanding extreme impacts. *WiRes. Clim. Change*, **5**(1), 113–128, doi:10.1002/wcc.252.

Lewis, S.C. and J. Mallela, 2018: A multifactor risk analysis of the record 2016 Great Barrier Reef bleaching. In *"Explaining Extreme Events of 2016 from a Climate Perspective"*. *Bull. Am. Meteorol. Soc.*, **99**(1), S144–S149, doi:10.1175/BAMS-D-17-0118.1.

Li, G., S.-P. Xie and Y. Du, 2016a: A Robust but Spurious Pattern of Climate Change in Model Projections over the Tropical Indian Ocean. *J. Clim.*, **29**(15), 5589–5608, doi:10.1175/jcli-d-15-0565.1.

Li, R.C.Y., W. Zhou, C.M. Shun and T.C. Lee, 2017: Change in destructiveness of landfalling tropical cyclones over China in recent decades. *J. Clim.*, **30**(9), 3367–3379, doi:10.1175/Jcli-D-16-0258.1.

Li, T. et al., 2010: Global warming shifts Pacific tropical cyclone location. *Geophys. Res. Lett.*, **37**(21), L21804, doi:10.1029/2010gl045124.

Li, X.C., S.P. Xie, S.T. Gille and C. Yoo, 2016b: Atlantic-induced pan-tropical climate change over the past three decades. *Nat. Clim. Change*, **6**(3), 275–279, doi:10.1038/NCLIMATE2840.

Lima, F.P. and D.S. Wethey, 2012: Three decades of high-resolution coastal sea surface temperatures reveal more than warming. *Nat. Commun.*, **3**, 704, doi:10.1038/ncomms1713.

Lin, N., K. Emanuel, M. Oppenheimer and E. Vanmarcke, 2012: Physically based assessment of hurricane surge threat under climate change. *Nat. Clim. Change*, **2**, 462, doi:10.1038/nclimate1389.

Link, L.E., 2010: The anatomy of a disaster, an overview of Hurricane Katrina and New Orleans. *Ocean Engineering*, **37**(1), 4–12, doi:10.1016/j.oceaneng.2009.09.002.

Link, P.M. and R.S. Tol, 2009: Economic impacts on key Barents Sea fisheries arising from changes in the strength of the Atlantic thermohaline circulation. *Global Environ. Change*, **19**(4), 422–433, doi:10.1016/j.gloenvcha.2009.07.007.

Link, P.M. and R.S. Tol, 2011: Estimation of the economic impact of temperature changes induced by a shutdown of the thermohaline circulation: an application of FUND. *Clim. Change*, **104**(2), 287–304, doi:10.1007/s10584-009-9796-7.

Linnerooth-Bayer, J. et al., 2019: Insurance as a response to Loss and Damage? In: *Loss and Damage from Climate Change: Concepts, Principles and Policy Options* [Mechler, R., L. Bouwer, J. Linnerooth-Bayer, T. Schinko and S. Surmiski (eds.)]. Springer, Heidelberg, Germany. [ISBN: 978-3-319-72025-8].

Lique, C. and M.D. Thomas, 2018: Latitudinal shift of the Atlantic Meridional Overturning Circulation source regions under a warming climate. *Nat. Clim. Change*, **8**(11), 1013–1020, doi:10.1038/s41558-018-0316-5.

Little, C.M. et al., 2015: Joint projections of US East Coast sea level and storm surge. *Nat. Clim. Change*, **5**(12), 1114–1120, doi:10.1038/nclimate2801.

Little, C.M., C.G. Piecuch and R.M. Ponte, 2017: On the relationship between the meridional overturning circulation, alongshore wind stress, and United States East Coast sea level in the Community Earth System Model Large Ensemble. *J. Geophys. Res.-Oceans*, **122**(6), 4554–4568, doi:10.1002/2017jc012713.

Liu, G. et al., 2018: Predicting Heat Stress to Inform Reef Management: NOAA Coral Reef Watch's 4-Month Coral Bleaching Outlook. *Front. Mar. Sci.*, **5**, 57, doi:10.3389/fmars.2018.00057.

Liu, G. et al., 2014a: Reef-Scale Thermal Stress Monitoring of Coral Ecosystems: New 5-km Global Products from NOAA Coral Reef Watch. *Remote Sens.*, **6**(11), 11579, doi:10.3390/rs61111579.

Liu, J. et al., 2012: Impact of declining Arctic sea ice on winter snowfall. *PNAS*, **109**(11), 4074, doi:10.1073/pnas.1114910109.

Liu, Q.Y., M. Feng, D.X. Wang and S. Wijffels, 2015a: Interannual variability of the Indonesian Throughflow transport: A revisit based on 30 year expendable bathythermograph data. *J. Geophys. Res.-Oceans*, **120**(12), 8270–8282, doi:10.1002/2015JC011351.

Liu, W., Z. Liu and E.C. Brady, 2014b: Why is the AMOC Monostable in Coupled General Circulation Models? *J. Clim.*, **27**(6), 2427–2443, doi:10.1175/jcli-d-13-00264.1.

Liu, W., J. Lu and S.-P. Xie, 2015b: Understanding the Indian Ocean response to double CO_2 forcing in a coupled model. *Ocean Dynam.*, **65**(7), 1037–1046, doi:10.1007/s10236-015-0854-6.

Liu, W., S.P. Xie, Z.Y. Liu and J. Zhu, 2017a: Overlooked possibility of a collapsed Atlantic Meridional Overturning Circulation in warming climate. *Sci. Adv.*, **3**(1), e1601666, doi:10.1126/sciadv.1601666.

Liu, Y. et al., 2017b: Recent enhancement of central Pacific El Niño variability relative to last eight centuries. *Nat. Commun.*, **8**, 15386, doi:10.1038/ncomms15386.

Lloyd, S.J. et al., 2016: Modelling the influences of climate change-associated sea level rise and socioeconomic development on future storm surge mortality. *Clim. Change*, **134**(3), 441–455, doi:10.1007/s10584-015-1376-4.

Lontzek, T.S., Y. Cai, K.L. Judd and T.M. Lenton, 2015: Stochastic integrated assessment of climate tipping points indicates the need for strict climate policy. *Nat. Clim. Change*, **5**(5), 441, doi:10.1038/Nclimate2570.

Loveday, B.R., P. Penven and C.J.C. Reason, 2015: Southern Annular Mode and westerly-wind-driven changes in Indian-Atlantic exchange mechanisms. *Geophys. Res. Lett.*, **42**(12), 4912–4921, doi:10.1002/2015GL064256.

Lozier, M.S. et al., 2017: Overturning in the Subpolar North Atlantic Program: A New International Ocean Observing System. *Bull. Am. Meterol. Soc.*, **98**(4), 737–752, doi:10.1175/bams-d-16-0057.1.

Lübbecke, J.F., J.V. Durgadoo and A. Biastoch, 2015: Contribution of increased Agulhas leakage to tropical Atlantic warming. *J. Clim.*, **28**(24), 9697–9706, doi:10.1175/JCLI-D-15-0258.1.

Lumbroso, D.M., N.R. Suckall, R.J. Nicholls and K.D. White, 2017: Enhancing resilience to coastal flooding from severe storms in the USA: international lessons. *Nat. Hazard. Earth Sys.*, **17**(8), 1357, doi:10.5194/nhess-17-1357-2017.

Luo, J.-J., W. Sasaki and Y. Masumoto, 2012: Indian Ocean warming modulates Pacific climate change. *PNAS*, **109**(46), 18701–18706, doi:10.1073/pnas.1210239109.

Lynch-Stieglitz, J., 2017: The Atlantic Meridional Overturning Circulation and Abrupt Climate Change. *Annu. Rev. Mar. Sci.*, **9**, 83–104, doi:10.1146/annurev-marine-010816-060415.

Lyth, A., A. Harwood, A.J. Hobday and J. McDonald, 2016: Place influences in framing and understanding climate change adaptation challenges. *Local Environment*, **21**(6), 730–751, doi:10.1080/13549839.2015.1015974.

MacKenzie, B.R. et al., 2014: A cascade of warming impacts brings bluefin tuna to Greenland waters. *Global Change Biol.*, **20**(8), 2484–2491, doi:10.1111/gcb.12597.

Majkut, J.D., J.L. Sarmiento and K.B. Rodgers, 2014: A growing oceanic carbon uptake: Results from an inversion study of surface pCO_2 data. *Global Biogeochem. Cy.*, **28**(4), 335–351, doi:10.1002/2013gb004585.

Makarim, S. et al., 2019: Previously unidentified Indonesian Throughflow pathways and freshening in the Indian Ocean during recent decades. *Sci. Rep.*, **9**(1), 7364, doi:10.1038/s41598-019-43841-z.

Mal, S., R.B. Singh, C. Huggel and A. Grover, 2018: Introducing Linkages Between Climate Change, Extreme Events, and Disaster Risk Reduction. In: *Climate Change, Extreme Events and Disaster Risk Reduction: Towards Sustainable Development Goals* [Mal, S., R.B. Singh and C. Huggel (eds.)]. Springer International Publishing, Cham, pp. 1–14. ISBN: 978-3-319-56468-5.

Malka, A., J.A. Krosnick and G. Langer, 2009: The association of knowledge with concern about global warming: Trusted information sources shape public thinking. *Risk Anal.*, **29**(5), 633–647, doi:10.1111/j.1539-6924.2009.01220.x.

Manganello, J.V. et al., 2014: Future Changes in the Western North Pacific Tropical Cyclone Activity Projected by a Multidecadal Simulation with a 16-km Global Atmospheric GCM. *J. Clim.*, **27**(20), 7622–7646, doi:10.1175/JCLI-D-13-00678.1.

Mann, M.E. et al., 2017: Influence of Anthropogenic Climate Change on Planetary Wave Resonance and Extreme Weather Events. *Sci. Rep.*, **7**, 45242, doi:10.1038/srep45242.

Manta, G. et al., 2018: The 2017 Record Marine Heatwave in the Southwestern Atlantic Shelf. *Geophys. Res. Lett.*, **45**(22), 12,449–12,456, doi:10.1029/2018GL081070.

Marengo, J.A. et al., 2011: The drought of 2010 in the context of historical droughts in the Amazon region. *Geophys. Res. Lett.*, **38**(12), L12703, doi:10.1029/2011GL047436.

Marino, E., 2011: Case 2: Flood Waters, Politics, and Relocating Home: One Story of Shishmaref, Alaska. In: *North by 2020: perspectives on Alaska's changing social-ecological systems* [Lovecraft, A.L. and H. Eicken (eds.)]. The University of Chicago Press, 1427 East 60th Street, Chicago, IL 60637 U.S.A., pp. 261–264. ISBN: 9781602231429.

6

Marino, E., 2012: The long history of environmental migration: Assessing vulnerability construction and obstacles to successful relocation in Shishmaref, Alaska. *Global Environ. Change*, **22**(2), 374–381, doi:10.1016/j.gloenvcha.2011.09.016.

Marotzke, J. and P.M. Forster, 2015: Forcing, feedback and internal variability in global temperature trends. *Nature*, **517**(7536), 565, doi:10.1038/nature14117.

Marshall, J. et al., 2014: The ocean's role in polar climate change: asymmetric Arctic and Antarctic responses to greenhouse gas and ozone forcing. *Philos. Trans. Royal Soc. A.*, **372**(2019), 20130040, doi:10.1098/rsta.2013.0040.

Marshall, M.I., L.S. Niehm, S.B. Sydnor and H.L. Schrank, 2015: Predicting small business demise after a natural disaster: an analysis of pre-existing conditions. *Nat. Hazards*, **79**(1), 331–354, doi:10.1007/s11069-015-1845-0.

Marzin, C. et al., 2013: Glacial fluctuations of the Indian monsoon and their relationship with North Atlantic climate: new data and modelling experiments. *Clim. Past*, **9**(5), 2135–2151, doi:10.5194/cp-9-2135-2013.

Marzocchi, W. et al., 2012: Basic principles of multi-risk assessment: a case study in Italy. *Nat. Hazards*, **62**(2), 551–573, doi:10.1007/s11069-012-0092-x.

Masson-Delmotte, V. et al., 2013: Information from Paleoclimate Archives. In: Climate Change 2013: The Physical Science Basis. Contribution of Working Group I to the Fifth Assessment Report of the Intergovernmental Panel on Climate Change [Stocker, T.F., D. Qin, G.-K. Plattner, M. Tignor, S.K. Allen, J. Boschung, A. Nauels, Y. Xia, V. Bex and P.M. Midgley (eds.)]. Cambridge University Press, Cambridge, United Kingdom and New York, NY, USA, 383–464.

Matei, D. et al., 2012: Multiyear prediction of monthly mean Atlantic meridional overturning circulation at 26.5 N. *Science*, **335**(6064), 76–79, doi:10.1126/science.1210299.

Matthews, T. and R. Potts, 2018: Planning for climigration: a framework for effective action. *Clim. Change*, **148**(4), 607–621, doi:10.1007/s10584-018-2205-3.

Mayhorn, C.B. and A.C. McLaughlin, 2014: Warning the world of extreme events: A global perspective on risk communication for natural and technological disaster. *Saf. Sci.*, **61**, 43–50, doi:10.1016/j.ssci.2012.04.014.

McAdam, J., 2011: Swimming against the tide: why a climate change displacement treaty is not the answer. *Int. J. Refug. Law*, **23**(1), 2–27, doi:10.1093/ijrl/eeq045.

McCabe, R.M. et al., 2016: An unprecedented coastwide toxic algal bloom linked to anomalous ocean conditions. *Geophys. Res. Lett.*, **43**(19), 10,366–10,376, doi:10.1002/2016gl070023.

McCarthy, G. et al., 2012: Observed interannual variability of the Atlantic meridional overturning circulation at 26.5 degrees N. *Geophys. Res. Lett.*, **39**, L19609, doi:10.1029/2012gl052933.

McCarthy, G.D. et al., 2015a: Ocean impact on decadal Atlantic climate variability revealed by sea level observations. *Nature*, **521**(7553), 508–510, doi:10.1038/nature14491.

McCarthy, G.D. et al., 2015b: Measuring the Atlantic Meridional Overturning Circulation at 26 degrees N. *Progress in Oceanography*, **130**, 91–111, doi:10.1016/j.pocean.2014.10.006.

McGregor, S. et al., 2018: Model tropical Atlantic biases underpin diminished Pacific decadal variability. *Nat. Clim. Change*, **8**(6), 493–498, doi:10.1038/s41558-018-0163-4.

McGregor, S. et al., 2013: Inferred changes in El Niño-Southern Oscillation variance over the past six centuries. *Clim. Past*, **9** 5), 2269, doi:10.5194/cp-9-2269-2013.

McGregor, S. et al., 2014: Recent Walker circulation strengthening and Pacific cooling amplified by Atlantic warming. *Nat. Clim. Change*, **4**(10), 888–892, doi:10.1038/NCLIMATE2330.

McInnes, K.L. et al., 2016: Application of a synthetic cyclone method for assessment of tropical cyclone storm tides in Samoa. *Nat. Hazards*, **80**(1), 425–444, doi:10.1007/s11069-015-1975-4.

McInnes, K.L. et al., 2014: Quantifying storm tide risk in Fiji due to climate variability and change. *Glob. Planet. Change*, **116**, 115–129, doi:10.1016/j.gloplacha.2014.02.004.

McNie, E.C., 2007: Reconciling the supply of scientific information with user demands: an analysis of the problem and review of the literature. *Environ. Sci. Policy*, **10**(1), 17–38, doi:10.1016/j.envsci.2006.10.004.

McTaggart-Cowan, R. et al., 2006: Analysis of Hurricane Catarina (2004). *Mon. Weather Rev.*, **134**(11), 3029–3053, doi:10.1175/mwr3330.1.

Mechler, R., 2016: Reviewing estimates of the economic efficiency of disaster risk management: opportunities and limitations of using risk-based cost–benefit analysis. *Nat. Hazards*, **81**(3), 2121–2147, doi:10.1007/s11069-016-2170-y.

Mechler, R. et al., 2019: *Loss and Damage from Climate Change: Concepts, Principles and Policy Options*. Springer, Heidelberg, Germany. ISBN: 978-3-319-72025-8.

Mecking, J.V., S.S. Drijfhout, L.C. Jackson and M.B. Andrews, 2017: The effect of model bias on Atlantic freshwater transport and implications for AMOC bi-stability. *Tellus A: Dynamic Meteorology and Oceanography*, **69**, 1299910, doi:10.1080/16000870.2017.1299910.

Mecking, J.V., S.S. Drijfhout, L.C. Jackson and T. Graham, 2016: Stable AMOC off state in an eddy-permitting coupled climate model. *Clim. Dynam.*, **47**(7–8), 2455–2470, doi:10.1007/s00382-016-2975-0.

Meehl, G.A. et al., 2019: Sustained ocean changes contributed to sudden Antarctic sea ice retreat in late 2016. *Nat. Commun.*, **10**(1), 14, doi:10.1038/s41467-018-07865-9.

Mei, W. and S.-P. Xie, 2016: Intensification of landfalling typhoons over the northwest Pacific since the late 1970s. *Nat. Geosci.*, **9**, 753, doi:10.1038/ngeo2792.

Meier, W.N. et al., 2014: Arctic sea ice in transformation: A review of recent observed changes and impacts on biology and human activity. *Rev. Geophys.*, **52**(3), 185–217, doi:10.1002/2013RG000431.

Meinen, C.S. et al., 2013: Temporal variability of the meridional overturning circulation at 34.5 degrees S: Results from two pilot boundary arrays in the South Atlantic. *J. Geophys. Res.-Oceans*, **118**(12), 6461–6478, doi:10.1002/2013jc009228.

Meinen, C.S. et al., 2018: Meridional Overturning Circulation transport variability at 34.5° S during 2009–2017: Baroclinic and barotropic flows and the dueling influence of the boundaries. *Geophys. Res. Lett.*, **45**(9), 4180–4188, doi:10.1029/2018GL077408.

Melet, A., R. Almar and B. Meyssignac, 2016: What dominates sea level at the coast: a case study for the Gulf of Guinea. *Ocean Dynam.*, **66**(5), 623–636, doi:10.1007/s10236-016-0942-2.

Mellin, C. et al., 2016: Marine protected areas increase resilience among coral reef communities. *Ecol. Lett.*, **19**(6), 629–637, doi:10.1111/ele.12598.

Menary, M.B. and L. Hermanson, 2018: Limits on determining the skill of North Atlantic Ocean decadal predictions. *Nat. Commun.*, **9**, 1694, doi:10.1038/s41467-018-04043-9.

Menary, M.B. and R.A. Wood, 2018: An anatomy of the projected North Atlantic warming hole in CMIP5 models. *Clim. Dynam.*, **50**(7–8), 3063–3080, doi:10.1007/s00382-017-3793-8.

Mengel, M., A. Levermann, C.-F. Schleussner and A. Born, 2012: Enhanced Atlantic subpolar gyre variability through baroclinic threshold in a coarse resolution model. *Earth Syst. Dynam.*, **3**(2), 189–197, doi:10.5194/esd-3-189-2012.

Mentaschi, L. et al., 2017: Global changes of extreme coastal wave energy fluxes triggered by intensified teleconnection patterns. *Geophys. Res. Lett.*, **44**(5), 2416–2426, doi:10.1002/2016GL072488.

Mercier, H. et al., 2015: Variability of the meridional overturning circulation at the Greenland-Portugal OVIDE section from 1993 to 2010. *Progress in Oceanography*, **132**, 250–261, doi:10.1016/j.pocean.2013.11.001.

Messmer, M., J.J. Gómez-Navarro and C.C. Raible, 2017: Sensitivity experiments on the response of Vb cyclones to sea surface temperature and soil moisture changes. *Earth Syst. Dynam.*, **8**(3), 477–493, doi:10.5194/esd-8-477-2017.

Miesner, A.K. and M.R. Payne, 2018: Oceanographic variability shapes the spawning distribution of blue whiting (Micromesistius poutassou). *Fish. Oceanogr.*, **27**(6), 623–638, doi:10.1111/fog.12382.

Mills, K. et al., 2013: Fisheries Management in a Changing Climate: Lessons From the 2012 Ocean Heat Wave in the Northwest Atlantic. *Oceanography*, **26**(2), 191–195, doi:10.5670/oceanog.2013.27.

Mishra, V., B.V. Smoliak, D.P. Lettenmaier and J.M. Wallace, 2012: A prominent pattern of year-to-year variability in Indian Summer Monsoon Rainfall. *PNAS*, **109**(19), 7213–7217, doi:10.1073/pnas.1119150109.

Moftakhari, H.R., A. AghaKouchak, B.F. Sanders and R.A. Matthew, 2017: Cumulative hazard: The case of nuisance flooding. *Earth's Future*, **5**(2), 214–223, doi:10.1002/2016EF000494.

Mogollon, R. and P. Calil, 2017: On the effects of ENSO on ocean biogeochemistry in the Northern Humboldt Current System (NHCS): A modeling study. *J. Marine Syst.*, **172**, 137–159, doi:10.1016/j.jmarsys.2017.03.011.

Mohino, E., N. Keenlyside and H. Pohlmann, 2016: Decadal prediction of Sahel rainfall: where does the skill (or lack thereof) come from? *Clim. Dynam.*, **47**(11), 3593–3612, doi:10.1007/s00382-016-3416-9.

Moloney, S., H. Fünfgeld and M. Granberg, 2017: *Local Action on Climate Change: Opportunities and Constraints*. Routledge, 2 Park Square, Milton Park, Abingdon, Oxon OX14 4RN, UK. ISBN: 9781138681521

Monerie, P.-A. et al., 2019: Effect of the Atlantic Multidecadal Variability on the Global Monsoon. *Geophys. Res. Lett.*, **46**(3), 1765–1775, doi:10.1029/2018gl080903.

Moon, I.-J., S.-H. Kim, P.J. Klotzbach and J.C.L. Chan, 2015a: Roles of interbasin frequency changes in the poleward shifts of the maximum intensity location of tropical cyclones. *Environ. Res. Lett.*, **10**(10), 104004, doi:10.1088/1748-9326/10/10/104004.

Moon, I.-J., S.-H. Kim and C. Wang, 2015b: El Niño and intense tropical cyclones. *Nature*, **526**(7575), E4–E5, doi:10.1038/nature15546.

Moore, G.W.K., K. Vage, R.S. Pickart and I.A. Renfrew, 2015: Decreasing intensity of open-ocean convection in the Greenland and Iceland seas. *Nat. Clim. Change*, **5**(9), 877–882, doi:10.1038/nclimate2688.

Mora, C., N.A.J. Graham and N. Nyström, 2016: Ecological limitations to the resilience of coral reefs. *Coral Reefs*, **35**(4), 1271–1280, doi:10.1007/s00338-016-1479-z.

Morchain, D., 2018: Rethinking the framing of climate change adaptation: knowledge, power, and politics. In: A Critical Approach to Climate Change Adaptation. Routledge, pp. 77–96. ISBN: 9781138056299.

Moreno-Chamarro, E. et al., 2017: Winter amplification of the European Little Ice Age cooling by the subpolar gyre. *Sci. Rep.*, **7**, 9981, doi:10.1038/s41598-017-07969-0.

Mori, A.S., 2016: Resilience in the Studies of Biodiversity–Ecosystem Functioning. *Trends Ecol. Evol.*, **35**(2), 87–90, doi:10.1016/j.tree.2015.12.010.

Morim, J. et al., 2018: On the concordance of 21st century wind-wave climate projections. *Glob. Planet. Change*, **167**, 160–171, doi:10.1016/j.gloplacha.2018.05.005.

Morrow, B.H., J.K. Lazo, J. Rhome and J. Feyen, 2015: Improving storm surge risk communication: Stakeholder perspectives. *Bull. Am. Meteor. Soc.*, **96**(1), 35–48, doi:10.1175/Bams-D-13-00197.1.

Moser, S.C. and M.T. Boykoff, 2013: *Successful adaptation to climate change: Linking science and policy in a rapidly changing world*. Routledge, 2 Park Square, Milton Park, Abingdon, Oxon OX14 4RN, UK. ISBN: 9780415524995.

Msadek, R. et al., 2014: Predicting a Decadal Shift in North Atlantic Climate Variability Using the GFDL Forecast System. *J. Clim.*, **27**(17), 6472–6496, doi:10.1175/jcli-d-13-00476.1.

Muis, S. et al., 2016: A global reanalysis of storm surges and extreme sea levels. *Nat. Commun.*, **7**, 11969, doi:10.1038/ncomms11969.

Mulitza, S. et al., 2017: Synchronous and proportional deglacial changes in Atlantic meridional overturning and northeast Brazilian precipitation. *Paleoceanography*, **32**(6), 622–633, doi:10.1002/2017PA003084.

Munich R.E., 2016: *2015 US natural catastrophe losses curbed by El Niño; brutal North American winter caused biggest insured losses*. Press Release, Munich Reinsurance America, Inc., Princeton, NJ, USA, 4 pp. www.munichre.com/site/mram-mobile/get/documents_E1831215213/mram/assetpool.munichreamerica.wrap/PDF/07Press/2015_natural_catastrophe_losses_US_010416.pdf. Accessed 2019/20/08.

Munro, D.R. et al., 2015: Recent evidence for a strengthening CO_2 sink in the Southern Ocean from carbonate system measurements in the Drake Passage (2002–2015). *Geophys. Res. Lett.*, **42**(18), 7623–7630, doi:10.1002/2015gl065194.

Murakami, H. et al., 2015: Investigating the influence of anthropogenic forcing and natural variability on the 2014 Hawaiian hurricane season. *Bull. Am. Meterol. Soc.*, **96**(12), S115–S119, doi:10.1175/Bams-D-15-00119.1.

Murakami, H. et al., 2017: Dominant role of subtropical Pacific warming in extreme eastern Pacific hurricane seasons: 2015 and the future. *J. Clim.*, **30**(1), 243–264, doi:10.1175/Jcli-D-16-0424.1.

Murakami, H. et al., 2012: Future changes in tropical cyclone activity projected by the new high-resolution MRI-AGCM. *J. Clim.*, **25**(9), 3237–3260, doi:10.1175/Jcli-D-11-00415.1.

Murphy, B.F., S.B. Power and S. McGree, 2014: The varied impacts of El Niño–Southern Oscillation on Pacific island climates. *J. Clim.*, **27**(11), 4015–4036, doi:10.1175/Jcli-D-13-00130.1.

Nakamura, J. et al., 2017: Western North Pacific Tropical Cyclone Model Tracks in Present and Future Climates. *J. Geophys. Res.-Atmos.*, **122**(18), 9721–9744, doi:10.1002/2017jd027007.

Nalau, J., J. Handmer and M. Dalesa, 2017: The Role and Capacity of Government in a Climate Crisis: Cyclone Pam in Vanuatu. In: *Climate Change Adaptation in Pacific Countries*. Springer Nature. Gewerbestrasse 11, 6330 Cham, Switzerland, pp. 151–161. ISBN: 978-3-319-50093-5.

Nalau, J. et al., 2016: The practice of integrating adaptation and disaster risk reduction in the south-west Pacific. *Clim. Dev.*, **8**(4), 365–375, doi:10.1080/17565529.2015.1064809.

Needham, H.F., B.D. Keim and D. Sathiaraj, 2015: A review of tropical cyclone-generated storm surges: Global data sources, observations, and impacts. *Rev. Geophys.*, **53**(2), 545–591, doi:10.1002/2014RG000477.

Nemakonde, L.D. and D. Van Niekerk, 2017: A normative model for integrating organisations for disaster risk reduction and climate change adaptation within SADC member states. *Disaster Prevention and Management: An International Journal*, **26**(3), 361–376, doi:10.1108/Dpm-03-2017-0066.

Newman, M. et al., 2018: The extreme 2015/16 El Nino, in the context of historical climate variability and change. In "*Explaining Extreme Events of 2016 from a Climate Perspective*". *Bull. Am. Meterol. Soc.*, **99**(1), S15–S20, doi:10.1175/bams-d-17-0116.1.

Nielsen, U.N. et al., 2012: The ecology of pulse events: insights from an extreme climatic event in a polar desert ecosystem. *Ecosphere*, **3**(2), 1–15, doi:10.1890/Es11-00325.1.

Nieves, V., J. Willis and W. Patzert, 2015: Recent hiatus caused by decadal shift in Indo-Pacific heating. *Science*, **349**(6247), 532–535, doi:10.1126/science.aaa4521.

Ñiquen, M. and M. Bouchon, 2004: Impact of El Niño events on pelagic fisheries in Peruvian waters. *Deep-Sea Res. Pt. II.*, **51**(6–9), 563–574, doi:10.1016/j.dsr2.2004.03.001.

Nordhaus, W.D., 1992: An optimal transition path for controlling greenhouse gases. *Science*, **258**(5086), 1315–1319, doi:10.1126/science.258.5086.1315.

Nordhaus, W.D., 2017: Revisiting the social cost of carbon. *PNAS*, **114**(7), 201609244, doi:10.1073/pnas.1609244114.

Noy, I., 2016: Tropical storms: the socio-economics of cyclones. *Nat. Clim. Change*, **6**(4), 343–345, doi:10.1038/nclimate2975.

Nunn, P.D. et al., 2016: Spirituality and attitudes towards Nature in the Pacific Islands: insights for enabling climate-change adaptation. *Clim. Change*, **136**(3–4), 477–493, doi:10.1007/s10584-016-1646-9.

6

Nyamwanza, A.M. et al., 2017: Contributions of decadal climate information in agriculture and food systems in east and southern Africa. *Clim. Change*, **143**(1–2), 115–128, doi:10.1007/s10584-017-1990-4.

Nye, J.A. et al., 2014: Ecosystem effects of the Atlantic Multidecadal Oscillation. *J. Marine Syst.*, **133**, 103–116, doi:10.1016/j.jmarsys.2013.02.006.

Oey, L.Y. and S. Chou, 2016: Evidence of rising and poleward shift of storm surge in western North Pacific in recent decades. *J. Geophys. Res.-Oceans*, **121**(7), 5181–5192, doi:10.1002/2016JC011777.

Ohman, M.D., N. Mantua, J. Keister, M. Garcia-Reyes and S. McClatchie 2017: ENSO impacts on ecosystem indicators in the California Current System., Woods Hole Oceanographic Institution. 266 Woods Hole Rd, MS #25, Woods Hole, MA 02543 USA. Available at: https://www.us-ocb.org/enso-impacts-on-ecosystem-indicators-in-the-california-current-system/ Accessed 2018/03/30.

Olita, A. et al., 2007: Effects of the 2003 European heatwave on the Central Mediterranean Sea: surface fluxes and the dynamical response. *Ocean Sci.*, **3**(2), 273–289, doi:10.5194/os-3-273-2007.

Oliver, E. C., 2019: Mean warming not variability drives marine heatwave trends. *Clim. Dynam.*, **53**(3-4), 1653–1659, doi:10.1007/s00382-019-04707-2.

Oliver, E.C.J. et al., 2017: The unprecedented 2015/16 Tasman Sea marine heatwave. *Nat. Commun.*, **8**, 16101, doi:10.1038/ncomms16101.

Oliver, E.C.J. et al., 2018a: Longer and more frequent marine heatwaves over the past century. *Nat. Commun.*, **9**(1), 1324, doi:10.1038/s41467-018-03732-9.

Oliver, E.C.J. and N.J. Holbrook, 2014: Extending our understanding of South Pacific gyre "spin-up": Modeling the East Australian Current in a future climate. *J. Geophys. Res.-Oceans*, **119**(5), 2788–2805, doi:10.1002/2013JC009591.

Oliver, E.C.J., T.J. O'Kane and N.J. Holbrook, 2015: Projected changes to Tasman Sea eddies in a future climate. *J. Geophys. Res.-Oceans*, **120**(11), 7150–7165, doi:10.1002/2015JC010993.

Oliver, E.C.J., S.E. Perkins-Kirkpatrick, N. J. Holbrook and N. L. Bindoff, 2018b: Anthropogenic and natural influences on record 2016 marine heat waves. In *"Explaining Extreme Events of 2016 from a Climate Perspective"*. *Bull. Am. Meterol. Soc.*, **99**(1), S44–S48, doi:10.1175/BAMS-ExplainingExtremeEvents2016.1.

Oliver, E.C.J., S.J. Wotherspoon, M.A. Chamberlain and N.J. Holbrook, 2014: Projected Tasman Sea Extremes in Sea Surface Temperature through the Twenty-First Century. *J. Clim.*, **27**(5), 1980–1998, doi:10.1175/jcli-d-13-00259.1.

Oltmanns, M., J. Karstensen and J. Fischer, 2018: Increased risk of a shutdown of ocean convection posed by warm North Atlantic summers. *Nat. Clim. Change*, **8**(4), 300–304, doi:10.1038/s41558-018-0105-1.

Oppenheimer, M. et al., 2014: Emergent risks and key vulnerabilities. In: Climate Change 2014: Impacts, Adaptation, and Vulnerability. Part A: Global and Sectoral Aspects. Contribution of Working Group II to the Fifth Assessment Report of the Intergovernmental Panel of Climate Change [Field, C.B., V.R. Barros, D.J. Dokken, K.J. Mach, M.D. Mastrandrea, T.E. Bilir, M. Chatterjee, K.L. Ebi, Y.O. Estrada, R.C. Genova, B. Girma, E.S. Kissel, A.N. Levy, S. MacCracken, P.R. Mastrandrea and L.L. White (eds.)]. Cambridge University Press, Cambridge, United Kingdom and New York, NY, USA, 1039–1099.

Organization of American States, 2014: *Mainstreaming disaster risk reduction and adaptation to climate change*. Department of Sustainable Development, Executive Secretariat for Integral Development, Washington, D.C., 39pp [Available at: http://www.oas.org/legal/english/gensec/exor1604_annex_a.pdf].

Osman, M. B. et al., 2019: Industrial-era decline in subarctic Atlantic productivity. *Nature*, **569**, 551–555, doi:10.1038/s41586-019-1181-8.

Pal, I., T. Ghosh and C. Ghosh, 2017: Institutional framework and administrative systems for effective disaster risk governance – Perspectives of 2013 Cyclone Phailin in India. *Int. J. Disast. Risk Re.*, **21**, 350–359, doi:10.1016/j.ijdrr.2017.01.002.

Palter, J.B., T.L. Frölicher, D. Paynter and J.G. John, 2018: Climate, ocean circulation, and sea level changes under stabilization and overshoot pathways to 1.5 K warming. *Earth Syst. Dynam.*, **9**, 817–828, doi:10.5194/esd-2017-105.

Pappenberger, F. et al., 2015: The monetary benefit of early flood warnings in Europe. *Environ. Sci. Policy*, **51**, 278–291, doi:10.1016/j.envsci.2015.04.016.

Park, D.-S.R. et al., 2017: Asymmetric response of tropical cyclone activity to global warming over the North Atlantic and western North Pacific from CMIP5 model projections. *Sci. Rep.*, **7**, 41354, doi:10.1038/srep41354.

Parsons, L. A. et al., 2014: Influence of the Atlantic Meridional Overturning Circulation on the monsoon rainfall and carbon balance of the American tropics. *Geophys. Res. Lett.*, **41**(1), 146–151, doi:10.1002/2013gl058454.

Pasch, R.J., A.B. Penny and R. Berg, 2018: *Hurricane Maria*. Tropical Cyclone Report, National Hurricane Center, Miami, USA., 48 pp. www.nhc.noaa.gov/data/tcr/AL152017_Maria.pdf. Accessed 20/08/2019.

Paton, D., 2007: Preparing for natural hazards: the role of community trust. *Disaster Prevention and Management: An International Journal*, **16**(3), 370–379, doi:10.1108/09653560710758323.

Paton, D., 2008: Risk communication and natural hazard mitigation: how trust influences its effectiveness. *IJGEnvl*, **8**(1–2), 2–16, doi:10.1504/IJGENVI.2008.017256.

Patterson, J.J. et al., 2018: Political feasibility of 1.5°C societal transformations: the role of social justice. *Curr. Opin. Env. Sust.*, **31**, 1–9, doi:10.1016/j.cosust.2017.11.002.

Paul, B.K., 2009: Why relatively fewer people died? The case of Bangladesh's Cyclone Sidr. *Nat. Hazards*, **50**(2), 289–304, doi:10.1007/s11069-008-9340-5.

Paul, S. et al., 2016: Weakening of Indian Summer Monsoon Rainfall due to Changes in Land Use Land Cover. *Sci. Rep.*, **6**, 32177, doi:10.1038/srep32177.

Payne, M.R. et al., 2012: The rise and fall of the NE Atlantic blue whiting *Micromesistus poutassou. Mar. Biol. Res.*, **8**(5–6), 475–487, doi:10.1080/17451000.2011.639778.

Payne, M.R. et al., 2017: Lessons from the First Generation of Marine Ecological Forecast Products. *Front. Mar. Sci.*, **4**, doi:10.3389/fmars.2017.00289.

Pearce, A.F. and M. Feng, 2013: The rise and fall of the "marine heat wave" off Western Australia during the summer of 2010/2011. *J. Marine Syst.*, **111–112**, 139–156, doi:10.1016/j.jmarsys.2012.10.009.

Pearce, T., J. Ford, A.C. Willox and B. Smit, 2015: Inuit traditional ecological knowledge (TEK), subsistence hunting and adaptation to climate change in the Canadian Arctic. *Arctic*, **68**(2), 233–245, doi:10.14430/arctic4475.

Peings, Y., G. Simpkins and G. Magnusdottir, 2016: Multidecadal fluctuations of the North Atlantic Ocean and feedback on the winter climate in CMIP5 control simulations. *J. Geophys. Res.-Atmos.*, **121**(6), 2571–2592, doi:10.1002/2015jd024107.

Penaflor, E.L. et al., 2009: Sea-surface temperature and thermal stress in the Coral Triangle's over the past two decades. *Coral Reefs*, **28**, 841–850, doi:10.1007/s00338-009-0522-8.

Pepler, A.S., L.V. Alexander, J.P. Evans and S.C. Sherwood, 2016a: The influence of local sea surface temperatures on Australian east coast cyclones. *J. Geophys. Res.-Atmos.*, **121**(22), 13,352–13,363, doi:10.1002/2016JD025495.

Pepler, A.S. et al., 2016b: Projected changes in east Australian midlatitude cyclones during the 21st century. *Geophys. Res. Lett.*, **43**(1), 334–340, doi:10.1002/2015GL067267.

Perkins-Kirkpatrick, S. E. et al., 2016: Natural hazards in Australia: heatwaves. *Clim. Change*, **139**(1), 101–114, doi:10.1007/s10584-016-1650-0.

Perl, T.M. et al., 1990: An outbreak of toxic encephalopathy caused by eating mussels contaminated with domoic acid. *N. Eng. J. Med.*, **322**(25), 1775–1780, doi:10.1056/NEJM199006213222504.

Persechino, A. et al., 2013: Decadal predictability of the Atlantic meridional overturning circulation and climate in the IPSL-CM5A-LR model. *Clim. Dynam.*, **40**(9–10), 2359–2380, doi:10.1007/s00382-012-1466-1.

Pershing, A.J. et al., 2015: Slow adaptation in the face of rapid warming leads to collapse of the Gulf of Maine cod fishery. *Science*, **350**(6262), 809–812, doi:10.1126/science.aac9819.

Pescaroli, G. and D. Alexander, 2018: Understanding compound, interconnected, interacting, and cascading risks: a holistic framework. *Risk Anal.*, **38**(11), 2245–2257, doi:10.1111/risa.13128.

Peterson, T.C., M.P. Hoerling, P.A. Stott and S. Herring, 2013: Explaining extreme events of 2012 from a climate perspective. *Bull. Am. Meterol. Soc.*, **94**(9), S1–S74, doi:10.1175/Bams-D-13-00085.1.

Peterson, T.C., P.A. Stott and S. Herring, 2012: Explaining extreme events of 2011 from a climate perspective. *Bull. Am. Meterol. Soc.*, **93**(7), 1041–1067, doi:10.1175/Bams-D-12-00021.1.

Peterson, W.T. et al., 2017: The pelagic ecosystem in the Northern California Current off Oregon during the 2014–2016 warm anomalies within the context of the past 20 years. *J. Geophys. Res.-Oceans*, **122**(9), 7267–7290, doi:10.1002/2017jc012952.

Petty, A.A. et al., 2017: Skillful spring forecasts of September Arctic sea ice extent using passive microwave sea ice observations. *Earth's Future*, **5**(2), 254–263, doi:10.1002/2016EF000495.

Petty, A.A. et al., 2018: The Arctic sea ice cover of 2016: a year of record-low highs and higher-than-expected lows. *The Cryosphere*, **12**(2), 433–452, doi:10.5194/tc-12-433-2018.

Pezza, A.B. and I. Simmonds, 2005: The first South Atlantic hurricane: Unprecedented blocking, low shear and climate change. *Geophys. Res. Lett.*, **32**(15), L15712, doi:10.1029/2005GL023390.

Pickering, M.D. et al., 2017: The impact of future sea level rise on the global tides. *Cont. Shelf Res.*, **142**(Supplement C), 50–68, doi:10.1016/j.csr.2017.02.004.

Plummer, R. et al., 2017: Is adaptive co-management delivering? Examining relationships between collaboration, learning and outcomes in UNESCO biosphere reserves. *Ecol. Econ.*, **140**, 79–88, doi:10.1016/j.ecolecon.2017.04.028.

Poli, P. et al., 2016: ERA-20C: An atmospheric reanalysis of the 20th century. *J. Clim.*, **29**, 4083–4097, doi:10.1175/JCLI-D-15-0556.1.

Poloczanska, E.S. et al., 2013: Global imprint of climate change on marine life. *Nat. Clim. Change*, **3**(10), 919–925, doi:10.1038/nclimate1958.

Pomeroy, R. et al., 2015: Status and Priority Capacity Needs for Local Compliance and Community-Supported Enforcement of Marine Resource Rules and Regulations in the Coral Triangle Region. *Coastal Management*, **43**(3), 301–328, doi:10.1080/08920753.2015.1030330.

Pörtner, H.-O., 2002: Climate variations and the physiological basis of temperature dependent biogeography: systemic to molecular hierarchy of thermal tolerance in animals. *Comp. Biochem. Phys. A.*, **132**(4), 739–761, doi:10.1016/S1095-6433(02)00045-4.

Pörtner, H.O. et al., 2014: Ocean systems. In: Climate Change 2014: Impacts, Adaptation, and Vulnerability. Part A: Global and Sectoral Aspects. Contribution of Working Group II to the Fifth Assessment Report of the Intergovernmental Panel of Climate Change [Field, C.B., V.R. Barros, D.J. Dokken, K.J. Mach, M.D. Mastrandrea, T.E. Bilir, M. Chatterjee, K.L. Ebi, Y.O. Estrada, R.C. Genova, B. Girma, E.S. Kissel, A.N. Levy, S. MacCracken, P.R. Mastrandrea and L.L. White (eds.)]. Cambridge University Press, Cambridge, United Kingdom and New York, NY, USA, 411–484.

Pörtner, H.O. and R. Knust, 2007: Climate change affects marine fishes through the oxygen limitation of thermal tolerance. *Science*, **315**(5808), 95–97, doi:10.1126/science.1135471.

Post, E. et al., 2013: Ecological Consequences of Sea-Ice Decline. *Science*, **341**(6145), 519, doi:10.1126/science.1235225.

Potemra, J.T. and N. Schneider, 2007: Interannual variations of the Indonesian throughflow. *J. Geophys. Res.-Oceans*, **112**, C05035, doi:10.1029/2006jc003808.

Power, S. et al., 2013: Robust twenty-first-century projections of El Nino and related precipitation variability. *Nature*, **502**(7472), 541–545, doi:10.1038/nature12580.

Power, S.B. and F.P. Delage, 2018: El Niño–Southern Oscillation and Associated Climatic Conditions around the World during the Latter Half of the Twenty-First Century. *J. Clim.*, **31**(15), 6189–6207, doi:10.1175/JCLI-D-18-0138.1.

Press, T., 2016: *Tasmanian Wilderness World Heritage Area Bushfire and Climate Change Report*. Department of Premier and Cabinet, Tasmanian Climate Change Office, GPO Box 123, Hobart TAS 7001, Australia, 156 pp. ISBN: 978 0 7246 5715 0.

Proshutinsky, A. et al., 2015: Arctic circulation regimes. *Philos. Trans. Royal Soc. A.*, **373**(2052), 20140160, doi:10.1098/rsta.2014.0160.

Pujiana, K., M.J. McPhaden, A.L. Gordon and A.M. Napitu, 2019: Unprecedented response of Indonesian throughflow to anomalous Indo-Pacific climatic forcing in 2016. *J. Geophys. Res.-Oceans*, **124**, 3737–3754, doi:10.1029/2018jc014574.

Putra, M.I.S., W. Widodo, B. Jatmiko and M. Mundilarto, 2018: The effectiveness of project based learning model to improve vocational skills on the vocational high school students. *Unnes Science Education Journal*, **7**(1), 35–49, doi:10.15294/usej.v7i1.19536.

Puy, M. et al., 2017: Influence of Westerly Wind Events stochasticity on El Niño amplitude: the case of 2014 vs. 2015. *Clim. Dynam.*, **52** 12), 7435–7454, doi:10.1007/s00382-017-3938-9.

Qian, C. et al., 2018: Human influence on the record-breaking cold event in January of 2016 in Eastern China. In "*Explaining Extreme Events of 2016 from a Climate Perspective*". *Bull. Am. Meterol. Soc.*, **99**(1), S118–S122, doi:10.1175/BAMS-ExplainingExtremeEvents2016.1.

Rahmstorf, S. et al., 2015: Exceptional twentieth-century slowdown in Atlantic Ocean overturning circulation. *Nat. Clim. Change*, **5**(5), 475–480, doi:10.1038/nclimate2554.

Rajeevan, M. et al., 2013: On the epochal variation of intensity of tropical cyclones in the Arabian Sea. *Atmos. Sci. Lett.*, **14**(4), 249–255, doi:10.1002/asl2.447.

Ramírez, I.J. and F. Briones, 2017: Understanding the El Niño Costero of 2017: The Definition Problem and Challenges of Climate Forecasting and Disaster Responses. *Int. J. Disast. Risk Sci.*, **8**(4), 489–492, doi:10.1007/s13753-017-0151-8.

Ranasinghe, R., D. Callaghan and M.J.F. Stive, 2012: Estimating coastal recession due to sea level rise: beyond the Bruun rule. *Clim. Change*, **110**(3), 561–574, doi:10.1007/s10584-011-0107-8.

Randerson, J.T. et al., 2015: Multicentury changes in ocean and land contributions to the climate-carbon feedback. *Global Biogeochem. Cy.*, **29**(6), 744–759, doi:10.1002/2014gb005079.

Ranger, N., T. Reeder and J. Lowe, 2013: Addressing 'deep' uncertainty over long-term climate in major infrastructure projects: four innovations of the Thames Estuary 2100 Project. *EURO Journal on Decision Processes*, **1**(3-4), 233–262, doi:10.1007/s40070-013-0014-5.

Rasquinho, O., J. Liu and D. Leong, 2013: Assessment on disaster risk reduction of tropical storm Washi. *Tropical Cyclone Research and Review*, **2**(3), 169–175, doi:10.6057/2013TCRR03.04.

Red Cross Climate Centre, 2016: *El Niño in Peru: Changing the paradigm, acting faster*. International Federation of Red Cross and Red Crescent Societies, 12 pp. http://climatecentre.org/downloads/files/NotaTecnicaFEN%20-%20Ingles%2020set2016.pdf. Accessed 2019/20/08.

Reguero, B.G., I.J. Losada and F.J. Méndez, 2019: A recent increase in global wave power as a consequence of oceanic warming. *Nat. Commun.*, **10**(1), 205–205, doi:10.1038/s41467-018-08066-0.

Reimer, J.J. et al., 2015: Sea surface temperature influence on terrestrial gross primary production along the Southern California current. *PLoS ONE*, **10**(4), e0125177, doi:10.1371/journal.pone.0125177.

Reintges, A., M. Latif and W. Park, 2017: Sub-decadal North Atlantic Oscillation variability in observations and the Kiel Climate Model. *Clim. Dynam.*, **48**(11), 3475–3487, doi:10.1007/s00382-016-3279-0.

Rexach, M. et al., 2017: *Hurricane Maria's Aftermath: Highlights of Available Government Assistance for Puerto Rico Residents*, Littler Mendelson. San Francisco, California, USA.. www.littler.com/files/hurricane_marias_aftermath_-_highlights_of_available_government_assistance_for_puerto_rico_residents.pdf Accessed 2018/04/19.

Reynolds, R.W. et al., 2007: Daily high-resolution-blended analyses for sea surface temperature. *J. Clim.*, **20**(22), 5473–5496, doi:10.1175/2007jcli1824.1.

Rhein, M. et al., 2013: Observations: Ocean. In: Climate Change 2013: The Physical Science Basis. Contribution of Working Group I to the Fifth Assessment Report of the Intergovernmental Panel on Climate Change [Stocker, T.F., D. Qin, G.-K. Plattner, M. Tignor, S.K. Allen, J. Boschung, A. Nauels, Y. Xia, V. Bex and P.M. Midgley (eds.)]. Cambridge University Press, Cambridge, United Kingdom and New York, NY, USA, 255-316.

Rhein, M. et al., 2017: Ventilation variability of Labrador SeaWater and its impact on oxygen and anthropogenic carbon: a review. *Philos. Trans. Royal Soc. A.*, **375**(2102), 20160321, doi:10.1098/rsta.2016.0321.

Rhodes, R.H. et al., 2015: Enhanced tropical methane production in response to iceberg discharge in the North Atlantic. *Science*, **348**(6238), 1016–1019, doi:10.1126/science.1262005.

Ridderinkhof, H. et al., 2010: Seasonal and interannual variability in the Mozambique Channel from moored current observations. *J. Geophys. Res.-Oceans*, **115**, C06010, doi:10.1029/2009JC005619.

Rimi, R.H., K. Huaustein, M.R. Allen and E.J. Barbour, 2018: Risks of pre-monsoon extreme rainfall events of Bangladesh: Is anthropogenic climate change playing a role? In *"Explaining Extremes of 2017 from a Climate Perspective"*. *Bull. Am. Meterol. Soc.*, **100**(1), S61–S66, doi:10.1175/BAMS-D-18-0135.1.

Ringeval, B. et al., 2013: Response of methane emissions from wetlands to the Last Glacial Maximum and an idealized Dansgaard-Oeschger climate event: insights from two models of different complexity. *Clim. Past*, **9**(1), 149–171, doi:10.5194/cp-9-149-2013.

Risser, M.D. and M.F. Wehner, 2017: Attributable Human-Induced Changes in the Likelihood and Magnitude of the Observed Extreme Precipitation during Hurricane Harvey. *Geophys. Res. Lett.*, **44**(24), 12,457–12,464, doi:10.1002/2017GL075888.

Rivera, C. and C. Wamsler, 2014: Integrating climate change adaptation, disaster risk reduction and urban planning: A review of Nicaraguan policies and regulations. *Int. J. Disast. Risk Re.*, **7**, 78–90, doi:10.1016/j.ijdrr.2013.12.008.

Roberts, M.J. et al., 2015: Tropical cyclones in the UPSCALE ensemble of high-resolution global climate models. *J. Clim.*, **28**(2), 574–596, doi:10.1175/Jcli-D-14-00131.1.

Robson, J., P. Ortega and R. Sutton, 2016: A reversal of climatic trends in the North Atlantic since 2005. *Nat. Geosci.*, **9**(7), 513–517, doi:10.1038/ngeo2727.

Robson, J.I. et al., 2018: Decadal prediction of the North Atlantic subpolar gyre in the HiGEM high-resolution climate model. *Clim. Dynam.*, **50**(3–4), 921–937, doi:10.1007/s00382-017-3649-2.

Robson, J.I., R.T. Sutton and D.M. Smith, 2012: Initialized decadal predictions of the rapid warming of the North Atlantic Ocean in the mid 1990s. *Geophys. Res. Lett.*, **39**(19), L19713, doi:10.1029/2012gl053370.

Robson, J.I., R.T. Sutton and D.M. Smith, 2013: Predictable Climate Impacts of the Decadal Changes in the Ocean in the 1990s. *J. Clim.*, **26**(17), 6329–6339, doi:10.1175/jcli-d-12-00827.1.

Rocha, J.C., G. Peterson, Ö. Bodin and S. Levin, 2018: Cascading regime shifts within and across scales. *Science*, **362**(6421), 1379–1383, doi:10.1126/science.aat7850.

Rogers, R.F. et al., 2017: Rewriting the Tropical Record Books: The Extraordinary Intensification of Hurricane Patricia (2015). *Bull. Am. Meterol. Soc.*, **98**(10), 2091–2112, doi:10.1175/bams-d-16-0039.1.

Rohini, P., M. Rajeevan and A. Srivastava, 2016: On the variability and increasing trends of heat waves over India. *Sci. Rep.*, **6**, 26153, doi:10.1038/srep26153.

Romanou, A., J. Marshall, M. Kelley and J. Scott, 2017: Role of the ocean's AMOC in setting the uptake efficiency of transient tracers. *Geophys. Res. Lett.*, **44**(11), 5590–5598, doi:10.1002/2017GL072972.

Rose, B.E.J. et al., 2014: The dependence of transient climate sensitivity and radiative feedbacks on the spatial pattern of ocean heat uptake. *Geophys. Res. Lett.*, **41**(3), 1071–1078, doi:10.1002/2013gl058955.

Rose, B.E.J. and L. Rayborn, 2016: The Effects of Ocean Heat Uptake on Transient Climate Sensitivity. *Curr. Clim.*, **2**(4), 190–201, doi:10.1007/s40641-016-0048-4.

Rosenzweig, C. and W. Solecki, 2014: Hurricane Sandy and adaptation pathways in New York: Lessons from a first-responder city. *Global Environ. Change*, **28**, 395–408, doi:10.1016/j.gloenvcha.2014.05.003.

Rosier, S. et al., 2015: Extreme Rainfall in Early July 2014 in Northland, New Zealand – Was There an Anthropogenic Influence? In *"Explaining Extreme Events of 2014 from a Climate Perspective"*. *Bull. Am. Meterol. Soc.*, **96**(12), S136–S140, doi:10.1175/BAMS-D-15-00105.1.

Rosinski, A. et al., 2017: *Coral bleaching recovery plan*. University of Hawai'i, Social Science Research Institute, 47 pp. https://dlnr.hawaii.gov/dar/files/2017/04/Coral_Bleaching_Recovery_Plan_final.pdf. Accessed 2019/20/08.

Rosselló, R., 2017: *Build Back Better Puerto Rico*. Request for Federal Assistance for Disaster Recovery, Puerto Rico, 94 pp. http://nlihc.org/sites/default/files/Build_Back_Better_PR_Request_94B.pdf. Accessed 2019/20/08.

Röthlisberger, M., S. Pfahl and O. Martius, 2016: Regional-scale jet waviness modulates the occurrence of midlatitude weather extremes. *Geophys. Res. Lett.*, **43**, 10,989–10,997, doi:10.1002/2016GL070944.

Roxy, M.K. et al., 2017: A threefold rise in widespread extreme rain events over central India. *Nat. Commun.*, **8**(1), 708, doi:10.1038/s41467-017-00744-9.

Roxy, M.K., K. Ritika, P. Terray and S. Masson, 2014: The curious case of Indian Ocean warming. *J. Clim.*, **27**(22), 8501–8509, doi:10.1175/jcli-d-14-00471.1.

Roxy, M.K. et al., 2015: Drying of Indian subcontinent by rapid Indian Ocean warming and a weakening land-sea thermal gradient. *Nat. Commun.*, **6**, 7423, doi:10.1038/ncomms8423.

Rudnick, D.L., K.D. Zaba, R.E. Todd and R.E. Davis, 2017: A climatology of the California Current System from a network of underwater gliders. *Progress in Oceanography*, **154**(Supplement C), 64–106, doi:10.1016/j.pocean.2017.03.002.

Rugenstein, M.A.A., K. Caldeira and R. Knutti, 2016: Dependence of global radiative feedbacks on evolving patterns of surface heat fluxes. *Geophys. Res. Lett.*, **43**(18), 9877–9885, doi:10.1002/2016gl070907.

Rugenstein, M.A.A. et al., 2013: Northern High-Latitude Heat Budget Decomposition and Transient Warming. *J. Clim.*, **26**(2), 609–621, doi:10.1175/jcli-d-11-00695.1.

Rustic, G.T., A. Koutavas, T.M. Marchitto and B.K. Linsley, 2015: Dynamical excitation of the tropical Pacific Ocean and ENSO variability by Little Ice Age cooling. *Science*, **350**(6267), aac9937, doi:10.1126/science.aac9937.

Saba, V.S. et al., 2016: Enhanced warming of the Northwest Atlantic Ocean under climate change. *J. Geophys. Res.-Oceans*, **121**(1), 118–132, doi:10.1002/2015jc011346.

Sacks, C.A., A.S. Kesselheim and M. Fralick, 2018: The shortage of normal saline in the wake of Hurricane Maria. *JAMA Intern. Med.*, **178**(7), 885–886, doi:10.1001/jamainternmed.2018.1936.

Saha, S.K., 2017: Cyclone Aila, livelihood stress, and migration: empirical evidence from coastal Bangladesh. *Disasters*, **41**(3), 505–526, doi:10.1111/disa.12214.

Sainsbury, N.C. et al., 2018: Changing storminess and global capture fisheries. *Nat. Clim. Change*, **8**(8), 655–659, doi:10.1038/s41558-018-0206-x.

Saleh, F. and M.P. Weinstein, 2016: The role of nature-based infrastructure (NBI) in coastal resiliency planning: a literature review. *J. Environ. Manage.*, **183**(Pt 3), 1088–1098, doi:10.1016/j.jenvman.2016.09.077.

Sanseverino, I. et al., 2016: *Algal bloom and its economic impact*, EUR 27905 EN Joint Research Centre, European Commission. 55 pp. www.matrixenvironment.com/2016_algae_bloom_and_economic_impact.pdf. Accessed 2019/20/08.

Santoso, A., M.J. McPhaden and W. Cai, 2017: The defining characteristics of ENSO extremes and the strong 2015/2016 El Niño. *Rev. Geophys.*, **55**(4), 1079–1129, doi:10.1002/2017rg000560.

Scaife, A.A. et al., 2017: The predictability of European winter 2015/2016. *Atmos. Sci. Lett.*, **18**, 38–44, doi:10.1002/asl/721.

Scannell, H.A. et al., 2016: Frequency of marine heatwaves in the North Atlantic and North Pacific since 1950. *Geophys. Res. Lett.*, **43**(5), 2069–2076, doi:10.1002/2015GL067308.

Schaeffer, A. and M. Roughan, 2017: Subsurface intensification of marine heatwaves off southeastern Australia: The role of stratification and local winds. *Geophys. Res. Lett.*, **44**(10), 5025–5033, doi:10.1002/2017gl073714.

Schipper, E.L.F. et al., 2016: Linking disaster risk reduction, climate change and development. *International Journal of Disaster Resilience in the Built Environment*, **7**(2), 216–228, doi:10.1108/Ijdrbe-03-2015-0014.

Schlegel, R.W. et al., 2017a: Predominant Atmospheric and Oceanic Patterns during Coastal Marine Heatwaves. *Front. Mar. Sci.*, **4**, 323, doi:10.3389/fmars.2017.00323.

Schlegel, R.W., E.C.J. Oliver, T. Wernberg and A.J. Smit, 2017b: Nearshore and offshore co-occurrence of marine heatwaves and cold-spells. *Progress in Oceanography*, **151**(Supplement C), 189–205, doi:10.1016/j.pocean.2017.01.004.

Schleussner, C.-F. et al., 2015: Indications for a North Atlantic ocean circulation regime shift at the onset of the Little Ice Age. *Clim. Dynam.*, **45**(11–12), 3623–3633, doi:10.1007/s00382-015-2561-x.

Schleussner, C.-F. and G. Feulner, 2013: A volcanically triggered regime shift in the subpolar North Atlantic Ocean as a possible origin of the Little Ice Age. *Clim. Past*, **9**(3), 1321–1330, doi:10.5194/cpd-8-6199-2012.

Schleussner, C.-F., A. Levermann and M. Meinshausen, 2014: Probabilistic projections of the Atlantic overturning. *Clim. Change*, **127**(3–4), 579–586, doi:10.1007/s10584-014-1265-2.

Schoenefeld, J.J. and M.R. McCauley, 2016: Local is not always better: the impact of climate information on values, behavior and policy support. *Journal of Environmental Studies and Sciences*, **6**(4), 724–732, doi:10.1007/s13412-015-0288-y.

Scolobig, A., 2017: Understanding Institutional Deadlocks in Disaster Risk Reduction: The Financial and Legal Risk Root Causes in Genova, Italy. *J. Extr. Even.*, **4**(02), 1750010, doi:10.1142/S2345737617500105.

Screen, J.A., C. Deser and L. Sun, 2015: Projected changes in regional climate extremes arising from Arctic sea ice loss. *Environ. Res. Lett.*, **10**(8), 084006, doi:10.1088/1748-9326/10/8/084006.

Screen, J.A. and I. Simmonds, 2013: Exploring links between Arctic amplification and mid-latitude weather. *Geophys. Res. Lett.*, **40**(5), 959–964, doi:10.1002/grl.50174.

Seager, R. et al., 2015: Causes of the 2011–14 California Drought. *J. Clim.*, **28**(18), 6997–7024, doi:10.1175/jcli-d-14-00860.1.

Seager, R. and I.R. Simpson, 2016: Western boundary currents and climate change. *J. Geophys. Res.-Oceans*, **121**(9), 7212–7214, doi:10.1002/2016JC012156.

Seidler, R. et al., 2018: Progress on integrating climate change adaptation and disaster risk reduction for sustainable development pathways in South Asia: Evidence from six research projects. *Int. J. Disast. Risk Re.*, **31**, 92–101, doi:10.1016/j.ijdrr.2018.04.023.

Sen Gupta, A. et al., 2016: Future changes to the Indonesian Throughflow and Pacific circulation: The differing role of wind and deep circulation changes. *Geophys. Res. Lett.*, **43**(4), 1669–1678, doi:10.1002/2016GL067757.

Send, U., M. Lankhorst and T. Kanzow, 2011: Observation of decadal change in the Atlantic meridional overturning circulation using 10 years of continuous transport data. *Geophys. Res. Lett.*, **38**(24), L24606, doi:10.1029/2011GL049801.

Sevellec, F., A.V. Fedorov and W. Liu, 2017: Arctic sea-ice decline weakens the Atlantic Meridional Overturning Circulation. *Nat. Clim. Change*, **7**(8), 604–610, doi:10.1038/nclimate3353.

Sgubin, G. et al., 2015: Multimodel analysis on the response of the AMOC under an increase of radiative forcing and its symmetrical reversal. *Clim. Dynam.*, **45**(5–6), 1429–1450, doi:10.1007/s00382-014-2391-2.

Sgubin, G. et al., 2017: Abrupt cooling over the North Atlantic in modern climate models. *Nat. Commun.*, **8**, 14375, doi:10.1038/ncomms14375.

Sharmila, S. and K.J.E. Walsh, 2018: Recent poleward shift of tropical cyclone formation linked to Hadley cell expansion. *Nat. Clim. Change*, **8**, 730–736, doi:10.1038/s41558-018-0227-5.

Shaw, T.A. et al., 2016: Storm track processes and the opposing influences of climate change. *Nat. Geosci.*, **9**, 656, doi:10.1038/ngeo2783.

Sheen, K.L. et al., 2017: Skilful prediction of Sahel summer rainfall on inter-annual and multi-year timescales. *Nat. Commun.*, **8**, 14966, doi:10.1038/ncomms14966.

Shepherd, T.G., 2016: A Common Framework for Approaches to Extreme Event Attribution. *Curr. Clim.*, **2**(1), 28–38, doi:10.1007/s40641-016-0033-y.

Sherwood, O.A. et al., 2011: Nutrient regime shift in the western North Atlantic indicated by compound-specific δ15N of deep-sea gorgonian corals. *PNAS*, **108**(3), 1011–1015, doi:10.1073/pnas.1004904108.

Shimizu, M. and A.L. Clark, 2015: Interconnected risks, cascading disasters and disaster management policy: a gap analysis. *Planet@ Risk*, **3**(2), Global Risk Forum GRF Davos, Promenade 35, CH-7270 Davos Platz, Switzerland 260–270. ISSN 2296-8172.

Shimura, T., N. Mori and H. Mase, 2015: Future Projection of Ocean Wave Climate: Analysis of SST Impacts on Wave Climate Changes in the Western North Pacific. *J. Clim.*, **28**(8), 3171–3190, doi:10.1175/jcli-d-14-00187.1.

Shiogama, H. et al., 2013: An event attribution of the 2010 drought in the South Amazon region using the MIROC5 model. *Atmos. Sci. Lett.*, **14**(3), 170–175, doi:10.1002/asl2.435.

Shuckburgh, E., D. Mitchell and P. Stott, 2017: Hurricanes Harvey, Irma and Maria: How natural were these 'natural disasters'? *Weather*, **72**(11), 353–354, doi:10.1002/wea.3190.

Simpson, T.W., 2012: What is trust? *Pac. Philos. Q.*, **93**(4), 550–569, doi:10.1111/j.1468-0114.2012.01438.x.

Smale, D.A. and T. Wernberg, 2013: Extreme climatic event drives range contraction of a habitat-forming species. *Proc. Royal Soc. B.*, **280**(1754), 20122829, doi:10.1098/rspb.2012.2829.

Smale, D.A. et al., 2019: Marine heatwaves threaten global biodiversity and the provision of ecosystem services. *Nat. Clim. Change*, **9**, 306–312, doi:10.1038/s41558-019-0412-1.

Smeed, D.A. et al., 2018: The North Atlantic Ocean Is in a State of Reduced Overturning. *Geophys. Res. Lett.*, **45**(3), 1527–1533, doi:10.1002/2017gl076350.

Smit, B. et al., 2001: Adaptation to climate change in the context of sustainable development and equity. In: Climate Change 2001: Impacts, Adaptation and Vulnerability. Contribution of Working Group II to the Third Assessment Report of the Intergovernmental Panel on Climate Change [McCarthy, J.J., O.F. Canziani, N.A. Leary, D.J. Dokken, and K.S. White, (eds)]. Cambridge University Press, UK, Cambridge, 877–912.

Smith, D.M. et al., 2013: Real-time multi-model decadal climate predictions. *Clim. Dynam.*, **41**(11–12), 2875–2888, doi:10.1007/s00382-012-1600-0.

Smith, M D., 2011: An ecological perspective on extreme climatic events: a synthetic definition and framework to guide future research. *J. Ecol.*, **99**(3), 656–663, doi:10.1111/j.1365-2745.2011.01798.x.

Sobel, A.H. et al., 2016: Human influence on tropical cyclone intensity. *Science*, **353**(6296), 242–246, doi:10.1126/science.aaf6574.

Speers, A.E., E.Y. Besedin, J.E. Palardy and C. Moore, 2016: Impacts of climate change and ocean acidification on coral reef fisheries: An integrated ecological–economic model. *Ecol. Econ.*, **128**, 33–43, doi:10.1016/j.ecolecon.2016.04.012.

Spillman, C.M. and A.J. Hobday, 2014: Dynamical seasonal ocean forecasts to aid salmon farm management in a climate hotspot. *Climate Risk Management*, **1**, 25–38, doi:10.1016/j.crm.2013.12.001.

Sprintall, J. and A. Revelard, 2014: The Indonesian Throughflow response to Indo-Pacific climate variability. *J. Geophys. Res.-Oceans*, **119**(2), 1161–1175, doi:10.1002/2013JC009533.

Stadtherr, L. et al., 2016: Record Balkan floods of 2014 linked to planetary wave resonance. *Sci. Adv.*, **2**(4), e1501428, doi:10.1126/sciadv.1501428.

6

Staten, P.W., J. Lu, K.M. Grise, S.M. Davis and T. Birner, 2018: Re-examining tropical expansion. *Nat. Clim. Change.*, **8**, 768–775, doi:10.1038/s41558-018-0246-2.

Stevenson, S. et al., 2010: ENSO model validation using wavelet probability analysis. *J. Clim.*, **23**(20), 5540–5547, doi:10.1175/2010jcli3609.1.

Stopa, J.E., F. Ardhuin and F. Girard-Ardhuin, 2016: Wave climate in the Arctic 1992–2014: seasonality and trends. *The Cryosphere*, **10**(4), 1605–1629, doi:10.5194/tc-10-1605-2016.

Studholme, J. and S. Gulev, 2018: Concurrent changes to Hadley circulation and the meridional distribution of tropical cyclones. *J. Clim.*, **31**, 4367–4389, doi:10.1175/JCLI-D-17-0852.1.

Strobl, E., 2012: The economic growth impact of natural disasters in developing countries: Evidence from hurricane strikes in the Central American and Caribbean regions. *J. Dev. Econ.*, **97**(1), 130–141, doi:10.1016/j.jdeveco.2010.12.002.

Stuart-Smith, R.D., C.J. Brown, D.M. Ceccarelli and G.J. Edgar, 2018: Ecosystem restructuring along the Great Barrier Reef following mass coral bleaching. *Nature*, **560**(7716), 92–96, doi:10.1038/s41586-018-0359-9.

Sugi, M., H. Murakami and K. Yoshida, 2017: Projection of future changes in the frequency of intense tropical cyclones. *Clim. Dynam.*, **49**(1–2), 619–632, doi:10.1007/s00382-016-3361-7.

Sun, Q. and C. Miao, 2018: Extreme rainfall (R20mm, Rx5day) in Yangtze-Huai, China in June-July 2016: The role of ENSO and anthropogenic climate change. In "*Explaining Extreme Events of 2016 from a Climate Perspective*". *Bull. Am. Meteorol. Soc.*, **99**(1), S102–S106, doi:10.1175/BAMS-D-17-0118.1.

Surminski, S., L.M. Bouwer and J. Linnerooth-Bayer, 2016: How insurance can support climate resilience. *Nat. Clim. Change*, **6**(4), 333–334, doi:10.1038/nclimate2979.

Susanto, R.D., A. Field, A.L. Gordon and T.R. Adi, 2012: Variability of Indonesian throughflow within Makassar Strait, 2004-2009. *J. Geophys. Res.-Oceans*, **117**, C09013, doi:10.1029/2012JC008096.

Susanto, R.D. and Y.T. Song, 2015: Indonesian throughflow proxy from satellite altimeters and gravimeters. *J. Geophys. Res.-Oceans*, **120**(4), 2844–2855, doi:10.1002/2014JC010382.

Susanto, R.D. et al., 2013: Observations of the Karimata Strait throughflow from December 2007 to November 2008. *Acta Oceanol. Sin.*, **32**(5), 1–6, doi:10.1007/s13131-013-0307-3.

Sutton, R.T., 2018: ESD Ideas: a simple proposal to improve the contribution of IPCC WGI to the assessment and communication of climate change risks. *Earth Syst. Dynam.*, **9**(4), 1155–1158, doi:10.5194/esd-9-1155-2018.

Sweet, W.V. et al., 2016: In tide's way: Southeast Florida's September 2015 sunny-day flood. In "*Explaining Extremes of 2012 from a Climate Perspective*". *Bull. Am. Meteorol. Soc.*, **97**(12), S25–S30, doi:10.1175/BAMS-D-13-00085.1.

Sweet, W.V. and J. Park, 2014: From the extreme to the mean: Acceleration and tipping points of coastal inundation from sea level rise. *Earth's Future*, **2**(12), 579–600, doi:10.1002/2014ef000272.

Sweet, W.V., C. Zervas, S. Gill and J. Park, 2013: Hurricane Sandy inundation probabilities today and tomorrow. In "*Explaining Extremes of 2012 from a Climate Perspective*". *Bull. Am. Meteorol. Soc.*, **94**(9), S17–S20, doi:10.1175/BAMS-D-13-00085.1.

Swingedouw, D. et al., 2017: Tentative reconstruction of the 1998–2012 hiatus in global temperature warming using the IPSL–CM5A–LR climate model. *C.R Geosci.*, **349**(8), 369–379, doi:10.1016/j.crte.2017.09.014.

Swingedouw, D. et al., 2013: Decadal fingerprints of freshwater discharge around Greenland in a multi-model ensemble. *Clim. Dynam.*, **41** (3-4), 695–720, doi:10.1007/s00382-012-1479-9.

Tàbara, J.D., J. Jäger, D. Mangalagiu and M. Grasso, 2019: Defining transformative climate science to address high-end climate change. *Reg. Environ. Change*, **19**(3), 807–818, doi:10.1007/s10113-018-1288-8.

Takagi, H. and M. Esteban, 2016: Statistics of tropical cyclone landfalls in the Philippines: unusual characteristics of 2013 Typhoon Haiyan. *Nat. Hazards*, **80**(1), 211–222, doi:10.1007/s11069-015-1965-6.

Takahashi, K. et al., 2018: The 2017 coastal El Niño. In "*State of the Climate in 2017*". *Bull. Am. Meteorol. Soc.*, **99**(8), S210–S211, doi:10.1175/2018BAMSStateoftheClimate.1.

Takahashi, K. and B. Dewitte, 2016: Strong and moderate nonlinear El Niño regimes. *Clim. Dynam.*, **46**(5–6), 1627–1645, doi:10.1007/s00382-015-2665-3.

Takayabu, I. et al., 2015: Climate change effects on the worst-case storm surge: a case study of Typhoon Haiyan. *Environ. Res. Lett.*, **10**(6), 064011, doi:10.1088/1748-9326/10/6/064011.

Tan, H. and R. Cai, 2018: What caused the record-breaking warming in East China Seas during August 2016? *Atmos. Sci. Lett.*, **19**(10), e853, doi:10.1002/asl.853.

Tang, Q., X. Zhang, X. Yang and J.A. Francis, 2013: Cold winter extremes in northern continents linked to Arctic sea ice loss. *Environ. Res. Lett.*, **8**(1), 014036, doi:10.1088/1748-9326/8/1/014036.

Tang, S. and S. Dessai, 2012: Usable science? The UK climate projections 2009 and decision support for adaptation planning. *Weather Clim. Soc.*, **4**(4), 300–313, doi:10.1175/WCAS-D-12-00028.1.

Tasmanian Climate Change Office, 2017: *Tasmanian Wilderness and World Heritage Area Bushfire and Climate Change Research Project: Tasmanian Government's Response*. Tasmanian Climate Change Office, Department of Premier and Cabinet, 49 pp. www.dpac.tas.gov.au/__data/assets/pdf_file/0015/361005/Tasmanian_Government_response_Final_Report_TWWHA_Bushfire_and_Climate_Change_Research_Project.pdf. Accessed 2019/20/08.

Taupo, T. and I. Noy, 2017: At the very edge of a storm: The impact of a distant cyclone on Atoll Islands. *Economics of Disasters and Climate Change*, **1**(2), 143–166, doi:10.1007/s41885-017-0011-4.

Tedeschi, R.G. and M. Collins, 2017: The influence of ENSO on South American precipitation: simulation and projection in CMIP5 models. *Int. J. Climatol.*, **37**(8), 3319–3339, doi:10.1002/joc.4919.

Tett, S.F.B. et al., 2018: Anthropogenic forcings and associated changes in fire risk in western North America and Australia during 2015/16. In "*Explaining Extreme Events of 2016 from a Climate Perspective*". *Bull. Am. Meteorol. Soc.*, **99** 1), S60–S64, doi:10.1175/BAMS-D-17-0118.1.

Tharakan, J., 2015: Indigenous knowledge systems—A rich appropriate technology resource. *African journal of science, technology, innovation and development*, **7**(1), 52–57, doi:10.1080/20421338.2014.987987.

The Government of the Commonwealth of Dominica, 2017: *Post-Disaster Needs Assesment: Hurricane Maria*, 20 pp. https://resilientcaribbean. caricom.org/wp-content/uploads/2017/11/DOMINICA-EXECUTIVE-SUMMARY.pdf]. Accessed 2019/20/08.

Thibodeau, B. et al., 2018: Last Century Warming Over the Canadian Atlantic Shelves Linked to Weak Atlantic Meridional Overturning Circulation. *Geophys. Res. Lett.*, **45**(22), 12376–12385, doi:10.1029/2018gl080083.

Thirumalai, K., P.N. DiNezio, Y. Okumura and C. Deser, 2017: Extreme temperatures in Southeast Asia caused by El Nino and worsened by global warming. *Nat. Commun.*, **8**, 15531, doi:10.1038/ncomms15531.

Thomas, A.R., 2015: *Resettlement in the Wake of Typhoon Haiyan in the Philippines: A Strategy to Mitigate Risk or a Risky Strategy?* The Brookings-LSE Project on Internal Displacement, The Brookings Institution, Washington, D.C., 29 pp. www.brookings.edu/wp-content/uploads/2016/06/Brookings-Planned-Relocations-Case-StudyAlice-Thomas-Philippines-case-study-June-2015.pdf. Accessed 2019/20/08.

Thompson, L.G. et al., 2017: Impacts of recent warming and the 2015/16 El Niño on Tropical Peruvian ice fields. *J. Geophys. Res.-Atmos.*, **122**(23), 12,688–12,701, doi:10.1002/2017JD026592.

Thompson, P.R., G.T. Mitchum, C. Vonesch and J.K. Li, 2013: Variability of Winter Storminess in the Eastern United States during the Twentieth Century from Tide Gauges. *J. Clim.*, **26**(23), 9713–9726, doi:10.1175/jcli-d-12-00561.1.

Thomson, J. et al., 2016: Emerging trends in the sea state of the Beaufort and Chukchi seas. *Ocean Model.*, **105**, 1–12, doi:10.1016/j.ocemod.2016.02.009.

6

Thornalley, D.J. et al., 2018: Anomalously weak Labrador Sea convection and Atlantic overturning during the past 150 years. *Nature*, **556**(7700), 227–230, doi:10.1038/s41586-018-0007-4.

Timmermans, B., D. Stone, M. Wehner and H. Krishnan, 2017: Impact of tropical cyclones on modeled extreme wind-wave climate. *Geophys. Res. Lett.*, **44**(3), 1393–1401, doi:10.1002/2016GL071681.

Toimil, A., I.J. Losada, P. Camus and P. Díaz-Simal, 2017: Managing coastal erosion under climate change at the regional scale. *Coast. Eng.*, **128**, 106–122, doi:10.1016/j.coastaleng.2017.08.004.

Tol, R.S.J. et al., 2006: Adaptation to five metres of sea level rise. *J. Risk Res.*, **9**(5), 467–482, doi:10.1080/13669870600717632.

Tommasi, D. et al., 2017a: Multi-Annual Climate Predictions for Fisheries: An Assessment of Skill of Sea Surface Temperature Forecasts for Large Marine Ecosystems. *Front. Mar. Sci.*, **4**, 201, doi:10.3389/fmars.2017.00201.

Tommasi, D. et al., 2017b: Managing living marine resources in a dynamic environment: The role of seasonal to decadal climate forecasts. *Progress in Oceanography*, **152**, 15–49, doi:10.1016/j.pocean.2016.12.011.

Toomey, M.R., J.P. Donnelly and J.D. Woodruff, 2013: Reconstructing mid-late Holocene cyclone variability in the Central Pacific using sedimentary records from Tahaa, French Polynesia. *Quaternary Sci. Rev.*, **77**, 181–189, doi:https://doi.org/10.1016/j.quascirev.2013.07.019.

Townley, C. and J.L. Garfield, 2013: Public Trust. In: Trust: analytic and applied perspectives. Rodopi, Amsterdam, Netherlands, **263**, 95–107. ISBN: 978-94-012-0941-0.

Tozier de la Poterie, A. and M.-A. Baudoin, 2015: From Yokohama to Sendai: Approaches to Participation in International Disaster Risk Reduction Frameworks. *Int. J. Disast. Risk Sci.*, **6**(2), 128–139, doi:10.1007/s13753-015-0053-6.

Trenberth, K.E., J.T. Fasullo and T.G. Shepherd, 2015: Attribution of climate extreme events. *Nat. Clim. Change*, **5**, 725, doi:10.1038/nclimate2657.

Trenberth, K.E. and D.J. Shea, 2006: Atlantic hurricanes and natural variability in 2005. *Geophys. Res. Lett.*, **33**(12), L12704, doi:10.1029/2006GL026894.

Trossman, D.S. et al., 2016: Large-scale ocean circulation-cloud interactions reduce the pace of transient climate change. *Geophys. Res. Lett.*, **43**(8), 3935–3943, doi:10.1002/2016GL067931.

Trusel, L.D. et al., 2018: Nonlinear rise in Greenland runoff in response to post-industrial Arctic warming. *Nature*, **564** (7734), 104–108, doi:10.1038/s41586-018-0752-4.

Tuleya, R.E. et al., 2016: Impact of upper-tropospheric temperature anomalies and vertical wind shear on tropical cyclone evolution using an idealized version of the operational GFDL hurricane model. *J. Atmos. Sci.*, **73**(10), 3803-3820, doi:10.1175/Jas-D-16-0045.1.

Turner, J. et al., 2017: Unprecedented springtime retreat of Antarctic sea ice in 2016. *Geophys. Res. Lett.*, **44**(13), 6868-6875, doi:10.1002/2017GL073656.

Ulrichs, M., R. Slater and C. Costella, 2019: Building resilience to climate risks through social protection: from individualised models to systemic transformation. *Disasters*, **43**(S3), S368–S387, doi:10.1111/disa.12339.

Ummenhofer, C.C., M. Kulüke and J.E. Tierney, 2018: Extremes in East African hydroclimate and links to Indo-Pacific variability on interannual to decadal timescales. *Clim. Dynam.*, **50**(7–8), 2971–2991, doi:10.1007/s00382-017-3786-7.

Ummenhofer, C.C. and G.A. Meehl, 2017: Extreme weather and climate events with ecological relevance: a review. *Philos. Trans. R. Soc. London. B.*, **372**(1723), 20160135, doi:10.1098/rstb.2016.0135.

UNISDR, 2015: *Sendai framework for disaster risk reduction 2015-2030*. World Conference on Disaster Risk Reduction, Sendai, Japan, 37 pp. www.unisdr.org/we/inform/publications/43291. Accessed 2019/20/08.

United Nations, 2017: *North coast of Perú Flash Appeal*. United Nations Office for the Coordination of Humanitarian Affairs, Geneva, Switzerland., 49 pp. https://reliefweb.int/report/peru/north-coast-peru-2017-flash-appeal-april. Accessed 2019/20/08.

UNSCAP, 2015: *El Niño 2105/2106: Impact outlook and policy implications*. Science and Policy Knowledge Series, 21 pp. www.unescap.org/sites/default/files/El%20Nino%20Advisory%20Note%20Dec%202015%20Final.pdf. Accessed 2019/20/08.

van den Honert, R.C. and J. McAneney, 2011: The 2011 Brisbane Floods: Causes, Impacts and Implications. *Water*, **3**(4), 1149, doi:10.3390/w3041149.

van den Hurk, B. et al., 2015: Analysis of a compounding surge and precipitation event in the Netherlands. *Environ. Res. Lett.*, **10**(3), 035001, doi:10.1088/1748-9326/10/3/035001.

van Hooidonk, R., J.A. Maynard and S. Planes, 2013: Temporary refugia for coral reefs in a warming world. *Nat. Clim. Change*, **3**, 508, doi:10.1038/nclimate1829.

van Oldenborgh, G.J. et al., 2017: Attribution of extreme rainfall from Hurricane Harvey, August 2017. *Environ. Res. Lett.*, **12**(12), 124009, doi:10.1088/1748-9326/aa9ef2.

Vance, T. et al., 2015: Interdecadal Pacific variability and eastern Australian megadroughts over the last millennium. *Geophys. Res. Lett.*, **42**(1), 129–137, doi:10.1002/2014GL062447.

Vanniere, B., A. Czaja and H.F. Dacre, 2017: Contribution of the cold sector of extratropical cyclones to mean state features over the Gulf Stream in winter. *Q.R. Roy. Meteorol. Soc.*, **143**(705), 1990–2000, doi:10.1002/qj.3058.

Varotsos, C.A., C. Tzanis and A.P. Cracknell, 2016: Precursory signals of the major El Nino Southern Oscillation events. *Theor. Appl. Climatol.*, **124**(3–4), 903–912, doi:10.1007/s00704-015-1464-4.

Vecchi, G.A., T.L. Delworth and B. Booth, 2017: Climate science: Origins of Atlantic decadal swings. *Nature*, **548**(7667), 284, doi:10.1038/nature23538.

Verlaan, M., S. De Kleermaeker and L. Buckman, 2015: GLOSSIS: Global storm surge forecasting and information System. In: *Australasian Coasts and Ports Conference 2015 [*22nd Australasian Coastal and Ocean Engineering Conference and the 15th Australasian Port and Harbour Conference, Auckland, New Zealand.. Engineers Australia and IPENZ, 229–234. ISBN: 9781922107794.

Veron, J.C.E.N. et al., 2011: The Coral Triangle. In: *Coral Reefs: An Ecosystem in Transition.* Springer, pp. 47–55.

Vignola, R., C. Kuzdas, I. Bolaños and K. Poveda, 2018: Hybrid governance for drought risk management: the case of the 2014/2015 El Niño in Costa Rica. *Int. J. Disast. Risk Re.*, **28**, 363–374, doi:10.1016/j.ijdrr.2018.03.011.

Vitousek, S. et al., 2017: Doubling of coastal flooding frequency within decades due to sea level rise. *Sci. Rep.*, **7**(1), 1399, doi:10.1038/s41598-017-01362-7.

von Känel, L., T.L. Frölicher and N. Gruber, 2017: Hiatus-like decades in the absence of equatorial Pacific cooling and accelerated global ocean heat uptake. *Geophys. Res. Lett.*, **44**(15), 7909–7918, doi:10.1002/2017GL073578.

Vousdoukas, M.I. et al., 2018: Climatic and socioeconomic controls of future coastal flood risk in Europe. *Nat. Clim. Change*, **8**(9), 776–780, doi:10.1038/s41558-018-0260-4.

Vousdoukas, M.I. et al., 2017: Extreme sea levels on the rise along Europe's coasts. *Earth's Future*, **5**(3), 304–323, doi:10.1002/2016ef000505.

Wadey, M., S. Brown, R.J. Nicholls and I. Haigh, 2017: Coastal flooding in the Maldives: an assessment of historic events and their implications. *Nat. Hazards*, **89**(1), 131–159, doi:10.1007/s11069-017-2957-5.

Wahl, T. et al., 2015: Increasing risk of compound flooding from storm surge and rainfall for major US cities. *Nat. Clim. Change*, **5**(12), 1093–1097, doi:10.1038/Nclimate2736.

Wahl, T., N.G. Plant and J.W. Long, 2016: Probabilistic assessment of erosion and flooding risk in the northern Gulf of Mexico. *J. Geophys. Res.-Oceans*, **121**(5), 3029–3043, doi:10.1002/2015JC011482.

Wallimann-Helmer, I., 2015: Justice for climate loss and damage. *Clim. Change*, **133**(3), 469–480, doi:10.1007/s10584-015-1483-2.

Wallimann-Helmer, I., 2016: Differentiating responsibilities for climate change adaptation. *Archiv für Rechts- und Sozialphilosphie (ARSP)*, **149**, 119–132, doi:10.5167/uzh-112531.

6

Wallimann-Helmer, I. et al., 2019: The ethical challenges in the context of climate loss and damage. In: *Loss and Damage from Climate Change*. Springer Nature Switzerland AG. Gewerbestrasse 11, 6330 Cham, Switzerland, pp. 39-62. ISBN: 978-3-319-72025-8.

Walsh, J.E. et al., 2017: The Exceptionally Warm Winter of 2015/16 in Alaska. *J. Clim.*, **30**(6), 2069–2088, doi:10.1175/jcli-d-16-0473.1.

Walsh, J.E. et al., 2018: The high latitude marine heat wave of 2016 and its impacts on Alaska. In *"Explaining Extreme Events of 2016 from a Climate Perspective"*. *Bull. Am. Meteorol. Soc.*, **99**(1), S39–S43, doi:10.1175/bams-d-17-0105.1.

Walsh, K.J.E. et al., 2016: Tropical cyclones and climate change. *WiRes. Clim. Change*, **7**(1), 65–89, doi:10.1002/wcc.371.

Wang, B. et al., 2013: Northern Hemisphere summer monsoon intensified by mega-El Niño/southern oscillation and Atlantic multidecadal oscillation. *PNAS*, **110**(14), 5347–5352, doi:10.1073/pnas.1219405110.

Wang, G. et al., 2017: Continued increase of extreme El Niño frequency long after 1.5°C warming stabilization. *Nat. Clim. Change*, **7**(8), 568–572, doi:10.1038/Nclimate3351.

Wang, S. and R. Toumi, 2016: On the relationship between hurricane cost and the integrated wind profile. *Environ. Res. Lett.*, **11**(11), 114005, doi:10.1088/1748-9326/11/11/114005.

Wang, S., X. Yuan and R. Wu, 2018: Attribution of the Persistent Spring–Summer Hot and Dry Extremes over Northeast China in 2017. In *"Explaining Extremes of 2017 from a Climate Perspective"*. *Bull. Am. Meteorol. Soc.*, **100**(1), S85–S90, doi:10.1175/BAMS-D-18-0135.1.

Wang, S.-Y., L. Hipps, R.R. Gillies and J.H. Yoon, 2014a: Probable causes of the abnormal ridge accompanying the 2013–2014 California drought: ENSO precursor and anthropogenic warming footprint. *Geophys. Res. Lett.*, **41**(9), 3220–3226, doi:10.1002/2014GL059748.

Wang, X. and H. Liu, 2016: PDO modulation of ENSO effect on tropical cyclone intensification in the western North Pacific. *Clim. Dynam.*, **46**(1–2), 15–28, doi:10.1007/s00382-615-2563-8.

Wang, Y. et al., 2014b: Assessing the effects of anthropogenic aerosols on Pacific storm track using a multiscale global climate model. *PNAS*, **111**(19), 6894–6899, doi:10.1073/pnas.1403364111.

Wang, Y. et al., 2019: Seasonal variation of water transport through the Karimata Strait. *Acta Oceanol. Sin.*, **38**(4), 47–57, doi:10.1007/s13131-018-1224-2.

Ward, P.J. et al., 2014: Strong influence of El Nino Southern Oscillation on flood risk around the world. *PNAS*, **111**(44), 15659–15664, doi:10.1073/pnas.1409822111.

Ward, P.J., M. Kummu and U. Lall, 2016: Flood frequencies and durations and their response to El Nino. *J. Hydrol.*, **539**, 358–378, doi:10.1016/j.jhydrol.2016.05.045.

Warner, K. and K. van der Geest, 2013: Loss and damage from climate change: local-level evidence from nine vulnerable countries. *Int. J. Global Warm.*, **5**(4), 367–386, doi:10.1504/Ijgw.2013.057289.

Watson, K.B. et al., 2016: Quantifying flood mitigation services: The economic value of Otter Creek wetlands and floodplains to Middlebury, VT. *Ecol. Econ.*, **130**, 16–24, doi:10.1016/j.ecolecon.2016.05.015.

Weaver, C.P. et al., 2013: Improving the contribution of climate model information to decision making: the value and demands of robust decision frameworks. *WiRes. Clim. Change*, **4**(1), 39–60, doi:10.1002/wcc.202.

Wehner, M. et al., 2015: Resolution Dependence of Future Tropical Cyclone Projections of CAM5.1 in the US CLIVAR Hurricane Working Group Idealized Configurations. *J. Clim.*, **28**(10), 3905–3925, doi:10.1175/JCLI-D-14-00311.1.

Wei, Z. et al., 2019: An overview of 10-year observation of the South China Sea branch of the Pacific to Indian Ocean throughflow at the Karimata Strait. *Acta Oceanol. Sin.*, **38**(4), 1–11, doi:10.1007/s13131-019-1410-x.

Weijerman, M. et al., 2015: An Integrated Coral Reef Ecosystem Model to Support Resource Management under a Changing Climate. *PLoS ONE*, **10**(12), e0144165, doi:10.1371/journal.pone.0144165.

Weijerman, M. et al., 2016: Atlantis Ecosystem Model Summit: Report from a workshop. *Ecol. Model.*, **335**, 35–38, doi:10.1016/j.ecolmodel.2016.05.007.

Weinkle, J., R. Maue and R. Pielke Jr, 2012: Historical global tropical cyclone landfalls. *J. Clim.*, **25**(13), 4729–4735, doi:10.1175/JCLI-D-11-00719.1.

Weller, E. et al., 2015: Human Contribution to the 2014 Record High Sea Surface Temperatures Over the Western Tropical And Northeast Pacific Ocean. *Bull. Am. Meteorol. Soc.*, **96**(12), S100–S104, doi:10.1175/bams-d-15-00055.1.

Wernberg, T. et al., 2016: Climate-driven regime shift of a temperate marine ecosystem. *Science*, **353**(6295), 169–172, doi:10.1126/science.aad8745.

Wernberg, T. et al., 2013: An extreme climatic event alters marine ecosystem structure in a global biodiversity hotspot. *Nat. Clim. Change*, **3**(1), 78–82, doi:10.1038/Nclimate1627.

WFP, FEWS NET, European Commission and FAO, 2017: *Persistent drought in Somalia leads to major food security crisis*. Joint Press Release, World Food Programme; Famine Early Warning Systems Network; European Commission; Food and Agriculture Organization of the United Nations, Rome, Italy, 6 pp. https://documents.wfp.org/stellent/groups/public/documents/ena/wfp290554.pdf?_ga=2.40555818.965568665.1558310041-1569466447.1558310041].

Whan, K. and F. Zwiers, 2017: The impact of ENSO and the NAO on extreme winter precipitation in North America in observations and regional climate models. *Clim. Dynam.*, **48**, 1401–1411, doi:10.1007/s00382-016-3148-x.

White, A.T. et al., 2014: Marine Protected Areas in the Coral Triangle: Progress, Issues, and Options. *Coast. Manage.*, **42**(2), 87–106, doi:10.1080/08920753.2014.878177.

Whitney, F.A., 2015: Anomalous winter winds decrease 2014 transition zone productivity in the NE Pacific. *Geophys. Res. Lett.*, **42**(2), 428–431, doi:10.1002/2014gl062634.

Winton, M. et al., 2013: Connecting Changing Ocean Circulation with Changing Climate. *J. Clim.*, **26**(7), 2268–2278, doi:10.1175/jcli-d-12-00296.1.

Winton, M., K. Takahashi and I.M. Held, 2010: Importance of Ocean Heat Uptake Efficacy to Transient Climate Change. *J. Clim.*, **23**(9), 2333–2344, doi:10.1175/2009jcli3139.1.

Wittenberg, A.T., 2009: Are historical records sufficient to constrain ENSO simulations? *Geophys. Res. Lett.*, **36**(12), L12702, doi:10.1029/2009GL038710.

WMO, 2016: Exceptionally strong El Niño has passed its peak, but impacts continue. Press release, World Meteorological Organization. Geneva, Switzerland. https://public.wmo.int/en/media/press-release/exceptionally-strong-el-niño-has-passed-its-peak-impacts-continue Accessed 2018/09/27.

Woollings, T. et al., 2018: Blocking and its Response to Climate Change. *Curr. Clim.*, **4**(3), 287–300, doi:10.1007/s40641-018-0108-z.

Woollings, T. et al., 2012: Response of the North Atlantic storm track to climate change shaped by ocean-atmosphere coupling. *Nat. Geosci.*, **5**(5), 313–317, doi:10.1038/ngeo1438.

Wouters, B. et al., 2013: Multiyear predictability of the North Atlantic subpolar gyre. *Geophys. Res. Lett.*, **40**(12), 3080–3084, doi:10.1002/grl.50585.

Wu, W. et al., 2018: Mapping Dependence Between Extreme Rainfall and Storm Surge. *J. Geophys. Res.-Oceans*, **123**(4), 2461–2474, doi:10.1002/2017jc013472.

Xie, S.-P., Y. Kosaka and Y.M. Okumura, 2016: Distinct energy budgets for anthropogenic and natural changes during global warming hiatus. *Nat. Geosci.*, **9**(1), 29–33, doi:10.1038/ngeo2581.

Yamada, Y. et al., 2017: Response of Tropical Cyclone Activity and Structure to Global Warming in a High-Resolution Global Nonhydrostatic Model. *J. Clim.*, **30**, 9703–9724, doi:10.1175/Jcli-D-17-0068.1.

Yamamoto, A. and J.B. Palter, 2016: The absence of an Atlantic imprint on the multidecadal variability of wintertime European temperature. *Nat. Commun.*, **7**, 10930, doi:10.1038/ncomms10930.

Yan, X., R. Zhang and T.R. Knutson, 2017: The role of Atlantic overturning circulation in the recent decline of Atlantic major hurricane frequency. *Nat. Commun.*, **8**(1), 1695, doi:10.1038/s41467-017-01377-8.

Yang, H. et al., 2016a: Intensification and poleward shift of subtropical western boundary currents in a warming climate. *J. Geophys. Res.-Oceans*, **121**(7), 4928–4945, doi:10.1002/2015JC011513.

Yang, Q. et al., 2016b: Recent increases in Arctic freshwater flux affects Labrador Sea convection and Atlantic overturning circulation. *Nat. Commun.*, **7**, 10525, doi:10.1038/ncomms10525.

Yashayaev, I. and J.W. Loder, 2017: Further intensification of deep convection in the Labrador Sea in 2016. *Geophys. Res. Lett.*, **44**(3), 1429–1438, doi:10.1002/2016gl071668.

Yeager, S.G. et al., 2018: Predicting near-term changes in the Earth System: A large ensemble of initialized decadal prediction simulations using the Community Earth System Model. *Bull. Am. Meteorol. Soc.*, **99**(9), 1867–1886, doi:10.1175/bams-d-17-0098.1.

Yeager, S.G., A.R. Karspeck and G. Danabasoglu, 2015: Predicted slowdown in the rate of Atlantic sea ice loss. *Geophys. Res. Lett.*, **42**(24), 10704–10713, doi:10.1002/2015GL065364.

Yeh, E.T., 2016: 'How can experience of local residents be "knowledge"?' Challenges in interdisciplinary climate change research. *Area*, **48**(1), 34–40, doi:10.1111/area.12189.

Yeh, S.W. et al., 2018: ENSO atmospheric teleconnections and their response to greenhouse gas forcing. *Rev. Geophys.*, **56**(1), 185–206, doi:10.1002/2017rg000568.

Yonekura, E. and T. Hall, 2014: ENSO Effect on East Asian Tropical Cyclone Landfall via Changes in Tracks and Genesis in a Statistical Model. *J. Appl. Meteorol. Clim.*, **53**(2), 406–420, doi:10.1175/JAMC-D-12-0240.1.

Yonson, R., I. Noy and J.C. Gaillard, 2018: The measurement of disaster risk: An example from tropical cyclones in the Philippines. *Review of Development Economics*, **22**(2), 736–765, doi:10.1111/rode.12365.

Yoshida, K. et al., 2017: Future Changes in Tropical Cyclone Activity in High-Resolution Large-Ensemble Simulations. *Geophys. Res. Lett.*, **44**(19), 9910–9917, doi:10.1002/2017GL075058.

Young, I.R. and A. Ribal, 2019: Multiplatform evaluation of global trends in wind speed and wave height. *Science*, **364**(6440), 548, doi:10.1126/science.aav9527.

Young, O.R., 2002: The institutional dimensions of environmental change: fit, interplay, and scale. The MIT Press. One Rogers Street, Cambridge, MA, USA. ISBN: 9780262240437.

Yuan, X., S. Wang and Z.Z. Hu, 2018: Do climate change and El Nino increase likelihood of Yangtze River extreme rainfall? In *"Explaining Extreme Events of 2016 from a Climate Perspective"*. *Bull. Am. Meteorol. Soc.*, **99**(1), S113–S117, doi:10.1175/BAMS-D-17-0118.1.

Zador, S. and E. Siddon, 2016: *Ecosystem Considerations 2016, Status of the Eastern Bering Sea Marine Ecosystem*. North Pacific Fishery Management Council, Anchorage, Alaska, USA, 210 pp. www.afsc.noaa.gov/REFM/Docs/2016/ecosysEBS.pdf. Accessed 2019/20/08.

Zhai, P. et al., 2016: The strong El Nino of 2015/16 and its dominant impacts on global and China's climate. *J. Meteorol. Res.*, **30**(3), 283–297, doi:10.1007/s13351-016-6101-3.

Zhan, R., 2017: Intensified mega-ENSO has increased the proportion of intense tropical cyclones over the western northwest Pacific since the late 1970s. *Geophys. Res. Lett.*, **44**(23), 11,959–11,966, doi:10.1002/2017glo75916.

Zhang, H. and Y. Guan, 2014: Impacts of four types of ENSO events on tropical cyclones making landfall over mainland china based on three best-track datasets. *Adv. Atmos. Sci.*, **31**(1), 154–164, doi:10.1007/s00376-013-2146-8.

Zhang, L. and K.B. Karnauskas, 2017: The role of tropical interbasin SST gradients in forcing Walker circulation trends. *J. Clim.*, **30**(2), 499–508, doi:10.1175/Jcli-D-16-0349.1.

Zhang, L., K.B. Karnauskas, J.P. Donnelly and K. Emanuel, 2017a: Response of the North Pacific Tropical Cyclone Climatology to Global Warming: Application of Dynamical Downscaling to CMIP5 Models. *J. Clim.*, **30**(4), 1233–1243, doi:10.1175/jcli-d-16-0496.1.

Zhang, R., 2017: On the persistence and coherence of subpolar sea surface temperature and salinity anomalies associated with the Atlantic multidecadal variability. *Geophys. Res. Lett.*, **44**(15), 7865–7875, doi:10.1002/2017gl074342.

Zhang, R. et al., 2013: Have aerosols caused the observed Atlantic multidecadal variability? *J. Atmos. Sci.*, **70**(4), 1135–1144, doi:10.1175/JAS-D-12-0331.1.

Zhang, R. and T.R. Knutson, 2013: The role of global climate change in the extreme low summer Arctic sea ice extent in 2012. In *"Explaining Extremes of 2012 from a Climate Perspective"*]. *Bull. Am. Meteorol. Soc.*, **94**(9), S23–S26, doi:10.1175/BAMS-D-13-00085.1.

Zhang, W. et al., 2016: The Pacific Meridional Mode and the Occurrence of Tropical Cyclones in the Western North Pacific. *J. Clim.*, **29**(1), 381–398, doi:10.1175/jcli-d-15-0282.1.

Zhang, X., J.A. Church, D. Monselesan and K.L. McInnes, 2017b: Sea level projections for the Australian region in the 21st century. *Geophys. Res. Lett.*, **44**(16), 8481–8491, doi:10.1002/2017gl074176.

Zhang, Y. et al., 2018: Strengthened Indonesian throughflow drives decadal warming in the Southern Indian Ocean. *Geophys. Res. Lett.*, **45**(12), 6167–6175, doi:10.1029/2018gl078265.

Zhang, Y., L. Zhang, S.P. Wang and J. Feng, 2017c: Drought events and their influence in summer of 2017 in China (in Chinese). *J. Arid. Meteorol.*, **35**, 899–905.

Zhao, H., X. Duan, G.B. Raga and P.J. Klotzbach, 2018: Changes in Characteristics of Rapidly Intensifying Western North Pacific Tropical Cyclones Related to Climate Regime Shifts. *J. Clim.*, **31**(19), 8163–8179, doi:10.1175/jcli-d-18-0029.1.

Zhou, C., M.D. Zelinka, A.E. Dessler and S.A. Klein, 2015: The relationship between interannual and long-term cloud feedbacks. *Geophys. Res. Lett.*, **42**(23), 10,463–10,469, doi:10.1002/2015GL066698.

Zhou, X. et al., 2016: Catastrophic drought in East Asian monsoon region during Heinrich event 1. *Quaternary Sci. Rev.*, **141**, 1–8, doi:10.1016/j.quascirev.2016.03.029.

Zickfeld, K., M. Eby and A.J. Weaver, 2008: Carbon-cycle feedbacks of changes in the Atlantic meridional overturning circulation under future atmospheric CO_2. *Global Biogeochem. Cy.*, **22**(3), GB3024, doi:10.1029/2007GB003118.

Zommers, Z. et al., 2017: Early Warning Systems for Disaster Risk Reduction Including Climate Change Adaptation. In: *The Routledge Handbook of Disaster Risk Reduction Including Climate Change Adaptation* [Kelman, I., J. Mercer and J.C. Gaillard (eds.)], London, pp. 428–443. ISBN: 9781315684260.

Zscheischler, J. and S.I. Seneviratne, 2017: Dependence of drivers affects risks associated with compound events. *Sci. Adv.*, **3**(6), e1700263, doi:10.1126/sciadv.1700263.

Zscheischler, J. et al., 2018: Future climate risk from compound events. *Nat. Clim. Change*, **8**, 469–477, doi:10.1038/s41558-018-0156-3.

Zuo, X.L. et al., 2015: Spatial and temporal variability of thermal stress to China's coral reefs in South China Sea. *Chinese Ge.*, **25**(2), 159–173, doi:10.1007/s11769-015-0741-6.

Zurcher, S. et al., 2013: Impact of an abrupt cooling event on interglacial methane emissions in northern peatlands. *Biogeosciences*, **10**(3), 1963–1981, doi:10.5194/bg-10-1963-2013.

6

Cross-Chapter Box 9: Integrative Cross-Chapter Box on Low-lying Islands and Coasts

Authors:

Alexandre K. Magnan (France), Matthias Garschagen (Germany), Jean-Pierre Gattuso (France), John E. Hay (Cook Islands/New Zealand), Nathalie Hilmi (Monaco/France), Elisabeth Holland (Fiji), Federico Isla (Argentina), Gary Kofinas (USA), Iñigo J. Losada (Spain), Jan Petzold (Germany), Beate Ratter (Germany), Ted Schuur (USA), Tammy Tabe (Fiji), Roderik van de Wal (Netherlands)

Review Editor:

Joy Pereira (Malaysia)

Chapter Scientist:

Jan Petzold (Germany)

This cross-chapter box should be cited as:

Magnan, A.K., M. Garschagen, J.-P. Gattuso, J.E. Hay, N. Hilmi, E. Holland, F. Isla, G. Kofinas, I.J. Losada, J. Petzold, B. Ratter, T.Schuur, T. Tabe, and R. van de Wal, 2019: Cross-Chapter Box 9: Integrative Cross-Chapter Box on Low-Lying Islands and Coasts. In: *IPCC Special Report on the Ocean and Cryosphere in a Changing Climate* [H.-O. Pörtner, D.C. Roberts, V. Masson-Delmotte, P. Zhai, M. Tignor, E. Poloczanska, K. Mintenbeck, A. Alegría, M. Nicolai, A. Okem, J. Petzold, B. Rama, N.M. Weyer (eds.)]. Cambridge University Press, Cambridge, UK and New York, NY, USA, pp. 657–674. https://doi.org/10.1017/9781009157964.009.

Executive Summary

Ocean and cryosphere changes already impact Low-Lying Islands and Coasts (LLIC), including Small Island Developing States (SIDS), with cascading and compounding risks. Disproportionately higher risks are expected in the course of the 21st century. Reinforcing the findings of the IPCC Special Report on Global Warming of 1.5°C, vulnerable human communities, especially those in coral reef environments and polar regions, may exceed adaptation limits well before the end of this century and even in a low greenhouse gas emission pathway (*high confidence*[1]). Depending on the effectiveness of 21st century mitigation and adaptation pathways under all emission scenarios, most of the low-lying regions around the world may face adaptation limits beyond 2100, due to the long-term commitment of sea level rise (*medium confidence*). LLIC host around 11% of the global population, generate about 14% of the global Gross Domestic Product and comprise many world cultural heritage sites. LLIC already experience climate-related ocean and cryosphere changes (*high confidence*), and they share both commonalities in their exposure and vulnerability to climate change (e.g., low elevation, human disturbances to terrestrial and marine ecosystems), and context-specificities (e.g., variable ecosystem climate sensitivities and risk perceptions by populations). Options to adapt to rising seas, e.g., range from hard engineering to ecosystem-based measures, and from securing current settings to relocating people, built assets and activities. Effective combinations of measures vary across geographies (cities and megacities, small islands, deltas and Arctic coasts), and reflect the scale of observed and projected impacts, ecosystems' and societies' adaptive capacity, and the existence of transformational governance (*high confidence*) {Sections 3.5.3, 4.4.2 to 4.4.5, 5.5.2, 6.8, 6.9, Cross-Chapter Box 2 in Chapter 1}.

Introduction

LLIC are already experiencing the impacts of climate-related changes to the ocean and cryosphere, for both extreme events and slow onset changes (Sections 4.3.3, 5.3.1 to 5.3.6, 6.2, 6.8, 6.9), due to their low elevation, narrow ecological zonation, climate sensitive ecosystems and natural resources, as well as increasing anthropogenic pressures (Sections 1.5, 4.3.2). High levels of impacts to coastal morphology, ecosystems and dependent human communities are detectable today and disproportionately higher risks are expected in the course of the 21st century (*medium evidence, high agreement*) (Sections 4.3.4, 5.3.7), even under a low emission pathway compatible with a 1.5°C global warming (Hoegh-Guldberg et al., 2018; IPCC, 2018). The magnitude of projected impacts (i.e., risks; Cross-Chapter Box 2 in Chapter 1) will depend on future greenhouse gas emissions and the associated climate changes, as well as on other drivers such as population movement into risk-prone areas and societal efforts to adapt.

LLIC include a wide diversity of systems (Figure CB9.1). Relevant regions occur on both islands and continents from the tropics to the poles, and support urban and rural societies from across the development spectrum (including SIDS and Least Developed Countries (LDCs)). LLIC host around 11% of the global population (Neumann et al., 2015), and generate about 14% of the global Gross Domestic Product (GDP) (Kummu et al., 2016). This integrative Cross-Chapter Box focuses on the array of challenges created by the melting of the cryosphere and the changing ocean, described throughout the report, to address societal risks, adaptation and the future habitability of LLIC.

---

[1] In this Report, the following summary terms are used to describe the available evidence: limited, medium, or robust; and for the degree of agreement: low, medium, or high. A level of confidence is expressed using five qualifiers: very low, low, medium, high, and very high, and typeset in italics, e.g., *medium confidence*. For a given evidence and agreement statement, different confidence levels can be assigned, but increasing levels of evidence and degrees of agreement are correlated with increasing confidence (see Section 1.9.2 and Figure 1.4 for more details).

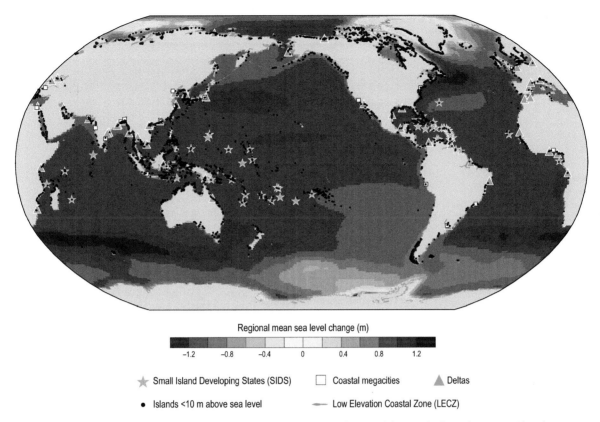

Regional mean sea level change (m)

−1.2   −0.8   −0.4   0   0.4   0.8   1.2

⭐ Small Island Developing States (SIDS)   ☐ Coastal megacities   ▲ Deltas

● Islands <10 m above sea level   —— Low Elevation Coastal Zone (LECZ)

Figure CB9.1 | The global distribution of low-lying islands and coasts (LLIC) particularly at risk from sea level rise. This map considers the Low Elevation Coastal Zone (elevation data from National Geophysical Data Center, 1999; LECZ, defined by McGranahan et al., 2007), islands with a maximum elevation of 10 m above sea level (Weigelt et al., 2013), Small Island Developing States (SIDS; UN-OHRLLS, n.d.), coastal megacities (cities with more than 10 million inhabitants, within 100 km from coast, and maximum 50 m above sea level; Pelling and Blackburn, 2013; UN-DESA, 2018) and deltas (Tessler et al., 2015). Regional sea level changes refer to projections under Representative Concentration Pathway (RCP)8.5 (2081–2100) (see Figure 4.8).

Drivers of Impacts and Risks

Climate-related hazards – LLIC are subject to the same climate-related hazards as other islands and coasts (overview in Wong et al., 2014), for both extreme events, for example, marine heat waves, tropical and extratropical storms, associated storm surges, and heavy precipitation; and slow onset changes, for example, retreat of glaciers and ice sheets, sea ice and permafrost thaw, sea level rise, and ocean warming and acidification (Sections 1.4, 2.2, 3.2 to 3.4, 4.2, 5.2, 6.2 to 6.6, Box 6.1). Table CB9.1 summarises the SROCC updates of these hazards, which often combine to explain part of observed climate impacts and projected risks. For example, accelerating sea level rise will combine with storm surges, tides and waves to generate to extreme sea level events that affect flooding (Section 4.3.3.2), shoreline changes (Section 4.3.3.3) and salinisation of soils, groundwater and surface waters (Section 4.3.3.4). Sea level rise will also combine with ocean warming to accelerate permafrost thawing in the Arctic (Sections 3.4.1.2, 3.4.2.2). Ocean acidification will combine with ocean warming and deoxygenation to impact benthic and pelagic organisms, associated ecosystems (e.g., coral reefs, oyster beds) and top predators, with subsequent impacts on species' abundance and distribution, and the ecosystem services benefiting human societies (Sections 4.3.3.5, 5.2.2, 5.3.1 to 5.3.6, 5.4.1, 6.4.2, 6.5.2, 6.6.2, 6.7.2, 6.8.2). Importantly, LLIC are at risk for multi-metre sea level rise projected post-2100 under Representative Concentration Pathway (RCP)8.5 and restricted to 1–2 m in 2300 under RCP2.6 (Section 4.2.3.5).

Table CB9.1 | Summary information on the critical climate-related drivers for low-lying islands and coasts, their trends due to climate change, and their main physical and ecosystem effects. Based on SROCC chapters and IPCC 5th Assessment Report (AR5). MSL is mean sea level, RCP is Representative Concentration Pathway, TC is tropical cyclone, ETC is extratropical cyclone, SLR is sea level rise, SST is sea surface temperature.

Climate-related driver	Physical/chemical effects	Observed trends	Projections	SROCC section
Global mean sea level (MSL)	Submergence, flood damage, erosion; saltwater intrusion; rising water tables/impeded drainage; ecosystem loss (and change)	Tide gauge records: *very likely* increase of 1.5 (1.1–1.9) mm yr^{-1} (1902–2010) and a total sea level rise of 0.16 (0.12–0.21) m Acceleration: with *high confidence* (–0.002–0.019) mm yr^{-2} over (1902–2010) Satellite altimetry: Global MSL of 3.0 mm yr^{-1} (2.4–3.6) over (1993–2015) Acceleration: with *high confidence* 0.084 (0.059–0.090) mm yr^{-2} over (1993–2015)	RCP2.6 (2046–2065): 0.24 (0.17–0.32) m RCP2.6 (2081–2100): 0.39 (0.26–0.53) m RCP2.6 (2100): 0.43 (0.29–0.59) m Rate of sea level rise (SLR) 4 (2–6) mm yr^{-1} in 2100 RCP4.5 (2046–2065): 0.26 (0.19–0.34) m RCP4.5 (2081–2100): 0.49 (0.34–0.64) m RCP4.5. (2100): 0.55 (0.39–0.72) m Rate of SLR 7 (4–9) mm yr^{-1} in 2100 RCP8.5 (2046–2065): 0.32 (0.23–0.40) m RCP8.5 (2081–2100): 0.71 (0.51–0.92) m RCP8.5 (2100): 0.84 (0.61–1.10) m Rate of SLR 15 (10–20) mm yr^{-1} in 2100	4.2.2.2 4.2.3.2
Regional sea level		Substantial regional variability at decadal at multi-decadal time scales due to changing winds, air-sea heat and freshwater fluxes and altered ocean circulation	Increased regional relative sea level with respect to AR5 nearly everywhere for RCP8.5 because of the increased Antarctic contribution (Figure 4.8)	4.2.2.4 4.2.3.2
Extreme sea levels		It is *very likely* that flood return period in low-lying areas has decreased over the past 20th century	*High confidence* in more frequently or yearly extreme sea level events which are currently rare (e.g., return period of 100 years) as a consequence of sea level rise at many locations for RCP8.5 by the end of the century (Figure 4.10) Even earlier and for RCP2.6 in locations where historical sea level variability (tides and storm surges) is small compared to projected sea level rise	4.2.3.4.1 4.2.3.4.3
Storms: tropical cyclones (TCs), extratropical cyclones (ETCs)	Storm surges and storm waves, coastal flooding, erosion; saltwater intrusion; rising water tables/impeded drainage; wetland loss (and change); coastal infrastructure damage and flood defense failure	TCs: Decreasing frequency of severe TCs in eastern Australia since the late 1800s; increase in frequency of moderately large US storm surge events since 1923; recent increase of extremely severe cyclonic storms over the Arabian Sea and intense TCs that make landfall in East and Southeast Asia in recent decades; increase in annual global proportion of hurricanes reaching Category 4 or 5 intensity in recent decades ETCs: *likely* poleward movement of circulation features but *low confidence* in intensity changes (AR5)	TCs: SLR will lead to higher storm surge levels for the TCs that do occur, assuming all other factors are unchanged (*high confidence*). *Medium confidence* that the proportion of TCs that reach Category 4 or 5 levels will increase, that the average intensity of TCs will increase (by roughly 1–10%, assuming a 2°C global temperature rise), and that average tropical cyclone precipitation rates (for a given storm) will increase by at least 7% per degree Celsius (SST) warming. *Low confidence* in how global TC frequency will change, although most studies project some decrease in global TC frequency ETCs: *Low confidence* in future changes in blocking and storm tracks in the northern hemisphere. The storm track projections for the southern hemisphere indicate an observed poleward contraction and a continued strengthening and southward contraction of storm tracks in the future (*medium confidence*)	6.3.1.1 6.3.1.2

Climate-related driver	Physical/chemical effects	Observed trends	Projections	SROCC section
Waves	Coastal erosion, overtopping and coastal flooding	Small increases in significant wave height globally and larger increases (5%) in extreme wave height, especially in the Southern Ocean (*medium confidence*). Global wave power has increased over the last six decades with marked spatial changes by oceans and long-term correlations with sea surface temperature (*low confidence*)	*High confidence* for projected increase of the mean significant wave height across the Southern Ocean, tropical eastern Pacific and Baltic Sea and for projected decrease of significant wave height over the North Atlantic and Mediterranean Sea. *Low confidence* in projections of significant wave height over the eastern north Pacific and Southern Indian and Atlantic Oceans. *Low confidence* in projected extreme significant wave height everywhere, except for the Southern Ocean (increase) and North Atlantic (decrease) (*high confidence*). Limited knowledge on projected wave period and direction	4.2.3.4.2 6.3.1.3
Sea surface temperature (SST)	Changes to stratification and circulation; reduced incidence of sea ice at higher latitudes; increased coral bleaching and mortality, poleward species migration; increased algal blooms	The ocean has warmed unabated, continuing the clear multi-decadal ocean warming trends documented in AR5. The 0–700 m layer of the ocean has warmed at rate of 5.31 ZJ yr^{-1} from 2005 to 2017. The long-term trend for 0–700 m layer has warmed 4.35 ZJ yr^{-1} from 1970 to 2017	For RCP8.5, the 0–2000 m layers of the ocean are projected to warm by a further 2150 ZJ (*very likely* range 1710 to 2790 ZJ) between 2017 and 2100. For RCP2.6, the 0–2000 m layers are projected to warm by 900 ZJ (*very likely* range 650 to 1340 ZJ) by 2100 (*) ZJ is Zettajoule	5.2.2.2.1
Marine heat waves		Have *very likely* doubled since 1980s	*Very likely* increase in frequency, duration, spatial extent and intensity, even under future low levels of warming	6.4.1
Freshwater inputs	Altered flood risk in coastal lowlands; altered water quality/salinity; altered fluvial sediment supply; altered circulation and nutrient supply	*Medium confidence* in a net declining trend in annual volume of freshwater input	*Medium confidence* for general increase in high latitude and wet tropics and decrease in other tropical regions	AR5
Ocean acidity	Increased CO$_2$ fertilization; decreased seawater pH and carbonate ion concentration (or 'ocean acidification')	*Virtually certain* that ocean surface water pH is declining by a *very likely* range 0.017 to 0.027 pH units per decade, since 1980, everywhere individual time-series observations exist	*High confidence* that the ocean will experience pH drops of between 0.1 (RCP2.6) or 0.3 (RCP8.5) pH units by 2100, with regional and local variability, exacerbated in polar regions	5.2.2.3
Sea ice and permafrost thaw	More storm surges, increasing ocean swells, coastal erosion	Permafrost temperatures have continued to increase to record high levels (*very high confidence*) Between 2007 and 2016, permafrost temperatures here increased 0.39°C ± 0.15°C in cold continuous zone permafrost and 0.20°C ± 0.10°C in warmer discontinuous zone permafrost. It is *very likely* that Arctic sea ice extent continues to decline in all months of the year; the strongest reductions in September (–12.8% ± 2.3% per decade; 1979–2018) are *likely* unprecedented in at least 1000 years. It is *virtually certain* that Arctic sea ice has thinned concurrent with a shift to younger ice: since 1979, the areal proportion of thick ice at least 5 years old has declined by approximately 90%	For stabilised global warming of 1.5°C, an approximately 1% chance of a given September being sea ice free at the end of century is projected; for stabilised warming at a 2°C increase, this rises to 10–35% (*high confidence*). The potential for reduced (further 5–10%) but stabilised Arctic autumn and spring snow extent by mid-century for RCP2.6 contrasts with continued loss under RCP8.5 (a further 15–25% reduction to end of century) (*high confidence*). Widespread disappearance of Arctic near-surface permafrost is projected to occur this century as a result of warming (*high confidence*). Near-surface permafrost area is projected to be reduced by 2–66% for RCP2.6 and 30–99% by 2100 under RCP8.5	3.2.1.1 Box 3.2 3.2.2 3.3.2 3.4.1 3.4.2

Anthropogenic drivers – Human factors play a major role in shaping exposure and vulnerability to climate-related changes in the Arctic, in temperate and tropical small islands, and in coastal urban areas (Sections 2.5.2, 4.3.2, Cross-Chapter Box 2 in Chapter 1). In the absence of major additional adaptation efforts compared to today (i.e., neither further significant action nor new types of actions), the anthropogenic drivers' contribution to climate change related risk will substantially increase (*high confidence*) (Section 4.3.4.2).

Highly context-specific territorial and societal dynamics have resulted in major changes at the coast, for instance the growing concentration of people and assets in risk prone coastal areas (Section 4.3.2.2), and the degradation of coastal ecosystem services such as coastal protection and healthy conditions for coastal fisheries and aquaculture (Section 4.3.2.3, 5.4.1.3, 5.4.2.2.2). Local drivers of exposure and vulnerability include, for example, coastal squeeze, inadequate land use planning, changes in construction modes, sand mining and unsustainable resource extraction (e.g., in the Comoros; Betzold and Mohamed, 2016; Ratter et al., 2016), as well as loss of Indigenous Knowledge and Local Knowledge (IK and LK; Cross-Chapter Box 4 in Chapter 1). For example, the loss of IK and LK-based practices and associated cultural heritage limits both the ability to recognise and respond to ocean and cryosphere related risk and the empowerment of local communities (*high confidence*) (Section 4.3.2.4.2). Population growth in medium-to-mega coastal cities is also of concern. For the year 2000, the Low Elevation Coastal Zones (LECZ, highest elevation up to 10 m above sea level) were estimated to host around 625 million people (Lichter et al., 2011; Neumann et al., 2015), with the vast majority (517 million) living in non-developed contexts. By 2100, the LECZ population may increase to as much as 1.14 billion under a Shared Socioeconomic Pathway (SSP) where countries focus on domestic, or even regional issues (SSP3; Jones and O'Neill, 2016). Poor planning can combine with coastal population growth and climate-related ocean change to create maladaptation (Juhola et al., 2016; Magnan et al., 2016).

Local factors drive – as well as are driven by – more regional processes such as extensive coastal urbanisation, human-induced sediment starvation (and implications on subsidence), degradation of vegetated coastal ecosystems (e.g., mangroves, coral reefs and salt-marshes), lack of long-term integrated planning, changing consumption modes, conflicting resource use and socioeconomic inequalities (*high confidence*), among others. These are vehicles of increasing exposure and vulnerability at multiple scales.

Observed and Projected Impacts on Geographies and Major Sectors
Coastal cities and megacities – Coastal cities, especially megacities with over 10 million inhabitants, are at serious risk from climate-related ocean and cryosphere changes (Abadie, 2018). Over half of today's global population lives in cities and megacities, many of which are located in LLIC, including New York City, Tokyo, Jakarta, Mumbai, Shanghai, Lagos and Cairo (Figure CB9.1). Without substantial adaptation interventions, and based on the compounding effects of future growth in population and assets, sea level rise and continued subsidence, future flood losses in the 136 largest coastal cities are projected to rise from 6 billion USD yr^{-1} at present to 1 trillion USD yr^{-1} in 2050 (Hallegatte et al., 2013; Sections 4.3.3.2 and 6.3.3). In addition to important impacts on coastal megacities and large port cities, small and mid-sized cities are also considered highly vulnerable because of fast growth rates and low political, human and financial capacities for risk reduction compared to larger cities (Birkmann et al., 2016; Box 4.2).

At a more local scale, and regardless of the size of the city, coastal property values and development will be affected by sea level changes, storms and other weather and climate-related hazards. Real estate values, and the cost and availability of insurance, will be impacted by actual and perceived flood risks (McNamara and Keeler, 2013; Section 5.4.2.3.1; Putra et al., 2015). Properties are also at risk of losing value due to coastal landscape degradation (McNamara and Keeler, 2013; Fu et al., 2016) and increasing risk aversion. The economic consequences manifest in declining rental incomes, business activities and local employment (Rubin and Hilton, 1996).

Coastal megacities are especially critical nodes for transboundary risks (Atteridge and Remling, 2018; Miller et al., 2018) as they contribute substantially to national economies and serve as a hub for global trade and transportation networks. The 2011 floods in Bangkok, for example, not only resulted in direct losses of 46.5 billion USD (World Bank, 2012; Haraguchi and Lall, 2015), but also in important effects on supply chains across the globe (Abe and Ye, 2013). Urbanisation could, however, also provide opportunities for risk reduction, given that cities are centres of innovation, political attention and private sector investments (Garschagen and Romero-Lankao, 2015).

Small islands – The extreme events occurring today, such as storms, tropical cyclones (TC), droughts, floods and marine heat waves (Herring et al., 2017), provide striking illustrations of the vulnerability of small island systems (*high confidence*) (Section 6.8.5, Box 4.2, Box 6.1). Societal dimensions can combine with climate changes, e.g., sea level rise, to amplify the impact of TCs, storm surge and ocean acidification in small islands contributing to loss and damage (Moser and Hart, 2015; Noy and Edmonds, 2016). For example, Category 5 TC Pam devastated Vanuatu in 2015 with 449.4 million USD in losses for an economy with a GDP of 758 million USD (Government of Vanuatu, 2015; Handmer and Iveson, 2017). Kiribati, Papua New Guinea, Solomon Islands and Tuvalu were all impacted by the TC Pam system (IFRC, 2018). In 2016, TC Winston caused 43 deaths in Fiji and losses of more than one third of the GDP (Government of Fiji, 2016; Cox et al., 2018). In 2017, Hurricanes Maria and Irma swept through 15 Caribbean countries, causing major damages and casualties across numerous islands. Rebuilding in three countries alone – Dominica, Barbuda and the British Virgin Islands – will cost an estimated 5 billion USD (UNDP, 2017). The Post-Disaster Needs Assessment for Dominica concluded that hurricane Maria resulted in total damages amounting to 226% of 2016 GDP (The Government of the Commonwealth of Dominica, 2017). In 2018, Category 4 TC Gita struck the Pacific islands of Eua and Tongatapu, impacting 80% of the population of Tonga through destruction of buildings, crops and infrastructure, and resulting in 165 million USD of losses with a national GDP of 461 million USD (Government of Tonga, 2018). Effective early warning systems, in some Caribbean islands, have reduced the impact (WMO, 2018). Projected changes in extreme weather include increased intensity of TCs with increased wind speed and rainfall, together with reduced translational speed creating greater destruction from individual storms and counteracting the decreased frequency of occurrence (Sections 6.3 and 6.8).

SIDS are home to 65 million people (UN-OHRLLS, 2015). More than 80% of small island residents live near the coast where flooding and coastal erosion already pose serious problems (Nurse et al., 2014) and since the IPCC 5th Assessment Report (AR5) and the Special Report on Global Warming of 1.5°C (SR1.5), there is consensus on the increasing threats to island sustainability in terms of land, soils and freshwater availability. As a result, there is growing concern that some island nations as a whole may become uninhabitable due to rising sea levels and climate change, with implications for relocation, sovereignty and statehood (Burkett, 2011; Gerrard and Wannier, 2013; Yamamoto and Esteban, 2014; Donner, 2015). For example, at the island scale, recent studies (e.g., on Roi-Namur Island, Marshall Islands; Storlazzi et al., 2018) estimate some atoll islands to become uninhabitable before the middle of the 21st century due to the exacerbation of wave-driven flooding by sea level rise, compromising soil fertility and the integrity of freshwater lenses (Cheriton et al., 2016). The literature also discusses the future of atoll island shoreline. Atoll islands are not 'static landforms' (*high confidence*) and they experience both erosion (Section 4.3.3.3) and accretion of land. In the Solomon Islands, where rates of sea level rise exceed the global average at 7–10 mm yr^{-1} (Becker et al., 2012), a study of 33 reef islands showed five vegetated islands had disappeared and six islands were concerned with severe shoreline erosion (Albert et al., 2016). In Micronesia, a study showed the disappearance of several reef islands, severe erosion in leeward reef edge islands and coastal expansion in mangrove areas (Nunn et al., 2017). In Tuvalu, with sea level rise of ~15 cm between 1971 and 2014, small islands decreased in land area while larger populated islands maintained or increased land area with the exception of the remote island of Nanumea (Kench et al., 2018). Positive shoreline and surface area changes over the recent decades to century have been observed for atoll islands in the Pacific and Indian oceans (McLean and Kench, 2015; Albert et al., 2016; Kench et al., 2018; Duvat, 2019). Out of 709 islands studied, 73.1% had stable surface area, 15.5% increased and 11.4% decreased in size over the last 40–70 years (Duvat, 2019). It has, however, been argued that the capacity of some atoll islands to maintain their land area by naturally adjusting to sea level rise could be reduced in the coming decades (*low evidence, high agreement*). Indeed, the projected combination of higher rates of sea level rise (Sections 4.2.3.2, 4.2.3.3 and 4.2.3.5), increased wave energy (Albert et al., 2016; see also Section 6.3), changes in storm wave direction (Harley et al., 2017), as well as the impacts of ocean warming and acidification on the reef system (Quataert et al., 2015; Hoegh-Guldberg et al., 2018), is expected to shift the balance towards more frequent flooding and increased erosion (Sections 4.3.3, 5.3.3).

Deltas – In a context of natural subsidence exacerbated by high human disturbances to sediment supply, for example, due to fresh water exploitation or damming and land use change upstream from the coast (Kondolf et al., 2014), marine flooding is already affecting deltas around the world (Brown et al., 2018; Section 4.3.3.4, Box 4.1). An estimated 260,000 km^2 of delta area have been temporarily submerged over the 1990s–2000s (Syvitski et al., 2009; Wong et al., 2014). The recurrence of El Niño associated floods in the San Juan River delta, Colombia, led to the relocation of several villages, including El Choncho, San Juan de la Costa, Charambira and Togoroma (Correa and Gonzalez, 2000). The intrusion of saline or brackish water due to relative sea level rise in combination with storm surges and natural and human-induced subsidence, results in increasing residual salinity, as already reported in the Delaware Estuary, USA (Ross et al., 2015), in the Ebro Delta, Spain (Genua-Olmedo et al., 2016) and in the Mekong Delta, Vietnam (Smajgl et al., 2015; Gugliotta et al., 2017). This affects livelihoods, for example., freshwater fish habitat in Bangladesh (Dasgupta et al., 2017; Section 4.3.3.4.2). Increased salinity limits drinking water supply (Wilbers et al., 2014), with associated repercussions for the abundance and toxicity of cholera vibrio (*Vibrio cholerae*) as shown in the Ganges Delta (Batabyal et al., 2014). Local agriculture is also at risk. Oilseed, sugarcane and jute cultivation have already ceased due to high salinity levels in coastal Bangladesh (Khanom, 2016) and dry-season crops are projected to decline over the next 15 to 45 years, especially in the Southwest (Kabir et al., 2018). In the Ebro delta, Spain, Genua-Olmedo et al. (2016) anticipate a decrease of the rice production index from 61.2% in 2010 to 33.8% by 2100 for a 1.8 m sea level rise scenario, far above the upper end of the RCP8.5 *likely*[2] range (Section 4.2.3.2, Table 4.3).

Arctic coasts – Climate-related ocean and cryosphere changes combine to negatively impact not only the economy and life-styles of the Arctic coastal communities, but also the local cultural identity, self-sufficiency, IK and LK and related skills (Lacher, 2015; Sections 3.4.3, 4.3.2.4.2). Changes in fish and seabird populations amplified by climate change have an impact on ecosystems and livelihoods in Arctic island communities such as in Norway's Lofoten archipelago (Dannevig and Hovelsrud, 2016; Kaltenborn et al., 2017). Another concern relates to coastal erosion, for example triggered by permafrost thaw (Günther et al., 2013; Jones et al., 2018), and which already affects 178 Alaskan communities, with 26 in a very critical situation, such as Newtok, Shishmaref, Kivilina and northwestern coastal communities on the Chukchi Sea (Bronen and Chapin III, 2013). Noteworthy, erosion does not affect all Arctic coastlines: many of them are located in areas that experience rapid glacial-isostatic adjustment (GIA) uplift (James et al., 2015; Forbes et al., 2018) and have low sensitivity to extreme sea levels and sea level rise. An additional factor unique to the Arctic coasts compared to other LLIC is the decrease in seasonal sea ice extent (Section 3.2.1, 4.3.4.2.1), that both reduces the physical protection of the land (Overeem et al., 2011; Fang et al., 2018), for example, from wave action, and allows for greater open water fetch producing stronger wind-generated waves in the open water (Lantuit et al., 2011). In combination with a decreased stability of permafrost – another specificity of polar regions (Romanovsky et al., 2010) – and sea level rise, seasonal sea ice extent reduction results in shoreline erosion (Gibbs and Richmond, 2017; Jones et al., 2018), with associated impacts on coastal settlements (Table 3.4). However, as mentioned above, local geomorphology and geology in the Arctic is as important as permafrost and sea ice extent for determining current and future erosion (Lantuit et al., 2011).

Risks to Arctic coasts will be reinforced by anthropogenic drivers originating in the recent decades of history (e.g., socioeconomic adjustments after government policies requiring children to attend school) which resulted in the construction of infrastructure in near-shore areas. While risk levels vary by village, in several cases infrastructure has been lost and subsistence use areas modified (Gorokhovich et al., 2013; Marino, 2015). More broadly, in the Arctic, 'indigenous peoples (…) have been pushed into marginalised territories that are more sensitive to climate impacts' (Ford et al., 2016: 350), with consequences in terms of undermining aspects of socio-cultural resilience.

Impacts on critical sectors and livelihoods – Economic impacts for LLIC are expected to be significant in the course of the century due to the convergence of the anticipated increase in the number of LECZ inhabitants (Jones and O'Neill, 2016; Merkens et al., 2016), the high dependency of societies on ocean and marine ecosystems and services (Section 5.4.1, 5.4.2), and increased detrimental effects of climate-related ocean and cryosphere changes on natural and human systems (*medium evidence, high agreement*) (Hsiang et al., 2017; United Nations, 2017). However, the degree of impacts on the economy and related dimensions – for example, on employment, livelihood, poverty, health (Kim et al., 2014; Weatherdon et al., 2016), well-being and food security (Sections 1.1 and 5.4.2, FAQ 1.2 in Chapter 1) and public budgets and investments – will vary across context-specific physical settings and exposure and vulnerability levels.

Considering a sea level rise scenario range of 25–123 cm – all RCPs; wider range of sea level rise scenarios than the *likely* range of AR5 but relatively consistent with the range of projections assessed in this report (Section 4.2.3.2) – and no adaptation, Hinkel et al. (2014) estimated annual losses from future marine flooding to amount to 0.3–9.3% of global GDP in 2100. Noteworthy, coastal protection will inevitably have economic costs (DiSegni and Shechter, 2013), whether it involves hard coastal protection (Hinkel et al., 2018), ecosystem-based approaches (Narayan et al., 2016; Pontee et al., 2016) or a combination of both (Schoonees et al., 2019). Coastal agriculture (e.g., rice crops; Smajgl et al., 2015; Genua-Olmedo et al., 2016), and fisheries and aquaculture will also be seriously impacted (Sections 4.3.3.6.1, 4.3.3.6.3, 5.4.1). For example, it is expected that the marine fisheries revenues of 89% of the world's fishing countries will be negatively affected by mid-century under RCP8.5 (Hilmi et al., 2015). The fact that more than 90% of the world's rural poor are located in the LECZ of 15 developing countries (Barbier, 2015) and that these regions are highly dependent on fish for their dietary consumption, raises a serious concern about future food security (FAO et al., 2017; Section 5.4.2.1.2). But not all regions are equally threatened,

---

[2]   In this Report, the following terms have been used to indicate the assessed likelihood of an outcome or a result: Virtually certain 99–100% probability, Very likely 90–100%, Likely 66–100%, About as likely as not 33–66%, Unlikely 0–33%, Very unlikely 0–10%, and Exceptionally unlikely 0–1%. Additional terms (Extremely likely: 95–100%, More likely than not >50–100%, and Extremely unlikely 0–5%) may also be used when appropriate. Assessed likelihood is typeset in italics, e.g., *very likely* (see Section 1.9.2 and Figure 1.4 for more details). This Report also uses the term '*likely* range' to indicate that the assessed likelihood of an outcome lies within the 17–83% probability range.

with Lam et al. (2016) estimating that the impacts on fisheries will be more important in SIDS, Africa and Southeast Asia. Cascading effects are also expected from risks to coral reefs and associated living resources, both on direct consumption by local communities and through disturbances to the broader food web chains (Sections 5.4.2, 6.5 and Box 6.1).

Coastal tourism could be affected in various ways by ocean- and cryosphere-related changes (Hoegh-Guldberg et al., 2018; Sections 4.3.3.6.2, 5.4.2.1.3). Coastal infrastructure and facilities, such as harbours and resorts (e.g., in Ghana; Sagoe-Addy and Appeaning Addo, 2013), are prone to storm waves. For coral reefs for recreational activities and tourism (especially diving and snorkelling), Chen et al. (2015) estimated that the global economic impact of the expected decline in reef coverage (between 6.6 and 27.6% under RCPs 2.6 and 8.5, respectively) will range from 1.9 to 12.0 billion USD yr^{-1}. The future appeal of tourism destinations will partly depend on sea surface temperature, including induced effects such as an increase in invasive species, e.g., jellyfishes (Burge et al., 2014; Weatherdon et al., 2016) and lion fish in the Northwest Atlantic, the Gulf of Mexico and the Caribbean (Albins, 2015; Johnston et al., 2015; Holdschlag and Ratter, 2016). It will also depend on how tourists and tourism developers perceive the risks induced by ocean-related changes (e.g., Shakeela et al., 2013; Davidson and Sahli, 2015). This will combine with the influence of changes in climatic conditions in tourists' areas of origin (Bujosa and Rosselló, 2013; Amelung and Nicholls, 2014; Hoegh-Guldberg et al., 2018) and of non-climatic components such as accommodation and travel prices. Importantly, estimating the effects on global-to-local tourism flows remains challenging (Rosselló-Nadal, 2014; Wong et al., 2014).

Recent studies provide further empirical evidence that people are rarely moving exclusively due to changes in ocean- and cryosphere-based conditions, and that migration as a result of disasters and increasing hazards strongly interact with other drivers, especially economic and political motivations (*high confidence*) (Kelman, 2015; Marino and Lazrus, 2015; Hamilton et al., 2016; Bettini, 2017; Stojanov et al., 2017; Perumal, 2018). While significantly higher risks of human displacement are expected in low-income LLIC, for example in Guatemala (Milan and Ruano, 2014) and Myanmar (Brakenridge et al., 2017), the issue also concerns developed countries. For example, Logan et al. (2016) show that people temporarily or permanently displaced by hurricanes in the Gulf Coast, USA, create a significant economic burden to tourism-dependent coastal cities and harbours.

Responses: Adaptation Strategies in Practice

A wide range of coastal adaptation measures are currently implemented in LLIC worldwide (Sections 1.6.2, 2.3.7, 3.5.2, 3.5.3, 4.4.3, 5.5.2, 6.9, Figure 1.2, Box 5.4), including the installation of major infrastructure such as armouring of coasts (e.g., seawalls, groynes, revetments and rip-raps), soft engineering (e.g., beach nourishment and dune restoration), reclamation works to build new lands seaward and upwards, ecosystem-based measures (e.g., vegetation planting and coral farming), community-based approaches (e.g., social networks, education campaigns and economic diversification) and institutional innovations (e.g., marine protected areas and evacuation plans). The effectiveness of the measures to reduce risks depends on both local context-specificities (Gattuso et al., 2018) and the magnitude and timing of local climate impacts. However, there is still a gap in on-the-ground evidence, good practices and guidelines to evaluate the observed and projected benefits of each type of measures applied in various contexts, for example, to decide whether nature-based options represent low- to no-regret solutions, or not a solution at all.

Protection with hard coastal defences is commonly used to prevent inundation from extreme water levels and wave overtopping (Section 4.4.2.2). In environments such as megacities, adequately engineered hard coastal defences are considered to be successful options and an efficient adaptation option in the long run (Hinkel et al., 2018). However, such measures can also lead to detrimental effects, such as erosion exacerbation by seawalls reflecting wave energy and jetties disrupting cross-shore sediment transport. Adaptation labelled measures 'may [thus] lead to increased risk of adverse climate-related outcomes, increased vulnerability to climate change, or diminished welfare' (Noble et al., 2014: 857) and therefore be maladaptive (Barnett and O'Neill, 2013; Juhola et al., 2016; Magnan et al., 2016). As a result, alternatives have emerged, such as ecosystem-based design measures including coconut fibre blankets (David et al., 2016), plantations of seagrass (Paul and Gillis, 2015), artificial reefs made from bio-rock materials (Beetham et al., 2017; Goreau and Prong, 2017) and bamboo breakwaters (David et al., 2016). While restoration operations are often rather associated to conservation practices, they can have co-benefits in terms of coastal protection services (Section 4.3.2.3). For example, soft protection systems used in 69 studies were found to exhibit effectiveness in reducing wave heights at 70% for coral reefs, 62–79% for salt marshes, 36% for seagrass meadows and 31% for mangroves (Narayan et al., 2016). Arguing that coral reefs can provide comparably higher wave attenuation benefits to artificial defences such as breakwaters, Ferrario et al. (2014) conclude that reef defences for reducing coastal hazards can be enhanced cost effectively on the order of 1/10th. Coral reefs are, however, at very high risk from climate change (Hoegh-Guldberg et al., 2018; Section 5.3.4), which challenges the duration of such benefits.

CCB9

Ecosystem-based measures, if applied place specific and adequate – for example, use of indigenous rather than exotic species (e.g., Duvat et al., 2016) – , are usually considered low-regret in that they can stabilise the coastal vegetation and protect against coastal hazards, while at the same time enhancing the adaptive capacity of natural ecosystems (*medium evidence, high agreement*) (Schoonees et al., 2019; Sections 2.3.3.4, 5.5.2, 6.9).

While human migration and relocation are expected to be a growing challenge for LLIC (*medium evidence, high agreement*) (Adger et al., 2014; Birk and Rasmussen, 2014; Milan and Ruano, 2014; Thomas, 2015; Sections 3.5.3.5, 4.4.2.4, 6.3.4, Table 3.4; Hajra et al., 2017; Stojanov et al., 2017), recent studies advocate for considering these options as adaptation to climate-related changes in the ocean and cryosphere (Shayegh et al., 2016; Allgood and McNamara, 2017; Hauer, 2017; Morrison, 2017; Perumal, 2018; Section 4.4.2.4). Such a view is, however, convoyed by discussions on related costs and impacts on the wellbeing of the people who are relocated (Null and Herzer Risi, 2016). Coastal retreat is underway in various LLIC around the world, for example, in Alaska and the US (Bronen, 2015; Ford et al., 2015; Logan et al., 2016; Hino et al., 2017), Guatemala (Milan and Ruano, 2014), Western Colombia (Correa and Gonzalez, 2000), the Caribbean (Apgar et al., 2015; Rivera-Collazo et al., 2015) and Vietnam (Collins et al., 2017). Noteworthy, environmentally-induced relocation is not necessarily new, for example, in the Pacific (Nunn, 2014; Boege, 2016). The Gilbertese people from Kiribati moved to the Solomon Islands during the 1950s–1960s, as a result of long periodic droughts and subsequent environmental degradation (Birk and Rasmussen, 2014; Albert et al., 2016; Tabe, 2016; Weber, 2016). In the Solomon Islands, the relocation of the Taro Township (Choiseul Province) as a result of rising sea level and coastal erosion is already underway (Haines and McGuire, 2014; Haines, 2016). In Fiji, the relocation of Vunidogoloa village as a result of sea level rise and coastal erosion was successfully carried out in 2014 (McNamara and Des Combes, 2015). In Alaska, some communities (e.g., Newtok) responded to changing environmental and livelihood conditions due to permafrost thaw with self-initiated relocation efforts. Subsequently, Alaska state funding has been allocated to assist them (Bronen, 2015; Hamilton et al., 2016). Conflict escalation is a serious concern in the resettlement areas, between newcomers and locals, or between different groups of newcomers, particularly under conditions of land scarcity, high population density and (perceived) inequality (Connell and Lutkehaus, 2017; Boege, 2018). The obstacles thus extend well beyond the cost of relocation itself because of the multi-dimensional impacts on people's lives. Relocation also concerns economic activities, as illustrated with shellfish aquaculture relocation in the west coast of the US due to ocean acidification-driven crises (Cooley et al., 2016).

For all interventions, adaptation is fully recognised as being a societal challenge, and not merely a question of technological solutions (*medium evidence, high agreement*) (Jones and Clark, 2014; McCubbin et al., 2015; Gerkensmeier and Ratter, 2018). Enhancing adaptation implies various sociopolitical and economic framings, coping capacities and cross-scale social and economic impacts (Sections 4.4.3, 4.4.5, Cross-Chapter Box 3 in Chapter 1). As a result, community-based decision making, sustainable spatial planning and new institutional arrangements gain increasing attention (Sections 4.4.4). Such approaches can involve working with local informal and formal institutions (Barron et al., 2012), enhancing risk ownership by communities through participative approaches (McEwen et al., 2017), establishing collaborative community networks (Hernández-González et al., 2016), and better integrating LLIC communities' IK and LK (see McMillen et al., 2014; Cross-Chapter Box 4 in Chapter 1). Small island communities, in particular, can strengthen their adaptive capacities by building on relatively high degrees of social capital, that is, dense social networks, collective action, reciprocity and relations of trust (Petzold and Ratter, 2015; Barnett and Waters, 2016; Petzold, 2016; Kelman, 2017; Section 4.3.2.4.3). The aim of all these approaches is both to facilitate the effective implementation of adaptive action, and create widespread acceptance of adaptation policies by stakeholders and local populations.

Participatory scenario building processes, collaborative landscape planning and co-design of ecosystem-based management for LLIC resilience are underway along with promising approaches to actively engage all levels of society in the exploration of future adaptation scenarios. Experiences are reported for the German North Sea coast (Karrasch et al., 2017), Tenerife Island in the Atlantic Ocean (Hernández-González et al., 2016) and Pacific island communities (Burnside-Lawry et al., 2017). While adaptation labelled measures currently applied 'on the ground' are mainly reactive and short-term, long-term approaches are emerging (Noble et al., 2014; Wong et al., 2014), as illustrated by the development of 'adaptation pathways' – that is, long-term adaptation strategies based upon decision cycles that, over time, explore and sequence a set of possible actions based on alternative external, uncertain developments (Haasnoot et al., 2013; Barnett et al., 2014; Wise et al., 2014; Werners et al., 2015; Hermans et al., 2017; Section 4.4.4.3.4). Key expected benefits are an improved consideration of both the evolving nature of vulnerability (Denton et al., 2014; Dilling et al., 2015; Duvat et al., 2017; Fawcett et al., 2017) and climate change uncertainty (O'Brien et al., 2012; Brown et al., 2014; Noble et al., 2014), as well as better anticipation of the risks of maladaptation (Magnan et al., 2016). Practical applications of adaptation pathways in LLIC are occurring, for example, in the Netherlands (Haasnoot et al., 2013), Indonesia (Butler et al., 2014), New York City (Rosenzweig and Solecki, 2014) and Singapore (Buurman and Babovic, 2017).

Conclusions

LLIC are particularly at risk from climate-related changes to the ocean and the cryosphere, whether they are urban or rural, continental or island, at any latitude and regardless of level of development (*high confidence*). Over the course of the 21st century, they are expected to experience both increasing risks (*high confidence*) and limits to ecological and societal adaptation (de Coninck et al., 2018; Djalante et al., 2018; Section 4.3.4.2, Figure 6.2, Figure CB9.2; Hoegh-Guldberg et al., 2018), which has the potential to significantly increase the level of loss and damage experienced by local coastal livelihoods (e.g., fishing, logistics or tourism) (Djalante et al., 2018). However, there are still important research gaps on residual risks and adaptation limits, given that these limits can be reached due to the intensity of the hazards and/or to the high vulnerability of a given system, and can be ecological, technological, economic, social, cultural, political or institutional. In addition, ocean and cryosphere changes have the potential to accumulate in compound events and cause cascades of impacts through economic, environmental and social processes (*medium evidence, high agreement*) (Sections 6.8.2 to 6.8.3, Box 6.1). This is the case when coastal flooding and riverine inundation occur together, for example, during the 2012 Superstorm Sandy in New York City, USA (Rosenzweig and Solecki, 2014); the 2014 cyclone Bejisa in Reunion Island, France (Duvat et al., 2016), and the 2017 Hurricane Harvey in Houston, USA (Emanuel, 2017). Cascade effects far beyond the extent of the original impacts bring the risk in LLIC of slowing down and reversing overall development achievements, particularly on poverty reduction (*low evidence, medium agreement*) (Hallegatte et al., 2016). Global time series analysis of risk and vulnerability trends show that many Pacific island states have fallen behind the global average in terms of progress made in the reduction of social vulnerability towards natural hazards over the past years (Feldmeyer et al., 2017). These findings may well be indicative of the situation for other LLIC (*medium confidence*) (Hay et al., 2019).

In addition, LLIC provide relevant illustrations of some of the IPCC Reasons for Concern (RFC) that describe potentially dangerous anthropogenic interference with the climate system (IPCC, 2014; IPCC, 2018). LLIC especially illustrate the risks to unique and threatened systems (RFC1), and risks associated with extreme weather and compound events (RFC2), and the uneven distribution of impacts (RFC3). Using this frame, O'Neill et al. (2017) estimate, for example, that the potential for coastal

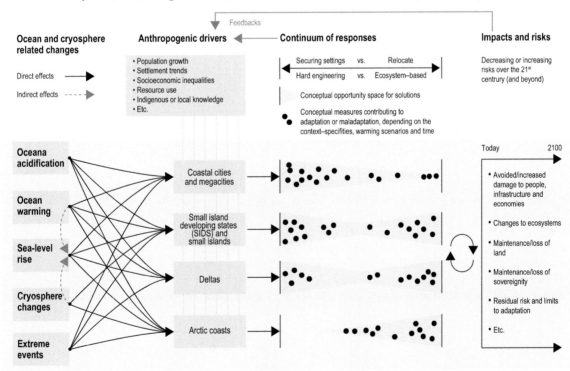

Figure CB9.2 | The storyline of risk for Low-Lying Islands and Coasts (LLIC). From left to right, this figure shows that ocean- and cryosphere-related changes (ocean acidification, ocean warming, sea level rise, etc.) will combine with anthropogenic drivers (population growth, settlement trends, socioeconomic inequalities, etc.) to explain impacts on various LLIC geographies (cities, islands, deltas, Arctic coasts). Depending on the combinations of responses (black dots; stylised representation of potential responses) along a continuum going from hard engineering to ecosystem-based approaches, and from securing current settings to relocation (light blue triangles), risks will increase or decrease in the coming decades. Some responses (black dots) will enhance either adaptation or maladaptation. SIDS is Small Island Developing States.

protection and ecosystem-based adaptation will reach significant limits by 2100 in the case of a 1 m rise in sea level, suggesting the need for research into the crossing of environmental and/or anthropogenic tipping points (Sections 6.2). The SROCC report confirms that high risk to various geographies (Arctic communities remote from regions of rapid positive glacial-isostatic adjustment, megacities, urban atoll islands and large tropical agricultural deltas) are to be expected before a 1 m rise in global mean sea level (Section 4.3.4.2.1). More broadly, this report suggests, first, that the drivers and timing of the future habitability of LLIC will vary from one case to another (Manley et al., 2016; Hay et al., 2018). Second, future storylines of risks will also critically depend on the multi-decadal effectiveness of coastal nations' and communities' responses (*medium evidence, high agreement*). This will, in turn, partly depend on transformation of risk management regimes in order to harness these potentials and shift course towards climate-resilient development pathways (*low evidence, high agreement*) (Solecki et al., 2017).

References

Abadie, L.M., 2018: Sea level damage risk with probabilistic weighting of IPCC scenarios: An application to major coastal cities. *Journal of Cleaner Production*, **175**, 582–598, doi:10.1016/j.jclepro.2017.11.069.

Abe, M. and L. Ye, 2013: Building Resilient Supply Chains against Natural Disasters: The Cases of Japan and Thailand. *Global Business Review*, **14** (4), 567–586, doi:10.1177/0972150913501606.

Adger, W.N. et al., 2014: Human security. In: *Climate Change 2014: Impacts, Adaptation, and Vulnerability. Part A: Global and Sectoral Aspects. Contribution of Working Group II to the Fifth Assessment Report of the Intergovernmental Panel of Climate Change* [Field, C.B., V.R. Barros, D.J. Dokken, K.J. Mach, M.D. Mastrandrea, T.E. Bilir, M. Chatterjee, K.L. Ebi, Y.O. Estrada, R.C. Genova, B. Girma, E.S. Kissel, A.N. Levy, S. MacCracken, P.R. Mastrandrea and L.L. White (eds.)]. Cambridge University Press, Cambridge, United Kingdom and New York, NY, USA, 755–791.

Albert, S. et al., 2016: Interactions between sea-level rise and wave exposure on reef island dynamics in the Solomon Islands. *Environmental Research Letters*, **11** (5), 054011.

Albins, M.A., 2015: Invasive Pacific lionfish Pterois volitans reduce abundance and species richness of native Bahamian coral-reef fishes. *Marine Ecology Progress Series*, **522**, 231–243, doi:10.3354/meps11159.

Allgood, L. and K.E. McNamara, 2017: Climate-induced migration: Exploring local perspectives in Kiribati. *Singapore Journal of Tropical Geography*, **38** (3), 370–385, doi:10.1111/sjtg.12202.

Amelung, B. and S. Nicholls, 2014: Implications of climate change for tourism in Australia. *Tourism Management*, **41**, 228–244, doi:10.1016/j.tourman.2013.10.002.

Apgar, M.J., W. Allen, K. Moore and J. Ataria, 2015: Understanding adaptation and transformation through indigenous practice: the case of the Guna of Panama. *Ecology and Society*, **20** (1), doi:10.5751/es-07314-200145.

Atteridge, A. and E. Remling, 2018: Is adaptation reducing vulnerability or redistributing it? *Wiley Interdisciplinary Reviews: Climate Change*, **9** (1), 1–16, doi:10.1002/wcc.500.

Barbier, E.B., 2015: Climate change impacts on rural poverty in low-elevation coastal zones. *Estuarine, Coastal and Shelf Science*, **165**, A1-A13, doi:10.1016/j.ecss.2015.05.035.

Barnett, J. et al., 2014: A local coastal adaptation pathway. *Nature Climate Change*, **4**, 1103–1108, doi:10.1038/nclimate2383.

Barnett, J. and S. O'Neill, 2013: Minimising the risk of maladaptation: a framework for analysis. In: *Climate Adaptation Futures* [Palutikof, J., S. Boulter, A. Ash, M. Stafford-Smith, M. Parry, M. Waschka and D. Guitart (eds.)]. Wiley-Blackwell, Chichester, 87–94, ISBN 9780470674963.

Barnett, J. and E. Waters, 2016: Rethinking the Vulnerability of Small Island States: Climate Change and Development in the Pacific Islands. In: *The Palgrave Handbook of International Development* [Grugel, J. and D. Hammett (eds.)]. Palgrave Macmillan UK, London, 731–748, ISBN 9781137427236.

Barron, S. et al., 2012: A Climate Change Adaptation Planning Process for Low-Lying, Communities Vulnerable to Sea Level Rise. *Sustainability*, **4** (9), 2176–2208, doi:10.3390/su4092176.

Batabyal, P. et al., 2014: Influence of hydrologic and anthropogenic factors on the abundance variability of enteropathogens in the Ganges estuary, a cholera endemic region. *Science of the Total Environment*, **472**, 154–61, doi:10.1016/j.scitotenv.2013.10.093.

Becker, M. et al., 2012: Sea level variations at tropical Pacific islands since 1950. *Global and Planetary Change*, **80–81**, 85–98, doi:10.1016/j.gloplacha.2011.09.004.

Beetham, E., P.S. Kench and S. Popinet, 2017: Future Reef Growth Can Mitigate Physical Impacts of Sea-Level Rise on Atoll Islands. *Earth's Future*, **5** (10), 1002–1014, doi:10.1002/2017ef000589.

Bettini, G., 2017: Where Next? Climate Change, Migration, and the (Bio) politics of Adaptation. *Global Policy*, **8** (S1), 33–39, doi:10.1111/1758-5899.12404.

Betzold, C. and I. Mohamed, 2016: Seawalls as a response to coastal erosion and flooding: a case study from Grande Comore, Comoros (West Indian Ocean). *Regional Environmental Change*, **17** (4), 1077–1087, doi:10.1007/s10113-016-1044-x.

Birk, T. and K. Rasmussen, 2014: Migration from atolls as climate change adaptation: Current practices, barriers and options in Solomon Islands. *Natural Resources Forum*, **38** (1), 1–13, doi:10.1111/1477-8947.12038.

Birkmann, J. et al., 2016: Boost resilience of small and mid-sized cities. *Nature*, **537** (7622), 605–8, doi:10.1038/537605a.

Boege, V., 2016: Climate Change and Planned Relocation in Oceania. *S&F Sicherheit und Frieden*, **34** (1), 60–65, doi:10.5771/0175-274X-2016-1-60.

Boege, V. 2018: *Climate change and conflict in Oceania – Challenges, responses and suggestions for a policy-relevant research agenda (Toda Peace Institute Policy Brief 17)*. Toda Peace Institute, Tokyo [Available at: http://toda.org/files/policy_briefs/T-PB-17_Volker%20Boege_Climate%20Change%20and%20Conflict%20in%20Oceania.pdf, accessed 18/10/2018].

Brakenridge, G.R. et al., 2017: Design with nature: Causation and avoidance of catastrophic flooding, Myanmar. *Earth-Science Reviews*, **165**, 81–109, doi:10.1016/j.earscirev.2016.12.009.

Bronen, R., 2015: Climate-induced community relocations: using integrated social-ecological assessments to foster adaptation and resilience. *Ecology and Society*, **20** (3), doi:10.5751/ES-07801-200336.

Bronen, R. and F.S. Chapin III, 2013: Adaptive governance and institutional strategies for climate-induced community relocations in Alaska. *Proceedings of the National Academy of Sciences*, **110** (23), 9320–5, doi:10.1073/pnas.1210508110.

Brown, S. et al., 2014: Shifting perspectives on coastal impacts and adaptation. *Nature Climate Change*, **4**, 752–755, doi:10.1038/nclimate2344.

Brown, S. et al., 2018: What are the implications of sea-level rise for a 1.5, 2 and 3 °C rise in global mean temperatures in the Ganges-Brahmaputra-Meghna and other vulnerable deltas? *Regional Environmental Change*, **18** (6), 1829–1842, doi:10.1007/s10113-018-1311-0.

Bujosa, A. and J. Rosselló, 2013: Climate change and summer mass tourism: the case of Spanish domestic tourism. *Climatic Change*, **117** (1), 363–375, doi:10.1007/s10584-012-0554-x.

Burge, C.A. et al., 2014: Climate change influences on marine infectious diseases: implications for management and society. *Annual Review of Marine Science*, **6**, 249–77, doi:10.1146/annurev-marine-010213-135029.

Burkett, M., 2011: The nation ex-situ: On climate change, deterritorialized nationhood and the postclimate era. *Climate Law*, **2**, 345–374.

Burnside-Lawry, J. et al., 2017: Communication, Collaboration and Advocacy: A Study of Participatory Action Research to Address Climate Change in the Pacific. *The International Journal of Climate Change: Impacts and Responses*, **9** (4), 11–33, doi:10.18848/1835-7156/CGP/v09i04/11-33.

Butler, J.R.A. et al., 2014: Framing the application of adaptation pathways for rural livelihoods and global change in eastern Indonesian islands. *Global Environmental Change*, **28**, 368–382, doi:10.1016/j.gloenvcha.2013.12.004.

Buurman, J. and V. Babovic, 2017: Adaptation Pathways and Real Options Analysis: An approach to deep uncertainty in climate change adaptation policies. *Policy and Society*, **35** (2), 137–150, doi:10.1016/j.polsoc.2016.05.002.

Chen, P.-Y., C.-C. Chen, L. Chu and B. McCarl, 2015: Evaluating the economic damage of climate change on global coral reefs. *Global Environmental Change*, **30**, 12–20, doi:10.1016/j.gloenvcha.2014.10.011.

CCB9

Cheriton, O.M., C.D. Storlazzi and K.J. Rosenberger, 2016: Observations of wave transformation over a fringing coral reef and the importance of low-frequency waves and offshore water levels to runup, overwash, and coastal flooding. *Journal of Geophysical Research: Oceans*, **121** (5), 3121–3140, doi:10.1002/2015jc011231.

Collins, N., S. Jones, T.H. Nguyen and P. Stanton, 2017: The contribution of human capital to a holistic response to climate change: learning from and for the Mekong Delta, Vietnam. *Asia Pacific Business Review*, **23** (2), 230–242, doi:10.1080/13602381.2017.1299449.

Connell, J. and N. Lutkehaus, 2017: Environmental Refugees? A tale of two resettlement projects in coastal Papua New Guinea. *Australian Geographer*, **48** (1), 79–95, doi:10.1080/00049182.2016.1267603.

Cooley, S.R., C.R. Ono, S. Melcer and J. Roberson, 2016: Community-Level Actions that Can Address Ocean Acidification. *Frontiers in Marine Science*, **2** (128), 1–12, doi:10.3389/fmars.2015.00128.

Correa, I.D. and J.L. Gonzalez, 2000: Coastal erosion and village relocation: a Colombian case study. *Ocean & Coastal Management*, **43** (1), 51–64, doi:10.1016/S0964-5691(99)00066-6.

Cox, J. et al., 2018: Disaster, Divine Judgment, and Original Sin: Christian Interpretations of Tropical Cyclone Winston and Climate Change in Fiji. *The Contemporary Pacific*, **30** (2), 380–410, doi:10.1353/cp.2018.0032.

Dannevig, H. and G.K. Hovelsrud, 2016: Understanding the need for adaptation in a natural resource dependent community in Northern Norway: issue salience, knowledge and values. *Climatic Change*, **135** (2), 261–275, doi:10.1007/s10584-015-1557-1.

Dasgupta, S., I. Sobhan and D. Wheeler, 2017: The impact of climate change and aquatic salinization on mangrove species in the Bangladesh Sundarbans. *Ambio*, **46** (6), 680–694, doi:10.1007/s13280-017-0911-0.

David, C.G., N. Schulz and T. Schlurmann, 2016: Assessing the Application Potential of Selected Ecosystem-Based, Low-Regret Coastal Protection Measures. In: *Ecosystem-Based Disaster Risk Reduction and Adaptation in Practice* [Renaud, F., K. Sudmeier-Rieux, M. Estrella and U. Nehren (eds.)]. Springer International Publishing, Cham, 457–482, ISBN 9783319436319.

Davidson, L. and M. Sahli, 2015: Foreign direct investment in tourism, poverty alleviation, and sustainable development: a review of the Gambian hotel sector. *Journal of Sustainable Tourism*, **23** (2), 167–187, doi:10.1080/09669582.2014.957210.

de Coninck, H. et al., 2018: Strengthening and implementing the global response. In: *Global Warming of 1.5°C. An IPCC Special Report on the impacts of global warming of 1.5°C above pre-industrial levels and related global greenhouse gas emission pathways, in the context of strengthening the global response to the threat of climate change, sustainable development, and efforts to eradicate poverty* [Masson-Delmotte, V., P. Zhai, H.O. Pörtner, D. Roberts, J. Skea, P.R. Shukla, A. Pirani, W. Moufouma-Okia, C. Péan, R. Pidcock, S. Connors, J.B.R. Matthews, Y. Chen, X. Zhou, M.I. Gomis, E. Lonnoy, T. Maycock, M. Tignor and T. Waterfield (eds.)]. In press.

Denton, F. et al., 2014: Climate-resilient pathways: adaptation, mitigation, and sustainable development. In: *Climate Change 2014: Impacts, Adaptation, and Vulnerability. Part A: Global and Sectoral Aspects. Contribution of Working Group II to the Fifth Assessment Report of the Intergovernmental Panel of Climate Change* [Field, C.B., V.R. Barros, D.J. Dokken, K.J. Mach, M.D. Mastrandrea, T.E. Bilir, M. Chatterjee, K.L. Ebi, Y.O. Estrada, R.C. Genova, B. Girma, E.S. Kissel, A.N. Levy, S. MacCracken, P.R. Mastrandrea and L.L. White (eds.)]. Cambridge University Press, Cambridge, United Kingdom and New York, NY, USA, 1101–1131.

Dilling, L. et al., 2015: The dynamics of vulnerability: why adapting to climate variability will not always prepare us for climate change. *Wiley Interdisciplinary Reviews: Climate Change*, **6** (4), 413–425, doi:10.1002/wcc.341.

DiSegni, D.M. and M. Shechter, 2013: Socioeconomic Aspects: Human Migrations, Tourism and Fisheries. In: *The Mediterranean Sea. Its history and present challenges* [Goffredo, S. and Z. Dubinski (eds.)]. Springer Netherlands, Dordrecht, 571–575, ISBN 9789400767034.

Djalante, R. et al., 2018: Cross-Chapter Box 12: Residual risks, limits to adaptation and loss and damage. In: *Global Warming of 1.5°C. An IPCC Special Report on the impacts of global warming of 1.5°C above pre-industrial levels and related global greenhouse gas emission pathways, in the context of strengthening the global response to the threat of climate change, sustainable development, and efforts to eradicate poverty* [Masson-Delmotte, V., P. Zhai, H.O. Pörtner, D. Roberts, J. Skea, P.R. Shukla, A. Pirani, W. Moufouma-Okia, C. Péan, R. Pidcock, S. Connors, J.B.R. Matthews, Y. Chen, X. Zhou, M.I. Gomis, E. Lonnoy, T. Maycock, M. Tignor and T. Waterfield (eds.)]. In press.

Donner, S.D., 2015: The legacy of migration in response to climate stress: learning from the Gilbertese resettlement in the Solomon Islands. *Natural Resources Forum*, **39** (3–4), 191–201, doi:10.1111/1477-8947.12082.

Duvat, V. et al., 2017: Trajectories of exposure and vulnerability of small islands to climate change. *Wiley Interdisciplinary Reviews: Climate Change*, **8** (6), e478, doi:10.1002/wcc.478.

Duvat, V.K.E., 2019: A global assessment of atoll island planform changes over the past decades. *Wiley Interdisciplinary Reviews: Climate Change*, **10**, e557, doi:10.1002/wcc.557.

Duvat, V.K.E. et al., 2016: Assessing the impacts of and resilience to Tropical Cyclone Bejisa, Reunion Island (Indian Ocean). *Natural Hazards*, **83** (1), 601–640, doi:10.1007/s11069-016-2338-5.

Emanuel, K., 2017: Assessing the present and future probability of Hurricane Harvey's rainfall. *Proceedings of the National Academy of Sciences*, **114** (48), 12681–12684, doi:10.1073/pnas.1716222114.

Fang, Z., P.T. Freeman, C.B. Field and K.J. Mach, 2018: Reduced sea ice protection period increases storm exposure in Kivalina, Alaska. *Arctic Science*, **4** (4), 1–13, doi:10.1139/as-2017-0024.

FAO et al., 2017: *The State of Food Security and Nutrition in the World 2017. Building resilience for peace and food security.* FAO, Rome, 117 pp. [Available at: http://www.fao.org/3/a-I7695e.pdf, accessed 17/04/2018].

Fawcett, D., T. Pearce, J.D. Ford and L. Archer, 2017: Operationalizing longitudinal approaches to climate change vulnerability assessment. *Global Environmental Change*, **45**, 79–88, doi:10.1016/j.gloenvcha.2017.05.002.

Feldmeyer, D., J. Birkmann and T. Welle, 2017: Development of Human Vulnerability 2012–2017. *Journal of Extreme Events*, **04** (04), doi:10.1142/s2345737618500057.

Ferrario, F. et al., 2014: The effectiveness of coral reefs for coastal hazard risk reduction and adaptation. *Nature Communications*, **5**, 3794, doi:10.1038/ncomms4794.

Forbes, D.L. et al., 2018: Coastal environments and drivers. In: *From Science to Policy in the Eastern Canadian Arctic: an Integrated Regional Impact Assessment (IRIS) of Climate Change and Modernization* [Bell, T. and T.M. Brown (eds.)]. ArcticNet, Québec, 210–249.

Ford, J.D. et al., 2016: Including indigenous knowledge and experience in IPCC assessment reports. *Nature Climate Change*, **6** (4), 349–353, doi:10.1038/Nclimate2954.

Ford, J.D., G. McDowell and T. Pearce, 2015: The adaptation challenge in the Arctic. *Nature Climate Change*, **5** (12), 1046–1053.

Fu, X., J. Song, B. Sun and Z.-R. Peng, 2016: "Living on the edge": Estimating the economic cost of sea level rise on coastal real estate in the Tampa Bay region, Florida. *Ocean & Coastal Management*, **133**, 11–17, doi:10.1016/j.ocecoaman.2016.09.009.

Garschagen, M. and P. Romero-Lankao, 2015: Exploring the relationships between urbanization trends and climate change vulnerability. *Climatic Change*, **133** (1), 37–52, doi:10.1007/s10584-013-0812-6.

Gattuso, J.-P. et al., 2018: Ocean solutions to address climate change and its effects on marine ecosystems. *Frontiers in Marine Sciences*, **5**, 1–18, doi:10.3389/fmars.2018.00337.

Genua-Olmedo, A., C. Alcaraz, N. Caiola and C. Ibáñez, 2016: Sea level rise impacts on rice production: The Ebro Delta as an example. *Science of the Total Environment*, **571**, 1200–1210, doi:10.1016/j.scitotenv.2016.07.136.

Gerkensmeier, B. and B.M.W. Ratter, 2018: Governing coastal risks as a social process – Facilitating integrative risk management by enhanced multi-stakeholder collaboration. *Environmental Science & Policy*, **80**, 144–151, doi:10.1016/j.envsci.2017.11.011.

Gerrard, M.B. and G.E. Wannier, Eds., 2013: *Threatened Island Nations: Legal Implications of Rising Seas and a Changing Climate*. Cambridge University Press, Cambridge, ISBN 9781107025769.

Gibbs, A.E. and B.M. Richmond, 2017: *National assessment of shoreline change – Summary statistics for updated vector shorelines and associated shoreline change data for the north coast of Alaska, U.S.-Canadian border to Icy Cape: U.S. Geological Survey Open-File Report 2017–1107*. U.S. Geological Survey, Reston, Virginia, 21 pp. [Available at: https://pubs.usgs.gov/of/2017/1107/ofr2017–1107.pdf, accessed 27/05/2019].

Goreau, T. and P. Prong, 2017: Biorock Electric Reefs Grow Back Severely Eroded Beaches in Months. *Journal of Marine Science and Engineering*, **5** (4), 48.

Gorokhovich, Y., A. Leiserowitz and D. Dugan, 2013: Integrating Coastal Vulnerability and Community-Based Subsistence Resource Mapping in Northwest Alaska. *Journal of Coastal Research*, **30** (1), 158–169, doi:10.2112/JCOASTRES-D-13-00001.1.

Government of Fiji, 2016: *Fiji. Post-Disaster Needs Assessment. Tropical Cyclone Winston, February 20, 2016*. Government of Fiji, Suva, Fiji [Available at: www.gfdrr.org/sites/default/files/publication/Post%20Disaster%20Needs%20Assessments%20CYCLONE%20WINSTON%20Fiji%202016%20(Online%20Version).pdf, accessed 17/04/2018].

Government of Tonga, 2018: *Post Disaster Rapid Assessment. Tropical Cyclone Gita*. Government of Tonga, Nuku'alofa [Available at: https://reliefweb.int/sites/reliefweb.int/files/resources/tonga-pdna-tc-gita-2018.pdf, accessed 18/10/2018].

Government of Vanuatu, 2015: *Vanuatu. Post-Disaster Needs Assessment. Tropical Cyclone Pam, March 2015*. Government of Vanuatu, Port Vila [Available at: https://reliefweb.int/sitesreliefweb.int/files/resources/vanuatu_pdna_cyclone_pam_2015.pdf, accessed 18/04/2018].

Gugliotta, M. et al., 2017: Process regime, salinity, morphological, and sedimentary trends along the fluvial to marine transition zone of the mixed-energy Mekong River delta, Vietnam. *Continental Shelf Research*, **147**, 7–26, doi:10.1016/j.csr.2017.03.001.

Günther, F. et al., 2013: Short- and long-term thermo-erosion of ice-rich permafrost coasts in the Laptev Sea region. *Biogeosciences*, **10** (6), 4297–4318, doi:10.5194/bg-10-4297-2013.

Haasnoot, M., J.H. Kwakkel, W.E. Walker and J. ter Maat, 2013: Dynamic adaptive policy pathways: A method for crafting robust decisions for a deeply uncertain world. *Global Environmental Change*, **23** (2), 485–498, doi:10.1016/j.gloenvcha.2012.12.006.

Haines, P., 2016: *Choiseul Bay Township Adaptation and Relocation Program, Choiseul Province, Solomon Islands. Case Study for CoastAdapt*. National Climate Change Adaptation Research Facility, Gold Coast [Available at: https://coastadapt.com.au/sites/default/files/case_studies/CSS3_Relocation_in_the_Solomon_Islands.pdf, accessed 20/08/2018].

Haines, P. and S. McGuire, 2014: Rising tides, razing capitals: A Solomon Islands approach to adaptation [online]. In: *Practical Responses to Climate Change Conference 2014*, Barton, Engineers Australia, 152–162.

Hajra, R. et al., 2017: Unravelling the association between the impact of natural hazards and household poverty: evidence from the Indian Sundarban delta. *Sustainability Science*, **12** (3), 453–464, doi:10.1007/s11625-016-0420-2.

Hallegatte, S. et al., 2016: *Shock Waves: Managing the Impacts of Climate Change on Poverty*. World Bank, Washington D.C. [Available at: https://openknowledge.worldbank.org/bitstream/handle/10986/22787/9781464806735.pdf?sequence=13&isAllowed=y, accessed 17/04/2018].

Hallegatte, S., C. Green, R.J. Nicholls and J. Corfee-Morlot, 2013: Future flood losses in major coastal cities. *Nature Climate Change*, **3** (9), 802–806, doi:10.1038/nclimate1979.

Hamilton, L.C. et al., 2016: Climigration? Population and climate change in Arctic Alaska. *Population and Environment*, **38** (2), 115–133, doi:10.1007/s11111-016-0259-6.

Handmer, J. and H. Iveson, 2017: Cyclone Pam in Vanuatu: Learning from the low death toll [online]. *The Australian Journal of Emergency Management*, **32** (2), 60–65.

Haraguchi, M. and U. Lall, 2015: Flood risks and impacts: A case study of Thailand's floods in 2011 and research questions for supply chain decision making. *International Journal of Disaster Risk Reduction*, **14** (3), 256–272, doi:10.1016/j.ijdrr.2014.09.005.

Harley, M.D. et al., 2017: Extreme coastal erosion enhanced by anomalous extratropical storm wave direction. *Scientific Reports*, **7** (1), 6033, doi:10.1038/s41598-017-05792-1.

Hauer, M.E., 2017: Migration induced by sea-level rise could reshape the US population landscape. *Nature Climate Change*, **7** (5), 321–325, doi:10.1038/nclimate3271.

Hay, J., V. Duvat and A.K. Magnan, 2019: Trends in Vulnerability to Climate-related Hazards in the Pacific: Research, Understanding and Implications. In: *The Oxford Handbook of Planning for Climate Change Hazards* [Pfeffer, W.T., J.B. Smith and K.L. Ebi (eds.)]. Oxford University Press, Oxford, 136.

Hay, J.E. et al., 2018: *Climate Change and Disaster Risk Reduction: Research Synthesis Report*. Submitted to the New Zealand Ministry of Foreign Affairs and Trade, Wellington, New Zealand [Available at: https://www.mfat.govt.nz/assets/Aid-Prog-docs/Research/Climate-Change-and-DRR-Synthesis-Report-Final-v6.4.pdf, accessed 17/04/2018].

Hermans, L.M., M. Haasnoot, J. ter Maat and J.H. Kwakkel, 2017: Designing monitoring arrangements for collaborative learning about adaptation pathways. *Environmental Science & Policy*, **69**, 29–38, doi:10.1016/j.envsci.2016.12.005.

Hernández-González, Y. et al., 2016: *Perspectives on contentions about climate change adaptation in the Canary Islands: A case study for Tenerife*. European Union, Luxembourg [Available at: http://publications.jrc.ec.europa.eu/repository/bitstream/JRC104349/lbna28340enn.pdf, accessed 17/04/2018].

Herring, S.C. et al., 2017: Explaining Extreme Events of 2016 from a Climate Perspective. *Bulletin of the American Meteorological Society*, **98** (12), S1–S157.

Hilmi, N. et al., Eds., 2015: *Bridging the Gap Between Ocean Acidification Impacts and Economic Valuation: Regional Impacts of Ocean Acidification on Fisheries and Aquaculture*. IUCN, Gland, Switzerland.

Hinkel, J. et al., 2018: The ability of societies to adapt to twenty-first-century sea-level rise. *Nature Climate Change*, **8** (7), 570–578, doi:10.1038/s41558-018-0176-z.

Hinkel, J. et al., 2014: Coastal flood damage and adaptation costs under 21st century sea-level rise. *Proceedings of the National Academy of Sciences*, **111** (9), 3292–3297, doi:10.1073/pnas.1222469111.

Hino, M., C.B. Field and K.J. Mach, 2017: Managed retreat as a response to natural hazard risk. *Nature Climate Change*, **7**, 364–370, doi:10.1038/nclimate3252.

Hoegh-Guldberg, O. et al., 2018: Impacts of 1.5°C global warming on natural and human systems. In: *Global Warming of 1.5°C. An IPCC Special Report on the impacts of global warming of 1.5°C above pre-industrial levels and related global greenhouse gas emission pathways, in the context of strengthening the global response to the threat of climate change, sustainable development, and efforts to eradicate poverty* [Masson-Delmotte, V., P. Zhai, H.O. Pörtner, D. Roberts, J. Skea, P.R. Shukla, A. Pirani, W. Moufouma-Okia, C. Péan, R. Pidcock, S. Connors, J.B.R. Matthews, Y. Chen, X. Zhou, M.I. Gomis, E. Lonnoy, T. Maycock, M. Tignor and T. Waterfield (eds.)]. In press.

Holdschlag, A. and B.M.W. Ratter, 2016: Sozial-ökologische Systemdynamik in der Panarchie. Adaptivität und Umweltwissen am Beispiel karibischer

CCB9

671

Small Island Developing States (SIDS). *Geographische Zeitschrift*, **104** (3), 183–211.

Hsiang, S. et al., 2017: Estimating economic damage from climate change in the United States. *Science*, **356** (6345), 1362–1369, doi:10.1126/science.aal4369.

IFRC, 2018: *Pacific Region: Tropical Cyclone Pam. Emergency Plan of Action Final Report*. International Federation of Red Cross and Red Crescent Societies [Available at: https://reliefweb.int/sites/reliefweb.int/files/resources/MDR55001efr.pdf, accessed 27/05/2019].

IPCC, 2014: *Climate Change 2014: Synthesis Report. Contribution of Working Groups I, II and III to the Fifth Assessment Report of the Intergovernmental Panel on Climate Change* [Core Writing Team, R.K. Pachauri and L.A. Meyer (eds.)]. IPCC, Geneva, Switzerland, 151 pp.

IPCC, 2018: Summary for Policymakers. In: *Global Warming of 1.5°C. An IPCC Special Report on the impacts of global warming of 1.5°C above pre-industrial levels and related global greenhouse gas emission pathways, in the context of strengthening the global response to the threat of climate change, sustainable development, and efforts to eradicate poverty* [Masson-Delmotte, V., P. Zhai, H.-O. Pörtner, D. Roberts, J. Skea, P.R. Shukla, A. Pirani, W. Moufouma-Okia, C. Péan, R. Pidcock, S. Connors, J.B.R. Matthews, Y. Chen, X. Zhou, M.I. Gomis, E. Lonnoy, T. Maycock, M. Tignor and T. Waterfield (eds.)]. World Meteorological Organization, Geneva, Switzerland.

James, T.S. et al., 2015: *Tabulated Values of Relative Sea-Level Projections in Canada and the Adjacent Mainland United States*. Geological Survey of Canada [Available at: http://ftp.maps.canada.ca/pub/nrcan_rncan/publications/ess_sst/297/297048/of_7942.pdf, accessed 27/05/2019].

Johnston, M.W., S.J. Purkis and R.E. Dodge, 2015: Measuring Bahamian lionfish impacts to marine ecological services using habitat equivalency analysis. *Marine Biology*, **162** (12), 2501–2512, doi:10.1007/s00227-015-2745-2.

Jones, B. and B.C. O'Neill, 2016: Spatially explicit global population scenarios consistent with the Shared Socioeconomic Pathways. *Environmental Research Letters*, **11** (8), doi:10.1088/1748-9326/11/8/084003.

Jones, B.M. et al., 2018: A decade of remotely sensed observations highlight complex processes linked to coastal permafrost bluff erosion in the Arctic. *Environmental Research Letters*, **13** (11), doi:10.1088/1748-9326/aae471.

Jones, N. and J.R.A. Clark, 2014: Social capital and the public acceptability of climate change adaptation policies: a case study in Romney Marsh, UK. *Climatic Change*, **123** (2), 133–145, doi:10.1007/s10584-013-1049-0.

Juhola, S., E. Glaas, B.-O. Linnér and T.-S. Neset, 2016: Redefining maladaptation. *Environmental Science & Policy*, **55**, 135–140, doi:10.1016/j.envsci.2015.09.014.

Kabir, M.J., D.S. Gaydon, R. Cramb and C.H. Roth, 2018: Bio-economic evaluation of cropping systems for saline coastal Bangladesh: I. Biophysical simulation in historical and future environments. *Agricultural Systems*, **162**, 107–122, doi:10.1016/j.agsy.2018.01.027.

Kaltenborn, B.P. et al., 2017: Ecosystem Services and Cultural Values as Building Blocks for 'The Good life'. A Case Study in the Community of Røst, Lofoten Islands, Norway. *Ecological Economics*, **140**, 166–176, doi:10.1016/j.ecolecon.2017.05.003.

Karrasch, L., M. Maier, M. Kleyer and T. Klenke, 2017: Collaborative Landscape Planning: Co-Design of Ecosystem-Based Land Management Scenarios. *Sustainability*, **9** (9), 1–15, doi:10.3390/su9091668.

Kelman, I., 2015: Difficult decisions: Migration from Small Island Developing States under climate change. *Earth's Future*, **3** (4), 133–142, doi:10.1002/2014ef000278.

Kelman, I., 2017: How can island communities deal with environmental hazards and hazard drivers, including climate change? *Environmental Conservation*, **44** (3), 244–253, doi:10.1017/S0376892917000042.

Kench, P.S., M.R. Ford and S.D. Owen, 2018: Patterns of island change and persistence offer alternate adaptation pathways for atoll nations. *Nature Communications*, **9** (1), 1–7, doi:10.1038/s41467-018-02954-1.

Khanom, T., 2016: Effect of salinity on food security in the context of interior coast of Bangladesh. *Ocean & Coastal Management*, **130**, 205–212, doi:10.1016/j.ocecoaman.2016.06.013.

Kim, I.N. et al., 2014: Chemical oceanography. Increasing anthropogenic nitrogen in the North Pacific Ocean. *Science*, **346** (6213), 1102–6, doi:10.1126/science.1258396.

Kondolf, G.M. et al., 2014: Sustainable sediment management in reservoirs and regulated rivers: Experiences from five continents. *Earth's Future*, **2** (5), 256–280, doi:10.1002/2013ef000184.

Kummu, M. et al., 2016: Over the hills and further away from coast: global geospatial patterns of human and environment over the 20th–21st centuries. *Environmental Research Letters*, **11** (3), 034010, doi:10.1088/1748-9326/11/3/034010.

Lacher, K., 2015: The island community of Chenega: Earthquake, tsunami, oil spill... What's next? *Global Environment*, **8** (1), 38–60, doi:10.3197/ge.2015.080103.

Lam, V.W., W.W. Cheung, G. Reygondeau and U.R. Sumaila, 2016: Projected change in global fisheries revenues under climate change. *Scientific Reports*, **6**, 32607, doi:10.1038/srep32607.

Lantuit, H. et al., 2011: The Arctic Coastal Dynamics Database: A New Classification Scheme and Statistics on Arctic Permafrost Coastlines. *Estuaries and Coasts*, **35** (2), 383–400, doi:10.1007/s12237-010-9362-6.

Lichter, M., A.T. Vafeidis, R.J. Nicholls and G. Kaiser, 2011: Exploring Data-Related Uncertainties in Analyses of Land Area and Population in the "Low-Elevation Coastal Zone" (LECZ). *Journal of Coastal Research*, **27** (4), 757–768, doi:10.2112/jcoastres-d-10-00072.1.

Logan, J.R., S. Issar and Z. Xu, 2016: Trapped in Place? Segmented Resilience to Hurricanes in the Gulf Coast, 1970–2005. *Demography*, **53** (5), 1511–1534, doi:10.1007/s13524-016-0496-4.

Magnan, A.K. et al., 2016: Addressing the risk of maladaptation to climate change. *Wiley Interdisciplinary Reviews: Climate Change*, **7** (5), 646–665, doi:10.1002/wcc.409.

Manley, M. et al., 2016: *Research and Analysis on Climate Change and Disaster Risk Reduction. Working Paper 1: Needs, Priorities and Opportunities Related to Climate Change Adaptation and Disaster Risk Reduction. Report to the New Zealand Ministry of Foreign Affairs and Trade*. Wellington, 128 pp. [Available at, accessed 18/04/2018].

Marino, E., 2015: *Fierce Climate, Sacred Ground: An Ethnography of Climate Change in Shishmaref, Alaska*. University of Alaska Press, Fairbanks, 122 pp., ISBN 9781602232662.

Marino, E. and H. Lazrus, 2015: Migration or Forced Displacement?: The Complex Choices of Climate Change and Disaster Migrants in Shishmaref, Alaska and Nanumea, Tuvalu. *Human Organization*, **74** (4), 341–350, doi:10.17730/0018-7259-74.4.341.

McCubbin, S., B. Smit and T. Pearce, 2015: Where does climate fit? Vulnerability to climate change in the context of multiple stressors in Funafuti, Tuvalu. *Global Environmental Change*, **30**, 43–55, doi:10.1016/j.gloenvcha.2014.10.007.

McEwen, L. et al., 2017: Sustainable flood memories, lay knowledges and the development of community resilience to future flood risk. *Transactions of the Institute of British Geographers*, **42** (1), 14–28, doi:10.1111/tran.12149.

McGranahan, G., D. Balk and B. Anderson, 2007: The rising tide: assessing the risks of climate change and human settlements in low elevation coastal zones. *Environment and Urbanization*, **19** (1), 17–37, doi:10.1177/0956247807076960.

McLean, R. and P. Kench, 2015: Destruction or persistence of coral atoll islands in the face of 20th and 21st century sea-level rise? *Wiley Interdisciplinary Reviews: Climate Change*, **6** (5), 445–463, doi:10.1002/wcc.350.

McMillen, H.L. et al., 2014: Small islands, valuable insights: systems of customary resource use and resilience to climate change in the Pacific. *Ecology and Society*, **19** (4), doi:10.5751/es-06937-190444.

McNamara, D.E. and A. Keeler, 2013: A coupled physical and economic model of the response of coastal real estate to climate risk. *Nature Climate Change*, 3 (6), 559–562, doi:10.1038/nclimate1826.

McNamara, K.E. and H.J. Des Combes, 2015: Planning for Community Relocations Due to Climate Change in Fiji. *International Journal of Disaster Risk Science*, 6 (3), 315–319, doi:10.1007/s13753-015-0065-2.

Merkens, J.-L., L. Reimann, J. Hinkel and A.T. Vafeidis, 2016: Gridded population projections for the coastal zone under the Shared Socioeconomic Pathways. *Global and Planetary Change*, **145**, 57–66, doi:10.1016/j.gloplacha.2016.08.009.

Milan, A. and S. Ruano, 2014: Rainfall variability, food insecurity and migration in Cabricán, Guatemala. *Climate and Development*, 6 (1), 61–68, doi:10.1080/17565529.2013.857589.

Miller, M.A., M. Douglass and M. Garschagen, Eds., 2018: *Crossing Borders. Governing Environmental Disasters in a Global Urban Age in Asia and the Pacific*. Springer Nature Singapore, Singapore, 288 pp., ISBN 9789811061257.

Morrison, K., 2017: The Role of Traditional Knowledge to Frame Understanding of Migration as Adaptation to the 'Slow Disaster' of Sea Level Rise in the South Pacific. In: *Identifying Emerging Issues in Disaster Risk Reduction, Migration, Climate Change and Sustainable Development* [Sudmeier-Rieux, K., M. Fernández, I. Penna, M. Jaboyedoff and J. Gaillard (eds.)]. Springer International Publishing, Cham, 249–266.

Moser, S.C. and J.A.F. Hart, 2015: The long arm of climate change: societal teleconnections and the future of climate change impacts studies. *Climatic Change*, **129** (1–2), 13–26, doi:10.1007/s10584-015-1328-z.

Narayan, S. et al., 2016: The Effectiveness, Costs and Coastal Protection Benefits of Natural and Nature-Based Defences. *PLoS One*, **11** (5), e0154735, doi:10.1371/journal.pone.0154735.

National Geophysical Data Center, 1999: Global Land One-kilometer Base Elevation (GLOBE) v.1.D. and P.K. Dunbar. National Geophysical Data Center, NOAA, Hastings, doi:10.7289/V52R3PMS. Neumann, B., A.T. Vafeidis, J. Zimmermann and R.J. Nicholls, 2015: Future coastal population growth and exposure to sea-level rise and coastal flooding--a global assessment. *PLoS One*, **10** (3), e0118571, doi:10.1371/journal.pone.0118571.

Noble, I.R. et al., 2014: Adaptation needs and options. In: *Climate Change 2014: Impacts, Adaptation, and Vulnerability. Part A: Global and Sectoral Aspects. Contribution of Working Group II to the Fifth Assessment Report of the Intergovernmental Panel of Climate Change* [Field, C.B., V.R. Barros, D.J. Dokken, K.J. Mach, M.D. Mastrandrea, T.E. Bilir, M. Chatterjee, K.L. Ebi, Y.O. Estrada, R.C. Genova, B. Girma, E.S. Kissel, A.N. Levy, S. MacCracken, P.R. Mastrandrea and L.L. White (eds.)]. Cambridge University Press, Cambridge, United Kingdom and New York, NY, USA, 833–868.

Noy, I. and C. Edmonds. 2016: *The economic and fiscal burdens of disasters in the Pacific (SEF Working Paper 25/2016)*. Victoria University of Wellington, Wellington [Available at: http://researcharchive.vuw.ac.nz/bitstream/handle/10063/5439/Working%20Paper.pdf?sequence=1, accessed 26/05/2019].

Null, S. and L. Herzer Risi, 2016: *Navigating Complexity: Climate, Migration, and Conflict in a Changing World. USAID Office of Conflict Management and Mitigation Discussion Paper November 2016*. United States Agency for International Development [Available at: https://www.wilsoncenter.org/sites/default/files/ecsp_navigating_complexity_web_1.pdf, accessed 21/09/2018].

Nunn, P.D., 2014: Geohazards and myths: ancient memories of rapid coastal change in the Asia-Pacific region and their value to future adaptation. *Geoscience Letters*, **1** (1), doi:10.1186/2196-4092-1-3.

Nunn, P.D., J. Runman, M. Falanruw and R. Kumar, 2017: Culturally grounded responses to coastal change on islands in the Federated States of Micronesia, northwest Pacific Ocean. *Regional Environmental Change*, **17** (4), 959–971, doi:10.1007/s10113-016-0950-2.

Nurse, L.A. et al., 2014: Small islands. In: *Climate Change 2014: Impacts, Adaptation, and Vulnerability. Part B: Regional Aspects. Contribution of Working Group II to the Fifth Assessment Report of the Intergovernmental Panel of Climate Change* [Barros, V.R., C.B. Field, D.J. Dokken, M.D. Mastrandrea, K.J. Mach, T.E. Bilir, M. Chatterjee, K.L. Ebi, Y.O. Estrada, R.C. Genova, B. Girma, E.S. Kissel, A.N. Levy, S. MacCracken, P.R. Mastrandrea and L.L. White (eds.)]. Cambridge University Press, Cambridge, United Kingdom and New York, NY, USA, 1613–1654.

O'Neill, B.C. et al., 2017: IPCC reasons for concern regarding climate change risks. *Nature Climate Change*, **7** (1), 28–37, doi:10.1038/nclimate3179.

O'Brien, K. et al., 2012: Toward a Sustainable and Resilient Future. In: *Managing the Risks of Extreme Events and Disasters to Advance Climate Change Adaptation. A Special Report of Working Groups I and II of the Intergovernmental Panel on Climate Change (IPCC)* [Field, C.B., V. Barros, T.F. Stocker, D. Qin, D.J. Dokken, K.L. Ebi, M.D. Mastrandrea, K.J. Mach, G.K. Plattner, S.K. Allen, M. Tignor and P.M. Midgley (eds.)]. Cambridge University Press, Cambridge, United Kingdom and New York, NY, USA, 437–486.

Overeem, I. et al., 2011: Sea ice loss enhances wave action at the Arctic coast. *Geophysical Research Letters*, **38** (17), 1–6, doi:10.1029/2011gl048681.

Paul, M. and L.G. Gillis, 2015: Let it flow: how does an underlying current affect wave propagation over a natural seagrass meadow? *Marine Ecology Progress Series*, **523**, 57–70, doi:10.3354/meps11162.

Pelling, M. and S. Blackburn, 2013: *Megacities and the Coast: Risk, Resilience, and Transformation*. Earthscan, Abingdon, 248 pp., ISBN 9780415815048.

Perumal, N., 2018: "The place where I live is where I belong": community perspectives on climate change and climate-related migration in the Pacific island nation of Vanuatu. *Island Studies Journal*, **13** (1), 45–64, doi:10.24043/isj.50.

Petzold, J., 2016: Limitations and opportunities of social capital for adaptation to climate change: A case study on the Isles of Scilly. *The Geographical Journal*, **182** (2), 123–134, doi:10.1111/geoj.12154.

Petzold, J. and B.M.W. Ratter, 2015: Climate change adaptation under a social capital approach – An analytical framework for small islands. *Ocean & Coastal Management*, **112**, 36–43, doi:10.1016/j.ocecoaman.2015.05.003.

Pontee, N., S. Narayan, M.W. Beck and A.H. Hosking, 2016: Nature-based solutions: lessons from around the world. *Proceedings of the Institution of Civil Engineers - Maritime Engineering*, **169** (1), 29–36, doi:10.1680/jmaen.15.00027.

Putra, H.C., H. Zhang and C. Andrews, 2015: Modeling Real Estate Market Responses to Climate Change in the Coastal Zone. *Journal of Artificial Societies and Social Simulation*, **18** (2), 18, doi:10.18564/jasss.2577.

Quataert, E. et al., 2015: The influence of coral reefs and climate change on wave-driven flooding of tropical coastlines. *Geophysical Research Letters*, **42** (15), 6407–6415, doi:10.1002/2015GL064861.

Ratter, B.M.W., J. Petzold and K.M. Sinane, 2016: Considering the locals: coastal construction and destruction in times of climate change on Anjouan, Comoros. *Natural Resources Forum*, **40** (3), 112–126, doi:10.1111/1477-8947.12102.

Rivera-Collazo, I. et al., 2015: Human adaptation strategies to abrupt climate change in Puerto Rico ca. 3.5 ka. *The Holocene*, **25** (4), 627–640, doi:10.1177/0959683614565951.

Romanovsky, V.E., S.L. Smith and H.H. Christiansen, 2010: Permafrost thermal state in the polar Northern Hemisphere during the international polar year 2007–2009: a synthesis. *Permafrost and Periglacial Processes*, **21** (2), 106–116, doi:10.1002/ppp.689.

Rosenzweig, C. and W. Solecki, 2014: Hurricane Sandy and adaptation pathways in New York: Lessons from a first-responder city. *Global Environmental Change*, **28**, 395–408, doi:10.1016/j.gloenvcha.2014.05.003.

Ross, A.C. et al., 2015: Sea-level rise and other influences on decadal-scale salinity variability in a coastal plain estuary. *Estuarine, Coastal and Shelf Science*, **157**, 79–92, doi:10.1016/j.ecss.2015.01.022.

Rosselló-Nadal, J., 2014: How to evaluate the effects of climate change on tourism. *Tourism Management*, **42**, 334–340, doi:10.1016/j.tourman.2013.11.006.

Rubin, B.M. and M.D. Hilton, 1996: Identifying the Local Economic Development Impacts of Global Climate Change. *Economic Development Quarterly*, **10** (3), 262–279, doi:10.1177/089124249601000306.

Sagoe-Addy, K. and K. Appeaning Addo, 2013: Effect of predicted sea level rise on tourism facilities along Ghana's Accra coast. *Journal of Coastal Conservation*, **17** (1), 155–166, doi:10.1007/s11852-012-0227-y.

Schoonees, T. et al., 2019: Hard Structures for Coastal Protection, Towards Greener Designs. *Estuaries and Coasts*, 1–12, doi:10.1007/s12237-019-00551-z.

Shakeela, A., S. Becken and N. Johnston, 2013: *Gaps and Disincentives that Exist in the Policies, Laws and Regulations which Act as Barriers to Investing in Climate Change Adaptation in the Tourism Sector of the Maldives. Final Project Report*. UNDP, Maldives [Available at: http://www.tourism.gov.mv/downloads/tap/2014/FINALReport_REVISED_UNDPMaldives_161213.pdf, accessed 18/04/2018].

Shayegh, S., J. Moreno-Cruz and K. Caldeira, 2016: Adapting to rates versus amounts of climate change: a case of adaptation to sea-level rise. *Environmental Research Letters*, **11** (10), doi:10.1088/1748-9326/11/10/104007.

Smajgl, A. et al., 2015: Responding to rising sea levels in the Mekong Delta. *Nature Climate Change*, **5** (2), 167–174, doi:10.1038/nclimate2469.

Solecki, W., M. Pelling and M. Garschagen, 2017: Transitions between risk management regimes in cities. *Ecology and Society*, **22** (2), doi:10.5751/ES-09102-220238.

Stojanov, R. et al., 2017: Local perceptions of climate change impacts and migration patterns in Malé, Maldives. *The Geographical Journal*, **183** (4), 370–385, doi:10.1111/geoj.12177.

Storlazzi, C.D. et al., 2018: Most atolls will be uninhabitable by the mid-21st century because of sea-level rise exacerbating wave-driven flooding. *Science Advances*, **4** (4), eaap9741, doi:10.1126/sciadv.aap9741.

Strauss, B.H., S. Kulp and A. Levermann, 2015: *Mapping Choices: Carbon, Climate, and Rising Seas, Our Global Legacy. Climate Central Research Report*. Climate Central, Princeton [Available at: http://sealevel.climatecentral.org/uploads/research/Global-Mapping-Choices-Report.pdf, accessed 18/04/2018].

Syvitski, J.P.M. et al., 2009: Sinking deltas due to human activities. *Nature Geoscience*, **2**, 681, doi:10.1038/ngeo629.

Tabe, T., 2016: Ngaira Kain Tari: We are people of the Sea. PhD Thesis. University of Bergen, Bergen, Norway.

Tessler, Z.D. et al., 2015: Profiling risk and sustainability in coastal deltas of the world. *Science*, **349** (6248), 638–43, doi:10.1126/science.aab3574.

The Government of the Commonwealth of Dominica, 2017: *Post-Disaster Needs Assessment Hurricane Maria September 18, 2017*. 161 pp. [Available at: https://reliefweb.int/sites/reliefweb.int/files/resources/dominica-pdna-maria.pdf, accessed 21/09/2018].

Thomas, A.R., 2015: *Resettlement in the Wake of Typhoon Haiyan in the Philippines: A Strategy to Mitigate Risk or a Risky Strategy?* The Brookings Institution, Washington D.C. [Available at: www.brookings.edu/wp-content/uploads/2016/06/Brookings-Planned-Relocations-Case-StudyAlice-Thomas-Philippines-case-study-June-2015.pdf, accessed 18/10/2018].

UN-DESA, 2018: *World Urbanization Prospects: The 2018 Revision, Online Edition*. United Nations Department of Economic and Social Affairs, New York [Available at: https://population.un.org/wup/Download/Files/WUP2018-F11a-30_Largest_Cities.xls, accessed 18/10/2018].

UN-OHRLLS. n.d.: About the Small Island Developing States. [Available at: http://unohrlls.org/about-sids/, accessed 22/06/2018].

UNDP, 2017: *Regional Overview: Impact of Hurricanes Irma and Maria*. UN Development Programme, New York [Available at: https://reliefweb.int/sites/reliefweb.int/files/resources/UNDP%20%20Regional%20Overview%20Impact%20of%20Hurricanes%20Irma%20and%20Maria.pdf, accessed 17/04/2018].

United Nations, Eds., 2017: *The First Global Integrated Marine Assessment. World Ocean Assessment I*. Cambridge University Press, Cambridge, 973 pp., ISBN 9781316510018.

Weatherdon, L.V. et al., 2016: Observed and Projected Impacts of Climate Change on Marine Fisheries, Aquaculture, Coastal Tourism, and Human Health: An Update. *Frontiers in Marine Science*, **3**, 1–21, doi:10.3389/fmars.2016.00048.

Weber, E., 2016: Only a pawn in their games? environmental (?) migration in Kiribati – past, present and future. *Die Erde*, **147** (2), 153–164, doi:10.12854/erde-147-11.

Weigelt, P., W. Jetz and H. Kreft, 2013: Bioclimatic and physical characterization of the world's islands. *Proceedings of the National Academy of Sciences*, **110** (38), 15307–12, doi:10.1073/pnas.1306309110.

Werners, S.E. et al., 2015: Turning points in climate change adaptation. *Ecology and Society*, **20** (4), doi:10.5751/ES-07403-200403.

Wilbers, G.J. et al., 2014: Spatial and temporal variability of surface water pollution in the Mekong Delta, Vietnam. *Science of the Total Environment*, **485–486**, 653–665, doi:10.1016/j.scitotenv.2014.03.049.

Wise, R.M. et al., 2014: Reconceptualising adaptation to climate change as part of pathways of change and response. *Global Environmental Change*, **28**, 325–336, doi:10.1016/j.gloenvcha.2013.12.002.

WMO, 2018: *Caribbean 2017 Hurricane Season an evidence-based assessment of the Early Warning System*. World Meteorological Organization, Geneva [Available at: https://library.wmo.int/index.php?lvl=notice_display&id=20700#.XJuRkK6nFaQ, accessed 01/05/2019].

Wong, P.P. et al., 2014: Coastal systems and low-lying areas. In: *Climate Change 2014: Impacts, Adaptation, and Vulnerability. Part A: Global and Sectoral Aspects. Contribution of Working Group II to the Fifth Assessment Report of the Intergovernmental Panel of Climate Change* [Field, C.B., V.R. Barros, D.J. Dokken, K.J. Mach, M.D. Mastrandrea, T.E. Bilir, M. Chatterjee, K.L. Ebi, Y.O. Estrada, R.C. Genova, B. Girma, E.S. Kissel, A.N. Levy, S. MacCracken, P.R. Mastrandrea and L.L. White (eds.)]. Cambridge University Press, Cambridge, United Kingdom and New York, NY, USA, 361–409.

World Bank, 2012: *Thai Flood 2011: Rapid Assessment for Resilient Recovery and Reconstruction Planning*. World Bank, Bangkok [Available at: https://openknowledge.worldbank.org/handle/10986/26862, accessed 18/04/2018].

Yamamoto, L. and M. Esteban, 2014: *Atoll Island States and International Law. Climate Change Displacement and Sovereignty*. Springer, Berlin and Heidelberg, 307 pp., ISBN 9783642381850.

Annexes

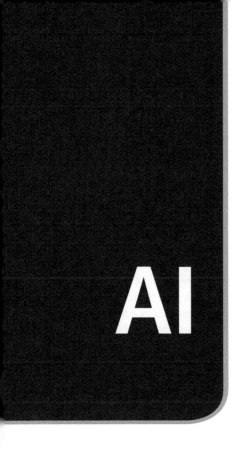

Annex I: Glossary

Coordinating Editor:
Nora M. Weyer (Germany)

Editorial team:
Miguel Cifuentes-Jara (Costa Rica), Thomas Frölicher (Switzerland), Miriam Jackson (Norway), Raphael M. Kudela (USA), Valérie Masson-Delmotte (France), J.B. Robin Matthews (United Kingdom), Katja Mintenbeck (Germany), Mônica M.C. Muelbert (Brazil), Hans-Otto Pörtner (Germany), Elvira S. Poloczanska (Australia/United Kingdom), Debra C. Roberts (South Africa), Renée van Diemen (Netherlands), Phil Williamson (United Kingdom), Panmao Zhai (China)

This annex should be cited as:
IPCC, 2019: Annex I: Glossary [Weyer, N.M. (ed.)]. In: *IPCC Special Report on the Ocean and Cryosphere in a Changing Climate* [H.-O. Pörtner, D.C. Roberts, V. Masson-Delmotte, P. Zhai, M. Tignor, E. Poloczanska, K. Mintenbeck, A. Alegría, M. Nicolai, A. Okem, J. Petzold, B. Rama, N.M. Weyer (eds.)]. Cambridge University Press, Cambridge, UK and New York, NY, USA, pp. 677–702. https://doi.org/10.1017/9781009157964.010.

Note: This Glossary defines some specific terms as the Lead Authors intend them to be interpreted in the context of this Special Report. *Blue italicisations* indicate terms defined in this Glossary. Note that subterms are in ***bold italics*** beneath main terms.

Ablation (of glaciers, ice sheets, or snow cover)

All processes that reduce the mass of a *glacier*, *ice sheet* or snow cover. The main processes are melting, and for glaciers also *calving* (or, when the glacier nourishes an *ice shelf*, *discharge of ice* across the *grounding line*), but other processes such as sublimation and loss of wind-blown snow can also contribute to ablation. Ablation also refers to the mass lost by any of these processes. See also *Mass balance/budget (of glaciers or ice sheets)*.

Abrupt climate change

Abrupt change refers to a change that is substantially faster than the rate of change in the recent history of the affected components of a system. Abrupt climate change refers to a large-scale change in the *climate system* that takes place over a few decades or less, persists (or is anticipated to persist) for at least a few decades, and causes substantial disruptions in human and natural systems. See also *Climate change*, *Human system*, *Natural systems* and *Tipping point*.

Accumulation (of glaciers, ice sheets or snow cover)

All processes that add to the mass of a *glacier*, an *ice sheet* or snow cover. The main process of accumulation is snowfall. Accumulation also includes deposition of hoar, freezing rain, other types of solid precipitation, gain of wind-blown snow, avalanching and basal accumulation (often beneath floating ice). See also *Avalanche* and *Mass balance/budget (of glaciers or ice sheets)*.

Active layer

Layer of ground above *permafrost* subject to annual thawing and freezing.

Adaptability

See *Adaptive capacity*.

Adaptation

In *human systems*, the process of adjustment to actual or expected *climate* and its effects, in order to moderate harm or exploit beneficial opportunities. In *natural systems*, the process of adjustment to actual climate and its effects; human intervention may facilitate adjustment to expected climate and its effects.

> ***Ecosystem-based Adaptation (EbA)*** The use of ecosystem management activities to increase the *resilience* and reduce the *vulnerability* of people and *ecosystems* to *climate change* (Campbell et al., 2009).

> ***Evolutionary adaptation*** The process whereby a species or population becomes better able to live in a changing environment, through the selection of heritable traits. Biologists usually distinguish evolutionary adaptation from acclimatisation, with the latter occurring within an organism's lifetime.

> ***Incremental adaptation*** Adaptation that maintains the essence and integrity of a system or process at a given scale (Park et al., 2012). In some cases, incremental adaptation can accrue to result in *transformational adaptation* (Tàbara et al., 2018; Termeer et al., 2017). Incremental adaptations to change in climate are understood as extensions of actions and behaviours that already reduce the losses or enhance the benefits of natural variations in *extreme weather/climate events*.

> ***Transformational adaptation*** Adaptation that changes the fundamental attributes of a *social-ecological system* in anticipation of *climate change* and its *impacts*; and adaptation responses that will be required in the face of a global failure to mitigate the causes of *anthropogenic* climate change and are characterised by system-wide change or changes across more than one system, by a focus on the future and long-term change, or by a direct questioning of the effectiveness of existing systems, social injustices and power imbalances.

> ***Adaptation limits*** The point at which an actor's objectives (or system needs) cannot be secured from intolerable risks through adaptive actions.

>> *Hard adaptation limit* – No adaptive actions are possible to avoid intolerable risks.

>> *Soft adaptation limit* – Options may exist but are currently not available to avoid intolerable risks through adaptive action.

See also *Adaptation options*, *Adaptive capacity*, *Justice*, *Maladaptive actions (Maladaptation)* and *Mitigation (of climate change)*.

Adaptation limits

See *Adaptation*.

Adaptation options

The array of strategies and measures that are available and appropriate for addressing *adaptation*. They include a wide range of actions that can be categorised as structural, institutional, ecological or behavioural. See also *Adaptive capacity* and *Maladaptive actions Maladaptation)*.

Adaptation pathways

See *Pathways*.

Adaptive capacity

The ability of systems, *institutions*, humans and other organisms to adjust to potential damage, to take advantage of opportunities or to respond to consequences (IPCC, 2014; MA, 2005). See also *Adaptation*.

Adaptive governance

See *Governance*.

Aerosol

A suspension of airborne solid or liquid particles, with a typical size between a few nanometres and 10 μm, that reside in the *atmosphere* for at least several hours. The term aerosol, which includes both the particles and the suspending gas, is often used in this Special Report in its plural form to mean 'aerosol particles'. Aerosols may be of either natural or *anthropogenic* origin. Aerosols can influence *climate* in several ways: directly through scattering and absorbing radiation, and indirectly by acting as cloud condensation nuclei or ice nuclei, modifying the optical properties and lifetime of clouds or upon deposition on snow or ice-covered surfaces thereby altering their *albedo* and contributing to climate feedback. Atmospheric aerosols, whether natural or anthropogenic, originate from two different pathways: emissions of primary particulate matter (PM), and formation of secondary PM from gaseous *precursors*. The bulk of aerosols are of natural origin. Some scientists use group labels that refer to the chemical composition, namely: sea salt, organic carbon, *black carbon (BC)*, mineral species (mainly desert dust), sulphate, nitrate and ammonium. These labels are, however, imperfect as aerosols combine particles to create complex mixtures. See also *Short-lived climate forcers (SLCFs)*.

Agreement

In this Special Report, the degree of agreement within the scientific body of knowledge on a particular finding is assessed based on multiple lines of *evidence* (e.g., mechanistic understanding, theory, data, models, expert judgement) and expressed qualitatively (Mastrandrea et al., 2010). See also *Confidence*, *Likelihood* and *Uncertainty*.

Albedo

The proportion of sunlight (solar radiation) reflected by a surface or object, often expressed as a percentage. Clouds, snow and ice usually have high albedo; soil surfaces cover the albedo range from high to low; vegetation in the dry season and/or in arid zones can have high albedo; whereas photosynthetically active vegetation and the ocean have low albedo. The Earth's planetary albedo changes mainly through varying cloudiness, snow, ice, leaf area and land cover changes.

Alien (non-native) species

An introduced species (alien species, exotic species, non-indigenous species, or non-native species) living outside its native distributional range, but which has arrived there by human activity, either deliberate or accidental. Non-native species can have various effects on and adversely affect the local *ecosystem*. See also *Endemic species* and *Invasive species*.

Anomaly

The deviation of a variable from its value averaged over a *reference period*.

Anthropogenic

Resulting from or produced by human activities. See also *Anthropogenic emissions*.

Anthropogenic emissions

Emissions of *greenhouse gases (GHGs)*, *precursors* of GHGs, and *aerosols*, caused by human activities. These activities include the burning of *fossil fuels*, *deforestation*, *land use* and land use changes (LULUC), livestock production, fertilisation, waste management, and industrial processes. See also *Anthropogenic*.

Anthropogenic subsidence

Downward motion of the *land* surface induced by anthropogenic *drivers* (e.g., loading, extraction of hydrocarbons and/or groundwater, drainage, mining activities) causing sediment compaction or subsidence/deformation of the sedimentary sequence, or oxidation of organic material, thereby leading to *relative sea level* rise. See also *Anthropogenic* and *Sea level change (sea level rise/sea level fall)*.

Atlantic Meridional Overturning Circulation (AMOC)

See *Meridional Overturning Circulation (MOC)*.

Atmosphere

The gaseous envelope surrounding the Earth, divided into five layers – the troposphere, which contains half of the Earth's atmosphere, the stratosphere, the mesosphere, the thermosphere, and the exosphere, which is the outer limit of the atmosphere. The dry atmosphere consists almost entirely of nitrogen (N_2, 78.1% volume mixing ratio) and oxygen (O_2, 20.9% volume mixing ratio), together with a number of trace gases, such as argon (Ar, 0.93% volume mixing ratio), helium (He) and radiatively active *greenhouse gases (GHG)* such as *carbon dioxide (CO_2*, 0.04% volume mixing ratio) and *ozone (O_3)*. In addition, the atmosphere contains the GHG water vapour (H_2O), whose amounts are highly variable but typically around 1% volume mixing ratio. The atmosphere also contains clouds and *aerosols*. See also *Climate system*, *Hydrological cycle*, *Methane (CH_4)* and *Radiative forcing*.

Atmosphere-ocean general circulation model (AOGCM)

See *Climate model*.

Attribution

See *Detection and attribution*.

Avalanche

A mass of snow, ice, earth or rocks, or a mixture of these, falling down a mountainside.

Benthos

The community of organisms living on the bottom or in sediments of a body of water (such as an *ocean*, river or lake). The ecological zone at the bottom of a body of water, including the sediment surface and some sub-surface layers, is known as the 'benthic zone'.

Biodiversity

or biological diversity means the variability among living organisms from all sources including, among other things, terrestrial, marine and other aquatic *ecosystems*, and the ecological complexes of which they are part; this includes diversity within species, between species and of ecosystems (UN, 1992). See also *Ecosystem service* and *Functional diversity*.

Biological (carbon) pump

A series of *ocean* processes through which inorganic carbon (as *carbon dioxide, CO_2*) is fixed as organic matter by photosynthesis in sunlit surface water and then transported to the ocean interior, and possibly the sediment, resulting in the storage of carbon. See also *Carbonate pump*, *Dissolved organic carbon (DOC) and particulate organic carbon (POC)*, *Microbial carbon pump* and *Solubility pump*.

Biomass
Organic material excluding the material that is fossilised or embedded in geological formations. Biomass may refer to the mass of organic matter in a specific area (ISO, 2014).

Black carbon (BC)
A relatively pure form of carbon, also known as soot, arising from the incomplete combustion of fossil fuels, biofuel and *biomass*. It only stays in the *atmosphere* for days or weeks. BC is a *climate* forcing agent with strong warming effect, both in the atmosphere and when deposited on snow or ice. See also *Aerosol*, *Albedo*, *Forcing* and *Short-lived climate forcers (SLCF)*.

Blue carbon
All biologically-driven carbon fluxes and storage in marine systems that are amenable to management can be considered as blue carbon. Coastal blue carbon focuses on rooted vegetation in the coastal zone, such as tidal marshes, mangroves and seagrasses. These *ecosystems* have high carbon burial rates on a per unit area basis and accumulate carbon in their soils and sediments. They provide many non-climatic benefits and can contribute to *ecosystem-based adaptation.* If degraded or lost, coastal blue carbon ecosystems are likely to release most of their carbon back to the *atmosphere*. There is current debate regarding the application of the blue carbon concept to other coastal and non-coastal processes and ecosystems, including the open *ocean*. See also *Carbon cycle*, *Coast*, *Ecosystem service* and *Sequestration*.

Calving (of glaciers or ice sheets)
The process of mechanical destruction of a mass of ice usually typical of marine-terminating *glaciers*; in the latter case, the ice calving (or breaking away) from the glacier edge can lead to the formation of *icebergs*. See also *Ice sheet* and *Marine ice cliff instability (MICI)*.

Carbonate pump
Ocean carbon fixation through the biological formation of carbonates, primarily by plankton that generate bio-mineral particles that sink to the *ocean* interior, and possibly the sediment. It is also called carbonate counter-pump, since the formation of calcium carbonate ($CaCO_3$) is accompanied by the release of *carbon dioxide (CO_2)* to surrounding water and subsequently to the *atmosphere*. See also *Biological (carbon) pump*, *Blue carbon*, *Dissolved organic carbon (DOC) and particulate organic carbon (POC)*, *Microbial carbon pump* and *Solubility pump*.

Carbon budget
refers to three concepts in the literature: (1) an assessment of *carbon cycle* sources and *sinks* on a global level, through the synthesis of *evidence* for *fossil-fuel* and cement emissions, *land use* change emissions, *ocean* and land *carbon dioxide (CO_2)* sinks, and the resulting atmospheric CO_2 growth rate. This is referred to as the global carbon budget; (2) the estimated cumulative amount of global CO_2 emissions that is estimated to limit global surface temperature to a given level above a *reference period*, taking into account global surface temperature contributions of other GHGs and climate forcers; (3) the distribution of the carbon budget defined under (2) to the regional, national, or sub-national level based on considerations of *equity*, costs or efficiency. See also *Atmosphere*, *Forcing* and *Land*.

Carbon cycle
The flow of carbon (in various forms, e.g., as *carbon dioxide (CO_2)*, carbon in *biomass*, and carbon dissolved in the *ocean* as carbonate and bicarbonate) through the *atmosphere*, hydrosphere, ocean, terrestrial and marine biosphere and lithosphere. In this Special Report, the reference unit for the global carbon cycle is GtCO2 or GtC (one Gigatonne = 1 Gt = 10^{15} grams; 1 GtC corresponds to 3.667 GtCO2). See also *Atmosphere*, *Blue carbon* and *Ocean acidification (OA)*.

Carbon dioxide (CO_2)
A naturally occurring gas, CO_2 is also a by-product of burning *fossil fuels* (such as oil, gas and coal), of burning *biomass*, of *land use* changes (LUC) and of industrial processes (e.g., cement production). It is the principal *anthropogenic* greenhouse gas (GHG) that affects the Earth's radiative balance. It is the reference gas against which other GHGs are measured and therefore has a Global Warming Potential (GWP) of 1. See also *Global warming*, *Greenhouse gas (GHG)*, *Land* and *Ocean acidification (OA)*.

Carbon dioxide removal (CDR)
Anthropogenic activities removing *carbon dioxide (CO_2)* from the *atmosphere* and durably storing it in geological, terrestrial, or *ocean* reservoirs, or in products. It includes existing and potential anthropogenic enhancement of biological or geochemical CO_2 *sinks* and direct air capture and storage but excludes natural CO_2 *uptake* not directly caused by human activities. See also *Greenhouse gas removal (GGR)*, *Mitigation (of climate change)* and *Negative emissions*.

Carbon price
The price for avoided or released *carbon dioxide (CO_2)* or CO2-equivalent emissions. This may refer to the rate of a carbon tax, or the price of emission permits. In many models that are used to assess the economic costs of *mitigation*, carbon prices are used as a proxy to represent the level of effort in mitigation policies.

Carbon sequestration
See *Sequestration*.

Carbon sink
See *Sink*.

Cascading impacts
from *extreme weather/climate events* occur when an extreme *hazard* generates a sequence of secondary events in natural and *human systems* that result in physical, natural, social or economic disruption, whereby the resulting impact is significantly larger than the initial impact. Cascading impacts are complex and multidimensional, and are associated more with the magnitude of *vulnerability* than with that of the hazard (modified from Pescaroli & Alexander, 2015). See also *Impacts (consequences, outcomes)*, *Natural systems* and *Risk*.

Climate
in a narrow sense is usually defined as the average weather – or more rigorously, as the statistical description in terms of the mean and variability of relevant quantities – over a period of time ranging from months to thousands or millions of years. The classical period for aver-

aging these variables is 30 years, as defined by the World Meteorological Organization (WMO). The relevant quantities are most often surface variables such as temperature, precipitation and wind. Climate in a wider sense is the state, including a statistical description, of the *climate system*.

Climate change

A change in the state of the *climate* that can be identified (e.g., by using statistical tests) by changes in the mean and/or the variability of its properties and that persists for an extended period, typically decades or longer. Climate change may be due to natural internal processes or external *forcings* such as modulations of the solar cycles, volcanic eruptions and persistent *anthropogenic* changes in the composition of the *atmosphere* or in *land use*. Note that the *United Nations Framework Convention on Climate Change (UNFCCC)*, in its Article 1, defines climate change as: 'a change of climate which is attributed directly or indirectly to human activity that alters the composition of the global atmosphere and which is in addition to natural climate variability observed over comparable time periods'. The UNFCCC thus makes a distinction between climate change attributable to human activities altering the atmospheric composition and *climate variability* attributable to natural causes. See also *Global warming*, *Ocean acidification (OA)* and *Detection and attribution*.

Climate extreme (extreme weather or climate event)

The occurrence of a value of a weather or *climate* variable above (or below) a threshold value near the upper (or lower) ends of the range of observed values of the variable. For simplicity, both extreme weather events and extreme climate events are referred to collectively as 'climate extremes'. See also *Extreme weather/ climate event*.

Climate feedback

An interaction in which a perturbation in one *climate* quantity causes a change in a second and the change in the second quantity ultimately leads to an additional change in the first. A negative feedback is one in which the initial perturbation is weakened by the changes it causes; a positive feedback is one in which the initial perturbation is enhanced. The initial perturbation can either be externally forced or arise as part of internal variability. See also *Climate variability* and *Forcing*.

Climate governance

See *Governance*.

Climate model

A qualitative or quantitative representation of the *climate system* based on the physical, chemical and biological properties of its components, their interactions and feedback processes and accounting for some of its known properties. The *climate* system can be represented by models of varying complexity; that is, for any one component or combination of components a spectrum or hierarchy of models can be identified, differing in such aspects as the number of spatial dimensions, the extent to which physical, chemical or biological processes are explicitly represented, or the level at which empirical parametrisations are involved. There is an evolution towards more complex models with interactive chemistry and biology scenarios. Climate models are applied as a research tool to study and simulate the climate and for operational purposes, including monthly, seasonal and interannual climate predictions. See also *Climate sensitivity* and *Earth system model (ESM)*.

Climate projection

Simulated response of the *climate system* to a scenario of future emissions or concentrations of *greenhouse gases (GHGs)* and *aerosols* and changes in *land use*, generally derived using *climate models*. Climate projections depend on an emission/concentration/ radiative forcing *scenario*, which is in turn based on assumptions concerning, for example, future socioeconomic and technological developments that may or may not be realised. See also *(Model) Ensemble*, *Projection* and *Radiative forcing*.

Climate-resilient development pathways (CRDPs)

Trajectories that strengthen *sustainable development* and efforts to eradicate *poverty* and reduce inequalities while promoting fair and cross-scalar *adaptation* to and *resilience* in a changing *climate*. They raise the ethics, *equity*, and *feasibility* aspects of the deep societal transformation needed to drastically reduce emissions to limit *global warming* (e.g., to well below 2°C) and achieve desirable and liveable futures and *well-being* for all. See also *Equality*.

Climate sensitivity

The change in the annual *global mean surface temperature (GMST)* in response to a change in the atmospheric *carbon dioxide (CO_2)* concentration or other radiative *forcing*.

Equilibrium climate sensitivity The equilibrium (steady state) change in the globally-averaged near-surface temperature following a doubling of the atmospheric CO_2 concentration from preindustrial conditions. Often estimated through experiments in atmosphere-ocean general circulation models (AOGCMs) where CO_2 levels are either quadrupled or doubled from *pre-industrial* levels and which are integrated for 100–200 years. A related quantity, the *climate feedback* parameter (unit: W m^{-2} °C^{-1}) refers to the top of *atmosphere* budget change per degree of globally-averaged near-surface temperature change. See also *Climate model* and *Global mean surface temperature (GMST)*.

Climate system

Global system consisting of five major components: the *atmosphere*, hydrosphere, *cryosphere*, lithosphere and biosphere, and the interactions between them. The climate system changes in time under the influence of its own internal dynamics and because of external *forcings* such as volcanic eruptions, solar variations, orbital forcing, and *anthropogenic* forcings such as the changing composition of the atmosphere and *land use* change.

Climate variability

Deviations of some *climate* variables from a given mean state (including the occurrence of extremes, etc.) at all spatial and temporal scales beyond that of individual weather events. Variability may be intrinsic, due to fluctuations of processes internal to the *climate system* (internal variability), or extrinsic, due to variations in natural or *anthropogenic* external *forcing* (forced variability).

Coast

The *land* near to the sea. The term 'coastal' can refer to that land (e.g., as in 'coastal communities'), or to that part of the marine environment that is strongly influenced by land-based processes. Thus, coastal seas are generally shallow and near-shore. The landward and seaward limits of the coastal zone are not consistently defined, neither scientifically nor legally. Thus, coastal waters can either be considered as equivalent to territorial waters (extending 12 nautical miles/22.2 km from mean low water), or to the full Exclusive Economic Zone, or to *shelf seas*, with less than 200 m water depth. See also *Ocean*, *Ocean deoxygenation* and *Sea level change (sea level rise/sea level fall)*.

Co-benefits

The positive effects that a policy or measure aimed at one objective might have on other objectives, thereby increasing the total benefits for society or the environment. Co-benefits are often subject to *uncertainty* and depend on local circumstances and implementation practices, among other factors. Co-benefits are also referred to as ancillary benefits. See also *Risk*.

Compound events

See *Compound weather/climate events*.

Compound weather/climate events

The combination of multiple *drivers* and/or *hazards* that contributes to societal and environmental *risk* (Zscheischler et al., 2018).

Compound risks

arise from the interaction of *hazards*, which may be characterised by single extreme events or multiple coincident or sequential events that interact with exposed systems or sectors. See also *Extreme weather/ climate event* and *Risk*.

Confidence

The robustness of a finding based on the type, amount, quality and consistency of *evidence* (e.g., mechanistic understanding, theory, data, models, expert judgment) and on the degree of *agreement* across multiple lines of evidence. In this Special Report, confidence is expressed qualitatively (Mastrandrea et al., 2010). See Section 1.8.3 for the list of confidence levels used. See also *Likelihood* and *Uncertainty*.

Coral reef

An underwater *ecosystem* characterised by structure-building stony corals. Warm water coral reefs occur in shallow seas, mostly in the tropics, with the corals (animals) containing algae (plants) that depend on light and relatively stable temperature conditions. Cold water coral reefs occur throughout the world, mostly at water depths of 50–500 m. In both kinds of reef, living corals frequently grow on older, dead material, predominantly made of calcium carbonate ($CaCO_3$). Both warm and cold-water coral reefs support high biodiversity of fish and other groups, and are considered to be especially vulnerable to *climate change*. See also *Ocean acidification (OA)*.

Cost-benefit analysis

Monetary assessment of all negative and positive impacts associated with a given action. Cost-benefit analysis enables comparison of different interventions, investments or strategies and reveal how a given investment or policy effort pays off for a particular person, company or country. Cost-benefit analyses representing society's point of view are important for *climate change* decision making, but there are difficulties in aggregating costs and benefits across different actors and across timescales. See also *Discounting*.

Cost-effectiveness

A measure of the cost at which a policy goal or outcome is achieved. The lower the cost, the greater the cost-effectiveness. See also *Private costs*.

Coupled Model Intercomparison Project (CMIP)

A *climate* modelling activity from the World Climate Research Programme (WCRP) which coordinates and archives *climate model* simulations based on shared model inputs by modelling groups from around the world. The CMIP3 multi-model data set includes projections using Special Report on Emissions Scenarios (SRES) scenarios. The CMIP5 data set includes projections using the *Representative Concentration Pathways (RCP)*. The CMIP6 phase involves a suite of common model experiments as well as an ensemble of CMIP-endorsed Model Intercomparison Projects (MIPs). See also *Climate projection*.

Cryosphere

The components of the Earth System at and below the *land* and *ocean* surface that are frozen, including snow cover, *glaciers*, *ice sheets*, *ice shelves*, *icebergs*, *sea ice*, lake ice, river ice, *permafrost* and seasonally *frozen ground*. See also *Climate system*.

Cultural services

See *Ecosystem services*.

Cumulative emissions

The total amount of emissions released over a specified period of time. See also *Carbon budget*.

Deep uncertainty

A situation of deep uncertainty exists when experts or stakeholders do not know or cannot agree on: (1) appropriate conceptual models that describe relationships among key driving forces in a system; (2) the probability distributions used to represent uncertainty about key variables and parameters; and/or (3) how to weigh and value desirable alternative outcomes (Lempert et al., 2003).

Deforestation

Conversion of *forest* to non-forest. [Note: For a discussion of the term forest and related terms such as afforestation, reforestation, and deforestation in the context of reporting and accounting Article 3.3 and 3.4 activities under the *Kyoto Protocol*, see 2013 Revised Supplementary Methods and Good Practice Guidance Arising from the Kyoto Protocol].

Detection

See *Detection and attribution*.

Detection and attribution

Detection of change is defined as the process of demonstrating that *climate* or a system affected by climate has changed in some defined statistical sense, without providing a reason for that change. An identified change is detected in observations if its *likelihood* of occurrence by chance due to internal variability alone is determined to be small, for example, <10%. Attribution is defined as the process of evaluating the relative contributions of multiple causal factors to a change or event with a formal assessment of *confidence*.

Developed/developing countries (Industrialised/developed/developing countries)

There is a diversity of approaches for categorizing countries on the basis of their level of development, and for defining terms such as industrialised, developed, or developing. Several categorisations are used in this Special Report. (1) In the United Nations (UN) system, there is no established convention for the designation of developed and developing countries or areas. (2) The UN Statistics Division specifies developed and developing regions based on common practice. In addition, specific countries are designated as least developed countries, landlocked developing countries, *small island developing states (SIDS)*, and transition economies. Many countries appear in more than one of these categories. (3) The World Bank uses income as the main criterion for classifying countries as low, lower middle, upper middle, and high income. (4) The UN Development Programme (UNDP) aggregates indicators for life expectancy, educational attainment, and income into a single composite Human Development Index (HDI) to classify countries as low, medium, high, or very high human development.

Development pathways

See *Pathways*.

Disaster

A 'serious disruption of the functioning of a community or a society at any scale due to hazardous events interacting with conditions of exposure, vulnerability and capacity, leading to one or more of the following: human, material, economic and environmental losses and impacts' (UNISDR, 2017). See also *Disaster risk management (DRM)*, *Exposure*, *Hazard*, *Risk* and *Vulnerability*.

Disaster risk management (DRM)

Processes for designing, implementing, and evaluating strategies, policies, and measures to improve the understanding of current and future disaster *risk*, foster *disaster* risk reduction and transfer, and promote continuous improvement in disaster preparedness, prevention and protection, response, and recovery practices, with the explicit purpose of increasing human security, *well-being*, quality of life, and *sustainable development (SD)*.

Discharge (of ice)

Rate of the flow of ice through a vertical section of a *glacier* perpendicular to the direction of the flow of ice. Often used to refer to the loss of mass at marine-terminating glacier fronts (mostly *calving* of *icebergs* and submarine melt), or to mass flowing across the *grounding line* of a floating *ice shelf*. See also *Mass balance/budget (of glaciers or ice sheets)*.

Discounting

A mathematical operation that aims to make monetary (or other) amounts received or expended at different times (years) comparable across time. The discounter uses a fixed or possibly time-varying discount rate from year to year that makes future value worth less today (if the discount rate is positive). The choice of discount rate(s) is debated as it is a judgement based on hidden and/or explicit values.

Discount rate

See *Discounting*.

Displacement

See *(Internal) Displacement (of humans)*.

Dissolved inorganic carbon

The combined total of different types of non-organic carbon in (seawater) solution, comprising carbonate (CO_3^{2-}), bicarbonate (HCO_3^-), carbonic acid (H_2CO_3) and *carbon dioxide (CO_2)*.

Dissolved organic carbon (DOC) and particulate organic carbon (POC)

Organic carbon types – for example, in the *ocean* – operationally separated by filtration. Filter pore size typically is 0.45 micrometres but may vary between 0.22 and 0.7 micrometres, with smaller carbon types in the solution (DOC) and larger carbon types (POC) being filtered out. In the global ocean, the ratio of DOC and POC is approximately 20:1. DOC can be further classified as labile DOC (LDOC) and refractory DOC (RDOC; also known as recalcitrant DOC). In the global ocean, DOC is mainly (>90%) comprised of RDOC. RDOC can be generated by *microbial carbon pump* processes, and is able to persist for hundreds to thousands of years due to its resistance to microbial decomposition. LDOC occurs mainly in surface seawaters and is readily available for biological utilisation or decomposition. See also *Carbon cycle*.

Downscaling

A method that derives local- to regional-scale (up to 100 km) information from larger-scale models or data analyses. Two main methods exist: dynamical downscaling and empirical/statistical downscaling. The dynamical method uses the output of regional *climate models*, global models with variable spatial resolution, or high-resolution global models. The empirical/statistical methods are based on observations and develop statistical relationships that link the large-scale atmospheric variables with local/regional *climate* variables. In all cases, the quality of the driving model remains an important limitation on quality of the downscaled information. The two methods can be combined, for example, applying empirical/statistical downscaling to the output of a regional climate model, consisting of a dynamical downscaling of a global climate model.

Driver

Any natural or human-induced factor that directly or indirectly causes a change in a system (adapted from MA, 2005). See also *Forcing*.

Drought

A period of abnormally dry weather long enough to cause a serious hydrological imbalance. Drought is a relative term; therefore any discussion in terms of precipitation deficit must refer to the particular precipitation-related activity that is under discussion. For example, shortage of precipitation during the growing season impinges on crop production or *ecosystem* function in general (due to *soil moisture* drought, also termed agricultural drought) and during the *runoff* and percolation season primarily affects water supplies (hydrological drought). Storage changes in soil moisture and groundwater are also affected by increases in actual evapotranspiration in addition to reductions in precipitation. A period with an abnormal precipitation deficit is defined as a meteorological drought. See also *Heatwave* and *Hydrological cycle*.

Early warning systems (EWS)

The set of technical and institutional capacities to forecast, predict, and communicate timely and meaningful warning information to enable individuals, communities, managed *ecosystems*, and organisations threatened by a *hazard* to prepare to act promptly and appropriately to reduce the possibility of harm or loss. Dependent upon context, EWS may draw upon scientific and/or *indigenous knowledge*, and other knowledge types. EWS are also considered for ecological applications, for example, conservation, where the organisation itself is not threatened by hazard but the ecosystem under conservation is (e.g., coral bleaching alerts), in agriculture (e.g., warnings of heavy rainfall, *drought*, ground frost and hailstorms) and in fisheries (e.g., warnings of storm, *storm surge* and tsunamis) (UNISDR 2009; IPCC, 2012a). See also *Disaster*, *Institutions*, *Local knowledge* and *Loss and Damage, and losses and damages*.

Earth system model (ESM)

A coupled *atmosphere–ocean* general circulation model (AOGCM) in which a representation of the *carbon cycle* is included, allowing for interactive calculation of atmospheric *carbon dioxide (CO₂)* or compatible emissions. Additional components (e.g., atmospheric chemistry, *ice sheets*, dynamic vegetation, nitrogen cycle, but also urban or crop models) may be included. See also *Climate model*.

Ecosystem

A functional unit consisting of living organisms, their non-living environment and the interactions within and between them. The components included in a given ecosystem and its spatial boundaries depend on the purpose for which the ecosystem is defined: in some cases they are relatively sharp, while in others they are diffuse. Ecosystem boundaries can change over time. Ecosystems are nested within other ecosystems and their scale can range from very small to the entire biosphere. In the current era, most ecosystems either contain people as key organisms, or are influenced by the effects of human activities in their environment. See also *Ecosystem services*.

Ecosystem-based adaptation (EbA)

See *Adaptation*.

Ecosystem services

Ecological processes or functions having monetary or non-monetary value to individuals or society at large. These are frequently classified as (1) supporting services such as productivity or *biodiversity* maintenance, (2) provisioning services such as food or fibre, (3) regulating services such as climate regulation or carbon *sequestration* and (4) cultural services such as tourism or spiritual and aesthetic appreciation. See also *Ecosystem* and *Nature's Contribution to People (NCP)*.

Elevation-dependent warming (EDW)

Characteristic of many regions where mountains are located, in which past and/or future surface air temperature changes vary neither uniformly nor linearly with elevation. In many cases, warming is enhanced within or above a certain elevation range.

El Niño-Southern Oscillation (ENSO)

The term El Niño was initially used to describe a warm-water current that periodically flows along the *coast* of Ecuador and Peru, disrupting the local fishery. It has since become identified with warming of the tropical Pacific Ocean east of the dateline. This oceanic event is associated with a fluctuation of a global-scale tropical and subtropical surface pressure pattern called the Southern Oscillation. This coupled atmosphere-ocean phenomenon, with preferred time scales of two to about seven years, is known as the El Niño-Southern Oscillation (ENSO). It is often measured by the surface pressure anomaly difference between Tahiti and Darwin and/or the *sea surface temperatures (SST)* in the central and eastern equatorial Pacific. During an ENSO event, the prevailing trade winds weaken, reducing upwelling and altering *ocean* currents such that the SSTs warm, further weakening the trade winds. This phenomenon has a great impact on the wind, SST and precipitation patterns in the tropical Pacific. It has climatic effects throughout the Pacific region and in many other parts of the world, through global *teleconnections*. The cold phase of ENSO is called La Niña. See also *Climate*.

Emission pathways

See *Pathways*.

Emission scenario

A plausible representation of the future development of emissions of substances that are radiatively active (e.g., *greenhouse gases*, or *aerosols*) based on a coherent and internally consistent set of assumptions about driving forces (such as demographic and socioeconomic development, technological change, energy and *land use*) and their key relationships. Concentration scenarios, derived from emission scenarios, are often used as input to a *climate model* to compute *climate projections*. See also *Driver*, *Forcing*, *Mitigation scenario*, *Radiative forcing*, *Representative concentration pathways (RCPs, under Pathways)*, *Shared socioeconomic pathways (SSPs, under Pathways)* and *Scenario*.

Endemic species

Plants and animals that are only found in one geographic region. See also *Alien (non-native) species*, *Ecosystem* and *Invasive species*.

Enhanced weathering

A proposed method to increase the natural rate of removal of *carbon dioxide (CO₂)* from the *atmosphere* using silicate and carbonate rocks. The active surface area of these minerals is increased by grinding, before they are actively added to soil, beaches or the open *ocean*. See also *Carbon dioxide removal (CDR)*, *Geoengineering* and *Sequestration*.

Ensemble

See *(Model) Ensemble*.

Equality

A principle that ascribes equal worth to all human beings, including equal opportunities, rights, and obligations, irrespective of origins.

Inequality Uneven opportunities and social positions, and processes of discrimination within a group or society, based on gender, class, ethnicity, age and (dis)ability, often produced by uneven development. Income inequality refers to gaps between highest and lowest income earners within a country and between countries.

See also *Equity*.

Equilibrium climate sensitivity

See *Climate sensitivity*.

Equity

The principle of being fair and impartial, and a basis for understanding how the *impacts* and responses to *climate change*, including costs and benefits, are distributed in and by society in more or less equal ways. Often aligned with ideas of *equality*, *fairness* and *justice* and applied with respect to equity in the responsibility for, and distribution of, *climate* impacts and policies across society, generations, and gender, and in the sense of who participates and controls the processes of decision making.

Distributive equity Equity in the consequences, outcomes, costs and benefits of actions or policies. In the case of climate change or climate policies for different people, places and countries, including equity aspects of sharing burdens and benefits for *mitigation* and *adaptation*.

Gender equity Equity between women and men with regard to their rights, resources and opportunities. In the case of climate change, gender equity recognises that women are often more vulnerable to the impacts of climate change and may be disadvantaged in the process and outcomes of climate policy.

Intergenerational equity Equity between generations that, in the context of climate change, acknowledges that the effects of past and present emissions, vulnerabilities and policies impose costs and benefits for people in the future and of different age groups.

Procedural equity Equity in the process of decision making including recognition and inclusiveness in participation, equal representation, bargaining power, voice and equitable access to knowledge and resources to participate.

Evidence

Data and information used in the scientific process to establish findings. In this Special Report, the degree of evidence reflects the amount, quality, and consistency of scientific/technical information on which the Lead Authors are basing their findings. See also *Agreement*, *Confidence*, *Likelihood* and *Uncertainty*.

Evolutionary adaptation

See *Adaptation*.

Exposure

The presence of people; *livelihoods*; species or *ecosystems*; environmental functions, services, and resources; infrastructure, or economic, social, or cultural assets in places and settings that could be adversely affected. See also *Hazard*, *Risk* and *Vulnerability*.

Extratropical cyclone

Any cyclonic-scale storm that is not a *tropical cyclone*. Usually refers to a middle- or high-latitude migratory storm system formed in regions of large horizontal temperature variations. Sometimes called extratropical storm or extratropical low.

Extreme event

See *Extreme weather/climate event*.

Extreme sea level

See *Storm surge*.

Extreme weather/climate event

An extreme weather event is an event that is rare at a particular place and time of year. Definitions of 'rare' vary, but an extreme weather event would normally be as rare as or rarer than the 10th or 90th percentile of a probability density function estimated from observations. By definition, the characteristics of what is called extreme weather may vary from place to place in an absolute sense. When a pattern of extreme weather persists for some time, such as a season, it may be classified as an extreme climate event, especially if it yields an average or total that is itself extreme (e.g., high temperature, *drought*, or total rainfall over a season). See also *Heat wave* and *Climate extreme (extreme weather or climate event)*.

Fairness

Impartial and just treatment without favouritism or discrimination in which each person is considered of equal worth with equal opportunity. See also *Equity* and *Equality*.

Feasibility

The degree to which climate goals and response options are considered possible and/or desirable. Feasibility depends on geophysical, ecological, technological, economic, social and *institutional* conditions for change. Conditions underpinning feasibility are dynamic, spatially variable, and may vary between different groups.

Economic feasibility An indicator of the benefits and costs of a climate adaptation or response, often expressed as a ratio of the two, used in order to judge whether it is possible or wise to proceed with the option.

Social and institutional feasibility Institutional feasibility has two key parts: (1) the extent of administrative workload, both for public authorities and for regulated entities, and (2) the extent to which the policy is viewed as legitimate, gains acceptance, is adopted, and is implemented.

Feedback

See *Climate feedback*.

Firn

Snow that has survived at least one *ablation* season but has not been transformed to *glacier* ice. Its pore space is at least partially interconnected, allowing air and water to circulate. Firn densities typically are 400–830 kg m^{-3}. See also *Cryosphere*.

Flood

The overflowing of the normal confines of a stream or other water body, or the accumulation of water over areas that are not normally submerged. Floods can be caused by unusually heavy rain, for example during storms and cyclones. Floods include river (fluvial) floods, flash floods, urban floods, rain (pluvial) floods, sewer floods, *coastal* floods and *glacial lake outburst floods (GLOFs)*. See also *Runoff*.

Food security

A situation that exists when all people, at all times, have physical, social and economic access to sufficient, safe and nutritious food that meets their dietary needs and food preferences for an active and healthy life (FAO, 2001). [Note: Whilst the term 'food security' explicitly includes nutrition within it 'dietary needs … for an active and healthy life', in the past the term has sometimes privileged the supply of energy, especially to the hungry. Thus, the term 'food and nutrition security' is often used (with the same definition as food security) to emphasise that the term food covers both energy and nutrition (FAO, 2009).] See also *Food system* and *Malnutrition*.

Food system

All the elements (environment, people, inputs, processes, infrastructures, *institutions*, etc.) and activities that relate to the production, processing, distribution, preparation and consumption of food, and the output of these activities, including socioeconomic and environmental outcomes (HLPE, 2017). [Note: Whilst there is a global food system (encompassing the totality of global production and consumption), each location's food system is unique, being defined by that place's mix of food produced locally, nationally, regionally or globally]. See also *Food security*.

Forcing

The *driver* of a change in the *climate system*, usually through an imbalance between the radiative energy received by and leaving the Earth's surface. See also *Radiative forcing* and *Short-lived climate forcers (SLCF)*.

Forest

A vegetation type dominated by trees. Many definitions of the term forest are in use throughout the world, reflecting wide differences in biogeophysical conditions, social structure and economics. [Note: For a discussion of the term forest and related terms such as afforestation, reforestation and *deforestation*, see the IPCC Special Report on Land Use, Land-Use Change, and Forestry (IPCC, 2000). See also information provided by the *United Nations Framework Convention on Climate Change (UNFCCC*, 2013) and the Report on Definitions and Methodological Options to Inventory Emissions from Direct Human-induced Degradation of Forests and Devegetation of Other Vegetation Types (IPCC, 2003)].

Fossil fuels

Carbon-based fuels from fossil hydrocarbon deposits, including coal, oil and natural gas.

Framework Convention on Climate Change

See *United Nations Framework Convention on Climate Change (UNFCCC)*.

Frozen ground

Soil or rock in which part or all of the pore water consists of ice. See also *Permafrost*.

Functional diversity

'The range and value of those species and organismal traits that influence *ecosystem* functioning' (Tilman 2001). See also *Biodiversity*.

General circulation model

See *Climate model*.

Geoengineering

A broad set of methods and technologies that aim to deliberately alter the *climate system* in order to alleviate the impacts of *climate change*. Most, but not all, methods seek to either (1) reduce the amount of absorbed solar energy in the climate system (solar radiation management, or *solar radiation modification, SRM*) or (2) increase net carbon *sinks* from the *atmosphere* at a scale sufficiently large to alter *climate* (i.e., *carbon dioxide removal, CDR*). Scale and intent are of central importance. Two key characteristics of geoengineering methods of particular concern are that they use or affect the climate system (e.g., atmosphere, *land* or *ocean*) globally or regionally and/or could have substantive unintended effects that cross national boundaries. Geoengineering is different from weather modification and ecological engineering, but the boundary can be unclear (IPCC, 2012b, p. 2). See also *Blue carbon*.

Glaciated

State of a surface that was covered by *glacier* ice in the past, but not at present.

Glacier

A perennial mass of ice, and possibly *firn* and snow, originating on the *land* surface by *accumulation* and compaction of snow and showing evidence of past or present flow. A glacier typically gains mass by accumulation of snow, and loses mass by *ablation*. Land ice masses of continental size (>50,000 km^2) are referred to as *ice sheets* (Cogley et al., 2011). See also *Calving (of glaciers or ice sheets)*, *Cryosphere*, *Grounding line* and *Mass balance/budget (of glaciers or ice sheets)*.

Glacial lake outburst flood (GLOF)/Glacier lake outburst

A sudden release of water from a *glacier* lake, including any of the following types – a glacier-dammed lake, a pro-glacial moraine-dammed lake or water that was stored within, under or on the glacier.

Global climate model

See *Climate model*.

Global mean surface temperature (GMST)

Estimated global average of near-surface air temperatures over *land* and *sea ice*, and *sea surface temperature (SST)* over ice-free *ocean* regions, with changes normally expressed as departures from a value over a specified *reference period*. When estimating changes in GMST, near-surface air temperatures over both land and oceans are also used.

Global warming

An increase in *global mean surface temperature (GMST)* averaged over a 30-year period, or the 30-year period centred on a particular year or decade, expressed relative to *pre-industrial* levels unless otherwise specified. For 30-year periods that span past and future years, the current multi-decadal warming trend is assumed to continue. See also *Climate change* and *Climate variability*.

Governance

In this Special Report, governance refers to the effort to establish, reaffirm or change formal and informal *institutions* at all scales to negotiate relationships, resolve social conflicts and realise mutual gains (Paavola, 2007; Williamson, 2000). It refers to how the economy and society are governed or regulated; and how collective interests are defined, reconciled and institutionalised (Peters and Pierre, 2001). Governance may be an act of governments (e.g., a government restricting resource use), non-governmental organisation (e.g., issuing green certification), private actors (e.g., resource users establishing rules or norms for restricting use of a common resource) or any combination of these. Governance does not only include establishing institutions such as laws or policies, but also their implementation, enforcement and monitoring. The term 'governance' is used in diverse and contested ways.

Adaptive governance An emerging term in the literature for the evolution of formal and informal institutions of governance that prioritise planning, implementation and evaluation of policy through iterative social learning; in the context of climate change, governance facilitating social learning to steer the use and protection of natural resources, and *ecosystem services*, particularly in situations of complexity and *uncertainty*.

Climate governance includes efforts to share the burden of emission reduction amongst countries, sectors and groups of society (*mitigation*), and to resolve conflicts involved in, or to realise mutual gains through, adapting to *climate change*.

Deliberative governance involves decision making through inclusive public conversation which allows opportunity for developing policy options through public discussion rather than collating individual preferences through voting or referenda (although the latter governance mechanisms can also be proceeded and legitimated by public deliberation processes).

Multi-level governance refers to the dispersion of governance across multiple levels of jurisdiction and decision making (Hooghe and Marks, 2003), including trans-regional and trans-national, regional, national and local levels. The concept emphasises that modern governance generally consists in, and is more flexible when there is, a vertical 'layering' of governance processes at different levels.

Participatory governance favours direct public engagement in decision and policy making using a variety of techniques such as referenda, community deliberation, citizen juries or participatory budgeting. The approach can be applied in formal and informal institutional contexts from national to local levels, but is usually associated with devolved decision making (Fung and Wright, 2003; Sarmiento and Tilly, 2018).

Polycentric governance involves multiple centres of decision making with overlapping jurisdictions. While the centres have some degree of autonomy, they also take each other into account, coordinating their actions and seeking to resolve conflicts (Carlisle and Gruby, 2017; Jordan et al., 2018; McGinnis and Ostrom, 2012).

Gravity Recovery And Climate Experiment (GRACE)

A pair of satellites to measure the Earth's gravity field anomalies from 2002 to 2017. These fields have been used, among other things, to study mass changes of the polar *ice sheets and glaciers*. See also *Marine ice sheet instability (MISI)* and *Mass balance/budget (of glaciers or ice sheets)*.

Green infrastructure

The interconnected set of natural and constructed ecological systems, green spaces and other landscape features. It includes planted and indigenous trees, wetlands, parks, green open spaces and original grassland and woodlands, as well as possible building and street level design interventions that incorporate vegetation. Green infrastructure provides services and functions in the same way as conventional infrastructure (Culwick and Bobbins, 2016). See also *Ecosystem* and *Ecosystem services*.

Greenhouse gases (GHG)

Gaseous constituents of the *atmosphere*, both natural and *anthropogenic*, that absorb and emit radiation at specific wavelengths within the spectrum of radiation emitted by the Earth's *ocean* and *land* surface, by the atmosphere itself, and by clouds. This property causes the greenhouse effect. Water vapour (H_2O), *carbon dioxide (CO_2)*, nitrous oxide (N_2O), *methane (CH_4)* and *ozone (O_3)* are the primary GHGs in the Earth's atmosphere. Human-made GHGs include sulphur hexafluoride (SF_6), hydrofluorocarbons (HFCs), chlorofluorocarbons (CFCs) and perfluorocarbons (PFCs); several of these are also O_3-depleting (and are regulated under the Montreal Protocol).

Greenhouse gas removal

Withdrawal of a *greenhouse gas (GHG)* and/or a *precursor* from the *atmosphere* by a *sink*. See also *Carbon dioxide removal (CDR)* and *Negative emissions*.

Gross domestic product (GDP)

The sum of gross value added, at purchasers' prices, by all resident and non-resident producers in the economy, plus any taxes and minus any subsidies not included in the value of the products in a country or a geographic region for a given period, normally one year. GDP is calculated without deducting for depreciation of fabricated assets or depletion and degradation of natural resources.

Grounding line
The junction between a *glacier* or *ice sheet* and an *ice shelf*; the place where ice starts to float. This junction normally occurs over a zone, rather than at a line.

Habitability
The ability of a place to support human life by providing protection from *hazards* which challenge human survival, and by assuring adequate space, food and freshwater.

Hazard
The potential occurrence of a natural or human-induced physical event or trend that may cause loss of life, injury, or other health *impacts*, as well as damage and loss to property, infrastructure, *livelihoods*, service provision, *ecosystems* and environmental resources. See also *Disaster*, *Exposure*, *Loss and Damage, and losses and damages*, *Risk* and *Vulnerability*.

Heat wave
A period of abnormally hot weather. Heat waves and warm spells have various and in some cases overlapping definitions. See also *Climate extreme (extreme weather or climate event)*, *Extreme weather event* and *Marine heatwave*.

Holocene
The current interglacial geological epoch, the second of two epochs within the Quaternary period, the preceding being the Pleistocene. The International Commission on Stratigraphy defines the start of the Holocene at 11,700 years before 2000 (ICS, 2019).

Human behaviour
The responses of persons or groups to a particular situation, here likely to relate to *climate change*. Human behaviour covers the range of actions by individuals, communities, organisations, governments and at the international level.

> **Adaptation behaviour** Human actions that directly or indirectly affect the *risks* of climate change *impacts*.

> **Mitigation behaviour** Human actions that directly or indirectly influence *mitigation*.

See also *Adaptation*.

Human mobility
The permanent or semi-permanent move by a person for at least one year and involving crossing an administrative, but not necessarily a national, border.

Human rights
Rights that are inherent to all human beings, universal, inalienable, and indivisible, typically expressed and guaranteed by law. They include the right to life, economic, social, and cultural rights, and the right to development and self-determination (UNOHCHR, 2018).

> **Procedural rights** Rights to a legal procedure to enforce *substantive rights*.

> **Substantive rights** Basic human rights, including the right to the substance of being human such as life itself, liberty and happiness.

See also *Equity*, *Equality*, *Justice* and *Well-being*.

Human security
A condition that is met when the vital core of human lives is protected, and when people have the freedom and capacity to live with dignity. In the context of *climate change*, the vital core of human lives includes the universal and culturally specific, material and non-material elements necessary for people to act on behalf of their interests and to live with dignity.

Human system
Any system in which human organisations and *institutions* play a major role. Often, but not always, the term is synonymous with society or social system. Systems such as agricultural systems, urban systems, political systems, technological systems and economic systems are all human systems in the sense applied in this report.

Hydrological cycle
The cycle in which water evaporates from the *ocean* and the *land* surface, is carried over the Earth in atmospheric circulation as water vapour, condenses to form clouds, precipitates over the ocean and land as rain or snow, which on land can be intercepted by trees and vegetation, potentially accumulating as snow or ice, provides *runoff* on the land surface, infiltrates into soils, recharges groundwater, discharges into streams, and ultimately, flows into the oceans as rivers, polar *glaciers* and *ice sheets*, from which it will eventually evaporate again. The various systems involved in the hydrological cycle are usually referred to as hydrological systems.

Iceberg
Large piece of freshwater ice broken off from a *glacier* or an *ice shelf* during *calving* and floating in open water (at least five metres height above sea level). Smaller pieces of floating ice known as 'bergy bits' (less than 5 metres above sea level) or 'growlers' (less than 2 metres above sea level) can originate from glaciers or ice shelves, or from the breaking up of a large iceberg. Icebergs can also be classified by shape, most commonly being either tabular (steep sides and a flat top) or non-tabular (varying shapes, with domes and spires) (NOAA, 2019). In lakes, icebergs can originate by breaking off shelf ice, which forms through freezing of a lake surface. See also *Calving (of glaciers or ice sheets)* and *Marine ice cliff instability (MICI)*.

Iceberg calving
See *Calving (of glaciers or ice sheets)*.

Ice core
A cylinder of ice drilled out of a *glacier* or *ice sheet* to gain information on past changes in *climate* and composition of the *atmosphere* preserved in the ice or in air trapped in ice.

Ice sheet
An ice body originating on *land* that covers an area of continental size, generally defined as covering >50,000 km^2, and that has formed over thousands of years through *accumulation* and compaction of

snow. An ice sheet flows outward from a high central ice plateau with a small average surface slope. The margins usually slope more steeply, and most ice is *discharged* through fast-flowing ice streams or outlet *glaciers*, often into the sea or into *ice shelves* floating on the sea. There are only two ice sheets in the modern world, one on Greenland and one on Antarctica. The latter is divided into the East Antarctic Ice Sheet (EAIS), the West Antarctic Ice Sheet (WAIS) and the Antarctic Peninsula ice sheet. During glacial periods, there were other ice sheets. See also *Ablation*, *Calving (of glaciers or ice sheets)*, *Grounding line*, *Hydrological cycle*, *Marine ice cliff instability (MICI)*, *Marine ice sheet instability (MISI)* and *Mass balance/budget (of glaciers or ice sheets)*.

Ice shelf

A floating slab of ice originating from *land* of considerable thickness extending from the *coast* (usually of great horizontal extent with a very gently sloping surface), resulting from the flow of *ice sheets*, initially formed by the accumulation of snow, and often filling embayments in the coastline of an ice sheet. Nearly all ice shelves are in Antarctica, where most of the ice *discharged* into the *ocean* flows via ice shelves. See also *Calving (of glaciers or ice sheets)*, *Glacier*, *Hydrological cycle*, *Marine ice cliff instability (MICI)* and *Marine ice sheet instability (MISI)*.

Ice stream

A stream of ice with strongly enhanced flow that is part of an *ice sheet*. It is often separated from surrounding ice by strongly sheared, crevassed margins. See also *Outlet glacier*.

Impacts (consequences, outcomes)

The consequences of realised *risks* on natural and *human systems*, where risks result from the interactions of climate-related *hazards* (including *extreme weather/climate events*), *exposure*, and *vulnerability*. Impacts generally refer to effects on lives, livelihoods, health and *well-being*, *ecosystems* and species, economic, social and cultural assets, services (including *ecosystem services*), and infrastructure. Impacts may be referred to as consequences or outcomes, and can be adverse or beneficial. See also *Adaptation*, *Loss and Damage, and losses and damages* and *Natural systems*.

Incremental adaptation

See *Adaptation*.

Indigenous knowledge (IK)

The understandings, skills and philosophies developed by societies with long histories of interaction with their natural surroundings. For many indigenous peoples, IK informs decision making about fundamental aspects of life, from day-to-day activities to longer term actions. This knowledge is integral to cultural complexes, which also encompass language, systems of classification, resource use practices, social interactions, values, ritual and spirituality. These distinctive ways of knowing are important facets of the world's cultural diversity (UNESCO, 2018). See also *Local knowledge (LK)*.

Industrial revolution

A period of rapid industrial growth with far-reaching social and economic consequences, beginning in Britain during the second half of the 18th century and spreading to Europe and later to other countries including the United States. The invention of the steam engine was an important trigger of this development. The industrial revolution marks the beginning of a strong increase in the use of *fossil fuels*, initially coal, and hence emission of *carbon dioxide (CO₂)*. See also *Pre-industrial*.

Inequality

See *Equality*.

Institutions

The 'prescriptions', that is the rules, norms and conventions, used by humans 'to organize all forms of repetitive and structured interactions including those within families, neighborhoods, markets, firms, sports leagues, churches, private associations, and governments at all scales' (Ostrom, 2005, p. 3). Institutions can be formal, such as laws and policies, or informal, such as traditions, customs, norms and conventions. Individuals and organisations, such as parliaments, regulatory agencies, private firms, and community bodies, develop and act in response to institutions and the incentives they frame. Institutions can guide, constrain and shape human interaction through direct control, incentives and processes of socialisation.

Integrated assessment

A method of analysis that combines results and models from the physical, biological, economic and social sciences and the interactions among these components in a consistent framework to evaluate the status and the consequences of environmental change and the policy responses to it.

(Internal) Displacement (of humans)

The involuntary movement, individually or collectively, of persons from their country or community, notably for reasons of armed conflict, civil unrest, or natural or man-made *disasters* (adapted from IOM, 2011). See also *Migration (of humans)* and *Planned relocation (of humans)*.

Internal variability

See *Climate variability*.

Invasive species

A species that is not native to a specific location or nearby, lacking natural controls, and has a tendency to rapidly increase in abundance, displacing native species. Invasive species may also damage the human economy or human health. See also *Alien (non-native) species*, *Ecosystem* and *Endemic species*.

Irreversibility

A perturbed state of a dynamical system is defined as irreversible on a given timescale if the recovery timescale from this state due to natural processes is significantly longer than the time it takes for the system to reach this perturbed state. In the context of this Special Report, the recovery time scale of interest is hundreds to thousands of years. See also *Tipping point*.

Justice

is concerned with ensuring that people get what is due to them setting out the moral or legal principles of *fairness* and *equity* in the way people are treated, often based on the ethics and values of society.

Climate justice Justice that links development and *human rights* to achieve a human-centred approach to addressing *climate change*, safeguarding the rights of the most vulnerable people and sharing the burdens and benefits of climate change and its impacts equitably and fairly (MRFJC, 2018).

Distributive justice Justice in the allocation of economic and non-economic costs and benefits across society.

Inter-generational justice Justice in the distribution of economic and non-economic costs and benefits across generations.

Procedural justice Justice in the way outcomes are brought about including who participates and is heard in the processes of decision making.

Social justice Just or fair relations within society that seek to address the distribution of wealth, access to resources, opportunity, and support according to principles of justice and fairness.

See also *Equity* and *Human rights*.

Kyoto Protocol

The Kyoto Protocol to the *United Nations Framework Convention on Climate Change (UNFCCC)* is an international treaty adopted in December 1997 in Kyoto, Japan, at the Third Session of the Conference of the Parties (COP3) to the UNFCCC. It contains legally binding commitments, in addition to those included in the UNFCCC. Countries included in Annex B of the Protocol (mostly OECD countries and countries with economies in transition) agreed to reduce their anthropogenic *greenhouse gas (GHG)* emissions (*carbon dioxide (CO$_2$)*, *methane (CH$_4$)*, nitrous oxide (N2O), hydrofluorocarbons (HFCs), perfluorocarbons (PFCs), and sulphur hexafluoride (SF6)) by at least 5% below 1990 levels in the first commitment period (2008–2012). The Kyoto Protocol entered into force on 16 February 2005 and as of May 2018 had 192 Parties (191 States and the European Union). A second commitment period was agreed in December 2012 at COP18, known as the Doha Amendment to the Kyoto Protocol, in which a new set of Parties committed to reduce GHG emissions by at least 18% below 1990 levels in the period from 2013 to 2020. However, as of May 2018, the Doha Amendment had not received sufficient ratifications to enter into force. See also *Anthropogenic* and *Paris Agreement*.

Labile dissolved organic carbon (LDOC)

See *Dissolved organic carbon (DOC)* and *particulate organic carbon (POC)*.

La Niña

See *El Niño-Southern Oscillation*.

Land

The terrestrial portion of the biosphere that comprises the natural resources (soil, near-surface air, vegetation and other biota, and water), the ecological processes, topography, and human settlements and infrastructure that operate within that system (FAO, 2007; UNCCD, 1994). See also *Ecosystem services* and *Land use*.

Land management

Sum of *land use* practices (e.g., sowing, fertilising, weeding, harvesting, thinning and clear-cutting) that take place within broader land use categories (Pongratz et al., 2018).

Land restoration

The process of assisting the recovery of *land* from a degraded state (IPBES, 2018; McDonald et al. 2015).

Land use

The total of arrangements, activities and inputs applied to a parcel of *land*. The term 'land use' is also used in the sense of the social and economic purposes for which land is managed (e.g., grazing, timber extraction, conservation and city dwelling). In national *greenhouse gas (GHG)* inventories, land use is classified according to the IPCC land use categories of *forest* land, cropland, grassland, wetlands, settlements, and other lands (see the 2006 IPCC Guidelines for National GHG Inventories for details). See also *Land management*.

Likelihood

The chance of a specific outcome occurring, where this might be estimated probabilistically. Likelihood is expressed in this Special Report using a standard terminology (Mastrandrea et al., 2010). See Section 1.9.2 in this Special Report for the list of likelihood qualifiers used. See also *Agreement*, *Evidence*, *Confidence* and *Uncertainty*.

Livelihood

The resources used and the activities undertaken in order for people to live. Livelihoods are usually determined by the entitlements and assets to which people have access. Such assets can be categorised as human, social, natural, physical, or financial.

Local knowledge (LK)

The understandings and skills developed by individuals and populations, specific to the places where they live. Local knowledge informs decision making about fundamental aspects of life, from day-to-day activities to longer term actions. This knowledge is a key element of the social and cultural systems which influence observations of and responses to *climate change*; it also informs *governance* decisions (UNESCO, 2018). See also *Indigenous knowledge (IK)*.

Local sea level change

Change in sea level relative to a datum (such as present-day mean sea level) at spatial scales smaller than 10km. See also *Regional sea level change* and *Sea level change (sea level rise/sea level fall)*.

Lock-in

A situation in which the future development of a system, including infrastructure, technologies, investments, *institutions* and behavioural norms, is determined or constrained ('locked in') by historic developments.

Loss and Damage, and losses and damages

Research has taken the term 'Loss and Damage' (capitalised letters) to refer to political debate under the *United Nations Framework Convention on Climate Change (UNFCCC)* following the establishment of the Warsaw Mechanism on Loss and Damage in 2013, which is to 'address

AI

loss and damage associated with impacts of climate change, including extreme events and slow onset events, in developing countries that are particularly vulnerable to the adverse effects of climate change.' The expression 'losses and damages' (lowercase letters) has been taken to refer broadly to harm from (observed) *impacts* and (projected) *risks* (Mechler et al., 2018).

Maladaptive actions (Maladaptation)

Actions that may lead to increased *risk* of adverse *climate*-related outcomes, including via increased *greenhouse gas (GHG)* emissions, increased *vulnerability* to *climate change*, or diminished welfare, now or in the future. Maladaptation is usually an unintended consequence. See also *Adaptation* and *Adaptive capacity*.

Marine heatwave

A period of extreme warm near-*sea surface temperature (SST)* that persists for days to months and can extend up to thousands of kilometres. See also *Climate extreme (extreme weather or climate event)*, *Extreme weather event* and *Heat wave*.

Marine ice cliff instability (MICI)

A hypothetic mechanism of an ice cliff failure. In case a marine-terminated *ice sheet* loses its buttressing *ice shelf*, an ice cliff can be exposed. If the exposed ice cliff is tall enough (about 800 m of the total height, or about 100 m of the above-water part), the stresses at the cliff face exceed the strength of the ice, and the cliff fails structurally in repeated *calving* events. See also *Iceberg* and *Marine ice sheet instability (MISI)*.

Marine ice sheet instability (MISI)

A mechanism of irreversible (on the decadal to centennial time scale) retreat of a *grounding line* for the marine-terminating *glaciers*, in case the glacier bed slopes towards the *ice sheet* interior. See also *Hydrological cycle*, *Ice shelf*, *Marine ice cliff instability (MICI)* and *Sea ice*.

Mass balance/budget (of glaciers or ice sheets)

Difference between the mass input (*accumulation*) and the mass loss (*ablation*) of an ice body (e.g., a glacier or ice sheet) over a stated time period, which is often a year or a season. Surface mass balance refers to the difference between surface accumulation and surface ablation. See also *Calving (of glaciers or ice sheets)* and *Discharge (of ice)*.

Measurement, reporting and verification (MRV)

Measurement 'Processes of data collection over time, providing basic datasets, including associated accuracy and precision, for the range of relevant variables. Possible data sources are field measurements, field observations, detection through remote sensing and interviews' (UN REDD, 2009).

Reporting 'The process of formal reporting of assessment results to the *UNFCCC*, according to predetermined formats and according to established standards, especially the Intergovernmental Panel on Climate Change (IPCC) Guidelines and GPG (Good Practice Guidance)' (UN REDD, 2009).

Verification 'The process of formal verification of reports, for example, the established approach to verify national communications and national inventory reports to the UNFCCC' (UN REDD, 2009).

Meridional Overturning Circulation (MOC)

Meridional (north-south) overturning circulation in the *ocean* quantified by zonal (east-west) sums of mass transports in depth or density layers. In the North Atlantic, away from the subpolar regions, the MOC (which is in principle an observable quantity) is often identified with the thermohaline circulation (THC), which is a conceptual and incomplete interpretation. It must be borne in mind that the MOC is also driven by wind, and can also include shallower overturning cells such as occur in the upper ocean in the tropics and subtropics, in which warm (light) waters moving poleward are transformed to slightly denser waters and subducted equatorward at deeper levels.

Atlantic Meridional Overturning Circulation (AMOC) The main current system in the South and North Atlantic Oceans. AMOC transports warm upper-ocean water northwards, and cold, deep water southwards, as part of the global ocean circulation system. Changes in the strength of AMOC can affect other components of the *climate system*.

Methane (CH$_4$)

One of the six *greenhouse gases (GHGs)* to be mitigated under the *Kyoto Protocol* and is the major component of natural gas and associated with all hydrocarbon fuels. Under future *global warming*, there is *risk* of increased methane emissions from thawing *permafrost*, coastal wetlands and sub-sea gas hydrates. See also *Mitigation*.

Microbial carbon pump

Microbial processes that transform organic carbon from rapidly-degradable states to biologically-unavailable forms, resulting in long-term carbon storage in the *ocean*. The unavailable states of organic carbon can be due to their intrinsic refractory nature, or to extremely low concentrations of each of the diverse individual molecules. The microbial carbon pump can take place at any depth in the water column and is the principal mechanism generating and sustaining *refractory dissolved organic carbon (RDOC)* in the ocean. See also *Biological (carbon) pump*, *Blue carbon* and *Dissolved organic carbon (DOC) and particulate organic carbon (POC)*.

Migrant

See *Migration (of humans)*.

Migration (of humans)

'Movement of a person or a group of persons, either across an international border, or within a State. It is a population movement, encompassing any kind of movement of people, whatever its length, composition and causes; it includes migration of refugees, displaced persons, economic *migrants*, and persons moving for other purposes, including family reunification' (IOM, 2018).

Migrant 'Any person who is moving or has moved across an international border or within a State away from his/her habitual place of residence, regardless of (1) the person's legal status; (2) whether the movement is voluntary or involuntary; (3) what the causes for the movement are; or (4) what the length of the stay is' (IOM, 2018).

See also *(Internal) Displacement (of humans)*.

Mitigation (of climate change)

A human intervention to reduce emissions or enhance the *sinks* of *greenhouse gases (GHG)*.

Mitigation measures In *climate* policy, mitigation measures are technologies, processes or practices that contribute to mitigation, for example renewable energy technologies, waste minimisation processes, public transport commuting practices. See also *Mitigation option*.

Mitigation option A technology or practice that reduces GHG emissions or enhances sinks.

Mitigation scenario A plausible description of the future that describes how the (studied) system responds to the implementation of mitigation policies and measures.

See also *Emission scenario*.

Mobility

See *Human mobility*.

(Model) Ensemble

A group of parallel model simulations characterising historical *climate* conditions, climate predictions, or *climate projections*. Variation of the results across the ensemble members may give an estimate of modelling-based *uncertainty*. Ensembles made with the same model but different initial conditions only characterise the uncertainty associated with internal climate variability, whereas multi-model ensembles including simulations by several models also include the impact of model differences. Perturbed parameter ensembles, in which model parameters are varied in a systematic manner, aim to assess the uncertainty resulting from internal model specifications within a single model. Remaining sources of uncertainty unaddressed with model ensembles are related to systematic model errors or biases, which may be assessed from systematic comparisons of model simulations with observations wherever available. See also *Projection*.

Monitoring and evaluation (M&E)

Mechanisms put in place at national to local scales to respectively monitor and evaluate efforts to reduce *greenhouse gas (GHG)* emissions and/or adapt to the *impacts* of *climate change* with the aim of systematically identifying, characterising and assessing progress over time. See also *Adaptation*.

Multi-level governance

See *Governance*.

Narratives (in the context of scenarios)

Qualitative descriptions of plausible future world evolutions, describing the characteristics, general logic and developments underlying a particular quantitative set of *scenarios*. Narratives are also referred to in the literature as 'storylines'. See also *Pathways*.

Nationally determined contributions (NDCs)

A term used under the *United Nations Framework Convention on Climate Change (UNFCCC)* whereby a country that has joined the *Paris Agreement* outlines its plans for reducing its emissions. Some countries' NDCs also address how they will adapt to climate change impacts, and what support they need from, or will provide to, other countries to adopt low-carbon pathways and to build climate resilience. According to Article 4 paragraph 2 of the Paris Agreement, each Party shall prepare, communicate and maintain successive NDCs that it intends to achieve.

Natural systems

The dynamic physical and biological components of the environment that would operate in the absence of human impacts. Most, if not all, natural systems are also now affected by human activities to some degree.

Nature's contributions to people (NCP)

'All the contributions, both positive and negative, of living nature (i.e., diversity of organisms, *ecosystems*, and their associated ecological and evolutionary processes) to the quality of life for people. Beneficial contributions from nature include such things as food provision, water purification, flood control, and artistic inspiration, whereas detrimental contributions include disease transmission and predation that damages people or their assets. Many NCP may be perceived as benefits or detriments depending on the cultural, temporal or spatial context' (Díaz et al., 2018). See also *Biodiversity* and *Ecosystem services*.

Near-surface permafrost

See *Permafrost*.

Negative emissions

Removal of *greenhouse gases (GHGs)* from the *atmosphere* by deliberate human activities, that is, in addition to the removal that would occur via natural *carbon cycle* processes. See also *Anthropogenic*, *Carbon dioxide removal (CDR)* and *Greenhouse gas removal (GGR)*.

Net-negative emissions

A situation of net-negative emissions is achieved when, as result of human activities, more *greenhouse gases (GHGs)* are removed from the *atmosphere* than are emitted into it. Where multiple GHGs are involved, the quantification of *negative emissions* depends on the *climate* metric chosen to compare emissions of different gases (such as *global warming* potential, global temperature change potential, and others, as well as the chosen time horizon). See also *Greenhouse gas removal (GGR)*, *Net-zero emissions* and *Net-zero CO$_2$ emissions*.

Net-zero CO$_2$ emissions

Net-zero *carbon dioxide (CO$_2$)* emissions are achieved when *anthropogenic* CO$_2$ emissions are balanced by anthropogenic CO$_2$ removals over a specified period. See also *Carbon dioxide removal (CDR)*, *Greenhouse gas removal (GGR)*, *Net zero emissions* and *Net negative emissions*.

Net-zero emissions

Net-zero emissions are achieved when *anthropogenic* emissions of *greenhouse gases (GHGs)* to the *atmosphere* are balanced by anthropogenic removals over a specified period. Where multiple GHGs are involved, the quantification of net-zero emissions depends on the climate metric chosen to compare emissions of different gases (such as *global warming* potential, global temperature change potential, and others, as well as the chosen time horizon). See also *Greenhouse gas removal (GGR)*, *Net-zero CO_2 emissions*, *Negative emissions* and *Net-negative emissions*.

Ocean

The interconnected body of saline water that covers 71% of the Earth's surface, contains 97% of the Earth's water and provides 99% of the Earth's biologically habitable space. It includes the Arctic, Atlantic, Indian, Pacific and Southern Oceans, as well as their marginal seas and coastal waters. See also *Blue carbon*, *Coast*, *Ocean acidification (OA)*, *Ocean deoxygenation* and *Southern Ocean*.

Ocean acidification (OA)

A reduction in the *pH* of the *ocean*, accompanied by other chemical changes (primarily in the levels of carbonate and bicarbonate ions), over an extended period, typically decades or longer, which is caused primarily by *uptake* of *carbon dioxide (CO_2)* from the *atmosphere*, but can also be caused by other chemical additions or subtractions from the ocean. *Anthropogenic* OA refers to the component of pH reduction that is caused by human activity (IPCC, 2011, p. 37). See also *Carbon cycle*, *Climate change* and *Global warming*.

Ocean deoxygenation

The loss of oxygen in the *ocean*. It results from ocean warming, which reduces oxygen solubility and increases oxygen consumption and stratification, thereby reducing the mixing of oxygen into the ocean interior. Deoxygenation can also be exacerbated by the addition of excess nutrients in the *coastal* zone.

Outburst flood

See *Glacial lake outburst flood (GLOF)/Glacier lake outburst*.

Outlet glaciers

A *glacier*, usually between rock walls, that is part of, and drains an *ice sheet*. See also *Ice stream* and *Hydrological cycle*.

Outflow

See *Discharge (of ice)*.

Overshoot

See *Temperature overshoot*.

Ozone (O_3)

The triatomic form of oxygen, and a gaseous *atmospheric* constituent. In the troposphere, O_3 is created both naturally and by photochemical reactions involving gases resulting from human activities (e.g., smog). Tropospheric O_3 acts as a *greenhouse gas (GHG)*. In the stratosphere, O_3 is created by the interaction between solar ultraviolet radiation and molecular oxygen (O_2). Stratospheric O_3 plays a dominant role in the stratospheric radiative balance. Its concentration is highest in the ozone layer. See also *Anthropogenic* and *Radiative forcing*.

Paris Agreement

The Paris Agreement under the *United Nations Framework Convention on Climate Change (UNFCCC)* was adopted in December 2015 in Paris, France, at the 21st session of the Conference of the Parties (COP) to the UNFCCC. The agreement, adopted by 196 Parties to the UNFCCC, entered into force on 4 November 2016 and as of May 2018 had 195 Signatories and was ratified by 177 Parties. One of the goals of the Paris Agreement is 'Holding the increase in the global average temperature to well below 2°C above pre-industrial levels and pursuing efforts to limit the temperature increase to 1.5°C above pre-industrial levels', recognising that this would significantly reduce the risks and impacts of climate change. Additionally, the Agreement aims to strengthen the ability of countries to deal with the impacts of climate change. The Paris Agreement is intended to become fully effective in 2020. See also *Kyoto Protocol* and *Nationally determined contributions (NDCs)*.

Participatory governance

See *Governance*.

Particulate organic carbon (POC)

See *Dissolved organic carbon (DOC) and particulate organic carbon (POC)*.

Pathways

The temporal evolution of natural and/or *human systems* towards a future state. Pathway concepts range from sets of quantitative and qualitative *scenarios* or *narratives* of potential futures to solution oriented decision making processes to achieve desirable societal goals. Pathway approaches typically focus on biophysical, techno-economic, and/or sociobehavioural trajectories and involve various dynamics, goals, and actors across different scales.

> ***Adaptation pathways*** A series of *adaptation* choices involving trade-offs between short-term and long-term goals and values. These are processes of deliberation to identify solutions that are meaningful to people in the context of their daily lives and to avoid potential *maladaptation*.

> ***Development pathways*** Trajectories based on an array of social, economic, cultural, technological, institutional and biophysical features that characterise the interactions between human and *natural systems* and outline visions for the future, at a particular scale. See also *Climate-resilient development pathways (CRDPs)* and *Human systems*.

> ***Emission pathways*** Modelled trajectories of global anthropogenic emissions over the 21st century are termed emission pathways. Emission pathways are classified by their temperature trajectory over the 21st century: pathways giving at least 50% probability based on current knowledge of limiting *global warming* to below 1.5°C are classified as 'no overshoot'; those limiting warming to below 1.6°C and returning to 1.5°C by 2100 are classified as '1.5°C limited overshoot'; while those exceeding 1.6°C but still returning

to 1.5°C by 2100 are classified as 'higher overshoot'. See also *Temperature overshoot*.

Representative concentration pathways (RCPs) *Scenarios* that include time series of emissions and concentrations of the full suite of *greenhouse gases (GHGs)* and *aerosols* and chemically active gases, as well as *land use/land* cover (Moss et al., 2008). The word 'representative' signifies that each RCP provides only one of many possible scenarios that would lead to the specific *radiative forcing* characteristics. The term 'pathway' emphasises the fact that not only the long-term concentration levels, but also the trajectory taken over time to reach that outcome are of interest (Moss et al., 2010). RCPs were used to develop *climate projections* in *Coupled Model Intercomparison Project* CMIP5.

> *RCP2.6:* One pathway where radiative forcing peaks at approximately 3 W m^{-2} and then declines to be limited at 2.6 W m^{-2} in 2100 (the corresponding Extended Concentration Pathway (ECP) assuming constant emissions after 2100).

> *RCP4.5 and RCP6.0:* Two intermediate stabilisation pathways in which radiative forcing is limited at approximately 4.5 W m^{-2} and 6.0 W m^{-2} in 2100 (the corresponding ECPs assuming constant concentrations after 2150).

> *RCP8.5:* One high pathway which leads to >8.5 W m^{-2} in 2100 (the corresponding ECP assuming constant emissions after 2100 until 2150 and constant concentrations after 2250). See also *Shared Socioeconomic Pathways (SSPs)*.

Shared socioeconomic pathways (SSPs) were developed to complement the *RCPs* with varying socioeconomic challenges to *adaptation* and *mitigation* (O'Neill et al., 2014). Based on five *narratives*, the SSPs describe alternative socioeconomic futures in the absence of *climate* policy intervention, comprising *sustainable development* (SSP1), regional rivalry (SSP3), *inequality* (SSP4), *fossil-fuelled* development (SSP5), and a middle-of-the-road development (SSP2) (O'Neill et al., 2017; Riahi et al., 2017). The combination of SSP-based socioeconomic *scenarios* and RCP-based *climate projections* provides an integrative frame for climate impact and policy analysis.

Sustainable development pathways (SDPs) Trajectories aimed at attaining the *Sustainable Development Goals (SDGs)* in the short term and the goals of *sustainable development* in the long term. In the context of *climate change*, such pathways denote trajectories that address social, environmental, and economic dimensions of sustainable development, *adaptation* and *mitigation*, and *transformation*, in a generic sense or from a particular methodological perspective such as *integrated assessment* models and *scenario* simulations.

See also *Emission scenario*, *Institution*, *Mitigation scenario* and *Natural Systems*.

Pelagic
The pelagic zone consists of the entire water column of the open *ocean*. It is subdivided into the 'epipelagic zone' (<200 m, the uppermost part of the ocean that receives enough sunlight to allow photosynthesis), the 'mesopelagic zone' (200–1000 m depth) and the 'bathypelagic zone' (>1000 m depth). The term 'pelagic' can also refer to organisms that live in the pelagic zone.

Permafrost
Ground (soil or rock, and included ice and organic material) that remains at or below 0°C for at least two consecutive years (Harris et al., 1988). Note that permafrost is defined via temperature rather than ice content and, in some instances, may be ice-free.

> *Near-surface permafrost* Permafrost within ~3–4 m of the ground surface. The depth is not precise, but describes what commonly is highly relevant for people and *ecosystems*. Deeper permafrost is often progressively less ice-rich and responds more slowly to warming than near-surface permafrost. Presence or absence of near-surface permafrost is not the only significant metric of permafrost change, and deeper permafrost may persist when near-surface permafrost is absent.

> *Permafrost degradation* Decrease in the thickness and/or areal extent of permafrost.

> *Permafrost thaw* Progressive loss of ground ice in permafrost, usually due to input of heat. Thaw can occur over decades to centuries over the entire depth of permafrost ground, with impacts occurring while thaw progresses. During thaw, temperature fluctuations are subdued because energy is transferred by phase change between ice and water. After the transition from permafrost to non-permafrost, ground can be described as thawed.

See also *Cryosphere* and *Frozen ground*.

Permafrost degradation
See *Permafrost*.

Permafrost thaw
See *Permafrost*.

pH
A dimensionless measure of the acidity of a solution given by its concentration of hydrogen ions (H$^+$). pH is measured on a logarithmic scale where pH = $-\log_{10}$(H$^+$). Thus, a pH decrease of 1 unit corresponds to a 10-fold increase in the concentration of H$^+$, or acidity. See also *Ocean acidification (OA)*.

Planned relocation (of humans)
A form of human mobility response in the face of sea level rise and related *impacts*. Planned relocation is typically initiated, supervised and implemented from national to local level and involves small communities and individual assets but may also involve large populations. Also termed resettlement, managed retreat, or managed realignment. See also *(Internal) Displacement (of humans)* and *Sea level change (sea level rise/sea level fall)*.

Plasticity

Change in organismal trait values in response to an environmental cue, and which does not require change in underlying DNA sequence.

Political economy

The set of interlinked relationships between people, the state, society and markets as defined by law, politics, economics, customs and power that determine the outcome of trade and transactions and the distribution of wealth in a country or economy.

Poverty

A complex concept with several definitions stemming from different schools of thought. It can refer to material circumstances (such as need, pattern of deprivation or limited resources), economic conditions (such as standard of living, *inequality* or economic position) and/or social relationships (such as social class, dependency, exclusion, lack of basic security or lack of entitlement). See also *Equality* and *Poverty eradication*.

Poverty eradication

A set of measures to end *poverty* in all its forms everywhere. See also *Sustainable Development Goals (SDGs)*.

Precursors

Atmospheric compounds that are not *greenhouse gases (GHGs)* or *aerosols*, but that have an effect on GHG or aerosol concentrations by taking part in physical or chemical processes regulating their production or destruction rates.

Pre-industrial

The multi-century period prior to the onset of large-scale industrial activity around 1750. In this Special Report, as in IPCC 2018a, the *reference period* 1850–1900 is used to approximate pre-industrial *global mean surface temperature*. See also *Industrial Revolution*.

Private costs

Costs carried by individuals, companies or other private entities that undertake an action, whereas social costs include additionally the external costs on the environment and on society as a whole. Quantitative estimates of both private and social costs may be incomplete, because of difficulties in measuring all relevant effect.

Primary production

The synthesis of organic compounds by plants and microbes, on *land* or in the *ocean*, primarily by photosynthesis using light and *carbon dioxide (CO_2)* as sources of energy and carbon, respectively. It can also occur through chemosynthesis, using chemical energy, for example, in deep sea vents.

> **Gross primary production (GPP)** The total amount of carbon fixed by photosynthesis over a specified time period.

> **Net primary production (NPP)** The amount of carbon accumulated through photosynthesis minus the amount lost by respiration over a specified time period.

Projection

A potential future evolution of a quantity or set of quantities, often computed with the aid of a model. Unlike predictions, projections are conditional on assumptions concerning, for example, future socioeconomic and technological developments that may or may not be realised. See also *Climate projection*, *(Model) ensemble*, *Scenario* and *Pathways*.

Radiative forcing

The change in the net, downward minus upward, radiative flux (expressed in W m^{-2}) at the tropopause or top of *atmosphere* due to a change in an external *driver* of *climate change*, such as a change in the concentration of *carbon dioxide (CO_2)*, the concentration of volcanic *aerosols* or in the output of the Sun. The traditional radiative *forcing* is computed with all tropospheric properties held fixed at their unperturbed values, and after allowing for stratospheric temperatures, if perturbed, to readjust to radiative-dynamical equilibrium. Radiative forcing is called instantaneous if no change in stratospheric temperature is accounted for. The radiative forcing once rapid adjustments are accounted for is termed the effective radiative forcing. Radiative forcing is not to be confused with cloud radiative forcing, which describes an unrelated measure of the impact of clouds on the radiative flux at the top of the atmosphere.

Reasons for concern (RFC)

Elements of a classification framework, first developed in the IPCC Third Assessment Report, which aims to facilitate judgments about what level of *climate change* may be dangerous (in the language of Article 2 of the *United Nations Framework Convention on Climate Change, UNFCCC*) by aggregating *risks* from various sectors, considering *hazards*, *exposures*, *vulnerabilities*, capacities to adapt, and the resulting *impacts*.

Reference period

The period relative to which *anomalies* are computed.

Refractory dissolved organic carbon (RDOC)

See *Dissolved organic carbon (DOC)* and *particulate organic carbon (POC)*.

Region

A relatively large-scale *land* or *ocean* area characterised by specific geographical and climatological features. The *climate* of a land-based region is affected by regional and local scale features like topography, *land use* characteristics and large water bodies, as well as remote influences from other regions, in addition to global climate conditions. The IPCC defines a set of standard regions for analyses of observed climate trends and climate model *projections* (see IPCC, 2018a, Figure 3.2; IPCC 2012a).

Regional sea level change

Change in sea level relative to a datum (such as present-day mean sea level) at spatial scales of about 100 km.

Relative sea level

Sea level measured by a tide gauge with respect to the land upon which it is situated. See also *Coast, Small Island Developing States*

(SIDS), *Local sea level change*, *Regional sea level change*, *Sea level change (sea level rise/sea level fall)*, *Steric sea level change* and *Anthropogenic subsidence*.

Relocation
See *Planned relocation (of humans)*.

Reporting
See *Measurement/Measurement, reporting and verification (MRV)*.

Representative concentration pathways (RCPs)
See *Pathways*.

Resettlement
See *Planned relocation (of humans)*.

Residual risk
The *risk* that remains following *adaptation* and risk reduction efforts.

Resilience
The capacity of interconnected social, economic and ecological systems to cope with a hazardous event, trend or disturbance, responding or reorganising in ways that maintain their essential function, identity and structure. Resilience is a positive attribute when it maintains capacity for *adaptation*, learning and/or *transformation* (Arctic Council, 2016). See also *Hazard*, *Risk* and *Vulnerability*.

Restoration
In environmental context, restoration involves human interventions to assist the recovery of an *ecosystem* that has been previously degraded, damaged or destroyed.

Risk
The potential for adverse consequences for human or ecological systems, recognising the diversity of values and objectives associated with such systems. In the context of *climate change*, risks can arise from potential *impacts* of climate change as well as human responses to climate change. Relevant adverse consequences include those on lives, *livelihoods*, health and *well-being*, economic, social and cultural assets and investments, infrastructure, services (including *ecosystem services*), *ecosystems* and species.

In the context of climate change impacts, risks result from dynamic interactions between climate-related *hazards* with the *exposure* and *vulnerability* of the affected human or ecological system to the hazards. Hazards, exposure and vulnerability may each be subject to *uncertainty* in terms of magnitude and *likelihood* of occurrence, and each may change over time and space due to socioeconomic changes and human decision making.

In the context of climate change responses, risks result from the potential for such responses not achieving the intended objective(s), or from potential trade-offs with, or negative side-effects on, other societal objectives, such as the *Sustainable Development Goals (SDGs)*. Risks can arise for example from uncertainty in implementation, effectiveness or outcomes of climate policy, climate-related investments, technology development or adoption, and system transitions.

See also *Adaptation*, *Human systems*, *Mitigation* and *Risk management*.

Risk assessment
The qualitative and/or quantitative scientific estimation of risks. See also *Risk*, *Risk management* and *Risk perception*.

Risk management
Plans, actions, strategies or policies to reduce the *likelihood* and/or magnitude of adverse potential consequences, based on assessed or perceived *risks*. See also *Risk assessment* and *Risk perception*.

Risk perception
The subjective judgment that people make about the characteristics and severity of a *risk*. See also *Risk assessment* and *Risk management*.

Runoff
The flow of water over the surface or through the subsurface, which typically originates from the part of liquid precipitation and/or snow-/ice-melt that does not evaporate, transpire or refreeze, and returns to water bodies. See also *Hydrological cycle*.

Scenario
A plausible description of how the future may develop based on a coherent and internally consistent set of assumptions about key driving forces (e.g., rate of technological change, prices) and relationships. Note that scenarios are neither predictions nor forecasts, but are used to provide a view of the implications of developments and actions. See also *Climate projection*, *Driver*, *Emission scenario*, *Mitigation scenario*, *(Model) ensemble*, *Pathways* and *Projection*.

Sea ice
Ice found at the sea surface that has originated from the freezing of seawater. Sea ice may be discontinuous pieces (ice floes) moved on the *ocean* surface by wind and currents (pack ice), or a motionless sheet attached to the *coast* (land-fast ice). Sea ice concentration is the fraction of the *ocean* covered by ice. Sea ice less than one year old is called first-year ice. Perennial ice is sea ice that survives at least one summer. It may be subdivided into second-year ice and multi-year ice, where multiyear ice has survived at least two summers. See also *Cryosphere*.

Sea level change (sea level rise/sea level fall)
Change to the height of sea level, both globally and locally (*relative sea level* change) at seasonal, annual, or longer time scales due to (1) a change in *ocean* volume as a result of a change in the mass of water in the ocean (e.g., due to melt of *glaciers* and *ice sheets*), (2) changes in ocean volume as a result of changes in ocean water density (e.g., expansion under warmer conditions), (3) changes in the shape of the ocean basins and changes in the Earth's gravitational and rotational fields, and (4) local subsidence or uplift of the *land*. Global *mean sea level* change resulting from change in the mass of the ocean is called barystatic. The amount of barystatic sea level change due to the addition or removal of a mass of water is called its *sea level equivalent (SLE)*. Sea level changes, both globally and locally, resulting from changes in water density are called steric. Density changes induced by temperature changes only are called thermosteric, while density changes induced by salinity changes are called halosteric. Barystatic and steric sea level changes do not

include the effect of changes in the shape of ocean basins induced by the change in the ocean mass and its distribution. See also *Anthropogenic subsidence*, *Local sea level change*, *Regional sea level change* and *Steric sea level change*.

Sea level equivalent (SLE)

The SLE of a mass of water, ice, or water vapour is that mass, converted to a volume using a density of 1000 kg m^{-3}, and divided by the present-day *ocean* surface area of 3.625×10^{14} m^2. Thus, 362.5 Gt of water mass added to the ocean correspond to 1 mm of global mean sea level rise. However, more accurate estimates of SLE must account for additional processes affecting mean sea level rise, such as shoreline migration, changes in ocean area, and for vertical land movements. See also *Sea level change (sea level rise/sea level fall)*.

Sea level rise (SLR)

See *Sea level change (sea level rise/sea level fall)*.

Sea surface temperature (SST)

The subsurface bulk temperature in the top few metres of the *ocean*, measured by ships, buoys, and drifters. From ships, measurements of water samples in buckets were mostly switched in the 1940s to samples from engine intake water. Satellite measurements of skin temperature (uppermost layer; a fraction of a millimetre thick) in the infrared or the top centimetre or so in the microwave are also used, but must be adjusted to be compatible with the bulk temperature. See also *Global mean surface temperature (GMST)*.

Sendai Framework for Disaster Risk Reduction

The Sendai Framework for Disaster Risk Reduction 2015–2030 outlines seven clear targets and four priorities for action to prevent new and reduce existing *disaster* risks. The voluntary, non-binding agreement recognises that the State has the primary role to reduce disaster *risk* but that responsibility should be shared with other stakeholders including local government, the private sector and other stakeholders, with the aim for the substantial reduction of disaster risk and losses in lives, *livelihoods* and health and in the economic, physical, social, cultural and environmental assets of persons, businesses, communities and countries.

Sequestration

The long-term removal of *carbon dioxide (CO$_2$)* or other forms of carbon from the *atmosphere*, with secure storage on climatically significant time scales (decadal to century). The period of storage needs to be known for climate modelling and carbon accounting purposes. See also *Blue carbon*, *Carbon dioxide removal (CDR)*, *Sink* and *Uptake*.

Shared socioeconomic pathways (SSPs)

See *Pathways*.

Shelf seas

Relatively shallow water covering the shelf of continents or around islands. The limit of shelf seas is conventionally considered as 200 m water depth at the continental shelf edge, where there is usually a steep slope to the deep *ocean* floor. During glacial periods, most shelf seas are lost since they become *land* as the build-up of *ice sheets* caused a decrease of global sea level. See also *Coasts*, *Glacier* and *Ice shelf*.

Short-lived climate forcers (SLCF)

A set of compounds that are primarily composed of those with short lifetimes in the *atmosphere* compared to well-mixed *greenhouse gases (GHGs)*, and are also referred to as near-term climate *forcers*. This set of compounds includes *methane (CH$_4$)*, which is also a well-mixed greenhouse gas, as well as *ozone* (O$_3$) and *aerosols*, or their *precursors*, and some halogenated species that are not well-mixed GHGs. These compounds do not accumulate in the atmosphere at decadal to centennial timescales, and so their effect on *climate* is predominantly in the first decade after their emission, although their changes can still induce long-term climate effects such as *sea level change*. Their effect can be cooling or warming. A subset of exclusively warming SLCFs is referred to as short-lived climate pollutants. See also *Forcing* and *Sea level change (sea level rise/sea level fall)*.

Sink

Any process, activity or mechanism which removes a *greenhouse gas (GHG)*, an *aerosol* or a *precursor* of a GHG from the *atmosphere* (*United Nations Framework Convention on Climate Change, UNFCCC*, Article 1.8). See also *Blue carbon, Sequestration* and *Uptake*.

Small Island Developing States (SIDS)

as recognised by the United Nations Office of the High Representative for the Least Developed Countries, Landlocked Developing Countries and Small Island Developing States, are a distinct group of developing countries facing specific social, economic and environmental vulnerabilities (UN-OHRLLS, 2011). They were recognised as a special case both for their environment and development at the Rio Earth Summit in Brazil in 1992. Fifty-eight countries and territories are presently classified as SIDS by the UN-OHRLLS, with 38 being UN member states and 20 being Non-UN-Members or Associate Members of the Regional Commissions (UN-OHRLLS, 2018).

Social costs

See *Private costs*.

Social-ecological system

An integrated system that includes human societies and *ecosystems*, in which humans are part of nature. The functions of such a system arise from the interactions and interdependence of the social and ecological subsystems. The system's structure is characterised by reciprocal feedbacks, emphasising that humans must be seen as a part of, not apart from, nature (Arctic Council, 2016; Berkes and Folke, 1998).

Social learning

A process of social interaction through which people learn new behaviours, capacities, values and attitudes.

Soil moisture

Water stored in the soil in liquid or frozen form. Root-zone soil moisture is of most relevance for plant activity. See also *Drought* and *Permafrost*.

Solar radiation management

See *Solar radiation modification (SRM)*.

Solar radiation modification (SRM)

The intentional modification of the Earth's shortwave radiative budget with the aim of reducing warming. Artificial injection of stratospheric *aerosols*, marine cloud brightening, and *land* surface *albedo* modification are examples of proposed SRM methods. SRM does not fall within the definitions of *mitigation* and *adaptation* (IPCC, 2012b, p. 2). Note that in the literature, SRM is also referred to as solar radiation management, or albedo enhancement. See also *Geoengineering*.

Solubility pump

A physicochemical process that transports dissolved inorganic carbon from the *ocean*'s surface to its interior. The solubility pump is primarily driven by the solubility of *carbon dioxide (CO_2)* (with more CO_2 dissolving in colder water) and the large-scale, thermohaline patterns of ocean circulation. See also *Biological (carbon) pump* and *Dissolved inorganic carbon*.

Source

Any process or activity which releases a *greenhouse gas (GHG)*, an *aerosol* or a *precursor* of a GHG into the *atmosphere* (*United Nations Framework Convention on Climate Change, UNFCCC*, Article 1.9). See also *Sink*.

Southern Ocean

The *ocean* region encircling Antarctica that connects the Atlantic, Indian and Pacific Oceans together, allowing inter-ocean exchange. This region is the main source of much of the deep water of the world's ocean and also provides the primary return pathway for this deep water to the surface (Marshall and Speer, 2012; Toggweiler and Samuels, 1995). The drawing up of deep waters and the subsequent transport into the ocean interior has major consequences for the global heat, nutrient, and carbon balances, as well as the Antarctic *cryosphere* and marine *ecosystems*.

Stabilisation (of GHG or CO_2-equivalent concentration)

A state in which the atmospheric concentration of one *greenhouse gas (GHG)* (e.g., *carbon dioxide, CO_2*) or of a CO_2-equivalent basket of GHGs (or a combination of GHGs and *aerosols*) remains constant over time. See also *Atmosphere*.

Steric sea level change

Change in sea level due to thermal expansion and salinity variations. Thermal expansion refers to the increase in volume (and decrease in density) that results from warming water. See also *Anthropogenic subsidence*, *Coast*, *Local sea level change*, *Regional sea level change*, *Relative sea level*, *Sea level change (sea level rise/sea level fall)* and *Small Island Developing States (SIDS)*.

Storm surge

The temporary increase, at a particular locality, in the height of the sea due to extreme meteorological conditions (low atmospheric pressure and/or strong winds). The storm surge is defined as being the excess above the level expected from the tidal variation alone at that time and place. See also *Extreme weather/climate event*.

Stratification

Process of forming of layers of (*ocean*) water with different properties such as salinity, density and temperature that act as barrier for water mixing. The strengthening of near-surface stratification generally results in warmer surface waters, decreased oxygen levels in deeper water, and intensification of *ocean acidification (OA)* in the upper ocean. See also *Ocean deoxygenation*.

Subsidence

See *Anthropogenic subsidence*.

Sustainability

involves ensuring the persistence of natural and *human systems*, implying the continuous functioning of *ecosystems*, the conservation of high *biodiversity*, the recycling of natural resources and, in the human sector, successful application of *justice* and *equity*. See also *Natural systems* and *Sustainable development (SD)*.

Sustainable development (SD)

Development that meets the needs of the present without compromising the ability of future generations to meet their own needs (WCED, 1987) and balances social, economic and environmental concerns. See also *Development pathways* (under *Pathways*), *Sustainability* and *Sustainable Development Goals (SDGs)*.

Sustainable Development Goals (SDGs)

The 17 global goals for development for all countries established by the United Nations through a participatory process and elaborated in the 2030 Agenda for *Sustainable Development* (UN, 2015), including ending *poverty* and hunger; ensuring health and *well-being*, education, gender *equality*, clean water and energy, and decent work; building and ensuring resilient and sustainable infrastructure, cities and consumption; reducing inequalities; protecting land and water *ecosystems*; promoting peace, *justice* and partnerships; and taking urgent action on *climate change*. See also *Resilience* and *Sustainability*.

Sustainable development pathways (SDPs)

See *Pathways*.

Teleconnection

A statistical association between *climate* variables at widely separated, geographically-fixed spatial locations. Teleconnections are caused by large spatial structures such as basin-wide coupled modes of *ocean-atmosphere* variability, Rossby wave-trains, mid-latitude jets, and storm tracks.

Temperature overshoot

The temporary exceedance of a specified level of *global warming*, such as 1.5°C. Overshoot implies a peak followed by a decline in global warming, achieved through *anthropogenic* removal of *carbon dioxide (CO_2)* exceeding remaining CO_2 emissions globally. See also *Carbon dioxide removal (CDR)* and *Emission pathways* (under *Pathways*).

Thermokarst

Processes, such as collapse, subsidence and erosion, by which characteristic landforms result from the thawing of ice-rich *permafrost* (Harris et al., 1988).

Time of Emergence (ToE)

Time when a specific *anthropogenic* signal related to *climate change* is statistically detected to emerge from the background noise of natural *climate variability* in a *reference period*, for a specific *region* (Hawkins and Sutton, 2012).

Tipping point

A level of change in system properties beyond which a system reorganises, often in a non-linear manner, and does not return to the initial state even if the *drivers* of the change are abated. For the *climate system*, the term refers to a critical threshold at which global or regional *climate* changes from one stable state to another stable state. Tipping points are also used when referring to *impact*: the term can imply that an impact tipping point is (about to be) reached in a natural or *human system*. See also *Abrupt climate change*, *Adaptation*, *Irreversibility* and *Natural Systems*.

Transformation

A change in the fundamental attributes of natural and *human systems*.

> ***Societal (social) transformation*** A profound and often deliberate shift initiated by communities toward sustainability, facilitated by changes in individual and collective values and behaviours, and a fairer balance of political, cultural and *institutional* power in society.

> ***Transformative change*** A system-wide change that requires more than technological change through consideration of social and economic factors that with technology can bring about rapid change at scale.

See also *Natural systems*.

Transformational adaptation

See *Adaptation*.

Transformative change

See *Transformation*.

Transition

The process of changing from one state or condition to another in a given period of time. Transition can be in individuals, firms, cities, *regions* and nations, and can be based on incremental or *transformative change*.

Tropical cyclone

The general term for a strong, cyclonic-scale disturbance that originates over tropical oceans. Distinguished from weaker systems (often named tropical disturbances or depressions) by exceeding a threshold wind speed. A tropical storm is a tropical cyclone with one-minute average surface winds between 18 and 32 m s^{-1}. Beyond 32 m s^{-1}, a tropical cyclone is called a hurricane, typhoon or cyclone, depending on geographic location. See also *Extratropical cyclone*.

Uncertainty

A state of incomplete knowledge that can result from a lack of information or from disagreement about what is known or even knowable. It may have many types of sources, from imprecision in the data to ambiguously defined concepts or terminology, incomplete understanding of critical processes, or uncertain *projections* of *human behaviour*. Uncertainty can therefore be represented by quantitative measures (e.g., a probability density function) or by qualitative statements (e.g., reflecting the judgment of a team of experts) (see IPCC, 2004; Mastrandrea et al., 2010; Moss and Schneider, 2000). See also *Agreement*, *Confidence*, *Deep Uncertainty* and *Likelihood*.

United Nations Framework Convention on Climate Change (UNFCCC)

The UNFCCC was adopted in May 1992 and opened for signature at the 1992 Earth Summit in Rio de Janeiro. It entered into force in March 1994 and as of May 2018 had 197 Parties (196 States and the European Union). The Convention's ultimate objective is the 'stabilisation of *greenhouse gas* concentrations in the *atmosphere* at a level that would prevent dangerous *anthropogenic* interference with the *climate system*'. The provisions of the Convention are pursued and implemented by two treaties: the *Kyoto Protocol* and the *Paris Agreement*.

Uptake

The transfer of substances (such as carbon) or energy (e.g., heat) from one compartment of a system to another; for example, in the Earth system from the *atmosphere* to the *ocean* or to the *land*. See also *Sequestration* and *Sink*.

Vulnerability

The propensity or predisposition to be adversely affected. Vulnerability encompasses a variety of concepts and elements including sensitivity or susceptibility to harm and lack of capacity to cope and adapt. See also *Adaptation*, *Exposure*, *Hazard* and *Risk*.

Water cycle

See *Hydrological cycle*.

Well-being

A state of existence that fulfils various human needs, including material living conditions and quality of life, as well as the ability to pursue one's goals, to thrive, and feel satisfied with one's life. Ecosystem well-being refers to the ability of *ecosystems* to maintain their diversity and quality. See also *Biodiversity*, *Climate-resilient development pathways (CRDPs)*, *Human rights* and *Sustainable Development Goals (SDGs)*.

References

Arctic Council (2016). *Arctic Resilience Report*. M. Carson and G. Peterson (eds). Stockholm Environment Institute and Stockholm Resilience Centre, Stockholm. [Available at: www.arctic-council.org/arr.]. Accessed: 2019/09/30.

Berkes, F. and C. Folke, 1998: *Linking Social and Ecological Systems: Management Practices and Social Mechanisms for Building Resilience*. Cambridge University Press, Cambridge, United Kingdom and New York, NY, USA, 459 pp. ISBN: 0-521-59140-6.

Campbell, A., Kapos, V., Scharlemann, J.P.W., Bubb, P., Chenery, A., Coad, L., Dickson, B., Doswald, N., Khan, M.S.I., Kershaw, F. and Rashid, M. 2009: *Review of the Literature on the Links between Biodiversity and Climate*

Change: Impacts, Adaptation and Mitigation. Secretariat of the Convention on Biological Diversity (CBD), Montreal. Technical Series No. 42, 124 pp. ISBN: 92-9225-135-X.

Carlisle, K. and R.L. Gruby, 2017: Polycentric Systems of Governance: A Theoretical Model for the Commons. *Policy Studies Journal*, **0(0)**. doi:10.1111/psj.12212.

Cogley, J.G., R. Hock, L.A. Rasmussen, A.A. Arendt, A. Bauder, R.J. Braithwaite, P. Jansson, G. Kaser, M. Möller, L. Nicholson and M. Zemp, 2011: *Glossary of Glacier Mass Balance and Related Terms*. IHP-VII Technical Documents in Hydrology No. 86, IACS Contribution No. 2, UNESCO-IHP, Paris. 114pp. [Available at: https://unesdoc.unesco.org/ark:/48223/pf0000192525.]. Accessed: 2019/09/30.

Culwick, C. and K. Bobbins, 2016: *A Framework for a Green Infrastructure Planning Approach in the Gauteng City–Region*. GCRO Research Report No. 04, Gauteng City–Region Observatory (GRCO), Johannesburg, South Africa, 127 pp.

Díaz, S. et al. 2018: Assessing Nature's Contributions to People. *Science*, **359(6373)**, 270–272. doi:10.1126/science.aap8826.

FAO, 2001: Glossary. In: *The State of Food Insecurity in the World 2001*. Food and Agriculture Organisation of the United Nations, Rome, Italy, pp. 49–50, ISBN 92-5-104628-X.

FAO, 2007: *Land evaluation: Towards a revised framework. Land and water discussion paper*. Food and Agriculture Organisation of the United Nations, Rome, Italy, 107 pp. ISSN 1729-0554.

FAO, 2009: *Declaration of the World Summit on Food Security. WSFS 2009/2*. Food and Agriculture Organisation of the United Nations, Rome, Italy, 7 pp. [Available at: www.fao.org/wsfs/wsfs-list-documents/en/]. Accessed: 2019/09/30.

Fung, A. and E.O. Wright (eds.), 2003: *Deepening Democracy: Institutional Innovations in Empowered Participatory Governance*. Verso, London, UK, 312 pp.

Harris, S.A., French, H.M., Heginbottom, J.A., Johnston, G.H., Ladanyi, B., Sego, D.C., van Everdingen, R.O., 1988: *Glossary of Permafrost and Related Ground-Ice Terms*. Technical Memorandum No. 142. Permafrost Subcommittee, Committee on Geotechnical Research, National Research Council of Canada. [Available at: https://ipa.arcticportal.org/publications/glossary]. Accessed: 2019/09/30.

Hawkins, E. and Sutton, R. 2012: Time of emergence of climate signals. *Geophys. Res. Lett.*, **39(1)**, 6 pp.

HLPE, 2017: *Nutrition and food systems*. A report by the High Level Panel of Experts on Food Security and Nutrition of the Committee on World Food Security, Rome. [Available at: www.fao.org/3/a-i7846e.pdf]. Accessed: 2019/09/30.

Hooghe, L., Marks, G., 2003: Unraveling the Central State, but How? Types of Multi-Level Governance. *Am. Political Sci. Rev.*, **97**, 233–243.

IOM, 2011: *Glossary on Migration. 2nd Edition*. [R. Perruchoud and J. Redpath-Cross (eds.)]. International Organization for Migration, 114 pp. ISSN: 1813-2278.

IOM, 2018: *Key Migration Terms*. International Organization for Migration (IOM). [Available at: www.iom.int/key–migration–terms]. Accessed: 2019/09/30.

ICS, 2019: *Formal subdivision of the Holocene Series/Epoch*. International Commission on Stratigraphy (ICS). [Available at: http://www.stratigraphy.org/index.php/ics-news-and-meetings/125-formal-subdivision-of-the-holocene-series-epoch]. Accessed: 2019/09/30.

IPBES, 2018: *The IPBES assessment report on land degradation and restoration*. [Montanarella, L., Scholes, R., and Brainich, A. (eds.)]. Secretariat of the Intergovernmental Science-Policy Platform on Biodiversity and Ecosystem services, Bonn, Germany, 744 pp.

IPCC, 2000: Land Use, Land-Use Change, and Forestry: A Special Report of the IPCC. [Watson, R.T., I.R. Noble, B. Bolin, N.H. Ravindranath, D.J. Verardo, and D.J. Dokken (eds.)]. Cambridge University Press, Cambridge, UK, 375 pp.

IPCC, 2003: Definitions and Methodological Options to Inventory Emissions from Direct Human–induced Degradation of Forests and Devegetation of Other Vegetation Types. [Penman, J., M. Gytarsky, T. Hiraishi, T. Krug, D. Kruger, R. Pipatti, L. Buendia, K. Miwa, T. Ngara, K. Tanabe, and F. Wagner (eds.)]. Institute for Global Environmental Strategies (IGES), Hayama, Kanagawa, Japan, 32 pp.

IPCC, 2004: *IPCC Workshop on Describing Scientific Uncertainties in Climate Change to Support Analysis of Risk of Options. Workshop Report*. Intergovernmental Panel on Climate Change (IPCC), Geneva, Switzerland, 138 pp.

IPCC, 2006: IPCC Guidelines for National Greenhouse Gas Inventories. [H.S. Eggleston, L. Buendia, K. Miwa, T. Ngara, K. Tanabe (eds)]. Institute for Global Environmental Strategies (IGES), Hayama, Kanagawa, Japan, 20 pp.

IPCC, 2011: Workshop Report of the Intergovernmental Panel on Climate Change Workshop on Impacts of Ocean Acidification on Marine Biology and Ecosystems. [Field, C.B., V. Barros, T.F. Stocker, D. Qin, K.J. Mach, G.-K. Plattner, M.D. Mastrandrea, M. Tignor, and K.L. Ebi (eds.)]. IPCC Working Group II Technical Support Unit, Carnegie Institution, Stanford, California, USA, 164 pp.

IPCC, 2012a: Managing the Risks of Extreme Events and Disasters to Advance Climate Change Adaptation. A Special Report of Working Groups I and II of the Intergovernmental Panel on Climate Change (IPCC). [Field, C.B., V. Barros, T.F. Stocker, D. Qin, D.J. Dokken, K.L. Ebi, M.D. Mastrandrea, K.J. Mach, G.-K. Plattner, S.K. Allen, M. Tignor, and P.M. Midgley (eds.)]. Cambridge University Press, Cambridge, UK and New York, NY, USA, 582 pp.

IPCC, 2012b: *Meeting Report of the Intergovernmental Panel on Climate Change Expert Meeting on Geoengineering*. IPCC Working Group III Technical Support Unit, Potsdam Institute for Climate Impact Research, Potsdam, Germany, 99 pp.

IPCC, 2014: Annex II: Glossary [Mach, K.J., S. Planton and C. von Stechow (eds.)]. In: Climate Change 2014: Synthesis Report. Contribution of Working Groups I, II and III to the Fifth Assessment Report of the Intergovernmental Panel on Climate Change [Core Writing Team, R.K. Pachauri and L.A. Meyer (eds.)]. IPCC, Geneva, Switzerland, pp. 117–130.

IPCC, 2018a: Hoegh-Guldberg, O., D. Jacob, M. Taylor, M. Bindi, S. Brown, I. Camilloni, A. Diedhiou, R. Djalante, K.L. Ebi, F. Engelbrecht, J. Guiot, Y. Hijioka, S. Mehrotra, A. Payne, S.I. Seneviratne, A. Thomas, R. Warren, and G. Zhou, 2018: Impacts of 1.5°C Global Warming on Natural and Human Systems. In: Global Warming of 1.5°C. An IPCC Special Report on the impacts of global warming of 1.5°C above pre-industrial levels and related global greenhouse gas emission pathways, in the context of strengthening the global response to the threat of climate change, sustainable development, and efforts to eradicate poverty [Masson-Delmotte, V., P. Zhai, H.-O. Pörtner, D. Roberts, J. Skea, P.R. Shukla, A. Pirani, W. Moufouma-Okia, C. Péan, R. Pidcock, S. Connors, J.B.R. Matthews, Y. Chen, X. Zhou, M.I. Gomis, E. Lonnoy, T. Maycock, M. Tignor, and T. Waterfield (eds.)]. In Press.

IPCC, 2018b: Annex I: Glossary [R. Matthews (ed.)]. In: Global warming of 1.5°C. An IPCC Special Report on the impacts of global warming of 1.5°C above pre-industrial levels and related global greenhouse gas emission pathways, in the context of strengthening the global response to the threat of climate change, sustainable development, and efforts to eradicate poverty [V. Masson-Delmotte, P. Zhai, H.O. Pörtner, D. Roberts, J. Skea, P.R. Shukla, A. Pirani, W. Moufouma-Okia, C. Péan, R. Pidcock, S. Connors, J.B.R. Matthews, Y. Chen, X. Zhou, M.I. Gomis, E. Lonnoy, T. Maycock, M. Tignor, T. Waterfield (eds.)]. In Press.

ISO, 2014: ISO 16559:2014(en) Solid biofuels – Terminology, definitions and descriptions. International Standards Organisation (ISO). [Available at: www.iso.org/obp/ui/#iso:std:iso:16559:ed-1:v1:en.]. Accessed: 2019/09/30.

Jordan, A., Huitema, D., Asselt, H. van, Forster, J., 2018. *Governing Climate Change: Polycentricity in Action?* Cambridge University Press. doi:10.1017/9781108284646.

Lempert, R.J., S.W. Popper and S.C. Bankes, 2003: *Shaping the Next One Hundred Years: New Methods for Quantitative, Long-Term Policy Analysis.* RAND Corporation, Santa Monica, CA, 186 pp. ISBN: 0-8330-3275-5.

MA, 2005: Appendix D: Glossary. In: *Ecosystems and Human Well-being: Current States and Trends. Findings of the Condition and Trends Working Group* [Hassan, R., R. Scholes, and N. Ash (eds.)]. Millennium Ecosystem Assessment. Island Press, Washington DC, USA, pp. 893–900. ISBN: 1-55963-227-5.

Marshall, J., and K. Speer, 2012: Closure of the meridional overturning circulation through Southern Ocean upwelling. *Nat. Geosci.,* **5**, 171–180, doi:10.1038/ngeo1391.

Mastrandrea, M.D. et al., 2010: *Guidance Note for Lead Authors of the IPCC Fifth Assessment Report on Consistent Treatment of Uncertainties.* Intergovernmental Panel on Climate Change (IPCC), Geneva, Switzerland, 6 pp.

McDonald, T., J. Jonson, and K.W. Dixon, 2016: National standards for the practice of ecological restoration in Australia. *Restor. Ecol.,* **24(S1)**, S4-S32. doi:10.1111/rec.12359.

McGinnis, M.D., Ostrom, E., 2012. Reflections on Vincent Ostrom, Public Administration, and Polycentricity. *Pub. Admin. Rev.,* **72**, 15–25. doi:10.1111/j.1540-6210.2011.02488.x.

Mechler, R., L.M. Bouwer, T. Schinko, S. Surminski, and J. Linnerooth-Bayer (eds.), in press: *Loss and Damage from Climate Change: Concepts, Methods and Policy Options.* Springer International Publishing, 561 pp.

Moss, R.H. and S.H. Schneider, 2000: Uncertainties in the IPCC TAR: Recommendations to Lead Authors for More Consistent Assessment and Reporting. In: *Guidance Papers on the Cross Cutting Issues of the Third Assessment Report of the IPCC* [Pachauri, R., T. Taniguchi, and K. Tanaka (eds.)]. Intergovernmental Panel on Climate Change (IPCC), Geneva, Switzerland, pp. 33–51.

Moss, R.H. et al., 2008: Towards New Scenarios for Analysis of Emissions, Climate Change, Impacts, and Response Strategies. Technical Summary. Intergovernmental Panel on Climate Change (IPCC), Geneva, Switzerland, 25 pp.

Moss, R.H. et al., 2010: The next generation of scenarios for climate change research and assessment. *Nature,* **463(7282)**, 747–756, doi:10.1038/nature08823.

MRFCJ, 2018: Principles of Climate Justice. Mary Robinson Foundation For Climate Justice (MRFCJ). Retrieved from: www.mrfcj.org/principles-of-climate-justice.

NOAA, 2019: *What is an iceberg?* National Oceanic and Atmospheric Administration. [Available at: https://oceanservice.noaa.gov/facts/iceberg.html]. Accessed: 2018/06/25.

O'Neill, B.C. et al., 2014: A new scenario framework for climate change research: the concept of shared socioeconomic pathways. *Clim. Change,* **122(3)**, 387–400, doi:10.1007/s10584-013-0905-2.

O'Neill, B.C. et al., 2017: The roads ahead: Narratives for shared socioeconomic pathways describing world futures in the 21st century. *Global Environ. Change,* **42**, 169–180, doi:10.1016j.gloenvcha.2015.01.004.

Ostrom, E., 2005. *Understanding institutional diversity.* Princeton University Press, Princeton, New Jersey, 355 pp. ISBN: 978-0-691-12207-6.

Paavola, J., 2007. Institutions and environmental governance: A reconceptualization. Ecological Economics **63**, 93–103, doi:10.1016/j.ecolecon.2006.09.026.

Park, S. E., N.A. Marshall, E. Jakku, A.M Dowd, S.M. Howden, E.K. Mendham and A. Fleming, 2012: Informing adaptation responses to climate change through theories of transformation. Global Environmental Change, **22(1)**, 115–126. doi:10.1016/j.gloenvcha.2011.10.003.

Pescaroli, G. and D. Alexander, 2015: A definition of cascading disasters and cascading effects: Going beyond the "toppling dominos" metaphor. In:

Planet@Risk, **3(1)**, 58–67, Davos: Global Risk Forum GRF Davos. [Available at: https://planet-risk.org/index.php/pr/article/view/208/355.]. Accessed: 2019/09/30.

Peters, B.G. and J. Pierre, 2001: Developments in intergovernmental relations: towards multi-level governance. *Policy & Politics,* **29(2)**, 131–135, doi:10.1332/0305573012501251.

Pongratz, J. et al., 2018: Models meet data: Challenges and opportunities in implementing land management in Earth system models. *Global Change Biol.,* **24(4)**, 1470–1487, doi: 10.1111/gcb.13988.

Riahi, K. et al., 2017: The Shared Socioeconomic Pathways and their energy, land use, and greenhouse gas emissions implications: An overview. *Global Environ. Change,* **42**, 153–168, doi:10.1016/j.gloenvcha.2016.05.009.

Sarmiento, H. and C. Tilly, 2018: Governance Lessons from Urban Informality. *Politics and Governance,* **6(1)**, 199–202, doi:10.17645/pag.v6i1.1169.

Tàbara, J.D., J. Jäger, D. Mangalagiu, and M. Grasso, 2018: Defining transformative climate science to address high-end climate change. *Reg. Environ. Change,* **19 (3)**, 807–818, doi:10.1007/s10113-018-1288-8.

Termeer, C.J.A.M., A. Dewulf, and G.R. Biesbroek, 2017: Transformational change: governance interventions for climate change adaptation from a continuous change perspective. *J. Environ. Plan. Manage.,* **60(4)**, 558–576, doi:10.1080/09640568.2016.1168288.

Tilman, D., 2001: Functional diversity. In: *Encyclopedia of Biodiversity* [Levin, S.A. (ed.)]. Academic Press, San Diego, CA, pp. 109–120. ISBN: 978-0-12-226865-6.

Toggweiler, J.R., and B. Samuels, 1995: Effect of Drake Passage on the global thermohaline circulation. *Deep-Sea Res. Pt. I,* **42**, 477–500, doi:10.1016/0967-0637(95)00012-U.

UN, 1992: Article 2: Use of Terms. In: *Convention on Biological Diversity.* United Nations (UN), pp. 3–4.

UN, 2015: *Transforming Our World: The 2030 Agenda for Sustainable Development.* A/RES/70/1, United Nations General Assembly (UNGA), New York, NY, USA, 35 pp.

UNCCD, 1994: United Nations Convention to Combat Desertification in countries experiencing serious drought and/or desertification, particularly in Africa. A/AC.241/27, United Nations General Assembly (UNGA), New York, NY, USA, 58 pp.

UNESCO, 2018: Local and Indigenous Knowledge Systems. United Nations Educational, Scientific and Cultural Organization (UNESCO). [Available at: www.unesco.org/new/en/natural-sciences/priority-areas/links/related-information/what-is-local-and-indigenous-knowledge]. Accessed: 2019/09/30.

UNFCCC, 2013: Reporting and accounting of LULUCF activities under the Kyoto Protocol. United Nations Framework Convention on Climatic Change (UNFCCC), Bonn, Germany. [Available at: http://unfccc.int/methods/lulucf/items/4129.php]. Accessed: 2019/09/30.

UNISDR, 2009: *2009 UNISDR Terminology on Disaster Risk Reduction.* United Nations International Strategy for Disaster Reduction (UNISDR), Geneva, Switzerland, 30 pp. [Available at: https://www.unisdr.org/we/inform/publications/7817.] Accessed: 2019/09/30.

UNISDR, 2017: Report of the open-ended intergovernmental expert working group on indicators and terminology relating to disaster risk reduction. UNISDR. [Available at: https://www.preventionweb.net/files/50683_oiewgreportenglish.pdf]. Accessed: 2019/09/30.

UNOHCHR, 2018: What are Human rights? UN Office of the High Commissioner for Human Rights (UNOHCHR). [Available at: www.ohchr.org/EN/Issues/Pages/whatarehumanrights.aspx]. Accessed: 2019/09/30.

UN-OHRLLS, 2011: *Small Island Developing States: Small Islands Big(ger) Stakes.* Office for the High Representative for the Least Developed Countries, Landlocked Developing Countries and Small Island Developing States (UN-OHRLLS), New York, NY, USA, 32 pp.

UN-OHRLLS, 2018: Small Island Developing States: Country profiles. Office for the High Representative for the Least Developed Countries, Landlocked Developing Countries and Small Island Developing States (UN-OHRLLS).

[Available at: http://unohrlls.org/about–sids/country–profiles]. Accessed: 2019/09/30.

UN–REDD, 2009: *Measurement, Assessment, Reporting and Verification (MARV): Issues and Options for REDD.* Draft Discussion Paper, United Nations Collaborative Programme on Reducing Emissions from Deforestation and Forest Degradation in Developing Countries (UN–REDD), Geneva, Switzerland, 12 pp.

WCED, 1987: *Our Common Future.* World Commission on Environment and Development (WCED), Geneva, Switzerland, 400 pp. doi:10.2307/2621529.

Williamson, O.E., 2000. The New Institutional Economics: Taking Stock, Looking Ahead. J. *Econ. Lit.,* **38**, 595–613.

Zscheischler, J., Westra, S., Hurk, B.J., Seneviratne, S.I., Ward, P.J., Pitman, A., AghaKouchak, A., Bresch, D.N., Leonard, M., Wahl, T. and Zhang, X., 2018: *Future climate risk from compound events. Nat. Clim. Change,* **8,** 469–477, doi: 10.1038/s41558-018-0156-3.

Annex II: Acronyms

This annex should be cited as:
IPCC, 2019: Annex II: Acronyms. In: *IPCC Special Report on the Ocean and Cryosphere in a Changing Climate* [H.-O. Pörtner, D.C. Roberts, V. Masson-Delmotte, P. Zhai, M. Tignor, E. Poloczanska, K. Mintenbeck, A. Alegría, M. Nicolai, A. Okem, J. Petzold, B. Rama, N.M. Weyer (eds.)]. Cambridge University Press, Cambridge, UK and New York, NY, USA, pp. 703–705.

a.s.l.	Above Sea Level		ESL	Extreme Sea Level
AABW	Antarctic Bottom Water		ESM	Earth System Model
ABNJ	Areas Beyond National Jurisdiction		ETC	Extratropical Cyclone
AC	Arctic Council		EWS	Early Warning System
ACC	Antarctic Circumpolar Current		FISHMIP	Fisheries and Marine Ecosystems Impact
AEWC	Alaska Eskimo Whaling Commission			Model Intercomparison Project
AIS	Antarctic Ice Sheet		GACS	Global Alliance of Continuous
ALT	Active-Layer Thickness			Plankton Recorder Surveys
AMO	Atlantic Multidecadal Oscillation		GCM	General Circulation Model/Global Climate Model
AMOC	Atlantic Meridional Overturning Circulation		GDP	Gross Domestic Product
AMV	Atlantic Multidecadal Variability		GGR	Greenhouse Gas Removal
AOGCM	Atmosphere-Ocean General Circulation Model		GHG	Greenhouse Gas
AP	Antarctic Peninsula		GIA	Glacio-Isostatic Adjustment
AR4	IPCC 4th Assessment Report		GIS	Greenland Ice Sheet
AR5	IPCC 5th Assessment Report		GLOF	Glacial Lake Outburst Flood
AR6	IPCC 6th Assessment Report		GMSL	Global Mean Sea Level
ASE	Amundsen Sea Embayment		GMST	Global Mean Surface Temperature
ASF	Animal Sourced Food		GPD	Generalised Pareto Distribution
ATS	Antarctic Treaty System		GPG	Good Practice Guidance
BC	Black Carbon		GPP	Gross Primary Production
BECCS	Bioenergy with Carbon Capture and Storage		GPS	Global Positioning System
CAO	Central Arctic Ocean		GRACE	Gravity Recovery and Climate Experiment
CbA	Community-based Adaptation		GRD	Gravity, Rotation, and Deformation
CBD	Convention on Biological Diversity		GSP	Gross State Product
CBM	Community-based Monitoring		GWP	Global Warming Potential
CCAMLR	Convention for the Conservation of		HAB	Harmful Algal Blooms
	Antarctic Marine Living Resources		HFC	Hydrofluorocarbons
CDR	Carbon Dioxide Removal		HSM	Historical Simulation
CFC	Chlorofluorocarbons		IAEA	International Atomic Energy Agency
CMIP	Coupled Model Intercomparison Project		ICC	Inuit Circumpolar Council
CMIP3	Coupled Model Intercomparison Project Phase 3		ICM	Integrated Coastal Management
CMIP5	Coupled Model Intercomparison Project Phase 5		ICZM	Integrated Coastal Zone Management
CMIP6	Coupled Model Intercomparison Project Phase 6		IK	Indigenous Knowledge
COP	Conference of Parties		IMO	International Maritime Organization
CoSMoS	Coastal Storm Modeling System		INGO	International Non-Governmental Organisation
COWCLIP	Coordinated Ocean Wave Climate Project		IOC-UNESCO	Intergovernmental Oceanographic Commission
CPR	Continuous Plankton Recorder			of the United Nations Educational, Scientific and
CRDP	Climate-Resilient Development Pathway			Cultural Organization
DDT	Dichlorodiphenyl-trichloroethane		IOD	Indian Ocean Dipole
DHW	Degree Heating Week		IPBES	Intergovernmental Platform on Biodiversity and
DIC	Dissolved Inorganic Carbon			Ecosystem Services assessments
DJF	December, January, February		IPO	Interdecadal Pacific Oscillation
DO	Dissolved Oxygen		IRC	Inuvialuit Regional Corporation
DOC	Dissolved Organic Carbon		ISR	Inuvialuit Settlement Region
DRM	Disaster Risk Management		ITCZ	Intertropical Convergence Zone
DSL	Deep Scattering Layers		ITF	Indonesia Throughflow
EAD	Expected Annual Damages		IWRM	Integrated Water Resource Management
EAIS	East Antarctic Ice Sheet		JJA	June, July, August
EbA	Ecosystem-based Adaptation		L&D	Loss and Damage
EBUS	Eastern Boundary Upwelling System		LDC	Least Developed Countries
ECMWF	European Centre for Medium-Range		LDOC	Labile Dissolved Organic Carbon
	Weather Forecasts		LECZ	Low Elevation Coastal Zone
ECP	Extended Concentration Pathway		LGM	Last Glacial Maximum
EDW	Elevation Dependent Warming		LIG	Last Interglacial
EEZ	Exclusive Economic Zone		LK	Local Knowledge
ENSO	El Niño-Southern Oscillation		LLIC	Low-Lying Islands and Coasts
ESA CCI	European Space Agency Climate Change Initiative		LMMA	Locally Managed Marine Protected Area

LUC	Land-Use Change		**RSL**	Relative Sea Level
LULUC	Land Use and Land-Use Change		**RSLR**	Relative Sea Level Rise
M&E	Monitoring and Evaluation		**SAM**	Southern Annular Mode
MAGT	Mean Annual Ground Temperature		**SAMOC**	South Atlantic Meridional Overturning Circulation
MCDA	Multi-Criteria Decision Analysis		**SD**	Sustainable Development
MHW	Marine Heatwave		**SDGs**	Sustainable Development Goals
MICI	Marine Ice Cliff Instability		**SDM**	Structured Decision Making
MIP	Model Intercomparison Project		**SDP**	Sustainable Development Pathway
MISI	Marine Ice Sheet Instability		**SH**	Southern Hemisphere
MOC	Meridional Overturning Circulation		**SIDS**	Small Island Developing States
MPA	Marine Protected Area		**SLCF**	Short-Lived Climate Forcers
mPWP	mid-Pliocene Warm Period		**SLE**	Sea Level Equivalent
MRV	Measurement, Reporting and Verification		**SLP**	Sea Level Pressure
MSL	Mean Sea Level		**SLR**	Sea Level Rise
MSY	Maximum Sustainable Yield		**SMB**	Surface Mass Balance
NADW	North Atlantic Deep Water		**SPG**	Subpolar Gyre
NAO	North Atlantic Oscillation		**SPM**	Summary for Policymakers
NAP	National Adaption Plan		**SR15**	IPCC Special Report on Global Warming of 1.5°C
NCEP	National Centers for Environmental Prediction		**SRCCL**	IPCC Special Report on Climate Change and Land
NCP	Nature's Contribution to People		**SRES**	IPCC Special Report on Emissions Scenarios
NDC	Nationally Determined Contributions		**SREX**	IPCC Special Report on Managing the
NGO	Non-Governmental Organisation			Risks of Extreme Events and Disasters
NOAA	National Oceanic and Atmospheric Administration			to Advance Climate Change Adaptation
NPP	Net Primary Production		**SRM**	Solar Radiation Modification
NPV	Net Present Value		**SROCC**	IPCC Special Report on the Ocean and
OA	Ocean Acidification			Cryosphere in a Changing Climate
OHC	Ocean Heat Content		**SRTM**	Shuttle Radar Topography Mission
OLS	Ordinary Least Square		**SSPs**	Shared Socioeconomic Pathways
OMZ	Oxygen Minimum Zone		**SST**	Sea Surface Temperature
PCBs	Polychlorinated Biphenyls		**SWE**	Snow Water Equivalent
PDO	Pacific Decadal Oscillation		**TC**	Tropical Cyclone
PFC	Perfluorocarbons		**THC**	Thermohaline Circulation
PIG	Pine Island Glacier		**ToE**	Time of Emergence
PISM	Parallel Ice Sheet Model		**UNCLOS**	United Nations Convention on the Law of the Sea
PM	Particulate Matter		**UNEP**	United Nations Environment Programme
POC	Particulate Organic Carbon		**UNFCCC**	United Nations Framework Convention on
POP	Persistent Organic Pollutants			Climate Change
ppmv	parts per million volume		**UQ**	Uncertainty Quantification
PSU	Practical Salinity Unit		**VLM**	Vertical Land Motion
RCM	Regional Climate Model		**WAIS**	West Antarctic Ice Sheet
RCP	Representative Concentration Pathway		**WBC**	Western Boundary Current
RDM	Robust Decision Making		**WGI**	IPCC Working Group I
RDOC	Refractory Dissolved Organic Carbon		**WGI**	IPCC Working Group I
RFC	Reason for Concern		**WGII**	IPCC Working Group II
RGI	Randolph Glacier Inventory		**WGMS**	World Glacial Monitoring Service
RITC	Rapid Intensification of Tropical Cyclones			

All

Annex III: Contributors to the IPCC Special Report on the Oceans and Cryosphere in a Changing Climate

This annex should be cited as:
IPCC, 2019: Annex III: Contributors to the IPCC Special Report on the Ocean and Cryosphere in a Changing Climate. In: *IPCC Special Report on the Ocean and Cryosphere in a Changing Climate* [H.-O. Pörtner, D.C. Roberts, V. Masson-Delmotte, P. Zhai, M. Tignor, E. Poloczanska, K. Mintenbeck, A. Alegría, M. Nicolai, A. Okem, J. Petzold, B. Rama, N.M. Weyer (eds.)]. Cambridge University Press, Cambridge, UK and New York, NY, USA, pp. 707–716.

Coordinating Lead Authors, Lead Authors, Review Editors, Contributing Authors and Chapter Scientists are listed alphabetically by surname.

ABD-ELGAWAD, Amro
Governmental, Academic & International Organisation
Egypt

ABDULLA, Amjad
Ministry of Environment and Energy
Maldives

ABE-OUCHI, Ayako
University of Tokyo
Japan

ABRAM, Nerilie
Australian National University
Australia

ACAR, Sevil
Boğaziçi University
Turkey

ADLER, Carolina
Mountain Research Initiative
Switzerland

ALAVA, Juan Jose
University of British Columbia
Ecuador/Canada

ALEGRIA, Andrés
IPCC WGII TSU
Alfred Wegener Institute
Germany/Honduras

ALLEN, Simon
University of Zürich
Switzerland

ALLISON, Eddie
University of Washington
USA

ANISIMOV, Oleg
State Hydrological Institute
Russian Federation

ARBIC, Brian
University of Michigan
USA

ARBLASTER, Julie
Monash University
Australia

ARENSON, Lukas
BGC Engineering Inc.
Canada

ARÍSTEGUI, Javier
Universidad Las Palmas de Gran Canaria
Spain

ARRIGO, Kevin
Stanford University
USA

AZETZU-SCOTT, Kumiko
Department of Fisheries and Oceans
Canada

BAKKER, Pepjin
Universität Bremen
Germany

BAMBRIDGE, Tamatoa
National Center for Scientific Research
French Polynesia

BANERJEE, Soumyadeep
International Centre for Integrated Mountain Development
Nepal

BARANGE, Manuel
Food and Agriculture Organisation
Italy

BARBER, David
University of Manitoba
Canada

BARR, Iestyn
Manchester Metropolitan University
United Kingdom

BARTSCH, Inka
Alfred Wegener Institute
Germany

BASSIS, Jeremy
University of Michigan
USA

BAUCH, Dorothea
GEOMAR
Germany

BERKES, Fikret
University of Manitoba
Canada

BIESBROEK, Robbert
Wageningen University
Netherlands

BINDOFF, Nathaniel Lee
University of Tasmania and Commonwealth Scientific and Industrial Research Organisation
Australia

BOPP, Laurent
CNRS
France

BÓRQUEZ, Roxana
King's College London
United Kingdom

BOUWER, Laurens
Climate Service Center Germany
Helmholtz-Zentrum Geesthacht
Germany

BOYD, Philip
University of Tasmania
Australia/United Kingdom

BRANDT, Angelika
Senckenberg
Germany

BROWN, Lee
University of Leeds
United Kingdom

BROWNING, Thomas
GEOMAR Helmholtz Centre for Ocean Research
Germany/United Kingdom

BRUGGEMAN, Jorn
Plymouth Marine Laboratory
Netherlands

BUCHANAN, Maya
Climate Central
USA

AIII

BUTENSCHÖN, Momme
Centro Euro-Mediterraneo sui
Cambiamenti Climatici
Germany

CÁCERES, Bolívar
Instituto Nacional de
Meteorología e Hidrología
Ecuador

CAI, Rongshuo
State Oceanic Administration of China
China

CAO, Bin
Lanzhou University
China

CAREY, Mark
University of Oregon
USA

CASSOTTA, Sandra
Aalborg University
Denmark

CHAPIN III, F. Stuart
University of Alaska Fairbanks
USA

CHÁVEZ, Francisco P.
Monterey Bay Aquarium Research Institute
USA

CHENG, Lijing
Chinese Academy of Sciences
China

CHEONG, So Min
University of Kansas
USA

CHEUNG, William (wai Lung)
The University of British Columbia
Canada

CHIDICHIMO, María Paz
Argentine Scientific Research Council/
Hydrographic Service/
University of Buenos Aires
Argentina

CHOWN, Steven
Monash University
Australia

CIFUENTES JARA, Miguel
Centro Agronómico Tropical de
Investigación y Enseñanza
Costa Rica

CINAR, Mine
Loyola Universuty Chicago
USA

COGLEY, Graham
Trent University
Canada

COLLINS, Matthew
University of Exeter
United Kingdom

COLLINS, Sinead
University of Edinburgh
United Kingdom

COLVIN, Rebecca
Australian National University
Australia

COOK, Alison
University of Ottawa
Canada

COSTA, Daniel P.
University of California Santa Cruz
USA

CRATE, Susie
George Mason University
USA

DANGENDORF, Sönke
University of Siegen
Germany

DAWSON, Jackie
University of Ottawa
Canada

DE CAMPOS, Ricardo Safra
University of Exeter
United Kingdom

DE MARFFY-MANTUANO, Annick
United Nations Office of Legal Affairs
Monaco

DE SHERBININ, Alex
Columbia University
USA

DECONTO, Robert
University of Massachusetts-Amherst
USA

DEFEO, Omar
Universidad de la República de Uruguay
Uruguay

DERKSEN, Chris
Environment and Climate Change Canada
Canada

DJOUNDOURIAN, Salpie
LAU Adnan Kassar School of Buiseness
Lebanon

DÖLL, Petra
University of Frankfurt
Germany

DOMINGUES, Catia
Antarctic Climate & Ecosystems
Cooperative Research Centre
Australia

DUNSE, Thorben
Universitetet i Oslo
Norway/Germany

DURAND, Axel
University of Tasmania
Australia

DUTTON, Andrea
University of Florida
USA

DUVAT, Virginie K.E.
Université de La Rochelle
France

EAKIN, C. Mark
National Oceanic and Atmospheric
Administration
USA

ECKERT, Nicolas
Irstea, Univ. Grenoble Alpes
France

EDDY, Tyler
University of South Carolina
Canada

EDWARDS, Tamsin
Kings College London
United Kingdom

AIII

709

EERKES-MEDRANO, Laura
University of Victoria
Canada

EIDE, Arne
University i Tromso
Norway

EKAYKIN, Alexey
Arctic and Antarctic Research Institute
Russian Federation

ELL-KANAYUK, Monica
Inuit Circumpolar Council
Canada

ELLIS, Bethany
The Australian National University
Australia

EMANUEL, Kerry
Massachusetts Institute of Technology
USA

ENDRES, Sonja
Max Planck Institute for Chemistry
Germany

ENOMOTO, Hiroyuki
National Institute of Polar Research
Japan

EPSTEIN, Howard
University of Virginia
USA

FARINOTTI, Daniel
ETH Zürich
Switzerland

FISCHLIN, Andreas
ETH Zürich
Switzerland

FLANNER, Mark
University of Michigan
USA

FLATO, Gregory
Environment and Climate Change Canada
Canada

FORBES, Bruce
University of Lapland
Finland

FORBES, Donald
Geological Survey of Canada
Canada

FORD, James
Univeristy of Leeds
United Kingdom

FORTES, Miguel D.
University of the Philippines
Philippians

FOX, Alan
University of Edinburgh
United Kingdom

FREDERIKSE, Thomas
Utrecht University
USA

FREE, Christopher
University of California Santa Barbara
USA

FRÖLICHER, Thomas
University of Bern
Switzerland

FYKE, Jeremy
Los Alamos National Laboratory
USA

GARSCHAGEN, Matthias
Ludwig-Maximilians-Universität München
Germany

GATTUSO, Jean-pierre
CNRS, Université Pierre et
Marie Curie and IDDRI
France

GEERTSEMA, Marten
University of Northern British Columbia
Canada

GERBER, Gemma
University of KwaZulu-Natal
South Africa

GHOSH, Tuhin
Jadavpur University
India

GLAVOVIC, Bruce
Massey University
New Zealand

GLAZOVSKY, Andrey
Russian Academy of Sciences
Russian Federation

GREBMEIER, Jacqueline
University of Maryland
USA

GREENE, Charles
Cornell University
USA

GROSE, Michael
Commonwealth Scientific and
Industrial Research Organisation
Australia

GROSSE, Guido
Alfred Wegener Institute
Germany

GRUBER, Nicolas
ETH Zürich
Switzerland

GRUBER, Stephan
Carleton University
Canada

GUINDER, Valeria Ana
Instituto Argentino de Oceanografía
Argentina

GUNN, Anne
Retired/Independent Consultant
Canada

GUPTA, Kapil
Indian Institute of Technology Bombay
India

HALLBERG, Robert
National Oceanic and Atmospheric
Administration
USA

HALLEGRAEFF, Gustaaf
Institute for Marine and Antarctic Studies
Australia

HARPER, Sherilee
University of Alberta
Canada

HARRISON, Matthew J.
National Oceanic and Atmospheric
Administration, Princeton University
USA

HAY, John
University of the South Pacific
Cook Islands

HE, Shengping
University of Bergen
Norway

HEMER, Mark
Commonwealth Scientific and
Industrial Research Organisation
Australia

HENNIGE, Sebastian
Sebastian Hennige
United Kingdom

HERNÁNDEZ GONZÁLEZ, Marcelino
Instituto de Ciencias del Mar
Cuba

HILMI, Nathalie Jeanne Marie
Centre Scientifique de Monaco
Monaco

HINDELL, Mark
Institute for Marine and Antarctic Studies
Australia

HINKEL, Jochen
Global Climate Forum
Germany

HIRABAYASHI, Yukiko
Shibaura Institute of Technology
Japan

HJORT, Jan
University of Oulu
Finland

HOBBS, Will
University of Tasmania
Australia

HOBERG, Eric P.
US National Parasite Collection
USA

HOCK, Regine
University of Alaska Fairbanks
USA

HODGSON-JOHNSTON, Indi
University of Tasmania
Australia

HOGG, Andrew
Australian National University
Australia

HOLLAND, David
New York University
USA

HOLLAND, Elisabeth
University of the South Pacific
Fiji

HOLLAND, Paul
British Antarctic Survey
United Kingdom

HOLLOWED, Anne
National Oceanic and Atmospheric
Administration
USA

HONSBERG, Martin
Tüv Süd
Germany

HOOD, Eran
University of Alaska Southeast
USA

HOPCROFT, Russell
University of Alaska Fairbanks
USA

HUNT, George
University of Washington
USA

HUNTINGTON, Henry
Independent
USA

HUSS, Matthias
ETH Zürich
Switzerland

ISLA, Federico
National Research Council of Argentina
Argentina

ITO, Taka
Georgia Tech
USA

JACKSON, Laura
Met Office
United Kingdom

JACKSON, Miriam
Norwegian Water Resources
and Energy Directorate
Norway

JACOT DES COMBES, Helene
University of the South Pacific
Fiji

JENKINS, Adrian
British Antarctic Survey
United Kingdom

JIAO, Nianzhi
State Key Laboratory of Marine
Environmental Science, Xiamen University
China

JIMENEZ ZAMORA, Elizabeth
Universidad Mayor de San Andres
Bolivia

KÄÄB, Andreas
University of Oslo
Norway

KAIRO, James Gitundu
Kenya Marine and Fisheries
Research Institute
Kenya

KAJTAR, Jules
University of Exeter
United Kingdom

KANG, Shichang
State Key Laboratory of Cryospheric
Science, Chinese Academy of Sciences
China

KARIM, Md Saiful
Queensland University of Technology
Australia

KASER, Georg
University of Innsbruck
Austria

KENNY, Tiff-Annie
University of Ottawa
Canada

AIII

KNUTSON, Thomas
National Oceanic and Atmospheric
Administration
USA

KOFINAS, Gary
University of Alaska Fairbanks
USA

KOLL, Roxy Mathew
Indian Institute of Tropical Meteorology
India

KOPP, Robert
Rutgers University
USA

KOTLARSKI, Sven
MeteoSwiss
Switzerland

KOVACS, Kit
Norwegian Polar Institute
Norway

KROEKER, Kristy
Univeristy of California Santa Cruz
USA

KUDELA, Raphael Martin
University of California Santa Cruz
USA

KÜNSTING, Martin
Freelance Graphics Designer
Germany

KUTUZOV, Stanislav
Russian Academy of Sciences
Russian Federation

KWIATKOWSKI, Lester
Institut Pierre-Simon-Laplace
France/United Kingdom

LAM, Vicky
The University of British Columbia
China/Canada

LAMBERT, Erwin
Utrecht University
Netherlands

LAUFKÖTTER, Charlotte
University of Bern
Switzerland

LAWRENCE, Judy
Victoria University of Wellington
New Zealand

LE COZANNET, Gonéri
Bureau de Recherches
Géologiques et Minières
France

LEBILLON, Philippe
The University of British Columbia
Canada

LEBRIS, Nadine
Laboratoire d'Ecogéochimie des
Environnements Benthiques
France

LEVIN, Lisa
University of California San Diego
USA

LJUBICIC, Gita
Carleton University
Canada

LÓPEZ MORENO, Juan Ignacio
Instituto Pirenaico de Ecologia
Spain

LORANTY, Michael
Colgate University
USA

LOSADA, Iñigo J.
University of Cantabria
Spain

LOTZE, Heike
Dalhousie University
Canada

LUNDQUIST, Jessica
University of Washington
USA

MACK, Michelle
North Arizona University
USA

MACKINNON, Jennifer
Scripps Institution of Oceanography
USA

MACKINTOSH, Andrew
Victoria University of Wellington
New Zealand

MAGNAN, Alexandre
Institute for Sustainable Development
and International Relations
France

MARCA, Hagenstad
Circle Economics
USA

MARSHALL, Nadine
Commonwealth Scientific and
Industrial Research Organisation
Australia

MARTEL, Patrick
University of KwaZulu-Natal
South Africa

MARZEION, Ben
University of Bremen
Germany

MASSON-DELMOTTE, Valerie
IPCC WGII Co-Chair
LSCE
France

MATHIAS COSTA MUELBERT, Mônica
PPGOB-IO/FURG
Brazil

MCDOWELL, Graham
University of British Columbia
Canada

MCINNES, Kathleen
Commonwealth Scientific and
Industrial Research Organisation
Oceans and Atmosphere
Australia

MCLEOD, Elizabeth
The Nature Conservancy
USA

MEIJERS, Andrew
British Antarctic Survey
United Kingdom/Australia

MELBOURNE-THOMAS, Jessica
Australian Antarctic Division
Australia

MELET, Angélique
LEGOS
France

MELTOFTE, Hans
Arhuus University
Denmark

MEREDITH, Michael
British Antarctic Survey
United Kingdom

MERRIFIELD, Mark
Scripps Institute of Oceanography
USA

MEYSSIGNAC, Benoit
Laboratoire d'Etudes en Géophysique
et Océanographie Spatiale
France

MILLS, L. Scott
University of Montana
USA

MILNER, Alexander
School of Geography Earth and
Environmental Sciences
United Kingdom

MINTENBECK, Katja
IPCC WGII TSU
Alfred Wegener Institute
Germany

MOLAU, Ulf
University of Gothenburg
Sweden

MOLINOS, Jorge García
University of Hokkaido
Japan/Spaim

MONTEIRO, Pedro
Council for Scientific and Industrial Research
South Africa

MORIN, Samuel
Météo-France – CNRS
France

MOSEMAN-VALTIERRA, Serena
The University of Rhode Island
USA

MOTAU, Andries
University of KwaZulu-Natal
South Africa

MOU, Cuicui
Lanzhou University
China

MUDRYK, Lawrence
Climate Research Division, Environment
and Climate Change Canada
Canada

MUKHERJI, Aditi
International Water Management Institute
India

MULSOW, Sandor
International Seabed Authority
Jamaica

MUTOMBO, Kana
CONGO MUDIMU
South Africa

NARAYAN, Siddharth
University of California Santa Cruz
USA

NEPAL, Santosh
International Centre for Integrated
Mountain Development
Nepal

NICHOLLS, Robert J.
University of Southhampton
United Kingdom

NICOLAI, Maike
IPCC WGII TSU
Alfred Wegener Institute
Germany

NÖTZLI, Jeannette
WSL Institute for Snow and
Avalanche Research SLF
Switzerland

NOY, Ilan
Victoria University of Wellington
New Zealand

NUTTALL, Mark
University of Alberta
Canada

O'DONOGHUE, Sean
University of KwaZulu-Natal and
EThekwini Municipality
South Africa

OKEM, Andrew
IPCC WGII TSU
University of KwaZulu-Natal
South Africa/Nigeria

OLIVER, Jamie
British Antarctic Survey
United Kingdom

OPPENHEIMER, Michael
Princeton University
USA

ORLOVE, Ben
Columbia University
USA

OSCHLIES, Andreas
GEOMAR Helmholtz Centre
for Ocean Research
Germany

OTTERSEN, Geir
Institute of Marine Research
Norway

OVERLAND, James
National Oceanic and Atmospheric
Administration
USA

OYINLOLA, Muhammed
University of British Columbia
Canada

PALAZZI, Elisa
National Research Council of Italy
Italy

PANDEY, Avash
International Centre for Integrated
Mountain Development
Nepal

PASCAL, Nicolas
Ecole Pratique des Hautes Etudes
France

PAYNE, Mark
Technical University of Denmark
Denmark

PECK, Victoria
British Antarctic Survey
United Kingdom

AIII

PEPIN, Nick
University of Portsmouth
United Kingdom

PEREIRA, Joy Jacqueline
Institute for Environment and Development
Malaysia

PETRASEK MACDONALD, Joanna
Inuit Circumpolar Council
Canada

PETZOLD, Jan
IPCC WGII TSU
Alfred Wegener Institute
Germany

PHILIP, Maxime
Blue Finance
France

PIERRE-MARIE, Lefeuvre
University of Oslo
Norway/France

PIRANI, Anna
IPCC WGI TSU,
Université Paris-Saclay/Abdus Salam
International Centre for Theoretical Physics
Italy/United Kingdom

POLOCZANSKA, Elvira S.
IPCC WGII TSU
Alfred Wegener Institute
Germany/United Kingdom

PÖRTNER, Hans-Otto
IPCC WGII Co-Chair
Alfred Wegener Institute
Germany

PRAKASH, Anjal
TERI School of Advanced Studies
India

PRITCHARD, Hamish
British Antarctic Survey, Natural
Environment Research Council
United Kingdom

PURCA CUICAPUSA, Sara Regina
Instituto del Mar del Perú
Peru

PURKEY, Sarah
Lamont Doherty Earth Observatory
USA

RANASINGHE, Roshanka
Deltares
Netherlands

RASUL, Golam
International Center for Integrated
Mountain Development
Nepal

RATHORE, Saurabh
Institute for Marine and Antarctic Studies
India

RATTER, Beate
University of Hamburg and
Institute of Coastal Research
Helmhotz Zentrum Geesthacht
Germany

REBELO, Xavier
University of Cape Town
South Africa

REID, Keith
CCAMLR
United Kingdom

RENAUD, Fabrice
University of Glasgow
United Kingdom

REYGONDEAU, Gabriel
University of British Columbia
France

RHEIN, Monika
University Bremen
Germany

RICE, Jake
Department of Fisheries and Oceans
Canada

RICHARDSON, Anthony
Commonwealth Scientific and
Industrial Research Organisation
Australia

RIEBESELL, Ulf
GEOMAR Helmholtz Centre
for Ocean Research
Germany

RINKEVICH, Baruch
National Institute of Oceanography
Israel

RIVERA-ARRIAGA, Evelia
University of Campeche
Mexico

RIXEN, Christian
WSL Institute for Snow and
Avalanche Research SLF
Switzerland

ROACH, Christopher
Laboratoire Oceanographie et
du Climat Experimentations et
Approches Numeriques LOCEAN2
France

ROBERTS, Debra
IPCC WGII Co-Chair
EThekwini Municipality
South Africa

ROBERTS, Murray
University of Edinburgh
United Kingdom

ROCKLÖV, Joacim
Umeå University
Sweden

ROMANOVSKY, Vladimir
University of Alaska Fairbanks
USA/Russian Federation

RUSSELL, Don E.
Yukon College
Canada

SAFA, Alain
Université Sofia Antipolis
France

SCHÄDEL, Christina
Northern Arizona University
USA/Switzerland

SCHMIDTKO, Sunke
GEOMAR Helmholtz Centre
for Ocean Research
Germany

SCHOEMAN, David
School of Science and Engineering,
University of the Sunshine Coast
Australia

SCHUUR, Ted
Northern Arizona University
USA

AIII

SEBESVARI, Zita
United Nations University
Germany

SEIBEL, Brad
University of South Florida
USA

SGUBIN, Giovanni
Université de Bordeaux
France

SHAHGEDANOVA, Maria
University of Reading
United Kingdom

SHEHABI, Manal
Oxford Institute for Energy Studies
United Kingdom

SHERPA, Pasang Yangjee
University of Washington
Nepal/USA

SHRESTHA SINGH, Mandira
International Centre for Integrated
Mountain Development
Nepal

SHURETY, Amy
University of KwaZulu-Natal
South Africa

SIMM, Jonathan
HR Wallingford
United Kingdom

SINGH, Gerald
University of British Columbia
Canada

SINISALO, Anna
International Centre for Integrated
Mountain Development
Nepal/Finland

SKILES, S. McKenzie
University of Utah
USA

SLOYAN, Bernadette
Commonwealth Scientific and
Industrial Research Organisation
Australia

SMEDSRUD, Lars H.
University of Bergen
Norway

SMIT, AJ
University of Western Cape
South Africa

SMITH, Matthew
Harvard University
USA

SOMMERKORN, Martin
World Wildlife Federation
Norway

STEFFEN, Konrad
Swiss Federal Research Institute WSL
Switzerland

STELTZER, Heidi
Fort Lewis College
USA

STROEVE, Julienne
NSIDC
Canada/USA

SUGA, Toshio
Graduate School of Science, Tohoku
University and RCGC, JAMSTEC
Japan

SUSANTO, Raden Dwi
University of Maryland
USA

SUTHERLAND, Catherine
University of KwaZulu-Natal
South Africa

SUTHERLAND, Michael
University of the West Indies
Trinidad and Tobago

SWINGEDOUW, Didier
CNRS-EPOC
France

TABE, Tammy
University of the South Pacific
Fiji

TAGLIABUE, Alessandro
University of Liverpool
United Kingdom

TIBIG, Lourdes
Climate Change Commission
Philippines

TIMMERMANS, Mary-Louise
Yale University
USA

TU, Nguyen Minh
United Nations University
Germany

TURETSKY, Merritt
University of Guelph
Canada

TURLEY, Carol
Plymouth Marine Laboratory
United Kingdom

TUYA, Fernando
University of Las Palmas de Gran Canaria
Spain

VAN DE WAL, Roderik
Utrecht University
Netherlands

VAN DEN BROEKE, Michiel
Utrecht University
Netherlands

VARGAS, Cristian
University of Concepción
Chile

VELICOGNA, Isabella
University of California Irvine
USA/Italy

VINCENT, Christian
University of Grenoble Alpes
France

VIVIROLI, Daniel
University of Zürich
Switzerland

VON SCHUCKMANN, Karina
Mercator Ocean
France

WABNITZ, Colette
University of British Columbia
Canada

WADHAM, Jemma
University of Bristol
United Kingdom

AIII

715

WALVOORD, Michelle
United States Geological Survey
USA

WANG, Gongjie
Chinese Academic of Sciences
China

WEYER, Nora M.
IPCC WGII TSU
Alfred Wegener Institute
Germany

WEYHENMEYER, Gesa
Uppsala University
Sweden

WHALEN, Caitlin
University of Washington
USA

WILLIAMS, Dee
United States Geological Survey
USA

WILLIAMSON, Phillip
Natural Environment Research Council
and University of East Anglia
United Kingdom

WIPFLI, Mark
University of Alaska Fairbanks
USA

WONG, Poh Poh
National University of Singapore
Singapore

WOODRUFF, Jon
University of Massachussets-Amherst
USA

WOUTERS, Bert
Utrecht University and Delft
University of Technology
Netherlands

XIAN, Siyuan
Accenture Strategy
China

XIAO, Cunde
Beijing Normal University
China

YANG, Daqing
Environment Canada
Canada

YASUNARI, Teppei J.
Hokkaido University
Japan

YOU, Qinglong
Fudan University
China

ZHAI, Panmao
IPCC WGI Co-Chair
China Meteorological Administration
China

ZHANG, Yangjiang
Chinese Academy of Sciences
China

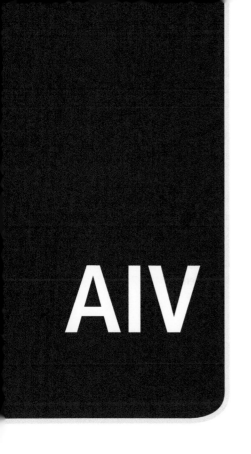

Annex IV: Expert Reviewers of the IPCC Special Report on the Oceans and Cryosphere in a Changing Climate

This annex should be cited as:

IPCC, 2019: Annex IV: Expert Reviewers of the IPCC Special Report on the Ocean and Cryosphere in a Changing Climate. In: *IPCC Special Report on the Ocean and Cryosphere in a Changing Climate* [H.-O. Pörtner, D.C. Roberts, V. Masson-Delmotte, P. Zhai, M. Tignor, E. Poloczanska, K. Mintenbeck, A. Alegría, M. Nicolai, A. Okem, J. Petzold, B. Rama, N.M. Weyer (eds.)]. Cambridge University Press, Cambridge, UK and New York, NY, USA, pp. 717–738.

ACKERMANN, Thomas
University of Applied Sciences
Germany

ADESALU, TAOFIKAT
University of Lagos
Nigeria

ADUSUMILLI, Susheel
Scripps Institution of Oceanography
USA

ADVANI, Nikhil
World Wildlife Fund
USA

AESCHBACH (NÉE VOLLWEILER), Nicole
Heidelberg University
Germany

AGGARWAL, Anubha
Delhi Technological University
India

AHMADUN, Fakhru'L-Razi
National Defense University of Malaysia
Malaysia

AHMED, Essam Hassan Mohamed
Climate Change and Sustainable
Development Expert
USA

ALAKKAT, Unnikrishnan
National Institute of Oceanogrphy
India

ALAVA, Juan Jose
University of British Columbia
Canada

ALEGRIA, Andrés
IPCC WGII TSU
Alfred Wegener Institute
Germany/Honduras

ALFREDSEN, Knut
Norwegian University of Science
and Technology
Norway

ALI, Meer
Florida State University
India

ALI, Syed Hammad
Glacier Monitoring and Research Center
Water and Power Development Authority
Pakistan

ALLAN, Richard
University of Reading
United Kingdom

ALLEN, Simon
University of Zurich
Switzerland

ALLEY, Richard B.
Pennsylvania State University
USA

ALPERT, Alice
US Department of State
USA

ANDRADE-VELAZQUEZ, Mercedes
Science and Technology Agency-
Global Change and Sustainability
Center in the Southeast
Mexico

ANDRELLO, Marco
Institut de Recherche pour le Développement
Canada

ANDRESEN, Camilla Snowman
Geological Survey of Denmark and Greenland
Denmark

ANICAMA DIAZ, Jahir
Pontificia Universidad Católica de Chile
Chile

ANORUO, Chukwuma
IMO State University Owerri
Nigeria

ANTIA, Effiom Edem
University Of Calabar
Nigeria

APPADOO, Chandani
University of Mauritius
Mauritius

ARBLASTER, Julie
Monash University
Australia

ARELLANO-TORRES, Elsa
Universidad Nacional Autónoma de México
Mexico

ARENSON, Lukas
BGC Engineering Inc./University of Manitoba
Canada

ARTHERN, Robert
British Antarctic Survey
United Kingdom

ATWOOD, Trisha
Utah State University
USA

AYALA, Alvaro
Centre for Advanced Studies in Arid Zones
Chile

AYYUB, Bilal
University of Maryland
USA

AZIZ, Danyal
Global Change Impact Studies Center
Pakistan

BABIKER, Mustafa
Saudi Aramco
Saudi Arabia

BAHRI, Tarub
Food and Agriculture Organization
Italy

BALAH, Belkacem
Département d'Hydraulique
Algeria

BALLINGER, Thomas
Texas State University
USA

BAMBER, Jonathan
University of Bristol
United Kingdom

BARANGE, Manuel
Food and Agriculture Organization
Italy

BARLETTA, Valentina R.
DTU Space
Denmark

AIV

BARLOW, Natasha
University of Leeds
United Kingdom

BARR, Iestyn
Manchester Metropolitan University
United Kingdom

BARRAZA, Francisco
Center for Climate and Resilience Research
Chile

BARRETT, Ko
IPCC Vice Chair
National Oceanic and Atmospheric
Administration
USA

BARRETT, Peter
Victoria University of Wellington
New Zealand

BARTLETT, Jesamine
University of Birmingham
United Kingdom

BARTSCH, Annett
Austrian Polar Research Institute
Austria

BARTSCH, Inka
Alfred Wegener Institute
Germany

BASANTES-SERRANO, Rubén
Escuela Politécnica Nacional
Ecuador

BAUMANN, Sabine
TUM
Germany

BEALL, Jacques
Surfrider Foundation Europe
France

BEDNARSEK, Nina
Southern California Coastal
Waters Research Project
USA

BEEVER, Erik
Montana State University and
U.S. Geological Survey
USA

BEHZAD, Layeghi
I.R. of Iran Meteorological Organization
Iran

BELL, Robert
National Institute of Water
& Atmospheric Research
New Zealand

BENKEBLIA, Noureddine
University of the West Indies
Jamaica

BENSON, Nsikak
Covenant University
Nigeria

BERNHARD, Luzi
Swiss Federal Research Institute WSL
Switzerland

BERTHIER, Etienne
LEGOS-CNRS
France

BESTER, Marthan
University of Pretoria
South Africa

BEVERIDGE, Malcolm
University of Stirling
United Kingdom

BEYLICH, Achim A.
Geomorphological Field Laboratory
Norway

BHUTIYANI, Mahendra
Defence R & D Organization
India

BIDDLE, Louise
University of Gothenburg
Sweden

BISHOP-WILLIAMS, Katherine
University of Guelph
Canada

BLARD, Pierre-Henri
CRPG, CNRS, Nancy France and Laboratoire
de Glaciologie ULB Bruxelles
France

BLOCKLEY, Ed
Met Office Hadley Centre
United Kingdom

BOCK, Christian
Alfred Wegener Institut
Germany

BODNER, Abigail
Brown University
USA

BOESCH, Donald
University of Maryland Center
for Environmental Sciences
USA

BOLCH, Tobias
University of Zurich
Germany

BONALDO, Davide
Institute of Marine Sciences,
National Research Council
Italy

BÖRGEL, Florian
Insitute for Baltic Sea Research
Germany

BORRAS-CHAVEZ, Renato
Pontificia Universidad Católica de Chile-Center
of Applied Ecology and Sustainability
Chile

BOUCHARD, Frédéric
Géosciences Paris Sud
France

BOUWER, Laurens
GERICS
Germany

BOX, Jason
Geological Survey of Denmark and Greenland
Denmark

BRANDT, Angelika
Senckenberg Research Institute
and Natural History Museum
Germany

BRASIER, Madeleine
University of Tasmania
Australia

BRENT, Kerryn
University of Tasmania
Australia

AIV

BRÖDER, Lisa
Vrije Universiteit
Netherlands

BRONSTERT, Axel
University of Potsdam
Germany

BROWN, Ross
Environment and Climate Change Canada
Canada

BRUN, Fanny
Université Grenoble Alpes
France

BURBA, George
LI-COR Biosciences/University of Nebraska
USA

BURGARD, Clara
Max Planck Institute for Meteorology
Germany

BURGER, Friedrich Anton
University of Bern
Switzerland

BURGHELEA, Carmen
University of Arizona
Romania

BURI, Pascal
University of Alaska Fairbanks
USA

BUTENSCHÖN, Momme
Centro Euro-Mediterraneo sui
Cambiamenti Climatici
Italy

CAESAR, Levke
Potsdam Institute of Climate Impact Research
Germany

CAI, Rongshuo
Third Institute of Oceanography,
Ministry of Natural Resource
China

CALDERON-AGUILERA, Luis Eduardo
Centro de Investigacion Cientifica
y de Educacion Superior de Ensenada
Mexico

CAMERON, Karen
Aberystwyth University
United Kingdom

CAMPBELL, Karley
University of Bristol
United Kingdom

CAMPBELL, Kristin
Institute for Governance and
Sustainable Development
USA

CAMPBELL, Seth
University of Maine
USA

CAPOBIANCO, Michele
CAPOBIANCO
Italy

CARL-FRIEDRICH, Schleussner
Climate Analytics
Germany

CARMAGNOLA, Carlo
Météo-France - CNRS, CNRM,
Centre d'Études de la Neige
France

CARRASCO, Jorge
Universidad de Magallanes
Chile

CARRIVICK, Jonathan
University of Leeds
United Kingdom

CARTER, Peter
Climate Emergency Institute
Canada

CASADO, Mathieu
Alfred Wegener Institute and APECS
Germany

CASEY, Michael
Germany

CASTANEDA MENA, Fátima Castaneda
Galileo University
Guatemala

CASTANEDA, Fátima
Galileo University
Guatemala

CASTRUCCI, Luca
Pacific Northwest National Laboratory
USA

CASWELL, Bryony
University of Hull
United Kingdom

CAVAN, Emma
University of Tasmania
Australia

CEARRETA, Alejandro
Universidad del País Vasco
Spain

CHAMBERS, robert
Institute of Development Studies
United Kingdom

CHAN, Kim Lian
Unversiti Malaysia SABAH
Malaysia

CHANDRA, Alvin
Department of Environment and Science
Australia

CHAPIN, F Stuart
University of Alaska Fairbanks
USA

CHARALAMPIDIS, Charalampos
Bavarian Academy of Sciences and Humanities
Germany

CHATURVEDI, Sanjay
Panjab University
India

CHEN, Wenting
Norwegian Institute for Water Research
Norway

CHEN, Xingrong
National Marine Environmental
Forecasting Center
China

CHENG, Lijing
Chinese Academy of Sciences
China

CHEUNG, Anson
Brown University
USA

CHEVALLIER, Matthieu
Météo France
France

CHOWN, Steven
Monash University and Scientific
Committee on Antarctic Research
Australia

CHRISTENSEN, Torben R.
Lund University/Aarhus University
Sweden

CHRISTIANEN, Susan
Extreme Design Lab
Iceland

CHUNG CHAI TSANG, Jean Bertrand
SSR Medical College
Mauritius

CHUNG, Ik Kyo
Pusan National University
Republic of Korea

CHURCH, John
University of New South Wales
Australia

CHUST, Guillem
AZTI
Spain

CIPOLLINI, Paolo
Telespazio VEGA UK for ESA Climate Office
United Kingdom

CLARK, Peter
Oregon State University
USA

CLARKE, William
Independent Researcher
Australia

CLAY, Jacinta
Brown University
USA

CLOSSET, Ivia
University of California Santa Barbara
USA

COBB, Kim
Georgia Tech
USA

COGLEY, J. Graham
Trent University
Canada

COLGAN, William
Geological Survey of Denmark and Greenland
Denmark

COLLEONI, Florence
Istituto Nazionale di Oceanografia
e Geofisica Sperimentale
Italy

COLLINS, Matthew
University of Exeter
United Kingdom

COMTE, Adrien
Université de Brest, IUEM
France

CONDE, Cecilia
National Autonomous University of Mexico
Mexico

CONSTABLE, Andrew
Australian Antarctic Division
Australia

COOK, Joseph
University of Sheffield
United Kingdom

COOLEY, Sarah
Ocean Conservancy
USA

CORNELIUS, Stephen
World Wildlife Federation
United Kingdom

COUGNON, Eva
University of Tasmania
Australia

CRAIG, Marlies
IPCC WGII TSU
University of KwaZulu-Natal
South Africa

CRAMER, Wolfgang
CNRS-IMBE
France

CRAWFORD, Alex
College of Wooster
USA

CRISE, Alessandro
OGS
Italy

CROOKALL, David
UCA
France

CUDENNEC, Christophe
Agrocampus Ouest
France

CUI, Peng
Chinese Academy of Sciences
China

CUNNINGHAM, Sari
University of Oslo
Norway

CVETKOVSKA, Marina
University of Ottawa
Canada

CZYZOWSKA-WISNIEWSKI, Elzbieta
University of Arizona, Pima Community College
USA

DADA, Olusegun A.
Federal University of Technology
Nigeria

DAGSSON WALDHAUSEROVA, Pavla
Agricultural University of Iceland
Iceland

DALEI, Narendra
University of Petroleum and Energy Studies
India

DANGENDORF, Sönke
University of Siegen
Germany

DANIELA, Coswig Kalikoski
Food and Agriculture Organization
Italy

DANSEREAU, Véronique
Nansen Environmental and
Remote Sensing Center
Norway

DARMARAKI, Sofia
Meteo France
France

AIV

DAS, Supriyo
Presidency University
India

DAVIES, Kirsten
Macquarie University
Australia

DAWSON, Jackie
University of Ottawa
Canada

DAX, Thomas
Federal Instittue for Less-fevoured
and Mountainous Areas
Austria

DE FONTAUBERT, Charlotte
The World Bank
USA

DE RAMON N'YEURT, Antoine
University of the South Pacific
Fiji

DEHECQ, Amaury
California Institute of Technology
USA

DEISSENBERG, Christophe
Aix-Marseille University
Luxembourg

DEMIROGLU, Osman Cenk
Umeå University
Sweden

DENG, Haijun
Fujian Normal University
China

DENIS ALLEMAND, Denis
Centre Scientifique de Monaco
Monaco

DESCAMPS, Sebastien
Norwegian Polar institute
Norway

DI MAURO, Biagio
University of Milano-Bicocca
Italy

DIAZ MOREJON, Cristobal Felix
Ministry of Science, Technology
and the Environment
Cuba

DIIWU, John
Government of Alberta
Canada

DIOP, E. Salif
Academy of Sciences
Senegal

DOCQUIER, David
Université Catholique de Louvain
Belgium

DOHERTY, Sarah
University of Washington
USA

DOLK, Michaela
Swiss Re
USA

DOVEY, Liz
Australian National University
Australia

DOW, Christine
University of Waterloo
Canada

DREYFUS, Gabrielle
Institute for Governance and
Sustainable Development
USA

DRIJFHOUT, Sybren
Royal Netherlands Meteorological Institute
Netherlands

DUMONT, Marie
Meteo-France CNRS, CNRM/CEN
France

DUPAR, Mairi
Overseas Development Institute
United Kingdom

DURGADOO, Jonathan
GEOMAR Helmholtz Centre
for Ocean Research Kiel
Germany

DUVIVIER, Alice
National Center for Atmospheric Research
USA

ECHEVESTE, Pedro
Instituto de Ciencias Naturales
Alexander von Humboldt
Chile

EDDEBBAR, Yassir
Scripps Institution of Oceanography
USA

EDDY, Tyler
University of British Columbia
Canada

EDWARDS, Charity
University of Melbourne
Australia

EDWARDS, Martin
Sir Alister Hardy Foundation for Ocean Science
United Kingdom

EDWARDS, Tamsin
King's College London
United Kingdom

EERKES-MEDRANO, Laura
University of Victoria
Canada

EGGERMONT, Hilde
Belgian Science Policy Office
Belgium

EHRENFELS, Benedikt
EAWAG & ETH Zurich
Switzerland

EINALI, Abbas
PMO
Iran

ELING, Lukas
Victoria University of Wellington
New Zealand

ELLIFF, Carla
Universidade Federal da Bahia
Brazil

ELOKA-EBOKA, Andrew
University of KwaZulu-Natal
South Africa

AIV

ENGELBRECHT, Francois
University of the Witwatersrand
South Africa

ENGLAND, Mark
Columbia University
United Kingdom

ENGSTRÖM, Erik
Swedish Meteorological and
Hydrological Institute
Sweden

EPSTEIN, Howard
University of Virginia
USA

ERLANIA, Erlania
Center for Fisheries Research
Indonesia

ERNANI DA SILVA, Carolina
TU Delft
Netherlands

ERWIN, Rottler
University of Potdam
Germany

FARIA, Sérgio Henrique
Basque Centre for Climate Change
Spain

FARINOTTI, Daniel
VAW ETH Zurich / WSL Birmensdorf
Switzerland

FATORIC, Sandra
North Carolina State University
USA

FEELY, Richard
National Oceanographic and
Atmospheric Administration
USA

FEHRENBACHER, Jennifer
Oregon State University
USA

FELDMAN, Daniel
Lawrence Berkeley National Laboratory
USA

FELIKSON, Denis
National Aeronautics and Space Administration
USA

FERNANDEZ PEREZ, Fiz
CSIC
Spain

FERNANDINO, Gerson
Red ProPlayas
Brazil

FETTWEIS, Xavier
University of Liège
Belgium

FIAMMA, Straneo
Scripps/University of California San Diego
USA

FIERZ, Charles
WSL Institute for Snow and
Avalanche Research SLF
Switzerland

FIRING, Yvonne
National Oceanography Centre
United Kingdom

FISCHLIN, Andreas
IPCC WGII Vice Chair
ETH Zurich
Switzerland

FIZ, Fernandez
CSIC
Spain

FLEMING, Sean
Oregon State University/ University of
British Columbia/ White Rabbit R&D LLC
USA

FLOWERS, Gwenn
Simon Fraser University
Canada

FOGGIN, Marc
University of Central Asia
Kyrgyzstan

FOGWILL, Christopher
Keele University
United Kingdom

FORD, James
McGill University
Canada

FORD, Victoria
Texas A&M University
USA

FORSBERG, Rene
Technical University of Denmark
Denmark

FÖRSTER, Kristian
Leibniz Universität Hannover
Germany

FOURNIER, Aimé
MIT
USA

FOX-KEMPER, Baylor
Brown University
USA

FRA.PALEO, Urbano
University of Extremadura
Spain

FRAJKA-WILLIAMS, Eleanor
National Oceanography Centre
United Kingdom

FRANS-JAN, Parmentier
University of Oslo
Norway

FREPPAZ, Michele
Università degli Studi di Torino
Italy

FROELICHER, Thomas
University of Bern
Switzerland

FUGLESTVEDT, Jan
Centre for International Climate
and Environmental Research
Norway

FÜSSEL, Hans-Martin
European Environment Agency
Denmark

GAGLIARDINI, Olivier
Université Grenoble Alpes
France

AIV

GAGNE, Karine
University of Guelph
Canada

GALBÁN-MALAGÓN, Cristóbal
Universidad Andrés Bello
Chile

GALEN MCKINLEY, Galen
Lamont Doherty Earth Observatory/
Columbia University
USA

GAN, Thian Yew
University of Alberta
Canada

GAO, Jing
Chinese Academy of Sciences
China

GARCIA-REYES, Marisol
Farallon Institute for Advanced
Ecosystem Research
USA

GARELICK, Sloane
Brown University
USA

GARNER, Andra
Rutgers University
USA

GARZA-GIL, M. Dolores
University of Vigo
Spain

GATTUSO, Jean-Pierre
CNRS-Sorbonne Université-Iddri
France

GELDSETZER, Torsten
University of Calgary
Canada

GERLAND, Sebastian
Norwegian Polar Institute
Norway

GIDDY, Isabelle Sindiswa
University Of Cape Town
South Africa

GIOLI, Giovanna
University of Edinburgh
United Kingdom

GLASER, Paul
University of Minnesota
USA

GOETZ, Scott
Northern Arizona University
USA

GOHEER, Muhammad
GCISC
Pakistan

GOLLEDGE, Nicholas
Victoria University of Wellington
New Zealand

GONG, Crystal
University of Alberta
Canada

GONZALEZ, Patrick
University of California
USA

GOUTTEVIN, Isabelle
Météo-France
France

GREGOIRE, Marilaure
MAST -ULiege
Belgium

GREGORY, Jonathan
University of Reading
United Kingdom

GREMION, Gwenaëlle
Université du Québec à Rimouski
Canada

GRIBBLE, Matthew
Emory University
USA

GROSSE, Guido
Alfred Wegener Institute
Germany

GRUBER, Nicolas
ETH Zürich
Switzerland

GRUBER, Stephan
Carleton University
Canada

GUARINO, Maria Vittoria
British Antarctic Survey
United Kingdom

GUILLAUME, Anne
Météo et Climat
France

GUILLOUX, Bleuenn Gaëlle
Cluster of Excellence the Future Ocean
Germany

GUILYARDI, Eric
LOCEAN/IPSL CNRS
France

GUNDERSEN, Hege
Norwegian Institute for Water Research
Norway

GUPTA, Mukesh
Institut de Ciencies del Mar
Spain

HAASNOOT, Marjolijn
Delft University of Technology
Netherlands

HAEBERLI, Wilfried
University of Zurich
Switzerland

HAINE, Thomas
Johns Hopkins University
USA

HALL-SPENCER, Jason
University of Plymouth/University of Tsukuba
United Kingdom/Japan

HAMILTON, Lawrence
University of New Hampshire
USA

HAN, In-Seong
National Institute of Fisheries Science
Republic of Korea

HANSMAN, Roberta
International Atomic Energy Agency
France

HAQUE, Md Enamul
University of Genova
Italy

AIV

HARDMAN-MOUNTFORD, Nicholas
Commonwealth Secretariat
United Kingdom

HASHMI, Danial
WAPDA
Pakistan

HAWARD, Marcus
Unversity of Tasmania
Australia

HAYASHI, Kentaro
National Agriculture and Food
Research Organization
Japan

HAYASHI, Masaki
University of Calgary
Canada

HAYMAN, Garry
Centre for Ecology & Hydrology
United Kingdom

HELENE, Frigstad
Norwegian Institute for Water Research
Norway

HERNÁNDEZ, Armand
ICTJA-CSIC
Spain

HEWITT, Helene
Met Office
United Kingdom

HEYD, Thomas
University of Victoria
Canada

HIDEKI, Kanamaru
Food and Agriculture Organization
Thailand

HIERONYMUS, Magnus
Swedish Meteorological and
Hydrological Institute
Sweden

HIRABAYASHI, Yukiko
Shibaura Institute of Technology
Japan

HJØLLO, Solfrid Sætre
Institute of Marine Research
Norway

HJORT, Jan
University of Oulu
Finland

HOCK, Regine
University of Alaska Fairbanks
USA

HOENISCH, Baerbel
University
USA

HOFFMANN, Dirk
Bolivian Mountain Institute
Germany

HOFSTEDE, Jacobus
Schleswig-Holstein Ministry for
Energy, Agriculture, Environment,
Nature and Digitization
Germany

HOLDING, Johnna
Aarhus University
Denmark

HOPWOOD, Mark
Geomar Helmholtz Centre for
Ocean Research Kiel
Germany

HOVLAND, Martin
University of Tromsø
Norway

HOWARD, William
Australian National University
Australia

HU, Zeng-Zhen
National Oceanographic and
Atmospheric Administration
USA

HUANG, Ping
Chinese Academy of Sciences
China

HUDSON, Thomas
British Antarctic Survey
United Kingdom

HUETTMANN, Falk
University of Alaska Fairbanks
USA

HUGGEL, Christian
University of Zurich
Switzerland

HUNT, George
University of Washington
USA

HUNTER, Nina
University of KwaZulu-Natal
South Africa

HUNTINGTON, Henry
Huntington Consulting
USA

HURLBERT, Margot
University of Regina
Canada

HUSS, Matthias
ETH Zürich
Switzerland

IBRAHIM, Zelina
Universiti Putra Malaysia
Malaysia

IMMERZEEL, Walter
Utrecht University
Netherlands

INSAROV, GRIGORY
Russian Academy of Sciences
Russian Federation

ISLAM, Akm Saiful
Bangladesh University of
Engineering and Technology
Bangladesh

ITKIN, Polona
Nansen Environmental and
Remote Sensing Center
Norway

ITO, Shin-Ichi
University of Tokyo
Japan

JACKSON, Lagipoiva Cherelle
National University of Samoa
Samoa

JACKSON, Laura
Met Office Hadley Centre
United Kingdom

JAFARI, Mostafa
National Macro Plan on Climate
Change Research/RIFR
Iran

JAHN, Alexandra
University of Colorado Boulder
USA

JAMIE, Shutler
University of Exeter
United Kingdom

JANSSEN, David
University of Bern
Switzerland

JENNINGS, Keith
University of Nevada, Reno and
The Desert Research Institute
USA

JEWETT, Elizabeth
National Oceanographic and
Atmospheric Administration
USA

JIE, Liu
University of Bergen
Norway

JOCHUMSEN, Kerstin
Federal Maritime and Hydrographic Agency
Germany

JOHN, Emeka
Georg-August-Universität Göttingen
Germany

JOHN, Jasmin
National Oceanographic and
Atmospheric Administration
USA

JOHNSON, Joanne
British Antarctic Survey
United Kingdom

JÖNSSON, Anette
Swedish Meteorological and
Hydrological Institute
Sweden

JOOS, Fortunat
University of Bern
Switzerland

JOSEY, Simon
National Oceanography Centre
United Kingdom

JOURDAIN, Nicolas
IGE, CNRS/Univ. Grenoble-Alpes
France

JOUZEL, Jean
CEA
France

JRRAR, Amna
Independent Researcher
Jordan

JUAN IGNACIO, López Moreno
CSIC
Spain

KÄÄB, Andreas
University of Oslo
Norway

KADIBI, Khadija
National Meteorological Direction
Morocco

KALEN, Ola
Swedish Meteorological
and Hydrological Institute
Sweden

KALÉN, Ola
Swedish Meteorological
and Hydrological Institute
Sweden

KANDASAMY, Kathiresan
Annamalai University
India

KANTH, Malin
Governmental Agency
Sweden

KAPNICK, Sarah
National Oceanographic and
Atmospheric Administration
USA

KAPSENBERG, Lydia
CSIC Institute of Marine Sciences
Spain

KASER, Georg
University of Innsbruck
Austria

KATSUMATA, Katsuro
JAMSTEC
Japan

KAWAMIYA, Michio
Japan Agency for Marine-Earth
Science and Technology
Japan

KENNEDY, Hilary
Bangor University
United Kingdom

KENNY, Tiff-Annie
Nereus Program
Canada

KERSTING, Diego
Freie Universität Berlin
Germany

KETTLES, Helen
Department of Conservation
New Zealand

KHAEMBA, Winnie
African Centre for Technology Studies
Kenya

KHALEEL, Zammath
Ministry of Environment
Maldives

KHESHGI, Haroon
ExxonMobil
USA

KIESSLING, Wolfgang
Friedrich-Alexander University
Erlangen-Nürnberg
Germany

KIM, Seong-Joong
Korea Polar Research Institute
Republic of Korea

KIM, Sung Yong
Korea Advanced Institute of
Science and Technology
Republic of Korea

AIV

KIMBLE, Melinda
United Nations Foundation
Syracuse University
USA

KING, Matt
University of Tasmania
Australia

KITOH, Akio
Japan Meteorological Business Support Center
Japan

KJELDSEN, Kristian Kjellerup
Geological Survey of Denmark and Greenland
Denmark

KNIEBUSCH, Madline
Leibniz Institute for Baltic Sea
research Warnemünde
Germany

KOBASHI, Takuro
Renewable Energy Institute
Japan

KOCHTITZKY, William
University of Maine
USA

KOHNERT, Katrin
GFZ German Research Centre for Geosciences
Germany

KOIVUROVA, Timo
Arctic Centre/University of Lapland
Finland

KONOVALOV, Vladimir
Institute of Geography
Russian Federation

KONYA, Keiko
Japan Agency for Mari-Earth
Science and Technology
Japan

KOPP, Robert
Rutgers University
USA

KOSZALKA, Inga
GEOMAR Helmholtz Centre for Ocean
Research Kiel & Kiel University
Germany

KOTLARSKI, Sven
Federal Office of Meteorology
and Climatology MeteoSwiss
Switzerland

KOUL, Vimal
Universität hamburg
Germany

KOURANTIDOU, Melina
University of Southern Denmark
Denmark

KOVEN, Charles
Lawrence Berkeley National Lab
USA

KRAUSE-JENSEN, Dorte
Aarhus University
Denmark

KRINNER, Gerhard
IGE/CNRS
France

KROGLUND, Marianne
Norwegian Environment Agency/Arctic
Monitoring and Assessment Programme
Norway

KRUEMMEL, Eva
ScienTissiME, Inuit Circumpolar Council
Canada

KUDELA, Raphael
University of California Santa Cruz
USA

KUSWARDANI, Anastasia Rita Tisiana Dwi
Ministry of Marine Affairs and Fisheries
Indonesia

KUWAE, Tomohiro
Port and Airport Research Institute
Japan

KWIATKOWSKI, Lester
Laboratoire de Météorologie Dynamique,
Institut Pierre-Simon Laplace
France

KYZIVAT, Ethan
Brown University
USA

LAGERLOEF, Gary
Earth and Space Research (Retired)
USA

LAM, Steven
University of Guelph
Canada

LANDOLFI, Angela
GEOMAR Helmholtz Centre
for Ocean Research
Germany

LANGE, Benjamin A.
Fisheries and Oceans Canada
Canada

LAVRILLIER, Alexandra
CEARC, OVSQ, Versailles University
France

LAW, Cliff
National Institute of Water &
Atmospheric Research
New Zealand

LAWRENCE, Judy
Victoria University of Wellington
New Zealand

LAYEGHI, Behzad
IRIMO
Iran

LE BARS, Dewi
KNMI
Netherlands

LE COZANNET, Goneri
BRGM
France

LE QUERE, Corinne
University of East Anglia
United Kingdom

LE TRAON, Pierre Yves
Mercator Ocean International
France

LE, Hoang Anh
Vietnam

LECLERC, Boris
Public Institution of the Ministry of Environment
France

AIV

LEE-SIM, Lim
Universiti Sains Malaysia
Malaysia

LEE, Brown
University of Leeds
United Kingdom

LEE, Sai Ming
Hong Kong Observatory
China

LEE, Suk Hui
Korea Marine Environment
Management Corporation
Republic of Korea

LEE, WON SANG
Korea Polar Research Institute
Republic of Korea

LEE, Yoo Kyung
Korea Polar Research Institute
Republic of Korea

LEGG, Sonya
Princeton University
USA

LEILA, Rashidian
Iran

LEMKE, Peter
Alfred Wegener Institute
Germany

LEVY, Joseph
Colgate University
USA

LEY, Debora
University of Oxford
Guatemala

LI, Hai
Ministry of Natural Resources
China

LI, Xichen
Chinese Academy of Sciences
China

LIANG, Yantao
Chinese Academy of Sciences
China

LIAO, Wenjie
Sichuan University
China

LIFANG, Chiang
University of California
USA

LIM, Lee-Sim
Universiti Sains Malaysia
Malaysia

LINDBÄCK, Katrin
Norwegian Polar Insitute
Norway

LING, Frank
University of Tokyo
Japan

LIPKA, Oxana
WWF Russia
Russian Federation

LIU, Jihua
Shandong University
China

LIU, Kexiu
National Marine Data and Information Service
China

LIU, Shiyin
Yunnan University
China

LIVINGSTON, Mary
Fisheries New Zealand, Ministry
for Primary Industries
New Zealand

LLANILLO, Pedro J.
University of Santiago de Chile
Chile

LOCKLEY, Andrew
University College London
United Kingdom

LOKMAN, Kees
University of British Columbia
Canada

LOMBARDI, Chiara
Italian National Agency for New
Technologies, Energy and Sustainable
Economic Development
Italy

LOPEZ-GASCA, Mariela
Venezuelan Research Institute
Venezuela

LORANTY, Michael
Colgate University
USA

LORENZ, William
University of Southern Queensland
Australia

LORENZONI, Laura
University of South Florida
USA

LOVEJOY, Connie
Université Laval
Canada

LOWTHER, Andrew
Norwegian Polar Institute
Norway

LUBANGO, Louis Mitondo
United Nations Economic
Commission for Africa
Ethiopia

LUCEY, Noelle
Smithsonian Tropical Research Institute
Panama

LUENING, Sebastian
Institute for Hydrography, Geoecology
and Climate Sciences
Portugal

LUKAS, Sven
Lunds Universitet
Sweden

LUPO, Anthony
University of Missouri
USA

ŁUSZCZUK, Michał
Maria Curie Skłodowska University
Poland

MACCRACKEN, Michael
Climate Institute
USA

MACDONALD, Joanna
Inuit Circumpolar Council Canada
Canada

AIV

MACDONALD, Robie
Fisheries and Oceans Canada
Canada

MADSEN, Kristine Skovgaard
Danish Meteorological Institute
Denmark

MAGNAN, Alexandre
IDDRI
France

MAHMOOD, Ali
University of Basrah
Iraq

MAHMUD, Mastura
Universiti Kebangsaan Malaysia
Malaysia

MAJID COOKE, Fadzilah
National University of Malaysia
Malaysia

MALATESTA, Stefano
University of Milano-Bicocca
Italy

MAŁECKI, Jakub
Adam Mickiewicz University in Poznań
Poland

MANTILLA-MELUK, Hugo
Unversidad del Quindío
Colombia

MARBAIX, Philippe
Université Catholique de Louvain
Belgium

MARCIL, Catherine
Institut des Sciences de la Mer
Canada

MARKUSZEWSKI, Piotr
Polish Academy of Sciences
Poland

MARTIN-VIDE, Javier
University of Barcelona
Spain

MARTINERIE, Patricia
Institut des Géosciences de
l'Environnement, CNRS
France

MARTÍNEZ FONTAINE, Consuelo
Université Paris-Saclay
France

MARTY, Christoph
SLF
Switzerland

MASSON-DELMOTTE, Valerie
IPCC WGII Co-Chair
LSCE
France

MASSONNET, François
Université Catholique de Louvain and
Barcelona Supercomputing Center
Belgium

MATT, King
University of Tasmania
Australia

MATTHEW, Collins
University of Exeter
United Kingdom

MAURITZEN, Cecilie
Norwegian Meteorological Institute
Norway

MAXIMILLIAN, Van Wyk de Vries
University of Minnesota
USA

MCCARTHY, Gerard
Maynooth University
Ireland

MCDOWELL, Graham
University of British Columbia
Canada

MCGEE, Jeffrey
University of Tasmania
Australia

MEDINA, Josep Ramon
Universitat Politècnica de València
Spain

MEINANDER, Outi
Finnish Meteorological Institute
Finland

MELET, Angelique
Mercator Ocean International
France

MÉMIN, Anthony
Université Côte d'Azur
France

MENARY, Matthew
Met Office Hadley Centre
United Kingdom

MENDOZA, Marcos
University of Mississippi
USA

MÉNÉGOZ, Martin
Institut des Géosciences de l'Environnement
France

MENGEL, Matthias
Potsdam Institute for Climate Impact Research
Germany

MENVIEL, Laurie
University of New South Wales
Australia

MIDGLEY, Pauline
Independent Consultant
Germany

MIGNOT, Juliette
IRD-IPSL/LOCEAN
France

MILINEVSKY, Gennadi
Taras Shevchenko University of Kyiv
Ukraine

MILLS, L. Scott
University of Montana
USA

MILNER, Alexander
University of Birmingham
United Kingdom

MINER, Kimberley
University of Maine
USA

MINTENBECK, Katja
IPCC WGII TSU
Alfred Wegener Institute
Germany

MISHRA, Anil
United Nations Educational,
Scientific and Cultural Organization
France

MIX, Alan
Oregon State University
USA

MONCKTON OF BRENCHLEY, Viscount
Science and Public Policy Institute
United Kingdom

MONTPETIT, Benoit
Environment and Climate Change Canada
Canada

MOORE, Robert
University of British Columbia
Canada

MOORE, Tommy
Northwest Indian Fisheries Commission
USA

MORA, Carla
IGOT
Portugal

MORAN, Kate
University of Victoria
Canada

MORELLE, Nathalie
Permanent Secretariat of the
Alpine Convention
Austria

MORIN, Samuel
Météo-France – CNRS
France

MOSER, Gleyci
Universidade Estadual do Rio de Janeiro
Brazil

MOTTRAM, Ruth
DMI
Denmark

MOUREY, Jacques
University Grenoble Alpes/University
Savoie Mont Blanc/CNRS
France

MRAK, Irena
Environmnetal Protection College
Slovenia

MSADEK, Rym
CNRS

France
MUELLER, Bennit
Institute of Ocean Sciences
Canada

MÜLLER, Marius
Federal University of Pernambuco
Brazil

MUÑOZ SOBRINO, Castor
Universidade de Vigo
Spain

MURATA, Akihiko
RCGC/JAMSTEC
Japan

MURPHY, Eugene
British Antarctic Survey
United Kingdom

NAM, Sunghyun
Seoul National University
Republic of Korea

NAUELS, Alexander
Climate Analytics
Germany

NAUGHTEN, Kaitlin
British Antarctic Survey
United Kingdom

NDIONE, Jacques-Andre
Centre de Suivi Ecologique
Senegal

NEBDI, Hamid
Chouaïb Doukkali University
Morocco

NELSON, Joanna
LandSea Science
USA

NEOGI, Suvadip
IPCC WGIII TSU
Ahmedabad University
India

NERILIE, Abram
Australian National University
Australia

NICHOLLS, Robert
University of Southampton
United Kingdom

NICOLAI, Maike
IPCC WGII TSU
Alfred Wegener Institute
Germany

NIENOW, Peter
University of Edinburgh
United Kingdom

NIFENECKER, Herve
Sauvons Le Climat
France

NIKAM, Jaee
UN Environment
India

NORTH, Michelle A.
University of KwaZulu-Natal
South Africa

NOTZ, Dirk
Max Planck Institute for Meteorology
Germany

NUGRAHA, Adi
Pacific Northwest National Laboratory
USA

NÜSSER, Marcus
Heidelberg University
Germany

OAKES, Robert
United Nations University Institute for
Environment and Human Security
United Kingdom

OGUTU, Geoffrey Evans Owino
Meteorological Services
Kenya

OKEM, Andrew
IPCC WGII TSU
University of KwaZulu-Natal
South Africa/Nigeria

OLIVA, Frank
Environment and Climate Change Canada
Canada

OLIVA, Marc
University of Barcelona
Spain

ONINK, Victor
University of Bern
Switzerland

ONO, Tsuneo
Japan Fisheries Research and
Education Agency
Japan

ORENSTEIN, Patrick
Brown University
USA

ORLOVE, Ben
Columbia University
USA

ORR, James
Climate & Environment Sciences
Lab/IPSL, CEA-CNRS-UVSQ
France

OSWALD SPRING, Úrsula
UNAM, National Autonomous
University of Mexico
Mexico

OTTERSEN, Geir
Institute of Marine Research
Norway

OTTO SIMONETT, otto
Zoï Environment Network
Switzerland

OTTO, Jan-Christoph
University of Salzburg
Germany

OUILLON, Sylvain
Institut de Recherche pour le Développement
France

OURBAK, Timothée
AFD
France

OVE HOEGH-GULDBERG, Ove
University of Queensland
Australia

OVERDUIN, Pier-Paul
Alfred Wegener Institute
Germany

OZSOY, Burcu
Istanbul Technical University
Polar Research Center
Turkey

PALMER, Matthew
Met Office Hadley Centre
United Kingdom

PALUPI, Listyati
Airlangga University
Indonesia

PARK, Jinsoon
Korea Maritime and Ocean University
Republic of Korea

PARK, Taehyun
Greenpeace East Asia
Republic of Korea

PATTYN, Frank
Université libre de Bruxelles
Belgium

PAUL, Frank
University of Zurich
Switzerland

PAULI, Harald
Austrian Academy of Sciences, University of
Natural Resources and Life Sciences Vienna
Austria

PAYNE, Mark
Technical University of Denmark
Denmark

PEARSON, Jenna
Brown University
USA

PEARSON, Pamela
International Cryosphere Climate Initiative
USA

PEBAYLE, Antoine
Ocean and Climate Platform
France

PELEJERO, Carles
ICREA and Institut de Ciències del Mar, CSIC
Spain

PENG, Ge
North Carolina State University/National
Centers for Environmental Information
USA

PENTZ, Brian
University of Toronto
Canada

PEREIRA, Christopher
Secretariat of the Convention
on Biological Diversity
Canada

PETER, Croot
National University of Ireland Galway
Ireland

PETER, Maria
Norwegian University of
Science and Technology
Norway

PETRASEK MACDONALD, Joanna
Inuit Circumpolar Council
Canada

PETRIE, Elizabeth
University of Glasgow
United Kingdom

PETZOLD, Jan
IPCC WGII TSU
Alfred Wegener Institute
Germany

PEZZOLI, Alessandro
Politecnico di Torino e Università di Torino
Italy

PFEFFER, Julia
The Australian National University
Australia

PIEPENBURG, Dieter
Alfred Wegener Institute
Germany

PIERCE, Ethan
Brown University
USA

PLANTON, Serge
Météo-France, CNRM
France

AIV

POHJOLA, Veijo
Uppsala University
Sweden

POITOU, Jean
Sauvons Le Climat
France

POLOCZANSKA, Elvira S.
IPCC WGII TSU
Alfred Wegener Institute
Germany/United Kingdom

POPE, James
British Antarctic Survey
United Kingdom

PÖRTNER, Hans-Otto
IPCC WGII Co-Chair
Alfred Wegener Institute
Germany

POULAIN, Florence
Food and Agriculture Organization
Italy

POUSSIN, Charlotte
University of Geneva and UNEP GRID
Switzerland

PRINZ, Rainer
University of Graz
Austria

QUEIROS, Ana
Plymouth Marine Laboratory
United Kingdom

QUEIRÓS, José Pedro
University of Coimbra
Portugal

RABATEL, Antoine
University of Grenoble
France

RABE, Benjamin
Alfred Wegener Institute
Germany

RABEHI, Walid
CTS/ASAL
Algeria

RABOUILLE, Sophie
CNRS
France

RACAULT, Marie-Fanny
Plymouth Marine Laboratory
United Kingdom

RAHMSTORF, Stefan
Potsdam Institute for Climate Impact Research
Germany

RAJAPAKSHE, Chamara
University of Maryland Baltimore County
Sri Lanka

RAMA, Bardhyl
IPCC WGII TSU
Alfred Wegener Institute
Germany/Kosovo

RAMAGE, Justine
Stockholm University
Sweden

RAMOS, Isabel
SENAMHI
Peru

RANZI, ROBERTO
University of Brescia
Italy

RASHIDIAN, Leila
OASC, IRIMO
Iran

RASSMANN, Jens
University of Liège
Belgium

RASUL, Golam
International Centre for Integrated
Mountain Development
Nepal

REBETEZ, Martine
WSL and University of Neuchatel
Switzerland

RECINOS RIVAS, Beatriz Margarita
University of Bremen
Germany

RECKIEN, Diana
Universiteit Twente
Germany

REESE, Ronja
Potsdam Institute for Climate Impact Research
Germany

REID, Keith
Commission for the Conservation
of Antarctic Marine Living Resources
Australia

RENNER, Angelika
Institute of Marine Research
Norway

RENWICK, James
Victorial University of Wellington
New Zealand

REUTEN, Christian
University of Calgary
Canada

RICHTER, Claudio
Alfred Wegener Institute/ University of Bremen
Germany

RICHTER, Nora
Brown University
USA

RIDING, Tim
New Zealand Ministry for the Environment
New Zealand

RIIHELÄ, Aku
Finnish Meteorological Institute
Finland

RIMI, Ruksana
University of Oxford
United Kingdom

RINTOUL, Stephen
Commonwealth Scientific and
Industrial Research Organisation
Australia

RIYAZ, Mahmood
Maldivian Coral Reef Society
Maldives

ROBERTS, Debra
IPCC WGII Co-Chair
EThekwini Municipality
South Africa

ROCKMAN, Marcy
International Council on Monuments and Sites
USA

ROHMER, Jeremy
Bureau de Recherches Géologiques et Minières
Finland

ROSE, Tseng
Air Force Institute of Technology
USA

ROSS, Nathan
Ministry of Foreign Affairs and Trade
New Zealand

RUBENSDOTTER, Lena
Geological Survey of Norway
Norway

RUIZ, Lucas
Instituto Argentino de Nivología,
Glaciología y Ciencias Ambientales
Argentina

RUMMUKAINEN, Markku
Swedish Meteorological and
Hydrological Institute
Sweden

RUNCIE, John
University of Sydney
Australia

RYABININ, Vladimir
Intergovernmental Oceanographic Commission
France

RYBSKI, Diego
Potsdam Institute for Climate Impact Research
Germany

RYKACZEWSKI, Ryan
University of South Carolina
USA

SAITO, Kazuyuki
JAMSTEC
Japan

SAKYA, Andi Eka
Agency for the Assessment and
Application of Technology
Indonesia

SALA, Hernan Edgardo
Argentine Antarctic Institute
Argentina

SALERNO, Franco
IRSA-CNR
Italy

SALLEE, Jean Baptiste
LOCEAN/CNRS
France

SAMELSON, Roger
Oregon State University
USA

SAMMIE, Buzzard
University College London
United Kingdom

SANDBERG SOERENSEN, Louise
DTU Space
Denmark

SANDER, Sylvia
IAEA-NAEL
France

SANE, Aakash
Brown University
USA

SANZ SANCHEZ, Maria Jose
Basque Centre for Climate Change
Spain

SATO, Kirk
Okinawa Institute of Science and Technology
Japan

SATOH, Masaki
University of Tokyo
Japan

SAVOSKUL, Oxana
Sri Lanka

SCHARFFENBERG, Martin
University of Hamburg
Germany

SCHAUWECKER, Simone
University of Geneva
Chile

SCHLEUSSNER, Carl-Friedrich
Climate Analytics
Germany

SCHLOGEL, Romy
ESA Climate Office
United Kingdom

SCHOEMAN, David
University of the Sunshnie Coast
Australia

SCHÖLD, Sofie
Swedish Meteorological and
Hydrological Institute
Sweden

SCHRAM, Julie
University of Oregon
USA

SCHRAMA, Ernst
Delft University of Technology
Netherlands

SCHREIBER, Erika
University of Colorado Boulder
USA

SCHROEDER, Katrin
CNR ISMAR
Italy

SCHUBACK, Nina
EPFL
Switzerland

SCHULER, Thomas Vikhamar
University of Oslo
Norway

SCHWEBEL, Michael
Temple University/The Johns
Hopkins University
USA

SCHWEIZER, Juerg
WSL Institute for Snow and
Avalanche Research
Switzerland

SEFERIAN, Roland
CNRM (Météo-France/CNRS)
France

SEIBERT, Petra
University of Natural Resources
and Life Sciences
Austria

SENSOY, Serhat
Turkish State Meteorological Service
Turkey

SERGI, Sara
LOCEAN-IPSL
France

SERGIENKO, Olga
Princeton University
USA

SHEA, Joseph
University of Northern British Columbia
Canada

SHIGEMITSU, Masahito
Japan Agency for Marine-Earth
Science and Technology
Japan

SHRESTHA, Arun
International Centre for Integrated
Mountain Devlopment
Nepal

SHRESTHA, Maheswor
Water and Energy Commission Secretariat
Nepal

SHUTLER, Jamie
University of Exeter
United Kingdom

SILLMANN, Jana
Centre for International Climate
and Environmental Research
Norway

SILVESTRE, Elizabeth
Universidad Catolica Santo
Toribio de Mogrovejo
Peru

SINGH, Nayanika
Ministry of Environment, Forest
and Climate Change
India

SINGH, Shalini
University of the South Pacific
Fiji

SINKLER, Emilie
University of Alaska Fairbanks
USA

SIORAK, Nicolas
Business Alliance for Climate Resilience
France

SISWANTO, Siswanto
The Agency for Meteorology,
Climatology, and Geophysics
Indonesia

SKEIE, Ragnhild Bieltvedt
Centre for International Climate
and Environmental Research
Norway

SLANGEN, Aimee
Royal Netherlands Institute for Sea Research
Netherlands

SMEDSRUD, Lars
University of Bergen
Norway

SMEED, David
National Oceanography Centre
United Kingdom

SMITH, Inga
University of Otago
New Zealand

SMITH, Sharon
Geological Survey of Canada
Natural Resources Canada
Canada

SMOLYANITSKY, Vasily
Arctic and Antarctic Research Institute
Russian Federation

SOMMARUGA, Ruben
University of Innsbruck
Austria

SOUZA, Alejandro
CINVESTAV
Mexico

SOYSA, Ramesh
PELCO Development Consultants Pvt. Ltd.
Sri Lanka

SPANDRE, Pierre
Institu National de Recherche en Sciences et
Technologies de l'Environnement et Agriculture
France

SPEER, Elizabeth
Natural Resources Defense Council
USA

SPEER, Lisa
Natural Resources Defense Council
USA

SPENCER, Thomas
University of Cambridge
United Kingdom

SPRING, Aaron
MPI-M
Germany

SROKOSZ, Meric
National Oceanography Centre
United Kingdom

ST. PIERRE, Kyra
University of Alberta
Canada

STENDEL, Martin
Danish Meteorological Institute
Denmark

STENMARK, Aurora
Norwegian Environment Agency
Norway

STOJANOV, Robert
Mendel University in Brno/
European University Institute
Czech Republic

STORCH, Daniela
Alfred Wegener Institute
Germany

STRANEO, Fiamma
Scripps/University of California
USA

STRAUSS, Sarah
University of Wyoming
USA

STRIEGEL, Sandra
University of Bern
Switzerland

STROBACH, Ehud
National Aeronautics and Space Administration
USA

STUDHOLME, Joshua
Russian Academy of Sciences
Russian Federation

SUGA, Toshio
Tohoku University/JAMSTEC
Japan

SULISTYAWATI, Sulistyawati
Ahmad Dahlan University
Indonesia

SULTAN, Maitham
Ministry of Science and Technology
Iraq

SUN, Jianqi
Chinese Academy of Sciences
China

SUN, Jun
Tianjin University of Science and Technology
China

SUNDBY, Svein
Institute of Marine Research
Norway

SUNGHYUN, Nam
Seoul National University
Republic of Korea

SURIANO, Zachary
University of Nebraska Omaha
USA

SUTTER, Johannes
Alfred Wegener Institute
Germany

SUTTON-GRIER, Ariana
The Nature Conservancy
USA

SUTTON, Adrienne
National Oceanographic and
Atmospheric Administraton
USA

SUZUKI, Kazuyoshi
Japan Agency for Marine-Earth
Science and Technology
Japan

SWART, Neil
Environment and Climate Change Canada
Canada

TAILLARDAT, Pierre
Université du Québec à Montréal
Canada

TAKAHASHI, Ken
Servicio Nacional de Meteorología
e Hidrología del Perú
Peru

TAKATA, Kumiko
National Institute for Environmental Studies
Japan

TALLEY, Lynne
University of California San Diego
USA

TAMURA, Makoto
Ibaraki University
Japan

TANG, Malcolm
University of Malaya
Malaysia

TAQUI, Muhammad
Comsats University Islamabad
Pakistan

TARASOV, Lev
Memorial University of Newfoundland
Canada

TAYLOR, Patrick
National Aeronautics and Space Administration
USA

TEDESCO, Letizia
Finnish Environment Institute
Finland

THAKUR, Praveen Kumar
Indian Institute of Remote Sensing
India

THALER, Thomas
University of Natural Resources
and Life Sciences
Austria

THIBERT, Emmanuel
Université Grenoble Alpes
France

THOMAS, Elizabeth
British Antarctic Survey
United Kingdom

THOMAS, Frank
University of the South Pacific
Fiji

THOMSEN, Soeren
Sorbonne University, IPSL, LOCEAN
France

THOR, Peter
Swedish Meteorological and
Hydrological Institute
Sweden

TIGNOR, Melinda
IPCC WGII TSU
Alfred Wegener Institute
Germany/USA

TINKER, Jonathan
Met Office Hadley Centre
United Kingdom

TIWARI, Pushp Raj
University of Hertfordshire
United Kingdom

TJERNSTRÖM, Michael
Stockholm University
Sweden

TOKARSKA, Katarzyna B.
Univerisity of Edinrbugh
United Kingdom

TOMANEK, Lars
California Polytechnic State University
USA

TOTIN, Edmond
National University of Agriculture
Benin

TOY, Suleyman
Atatürk University
Turkey

TREWIN, Blair
Australian Bureau of Meteorology
Australia

TROVATO, Maria Rosa
University of Catania
Italy

TRUFFER, Martin
University of Alaska Fairbanks
USA

TSANI, Stella
International Centre for Research
on the Environment and the Economy
Greece

TURCO, Marco
Barcelona Supercomputing Center
Spain

TURNER, Kate
University of Alaska Fairbanks
USA

AIV

ULTEE, Elizabeth
Massachusetts Institute of Technology
USA

UOTILA, Petteri
University of Helsinki
Finland

UYSAL, Irfan
Ministry of Agriculture & Forestry
Turkey

VACHAUD, Georges
CNRS
France

VALLOT, Dorothée
Swedish Meteorological and
Hydrological Institute
Sweden

VAN DE WAL, Roderik
Utrecht University
Netherlands

VAN DEN BROEKE, Michiel
Utrecht University
Netherlands

VAN DEN HEUVEL, Floortje
Ecole Polytechnique Federale de Lausanne
Switzerland

VAN DER LAAN, Larissa
Bavarian Academy of Sciences and Humanities
Germany

VAN DONGEN, Eef
ETH Zurich
Switzerland

VAN MEERBEECK, Cedric
Caribbean Institute for
Meteorology and Hydrology
Barbados

VAN WYCHEN, Wesley
University of Waterloo
Canada

VAN YPERSELE, Jean-Pascal
Université Catholique de Louvain
Belgium

VARADE, Divyesh
Indian Institute of Technology Kanpur
India

VERFAILLIE, Deborah
Barcelona Supercomputing Center
Spain

VIEIRA, Gonçalo
Universidade de Lisboa
Portugal

VIJAY, Saurabh
Technical University of Denmark
Denmark

VINCENT, Christian
University of Grenoble/CNRS
France

VIVIAN, Christopher
Centre for Environment, Fisheries
and Aquaculture Science (Retired)
United Kingdom

VLADU, Iulian Florin
United Nations Framework
Convention on Climate Change
Germany

VON SCHUCKMANN, Karina
Mercator Ocean International
France

VOORTMAN, Hessel
Arcadis
Netherlands

VOUSDOUKAS, Michail
EC Joint Research Centre
Italy

WAGNER, Thomas
National Aeronautics and Space Administration
USA

WAGNER, Till
University of North Carolina Wilmington
USA

WAGNON, Patrick
IGE-IRD
France

WAHL, Thomas
University of Central Florida
USA

WALKER, Scott
Northwest Vista College
USA

WANG, Chunzai
Chinese Academy of Sciences
China

WANG, Dongxiao
South China Sea Institute of Oceanology
China

WANG, Feiteng
Chinese Academy of Sciences
China

WANG, Junye
Athabasca University
Canada

WANG, Pengling
China Meterological Administration
China

WANG, XIAOMING
Chinese Academy of Science
Australia

WANG, Xiujun
Beijing Normal University
China

WANG, Zhaomin
Hohai University
China

WARREN, Stephen
University of Washington
USA

WARRICK, Olivia
Red Cross Red Crescent Climate Centre
New Zealand

WATSON, Phil
US Coastal Education and
Research Foundation
Australia

WEATHERHEAD, Elizabeth
Jupiter Intelligence; University
of Colorado (Retired)
USA

WEBBER, Ben
University of East Anglia
United Kingdom

WEISSENBERGER, Sebastian
Université du Québec à Montréal
Canada

AIV

WEN, Jiahong
Shanghai Normal University
China

WEPKING, Carl
Colorado State University
USA

WESTER, Philippus
International Centre for Integrated
Mountain Development
Netherlands

WEYER, Nora M.
IPCC WGII TSU
Alfred Wegener Institute
Germany

WHITE, Dave
Climate Change Truth Inc.
USA

WHITE, Rehema
University of St Andrews
United Kingdom

WHITEHOUSE, Pippa
Durham University
United Kingdom

WILLIAMS, Dee
United States Geological Survey
USA

WILLIAMS, Emily
University of California Santa Barbara
USA

WINBERG VON FRIESEN, Lisa
IVL Swedish Environmental Research Institute
Sweden

WINIGER, Patrik
Independent Expert
Netherlands

WINKLER, Manuela
University of Natural Resources
and Life Sciences
Austria

WOLFF, Eric
University of Cambridge
United Kingdom

WONG, Poh Poh
University of Adelaide
Singapore

WOOD, Thomas
Retired
USA

WRIGHT, Jeneva
East Carolina University
USA

WU, Bingyi
Fudan University
China

WU, Mengxi
Brown University
USA

WU, Renguang
Chinese Academy of Sciences
China

WU, Shaohong
Chinese Academy of Sciences
China

WUITE, Jan
Environmental Earth Observation
Information Technology GmbH
Austria

YAMANOUCHI, Takashi
National Institute of Polar Research
Japan

YANG, Handa
University of California
USA

YANG, Y. Jeffrey
United States Environmental Protection Agency
USA

YATES, Katherine
University of Salford
United Kingdom

YETTELLA, Vineel
University of Colorado Boulder
USA

YIN, Baoshu
Chinese Academy of Sciences
China

YIN, Yixing
Nanjing University of Information
Science and Technology
China

YOUNG, Gillian
British Antarctic Survey
United Kingdom

YOUNG, Tun Jan
Taroko National Park
United Kingdom

YOUNGFLESH, Casey
Stony Brook University
USA

YUMRUKTEPE, Veli Caglar
Nansen Environmental and
Remore Sensing Center
Norway

ZAELKE, Durwood
Institute for Governance and
Sustainable Development
USA

ZAFAR, Qudsia
Global Change Impact Studies Centre
Pakistan

ZAITON IBRAHIM, Zelina
Universiti Putra Malaysia
Malaysia

ZAREIAN, Mohammad Javad
Ministry of Energy
Iran

ZARGARLELLAHI, Hanieh
Geological Survey of Iran
Iran

ZEKOLLARI, Harry
ETH Zürich
Switzerland

ZHAI, Panmao
IPCC WGI Co-Chair
Meteorological Administration
China

ZHANG, Fan
Chinese Academy of Sciences
China

AIV

ZHANG, Rui
University of British Columbia
Canada

ZHOU, Botao
Nanjing University of Information
Science and Technology
China

ZINKE, Jens
Freie Universität Berlin
Germany

ZIVERI, Patrizia
ICREA-ICTA UAB
Spain

ZIVIAN, Anna
Ocean Conservancy
USA

ZOLKOS, Scott
University of Alberta
USA

ZUO, Juncheng
Zhejiang Ocean University
China

AIV

Index

This index should be cited as:
IPCC, 2019: Index. In: *IPCC Special Report on the Ocean and Cryosphere in a Changing Climate* [H.-O. Pörtner, D.C. Roberts, V. Masson-Delmotte, P. Zhai, M. Tignor, E. Poloczanska, K. Mintenbeck, A. Alegría, M. Nicolai, A. Okem, J. Petzold, B. Rama, N.M. Weyer (eds.)]. Cambridge University Press, Cambridge, UK and New York, NY, USA, pp. 739–755.

Index

Index

Index

Index

H

Index

Index

Index